The New Public Health

Third Edition

ELSEVIER *science & technology books*

Companion Web Site:

http://booksite.elsevier.com/9780124157668

The New Public Health, Third Edition
By Theodore H. Tulchinsky and Elena A. Varavikova

Resources for Students and Professors:

- Instructor's Guide
- Glossary of Terms
- Chapter headings and content, abstracts, key words
- Student review, example questions and extended bibliography
- Links to appropriate websites
- Power Points of author key lectures
- Selected case studies

TOOLS FOR ALL YOUR TEACHING NEEDS
textbooks.elsevier.com

ACADEMIC PRESS

The New Public Health

Third Edition

Theodore H. Tulchinsky MD, MPH
Braun School of Public Health,
Hebrew University–Hadassah, Ein Karem, Jerusalem, Israel

and

Elena A. Varavikova MD, MPH, PhD
Research Institute for Public Health Organization and Information (CNIIOIZ),
Moscow, Russian Federation

With **Joan D. Bickford, MSN**
Former Chief Public Health Nurse, Province of Manitoba, Canada

Foreword By **Jonathan Fielding, MD, MPH**
University of Los Angeles (UCLA), California

AMSTERDAM • BOSTON • HEIDELBERG • LONDON • NEW YORK • OXFORD • PARIS
SAN DIEGO • SAN FRANCISCO • SINGAPORE • SYDNEY • TOKYO

Academic Press is an imprint of Elsevier

Academic Press is an imprint of Elsevier
32 Jamestown Road, London NW1 7BY, UK
225 Wyman Street, Waltham, MA 02451, USA
525 B Street, Suite 1800, San Diego, CA 92101-4495, USA

Recommended Citation: Tulchinsky TH, Varavikova EA. *The New Public Health, Third Edition.* Elsevier, Academic Press, San Diego, 2014.

British Library Cataloguing-in-Publication Data
A catalogue record for this book is available from the British Library

Library of Congress Cataloging-in-Publication Data
A catalog record for this book is available from the Library of Congress

ISBN: 978-0-12-415766-8

For information on all Academic Press publications
visit our website at elsevierdirect.com

Typeset by TNQ Books and Journals
www.tnq.co.in

Printed and bound in United States of America

14 15 16 17   10 9 8 7 6 5 4 3 2 1

Working together
to grow libraries in
developing countries

www.elsevier.com • www.bookaid.org

The authors wish to make the following dedications:

Theodore Tulchinsky – to my wife and family: Joan, our children (Daniel, Joel, and Karen), and their children (Guy, Noa, Amir, Ariel, Jonathan, and Or), as well as the many family, friends, colleagues, organizations, and students who encouraged us in this work.

Elena Varavikova – to my family Inna, Tatyana and Natalya, to the great friends and colleagues: Dean David Carpenter, David Momrow, Dr Harvey Bernard, Thaisa and Charles Beach, Irina Volguina, Professor Michael Reich, Dr Halfdan T. Mahler, Dr Margaret Chan, Professor Tamara Maximova, Professor Vladimir Starodubov, and many wonderful people who inspired and helped through the life of this book.

We jointly dedicate this book to all those who sustained us and contributed to its development and its translations into many languages. We also want to add a dedication to all public health workers, current and future, and to the improved health of all, particularly the poor and needy in all parts of the globe.

Contents

Contents

10. Organization of Public Health Systems

Contents

15. Health Technology, Quality, Law, and Ethics

16. Global Health

Theodore H. Tulchinsky MD, MPH

Braun School of Public Health, Hebrew University-Hadassah, Jerusalem

Theodore Herzl (Ted) Tulchinsky was born and raised in Brantford, Ontario, Canada, and studied medicine at the University of Toronto. He began his medical career as a physician in support of Saskachewan's pioneer Medicare program, and a general practitioner in community clinics. He then trained in internal medicine and obtained a MPH at Yale University, (USA).

He served the Province of Manitoba Department of Health and Social Development as Associate Deputy Minister and then Deputy Minister during the 1970s, managing the department of over 5,000 staff. He initiated and led the development of Manitoba's province-wide home care program, rural district health systems, mental health reform, and urban community clinics, as well as leading in controlling an epidemic of western equine encephalitis.

In 1976 Dr Tulchinsky and his family moved to Israel. He joined the Ministry of Health and served as Director of Public Health and from 1981 to 1994 as Ministry of Health Coordinator of Health Services for the West Bank and Gaza, with oversight responsibility for immunization, nutrition, sanitation, maternal and child health, community health workers, hospital development, and other topics for the Palestinian population. He has been active in developing programs in vaccination, maternal and child health, prevention of anemia of infancy, fortification of foods to prevent micronutrient deficiency conditions in Israel and the Palestinian Authority. During the 1985–1992 period he served as a visiting consultant to the New York State Department of Health in Albany New York.

Since 1981, he has taught at the Braun School of Public Health and Community Medicine at the Hebrew University–Hadassah in Jerusalem. His courses include Organization of Public Health Services and a seminar course on the Risk Approach in the International Master of Public Health Program for students from low-income countries around the world. He served as public health

consultant for a World Bank mission to the Russian Federation from 1992 to 1995 and was a Fulbright Scholar during a one-year sabbatical at the University of California, Los Angeles (UCLA) School of Public Health. He has served as a consultant and visiting lecturer on developing new schools of public health in Moldova, Albania, Macedonia, Mongolia, Russia, and Kazakhstan, and as a member of the Executive Board of the European Association of Schools of Public Health (ASPHER).

With coauthor Elena Varavikova, the first edition of *The New Public Health: An Introduction for the 21st Century* was developed and published in 1998 and translated into many languages; the second edition, in 2009, was translated into the Georgian language.

In 2007–2008, he led in the preparation of the Working Paper on Nutrition for the Israel Ministry of Health, Health Israel 2020 Project, and in activities of the Ministry of Health in fortification of flour, milk, and salt with essential micronutrients. Since 2008 he has served as Deputy Editor of the online journal *Public Health Reviews*, based in the French School of Public Health (EHESP) in Paris, France, being responsible for issues of the new journal on The New Public Health, Public Health Education, Mental Health and Public Health Ethics. He is currently leading in the organization of bachelor's programs in public health, health promotion, nursing, and nutrition in Ashkelon College in Ashkelon, Israel.

Awards and Honors

Member of the Israeli delegation to the 1978 Alma-Ata Conference on Primary Health Care, Alma-Ata, Kazakhstan, and participated in the 30th anniversary conference, Almaty, in October 2008.

Professor Honorarium, Faculty of Health Sciences, University of Mongolia: "For your significant contributions in developing and strengthening the public health sector and medical research of Mongolia, you will forever be held in the highest regards at HSUM", 2005.

Honorable Doctor of Tbilisi State Medical University (Georgia) Academic Council award of the title in the Department of Public Health, December 2006.

Professor Honoris Causa, Saints Cyril and Methodius University, Macedonia, Faculty of Medicine, May 2008.

Andreas Stampar Medal awarded by the European Association of Schools of Public Health, (ASPHER), November 2008, "for distinguished contribution to development of public health in Europe".

Publications

Dr Tulchinsky has 88 publications in peer-reviewed journals, and 32 books, chapters, and other publications. Topics include poliomyelitis, measles and tetanus immunization programs, vitamin K deficiency bleeding disorder, iron-deficiency anemia, other micronutrient deficiency disorders and their prevention and monitoring, child growth, asbestosis and other occupational diseases, non-communicable diseases, evaluation of health systems, public health education, preventing birth defects, community health workers, public health ethics, and public health education.

Contact: tulchinskyted@hotmail.com; tedt@hadassah.org.il

Elena A. Varavikova MD, MPH, PhD
Federal Research Institute for Public Health Organization and Information, Moscow, Russian Federation

Elena A. Varavikova was born in Alma-Ata, Kazakhstan, USSR, and studied medicine and public health at the Moscow Medical Academy named after I. M. Sechenov. She completed her PhD in Moscow, an MPH degree at the School of Public Health at the State University of New York in Albany, and postdoctoral studies at the Takemi Program of Harvard University School of Public Health. She served as Chief of the Unit for Monitoring of Health and Preventable Deaths, Research Institute for Public Health Organization and Information, Russia, as well as Associate Professor of Public Health at the Moscow Medical Academy.

She was recruited to the World Health Organization (WHO) in 2002, where she has worked on strengthening regional health systems, policy analysis, evaluation of the health services outcomes, studies of avoidable death, regional inequality in health services and human resources. She participated in preparing National White Papers on health and various national reports, and in the development of the computerized mortality database.

Dr Varavikova has worked on the supervision and management of grants and projects from OSI–Soros Foundation in the development and management of health programs in Russia (1995–2002). She was involved in the development and strengthening of the schools of public health in the former soviet countries and regions of Russia. She was also a consultant in health policy for the World Bank, UNICEF, UNDP, and UNHCR.

During more than four years at the WHO headquarters in Geneva, her area of work included projects on avoidable conditions and death; strategies for improvement in health services coverage; human resources for health, quality improvement in public health education, mapping of the School of Public Health and related institutions around the world; and patient safety (Regional and National Patient Safety Programme Development). She has participated in different committees and working groups, including "Future of the WHO-EURO in 2020" and prevention of non-communicable diseases.

After her affiliation as a scientist for the WHO, Dr Varavikova returned to Russia to work in the Ministry of Health and Social Development in the Department for International Collaboration, and later in the Federal Agency for High-Tech Medical Care, where she was a State Adviser for the Russian Federation. She has managed and participated in a number of projects in many countries and has published on public health topics including health policy, population health, future studies and globalization, health technology assessment, and professional education. She is now a leading researcher on the issues of medical science management, prevention of fetal alcohol syndrome, and medical diplomacy.

Improving quality of life and promoting health for all are among the main goals of modern public health, and improvement in the health of nations is dependent on effective public health endeavors. No field of endeavor is more multifaceted than public health, because it includes the wide range of individual and collective efforts to improve health, forestall disease, and protect health. To adequately cover this field requires many different types of expertise across a range of professionals and up-to-date knowledge ranging widely from genetics to biostatistics and health services organization and management, and from environmental epidemiology to sociology and demography.

That is why it is so rare to see a comprehensive treatment of public health written by only two authors. Yet Ted Tulchinsky and Elena Varavikova have responded to this challenge with a volume that covers the public health waterfront, is eminently readable, and provides a valuable resource for both students and practicing public health professionals. One advantage of this economy of authors is a consistent tone, perspective, and extent of references.

Originally intended for use as the introductory and core textbook for the new Schools of Public Health being developed in the Newly Independent States, it is now used in Israel, Eastern Europe, Central Asia, and the USA. It is also appropriate as an introductory text for the increasing number of students studying public health in undergraduate programs. The previous editions of this work have been translated into many languages. Much of the information in this book derives from the experiences in Western Europe and the USA, and it is being successfully used in departments and schools of public health and other related institutions.

This book assumes limited prior knowledge, providing definitions for key terms. Yet it can also serve as a reference for the experienced public health generalist and for those in other disciplines for whom public health knowledge is advantageous.

The chapters on the history of public health and its expanding definition are an excellent introduction to the underlying concepts and the evolution of our understanding of health determinants and disease and injury control opportunities. A significant portion of the volume treats health care systems, both generically as a set of complex organizations whose stewardship requires strategic thinking, planning, and operational expertise, and more specifically as to how these systems operate in different social, cultural, and economic environments.

The title *The New Public Health* is apt. Today, public health is defined broadly, understanding that what we do in the health sector is insufficient to produce maximum attainable health, but must be complemented by attention to how decisions in other sectors affect the public's health. Is there any doubt that decisions about taxation policy, educational systems, transportation, urban planning, or agricultural subsidies affect health? Also, the New Public Health emphasizes the importance of reducing preventable health disparities within and between populations. Furthermore, governmental public health needs many partners to fulfill its mission – schools, religious institutions, employers, and other agencies. While these concepts are not entirely new, they have been rediscovered and emphasized to a much greater extent than was common during the twentieth century. We also have much new evidence about which programs and policies can effectively improve health. Recent research has aided our understanding of the basic mechanisms influencing health and disease, and the interventions to prevent and reduce disease processes and injuries. And we have standardized methods to assess bodies of evidence that increase our confidence in the reproducibility of results of policies and programs.

In short, this work is approachable, instructive, practical, and timely. Its use will help to assure that public health professionals in all nations where it is read are current and well informed about the New Public Health.

Jonathan Fielding, MD, MPH
2 February 2013

Dr Fielding is a Distinguished Professor of Health Services and Pediatrics at University of California, Los Angeles (UCLA) Fielding School of Public Health. He serves as Director of Public Health and Health Officer for Los Angeles County, where he is responsible for the full range of public health activities for 10 million county residents. Dr Fielding teaches Determinants of Health and participates as faculty lecturer in several other courses. He received his MD, MA (History of Science), and MPH from Harvard University, and his MBA from the Wharton School of Business Administration. His areas of expertise include systematic assessment of the best evidence of effective population-oriented interventions, the development of clinical preventive services guidelines, economics and financing of prevention and health promotion for children, adults and families in community, clinical and occupational settings. As the founding Co-Director of the Center for Health Enhancement, Education and Research at UCLA, he directed the first comprehensive university-based center to focus on clinical and worksite prevention opportunities. He formerly served as Founding Board Member, Chairman of the Board, and member of the Executive Committee of The California Wellness Foundation, the largest US Foundation devoted to disease prevention and health promotion. He was a founding member of the US Preventive Services Task Force and chairs the US Community Preventive Services Task Force. He is a past President of the American College of Preventive Medicine. Dr Fielding's awards include the UCLA Medal, the Fries Prize for contributions to improving health, the Sigerist and Roemer prizes from the American Public Health Association, the Porter Prize for his national impact on improving the lives of Americans, and membership in the National Academy of Sciences Institute of Medicine. He is the author of over 200 original scientific articles and chapters, and the long-time Editor of *Annual Review of Public Health*. He is also a founding member of the Editorial Board of *Public Health Reviews*.

Classical Public Health

New Public Health

LOOKING BACK HELPS IN LOOKING AHEAD

"It's déjà vu all over again"

Famously attributed to Lawrence Peter "Yogi" Berra, famed catcher for the New York Yankees baseball team, noted for pithy and wise humor, such as: "You can observe a lot by watching".

We greatly appreciate the warm reception that the first (1999) and second (2008) English editions of this textbook has received from students, teachers, and practitioners of public health in many countries over the past 14 years. It has also been well accepted in translated editions in Russian, Bulgarian, Albanian, Moldovan, Romanian, Macedonian, Uzbek, and Mongolian languages. The most recent translation was released in the Georgian language in 2013.

This book is used not only in introductory courses in public health at bachelor's and master's levels but also as a general review for PhD students coming to public health from different disciplines, in North America, Europe, and many other countries. It has also been frequently recommended for use as a desk reference for practitioners. This acceptance has been gratifying and we hope this third edition will also be widely used as a working tool in existing and new programs in professional and vocational public health education.

Looking ahead, we should remember where we as public health educators and practitioners have come from. In the early 1960s, no one could have predicted the path ahead in public health. It was a time when it was widely thought that infectious diseases were soon to be completely controlled or eradicated. At that time, the pandemic of coronary heart diseases was increasing as the leading cause of death in the industrialized world, as it is now becoming in developing countries. Pioneering epidemiological studies of that time, such as the Framingham Heart Study conducted from 1948 to the present time and many like it, provided breakthrough knowledge. The new term "risk factors" identified "preventable causes" that needed to be addressed by public health and clinical medicine.

In 1964, the US Surgeon General's Report on Smoking brought together a vast literature on the health effects of smoking, launching a struggle which has been a major public health success but continues as a challenge to the present time. In the early 1970s, key policy analyses such as the Lalonde Report in Canada linked the importance of environment, genetics, and lifestyle as well as medical care in determining health status. This opened the path to the Ottawa Charter on Health Promotion, moving public health to a major new professional sphere and application.

The US Surgeon General's adoption of health targets in "Healthy People" and the Alma-Ata Declaration of 1978 placing emphasis on "health for all" and community-based interventions set new directions for public health action. These analyses provided the conceptual infrastructure for a continuously developing New Public Health.

New concepts and applications of public health and medical interventions adopted in the following years were associated with dramatic reductions in mortality from coronary heart disease, at its peak in the mid-1960s, and from stroke and more recently from cancer. When the HIV pandemic came out of the blue in the 1980s, no one could have predicted that this horrific disease would largely be brought under control within a decade by a combination of health promotion and later antiretroviral treatments. The great achievements of public health and medical sciences have given the world substantial gains in life expectancy and freedom from many historic diseases and debilitating conditions. It would have been difficult to foresee that viruses causing cancers would come to be preventable not only by lifestyle changes but also by screening and early intervention, as well as by new vaccines for hepatitis B and later for human papillomavirus, and treatment of *Helicobacter pylori* to prevent chronic peptic ulcer diseases and gastric cancer, or the enormous impact of hypertension control, use of statins, and smoking cessation in reducing the cardiovascular pandemic mortality rates.

There has been a rapid decline in cardiovascular mortality since the 1960s from reduction in smoking, limiting alcohol use, healthier diet, and exercise, in lowering cholesterol levels, as well as in improvements in medical treatment and access to it. Tobacco control has made tremendous strides forward even in the face of powerful opposition from the giant tobacco agroindustry. We are continuing to see great improvements in access to safe water, food, and sanitation, and in malaria and tuberculosis prevention and control. Globally, the 1960s saw the gradual eradication of smallpox and in subsequent decades the growing control and near-eradication of poliomyelitis, with other great advances in lesser known achievements in the control of leprosy, onchocerciasis, and filariasis, diseases that drained the energy, vision, and health of millions in tropical countries. Road safety improvements have reduced injuries and deaths, and suicide rates have also fallen in many countries.

Public health worked to become better prepared to face health threats after the 9/11 terrorist attacks in the USA, and natural disasters such as Hurricane Katrina, as well as actual or threatened pandemics from SARS and H1N1 and the newly appearing potential pandemic threat of H7N9. Having learned from past successes and errors and with more effective tools including organization and training, communities and countries are better able to cope. The Millennium Development Goals (MDGs) of 2001–2015 have been substantially but variably successful, achieving major reductions in child mortality and in vaccine-preventable diseases. There have been setbacks as well as accomplishments in the return of once controlled diseases such as measles, pertussis, and diphtheria. We in public health have learned many lessons that will be applied, we hope, in this decade and beyond 2020. We have come a long way and have a long way to go; we have every reason to face our challenges with confidence and energetic commitment with continuous learning and practice standards.

Are the struggles against poverty, disease, and premature death due to preventable diseases over? Of course not, but looking ahead we see progress in achieving the MDGs set globally in 2001 for the target year 2015. Current reviews show uneven progress in the three MDGs directly related to health; we may not able to reach the stated targets by the target year of 2015, but the global health community should take heart from achievements, even if there are limitations to the achievements and more work lies ahead. New health technologies, assuring access to care for all, eliminating inequalities, economizing, and reducing waste and risk to patient safety and quality of health care are all part of the challenges that face us during social and economic crises, terrorism, conflicts and disasters, climate change, drinking water shortages, incitement to genocide, and many other events affecting current and future global and local population health.

The long-standing public health challenges such as tuberculosis, malaria, diarrheal, environmental, and sexually transmitted diseases, antibiotic resistance, mental health, dementias, diabetes, and obesity remain important, and new challenges lie ahead. Immunization, even while preventing millions of deaths, has faced public resistance and even opposition based on misinformation and fraudulent research quickly adopted by internet players, so that diseases thought to have been controlled, such as measles, pertussis, and diphtheria, are being seen commonly again. At the same time, hopes for new advances in diagnostics, therapeutics, prevention, and health promotion will reduce illness and premature deaths and reduce the inequalities that trouble all regions and nations of the globe.

This book evolved from many years of teaching the principles of health organization to students of public health from Africa, Latin America, the Caribbean, Asia, the USA, Eastern Europe, and Russia, as well as from the practice of public health in a wide variety of international settings, Because of globalization, migration, and the rapidly changing context of public health, we concluded from this experience that there was a need for a new textbook of public health that both provides a basis in the classic knowledge and achievements of public health, and brings current thinking in the broad base to new students and veteran practitioners with an international orientation.

We draw upon ancient traditions from Biblical Mosaic and Greco-Roman societies with belief systems of Sanctity of Human Life (*Pikuach nefesh*), Improve the World (*Tikkun olam*) and Healthy Mind–Healthy Body (*Mens sana in corpore sano*) together with modern applications of social solidarity and human rights that link between individual and community responsibility for health. Organizational philosophies of health as a right and scientific advances provide the basis for the scientific and ethical approaches of the New Public Health. The New Public Health is a synthesis of classical public health with evolving modern public health and standards of preventive medicine and social policy. Both society and individuals have rights and responsibilities in promoting and maintaining health and the quality of life, as well as preventing disease and premature death, with equity and application of best practices and policies gained from science and practice.

The New Public Health is a cumulative philosophy of saving lives and improving health by a wide variety of professions and methods based on scientific achievements in the context of societal responsibility for the health and well-being of the population. The New Public Health is a composite of social policy, law, and ethics, with integration of social, behavioral, economic, management, and biological sciences. It is an intersectoral and interdisciplinary application of social policy, health promotion, preventive, and curative health services, all of which are vital to sustain and improve health for individuals and populations. We hope that this edition will help enable students, teachers, practitioners and policy makers to understand this complexity and to apply it as their profession, work, avocation, and dedication.

T. H. Tulchinsky MD MPH
E. A. Varavikova MD MPH PhD
15 January 2014

REFERENCES

Centers for Diseases Control and Prevention, 2011. Ten great public health achievements – United States, 2001–2010. MMWR Morb. Mortal Wkly. Rep. 60, 619–623. Available at: http://www.cdc.gov/mmwr/preview/mmwrhtml/mm6019a5.htm (accessed 15.01.14).

Centers for Diseases Control and Prevention, 2011. Ten great public health achievements – worldwide, 2001–2010. 60, 614–618. Available at: MMWR Morb. Mortal Wkly Rep. 60, 814–818 Available at:. http://www.cdc.gov/mmwr/preview/mmwrhtml/mm6019a5.htm (accessed 15.01.14).

Surgeon General of the Public Health Service, 1979. US Department of Health, Education, and Welfare. Healthy People: The Surgeon General's report on health promotion and disease prevention. US Government Printing Office, Washington, DC. Available at: http://profiles.nlm.nih.gov/ps/access/NNBBGK.pdf (accessed 15/01.14).

Tulchinsky, T.H., Varavikova, E.A., 2010. What is the "New Public Health"? Public Health Rev. 32, 25–53. Available at: http://www.publichealthreviews.eu/show/f/23 (accessed 15.01.14).

United Nations, 2012. Millennium Development Goals Report 2012. UN Department of Economics and Social Affairs, New York. Available at: http://www.un.org/en/development/desa/publications/mdg-report-2012.html (accessed 05.03.13).

Acknowledgments

The first two editions of this textbook were published in 1999/2000, and 2008, and now the third in 2014. We wish to express our gratitude to the many persons who contributed to this process, including many friends and colleagues at the Open Society Institute of New York, the American Joint Distribution Committee, the World Bank, the State University of New York, Albany School of Public Health, the Takemi Program in International Health in the Harvard School of Public Health, the World Health Organization, the University of California Los Angeles (UCLA) School of Public Health, the Braun School of Public Health, Hebrew University-Hadassah, Jerusalem, the IM Sechenov Moscow Medical Academy, Academic Press and Elsevier, as well as to many students, faculty colleagues and supporters mentioned in the acknowledgement of all editions.

We owe a very special gratitude to Joan Bickford MSN, Winnipeg, Manitoba for her steadfast, thorough and consistent support throughout the preparation of the second and the third edition and to Amy Ovadia and Britt Das for their editorial support.

We also wish to extend our warm thanks to the following colleagues who helped in the editing and updating of this book: Britt Dash, Ksenia Kubasova, Amy Ovadia, Sheeba Qureshi, Maureen Malowany, Anders Foldpsang, Rajesh Rai, Rilwan Raji, Miguel Reina, Francisco Sarmiento, Ahyan Shandilya, and Vineet Srivastava.

In this edition we include boxes prepared by colleagues and friends including: Chris Birt, Mayer Brezis, Mauro Cibin, George DiFerdinando, Joseph Dorsey, Anders Foldpsang, Rene Galera, Gary Ginsberg, Selena Gray, Ina Hinnenthal, Dena Jaffe, Shimelis Kitancho, Eli Rosenberg, Eli Richter, Martin McKee, Nicola Nante, Ellen Nolte, Amy Ovadia, Ora Paltiel, Walter Ricciardi and Valerie Saatchi.

Our warm appreciations also go to Nancy Maragioglio (Senior Acquisitions Editor) and Carrie Bolger (Editorial Project Manager) of Elsevier who have worked with us closely throughout the process with continuous support and efficient management.

Of course, we could not have developed this edition without the encouragement and very constructive input of family, friends, students and colleagues. We are grateful for their support and contributions to the international flavor of the book. The common goal is to improve health knowledge.

For all help we are eternally grateful. The final responsibility is, of course, with the authors.

TH Tulchinsky, EA Varavikova
15 January 2014

Chapter 1

A History of Public Health

Learning Objectives

Upon completion of this chapter, the student should be able to:

1. Identify major historical trends and concepts of public health, and their relationship to the individual and the community;
2. Address health issues within a historical perspective;
3. Apply experience from the past to address present and new health problems.

INTRODUCTION

The history of public health is derived from many historical ideas, trial and error, the development of basic sciences, technology, and epidemiology. In the modern era, James Lind's clinical trial of various dietary treatments of British sailors with scurvy in 1756 and Edward Jenner's 1796 discovery that cowpox vaccination prevents smallpox have modern-day applications as the science and practices of nutrition and immunization are crucial influences on health among the populations of developing and developed countries.

History provides a perspective to develop an understanding of health problems of communities and how to cope with them. We visualize through the eyes of the past how societies conceptualized and dealt with disease. All societies must face the realities of disease and death, and develop concepts and methods to manage them. These strategies evolved from scientific knowledge and trial and error, but are associated with cultural and societal conditions, beliefs and practices that are important in determining health status and curative and preventive interventions to improve health.

The history of public health is a story of the search for effective means of securing health and preventing disease in the population. Epidemic and endemic infectious disease stimulated thought and innovation in disease prevention on a pragmatic basis, often before the causation was established scientifically. The prevention of disease in populations revolves around defining diseases, measuring their occurrence, and seeking effective interventions.

Public health evolved through trial and error and with expanding scientific medical knowledge, at times controversial, often stimulated by war and natural disasters. The need for organized health protection grew as part of the development of community life, and in particular, urbanization and social reforms. Religious and societal beliefs influenced approaches to explaining and attempting to control communicable disease by sanitation, town planning, and provision of medical care. Religions and social systems have also viewed scientific investigation and the spread of knowledge as threatening, resulting in inhibition of developments in public health, with modern examples of opposition to birth control, immunization, and food fortification.

Scientific controversies, such as the contagionist and anticontagionist disputations during the nineteenth century and opposition to social reform movements, were ferocious and resulted in long delays in adoption of the available scientific knowledge. Such debates continued into the twentieth and still continue into the twenty-first century with a melding of methodologies proven to be interactive incorporating the social sciences, health promotion, and translational sciences bringing the best available evidence of science and practice together for greater effectiveness in policy development for individual and population health practices.

Modern society in high, medium and low income countries still faces the ancient scourges of communicable diseases, but also the modern pandemics of cardiovascular disease, cancers, mental illness, and trauma. The emergence of acquired immunodeficiency syndrome (AIDS), severe acute respiratory syndrome (SARS), avian influenza, and drug-resistant microorganisms forces us to seek new ways of preventing their potentially serious consequences to society. Threats to health in a world facing severe climate and ecological change pose harsh and potentially devastating consequences for society.

The evolution of public health is a continuing process; pathogens change, as do the environment and the host. In order to face the challenges ahead, it is important to have an understanding of the past. Although there is much in this age that is new, many of the current debates and arguments in public health are echoes of the past. Experience from the

The New Public Health. http://dx.doi.org/10.1016/B978-0-12-415766-8.00001-X

past is a vital tool in the formulation of health policy. An understanding of the evolution and context of those challenges and innovative ideas can help us to navigate the public health world of today and the future.

PREHISTORIC SOCIETIES

The Paleolithic Age is the earliest stage of human development where organized societal structures are known to have existed. These social structures consisted of people living in bands which survived by hunting and gathering food. There is evidence of the use of fire going back some 230,000 years, and increasing sophistication of stone tools, jewelry, cave paintings, and religious symbols during this period. Modern humans evolved from *Homo sapiens*, probably originating in Africa and the Middle East about 90,000 years ago, and appearing in Europe during the Ice Age period (40,000–35,000 BCE). During this time, humanity spread over all major land masses following the retreating glaciers of the last Ice Age at 11,000–8000 BCE.

A Mesolithic Age or transitional phase of evolution from hunter–gatherer societies into the Neolithic Age of food-raising societies occurred during different periods in various parts of the world, first in the Middle East from 9000 to 8000 BCE onward, reaching Europe about 3000 BCE. The change from hunting, fishing, and gathering modes of survival to agriculture was first evidenced by domestication of animals and then the growing of grain and root crops, and vegetables. Associated skills, including food storage and cooking, pottery, basket weaving, ovens, smelting, and trade, led to improved survival techniques and population growth gradually spread throughout the world.

Communal habitation became essential to adaptation to changing environmental conditions and hazards, allowing population growth and geographic expansion. At each stage of human biological, technological, and social evolution, humans coexisted with diseases associated with the environment and living patterns, seeking herbal and mystical treatments for the maladies. People called on the supernatural and magic to appease these forces and prevent plagues, famines, and disasters. Shamans or witch doctors attempted to remove harm by magical or religious practices along with herbal treatments acquired through trial and error. Life expectancy in prehistoric times was 25–30 years, with men living longer than women, probably due to malnutrition and maternity-related causes.

As human society evolved, technologically, culturally, and biologically, nutrition and exposure to communicable and non-infectious disease changed. Social organization led to innovations in tools and skills for hunting, clothing, shelter, fire for warmth and cooking, food for use and storage, burial of the dead, and removal of waste products from living areas. Adaptation of human society to the environment has been and remains a central issue in population

health. This is a recurrent theme in the development of public health, with resilience in facing daunting new challenges of adaptation and balance with the environment.

THE ANCIENT WORLD

The development of agriculture served growing populations unable to survive solely on hunting, gathering, crafts and trading, stimulating the organization of more complex societies able to share production and irrigation systems, in response to disease, malnutrition, and stunted growth. Division of labor, trade, commerce, and government was associated with the development of urban societies. Population growth and communal living led to improved standards of living but also created new health hazards, including the spread of diseases. As in our time, these challenges required community action to prevent disease and promote survival.

In the first civilizations, mystical beliefs, divination, and shamanism coexisted with practical knowledge of herbal medicines, midwifery, management of wounds or broken bones, and trepanation to remove "evil spirits". All were part of communal life with variations in historical and cultural development. The advent of writing led to medical documentation. Requirements of medical conduct were spelled out as part of the general legal code of Hammurabi in Mesopotamia (*c.* 1700 BCE). This code included regulation of physician fees, with punishment for treatment failure, which set a legal basis for the subsequent secular practice of medicine. Many of the main traditions of medicine were based on magic or derived from religion. Medical practice was often based on belief in the supernatural, and healers were believed to have a religious calling. Training of medical practitioners, regulation of their practice, and ethical standards evolved in a number of ancient societies. In general, physicians were regulated by specific schools that acted as trade guilds, often with many competing schools based on differing gods, methods, and mystical beliefs.

Some cultures equated cleanliness with godliness and associated hygiene with religious beliefs and practices. Chinese, Egyptian, Hebrew, Indian, and Incan societies all provided sanitary amenities as part of the religious belief system and took measures to provide water, sewerage, and drainage systems. These measures allowed for successful urban settlement and reinforced the beliefs upon which such practices were based. Technical achievements in providing hygiene at the community level slowly coevolved with urban society.

Chinese practice in the twenty-first to eleventh centuries BCE included digging wells for drinking water; from the eleventh to the seventh centuries this included the use of protective measures for drinking water and destruction of rats and rabid animals. In the second century BCE, Chinese communities were using sewers and latrines. The basic concept of health was that of countervailing forces between the principles of yin (female) and yang (male), with an emphasis on a balanced lifestyle. Medical care emphasized diet, herbal medicine, hygiene, massage, and acupuncture.

Ancient cities in India were planned with building codes, street paving, and covered sewer drains built of bricks and mortar. Indian medicine originated in herbalism associated with gods. Between 800 and 200 BCE, Ayurvedic medicine developed, and with it, medical schools and "public hospitals". The ancient Indian way of medical practice, Ayurveda, is the Sanskrit translation of "knowledge of life". Primarily originating in the Indus Valley, the golden age of ancient Indian medicine began approximately 800 BCE. Personal hygiene, sanitation, and water supply engineering were emphasized in the laws of Manu. Pioneering physicians, supported by Buddhist kings, developed the use of drugs and surgery, and established schools of medicine and public hospitals as part of state medicine. Indian medicine played a leading role throughout Asia between 800 BCE and 400 CE, when major texts on medicine and surgery were written. Among the most valued pieces of ancient Indian writings are those created by Sushruta, a surgeon, and Charaka, a physician, both prominent teachers who ran prestigious schools of medicine. These writings contribute to validating ancient India's medical history. According to some historians, the teachings of Sushruta and Charaka were passed along to the Romans and Greeks. Despite these advanced medical teachings, with the Mogul invasion of 600 CE, state support declined, and with it, Indian medicine.

In addition to the ancient Indian medical texts, there is evidence of several ancient Egyptian texts, dating from the years around 1900 BCE. The Kahoun Papyrus, from 1950 BCE, the most ancient scroll, includes three parts: human medicine, veterinary science, and mathematics.

Ancient Egyptian intensive agriculture and irrigation practices were associated with widespread parasitic disease. The cities had stone masonry gutters for drainage, and personal hygiene was highly emphasized. Egyptian medicine developed surgical skills and organization of medical care, including specialization and training that greatly influenced the development of Greek medicine. The Ebers Papyrus, written 3400 years ago, gives an extensive description of Egyptian medical science, including the isolation of infected surgical patients. It is recognized as the most extensive and significant of all the known papyri, given the physiological knowledge uncovered. While the first section of the Ebers Papyrus revolves around divine origin and the strength of magic, the latter portions discuss the treatment of medical conditions including digestive diseases, eye diseases, and skin problems. Fractures and painful limbs are also described. The papyrus includes a treatise on the heart and vessels, standing out as only one of many covering anatomy and physiology. The last portion of this important papyrus focuses on surgery, in particular, tumors and abscesses.

The Hebrew Mosaic law of the five Books of Moses (c. 1000 BCE) stressed prevention of disease through regulation of personal and community hygiene, reproductive and maternal health, isolation of lepers and other "unclean conditions", and family and personal sexual conduct as part of religious practice. It also laid a basis for medical and public health jurisprudence. Personal and community responsibility for health included a mandatory day of rest, limits on slavery and guarantees of the rights of slaves and workers, protection of water supplies, sanitation of communities and camps, waste disposal, and food protection, all codified in detailed religious obligations. Food regulation prevented use of diseased or unclean animals, and prescribed methods of slaughter improved the possibility of preservation of the meat. While there was an element of viewing illness as a punishment for sin, there was also an ethical and social stress on the value of human life with an obligation to seek and provide care. The Talmudic interpretation of biblical law is the concept of sanctity of human life (*Pikuah Nefesh*); the saving of a single human life was considered "as if one saved the whole world", which has been given overriding religious and social roles in community life. A second principle from this source is improving the quality of life on Earth (*Tikkun Olam*). In this tradition, there is an ethical imperative to achieve a better earthly life for all. The Mosaic Law, which forms the basis for Judaism, Christianity, and Islam, codified health behaviors for the individual and for society. These found secular versions in Humanism over recent centuries, which have continued into the modern era as basic concepts in societal values and in practical application in environmental and social hygiene.

In Cretan and Minoan societies, climate and environment were recognized as playing a role in disease causation. Malaria was related to swampy and lowland areas, and prevention involved planning the location of settlements. Ancient Greece placed high emphasis on healthful living habits in terms of personal hygiene, nutrition, physical fitness, and community sanitation. Hippocrates articulated the clinical methods of observation and documentation and a code of ethics of medical practice. He articulated the relationship between disease patterns and the natural environment (air, water, and places), which dominated epidemiological thinking until the nineteenth century. Preservation of health was seen as a balance of forces: exercise and rest, nutrition and excretion, and recognizing the importance of age and sex variables in health needs. Disease was seen as having inevitable natural causation, and medical care was valued, with the city-states providing free medical services for the poor and for slaves. City officials were appointed to look after public drains and water supply, providing organized sanitary and public health services. Hippocrates gave medicine a rudimentary, scientific, and ethical spirit which lasts to the present time.

Ancient Rome adopted much of the Greek philosophy and experience concerning health matters, with high levels of achievement and new innovations in the development of public health. The Romans were extremely skilled in engineering of water supply, sewerage and drainage systems, public baths and latrines, town planning, sanitation of military encampments, and medical care. Roman law also regulated businesses and medical practice. The influence of the Roman Empire resulted in the transfer of these ideas throughout much of Europe and the Middle East. Rome itself had access to clean water via 10 aqueducts supplying ample water for the citizens. Rome also built public drains. By the early first century BCE, the aqueducts made available 600–900 liters per person per day of household water from mountains. Marshlands were drained to reduce endemic malaria. Public baths were built to serve the poor, and fountains were built in private homes for the wealthy. Streets were paved, and organized garbage disposal served the cities.

Roman military medicine included well-designed sanitation systems, food supplies, and surgical services. Roman medicine, based on mystical beliefs and religious rites, with slaves as physicians, developed partly from Greek physicians who brought their skills and knowledge to Rome after the destruction of Corinth in 146 BCE. Training as apprentices, Roman physicians achieved a highly respected role in society. Hospitals and municipal doctors were employed by Roman cities to provide free care to the poor and the slaves, but physicians also engaged in private practice, mostly on retainers to families. Occupational health was described with measures to reduce known risks such as lead exposure, particularly in mining. Commercial weights and measures were standardized and supervised. Rome made important contributions to the public health tradition of sanitation, urban planning, and organized medical care. Galen, Rome's leading physician, perpetuated the fame of Hippocrates through his medical writings, basing medical assessment on the four humors (sanguine, phlegmatic, choleric, and melancholic). These ideas dominated European medical thought for nearly 1500 years until the advent of modern science.

THE EARLY MEDIEVAL PERIOD (FIFTH TO TENTH CENTURIES CE)

The Roman Empire disappeared as an organized entity following the sacking of Rome in the fifth century CE. The eastern empire survived in Constantinople, with a highly centralized government. Later conquered by the Muslims, it provided continuity for Greek and Roman teachings in health. The western empire integrated Christian and pagan cultures, which viewed disease as punishment for sin. Possession by the devil and witchcraft were accepted as causes of disease. Prayer, penitence, and exorcising witches were accepted means of dealing with health problems. The ensuing period of history was dominated in health, as in all other spheres of human life, by the Christian doctrine institutionalized by the Church. The secular political structure was dominated by feudalism and serfdom, associated with a strong military landowning class in Europe.

Church interpretation of disease was related to original or acquired sin. Humanity's destiny was to suffer on Earth and hope for a better life in heaven. The appropriate intervention in this philosophy was to provide comfort and care through the charity of church institutions. The idea of prevention was seen as interfering with the will of God. Monasteries with well-developed sanitary facilities were located on major travel routes and provided hospices for travelers. The monasteries were the sole centers of learning and for medical care. They emphasized the tradition of care of the sick and the poor as a charitable duty of the righteous and initiated hospitals. These institutions provided care and support for the poor, and made efforts to cope with epidemic and endemic disease.

Most physicians were monks guided by Church doctrine and ethics. Medical scholarship was based primarily on the teachings of Galen (131 CE), sustained in Muslim centers of medical learning and later brought to Europe with the return of the Crusades, whose teachings provided the basis of medical teaching until the fifteenth century. Education and knowledge were under clerical dominance. Scholasticism, or the study of what was already written, stultified the development of descriptive or experimental science. The largely rural population of the European medieval world lived with poor nutrition, education, housing, sanitary, and hygienic conditions. Endemic and epidemic diseases resulted in high infant, child, and adult mortality. Commonly, 75 percent of newborns died before the age of five. Maternal mortality was high. Leprosy, malaria, measles, and smallpox were established endemic diseases, along with many other less well-documented infectious diseases.

Between the seventh and tenth centuries, outside the area of Church domination, Muslim medicine flourished under Islamic rule primarily in Persia, Central Asia and later Baghdad and Cairo. Famous physicians, including the Persian Rhazes (850–c. 932) and the outstanding Islamic Bukhara-born philosopher and physician Ibn Sinna (Avicenna, 980–1037), translated and adapted ancient Greek and Mosaic teachings, adding clinical skills developed in medical academies and hospitals. Piped water supplies were documented in Cairo in the ninth century. Great medical academies were established, including one in Muslim-conquered Spain at Cordova. The Cordova Medical Academy was a principal center for medical knowledge and scholarship prior to the expulsion of Muslims and Jews from Spain in 1492, and the Inquisition. The Academy helped to stimulate European medical thinking and the beginnings of western medical science in anatomy, physiology, and descriptive clinical medicine.

THE LATE MEDIEVAL PERIOD (ELEVENTH TO FIFTEENTH CENTURIES)

In the later feudal period, ancient Hebraic and Greco-Roman concepts of health were preserved and flourished in the Muslim Empire. The twelfth-century Jewish rabbi–philosopher–physician Moses Maimonides (Rambam), who trained in Cordova and was expelled to Cairo, helped to synthesize Roman, Greek, and Arabic medicine with Mosaic concepts of communicable disease isolation and sanitation.

Monastery hospitals were established between the eighth and twelfth centuries to provide charity and care to ease the suffering of the sick and dying. Monasteries provided centers of literacy, medical care, and the ethic of caring for the sick patient as an act of charity. The monastery hospitals (described in eleventh-century Russia) were gradually supplanted by municipal, voluntary, and guild hospitals developed in the twelfth to sixteenth centuries. By the fifteenth century, Britain had 750 hospitals. Medical care insurance was provided by guilds to its members and their families. Hospitals employed doctors, and the wealthy had access to private doctors.

In the early Middle Ages, most physicians in Europe were monks, and the medical literature was compiled from ancient sources. In 1131 and 1215, Papal rulings increasingly restricted clerics from doing medical work, thus promoting secular medical practice. In 1224, Emperor Frederick II of Sicily published decrees regulating medical practice, establishing licensing requirements: medical training (3 years of philosophy, 5 years of medicine), 1 year of supervised practice, then examination followed by licensure. Similar ordinances were published in Spain in 1238 and in Germany in 1347.

The Crusades (1096–1270 CE) exposed Europe to Arabic medical concepts, as well as leprosy. The Hospitallers, a religious order of knights, developed hospitals in Rhodes, Malta, and London to serve returning pilgrims and crusaders. The Muslim world had hospitals, such as Al Mansour in Cairo, available to all as a service provided by the governate. Increasing contact between the Crusaders and the Muslims through war, conquest, cohabitation, and trade introduced Arabic culture and diseases, and revised ancient knowledge of medicine and hygiene.

Leprosy became a widespread disease in Europe, particularly among the poor, during the early Middle Ages, but the problem was severely accentuated during and following the Crusades, reaching a peak during the thirteenth to fourteenth centuries. Isolation in leprosaria was common in Europe. In France alone, there were 2000 leprosaria in the fourteenth century. This disease has caused massive suffering and although leprosy still exists in tropical countries it is gradually disappearing globally. The development of modern antimicrobials has cured millions of leprosy (Hansen's disease) cases; with early case finding and multidrug therapy, this disease and its disabling and deadly effects are now largely a matter of history.

As rural serfdom and feudalism declined in Western Europe, cities developed with crowded and unsanitary conditions. Towns and cities were allowed to develop in Europe with royal charters for self-government, primarily located at the sites of former Roman settlements and at river crossings related to trade routes. The Church provided stability in society, but repressed new ideas and imposed its authority particularly via the Inquisition. Established by Pope Gregory in 1231, the Inquisition was renewed and intensified, especially in Spain in 1478 by Pope Sixtus IV, to exterminate heretics, Jews, and anyone seen as a challenge to the accepted Papal dogmas.

Universities established under royal charters in Paris, Bologna, Padua, Naples, Oxford, Cambridge, and others provided a haven for scholarship outside the realm of the Church. In the twelfth and thirteenth centuries there was a burst of creativity in Europe, with inventions including the compass, the mechanical clock, and the loom, with a surge in use of the waterwheel and the windmill. Physical and intellectual exploration opened up with the travels of Marco Polo and the writings of Thomas Aquinas, Roger Bacon, and Dante. Trade, commerce, and travel flourished.

Medical education was widespread in institutions of higher education in many parts of Muslim and other societies. Medical schools in Europe evolved in Salerno, Italy, in the tenth century and in universities throughout Europe in the eleventh to fifteenth centuries: in Paris (1110), Bologna (1158), Oxford (1167), Montpellier (1181), Cambridge (1209), Padua (1222), Toulouse (1233), Seville (1254), Prague (1348), Krakow (1364), Vienna (1365), Heidelberg (1386), Glasgow (1451), Basel (1460), and Copenhagen (1478). By the end of the fifteenth century there were around 80 universities in Europe. Printed books opened a new potential for secular as well as religious education. Physicians, recruited from the new middle class, were trained in scholastic traditions based on translations of Arabic literature and the ancient Roman and Greek texts, mainly Aristotle, Hippocrates, and Galen, but with some more current texts, mainly written by Arab and Jewish physicians.

Growth exacerbated public health problems in the newly walled commercial and industrial towns, leading to eventual emergencies which demanded solutions. Rapidly growing medieval towns lacked systems of sewers or water pipes. Garbage and human waste were thrown into the streets. Houses were made of wood, mud, and dung. Rats, lice, and fleas flourished in the rushes or straw used on the clay floors of people's houses.

Crowding, poor nutrition and sanitation, lack of adequate water sources and drainage, unpaved streets, keeping of animals in towns, and lack of organized waste disposal created conditions for widespread infectious diseases.

Municipalities developed protected water sites (cisterns, wells, and springs) and public fountains with municipal regulation and supervision. Piped community water supplies were developed in Dublin, Basel, and Bruges (Belgium) in the thirteenth century. Between the eleventh and fifteenth centuries, Novgorod in Russia used clay and wooden pipes for water supplies, and municipal bath houses were available.

Medical care was still largely oriented towards symptom relief, with few curative resources to draw upon. Traditional folk medicine survived especially in rural areas, but was suppressed by the Church as witchcraft. Physicians provided services for those able to pay, but medical knowledge was a mix of pragmatism and mysticism, and there was a sheer lack of scientific knowledge. Conditions were ripe for vast epidemics of smallpox, cholera, measles, and other epidemic diseases, fanned by the debased conditions of life and chronic banditry, warfare, and famines raging throughout Europe, such as during the English invasion of France during the Hundred Years War (1337–1453).

The Black Death, mainly pneumonic and bubonic plague due to *Yersinia pestis* infection transmitted by fleas on rodents, was brought from the steppes of Central Asia to Europe with the Mongol invasions, and then transmitted via extensive trade routes throughout Europe by sea and overland. The Black Death was also introduced to China with Mongol invasions, bringing tremendous mortality, halving the population of China between 1200 and 1400 CE. Between the eleventh and thirteenth centuries, during the Mongol–Tatar conquests, many widespread epidemics, including plague, were recorded in Rus (now Russia). The plagues traveled rapidly with armies and caravan traders, and later by ship as world trade expanded in the fourteenth to fifteenth centuries (Box 1.1). The plague ravaged most of Europe between 1346 and 1350, killing between 24 and 50 million people, approximately one-third of the population, and leaving vast areas of Europe sparsely populated. Despite local efforts to prevent disease by quarantine and isolation of the sick, the disease devastated whole communities.

Fear of a new and deadly disease, lack of knowledge, speculation, and rumor led to countermeasures which often exacerbated the spread of epidemics (as seen in the last decades of the twentieth century, and in the twenty-first century, with SARS and pandemic H1N1 influenza). In Western Europe, public and religious ceremonies and burials were promoted by religious and civil authorities, which increased contact with infected people. The misconception that cats were the cause of plague led to their slaughter; however, they could have helped to stem the tide of disease brought by rats and their fleas to humans. Hygienic practices limited the spread of plague in Jewish ghettos, leading to the Jews being blamed for the plague's spread, and widespread massacres, especially in Germany and Central Europe.

Seaport cities in the fourteenth century began to apply the biblical injunction to separate lepers by keeping ships coming from places with the plague waiting in remote parts of the harbor, initially for 30 days (*treutina*), then for 40 days (*quarantina*) (Ragusa in 1465, and Venice in 1485), establishing the public health act of quarantine as a government measure, which on a pragmatic basis was found to reduce the chance of entry of the plague. Towns along major overland trading routes in Russia took measures to restrict movement in homes, streets, and entire towns during epidemics. In sixteenth-century Russia, Novgorod banned public funerals during plague epidemics, and in the seventeenth century, Czar Boris Godunov banned trade, prohibited religious and other ceremonies, and instituted quarantine-type measures. All over Europe, municipal efforts to enforce isolation broke down as crowds gathered and were uncontrolled by inadequate police forces and public health. In 1630, all officers of the Board of Health of Florence, Italy, were excommunicated because of efforts to prevent spread of the contagion by isolation of cases, thereby interfering with religious ceremonies to assuage God's wrath through appeals to divine providence.

The plague continued to strike, with epidemics in London in 1665, Marseille in 1720, Moscow in 1771, and Russia, India, and the Middle East through the nineteenth

---

BOX 1.1 "This is the End of the World": The Black Death

"Rumors of a terrible plague supposedly arising in China and spreading through Tartary (Central Asia) to India and Persia, Mesopotamia, Syria, Egypt and all of Asia Minor had reached Europe in 1346. They told of a death toll so devastating that all of India was said to be depopulated, whole territories covered by dead bodies, other areas with no one left alive. As added up by Pope Clement VI at Avignon, the total of reported dead reached 23,840,000. In the absence of a concept of contagion, no serious alarm was felt in Europe until the trading ships brought their black burden of pestilence into Messina while other infected ships from the Levant carried it to Genoa and Venice. By January 1348 it penetrated France via Marseille, and North Africa via Tunis. Ship-borne along coasts and navigable rivers, it spread westward from Marseille through the ports of Languedoc to Spain and northward up the Rhone to Avignon, where it arrived in March. It reached Narbonne, Montpellier, Carcassone, and Toulouse between February and May, and at the same time in Italy spread to Rome and Florence and their hinterlands. Between June and August it reached Bordeaux, Lyon, and Paris, spread to Burgundy and Normandy into southern England. From Italy during the summer it crossed the Alps into Switzerland and reached eastward to Hungary. In a given area the plague accomplished its kill within four to six months and then faded, except in the larger cities, where, rooting into the close-quartered population, it abated during the winter, only to appear in spring and rage for another six months."

Source: *Tuchman BW. A distant mirror: the calamitous fourteenth century. New York: Alfred A. Knopf; 1978.*

century. Furthermore, the plague continued into the twentieth century with epidemics in Australia (1900), China (1911), Egypt (1940), and India (1995). (See *The Plague*, a historical novel by Albert Camus.) The disease is endemic in rodents in many parts of the world, including the USA; however, modern sanitation, pest control, and antibiotic treatment have greatly reduced the potential for large-scale plague epidemics.

Guilds organized to protect the economic interests of traders and skilled craftsmen, and limited competition by regulating training and entry requirements of new members. They also placed high priority on mutual benefit funds to provide financial assistance and other benefits for illness, death, widows and orphans, and medical care, as well as burial benefits for members and their families. The guilds wielded strong political power during the late Middle Ages. These brotherhoods provided a tradition later expressed in the mutual benefit or friendly societies, sick funds, and insurance for health care based on employment groups. This tradition has continued in western countries, where labor unions are among the leading advocates for the health of workers and their families.

The fourteenth century saw a devastation of the population of Europe by plague, wars, and the breakdown of feudal society. It also set the stage for the agricultural revolution and later the industrial revolution. The period following the Black Death was innovative and dynamic. Shortages of farm laborers led to innovations in agriculture. Enclosures of common grazing land reduced the spread of disease among animals, increased field crop productivity, and improved sheep farming, leading to the development of the wool and textile industries and the search for energy sources, industrialization, and international markets.

THE RENAISSANCE (1400–1600s)

Commerce, industry, trade, merchant fleets, and voyages of discovery to seek new markets led to the development of a moneyed middle class and wealthy cities. During this period, mines, foundries, and industrial plants flourished, creating new goods and wealth. Partly as a result of the trade generated and the increased movement of goods and people, vast epidemics of syphilis, typhus, smallpox, measles, and the plague continued to spread across Europe. Malaria was still widespread throughout Europe. Rickets, scarlet fever, and scurvy, particularly among sailors, were rampant. Pollution and crowding in industrial areas resulted in centuries-long epidemics of environmental disease, particularly among the urban working class.

A virulent form of syphilis, allegedly brought back from America by the crews of Columbus, spread rapidly throughout Europe between 1495 and 1503, when it was first described by Girolamo Fracastoro (1478–1553). Control measures tried in various cities included examination and registration of prostitutes, closure of communal bath houses, isolation in special hospitals, reporting of disease, and expulsion of sick prostitutes or strangers. The disease gradually decreased in virulence, but it lingers as a diminishing public health problem to the present time.

The Ottoman conquest of Constantinople in 1453 resulted in the westward movement of many Greek thinkers and the end of the Hundred Years War brought stability to north-west Europe. In Europe, the growth of cities with commerce and industrialization and the massive influx of the rural poor brought the focus of public health needs to the doorsteps of municipal governments. The breakdown of feudalism, the decline of the monasteries, and the land enclosures dispossessed the rural poor. Municipal and voluntary organizations increasingly developed hospitals, replacing those previously run by monastic orders. In 1601, the British Elizabethan Poor Laws defined the local parish government as being responsible for the health and social well-being of the poor, a system later brought to the New World by British colonists. Municipal control of sanitation was weak. Each citizen was in theory held responsible for cleaning his part of the street, but hygienic standards were low, with animal and human waste freely accumulating.

During the Renaissance, the sciences of anatomy, physiology, chemistry, microscopy, and clinical medicine opened medicine to a scientific base. Medical schools in universities developed affiliations with hospitals, promoting clinical observation with increasing precision in the description of disease. The contagion theory of disease, described in 1546 by Fracastorus and later the German–Swiss physician Paracelsus (Phillipus von Hohenheim, 1493–1541), including the terms *infection* and *disinfection*, was contrary to the until-then sacrosanct miasma teachings of Galen.

In Russia, Czar Ivan IV (Ivan the Terrible) (1530–1584) in the sixteenth century arranged to hire a court physician of Queen Elizabeth I, who brought with him to Moscow a group of physicians and pharmacists to serve the court. The Russian army had a tradition of regimental doctors. In the mid-seventeenth century, the czarist administration developed pharmacies in major centers throughout the country for military and civilian needs, and established a State Pharmacy Department to control pharmacies and medications, education of doctors, military medicine, quarantine, forensic medicine, and medical libraries. Government revenues from manufacturing, sale, and promotion of vodka provided for these services. Preparation of military doctors (*Lekars*) with 5–7 years of training was instituted in 1654. Hospitals were mainly provided by monasteries, serving both civilian and military needs. In 1682, the first civic hospital was opened in Moscow, and in the same year, two hospitals were opened, also in Moscow, by the central government for the care of patients and training of *Lekars*.

From 1538, parish registers of christenings and burials were published in England as weekly and annual abstracts, known as the *Bills of Mortality*. Beginning in 1629, national annual Bills of Mortality included tabulation of death by cause. On the basis of the Bills of Mortality, novelist Daniel Defoe described the plague epidemic of London of 1665 over 60 years later (*A Journal of the Plague Year*, Daniel Defoe, 1722).

In England in 1662, John Graunt published *Natural and Political Observations Upon the Bills of Mortality*. He compiled and interpreted mortality figures by inductive reasoning, demonstrating the regularity of certain social and vital phenomena. He showed statistical relationships between mortality and living conditions. Graunt's work was important because it was the first instance of statistical analysis of mortality data, providing a foundation for the use of health statistics in the planning of health services. It established the sciences of demography and vital statistics and methods of analysis, providing basic measurements for health status evaluation with mortality rates by age, sex, and location. Also in 1662, William Petty took the first census in Ireland. In addition, he studied statistics on the supply of physicians and hospitals.

ENLIGHTENMENT, SCIENCE, AND REVOLUTION (1600s–1800s)

The Enlightenment, a dynamic period of social, economic, and political thought, provided great impetus for political and social emancipation and rapid advances in science and agriculture, technology, and industrial power. Changes in many spheres of life were exemplified by the American and French Revolutions, along with the economic theory of Adam Smith (author of *The Wealth of Nations*), which developed the political and economic rights of the individual.

In this influential and notable era, it became evident that advanced ideas and new ways of thinking could materialize into practical, tangible objects. This is exemplified by the development of microscopy, invented in 1676, as a tool that provided a method for the study of microorganisms (Box 1.2).

In the seventeenth century, the great medical centers were located in Leyden, Paris, and Montpelier. Bernardino Ramazzini published the first modern comprehensive treatise on occupational diseases in 1700.

In Russia, Peter the Great (1682–1725) initiated political, cultural, and health reforms. He sent young aristocrats to study sciences and technology, including medicine, in Western Europe. He established the first hospital-based medical school in St. Petersburg and subsequently in other centers as well, mainly to train military doctors. He established the Anatomical Museum of the Imperial Academy of Sciences in St. Petersburg in 1717, and initiated a census of males for military service in 1722. In 1724, V. N. Tatishev carried out a survey by questionnaire of all regions of the Russian empire regarding epidemic disease and methods of treatment.

BOX 1.2 The Invention of the Microscope

Of the many important medical and scientific discoveries, the creation of the microscope provides a crucial tool for the development of modern science applied to biological and medical progress. It has influenced the way in which scientists study, identify, diagnose, treat, and prevent diseases that have so greatly plagued and limited human life in the past.

The first compound microscope was created by Zacharias Janson and his father, Dutch spectacle-makers who experimented with lenses in 1595. They placed lenses in a tube, and noted that the object examined looked substantially enlarged. Robert Hooke (1635–1703) in England and Jan Swammerdam in the Netherlands built compound microscopes and made important discoveries with them. Hooke's book *Micrographia*, published in 1665, showed his compound microscope and illumination system, one of the best such microscopes of his time, and demonstrated at the Royal Society's meetings, with observations of insects, sponges, plant cells, fossils, and bird feathers.

However, credit for invention of the microscope and its medical use is given to Anton van Leeuwenhoek (1632–1723), a Dutch scientist and draper, who attained great success by creating efficient, better functioning lenses. His skill in grinding and polishing lenses provided remarkably high magnifying power. He was the first to see and describe bacteria (1674), yeast plants, the teeming life in a drop of water, and the circulation of blood corpuscles in capillaries. In 1678, after Leeuwenhoek had written to the Royal Society with a report of discovering "little animals" – bacteria and protozoa – Hooke was asked by the Society to confirm Leeuwenhoek's findings. He did so, paving the way for the wide acceptance of Leeuwenhoek's discoveries.

The initial scientific discoveries founded upon microscopy pertained to the circulating blood, microbiological organisms, and tissue cellular structure. As models of microscopes advanced, new capabilities were made possible so that more minute samples could be investigated for vital discoveries throughout the microbiological revolution. Subsequently, cellular structure opened up for scientific research with further advances such as the electron microscope.

Observations and new discoveries made through use of a microscope have shaped how we view disease, cellular processes, microorganisms, and the building blocks of life. From scientists investigating nerve cell function, to Koch studying bacilli responsible for tuberculosis infection, and Pasteur observing microbes as foreign organisms, the microscope has provided one of the key technological contributions to medical and health sciences of all time.

Sources: *Nobelprize.org. From thrilling toy to important tool [updated 2012]. Stockholm: Nobel Media. Available at: http://www.nobelprize. org/educational/physics/microscopes/discoveries/ [Accessed 10 August 2012].*
History of the microscope. UK: History of the Microscope [updated 2012]. Available at: http://www.history-of-the-microscope.org/terms.php [Accessed 10 August 2012].

Improvements in agriculture created greater productivity and better nutrition. These were associated with higher birth rates and falling death rates, leading to rapid population growth. The agricultural revolution during the sixteenth and seventeenth centuries, based on mechanization and larger land units of production with less labor, was associated with rural depopulation and provided excess workers to staff the factories, mines, ships, home construction, and shops of the industrial revolution. Other significant achievements of the agricultural revolution included expanding commerce and nourishing a growing middle class. Exploration and colonization provided the expansion of markets that fueled the industrial revolution, and stimulated the growth of science, technology, and wealth.

Colonization also contributed to the agricultural revolution through the introduction of new crops from the Americas, including the potato, the tomato, peppers, and maize. Thus, in addition to the new crops, animal husbandry, improved land use, and farm machinery all contributed to a general improvement in food security and nutrition. This was supplemented by increasing availability of cod from the Grand Banks of the Atlantic, adding protein to the common diet.

Industrialized urban centers grew rapidly. Crowded cities were ill-equipped to house and provide services for the growing working class. Urban areas suffered from crowding, poor housing, sanitation, poor nutrition, and harsh working conditions, which together produced appalling health conditions. During this period, documentation and statistical analysis developed in various forms, becoming the basis for social sciences including demography and epidemiology. Intellectual movements of the eighteenth century defined the rights of man and gave rise to revolutionary movements to promote liberty and release from tyrannical rule, as in the American and French Revolutions of 1775 and 1789, respectively. Following the final defeat of Napoleon at Waterloo in 1815, conservative governments were faced with strong middle-class movements for reform of social conditions, with important implications for health.

Eighteenth-Century Reforms

The period of enlightenment and reason was led by philosophers John Locke, Diderot, Voltaire, Rousseau, and others. These men produced a new approach to science and knowledge derived from observations and systematic testing and philosophical debate of ideas as opposed to instinctive or innate knowledge as the basis for human progress. The newness of the enlightenment was the idea of *progress*, *Sapere Aude* [dare to know and "have courage to use your own understanding!"], as the motto of enlightenment. The idea of the rights of man contributed to the American and French Revolutions, but also to a widening belief that society was obliged to serve all rather than just the privileged.

This had a profound impact on approaches to health and societal issues.

The late eighteenth century was a period of growth and development of clinical medicine, surgery, and therapeutics, as well as of the sciences of chemistry, physics, physiology, and anatomy. From the 1750s onward, voluntary hospitals were established in major urban centers in Britain, America, and Eurasia. Medical–social reform involving hospitals, prisons, and lazarettos (leprosy hospitals) in Britain, led by John Howard (*On the State of Prisons,* published in 1777), produced substantive improvements in these institutions. Following the French Revolution, Philippe Pinel (1745–1826) was instrumental in the development of a more humane psychological approach to the custody and care of psychiatric patients. He fostered reform of insane asylums by removing the chains from patients at the Bicetre Mental Hospital and later Hospice de la Salpêtrière near Paris. Pinel made notable contributions to the classification of mental disorders, and he is often identified as the "father of modern psychiatry". Reforms in this field were also carried out in Britain by the Society of Friends (the Quakers), who built the York Retreat, providing humane care as an alternative to the inhuman conditions of the York Asylum.

In 1700, Bernardino Ramazzini (1633–1714) published a monumental piece on occupational diseases (*Diseases of Workers*), applying epidemiological principles and highlighting specific health hazards. These occupational risks included exposure to chemicals, dust, and metals, as well as musculoskeletal injury from unnatural postures and repetitive or violent motions. In his publication, he described other various disease-causative agents encountered by workers in 52 major occupations. Considered to be the "father of occupational medicine", Ramazzini established the basis for this field, although progress in the application of his views was slow. Despite the reluctance to apply his beliefs, the latter part of the century fostered interest in the health of sailors and soldiers, which led to important developments in military and naval medicine. Studies of prevalent diseases were carried out by pioneering physicians among workers in various trades, such as metalworkers, bakers, shoemakers, and hatmakers. Deeper understanding of these trades and the risks involved allowed for the identification of causative agents, and thus methods of prevention. The observational studies of Percivall Pott (1714–1789) identified scrotal cancer as an occupational hazard of chimney sweeps (1775). In 1767, George Baker (1722–1809) studied Devonshire colic, acquired from lead poisoning in cider production. Each of these and other similar studies helped to lay the basis for the development of investigative epidemiology.

Pioneers and supporting movements successfully agitated for reform in Britain through the parliamentary system. The anti-gin movement, aided by the popular newspapers (the "Penny Press") and the brilliant engravings of William Hogarth (1697–1764), helped to bring about legal, social, and police reforms in English townships. The reform spirit also produced an effective antislavery movement led by Protestant Christian churches, which goaded the British government to ban slavery in 1797 and the slave trade in 1807. This was achieved using the Royal Navy to sweep the slave trade from the seas during the early part of the nineteenth century.

Applied Epidemiology

Scurvy (the Black Death of the Sea) was a major health problem among sailors during long voyages. In 1498, Vasco da Gama (1460s–1524) lost 55 crewmen to scurvy during his voyages. Moreover, in 1535, Jacques Cartier's (1491–1557) crew suffered severely from scurvy on his voyage of discovery to Canada. During the sixteenth century, Dutch sailors knew of the value of fresh vegetables and citrus fruit in preventing scurvy.

Samuel Purchas (1577–1626) in 1601 and John Woodall (a British naval doctor, 1570–1643) in 1617 recommended the use of lemons and oranges in the treatment of scurvy, but this was not widely practiced. During the seventeenth to eighteenth centuries, Russian military practice included antiscorbutic preparations, and the use of sauerkraut for this purpose became common in European armies. Scurvy was a major cause of sickness and death among sailors when supplies of fruit and vegetables ran out, thus significantly limiting long voyages and contributing to frequent mutinies at sea.

Conditions for sailors in the British navy improved following the explorations of Captain James Cook during the period 1766–1779. As mentioned in Box 1.3, a British naval squadron of seven ships and nearly 2000 men led by Commodore George Anson left Plymouth to circumnavigate the globe in 1740–1744. The squadron returned to England comprised of only one ship and 145 men, after losing the majority of the crews to scurvy. In 1747, James Lind carried out his pioneering epidemiological investigation on scurvy among sailors on long voyages. His work led to the adoption of lemon or lime juice as a routine nutrition supplement for British sailors some 50 years later. Vitamins were not understood or isolated until almost 150 years later; however, Lind's scientific technique of hypothesis formulation, study design, careful observation and testing, followed by documentation and publication, was exceptional and monumental. This established the investigation of nutrition in public health in what is now recognized as the first clinical trial and epidemiological investigation.

This discovery was adopted by progressive sea captains and aided Captain Cook in his voyages of discovery in the South Pacific in 1768–1771. In 1795, the Royal Navy adopted routine issuance of lime juice to sailors to prevent scurvy. This measure effectively doubled the fighting strength of the Royal Navy by extending the capacity to remain longer at sea with a healthy crew. This was crucial

---

BOX 1.3 James Lind and Scurvy, 1747

Captain James Lind (1716–1794), a physician serving in Britain's Royal Navy, developed a hypothesis explaining the cause of scurvy, founded upon clinical observations in what is currently regarded as the first clinical epidemiological study. It was the tragedy of Admiral Anson's expedition of circumnavigation, with the deaths of 380 men out of a crew of 510 on one of his ships, which led to Lind's interest in investigating scurvy.

In May 1747, on HMS *Salisbury*, Lind conducted his study by treating 12 sailors who had fallen sick to scurvy. He gave each sailor one of six different dietary regimens. The two sailors who were fed oranges and lemons recovered from their illness and were fit for duty within 6 days. This is in contrast to all of the other sailors, who were given different treatments, and consequently, remained sick. Lind concluded that citrus fruits would treat and prevent scurvy. In 1753, he published *A treatise of the scurvy: in three parts. Containing an inquiry into the nature, causes and cure of that disease together with a critical and chronological view of what has been published on the subject.*

Lind reported: "Scurvy began to rage after being a month or six weeks at sea … the water on board … was uncommonly sweet and good [and] provisions such as could afford no suspicion … yet, at the expiration of ten weeks, we brought into Plymouth 80 men, out of a complement of 350, more or less afflicted with the diseases". Captain Lind observed during his experiment that: "the most sudden and visible good effects were perceived from the use of oranges and lemons", and that in a short time this group was fit for duty, whereas all the other groups remained ill. He concluded that: "experience indeed sufficiently shows that as green or fresh vegetables with ripe fruit were the best remedies for it [i.e., scurvy], so they prove the most effectual preservatives against it", and that oranges are "the most effectual preservatives against the distemper".

Scurvy was eliminated in the Royal Navy by the end of the eighteenth century, but continued to plague merchant seaman during most of the nineteenth century until compulsory lime juice was imposed and steam ships led to shortened voyage times.

Sources: *Carpenter KJ. The history of scurvy and vitamin C. Cambridge: Cambridge University Press; 1986.*
Rosen G. A history of public health. Expanded edition. Baltimore, MD: Johns Hopkins University Press; 1993.
Trohler U. James Lind and scurvy, 1747–1795. Bern: James Lind Library. Available at: http://www.jameslindlibrary.org/illustrating/articles/james-lind-and-scurvy-1747-to-1795 [Accessed 10 August 2012].
Cook GC. Scurvy in the British mercantile marine in the 19th century, and the contribution of the Seamen's Hospital Society. Postgrad Med J 2004;80:224–9. Available at: http://pmj.bmj.com/content/80/942/224. long [Accessed 10 August 2012].
Bartholomew M. James Lind's Treatise of the scurvy (1753). Postgrad Med J 2002;78:695–6. Available at: http://pmj.bmj.com/content/78/925/695. long [Accessed 10 August 2012].

---

during the Napoleonic wars of 1797–1814, so that "Lind as much as Nelson, broke the power of Napoleon". Lind also instigated reforms in living conditions for sailors, thus contributing to improvements in their health and fitness and the functioning of the fleet. The inquiries following the

Spithead mutiny of 1797 led to the adoption of Lind's nutrition and health recommendations in the same year. In 1798, the USA developed the Marine Hospitals Service for treatment and quarantine of sailors, which later became the US Public Health Service.

Jenner and Vaccination

Smallpox, a devastating and disfiguring epidemic disease, ravaged all parts of the world and had been recognized since the third century BCE. Described first by Rhazes in the tenth century, the disease was confused with measles and was widespread in Asia, the Middle East, and Europe during the Middle Ages. It was a designated cause of death in the Bills of Mortality in 1629 in London. Epidemics of smallpox occurred throughout the seventeenth to eighteenth and into the nineteenth centuries. Primarily a disease of childhood, mortality rates were 25 to 40 percent or more and the disease was characterized by disfiguring sequelae.

Smallpox was a key factor in the near elimination of the Aztec and other societies in Central and South America following the Spanish invasion. Traditions of prevention of this disease by inoculation or transmission of the disease to healthy people to prevent them from acquiring a more virulent form during epidemics were reported in ancient China. This practice, called variolation, was first brought to England in 1721 by Lady Mary Montagu, wife of the British ambassador to Constantinople, where it was common practice. It was widely adopted in England in the mid-eighteenth century, when the disease affected millions of people in Europe. Catherine the Great of Russia had her son inoculated by variolation by a leading English practitioner.

Edward Jenner (1749–1823) was the first to use vaccination with cowpox to prevent smallpox in 1796 (Box 1.4), initiating one of the most dramatically successful endeavors of public health. This revolutionary experiment culminated in the eventual eradication of this dreaded killing and disfiguring disease some 200 years later. In 1800, vaccination was adopted by the British armed forces, and the practice spread to Europe, the Americas, and the British Empire. Denmark made vaccination mandatory in the early nineteenth century and soon eradicated smallpox locally. Despite some professional opposition, the practice spread rapidly from the upper classes and voluntary groups to the common people as a result of the fear of becoming infected with smallpox. Vaccination later became compulsory in many countries, leading to the ultimate public health achievement: global eradication of smallpox in the late twentieth century.

FOUNDATIONS OF HEALTH STATISTICS AND EPIDEMIOLOGY

Registration of births and deaths, originating in ancient societies, Egypt, China, India, Greece, and Rome, was used for tax purposes as well as the determination of potential military manpower. Birth and death rates form the foundation of demography, which is fundamental to epidemiology, a discipline which utilizes demography, sociology, and statistics. Churches maintained registries of births and deaths, and compulsory registration with local government was adopted in the UK in 1853.

Statistical and epidemiological methods emerged in the early seventeenth century with inductive reasoning put forward by Francis Bacon and applied by Robert Boyle in chemistry, Isaac Newton in physics, William Petty in

---

BOX 1.4 Jenner and Smallpox

Variolation, or the exposure of people to the pustular matter of cases of smallpox, was originally documented in ancient China in 320 CE. This practice was used widely in the eighteenth century as a lucrative medical practice, and was a powerful tool in protecting armies from the ravages of smallpox. Variolation was made mandatory by George Washington in the Continental Army during the American Revolution.

In 1796, Edward Jenner (1749–1823), a country physician in Gloucestershire, England, investigated local folklore that milkmaids were immune to smallpox because of their exposure to cowpox. He took matter from a cowpox pustule on a milkmaid, Sarah Nelmes, and applied it with scratches to the skin of a young boy named James Phipps. This inoculated the boy with smallpox, and in turn, he did not develop the disease. Jenner's 1798 publication, *An enquiry into the causes and effects of the variolae vaccina*, described his widescale vaccination and its successful protection against smallpox. Jenner prophesied that "the annihilation of the smallpox, the most dreadful scourge of the human species, must be the result of this practice". He then developed vaccination as a method to replace variolation.

Opposition to vaccination was intense, and Jenner's contribution was ignored by the scientific and medical establishment of the day, but he was later rewarded by Parliament. Vaccination was adopted as a universal practice by the British military in 1800 and by Denmark in 1803. A critical public health tool, vaccination became an increasingly widespread practice during the nineteenth century. In 1977, the last case of smallpox was identified; thus, smallpox eradication was declared by the World Health Organization in 1980.

Remaining stocks of the virus in the USA and Russia were to be destroyed in 1999. However, this was delayed, and following the 9/11 attack on the Twin Towers in New York City, the threat of bioterrorism was taken seriously, including the possibility of use of smallpox. Consequently, vaccination was reinstated for "first responders" including fire, police, and hospital staff in the USA and other countries.

Sources: *Riedel S. Edward Jenner and the history of smallpox and vaccination. Proc Bayl Univ Med Cent 2005;18:21–5. Available at: http://www.ncbi.nlm.nih.gov/pmc/articles/PMC1200696/ [Accessed 10 August 2012].*
Centers for Disease Control and Prevention. Smallpox [updated 6 February 2007]. Atlanta, GA: CDC. Available at: http://www.bt.cdc.gov/agent/smallpox/ [Accessed 18 July 2012].

economics, and John Graunt in demography. Bacon's writing inspired a whole generation of scientists in different fields and led to the founding of the Royal Society of London in 1660.

In Russia, in 1722, Peter the Great began a system of registration of births of male infants for military purposes. In 1755, Mikhail Vasilyevich Lomonosov (1711–1765) was central in establishing the study of demography, carrying out surveys and studies of birth statistics, infant mortality, quality of medical care, alcoholism, and workers' health. He brought the results of these studies to the government's attention, which led to improved training of doctors and midwives, as well as epidemic control measures. Lomonosov also helped to set up the medical faculty of Moscow University (1765).

Daniel Bernoulli (1700–1782), a member of a European family of mathematicians, constructed life tables based on available data showing that variolation against smallpox conferred lifelong immunity and vaccination at birth increased life expectancy. Following the French Revolution, health statistics flourished in the mid-nineteenth century in the work of Pierre Charles Alexandre Louis (1787–1872), who is considered the founder of modern epidemiology. Louis conducted several important observational studies, including one showing that bloodletting, then a common form of therapy, was ineffective. The importance of Louis's studies was demonstrated by the decline in this harmful practice. His students included Marc D'Epigne in France, William Farr in Britain, and others in the USA who became the pioneers in spreading *la méthode numérique* in medicine. *The Lancet*, one of the oldest, best known, and most respected medical journals, was founded in 1823 by Thomas Wakely, an English surgeon, and its creation played an important role in promoting statistical analysis in medical sciences.

Health statistics for social and public health reform took an important place in the work of Edwin Chadwick (1800–1890), Lemuel Shattuck (1793–1859), and Florence Nightingale (1820–1910). Recognizing the extraordinary significance of accurate statistical information in health planning and disease prevention, Edwin Chadwick's work led to legislation establishing the Registrar-General's Office of Britain in 1836. William Farr became its director-general and placed the focus of the office on public health. Farr's analysis of mortality in Liverpool, for example, showed that barely half of its native-born lived to their sixth birthday, whereas in England, the overall median age at death was 45 years. As a result, Parliament passed the Liverpool Sanitary Act of 1846, creating a legislated sanitary code, a medical officer of health position, and a local health authority.

In 1842 in Boston, Massachusetts, Lemuel Shattuck initiated a statewide registration of vital statistics, which later became a model elsewhere in the USA. His report was a landmark in the evolution of public health administration and planning. It provided a detailed account of data collection by age, sex, race, and occupation, and uniform nomenclature for causes of diseases and death. He emphasized the importance of a routine system for exchanging data and information. The London Epidemiological Society, founded in 1850, became an active investigative and lobbying group for public health action. Its work on smallpox led to the passage of the Vaccination Act of 1853, establishing compulsory vaccination in the UK.

In the later part of the nineteenth century, Florence Nightingale highlighted the value of a hospital discharge information system. She promoted the collection and use of statistics that could be derived from the records of patients treated in hospitals. Her work led to marked improvements in the management and design of hospitals, military medicine, and nursing as a profession.

SOCIAL REFORM AND THE SANITARY MOVEMENT (1830–1875)

Following the English Civil War in 1646, veterans of the Parliamentary Army called on the government to provide free schools and free medical care throughout the country as part of democratic reform. However, they failed to sustain interest or gain support for their revolutionary ideas amidst postwar religious conflicts and restoration of the monarchy.

In Russia, the role of the state in health was promoted following initiatives of Peter the Great to introduce western medicine to the country. During the rule of Catherine the Great, under the supervision of Count Orlov, an epidemic of plague in Moscow (1771–1772) was suppressed by incentive payments to bring the sick for care. In 1784, a Russian physician, I. L. Danilevsky, defended a doctoral dissertation on "Government power – the best doctor". In the eighteenth and nineteenth centuries, reform movements promoted health initiatives by government. Although these movements were suppressed (the Decembrists, 1825–1830) and liberal reform steps reversed, their ideas influenced later reforms in Russia.

Following the revolution in France, the Constituent Assembly established a Health Commission. A national assistance program for indigents was established. Steps were taken to strengthen the Bureaux de Sante (Offices of Health) of municipalities which had previously dealt primarily with epidemics. In 1802, the Paris Bureau addressed a wide range of public health concerns, including sanitation, food control, health statistics, occupational health, first aid, and medical care issues. The other major cities of France followed with similar programs over the next 20 years, and in 1848 a central national health authority was established. Child welfare services were also developed in France in the middle part of the nineteenth century. The reporting of vital statistics became reliable in the German states and even more so in France, fostering the development of epidemiological analysis of causes of death.

The governmental approach to public health was articulated by Johann Peter Franck for the Germanic states in his monumental series of books, *A Complete System of Medical Police* (1779–1817). This text explained the government's role in states with strong central governments and how to achieve health reform through administrative action. State regulations were to govern public health and personal health practices including marriage, procreation, and pregnancy. He promoted dental care, rest following obstetric delivery, maternity benefits, school health, food hygiene, housing standards, sanitation, sewage disposal, and clean water supplies. In this system, municipal authorities were responsible for keeping cities and towns clean and for monitoring vital statistics, military medicine, venereal diseases, hospitals, and communicable disease.

This system emphasized a strong, even authoritarian role of the state in promoting public health, including provision of prepaid medical care. It was a comprehensive and coherent approach to public health, emphasizing the key roles of municipal and higher levels of government. This work was influential in Russia, where Franck spent the years 1805–1807 as director of the St. Petersburg Medical Academy. Because of its primary reliance on authoritarian governmental roles, however, this approach was resisted in most western countries, especially following the collapse of absolutist government ideas following the Napoleonic period.

Municipal (voluntary) boards of health were established in some British and American cities in the late eighteenth and early nineteenth centuries. A Central Board of Health was established in Britain in 1805, primarily to govern quarantine regulations to prevent the entry of yellow fever and cholera into the country. Town life improved as sanitation, paving, lighting, sewers, iron water pipes, and water filtration were introduced. Despite the progress, organization for the development of such services was inadequate. Multiple agencies and private water companies provided unsupervised and overlapping services. London City Corporation had nearly 100 paving, lighting, and cleansing boards, 172 welfare boards, and numerous other health-related authorities in 1830. These were later consolidated into the London Board of Works in 1855.

In Great Britain, early nineteenth-century reforms were stimulated by the Philosophic Radicals led by Jeremy Bentham, who advocated dealing with public problems in a rational and scientific way, initiating a reform movement utilizing parliamentary, legal, and educational means. Economic and social philosophers in Britain, including Adam Smith and Jeremy Bentham, argued for liberalism, rationalism, free trade, political rights, and social reform, all contributing to "the greatest good for the greatest number". Labor law reforms (the Mines and the Factory Acts) banned children and women from underground work in the mines and regulated reduction in the workday to 10 hours. These reforms were adopted by the British Parliament in the 1830s to 1840s. The spread of railroads and steamships, the Penny Post (1840), and telegraphs (1846), combined with growing literacy and compulsory primary education introduced in Britain in 1876, dramatically altered local and world communication.

The British Poor Law Amendment Act of 1834 replaced the old Elizabethan Poor Laws, shifting responsibility for welfare of the poor from the local parish to the central government's Poor Law Commission. The parishes were unable to cope with the needs of the rural poor, whose condition was deteriorating with loss of land rights due to agricultural innovations and enclosures. Losing strength, the old system was breaking down, and the new industrialization needed workers, miners, sailors, and soldiers. The new conditions forced the poor to move from rural areas to the growing industrial towns. The urban poor suffered or were forced into workhouses while resistance to reform led to more radicalization and unsuccessful revolution, followed by deep political conservatism.

Deteriorating housing, sanitation, and work conditions in Britain in the 1830s resulted in rising mortality rates recorded in the Bills of Mortality. Industrial cities like Manchester (1795) had established voluntary boards of health, but they lacked the authority to alter fundamental conditions to control epidemics and urban decay. The boards of health were unable to deal with sewage, garbage, animal control, crowded slum housing, privies, adulterated foods and medicines, industrial polluters, or other social or environmental risk sources. Legislation in the 1830s in Britain and Canada improved the ability of municipalities and boards of health to cope with oversight of community water supplies and sanitation.

Under pressure from reformists and the Health of Towns Association, the British government commissioned Edwin Chadwick to undertake a study, which led to the *Report on the Sanitary Conditions of the Labouring Population of Great Britain* (1842), resulting in a further series of reforms through the Poor Law Commission (Box 1.5). The British Parliament passed the Health of Towns Act and the Public Health Act of 1848. This established the General Board of Health, mainly to ensure the safety of community water supplies and drainage, establishing municipal boards of health in the major cities and rural local authorities, along with housing legislation and other reforms. Despite setbacks due to reaction to these developments, the basis was laid for the "sanitary revolution", dealing with urban sanitation and health conditions, as well as cholera, typhoid, and tuberculosis (TB) control.

In 1850, the Massachusetts Sanitary Commission, chaired by Lemuel Shattuck, was established to look into similar conditions in the state. Boards of health established

BOX 1.5 Edwin Chadwick (1800–1900), Social Reform, and the Miasma Theory

Edwin Chadwick, a Manchester lawyer interested in political and social reform, was a leader in the reform movement in Great Britain. In 1832 he was appointed to a Royal Commission to investigate the revision of the Elizabethan Poor Laws, in effect since 1601, leading to the Poor Law Amendment Act of 1834.

In the 1830s a series of epidemics of cholera, typhoid, and influenza prompted the government to launch an investigation of sanitation. Chadwick, a strong believer in miasma theories, was convinced that measures such as cleaning, drainage and ventilation would make people healthier and thus less dependent on welfare. He was appointed to lead the inquiry and produced the report *The Sanitary Conditions of the Labouring Population* (1842). In this report, Chadwick used quantitative methods to show a direct link between poor living conditions and disease and life expectancy. This report inspired major efforts by local authorities to improve sanitation, and led to parliamentary adoption of the Public Health Act of 1848 establishing a General Board of Health, which Chadwick led. He was later forced from office and the national Board of Health was abolished; however, local Boards of Health at the municipal and county levels were developed to implement sanitary reform.

As an advocate of the miasma theory, he was sidelined by the growing strength of the rival germ theory advocates, whose scientific base was growing rapidly in the late nineteenth century. However, Chadwick's impact on promotion of sanitation and emphasis on the link between poverty and disease gave him a place among the most important pioneers in public health in the nineteenth century. While the miasma and germ theory advocates struggled bitterly for dominance for many years, each played a key role in modern public health.

In recent years, the association of poor sanitation, poverty, and adverse social conditions with health risk has re-emerged as being of central importance. Both the biomedical and the social hygiene models are seen to be interactive in the Inequalities in Health movement of the twenty-first century, recognizing the multidimensionality of societal and medical interaction to improve health and quality of life.

Sources: *Science Museum. Edwin Chadwick 1800–1900. London: Science Museum.* Available at: http://www.sciencemuseum.org.uk/broughttolife/people/edwinchadwick.aspx [Accessed 11 August 2012]. *Rosen G. A history of public health. Expanded edition. Baltimore, MD: Johns Hopkins University Press; 1993.*

earlier in the century became efficiently organized and effective in sanitary reform in the USA. The report of that committee has become a classic public health document. Reissued in the 1970s, it remains a useful model for a comprehensive approach to public health.

The Chadwick Report in Great Britain (1842) and the Shattuck Report in Massachusetts (1850) promoted the concept of municipal boards of health based on public health law with a public mandate to supervise and regulate community sanitation. This included urban planning, zoning, restriction of animals and industry in residential areas, and regulation of working conditions, setting the basis of public health infrastructure in the English-speaking world and beyond for the next century.

The interaction between sanitation and social hygiene was a theme promoted by Rudolf Virchow, the founder of cellular pathology and a social–medical philosopher. Virchow was a leading German physician in the mid-nineteenth century. He promoted the ideas of observation, hypothesis, and experimentation, helping to establish the scientific approach to medical issues. He was a social activist and linked the health of the people to social and economic conditions, emphasizing the need for political solutions. Virchow played an important part in the 1848 revolutions in Central and Western Europe, in the same year as the publication of the *Communist Manifesto* by Karl Marx. These all contributed to growing pressure on governments by workers' and political movements to promote better living, working, and health conditions in the 1870s. Virchow was an avowed anticontagionist, and his emphasis on and advances made in the social, economic, and political environment were as much a factor in public health progress as the bacteriological discoveries.

The Massachusetts State Board of Health was established in 1869, and in the same year a Royal Sanitary Commission was appointed in the UK. The American Public Health Association (APHA), established in 1872, served as a professional educational and lobbying group to promote the interests of public health in the USA, often successfully prodding federal, state, and local governments to act in the public interests in this field. The APHA definition of appropriate services at each level of government continues to set standards and guidelines for local health authorities. The organization of local, state, and national public health activities over the twentieth century in the USA owes much to the professional leadership and lobbying skills of the APHA.

Max von Pettenkoffer in 1873 studied the high mortality rates of Munich, comparing them to rapidly declining rates in London. His public lectures on the value of health to a city led to sanitary reforms, which were being achieved in Berlin at the same time under Virchow's leadership. Pettenkoffer introduced laboratory analysis to public health practice and established the first academic chair in hygiene and public health, emphasizing the scientific basis for public health; he is considered to be the first professor of experimental hygiene. A strongly outspoken anticontagionist until the beginning of the twentieth century, Pettenkoffer promoted the concept of the value of a healthy city, stressing that health is the result of a number of factors, and that public health is a community concern since the measures taken to help those in need benefit the entire community.

Social Security

In 1861, Russia freed the serfs and returned independence to universities. Departments of hygiene were established in the university medical schools in the 1860s and 1870s to train future hygienists, and to carry out studies of sanitary and health conditions in manufacturing industries. F. F. Erisman, a pioneer in sanitary research in Russia, promoted the connection between experimental science, social hygiene, and medicine, and he established a school of hygiene in 1890, later closed by the czarist government. In 1864, the government initiated the Zemstvos system of providing medical care in rural areas as a governmental program. These health reforms were implemented in 34 of 78 regions of Russia, before the Revolution. Prior to these reforms, medical services in rural areas were practically non-existent. Epidemics and the high mortality of the working population induced the nobility and new manufacturers in rural towns to promote Zemstvos' public medical services. In rural areas previously served by doctors based in the towns traveling to the villages, local hospitals and delivery homes were established. The Russian medical profession largely supported free public medical care as a fundamental right.

In 1883, Otto von Bismarck (1815–1898), Chancellor of Germany, introduced legislation providing mandatory insurance for injury and illness for workers in industrial plants, and survivor benefits. In 1883, he introduced social insurance for health care of workers and their families, based on mandatory payments from workers' salaries and employer contributions. In the UK in 1911, Chancellor of the Exchequer David Lloyd-George established compulsory insurance for workers and their families for medical care for general practitioner services based on capitation payment. This was followed by similar programs in Russia in 1912 and in virtually all Central and Western European countries by the 1930s. In 1918, Vladimir Lenin (1870–1924) established the state-operated health program, named after its founder Nikolai Semashko (1874–1949), bringing health care to the vast reaches of the Soviet Union. These programs were based on wide recognition of the principle of social solidarity, with governmental responsibility for health of the population being established in virtually all developed countries by the 1960s (see Chapter 13). In the USA, pensions were established for Civil War veterans, widows and orphans, and were made a national social security system only in 1935. Health care insurance was developed through trade unions, and only extended to governmental medical care insurance for the elderly and the poor in 1965.

During the industrial revolution, the harsh conditions in the urban industrial and mining centers of Europe led to efforts in social reform preceding and contributing to sanitary reform. These changes occurred well before the germ theory of disease causation was proven and the science of microbiology was established. Pioneering breakthroughs were made based on trial and error, challenging then established dogmas and producing the sanitary revolution, still unfinished and perhaps the most basic of the foundations of public health.

The issues of universal access to health care and especially prevention are challenges to public health in the twenty-first century. In many of the industrialized countries with universal health care, social inequalities still exist, with gaps between rich and poor, urban and rural, minority groups and other groups at special risk. These are discussed in following chapters of this book. In the USA, the struggle to achieve universal health care is an ongoing political issue in the second decade of the twenty-first century and is still unresolved. In the countries in transition from the Soviet system of health protection, an epidemiological shift occurred but the health system has been slow to respond. In the developing countries universal access to health care is still a distant dream.

Snow on Cholera

The great cholera pandemics originated in India between 1825 and 1854 and spread via increasingly rapid transportation to Europe and North America. Moscow lost some 33,000 people in the cholera epidemic of 1829, which recurred in 1830–1831. In Paris, the 1832 cholera epidemic killed over 18,000 people (just over 2 percent of the population) in 6 months.

Between 1848 and 1854, a series of outbreaks of cholera occurred in London with large-scale loss of life. The highest rates were in areas of the city where two water companies supplied homes with overlapping water mains. One of these (the Lambeth Company) then moved its water intake to a less polluted part of the River Thames, while the Southwark and Vauxhall company left its intake in a part of the river heavily polluted with sewage. John Snow, a founding member of the London Epidemiological Society and anesthetist to Queen Victoria, investigated an outbreak of cholera in Soho from August to September 1854, in the area adjacent to Broad Street. He traced some 500 cholera deaths occurring in a 10-day period. Cases either lived close to or used the Broad Street pump for drinking water. He determined that brewery workers and poorhouse residents in the area, using uncontaminated wells, escaped the epidemic. Snow concluded that the Broad Street pump was probably contaminated. He persuaded the authorities to remove the handle from the pump, and the already subsiding epidemic disappeared within a few days.

During September to October, 1854, Snow investigated another outbreak, again suspecting water transmission. He identified cases of mortality from cholera by their place of residence and which water company supplied the home (Table 1.1). Snow calculated the cholera rates in a 4-week

TABLE 1.1 Deaths from Cholera Epidemic in Districts of London Supplied by Two Water Companies, 7 Weeks, 1854

Water Supply Company	Number of Houses	Deaths from Cholera	Cholera Deaths/10,000 Houses
Southwark and Vauxhall	40,046	1,263	315
Lambeth	26,107	98	37
Rest of London	256,423	1,422	59

Sources: Snow J. On the mode of transmission of cholera. 1854. In: Snow on cholera: a reprint of two papers. New York: Commonwealth Fund; 1936.
Sack DA, Sack RB, Nair GB, Siddique AK. Cholera. Lancet 2004;363:223–33. Available at: http://www.thelancet.com/journals/lancet/article/PIIS0140-6736(03)15328-7/abstract [Accessed 11 August 2012].

period in homes supplied by each of the two companies. Homes supplied by the Southwark and Vauxhall Water Company were affected by high cholera death rates while adjacent homes supplied by the Lambeth Company had rates lower than the rest of London. This observation provided overwhelming epidemiological support for Snow's hypothesis that the cholera epidemic source was the contaminated water from the River Thames, distributed to homes in a large area of south London. The risk to local residents of becoming infected and falling ill with cholera was dependent upon the specific water company as well as the pump utilized.

This investigation, with a study and control group occurring in an actual disease outbreak, strengthened the case of the germ theory supporters, who were still opposed by strong proponents of miasmatic theories. It also led to legislation mandating filtration of water companies' supplies in 1857. *Vibrio cholerae* was not isolated until 1883, during an investigation of waterborne cholera outbreaks in Egypt by Robert Koch. Snow's work on cholera has become one of the classic epidemiological investigations, studied to this day for its scientific imagination and thoroughness, despite preceding the discovery of the causative organism by nearly 30 years.

A landmark case, Snow's work on cholera stimulated more investigation of causes of enteric diseases. William Budd (1811–1880), a physician at the Bristol Royal Infirmary, carried out a number of epidemiological investigations of typhoid fever in the 1850s, finding waterborne episodes of the disease. He investigated an outbreak in 1853 in Cowbridge, a small Welsh village, where a ball attracted 140 participants from surrounding counties. Almost immediately afterwards, many of those attending the ball became sick with typhoid fever. He found that a person with typhoid had been at the location some days before, and that his excreta had been disposed of near a well, from which water was drawn for the ball. Budd then concluded that water was the vehicle of transmission of the disease. He investigated other outbreaks and summarized his reports in *Typhoid Fever: Its Nature, Mode of Transmission and Prevention*, published in 1873, which is a classic work on waterborne transmission of enteric disease. These investigations were very valuable, as they contributed to the movement to disinfect public water systems on a preventive basis.

The brilliant epidemiological studies of Snow and Budd set a new direction in epidemiology and public health practice, not only with waterborne disease. They established a standard for investigation of the distribution of disease in populations with the purpose of finding a way to interrupt the transmission of disease. Improved sanitation and water safety, developed in urban and rural population centers, contributed greatly to improved survival and a reduction in cholera and typhoid epidemics. However, globally, waterborne disease remains a major cause of morbidity and mortality, especially among children living in poverty.

The Germ and Miasma Theories

Until the early and middle parts of the nineteenth century, the causation of disease was hotly debated. The miasma theory, holding that disease was the result of environmental emanations or miasmas, went back to Greek and Roman medicine, and Hippocrates' treatise *On Air, Water, and Places*. Miasmists believed that disease was caused by infectious mists or noxious vapors emanating from filth in the towns and that the method of prevention of infectious diseases was to establish sanitary measures to clean the streets of garbage, sewage, animal carcasses, and wastes that were features of urban living. This provided the basis for the Sanitary Movement, with great benefit to improving health conditions. The miasma theory had strong proponents well into the later part of the nineteenth century.

The contagion or germ theory gained ground, despite the lack of scientific proof, on the basis of biblical and Middle Ages' experience with isolation of lepers and quarantine of other infectious conditions. In 1546, Fracastoro (1478–1553) published *De Contagione*, a treatise on microbiological organisms as the case of specific diseases. The germ theory was strengthened by the work of Antony van Leeuwenhoek (1632–1723), who invented the microscope in 1676. The invention of this apparatus is considered to be a groundbreaking discovery, a watershed in the history of science. His research showing small microorganisms led to his recognition as a Fellow of the Royal Society of England in 1680. The germ theorists believed that microbes, such as those described by van Leeuwenhoek, were the cause of diseases which could be transmitted from person to person or by contact with sewage or contaminated water.

Major contributions to resolving this issue came from the epidemiological studies of Snow and Budd in the 1850s, proving waterborne transmission of cholera and typhoid, suggesting that if the disease was not from a miasma source then it was due to germ (contagion) sources. The classic study of a measles epidemic in the remote Faroe Islands by Peter Panum in 1846 clearly showed person-to-person transmission of this disease, its incubation period, and the lifelong natural immunity that exposure gives (Box 1.6). The dispute continued, with miasmists or sanitationists and germ theorists arguing with equal vehemence.

While the science of the issue was debated until the end of the nineteenth century, the practical application of sanitary reform was promoted by both theories. Increasing attention was paid to sewage and water safety, and the removal of waste products by organized municipal activities was adopted in European and North American cities. The sanitary revolution proceeded while the debates raged and solid scientific proof of the germ theory accumulated, primarily in the 1880s. Fear of cholera stimulated New York City to establish a Board of Health in 1866. In the city of Hamburg, Germany, a Board of Health was established in 1892 only after a cholera epidemic attacked the city, while neighboring Altona remained cholera free because it had established a water filtration plant.

The specific causation of disease (the germ theory) has been a vital part of the development of public health. The bacteriological revolution (see later section entitled "The Bacteriological Revolution"), led by the work of Louis Pasteur and Robert Koch, provided enormous benefit to medicine and public health. Those who argued that disease is environmental in origin (the miasma theory), however, also contributed to public health because of their recognition of the importance of social or other environmental factors, such as poor sanitation and housing conditions or nutritional status, all of which increase susceptibility to specific agents of disease, or to the severity of disease.

HOSPITAL REFORM

Hospitals developed by monasteries as charitable services were supplanted by voluntary or municipal hospitals mainly for the poor during and after the Renaissance. Reforms in hospital care evolved along with the sanitary revolution. In eighteenth-century Europe, hospitals that were operated by religious orders of nuns and by municipal or charitable organizations were dangerous cesspools of pestilence. The dangers arose from lack of knowledge about and practice of basic hygiene for infection control, the concentration of patients with highly communicable diseases, and transmission of disease by medical and other staff.

Reforms in hospitals in England were stimulated by the reports of John Howard in the late eighteenth century, becoming part of wider social reform in the early part of the

---

BOX 1.6 Panum on Measles in the Faroe Islands, 1846

Peter Ludwig Panum (1820–1885), a 26-year-old newly graduated medical doctor from the University of Copenhagen, was sent to the Faroe Islands by the Danish government in 1846 to investigate an outbreak of measles. On the islands, located in the far reaches of the North Atlantic, there was no documentation of measles since 1781. During the 1846 epidemic, approximately 6000 of the 7782 islanders were stricken with measles and 102 of them died of the disease or its sequelae. Panum visited all isolated corners of the islands, tracing the chain of transmission of the disease from location to location, and the immunity of those exposed during the epidemic.

From his well-documented observations he concluded, contrary to the prevailing opinion, that measles is a contagious disease spread from person to person, and that one attack gives lifelong immunity. His superb report clearly demonstrated the contagious nature of the disease and its incubation period. It also proved that measles is not a disease of "spontaneous generation", nor is it generally dispersed in the atmosphere and spread as a "miasma", giving strength to and providing evidence for the germ theory.

Since the 1960s, the availability of an inexpensive, highly effective, and safe vaccine has led to the elimination of domestic circulation of the virus in many countries, yet measles remains a serious global health problem in 2013. An estimated 250,000 children died of this highly contagious disease in 2006.

A European-wide measles epidemic in 2010–2012 had over 50,000 cases in 2011 alone. As a result of imported cases and local spread, outbreaks of measles are occurring in countries that were thought to be measles free, including countries in North and South America. Measles elimination is possible with the two-dose policy with current vaccines, if immunization is given high priority with catch-up campaigns for vulnerable age groups and travelers, and pursued with determined national and international efforts.

Sources: *Emerson H. Panum on measles: observations made during the epidemic of measles on the Faroe Islands in the year 1846 (A translation from the Danish). Am J Public Health Nations Health 1940;30:1245–6. Available at: http://www.ncbi.nlm.nih.gov/pmc/articles/PMC1530953/ [Accessed 11 August 2012].*
Rosen G. A history of public health. Expanded edition. Baltimore, MD: Johns Hopkins University Press; 1993.
World Health Organization. Measles fact sheet [updated April 2012]. Geneva: WHO. Available at: http://www.who.int/mediacentre/factsheets/fs286/en/ [Accessed 11 August 2012].

---

nineteenth century. Professional reform in hospital organization and care started in the latter half of the nineteenth century under the influence of Florence Nightingale, Oliver Wendel Holmes, and Ignaz Semmelweiss. Clinical–epidemiological studies of "antiseptic principles" provided a new, scientific approach to improvement in health care.

In the 1840s, puerperal fever was a major cause of death in childbirth, and consequently, was the subject of investigation by Holmes in the USA, who argued that it was due to a

BOX 1.7 Crede and Prevention of Gonococcal Ophthalmia Neonatorum

Gonorrhea was common in all levels of society in nineteenth century Europe. Ophthalmic infection of newborns was a widespread cause of infection, scarring, and blindness. Carl Franz Crede (1819–1892), professor of obstetrics at the University of Leipzig, attempted to treat neonatal gonococcal ophthalmic infection with many medications. Crede discovered the use of silver nitrate as a treatment and introduced its use as a preventive measure during the period 1854–1860, with astonishing success.

The prophylactic use of silver nitrate spread rapidly hospital by hospital. However, owing to widespread medical opposition to this innovation, decades passed before it was mandated widely. It was only in 1879 that the gonococcus organism was discovered by Albert Ludwig Neisser (1855–1916). Estimates of children saved from blindness by this procedure in Europe during the nineteenth century are as high as one million.

Sources: *Schmidt A. Gonorrheal ophthalmia neonatorum: historic impact of Crede's eye prophylaxis. Pediatr Infect Dis Revisited 2007;95–115. Available at: http://www.springerlink.com/content/xtu8475716207264/ [Accessed 11 August 2012].*
Dunn PM. Perinatal lessons from the past: Sir Norman Gregg, ChM, MC, of Sydney (1892–1966) and rubella embryopathy. Arch Dis Child Fetal Neonatal Ed 2007;92:F513–4. Available at: http://www.ncbi.nlm.nih.gov/pubmed/17951553 [Accessed 11 August 2012].

contagion. In 1846, Semmelweiss (1818–1865), a Hungarian obstetrician at the Vienna Lying-In Hospital, suspected that the deaths from puerperal fever were the result of contamination on the hands of physicians, who transmitted autopsy material to living patients. He showed that death rates among women attended by medical personnel were two to five times the rates among those attended by midwives. By requiring doctors and medical students to soak their hands in chlorinated lime after autopsies, he reduced the mortality rates among the medically attended women to the rate of the midwife-attended group.

Semmelweiss's work, although carefully documented, was slow to be accepted by the medical community, taking some 40 years for general adoption. His pioneering investigation of childbed fever (streptococcal infection in childbirth) in Vienna contributed to improvements in obstetrics and a reduction in maternal mortality. In the 1850s, prevention of blindness in newborns by prophylactic use of silver nitrate eyedrops, developed by Carl Crede (1819–1892) in Leipzig, spread rapidly through the medical world (Box 1.7). This practice continues to be a standard in the prevention of ophthalmia neonatorum.

Florence Nightingale's momentous work in nursing and hospital administration in the Crimean War (1854–1856) established the professions of nursing and modern hospital administration. In the 1860s, she emphasized the importance of the "Poor Laws", including workhouse reform and training special district nurses for care of the sick and poor at home. Nightingale's subsequent long and successful campaigns to raise standards of military medicine, hospital planning, supply services and management, hospital statistics, and community health nursing were outstanding contributions to the development of modern, organized health care and antisepsis.

Despite all of the cumulative progress over the past 150 years, including the advent of sterile techniques and antibiotics, hospital-acquired infection remains a serious public health problem today. This major medical challenge is exacerbated by multidrug-resistant organisms and a persistent failure of regular hand washing between patient care by doctors and nurses. It is also complicated by antiseptic measures needed for invasive procedures such as central venous and bladder catheters.

THE BACTERIOLOGICAL REVOLUTION

In the third quarter of the nineteenth century, the sanitary movement rapidly spread through the cities of Europe, America and elsewhere with demonstrable success in reducing disease in areas served by sewage drains, improved water supplies, street paving, and waste removal. At the same time, innovations occurred in hospitals, stressing hygiene and professionalization of nursing and administration. These were accompanied by breakthroughs in establishing scientific and practical applications of bacteriology and immunology.

Pasteur, Cohn, Koch, and Lister

Louis Pasteur (1822–1895), a French professor of chemistry, serves as one of the most notable figures in scientific history. One of his many groundbreaking achievements involves the science of immunology, through his work with vaccines (Box 1.8).

Rabies was widely feared as a disease transmitted to humans through the bites of infected animals, and was universally fatal. Pasteur reasoned that the disease affected the nervous system and was transmitted in saliva. He injected material from infected animals, attenuated to produce protective antibodies but not the disease. In 1885, a 14-year-old boy from Alsace was severely bitten by a rabid dog. Local physicians agreed that because death was certain, Pasteur, a chemist and not a physician, be allowed to treat the boy with a course of immunization. The boy, Joseph Meister, survived, and similar cases were brought to Pasteur and successfully immunized. Pasteur was criticized in medical circles, but both the general public and scientific circles soon recognized his enormous contribution to public health.

Ferdinand Julius Cohn (1828–1898), professor of botany at Breslau University, developed and systematized the science of bacteriology using morphology, staining, and media characteristics of microorganisms, and trained a key

BOX 1.8 Louis Pasteur, the Pioneer of Pasteurization, Microbes & Vaccines

Louis Pasteur was a French chemist and biologist who proved the germ theory of disease and invented the process of pasteurization. He brilliantly developed the basis for modern bacteriology as a cornerstone of public health, establishing scientific, experimental proof for the germ theory with his demonstration in 1854 of anaerobic microbial fermentation.

Pasteur was asked to investigate the threatened destruction of the French silk industry by epidemics destroying the silkworms. His analysis pointed to living microorganisms causing the disease. Consequently, he devised new growing conditions, which eliminated the problem.

This, in turn, raised both scientific and industrial interest in the germ theory. In the period 1856–1860, he showed how to prevent wine spoilage due to contamination from foreign organisms. His microscopic work helped him to observe that certain liquids went rancid owing to the multiplication of minute organisms in wine, then beer and milk. With more investigation, he saw that these microorganisms could be destroyed by heating the liquid, an important process later termed "pasteurization". As with the other liquids he worked with, he heated the wine to a certain temperature before bottling it, thus killing the undesired ferments.

When he published the concept that the very microorganisms that contaminated the liquids also floated in the air, it was met with ridicule by the medical establishment. The germ theory is the idea that certain microorganisms are responsible for causing many specific diseases. Monumental and groundbreaking for its time, the germ theory drastically influenced the way in which medicine was practiced. Pasteur's work confirmed previous awareness of the existence of germs and development of the germ theory, so that the discoveries of many scientists were retrospectively recognized as contributing to the understanding of the germ theory.

Pasteur moved on to studying solid compounds through experimental trials, demonstrating that microbes were the reason behind the decay of meat. He was confident that this concept explained the development of disease, arguing that the multiplication of germs leads to a specific disease. This was a very significant realization, as it meant that microbes not only affected beer, milk, and various foods, but had the potential to affect humans as well.

He succeeded in producing vaccines through attenuation, or weakening an organism's strength by passing it successively through animals, recovering it, and retransmitting it to other animals. He calculated that if a vaccine can be developed for smallpox, then one can be created for all diseases. In collaboration with physician Charles Chamberland, he inoculated chickens with chicken cholera germs taken from an old culture. Following this, using cholera germs from a new, fresh culture, they inoculated two groups of chickens: those that had previously been inoculated with the old culture, and those that had not. The chickens initially inoculated with the old sample survived, while the other group of chickens did not, demonstrating the initial inoculation of the old culture to be successful in producing immunity to chicken cholera.

This experiment illustrated the principles of vaccination; Pasteur's explanation for it surrounded the idea of weaker germs creating a form of defense and establishing protection to fight the stronger, more potent germs later presented in the fresh sample. In 1883, he produced a similar protective vaccine for swine erysipelas, and then in 1884–1885, a vaccine for rabies.

Louis Pasteur was a brilliant scientific pioneer whose many outstanding achievements greatly contributed to the advance of medical sciences and public health.

Sources: *History Learning Site. Louis Pasteur [updated 2012]. UK: History Learning Site. Available at: http://www.historylearningsite.co.uk/louis_pasteur.htm [Accessed 11 August 2012].*
Science Museum. Louis Pasteur (1822–1895). London: Science Museum. Available at: http://www.sciencemuseum.org.uk/broughttolife/people/louispasteur.aspx [Accessed 11 August 2012].

generation of microbiological investigators. One student, Robert Koch (1843–1910), a German rural district medical officer, investigated anthrax using mice inoculated with blood from sick cattle, with transmission of the disease for more than 20 generations. He developed basic bacteriological techniques including methods of culturing and staining bacteria. He demonstrated the organism causing anthrax, recovered it from sick animals, and passed it through several generations of animals, proving the transmission of specific disease by specific microorganisms.

In 1882, Koch cultured and demonstrated the tubercle bacillus. He then headed the German Cholera Commission visiting Egypt and India in 1883, isolating and identifying *Vibrio cholerae* (Nobel Prize 1905). He demonstrated the efficacy of water filtration in preventing transmission of enteric disease including cholera. His development of "Koch's postulates", or criteria for causation of a disease by a specific organism that produced scientific evidence and substantiated the germ theory, was a long-lasting contribution to the science of medicine. He was awarded the Nobel Prize in Physiology or Medicine in 1905. In 1883, Koch, adapting postulates on causation of disease from clinician–pathologist Jacob Henle (1809–1885), established criteria for attribution of causation of a disease to a particular parasite or agent (Box 1.9). These were fundamental to the establishment of the science of bacteriology and the relationship of microorganisms to disease causation.

The Koch–Henle postulates serve as guidelines – in their pure form were later seen as too rigid, and would limit identification of the causes of many diseases – but they were important in establishing germ theory and the scientific basis of bacteriology, dispelling the many other theories of disease that were still widespread in the late nineteenth century. These postulates served as guidelines for evidence of causation, but had limitations in that not all microbiological agents can be grown in pure culture, some

BOX 1.9 Koch–Henle Postulates on Microorganisms as the Cause of Disease

1. The organism (agent) must be shown to be present in every case of the disease by isolation in pure culture.
2. The agent should not be found in cases of any other disease.
3. Once isolated, the agent should be grown in a series of cultures, and then must be capable of reproducing the disease in experimental animals.
4. The agent must then be recovered from the disease produced in experimental animals.

These criteria for disease causation by bacteria have been modified to accommodate new scientific knowledge, newly identified organism (e.g., viruses), asymptomatic infections, host and environmental and other factors. However, these postulates were important as basic tools in developing public health science.

Sources: *Last JM. A dictionary of epidemiology. 4th ed. New York: Oxford University Press; 2001.*
Last JM. A dictionary of public health. Oxford: Oxford University Press; 2006.
MedicineNet. Definition of Koch's postulates [updated 10 October 1998]. New York: MedicineNet. Available at: http://www.medterms.com/script/main/art.asp?articlekey=7105 [Accessed 11 August 2012].
Frederichs DN, Relman DA. Sequence-based identification of microbial pathogens: a reconsideration of Koch's postulates. Clin Microbiol Rev 1996;9:18–33. Available at: http://www.ncbi.nlm.nih.gov/pmc/articles/PMC172879/pdf/090018.pdf [Accessed 11 August 2012].

BOX 1.10 Gregor Mendel: The Father of Genetics

Gregor Mendel (1822–1884) was an Augustinian monk in Brno (now in the Czech Republic) who was sent to study at the University of Vienna. He carried out botanical experimentation stimulated by a wide demand for knowledge of plant heredity. The introduction of new plants by explorations overseas since 1500 brought many new species of vegetables, fruit, and flora to Europe, while improved transportation and the rise of cities in the industrial revolution brought a need for improved agricultural production.

Mendel studied variation in plant height (tall or short) and seed color (green or yellow) in the monastery's experimental garden, focusing on 29,000 pea plants. He published his findings on the occurrence of paired elementary units of heredity, now known as important principles of genetics. Using his extensive data, Mendel was successful in demonstrating the concept that genes obey simple statistical laws.

His experiments led him to two generalizations, the Law of Segregation and the Law of Independent Assortment, later known as Mendel's Laws of Inheritance. His studies were published with little impact at the time, but were later recognized as the basics of genetic studies and Mendel posthumously became known as the "father of modern genetics".

Sources: *Nobelprize.org. The Nobel Prize in Physiology or Medicine 1962: Francis Crick, James Watson, Maurice Wilkins. Sweden: Nobelprize.org. Available at: http://www.nobelprize.org/nobel_prizes/medicine/laureates/1962/watson.html [Accessed 11 August 2012].*
Human Genome Project information. History of the Human Genome Project. Available at: http://www.ornl.gov/sci/techresources/Human_Genome/project/hgp.shtml [Accessed 2 August 2012].
Johns Hopkins School of Public Health. Genetically engineered bacteria prevent mosquitoes from transmitting malaria. Available at: http://www.jhsph.edu/news/news-releases/2012/jacobs_lorena_bacteria.html [Accessed 2 August 2012].

organisms undergo antigenic drift or change in antigenicity, and some organisms have no animal host. Koch's postulates were later adapted by Evans (1976) to include non-infectious disease-causing agents, such as cholesterol, following the changing emphasis in epidemiology of non-infectious diseases.

In the mid-1860s, Joseph Lister in Edinburgh, under the influence of Pasteur's work and with students of Semmelweiss, developed a theory of "antisepsis". His 1865 publication *On the Antiseptic Principle in the Practice of Surgery* described the use of carbolic acid to spray operating theaters and to cleanse surgical wounds, applying the germ theory with great benefit to surgical outcomes. Lister's work on chemical disinfection for surgery in 1865 was a pragmatic development and a major advance in surgical practice; it was also an important contribution to establishing the germ theory in nineteenth-century medicine.

The Basis of Genetics

The introduction of new plants from explorations overseas since 1500 brought many new species of vegetables, fruit, and flora to Europe, while improved transportation and the rise of cities in the industrial revolution brought a need and demand for improved agricultural production. This led to the wide practice of animal and plant breeding, but the basic science of genetics was lacking.

Gregor Mendel, an Augustinian monk, carried out botanical experimentation stimulated by a wide demand for knowledge of plant heredity. He studied variation in pea plants in terms of their characteristics and discovered that genes obey simple statistical laws (Box 1.10). His work was later defined as Mendel's laws of inheritance. These were recognized as the basics of genetic studies and Mendel became known as the "father of modern genetics".

Later in the nineteenth century, the discovery of chromosomes gave new life to genetic studies on human health. During the twentieth century many genetically determined diseases were identified and practical solutions were worked out to test and counsel families. These achievements allowed hereditary diseases such as Tay Sachs and thalassemia to be reduced or eliminated by public health programs of screening, education, and individual counseling.

The mid-twentieth century brought on Francis Crick, James Watson, and Maurice Wilkins's discovery of the famous "double helix" molecular structure of nucleic acid and its significance for information transfer in living material (DNA, Nobel Prize 1962). This was an enormous step forward in the field of genetics and provided the basis for the Human Genome Project (HGP), which defined thousands of human genes. The HGP was an international 13-year effort, begun in 1990 and completed in 2003. Its purpose was to discover all of the estimated 20,000–25,000 human genes in order to make them accessible for further biological study, and to determine the complete sequence of the 3 billion DNA subunits (bases in the human genome).

The application of these new sciences opened the fields of genetic screening and counseling, and drug development for specific genome-associated diseases, creating enormous potential for risk factor identification and targeted medical interventions. The applications will change medicine and public health in the coming decades in diverse fields of agriculture, biology, medicine, nutrition, and public health for the prevention and control of birth defects, cancer, and degenerative, infectious, allergic, and other diseases.

Vectorborne Disease

Studies of disease transmission defined the importance of carriers (i.e., those who can transmit a disease without showing clinical symptoms) in the transmission of diphtheria, typhoid, and meningitis, and promoted studies of diseases borne by intermediate hosts or vectors. Parasitic diseases of animals and humans, including Guinea worm disease, tapeworms, filariasis, and veterinary parasitic diseases such as Texas cattle fever, were investigated in many centers during the nineteenth century. David Bruce (1855–1931) demonstrated transmission of nagana (animal African trypanosomiasis), a disease of cattle and horses in Zululand, South Africa, in 1894–1895. Nagana is caused by a trypanosome parasite transmitted by the tsetse fly, and its study led to the use of environmental methods of control to halt disease transmission. Alexandre Yersin (1863–1943) and Shibasaburo Kitasato (1853–1931) discovered the plague bacillus in 1894, and in 1898 French epidemiologist P. L. Simmond demonstrated that the plague was a disease of rats spread by fleas to humans.

Malarial parasites were identified by French army surgeon Alphonse Laveran (1845–1922, Nobel Prize 1907) in Algeria in 1880. He referred to the organism as *Oscillaria malariae*, and as with many major public health discoveries, his was initially met with doubt. Laveran persisted, and after ending his military career he continued his research, producing irrefutable evidence that was later validated by other experts. Moreover, other nineteenth-century investigators suspected mosquitoes as the method of transmission, and in 1897, Ronald Ross (1857–1932, Nobel Prize 1902), a British army doctor in India, Patrick Manson (1844–1922) in England, and Benvenuto Grassi in Rome demonstrated transmission of malaria by the *Anopheles* mosquito. Yellow fever, probably imported by the slave trade from Africa, was endemic in the southern USA but spread to northern cities

---

BOX 1.11 Havana and Panama: Control of Yellow Fever and Malaria, 1901–1906

The United States Army Commission on Yellow Fever, led by Walter Reed (1851–1902), a military physician, organized a team in Cuba in 1900, with physicians Carlos Finlay (1833–1915) and Jesse Lazear (1866–1900), to test hypotheses regarding yellow fever transmission. Working with volunteers, he demonstrated transmission of the disease from person to person by the specific mosquito *Stegomyia fasciata*. The Commission accepted that "the mosquito acts as the intermediate host for the parasite of Yellow Fever".

Another US army doctor, William Gorgas (1845–1920), applied the new knowledge of transmission of yellow fever and the life cycle of the vector mosquito. He organized a campaign to control the transmission of yellow fever in Havana, isolating clinical cases from mosquitoes and eliminating the breeding places for *Stegomyia* with Mosquito Brigades.

Yellow fever was eradicated in Havana within 8 months. This demonstrated the possibility for control of other mosquito-borne diseases, principally malaria with its specific vector, *Anopheles*. Gorgas then successfully applied mosquito control to prevent both yellow fever and malaria between 1904 and 1906, permitting construction of the Panama Canal.

Sources: *McCullough D. The path between the seas: the creation of the Panama Canal 1870–1914. New York: Touchstone; 1977.*
Harvard University Open Collections Program. Contagion historical views. William Gorgas 1854–1920. Available at: http://ocp.hul.harvard.edu/contagion/gorgas.html [Accessed 12 August 2012].

---

in the late eighteenth century. An outbreak in Philadelphia in 1798 killed nearly 8 percent of the population. Outbreaks in New York killed 732 people in 1795, 2086 in 1798, and 606 in 1803. The Caribbean and Central America were endemic with both yellow fever and malaria.

The conquest of yellow fever also contributed to the germ or contagion theory becoming established and accepted over the miasma theory, when the work of Cuban physician Carlos Finlay was confirmed by Walter Reed in 1901. His studies in Cuba proved the mosquito-borne nature of the disease as a transmissible disease via an intermediate host (vector), but which was not contagious between humans.

William Gorgas applied this to vector control activities and protection of sick people from contact with mosquitoes, resulting in an eradication of yellow fever in Havana within 8 months, and in the Panama Canal Zone within 16 months (Box 1.11). This work showed a potential for the control of vectorborne disease, which has had important success in the control of many tropical diseases, including yellow fever and, currently, Guinea worm disease and onchocerciasis. Despite being well controlled in many parts of the world, there has been a resurgence of malaria in many tropical countries since the 1960s.

MICROBIOLOGY AND IMMUNOLOGY

In Russia in 1883, Ilya Ilyich Mechnikov (1845–1916) described phagocytosis, a process in which white cells in the blood surround and destroy bacteria; his elaboration of the processes of inflammation and humoral and cellular response led to a joint Nobel Prize in 1908 with Paul Ehrlich (1854–1915). Other investigators searched for the bactericidal or immunological properties of blood that enabled cell-free blood or serum to destroy bacteria. This work greatly strengthened the scientific basis for bacteriology and immunology.

Pasteur's co-workers, Emile Roux (1853–1933) and Alexandre Yersin, isolated and grew the causative organism for diphtheria and suggested that the organism produced a poison or toxin which, in turn, caused the lethal effects of the disease. In Berlin in 1890, Karl Fraenkel published his work showing that inoculating guinea pigs with attenuated diphtheria organism could produce immunity. At the same time, also in Germany, Emile Behring (1854–1917) and Japanese co-worker Shibasaburo Kitasato produced evidence of immunity to tetanus bacilli in rabbits and mice. Behring also developed a protective immunization against diphtheria in humans with active immunization, as well as an antitoxin for passive immunization of an already infected person (Nobel Prize 1901). By 1894, diphtheria antitoxin was ready for general use. The isolation and identification of new disease-causing organisms proceeded rapidly in the last decades of the nineteenth century. The diphtheria organism was discovered in 1885 by Edwin Klebs (1834–1913) and Friedrich Loeffler (1852–1915), both students of Koch. A diphtheria vaccine was developed in 1912, leading to the control of this disease in many parts of the world. Between 1876 and 1898, many pathogenic organisms were identified, providing a strong foundation for advances in vaccine development.

During the last quarter of the nineteenth century, it was apparent that inoculation of attenuated microorganisms could produce protection through active immunization of a host by generating antibodies to that organism. This, in turn, would protect the individual when exposed to the virulent (wild) organism. Passive immunization could be achieved in an already infected person by injecting the serum of animals infected with attenuated organisms. The serum from that animal helps to counter effects of the toxins produced by an invading organism. Pasteur's vaccines were followed by those of Waldemar Haffkine (1860–1930), a bacteriologist working in India. A remarkable figure, he was the first microbiologist to develop and use vaccines against cholera and bubonic plague, after testing them on himself. Other pioneering achievements include those of Richard Pfeiffer (1858–1945), a bacteriologist studying under Robert Koch and Sir Almoth Wright (1861–1947), known for his work in co-developing the typhoid vaccine. Albert Calmette (1863–1933), a bacteriologist, and Camille Guérin (1872–1961) a veterinarian, worked together, examining the intestinal route of TB. Through persistence and the constant replanting of bacterial cultures, the two developed the TB vaccine, called BCG (bacille Calmette–Guérin), named after the remarkable pair of researchers. Further contributory achievements include those of Theobald Smith (1859–1934), a pathologist most recognized for his research in Texas cattle fever, and Max Theiler (1899–1972, Nobel Prize 1951), a prominent South African physician successful in discovering the vaccine against yellow fever.

The twentieth century produced a flowering of immunology in the prevention of important diseases in animals and in humans based on the pioneering work of Jenner, Pasteur, Koch, and those who followed. Many major childhood infectious diseases have been controlled by immunization, one of the outstanding achievements of twentieth-century public health. The success of vaccines and other infection control in preventing cancers such as of the liver (hepatitis B vaccine) and the cervix (human papillomavirus, HPV), and gastric cancer (*Helicobacter pylori*) marked the beginning of an important new stage in public health.

Poliomyelitis

Poliomyelitis was endemic in most parts of the world prior to World War II, causing widespread crippling of infants and children, hence its common name of "infantile paralysis". The most famous polio patient was Franklin Delano Roosevelt, crippled by polio in his early thirties, who went on to become president of the USA. Massive epidemics of poliomyelitis during the 1940s and 1950s affected thousands of North American children and young adults. Consequently, national hysteria and fear of this disease ensued because of its crippling and killing power. In 1952, 52,000 cases of poliomyelitis were reported in the USA, bringing a national response and support for the "March of Dimes" Infantile Paralysis Association for research and field vaccine trials.

Based on the development of methods for isolating and growing the virus by John Enders and colleagues, Jonas Salk developed an inactivated vaccine in 1955 and Albert Sabin a live attenuated vaccine in 1961. Salk's field trial proved the safety and efficacy of his vaccine in preventing poliomyelitis. Sabin's vaccine proved to be cheaper and easier to use on a mass basis and is still the mainstay of polio eradication worldwide (Box 1.12). The conquest of this dreaded, disabling, and disfiguring disease has provided one of the most dramatic achievements of public health in the twentieth and early twenty-first centuries. Despite setbacks and economic recession limiting donor funding, there are good prospects for the elimination of poliomyelitis by 2015 or soon thereafter.

BOX 1.12 Salk and Sabin Vaccines and Poliomyelitis Eradication

In the early 1950s, John Enders (1897–1985) and colleagues developed methods of growing polio virus in laboratory conditions, for which they were ultimately awarded a Nobel Prize. At the University of Pittsburgh, physician and epidemiologist Jonas Salk (1914–1995) developed the first inactivated (killed) vaccine under sponsorship of a large voluntary organization (The March of Dimes), which mobilized the resources to fight this much dreaded disease. In 1954, Salk successfully completed the largest field trial ever, involving 1.8 million children. The vaccine was rapidly licensed and quickly developed and distributed in North America and Europe, interrupting the epidemic cycle and rapidly reducing polio incidence to low levels.

Albert Sabin (1906–1994) at the University of Cincinnati developed a live, attenuated vaccine given orally (OPV), which was approved by the FDA in 1961. This vaccine was an immediate success as it has many advantages: it is administered easily, spreads its benefits to non-immunized people, and is inexpensive. It became the vaccine of choice and was used widely, thus reducing polio to a negligible disease in most developed countries within a few years. Sabin also pioneered application of OPV through national immunization days in South America, which contributed to the eradication of polio in that region, and more recently, in many other continents and countries.

In 1987, the World Health Assembly of the WHO declared the target of eradication of poliomyelitis by the year 2000. With the help of international and national commitment, the Americas were declared polio free in 1990. A worldwide campaign has been conducted with improved routine immunization coverage, mass immunization days, and localized control measures, resulting in a widening of successful eradication across all continents.

By the end of 2010, polio was still active in four countries, with a substantial outbreak in Tajikistan. Total polio cases worldwide declined from 650 cases in 2011 to 97 cases up to 11 July 2012. In 2012, polio remains endemic in Afghanistan, Nigeria, Pakistan, the Congo (DRC), and Chad, while India seems to have been polio free in 2011. Eradication of the natural transmission of the disease is anticipated by 2015, and an end-stage strategy recommended by the WHO will include a combined use of OPV and IPV.

Salk vaccine has been adopted by most western countries since 2006, but OPV continues to be used as the basic immunization for polio in most parts of the world, both as routine immunization and in mass immunization campaigns.

In 2012, a policy shift in use of polio vaccines occurred: "The polio eradication endgame plan is to switch from the trivalent oral polio vaccine, currently the vaccine of choice in most countries, to two vaccines: a new bivalent oral polio vaccine for routine immunization backed up by judicious use of inactivated polio vaccine (IPV)".

Eradication is within sight and the end-game strategy is important. In 2013, spread of Wild Polio Virus type 1 in the sewage in highly immunized Israel suggested that IPV alone may not prevent recurrence of polio spread.

Sources: *Tulchinsky TH, Goldblum N. Poliomyelitis immunization [letter to the editor]. N Engl J Med 2001;344:61–2. Available at: http://www.ncbi.nlm. nih.gov/pubmed/11187114 [Accessed 11 August 2012].*
Global Polio Eradication Initiative. Polio this week – as of 8 August 2012 [updated August 2012]. Geneva: Global Polio Eradication Initiative. Available at: http://www.polioeradication.org/Dataandmonitoring/Poliothisweek.aspx [Accessed 11 August 2012].
Aylward B. Ending polio, one type at a time. Bull World Health Organ 2012;90:482–3. Available at: http://www.who.int/bulletin/volumes/90/7/ 12-020712/en/index.html [Accessed 6 August 2012].

Advances in Prevention and Treatment of Infectious Diseases

Treatment of infectious diseases has also played a vital part in reducing the toll of disease and limiting its spread. Paul Ehrlich (1854–1915), seeking a "magic bullet", discovered an effective antimicrobial agent for syphilis (Salvarsan) and was awarded the Nobel Prize in 1908, jointly with Ilya Ilych Metchnikov (1845–1916) for the discovery of cellular immunity. In 1928, Alexander Fleming discovered the antibiotic quality of the momoldicillium, and in 1935 the first sulpsulfa(Prontosil) was discovered by German chemist Gerhard Domakh (1895-1964; 1939 Nobel Prize) followed by streptomycin by Selman Waksman (1888-1973) (Nobel Prizes 1945 and 1952). These and later generations of antibiotics have proven powerful tools in the treatment of infectious diseases.

Antibiotics and vaccines, along with improved nutrition, general health, and social welfare, led to dramatic reductions in infectious disease morbidity and mortality. As a result, optimistic forecasts of the conquest of communicable disease led to widespread complacency in the medical and research communities by the late twentieth century. In the 1990s, organisms resistant to available antibiotics constituted a major problem for public health and health care systems. Resistant organisms are now evolving as quickly as newer generation antimicrobials can be developed, threatening a return of diseases once thought to be under control. The pandemic of AIDS and other emerging and re-emerging diseases like SARS will require new strategies in treatment and prevention including new vaccines, antibiotics, chemotherapeutic agents, and risk reduction through community education.

Infections and Chronic Diseases

Since World War II, advances in immunology as applied to public health, eradication of smallpox, and near eradication of polio as well as the control and in some cases potential eradication of diphtheria, pertussis, tetanus, poliomyelitis, measles, mumps, rubella, and more recently hepatitis B and *Haemophilus influenzae* type b. The advent of immunizations to prevent infectious disease, and even potentially eradicate it, heralds a whole new area of endeavor for the vaccine field. The future in this field is promising and will play a central role in public health well into the twenty-first century.

In the past several decades, evidence for the long-held association between microbiological agents and cancer has accelerated and has now reached the point where effective treatment with antibiotics (for *H. pylori*) is associated with the cure of chronic peptic ulcer diseases. This, in turn, is associated with the decline in gastric cancer seen in developed countries.

The advent of hepatitis B vaccine gives real hope to the goal of reducing liver cancer, which is the third leading cause of cancer mortality globally, causing 695,000 deaths out of a total of 7.6 million cancer deaths in 2008. Hepatitis B affects 2 billion people worldwide, with 600,000 deaths from cirrhosis and cancer, while the global prevalence of hepatitis C is reported as 150 million people, with 350,000 deaths from cirrhosis and cancer (WHO, 2012, 2013). The recent discovery of Human Papilloma Virus (HPV) and the production of safe and effective vaccines provide hope for reducing even further cancer of the cervix, which has already been greatly reduced in most developed countries by Pap smear screening, and other cancers. The discovery of *H. pylori* as the main cause of chronic peptic ulcer diseases and gastric cancer offers hope for new breakthroughs in infectious origins of non-communicable disease which can be controlled or prevented by biomedical as well as behavioral changes. This breakthrough provides encouragement that infectious causes of other chronic diseases may lead to the development of new vaccines or antimicrobials for diseases long thought to be of genetic or environmental origin; the successes and great potential of public health are evident as ways are developed to continue the reduction in mortality and morbidity from chronic diseases.

MATERNAL AND CHILD HEALTH

Preventive care for the special health needs of women and children developed as a result of public concerns in the late nineteenth century. Public awareness of severe working conditions, especially for women and children, grew to encompass the health effects of poverty, poor living conditions and general hygiene, home deliveries, lack of prenatal care, and poor nutrition.

Preventive care as a service separate from curative medical services for women and children was initiated in the unsanitary urban slums of industrial cities in nineteenth-century France, in the form of "milk stations" (*gouttes de lait*). The plan was later expanded to a complete child welfare effort, especially promoting breastfeeding and a clean supply of milk to children, which had dramatic effects in reducing infant deaths.

The concept of child health spread to other parts of Europe and the USA with the development of pediatrics as a specialty and an emphasis on appropriate child nutrition.

Henry Koplik (1858–1927) in 1889 and Nathan Strauss (1848–1931) in 1893 promoted centers to provide safe milk to pregnant women and children in the slums of New York City in order to combat summer diarrhea. The Henry Street Mission, serving poor immigrant areas, developed the model of visiting nurses and public milk stations. The concept of the "milk station", combined with home visits, was pioneered by Lillian Wald (1867–1940), who coined the term *district nurse* or *public health nurse*. This became the basis for public prenatal, postnatal, and well-child care, as well as school health supervision. Visiting Nursing Associations (VNAs) gradually developed throughout the USA to provide such services. Physicians' services in the USA were mainly provided on a fee-for-service basis for those able to pay, with charitable services in large city hospitals. The concept of direct provision of care to those in need by local authorities and by voluntary charitable associations, with separation between preventive and curative services, is still a model of health care in many countries. Maternal mortality at the beginning of the twentieth century was at levels current in developing countries today. Since the 1920s, the maternal mortality rates drastically declined in the USA owing to improved access to professional prenatal care and delivery (Figure 1.1).

In Jerusalem in 1902, Shaare Zedek Hospital kept cows to provide safe milk for infants and pregnant women. In 1911, two public health nurses came from New York to Jerusalem to establish milk stations (*Tipat Halav*, "drop of milk") for poor pregnant women and children. This model became the standard method of Maternal and Child Health (MCH) provision throughout Israel, operating in parallel to the Sick Funds which provide medical care. The separation between preventive and curative services persists to the present, and is sustained by the Israeli national government's obligation to ensure basic preventive care to all, regardless of ethnicity, gender, prior conditions or ability to pay.

In the Soviet Union, institution of the state health plan in 1918 by Nikolai Semashko emphasized maternal and child health, along with epidemic and communicable disease control. All services were provided free as a state responsibility through an expanding network of polyclinics and prenatal and child care centers, including preventive check-ups, home visits, and vaccinations and other services. Infant mortality declined rapidly even in the Asian republics with previously poor health conditions.

During the 1990s, the USA was having difficulty in immunizing children in areas of high poverty in urban centers. Immunization was adopted as part of the USA's excellent Women, Infants, and Children (WIC) food support program for poor pregnant women and children. WIC's inclusion of immunization contributed to much higher coverage levels being achieved than in previous years.

The emphasis placed on maternal and child health continues to be a keystone and a major pillar of public health.

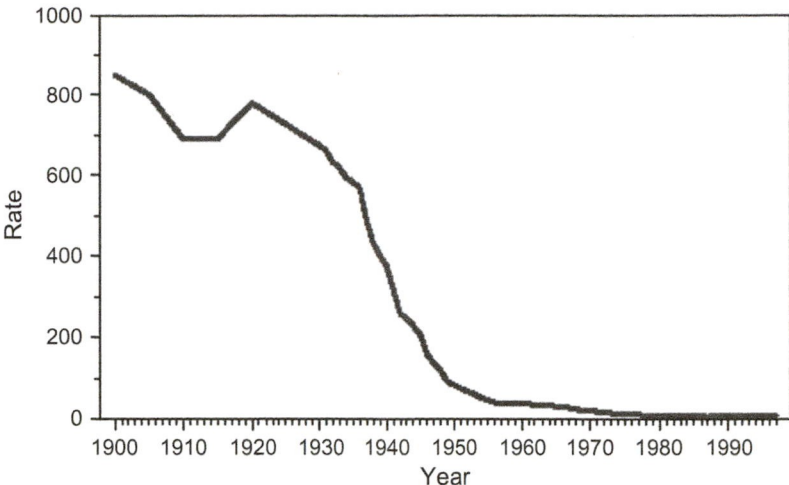

FIGURE 1.1   Maternal mortality per 100,000 live births, USA, 1900–1997. *Source: Centers for Disease Control and Prevention. Achievements in public health, 1900–1999: healthier mothers and babies. MMWR Morb Mortal Wkly Rep 1999;48:849–88. Available at: http://www.cdc.gov/mmwr/preview/mmwrhtml/mm4838a2.htm [Accessed 11 August 2012].*

Central to care of children and women in relation to fertility is the application of what later came to be called the "risk approach", where attention is focused on designing health programs for the most vulnerable groups in the population.

Maternal mortality remains a major public health issue globally, as currently approximately 800 women lose their lives to pregnancy or birth-related complications each day. Moreover, there is widespread serious morbidity due to anemia, susceptibility to infections and vesicovaginal fistulae causing drastic health and social effects. The 2001 Millennium Development Goals (MDGs) included reducing maternal mortality by one-third; this target has been met in some countries but largely failed in sub-Saharan Africa and in India. Child marriages and pregnancies, low levels of education, and lack of access to professional prenatal and delivery care are still widely prevalent causative conditions.

NUTRITION IN PUBLIC HEALTH

As infectious disease control and later maternal and child health became public health issues in the eighteenth to nineteenth centuries, nutrition gained recognition from the work of pioneers such as James Lind (see the earlier section entitled "Applied Epidemiology"). In 1882, Kanehiro Takaki (1849–1920), surgeon-general of the Japanese navy, reduced the incidence of beriberi among naval crews by adding meat and vegetables to their diet of rice. In 1900, Christiaan Eijkman (1858–1930), a Dutch medical officer in the East Indies, found that inmates of prison camps who ate polished rice developed beriberi, while those eating whole rice did not. He also produced beriberi experimentally in fowls on a diet of polished rice, thus establishing the etiology of the disease as a deficiency condition and fulfilling a nutritional epidemiological hypothesis. Eijkman

was awarded the Nobel Prize in Physiology or Medicine in 1929.

In the USA, the pioneering Pure Food and Drug Act was passed in 1906, stimulated by journalistic exposures of conditions in the food industry and Upton Sinclair's famous 1906 novel *The Jungle*. The legislation established federal authority in food and labeling standards, originally for interstate commerce, but later for the entire country. This provided for a federal regulatory agency and regulations for food standards. The Food and Drug Administration (FDA) has pioneered nutritional standards now used throughout the world.

In the early part of the twentieth century, the United States Department of Agriculture (USDA) supported "land grant colleges" and extension services in rural counties to promote agricultural improvement and good nutrition in poor agricultural areas of the country. These services, along with local women's organizations, helped to create a mass movement to improve nutrition. Actions included canning surplus foods, house gardening, home poultry production, home nursing, furniture refinishing, and other skills that helped farm families to survive the years of economic depression and drought. These, in turn, promoted better nutrition through education and community participation.

In 1911, the chemical nature of vitamin D was discovered, and a year later, Kasimir Funk (1884–1967) coined the term vitamin ("vital amine"). Goldberger's work established the dietary causes of pellagra, and in 1928 he discovered the pellagra-preventing factor in yeast (Box 1.13). In 1916, US investigators defined fat-soluble vitamin A and water-soluble vitamin B; the latter was later shown to be more than one factor. In 1922, Elmer McCollum (1879–1967) identified vitamin D in cod liver oil, which became a staple in child care for many decades. Rickets, still common in

BOX 1.13 Goldberger on Pellagra

"Mal de la rosa", first described in Spain by Casal in 1735, was common in northern Italy. In 1771, Frappolli described "pelle agra" or farmers' skin, often mistaken for leprosy. Pellagra was a common disorder among poor farm families whose diet consisted mainly of corn flour. Pellagra is characterized by the "4 Ds": diarrhea, dermatitis, dementia, and death. In 1818, Hameau described a widespread skin disease among poor farmers in southern France. Theophile Roussel (1816–1903) investigated and concluded that pellagra was endemic and due to poverty rather than a diet heavy in corn. The French Department of Agriculture implemented his recommended reforms, raising standards of living among poor farmers by encouraging them to grow wheat and potatoes instead of corn. Consequently, the disease disappeared by the beginning of the twentieth century. Similar measures in Italy reduced the growth of corn, and here too the disease disappeared. Lambrozo, in Verona, Italy, reported many cases of pellagra among mental hospital patients, also concluding that it was due to toxic material in corn. At the beginning of the twentieth century, the corn theory was less accepted and the common view was that pellagra was an infectious disease. British investigator L. V. Sambon, discoverer of the role of the tsetse fly in trypanosomiasis in 1910, took the view that the disease was mosquito borne.

Pellagra was first reported in the USA in 1906 as an epidemic in a mental hospital in Alabama. In the first decades of the twentieth century, pellagra was considered the leading public health problem in the southern USA, where poverty was rampant, and generally believed to be infectious in origin. Between 1907 and 1940, approximately three million Americans contracted pellagra and 100,000 of them died.

In 1913, Joseph Goldberger (1874–1929) of the US Public Health Service was appointed by the Surgeon General to investigate the pellagra epidemic in the southern USA. Goldberger had previously worked on yellow fever, dengue, and typhus. He visited psychiatric hospitals and orphanages with endemic pellagra and was struck by the observation that the staff was not affected, suggesting that the disease was not infectious. He recognized that non-transmission from patients to staff suggested a non-infectious cause. He postulated that it may have been due to the diet. In one mental hospital, he eliminated pellagra by adding milk and eggs to the diet and concluded that the disease was due to a lack of vitamins and preventable by a change in diet alone. He went on to establish the nutritional basis of this disease: a landmark in nutritional epidemiology in public health.

In 1999, the Centers for Disease Control and Prevention reported:

"Pellagra is a good example of the translation of scientific understanding to public health action to prevent nutritional deficiency. Pellagra, a classic dietary deficiency disease caused by insufficient niacin, was noted in the South after the Civil War. Then considered infectious, it was known as the disease of the four Ds: diarrhea, dermatitis, dementia, and death. The first outbreak was reported in 1907. In 1909, more than 1000 cases were estimated based on reports from 13 states. One year later, approximately 3000 cases were suspected nationwide based on estimates from 30 states and the District of Columbia. By the end of 1911, pellagra had been reported in all but nine states, and prevalence estimates had increased nearly ninefold.

During 1906–1940, approximately 3 million cases and approximately 100,000 deaths were attributed to pellagra. From 1914 until his death in 1929, Joseph Goldberger, a Public Health Service physician, conducted groundbreaking studies that demonstrated that pellagra was not infectious but was associated with poverty and poor diet. Despite compelling evidence, his hypothesis remained controversial and unconfirmed until 1937. The near elimination of pellagra by the end of the 1940s has been attributed to improved diet and health associated with economic recovery during the 1940s and to the enrichment of flour with niacin. Today, most physicians in the United States have never seen pellagra although outbreaks continue to occur, particularly among refugees and during emergencies in developing countries.

The growth of publicly funded nutrition programs was accelerated during the early 1940s because of reports that 25% of draftees showed evidence of past or present malnutrition; a frequent cause of rejection from military service was tooth decay or loss. In 1941, President Franklin D. Roosevelt convened the National Nutrition Conference for Defense, which led to the first recommended dietary allowances of nutrients, and resulted in issuance of War Order Number One, a program to enrich wheat flour with vitamins and iron. In 1998, the most recent food fortification program was initiated; folic acid, a water-soluble vitamin, was added to cereal and grain products to prevent neural tube defects."

Sources: *Centers for Disease Control and Prevention. Ten great public health achievements – United States, 1900–1999. MMWR Morb Mortal Wkly Rep 1999;48:241–3. Available at: http://www.cdc.gov/mmwr/preview/mmwrhtml/00056796.htm [Accessed 11 August 2012].*
Office of NIH History. Dr Joseph Goldberger and the war on pellagra: Goldberger vs. the south. Bethesda, MD: Office of NIH History. Available at: http://history.nih.gov/exhibits/goldberger/docs/south_6.htm [Accessed 11 August 2012].

industrialized countries prior to World War II and into the 1950s, virtually disappeared following fortification of milk with vitamin D. In the period 1931–1937, fluoride in drinking water was found to prevent tooth decay, in 1932 vitamin C was isolated from lemon juice.

Iodization of salt to prevent iodine deficiency disorders (IDD) has been one of the greatest successes, and failures, of twentieth-century public health. From studies in Zurich and in the USA, the efficacy of iodine supplements

in preventing goiter was demonstrated. In 1924 Morton's iodized salt became a national standard in the USA; it served as an early and noble example of voluntary public health action by private industry. In 1941, the USA initiated mandatory fortification of "iodine-enriched" salt, as well as flour (iron, vitamins B and later folic acid) and milk (with vitamin D). In Canada, iodized salt became mandatory in 1979 along with other vitamin and mineral fortification of bread and milk (see Chapter 8). Prevention of IDD by salt iodization has become an important goal in international health, and progress is being made towards universal iodization of salt in many countries where goiter, cretinism, and iodine deficiency are still endemic. In 1998, folic acid fortification of flour for prevention of birth defects was adopted in the USA, Canada, and Chile, and by 2013 had been adopted in over 60 other countries.

The international movement to promote proper nutrition is vital to reduce the toll of the malnutrition–infection cycle in developing countries. No less important is prevention of non-communicable diseases associated with overnutrition, including cardiovascular diseases, diabetes, and some cancers in industrialized nations. Nutrition is a key issue in the New Public Health, with international movements to eradicate vitamin and mineral (micronutrient) deficiency conditions, all of which are important, widespread, and preventable.

MILITARY MEDICINE

Professional armies evolved with urban civilizations and developed in the ancient world from about 4000 BCE. Since organized conflict began, armies have had to deal with the health of soldiers as well as treatment of the wounded. Injunctions on military and civilian camp placement and sanitation were clearly spelled out in the Bible (Old Testament). Roman armies excelled at construction of camps with care and concern for hygienic conditions, food, and medical services for the soldiers. Throughout history, examples of the defeat of armies by disease and lack of support services prove the need for paying serious attention to the health and care of the soldier. Studies of casualties of war in major conflicts contribute not only to military medicine but to knowledge of the care of civilian populations in natural or human-made disasters.

As the armies and weapons became increasingly powerful, the care of the sick and wounded became more complex. Military medicine perfected knowledge and skills in taking care of wounded on the battlefield and preventing loss of life. Epidemics in armies have killed more troops than weapons have; thus, treating and preventing disease is an integral part of military medicine. Many medical discoveries have been implemented in the army and later in civil society, including surgery, vaccination, antibiotics, and nutrition.

The Roman Empire developed military medicine as its professional armies spread across the known world. The Roman army included physicians to provide medical care for the legions, beginning with ensuring that only the best (and most intelligent) candidates were recruited. Once in service, the military medical corps strove to ensure the general health of the soldier by a continuous emphasis on hygiene. The design of legion fortifications and encampments ensured a healthy environment for the troops.

Following the destruction of Rome and later the eastern empire, the Roman military medical tradition disappeared. Military medicine during the Middle Ages was relatively primitive.

Jean Henri Dunant (1828–1910), a young Swiss businessman, arrived in Solferino, Italy, on the evening of the battle fought on 24 June 1859, between the French–Sardinian allies against the Austrian army. Some 38,000 injured, dying, and dead soldiers remained on the battlefield, with little attempt to provide care. Dunant took the initiative to organize volunteers from the local civilian population, especially the women and girls, to provide assistance to the injured and sick soldiers. He organized the purchase of needed materials and helped to erect makeshift hospitals, providing care for all without regard to their affiliation in the conflict. He worked alongside volunteer doctors and Austrian doctors captured by the French.

After returning to Geneva, Dunant published a book about his experience, *A Memory of Solferino*, describing the battle, its costs, and the chaos afterwards. He proposed that in the future a neutral organization should be established to provide care to wounded soldiers. His work led to the First Geneva Convention on the treatment of non-combatants and prisoners of war, now ratified by 194 countries. Moreover, his experiences and achievements resulted in the foundation of the International Committee of the Red Cross in 1863. Dunant was awarded the first Nobel Prize for Peace in 1901.

The Crimean War was a medical disaster for the British Army, characterized by higher mortality from disease than from battle, largely due to poorly organized sanitation, supply, and medical services. Mortality rates among British amputees in the Scutari Hospital averaged nearly 30 percent. Of every 100 men in the French forces admitted to military hospitals, 42 percent died – a hospital mortality rate equivalent to that of the Middle Ages. In November 1854, after the battle of Balaklava, Florence Nightingale and her 18 trained nurses introduced basic standards of hygiene, nutrition, sanitation, and administration in the British military hospitals. Upon her arrival, Nightingale reported a hospital mortality rate at Scutari of 44 percent. As a result of her efforts, the rate dropped remarkably, to 2 percent by the end of the war.

Nightingale's work made an enormous contribution to the knowledge and practice of hospital organization and management. On the opposing side of the same Crimean War, Nikolai Perogov (1810–1881), a military surgeon in the Russian czarist army, developed rectal anesthesia for field surgery. He also established triage of the wounded by degree of severity, as well as hygiene of wounds. Perogov defined improved systems for the management of the wounded in war theaters, which had applicability in civilian hospitals. The French army in World War I further developed the triage system of casualty clearance, now used worldwide in military and disaster situations, in public health emergencies, and in hospital emergency rooms.

Nutrition of sailors on long sea voyages and Lind's classic epidemiological study of scurvy, followed a century later by the work on beriberi, were crucial steps in identifying the importance of nutrition to public health. Bismarck's establishment of national health insurance and other benefits for workers was partly based on the need to improve the health of the general population in order to build mass armies of healthy conscripts (see Chapter 13). During conscription to the US Army in World War I, the high rates of rejection of draftees who were considered medically unfit for military service raised concerns over national health standards. Finding high rates of goiter in the draftees to the US Army led to efforts to identify high-risk areas and to reduce iodine deficiency in the civilian population by iodization of salt.

In the wake of World War I, a massive pandemic of influenza killed some 20 million people. The "Spanish flu" pandemic lasted from 1918 to 1919. Current estimates reveal that 50–100 million people worldwide died in this pandemic, described as the "greatest medical holocaust in history" as it may have killed as many people as the Black Death. The Spanish flu, closely following the huge losses of World War I, was to a large degree spread in the close quarters of army camps and by the mass movement of troops, with a high percentage of the deaths occurring among young men, the group most affected by the war itself.

Epidemics of louse-borne typhus in Russia, following the war and the Russian Revolution, contributed to the chaos of the period and prompted Lenin's famous statement, "Either socialism will conquer the louse, or the louse will conquer socialism". In World War II, sulfa drugs, antimalarials, and antibiotics made enormous contributions to the Allied war effort and later to general health care and preserving the health of the population. Lessons learned in war for protection of soldiers from disease and treatment of burns, crash injuries, amputations, battle fatigue, post-traumatic stress disorder, and many other forms of trauma were brought back to civilian health systems. Much of modern medical technology was first developed for or tested by the military. As an example, sonar radio wave mechanisms developed to detect submarines were adapted after World War II as ultrasound, now a standard non-invasive instrument in medical care.

In the twentieth century, two world wars, with enormous loss of human life not only of combatants but also of civilians, the eugenics movement and its downstream effects of the mass murder of handicapped people and genocide in the Holocaust, brought wider challenges to public health ethics. Not only had the destructiveness of war increased enormously with chemical, biological, and nuclear weapons of mass destruction, but medical and public health ethics were to have severe consequences on the public acceptance of public health as a profession and in civil society. The Nuremberg Trials addressed the Holocaust and unethical medical experimentation by the Nazi military on civilian and military prisoners. The infamous Tuskegee experiment of untreated syphilis in African American men in the USA in the period 1932–1972 brought these issues into the open arena of public debate. The International Declaration of Human Rights (1948), Convention on the Prevention and Punishment of the Crime of Genocide (1948) and the Helsinki Declaration (1964) set new standards for medical and research ethics, with important implications for public health (see Chapter 15).

The brutalities of wars against civilian populations have tragically recurred even near the end of the twentieth century in genocidal warfare in Iraq, Rwanda, the former Yugoslavia and in Syria (2011-2013). Those tragedies produced massive casualties and numerous refugees, with resultant public health crises requiring intervention by local and international health agencies. International and national public health agencies have a major responsibility for preventing the mass tragedies of the twentieth century from recurring, perhaps on a larger scale in the twenty-first century, with the spread of the potential for chemical, biological, and nuclear terrorism.

GLOBALIZATION OF HEALTH

Cooperation in health has been a part of international diplomacy from the first international conference on cholera in 1851 in Cairo to the health organization of the League of Nations after World War I, and further, into modern times. Following World War II, international health began to promote the widespread application of public health technology, such as immunization, to developing countries. Established in 1948, the World Health Organization's (WHO's) constitution was signed in 1946, with the charter defining health as "the complete state of physical, social and mental well-being, and not merely the absence of disease".

The tradition of international cooperation continues by organizations such as the WHO, the International Red Cross/Red Crescent (IRC), the United Nations Children's Fund (UNICEF), and a myriad of others. Under the leadership of the WHO, the eradication of smallpox was achieved by 1977, through united action, demonstrating that major threats to health could be controlled through international cooperation. The potential eradication of polio further demonstrates this principle.

The international spread of disease has taken enormous tolls on human life globally and the threat continues in the twenty-first century. Public health threats can emerge and spread rapidly, as seen with the human immunodeficiency virus (HIV) pandemic since the 1980s and SARS in 2003. More recently, concerns have grown over potentially devastating pandemic influenza, such as the H5N1 virus strain known as avian influenza.

Chronic diseases, including mental health, characterized the most common causes of mortality and disability in the industrialized countries in the latter half of the twentieth century. They are now also predominant in most developing countries with growing per capita incomes and middle-class communities. Obesity, diabetes, heart disease, and cancer are among the leading causes of morbidity and mortality in the modern world. The toll of violence, often overlooked as a public health problem, cannot be overstated. Many other factors affect health globally, including environmental degradation as it relates to global warming, accumulation of toxic wastes, acid rain, nuclear accidents, loss of biodiversity with the destruction of natural ecosystems such as the Amazon basin rainforests, and the human tragedy of chronic poverty of many people in developing countries and some even in the wealthiest of countries. Global health issues and social disparities are by their very nature beyond the capacity of individual or even groups of countries to solve. They require organized common efforts of governments, international agencies, and non-governmental organizations, cooperating with each other, with industry, and with the media to bring about change and reduce the common hazards caused by abuse of the environment and social gaps.

Tobacco is the leading preventable risk factor for premature mortality worldwide. Estimates of 5 million annual deaths attributable to smoking do not adequately measure the impact of the tobacco pandemic. Tobacco use contributes to and exists as a comorbidity factor in a very wide spectrum of diseases and has justly become a major issue of public health. Over half of the estimated 650 million people currently smoking will die from effects of this addiction and if current smoking patterns continue, it will cause more than 10 million deaths yearly by 2020. Recognition of tobacco control is gaining the support of researchers, medical professionals, politicians, and communities around the world. The Centers for Disease Control and Prevention (CDC) sees recognition of tobacco as a public health hazard as one of the greatest public health achievements of the twentieth century in the USA. The WHO's Framework Convention on Tobacco Control, adopted by the 56th World Health Assembly in 2003, placed elimination of tobacco use as one of the greatest public health challenges, but one far from fruition.

Bringing health care to all is as great a challenge as feeding a rapidly growing global population. Successes in the eradication of smallpox and the control of many other diseases by public health measures show the potential for concerted international cooperation and action targeted at specific objectives that reduce disease and suffering.

EPIDEMIOLOGICAL TRANSITION

As societies evolve, so do patterns of disease. These changes are partly the result of public health and medical care but just as surely are due to improved standards of living, nutrition, housing, and economic security, as well as changes in fertility and other family and social factors. As disease patterns change, so do appropriate strategies for intervention.

During the first half of the twentieth century, infectious diseases predominated as causes of death even in the developed countries. Since World War II, a major shift in epidemiological patterns has taken place in the industrialized countries, with the decline in infectious diseases and the growth in non-infectious diseases as causes of death. Increases in longevity have occurred primarily from declining infant and child mortality, improved nutrition, control of vaccine-preventable diseases, and the advent of antibiotics for treatment of acute infectious diseases. The rising incidence of cardiovascular diseases and cancer affects primarily older people, leading to a growing emphasis in epidemiological investigations on causative risk factors for these non-infectious diseases.

Studies on the distribution of non-infectious diseases in specific groups go back many centuries to when the Romans reported excess death rates among specific occupational groups. These studies were updated by Ramazzini in the early eighteenth century. As noted earlier, in eighteenth-century London, Percivall Pott documented that cancer of the scrotum was more common among chimney sweeps than in the general population. Nutritional epidemiological studies, from Lind on scurvy among sailors in 1747 to Goldberger on pellagra in the southern USA in 1914, focused on nutritional causes of non-infectious diseases in public health.

Observational epidemiological studies of "natural experiments" produced enormously important data in the early 1950s, when pioneering investigators in the UK, Richard Doll (1912–2005), Austin Bradford Hill (1897–1991), and Richard Peto (1943–present), demonstrated a relationship between tobacco use and lung cancer. They followed the mortality patterns of British patients from different causes, especially lung cancer. They found that mortality rates from lung cancer were 10 times higher in smokers than in non-smokers. Epidemiological studies pointing to the relationship of diet and hypertension with cardiovascular diseases also provided critically important material for public health policy. This raised public concern and consciousness in western countries of the impact and influence of lifestyle on public health. In this new era of public health, the complementary relationship between miasma and germ theories is recognized (Box 1.14). These issues are discussed subsequently throughout this book.

In the mid-twentieth century, while communicable diseases were coming under control, risks related to modern living developed. These include cardiovascular disease, trauma, cancer, and other chronic diseases, which have become the predominant causes of premature death, hospitalization, and disability. These conditions are more complex than the infectious diseases, both in causation and in the means of prevention. Despite the complexity and associated challenges, public health interventions have shown surprising success in combating this set of mortality patterns, with a combination of improved medical care and activities under the general title of health promotion.

BOX 1.14 Complementarities of the Miasma and Germ Theories

In the mid-nineteenth century, the miasma theory (i.e., the concept that airborne vapors or "miasmata" caused most diseases) competed with the germ theory (i.e., specific microorganisms cause specific diseases). The latter gained pre-eminence among scientists and biological sciences, yet the miasma theory was the basis for action by sanitary reformers. Miasma explained why cholera and other diseases were epidemic in places where the undrained sewage water was foul smelling. Their endeavors led to improved sanitation systems, which resulted in decreased episodes of cholera. The connection between dirtiness and diseases led to public health reforms and encouraged environmental and personal cleanliness. The miasma theory was consistent with the observations that disease was associated with poor sanitation and foul odors, and that sanitary improvements were successful in reducing disease.

Echoes of these two theories continue to compete even today. Applied to chronic disease related to toxins (e.g., smoking) or nutritional indicators (e.g., blood lipids and micronutrient deficiency conditions), the environmental focus of the sanitary reformers is reflected in the idea that environmental and social conditions are the main factors in these diseases. This is in contrast to the more biomedically oriented approach of the intellectual descendants of germ theory of infectious diseases. Clearly, both are operative, with improved infectious disease control and environmental and social conditions all contributing to improved longevity and reduced burden of many diseases. However, large gaps remain between rich and poor as a result of differential social, economic, and cultural differences.

In 2007, the *British Medical Journal* conducted an opinion survey of the most important medical innovations of all time. The clear winner was the "sanitary revolution as greatest medical advance since 1840". Major problems remain today, as millions of people die annually from lack of modern sanitation of safe drinking water and poor sewage and solid waste disposal. Other large-scale killers are infectious diseases for which highly effective vaccines or other management tools exist.

Even in developed countries, management of infectious diseases is still a major public health issue, and includes the annual loss of life from influenza, pneumonia, medically related diseases such as multidrug-resistant tuberculosis, and the rise of drug-resistant organisms, which were easily controlled by antibiotics a short generation ago. Biomedical and socioecological approaches will need to be combined in the future to address these major issues.

Sources: Ferriman A. BMJ readers choose the "sanitary revolution" as greatest medical advance since 1840. BMJ 2007;334:111. Available at: http://www.bmj.com//content/334/7585/111.2?variant=f ull-text [Accessed 11 August 2012].
Mackenbach JP. Sanitation: pragmatism works. BMJ 2007;334: s17. Available at: http://www.bmj.com/content/334/suppl_1/s17.full [Accessed 11 August 2012].

At the beginning of the twenty-first century, the need to link public health with clinical medical care and organization of services became increasingly apparent. The decline in coronary heart disease mortality has been accompanied by a slow increase in morbidity, and recent epidemiological evidence shows new risk factors not directly related to lifestyle, but requiring longitudinal preventive care to avert early recurrence and preventable premature death. Progress continues into the twenty-first century as new challenges arise.

ACHIEVEMENTS OF PUBLIC HEALTH IN THE TWENTIETH CENTURY

The foundations of public health organization were established in the second half of the nineteenth and first half of the twentieth centuries. Water sanitation, waste removal, and food control developed at municipal and higher levels of government, the establishment of organized local public health offices with state and federal grants, and improved vaccination technology all contributed to the control of communicable diseases. Organized public health services implemented the regulatory and service components of public health in developed countries, with national standards for food and drug safety, state licensing, and discipline in the health professions.

At the beginning of the twentieth century, there were few effective medical treatments for disease, but improved public health standards resulted in reduced mortality and increased longevity. As medical technology improved following World War II with antibiotics, antihypertensives, and psychotherapeutic drugs, much focus shifted to curative medical care, with a widening chasm between public health and medicine. In our time, a new interest in the commonality between the two is emerging, as new methods of organizing and financing health care develop. The aim is to contain the rising costs of health care, while simultaneously increasing the utilization of preventive medicine.

National and state efforts to promote public health during the twentieth century widened in scope of activities and financing programs. This required linkage between governmental and non-governmental activities for effective public health services. Dramatic scientific innovations brought vaccines and antibiotics that, in conjunction with improved nutrition and living standards, helped to control infectious disease as the major cause of death. In the developed countries, the advent of national or voluntary health insurance on a wide scale opened access to health care to all of the population.

The modern era of public health from the 1960s to today has brought a new focus on non-infectious disease epidemiology and prevention. Important epidemiological studies of the impact of diet and smoking on cardiovascular diseases, and smoking on lung cancer, identified crucial

BOX 1.15 Ten Great Achievements of Public Health in the USA in the Twentieth Century

During the twentieth century, health and life expectancy in the USA improved dramatically. Since 1900, average lifespan has lengthened by over 30 years, and 25 years of this gain is attributable to advances in public health. *Morbidity and Mortality Weekly Report* (*MMWR*) profiled 10 public health achievements in a series of reports published in 1999. These reflect similar public health achievements in many industrialized countries:

- Control of infectious disease
- Vaccination
- Motor vehicle safety
- Safer workplaces
- Decline in deaths from coronary heart disease, strokes
- Safer and healthier foods
- Healthier mothers and babies
- Family planning
- Fluoridation of drinking water
- Recognition of tobacco as a health hazard.

Source: *Centers for Disease Control and Prevention. Ten great public health achievements – United States, 1900–1999. MMWR Morb Mortal Wkly Rep 1999;48:241–3. Available at: http://www.cdc.gov/mmwr/preview/mmwrhtml/00056796.htm [Accessed 11 August 2012].*

preventable risks factors for chronic disease. As a result of these and similar studies of disease and injury related to the environment, modern public health has, through health promotion and consumer advocacy, played a significant role in mortality and morbidity reduction for a spectrum of diseases. For prevention of premature disease and death, more comprehensive approaches will be needed by public health and health care providers than have been developed to date.

The twentieth century saw great achievements in public health in the industrialized countries, indeed throughout the world. The CDC reviewed these achievements in a series of publications in 1999, which represent the potential for public health and, if not a universal "gold standard", at least a well-documented set of achievements and potential for public health everywhere (Box 1.15).

The dream of international and national health agencies to achieve *Health for All* faces serious obstacles of inequities, lack of resources, distortions with overdevelopment of some services at the expense of others, and competing priorities. Managing health care to use resources more effectively is now a concern of every health professional. At the same time, public expectations are high for unlimited access to care, including the specialized and highly technical services that can overwhelm the available budgetary and personnel resources. All nations, wealthy or poor, face the problem of managing limited resources, exacerbated since 2008 by a severe global recession. How that will be achieved is part of the challenge we discuss as the New Public Health.

CREATING AND MANAGING HEALTH SYSTEMS

Provision of medical care to the entire population is one of the great challenges of public health. Governments as insurers, providers, or regulators of health care have broad responsibilities for health of the people. As will be discussed in subsequent chapters, nations have many reasons to ensure health for all, just as they promote universal education and literacy. National interests in the late nineteenth and early twentieth centuries were defined to include having healthy populations, especially of workers and soldiers, and for national prestige. Responsibility for the health of a nation included measures for the prevention of disease, but also financing and prepayment for medical and hospital care. National policies gradually took on measures to promote health, structures to evaluate the health of the nation, and modification of policies to keep up with changing needs.

The health of a population requires that people have access to medical and hospital services as well as preventive care, a healthy environment, and a health promotion policy orientation. Greek and Roman cities appointed doctors to provide free care for the poor and the slaves. Medieval guilds provided free medical services to their members. In 1883, Germany introduced compulsory national health insurance to ensure healthy workers and army recruits, which would provide a political advantage. In 1911, Britain's Chancellor of the Exchequer, Lloyd-George, instituted the National Insurance Act, providing compulsory health insurance for workers and their families. In 1918, following the October Revolution, the Soviet Union created a comprehensive state-operated health system with an emphasis on prevention, providing free comprehensive care in all parts of the country.

During the 1920s, national health insurance was expanded in many countries in Europe. Following the Great Depression of the 1930s and hopes raised by the Allied victory in World War II, important social and health legislation was enacted to provide health care to the populations of Britain, Canada, and the USA. In Britain the welfare state including the National Health Service (NHS) was developed by the Labour Government. In Canada, a more gradual development took place in the period from 1946 to 1971, including the establishment of national pensions and a national health insurance program. In the USA, social legislation has been slow in coming following the defeat of national health insurance legislation in Congress in 1946 and long-standing ideological opposition to "socialized medicine". In 1965, however, universal coverage of the population over age 65 (Medicare) was instituted and coverage for the poor under Medicaid soon followed. Inadequate coverage of workers and low-income American families is still a serious problem, although health reforms initiated by President Barak Obama in 2009 and approved by Congress in 2010 promise to extend health coverage to millions of Americans in the coming years. The Affordable Care Act is politically controversial; however, when challenged in the Supreme Court it was ruled as within constitutional limits. It will add many millions of Americans to health insurance but not yet to universal health care, which is the gold standard in all other developed countries. In the first decade of the twenty-first century, virtually every country has in principle undertaken responsibility to provide for the health and social well-being of its population, but the gaps and inequities in their achievements are wide. The term *health systems* may imply a formalized structure or a network of functions that work together to meet the needs of a population through health insurance or health service systems. Private health insurance is still the dominant mode in the USA, but the elderly and the poor are covered by government health insurance (see Chapter 13). The American public health community is currently seeking means to achieve universal health coverage. Prepayment for health is financed through general tax revenues in many countries, and in others through payment by workers and employers to social security systems. Both developed and developing countries are involved in financing health care, as well as research and training of health professionals.

Industrialized countries share increasing concerns of cost escalation, with health expenditure costs impacting general economic growth. While health care is a large-scale employer in all developed countries, high and rising expenditures for health, reaching over 16 percent of gross domestic product in the USA, and around 10 percent in many other western countries, is a major factor in stimulating health care reform. Many countries are struggling to keep up with the rising costs of technology and competition from other social needs, such as education, employment, and social welfare, all of which are important for national health and well-being. Some economic theories allocate no economic value to a person except as an employee and a consumer. Liberal and social democratic political philosophies advocate an ethical concern and societal responsibility for health. Both approaches now concur that health has social and economic value. The very success of public health has produced a large increase in longevity, thus increasing the percentage of elderly people in the population, raising ethical and economic questions regarding improved preventive, therapeutic and diagnostic techniques, health care consumption, allocation of services, and social support systems.

For developing countries, providing health care for the entire population is a distant dream. Limited resources and overspending on high-technology facilities in larger cities leave little funding for primary care for the rural and urban poor. Despite this, there has been real progress in implementing fundamental services such as immunization and prenatal care. Still, millions of preventable deaths occur annually because of a lack of basic primary care programs.

CHALLENGES IN THE TWENTY-FIRST CENTURY

Dramatic events in the late decades of the twentieth century and the first decade of the twenty-first have deeply affected public health. Since the 1960s, the capacity of public health has widened, with local, national, and global perspectives. The public sector has come to work with private sector influences on activities such as the global eradication of polio and the control of AIDS and malaria in sub-Saharan Africa. New attention has come to focus on many unmet human rights issues such as slave trafficking and genocide, which are still inadequately addressed in the international development agenda. Economic globalization is increasingly associated with health and human rights in academic and public discourse. They have met partial success and some failures in achieving global health goals in the context of political, economic, and public health values. The global recession that began in 2008 will undoubtedly have a negative impact on such global health goals, but the clear advantage of cost-effective prevention interventions will become more and more apparent.

The understanding of the enormous impact of smoking reduction, healthier diets, improved road safety, improved medical care for hypertension, and long-term management of chronic conditions, mainly in the developed countries, is now being recognized as essential for developing countries at all stages of the development process. Public health achievements are reaching mid-level developing countries and the emerging economies such as Brazil, India, China, South Africa, and South-East Asia. Despite powerful economic growth, China and India especially still remain poor for the majority of their citizens. The countries lagging behind in sub-Saharan Africa still suffer from weak infrastructure, corruption, and political instability, which prevent the use of their natural resources to build modern economies and civil structures of public health. Political instability and inadequate infrastructure and the mixed effects of globalization and recession also negatively influence public health.

Scientific advances in the discovery of causes of some chronic diseases in the late twentieth century are coming as public health advances in the prevention of birth defects (by folic acid consumption before and during pregnancy), cancers of the stomach (from *H. pylori* infection), liver (from hepatitis B infection), and cervix (from HPV infection). These remarkable discoveries hold out hope for further advances in immunology in the years ahead. Similarly, advances in knowledge of nutrition, genetics, nanotechnology and chronic disease are already contributing to reductions in cardiovascular disease and cancer mortality, and perhaps incidence, with more benefits to come in the years ahead. Success in reducing mortality from HIV as a result of the wider availability and lower costs of antiretroviral drugs has opened a new phase in HIV control by reducing transmission, such as from mothers to infants, but also for preventive care after exposure and prevention of infection in high-risk groups. Here again, economic recession may hinder the sustainability of such programs, such as in preventing maternal–fetal HIV transmission.

At the same time, malaria control is disappointingly difficult in many countries where it is still a major cause of childhood mortality; the search for an effective vaccine is one of the great challenges so far unsuccessfully met in immunology. Drug-resistant diseases are also an increasing challenge awaiting scientific advances.

The first decade of the new century saw the enormous effects of terrorism, such as the 9/11 attack in New York and the 7/7 attack in London, and these effects are not limited to the direct deaths and casualties. Major wars and chronic low-grade warfare in Afghanistan and Pakistan have serious consequences not only in military and civilian casualties, but in financial burden and economic downturn. Recurrent civil wars such as those plaguing some African countries have tragic local effects but also hinder the adequate development of public health infrastructure and public policy. Disasters such as the Thailand tsunami and floods in Pakistan and China in 2010, the enormous earthquake in Haiti and the Japanese tsunami of 2011 caused huge devastation, loss of life, and the subsequent effects of homelessness, poor sanitation, and consequent disease and death.

The world now faces a long-term threat of weapons of mass destruction in the hands of fanatically religious, transnational groups who may promote the use of such weapons in genocidal acts following years of incitement, a known precursor to genocide. The public health consequences may be devastating and result in new genocides, perhaps on an unimaginable scale, with weapons of mass destruction available to radical regimes and terrorist organizations.

Climate change has serious and far-reaching health implications, now and in the future. It is already changing the distribution of some infectious disease vectors; of death, disease, and injury from heat waves; of floods, storms, fires, and droughts. Climate change may also cause social disruption, economic decline, and displacement of populations, all of which may impact health substantially, affecting especially the very young and the elderly, the physically and mentally disabled, the poor and economically disadvantaged, and other marginalized groups. Local public health agencies are tasked with showing leadership in dealing with the effects of disasters on this issue. Budget and labor cuts in a time of recession can seriously jeopardize emergency preparedness planning and preparation.

Great challenges and opportunities lie ahead for public health, including access to future scientific and health technology achievements in genomics, nanotechnologies, and other scientific advances in vector control, vaccines, cancer prevention and management, and coping with diseases associated with aging and mental illness. The economics of health care will be challenged by rising costs, economic recession, and still prevalent inequalities in health status. The economic crisis building since 2008 will push more people into poverty and unemployment while reducing spending on public services, including public health and social support systems. Privatization of health care and public health services will reduce access to needed care.

Migration, civil strife, and threats of genocide will continue to challenge human rights and public health. Natural and human-made disasters now and in the future will cause large-scale loss of life and social disruption. Yet health technology and the widening use of currently available preventive measures will increase longevity in more and more countries. The translation of scientific and technical capacity to meet human needs will continue to be a challenge for civil societies and public health. Human resources for health continues to be a serious challenge for health systems, and with aging societies, new health disciplines such as community health workers will be needed to meet human needs. These and many other issues in public health will raise important ethical questions and challenges for health promotion in mandating risk reduction, such as in smoking and dealing with apathy or opposition to essential programs of public health protection such as immunization for population as well as individual health. Inequalities will perhaps be the greatest ethical challenge to face.

History has shown that societies can achieve greater longevity and a healthier quality of life by the application of public health and health promotion measures. The achievements of the past can be equaled in the future with the brilliant achievements of individual and systems research and the application of findings. Many innovations face apathy and serious resistance, with a high cost in unnecessary morbidity and mortality. These and related issues will be discussed in the chapters that follow and in recommended readings, and in specialized courses in the broad context of public health.

SUMMARY

The history of public health is directly related to the evolution of thinking about health. Ancient societies in one way or another realized the connection between sanitation and health and the role of personal hygiene, nutrition, and fitness. The sanctity of human life (*Pikuah Nefesh*), which established an overriding human responsibility to save life, and improving the world (*Tikkun Olam*), were both derived from Mosaic Law from 1500 BCE. The scientific and ethical basis of medicine was also influenced largely by the teachings of Hippocrates in the fourth century BCE. Sanitation, hygiene, good nutrition, and physical fitness all had roots in ancient societies, including the obligations of the society to provide care for the poor. These ethical foundations support efforts to preserve life even at the expense of other religious or civil ordinances.

Social and religious systems linked disease to sin and punishment by higher powers. Moreover, they viewed investigation or intervention by society (except for relief of pain and suffering) as interference with God's will. Childbirth was associated with pain, disease, and frequent death as a general concept of "in sorrow shall you bring forth children". Health care was seen as a religious charitable responsibility to ease the suffering of sinners. In the modern world, the effectiveness of medical and preventive care is widely accepted, yet religious and cultural issues are still expressed as limitations in funding for family planning, birth-related care, and many other public health issues.

The clear need and responsibility of society to protect itself by preventing the entry or transmission of infectious diseases was driven home by pandemics of leprosy, plague, syphilis, smallpox, measles, and other communicable diseases which occurred in the Middle Ages. The diseases themselves evolved and pragmatic measures were gradually found to control their spread, including the isolation of lepers, quarantine of ships, and closure of public bath houses. Epidemiological investigations of cholera, typhoid, occupational diseases, and nutritional deficiency disorders in the eighteenth and nineteenth centuries began to show causal relationships and effective methods of intervention before scientific proof of causation was established.

Public health practice continued to evolve on a pragmatic basis, often before full scientific basis of the causation of many diseases had been worked out. Public health organizations to ensure basic community sanitation and other modalities of prevention evolved through the development of local health authorities, fostered, financed, and supervised by civic, state, or provincial and national health authorities, as governments became increasingly involved in health issues. Pioneers such as Lillian Wald brought public health nursing to the homes of the sick and poor immigrants in New York City at the beginning of the twentieth century. Through Wald's efforts, we now understand that the application of outreach, home care, and preventive work by community health workers of all kinds is part of the complex of modern public health.

Freeing human thought from restrictive dogmas that limited the scientific exploration of health and disease fostered the search for the natural causation of disease. During the Enlightenment, scientific inquiry emerged and natural philosophy was strongly tied to spiritual and religious motives. This was of paramount importance in seeking interventions and preventive activities. This concept, first articulated in ancient Greek medicine, provided the basis for clinical and scientific observations leading to the successes of public health over the past two centuries. The epidemiological method led to public health interventions before the biological basis of disease was determined. Sanitation to prevent disease was accepted in many ancient societies, and codified in some as part of civil and religious obligations. Lind's investigation of scurvy, Jenner's discovery of vaccination to prevent smallpox, and Snow's investigation of cholera in London demonstrated the investigation of disease using modern scientific epidemiological methods. Their results were eventually accepted despite a lack of contemporary biochemical or bacteriological proof. Their remarkable contributions helped to formulate the core methodology of public health.

Public health has developed through pioneering epidemiological studies, devising forms of preventive medicine, and community health promotion. Reforms pioneered in many areas, from the abolition of slavery and serfdom to the provision of state-legislated health insurance, have all improved the health and well-being of the general population. In the last years of the twentieth century, the relationship between health and social and economic development gained recognition internationally. The twentieth century has seen a dramatic expansion of the scientific basis for medicine and public health. Immunology, microbiology, pharmacology, toxicology, and epidemiology have provided powerful tools and resulted in improved health status of populations. New medical knowledge and technology have come to be available to the general public in many countries in the industrialized world through the advent of national health insurance. In this century, virtually all industrialized countries established systems of ensuring access to care for their whole population as essential for the health of both the individual and the collective.

Major historical concepts have had profound effects on the development of public health. The idea of sickness as punishment for sin prevented attempts to control disease over many centuries. This mentality persists in modern times through "blaming the victim". AIDS patients are seen as deserving their fate because of their behavior; workers are believed to become injured because of their own negligence; and the obese person and the smoker are believed to deserve their illnesses because of weakness in the way they conduct their lives. The sanctity of human life, improving the world, and human rights are fundamental to the ethics and values of public health, as is charity in care, in which there is a societal and professional responsibility for kindness and relief of suffering. Ethical controversies pitting individual rights versus community benefit are still ongoing in many diverse areas such as universal health insurance, food fortification, fluoridation of water supplies, managed care, reproductive health, cost–benefit analysis, euthanasia, the care of sex workers and prisoners, and many others.

Acceptance of the right to health for all by the founders of the United Nations and the WHO added a universal element to the mission of public health. This concept was embodied in the constitution of the WHO and given more concrete form in the *Health for All* concept of Alma-Ata, which emphasized the right of health care for everyone and the responsibility of governments to ensure that right. This concept also articulates the primary importance of prevention and primary care, which became a vital issue in competition for resources between public health and hospital-oriented health care.

The lessons of history are important in public health. Basic issues of public health need to be revived because new challenges for health appear and old ones re-emerge. The philosophical and ethical basis of modern public health is a belief in the inherent worth of the individual and his or her human right to a safe and healthful environment. The health and well-being of the individual and the community are interdependent. Investment in health, as in education, is a contributor to economic growth, as healthy and educated individuals contribute to a creative and economically productive society.

Globalization in health has entered a new phase in the twenty-first century with the articulation of the MDGs in 2001, accepted by virtually all countries in the world as a basis for poverty reduction, improved education and health standards, especially for women and children, and in the control of HIV, malaria, and other diseases. The progress made up to 2013 has been substantial in many regions, with reduction in rates of extreme poverty, enrolling children into primary schools, addressing AIDS, malaria and child health, and a good likelihood of reaching the target for access to clean drinking water. Despite this progress, progress in sub-Saharan Africa has been very slow, with no advances at all in addressing maternal mortality rates.

The New Public Health emerges from the experience of history. Organized activity to prevent disease and promote health had to be relearned from the ancient and postindustrial revolution worlds. Over the coming decades of the twenty-first century, we must learn from a wider framework how to use all health modalities, including clinical and prevention-oriented services, health promotion, and proactive efforts in the public and private sectors to effectively and economically preserve, protect, and promote the health of individuals and of greater society. The New Public Health, as public health did in the past, faces ethical issues that relate to health expenditures, priorities, and social philosophy. Throughout the course of this book, we discuss these issues and attempt to illustrate a balanced, modern approach towards the New Public Health.

HISTORICAL MARKERS

3000 BCE	Dawn of Sumerian, Egyptian, and Minoan cultures – drains, flush toilets
2000 BCE	Indus valley – urban society with sanitation facilities
1700 BCE	The Code of Hammurabi – rules governing medical practice
1500 BCE	Mosaic Law – personal, food, and camp hygiene, segregating lepers, overriding duty of sanctity of human life (Pikuah Nefesh) and improving the world (Tikkun Olam) as religious imperatives
400 BCE	Greece – personal hygiene, fitness, nutrition, sanitation, municipal doctors, occupational health; Hippocrates – clinical and epidemic observation and environmental health
500 BCE to 500 CE	Rome – aqueducts, baths, sanitation, municipal planning, and sanitation services, public baths, municipal doctors, military, and occupational health
170 CE	Galen – physiology, anatomy, humors dominated western medicine until 1500 CE
500–1000	Europe – destruction of Roman society and the rise of Christianity; sickness as punishment for sin; mortification of the flesh, prayer, fasting, and faith as therapy; poor nutrition and hygiene, pandemics; antiscience; care of the sick as religious duty
700–1200	Islam – preservation of ancient health knowledge, schools of medicine, Arab–Jewish medical advances (Ibn Sinna and Maimonides)
1000+	Universities and hospitals in Middle East and Europe
1000+	Rise of cities, trade, and commerce, craft guilds, municipal hospitals
1096–1272	Crusades – contact with Arabic medicine, hospital orders of knights, leprosy
1268	Roger Bacon publishes treatise on use of eyeglasses to improve vision
1348	Venice – board of health and quarantine established
1348–1350	Black Death – origins in Asia, spread by armies of Genghis Khan, world pandemic kills 60 million in fourteenth century, one-third to one-half of the population of Europe
1300	Pandemics – bubonic plague, smallpox, leprosy, diphtheria, typhoid, measles, influenza, tuberculosis, anthrax, trachoma, scabies, and others until eighteenth century

1400–1600s	Renaissance and Enlightenment, decline of feudalism, rise of urban middle class, trade, commerce, exploration, new technology, printing, arts, science, anatomy, microscopy, physiology, surgery, clinical medicine, hospitals (religious, municipal, voluntary)
1518	Royal College of Physicians founded in London
1532	Bills of Mortality published
1546	Girolamo Fracastorus publishes De Contagione – the germ theory
1562–1601	Elizabethan Poor Laws – responsibility for the poor on local government
1628	William Harvey publishes findings on circulation of the blood
1629	London Bills of Mortality specify causes of death
1639	Massachusetts law requires recording of births and deaths
1660s	Leyden University strengthens anatomical education
1661	John Graunt founds medical statistics
1661	Rene Descartes publishes first treatise on physiology
1662	Royal Society of London founded by Francis Bacon
1665	Great Plague of London
1673	Antony van Leeuwenhoek – microscope, observes sperm and bacteria
1667	Pandemics of smallpox in London; pandemic of malaria in Europe
1687	William Petty publishes Essays in Political Arithmetic
1700	Bernardino Ramazzini publishes compendium of occupational diseases
1701	London – 75% of newborns die before fifth birthday
1701	Variolation against smallpox practiced in Constantinople, isolation practiced in Massachusetts
1710	English Quarantine Act
1720+	London – voluntary teaching in hospitals; Guy's, Westminster
1721	Lady Mary Montagu introduces inoculation for smallpox to England
1730	Science and scientific medicine; Rights of Man, encyclopedias, agricultural and industrial revolutions, population growth – high birth rates, falling death rates
1733	Obstetrical forceps invented
1733	Stephen Hales measures blood pressure
1747	James Lind – case–control study of scurvy in sailors
1750	British naval hospitals established
1750	John Hunter establishes modern surgical practice and teaching
1752	William Smellie publishes textbook of midwifery
1762	Jean Jacques Rousseau publishes the Social Contract
1775	Percivall Pott investigates scrotal cancer in chimney sweeps
1777	John Howard promotes prison and hospital reform in England
1779	Johann Frank promotes Medical Police in Germany
1785	William Withering – discovers foxglove (Digitalis) treatment of dropsy
1788	UK legislation to protect boys employed as chimney sweeps
1796	Edward Jenner – vaccinates 24 children against smallpox from milkmaid's cowpox pustules
1796	British Admiralty adopts daily issue of lime juice for sailors at sea to prevent scurvy
1797	Massachusetts legislation permitting local boards of health
1798	Philippe Pinel removes chains from insane in Bicetre Asylum in France
1798	President John Adams signs law for care of sick and injured seamen, establishing marine hospital service, later becoming US Public Health Service (1912)
1800	Britain and US establish Municipal Boards of Health
1800	Vaccination adopted by British army and navy
1800	Adam Smith, Jeremy Bentham – economic, social philosophers
1801	Vaccination mandatory in Denmark, local eradication of smallpox
1801	First national census, UK
1804	Modern chemistry established – Humphrey Davey, John Dalton
1807	UK Abolition Act – mandates eradication of international slave trade enforced by the Royal Navy

1827	Carl von Baer in St. Petersburg establishes science of embryology
1834	UK Poor Law Amendment Act documents harsh state of urban working class in the USA
1837	UK National Vaccination
1830s–1840s	Sanitary and social reform, growth of science; voluntary societies for reform, boards of health, mines and factory acts – improving working conditions
1842	Edwin Chadwick – UK Poor Law Commission on Sanitary Conditions of the Labouring Population of Great Britain – links poverty and disease
1844	Horace Wells – anesthesia in dentistry, then surgery in the USA
1848	UK Parliament passes Public Health Act establishing the General Board of Health
1850	Massachusetts – Shattuck Report of Sanitary Commission
1852	Adolph Chatin uses iodine for prophylaxis of goiter
1854	John Snow – waterborne cholera in London: the Broad Street pump
1854	Florence Nightingale, modern nursing and hospital reform – Crimean War
1855	London – mandatory filtration of water supplies and consolidation of sanitation authorities
1858	Louis Pasteur proves no spontaneous generation of life
1858	Rudolf Virchow publishes *Cellular Pathology*; pioneer in political–social health context
1858	Public Health and Local Government Act and Medical Act in UK – local health authorities and national licensing of physicians
1859	Charles Darwin publishes *On the Origin of Species*
1861	Emancipation of the serfs in Russia
1861	Ignaz Semmelweiss publishes *The Cause, Concept and Prophylaxis of Puerperal Fever*
1862	Louis Pasteur publishes findings on microbial causes of disease
1862	Florence Nightingale founds St. Thomas' Hospital School of Nursing
1862	Sanitary Commission during US Civil War
1862	Emancipation of slaves in the USA
1864	Boston bans use of milk from diseased cows
1864	Russia – rural health as tax-supported local service through Zemstvos

1864	First International Geneva Convention and founding of International Committee of the Red Cross
1866	Gregor Johann Mendel, a Czech monk, publishes basic laws of heredity establishing the scientific basis of genetics
1867	Joseph Lister describes use of carbolic spray for antisepsis
1869	Dimitri Ivanovitch Mendeleev – periodic table
1872	American Public Health Association founded
1872	Milk stations established in New York immigrant slums
1876	Robert Koch discovers anthrax bacillus
1876	Neisser discovers Gonococcus organism
1879	US National Board of Health established
1879	US Food and Drug Administration established
1880	Typhoid bacillus discovered (Laveran); leprosy organism (Hansen); malaria organism (Laveran)
1882	Robert Koch discovers the tuberculosis organism, tubercle bacillus
1883	Otto von Bismarck introduces social security with workmen's compensation, national health insurance for workers and their families in Germany
1883	Robert Koch discovers bacillus of cholera
1883	Louis Pasteur vaccinates against anthrax
1885	Kanehiro Takaki of the Japanese navy describes beriberi; recommends diet change eliminating the sailor's disease
1884	Diphtheria, Staphylococcus, Streptococcus, tetanus organisms identified
1885	Pasteur develops rabies vaccine; Escherich discovers coli bacillus
1886	Karl Fraenkel discovers the Pneumococcus organism
1887	Malta fever or brucellosis (Bruce) and chancroid (Ducrey) organisms identified
1887	US National Institutes of Health founded
1892	Gas gangrene organism discovered by Welch and Nuttal
1893	Lillian Wald organizes Henry Street Mission and the Visiting Nurses Association of New York for care of the poor and disabled in their own homes
1894	Plague organism discovered (Yersin, Kitasato); botulism organism (Van Ermengem)
1895	Louis Pasteur develops vaccine for rabies

1895	Wilhelm Roentgen – discovers electromagnetic waves (X-rays) for diagnostic imaging
1895	Emil von Behring develops diphtheria vaccine (Nobel Prize 1901)
1897	Edmond Nocard develops antitetanus serum (ATS) for passive immunity
1897	London School of Hygiene and Tropical Medicine founded
1897	Felix Hoffman – synthesizes acetylsalicylic acid (aspirin)
1904	Ivan Petrovitch wins Nobel Prize for work in conditioned reflexes, neurophysiology
1905	Abraham Flexner – major report on medical education in the USA
1905	Workman's Compensation Acts in Canada
1906	US Pure Food and Drug Act passed by Congress
1910	Paul Ehrlich – chemotherapy use of arsenical salvarsan for treatment of syphilis
1911	Lloyd-George, UK compulsory health insurance for workers
1911	Kasimir Funk investigates "vital amines" and names them vitamins
1912	Health insurance for industrial workers in Russia
1912	US Children's Bureau and US Public Health Service established
1914	Joseph Goldberger of US Public Health Service investigates cause and prevention of pellagra
1915	Johns Hopkins and Harvard Schools of Public Health founded
1915	Tetanus prophylaxis and antitoxin for gas gangrene
1918–1919	Pandemic of Spanish flu (influenza) kills some 20 million people
1918	Nikolai Semashko introduces USSR national health plan
1921	Frederick Banting and Charles Best discover insulin in Toronto (Nobel Prize 1923)
1923	Health Organization of League of Nations established
1924	David Cowie promotes widespread ionization of salt in the USA; Morton's iodized salt popular in North America
1924	Tetanus toxoid vaccine developed
1926	Pertussis vaccine developed
1928	Alexander Fleming discovers penicillin (Nobel Prize 1945)
1928	George Papanicolaou develops Pap smear for early detection of cancer of cervix
1929–1936	The Great Depression – widespread economic collapse, unemployment, poverty, and social distress in industrialized countries
1930	US Food and Drug Administration established
1935	President Roosevelt – Social Security Act and the New Deal in the USA
1939	UK National Hospital Service – wartime nationalization of hospitals
1940	Charles Drew describes storage and use of blood plasma for transfusion
1941	Norman Gregg reports rubella in pregnancy causing congenital anomalies
1941	President Roosevelt initiates food fortification in the USA, adopted in Canada and UK
1942	William Beveridge Report in the UK – the "Welfare State"
1942	USA establishes National Centers for Disease Control and Emergency Maternity and Infant Care for families of servicemen
1939–1945	World War II, with catastrophic military and civilian loss of life, wartime emergency medical structure; Nazi Holocaust of 6 million Jews and many others
1945	Diphtheria, pertussis, tetanus (DPT) vaccine developed
1945	Trial of fluoridation of community water supplies, Grand Rapids MI; Newburgh, NY; and Brantford, Ontario
1946	World Health Organization founded
1946	National health insurance defeated in US Congress
1946	US Communicable Disease Center (CDC) established in Atlanta; later called the Centers for Disease Control and Prevention
1946	US Congress Hill–Burton Act supports local hospital construction up to 4.5 beds per 1000 population
1946	Tommy Douglas – Saskatchewan provincial hospital insurance plan
1947	Nuremberg Doctors Trial of Nazi crimes against humanity
1948	International Declaration of Human Rights
1948	UK establishes National Health Service
1950	CDC establishes the Epidemiological Intelligence Service (EIS)
1953	James Watson and Francis Crick discover the double helix structure of DNA (Nobel Prize 1962)

1954	Framingham study of heart disease risk factors
1954	Richard Doll reports on link between smoking and lung cancer
1954	Jonas Salk's inactivated poliomyelitis vaccine licensed
1955	Michael Buonocore develops dental sealants
1956	Gregory Pincus reports first successful trials of birth control pills
1960	Albert Sabin – live poliomyelitis vaccine licensed
1961	American Academy of Pediatrics recommends routine vitamin K for all newborns
1961	CDC publishes *Morbidity and Mortality Weekly Report (MMWR)*
1963	Measles vaccine licensed
1964	US Surgeon General's Report on Smoking (Luther Terry)
1965	The USA enacts Medicare for the elderly, Medicaid for the poor
1966	US National Traffic and Motor Vehicle Safety Act
1967	Mumps vaccine licensed
1970	Rubella vaccine licensed
1971	Canada has universal health insurance in all provinces
1971	US National Center for Health Statistics conducts the first National Health and Nutrition Examination Survey (NHANES) to capture the health status of Americans.
1972	US Stanford Three Community Study starts (later The Stanford Five-City Project); a 23% reduction in coronary heart disease risk by community-based interventions to change lifestyle risk factors – physical activity, dietary habits, and tobacco use
1972	Finland's North Karelia Project begins, to prevent cardiovascular disease; cardiovascular mortality rates for men aged between 35 and 64 years decreased by 57% from 1970 to 1992
1973	*MMWR* reports that lead emissions in a residential area constitute a public health threat
1974	Marc Lalonde New Perspectives on the Health of Canadians
1977	WHO adopts Health for All by the Year 2000
1977	Last known outbreak of smallpox reported in Somalia
1977	Framingham study shows effects of triglycerides and LDL- and HDL-cholesterol on heart disease

1978	Alma-Ata Conference on Primary Health Care
1978	Hepatitis B vaccine licensed
1979	Canada adopts mandatory vitamin/mineral enrichment of foods
1979	WHO declares eradication of smallpox achieved
1981	AIDS – first recognition of cases of acquired immunodeficiency syndrome
1983	CDC – Violence Epidemiology Branch to apply prevention strategies to child abuse, homicide, and suicide
1984	Harald zur Haisen discovers link between human papillomavirus and cancer of cervix (Nobel Prize 2008)
1985	WHO European Region Health Targets
1985	*Haemophilus influenzae* b (Hib) vaccine licensed by FDA
1985	Luc Montaignier publishes genetic sequence of HIV (with Francoise Barre-Sinoussi, Nobel Prize 2008)
1986	First coronary stent implanted by Jacques Puel and Ulrich Sigwart in France
1988	American College of Obstetricians and Gynecologists recommends annual Pap smears for all women
1988	Framingham study shows isolated systolic hypertension linked to increase risk of heart disease
1988	Framingham study shows cigarette smoking increases risk of stroke
1989	WHO targets eradication of polio by the year 2000
1989	Warren and Marshall discover *Helicobacter pylori* as treatable cause of peptic ulcers (Nobel Prize 2005)
1989	International Convention on the Rights of the Child
1990	World Summit on Children, New York
1990	World Conference on Education for All, Jomtien, Thailand
1990	W. F. Anderson performs first successful gene therapy
1990	Newly emerging and re-emerging diseases (HIV, Marburg, Ebola, cholera, BSE, TB) and multidrug-resistant organisms
1991	Folic acid proven to prevent neural tube defects
1992	United Nations Conference on Environment and Development, Rio de Janeiro

1992	International Conference on Nutrition
1992	The Victoria Declaration in Canada on Heart Health affirms that CVD is largely preventable, that scientific knowledge exists to eliminate most CVD, and that public health infrastructure and capacity to address prevention are lacking
1993	World Conference on Human Rights, Vienna
1993	World Development Report: Investing in Health published by World Bank
1993	Russian Federation approves compulsory national health insurance
1994	International Conference on Population and Development, Cairo
1994	Clinton National Health Insurance plan defeated in US Congress
1995	World Summit for Social Development, Copenhagen
1995	United Nations Fourth World Conference on Women, Beijing
1996	Second United Nations Conference on Human Settlement (Habitat II), Istanbul
1996	Explosive growth of managed care plan coverage in the USA
1997	Legal action for damages against tobacco companies for costs of health effects of smoking, 33 states in the USA and other countries
1997	US President Clinton apologizes for Tuskegee study of syphilis among black American men (1932–1972)
1998	US President Clinton proposes legislation on patients' rights in managed care
1998	FDA approves rotavirus vaccine
1998	WHO Health for All in the Twenty-First Century adopted
1998	US National Academy of Sciences recommends routine vitamin supplements for adults
1998	Bologna Declaration on postgraduate education in Europe adopts BA, MA, and PhD levels
1998	The USA, Canada, and Chile adopt mandatory fortification of flour with folic acid to prevent birth defects
1999	US Congress passes legislation regulating patients' rights in managed care
1999	Master Settlement Agreement between US states and tobacco companies for $206 billion for Medicaid damages
1999	MMWR publishes Ten Great Public Health Achievements – United States, 1900–1999
2000	The entire human genome is mapped

2000	WHO 53rd World Health Assembly endorses global strategy for non-communicable disease (NCD) prevention and control, with monitoring, preventing, and managing major NCDs with common risk factors and determinants: cardiovascular disease, cancer, diabetes, and chronic respiratory disease
2001	9/11 Terrorism and mass casualties in destruction by Islamic terrorists of Twin Towers in New York City
2001	Anthrax bioterrorism threats in USA
2001	Millennium Development Goals proposed by the United Nations accepted by most member states as global effort to reduce poverty, and improve education and health in poor countries
2003	SARS epidemic in China reaches Toronto; 8098 total cases with 774 deaths
2003	WHO's Framework Convention on Tobacco Control adopted by the 56th World Health Assembly
2004	Tsunami and mass casualties in South-East Asia
2004	WHO Global Strategy on Diet, Physical Activity and Health endorsed by World Health Assembly
2005	Hurricanes Katrina and Rita cause widespread devastation and mass casualties
2005	Bangkok Charter for Health Promotion in a globalized world
2005	International Health Regulations promoted by WHO adopted by 194 countries
2006	Bird flu of H5N1 virus threatens world pandemic
2006	Human papillomavirus (HPV) vaccine approved by FDA for prevention of cervical cancer
2006	Medicare Part D prescription drug plan for seniors instituted in the USA
2007	HPV vaccine in wide use for preteen girls in industrialized countries
2008	China – milk products deliberately contaminated with melanine; over 14,000 hospitalized
2008	Commission on Social Determinants of Health reveals the appalling levels of health inequality resulting in premature deaths and stunted lives
2008	Global tuberculosis control – progress to control the TB epidemic slowed in 2006
2009	Creuzfeldt–Jakob disease – outbreaks of BSE in animals in several countries

2009	WHO and UNICEF launch the Global Action Plan for the prevention and control of pneumonia (GAPP) in 65 countries to prevent up to 5.3 million child deaths from pneumonia by 2015
2009	World malaria report 2009 – reduced impact of malaria needed to achieve Millennium Development Goals; 243–311 million malaria cases worldwide and 863,000 to 1 million early deaths per year, almost all in the poorest countries
2009	H1N1 pandemic announced by WHO
2010	Haiti suffers 7.0 magnitude earthquake with massive loss of life and displacement; many deaths from cholera
2010	Massive floods in Pakistan and China: Pakistan's flood crisis affects over 215 million people, with 6 million needing life-saving humanitarian and health care; in China more than 400 million
2010	Millennium Development Goals 2010 status report – progress in some regions, but in sub-Saharan Africa goals will not be achieved by 2015
2010	US Congress enacts President Obama's Patient Protection and Affordable Care Act (PPACA or "Obamacare") to extend health insurance coverage to millions of uninsured Americans
2011	HPV vaccine recommended by US CDC for boys as well as girls
2012	US Supreme Court upholds legality of PPACA

Source: Deutsche Welle Focus: Millennium Development Goals [updated 21 September 2012]. Bonn: Deutsche Welle. Available at: http://www.dw-world.de/dw/article/0,,6003071,00.html [Accessed 17 July 2012].

NOTE

For a complete bibliography and guidance for student reviews and expected competencies please see companion web site at http://booksite.elsevier.com/9780124157668

BIBLIOGRAPHY

Carter, K.C., 1991. The development of Pasteur's concept of disease causation and the emergence of specific causes in nineteenth-century medicine. Bull. Hist. Med. 65, 528–548. Available at: http://www.ncbi.nlm.nih.gov/pubmed/1802317 (accessed 11.08.12).

Centre for History in Public Health. London: London School of Hygiene and Tropical Medicine. Available at: http://history.lshtm.ac.uk/ (accessed 11.08.12).

Centers for Disease Control and Prevention, 1999. Achievements in public health, 1900–1999: control of infectious diseases. MMWR Morb. Mortal. Wkly. Rep. 48, 621–629. Available at: http://www.cdc.gov/mmwr/preview/mmwrhtml/mm4829a1.htm (accessed 11.08.12).

Centers for Disease Control and Prevention, 1999. Achievements in public health, 1900–1999: decline in deaths from heart disease and stroke – United States, 1900–1999. MMWR Morb. Mortal. Wkly. Rep. 48, 649–656. Available at: http://www.cdc.gov/mmwr/preview/mmwrhtml/mm4830a1.htm (accessed 11.08.12).

Centers for Disease Control and Prevention, 1999. Achievements in public health, 1900–1999: family planning. MMWR Morb. Mortal. Wkly. Rep. 48, 1073–1080. Available at: http://www.cdc.gov/mmwr/preview/mmwrhtml/mm4847a1.htm (accessed 11.08.12).

Centers for Disease Control and Prevention, 1999. Achievements in public health, 1900–1999: fluoridation of drinking water to prevent dental caries. MMWR Morb. Mortal. Wkly. Rep. 48, 933–940. Available at: http://www.cdc.gov/mmwr/preview/mmwrhtml/mm4841a1.htm (accessed 11.08.12).

Centers for Disease Control and Prevention, 1999. Achievements in public health, 1900–1999: healthier mothers and babies. MMWR Morb. Mortal. Wkly. Rep. 48, 849–858. Available at: http://www.cdc.gov/mmwr/preview/mmwrhtml/mm4838a2.htm (accessed 11.08.12).

Centers for Disease Control and Prevention, 1999. Achievements in public health, 1900–1999: impact of vaccines universally recommended for children – United States, 1990–1998. MMWR Morb. Mortal. Wkly. Rep. 48, 243–248. Available at: http://www.cdc.gov/mmwr/preview/mmwrhtml/00056803.htm (accessed 11.08.12).

Centers for Disease Control and Prevention, 1999. Achievements in public health, 1900–1999: improvements in workplace safety – United States, 1900–1999. MMWR Morb. Mortal. Wkly. Rep. 48, 461–469. Available at: http://www.cdc.gov/mmwr/preview/mmwrhtml/mm4822a1.htm (accessed 11.08.12).

Centers for Disease Control and Prevention, 1999. Achievements in public health, 1900–1999: motor-vehicle safety: a 20th century public health achievement. MMWR Morb. Mortal. Wkly. Rep. 48, 369–374. Available at: http://www.cdc.gov/mmwr/preview/mmwrhtml/mm4818a1.htm (accessed 11.08.12).

Centers for Disease Control and Prevention, 1999. Achievements in public health, 1900–1999: safer and healthier foods. MMWR Morb. Mortal. Wkly. Rep. 48, 905–913. Available at: http://www.cdc.gov/mmwr/preview/mmwrhtml/mm4840a1.htm (accessed 11.08.12).

Centers for Disease Control and Prevention, 1999. Achievements in public health, 1900–1999: tobacco use – United States, 1900–1999. MMWR Morb. Mortal. Wkly. Rep. 48, 986–993. Available at: http://www.cdc.gov/mmwr/preview/mmwrhtml/mm4843a2.htm (accessed 11.08.12).

Centers for Disease Control and Prevention, 2011. Ten great public health achievements – United States, 2001–2010. MMWR Morb. Mortal. Wkly. Rep. 60, 619–623. Available at: http://www.cdc.gov/mmwr/preview/mmwrhtml/mm6019a5.htm.

Centers for Disease Control and Prevention, 2011. Ten great public health achievements – worldwide, 2001–2010. MMWR Morb. Mortal. Wkly. Rep. 60, 814–818. Available at: http://www.cdc.gov/mmwr/preview/mmwrhtml/mm6024a4.htm (accessed 11.08.12).

Centers for Disease Control and Prevention, CDC timeline. [updated 7 July 2011]. Atlanta, GA: CDC. Available at: http://www.cdc.gov/about/history/timeline.htm (accessed 11.08.12).

College of Physicians of Philadelphia, Jesse Lazear. [updated 2012]. Philadelphia, PA: College of Physicians of Philadelphia. Available at: http://www.historyofvaccines.org/content/jesse-lazear (accessed 11.08.12).

Cook, G.C., 2004. Scurvy in the British mercantile marine in the 19th century, and the contribution of the Seamen's Hospital Society. Postgrad Med. J. 80, 224–229. Available at: http://pmj.bmj.com/content/80/942/224.long (accessed 10.08.12).

Ferriman, A., 2007. BMJ readers choose the "sanitary revolution" as greatest medical advance since 1840. BMJ 334, 111. Available at: http://www.bmj.com//content/334/7585/111.2?variant=full-text (accessed 11.08.12).

Grob, G.N., 1985. The origins of American psychiatric epidemiology. Am. J. Public Health 75, 229–236. Available at: http://ajph.aphapublications.org/doi/pdf/10.2105/AJPH.75.3.229 (accessed 10.08.12).

History Learning Site, updated 2012. Louis Pasteur. History Learning Site, UK. Available at: http://www.historylearningsite.co.uk/louis_pasteur.htm (accessed 11.08.12).

Jahan, S., 2010. The MDGs beyond 2015. IDS Bulletin 41, 51059. Available at: http://onlinelibrary.wiley.com/doi/10.1111/j.1759-5436.2010.00104.x/abstract (accessed 11.08.12).

Mackay, J., Mensah, G.A., updated 2012. The atlas of heart disease and stroke. Milestones in heart and vascular disorders. WHO, Geneva. Available at: http://www.who.int/cardiovascular_diseases/resources/atlas/en/ (accessed 10.08.12).

Office of NIH History, Dr Joseph Goldberger and the war on pellagra: Goldberger vs. the south. Bethesda, MD: Office of NIH History. Available at: http://history.nih.gov/exhibits/goldberger/docs/south_6.htm (accessed 11.08.12).

Plotkin, S.A., Orenstein, W.A., Offit, P.A., 2008. Vaccines, fifth ed. Saunders/Elsevier, Philadelphia, PA.

Program in the History of Public Health and Medicine. New York: Columbia University. Available at: http://www.cumc.columbia.edu/dept/hphm/course.htm. (accessed 11.08.12). Science Museum. Edwin Chadwick 1800–1900. London: Science Museum. Available at: http://www.sciencemuseum.org.uk/broughttolife/people/edwinchadwick.aspx(accessed 11.08.12).

Public Health Infolinks. Atlanta, GA: Rollins School Public Health Emory University. Available at: http://www.sph.emory.edu/cms/academic_programs/research/phi_links.html (accessed 11.08.12).

Rajakumar, K., Greenspan, S.L., Thomas, S.B., Holick, M.F., 2007. Solar ultraviolet radiation and vitamin D: a historical perspective. Am. J. Public Health 97, 1746–1754. Available at: http://www.ncbi.nlm.nih.gov/pubmed/17761571 (accessed 11.08.12).

Rosen, G., 1993. A history of public health. Expanded edition. Johns Hopkins University Press, Baltimore, MD.

Science Museum, Edwin Chadwick 1800–1900. London: Science Museum. Available at: http://www.sciencemuseum.org.uk/broughttolife/people/edwinchadwick.aspx (accessed 11.08.12).

Science Museum, 1822–1895. Louis Pasteur. Science Museum, London. Available at: http://www.sciencemuseum.org.uk/broughttolife/people/louispasteur.aspx (accessed 11.08.12).

Scrimshaw, N.S., 2007. Fifty-five-year personal experience with human nutrition worldwide. Annu. Rev. Nutr. 27, 1–18. Available at: http://www.ncbi.nlm.nih.gov/pubmed/17506667 (accessed 11.08.12).

Steckler, A., McLeroy, K.R., Holtzman, D., Godfrey, H., 2010. Hochbaum (1916–1999): from social psychology to health behavior and health education. Am. J. Public Health 100, 1864. Available at: http://www.ncbi.nlm.nih.gov/pubmed/20724676 (accessed 11.08.12).

Tarantola, D., 2008. A perspective on the history of health and human rights: from the Cold War to the Gold War. J. Public Health Policy 29, 42–53. Available at: http://www.ncbi.nlm.nih.gov/pubmed/18368018 (accessed 11.08.12).

Trohler, U., James Lind and scurvy, 1747–1795 . Bern: James Lind Library. Available at: http://www.jameslindlibrary.org/illustrating/articles/james-lind-and-scurvy-1747-to-1795 (accessed 10.08.12).

Tuchman, B.W., 1978. A distant mirror: the calamitous fourteenth century. Alfred A. Knopf, New York.

US Department of Energy, History of the Human Genome Project. [updated 4 June 2012]. Washington, DC: Human Genome Program, US Department of Energy. Available at: http://www.ornl.gov/sci/techresources/Human_Genome/project/hgp.shtml (accessed 11.08.12).

US Department of Health and Human Services, updated 5 January 2012. Images from the history of the US Public Health Service. Available at:. Department of Health and Human Services, Bethesda, MD. http://www.nlm.nih. gov/exhibition/phs_history/contents.html (accessed 11.08.12).

Chapter 2

Expanding the Concept of Public Health

Learning Objectives

Upon completion of this chapter, the student should be able to:

1. Describe basic historical concepts of public health;
2. Analyze these concepts and their applicability to current and newly emerging public health problems;
3. Discuss the principles and component elements of the New Public Health.

INTRODUCTION

The development of public health from its ancient and recent roots, especially in the past several centuries, is a continuing process, with evolutionary and sometimes dramatic leaps forward, and important continuing and new challenges for personal and population health and well-being. Everything in the New Public Health is about preventing avoidable disease, injuries, disabilities, and death while promoting and maximizing a healthy environment and optimal conditions for current and future generations. Thus, the New Public Health addresses overall health policy, resource allocation, as well as the organization, management, and provision of medical care and of health systems in general within a framework of overall social policy and in a community, state, national, transnational, and global context.

The study of history (see Chapter 1) helps us to understand the process of change, to define where we came from and where we are going. It is vital to recognize and understand change in order to deal with radical transformations in direction that occur as a result of changing demography and epidemiology, new science, evolving best practices in public health and clinical medicine, and above all inequalities in health resulting from societal system failures and social and economic factors. Health needs will continue to develop in the context of environmental, demographic and societal adjustments, with knowledge gained from social and physical sciences, practice, and economics. For the coming generations, this is about not only the quality of life, but the survival of society itself.

CONCEPTS OF PUBLIC HEALTH

Over the past century there have been many definitions of public health and health for all. Mostly they represent visions and ideals of societal and global aspirations. This chapter examines the very base of the New Public Health, which encompasses the classic issues of public health with recognition of the advances made in health promotion and the management of health care systems as integral components of societal efforts to improve the health of populations and of individuals. What follows in succeeding chapters will address the major concepts leading to modern and comprehensive elements of public health. Inevitably, concepts of public health continue to evolve and to develop both as a philosophy and as a structured discipline. As a professional field, public health requires specialists trained with knowledge and appreciation of its evolution, scientific advances, concepts, and best practices, old and modern. It demands sophisticated professional and managerial skills, the ability to address a problem, reasoning to define the issues, and to advocate, initiate, develop, and implement new and revised programs. It calls for profoundly humanistic values and a sense of responsibility towards protecting and improving the health of communities and every individual. In the twenty-first century, this set of values was well expressed in the Human Development Index agreed to by 160 nations (Box 2.1).

Public health is a multidimensional field and therefore multidisciplinary in its workforce and organizational needs. It is based on scientific advances and application of best practices as they evolve, and includes many concepts, including holistic health, first established in ancient times.

BOX 2.1 Human Development Index 2010

"People are the real wealth of a nation". With these words the 1990 *Human Development Report (HDR)* began a forceful case for a new approach to thinking about development. That the objective of development should be to create an enabling environment for people to enjoy long, healthy and creative lives may appear self-evident today. But that has not always been the case. A central objective of the *HDR* for the past 20 years has been to emphasize that development is primarily and fundamentally about people.

Source: *Overview: Human Development Report 2010. 20th anniversary edition. The real wealth of nations: pathways to human development.* New York: United Nations Development Programme; 2010. Available at: *http://hdr.undp.org/en/media/HDR_2010_EN_Complete_reprint.pdf* [Accessed 16 January 2011].

The New Public Health. http://dx.doi.org/10.1016/B978-0-12-415766-8.00002-1

The discussion will return to the diversity of public health throughout this chapter and book many times.

In previous centuries, public health was seen primarily as a discipline which studies and implements measures for control of communicable diseases, primarily by sanitation and vaccination. The sanitary revolution, which preceded the development of modern bacteriology, made an enormous contribution to improved health, but many other societal factors including improved nutrition, education, and housing were no less important for population health. Maternal and child health, occupational health, and many other aspects of a growing public health network of activities played important roles, as have the physical and social environment and personal habits of living in determining health status. In recent decades recognition of the importance of women's health and health inequalities associated with many high-risk groups in the population have seen both successes and failures in addressing their challenges. Male health issues have received less attention, apart from issues associated with specific diseases, or those of healthy military personnel.

The scope of public health has changed along with growth of the medical, social, and public health sciences, public expectations, and practical experience. Taken together, these have all contributed to changes in the concepts and causes of disease. Health systems that fail to adjust to changes in fundamental concepts of public health suffer from immense inequity and burdens of preventable disease, disability, and death. This chapter examines expanding concepts of public health, leading to the development of a New Public Health.

Public health has evolved as a multidisciplinary field that includes the use of basic and applied science, education, social sciences, economics, management, and communication skills to promote the welfare of the individual and the community. It is greater than the sum of its component elements and includes the art and politics of the funding and coordination of the wide diversity of community and individual health services.

The concept of the interdependence of health in body and in mind has ancient origins. They continue to be fundamental to individuals and societies, and part of the fundamental rights of all humans to have knowledge of healthful lifestyles and to have access to those measures of good health that society alone is able to provide, such as immunization programs, food and drug safety and quality standards, environmental and occupational health, and universal access to high-quality primary and specialty medical and other vital health services. This holistic view of balance and equilibrium may be a renaissance of classical Greek and biblical traditions, applied with the broad new knowledge and experience of public health and medical care of the nineteenth, twentieth, and the early years of the twenty-first centuries as change continues to challenge our capacity to adapt.

The competing nineteenth-century germ and miasma theories of biological and environmental causation of illness each contributed to the development of sanitation, hygiene, immunization, and understanding of the biological and social determinants of disease and health. They come together in the twenty-first century encompassed in a holistic New Public Health addressing individual and population health needs. Medicine and public health professionals both engage in organization and in direct care services. These all necessitate an understanding of the issues that are included in the New Public Health, how they evolved, interact, are put together in organizations, and are financed and operated in various parts of the world in order to understand changes going on before our eyes.

Great success has been achieved in reducing the burden of disease with tools and concepts currently at our disposal. The idea that this is an entitlement for everyone was articulated in the Health for All concept of Alma-Ata in 1978. The health promotion movement emerged in the 1970s and showed dramatically effective results in managing the new human immunodeficiency virus (HIV) pandemic and in tackling smoking and other risk factors for non-communicable diseases (NCDs). A Health in All policy concept emerged in 2006 promoting the concept that health should be a basic component of all public and private policies to achieve the full potential of public health and eliminate inequalities associated with social and economic conditions.

Profound changes are taking place in the world population, and public health is crucial to respond accordingly: mass migration to the cities, fewer children, extended life expectancy, and the increase in the population of older people who are subject to more chronic diseases and disabilities in a changing physical, social, and economic climate. Health systems are challenged with continuing development of new medical technologies and related reforms in clinical practice, while experiencing strong influences of pharmaceuticals and the medicalization of health, with prevention and health promotion less central in priorities and resource allocation.

Globalization of health has many meanings: international trade, improving global communications, and economic changes with increasing flows of goods, services, and people. Ecological and climate change bring droughts, hurricanes, Arctic meltdown, and rising sea levels. Globalization also has political effects, with water and food shortages, terrorism, and economic distress affecting billions of people. In terms of health, disease can spread from one part of the world to others, as in pandemics or in a quiet spread such as that of West Nile Fever moving from its original Middle Eastern natural habitat to the Americas and Europe, or severe acute respiratory syndrome (SARS), which spread with lightning speed from Chinese villages to metropolitan cities such as Toronto, Canada. It can also mean that the NCDs characteristic of the industrialized countries are now recognized as the leading causes of death in low- and middle-income countries, associated with diet, activity levels, and smoking, which are themselves pandemic risk factors.

The potential for global action in health can also be dramatic. The eradication of smallpox was a stunning victory for public health. The campaign to eradicate poliomyelitis is succeeding even though the end-stage is fraught with setbacks, and measles elimination has turned out to be more of a challenge than was anticipated a decade ago, with resurgence in countries thought to have it under control. Global health policies have also made the achievements of public–private partnerships of great importance, particularly in vaccination and acquired immunodeficiency syndrome (AIDS) control programs. There have been failures as well, with very limited progress in human resources development of the public health workforce in low-income countries.

The New Public Health is necessarily comprehensive in scope and it will continue to evolve as new technologies and scientific discoveries – biological, genetic, and sociological – reveal more methods of disease control and health promotion. It relates to or encompasses all community and individual activities directed towards improving the environment for health, reducing factors that contribute to the burden of disease, and fostering those factors that relate directly to improved health. Its programs range broadly from immunization, health promotion, and child care, to food labeling and fortification, as well as to the assurance of well-managed, accessible health care services. A strong public health system should have adequate preparedness for natural and human-made disasters, as seen in the recent tsunamis, hurricanes, biological or other attacks by terrorists, wars, conflicts, and genocidal terrorism (Box 2.2).

The concepts of health promotion and disease prevention are essential and fundamental elements of the New Public Health. Parallel scientific advances in molecular biology, genetics and pharmacogenomics, imaging, information technology, computerization, biotechnology, and nanotechnology hold great promise for improving the productivity of the health care system. Advances in technology with more effective and less expensive drug and vaccine development, with improved safety and effectiveness, and fewer adverse reactions, will over time greatly increase efficiency in prevention and treatment modalities.

The New Public Health is important as a conceptual base for training and practice of public health. It links classical topics of public health with adaptation in the organization and financing of personal health services. It involves a changed paradigm of public health to incorporate new advances in political, economic, and social sciences. Failure at the political level to appreciate the role of public health in disease control holds back many societies in economic and social development. At the same time, organized public health systems need to work to reduce inequities between and inside countries to ensure equal access to care. It also demands special attention through health promotion activities of all kinds at national and local societal levels to provide access for groups with special risks

---

BOX 2.2 The New Public Health: Mission and Methods

The mission of the NPH is to maximize human health and well-being for individuals and communities, nationally and globally.

The methods with which the NPH works to achieve this are in keeping with recognized international best practices and scientific advances:

1. *Societal commitment* and sustained efforts to maximize quality of life and health, economic growth with equity for all (Health for All and Health in All).
2. *Collaboration* between international, national, state, and local health authorities working with public and private sectors to promote health awareness and activities essential for population health.
3. *Health promotion* of knowledge, attitudes, and practices, including legislation and regulation to protect, maintain, and advance individual and community health.
4. *Universal access* to services for prevention and treatment of illness and disability, and promotion of maximum rehabilitation.
5. *Environmental, biological, occupational, social, and economic factors* that endanger health and human life, addressing:
   (a) physical and mental illness, diseases and infirmity, trauma and injuries
   (b) local and global sanitation and environmental ecology
   (c) healthful nutrition and food security including availability, quality, safety, access, and affordability of food products
   (d) disasters, natural and human-made, including war, terrorism, and genocide
   (e) population groups at special risk and with specific health needs.
6. *Promoting links between health protection and personal health services through health policies and health systems management*, recognizing economic and quality standards of medical, hospital, and other professional care in health of individuals and populations.
7. *Training of professional public health workforces* and education of all health workers in the principles of ethical best practices of public health and health systems.
8. *Research and promotion of current best practices*: wide application of current international best practices and standards.
9. *Mobilizing the best available evidence* from local and international scientific and epidemiological studies and best practices recognized as contributing to the overall goal.
10. *Maintaining and promoting equity* for individual and community rights to health with high professional and ethical standards.

---

and needs to medical and community health care with the currently available and newly developing knowledge and technologies.

The great gap between available capabilities to prevent and treat disease and actually reaching all in need is still the source of great international and internal national inequities. These inequities exist not only between developed and developing countries, but also within transition countries, mid-level developing countries, and those newly emerging with rapid economic development. The historical experience of public health will help to develop the applications of existing and new knowledge and societal commitment to social solidarity in implementation of the new discoveries for every member of the society, despite socioeconomic, ethnic, or other differences.

Political will and leadership in health, adequate financing, and organization systems in the health setting are crucial to furthering health as an objective with defined targets, supported by well-trained staff for planning, management, and monitoring the population health and functioning of health systems. Political leadership and professional support are both indispensable in a world of limited resources, with high public expectations and the growing possibilities of effectiveness of public health programs. Well-developed information and knowledge management systems are required to provide the feedback and information needed for good management. It includes responsibilities and coordination at all levels of

government. Non-governmental organizations (NGOs) and participation of a well-informed media and strong professional and consumer organizations also have significant roles in furthering population health. No less important are clear designations of responsibilities of the individual for his or her own health, and of the provider of care for humane, high-quality professional care. The complexities and interacting factors are suggested in Figure 2.1, with the classic host–agent–environment triad.

EVOLUTION OF PUBLIC HEALTH

Many changes have signaled a need for transformation towards the New Public Health. Religion, although still a major political and policy-making force in many countries, is no longer the central organizing power in most societies. Organized societies have evolved from large extended families and tribes to rural societies, cities, regions, and national governments. With the growth of industrialized urban communities, rapid transport, and extensive trade and commerce in multinational economic systems, the health of individuals and communities has become more than just a personal, family, and/or local problem. An individual is not

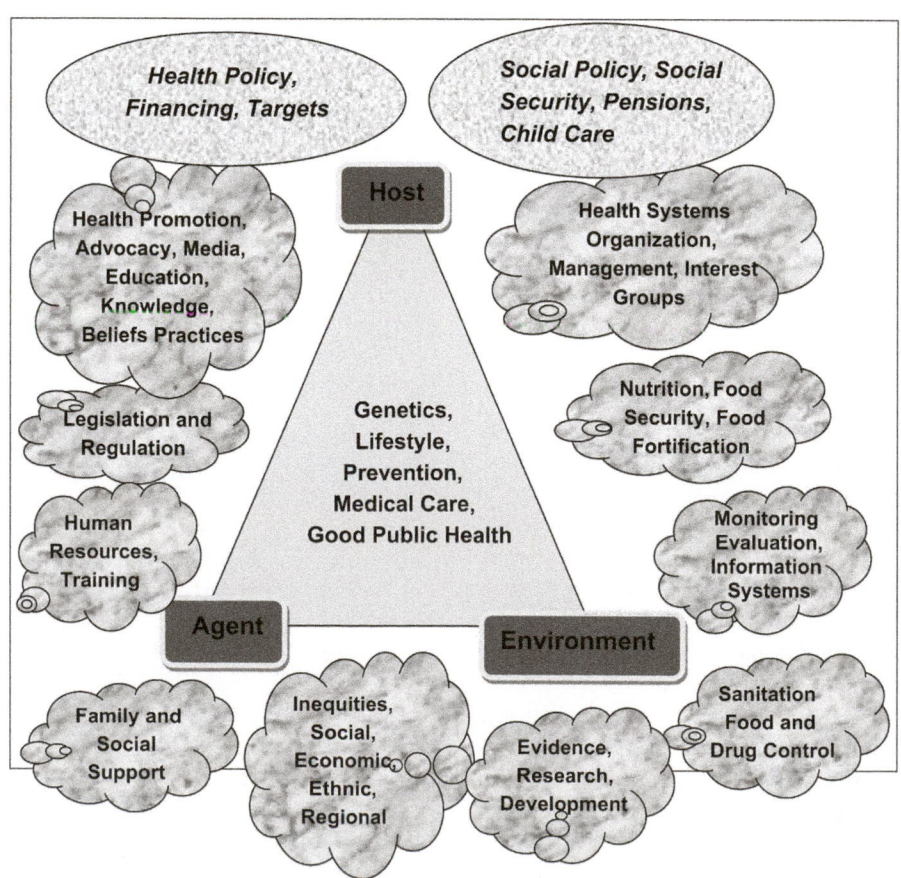

FIGURE 2.1   The New Public Health.

only a citizen of the village, city, or country in which he or she lives, but a citizen of a "global village".

The agricultural revolutions and international explorations of the fifteenth to seventeenth centuries that increased food supply and diversity were followed only much later by knowledge of nutrition as a public health issue. The scientific revolution of the seventeenth to nineteenth centuries provided the basics to describe and analyze the spread of disease and the poisonous effects of the industrial revolution, including crowded living conditions and pollution of the environment with serious ecological damage. In the latter part of the twentieth century, a new agricultural "green revolution" had a great impact in reducing human deprivation internationally, yet the full benefits of healthier societies are yet to be realized in the large populations living in abject poverty in sub-Saharan Africa, South-East Asia, and other parts of the world. Global water shortages can be addressed with new methods of irrigation, water conservation and the application of genetic sciences to food production, and issues of economics and food security are of great importance to a still growing world population with limited supplies. Further, food production capacity can and must be enlarged to meet current food insecurity, rising expectations of developing nations, and population growth. The sciences of agriculture-related fields, including genetic sciences and practical technology, will be vital to human progress in the coming decades.

These and other societal changes discussed in Chapter 1 have enabled public health to expand its potential and horizons, while developing its pragmatic and scientific base. Organized public health in the twentieth century proved effective in reducing the burden of infectious diseases and has contributed to improved quality of life and longevity by many years. In the last half-century, chronic diseases have become the primary causes of morbidity and mortality in the developed countries and increasingly in developing countries. Growing scientific and epidemiological knowledge increases the capacity to deal with these diseases. Many aspects of public health can only be influenced by the behavior of and risks to the health of individuals. These require interventions that are more complex and relate to societal, environmental, and community standards and expectations as much as to personal lifestyle. The dividing line between communicable and non-communicable diseases changes over time. Scientific advances have shown the causation of chronic conditions by infectious agents and their prevention by curing the infection, as in *Helicobacter pylori* and peptic ulcers, and in prevention of cancer of the liver and cervix by immunization for hepatitis B and human papillomavirus (HPV), respectively.

Chronic diseases have come to the center stage in the "epidemiological transition", as infectious diseases came under increasing control. This, in part, has created a need for reform in the funding and management of health systems due to rapidly rising costs, aging of the population, the rise of obesity and diabetes and other chronic conditions, mushrooming therapeutic technology, and expanding capacity to deal with public health emergencies. Reform is also needed in international assistance to help less developed nations build the essential infrastructure to sustain public health in the struggle to combat AIDS, malaria, tuberculosis (TB), and the major causes of preventable infant, childhood, and motherhood-related deaths.

The nearly universal recognition of the rights of people to have access to health care of acceptable quality by international standards is a challenge of political will and leadership backed up by adequate staffing with public health-trained staff and organizations. The challenges of the current global economic crisis are impacting social and health systems around the world. The interconnectedness of managing health systems is part of the New Public Health. Setting the priorities and allocating resources to address these challenges requires public health training and orientation of the professionals and institutions participating in the policy, management, and economics of health systems. Conversely, those who manage such institutions are recognizing the need for a wide background in public health training in order to fulfill their responsibilities effectively. Concepts such as objectives, targets, priorities, cost-effectiveness, and evaluation have become part of the New Public Health agenda. An understanding of how these concepts evolved will help the future health provider or manager to cope with the complexities of mixing science, humanity, and effective management of resources to achieve higher standards of health, and to cope with new issues as they develop in the broad scope of the New Public Health for the twenty-first century, in what Breslow called the "Third Public Health Era" of long and healthy quality of life (Box 2.3).

HEALTH AND DISEASE

Health can be defined from many perspectives, ranging from statistics on mortality, life expectancy, and morbidity rates to idealized versions of human and societal perfection, as in the World Health Organization's (WHO's) founding charter. The

---

BOX 2.3 Breslow: The Continuing Epidemiological Transition

First Public Health Era – the control of communicable diseases.
Second Public Health Era – the rise and fall of chronic diseases.
Third Public Health Era – the development of long and high-quality life.

Source: Breslow L. 2006. Health measurement in the third era of health. *Am J Public Health* 2006;96:17–9.

preamble to the constitution of the WHO, as adopted by the International Health Conference in New York in 1946 and signed by the representatives of 61 states, entered into force on 7 April 1948, with the widely cited definition: "Health is a state of complete physical, mental and social well-being and not merely the absence of disease or infirmity". This definition is still important conceptually as an ideal accepted as fundamental to public policy over the years.

A more operational definition of health is a state of equilibrium of the person with the biological, physical, and social environment, with the object of maximum functional capability. Health is thus seen as a state characterized by anatomical, physiological, and psychological integrity, and an optimal functional capability in the family, work, and societal roles (including coping with associated stresses), a feeling of well-being, and freedom from risk of disease and premature death. Deviances in health are referred to as unhealthy and constitute a disease nomenclature.

There are many interrelated factors in disease and in their management through what is now called risk reduction. In 1878, Claude Bernard described the phenomenon of adaptation and adjustment of the internal milieu of the living organism to physiological processes. This concept is fundamental to medicine. It is also central to public health because understanding the spectrum of events and factors between health and disease is basic to the identification of contributory factors affecting the balance towards health, and to seeking the points of potential intervention to reverse the imbalance.

As described in Chapter 1, from the time of Hippocrates and Galen, diseases were thought to be due to humors and miasma or emanations from the environment. This was termed the miasma theory, and while without a direct scientific explanation, it was acted upon in the early to mid-nineteenth century and promoted by leading public health theorists including Florence Nightingale, with practical and successful measures to improve sanitation, housing, and social conditions, and having important results in improving health conditions. The competing germ theory developed by pioneering nineteenth-century epidemiologists (Panum, Snow, and Budd), scientists (Pasteur, Cohn, and Koch), and practitioners (Lister and Semmelweiss) led to the science of bacteriology and a revolution in practical public health measures. The combined application of the germ (agent–host–environment) and miasma theories (social and sanitary environment) has been the basis of classic public health, with enormous benefits in the control of infectious and other diseases or harmful conditions.

The revolutionary changes occurring since the 1960s have brought about a decline in cardiovascular and cancer mortality, and conceptual changes such as Health for All and Health in All to bring health issues to all policies at both governmental and individual levels. The concepts of public health advanced with the 1974 Marc Lalonde Health Field Concept (New Perspectives on the Health of Canadians,

1974), stating that health was the result of the physical and social environment, lifestyle and personal habits, genetics, as well as organization and provision of medical care.

The Lalonde report was a key concept leading to ideas advanced at the Alma-Ata conference on primary care held in 1978 and more explicitly in the development of the basis for health promotion as articulated in the Ottawa Charter of 1986 on Health Promotion. This marked the beginning of a whole new aspect of public health, which proved itself in addressing with considerable success the epidemic of HIV and cardiovascular diseases.

In the USA, the Surgeon General's reports of 1964 on smoking and health, and of 1979 defining health targets as national policy promoted the incorporation of "management by objectives" from the business world applied to the health sector (see Chapter 12). This led to Healthy People USA 2000 and later versions, and the United Nations (UN) Millennium Development Goals (MDGs), aimed primarily at the middle- and low-income countries (Box 2.4). The identification of infectious causes of cancers of the liver and cervix established a new paradigm in epidemiology, and genetic epidemiology has important potential for public health and clinical medicine.

BOX 2.4 Health Policy Evolution Since the 1960s

- Risk factors for chronic disease – 1960s
- Social and behavioral sciences – social epidemiology – from 1960s
- Health Field Concept (Lalonde – Health Field Concept) – 1974
- Health for All (Alma Ata Declaration) – 1978
- Declining mortality from stroke, CHD, trauma – from the 1960s
- Health promotion – from the 1980s
- Advances in drugs, vaccines and diagnostics
- Control of infectious diseases – antibiotics, vaccines, eradication of smallpox, advances in control of poliomyelitis, measles, and other childhood diseases
- Rapid increase in costs of care: health system reform
- Health targets, e.g., Healthy People 2000 and 2010
- Health in All Policies – from 1990s
- Screening, nutrition, and immunization to prevent cancers
- Inequalities in health – from 2000
- Health systems reform – universal coverage and new organizational systems
- Identification of infectious causes of non-communicable diseases – from 1990s
- Genetic epidemiology – from 1990s
- Millennium Development Goals – 2001–2015
- Recognition of non-communicable diseases as central issues in low- and medium-income as well as high-income countries
- Health target planning for Healthy People 2020 in the USA and Health 2020 in Europe

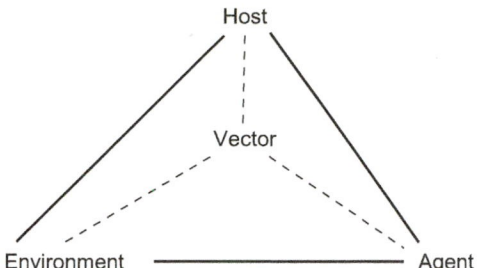

FIGURE 2.2   The host–agent–environment paradigm.

The Host–Agent–Environment Paradigm

In the basic host–agent–environment paradigm, a harmful agent comes through a sympathetic environment into contact with a susceptible host, causing a specific disease. This idea dominated public health thinking until the mid-twentieth century. The host is the person who has or is at risk for a specific disease. The agent is the organism or direct cause of the disease. The environment includes the external factors which influence the host, his or her susceptibility to the agent, and the vector which transmits or carries the agent to the host from the environment. This explains the causation and transmission of many diseases. This paradigm (Figure 2.2), in effect, joins together the contagion and miasma theories of disease causation. A specific agent, a method of transmission, and a susceptible host are involved in an interaction, which are central to the infectivity or severity of the disease. The environment can provide the carrier or vector of an infective (or toxic) agent, and it also contributes factors to host susceptibility; for example, unemployment, poverty, or low education level.

The Expanded Host–Agent–Environment Paradigm

The expanded host–agent–environment paradigm widens the definition of each of the three components (Figure 2.3), in relation to both acute infectious and chronic non-infectious disease epidemiology. In the latter half of the twentieth century, this expanded host–agent–environment paradigm took on added importance in dealing with the complex of factors related to chronic diseases, now the leading causes of disease and premature mortality in the developed world, and increasingly in developing countries.

Interventions to change host, environmental, or agent factors are the essence of public health. In infectious disease control, the biological agent may be removed by pasteurization of food products or filtration and disinfection (chlorination) of water supplies to prevent transmission of waterborne disease. The host may be altered by immunization to provide immunity to a specific infective organism. The environment may be changed to prevent transmission by destroying the vector or its reservoir of the disease.

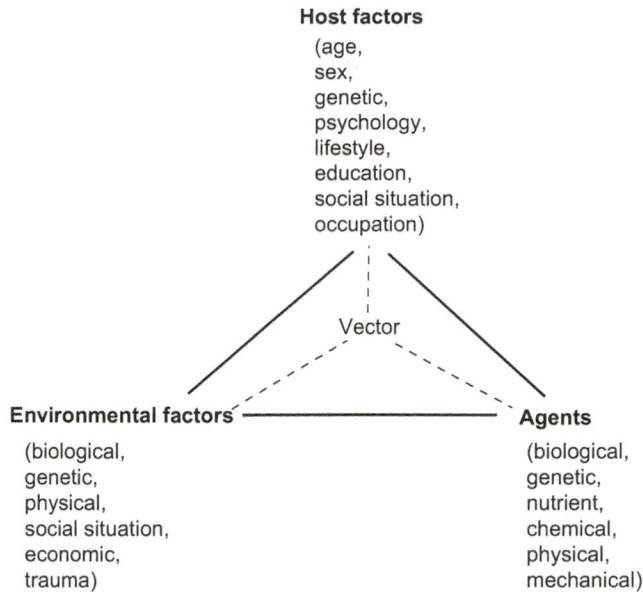

FIGURE 2.3   The expanded host–agent–environment paradigm.

A combination of these interventions can be used against a specific risk factor, toxic or nutritional deficiency, infectious organism, or disease process.

Vaccine-preventable diseases may require both routine and special activities to boost herd immunity to protect the individual and the community. For other infectious diseases for which there is no vaccine (e.g., malaria), control involves a broad range of activities including case finding and treatment to improve the individual's health and to reduce the reservoir of the disease in the population, and other measures such as bed nets to reduce exposure of the host to vector mosquitoes, as well as vector control to reduce the mosquito population. TB control requires not only case finding and treatment, but understanding the contributory factors of social conditions, diseases with TB as a secondary condition (substance abuse and AIDS), agent resistance to treatment, and the inability of patients or carriers to complete treatment without supervision. Sexually transmitted infections (STIs) which are not controllable by vaccines require a combination of personal behavior change, health education, medical care, and skilled epidemiology.

With non-infectious diseases, intervention is even more complex, involving human behavior factors and a wide range of legal, administrative, and educational issues. There may be multiple risk factors, which have a compounding effect in disease causation, and they may be harder to alter than infectious diseases factors. For example, smoking in and of itself is a risk factor for lung cancer, but exposure to asbestos fibers has a compounding effect. Preventing exposure to the compounding variables may be easier than smoking cessation. Reducing trauma morbidity and mortality is equally problematic.

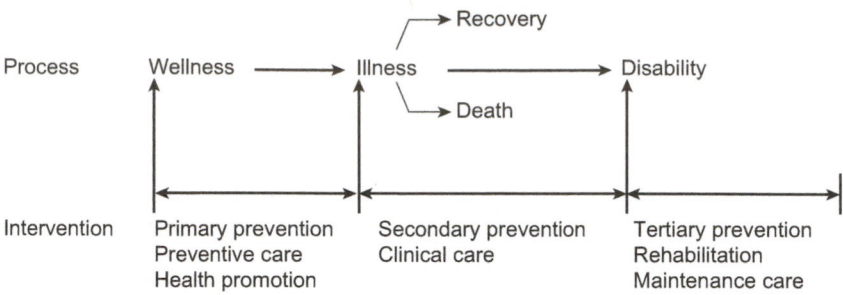

FIGURE 2.4 The process of disease and intervention.

The identification of a single specific cause of a disease is of great scientific and practical value in modern public health, enabling such direct interventions as the use of vaccines or antibiotics to protect or treat individuals from infection by a causative organism, toxin, deficiency condition, or social factor. The cumulative effects of several contributing or risk factors in disease causation are also of great significance in many disease processes, in relation to infectious diseases such as nutritional status as for chronic diseases such as the cardiovascular group.

The health of an individual is affected by risk factors intrinsic to that person as well as by external factors. Intrinsic factors include the biological ones that the individual inherits and those life habits he or she acquires, such as smoking, overeating, or engaging in other high-risk behaviors. External factors affecting individual health include the environment, the socioeconomic and psychological state of the person, the family, and the society in which he or she lives. Education, culture, and religion are also contributory factors to individual and community health.

There are factors that relate to health of the individual in which the society or the community can play a direct role. One of these is provision of medical care. Another is to ensure that the environment and community services include safety factors that reduce the chance of injury and disease, or include protective measures; for example, fluoridation of a community water supply to improve dental health, and seat-belt or helmet laws to reduce motor vehicle injury and death. These modifying factors may affect the response of the individual or the spread of an epidemic (see Chapter 3). An epidemic may also include chronic disease, because common risk factors may cause an excess of cases in a susceptible population group, in comparison to the situation before the risk factor appeared, or in comparison to a group not exposed to the risk factor. These include rapid changes or "epidemics" in such conditions as type 2 diabetes, asthma, cardiovascular diseases, trauma, and other non-infectious disorders.

THE NATURAL HISTORY OF DISEASE

Disease is a dynamic process, not only of causation, but also of incubation or gradual development, severity, and

the effects of interventions intended to modify outcome. Knowledge of the natural history of disease is fundamental to understanding where and with what means intervention can have the greatest chance for successful interruption or change in the disease process for the patient, family, or community.

The natural history of a disease is the course of that disease from beginning to end. This includes the factors that relate to its initiation; its clinical course leading up to resolution, cure, continuation, or long-term sequelae (further stages or complications of a disease); and environmental or intrinsic (genetic or lifestyle) factors and their effects at all stages of the disease. The effects of intervention at any stage of the disease are part of the disease process (Figure 2.4).

As discussed above, disease occurs in an individual when agent, host, and environment interact to create adverse conditions of health. The agent may be an infectious organism, a chemical exposure, a genetic defect, or a deficiency condition. A form of individual or social behavior, such as reckless driving or risky sexual behavior, may lead to injury or disease. The host may be immune or susceptible as a result of many contributing social and environmental factors. The environment includes the vector, which may be a malaria-bearing mosquito, a contaminated needle shared by drug users, lead-contaminated paint, or an abusive family situation.

Assuming a natural state of "wellness" – i.e., optimal health or a sense of well-being, function, and absence of disease – a disease process may begin with the onset of a disease, infectious or non-infectious, following a somewhat characteristic pattern of "incubation" described by clinicians and epidemiologists. Preclinical or predisposing events may be detected by a clinical history, with determination of risk including possible exposure or presence of other risk factors. Interventions, before and during the process, are intended to affect the later course of the disease.

The clinical course of a disease, or its laboratory or radiological findings, may be altered by medical or public health intervention, leading to the resolution or continuation of the disease with fewer or less severe secondary sequelae. Thus, the intervention becomes part of the natural history of the disease. The natural history of an infectious disease in a population will be affected by the extent of prior vaccination or previous exposure in the community. Diseases particular to children are often so because the adult population is immune from previous exposure or vaccinations. Measles and diphtheria, primarily childhood diseases, now affect adults to a large extent because they are less protected by naturally acquired immunity or are vulnerable when their immunity wanes naturally or as a result of inadequate vaccination in childhood.

In chronic disease management, high costs to the patient and the health system accrue where preventive services or management are inadequate, not yet available, or inaccessible or where there is a failure to apply the necessary interventions. The progress of diabetes to severe complications such as cardiovascular, renal, and ocular disease is delayed or reduced by good management of the condition, with a combination of smoking cessation, diet, exercise, and medications with good medical supervision. The patient with advanced chronic obstructive pulmonary disease or congestive heart failure may be managed well and remain stable with smoking avoidance, careful management of medications, immunizations against influenza and pneumonia, and other prevention-oriented care needs. Where these are not applied or if they fail, the patient may require long and expensive medical and hospital care. Failure to provide adequate supportive care will show up in ways that are more costly to the health system and will prove more life-threatening to the patient. The goal is to avoid where possible the necessity for tertiary care, substituting tertiary prevention, i.e., supportive rehabilitation to maximum personal function and maintaining a stable functional status.

As in an individual, the phenomenon of a disease in a population may follow a course in which many factors interplay, and where interventions affect the natural course of the disease. The epidemiological patterns of an infectious disease can be assessed in their occurrence in the population or their mortality rates, just as they can for individual cases. The classic mid-nineteenth-century description of measles in the Faroe Islands by Panum showed the transmission and the epidemic nature of the disease as well as the protective effect of acquired immunity (see Chapter 1). Similar, more recent breakthroughs in medical, epidemiological, biological, and social sciences have produced enormous benefit for humankind as discussed throughout this text, with some examples. These include the eradication of smallpox and in the coming years, poliomyelitis, measles, leprosy, and other dreaded diseases known for millennia; the near-elimination of rheumatic heart disease and peptic ulcers in the industrialized countries; vast reduction in mortality from stroke and coronary heart disease (CHD); and vaccines (against hepatitis B and HPV) for the prevention of cancers. These and other great achievements of the twentieth and early part of the twenty-first centuries hold great promise for humankind in the coming decades, but great challenges lie ahead as well. The biggest challenge is to bring the benefits of known public health capacity to the poorest population of each country and the poorest populations globally. In developed countries a major challenge is to renew efforts of public health capacity to bear on prevention of chronic conditions such as diabetes and obesity, considered to be at pandemic proportions; and the individual and societal effects of mental diseases.

In public health today, fears of a pandemic of avian influenza are based on transmission of avian or other animal-borne (zoonotic) prions or viruses to humans and then their adaptation permitting human-to-human spread. With large numbers of people living in close contact with many animals (wild and domestic fowl), such as in China and South-East Asia, and rapid transportation around the world, the potential for global spread of disease is almost without historical precedent. Indeed, many human infectious diseases are zoonotic in origin and transferred from natural wildlife reservoirs to humans either directly or via domestic or other wild animals, such as from birds to chickens to humans in avian influenza. Monitoring or immunization of domestic animals requires a combination of multidisciplinary zoonotic disease management strategies, public education and awareness, and veterinary public health monitoring and control. Rift Valley Fever, equine encephalitis, and more recently SARS and avian influenza associated with bird-borne viral disease which can affect humans, each show the terrible dangers of pandemic diseases. Ebola virus is probably sustained between outbreaks among fruit bats, or as recently suggested wild or domestic pigs, and may become a major threat to public health as human case fatality rates decline, meaning that patients and carriers, or genetic drift of the virus with possible airborne transmission, may spread this deadly disease more widely than in the past (see Chapter 4).

SOCIETY AND HEALTH

The health of populations, like the health of individuals, depends on societal factors no less than on genetics, personal risk factors, and medical services. Social inequalities in health have been understood and documented in public health over the centuries. The Chadwick and Shattuck reports of 1840–1850 documented the relationship of poverty and bad sanitation, housing, and working conditions with high mortality, and ushered in the idea of social epidemiology. Political and social ideologies thought that the welfare state, including universal health care systems of one type or another, would eliminate social and geographic differences in health status and this is in large part true.

Health Insurance Systems

From the introduction of compulsory health insurance in Germany in the 1880s to the failed attempt in the USA at national health insurance in 1995 (see Chapters 1, 10 and 13) and the more recent achievements of US President Obama in 2010–2011, social reforms to deal with inequalities in health have focused on improving access to medical and hospital care. Almost all industrialized countries have developed such systems, and the contribution of these programs to improve health status has been an important part of social progress, especially since World War II.

But even in societies with universal access to health care, people of lower socioeconomic status (SES) suffer higher rates of morbidity and mortality from a wide variety of diseases. The Black Report (Douglas Black) in the UK in the early 1980s pointed out that the class V population (unskilled laborers) had twice the total and specific mortality rates of the class I population (professional and business) for virtually all disease categories, ranging from infant mortality to death from cancer. The report was shocking because all Britons have had access to the comprehensive National Health Service (NHS) since its inception in 1948, with access to a complete range of services at no cost at time of service, close relations to their general practitioners, and good access to specialty services. These findings initiated reappraisals of the social factors that had previously been regarded as the academic interests of medical sociologists and anthropologists and marginal to medical care. More recent studies and reviews of regional, ethnic, and socioeconomic differentials in patterns of health care access, morbidity, and mortality indicate that health inequities are present in all societies including the UK, the USA, and others, even with universal health insurance or services.

Health Promotion

The Ottawa Charter on Health Promotion in 1986 placed a new paradigm before the world health community that recognized social and political factors as no less important ion health that traditional medical and sanitary public health measures. These concepts helped the world health community to cope with new problems such as HIV/AIDS – for which there was neither a medical cure nor a vaccine to prevent the disease. Its control came to depend in the initial decades almost entirely on education and change in lifestyles, until the advent of the antiretroviral drugs in the 1990s. There is still no viable vaccine.

Although the epidemiology of cardiovascular disease shows the direct relationship of the now classic risk factors of stress, smoking, poor diet, and physical inactivity, differences in mortality from cardiovascular disease between different classes among British civil servants are not entirely explainable by these factors. The differences are also affected by social and economic issues that may relate to the psychological needs of the individual, such as the degree of control people have over their own lives. Blue-collar workers have less control over their lives, their working life in

particular, than their white-collar counterparts, and have higher rates of CHD mortality than higher social classes. Other work shows the effects of migration, unemployment, drastic social and political change, and binge drinking, along with protective effects of healthy lifestyle, religiosity, and family support systems in cardiovascular diseases.

Social conditions affect disease distribution in all societies. In the USA and Western Europe, TB has re-emerged as a significant public health problem in urban areas partly because of high-risk population groups, owing to poverty and alienation from society, as in the cases of homelessness, drug abuse, and HIV infection. In countries of Eastern Europe and the former Soviet Union, the recent rise in TB incidence has resulted from various social and economic factors in the early 1990s, including the large-scale release of prisoners. In both cases, diagnosis and prescription of medication are inadequate, and the community at large becomes at risk because of the development of antibiotic-resistant strains of tubercle bacillus readily spread by inadequately treated carriers, acting as human vectors.

Studies of SES and health are applicable and valuable in many settings. In Alameda County, California, differences in mortality between black and white population groups in terms of survival from cancer became insignificant when controlled for social class. A 30-year follow-up study of the county population reported that low-income families in California are more likely than those on a higher income to have physical and mental problems that interfere with daily life, contributing to further impoverishment.

Studies of the association between indicators of SES and recent screening in the USA, Australia, Finland, and elsewhere showed that lower SES women use less preventive care such as Papanicolaou (Pap) smears for cervical cancer than women of higher SES, despite having greater risk for cervical cancer. Many factors in SES inequalities are involved, including transportation and access to primary care, differences in health insurance coverage, educational levels, poverty, high-risk behaviors, social and emotional distress, feeling a lack of control over one's own life, employment, occupation, and inadequate family or community social support systems. Many barriers exist owing to difficulties in access and the lack of availability of free or low-cost medical care, and the absence or limitations of health insurance is a further factor in the socioeconomic gradient.

Social Determinants of Health

The recognition that health and disease are influenced by many factors, including social inequalities, plays a fundamental role in the New Public Health paradigm. Health care systems need to take into account economic, social, physical, and psychological factors that otherwise will limit the effectiveness of even the best medical care. The health system includes access to competent and responsible primary care as well as by the wider health system, including health promotion, specific prevention and population-based health protection. The paradigm of the host–agent–environment triad (Figures 2.2 and 2.3) is profoundly affected by the wider context. The sociopolitical environment and organized efforts at intervention affect the epidemiological and clinical course of disease of the individual. Medical care is essential, as is public health, but the persistent health inequities seen in most regions and countries require societal attention. Success or failure in improving the conditions of life for the poor, and other vulnerable "risk groups", affect national or regional health status and health system performance. The health system is meant to reduce the occurrence or bad outcome of disease, either directly by primary prevention or treatment as secondary prevention or by maximum rehabilitation as tertiary prevention, or equally important indirectly by reducing community or individual risk factors.

The World Health Organization Commission on Inequities (2007; see Box 2.5) states:

"The gross inequalities in health that we see within and between countries present a challenge to the world. That there should be a spread of life expectancy of 48 years among countries and 20 years or more within countries is not inevitable. A burgeoning volume of research identifies social factors at the root of much of these inequalities in health. Social determinants are relevant to communicable and non-communicable disease alike. Health status, therefore, should be of concern to policy makers in every sector, not solely those involved in health policy. As a response to this global challenge, The WHO Commission on Social Determinants of Health reviewed the evidence, raised societal debate, and recommend policies with the goal of improving health of the world's most vulnerable people. A major challenge put forward by the Commission is turning public health knowledge into political action."

The effects of social conditions on health can be partly offset by interventions intended to promote healthful conditions; for example, improved sanitation, or through good-quality primary and secondary health services, used efficiently and effectively made available to all. The approaches to preventing disease or its complications may require physical changes in the environment, such as removal of the Broad Street pump handle to stop the cholera epidemic in London, or altering diets as in Goldberger's work on pellagra. Some of the great successes of public health have been and continue to be low technology. Examples, among many others, include insecticide-impregnated bednets and other vector control measures, oral rehydration solutions, treatment and cure of peptic ulcers, exercise and diet to reduce obesity, hand washing in hospitals (and other health facilities), community health workers, and condoms and circumcision for the prevention of STIs, including HIV and cancer of the cervix.

The societal context in terms of employment, social security, female education, recreation, family income, cost of living, housing, and homelessness is relevant to the health status of a population. Income distribution in a wealthy country may leave a wide gap between the upper and lower socioeconomic groups, which affects health status. The media have great power to sway public perception of health issues by choosing what to publish and the context in which to present information to society. Modern media may influence an individual's tendency to overestimate the risk of some health issues while underestimating the risk of others, ultimately influencing health choices, such as occurred with public concern regarding false claims of an association between the measles–mumps–rubella (MMR) vaccine and autism in the UK (see the Wakefield effect, Chapter 4). The New Public Health has an intrinsic responsibility for advocacy of improved societal conditions in its mission to promote optimal community health.

MODES OF PREVENTION

An ultimate goal of public health is to improve health and to prevent widespread disease occurrence in the population and in an individual. The methods of achieving this are wide and varied. When an objective has been defined in

---

BOX 2.5 Social Determinants of Health

"Social justice is a matter of life and death. It affects the way people live, their consequent chance of illness, and their risk of premature death. We watch in wonder as life expectancy and good health continue to increase in parts of the world and in alarm as they fail to improve in others. A girl born today can expect to live for more than 80 years if she is born in some countries – but less than 45 years if she is born in others. Within countries there are dramatic differences in health that are closely linked with degrees of social disadvantage. Differences of this magnitude, within and between countries, simply should never happen. These inequities in health, avoidable health inequalities arise because of the circumstances in which people grow, live, work, and age, and the systems put in place to deal with illness. The conditions in which people live and die are, in turn, shaped by political, social, and economic forces. Social and economic policies have a determining impact on whether a child can grow and develop to its full potential and live a flourishing life, or whether its life will be blighted. Increasingly the nature of the health problems rich and poor countries have to solve are converging. The development of a society, rich or poor, can be judged by the quality of its population's health, how fairly health is distributed across the social spectrum, and the degree of protection provided from disadvantage as a result of ill-health."

Source: *World Health Organization. Commission on social determinants of health: closing the gap in a generation: health equity through action on the social determinants of health. Geneva: WHO; 2008. Available at: http://whqlibdoc.who.int/publications/2008/9789241563703_eng.pdf [Accessed 21 November 2012].*

BOX 2.6 Modes of Prevention

- *Health promotion* – fostering national, community, and individual knowledge, attitudes, practices, policies, and standards conducive to good health; promoting legislative, social, or environmental conditions; promoting knowledge and practices for self-care that reduce individual and community risk; and creating a healthful environment. It is directed toward action on the determinants of health.
- *Health protection* – activities of official health departments or other agencies empowered to supervise and regulate food hygiene, community and recreational water safety, environmental sanitation, occupational health, drug safety, road safety, emergency preparedness, and many other activities to eliminate or reduce as much as possible risks of adverse consequences to health.
- *Primary prevention* – preventing a disease from occurring, e.g., vaccination to prevent infectious diseases, advice to stop smoking to prevent lung cancer.
- *Secondary prevention* – making an early diagnosis and giving prompt and effective treatment to stop progress or shorten the duration and prevent complications from an already existing disease process, e.g., screening for hypertension or cancer of cervix and colorectal cancer for early case finding, early care and better outcomes.
- *Tertiary prevention* – stopping progress of an already occurring disease, and preventing complications, e.g., in managing diabetes and hypertension to prevent complications; restoring and maintaining optimal function once the disease process has stabilized, e.g., promoting functional rehabilitation after stroke and myocardial infarction with long-term follow-up care.

Source: *Adapted from Last JM. A dictionary of public health. New York: Oxford University Press; 2007.*

---

preventing disease, the next step is to identify suitable and feasible methods of achieving it, or a strategy with tactical objectives. This determines the method of operation, course of action, and resources needed to carry it out. The methods of public health are categorized as health promotion, and primary, secondary, and tertiary prevention (Box 2.6).

Health Promotion

Health promotion is the process of enabling people and communities to increase control over factors that influence their health, and thereby to improve their health (adapted from the Ottawa Charter of Health Promotion, 1986; Box 2.7). Health promotion is a guiding concept involving activities intended to enhance individual and community health and well-being (Box 2.8). It seeks to increase involvement and control by the individual and the community in their own health. It acts to improve health and social welfare, and to reduce specific determinants of diseases and risk factors that adversely affect the health, well-being, and productive capacities of an individual or society, setting targets based on the size of the problem but also the feasibility of successful intervention, in a cost-effective way. This can be through direct contact with the patient or risk group, or act indirectly through changes in the environment, legislation, or public policy. Control of AIDS relies on an array of interventions that promote change in sexual behavior and other contributory risks such as sharing of needles among drug users, screening of blood supply, safe hygienic practices in health care settings, and education of groups at risk such as teenagers, sex workers, migrant workers, and many others. Control of AIDS is also a clinical problem in that patients need antiretroviral therapy (ART), but this becomes a management and policy issue for making these drugs available and at an affordable price for the poor countries most affected.

This is an example of the challenge and effectiveness of health promotion and the New Public Health.

Health promotion is a key element of the New Public Health and is applicable in the community, the clinic or hospital, and in all other service settings. Some health promotion activities are government legislative and

---

BOX 2.7 The Ottawa Charter: Health Promotion

Health promotion (HP) is the process of enabling people to increase control over, and to improve their health.

HP represents a comprehensive social and political process, and not only embraces actions directed at strengthening the skills and capabilities of individuals.

HP also undertakes action directed towards changing social, environmental, and economic conditions so as to alleviate their impact on public and individual health.

Health promotion is the process of enabling people to increase control over the *determinants of health* and thereby improve their *health*. Participation is essential to sustain health promotion action.

The *Ottawa Charter* identifies three basic strategies for health promotion. These are *advocacy* for health to create the essential conditions for health indicated above; *enabling* all people to achieve their full health potential; and *mediating* between the different interests in society in the pursuit of health.

These strategies are supported by five priority action areas as outlined in the Ottawa Charter for health promotion:

- Build healthy public policy.
- Create supportive environments for health.
- Strengthen community action for health.
- Develop personal skills.
- Reorient health services.

Source: *Ottawa Charter for Health Promotion. Geneva: WHO; 1986. Available at: http://www.who.int/hpr/NPH/docs/hp_glossary_en.pdf [Accessed 21 November 2012].*

BOX 2.8 Elements of Health Promotion

1. Address the population as a whole in health-related issues, in everyday life as well as people at risk for specific diseases.
2. Direct action to risk factors or causes of illness or death.
3. Undertake activist approach to seek out and remedy risk factors in the community that adversely affect health.
4. Promote factors that contribute to a better condition of health of the population.
5. Initiate actions against health hazards, including communication, education, legislation, fiscal measures, organizational change, community development, and spontaneous local activities.
6. Involve public participation in defining problems and deciding on action.
7. Advocate relevant environmental, health, and social policy.
8. Encourage health professional participation in health education and health advocacy.
9. Advocate for health based on human rights and solidarity.
10. Invest in sustainable policies, actions, and infrastructure to address the determinants of health.
11. Build capacity for policy development, leadership, health promotion practice, knowledge transfer and research, and health literacy.
12. Regulate and legislate to ensure a high level of protection from harm and enable equal opportunity for health and well-being for all people.
13. Partner and build alliances with public, private, nongovernmental, and international organizations and civil society to create sustainable actions.
14. Make the promotion of health central to the global development agenda.

Source: *Adapted from World Health Organization. Ottawa Charter for Health Promotion. Geneva: WHO; 1986; Bangkok Charter for Health Promotion in a Globalized World, 2005; other conferences are available at: http://www.who.int/healthpromotion/conferences/en/ [Accessed 25 May 2012].*

regulatory interventions such as mandating the use of seat belts in cars, requiring that children be immunized to attend school, declaring that certain basic foods must have essential minerals and vitamins added to prevent nutritional deficiency disorders in vulnerable population groups, and mandating that all newborns should be given prophylactic vitamin K to prevent hemorrhagic disease of the newborn. Setting food and drug standards and raising taxes on cigarettes and alcohol to reduce their consumption are also part of health promotion. Promoting a healthy lifestyle is a major known obesity-preventive activity. Health promotion is provided by organizations and people with varied professional backgrounds working towards common goals of improvement in the health and quality of individual and community life. Initiatives may come from government with dedicated allocation of funds to address specific health issues, from donors, or from advocacy or

community groups or individuals to promote a specific or general cause in health.

Raising awareness to inform and motivate people about their own health and lifestyle factors that might put them at risk requires teaching young people about the dangers of sexually transmitted diseases, smoking, and alcohol abuse to reduce risks associated with their social behavior. It might include disseminating information on healthy nutrition; for example, the need for folic acid supplements for women of childbearing age and multiple vitamins for elderly, as well as the elements of a healthy diet, compliance with immunization recommendations, compliance with screening programs, and many others. Community and peer group attitudes and standards affect individual behavior. Health promotion endeavors to create a climate of knowledge, attitudes, beliefs, and practices that are associated with better health outcomes.

International conferences following on from the Ottawa Charter were held in Adelaide in 1988, Sundsvall in 1991, Jakarta in 1999, Mexico in 2000, Bangkok in 2005, and Nairobi in 2009. The principles of health promotion have been reiterated and have influenced public policy regarding public health as well as the private sector.

Health promotion has a track record of proven success in numerous public health issues where a biomedical solution was not available. The HIV/AIDS pandemic from the 1980s until the late 1990s had no medical treatment and control measures relied on screening, education, lifestyle changes, and supportive care. Health promotion brought forward multiple interventions, from condom use and distribution, to needle exchanges for intravenous drug users, to male circumcision in high-prevalence African countries. Medical treatment was severely limited until ART was developed.

The success of ART also depends on a strong element of health promotion in widening the access to treatment and the success of medications to reduce transmission, most remarkably in reducing maternal–fetal transmission (see Chapter 4). Similarly, in the battle against cardiovascular diseases, health promotion was an instrumental factor in raising public awareness of the importance of management of hypertension and smoking reduction, dietary restraint, and physical exercise. The success of massive reductions in stroke and CHD mortality is as much the result of health promotion as of improved medical care (see Chapter 5).

Health Protection

The character of public health carries with it a "good cop, bad cop" dichotomy. The "good cop" is persuasive and educational trying to convince people to do the right thing in looking after their own health: diet, exercise, smoking cessation, and others. On the other side, the "bad cop" role is regulatory and punitive. Public health has a serious responsibility and role in the enforcement of laws and regulation to protect the public health. Some of these are restrictive of individual rights that may damage other people or are requirements based on strong evidence of benefits to population health. Readily accepted are food and drug standards, such as pasteurization of milk, and iodization of salt; requirements to drive on the right-hand side of the road (except in some countries such as the UK), to wear seat belts and for motorcyclists to wear safety helmets; and not smoking in public places.

Enforcement of these and similar statutory or regulatory requirements is vital in a civil society to protect the public from health hazards and to protect people from harm and exploitation by unscrupulous manufacturers and marketing. Cigarette advertising and sponsorship of sports events by tobacco companies are banned in most upper income countries. The use of transfats in food manufacturing and baking is now banned and salt reduction is being promoted and even mandated in many US local authorities to reduce cardiovascular disease. Advertising of unhealthy snack foods on children's television programs and during child-watching hours is commonly restricted. Banning high-sugar soda drink distribution in schools is a successful intervention to reduce the current child obesity epidemic. Melamine use in milk powders and baby formulas, which caused widespread illness and death of infants in China, is now banned and a punishable offence for manufacture or distribution in China and worldwide.

Examples of this aspect of public health are mentioned throughout this text, especially in Chapters 8 and 9 on nutrition, and environmental and occupational health, respectively. The regulatory enforcement function of public health is sometimes controversial and portrayed as interference with individual liberty. Fluoridation of community water supplies is an example where aggressive lobby groups opposing this safe and effective public health measure are still common. This is discussed in Chapter 7.

Equally important is the public health policy issue of resource allocation and taxation for health purposes. Taxation is an unpopular measure that governments must employ and enforce in order to do the public's business. The debate over the Patient Protection and Affordable Care Act 2010 (PPACA or "Obamacare"), discussed elsewhere in this and other chapters, shows how bitter the arguments can become, yet the goal of equality of access to health care cannot be denied as a public good, demonstrably contributing to the health of the nation.

Primary Prevention

Primary prevention refers to those activities that are undertaken to prevent disease or injury from occurring at all. Primary prevention works with both the individual and the community. It may be directed at the host to increase resistance to the agent (such as in immunization or cessation of smoking), or at environmental activities to reduce conditions favorable to the vector for a biological agent, such as mosquito vectors of malaria or dengue fever. Landmark examples include the treatment and prevention of scurvy among sailors based on James Lind's findings in a classic clinical epidemiological study in 1747, and John Snow's removal of the handle from the Broad Street pump to stop a cholera epidemic in London in 1854 (see Chapter 1).

Primary prevention includes elements of health protection such as ensuring water, food and drug, and workplace safety; chlorination of drinking water to prevent transmission of waterborne enteric diseases; pasteurization of milk to prevent gastrointestinal diseases; mandating wearing seat belts in motor vehicles to prevent serious injury and death in road crashes; and reducing the availability of firearms to reduce injury and death from intentional, accidental, or random violence. It also includes direct measures to prevent diseases, such as immunization to prevent polio, tetanus, pertussis, and diphtheria.

Health promotion and health protection blend together as a group of activities that reduce risk factors and diseases through many forms of intervention such as changing smoking legislation or preventing birth defects by fortification of flour with folic acid. Prevention of HIV transmission by needle exchange for intravenous drug users, promoting condom usage, and promoting male circumcision in Africa, and the distribution of condoms and clean needles for HIV-positive drug users are recent examples of primary prevention associated with health promotion programs.

Primary prevention also includes activities within the health system that can lead to better health. This may mean, for example, setting standards and to reduce hospital infections, and ensuring that doctors not only are informed of appropriate immunization practices and modern prenatal care or screening programs for cancer of the cervix, colon, and breast, but also are aware of their vital role in preventing cardiovascular and other non-communicable diseases. In this role, the health care provider serves as a teacher and guide, as well as a diagnostician and therapist. Like health promotion, primary prevention does not depend on health care providers alone; health promotion works to increase individual and community consciousness of self-care, mainly by raising awareness and information levels and empowering the individual and the community to improve self-care, to reduce risk factors, and to live healthier lifestyles.

Secondary Prevention

Secondary prevention is early diagnosis and management to prevent complications from a disease. Public health interventions to prevent the spread of disease include the identification of sources of the disease and the implementation of steps to stop it, as shown in Snow's closure of the Broad Street pump. Secondary prevention includes steps to isolate cases and treat or immunize contacts so as to prevent further cases of meningitis or measles, for example, in outbreaks. For current epidemics such as HIV/AIDS, primary prevention is largely based on education, abstinence from any and certainly risky sexual behavior, circumcision, and treatment of patients in order to improve their health and to reduce the risk of spread of HIV. For high-risk groups such as intravenous drug users, needle-exchange programs reduce the risk of spread of HIV, and hepatitis B and C. Distribution of condoms to teenagers, military personnel, truck drivers, and commercial sex workers helps to prevent the spread of STIs and AIDS in schools and colleges, as well as among the military. The promotion of circumcision is shown to be effective in reducing the transmission of HIV and of HPV (the causative organism for cancer of the cervix).

All health care providers have a role in secondary prevention; for example, in preventing strokes by early identification and adequate care of hypertension. The child who has an untreated streptococcal infection of the throat may develop complications which are serious and potentially life-threatening, including rheumatic fever, rheumatic valvular heart disease, and glomerulonephritis. A patient found to have elevated blood pressure should be advised about continuing management by appropriate diet and weight loss if obese, regular physical exercise, and long-term medication with regular follow-up by a health provider in order to reduce the risk of stroke and other complications. In the case of injury, competent emergency care, safe transportation, and good trauma care may reduce the chance of death and/or permanent handicap. Screening and high-quality care in the community prevent complications of diabetes, including heart, kidney, eye, and peripheral vascular disease. They can also prevent hospitalizations, amputations, and strokes, thus lengthening and improving the quality of life. Health care systems need to be actively engaged in secondary prevention, not only as individual doctors' services, but also as organized systems of care.

Public health also has a strong interest in promoting high-quality care in secondary and tertiary care hospital centers in such areas of treatment as acute myocardial infarction, stroke, and injury in order to prevent irreversible damage. Measures include quality of care reviews to promote adequate long-term postmyocardial infarction care with aspirin and beta-blockers or other medication to prevent or delay recurrence and second or third myocardial infarctions. The role of high-quality transportation and care in emergency facilities of hospitals in public health is vital to prevent long-term damage and disability; thus, cardiac care systems including publicly available defibrillators, catheterization, the use of stents, and bypass procedures are important elements of health care policy and resource allocation, which should be accessible not only in capital cities but also to regional populations.

Tertiary Prevention

Tertiary prevention involves activities directed at the host or patient, but also at the social and physical environment in order to promote rehabilitation, restoration, and maintenance of maximum function after the disease and its complications have stabilized. The person who has undergone a cerebrovascular accident or trauma will reach a stage where active rehabilitation can help to restore lost functions and prevent recurrence or further complications. The public health system has a direct role in the promotion of disability-friendly legislation and standards of building, housing, and support services for chronically ill, handicapped, and elderly people. This role also involves working with many governmental social and educational departments, but also with advocacy groups, NGOs, and families. It may also include the promotion of disability-friendly workplaces and social service centers.

Treatment for conditions such as myocardial infarction or a fractured hip now includes early rehabilitation in order to promote early and maximum recovery with restoration to optimal function. The provision of a wheelchair, walkers, modifications to the home such as special toilet facilities, doors, and ramps, along with transportation services for paraplegics are often the most vital factors in rehabilitation. Public health agencies work with groups in the community concerned with promoting help for specific categories of risk group, disease, or disability to reduce discrimination. Community action is often needed to eliminate financial, physical, or social barriers, promote community awareness, and finance special equipment or other needs of these groups. Close follow-up and management of chronic disease, physical and mental, require home care and ensuring an appropriate medical regimen including drugs, diet, exercise, and support services. The follow-up of chronically ill people to supervise the taking of medications, monitor changes, and support them in maximizing their independent capacity in activities of daily living is an essential element of the New Public Health.

DEMOGRAPHIC AND EPIDEMIOLOGICAL TRANSITION

Public health uses a population approach to achieve many of its objectives. This requires defining the population, including trends of change in the age and gender distribution of the population, fertility and birth rates, spread of disease and disability, mortality, marriage and migration, and socioeconomic factors. The reduction of infectious disease as the major cause of mortality, increased longevity coupled with declining fertility rates, resulted in changes in the age composition, or a demographic transition. Demographic changes, such as fertility and mortality patterns, are important factors in changing the age distribution of the population, resulting in a greater proportion of people surviving to older ages. Declining infant mortality, increasing educational levels of women, the availability of birth control, and other social and economic factors lead to changes in fertility patterns and the demographic transition – an aging of the population – with important effects on health service needs.

The age and gender distribution of a population affects and is affected by patterns of disease. Change in epidemiological patterns, or an epidemiological shift, is a change in predominant patterns of morbidity and mortality. The transition of infectious diseases becoming less prominent as causes of morbidity and mortality and being replaced by chronic and non-infectious diseases has occurred in both developed and developing countries. The decline in mortality from chronic diseases, such as cardiovascular disease, represents a new stage of epidemiological transition, creating an aging population with higher standards of health but also long-term community support and care needs. Monitoring and responding to these changes are fundamental responsibilities of public health, and a readiness to react to new, local, or generalized changes in epidemiological patterns is vital to the New Public Health.

Societies are not totally homogeneous in ethnic composition, levels of affluence, or other social markers. On one hand, a society classified as developing may have substantial numbers of people with incomes that promote overnutrition and obesity, so that disease patterns may include increasing prevalence of diseases of excesses, such as diabetes. On the other hand, affluent societies include population groups with disease patterns of poverty, including poor nutrition and low birth-weight babies.

A further stage of epidemiological transition has been occurring in the industrialized countries since the 1960s, with dramatic reductions in mortality from CHD, stroke and, to a lesser extent, trauma. The interpretation of this epidemiological transition is still not perfectly clear. How it occurred in the industrialized western countries but not in those of the former Soviet Union is a question whose answer is vital to the future of health in Russia and some countries of Eastern Europe. Developing countries must also prepare to cope with increasing epidemics of non-infectious diseases, and all countries face renewed challenges from infectious diseases with antibiotic resistance or newly appearing infectious agents posing major public health threats.

Demographic change in a country may reflect social and political decisions and health system priorities from decades before. Russia's rapid population decline since the 1990s, China's gender imbalance with a shortage of millions of young women, Egypt's rapid population growth outstripping economic capacity, and many other examples indicate the severity and societal importance of capacity to analyze and formulate public health and social policies to address such fundamental sociopolitical issues.

Aging of the population is now the norm in most developed countries as a result of low birth and declining mortality rates. This change in the age distribution of a population has many associated social and economic issues as to the future of social welfare with a declining age cohort to provide the workforce. The aging population requires pension and health care support which make demands of social security systems that will depend on economic growth with a declining workforce. In times of economic stress, as in Europe, this situation is made more difficult by longstanding short working weeks, early pension ages, and high social benefits. However, this results in unemployment among young people in particular and social conflict. The interaction of increasing life expectancy and a declining workforce is a fundamental problem in the high-income countries. This imbalance may be resolved in part through productivity gains and switching of primary production to countries with large still underutilized workforces, while employment in the developed countries will depend on service industries including health and the economic growth generated by higher technology and intellectual property and service industries.

INTERDEPENDENCE OF HEALTH SERVICES

The challenge of keeping populations and individuals healthy is reflected in modern health services. Each component of a health service may have developed with different historical emphases, operating independently as a separate service under different administrative auspices and funding systems, competing for limited health care resources. In this situation, preventive community care receives less attention and resources than more costly treatment services. Figure 2.5 suggests a set of health services in an interactive relationship to serve a community or defined population, but the emphasis should be on the interdependence of these services with one other and with the comprehensive network in order to achieve effective use of resources and a balanced set of services for the patient, the client or patient population, and the community.

Clinical medicine and public health each play major roles in primary, secondary, and tertiary prevention. Each may function separately in their roles in the community, but optimal success lies in their integrated efforts. Allocation of resources should promote management and planning practices to assist this integration. There is a functional interdependence of all elements of health care serving a definable population. The patient should be the central figure in the continuum or complex of services available. Effectiveness in use of resources means that providing the service most appropriate for meeting the individual's or group's needs at a point in time are those that should be applied. This is the central concept in currently developing innovations in health care delivery in the USA with organizations using terms such as patient centered medical home, accountable care organizations (ACOs), and population health management systems, which are being promoted in the Obamacare

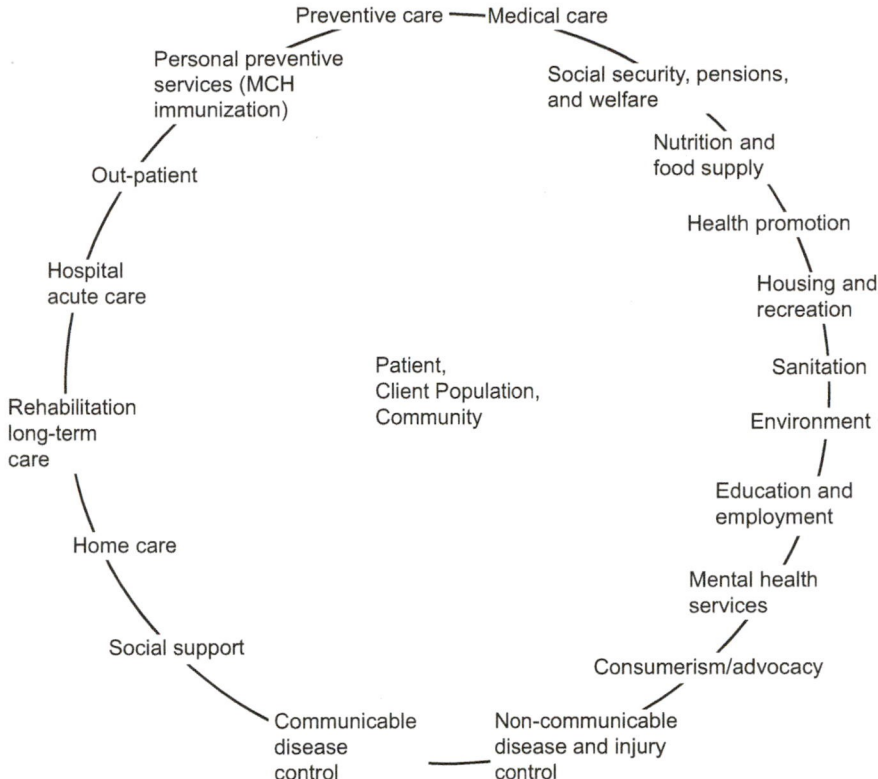

FIGURE 2.5 **Community health as a network of services, regulation, and health promotion serving a defined population.**

health reforms now in process (see Chapter 10) (Shortell et al., 2010).

Separate organization and financing of services place barriers to appropriate provision of services for both the community and the individual patient. The interdependence of services is a challenge in health care organizations for the future. Where there is competition for limited resources, pressures for tertiary services often receive priority over programs to prevent children from dying of preventable diseases. Public health must be seen in the context of all health care and must play an influential role in promoting prevention at all levels. Clinical services need public health in order to provide prevention and community health services that reduce the burden of disease, disability, and dependence on the institutional setting.

DEFINING PUBLIC HEALTH

Health was traditionally thought of as a state of absence of disease, pain, or disability, but has gradually been expanded to include physical, mental, and societal well-being. In 1920, C. E. A. Winslow, professor of public health at Yale University, defined public health as follows:

"Public health is the Science and Art of (1) preventing disease, (2) prolonging life, and (3) promoting health and efficiency through organized community effort for:

(a) the sanitation of the environment,

(b) the control of communicable infections,

(c) the education of the individual in personal hygiene,

(d) the organization of medical and nursing services for the early diagnosis and preventive treatment of disease, and

(e) the development of social machinery to ensure everyone a standard of living adequate for the maintenance of health, so organizing these benefits as to enable every citizen to enjoy his birthright of health and longevity."

(Quoted in Institute of Medicine. The future of public health. Washington, DC: National Academy Press; 1988)

Winslow's far-reaching definition remains a valid framework but is unfulfilled when clinical medicine and public health have financing and management barriers between them. In many countries, isolation from the financing and provision of medical and nursing care services left public health with the task of meeting the health needs of the indigent and underserved population groups with inadequate resources and recognition. Health insurance organizations for medical and hospital care have in recent years been more open to incorporating evidence-based preventive care, but the organization of public health has lacked the same level of attention. In some countries, the limitations have been conceptual in that public health was defined primarily in terms of control of infectious, environmental, and occupational diseases.

A more recent and widely used definition is:

"Public health is the science and art of preventing disease, prolonging life, and promoting health through the organized efforts of society."

This definition, coined in 1988 in the Public Health in England report by Sir Donald Acheson, reflects the broad focus of modern public health.

Terms such as *social hygiene*, *preventive medicine*, *community medicine*, and *social medicine* have been used to denote public health practice over the past century. Preventive medicine is the application of preventive measures by clinical practitioners combining some elements of public health with clinical practice relating to individual patients. Preventive medicine defines medical or clinical personal preventive care, with stress on risk groups in the community and national efforts for health promotion. The focus is on the health of defined populations to promote health and well-being using evidence-based guidelines for cost-effective preventive measures. Measures emphasized include screening and follow-up of chronic illnesses, and immunization programs; for example, influenza and pneumococcal pneumonia vaccines are used by people who are vulnerable because of their age, chronic diseases, or risk of exposure, such as medical and nursing personnel and those providing other personal clinical services. Clinical medicine also deals in the area of prevention in the management of patients with hypertension or diabetes, and in doing so prevents the serious complications of these diseases.

Social Medicine and Community Health

Social medicine is also primarily a medical specialty which looks at illness in an individual in the family and social context, but lacks the environmental and regulatory and organized health promotion functions of public health. Community health implies a local form of health intervention, whereas public health more clearly implies a global approach, which includes action at the international, national, state, and local levels. Some issues in health can be dealt with at the individual, family, or community level; others require global strategies and intervention programs with regional, national, or international collaboration and leadership.

The social medicine movement originated to address the harsh conditions of the working population during the industrial revolution in mid-nineteenth-century Europe. An eminent pioneer in cellular pathology, Rudolph Virchow provided leadership in social medicine powered by the revolutionary movements of 1848, and subsequent social democrat political movements. Their concern focused on harsh living and health conditions among the urban poor working class and neglectful political norms of the time. Social medicine also developed as an academic discipline and advocacy orientation by providing statistical evidence showing, as in various governmental reports in the mid-nineteenth century, that poverty among the working class was associated with short life expectancy and that social conditions were key factors in the health of populations and individuals. This movement provided the basis for departments in medical faculties and public health education throughout the world stressing the close relationship between political priorities and health status. This continued in the twentieth century and in the USA found expression in pioneering work since the 1940s at Montefiore Hospital in New York and with Victor Sidel, founding leader of the community health center movement the USA from the 1960s.

In the twenty-first century this movement continues to emphasize relationships between politics, society, disease, and medicine, and forms of medical practice derived from it, as enunciated by prominent advocates such as Harvard-based Paul Farmer in Haiti, Russia and Rwanda, and in the UK by Martin McKee and others (Nolte and McKee, 2008). Similar concepts are current in the USA under headings such as family medicine, preventive medicine, and social medicine. This movement has also influenced Sir Michael Marmot and others in the World Health Commission of Health Inequalities of 2008, with a strong influence on the UN initiative to promote MDGs, whose first objective is poverty reduction (Commission on Inequalities report 2010). Application of the idea of poverty reduction as a method of reducing health inequalities has been successful recently in a large field trial in Brazil showing greater reduction in child mortality where cash bonuses were awarded by municipalities for the poor families than that observed in other similar communities (Rasella, 2013).

In the USA, this movement is supported by increased health insurance coverage for the working poor, with funding for preventive care and incentives for community health centers in the Obamacare plan of 2010 for implementation in the coming years to provide care for uninsured and underserved populations, particularly in urban and rural poverty areas. The political aspect of social medicine is the formulation of and support for national initiatives to widen health care coverage to the 16 percent of the US population who are still uninsured, and to protect those who are arbitrarily excluded owing to previous illnesses, caps on coverage allowed, and other exploitative measures taken by private insurance that frequently deny Americans access to the high levels of health care available in the country.

Social Hygiene, Eugenics, and Corruption of Public Health Concepts

The ethical base of public health in Europe evolved in the context of its successes in the nineteenth and early twentieth centuries along with ideas of social progress. But the twentieth century was also replete with extremism and wide-scale abuse of human rights, with mass executions, deportations, and starvation as official policy in fascist and Stalinist regimes. Eugenics, a pseudoscience popularized in the early decades of the twentieth century, promoted social policies meant to improve the hereditary qualities of a race by methods such as sterilization of mentally handicapped people.

The "social and racial hygiene" of the eugenics movements led to the medicalization of sterilization in the USA and other countries. This was adopted and extended in Nazi Germany to a policy of murder, first of the mentally and physically handicapped and then of "racial inferiors". These eugenics theories were widely accepted in the medical community in Germany, then used by the Nazi regime to justify medically supervised killing of hundreds of thousands of helpless, incapacitated individuals. This practice was linked to wider genocide and the Holocaust, with the brutalization and industrialized murder of over 6 million Jews and 6 million other people, and corrupt medical experimentation on prisoners. Following World War II, the ethics of medical experimentation (and public health) were codified in the Nuremberg Code and Universal Declaration of Human Rights based on lessons learned from these and other atrocities inflicted on civilian populations (see Chapter 15).

Threats of genocide, ethnic cleansing, and terrorism are still present on the world stage, often justified by current warped versions of racial hygienic theories. Genocidal incitement and actual genocide and terrorism have recurred in the last decades of the twentieth century and into the twenty-first century in the former Yugoslav republics, Africa (Rwanda and Darfur), south Asia, and elsewhere. Terrorism against civilians has become a worldwide phenomenon with threats of biological and chemical agents, and potentially with nuclear capacity. Asymmetrical warfare of insurgencies which use innocent civilians for cover, as with other forms of warfare, carries with it grave dangers to public health, human rights, and international stability, as seen in the twenty-first century in South Sudan, Darfur, DR Congo, Chechnya, Iraq, Afghanistan, and Pakistan.

Medical Ecology

In 1961, Kerr White and colleagues defined medical ecology as population-based research providing the foundation for management of health care quality. This concept stresses a population approach, including those not attending and those using health services. This concept was based on previous work on quality of care, randomized clinical trials, medical audit, and structure–process–outcome research. It also addressed health care quality and management.

These themes influenced medical research by stressing the population from which clinical cases emerge as well as public health research with clinical outcome measures, themes that recur in the development of health services research and, later, evidence-based medicine. This led to the development of the Agency for Health Care Policy and

Research and Development in the US Department of Health and Human Services and evidence-based practice centers to synthesize fundamental knowledge for the development of information for decision-making tools such as clinical guidelines, algorithms, or pathways. Clinical guidelines and recommended best practices have become part of the New Public Health to promote quality of patient care and public health programming. These can include recommended standards; for example, follow-up care of the postmyocardial infarction patient, an internationally recommended immunization schedule, recommended dietary intake or food fortification standards, and mandatory vitamin K and eye care for all newborns and many others (see Chapter 15).

Community-Oriented Primary Care

Community-oriented primary care (COPC) is an approach to primary health care that links community epidemiology and appropriate primary care, using proactive responses to the priority needs identified. COPC, originally pioneered in South Africa and Israel by Sidney and Emily Kark and colleagues in the 1950s and 1960s, stresses medical services in the community which need to be adapted to the needs of the population as defined by epidemiological analysis. COPC involves community outreach and education, as well as clinical preventive and treatment services.

COPC focuses on community epidemiology and an active problem-solving approach. This differs from national or larger scale planning that sometimes loses sight of the local nature of health problems or risk factors. COPC combines clinical and epidemiological skills, defines needed interventions, and promotes community involvement and access to health care. It is based on linkages between the different elements of a comprehensive basket of services along with attention to the social and physical environment. A multidisciplinary team and outreach services are important for the program, and community development is part of the process.

In the USA, the COPC concept has influenced health care planning for poor areas, especially provision of federally funded community health centers in attempts to provide health care for the underserved since the 1960s. In more recent years, COPC has gained wider acceptance in the USA, where it is associated with family physician training and community health planning based on the risk approach and "managed care" systems. Indeed, the three approaches are mutually complementary (Box 2.9). As the emphasis on health care reform in the late 1990s moved towards managed care, the principles of COPC were and will continue to be important in promoting health and primary prevention in all its modalities, as well as tertiary prevention with follow-up and maintenance of the health of the chronically ill.

COPC stresses that all aspects of health care have moved towards prevention based on measurable health issues in the community. Through either formal or informal linkages

BOX 2.9 Features of Community-Oriented Primary Health Care (COPC)

- Essential features:
  - Clinical and epidemiological skills
  - A defined population
  - Defined programs to address community health issues
  - Community involvement
  - Accessibility to health care – reducing geographic, fiscal, social, and cultural barriers.
- Desirable features:
  - Integration/coordination of curative, rehabilitative, preventive, and promotive care
  - Comprehensive approach extending to behavioral, social, and environmental factors
  - Multidisciplinary team
  - Mobility and outreach
  - Community development.

Sources: *Tollman S. Community oriented primary care: origins, evolution, applications. Soc Sci Med 1991;32:633–42.*
Epstein L, Gofin J, Gofin R, Neumark Y. The Jerusalem experience: three decades of service, research, and training in community-oriented primary care. Am J Public Health 2002;92:1717–21.
Gofin J, Gofin R. Essentials of global community health. American Public Health Association; 2010.

BOX 2.10 Definitions of Health and Mission of the World Health Organization

The World Health Organization defines health as "a state of complete physical, mental and social well-being, not merely the absence of disease or infirmity" (WHO Constitution, 1948).

In 1978 at the Alma-Ata Conference on Primary Health Care, the WHO related health to "social and economic productivity in setting as a target the attainment by all the people of the world of a level of health that will permit them to lead a socially and economically productive life". Three general programs of work for the periods 1984–1989, 1990–1995, and 1996–2001 were formulated as the basis of national and international activity to promote health.

In 1995, the WHO, recognizing changing world conditions of demography, epidemiology, environment, and political and economic status, addressed the unmet needs of developing countries and health management needs in the industrialized countries, calling for international commitment to "attain targets that will make significant progress towards improving equity and ensuring sustainable health development".

The 1999 object of the WHO is restated as "the attainment by all peoples of the highest possible level of health" as defined in the WHO constitution, by a wide range of functions in promoting technical cooperation, assisting governments, and providing technical assistance, international cooperation, and standards.

Source: *World Health Organization. New challenges for public health: report of an international meeting. Geneva: WHO; 1996. Available at: www.//who.org/aboutwho [Accessed 21 November 2012].*

between health services, the elements of COPC are part of the daily work of health care providers and community services systems. The US Institute of Medicine issued the Report on Primary Care in 1995, defining *primary care* as "the provision of integrated, accessible health care services by clinicians who are accountable for addressing the majority of personal health care needs, developing a sustained partnership with patients and practicing in the context of the family and the community". This formulation was criticized by the American Public Health Association (APHA) as lacking a public health perspective and failing to take into account both the individual and the community health approaches. COPC tries to bridge this gap between the perspectives of primary care and public health.

The community, whether local, regional, or national, is the site of action for many public health interventions. Moreover, understanding the characteristics of the community is vital to a successful community-oriented approach. By the 1980s, new patterns of public health began to emerge, including all measures used to improve the health of the community, and at the same time working to protect and promote the health of the individual. The range of activities to achieve these general goals is very wide, including individual patient care systems and the community-wide activities that affect the health and well-being of the individual. These include the financing and management of health systems, evaluation of the health status of the population, and measures to improve the quality of health care. They place reliance on health promotion activities to change environmental risk factors for disease and death. They promote integrative and multisectoral approaches and the international health teamwork required for global progress in health.

THE WORLD HEALTH ORGANIZATION'S DEFINITION OF HEALTH

The definition of health in the charter of the WHO as a complete state of physical, mental, and social well-being had a ring of utopianism and irrelevance to states struggling to provide even minimal care in severely adverse political, economic, social, and environmental conditions (Box 2.10). In 1977, a more modest goal was set for attainment of a level of health compatible with maximum feasible social and economic productivity. One needs to recognize that health and disease are on a dynamic continuum that affects everyone. The mission for public health is to use a wide range of methods to prevent disease and premature death, and improve quality of life for the benefit of individuals and the community.

In the 1960s, most industrialized countries were concentrating energies and financing in health care on providing access to medical and hospital services through national insurance schemes. Developing countries were often spending scarce resources trying to emulate this trend. The WHO was concentrating on categorical programs, such as eradication of smallpox and malaria, as well as the Expanded Program of Immunization and similar specific efforts. At the same time, there was a growing concern that developing countries were placing too much emphasis and expenditure on curative services and not enough on prevention and primary care.

Alma-Ata: Health for All

The World Health Assembly (WHA) in 1977 endorsed the primary care approach under the banner of "Health for All by the Year 2000" (HFA 2000). This was a landmark decision and has had important practical results. The WHO and the United Nations Children's Fund (UNICEF) sponsored a seminal conference held in Alma-Ata, in the USSR (Kazakhstan), in 1978, which was convened to refocus health policy on primary care. The Alma-Ata Declaration stated that health is a basic human right, and that governments are responsible to assure that right for their citizens and to develop appropriate strategies to fulfill this promise. This proposition has come to be increasingly accepted in the international community. The conference stressed the right and duty of people to participate in the planning and implementation of their health care. It advocated the use of scientifically, socially, and economically sound technology. Joint action through intersectoral cooperation was also emphasized.

The Alma-Ata Declaration focused on primary health care as the appropriate method of assuming adequate access to health care for all (Box 2.11). Many countries have gradually come to accept the notion of placing priority on primary care, resisting the temptation to spend high percentages of health care resources on high-tech and costly medicine. Spreading these same resources into highly cost-effective primary care, such as immunization and nutrition programs, provides greater benefit to individuals and to society as a whole.

Alma-Ata provided a new sense of direction for health policy, applicable to developing countries and in a different way than the approaches of the developed countries. During the 1980s, the Health for All concept influenced national health policies in the developing countries with signs of progress in immunization coverage, for example, but the initiative was diluted as an unintended consequence by more categorical programs such as eradication of poliomyelitis. For example, developing countries have accepted immunization and diarrheal disease control as high-priority issues and achieved remarkable success in raising immunization coverage from some 10 percent to over 75 percent in just a decade.

> **BOX 2.11 Declaration of Alma-Ata, 1978: A Summary of Primary Health Care (PHC)**
>
> 1. Reaffirms that health is a state of complete physical, mental, and social well-being, and not merely the absence of disease or infirmity, and is a fundamental human right.
> 2. Existing gross inequalities in the health status of the people, particularly between developed and developing countries as well as within countries, are of common concern to all countries.
> 3. Governments have a responsibility for the health of their people. The people have the right and duty to participate in planning and implementation of their health care.
> 4. A main social target is the attainment, by all peoples of the world by the year 2000, of a level of health that will permit them to lead a socially and economically productive life.
> 5. PHC is essential health care based on practical, scientifically sound, and socially acceptable methods and technology.
> 6. It is the first level of contact of individuals, the family, and the national health system bringing health care as close as possible to where people live and work, as the first element of a continuing health care process.
> 7. PHC evolves from the conditions and characteristics of the country and its communities, based on the application of social, biomedical, and health services research and public health experience.
> 8. PHC addresses the main health problems in the community, providing promotive, preventive, curative, and rehabilitative services accordingly.
> 9. PHC includes the following:
>   (a) Education concerning prevailing health problems and methods of preventing and controlling them
>   (b) Promotion of food supply and proper nutrition
>   (c) Adequate supply of safe water and basic sanitation
>   (d) Maternal and child health care, including family planning
>   (e) Immunization against the major infectious diseases
>   (f) Prevention of locally endemic diseases
>   (g) Appropriate treatment of common diseases and injuries
>   (h) The provision of essential drugs
>   (i) Relies on all health workers … to work as a health team.
> 10. All governments should formulate national health policies, strategies and plans, mobilize political will and resources, used rationally, to ensure PHC for all people.
>
> **Source:** *World Health Organization. Declaration of Alma-Ata: International Conference on Primary Health Care, Alma-Ata, USSR, 6–12 September 1978. Available at: http://www.who.int/hpr/NPH/docs/declaration_almaata.pdf [Accessed 21 November 2012].*

Developed countries addressed these principles in different ways. In these countries, the concept of primary health care led directly to important conceptual developments in health. National health targets and guidelines are now common in many countries and are integral parts of national health planning. Reforms of the NHS – for example, as discussed in Chapter 13, remuneration increases for family physicians and encouraging group practice with public health nursing support – have become widespread in the UK. Leading health maintenance organizations, such as Kaiser Permanente in the USA and district health systems in Canada, have emphasized integrated approaches to health care for registered or geographically defined populations (see Chapters 11–13). This approach is becoming common in the USA in ACOs, which will be fostered by the 2010 Obamacare legislation (PPACA). This systematic approach to individual and community health is an integral part of the New Public Health.

The interactions among community public health, personal health services, and health-related behavior, including their management, are the essence of the New Public Health. How the health system is organized and managed affects the health of the individual and the population, as does the quality of providers. Health information systems with epidemiological, economic, and sociodemographic analysis are vital to monitor health status and allow for changing priorities and management. Well-qualified personnel are essential to provide services, manage the system, and carry out relevant research and health policy analysis. Diffusion of data, health information, and responsibility helps to provide a responsive and comprehensive approach to meet the health needs of the individual and community. The physical, social, economic, and political environments are all important determinants of the health status of the population and the individual. Joint action (intersectoral cooperation) between public and non-governmental or community-based organizations is needed to achieve the well-being of the individual in a healthy society.

In the 1980s and 1990s, these ideas contributed to an evolving New Public Health, spurred on by epidemiological changes, health economics, the development of managed care linking health systems, and prepayment. Knowledge and self-care skills, as well as community action to reduce health risks, are no less important in this than the roles of medical practitioners and institutional care. All are parts of a coherent holistic approach to health.

SELECTIVE PRIMARY CARE

The concept of selective primary care, articulated in 1979 by Walsh and Warren, addresses the needs of developing countries to select those interventions on a broad scale that would have the greatest positive impact on health, taking into account limited resources such as money, facilities, and human resources.

The term *selective primary care* is meant to define national priorities that are based not on the greatest causes of morbidity or mortality, but on common conditions of epidemiological importance for which there are effective

and simple preventive measures. Throughout health planning, there is an implicit or explicit selection of priorities for allocation of resources. Even in primary care, selection of targets is a part of the process of resource allocation. In modern public health, this process is more explicit. A country with limited resources and a high birth rate will emphasize maternal and child health before investing in geriatric care.

This concept has become part of the microeconomics of health care and technology assessment, discussed in Chapters 11 and 15, respectively, and is used widely in setting priorities and resource allocation. In developing countries, cost-effective primary care interventions have been articulated by many international organizations, including iodization of salt, use of oral rehydration therapy (ORT) for diarrheal diseases, vitamin A supplementation for all children, expanded programs of immunization, and others that have the potential for saving hundreds of thousands of lives yearly at low cost. In developed countries, health promotions targeted to reduce accidents and risk factors such as smoking, high-fat diets, and lack of exercise for cardiovascular diseases are low-cost public health interventions that save lives and reduce the use of hospital care.

Targeting specific diseases is essential for efforts to control TB or eradicate polio, but at the same time, development of a comprehensive primary care infrastructure is equally or even more important than the single-disease approach. Some disease entities such as HIV/AIDS attract donor funding more readily than basic infrastructure services such as immunization, and this can sometimes be detrimental to addressing the overall health needs of the population and other neglected but also important diseases.

THE RISK APPROACH

The risk approach selects population groups on the basis of risk and helps to determine interventional priorities to reduce morbidity and mortality. The measure of health risk is taken as a proxy for need, so that the risk approach provides something for all, but more for those in need, in proportion to that need. In epidemiological terms, these are people with higher relative risk or attributed risk.

Some groups in the general population are at higher risk than others for specific conditions. The Expanded Programme on Immunization (EPI), Control of Diarrhoeal Diseases (CDD), and Acute Respiratory Disease (ARD) programs of the WHO are risk approaches to tackling fundamental public health problems of children in developing countries.

Public health places considerable emphasis on maternal and child health because these are vulnerable periods in life for specific health problems. Pregnancy care is based on a basic level of care for all, with continuous assessment of risk factors that require a higher intensity of follow-up. Prenatal care helps to identify factors that increase the risk for the pregnant woman or her fetus/newborn. Efforts directed towards these special risk groups have the potential to reduce morbidity and mortality. High-risk case identification, assessment, and management are vital to a successful maternal care program.

Similarly, routine infant care is designed not only to promote the health of infants, but also to find the earliest possible indications of deviation and the need for further assessment and intervention to prevent a worsening of the condition. Low birth-weight babies are at greater risk for many short- and long-term hazards and should be given special treatment. All babies are routinely screened for birth defects or congenital conditions such as hypothyroidism, phenylketonuria, and other metabolic and hematological diseases. Screening must be followed by investigating and treating those found to have a clinical deficiency. This is an important element of infant care because infancy itself is a risk factor.

As will be discussed in Chapters 6 and 7 and others, epidemiology has come to focus on the risk approach with screening based on known genetic, social, nutritional, environmental, occupational, behavioral, or other factors contributing to the risk for disease. The risk approach has the advantage of specificity and is often used to initiate new programs directed at special categories of need. This approach can lead to narrow and somewhat rigid programs that may be difficult to integrate into a more general or comprehensive approach, but until universal programs can be achieved, selective targeted approaches are justifiable. Indeed, even with universal health coverage, it is still important to address the health needs or issues of groups at special risk.

Working to achieve defined targets means making difficult choices. The supply and utilization of some services will limit availability for other services. There is an interaction, sometimes positive, sometimes negative, between competing needs and the health status of a population.

THE CASE FOR ACTION

Public health identifies needs by measuring and comparing the incidence or prevalence of the condition in a defined population with that in other comparable population groups and defines targets to reduce or eliminate the risk of disease. It determines ways of intervening in the natural epidemiology of the disease, and develops a program to reduce or even eliminate the disease. It also assesses the outcomes in terms of reduced morbidity and mortality, as well as the economic justification in cost-effectiveness analysis to establish its value in health priorities.

Because of the interdependence of health services, as well as the total financial burden of health care, it is essential to look at the costs of providing health care, and how

resources should be allocated to achieve the best results possible. Health economics has become a fundamental methodology in policy determination. The costs of health care, the supply of services, the needs for health care or other health-promoting interventions, and effective means of using resources to meet goals are fundamental in the New Public Health. It is possible to err widely in health planning if one set of factors is overemphasized or underemphasized. Excessive supply of one service diminishes the availability of resources for other needed investments in health. If diseases are not prevented or their sequelae not well managed, patients must use costly health care services and are unable to perform their normal social functions such as learning at school or performing at work. Lack of investment in health promotion and primary prevention creates a larger reliance on institutional care, driving health costs upwards, and restricting flexibility in meeting patients' needs. The interaction of supply and demand for health services is an important determinant of the political economy of health care. Health and its place in national priorities are determined by the social-political philosophy and resource allocation of a government.

The case for action, or the justification for a public health intervention, is a complex of epidemiological, economic, and public policy factors (Table 2.1). Each disease or group of diseases requires its own case for action. The justification for public health intervention requires sufficient evidence of the incidence and prevalence of the disease (see Chapter 3).

Evidence-based public health takes into account the effectiveness and safety of an intervention; risk factors; safe means at hand to intervene; the human, social, and economic cost of the disease; political factors; and a policy decision as to the priority of the problem. This often depends on subjective factors, such as the guiding philosophy of the health system and the way it allocates resources.

Some interventions are so well established that no new justification is required to make the case, and the only question is how to do it most effectively. For example, infant vaccination is a cost-effective and cost-beneficial program for the protection of the individual child and the population as a whole. Whether provided as a public service or as a clinical preventive measure by a private medical practitioner, it is in the interest of public health that all children be immunized.

An outbreak of diarrheal disease in a kindergarten presents an obvious case for action, and a public health system must respond on an emergency basis, with selection of the most suitable mode of intervention. The considerations in developing a case for action are outlined above. Need is based on clinical and epidemiological evidence, but also on the importance of an intervention in the eyes of the public. The technology available, its effectiveness and safety, and accumulated experience are important in the equation, as are the acceptability and affordability of appropriate interventions. The precedents for use of an intervention are also important.

TABLE 2.1 The Case for Action: Factors in Justifying Public Health Interventions

Ethics and Potential	Issues
The right to health	Public expectation and social norms
Public advocacy	Concerned groups, the media, an individual
Need – epidemiological and clinical	Morbidity, mortality, functional disability, physiological indicators
Available technology	Documented effectiveness, safety, experience, acceptability, affordability
Precedent – "state of the art"	Good public health practice; standards from leading centers of excellence, not necessarily consensus
Research and cumulative evidence	New evidence from research and practice should be published in peer-reviewed journals and made accessible to all policy makers and practitioners
Teaching public health	All health professionals should have broad introductory courses in public health
Legal constraints and liability	Law and court decisions, providers, managed care, and governments
Costs and benefits	Direct cost to health system; indirect cost to the individual, family, and society
Acceptability	Media and public opinion
Leadership	Political, professional, the media, advocacy groups, public opinion
Quality of life	Optimizing human potential in healthy communities

On epidemiological evidence, if the preventive practice has been seen to provide reduction in risk for the individual and for the population, then there is good reason to implement it. The costs, risks and benefits must be examined as part of the justification to help in the selection of health priorities.

Health systems research examines the efficiency of health care and promotes improved efficiency and effective use of resources. This is a vital function in determining how best to use resources and meet current health needs. Past emphasis on hospital care at the expense of less development of primary care and prevention is still a common issue, particularly in former Soviet and developing countries, where a high percentage of total health expenditure goes to acute hospital care with long length of stay, with smaller allocation to preventive and community health care. The result of this imbalance is high mortality from preventable diseases.

New drugs, vaccines, and medical equipment are continually becoming available, and each new addition needs to be examined among the national health priorities. Sometimes, owing to cost, a country cannot afford to add a new vaccine to the routine. However, when there is good evidence for efficacy and safety of new vaccines, drugs, diagnostic methods or other innovations, it could be applied for those at greatest risk. Although there are ethical issues involved, it may be necessary to advise parents or family members to purchase the vaccine independently. Clearly, recommending individual purchase of a vaccine is counter to the principle of equity and solidarity, benefiting middle-class families, and providing a poor basis of data for evaluation of the vaccine and its target disease. On the other hand, failure to advise parents of potential benefits to their children creates other ethical problems, but may increase public pressure and insurance system acceptance of new methods, e.g., varicella and HPV vaccines.

Mass screening programs involving complete physical examinations have not been found to be cost-effective or to significantly reduce disease. In the 1950s and 1960s, routine general health examinations were promoted as an effective method of finding disease early. Since the late 1970s, a selective and specific approach to screening has become widely accepted. This involves defining risk categories for specific diseases and bearing in mind the potential for remedial action. Early case finding of colon cancer by routine fecal blood testing and colonoscopy has been found to be effective, and Pap smear testing to discover cancer of the cervix is timed according to risk category. Screening for colorectal cancer is essential for modern health programs and has been adopted by most industrialized countries. Outreach programs by visits, telephones, emails or other modern methods of communication are important to contact non-attenders to promote utilization, and have been shown to increase compliance with proven effective measures. These programs are important for screening, follow-up, and maintenance of treatment for hypertension, diabetes, and other conditions requiring long-term management. Screening technology is changing and often the subject of intense debate as such programs are costly and their cost-effectiveness is an important matter for policy making: screening for lung cancer is becoming a feasible and effective matter for high-risk groups, whereas breast cancer screening frequency is now in dispute; while nanotechnology and bioengineering promises new methods for cancer screening.

The factor of contribution to quality of life should be considered. A vaccine for varicella is justified partly for the prevention of deaths or illness from chickenpox. A stronger argument is often based on the fact that this is a disease that causes moderate illness in children for up to 2 weeks and may require parents to stay home with the child, resulting in economic loss to the parent and society. The fact that this vaccination prevents the occurrence of herpes zoster or shingles later in life may also be a justification. Widespread adoption of hepatitis B vaccine is justified on the grounds that it prevents cancer of the liver, liver cirrhosis, and hepatic failure in a high percentage of the population affected.

How many cases of a disease are enough to justify an intervention? One or several cases of some diseases, such as poliomyelitis, may be considered an epidemic in that each case constitutes or is an indicator of a wider threat. A single case of polio suggests that another 1000 persons are infected but have not developed a recognized clinical condition. Such a case constitutes a public health emergency, and forceful organization to meet a crisis is needed. Current standards are such that even one case of measles imported into a population free of the disease may cause a large outbreak, as occurred in the UK, France, and Israel during 2007 through 2013, by contacts on an aircraft, at family gatherings, or even in medical settings. A measles epidemic indicates a failure of public health policy and practice. Screening for some cancers, such as cervix and colon, is cost effective. Screening of all newborns for congenital disorders is important because each case discovered early and treated effectively saves a lifetime of care for serious disability.

Assessing a public health intervention to prevent the disease or reduce its impact requires measurement of the disease in the population and its economic impact. There is no simple formula to justify a particular intervention, but the cost–benefit approach is now commonly required to make such a case for action. Sometimes public opinion and political leadership may oppose the views of the professional community, or may impose limitations of policy or funds that prevent its implementation. Conversely, professional groups may press for additional resources that compete for limited resources available to provide other needed health activities. Both the professionals of the health system and the general public need full access to health-related information to take part in such debates in a constructive way. To maintain progress, a system must examine new technologies and justify their adoption or rejection (see Chapter 15).

POLITICAL ECONOMY AND HEALTH

The association between health and political issues was emphasized by European innovators such as Rudolf Virchow (and in Great Britain by Edwin Chadwick; see Chapter 1) in the mid-nineteenth century, when the conditions of the working population were such that epidemic diseases were rife and mortality was high, especially in the crowded slums of the industrial revolution. The same observations led Bismarck in Germany to introduce early forms of social

insurance for the health of workers and their families in the 1880s, and to Britain's 1911 national health insurance, also for workers and families. The role of government in providing universal access to health care was a struggle in individual countries during the twentieth century and lasting into the second decade of the twenty-first century (e.g. President Obama's Affordable Health Care Act of 2010).

As the concept of public health has evolved, and the cost effectiveness of medical care has improved through scientific and technological advances, societies have identified health as a legitimate area of activity for collective bargaining and government. With this process, the need to manage health care resources has become more clearly defined as a public responsibility. In industrialized countries, each with very different political make-up, national responsibility for universal access to health has become part of the social ethos. With that, the financing and managing of health services have developed into part of a broad concept of public health, and economics, planning, and management have come to be part of the New Public Health (discussed in Chapters 10–13).

Social, ethical, and political philosophies have profound effects on policy decisions including allocation of public monies and resources. Investment in public health is now recognized as an integral part of socioeconomic development. Governments are major suppliers of funds and leadership in health infrastructure development, provision of health services, and health payment systems. They also play a central role in the development of health promotion and regulation of the environment, food, and drugs essential for community health.

In liberal social democracies, the individual is deemed to have a right to health care. The state accepts responsibility to ensure availability, accessibility, and quality of care. In many developed countries, government has also taken responsibility to arrange funding and services that are equitably accessible and of high quality. Health care financing may involve taxation, allocation, or special mandatory requirements on employers to pay for health insurance. Services may be provided by a state-financed and -regulated service or through NGOs and/or private service mechanisms. These systems allocate between 6 percent and 14 percent of gross national product (GNP) to health services, with some governments funding over 80 percent of health expenditure; for example, Canada and the UK.

In communist states, the state organizes all aspects of health care with the philosophy that every citizen is entitled to equity in access to health services. The state health system manages research, staff training, and service delivery, even if operational aspects are decentralized to local health authorities. This model applied primarily to the Soviet model of health services. These systems, except for Cuba, placed financing of health low on the national priority, with funding less than 4 percent of GNP. In the shift to market economies in the 1990s, some former Socialist countries, such as Russia, are struggling with poor health status and a difficult shift from a strongly centralized health system to a decentralized system with diffusion of powers and responsibilities. Promotion of market concepts in former Soviet countries has reduced access to care and created a serious dilemma for their governments.

Former colonial countries, independent since the 1950s and 1960s, largely carried on the governmental health structures established in the colonial times. Most developing countries have given health a relatively low place in budgetary allotment, with expenditures under 3 percent of GNP. Since the 1980s, there has been a trend in developing countries towards decentralization of health services and greater roles for NGOs, and the development of health insurance. Some countries, influenced by medical concepts of their former colonial master countries, fostered the development of specialty medicine in the major centers with little emphasis on the rural majority population. Soviet influence in many ex-colonial countries promoted state-operated systems. The WHO promoted primary care, but the allocations favored city-based specialty care. Israel, as an ex-colony, adapted British ideas of public health together with Central European sick funds and maternal and child health as major streams of development until the mid-1990s.

A growing new conservatism in the 1980s and 1990s in the industrialized countries is a restatement of old values in which market economics and individualistic social values are placed above concepts of the "common good" of liberalism and socialism in its various forms. In the more extreme forms of this concept, the individual is responsible for his or her own health, including payment, and has a choice of health care providers that will respond with high-quality personalized care.

Market forces, meaning competition in financing and provision of health services with rationing of services, based on fees or private insurance and willingness and ability to pay, have become part of the ideology of the new conservatism. It is assumed that the patient (i.e., the consumer) will select the best service for his or her need, while the provider best able to meet consumer expectations will thrive. In its purest form, the state has no role in providing or financing of health services except those directly related to community protection and promotion of a healthful environment without interfering with individual choices. The state ensures that there are sufficient health care providers and allows market forces to determine the prices and distribution of services with minimal regulation. The USA retains this orientation in a highly modified form, with 86 percent of the population covered by some form of private or public insurance systems (see Chapters 10 and 13).

Modified market forces in health care are part of health reforms in many countries as they seek not only to ensure quality health care for all but also to constrain costs. A free market in health care is costly and ultimately inefficient because it encourages inflation of provider incomes or budgets and increasing utilization of highly technical services. Further, even in the most free market societies, the economy of health care is highly influenced by many factors outside the control of the consumer and provider. The total national health expenditure in the USA rose rapidly until reaching over 17.7 percent of gross domestic product (GDP) in 2011, the highest of any country, despite serious deficiencies for those without any or with very inadequate health insurance (in total more than 30 percent of the population). This figure compares to some 11.2 percent of GDP in Canada, which has universal health insurance under public administration. Following the 1994 defeat of President Clinton's national health program, the conservative Congress and the business community took steps to expand managed care in order to control costs, resulting in a revolution in health care in the USA (see Chapters 11 and 13). In the 2011–2019 decade health expenditure in the USA is expected to rise to 19.6 percent of GDP, partly owing to increased population coverage with implementation of the PPACA (Obamacare).

Reforms are being implemented in many "socialized" health systems. These may be through incentives to promote achievement of performance indicators, such as full immunization coverage. Others are using control of supply, such as hospital beds or licensed physicians, as methods of reducing overutilization of services that generates increasing costs. Market mechanisms in health are aimed not only at the individual but also at the provider. Incentive payment systems must work to protect the patient's legitimate needs, and conversely incentives that might reduce quality of care should be avoided. Fee-for-service promotes high rates of services such as surgery. Increasing private practice and user fees can adversely affect middle- and low-income groups, as well as employers, by raising the costs of health insurance. Managed care systems, with restraints on fee-for-service medical practice, have emerged as a positive response to the market approach. Incentive systems in payments for services may be altered by government or insurance agencies in order to promote rational use of services, such as reduction of hospital stays. The free market approach is affecting planning of health insurance systems in previously highly centralized health systems in developing countries as well as the redevelopment of health systems in former Soviet countries.

Despite political differences, reform of health systems has become a common factor in virtually all health systems since the 1990s, as each government searches for cost-effectiveness, quality of care, and universality of coverage. The new paradigm of health care reform sees the convergence of different systems to common principles. National responsibility for health goals and health promotion leads to national financing of health care with regional and managed care systems. Most developed countries have long since adopted national health insurance or service systems. Some governments may, as in the USA, insure only the highest risk groups such as the elderly and the poor, leaving the working and middle classes to seek private insurers. The nature and direction of health care reform affecting coverage of the population are of central importance in the New Public Health because of its effects on allocation of resources and on the health of the population.

The effects of the economic crisis in the USA are being felt worldwide. While the downturn has largely occurred in wealthier nations, the poor in low-income countries will be among those affected. Past economic downturns have been followed by substantial drops in foreign aid to developing countries.

SOCIAL, BEHAVIORAL, AND POLITICAL SCIENCES

As public health gained from sanitary and other control measures for infectious diseases, along with mother and child care, nutrition, and environmental and occupational health, it also gained strength and applicability from advances in the social and behavioral sciences. Social Darwinism, a political philosophy that assumed "survival of the fittest" and no intervention of the sate to alleviate this assumption, was popular in the early nineteenth century but became unacceptable in industrialized countries, which adopted social policies to alleviate the worst conditions of poverty, unemployment, poor education, and other societal ills.

The political approach to focusing on health and poverty is associated with Jeremy Bentham in Britain in the late eighteenth century, who promoted social and political reform and "the greatest good for the greatest number", or utilitarianism. Rudolf Virchow, an eminent pathologist and a leader in recognizing ill-health and poverty as cause and effect, called for political action to create better conditions for the poor and working-class population.

The struggle for a social contract was promoted by pioneer reformists such as Edwin Chadwick (*General Report on the Sanitary Condition of the Labouring Population of Great Britain*, 1842), who later became the first head of the Board of Health in Britain, and Lemuel Shattuck (*Report of a General Plan for the Promotion of Public and Personal Health*, 1850). Shattuck was the organizer and first president of the American Statistical Association.

The social sciences have become fundamental to public health, with a range of disciplines including vital statistics and demography (seventeenth century), economics and politics (nineteenth century), sociology (twentieth century), history, anthropology, and others, which provide collectively important elements of epidemiology of crucial significance for survey methods and qualitative research (see Chapter 3). These advances contributed greatly to the development of methods of studying diseases and risk factors in a population and are still highly relevant to addressing inequalities in health.

HEALTH AND DEVELOPMENT

Individuals in good health are better able to study and learn, and be more productive in their work. Improvements in the standard of living have long been known to contribute to improved public health; however, the converse has not always been recognized. Investment in health care was not considered a high priority in many countries where economic considerations directed investment to the "productive" sectors such as manufacturing and large-scale infrastructure projects, such as hydroelectric dams.

Whether health is a contributor to economic development or a drain on societies' resources has been a fundamental debate between socially and market-oriented advocates. Classic economic theory, both free enterprise and communist, has tended to regard health as a drain on economies, distracting investment needed for economic growth. As a result, in many countries health has been given low priority in budgetary allocation, even when the major source of financing is governmental. This belief among economists and banking institutions prevented loans for health development on the grounds that such funds should focus on creating jobs and better incomes, before investing in health infrastructure. Consequently, the development of health care has been hampered.

A socially oriented approach sees investment in health as necessary for the protection and development of "human capital", just as investment in education is needed for the long-term benefit of the economy of a country. In 1993, the World Bank's World Development Report: Investing in Health articulated a new approach to economics in which health, along with education and social development, is seen as an essential precondition for and contributor to economic development. While many in the health field have long recognized the importance of health for social and economic improvement, its adoption by leading international development banking may mark a turning point for investment in developing nations, so that health may be a contender for increased development loans.

The concept of an essential package of services discussed in that report establishes priorities in low- and middle-income countries for efficient use of resources based on the burden of disease and cost-effectiveness analysis of services. It includes both preventive and curative services targeted to specific health problems. It also recommends support for comprehensive primary care, such as for children, and infrastructure development including maternity and hospital care, medical and nursing outreach services, and community action to improve sanitation and safe water supplies.

Reorientation of government spending on health is increasingly being adopted, as in the UK, to improve equity in access for the poor and other neglected sectors or regions of society with added funding for relatively deprived areas to improve primary care services. Differential capitation funding as a form of affirmative action to provide for high-needs populations is a useful concept in public health terms to address the inequities still prevalent in many countries.

HEALTH SYSTEMS: THE CASE FOR REFORM

As medical care has gradually become more involved in prevention, and as it has moved into the era of managed care, the gap between public health and clinical medicine has narrowed. As noted above, many countries are engaged in reforms in their health care systems. The motivation is largely derived from the need for cost containment, but also to extend health care coverage to underserved parts of the population. Countries without universal health care still have serious inequities in distribution of or access to services, and may seek reform to reduce those inequities, perhaps under political pressures to improve the provision of services. Incentives for reform are needed to address regional inequities, and preserving or developing universal access and quality of care, but also on inequities in health between the rich and the poor countries and within even the wealthy countries.

In some settings, a health system may fail to keep pace with developments in prevention and in clinical medicine. Some countries have overdeveloped medical and hospital care, neglecting important initiatives to reduce the risk of disease. The process of reform requires setting standards to measure health status and the balance of services to optimize health. A health service can set a target of immunizing 95 percent of infants with a national immunization schedule, but requires a system to monitor performance and incentives for changes.

A health system may also have failed to adapt to changing needs of the population through lack, or misuse, of health information and monitoring systems. As a result, the system may err seriously in its allocation of resources, with excessive emphasis on hospital care and insufficient attention to primary and preventive care. All health services should have mechanisms for correctly gathering and analyzing needed data for monitoring the incidence of disease and other health indicators, such as hospital utilization, ambulatory care, and preventive care patterns. For example, the UK's NHS periodically undertakes a restructuring process of parts of the system to improve the efficiency of service. This involves organizational changes and decentralization with regional allocation of resources (see Chapter 13).

Health systems are under pressures of changing demographic and epidemiological patterns as well as public expectations, rising costs of new technology, financing, and organizational change. New problems must be continually addressed with selection of priority issues and the most effective methods chosen. Reforms may create unanticipated problems, such as professional or public dissatisfaction, which must be evaluated, monitored, and addressed as part of the evolution of public health.

ADVOCACY AND CONSUMERISM

Literacy, freedom of the press, and increasing public concern for social and health issues have contributed to the development of public health. The British medical community lobbied for restrictions on the sale of gin in the 1780s in order to reduce the damage that it caused to the working class. In the late eighteenth and the nineteenth centuries, reforms in society and sanitation were largely the result of strongly organized advocacy groups influencing public opinion through the press. Such pressure stimulated governments to act in regulating the working conditions of mines and factories. Abolition of the slave trade and its suppression by the British navy in the early nineteenth century resulted from successful advocacy groups and their effects on public opinion through the press. Vaccination against smallpox was promoted by privately organized citizen groups, until later taken up by local and national government authorities.

Advocacy consists of activities of individuals or groups publicly pleading for, supporting, espousing, or recommending a cause or course of action. The advocacy role of reform movements in the nineteenth century was the basis of the development of modern organized public health. Campaigns ranged from the reform of mental hospitals, nutrition for sailors to prevent scurvy and beriberi, and labor laws to improve working conditions for women and children in particular, to the promotion of universal education and improved living conditions for the working population. Reforms on these and other issues resulted from the stirring of the public consciousness by advocacy groups and the public media, all of which generated political decisions in parliaments (Box 2.12). Such reforms were in large part motivated by fear of revolution throughout Europe in the mid-nineteenth century and the early part of the twentieth century.

Trade unions, and before them medieval guilds, fought to improve hours, safety, and conditions of work, as well as social and health benefits for their members. In the USA, collective bargaining through trade unions achieved wage increases and widespread coverage of the working population under voluntary health insurance. Unions and some industries pioneered prepaid group practice, the predecessor of health maintenance organizations and managed care or the more recent ACOs (see Chapters 10 and 13).

Through raising public consciousness on many issues, advocacy groups pressure governments to enact legislation to restrict smoking in public places, prohibit tobacco advertising, and mandate the use of bicycle helmets. Advocacy groups play an important role in advancing health based on disease groups, such as cancer, multiple sclerosis, and thalassemia, or advancing health issues, such as the organizations promoting breastfeeding, environmental improvement, or smoking

BOX 2.12 The Plimsoll Line

Political activism for reform in nineteenth-century Britain led to banning and suppressing the slave trade, improvements in working conditions for miners and factory workers, and other major political reforms.

In keeping with this tradition, Samuel Plimsoll (1824–1898), British Member of Parliament elected for Derby in 1868, conducted a solo campaign for the safety of seamen. His book, *Our Seamen*, described ships sent to sea so heavily laden with coal and iron that their decks were awash. Seriously overloaded ships, deliberately sent to sea by unscrupulous owners, frequently capsized, drowning many crew members, with the owners collecting inflated insurance fees.

Overloading was the major cause of wrecks and thousands of deaths in the British shipping industry. Plimsoll pleaded for mandatory Load-Line Certificate markers to be issued to each ship to prevent any ships putting to sea when the marker was not clearly visible. Powerful shipping interests fought him every inch of the way, but he succeeded in having a Royal Commission established, leading to an Act of Parliament mandating the "Plimsoll Line", the safe carrying capacity of cargo ships. This regulation was adopted by the US Bureau of Shipping as the Load Line Act in 1929 and is now standard practice worldwide.

reduction. Some organizations finance services or facilities not usually provided within insured health programs. Such organizations, which can number in the hundreds in a country, advocate the importance of their special concern and play an important role in innovation and meeting community health needs. Advocacy groups, including trade unions, professional groups, women's groups, self-help groups, and many others, focus on specific issues and have made major contributions to advancing the New Public Health.

Professional Advocacy and Resistance

The history of public health is replete with pioneers whose discoveries led to strong opposition and sometimes violent rejection by conservative elements and vested interests in medical, public, or political circles. Opposition to Jennerian vaccination, the rejection of Semmelweiss by colleagues in Vienna, and the contemporary opposition to the work of great pioneers in public health such as Pasteur, Florence Nightingale, and many others may deter or delay implementation of other innovators and new breakthroughs in preventing disease. Although opposition to Jenner's vaccination lasted well into the late nineteenth century in some areas, its supporters gradually gained ascendancy, ultimately leading to the global eradication of smallpox. These and other pioneers led the way to improved health, often after bitter controversy on topics later accepted and which, in retrospect, seem to be obvious.

Advocacy has sometimes had the support of the medical profession but elicited a slow response from public authorities. David Marine of the Cleveland Clinic and David Cowie, professor of pediatrics at the University of Michigan, proposed the prevention of goiter by iodization of salt. Marine carried out a series of studies in fish, and then in a controlled clinical trial among schoolgirls in 1917–1919, with startlingly positive results in reducing the prevalence of goiter. Cowie campaigned for the iodization of salt, with support from the medical profession. In 1924, he convinced a private manufacturer to produce Morton's iodized salt, which rapidly became popular throughout North America. Similarly, iodized salt came to be used in many parts of Europe, mostly without governmental support or legislation. Iodine-deficiency disorders (IDDs) remain a widespread condition, estimated to have affected 2 billion people worldwide in 2013. The target of international eradication of IDDs by 2000 was set at the World Summit for Children in 1990, and the WHO called for universal iodization of salt in 1994. By 2008, nearly 70 percent of households in developing countries consumed adequately iodized salt. China and Nigeria, have had great success in recent years with mandatory salt fortification in increasing iodization rates, in China from 39 percent to 95 percent in 10 years. But the problem is not yet gone and even in Europe there is inadequate standardization of iodine levels and population follow-up despite decades of work on the problem.

Professional organizations have contributed to promoting causes such as children's and women's health, and environmental and occupational health. The American Academy of Pediatrics has contributed to establishing and promoting high standards of care for infants and children in the USA, and to child health internationally. Hospital accreditation has been used for decades in the USA, Canada, and more recently in Australia and the UK. It has helped to raise standards of health facilities and care by carrying out systematic peer review of hospitals, nursing homes, primary care facilities, and mental hospitals, as well as ambulatory care centers and public health agencies (see Chapter 15).

Public health needs to be aware of negative advocacy, sometimes based on professional conservatism or economic self-interest. Professional organizations can also serve as advocates of the status quo in the face of change. Opposition by the American Medical Association (AMA) and the health insurance industry to national health insurance in the USA has been strong and successful for many decades. The passage of the PPACA has been achieved despite widespread political and public opposition, yet was sustained in the US Supreme Court and is gaining widening popular support as the added value to millions of formerly uninsured Americans becomes clear. In some cases, the vested interest of one profession may block the legitimate development of others, such as when ophthalmologists lobbied successfully against the development of optometry, now widely accepted as a legitimate profession.

Jenner's discovery of vaccination with cowpox to prevent smallpox was adopted rapidly and widely. However, intense opposition by organized groups of antivaccinationists, often led by those opposed to government intervention in health issues and supported by doctors with lucrative variolation practices, delayed the implementation of smallpox vaccination for many decades. Ultimately, smallpox was eradicated in 1972, owing to a global campaign initiated by the WHO. Opposition to legislated restrictions on private ownership of assault weapons and handguns is intense in the USA, led by well-organized, well-funded, and politically powerful lobby groups, despite the amount of morbidity and mortality due to gun-associated violent acts (see Chapter 5).

Fluoridation of drinking water is the most effective public health measure for preventing dental caries, but it is still widely opposed, and in some places the legislation has been rescinded even after implementation, by well-organized antifluoridation campaigns. Opposition to fluoridation of community water supplies is widespread, and effective lobbying internationally has slowed but has not stopped progress (see Chapter 6). Despite the life-saving value of immunization, opposition still exists in 2013 and harms public health protection. Opposition has slowed progress in poliomyelitis eradication; for example, radical Islamists killed polio workers in northern Nigeria in 2012, one of the last three countries with endemic poliomyelitis. Resistance to immunization in the 1980s has resulted in the recurrence of pertussis and diphtheria and a very large epidemic of measles across Western Europe, including the UK, with further spread to the western hemisphere in 2010–2013 (see Chapter 4).

Progress may be blocked where all decisions are made in closed discussions, not subject to open scrutiny and debate. Public health personnel working in the civil service of organized systems of government may not be at liberty to promote public health causes. However, professional organizations may then serve as forums for the essential professional and public debate needed for progress in the field. Professional organizations such as the APHA provide effective lobbying for the interests of public health programs and can have an important impact on public policy. In mid-1996, efforts by the Secretary of Health and Human Services in the USA brought together leaders of public health with representatives of the AMA and academic medical centers to try to find areas of common interest and willingness to promote the health of the population. In Europe too, increasing cooperation between public health organizations is stimulating debate on issues of transnational importance across the region, which, for example, has a wide diversity of standards on immunization practices and food policies.

Public advocacy has played an especially important role in focusing attention on ecological issues (Box 2.13). In 1995, Greenpeace, an international environmental activist group, fought to prevent the dumping of an oil rig in the

---

BOX 2.13 An "Enemy of the People"

Advocacy is a function in public health that has been important in promoting advances in the field, and one that sometimes places the advocate in conflict with established patterns and organizations. One of the classic descriptions of this function is in Henrik Ibsen's play *An Enemy of the People*, in which the hero, a young doctor, Thomas Stockmann, discovers that the water in his community is contaminated. This knowledge is suppressed by the town's leadership, led by his brother the mayor, because it would adversely affect plans to develop a tourist industry of baths in their small Norwegian town in the late nineteenth century.

The young doctor is taunted and abused by the townspeople and driven from the town, having been declared an "enemy of the people" and a potential risk. The allegory is a tribute to the man of principle who stands against the hysteria of the crowd. The term also took on a far more sinister and dangerous meaning in George Orwell's novel *1984* and in totalitarian regimes of the 1930s to the present time.

North Sea and forced a major oil company to find another solution that would be less damaging to the environment. An explosion on an oil rig in the Gulf of Mexico in 2010 led to enormous ecological and economic damage as well as loss of life. Damages levied on the responsible company (British Petroleum) amount to some $4.5 billion dollars and several criminal negligence charges are pending. Greenpeace also continued its efforts to stop the renewal of testing of atomic bombs by France in the South Pacific.

International protests led to the cessation of almost all testing of nuclear weapons. International concern over global warming has led to growing efforts to stem the tide of air pollution from fossil fuels, coal-burning electrical production, and other manifestations of carbon dioxide and toxic contamination of the environment. Progress is far from certain as newly enriched countries such as China and India follow the rising consumption patterns of western countries. Public advocacy and rejection of wanton destruction of the global ecology may be the only way to prod consumers, governments, and corporate entities such as the energy and transportation industries to change direction. The pace of change from fossil fuels is slow but has captured public attention, and private companies are seeking more fuel efficiency in vehicles and electrical power production, mainly though the use of natural gas instead of fuel oil and coal for electricity production or better still by wind and solar energy. The search for "green solutions" to the global warming crisis has become increasingly dynamic, with governments, the private sector, and the general public keenly aware of the importance of the effort and the dangers of failure.

In the latter part of the twentieth century and the early twenty-first century, prominent international personalities and entertainers have taken up causes such as the removal of land mines in war-torn countries, illiteracy in disadvantaged populations, and funding for antiretroviral drugs for African countries to reduce maternal–fetal transmission of HIV and to provide care for the large numbers of cases of AIDS devastating many countries of sub-Saharan Africa. Rotary International has played a key role in polio eradication efforts globally. The public–private consortium Global Alliance for Vaccines and Immunization (GAVI) has been instrumental in promoting immunization in recent years, with participation by the WHO, UNICEF, the World Bank, the Gates Foundation, vaccine manufacturers, and others. This has had an important impact on extending immunization to protect and save the lives of millions of children in deprived countries not yet able to provide fundamental prevention programs such as immunization at adequate levels. GAVI has brought vaccines to low-income countries around the world, such as rotavirus vaccine, pentavalent vaccine in Myanmar, and pneumococcal vaccine for children in 15 countries in sub-Saharan Africa, including DR Congo. The Bill & Melinda Gates Foundation pledged US $750 million in 1999 to establish GAVI, with US $75 million per year and US $1 billion in 2010 to promote the Decade of Vaccines.

International conferences help to create a worldwide climate of advocacy for health issues. International sanitary conferences in the nineteenth century were convened in response to the cholera epidemics. International conferences continue in the twenty-first century to serve as venues for advocacy on a global scale, bringing forward issues in public health that are beyond the scope of individual nations. The WHO, UNICEF, and other international organizations perform this role on a continuing basis (see Chapter 16). Criticisms of this approach have focused on the lack of similar effort or donors to address NCDs, weak public health infrastructure, and that this frees national governments from responsibility to care for their own children. No one can question, however, that this kind of endeavor has saved countless lives and needs the backing of other aid donors and national government participation.

Consumerism

Consumerism is a movement that promotes the interests of the purchaser of goods or services. In the 1960s, a new form of consumer advocacy emerged from the civil rights and antiwar movement in the USA. Concern was focused on the environment, occupational health, and the rights of the consumer. Rachel Carson stimulated concern by dramatizing the effects of DDT on wildlife and the environment but inadvertently jeopardized anti-malarial efforts in many countries. This period gave rise to environmental advocacy efforts worldwide, and a political movement, the Greens, in Western Europe.

Ralph Nader showed the power of the advocate or "whistle-blower" who publicizes health hazards to stimulate active public debate on a host of issues related to the public well-being. Nader, a consumer advocate lawyer, developed a strategy for fighting against business and government activities and products which endangered public health and safety. His 1965 book *Unsafe at Any Speed* took issue with the US automobile industry for emphasizing profit and style over safety, and led to the enactment of the National Traffic and Motor Safety Act of 1966, establishing safety standards for new cars. This was followed by a series of enactments including design and emission standards and seat-belt regulations. Nader's work continues to promote consumer interests in a wide variety of fields, including the meat and poultry industries, and coal mining, and promotes greater government regulatory powers regarding pesticide usage, food additives, consumer protection laws, rights to knowledge of contents, and safety standards.

Consumerism has become an integral part of free market economies, and the educated consumer does influence the quality, content, and price of products. Greater awareness of nutrition in health has influenced food manufacturers to improve packaging, content labeling, enrichment with vitamins and minerals, and advertisement to promote those values. Low-fat dietary products are available because of an increasingly sophisticated public concerned over dietary factors in cardiovascular diseases. The same process occurred in safe toys and clothing for children, automobile safety features such as mandatory use of car seats for infants, and other innovations that quickly became industry standards in the industrialized world. Dangerous practices such as the use of lead paint in toys and melamine contamination of milk products from China capture the public attention quickly and remind public health authorities of the importance of continuous alertness to potential hazards.

Consumerism can also be exploited by pharmaceutical companies with negative impacts on the health system, especially in the advertising of health products which leads to unnecessary visits to health providers and pressure for approval to obtain the product. The Internet has provided people with access to a vast array of information and opinion, and to current literature otherwise unavailable because of the often inadequate library resources of medical and other health professionals. The very freedom of information the Internet allows, however, also provides a vehicle for extremist and fringe groups to promote disinformation such as "vaccination causes autism, fluoridation causes cancer", which can cause considerable difficulties for basic public health programs or lead to self-diagnosis of conditions, with often disastrous consequences.

Advocacy and voluntarism go hand in hand. Voluntarism takes many forms, including raising funds for the development of services or operating services needed in the community. It may take the form of fund-raising to build clinics or hospitals in the community, or to provide medical equipment for elderly or handicapped people; or retirees and teenagers working as hospital volunteers to provide services that are not available through paid staff, and to provide a sense of community caring for the sick in the best traditions of religious or municipal concerns. This can also be extended to prevention, as in support for immunization programs, assistance for the handicapped and elderly in transportation, Meals-on-Wheels, and many other services that may not be included in the "basket of services" provided by the state, health insurance, or public health services.

Community involvement can take many forms, and so can voluntarism. The pioneering role of women's organizations in promoting literacy, health services, and nutrition in North America during the latter part of the nineteenth and the early twentieth centuries profoundly affected the health of the population. The advocacy function is enhanced when an organization mobilizes voluntary activity and funds to promote changes or needed services, sometimes forcing official health agencies or insurance systems to revise their attitudes and programs to meet these needs.

THE HEALTH FIELD CONCEPT

By the early 1970s, Canada's system of federally supported provincial health insurance plans covered all of the country. The federal Minister of Health, Marc Lalonde, initiated a review of the national health situation, in view of concern over the rapidly increasing costs of health care. This led to articulation of the "Health Field Concept" in 1974, which defined health as a result of four major factors: human biology, environment, behavior, and health care organization (Box 2.14). Lifestyle and environmental factors were seen as important contributors to the morbidity and mortality in modern societies. This concept gained wide acceptance, promoting new initiatives that emphasized health promotion in response to environmental and lifestyle factors. Conversely, reliance primarily on medical care to solve all health problems could be

---

BOX 2.14 The Health Field Concept – Marc Lalonde

Definition
Health is a result of factors associated with genetic inheritance, the environment, and personal lifestyle and of medical care. Promotion of healthy lifestyles can improve health and reduce the need for medical care.

Elements
- Genetic and biological factors
- Behavioral and attitudinal factors (lifestyle)
- Environment, including economic, social, cultural, and physical factors
- The organization of health care systems.

Source: Lalonde M. A new perspective on the health of Canadians: a working document. Ottawa: Information Canada; 1974. Available at: http://www.hc-sc.gc.ca/hcs-sss/com/fed/lalonde-eng.php [Accessed 25 May 2012].

---

counterproductive. This concept was a fundamental contributor to the idea of health promotion later articulated in the Ottawa Declaration, discussed below.

The health field concept came at a time when many epidemiological studies were identifying risk factors for cardiovascular diseases and cancers that related to personal habits, such as diet, exercise, and smoking. The concept advocated that public policy needed to address individual lifestyle as part of the overall effort to improve health status. As a result, the Canadian federal government established health promotion as a new activity. This quickly spread to many other jurisdictions and gained wide acceptance in many industrialized countries.

Concern was expressed that this concept could become a justification for a "blame the victim" approach, in which those ill with a disease related to personal lifestyles, such as smokers or AIDS patients, are seen as having chosen to contract the disease. Such a patient might then be considered not to be entitled to all benefits of insurance or care that others may receive. The result may be a restrictive approach to care and treatment that would be unethical in the public health tradition and probably illegal in western jurisprudence. This concept was also used to justify withdrawal from federal commitments in cost sharing and escape from facing controversial health reform in the national health insurance program.

THE VALUE OF MEDICAL CARE IN PUBLIC HEALTH

During the 1960s and 1970s, outspoken critics of health care systems, such as Ivan Illytch, questioned the value of medical care for the health of the public. This became a widely discussed, somewhat nihilistic, view towards medical care, and was influential in promoting skepticism regarding the value of the biomedical mode of health care, and antagonism towards the medical profession.

In 1976, Thomas McKeown presented a historical–epidemiological analysis showing that up to the 1950s, medical care had only a limited impact on mortality rates, although improvements in surgery and obstetrics were notable. He showed that crude death rates in England averaged about 30 per 1000 population from 1541 to 1750, declining steeply to 22 per 1000 in 1851, 15 per 1000 in 1901, and 12 per 1000 in 1951, when medical care became truly effective. McKeown concluded that much of the improvement in health status over the past several centuries was due to reduced mortality from infectious diseases. This he related to limitation of family size, increased food supplies, improved nutrition and sanitation, specific preventive and therapeutic measures, and overall gains in quality of life for growing elements of the population. He cautioned against placing excessive reliance for health on medical care, much of which was of unproved effectiveness.

This skepticism of the biomedical model of health care was part of wider antiestablishment feelings of the 1960s and 1970s in North America. In 1984, Milton Roemer pointed out that the advent of vaccines, antibiotics, antihypertensives, and other medications contributed to great improvements in infant and child care, and in the management of infectious diseases, hypertension, diabetes, and other conditions. Therapeutic gains continue to arrive from teaching centers around the world. Vaccine, pharmaceutical, and diagnostic equipment manufacturers continue to provide important innovations that have major benefits, but also raise the cost of health care. The latter issue is one which has stimulated the search for reforms, and search for lower cost technologies such as in treatment of hepatitis C patients, a huge international public health issue.

The value of medical care to public health and vice versa has not always been clear, either to public health personnel or to clinicians. The achievements of modern public health in controlling infectious diseases, and even more so in reducing the mortality and morbidity associated with chronic diseases such as stroke and CHD, were in reality a shared achievement between clinical medicine and public health (see Chapter 5).

Preventive medicine has become part of all medical practice, with disease prevention through early diagnosis and health promotion through individual and community-focused activities. Risk factor evaluation determines appropriate screening and individual and community-based interventions. Medical care is crucial in controlling hypertension and in reducing the complications and mortality from CHD. New modalities of treatment are reducing death rates from first time acute myocardial infarctions. Better management of diabetes prevents the early onset of complications. At the same time, the contribution of public health to improving outcomes of medical care is equally important. Control of the vaccine-preventable diseases, improved nutrition, and preparation for motherhood contribute to improved maternal and infant outcomes. Promotions of reduced exposure to risk factors for chronic disease are a task shared by public health and clinical medical services. Both clinical medicine and public health contribute to improved health status. They are interdependent and rely on funding systems for recognition as part of the New Public Health.

HEALTH TARGETS

During the 1950s, many new management concepts emerged in the business community, such as "management by objective", a concept developed by Peter Drucker at General Motors, with variants such as "zero-based budgeting" developed in the US Department of Defense (see Chapter 12). They focused the activities of an organization and its budget on targets, rather than on previous allocation of resources. These concepts were applied in other spheres, but they influenced thinking in health, whose professionals were seeking new ways to approach health planning. The logical application was to define health targets and to promote the efficient use of resources to achieve those targets. This occurred in the USA and soon afterwards in the WHO European Region. In both cases, a wide-scale process of discussion and consensus building was used before reaching definitive targets. This process contributed to the adoption of the targets by many countries in Europe as well as by states and many professional and consumer organizations. The USA developed national health objectives in 1979 for the year 1990 and subsequently for the year 2000, with monitoring of progress in their achievement and development of further targets for 2010 and now for 2020. Beginning in 1987, state health profiles are prepared by the Epidemiology Program Office of the Centers for Disease Control and Prevention based on 18 health indicators recommended by a consensus panel representing public health associations and organizations.

The eight MDGs adopted by the UN in 2000 include halving extreme poverty, reducing child mortality by two-thirds, improving maternal health, halting the spread of HIV/AIDS, malaria, and other diseases, and providing universal primary education, all by the target date of 2015. The MDGs form a common blueprint agreed to by all countries and the world's leading development institutions. The process has galvanized unprecedented efforts to meet the needs of the world's poorest, yet 2008 reviews of progress indicate that most developing nations will not meet the targets at current rates of progress.

The United Nations Development Programme (UNDP) Global Partnership for Development 2012 report on the MDGs states that if the national development strategies and initiatives are supported by international development partners, the goals can be achieved by 2015. The MDGs were adopted by over 120 nations and provided guidance for national policies and for international aid agencies. The focus was on middle- and low-income countries and their achievements have been considerable but variable (see Box 2.15 and Chapter 16). As of July 2012, extreme poverty was falling in every region, the poverty reduction target had been met, the world had met the target of halving the proportion of people without access to improved sources of water, and the world had achieved parity in primary education between girls and boys.

Further progress will require sustained political commitment to develop the primary care infrastructure: improved reporting and epidemiological monitoring, consultative mechanisms, and consensus by international agencies, national governments, and non-governmental agencies. The achievement of the targets will also require sustained international support and national commitment with all the difficulties of a time of economic recession. Nevertheless, defining a target is crucial to the process.

BOX 2.15 Achievements of Millennium Development Goals 2000–2012

- MDG1. *Eradicate extreme hunger and poverty* – The number of people living in extreme poverty and the poverty rates fell in every developing region, including in sub-Saharan Africa, where rates are highest. In the developing regions, the proportion of people living on less than $1.25 a day fell from 47 percent in 1990 to 24 percent in 2008. In 2008, about 110 million fewer people than in 2005 lived in conditions of extreme poverty.
- MDG2. *Achieve universal primary education* – In the developing regions, the net enrolment rate for children of primary school age rose from 82 percent to 90 percent between 1999 and 2010, mainly between 1999 and 2004, but has leveled off since.
- MDG3. *Promote gender equality and empower women* – By the end of January 2012, women accounted for 19.7 percent of parliamentarians worldwide. This amounts to nearly a 75 percent increase since 1995, when women held 11.3 percent of seats worldwide, and a 44 percent increase over the 2000 level.
- MDG4. *Reduce child mortality* – Progress on child mortality is gaining momentum. The target is to reduce by two-thirds, between 1990 and 2015, the under-5-year-old mortality rate, from 93 children of every 1000 dying to 31 of every 1000. Child deaths are falling, but much more needs to be done in order to reach the development goal. Revitalizing efforts against pneumonia and diarrhea, while bolstering nutrition, could save millions of children.
- MDG5. *Improve maternal health* – Maternal mortality has nearly halved since 1990, but levels are far removed from the 2015 target. The targets for improving maternal health include reducing by three-quarters the maternal mortality ratio and achieve universal access to reproductive health. Poverty and lack of education perpetuate high adolescent birth rates. Inadequate funding for family planning is a major failure in fulfilling commitments to improving women's reproductive health.
- MDG6. *Combat HIV/AIDS, malaria, tuberculosis, and other diseases* – More people than ever are living with HIV owing to fewer AIDS-related deaths and the continued large number of new infections. In 2011, an estimated 34.2 million were living with HIV, up 17 percent from 2001. This persistent increase reflects the continued large number of new infections along with a significant expansion of access to life-saving antiretroviral therapy, especially in more recent years.
- MDG7. *Ensure environmental sustainability* – The unparalleled success of the Montreal Protocol shows that action on climate change is within grasp. The 25th anniversary of the Montreal Protocol on Substances that Deplete the Ozone Layer, in 2012, had many achievements to celebrate. Most notably, there has been a reduction of over 98 percent in the consumption of ozone-depleting substances. Further, because most of these substances are also potent greenhouse gases, the Montreal Protocol has contributed significantly to the protection of the global climate system. The reductions achieved to date leave hydrochlorofluorocarbons (HCFCs) as the largest group of substances remaining to be phased out.
- MDG8. *A global partnership for development* – Core development aid fell in real terms for the first time in more than a decade, as donor countries faced fiscal constraints. In 2011, net aid disbursements amounted to $133.5 billion, representing 0.31 percent of developed countries' combined national income. While constituting an increase in absolute dollars, this was a 2.7 percent drop in real terms over 2010. If debt relief and humanitarian aid are excluded, bilateral aid for development programmes and projects fell by 4.5 percent in real terms.

Source: *United Nations Development Programme. The Millennium Development Goals Report 2012. Summary 2 July 2012. Available at: http:// www.undp.org/content/undp/en/home/librarypage/mdg/the-millennium-development-goals-report-2012/ [Accessed 21 November 2012]. The full report is available at: http://www.undp.org/content/dam/undp/library/ MDG/english/The_MDG_Report_2012.pdf [Accessed 21 November 2012].*

There are encouraging signs that national governments are influenced by the general movement to place greater emphasis on resource allocation and planning on primary care to achieve internationally recognized goals and targets. The successful elimination of smallpox, rising immunization coverage in the developing countries, and increasing implementation of salt iodization have shown that such goals are achievable.

US Health Targets

While the USA has not succeeded in developing universal health care access, it has a strong tradition of public health and health advocacy. Federal, state, and local health authorities have worked out cooperative arrangements for financing and supervising public health and other services. With growing recognition in the 1970s that medical services alone

would not achieve better health results, health policy leadership in the federal government formulated a new approach, in the form of developing specific health targets for the nation.

In 1979, the Surgeon General of the USA published the Report on Health Promotion and Disease Prevention (Healthy People). This document set five overall health goals for each of the major age groups for the year 1990, accompanied by 226 specific health objectives. New targets for the year 2000 were developed in three broad areas: to increase healthy lifespans, to reduce health disparities, and to achieve access to preventive health care for all Americans. These broad goals are supported by 297 specific targets in 22 health priority areas, each one divided into four major categories: health promotion, health protection, preventive services, and surveillance systems. This set the public health agenda on the basis of measurable indicators that can be assessed year by year.

Leading health indicators selected for 2010 incorporate the original 467 objectives in Healthy People 2010, which served as a basis for planning public health activities for many state and community health initiatives. For each of the leading health indicators, specific objectives and subobjectives derived from Healthy People 2010 are used to monitor progress. The specific objectives set for Healthy People 2020 are listed in Box 2.16. Thirteen new topic areas are listed for 2020, such as older adults, genomics, dementias, and social determinants of health. These provide guidelines for national, state, and local public health agencies as well as insurance providers, primary care services, and health promotion advocates. A key issue will be in reducing regional, ethnic, and socioeconomic health disparities.

The process of working towards health targets in the USA has moved down from the federal level of government to the state and local levels. Professional organizations, NGOs, as well as community and fraternal organizations are also involved. The states are encouraged to prepare their own targets and implementation plans as a condition for federal grants, and many states require county health departments to prepare local profiles and targets.

Diffusion of this approach encourages state and local initiatives to meet measurable program targets. It also sets a different agenda for local prestige in competitive terms, with less emphasis on the size of the local hospital or other agencies than on having the lowest infant mortality or the least infectious disease among neighboring local authorities.

European Health Targets

The WHO European Region document "Health 21 – Health for All in the 21st Century" addresses health in the twenty-first century, with 21 principles and objectives for improving the health of Europeans, within and between countries of Europe. The Health 21 targets include:

1. Closing the health gap between countries.
2. Closing the health gap within countries.
3. A healthy start in life (supportive family policies).
4. Health of young people (policies to reduce child abuse, accidents, drug use, and unwanted pregnancies).
5. Healthy aging (policies to improve health, self-esteem, and independence before dependence emerges).
6. Improving mental health.
7. Reducing communicable diseases.
8. Reducing non-communicable diseases.
9. Reducing injury from violence and accidents.
10. A healthy and safe physical environment.
11. Healthier living (fiscal, agricultural, and retail policies that increase the availability of and access to and consumption of vegetables and fruits).
12. Reducing harm from alcohol, drugs, and tobacco.
13. A settings approach to health action (homes should be designed and built in a manner conducive to sustainable health and the environment).

BOX 2.16 Healthy People 2020 Objective Topic Areas

- Access to Health Services
- Adolescent Health
- Arthritis, Osteoporosis, and Chronic Back Conditions
- Blood Disorders and Blood Safety
- Cancer
- Chronic Kidney Disease
- Dementias, including Alzheimer's Disease
- Diabetes
- Disability and Health
- Early and Middle Childhood
- Educational and Community-Based Programs
- Environmental Health
- Family Planning
- Food Safety
- Genomics
- Global Health
- Health Communication and Health Information Technology
- Healthcare-Associated Infections
- Health-Related Quality of Life and Well-Being
- Hearing and Other Sensory or Communication Disorders
- Heart Disease and Stroke
- HIV
- Immunization and Infectious Diseases
- Injury and Violence Prevention
- Lesbian, Gay, Bisexual, and Transgender Health
- Maternal, Infant, and Child Health
- Medical Product Safety
- Mental Health and Mental Disorders
- Nutrition and Weight Status
- Occupational Safety and Health
- Older Adults
- Oral Health
- Physical Activity
- Preparedness
- Public Health Infrastructure
- Respiratory Diseases
- Sexually Transmitted Diseases
- Sleep Health
- Social Determinants of Health
- Substance Abuse
- Tobacco Use
- Vision

   Note: Under each objective topic areas, specific objectives are defined.

Source: *Healthy People 2020. Available at: http://www.healthy-people.gov/2020/topicsobjectives2020/overview.aspx?topicid=15 and http://www.healthypeople.gov/2020/topicsobjectives2020/pdfs/HP2020objectives.pdf [Accessed 21 November 2012].*

14. Multisectoral responsibility for health.
15. An integrated health sector and much stronger emphasis on primary care.
16. Managing for quality of care using the European health for all indicators to focus on outcomes and compare the effectiveness of different inputs.

17. Equitable and sustainable funding of health services.
18. Developing human resources (educational programs for providers and managers based on the principles of the Health for All policy).
19. Research and knowledge: health programs based on scientific evidence.
20. Mobilizing partners for health (engaging the media/ television/Internet).
21. Policies and strategies for Health for All – national, targeted policies based on Health for All.

A 2010–2012 review has been commissioned by the European Office of the WHO to assess inequalities in the social determinants of health. While health has improved there are still significant inequalities. Factors include variance in local, regional, national, and global economic forces. The European Union and the European Region of WHO are both working on health targets for the year 2020.

UK Health Targets

There are competing demands in society for expenditure by the government, and therefore making the best use of resources – money and people – is an important objective. The UK has devolved many of the responsibilities to the constituent countries (England, Wales, Scotland, and Northern Ireland) within an overall national framework (Box 2.17).

BOX 2.17 UK National Health Service Outcomes Framework 2013–2014 and Public Health Service Outcomes Framework 2013–2016

The UK National Health Service (NHS) has semi-autonomous units in England, Scotland, Wales, and Northern Ireland. They are funded from the central UK NHS but with autonomy within national guidelines. The NHS has defined national health outcomes for improvements grouped around five domains, each comprised of key indicators aimed at improving health with reducing inequalities.

NHS Domains
- Preventing people from dying prematurely from causes amenable to health care for all ages:
  - the target diseases include cardiovascular, respiratory, and liver diseases, and cancer (with focus on cancer of breast, lung, and colorectal cancer)
  - reducing premature death in people with serious mental illnesses
  - reducing infant mortality, neonatal mortality, still births, and deaths in young children
  - increasing 5-year survival for children with cancer.
- Enhancing quality of life for people with long-term conditions:
  - ensuring people feel supported to manage their condition
  - improving functional ability in people with long-term conditions
  - reducing time spent in hospital by people with long-term conditions
  - ensuring quality of life for caregivers
  - enhancing quality of life for people with mental illness.
- Helping people to recover from episodes of ill-health or injury:
  - improving outcomes from planned treatments
  - preventing lower respiratory tract infections in children from becoming serious
  - improving recoveries from injury and traumas
  - improving recovery from stroke
  - improving recovery from fragility fracture
  - helping older people to recover their independence after illness or injury.

- Ensuring that people have a positive experience of care:
  - improving people's experience of outpatient care
  - improved hospital responsiveness to personal needs
  - improving people's experience with accident and emergency services
  - improving women's and their families' experiences with maternity services
  - improving the experience of care for people at the end of their lives
  - improving experiences of health care for people with mental illness
  - improving children's and young people's experience of healthcare
  - improving people's experience of integrated care.
- Treating and caring for people in a safe environment and protecting them from avoidable harm:
  - reducing the incidence of avoidable harm
  - improving the safety of maternity services
  - delivering safe care to children in acute settings.

Public Health Domains
- Determinants of health; improve wider factors that affect health and well-being, and health inequalities.
- Health improvement; help people to live healthy lifestyles, healthy choices, reduce health inequalities, protection from major incidents and other threats, while reducing health inequalities.
- Health care, public health and preventing premature mortality; reduce the numbers of people living with preventable ill-health and people dying prematurely, while reducing the gap between communities.

Source: *UK Department of Health. Available at: https://www.gov.uk/government/organisations/department-of-health/about#our-priorities, https://www.gov.uk/government/uploads/system/uploads/attachment_data/file/193619/Improving-outcomes-and-supporting-transparency-part-1A.pdf. pdf, and https://www.gov.uk/government/uploads/system/uploads/attachment_data/file/127106/121109-NHS-Outcomes-Framework-2013-14.pdf. pdf [Accessed 24 June 2013].*

INDIVIDUAL AND COMMUNITY PARTICIPATION IN HEALTH

National policy in health ultimately relates to health of the individual. The various concepts outlined in the health field concept, community-oriented primary health care, health targets, and effective management of health systems, can only be effective if the individual and his or her community are knowledgeable participants in seeking solutions. Involving the individual in his or her own health status requires raising levels of awareness, knowledge, and action. The methods used to achieve these goals include health counseling, health education, and health promotion (Figure 2.6).

Health counseling has always been a part of health care between the doctor or nurse and the patient. It raises levels of awareness of health issues of the individual patient. Health education has long been part of public health, dealing with promoting consciousness of health issues in selected target population groups. Health promotion incorporates the work of health education but takes health issues to the policy level of government and involves all levels of government and NGOs in a more comprehensive approach to a healthier environment and personal lifestyles.

Health counseling, health education, and health promotion are among the most cost-effective interventions for improving the health of the public. While costs of health care are rising rapidly, demands to control cost increases should lead to greater emphasis on prevention, and adoption of health education and promotion as an integral part of modern life. This should be carried out in schools, the workplace, the community, commercial locations (e.g., shopping centers), and recreation centers, and in the political agenda.

Psychologist Abraham Maslow described a hierarchy of needs of human beings. Every human has basic requirements including physiological needs of safety, water, food, warmth, and shelter. Higher levels of needs include recognition, community, and self-fulfillment. These insights supported observations of efficiency studies such as those of Elton Mayo in the famous Hawthorne effect in the 1920s, showing that workers increased productivity when acknowledged by management in the objectives of the organization (see Chapter 12). In health terms, these translate into factors that motivate people to positive health activities when all barriers to health care are reduced.

Modern public health faces the problem of motivating people to change behavior; sometimes this requires legislation, enforcement, and penalties for failure to comply, such as in mandating car seat-belt use. In other circumstances it requires sustained performance by the individual, such as the use of condoms to reduce the risk of STI and/or HIV transmission. Over time, this has been developed into a concept known as knowledge, attitudes, beliefs, and practices (KABP), a measurable complex that cumulatively affects health behavior (see Chapter 3). There is often a divergence between knowledge and practice; for example, the knowledge of the importance of safe driving, yet not putting this into practice. This concept is sometimes referred to as the "KABP gap".

The health belief model has been a basis for health education programs, whereby a person's readiness to take action for health stems from a perceived threat of disease, a recognition of susceptibility to disease and its potential severity, and the value of health. Action by an individual may be triggered by concern and by knowledge. Barriers to appropriate action may be psychological, financial, or physical, including fear, time loss, and inconvenience. Spurring action to avoid risk to health is one of the fundamental goals in modern health care. The health belief model is important in defining any health intervention in that it addresses the emotional, intellectual, and other barriers to taking steps to prevent or treat disease.

Health awareness at the community and individual levels depends on basic education levels. Mothers in developing countries with primary or secondary school education are more successful in infant and child care than less educated women. Agricultural and health extension services reaching out to poor and uneducated farm families in North America in the 1920s were able to raise consciousness of safe self-health practices and good nutrition, and when this was supplemented by basic health education in schools, generational differences could be seen in levels of awareness of the importance of balanced nutrition. Secondary prevention with diabetics and patients with CHD hinges on education and awareness of nutritional and physical activity patterns needed to prevent or delay a subsequent myocardial infarction.

Founding of Health Promotion: The Ottawa Charter of 1986

The WHO sponsored the First International Conference on Health Promotion held in Ottawa, Canada, in 1986 (Figure 2.7). The resulting Ottawa Charter defined health promotion and set out five key areas of action: building healthy public policy, creating supportive environments, strengthening community action, developing personal skills, and reorienting health services. The Ottawa Charter called on all countries to put health on the agenda of policy makers in all sectors and at all levels, directing them to be aware

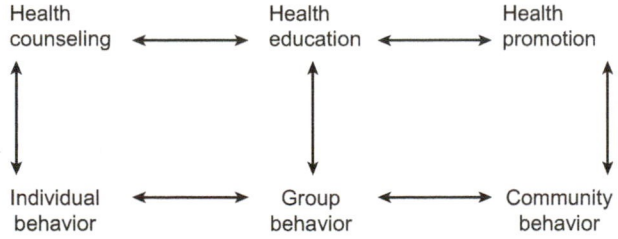

FIGURE 2.6   Health counseling–health education–health promotion.

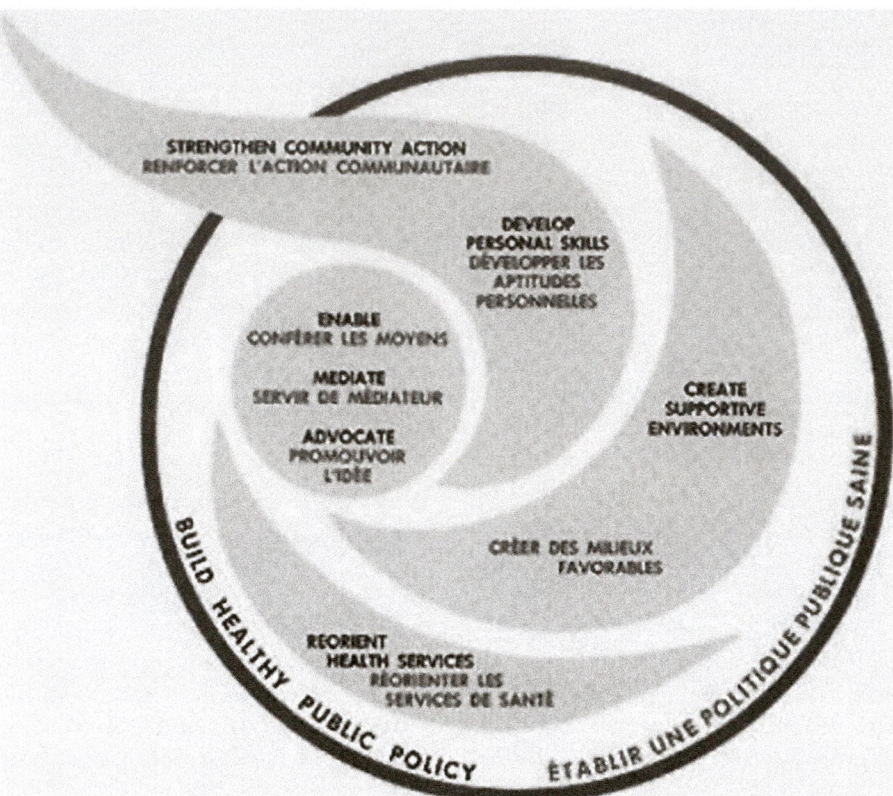

FIGURE 2.7   Logo of the First International Conference on Health Promotion, Ottawa, 21 November 1986: the Ottawa Charter 1986.

of the health consequences of their decisions and to accept responsibility for health. Health promotion policy combines diverse but complementary approaches, including legislation, fiscal measures, taxation, and organizational change. It is a coordinated action that leads to health, income, and social policies that foster greater equity. Joint action contributes to ensuring safer and healthier goods and services, healthier public services, and cleaner, more enjoyable environments. Health promotion policies require the identification of obstacles to the adoption of healthy public policies in non-health sectors, and ways of removing them.

Built on progress made from the Declaration on Primary Health Care at Alma-Ata, the aim was to make the healthier choice the easier choice for policy makers as well. The logo of the Ottawa Charter has been maintained by the WHO as the symbol and logo of health promotion. Health promotion represents activities to enhance and embed the concept of building healthy public policy through:

- building healthy public policy in all sectors and levels of government and society
- enhancing both self help and social support
- developing personal skills through information and education for health
- enabling, mediating, and advocating healthy public policy in all spheres

- creating supportive environments of mutual help and conservation of the natural environment
- reorienting health services beyond providing clinical curative services with linkage to broader social, political, economic, and physical environmental components.

(Adapted from Ottawa Charter; Health and Welfare Canada and World Health Organization, 1986)

State and Community Models of Health Promotion

An effective approach to health promotion was developed in Australia where, in the state of Victoria, revenue from a cigarette tax has been set aside for health promotion purposes. This has the effect of discouraging smoking, and at the same time finances health promotion activities and provides a focus for health advocacy in terms of promoting cessation of cigarette advertising at sports events or on television. It also allows for assistance to community groups and local authorities to develop health promotion activities at the workplace, in schools, and at places of recreation. Health activity in the workplace involves reduction of work hazards as well as promotion of a healthy diet and physical fitness, and avoidance of risk factors such as smoking and alcohol abuse.

In the Australian model, health promotion is not only the persuasion of people to change their life habits; it also

TABLE 2.2 Core Themes of Phase V of the World Health Organization European Healthy Cities Network

The choice of core themes offers the opportunity to work on priority urban health issues that are relevant to all European cities. Topics that are of particular concern to individual cities and/or are challenging and cutting edge for innovative public health action are especially emphasized. Healthy Cities encourages and supports experimentation with new ideas by developing concepts and implementing them in diverse organizational contexts.

1. Creating caring and supportive environments	A healthy city is a city for all its citizens: inclusive, supportive, sensitive and responsive to their diverse needs and expectations.
2. Healthy living	A healthy city provides conditions and opportunities that encourage, enable and support healthy lifestyles for people of all social groups and ages.
3. Healthy urban environment and design	A healthy city offers a physical and built environment that encourages, enables and supports health, recreation and well-being, safety, social interaction, accessibility and mobility, a sense of pride and cultural identity and is responsive to the needs of all its citizens.

World Health Organization. WHO European Healthy Cities Network: goals and requirements Phase V (2009–2013). Available at: http://www.euro.who.int/__data/assets/pdf_file/0009/100989/E92260.pdf [Accessed 19 November 2012].
Sources: World Health Organization. Zagreb Declaration of Healthy Cities: Health and health equity in all local policies. Copenhagen: WHO Regional Office for Europe; 2009. Available at: http://www.euro.who.int/__data/assets/pdf_file/0015/101076/E92343.pdf [Accessed 19 November 2012].

involves legislation and enforcement towards environmental changes that promote health. For example, this involves mandatory filtration, chlorination, and fluoridation for community water supplies, vitamin and mineral enrichment of basic foods. Primary Care alliances of service providers are organized including hospitals, community health services serving a sub-district population for more efficient and comprehensive care. These are at the level of national or state policy, and are vital to a health promotion program and local community action.

Community-based programs to reduce chronic disease using the concept of community-wide health promotion have developed in a wide variety of settings. Such a program to reduce risk factors for cardiovascular disease was pioneered in the North Karelia Project in Finland. This project was initiated as a result of pressures from the affected population of the province, which was aware of the high incidence of mortality from heart disease. Finland had the highest rates of CHD in the world and in the rural area of North Karelia the rate was even higher than the national average. The project was a regional effort involving all levels of society, including official and voluntary organizations, to try to reduce risk factors for CHD. After 15 years of follow-up, there was a substantial decline in mortality with a similar decline in a neighboring province taken for comparison, although the decline began earlier in North Karelia.

In many areas where health promotion has been attempted as a strategy, community-wide activity has developed with participation of NGOs or any valid community group as initiators or participants. Healthy Heart programs have developed widely with health fairs, sponsored by charitable or fraternal societies, schools, or church groups, to provide a focus for leadership in program development. A wider approach to addressing health problems in the community has developed into an international movement of "Healthy Cities".

Healthy Cities/Towns/Municipalities

Following deliberations of the Health of Towns Commission chaired by Edwin Chadwick, the Health of Towns Association was founded in 1844 by Southwood Smith, a prominent reform leader of the sanitary movement, to advocate change to reduce the terrible living conditions of much of the population of cities in the UK. The association established branches in many cities and promoted sanitary legislation and public awareness of the "sanitary idea" that overcrowding, inadequate sanitation, and absence of safe water and food created the conditions under which epidemic disease could thrive.

In the 1980s, Iona Kickbush, Trevor Hancock, and others promoted renewal of the idea that local authorities have a responsibility to build health issues into their planning and development processes. This "Healthy Cities" approach promotes urban community action on a broad front of health promotion issues (Table 2.2). Activities include environmental projects (such as recycling of waste products), improved recreational facilities for young people to reduce violence and drug abuse, health fairs to promote health awareness, and screening programs for hypertension, breast cancer, and other diseases. It combines health promotion with consumerism and returns to the tradition of local public health action and advocacy.

The municipality, in conjunction with many NGOs, develops a consultative process and program development approach to improving the physical and social life of the urban environment and the health of the population. In 1995, the Healthy Cities movement involved 18 countries with 375 cities in Europe, Canada, the USA, the UK, South America, Israel, and Australia, an increase from 18 cities in 1986. The model now extends to small municipalities, often with populations of fewer than 10,000. Networks of healthy cities are the backbone of the movement, with more than 1400 member towns and cities across Europe.

A typical healthy city has a population in the multiple thousands, often multilingual, with an average middle-class income. A Healthy Cities project builds a coalition of municipal and voluntary groups working together in a continuing effort to improve quality of service, facilities, and living environment. The city is divided into neighborhoods, engaged in a wide range of activities fostered by the project. Municipalities have traditionally had a leading role in sanitation, safe water supply, building and zoning laws and regulation, and many other responsibilities in public health (see Chapter 10). The Healthy Cities or Communities movement has elevated this to a higher level with policies to promote health in all actions. Some examples are listed of municipal, advocacy group, and higher governmental activities for healthier city environments:

- improved public transport
- developing sanitation, water supply, and waste disposal in urban slums
- traffic circles, crosswalks, and road bumps to slow urban vehicle traffic and improve pedestrian safety
- nuisance abatement in local quarries
- encouraging business enterprise for healthy food markets and local services in low-income areas
- parks, tree planting, and home gardens in poor and low-income neighborhoods and schools
- physical and security improvements to primary and secondary schools
- monitoring neighborhood profiles of social and crime indicators
- promoting preschool, primary and secondary school improvement
- cooperative housing for low-income families
- intercultural communication
- recreational facilities for young people
- restoration of neglected sites – green spaces
- extension of public parks, with improved walking and bicycle trails
- youth and community activities
- reduction of drug and crime environment through community policing and social support
- safe houses for battered women and homeless people
- community centers for older adults
- encouraging private enterprises and individuals in environmentally friendly planning, building, energy supply and conservation
- developing and supporting community health centers in urban slums
- reducing barriers for physically disabled people in public spaces and buildings
- reducing the stigma associated with mental illness and handicaps.

Working with senior levels of government, other departments in the municipalities, religious organizations, private donors, and the NGO sector to innovate and especially to improve conditions in poverty-afflicted areas of cities is a vital role for health-oriented local political leadership.

HUMAN ECOLOGY AND HEALTH PROMOTION

Human ecology, a term introduced in the 1920s and revived in the 1970s, attempted to apply theory from plant and animal life to human communities. It evolved as a branch of demography, sociology, and anthropology, addressing the social and cultural contexts of disease, health risks, and human behavior. Human ecology addresses the interaction of humans with and adaptation to their social and physical environment.

Parallel subdisciplines of social, community, and environmental psychology, medical sociology, anthropology, and other social sciences contributed to the development of this academic field with wide applications in health-related issues. This led to the incorporation of qualitative research methods alongside the quantitative research methods traditionally emphasized in public health, providing crucial insights into many public health issues where human behavior is a key risk factor.

Health education developed as a discipline and function within public health systems in school health, rural nutrition, military medicine, occupational health, and many other aspects of preventive-oriented health care, and is discussed in later chapters of this text. Directed at behavior modification through information and raising awareness of consequences of risk behavior, this has become a longstanding and major element of public health practice in recent times, being almost the only effective tool to fight the epidemic of HIV and the rising epidemic of obesity and diabetes.

Health promotion as an idea evolved, in part, from Marc Lalonde's Health Field Concepts and from growing realization in the 1970s that access to medical care was necessary but not sufficient to improve the health of a population. The integration of the health behavior model, social ecological approach, environmental enhancement, or social engineering formed the basis of the social ecology approach to defining and addressing health issues (Table 2.3).

Individual behavior depends on many surrounding factors, while community health also relies on the individual; the two cannot be isolated from one another. The ecological perspective in health promotion works towards changing people's behavior to enhance health. It takes into account factors not related to individual behavior, which are determined by the political, social, and economic environment. It applies broad community, regional, or national approaches that are needed to address severe public health problems, such as controlling HIV infection, TB, malnutrition, STIs, cardiovascular disorders, violence and trauma, and cancer.

TABLE 2.3 Health Promotion Approaches: Behavior Modification, Environment Enhancement, and Social Ecology

Health Behavior Model Change and Lifestyle Modification	Social Ecological Approaches	Environment Enhancement: National, Municipal, and Community Based
Behavior modification	Cultural change models of health	Universal access to health care
Social learning theory	Biopsychosocial models of health	Environmental health
Health belief model	Stressful life events	Industrial hygiene
Theory of reasoned action	Ecology of human development	Social security
Theory of planned behavior	Public health psychology	Societal support
Risk perception theory	Medical sociology	Community organization
Fear arousal	Ethnography	Ergonomics/human factors
Protection–motivation theory	Social epidemiology	Health monitoring epidemiology
Health communications	Social ecology of health	Urban planning, architecture
Mass media	Community health promotion	Regulation of housing, zoning
	Public policy initiatives	Injury and disaster control
	Healthy communities	Food and drug control
		Nutrition and food fortification

American Public Health Association. 10 Essential public health services, 2012. Available at: http://www.apha.org/programs/standards/performancestandards-program/reseexxentialservices.htm [Accessed 17 November 2012].
Source: Modified from Stokols D. Translating social ecological theory into guidelines for community health promotion. Am J Health Promot 1996;10:282–98.

DEFINING PUBLIC HEALTH STANDARDS

The APHA's formulation of the public health role in 1995, entitled The Future of Public Health in America, was presented at the annual meeting in 1996. The APHA periodically revises standards and guidelines for organized public health services provided by federal, state, and local governments (Table 2.4). These reflect the profession of public health as envisioned in the USA where access to medical care is limited for large numbers of the population because of a lack of universal health insurance. Public health in the USA has been very innovative in determining risk groups in need of special care and finding direct and indirect methods of meeting those needs.

European countries such as Finland have called for setting public health into all public policy, which reflects the vital role that local and county governments can play in developing health-oriented policies. These include policies in housing, recreation, regulation of industrial pollution, road safety, promotion of smoke-free environments, bicycle paths, health impact assessment, and many other applications of health principles in public policy.

INTEGRATIVE APPROACHES TO PUBLIC HEALTH

Public health involves both direct and indirect approaches. Direct measures in public health include immunization of children, modern birth control, and chronic disease case finding – hypertension, diabetes, and cancer. Indirect methods used in public health protect the individual by community-wide means, such as raising standards of environmental safety, ensuring a safe water supply, sewage disposal, and improved nutrition (Box 2.18).

In public health practice, the direct and indirect pproaches are both relevant. To reduce morbidity and mortality from diarrheal diseases requires an adequate supply of safe water and waste disposal, and also education of the individual in hygiene and the mother in use of ORT, and rotavirus vaccination of all children. The targets of public health action therefore include the individual, family, community, region, or nation, as well as a functioning and health system adopting current best practices for health care and health protection.

The targets for protection in infectious disease control are both the individual and the total group at risk. For vaccine-preventable diseases, immunization protects the individual but also has an indirect effect by reducing the risk even for non-immunized persons. In control of some diseases, individual case finding and management reduce risk of the disease in others and the community. For example, TB requires case finding and adequate care among high-risk groups as a key to community control. In malaria control, case finding and treatment are essential together with environmental action to reduce the vector population, to prevent transmission of the organism by the mosquito to a new host.

Control of NCDs, where there is no vaccine for mass application, depends on the knowledge, attitudes, beliefs, and practices of individuals at risk. In this case, the social

TABLE 2.4 Vision, Mission, and Standards for Public Health Services, American Public Health Association

Vision	Excellence in Public Health Practice defined by recognized Performance Standards
Mission	To improve the quality of public health practice and performance of public health systems
Goals	1. Provide performance standards for public health systems and encourage their widespread use
	2. Encourage and leverage national, state, and local partnerships to build a stronger foundation for public health preparedness
	3. Promote continuous quality improvement of public health systems
	4. Strengthen the science base for public health practice improvement
Essential services	1. *Monitor* health status to identify community health problems
	2. *Diagnose and investigate* health problems and health hazards in the community
	3. *Inform, educate, and empower* people about health issues
	4. *Mobilize* community partnerships to identify and solve health problems
	5. *Develop policies and plans* that support individual and community health efforts
	6. *Enforce* laws and regulations that protect health and ensure safety
	7. *Link* people to needed personal health services and assure the provision of health care when otherwise unavailable
	8. *Assure* a competent public health and personal health care workforce
	9. *Evaluate* effectiveness, accessibility, and quality of personal and population-based health services
	10. *Research* for new insights and innovative solutions to health problems

Centers for Disease Control and Prevention. Ten great public health achievements – United States, 2001–2010. MMWR Morb Mortal Wkly Rep 2011;60:619–23.
Sources: American Public Health Association. Vision, mission and goals [posted 23 July 2012]. Available at: http://www.apha.org/programs/standards/performancestandardsprogram/resmission.htm [Accessed 17 November 2012].

context is of importance, as is the quality of care to which the individual has access. Control and prevention of non-infectious diseases involve strategies using individual and population-based methods. Individual or clinical measures include professional advice on how best to reduce the risk of the disease by early diagnosis and implementation of appropriate therapy. Population-based measures involve indirect measures with government action banning cigarette advertising, or direct taxation on cigarettes. Mandating food quality standards, such as limiting the fat content of meat, and requiring food labeling laws are part of the control of cardiovascular diseases.

The way individuals act is central to the objective of reducing disease, because many non-infectious diseases are dependent on behavioral risk factors of the individual's choosing. Changing the behavior of the individual means addressing the way a person sees his or her own needs. This can be influenced by the provision of information, but how someone sees his or her own needs is more complex than that. An individual may define needs differently from the society or the health system. Reducing smoking among women may be difficult to achieve if smoking is thought to reduce appetite and food intake, given the social message that "slim is beautiful". Reducing smoking among young people is similarly difficult if smoking is seen as fashionable and diseases such as lung cancer seem very remote.

Recognizing how individuals define needs helps the health system to design programs that influence behavior that is associated with disease.

Public health has become linked to wider issues as health care systems are reformed to take on both individual and population-based approaches. Public health and mainstream medicine have found increasingly common ground in addressing the issues of chronic disease, growing attention to health promotion, and economics-driven health care reform. At the same time, the social ecology approaches have shown success in slowing major causes of disease, including heart disease and AIDS, and the biomedical sciences have provided major new technology for preventing major health problems, including cancer, heart disease, genetic disorders, and infectious diseases.

Technological innovations unheard of just a few years ago are now commonplace, in some cases driving up costs of care and in others replacing older and less effective care. At the same time, resistance of important pathogenic microorganisms to antibiotics and pesticides is producing new challenges from diseases once thought to be under control, and newly emerging infectious diseases challenge the entire health community. New generations of antibiotics, antidepressants, antihypertensive medications, and other treatment methods are changing the way many conditions are treated. Research and development in the biomedical

---

BOX 2.18 Why Health Systems Matter to the Social Determinants of Health Inequity

General Population Benefits

Health systems offer general population benefits that go beyond preventing and treating illness. Appropriately designed and managed, they:

- Provide a vehicle to improve people's lives, protecting them from the vulnerability of sickness, generating a sense of life security, and building common purpose within society
- Ensure that all population groups are included in the processes and benefits of socioeconomic development
- Generate the political support needed to sustain them over time.

Promote Health Equity

Health systems promote health equity when their design and management specifically consider the circumstances and needs of socially disadvantaged and marginalized populations, including women, the poor, and groups who experience stigma and discrimination, enabling social action by these groups and the civil society organizations supporting them.

Contribute to Achieving the Millennium Development Goals

Health systems can, when appropriately designed and managed, contribute to achieving the Millennium Development Goals.

Source: Gilson L, Doherty J, Loewenson R, Francis V. Final report – Knowledge Network on Health Systems – June 2007. WHO Commission on Social Determinants of Health (CSDOH); 2007. Available at: http://www.who.int/social_determinants/resources/csdh_media/hskn_final_2007_en.pdf [Accessed 21 November 2012].

---

sciences are providing means of prevention and treatment that profoundly affect disease patterns where they are effectively applied.

The technological and organizational revolutions in health care are accompanied by many ethical, economic, and legal dilemmas. The choices in health care include heart transplantation, an expensive life-saving procedure, which may compete with provision of funds and labor resources for immunizations for poor children or for health promotion to reduce smoking and other risk factors for chronic disease. New means of detecting and treating acute conditions such as myocardial infarction and peptic ulcers are reducing hospital stays, and improving long-term survival and quality of life. Imaging technology has been an important development in medicine since the advent of X-rays in the early twentieth century. Technology has forged ahead with high-technology instruments and procedures, new medication, genetic engineering, and important low-technology gains such as impregnated bed nets, simplified tests for HIV and TB, and many other "game changers". New technologies that can enable lower cost diagnostic devices, electronic transmission, and distant reading of transmitted imaging all open up possibilities for advanced diagnostic

capacities in rural and less developed countries and communities. Molecular biology has provided methods of identifying and tracking movement of viruses such as polio and measles from place to place, greatly expanding the potential for appropriate intervention.

The choices in resource allocation can be difficult. In part, these add political commitment to improve health, competent professionally trained public health personnel, the public's level of health information, and legal protection, whether through individuals, advocacy, or regulatory approaches for patients' rights. These are factors in a widening methodology of public health.

ACHIEVEMENTS OF PUBLIC HEALTH

The Centers for Disease Control and Prevention (Morbidity and Mortality Weekly Report) in 1999 summarized 10 great achievements of public health in the USA, with an extension of the lifespan by over 30 years and improvements in many measures of quality of life. They were updated in a similar summary report in 2011, showing continuous progress, and a global version which was also encouraging in its scope of progress (Table 2.5). These achievements were also seen in all developed countries over the past century and are beginning to be seen in developing countries as well. They reflect a successful application of a broad approach to prevention and health promotion along with improved medical care and growing access to its benefits. In the past several decades alone, major new innovations are leading to greater control of cardiovascular disease, cancer prevention, and many other improvements to health affecting hundreds of millions of people. A similar 2011 report by the CDC shows global progress in the first decade of the twenty-first century, while MDG reports show progress on all eight target topics, although not at uniformly satisfactory rates. These achievements are discussed throughout this text.

This successful track record is very much at the center of a New Public Health involving a wide range of programs and activities, shown to be feasible and benefiting from continuing advances in science and understanding of social and management issues affecting health care systems worldwide.

THE FUTURE OF PUBLIC HEALTH

Public health issues have received new recognition in recent years because of a number of factors, including a growing understanding among the populace at different levels in different countries that health behavior is a factor in health status and that public health is vital for protection against natural or human-made disasters. The challenges are also increasingly understood: preparation for bioterrorism, avian influenza, rising rates of diabetes and obesity, high mortality rates from cancer, and a wish for prevention to be effective.

TABLE 2.5 Great Achievements of Public Health in the USA, 1900–1999 and 2001–2010

1900–1999	2001–2010
Vaccination	Vaccine-preventable diseases
Motor-vehicle safety	Prevention and control of infectious diseases
Safer workplaces	Tobacco control
Control of infectious diseases	Maternal and infant health
Decline in deaths from coronary heart disease and stroke	Motor vehicle safety
Safer and healthier foods	Cardiovascular disease prevention
Healthier mothers and babies	Occupational safety
Family planning	Cancer prevention
Fluoridation of drinking water	Childhood lead poisoning prevention
Recognition of tobacco use as a health hazard	Public health preparedness and response

Sources: Centers for Disease Control and Prevention. Ten great public health achievements – United States, 1900–1999. MMWR Morb Mortal Wkly Rep 1999;48:1141–7.

The MDGs selected by the UN in 2000 have eight global targets for the year 2015, including four directly related to public health (discussed above, Box 2.15). These are a recognition and a challenge to the international community and public health as a profession and as organized systems. Formal education in newly developing schools of public health is increasing in Europe, including many countries of Eastern Europe, and beginning to develop in India and sub-Saharan Africa. But there is delay in establishing centers of postgraduate education and research in many developing countries which are concentrating their educational resources on training physicians. Many physicians from developing nations are moving to the developed countries, which have become dependent on these countries for a significant part of their supply of medical doctors. Progress in implementation of the MDGs is mixed in sub-Saharan Africa, making some progress in immunization, but falling back on other goals. Proposals to renew global health targets following the 2015 end-stage of the MDG health goals will need to add a focus on NCDs, which account for 60 percent of global deaths, including 8.1 million premature deaths below the age of 60 (UNDP).

Economic growth has been hampered by the global recession since 2008, which will affect continued progress with many other factors of changing population dynamics, the economics of prevention versus expensive treatment costs, and the high costs of health care. Environmental degradation with high levels of carbon dioxide contamination is a growing concern, with disastrous global warming and consequent effects of drought, flooding, hurricane, and elevated particulate matter-induced asthma and effects on cardiovascular disease. The potential for the development of basic and medical sciences in genetics, nanotechnology, and molecular biology shows enormous promise for health benefits as yet unimagined.

At the same time, the effectiveness of health promotion has shown dramatic successes in reducing the toll of AIDS, reducing smoking, and increasing consciousness of nutrition and physical fitness in the population, and of the tragic effects of poverty and poor education on health status. The ethics of public health issues are complex and changing with awareness that failure to act on strong evidence-based policies is itself ethically problematic. The future of public health is not as a solo professional sector; it is at the heart of health systems, without which societies are open to chronic and infectious diseases that are preventable, affecting the society as a whole in economic and development matters.

There is an expanding role of private donors in global health efforts, such as the Rotary Club and the polio eradication program, GAVI with immunization and bed-nets in sub-Saharan Africa, and bilateral donor countries' help in reducing the toll of AIDS in sub-Saharan Africa.

THE NEW PUBLIC HEALTH

The New Public Health has emerged as a concept to meet a whole new set of conditions, associated with increasing longevity and aging of the population, with the post-World War II baby-boom generation reaching the over-65 age group facing the growing importance of chronic diseases. Inequalities in health exist in and between affluent and developing societies, as well as within countries, even those having advanced health care systems. Regional inequalities are seen across the European Region in an east–west gradient and globally a north–south divide of extremes of inequality. The global environmental and ecological degradation and pollution of air and water present grave challenges for developed and developing countries worldwide. Yet optimism can be derived from proven track records of success in public health measures that have already been implemented. Many of the underlying factors are amenable to prevention through social, environmental, or behavioral change and effective use of medical care.

The New Public Health idea has evolved since Alma-Ata, which articulated the concept of Health for All, followed by a trend in the late 1970s to Health in All policies and establishing health targets as a basis for health planning. During the late 1980s and early 1990s, the debate on the future of public health in the Americas intensified as health professionals looked for new models and approaches to public health research, training, and practice. This debate helped to redefine traditional approaches of social, community, and preventive medicine. The search for the "new" in public health continued with a return to the Health for All concept of Alma-Ata (renewed in 2008) and a growing realization that the health of both the individual and the society involves the management of personal care services and community prevention, with a comprehensive approach taking advantage of advancing technology and experience of best practices globally.

The New Public Health is an extension of the traditional public health. It describes organized efforts of society to develop healthy public policies: to promote health, to prevent disease, and to foster social equity within a framework of sustainable development. A new, revitalized public health must continue to fulfill the traditional functions of sanitation, protection, and related regulatory activities, but in addition to its expanded functions. It is a widened philosophy and practical application of many different methods of addressing health, and preventing disease and avoidable death. It necessarily addresses inequities so that programs need to meet special needs of different groups in the population according to best standards, limited resources, and population needs. It is proactive and advocates interventions within legal and ethical limits to promote health as a value in and of itself and as an economic gain for society as well for its individual members.

The New Public Health is a comprehensive approach to protecting and promoting the health status of the individual and the society, based on a balance of sanitary, environmental, health promotion, personal, and community-oriented preventive services, coordinated with a wide range of curative, rehabilitative, and long-term care services. It evolves with new science, technology, and knowledge of human and systems behavior to maximize health gains for the individual and the population.

The New Public Health requires an organized context of national, regional, and local governmental and non-governmental programs with the object of creating healthful social, nutritional, and physical environmental conditions. The content, quality, organization, and management of component services and programs are all vital to its successful implementation.

Whether managed in a diffused or centralized structure, the New Public Health requires a systems approach acting towards achievement of defined objectives and specified targets. The New Public Health works through many channels to promote better health. These include all levels of government and parallel ministries; groups promoting advocacy, academic, professional, and consumer interests; private and public enterprises; insurance, pharmaceutical, and medical products industries; the farming and food industries; media, entertainment, and sports industries; legislative and law enforcement agencies; and others.

The New Public Health is based on responsibility and accountability for defined populations in which financial systems promote achievement of these targets through effective and efficient management, and cost-effective use of financial, human, and other resources. It requires continuous monitoring of epidemiological, economic, and social aspects of health status as an integral part of the process of management, evaluation, and planning for improved health.

The New Public Health provides a framework for industrialized and developing countries, as well as countries in political–economic transition such as those of the former Soviet system. They are at different stages of economic, epidemiological, and sociopolitical development, each attempting to ensure adequate health for its population with limited resources. The challenges are many, and affect all countries with differing balances, but there is a common need to seek better survival and quality of life for their citizens (Table 2.6).

SUMMARY

The object of public health, like that of clinical medicine, is better health for the individual and for society. Public health works to achieve this through indirect methods, such as by improving the environment, or through direct means such as preventive care for mothers and infants or other at-risk groups. Clinical care focuses directly on the individual patient, mostly at the time of illness. But the health of the individual depends on the health promotion and social programs of the society, just as the well-being of a society depends on the health of its citizens. The New Public Health consists of a wide range of programs and activities that link individual and societal health.

The "old" public health was concerned largely with the consequences of unhealthy settlements and with safety of food, air, and water. It also targeted the infectious, toxic, and traumatic causes of death, which predominated among young people and were associated with poverty.

A summary of the great achievements of public health in the twentieth and in the early twenty-first century in the industrialized world is included in Chapter 1 and throughout this text. These achievements are reflective of public health gains throughout the industrialized world and are

TABLE 2.6 Origins and Synthesis of the New Public Health

Classical Public Health	Social Ecology	Biomedical Care	Organization and Financing
To End of Nineteenth Century			
Food and personal hygiene	Church and serfdom	Basic sciences	Private payment for rich
Settlement health	Renaissance	Clinical sciences	Municipal doctors for poor
Quarantine	Agricultural revolution	Medical education	Charity, church, voluntary hospital care
Nutrition/fitness	Improved nutrition	Hospitals: church, municipal, voluntary, university	Guilds, mutual benefit, friendly societies for medical, pensions, burial benefits
Vital statistics	Rise of cities	Specialization	National health insurance for workers
Epidemiology	Rights of man	Therapeutics	Sick funds and voluntary health insurance
Miasma theory	Industrial revolution	Antisepsis	Sanitation
Municipal organization	Labor laws		Municipal sanitation
Bacteriology, germ theory	Universal education		
Vaccines, immunology	Social reform	Vaccines, antitoxins	
Control of infectious diseases	Political revolution	Antibiotics	
Maternal and child health	Information revolution		Federla, state and local authorities
To the 1980s			
Epidemiological transition	Aging of population	Advancing medical sciences	Collective bargaining health benefits
Declining mortality and birth rates, aging of population	Rising expectations	Clinical specialization	Government responsibility
Demographic transition	Lifestyle and risk factors	Diagnostics, imaging, laboratory technology	National health insurance or national health service
Decreasing infectious disease	Social inequities	Therapeutics, antibiotics, antihypertensives, cardiac, psychotropic drugs	Rising costs of health care
Increase in non-infectious disease	Social security	Preventive medicine	Imbalance of hospital and primary care
International health	The welfare state	Home care	Health maintenance organizations
Eradication of smallpox	Governmental responsibility for health	Long-term care	Cost–benefit evaluation
Alma-Ata	Advocacy	Hospital versus community care	Rationalization
Ottawa Charter	Health promotion	Ambulatory surgery	Reforms
1980 and Beyond: The New Public Health			
Policy coordination	National health policy	University medical schools	National health targets
Evaluation of health status	Resource allocation	Postgraduate education	Decentralization/diffusion of implementation
Health promotion	Economic development	Health management training	District health systems
Regulation of food, drugs, water, worksite, toxic agents, trauma,	Social context	Peer review systems	Managed care systems (HMOs)
Communicable disease control	Social security	Accreditation	Modified market mechanisms, regulation of supply, incentives, fee control, competition, managed care
Control non-communicable diseases	Ecology and environment	Quality of care (TQM)	Management accountability
Reduce risk factors	Nutrition and food policy	Targeted research	Economic assessment

TABLE 2.6 Origins and Synthesis of the New Public Health—cont'd

Classical Public Health	Social Ecology	Biomedical Care	Organization and Financing
Special needs groups	Healthy public policy	Hospital/community care, long-term care, home care, elderly housing, community services	Integrated health systems
Mental health	Healthy communities	Integrated health systems	
Dental health	Intersectoral cooperation	Managed care systems	
Health information systems	Advocacy	Ethical issues	
Epidemiology in planning and management	Voluntarism	Evidence-based medicine	
	Community participation	Cost-effectiveness analysis	

beginning to affect the health situation in countries in transition from the socialist period. Countries emerging from developing status are also showing signs of mixed progress in the dual burden of infectious and maternal/child health issues, along with growing exposure to the chronic diseases of developed nations such as cardiovascular diseases, obesity, and diabetes.

The New Public Health synthesizes traditional public health with management of personal services and community action for a holistic approach. Evaluation of cost-effective public health and medical interventions to reduce the burden of disease also contributes to the need to seek and apply new approaches to health. The New Public Health will continue to evolve as a framework drawing on new ideas, science, technology, and experiences in public health throughout the world. It must address the growing recognition of social inequality in health, even in developed countries with universal health programs with improved education and social support systems.

NOTE

For a complete bibliography and guidance for student reviews and expected competencies please see companion web site at http://booksite.elsevier.com/9780124157668

BIBLIOGRAPHY

Allen, L., de Benoist, B., Dary, O., Hurrell, R., 2006. Guidelines on food fortification with micronutrients. WHO, Geneva.

Alliance for Health Policy and Systems Research. Systematic reviews in health policy, September 2009. Available at: http://www.who.int/alliance-hpsr/resources/alliancehpsr_briefingnote4.pdf (accessed 18.11.12).

American Public Health Association, 10 essential public health services. Available at: http://www.apha.org/programs/standards/performancestandardsprogram/resexxentialservices.htm (accessed 19.11.12).

American Public Health Association. Healthy communities, 2000. model standards for community attainment of the year 2000 national health objectives, third ed. APHA, Washington, DC 1991.

Bobak, M., Murphy, M., Rose, R., Marmot, M., 2003. Determinants of adult mortality in Russia: estimates from sibling data. Epidemiology 14, 603–611.

Centers for Disease Control and Prevention, 2011. Ten great public health achievements – United States, 2001–2010. MMWR Morb. Mortal. Wkly. Rep. 60, 619–623.

Centers for Disease Control and Prevention, 2011. Ten great public health achievements – worldwide, 2001–2010. MMWR Morb. Mortal. Wkly. Rep. 60, 814–818.

College of Physicians and Surgeons of Philadelphia, History of anti-vaccination movements. Available at: http://www.historyofvaccines.org/content/articles/history-anti-vaccination-movements (accessed 21.06.13).

Commission on Social Determinants and Health. Closing the gap in a generation: health equity through action on the social determinants of health. Final report of the Commission on Social Determinants of Health. Geneva: WHO; 2008. Available at: http://whqlibdoc.who.int/publications/2008/9789241563703_eng.pdf (accessed 10.12.10).

Fries, J.F., 2000. Compression of morbidity in the elderly. Vaccine 18, 1584–1589.

Gebbie, K., Rosenstock, L., Hernandez, L.M. (Eds.), 2003. Institute of Medicine. Who will keep the public healthy? Educating public health professionals for the 21st century. National Academies Press, Washington, DC. Available at: http://books.nap.edu/openbook.php?record_id=10542. (accessed 6.12.10).

Global Alliance for Vaccine and Immunization (GAVI), http://www.gavialliance.org/about/partners/bmgf/ (accessed 21.11.12).

Halpin, H.A., Morales-Suarez-Varela, M., Martin-Moreno, J.M., 2010. Chronic disease prevention and the new public health. Public Health Rev. 32, 120–154.

Hancock, T., 1993. The evolution, impact and significance of healthy cities/healthy communities. J. Public Health Policy 14, 5–18.

Health and Welfare Canada, World Health Organization. Ottawa Charter for health promotion: an international conference on health promotion, Ottawa, Canada, 1986. Available at: http://www.who.int/hpr/NPH/docs/ottawa_charter_hp.pdf (accessed 23.04.11).

Healthy People. 2020. Available at: http://www.healthypeople.gov/hp2020/ (accessed 12.11.12).

Holtzman, D., Neumann, M., Sumartojo, E., Lansky, A., 2006. Behavioral and social sciences and public health at CDC. MMWR Morb. Mortal. Wkly. Rep. 55 (Suppl. 2), 14–16.

Kimmo, L., Ollila, E., Pena, S., Wismar, H., Cook, S. (Eds.), 2013. Health in All policies: seizing opportunities, implementing policies. Ministry of Social Affairs and Health, Finland. Available at: http://www.euro.who.int/__data/assets/pdf_file/0007/188809/Health-in-All-Policies-final.pdf. (accessed 19.06.13).

Lalonde, M., 1974. New perspectives on the health of Canadians: a working document. Ottawa.

Lalonde, M., 2002. New perspective on the health of Canadians: 28 years later. Rev. Panam. Salud Publica/Pan. Am. J. Public Health 12, 149–152. Available at: http://www.bvsde.paho.org/bvsacd/cd51/heroes.pdf (accessed 6.12.10).

Lawrence, W., Green, L.W., Fielding, J., 2011. The US Healthy People initiative: its genesis and its sustainability. Annu. Rev. Public Health 32, 451–470.

Levi, F., Chatemoud, L., Bertuccio, P., Lucchini, F., Nigri, E., La Vecchia, C., 2009. Mortality from cardiovascular and cerebrovascular diseases in Europe and other areas of the world: an update. Eur. J. Cardiovasc. Prev. Rehabil. 16, 333–350.

Marmot, M., 2010. Fair society, healthy lives: the Marmot Review: Executive Summary. Strategic review of health inequalities in England post. Department of Health, London.

Nutbeam, D., 1998. Health promotion glossary: Foreword: moving toward a New Public Health. Health Promot. Int. 13, 1–16.

Organisation for Economic Co-operation and Development, 2012. Health policy and data. OECD health data. Available at: http://www.oecd.org/health/healthpoliciesanddata/oecdhealthdata2012.htm. (accessed 18.11.12).

Sachs, J.D., 2007. Primary care (extended version): ten key actions could globally ensure a basic human right at almost unnoticeable cost. Sci. Am. (December). Available at: http://www.scientificamerican.com/article.cfm?id=primary-health-for-all-extended (accessed 16.11.12).

Stuckler, D., Basu, S., McKee, M., 2010. Public health in Europe: power, politics, and where next. Public Health Rev. 32, 213–242.

Suhrcke, M., Rocco, L., McKee, M., 2007. Health: a vital investment for economic development in Eastern Europe and Central Asia. European Observatory on Health Systems and Policies. WHO, European Regional Office, Copenhagen.

Tulchinsky, T.H., 2010. It is not just the Broad Street pump. J. Public Health 32, 134–135.

Tulchinsky, T.H., Varavikova, E.A., 1996. Addressing the epidemiologic transition in the former Soviet Union: strategies for health systems and public health reform in Russia. Am. J. Public Health 86, 313–320.

Tulchinsky, T.H., Varavikova, E.A., 2010. What is the "New Public Health"? Public Health Rev. 32, 25–53.

United Nations. Millenium Development Goals: 2013 progress chart. Available at: http://www.un.org/millenniumgoals/pdf/report-2013/2013_progress_english.pdf (accessed 8.11.2013).

United Nations Development Programme, Millennium Development Goals. Eight goals for 2015. Available at: http://www.undp.org/content/undp/en/home/mdgoverview.html (accessed 16.11.12.).

US Department of Health and Human Services, Healthy People 2020. Available at: http://www.healthypeople.gov/2020/about/default.aspx (accessed 18.11.12).

US Department of Health, Education, and Welfare. Healthy People. The Surgeon General's report on health promotion and disease prevention. Washington, DC: US DHEW; 1979. Available at: http://profiles.nlm.nih.gov/NN/B/B/G/K/_/nnbbgk.pdf (accessed 19.11.12).

Waage, J., Banerji, R., Campbel, O., Chirwa, E., Collender, G., Dieltiens, V., et al., 2010. The Millennium Development Goals: a cross-sectoral analysis and principles for goal-setting after 2015. Lancet 376, 991–1023. Available at: http://www.thelancet.com/journals/lancet/article/PIIS0140-6736(10)61196-8/fulltext#article_upsell (accessed 29.07.13).

Walsh, J.A., Warren, K.S. Selective primary health care: an interim strategy for disease control in developing countries. N. Engl. J. Med. 1979:301 (18):967–974. Available at: http://www.ais.up.ac.za/med/pcm870/interimstrategy.PDF (accessed 5.11.2013).

World Health Organization. Declaration of Alma-Ata. International conference on primary health care, Alma-Ata, USSR, 6–12 September 1978. Geneva: WHO. Available at: http://www.who.int/hpr/NPH/docs/declaration_almaata.pdf (accessed 6.12.10).

World Health Organization. Healthy Cities networks across the WHO, European Region. Available at: http://www.euro.who.int/en/what-we-do/health-topics/environmental-health/urban-health/activities/healthy-cities (accessed 18.11.12).

World Health Organization. Preamble to the Constitution of the World Health Organization as adopted by the International Health Conference, New York, June 19–22, 1946; signed on July 22, 1946, by the representatives of 61 States (Official Records of the World Health Organization, no. 2, p. 100) and entered into force on April 7, 1948. Available at: http://www.who.int/about/definition/en/print.html (accessed 14.02.08).

World Health Organization, 1999. Regional Office for Europe. Health 21 – Health for All in the 21st Century. WHO Regional Office for Europe, Copenhagen.

World Health Organization, 2012. Regional Office for Europe. WHO European healthy cities network. Available at:. WHO Regional Office for Europe, Copenhagen. http://www.healthypeople.gov/2020/about/new2020.aspx. (accessed 21.11.12).

Chapter 3

Measuring, Monitoring, and Evaluating the Health of a Population

Learning Objectives

Upon completion of this chapter, the student should be able to:

1. Explain the basic terms and concepts of epidemiology;
2. Interpret health data from an epidemiological viewpoint;
3. Apply a systemic approach in evaluating the health status of a population.

INTRODUCTION

The history of health, health concepts, and scientific developments has been discussed in previous chapters. Measuring the health of populations is fundamental to improving their health status. Traditionally, public health deals with the health of populations, while the New Public Health deals with the health of both individuals and population groups. This chapter discusses how measurements are used to describe, analyze, prescribe, and justify interventions to protect and improve the health of populations and of individuals, and to monitor the outcomes of interventions.

The public health professional working with individual and community health needs to acquire the knowledge and skills necessary to measure and interpret the factors that relate to disease and health, both in the individual and in population groups. Demography and epidemiology are the basis of health information systems, but the social and basic medical sciences are also vitally important in understanding public health, providing an expanding array of health status indicators and measures of the impact of interventions.

Demography deals with the recording of the characteristics and trends of a population over time. The field has broadened to include social demography, which has a broader focus on economic, social, cultural, and biological factors, an important aspect of the New Public Health because of the vital role that risk factors, which are deeply affected by social conditions, play in health protection and disease prevention. Epidemiology measures the distribution, causes, control, and outcomes of disease in population

groups. It provides the basic tools for quantification of the extent of disease, its patterns of change, and associated risk factors. Epidemiology also provides basic information needed for planning, evaluating, and managing health services. Other disciplines provide additional information and insights needed for community and national health assessment. These include the social sciences (sociology, psychology, anthropology, and economics), as well as clinical fields such as pediatrics and geriatrics, and basic sciences such as microbiology, immunology, and genetics.

This chapter is an introduction to epidemiology and health information systems intended to familiarize the student with basic terms, concepts, and methods. The scope of this text does not lend itself to detailed discussion of biostatistics and epidemiological methods, but instead focuses on the basic ideas and their relevance to the New Public Health. This chapter is meant to provide a general overview of the role of epidemiology and health information systems in the context of the New Public Health; it cannot serve as an authoritative, detailed text on the subject. Specialized texts and reviews, such as the Centers for Disease Control and Prevention's (CDC's) Health Disparities and Inequalities Report – United States 2011, are listed in the bibliography at the end of this chapter and on the companion web site (http://booksite.elsevier.com/9780124157668).

DEMOGRAPHY

Demography is "the study of populations, especially with reference to size and density, fertility, mortality, growth, age distribution, migration, and vital statistics and the interaction of all these with social and economic conditions" (Last, 2001). Demography is based on vital statistics reporting and special surveys of population size and density; it measures trends over time. It includes indices such as fertility, birth, and death rates; rural–urban residential patterns; marriage and divorce rates and migrations; and their interaction with social and economic conditions. Since public health deals with disease as it occurs in the population, the definition of populations and their characteristics is fundamental.

The New Public Health. http://dx.doi.org/10.1016/B978-0-12-415766-8.00003-3

Vital statistics include births; deaths; and population by age, gender, location of residence, marital status, socioeconomic status (SES), and migration. Birth data are derived from mandatory reporting of births and mortality data from compulsory death certificates. Other sources of data include population registries, including marriage/divorce, adoption, emigration, and immigration, residential patterns, as well as census data, economic and labor force statistics, and data from special household surveys conducted by home visits, telephone, or electronic media methods.

A census is a survey covering the entire population of a defined geographic, political, or administrative entity. It is an enumeration of the population, recording the identity of all people in every residence at a specified time. The census provides important information on all members of the household, including age, date of birth, gender, occupation, national origin, marital status, income, relation to head of the household, literacy, education level, and health status (e.g., permanent disabling conditions). The census also covers residents of health and social facilities such as nursing homes or similar care facilities. Other information on the home and its facilities may be included. A census may assign people according to their location at the time of the enumeration (*de facto*) or to the usual place of residence (*de jure*). A census tract is the smallest geographic area for which census data are aggregated and published. Data for larger geographic areas (metropolitan/regional statistical areas) are also published. More extensive data may be collected for representative samples of the population. These surveys are carried out over a period of years by a specialized national agency (e.g., Bureau of the Census in the USA and the Central Bureau of Statistics, Office of Population, Censuses and Surveys in the UK).

Census data are published in multiple-volume series with availability for research on computer disks, CD-ROMs, and the Internet. Intercensus surveys are systematically collected information sets, without prior hypothesis, usually by questionnaires with questions carefully composed and tested for validity and consistency (Last, 2007). They may include interviews, biological samples and physical examination. An outstanding example is the US National Health and Nutrition Examination Surveys (NHANES) conducted by the US Center for Health Statistics. These are carried out to determine trends in important economic or demographic data such as individual and family incomes, nutrition, employment, and other social indicators. Such a complex and costly process can never be 100 percent accurate, but great care is taken to maximize response and standardization in interview methods and processing to ensure precision.

Despite its limitations, the census is accepted as the basis of statistical definition of a population. It is well established in developed countries, but is problematic in developing countries where birth and death registration may be inadequate, requiring community-based registration systems. In the Scandinavian countries, population censuses have been replaced by continuously updated databases containing information about all inhabitants, who are assigned a personal identification number at birth or upon immigration.

Demographic transition is a long-term trend of declining birth and death rates, resulting in substantive change in the age distribution of a population. Population age and gender distribution is mainly affected by birth and death rates, as well as other factors such as migration, economics, war, political and social change, famine, or natural disasters. *Biodemography*, the study of the senescent process, focuses on aspects such as the length of life, the length of healthy life, and the limits to the lifespan. Economic development has a profound effect on population patterns, and demographic transition may be characterized by the following stages:

1. *Traditional* – high and balanced birth and death rates.
2. *Transitional* – falling death rates and sustained birth rates.
3. *Low stationary* – low and balanced birth and death rates.
4. *Graying of the population* – increased proportion of elderly people as a result of decreasing birth and death rates, and increasing life expectancy.
5. *Regression* – low birth rates, migration, or increasing death rates among young adults due to trauma, acquired immunodeficiency syndrome (AIDS), early cardiovascular disease (CVD) mortality, or war can result in a steady or declining population (i.e., demographic regression).

Fertility, mortality, disease patterns, and migration are the major influences on this transition within the population. The many factors that affect fertility decline and increasing longevity are outlined in Box 3.1. Education of women, urbanization, improved hygiene and preventive care, economic improvement with better living conditions, and declining mortality of infants and children are the major factors. This is an important issue in developing countries where high fertility rates and declining mortality of children contribute to rapid population growth and poverty.

Birth rates in the industrialized countries have fallen over the past half-century and are continuing to fall in many countries to levels below rates needed to sustain or maintain population size and age distribution. This contributes to aging of the population, with important economic and societal effects. Economic prosperity, efficient and easily available methods of birth control, and greater education and work opportunities for women in the workforce are major factors in choices made in terms of the number of children a woman wishes to have, and her right to determine the number and spacing of pregnancies. In some countries, access to prenatal diagnosis of the gender of the fetus has resulted

BOX 3.1 Factors in Fertility Decline and Increasing Longevity

Factors in Fertility Decline

- Education, especially of women.
- Decreasing infant and child mortality, reducing pressure for more children to ensure survivors.
- Economic development, improved standards of living, rising expectations and family income levels.
- Urbanization – family needs and resources change compared to rural society.
- Birth control methods – safe, inexpensive, supply, accessibility, and knowledge.
- Government policy promoting fertility control as a health measure.
- Mass media can raise awareness of birth control, and aspiration to higher standards of living.
- Health system development and improved access to medical care.
- Changing economic status, social role, and self-image of women.
- Changing social, religious, political and ideological values.

Factors in Increasing Longevity

- Increasing family income, education level and standards of living.
- Improved nutrition including improved food supply, distribution, quality, and nutritional knowledge.
- Control of infectious diseases.
- Reduction in non-infectious disease mortality.
- Adequacy of safe food and water, sewage and garbage disposal, adequate housing conditions.
- Disease prevention, reducing risk factors, promoting healthy lifestyle.
- Medical care services with improved access and quality.
- Health promotion and education activities of the society, community, and individual.
- Social security systems, child allowances, pensions, unemployment insurance, national health insurance.
- Improved conditions of employment and recreation, economic and social well-being.

BOX 3.2 Commonly Used Fertility Rates

- *Crude birth rate* (CBR) – the number of live births in a population over a given period, usually one calendar year, divided by the midyear population of the same jurisdiction, multiplied by 1000.
- *Total fertility rate* (TFR) – the average number of children that a woman would bear if all women lived to the end of their childbearing years and bore children according to age-specific fertility rates; most accurately answering the question "how many children does a woman have, on average?"

Source: *Modified from Last JM, editor. A dictionary of public health. New York: Oxford University Press; 2007.*

complex issue influenced by cultural, social, economic, religious, and even political factors. Although economic prosperity may initially promote higher birth rates, increases in education levels and economic prospects, as well as in survival of those born, are generally related to reduced birth rates and natural population growth (Box 3.2). Changes in the status of women, and sexual and reproductive health standards and methods have contributed to changing birth patterns and expectations of family size in evolving societies. In recent decades, new medical advances have led to in vitro fertilization methods becoming widely available in upper- and middle-income countries; these are now an option in some instances of infertility, as is surrogate motherhood.

Population Pyramid

A population pyramid provides a graphic display of the percentage of men and women in each age group in a total population (Figures 3.1 and 3.2). A wide population base and a high birth rate in a country or region result in a large percentage of its population being under 15 years of age; when accompanied by limited economic resources, this is a formula for continued poverty. A population pyramid with a narrow base (i.e., few young people) and a growing elderly population will have a smaller workforce to provide the economic base for the "dependent age" population (i.e., both the young and the old). Aging of the population represents an increase in the over-65 population to some 13 percent of the population (Figure 3.3).

With a smaller working-age population to support these social costs of dependent subgroups, adverse economic consequences may prejudice costly pension and health services for the population. Other factors may also affect the population pyramid; for example, the loss of a large number of people during wartime. This loss affects a particular age–gender group as well as fertility patterns during and after the war; for example, the postwar "baby boom" after World War II. With aging of the population in many countries due

in wide-scale abortion of females because of birth policies, with parental preference for male children in China and India as examples. This is resulting in a major numerical deficiency of young women in the population with many attendant social and political effects. Reduced fertility and mortality, as in Japan and many countries in Western Europe, also have many societal and economic consequences, as a smaller workforce has to maintain a higher elderly population dependent on social security benefits.

Fertility

Fertility is the bearing of living children and is clearly determined by more than biological potential. Fertility is a

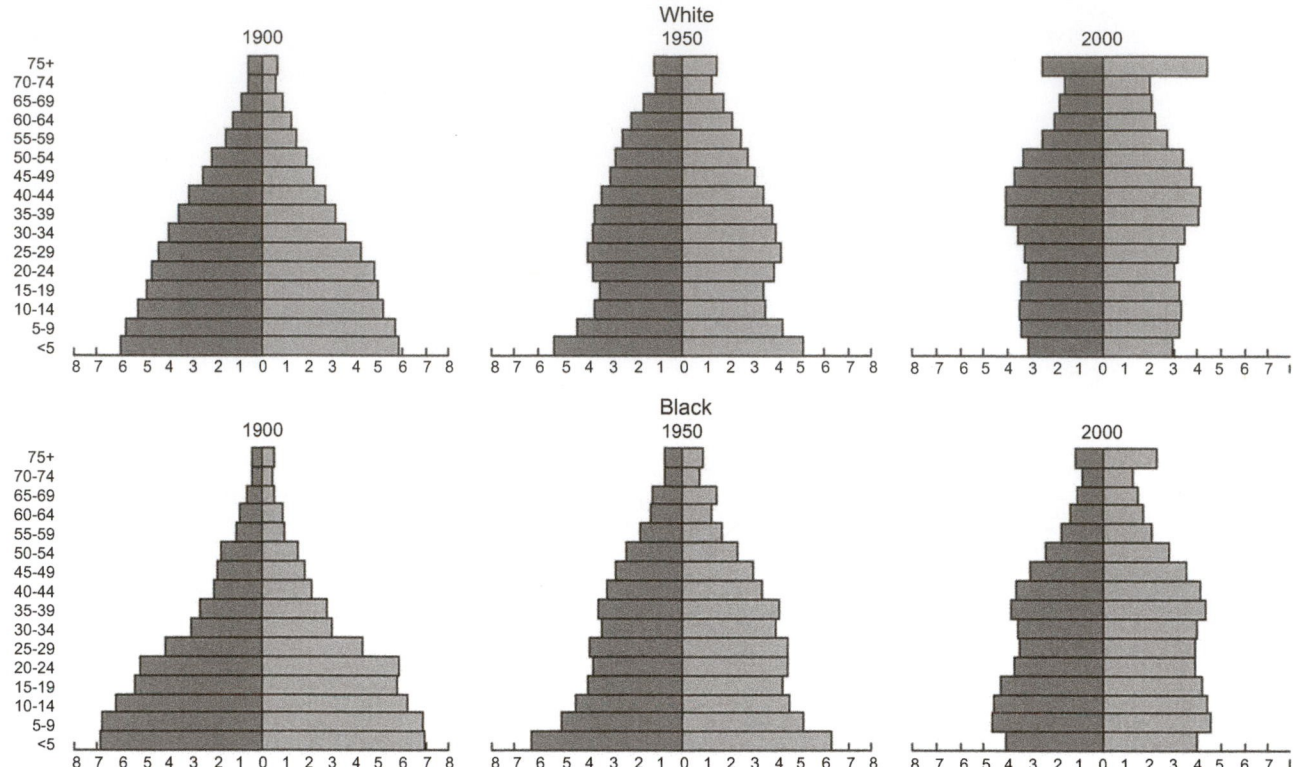

FIGURE 3.1   Population pyramids for the USA, 1900, 1950, and 2000, by gender for white and black populations. Note: bars (left) = male; bars (right) = female. *Source: Hobbs F, Stoops N. US Census Bureau: Census 2000 special reports, Series CENSR-4, Demographic trends in the 20th century. Washington, DC: US Government Printing Office; 2002.*

to low birth rates and increasing longevity, the concept of dependent population groups of those under the age of 15 and those over 65 as a percentage of the total population is increasingly relevant to social and economic planning.

LIFE EXPECTANCY

Life expectancy is an important health status indicator based on the average number of years a person at a given age may be expected to live given current mortality rates. Life expectancy can be measured at birth (age 0), which is most commonly used for national and international comparisons (Table 3.1).

Life expectancy is also reported at other specific ages, representing expected survival time once a person has reached that age; for example, at age 15, 60, or 75 by gender and by ethnic group, or by specific medical conditions such as cancer of the colon, myocardial infarction, and others.

Between 1970 and 2009, life expectancy for people aged 65 by gender and race was similar (Figure 3.4). However, variation in life expectancy at birth by gender and race remains constant. Looking at the years 1900–2000, life expectancy at birth in the USA increased dramatically in the first half of the century, reflecting mainly the reduction in infectious diseases and adverse conditions of maternity

and infancy. The second half of the century was characterized by an increase and then a decrease in CVD as a cause of mortality and an increase in cancer and trauma-related deaths, so that life expectancy increased, but at a lower rate than in the earlier period.

Life expectancy at birth increased dramatically in the USA from 47.3 years in 1900 to 68.2 years in 1950. Since the 1950s, life expectancy increased to 73.7 years in 1980, and to 78.3 years in 2011. In 2011, male life expectancy was 76.2 years and female life expectancy 81.0 years (Hoypert and Xu, National Vital Statistics Report, 2012). From 1900 to 1999, the average lifespan of people in the USA lengthened by more than 30 years; 25 years of this gain can be attributed to advances in public health. Life expectancy (76.8 years in 2000) at birth among US residents increased by 62 percent during the twentieth century with great improvements in population health status at all stages of life, and this process of decline in death rates is continuing in the twenty-first century (Ten Great Public Health Achievements – United States, 2001–2010, MMWR, 2011).

Country ranking by estimated life expectancy in 2012 is shown in Table 3.2, based on vital statistics and the CIA *World Factbook.* The USA ranks low among in life expectancy, in 2012 coming in 51st, ranking below many countries with lower per capita gross domestic product (GDP)

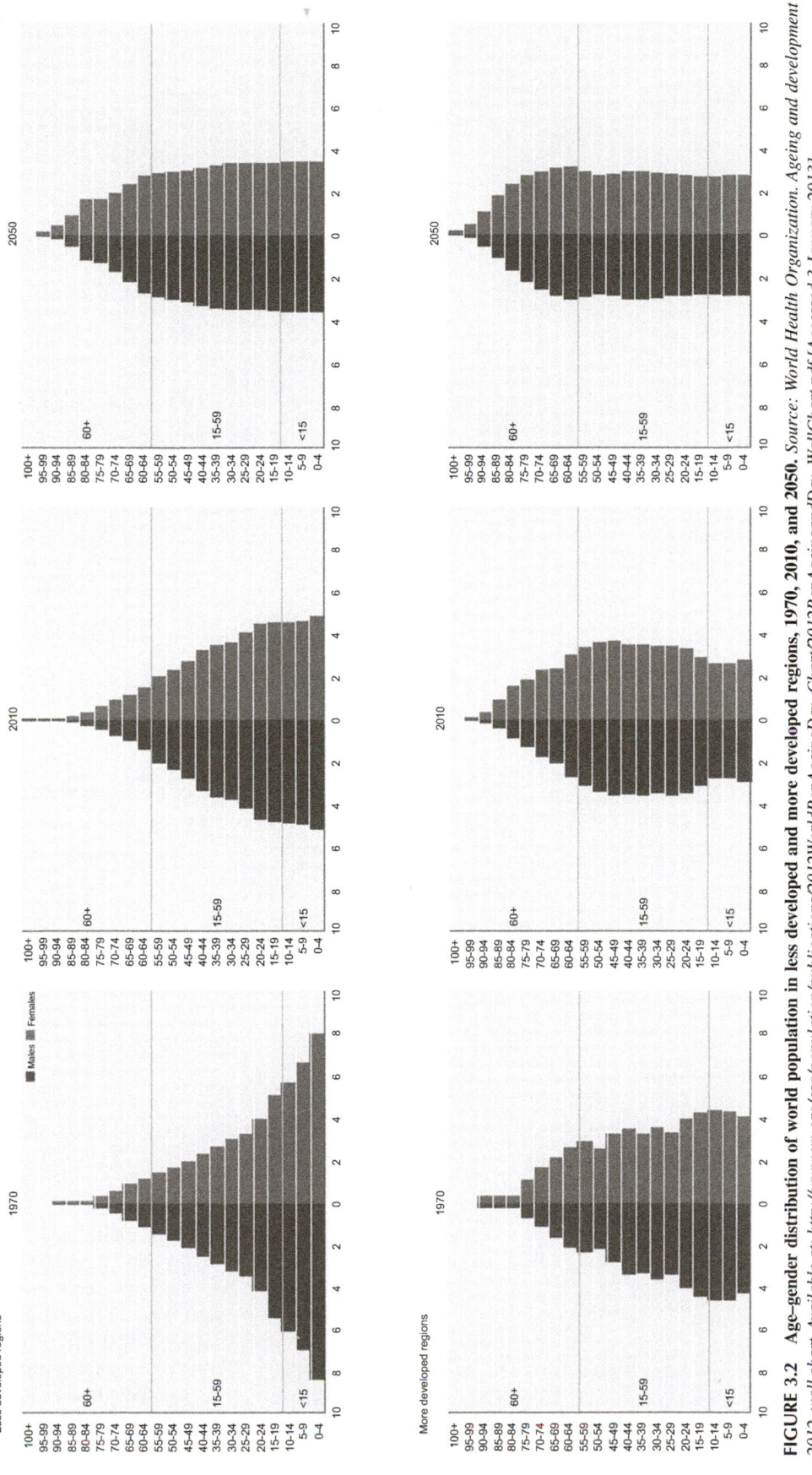

FIGURE 3.2 Age–gender distribution of world population in less developed and more developed regions, 1970, 2010, and 2050. *Source: World Health Organization. Ageing and development 2012, wall chart. Available at: http://www.un.org/esa/population/publications/2012WorldPopAgeingDev_Chart/2012PopAgeingandDev_WallChart.pdf [Accessed 3 January 2013].*

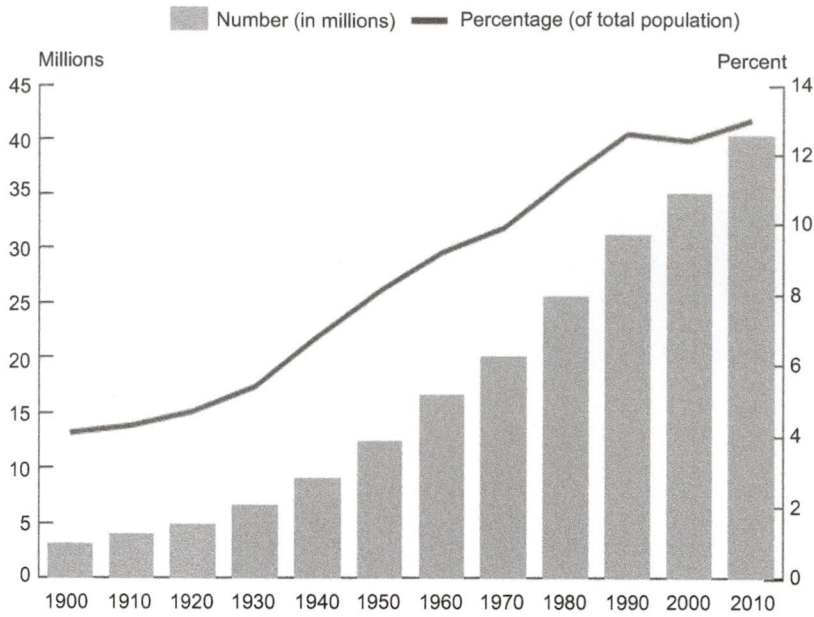

FIGURE 3.3 **Population over age 65, USA, 1900–2010.** *Source: US Census Bureau. The older population: 2010 Census Briefs. Decennial census of population, 1900–2000; 2010 census Summary File 1. Available at: http://www.census.gov/prod/cen2010/briefs/c2010br-09.pdf [Accessed 3 January 2013].*

TABLE 3.1 Life Expectancy at Birth in Years for Selected Organisation for Economic Co-operation and Development Countries and Russia, 1970–2009

	1970	1980	1990	2000	2009
Canada	72.8	75.3	77.6	79.0	80.7[a]
Denmark	73.3	74.3	74.9	76.8	79.0
Finland	70.8	73.6	75.0	77.7	80.0
France	72.2	74.3	76.8	79.0	81.0
Germany	70.5	72.9	75.3	78.2	80.3
Ireland	71.2	72.8	74.9	76.6	80.0
Israel	71.8	73.9	77.5	78.8	81.6
Japan	72.0	76.1	78.9	81.2	83.0
Korea	62.1	65.9	71.4	78.0	80.3
Netherlands	73.7	75.8	77.0	78.0	80.6
New Zealand	71.5	73.2	75.5	78.3	80.8
Russian Federation	68.3	67.3	69.0	65.7	68.7
Sweden	74.7	75.8	77.6	79.7	81.4
UK	71.8	73.2	75.7	77.8	80.4
USA	70.9	73.7	75.3	76.8	78.2
OECD average	70.5	72.6	74.9	77.1	79.5

Note: [a] 2007 data.
Source: Organisation for Economic Co-operation and Development. OECD factbook 2011–2012: economic, environmental, and social statistics. OECD Publishing; 2012. Available at: http://www.oecd-ilibrary.org/economics/oecd-factbook-2011-2012_factbook-2011-en [Accessed 3 January 2013].

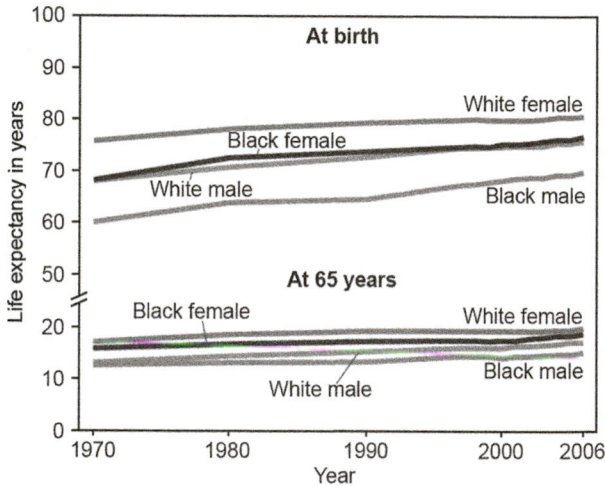

FIGURE 3.4 **Life expectancy at birth and age 65, by gender and ethnicity, USA, 1970–2006.** *Source: US Department of Health and Human Services. Health United States, 2009 with chartbook on health of Americans (Figure 16).*

and other markers of prosperity and civil society. This is in contrast with the high ranking in Human Development Index (HDI) measures and the high level of expenditure on health. The reasons for this are debated: most attribute it to a lack of universal health coverage, but the regional variations in mortality rates seen in the USA suggest that the key issues include differences in diet and life habits. Life expectancies are as much as 5 years longer in other high-income countries such as Singapore, Japan, Israel, Scandinavia,

TABLE 3.2 Country Ranking by Estimated Life Expectancy, selected countries 2012

Rank	Country	Years	Rank	Country	Years
3	Japan[a]	83.9	74	Uruguay	76.4
4	Singapore	83.8	78	Poland	76.3
9	Australia	81.9	89	Macedonia	75.4
10	Italy	81.8	90	West Bank	75.2
12	Canada[a]	81.5	93	Hungary	75.0
14	France	81.5	96	China[a]	74.8
15	Spain	81.3	98	Colombia[a]	74.8
16	Sweden[a]	81.2	99	Algeria	74.7
17	Switzerland	81.2	108	Saudi Arabia	74.4
19	Israel[a]	81.1	109	Romania	74.2
21	Netherlands[a]	81.0	110	Gaza Strip	74.1
25	New Zealand	80.7	111	Venezuela	74.0
26	Ireland	80.3	122	Egypt	72.9
27	Norway[a]	80.3	124	Brazil	72.8
28	Germany[a]	80.2	125	Turkey	72.8
29	Jordan	80.2	129	Vietnam	72.4
30	UK[a]	80.2	133	Philippines	71.9
31	Greece	80.1	147	Iran	70.4
36	European Union[a]	79.8	149	Kazakhstan	69.6
38	Belgium	79.7	155	Mongolia	68.6
40	Finland[a]	79.4	162	India	67.1
41	Korea, South	79.3	164	Russia[a]	66.5
48	Denmark[a]	78.8	166	Pakistan	66.4
49	Portugal	78.7	177	Kenya	63.1
51	USA[a]	78.5	192	Ghana	58.6
54	Chile	78.1	196	Ethiopia	56.6
60	Cuba	77.9	205	Uganda	53.5
62	Albania	77.6	212	Nigeria[a]	52.0
65	Czech Republic	77.4	218	Afghanistan	49.7
69	Argentina	77.1	220	South Africa	49.4

Note: [a]Discussed in Chapter 13.
Source: CIA. The world factbook. Country comparison, life expectancy at birth. Available at: https://www.cia.gov/library/publications/the-world-factbook/rankorder/2102rank.html [Accessed 7 January 2012].

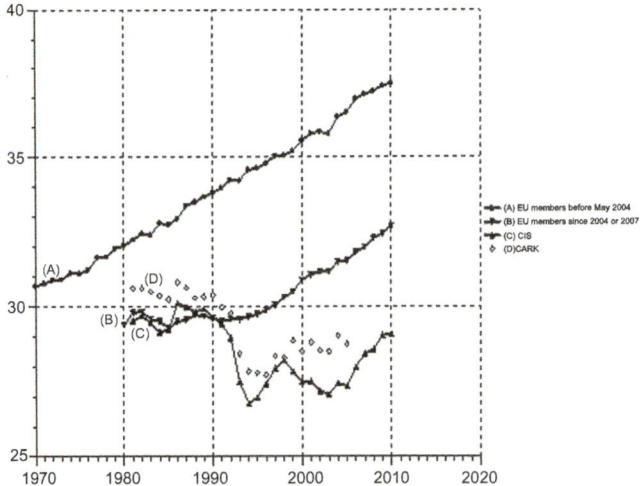

FIGURE 3.5   Life expectancy at age 45 years, European Region, 1970–2010. Note: CARK=Central Asian Republics; Old EU=members of the European Union before 2004; New EU=members of the European Union after 2004; CIS=Commonwealth of Independent States (Russia, Ukraine, Byelorussia). *Source: Health for All Database, WHO European Region, August 2012.*

in a population, such as people with breast cancer. This is important in clinical epidemiology where studies of effectiveness of specific interventions are assessed. Life expectancy is quite different for males and females; thus, gender is an important factor in the assessment of disease prevalence and also in the effectiveness of interventions.

Demography is becoming a major political and social issue in countries where demographic transition is resulting in major shifts in population make-up, and less severely in Western European countries. Russia is experiencing a major reduction in population, with low birth rates and low life expectancy. In the late 1970s, China implemented a "one child per family" policy, and a preference for males means that the country now has a major gender imbalance, with excess males and a deficit in the female population. Developing countries with high birth rates are experiencing population growth exceeding economic growth capacity. The USA has the benefit of steadily improving life expectancy and high immigration rates to offset low birth rates. Japan and many European countries with very high life expectancy and low birth rates face declining and aging populations. These population transitions have important political and economic implications in every country and in regions within countries (see Chapter 13).

International migration has important demographic, economic, social, cultural, political, and health implications for the migrants original and adopted countries. The World Bank estimates that 215 million people, or 3 percent of the world's population, are living in countries other than their home countries. Internal migration has even more importance for the development of many countries, with huge transfers of the rural population to urban settings. Developing countries, where more than 80 per cent of the world's

Canada, and the UK than for the US population. Figure 3.5 shows life expectancy at the age of 45 since 1970: it has risen steadily with variation between western countries, including Western European countries, with the Central and Eastern European countries also rising, but the countries of the former Soviet Union lagging well behind.

Life expectancy is also used in chronic disease epidemiology to summarize patterns of mortality and survival

population live, are a significant source of international migration to industrialized countries as people search for better opportunities in more developed economies and for political freedom in stable civil societies. In aging western societies, migration provides young workers to sustain jobs that local educated young people avoid. High birth rates in poor countries and stagnant economies offer little opportunity to young adults in many developing countries.

Migration is a modest but complex and important factor in demography. Vastly differing birth rates, together with increased life expectancy in most regions, are important factors in regional differences in population growth and aging which affect the supply of labor. As the twentieth century drew to a close, the rate of global population growth began to fall, due primarily to continuing declines in population replacement and declining fertility rate, but partially offset by longevity and an aging population.

Population projection is fraught with many uncertainties such as fertility rates, death rates, life expectancy, and economic, cultural, and political factors. Demographic research explores the potential for further extensions of the lifespan in mathematical, evolutionary, and empirical contexts, as well as economic transfers, both public and private, between age groups and social inequalities in terms of poverty, prosperity, economic growth, and lifetime choices. Migration changes traditional demographics and ethnicity in many countries along with differing religious, cultural, political, and fertility patterns. In a spatially and socially mobile world, civil and human rights issues arise with security and threats to public order are frequent. Passions of nationalism can emerge with incitement and actual events of ethnic cleansing, even genocide, and birth policies promoting gender selection or ethnic preference.

World population growth is uneven, as high- and medium-income countries have reduced their birth rates to near or below population replacement levels, while low-income countries continue with high birth rates, so that world population growth will continue to levels that will challenge the provision of basics such as water, food, and economic development. The United Nations (UN) world population projection for 1950–2050 is shown in Figure 3.6.

EPIDEMIOLOGY

Health care providers are generally oriented towards individual patient assessment and care. However, every health worker, especially the specialized clinician, must have a basic understanding that disease is not an event isolated to an individual, but affects population groups and communities alike, and vice versa. Many epidemics are first identified

---

BOX 3.3 Goals, Methods, Ethics and Challenges of Epidemiology

Goals

- To eliminate, contain or reduce health problems and related consequences.
- To prevent the occurrence or recurrence of problems.

Methods

- Describe the distribution and size of disease problems in human populations.
- Identify etiological (i.e., the cause of disease) processes and factors involved in the pathogenesis of disease.
- Provide data essential to the planning, implementation, and assessment of services for the prevention, control, and treatment of disease and to establish priorities among these services.

Ethical Principles

The Helsinki Declaration is primarily concerned with experimental designs in clinical research, and does not cover many of the observational designs used so often in public health enquiry. There are four general ethical principles for research: autonomy (respect for individual rights), beneficence (do good), non-maleficence (do no harm), and justice.

Current Challenges of Epidemiology

Addressing health inequities and promoting health equity in all fields of disease and health, including: injury epidemiology, occupational health, infectious diseases, chronic diseases, maternal and child health, surveillance and field epidemiology, mental health, violence (from self-directed, e.g., suicide, to interpersonal to structural), psychoactive substance use (including tobacco), and measures of subjective health. Attention will be given to epidemiology's theoretical frameworks and emphasizing knowledge translation, from epidemiology to health systems, and policy.

Sources: *International Epidemiologic Association. http://www.dundee. ac.uk/iea/GEP07.htm [Accessed 8 July 2007].*
Monsour BB, Johnston JM, Hennessy TW, Schmidt MI, Krieger N. Visions for the 20th International Epidemiological Association's World Congress of Epidemiology (WCE 2014). Public Health 2012;126:274–6. http:// dx.doi.org/10.1016/j.puhe.2011.12.015.

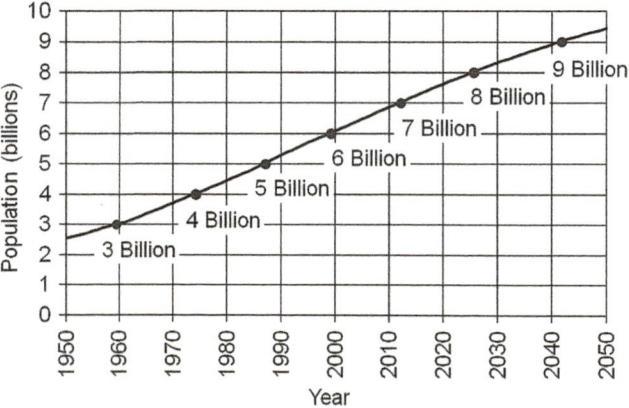

FIGURE 3.6 World population, 1950–2050. *Source: US Census Bureau. International programs. International data base, June 2011 update. Available at: http://www.census.gov/population/international/ data/idb/worldpopgraph.php [Accessed 5 January 2012].*

by "index cases" being reported to public health authorities who begin to put together a picture of moving events. The human immunodeficiency virus (HIV) epidemic was first reported with a small number of cases in New York City [reported in CDC's *Morbidity and Mortality Weekly Report* (MMWR)], soon followed by a new group of cases in San Francisco and the rapid spread to become a pandemic globally costing millions of lives. The epidemic of severe acute respiratory syndrome (SARS) crossed borders and continents, spreading from China to Canada in a matter of days, and closing down the city of Toronto for many days.

The importance of monitoring disaster events is exemplified by the British Petroleum oil spill in the Gulf of Mexico in 2011. A recent epidemic of fungal meningitis was suspected by a hospital clinician and followed by public health authorities in Tennessee, who found the source to be contaminated medication produced in Massachusetts, leading to the US Food and Drug Administration (FDA) establishing new regulations on pharmaceutical company safety standards (see Chapter 15). The clinician must be aware of the potential for epidemic and pandemic disease as well as the risk factors for a non-communicable disease (NCD), such as CVD, affecting the individual patient, in order to determine management over the long term.

Epidemiology is the study of health events in a population. The goals and methods of epidemiology incorporate ethical principles consistent with the Helsinki Declaration (Box 3.3). Its purpose is to help understand disease processes and outcomes, determine factors in causation, assess the effectiveness of interventions, and provide direction for medical or public health interventions. The distribution and determinants of health-related states, conditions, or events in defined populations are important in the identification of potential interventions and priorities to control health problems and reduce "avoidable or "amenable" deaths (Box 3.4). Methods include surveillance, observation, and hypothesis generation and testing in analytical research and experiments. Health events occur in population groups and the study of epidemiology requires definition of the events and the population studied. Specified populations are those with common, identifiable characteristics that can be quantified, such as gender, age, ethnicity, and region of residence. Potential determinants include physical, biological, social, cultural, economic, environmental, and psychological and behavioral factors.

BOX 3.4 Amenable Mortality as a Public Health Outcome Indicator

One challenge of epidemiology is to measure the contribution of health care to population health outcomes with precision, given the often multifactorial nature of many outcomes. The concept of "amenable" or avoidable mortality is one approach shown to provide a useful approximation. Using this concept, previous work illustrated how health care impacted positively on population health in many industrialized countries during the 1980s and 1990s. However, the pace of change has differed among countries and over time.

Recent work demonstrated, for example, that progress in the USA on this indicator was lagging behind other industrialized countries. The USA had a higher rate than European countries, that is, a higher rate of deaths from conditions such as diabetes or acute infection that could potentially have been treated with timely and effective care. In 2007, for example, US rates of such deaths were almost twice those in France, which had the lowest rates of the countries studied.

The USA spends an average of nearly $8000 a year per person on health care – roughly double the average in Western European countries. Yet Americans die sooner and also experience poorer health throughout life than people in many other countries. While the USA enjoyed steady increases in the length of life during the twentieth century in particular, in the latter part it fell increasingly behind other high-income countries.

This can be illustrated by recent trend in amenable mortality, comparing the USA with France, Germany, and the UK. Thus, between 1999 and 2007, the rate of potentially preventable deaths among men under the age of 75 fell by 18.5 percent in the USA compared to a 37 percent decline in the UK,

a 28 percent decline in France, and a 24 percent decline in Germany. For women, the rates fell by 17.5 percent in the USA, compared to 32 percent in the UK, and 23 percent in both France and Germany.

The lag in improvement was most notable among American men and women under the age of 65. These individuals are more likely to be uninsured than are Americans over 65, who are eligible for Medicare. The observed differences are not inevitable, however, and there are regional variations, as in all countries. For example, the state of Minnesota achieved outcomes on a par with those found in many European countries and an amenable mortality rate less than half that of Mississippi or the District of Columbia.

Evidence indicates that these outcomes were achieved with patients receiving care that meets best practice guidelines and preventive care to reduce unneeded hospitalization. These findings underscore the importance of improving access to timely and effective health care in the USA. Amenable or avoidable mortality is an important tool for epidemiological monitoring and comparisons of population health between countries and between regions within countries.

Sources: *Nolte E, RAND Europe; McKee M, London School of Hygiene and Tropical Medicine. Personal communication; January 2013.*
Nolte E, McKee M. Does healthcare save lives? Avoidable mortality revisited. London: Nuffield Trust; 2004.
Nolte E, McKee M. In amenable mortality – deaths avoidable through health care – progress in the US lags that of three European countries. Health Aff 2012;31:2114–22.
Woolf SH, Aron L, editors. Shorter lives, poorer health. Washington, DC: National Academies Press; 2013.

BOX 3.5 Health Determinants and Measures of the Individual and Community

Factors

- *Biology* – age, gender, genetics
- *Geography* – urban, rural, climate, nomadic
- *Economics* – GDP per capita, family income, unemployment, living standards, poverty levels
- *Social security* – pensions for the elderly, disability and chronic illness pensions
- *Cultural, religious, and economic factors*
- *Education* – literacy, gender differences, higher education
- *Lifestyle, personal habits* – diet, smoking, exercise, drug use, risky sexual habits
- *Occupation* – injuries, toxic exposures, mental and physical stress
- *Environment* – exposure to toxins, air pollutants, carcinogens, infectious agents
- *Societal factors and physical urban and rural environment*
- *Nutrition* – food security and safety, diet, cost, quality with fortification with essential nutrients
- *Health services and insurance* – accessibility, quality of care, comprehensiveness, organization, financing
- *Public health infrastructure and policies*
- *Family and social support* – stability and family function.

Measures

- *Demography* – births, deaths, marriages, divorces, migration
- *Infrastructure* – safe water, food, air, solid waste disposal, transport measures
- *Health insurance* – coverage, comprehensiveness
- *Resources* – hospital beds and medical personnel per 1000 population, and their distribution
- *Process* – utilization, immunization, hospitalization rates
- *Outcomes*:
  - mortality: by age, gender, cause
  - morbidity: by cause, time, place, common exposure, nutritional micronutrient deficiencies
  - physiological indicators: growth and development, body mass index
  - functions of daily living and disability
- *Quality measures* – accreditation, peer review, quality improvement
- *Knowledge, attitudes, beliefs, practices*
- *Satisfaction and self-assessment*
- *Costs and benefits.*

Variables are "any attribute, phenomenon, or event that can have different values" (Last, 2001). They include all the physical, biological, social, cultural, environmental, economic, psychological, and behavioral factors that influence health. Health-related states and events include diseases, causes of death, behavior such as use of tobacco, compliance with preventive regimens, and provision and use of health services.

Distribution includes analysis by time (e.g., month, season, time of day), place, identifying individuals or groups of people affected by common events such as foodborne disease at a festival, on a cruise ship, or in a workplace, nursing home, hospital, or school. Health status monitoring covers a large range of health-related states and events including diseases, handicapping conditions, causes of death, fertility and fecundity, birth defects, growth and development in childhood, health-related behavior (e.g., use of tobacco), compliance with public health intervention (e.g., immunizations), and access to and use of health services (Box 3.5).

Avoidable mortality includes deaths for diseases that are totally or largely preventable by public health and clinical care, such as measles, or lung cancer from smoking. Amenable mortality is death from a disease that can be managed with prolongation of life, such as diabetes or hypertension. Box 3.4 addresses the changes in amenable mortality comparing countries to indicate the effectiveness of their health and social systems.

Epidemiological studies may include descriptive studies of routinely or ad hoc reported and collected data on mortality, morbidity, and related factors. They focus on the distribution of disease or risk factor by time, place, and person characteristics, and form a crucial basis for public health activities and evaluation. Analytical epidemiological studies are based on hypothesis testing and include observational studies such as cross-sectional, case–control, and cohort studies, as well as intervention studies, including clinical and program trials. They focus on exposures and outcomes, and attempt to determine their associations. An interpretation of this wide range of data sources is shown in Figure 3.7. CDC's vision for public health surveillance for the twenty-first century represents the broad scope of information and professions involved in population health monitoring (CDC, 2012).

Classically, the clinician diagnoses and treats a patient who presents for medical care, including remedial and preventive care. Community public health workers focus on health protection and preventive care for the population. Epidemiologists study the health of a defined population in partnership with the many other disciplines represented in Figure 3.7, including geneticists, microbiologists, information specialists, statisticians, economists, social scientists, and others. This provides a strong base for assessment of the need for preventive action. Epidemiologists also evaluate the effects of preventive or treatment measures and share the need to understand risk factors and the natural process of disease. Epidemiology studies a particular disease in a population, taking into account factors such as age, gender, ethnicity, exposure to known or suspected risk factors, and socioeconomic patterns, as well as the effect of various interventions. This study is undertaken to understand the natural history of disease and related diagnostic criteria, to identify risk groups and relevant target groups for intervention and, accordingly, appropriate methods of prevention or

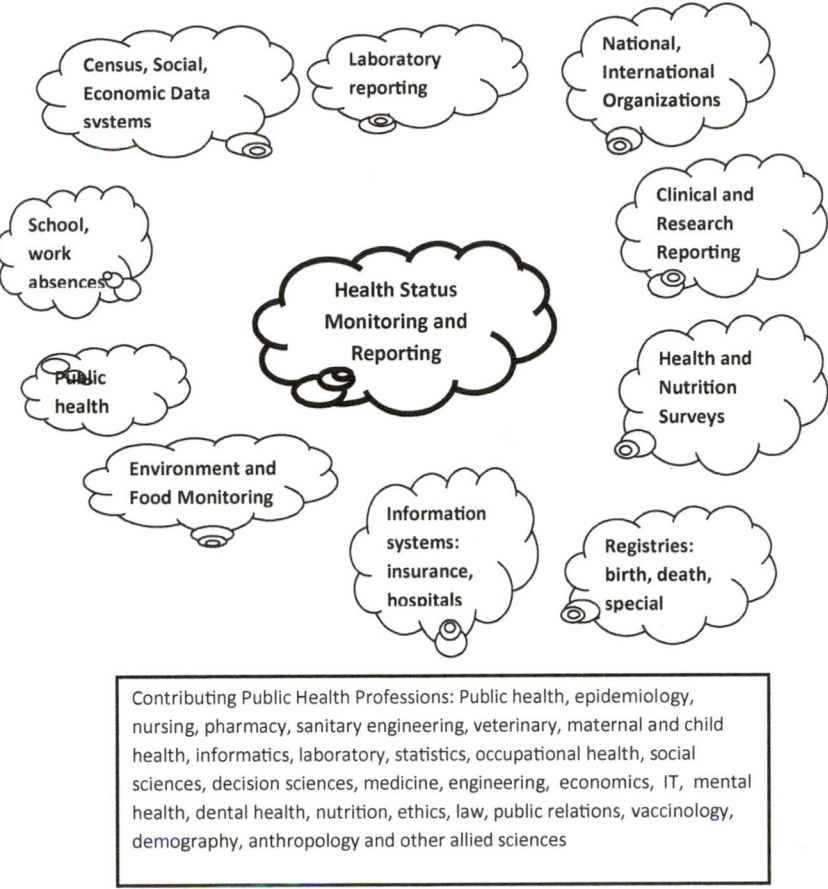

FIGURE 3.7   **Centers for Disease Control and Prevention model of information systems and professions in public health monitoring.** *Adapted from Lee LM, Thacker SB. CDC's vision for public health surveillance in the 21st century. The cornerstone of public health practice: public health surveillance, 1961–2011. MMWR Morb Mort Wkly Rep Suppl 2011;60(04):15–21.*

management, outcomes to be expected, and the costs and benefits of the different methods of control in addition to aspects relevant to ethical assessment.

Clinicians and epidemiologists depend on each other, need to collaborate with professionals from other fields, such as health economics and management, and require documented experience of interventions to improve care and efficient use of resources. Reliance is also placed on interaction with the various disciplines within public health, health policy, health systems management, and clinical medicine. Difficult choices in public policy regarding allocation of resources must be made with many factors in mind, including the epidemiology of the condition, cost-effectiveness of intervention, and ethical questions.

In the nineteenth to twentieth centuries, a profound transition occurred in the industrialized countries as the diseases of "pestilence and famine" waned, and chronic diseases became the leading causes of death. Many of these were associated with human-caused environmental problems and personal lifestyle. This epidemiological transition took place, initially primarily because of the cumulative effects of successful public health activities such as environmental sanitation and food safety, and later through communicable disease control with the success of vaccines and antibiotics in reducing the major diseases of childhood, and improvements in living conditions. A "second era" in public health occurred during the latter half of the twentieth century with the rise and fall of chronic disease in the industrialized countries, but this era is still a great challenge in countries in transition (e.g., former Soviet countries, especially in the Russian Federation, Ukraine, and Central Asian Republics), and increasingly in developing countries as well. Now a "third era" of health has arrived, with people living well into their seventies and eighties, often not only free from serious morbidity but leading vibrant and active lives, requiring a reorientation of personal perspectives as well as adjustments in the community and the health system (Breslow, 2006). In the second decade of the twenty-first century people in their nineties and even some centenarians are living healthful active lives.

During the 1950s and 1960s, rising standards of living in the industrialized world were associated with increases in NCDs, including CVDs, malignancies associated with smoking, other "lifestyle" diseases, and trauma associated with industrialization, violence, self-injury, and motor vehicle accidents. This transition is playing an important role in the disease patterns of developing countries as they urbanize and the middle class grows. The Global Burden of Diseases, Injuries, and Risk Factors Study 2010 (GBD 2010), sponsored by the Bill and Melinda Gates Foundation, launched in spring 2007, is the most comprehensive effort since the GBD began in 1990. The new GBD will produce complete and comparable estimates of the burden of diseases, injuries, and risk factors for 21 regions of the globe for the years 1990, 2005, and 2010. The study is the collective work of a large community of experts and leaders in epidemiology and other areas of public health research from around the world. The main methods and findings from the study are published in *The Lancet* in a series of seven papers, commentaries, and accompanying material totaling over 2300 pages in length. This database will be important to develop strategies for global health to follow on from the Millennium Development Goals (MDGs) of 2001–2015. Table 3.3 shows the leading causes of death globally in 2008, with ischemic heart disease, stroke, and other CVDs far ahead of other causes. NCDs are discussed in depth in Chapter 5.

Since the 1960s, a new and equally profound epidemiological transition has occurred with the decline of heart disease, stroke, and trauma as causes of death, in the industrialized world but also increasingly in the developing countries. This decline has contributed to increasing longevity. Greater health consciousness and self-care, improved social security for the elderly and disabled and vulnerable adults, and advances in medical care have contributed to this phenomenon.

In the early 1980s, a dramatic new epidemiological challenge appeared with the advent of a pandemic of HIV infection and a return of diseases thought to have been under control. Potentially dangerous infectious diseases can be transmitted far from their original habitat with the rapid transportation and movement of populations, including migrants, tourists, and other travelers around the globe. Other infectious diseases are becoming resistant to available treatments, and multidrug-resistant (MDR) infectious diseases, especially tuberculosis (TB), are emerging.

TABLE 3.3 Leading Causes of Death Worldwide, 2008

Country Groupings	World Total		Low-Income Countries		Middle-Income Countries		High-Income Countries	
Diseases	Deaths (millions)	Deaths (% total)	Deaths (millions)	Deaths (% total)	Deaths (millions)	Deaths (% total)	Deaths (millions)	Deaths (% total)
Ischemic heart disease	7.25	12.8	0.57	6.1	5.27	13.7	1.42	15.6
Stroke and other cardiovascular disease	6.15	10.8	0.45	4.9	4.91	12.8	0.79	8.7
Lower respiratory infections	3.46	6.1	1.05	11.3	2.07	5.4	0.35	3.8
Chronic obstructive pulmonary disease	3.28	5.8	NA	NA	2.79	7.2	0.32	3.5
Diarrheal diseases	2.46	4.3	0.76	8.2	1.68	4.4	NA	NA
HIV/AIDS	1.78	3.1	0.72	7.8	1.03	2.7	NA	NA
Trachea, bronchus, lung cancers	1.39	2.4	NA	NA	NA	NA		
Breast cancer	NA	NA	NA	NA	NA	NA	0.17	1.9
Tuberculosis	1.34	2.4	0.40	4.3	0.93	2.4		
Diabetes mellitus	1.26	2.2	NA	NA	0.87	2.3	0.24	2.6
Road traffic accidents	1.21	2.1	NA	NA	0.94	2.4	NA	NA
Prematurity, low birth weight, birth asphyxia and trauma, neonatal infections	NA	NA	0.81	8.7	NA	NA	NA	NA
Alzheimer's and dementias	NA	NA	NA	NA	NA	NA	0.37	4.1
Hypertensive heart disease	NA	NA	NA	NA	0.83	2.2	0.21	2.3

Note: HIV=human immunodeficiency virus; AIDS=acquired immunodeficiency syndrome; NA=not indicated.
Source: World Health Organization. The top 10 causes of death. Fact sheet no. 310 [updated June 2011]. Geneva: WHO. Available at: http://www.who.int/mediacentre/factsheets/fs310/en/index.html [Accessed 8 January 2013].

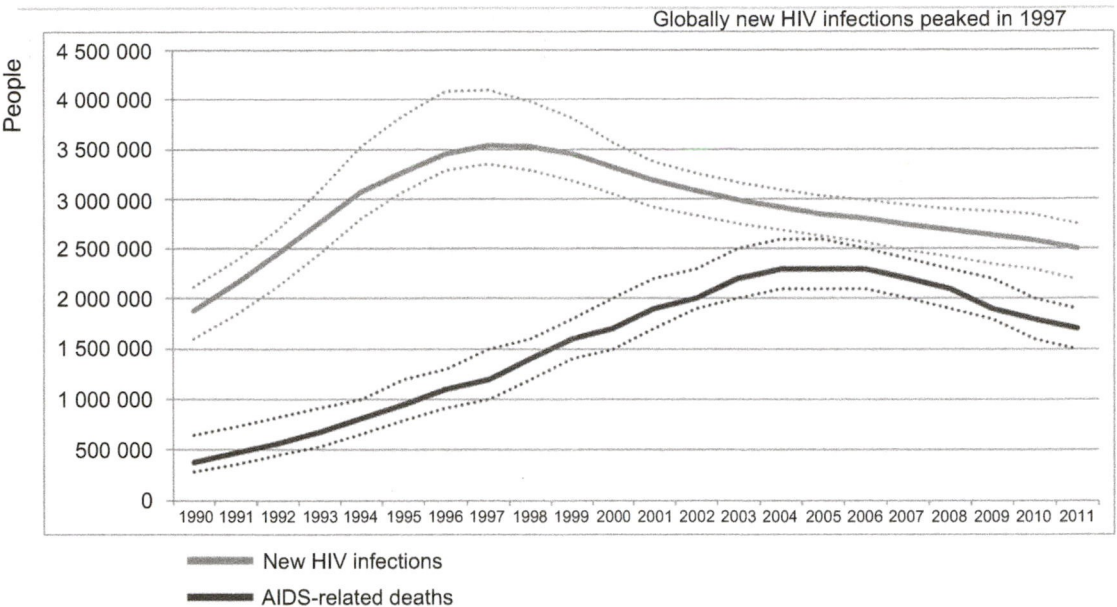

FIGURE 3.8   Global new HIV infections and AIDS-related deaths. *Source: World Health Organization. Core slides HIV/AIDS, 2012. Available at: http://www.who.int/hiv/data/en/ [Accessed 4 January 2013].*

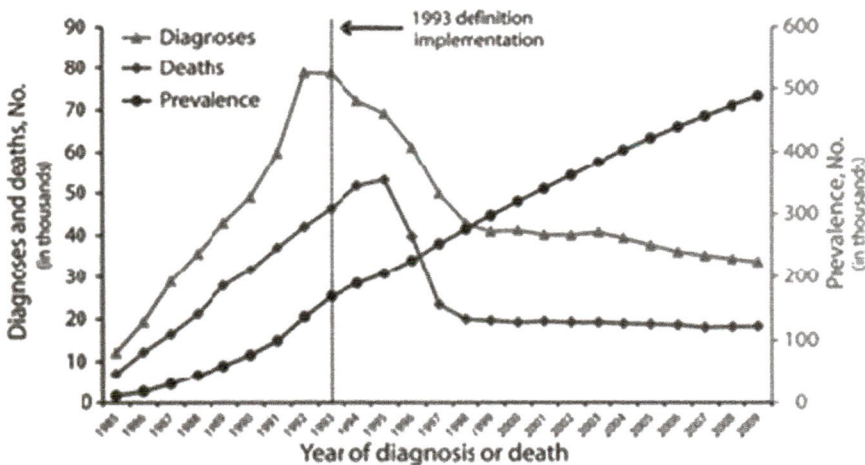

FIGURE 3.9   AIDS diagnoses, deaths and people living with AIDS, USA, 1985–2009. *Source: Centers for Disease Control and Prevention. HIV surveillance report. Diagnoses of HIV infection and AIDS in the USA and dependent areas, 2008, Vol. 20. Available at: http://www.cdc.gov/hiv/topics/surveillance/resources/slides/trends/index.htm [Accessed 3 January 2012].*

The HIV/AIDS epidemic (see Chapter 4) created a new and deadly situation with a worldwide pandemic. The epidemic rose to vast proportions in the 1990s but appears to have peaked and fallen since 2000, although it is still spreading in Central and Eastern Europe. Globally, 34 million people are living with HIV, including 3.3 million children, and 1.7 million people died of AIDS-related illnesses in 2011 (Figure 3.8). There are an estimated 2.5 million new infections globally per year, but this is a reduction of over 20 percent since 2001. Antiretroviral therapy (ART) is holding down the epidemic through treatment and preventing further spread. Progress in reducing maternal–infant HIV transmission by ART drugs, now used widely in Africa by international donor

agencies, has been a very impressive achievement. This experience gives cause for cautious optimism, but economic recession may slow this process.

The transmission of HIV is still high among some groups in the USA (see Chapter 4), where the number of people living with HIV is estimated at 1.1 million. Most transmission is via male-to-male sexual contact. An estimated 20 percent of infected people are undiagnosed and therefore not in treatment and still transmitting the diseases (Figure 3.9). In countries of Eastern Europe and the former Soviet Union, the disease is still primarily among intravenous drug users and the official agencies are not actively promoting ART or other control measures, so the epidemic has yet to run its course.

Partial control of the spread of HIV was achieved in the industrialized countries through scientific achievements and the application of public health measures. AIDS has had enormous effects on the need for trained infectious disease clinicians, epidemiologists, and virologists to provide the care and to carry out the monitoring and research that will be needed to control this devastating pandemic. Hope for a vaccine is still unfulfilled but trial and error have produced a promising start to control with education, condom use, circumcision, and the revolutionary ART that has saved so many lives. This work has involved partnerships and cooperative activities among governments, international organizations, bilateral aid agencies, non-governmental organizations (NGOs), and private agencies to establish screening, education, risk-reduction programs, prophylaxis, and treatment with ART to improve clinical care and prevent transmission of the virus.

"Newly emerging" diseases are a notable threat to the gains made in the health status of the industrialized world, and an even greater threat to the struggling health systems of developing countries (see Chapters 4 and 16). However, the chief threat to the public's health remains the massive deprivation in developing countries and poverty still present in the industrialized countries. Newly emerging diseases present a growing challenge to public health, with new disease entities (AIDS, Ebola, SARS, avian influenza, H1N1) along with renewed threats from diseases present for centuries, and multi-drug resistant (MDR) cases of TB, methicillin-resistant *Staphylococcus aureus* (MRSA), and others from abuse of antibiotics and molecular shifts in the organisms. Even diseases thought to have been brought under control by vaccines, such as pertussis and measles, have reappeared in imported and localized epidemic forms.

In the 1990s, there were new breakthroughs in the epidemiology of infections causing highly prevalent chronic diseases. A new infectious agent, the prion, was identified by Stanley Pruziner (Nobel Prize 1997) as transmitting Creutzfeld–Jakob disease, a serious degenerative and fatal neurological disorder. This experience resulted in a closer relationship between veterinary public health and those with responsibility for a variety of agricultural products used for animals. This aspect of veterinary public health is coming into greater focus. A new bacterium first identified in the 1980s, *Helicobacter pylori*, was shown to be the cause of peptic ulcers and cancer of the stomach (B. J. Marshall and J. R. Warren, Nobel Prize 2005). The previously known relationship of hepatitis B to cancer of the liver and chronic cirrhosis took on new importance as an effective and inexpensive vaccine became available. Furthermore, nutritional deficiencies were found to be cofactors in a variety of diseases.

In the first years of the twenty-first century, human papillomavirus (HPV), a sexually transmitted virus, was identified as the cause of cancer of the cervix. An effective vaccine was approved in the USA by the Food and Drug Administration (FDA) in 2006 and is already being used in the industrialized countries, but is too costly for developing countries where it is most needed. It provides the means to control and possibly eliminate one of the leading causes of cancer in women worldwide, but will need to be used along with Papanicolaou (Pap) smear screening for many years to come.

Such breakthroughs in medical science and public health practice demonstrate the vital importance of combining epidemiological and clinical investigations to confirm these relationships and to seek out preventive mechanisms.

SOCIAL EPIDEMIOLOGY

Epidemiology has evolved from its origins as a factor in sanitary statistics in the first half of the nineteenth century, as exemplified in the political arithmetic and vital statistics of Farr and the social statistics of Chadwick and Shattuck. It helped to foster the sanitary movement and public health benefits through the development of drains, sewage systems, and community sanitation. In the late nineteenth century through the first half of the twentieth century, epidemiology was associated with the germ theory of single agents relating to one specific disease, and public health activities focused on interruption of transmission or primary prevention through vaccinations. In the latter half of the twentieth century, chronic disease epidemiology showed associations among multiple risk factors and outcomes, without full understanding of the intervening factors or pathogenesis. Some landmarks of epidemiology are shown in Table 3.4. They are further discussed in Chapters 1, 4, 5, 8, and 13.

Chronic disease epidemiology led to health promotion as a key approach in risk-control public health measures, affecting lifestyle (diet, exercise, smoking), products (food, guns, cars), and environment (pollution, passive smoking). A new era of epidemiology is emerging in the twenty-first century in which organization, information, and application of biomedical technology are vital in population health. A wider, multidisciplinary approach is taken, in which statisticians, economists, social scientists, health systems managers, and epidemiologists bring different skills to a more complex paradigm of public health.

Social inequalities in morbidity and mortality have been a major field of interest in epidemiological studies for many years. A study of late-stage diagnosis of colorectal cancer in New York State showed that women and African Americans were more likely to have late-stage cancer than men and whites. Individuals living in areas of low SES were significantly more likely to be diagnosed at a later stage than those living in higher SES areas. Similar patterns of socioeconomic disparity in mortality have been shown among men in the state of São Paulo, Brazil, with the poor having three times greater rates of mortality than the wealthy minority. In the UK, regional differences in mortality patterns are closely linked to socioeconomic conditions, with poverty

TABLE 3.4 Selected Landmarks in Epidemiology

Vital Statistics and Social Epidemiology		Non-Infectious Disease Epidemiology	
1662	Graunt publishes *Natural and Political Observations Made upon the Bills of Mortality*	1747	Lind demonstrates prevention of scurvy by citrus fruits
1836	Registrar General's Office established by UK Parliament	1775	Pott shows high rate cancer of scrotum in chimney sweeps
1842	Chadwick: *Report on the Sanitary Condition of the Labouring Population of Great Britain*	1914	Goldberger demonstrates nutritional cause of pellagra
1848	Virchow: "medicine is a social science"	1950	Doll and Hill relate cigarette smoking to lung cancer
1858	Simon maps mortality by district in relation to social and environmental conditions	1954	Framingham study reports on heart disease risk factors
1974	LaLonde: *New Perspectives on the Health of Canadians* – lifestyle, genetics, environment, and medical care key health factors	1960s	US Surgeon General's Report on Smoking and Health. Decreasing mortality from cardiovascular diseases, trauma
1982	Black Report: social class differences in mortality in the UK	1980s	Infections as causes of chronic diseases; *Helicobacter pylori* causing peptic ulcers and cancer of stomach
1986	Ottawa Charter on Health Promotion	1980s	Advances in cardiovascular epidemiology and successful preventive and treatment interventions
1995	Beijing Conference on Women, empowerment for health of women and children	1990s	Vaccines for hepatitis B to prevent cancer of liver. Health promotion plays increasing role in public health
2001	UN Millennium Development Goals (MDGs) and Human Development Index (HDI)	2006	Human papillomavirus vaccines to prevent cancer of cervix
Infectious Disease Epidemiology		**Health Policy Epidemiology**	
1796	Jenner uses cowpox to vaccinate against smallpox	1883	Bismarck initiates workers' compensation and national health insurance
1854	Snow identifies and interrupts water transmission of cholera in London	1917	Semashko establishes Soviet state health system
1882	Koch discovers tubercle bacillus and cholera. Koch–Henle postulates on causation of disease	1948	UK establishes National Health Service
1920–2000	Sanitation, vaccines and antibiotics control many infectious diseases	1961	Canada's provincial Medicare plans advance
1980	Eradication of smallpox declared achieved (WHO)	1965	US Medicare and Medicaid amendments to Social Security Act of 1935
1980s	HIV and other newly emerging or resurging infectious diseases	1978	Declaration on Alma-Ata and Health for All 2000
1990s	Vaccines for hepatitis B prevent cancer of liver	1979	US Surgeon General: Health People, health targets
2000s	Antiretroviral therapy for HIV is a dramatic success. Progress in elimination of yaws, poliomyelitis, leprosy, dracunculiasis, measles, mumps, and rubella being achieved	1990s	Health promotion plays central role in HIV management, tobacco control, CVD risk factor reduction
2000s	Terrorism, potential bioterrorism	1990s	Managed care expansion in the USA
2000s	SARS, avian flu, multidrug-resistant organisms. Control of *Helicobacter pylori*, chronic peptic ulcer disease and cancer of stomach	2000s	Health reforms in Central and Eastern Europe, Commonwealth of Independent States, central Asia, and emerging developing countries
2007–2012	H1N1 avian influenza spread and threatened pandemic	2008	Recession in USA and Europe
2010–2013	Measles epidemic in Europe; diphtheria and pertussis return in USA, Canada, and Europe	2010	Affordable Care Act in USA ("Obamacare") to add millions of Americans to health insurance; reforms in health care

Note: See Chapter 1 timeline.
Source: Adapted from Susser M, Susser E. Choosing a future for epidemiology: eras and paradigms. Am J Public Health 1996;86:668–73.

and its associated conditions as key variables despite universal access to the National Health Service (NHS). In contrast, a study in Denmark of regional and social class variation in relative risk of death showed little social variation except for people with no known address. Social inequities in health occur in virtually all societies, even those with "universal" access to health care, including the USA, the UK, Israel, and many others, with differences in physical access to care; differences in lifestyle and risk factors; socioeconomic conditions, and knowledge, attitudes, and practices related to health and health care.

Social epidemiology in some senses reflects the nineteenth-century traditions of Virchow, Chadwick, Shattuck, and Farr (see Chapter 1), and a return to the "miasma theory" of disease, in which health of populations is largely determined by environmental factors of society, and that to understand causation of disease it is essential to understand its historical and social context. This social epidemiology necessarily incorporates qualitative methodologies based on the social sciences in addition to the quantitative epidemiological tools of measuring associations between exposure and disease in individuals or groups. The New Public Health integrates the qualitative and quantitative methods with management sciences based on successful applications of all these modalities to public health issues over the past century and more.

EPIDEMIOLOGY IN BUILDING HEALTH POLICY

Epidemiology, originally seen as the study of epidemics, evolved rapidly in the latter part of the nineteenth century with the growth of bacteriology and the physical sciences. "Epidemiology burgeoned as an increasingly rigorous science based on observation, inference, and experimentation, and around the middle of the twentieth century with development of methods, notably case control and cohort studies to investigate non-communicable diseases such as coronary heart disease and cancer, and randomized control trials to evaluate therapeutic and preventive regimens aimed at control of the conditions" (Last, 2007) (see Chapter 5).

Empirical documentation increased in Europe in the middle of the nineteenth century, concerning health, disease, and mortality, in the form of death certificates and mortality registers. The most influential events in establishing infectious disease epidemiology were the discovery by Edward Jenner in 1797 that vaccination with cowpox could prevent smallpox, Peter Panum's description of the spread of a measles epidemic in the Faroe Islands in 1846, and John Snow's and William Farr's successful analyses of the London cholera epidemic of the 1850s. James Lind's classic case-controlled experiment showing the nutritional cause of scurvy in 1847 opened up the field of nutritional epidemiology; and the work of Edwin Chadwick and William Farr concerning social inequity in mortality and on classification of causes of death initiated social epidemiology. Farr observed that

"Hunger destroys a much higher proportion than is indicated by the registers in this and every other country, but its effects, like the effects if excess, are generally manifested indirectly in the production of disease of various kinds" (Whitehead, 2000). Ramazzini's work in occupational health and Percivall Pott's identification of cancer of the scrotum among chimney sweeps opened up new areas of epidemiological investigation. Ignaz Semmelweis improved maternal health by documenting the causes of high neonatal mortality in a maternity ward, identifying a lack of hand washing by medical doctors as the culprit. These pioneers set the stage for infectious disease, NCD, occupational, women's health, maternity, and social epidemiology (see Chapter 1).

Until a few decades ago, epidemiology was considered the dominating, central discipline of public health. The 1980s and 1990s brought increasing acknowledgement that public health epidemiology – with the probability concept – must interact with qualitative methodology, and public health must integrate a variety of disciplines, including sociology, social psychology, health economics, environmental health, systems analysis, and political science. Selected landmarks in the development of epidemiology from the early twentieth century are highlighted in Table 3.4.

Epidemiology proved itself in enormously successful interventions for public health in the first half of the nineteenth century. The golden period of infectious disease epidemiology in the late nineteenth and first half of the twentieth centuries established the basis for control of communicable disease, a revolution still in process. During the mid-twentieth century, the development of non-infectious epidemiology and social epidemiology provided the basis for health promotion and lifestyle changes contributing to reduced morbidity and mortality from CVD and the potential for control of cancer, trauma, and other non-communicable conditions. In the case of HIV/AIDS, health promotion was the only tool available until the antiretroviral drugs became available, but the hoped-for vaccine is still in the future. There is an important role for controlled trials for preventive modalities and treatment in relation to chronic disease (mammography, hormone therapy, and many others). Highlights of the development of modern epidemiological methods are discussed in Box 3.6.

The fundamentals of epidemiology are as vital for the student of health sciences as are the study of bacteriology, biochemistry, or surgery. It is equally important that health planners, economists, and others concerned with the policy aspects of health be conversant with epidemiology. This is so that they understand the need to adapt health services to changes occurring in the epidemiological and technological aspects in health and in society, as well as the application of data from studies to the changing needs for health care.

David Sackett, one of the founders of evidence-based medicine (EBM), defined it as "the conscientious, explicit, and judicious use of current best evidence in making decisions about the care of individual patients". Medical students and students in medically related programs must be

BOX 3.6 Highlights of the Development of Modern Epidemiological Methods

Early twentieth century theoretical developments of epidemiology included contributions of Ronald Fisher and others from the 1920s onwards creating the foundations for modern statistical science. Multivariate analysis methods enabled epidemiological concurrent analysis of various potential health risk factors. Karl Popper's work on the logic of scientific discovery, published in German in the 1930s, reached wide recognition after publication in English in 1959. It provided a philosophical theory of science and a basis for academic epidemiology.

Since the mid-twentieth century, epidemiological activity has grown gradually and substantially, so that today's epidemiological research production is breathtaking compared to the situation just a few decades ago. Cancer, tuberculosis, birth defect, and heart and other disease registries have been developed, which are the data basis for epidemiological analysis. Population cohorts were established and followed for years. Cancer and cardiovascular epidemiology developed to become major fields. The etiologies of most large disease categories were a productive sphere of epidemiological methods, including mental disease prevalence studies (see Chapters 4–9).

An early breakthrough in modern epidemiology was the 1954 publication of the British Doctors' study by Doll and Hill, showing a strong association between smoking and lung cancer. In the first part of the 1980s, the interaction between developments in immunology and developments in epidemiological methods and in biostatistics resulted in the identification of HIV as the infectious cause of AIDS.

Major theoretical, methodological, practical, and organizational developments have taken place in the last half-century. In 1960, MacMahon, Pugh, and Ipsen published their ground-breaking theory, the "Web of Causation", moving epidemiology from monocausality to multicausality, so that it came work with other sciences in etiological studies. In 1964, Bradford Hill published his 10 criteria for causality. Increasing development of the logical foundation of epidemiological designs, as well as theoretical development, included more precise and consistent validity and bias concepts, such as:

- cohort designs ("prospective"), fixed and open, dynamic
- case–control ("retrospective"), case–referent, case–base designs
- quasi-experimental designs
- experimental designs and randomized controlled population experiments, inspired by the work of Cochrane.

Interaction between the development of theory, principles and methods in epidemiology and biostatistics increased, especially from the 1970s onwards. Discriminant analysis and other types of discrete analysis of central importance for epidemiological thought and documentation were developed, e.g., multiple logistic regression, multilevel logistic regression, Cox regression, Poisson regression, longitudinal analysis, structured equation modeling, Markov chains and processes, and multidimensional methods for forecasting.

Clinical epidemiology developed as a discipline in the last part of the twentieth century, applying population-epidemiological methods to patient populations. Modern technology opens new perspectives in epidemiology, including genetic epidemiology, with the use of disease, occupation, and population registries linking individual data on health and health services consumption with social conditions (e.g., education, occupation, employment, and family structure), economy, and residence. With lifelong follow-up, this makes possible large-scale, multidimensional, observational population (and patient) studies, which are suitable for the study of rare diseases – as well as quasi-experimental estimation of effects of structural and dynamic interventions in health systems and other systems.

The World Health Organization pioneered multiple-country databases, such as WHO Europe's Health for All Database (HFA-DB), publicly accessible on the Internet, which includes major health indicators for European countries since the 1970s. Such databases are vital tools for education and policy analysis to inform national and international health policies.

Sources: *Foldspang A, Aarhus University. Personal communication.*
Saracci R. Introducing the history of epidemiology. Chapter 1. In: Olsen J, Saracci R, Trichopoulos D, editors. Teaching epidemiology. Oxford: Oxford University Press; 2010.
Fisher R. Statistical methods for research workers. Edinburgh: Oliver and Boyd; 1925.
Popper KR. The logic of scientific discovery. London: Hutchinson & Co.; 1959.
Doll R, Hill AB. Lung cancer and other causes of death in relation to smoking; a second report on the mortality of British doctors. BMJ 1956;ii:1071–81.
Pugh B, Ipsen TF, MacMahon H. Epidemiologic methods. Boston, MA: Little, Brown & Company; 1960.
Bradford Hill A. The environment and disease: association or causation? Proc R Soc Med 1965;58:295–300.

exposed to the sources of evidence and therefore require grounding in search techniques, medical databases, and the structure and function of the medical library (both the physical and the virtual entities).

Medical and public health decision-making requires skills in communication, information retrieval, and formulating and answering focused clinical questions. Medical students need increasing exposure to the principles of EBM and methodological training in epidemiology and biostatistics in order to cope with the explosion of medical information and to appraise, interpret, and perform clinical

investigations and research. The principles applied to teaching these subjects to medical students, and to students in Master's programs who will pursue careers in public health or clinical research are outlined in Box 3.7.

The combination of a Doctor of Medicine (MD) with a Master of Public Health (MPH) program as a joint degree is becoming more widespread as a means of educating future physicians towards applying public health principles to clinical work and research. The core curriculum for MPH programs is discussed in Chapter 14. Sound methodological training provides the basis of evidence-based public health practice,

BOX 3.7 Syllabus in Epidemiology, Clinical Epidemiology and Biostatistics for Medical Schools and Master's Degree Students

Medical School Education
- Measures of disease frequency, morbidity, and mortality
- Rates and standardization
- Morbidity and mortality
- Research design I: Cohort studies
- Measures of association
- Statistical inference
- Research design II: Case–control studies
- Sample size
- Occupational and environmental epidemiology
- Clinical trials
- Analysis of clinical trials (multivariate models)
- Survival analysis
- Diagnostic tests
- Screening
- Meta-analysis
- Evidence-based medicine
- Causal and non-causal associations; bias in research
- Preventive medicine
- Workshop in preventive medicine (e.g., smoking cessation)
- Small group critical appraisal sessions: prognosis; therapy; prevention
- Final written examination: short answers, multiple choice based on lectures and journal articles

Master's in Clinical Epidemiology
- Basic courses: as in column 1
- Principles and uses of epidemiology
- Survey methods
- Basic statistics
- Interpretation of epidemiological data
- Statistical analysis of rate and proportions
- Clinical trials
- Logistic regression
- Survival analysis
- Seminar in clinical epidemiology
- Health economics and economic assessments
- Advanced topics in epidemiology
- Examinations
- Thesis

Sources: Paltiel O, Hebrew University; Brezis M, Hadassah Hospital; Cohen MJ, Hadassah Medical Center, Jerusalem. Personal communication. Sackett DL, Strauss SE, Richardson WS, Rosenberg W, Haynes RB. Evidence-based medicine. How to practice and teach EBM. 2nd ed. London: Churchill Livingstone; 2000.
Paltiel O, Brezis M, Lahad M. The principles for planning the teaching of evidence-based medicine/clinical epidemiology in an MPH and for medical students. Public Health Rev 2002;30:261–70.

in health promotion or health administration and its application in clinical medicine. Life-long learning provides essential knowledge for scientific reasoning, and the ability to evaluate the literature critically, skills essential for physicians and other health workers to analyze and incorporate new information.

Epidemiology and demography are necessary, but not sufficient alone, for the determination of health policy. Other factors include the funds, human resources, and facilities availability, and their utilization, community attitudes, and political will. Epidemiology, health care financing, and resource allocation relate to supply and demand, and ultimately to policy. These are all issues of great importance to the management of health systems and addressing the changing needs of an aging population with growing obesity, diabetes, and other chronic diseases, while infectious diseases continue to play an important role in population health. They are also of importance in addressing issues of inequalities in health even in countries with universal access through national health insurance. In the USA, without universal health insurance, a serious gap in social policy has many downstream effects perpetuating social and ethnic disparities in health.

The multidisciplinarity of epidemiology is also essential to the formulation of policy and operation of health systems. It is essential for the smooth functioning of a health system, as a method of analysis, and as a monitoring tool. Assessment and monitoring of the health status of a population are, by their very nature, multifactorial. Preliminary and, perhaps, impressionistic reading of the situation makes use of data available from routine sources and serves to generate hypotheses for testing. Evaluation is a more formal and systematic approach in determining the quality of the health of a population as objectively as possible. All evaluations need to look at the input, process, and output of a system. The epidemiological method is applied to measurements (indicators) of inputs (resources) of a health system, the process (manner) of their utilization, and outcomes of care (indicators of morbidity, mortality, or functional status of a population).

Analysis of these complex factors provides the intelligence or feedback for managing the broad scope of public health. The New Public Health integrates assessment, evaluation, and epidemiological analysis with the organization, supply of health care, and other factors relating to the health of the community as a whole. These disciplines provide vital material to link population health needs and the use of resources.

Epidemiology is addressing new challenges of social equity in health; this has become an important part of modern epidemiology with the growing understanding that social conditions and cultural background are key factors not only in disease incidence and prevalence, but in access to health care, both preventive and curative. While the epidemiological identification of health inequality/inequity has been important in identifying the extent and severity of the issues, it is in the utilization of these data in policy making and action where the real challenge lies. Gender, sexual preference and behavior, ethnicity, place of residence, income, family status, religion and religiosity, social connectivity, occupation, and education are all part of the health–sickness spectrum.

DEFINITIONS AND METHODS OF EPIDEMIOLOGY

Rates and Ratios

Measuring the extent of a disease (or risk factor) in a population relates known cases to a population base, and is expressed as a rate or a risk. These rates can be standardized for age and gender for comparisons (see below). Comparing the extent of a disease or a risk factor among population groups can be expressed as a ratio (or a relative risk). The *risk group* may be the entire population defined by a geographic area, an occupational group, a school, a health service, an insurance system, or any other specified groups of people such as defined by occupation, place of work, or lifestyle. The population may also be people who share a risk factor for disease, such as smokers, substance abusers, sex workers, or people attending a celebration who eat certain foods that may be the common source of a disease outbreak.

Incidence of a disease is a measure of the new events or cases which can be stated as numbers or as rates. *Prevalence* is the term used for new and pre-existing cases of a specific condition. The denominator is a fixed population about which information is available on the condition under study for each individual within that population. Identification of cumulative case incidence and prevalence may be complex, with people entering and leaving the study population (representing the denominator), but risks may be estimated.

Rates indicate the occurrence of a phenomenon, such as the occurrence of a health event, in a defined population, in a given period. The components of a rate are the numerator (A) defining the number of cases of a specified condition, over the denominator (B) defining in a specified time-frame in which the events occurred in a defined population, place, region, or country. A multiplier to convert the fraction to a decimal number may be used for convenient comparisons between the frequencies of the event in different population groups.

Crude rates are summary rates based on the actual number of events (e.g., births or deaths) reported in a total population in a given period. Cause-specific rates measure specified conditions (e.g., TB) occurring in the total population or in a designated population group (e.g., age–gender groups) in a specified period. The population used for annual rate calculations is usually estimated at 1 July of that year or may use an average for the entire year.

Cumulative incidence of reported cases of the disease under study that occur in a defined population group may be followed over a period of time to allow the identification of incidence of new cases, such as in mesothelioma among a group of people exposed to asbestos at work or in the community (Box 3.8).

Rates may be crude or specified by age, gender, or other characteristics. Ways of computing them are identical in principle. Furthermore, rates may be compared between two or more populations or population groups, using classical tools of the two techniques of standardization. There are many other ways of comparing rates, such as performing multivariate regression for discrete data (e.g., logistic regression).

Defining the population at risk is a crucial aspect of any epidemiological study and is subject to common errors. Defining the number of cases of a disease or the risk factors being studied is essential to provide the numerator of the rate or ratio. This is also difficult because not all cases of a disease may be reported at the same time or at all. Outreach may be needed to contact exposed people who do not report ill, such as passengers on an aircraft in which a person with a serious infectious disease (e.g., drug-resistant TB, measles, or Ebola virus) may have traveled.

The numerator may be an underestimation of the true value in the population. This problem can occur with common infectious diseases (e.g., mumps, rubella) or where many cases of disease are not clinically diagnosed and therefore go unreported. The same applies in chronic diseases (e.g., hypertension or diabetes mellitus) for many reasons, including non-presentation to the medical system of asymptomatic cases, unclear case definition, and medical error. There may be discrepancies in reporting, such as in coronary heart disease where symptoms differ quite significantly between men and women, or when access to care varies between people in different socioeconomic groups.

A *proportion* is a ratio where the numerator is included in the denominator population, such as describing the number of cases found in a given population, or the proportion of people with a certain attribute or risk factor within the defined population; for example, the proportion of smokers within a certain community.

When cases are relatively rare, an approximation can be made using the total population (including both the disease free and the cases) as the numerator. In such conditions, the odds ratio may serve as a good estimate of the relative risk.

Measures of Disease Occurrence or Morbidity

A measure of disease occurrence or morbidity is a departure, subjective or objective, from a state of physiological and psychological well-being or normal function. It can be measured as the number of people who are ill, periods or spells of illness, or duration of illnesses (days, weeks, months). Morbidity is also described in terms of frequency or severity, including indicators of deaths, disease, disability, and risk factors related to health outcomes. Disability or incapacity rates measure the extent of long-term reduction of a person's capacity to function in society. These measures are also related to disability-adjusted life years (DALYs) and quality-adjusted life years (QALYs) combining morbidity, impact on longevity, and quality of life or disability (see Chapter 11) (Box 3.9).

BOX 3.8 Commonly Used Mortality Rates and Ratios

- *Crude death rate* (CDR) – the number of deaths from all or a specific cause (*A*) per 1000 population (*B*) in a given year $= A/(B \times 1000)$ (total deaths/average population $\times 1000$, or per 10,000).
- *Age-specific mortality rate* – the number of deaths of people in the specified age group per 1000 live population in that age group over a period of time, usually a year for all causes or for specific causes.
- *Cause-specific mortality rate* – the number of deaths from a specific cause per 100,000 live population (estimated on 1 July of the given year); e.g., annual number of deaths from lung cancer in a given year $= 400$ in a population of 1 million $= 400/1,000,000 = 40$ lung cancer deaths per 100,000 population.
- *Case fatality rate* (CFR) – The number of deaths from a specified cause during a given period over the number of diagnosed cases of that disease during the same period $\times 100$; e.g., 10 deaths from measles among 5000 cases is a CFR of $(10/5000) \times 100 = 0.2\%$.
- *Proportional mortality rate* (PMR) (for a specific cause) – the number of deaths from that cause in a specified period over the total number of deaths in that population in the same period $\times 100$; e.g., 25 deaths from motor vehicle accidents/1000 total deaths from all causes $\times 100 = 2.5\%$ (the denominator includes the numerator).
- *Standardized mortality rate or ratio* (SMR) – the ratio of the number of deaths from a specified condition observed in a study population over the number that would be expected if the study population had the same specific rates as the standard population $\times 100$. There is both indirect standardization and direct standardization.

- *Risk* – the measure of estimated probability that an event will occur. Analysis of risk is based on review of the evidence related to a particular risk or group of risks; this may be due to an agent, e.g., toxic, biological, radiological, nutritional (deficiency or excess), behavioral (smoking, risk taking, lack of exercise), stress, alcohol and drug abuse, social deprivation, and others.
- *Hazard identification and quantification* – to determine the extent or degree of exposure of the exposed population to a toxin, carcinogens, air pollutants, alcohol and drug abuse, driving habits, gun exposure, and other risk factors.
- *Relative risk* or *risk ratio* (RR) – the ratio of the risk of a disease (or death or other exposure outcome) among those exposed to the agent or risk factor relative to those not exposed; RR also defined as relative cumulative incidence rate among those exposed to the same cumulative incidence among the non-exposed; in analysis of results in a case–control study this is often expressed as the *odds ratio* (OR).
- *Risk characterization* – to quantify exposure, dose–effect, and dose–response relationships.
- *Risk estimation* – to assemble the relevant data, to define the risk level of the exposed population, leading to quantification of the estimate of the numbers in the population at risk to be affected by the exposure.

Source: *Modified from Last JM, editor. A dictionary of public health. New York: Oxford University Press; 2007.*

Morbidity data are derived from reported communicable diseases or chronic, genetic, and other conditions for which there are established, recognized reporting systems and registries, which are usually operated by ministries or departments of health for the population of their jurisdiction. Databases are provided for monitoring and providing direction for etiological studies and for priorities and avenues for intervention to control the spread of disease. Morbidity is measured by incidence and prevalence rates, as well as severity and duration, although these are not usually available on routine reporting and may require special investigation. Incidence is more useful for acute conditions, whereas prevalence is more important in measuring chronic disease and assessing the long-term impact of a disease.

Latency is the period between exposure to a disease-causing agent and the appearance or manifestation of the disease. For an infectious disease, it is called the *incubation period*. A disease may appear clinically days, weeks, months, or even years after exposure to the causative agent, whether it is microbiological, toxic, carcinogenic, or traumatic.

An attack rate is a specific incidence rate expressed as the percentage of the exposed population suffering from the disease. When the population is at risk for a limited period, such as during an epidemic, the study period can readily encompass the entire epidemic. The attack rate gives a measure of the extent of the epidemic and may provide information needed to control it. For example, if an epidemic of measles spreads from the initial, or index, cases with an increasing attack rate among the exposed population, a change in vaccination tactics and control measures may be needed in order to avoid rapid spread to other vulnerable groups.

Incidence rates measure the frequency of health-related events in a certain population during a specified period. The denominator for incidence rates is defined as the "population at risk", in which the studied events may occur. For example, the incidence rate for breast cancer in a certain region will be the number of new cases diagnosed over a 1-year period, divided by the total number of women in that region. An attack rate is the cumulative incidence of infectious cases in a group, observed over a period during an epidemic, either by identification of cases or by seroepidemiology (Last, 2001).

BOX 3.9 Measures of Frequency of Disease in a Population

$$Rate = \frac{Number\ of\ cases\ in\ a\ given\ time\ period}{Population\ at\ risk\ in\ the\ same\ time\ period} \times N^{th}$$

where N^{th} = 100, 1000, 10,000, 100,000, or 1,000,000; Period: usually = 1 year; Population: mid-year (1 July) estimate.

- *Incidence rate* – the rate at which new health-related events occur in a population. The numerator is the number of new events occurring in a defined period (usually 1 year); the denominator is the population at risk of experiencing the event during this period.
- *Prevalence* – the total number of all individuals who have an attribute or a given disease or condition at a point in time or a designated period. The *prevalence rate* (or rather the *prevalence proportion*) is the number of individuals with the attribute divided by the population at risk, at that point in time (*point prevalence*). Period prevalence usually is defined as the sum of (a) the point prevalence at the beginning of the period, plus (b) the cumulative incidence during the period.
- *Attack rate* – the cumulative number of cases of a specified disease among the population known to be exposed to that disease over a defined period.

Source: *Modified from Last JM, editor. A dictionary of public health. New York: Oxford University Press; 2007.*

There are several ways to define the denominator for incidence rates:

- *Ordinary incidence rate* is used when calculating incidence rates in a changing population; for instance, where there is a natural movement in and out of the studied population (due to births, deaths, and migration). In that case, the average size of the population in the specified period is used as the denominator, usually including both the "population at risk" and cases already with the disease (prevalence). Although only the "population at risk" should theoretically be included, such an approximation is often made. The reason for this is that when the condition is relatively rare, the influence that prevalence cases will have over the denominator can be considered negligible. In addition, the information about prevalence cases is often not available.
- *Cumulative incidence rate* is usually calculated in longitudinal epidemiological studies. When a cohort (a group of people), initially free from the disease, is being followed during a certain period, incidence cases can be identified as they occur. The sum of those incidence cases is referred to as "cumulative incidence". Here, the denominator includes only the "population at risk"; therefore, cumulative incidence may also be termed risk of the condition (Abramson, 2001).

- *Person–time incidence rate or incidence density* is usually used in follow-up studies in which individuals are "at risk" or may be followed up during different periods. In this case, the total number of events is divided by the sum of all subjects' periods at risk, measured, for example, in years, months, or days. In order to calculate the denominator, each individual's "period at risk" must be calculated, measuring the time from the beginning of the follow-up until withdrawal from the study (due to either occurrence of the condition under study, or "censoring"; i.e., any other reason causing cessation of follow-up).

Prevalence Rates

Unlike incidence (indication of occurrence), prevalence is the measure of the total existing situation of a health-related condition or risk factor, including old and new cases. A prevalence rate measures the proportion of individuals having that condition within a defined population group at or during a specified time. Several measures of prevalence rates exist:

- *Point prevalence* – the proportion of people with the condition being studied at a certain point in time is divided by the size of the group or population. Point prevalence is influenced by the incidence rate of the condition, as well as its mean duration up to death or recovery.
- *Period prevalence* – the proportion of people who developed the condition before and during the specified period. The denominator includes all the individuals who have or had the condition during the defined period, including those who left, died, or recovered during that period. It allows comparison over time with the same or other population groups. Thus, morbidity from a specific condition during one year can be compared to previous years, weeks, or months, and between countries or regions in a country.
- *Lifetime prevalence* – the proportion of people who have had the condition at any time during their lives; for instance, those who have or had the condition divided by the total population.

The prevalence rate is calculated on the basis of the number of cases and the number of people exposed, and may be compared to the non-exposed population. Estimation of case prevalence in an exposed population may be underreported if insufficient time has elapsed for a disease with a long latency period. An example of period prevalence is the number of cases of cancer among people exposed to a carcinogenic agent in the past; for example, mesothelioma cases occurring in a former asbestos worker population over a 30-year latency period following exposure.

Measures of Mortality

A death rate (mortality rate) is the incidence rate that measures the frequency of deaths over a given period in a defined population. Mortality rates may be standardized to allow comparability between population groups and may be specific to defined diseases or conditions. Modern epidemiology originated in studies of mortality derived from the Bills of Mortality (publication of deaths by location and cause) in the UK by John Graunt in 1662. Mortality data are based on the mandatory reporting of all deaths (see Chapter 1).

Death certificates are mandatory in most countries and must be signed by a licensed physician before the body can be buried or cremated and before insurance payment or inheritance can occur. The contents of the death certificate are important because the medically certified cause of death is the basis for mortality statistics. Personal data include the age, gender, ethnicity, place of residence, and other variables such as occupation and injury. Completeness of reporting, accuracy of diagnosis, and coding of causes of death may limit the conclusions that can be drawn from such data. In practice, however, the data reported in large disease categories are an acceptable guide to actual events.

Analysis of causes of death may take into account more than one diagnosis so as to determine the underlying causes of death such as diabetes. This seems straightforward, but standardization of reporting causes of death is far from simple. Doctors who fill in the form may vary in their perception of diagnosis and the difference between immediate and underlying cause of death. In developing countries, data from death certificates may not be available and determination of leading causes of death may have to be studied by "verbal autopsies" conducted as part of community surveys.

A standard national death certificate is vital for public health as it provides basic information needed for demographic and epidemiological purposes. Box 3.10 (see companion web site at http://booksite.elsevier.com/9780124157668) presents the data required in a standard death certificate as modified in 2003 in the USA, although the format may vary from country to country.

Causes of death recorded on the death certificate include the immediate cause of death (e.g., cardiac arrest); the second and third lines include contributing conditions (e.g., acute myocardial infarction and congestive heart failure); and the fourth line is the underlying cause (e.g., coronary heart disease). The death certificate is filed with a public registry office and forwarded to a vital records office where the causes of death are recorded by a registrar trained to federal standards to interpret and code medical diagnoses, according to the 10th revision of the International Classification of Diseases (ICD-10), adopted by the World Health Organization (WHO) in 1990.

Overall patterns of mortality are examined by age, gender, and ethnic group, and by cause of death. Mortality trends will be discussed under communicable and non-communicable disease in Chapters 4 and 5. National mortality trends give vital information on disease and changing epidemiological patterns, allowing for regional and international comparisons, and help to define health programs and targets (Box 3.8).

Mortality patterns can be studied in a particular year or over time. A cohort is usually a group of people born in a particular year, but it can be any defined group being followed epidemiologically. Cohorts of people born in particular years can be followed to observe and compare mortality patterns. With suitable age standardization, the mortality patterns of men born, for example, in 1900, 1920, 1940, and 1990 can be compared with each other.

Mortality statistics are fundamental to epidemiology and provide some of the most reliable data available. Epidemiological analysis of mortality data depends on the registration of deaths with basic demographic data and causation of death as recorded by the physician certifying it. Total, age-specific, and gender-specific mortality are usually calculated on an annual basis, with the mid-year population as the denominator. This provides crude, age-specific, cause-specific, and proportional mortality rates from which standardized mortality rates or ratios (SMRs) are calculated. Case fatality rates (CFRs) relate mortality from a cause to the incidence or prevalence of that disease.

Changes in mortality patterns may occur as a result of a number of factors affecting the outcome of a disease, such as changes in socioeconomic conditions, disease prevention, or methods of treatment. Diagnostic criteria or accuracy of death certificates may also change over time. Thus, a change in mortality may reflect a change in incidence of the disease or CFRs related to treatment methods and access to care, or changes in the definition or classification of diseases.

Social Classification

The British Registrar General's Classification of Occupations was established in 1911 and is updated every 10 years (Box 3.11). It is easy to use and provides an excellent demographic and epidemiological tool that has been used in many studies of disease outcomes. It can help to illustrate the different health experiences of the various social classes, even within the universal NHS. It has become part of the database of vital statistics and morbidity patterns in the UK.

Many other classifications are used for research purposes in the UK and in other countries, addressing other issues in social inequality such as unemployment. Other proxy indicators for social class or social inequality include number of siblings, infant and maternal mortality, single parenthood, and many others. The UK uses social indicators of deprivation to classify counties and collate them with health outcome data.

The USA and most other western countries do not have social class data recorded on death certificates and therefore proxy measures of social classification are used, such

BOX 3.11 British Occupational Based Social Class

- Class I – professional and business occupations (e.g., physician, banker)
- Class II – intermediate occupations (e.g., schoolteacher, storekeeper)
- Class III – non-manual occupations (e.g., clerk)
- Class III – manual skilled occupations (e.g., foreman)
- Class IV – partly skilled occupations (e.g., salesperson, factory worker)
- Class V – unskilled occupations (e.g., porter, waiter)

Note: Before 1990 known as the British Registrar General's Classification of Occupations.

Source: *Galobardes B, Shaw M, Lawlor DA, Lynch JW, Smith GD. Indicators of socioeconomic position (part 2). J Epidemiol Community Health 2006;60:95–101. http://dx.doi.org/10.1136/jech.2004.028092 PMCID: PMC2566160. Available at: http://www.ncbi.nlm.nih.gov/pmc/articles/PMC2566160/ [Accessed 8 January 2013].*

as ethnicity, national origin, education, and poverty levels. In the USA, race is recorded on death certificates and these mortality data can be analyzed by racial groups including Native American or Alaskan Native, Asian or Pacific Islander, Black, Hispanic, and White. Education level and occupation are also recorded, but mortality data are generally presented by racial group, not social indicators.

The interrelationship between ethnicity and disease or mortality often masks other socioeconomic factors, such as higher levels of poverty or reduced access to medical care among African American and Hispanic groups in the USA or immigrant groups in European countries. Because there are wide variations in socioeconomic and educational levels within ethnic or racial groups, and many confounding factors in ethnicity or race that may affect disease patterns, data classified in this way should be interpreted carefully.

Social class is increasingly identified as a major variable in health status. It serves as a proxy measure for many health-related issues, such as nutrition, access to care, and dependence on occupations with hazards, with little opportunity for personal development, or lacking security. Social class variations in health status exist even where universal access health systems are well established, in countries with universal health insurance or services; for example, the UK, Sweden, and Israel. However, social differences are less pronounced in the Nordic countries where social gaps are generally narrower than in countries with less developed social welfare systems. These systems are, however, coming under pressure from immigrant and migrant worker populations, which have become significant both demographically and with regard to health issues.

SENTINEL EVENTS

The US Joint Commission defines a sentinel event as: "an unexpected occurrence involving death or serious physical or psychological injury, or the risk thereof". The Commission includes a review of organizations' activities in response to sentinel events in its accreditation process, including all full accreditation surveys and random unannounced surveys (Joint Commission, 2013). Health facility errors and adverse events are considered major causes of preventable deaths and high costs to the health system.

Sentinel events include the suicide of a patient who was supposed to be under constant surveillance, unexpected maternal or infant deaths, infant abduction or discharge to the wrong family, rape, hemolytic transfusion reaction due to mismatched blood, surgical deaths, surgery on the wrong patient or wrong body part, or an instrument left in a patient during surgery. They can also include unusual rates of infection, poor surgical outcomes, medication errors, infections occurring in a hospital that may jeopardize patients' health, and many others. Such events occur frequently enough to pose both health risks and an economic burden to the hospital, the insurer, and of course, the patient.

Sentinel events are taken as measures of problems in a health care process. They are events, such as avoidable deaths, which should be uncommon if all goes well with preventive and curative care, and acceptable standards are in place. Avoidable deaths will vary according to the state of health development of a country, and each country may define its own sentinel events for review. The response to such events and preventive measures to reduce such errors are part of quality assessment in hospitals, nursing homes, and other health care programs.

In infectious disease epidemiology, the index case is the first case or group of cases of a condition that come to attention, providing the first clues in an outbreak or epidemic. In non-infectious conditions, the sentinel event may be a death, where the investigation of the circumstances may help in understanding the process of the disease or the care that was received. One case of clinical poliomyelitis represents up to 1000 people infected with the virus, which can develop in the gut of a person protected from the disease by the killed vaccine. Several epidemiologically linked cases of measles in a country previously free from the disease must be considered sentinel events that should not happen, and their investigation may show errors of omission or commission which explain the event and point to a need for remedial action or a change in policy.

Reporting and data systems should be arranged to indicate avoidable deaths from vital records or hospital discharge information systems. Comparison between areas may also include avoidable deaths as a health status indicator. There are selected conditions that are generally preventable or treatable, and therefore warrant investigation when they occur. Maternal deaths (i.e., deaths associated in time and related to pregnancy or the postpartum period), deaths within 24 hours of hospital admission, or deaths following surgery (usually within 48 hours) are examples of sentinel events which are uncommon and should always be investigated. Deaths from appendicitis or appendectomy, tonsillectomy, hysterectomy, tubal ligation, or other elective surgical procedures should be investigated as sentinel avoidable deaths until other explanatory factors are found. Nosocomial (hospital-acquired) infections are a major cause of mortality, increased length of hospital stay, and health care expenditures, and require an active program of surveillance and prevention within the care setting.

With the advent of newly emerging frightening diseases such as SARS, Ebola, and avian flu, the development of rapid reporting of cases of suspect infectious disease takes on a new urgency. The situation is even more worrying with the potential for bioterrorism in the twenty-first century, raising specters not seen before in modern public health. Hospital emergency rooms and doctors' offices in the community become front-line monitoring sites for such disease, and depend on current information on possible symptoms or forms of presentation of a disease even in the earliest stages of its development. The identification of index cases and sentinel events is crucial to the functions of public health, especially with emergence of newly identified diseases, such as Legionnaire's disease, Ebola and Marburg viruses, and many others which sometimes move from their natural habitat via travelers and can become entrenched and even endemic in new environments, as has happened with West Nile fever, Rift Valley fever, Chikungunya, Lyme disease, dengue, and others. It is fundamental to detect and localize outbreaks of such highly dangerous infectious diseases.

THE BURDEN OF DISEASE

Burden of disease refers to the combined measurement of mortality and nonfatal health outcomes. The assessment of burden of disease serves to design, test, and implement methodologies to aid in setting priorities for the effective allocation of health resources. The challenge is to develop valid, reliable, comparable, and comprehensive measures of population health and comparative assessments of the burden of diseases, injuries, and risk factors. This assessment can then be linked with the investigation of costs, efficacy, and effectiveness of major health interventions to establish appropriate cost-effectiveness estimations, which should be a major tool in policy design and decision making.

The burden of disease is an important epidemiological research instrument. This approach recognizes that social and other factors contribute to diseases which are multifactorial in origin. These estimations, combining economic and epidemiological data, use the DALY as the unit of measurement of the burden of disease, representing the loss of 1 year of "healthy" life.

YEARS OF POTENTIAL LIFE LOST

Years of potential life lost (YPLL) are calculated based on age-specific rates of mortality or disability. They provide a refinement in epidemiology which has added important new perspectives in the analysis of specific problems. The leading causes of death in the USA, as in most developed countries, are coronary heart disease, cancer, and stroke. However, when the data are examined from the point of view of YPLL, trauma (unintentional injuries, homicides, and suicides) becomes the leading cause of death.

YPLL is a better reflection of the impact of diseases on a society than other mortality rates because it is age related, showing the relative impact of early mortality, which should be taken into account when determining national health priorities. Trends in YPLL for the years 1980–1996 are shown in Table 3.5 (see companion web site at http://booksite.elsevier.com/9780124157668). There was a large drop in YPLL for total mortality of most specific causes. There was also a substantial decline in YPLL for total and some categories from 1995 to 1996, especially in HIV, suicide, and homicide deaths.

Qualitative Measures of Morbidity and Mortality

QALYs and DALYs are calculations of morbidity introduced in the international health literature (Box 3.12). They serve as statistical measures of the burden of disease, allowing for international comparisons. Other terms used include disability-free life expectancy (DFLE) and health expectancy, which are both measures of mortality, morbidity, and impairment or disability. Burden of disease measures are used to assess the cost-effectiveness of specific interventions (see Glossary).

The World Bank calculates the variation in burden of disease between demographic regions, varying from nearly 600 DALYs lost per 1000 population in sub-Saharan African countries to approximately 120 per 1000 in the industrialized countries. These measures are used in economic analyses of health status, helping to focus on outcome measures to justify resource allocation by comparing benefits in terms of reduced mortality and morbidity.

MEASUREMENT

Epidemiology and public health are dependent on quantitative and qualitative observations to establish relationships and possible points of intervention. Therefore, an appreciation of the methods of handling statistics and their interpretation is fundamental. A complete presentation of this field is beyond the scope of this text; however, some general concepts are important to establish.

BOX 3.12   Measures of the Burden of Disease

- *Potential years of life lost* (PYLL) – PYLL is a measure of the relative impact of various diseases and lethal forces on society. It highlights the loss to society as a result of youthful or early deaths. The figure for PYLL due to a particular cause is the sum, over all people dying of that cause, of the years that these people would have lived had they reached a specified age.

- *Disability-adjusted life year* (DALY) – DALYs are units for measuring the global burden of disease and the effectiveness of health interventions and changes in living conditions. The DALY is a summary measure of population health. DALYs are calculated as the present value of future years of disability-free life that are lost as a result of premature death or disability occurring in a particular year.

  DALYs are calculated by a formula that includes five main components: the duration of time lost due to a death at each age, disability weights, age weights, time preference (expressed as a discounting function), and the integration of health measures among a population.

  DALYs for a disease or health condition are calculated as the sum of the years of life lost (YLL) due to premature mortality in the population and the years of healthy life lost from some degree of disability (YLD) for incident cases of the health condition.

  For detailed procedures on how to calculate DALYs, see WHO (2001, 2010).

- *Quality-adjusted life year* (QALY) – QALYs are an adjustment or reduction of life expectancy reflecting chronic conditions, disability, or handicap, derived from survey, hospital discharge, or other data. Numerical weighting of severity of disability is established on the basis of patient and health professional judgment.

Sources: *Last JM, editor. A dictionary of public health. New York: Oxford University Press; 2007.*
Harvard School of Public Health, Burden of Disease Unit. http://www.hsph.harvard.edu/organizations/bdu [Accessed 21 April 2008].
Murray CJ. Quantifying the burden of disease: the technical basis for disability-adjusted life years. Bull World Health Organ 1994;72:429–45.
World Health Organization. Global burden of disease. Available at: http://www.who.int/topics/global_burden_of_disease/en/ [Accessed 2 January 2013].
World Health Organization. Global health risks: mortality and burden of disease attributable to selected major risks. Geneva: WHO; 2009. Available at: http://www.who.int/healthinfo/global_burden_disease/global_health_risks/en/index.html [Accessed 6 January 2013].
World Health Organization. Health statistics and information systems. Metrics: disability-adjusted life years (DALY). Quantifying the burden of diseases from mortality and morbidity. Geneva: WHO; 2010. Available at: http://www.who.int/healthinfo/global_burden_disease/metrics_daly/en/ [Accessed 1 February 2013].
World Health Organization. National burden of disease studies: a practical guide. Edition 2.0. Geneva: WHO; October 2001. Available at: http://www.who.int/healthinfo/nationalburdenofdiseasemanual.pdf [Accessed 1 February 2013].
World Health Organization. The top ten causes of death. Available at: http://www.who.int/mediacentre/factsheets/fs310/en/index.html [Accessed 2 January 2013].

Routine data sets and their analysis can provide vital information for state- and county-level health agencies as well as to members of the health professions and the public at large. New York State has developed a remarkably ambitious and highly developed Community Health Data Set of the state by county, with many indicators (see Box 3.13 on companion web site at http://booksite.elsevier.com/9780124157668).

Research and Survey Methods

The scope and depth of research methods and the many other quantitative and qualitative sciences related to conducting investigations of health and disease in population groups are now important elements of training in public health. This area of public health is basic not only for research but also in reading the literature of a dynamic field such as public health, and in the design of policies and intervention programs, resource allocation, and the management of health systems. Research and surveys are integral parts of public health practice, and especially of academic public health. Familiarity with their basic principles is an important part of the preparation of public health professionals and a responsibility of academic centers training the public health workforce.

A thorough review of the peer-reviewed literature is a prerequisite for development of a study, requiring skills in the use of Internet search engines such as PubMed and Medline, as well as important sources such as the CDC Atlanta, the WHO, and other respected professional bodies. Organized literature reviews are called Cochrane Reviews, after the leading British epidemiologist Archie Cochrane, using meta-analysis. This is a formal method of review and analysis of multiple studies of a causal relationship of a therapeutic or preventive measure that yields a quantitative aggregate summary of all results. Meta-analysis includes selection of studies of similar design, mostly of randomized controlled trials (RCTs), pooling of the data to make a larger sample. This increases the chance that any change and comparison of study and control groups would be statistically significant, but also based on critical analysis and selection of those studies meeting acceptable criteria of methodology. A 2007 study reported in *The Lancet* on meta-analysis of previous studies showed a significant benefit of folic acid supplementation in reducing the incidence and severity of stroke, whereas individual studies were equivocal or showed change that was not statistically significant.

The formulation of a study question and its hypothesis includes defining its purposes and objectives. This leads to basic study design, definition and selection of the study population, sample selection, and selection of variables to be measured. A study is dependent on funding and the presentation of the proposal is crucial for success. The study design requires development and testing of survey instruments, organization of the study team, and collection of data. Assessment of reliability and validity of the data is a key part of preparing it for analysis. Training in research methods is thus integral to studies of epidemiology and descriptive and inferential statistics.

Qualitative methods, including quantitative measures used in the social and behavioral sciences, are also important in public health, with health behavior as a basis for "lifestyle" or personal choices. These methods are also applied to societal conditions, cultural, socioeconomic and geographic factors, and support systems, which are all related to fundamental risk factors for some diseases and their severity, access to health care, and health outcomes. They are also related to organizational systems, management of health systems, economics, and professional interactions.

In these areas, the applicable social sciences include sociology, psychology, anthropology, political science, organization theory, and information technology. "Social marketing" is based on the study of human behavior and how to change it. Public health campaigns against risk factors such as smoking or high-risk sexual behavior depend on such knowledge of awareness, attitudes, behavior, and practices. Qualitative studies are more exploratory and developmental in pursuit of non-numerical aspects of the study question, and relate to attitudes, concerns, fears, and social aspects of study questions crucial to success in public health. Examples include studies of teen pregnancies, parental concerns regarding new vaccines, sexual practices such as condom use, interfamilial relationships and their impact on risk behavior and antisocial behavior, and smoking-related issues. Epidemiological and qualitative studies can be complementary to each other, providing important scientific evidence related to real public health issues of national and international importance.

Interpretation of statistical events requires a familiarity with methods of gathering and processing basic information. Statistics is "the science and art of collecting, summarizing, and analyzing data that are subject to random variation" (Last, 2001). Biostatistics is the application of statistics to biological problems.

Variables

A variable is any factor being studied which is considered to affect health status and which can be measured. It may be an attribute, a phenomenon, or an event that can have different values, such as age, gender, SES, behavior, other disease conditions, characteristics of the health care system, or exposure to a toxic or an infectious agent. A dependent variable is the outcome being studied. An independent variable is the characteristic being observed or measured which is hypothesized to cause or contribute to the event or outcome being studied, but is not itself influenced by that event. For instance, in the study of the association between smoking

and coronary heart disease, smoking (described as the average number of cigarettes smoked per day, for example) is the independent variable, or the exposure. Coronary heart disease is the dependent variable, or the outcome.

The Null Hypothesis

The null hypothesis is the assumption that one variable has no association with another variable, and that two or more populations being studied do not differ from one another. A statistical test is used to decide whether the null hypothesis may be rejected or accepted; that is, the probability that any differences observed may be due to chance alone and not indicative of a real difference.

If the probability of chance alone explaining the observed differences is very low then the null hypothesis may be rejected, suggesting that the studied association or difference may actually exist. The definition of the threshold for "low probability" depends upon the decision of the level of significance required. Statistical testing thus provides the basis for inference or decisions regarding the results of a study as statistically significant and to what degree.

Confounders

A confounding variable (confounder) is a factor other than the one being studied that is associated both with the disease (dependent variable) and with the factor being studied (independent variable). A confounding variable may distort or mask the effects of another variable on the disease in question. For example, a hypothesis that coffee drinkers have more heart disease than non-coffee drinkers may be influenced by another factor (Figure 3.10). Coffee drinkers may smoke more cigarettes than non-coffee drinkers, so smoking is a confounding variable in the study of the association between coffee drinking and heart disease. The increase in heart disease may be due to the smoking and not the coffee. More recent studies have shown coffee drinking to have substantial benefit in heart health and in the prevention of dementia.

In public health, researchers are often limited to observational studies to find evidence of causal relations. Experimental studies may not be possible for many technical, ethical, financial, or other reasons. The proper causal interpretation of the relations from carefully developed epidemiological studies is vital to the development of effective measures of prevention.

Sampling

The majority of epidemiological studies cannot collect information about every individual in the target population (the general population of a country or a region, or a defined group of people). Therefore, a sample, which is

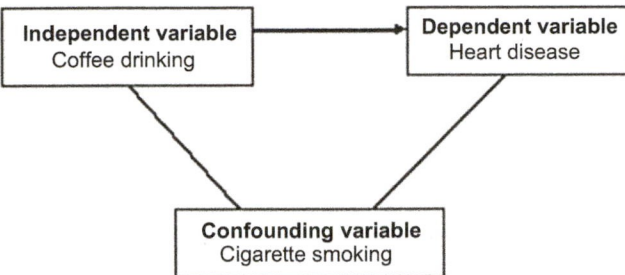

FIGURE 3.10   Independent, dependent, and confounding variables in a study.

chosen from that target population, is defined and used as the study population, for which all the required information is collected. The appropriate choice of the study population is crucial to ensure that the results obtained from the study can later be generalized to the general population. Therefore, a sample must be selected randomly, representative of the general population, and of sufficient size as to increase the likelihood (or probability) that the results obtained from the sample are close enough to the actual situation in the general population (i.e., where the level of significance of a statistical test is acceptable) (Box 3.14).

Random Sampling

A distinction should be made between sampling and randomization, which is the allocation of two or more groups to different interventions.

The main sampling methods are described as follows.

- *Simple randomization* – when all individuals in the population have an equal chance of being selected, the group is known as a *random sample*. This is often achieved by assigning each person in the group a number and then selecting the sample from a table of random numbers until the desired sample size is reached.
- *Systematic randomization* – every *n*th unit is selected.
- *Stratified randomization* – the population is divided into strata (subgroups) and simple randomization is applied within subgroups. For example, if 20 percent of the population is in the age group 40–59 and 20 percent of the sample comes from this age group, and similarly for other age groups, then all strata are fairly represented with regard to numbers of people in the sample.
- *Cluster sampling* – a population is non-randomly divided into subgroups (such as households, schools in a city, or classes in a school) and clusters (subgroups) are randomly selected.
- *Multistep sampling* – groups are randomly selected, and then individuals within groups are chosen.

A non-random sample is one in which a form of bias is introduced into the sampling process. For example, a

---

BOX 3.14 Sample Size

Principles

- Samples are drawn to represent a population and the larger the number of samples and their sizes, the higher the probability that their average value (of the parameter under study) is equal to the value in the population.
- Because sample sizes are limited, sampling error (i.e., the probable difference between the value in the sample and in the population) must always be taken into account.
- The size of the sampling error is affected by the size of the sample drawn, and by other factors, some of which are called biases. Increasing the sample size decreases the size of the sampling error, unless there is a selection bias, in which case increasing the size will sustain the sampling error.
- The principles of sampling rest on the assumption that samples are randomly obtained.

Factors in the Calculation of Sample Size

- *Type 1 error* – the risk of a false positive result (α) (i.e., the chance of detecting a statistically significant difference when there is no real difference).
- *Type 2 error* – the risk of a false negative result (β) (i.e., the likelihood of not detecting a significant difference when there really is a difference that is greater than the specified threshold).

The power of a study is its ability to demonstrate an association if one exists. It is determined by several factors, including the frequency of the condition under study, the magnitude of the effect, the study design, and sample size. It is defined as the chance of not getting a false-negative result and is equal to $1 - \beta$ (type 2 error).

Calculation of sample size is beyond the scope of this text and is found in free computer programs, including those in the Sources below.

Sources: *Last JM, editor. A dictionary of public health. New York: Oxford University Press; 2007.*
Centers for Disease Control and Prevention. Epi Info 7 [updated December 2012]. Available at: http://wwwn.cdc.gov/epiinfo/7/index.htm [Accessed 3 January 2013]
Abramson JH. WINPEPI (PEPI-for-Windows) [new version posted 17 December 2012]. Available at: http://www.brixtonhealth.com/pepi4windows.html [Accessed 3 January 2013].

---

convenience sample is a group of people who are readily accessible, such as volunteer blood donors or people who appear at a health fair for blood pressure examination. The bias in such samples is that there may be a self-selection process not representative of the total population. A selection of a group at special risk, for example, might entail choosing districts with known low immunization coverage in order to attempt to determine the reasons for this. Such a study would then be applicable to those districts and, although not generalizable to the total population, could provide valuable information affecting the immunization program.

Conclusions based on sample results may be attributable to the population from which the sample is taken. Extrapolation to the total population or a different population is a judgment, which may be justified but must be qualified by description of the sampling methods and the potential biases with appropriate statistical testing used. Despite these limitations, careful sampling is essential for assessing a particular characteristic in a larger population and should give results that are reproducible by other investigators.

NORMAL DISTRIBUTION

The evaluation of certain characteristics in a population group is based on the assumption of normal distribution (nutrition, height, weight). A normal distribution is continuous and symmetrical about a mid-point. It is often described as a bell-shaped frequency distribution of observations. A normal distribution has upper and lower values that may extend to infinity, but it has an arithmetic mean, mode, and median from a central point (Figure 3.11).

Mean, median, and mode are measures of central tendency in a group of numbers (Box 3.15). The symmetrical bell-shaped (Gaussian) curve represents the normal distribution of biological characteristics, such as heart rate, height, weight, or blood pressure in a normal population group. In such a distribution, approximately two-thirds of the observations fall within one standard deviation and approximately 95 percent fall within two standard deviations of the mean.

Normality may be defined in several senses. It is a range of variation in a given population, within two standard deviations below and above the mean, or between specified percentiles, for example, the 10th and 90th of the distribution. *Normal* also refers to the limits of a range of a test or measurement and is an indication of the finding being conducive to good health.

Deciding when a group of observations is "normal" or "abnormal" requires defining cut-off points, both in clinical medicine and in epidemiology. In clinical medicine, deciding what is a normal blood pressure, cholesterol level, or growth of a child is based on norms determined from a large number of observations of what is assumed to be a "normal" population. For example, growth patterns

of children used as an international standard are based on data derived from a white, middle-class American population (see Chapter 6).

STANDARDIZATION OF RATES

Age structures of populations vary widely in countries around the world. For comparisons, it is important to standardize rates. After many years of examining alternative standard populations, standardization now uses the WHO standard population based on world average population estimated between 2000 and 2025 (Ahmad et al., WHO, 2001).

The age-adjusted death rate is the number of deaths per 1000 people of a specified population during 1 year, with the rate adjusted to prevent distortion by the age composition of the population. A standard population is used for determining this rate.

BOX 3.15 Summarizing a Group of Numbers

- *Mean* – the average value of the observations, i.e., the sum of values of the observations divided by the number of observations.
- *Median* – the midpoint value to which half of the observations are equal or less, and half are equal or greater. It is the middle observation when a set of observation numbers is arranged in order of increasing value.
- *Mode* – the most frequently occurring value in a set of observations. In a normal distribution, the mean, median, and mode are all equal to one another.
- *Standard deviation* – the common method of summary of how widely spread or dispersed the observed values are from the mean of the observations.
- *Confidence interval* – the range or interval within which the true value of a variable, such as mean, proportion, or rate, lies at a specified degree of probability (e.g., 95% or 99%). It indicates how precisely the results of an analysis based on a sample approaches the true value of the rate in the population which the sample is meant to represent.

Sources: *Adapted from Last JM, editor. A dictionary of epidemiology. 4th ed. New York: Oxford University Press; 2001.*
Last JM, editor. A dictionary of public health. New York: Oxford University Press; 2007.

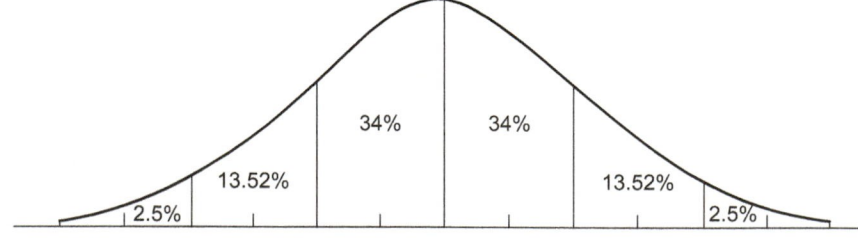

FIGURE 3.11   Normal distribution. *Source: Last JM, editor. A dictionary of epidemiology. 4th ed. New York: Oxford University Press; 2001.*

Standardization of rates is important in comparing data between populations of different age and gender distribution and to remove, as far as possible, the effects of confounders in epidemiological studies. Comparing mortality rates in one country, for example, will require using a standard population such as that of the USA in 1940 to compare mortality in 1940 with subsequent rates, thus removing the effects of changes in the age and gender composition of the populations. Without standardization of the population, the age–gender changes would act as confounders when describing distributions or comparing mortality or disease incidence between two or more designated groups.

Standardization uses a "standard population" selected to adjust for differences in the distribution of the relevant variables between groups being compared, or between the sample used in a study and the population it was chosen from. The standard population in this procedure is one in which the age and gender composition is known and therefore is used as a benchmark to compare rates for a number of different population groups. Comparisons between different states in the USA or countries in Europe would use a US, European, or world population distribution. Standardization can be done by direct or indirect methods.

It is important to note that although age and gender distributions often act as confounders, many other variables may affect the outcome being measured, depending on the study. For example, smoking and coffee drinking may act as confounders when studying the association between physical activity and coronary heart disease. If such a confounding effect is present and identified, then the study analysis must control for the confounder in order to correctly assess the main study variables.

Direct Method of Standardization

The direct method of standardization is used when age–gender-specific mortality rates are known for the populations being compared to a standard population. These rates are then applied to the standard population to calculate the *expected numbers of deaths for each group* in the population, as if its composition (with respect to the variable being standardized for) were the same as in the standard population. They are then summed and divided by the total standard population to give a summary adjusted rate. Standardized death rates can be calculated for particular diseases. For example, if one is comparing lung cancer death rates in a number of countries to see whether there are differences that might be attributed to external factors such as air pollution patterns, the data for each city can be compared using standardized (cause-specific) mortality rates.

The direct standardization of rates is an important method of comparing mortality patterns between cities, districts, regions, and countries. Table 3.6 shows mortality rates from a range of countries with very different age distributions comparing mortality from NCDs in 2004. With age standardization, countries can be compared with key causes of mortality which illustrate the high rates of mortality from NCDs in developing and mid-level income countries. Age adjustment of rates is important for time trends and comparisons between countries and regions within countries.

Indirect Method of Standardization

When age–gender-specific mortality rates for the study population are not available or if the numbers in some age groups are too small, the indirect method of standardization is used. This method uses known age–gender-specific rates from a standard population to calculate the expected number of the same health event for the population being studied, given that population's distribution (Box 3.16). The expected number of deaths or cases thus calculated is then compared to the actually observed number of deaths or

TABLE 3.6 Age-Standardized Mortality from Non-Communicable Diseases (NCDs), Rates per 100,000 Population, Selected Countries, 2004

	Total NCD Mortality	CVD	Cancer	Injury
Canada	374	131	135	33
Egypt	891	515	81	36
Ethiopia	817	384	142	105
France	387	123	154	45
India	713	382	100	116
Israel	368	121	121	29
Japan	284	103	120	39
Russia	904	645	142	218
Sweden	372	171	115	32
UK	441	175	147	26
USA	450	179	133	50

Note: Rates are age-standardized to WHO's world standard population. CVD = cardiovascular disease.
Sources: World Health Organization. World health statistics, 2009. Table 2. Geneva: WHO; 2009. Available at: http://www.who.int/gho/publications/world_health_statistics/EN_WHS09_Full.pdf and http://www.who.int/whosis/whostat/EN_WHS09_Table2.pdf [Accessed 9 January 2013].
Ahmad OB, Boschi-Pinto C, Lopez AD, Murray CJL, Lozano R, Inoue M. Age standardization of rates: a new WHO standard. GPE discussion paper series no. 31. Geneva: WHO; 2001. Available at: http://www.who.int/healthinfo/paper31 [Accessed 1 February 2013].

BOX 3.16 Standardized Mortality Rates (SMRs) and Standardized Incidence Ratios (SIRs)

$$SMR\,(SIR) = O/E = \frac{Observed\ deaths\,(cases)}{Expected\ deaths\,(cases)} \times 100$$

cases. The ratio of observed to expected is then multiplied by 100 (or another decimal multiplier) to give the standardized mortality ratio (SMR), which now shows the comparisons free from confounding factors such as different age distributions.

The SMR thus allows for comparison of one national, regional, or other defined population group where the age–gender-specific rates are not available, to a selected standard population for which these specific rates are known. This same method is also used to calculate morbidity as standardized incidence ratios (SIRs) or other health-related observations.

Comparing mortality or morbidity rates in European countries is made accessible to all by the Health for All database. This compares rates of mortality, morbidity, health resources, utilization, lifestyle, and others. Data for all countries are standardized to the European population standard, so the reported rates are comparable.

Standardized mortality (incidence) ratios (SMRs or SIRs) are therefore the crude rate or the total number of deaths or cases occurring in the study group, compared to the expected number of deaths if that population had experienced the same death (or incidence) rates as the standard population. The standard population provides a strong base of comparison as it is larger in size, with less likelihood of random variation.

SMRs can be calculated for a specific population group at special risk and compared to a standard population to see whether it is vulnerable to higher rates. A group of people who have been employed in a certain industry and exposed to asbestos may, after a long latency period, develop mesothelioma. The SMR for a population of former asbestos workers in a 25-year follow-up study is seen in Table 3.7. Studies in the USA, the UK, and Italy followed the expected burden of mesothelioma calculated based on exposed population and degree of exposure, to document and project future expected deaths from this highly specific asbestos-induced disease.

In the UK, the SMR is used as the adjustment factor for allocation of funds to district health authorities. Following a lengthy examination of many alternatives, the SMR was believed to incorporate many variables affecting health, including age, gender, and socioeconomic and environmental factors. Populations living in areas with higher than expected mortality may have more disease or higher case fatality rates (CFRs) resulting from a greater prevalence of risk factors (genetic, environmental, and/or socioeconomic). Excess mortality may also be due to less access to or poorer quality of health care. Extra resources are made available on this basis to deal with the poorer health status of the population. This is a practical method of addressing regional differences in health, providing a high degree of equity in resource allocation. It takes into account greater need in some areas than in others. The SMR applies epidemiological methods to improve management practice in health.

POTENTIAL ERRORS IN MEASUREMENT

Data must be assessed as to their validity and reliability. They should also be considered for their biological plausibility (Box 3.17). These all affect the degree to which inferences can be made and generalizations drawn from the study sample.

Reliability

Reproducibility or reliability is the degree of stability of the data when the measurement is repeated under similar

---

TABLE 3.7 Mesothelioma Death Rates Among Former Asbestos Workers in Israel, 1950–1990

Study group (n)	4401
No. of mesothelioma deaths in study group	26
Expected deaths from national population rates	0.12
Standardized mortality rate (SMR)	26/0.12 = 216.7

Note: Expected deaths derived from applying age-specific mesothelioma mortality rates of total population of Israel to the study group.
Sources: Tulchinsky TH, Ginsberg GM, Shihab S, Goldberg E, Laster R. Mesothelioma mortality among former asbestos-cement workers in Israel, 1953–1990. Isr J Med Sci 1992;28:542–7.
Hodgson JT, McElvenny DM, Darnton AJ, Price MJ, Peto J. The expected burden of mesothelioma mortality in Great Britain from 2002 to 2050. Br J Cancer 2005;92;587–93.

BOX 3.17 Observation Measurement Issues in Epidemiology

- *Validity* – the degree to which a measure actually measures what it claims to measure.
- *Accuracy* – the extent to which a measure conforms to or agrees with the true value.
- *Precision* – the quality of being sharply defined.
- *Reliability, reproducibility* – the stability seen when a measure is repeated under similar conditions.
- *Instrumental error* – this includes all sources of variation inherent in the test itself.
- *Digit preference* – a consistent bias by observer rounding of numbers (e.g., to the nearest whole number).
- *Interobserver variation* – differences in observation between different observers of the same phenomenon.
- *Individual observer variation* – the same observer may record the same observation differently owing to changes within the observer, not the observed.
- *Bias* – an effect or inference that departs systematically from the true value.
- *Spurious* – an apparent but not genuine epidemiological relationship.

Source: Adapted from Last JM, editor. A dictionary of epidemiology. 4th ed. New York: Oxford University Press; 2001.

conditions. If the findings of two researchers carrying out the same test (such as the measurement of blood pressure) are very close, the observations show a high degree of interobserver reproducibility. However, it is common in medicine that even relatively objective measurements by different observers, such as radiologists' readings of the same X-ray or cardiologists' readings of the same cardiogram, show high degrees of variability. Instrument standardization, observer training in common standards, and standardization of recording observations are needed to ensure acceptable standards of reliability in any data set. Measuring the same patient at different times can

produce different results (as in measuring blood pressure or blood sugar), such that standardization of conditions of recording or timing the test is essential to ensure comparable data. Standardization of a test requires, as part of quality control, sending samples tested in one laboratory to a reference laboratory to see whether the test results are the same.

There are three main types of bias: selection, information, and analytical bias. All other types are subtypes of these. It is worth noting that bias is a dynamic concept; that is, if no conclusion is drawn, there is no bias. This means that bias cannot be defined only based on material aspects. It is important to minimize sources of bias (Box 3.18).

Validity

Validity refers to the degree that a measurement actually measures what it claims to measure. This includes the representativeness of the sample and the nature of the population from which the sample is taken. It includes the nature of the phenomenon being tested and whether the sampling method takes it into account, such as when a condition changes with age, does the sample take that into account, or whether the content of the testing, such as a questionnaire, truly reflects the nature of the phenomenon being studied. A set of findings from a study using white middle-class males or US nurses as subjects may not be generalizable to females or males of other ethnic or socioeconomic status, or populations with different sociocultural environments.

SCREENING FOR DISEASE

Screening for disease may be carried out on a mass basis of a whole population, as was commonly done in the past for TB. When done with a number of tests it is called *multiphasic screening*. Screening may target a group at special risk, such as blood lead screening among workers exposed to lead at their place of work or children living in the vicinity of a plant using lead.

Screening is an essential part of patient care when the caregiver routinely tests, for example, blood pressure, blood sugar, or blood lipids. Hypertension is common and undiagnosed in a high percentage of affected persons, with serious long-term effects such as strokes and other vascular diseases. Blood pressure testing is a simple procedure that should be carried out in all possible health visit situations to find those cases for whom preventive care programs can be life saving.

The *accuracy* of a test is usually measured in terms of sensitivity and specificity. Targeted screening may be required by law, as in the case of newborn screening for phenylketonuria (PKU), hypothyroidism, and other congenital disorders. The value of the screening test is defined with regard to its degree of sensitivity and specificity, as well as its costs and benefits for screening or not screening.

BOX 3.18 Sources of Bias

The reliability of a data set may be compromised by systematic biases in the data collection or processing. Such biases include the following:

- *Assumption bias* – errors from faulty logic, premises, or assumptions on which the study is based.
- *Response bias* – systematic error due to differences between those who choose or volunteer for a study as compared to those who do not.
- *Selection bias* – error due to inclusion of those who appear and are included in a study, leaving out those who did not arrive because they had died, were cured without care, were not interested, and so forth.
- *Sampling bias* – error when sampling methodology does not ensure that all members of the reference population have a known and equal chance of being selected for the sample.
- *Observer bias* – error due to differences between observers; may be between observers (interobserver) or by the same observer on different occasions (intraobserver).
- *Detection bias* – systematic error due to faulty methods of diagnosis or verification of cases in a survey.
- *Design bias* – systematic bias due to faulty design of the study.
- *Information bias* – flaws in measuring exposure or outcome resulting in data being not comparable.
- *Measuring instrument bias* – faulty calibration, inaccurate measuring instruments, contaminated reagents, incorrect dilutions/mixing of reagents, flawed questionnaire.
- *Interviewer bias* – conscious or subconscious selection in gathering of data.
- *Reporting bias* – self-report selective reporting, suppressing, or exaggerating of information; e.g., history of STIs.
- *Publication bias* – editors and reviewers prefer positive results so that a distorted perception of an issue may occur.
- *Bias due to withdrawals* – loss of cases from the sample by withdrawal or non-appearance in follow-up.
- *Ascertainment bias* – error due to the type of patients seen by the observer, or in the diagnostic process affected by the culture, customs, or idiosyncrasies of the provider of care.

Source: *Adapted from Last JM, editor. A dictionary of epidemiology. 4th ed. New York: Oxford University Press; 2001.*

TABLE 3.8 Screening Tests: Validity, Sensitivity, and Specificity

Screening Test	Disease Present	Disease Absent	Total
Test positive	True positive (A)	False positive (B)	A + B
Test negative	False negative (C)	True negative (D)	C + D
Total	A + B	B + D	A + B + C + D

Note: Sensitivity = TP/TP + FN.
Specificity = TN/TN + FP.
Positive predictive value = TP/TP + FP.
Negative predictive value = TN/TN + FN.
True-positive rate (TP)
True-negative rate (TN)
False negatives (FN)
False positive (FP)

Sensitivity is the proportion of truly diseased people in the screened population who are identified as such by a screening test, and is sometimes called the *true-positive rate* (TP). *Specificity* is the proportion of truly non-diseased people who are identified as not having the disease; that is, it measures the probability of correctly identifying a non-diseased person with a screening test, or the *true-negative rate* (TN). A test that produces too many false positives or false negatives is not valid (Table 3.8).

Screening for cancer of cervix is still a life-saving procedure even though an effective vaccine against HPV is now being used. The interval and age of onset of testing are revised periodically, but the Pap smear test of the cervix has proven its value over many years in many countries. Breast cancer screening with mammography is somewhat controversial but still recommended regularly by many professional organizations. Screening for colorectal cancer is now accepted as essential for all people over the age of 50 at intervals of 5–6 years (see Chapter 5).

False negatives (FN) occur when a negative laboratory result appears in a person who has the condition for which the test is being conducted. The condition is present but does not show up on the initial screening test or data set. If screening for PKU is done too soon after birth, some cases may be missed and will only appear later. False negatives can compromise the effectiveness of the screening program.

False-positive (FP) results are those cases in which a positive laboratory result occurs in a person without the condition for which the test is being conducted. Not everyone with an isolated elevated reading of blood pressure has true hypertension. False-positive results must be checked because they cannot be excluded without confirmation by more specific testing, such as repeated blood pressure readings. Precision is the quality of sharp definition of the test. If a laboratory test for environmental contamination is accurate to parts per billion as compared to parts per million, then the precision is enhanced.

Screening for disease and risk factors is a common and necessary part of public health. In order to be valuable, screening requires a valid test and a significant condition with a high prevalence in the population. Screening for breast cancer, carcinoma of the cervix, and many other conditions is part of the armamentarium of public health and contributes to lowering mortality and improving survival rates for these diseases. Screening of newborns is important for conditions that are serious and treatable but uncommon (e.g., PKU) and those that are more common and treatable (e.g., congenital hypothyroidism) (see Chapter 6). PKU is manageable with a strict diet to prevent serious consequences of the abnormal biochemical condition. Screening for these and other birth disorders, cancers of the cervix and colon, and many other conditions is now accepted in standard clinical guidelines, while screening for breast cancer is recommended but is under review as to its cost-effectiveness.

EPIDEMIOLOGICAL STUDIES

Epidemiological methods of study are important, not only to define the extent of disease in the population, but also to look for specific risk or causal factors for the disease. Epidemiological studies permit analysis of a risk factor, a variable, or an intervention (such as a new vaccine or drug), allowing the testing of new hypotheses and innovations in medicine and public health.

Epidemiological studies are classified as observational or experimental (Figure 3.12). No intervention is made in an observational study, whereas an experimental study involves interventions.

Observational Studies

Observational studies are those where the population is studied, but nature is allowed to take its course. They may be descriptive or analytical. Descriptive studies are limited to describing the occurrence of a disease in a population, which is often the first step in investigation, as it may provide clues for more in-depth investigation. Analytical studies go further by looking for specific variables that may be causally associated with the disease.

Descriptive Epidemiology

Descriptive epidemiology uses observational studies of the distribution of disease in terms of person, place, and time. The study describes the distribution of a set of variables, without regard to causal or other hypotheses. Personal factors include age, gender, SES, educational level, ethnicity, and occupation. The place of occurrence can be defined by natural or political boundaries, and can also include such variables as location of residence, work, school, or

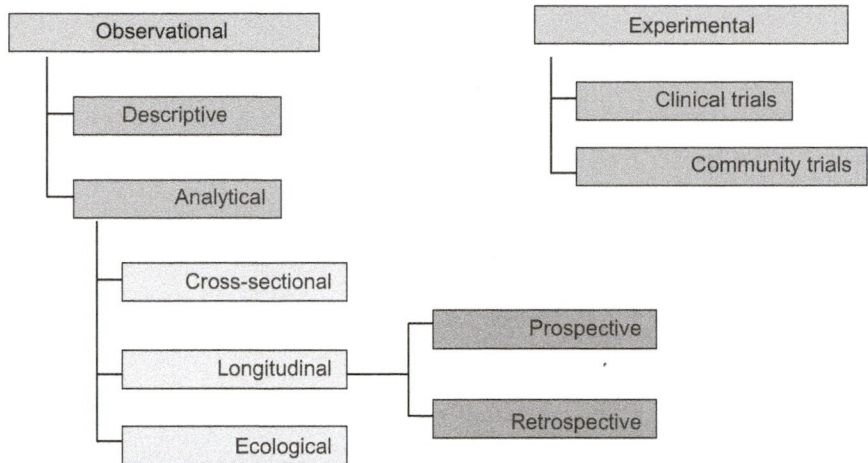

FIGURE 3.12   Classification of epidemiological studies. *Source: Abramson ZH, Abramson JH. Research methods in community medicine: surveys, epidemiological research, programme evaluation, clinical trials. 6th ed. Indianapolis, IN: Wiley; 2008.*

recreation. Time factors include time trends, which are generally divided into three types:

- *secular trends* – long-term variations
- *cyclic changes* – periodic fluctuations on an annual or other basis
- *epidemic disease outbreaks* – short-term fluctuations.

Time trends contribute to our understanding of the natural history of epidemics of acute infectious diseases such as measles or waterborne disease, as well as NCDs such as stroke or cancer. The study of a "natural experiment" when a public health situation is occurring may provide valuable experience and hypothesis for further investigation. Epidemiology also examines the frequency and distribution of potential health indicators and health-related events (such as smoking).

Natural experiments allow observation of the effects of events not in the control of the observer. Increases in legal speed limits are associated with increases in the incidence and severity of traffic collisions, and deaths as velocity increases. Fluoridation of community water supplies is associated with reductions in dental caries and poor dental health of children and elderly people. The observations are consistent and show a strong association, but are denied by ideologues as "not proven".

Laws on smoking restrictions are important in reducing this previously highly prevalent risk factor for CVD and cancer, but this cannot be demonstrated by usual epidemiological methods. Yet it was the good epidemiology that provided the strong association between smoking and these diseases which justified the legal restrictions imposed by many civil societies. Natural experiments are an important part of the evidence base for public health where in many circumstances more definitive epidemiological research methods such as RCTs are neither feasible nor ethical.

Analytical Studies

Analytical studies are concerned with establishing causes or contributory risk factors to disease, including social, economic, psychological, or political conditions that impinge on health. They help to define programs to intervene in order to reduce the burden of disease in the population. Analytical epidemiology has made vital contributions to modern medicine through identification of key risk factors, such as higher rates of lung cancer among smokers and higher rates of stroke among people with hypertension. Analytical studies may include cross-sectional (or prevalence), as well as retrospective, prospective, and ecological studies.

Analytical studies may be individual or group-based studies. Individual-based epidemiological studies collect information about individuals, and both the exposure and the outcome status should be known for each individual within the study. An ecological study is one in which the units of analysis are populations or groups of a population rather than individuals.

Ecological Studies

Ecological studies, also known as *group-based studies*, compare the mean (or summary) values of exposures and outcomes of different population groups. For example, a study analyzing the association between the GDP of different countries and the prevalence of malnutrition in those countries is an ecological study. However, conclusions from ecological studies should be drawn carefully, as the mean values may not be truly representative of the actual situation, and furthermore, because exposures and outcomes for individuals are not established. A group may have a high prevalence of a specific exposure (e.g., oral contraceptive use among women) and an outcome (e.g., prevalence of heart disease), but it is not known whether the individuals with a positive exposure status are also those with a positive outcome. Drawing a conclusion from this apparent relationship is a bias termed an *ecological fallacy*. The association between aggregated variables based on group characteristics does not necessarily represent the association at the individual level.

Studies showing an apparent correlation between quality of drinking water and mortality rates from heart disease have not been substantiated as indicating a "cause–effect" relationship. It would be an inappropriate conclusion (ecological fallacy) to infer from this finding alone that exposure to water of a particular level of hardness necessarily influences an individual's chances of developing or dying of heart disease.

However, ecological studies are important for population health monitoring and for generation of hypotheses for further investigation and intervention. For example, comparison of SMRs for disease categories from routine mortality sets can identify regions with high rates of a specific disease, such as lung cancer or diabetes-related conditions, or motor vehicle accidents, which require follow-up, investigation, and possibly intervention even before more complete epidemiological studies can be carried out. Studies have shown higher rates of CVD mortality for African Americans compared to whites in the USA. However, further analysis shows that there are gradients for cardiovascular mortality for both whites and African Americans according to median family income, such that SES emerges as a more important factor than race.

Cross-Sectional Studies or Prevalence Studies

These studies examine the relationship between specific diseases and health-related factors as they exist in individuals in a population at a particular time. The population may then be divided into subgroups, with and without the disease, and the characteristics of each member of each group analyzed for different variables; for example, age, gender, region of residence, occupation, and social class. Comparisons of these variables may indicate a higher risk for disease in one population group than in an otherwise similar comparable population.

This type of study is relatively simple and easy to perform. However, it has some serious drawbacks resulting from the simultaneous measurement of both exposure and outcome. When investigating two variables (a presumed exposure and presumed outcome) it may be impossible to determine which one is the exposure and which is the outcome, as there is no information about a time relationship. For example, a cross-sectional study of body mass index (BMI) and blood pressure may find that high BMI correlates with high blood pressure, but will not be able to indicate whether people with high BMI had an increase in blood pressure or if people with high blood pressure became fatter. A cross-sectional study may fail to produce valuable information where the main studied exposure or outcome is only present during a short period. If the exposure is short, the recovery from the outcome condition is rapid, or its case fatality is high, it is unlikely that their assessment at one point in time will actually reflect all the exposures and outcomes.

Case–Control Studies

Case–control studies are observational studies of people with the disease (or other outcome variable of interest) and a suitable control group of people without the disease. These studies are retrospective, taking a known outcome status (e.g., disease status) and looking at the exposure status. They compare two similar population groups for their exposure outcomes, one with the disease or condition and the other without. An example is the study of the occurrence of a serious upper limb defect (phocomelia) in children born in Germany in the late 1950s, which showed that of those born with this defect, 41 out of 46 mothers had taken the medication thalidomide as an antinausea pill promoted for use during pregnancy, whereas none of the 300 control mothers with normal children had done so. This study was confirmed by studies in other European countries which had licensed thalidomide, which led to the FDA stopping approval for this drug in the USA, and later to its being banned in countries where it was already in use. Case–control studies are defined as retrospective (defining the outcome status and then looking at the exposure).

The odds ratio is commonly used to summarize findings of case–control studies. It is a ratio of the odds of exposure among cases to the odds of exposure among controls. Case–control studies may be vital to define the differences between the sick and the control groups in an epidemic or outbreak situation.

Case–control studies are ideal for the study of rare disease or conditions that are slow to evolve, as they permit the assembly of a group of cases of appropriate size for analysis, without requiring an extremely large study population. This presents an important advantage as it reduces the cost and time necessary for the study of such conditions. However, a case–control study is prone to various sources of bias, notably "recall bias", where people with (or without) a studied outcome may tend to better remember their exposure status. For example, a study of environmental exposures during pregnancy and fetus malformation may reveal a higher proportion of exposures among women who had an affected fetus because they were more aware and recalled all potential hazards that may have caused the severe adverse outcome of their pregnancies.

Cohort Studies

Cohort studies are also referred to as prospective, longitudinal, or follow-up studies. They examine a population that is initially free from the disease, dividing the population into subgroups according to exposure to a potential risk factor. Such studies can yield the magnitude of risk or incidence rates of the disease under study. The relative risk (a ratio of risk of disease in the two groups; i.e., exposed and non-exposed) can then be calculated. Where risk cannot be determined, the rates of disease for each group (exposed and non-exposed) are determined and the rate ratio may be determined for comparison of risk.

The main disadvantages of cohort studies are the follow-up time they require (during which people may leave the study or be lost to follow-up, an important source of bias) and the relatively large study populations needed to ensure the appearance of a sufficient number of cases for analysis. Long follow-up and large samples usually imply high costs, and make cohort studies less suitable for the investigation of rare diseases or conditions that develop slowly. However, cohort studies present many advantages in terms of the reliability of the information collected, as all exposures may be assessed by the investigators at the beginning of the study and outcomes are identified as they appear during the study period, so that there is no doubt about time relationships.

Cohort (prospective) studies permit the observation of many outcomes from long-term follow-up of a selected population to ascertain morbidity and mortality data not readily available in the general population reporting systems. The British doctors' smoking habits study initiated by Richard Doll and Bradford Hill was carried on from 1951 to 2001, showing the harmful effects of smoking in terms of lung cancer, coronary heart disease, and early mortality. The Framingham Study, initiated in 1949, has provided a wealth of epidemiological information on risk factors for CVD in the population of Framingham, Massachusetts (see Chapter 5). Many epidemiological prospective studies follow selected population groups, such as the long-term prospective study of nurses conducted by Walter Willett at the Harvard School of Public Health, providing a major source of new information on the health of women. This is the largest cohort study of women, tracking over 120,000 nurses since 1976.

Retrospective or historical cohort analysis looks back at earlier records of groups with a specific disease and their earlier life experience. Factors such as smoking, birth weight, obesity, hypertension, or exposure to toxic substances (e.g., asbestos) are analyzed in relation to current morbidity and mortality from lung cancer, coronary heart disease, diabetes, and mesothelioma.

Observational studies of particular population groups have provided important public health advances over the past 50 years. A natural experiment is a situation in which naturally occurring circumstances result in two similar population groups, one exposed to a supposed causal factor and one not exposed as a study or control group. The term *natural experiment* is derived from John Snow's 1850 study of Londoners exposed to drinking water supplied by two different water companies, one group having high rates of cholera and the other low rates. This term is currently used in investigating epidemiological events, regarding each event as a unique situation for which relevant factors need to be defined and, to the extent possible, linked to the disease.

A cohort study of 68,444 adults exposed to the 9/11 terrorist destruction of the New York World Trade Center was made up of lower Manhattan residents, area workers, and passers-by enrolled in the World Trade Center Health Registry. The cohort was followed for 5–6 years and showed morbidity from post-traumatic stress syndrome and lower respiratory symptoms. The study showed that: "respiratory and mental illness are closely linked in individuals exposed to 9/11 and should be considered jointly in public health outreach and treatment programs" (Stellman SD, personal communication; Nair et al., 2012).

Experimental Epidemiology

Experimental studies are studies of conditions under the direct control of the investigator, conducted as closely as possible to a laboratory experiment. Experimental epidemiology involves changing a variable and measuring the effect in one or more population groups. Clinical epidemiology applies experimental epidemiological research methods to clinical problems and practice. It includes promoting the use of epidemiological knowledge in the clinical care of individual patients. Clinical epidemiology also contributes knowledge to the planning and operation of health care systems and clinical and community trials.

Controlled Trials

Controlled trials are epidemiological experiments designed to study an intervention (preventive or therapeutic). A clinical trial requires a random method of allocating the cases to the experimental or the control group, and then both groups are observed for change over time in relation to the condition being studied. Assignment to the treatment or the control group is by random selection. If the people in both the test and control groups do not know which group they are in, the study is called *blind*. If, in addition, the people judging the outcome are not aware whether the person tested is in the test or control group, the trial is called *double blind*. Furthermore, if those analyzing the data also do not know who was in each group, the study may be called *triple blind*. Blinding helps to avoid various biases which limit the value of a study. If the difference in outcomes is statistically significant for the control group and the treatment group, then the treatment is deemed to have been effective.

Although RCTs are considered the gold standard in clinical epidemiology, they are often not available for important policy issues and would be unethical to conduct because denying the benefits of a known positive intervention would be unacceptable. They are also often difficult for policy-making generalization because of inherent limitations in the methodological limitations and resources available for the study.

Field Trials

Field trials follow people who are disease free in two groups, one with and one without a specific intervention, to determine whether the intervention affects the risk of developing the disease. They are often used to test new vaccines in susceptible populations. The field trial conducted by Jonas Salk of inactivated poliomyelitis vaccine in 1956 demonstrated its protective effect and safety in some 1.5 million American children, and the vaccine was subsequently adopted throughout the world. Field trials are part of the process of approval for new vaccines and medications.

There are many ethical traps in conducting such trials in developing countries without adequate transmission of information to subjects in field trials. Serious ethical breaches in such experiments are discussed in Chapters 4 and 15: the Tuskegee and Guatemala experiments with syphilis in the 1930s and 1940s stand as important warnings to overzealous research with inadequate protection and ethical clearance, as now required according to the Helsinki Declaration and more recent iterations of ethical standards in epidemiological research.

Community Trials

Community trials are conducted on whole communities to measure the effect of a risk factor or intervention. They cannot easily be randomized because the entire community is selected, and it may not be possible to isolate the community from changes going on in the general population. Community-based heart disease prevention programs have been undertaken in many settings, such as in North Karelia, Finland; in the USA, such as the Minnesota Heart Health Project, Pawtucket Heart Health Project in Rhode Island (CHAD project); in Kiryat Yovel, Jerusalem, Israel; and many others. These are difficult to evaluate, with a conflict between experimental design and community realities. Regional programs for prevention of heart disease cannot be isolated from time trends in the surrounding communities, limiting the interpretation of measured outcomes. Nevertheless, community trials are necessary in evaluating health interventions to reduce risks or adverse health outcomes. They often rely on performance or utilization indicators as proxies. For example, a village health worker program may lead to earlier and more frequent use of prenatal care or immunization coverage, but measurement of outcome variables may be hard to establish in field conditions, mainly because of a lack of reliable data.

ESTABLISHING CAUSAL RELATIONSHIPS

Classically, the search for causation in medicine and in public health is for the agent–host–vector relationship, with the agent being a specific causative organism. In infectious disease epidemiology, this has provided the scientific basis for immunology and control of vaccine-preventable diseases, and for sanitation to prevent transmission of foodborne and waterborne diseases. Criteria for attributing causation for communicable disease were established in the nineteenth century by Jacob Henle and Robert Koch (Box 3.19).

Criteria for causation include the strength of the association, biological plausibility, consistency with other investigations, and dose–response relationship. Biological plausibility is a test of the plausibility of a causal association based on existing biological or medical knowledge. Consistency with other investigations means that the findings are similar to those of other studies. The dose–response relationship is that in which a change in amount, intensity, or duration of exposure is associated with a change (increase or decrease) in a specified outcome.

Even in infectious disease control, the public health reality is often more complex than the single-causation model. TB deaths fell during the nineteenth century, presumably due to improved nutrition and living conditions, and were further reduced in the early part of the twentieth century before the antibiotic era by a combination of improved nutrition and symptomatic treatment. Mortality from measles dropped dramatically despite its endemicity (the continuing presence of a disease in a given geographic area) prior to the successful vaccine introduced in the 1960s. This can be attributed to rising standards of living and improved means of treatment of complications. Even today, the mortality rate from measles is seen to be affected by improving the nutrition of children and by vitamin A supplementation.

For NCDs, causation is even more clearly multifactorial, and a risk factor for one disease may also be a contributor to increased risk for another disease. Diet has been established as a major risk factor for coronary heart disease, as well as diabetes and hypertension. Diabetes is a major risk factor for coronary heart disease, stroke, renal, eye, and peripheral vascular disease. Nutrition is an important contributor to certain cancers, so that the multiple-factor causation of disease cannot be ignored.

Risk factors for disease are those aspects of personal behavior or lifestyle, occupational or environmental exposure, social and economic conditions, and inborn or inherited characteristics which, on the basis of epidemiological evidence, are known to be associated with health-related conditions considered important to prevent. Non-infectious diseases are often related to and exacerbated by a number of risk factors, so that measurement of the prevalence of risk factors, or intervening variables, is important to epidemiological assessment of the future risk of such diseases. The

BOX 3.19 Henle–Koch Postulates on Microorganisms as the Cause of Disease

- The organism (agent) must be shown to be present in every case of the disease and must be isolated, cultured, and identified.
- The organism must produce the disease when a pure culture is given to a susceptible animal.
- The organism must be recoverable from the animal.

Source: *Last JM, editor. A dictionary of epidemiology. 4th ed. New York: Oxford University Press, 2001.*

BOX 3.20 Criteria for Causation in Chronic Disease: The Evans Postulates

- Prevalence of the disease should be significantly higher in those exposed to the hypothesized cause than in controls not so exposed.
- Exposure to the hypothesized cause should be more frequent among those with the disease than in controls without the disease, when all other risk factors are held constant.
- Incidence of the disease should be significantly higher in those exposed to the hypothesized cause than in controls not so exposed, as shown by prospective studies.
- The disease should follow exposure to the hypothesized causative agent with a normal or log-normal distribution of incubation periods.
- A spectrum of host responses should follow exposure to the hypothesized agent along a logical biological gradient from mild to severe.
- A measurable host response following exposure to the hypothesized cause should have a high probability of appearing in those lacking this before exposure (e.g., antibody, cancer cell) or should increase in magnitude if present before exposure. This response pattern should occur infrequently in people not so exposed.
- Experimental reproduction of the disease should occur more frequently in animals or humans appropriately exposed to the hypothesized cause than in those not so exposed; this exposure may be deliberate in volunteers, experimentally induced in the laboratory, or may represent a regulation of a natural exposure.
- Elimination or modification of the hypothesized cause should decrease the incidence of the disease (e.g., attenuation of a virus, removal of tar from cigarettes).
- Prevention or modification of the host's response on exposure to the hypothesized cause should decrease or eliminate the disease (e.g., immunization, drugs to lower cholesterol, specific lymphocyte transfer factor in cancer).
- All of the relationships and findings should make biological and epidemiological sense.

Sources: *Evans AS. Causation and disease: the Henle–Koch postulates revisited. Yale J Biol Med 1976;49:175–95.*
Porta M, Greenland S, Last JM, editors. International Epidemiological Association. A dictionary of epidemiology. 5th ed. New York: Oxford University Press; 2008.

prevalence of smoking, as an example, may serve as an indicator of the future potential of lung cancer and CVD. BMI, blood pressure, and serum cholesterol levels measured in the community serve as indicators of risk for coronary heart disease (Box 3.20). These measurements indicate individual and community risk, and the potential effectiveness of health promotion programs.

NOTIFICATION OF DISEASES

Morbidity data are reported by doctors, usually based on compulsory reporting of specific infectious and non-infectious diseases. Some diseases such as plague, cholera, yellow fever, louseborne typhus, and louseborne relapsing fever are notifiable by international convention. Locally endemic diseases are notifiable under national and also state/provincial public health laws in order to monitor their prevalence and the impact of public health measures (see Chapter 4). Additional diseases reported routinely in other countries include waterborne and foodborne disease, chemical poisonings, botulism, leishmaniasis, septicemia, *Chlamydia trachomatis* (genital), gonococcal ophthalmia, and listeriosis. Other diseases or health events may be added to routine reporting (or to special surveys) according to endemic environmental conditions. Reporting of infectious diseases is one of the most important foundations of public health practice.

SPECIAL REGISTRIES AND REPORTING SYSTEMS

Special registries are used to establish a basis for the epidemiological study of vital health events pertinent to the population and clinical states important to population health. These include mandatory reporting and special registries and surveys. They are vital for monitoring the health of a population and providing epidemiological information to guide health policy, whether it is for an acute infectious disease challenge or a long-term chronic disease problem such as CVD or diabetes. The range of such reporting systems is necessarily very wide (Table 3.9), with recent additions including mandatory reporting of child and elder abuse.

Priorities may vary from country to country, but the basic registry needs in health care include a range of conditions, including infectious diseases, cancer, birth defects, and hospital discharge information systems. Data from cancer, birth defect, and low birth weight registries can provide valuable clues about environmental exposures of public health importance.

Ideally, disease registries and reporting systems should be coordinated into unified health information systems. The USA has an effective network of such reporting systems, such as the Census Bureau, the Department of Health and

TABLE 3.9 Public Health Mandatory or Voluntary Reporting and Registries

Mandatory	Special Registries or Surveys
Vital events: birth, death, marriages, and divorces	Cancer registries
Notifiable infectious diseases, including STIs, HIV, and TB (see Chapter 4)	Chronic diseases registries
Birth weight and condition (Apgar score)	Neurological disorders registries
Birth defect registries	Diabetes registries
Congenital screening for PKU, hypothyroidism	Coronary heart disease
Abortions and other pregnancy events	Thalassemia
Hospital discharge information systems	Sickle cell disease
Battered children, partners/spouses	Mental illness – psychiatric conditions
Domestic violence and elder abuse	Nutritional status indicators surveys, e.g., NHANES
Motor vehicle accident injuries	Growth and development indicators
Air and water quality monitoring	Blind and partially sighted people
Environmental hazards and monitoring	Deaf and hearing impaired
Occupational safety and health hazards	Disability surveys
Animal disease monitoring	At-risk workers' groups
Vaccine and drug reactions	Behavioral risk factors surveys
Hospital infections and incident reports	Internet and news media obituaries
Poison control centers	Influenza – sentinel reporting centers
Injuries, trauma	Autism registries
Workers' compensation	Alzheimer's and other dementias
School absence	Toxic substance and poison control centers
Public health laboratories	Hazardous waste sites
Social security: Medicare, Medicaid, special categories (e.g., end-stage renal disease patients)	Psychiatric/mental health
Hospital discharge information systems	Cancer, leukemia, lymphoma, and transplant registries
Blood bank	Cystic fibrosis registries
Public health laboratories	Self-rated health status surveys
Veterinary public health surveillance	Sentinel sites for influenza reporting
Animal reservoirs and health	Behavioral risk factors surveys (e.g., smoking, teen pregnancies, car seat belt use)
Vaccine and drug reactions	Nutritional surveys (e.g., NHANES)
Hospital (nosocomial) infections	Growth and development indicators
Injuries	Health insurance systems utilization
Poisonings (e.g., poison control centers)	Performance indicators (e.g., GP immunization and preventive service coverage rates, hospital utilization)
Violence and trauma (i.e., emergency services)	

Note: STI=sexually transmitted infection; HIV=human immunodeficiency virus; TB=tuberculosis; PKU=phenylketonuria; NHANES=National Health and Nutrition Examination Survey; GP=general practitioner.
Roush S, Birkhead G, Koo D, Cobb A, Fleming D. Mandatory reporting of diseases and conditions by health care professionals and laboratories. JAMA 1999;282:164–70.
New York State Department of Health. Chronic Disease Registries; 1999. Available at: http://www.health.ny.gov/diseases/chronic/diseaser.htm [Accessed 11 January 2013].
Sources: Adapted from Declich S, Carter AO. Public health surveillance: historical origins, methods and evaluation. Bull World Health Organ 1994;72: 285–304.

Human Services, state health departments, and the CDC, which has a variety of surveillance systems and a regular weekly publication with periodic special reports on special surveys and routine reports of disease incidence and prevalence. Individual identification numbers, such as Social Security numbers, for each member of the population enable the use of data from related special registries. However, protective measures must be in place to ensure privacy and to prevent the misuse of these data for unethical purposes. Safeguard mechanisms can be built into data systems to protect the privacy of the individual. This is a particular problem in the USA, which has a large unregistered immigrant population, many of whom receive services from public programs, but who may be put in jeopardy by the threat of possible deportation by federal immigration authorities.

Linkages among data sets allow important epidemiological correlations to be studied. For example, linking data sets for cancer registries, vital records, pollution indicators, and hospital discharge information systems may enhance the investigation of specific medical conditions, such as monitoring longevity and hospital use for childhood cancer. Such links may also be used to compare morbidity and mortality patterns for specific conditions by comparing hospitalizations with mortality patterns.

A study by the Department of Health, based on an observation from routine death reports of 32 infant deaths in New York State over a 10 year period, found that none of the 24 hospitals where these deaths occurred had standing orders for vitamin K injection at birth, as recommended by the American Academy of Pediatrics since 1961. The Commissioner of Health initiated a decision by the State Board of Health for the adoption of mandatory vitamin K by injection as a routine for newborns. This was gradually adopted by all states and there are now zero deaths in the USA from Vitamin K Deficiency Bleeding (VKDB) previously known as Hemmorhagic Disease of the Newborn (HDN).

Mandatory care in most states now includes vitamin K along with antibiotic eye care and heel blood for newborn screening for phenylketonuria (PKU), congenital hypothyroidism, sickle-cell anemia, and many other inborn errors of metabolism to prevent Vitamin K Deficiency Bleeding (VKDB) (see Chapter 6 and Box 3.21). Birth defect registries are very important as there are many preventive interventions that can reduce birth defects, such as folic acid fortification, reduction of low birth weight in newborns, and intervention in cases of social deprivation associated with low education and social support for young single mothers. Monitoring the incidence of new cases and rates will help in evaluation of the effectiveness of interventions such as folic acid supplements before pregnancy and fortification of flour with folic acid (see Chapters 6 and 8).

The importance of records linkage may also be demonstrated by the following epidemiological example. Mortality

---

BOX 3.21 Identification of Vitamin K Deficiency Bleeding (VKDB) in a Review of Vital Records and Follow-Up Study in New York State

Studies of vital statistics registries may raise epidemiological questions or hypotheses for further investigation. Special surveys become important as the follow-up to initial findings. Intervention can then be planned on the basis of these investigations. An example review of vital statistics in New York State (1987) showed 32 infant deaths reported during the 1980s attributed to Vitamin K Deficiency Bleeding (VKDB), then called Hemorrhagic Disease of Newborne (HDN), a disease preventable by prophylactic vitamin K injections of newborns.

A study of the State Hospital Discharge Information system showed a substantial number of hospital discharges with the diagnosis of HDN (first to fourth diagnosis) during the same period. A case record review conducted of infant deaths with VKDB, then known as Hemorrhagic Disease of Newborn (HDN) as a diagnosis (first to fourth diagnosis). Two-thirds of the cases did not receive vitamin K at all, or not until after bleeding had already begun. None of the 22 hospitals in which these cases occurred had standing orders for vitamin K injections for newborns. Up to that time, five states had mandatory vitamin K requirements for newborns and is standard practice since first recommended by the American Academy of Pediatrics in 1961.

As a result, the New York State Department of Health adopted mandatory vitamin K prophylaxis for newborns. Record linkage between hospitalization data and the individual cases would have made such a study more readily achievable. This study led to adoption of mandatory vitamin K injection for all infants in New York State and subsequently in all US states. No cases of mortality from this condition were reported in 2011 and 2012. In 2013, 4 cases of late VKDB were reported in a childrens' hospital in Tennessee due to mothers refusal to give vitamin K to their newborns. Three of these children had intracranial hemorrhages. Vitamin K is not standard international recommended care for newborns.

Source: *Tulchinsky TH, Patton MM, Randolph LA, Meyer MR, Linden JV. Mandating vitamin K prophylaxis for newborns in New York State. Am J Public Health 1993;83:1166–8.*
Zipursky A. Prevention of vitamin K deficiency bleeding in newborns. Br J Haematol 1999;104:430–7.

---

from CVD has fallen dramatically in industrialized countries since its peak in the early 1960s. This decrease can be attributed to many factors, including changes in nutrition, smoking, and other risk factors, but also to improved medical care for hypertension and for acute coronary events, as well as long-term cardiac rehabilitation and care. The prevalence of the basic disease process may not have declined, but primary and secondary prevention is much improved. Studies linking hospitalization patterns with preventive action such as smoking education laws and CFRs for CVDs are helping to provide support for prevention and new modalities of care.

DISEASE CLASSIFICATION

Because comparative statistics are vital in monitoring the health status of a population, it has been essential to develop internationally accepted standard nomenclature and a coding system in order to minimize differences in classification. The Bills of Mortality used in the seventeenth century defined 17 categories. Classification of disease by anatomic site or body system was initiated by William Farr at the Second International Statistical Congress in Paris in 1855.

After World War I, the League of Nations supervised revisions of the *International Classification of Diseases* (ICD), and since the 1948 sixth revision, the ICD has been updated at approximately 10-year intervals by the WHO. The tenth revision of the *International Classification of Diseases* (ICD-10) came into general use in 1993. The classification is broken down into many subcategories with coding to indicate precise disease and procedure groups (see Table 3.10 on the companion web site at http://booksite.elsevier.com/9780124157668). Similarly, a classification of mental health disorders has been developed (see Chapter 7).

HOSPITAL DISCHARGE INFORMATION

Admission to a hospital is a major medical event, no less important from an epidemiological point of view than the reporting of a death or an infectious disease. A hospital discharge data system is an informational, planning, budgeting, epidemiological, and quality control tool in modern health care. It involves gathering a basic data set on all hospital discharges, input of data into a central file on a regular basis, and processing the data for administrative and epidemiological purposes. This process requires a basic data retrieval form for all hospitalized patients and a system of reporting and analysis, preferably with computerized data retrieval.

Hospital statistics were originally promoted by Florence Nightingale in the nineteenth century as essential to improve outcomes of care. The Uniform Hospital Discharge Information System (UHDIS) evolved as a result of the growing recognition of the importance of hospital utilization in the economics of health care (Box 3.22). Introduced in the 1960s by the US National Center for Health Statistics (NCHS), it provided the basis for the development of diagnosis-related groups (DRGs), which have become the

---

BOX 3.22 Hospital Discharge Information Systems

- *Planning* – organizing based on admission and surgical rates, utilization by age and gender, diagnosis, length of stay, and "small area analysis" which compares practice patterns and use or excess and waste of resources; search for new methods to promote patient flow to alternative care facilities (e.g., minimal supervisory residential care, ambulatory, or home care).
- *Case-mix analysis* – make-up of the hospital case load, looking for common diagnoses or rare events which might be of epidemiological significance, or may have administrative and quality control importance. Case mix has become part of payment systems for hospital care in the USA and other countries.
- *Budgeting* – planning within the hospital and in relation to referral sources based on utilization patterns by diagnosis and department.
- *Quality of care monitoring* – determination of aberrant practice, complications, or outcomes (e.g., excess surgical rates, infections, mortality). Organisation for Economic Development and Cooperation (OECD) includes many measures of hospitalization as quality of care measures, including in-hospital case fatality rates for myocardial infarction, strokes, and cancer of the colon, and avoidable hospital admissions for asthma and asthma mortality rates.
- *Epidemiology* – tracing and mapping epidemics of communicable diseases and identifying localizations and sources; using "tracer conditions" to pick out medically and epidemiologically significant events such as strokes or diabetes mellitus; supplementing international, national, or regional mortality data.

- *Research* – through case finding of particular clinical events which may then be analyzed for related variables (e.g., incidence of coronary heart disease to compare with mortality patterns, intracranial hemorrhages, and administration of prophylactic vitamin K to newborns, or follow-up of patients with coronary artery bypass procedures).
- *Linkage with other registries* – linkage with death records, cancer, birth defects, or other special disease registries; relating hospitalization events to special disease registries, such as birth defects, cystic fibrosis, asbestosis, and mesothelioma; supplementing a cancer registry.
- *Economic analysis* – this is an essential aspect of modern health care and the use of hospital care and its alternatives, central to health economics; linked data from various registries and hospitalization data can provide data for important cost-effectiveness and other economic planning models.

Sources: *Dennison C, Pokras R. Design and operation of the National Hospital Discharge Survey: 1988 redesign. Vital Health Stat 2000;1(39). Organisation for Economic Co-operation and Development. Health policies and data: OECD health data 2012 – frequently requested data. October 2012. Available at: http://www.oecd.org/els/healthpoliciesanddata/oecd-healthdata2012-frequentlyrequesteddata.htm [Accessed 11 January 2013]. Centers for Disease Control and Prevention/National Center for Health Statistics. National Hospital Discharge Survey. 16 October 2012. Available at: http://www.cdc.gov/nchs/nhds.htm [Accessed 11 January 2013]. Department of Health and Human Services, Centers for Medicare & Medicaid Services. Federal Register, 11 May 2012. 42 CFR Parts 412, 413, 424, et al. Medicare Program; Hospital inpatient prospective payment systems for acute care hospitals and the long-term care hospital prospective payment system and fiscal year 2013 rates; Hospitals' resident caps for graduate medical education payment purposes; Quality reporting requirements for specific providers and for ambulatory surgical centers; Proposed rule; 77(92):1–324.*

major mode of payment for hospitals in the USA and in some other countries since the 1980s. Use of the ICD allows for comparisons among data sets, regions, and countries. The National Hospital Discharge Survey (NHDS) was conducted annually from 1965 to 2010, using a national probability survey of 500 and later 239 US hospitals. It provides information on characteristics of inpatients discharged from acute-care short-stay hospitals to examine important topics of interest in public health (NHDS/NHCS, 2012). A central governmental professional unit is needed at the state level to plan, train, and supervise data retrieval, and to process and interpret the output data. Data provided by all hospitals provide a complete picture of the entire population using all hospital services, rather than just those services provided by an individual hospital in the region. This is necessary, as people residing in a hospital catchment area may be hospitalized in another region by referral or for emergency care.

Developing countries need assistance in developing basic registration systems of births, deaths, and other vital events. The WHO estimates that tens of millions of such events occur annually without registration or reporting. At the same time, the understaffed primary care services compile daily records with large amounts of indigestible data on ambulatory care utilization. Scarce financial and personnel resources should instead be focused on more significant and higher quality data associated with hospitalizations. Fewer centers are involved in hospital care than in ambulatory care, so that data retrieval is easier to control. Most importantly, the less common event of hospitalization is medically and epidemiologically more significant because it consumes 40–75 percent of health care financing. A UHDIS may be seen as a priority information system after the reporting of infectious diseases, mortality, cancer, and birth defects.

The three primary users of information flow in a hospital information system are clinical medicine, epidemiology, and managerial services. However, much of the development of information systems in recent years has been for managerial purposes. Good data should be easy to interpret for managers and clinicians alike. This requires informatics staff (knowledgeable of modern technology) to tailor the data reporting method so that the manager and others can analyze the data for their needs. The data should be in a manageable format and training should be provided for users of the system.

Hospital discharge provides a basis for epidemiological monitoring and control of diseases and simple research information. Analysis of hospital discharge data, especially mortality, surgical complications, and excessive length of stay, provides important indicators of efficiency and quality of care. Interregional variations in hospital utilization provide a clear premise for designing and implementing policies. With the increasing use of surgery, cancer care, and other medical care on an outpatient basis or with endoscopic methods, long lengths of stay in hospital are unjustified

from both the patient welfare and economic points of view, which are important to the health insurance or health service system (Box 3.22).

Hospital discharge data studies permit case-mix studies, reveal trends in care patterns and patient safety conditions, and provide a basis for peer review within a hospital and between hospitals. They provide material for analysis and policy formation at the clinical level, as well as for hospital management and planning; for example, in the development of ambulatory care, reducing admissions and length of stay for services better provided on an outpatient basis.

The number of hospitalizations is reducing over the years, with rates varying by age group. Limitations of the data include factors such as lack of standardization of diagnostic criteria. Some patients do not reach a hospital, for economic or other reasons; they may have transportation problems, or may have died prior to admission. Others may be unaware of the existence of some health services or are simply afraid of them. Moreover, the denominator for rates is missing because the hospitals may not have a defined catchment population. Nevertheless, hospital discharge information is an important tool for planning, monitoring, and evaluating health services (Box 3.22).

Vast numbers of people use ambulatory care, generating too large a data set for effective monitoring. The number of ambulatory care visits may range from four to 10 per person per year, depending on the country. Ambulatory care data are of poorer quality because they are usually in broad categories of diagnosis, such as musculoskeletal and respiratory complaints, which comprise the bulk of visits. Ambulatory care can be monitored selectively through sampling or monitoring of representative sentinel centers to provide examples for wider replication. Specific components of ambulatory care should be monitored, such as infants and school-age children receiving immunizations, attendance for prenatal care, birth control services, screening for hypertension and diabetes, or breast cancer screening, as particular health goals. With increasing trends for ambulatory care surgery and medical care, linkage of such data with inpatient care is needed to ensure continuity of comparisons with previous patterns of care.

HEALTH INFORMATION SYSTEMS (INFORMATICS)

Information is needed for the management of any health system. It is vital to establishing objectives, developing programs, and managing the use of resources. Modern information technology, or informatics, provides the tools for analysis and policy formation to adjust the service. Informatics is as much a part of health care as the cardiograph or ultrasound machines. It provides the feedback, "imaging", or cybernetics potential for management.

Dissemination of information is no less vital than its collection or interpretation in central offices. Reporting of vital data is meaningless unless the data are processed and fed back to the service system in a regular, timely, and usable fashion or, in current computer terminology, in a user-friendly manner. Modern health information monitors the operation of a health care system, including component parts such as objects (hospital buildings), people (health personnel), services, policy (equity), finance, organization, administration, regulation, quality assurance, and health promotion. The component parts interact to support the system as a whole. Interaction is made possible through information and communications technology and driven by financing and organizational imperatives.

Health care services are a source of increasing expense to governments and individuals. As a result, governments throughout the world are recognizing the importance of health information for effective health services management and planning. The requirement for public accountability has led to the design of policies to ensure appropriate quantity, quality, and effectiveness of care with the best use of resources. This has created substantial requirements for information.

Public health informatics is the systematic application of information and computer science to clinical and public health practice, research, and learning. It includes the use of computerized medical and hospital records, the use of clinical and preventive care guidelines, and disease registry information retrieval.

Each country must develop its own health information system and uniform health information systems, such as that developed by the WHO European Region (Box 3.23). This system provides a timely (current or real-time) spectrum of vital statistics, demography, and key outcome measures, as well as data on health care resources and utilization. Each country should provide local, district, community or municipal, and regional health profiles. This information should be widely distributed and available for analysis and discussion to the media, the public, and health professionals. Data are of little value if locked away and unavailable for regular circulation and dissemination to a wide audience, who require this information in order to make an informed contribution to policy analysis and formation.

Precision is limited by the quality of the data, but even limited data are extremely important in epidemiology and for health planning. Some infectious diseases are reported less stringently than others, partly because of lesser concern by physicians, but also because the clinical presentation may be atypical, or some cases may be entirely subclinical. A clinical case of poliomyelitis may represent 100 subclinical cases. Many infectious diseases of public health importance (e.g., measles, rubella) are underreported because non-immunized, vulnerable children may not be brought to medical care despite mandatory reporting requirements, while some reported cases are unconfirmed by laboratory evidence. Nevertheless,

---

BOX 3.23 Functions of Health Information Systems

- *Monitoring* – of the health status of a population.
- *Comparisons* – using historical, regional, national, or international patterns and standards.
- *Assessment* – an overview of the health status of a population based on available data, the professional literature, field visits, and interviews with key health personnel and community representatives.
- *Evaluation* – monitoring use of resources, performance, and outcomes of programs as part of total quality management.
- *Prediction* – using current data to predict trends in disease ("modeling") and utilization patterns, costs, potential outcomes, program planning, policy formulation, and setting priorities.
- *Explanation* – data to understand disease patterns, risk factors, and service utilization of a population of a district and determine causal relations, or need for intervention.
- *Planning* – data are needed for planning responses to public health problems and monitoring the outcomes of interventions.
- *Payment systems* – diagnosis-related groups (DRGs) and case-mix systems of payment are now used widely in the USA and elsewhere to provide incentives for efficiency in care and short stay in hospital. This requires good home care and ambulatory care in hospital and in primary care settings.
- *Adaptation* – as new technologies (e.g., laparoscopy and robotic surgery) increase the effectiveness of care, hospital care patterns change; as science advances (e.g., discovery of *Helicobacter* as the cause of chronic peptic ulcer disease), much of the surgery done in previous decades is no longer performed.
- *Quality improvement* – early response to index cases of infectious diseases provides information critical to rapid response and management of longstanding diseases that recur, e.g., diphtheria after decades of its control, or new entities such as HIV in the 1980s, and many examples since. Patients in health care facilities are at risk of serious hospital-acquired infections or human error which cost many lives each year and prolong hospital stay. Monitoring and preventive systems help to avert these issues.

---

reported cases are the basis for monitoring and policy formation. Awareness of the direction and magnitude of errors will enable the user to determine the validity of the data.

Making health information data available on a routine basis to providers and managers of services helps to promote an awareness of the overall operation of the health system in which they are involved. Information provides the basis for accountability, which implies that the provider of care or the manager of a health system is responsible for and must report on the results of his or her work, including unintended outcomes. Any system of service requires a system of

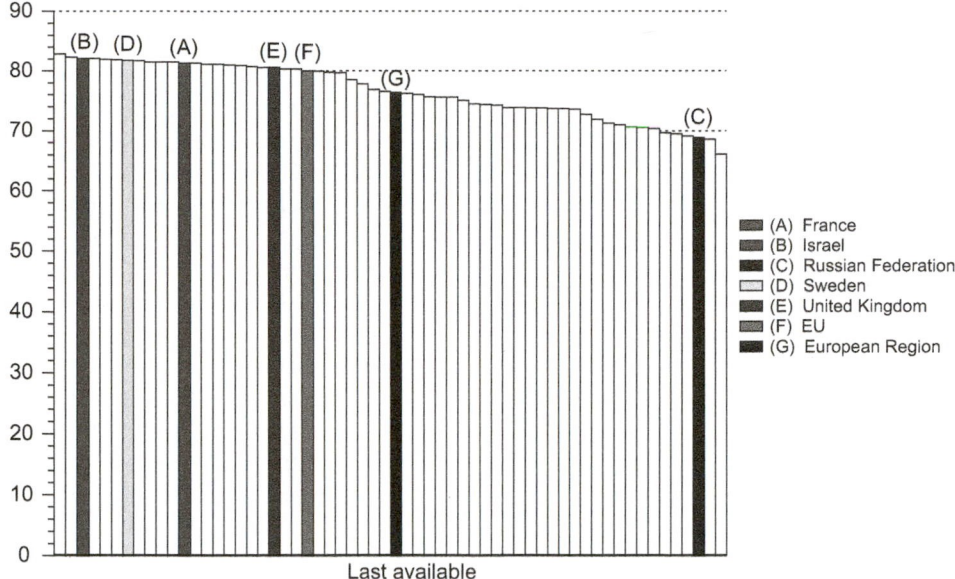

FIGURE 3.13   Life expectancy at birth, European Region, 1970 to 2010–2011. *Source: World Health Organization, European Region. Health for All database; July 2012. Available at: http://data.euro.who.int/hfadb/ [Accessed 2 February 2013].*

accountability in order to maintain standards and to provide the consumer with an assurance of quality care.

In a centrally managed system, reporting of services provided is part of the chain of command. In a decentralized system, such data may be derived from billing patterns from hospitals or physician payments. They are then transferred to the higher levels of the health service administration and used for decision making and planning. Those who provide the data should be informed of the outcome, including resultant operational decisions.

The United Nations Statistics Division is the primary agency responsible at the international level for collecting official national statistics related to population size and structure, birth, death, migration and social concerns globally, with updated country reports. The WHO Technical Committee on Information Systems emphasizes that the more active and innovative a health policy, the greater the need for information. Data collection and processing require planning, training, and continuing monitoring. While massive data banks are not helpful, well-selected and widely available information systems targeted to vital events in the health process can promote flexibility and relevance in the planning of health services. Other international organizations maintain vital statistics and socioeconomic data systems with regular reporting available online; these include the United Nations Children's Fund (UNICEF), the Organisation for Economic Co-operation and Development (OECD), the United Nations Development Programme (UNDP), US Census Bureau International, the European Union (EU), and regional offices of the WHO (Africa, the Americas, South-East Asia, Europe, Eastern Mediterranean, and Western Pacific).

WHO European Region Health for All Database

The WHO European Region makes available an outstanding database as a free service. It provides some 500 health indicators for all countries in the European region of WHO and is updated twice yearly. It can be accessed at http://www.euro.who.int/ under Publications and Data. It can be downloaded to a computer and unzipped to provide continuous access to up-to-date data on demographics, mortality, morbidity, lifestyle, resources, and utilization data, and presented as time trends or single-year comparisons of all countries in the region or as a single-year map. It is excellent for teaching purposes and the graphs and data can be downloaded to Microsoft PowerPoint or Word documents. An example is shown in Figure 3.13, which compares life expectancy at birth in 2010–2011 in all countries in the European Region of WHO and indicates selected countries (France, Israel, Russia, Sweden, the UK, EU member states, and the European average).

SURVEILLANCE, REPORTING, AND PUBLICATION

The publication and wide distribution of weekly summaries of specified reportable diseases are essential to maintain the viability of reporting and promote meaningful use of the data (Box 3.24). The CDC of the US Public Health Service publishes and widely distributes the *Morbidity and Mortality Weekly Report* (MMWR), reporting on national and international epidemiological events through surveys and special reports. The weekly report is supplemented by in-depth special reviews of important public health topics.

BOX 3.24 Factors Affecting the Value of Data

- *Relevancy* – Are the right data being collected? Are some data no longer useful?
- *Coverage* – Do the data help to identify high-risk groups?
- *Quality* – How good do data need to be to be useful? Limitations of data are a factor in decision making.
- *Acceptability* – Are the data collected acceptable in terms of design, cost, and ethical standards?
- *Timeliness* – How recent are the data? How long a time-series is needed to show temporal patterns?
- *Accessibility* – Are the data available to those who need them? Are the data suitable for publication? Are they published and distributed on the Internet and hard copy?
- *Usability* – Are the data in a usable format? Are they presented in a user-friendly manner (i.e., easy to access and use for non-specialists)? Can one generate summaries, graphs, and tabulations from the data?
- *Cost* – What does it cost to collect and process the data? Are the data available to students and researchers without prohibitive cost?
- *Validity* – To what extent do the data relate to the issue of concern?
- *Specificity and sensitivity* – Were the raw data collected using accurate measures (i.e., measures with a high capacity of detecting actual cases and determining non-cases as such)?
- *Data aggregation and reporting* – Are data reported by disease, category of service, social indicators, and region of residence? What is the population at risk?
- *Biological plausibility* – Is the observed or presumed causal association compatible with existing biological and medical science? Can it be explained from a biological perspective?
- *Equity* – Do the data show interregional and social class variation and inequity?
- *Dissemination* – Information obtained, collated, and analyzed must be organized and available to those who report the raw data, who need data to monitor health status, and who plan health services and health promotion needs of the population.

Source: *Last JM. A dictionary of public health. New York: Oxford University Press; 2007.*

The Department of Health and Human Services and Census Bureau publish frequent topical reports, as do the Agency for Healthcare Research and Quality and other agencies of government. The WHO publishes the *Weekly Epidemiological Record* (WER), which reports global, regional, and country epidemiological events and offers highly professional reviews of selected topics of infectious diseases internationally. *Eurosurveillance* is published by the European Center for Disease Prevention and Control, based in Stockholm and sponsored by the EU, and monitors infectious disease events in the EU and potential candidate countries. The OECD, UNICEF, and UNDP publish annual reports of high importance for the field of public health. The UNDP annual reports on progress in MDGs overall and by country are an important source of health-related data.

The UK Health Protection Agency publishes regular reports on infectious diseases and a wide variety of environmental and other publications of public health importance. Ministries of health often use online reporting and publication to keep the flow of information available to public health practitioners. The Public Health Agency of Canada publishes *Canada Communicable Disease Report*, as well as *Chronic Diseases and Injuries in Canada* for non-infectious diseases and related laboratory findings, in addition to Statistics Canada publishing annual updates on important economic, environmental health, and other databases. Many countries publish similar bulletins and annual reports vital to following trends in health status of their populations.

Reporting systems and publication of the data are both vital to epidemiological monitoring of infectious and non-infectious disease trends. Regular circulation to field personnel increases the sense of awareness and participation in epidemiological monitoring and shows that the reporting is put to good use. Awareness of the reported data helps local health providers and managers in managing their services more effectively (Box 3.25).

Providing ready access to historical and current data as the events unfold is vital to promote a sense of involvement and challenge for the achievement of goals, such as high coverage of immunization and rapid control of disease outbreaks. Linking data sets such as for hospitalization and ambulatory care with mortality data provides important material for studies of the health impact of interventions with comparison groups. One challenge in managing health systems is to monitor population health by linking multiple factors. Studying the impact of health promotion activities such as those of community health workers can provide a rationale for introducing new approaches to community health to improve patient education for diabetics, smokers, or young people at risk for intravenous drug use and suicide. Macrostudies into natural changes in the socio-economic and physical environment include investigations assessing the impact of economic change on air pollution in California over the period 1980–2000 by linking multiple data sets (Davis, 2012), and monitoring complications from influenza vaccinations by studying Medicare claims (Burwen et al., 2012). Internet surveys of physicians can help researchers to understand doctors' attitudes to immunization for influenza or managing hypertension and help to elucidate quality of care with outcome data.

The Internet is clearly an essential tool for public health, for reporting and obtaining data, and for access to the world literature. Many resources such as the MMWR, WER, and *Eurosurveillance*, as well as major journals, are available online free, at least as abstracts and as full articles for some publications (Public Health Reviews at www.publichealthreviews.eu). CDC

BOX 3.25 Evidence- and Best Practice-Based Public Health

Evidence-based evaluation of policies to improve health and reduce inequalities, prioritization, and providing resources for these policies requires four basic types of information:

- a detailed assessment of the magnitude and impact of health problems in the population, including information on the causes of loss of health in the population in terms of both diseases and injury, and risk factors or broader determinants
- information on health expenditure and health infrastructure (a national system of health accounts) detailing the availability of resources for health improvement and what the resources are currently used for
- information on the cost-effectiveness of available technologies and strategies for improving health
- information on inequalities in health status, health determinants, and access to and use of health services (including both prevention and treatment services).

Performance-based measures have become essential elements of public health policy and implementation strategies. These are generally based on professional consensus criteria determined by Delphi methods of consultation. These may be translated into "gold standards" and health targets. They may be used for performance monitoring and indeed administrative payment systems to encourage their complete implementation. Examples of performance indicators for payment include immunization coverage, Pap smears, and mammograms for patients registered with British general practitioners.

The concept of health targets has become an essential element of US public health policy with *Healthy People 2020* at the federal level with state compliance with such measures. When reviewing policy issues in public health, currently accepted practices used in other countries with recognized stature in this field should be taken into account, as well as recommendations by respected international health agencies such as WHO, UNICEF, and others.

Source: *Brownson RC, Fielding JE, Maylahn CM. Evidence-based public health: a fundamental concept for public health practice. Annu Rev Public Health 2009;30:175–201. http://dx.doi.org/10.1146/annurev.publhealth.031308.100134.*

BOX 3.26 Assessing the Health Status of the Individual

- Current chief complaint
- Personal data – age, gender, ethnicity, education, marital status, children, living situation
- Occupational history
- Family history
- Personal history
- Functional inquiry – systems review
- Summary of risk factors – family history, hypertension, diabetes, smoking, sedentary lifestyle, high-fat diet, occupation, alcohol use, stress, other
- History of the present illness
- Physical examination
- Differential diagnosis
- Other medical problems
- Investigation: laboratory, cardiographic, imaging, other
- Presumptive or working diagnosis
- Treatment and its effects
- Definitive diagnosis
- Management of other medical problems
- Follow-up management and monitoring
- Counseling regarding long-term health needs

ASSESSING THE HEALTH OF THE INDIVIDUAL

Physicians and other health professionals are trained to assess the health of the individual patient seeking care (Box 3.26). This involves more than dealing with the chief complaint, requiring a history of the present illness, as well as a wider review of body functions, family and occupational history, physical examination, and laboratory and imaging tests.

Defining a differential diagnosis and treatment for a presumptive diagnosis allows for follow-up of a patient to observe the course of the disease, the outcomes of diagnostic tests, and the effects of intervention. Caregivers must take into account the effects of the process on the patient, the family, and the community. Providers must also be concerned about costs of care, alternative methods of looking after the patient to meet changing needs, and promoting early and maximum recovery. Continuous monitoring and re-evaluation are key parts of the process. There are many parallels in care of the individual and care of the population.

ASSESSMENT OF POPULATION HEALTH

Health service administration is being increasingly decentralized in many countries, and the concept of healthy cities/municipalities is becoming more widespread. These developments have increased the need for and value of health profiles at the community, county, and district levels. This type of health profile provides management with regular monitoring of the health situation, including resources, utilization, morbidity, and mortality. This application of

provides regular and special reports, as do WHO, UNICEF, UNDP, EU, OECD, and other international agencies. These are available online, free of charge. Newsgroups enable convenient and immediate discussions by professionals on particular topics, such as Promed for almost daily current infectious disease reporting from around the world (see Chapter 4). Similarly, the Internet permits literature searches and access to interest groups on virtually any topic in health. This allows people to be in contact with and to obtain support from many others in their field. The WHO home page (http://www.who.int/en/) provides access to its component departments and regional offices.

modern health informatics at a community level does not require advanced computer capacity or skills. Annual reports in a standard format using all existing data sources can be brought together in a user-friendly manner to provide valuable health status monitoring.

District or community health information systems increase the potential for local health authorities and communities to have greater power in determining local health policy. National health authorities need to provide guidance on health targets and resources that may be used flexibly to meet local needs. But supervision and regulation by national health authorities are essential to ensure that resources are well used and that targets are being met, as well as to reduce inequalities between regions.

The WHO European Region has developed a user-friendly computer program for 1000 health indicators, including sociodemographic, mortality, morbidity, health resources, utilization, and lifestyle indicators. These can readily be produced in tabular or graphic form with time trends and mapping capability. The program is accessible free of charge to anyone with a personal computer, Internet access, and modest computer skills via http://www.euro.who.int/hfadb.

As with individual health assessment, evaluation of the health status of a population is based on the accumulation of a portfolio of observations and data from a variety of sources and their interpretation, with comparisons to international, national, or regional patterns or standards. Community health assessment (CHA) begins with identification of the main health problems or chief complaints as understood by key health professionals and the community, or from regular community health profiles.

Information should be derived about the community's SES, the resources available for health care, how they are distributed, and how services are utilized, as well as morbidity, mortality, and other "outcome" measures that help to describe or compare health status (Table 3.11). Health measures include how care is provided, how it governs or monitors itself, and how the system is accountable for its component services. The knowledge, attitudes, beliefs, and practices (KABP) of the people and health providers and the way in which society addresses risk factors for ill-health may also be important determinants of health status.

Gathering the data necessary for monitoring the system itself should be part of the standard functions of a health system. This provides for accountability in use of public resources and maintains a self-correcting feature of the system. CHAs help to point out health risk factors at the population level, and if carried out in a timely and regular fashion, changes can be made without inordinately long waiting periods and without any unnecessary increase in morbidity or mortality.

The CHA is part of the health planning process; it may be designed to monitor the impact of an intervention program meant to deal with a particular health problem, such as coronary heart disease, or a set of risk factors for disease, such as smoking. The CHA is also part of program evaluation, especially in community trials, with an evaluation protocol based on a multiphasic approach and data from many sources.

Defining the Population

The population served by a health system must be defined in terms of age and gender distribution. This is one of the key factors in the planning of health care services, as different age groups have different needs. Women, children, and the elderly utilize more health services and institutional care than the population in general. The demographic pyramid is an excellent graphic summary of the population distribution. The health status of elderly people is affected by the major chronic diseases and the associated disability and mortality patterns. While increasing longevity is associated with a healthier elderly population, the demand for care still grows with age. The elderly, and increasingly the very elderly (those over 85), are high users of health services, including institutional care in hospitals and long-term facilities.

Socioeconomic Status

Health is affected by standards of living and therefore analysis of income and its distribution is a component of the process of assessing the health status of a population. The national average income is often represented by the gross national product (GNP) or gross domestic product (GDP) per capita; for instance, the average of the total production of goods and services of a nation. Real income may vary by state or district, ethnic group, educational levels, gender, or family size. These and many other factors may affect the distribution of wealth in the population.

Living conditions as reflected in housing standards, density of housing, and crowding (people per room or per square meter) are dependent on family income. Services, such as electricity, running water, indoor toilet and bathing facilities, as well as other service facilities in the home (e.g., refrigerators, toilets, baths, stoves, central heating and air conditioning), are also important measures of health-related socioeconomic conditions. Adverse economic conditions prejudice health status in measurable ways. In developing countries, the poverty–disease–malnutrition cycle affects children, women, and the elderly predominantly, reducing potential for economic growth. Even in industrialized countries, there is unevenness in the patterns of income and of health status; the health status of the upper social class is much better than that of the unskilled workers for many health indicators. Where there are large gaps between the rich and the poor, such as in the USA, there is poorer health status than in countries with smaller social gaps, such as Japan and the Scandinavian countries.

Educational level of parents is an important factor in family health. In the case of the father in a family, level of

TABLE 3.11 Evaluation of Population Health of a Community, District, State, or Country

Factor	Topics	Example Indicators
Geography	Climate, topography, density, urban/rural	Tropical, temperate, mountainous, desert, distance from medical facilities
Demography	Vital statistics	Population size, age/gender, urban/rural
Socioeconomic	Ethnic, cultural, religious practices Community, family economic status	Per capita and family income, education, literacy (women), employment, religious affiliation, social attitudes, occupations
Nutrition	Supply, affordability, use of major food groups Food safety and quality Food fortification	Undernutrition and overnutrition Risk group identification Monitoring child growth patterns, anemia
Environment and occupational	Water, air, waste and sewage disposal, toxic wastes, radiological hazards Industrial or agricultural toxic materials	Ambient air pollutants, bacteriological and chemical qualities of community and recreational water, radiation and radon levels, lead levels in soil, water
Public health infrastructure	Organization, training and deployment of public health functions and personnel	Legal and regulatory functions Schools of public health Research capacity in epidemiology, public health
Health care system	Organization Prepaid coverage Finance total and internal allocation	Decentralized administration and finances Integration of local services and finances Total resources; % GNP and per capita (US $) spent on health care; % population with full, partial, or no health benefit insurance
Health resources	Expenditures per capita Hospital beds per capita Long-term care facilities Clinics Personnel, doctors, nurses per capita	Expenditure by type of service, preventive, curative, hospital Acute care beds per 1000 population Special hospital beds per 1000 population Long-term care facilities per 1000 population Doctors and nurses per 1000 population
Community and home care	Post and pre hospital care at home Outreach services to chronically ill Day centers for elderly and handicapped Patient guidance for individuals and groups	Diabetics
Health care utilization	Hospitals, general, chronic, and mental Ambulatory care Preventive services	Admissions and days of care per 1000 population Physician visits per person per year Immunization coverage at age 2 years Ambulatory surgery, home care measures
Process (quality) of care	Professional care standards Accreditation by external agency Peer-review mechanisms Records review Mortality case review Clinical guidelines	Criteria for surgery, second opinion Immunization and child health monitoring rates Correction of deficiencies from accreditation Departmental reviews of caesarean, infection rates Maternal and infant mortality case by case reviews Computerized medical records
Health outcomes	Morbidity Mortality Functional/physiological status "Tracer conditions" – common, treatable or preventable diseases to indicate system failure	Infectious and chronic disease incidence/prevalence Infant, child, maternal, age–gender-specific mortality rates by cause, cardiovascular disease, trauma Anemia of infancy, pregnancy, blood lead levels Lower limb amputation rates
Costs and benefits	Examine specific diseases, procedures, services or health promotion	Cost–benefit of second dose of measles vaccine, bicycle helmets, air bags in cars, antismoking campaigns, e.g., smoking among teenagers
Knowledge, attitude, beliefs, practices (KABP)	General population Risk groups Patients Patients' families Health providers	Diet, smoking, eating, moderate alcohol use, exercise Diabetes, hypertension Birth control, rights of women AIDS/STI-related issues

Note: GNP = gross national product; AIDS = acquired immunodeficiency syndrome; STI = sexually transmitted infection.

education is often a direct determinant of income. In the case of the mother, education relates to income, but even more strongly to successful health care of infants and children. Mothers with higher levels of education, as measured by years of school attendance, are more likely to absorb new knowledge regarding self-care in pregnancy and care of the infant in areas such as nutrition, immunization, and routine baby care. Better educated women tend to have fewer pregnancies, not only because of knowledge of the need for and methods of birth control, but also because of greater self-awareness and different life goals. Ethnic, cultural, political, and religious beliefs and practices have important implications for health, in such areas as the status of women, mental health, family structure, nutrition, substance use and abuse, and birth control and abortion. These beliefs and practices can affect attitudes towards issues such as national health insurance and the funding of health care.

Studies on regional variation in health indicators in the UK show large differences between deprived and non-deprived regions of the country, and between Scotland and northern England on the one hand and southern regions of England on the other. Figure 3.14 shows a comparison of standardized values for life expectancy at birth, mortality, cancer incidence, "limiting illnesses", current smokers, alcohol consumption, childhood obesity and drug use for men and women for three relatively deprived northern regions of England compared to the English average.

Nutrition

Appropriate nutrition, overnutrition, and undernutrition are fundamental determinants of the health of a population.

Overnutrition and obesity place a heavy burden of morbidity and mortality on the health system, with such diseases as diabetes, coronary heart disease, hypertension, and stroke, and their complications. Undernutrition in the form of gross malnutrition is rare in the industrialized countries, but extremely common in many developing nations. In all societies there are groups at risk for overt or subclinical malnutrition, such as iron-deficiency anemia, iodine deficiency, vitamin D and osteoporosis, and other essential minerals or vitamins. A society that acts to prevent malnutrition in vulnerable groups is acting on behalf of the vulnerable groups in the population and indicates the well-being of that society. Public health and economic measures to promote good quality of food and its accessibility to the population, fortification of basic foods, school lunch programs, and meal services for the elderly and chronically ill are health promotion programs that show the level of organized community responsibility for its members (see Chapter 8).

Special surveys, such as low birth weight or nutritional status conditions, are needed to provide nutrition status data. Monitoring of nutrition status, discussed in detail in Chapter 8, is of fundamental importance to population health evaluation. Periodic large-scale national surveys, such as the NHANES, initiated in 1971 in the USA, provide meaningful information on nutrition status in the country. Within the USA, the surveys provide vital information for adjusting recommended dietary allowances and national, state, or local nutrition programs. This information is of great importance for the food industry, which is obliged to follow federal government standards of labeling and content of packaged and processed foods.

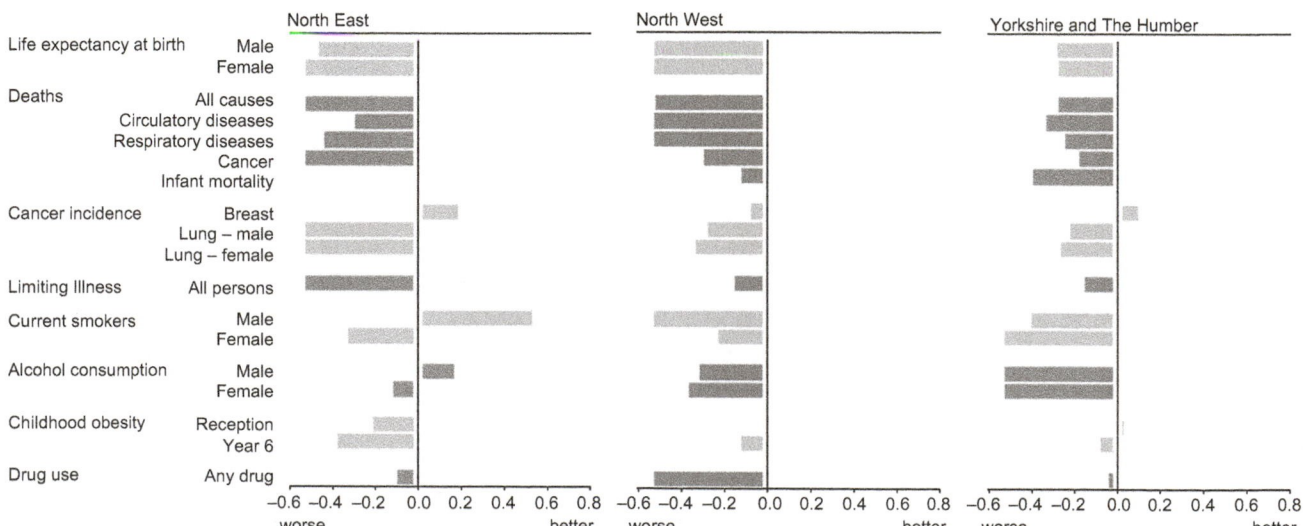

FIGURE 3.14   Health indicators for three deprived regions of England compared to England average, 2006–2008. *Source: Ellis A, Fry R. United Kingdom. Office for National Statistics. Regional trends, no. 42, 2010 edition – Regional health inequalities in England 8 Jun 2010. Available at: http:// www.ons.gov.uk/ons/search/index.html?content-type=Article&pubdateRangeType=allDates&newquery=stroke+mortality+by+regions&pageSize=50& applyFilters=true [Accessed 3 January 2013].*

Environment and Occupation

Safety of community water, management of solid and toxic wastes, air and noise pollution, and ambient air standards are all factors in the health of the community. Organized public health has traditionally focused on these issues, but they remain public policy issues in virtually all countries and internationally. Healthy societies are dealing with these issues with a very high degree of public awareness, sometimes overcoming strong economic interest groups to force improved attention to the environment by governments, communities, and businesses.

Environment includes housing, recreation, schools, businesses, parks, urban and rural planning, and many other aspects of community life that are addressed in "healthy community" initiatives. Employment of children and work in hazardous industries are health issues. Societies that tolerate toxic and dangerous work settings create health hazards that are preventable, but costly to treat. Unemployment, job insecurity, loss of health insurance with change of employer, job-related injury or disease, and low income levels for many workers all contribute to poor health (see Chapter 9). Where health insurance is related to employment, as in the USA, health protection can be a major factor in relation to losing or changing place of employment.

The development of the New Public Health has moved national agendas and local authorities with major roles in improving the health of populations. The idea of community diagnosis and community-oriented primary care has played an important part in this process. It is of vital importance in developing countries where the infrastructure for prevention and primary care remain weak. In countries in transition from the Soviet system of health care, reform should address the imbalance between excessive expenditure on hospitalization and inadequate development of primary care and health promotion. Countries in transition should address high rates of mortality from CVD and trauma (see Chapters 11 and 13).

HEALTH CARE FINANCING AND ORGANIZATION

The way in which a nation finances and organizes health care is an important aspect of health status evaluation. Where there is universal coverage of the population, either through health insurance or through a state-operated health care program, the population in principle has equity in access to care. Financial access, however, does not guarantee actual access because the distribution and supply of services are important variables in utilization. Financing and organization of health services are related issues, discussed in Chapters 10–13, that must be recognized as part of the process of assessing the health status of the population of a country or region. Assurance of access to medical and hospital care does not necessarily ensure that appropriate or effective services are provided.

How services link facilities of different levels of intensity of care and costs is a basic issue in health reform in many countries. The way in which preventive care is provided to special groups in the population (such as infants, children, adults, the elderly, and the chronically ill) and how these services fit together as a holistic entity, interacting to serve the community, are important in determining the status of health and health costs of a community or a country.

Health Care Resources

While overall expenditures for health are important determinants of the level of health care available, no less important is how these resources are spent; that is, what the internal financial allocation is within total health. The major resources for health care are in primary care services, hospitals, and long-term care facilities. All countries have limited health financial resources for health expenditures, so that to a great extent one aspect of the health system can only grow at the expense of another.

Hospitals are the largest segment of the health care system in terms of expenditures and may consume more than 50 percent of total expenditures. The supply of hospital beds is, therefore, a central factor in the health care economy. The number of hospital beds per 1000 population is a key indicator for health economics. The hospital bed-to-population ratio varies widely, from 2.5 to 16 care beds per 1000 in OECD countries, with most countries having reduced hospital bed supplies rapidly since the 1980s.

Age distribution of the population affects morbidity and therefore hospitalizations; countries with a high percentage of elderly people may need more hospital facilities, as well as alternative care services, such as home care and long-term institutional care services. Innovations in health care organization are influencing health planning, with many developed countries reducing acute care hospital admissions and length of stay by a variety of incentive and management systems (see Chapters 11–13). Health planning requires facing up to political and other pressures to sustain or even increase levels of hospital bed-to-population ratios beyond real need, at the expense of other more appropriate alternative services. The absence of organized home care programs is an indicator of inadequate planning to address the needs of the elderly and chronically ill in a society.

The ratio of medical doctors per 10,000 population also varies widely. A high ratio may indicate an overpopulation of specialists and a lack of primary care services, while a low ratio may indicate a need for training more physicians. Countries in Eastern Europe have high doctor-to-population ratios and lower ratings on health status indicators (such as SMRs for trauma) than countries with fewer doctors. Nurse-to-population ratios are also equally variable, but typically, many countries that have high levels of physician-to-population ratios have relatively low numbers of nursing personnel. The number of nurses registered to practice often overstates the actual supply because many nurses never practice following graduation, work only part time, or stay in the profession for only a short period.

Excessive supply of medical doctors, inequitable distribution, relative shortages of nurses, inefficient development of community health programs, and inefficient use of community health workers are important issues in many countries (see Chapter 15). These all have economic and health outcome implications, requiring continuous review and reassessment in each country, and application of lessons learned from other countries.

The organization of health services, discussed in Chapters 10–12, is an important factor in the efficiency and quality of care. Community health services are a hallmark of provision of primary care to address population health needs, while many health systems in the past especially emphasized hospital and other institutional care in their norms and financial incentives.

Utilization of Services

Rapid cost increases have fostered a search for efficient ways of organizing and financing health services. In the USA, the development of the DRG method of payment for hospital services has reduced hospital length of stay. Health maintenance organizations (HMOs) have been successful in providing comprehensive care with less hospitalization and fewer hospital beds than traditional fee-for-service practice. Policy makers and the business community have therefore begun to focus on "managed care" systems to meet the need to extend insurance coverage and to control costs.

While supply of services is important, actual utilization patterns are also a valuable part of the overall evaluation program. Hospital care is a key issue because of its dominance in the economics of health care. Monitoring hospital performance indicators can play an important role in determining the effective functioning of the health care system.

Surgical and other procedure rates are continuing issues in health systems management. For instance, age-standardized hysterectomy rates varied widely among Canadian provinces in 2010, from 512 per 100,000 in Prince Edward Island to 311 per 100,000 in British Colombia, and varied by a factor of 4 within Ontario on a county-to-county basis (2008–2009). A study of this phenomenon indicates that if all provinces achieved the hysterectomy rates of British Columbia, there would be 3700 fewer hysterectomies with a cost saving of $19 million per year (Canadian Institute for Health Information, 2010). A study conducted in Saskatchewan showed that the introduction of mandatory second opinions resulted in dramatic reductions in hysterectomy rates. Appendectomy rates in Germany are up to three times

higher than those in other countries, with no epidemiological explanation.

Studies abound in the USA showing differential utilization of health services by African American and white populations for coronary heart bypass procedures, for localized compared to radical surgery for lumps in the breast, and for mammography and other services currently considered to be of benefit to the patient. These differences generally are primarily due to differences in health insurance coverage, but other socioeconomic or ethnic variables may also be responsible. Excess surgical procedures, for example, caesarean sections, are a widespread problem in countries where fee-for-service is the method of payment, but the amount of surgery is also related to the number of surgeons and fee-for-service payments.

Health Care Outcomes

While it is clear that health status is affected by many social and economic factors, the general state of the country's health is often described by epidemiological indicators, such as mortality and morbidity rates as indicators of health status. Epidemiological information on communicable and non-communicable diseases helps to determine a potential for intervention and alteration of the natural history of the disease.

Outcomes can include morbidity, mortality, and physiological and functional measures (Box 3.27). They may also include measures of self-assessment of health status; risk behavior such as smoking or engaging in unsafe sexual practices; or knowledge, attitude, and beliefs of health-related issues. These measures may be part of the evaluation of the health status of a population or a program meant to cause change.

Outcome indicators include a variety of measures from routine data sources and special surveys. DALYs and QALYs (described earlier) attempt to quantify mortality and quality of life measures for comparisons and for analysis of specific interventions. In addition, physiological or functional indicators such as activities of daily living measure patient performance. Special surveys for clinical signs of undernutrition such as anthropometric measures (growth and body size) should be supplemented by biochemical-level and hematological surveys to establish patterns of undernutrition. Special surveys of nutrition status and disability, school performance, and other indicators of functional status are important aspects of health status evaluation (see Chapter 8).

Quality of Care

The quality of care (see Chapter 11) is part of evaluation of health in any population. Assessment of how available funds are spent to address the health problems specific to that

BOX 3.27 Outcome Indicators of Health Status of a Population

Outcome is a variable with a value which varies according to the outcome or the effectiveness of an intervention (Last, 2007), taking into account independent variables, such as more general changes occurring in the same time-frame. Examples include the following.

Mortality-related indicators
- Infant and child mortality rates (IMRs)
- Maternal mortality rates (MMRs)
- Crude mortality rates (CMRs)
- Age-specific mortality rates
- Cause-specific mortality rates – infectious, non-infectious
- Case fatality rates as a measure of the success of medical care
- Life expectancy (LE) at ages 0, 1, 65, and other ages
- Standardized mortality rates (SMRs) – total specific
- Years of potential life lost (YPLL) – a measure of the impact of mortality on different age groups to reflect relative impact of diseases or conditions on the population
- Quality-adjusted life years (QALYs) – an adjustment of life expectancy by inclusion of chronic conditions with impairment, disability, or handicap
- Disability-adjusted life years (DALYs) – a measurement based on adjustment of life expectancy and includes the estimated effect of long-term disability.

Morbidity outcome indicators
- Incidence of vaccine-preventable disease
- Incidence of waterborne disease
- Incidence of foodborne disease

- Incidence/prevalence of tuberculosis
- Incidence/prevalence of STIs/AIDS
- Incidence of malaria, other tropical diseases
- Prevalence of non-infectious diseases – cardiovascular diseases, diabetes, cancer, trauma
- Prevalence of disabling conditions
- Prevalence of risk factors.

Behavioral indicators
- Knowledge, attitudes, beliefs, practices regarding risk factors – smoking, alcohol and drug use; unsafe sexual practices; high-risk behavior regarding motor vehicles, violence, drug use, suicide
- Compliance with immunization, preventive care, medical treatment and advice, physical fitness, suitable weight.

Physiological indicators
- Nutritional status – growth patterns of infants and children; body mass index of adults; dietary patterns
- Hematological and biochemical indicators (blood sugar; cholesterol; lipids; vitamins A, B, C, D); anemia among infants; children, and women; iodine status; environment.

Functional indicators
- Work and school absence
- Psychomotor function
- Work capacity
- School performance
- Fitness test performance
- Activities of daily living (ADLs)
- Cognitive capacity.

population is part of the CHA. The findings of such evaluations are meant to affect resource allocation and address unmet needs. Health care is increasingly being evaluated by managers of health insurance programs, whether as health maintenance organizations or veterans' health services and Medicare of the US federal government, or by international organizations (such as WHO, UNICEF, OECD, and UNDP), seeing health as an economic investment, and international comparisons, as in the Human Development Index (HDI) and Health for All database. Data systems for epidemiological studies and for population health monitoring include the most basic reporting systems of infectious diseases, vital statistics, and special disease registries such as birth defect registries, special surveys such as NHANES on nutrition status (see Chapter 8), and hospitalizations as seminal health events or "tracer conditions" to provide vital material to study and compare the effectiveness of health systems, and indeed individual provider performance.

Other important indicators of quality health systems include health system responsiveness and patient or population satisfaction. Responsiveness is a measure of ease of access and comfort level of clients with "consumer-friendly"

and psychologically supportive facilities and staff for the population served.

Practices in prescription drug use may indicate utilization much beyond accepted clinical guidelines, as in the use of proton pump inhibitors (PPIs) in the treatment of acid-related dyspepsia and peptic ulcers by the UK NHS. These drugs are important but overused, according to National Institute for Health and Care Excellence (NICE) standards in the UK (see Chapter 15): expenditure on PPIs by the NHS was estimated at €595 million (euros) in England in 2006 and €4.5 billion in the USA in 2009 on one PPI, whereas less costly methods are just as or more effective (Cahir et al., 2012). Such analysis of data sets on prescription drug use is of great importance to the economic survival of health systems, permitting limited resources to be used to better effect for unmet health needs.

Self-Assessment of Health

Data on self-assessment of health are used along with household expenditure and nutrition surveys to provide information on the health-related experiences of selected samples of the population, sometimes by household interviews and by telephone surveys. These may yield estimates of poverty, illness, or inequality for small areas for which no or few other data are available. Reliability of recall and reporting is limiting, but this method does provide important information that cannot be measured in other ways. Health surveys are vital to monitoring population health and self-assessment is an important component of ongoing monitoring, and to measure inequalities within a health system.

Costs and Benefits

Analysis of costs and benefits is reviewed in more detail in Chapter 11 on economics and health policy, and will be mentioned here only briefly. Evaluation of the health status of a population requires examination of the choices made in resource allocation in a particular geographic area. This is of concern not only to the planner, but also to the provider of health care and to the public. If priorities in resource allocation promote highly technological medicine, then primary care may lag behind in resources, and the health status of the population may be compromised. Cost–benefit analyses can contribute to establishing priorities within a health care system (see Chapter 11).

Effects of Intervention

The adoption of *Haemophilus influenzae* vaccine for infant immunization will result in an almost immediate drop in *H. influenzae* meningitis and pneumonia, in the same way as adoption of a two-dose policy for measles vaccination will lead to a very rapid reduction in measles morbidity and mortality. Other interventions in public health affect an epidemic curve more slowly, as smoking reduction actions lead to reduced hospitalization and mortality from coronary heart disease.

Many interventions in preventive medicine and public health are complementary, so that a doctor's advice to quit smoking and antismoking legislation mutually reinforce the same message. The natural history of disease is affected by many sociological and economic factors as well as medical or public health interventions. The dramatic reduction in coronary heart disease mortality, but not necessarily morbidity, is attributable to improved medical care, preventive medical care, and wider public health activities related to improving knowledge, attitudes, and practices for lifestyle change. These themes were discussed in Chapters 1 and 2, and will recur in coming chapters of this book as part of the continuously evolving New Public Health.

Qualitative and Quantitative Research Methods

Public health research capacity is important to investigate how diseases are generated by causative agents, and in the context of contributory factors, how the social, physical, or policy environment influences people's perceptions and behavior. Research methods in epidemiology rely on quantitative studies based on centuries of population data analysis.

Quantitative studies are important for new epidemiological and clinical research. They are the basis for analysis of routinely collected health information such as births, mortality and morbidity rates, and associated factors. They also investigate the utilization of health services, such as short- and long-term hospitalization by cause, and many others such as registries of birth defects, cancer, diabetes, asthma, neurological disorders and other diseases, and socioeconomic data.

Quantitative research uses questionnaires and surveys, including telephone and electronic mail surveys, to provide objective evidence of population health, and its associated factors such as nutrition, smoking, diet, physical activity, self-defined health status, activities of daily living, and many other measures of health and social well-being. Some surveys study age, gender, and ethnic groups for biological factors by, for example, BMI, micronutrient levels (e.g., vitamin D), blood lipid levels, and dietary intake. These are basic to monitoring population disease as cornerstones of public health.

Quantitative research yield data analyzed as rates, proportions, associations, and multifactorial correlations. Quantitative surveys emphasize structure, consistency, precisely worded questions, and analysis methods to quantify experiences and produce measurable outcomes. Quantitative studies generate or use existing databases for analysis which can aid understanding and add precision in evidence of disease risk factors that have become part of modern epidemiology and public health, such as smoking and cholesterol, the reduction of which has led to declines in cardiovascular and cancer mortality (see Chapter 5).

Qualitative research methodologies developed by social sciences are valuable in the direct observation of behavior and attitudes, and have been especially important in exploring issues related to human sexuality, strategies for managing complex public health issues such as the AIDS pandemic, malaria control, and many other public health challenges.

Qualitative research is increasingly related to health issues. The social sciences (psychology, sociology, and anthropology) are important in studying human behavior and the societies in which they live, but with increasing difficulty in trying to explain human behavior in quantifiable, measurable terms alone. Although qualitative research also starts with research questions, these may change with the experience of addressing people in an open fashion in their own communities. This helps to generate knowledge of social influences and processes by understanding what they mean to people. Qualitative research methods are valuable for exploration, with open-ended collection of information by questionnaires, interviews, and focus groups to develop hypotheses for further study using quantitative methods. These types of study supplement quantitative research, or provide new hypotheses and issues for quantitative research to provide important information on the policy alternatives for decision making, and to modify intervention programs.

Clinical observation and analysis is a form of qualitative methodology with exploratory epidemiology that has contributed greatly to development of the field. The observations of Peter Panum of measles in the Faroe Islands in the 1840s made an enormous contribution to infectious disease epidemiology. Observations of a large number of cases of infant cataracts by Australian ophthalmologist Norman Gregg in 1941 led to the discovery of rubella syndrome. The observation in 1979 by pathologist Robin Warren in Adelaide, Australia, of small, curved, organism-like objects in crypts of gastric biopsy specimens led to the discovery of *Helicobacter pylori* as the cause of chronic peptic ulcer disease in the early 1980s, and the Nobel Prize in 2005 (see Chapter 1). In the early years of the HIV/AIDS pandemic, qualitative research provided clues for educational and behavioral interventions that were the only tools available until the advent of ART in the 1990s.

Quantitative and qualitative methods, in principle, both start with a research question as a study hypothesis but differ in their methods of data collection, analysis, and interpretation. Researchers working with behavioral aspects of health serve to generate hypotheses or modifications for quantitative studies or trial interventions. Qualitative researchers should be familiar with methods of quantitative research, and vice versa. In the era of webs of causation, with multiple factors in play, quantitative research provides greater precision and statistical strength to determine causal relationships. Qualitative research provides valuable exploration to elucidate questions which can add to our understanding of the epidemiology of a multifactorial causation, especially regarding compliance with best practices. Both methods are vital to progress in public health (Table 3.12).

The emphasis in qualitative research is on exploration. It relies on the synergy between design and discovery, and thus is valuable for program evaluation. This research helps investigators to elucidate and understand how the social, physical, or policy environment influences people's perceptions and behavior. It does this by focusing on both verbal and non-verbal language using an unstructured interview format so that participants can answer for as long and as openly as they choose.

Important clues to public health issues can be revealed by talking to people. For example, studies on the use of low-cost insecticide-treated nets (ITNs) to prevent malaria in sub-Saharan Africa showed cultural and beliefs to be important in their uptake, including information on their benefit, seasonality of use, and many other factors that could only be determined by interviews and community participation focus groups (Binka and Adongo, 1997). Another study report on this issue (Alaii et al., 2003) states:

"… findings from our anthropologic studies early in the trial indicated that the study population would accept and use ITNs. After introduction, an array of social and cultural issues associated with the ITN studies became apparent. While the majority of these problems could be addressed during the trial they illustrate the shifting roles of communication, time, and the social system in the diffusion process. Individuals seek information at various stages of the diffusion process to decrease uncertainty about its expected consequences. The decision leads to either rejection or adoption of the innovation and success or failure of the intervention."

In another example, research focusing on high birth rates among indigenous adolescent women in rural Mexico would require quantitative surveys to provide relevant data such as the percentage of women pregnant in the age groups 15–17 and 17–19, the probability that a woman will use a contraceptive method, frequency of abortions, or the risk of her dying from pregnancy. Qualitative research would be able to elucidate factors such as misinformation regarding contraception, parental or partner opinions about adolescent pregnancy, and beliefs and problems regarding accessing prenatal and postnatal care. Qualitative research methods can operate independently or complement quantitative instruments by either proceeding or preceding them, depending on the study goals.

Qualitative research is guided by the research problem and community responses in less formal questionnaires or discussion with community residents and key people, which can fuel further research questions. A conceptual framework is often applied to keep the research directed and dictates the combination of questions asked such as ones based on experiences, behaviors, opinions, values, concerns, or knowledge. Qualitative research should be dynamic, using questions and approaches that evolve as new insights are gained. Approaches to data collection can take the form of words, images, and observations; observation, in-depth interviews, and focus groups are the fundamental approaches to qualitative research. Other methods, such as documentary research and videotaping, can also play an important role in gaining participants' perspectives.

Entering the community by acknowledging and consulting with "gatekeepers" or leaders of the potential research site population helps in accessing members of the community. It also facilitates follow-up, such as identifying local people to work with, presenting oneself and the research to key stakeholders, and recruiting participants. Researchers often visit common meeting places, chat with potential participants, and then select a sample purposively based on readiness of individuals to participate, as well as their demographic characteristics to represent a defined subgroup. Sampling can be varied and, depending on strategy,

TABLE 3.12 Quantitative Versus Qualitative Research

Quantitative Research	Qualitative Research
Methodological Approaches	
Define the issue to be examined – case for action	Define the issue to be examined – case for action
Theory or question driven	Theory or question driven
Deductive process to test prespecified concepts, constructs, and hypothesis that make up a theory	Inductive process of observation to formulate a theory, or hypotheses
Objective in observing effects (interpreted by researchers) of a program, problem or condition	Interviews and focus groups use semi-structured but open-ended questions/formats
Sampling representative of population size, composition, randomization crucial	Describes a problem or condition from the point of view of those experiencing it
Surveys, structured interviews, observations, and reviews of records or documents numeric information	Time expenditure lighter on the planning end and heavier during the analysis phase
Fixed response options use numbers to define relationships via closed-ended answers, experimental, empirical means	Sample size and composition less formal, structured for exploration
Data collection: surveys with closed answers	Interpretive "experience near"
Statistical tests used for analysis	Sampling – selection of sample of people with direct familiarity with the population and research question
Specificity and reliability key issues	Text-based, and not numerical
Analysis: turning beliefs, behaviors, or attitudes into numbers to support hypotheses	Analysis of observed interactions, behaviors, and attitudes
Conclusions in keeping with the findings, limitations and plausibility given the literature and knowledge of the topic	No statistical tests
	In-depth information on fewer participants
	Conclusions in keeping with the findings, limitations of the study, more research questions indicated, and policy implications
Research Questions	
Precisely worded questions, structured response options	Interview skills require well-trained personnel
Aim to quantify information/data and produce measurable outcomes	Unstructured or semi-structured response options, room for follow-up questions
Structured by hypothesis	Aim to explore and gain insight into behavior and perceptions
Less in-depth but more breadth of information across appropriate sample size	How people interpret and experience their interactions and perceptions and/or attitudes
Place emphasis on structure	Open-ended or semi-fixed structure: discovery and exploration, synergy between design and discovery
Statistical tests used for analysis	Methods include focus groups, in-depth interviews, and reviews of documents for types of themes
Can be valid and reliable: largely depends on sample, measurement device, or instrument used	Can be valid and reliable: largely depends on skill and rigor of the researcher
Time and cost expenditure heavier on the planning phase and lighter on the analysis phase	More in-depth information on a few cases
Reliability, uniformity, objectivity, and freedom from bias are paramount	Less generalizable
More generalizable	Generate hypotheses for future research or policy decisions
Generate further research and policy guidelines, standards	

Centers for Disease Control and Prevention. CDCynergy "Lite". Evaluation. Available at: http://www.cdc.gov/healthcommunication/cdcynergy/evaluation. html [Accessed 3 January 2012].
US Department of Energy. Differences between qualitative and quantitative research methods. Available at: http://www.orau.gov/cdcynergy/soc2web/Content/phase05/phase05_step03_deeper_qualitative_and_quantitative.htm [Accessed 3 January 2012].
Sources: Feldman B. Personal communication; 2007.

may select homogeneous, heterogeneous, extreme, or typical participants. Pilot testing often follows to assess how well the objectives of the study are fulfilled, and provides the opportunity to circumvent any constraints and obstacles before study initiation.

One-to-one, or in-depth interviewing allows participants to play an active role in determining the direction of the interview. Questions follow the flow of conversation and the interview has a conversational quality. The interviews can take the form of unstructured informal conversations, or can be semi-structured or structured. They generate empirical data as participants talk freely about their experiences and beliefs. This is an effective approach when inquiring about sensitive information and when assessing an individual's opinions and perceptions rather than understanding community norms and customs. In-depth interviews can highlight the differences between individuals, elicit detailed information, and also provide a forum for follow-up questions.

In the 1960s, the NHANES began to study the US population health and nutrition behaviors and the links between dietary habits and NCDs. By the 1980s, epidemiological evidence showed that personal health behavior was a major risk for premature morbidity and mortality from many diseases including lung cancer, CVD, and HIV, and health promotion became an established part of public health. In 1984, the CDC established behavioral surveys with standard questionnaires administered through telephone surveys to monitor established risk factors in 15 states of the USA. These surveys supplement other important epidemiological monitoring systems, such as vital statistics, disease registries, and health systems monitoring, with counterparts in other countries.

FROM HEALTH INFORMATION TO KNOWLEDGE TO POLICY

Internal review boards (IRBs) are research monitoring bodies or committees, sometimes called Helsinki Committees, whose approval is required for research funding and publication purposes. IRBs require that all precautions are taken so that participants are not exposed to harm by the study, and that the project is scientifically sound. They also require that follow-up care is provided with referrals, that a researcher/practitioner is clear about his or her role boundaries, and that appropriate information and support are available.

Consent requires that participants are informed that research is not therapeutic. Some situations do not require consent when it is made clear that participants understand the study. Confidentiality must be maintained (e.g., the secure storage of tapes and transcripts), using as few details about participants as possible. This is to prevent anxiety and distress, exploitation, misrepresentation, and identification of participants in published papers. Validation for respondents refers to the process whereby researchers review the results of the study with the participants before the findings are published.

Information is the basis for planning, organizing, managing, and providing high-quality care. The process begins with basic vital statistics and the epidemiology of infectious and non-infectious diseases to identify and quantify the health needs of the population. It extends into health information systems to manage and monitor the functioning of the health care system. Surveillance of health events at national, regional, and community levels depends on building information systems and linking data to provide community health profiles. This process is fundamental to monitoring and managing health systems. It requires clear policy to ensure that information systems do not exist to serve only those who process the data at national levels, but are returned to the community level and linked with other data sources in readily usable formats (Box 3.28).

SUMMARY

Epidemiology and related sciences have made enormous contributions to defining the causes of disease and articulating their risk factors, and translating them into effective public health policy saving millions of lives. Information is widely available in the form of health statistics and published data of all kinds, today more than ever on the Internet. The sophisticated methods and data sets available provide a wide array of information allowing the continuous development of information technology and monitoring systems for health policy and the management of health facilities and health systems.

Health policy formulation requires seeking the appropriate information and making intelligent use of it. Educating health workers in coordinating information and streamlining data will help them to understand the relevance and impact of their actions. Information systems and the flow of properly organized and disseminated data are vital for management. They are as important to the functioning of the system as an intelligence service is to a military operation. The vast and expensive mechanism of a health service operates in the dark without a continuously monitoring information system and applied research methods of epidemiology.

Translation of knowledge into practice in many cases moves with glacial speed. Delayed implementation of established preventive interventions such as weight loss and prescription of beta-blockers and antihypertensive medications costs many needless premature deaths. These practices no longer require research to demonstrate efficacy and effectiveness; what is at issue is how to ensure that they reach all those in need. The vast majority of cardiovascular deaths could be eliminated through measures that have already been demonstrated in etiological studies (Ness, 2013).

Throughout the world, health care systems are under critical scrutiny because of concerns over costs, accessibility, appropriateness, quality, and outcomes of care. The

BOX 3.28 Evidence-Based Public Health and the Burden of Proof

The Hippocratic Oath specifies: do good and do no harm. It has found expression in the precautionary principle, a contemporary redefinition of Bradford Hill's case for action; when in doubt about the possible presence of a hazard, the burden of proof is shifted from showing presence of risk to showing total absence of risk. This creates a dilemma in public health and in clinical medicine suggesting that the normal evidence required for action is without validity. It implies that any possible risk of an intervention outweighs the risk of non-intervention.

Great care is warranted when introducing new public health interventions, but the weight of evidence must include not only epidemiological studies but policies derived from Delphi consultative procedures and successful experience of the intervention in large population groups over long periods, without substantive evidence of harmful effect.

A balance between the precautionary principle, the experience of "good public health practice" and epidemiological evidence is often a delicate judgment, but is nonetheless essential for policy in this field. Last's definition of *evidence-based public health* is wise: "application of best available experience in setting public health policies and priorities. The evidence comes from official vital and health statistics and from peer reviewed publications in epidemiology, sociology, economics and other relevant disciplines".

Failure to act on best practices and cumulative evidence can be an ethical and indeed a legal problem (see Chapter 15), where inordinate delay in implementing scientific and practical positive experience with public health interventions can allow serious morbidity and mortality to go unchecked when they are preventable.

The time lag between adequate scientific evidence and positive experience with good public health practices can be very long, and measures that can save or improve the quality of life for large numbers of people are delayed in implementation due to lack of political motivation, priorities, and active or passive resistance by professional or lobby groups with other agendas.

Delays in the adoption of a two-dose policy for measles vaccination and slow implementation in some developing countries have cost millions of lives. The implementation of folic acid fortification of flour has been slow, despite overwhelming evidence and positive experience in over 60 countries showing that folic acid fortification prevents birth defects and late pregnancy terminations with low cost and great safety. The banning of DDT in the 1960s due to legitimate environmental concerns without replacement of equally effective insecticides contributed to the resurgence of malaria, again costing millions of lives. Keeping up with scientific and best public health practices is an important responsibility of public health in balance with due precaution.

As Brownson et al. (2009) point out, "An array of effective interventions is now available from numerous sources including the *Guide to Community Preventive Services*, the *Guide to Clinical Preventive Services*, *Cancer Control PLANET*, and the *National Registry of Evidence-based Programs and Practices*: Second, to translate science to practice, we need to marry information on evidence-based interventions from the peer-reviewed literature with the realities of a specific real-world environment. Finally, wide-scale dissemination of interventions of proven effectiveness must occur more consistently at state and local levels."

Jacobs et al. (2012) address the "free online resources in the following topic areas: training and planning tools, US health surveillance, policy tracking and surveillance, systematic reviews and evidence-based guidelines, economic evaluation, and gray literature. Key elements of EBPH are engaging the community in assessment and decision making; using data and information systems systematically; making decisions on the basis of the best available peer-reviewed evidence (both quantitative and qualitative); applying program-planning frameworks (often based in health-behavior theory); conducting sound evaluation; and disseminating what is learned."

Sources: *Last JM. A dictionary of public health. New York: Oxford University Press; 2007.*
Coughlin SS, Barker A, Dawson A. Ethics and scientific integrity in public health, epidemiological and clinical research. Public Health Rev 2012;34: Epub ahead of print. Available at: www.publichealthreviews.eu [Accessed 10 January 2013].
Brownson RC, Fielding JE, Maylahn CM. Evidence-based public health: a fundamental concept for public health practice. Annu Rev Public Health 2009;30:175–201. http://dx.doi.org/10.1146/annurev.publhealth.031308.100134.
Jacobs JA, Jones E, Gabella BA, Spring B, Brownson RC. Tools for implementing an evidence-based approach in public health practice. Prev Chronic Dis 2012;9:110324. doi: http://dx.doi.org/10.5888/pcd9.110324. Available at: http://www.cdc.gov/pcd/issues/2012/11_0324.htm

effectiveness of a health system is frequently on the political agenda. Quality assurance and accountability are critical in the operation of any health system. Health expenditures must be increasingly justified in terms of their need and cost-effectiveness, policy formulation, strategies, and priorities, taking into account economic, sociological, and political factors.

Curbing the soaring costs of health care is a necessity and not a matter of choice for governments and individuals if the WHO policy of Health for All is to be achieved. One means of reaching the goals and objectives of this policy is to develop an efficient health information system. Knowing the population, the epidemiological patterns of its diseases, and its health care services and utilization, are all part of the monitoring and feedback systems essential to allow the health system to evaluate health status and to keep pace with changes. They are therefore essential elements of the New Public Health.

NOTE

For a complete bibliography and guidance for student reviews and expected competencies please see companion web site at http://booksite.elsevier.com/9780124157668

BIBLIOGRAPHY

Abramson, Z.H., Abramson, J.H., 2008. Research methods in community medicine: surveys, epidemiological research, programme evaluation, clinical trials, sixth ed. Wiley, Indianapolis, IA.

American College of Epidemiology, http://acepidemiology.org/ (accessed 06.01.13).

American Public Health Association, http://www.alpha.org (accessed 03.01.13).

Breslow, L., 2006. Health measurement in the third era of health. Am. J. Public Health 96, 17–19.

Brownson, R.C., Fielding, J.E., Maylahn, C.M., 2009. Evidence-based public health: a fundamental concept for public health practice. Annu. Rev. Public Health 30, 175–201. http://dx.doi.org/10.1146/annurev.publhealth.031308.100134.

Cahir, C., Fahey, T., Tilson, L., Teljeur, C., Bennett, K., 2012. Proton pump inhibitors: potential cost reductions by applying prescribing guidelines. BMC Health Serv. Res. (12), 408. Available at: http://www.ncbi.nlm.nih.gov/pmc/articles/PMC3529111/ (accessed 08.08.13).

Canadian Institute for Health Information, 2010. Health Care in Canada 2010. CIHI, Ottawa. Available at: https://secure.cihi.ca/free_products/HCIC_2010_Web_e.pdf (accessed 08.08.13).

Centers for Disease Control and Prevention, http://www.cdc.gov/ (accessed 03.01.13).

Centers for Disease Control and Prevention, 2008. National Center for Chronic Disease Prevention and Health Promotion. Behavioral Risk Factor Surveillance System – BRFSS history. CDC, Atlanta, GA. Available at: http://www.cdc.gov/brfss/history.htm (accessed 03.01.13).

Centers for Disease Control and Prevention, National Health and Nutrition Examination Survey (NHANES). Available at: http://www.cdc.gov/nchs/nhanes.htm (accessed 03.01.12).

Centers for Disease Control and Prevention. Public Health Then and Now: Celebrating 50 Years of *MMWR* at CDC. Available at: 1961-2011. MMWR Morb Mort Wkly Rep Suppl 2011;60(04):15-21. In: http://www.cdc.gov/mmwr/preview/ind2011_su.html (accessed 20 November 2013).

Centers for Disease Control and Prevention. CDC's vision for public health surveillance in the 21st century. MMWR Morb Mort Wkly Rep Suppl 2012;61:1-40. Available at: http://www.cdc.gov/mmwr/pdf/other/su6103.pdf (accessed 20 November 2013).

Centers for Disease Control and Prevention. Notes from the Field: Late Vitamin K Deficiency Bleeding in Infants Whose Parents Declined Vitamin K Prophylaxis — Tennessee, 2013. Available at: http://www.cdc.gov/mmwr/preview/mmwrhtml/mm6245a4.htm (accessed 18 November 2013).

Coughlin, S.S., Barker, A., Dawson, A., 2012. Ethics and scientific integrity in public health, epidemiological and clinical research. Public Health Rev., 34. Epub ahead of print. Available at: www.publichealthreviews.eu (accessed 10.01.13).

Ellis A, Fry R. United Kingdom. Office for National Statistics. Regional Trends, No. 42, 2010 Edition – Regional health inequalities in England 8 Jun 2010. Available at: http://www.ons.gov.uk/ons/search/index.html?content-type=Article&pubdateRangeType=allDates&newquery=stroke+mortality+by+regions&pageSize=50&applyFilters=true and http://www.ons.gov.uk/ons/rel/regional-trends/regional-trends/no-42-2010-edition/regional-health-inequalites-in-england-and-wales.pdf (accessed 03.01.13).

Eurostat, Causes of death_statistics. Available at: http://epp.eurostat.ec.europa.eu/statistics_explained/index.php/Causes_of_death_statistics (accessed 09.01.13).

Epidemiology, Gordis L., 2008. fourth ed. WB Saunders, Philadelphia, PA.

Green, L., Ottoson, J., Hiatt, R., Garcia, C., 2009. Diffusion, dissemination and implementation of evidence based public health. Annu. Rev. Public Health 30, 151–174.

Health Metrics Network, http://www.who.int/healthmetrics/en/ (accessed 03.01.13).

International Epidemiologic Association, November 2007. Good epidemiologic practice: IEA guidelines for epidemiologic practice (GEP). Available at: http://webcast.hrsa.gov/conferences/mchb/mchepi_2009/communicating_research/Ethical_guidelines/IEA_guidelines.pdf. (accessed 22.12.12).

International Society for Infectious Diseases, Promed-mail. Available at: http://www.promedmail.org/ (accessed 09.01.13).

Institute of Medicine, 2002. The future of the public's health in the 21st century. National Academies Press, Washington, DC.

Joint Commission, 2013. Sentinel events. . Available at: http://www.joint-commission.org/sentinel_event.aspx. (accessed 10.01.13).

Kozak, L.J., DeFrances, C.J., Hall, M.J., 2006. National hospital discharge survey: 2004 annual summary with detailed diagnosis and procedure data. Vital Health Stat. 13, 1–209.

Last, J.M. (Ed.), 2007. A dictionary of public health. Oxford University Press, New York.

Lee, L.M., Thacker, S.B., 2011. The cornerstone of public health practice: public health surveillance, 1961–2011. MMWR Morb. Mort. Wkly. Rep. Suppl. 60 (04), 15–21. Available at: http://www.cdc.gov/mmwr/preview/mmwrhtml/su6004a4.htm?s_cid=su6004a4_w (accessed 03.01.13).

MacDonald, P.D.M., 2011. Methods in field epidemiology. Burlington, MA: America Public Health Association Press/Jones and Bartlett Learning; 2012. Available at:. World Health Organization. World health statistics, 2011, Geneva: WHO. (accessed 03.01.13) http://www.who.int/whosis/whostat/EN_WHS2011_Full.pdf.

Martinson, M.L., 2012. Income inequality in health at all ages: a comparison of the United States and England. Am. J. Public Health 102, 2049–2056.

Murray, C.J.L., Richards, M.A., Newton, J.N., Fenton, K.A., Anderson, H.R., Atkinson, C., et al., 2013. UK health performance: findings of the Global Burden of Disease Study 2010. Lancet 381, 997–1020.

National Center for Health Statistics, 2011. Health, United States. with special feature on socioeconomic status and health. Department of Health and Human Services, Hyattsville, MD 2012.

Needleman, J., Buerhaus, P., Pankratz, S.V., Leibson, C.V., Steven, S.R., 2011. Nurse staffing and inpatient hospital mortality. N. Engl. J. Med. 364 (11), 1037–1045.

Nolte, E., McKee, M., 2012. In amenable mortality – deaths avoidable through health care – progress in the US lags that of three European countries. Health Aff. 31, 2114–2122.

Organisation for Economic Co-operation and Development, 2011. Health policies and data: health at a glance. Available at: http://www.oecd.org/health/healthpoliciesanddata/healthataglance2011.htm. (accessed 08.12.12).

Organisation for Economic Co-operation and Development, OECD health data 2012 – frequently requested data. Available at: http://www.oecd.org/els/healthpoliciesanddata/oecdhealthdata2012-frequentlyrequesteddata.htm (accessed 03.01.12).

Tulchinsky, T.H., Ginsberg, G.M., Ishovitz, J., Shihab, S., Fischbein, A., Richter, E.D., 1998. Cancer in ex-asbestos cement workers in Israel, 1953–1992. Am. J. Ind. Med. 35, 1–8.

World Health Organization, The global burden of disease: 2004 update. Available at: http://www.who.int/healthinfo/global_burden_disease/GBD_report_2004update_full.pdf (accessed 09.01.13).

World Health Organization, 2011. World health statistics, 2011. WHO, Geneva. Available at: http://www.who.int/whosis/whostat/EN_WHS2011_Full.pdf (accessed 03.01.13).

World Health Organization, January 2013. European Region. Health for All data set. WHO, Copenhagen http://data.euro.who.int/hfadb/ (accessed 13.06.13).

Chapter 4

Communicable Diseases

Learning Objectives

Upon completion of this chapter, the student should:

1. Be familiar with the major communicable diseases, their modes of transmission, and their prevention;
2. Be familiar with measures for the investigation, control, and/or eradication of the major communicable diseases;
3. Understand the New Public Health approach integrating environmental, vaccination, clinical care, and health systems management for control of infectious diseases, and be able to apply this to changing disease patterns and problems.

INTRODUCTION

Despite enormous advances in medical sciences and their applications in public health, infectious diseases remain a central challenge for public health in the twenty-first century. Of particular concern are human immunodeficiency virus/acquired immunodeficiency syndrome (HIV/AIDS), tuberculosis (TB), malaria, severe acute respiratory syndrome (SARS), avian flu, and antibiotic-resistant infections (superbugs), among others. Globalization has facilitated the spread of many infectious agents to all corners of the globe. Mass travel, economic globalization, and climate change, along with accelerating urbanization of human populations, are causing environmental disruption, including global warming. There are and will be more consequences in international transmission of infectious diseases than are now known, in humans and wildlife, as well as domestic animals in the food chain.

This chapter describes communicable and infectious diseases and programs for their prevention, control, elimination, and eradication. Eradication refers to the total elimination of the organism from nature; elimination designates a stop in the circulation of the organism locally; and control means reducing the disease as a public health problem. All of these require a systems approach using available resources, conducting research, and effectively mobilizing environmental measures. This must be carried out in conjunction with strengthening primary care and the overarching health care framework. Rapid transportation and communication make a virus outbreak in any part of the world an international concern, both for health professionals and for the general public. With rapid changes in our understanding of basic sciences, and in vaccine research, production, and

associated measures, it is incumbent upon all medical and allied professionals, educators, policy makers, and students entering the field to have a working understanding of the exciting and dynamic advances. Here, the interesting overlap between communicable diseases and non-communicable diseases (NCDs) becomes evident and imperative. Furthermore, it is crucial to be aware of the social environment in which the risk and exposure of vulnerable groups and individuals cause greatly varying degrees of morbidity and mortality in populations.

The material presented in this chapter is intended to provide an introduction to the student or a review for the public health practitioner, with an emphasis on the applied aspects of communicable disease control. The authors have relied for the content of this chapter on several standard references, especially Heymann's *Control of Communicable Diseases Manual*, 19th edition (2008), *WHO Vaccine Preventable Diseases Monitoring System: 2007 Global Summary*, and *Jawetz, Melnick and Adelberg's Medical Microbiology*, 26th edition (2012). Regular access to the Centers for Disease Control and Prevention (CDC) publication *Morbidity and Mortality Weekly Report* (MMWR), the European Union's *Eurosurveillance*, and the World Health Organization's (WHO's) *Weekly Epidemiological Record* (WER) provides continuing sources of information on communicable diseases, and these are available free online. ProMed, a highly effective Harvard University-based website, is a frequently updated source of current infectious disease outbreaks around the world. The authors have also relied on electronic sources such as PubMed, the American Academy of Pediatrics, WHO, and United Nations Children's Fund (UNICEF) websites, as well as library-access journals. The references listed will augment the limited discussion possible in this text. A recommended standard reference text is Plotkin, Orenstein, and Offit's *Vaccines*, 5th edition (2008).

Each disease has its own characteristic organism and natural history from onset to resolution. Many infectious diseases may remain at a presymptomatic or subclinical stage without progressing to clinical symptoms and signs, but may be transmissible to other people. Even a subclinical disease may cause an immunological effect, producing immunity. The drama of infectious disease is exemplified in the tragic event of the plague in the fourteenth century and its periodic recurrence, as in the epidemic of 1665 in London, described by Daniel Defoe (Box 4.1).

The New Public Health. http://dx.doi.org/10.1016/B978-0-12-415766-8.00004-5

PUBLIC HEALTH AND THE CONTROL OF COMMUNICABLE DISEASE

Organized public health emerged out of the sanitation movement of the mid-nineteenth century, which sought to reduce the environmental and social factors in communicable disease (Box 4.2). Traditionally, the prevention and control of communicable diseases have been accomplished by sanitation, safe water and food supply, isolation, and immunization.

The potential for infectious disease to disturb or destroy human life still exists today, especially in low-income countries, but can also pose serious challenges in the high-income countries. This threat may increase as infectious diseases evolve and escape current human-developed control mechanisms. The spread of the plague throughout Europe and Asia in the fourteenth century caused death and social destruction to an estimated one-third to half of the population of Europe, and is long embedded in the folk culture of the western world. The 1918–1919 Spanish (swine) influenza pandemic may have affected one-third of the world's population; it was very severe and carried a case fatality rate of over 2.5 percent, particularly hitting young adults, resulting in between 50 and possibly as high as 100 million deaths, and killing more young men than died in World War I (Tautenberger, 2006).

Other pandemics that have caused massive recurring devastation, such as smallpox, TB, syphilis, measles, cholera, and influenza, show the explosive potential and epidemic nature of infectious diseases. Some of these diseases have been brought under control and some may be eliminated as public health problems; however, new or recurrent communicable diseases continue to emerge. The spread of AIDS since the 1980s, ongoing cholera epidemics in Asia, Africa, and South America, diphtheria in the former Soviet Union in the 1990s, measles in Western Europe in 2010–2012, and diphtheria and pertussis in many western countries in 2011–2013, all remind us why communicable disease control remains one of the major responsibilities of public health.

The miasma theory (i.e., environment–host) and germ theory (microbiological agent–host) were bitterly contested by their proponents in the nineteenth century. Both have contributed to great achievements in the control of communicable disease in the first half of the twentieth century. The emergence of the germ theory in the late nineteenth century led to the sciences of bacteriology and immunology, growing out of the work of Jenner, Pasteur, Koch, Lister, and many others (see Chapter 1). The control of vaccine-preventable diseases (VPDs) has been a boon to humankind, saving countless lives and providing a cornerstone for public health. Despite this, millions of children still die annually from preventable or readily treatable diseases, such as respiratory infections, diarrheal diseases, and measles. Infectious diseases of childhood are still tragically undercontrolled internationally. Infectious diseases also undermine the health of other vulnerable groups in the population, such as the elderly, the socially and economically most disadvantaged, and the chronically ill, thereby playing a major role in the economics of health care.

Great strides have been made in the control of communicable diseases through public health successes in environmental sanitation, safe foods, vaccination, and antibiotics in the USA (Figure 4.1), as in other industrialized countries. However, the field of infectious disease continues to be dynamic and challenging. Emerging infectious disease threats from new diseases not previously identified, such as HIV and SARS, and new variants of old diseases with resistance to current methods of treatment, together provide great challenges to public health. Increasing resistance to therapeutic agents augments the need for new strategies and

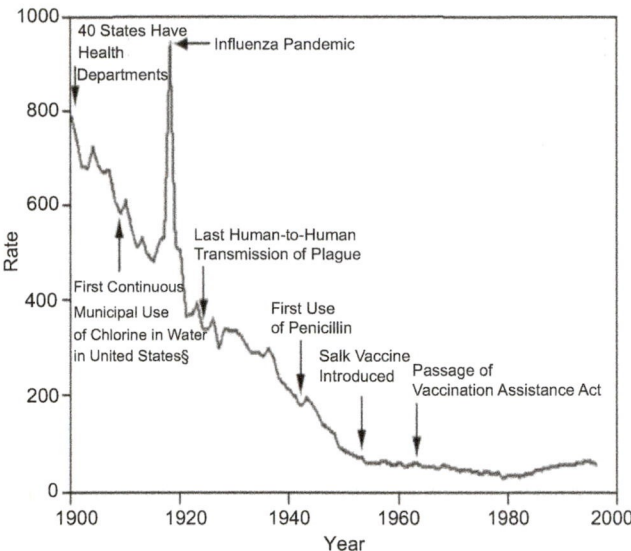

FIGURE 4.1   Crude death rate from infectious diseases, USA, 1900–1999. *Source: Centers for Disease Control and Prevention. Achievements in public health, 1900–1999: control of infectious diseases. MMWR Morb Mortal Wkly Rep 1999;48:621–9. Available at: http://fis.org/public/1900–1999-Infection.pdf [Accessed 18 January 2013].*

coordination between public health and clinical services. Understanding the principles and methods of communicable disease control and eradication is important for all health providers and public health personnel.

THE NATURE OF COMMUNICABLE DISEASE

An infectious disease may or may not be clinically manifested, and therefore it is possible for a person to carry the disease agent without having clinical illness. Acute infectious diseases are intense or short term, but may have long-term sequelae of great public health importance, such as poststreptococcal glomerulonephritis or rheumatic heart disease. Other infectious diseases are chronic with their own long-term effects, such as HIV infection or peptic ulcers. Infections may have both short-term and long-term morbidity, as with viral hepatitis infections. The importance of infectious disease prevention in a global context is shown in Box 4.3. The stages and context of infectious disease include:

- exposure and infection
- presymptomatic/prodromal stage
- non-manifest or subclinical disease
- clinically manifest disease and its progression
- effects of diseases (resolution, recovery, remission, relapse, suprainfection, or death)
- cause of and comorbidity with NCDs
- long-term consequences or sequelae
- social, economic, and environmental factors of communicable diseases.

A modern example of the recurrence of an infectious disease is measles in 2010–2011 in Europe and continuing in Africa. Diseases once localized to specific parts of the world, such as Dengue, West Nile Fever, Lyme disease, Chikungunya, and Rift Valley Fever (RVF), are emerging and spreading in locations far from their normal habitat, and in some cases these diseases may become endemic (Box 4.4); this means that after importation they may spread locally and become "resident" or endemic and transmit locally.

BOX 4.4 Examples of Infectious Diseases, Unusual Health Events, and Newly Discovered Pathogens, Worldwide, 2000–2011

2000 – Outbreak of Rift Valley Fever in Saudi Arabia and Yemen, the first reported cases of the disease outside the African continent

2000 – First detection of carbapenem resistance among the common Gram-negative bacteria Enterobacteriaceae (*Klebsiella pneumoniae*)

2001 – Intentionally caused anthrax in the USA

2001 – Identified new respiratory virus in the Netherlands; human metapneumovirus, among children with respiratory infections

2002 – First detection of *Staphylococcus aureus* bacteria completely resistant to vancomycin

2002 – Outbreak of multidrug-resistant *Salmonella* Newport in the USA

2002 – Norovirus infection of cruise ships entering US ports

2003 – Global outbreak of severe acute respiratory syndrome (SARS) caused by a previously unknown coronavirus

2003 – Identification of a new, hypervirulent strain of *Clostridium difficile* as the cause of hospital outbreaks of gastrointestinal illness in the USA and Canada

2003 – Re-emergence of avian influenza A (H5N1) in South-East Asia, and outbreaks in Africa

2005 – Outbreak of Marburg hemorrhagic fever in Angola

2005 – Identification in Sweden of a new virus, human bocavirus, among children hospitalized with acute respiratory infections

2006 – Outbreak of Rift Valley fever in Kenya

2007 – Outbreak of Ebola hemorrhagic fever in the Democratic Republic of the Congo

2007 – Outbreak of Nipah virus encephalitis in Bangladesh

2007 – First detection in Italy of mosquito-borne transmission of Chikungunya fever, previously detected only in parts of Africa and South and South-East Asia

2007 – Discovery in Thailand of a new human species of *Bartonella*, an insect-borne bacterium that multiplies inside red blood cells causing fever, fatigue, muscle pain, headache, and rash

2007 – Outbreak of hemorrhagic fever in Uganda caused by a new strain of Ebola, Bundibugyo Ebola virus

2007 – Outbreak of Marburg hemorrhagic fever in Uganda

2008 – Ebola-like outbreak in Zambia due to a previously unknown virus, Lujo hemorrhagic fever virus

2008 – Australia - isolation of arenavirus, a new virus, transmitted by transplantation causing lymphocytic choriomeningitis

2008 – Increasing outbreaks and international spread of carbapenem-resistant Enterobacteriaceae, and first detection of New Delhi metallo-beta-lactamose (NDM-1), a genetic element that can confer such resistance

2009–10 – Imported and local transmission outbreaks of Dengue recently in Florida, Hawaii, and Texas

2009–10 – Influenza pandemic caused by a new influenza strain, influenza A (H1N1)

2010 – Outbreaks of cholera in Haiti

2011 – Outbreak of Shiga toxin-producing *Escherichia coli* 0104:H4 (STEC 0104:H4) infections in Germany

2011 – New strain of gonorrhea resistant to antibiotics found

2012–13 – New strain of Coronavirus MERS-coV causing severe respiratory illness in Saudi Arabia and Middle East; local and visitor cases; over 40 percent mortality

Source: *Centers for Disease Control and Prevention. A CDC framework for preventing infectious diseases: sustaining the essentials and innovating for the future. Atlanta, GA: CDC; 2011. Available at: http://www.cdc.gov/oid/framework.html [Accessed 18 January 2013].*

HOST–AGENT–ENVIRONMENT TRIAD

The host–agent–environment triad, discussed in Chapter 2, is fundamental to the success of understanding the transmission of infectious diseases and their control, including well-known diseases, those changing their patterns, and those newly emerging or escaping current methods of control. Infection occurs when the organism successfully invades the host's body, where it multiplies and produces an illness.

Hosts are people, or other living animals, including birds and arthropods, which provide a place for growth and sustenance to an infectious agent under natural conditions. Some organisms, such as protozoa or helminths, may pass through successive stages of their life cycle in different hosts, but the definitive host is the one in which the organism passes its sexual stage. The intermediate host is where the parasite passes the larval or asexual stage. A transport host is a carrier in which the organism remains alive, but does not develop.

An agent of an infectious disease is necessary but not always sufficient to cause a disease or disorder. The infective dose is the quantity of the organism needed to cause clinical disease. A disease may have a single agent as a cause, or it may occur as a result of the agent in company with contributory factors, such as in socially deprived and undernourished populations or among immunocompromised people who are vulnerable to the spread and development of the disease. A disease may be present in an infected person in a dormant form, such as TB, or a preclinical stage, such as poliomyelitis (polio) or HIV, without clinical paralytic disease in the case of polio or before clinical AIDS appears in the case of HIV. The virulence or pathogenicity of an infective agent is the capacity of an infectious agent to enter the host, replicate, damage tissue, and cause disease. Virulence describes the severity of clinical disease and may vary among serotypes or strains of the same agent.

The environment provides a reservoir for the organism and the mode of transmission by which the organism reaches a new host. The reservoir is the natural habitat where an infectious agent lives and multiplies, from which it can be transmitted directly or indirectly to a new host. Reservoirs may be in people, animals, arthropods, avians, reptiles, plants, soil, or substances in which an organism normally lives and multiplies, and on which it depends for survival or in which it survives in a dormant form. A fomite is an inanimate object contaminated with infectious material which may transmit disease, such as improperly sanitized medical equipment.

Contacts are people or animals that have been in some form of association with an infected person, animal, fomite, or environment that may provide a risk for acquiring and transmitting the infective agent. People or animals that harbor a specific infectious agent, often in the absence of discernible clinical disease, and who serve as a source of infection or contamination of food, water, or other materials, are carriers. A carrier may have an unapparent infection (a healthy carrier) or may be in the incubation or convalescent stage of the infection.

The emergence of health facility-acquired infections, often from catheters or other invasive devices, has become a major issue in all health systems with multidrug-resistant (MDR) organisms. Infection control and isolation techniques are vital to control such infections, which in individual patients cause overwhelming complications and in multiple patients can spread in epidemic form, such as in nurseries, nursing homes, or even in well-respected hospitals.

CLASSIFICATIONS OF COMMUNICABLE DISEASES

Communicable diseases may be classified by a variety of methods: by clinical syndrome, mode of transmission, methods of prevention (e.g., vaccine preventable), or by major organism classification, that is, viral, bacterial, fungal, and parasitic disease.

A virus is a nucleic acid molecule (RNA or DNA) encapsulated in a protein coat or capsid. The virus is not a complete cell and can only replicate inside a living cell. The capsid may have a protective lipid-containing envelope. The capsid and envelope facilitate attachment and penetration into host cells, and often contain virulence factors. Inside the host cell, the nucleic acid molecule utilizes cellular proteins and processes for virus replication. Prions – discovered in recent years (Stanley Prusiner, Nobel Prize 1997) – are proteins, which can induce disease. As infectious agents, prions cause a number of degenerative central nervous system (CNS) diseases, including spongiform encephalopathy in livestock (mad cow disease and scrapie in sheep) and humans [variant Creutzfeldt–Jakob Disease (vCJD)].

Bacteria are unicellular organisms that reproduce sexually or asexually and can exist in an environment with oxygen (aerobic) or in a situation lacking oxygen (anaerobic). Some may enter a dormant state and form spores where they are protected from the environment and may remain viable for years. Bacteria have a nucleus of chromosomal DNA material within a membrane surrounded by cytoplasm, itself usually enclosed by a cellular membrane. Bacteria are classified by morphology and growth conditions, including coloration under Gram stain (Gram-negative or Gram-positive), microscopic morphology, immunological (antigen) or molecular (DNA) markers, or by the diseases they may cause. Bacteria include both indigenous flora (normal resident) bacteria and pathogenic (disease-causing) bacteria. Pathogenic bacteria cause disease by invading, overcoming natural or acquired resistance, and multiplying in the body. Bacteria may produce a toxin or poison that can affect a body site distant from where the bacterial replication occurs, such as in tetanus. Bacteria may also initiate an excessive immune response, producing damage to other body tissues away from the site of infection (e.g., acute rheumatic fever and glomerulonephritis).

Mycoses are infections caused by molds and yeasts. Clinical manifestations of fungal disease range from relatively mild superficial infection to systemic, life-threatening conditions. Immunocompromised individuals are at elevated risk. *Cryptococcus*, *Candida*, *Aspergillus*, and *Mucor* molds or fungi are among the leading causes of morbidity in HIV-positive patients and among immunosuppressed populations, such as those receiving chemotherapy and radiation for cancer treatment. *Pneumocystis jiroveci* (formerly *P. carinii*), once thought to be a protozoan, is now classified a fungus, based on genetic analysis. Common dermatophytic infections, known as tinea, are caused by fungi invading the hair, skin, or nails, and they occur in nearly all living organisms. A case report on contaminated drugs used for treating neurological conditions leading to a fungal meningitis outbreak in the USA in 2012 is reported in Chapter 15, including index case follow-up and epidemic investigation.

Parasitology studies protozoa, helminths, and arthropods that live within, on, or at the expense of a host. Protozoa include oxygen-producing, unicellular organisms such as the flagellates *Giardia* and *Trichomonas*, and amoebae such as *Entamoeba*, in enteric and gynecological disorders. Sporozoa are parasites with complex life cycles in different hosts, such as *Cryptosporidium* or malarial parasites. Helminths are worms that infest humans, especially in places with poor sanitation and tropical areas. Arthropods, the most numerous of animal species, include lice, fleas, sandflies, blackflies, and ticks, and they serve as important disease vectors. Arthropods can live on the body's surface (ectoparasites) and transmit bacterial, viral, rickettsial, and other diseases. They are also capable of fecal–oral transmission, for instance, in cases of *Shigella* and *Escherichia coli*, in or via biological effects within the host such as in malaria. This group constitutes among the most important public health threats globally and their control is a continuing public health challenge.

MODES OF TRANSMISSION OF DISEASE

Transmission of diseases occurs via the spread of an infectious agent from a source or reservoir to a person (Table 4.1). Direct transmission from one host to another occurs during touching, biting, kissing, sexual intercourse, projection via droplets, as in sneezing, coughing, or spitting, or by entry through the skin. Indirect transmission includes aerosols of long-lasting suspended particles in air; this can be central among passengers on an aircraft being exposed to a TB patient or a measles carrier. Fecal–oral transmission such as foodborne and waterborne can occur easily, as well as by poor hygienic conditions with fomites, such as soiled clothes, handkerchiefs, toys, or other objects. Waterborne and foodborne diseases are still among the most common causes of death, particularly of children in low-income countries with poor infrastructure for water and sewage management. This situation is exacerbated by crowded and unhygienic housing and by the effects of poverty on sanitation, nutrition, and access to health care. Transmission in medical settings is common, yet preventable by hand washing and consistent use of sterile techniques and cleaning procedures. The use of face masks, hand washing, gloves, gowns, and tissues when sneezing are all vital in implementing practices to reduce the spread of hospital-acquired infections and influenza. Promoting these simple measures is of utmost importance as hospital-acquired infections represent a major cause of morbidity and result in extended hospital stay, thus significantly elevating economic costs. Sterile practices among health care and hospital workers should certainly be made a priority. The use of bed nets and vector control for malaria are among the most effective ways of reducing the burden of this highly dangerous disease.

Vectorborne diseases are transmitted via crawling or flying insects, in some cases with multiplication and development of the organism in the vector, as in malaria. The subsequent transmission to humans is by injection of salivary gland fluid during biting or by deposition of feces, urine, or other material capable of penetrating the skin through a bite wound or other trauma. Transmission may occur with insects as a transport mechanism, as in *Shigella* on the legs of a housefly.

Airborne transmission occurs indirectly via infective organisms in small aerosols that may remain suspended for long periods and which easily enter the respiratory tract. Viruses such as influenza, the common cold, and measles can be transmitted in this way. Particles of dust may spread organisms from soil, clothing, or bedding.

TABLE 4.1 Classification of Infectious Diseases by Principal Modes of Transmission

Method of Transmission	Examples
Airborne (droplet and aerosols)	Viral exanthems (measles), streptococcal diseases, various upper and lower respiratory tract diseases, tuberculosis, Legionnaire's disease, influenza (seasonal and H1N1), SARS, measles, mumps, rubella
Physical contact	Leprosy, impetigo, scabies, anthrax
Sexual contact	HIV, syphilis, gonorrhea, herpes genitalis, hepatitis B, chlamydia, human papillomavirus
Blood and blood products	HIV, hepatitis B, hepatitis C
Fecal–oral	Hepatitis A, poliovirus, enteroviruses, *Shigella*, rotavirus, adenoviruses, typhoid
Foodborne	*Salmonella, Escherichia coli, Helicobacter pylori, Campylobacter, Listeria*
Waterborne	Cholera, *Giardia, Cryptosporidium, Helicobacter pylori*
Transcutaneous	Vectorborne via insects (arthropods): malaria, viral hemorrhagic fevers, schistosomiasis, plague Animal bite (zoonoses): rabies Self-injected (illicit drug users): HIV, hepatitis B
Congenital maternal–fetal	Congenital rubella syndrome, congenital syphilis, gonorrheal ophthalmia, cytomegalovirus, HIV, rubella, syphilis, hepatitis B and C, gonorrhea, chlamydia
Health care associated	Transmission: hospital, long-term care facilities, community surgical centers, and community-acquired *Klebsiella pneumoniae, Clostridium difficile, Staphylococcus aureus* including methicillin-resistant organisms (MRSA), HIV, hepatitis B, hepatitis C, fungal infections, central venous line-, ventilator-, and catheter-associated pneumonia, surgical site infections

Note: SARS = severe acute respiratory syndrome; HIV = human immunodeficiency virus; MRSA = methicillin-resistant *Staphylococcus aureus*.

Vertical transmission occurs from one generation to another or from one stage of the insect life cycle to another stage. Maternal–infant transmission occurs during pregnancy (transplacental), or during delivery (as in gonorrhea) or breastfeeding (e.g., HIV), with transfer of infectious agents from mother to fetus or newborn.

Many types of invasive devices and procedures are used to treat patients in hospitals, long-term care facilities, and surgical centers, and in the community. Healthcare-associated infections (HAIs) are complications with devices including catheters, ventilators, central venous lines causing

BOX 4.5 Terms in Immunology of Infectious Diseases

- *Infectious agent* – a pathogenic organism (e.g., virus, bacterium, rickettsia, fungus, protozoa, helminth, pollen, or chemical) is one capable of producing infection or an infectious disease in humans, animals, and plants.
- *Infection* – the process of entry, development, and proliferation of an infectious agent in the body tissue of a living organism overcoming the host's defense mechanisms, resulting in a non-apparent or clinically manifest disease.
- *Antigen* – any substance (e.g., protein, polysaccharide) which causes the immune system to produce antibodies against it. An antigen may be introduced into the host by invasion of an infectious agent, inhalation, ingestion, or through the skin, wounds, or via transplantation; antigens are also produced by immunization with a live or inactivated organism, or its components.
- *Antibody* – a protein molecule produced by the body's immune system in response to a harmful foreign substance (an antigen) or acquired by passive transfer. Antibodies bind to the specific antigen that elicits their production, causing the infective agent to be susceptible to immune defense mechanisms against infections (e.g., humoral and cellular). Antibodies may also form against one's own tissue, producing an autoimmune disorder.
- *Innate immunity* – includes the cough reflex, skin, mucus, and stomach acidity as barriers which protect the body against infection.
- *Acquired immunity* – developed as result of natural exposure or deliberate exposure by immunization to an

infectious agent or its antigenic components which protects against later exposure to the active live agent.
- *Passive immunity* – the transfer of antibodies produced in another body, such as maternal antibodies transferred from mother to her fetus via the placenta to provide protection for part of the first year of life, or by antiserum or antitoxin of antibodies from another person to give short-term protection against an antigen such as serum globulin against hepatitis infection or tetanus antiserum.
- *Immunoglobulins* – molecules produced by plasma cells in response to an antigen challenge and are present in blood or other body fluids. There are five major classes (IgG, IgM, IgA, IgD, and IgE) and subclasses based on molecular weight. They can cross from a mother to fetus in utero to provide passive immunity to the fetus.
- *Antisera or antitoxin* – materials prepared in animals for use in passive immunization against infection or toxins.
- *Cellular immunity (cell-mediated immunity)* – immunity acquired with T lymphocyte cells producing chemicals which activate natural killer cells (macrophages).
- *Herd immunity* – resistance of a group to an infectious disease when a large percentage of the population at risk is immune through previous exposure to the disease or by immunization.

Sources: *Last JM. Dictionary of public health. New York: Oxford University Press; 2007.*
US National Library of Medicine. Medlineplus. Immune responses. Available at: http://www.nlm.nih.gov/medlineplus/ency/article/000821.htm
Free Medical Dictionary. Available at: http://medical-dictionary.thefreedictionary.com/cell-mediated+immunity [Accessed 18 January 2013].

bloodstream infections, catheter-associated urinary tract infections, and ventilator-associated pneumonia, together accounting for roughly two-thirds of all HAIs, or at surgery sites. *Clostridium difficile* can cause gastrointestinal infection; patients can be exposed to this bacterium through contaminated surfaces or the spores can be transferred from other people's unclean hands. Methicillin-resistant *Staphylococcus aureus* (MRSA) is a widespread skin contaminant especially of concern in hospitals. CDC monitors and promotes preventive procedures as these infections are an important threat to patient safety (CDC, 2012).

IMMUNITY

Resistance to infectious diseases is related to many host and environmental factors, including age, gender, pregnancy, nutrition, trauma, fatigue, living and socioeconomic conditions, and emotional status. Good nutritional status has a protective effect and bolsters immune competency. Vitamin A supplements reduce complication rates of measles and enteric infections. TB may be present in an individual person whose resistance is sufficient to prevent clinical disease, but the infected person (with or without symptoms)

may be a carrier of an organism which can be transmitted to another or cause clinical disease if the person's susceptibility is reduced (Box 4.5).

The body is protected by physical barriers, e.g., skin and stomach acidity, against the entry of foreign organisms. The body also has passive immunity and acquired immunity both cellular and in humoral components of blood. Immunity is the means by which the body recognizes and resists infection resulting from the presence of specific foreign antigens on the surface of bacteria, viruses, fungi, or other toxins, chemicals, drugs or foreign objects, e.g., a splinter, which may be harmful. The immune system recognizes and acts to destroy or contain the dangerous agent. Humoral (blood) immunity is activated by B lymphocytes, which produce antibodies, complement proteins, or cells that act on the microorganism associated with a specific disease, toxin, or foreign body. The body also reacts to infective antigens with cellular responses, including those that directly defend against invading organisms and other cells which produce antibodies. T lymphocytes attack antigens directly and assist with chemicals controlling the immune response (cytokines). Inflammation attracts white cells (macrophages and neutrophils) which act as phagocytes to kill germs and dead or damaged cells.

Immunity can be acquired by response to an organism or its antigenic components which, when introduced into a person's body, produce a natural protective immunity. Passive immunity is temporary, by the passage of preformed antibody from mother to infant via the placenta and breast milk, or by injection of preformed immunoglobulins. Active immunization introduces effective killed or attenuated organisms, or parts of organisms as antigens into the body, which responds by producing antibodies. This enables public health systems to prevent millions of deaths from communicable disease and provides hope for more success in the future as the sciences of microbiology, vaccinology, and public health practice advance.

SURVEILLANCE

Surveillance of disease is the continuous scrutiny of all aspects of the occurrence and spread of a disease pertinent to effective control of that disease. Maintaining ongoing surveillance is one of the basic duties of a public health system, and is vital to the control of communicable disease, providing the essential data for tracking of disease, planning interventions, and responding to future disease challenges. Surveillance of infectious disease incidence relies on reports of notifiable diseases by physicians, supplemented by individual and summary reports of public health laboratories. Such a system must concern itself with the completeness and quality of reporting and potential errors and artifacts. Quality is maintained by seeking clinical and laboratory support to confirm first reports. Completeness, rapidity, and quality of reporting by physicians and laboratories should be emphasized in undergraduate and postgraduate medical education. Enforcement of legal sanctions may be needed where standards are not met. Surveillance of infectious diseases includes the elements listed in Box 4.6.

Epidemiological monitoring based on individual and aggregated reports of infectious diseases provide data vital to planning interventions at the community level or for individual patients, along with other information sources, such as hospital discharge data and monitoring of sentinel centers. These may be specific medical or community sites that are representative of the population and are able to provide good levels of reporting to monitor an area or population group. A sentinel center can be a pediatric practice site, a hospital emergency room, or another location that will provide a "finger on the pulse" to assess suspicious changes occurring in the community. It can also include monitoring in a location previously known for disease transmission, such as Hong Kong in relation to influenza typing, for vaccine planning, production, and distribution.

BOX 4.6 Infectious Disease Surveillance and Monitoring Tools

Reporting Systems
- Public health departments – local, state, and national
- Mandatory morbidity reports from doctors or clinics to public health offices
- Mortality reports from attending doctors to vital records
- Births including birth defects reporting, e.g., rubella syndrome, congenital syphilis
- Reports from selected sentinel centers, e.g., hospital emergency rooms, pediatric centers
- Hospital discharge information systems
- Hospital infection and incident reports
- Special field investigations of epidemics or individual cases
- Laboratory monitoring of infectious agents in population samples
- Data on supply, use, and side-effects of vaccines, toxoids, and immunoglobulins
- Data on vector monitoring and control activities such as insecticide use
- Immunity levels in samples of the population at risk
- Review of current literature on the disease
- Epidemiological and clinical reports from other jurisdictions
- School and work attendance/absence

Publications
- Centers for Disease Control and Prevention. *Morbidity and Mortality Weekly Report* (*MMWR*) and special reports. Free email subscription available at: http://www.cdc.gov/mmwr/mmwrsubscribe.html

- Centers for Disease Control and Prevention. National Notifiable Diseases Surveillance System (NNSDSS); December 2012. Available at: http://www.cdc.gov/osels/ph_surveillance/nndss/phs/infdis2011.htm
- Centers for Disease Control and Prevention. Emergency preparedness and response. Available at: http://emergency.cdc.gov/han/
- World Health Organization. *Weekly Epidemiological Record* (*WER*). Free subscription available at: http://www.who.int/wer/en/
- European Union. European Center for Disease Control (ECDC). *Eurosurveillance*. Free subscription available at: http://www.eurosurveillance.org/public/Subscribe/Subscribe.aspx
- International Society for Infectious Diseases. Promed.mail. Available at: http://www.promedmail.org/
- *Emerging Infectious Diseases* – CDC journal. Available at: http://wwwnc.cdc.gov/eid/
- *Promed Digest*. Excellent early warning and updating daily digest of infectious disease information from around the world. Free subscription available at: majordomo@promedmail.org
- Electronic media – websites, Facebook, Twitter activity.
- Public Health Agency of Canada. Available at: http://www.phac-aspc.gc.ca/id-mi/

Epidemiological analysis provided by government public health agencies should be published weekly, monthly, and annually, and distributed to a wide audience of public health and health-related professionals throughout the country. Feedback is vital in order to promote involvement and improved quality of data, as well as to allow evaluation of local situations in comparison to other areas. In a federal system of government, national agencies report regularly on all state or provincial health patterns. State or provincial health authorities provide data to the counties and cities in their jurisdictions. Such data should also be readily available to researchers in other government agencies and academic settings for further research and analysis.

Notifiable diseases are those that a physician is legally required to report to state or local public health officials. Notification is mandatory because of the degree of contagiousness, severity, frequency, or other public health importance of these diseases (Table 4.2). Public health laboratory services provide validation of clinical and epidemiological reports. They also provide day-to-day supervision of public

TABLE 4.2 Notifiable Infectious Diseases in the USA, 2011

Anthrax	Mumps
Arboviral disease	Pertussis (whooping cough)
Babesiosis	Plague
Botulism	Poliomyelitis, paralytic or non-paralytic
Brucellosis (undulant fever)	Psittacosis
Chancroid	Q-fever
Chlamydia trachomatis, genital infection	Rabies (animal and human)
Cholera	Rocky Mountain spotted fever
Coccidiomycosis	Rubella and rubella congenital syndrome
Cryptosporidiosis	Rubella congenital syndrome
Cyclosporiasis	Salmonellosis
Diphtheria	Severe acute respiratory syndrome-associated coronavirus (SARS-CoV)
Escherichia coli, Shiga toxin-producing coli	Shigellosis
Erlichiosis	Smallpox
Giardiasis	Streptococcal disease, invasive group A
Gonorrhea	Streptococcal pneumonia, pediatric or drug-resistant invasive
Haemophilus influenzae, invasive disease	Streptococcal toxic shock syndrome
Hansen's disease (leprosy)	Syphilis (primary, secondary, latent, late, congenital)
Hantavirus pulmonary syndrome	Tetanus
Hemolytic uremic syndrome (postdiarrhea)	Toxic shock syndrome, streptococcal and non-streptococcal
Hepatitis, viral A, B, C, others	Trichinellosis
HIV[a] type 1, 2, 3 (AIDS), and HIV type unknown	Tuberculosis
Influenza, pediatric mortality or novel influenza A	Tularemia
Legionnellosis	Typhoid fever
Lyme disease	Vancomycin-resistant or intermediate *Staphylococcus aureus*
Malaria	Varicella
Measles	Vibriosis (non-cholera)
Meningococcal disease	Yellow fever

Note: [a]*Human immunodeficiency virus (HIV) is now classified by stages and includes acquired immunodeficiency syndrome (AIDS). Babesiosis is a tick-borne disease. Other diseases for which individual local or state authority monitoring may be required include amebiasis, meningitis (aseptic and other bacterial), campylobacteriosis, dengue fever, genital herpes, genital warts, granuloma inguinale, leptospirosis, listeriosis, lymphogranuloma venereum, mucopurulent cervicitis, non-gonococcal urethritis, pelvic inflammatory disease, and poststreptococcal disease.*
Sources: Centers for Disease Control and Prevention. December 2012. National Notifiable Diseases Surveillance System (NNSDSS). Available at: http://www.cdc.gov/osels/ph_surveillance/nndss/phs/infdis2011.htm; Centers for Disease Control and Prevention 2012, case definitions: nationally notifiable conditions infectious and non-infectious case. Atlanta, GA: CDC; 2012. Available at: http://wwwn.cdc.gov/nndss/document/2012_Case%20Definitions.pdf [Accessed 18 January 2013].

health conditions, and can monitor communicable disease and vaccine efficacy and coverage. In addition, they support standards of clinical laboratories in biochemistry, microbiology, and genetic screening. Reference laboratories are specialized central facilities usually operated by the public health at higher levels of government (i.e., state or federal). They enable public health authorities to monitor and validate the work of other laboratories and may be assisted by specialized faculties in teaching centers.

With newly emerging diseases capable of spreading far from their previously known habitat, and the threats of pandemics such as SARS and, of greater concern, avian influenza, surveillance for human and animal disease is crucial to the societies we live in, including the global society. The first diagnosis of a strange new disease entity may lead to its identification and practical measures to halt its spread. When signs point to anticipated or surprise epidemics and pandemics, and when the real threat of bioterrorism emerges, multisectoral preparation and training are of utmost importance.

HEALTHCARE-ASSOCIATED INFECTIONS

Healthcare-associated infections (HAIs) are among the leading communicable and preventable causes of morbidity and mortality throughout the world. Nosocomial infections are those wherein a patient is exposed to and contracts disease while hospitalized or in another care facility. While great strides have been made in hospital sanitation, HAI still occurs in as many as 10 percent of admissions in developed countries. Recent CDC estimates place the number of nosocomial infections in the USA for 2002 at 1.7 million, a higher incidence than any notifiable disease. With a case mortality of nearly 6 percent, HAIs are also among the most deadly. Although progress has been made in HAI prevention, the organisms implicated are becoming resistant to conventional therapy.

MRSA is among the most virulent and treatment-resistant bacteria, now accounting for over 50 percent of wound infections in many hospitals. Rare reports of vancomycin-resistant *Staphylococcus aureus* (VRSA) cause alarm, proving that antibiotic resistance has transferred from other species. Treatment options for VRSA and vancomycin-resistant *Enterococcus* species are extremely limited, with major concern that these organisms could spread or become resistant to the few known effective therapies. The increasing number of immunodeficient patients has increased the importance of prevention of nosocomial infections (Box 4.7).

Where standards of infection control are deficient or lacking in both developed and developing countries, hospital patients and staff are vulnerable to serious infection. Of note, TB and hepatitis B exposure is common among health care workers, but preventable through airborne precautions and vaccination, respectively. In developing countries, deadly emerging viruses, such as avian influenza H5N1 and

---

BOX 4.7 Prevention of Health Care Facility-Associated Infections: Key Elements of Standard Precautions

- Commitment of facility management to infection monitoring and prevention
- Continuous quality improvement on this topic
- Sterile technique training and supervision
- Surgical suite sterility procedures and monitoring
- Hand hygiene as key to improvement throughout a facility
- Gloves, gowns, masks
- Safe injection practices and needle safety
- Respiratory hygiene and cough etiquette
- Environmental cleaning
- Linens and waste disposal
- Patient care equipment
- Dialysis equipment and space
- Catheter care sterility and regular change
- Respirator/ventilator sterility and regular change
- Central venous line sterile maintenance
- Colonoscopy and other invasive device sterility
- Reporting and follow-up

Sources: *Adapted from: World Health Organization. WHO guidelines on hand hygiene in health care. Geneva: WHO; 2009. Available at: http://whqlibdoc.who.int/publications/2009/9789241597906_eng.pdf [Accessed 18 January 2013].*
Centers for Disease Control and Prevention. Hand hygiene in health care settings: guidelines. Available at: http://www.cdc.gov/handhygiene/ [Accessed 18 January 2013].
Centers for Disease Control and Prevention. CDC definitions of nosocomial infections. Available at: http://health.utah.gov/epi/diseases/legionella/plan/cdcdefsnosocomial%20infection.pdf
Allegranzi B, Pittet D. Role of hand hygiene in healthcare-associated infection prevention. J Hosp Infect 2009;73:305–15. Available at: http://www.ncbi.nlm.nih.gov/pubmed/19720430 [Accessed 18 January 2013].

Ebola viruses, infect nursing, medical, and other staff as secondary cases.

A great obstacle in quantifying the impact of HAI is the lack of uniform and clear case definitions, as well as reliance, in most countries, on voluntary reporting by institutions. While many recommendations have been made, notably by the Society for Healthcare Epidemiology of America, no uniform regulations have been established to mandate reporting of HAIs. However, much work has been focused on prevention. Standard Precautions (formerly known as Universal Precautions) are a set of basic practices by which health care workers may reduce the spread of nosocomial infection among patients, visitors, and staff, as well as protect health workers from occupationally acquired disease. These include adequate hand-washing hygiene and use of protective barriers suited to specific risks. Expanded precautions and mandatory use of organism-specific clinical guidelines are necessary procedures in many health care institutions as protective measures.

The 2007 CDC and Healthcare Infection Control Practices Advisory Committee (HICPAC) guidelines provide recommendations applicable to all settings. In 2011 the CDC published evidence-based guidelines for minimum prevention expectations for safe ambulatory care settings. Organizational policy must be established for each institution by an integrated and authoritative department of infection control and epidemiology (CDC Guide to Infection Prevention, 2011).

In the USA, approximately one out of every 20 hospitalized patients will contract an HAI. Costs of HAI to US hospitals range from US$28.4 to US$45 billion (2007 dollars). With 20 percent of infections preventable, potential cost savings range from an estimated low of US$5.7 to US$6.8 billion annually; with 70 percent of infections preventable, cost savings range from US$25.0 to US$31.5 billion (CDC 2009 and Public Health Reports 2007).

As illustrated, the cost of nosocomial infections serves as a major consideration in planning health budgets. Reducing the risk of HAIs justifies substantial expenditure for hospital epidemiology and infection control activities. With the diagnosis-related group (DRG) payment system for hospital care (classified by diagnosis rather than by days of stay), the effective manager has a major incentive to minimize the risk of nosocomial infections to improve patient care. Infections can greatly prolong hospital stay, increasing serious complications, patient dissatisfaction, and health care costs.

The US Agency for Healthcare Research and Quality patient safety program initiated a program to reduce central line-associated bloodstream infections (CLABSIs) in newborns. CLABSIs are health care-associated infections of central vein or artery catheters, especially in premature low birth-weight babies. These catheters may be in place for long periods to provide fluids, nutrients, and medications, but they are readily subject to infections that seriously harm or kill infants or adults.

Neonatal intensive care units (NICUs) participating in this project included 100 NICUs in nine states caring for 8400 newborns. The project was to improve the safety of procedures of care of these infants with intravenous bloodstream infections, adopting safe practices and guidelines provided by the CDC. As a result, the program reduced in-hospital infections by 58 percent in less than a year and relied on the program's prevention practice checklists and better communication to prevent an estimated 131 infections and up to 41 deaths and to avoid more than US$2 million in health care costs (AHQR, 2013).

Patient safety and prevention of infections are long-standing issues in health care, going back to Florence Nightingale at Scutari Hospital in the Crimea, Ignaz Semmelweiss in Vienna, and Joseph Lister in Glasgow in the nineteenth century (see Chapter 1), but they remain vital issues in health care management economics and epidemiology.

ENDEMIC AND EPIDEMIC DISEASE

An endemic disease is the continuous usual presence of a disease or infectious agent in a given geographic area or population group. Hyperendemic means a state of persistence of high levels of incidence of the disease. Holoendemic means that the disease appears early in life and affects most of the population, as in malaria or hepatitis A and B in some regions.

An epidemic takes place in a community or region when the occurrence of a number of cases of an illness is in excess of the usual or expected number of cases, or health-related behaviors (e.g., smoking) or events (e.g., road traffic injuries). The number of cases constituting an epidemic varies with the disease. A number of factors such as previous epidemiological patterns of the disease, time and place of the occurrence, and the population involved, must be taken into account. A single case of a disease long absent from an area, such as polio, constitutes an epidemic. Therefore, it is a public health emergency, as one clinical case may represent as many as 1000 carriers with non-paralytic or subclinical polio. If two to three or more cases of any unusual disease locally are linked in time and place, this may be considered sufficient evidence of transmission and presumed to be an epidemic. Moreover, a pandemic refers to the occurrence of a disease on a wide scale over an expansive area, crossing international boundaries and affecting a large proportion of the world.

Epidemic Investigation

Each epidemic should be regarded as a unique natural experiment. The investigation of an epidemic requires preparation and field investigation in conjunction with local health and other relevant authorities. Verification of cases and the scope of the epidemic will require case definition and laboratory confirmation. Tabulation of known cases according to time, place, person and potential common source is important for immediate control measures and formulation of the hypothesis as to the nature of the epidemic. An epidemic curve is a graphic plotting of the distribution of cases by the time of onset or reporting, which gives a picture of the timing, spread, and extent of the disease from the time of the initial index cases and the secondary spread.

Epidemic investigation requires a series of steps. It starts with confirmation of the initial report and preliminary investigation, defining who is affected, determining the nature of the illness and confirming the clinical diagnosis, and recording when and where the first (index) and follow-up (secondary) cases occurred, as well as how the disease was transmitted. Samples are taken from index case patients (e.g., blood, feces, throat swabs) as well as from possible reservoirs (e.g., food, water, sewage, environment). A working hypothesis is established based on the first findings, taking into account all plausible explanations. The epidemic pattern is studied, establishing common sources or risk factors, such as food, water, contact, and environment, and drawing a timeline of cases to define the epidemic curve.

The number of individuals who are ill (the numerator) and the population at risk (the denominator) establish the attack rate; namely, the percentage of sick among those exposed to the common factor or common source. Questions arise to determine a reasonable explanation of the occurrence: Is there a previous pattern, with the present episode a recurrence or new event? Consultation with colleagues and the literature helps to establish both biological and epidemiological plausibility. What steps are needed to prevent spread and recurrence of the disease? Coordination with relevant health and other officials and providers is required to establish surveillance and control systems, to document and distribute reports, and to respond to the public's right to know.

The first reports of excess cases may come from a medical clinic or hospital. The initial (sentinel or index) cases provide the first clues that may point to a common source. Investigation of an epidemic is designed to quickly elucidate the cause and points of potential intervention to stop its continuation. This requires skilled investigation and interpretation. The term "epidemiological investigation" means a broad review of all evidence related to a topic, not just one epidemic or outbreak. Epidemiological investigations have defined many public health problems. Steps in epidemiological investigation are shown in Box 4.8. Rubella syndrome, Legionnaire's disease, AIDS, Lyme disease, and hantavirus diseases were first identified clinically when unusually large numbers of cases appeared with common features. The suspicions that were raised led to a search for causes and the identification of control methods.

A working hypothesis of the nature of an epidemic is developed based on the initial assessment, the type of presentation, the condition involved, and previous local, regional, national, and international experience. The hypothesis provides the basis for further investigation, control measures, and planning additional clinical and laboratory studies. Surveillance will then monitor the effectiveness of control measures. Communication of findings to local, regional, national, and international health reporting systems is important for sharing the knowledge with other potential support groups or other areas where similar epidemics may occur.

The CDC, originally organized in 1946 as the Office for Malaria Control in War Areas, is part of the US Public Health Service. As of 1993, the CDC had a budget of US$1.5 billion, and its 7300 employees included epidemiologists, microbiologists, and many other professionals. By 2007, the CDC budget had reached US$9 billion, employing 8467 individuals. The CDC includes national centers for environmental health and injury control, chronic disease prevention and health promotion, infectious diseases, prevention services, health statistics, occupational safety and health, and international health. Recently, however, budget reductions have imposed limits of capacity in such areas as overseas work. In 2010, the budget reached nearly US$10.9 billion, and CDC employed over 14,000 people in 2011. The key increases for 2012 were support for the prevention and control of infectious diseases [HIV/AIDS, other sexually transmitted infections (STIs), global polio eradication, Strategic National Stockpile] and chronic disease prevention and health promotion (CDC, 2011).

The Epidemic Intelligence Service (EIS) of the CDC in the USA is an excellent model for the organization of the national control of communicable diseases. Clinicians are trained to carry out epidemiological investigations as part of training to become public health professionals. EIS officers are assigned to state health departments, other public health units, and research centers as part of their training, carrying out epidemic investigation and special tasks in disease control. The CDC, in cooperation with the WHO, has developed a personal computer program to support field epidemiology, including epidemic investigations (EPI-INFO). It can be accessed and downloaded free of charge from the Internet. This program should be adopted widely in order to improve field investigations, to encourage reporting in real time, and to develop high standards in this discipline.

Any unusual increase in disease incidence should be investigated. The intensity and effort of the investigation are dependent on the severity of the disease, the number of people affected, the potential for the disease to spread, and the effectiveness of available countermeasures. The fundamental objective of syndromic surveillance is to identify illness clusters early, before diagnoses are confirmed and reported to public health agencies, and to mobilize a rapid response, thereby reducing morbidity and mortality. Epidemic curves for people with earliest symptom onset and those with severe illness can be depicted graphically.

Public health surveillance systems for early outbreak detection include early warning systems, prodrome surveillance, outbreak detection systems, information system-based sentinel surveillance, biosurveillance systems, health indicator surveillance, and symptom-based surveillance. The term "syndromic surveillance" is generally used and meant for early detection of potentially serious events such as terrorism with biological or chemical agents (Pavlin, 2003; Henning, 2004). A 2012 epidemiological event of an outbreak of fungal meningitis due to contaminated medication used for back pain relief (see Chapter 15) was detected by an alert clinician and followed by traditional "shoe leather epidemiology" and by syndrome surveillance with rapid conclusions, withdrawal of the offending drug mix, and onsite investigation, followed by legal action (CDC, 2013).

The CDC's epidemiological data are published in the excellent *Morbidity and Mortality Weekly Report* (*MMWR*), which contains articles on diseases of current interest and may be subscribed to as a free resource available from the Internet. The publication includes special summaries of

BOX 4.8 Steps of Epidemic Investigation

1. Awareness of clinicians; provide reports of infectious events to the local public health offices.
2. Surveillance monitoring of school attendance, Internet reporting and reporting from sentinel pediatric and family physician offices.
3. Identify index cases, possible causes, and contacts.
4. "Syndromic surveillance".
5. Confirm the existence of an epidemic, numbers of cases, common factors and exposures, and geographic spread.
6. Orient to person, place, and time of events.
7. Consult with relevant clinicians, and local, state, or national authorities.
8. Use medical data sources routinely collected for other purposes (e.g., emergency room logs).
9. Use non-clinical data (e.g., pharmacy sales, school absenteeism) to trigger alert.
10. Verify the diagnosis: clinical and laboratory.
11. Develop a case definition.
12. Thoroughly review relevant literature.
13. Develop a case report form.
14. Estimate the numbers from count of index cases and approximation analysis.
15. Orient data: time, place, person, common exposures.
16. Analyze the data: agent, transmission, vector, and host.
17. Use CDC's free Epi Info program. Available at: http://wwwn.cdc.gov/epiinfo/html/downloads.htm (last updated 13 February 2013).
18. Develop a hypothesis.
19. Test the hypothesis, e.g. with case controlled studies matching sick people with similar non-sick people.
20. Plan and institute control and prevent further cases.
21. Take legal action to withdraw drugs, food, water, or other causative factors.
22. Ensure compliance and evaluate the implemented measures.
23. Establish or improve the public health surveillance.
24. Write reports to home office and local authorities.
25. Keep local, state and federal officials, health providers and the public informed.
26. Plan and conduct additional studies.
27. Use the experience for teaching purposes.
28. Publish in peer-reviewed journals.
29. Follow-up on the outbreak and the literature.
30. Go home to your family and tell them all about it.

Sources: *Adapted from: Brachmen PS, Thacker SB. Evolution of epidemic investigations and field epidemiology during the MMWR era at CDC – 1961–2011. Public health then and now: celebrating 50 years of MMWR at CDC. MMWR Morb Mortal Wkly Rep Suppl. 2011;60(Suppl.);22–6 [Accessed 22 January 2013].*
Pavlin J. Investigation of disease outbreaks detected by "syndromic" surveillance systems. J Urban Health: Bull N Y Acad Med 2003;80(2 Suppl. 1);i107–14.
Henning KJ. Overview of syndromic surveillance: what is syndromic surveillance? MMWR Morb Mortal Wkly Rep 2004;53(Suppl.);5–11.

reportable infectious diseases as well as NCDs of epidemiological interest, comprehensive reviews of the literature, and recent investigative work by the CDC and other reputable health organizations. In 1999, *MMWR* published a review of "Ten great achievements of public health in the United States in the twentieth century", which included control of communicable disease and VPDs, as well as improvements in public health organization, occupational health, maternal and child health, motor vehicle accidents, tobacco control, reduction in cardiovascular disease mortality and motor vehicle deaths and injuries, and fluoridation of community water supplies (see Chapter 2). In 2011, *MMWR* published a special report on advances in the first decade of the twenty-first century in the USA and one on advances in public health globally.

CONTROL OF COMMUNICABLE DISEASES

An infectious disease is an event affecting an individual; however, it is transmissible to others, and therefore infection control requires both individual and community measures. Control of a disease comprises reduction in its incidence, prevalence, morbidity, and mortality. Elimination of a disease in a specified geographic area may be achieved as a result of intervention programs such as individual protection against tetanus; elimination of infections such as measles requires a halt in the circulation of the organism. Eradication of a disease is the reduction to zero of naturally occurring incidence, such as with smallpox. Extinction means that a specific organism no longer exists in nature or in laboratories.

Public health applies a wide variety of tools for the prevention of infectious diseases and their transmission, including activities ranging from filtration and disinfection of community drinking water to environmental vector control, pasteurization of milk, and immunization programs (Table 4.3). No less important are organized programs to promote self-protection, case finding, and effective treatment of infections to stop their spread to other susceptible people (e.g., HIV, STIs, TB, and malaria). Planning measures to control and eradicate specific communicable diseases is one of the principal activities of public health and remains so for the twenty-first century.

Treatment

Treating an infection once it has occurred is vital to the control of a communicable disease. Each person infected may become a vector and continue the chain of transmission. Successful treatment of the infected person reduces the potential for an uninfected contact person to acquire the infection. Bacteriostatic agents or drugs such as sulfonamides inhibit growth or stop replication of the organism, allowing normal body defenses to overcome the organism. Bactericidal drugs such as penicillin act to kill pathogenic organisms.

TABLE 4.3 Methods of Prevention and Control of Infectious Diseases by Type of Organism

Control of Major Infectious Diseases	Viruses	Bacteria	Parasites
Vaccination – pre-exposure to protect individuals and the community (herd immunity); post-exposure for individual protection (e.g., for rabies following animal bite, or contact after exposure to measles cases); or immunization of animals to prevent infected meat or milk transfer of disease to humans (e.g., brucellosis)	Rabies, polio, measles, rubella, mumps, hepatitis B, influenza, varicella, hepatitis A, HPV	Diphtheria, pertussis, tetanus, tuberculosis, anthrax, brucella, pneumococcal pneumonia, *Haemophilus influenzae* type b	Neglected tropical diseases prevention programs with mass treatment and case finding
Environmental measures – water and sewage control (e.g., chlorination of water to reduce burden of gastroenteric disease), vector control, antimosquito control measures (draining pooled water, larvicides, insecticides, repellents, protective bed nets and clothing)	Hepatitis A, rotaviruses, polio, arboviruses, tick- and mosquito-borne viruses	*Salmonella*, *Shigella*, cholera, Legionnaire's disease, *Escherichia coli*, *Helicobacter pylori*	Malaria, onchocerciasis, dracunculiasis, schistosomiasis, elephantiasis, worms
Education/social/behavioral measures – to promote self-care and self-protection to reduce risk (e.g., safe sexual practices to prevent STIs and HIV), needle exchange, condom distribution among risk groups	HIV, HPV, hepatitis B and C	Diarrheal diseases, syphilis, gonorrhea, chancroid	Malaria, scabies, onchocerciasis, dracunculiasis
Animal and food control – to reduce transmission by pasteurization of milk, veterinary supervision of animal husbandry, meat production and distribution, food hygiene and safety measures, radiation of food	Rabies vaccination of domestic and wild animals; suspect case management	Brucellosis, coliforms, salmonellosis, shigellosis	Tapeworms
Case finding and treatment – to cure or prevent transmission and reduce the carrier population (e.g., clinical diagnosis, laboratory confirmation with blood, sputum, immunological tests of animals, vectors, and human diseases	Rabies, herpes, cytomegalovirus, HIV, hepatitis C	TB, STIs, rheumatic fever	
Helicobacter pylori – chronic peptic ulcer diseases	Malaria, worms, dracunculiasis, leprosy, onchocerciasis, schistosomiasis		
Occupational measures – to protect people exposed at place of work (e.g., immunization of food handlers, health care and child care workers)	HIV, hepatitis A and B, measles, rubella, arboviruses	Brucellosis, tuberculosis, anthrax	Hydatid cyst, trichinosis

Note: STI = sexually transmitted infection; HIV = human immunodeficiency virus; HPV = human papillomavirus; TB = tuberculosis.

Traditional medical emphasis on single antibiotics has changed to the use of multiple drug combinations for tuberculosis (TB) and more recently for hospital-acquired infections. Antibiotics have made enormous contributions to clinical medicine and public health. However, pathogenic organisms are able to adapt or mutate and develop resistance to antibiotics, resulting in drug resistance. Wide-scale use of antibiotics has led to increasing incidence of resistant organisms. Multidrug resistance constitutes one of the major public health challenges in the twenty-first century.

Antiviral agents (e.g., ribavirin) are important additions to medical treatment potential, as are "cocktails" of antiviral agents for the management of HIV infection, known as highly active antiretroviral treatment (HAART). Prudent antibiotic use requires the attention of clinicians and their teachers as well as the public health community and health

care managers, representing the interaction of health issues across a broad spectrum of services.

Methods of Prevention

Organized public health services are responsible for advocating legislation and for regulating and monitoring programs to prevent infectious disease occurrence and transmission. They function to educate the population in measures to reduce or prevent the spread of disease.

Health promotion is one of the most essential instruments of infectious disease management. It promotes compliance and community support of preventive measures, including personal hygiene and safe handling of water, milk, and food supplies. Health education is the major method of prevention of STIs. Each of the infectious diseases or classifications of infectious diseases has one or more preventive or control approaches (Table 4.3). These may involve the coordinated intervention of different disciplines and modalities, including epidemiological monitoring, laboratory confirmation, environmental measures, immunization, and health education, all of which require teamwork and organized collaboration.

Remarkable progress has been made in infectious disease control by clinical, public health, and societal measures since 1900 in the industrialized countries, and since the 1970s in the developing world. This progress is attributable to a variety of factors, including organized public health services, the rapid development and wide use of new and improved vaccines and antibiotics, better access to health care, and improved sanitation, living conditions, and nutrition. Triumphs have been achieved in the eradication of smallpox and in the increasing control of other VPDs. Despite the great advances, major challenges persist, such as TB, STIs, malaria, HIV, child mortality from respiratory infections and diarrheal diseases, an increase in MDR organisms, and the rise of NCDs in low- and medium-income countries.

VACCINE-PREVENTABLE DISEASES

Vaccines are one of the most important and indispensable tools of public health in the control of infectious diseases, particularly for child health. VPDs are diseases preventable by currently available vaccines (Table 4.4). The term *vaccine* is derived from the use of cowpox (vaccinia virus) to stimulate immunity to smallpox, first demonstrated by Jenner in 1796 (see Chapter 1), and is generally used for all immunizing agents.

According to the trend tables published by the CDC's *Health United States, 2010*, since 2000 there have been no reported cases of polio in the USA. While the table (Table 4.5) illustrates zero cases for the years prior to 2000, it represents the number of new cases per 100,000 population; the rate is greater than zero but less than 0.005, and there have been too few cases to count. Thus, in 2000 and the years after, no polio cases have been documented, and since 1950, the incidence has continuously declined each year. The success of the polio vaccine is indicative of what can be achieved with the implementation of a comprehensive program.

As discussed above, the body responds to invasion by disease-causing organisms by antigen–antibody reactions and cellular responses. Together, these act to restrain or destroy the disease-causing potential. Strengthening this defense mechanism is possible through immunization. Vaccines are suspensions of live or killed microorganisms or the antigenic portion of those agents presented to a potential host to induce immunity to prevent the specific disease caused by that organism. The preparation of vaccines uses different techniques, as seen in Box 4.9.

The process of immunization (vaccination) increases host resistance to specific microorganisms to prevent them from causing disease. Doing so induces primary and secondary responses in the human or animal body:

TABLE 4.4 Vaccine-Preventable Diseases and Year of Vaccine Development/Licensure, USA, 1798–2012

Vaccine	Year Vaccine Developed or Licensed
Smallpox[a]	1798[b]
Rabies	1885[b]
Typhoid	1896[b]
Cholera	1896[b]
Plague	1897[b]
Diphtheria[a]	1923[b]
Pertussis[a]	1926[b]
Tetanus[a]	1927[b]
Tuberculosis	1927[b]
Influenza	1945[c]
Yellow fever	1953[c]
Poliomyelitis[a]	1955[c]
Measles[a]	1963[c]
Mumps[a]	1967[c]
Rubella[a]	1969[c]
Anthrax	1970[c]
Meningitis	1975[c]
Pneumonia	1977[c]
Adenovirus	1980[c]
Hepatitis B[a] (HBV)	1981[c]
Haemophilus influenzae type b[a] (Hib)	1985[c]
Japanese encephalitis	1992[c]
Hepatitis A (HAV)	1995[c]
Varicella[a]	1995[c]
Lyme disease[a]	1998[c]
Rotavirus[a]	1998[c]
Rotavirus, quadrivalent meningococcal conjugate, herpes zoster, pneumococcal conjugate, and influenza; human papillomavirus vaccines; tetanus, diphtheria, and acellular pertussis vaccine for adults and adolescents introduced. Total number of VPDs targeted by the US routine immunization policy in 2012 is 17	2000–2010[d]

Note: [a]Vaccine recommended for universal use in US children. Routine vaccination for smallpox was ended in 1971. [b]Vaccine developed (i.e., first published results of vaccine usage). [c]Vaccine licensed for use in USA. [d]Vaccine added to the US list of vaccines recommended for routine use in all or for selected population groups.
Sources: Centers for Disease Control and Prevention. Achievements in public health, 1900–1999. Impact of vaccines universally recommended for children – United States, 1990–1998. MMWR Morb Mortal Wkly Rep 1999;48:243–8. Available at: http://www.cdc.gov/mmwr/preview/mmwrhtml/00056803.htm [Accessed 6 February 2013]. Centers for Disease Control and Prevention. Ten great public health achievements – United States, 2001–2010. MMWR Morb Mortal Wkly Rep 2011;60:619–23. Available at: http://www.cdc.gov/mmwr/preview/mmwrhtml/mm6019a5.htm [Accessed 6 February 2013]. College of Physicians and Surgeons of Philadelphia. The history of vaccines. Timelines overview. Posted 2013. Available at: http://www.historyofvaccines.org/ [Accessed 6 February 2013].

TABLE 4.5 Annual Incidence of Selected Vaccine-Preventable Infectious Diseases, Selected Years, USA, 1970–2008, Rates per 100,000 Population

Disease	1970	1980	1990	2000	2004	2006	2008
Diphtheria	0.2	0	0	0	0	0	0
Pertussis	2.1	0.8	1.8	2.9	8.9	5.3	4.4
Poliomyelitis	0	0	0	0	0	0	0
Measles	23.2	6.0	11.2	0.03	0.04	0.02	0.05
Mumps	55.6	3.9	2.2	0.08	0.13	2.2	0.15
Rubella	27.8	1.7	0.5	0.06	0	0	0.01
Hepatitis A	27.8	12.8	12.6	4.9	2.0	1.2	0.86
Hepatitis B	4.1	8.4	8.5	3.0	2.1	1.6	1.3

Sources: National Center for Health Statistics. Health, United States, 2010 with special feature on death and dying. Hyattsville, MD: National Center for Health Statistics; 2011. Available at: http://www.cdc.gov/nchs/data/hus/hus10.pdf [Accessed 31 January 2013].
National Center for Health Statistics. Health, United States, 2006 with chartbook on trends in the health of Americans. Hyattsville, MD: National Center for Health Statistics; 2006. Available at: http://www.cdc.gov/nchs/data/hus/hus06.pdf [Accessed 31 January 2013].
National Center for Health Statistics. Health, United States, 1998 with socioeconomic status and health chartbook. Hyattsville, MD: National Center for Health Statistics; 1998. Available at: http://www.cdc.gov/nchs/data/hus/hus98.pdf [Accessed 31 January 2013].

BOX 4.9 Immunizing Agents and Processes

- *Live attenuated organisms* – have been passed repeatedly in tissue culture or chick embryos so that they have lost their capacity to cause disease but retain an ability to induce antibody response. Examples: polio (Sabin), measles, rubella, mumps, yellow fever, BCG, typhoid, plague.
- *Inactivated or killed organisms* – have been killed by heat or chemicals but retain an ability to induce antibody response; they are generally safe but less efficacious than live vaccines and require multiple doses. Examples: polio (Salk), influenza, rabies, Japanese encephalitis.
- *Cellular fractions* – usually of a polysaccharide fraction of the cell wall of a disease-causing organism. Examples: pneumococcal pneumonia, meningococcal meningitis.
- *Recombinant vaccines* – produced by recombinant DNA methods in which specific DNA sequences are inserted by molecular engineering techniques, such as DNA sequences spliced to vaccinia virus grown in cell culture to produce influenza and hepatitis B vaccines.
- *Toxoids or antisera* – modified toxins are made non-toxic to stimulate formation of an antitoxin. Examples: tetanus, diphtheria, botulism, gas gangrene.
- *Immune globulins* – antibody-containing solutions derived from immunized animals or human blood plasma, used primarily for short-term passive immunization, e.g., rabies, IgG globulin for immunocompromised people.
- *Antitoxin* – an antibody derived from serum of animals after stimulation with specific antigens and used to provide passive immunity. Examples: tetanus, snake and scorpion venom.

Sources: *Brooks GE, Carroll KC, Butel JS, Morse SA, Mietzner TA. Jawetz, Melnick and Adelberg's medical microbiology. 26th ed. New York: McGraw-Hill; 2012.*
Fauci AS, Eugene Braunwald E, Kasper DL, Hauser SL, Longo DL, Jameson JL, Loscalzo J. Harrison's textbook of internal medicine. 17th ed. New York: McGraw-Hill; 2008.

- *Primary response* – occurs on first exposure to an antigen. After a lag or latent period of 3–14 days (depending on the antigen), specific antibodies appear in the blood. Antibody production ceases after several weeks but memory cells that can recognize the antigen and respond to it remain ready to respond to a further challenge by the same antigen.
- *Secondary (booster) response* – the response to a second and subsequent exposure to an antigen. This lag period is shorter than the primary response, with the peak being higher and lasting longer. The antibodies produced have a higher affinity for the antigen, and a much smaller dose of the antigen is required to initiate a response. Booster doses of vaccines are used to activate memory cells to strengthen immunity.
- *Immunological memory* – exists even when circulating antibodies are insufficient to protect against the antigen. When the body is exposed to the same antigen again, it responds by rapidly producing high levels of antibody to destroy the antigen before it can replicate and cause disease.

Thus, immunization protects susceptible individuals from communicable disease by administration of a living modified agent, a subunit of the agent, a suspension of killed organisms, or an inactivated toxin (Box 4.9) to stimulate development of antibodies to that agent. In disease control, individual immunity may also protect another individual.

- *Herd immunity* – occurs when sufficient numbers of people are protected (naturally or by immunization) against a specific infectious disease, reducing circulation of the organism, and thereby lowering the chance of an unprotected person becoming infected. Each pathogen has different characteristics of infectivity, and therefore different levels of herd immunity are required to protect the non-immune individual who may not have been immunized or who is immunocompromised, or whose immunity from vaccination may have waned.

Immunization Coverage

The critical proportion of a population that must be immunized in order to interrupt local circulation of the organism varies from disease to disease. Eradication of smallpox was achieved with approximately 80 percent world coverage, followed by concentration on new case findings and immunization of contacts and surrounding communities.

For highly infectious diseases such as measles, immunization coverage of over 95 percent is required in order to achieve local eradication.

Immunization coverage in a community must be monitored to gauge the extent of protection and need for program modification to achieve targets of disease control. Immunization coverage is expressed as a proportion in which the numerator is the number of people in the target group immunized at a specific age, and the denominator is the number of people in the target cohort who should have been immunized according to the accepted standard:

$$\text{Vaccine coverage} = \frac{\text{No. of persons immunized in specific age group}}{\text{No. of persons in the age group during that year}} \times 100$$

Immunization coverage in the USA is regularly monitored by the National Immunization Survey, a telephone-based questionnaire of households from all 50 states, as well as selected areas at high risk for inadequate levels of vaccination. An initial telephone survey is followed by confirmation, where possible, from documentation from the parents or health care providers. The childhood immunization survey for 2006, for instance, examined children aged 19–35 months. The results reveal 85 percent of US children having received four or more (4+) doses of diphtheria–tetanus–acellular pertussis (DTaP), 93 percent with three or more (3+) doses of oral or injected polio vaccine, and 93 percent with three or more (3+) doses of *Haemophilus influenzae* type b (Hib). Hepatitis B coverage (3+) greatly increased to 93 percent, while institution of pneumococcal (3+) and varicella (1+) vaccination policies has rapidly achieved 87 percent and 89 percent, respectively. Despite these gains, only 77 percent of children received all vaccinations at the recommended ages.

Present technology allows for control or eradication of important infectious diseases that still cause millions of deaths globally each year. Other important infectious diseases are still not subject to vaccine control owing to difficulties in their development. In some cases, a microorganism can mutate with changes. Viruses can undergo antigenic shifts in their molecular structure, producing completely new subtypes of the organism. Hosts previously exposed to other strains may have little or no immunity to the new strains.

Antigenic drift refers to relatively minor antigenic changes which occur in viruses, and is responsible for frequent epidemics. Antigenic shift is believed to explain the occurrence of new strains of influenza virus, necessitating annual reformulation of the influenza vaccine. New variants of poliovirus strains are similar enough to three main types that immunity to one strain is carried over to the new strain. Molecular epidemiology is a powerful genetic technique used to determine geographic origin, permitting tracking of the spread of infectious organisms and epidemics.

The trend for increasing the number of vaccines included in a "cocktail" of vaccines (i.e., combination of more than one vaccine) has many advantages in lowering costs and reducing the number of visits and injections, thus increasing convenience and compliance by the public. There are virtually no contraindications to the use of multiple antigens simultaneously. Examples of vaccine cocktails include DTaP in combinations with Hib, polio, varicella, or measles–mumps–rubella (MMR) vaccines. The term DTaP is used for the combination which includes acellular pertussis vaccine. It is more expensive but with reduced complications and is used in the USA and other high-income countries (see website material).

Interventions in the form of effective vaccination save millions of lives each year and immunization coverage contributes to the improved health of countless children and adults globally. Vaccination is accepted as one of the most cost-effective health interventions currently available. Continuous policy review is needed regarding allocation of adequate resources, logistical organization, and continued scientific effort to seek effective, safe, and inexpensive vaccines for other major diseases such as malaria and HIV. Molecular technology, producing recombinant vaccines, such as those for hepatitis A and B, holds promise for important vaccine breakthroughs in the decades ahead.

The introduction of new vaccines such as for rotavirus and its potential to reduce child morbidity and mortality is an ongoing challenge pertaining to strengthening immunization programs and prioritizing effective strategies. It is especially difficult for low-income countries to meet the vaccine targets and Millennium Development Goals (MDGs), particularly in the area of reducing child mortality.

CDC reports indicate that worldwide, vaccination prevents over 2 million childhood deaths from VPDs each year (2004 estimate). Globally, over 130 million children are born annually, all needing immunization. Annual routine immunization coverage levels included 83 percent of the world's children for three doses of diphtheria–tetanus–pertussis vaccine (DTP3) in 2012, as compared with 20% in 1980 and 73% in 2000 respectively); in 2012, 84 percent had a least one dose of a measles-containing vaccine. However, this means that approximately 27 million infants remained unvaccinated. Hepatitis B coverage in 2012 was 79 percent and Hib vaccine coverage 45 percent. Further data indicate that one in six children is not vaccinated against TB. Globally, 79 percent of infants receive the complete dose of hepatitis B immunizations, and only 45 percent are vaccinated against Hib disease. Thus, the remaining infants are left unprotected from serious diseases (WHO, 2013).

The negative impact and the problems that ensue owing to these significant gaps in immunization are tremendous. In 2004, approximately 1.4 million children under 5 years old died from the six major VPDs; an additional 1.1 million succumbed to pneumococcal disease and rotavirus and consequently died. The vaccines protecting against these two diseases are available and accessible in the USA but rarely administered in developing countries. While some medical issues may be inevitable (such as genetic disorders), VPDs are preventable; thus, these deaths and serious morbidities could have, and should have, been averted.

Since 1991, polio eradication and measles elimination programs have received significant financial and professional support from the CDC. CDC's annual investment dedicated to global immunization has increased from slightly over US$3 million to US$140 million in 2006. A US federal agency, the CDC has substantially expanded its influence in the area of global immunization. It plays a key role in the establishment and initiation of instrumental vaccine initiatives, such as the Global Immunization Vision and Strategy for 2006–2015 (GIVS) and the Global Alliance for Vaccines and Immunization (GAVI). Moreover, the CDC supports the research, development, and evaluation needed to create new vaccines, particularly those required to protect against HIV, TB, and malaria, the greatest causes of mortality in developing countries.

In the developing countries, immunization averts some 3 million child deaths each year. Internationally, substantial progress was made in controlling VPDs in the 1980s. Towards the end of the 1970s, fewer than 10 percent of children worldwide were receiving immunizations. Collaboration between the WHO, UNICEF and other international bodies allowed for the promotion of the Expanded Programme on Immunization (EPI) and the target of achieving 80 percent uptake by 1990. In 2007, the WHO reported that the following vaccines reached 80–90 percent coverage in 2006: DTP3, polio (three doses), and measles.

Estimated global DTP3 coverage in the 193 WHO member states increased from 74 percent in 2000 to 82 percent in 2009; this reflects vaccination of 107.1 million infants with DTP3 vaccine in 2009 (14.6 million more than in 2000).

Changes in coverage varied by geographic region, and the overall increase mainly was attributed to improvements in vaccination coverage in the African (+16 percent), Eastern Mediterranean (+12 percent), and Western Pacific (+10 percent) WHO regions.

National DTP3 coverage of at least 90 percent was reported by 122 countries (63 percent), but only 48 (25 percent) reported 80 percent coverage or higher in all districts, and only 55 percent of low-income countries are on track to achieve 90 percent coverage by 2015 (UNICEF, unpublished data, 2010). During 2007–2009, 149 countries (77 percent) had sustained DTP3 coverage of 80 percent or higher. However, coverage in 2009 was less than 80 percent in 36 countries (19 percent), and six countries failed to achieve even 50 percent DTP3 coverage. Globally, 23.2 million children worldwide did not receive three doses of DTP vaccine during the first year of life in 2009; 70 percent live in 10 countries and approximately half live in India (37 percent) and Nigeria (14 percent).

From 2000 to 2012, estimated global measles-containing vaccine (MCV1) coverage increased from 71 to 84 percent, and 136 countries (70 percent) added a second MCV dose to their routine vaccination schedules. Three-dose coverage with hepatitis B vaccine (HepB3) increased from 30 to 79 percent during this period, and three-dose coverage with Hib vaccine (Hib3) increased from 13 to 45 percent.

In countries where Hib vaccine had been introduced, Hib3 coverage was similar to DTP3 coverage; however, an increase in global coverage did not occur because several large countries (e.g., China, India, Indonesia, and Nigeria) had not yet introduced Hib vaccine. Global immunization coverage has increased in all WHO member states (Figures 4.2 and 4.3).

While the three-dose Hib vaccine reached 90 percent of those residing in the Americas, coverage was much lower in the European Region (44 percent). Even lower and more disappointing was the 24 percent uptake in the African Region of the WHO. A more positive outcome demonstrates international increases in coverage of other vaccines.

Bacille Calmette–Guérin (BCG) uptake rose from 31 to 89 percent and polio with oral poliomyelitis vaccine (OPV, three doses) increased from 24 to 85 percent. Furthermore, tetanus toxoid vaccine coverage among pregnant women rose from 14 to 57 percent. Despite this, recent drops in uptake have taken place in many countries, most notable in Sudan, Myanmar (Burma), and other areas struggling with violent conflicts.

The number of global polio cases has been reduced by more than 99.8 percent since 1988, including the prevention of 5 million cases of paralysis and of more than 250,000 deaths. This reduction in cases is a remarkable achievement of many agencies (WHO, UNICEF, GAVI, AID, CDC, Rotary International, and many others). By 2010, only four countries remain endemic for polio, the

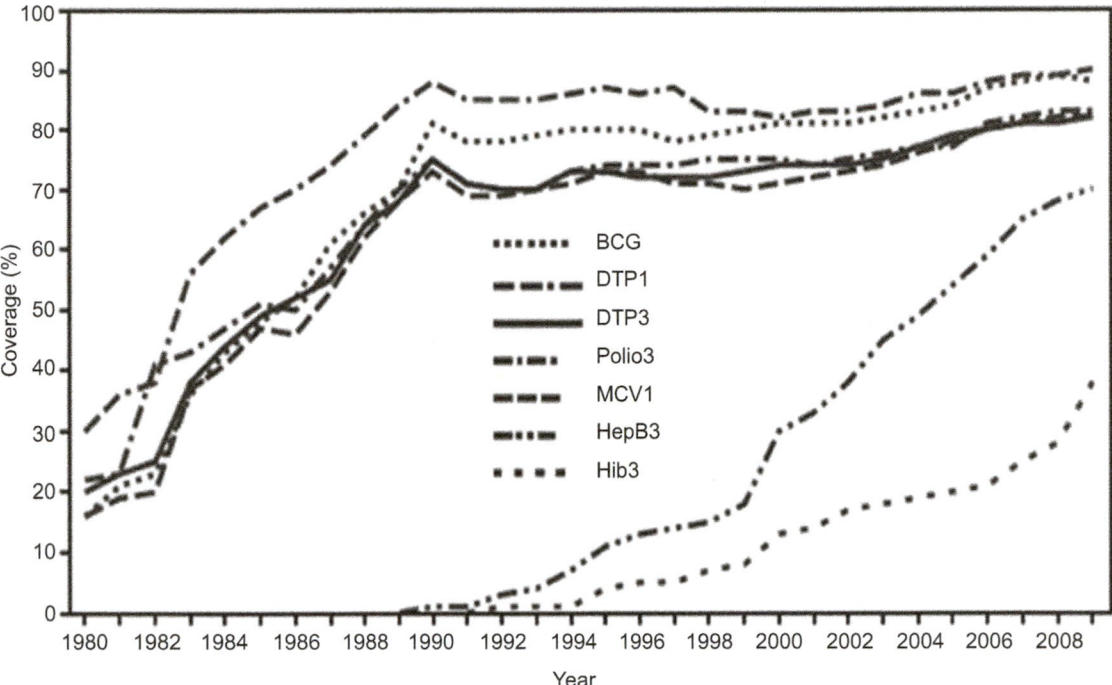

FIGURE 4.2   Estimated global vaccination coverage by vaccine type in children by age 12 months, 1980–2009. Note: BCG = bacille Calmette–Guérin; DTP1 = one dose of diphtheria–tetanus–pertussis vaccine; DTP3 = three doses of diphtheria–tetanus–pertussis vaccine; Polio3 = three doses of polio vaccine; MCV1 = one dose of measles-containing vaccine; HepB3 = three doses of hepatitis B vaccine; Hib3 = three doses of *Haemophilus Influenzae* type b vaccine. *Source: Centers for Disease Control and Prevention. Global routine vaccination coverage, 2009. MMWR Morb Mortal Wkly Rep 2010;59:1367–71. Available at: http://www.cdc.gov/mmwr/preview/mmwrhtml/mm5942a3.htm#fig1. See also Global vaccine coverage 2011. MMWR Morb Mortal Wkly Rep 2012;61:883–5. Available at: http://www.cdc.gov/mmwr/preview/mmwrhtml/mm6143a5.htm [Accessed 21 January 2013].*

fewest ever recorded, and in 2012 this was reduced to three countries, as India had no cases.

Endemic measles has remained eliminated from the western hemisphere since 2002, and no importations from Latin America have occurred in the USA since 2000 (in 1990, more than 90 percent of measles importations into the USA were from Latin America). The Measles Initiative has reduced measles deaths by 60 percent in Africa from 1999 to 2004. Global measles deaths have dropped by 78 percent, falling from 733,000 deaths in 2000 to 164,000 in 2008.

With technical support from the WHO, UNICEF, CDC, and other partners, global immunization coverage for DTP3 has increased from 20 percent in 1980 to 82 percent in 2009.

The challenge remains to achieve control or eradication of VPDs, thus saving millions more lives. Part of Health for All (HFA) stresses the EPI approach, which includes immunization against diphtheria, pertussis, tetanus, polio, measles, and TB. In 2009, of the 23.2 million children who did not receive three doses of DTP vaccine during the first year of life, 70 percent lived in 10 countries, with approximately half residing in India (37 percent) and Nigeria (14 percent). A further challenge in eradicating VPDs is overcoming the efforts of segments of the population who are averse to immunization due to transitory side-effects or immunization as a perceived cause of other morbidity (Box 4.10).

An extended form of this is the EPI-plus program, which combines EPI with immunization against hepatitis B and yellow fever and, where appropriate, supplementation with vitamin A and iodine. The success in the international eradication of smallpox has been followed with major progress towards the eradication of polio, measles, and other important infectious diseases.

Diphtheria

Diphtheria is an acute bacterial disease of the tonsils, nasopharynx, and larynx caused by the organism *Corynebacterium diphtheriae*. It occurs in colder months in temperate climates where the organism is present in human hosts and is spread by contact with patients or carriers. Most typically transmitted via respiratory droplets, it has an incubation period of 2–5 days. In the past, this was primarily an infection of children and was a major contributor to child mortality in the prevaccine and preantibiotic eras. Diphtheria has been virtually eliminated in countries with well-established immunization programs.

In the 1980s, an outbreak of diphtheria occurred in the countries of the former Soviet Union among people over the age of 15. It reached epidemic proportions in the 1990s, with 140,000 cases (1991–1995), and 1100 deaths in 1994 in Russia alone. This indicates a failure of the vaccination

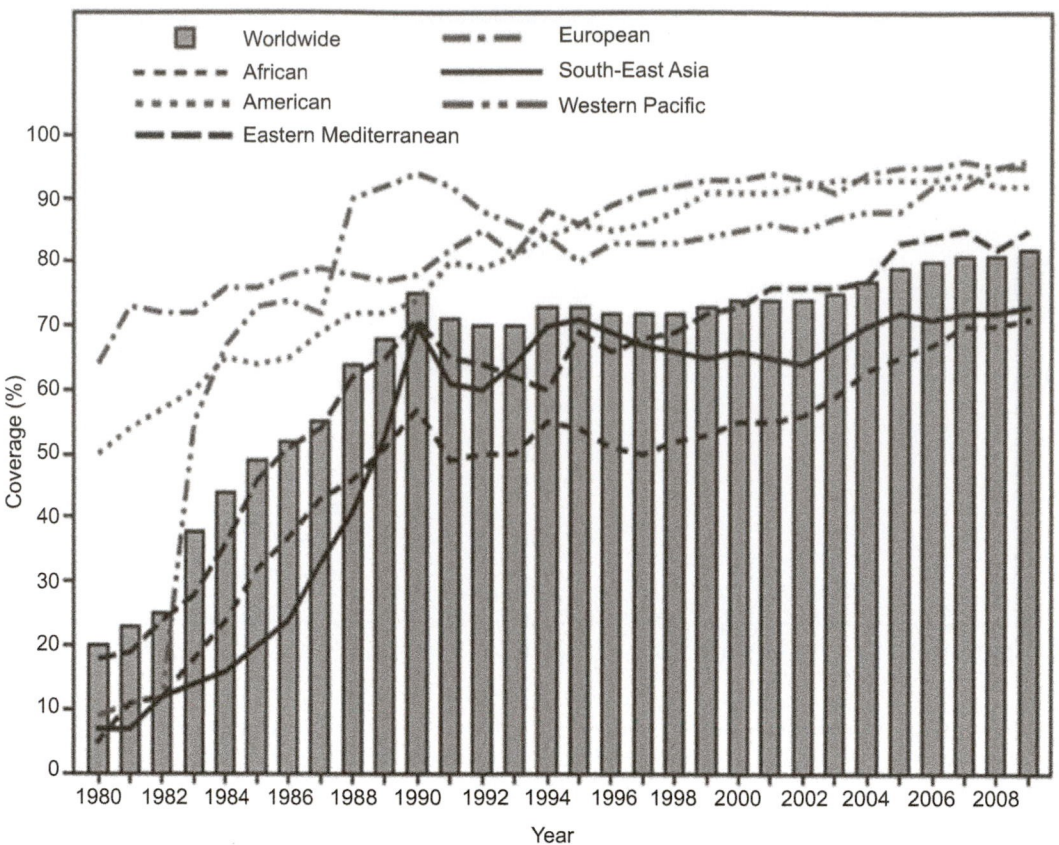

FIGURE 4.3 Estimated vaccination coverage with three doses of diphtheria–tetanus–pertussis vaccine among infants, worldwide and by region 1980–2009. *Source: Centers for Disease Control and Prevention. Global routine vaccination coverage, 2009. MMWR Morb Mortal Wkly Rep 2010;59:1367–71. Available at: http://www.cdc.gov/mmwr/preview/mmwrhtml/mm5942a3.htm [Accessed 1 February 2013].*

program in several respects: it used only three doses of DTP or DTaP for infants, no boosters were given at school age or subsequently, the efficacy of diphtheria vaccine may have been low, and coverage was below 80 percent. Cases have declined from 1409 in 1998 to 909 in 2001 and gradually further declined to 353 in 2005 and nine in 2010 (WHO–UNICEF, 2011).

Efforts to control the epidemic included mass vaccination campaigns for people over 3 years of age with a single dose of DT (diphtheria and tetanus) and increasing coverage of routine DTP vaccines to four doses by the age of 2 years. The epidemic and its control measures have led to improved coverage with DT for those over 18 years, and 93 percent coverage among children aged 12–23 months. By 2010 coverage in Russia reached 97–98 percent.

The WHO recommends three doses of DTP in the first year of life and a booster at primary school entry, as well as at enrollment at college, military, or other organized settings. This is considered by many to be insufficient to produce long-lasting immunity. The USA and other industrialized countries use a four-dose schedule and recommend periodic boosters with DT for adults.

Pertussis

Pertussis is an acute bacterial disease of the respiratory tract caused by the bacillus *Bordetella pertussis*. After an initial cold-like (catarrhal) stage, the patient develops a severe cough which comes in spasms (paroxysms). The disease can last for 1–2 months. The paroxysms can become violent and may be followed by a characteristic crowing or high-pitched inspiratory whooping sound, followed by expulsion of tenacious clear sputum, often followed by vomiting. In poorly immunized populations and malnourished people, pneumonia often follows, and death is common.

Pertussis declined dramatically in the industrialized countries as a result of widespread coverage with DTP. However, because the pertussis component of early vaccines caused rare reactions, many physicians and parents avoided its use, instead opting for DT alone and leaving children susceptible to infection. During the 1970s in the UK, many physicians recommended against vaccination with DTP. As a result, pertussis incidence increased, with substantial mortality rates. This led to a reappraisal of the immunization program, with institution of incentive payments to general practitioners for completion of vaccination schedules. As a result of these

BOX 4.10 Anti-Vaccination Movements

Although vaccination is recognized as one of the 10 great achievements of public health of the twentieth century, ever since its origins with Jenner's work in 1796, anti-vaccination movements have been active in opposing this indispensible, life-saving tool. This was highlighted in the mid-nineteenth century by anti-vaccination leagues.

Opposition centered on provocative fears of the inefficiency of vaccines, their side-effects, or the principle that the state is interfering with individual freedoms by promoting or mandating vaccination. Opposition was in part promoted by doctors who made good fees from variolation, i.e., passing actual smallpox to young children to give immunity, a dangerous procedure. In Britain and the USA various acts mandating smallpox vaccination in the nineteenth century were bitterly opposed, and modifications of laws allowed for conscientious objectors to opt out of the required immunization.

In the late twentieth century this took on a new form when objections to pertussis vaccine became frequent on the grounds of common side-effects. This reduced compliance with DTP vaccination in the UK, and in consequence, an upsurge in pertussis cases was seen in 2011–2012 in babies of young mothers who themselves may not have been immunized as children.

In 1998, an infamous controversy provoked by a Dr Andrew Wakefield in the UK created a storm of opposition toward the MMR vaccine owing to an alleged causative connection with autism. This alleged association was published in the prestigious medical journal *The Lancet*. The study was later refuted and ultimately proven fraudulent. Wakefield lost his UK medical license for serious misconduct (General Medical Council Great Britain 2010). His fictitious findings and fraudulent science allegedly motivated for economic gains from legal actions against MMR producers did great harm to vital immunization coverage (BMJ editorial, 2011). Despite legal action against Wakefield, considerable damage was done to public opinion, and compliance with MMR vaccination reduced in the UK; it gradually recovered from low 80 percent to over 90 percent. *The Lancet* eventually retracted the paper in 2010, 12 years after initial publication.

A measles epidemic in Europe in 2010–2012 and its subsequent spread to North and South America show the fragility of the immunization system globally even in the advanced countries. Opposition to vaccination is still a major obstacle in achieving the full benefits of vaccines in the USA and other countries. Opposition has hindered efforts to eradicate poliomyelitis in Nigeria and in Pakistan owing to beliefs in some uneducated regions that the vaccine is meant to sterilize girls. Other controversies lurk around the adoption of human papillomavirus vaccine on the grounds that immunization of girls with HPV will encourage early sexual activity.

The H1N1 influenza pandemic of 2009 showed the extent of public concern or apathy to vaccination when the uptake of a vaccine made freely available to avert an international crisis was largely ignored by the general public in most countries. Furthermore, the vaccine was refused by some medical and nursing personnel, a population at high risk of contracting the infection during a pandemic, and thus at risk of transmitting the disease to other patients as well as their own families and contacts.

Vaccination is a pillar of modern public health, saving millions of lives and with potential for millions more to be saved, as medical sciences bring more vaccines into general use. Public apathy and unrelenting opposition by anti-vaccinationists are substantial obstacles to the wider prevention of diseases. Health promotion is crucial to increase support for vaccination by the general public, in order to control very important diseases. New vaccines for HIV, malaria, *Helicobacter pylori*, sexually transmitted diseases, and cancers will provide both opportunities and challenges for public health to disseminate its message widely.

Sources: *Amy Ovadia MPH. Personal communication; December 2012.*
Centers for Disease Control and Prevention. Achievements in public health, 1900–1999: impact of vaccines universally recommended for children – United States, 1990–1998. MMWR Morb Mortal Wkly Rep 1999;48;243–8. Available at: http://www.cdc.gov/mmwr/preview/mmwrhtml/00056803. htm [Accessed 30 January 2013].
Editorial. The fraud behind the MMR scare. BMJ 2011;342:c7452 http:// www.bmj.com/content/342/bmj [Accessed 19 November 2013].c7452
General Medical Council. Date: 24 May 2010. Dr Andrew Jeremy WAKEFIELD Determination on Serious Professional Misconduct (SPM) and sanction: Available at: http://www.gmc-uk.org/Wakefield_SPM_and_SANCTION. pdf_32595267.pdf [Accessed 10 November 2013].
Philadelphia College of Physicians and Surgeons. A history of anti-vaccination movements. Available at: http://www.historyofvaccines.org/content/articles/ history-anti-vaccination-movements [Accessed 23 November 2011].

measures, vaccination coverage, with resulting pertussis control, improved dramatically in the UK. A new acellular vaccine is now in widespread use and will be safer with fewer and less severe reactions in infants, increasing the potential for improved confidence and support for routine vaccination. The new vaccine is used in the USA and other industrialized countries, and forms part of the US recommended vaccination schedule. Although most Western European countries are advanced in the use of vaccines, there is no Europe-wide equivalent of the CDC-recommended immunization schedule for the region, which will be coming up for discussion in European Union (EU) health forums.

The CDC reports that estimates of childhood vaccination coverage in the USA with at least doses of pertussis-containing vaccine have exceeded 90 percent since 1993. However, reported pertussis cases increased from a historic low of 1010 cases in 1976 to 11,647 in 2003, with a substantial increase in reported cases among adolescents, who become susceptible to pertussis approximately 6–10 years after their childhood vaccination. This increase is attributed to waning immunity and lack of booster doses. Consequently, booster doses in adolescence are now recommended. However, reported cases reached 25,827 and 25,616 in 2004 and 2005, respectively. Cases then declined to 15,632 and 10,454 in 2006 and 2007 but again rose to 13,213 in 2008, and 27,550 reported cases in 2012 (and an unknown number of unreported cases). Despite DTP, three-dose immunization has remained at 95–96 percent since

2003, but pertussis remains endemic and careful protection is required to prevent disease and deaths from pertussis.

Pertussis continues to be a public health threat and recurs wherever there is inadequate immunization in infancy. In addition, recent epidemics have been noted in adults who have lost childhood immunity. While the disease generally follows a milder course in healthy adults, concerns have been raised over adults serving as reservoirs for infection of children and immunocompromised individuals. To eliminate this risk, pertussis booster vaccination is recommended during adolescence and once again in adulthood and also during pregnancy to protect the newborn until routine infant immunization offers full protection. Outbreaks in schools, kindergartens, and hospitals require booster TdaP doses of vaccine.

Tetanus

Tetanus is an acute disease caused by an exotoxin of the tetanus bacillus (*Clostridium tetani*) which grows anaerobically at the site of an injury. The bacillus is universally present in the environment and enters the human body via penetrating injuries. Following an incubation period of 3–21 days, it causes an acute condition of painful muscular contractions. Unless there is modern medical care available, patients are at risk of high case fatality rates of 30–90 percent (highest in infants and elderly people).

Antitetanus serum (ATS) was discovered in 1890, and during World War I, ATS contributed to saving the lives of many thousands of wounded soldiers. Tetanus toxoid was developed in 1993. Owing to the organism's universal presence in the environment, it cannot be eradicated. However, the disease can be controlled by effective immunization of every child during infancy and school age. Adults should receive routine boosters of tetanus toxoid once every decade.

Newborns are infected by tetanus spores (tetanus neonatorum) where unsanitary conditions or practices are present. Infections can occur when traditional birth attendants at home deliveries use unclean instruments to sever the umbilical cord, or dress the severed cord with contaminated material. Tetanus neonatorum remains a serious public health problem in developing countries. Immunization of pregnant women and women of childbearing age is reducing the problem by conferring passive immunity to the newborn. The training of traditional birth attendants in hygienic practices and the use of medically supervised birth centers for delivery also contribute to the reduction of the incidence of tetanus neonatorum.

Elimination of tetanus neonatorum was made a health target by the World Summit of Children in 1990. In that year, the number of deaths from neonatal tetanus was reported by the WHO as 25,293 infants worldwide, declining to 8376 in 2006 (112 countries reporting). Immunization of pregnant women increased from under 20 percent in 1984 to 69 percent in 2006. In 2008, maternal and neonatal tetanus (MNT) still occurred in 46 countries; however, progress continues and at the end of 2010, the goal for elimination had not been reached

by 39 countries. The WHO reported the approximate number of cases for 2008 to be an estimated 59,000 newborn deaths caused by MNT (a 92 percent reduction from the late 1980s).

Tetanus cases have declined dramatically in the USA, but the disease still occurs, mainly among older adults. According to the CDC, during 1990–2001 a total of 534 cases of tetanus were reported; 301 (56 percent) cases occurred among adults aged 19–64 years and 201 (38 percent) among adults aged 65 years or older. Data from a national population-based serosurvey indicated that the prevalence of immunity to tetanus was over 80 percent among adults aged 20–39 years; however, this declined with increasing age. These figures support current recommendations to give booster doses of tetanus (with diphtheria) vaccine for adolescents and adults every 10 years.

Poliomyelitis

Poliovirus infection may be asymptomatic or cause an acute non-specific febrile illness. It may reach more severe forms of aseptic meningitis and acute flaccid paralysis (AFP) with long-term residual paralysis or death during the acute phase. Polio is transmitted mainly by direct person-to-person contact, but also via sewage contamination. Large-scale epidemics of disease, with attendant paralysis and death, occurred in industrialized countries in the 1940s and 1950s, engendering widespread fear and panic and thousands of clinical cases of "infantile paralysis".

Growth of the poliovirus in tissue culture, by John Enders and colleagues in 1949, led to the development and wide-scale testing of the first inactivated (killed) polio vaccine by Jonas Salk in the mid-1950s. This achievement and the largest clinical trial ever conducted up to that point helped to build great hopes and outstanding success in the control of this much feared disease, making Salk a national and global hero. Albert Sabin's development of the live attenuated OPV, licensed in 1960, added a major new dimension to polio control owing to its effectiveness, low cost, and ease of administration. The two vaccines in their more modern forms, enhanced strength inactivated poliomyelitis vaccine (eIPV), and triple oral poliomyelitis vaccine (TOPV), have been used in different settings with great success.

OPV induces both humoral and cellular (including intestinal) immunity. The presence of OPV in the environment by contact with immunized infants and via excreta of immunized people in the sewage gives a booster effect in the community. Immunization using OPV, in both routine practice and on national immunization days (NIDs), has proven effective in dramatically reducing polio and circulation of the wild virus in many parts of the world. Use of the eIPV produces early and high levels of circulating antibodies, as well as protecting against the vaccine-associated disease.

In rare cases, OPV can cause vaccine-associated paralytic poliomyelitis (VAPP), with a risk of one case per 520,000 with initial doses, and one case per over 12 million with subsequent doses. Approximately eight to 10 cases of VAPP occurred annually in the USA during the 1990s following the elimination of natural transmission. The CDC changed its recommendations to inactivated polio vaccine (IPV) use in 1999, out of concern that the risk of VAPP would outweigh the risk of local wild polio from imported cases. Many developed countries have followed suit. While this eliminates the risk for VAPP, concerns have been raised that herd immunity may be reduced owing to the shorter memory and lower intestinal immunity noted with IPV use.

Controversy as to the relative advantages of each vaccine continues. The OPV program of mass repeated vaccination in the control of polio in the Americas established the primacy of OPV in practical public health, and the momentum to eradicate polio is building. OPV requires multiple doses to achieve protective antibody levels. Where there are many enteroviruses in the environment, interference in the uptake of OPV may result in cases of paralytic polio among people who have received three or even four doses of adequate OPV. The use of IPV as initial protection eliminates this problem. During the 1970s and 1980s, a combined approach bolstering IPV immunity with OPV boosters showed promise in Gaza and Israel, where natural poliovirus was eradicated. Although the sequential use of IPV and OPV was adopted in the routine infant immunization program in the USA in 1997, since 2000 programs have used IPV alone. IPV has been adopted as the exclusive polio vaccine in most of the industrialized countries, while developing countries continue to rely on the less costly and easier to administer OPV. Mop-up campaigns using monovalent OPV (type 1) in still endemic areas.

There are concerns that exclusive use of either vaccine alone will not lead to the desired goal of eradication of polio. In 1988, the global polio eradication initiative was launched and progress since then has been impressive. Global coverage of infants with three doses of OPV reached 85 percent in 2005 compared with 83 percent in 1995 (UNICEF). During this period, OPV coverage increased in the African Region of the WHO from 51 percent in 2000 to 75–80 percent (2006), but since then has fallen slightly. NIDs are conducted in many countries throughout the world, achieving coverage of over 400 million children annually. Mop-up operations to reinforce coverage of children in still endemic areas are proceeding, along with increased emphasis on AFP monitoring. The number of global polio cases has been reduced by more than 99.8 percent from 1988 to 2009, including the prevention of five million cases of paralysis and more than 250,000 deaths.

With continued national and international emphasis, and the support of the WHO, Rotary International, UNICEF, GAVI, and donor countries, there is real prospect of a world without polio. India had no cases in 2011–2012, while Nigeria, Pakistan, and Afghanistan continue to have endemic wild poliovirus cases. In 2012, a total of 223 polio cases were reported from five countries: Afghanistan, Chad, Niger, Nigeria, and Pakistan. In 2013 (up to 5/11/2013) 328 polio cases were reported from: Afghanistan, Cameroon, Ethiopia,

Kenya, Nigeria, Pakistan, Somalia, and Syrian Arab Republic. The end of polio eradication is in sight, but great care must be taken in the end-stage strategy of continued vaccination in the decade ahead at least. In 2012, the murder of polio workers by radical Islamic Taliban fighters in Pakistan cast a pall over the program, but the vaccination program continues and will succeed (WHO Polio Eradication Initiative, 2013).

WHO has a policy to promote conversion from OPV to IPV-only policies. This needs to be reconsidered. In 2013 routine sewage monitoring in Israel revealed widely dispersed WPV was detected in 87 of 220 samples from 79 sites at 26 locations of southern and central parts of the country. There were no clinical cases in the population with 95 percent IPV immunization coverage but 42 healthy WPV1 carriers were found (4.4% of those sampled). In contrast, the West Bank and Gaza with a combined IPV- OPV immunization program since 1978 and over 95 percent coverage and no cases since 1992 had in 2013 sewage sampling identified only 4 positive samples of WPV1. This constitutes a "natural experiment" comparing two adjacent interacting jurisdictions: one with a combined OPV and IPV and one with IVP comparing the effectiveness of the two systems for preventing entry of WPV into an area. This experience should be considered of great importance in determining appropriate end-stage policy for the global polio eradication strategy. IVP alone is insufficient protection against spread of WPV from an unknown source even to well immunized and high level sanitation countries. A combined OPV/IPV policy is recommended (Tulchinsky et al., 2013).

Measles

Measles is an acute disease caused by a virus of the *Paramyxovirus* family. It is highly infectious with a very high ratio of clinical to subclinical cases (99 to 1). Measles has a characteristic clinical presentation with fever, rhinorrhea, white spots (Koplik spots) on the membranes of the mouth, and a red blotchy rash appearing on day 3–7, lasting for 4–7 days. Mortality rates are high in young children with compromised nutritional status, especially vitamin A deficiency.

The measles virus evolved from a virus disease of cattle (rinderpest) some 3000–5000 years ago, becoming an important disease of humans with high mortality rates in debilitated, poorly nourished children, and significant mortality and morbidity even in industrialized countries. In the prevaccine era, measles was endemic worldwide, and it remains a major childhood infectious disease today.

Single-dose immunization failed to meet control or eradication requirements even in the most developed parts of the world. A live vaccine, licensed in 1963, was later replaced by a more effective and heat-stable vaccine, but still with a primary vaccination failure rate (i.e., fails to produce protective antibodies) of 4–8 percent, and secondary failure rate (i.e., produces antibodies but protection is lost over time) of 4 percent. A two-dose policy incorporates a booster dose, usually at school age, in addition to maximum feasible infant coverage of children in the 9–15 month period (timing varies in different countries). Catch-up campaigns among school-age children should be carried out until the routine two-dose policy has time to take full effect. Nearly universal primary education in developing countries offers an opportunity for mass coverage of school-age children with a second dose of measles vaccine and a resulting increase in herd immunity to reduce the transmission of the virus. The two-dose policy adopted in many countries should be supplemented with catch-up campaigns in schools to provide the booster effect for those previously immunized and to cover those previously unimmunized, especially in developing countries. Figure 4.4 illustrates the annual number of reported measles cases in the USA and, most notably, the dramatic drop in cases following the introduction of the first measles vaccine in 1963.

The CDC declares that domestic transmission in the USA has been interrupted and that most localized outbreaks were traceable to imported cases. South America and the Caribbean countries are now considered free from indigenous measles, based on their successful use of NIDs, although a large epidemic occurred in 1999 in Brazil. Eradication of measles is feasible in the second decade of this century, if a two-dose policy is used and sustained with high priority globally, supplemented by catch-up campaigns for older children and young adults, and outbreak control.

Measles eradication is one of the central targets on the global public health community agenda, with emphasis placed on reducing mortality and then on gradual eradication of the disease (Box 4.11). The relationship between measles vaccination and cases is shown in Figure 4.4. In 2010, there were 139,300 measles deaths worldwide, which translates to almost 380 deaths daily, or 15 deaths per hour. Europe has experienced large waves of measles in the 1990s and especially between 2010 and 2012. Globally, measles vaccination led to a reduction of 74 percent in measles mortality between 2000 and 2010. From January to November 2012, France, Italy, Romania, Spain, and the UK accounted for 87 per cent of all reported cases, and there were no measles-related deaths, but seven cases were complicated by acute measles encephalitis. This was lower than the 2011 level, but measles transmission continues across Western and Central Europe. Much of the measles epidemic in Europe is school related, occurring among the underimmunized age group 10–16 years. England reports an estimated 330,000 underimmunized children in this age group (Ramsay, 2013). The trend of confirmed measles cases in England from 2009 to 2013 is shown in Figure 4.5.

In 2000, 72 percent of the world population received the first dose of measles-containing vaccine. Ten years later, total coverage increased to 85 percent. In 2000, measles incidence (cases per million population) amounted to 838 in the African Region of the WHO. In 2010, measles incidence

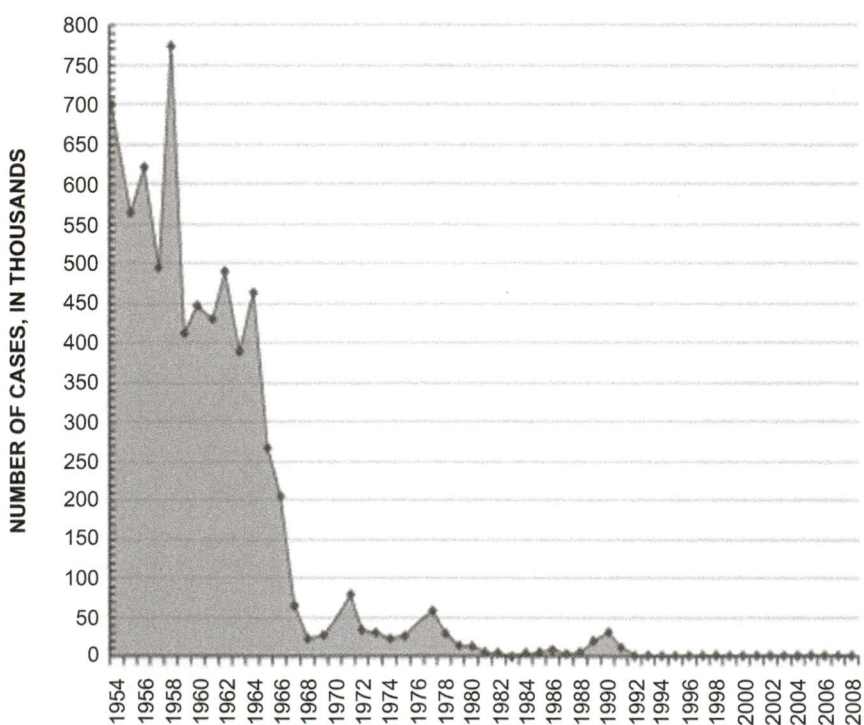

FIGURE 4.4 **Reported measles cases per year, USA, 1954–2008.** *Source: College of Physicians of Philadelphia. The history of vaccines. Available at: http://www.historyofvaccines.org/content/graph-us-measles-cases [Accessed 18 January 2013]. Hall-Baker PA, Nieves E Jr, Jajosky RA, Adams DA, Sharp P, Anderson WJ, et al. Summary of notifiable diseases, United States, 2008. MMWR Morb Mortal Wkly Rep 2010;57:1–94. Available at: http://www. cdc.gov/mmwr/preview/mmwrhtml/mm5754a1.htm [Accessed 18 January 2013].*

dropped to 238 (per million population), thus representing a 72 percent decline in this period. Furthermore, in the African Region, 90 percent of children under the age of five mostly die from complications such as severe diarrhea, pneumonia, and encephalitis. Over 95 percent of deaths occur in low-income countries with weak health systems. In 2011, large outbreaks were reported by the Democratic Republic of the Congo (DRC) with over 103,000 cases, Nigeria with 17,428 cases, Zambia with 5397 cases, and Ethiopia with 2902 cases. The WHO regional office in DRC reported 1100 measles-associated deaths in 2011 in that country.

International transmission of the virus in carriers has led to importation and subsequent epidemics even in countries thought to have achieved local eradication. This is exemplified by outbreaks in 2006–2008 in the UK, Switzerland (2250 cases), Austria, France, Italy, and other countries. Israel had an epidemic of over 1200 cases in 2007–2008 following an imported case. In July 2008, the Health Protection Agency in the UK declared measles to be endemic for the first time in 14 years, owing to a decade of poor coverage with the measles vaccine. The USA had an annual average of 64 cases during 2000–2007, but an increase in 2008. Spikes in measles cases

BOX 4.11 Measles Control and Its Return (2010–2013)

Measles is a highly infectious viral disease that was responsible for the death of nearly 2 million children globally per year in the 1960s, before the advent of measles vaccine. Gradual adoption of the measles vaccine and growing coverage worldwide reduced global mortality to fewer than 200,000 deaths by 2010. Measles vaccination has been one of the success stories of public health; however, the picture is still mixed, and hopes for measles eradication have been put off until 2015.

Measles vaccine is one of the most immunogenic and safest of all the available vaccines. Two doses, with over 95 percent coverage, and catch-up of older teens and young adults is required in order to ensure full coverage, long-lasting individual protection, and herd immunity. The two-dose policy was adopted late in many countries; therefore, it took time for the circulation of the virus to appear to be under control. While measles cases and deaths were drastically reduced, large pockets of susceptible populations remained vulnerable.

Between 1989 and 1991, measles in the USA resulted in more than 100 deaths, with over 55,000 cases reported. This was followed by a return of subacute sclerosing panencephalitis (SSPE), a rare but fatal neurological complication of measles in children – a condition which had largely disappeared after the vaccine became widely used (CDC, MMWR, 2011).

In the 1990s in the USA and in 2000 in the Americas as a whole, measles was declared to have been eliminated. However, importation of the virus continued at low levels.

Late in the first decade of the twenty-first century, a number of large-scale outbreaks of measles occurred in countries that had not experienced the disease for many years, as a result of laxity and resistance to immunization. Complacency among parents, and medical and public health practitioners allowed outbreaks to continue due to a weak response to targeted and large-scale strategic immunization efforts.

A decline in coverage with the MMR vaccine in the UK in the late 1990s resulted from widespread publicity and concern over a fraudulent and disproven allegation of an association with autism (the "Wakefield effect"). This led to a reduction in MMR coverage in the UK and large-scale outbreaks, with a return to endemicity of the disease as the virus circulated, finding susceptible children who had not been not immunized and causing disease even among adequately immunized children.

Many children were left unimmunized and, consequently, unprotected, as a result of parents' fears of complications of

the MMR vaccine. Other children who were not immunized included those with chronic diseases such as HIV and leukemia, many of whom are unable to receive the vaccine owing to an increased risk of potential complications. Thus, a lack of appropriate levels of herd immunity left these people vulnerable. Unimmunized infants, older children, and young adults are also susceptible to contracting this highly contagious disease.

In sub-Saharan Africa, vaccination with one dose of a measles-containing vaccine reached 83 percent in 2009. In 2011, Nigeria reported 30,000 cases with 122 deaths; the Democratic Republic of the Congo reported 16,000 cases and 107 deaths between January and February 2011 alone. In 2012 and 2013, measles epidemics have been occurring throughout the UK including England, Wales, and Scotland. In the first quarter of 2013, 587 cases of measles were reported in England, mostly in the north-west and north-east regions, among infants before immunization age and among 7–16-year-olds.

Control, elimination, and potential eradication of the disease will require more years of intense effort to raise basic coverage with two doses of the vaccine as well as catch-up campaigns. New strategies are needed to influence global public perception of the vital importance of immunization and its safety. The current WHO target is to achieve measles eradication by 2015. In 2012, the WHO launched a new initiative combining measles and rubella vaccination with three doses of measles-containing vaccine, and promoted MMR so that rubella and mumps and their complications can also be eliminated.

Sources: Centers for Disease Control and Prevention. Measles – United States, January–May 20, 2011. MMWR Morb Mortal Wkly Rep 2011;60:666–8. Available at: http://www.cdc.gov/mmwr/preview/mmwrhtml/mm6020a7.htm
World Health Organization. Measles outbreaks and progress towards meeting measles pre-elimination goals. Wkly Epidemiol Rec WER 2011;86:129–40. Available at: http://www.who.int/wer/2011/wer8614.pdf
Johnson TD. Measles cases abroad linked to increase in cases in the US. The Nation's Health. September 2011. Available at: http://thenationshealth.aphapublications.org/content/41/7/1.2.full [Accessed 20 October 2011].
Public Health England. Measles cases in England: January to March 2013. Available at: http://www.hpa.org.uk/webc/HPAwebFile/HPAweb_C/1317138802384 [Accessed 8 August 2013].
World Health Organization. Increased transmission and outbreaks of measles, European Region, 2011. Wkly Epidemiol Rec WER 2011;86:559–64.
World Health Organization. Global measles and rubella strategic plan: 2012–2020. Geneva: WHO; 2012. Available at: http://www.who.int/immunization/newsroom/Measles_Rubella_StrategicPlan_2012_2020.pdf [Accessed 18 January 2013].

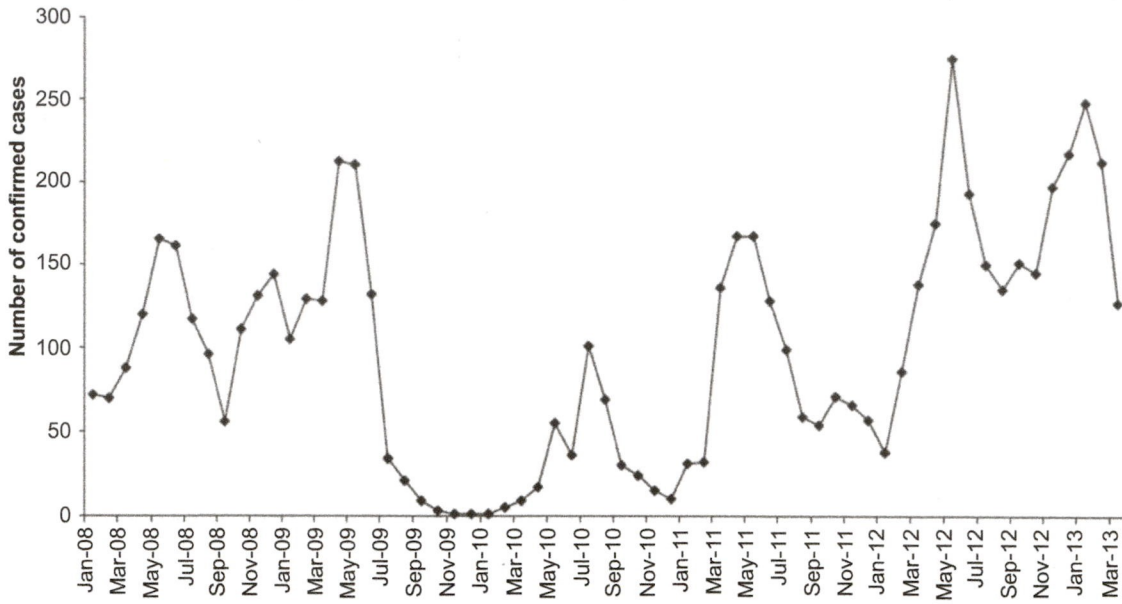

FIGURE 4.5 Measles cases, England, 2008–2013. *Source: Ramsay M. Measles in England 2012 and 2013. Public Health England; 2013. Available at: https://www.gov.uk/government/uploads/system/uploads/attachment_data/file/192611/Presentation_by_Mary_Ramsay_-_Measles_in_ England_2012___2013.pdf or http://www.hpa.org.uk/HPAwebHome/*

also occurred in the USA as a result of importations, in 2006, and again in 2008 and in 2011–2012.

In 2010–2012, there was a widespread epidemic of over 50,000 reported cases. The largest burden of disease fell on Western Europe, with the WHO European Region reporting 26,025 cases up to October 2011. In 2012, 115 measles outbreaks were reported from 36 countries with many thousands of cases in France, the UK, Germany, Bulgaria, Italy, and Switzerland, as well as Africa and Asia. Following this string of measles epidemics, the virus has spread outside the region, leading to hundreds of imported and secondary spread cases in the USA and Canada.

A dramatic drop occurred in US measles cases following the introduction of the first measles vaccine in 1963. The WHO has promoted measles vaccination in campaigns combined with other life-saving interventions such as bed nets to protect against malaria, deworming medicine, and vitamin A supplements to make use of the contact occasion to reduce child death rates, in keeping with the MDGs between 1990 and 2015. Elimination of measles as a public health problem, and even eradication, are feasible goals in the second decade of this century. This combination of interventions is critical in attaining the MDG target of reducing child mortality by 2015. The MDGs are a crucial and sometimes underrecognized issue, but this topic deserves to be one of the highest professional and political priorities of international and national donor and public health agencies, as well as national governments.

The WHO strategy of partnership with national governments and non-governmental organizations (NGOs) such as the Measles Initiative, GAVI, and others, includes:

- provision of one dose of measles vaccine for all infants via routine health services

- a second dose for children through mass vaccination campaigns
- effective surveillance for measles
- enhanced care, including the provision of supplemental vitamin A.

The number of countries implementing a two-dose policy has increased sharply from 97 (50 percent) to 141 (73 percent) in 2011, while coverage with MCV1 increased from 72 to 84 percent. The reported incidence rates declined by 65 percent from 146 to 52 per million population; deaths declined 71 percent from 548,000 to 158,000.

With a goal of eradicating measles by 2015, the European Region of the WHO undertook a study to review progress towards the goal. Low vaccination rates occur among hard-to-reach population groups, including migrants (employed but not citizens) and members of particular ethnic groups. Across Europe there is inadequate information to monitor vaccination coverage in these groups. The results of the study show that achieving measles eradication requires the collective efforts of policy makers and health providers.

Mumps

Mumps is an acute viral disease characterized by fever, swelling, and tenderness usually of the parotid glands, but also other glands. The incubation period ranges between 12 and 25 days. Orchitis, or inflammation of the testicles, occurs in 20–30 percent of postpubertal males and oophoritis, or inflammation of the ovaries, in 5 percent of postpubertal females. Sterility is an extremely rare result of mumps. CNS involvement can occur in the form of aseptic meningitis, almost always without sequelae. Encephalitis is reported in 1–2 per

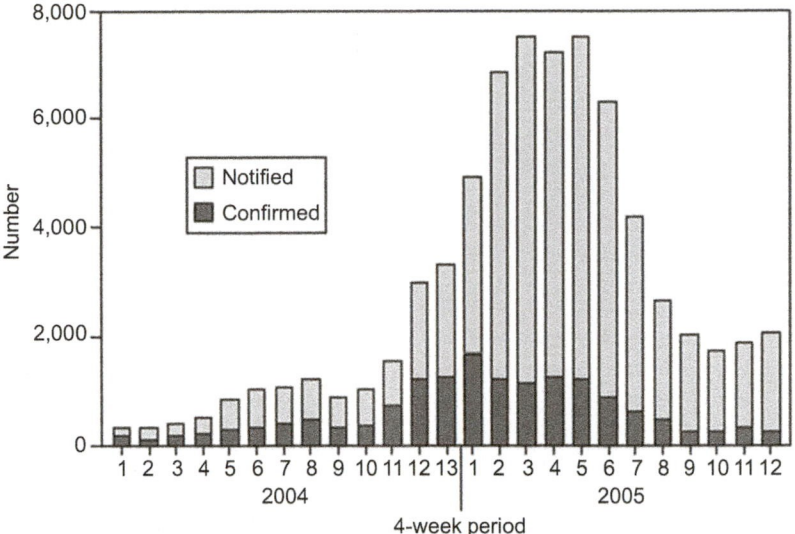

FIGURE 4.6   Epidemic curve of mumps epidemic, UK, 2004–2005. *Source: Centers for Disease Control and Prevention. Mumps epidemic – United Kingdom, 2004–2005. MMWR Morb Mortal Wkly Rep 2006;55:173–5.*

10,000 cases with an overall case fatality rate of 0.01 percent. Pancreatitis, neuritis, nerve deafness, mastitis, nephritis, thyroiditis, and pericarditis, although rare, may occur. Most people born before 1957 are immune to the disease, because of the nearly universal exposure to the disease prior to that year.

The live attenuated vaccine introduced in the USA in 1967 is available as a single vaccine or in combination with measles and rubella as the MMR vaccine. It provides long-lasting immunity in 95 percent of cases. Mumps vaccine is now recommended in a two-dose policy with the first dose of MMR given between 12 and 15 months of age and a second dose given either at school entry or in early adolescence. MMR in two doses is now standard policy in the USA, Sweden, the Netherlands, Canada, Israel, and other countries. The incidence of mumps overall has declined rapidly over the years, with routine MMR vaccine in two doses as the main recommended routine; however, it still remains a threat.

During 2004–2005, the UK experienced a nationwide epidemic of mumps, which peaked during 2005 when over 56,000 cases were reported in England and Wales, mostly in people aged 15–24 years, and most of whom had not been eligible for routine mumps vaccination. Figure 4.6 shows the epidemic curve during the period 2004–2005, as published in *MMWR*. The episode can be traced back to the period of controversy surrounding the use of MMR vaccine and increased susceptibility in a partially immunized population in the age group, a group which probably received only one dose of the vaccine, if at all.

Poland also experienced a major outbreak of mumps in 2005–2006, mostly among children aged 5–9 years. The UK had more than 100,000 mumps cases in 2004–2005; the USA had 4000 cases in a Midwest outbreak in 2006. Canada reported more than 450 cases of mumps among university students in the spring of 2007 (WHO, US CDC, Health Canada). In the summers of 2004 and 2005, 39 patients with mumps were hospitalized in Crete and Greece, and almost all were young tourists from Britain. The disease spread among the Greek population as well, in which six cases were reported. Outbreaks in the Netherlands and Canada in the 2006–2008 period were linked to fundamentalist Christian religious groups that refused immunization on religious grounds.

Many countries in Europe still do not use MMR or a two-dose policy; therefore, they are vulnerable to mumps along with rubella outbreaks. MMR vaccination should be adopted as an international standard, with two doses for all children and catch-up for school-age children. Local eradication of this disease is important and should be part of a basic international immunization program. This is an ongoing challenge in the European Region, which lacks a harmonized immunization program.

Rubella

Rubella (German measles) is generally a mild viral disease with lymphadenopathy and a diffuse, raised red rash. Low-grade fever, malaise, coryza, and lymphadenopathy characterize the prodromal period. The incubation period typically lasts for 16–18 days. Differentiation from scarlet fever, measles, or other febrile diseases with rash may require laboratory testing and recovery of the virus from nasopharyngeal, blood, stool, and urine specimens (Box 4.12).

Congenital rubella syndrome (CRS) occurs with single or multiple congenital anomalies including deafness, cataracts, microophthalmia, congenital glaucoma, microcephaly, meningoencephalitis, congenital heart defects, and others. Insulin-dependent diabetes is suspected as a late sequela of congenital rubella. Each case of CRS is estimated to cost some US$250,000 in health care during the patient's lifetime.

Moderate and severe cases are recognizable at birth, but mild cases may not be detected for months or years after

BOX 4.12 Discovery of Rubella Syndrome

In 1942, Norman Gregg, an Australian ophthalmologist, observed an epidemic of cases of congenital cataract in newborns and other birth defects associated with a history of rubella in the mother, during the first trimester of pregnancy. Subsequent investigation demonstrated that intrauterine death, spontaneous abortion, and birth defects including congenital heart disease and deafness occur commonly when rubella occurs early in pregnancy.

For this discovery, Gregg was knighted in 1953 and received many other honors. With the later development of rubella vaccine, its inclusion in MMR vaccine with measles and mumps, and its widespread use around the world, this cause of birth defects has been gradually reduced. The WHO in 2012 made rubella part of its disease eradication program, together with measles, to prevent rubella syndrome globally.

Sources: Dunn PM. Perinatal lessons from the past: Sir Norman Gregg, ChM, MC, of Sydney (1892–1966) and rubella embryopathy. Arch Dis Child Fetal Neonatal Ed 2007;92(6):F513–4. Available at: http://www.ncbi.nlm.nih.gov/pmc/articles/PMC2675410/ [Accessed 18 January 2013].
World Health Organization. Rubella vaccines: WHO position paper. Wkly Epidemiol Rec 2011;86:301–316. Available at: http://www.who.int/wer/2011/wer8629.pdf [Accessed 18 January 2013].
Centers for Disease Control and Prevention. Vaccines: rubella. Atlanta, GA: CDC; 2011. Available at: http://www.cdc.gov/vaccines/vpd-vac/rubella/in-short-adult.htm [Accessed 3 February 2013].
Plotkin SA, Orenstein WA, Offit PA, editors. Vaccines. 5th ed. Philadelphia, PA: Saunders; 2008.
World Health Organization. Global measles and rubella strategic plan: 2012–2020. Geneva: WHO; 2012. Available at: http://www.who.int/immunization/newsroom/Measles_Rubella_StrategicPlan_2012_2020.pdf [Accessed 18 January 2013].

birth. The WHO estimates that in 1996, CRS occurred in 22,000 babies in Africa, 46,000 in South-East Asia, and 13,000 in the Western Pacific Region, and that similar rates were occurring in 2008, since few of the countries in these regions have introduced rubella vaccine.

Prior to the availability of the attenuated live rubella vaccine in 1969, the disease was universally endemic, with epidemics or peak incidence every 6–9 years. In unvaccinated populations, rubella is primarily a disease of childhood. In areas where children are well vaccinated, adolescent and young adult infection is more apparent, with epidemics in institutions and colleges, and among military personnel.

A sharp reduction in rubella cases was seen in the USA following introduction of the vaccine in 1970; however, rates increased in 1978, following rubella epidemics in 1976–1978. A further reduction in cases was followed by a sharp upswing of rubella and CRS in 1988–1990. An outbreak of rubella among the Amish in the USA, who refuse immunization on religious grounds, resulted in seven cases of CRS in 1991. It is now thought that vaccination of sufficient numbers in the USA reduced circulation of the virus and protected the most vulnerable groups in the population. Most industrialized countries adopted MMR in the 1990s and, subsequently, a two-dose policy. Rubella and CRS incidence dropped

dramatically. Controversy in the UK in the early 2000s led to reduced MMR usage and an increase in cases of measles and rubella (see Box 4.11). This was subsequently improved by providing incentive payments for general practitioners, with 100 percent age-specific immunization coverage.

Some parts of Europe failed to adopt MMR vaccine use and have suffered recurrent outbreaks of these diseases. A number of outbreaks were reported in 2005–2007. Poland reported 7946 cases of rubella in 2005 (20.8 per 100,000 population), an increase of 64 percent compared to 2004. MMR was added to the routine immunization schedule at the end of 2003. In 2003, Italy approved a national plan for the elimination of measles and congenital rubella, with the aim of reducing and maintaining the incidence of CRS to less than one case per 100,000 live births by 2007.

Despite the common program recommended by the European Region of the WHO, childhood immunization programs vary widely across the EU and within some countries where regional autonomy for immunization is permitted. Coverage with a first dose of a measles-containing vaccine was 82 percent in the UK, 84 percent in Ireland, 87 percent in France, and 88 percent in Belgium in 2005, all well below herd immunity levels required for measles control (Venice Project 2007). This leaves each country to develop its own program, without guidelines for countries in transition from the socialist period, operating with high coverage rates. Thus, it allows for obsolescent immunization practices which only very slowly adopt current best practices from western standards. Many countries have not yet adopted MMR. The WHO declares the eradication of measles and rubella to be of higher priority than mumps and, furthermore, suggests that the combination MMR vaccine be used.

In the past, the immunization policy for rubella in some countries was to vaccinate schoolgirls at the age of 12 and women after pregnancy to provide protection throughout the period of fertility. The current approach is to give a routine dose of MMR in early childhood, followed by a second dose in early school age to reduce the pool of susceptible people. Women of reproductive age should be tested to confirm immunity prior to pregnancy and receive the immunization if not already immune. Should a woman become infected during pregnancy, termination of the pregnancy is no longer recommended; rather, the infection is now managed with hyperimmune globulin.

The infection of pregnant women during their first trimester of pregnancy is the primary public health implication of rubella. The emotional and financial burden of CRS, including the cost of treatment of its congenital defects, makes this vaccination program cost-effective and critical. Its inclusion in a modern immunization program is fully justified. Elimination of CRS should be one of the primary goals of a program for prevention of VPDs in developed and developing countries. Adoption of MMR and the two-dose policy will gradually lead to the eradication of rubella and rubella syndrome.

Since the WHO recommended the universal use of rubella vaccine in childhood, immunization coverage has become more widespread: it increased from 83 countries in 1996 to 130 of the 193 WHO member states in 2009. Rubella-containing vaccines (RCVs) are used in national immunization schedules in two of 46 countries in the African Region, all 35 in the Region of the Americas, four of 11 countries in South-East Asia, all 53 in the European Region, 15 of 21 in the Eastern Mediterranean Region, and 21 of 27 countries in the Western Pacific Region (WHO WER, 2010).

Viral Hepatitis

Viral hepatitis is a group of diseases of increasing public health importance owing to its large-scale worldwide prevalence, its serious consequences, and our increasing ability to take preventive action. Viral hepatic infectious diseases each have specific etiological, clinical, epidemiological, serological, and pathological characteristics. They have important short- and long-term sequelae. Vaccine development is of high priority for their control and ultimate eradication. Beginning in 2009, CDC requires US-reported cases of acute and chronic viral hepatitis to meet Council of State and Territorial Epidemiologists (CSTE)-defined clinical and laboratory criteria. The first official World Hepatitis Day was held on 28 July 2011.

Hepatitis A

Hepatitis A virus (HAV) is mainly transmitted by the fecal–oral route. Clinical severity varies from a mild illness of 1–2 weeks to a debilitating illness persisting for several months. The norm is complete recovery within 9 weeks, but a fulminating or even fatal hepatitis can occur. Severity of the disease worsens with increasing age. HAV is endemic worldwide.

Improving sanitation lowers childhood exposure to HAV, but infection among adults typically leads to more severe clinical symptoms. It is now prevalent particularly in people from industrialized countries who are exposed to environments characterized by poor hygiene or contaminated food products. HAV also occurs among young adults when traveling to areas where the disease is endemic. Common source outbreaks occur in school-age children and young adults from case contact or from food contaminated by infected handlers. Hepatitis A may be a serious public health problem in a disaster situation.

Prevention involves improving personal and community hygiene, with safe chlorinated water and proper food handling. Short-term risk of infection for people exposed to HAV may be reduced with prompt administration of HAV immune globulin. Hepatitis A vaccine is now recommended for all children over 12 months of age, as well as for people traveling to endemic areas or at increased risk of exposure or morbidity. CDC reports that 33 percent of the US population were ever infected with HAV, but there is no chronic carrier state. HAV immunization is being adopted for routine prevention programs in some countries, including the USA; it is used for pre-exposure prevention; however, immune globulin is still used for postexposure protection. As the costs of the vaccine begin to fall, its widespread routine use may be recommended.

Hepatitis B

Hepatitis B virus (HBV) was once thought to be transmitted only by injections of blood or blood products. It is now known to be present in all body fluids and easily transmissible by household and sexual contact, perinatal spread from mother to newborn, and between toddlers. In contrast to HAV, it is not typically spread by the fecal–oral route.

HBV is endemic worldwide and is especially prevalent in developing countries. Carrier status with persistent viremia is estimated by CDC to be 1.25 million in the USA, with 4.9 percent of the population ever infected. Carrier rates are 5–8 percent in sub-Saharan Africa but between 8 and 15 percent of babies become infected in some parts of the world, thus routine immunization is recommended. Carriers have detectable levels of HBsAg, the surface antigen (i.e., Australian antigen), in their blood. In 2006, an estimated 27 percent of newborns worldwide received a dose of hepatitis B vaccine and 69 percent of the 2008 birth cohort had three doses of hepatitis B vaccine. By 2011, the hepatitis B vaccine was an integral part of national immunization programs in 179 countries (WHO, 2013).

Transmission from mother to child and between children by unsafe injections is common, as is infection via sexual contact. It is crucial for the following high-risk groups in developed countries to be immunized: health care workers, intravenous drug users, men who have sex with men, people with high numbers of sexual partners, those receiving tattoos, body piercing, or acupuncture treatments, and residents or staff of institutions such as group homes and prisons. Immunocompromised and hemodialysis patients are commonly carriers of HBV. HBV may also be spread in a health system by use of inadequately sterilized reusable syringes, as has occurred in China and the former Soviet Union. Transmission is reduced by screening blood and blood products for HBsAg. Strict technique for handling blood and body fluids in health care settings is absolutely essential and greatly reduces the risk of transmission.

HBV is clinically recognizable in less than 10 percent of infected children but is apparent in 30–50 percent of infected adults. Clinically, HBV has an insidious onset with anorexia, abdominal discomfort, nausea, vomiting, and jaundice. The disease can vary in severity from subclinical, through very mild, to fulminating liver necrosis and death. It is a major cause of primary liver cancer, chronic liver disease, and liver failure, all devastating to health and expensive to treat.

HBV is considered to be the cause of 60 percent of primary cancer of the liver in the world and the most common carcinogen, second to cigarette smoking. The WHO estimates that more than 2 billion people alive today have been infected with HBV. It is also estimated that 350 million people are chronic carriers of HBV, with an estimated 1–1.5 million deaths per year from cirrhosis or primary liver cancer. This makes control of hepatitis B a vital issue in the revision of health priorities in many countries.

Strict discipline in blood banks and testing of all blood donations for HBV, as well as HIV and hepatitis C, is mandatory, with destruction of those donations that are detected as positive. Contacts should be immunized following exposure with HBV immunoglobulin and HBV vaccine. The inexpensive recombinant HBV vaccine should be adopted by all countries and included in routine vaccination of infants. Catch-up immunization for older children is also desirable. Immunization programs should include those exposed in prison or at work, such as in health care facilities, as well as sex workers and adults in group settings. HBV immunization has been included in the WHO's EPI-plus expanded program of immunization. In 1992, the World Health Assembly (WHA) passed a resolution to recommend global vaccination against hepatitis B, then used in 31 countries. By 2011, 179 countries had adopted this vaccine in their national infant immunization programs, and 93 member states have introduced the hepatitis B birth dose, protecting millions of children from chronic infection with HBV (WHO, 2013).

Hepatitis C

First identified in 1989, and previously known as non-A, non-B hepatitis, hepatitis C has an insidious onset characterized by jaundice, fatigue, abdominal pain, nausea, and vomiting. Hepatitis C virus (HCV) may cause mild to moderate illness; however, chronicity is common, progressing to cirrhosis and liver failure. The WHO estimates that 150 million people are chronically infected with HCV and 3–4 million are newly infected globally each year. More than 350,000 people die from hepatitis C-related liver diseases every year.

High rates of HCV are reported in Egypt (22 percent of the population), and more than 3 percent of the population in China and Pakistan are infected. The WHO reports identify contaminated equipment in the administration of injections as a primary factor in transmission (WHO, 2010). The CDC estimates that 3.2 million Americans are chronically infected with HCV, with an estimated 12,000–15,000 resulting deaths per annum (CDC, 2012). Furthermore, HCV is the main cause of illness necessitating liver transplants.

The virus is transmitted most commonly in blood products, but also among injecting drug users; 90 percent of intravenous drug users (IVDUs) were HCV positive in a Vancouver study in 1998. HCV also poses risk for health workers. The disease may also occur in dialysis centers and other medical situations. Person-to-person spread is unclear. Prevention of transmission includes routine testing of blood donations and antiviral treatment of blood products. For IVDUs, needle-exchange programs and hygiene education are of value. In 1998, the WHO declared hepatitis C a major public health crisis, stressing that this "silent epidemic" is being neglected and that screening of blood products is vital to reduce transmission of this disease, as for HIV.

HCV is a major cause of liver cirrhosis, end-stage liver disease, and hepatocellular carcinoma. The virus is primarily transmitted parenterally. No vaccine is available at present, but research is currently under way to find this "holy grail". The genetic diversity of the virus makes it difficult to find the correct antigen for effective antibody production, so the virus evades the host immune response and, in turn, poses a serious challenge for developing an effective vaccine. The incidence reported in the CDC is estimated to be about half of number of true new infections with hepatitis C, which often has mild symptoms (Figure 4.7).

Significant advances have been made in the treatment of chronic HCV infection. The combination of pegylated interferon and ribavirin, used since 2001, has been enhanced by the addition of one of two new drugs – boceprevir or telaprevir – with improved results over the previous two-drug therapy, including for those who had been previously treated and failed to clear the virus. Treatment is expensive and carries significant adverse effects, with symptoms of weakness and depression, but the new triple combination reduces the duration of treatment and improves success rates.

Treatment is needed in most cases of hepatitis C carrier status and is needed if there are signs of liver damage to prevent HCV-associated cirrhosis and liver cancer. Prevention of transmission is primarily addressed towards intravenous drug users but screening is recommended for adults including those in the 45-plus age groups. Current research is showing improved methods of multi drug treatment with improved success rates and possibility of eliminating interferon with a reduction in costs and far fewer side-effects as early as during 2013. Developing countries have high levels of this infection but limited resources to control it until a vaccine is developed.

Hepatitis D

Hepatitis D virus (HDV), also known as delta hepatitis, may be self-limiting or progress to chronic hepatitis. It is caused by a virus-like particle which requires HBV to reproduce. HDV infects cells along with HBV as a coinfection or in chronic carriers of HBV. HDV occurs worldwide in the same groups at risk for HBV. This virus has been the source of epidemics and is endemic in South America and Africa, as well as among drug users. Prevention is achieved by measures similar to those for HBV. Management of HDV is by passive immunity with immunoglobulin for contacts and high-risk groups, and should include HBV vaccination as the diseases often coincide. There is currently no vaccine for HDV.

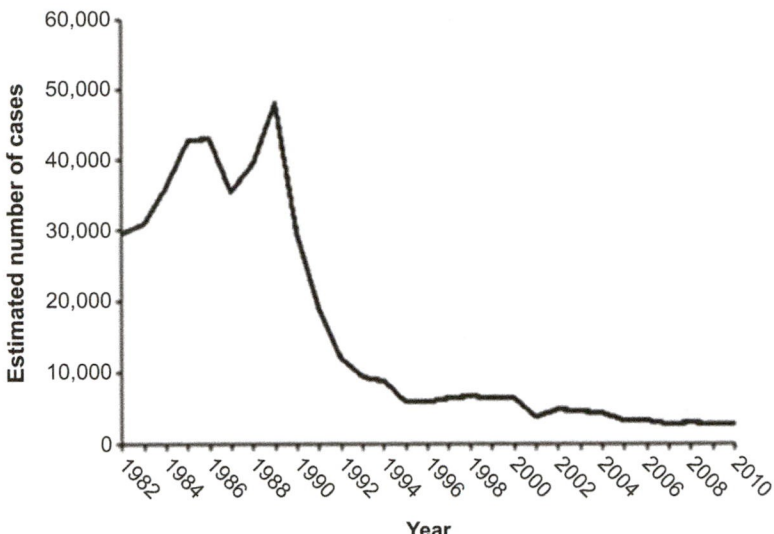

FIGURE 4.7 Number of new reported cases of acute hepatitis C, USA, 1982–2010. *Source: Centers for Disease Control and Prevention. Viral hepatitis statistics and surveillance: disease burden from viral hepatitis A, B, and C in the United States [updated 22 June 2012]. Available at: http://www.cdc. gov/hepatitis/Statistics/index.htm [Accessed 21 January 2013].*

Hepatitis E

Hepatitis E virus (HEV) has an epidemiological and clinical course similar to that of HAV, with an incubation period of 15–64 days. There is no evidence of a chronic form of HEV. One striking characteristic of HEV is the high mortality rate among pregnant women. Infections typically result via waterborne epidemics, or as sporadic cases in areas with poor hygiene, spread via the fecal–oral route. In disaster situations with crowding and poor sanitary conditions, HEV presents a major hazard. Prevention is by safe management of water supplies and sanitation. Treatment is supportive and aimed at the symptoms; passive immunization is not helpful and no vaccine is currently available.

Haemophilus influenzae *Type b*

The Hib bacterium causes meningitis, pneumonia, and other serious infections in children. Before the introduction of effective vaccines, as many as one in 200 children developed invasive Hib infection. Two-thirds of these developed Hib meningitis, with a case fatality rate of 2–5 percent. Long-term sequelae such as hearing impairment and neurological deficits occurred in 15–30 percent of survivors.

The first licensed Hib vaccine (1985) was based on capsular material from the bacterium. Extensive clinical trials conducted in Finland demonstrated a high degree of efficacy, but less impressive results were in seen in postmarketing efficacy studies. By 1989, a conjugate vaccine based on an additional protein cell capsular factor capable of enhancing the immunological response was introduced. Several conjugate vaccines are now available.

Since the introduction of Hib vaccine into the US recommended immunization schedule, bacterial meningitis cases were dramatically reduced by 55 percent; however, this may have also been an effect of the widening use of pneumococcal pneumonia vaccine. In recent years, the number of cases of Hib infection has increased in adults, especially in elderly or immunocompromised people.

The conjugate vaccines are now combined with DTaP as their schedule is simultaneous with that of DTaP. Hib vaccine has been found to be cost-effective, despite its costing as much as all the basic vaccines combined (i.e., DTP, OPV, MMR, and HBV vaccine). For this reason, its use thus far has been mainly limited to industrialized countries, although it is spreading to many developing countries with support from GAVI. As of 2009, Hib vaccine had been introduced into routine vaccination in 169 countries (WHO Global Advisory Committee on Vaccine Safety, 2013).

The vaccine is a valuable addition to the immunological armamentarium. It has shown dramatic results in local eradication of this serious early childhood infection in a number of European countries and a sharp reduction in the USA. The price of the vaccine has also fallen dramatically since the mid-1990s.

Influenza

Influenza is an acute viral respiratory illness characterized by fever, headache, myalgia, prostration, and cough. Transmission is rapid by close contact with infected individuals and by airborne particles with an incubation period of 1–5 days. It is generally mild and self-limiting, with recovery in 2–7 days. However, in certain population groups, such as young children, and elderly and chronically ill people, infection can lead to severe sequelae. Gastrointestinal symptoms commonly occur in children. During epidemics, mortality rates from respiratory diseases increase because of the large numbers of people affected, although the case fatality rates are generally low.

Over the past century or so, influenza pandemics have occurred in 1889, 1918, 1957, 1968, and 2009, while epidemics have presented themselves as annual events. The influenza pandemic of 1918 caused millions of deaths among young adults, by some estimates killing more than had died in World War I: the pandemic killed nearly 50 million people worldwide and was characterized by an atypical mortality curve. Influenza typically affects the very old and the very young. The principal group suffering from the 1918 pandemic was young men between the ages of 30 and 60 years, many in army training camps, as well as in the general population. Fear of recurrence of this pandemic led the CDC to launch a massive immunization program in the USA in 1976 to prevent swine flu (the virus was a strain antigenically similar to that of the 1918 pandemic influenza) from spreading from an isolated outbreak in an army camp. The effort was stopped after millions of people were immunized with an urgently produced vaccine when serious reactions occurred (Guillain–Barré syndrome, a type of paralysis) and when no further cases of swine flu were seen. This example demonstrates the difficulty in extrapolating scenarios from a historical experience.

In the 2009 pandemic H1N1 outbreak, the per capita risk factor for hospitalization was highest among children aged under 5 years. The highest risk of death was in the age group 50 and over. Following the pandemic, the WHO defined standards for reporting fatal influenza-related cases. A similar approach has been developed for severe respiratory infections. Assessing the problems with this pandemic, it became evident that many countries lack surveillance systems to monitor outbreaks over an extended period. Respiratory illness is the second leading cause of death in many low- and middle-income countries, and thus justifies directing sufficient resources towards management and prevention. Current best practices for universal immunization annually for influenza for all age groups have been shown to reduce serious complications, death, and the high costs of medical care. It is especially important for health workers and other caregivers (teachers, kindergarten workers, and others whose work entails meeting the public) to have up-to-date influenza vaccination. If there are shortages of vaccine, efforts should focus on those listed in Box 4.13. As shown in Box 4.14, vaccination is especially important to prevent influenza during pregnancy.

A CDC report in 2012 indicated that nearly one-third of surveyed health workers were not immunized for influenza because of concern over the vaccine's efficacy and side-effects and doubts as to whether it was needed (MMWR, 2012). Unvaccinated health workers put their patients and their families at risk. Current recommendations are for annual influenza vaccination for all people aged 6 months

> **BOX 4.13 High-Risk Groups Recommended for Annual Influenza Vaccination**
>
> - Children between 6 months and 59 months of age
> - Pregnant women
> - All adults aged 50 years and older
> - Pregnant women or women planning pregnancy
> - Adults and children with chronic medical conditions
> - People with chronic health conditions (e.g., lung, cardiovascular, renal, hepatic, or neurological disorders; immunosuppression, e.g., medical or HIV related)
> - Residents of long-term care facilities and nursing homes
> - People in contact with high-risk individuals or populations
> - Household contacts of children aged <5 years and adults aged >50 years and with people with chronic health conditions
> - Caregivers and contacts of infants and at-risk children
>
> *Source: Centers for Disease Control and Prevention. Seasonal influenza (flu). Who should get vaccinated against influenza? Available at: http://www.cdc.gov/flu/protect/whoshouldvax.htm [Accessed 21 January 2013].*

or older and all health care providers. Other groups, such as immunosuppressed patients and those receiving chronic aspirin therapy, should obtain medical advice regarding influenza risk and vaccination. People with allergy to previous flu shot, eggs, or other vaccine components, or with a history or risk for Guillain–Barré syndrome, may not be candidates for vaccination and should obtain medical advice.

In recent years, concern has again risen surrounding the likelihood of virulent influenza pandemics. Particularly noteworthy is the influenza A H5N1 strain, known as avian influenza. The WHO reports that from 2003 to July 2013, the number of confirmed human cases was 633 with 377 deaths. Although relatively few human-to-human transmissions have been documented, the first such transmission was reported in July/August 2013, and this virus has rapidly spread among wild and domestic bird populations throughout Asia and much of the world. People exposed to and in contact with infected birds or poultry are at risk for severe disease, with over 60 percent case mortality. A minor mutation or genetic conjugation with a known human strain could result in a virus as deadly and contagious as the swine flu of 1918. It is estimated that up to 1.9 million people in the USA could die if such an outbreak occurs. Extensive international plans have been developed for intervention should a virulent influenza pandemic occur. These include several vaccines with specificity to known virus strains. As many of the most devastating global communicable disease emergencies of recent centuries have been associated with highly pathogenic respiratory viruses, health systems and emergency plans must be prepared in case of a pandemic. Active surveillance using sentinel chicken flocks now under

BOX 4.14 Influenza Vaccine for Pregnant Women

Preventing influenza during pregnancy is an important part of prenatal care. Influenza can cause severe illness among pregnant women, placing them at high risk owing to changes to the immune system, lungs, and heart during pregnancy. Pregnant women suffered excess mortality in previous influenza pandemics, such as in 1918, 1957, and 2009.

To protect her health, as well as that of her unborn baby, a pregnant woman is strongly advised to receive an influenza vaccine, regardless of the trimester. The CDC Advisory Committee on Immunization Practices as well as the American College of Obstetricians and Gynecologists support this recommendation. The inactivated influenza vaccine has proven to be safe, with no adverse effects in pregnant women or their infants. The vaccine is not designed for infants aged <6 months, so maternal immunity is key to protecting newborns.

CDC surveyed vaccination uptake among pregnant women in the USA during the 2011–2012 influenza season. Of 1660 survey participants, only 47 percent had received influenza vaccination; 9.9 percent before, 36.5 percent during, and <1 percent after pregnancy. Health care and public health professionals play a key role in educating women on the importance of the vaccine and to increase influenza vaccine uptake.

Sources: *Centers for Disease Control and Prevention. Influenza vaccination coverage among pregnant women – 2011–12 influenza season, United States. MMWR Morb Mortal Wkly Rep 2012;61;758–63. Available at: http://www.cdc.gov/mmwr/preview/mmwrhtml/mm6138a2.htm [Accessed 21 January 2013].*
Centers for Disease Control and Prevention. Pregnant women need a flu shot. November 2012. Available at: http://www.cdc.gov/features/pregnancyandflu/
American College of Obstetrics and Gynecology. Influenza vaccination during pregnancy [posted October 2010]. Available at: http://www.acog.org/Resources_And_Publications/Committee_Opinions/Committee_on_Obstetric_Practice/Influenza_Vaccination_During_Pregnancy [Accessed 21 January 2013].

observation for West Nile Fever could be used to provide early warning of the entry of the bird-borne disease into a specific region. This, in turn, could help to trigger the activation of response mechanisms.

Each year, epidemiological services of the WHO and collaborating centers such as the CDC recommend which strains should be used in vaccine preparation for use among susceptible population groups. These vaccines are prepared with the current anticipated epidemic strains. The three main types of influenza (A, B, and C) have different epidemiological characteristics. Type A and its subtypes, which are subject to antigenic shift (abrupt major change), are associated with widespread epidemics and pandemics. Type B undergoes antigenic drift (small changes over time) and is associated with less widespread epidemics. Influenza type C is even more localized.

Active immunization against the prevailing wild strain of influenza virus produces a 70–80 percent level of protection in high-risk groups. The benefits of annual immunization outweigh the costs, and it has proven to be effective in reducing cases of influenza and its secondary complications, such as pneumonia and death from respiratory complications in high-risk groups.

Avian (H5N1) influenza is a threat to the world's population because of its potential to become a pandemic on the scale of the 1917–1918 influenza epidemic. It is a bird-borne zoonotic disease, so far affecting fowl such as chickens and turkeys contacted by infected wild fowl. Sensitive and robust surveillance measures are required to detect any evidence that the virus has changed and acquired the ability to be transmittable between humans. Surveillance is largely passive in relying on reports of infected wild and domestic fowl and, most importantly, human cases. The major concern is to detect human-to-human transmission, which would threaten to transform this disease into a local, regional, and world pandemic in a matter of months.

International efforts to improve national and local capacities in surveillance and response to this threat are vital to review the scale of the threat should the leap from animal-to-human to human-to-human transmission occur. An integral part of the pandemic planning response in the UK was established in 2005 with the National H5 Laboratory Network, capable of rapidly and accurately identifying potential human H5N1 infections in all regions of the UK and the Republic of Ireland.

The CDC relies on seven systems for national influenza surveillance, four of which operate year-round: the WHO and the National Respiratory and Enteric Virus Surveillance System (NREVSS) collaborating laboratory systems; the US Influenza Sentinel Provider Surveillance System; the 122 Cities Mortality Reporting System; and a national surveillance system that records pediatric deaths associated with laboratory-confirmed influenza.

The H1N1 pandemic control efforts (2009) led by the WHO had broad support from national governments, but there was widespread skepticism and apathy among health care providers and the general public. Response rates to immunization were low even among first responders and hospital personnel. There is a lingering controversy regarding alleged overreaction and conflicts of interests; however, preparation for potential pandemics was advanced. More attention must be paid to public information campaigns in the event of future pandemic threats as well as for annual seasonal influenza.

Pneumococcal Disease

Pneumococcal diseases, which are caused by *Streptococcus pneumoniae*, include pneumonia, meningitis, and otitis media. Together these constitute the world's leading cause of vaccine-preventable child mortality, as over 1 million children die from pneumococcal diseases each year. The 23 capsular types of pneumococci selected out of 83 known types of the organism for the conjugate vaccine (PPV23) are responsible for 88 percent of pneumococcal pneumonia cases and 10–25 percent of all pneumonia cases in the USA.

This vaccine has been found to be cost-effective for high-risk groups, including people with chronic disease, HIV carriers, patients whose spleens have been removed, the elderly, and those with immunosuppressive conditions. It should be included in preventive-oriented health programs, especially for long-term care of the chronically ill. In addition, seven-valent conjugate vaccines (PCV7 and PCV13) are now recommended for routine childhood immunization for children under 2 years of age, the highest risk age group for pneumococcal disease mortality. The WHO and CDC recommend PCV7 routinely for children under 2 years old and PPV23 for adults over 65 years of age. Moreover, others at risk for respiratory disease or pneumococcal infection should be vaccinated with PCV23. Pneumococcal conjugate vaccine is now recommended for routine childhood immunization in the USA and globally by the WHO. By 2010, pneumococcal conjugate vaccine was in routine use for childhood vaccination in 55 countries covering 42 percent of the world's child population. It is now also recommended for routine use among elderly and chronically ill people.

Varicella (Chickenpox, Shingles, Herpes Zoster)

Varicella is an acute, generalized viral disease caused by the varicella zoster virus (VZV). Despite varicella's reputation as an innocuous disease of childhood, patients can become quite ill. A mild fever and characteristic generalized red rash last for a few hours, followed by vesicles occurring in successive crops over various areas of the body. Affected areas may include the membranes of the eyes, mouth, and respiratory tract. The disease may be so mild as to escape observation or may be quite severe, especially in adults. Death can occur from viral pneumonia in adults and sepsis or encephalitis in children. Neonates whose mothers develop the disease within 2 days of delivery are at increased risk, with a case fatality rate of up to 30 percent.

Long-term sequelae include herpes zoster or shingles with a severely painful, vesicular rash along the distribution of sensory nerves, which can last for months. Its occurrence increases with age and it is primarily seen in elderly people. It can, however, occur in immunocompromised children (especially those on cancer chemotherapy), AIDS patients, and others. Some 15 percent of a population will experience herpes zoster during their lifetime. Reye's syndrome is an increasingly rare but serious complication of varicella or influenza type b. It occurs in children and affects the liver and CNS. Congenital varicella syndrome with birth defects similar to CRS has been identified, emphasizing the importance of effective immunization against VZV. Varicella

vaccine is now recommended for routine immunization at 12–18 months of age in the USA, with catch-up for non-immunized children and adults, especially non-pregnant women of childbearing age. To maintain immunity in adolescence and adulthood, booster vaccinations after 13 years of age and again after 50 years of age are effective in those who have no history of VZV infection or evidence of immunity. Varicella vaccine is likely to be added to a "cocktail vaccine" containing DTP, polio (IPV), and Hib.

Meningococcal Meningitis

Meningococcal meningitis, caused by the bacterium *Neisseria meningitides*, is characterized by headache, fever, neck stiffness, delirium, coma, and/or convulsions. The incubation period is 2–10 days. It has a case fatality rate of 5–15 percent if treated early and adequately, but rises to 50 percent in the absence of treatment. There are several important strains (A, B, C, X, Y, and Z). Serogroups A and C are the main causes of epidemics, with B causing sporadic cases and local outbreaks. Transmission is by direct contact and droplet spread.

Meningitis (group A) is common in sub-Saharan African countries, but epidemics have occurred worldwide. During epidemics, children, teenagers, and young adults are the most severely affected. In developed countries, outbreaks occur most frequently in military and college student populations. In 1997, meningococcal meningitis spread widely in the "meningitis belt" in Central Africa. In the 2009 epidemic season, 14 African countries implementing enhanced surveillance reported 88,199 suspected cases, including 5352 deaths, the largest number since an epidemic in 1996.

Epidemic control is achieved by mass chemoprophylaxis with antibiotics (e.g., rifampin or sulfa drugs) among case contacts, although the emergence of resistant strains is a concern. Vaccines against serotypes A and C (bivalent) or A, C, Y, and W-135 are available. Their use is effective in epidemic control and prevention in institutions and among military recruits, especially for A and C serogroups. Recommendations are to immunize using the tetravalent meningococcal conjugate vaccine (MCV4) during the preadolescent years, so that immunity is established before residential education or military service.

VACCINE REGULATION, SCHEDULES, AND PROGRAMS

Vaccine preventable diseases (VPDs) are still among the leading causes of death in developing countries; many mid-level developing or transition countries are not using the full potential of vaccines currently available to protect their children. VPDs are a fundamental aspect of public health not only because of the success achieved in saving millions of lives, but in the enormous potential for future

TABLE 4.6 Estimated Number of Deaths in 2002 from Vaccine-Preventable Diseases Among Children < 5 Years in 2002–2004, Diphtheria–Tetanus–Pertussis (DTP) Vaccine Coverage, and Numbers of Unreached Infants and Incompletely Vaccinated Infants by WHO Region

WHO Region	No. of Deaths	% Coverage with 1 Dose of DTP	No. of Unreached Infants[a]	% Coverage with 3 Doses of DTP	No. of Incompletely Vaccinated Infants[b]
African	1,113,000	78	5,607,000	66	3,048,000
American	44,000	96	562,000	92	659,000
Eastern Mediter-ranean	353,000	86	1,948,000	78	1,186,000
European	32,000	96	458,000	94	158,000
South-East Asian	757,000	77	8,082,000	69	2,959,000
Western Pacific	251,000	96	1,051,000	90	1,302,000
Total	2,550,000	86	17,708,000	78	9,312,000

Note: [a]Number of surviving infants who did not receive one dose of DTP, calculated on the basis of WHO/UNICEF estimates of vaccination coverage with one dose of DTP and estimates of surviving infants from World Population Prospects: the 2004 revision. [b]Number of surviving infants who did not receive three doses of DTP; unvaccinated infants were excluded. World Health Organization Annual Report.
Source: Centers for Disease Control and Prevention. Vaccine preventable deaths and the global immunization vision and strategy 2006–2015. MMWR Morb Mortal Wkly Rep 2006. Available at: http://www.cdc.gov/mmwr/preview/mmwrhtml/mm5518a4.htm [Accessed 18 January 2013].

developments that may make equally valuable contributions to the length and quality of life. Despite this, the potential of even currently available vaccines is not yet fully realized and traditional practices mean that many countries are slow to adopt the newer vaccines and their great life-saving capacity. Table 4.6 shows the number of deaths from VPDs in relation to level of DTP vaccine coverage, classified by WHO region.

ESSENTIALS OF AN IMMUNIZATION PROGRAM

Vaccination is one of the key modalities of primary prevention and one of the principal cornerstones of public health. Immunization is cost-effective and prevents wide-scale disease and death, with high levels of safety. Despite the general consensus in public health regarding the central role of vaccination, there are many areas of controversy and unfulfilled expectations.

A vaccination program should aim at 95 percent or higher coverage at appropriate times, including infants, schoolchildren, and adults. Immunization policy should be adapted from current international standards applying the best available program to national circumstances and financial capacities. Public health personnel with expertise in VPD control are needed to advise ministries of health and the practicing pediatric community on current issues in vaccination. Furthermore, they are needed to monitor the implementation and evolution of control programs. Controversies and changing views are common to immunization policy, and therefore discussions must be conducted on a continuing basis. Policy should be under continuing review

by a government-appointed national immunization advisory committee, including professionals from public health, academia, immunology, laboratory sciences, economics, and relevant clinical fields.

The WHO and UNICEF monitor global mortality from vaccine-preventable deaths and vaccination coverage as well as the introduction of new vaccines in routine immunization programs by countries around the world. Table 4.7 shows features of progress and remaining challenges in the prevention of VPDs.

Complacency is dangerous in vaccination program implementation. In 2000, the CDC declared measles eliminated in the USA. But in 2011, 222 Americans contracted the disease, the most cases the government health agency had seen in 15 years (CDC MMWR 2012). Measles is one of the most contagious VPDs. Most of the cases between 2010 and 2012 were among people who had not been vaccinated, but a significant number had received one or even two doses of measles-containing vaccine. Cases in North and South America have occurred after importation among people traveling in Europe or visitors from Europe, where there have been very large epidemics. France reported 14,966 measles cases in 2011 and six deaths. The UK has an ongoing epidemic of measles which often affects babies too young to receive the vaccine, who are vulnerable to severe cases. Measles is a highly infectious and dangerous disease with important complications, including pneumonia, encephalitis, and even death. Infants before the age of routine vaccination, people with compromised immune systems, including the elderly and those with HIV or types of cancer, are at risk. Mumps has also seen a resurgence in the USA, although the circumstances surrounding the spike differ from measles.

TABLE 4.7 Estimated Global Vaccine-Preventable Disease (VPD) Deaths and Vaccination Coverage, 2010–2011

Child deaths <5 years	2008: total 8.8 million child deaths, with 1.5 million (17%) from VPDs; 2010: 5.2 million deaths, 29% from VPDs, including *Haemophilus influenzae* type b: 199,000; pertussis: 195,000; measles: 118,000; neonatal tetanus: 59,000; tetanus (non-neonatal): 2000; pneumococcal disease: 476,000; rotavirus: 453,000
Immunization coverage	Averts 2–3 million deaths per year globally Countries with >80% coverage increased from 158 in 2010 to 162 in 2011
Unprotected children	2010: 21.1 million 2011: 22.4 million
Diphtheria–tetanus–pertussis coverage	1990: 75%; 2011: 83% (107 million children)
Polio coverage	1990: coverage with three doses 75%; 2011: 84%
Hepatitis B vaccine	Used in 180 countries – global coverage 75%
Haemophilus influenza b vaccine	Used in 177 countries at end of 2011; global coverage with three doses 43% in 2011
Rubella vaccine	Used in 130 countries in 2011; up from 83 countries in 1996
Measles vaccine	Vaccine since 1960s; two-dose policy since 1980s; declaration of measles elimination in the Americas in 2000; reduction of global mortality from 1 million to <200,000; return of measles in 2006–2012 in Europe and in the Americas by importation
Mumps vaccine	Used in 120 countries by end of 2011; mumps epidemics
Yellow fever vaccine	Use in 36 out of 48 countries with endemic yellow fever in Africa and the Americas
Maternal and neonatal tetanus (MNT)	Vaccine in routine prenatal care in over 100 countries by 2011; neonatal tetanus remains major cause of neonatal mortality in Africa and Asia
Pneumococcal vaccine	Used in 72 countries by end of 2011
Rotavirus vaccine	Used in 31 countries by end of 2011; up from 28 countries in 2010
Human papillomavirus vaccine	Used in 43 countries by end of 2011; up from 33 countries in 2010

Source: World Health Organization/UNICEF. Global immunization data. October 2012. Available at: http://www.who.int/immunization_monitoring/Global_Immunization_Data.pdf and http://www.who.int/immunization_monitoring/data/en/ [Accessed 23 January 2013].

Diphtheria and pertussis are "back in town". An undercurrent of anti-vaccinationism is quite common and ready to spring open in response to any willful unfounded damaging reports, as generated by an antipertussis movement in the 1980s in the UK and elsewhere, and in the aftermath of the 1998 MMR "Wakefield effect" alleging a causal relationship between MMR vaccine and autism (see Box 4.11). The latter was proven to be fraudulent, yet damaging, and an overreaction to charges of mercury poisoning due to the use of thiomersal as a preservative. The latter caused a switch in vaccine production to single-dose vials which carry a large increase in costs, making vaccines even less affordable in low-income countries.

Because of an increase in anti-vaccination attitudes many mothers are avoiding or delaying immunization, putting their children and others at risk. The US state of Washington is considering making the "opt-out clauses" for mandatory vaccination for school entry more stringent by requiring written statements from parents and their doctors. Opposition to immunization is widespread on the Internet among a generation who never experienced the horrors of children dying from pertussis, diphtheria, and measles, and being crippled by polio. Public health has a responsibility to work steadily with communities and advocacy groups to change the climate of public opinion towards vaccination.

The schedule shown in Table 4.8 reveals the recommended and optimal ages for routine administration of currently licensed immunizations in the USA, targeted for children from birth until the age of 6 years. Doses not administered at the suggested age should be given at a subsequent doctor's visit, when indicated. The 2011 immunization schedule, which has been approved by the Advisory Committee on Immunization Practices, the American Academy of Pediatrics, and the American Academy of Family Physicians, also outlines a range of ages to administer specific vaccines for certain high-risk groups.

Vaccine supply should be adequate and continuous. Supplies should be ordered from known manufacturers meeting international standards of good practice. All batches should be tested for safety and efficacy before being released for use. There should be a sufficient and continuously monitored cold chain to protect against high temperatures for heat-labile vaccines, sera, and other active biological preparations. The cold chain should include all stages of storage, transport, and maintenance at the site of usage. Only disposable syringes should be used in vaccination programs to prevent any possible transmission of blood-borne infection.

A vaccination program depends on a readily available service with no barriers or unnecessary prerequisites, free

TABLE 4.8 Recommended Immunization Schedule for Children Aged 0–6 Years, USA, 2012

Vaccine	Birth	1 month	2 months	4 months	6 months	9 months	12 months	15 months	18 months	19–23 months	2–3 years	4–6 years
Hepatitis B[1]	HepB	HepB			HepB							
Rotavirus[2]			RV	RV	RV[2]							
Diphtheria–tetanus–pertussis[3]			DTaP	DTaP	DTaP		See fn 3	DTaP				DTaP
Haemophilus influenzae type b[4]			Hib	Hib	Hib[4]		Hib					
Pneumococcal[5]			PCV	PCV	PCV		PCV				PPSV	
Inactivated poliovirus[6a]			IPV	IPV	IPV							IPV
Influenza[7]					Influenza (yearly)							
Measles, mumps, rubella[8]							MMR		See fn 8			MMR
Varicella[9]							VAR		See fn 9			VAR
Hepatitis A[10]							Dose 1[10]				HepA series	
Meningococcal[11]						MCV4; see fn 11						

Note: Detailed footnotes [1–11] and catch-up schedule for those who fall behind or start late are available at the site noted below.

Light shading = range of recommended ages for all children; dark shading = range of recommended ages for certain high-risk groups; medium shading: range of recommended ages for all children and certain high-risk groups.

HepB = hepatitis B; RV = rotavirus; DPT or DPaT = diphtheria, pertussis, tetanus; preferably the acellular preparation (DTaP) and tetanus toxoid (DPT4 can be given at age 12 months if 6 months has elapsed since previous DPT); Hib = *Haemophilus influenzae* type b; PCV = pneumococcal conjugate vaccine; PPSV = pneumococcal polysaccharide vaccine; IPV = inactivated polio vaccine; MMR = measles, mumps, rubella; VAR = varicella zoster virus; MCV4 = tetravalent meningococcal conjugate vaccine.

[a]*During 1999, the recommendation for poliovirus was changed to three doses of IPV in infancy.*

Source: Centers for Disease Control and Prevention. Immunization schedules: Birth-18 Years & "Catch-up" Immunization Schedules United States, 2013: Details For Health Care Professionals. Available at: http://www.cdc.gov/vaccines/schedules/hcp/child-adolescent.html (accessed 19 November 2013) and Centers for Disease Control and Prevention Advisory Committee on Immunization Practices (ACIP) Recommended Immunization Schedule for Persons Aged 0 Through 18 Years — United States, 2013 Morbid Mortal Wkly Rep MMWR Supplements. 2013 / 62(01);2-8. Available at: http://www.cdc.gov/mmwr/preview/mmwrhtml/su6201a2.htm (accessed 19 November 2013).

to parents or with a minimum fee, to administer vaccines in disposable syringes by properly trained individuals using patient-oriented and community-oriented approaches. Ongoing education and training on current immunization practices are needed. Incentive payments by ensuring agency or managed care systems promote complete, on-time coverage. All clinical encounters should be used to screen, immunize, and educate parents and guardians.

Contraindications to vaccination are very few; vaccines may be given even during mild illness with or without fever, during antibiotic therapy, during convalescence from illness, following recent exposure to an infectious disease, and to people having a history of mild to moderate local reactions, convulsions, or a family history of sudden infant death syndrome (SIDS). Simultaneous administration of vaccines and vaccine "cocktails" reduces the number of visits and thereby improves coverage; there are no known cases of interference between vaccine antigens.

Accurate, complete recording with computerization of records and automatic reminders helps to promote compliance, as does co-scheduling of immunization appointments with other services. Adverse events should be reported promptly, accurately, and thoroughly. A tracking system should operate with reminders of upcoming or overdue immunizations; various forms of communication such as mail, telephone, and home visits should be implemented, especially for high-risk families. Semiannual audits should be carried out to assess coverage and review patient records in the population served to determine the percentage of children covered by their second birthday. Tracking should identify children needing completion of the immunization schedule and assess the quality of documentation. It is important to maintain up-to-date, easily retrievable medical protocols where vaccines are administered, noting vaccine dosage, contraindications, and management of adverse events.

For example, Manitoba, Canada, established the population-based electronic Manitoba Immunization Monitoring System (MIMS) in 1988 for all Manitobans registered with the government public health insurance program. Routine immunizations provided by physicians and public health staff are included. Infants are registered at birth and the family is contacted to initiate immunization. Reminder letters are distributed through the public health offices. There is active follow-up of children to ensure completion of immunization according to the recommended schedule (MIMS Annual Report, 2011).

All health care providers should be trained in the education, promotion, and management of immunization policy. Health education should target parents as well as the general public. Monitoring of vaccines used and children immunized, individually and by category of vaccination, can be facilitated by computerization of immunization records, or regular manual review of child care records. Where immunization is carried out by physicians in private practice, as in the USA, coverage is determined by periodic surveys.

Regulation of Vaccines

Inspection of vaccines for safety, purity, potency, and standards is part of the public health regulatory function. Vaccines are defined as biological products and are therefore subject to regulation by national health authorities. In the USA, this comes under the legislative authority of the Public Health Service Act, as well as the Food, Drug and Cosmetics Act, with applicable regulations in the Code of Federal Regulations. The specific federal agency empowered to carry out this regulatory function is the Center for Drugs and Biologics of the Federal Food and Drug Administration.

In the past, litigation regarding adverse effects of vaccines has led to inflation of legal costs as well as efforts to limit court settlements. The US federal government enacted the Child Vaccine Injury Act of 1988, establishing the National Vaccine Injury Compensation Program (NVICP). This legislation requires providers to document vaccines administered and to report on complications or reactions. It was intended to pay benefits to people injured by vaccines faster and by means of a less expensive procedure than a civil suit for resolving claims. Using this no-fault system, petitioners do not need to prove that manufacturers or vaccine providers were at fault. They must only prove that the vaccine is related to the injury in order to receive compensation. The vaccines covered by this legislation include Hib, HAV, HBV, human papillomavirus

(HPV), influenza, meningococcal, pneumococcal, rotavirus, and VZV vaccines, and DTaP/TdaP, MMR, OPV, and IPV.

Adverse effects are documented through the US federal Vaccine Adverse Event Reporting System (VAERS). Created in 1990, it is a national passive reporting system that receives statements on side-effects and adverse events related to US licensed vaccines. Data obtained from VAERS are used to determine patient risk factors, assess increases in known adverse events, and identify new or rare harmful effects. An estimated 30,000 VAERS reports are filed per year, of which 10–15 percent are identified as serious. This degree of classification refers to a patient who received the vaccine and experienced permanent disability, hospitalization, a fatal illness, or death. Reports may be made by anyone, including health practitioners, manufacturers, and vaccine recipients and their guardians. Important information is obtained, but the data are limited so that determining whether the vaccine truly caused an adverse effect cannot be established from a VAERS report. A compensation system was established in the USA in 1988 for injury suffered as a result of vaccines. It includes a compensation system for such injuries as well as ensuring adequate supplies of recommended vaccines (HRSA National Vaccine Compensation Administration Program, 2012).

Newly recommended vaccines for children and adolescents have nearly doubled in number since 2000, and the cost of fully vaccinating a child has increased dramatically in the past decade. Funding of the extensive recommended schedule is a problem in all countries where this is provided as a public health service or where it is covered by health insurance. In the USA, with a lack of health insurance for some 15 percent of the population and low levels of coverage for another 15 percent, lack of coverage for immunization poses a significant problem. Many of the poorest children are covered under the Special Supplemental Nutrition Program for Women, Infants, and Children (WIC); however, many in the working poor population lack access. Providing universal coverage for children remains to be resolved and is crucial to meeting the international standards of developed nations; it is an issue of debate in current political struggles for national health insurance in the USA.

Vaccine Development

The development of vaccines, from Jenner in the eighteenth century to the advent of recombinant hepatitis B vaccine in 1987, stands as one of the pillars of public health and, consequently, vaccines have saved innumerable lives. A giant in the field of vaccinology was Maurice Hillman who, between 1944 and 1995, was the outstanding scientific leader in the development of most of the basic vaccines in use today (Box 4.15).

Vaccines for viral infections in humans for HIV, respiratory syncytial virus, Epstein–Barr virus, dengue fever, and hantavirus are the subjects of intense research with genetic

BOX 4.15 Maurice Hilleman: Creating Vaccines that Changed the World

Maurice R. Hilleman PhD (born 1919, Miles City, Montana, died Philadelphia 2005) was the lead microbiologist at Merck Company during the 1950s to the 1980s. Hilleman and his team created more than 36 human and animal vaccines, including measles, mumps, chickenpox, rubella, hepatitis A, hepatitis B, pneumonia, meningitis, pandemic influenza, and other important diseases.

His role in their development included the laboratory work, as well as scientific and administrative leadership, with fundamental breakthroughs leading to development of over 40 experimental and licensed animal and human vaccines. He characterized antigens and isolated them, and performed the basic and process research and the clinical studies, all the way through the manufacturing process.

In 1944 he developed a vaccine against Japanese B encephalitis, to immunize soldiers on the Pacific front of World War II. He developed the vaccine to fight influenza of the 1957 pandemic Hong Kong flu. He also developed most of the 14 vaccines currently recommended for childhood routine immunization:

- Attenuated measles vaccine, 1962
- Mumps vaccine, 1967 – strain from throat swabs from and named after his daughter Jeryl Lyn at age 5
- Rubella vaccine, 1969
- Measles, mumps, rubella vaccine, 1971
- Meningococcal polysaccharide vaccines, 1974
- Polysaccharide vaccine protecting against 14 types of pneumococcal bacteria in 1977 (with Robert Austrian) – later 23 types
- Hepatitis B, 1981, first viral subunit
- Hepatitis B recombinant vaccine
- Hepatitis A vaccine, 1995
- Chickenpox vaccine, 1995
- *Haemophilus influenzae* b
- Varicella vaccine, from Japanese strain, 1995.

Hilleman's contributions in the fields of virology, epidemiology, immunology, cancer research, and vaccinology led to many professional awards for lifetime achievement. Despite this, he never received a much deserved Nobel Prize or reached the level of the public or professional recognition given to Pasteur, Salk, or Sabin.

Maurice Hilleman's work led to more human and animal vaccines than any other scientist. His work saved tens of millions of lives, extending human life expectancy, improving economies, and changing the world.

Sources: *Dove A. Maurice Hilleman, 2005. Nat Med 2005;11(Suppl 2). http://dx.doi.org/10.1038/nm1223. Available at: http://www.nature.com/nm/journal/v11/n4s/full/nm1223.html [Accessed 21 January 2013]. Obituary. Maurice Hilleman. BMJ 2005;330:1028. Available at: http://www.historyofvaccines.org/content/articles/vaccine-development-licensing-events [Accessed 21 January 2013].*

approaches using recombinant techniques. The potential for the future of vaccines will be greatly influenced by scientific advances in genetic and molecular technology. Moreover, there is potential for the development of vaccines attached to bacteria or protein in plants, which may be given in combination against an increasing range of organisms and toxins.

Recombinant DNA technology has revolutionized basic and biomedical research since the 1970s. The industry of biotechnology has produced important diagnostic tests, such as for HIV, with great potential for vaccine development. Traditional whole-organism vaccines, alive or killed, may contain toxic products that may cause mild to severe reactions. Subunit vaccines are prepared from components of a whole organism. This avoids the use of live organisms that can cause the disease or create toxic products that cause reactions. Subunit vaccines traditionally prepared by inactivation of partially purified toxins are costly, difficult to prepare, and weakly immunogenic. Recombinant techniques are an important development for the production of new whole-cell or subunit vaccines that are safe and inexpensive, and produce more antibodies than other approaches. Their potential contribution to the future of immunology is enormous.

Molecular biology and genetic engineering have made it feasible to create new, improved, and less costly vaccines. New vaccines should be inexpensive, easily administered, capable of being stored and transported without refrigeration, and given orally. The search for inexpensive and effective vaccines for groups of viruses causing diarrheal diseases led to development of the rotavirus vaccine. Some "edible" research focuses on the genetic programming of plants to produce vaccines and DNA. Vaccine manufacturers, who spend enormous sums of money and years of research on new products, are more inclined to work on vaccines that will bring great financial rewards for the company and are critical to the local health care community. This has led to less effort being expended on developing vaccines for diseases such as malaria, which affect primarily the developing world. Industry plays a crucial role in continued progress in the field. Therefore, principles and guidelines must be created to establish incentives for research, development, and application of vaccine technology from a global perspective.

CONNECTIONS BETWEEN INFECTIOUS AND CHRONIC DISEASES

Advances in science have opened an entirely new front in public health with prevention of major chronic diseases by infection control. These advances include vaccines to prevent HPV (the major cause of cervical cancer) and hepatitis B, as well as control of peptic ulcer diseases caused by the nearly universally present *Helicobacter pylori*, a major cause of cancer of the stomach. Making links between infectious and chronic diseases has been one of the major

TABLE 4.9 Linkages Between Infectious and Chronic Diseases

Infectious Diseases Causing Chronic Diseases	Hepatitis B (HBV)	Cirrhosis and Cancer of Liver
	Hepatitis C (HCV)	Cirrhosis and cancer of liver
	Human papillomavirus (HPV)	Cancer of cervix and oropharynx, anal warts
	Helicobacter pylori	Peptic ulcer disease, gastric cancer
	HIV	AIDS, tuberculosis, cancer
	Lyme disease	Arthritis
	Gonorrhea	Syphilis
	Chlamydia	Neurological degenerative disease
	Streptococcal infection	Neonatal blindness
		Rheumatic heart disease, glomerulonephritis
Vaccines, antibiotics, and antiviral therapy to prevent infectious diseases from becoming cause of chronic disease	Influenza vaccination	Seasonal and pandemic
	HBV vaccination	Liver cirrhosis and cancer
	Hepatitis C treatment	Liver cirrhosis and cancer
	HPV vaccine: girls, women, boys	Cancer of cervix
	Helicobacter pylori	Cancer of stomach
	Syphilis	Neurological degenerative disease
	Gonorrhea	Sterility
	Streptococcal infections	Rheumatic fever, glomerulonephritis
Infectious diseases increasing risk and severity of chronic conditions, serious complications, and death	Community- and health facility-acquired infections	Sepsis, blood infections, *Staphylococcus aureus* and methicillin-resistant *Staphylococcus aureus* (MRSA)
	Heart diseases, diabetes, obesity	Influenza (seasonal and pandemic), pneumonia, urinary tract infections,
	Chronic liver and lung diseases	West Nile Fever, Legionnaire's and Lyme Disease and many respiratory infections, e.g., SARS, Middle East respiratory syndrome coronavirus (MERS-Cov)
	AIDS, mental depression	Tuberculosis, HIV, TB, Hepatitis C and B, Drug use, schizophrenia and depression

Source: Adapted from: Centers for Disease Control and Prevention. A CDC framework for preventing infectious diseases: sustaining the essentials and innovating for the future. Atlanta, GA: CDC; October 2011. Available at: http://www.cdc.gov/oid/docs/ID-Framework.pdf [Accessed 21 January 2013].

advances in public health and clinical medicine over the past several decades, as shown in Table 4.9, which uses data from a CDC review of infectious disease control in 2011.

CONTROL AND ERADICATION OF INFECTIOUS DISEASES

Control and eradication of infectious diseases is of national and international concern, particularly with diseases that may be used in bioterrorism or spread quickly on a worldwide basis, requiring strengthened biosafety, biosecurity, and readiness for outbreaks of dangerous and emerging pathogens. The Global Outbreak Alert and Response Network (GOARN) (Box 4.16), coordinated by the WHO on behalf of member countries, is a technical collaboration of existing institutions and networks to pool human and technical resources for rapid identification, confirmation, and response to disease outbreaks of international importance. GOARN provides an operational framework to link expertise and skills needed to keep the international community constantly alert to the threat of outbreaks and ready to respond.

Since the eradication of smallpox, much attention has focused on the possibility of similarly eradicating other diseases, and a list of potential candidates has emerged. Some of the diseases have been abandoned owing to practical difficulties with current technology as well as management, funding, and human resource limitations. Diseases that have been under discussion for eradication have included TB, measles, and polio; some tropical diseases, such as malaria, leprosy, onchocerciasis, filariasis, and dracunculiasis; and some non-infectious conditions of public health importance. In the past, eradication has encompassed multiple definitions, such as extinction of a disease pathogen, the achievement of a situation in which no further cases of a disease occur anywhere, or control of an infection whereby transmission has ceased in a specified area. The definitions in Box 4.17 were delineated by WHO consultant Walter Dowdle in a 1999 MMWR report, *The principles of disease elimination and eradication*. Although developed in reference to infectious disease, the definitions describing control and elimination are relevant to non-infectious diseases as well.

Reducing epidemics of infectious diseases, through control in selected areas or target groups, contributes to the ultimate goal of eradication of the disease. Local elimination can be achieved where domestic circulation of an organism is interrupted, with cases occurring from importation only. This requires a strong, sustained immunization program with adaptation to address the importation of carriers and changing epidemiological patterns.

BOX 4.16 The Global Outbreak Alert and Response Network (GOARN)

Partners meeting in Geneva in April 2000 brought together representatives of technical institutions, organizations, and networks in global epidemic surveillance and response, to discuss "Global Outbreak Alert and Response". Participants identified the need for a global network, building on new and existing partnerships, to strengthen biosafety and biosecurity to deal with global threats of epidemic-prone and emerging diseases. A steering committee of network partners is guiding development of the network.

WHO Geneva coordinates the international outbreak response of the Network as part of its Global Alert and Response operations. Protocols for network structure, operations, and communications have been developed to improve coordination between partners. The Network focuses technical and operational resources from scientific institutions in member states, medical and surveillance initiatives, regional technical networks, networks of laboratories, United Nations organizations (e.g., UNICEF, UNHCR), the Red Cross (International Committee of the Red Cross, International Federation of Red Cross and Red Crescent Societies and national societies), international humanitarian non-governmental organizations (e.g., Médecins sans Frontières, International Rescue Committee, Merlin, and Epicentre), and other technical institutions, networks, and organizations with capacity to contribute to international outbreak alerts and responses.

Agreed standards to international epidemic responses were developed as guiding principles for International Outbreak Alert and Response and operational protocols. These help to standardize epidemiological, laboratory, clinical management, research, communications, logistics support, security, evacuation, and communications systems. The guiding principles aim to improve the coordination of international assistance in support of local efforts by partners in the Network.

Emerging infectious diseases have been a growing concern of the global public health community for generations, but awareness and action have grown since the 1990s with more and more examples of the movement of diseases and their vectors to locations other than their natural habitat. Rift Valley fever (RVF) is one such disease which has moved to wider locations as environmental conditions permit. RVF could reach the USA via livestock, much as the Schmallenberg virus has recently reached European farm animals. International cooperation is vital to anticipate and attempt to halt the spread of such diseases, which can do very great damage to population health.

Examples to 25 February 2013:

- Novel coronavirus infection in the UK – January 2013.
- Poliovirus in environmental samples in Egypt – December 2012. WPV1 with polymerase chain reaction evidence of Pakistani origin. In 2013, WPV found in sewage samples across Israel and in human stool samples indicating vulnerability of highly immunized population using IPV only (see polio section)
- Measles epidemics across Europe, sub Saharan Africa and outbreaks in the Americas from imported cases 2010–2013.
- Avian influenza – the Ministry of Health in Cambodia reported five new human cases of avian influenza that were confirmed positive for the H5N1 virus in January 2013.
- Yellow fever in Chad – 14 February 2013.
- MERSCov Middle east respiratory syndrome (MERSCov) severe respiratory disease in Saudi Arabia and spreading; probably of camel origin 2013.

Sources: *World Health Organization. Global Outbreak Alert & Response Network. Available at: http://www.who.int/csr/outbreaknetwork/en/ [Accessed 12 February 2013].*
Hartley DM, Rinderknecht JL, Nipp TL, Clarke NP, Snowder GD, National Center for Foreign Animal and Zoonotic Disease Defense Advisory Group. Potential effects of Rift Valley fever in the United States. Emerg Infect Dis [serial on the Internet]; August 2011. Available at: http://dx.doi.org/10.3201/eid1708.101088 [Accessed 12 February 2013].
Emerging Infectious Diseases. Potential effects of Rift Valley fever in the United States. EID online: http://wwwnc.cdc.gov/eid/article/17/8/10-1088_article.htm#suggestedcitation [Accessed 12 February 2013].
World Health Organization. Novel coronavirus infection – update. Available at: http://www.who.int/csr/don/2013_02_11b/en/index.html
World Health Organization. Poliovirus detected from environmental samples in Egypt. Available at: http://www.who.int/csr/don/2013_02_11/en/index.html

Success in Eradicating Smallpox

Smallpox is characterized as one of the major pandemic diseases of the Middle Ages, and its recorded history goes back to antiquity. Prevention of smallpox was discussed in ancient China by Ho Kung (*c.* 320 CE), and inoculation against the disease was practiced there from the eleventh century CE. Prevention was carried out by nasal inhalation of powdered dried smallpox scabs. Exposure of children to smallpox when the mortality rate was lowest assumed a weakened form of the disease, and it was observed that a person could only have smallpox once in a lifetime. Isolation and quarantine were widely practiced in Europe during the sixteenth and seventeenth centuries.

Variolation was the practice of inoculating youngsters with material from scabs of pustules from mild cases of smallpox, with the hope that they would develop a mild form of the disease. Although this practice was associated with substantial mortality, it was widely adopted because mortality from variolation was well below that of smallpox acquired during epidemics. Introduced into England in 1721 (see Chapter 1), it was commonly practiced as a lucrative medical specialty during the eighteenth century. In the 1720s, variolation was also introduced into the American colonies, Russia, and subsequently Sweden and Denmark.

Despite all efforts, in the early eighteenth century, smallpox was a leading cause of death in all age groups. Towards the end of the eighteenth century, an estimated 400,000 people died annually from smallpox in Europe. Vaccination, or the use of cowpox vaccinia virus to protect against smallpox, was initiated late in the eighteenth century. In 1774, a cattle breeder in

BOX 4.17 Control and Elimination of Infectious Diseases

- *Control* – reduction of disease incidence, prevalence, morbidity, or mortality to a locally acceptable level as a result of deliberate efforts; continued intervention measures are required to maintain reduction. Example: diarrheal diseases.

- *Elimination of disease* – reduction to zero of the incidence of a specified disease in a defined geographic area as a result of deliberate efforts; continued intervention measures are required. Example: neonatal tetanus.

- *Elimination of infections* – reduction to zero of the incidence of infection caused by a specific agent in a defined geographic area as a result of deliberate efforts; continued measures to prevent re-establishment of transmission are required. Example: measles, poliomyelitis.

- *Eradication* – permanent reduction to zero of the worldwide incidence of infection caused by a specific agent as a result of deliberate efforts; intervention measures are no longer needed. Example: smallpox.

- *Extinction* – the specific infections agent no longer exists in nature or in the laboratory. Example: none.

Source: *Dowdle W. Principles of disease elimination and eradication. MMWR Morb Mortal Wkly Rep 1999;48(SU01):23–7. Available at: http://www.cdc.gov/mmwr/preview/mmwrhtml/su48a7.htm [Accessed 21 January 2013].*

Yorkshire, England, inoculated his wife and two children with cowpox to protect them during a smallpox epidemic. In 1796, Edward Jenner, an English rural general practitioner, experimented with inoculation from a milkmaid's cowpox pustule to a healthy youngster, who subsequently proved resistant to smallpox by variolation (see Chapter 1). Vaccination, or at the time, the deliberate inoculation of cowpox material, was slow to be adopted universally, but by 1801, over 100,000 people in England had been vaccinated. Vaccination gathered support in the nineteenth century in military establishments and in some countries that adopted it universally.

Opposition to vaccination remained strong for nearly a century based on religious grounds and observed failures of vaccination to give lifelong immunity, and because it was seen as an infringement by the state on the rights of the individual. Often the protest was led by medical variolationists whose medical practice and large incomes were threatened by the mass movement to vaccination. Resistance was also offered by "sanitarians" who opposed the germ theory and believed that cleanliness was the best method of prevention. Universal vaccination was increasingly adopted in Europe and America in the early nineteenth century and eradication of smallpox in developed countries was achieved by the mid-twentieth century.

In 1958, the Soviet Union proposed a program to the WHA to eradicate smallpox internationally, and subsequently donated 140 million doses of vaccine per year as part of the 250 million needed to promote vaccination of at least 80 percent of the world population. In 1967, the WHO adopted a target for the eradication of smallpox. The program included a massive increase in coverage to reduce the circulation of the virus through person-to-person contact. Where smallpox was endemic, with a substantial number of unvaccinated people, the aim of the mass vaccination phase was 80 percent coverage.

Increasing vaccination coverage in developing countries reduced the disease to periodic and increasingly localized outbreaks. In 1967, 33 countries were considered endemic for smallpox, and another 11 experienced importation of cases. By 1970, the number of endemic countries fell to 17, and by 1973 only six countries were still endemic, including India, Pakistan, Bangladesh, and Nepal. In these countries, a new strategy was needed, based on a search for cases and vaccination of all contacts, working with a case incidence below five per 100,000. The program then moved into the consolidation phase, with emphasis on the vaccination of newborns and new arrivals. Surveillance and case detection were improved with case-contact or risk-group vaccination. The maintenance phase began when surveillance and reporting were switched to the national or regional health service with intensive follow-up of any suspect case. The mass epidemic era had been controlled by mass vaccination, reducing the total burden of the disease, but eradication required the isolation of individual cases with vaccination of potential contacts.

Technical innovations greatly eased the problems associated with mass vaccination worldwide. During the 1920s, there was wide variation in sources of the smallpox vaccine. In the 1930s, efforts to standardize and further attenuate the strains used reduced complication rates from vaccinations. The development of lyophilization (freeze-drying) of the vaccine in England in the 1950s made a heat-stable vaccine that could be effective in tropical field conditions in developing countries. The invention of the bifurcated needle (by Rubin in 1961) allowed for easier and more widespread vaccination by less trained personnel in remote areas. The net result of these innovations was increased world coverage and a reduction in the spread of the disease. Smallpox became more and more confined by increasing herd immunity, thus allowing transition to the phase of monitoring and isolation of individual cases.

In 1977 the last case of smallpox was identified in Somalia, and in 1980 the WHO declared the disease eradicated. No subsequent cases have been found except for several associated with a laboratory accident in the UK in 1978. The cost of the smallpox eradication program was US$112 million or US$8 million per year. Globally, savings from vaccination were estimated to be in the order of tens of billions of US dollars of direct savings. Malaria is estimated to cost sub-Saharan African countries US$100 billion worth of lost annual gross domestic product (GDP), much of which can be saved by low-cost interventions such as insecticide-impregnated bed nets and vector control, and, it is hoped, with vaccines now in development.

The WHA recommended destruction of the last two remaining stocks of the smallpox virus in Atlanta, Georgia (USA) and Moscow (Russia) in 1999. This was delayed in 1999 owing to concern that illegal stocks may be held by some states or terrorists for potential use as weapons of mass destruction. Destruction of the stocks was also postponed owing to concern regarding the appearance of monkeypox, and a wish to use the virus for further research. Today, virus stocks are handled only in select laboratories with high security. In addition, emergency plans have been developed, including the immunization of key health workers to limit the extent of a bioterror-engendered epidemic.

Eradication of Poliomyelitis

Given the success in eradicating smallpox, the WHO in 1988 established a target for the eradication of polio. Although polio epidemics continue, largely in countries with limited access to public health services, the burden of disease worldwide has been greatly reduced. At the initiation of the polio eradication campaign, 350,000 cases of childhood paralysis were attributed to polio in 125 countries. By 2009, this number was reduced to 1604 cases. Thus, since 1988, the number of polio cases has fallen by over 99 percent. Only three countries have never achieved wild-poliovirus interruption, and remain endemic for polio: Afghanistan, Nigeria, and Pakistan. However, in 2009–2010, 23 countries that were previously polio free were reinfected, including Angola, Chad, and DRC. Support from member countries and international organizations and agencies such as UNICEF, GAVI, the Gates Foundation, and Rotary International has led to wide-scale increases in immunization coverage throughout the world. The WHO promotes use of OPV as part of routine infant immunization on National Immunization Days (NIDs) or supplementary immunization activities (SIAs). This strategy has been successful in the Americas, Europe, and China, but several countries remain problematic. In 2011, India experienced far fewer cases than in 2010 as a result of a massive commitment to supplementary immunization activities (SIAs) and a slow improvement in basic immunization coverage in key problematic states where polio remains endemic at low levels.

Eradication of wild polio will require flexibility in vaccination strategies and may necessitate the combined approach, using OPV (Sabin, attenuated live vaccine) and IPV (Salk, inactivated polio vaccine), as adopted in the USA between 1997 and 2000 to prevent vaccine-associated clinical cases (i.e., VAPP). The USA has since switched to an IPV-only vaccination policy, which has been adopted by most of the industrialized countries but is impractical for developing nations because of high cost and less than needed immunization coverage.

Currently, IPV is largely used only in countries where interruption of wild poliovirus has occurred, but lower intestinal immunity from IPV may be a risk for imported polio. The combination of OPV and IPV may be needed where enteric disease is common and leads to interference in OPV uptake, especially in tropical areas where endemic poliovirus and diarrheal diseases are still found. Polio made a resurgence in 2004 in Nigeria, and in 2005 in a number of countries previously thought to be under control. The use of OPV has been put in doubt by recent decisions in the industrialized countries to follow the US example of IPV only. The developing countries will need to rely on OPV in the coming years because of the high cost and limited supply of IPV.

Future Candidates for Disease Eradication

Malaria

Eradication of malaria was thought to be possible in the 1950s when major gains were seen in malaria control by aggressive environmental control, with case finding and management. However, poorly sustained vector control, banning of DDT, and lack of development of an effective vaccine have been major obstacles. Malaria control suffered serious setbacks because of a failure of political resolve and continuity needed in the support for necessary programs.

In the 1950s great progress was made largely using dichlorodiphenyltrichloroethane (DDT) in eliminating malaria in the Caribbean, the Balkans, and many countries in the Pacific Region. However, widespread use of DDT as an agricultural pesticide reduced its effectiveness as resistance developed in mosquito populations. A movement against the use of DDT was stimulated by the publication of *Silent Spring* by Rachel Carson in 1962, which raised concerns over the environmental and long-term effects of the large-scale use of DDT. In 1972, DDT was banned in the USA, and subsequently for agricultural use worldwide.

Malaria control efforts were not sustained in many countries, and a dreadful comeback of the disease occurred in Africa and Asia in the 1980s. The emergence of mosquitoes resistant to insecticides, and strains of the parasite resistant to antimalarial drugs have made control even more difficult, expensive, and cumbersome.

Renewed efforts in malaria control required new approaches with case finding and treatment as well as vector control. The use of community health workers (CHWs) in small villages in highly endemic regions of Colombia resulted in a major drop in malaria mortality during the 1990s. The CHWs investigate suspect cases by taking clinical histories and blood smears. A presumptive diagnosis is made clinically or by local examination of blood smears. Therapy is instituted rapidly and the patient is followed up. Quality control monitoring shows high levels of accuracy in the reading of slides compared to professional laboratories.

The total ban on DDT is now seen as an overreaction to legitimate concerns over its persistence in nature. Banning of its widespread use by international convention had a critically disabling effect on national malaria control programs, contributing to a relaxation of efforts to control the disease and its comeback on a massive scale in those countries least able to adapt with the few and expensive alternative insecticides. In 2006, nearly 30 years after DDT was phased out for indoor spraying to protect residents from malaria, the WHO announced that DDT would once again play a key role in household spraying to fight the disease. DDT remains a key element for vector control, a vital aspect of malaria control.

In 2010, there were an estimated 219 million cases of malaria (uncertainty range of 154–289 million) globally, with some 660,000 deaths (490–836,000). Malaria mortality rates have fallen by more than 25 percent globally since 2000, and by 33 percent in the WHO African Region. Most deaths occur among children in Africa (WHO, World Malaria Report 2011 and Malaria fact sheet no. 94, January 2013).

DDT for malaria control through indoor spraying of homes is once again widely used in endemic areas. Since 2006, the WHO has recommended the use of insecticide-impregnated bed nets and limited uses of DDT for internal residual spraying to protect homes and reduce the risk of infection to children. Vector control by removing still water sites, particularly near residences, is a crucial part of malaria control, along with early case finding, bed nets, and multidrug treatment.

The WHO's rollback malaria program has set ambitious targets for the reduction of malaria deaths by the year 2015. Real progress has been made in international donor and national governmental support, in the widespread use of insecticide-impregnated bed nets and indoor residual spraying, as well as outdoor vector control by reducing still waters where mosquito vectors breed, the clinical management of cases, and early diagnosis and treatment with effective combination therapy. In 2010, there were just under 700,000 deaths from malaria, mainly among children and mostly located in sub-Saharan Africa. The WHO has launched the "T3: Test. Treat. Track" initiative, urging endemic countries and stakeholders to scale up diagnostic testing, treatment, and surveillance for malaria. The potential for increased control of malaria has been demonstrated by a complex of factors that depend on international donor funding, which will be problematic in times of recession.

Measles and Rubella

Before the development of the measles vaccine, this disease affected most children, causing an estimated 2 million deaths globally and tens of thousands of cases of blindness. In the late 1970s, widening use of the successful measles vaccine, with local elimination of the disease in the Americas and Europe, brought measles deaths down to 535,000 in 2000 and to 240,000 in 2010. Rubella remains a threat to pregnancies, with 100,000 children born with CRS yearly. Measles combined with rubella and mumps vaccine if given in two doses can protect children and adults from these highly contagious diseases. Controversial claims against the MMR vaccine, which were later proven to be fraudulent, promoted a loss of confidence in measles vaccination in the UK and other countries.

Optimism over the potential for the eradication of measles led to an initial target date of 2010 being set, which was later moved to 2015. Measles eradication was set back owing to low levels of immunization coverage in many developing countries and breakthrough epidemics occurring in the USA, Canada, and many other countries during the 1980s and early 1990s. Despite this, regional eradication was achieved combining the two-dose policy with catch-up campaigns for older children or on NIDs, as demonstrated in the Caribbean countries. In the years 2010–2012, large-scale epidemics were spreading across Western Europe, with export of cases by travelers and visitors to the Americas and elsewhere.

In 2012, the WHO launched Strategic Plan 2012–2020 to promote measles and rubella elimination, applying lessons learned from polio eradication and from the measles experience to date. This needs to be strengthened by routine immunization of children with two doses of measles- and rubella-containing vaccines, plus supplemental campaigns to reach out to the immunized, backed by improved epidemiological monitoring and laboratory capacity surveillance capacity along with improved supply- and cold-chain efforts. This program has support from many international, national, and charitable organization donor agencies with successful experience in controlling VPDs.

Criteria for Eradicable Diseases

A decade after the eradication of smallpox was achieved, the International Task Force for Disease Eradication (ITFDE) was established to systematically evaluate the potential for global eradication of candidate diseases. Its goals were to identify specific barriers to the eradication of these diseases that might be surmountable, and thus promote eradication efforts. Selecting diseases for eradication is not purely a professional issue of resources such as vaccines and human resources, organization, and financing. It is also a matter of political will and perception of the burden of disease, and thus triggers many controversies. The CDC has published criteria for the selection of disease for eradication, as shown in Box 4.18.

NCDs are also included in the discussion of eradicability, but this is oriented towards elimination as public health problems with specified targets, such as eliminating iodine deficiency conditions together with measuring urinary

BOX 4.18 Criteria for Assessing Eradicability of
Diseases, International Task Force for Disease
Eradication (ITFDE)

- Scientific feasibility
- Epidemiological vulnerability: existence of non-human
  reservoir, ease of spread, natural cyclic decline in prevalence, naturally induced immunity, ease of diagnosis,
  duration of any relapse potential
- Effective practical intervention available: vaccine or
  other primary preventive or curative treatment, and
  means of eliminating vector. Intervention should be safe,
  inexpensive, long-lasting, and easily used in the field
- Demonstrated feasibility of elimination in specific locations: documented elimination from island or other geographic unit
- Political will/popular support
- Perceived burden of the disease: morbidity, mortality,
  disability, extent, other effects, true burden may not be
  perceived, the reverse of benefits expected to accrue
  from eradication, and costs of care in developed and
  developing countries
- Expected cost of eradication, particularly in relation to
  perceived burden of disease
- Synergy of implementation with other programs, other
  eradication efforts
- Reasons for eradication versus control

Sources: Centers for Disease Control and Prevention. World Health
Organization: Update international task force for disease eradication,
1990 and 1991. MMWR Morb Mortal Wkly Rep 1992;41:40–2. Available
at: http://www.cdc.gov/mmwr/preview/mmwrhtml/00015970.htm
[Accessed 2 February 2013].
Centers for Disease Control and Prevention. Recommendations of the
international task force for disease eradication. MMWR Morb Mortal
Wkly Rep 1993;42(RR-16):1–39. Available at: http://www.cartercenter.
org/documents/1184.pdf [Accessed 4 February 2013].
Dowdle WR. The principles of disease elimination and eradication. Bull
World Health Organ 1998;76(Suppl. 2):23–5.

iodine levels in surveys of schoolchildren and pregnant
women. In many developing countries, with the support of
international organizations and donors, mass immunization
days are held to reach those who did not receive the routine
vaccinations of childhood. These immunizations are often
coupled with micronutrient supplements such as vitamin
A, iron, zinc, and in some cases multivitamins as drops or
sprinkles to be added to regular foods.

The subject of eradication versus control of infectious
diseases is of central public health importance as technology expands the armamentarium of immunization and vector control into the twenty-first century (Table 4.10). The
control of epidemics, followed by interruption of transmission and ultimately eradication, will save countless lives
and prevent serious damage to children throughout the
world. The smallpox achievement, momentous in itself,
points to the potential for the eradication of other deadly
diseases. The skillful use of existing and new technology is

an important priority in the New Public Health. Flexibility
and adaptability are as vital as resources and personnel.

Health targets in the field of infectious disease control
for the twenty-first century, selected by the WHO, include
the following for control or eradication:

- Chagas disease
- neonatal tetanus
- leprosy
- measles – eradication by 2020
- trachoma – eradication by 2020
- reversing the current trend of increasing TB and HIV/
  AIDS.

The Neglected Tropical Diseases

Table 4.11 shows the causes, impact, and strategies for control of neglected tropical diseases (NTDs) as defined by a
2013 report of the US Institute of Medicine based on Hotez
et al. (2009).

Many of these campaigns have already successfully
achieved their interim goals. Primary targets for eradication, such as polio and measles, have proven problematic
and may require changes in immunization tactics, but are
achievable by 2020. Progress in the control of leprosy,
onchocerciasis, filariasis, and dracunculiasis has been
impressive. The coming years hold hope for breakthroughs
and elimination of many preventable diseases.

In 2005, the Department for Control of Neglected Tropical Diseases was established at the WHO headquarters.
Eradication is defined as the interruption of transmission
worldwide and zero cases in each country to be certified
as free of disease, as is the case for dracunculiasis (guinea
worm) in many countries.

A WHO review and roadmap for achieving goals in
NTDs noted progress made in reducing the NTDs up to
2009, with preventive chemotherapy for schistosomiasis,
soil-transmitted helminths, lymphatic filariasis, onchocerciasis, and trachoma. Most progress has been made
for onchocerciasis, with chemotherapy reaching nearly 60
percent of the population in need by 2009. Filariasis chemoprophylaxis reached nearly 42 percent of the population
in need in 2007, but declined subsequently. For trachoma,
schistosomiasis, and helminthiasis the level of coverage
with chemotherapy was low (reaching 13, 8.3, and 31 percent, respectively). Progress in the adoption of intensified
disease management has been uneven.

Yaws chemotherapy was made much easier when oral
azithromycin was found to be as effective as injected benza-
thine penicillin, since it can be administered by CHWs and
reach the population in need much more readily. Onchocerciasis eradication efforts in 2002 concluded that most probable elimination of the disease could be achieved in the WHO
Region of the Americas, but not yet in the African Region
given the existing tools. Achievements to date should be

TABLE 4.10 Potential Candidates for Disease Control and Eradication

Organism	Control – Elimination as a Public Health Problem	Eradicable – Regional/Global
Bacterial disease	Pertussis Neonatal tetanus Congenital syphilis Trachoma Tuberculosis Leprosy	Diphtheria *Haemophilus influenzae* type b
Viral disease	Hepatitis B Hepatitis A Yellow fever Rabies	Poliomyelitis Measles Rubella Mumps
Parasitic disease	Trachoma blindness – global elimination Yaws – control of treponematoses Leprosy – elimination as a public health problem Chagas disease – control and elimination Leishmaniasis, visceral – control Filariasis – elimination as a public health problem Onchocerciasis river blindness prevention Schistosomiasis – control Helminthic infestation – soil transmitted – control Cysticercosis	Dracunculiasis (Guinea worm) Echinococcus Teniasis
Non-infectious disease	Lead poisoning Silicosis Protein energy malnutrition Micronutrient malnutrition Iodine deficiency Vitamin A deficiency Folic acid deficiency Iron deficiency	

Sources: Goodman RA, Foster KL, Trowbridge FL, Figueroa JP, editors. Global disease elimination and eradication as public health strategies. MMWR Morb Mortal Wkly Rep 1999;48(Suppl):1–309. Available at: ftp://ftp.cdc.gov/pub/Publications/mmwr/other/suppl48.pdf [Accessed 23 January 2013].
Aylward B, Hennessey KA, Zagaria N, Olivé J-M, Cochi S. When is a disease eradicable? 100 years of lessons learned. Am J Public Health 2000;90:1515–20. Available at: http://ajph.aphapublications.org/doi/pdf/10.2105/AJPH.90.10.1515 [Accessed 23 January 2013].
World Health Organization. Accelerating work to overcome the global impact of neglected tropical diseases – a roadmap for implementation. Geneva: WHO; 2011. Available at: http://www.who.int/neglected_diseases/NTD_RoadMap_2012_Fullversion.pdf [Accessed 28 January 2013].

preserved and built upon by continued cooperative efforts of the WHO, World Bank, United Nations Development Programme (UNDP), and others. The struggle to eliminate and potentially eradicate important diseases, such as was achieved with smallpox, will require many years of strong political and funding support as well as a strong cadre of a public health workforce with new scientific breakthroughs (such as malaria and HIV vaccines). The movement, even when only partially successful, is saving millions of lives and improving the quality of life for many.

Tuberculosis

TB is caused by a group of organisms including *Mycobacterium tuberculosis* in humans and *M. bovis* in cattle. The disease is primarily found in humans, but it is also a disease of cattle and occasionally other primates in certain regions of the world. It is transmitted via airborne droplet nuclei from people with pulmonary or laryngeal TB through coughing, sneezing, talking, or singing. The initial infection may go unnoticed, but tuberculin sensitivity appears within a few weeks. About 95 percent of those infected enter a latent phase with a lifelong risk of reactivation. Approximately 5 percent go from initial infection to pulmonary TB. Less commonly, the infection develops as extrapulmonary TB, involving meninges, lymph nodes, pleura, pericardium, bones, kidneys, or other organs.

Untreated, about half of the patients with active TB will die of the disease within 2 years, but modern chemotherapy almost always results in a cure. Pulmonary TB symptoms include cough and weight loss, with clinical findings on chest examination and confirmation by findings of tubercle bacilli in stained smears of sputum and, if possible, growth of the organism on culture media, and changes in the chest X-ray. TB affects people in their adult working years, with 80–90 percent of cases in people between the ages of 15 and 49. Its devastating effects on the workforce and economic development contribute to a high cost-effectiveness for TB control.

TABLE 4.11 Neglected Tropical Diseases: Causes, Impact, and Strategies for their Control

	Disability-Adjusted Life Years	Deaths	Approximate Global Prevalence	Approaches to Control
High-prevalence diseases	14.9–52.1 million	24,000–415,000	1.0–1.2 billion	MDA with rapid effect package
Hookworm infection	1.8–22.1 million	3000–65,000	600 million	MDA with rapid effect package or albendazole
Ascariasis	1.2–10.5 million	3000–60,000	800 million	MDA with rapid effect package or albendazole or mebendazole
Trichuriasis	1.6–6.4 million	3000–10,000	600 million	MDA with rapid effect package or albendazole or mebendazole
Lymphatic filariasis	5.8 million	<500	120 million	MDA with rapid effect package or diethylcarbamazine + albendazole or ivermectin + albendazole
Schistosomiasis	1.7–4.5 million	15,000–280,000	200 million	MDA with rapid effect package or praziquantiel
Trachoma	2.3 million	<500	84 million	SAFE strategy with azethromycin
Onchocerciasis	0.7 million	<500	37 million	MDA with rapid effect package or ivermectin
Vectorborne protozoan and viral diseases	5.0 million	132,000	70 million	Integrated vector management or case detection and management or both
Dengue fever	0.7 million	19,000	50 million	Integrated vector management
Leishmaniasis	2.1 million	51,000	12 million	Case detection and management and integrated vector management
Chagas disease	0.7 million	14,000	8–9 million	Integrated vector management
Human African trypanosomiasis	1.5 million	48,000	<0.1 million	Case detection and management, and tsetse control

Note: MDA = mass drug administration by community health workers.
Sources: Choffnes ER, Relman DA, Institute of Medicine. Forum on microbial threats: the causes and impacts of neglected tropical and zoonotic diseases: opportunities for integrated prevention strategies. Washington, DC: Institute of Medicine, National Academies Press; 2011. Available at: www.nap.edu/catalog.php?record_id=13087 [Accessed 25 February 2013].
Hotez PJ, Fenwick L, Svioli L, Molyneux DH. Rescuing the bottom billion through control of neglected tropical diseases. Lancet 2009;373:1570–5.
World Health Organization. Working to overcome the global impact of neglected tropical diseases: Update 2011. Geneva: WHO; 2011. Available at: http://www.who.int/neglected_diseases/2010report/WHO_NTD_report_update_2011.pdf [Accessed 22 January 2013].

Nearly one-third of the world's population is infected with TB. In 2010, there were over 8.8 million incident cases, and nearly 1.1 million deaths and an additional 0.35 million deaths among HIV-positive people. During 2005, new cases of TB included 3 million in South-East Asia and 2.5 million in Africa, where HIV disease has become the leading comorbidity and risk for TB mortality. Approximately 13 percent of TB cases occur among HIV-positive individuals. Throughout the period between 1990 and 1999, the WHO estimates that there were 88 million new cases of TB, of which 8 million cases were in association with HIV infection. During the 1990s, an estimated 30 million people died of TB, including 2.9 million with HIV infection.

TB has also increased in the USA and several European countries for the first time in many decades. Unrealistic expectations can lead to inappropriate assessments and policy when confounding factors alter the epidemiological course of events. Such is the case with TB, where control

and eradication have receded from the picture. This deadly disease has returned to developed countries, partly in association with the HIV infection and multidrug-resistant strains (MDR), as well as homelessness, rising prison populations, poverty, and other deleterious social conditions. Directly observed therapy is an important recent breakthrough, making more effective use of the available technology, and this strategy will certainly play a major role in TB control in the twenty-first century.

The 2008 Global Tuberculosis Control Report noted 9.2 million new cases in 2006, including 400,000 new cases of multidrug-resistant tuberculosis (MDR-TB); approximately 1.5 million people died of TB in 2006. The incidence rate (per 100,000 population) has been declining annually by 1.3 percent since 2002. Sustaining this trend means reaching and attaining the MDG target for 2015. The WHO Report on Global Tuberculosis Control indicates that in 2012, there were approximately 8.8 million incident cases of TB.

BOX 4.19 Principal Issues of Control of Tuberculosis

- Political commitment and funding
- Public–private partnership
- Professional commitment
- Societal changes to reduce risks such as homelessness, malnutrition, and social isolation
- Reduce comorbidities of concomitant diseases, e.g., HIV
- Identify, treat, and follow up people with clinically active TB
- Educate health care providers on suspicion of TB and investigation of suspects
- Diagnostic methods: clinical suspicion, sputum smear for bacteriological examination, tuberculin skin testing, chest radiograph
- Outreach and case finding and investigation programs in high-risk groups, e.g., homeless people, HIV carriers, prisoners, and intravenous drug users
- Contact investigation
- Isolation techniques only during initial therapy
- Treatment, mainly ambulatory, of people with clinically active TB
- Investigation and treatment of contacts
- Directly observed treatment, short course (DOTS) to reinforce compliance
- Environmental control in treatment settings to reduce droplet infection

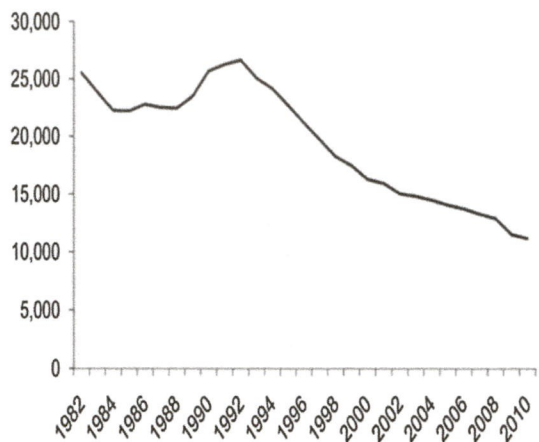

FIGURE 4.8 Reported number of tuberculosis cases, USA, 1982–2010. *Source: Centers for Disease Control and Prevention. Reported tuberculosis in the United States, 1982–2010. Available at: http://www. cdc.gov/Features/dsTB2010Data/ [Accessed 3 December 2012].*

Contrary to previous global reports, which suggested that TB cases would rise slowly, the absolute number of cases has, in fact, been declining since 2006. Moreover, progress is on the horizon, as findings demonstrate that all six WHO regions are on track for attaining the MDG established for 2015.

A new and dangerous period for TB resurgence has resulted from parallel epidemiological events, specifically the advent of HIV infection and the occurrence of MDR-TB. MDR-TB refers to organisms resistant at least to both isoniazid and rifampin, two mainstays of TB treatment. MDR-TB can have a case fatality rate as high as 70 percent. Since HIV reduces cellular immunity, people with latent TB have a high risk of activation of the disease. It is estimated that HIV-negative people have a 5–10 percent lifetime risk of contracting TB; HIV-positive people have a risk of 10 percent per year of developing clinical TB (Box 4.19).

The incidence of TB in the USA decreased steadily until 1985, increased in 1990, and since the early 1990s has been continuously declining (Figure 4.8). From 1986 to 1992, there was an excess of 51,600 cases over the expected rate if the previous decline in case incidence had continued. This sudden rise in TB was largely due to the HIV/AIDS epidemic and the emergence of MDR-TB, as well as a greater concentration among immigrants from areas of higher TB incidence, drug abusers, homeless people, and those with limited access to health care. This is particularly true in New York City, where MDR-TB has appeared in outbreaks

among prison inmates and hospital staff. Since 2000, the majority of cases have been among people born outside the USA. Although case rates have declined in all age groups, the burden of disease is highest among older people.

The resurgence of TB cases peaked in 1992, but since then the number of cases reported annually has declined by 58 percent (to 2009). The incidence of TB in the USA has been reduced because of strong TB control programs that promptly identified people with TB and ensured completion of appropriate therapy. Aggressive staff training, outreach, and case management approaches were vital to this success. Rising rates among recent immigrants, the continued challenge of HIV/AIDS, and coincidental transmission of hepatitis A, B, and C among drug users and marginal population groups demonstrate a need for continued support for TB. Primary multidrug resistance increased from 1.1 percent of reported cases in 2009 to 1.2 percent in 2010.

Bacille Calmette–Guérin (BCG) is an attenuated strain of the tubercle bacillus developed in the early 1920s at the Pasteur Institute in Paris. It was and still is used widely as a vaccination to prevent TB, especially in high-incidence areas. It induces tuberculin sensitivity or an antigen–antibody reaction in which the antibodies produced may be somewhat protective against the tubercle bacillus in 90 percent of vaccinees. There are now several strains of BCG. Although support for its general use is contradictory, there is evidence from case–control and contact studies of positive protection against TB meningitis and disseminated TB in children under the age of 5. In some developed, low-incidence countries, it is used selectively rather than routinely. It may also be used in asymptomatic HIV-positive people or other high-risk groups.

The BCG vaccine for TB remains controversial. While used widely internationally, in the USA and other industrialized countries, it is thought to hinder rather than help in the fight against TB. This concern is based on the usefulness of tuberculin testing for diagnosis of the disease. Where BCG has been administered, the diagnostic value of tuberculin testing is reduced, especially in the period soon after BCG is used. Studies showing equivocal benefit of BCG in preventing TB have added to the dispute and uncertainty. While those in the field in the USA continue to oppose the use of BCG, internationally it is still felt to be of benefit in preventing TB, primarily in children. Currently, the WHO recommends the use of a single dose of BCG as close to birth as possible as part of the EPI. However, there is concern that BCG given to immunocompromised people can be dangerous.

A 1994 meta-analysis of the literature on BCG, carried out by the Technology Assessment Group at Harvard School of Public Health, concluded that BCG vaccine significantly reduces the risk of TB, by 50 percent on average. Protection is observed across many populations, study designs, and forms of TB. Age at vaccination does not affect the efficacy of BCG. Protection against TB death, meningitis, and disseminated disease is higher than for total TB cases, although this result may reflect reduced error in disease classification rather than greater BCG efficacy.

Limitations of current chemotherapy and the only available vaccine, BCG, in the fight against TB make the continued search for new vaccines and therapeutics vital, possibly aided by new methods in the design of vaccines and drugs. However, the struggle is now best fought using the directly observed treatment, short-course (DOTS) strategy, improved diagnostic methods, and poverty alleviation and nutritional improvements in vulnerable population groups.

Directly Observed Therapy Short Course (DOTS)

DOTS is a case management strategy adopted in 1993 by the WHO to improve the effectiveness of compliance with treatment of TB and reduce the increasing global burden of the disease, especially in developing countries, but also in vulnerable population groups in developed countries. The five elements of the DOTS strategy are sustainable government commitment, quality assurance of sputum microscopy, standardized short-course treatment (including DOTS), regular supply of drugs, and establishment of reporting and recording systems (Box 4.20). The strategy of DOTS uses CHWs to visit the patient, and observe him or her taking the various medications, providing incentive, support, and moral coercion to complete the needed 6–8 month therapy. DOTS has been shown to cure up to 95 percent of cases, at a cost of as little as US$11 over the period of treatment per patient. It is one of the few hopes of containing the current TB pandemic.

The goal of this approach is to reduce TB morbidity and mortality and the chance of *M. tuberculosis* developing resistance to primary treatment drugs. Target goals of TB control

---

BOX 4.20 Control of Tuberculosis Using Directly Observed Therapy, Short Course (DOTS)

DOTS remains at the heart of the Stop TB Strategy. The following are crucial components of a national TB program as an internationally recommended strategy for TB control that has been recognized as a highly efficient and cost-effective strategy. DOTS comprises five components:

1. Political commitment with increased and sustained financing – legislation, planning, human resources, training.
2. Case detection through quality-assured bacteriology – strengthening TB laboratories, drug resistance surveillance.
3. Standard treatment with supervision and patient support – TB treatment and program management guidelines, International Standards of TB Care (ISTC), PPM, Practical Approach to Lung Health (PAL), community–patient involvement.
4. An effective drug supply and management system – availability of TB drugs, TB drug management, Global Drug Facility (GDF), Green Light Committee; regular, uninterrupted supply of high-quality anti-TB drugs.
5. Monitoring and evaluation system and impact measurement.

Sources: *World Health Organization. Tuberculosis (TB). Pursue high-quality DOTS expansion and enhancement. Available at: http://www.who.int/tb/dots/en/ and http://www.who.int/tb/dots/whatisdots/en/index.html [Accessed 27 January 2013].*

---

adopted in 1991 by the WHA include at least a 70 percent detection rate of the estimated incidence of sputum smear-positive pulmonary tuberculosis (PTB+) and a cure rate of 85 percent or higher for newly detected PTB+ cases. The 85 percent or higher cure rate was adopted on the basis of accumulated experience in Africa and certain districts of China.

Performance indicators surrounding the DOTS program use the proportion detected of PTB+, which is the most infectious form of TB. PTB+ is associated with high mortality and is the most effective form of TB to use for bacteriological monitoring of treatment progress. The proportion of newly detected PTB+ cases among the total number of adults with PTB reflects the proper application of diagnostic criteria. In countries with a medium or high TB burden, when necessary laboratory resources are available and sputum smears for microscopy are administered to TB patients, PTB+ accounts for more than 50 percent of all TB cases and over 65 percent of new PTB cases in adults. Achieving a high (i.e., ≥85 percent) cure rate for PTB+ is a critical priority for TB control programs. Failure to achieve this rate results in continued infectiousness and possible development of MDR-TB, characterized by resistance to at least isoniazid and rifampin.

Even under adverse conditions, DOTS produces excellent results. It is one of the most cost-effective health interventions combining public health and clinical medical approaches. It proves most efficacious among patients in poor self-care settings, such as homeless people, drug users, and those with AIDS.

In 2006, the WHO rededicated itself to TB control with the "Stop TB Strategy" for control of TB over the next decade. The plan calls for new guidelines for control, new aid funds for developing countries, and enlistment of NGOs to assist in the fight. The new guidelines stress short-term chemotherapy in well-managed programs of DOTS, emphasizing strict compliance with therapy for infectious cases with a goal of an 85 percent cure rate. The primary goals of the Stop TB Strategy are to reduce TB incidence and mortality by 50 percent by 2015, relative to 1990 rates, and to eliminate TB as a public health problem by 2050.

TB control remains feasible with current medical and public health methods. Deterioration in its control should not lead to despair and passivity. The recent trend towards successful control by DOTS, despite the growing problem of MDR-TB, suggests that control and gradual reduction can be achieved by an activist, community outreach approach. In 2006, the WHO reaffirmed TB control as one of its major priorities, expressing grave concern that the MDR organism, now widespread in countries of Asia, Eastern Europe, and the former Soviet Union, may spread the disease much more widely. The disease constitutes one of the great challenges to public health. Extremely drug-resistant tuberculosis (XDR-TB) has become a central concern in addressing the current TB epidemic and is part of a WHO-led strategy in this field.

To manage a high and increasing burden of TB in Kazakhstan, in 1998 the Ministry of Health adopted and implemented a new National Tuberculosis Program, the objectives and target goals of which are in accord with the DOTS strategy. Primary health care physicians and TB specialists have received training in case detection, and laboratories have been equipped with binocular microscopes. Unfavorable treatment outcomes for new TB+ cases were associated with alcohol abuse, homelessness, previous incarceration, unemployment, being male, and urban residence.

The epidemic curve peaked in 1998 and has since been in continuous decline. Treatment of MDR-TB is costly and complex, but has become an essential part of international TB work. The standardized mortality rates from TB for Kazakhstan, the Commonwealth of Independent States, the Central Asian Republics (Uzbekistan, etc.), the old EU countries and the new EU (Eastern European) countries are shown in Figure 4.9.

Streptococcal Diseases

Acute infectious diseases caused by group A streptococci include streptococcal sore throat, scarlet fever, puerperal fever, septicemia, erysipelas, cellulitis, mastoiditis, otitis media, pneumonia, peritonsillitis (quinsy), wound infections, toxic shock syndrome, and fasciitis, the "flesh-eating bacteria". *Streptococcus pyogenes* group A includes some 80 serologically distinct types which vary in geographic

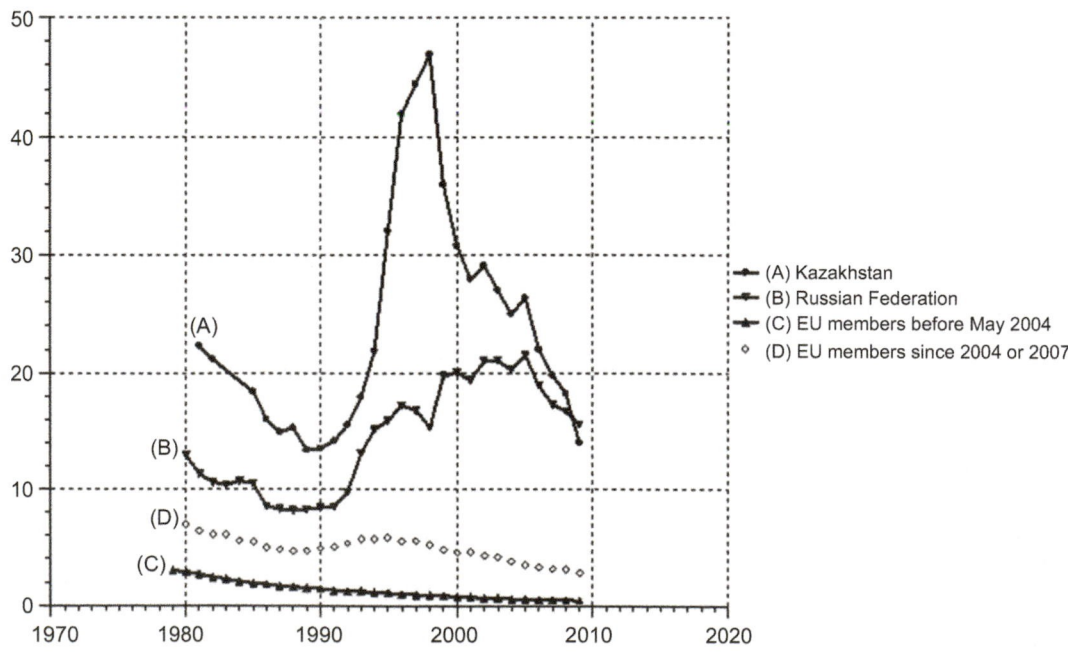

FIGURE 4.9 **Standardized death rates (per 100,000 population) from tuberculosis, Kazakhstan and selected European regional groupings, 1970–2010.** *Source: World Health Organization European Region. Health for All data base. August 2012. Available at: www.who.euro.HFADB [Accessed 28 January 2013].*

location and clinical significance. Transmission is by droplet, person-to-person direct contact, or food infected by carriers. Important complications from a public health point of view include acute rheumatic fever and acute glomerulonephritis, as well as skin infections and pneumonia.

Acute rheumatic fever (ARF) is a complication of *Streptococcus* A infection that has virtually disappeared from industrialized countries as a result of improved standards of living and antibiotic therapy. Mortality rates from rheumatic fever and rheumatic heart disease have declined steadily over the past three decades, largely due to the increased availability and use of antibiotics. In developing nations and lower socioeconomic areas where rheumatic fever is more prevalent, ARF is a major cause of death and disability in children and adolescents. Moreover, outbreaks were recorded in the USA in 1985, and an increasing number of cases has been seen since 1990. In developing countries, rheumatic fever remains a serious public health problem affecting school-age children, particularly those in crowded living arrangements. Long-term sequelae include disease of the mitral and aortic heart valves, which require cardiac care and surgery for repair or replacement with artificial valves.

Acute glomerulonephritis is a reaction to toxins of the streptococcal infection in the kidney tissue. It can result in long-term kidney failure and the need for dialysis or kidney transplantation. This disease has become far less common in the industrialized countries, but remains a public health problem in developing countries.

Group B streptococci (GBS) are related organisms. They commonly colonize the reproductive tract of women of reproductive age, and are the leading cause of meningitis in newborn infants. As with other strains of beta-hemolytic streptococci, treatment with penicillin (or appropriate therapy for allergic patients) is effective. Women should be screened for GBS at 35–37 weeks of pregnancy and treated during labor and delivery. If screening tests are unavailable, the risk of infection is high, thus recommendations are to treat prophylactically.

The streptococcal diseases are controllable by early diagnosis and treatment with antibiotics. This is a major function of primary care systems. Recent increases in rheumatic fever may herald a return of the problem, perhaps due to inadequate access to primary care in the USA for large sectors of the population, along with crowding and possibly poor access to medical care due to a lack of or inadequate health insurance.

Where access to primary care services is limited, infections with streptococci can result in a heavy burden of chronic heart and kidney disease with substantial health, emotional, and financial tolls. Measures to improve access to care and public information are needed to ensure rapid and effective care to prevent chronic and costly conditions.

Zoonoses

Zoonoses are infectious diseases transmissible from vertebrate animals to humans. Common examples of zoonoses of public health importance in non-industrialized countries include brucellosis and rabies. In industrialized countries, salmonellosis, mad cow disease, and influenza have reinforced the importance of relationships between animal and human health. Strong cooperation between public health and veterinary public health authorities is required to monitor and to prevent such diseases. Zoonoses have been described and recognized over many centuries. They involve several types of agents: bacteria, parasites, viruses, and unconventional agents. Bacterial organisms transmitted by animals include salmonellosis and campylobacteriosis, anthrax, brucellosis, *E. coli*, leptospirosis, plague, shigellosis, and tularemia.

Viruses transmitted by animals include rabies, which is a disease of carnivores and bats mainly transmissible to humans by bites. Almost all people severely exposed to rabid animals will die if not treated. An estimated 55,000 people, predominantly children, die of this disease in the world every year. Control measures focus on immunization of domestic animals and household pets. Infected dog-bite transmission is responsible for most human deaths.

Other viral zoonoses include avian influenza, Crimean–Congo hemorrhagic fever, Ebola, and RVF. Bovine spongiform encephalopathy (BSE) is thought to be the cause of variant Creutzfeldt–Jakob disease (vCJD), which is a neurological disease different from CJD, leading to death in humans.

Other important zoonoses are brucellosis and echinococcosis/hydatidosis. The class of zoonoses still represents significant and often neglected public health threats, affecting hundreds of thousands of people, particularly in developing countries. Schmallenberg virus, which affects cattle, is moving into new countries in Europe (for example, it was recently imported from France into Poland), indicating once again that disease agent transmission is a potent issue in public health affecting animals. This is of great economic importance in itself but in time it can also affect humans. Despite the heavy burden they represent and their potential to transfer to human diseases, many zoonotic diseases are preventable with professional veterinary public health measures, which are an essential part of general public health.

Brucellosis

Brucellosis is a disease occurring in cattle (*Brucella abortus*), dogs (*B. canis*), goats and sheep (*B. melitensis*), and pigs (*B. suis*). Humans are affected mainly through ingestion of contaminated milk products, by contact, or by inhalation. Brucellosis (also known as relapsing, undulant, Malta, or Mediterranean fever) is a systemic bacterial disease of acute or insidious onset characterized by fever, headache, weakness, sweating, chills, arthralgia, depression, weight loss, and generalized malaise. Transmission can occur as a result of contact with tissues, blood, urine, vaginal discharges; however, brucellosis is predominantly spread by ingestion of raw milk and dairy products from infected animals. The disease may last from a few days to a year or more. Complications include osteoarthritis and relapses. Case fatality is under 2 percent, but disability is common and can be pronounced.

The disease is primarily seen in Mediterranean countries, the Middle East, India, central Asia, and Central and South America. Brucellosis occurs primarily as an occupational disease of people working with and in contact with tissues, blood, and urine of infected animals, especially goats and sheep. It is an occupational hazard for farmers, veterinarians, packing house workers, butchers, tanners, and laboratory workers. It is also transmitted to consumers of unpasteurized milk from infected animals. Because animal vectors include wild animals, eradication of the disease is virtually impossible. Diagnosis is confirmed by laboratory findings of the organism in blood or other tissue samples, or with rising antibody titers in the blood, with confirmation by blood cultures.

Clinical cases are treated with antibiotics. Epidemiological investigation may help to track down contaminated animal flocks. Routine immunization of animals, monitoring of animals in high-risk areas, quarantining sick animals, destroying infected animals, and pasteurizing milk and milk products all serve as important actions to prevent spread of the disease. Control measures include educating farmers and the public not to use unpasteurized milk. Individuals who work with animals (cattle, swine, goats, sheep, dogs, coyotes) should take special precautions when handling animal carcasses and materials. Testing animals, destroying carriers, and enforcing mandatory pasteurization will restrict the spread of the disease. This is an economic as well as a public health problem, requiring full cooperation between ministries of health and of agriculture.

Rabies

Rabies is primarily a disease of animals, with a variety of wild animals serving as a reservoir for the disease, including foxes, wolves, bats, skunks, and raccoons; these wild animals may infect domestic animals such as dogs, cats, and farm animals. Animal bites break the skin or mucous membrane, thus allowing entry of the virus from the infected saliva into the bloodstream. The incubation period of the virus is 2–8 weeks; it can be as long as several years or as short as 5 days. Accordingly, postexposure preventive treatment is a public health emergency.

The clinical disease often begins with a feeling of apprehension, headache, and pyrexia, followed by muscle spasms, acute encephalitis, and death. Both fear of water ("hydrophobia") and fear of swallowing are characteristics of the disease. Rabies is almost always fatal within a week of onset of symptoms. There is no effective treatment, thus control relies on vaccination of animals, rapid prophylaxis of exposed people, and prevention of contact with biting and scratching animals. The disease is estimated to cause 30,000 deaths annually, primarily in developing countries. It is uncommon in developed countries.

Rabies control focuses on prevention in humans, domestic animals, and wildlife. Prevention in humans is based on pre-exposure prophylaxis for groups at risk (e.g., veterinarians, zoo workers) and postexposure immune globulin and vaccine administration for people bitten by potentially rabid animals. Because reducing exposure of pets to wild animals is difficult, immunization of domestic animals is one of the most important preventive measures. Prevention in domestic animals is by mandatory immunization of household pets. All domestic animals should be immunized at 3 months of age and revaccinated according to veterinary instructions.

Prevention in wild animals to reduce the reservoir is successful in achieving local eradication in settings where re-entry from neighboring settings is limited. Since 1978, the use of oral rabies immunization has been successful in reducing the population of wild animals infected by the rabies virus. Rabies eradication efforts, using aerial distribution of baits containing fox rabies vaccine in affected areas of Belgium, France, Germany, Italy, and Luxembourg, have been under way since 1989. The number of rabies cases in these affected areas has declined by some 70 percent. Switzerland is now virtually rabies free because of this vaccination program. However, the WHO Collaborating Centre for Rabies Surveillance and Research reports that rabies cases in 2012 increased by more than 300, to a total of 6185 cases, mainly due to the large increase in cases in Poland.

The potential exists for local eradication, especially on islands or in partially restricted areas with limited possibilities for wild animal entry. Livestock need not be routinely immunized against rabies, except in high-risk areas. In regions in which bats are major reservoirs of the disease, as in the USA, eradication is not currently feasible (WHO Collaborating Center for Rabies Surveillance and Research, 2012).

Salmonella

Salmonella, discussed later in this chapter under the classification of diarrheal diseases, is one of the most common infectious diseases among animals. It is easily spread to humans via poultry, meat, eggs, and dairy products. Transmission may also occur from contact with infected animals, particularly reptile pets. Specific antigenic types are associated with foodborne transmission to humans, causing generalized illness and gastroenteritis. The severity of the disease varies widely, and salmonella can be devastating among vulnerable populations, such as young children and elderly or immunocompromised people. Epidemiological investigation of common food source outbreaks may uncover hazardous food handling practices. Laboratory confirmation or serotypes help in monitoring the disease. Prevention is achieved by maintaining high standards of food hygiene in processing, inspection and regulation, food handling practices, and hygiene education.

Anthrax

Bacillus anthracis causes a bacterial infection in herbivorous animals and its spores contaminate soil worldwide. It predominantly affects humans exposed in occupational settings. Transmission is cutaneous by contact, gastrointestinal by ingestion, or respiratory by inhalation. In recent decades it has gained attention as a highly potent agent for germ warfare or terrorism, as in Iraq in 1997. In 2001, anthrax was used as a bioterror agent against the USA. Twenty-two people were infected, with a 50 percent case mortality rate.

Although most *B. anthracis* strains are susceptible to common antibiotics, concern over the possible existence of weaponized, antibiotic-resistant anthrax has prompted extensive planning to counter the possibility of terrorist or other attacks. Limited supplies of vaccine are available; however, in the absence of an epidemic, its use is only justified for veterinarians, key public health workers, soldiers, and laboratory personnel with a higher risk of exposure.

Creutzfeldt–Jakob Disease

Creutzfeldt–Jakob Disease (CJD) is a rare degenerative disease of the CNS linked to consumption of beef from cattle infected with BSE. It is transmitted by prion proteins in animal feed prepared from contaminated animal material and in transplanted organs. This disease was identified in the UK linked to infected cattle. This noteworthy case led to a 1997 ban on British beef in many parts of the world as well as the slaughter of large numbers of potentially contaminated animals. Between 2000 and 2011, the UK accumulated a total of 175 reported cases of vCJD, and there were 49 in other countries, but few cases of the disease have been reported since 2000. This disease is still under study with surveillance and follow-up by veterinary services among cattle, and under public health watch for human cases.

Other Major Zoonotic Diseases

The tapeworm causing diphyllobothriasis (*Diphyllobothrium latum*) is widespread in North American freshwater fish, passing from crustacean to fish to humans by consumption of raw freshwater fish. It is especially common among Inuit peoples and may be asymptomatic or cause severe general and abdominal disorders. Food hygiene (freezing and cooking of meat) is recommended; treatment is by anthelmintics.

Leptospiroses are a group of zoonotic bacterial diseases found worldwide in rats, raccoons, and domestic animals. They affect farmers, sewer workers, dairy and abattoir workers, veterinarians, military personnel, and miners with transmission via exposure to or ingestion of urine-contaminated water or tissues of infected animals. Disease is often asymptomatic or mild, but may cause generalized illness like influenza, meningitis, or encephalitis. Prevention requires education of the public in self-protection and immunization of workers in hazardous occupations, along with immunization and segregation of domestic animals and control of wild animals.

VECTORBORNE DISEASES

Vectorborne diseases are a group of diseases in which the infectious agent is transmitted to humans by crawling or flying insects. The vector is the intermediary between the reservoir and the host. Both the vector and the host may be affected by climatic conditions; mosquitoes thrive in warm, wet weather, and are suppressed by cold weather; and humans may wear less protective clothing in warm weather.

Malaria

The only important reservoir of malaria is humans. Its mode of transmission is from person to person via the bite of an infected female *Anopheles* mosquito (Ronald Ross, Nobel Prize 1902). The causative organism is a single-cell parasite with four species: *Plasmodium vivax*, *P. malariae*, *P. falciparum*, and *P. ovale*. Clinical symptoms are produced by the parasite invading and destroying red blood cells. The incubation period is approximately 12–30 days, depending on the specific *Plasmodium* involved. Some strains of *P. vivax* may have a protracted incubation period of 8–10 months, and even longer for *P. ovale*. The disease can also be transmitted through infected blood transfusions. Confirmation of diagnosis is by demonstrating malaria parasites on blood smears.

Falciparum malaria, the most serious form, presents with fever, chills, sweats, and headache. It may progress to jaundice, bleeding disorders, shock, renal or liver failure, encephalopathy, coma, and death; prompt treatment is essential. Case fatality rates in untreated children and adults are above 10 percent. An untreated attack may last for 18 months. Other forms of malaria may present as a nonspecific fever. Relapse of *P. ovale* malaria may occur up to 5 years after initial infection; malaria may persist in chronic form for up to 50 years.

Malaria control advanced during the 1940s to 1960s through improved chloroquine treatment and the use of

DDT for vector control, with optimism for eradication of the disease. However, control regressed in many developing countries as allocations for environmental control, case finding, and treatment were reduced. Moreover, the world saw an increase in drug resistance, thus this disease is now recognized as a central public health problem globally. The need for a vaccine for malaria control is now more apparent than ever.

Globally, 225 million people are infected with malaria each year. Of the hundreds of thousands of deaths resulting from malaria, more than two-thirds are concentrated in sub-Saharan Africa, and a large proportion strike children. Up to 50 percent of health expenditures are attributed to treatment of malaria patients. Large areas, particularly in forest or savannah regions with high rainfall, are holoendemic; this means that nearly all individuals residing in the region are infected. At higher altitudes, endemicity is lower, but epidemics do occur. Chloroquine-resistant *P. falciparum* has spread throughout Africa, accompanied by an increasing incidence of severe clinical forms of the disease. The World Bank estimates that 11 percent of all disability-adjusted life years (DALYs) lost per year in sub-Saharan Africa are from malaria, which places a heavy economic burden on a country's health system.

In the Americas, the number of cases detected has risen every year since 1974, and the WHO estimated there to have been 2.2–2.5 million cases in 1991. The nine most endemic countries in the Americas achieved a 60 percent reduction in malaria mortality between 1994 and 1997. In 2002, CDC reported that of the 1337 malaria cases in the USA, all but five were imported, i.e., acquired in malaria-endemic countries.

Malaria kills more nearly 800,000 people annually and infects 200–300 million (WHO, 2011). Trends show an apparent increase in the number of global malaria cases, from 233 million to 244 million from 2000 to 2005, respectively. This figure then fell to 225 million in 2009, and malaria deaths dropped from 985,000 in 2000 to 781,000 in 2009.

Sub-Saharan Africa is by far the hardest hit region, with approximately 90 percent of the deaths. Among the most vulnerable and afflicted are children: a child dies from malaria every 45 seconds. An enormous burden causing unspeakable harm, malaria is responsible for approximately 20 percent of all child mortality in Africa. Moreover, there is an increase in strains resistant to the major available drugs and of mosquitoes resistant to the insecticides in use.

Vector control, case finding, and treatment remain the mainstay of control. Use of insecticide-impregnated bed nets and curtains, residual house spraying, and strengthened vector control activities are important, as are early diagnosis and carefully monitored treatment with evaluating for resistance. Control of malaria will ultimately depend on a safe, effective, and inexpensive vaccine (Box 4.21). Attempts to develop a malaria vaccine have been unsuccessful to date owing to the large number of genetic types of *P. falciparum* even in localized areas. Twenty-three prospective

---

BOX 4.21 The Long Wait for a Malaria Vaccine

The World Health Organization has indicated that it could recommend the first malaria vaccine, RTS,S/AS01 *Plasmodium falciparum* vaccine, for use in some African countries as early as 2015; the full phase III trial results will be available in 2014. The vaccine has been developed by a public–private partnership primarily for use in infants and young children in sub-Saharan Africa. It is a hybrid construct of the hepatitis B surface antigen fused with a recombinant antigen derived from part of the malaria circum-sporozoite protein. This is the protein coat of the sporozoite, the parasite stage inoculated by the feeding anopheline mosquito; it then invades liver cells and multiplies there before entering the bloodstream. Many other potential malaria vaccines are in various stages of development, but the RTS,S/AS01 vaccine is closer to registration and potential deployment than the others (White, 2011).

Malaria vaccine development depends on financial support of major drug companies and donors such as the Gates Foundation. Great improvements have been achieved using existing methods of vector and transmission control prevention such as insecticide treated bed nets, use of DDT for targeted use in indoor spraying, reduction of still water breeding sites for carrier mosquitoes, *seasonal chemo-prevention and artesunate for severe cases to reduce mortality (White, 2011).*

In 2013, GlaxoSmithKline in the UK is seeking approval of a vaccine which has shown encouraging results and safety in a large field trial in Africa with 25% reduction in infant malaria cases in Phase III clinical trials.

A safe, effective, and affordable malaria vaccine will have an enormous impact on the world. As the tenacious scientists, researchers, and investigators persist, the rest of the world anticipates the day this vaccine is announced: it will be like the day the first man landed on the moon. The WHO is optimistic that a safe and effective new vaccine is within sight.

Sources: *White NJ. A vaccine for malaria [editorial]. N Engl J Med 2011;10. 1056/nejme1111777.*
RTS Clinical Trials Partnership. First results of phase 3 trial of RTS,S/AS01 malaria vaccine in African children. N Engl J Med 2011;365:1863–75 (10.1056/NEJMoa1102287).
World Health Organization. Initiative for Vaccine Research (IVR). June 2013. http://www.who.int/vaccine_research/links/Rainbow/en/ [Accessed 22 October 2013].
World Health Organization. WHO reports: Immunization, vaccines and biologicals. Promising results from clinical trial of advanced malaria vaccine candidate. Available at: http://www.who.int/immunization/newsroom/newsstory_malaria_vaccine_trial_results/en/index.html [Accessed 14 February 2013].

P. falciparum vaccines are currently in clinical trials, with some reported effectiveness. Research into vaccines for malaria has also been hampered by the fact that it is a relatively low priority for vaccine manufacturers because of the minimal potential for financial benefit. Because of increasing drug resistance, research on malaria has concentrated on the pharmacological aspects of the disease. Effective control of malaria will require both new drugs for resistant infections and primary prevention through vector control. Larvicides are used with some success, including bacteria which destroy larvae and genetic modification of mosquitoes to reduce their fertility, with hopes for eventual vaccine development.

In 1998, the WHO initiated a campaign to "Roll Back Malaria" and maintain the dream of eradication in the future; malaria is included in MDG6 and Rollback Malaria for the period 2006–2015. Effective low-technology interventions include community-based case finding, early treatment with good-quality insecticides, and vector control. The use of CHWs and widespread provision of insecticide-impregnated bed nets in endemic areas has shown promising results. Between 2008 and 2010 insecticide-treated nets protected 578 million people (10 percent of the at-risk population). DDT was banned in many countries; however, it is recommended by the WHO for limited uses such as in-house spraying but not for wide environmental insecticide use. Local control and even eradication can be achieved with currently available technology. This requires the integration of public health and clinical approaches with strong political commitment internationally and nationally in the affected countries. After decades of work in many centers, in 2013 there are hopeful signs of an effective vaccine for malaria emerging, but this will take more years of research and clinical trials to become an effective instrument for control of this still deadly disease.

Rickettsial Infections

The Rickettsiae are obligate parasites; they can only replicate in living cells, but otherwise they have characteristics of bacteria. They represent a class of clinically similar diseases, usually characterized by severe headache, fever, myalgia, rash, and capillary bleeding causing damage to brain, lungs, kidneys, and heart. Identification is by serological testing for antibodies, but the organisms can also be cultured in laboratory animals, embryonic eggs, or cell cultures. The organisms are transmitted by arthropod vectors such as lice, fleas, ticks, and mites. The diseases caused millions of deaths during war and famine periods before the advent of antibiotics.

This group of diseases exists in nature in ways that make them impossible to eradicate; however, clinical diagnosis, host protection, and vector control can help to reduce the burden of disease and control any outbreaks that may occur.

Public education regarding self-protection, appropriate clothing, tick removal, and localized control measures such as spraying and habitat modification are effective.

Epidemic typhus, first identified in 1836, is due to *Rickettsia prowazekii*. Spread primarily by the body louse, typhus was the cause of an estimated 3 million deaths, especially during war and famine, in Poland and the Soviet Union from 1915 to 1922. Untreated, the fatality rate ranges from 5 to 40 percent. Typhus responds well to antibiotics. It is currently largely confined to endemic foci in Central Africa, Central Asia, Eastern Europe, and South America. It is preventable by hygiene, anti-mosquito measures, judicious use of pesticides such as DDT and lindane, bed nets, vector control, and protective clothing. A vaccine is available for exposed laboratory personnel.

Murine typhus is a mild form of typhus due to *Rickettsia typhi*, which is found worldwide and spread in rodent reservoirs. Scrub typhus, also known as Tsutsugamushi or Japanese river fever, is located throughout the Far East and the Pacific islands, and represented a serious health problem for US armed forces in the Pacific during World War II. It is spread by *Rickettsia tsutsugamushi* and has a wide variation in case fatality according to region, organism, and age of the patient.

Rocky Mountain spotted fever is a well-known and deadly form of tick-borne typhus due to *Rickettsia rickettsii*, occurring in western North America, Europe, and Asia. Q fever is a tick-borne disease caused by *Coxiella burnetii* and is worldwide in distribution, usually associated with farm workers, in both acute and chronic forms. Regular antitick spraying of sheep, cows, and goats helps to protect exposed workers. Protective clothing and regular removal of body ticks help to protect exposed people.

Arboviruses (Arthropod-Borne Viral Diseases)

Arthropod-borne viral diseases are caused by a diverse group of viruses which are transmitted between vertebrate animals (often farm animals or small rodents) and people by the bite of blood-feeding vectors such as mosquitoes, ticks, and sandflies. Transmission also occurs through direct contact with infected animal carcasses. Usually the viruses have the capacity to multiply in the salivary glands of the vector, but some are carried mechanically in their mouth parts.

These viruses cause acute CNS infections (meningoencephalitis), myocarditis, or undifferentiated viral illnesses with polyarthritis and rashes, or severe hemorrhagic febrile illnesses. The most important of the arbovirus diseases are yellow fever, dengue, and a wide group of encephalitic diseases, such as eastern and western equine encephalitides, Japanese encephalitis, Murray Valley encephalitis, and another group which includes West Nile fever, and others with exotic names like Rift Valley fever, Chikungunya, African swine fever, Crimean–Congo hemorrhagic fever, Powassan virus, Ppataci fever, and many others. Arbovirus diseases are often asymptomatic in vertebrates but may be severe in humans.

Venezuelan, Japanese, and Murray Valley encephalitides

Over 250 antigenetically distinct arboviruses are associated with disease in humans, varying from benign fevers of short duration to severe hemorrhagic fevers. Each has a characteristic historic and geographic location, and vector-specific clinical and virological characteristics. They can spread globally via travelers and become endemic in new regions, as ecological conditions and mass travel enable disease transfer and endemicity in newly suitable environments with increased vector presence, such as the appropriate mosquito population.

They are of international public health importance because of the potential for spread via natural phenomena and modern rapid transportation of vectors and people incubating or ill with the disease, with potential for further spreading at the point of destination. Key preventive measures include vector control to reduce mosquito breeding, mosquito nets, and individual protection with protective clothing and the use of insect repellents against mosquitoes and ticks.

Encephalitides

Arboviruses are responsible for a large number of encephalitic diseases characterized by mode of transmission and geographic area. Mosquito-borne arboviruses causing encephalitis include eastern and western equine, Venezuelan, Japanese, and Murray Valley encephalitides. Japanese encephalitis is caused by a mosquito-borne arbovirus found in Asia and is associated with rice-growing areas. It is characterized by headache, fever, convulsions, and paralysis, with fatality rates in severe cases as high as 60 percent. A currently available vaccine is used routinely in endemic areas (Japan, Korea, Thailand, India, and Taiwan) and for people traveling to infected areas. Tick-borne arboviruses causing encephalitis include the Powassan virus, which occurs sporadically in the USA and Canada. Tick-borne encephalitis is endemic in Eastern Europe, Scandinavia, and the former Soviet Union.

West Nile Virus

West Nile Virus (WNV) was first discovered in Uganda in 1937 and in the Nile Delta in 1953. Outbreaks subsequently occurred in the 1990s in Algeria and in Romania, and in Israel in 1997. Prior to 1997, the virus was not recognized

as pathogenic for birds; however, scientists' understanding of the disease changed, as during this time in Israel, birds of various species died from a more virulent strain of the virus; these infected birds showed signs of paralysis and encephalitis.

WNV is a potentially severe disease typically spread by the bites of infected mosquitoes, which become infected by feeding on birds that carry the virus. *Culex pipiens* mosquitoes are recognized as the primary vectors of the virus. The virus remains in mosquito populations via vertical transmission, or transmission from adults to eggs; birds remain the reservoir hosts of WNV. Thus, the virus survives in nature through a mosquito–bird–mosquito cycle. Humans are at risk of acquiring the virus though the bite of infected mosquitoes. WNV can also rarely be spread through organ transplants, blood transfusions, breastfeeding and pregnancy.

WNV may cause deadly neurological disease in humans, although 80 percent of those infected with the virus show no symptoms at all. Despite this, there is no way to determine whether a person will develop an illness or not. Up to 20 percent of those who are infected will develop West Nile fever and show symptoms such as headache, body ache, nausea, swollen lymph nodes, and skin rash. The duration of these symptoms ranges from a few days to many weeks. A much smaller proportion (approximately one out of 150) of individuals infected with WNV represent a subset of people whose condition will advance, causing them to become severely ill. Corresponding symptoms include high fever, stiff neck, disorientation, coma, tremors, convulsions, loss of vision, numbness, and paralysis. The neurological effects may become permanent. Most commonly, symptoms begin to affect people approximately 3–14 days following the bite of an infected mosquito.

The disease has become endemic in many regions of the world (Box 4.22) and no WNV vaccine currently exists for humans. There is no specific treatment for WNV infection. The mild symptoms experienced by patients will pass by themselves, but patients experiencing the more severe symptoms and neuroinvasive WNV typically need to be hospitalized. Treatment is supportive and includes administration of intravenous fluids, respiratory assistance, and prevention of further infections. Individuals over the age of 50 incur the highest risk of WNV complications and death. Moreover, chronic kidney disease may occur in convalescent patients.

Many of the control measures revolve around vector control, which includes draining still waters where mosquitoes can breed, spraying with antimosquito compounds, and repairing window and door screens. Furthermore, it is very important that people use mosquito repellents and wear long sleeves and long pants (trousers) outdoors, particularly at dusk and dawn, when mosquitoes are most active. Laboratory confirmation is by serological testing. Monitoring the serology of dead birds or horses which are

BOX 4.22 West Nile Virus Outbreak 2012

West Nile Virus (WNV) is a potentially fatal neurological infection transmitted to people as well as horses and other mammals by infected mosquitoes and maintained in nature by transmission by birds through mosquitoes. The disease was first identified in the West Nile region of Uganda in 1937 and is now considered endemic in Africa, the Middle East, North America, Europe including Russia, the Mediterranean and Middle Eastern countries, and Australasia. Most cases are subclinical or with mild flu-like symptoms and rash. The virus is also reported in parts of Europe, e.g., north-eastern Italy and Serbia (37 cases with three deaths). Greece, Israel, Romania, Russia, and the USA represent countries which have experienced the largest outbreaks.

West Nile fever was first seen in the USA in 1999 in New York City. US populations are seeing a major resurgence of the virus, as the number of infected individuals has increased sharply, particularly in August 2012. The USA has now seen animal and human cases in nearly every state.

In 2012 CDC reported the USA to be "in the midst of one of the largest West Nile virus outbreaks ever seen in the United States". As of 11 December 2012, WNV was reported in 48 states, with 5387 cases including 243 deaths. Most cases (80 percent) were reported in 13 states, but one-third of cases were concentrated in Texas. In 2013, there were 2,059 reported cases (to 29 October) in the USA, of which 49 percent were neuroinvasive, mostly in the north east coast, the Midwest, California and Texas.

The time and place of outbreaks are affected by weather, the flight pattern of birds that sustain the virus, the number of mosquitoes that spread the virus, and human behavior. Atypically mild winters and hot summers experienced by many states may promote conditions for the spread of the virus to humans. International transportation of the WNV by bird vectors has led to importation and establishment of vectorborne pathogens outside their usual habitat. This is a serious global threat which may worsen as climatic changes favor vector proliferation.

Sources: *Centers for Disease Control and Prevention. West Nile virus. Available at: http://www.cdc.gov/ncidod/dvbid/westnile/index.htm [Accessed 6 February 2013].*
Barzon L, Pacenti M, Franchin E, Martello T, Lavezzo E, Squarzon L, et al. Clinical and virological findings in the ongoing outbreak of West Nile virus Livenza strain in northern Italy, July to September 2012. Euro Surveill 2012;17(36):pii=20260.
Centers for Disease Control and Prevention. West Nile Virus. Available at: http://www.cdc.gov/westnile/statsMaps/preliminaryMapsdata/index.html [Accessed 2 November 2013].
World Health Organization. West Nile virus. Fact sheet no. 354; July 2011. Available at: http://www.who.int/mediacentre/factsheets/fs354/en/index.html [Accessed 6 February 2013].

frequently infected via the mosquito vector may give the first indication of local appearance of the disease. In addition, the use of strategically located sentinel chicken flocks has been very effective in determining the geographic distribution of WNV and predicting local risks for infection. When birds test positive in a new area, health care providers are alerted to the signs and symptoms of WNV, increasing the effectiveness of surveillance, early intervention, and prevention. This highly successful model may potentially be applied to other zoonotic diseases.

Chikungunya

Chikungunya fever is a disease caused by a virus (alphavirus, in the family Togaviridae) spread by the bite of infected mosquitoes. Chikungunya fever has no specific treatment and care is symptomatic but the disease is not usually fatal. It is mainly located in over 40 countries in Africa, India, and South-East Asia, causing a severe dengue-like illness that is mostly non-fatal. Insect repellents, appropriate clothing, and staying in areas with screens are standard protective measures.

The disease has spread to Europe with outbreaks in France and later Italy following importation from India by a single traveler. This disease is common in former French colonies (e.g., Réunion) and it has become common in parts of France along with its major vectors *Aedes aegypti* and *A. albopictus*.

Because of outbreaks of Chikungunya fever such as that in Italy in 2007, with over 197 cases, and in France in 2010, with 76 suspected cases reported and 32 confirmed by laboratory analysis, two of which were indigenous cases, concerns have been raised that it may become endemic in Europe, especially with climatic warming and laxity in mosquito control measures.

Rift Valley Fever

Rift Valley fever (RVF) is a hemorrhagic fever whose virus (*Phlebotomus* genus, Bunyaviridae family) is spread by mosquitoes and other insect vectors. The virus mainly affects ruminant animals (e.g., cattle, sheep, and goats), resulting in hemorrhage, abortion, and death: it causes universal abortion in ewes and a high percentage of death in lambs. It also affects humans who have been in direct contact with the meat or blood of affected animals. RVF virus causes a generalized illness in humans, and can advance to encephalitis, hemorrhage, retinitis and retinal hemorrhage leading to partial or total blindness, and death (1–2 percent).

The virus's normal habitat is in the Rift Valley of East Africa, often spreading to southern Africa, depending on climatic conditions. The primary reservoir and vector is the *Aedes* mosquito. RVF was first identified in Kenya, near the great Syrian–African rift which stretches from South-East Africa to the Nile valley and the Red Sea, along the Arava and Jordan valleys, up to the Bekah Valley in Lebanon and Syria.

The disease has been known since the 1930s as a health hazard for cattle and sheep. Veterinary services were well developed in British Rhodesia and South Africa. An effective veterinary vaccine was produced to protect domestic animals. A vaccine for humans was developed by the US Army at Fort Dietrich for biological warfare defense, but in limited supply.

An unusual spread of RVF northward to the Sudan and along the Aswan Dam reservoir to Egypt in 1977–1978 caused hundreds of thousands of animal deaths, as well as 18,000 human cases and 598 deaths. RVF appeared again in Egypt in 1993. This disease is suspected to have been one of the 10 plagues of Egypt leading to the exodus of the Children of Israel from Egypt during pharaonic–biblical times.

Preventive measures taken in Israel to prevent entry of RVF included the immunization of 1.5 million domestic animals, from dairy cattle to Bedouin sheep and goat flocks in Israel, the West Bank, Gaza, and the Sinai Peninsula, to create a *cordon sanitaire* of protected animals. Laboratories were prepared to identify any suspected animal or human cases. Public health and veterinary staff of the Ministry of Agriculture were given training and guidelines to handle a public health emergency alert situation, lest the RVF virus become established in the country and endanger animals and humans alike, with possible spread throughout the region and into Europe.

In 1997, an outbreak of RVF occurred in Kenya, initially thought to be anthrax, with hundreds of cases and dozens of deaths, related to an abnormal rainy season and vector conditions. Satellite monitoring of rainfall and vegetation is being used to predict epidemics in Kenya and surrounding countries. Animal immunization, monitoring, vector control, and reduced contact with infected animals can limit the spread of this disease.

RVF has reappeared in the Middle East in Yemen and Saudi Arabia since 2000 and may have become endemic in the region. Box 4.23 summarizes unexpected RVF outbreaks in Mauritania in 2010 and 2012.

Renewed interest in RVF stems from its movement away from traditional habitats to the Middle East and potentially to Europe. This disease is of great economic importance because of its potential impact on animal husbandry. It is also a threat to human public health and its spread to new pastures in Africa and Europe may accelerate with changing climatic conditions. Imported suspected cases have been reported in France, in a tourist to Zimbabwe, and in French soldiers returning from duty in Chad.

RVF has recently reappeared in East Africa, including Sudan, the Nile Valley, and countries near the Indian Ocean. RVF is very active and sensitive to climate and other environmental as well as socioeconomic changes, such as occurred with the Nasser Dam in the 1970s. Ecological changes and growing human populations with increased demand for meat promote greater movements of livestock, controlled and uncontrolled. This increases the risk of spread of RVF in the Mediterranean basin, Central Europe, and the Middle East (Chevalier et al., 2010), as has occurred with other arboviruses. This provides an incentive for intensive research into the vaccinology of RVF virus (Dar et al., 2013). RVF was in the past considered a potentially seriously disruptive biological warfare agent. The potential methods and effects of deliberate or inadvertent introduction of RVF

BOX 4.23 Rift Valley Fever in Mauritania, 2010 and 2012

On 4 October 2012, the Mauritanian Ministry of Health announced an outbreak of Rift Valley fever (RVF). The first (index) case was identified in mid-September, and 1 month later 24 cases, including 13 deaths, were reported. By the end of October, the number of RVF cases had risen to 34 and 17 deaths in six regions of the country. Laboratory tests on human and human samples with enzyme-linked immunosorbent assay and polymerase chain reaction at the National Veterinary Research Laboratory confirmed that RVF virus was circulating in various regions.

Mauritania had a previous unanticipated outbreak in September and October 2010, after exceptional amounts of rainfall. The formation of large ponds resulted in oases in the Saharan region of Adrar. Atypical growth of vegetation drew in shepherds from distant regions, and provided optimal conditions for mosquito growth in masses, mostly *Culex* and *Anopheles*. In the weeks following the heavy rainfall, major outbreaks of both malaria and RVF occurred in many oases.

The first documented case in livestock was an infected camel. A herdsman slaughtered the animal and ate the uncooked meat, sharing it with others. A few days later, those who had consumed the camel meat had severe intestinal and hemorrhagic symptoms and died. Laboratory tests confirmed RVF virus.

By the end of 2010, 63 human cases including 13 deaths due to RVF had been confirmed, although the true RVF morbidity and mortality was likely to be much higher in these remote regions with unrecorded cases and deaths.

RVF outbreaks have long afflicted regions in Kenya, where the virus is endemic, and have moved to new regions further north, including Sudan, Egypt (1976–1977), and more recently Yemen, Aden, and Saudi Arabia (2000). Camels and small rodents are thought to have served a crucial role in transmission of RVF from northern Sudan to southern Egypt in 1977.

Sources: World Health Organization. Rift Valley fever in Mauretania, 1 November 2012. Available at: http://www.who.int/csr/don/2012_11_01/en/index.html
Chevalier V, Pepin M, Plee L, Lancelot R. Rift Valley fever – a threat for Europe? Euro Surveill 2010;15(10):pii=19506. Available at: http://www.eurosurveillance.org/ViewArticle.aspx?ArticleId=19506 [Accessed 3 January 2013].
Hartley DM, Rinderknecht JL, Nipp TL, Clarke NP, Snowder GD, National Center for Foreign Animal and Zoonotic Disease Defense Advisory Group. Potential effects of Rift Valley fever in the United States. Emerg Infect Dis [serial on the Internet]. August 2011. Available at: http://dx.doi.org/10.3201/eid1708.101088 [Accessed 8 February 2013].

into the USA via international movement of livestock or as a terrorist act come under the watchful eye of the CDC and security agencies (Hartley et al., 2011).

Hemorrhagic Fevers

Arboviruses can also cause hemorrhagic fevers, which are acute febrile illnesses. They are characterized by extensive hemorrhagic phenomena (internal and external), liver damage, shock, and often high mortality rates. The potential for international transmission is high.

Yellow Fever

Similar to a number of other infectious diseases examined in this chapter, yellow fever is transmitted via infected mosquitoes. It is characterized as an acute viral disease of short duration and varying severity with jaundice, hence the name "yellow" fever. The mosquito is the chief vector of yellow fever, which transmits the virus from host to host, typically from monkeys to humans.

Once the virus is transmitted to a human, it typically incubates for 3–6 days. Subsequently, the infection can follow one of two pathways. It can enter an acute phase, in which typical symptoms include fever, muscle pain, backache, headache, and nausea. For the majority of patients, their condition improves and these symptoms are gone a few days later.

The other possible route the infection can take represents a more toxic phase; generally 15 percent of patients infected with the yellow fever virus enter this phase as well. Multiple body systems are affected, the high fever reappears, and the patient is suddenly struck with jaundice, severe abdominal pain, and vomiting. As the disease advances, bleeding from the mouth, nose, eyes, or stomach may occur. Kidney function may weaken, causing an emergency situation. Approximately 50 percent of patients who experience this toxic phase die within 10–14 days; the other 50 percent recuperate without suffering from major organ damage.

The case fatality rate is 5 percent in endemic areas, but may be as high as 50 percent in non-endemic areas and during epidemics. It has caused major epidemics in the Americas in the past, but was successfully controlled by elimination of the vector, *A. aegypti*. As no specific treatment for yellow fever currently exists, supportive care is effective in treating dehydration and fever. While this form of care can be successful in improving the condition of severely ill patients, it is unlikely to be available in poor, low-resource regions.

Undoubtedly, vaccination is the most significant, effective preventive measure against yellow fever. To effectively prevent outbreaks in affected areas, experts agree that vaccination uptake must reach at least 60–80 percent of the population at risk. To improve vaccination rates throughout endemic areas, preventive vaccination can be implemented via routine infant immunizations as well as one-off mass campaigns. Also recommended for those traveling to infected areas, the live attenuated yellow fever vaccine is regarded as safe and affordable. One single dose offers protection from yellow fever for 30–35 years, and perhaps for life. Other important preventive measures include mosquito control and epidemic preparedness, which refers to rapid detection and blood tests for yellow fever capable of being carried out at a national laboratory.

Furthermore, determining the mode of transmission and vector control of yellow fever played a major role in the development of public health (see Chapter 1). The WHO reports that annually there are 200,000 cases of yellow fever, resulting in 30,000 deaths globally. Originally imported to the Americas from Africa, yellow fever is endemic in 45 countries in Africa and Latin America. In Africa, over 500 million individuals reside in 32 endemic countries.

Other populations are at risk in 13 countries in Latin America; Ecuador, Brazil, Bolivia, Columbia, and Peru represent the countries at highest risk of yellow fever. Cases of yellow fever have been increasing over the past two decades owing to declining population immunity to infection in conjunction with environmental factors such as deforestation, urbanization, and global travel.

Dengue and Severe Dengue

Dengue fever is a serious influenza-like illness, capable of advancing and leading to a fatal complication called severe dengue (formerly referred to as dengue hemorrhagic fever). An acute sudden-onset viral disease, dengue is transmitted by *Aedes* mosquitoes; humans contract the disease via bites of infected female mosquitoes. Unlike most other mosquitoes, *A. aegypti* feeds during the day: early morning and evening time before sunset account for its peak biting times. Dengue surveillance shows that dengue is occurring more frequently, mostly in tropical and subtropical climates (Box 4.24), with a dramatic increase globally as shown in Figure 4.10.

Dengue fever is characterized by 3–5 days of high fever, intense headache, myalgia, arthralgia, nausea, vomiting, and rash. Severe dengue is a lethal complication that may develop as a result of dengue fever. It may occur due to any of the following conditions: fluid accumulation, plasma leakage, respiratory problems, severe bleeding, or organ impairment. Warning signs that a person has developed severe dengue include continuous vomiting, rapid breathing, severe abdominal pain, bleeding gums, and the presence of blood in vomit. According to the WHO, these warning signs arise 3–7 days after the initial symptoms and they coincide with a reduction in fever. Medical treatment with fluid replacement infusions is essential to avoid further illness and death, as the next 1–2 days of this period can be fatal. To date, no particular treatment for dengue exists; however, for severe dengue the supportive medical treatment can drastically reduce mortality rates from over 20 percent to below 1 percent.

Scientists have recognized four particular, closely associated serotypes of the virus that lead to dengue. Thus, patients recovering from an infection by one serotype acquire a lifelong immunity protecting against that specific one. Nevertheless, following recovery, immunity against the other three serotypes is incomplete and temporary. Further infections by more than one serotype raise a patient's risk of acquiring severe dengue. Similar to other vectorborne

BOX 4.24 Dengue and Its Global Burden

Dengue is a mosquito-borne virus that is now endemic in over 100 countries, with a significant increase in the global incidence in recent decades. Found mostly in tropical and subtropical climates, dengue predominantly affects urban and semi-urban regions. Severe dengue was seen in the USA in the nineteenth century: in Charleston, South Carolina, in 1828; Savannah, Georgia, in 1850; and 16,000 cases in Austin, Texas, in 1885; further outbreaks occurred in the 1920s, and "breakbone fever", as it was called, affected almost 10 percent of the population of Miami, Florida, in the 1940s. After World War II, mosquito control with DDT reduced outbreaks in the USA.

Dengue virus is thought to have transferred from monkeys to humans hundreds of years ago, but remained localized geographically until it emerged as a worldwide problem in the 1950s, probably as a result of mass movement of goods and people with mosquitoes, when epidemics were seen in the Philippines and Thailand. From the 1980s, large numbers of cases began to appear in the Caribbean and Latin America, following declines in vector control programs from the 1970s. Dengue is now a global public health problem in Asia, the Pacific, the Americas, and Caribbean countries. Severe dengue represents a major cause of hospitalization, morbidity, and mortality, particularly among children living in these regions.

The WHO reports that more than 2.5 billion individuals, or 40 percent of the global population, are currently at risk for developing dengue fever. Estimates suggest that 50–100 million cases of dengue occur annually worldwide. The substantial increase in global evidence is illustrated in Figure 4.10. Dengue stands as the world's most significant mosquito-borne viral disease. Incidence is currently 30 times higher than that of 50 years ago. Furthermore, prior to 1970, severe dengue epidemics occurred in only nine countries. In 2012, dengue was endemic in over 100 countries in Africa, the Americas, and the Eastern Mediterranean; however, South-East Asian and Western Pacific countries are most dramatically affected.

Dengue thrives in tropical and subtropical areas of developing countries owing to environmental conditions that promote mosquito multiplication and viral transmission by *Aedes*

aegypti. Conditions in which mosquitoes flourish include rapid population growth, movement between rural and urban regions, and water being stored in containers close to homes owing to poor infrastructure with insufficient or inconsistent water supply. A high volume of solid waste also contributes to mosquito concentration and reproduction, as neglected and discarded objects contain water which serves as a larval habitat and breeding ground.

In addition to the increase in incidence and spread of endemic countries, the WHO warns of the potential for explosive outbreaks. There is now risk of potential dengue fever outbreaks occurring in Europe, as local transmission was officially detected in both Croatia and France in 2010. Moreover, imported cases have been recorded in three more countries in Europe. Most dengue cases in the USA were acquired abroad, but outbreaks have been reported in Florida, Texas, Puerto Rico, Alaska, and Hawaii in recent years. As a result of the rapid spread of the disease, the WHO has created a central data management system, DengueNet, as a standardized epidemiological system that can effectively monitor dengue.

Approximately half a million patients infected with severe dengue fever require hospitalization annually and an estimated 2.5 percent of those infected do not survive. Globalization, international commercial trade, the rise in air travel, and in many high-risk areas the lack of efficient vector control methods, all play major roles in the global burden of dengue and severe dengue. The global resurgence of dengue and the development of hyperendemicity in many regions reflect a level of concern that necessitates action and makes this disease a major public health challenge. In 2013, Thailand reported 7033 cases and five deaths in the first 5 weeks of the year: fives time higher than during the same period in 2012.

Sources: *World Health Organization. Dengue and severe dengue. Fact sheet no. 117; November 2012. Available at: http://www.who.int/mediacentre/factsheets/fs117/en/*
World Health Organization. Global Alert and Response (GAR): impact of dengue. WHO; 2013. Available at: http://www.who.int/csr/disease/dengue/impact/en/
Centers for Disease Control and Prevention. Dengue. Available at: http://www.cdc.gov/dengue/ [All accessed 25 February 2013].

diseases, prevention of dengue relies on human behavioral practices that limit contact with mosquitoes. These measures include disposing of solid waste appropriately, consistently cleaning water storage containers, and using household protective measures such as window screens, vaporizers, and insecticides. Epidemics of dengue fever can be explosive; however, as mentioned, adequate treatment can greatly reduce the number of deaths.

Other Hemorrhagic Fevers

Lassa Fever

Lassa fever is an acute viral hemorrhagic fever caused by the Lassa virus, first isolated in Lassa, Nigeria, in 1969. It is widely distributed in West Africa and is responsible

for 200,000–400,000 cases and 5000 deaths annually. It is spread by direct contact with blood, urine, or secretions of infected rodents as well as by direct person-to-person contact in hospital and laboratory settings. The disease is characterized by a persistent or spiking fever for 2–4 weeks, and may include severe hypotension, shock, and hemorrhaging. The case fatality rate is 15 percent. A global network of WHO Collaborating Centers in many parts of the world works together and with WHO member states to investigate and manage outbreaks (WHO, 2012).

Marburg Disease

Marburg disease is a rare viral disease causing a hemorrhagic fever. It was discovered in 1967 in an epidemic in Marburg, Germany, and Serbia, following exposure to African green

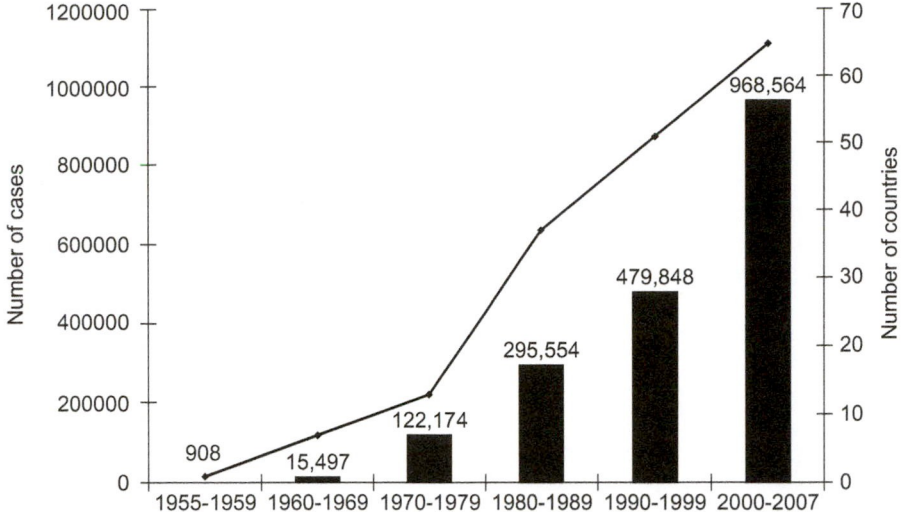

FIGURE 4.10   Dengue and severe dengue fever, globally, 1955–2007. *Source: WHO Global Alert and Response: impact of dengue. WHO; 2012. Available at: http://www.who.int/csr/disease/dengue/impact/en/ [Accessed 12 January 2013].*

monkeys, with 32 cases and seven deaths among laboratory workers and their families and caregivers. It has a sudden onset of generalized illness, malaise, fever, myalgia, headache, diarrhea, vomiting, rash, and hemorrhage. Person-to-person transmission occurs via blood, secretions, organs, and semen. Confirmed cases have been identified in a number of countries in Africa. The virus is carried by the fruit bat, which is widely distributed globally. Case fatality rates vary between 23 and 90 percent. Treatment is symptomatic support, and prevention is mainly focused on guidelines for the safety of health workers in African hospitals to prevent transfer of the disease to health workers.

Ebola Fever

Ebola is a virus that leads to severe viral hemorrhagic fever outbreaks, characterized by a particularly high case fatality rate of up to 90 percent. It is described by the WHO as "one of the most virulent viral diseases known to humankind". As an acute viral infection, Ebola hemorrhagic fever (EHF) typically leads to a sudden onset of fever, weakness, and extreme muscle pain and headache. Subsequent major symptoms include vomiting, diarrhea, and complications pertaining to liver and kidney functions, often with both internal and external bleeding. Laboratory results indicate low levels of white blood cells and platelets and elevated liver enzyme levels.

Ebola virus initially affects humans through contact with wild animals, specifically their secretions, organs, and blood. Fruit bats originating from the Pteropodidae family are recognized as the natural host of the virus; specifically in Africa, infections have occurred through contact with infected chimpanzees, gorillas, monkeys, fruit bats, forest antelopes, and porcupines. Typically, contact with these animals, which have been found dead or infected in the rainforest, leads to further transmission. The infection

subsequently enters a community through human-to-human transmission. Humans are capable of transmitting the virus to others when it is present in body fluids, i.e., blood and secretions. The incubation period ranges from 2 to 21 days and the case fatality varies depending on the outbreak, from 25 to 90 percent. A common route of transmission is through health workers treating Ebola patients; this carries a high risk if proper control measures are not taken and sufficient barrier nursing procedures not followed. Moreover, widespread African practices of the washing and displaying of dead bodies facilitate the spread of this deadly hemorrhagic fever. This can be exemplified through burial ceremonies in which mourners have direct contact with a dead body.

Since the virus was discovered, an estimated 1850 cases with more than 1200 deaths have occurred (Box 4.25). While there is no treatment or vaccine for this virus, it is crucial that patients receive intensive supportive care, as many are severely dehydrated and require oral rehydration or intravenous fluids. Prevention is critical and should be implemented through various measures, such as routine disinfection of monkey farms; detergents such as sodium hypochlorite are capable of inactivating Ebola virus. Prevention of the virus in humans should revolve around educating community members on the risk factors of acquiring the infection as well as protective methods, such as the use of gloves. This is absolutely imperative to all, but especially for health care workers, laboratory workers, and those caring for ill patients at home. Educational messages must be spread to populations afflicted by Ebola, providing information about the disease itself as well as methods to control an outbreak. Measures include practicing timely and safe burials of community members who have died from Ebola. This disease is considered highly dangerous unless outbreaks are effectively controlled. Once identified, an Ebola epidemic becomes an international emergency; public health workers

BOX 4.25 Ebola Hemorrhagic Fever Outbreaks

Outbreaks of Ebola hemorrhagic fever (EHF) predominantly affect populations in secluded villages in West and Central Africa, particularly those close to tropical rainforests. EHF cases were first documented in 1976 in two concurrent outbreaks, one occurring in Sudan and the other in the Democratic Republic of the Congo (DRC). Both were characterized as major and devastating outbreaks. Of the 284 cases in the Sudan outbreak, 151 deaths occurred, representing a case fatality rate of 53 percent. Even more disastrously, in the outbreak occurring in DRC, 318 fell ill and 280 died, a case fatality rate of 88 percent.

The nature of these outbreaks is not the same, as they were caused by two different virus subtypes. In total, there are five recognized Ebola virus subtypes; each has been named based upon the area in which it was initially identified in an Ebola outbreak.

Uganda experienced an outbreak in 2007 of a new strain of EHF (Bundibugyo) with a lower case fatality rate (17 of 43 cases or a 40 percent case fatality rate). In August 2012 an outbreak of EHF occurred in Kibaale district with 24 probable and confirmed cases, including 16 deaths. Of the total, 10 cases were confirmed through the Uganda Virus Research Institute in Entebbe. Some suspect cases had negative laboratory results, so 43 individuals were treated symptomatically then released from the isolation facility. They received counseling by psychosocial teams with guidance on reintroduction to the community. Even patients confirmed negative for the EHF virus need counseling to mitigate fears, diminish stigma, and improve their acceptance back into society. All contacts of probable and confirmed cases are monitored and evaluated daily for 21 days. Red Cross volunteers and village health teams created social mobilization teams, visiting households in the most afflicted regions of the district and communicating health awareness messages. As a precaution, countries neighboring Uganda are working to improve surveillance of the Ebola virus.

Another Ebola outbreak in DRC began on 3 September 2012, with 28 cases including 14 deaths in the Haut-Uele district, in Orientale Province. Of these, eight were confirmed, six considered probable, and the remaining 14 are suspect cases. Fatal cases in one of the two affected health zones include three health care providers. As of 24 September 2012, the number of documented Ebola cases in DRC has increased to 51; of these, 19 were laboratory confirmed and 32 probable cases; of the total, 20 cases were fatal.

Epidemiologists and logisticians from the WHO Regional Office for Africa as well as from WHO headquarters and from the CDC assisted the Ministry of Health in the outbreak response, social mobilization, active case finding, contact tracing, improved surveillance and monitoring, and case management with strengthened infection control.

The recent and simultaneous outbreaks in Uganda and DRC are considered to be epidemiologically unrelated and no specific travel warnings have been issued. However, the lower case fatality rates seen recently may leave more survivors to transmit this alarming disease.

Sources: *World Health Organization. Global Alert and Response. Ebola haemorrhagic fever. Available at: http://www.who.int/csr/disease/ebola/en/index.html [Accessed 8 February 2013].*
Shoemaker T, MacNeil A, Balinandi S, Campbell S, Wamala JF, McMullan LK, et al. Reemerging Sudan Ebola virus disease in Uganda, 2011. Emerg Infect Dis 2012. Available at: http://dx.doi.org/10.3201/eid1809.111536 [Accessed 2 September 2012].
Source: World Health Organization. CSR. Disease outbreak news; 27 September 2012 [edited]. Available at: http://www.who.int/csr/don/2012_09_27/en/index.html

from across the world are involved in control and intervention through WHO- and CDC-directed projects.

Lyme Disease

Lyme disease is characterized by the presence of a rash, as well as musculoskeletal, neurological, and cardiovascular symptoms. Confirmation is by laboratory investigation. It is the most common vectorborne disease in the USA. It primarily affects children in the 5–14-year age group and adults aged 35–55. Lyme disease is preventable by avoiding contact with ticks and by applying insect repellent.

Lyme disease infects some 24,000 Americans per year but the true incidence may be three times higher or more. Risk is highest in the north-east, north-central, and mid-Atlantic regions (Box 4.26). Although localized in 13 states, Lyme disease ranked sixth on a list of the leading US nationally notifiable diseases in 2011. Several US manufacturers have developed vaccines but these are not used owing to concerns about adverse event reporting and tracking.

NEGLECTED TROPICAL DISEASES

NTDs are a group of 17 parasitic diseases high rates of associated mortality and morbidity mostly in tropical countries. WHO estimates they affect nearly 1 billion persons in 77 countries, and CDC estimates that NTDs kill an estimated 534,000 people worldwide annually, mainly in 149 countries and territories with high burdens of years of life lost due to premature disability and death. Innovative and Intensified Disease Management (IDM) involves surveillance, capacity building, advocacy and research (Table 4.11). In 2012, the NTD goals and commitments of the London Declaration on Neglected Tropical Diseases was agreed to by key public and private stakeholders.

Medically important parasites are animals that live in, take nourishment from, and thrive in the body of a host; which may or may not harm the host, but never bring benefit. They include unicellular organisms such as protozoa (malaria, *Giardia*, amebiasis, and *Cryptosporidium*), and helminths (worms), which are categorized as nematodes, cestodes, and trematodes (Box 4.27).

BOX 4.26 Lyme Disease

In the mid-1970s, a mother of two young boys who were recently diagnosed with arthritis in the town of Old Lyme, Connecticut, conducted a private investigation among other town residents. She mapped each of the six arthritis cases in the town – cases which had occurred in a short time span among boys living in close proximity. Her findings suggested that this syndrome of "juvenile rheumatoid arthritis" was perhaps connected with the boys playing in the woods. She presented her data to the head of rheumatology at Yale Medical School in New Haven, who investigated this "cluster of a new disease entity". Some parents reported that their sons had experienced tick bites and a rash before onset of the arthritis. A tick-borne, spiral-shaped bacterium, a spirochete, *Borrelia burgdorferi*, was identified as the organism, and *Ixodes* ticks were shown to be the vector. Cases respond well to antibiotic therapy.

Lyme disease has been identified in many parts of North America, Europe, the former Soviet Union, China, and Japan. Personal protection by clothing and insect repellent applications to protect from tick bites, especially in forest areas, and environmental modification are important to limit spread of the disease.

Sources: *Centers for Disease Control and Prevention. Notice to readers, recommendations for test performance and interpretation from the second national conference on serologic diagnosis of Lyme disease. MMWR Morb Mortal Wkly Rep 1995;44:590–1. Available at: http://www.cdc.gov/mmwr/preview/mmwrhtml/00038469.htm*
Centers for Disease Control and Prevention. Lyme disease [updated 1 October 2012]. Available at: http://www.cdc.gov/lyme/ [Accessed 28 January 2013].
Bacon RM, Kugeler KJ, Mead PS. Surveillance for Lyme disease – United States, 1992–2006. MMWR Surveill Summ 2008;57(SS10):1–9. Available at: http://www.cdc.gov/mmwr/preview/mmwrhtml/ss5710a1.htm [Accessed 15 February 2013].

Box 4.28 discusses NTDs and their presence in the USA, predominantly among impoverished populations. Moreover, parasitic diseases such as malaria are among the most common causes of illness and death in the world. Milder illnesses such as giardiasis and trichomoniasis cause widespread morbidity. Intestinal infestations with worms may cause severe complications, although they commonly cause chronic low-grade symptomatology and iron-deficiency anemia. Deworming every 6 months has become an effective part of the Expanded Programme of Immunization (EPI-plus) along with vitamin A supplementation and insecticide-impregnated bed nets for children. The paradigm shift in global approaches to control of diseases originating at the animal–human interface requires strong support from highly professional veterinary public health.

Echinococcosis

Echinococcosis (hydatid cyst disease) is infection with *Echinococcus granulosus*, a small tapeworm commonly found in dogs. The tapeworm forms unilocular (single, non-compartmental) cysts in the host, primarily in the liver and lungs, but they can also grow in the kidney, spleen, CNS, or bones. Cysts, which may grow up to 10 cm in size, may be asymptomatic or, if untreated, may cause severe symptoms and even death. This parasite is common where dogs are used with herd-grazing animals and also have intimate contact with humans.

The Middle East, Greece, Sardinia, North Africa, and South America are endemic areas, as are a few areas in the USA and Canada. The human disease has been eliminated in Cyprus and Australia. While the dog is the major host, intermediate hosts include sheep, cattle, pigs, horses, moose, and wolves. Preventive measures include education in food and animal contact hygiene, destroying wild and stray dogs, and keeping dogs away from the viscera of slaughtered animals.

A similar, but multilocular, cystic hydatid disease is widely found in wild animal hosts in areas of the northern hemisphere, including Central Europe, the former Soviet Union, Japan, Alaska, Canada, and the north–central USA. Another echinococcal disease (*Echinococcus vogeli*) is found in South America, where its natural host is the bush dog and its intermediate host is the rat. The domestic dog also serves as a source of human infection.

Surgical resection is not always successful, and long-term medical treatment may be required. Control is through awareness and hygiene as well as the control of wild animals that come in contact with humans and domestic animals. Control may require cooperation between neighboring countries.

Tapeworm

Tapeworm infestation (taeniasis) is common in tropical countries where hygienic standards are low. Beef (*Taenia saginata*) and pork (*T. solium*) tapeworms are common where animals are fed with water or food exposed to human feces. *Taenia solium* is especially deadly; delay in diagnosis and treatment may lead to severe disease, including neurological cysticercosis. In developing countries, infection is associated with pork consumption, while in the USA, several epidemics have occurred from eating carnivorous game animals such as mountain lions and bears. Freezing or cooking meat, especially that of pigs and carnivorous mammals, is essential to destroy the tapeworm. Fish tapeworm (*Diphyllobothrium latum*) is common in populations living primarily on uncooked fish, such as Inuit, Eastern European, and Scandinavian. These tapeworms are usually associated with northern climates.

Toddlers are especially susceptible to dog tapeworm (*Dipylidium caninum*), which is present worldwide, and domestic pets are often the source of fecal–oral transmission of the eggs. The disease is usually asymptomatic. Similarly, dwarf tapeworm (*Hymenolepis nana*) is transmitted through fecal–oral contamination from person to person, or via contaminated food or water. Rat tapeworm (*Hymenolepis diminuta*) also mostly affects young children.

BOX 4.27 Control of Neglected Tropical Diseases

WHO reports that at least 1 billion people suffer from one or more of the 17 major diseases which affect the poorest, most vulnerable people mainly in tropical and subtropical areas of the world. Some diseases affect individuals throughout their lives, causing a high degree of morbidity, social stigmatization and abuse. The diseases include: Buruli ulcer, cysticercosis, dracunculiasis (guinea-worm disease), foodborne trematode infections (such as fascioliasis), hydatidosis, leishmaniasis, lymphatic filariasis, onchocerciasis, schistosomiasis, soil-transmitted helminthiasis, trachoma, trypanosomiasis, and vectorborne diseases including dengue fever. Several other conditions included in this category are: mycetoma, pdoconiosis, scabies, snakebite and strongyloidiasis.

For a large group of these diseases – mainly helminthic infections, effective, inexpensive or donated drugs are available for prevention and cure which when used on a large scale, are able to reduce and locally eradicate these diseases. Leprosy treatment with effective antibiotics is leading to its elimination, although recent local increases are seen in Cambodia of this disabling disease. Cost-effective treatment for yaws can lead to elimination and final eradication of this debilitating disease which causes severe deformation. In the case of blinding trachoma, the use of the recommended strategy (SAFE) of an effective antibiotic is leading to its local eradication where applied vigorously.

A second group of NTDs are diseases for which the main option currently available is systematic case-finding and early case management. These include Buruli ulcer, Chagas disease, and other diarrheal diseases, human African trypanosomiasis, and leishmaniasis. Simple diagnostic tools and safe and effective treatment regimens need to be developed urgently for some of these diseases. However, even for these infections, systematic use of the present, albeit imperfect tools at an early stage can dramatically reduce mortality and morbidity. For others, vector control tools are available and present the main method of transmission control, as in the case of Chagas disease.

There are examples of great successes in the fight against both of these groups of NTDs. Since 1985, cases of leprosy have been reduced by over 90 percent to 213,000 in 2009 with prevention of disabilities in 1-2 million people through use of multi-drug

therapy. Before the start of the Guinea-worm Eradication Programme in the early 1980s, an estimated 3.5 million people in 20 endemic countries were infected with the disease. The number of reported cases declined from some 10,000 in 2005 to 3,190 in 4 endemic countries (Ethiopia, Ghana, Mali and Sudan) in 2009 and the program is moving towards eradication in 2015. Onchocerciasis control measures have freed more than 25 million hectares of previously onchocerciasis-infected land available for resettlement and agricultural cultivation, thereby improving development prospects in Africa and Latin America. Increased awareness and advocacy are needed to draw attention to the realistic prospect of reducing the negative impact of NTDs on the health and social and economic well-being of affected communities, (WHO, 2010) with progress toward final elimination. Large-scale, regular treatment plays a central role in the control of many NTDs such as filariasis, onchocerciasis, schistosomiasis, and soil-transmitted nematode infections. For example, regular prophylactic chemotherapy against intestinal worms reduces mortality and morbidity in preschool children, improves the nutritional status and academic performance of schoolchildren, and improves the health and well-being of pregnant women and their babies.

In 2013, WHO is targeting eradication of dracunculiasis by 2015 and that of yaws by 2020. Worldwide, 149 countries and territories are affected by at least one neglected tropical disease (NTD). This includes setting six targets for the elimination of five neglected tropical diseases by 2015; a further 10 elimination targets are set for 2020 either globally or in selected geographical areas for nine neglected tropical diseases. Targets are also set for intensified control of dengue, Buruli ulcer, cutaneous leishmaniasis, selected zoonoses and helminthiases. This will require close cooperation between international agencies, national governments and international donors. With adequate support great progress can be made in the coming decade on these debilitating and often lethal conditions.

Source: *World Health Organization. Neglected tropical diseases. Available at: http://www.who.int/neglected_diseases/2010report/en/index.html [Accessed 26 January 2013] and WHO. http://www.who.int/neglected_diseases/EB_resolution_2013/en/index.html [Accessed 24 October 2013].*

Onchocerciasis

Onchocerciasis, known as river blindness, is caused by a parasitic worm capable of producing millions of larvae which move through the body causing intense itching, debilitation, and eventually blindness. The disease is spread via the bites of blackflies belonging to the genus *Simulium*; the blackflies transmit the larva from infected individuals to those uninfected.

Leading to visual impairment and debilitating skin disease, onchocerciasis is primarily located in sub-Saharan Africa and in Latin America. By December 2009, over 112,000 million individuals in Africa were at risk for onchocerciasis. Approximately 70 percent of this population lives in only five of the 24 countries in which this disease is endemic: Nigeria, DRC, Cameroon, Ethiopia, and Sudan. Earlier the same year, 73.7 percent of the population at risk received ivermectin treatments and, consequently, it

is estimated that the burden of this disease has dropped significantly. Control is by a combination of activities including environmental control by larvicidal sprays to reduce the vector population, protection of potential hosts by protective clothing and insect repellents, and case treatment.

A WHO-initiated program for onchocerciasis control started in 1974 and is sponsored by four international agencies: the Food and Agriculture Organization (FAO), UNDP, World Bank, and WHO. It covers 11 countries in sub-Saharan Africa, focusing on control of the blackfly by destroying its larvae, mainly via insecticides sprayed from the air. The Vision 2020 program of the WHO aims for control of river blindness by the year 2020. The program has been successful in protecting some 30 million people and helping 1.5 million infected people to recover from this disease.

The WHO estimates that the program prevented 500,000 cases of blindness by 2000 and has freed 25 million hectares

BOX 4.28 Neglected Tropical Diseases and Poverty in the USA

Public health continues to face the problems of parasitic diseases in the developing world but, increasingly, parasitic diseases are being recognized in industrialized countries as well. Giardiasis and *Cryptosporidium* infections in water and other outbreaks have occurred in the USA.

The First National Summit on Neglected Infections of Poverty in the USA held in Washington DC in 2009 was attended by public health experts, public policy leaders, and government officials. A follow-up workshop on the "neglected infections of poverty" (NIPs) and neglected tropical diseases (NTDs) was held in Washington in 2010.

Globally, there is no common definition of NTDs. NIPs, which include a range of NTDs, occur among poor groups living in wealthy countries such as the USA and Canada, and in Europe. In the USA, these are found in areas of the Mississippi Delta, post-Katrina Louisiana, along the border of Mexico, and in the Appalachians. These are largely made up of minority groups including African Americans, Hispanics, and Native Americans. The major US NTDs include dengue fever, toxocariasis, and Chagas disease.

There are an estimated 300,000 cases of Chagas in the USA with 30,000–45,000 cardiomyopathy cases and 63–315 congenital infections annually (Bern and Montgomery, 2009). The disability-adjusted life years (DALYs) associated with these NTDs in the USA are similar to those of HIV/AIDS, malaria, or tuberculosis in low- and middle-income countries.

NIPs and NTDs have a low profile and status in public health priority, often with poor statistics, but are well known in the public health community. Surveillance and monitoring to provide more precise data on morbidity, mortality, and transmission rates need strengthening. NIPs and NTDs are frequently unrecognized or misdiagnosed in the USA often because the people do not or cannot seek health care or the health care providers have not had training in these diseases. Led by Congressman Hank Johnson Jr, Georgia introduced legislation, the Neglected Infections of Impoverished Americans Act 2010 (HR 5896), for the Department of Health and Human Services to collect additional information on these "neglected" diseases.

Sources: *Editorial. Fighting neglected tropical diseases in the southern United States. BMJ 2012;345:e6112. Available at: http://www.bmj.com/content/345/bmj.e6112 [Accessed 23 February 2013].
Institute of Medicine. The causes and impacts of neglected tropical and zoonotic diseases: opportunities for integrated intervention strategies. Washington, DC: National Academies Press; 2011. Available at: http://www.ncbi.nlm.nih.gov/books/NBK62507/ [Accessed 23 February 2013].
Hotez PJ. Neglected infections of poverty in the United States of America. PLoS Negl Trop Dis 2008;2(6):e256. http://dx.doi.org/10.1371/journal.pntd.0000256. Available at: http://www.plosntds.org/article/info:doi/10.1371/journal.pntd.0000256#pntd-0000256-t003 [Accessed 23 February 2013].
Centers for Disease Control and Prevention. Emerging infectious diseases national summit on neglected infections of poverty in the United States. Available at: http://wwwnc.cdc.gov/eid/article/16/5/09-1863_article.htm [Accessed 23 February 2013].
Bern C, Montgomery SP. An estimate of the burden of Chagas disease in the United States. Clin Infect Dis 2009;49(5):e52–4. Available at: http://cid.oxfordjournals.org/content/49/5/e52.long [Accessed 24 February 2013].*

of land for resettlement and cultivation. The program cost US$570 million. This investment is considered by the World Bank to have a return of 16–28 percent in terms of large-scale land reuse and improved output of the population.

A WHO program, the African Program for Onchocerciasis Control (APOC), established in 1996, includes ivermectin and selective vector control efforts by spraying for the blackfly. This program involves 30 countries in Africa and six in a similar program in South America. APOC has utilized rapid epidemiological mapping techniques to determine populations at risk, and thus provide community-directed treatment. Mapping strategies allow experts to understand levels of endemicity of each area. Moreover, while onchocerciasis is unlikely to be eliminated from Africa in the foreseeable future, emphasis on interrupting transmission of the disease is vital and certainly making progress.

Dracunculiasis

Dracunculiasis (Guinea worm disease) is a parasitic disease of great public health importance in India, Pakistan, and Central and West Africa. It is an infection of the subcutaneous and deeper tissues caused by a large (60 cm) nematode roundworm, *Dracuculiasis medenisis*; it typically affects the lower extremities, causing pain and disability. This

disease is an infestation of the body via contaminated drinking water that contains the larvae of the parasite.

The parasite grows and mates in the intestine and over a year after infection, begins to emerge from the body via painful skin lesions. Most commonly, the victim bathes in stagnant waters to soothe the pain of the lesion. As a result, the worm, which may be up to 90 cm in length, emerges from the ulcer and releases new larvae to the water source. The larvae are ingested by water fleas which then can transmit the disease to new victims who drink the contaminated water.

Prevention is based on improving the safety of water supplies and preventing contamination by infected people. Education of people in endemic areas cautions individuals not to enter water sources. Promotion of filtering drinking water to reduce transmission is also essential. Insecticides remove the crustaceans and chlorine kills the larvae and crustaceans, which prologue larval infectivity. Box 4.29 discusses progress towards eradicating dracunciliasis.

Schistosomiasis

Schistosomiasis is a parasitic infection caused by the trematode (blood fluke) and transmitted from person to person via an intermediate host, the snail. It is endemic in 74 countries in Africa, South America, the Caribbean, and Asia. There

BOX 4.29 Progress Towards Global Eradication of Dracunculiasis

Dracunculiasis was traditionally endemic in the belt from West Africa through the Middle East to India and Central Asia. It was successfully eliminated from Central Asia and Iran, and has disappeared from the Middle East and some African countries (Gambia and Guinea).

Worldwide prevalence was reduced from 12 million cases in 1980 to 3 million in 1990 and 3,190 cases in 2009 with 187 countries certified as free of the disease by WHO. India's reported cases fell from 17,000 in 1987 to 900 in 1992, and the country was free from transmission in 1997. Similarly, formerly high-prevalence countries such as Kenya reported no cases in 1997. Major progress can be attributed to the World Health Organization's (WHO's) forceful promotion of eradicating dracunculiasis. In 1986 the World Health Assembly called for the elimination of dracunculiasis; at that time there were an estimated 3.5 million cases of the debilitating disease in 21 countries in Africa and Asia.

A Guinea Worm Eradication program was initiated leading to a great reduction in cases. Several target dates were set and not reached. However, the campaign was strengthened by strong support from non-governmental organizations (NGOs), and increasing cooperation from the governments of affected countries. Ghana was declared Guinea worm free in 2011. In 2012, the remaining countries with cases were South Sudan, Mali, Ethiopia, and Chad. The WHO reported 396 cases in 2012, from the beginning of the year to 30 June 2012, compared with 807 cases for the same period in 2011. A total of 1058 cases was documented for whole of 2011, a significant improvement from 2010 with 1797 cases. Transmission has been limited to four countries in 2011, compared with 20 countries in 1990.

As there is no vaccine or curative treatment, control of this formidable disease relies entirely on people screening water for household use and as well as case finding. Prevention is key and emphasis must be on education and change in human behavior, including preventing people from bathing in drinking water sources and strongly promoting the filtration of water from the source before use. The WHO, UNICEF, the Carter Center, the US Centers for Disease Control and Prevention, the Gates Foundation, and the UK government have all collaborated with affected countries in joint efforts to prevent and contain this disease.

The WHO International Commission for the Certification of Dracunculiasis Eradication sends teams to previously endemic countries to assess their progress. The great advances made to date suggest that with continuing political support in affected countries, eradication of dracunculiasis may be achieved in the coming few years. The success of this massive international effort to eradicate dracunculiasis relies on education and on village health workers, who are the backbone of the program. This is seen in evaluation measures demonstrating a marked reductions in cases where village health workers have been mobilized. Some experts believe that dracunculiasis eradication may be achieved before polio eradication, despite polio's two remarkably successful vaccines and massive global efforts. Extraordinary success can be achieved through the power of organized health promotion, supported by the affected nations, international agencies, and NGO participation.

Sources: *Centers for Disease Control and Prevention. Progress toward global eradication of dracunculiasis, January 2010–June 2011. MMWR Morb Mortal Wkly Rep 2011;60:1450–3. Available at: http://www.cdc.gov/mmwr/preview/mmwrhtml/mm6042a2.htm [Accessed 15 February 2013].*
World Health Organization. Dracunculiasis [posted 24 August 2012]. Available at: http://www.who.int/dracunculiasis/en/ [Accessed 15 February 2013].
Carter Center. Guinea worm disease eradication. Available at: http://www.cartercenter.org/health/guinea_worm/mini_site/index.html [Accessed 15 February 2013].

are an estimated 200 million people infected worldwide and more than 600 million at risk for the disease. The clinical symptoms include fever, nausea, vomiting, abdominal pain, diarrhea, and hematuria. The organisms *Schistosoma mansoni* and *S. japonicum* cause intestinal and hepatic symptoms, including diarrhea and abdominal pain. *Schistosoma haematobium* affects the genitourinary tract, causing chronic cystitis and pyelonephritis, with a high risk for bladder cancer, the ninth most common cause of cancer deaths globally.

A recently identified species, *Schistosoma intercalatum*, is genetically unique, but thought to cause both intestinal and genitourinary disease. *Schistosoma intercalatum* is largely identified in inhabitants and immigrants from western Africa. Infection by all schistosomes is acquired by skin contact with freshwater containing contaminated snails. The cercariae of the organism penetrate the skin, and in the human host it matures into an adult worm which mates and produces eggs. The eggs are disseminated to other parts of the body from the worm's location in the veins surrounding the bladder or the intestines, and may result in neurological symptoms.

Eggs may be detected under microscopic examination of urine and stools. Sensitive serological tests are also available. Treatment is effective against all three major species of schistosomiasis. Eradication of the disease can be achieved with the use of irrigation canals, prevention of contamination of water sources by urine and feces of infected people, treatment of infected people, destruction of snails, and health education in affected areas. People exposed to freshwater lakes, streams, and rivers in endemic areas should be warned of the danger of infection. Mass chemotherapy in communities at risk and improved water and sanitation facilities are resulting in improved control of this disease.

Leishmaniasis

Leishmaniasis causes both cutaneous and visceral disease. The cutaneous form is a chronic ulcer of the skin, called by various names (e.g., rose of Jericho, oriental sore, and Aleppo boil). It is caused by *Leishmania tropica*, *L. brasiliensis*, *L. mexicana*, or the *L. donovani* complex. This chronic ulcer may last from weeks to more than a year. Diagnosis is by biopsy, culture, and serological tests. The organism multiplies in the gut of sandflies (*Phlebotomus* and *Lutzomi*) and is transmitted to humans, dogs, and rodents through bites. The parasites may remain in the untreated lesion for 5–24 months, and the lesion does not heal until the parasites are eliminated.

Prevention is through limiting exposure to the phlebotomines and reducing the sandfly population by environmental control measures. Insecticide use near breeding places and homes has been successful in destroying the vector sandflies in their breeding places. Case detection and treatment reduce the incidence of new cases. There is no vaccine, and treatment is with specific antimonials and antibiotics.

Visceral leishmaniasis (kala azar) is a chronic systemic disease in which the parasite multiplies in the cells of the host's visceral organs. The disease is characterized by fever, the enlargement of the liver and spleen, lymphadenopathy, anemia, leukopenia, and progressive weakness and emaciation. Diagnosis is by culture of the organism from biopsy or aspirated material, or by demonstration of intracellular (Leishman–Donovan) bodies in stained smears from bone marrow, spleen, liver, or blood.

Kala azar is a rural disease occurring in the Indian subcontinent, China, the southern republics of the former USSR, the Middle East, Latin America, and sub-Saharan Africa. Some 90 percent of visceral leishmaniasis occurs in six countries: India, Bangladesh, Sudan, South Sudan, Ethiopia, and Brazil. It usually occurs as scattered cases among infants, children, and adolescents. Cutaneous leishmaniasis is more widespread and may be increasing with global climate change. Transmission is by the bite of the infected sandfly, with an incubation period of 2–4 months. There is no vaccine, but specific treatment is effective and environmental control measures, such as the use of antimalarial insecticides, reduce the disease prevalence. In localities where the dog population has been reduced, the disease is less prevalent. Estimates of deaths vary between 20,000 and 40,000 per year (Alvar, 2012).

Trypanosomiasis

African Trypanosomiasis (Sleeping Sickness)

Sleeping sickness is a fatal degenerative neurological disease caused by *Trypanosoma brucei*, transmitted by the tsetse fly, primarily in the African savannahs, affecting cattle and humans. Subspecies are known to cause both acute and chronic forms of sleeping sickness. Some 55 million people are at risk in sub-Saharan Africa. Between 1998 and 2004, renewed surveillance and control reduced the incidence of African trypanosomiasis from 38,000 to approximately 18,000. Prevention depends on vector control and effective treatment of human cases. A dramatic reduction has occurred since 2004 (17,600 cases), with 9878 cases in 2009. The drop below 10,000 is a first time achievement in 50 years.

Chagas Disease (American Trypanosomiasis)

Chagas disease is a chronic vector- and blood transfusion-borne parasitic disease (*Trypanosoma cruzi*) which causes significant disability and death. Globally, some 10 million people are infected with *T. cruzi*, the protozoan parasite that leads to Chagas disease. The majority of infected individuals reside in Latin America. The WHO estimates that 7–8 million people worldwide are infected with Chagas disease, with approximately 10,000 deaths (2008).

Once found mainly in Latin America, in recent decades increasing cases have been identified in the USA, Canada, and other continents including some European and Western Pacific countries. These cases can be explained by migration and mobility patterns between populations in Latin America and the rest of the world. Thus, the demographics of this disease have changed and its presence has expanded to other continents.

The disease manifests itself in two phases. Within the first, acute phase, parasites circulate throughout the blood. Typically symptoms are mild, but may consist of fever, muscle pain, trouble breathing, and enlarged lymph glands. By phase two, characterized as the chronic phase, parasites have traveled through the body and, by this time, are hidden in the heart and digestive muscle. About 30 percent of affected people develop severe heart disease. After years have passed, heart failure or sudden death may occur as a result of the deterioration of the heart muscle.

While vaccine development is not likely owing to the ability of trypanosome antigens to cause autoimmunity and rapid immunological drift of the organism, two drugs have been developed which show effectiveness in limiting early chronic disease. Brazil achieved elimination of transmission in 1998, after Uruguay (1996) and Venezuela (1997), and followed by Argentina (1999). While the initial WHO elimination goal by 2010 proved unfeasible, efforts continue to dramatically reduce the incidence of *T. cruzi* infection.

Control is difficult, but vector control by ecological and insecticide measures is the most effective measure of reducing the animal host and vector insect population in its habitat. Other measures emphasize educating those residing in high-risk areas in prevention by clothing, bed nets, and repellents. Chemotherapy is utilized for case management. Blood screening to prevent transmission via transfusion or transplantation is also vital.

Other Parasitic Diseases

Amebiasis

Amebiasis is an infection with a protozoan parasite (*Entamoeba histolytica*) which exists as an infective cyst. Infestation may be asymptomatic or cause acute, severe diarrhea with blood and mucus, alternating with constipation.

Entamoeba histolytica infection sometimes results in invasive abdominal infestation, severe liver disease, and death.

Amebic colitis can be confused with ulcerative colitis. Diagnosis is by microscopic examination of fresh fecal specimens showing trophozoites or cysts. Transmission is generally via ingestion of fecal-contaminated food or water containing cysts, or by oral–anal sexual practices. Amebiasis is found worldwide. Sand filtration of community water supplies removes nearly all cysts. Suspect water should be boiled. Education regarding hygienic practices with safe food and water handling and disposal of human feces is the basis for control.

Ascariasis

Ascariasis is infestation of the small intestine with the roundworm *Ascaris lumbricoides*, which may appear in the stool, occasionally the nose or mouth, or may be coughed up from lung infestation. The roundworm is very common in tropical countries, where infestation may reach or exceed 50 percent of the population. Children aged 3–8 years are especially susceptible. Infestation can cause pulmonary symptoms and frequently contributes to malnutrition, especially iron-deficiency anemia. Transmission is by ingestion of infective eggs, common among children playing in contaminated areas, or via the ingestion of uncooked products of infected soil. Eggs may remain viable in the soil for years. Vermox and other treatments are effective. Prevention is through education, adequate sanitary facilities for excretion, and improved hygienic practices, especially with food. Use of human feces for fertilizer, even after partial treatment, may spread the infestation. Mass treatment is indicated in high-prevalence communities.

Pinworm Disease (Enterobiasis)

Pinworm disease (oxyuriasis) is common worldwide in all socioeconomic classes; however, it is more widespread among people living in crowded and unsanitary conditions. The *Enterobius vermicularis* infestation of the intestine may be asymptomatic or may cause severe perianal itching or vulvovaginitis. It primarily affects schoolchildren and preschoolers. More severe complications may occur. Adult worms may be seen visually or identified by microscopic examination of stool specimens or perianal swabs. Transmission is by the fecal–oral ingestion of eggs. The larvae grow in the small intestine and upper colon. Prevention is by educating the public regarding hygiene and adequate sanitary facilities, as well as by treating cases and investigating contacts. Treatment is the same as for ascariasis. Likewise, mass treatment is indicated in high-prevalence communities.

Ectoparasites

Ectoparasites include scabies (*Sarcoptes scabiei*), the common bed bug (*Cimex lectularius*), fleas, and lice, including the body louse (*Pediculus humanis*), pubic louse (*Phthirius pubis*), and head louse (*Pediculus humanus capitis*). Their severity ranges from nuisance value to serious public health hazard. Head lice are common in schoolchildren worldwide and are mainly a distressing nuisance. The body louse serves as a vector for epidemic typhus, trench fever, and louse-borne relapsing fever. In disaster situations, disinfection and hygienic practices may be essential to prevent epidemic typhus. The flea plays an important role in the spread of the plague by transmitting the organism from rats to humans. Control of rats has reduced the flea population; however, during war and disasters, rat and flea populations may thrive. Scabies, which is caused by a mite, is common worldwide and transmitted from person to person. The mite burrows under the skin and causes intense itching. All of these ectoparasites are preventable by proper hygiene and the treatment of cases. The spread of these diseases is rapid and therefore warrants immediate attention in school health and public health policy.

LEGIONNAIRE'S DISEASE

Legionnaire's disease (legionnellosis) is an acute bacterial disease caused by Legionnellae, a Gram-negative group of bacilli, with 35 species and many serological groups. The first documented case was reported in the USA in 1947, and the first disease outbreak was reported in the USA in 1976 among participants of a veterans' convention in Philadelphia. General malaise, anorexia, myalgia, and headache are followed by fever, cough, abdominal pain, and diarrhea. Pneumonia followed by respiratory failure may ensue. The case fatality rate can be as high as 40 percent of hospitalized cases. A milder, non-pneumonic form of the disease (Pontiac fever) is associated with virtually no mortality.

The organism is found in water reservoirs and is transmitted through heating, cooling, and air-conditioning systems, as well as from tap water, showers, saunas, and jacuzzi baths. The disease has been reported worldwide. Significant epidemics have occurred on cruise ships, where insufficient air-conditioning sanitation and an older, more susceptible clientele are a dangerous combination. Prevention requires the cleaning of water towers and cooling systems, including whirlpool spas. Hyperchlorination of water systems and the replacement of filters are required where cases and/or organisms have been identified. Antibiotic treatment with erythromycin is effective.

LEPROSY

Leprosy (Hansen's disease) was widely prevalent in Europe and Mediterranean countries for many centuries, with some 19,000 leprosaria in the year 1300. The disease was largely wiped out during the Black Death in the fourteenth century, but maintained in endemic form until the twentieth century. Leprosy is a chronic bacterial infection of the skin, peripheral nerves, and upper airway. In the lepromatous form, there is diffuse infiltration of the skin nodules and macules, usually bilateral and extensive. The tuberculoid form of the disease is characterized by clearly demarcated skin lesions with peripheral nerve involvement. Diagnosis is based on clinical examination of the skin and signs of peripheral nerve damage, skin scrapings, and skin biopsy.

Transmission of the *Mycobacterium leprae* organism is by close contact from person to person, with incubation periods of between 9 months and 20 years (average of 4–8 years). Rifampin and other medications allow the patient to become non-infectious in a short time, thus ambulatory treatment is possible. Multidrug therapy (MDT) has been shown to be highly effective in combating the disease, with a very low relapse rate. Treatment with MDT ensures that the bacillus does not develop drug resistance. The increase has been associated with improved case finding. BCG may be useful in reducing tuberculoid leprosy among contacts. Investigation of contacts over 5 years is recommended.

The disease has been eliminated from 119 out of 122 countries in which leprosy was previously recognized and perceived as a public health problem in 1985. In the last two decades over 14 million people suffering from leprosy have been cured.

Despite this progress, leprosy is still present in 11 countries in South-East Asia (including India), in sub-Saharan Africa, the Middle East (Sudan, Egypt, and Iran), and in some parts of Latin America (namely Mexico and Colombia), with isolated cases in the USA. However, world prevalence has declined from 10.5 million cases in 1980, through 5.5 million in 1990, to fewer than 300,000 in 2005 declining to 228,474 cases in 2010 and increasing slightly to 232,857 cases in 2012 (WER, 2013).

The WHO aimed to eliminate leprosy as a public health problem by 2000, defined as prevalence of fewer than one per 10,000 population, or fewer than 300,000 cases. The goal was met and this achievement has served as a major historic event in public health. Extensive use of MDT has contributed to a marked reduction in leprosy. In addition, the implementation of national and subnational campaigns in endemic countries has strengthened the control of leprosy. Emphasis is increasingly placed on incorporating primary leprosy medical services into existing general health services; this approach has successfully simplified diagnosis and treatment of the disease. The goals for 2011–2015 are to continue reducing the disease burden of leprosy where the disease is still endemic. This is achievable with early diagnosis and treatment with MDT. Numbers of new cases of the disease have declined (Figure 4.11), but it remains a serious problem in some countries; those with the largest burden of new cases of leprosy in 2010 were India (126,800), Brazil (34,894), and Indonesia (17,012).

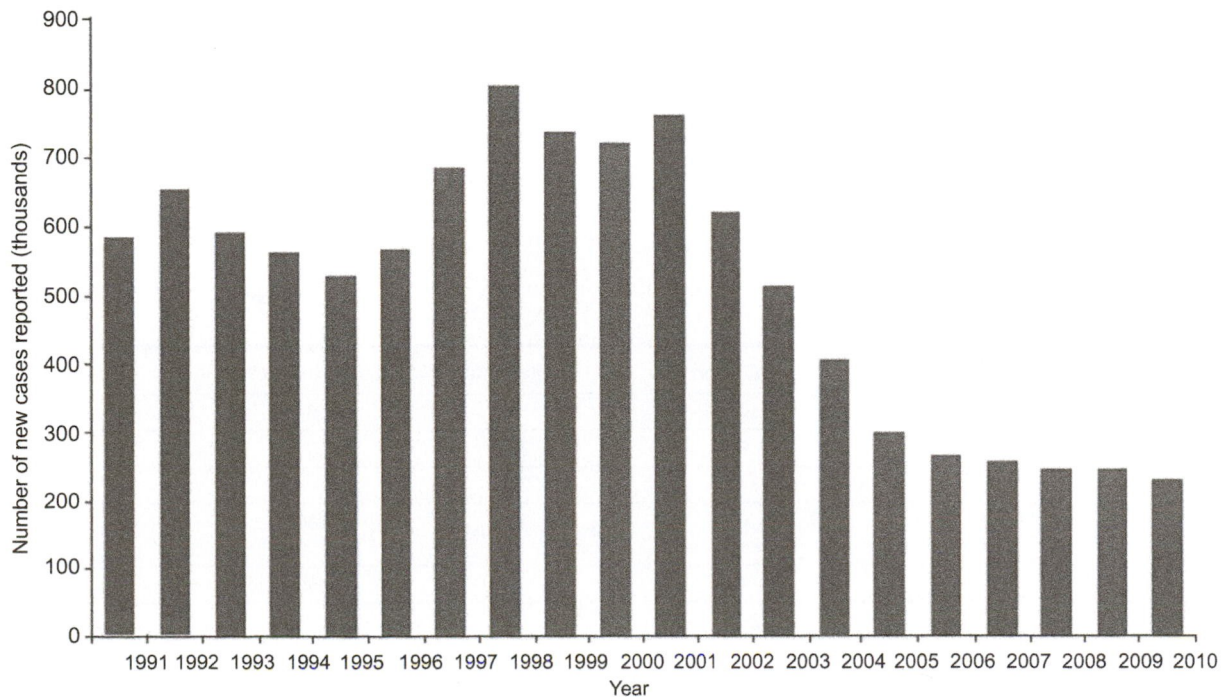

FIGURE 4.11   Number of new reported cases of leprosy, globally, 1991–2010. *Source: World Health Organization. First WHO report on neglected tropical diseases update 2011. Available at: http://www.who.int/neglected_diseases/en/ [Accessed 25 January 2013].*

SEXUALLY TRANSMITTED INFECTIONS

STIs are widespread internationally and continuing to increase, with an estimated 440 million new cases of curable infections (excluding HIV/AIDS) per year. The global burden of STIs is enormous (Tables 4.12 and 4.13). Scientists and infectious disease experts have identified more than 30 viral, parasitic, and bacterial pathogens that are transmitted sexually. CDC estimates that there are some 20 million new STIs in the USA each year, particularly among young adults, and a prevalence of 100 million cases. The annual health care costs of these illnesses are estimated to be in the order of US$16 billion. Mostly these do not cause serious harm if diagnosed and treated early, but many lead on to life-threatening conditions, e.g., HIV, HBV, syphilis, HPV, gonorrhea (now including antibiotic-resistant strains), chancroid, chlamydia, herpes simplex, and trichomonas (CDC, 2013).

This class of infections represents a substantial global cause of infertility, acute illness and disability, as well as further medical and psychological issues to many. Moreover, many STIs detrimentally affect pregnancy, leading to spontaneous abortions, stillbirths, or preterm deliveries. They can cause neonatal infections and, depending on the infection, may result in blindness, chronic respiratory disease, or herpes encephalitis. Accordingly, the public health and social consequences are devastating in many countries. Women, specifically adolescents, are most vulnerable to STIs. Many cases are asymptomatic, thus easing transmission of the disease. Populations residing in urban regions, people in low socioeconomic groups, and those involved in prostitution and drug use represent the highest risk of infection.

Since STIs, especially in women, may be asymptomatic, it is common for severe sequelae to occur before patients seek care. Infection by one STI increases the risk of infection by other diseases in this group. Prompt diagnosis and treatment can have enormous benefits for infant health and the overall health of a population. Coverage of services that prevent mother-to-child transmission of infections reached 45 percent in 2008; a major factor in saving infants' lives, UNAIDS estimates that transmission can be reduced by 30–35 percent, compared to a reduction of 1–2 percent without preventive services. Globally, an estimated 200,000 infant HIV infections were averted over a 12 year period, between 1996 and 2008 (Figure 4.12).

Syphilis

Syphilis is caused by the spirochete *Treponema pallidum*. After an incubation period of 10–90 days (mean of 21 days), primary syphilis develops as a painless ulcer or chancre on the penis, cervix, nose, mouth, or anus, lasting 4–6 weeks. The patient may first present with secondary syphilis 6–8 weeks (up to 12 weeks) after infection with a general rash and malaise, fever, hair loss, arthritis, and jaundice. These symptoms spontaneously disappear within weeks or up to 12 months later. Tertiary syphilis may appear 5–20 years after initial infection. Complications of tertiary syphilis

TABLE 4.12 Sexually Transmitted Infections

Bacteria	Gonorrhea (*Neisseria gonorrhoeae*) Chlamydia (*Chlamydia trachomatis*) Syphilis (*Treponema pallidum*) Chancroid (*Haemophilus ducreyi*)
Viruses	Genital warts, cervical and other cancers (human papillomavirus, HPV) Genital herpes (herpes simplex virus) Hepatitis B (hepatitis B virus)
Parasites	Trichomoniasis (*Trichomonas vaginalis*) Pubic lice (*Phthirus pubis*)

Source: Adapted from Global incidence and prevalence of four curable sexually transmitted infections (STIs): new estimates from WHO. PowerPoint presentation created by George Schmid, HIV/AIDS Department. Geneva: WHO; March, 2009. Available at: www.hivsurveillance2009.org/docs/session_iiia/pres5.ppt [Accessed 28 January 2013].

TABLE 4.13 Global Incidence of Sexually Transmitted Infections Among Men and Women Aged 15–49 Years by WHO Region, 2005 (Estimated Cases, Millions)

Region	Chlamydia	Gonorrhea	Syphilis	Trichomonas	Total
African	10.0	17.5	3.4	78.8	109.7
Americas	22.4	9.5	2.4	54.9	70.80
South-East Asian	6.6	22.7	2.9	54.9	89.20
Eastern Mediterranean	5.7	6.5	0.6	12.60	25.40
European	15.2	4.6	0.3	24.50	44.60
Western Pacific	41.6	26.9	1.1	39.10	108.70
Global total	101.5	87.7	10.6	248.5	448.3

Source: World Health Organization. Prevalence and incidence of selected sexually transmitted infections: methods and results used by WHO to generate 2005 estimates. Geneva: WHO; 2011. Available at: http://whqlibdoc.who.int/publications/2011/9789241502450_eng.pdf [Accessed 24 January 2013].

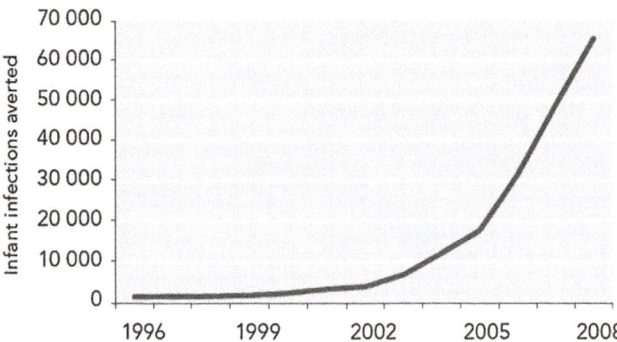

FIGURE 4.12   Global number of infant HIV infections prevented by antiretroviral prophylaxis to HIV-positive pregnant women, 1996–2008.
Source: Joint United Nations Programme on HIV/AIDS; World Health Organization. AIDS epidemic update: December 2009. Geneva: UNAIDS and WHO; 2009. Available at: http://data.unaids.org/pub/report/2009/ jc1700_epi_update_2009_en.pdf [Accessed 27 January 2013].

include catastrophic cardiovascular and CNS conditions. Early antibiotic treatment is highly effective when given in a large initial dose, but longer term therapy may be needed if treatment is delayed.

Gonorrhea

Gonorrhea is caused by the bacterium *Neisseria gonorrhoeae*. Worldwide, an estimated 62 million people are infected with gonorrhea annually. The incubation period is 1–14 days, and gonorrhea is often associated with concurrent chlamydia infection. In women, gonorrhea may be asymptomatic or it may cause vaginal discharge, pain on urination, bleeding on intercourse, or lower abdominal pain.

Untreated, it can lead to sterility. In men, gonorrhea causes urethral discharge and painful urination. Treatment with antibiotics ends infectivity, but untreated cases can remain infectious for months. Drug resistance to penicillin, tetracycline, and quinolones has emerged in many countries, thus more expensive and often unavailable drugs are necessary for treatment. Prevention of gonococcal eye infection in newborns is based on routine use of antibiotic ointments in the eyes of newborns (see Chapter 1).

While the control of STIs is an integral part of public health, work in this field must be carried out appropriately and delicately. This philosophy was certainly not applied to a study into STIs conducted in Guatemala (described in Box 4.30); on the contrary, study participants experienced a gross violation of human rights and ethical standards. Refer to Chapter 15 for more ethics in the framework of public health.

Chancroid

Chancroid is an infection caused by *Haemophilus ducreyi*. In women, chancroid may cause a painful, irregular ulcer

near the vagina, resulting in pain during intercourse, urination, and defecation; however, it may be asymptomatic. In men it causes a painful, irregular ulcer on the penis. The incubation period is usually 3–5 days, but may last up to 14 days. An individual is infectious as long as there are ulcers, usually 1–3 months. Treatment is by erythromycin or azithromycin.

Herpes Simplex

Herpes simplex is caused by herpes simplex virus types 1 and 2 and has an incubation period of 2–12 days. Genital herpes causes painful blisters around the mouth, vagina, penis, or anus. The genital lesions are infectious for 7–12 days. Among severe cases, herpes may lead to CNS meningoencephalitis infection. It can be transmitted to newborns during vaginal delivery, causing infection, encephalitis, and death. Caesarean delivery is therefore necessary when a mother is infected. Antiviral drugs are used in treatment, orally, topically, or intravenously. In 2003 the first systematic review was undertaken to establish the estimated global incidence and prevalence of herpes simplex virus (HSV) type 2. The 2003 estimated global prevalence among 15–49-year-olds was 535.5 million (16 percent of the world's population). In the same year, the estimated number of new HSV type 2 infections was 23.6 million.

Researchers indicate challenges associated with inadequate availability of data; however, they have recognized some general trends. For instance, there are more cases of HSV type 2 among women than men. Among the 535.5 million cases affecting 15–49-year-olds, women comprise approximately 315 million cases, and men the remaining 221 million infections. An additional trend is that the number of infected individuals directly increases with age. Data on prevalence rates of HSV type 2 in the USA indicate that 40–60 million individuals are infected. The estimated annual incidence is 1.2 million infections; the prevalence rate specifically among the 30–40-year-old age group is approximately 30 percent. Moreover, there is now plenty of evidence indicating that HSV type 2 is a major cofactor of HIV infection.

Chlamydia

Chlamydia, caused by *Chlamydia trachomatis*, is the second most common STI after HPV in the USA. The reported incidence increased to nearly 2.9 million in 2008. Despite this, underreporting is a major problem and actual incidence is estimated at more than twice the figure reported. In women, it usually presents asymptomatically, but may cause vaginal discharge, spotting, pain on urination, lower abdominal pain, and pelvic inflammatory disease (PID). In newborns, chlamydia may cause eye and respiratory infections. In men, chlamydia causes urethral discharge and pain

BOX 4.30 Unethical Medical Research on Syphilis in Guatemala

Informed consent, ethical standards, and regulation compliance may seem mindless, tedious, and perfunctory to some; to others, they may determine life or death, health, or illness. In the shameful medical experiment discussed below, US medical researchers conducted unscrupulous research victimizing their Guatemalan study participants.

Between 1946 and 1948, American researchers from the US Public Health Service conducted a study in Guatemala on 1300 prisoners, commercial sex workers, psychiatric patients, and soldiers. The group and local employees carried out experiments in which they deliberately infected these populations with sexually transmitted diseases (STDs). Consent was not obtained and the researchers made great effort to keep these acts a secret. Secrecy was critical in order to receive funding from senior authorities, who clearly should have never allowed these experiments to be carried out.

Funded by the US National Institutes of Health, the aim of the study was to determine new methods of preventing STDs, such as gonorrhea, syphilis, and chancroid. The researchers infected female sex workers with gonorrhea or syphilis; they subsequently allowed them to engage in unprotected sex with the other populations of their study, namely soldiers or prisoners. Consequently, some of these men contracted the STDs. The doctors then chose to directly inoculate the soldiers, inmates, and psychiatric patients, by injecting gonorrhea into the subjects' urethras; syphilis was transmitted both via skin injections and by exposing the penis to infectious agents.

Analysis of their own internal communications revealed that a few years before the Guatemalan experiment, the medical researchers attained informed consent prior to conducting research on prisoners in Terre Haute, Indiana. The researchers were fully aware of their obligation to obtain informed consent, but undertook the study with a conscious decision to skip this fundamental step, resulting in serious harm and immense humiliation.

This research has similarities with the infamous Tuskegee trial, which was primarily unethical for allowing syphilis-infected study subjects to remain untreated. This is partially because the director of the research conducted in Guatemala,

John Cutler, a US Public Health Service medical officer, also served as an investigator in the Tuskegee experiments. Moreover, further evidence indicates that the Surgeon General at that time, General Thomas Parran Jr, knew of the nature of the experiments Cutler and his colleagues were conducting.

An official apology was subsequently rendered to the government of Guatemala. The US Presidential Commission on Bioethics has highlighted this case as an awakener to ensure that something of this magnitude, grossly violating the rights of study participants, is never repeated. Current regulations dictate rules for medical research on human participants. These regulations are indispensable requirements for researchers and funding agencies, and limitations for publication. Implementation of and appropriate compliance with the regulations are equally important. Ignoring this protection for study participants in research in another country does not absolve anyone from responsibility and compliance towards ethical standards.

Deception was a major factor in this study of participants recruited from vulnerable populations. This notorious experiment undermined human rights, medicine, ethics, and the principle of protecting patients. Effort to determine new treatments and advances in medicine with the intention of helping patients must be conducted according to internationally accepted rules of procedure, such as those of the Helsinki Declaration of 1964 and its subsequent formulations (see Chapter 15).

Sources: Reverby SM. Ethical failures and history lessons: the US Public Health Service research studies in Tuskegee and Guatemala. Public Health Rev 2012;34. Epub ahead of print. Available at: www.publichealthreviews. eu [Accessed 28 January 2013].
US Department of Health and Human Services. Fact sheet on the 1946–1948 US Public Health Service Sexually Transmitted Diseases (STD) Inoculation Study. US Public Health Service sexually transmitted disease inoculation study of 1946–48. Available at: http://www.hhs.gov/1946inoculationstudy/factsheet.html [Accessed 10 January 2013].
Frieden TR, Collins FS. Intentional infection of vulnerable populations in 1946–1948: another tragic history lesson. JAMA 2010;304:2063–4. http://dx.doi.org/10.1001/jama.2010.1554. Available at: ama.jamanetwork.com/article.aspx?articleid=186859 [Accessed 10 January 2013].
US Presidential Commission on Bioethical Issues. Moral science: protecting participants in human subjects research. Washington, DC: US Presidential Commission for the Study of Bioethical Issues; December 2011. Available at: www.bioethics.gov [Accessed 10 January 2013].

on urination. The incubation period is 7–21 days and the infectious period is unknown. Treatment for chlamydia is doxycycline, azithromycin, or erythromycin. Because co-transmission with gonorrhea is extremely common, CDC recommends treatment for both diseases when either is confirmed. Chlamydia infection, not necessarily venereal in transmission, may be transmitted to the newborns of infected mothers.

In the past, *Chlamydia pneumoniae* has been suspected as a possible cause of or contributor to coronary heart disease. This correlation has been under much investigation, leading to the following findings. Chlamydia (*C. pneumoniae*) infection is known to be a widespread chronic risk factor for coronary artery disease. Intra-arterial

infection, among other contributing etiologies, contributes to plaque formation, thromboembolic occlusion of arteries, and myocardial infarction. While antibiotic treatment of chlamydia as a preventive measure for heart disease has not been used, this could potentially reduce the burden of the leading worldwide cause of death at a relatively low cost.

The WHO reports that *Chlamydia trachomatis* causes more cases of STIs than any other bacterial pathogen, thus it represents a worldwide public health challenge, exacerbated by the infection's silent, asymptomatic nature among many, leading to an ease and unawareness of transmission, affecting both men and women. Vaccine development is underway.

Trachoma

Trachoma, an infectious disease of the eye, is currently responsible for approximately 3 percent of blindness in the world. Although not classified as an STI, it is placed in this section because its causative organism is *Chlamydia trachomatis*, the same as a bacterium that causes chlamydia, one of the most common STIs. The organism is transmitted through contact with eye discharge of an infected individual, typically by use of household items such as handkerchiefs and washcloths; the bacterium can is also transmitted by flies that have landed on the eyes or nose of someone infected. If this infection is not treated and further trachoma infections occur, it can cause major scarring in the inside of the eyelid. As a result, the eyelid may turn inward and, consequently, the eyelashes can scratch the cornea. This becomes painful and disabling as the permanent corneal damage can result in irreversible blindness.

Trachoma is recognized as the top cause of preventable blindness due to an infection. It is most easily transmitted in regions with insufficient access to water and inadequate sanitation. Affecting the most marginalized populations, trachoma is common in poor rural areas of Central America, Brazil, Africa, parts of Asia, and some countries in the Eastern Mediterranean. The WHO estimates that approximately 2.2 million people are visually impaired as a result of trachoma, and 1.2 million cases have advanced to blindness. Once considered endemic in most countries, more recent reports indicate trachoma endemicity in 57 countries. Hygiene, vector control, and treatment with antibiotic eye ointments or simple surgery for scarring of eyelids and inturned eyelashes are important in preventing blindness. Moreover, a new drug, azithromycin, is effective in curing the disease. The WHO is promoting a program dedicated to the global elimination of trachoma in endemic areas by 2020. The comprehensive intervention, known as SAFE, focuses on four major elements: surgery (to correct advanced stages), antibiotics, facial cleanliness, and environmental improvements pertaining to water and sanitation.

Trichomoniasis

Trichomoniasis is caused by the protozoan parasite *Trichomonas vaginalis*. The incubation period is 4–20 days (mean 7 days). In women, trichomoniasis may be asymptomatic or may cause a frothy vaginal discharge with foul odor, and painful urination and intercourse. In men, the disease is usually mild, causing pain on urination or itching or irritation inside the penis. Treatment is by a single dose of antibiotics, metronidazole or tinidazole, taken orally. These antibiotics can cure trichomoniasis; without treatment, the disease may persist and remain infectious for years. CDC reports that in the USA approximately 3.7 million people are infected with trichomoniasis; however, only 30 percent develop symptoms of the infection. It is possible for those infected but experiencing no symptoms to transmit the infection to others. Moreover, approximately one in five people become reinfected within 3 months after taking the antibiotics.

Human Papillomavirus

HPV, endemic throughout the world, is now known as the leading cause of cervical neoplasia and cancer of the cervix. Silent among many people, the majority of HPV infections cause no symptoms. Despite this, genital HPV infections that are persistent can lead to the development of cervical cancer. Nearly all cases of cervical cancer (99 percent) are associated with a genital HPV infection, which represents the most prevalent viral infection affecting the reproductive tract. HPV includes many types associated with venereal warts (condylomas).

Scientists have identified over 40 types of HPV that infect the genital area in males and females. Among 90 percent of cases, the body's immune system manages to clear HPV naturally, within 2 years. Despite this, there is no way of determining which HPV cases will lead to cancer, genital warts, or other medical problems. Screening tests are integral in preventing cervical cancer or detecting it early. They include the HPV test, which detects the virus that can cause cervical cell changes, and the Papanicolaou (Pap) smear, which detects precancerous lesions as well as cell changes on the cervix.

An effective vaccine against the most common carcinogenic strains is now available and recommended for young women to prevent cervical cancer, a breakthrough of enormous importance as this is one of the leading causes of cancer mortality in women. HPV vaccines are administered in three doses over a period of 6 months. Although originally developed for females, an HPV vaccine has also been created to protect males from the most common types of the virus.

The prevention of cervical cancer by vaccination and Pap smear screening is a major advance in public health, along with the prevention of liver cancer by hepatitis B immunization. Circumcision is now recommended by the WHO for primary prevention of transmission of HPV (see Chapters 5 and 6).

Control of Sexually Transmitted Infections

In areas where a full range of diagnostic services is lacking, a "syndromic approach" is recommended for the control of STIs. The diagnosis is based on a group of symptoms and treatment on a protocol addressing all the diseases that could possibly cause those symptoms, without expensive laboratory tests and repeated visits. Early treatment without laboratory confirmation helps to cure people who might not return for follow-up. Moreover, it may place them in a noninfective stage so that even without follow-up care, they will not transmit the disease. STI incidence per 100,000

TABLE 4.14 Reportable Sexually Transmitted Infections, USA, Selected Years, 1950–2009: New Cases per 100,000 Population

Infection	1950	1960	1970	1980	1990	2000	2007	2008	2009
Syphilis	147	69	45	30	54	11	14	15	15
Gonorrhea	193	145	294	442	276	129	118	111	99
Chlamydia	NA	NA	NA	NA	160	251	367	398	409

Note: The figures listed for syphilis include all three stages of the disease as well as congenital syphilis. Rates are cases per 100,000 population, rounded. Prior to 1994, Chlamydia was not notifiable. Between 1994 and 1999, cases for New York were only reported by New York City. Only data beginning in 2000 onward represent cases for the entire state.
NA=not available.
Source: Centers for Disease Control and Prevention. Health United States, 2011. Trend table 44. Available at: http://www.cdc.gov/nchs/data/hus/hus11.pdf [Accessed 3 June 2013].

population from 1950 to 2009 is shown in Table 4.14. The decline in syphilis and gonorrhea from 1950 to 1990 was followed by a leveling or slow decline in the 2001–2009 period, while the increase in chlamydia is probably due to greater awareness and diagnosis.

Screening in prenatal and family planning clinics, prison medical services, and clinics serving sex workers, homosexuals, or other potential risk groups will detect subclinical cases of various STIs. Treatment can be carried out inexpensively and immediately. For instance, the screening test for syphilis costs US$0.10 and the treatment with benzathine penicillin injection costs about US$0.40. Partner notification is a controversial issue, but may be needed to identify contacts who may be the source of transmission to others.

Control of STIs through a syndrome approach based on primary care providers is being promoted by the WHO. Health education directed towards high-risk target groups and marginalized populations is essential. Providing easy and cost-free access to acceptable, non-threatening treatment is vital in promoting the early treatment of cases and thereby reducing the risk of transmission.

Promoting prevention through the use of condoms and/or monogamy requires long-term educational efforts that are now fostered by the HIV/AIDS pandemic. Increased use of condoms for HIV prevention is associated with a reduced risk of other STIs. Training medical care providers in STI awareness should be stressed in undergraduate and continuing educational efforts, including personal protection as caregivers.

HIV/AIDS

HIV/AIDS has captured world attention since it was first identified in the early 1980s. HIV is a retrovirus that infects various cells of the immune system, and also affects the CNS. Two types have been identified: HIV1, worldwide in distribution; and the less pathogenic HIV2, found mainly in West Africa. HIV is transmitted by sexual contact,

exposure to blood and blood products, perinatally, and via breast milk. The period of communicability is unknown, but studies indicate that infectiousness is high, both during the initial period after infection and later in the disease. Antibodies to HIV usually appear within 1–3 months.

Within several weeks to months of the infection, many people develop an acute self-limiting flu-like syndrome. They may then be free from any signs or symptoms for months to more than 10 years. Onset of illness is usually insidious with non-specific symptoms, including sweats, diarrhea, weight loss, and fatigue. AIDS represents the later clinical stage of HIV infection. According to the revised CDC case definition (1993), AIDS involves any one or more of the following: low CD4 count, severe systemic symptoms, opportunistic infections such as *Pneumocystis* pneumonia or TB, aggressive cancers such as Kaposi's sarcoma or lymphoma, and/or neurological manifestations, including dementia and neuropathy. The WHO case definition is more clinically oriented, relying less on often unavailable laboratory diagnoses for indicator diseases.

This pandemic presented lessons of public health and hygiene that had been forgotten in a smug confidence and reliance on antimicrobial therapy and vaccines that were assumed to be capable of defeating all infectious diseases. Regrettably this is not the case, and the HIV/AIDS experience showed the price of negligence in infectious disease control of STIs. With no vaccine yet on the horizon, the prospects for this disease are grim and its spread certain until an effective vaccine can be developed. However, the pattern of mortality in the USA is shown in Figure 4.13 from CDC, indicating modest success in the changing potential for prolonging survival, improving quality of life, and reducing transmission. Active public health measures include education on AIDS prevention and condom promotion, and effective medical care based on antiretroviral therapy (ART). Other measures include promoting early diagnosis and treatment for TB, with special attention to MDR-TB, and other opportunistic infections, as well as nutritional supplementation and general care, which are also gaining ground as preventive measures in sub-Saharan Africa.

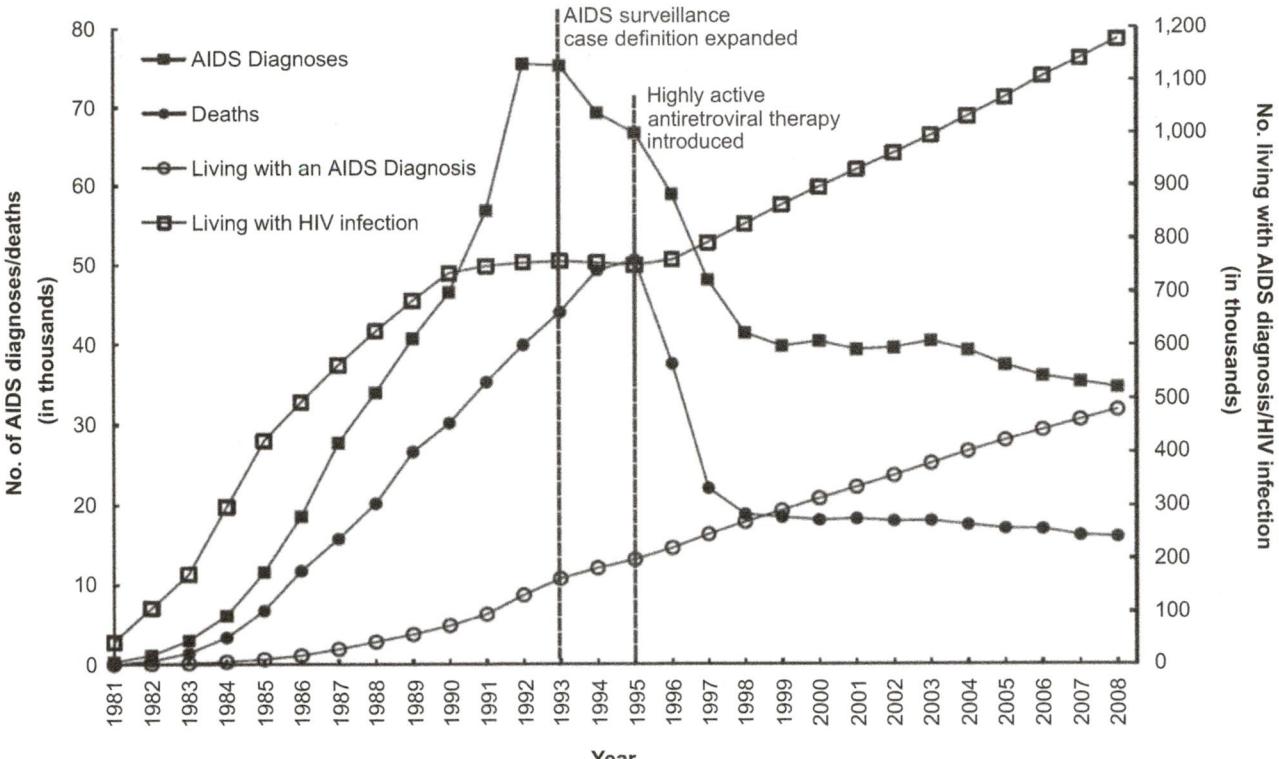

FIGURE 4.13 AIDS diagnosis and cases and estimated number of people aged ≥ 13 years, USA, 1981–2008. *Source: Centers for Disease Control and Prevention. HIV in the US: an overview. Atlanta GA: CDC; August 2011. Available at: http://www.cdc.gov/hiv/topics/surveillance/resources/factsheets/pdf/HIV-US-overview.pdf [Accessed 28 January 2013].*

The graph in Figure 4.13 shows important trends in AIDS diagnosis and prevalence in the USA over several years. The numbers of both AIDS diagnoses and deaths have fallen significantly, beginning around 1993 and 1995. When examining the number of Americans living with HIV infection or an AIDS diagnosis, the graph shows a gradual increase beginning in the early to mid-1990s. While the HIV prevalence is higher than ever before, this echoes a trend that can be described as a higher number of new infections than number of people who die of HIV/AIDS yearly. Thus, effective medical care and ART have contributed immensely to allowing those infected with HIV to live longer. Likewise, ART has allowed for decreased mortality due to HIV/AIDS.

According to the CDC, in the USA, by the end of 2008, nearly 1.2 million adults and adolescents were living with HIV. When compared to the estimate for 2006, this figure signifies a 7 percent increase. Another important trend discovered by researchers is that the majority of those infected with HIV did not transmit the virus to others in recent years. Experts estimate that in the USA in 2006, of every 100 people living with HIV, there were five transmissions. Thus, at least 95 percent of HIV-positive people did not transmit the virus to otherwise healthy individuals that year. This represents a significant (89 percent) decline in the approximate

rate of HIV transmission since the mid-1980s. The reduction in transmission is likely to be attributable to successful preventive measures, improved testing, and highly effective treatment.

AIDS was first recognized clinically in 1981 in Los Angeles and New York (Box 4.31). By mid-1982 it was considered an epidemic in those and other US cities. It was primarily seen among men who have sex with men and recipients of blood products. After initial errors, testing of blood and blood products became standard and subsequently closed off this method of transmission. Transmission has changed markedly since the initial onslaught of the disease, with needle-sharing among intravenous drug users, heterosexual activity, and maternal–fetal transmission becoming major sources of infection. Comorbidity with other STIs apparently increases HIV infectivity and, furthermore, may have helped to convert the epidemiology to a greater degree of heterosexual transmission. The disease grew exponentially in the USA but the incidence of new cases has declined since 1993. AIDS represents a major public health problem in most developed and developing countries, reaching catastrophic proportions in some sub-Saharan African countries, affecting up to 30 percent or more of some populations.

HIV-related deaths were the eighth leading cause of all deaths in 1993 in the USA, the leading cause among men

BOX 4.31 HIV/AIDS Retrospective, 1981–2009

Acquired immunodeficiency syndrome (AIDS) was first reported as a clinical entity in 1981 by Dr Michael Gottlieb at the University of California, Los Angeles (UCLA) Hospital. He reported in *Morbidity and Mortality Weekly Report* and later the *New England Journal of Medicine* on five cases of *Pneumocystis carinii* (now *P. jiroveci*) pneumonia with cytomegaloviremia (CMV) among young male homosexuals with evidence of immune deficiency. A few weeks later, Dr Alvin Friedman-Kien reported 26 cases of Kaposi's sarcoma in gay men from New York and California.

In 1983, heterosexual infection and maternal–fetal transmission were identified, and the human immunodeficiency virus (HIV) was isolated from an AIDS patient. By the mid-1980s HIV/AIDS had become a global pandemic. In 1985, the first HIV test was approved and used for testing blood donations, and public recognition of the pandemic was growing. Zidovudine (AZT) treatment became available in 1987. In 1992, the FDA licensed the first rapid HIV test. During 1996–1999, a treatment cocktail of antiretroviral therapy was approved to reduce viral loads, and death rates dropped by 40 percent in the USA. A safer and more effective combination of drugs was used in 1998–2000.

The Global Fund was established in 2001–2002 to fight HIV/AIDS, but treatment remained unavailable to the over 4 million HIV carriers in sub-Saharan Africa and AIDS was the leading cause of death among people in the age group 15–59 years. In 2003–2005, President George W. Bush provided US$15 billion in aid to HIV/AIDS treatment in 15 heavily affected countries. In 2006–2007, UNAIDS recommended adult male circumcision after evidence of its effectiveness in reducing HIV transmission in high-prevalence areas; HIV vaccine trials failed, but new candidate vaccines were being developed. In 2008, the Centers for Disease Control and Prevention indicated that there were more cases than previously thought, with 1.1 million Americans being affected, while the number of new cases again increased rapidly.

Over 30 million people globally, including an estimated 1.1 million Americans, have died from AIDS since 1981. UNAIDS estimated that the epidemic peaked in 1996 at 3.5 million new infections, and deaths peaked at 2.2 million in 2004, even though more than half of those who need treatment do not receive it.

The 2008 Nobel Prize for Medicine was awarded to Luc Montagne and Francoise Barre-Sinoussi from the Pasteur Institute for the discovery of HIV. In total, 33 million people were living with HIV in 2008, but death rates were dropping. Since then, AIDS has receded from public view but new case rates are still climbing, mainly among men who have sex with men, but still over 30 percent of new infections are acquired heterosexually. HIV vaccine studies are showing promise and there is growing optimism that an effective vaccine will be developed.

Sources: *Gottlieb MS. Pneumocystis carinii pneumonia and mucosal candidiasis in previously healthy homosexual men: evidence of a new acquired cellular immunodeficiency. N Engl J Med 1981;305:1425–31.*
Centers for Disease Control and Prevention. Kaposi's sarcoma and pneumocystis pneumonia among homosexual men – New York City and California. MMWR Morb Mortal Wkly Rep 1981;30:305–8.
Global Fund: Fighting AIDS, tuberculosis and malaria. Available at: http://www.theglobalfund.org/en/about/diseases/ [Accessed 27 January 2013].
Nobelprize.org. The Nobel Prize in Physiology or Medicine. Harald zur Hausen, Francoise Barre-Sinoussi, Luc Montagnier. Available at: http://www.nobelprize.org/nobel_prizes/medicine/laureates/2008/ [Accessed 28 January 2013].
AIDS retrospective slide show: a pictorial timeline of the HIV/AIDS pandemic. Available at: http://www.webmd.com/hiv-aids/ss/slideshow-aids-retrospective [Accessed 28 January 2013].

aged 25–44, and the fourth leading cause for women in this age group. By 2005, AIDS had been diagnosed in 984,000 people and 550,000 had died. At the end of 2003, it was estimated that up to 1.1 million people were HIV infected in the USA. Up to 30 percent of these individuals may be unaware of their HIV status. In 2005, 42,000 new diagnoses were reported. The rate of AIDS diagnoses, in 2009, was estimated at 11.2 per 100,000 population.

Globally, deaths from AIDS totaled 2.8 million in 2005, with an estimated 11.7 million people having died from this pandemic up to 1997. In 2005, there were an estimated 4.1 million new cases. HIV/AIDS is the fourth largest killer in the world, and the leading cause of death in sub-Saharan Africa. However, owing to implementation of coordinated control programs, it is believed that the pandemic expansion peaked in the late 1990s. The WHO aims to reverse the increase in HIV infection by 2015. With increased attention, training, funding, and resources, this may be possible. Globally in 2008, deaths totaled 1.8 million and new cases 2.6 million. Moreover, there were 33.3 million (2.5 million children) living with AIDS. The 2008 Report on Global AIDS Estimates indicates that globally, in 2007 there were approximately 15 million orphans due to AIDS. In this study, orphans are defined as children between the ages of 0 and 17 who have lost one or both parents to AIDS. Of the 15 million orphans worldwide, 11.6 million live in sub-Saharan Africa. These figures can be compared with those of 2001, in which globally, there were an estimated 8 million orphans due to AIDS. Of these, 6.5 million reside in sub-Saharan Africa.

The declining incidence of new cases in industrialized nations may be the result of greater awareness of the disease and methods of prevention of transmission. Improving early diagnosis and access to care, especially the combined therapy programs that are very effective in delaying onset of symptoms, are important facets of public health management of the AIDS crisis. In developed countries, highly active antiretroviral therapy (HAART) has been successful in substantially reducing disease advancement to AIDS. Thus, this form of treatment has converted HIV/AIDS from a fatal illness to a fairly manageable chronic disease. While this is, of course, a favorable outcome, HIV treatment does not come without major challenges. It typically consists of a cocktail of multiple drugs, and it functions as a lifelong therapy. Adherence is crucial for treatment be effective, for increased virological control, and in preventing drug resistance. Regimens are complex and many patients experience serious side-effects. A central challenge associated with treatment is access and availability to marginalized populations, particularly in poor, less developed countries.

In 1987, the first phase I trial of an HIV vaccine was carried out in the USA. Phase II and III trials have also been conducted, and while plenty of research is dedicated to this potential form of prevention, scientists have not been successful in developing a safe, effective, and affordable HIV vaccine. Until an effective vaccine is available, preventive reliance will continue to be on behavior risk reduction and other prevention strategies such as needle and condom distribution among high-risk population groups.

Throughout the world, HIV continues to spread rapidly, especially in poor countries in Africa, Asia, and South and Central America. The United Nations reports that 90 percent of people living with HIV/AIDS are in developing countries, where transmission largely occurs through heterosexual contact. Every day, more than 8500 people, including 1000 children, are infected. In Thailand, one person in 50 is now infected. In sub-Saharan Africa more than one in 40 is infected, and in some cities as many as one in three people carry the virus. Estimations of new infections per year in sub-Saharan Africa range from 1 to 2 million people, while in Asia there are 1.2–3.5 million new infected people per year. Lessons are still being learned from the AIDS pandemic. Furthermore, the prevalence of those living with the HIV is three times that of 1990. Between 2003 and 2008, there was a 10-fold increase in access to antiviral drugs among low- and middle-income countries. The explosive spread of this infection, from an estimated 100,000 people in 1980 to an anticipated 40 million people HIV infected, shows that the world is still vulnerable to pandemics of emerging infectious diseases. Enormous movements of tourists, businesspeople, truck drivers, migrants, soldiers, and refugees promote the spread of such diseases. Widespread sexual exchange, transfusion of blood products, and illicit drug use can all promote the international potential for pandemics. War and massive refugee situations promote rape and prostitution, worsening the AIDS situation in some settings in Africa.

The HIV pandemic has spread throughout the world. However, there is the somewhat hopeful indication that the rate of increase has slowed in the USA. This may be a reflection of a number of factors including higher levels of self-protective behavior, the most susceptible population groups having already been affected, and the spread into the general population at a slower rate. The slowdown may yet be only a lull in the storm, as heterosexual contact becomes a more important mode of transmission and male-to-male transmission is increasing, especially among black homosexual Americans.

The 2013 UNAIDS reports major progress in control of the world pandemic of HIV/AIDS, in part due to designation of the Millennium Development Goal 6 of halting and reversing the spread of AIDS and providing ART treatment for all those who need it. An estimated 2.5 million people became HIV positive in 2011, with 25 countries showing a drop in new HIV infections of 50 percent or more; half of this reduction was among newborns as a result of antiretroviral treatment of HIV-positive mothers during pregnancy. In 2011, over 8 million people were receiving ART management for HIV, an increase of over 60 percent from 2009, and this includes an increase in the number of people receiving ART in low- and medium-income countries. However, 7 million who need ART do not receive it, and new HIV infection rates are rising in the Middle East, North Africa, Eastern Europe, and Central Asia.

Combinations of several drugs from among a number of antiretroviral medications are showing promise in suppressing HIV in infected people. At a current annual price of nearly US$20,000 per patient, these sums are well beyond the capacity of most developing countries. The development of methods for measuring the HIV viral load has allowed for better evaluation of potential therapies and monitoring of patients receiving therapy. In developed countries, transmission by blood products has been largely controlled by screening tests, transmission among homosexuals has been reduced by safe-sex practices, and transmission to newborns has been reduced by recent therapeutic advances, specifically prevention of mother-to-child transmission (PMTCT). Safe-sex practices and condom use may have helped in reducing heterosexual transmission. Further advances in therapy and preventive measures based on a vaccine are expected over the next decade.

The HIV/AIDS pandemic is one of the great public health challenges of the twenty-first century for a myriad of reasons. It is a complex disease involving strong cultural elements, and challenging factors include its international spread, its sexual and other modes of transmission, the associated stigma, its devastating and costly clinical effects, and its impact on parallel diseases such as TB, respiratory infections, and cancer. The cost of care for the AIDS patient can be exceedingly high. Programs needing strengthening include home care with home health aides or CHWs to encourage compliance with treatment, along with adequate nutrition and self-care, as well as mutual help among HIV carriers and AIDS patients. Adding to the complexity, difficult ethical issues associated with AIDS arise. Important matters needing support include improved screening of pregnant women and newborns; partner notification, reporting, and contact tracing; and financing the cost of care. The AIDS pandemic is not by any means over or "under

TABLE 4.15 Major Organisms Associated with Diarrheal Diseases

Classification	Organisms
Bacteria	*Salmonella, Shigella, Escherichia coli, Vibrio cholerae, Bacillus cereus, Campylobacter jejuni*
Virus	Enteroviruses, rotaviruses, adenoviruses, astroviruses, calciviruses, coronaviruses, small round virus group, Norwalk group
Parasites/protozoa	*Schistosoma, Giardia lamblia, Cryptosporidium, Amoeba histolytica*

control", but progress has been made and there is hope for the "magic bullet" of an effective and inexpensive vaccine.

DIARRHEAL DISEASES

Diarrheal diseases are the leading cause of child mortality in the world. They are caused by a wide variety of bacteria, parasites, and viruses (Table 4.15) infecting the intestinal tract. They cause secretion of fluids and dissolved salts into the gut with mild to severe or fatal complications.

In developing countries, diarrheal diseases account for half of all morbidity and a quarter of all mortality. Diarrhea itself does not cause death, but the dehydration resulting from fluid and electrolyte loss is one of the most common causes of death in children worldwide. Deaths from dehydration can be prevented by the use of oral rehydration therapy (ORT), an inexpensive and simple method of intervention easily used by a non-medical primary care worker and by the mother of the child as a home intervention. In 1983, diarrheal diseases were the cause of almost 4 million child deaths, but by 1996 this figure had declined to 2.4 million, largely owing to the increased use of ORT.

Diarrheal diseases are transmitted by water, food, and directly from person to person via fecal–oral contamination. Diarrheal diseases occur in epidemic levels in situations of food poisoning or contaminated water sources; they can also be present at high levels when common source contamination is not found. Contamination of drinking water by sewage and poor management of water supplies are also major causes of diarrheal disease. A dangerous practice, the use of sewage for the irrigation of vegetables is a common cause of diarrheal disease in many areas.

Salmonella

Salmonella are a group of bacterial organisms causing acute gastroenteritis, associated with generalized illness including headache, fever, abdominal pains, and dehydration. There are over 2000 serotypes of *Salmonella*, many of which are pathogenic in humans, the most common of which are *Salmonella typhimurium*, *S. enteritidis*, and *S. typhi*. Transmission is by ingestion of the organisms in food, derived from fecal material from animal or human contamination. Common sources include raw or uncooked eggs, raw milk, meat, poultry and its products, as well as pet turtles or chicks. Fecal–oral transmission from person to person is common. Prevention is in safe animal and food handling, refrigeration, sanitary preparation and storage, protection against rodent and insect contamination, and the use of sterile techniques during patient care. Antibiotics rarely affect disease progression and may lead to increased carrier rates and produce resistant strains; therefore, only symptomatic and supportive treatment is recommended, except in systemic and life-threatening cases.

Salmonella typhi causes typhoid fever and according to the WHO kills some 500,000 people per year, while seriously affecting millions of others. While treatable by ampicillin and fluid replacement, the antibiotics are becoming less effective. Two vaccines are currently available and are used in high-risk areas.

Shigella

Shigella are a group of bacteria that are pathogenic in humans. The infectious dose of *Shigella* is among the lowest of all pathogens; fewer than 10 organisms are sufficient to cause disease within four groups: type A (*Shigella dysenteriae*), type B (*S. flexneri*), type C (*S. boydii*), and type D (*S. sonnei*). Types A, B, and C are each further divided into a total of 40 serotypes. *Shigella* are transmitted by direct or indirect fecal–oral methods from a patient or carrier, and illness follows ingestion of even a few organisms. Flies can transmit the organism, and in non-refrigerated foods the organism may multiply to an infectious dose. Control is in hygienic practices and in the safe handling of water and food. *Shigella* is a common cause of waterborne disease outbreaks where water supplies are contaminated and not treated adequately. *Shigella* bacteria can contaminate community water sources, local surface water, and recreational waters such as streams and swimming pools, via human, animal, or sewage sources, causing large waterborne disease outbreaks.

Escherichia coli

Escherichia coli bacteria are common fecal contaminants of inadequately prepared and cooked food. Particularly virulent strains such as O157:H17 can cause explosive outbreaks of severe (enterohemorrhagic) diarrheal disease with a hemolytic–uremic syndrome and death, as occurred in Japan in 1998 with cases and deaths due to a foodborne epidemic. Sporadic, but significant epidemics occur often, mostly in developed countries where food processing and transport are common. Other milder strains cause traveler's diarrhea and nursery infections. Inadequately cooked hamburgers, unpasteurized milk, and other food vectors are discussed in Chapter 8. Foodborne disease can occur anywhere, including in developed countries, as in the case of contaminated lettuce from California in 2007 and leafy vegetables grown in Germany and exported all over Europe in 2011.

Cholera

Cholera is an acute bacterial enteric disease caused by *Vibrio cholerae*. It is characterized as causing sudden-onset, profuse, painless watery stools, occasional vomiting, and if untreated, rapid dehydration, circulatory collapse, and death. Similar disease may be caused by other "cholerogenic" species of *Vibrio*. Asymptomatic infection or carrier status, and mild cases are common. In severe, untreated cases, mortality is over 50 percent, but with adequate treatment, mortality is under 1 percent. Diagnosis is based on clinical signs, epidemiology, serology, and bacteriological confirmation by culture. The two types of cholera are the classic and el Tor (with Inaba and Ogawa serotypes).

In 1991, a large-scale epidemic of cholera spread through much of South America. It was imported via a Chinese freighter, whose sewage contaminated shellfish in Lima harbor in Peru (Box 4.32). Since 1991, epidemics in South America, south Asia, and Iraq have caused hundreds of thousands of cases and, consequently, thousands of deaths. Haiti experienced its first outbreak of cholera in decades following the 2010 earthquake.

According to the WHO, the number of documented cases of cholera continues to increase. In 2011 alone, nearly 590,000 cases were reported from 58 countries, including some of the most severe cases and over 7800 deaths. Experts understand that these figures do not represent the actual cases and deaths. As a result of limitations in surveillance in conjunction with concern over potential trade and travel sanctions, many more cases have been unaccounted for and undocumented. Furthermore, discrepancies occur owing to inconsistencies in case definitions and a lack of agreed upon vocabulary. A more accurate depiction of the burden of cholera is 3–5 million cases, with 100,000–120,000 deaths per year.

Prevention requires sanitation, particularly the chlorination of drinking water, prohibiting the use of raw sewage for the irrigation of vegetable crops, and high standards of community, food, and personal hygiene. Crucial treatment is prompt fluid therapy with electrolytes in large volume to replace all fluid loss with ORT. Using this form of treatment can successfully treat up to 80 percent of cholera cases. Tetracycline shortens the duration of the disease, and chemoprophylaxis for contacts following stool samples may help in reducing its spread. A vaccine is available; however, it has no value in the prevention of outbreaks.

Cholera and its burden on a country provide meaningful information, as this disease serves as a chief indicator of inadequate social development. It persists as a major

public health challenge in developing countries that lack fundamental infrastructure capable of providing clean, safe water. Owing to unsanitary living conditions, these populations and communities are at high risk of major cholera outbreaks as well as other diarrheal diseases.

Viral Gastroenteritis

Viral gastroenteritis can occur in sporadic or epidemic forms in infants, children, and adults. Some viruses, such as the

---

BOX 4.32 Cholera Pandemics in South America, 1991–1998, and Haiti, 2010–2012

In the 1980s, Peruvian officials stopped the chlorination of community water supplies because of concern over possible carcinogenic effects of trihalomethanes, a view encouraged by officials of the US Environmental Protection Agency (EPA) and the US Public Health Service. In January 1991, a Chinese freighter arrived in Lima, Peru, and dumped bilge (sewage) in the harbor, apparently contaminating local shellfish. Raw shellfish is a popular local delicacy (ceviche) and is associated with cases of cholera seen in local hospitals.

Contamination of local water supplies from sewage resulted in an exponential increase in cases, and by the end of 1992, the Pan American Health Organization (PAHO) reported an epidemic of 391,000 cases and 4002 deaths. The epidemic spread to 21 countries, and in 1992 there were a further 339,000 cases and 2321 deaths spreading over much of South America, continuing in 1999.

In the USA, 102 cases of cholera were reported in 1992; of these, 75 cases and one death were among passengers of an airplane flying from South America to Los Angeles in which contaminated seafood was served. In 1993, 91 cases of cholera were reported in the USA, though unrelated to international travel. These occurred mostly among people consuming shellfish from the Gulf coast with a strain of cholera similar to the South American strain, also possibly introduced in a ship's ballast. Cholera organisms are reported in harbor waters in other parts of the USA.

Haiti has experienced recurring cholera epidemics following overwhelming damage and loss of life in an earthquake. This event killed over 220,000 people and led to the displacement of 1.3 million people, with enormous damage to an already poor sanitary infrastructure. The cholera epidemics that occurred in 2010 and 2012 caused a reported 635,980 cases and 7912 deaths.

Sources: *Anderson C. Cholera epidemic traced to risk miscalculation. Nature 1991;354:255.*
Centers for Disease Control and Prevention. Update cholera – western hemisphere, 1992. MMWR Morb Mortal Wkly Rep 1993;42:89–91.
Centers for Disease Control and Prevention. Isolation of Vibrio cholerae O1 from oysters – Mobile Bay, 1991–1992. MMWR Morb Mortal Wkly Rep 1993;42:91–3.
Centers for Disease Control and Prevention. Outbreak notice: cholera in Haiti. Posted 29 January 2013 at http://globalhealth.kff.org/Daily-Reports/2011/October/03/GH-100311-Haiti-Cholera-Death-Toll.aspx [Accessed 29 January 2013].

rotaviruses and enteric adenoviruses, affect mainly infants and young children; they may be severe enough to require hospitalization for dehydration. Others, such as Norwalk and Norwalk-like viruses, affect older children and adults in self-limited acute gastroenteritis in family, institution, or community outbreaks.

Rotaviruses

Rotaviruses cause acute gastroenteritis in infants and young children, characterized by fever and vomiting, followed by watery diarrhea and occasionally severe dehydration and death if not adequately treated. Diagnosis is by examination of stool or rectal swabs with commercial immunological kits. In both developed and developing countries, rotavirus is the cause of about one-third of all hospitalized cases for diarrheal diseases in infants and children up to the age of 5. Most children in developing countries experience this disease by the age of 4 years, with the majority of cases occurring between 6 and 24 months. In developing countries, rotaviruses are estimated to cause over 1 million deaths per year. The virus is found in temperate climates in the cooler months and in tropical countries throughout the year. Breastfeeding does not prevent the disease but may reduce its severity. ORT is the key treatment.

Rotavirus represents the most common trigger of severe diarrheal disease affecting infants and young children worldwide. It is estimated that this class of viruses causes 527,000 deaths annually; over 85 percent of these deaths occur in low-resource countries in Africa and Asia. Each year over 2 million of these infants and children require hospitalization and supportive care to combat severe dehydration.

A live attenuated vaccine was approved by the FDA in 1998 and adopted in the 1999 US recommended routine vaccination programs for infants. In 2009 the WHO recommended rotavirus vaccine be included in all national immunization programs. There is strong evidence of herd immunity, as fewer children and adults are hospitalized following the introduction of infant vaccination. The vaccine for rotavirus is considered the single form of prevention recognized as having the strongest impact on reducing new, severe cases. In the USA the burden of rotavirus disease as a cause of hospitalization is greater than previously known. Furthermore, rotavirus as a cause of hospitalization for gastroenteritis has declined significantly over the years since the introduction of the vaccine (Figure 4.14).

Adenoviruses

Adenoviruses, Norwalk, and a variety of other viruses (including astrovirus, calcivirus, and other groups) cause sporadic acute gastroenteritis worldwide, mostly in outbreaks. Spread occurs via the fecal–oral route, often in hospital or other communal settings, with secondary spread among family contacts. Foodborne and waterborne transmission are both likely, and can represent serious problems in disaster situations. No vaccines are available. Management is with fluid replacement and hygienic measures to prevent secondary spread.

Parasitic Gastroenteritis

Giardiasis

Giardiasis (caused by *Giardia lamblia*) is a protozoan parasitic infection of the upper small intestine. It is usually

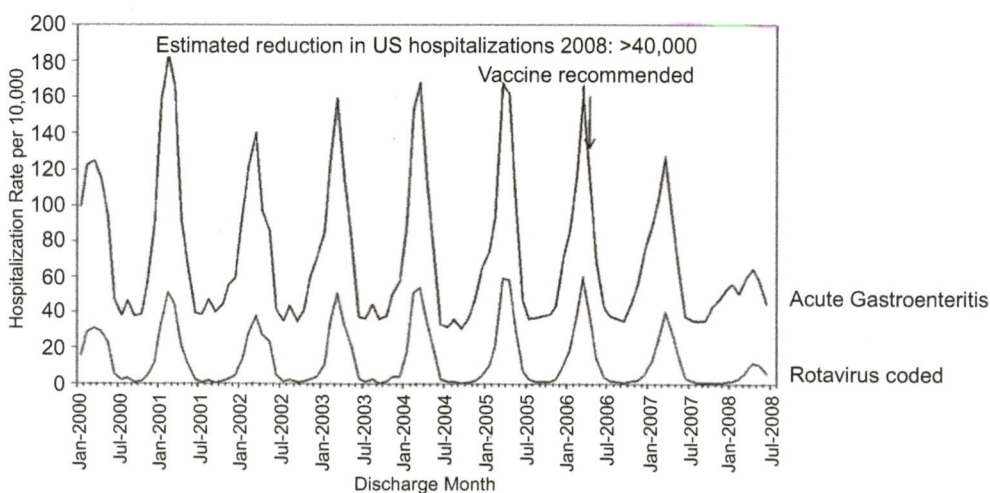

FIGURE 4.14   Acute gastroenteritis hospitalizations of children up to age 5, USA, 2000–2008, before and after routine use of the rotavirus vaccine. *Source: Curns AT, Steiner CA, Barrett M, Hiunter K, Wilson E, Parashar UD. Reduction in acute gastroenteritis hospitalizations among US children after introduction of rotavirus vaccine: analysis of hospital discharge data from 18 US states. J Infect Dis 2010;201:1617–24. Available at: http://jid.oxfordjournals.org/content/201/11/1617.full [Accessed 29 January 2013].*

asymptomatic, but is sometimes associated with chronic diarrhea; abdominal cramps; bloating; frequent, loose, greasy stools; fatigue; and weight loss. Malabsorption of fats and vitamins may lead to malnutrition. Diagnosis is by the presence of cysts or other forms of the organism in stools or duodenal fluid, or in intestinal mucosa from a biopsy. This disease is prevalent worldwide and affects mostly children. Like many other diarrheal diseases, it is spread in areas of poor sanitation, in preschool settings and swimming pools. Giardiasis is of increasing importance as a secondary infection among immunocompromised patients, especially those with AIDS.

Waterborne *Giardia* was recognized as a serious problem in the USA in the 1980s and 1990s, as the protozoa are not readily inactivated by chlorine; rather, adequate filtration before chlorination is required. Person-to-person transmission in day-care centers is common, as is transmission by unfiltered stream or lake water where contamination by human or animal feces is to be expected. An asymptomatic carrier state is common. Prevention relies on careful hygiene in settings such as day-care centers, filtration of public water supplies, and the boiling of water in emergency situations.

Cryptosporidium

Cryptosporidium parvum is a parasitic infection of the gastrointestinal tract in humans, small and large mammals, and other vertebrates. Infection may be asymptomatic or cause a profuse, watery diarrhea, abdominal cramps, general malaise, fever, anorexia, nausea, and vomiting. In immunosuppressed patients, such as people with AIDS, it can be a serious problem. The disease is most common in children under 2 years of age and those in close contact with them; it is also common among homosexual men. Diagnosis is by identification of the *Cryptosporidium* organism cysts in stools. The disease is present worldwide. In Europe and the USA, the organism has been found in 1–4.5 percent of individuals sampled. Spread is common by person-to-person contact via fecal–oral contamination, especially in such settings as day-care centers. Raw milk as well as waterborne outbreaks have also been identified in recent years. A large waterborne disease outbreak due to *Cryptosporidium* occurred in Milwaukee in 1986, as described in Chapter 9. Management is by rehydration and prevention is achieved by careful hygiene in food handling and water safety.

Helicobacter pylori

Helicobacter pylori, first identified in 1982, is a bacterium causally linked to gastrointestinal ulcers and gastritis, contributing to high rates of chronic peptic ulcer disease and to gastric cancer (Chapter 5). It is an important example of the causative link between infection and a group of chronic diseases. The discoverers of this link were Robin Warren and Barry Marshal in Australia, later rewarded with the Nobel Prize in Medicine in 2005. The diagnosis and management of chronic peptic ulcer revolutionized medical and surgical practices with a simple cure for this disease group. It had enormous implications for prevention of cancer of the stomach and chronic peptic ulcers, which filled medical and surgical wards of hospitals until the 1990s at great cost and consumed large amounts of medical resources.

A Program Approach to Diarrheal Disease Control

The control of diarrheal diseases requires a comprehensive program involving a wide range of activities, including good management of food and water supplies, education in hygiene and, particularly where morbidity and mortality are high, education in the use of ORT.

ORT is considered by UNICEF and WHO to have resulted in the saving of 1 million lives each year in the 1990s. Proper management of an episode of diarrhea by ORT (Table 4.16), along with continued feeding, not only saves the child from dehydration and immediate death, but also contributes to early restoration of nutritional adequacy, sparing the child the prolonged effects of malnutrition.

The World Summit for Children (WSC) in 1990 called for a reduction in child deaths from diarrheal diseases by one-third and malnutrition by one-half, with emphasis on the widest possible availability of, education in, and use of ORT. This requires a programmatic approach. Public health leadership must train primary care doctors, pediatricians,

---

TABLE 4.16 WHO Formula for Oral Rehydration Therapy

Ingredients	Amount (g/l)	Ions	Concentration (mmol/l)
Sodium chloride, NaCl	3.5	Sodium	90
Trisodium citrate, dihydrate, or sodium bicarbonate, NaHCO₃	2.9 or 2.5	Citrate[a]	20 citrate[b]
Potassium chloride, KCl	1.5	Potassium	10 of potassium, 80 of chloride
Glucose (anhydrous)	20.0	Glucose	111

Note: [a]Or 2.5 g sodium bicarbonate.
Sources: World Health Organization. Readings on diarrhoea: student manual. Geneva: WHO; 1992.
Heymann DL, editor. Control of communicable diseases manual. 18th ed. Washington, DC: American Public Health Association; 2004.

pharmacists, drug manufacturers, and primary care health workers of all kinds in ORT principles and usage. They must be supported by the widest possible publicity to raise awareness among parents.

ORT is an essential public health modality in developed countries as well as in developing countries. Although diarrheal disease does not cause death as frequently in developed countries, it represents a significant factor in infant and child health. Even under the most optimal conditions, it can cause setbacks in the nutritional state and physical development of a child. Use of ORT does not prevent the disease (i.e., it is not a primary prevention); however, it is an excellent form of secondary prevention, by preventing complications from diarrhea. Accordingly, it should be available in every home for symptomatic treatment of diarrheal diseases.

An adaptation of ORT has found its place in popular culture in the USA. A form of ORT, marketed as "sports drinks", is used in sports where athletes lose large quantities of water and salts in sweat and insensible loss from the respiratory tract. The wider application of the principles of ORT for use in adults in dry, hot climates and in adults under severe physical exertion with inadequate fluid and salt intake situations requires further exploration.

Management of diarrheal diseases should be part of a wider approach to child nutrition. The child who goes through an episode of diarrheal disease may falter in growth and development. Supportive measures may be needed following as well as during the episode. These involve providing primary care services that are attuned to monitoring individual infant and child growth. Growth monitoring surveillance is important to assess the health status of the individual child and the child population. Supplementation of infant feeding with vitamins A and D, and iron to prevent anemia, is important for routine infant and child care, and more so for conditions affecting total nutrition such as diarrheal diseases.

ACUTE RESPIRATORY INFECTIONS

In the developing world, respiratory infections account for over one-quarter of all deaths and illnesses in children. As diarrheal disease deaths are reduced, the major cause of death among infants in developing countries is becoming acute respiratory infections (ARIs). In industrialized countries, ARIs are important for their potentially devastating effects on elderly and chronically ill people. They are also the major cause of morbidity in infants in developed countries, causing much anxiety to parents even in areas with adequate living conditions. Cigarette smoking, chronic bronchitis, poorly controlled diabetes or congestive heart failure, and chronic liver and kidney disease increase susceptibility to ARIs. ARIs place a heavy burden on health care systems and individual families. Improved methods of management of such chronic diseases are needed to reduce

the associated toll of morbidity and mortality, and the considerable expenses of health care.

ARIs are due to a broad range of viral and bacterial infections. Secondary bacterial infections progress to pneumonia with mortality rates of 10–20 percent. Acute viral respiratory diseases include those affecting the upper respiratory tract, such as acute viral rhinitis, pharyngitis, and laryngitis, as well as those affecting the lower respiratory tract, such as tracheobronchitis, bronchitis, bronchiolitis, and pneumonia. ARIs are frequently associated with VPDs, including measles, varicella, and influenza. They are caused by a large number of viruses, producing a wide spectrum of acute respiratory illnesses. Some organisms affect any part of the respiratory tract, while others affect specific parts, and all predispose to bacterial secondary infection. While children and the elderly are especially susceptible to morbidity and mortality from acute respiratory disease, the vast numbers of respiratory illnesses among adults cause large-scale economic loss due to absence from work.

Bacterial agents causing upper respiratory tract infection include group A *Streptococcus*, *Mycoplasma pneumoniae*, pertussis, and parapertussis. Pneumonia, or acute bacterial infection of the lower respiratory tract and lung tissue, may be due to pneumococcal infection with *Streptococcus pneumoniae*. There are 83 known types of this organism, distinguished by capsule characteristics; 23 account for 88 percent of pneumococcal infections in the USA. An excellent polyvalent vaccine based on these types is available for high-risk groups such as elderly people, immunodeficient patients, and people with chronic heart, lung, liver, or blood disorders, or diabetes.

Opportunistic infections attack the chronically ill, especially those with compromised immune systems, often with life-threatening ARIs. Mycoplasma (primary atypical pneumonia) is a lower respiratory tract infection which sometimes progresses to pneumonia. TB and *Pneumocystis jiroveci* are especially problematic for AIDS patients. Other organisms causing pneumonia include *Chlamydia pneumoniae*, *Haemophilus influenzae*, *Klebsiella pneumoniae*, *E. coli*, *Staphylococcus*, rickettsia (Q fever), and *Legionella*. Parasitic infestation of the lungs may occur with nematodes (e.g., ascariasis). Fungal infections of the lungs may be caused by aspergillosis, histoplasmosis, and coccidiomycosis, often as a complication of antibiotic therapy.

Access to primary care and early institution of treatment are vital in controlling excess mortality from ARIs. In developed countries, ARIs as contributors to infant deaths are largely a problem in minority and deprived population groups. Because these groups contribute disproportionately to childhood mortality, infant mortality reduction has been slower in countries such as the USA and Russia than in other industrialized countries. The continuing gap in mortality rates between white and African American children in the USA can, to a large extent, be attributed to ARIs and less access to organized primary care. Children are brought to emergency rooms for care when the disease process is already advanced and more dangerous than had it been attended to professionally earlier in the process. Many field trials of ARI prevention programs have proved successful, involving parent education and training of primary care workers in early assessment and, if necessary, initiation of treatment. This needs field testing in multiple settings.

Reliance on vaccines to prevent respiratory infectious diseases is not currently feasible. ARIs are caused by a very wide spectrum of viruses, and the development of vaccines in this field has been slow and limited. The vaccine for pneumococcal pneumonia has been an important breakthrough, but it is still inadequately utilized by the chronically ill because of its limitations, costs, and lack of sufficient political and public awareness. Furthermore, it is too expensive for developing countries. This vaccine is recommended for infants in the USA and many industrial nations and recommended by the WHO for developing countries, but has yet to be widely applied in the latter. Improvements in bacterial and viral vaccine development will potentially help to reduce the burden of ARIs. A programmatic approach with clinical guidelines and education of family and caregivers is currently the only feasible way to reduce the still enormous burden of morbidity and mortality from ARIs among young and elderly people.

INEQUALITIES IN CONTROL OF COMMUNICABLE DISEASES

As in other fields of public health, there are wide variations or inequities between and within countries pertaining to the control of communicable diseases. The differences between the industrialized countries and the developing countries are enormous. The gaps are not only in coverage, but also in the content of the immunization programs. For instance, adoption of Hib vaccine is increasing; however, the decade-long gap from availability to widespread global usage results in many preventable deaths. Similarly, the lag in adoption of pneumococcal pneumonia and rotavirus vaccines will prolong the time it takes to achieve the MDGs of reducing child mortality in very many countries.

Even in the European Region, there are wide differences between groups of countries, as seen in Figure 4.15 comparing standardized mortality rates (3-year moving averages) for infectious and parasitic disease between long-standing members of the EU (such as France, Germany, and the UK) with the new members (since 2004, such as Hungary, Poland, and other countries of Eastern Europe), those of the Commonwealth of Independent States (e.g., Ukraine and Russia), and finally the Central Asian Republics (e.g., Kazakhstan, Uzbekistan, and Tajikistan). The trends show low and stable rates in the countries of Western and Eastern Europe, high and falling rates in Central Asia, but high and rising rates in the key countries of the former Soviet Union. While there may be artifacts of reporting, the trends

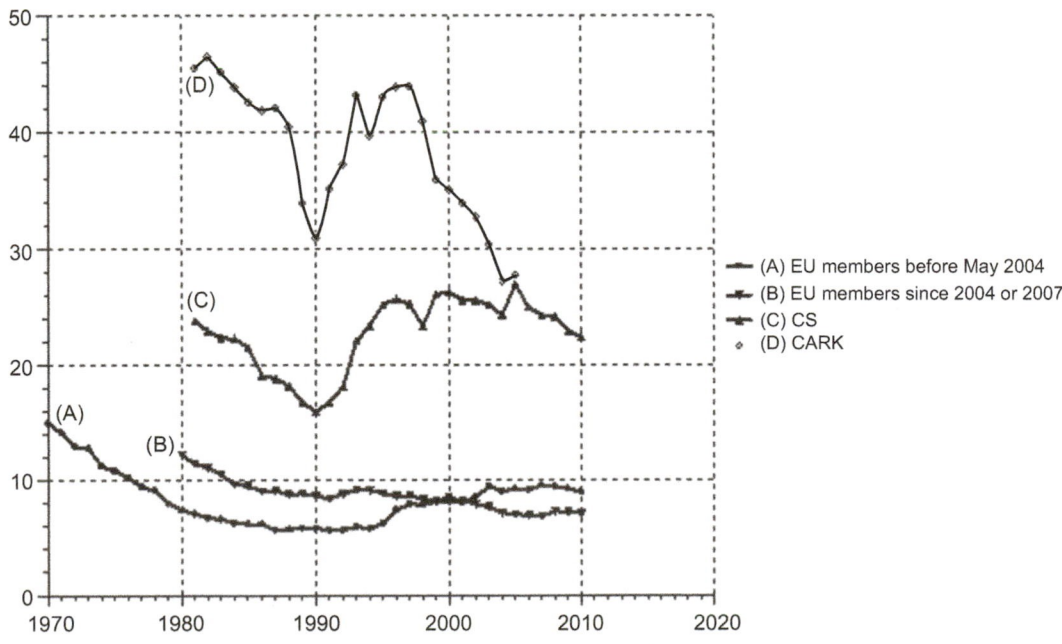

FIGURE 4.15   Standardized death rates (per 100,000 population, standardized to European population) from infectious and parasitic diseases, selected regions of Europe, 1970–2005. Note: CIS=Commonwealth of Independent States (Russia, Ukraine, etc.); CARK=Central Asian Republics (Kazakhstan, Uzbekistan, Tajikistan, Kirgizstan). *Source: World Health Organization, European Regional Office. Health for All database. August 2012. Available at: www.who/dk/hfadb [Accessed 28 January 2013].*

are thought to be accurate, and are likely to be related to many factors such as water and food safety, obsolescent immunization programs, TB and HIV control, and a myriad of other determinants. Comparisons within countries also reveal social and regional disparities, which constitute failings of public health systems. Inexcusably and unreasonably wide gaps indicate that communicable diseases must remain a key part of the modern public health agenda.

The European Region, which includes all of these groups of countries such as the EU, does not have a standard or harmonized immunization schedule; thus, each country follows its own patterns. Western European countries are generally up to date with the content of their immunization programs and with high coverage, but this is not uniformly so even in this group, and certainly not in the European Region as a whole. The countries of the former Soviet Union are gradually updating their immunization schedules but remain largely at least a decade behind. The WHO's advisory committee system on immunization has been updating its recommendations rapidly in recent years for hepatitis B, and more recently Hib and pneumococcal pneumonia, and adopting of the two-dose policy of MMR vaccination. As new vaccines become available transitional and developing countries will need support to expand their programs of immunization, a key part of the drive to attain the MDGs of reduced child mortality and control of infectious diseases (malaria, HIV, and others).

COMMUNICABLE DISEASE CONTROL IN THE NEW PUBLIC HEALTH

The success of sanitation, vaccines, and antibiotics led many to assume that all infectious diseases would sooner or later succumb to public health and medical technology. Unfortunately, this is a premature and even dangerous assumption. Despite the long-standing availability of an effective and inexpensive vaccine, the persistence of measles as a major killer of 1 million children per year represents a failure in effective use of both the vaccine and the health system. The resurgence of TB and malaria has led to new strategies, such as managed or directly observed care, with CHWs ensuring compliance needed to render the patient non-infectious to others and, similarly, to reduce the pool of carriers of the disease.

Successes achieved in reducing polio, measles, dracunculiasis, onchocerciasis, and other diseases to the level of local or global eradication have raised hopes for similar success in other fields. But there are many infectious diseases of importance in developed and developing countries where existing technologies are not fully utilized. ORT is one of the most cost-effective methods of preventing excess mortality from ordinary diarrheal diseases, yet it is not utilized on a sufficient scale.

Biases in the financing and management of medical insurance programs can result in underutilization of available effective vaccines. Hospital-based infections cause large-scale increases in lengths of stay and expenditures, although the application of epidemiological investigation and improved quality in hospital practices could reduce this burden. Control of the spread of AIDS using combined medical therapies is not financially or logistically possible in many countries, but education for safe sex is effective. CHW programs can greatly enhance TB, malaria, and STI control; or in AIDS care, promote prevention and appropriate treatment.

The link between infectious disease and non-communicable disease (NCDs) has become a major new development of public health in the past several decades. Vaccines for hepatitis B and HPV to prevent cancer of the liver and of the cervix will save countless lives. The discovery of *H. pylori* as the cause of chronic peptic ulcer disease and gastric cancer provided an easily diagnosed and treated infection which will also save many lives and reduce hospitalization for diseases that formerly filled the medical and surgical wards of hospitals. The potential for new discoveries of infectious and genetic factors in disease is augmented by a growing understanding of the importance of social and economic factors in all diseases. The advent of health promotion awareness became evident when the HIV/AIDS pandemic struck, for which there were no medical answers except palliative care, but human behavior was the vital link to management and remains so even since excellent medical treatments have become available.

In the industrialized and mid-level developing countries, epidemiological and demographic shifts have created new challenges in infectious disease control. The prevention and early treatment of infectious disease among the chronically ill and the elderly is not only a medical issue, it is also an economic one. Patients with chronic obstructive pulmonary disease (COPD), chronic liver or kidney disease, or congestive heart failure are at high risk of developing an infectious disease followed by prolonged hospitalization.

SUMMARY

Public health has addressed, and will continue to stress, the subject of communicable disease as one of its key issues in protecting individual and population health. Methods of intervention include classic public health through sanitation, safe water and foods, immunization, and well beyond that into nutrition, education, case finding, treatment, and changing human behavior. The knowledge, attitudes, beliefs, and practices of policy makers, health care providers, and parents are as important in the success of communicable disease control as the technology available and methods of financing health systems. Together, these encompass the broad programmatic approach of the New Public Health to the control of communicable diseases.

In a world of rapid international transport and contact between populations, systems are needed to monitor the potentially explosive spread of pathogens that may be transferred from their normal habitat. The potential for the international spread of new or re-emerged, reinvigorated infectious diseases constitutes a threat to humankind akin to ecological and other anthropogenic disasters.

The eradication of smallpox has paved the way for the eradication of polio, and perhaps measles, in the foreseeable future. New vaccines are showing the capacity to reduce important morbidity from rubella syndrome, mumps, meningitis, and hepatitis. Other new vaccines on the horizon will continue the immunological revolution into the twenty-first century.

As the triumphs of control or elimination of infectious diseases of children continue, the scourge of HIV infection continues, with distressingly slow progress in the development of an effective vaccine or cure for the disease it engenders. Partly as a result of HIV/AIDS, TB staged a comeback in many countries where it was thought to be merely a residual problem. At the same time an old/new method of intervention using directly observed short-term therapy has shown great success in controlling the TB epidemic. The resurgence of TB is dangerous in that MDR-TB has become a widespread problem. This issue highlights the difficulty of keeping ahead of drug resistance in the search for new generations of antibiotics, posing a difficult challenge for the pharmaceutical industry, basic scientists, and public health workers.

The burden of infectious diseases has appeared to recede as the predominant public health problem in the developed countries. Despite this, new challenges of emerging infectious diseases have come to the fore in public health, and communicable disease remains a dominating problem in the developing countries. With increases in longevity and the increased importance of chronic disease in the health status of the industrial and mid-level developing nations, the effects of infectious disease on the care of elderly and chronically ill people are of great importance in the New Public Health. Long-term management of chronic disease needs to address the care of vulnerable groups, promoting the use of existing vaccines and antibiotics. Most important is the development of health systems that provide close monitoring of groups at special risk for infectious disease, especially patients with chronic diseases, the immunocompromised, and the elderly. The combination of traditional public health with direct medical care needed for effective control and eradication of communicable diseases is an essential element of the New Public Health. The challenges include applying a comprehensive approach and managing resources to define and reach achievable targets in communicable disease control.

Control of communicable diseases is one of the fundamental pillars of public health. The new capacities of vaccines and other methods of control develop slowly, and the advent of effective vaccines for HIV, malaria, and TB will bring untold benefit to the global community. The challenges of natural dispersion of communicable disease can be made

more threatening because of the advent of bioterrorism and the emergence of new diseases or the spread of those previously localized in a less mobile, less globalized world. The challenges, the potential for harm, and the benefits that can be achieved in this aspect of public health are enormous.

NOTE

For an abstract, guidance for student review and expected competencies as well as an extended bibliography, please see companion website at http://booksite.elsevier.com/9780124157668

BIBLIOGRAPHY

Alvar, J., Vélez, I.D., Bern, C., Herrero, M., Desjeux, P., Cano, J., et al., 2012. Leishmaniasis worldwide and global estimates of its incidence. PLoS ONE 7 (5), e35671. Available at: http://www.plosone.org/article/info%3Adoi%2F10.1371%2Fjournal.pone.0035671 (accessed 04.02.13).

Aylward, B., Hennessey, K.A., Zagaria, N., Olivé, J.-M., Cochi, S., 2000. When is a disease eradicable? 100 years of lessons learned. Am. J. Public Health 90, 1515–1520. Available at: http://ajph.aphapublications.org/doi/pdf/10.2105/AJPH.90.10.1515 (accessed 04.02.13).

Bern, C., Montgomery, S.P., 2009. An estimate of the burden of Chagas disease in the United States. Clin. Infect. Dis. 49 (5), e52–e54. Available at: http://cid.oxfordjournals.org/content/49/5/e52.long (accessed 24.02.13).

Berger, S.A. Leprosy: Global status, 2013. Gideon e-books. Available at: http://www.gideononline.com/ebooks/disease/leprosy-global-status/ (accessed 16.11.13).

Brachmen, P.S., Thacker, S.B., 2011. Evolution of epidemic investigations and field epidemiology during the MMWR era at CDC – 1961–2011. MMWR. Surveill. Summ. 60 (Suppl. 4), 22–26. Available at: http://www.cdc.gov/mmwr/preview/mmwrhtml/su6004a5.htm (accessed 22.01.13).

Brooks, G.F., Carroll, K.C., Butel, J.S., Morse, S.A., 2012. Jawetz, Melnick and Adelberg's medical microbiology, twentysixth ed. McGraw-Hill, New York.

Carter Center, Guinea worm disease eradication. Atlanta, GA: Carter Center. Available at: http://www.cartercenter.org/health/guinea_worm/mini_site/index.html (accessed 17.02.13).

Carter Center, International Task Force for Disease Eradication. Available at: http://www.cartercenter.org/health/itfde/index.html (accessed 20.02.13).

Centers for Disease Control and Prevention. Ten great public health achievements – United States, 2001–2010. MMWR Morb Mortal Wkly Rep 2011;60:619–23. Available at: http://www.cdc.gov/mmwr/preview/mmwrhtml/mm6019a5.htm (accessed 30.01.13).

Centers for Disease Control and Prevention, 2004. 150th anniversary of John Snow and the pump handle. MMWR. Morb. Mortal. Wkly. Rep. 53, 783. Available at: http://www.cdc.gov/mmwr/preview/mmwrhtml/mm5334a1.htm (accessed 30.01.13).

Centers for Disease Control and Prevention, 2011. A CDC framework for preventing infectious diseases: sustaining the essentials and innovating for the future. CDC, Atlanta, GA. Available at: http://www.cdc.gov/oid/docs/ID-Framework.pdf (accessed 04.02.13).

Centers for Disease Control and Prevention, 1999. Achievements in public health, 1900–1999: control of infectious diseases. MMWR. Morb. Mortal. Wkly. Rep. 48, 621–629. Available at: http://www.cdc.gov/mmwr/PDF/wk/mm4829.pdf (accessed 04.02.13).

Centers for Disease Control and Prevention, 1999. Achievements in public health, 1900–1999: impact of vaccines universally recommended for children – United States, 1990–1998. MMWR. Morb. Mortal. Wkly. Rep. 48, 243–248.

Centers for Disease Control and Prevention, February 2013. CDC fact sheet. Incidence, Prevalence, and cost of sexually transmitted, infections in the United States. Available at: http://www.cdc.gov/std/stats/sti-estimates-fact-sheet-feb-2013.pdf (accessed 08.10.13).

Centers for Disease Control and Prevention, Chikungunya fever. [updated 22 May 2012]. Available at: http://www.cdc.gov/chikungunya/ (accessed 26.01.13).

Centers for Disease Control and Prevention, Emerging infectious diseases national summit on neglected infections of poverty in the United States. Available at: http://wwwnc.cdc.gov/eid/article/16/5/09-1863_article.htm (accessed 23.02.13).

Centers for Disease Control and Prevention. Health disparities and inequality report, United States, 2011. MMWR. Morb. Mortal. Wkly. Rep. 2011;60(Suppl):1–114. Available at: http://www.cdc.gov/mmwr/pdf/other/su6001.pdf (accessed 30.01.13).

Centers for Disease Control and Prevention, 2011 August. HIV in the United States: an overview. Available at: http://www.cdc.gov/hiv/topics/surveillance/resources/factsheets/pdf/HIV-US-overview.pdf. (accessed 04.02.13).

Centers for Disease Control and Prevention, Human papillomavirus (HPV), signs and symptoms. [updated 5 February 2013]. Available at: http://www.cdc.gov/hpv/Signs-Symptoms.html (accessed 14.02.13).

Centers for Disease Control and Prevention, 1998. Impact of the sequential IPV/OPV schedule on vaccination coverage – United States, 1997. MMWR. Morb. Mortal. Wkly. Rep. 47, 1017–1019. Available at: http://www.cdc.gov/mmwr/preview/mmwrhtml/00055785.htm (accessed 04.02.13).

Centers for Disease Control and Prevention, 2012. Measles – United States, 2011. MMWR. Morb. Mortal. Wkly. Rep. 61, 253–257. Available at: http://www.cdc.gov/mmwr/preview/mmwrhtml/mm6115a1.htm (accessed 25.01.13).

Centers for Disease Control and Prevention, Multistate fungal meningitis outbreak investigation. [updated 23 January 2013]. Available at: http://www.cdc.gov/hai/outbreaks/currentsituation/ (accessed 04.02.13).

Centers for Disease Control and Prevention, Pertussis (whooping cough) outbreaks. [updated 19 February 2013]. Available at: http://www.cdc.gov/pertussis/outbreaks.html (accessed 04.02.13).

Centers for Disease Control and Prevention, 2011. Progress toward global eradication of dracunculiasis, January 2010–June 2011. MMWR. Morb. Mortal. Wkly. Rep. 60, 1450–1453. Available at: http://www.cdc.gov/mmwr/preview/mmwrhtml/mm6042a2.htm (accessed 17.02.13).

Centers for Disease Control and Prevention, 2011. Public health then and now: celebrating 50 years of MMWR at CDC. MMWR. Morb. Mortal. Wkly. Rep. 60 (Suppl.). Available at: http://www.cdc.gov/mmwr/pdf/other/su6004.pdf (accessed 22.01.13).

Centers for Disease Control and Prevention, 2013. Recommended immunization schedule for persons age 0 through 18 years – United States. Available at: http://www.cdc.gov/vaccines/schedules/hcp/imz/child-adolescent.html. (accessed 20.02.13).

Centers for Disease Control and Prevention, Atlanta, GA: CDC. Rotavirus. Available at: http://www.cdc.gov/vaccines/pubs/pinkbook/downloads/rota.pdf (accessed 04.02.13).

Centers for Disease Control and Prevention, 2010. Sexually transmitted diseases treatment guidelines 2010. MMWR. Morb. Mortal. Wkly. Rep. 59 (RR-12). Available at: http://www.cdc.gov/mmwr/preview/mmwrhtml/rr5912a1.htm?s_cid=rr5912a1_e (accessed 04.02.13).

Centers for Disease Control and Prevention, 2011. Ten great public health achievements – United States, 2001–2010. MMWR. Morb. Mortal. Wkly. Rep. 60, 619–623. Available at: http://www.cdc.gov/mmwr/preview/mmwrhtml/mm6019a5.htm (accessed 12.08.13).

Centers for Disease Control and Prevention, 2011. Ten great public health achievements – worldwide, 2001–2010. MMWR. Morb. Mortal. Wkly. Rep. 60, 814–818. Available at: http://www.cdc.gov/mmwr/preview/mmwrhtml/mm6024a4.htm. (accessed 12.08.13).

Centers for Disease Control and Prevention, 2012. The ABCs of hepatitis. Available at: http://www.cdc.gov/hepatitis/Resources/Professionals/PDFs/ABCTable.pdf.

Centers for Disease Control and Prevention, 11 July 2013. Vaccines & immunizations Glossary/acronyms. Available at: http://www.cdc.gov/vaccines/about/terms/vacc-abbrev.htm. (accessed 08.08.13).

Centers for Disease Control and Prevention, 2012. Viral hepatitis statistics & surveillance. Available at: http://www.cdc.gov/hepatitis/statistics/ (accessed 22.10.13).

Centers for Disease Control and Prevention. West Nile virus, statistics, surveillance, and control: West Nile virus (WNV) human infections reported to ArboNET, by state, United States, 2012 (as of December 11, 2012). Available at: http://www.cdc.gov/ncidod/dvbid/westnile/surv&controlCaseCount12_detailed.htm (accessed 20.02.13).

Centers for Disease Control and Prevention. Immunization schedules: Birth-18 Years & "Catch-up" Immunization Schedules United States, 2013: Details For Health Care Professionals. Available at: http://www.cdc.gov/vaccines/schedules/hcp/child-adolescent.html (accessed 19.11.13).

Centers for Disease Control and Prevention Advisory Committee on Immunization Practices (ACIP) Recommended Immunization Schedule for Persons Aged 0 Through 18 Years — United States, 2013. MMWR. Morb. Mortal. Wkly. Rep. Suppl. 2013; 62(1):2–8. Available at: http://www.cdc.gov/mmwr/preview/mmwrhtml/su6201a2.htm (accessed 19.11.13).

Chevalier, V., Pepin, M., Plee, L., Lancelot, R., 2010. Rift Valley fever – a threat for Europe? Euro. Surveill. 15 (10). pii=19506. Available at: http://www.eurosurveillance.org/ViewArticle.aspx?ArticleId=19506 (accessed 03.01.13).

College of Physicians of Philadelphia, 2013. History of anti-vaccination movements. Vaccines. Available at: http://www.historyofvaccines.org/content/articles/history-anti-vaccination-movements (accessed 23.11.11).

College of Physicians of Philadelphia, Vaccines: vaccine development and licensing events. Available at: http://www.historyofvaccines.org/content/articles/vaccine-development-licensing-events (accessed 15.06.11).

College of Physicians and Surgeons of Philadelphia, 2013. Vaccines: the history of Lyme disease. Available at: http://www.historyofvaccines.org/content/articles/history-lyme-disease-vaccine. (accessed 24.01.13).

Dar, O., McIntyre, S., Hogarth, S., Heymann, D., 2013. Rift Valley fever and a new paradigm of research and development for zoonotic disease control. Emerg. Infect. Dis. 19, 189–193. Available at: http://wwwnc.cdc.gov/eid/article/19/2/pdfs/12-0941.pdf (accessed 04.02.13).

Dowdle, W.R., 1999. The principles of disease elimination and eradication. MMWR. Morb. Mortal. Wkly. Rep. 48, 23–27. Available at: http://www.cdc.gov/mmwr/preview/mmwrhtml/su48a7.htm (accessed 04.02.13).

European Creutzfeldt Jakob Disease Surveillance Network (EUROCJD), June 2012. EuroCJD surveillance data: vCJD cases worldwide 28. Available at: http://www.eurocjd.ed.ac.uk/surveillance%20data%204.htm (accessed 04.02.13).

Flahault, A., Zylberman, P., 2010. Influenza pandemics: past, present and future challenges. Public Health Rev. 32, 319–340. Available at: http://www.publichealthreviews.eu/show/f/37 (accessed 04.02.13).

General Medical Council Great Britain. Date: 24, May 2010. Dr Andrew Jeremy WAKEFIELD Determination on Serious Professional Misconduct (SPM) and sanction. Available at: http://www.gmc-uk.org/Wakefield_SPM_and_SANCTION.pdf_32595267.pdf (accessed 10.11.13).

Global Polio Eradication Initiative, Infected countries. Available at: http://www.polioeradication.org/Infectedcountries.aspx (accessed 18.01.13).

Goldblum, N., Gerichter, ChB., Tulchinsky, T.H., Melnick, J.L., 1994. Poliomyelitis control in Israel, the West Bank and Gaza Strip 1948–1993: changing strategies with the goal of eradication in an endemic area. Bull. World Health Organ. 72, 783–796. Available at: http://www.ncbi.nlm.nih.gov/pmc/articles/PMC2486552/ (accessed 15.02.13).

Goodman RA, Foster KL, Trowbridge FL, Figueroa JP, editors. Global disease elimination and eradication as public health strategies. MMWR Morb. Mortal. Wkly. Rep. 1999;48(Suppl):1–309. Available at: ftp://ftp.cdc.gov/pub/Publications/mmwr/other/suppl48.pdf (accessed 23.01.13).

Government of Manitoba, 2011. Health. Manitoba Immunization Monitoring System (MIMS). Annual Report - 2010. Available at: http://www.gov.mb.ca/health/publichealth/surveillance/reports.html.

Guide to Community Preventive Services, Targeted vaccinations: multiple interventions implemented in combination. [updated 3 December 2011]. Available at: http://www.thecommunityguide.org/vaccines/targeted/multi_combination.html (accessed 23.02.13).

Hartley, D.M., Rinderknecht, J.L., Nipp, T.L., Clarke, N.P., Snowder, G.D., 2011. National Center for Foreign Animal and Zoonotic Disease Defense Advisory Group. Potential effects of Rift Valley fever in the United States. Emerg. Infect. Dis. 17 (8). Available at: http://wwwnc.cdc.gov/eid/article/17/8/10-1088_article.htm (accessed 04.02.13).

Heymann, D.L., 2008. Control of communicable diseases manual, nineteenth ed. American Public Health Association, Washington, DC.

Hopkins, D.R., 2013. Global health: disease eradication. N. Engl. J. Med. 368, 54–63. Available at: http://www.nejm.org/doi/full/10.1056/NEJMra1200391 (accessed 20.02.13).

Hotez, P.J., Fenwick, L., Svioli, L., Molyneux, D.H., 2009. Rescuing the bottom billion through control of neglected tropical diseases. Lancet 373, 1570–1575.

Institute of Medicine, 2011. The causes and impacts of neglected tropical and zoonotic diseases: opportunities for integrated intervention strategies. National Academies Press, Washington, DC. Available at: http://www.ncbi.nlm.nih.gov/books/NBK62507/ (accessed 23.02.13).

London Declaration on Neglected Tropical Diseases, 2012. Uniting to combat neglected tropical diseases. Ending the neglect and reaching 2020 goals. Available at: http://www.unitingtocombatntds.org/downloads/press/london_declaration_on_ntds.pdf (accessed 30.07.13).

Plotkin, S.A., Orenstein, W.A., Offit, P.A. (Eds.), 2008. Vaccines. fifth ed. Saunders, Philadelphia, PA.

Reverby, S.M., 2012. Ethical failures and history lessons: the US Public Health Service research studies in Tuskegee and Guatemala. Public Health Rev. 34. epub ahead of print. Available at: www.publichealthreviews.eu (accessed 28.01.13).

Roll Back Malaria Partnership, January 2013. AMP zeros in on continued net distribution in malaria-endemic countries 31. Available at: http://www.rbm.who.int/globaladvocacy/eventsarchive2013.html (accessed 04.02.13).

Schneider, D., Evering-Watley, M., Walke, H., Bloland, P.B., 2011. Training the global public health workforce through applied epidemiology training programs: CDC's experience. Public Health Rev. 33, 190–203. Available at: http://www.publichealthreviews.eu/upload/pdf_files/9/Schneider.pdf (accessed 31.01.13).

Suerbaum, S., Michetti, P., 2002. *Helicobacter pylori* infection. N. Engl. J. Med. 347, 1175–1186. Available at: http://www.nejm.org/doi/full/10.1056/NEJMra020542 (accessed 04.02.13).

Tulchinsky, T.H., Ginsberg, G.M., Abed, Y., Angeles, M.T., Akukwe, C., Bonn, J., 1993. Measles control in developing and developed countries: the case for a two-dose policy. Bull. World Health Organ. 71, 93–103. Available at: http://www.ncbi.nlm.nih.gov/pmc/articles/PMC2393424/ (accessed 04.02.13).

Tulchinsky, T.H., Ramlawi, A., Abdeen, Z., Grotto, I., Flahault, A., 2013. Polio lessons 2013: Israel, the West Bank, and Gaza. Lancet 382 (9905), 211–212. Available at: http://www.thelancet.com/journals/lancet/issue/current (accessed 16.11.13).

Tulchinsky, T.H., Ramlawi, A., Abdeen, Z., Grotto, I., Flahault, A., 2013. Polio lessons 2013: Israel, the West Bank, and Gaza. Lancet. 382 (9905). Available at: http://www.thelancet.com/journals/lancet/issue/current (accessed 18.11.13).

UNAIDS, 2012. Global fact sheet. World AIDS Day. Available at: http://www.unaids.org/en/media/unaids/contentassets/documents/epidemiology/2012/gr2012/20121120_FactSheet_Global_en.pdf. (accessed 04.02.13).

UNAIDS, 2013. Global Report UNAIDS Report on the global AIDS epidemic. Available at: http://www.unaids.org/en/media/unaids/contentassets/documents/epidemiology/2013/gr2013/UNAIDS_Global_Report_2013_en.pdf (accessed 04.11.13).

Wise, J., 2006. Demand for male circumcision rises in a bid to prevent HIV. Bull. World Health Organ. 84, 509–511. Available at: http://www.ncbi.nlm.nih.gov/pmc/articles/PMC2627386/ (accessed 04.02.13).

World Health Organization, 2011–2012 Tuberculosis global facts. Available at: http://www.who.int/tb/publications/2011/factsheet_tb_2011.pdf (accessed 04.02.13).

World Health Organization, Chagas disease (American trypanosomiasis). Fact sheet no. 340. [updated March 2013]. Available at: http://www.who.int/mediacentre/factsheets/fs340/en/index.html (accessed 08.08.13).

World Health Organization, Cholera. Fact sheet no. 107 [updated July 2012]. Available at: http://www.who.int/mediacentre/factsheets/fs107/en/index.html (accessed 04.02.13).

World Health Organization, 2010. Controlling rubella and preventing congenital rubella syndrome – global progress, 2009. Wkly. Epidemiol. Rec. 85, 413–424. Available at: http://www.who.int/wer/2010/wer8542.pdf (accessed 04.02.13).

World Health Organization, 2010. First WHO report on neglected tropical diseases: working to overcome the global impact of neglected tropical diseases. Available at: http://whqlibdoc.who.int/publications/2010/9789241564090_eng.pdf.

World Health Organization. Cumulative number of confirmed human cases for avian influenza A(H5N1) reported to WHO, 2003–2013; July 2013. Available at: www.who.int/influenza/human_animal_interface/EN_GIP_20130705CumulativeNumberH5N1cases_2.pdf (accessed 08.08.13).

World Health Organization, November 2012. Dracunculiasis. Posted 5. Available at: http://www.who.int/dracunculiasis/en/ (accessed 16.02.13).

World Health Organization, August 2012. Ebola haemorrhagic fever. Fact sheet no. 103. Available at: http://www.who.int/mediacentre/factsheets/fs103/en/. (accessed 04.02.13).

World Health Organization. Global Advisory Committee on Vaccine Safety, report of meeting held 12–13 June 2013. Available at: http://www.who.int/vaccine_safety/committee/reports/Jun_2013/en/. (accessed 22.10.13).

World Health Organization, 14 August 2012. Global Alert and Response (GAR): Ebola in Uganda – update. Available at: http://www.who.int/csr/don/2012_08_14/en/index.html. (accessed 08.09.12).

World Health Organization, Global Alert and Response (GAR): Hepatitis C. Available at: http://www.who.int/csr/disease/hepatitis/whocdscsrlyo2003/en/index4.html (accessed 22.10.13).

World Health Organization, 2011. Hepatitis C. Wkly. Epidemiol. Rec. 86, 445–456. Available at: http://www.who.int/wer/2011/wer8641.pdf (accessed 04.02.13).

World Health Organization, April 2013. Immunization coverage. Fact sheet no. 378. Available at: http://www.who.int/mediacentre/factsheets/fs378/en/. (accessed 08 08.13).

World Health Organization, 2013. Initiative for Vaccine Research (IVR). Sexually transmitted diseases. Chlamydia trachomatis. Available at: http://www.who.int/vaccine_research/diseases/soa_std/en/index1.html. (accessed 08.08.13).

World Health Organization, 2013. Leishmaniasis. . Available at: http://www.who.int/leishmaniasis/en/. (accessed 12.08.13).

World Health Organization, 2013. Global leprosy: update on the 2012 situation. WER 35 (88), 365–380. Available at: http://www.who.int/wer/2013/wer8835.pdf (accessed 19.11.13).

World Health Organization, September 2012. Leprosy. Fact sheet no. 101. Available at: http://www.who.int/mediacentre/factsheets/fs101/en/index.html. (accessed 04.02.13).

World Health Organization, MDG 6: combat HIV/AIDS, malaria, and other diseases. [updated February 2013]. Available at: http://www.who.int/topics/millennium_development_goals/diseases/en/index.html (accessed 20.02.13).

World Health Organization/UNICEF. WHO vaccine-preventable disease monitoring system, 2013 global summary. Available at: http://www.who.int/immunization_monitoring/data/gs_gloprofile.pdf. (accessed 19.11.13).

World Health Organization, February 2013. Measles. Fact sheet no. 286; Available at: http://www.who.int/mediacentre/factsheets/fs286/en/. (accessed 04.02.13).

World Health Organization. Viral hepatitis statistics and surveillance. [Updated 19 August 2013]. Available at: www.cdc.gov/hepatitis/statistics/ (accessed 22.10.13).

World Health Organization, 2011. Meeting of the International Task Force for Disease Eradication, April 2011. Wkly. Epidemiol. Rec. 86, 341–352. Available at: http://www.who.int/wer/2011/wer8632.pdf (accessed 04.02.13).

World Health Organization, New and underutilized vaccines: Rotavirus. [updated October 2011]. Available at: http://www.who.int/nuvi/rotavirus/en/ (accessed 04.02.13).

World Health Organization, 2007. Pneumococcal conjugate vaccine for childhood immunization – WHO position paper. Wkly. Epidemiol. Rec. 82, 93–104. Available at: http://www.who.int/wer/2007/wer8212/en/index.html (accessed 04.02.13).

World Health Organization, Rabies Bulletin Europe. Available at: http://www.who-rabies-bulletin.org/Queries/Surveillance.aspx?Issue=2012_3 (accessed 08.08.13).

World Health Organization, 2012. Variant Creutzfeldt–Jakob disease. Fact sheet no. 180; revised February. Available at: http://www.who.int/mediacentre/factsheets/fs180/en/. (accessed 15.02.13).

World Health Organization, Water sanitation health: facts and figures on water quality and health. Available at: http://www.who.int/water_sanitation_health/facts_figures/en/index.html (accessed 04.02.13).

World Health Organization, July 2011. West Nile Virus. Fact sheet no. 354. Available at: http://www.who.int/mediacentre/factsheets/fs354/en/index.html. (accessed 04.02.13).

World Health Organization. Neglected tropical diseases. Available at: http://www.who.int/neglected_diseases/2010report/en/index.html (accessed 26.01.13) and WHO. http://www.who.int/neglected_diseases/EB_resolution_2013/en/index.html (accessed 24.10.13).

World Health Organization, 2009. WHO guidelines on hand hygiene in health care, first global patient safety challenge, clean care is safer care. Available at: WHO, Genevahttp://whqlibdoc.who.int/publications/2009/9789241597906_eng.pdf.

World Health Organization. WHO Report 2011: global tuberculosis control. Geneva: WHO; 2011. Available at: http://www.who.int/tb/publications/global_report/2011/gtbr11_full.pdf (accessed 04.02.13).

World Health Organization, 2011. Working to overcome the global impact of neglected tropical disease, update 2011. WHO, Geneva. Available at: http://www.who.int/neglected_diseases/2010report/WHO_NTD_report_update_2011.pdf (accessed 04.02.13).

World Health Organization. World Malaria Report 2011: Executive summary. Geneva: WHO; 2011. Available at: http://www.who.int/malaria/world_malaria_report_2011/wmr2011_summary_keypoints.pdf (accessed 04.02.13).

World Health Organization European Region, July 2013. European Health for All Data Base. Available at: http://www.euro.who.int/en/data-and-evidence/databases/european-health-for-all-database-hfa-db (accessed 22.09.13).

Non-Communicable Diseases and Conditions

INTRODUCTION

Disease and health conditions are classified to ease efforts in monitoring, controlling, preventing, and treating illness. The distinction between communicable and non-communicable diseases changes over time with growing knowledge of causation and risk factors, as discussed in previous chapters. Communicable diseases are those caused by specific pathogens and may be transmitted from an infected to an uninfected host, but this process is influenced by the physical, social, and economic environment. Non-communicable diseases (NCDs) are mostly due to societal conditions and lifestyle habits such as poor nutrition, smoking, excess alcohol, and lack of physical exercise. Degenerative, genetic, hereditary, and environmental conditions are also important factors. Some conditions are caused by infectious diseases, such as acquired immunodeficiency syndrome (AIDS) due to human immunodeficiency virus (HIV) infection, peptic ulcers being due to *Helicobacter pylori*, and chronic liver disease due to hepatitis B and C viruses. Chronic conditions may have a multiple-factor model of causation, as exemplified by cardiovascular diseases (CVDs) and diabetes, with the interactions of genetic tendencies and socioeconomic and behavioral factors. The World Health Organization (WHO) estimates that 150 million people suffer from catastrophic health care costs each year due to NCDs, with the problems being more frequent and severe in low- and middle-income countries.

Infections such as by *H. pylori* are causes of chronic peptic ulcer diseases and gastric cancer and their long-term chronic disease manifestations. Other infectious diseases including HIV/AIDS and hepatitis B and C produce chronic conditions and their prevention and treatment greatly improves the chance of survival and reduces the risk of onward transmission of infection, such as maternal–fetal transmission. Human papillomavirus (HPV) infection is causative of cervical, oral, and anogenital cancers, and therefore the HPV vaccine is now recommended for girls and for boys. Poor oral health with gum infection is considered to be a risk factor for CVD.

Disease causation is complex with many contributing factors. Rarely is a single necessary factor sufficient by itself to produce disease. The germ theory of a single causation of disease proved to be an enormously potent basis for the development of medical science and public health. However, environmental factors of the physical, social, economic, and cultural environment have emerged as important modern applications of the miasma theory. Both are important for understanding modern public health, with risk factors, social inequalities, and directed interventions.

Globally, the leading causes of death are NCDs, with 59.7 percent of total deaths (Table 5.1). For low-income countries, lower respiratory tract infections, diarrheal disease, HIV/AIDS, malaria, tuberculosis, low birth weight, and birth trauma are among the top 10 causes of death, but ischemic heart disease and stroke rank as the fourth and sixth leading causes. In middle-income countries, ischemic heart disease, stroke, and chronic obstructive pulmonary disease (COPD) are the leading causes of death. The low- and medium-income countries suffer the double burden of high morbidity and mortality from infectious diseases as well as NCDs. In high-income countries NCDs are the leading causes of death. Among them are ischemic heart disease, stroke, cancers, Alzheimer's disease, COPD, colon and rectal cancer, diabetes mellitus, hypertensive heart disease, and breast cancer – all with high potential for preventive care and public health measures. The leading causes of death in the USA over the period 1970–2008 are shown in Table 5.1.

The New Public Health. http://dx.doi.org/10.1016/B978-0-12-415766-8.00005-7

TABLE 5.1 Leading Causes of Death (Standardized Rates per 100,000 Population), USA, 1970–2008

Cause of Death	1970	1980	1990	2000	2005	2008
All causes	1222.6	1039.1	938.7	869.0	798.8	758.3
Diseases of heart	492.7	412.1	321.8	257.6	211.1	186.5
Ischemic heart disease	–	345.2	249.6	186.8	144.4	122.7
Cerebrovascular diseases	147.7	96.2	65.3	60.9	46.6	40.7
Malignant neoplasms.	198.6	207.9	216.0	199.6	183.8	175.3
Trachea, bronchus, and lung	37.1	49.9	59.3	56.1	52.6	49.5
Colon, rectum, and anus	28.9	27.4	24.5	20.8	17.5	16.4
Chronic lower respiratory diseases	–	28.3	37.2	44.2	43.2	44.0
Influenza and pneumonia	41.7	31.4	36.8	23.7	20.3	16.9
Chronic liver disease and cirrhosis	17.8	15.1	11.1	9.5	9.0	9.2
Diabetes mellitus	24.3	18.1	20.7	25.0	24.6	21.8
Alzheimer's disease	–	–	–	18.1	22.9	24.4
Human immunodeficiency virus (HIV) disease	–	–	10.2	5.2	4.2	3.7
Unintentional injuries	60.1	46.4	36.3	34.9	39.1	40.0
Motor vehicle-related injuries	27.6	22.3	18.5	15.4	15.2	12.9
Poisoning	2.8	1.9	2.3	4.5	7.9	10.2
Suicide	13.1	12.2	12.5	10.4	10.9	11.6
Homicide	8.8	10.4	9.4	5.9	6.1	5.9

Note: Data from death certificates; numbers rounded.
Source: National Center for Health Statistics. Health, United States, 2011: with special feature on socioeconomic status and health. Table 24. Hyattsville, MD: US DHHS, NCHS; 2011. Available at: http://www.cdc.gov/nchs/data/hus/hus11.pdf#listtables [Accessed 27 October 2012].

The relative effect of different causes of mortality on years of potential life lost (YPLL) (seen in Table 5.2) show that injuries, suicide, and homicide are prominent because they affect young people primarily, with many implications for productivity of society. Cancer is in second place as it affects people at earlier ages than heart disease. Congenital anomalies and perinatal period deaths also affect this measure of mortality because of the loss of many more years of potential life than conditions appearing later in life.

Identification of many risk factors and methods of early detection have increased the potential for interventions to lower the prevalence of these diseases and their complications. As a result of these changes, and improved capacity for successful health promotion and disease prevention, the scope of public health has broadened. In the high- and medium-income countries, NCDs are the major issues of public health, although sanitation, food safety, and infectious disease control are still potent issues. In lower income countries the additional burden of infectious diseases, maternal and child health, and nutritional challenges increase the vulnerability to all diseases and long-term effects. Other fundamentals of public health are at the same time facing the association of poverty with NCDs, which now serve as major causes of mortality. Despite the many

challenges, there is great potential for socially oriented health promotion along with individual and population primary, secondary, and tertiary prevention to reduce mortality and disability as well as the burden of disease and disability. The New Public Health stresses all aspects of prevention and care and is therefore increasingly required as a broad approach to this wide group of conditions and the needs associated with them.

This chapter examines the major chronic diseases and their effects on the health of the population. It also looks at the risk factors that contribute to these diseases and at the interventions needed to reduce their prevalence in the population.

THE RISE AND SPREAD OF CHRONIC DISEASE

The gradual shift in predominance from infectious diseases to NCDs is often referred to as the "epidemiological transition" which has occurred gradually over the twentieth century with sanitation, food safety, and other basic public health measures. Vaccines and antibiotics, along with improved living standards, sanitation, nutrition, and safe water, brought about a reduction in mortality rates from

TABLE 5.2 Years of Potential Life Lost (YPLL) Before Age 65, 2009, USA

Cause of Death	YPLL 2000	%	YPLL 2010	%
All causes	11,261,211	100	11,043,870	100
Unintentional injury	2,022,483	18.0	2,083,297	18.9
Malignant neoplasm	1,866,815	16.6	1,843,612	16.7
Heart disease	1,376,937	12.2	1,348,612	12.2
Perinatal period	913,066	8.1	786,472	7.1
Suicide	635,028	5.6	764,776	6.9
Homicide	551,612	4.9	522,701	4.7
Congenital abnormalities	490,687	4.4	439,731	4.0
HIV	320,582	2.8		
Liver disease	233,500	2.1	263,317	2.4
Cerebrovascular	248,446	2.2	230,587	2.1
Diabetes mellitus	NA	NA	216,229	2.0
All others	2,602,055	23.1	2,544,274	23.0

Note: Both genders, all ages, all deaths.
Categories were revised in 2010, and the category "All others" now includes human immunodeficiency virus (HIV) data for 2010.
Source: Centers for Disease Control and Prevention. Injury prevention and control: data and statistics (WISQARS). Years of potential life lost (YPLL) [updated 17 September 2012]. Atlanta, GA: CDC. Available at: http://www.cdc.gov/injury/wisqars/years_potential.html [Accessed 24 October 2012].

TABLE 5.3 Leading Causes of Death Worldwide, 2008

Disease	Deaths (Millions)	Deaths (% of Total)
Ischemic heart disease	7.25	12.8
Stroke and other cardiovascular disease	6.15	10.8
Lower respiratory infections	3.46	6.1
Chronic obstructive pulmonary disease	3.28	5.8
Diarrheal diseases	2.46	4.3
HIV/AIDS	1.78	3.1
Trachea, bronchus, lung cancers	1.39	2.4
Tuberculosis	1.34	2.4
Diabetes mellitus	1.26	2.2
Road traffic accidents	1.21	2.1

Source: World Health Organization. The top 10 causes of death. fact sheet no. 310. Geneva: WHO [updated June 2011]. Available at: http://www.who.int/mediacentre/factsheets/fs310/en/index.html [Accessed 24 October 2012].

infectious diseases with increasing in life expectancy. Infectious diseases, while still important, are no longer the primary concerns in public health in the developed countries, and similar trends are appearing in the developing countries.

The shift in predominance of infectious to non-infectious diseases in the USA is seen in Table 5.1, which shows the changes occurring in leading causes of death over the period 1970–2008. This reflects similar changes that have occurred in all the developed countries over the past half-century.

Of the 57 million total deaths globally in 2008, 36 million, or 63 percent, were due to NCDs, which represent the most frequent causes of death in most countries, except in Africa. But even in Africa, NCDs are projected to be the most common cause of death by 2030. CVD, diabetes, cancer, and chronic respiratory diseases are the principal NCDs. In the European Region, these four groups of diseases cause nearly 86 percent of deaths and 77 percent of the disease burden.

Table 5.3 shows the global pattern of causes of death, with non-infectious diseases the leading causes of death. In developing countries there is a double burden of high levels of both infectious and NCD morbidity and mortality.

CVDs, principally coronary heart disease and stroke, are the leading cause of death, with more than 13 million deaths worldwide; this figure is predicted to decrease in high income countries whilst increasing steadily in mid- and low-income countries. The 10 leading causes of death worldwide in 2008 are shown in Table 5.3. More than 36 percent of deaths are caused by the principal chronic diseases of CVD, diabetes mellitus, cancer, and chronic respiratory diseases. With 3.6 percent of world deaths, AIDS/HIV survival time has increased to 11 years with advances and success in management of the disease and is thus beginning to show features of chronic disease. Globally in 2002, of the 54 percent of years of life lost, 33 percent were due to NCD and a further 13 percent to injuries. Even in low-income countries 30 percent of years of life lost were from NCD and injuries (WHO statistics, 2008).

Chronic diseases as the leading cause of morbidity and mortality are associated with a number of demographic and epidemiological factors. First, the decline in infectious disease mortality has resulted in greater longevity, increasing the number of people surviving to ages when cancer and heart disease are more common. Second, changes in lifestyle such as smoking, lack of exercise, diets rich in unhealthy fats and sugars, and risk-taking behavior, have increased risk factors, thus influencing CVD and cancer to become the leading causes of disease, disability, and death. Third, trauma and chronic diseases are major contributors to rising costs of health care. Fourth, public health experience and new scientific knowledge are leading to new forms of prevention and medical treatment that are reducing the burden of disease and disability from chronic conditions. Poverty and illiteracy are increasingly recognized risk factors for NCDs in an increasing globally mobile population.

Figure 5.1 shows shifts in leading causes of death in the world between 2004 and 2030 (predicted). The highest ranked causes remain ischemic heart disease and stroke. Tuberculosis (TB) deaths are predicted to drop from the seventh position to the 20th, while diarrheal disease deaths are expected to fall dramatically from the fifth to the 23rd position. Death from cirrhosis of the liver remains stable, ranked 18, while diseases and factors causing death which are predicted to move up the scale include road traffic accidents, diabetes mellitus, violence, hypertensive heart disease, and cancers of the breast, esophagus, colon and rectum, and stomach.

Table 5.4 shows the standardized rates of mortality (SMR) for the leading NCDs and all causes of mortality in 2008 in all six geographic regions of the WHO (African, the Americas, South-East Asian, European, Eastern Mediterranean, and Western Pacific). It also shows countries grouped by income levels. NCDs are the leading cause not only of mortality, but also of the total burden of disease.

Causes of important chronic conditions are being identified with interventions and treatment that will further alter disease patterns in the years ahead. Prevention methods include primary prevention by health promotion such as immunization, smoking cessation, dietary improvement, and physical activity. Secondary prevention is achieved through screening and early discovery of potentially serious disease, which can be successfully treated. Tertiary prevention promotes long-term survival and best possible function. The secondary and tertiary levels of prevention rely on access to well-trained health care providers who play important and measurable roles in reducing the burden of disease.

As expressed in a series of papers in *The Lancet* in 2007, a global program goal for prevention and control of chronic diseases could avert 36 million deaths by 2015 with major economic benefits, using existing interventions. The evidence suggests that this goal is possible and realistic with interventions directed towards whole populations and individuals at high risk: "The total yearly cost of the interventions in 23 low-income and middle-income countries is about US$5.8 billion" (2005). This calls for "a serious and sustained worldwide effort to prevent and control

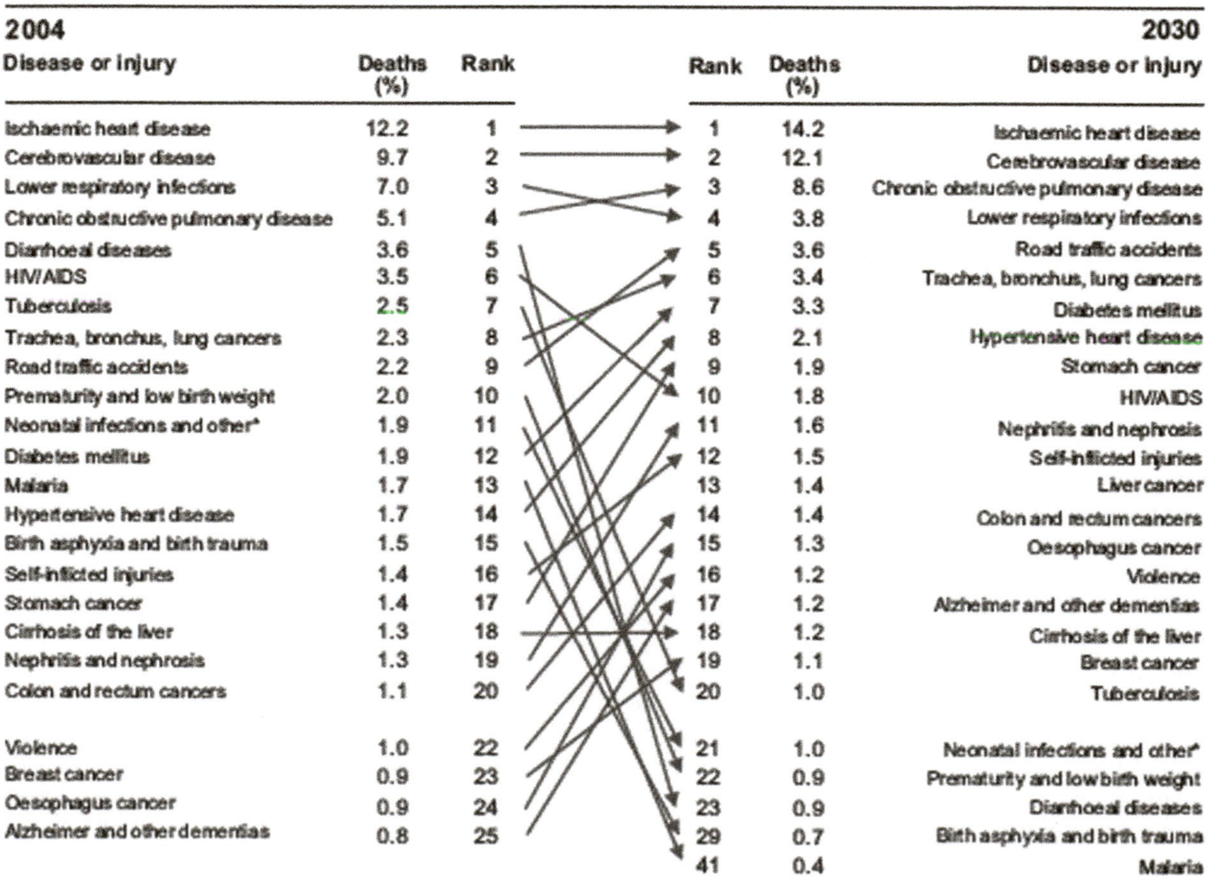

FIGURE 5.1   Rank order of leading causes of death worldwide, 2004 and 2030 predicted. *Source: World Health Organization. World health statistics 2008. Geneva: WHO; 2008. Available at: http://www.who.int/whosis/whostat/EN_WHS08_Full.pdf [Accessed 21 October 2012.].*

TABLE 5.4 Age-Standardized Mortality Rates by WHO Regions (Geographic and Income Groups), by Cause, per 100,000 population, 2008

WHO Region	Communicable	Non-Communicable	Cancer	CVD/Diabetes	Injuries	All Causes
African	798	779	147	382	107	1716
Americas	72	455	136	169	63	532
South-East Asian	334	676	125	322	101	987
European	51	532	166	238	63	626
Eastern Mediterranean	254	706	127	344	91	881
Western Pacific	74	534	168	184	64	545
Income level						
Low	636	757	154	375	124	1354
Lower middle	233	658	150	273	82	808
Upper middle	125	608	158	295	81	805
High	31	380	141	104	41	376
Global	230	573	150	245	78	764

Source: World Health Organization. World health statistics 2012. Geneva: WHO; 2012. Available at: http://www.who.int/healthinfo/EN_WHS2012_Full.pdf [Accessed 22 October 2012].

chronic diseases in the context of a general strengthening of health systems. Urgent action is called for by the WHO, the World Bank, regional banks and development agencies, foundations, national governments, civil society, non-governmental organizations, the private sector including the pharmaceutical industry, and academics". A Chronic Disease Action Group was established to encourage, support, and monitor action on the implementation of evidence-based efforts to promote global, regional, and national action to prevent and control chronic diseases (Beaglehole et al., 2007). Tobacco is considered a major cause of several of the world's top killer diseases, including CVD, COPD, and lung cancer. Smoking is often the hidden cause of the disease recorded as responsible for death for almost one in 10 adults worldwide. In the USA, as in all the high-income developed countries, death rates for heart disease and cancer have declined significantly since the 1960s (by 30.8 percent and 11.9 percent, respectively) (Figure 5.2).

ENVIRONMENT AND LIFESTYLE

During the twentieth century, the "diseases of modern life", or non-infectious conditions as they were commonly called, became the leading causes of morbidity and mortality in developed countries. This epidemiological transition is also occurring in many low- and middle-income countries (Table 5.4).

Causation in chronic diseases is multifactorial, and prevention must take into account many contributory risk factors. Tobacco use, unhealthy diet, low physical activity, and harmful use of alcohol are major behavioral risks that are

pervasive factors of economic transition, rapid urbanization, and twenty-first century lifestyles, all which are ongoing challenges in this millennium. Despite the complexity, for reasons such as control of hypertension and hyperlipidemia, and other factors not well understood, dramatic success has been achieved in reducing stroke and heart disease death rates in many countries over the past 20 years. Cancer and trauma death rates, key elements of non-infectious disease patterns, have also fallen owing to reduced smoking and improved road safety. However, they are far from reaching levels in countries with "best practices" in risk factor reduction and screening programs.

GLOBAL BURDEN OF MORTALITY

Chronic conditions are not only the major causes of death in western countries, but also increasingly recognized as a major cause of mortality in the developing countries (Box 5.1). They pose far greater challenges for these countries, which face a heavy burden of mortality from communicable diseases, diseases of nutrition, and maternal and child health complications. Thus, the combination of diseases places very great demands on weak health systems. A major global health crisis exists owing to rising levels of CVD and diabetes in developing countries, affecting the working population with serious economic consequences. The First Global Ministerial Conference on Healthy Lifestyles held in Moscow in April 2011 stated that the WHO estimates that two-thirds of all deaths in the world today are the result of chronic diseases, thus responsible for 53 million deaths each year. Of these, 9–10 million premature deaths are under 60 years of age, with serious

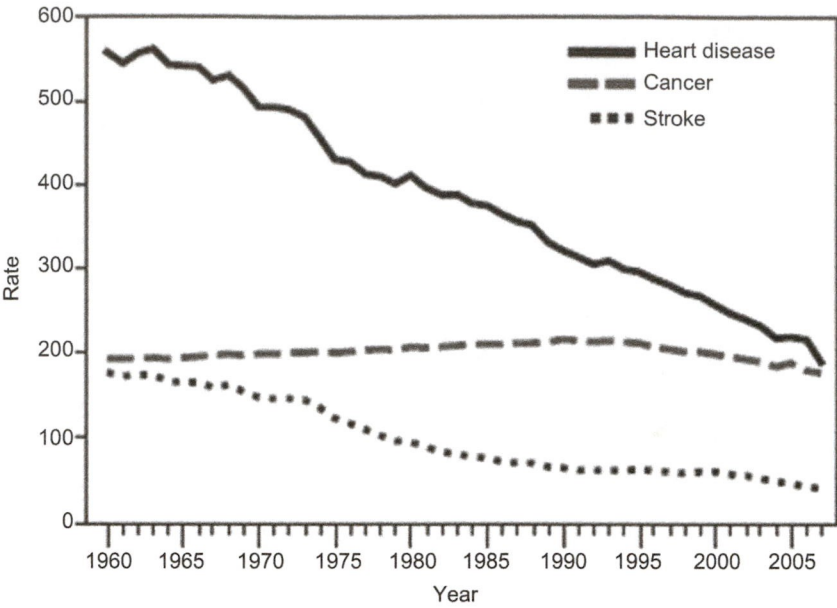

FIGURE 5.2   Age-adjusted death rates for heart disease, cancer and stroke, USA, 1960–2007. *Source: Centers for Disease Control and Prevention. Public health then and now: celebrating 50 years of MMWR at CDC. MMWR Morb Mortal Wkly Rep 2011;60:1–112. Available at: http://www.cdc.gov/ mmwr/pdf/other/su6004.pdf [Accessed 24 October 2012].*

losses to a developing or burgeoning economy. In Europe, 85 percent of all deaths are the result of chronic diseases.

CVDs – coronary heart disease (CHD), stroke, and renal failure – are the leading causes of death and disability in the world. Heart disease and stroke kill 13.5 million people every year, compared to 3 million due to HIV/AIDS, with 80 percent of these deaths occurring in low- and middle-income countries.

A WHO/Centers for Disease Control and Prevention (CDC) Joint Atlas of Heart Disease and Stroke (2004) attributes "many of these deaths to tobacco smoking, which increases the risk of dying from coronary heart disease and cerebrovascular disease 2–3 fold. Physical inactivity and unhealthy diet are other main risk factors which increase individual risks to cardiovascular diseases. One of the strategies to respond to the challenges to population health and well-being due to the global epidemic of heart attack and stroke is to provide actionable information for development and implementation of appropriate policies".

Globally, coronary heart disease, cancer and cerebrovascular diseases are expected to continue increasing up to 2030, with respiratory disease remaining a leading cause but expected to decline. The lessons learned from the decline in coronary heart and cerebrovacsular disease mortality seen in the upper income countries since the mid-1960s offer hope to reduce the current increase being seen in middle and lower income countries. HIV and tuberculosis mortality is also expected to decline. Road traffic accident mortality is predicted to continue to rise (Figures 5.1 and 5.3).

NCDs as a cause of mortality are projected to remain a major cause of death in low- and middle-income countries over the next two decades (Figure 5.4). The WHO reports that NCDs affect the poorest people in these countries, particularly low-resource countries, with a major contribution to deepening poverty by illness and loss of work capacity. CVD places a heavy burden on the economies of low- and middle-income countries, as premature deaths from heart disease, stroke, and diabetes are expected to hamper economic growth and reduce gross domestic product (GDP) by between 1 and 5 percent in low- and middle-income countries.

The WHO reports CVD as the leading NCD globally; nearly half of the 36 million deaths due to NCDs are caused by CVD. In 2008, 9 million people died of NCDs prematurely, before the age of 60. The vast majority (some 8 million) of these premature deaths occurred in low- and middle-income countries (WHO Global Atlas on CVD Prevention and Control, 2011). Often designated as diseases of "modern living", it is now understood that NCDs such as cardiovascular diseases affect poor people much more than upper income groups and countries (Figure 5.5). The increasing prominence of NCDs in middle- and low-income countries where dietary and lifestyle changes coexist with extreme poverty is one of the major challenges for public health in the coming decades. NCDs are etiologically due to factors such as smoking, lack of physical exercise, and a diet rich in animal fats, collectively known as "lifestyle" risk factors, but these factors weigh more heavily in conditions of poverty. They are

BOX 5.1 Global Burden of Non-Communicable Diseases

- NCDs are the leading cause of death globally, killing more people than all other causes combined, i.e., 36 million of a total of 57 million deaths in 2008, and one-quarter of these occurring before age 60.
- 80 percent of deaths from NCDs occur in developing countries.
- Currently known cost-effective and feasible interventions have proven to be effective in averting much of the human and social impact of NCDs.
- Main NCDs: cardiovascular diseases, cancer, diabetes, chronic lung disease.
- Main risk factors: diet, smoking, lack of physical activity, excess use of alcohol
- Associated factors: poverty, economic transition, rapid urbanization, twenty-first century lifestyle.
- Poverty predisposes people to increased risk factors.
- Population-wide interventions are cost-effective and may generate revenues, e.g.:
  - increased taxes on alcohol and cigarettes
  - regulation of smoking in public locations
  - regulation of fat and salt content of manufactured or processed foods (banning of trans fats)
  - food fortification with essential vitamin and minerals that are insufficiently available in regular diets for health.
- Tobacco control applied in only 10 percent of world population settings owing to a lack of political leadership to apply the WHO Framework Convention on Tobacco Control. This is due to lack of advocacy and strength of opposition from vested interests.
- Improved health care can reduce many of the NCDs and delay their most serious effects:
  - screening for specific cancers such as cervix, breast, and colon
  - screening and management for hypertension and diabetes
  - immunizations for pneumonia, influenza, rubella, hepatitis B, and human papillomavirus
  - counseling: diet, exercise.

Source: *World Health Organization. Global status report on non-communicable diseases 2010. Geneva: WHO; 2011. Available at: http://www.who.int/nmh/publications/ncd_report2010/en/ [Accessed 21 October 2012].*

amenable to change by the individual at risk and through community interventions and health promotion activities. Secondary prevention, which emphasizes early diagnosis and rapid intervention, has proven to be important for saving lives and for extending longevity and quality of life in population health. Perhaps the most important lesson in public health since the 1960s is that these risk factors and diseases can be reduced dramatically by suitable health promotion, public health and clinical interventions.

RISK FACTORS, HEALTH PROMOTION, AND PREVENTION OF NON-COMMUNICABLE DISEASES

The criteria for causation in infectious disease, discussed in Chapter 1, are known as the Koch–Henle–Evans postulates. The criteria for causation in chronic disease, called the Evans criteria, are outlined in Box 5.2. They articulate the relationship of predisposing or risk factors to the causation of chronic disease, and are important in analyzing the relative contribution of factors to disease in a population.

The most common, costly, and preventable health problems are chronic diseases, principally heart disease, stroke, cancer, diabetes, and injury. In the USA, chronic diseases are responsible for seven out of 10 deaths and associated morbidity. Nearly 19 million Americans have activity limitations. About one-third of adults and one-fifth of young people are obese. Diabetes is the leading cause of blindness, kidney failure, and lower limb amputations among adults. Lifestyle factors of physical inactivity, poor nutrition, tobacco usage, and excessive alcohol consumption are responsible for many of the effects of chronic disease illness and death. Medically manageable conditions are a major factor in high morbidity and mortality from NCDs; an estimated 45 percent of cardiovascular deaths can be attributed to elevated blood pressure, 16 percent to high cholesterol, and 13 percent to high blood sugar.

The challenges of addressing these issues are for each individual, family member, caregiver, insurer, school, academic institution, food producer and marketer, employer, union and professional organization, government and non-governmental organization (NGO) concerned with societal responsibility. Social policies and public health interventions can be very effective in reducing these risk factors, as seen in smoking bans and reduction. Poverty, and education among the poor, is one of the central issues because of its pervasive effects on social skills and lifestyle issues.

The eight key risk factors – alcohol use, tobacco use, high blood pressure, high body mass index (BMI), high cholesterol, high blood glucose, low fruit and vegetable intake, and physical inactivity – account for 61 percent of loss of healthy life years from CVD and 61 percent of cardiovascular deaths. These risk factors are linked to social and economic conditions in which poverty is a crucial factor. Typically, poor people smoke more, and have lower quality diets because of lack of opportunities for choice. These factors, alone or in combination, are implicated in many leading cancers. Worldwide, tobacco use causes 71 percent of lung cancer deaths. Tobacco, physical inactivity, and overweight are quickly replacing the traditional risks and leading to a double burden in low- and middle-income countries. Estimates of the prevalence of risk factors among Americans are shown in Figure 5.6.

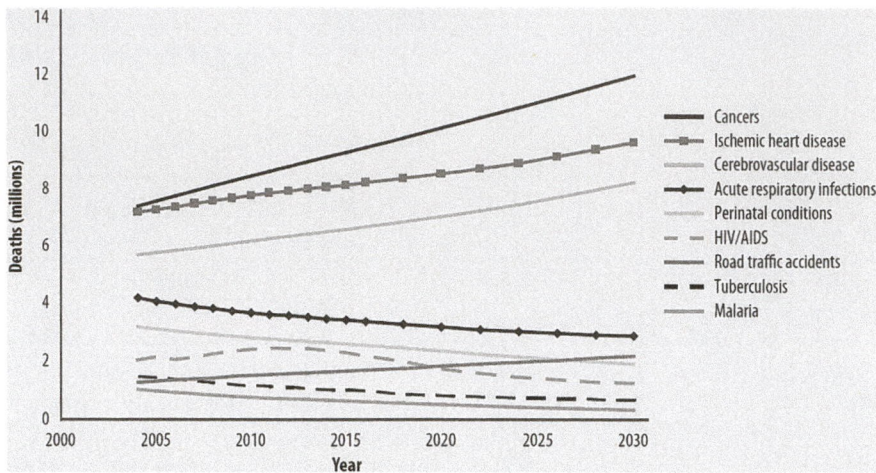

FIGURE 5.3 **Projected global deaths by selected causes, 2004–2030.** *Source: World Health Organization. The global burden of disease: 2004 update. Geneva: WHO; 2008. Available at: http://www.who.int/healthinfo/global_burden_disease/GBD_report_2004update_full.pdf [Accessed 21 October 2012].*

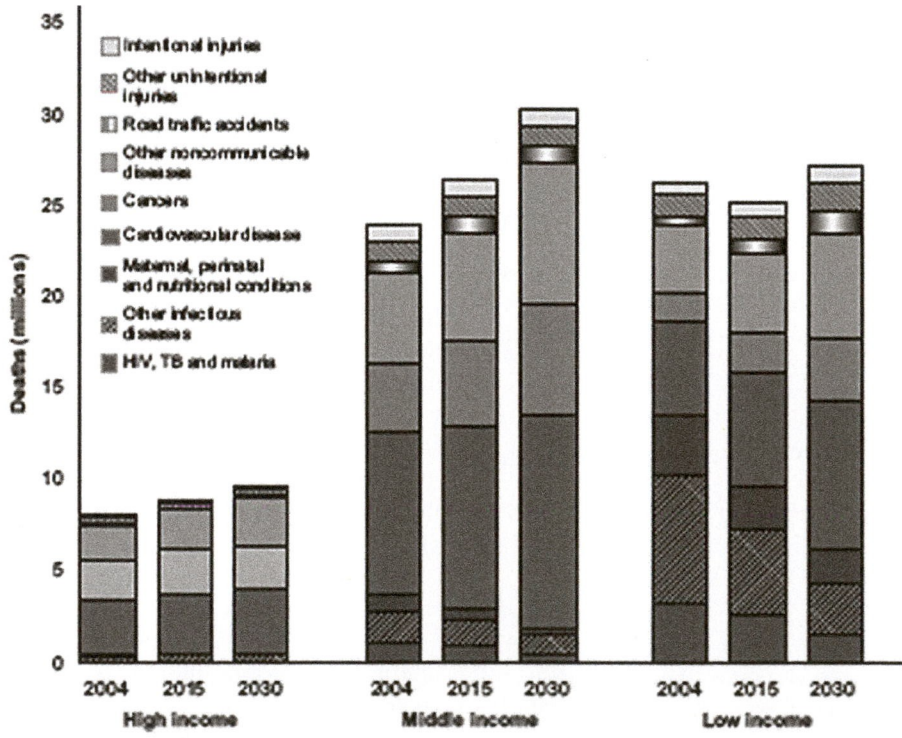

FIGURE 5.4 **Projected causes of death for high-, middle-, and low-income countries, 2004–2030.** *Sources: World Health Organization. WHO's statistical information system. Available at: http://www.who.int/statistics, http://www.who.int/gho/publications/world_health_statistics/EN_WHS2011_Part2. pdf World Health Organization. World health statistics 2011. Geneva: WHO; 2011. Available at: http://www.who.int/whosis/whostat/EN_WHS2011_TOC. pdf [Accessed 21 October 2012]. World Health Organization. World health statistics 2008. Geneva: WHO; 2008. Available at: http://www.who.int/whosis/ whostat/EN_WHS08_Full.pdf [Accessed 21 October 2012.].*

Atherosclerosis can begin in childhood, and as part of the aging process, along with genetic factors, leads to the end result of heart attacks and strokes and is thought to be associated with other degenerative diseases such as diabetes and dementias. Moreover, infections, environmental and behavioral exposures are causes in more than three-quarters of deaths from mouth, liver, lung, and cervical cancer. Environmental degradation

such as air pollution is linked to increases in cardiovascular morbidity and mortality, as well as in chronic pulmonary diseases such as asthma and COPD, and lead poisoning.

Mental disorders, especially depression, are among the leading worldwide causes of life years lived with disability, including lifetime prevalence among the adult population of 46 percent for any of the following disorders: anxiety (29 percent),

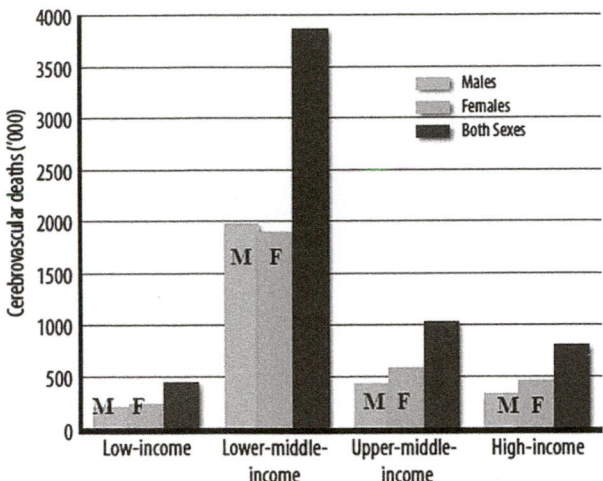

FIGURE 5.5 Total deaths due to cerebrovascular disease by World Bank income groups, 2008. *Source: Mendis S, Puska P, Norrving B, editors. Global atlas on cardiovascular disease prevention and control. Geneva: WHO; 2011. Available at: http://whqlibdoc.who.int/publications/2011/9789241564373_eng.pdf [Accessed 22 October 2012].*

impulse control (25 percent), mood (21 percent), and substance abuse (15 percent). The interaction between chronic illness and depression is profound and complicated by side-effects of drugs used to manage mental illnesses such as schizophrenia and bipolar disorders, promoting passivity, smoking, and a lack of exercise, socialization, and healthful nutrition. Depression is itself a risk factor for worsening of prognosis for patients with chronic illness, such as CHD and heart failure, as these conditions can engender a depressive response which contributes to a weakening of the physical condition.

Risk factors in NCDs, addressed in Table 5.5, have been identified by critically important epidemiological studies over the past six decades. The classic epidemiological work of Jeremy Morris showed the difference in risk for CHD between London bus drivers and bus conductors. The conductors, who are required to climb bus stairs and experience less stress as part of their job, had lower rates of CHD mortality than the sedentary, high-tension drivers. This study was followed by equally classic work of the Whitehall studies of British civil servants by Michael Marmot and others, showing marked differences in the morbidity and mortality patterns of those with higher status, income, and control over their own lives, and the protective effect of physical exercise. The causal association of smoking with lung cancer was shown in the 1950s by Richard Doll and Bradford Hill, and consequently became a public health issue with the publication of the US Surgeon General's Report on Smoking in 1964. The work of Staler and many epidemiologists, as well as the famous Framingham Heart Study and the North Karelia project in Finland, brought lipid control to the forefront of public health endeavors in the 1960s (see Chapter 1).

During the period 2005–2008, overall 37 percent of the US population reported having two or more of the risk

BOX 5.2 Criteria for Causation in Chronic Disease – Evans' Postulates

1. Prevalence of the disease should be significantly higher in those exposed to the hypothesized cause than in controls not so exposed.
2. Exposure to the hypothesized cause should be more frequent among those with the disease than in controls without the disease, when all other risk factors are held constant.
3. Incidence of the disease should be significantly higher in those exposed to the hypothesized cause than in controls not so exposed, as shown by prospective studies.
4. The disease should follow exposure to the hypothesized causative agent with a normal or log-normal distribution of incubation periods.
5. A spectrum of host responses should follow exposure to the hypothesized agent along a logical biological gradient from mild to severe.
6. A measurable host response following exposure to the hypothesized cause should have a high probability of appearing in those lacking this before exposure (e.g., antibody, cancer cell) or should increase in magnitude if present before exposure. This response pattern should occur infrequently in persons not so exposed.
7. Experimental reproduction of the disease should occur more frequently in animals or humans appropriately exposed to the hypothesized cause than in those not so exposed; this exposure may be deliberate in volunteers, experimentally induced in the laboratory, or may represent a regulation of a natural exposure.
8. Elimination or modification of the hypothesized cause should decrease the incidence of the disease (e.g., attenuation of a virus, removal of tar from cigarettes).
9. Prevention or modification of the host's response on exposure to the hypothesized cause should decrease or eliminate the disease (e.g., immunization, drugs to lower cholesterol, specific lymphocyte transfer factor in cancer).
10. All of the relationships and findings should make biological and epidemiological sense.

Sources: Evans AS. Causation and disease: the Henle–Koch postulates revisited. Yale J Biol Med 1976;49:175–95. Available at: http://www.ncbi.nlm.nih.gov/pmc/articles/PMC2595276/pdf/yjbm00143-0072.pdf [Accessed 21 October 2012].
Last JM. A dictionary of epidemiology. 4th ed. New York: Oxford University Press; 2001.

factors. CDC attributes the distribution of individual risk factors as follows:

- inactivity – 53 percent
- obesity – 34 percent
- high blood pressure – 32 percent
- smoking – 21 percent
- high cholesterol – 15 percent
- diabetes – 11 percent.

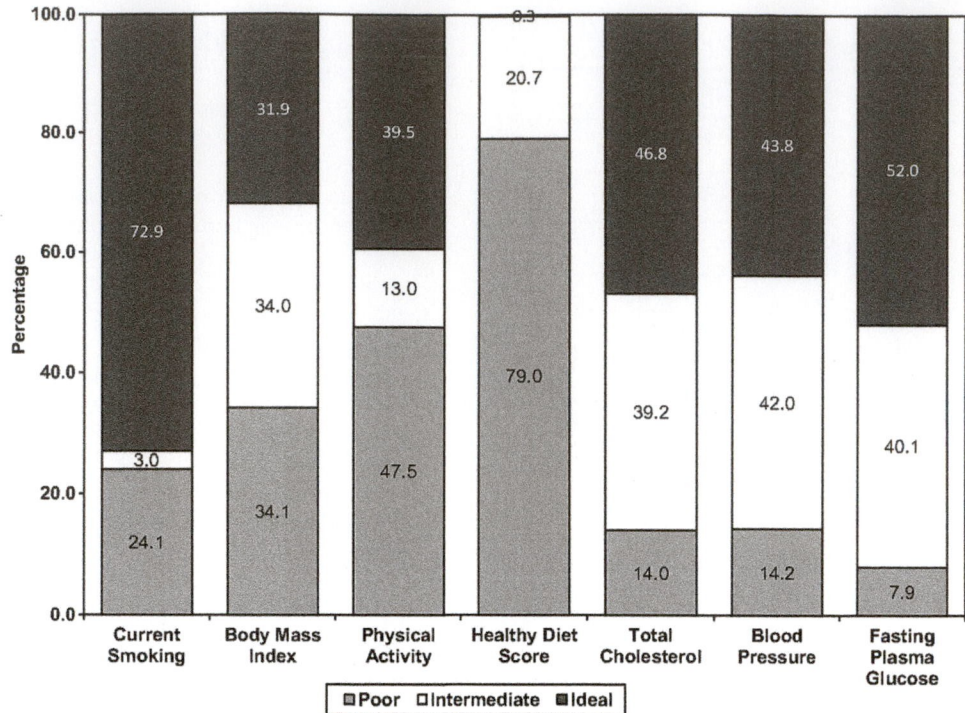

FIGURE 5.6 Risk factor prevalence in US adults in NHANES 2007–2008. Note: Age-standardized prevalence estimates for poor, intermediate, and ideal cardiovascular health for each of the seven metrics of cardiovascular health in the American Heart Association 2020 goals, among US adults aged ≥ 20 years, National Health and Nutrition Examination Survey (NHANES) 2007–2008 (available data as of 1 June 2011). *Source: Roger V, Go A, Lloyd-Jones D, Benjamin E, Berry J, Borden W, et al. American Heart Association Statistical Update: Heart disease and stroke statistics – 2012 update. Circulation 2012;125:e2–220. Available at: http://circ.ahajournals.org/content/125/1/e2.full.pdf+html [Accessed 22 October 2012].*

There have been very important reductions in mortality from CVD in recent decades. However, the increased obesity and diabetes prevalence in both high- and low-income countries serves as a cautionary sign that needs full attention by public health professionals and by policy makers responsible for directing and allocating resources in national health systems (Box 5.3).

Tobacco

Tobacco, principally cigarette smoking, is a leading health risk, accounting for 11 percent of the disease burden and 18 percent of deaths in high-income countries. Cigarette smoking is the most common form of using smoking tobacco and most data rely on self-reported usage. However, tobacco as a health hazard also includes secondary exposure to smoke, use of both smokeless and smoking tobacco, and the risk of exposure to tobacco through harvesting tobacco, which is less often thought about.

Globally, tobacco usage among men is 36 percent and women 8 percent, with an increasing predominance of smoking in low- and middle-income countries. Ten countries, including highly developed countries such as Japan, Germany, and the USA, along with China, India, Brazil, Indonesia, the Russian Federation, Bangladesh, and Turkey,

as well as being home to 58 percent of the world's population, account for nearly two-thirds of all smokers worldwide. In addition, one-third of adults in Eastern and Central Europe are smokers. Tobacco control measures reach only 5 percent of the world's population, while half of all countries do not implement any of the proven and effective tobacco control recommendations.

Smoking was identified as a risk factor in US studies by Ernst Wynder and others in the 1940s and 1950s. In the famous prospective studies of British doctors (1951–2001), epidemiologists Richard Doll, Austin Bradford Hill, and Richard Peto showed a strong relationship between smoking and poor health outcomes. By 1954 this study showed highly convincing statistical evidence of smoking as a major risk factor for lung cancer. In the 1950s, 35-year-old male cigarette smokers were shown to have less chance of surviving to age 65 (73 percent) than non-smokers (85 percent) and ex-smokers (81 percent). The classic 1964 US Surgeon General's Report on Smoking and Health presented a forceful summary of the hundreds of published studies up to that time, and concluded that cigarette smoking was a major health hazard and cause of lung cancer, CHD, chronic lung disease, and stroke.

Subsequent Surgeon General's Reports in 1983 and 1984 attributed 30 percent of CHD deaths and 80–90 percent of COPD deaths to smoking. For these reasons, smoking

TABLE 5.5 Chronic Disease Risk Factors

Age, gender, ethnicity	CVD, cancer, COPD, and liver disease increase with aging but tend to occur later in women than in men
Low socioeconomic status	Poverty, unemployment, poor housing, disorganized and dangerous neighborhoods, lack of access to medical care and to healthful foods are all associated with obesity, smoking, and alcohol abuse, and thus fundamental to social inequalities in NCD morbidity and mortality
Tobacco use	Of 1 billion smokers globally, 6 million people die from tobacco use and exposure each year; this accounts for 6% of all female and 12% of male deaths in the world. Of these, over 600,000 are attributable to second-hand smoke exposure among non-smokers and more than 5 million deaths are attributed to direct tobacco use
Alcohol use	Direct relationship between higher levels of alcohol consumption and rising risk of some cancers, liver diseases, and CVDs. The relationship between alcohol consumption and cardiovascular diseases is complex and highly dependent on amount and pattern of consumption
Hypertension	Worldwide, raised blood pressure causes 7.5 million deaths, 12.8% of the total of all annual deaths. This accounts for 57 million or 3.7% of total DALYs. Hypertension is a major risk factor for all CVDS: coronary heart disease, ischemic and hemorrhagic stroke, heart failure, peripheral vascular disease, renal impairment, retinal hemorrhage and visual impairment. The risk of CVDs doubles for each increment increase of 20/10 mmHg of blood pressure, from as low as 115/75 mmHg
Diet	Lack of a healthful diet is a major contributor to NCDs: high salt and fat intake, low fruit and vegetable consumption, and vitamin D deficiency are major contributors to high rates of NCDs
Low fruit and vegetable consumption	Cancer and CVDs are attributable to low fruit and vegetable consumption. Adequate consumption of fruit and vegetables reduces risk for CVDs, lung, stomach, colorectal, and other cancers
High salt intake	Dietary salt consumed is an important determinant of blood pressure levels and CVD risk. Average salt intake of <5 g per person per day is recommended for prevention of CVD. Decreasing dietary salt intake from the current global levels of 9–12 g per day to the recommended level will have a major impact on reducing blood pressure and CVD. This requires voluntary and regulatory changes in the food industry and public education
Physical inactivity	Insufficient physical activity is the fourth leading risk factor for mortality. Approximately 3.2 million deaths and 32.1 million DALYs (representing 2.1% of global DALYs) each year are attributable to insufficient physical activity
Obesity	Worldwide, 2.8 million people die each year as a result of overweight and obesity with an estimated 35.8 million (2.3%) of global DALYs. Adverse metabolic effects affect blood pressure, cholesterol, triglycerides, and insulin resistance; risks of coronary heart disease, ischemic stroke, and type 2 diabetes mellitus rise steadily with increasing body mass index
Stress	Stress associated with poverty and poor control over one's life contributes to high rates of NCDs
Environmental exposure	Exposure to NOx, particulate matter of outdoor and indoor air pollution, asbestos, benzene, and contaminants such as arsenic. Ionizing radiation increases the risk for several cancer types. Diagnostic X-rays were estimated to contribute between 0.5 and 3% to the overall cancer burden in high-income countries; residential radon has been estimated to cause 2% of cancer deaths in Europe. Excess exposure to solar radiation and ultraviolet tanning devices is a cancer risk
Occupation	Some 50 occupational exposure circumstances are carcinogenic to humans. In the UK, an overall 5% of cancers estimated to be attributable to occupation; this is likely to be higher in countries with lower standards of worker protection. Risk related to radon is high in miners. Injury and disability are associated with many occupations such as mining, lumber work, fishing, and others
Physiological and mental stress	Depression, poor self-care, low self-esteem, and low level of control over life factors; physical and mental stress contribute to high-risk behavior such as smoking, poor nutrition, and social isolation (e.g., unemployment, post-traumatic stress syndrome, loss of home, marital distress or breakdown, divorce, death in family); dehydration may trigger cardiac or stroke events

Note: CVD = cardiovascular disease; COPD = chronic obstructive pulmonary disease; NCD = non-communicable disease; NOx = Nitrogen oxide; DALY = disability-adjusted life year.
Sources: World Health Organization. Global status report on non-communicable diseases 2010. Description of the global burden of NCDs, their risk factors and determinants. Geneva: WHO; April 2011. Available at: http://www.who.int/nmh/publications/ncd_report2010/en/ [Accessed 21 October 2012].
Cardiovascular disease risk factors. Geneva: World Heart Federation. Available at: http://www.world-heart federation.org/cardiovascular-health/cardiovascular-disease-risk-factors/stress/ [Accessed 22 October 2012].
World Health Organization. Joint effects of risk factors. Geneva: WHO. Available at: http://www.who.int/healthinfo/global_burden_disease/GlobalHealthRisks_report_part3.pdf [Accessed 22 October 2012].

reduction has become one of the pillars of modern public health. Strong advocacy group activity with legislation banning cigarette advertising and smoking in more and more public places helped to raise consciousness of the hazards of smoking.

Cigarette smoking declined in the USA from 55 percent of men in 1955 to just over 30 percent in 1995; this was recognized as one of the great achievements of public health of the twentieth century in the USA and continuing in the first decade of the twenty-first century (MMWR 1999 and

BOX 5.3 Obesity in the USA and Globally

"Two-thirds of adults and almost one-third of children in the United States are overweight or obese. This epidemic of excess weight is associated with major causes of chronic disease, disability, and death. Obesity-related illness is estimated to carry an annual cost of US$190.2 billion.

The causes of obesity are multi-factorial, including everything from cultural norms to the availability of sidewalks or affordable foods. The broad societal changes that are needed to prevent obesity will inevitably affect eating and activity environments and settings for all ages." (Glickman, 2012)

"The global epidemic of overweight and obesity – "globesity" – is rapidly becoming a major public health problem in many parts of the world. Paradoxically coexisting with undernutrition in developing countries, the increasing prevalence of overweight and obesity is associated with many diet-related chronic diseases including diabetes mellitus, cardiovascular disease, stroke, hypertension and certain cancers." (WHO, 2012).

Sources: *Glickman D, Parker, L, Sim L, Del Valle Cook H, Miller E. Accelerating progress in obesity prevention, solving the weight of the nation. Washington, DC: National Academies Press; 2012. Available at: http://books.nap.edu/openbook.php?record_id=13275 [Accessed 22 October 2012].*
World Health Organization. Global database on body mass index: an interactive surveillance tool for monitoring nutrition transition. Posted 15 May 2012. Available at: http://apps.who.int/bmi/index.jsp [Accessed 22 September 2012].

2011). Smoking among men further declined from 28 percent in 1990 to 23 percent in 2009, and among women from 23 percent to 18 percent; among teenagers smoking rates fell from 28 percent to 19–20 percent (Health United States, 2010). However, smoking rates in the USA have stabilized since 2004 (men 23.5 percent, women 17.9 percent in 2010) (Figure 5.7). As a deadly habit, smoking is more prominent among less educated people and it varies by region. In 2009, one in five high-school students smoked. Since 2009, the Smoking Prevention and Tobacco Control Act has provided the US Food and Drug Administration (FDA) with the authority to regulate the manufacturing, marketing, and distribution of tobacco products.

Primary prevention includes community-based programs designed to change life habits as a method of reducing the risks of CHD. They have become accepted components of public health. Studies of community intervention programs such as the Minnesota Five Cities and the Stanford projects showed a small direct response when the target groups were broadly defined; however, other studies focusing on specific target groups, such as prevention of teenage smoking, proved to be more effective. Social behavior studies, economic status indicators, educational levels, ethnicity, and other factors all relate to risk behavior and help to define high-risk target groups and methods of approach. They also provide insight on the effect of social pressures for engaging in risk behaviors, as in dangerous driving, smoking, and binge drinking.

Governments that benefit from taxes on cigarettes, and manufacturers that profit from this market, can and do directly or indirectly promote smoking, as is the case for alcohol. Where governments permit advertising of alcohol and cigarettes in the media or fail to restrict tobacco use in public places, there is a tacit approval of the practice. Manufacturers promote sales of their products among the young and the poor; they target developing countries where governmental and NGO defense mechanisms and

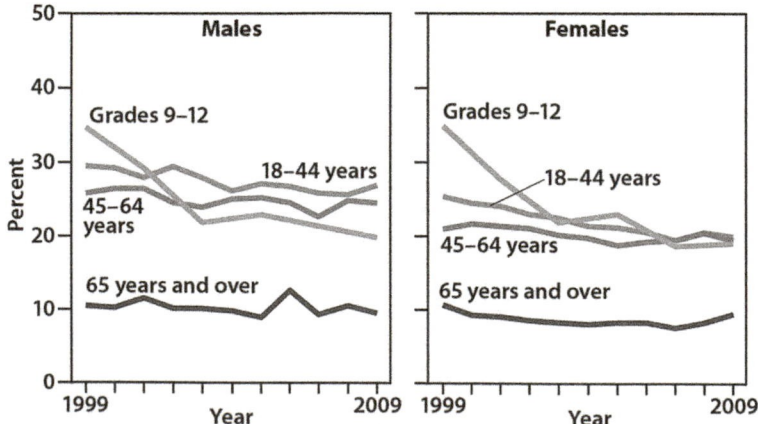

FIGURE 5.7 Cigarette smoking by age group, USA, 1999–2009. *Source: Centers for Disease Control and Prevention. Health, United States, 2010: with special feature on death and dying. Hyattsville, MD: US DHHS, NCHS; February 2011. Available at: http://www.ncbi.nlm.nih.gov/books/NBK54380/#morbidity.s3 and http://www.ncbi.nlm.nih.gov/books/NBK54380 [Accessed 21 October 2012].*

public awareness of the health risks of smoking are lower than in high-income countries with largely middle-class populations.

While governments profit from taxes derived from tobacco and alcohol sales and their consumption, there is a major conflict of interest since illness caused by these health hazards is also a heavy burden on governments owing to their major responsibilities for much or most of health costs and thus an overall liability for government treasuries. Most governments in upper income countries have taken action to raise cigarettes taxes, and ban or limit smoking advertising, smoking in public places, and sales to young people. A 1999–2005 tobacco survey of young people in 141 countries found that 80 percent of respondents have seen tobacco advertisements and 12 percent have been offered free cigarettes, thus showing the influence of tobacco advertising and promotion worldwide. While some young people are smokers, many others use tobacco products other than cigarettes. Telehealth is a specific form of disease management that hold promise for improving the self-care abilities of those most at risk for the principal NCDs as a component of health care encounters.

Alcohol and Other Drugs

Harmful alcohol usage was responsible for 2.5 million deaths globally in 2004 and considered responsible for 4.5 percent of disability-adjusted life years (DALYs), the measurement of global burden of disease. Worldwide, the WHO estimates that 14 percent of deaths caused by alcohol are due to CVD and diabetes, and 50 percent due to liver cirrhosis, and cancer of the liver and breast. In the USA just over 50.9 percent of adults 18 years of age and over are current regular drinkers (at least 12 drinks in the past year) and 13.6 percent are current infrequent drinkers (one to 11 drinks in the past year). In 2009, the number of alcoholic liver disease

deaths, 15,183, and alcohol-induced deaths, excluding accidents and homicides, totaled 24,518 (CDC, Alcohol use fact sheet, 2012).

Physical Activity

Lack of physical activity is the cause of 3.2 million deaths and 32.1 million DALYs around the world. Those who have insufficient physical activity have an estimated 20–30 percent increased risk of death from any cause, compared to physically active people. Participation in physical activity equivalent to an average of 30 minutes daily is thought to reduce the risk of ischemic heart disease by 30 percent and diabetes by 27 percent. The Americas and Eastern Mediterranean regions of the WHO have the highest prevalence of insufficient physical activity. In the USA, adult participation in physical activity (Figure 5.8) increased slightly from 15 to 19 percent between 1999 and 2009, with participation highest among males in the 18–44 age group. Arthritis, heart disease, depression, and diabetes are important factors contributing to a lack of physical activity (Table 5.6). A major challenge in people of all ages is motivation, as well as the presence of conditions that may make it difficult to engage in regular moderate exercise, particularly within the limits of chronic illness.

Hypertension

The prevalence of hypertension and prehypertension increased steadily between 1999 and 2006. Prehypertension is a major risk factor in the age group 18–29 years, leading to established hypertension in the 20–39 age group (Figure 5.9). Hypertension is a highly significant risk factor for cerebrovascular disease, coronary artery disease, congestive heart failure, atrial fibrillation, renal failure, peripheral vascular disease, dementia, and erectile dysfunction.

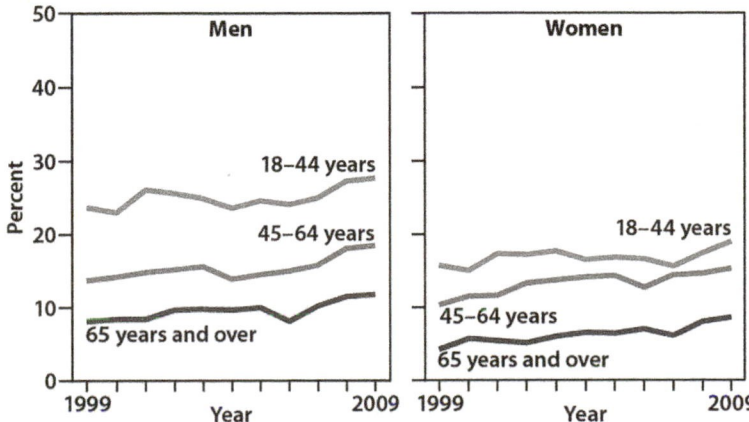

FIGURE 5.8   Adults' participation in physical exercise, by age group, USA, 1999–2009. *Source: Centers for Disease Control and Prevention. Health, United States, 2010: with special feature on death and dying. Hyattsville, MD: US DHHS, NCHS; February 2011. Available at: http://www.ncbi. nlm.nih.gov/books/NBK54380/#morbidity.s3 and http://www.ncbi.nlm.nih.gov/books/NBK54380 [Accessed 21 October 2012].*

TABLE 5.6 Prevalence of Leading Chronic Conditions Causing Limitation of Activity, USA, 2010

Chronic Condition	Prevalence (Millions)
Back/neck condition	7.5
Arthritis/rheumatism	6.8
Heart condition	4.2
Depression/anxiety	4.0
Musculoskeletal condition	3.8
Diabetes	3.6
Hypertension	3.6
Nervous system problem	3.3
Lung/breathing problem	3.1
Fracture/bone/joint injury	2.8
Vision problem	2.3
Stroke	1.7
Cancer	1.5

Source: National Heart Lung and Blood Institute. Morbidity and mortality: 2012 chartbook on cardiovascular, lung and blood diseases. Bethesda, MD: National Heart Lung and Blood Institute; February 2012. Available at: http://www.nhlbi.nih.gov/resources/docs/2012_ChartBook_508.pdf [Accessed 22 October 2012].

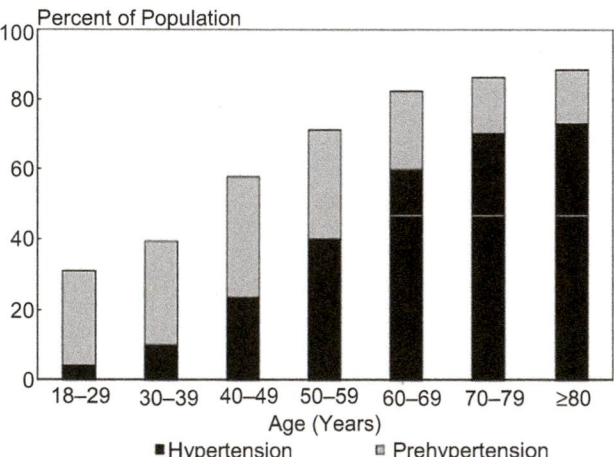

FIGURE 5.9 Prevalence of hypertension and prehypertension by age, USA, 1999–2006. Note: Hypertension = systolic blood pressure (SBP) ≥140 mmHg, or diastolic blood pressure (DBP) ≥90 mmHg, or on medication; prehypertension = SBP ≥120–139 mmHg or DBP 80–89 mmHg. *Source: Centers for Disease Control and Prevention. Health, United States, 2010: with special feature on death and dying. Hyattsville, MD: US DHHS, NCHS; February 2011. Available at: http://www.ncbi.nlm.nih.gov/books/NBK54380/#morbidity.s3 and http://www.ncbi.nlm.nih.gov/books/NBK54380 [Accessed 21 October 2012].*

High blood pressure causes an estimated 7.5 million (13 percent) deaths globally each year. A meta-analysis of data on 1 million adults showed that in the age group 40–69 years each 10 mmHg is associated with more than a two-fold difference in the stroke death rate across all income groups.

Overall, the proportional differences have about half the variation at age 80–89; the absolute differences in risk are greater at old age. These associations are similar for men and women. The study concluded that the average of systolic and diastolic blood pressure is a more informative, and pulse pressure is a much less informative predictor of vascular death. The usual blood pressure is strongly and directly related to vascular (and overall) mortality, and no evidence was found of a threshold down to at least 115/75 mmHg (Prospective Studies Collaboration, Lancet, 2002).

Because of tremendous advances in the prevention, diagnosis, and treatment of CVD, death rates have declined by more than 50 percent – from CHD since the peak in the 1960s and from stroke over a longer period. People in the high-income countries are living longer, healthier lives. The potential for life saving is high, through close supervision of hypertension in its association with serious cardiovascular consequences. Despite this progress, many people are not attached to the health system by insurance or are unaware of the dangers of this condition. In the USA nearly 90 percent of adults with uncontrolled hypertension have access to regular medical care and health insurance at an annual cost of US$131 billion.

Figure 5.10 indicates that a high proportion (30.4 percent) of adults in the USA have hypertension. Perhaps the most alarming and concerning aspect of this figure is that, of the millions of people with hypertension, a larger proportion have *uncontrolled* hypertension. This may be due to the fact that they are unaware of their condition, which is asymptomatic, or they are aware but have not received treatment. As a simple measure of medical care, the primary care physician should play a proactive role in screening and educating patients with even slight elevation of blood pressure to address lifestyle and dietary issues before the process leads to these serious complications. The health community, including public health, clinical practitioners, and health insurance systems, needs to promote awareness of this health problem. They must take action to ensure early diagnosis and management, which starts with smoking cessation, increased physical activity, and moderation in dietary practices, with high vegetable consumption and avoidance of weight gain.

Diet and Obesity

Overweight and obesity have reached epidemic and pandemic levels around the world, in both high- and low-income countries. Most worrisome is the overweight and obesity in children and young people. Daily dietary habits are key factors in causation of major NCDs. The WHO estimates that in 2005 over 1 billion people were overweight and 300 million obese, and by 2015 1.5 billion people worldwide will be overweight or obese. Excessive caloric, cholesterol, and salt consumption, and low levels of physical activity

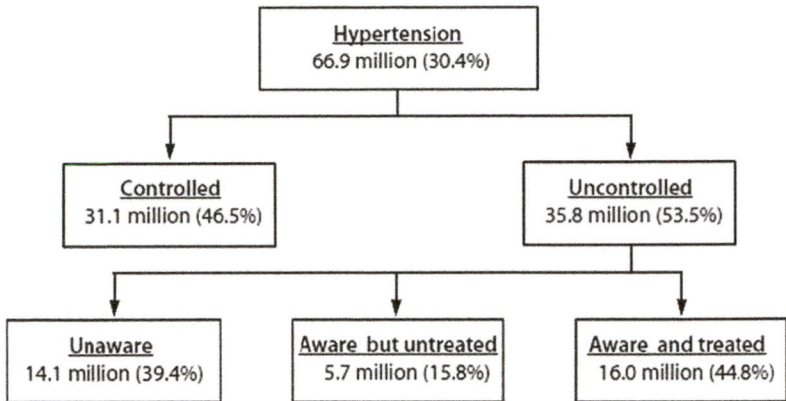

FIGURE 5.10   Number and percentage of adults aged 18 years and over who had hypertension, who had controlled or uncontrolled hypertension, and who were aware and/or pharmacologically treated for hypertension among those with uncontrolled hypertension – National Health and Nutrition Examination Survey (NHANES), USA, 2003–2010. *Source: Centers for Disease Control and Prevention. Vital signs: awareness and treatment of uncontrolled hypertension among adults – USA, 2003–2010. MMWR Morb Mortal Wkly Rep 2012;6:703–9. Available at: http://www.cdc. gov/mmwr/preview/mmwrhtml/mm6135a3.htm?s_cid=mm6135a3_w [Accessed 22 October 2012].*

are major factors in the high prevalence of CVD worldwide. Low consumption of fruit and vegetables is considered to be responsible for 1.7 million deaths (2.8 percent) worldwide from cardiovascular events or cancer.

There is evidence that the "Mediterranean diet" has positive effects on health status and NCDs. This diet is characterized by high consumption of olive oil, legumes, unrefined cereals, fruits and vegetables, and fish. Food sources that should be consumed in moderate to low amounts include dairy products (mostly cheese and yogurt), meat and meat products, with moderate wine consumption.

The UK has implemented population-wide measures to contain salt intake (Box 5.4). American levels of salt intake are very high, with 85 percent of men and 75 percent of women exceeding the recommended upper limits. Dietary issues are common to deaths and DALYs attributable to leading risk factors for NCDs, especially cardiovascular diseases, as shown in Figure 5.11.

Health problems associated with underweight, including anorexia and bulimia nervosa, pose a challenge in many countries. At the other end of the spectrum, overweight and obesity are problematic for a myriad of other reasons. Overweight and obesity among children and adults are developing into a global epidemic in developed and in developing countries. Despite the declining death rates for stroke, in the USA there has been an increase in the proportion of all strokes in the under-55 age group, from 12.9 percent in 1993–1994 to 18.6 percent in 2005. This change over time is probably related to an increase in risk factors for stroke among younger people including hypertension, diabetes with obesity, and high cholesterol (the metabolic syndrome). The steady increase in obesity among adults is seen in Figure 5.12 and in children in Figure 5.13, indicating that the cardiovascular problems of middle age are likely to increase in the coming decades unless major changes occur to reduce these risk factors.

BOX 5.4 Cost-Effective Policy: UK Salt Reduction Program

The UK salt reduction program, begun in 2003, has worked with industry to reduce levels of salt in food, raise consumer awareness and improve food labeling. Average intake was 9.5 g/day in 2000–2001, which was considerably above the recommended national level of no more than 6 g/day for adults.

Voluntary salt reduction targets were set, and industry made public commitments to work to reduce the amount of salt in food products. Public awareness campaigns about health issues, recommended salt intakes and consumer advice were promoted between 2004 and 2010.

Levels of salt in foods have been reduced in some products by up to 55 percent, with significant reductions in those food categories contributing most salt to the diet. Consumer awareness of the 6 g/day message increased 10-fold, and the number of people who say they make a special effort to reduce their intake has doubled. By 2008, average salt intake declined by 0.9 g, thus amounting to 8.6 g/day. This level is estimated to prevent more than 6000 premature deaths and save £1.5 billion every year in health care and other costs, dramatically more than the cost of running the salt reduction program.

Source: *World Health Organization. Global status report on non-communicable diseases 2010. Geneva: WHO; April 2011. Available at: http://www.who.int/nmh/publications/ncd_report2010/en/ [Accessed 21 October 2012].*

Mental Illness

Psychiatric disorders such as depression are a leading cause of disability, also associated with other leading NCDs such as cardiovascular conditions and diabetes. Depressive, substance abuse, and mood disorders affect nearly half of American adults and are also leading causes of life lived

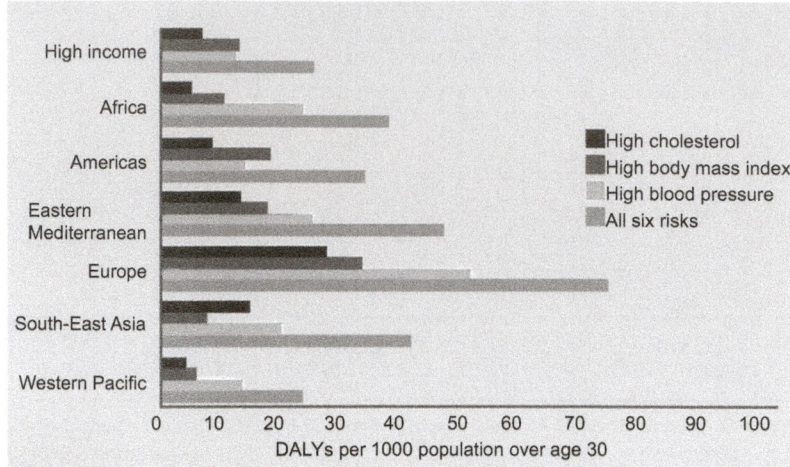

FIGURE 5.11   Diet-related disability-adjusted life years (DALYs) for risk factors by WHO region. *Source: World Health Organization. Global health risks: mortality and burden of disease attributable to selected major risks. Geneva: WHO; 2009. Available at: http://www.who.int/healthinfo/ global_burden_disease/GlobalHealthRisks_report_full.pdf [Accessed 21 October 2012].*

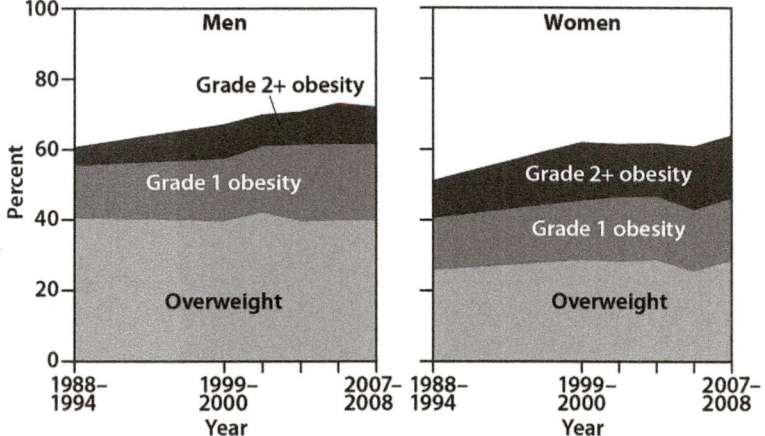

FIGURE 5.12   Overweight and obesity, adults 20 years and over by gender, USA, 1988–2008. Note: Overweight=body mass index (BMI) ≥25 but <30; grade 1 obesity=BMI ≥30 but <35; grade 2+ obesity=BMI ≥35. *Source: Centers for Disease Control and Prevention. Health, United States, 2010: with special feature on death and dying. Hyattsville, MD: US DHHS, NCHS; February 2011. Available at: http://www.ncbi.nlm.nih.gov/books/ NBK54380/#morbidity.s3 and http://www.ncbi.nlm.nih.gov/books/NBK54380 [Accessed 21 October 2012].*

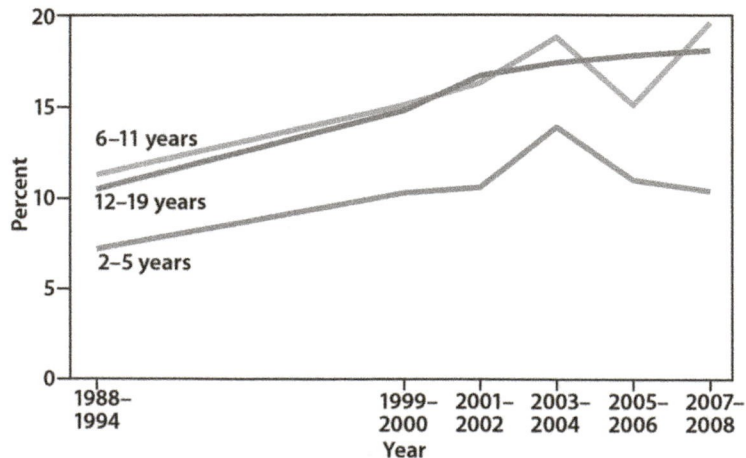

FIGURE 5.13   Obesity among children, by age, USA, 1988–2008. Note: Obesity=body mass index for age and gender ≥95th percentile of CDC growth charts. *Source: Centers for Disease Control and Prevention. Health, United States, 2010: with special feature on death and dying. Hyattsville, MD: US DHHS, NCHS; February 2011. Available at: http://www.ncbi.nlm.nih.gov/books/NBK54380/#morbidity.s3 and http://www.ncbi.nlm.nih.gov/ books/NBK54380 [Accessed 21 October 2012].*

TABLE 5.7 Cardiovascular Risk Factors in Mental Disorders

Modifiable Risk Factor	Estimated Prevalence and Relative Risk (RR)	
	Schizophrenia	Bipolar Disorder
Obesity	45–55%, 1.5–2× RR	26%
Smoking	50–80%, 2–3 × RR	55%
Diabetes	10–14%, 2 × RR	10%
Hypertension	≥18%	15%
Dyslipidemia	Up to 5 × RR	

Source: Parks J, Svendson D, Singer P, Foti ME, Mauer B, editors. Morbidity and mortality in people with serious mental illness. Alexandria, VA: National Association of Mental Health Programs Directors, Medical Directors Council; October 2006. Available at: http://www.dsamh.utah.gov/docs/mortality-morbidity_nasmhpd.pdf [Accessed 22 October 2012].

with disability (see Chapter 7). Depression is also associated with impaired cognitive functions in older people. The lifespan of people with major mental illness is about 25 years shorter owing to preventable conditions such as diabetes. The relationship functions in both directions: mental illness contributes to NCDs and, likewise, NCDs contribute to mental health dysfunction. The public health surveillance system is gradually adopting an integrative approach between public health and mental health to incorporate effective interventions (Institute of Medicine, 2012).

People with comorbid conditions such as heart disease and diabetes are at increased risk for developing depression. This may result in poor adherence to treatment regimens with exacerbation of the underlying chronic illness. Conversely, people with chronic mental illness are at higher risk than their age- and gender-matched counterparts for the development of NCDs. Table 5.7 demonstrates the high prevalence of risk factors for NCDs among people suffering from schizophrenia and bipolar disorder. Some of these are related to the basic condition and some are affected by medications for their treatment. The physical, organizational, financial, and customary separation of mental health care from general primary care and public health contributes to the failure to address the long-term NCD-related health issues for mental health patients. The emotional and behavioral effects of chronic diseases are important elements of comprehensive approaches to NCD prevention activities of clinical care and public health.

Genetic, Infectious, and Non-Communicable Diseases

Genetic markers of susceptibility for NCDs include those related to cancers, diabetes, CVD, asthma, and many others. While these may be independent factors, some may be interactive with environmental, nutritional, or other factors including smoking, diet, and parental age at time of conception and pregnancy. Genetic mutations may produce dysfunctional gene behavior at the cellular level to predispose an individual to these NCDs. Advances in science and medical research aim to provide future methods of intervention to inhibit the development of the undesired disease or play a role in treatment with new forms of gene therapy. Many congenital diseases are discovered by screening at birth and some, such as phenylketonuria and congenital hypothyroidism, should be screened in all newborns, as they are treatable. Many other inherited genetic disorders are included in newborn screening to provide guidance for management and genetic counseling for conditions such as Tay–Sachs disease and thalassemia, with successful elimination of diseases in many regions of the world (see Chapter 6).

Newer genetic scientific discoveries and technologies will enlarge clinical capacity with new methods of diagnosis and treatment, which will include widened applications of stem cell therapy in NCDs such as cardiovascular and neurological disorders. A small percentage of breast cancer cases (5–10 percent) can be linked to gene mutations inherited from the mother or father. These mutations increase a woman's risk of developing breast cancer by up to 80 percent during her lifetime, and at a younger age. Ovarian cancer risk is also associated with these genetic mutations.

Infections as causes of chronic diseases are of great importance for public health because such associations can lead to new treatments or preventive measures. Some of these associations are well established, while others are reported but still lack sufficient evidence. The search for vaccines could replicate the success of immunization in control of the acute infectious diseases of childhood, such as shown in hepatitis B as a major cause of liver cancer and human papillomavirus (HPV) in treatable causal relationships of *H. pylori* and chronic peptic ulcer disease and gastric cancer.

The number of established and proposed relationships between certain organisms and chronic diseases is growing. The relationship of hepatitis B with chronic hepatitis, cirrhosis, and hepatic carcinoma provides the justification for wide-scale immunization to protect individuals, especially those in developing countries who are at increased risk for hepatitis B infection. Hepatitis C also causes cirrhosis and cancer of the liver, and while there is still no vaccine, there are treatments that can eliminate the chronic infection in more than half of cases, depending on the subtype of the virus. HPV (mainly types 16 and 18) is a cause of cervical carcinoma. Screening for cancer of the cervix, with the Papanicolaou (Pap) smear, is an important public health modality; however, education in hygienic practices and the control of sexually transmitted infections (STIs) are also crucial in reducing the spread of these organisms (Box 5.5).

An HPV vaccine based on genetic engineering technology is now widely used for prevention of cancer of the cervix, with the ultimate potential of eradicating this disease,

BOX 5.5 Interaction of Infectious and Chronic Diseases

Infectious Causes of NCDs

- Helicobacter pylori – peptic ulcer disease, gastric cancer
- *Human papillomavirus (HPV)* – cancer of cervix, otolaryngeal and anogenital cancers
- *Human immunodeficiency virus (HIV)* – acquired immunodeficiency syndrome (AIDS), cofactor in reducing immune response, tuberculosis, Kaposi's sarcoma
- *Hepatitis B (HBV) and hepatitis C (HCV)* – chronic cirrhosis and liver cancer
- *Sexually transmitted infections (STIs)* – sterility
- *Syphilis* – neurological disorders and valvular heart disease
- *Chlamydia infection* – coronary heart disease and stroke (hypothesized)
- *Streptococcal infection* – glomerulonephritis
- *Acute rheumatic fever, streptococcal A infection* – rheumatic valvular heart disease, glomerulonephritis
- *Schistosoma haematobium* (schistosomiasis) – bladder cancer in developing countries
- *Epstein–Barr virus* – Hodgkin's lymphoma, Burkitt's lymphoma in Africa
- *Measles* – subacute sclerosing panencephalitis (SSPE), Crohn's disease (hypothesized)
- *Varicella* – herpes zoster
- *Prions* – bovine spongiform encephalopathy (BSE), Creutzfeldt–Jakob disease (CJD)
- *Fungal infection* – polycystic kidneys (hypothesized)
- *Borna disease virus (BDV)* – schizophrenia (hypothesized)
- *Hantavirus* – hypertensive renal disease (hypothesized)

Prevention by Immunization, Screening, Antibiotic Treatment, and Environmental Measures

- *Hepatitis B*
- *HPV*
- *Helicobacter pylori* – treatment of peptic ulcer disease, improved water safety

Comorbidity

- HIV with tuberculosis and Kaposi's sarcoma
- Chronic illness with depression, heart disease, chronic lung disease, and others
- Susceptibility to intercurrent infections such as influenza and pandemic H1N1
- Susceptibility to hospital infections

Sources: Adapted from Centers for Disease Control and Prevention. A CDC framework for preventing infectious diseases: sustaining the essentials and innovating for the future. Atlanta, GA: CDC; 2011. Available at: http://www.cdc.gov/oid/docs/ID-Framework.pdf [Accessed 21 October 2012]. De Martel C, Ferlay J, Franceschi S, Vignat J, Bray F, Forman D, et al. Global burden of cancers attributable to infections in 2008: a review and synthetic analysis. Lancet Oncol 2012;13:607–15. Available at: http://www.thelancet.com/journals/lanonc/article/PIIS1470-2045(12)70137-7/fulltext [Accessed 22 October 2012].

for control of this deadly disease in the coming decade and beyond. Varicella virus is associated with herpes zoster and postherpetic neuralgia. The varicella vaccine, long recommended for routine childhood immunization, may in time eliminate this problem, but length of immunity may require adult booster doses of varicella for preventing herpes zoster, a painful and severely discomfiting burden for older people.

Similarly, acute STIs can have long-term sequelae such as sterility, neurological disorders, and cancer. Untreated syphilis can lead to long-term neurological deterioration and valvular heart disease. Examples of established relationships and others that are still unproven hypotheses are shown in Box 5.5. The number of examples of infectious diseases causing chronic disease will increase with development of the biological sciences, as in the example of the discovery of prions as transmitters of such diseases as scrapie in sheep, bovine spongiform encephalopathy (BSE) in cattle, and Creutzfeldt–Jakob disease (CJD) in humans.

Since the 1990s, *Chlamydia pneumoniae* infection has been a suspected cause of CHD, providing fresh impetus to the search for new approaches to prevent the leading cause of death in industrialized and many developing countries. Initial reports from Finland, the UK, and Italy were followed by supportive studies in the USA. The association, however, has not been substantiated according to current criteria for causation of the Koch–Henle postulates. However, the association of elevated C-reactive protein (CRP) suggests that an inflammatory process is involved in atherosclerosis and its expression as CVD.

Cardiac disorders are a manifestation of chronic Chagas disease, which is known to cause sudden death or heart failure due to destruction of heart muscle. Largely as a result of migration from endemic countries, Chagas disease is increasingly being recognized in Canada, the USA, Europe, and Western Pacific countries (WHO Global Atlas on Cardiovascular Disease Prevention and Control).

Bacterial infections can also have long-term sequelae. Examples include peptic ulcer disease and chronic gastritis from the bacterium *H. pylori* (Box 5.6). In 1991, four case–control studies quantified and confirmed that *H. pylori* is associated with stomach cancer. They included the demonstration of antibodies and growth of the organism before and after diagnosis of cancer. One study showed a relation between severity of infection and risk of stomach cancer. Currently, the association of *H. pylori* infection with cancer of the stomach and peptic ulcer and its complications is accepted, with clear evidence of a dramatic reduction in cancer of the stomach and in surgical need for peptic ulcer treatment in recent decades.

Peptic ulcer diseases, comprising gastric and peptic ulcers, esophagitis and their chronic debilitating effects used to be among the most common medical conditions, requiring medical and surgical hospitalization for chronicity and complications such as life-threatening bleeding

in conjunction with Pap smear programs. HPV vaccine for young girls and women, and more recently for boys as well, gives hope to reducing HPV infection and transmission, but maintenance of screening programs with Pap smears is vital

BOX 5.6 *Helicobacter pylori*, Peptic Ulcers, and Stomach Cancer

Helicobacter (Campylobacter) pylori, first reported in *The Lancet* in 1984 as a curved bacterium associated with gastritis, is present in more than half the world's population. Once acquired, it persists and can cause serious disease decades after infection, acting as a "microbial parasite". The organism was shown in the pylorus of the gastric aspirates of 58 of 100 patients with peptic ulcer by Australian physicians Robin Warren and Barry Marshall. Warren discovered the organism, and Marshall experimentally ingested the organism himself and consequently developed nausea, stomach pain, and foul breath. They suggested that this organism might be the cause of peptic ulcer disorders and possibly stomach cancer. Treatment of peptic ulcer, after gastroscopy, biopsy, and demonstration of the organism, followed by a short course of antibiotics is now standard treatment.

Peptic ulcer disease (PUD) has become far less common in the industrialized countries, probably in relation to improved diagnosis and treatment sanitation. In the USA, age-adjusted hospitalization rates for PUD decreased by 21 percent from 1998 to 2005 (from 71.1 to 56.5 per 100,000 in 2005), and discharge diagnosis of *H. pylori* infection decreased by 47 percent (from 35.9 to 19.2 per 100,000 population) in the same period.

The introduction of powerful hydrogen ion antagonists in 1977 contributed to further decline in hospitalization rates from peptic ulcers. Many factors were considered important in causation for this disease, including stress, occupation, genetic tendencies, alcohol, smoking, coffee, and aspirin.

An ecological study in the USA showed high correlation between areas with high prevalence of the organism and high prevalence of cancer of the stomach. In 1998 *H. pylori* was reported to have been found in surface water, suggesting a major reservoir and method of transmission of this organism. Chlorination apparently kills the organism.

Gastric cancer, once the leading cancer in males, declined during the 1960s to 1980s worldwide, especially in the industrialized countries. Marshall and Warren were awarded the Nobel Prize for Physiology and Medicine in 2005 for this achievement. Finding the cause and cure for peptic ulcer has radically changed the pattern of stomach cancer epidemiology and the nature of surgical practice worldwide. Community and environmental aspects of *H. pylori* and its associated NCDs raise many issues of prevention of infection by water treatment and hopefully in the near future, a safe, effective, and inexpensive vaccine to protect the population from this widespread infection.

Sources: *Press Release: The 2005 Nobel Prize in Physiology or Medicine. Stockholm: Nobelprize.org. Available at: http://www.nobelprize.org/nobel_prizes/medicine/laureates/2005/press.html [Accessed 22 October 2012].*
Uemura N, Okamoto S, Yamamoto S, Matsumura N, Yamaguchi S, Yamakido M, et al. Helicobacter pylori infection and the development of gastric cancer. N Engl J Med 2001;345:784–9. Available at: http://www.nejm.org/toc/nejm/345/11 or http://www.nejm.org/doi/full/10.1056/NEJMoa001999 [Accessed 22 October 2012].
Feinstein LB, Holman RC, Yorita Christensen KL, Steiner CA, Swerdlow DL. Trends in hospitalizations for peptic ulcer disease, United States, 1998–2005. Emerg Infect Dis 2010;16:1410–8. Available at: http://wwwnc.cdc.gov/eid/article/16/9/pdfs/09-1126.pdf [Accessed 22 October 2012].
Centers for Disease Control and Prevention. Helicobacter and peptic ulcer diseases [updated 28 September 2006]. Atlanta, GA: CDC. Available at: http://www.cdc.gov/ulcer/history.htm [Accessed 22 October 2012].
Helicobacter pylori and cancer [updated 16 November 2011]. Bethesda, MD: National Cancer Institute. Available at: http://www.cancer.gov/cancer-topics/factsheet/Risk/h-pylori-cancer [Accessed 22 October 2012].
World Health Organization. Cancer. Fact sheet no. 297 [updated February 2012]. Geneva: WHO. Available at: http://www.who.int/mediacentre/factsheets/fs297/en/ [Accessed 22 October 2012].

ulcers. The work of Warren and Marshall in Western Australia showed *H. pylori* to be the cause of chronic peptic ulcer diseases (Box 5.6). This established strong support for the link between infectious diseases and NCDs, including gastric cancer. Gastric cancer, a major cause of death globally, declined during the 1960s to 1980s worldwide, especially in the industrialized countries. However, it remains one of the leading causes of death worldwide. The WHO reports that cancer deaths in 2008 totaled 7.6 million, with cancer of the stomach ranked as the second most common cause. The main types of cancer mortality were due to lung (1.4 million deaths), stomach (740,000 deaths), liver (700,000 deaths), colorectal (610,000 deaths), and breast cancer (460,000 deaths).

Helicobacter pylori is the major cause of peptic ulcer diease, now a treatable condition, thus giving impetus to the idea of infections as causes of chronic diseases (see Chapter 4). Standard medical treatment of gastric and duodenal ulcers now uses breath tests, which indicate the presence of *H. pylori*, as well as treatment with inexpensive antibiotics. This has further reduced hospitalizations and surgery for chronic peptic ulcer complications, as well as expensive long-term drug treatment and the suffering of patients with chronic peptic ulcers.

The continuing search for vaccines against common infectious diseases, especially those associated with chronic conditions, will have an important impact on public health in the coming years, as has happened in the past several decades.

Liver disease and cirrhosis combined were the fifth leading cause of death in the age group 45–64 in the USA in 2009 (12th in 2010), also serving as a major contributory cause of morbidity, leading to 1 percent of all hospital admissions. Patients with liver failure are high users of hospital care, mainly because of serious complications such as gastrointestinal bleeding from esophageal varices. This is a group of diseases related to chronic alcohol consumption and chronic viral hepatitis infection (mainly hepatitis B and C). The risk of cirrhosis is high among long-term heavy users of alcohol and is related to amounts consumed daily. Other nutritional factors, such as vitamin B deficiency, may be secondary contributory factors. The death rate from chronic liver disease and cirrhosis was 10.0 per 100,000 in 2009, and nearly twice as high in men as in women (13.1 deaths per 100,000 and 6.9 per 100,000, age-adjusted, respectively). Death rates (age adjusted) in the USA for chronic liver disease and cirrhosis increased by 3.3 percent between 2009 and 2010.

Hepatitis B infection, transmitted in blood products, body fluids, and household contacts, is common in many countries. An estimated 2 billion people are infected, along with 350 million carriers who are at high risk for cirrhosis of the liver and primary liver cancer. Hepatitis C prevalence worldwide is reported by the WHO as 170 million cases, with infection rates varying from under 1 percent in Canada, Australia, and parts of Western Europe to 1–2.4 percent in the USA and much of Europe, India, and most of South America. Hepatitis C rates are up to 10 percent of the population in China, much of Africa, and South America. Aflatoxin exposure in foods is a major contributor to chronic liver disease in developing countries. Liver disease varies widely by country; this is likely to be due to a combination of widespread hepatitis, high alcohol intake, and nutritional and environmental factors such as pesticide exposure. Rates of mortality from chronic liver disease vary widely; Romanian rates are staying very high while rates in Italy and France are falling steadily and rates in Israel remain low, which could possibly be explained primarily by alcohol consumption, nutritional differences, and hepatitis prevalence (Figure 5.14).

Liver cancer and liver cirrhosis are major public health problems, with hepatitis B being the cause of 60–80 percent of primary liver cancer, especially in developing countries of sub-Saharan Africa, East and South-East Asia, and the Pacific basin. About 2 billion people globally have been infected with the hepatitis B virus. Out of these, more than 360 million have chronic, lifelong infections. Those chronically infected are at high risk of death from cirrhosis of the liver and liver cancer, diseases that kill about 1 million people each year. Therefore, the WHO recommends prevention by inclusion of hepatitis B vaccine in routine infant vaccination programs and catch-up immunization of other age groups.

Hepatitis C virus, discovered in 1988, is estimated to affect 170 million people globally, with 3–4 million people being newly infected each year. Chronic carriers are also at risk for liver cirrhosis and primary liver cancer. Screening of blood and blood products for hepatitis C should be standard practice worldwide. The virus is commonly spread through unsanitary intravenous drug usage. There is no vaccine for hepatitis C. Prevention of cirrhosis focuses on reducing the daily consumption of alcohol and promoting universal immunization against hepatitis B. Needle exchange programs reduce transmission of hepatitis among intravenous drug users. Treatment of hepatitis C virus with interferon and antiviral drugs is expensive and lengthy, with a heavy burden of side-effects, and has limited success. The recent addition of combinations with newer antiviral medication has improved the success rate, but it still presents an onerous burden on patients and on the health system. No vaccine is yet in the offing.

Alcohol consumption in moderation is beneficial, but long-term high levels of consumption, such as binge drinking, bring health hazards in chronic liver disease. Environmental and occupational exposures to pesticides and other toxic chemicals are also factors in chronic liver disease, which is a major consumer of health care resources. Liver transplantation is effective for some in developed countries; however, a lack of donors limits the effectiveness of this modality, as does its high cost.

THE SOCIAL AND ECONOMIC BURDEN OF NON-COMMUNICABLE DISEASES

Chronic conditions place a heavy burden on the individual, the family, and society as a whole in terms of morbidity and mortality as well as in health costs. Measurement of the burden of disease (discussed in Chapter 3) is a fundamental responsibility of public health agencies. Health expenditures as a percentage of Gross Domestic Product (GDP) comparing high- and medium-income countries (see Chapter 11) vary widely. The economic burden for health care of the population, whether funded by national responsibility, as in most industrialized countries (see Chapters 10–13), or by a mix of public and private expenditures, is an important factor in national economies. As chronic disease and aging of the population both rise, so will the burden of disease become increasingly important in economic as well as in health terms.

The cost of individual disease groups can be enormous. The American Heart Association reports the cost of CVD in the USA in 2010 at an estimated US$444.0 billion (representing $1 of every $6 spent on health care). This figure includes direct costs of health services such as physicians and other professionals, hospital and nursing home services, medications, home health care and other medical durables, as well as lost productivity resulting from morbidity and mortality (indirect costs). With aging of the population the economic burden of caring for these diseases will grow rapidly, thus prevention is essential in reducing this financial burden on society and on individual families. As a comparison, in 2010 the estimated cost of all cancers was US$263.8 billion (US$102.8 billion in direct costs, US$20.9 billion in indirect morbidity costs, and US$140.1 billion in indirect costs associated with premature mortality). Between 1990 and 2007 death rates in the USA declined by 22 percent in men and 14 percent in women, equivalent to avoiding nearly 900,000 cancer deaths.

Less quantitative and more difficult to calculate is the burden of disease on the individual, the family, and the community. Traditional measures of morbidity and mortality are supplemented by quality-adjusted life years (QALYs) and disability-adjusted life years (DALYs), as seen in Chapter 3, but the physical and emotional burden of caring for someone is almost impossible to determine quantitatively. The burden of chronic disease on the individual is reflected in his or her ability to function in the normal activities of daily living (ADLs). The level of function of a person with a chronic condition is measured by his or her ability to perform ADLs, as seen in Chapter 6.

Chronic conditions often result in disabilities that impede capabilities in normal daily functions or activities. ADLs measure the degree of independent capacity the patient has pertaining to personal care, household management, and socializing. These measures help to determine the level and amount of home care required or the type of facility necessary for the patient. While ADL measures the function of a patient, it does not address the emotional, physical, and financial stress on the caregiver in a family. Self-perception of health status is important and is part of periodic surveys conducted by the US National Center for Health Statistics Summary Health Statistics for the US Population: National Health Interview Survey (NHIS), 2006 Series.

It is estimated that 75 percent of total health care costs can be attributed to patients with one or more chronic conditions which can be prevented. More than one-third of Americans have at least one type of CVD – 935,000 have CHD and 795,000 have had a stroke – with total costs of nearly US$300 billion (Figure 5.15). Of the 68 million with high blood pressure and the 71 million with high cholesterol, more than half represent uncontrolled cases. In many health care systems, the compensation structure for services focuses more heavily on diagnosis and treatment than on other health promotion and preventive measures, which serve as key components in containing health care service costs.

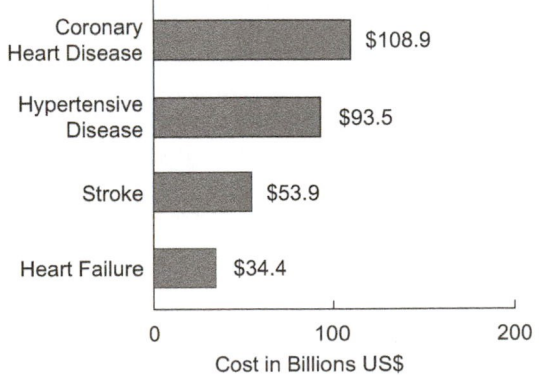

FIGURE 5.15 Estimated costs for cardiovascular diseases, USA, 2010. *Source: Centers for Disease Control and Prevention. Heart disease and stroke prevention – addressing the nation's leading killers: at a glance 2011. Atlanta, GA: CDC; 2011. Available at: http://www.cdc.gov/chronicdisease/resources/publications/aag/pdf/2011/Heart-Disease-and-Stroke-AAG-2011.pdf [Accessed 22 October 2012].*

HEALTH PROMOTION AND DISEASE PREVENTION

Health promotion and prevention programs built around evidence-based strategies are vital to contain risk factors and reduce the toll of NCDs. Population- and community-based strategies aim at increasing healthy behaviors, reducing the disease burden and lowering death rates. DALYs are an indicator of the impact of the premature avoidable death on society. There is increasing awareness of the effectiveness and success of efforts that focus on the common, parallel risk factors associated with the leading NCDs, in contrast to a vertical method of approaching each disease individually. Leadership is required at all levels, especially national government. In the USA, the Healthy People 2020 framework provides for all stakeholders in health issues to work towards defined targets, and this has a "trickle-down" effect reaching state and local government, health insurance agencies, health care providers, and individuals.

In Europe the term "Health in All Policies" has become widely accepted in trying to spread the tasks of public health among all governments (national, state, and municipal), as well as among commercial and charitable organizations and across all social agencies and activities. This approach also places stress on the role of the individual in protecting his or her own health with positive health behaviors and attention to preventive care activities such as immunization, healthful diet, and physical activity, with less risk taking such as smoking, alcohol overuse, and unsafe driving and pedestrian behavior. The role of municipal governments in urban planning for healthy urban environments is a key component of this, as are state and federal government promotion, regulation, and supportive programs for population and individual health behaviors.

HYPERTENSION AND CARDIOVASCULAR DISEASES

Cardiovascular disease (CVD) refers to a group of diseases of the heart and blood vessels, including coronary or ischemic heart disease, hypertension, and cerebrovascular disease (stroke). These diseases are associated with atherosclerosis, excess fats in the diet and lipids in the body, and often with impairment of endocrine functions related to glucose metabolism and diabetes mellitus.

CHD occurs when the arteries that supply the muscles of the heart become narrow and restricted with fatty plaques, leading to possible blockages. The coronary arteries may become blocked with thromboses or clots, thus cutting off the blood and oxygen supply (ischemia, i.e., ischemic heart disease) and leading to death (necrosis) of heart muscle, or other severe episodes, such as acute myocardial infarctions (AMIs), also called heart attacks or coronary incidents. Aside from these potential outcomes, milder forms may present with angina pectoris, or chest pain on exertion.

The famous longitudinal heart study in Framingham, Massachusetts (1948), provided important epidemiological data showing that hypertension, smoking, and elevated cholesterol are all associated with increased risk of CVD (Box 5.7). The Framingham study pioneered the epidemiological approach to gain insight into causes of CVD. This prospective cohort study was developed to quantify risks for these diseases, in terms of both absolute and relative risk. Observations made in this study have led to causal inferences (e.g., elevated blood pressure with increased risk of stroke).

Subsequent studies have elaborated on the Framingham findings. High blood pressure (hypertension) refers to elevated levels of systolic and/or diastolic blood pressure and is associated with an increased risk of morbidity and mortality from myocardial infarction, stroke, and renal disease. Concepts of normality have been replaced with guidelines for "optimal" values of blood lipid and blood pressure for long-term freedom from CVD. The atherogenic potential for serum total cholesterol was shown to be derived from the low-density lipoprotein cholesterol (LDL-C) fraction, which is positively related to CHD incidence. High-density lipoprotein cholesterol (HDL-C) is inversely related to CHD, as its function is to remove cholesterol from tissues. The risk of CHD is independently related to each of these lipoprotein fractions. Therefore, the ratio of total to HDL-C is an efficient lipid risk profile.

CVD is the leading cause of death worldwide, with 17.3 million deaths from coronary and vascular diseases (30 percent of total global mortality) in 2008 (7.3 million from

---

BOX 5.7 Timeline of Milestones from the Framingham Heart Study

1948 – Start of the Framingham Heart Study

1956 – Findings on progression of rheumatic heart disease reported

1960 – Cigarette smoking found to increase the risk of heart disease

1961 – Cholesterol level, blood pressure, and electrocardiogram abnormalities found to increase the risk of heart disease

1965 – First Framingham Heart Study report on stroke

1967 – Physical activity found to reduce the risk of heart disease, and obesity to increase the risk of heart disease

1970 – High blood pressure found to increase the risk of stroke

1974 – Diabetes and its complications associated with development of cardiovascular disease

1976 – Menopause found to increase the risk of heart disease

1977 – Effects of triglycerides and LDL and HDL cholesterol described

1978 – Psychosocial factors found to affect heart disease, and atrial fibrillation (heart beats irregularly) to increase the risk of stroke

1981 – Major report on relationship of diet and heart disease; filter cigarettes give no protection against coronary heart disease

1986 – First report on dementia related to vascular diseases

1987 – High blood cholesterol levels found to correlate directly with risk of death in young men, fibrinogen increases the risk of heart disease; estrogen replacement therapy found to reduce risk of hip fractures in postmenopausal women

1988 – High HDL cholesterol found to reduce risk of death; association of type "A" behavior with heart disease reported; isolated systolic hypertension found to increase risk of heart disease; cigarette smoking found to increase risk of stroke

1990 – Homocysteine (an amino acid) suggested as possible risk factor for heart disease

1993 – Mild isolated systolic hypertension shown to increase risk of heart disease

1993 – Major report predicts survival after diagnosis of heart failure

1994 – Enlarged left ventricle shown to increase the risk of stroke; lipoprotein A found as possible risk factor for heart disease

1994 – Risk factors for atrial fibrillation described; apolipoprotein E found to be possible risk factor for heart disease; first Framingham report on diastolic heart failure published

1995 – OMNI Study of Minorities started (applicability of Framingham findings to many ethnic groups)

1996 – Progression from hypertension to heart failure described

1997 – Cumulative effects of smoking and high cholesterol on the risk for atherosclerosis reported; impact of enlarged left ventricle and risk for heart failure in asymptomatic individuals investigated

1998 – New risk prediction formulae calculated risk for developing coronary disease over the next 10 years; a gene (angiotensin-converting enzyme deletion/insertion polymorphism) associated with hypertension in men

2002 – NEJM report linked body mass index (BMI) and obesity with increased risk of heart failure; high BMI shown as independent risk factor; third generation study enrolled 3900 grandchildren of the Framingham Heart Study's original enrollees

2003 – Offspring-based study published relating likelihood of heart attack three times greater in individuals with common genetic variation in an estrogen receptor

2004 – Demonstration that having a parent with a cardiovascular disease history doubles personal risk of the disease

2005 – Offspring study reported that an increase of up to 45 percent for risk of heart attack, stroke, or arterial disease may occur in middle-aged people with a sibling who suffered a similar cardiovascular event.

Source: A timeline of milestones from the Framingham Heart Study. Framingham, MA: Framingham Heart Study. Available at: http://www.framingham.com/heart/timeline.htm [Accessed 21 October 2012].

CHD and 6.2 million from stroke). The WHO estimates that by 2030 cardiovascular deaths will reach 23.6 million, with ischemic and cerebrovascular disease the two leading causes of death, and hypertensive heart disease deaths ranking 8th compared to 14th in 2004. Factors in the 2030 rankings will be related to population growth, population aging, and epidemiological changes that should be influenced by public health interventions and personal life habit changes. An important portion of these cases can be attributed to behavioral factors, such as those that can often be modified, including cigarette smoking, physical inactivity, and unhealthy diet. The combination of direct and indirect costs associated with CHD was estimated to be US$165 billion in 2009.

The global risk for CVD includes many risk factors and is therefore complex to control. One of the results of research based on the Framingham study is a predictive model, valid with adaptations for all societies. This allows not only the assessment of individual risk, but also a population's risk and preventable factors to be addressed. These evaluations are essential when planning and implementing health promotion programs, which are vital to address the global burden of CVD worldwide. CVD and other NCDs contribute to poverty and are in large measure associated with poverty and limited choices in life. The most affected are poor populations in low- and middle-income countries, where heart disease, stroke and diabetes reduce GDP by 1–5 percent.

In 1965, the classic Alameda County study of Lester Breslow and colleagues followed a cohort of nearly 7000 people residing in the county, which had a population of 1 million. Researchers identified seven health practices that were associated with reduced mortality and disability patterns. The following are the health practices that were assessed: excessive alcohol intake, cigarette smoking, obesity, sleeping less or more than 7–8 hours/day, physical inactivity, eating between meals, and not eating breakfast. Adjusting for age, gender, health status, and social networks, the occurrence of disability was half as great for those with good health practices as compared to those with poor practices. These patterns were prevalent in the cohort that followed in the 1960s and 1970s. The Alameda County studies reported socioeconomic, morbidity, and mortality gradients as well as the effectiveness of lifestyle changes, primarily in reducing the complications of chronic diseases and in prolonging life. The WHO MONICA project contributed to the development of cardiovascular epidemiology in an international context, which was highly influential in promoting a shift in health policies towards primary prevention (Box 5.8).

These and many other studies showed strong associations between social relationships and CVD, more so for men than for women. These findings have been confirmed in many studies and the association of CVD with poverty

---

BOX 5.8 The MONICA Project (Multinational Monitoring of Trends and Determinants in Cardiovascular Disease)

Following the end of World War II, coronary heart disease (CHD) assumed epidemic proportions in western countries. CHD mortality began to decline in the USA in the early 1960s and this pattern was followed later in many western countries, after peaking in 1968. In 1978, the National Heart, Lung, and Blood Institute of the NIH organized the Bethesda conference on CHD mortality to assess whether prevention or improved acute coronary care was responsible for the decline in age-specific CHD mortality rates. There was no clear answer to these questions, but the WHO took the leadership and organized the MONICA project as an international epidemiological investigation system to assess trends and determinants of cardiovascular mortality, incidence, and case fatality from the mid-1980s to the mid-1990s. The study included 38 population centers in 21 countries worldwide.

Altogether, some 13 million people were monitored over a 10-year period, 166,000 myocardial infarction patients were registered, and more than 300,000 men and women were sampled and examined for cardiovascular risk factors and many other health data. In western countries, where the CHD mortality declined on average 2–3 percent annually, two-thirds of this decline could be explained by a decline in CHD incidence and one-third by a decline in CHD case fatality. When the trends were examined in relation to changes in risk factors and CHD event rates in men over a period of 10 years in all MONICA populations, the greatest contribution to the observed decline was decreased smoking; however, reduction of hypertension levels also contributed.

The MONICA project was in company with many similar studies worldwide, the Seven Countries Study including Finland (high), the USA, the Netherlands, Italy, Yugoslavia, Greece, Japan (low), and the Framingham Heart Study, all contributing to the cumulative development of chronic disease epidemiology and to prevention of cardiovascular diseases.

Sources: *Tunstall-Pedoe H, Connaghan J, Woodward M, Tolonen H, Kuulasmaa K, for the WHO MONICA project. Pattern of declining blood pressure across replicate population surveys of the WHO MONICA project, mid-1980s to mid-1990s, and the role of medication. BMJ 2006;332:629–35. Available at: http://www.bmj.com/content/332/7542/629 [Accessed 22 October 2012].*
Luepker RV. WHO MONICA project: what have we learned and where to go from here? Public Health Rev 2011;33:373–96. Available at: http://www.publichealthreviews.eu/upload/pdf_files/10/00_Luepker.pdf [Accessed 22 October 2012].

---

provides an important basis for planning interventions in targeted groups and for those individuals at highest risk.

The early epidemiological work on CVD including controlled trials led to community-based intervention trials on CVD, such as the Minnesota Heart Health Program, and the North Karelia project in Finland. The Finnish project demonstrated a dramatic effect in reducing CVD mortality in a region which exhibited the highest mortality rates in the country. This was achieved by adopting a health promotion orientation of public health. Replication of health promotion activities and interventions showed similar dramatic effects across the country.

Many studies also showed the importance of social determinants in addition to the classic risk factors of smoking, elevated cholesterol, blood pressure, and physical inactivity. The social factors include fundamental elements, such as the social environment during infancy, level of education, income, economic and social policies, and education. Social risk factors include the work setting context (employment, job instability, and working conditions), social relationships and isolation, geographic environment, "ethnicity", and above all poverty. The UK Whitehall II study showed the association between job stress and, more importantly, lack of control in the job setting, with a dose–effect relationship between exposure to job strain and the onset of CHD. The results indicated that longer periods of exposure are associated with higher incidence, and lower grade civil servants were subject to higher rates of CVD than higher grades. Moreover, ethnicity and race have been commonly associated with CVD mortality; however, much of this variance among the same ethnic groups is found to be related to social class.

Forsdahl in 1977 and Barker in 1986 showed that early disadvantages in life (low birth weight, high risk of infant mortality, child malnutrition) are predictors of future early CVD risk. The chain of causality thus links many societal (such as income, level of education, economic and social policy, education policy) and individual coping with biomedical risk factors as "fundamental" to the origin of many diseases. Rose proposed a "high-risk strategy to prevention", targeting individuals with a high level of cardiovascular risk factor, along with a population-based strategy, aiming to shift the distribution curve of these risk factors by reducing average millimeters of blood pressure, millimoles of cholesterol, or serum sodium levels. This strategy would bring minimal benefit to the individual but appreciable gain to the population in terms of mortality.

Heart disease death rates in the USA vary by race and gender, with the highest among black males (Figure 5.16). "Since 1979, age-adjusted rates of death from heart disease declined significantly among blacks and whites for both men and women. Death rates remain highest for black males and lowest for white females, although differences by race and sex have narrowed in recent years. From 2005 to 2006, rates of death from heart disease declined 7.4 percent for black females, 5.8 percent for white females, 5.4 percent for white males, and 3.8 percent for black males" (CDC, 2008).

Other diseases of the heart include problems associated with the heart muscle, as well as rheumatic diseases, characterized by damage to valves of the heart. Depending on the extent of heart muscle damage, a patient may go into congestive heart failure (CHF) owing to weakened function of the heart as a pump, resulting in congestion of fluids in the lungs and other tissues.

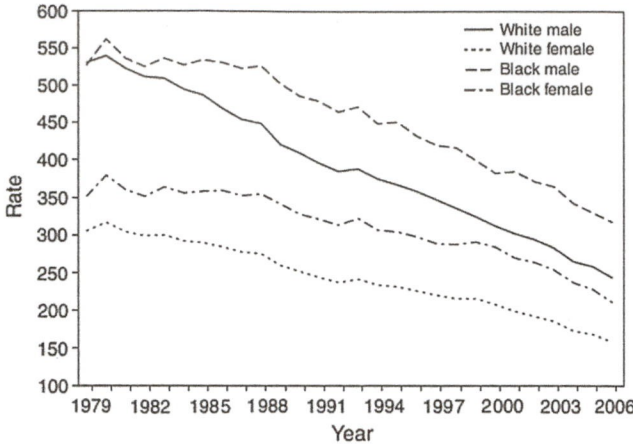

FIGURE 5.16 Age-adjusted rates of death per 100,000 population from heart disease by race and gender, USA, 1979–2006. Note: Data for 2006 are preliminary. *Sources: Centers for Disease Control and Prevention. Quick stats: age-adjusted rates of death from heart disease by race and sex, United States, 1979–2006. MMWR Morb Mortal Wkly Rep 2008;57:779. Available at: http://www.cdc.gov/mmwr/preview/mmwrhtml/mm5728a6.htm [Accessed 22 October 2012]. Heron MP, Hoyert DL, Xu JQ, Scott C, Tejada-Vera B. Deaths: preliminary data for 2006. Natl Vital Stat Rep 2008;56:1–52. Available at: http://www.cdc.gov/nchs/data/nvsr/nvsr56/nvsr56_16.pdf [Accessed 22 October 2012].*

Difficulties or disruption in the heart rhythm can occur, with sudden stoppage of the heart, possibly leading to death. Although some can be fatal, most arrhythmias are harmless and are manageable by medication or pacemakers to augment the heart's electrical system and provide an electrical stimulus when the heart rhythm slows to dangerous levels.

Cerebrovascular accidents (CVAs), or strokes, are incidents in which blood flow to a portion of the brain is blocked. Thus, if blood vessels supplying the brain do not properly deliver blood and oxygen, this results in the death of areas of brain tissue. Consequently, death or disability from loss of central nervous system function may occur. If a stroke occurs on the dominant side of the brain, for those with damage to the left side for right-handed people, depending on the area of permanent damage after recovery, there are varying degrees of motor and mental limitations. Partial occlusion may cause transient ischemic attacks (TIAs) resulting in brief loss of motor and mental function. In both Acute Myocardial Infarctions (AMIs) and cerebrovascular accidents (CVAs or stroke), the area of dead tissue will be surrounded by tissue that is inflamed and damaged. Treatment is intended to restore blood flow and minimize the inflammation or swelling (edema) and fibrosis with permanent loss of function. Immediate care is vital to minimize damage: this is imperative, as it has a direct effect on maximizing recovery.

Precise measurement of the prevalence of these diseases through epidemiologically sound, community-based prevalence studies is difficult because of a lack of standard diagnostic criteria and terminology, as well as limited resources. Measures such as mortality rates and hospitalization data may not give true prevalence rates; however, they do provide important time trends and comparisons between areas that are valid for planning public health interventions. CVDs have common pathophysiological features with diabetes, related to nutrition, exercise, and other lifestyle factors. Despite common features in the basic pathology, cerebrovascular and coronary heart diseases have important clinical differences in risk groups and clinical manifestations, as well as in the screening, measurement, and interventions required. There is growing awareness that the initial focus on CHD and stroke should be expanded to and integrated with other NCDs, with a strategic focus on underserved populations to reduce the burden of this entire class of diseases, and not only CVD.

Mortality from CHD and cerebrovascular disease increased in most western countries from the 1920s, reaching its peak in the 1950s. CHD mortality began to decline some years later than cerebrovascular disease and continues to decline. Cerebrovascular disease has continuously dropped since the start of the twentieth century, but the rate of decline has varied since 1990. CHD death rates vary widely between countries and by income group. In low- and middle-income countries heart disease and stroke are, respectively, the third and sixth leading causes of mortality among women. Table 5.8 compares the average annual percentage decline in coronary death rates among males in selected countries. In 2007, the USA ranked sixth highest for male CVD deaths and fourth highest for female CVD deaths among 12 industrialized countries.

There was a dramatic decline in death rates from cardiovascular and cerebrovascular diseases in the USA from 1950 to 2008, but less so than in many other industrialized countries. Table 5.9 shows the percentage change in death rates by cause for the USA for all causes and for CVD between 1968 and 2008. Age-adjusted death rates for ischemic heart disease declined in the USA by 59 percent from 1950 to 2002, and stroke mortality by 69 percent, while mortality from all causes fell by 26 percent, but the rate of decline has slowed since 2002. The rates of decline in CVD, however, were not uniform in all regions or population groups. The average annual decline over this period was 3 percent for women and 3.8 percent for men. Of the total deaths in 2007, 25 percent were from diseases of the heart. In the USA, rates of heart disease and cerebrovascular disease mortality for men and women, African Americans and whites have all declined, but at varying rates (Table 5.10). The costs attributable to CVD in the USA reached more than US$500 billion in 2010.

Despite the decline in CHD in most western countries, CHD is still the largest cause of death in the industrialized nations and increasingly in developing countries. From 2006 to 2010, age-adjusted CHD prevalence in the USA declined overall from 6.7 percent to 6.0 percent. Similar

TABLE 5.8 Change in Age-Adjusted Death Rates[a] for Coronary Heart Disease by Country and Gender, Ages 35–74 Years, 1999–2009[b]

Males		Females	
Country	Average Annual Percent Change[c]	Country	Average Annual Percent Change[c]
Denmark (1999–2006)	−8.8	Netherlands (1999–2009)	−9.0
Netherlands (1999–2009)	−8.6	Denmark (1999–2006)	−9.0
Norway (1999–2009)	−7.4	Norway (1999–2009)	−8.3
UK[d] (2001–2009)	−6.5	UK[d] (2001–2009)	−8.0
Germany (1999–2006)	−5.7	Germany (1999–2006)	−7.2
Poland (1999–2008)	−5.1	Finland (1999–2009)	−6.5
France (2000–2007)	−5.1	Poland (1999–2008)	−6.1
Finland (1999–2009)	−4.9	France (2000–2007)	−6.0
USA (1999–2008)	−4.6	Spain (1999–2009)	−5.2
Spain (1999–2008)	−4.4	USA (1999–2008)	−5.1
Czech Republic (1999–2009)	−4.3	Czech Republic (1999–2009)	−4.9
Romania (1999–2009)	−2.5	Romania (1999–2009)	−4.1
Hungary (1999–2009)	−1.8	Japan (1999–2009)	−3.3
Japan (1999–2009)	−1.8	Hungary (1999–2009)	−2.9
Republic of Korea (1999–2006)	−1.1	Republic of Korea (1999–2006)	−1.7

Note: [a] Age adjusted to European standard population.
[b] Data for years indicated in parentheses.
[c] Based on a log linear regression of the actual rates.
[d] Death rate is for the UK, not just England and Wales.
Source: National Institutes of Health. National Heart Lung and Blood Institute. Morbidity and mortality: 2012 chartbook on cardiovascular, lung, and blood diseases. Tables 3.35 and 3.36. Bethesda, MD: National Heart Lung and Blood Institute; February 2012. Available at: http://www.nhlbi.nih.gov/resources/docs/2012_ChartBook_508.pdf [Accessed 6 August 2013].

declines were observed across age group, gender, and education categories. In the USA in 2008, over 616,000 people died of heart disease, 405,309 from CHD. Each year some 785,000 people experience a first heart attack and another 47,000 have previously had one or more heart attacks. In 2010, CHD alone was projected to cost the USA nearly US$109 billion, including treatment and medication costs and lost productivity.

TABLE 5.9 Average Annual Percentage Change in Age-Adjusted Death Rates or All Causes and Cardiovascular Diseases, USA, 1968–2008

Years	All Causes	Total CVD[a]	CHD	Stroke	Other CVD	All Other Causes
1968–1978	−2.2	−3.6	−2.9	−4.2	−6.7	−0.7
1979–1988	−0.6	−2.2	−2.9	−3.7	0.9	1.0
1989–1998	−0.9	−1.8	−2.8	−0.9	−0.1	−0.1
1999–2008	−1.8	−4.2	−5.3	−5.0	−1.7	−0.4

Note: [a] Excludes congenital malformations of the circulatory system.
Source: National Heart Lung and Blood Institute. Morbidity and mortality: 2012 chartbook on cardiovascular, lung and blood diseases. Bethesda, MD: National Heart Lung and Blood Institute; February 2012. Available at: http://www.nhlbi.nih.gov/resources/docs/2012_ChartBook_508.pdf [Accessed 22 October 2012].

TABLE 5.10 Average Annual Percent Change in Age-Adjusted Death Rates for All Causes and Cardiovascular Diseases by Race and Sex, USA, 1999–2008

Cause of Death	All Causes	Black Male	White Male	Black Female	White Female
All causes	−1.8	−2.4	−2.0	−2.1	−1.6
Total CVD[a]	−4.2	−3.4	−4.3	−4.0	−4.3
Heart disease	−4.0	−3.5	−4.0	−4.2	−4.2
CHD	−5.2	−4.7	−5.0	−5.6	−5.6
Stroke	−4.9	−4.2	−5.3	−4.4	−4.9
Non-CVD	−0.4	−1.7	−0.7	−0.7	0.1

Note: [a] Excludes congenital malformations of the circulatory system.
Source: National Heart Lung and Blood Institute. Morbidity and mortality: 2012 chartbook on cardiovascular, lung and blood diseases. Bethesda, MD: National Heart Lung and Blood Institute; February 2012. Available at: http://www.nhlbi.nih.gov/resources/docs/2012_ChartBook_508.pdf [Accessed 22 October 2012].

Regional variation in CHD among adults in the USA is shown in Figure 5.17. In 2010, CHD prevalence ranged from the lowest rate of 3.7 percent in Hawaii to 8.2 percent in Kentucky, with the greatest regional prevalence observed in the southern states. CHD mortality rates are declining more rapidly in several of these states than nationally.

Of the more than 90 percent of the adult American population with CVD risk factors, more than one-third have high rates for at least three of the factors. With costs in excess of US$300 billion, a renewed emphasis should be of utmost priority among adults aged 30 years and older. Furthermore, prevention of these risk factors in the child population is imperative to achieve longer term results.

Mortality rates from CVD vary widely between countries, as well as between regions of a country, for men and women, for ethnic and socioeconomic groups, and for different periods in the same country. Some countries experienced the peak of mortality from CVD in the early to mid-1950s, followed by a dramatic and sustained decline. In others, the peak was reached in the mid- to late 1960s or early 1970s, followed by a more moderate rate of decline (Figure 5.18). Hospitalization and long-term care usage patterns have been modified with active preventive programs.

Comparisons between countries can be helpful in the evaluation of public health needs and priorities. The experience of one country or region is not necessarily directly applicable to another, but the trends in CVD mortality are now well established with declines of more than 40–50 percent in many countries. All countries or regions within a country that have persistent high rates should review their program priorities in public health. Figure 5.19 shows standardized death rates for groups of countries in the WHO European Region. More specifically, Finland has been a country with very high rates of cardiovascular mortality, but since the 1970s it experienced a sharp decline, as did Israel and, to a lesser degree, Sweden. Denmark and the UK did not show reductions until the late 1970s, possibly due to later adoption of new innovations, new treatments for acute coronary syndromes (e.g., insertion of stents), and health promotion with an emphasis on healthy diets, exercise, and smoking reduction.

Hypertension, labile or fixed, systolic or diastolic, mild or severe, for any age group or gender, is an independent contributor for CHD. Moreover, glucose intolerance or diabetes is an important risk factor for CHD. A familial history of CHD also confers excess risk, as does smoking, physical inactivity, and a diet high in fat. Multivariate analysis has

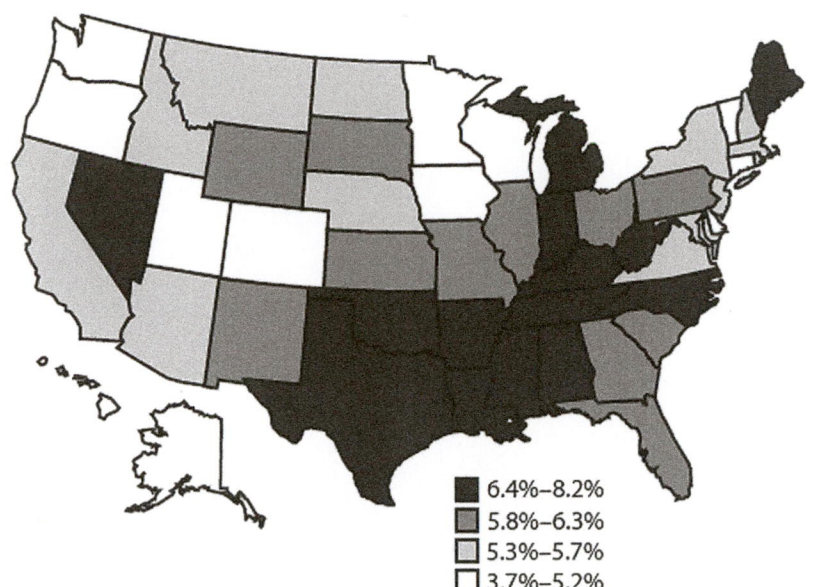

6.4%–8.2%
5.8%–6.3%
5.3%–5.7%
3.7%–5.2%

FIGURE 5.17   Age-adjusted prevalence of coronary heart disease among adults: behavioral risk factor surveillance system, USA, 2010.
Source: Centers for Disease Control and Prevention. Prevalence of coronary heart disease – United States, 2006–2010. MMWR Morb Mortal Wkly Rep 2011;60:1377–8. Available at: http://www.cdc.gov/mmwr/preview/mmwrhtml/mm6040a1.htm [Accessed 22 October 2012].

established risk factors for intervention programs. Hypertension is more common, less controlled, and less well managed among the poor, African Americans, Hispanics, and Native Americans. This condition is also common among native peoples in other parts of North America, Australasia, and the South Sea Islands.

Good quality medical care as early as possible during and after an AMI has the potential to reduce case fatality rates. Secondary prevention after a first AMI can reduce the risk or delay repeat AMIs and therefore increase long-term survival. Primary prevention to reduce risk factors remains an important aspect of reducing the burden of CVD. A review of the literature and computer modeling of the experience in the USA, published in 1997, attributed less than one-third of the reduction of mortality rates between 1980 and 1990 to primary prevention, while improved treatment accounted for half the reduction; secondary prevention measures, such as routine use of aspirin and beta-blockers following AMI, accounted for the rest. Results of a 2008 study by Young and colleagues concluded that between 1980 and 2000, 50 percent of the fall in CHD mortality was due to declines in risk factors, with primary prevention substantially reducing mortality, as compared to secondary prevention.

Diffusion of medical interventions, discussed in Chapter 15, is sometimes seen as too rapid and too costly. A survey in the late 1990s demonstrated that use of simple, low-cost medical technology, such as aspirin, statins, and beta-blockers, all proven to be highly effective in reducing CHD risk or delaying second AMIs, was not adopted by a majority of practitioners in the USA at that time. However, these forms of treatment are now in common usage, in conjunction with new therapies. Many clinical trials have shown the effectiveness of these medications in lowering cholesterol levels, resulting in a reduction of mortality rates from coronary events, CHF, and stroke. Elevated cholesterol levels in American adults declined from 33.0 percent in 1994–1995 to 16.3 percent in 2008, when 25 percent of adults were taking statin medications. However, when using a key informant approach of leading local physicians, the percentage of local primary care doctors using these secondary prevention medications rose sharply.

The role of national professional organizations such as the American Heart Association plays a crucial role in promoting awareness and translating the application of current "best practices". CDC reports that despite the gains made, an estimated 71 million Americans, 33.5 percent of the adult population, has high LDL-C levels and only one-third of them is well controlled. A more positive outcome is that the proportion of adults with high LDL-C who were being treated increased from 28.4 percent to 48.1 percent between the 1999–2002 and 2005–2008 study periods. Among adults with high LDL-C, the prevalence of LDL-C control increased from 14.6 percent to 33.2 percent between the periods. Accordingly, the overall population prevalence of high LDL-C did not change significantly from 1999–2002 (34.5 percent) to 2005–2008 (33.5 percent) (CDC, MMWR, 2011).

Priority should be given to reaching out to undertreated populations who are at risk of premature death from preventable diseases, as well as to persistent smokers, and inactive and overweight people. Physicians and all other health workers need to be advocates for healthful practices for their patients and the general public.

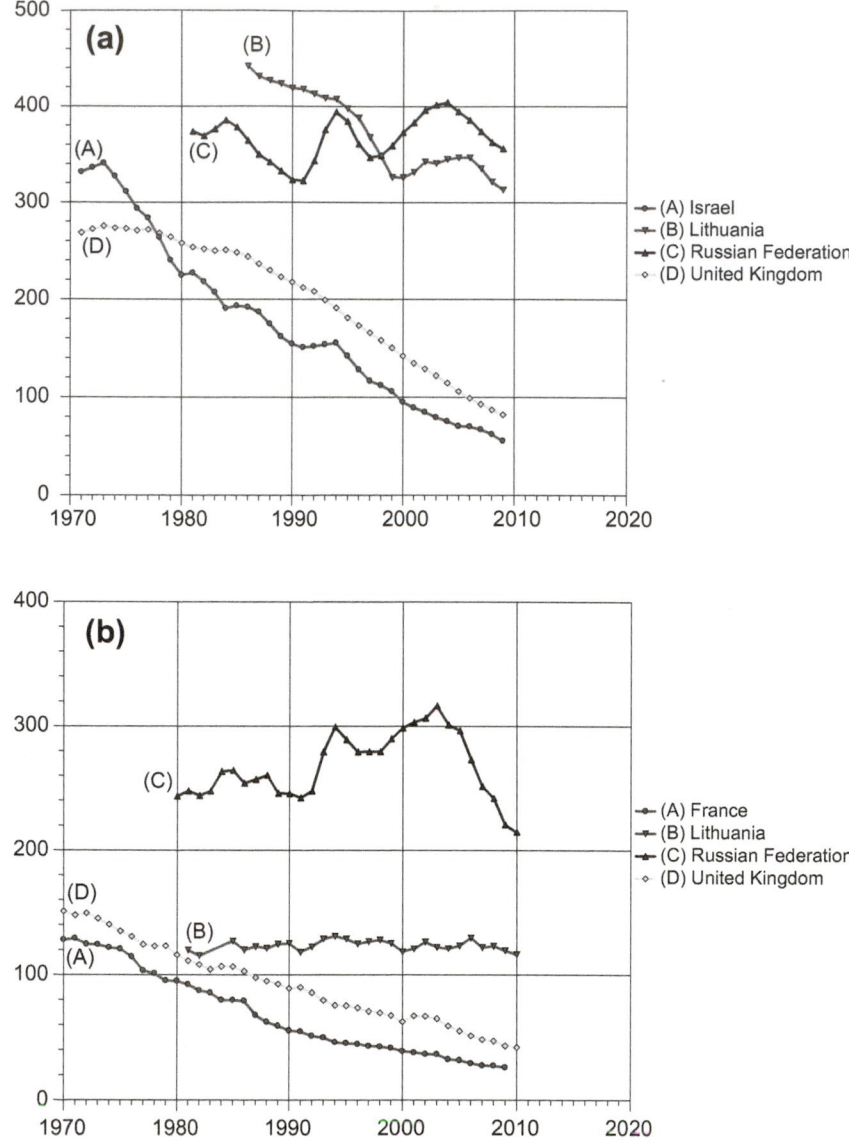

FIGURE 5.18 **Trends in (a) ischemic heart disease mortality rates and (b) cerebrovascular disease mortality rates, selected countries, 1970–2009, all ages (per 100,000 population).** *Source: World Health Organization, European Region. Health for All database. Copenhagen: WHO Regional Office for Europe; January 2013. Available at: http://data.euro.who.int/hfadb/ [Accessed 4 August 2013].*

Stroke risk factors include cardiac disease, atrial fibrillation, systolic hypertension, left ventricular hypertrophy, diabetes, high consumption of dietary fat and sugar, cigarette smoking, family history of early strokes, and low socioeconomic status, as well as previous stroke or transient ischemic episodes. Reduction of stroke deaths is dependent on the detection and management of hypertension and its control by changes in lifestyle, as well as supportive medication with long-term management and follow-up. Where stroke mortality is high, public health and medical services need to cooperate in developing education, screening, and management programs to reduce risk factors.

The long-term benefits of public health, community awareness, and improved medical care for management

of elevated cholesterol can be seen in the reduction in cholesterol levels in the US population, as determined in the National Health and Nutrition Examination Survey (NHANES) (see Chapter 8) since the 1980s (Figure 5.20).

With declining mortality, the long-term problem of CHF resulting from myocardial infarction and hypertension is increasing. CHF affects 5.8 million people in the USA, with 670,000 new cases diagnosed each year. CHF primarily affects the elderly. The aging population and prolonging of the lives of cardiac patients by modern and improved medical treatment has led to the increased incidence of this condition, reaching nearly 10 in 1000 people over the age of 65. CHF is the primary diagnosis for 3 million ambulatory care

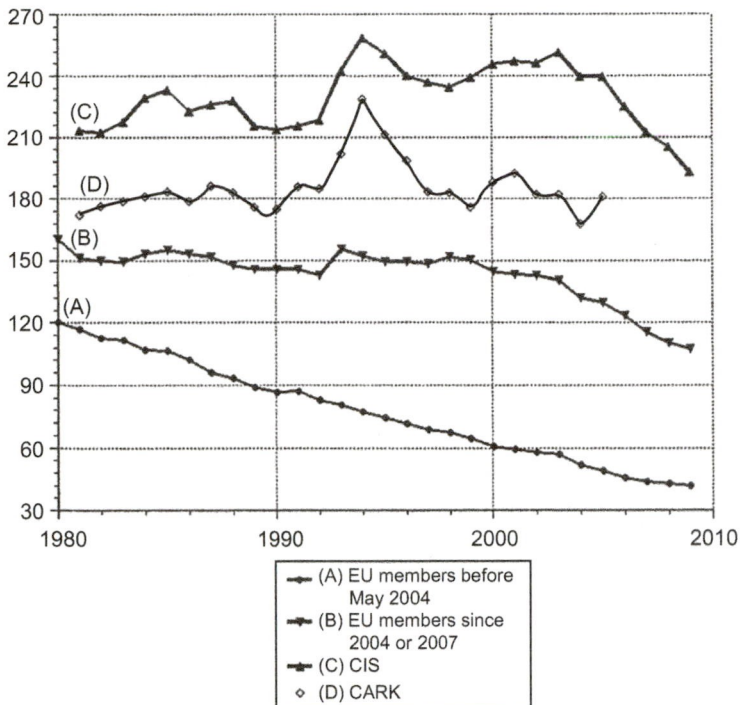

FIGURE 5.19   Standardized mortality rates (per 100,000 population) from cerebrovascular diseases, selected European countries, 1980–2010. Note: CARK=Countries of Central Asia; CIS=Russian Federation, Ukraine, and other members of Commonwealth of Independent States; EU=European Union. Three-year moving averages. *Source: World Health Organization European Region. Health for All database (HFA-DB). Copenhagen: WHO Regional Office for Europe; January 2012. Available at: http://data.euro.who.int/hfadb/ [Accessed 22 October 2012].*

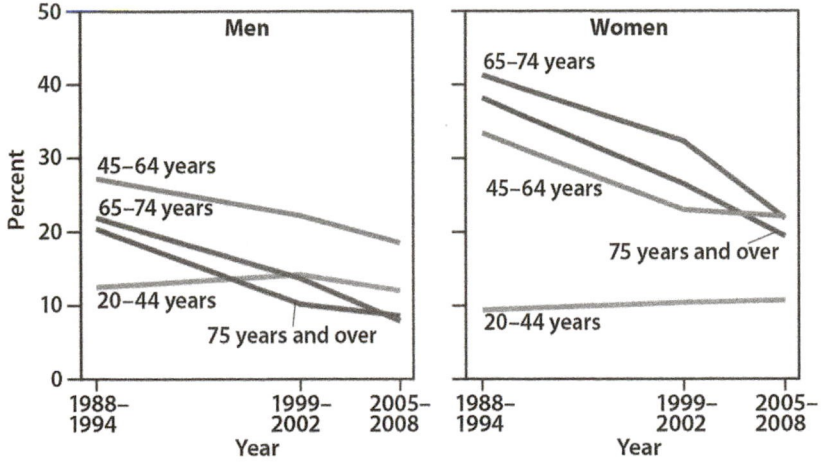

FIGURE 5.20   High-cholesterol adults, by gender, USA, 1988–2008. Note: High cholesterol=serum total cholesterol of ≥240 mg/dl. *Source: Centers for Disease Control and Prevention. Health, United States, 2010: with special feature on death and dying. High serum total cholesterol level. Hyattsville, MD: US DHHS, NCHS; February 2011. Available at: http://www.ncbi.nlm.nih.gov/books/NBK54380/#morbidity.s3 and http://www.ncbi.nlm. nih.gov/books/NBK54380 [Accessed 21 October 2012].*

and emergency department hospital visits and well over 1 million hospitalizations in the USA annually.

The yearly incidence of new cases of CHF is 550,000, and it is responsible for nearly 658,000 visits to hospital emergency rooms, accounting for 20 percent of total CHF-specific ambulatory care delivered each year in the USA. Nearly all of these visits require intervention, often with

lengthy hospitalizations. As the population ages, secondary prevention improves and treatment of acute coronary events advances, CHF is becoming a more common condition requiring attention not only from medical practitioners, but also from pharmacists, dieticians, nurse practitioners, home care workers, and others (see Chapter 14). Education in self-management is of vital importance and includes the following practices: smoking cessation, balanced diet, sodium and fluid restriction, daily exposure to sunlight, vitamin supplements, regular medical follow-up and immunizations, observations of fatigue, weight gain or loss, depression, isolation, family and neighbor support, and socialization. If managed consistently, these factors can prevent frequent and costly hospitalization and complications that hasten death. Hospital care should include education in the "teachable moment" for the patient, family, and caregivers on self-care upon returning home in order to prevent frequent relapses of acute CHF.

The following outlines the prevalence of comorbidities in patients with CHF: hypertension 73 percent, coronary artery disease 57 percent, diabetes mellitus 44 percent, and renal insufficiency 30 percent. An alarming and quite startling figure is that about 50 percent of those diagnosed with CHF die within 5 years. However, more adolescents with CHF are surviving to adulthood; but nearly one-third of adolescents and young adults with transplants do not adhere to medication regimens. Strong support networks are important in improving self-care (Box 5.9). Mortality rates in patients with CHF remain high; in 2007, there were 277,193 deaths from CHF as a primary and underlying cause. In addition, costs associated with CHF are extremely high. CHF is the most common Medicare diagnosis-related group (DRG), and more Medicare dollars are spent for diagnosis and treatment of CHF than for any other diagnosis. In 2009, the estimated total direct and indirect cost of CHF was US$37.2 billion, rising in 2010 to US$39.2 billion. The total direct and indirect cost of CVD and stroke in the USA for 2010 is estimated at US$503.2 billion.

The reduction in stroke and CHD mortality was one of the great achievements of public health in the twentieth century. It was one of the key factors in extension of life expectancy by more than 30 years over the century, largely due to public health interventions and important advances in clinical medicine.

The age-adjusted death rates for CHD and stroke between 2000 and 2009 show a continuing downward trend in the USA as in all upper income countries. In 1921, stroke was the leading cause of death and by 1938 it had fallen to third place. In 2009, with a death rate of 42.2 per 100,000, stroke was the fourth leading cause of death in the USA. The age-adjusted CHD death rate declined from 195 per 100,000 in 2000 to 126 per 100,000 in 2009. Estimated total direct and indirect costs in 2009 were US$475 billion in the USA.

The decline in CVD mortality common in the industrialized countries over the past 30–40 years has been attributed to many factors, without precise evidence of the relative importance of each factor. The decline in mortality does not necessarily indicate a decline in prevalence or severity (case-fatality rates) of the disease. It is more likely to be the result of reduced severity of risk factors and improved

> **BOX 5.9 Self-Care Behaviors Recommended for Patients With Congestive Heart Failure**
>
> - Maintain a healthful diet – fruit, vegetables, protein, and carbohydrates in balance.
> - Half an hour in the sun daily.
> - Daily physical activity to maintain physical fitness.
> - Cease all tobacco use and avoid exposure to second-hand smoke.
> - Avoid excessive fluid intake.
> - Restrict dietary sodium.
> - Restrict alcohol intake.
> - Use low-fat, low-salt foods.
> - Take multivitamins daily.
> - Develop a system for taking all medications as prescribed.
> - Visit your health care provider at regular intervals.
> - Maintain current immunizations, especially influenza (annual) and *Streptococcus pneumonia* (periodic).
> - Monitor for an unexpected decline in body weight.
> - Monitor for unexpected increase in body weight.
> - Monitor for signs/symptoms of shortness of breath, swelling, fatigue, and other indicators of worsening CHF.
> - Avoid other recreational toxins, especially cocaine.
> - Do not ignore emotional distress, especially depression and anxiety.
> - Seek treatment early.
> - Tell your provider about sleep disturbances.
> - Talk to a pharmacist or other provider before using herbal medicines.
> - If diabetic, achieve diabetes mellitus treatment goals.
>
> **Source:** *Riegel B, Moser DK, Anker SD, Appel LJ, Dunbar SB, Grady KL, et al., on behalf of American Heart Association Council on Cardiovascular Nursing; American Heart Association Council on Cardiovascular Nursing; American Heart Association Council on Clinical Cardiology; American Heart Association Council on Nutrition, Physical Activity, and Metabolism; American Heart Association Interdisciplinary Council on Quality of Care and Outcomes Research. State of the science: promoting self-care in persons with heart failure: a scientific statement from the American Heart Association. Circulation 2009;120:1141–63. Available at: http://circ.ahajournals.org/content/120/12/1141.full [Accessed 22 October 2012].*

access to and quality of care, with secondary and tertiary prevention including better medications, resuscitation, and emergency care, thus delaying mortality. Higher standards of living, leisure and recreation, greater awareness of healthful nutrition and availability of appropriate foods at reasonable cost, and wider community and individual awareness have all played a role in the reduction in mortality from CVD.

Future challenges in this field will include improved management of hypertension, further smoking reduction, healthful food regulation and dietary practices, and increasing physical fitness. Clinical medicine will also see new innovations based on improved medications for lipid reduction, better diabetes and hypertension control, and stem cell innovations to promote heart muscle recovery after myocardial infarctions. Access to care for all is essential, but not sufficient to reach high-risk groups, and appropriate outreach and follow-up methods will be enabled by modern information technology and cell phone communication (CDC, MMWR: Ten great public health achievements – United States, 2001–2010). Figure 5.21 (see companion web site at http://booksite.elsevier.com/9780124157668) gives a diagrammatic presentation of the interaction of key factors in ischemic heart disease: age, gender, and socioeconomic factors play a key role; physical activity, fat intake, body weight, alcohol, and smoking are factors that the individual and the society can modify; blood pressure, cholesterol, and diabetes management are the essential role of primary care practitioners. The potential for saving lives and costly medical care, and reaching healthy older ages is powerfully present in addressing these factors.

The public health implications of these alternative explanations are substantial. Current data do not allow for clear distinctions of the contribution of each factor to the decline in mortality. Prudent public health would continue to place stress on all of these and attempt to strengthen the trend, particularly in state or local areas where higher than average rates prevail. Prevention of NCDs includes organized efforts of primary prevention (e.g., education, smoking cessation), secondary prevention (e.g., screening), and tertiary prevention (e.g., emergency medical services).

Implementation of primary and secondary preventive activities to reduce the burden of CVD involves building institutional support, educating the public, community-based risk factor reduction activities, a healthy working environment, information systems to monitor morbidity, and a well-informed medical community. The US Institute of Medicine addressed the fragmented US emergency care system and, accordingly, called for creation of coordinated, regionalized, and accountable emergency care systems that include protocols for the treatment, triage, and transport of prehospital patients. Medical interventions include identification and aggressive treatment of hypertension and diabetes, counseling to promote lifestyle modifications, and a supportive psychosocial environment. These are both individual and population-wide issues.

Widespread use of a multidrug regimen for the prevention of CVD (i.e., statins, aspirin, and two blood-pressure-lowering medicines) has been proposed, stating that over a 10-year period, this regimen may avert some 18 million deaths from CVD. This regimen would largely benefit those below 70 years of age (Lim et al., 2007). The medical care provider is involved in screening, treatment, and advising patients of the importance of reducing risk factors both before and after the onset of symptoms. The client must take personal responsibility for many aspects of prevention, such as involvement in screening, sustaining management of high blood pressure, and reducing elevated blood lipids (Box 5.10).

BOX 5.10 Notes on Hypertension (HBP) in the USA

Prevalence and Costs
- 31% of Americans have hypertension (1/3 adults).
- Another 30% have pre-hypertension.
- Hypertension is a key risk factor for stroke, coronary heart disease, congestive heart failure, kidney, eye disease, peripheral vascular disease and other complications.
- About half of people with HBP have hypertension under control.
- 348,000 US deaths included hypertension as the primary or secondary cause.
- HBP costs the US nation $47.5 billion USD per year.
- Major global health issue.

Definitions
- Pre-hypertension Systolic <120; diastolic <80 mm.
- At risk Systolic 120-139 Diastolic 80-89.
- High risk Systolic 140 mm plus Diastolic 90 mm plus.
- Framingham Heart study showed that a 2-mmHg reduction in blood pressure would result in 14% reduction in the risk of stroke and transient ischemic attacks, and a 6% reduction in risk of coronary heart disease.

Prevention
- Check BP (20% of US adults with HBP don't know they have it).
- 70 % of US adults with HBP use medications to treat it.

Public health:
- Promotion of awareness by health workers and the public;
- Limit salt content of produced foods, along with banning transfats.

Self care:
- Smoking cessation.
- Low salt intake - reduce from 3300 mg to 2,399 mgm per day.
- Regular exercise – ½ hour per day.
- Weight control/loss.
- Diet high in vegetables, nuts, low in fat, manufactured foods and meat.
- Medications as prescribed – thiazides, ACE inhibitors, calcium.
- Regular BP checkups.

Sources: *NIH. National High Blood Pressure Education Program. Prevention, Detection, Evaluation, and Treatment of High Blood Pressure: The Seventh Report of the Joint National Committee on JNC 2003. Available at: http://www.nhlbi.nih.gov/guidelines/hypertension/express.pdf (accessed 20.11.2013).*
Centers for Disease Control and Prevention. High Blood Pressure. America's High Blood Pressure Burden. March 20, 2013. Available at: http://www.cdc.gov/bloodpressure/facts.htm (accessed 21.11.2013).
Rafey MA. Cleveland Clinic Center for Continuing Education. Hypertension. 2013. Available at: http://www.clevelandclinicmeded.com/medicalpubs/diseasemanagement/nephrology/arterial-hypertension/ (accessed 21.11.2013).

A conference on heart health in Victoria, British Columbia, in 1992 called on all government, community, and private agencies to join forces in eliminating this modern epidemic by adopting new policies, making regulatory changes, and implementing disease prevention programs directed at entire populations. Health personnel were encouraged to work on health promotion with the community, including the media, the education system, the social sciences, professional associations, government agencies, the private sector, international and private voluntary organizations, and community health coalitions (Box 5.11).

Numerous countries have recognized the importance of lifestyle factors and their impact on the health status of the population, and notably on the occurrence of heart diseases, representing a major burden of disease on a society. In order to facilitate the implementation of intervention programs to reduce lifestyle-related risk factors, the international Countrywide Integrated Non-communicable Diseases Intervention (CINDI) program was initiated by the WHO in 1990. CINDI addresses the wide gap between Eastern and Western Europe in the prevalence of chronic disease. The member states collaborate on the implementation of an integrated approach to chronic disease prevention. The objectives are to promote measures for integrated disease prevention and health promotion in an effort to reduce morbidity, with intercountry collaboration in 29 countries.

Countries with very high rates of CVD should develop aggressive intersectoral intervention programs at the national and community levels. Concerted national efforts of the ministry of health, other relevant ministries, the media, NGOs, food manufacturers, and local communities are needed to reverse the high rates of CVD in some US states and Northern and Eastern Europe. Legislation and litigation against tobacco manufacturers may affect the sale and pattern of smoking, but the problem is a major public health issue.

This topic should also be placed high on the national agenda in the developing countries. It is particularly relevant to those at the midlevel of development, where an epidemiological transition is under way and CVDs are emerging as the leading cause of death. In developing countries, the problem

---

BOX 5.11 Victoria Declaration on Heart Health: Conference Recommendations for Preventing Heart Disease, 1992

"Recognizing that both scientific knowledge and widely tested methods exist to prevent most cardiovascular disease, the Advisory Board of the International Conference on Heart Health calls upon
- *health, media, education and social science professionals*
- *and their associations*
- *the scientific research community*
- *government agencies concerned with health, education, trade, commerce and agriculture*
- *the private sector*
- *international organizations and agencies concerned with health and economic development*
- *community health coalitions*
- *voluntary health organizations*
- *employers and their organizations*
  to join forces in eliminating this modern epidemic by adopting new policies, making regulatory changes and implementing health promotion and disease prevention programs directed at entire populations."

A programmatic approach on prevention of cardiovascular disease includes the following:

1. Education: educate the public, health providers, community groups, and governments in risk factor reduction.
2. Food policy: reduce fat content of milk and meat products and reduce salt content of processed foods, working with ministries of agriculture, industry, and commerce, as well as with producers and manufacturers.
3. Reduce smoking: increase cost of cigarettes through taxation, ban advertising, ban smoking in work and public places, devote some revenue from cigarette taxes to health promotion and education against smoking.
4. Promote physical exercise: promote personal and community attitudes and facilities encouraging participation in regular physical activity.
5. Reduce obesity: encourage individual and community-based health promotion.
6. Community-based initiatives: promote healthy lifestyle, including smoking cessation and fitness promotion, raise consciousness of health self-care issues, teach cardiopulmonary resuscitation (CPR).
7. Medical care: promote primary prevention techniques including screening for risk factors, management of hypertension, and patient counseling for risk factor reduction and stress management.
8. Screening: screen for risk factors such as diabetes, elevated blood lipids, and hypertension, and counsel as to findings and implications.
9. Emergency and hospital care: reduce case fatality rates, perform CPR, transfer rapidly to designated medical centers with intensive care with current standards of antithrombotic agents (aspirin, streptokinase, or others) at district hospitals and ballooning, stents, or coronary artery bypass procedures at referral hospitals.
10. Rehabilitation: promote maximum recovery and function at work and in personal life, adopt preventive approaches to stop the pathological process.

The Victoria Conference on Heart Health was followed by similar conferences in Singapore in 2000 and the Biannual World Congress of Cardiology was held in Barcelona in 2006, bringing the latest scientific information on heart disease to cardiologists internationally.

Source: *International Heart Health Society. The Victoria declaration on heart health. Ottawa: Health and Welfare, Canada; 1992. Available at: http://www.internationalhearthealth.org/Publications/victoria_eng_1992.pdf [Accessed 22 October 2012].*

of CVD is hidden beneath more acute, more immediate issues of high morbidity and mortality from infectious diseases, nutritional challenges, and conditions associated with maternal and child health. Age-specific mortality rates from low- and middle-income countries are drastically higher than in most European countries (see Chapter 3), and CVD mortality is now the principal cause of death in many developing countries. Smoking is promoted for commercial reasons in low-resource countries, without countervailing health hazard warning labels, limits on advertising, and other restrictive legislation. This warrants attention by health authorities in developing countries and application of lessons learned in the industrialized countries over the past 30 years.

The global pandemic of CVD is made up of a number of trends. One is a major decline in mortality in the industrialized countries, while stroke and CHD mortality are increasing in developing countries. The downward trend of the past four decades due to health promotion success in reduced smoking and improved treatments in the western countries may not be sustained. The very high rates of cardiovascular mortality in Russia in recent decades have declined since 2005, but are still three times higher than Western European levels. There is growing mortality from stroke in countries of the WHO Western Pacific Region, including China, with high rates of smoking, in association with growing wealth and dietary change, which are also related to a widespread increase in diabetes and obesity.

DIABETES MELLITUS

Globally, health expenditure for diabetes care in 2011 was US$465 billion, equivalent to 11 percent of total health spending. This enormous figure is predicted to rise to US$595 billion by 2030, posing a major challenge for the prevention and management of diabetes. In 2004, of the approximately 3.4 million people who died from consequences of high blood sugar, more than 80 percent were in low-and middle-income countries. In 2004, an estimated 3.4 million people died from consequences of high fasting blood sugar and is projected to be the 7th leading cause of death in 2030. More than 80% of diabetes deaths occur in low- and middle-income countries.

The International Diabetes Federation reports the incidence of diabetes in 2011 as more than 366 million worldwide, and a further 280 million individuals are at risk of developing the disease. The Federation estimates that by 2030 the incidence will reach 552 million (one adult in every 10) and 308 million at high risk. The number of Americans with diagnosed diabetes is projected to increase 165 percent, from 11 million in 2000 (prevalence of 4.0 percent) to 29 million in 2050 (prevalence of 7.2 percent). Much of the increase in cases will be attributable to changes in demographic composition and to population growth; however, one-third will be due to increasing prevalence rates (World Diabetes Federation Global Diabetes Plan 2011–2020, 2011, Brussels).

Diabetes mellitus is a common chronic condition with disturbed carbohydrate metabolism, resulting from deficiency in the production of insulin in the pancreas or impaired function of the insulin receptors. This leads to increased levels of blood glucose. In the USA, diabetes has been ranked among the 10 leading causes of death since the 1930s; it was the seventh leading cause in 2007, with over 73,000 deaths directly attributed to the disease. It affected some 25.8 million Americans, or 8.3 percent of the population in 2010, of whom 18.8 million were diagnosed and an estimated 7 million undiagnosed and unaware they are diabetic (Figure 5.22). It is a major disease in its own right and an important risk factor for CVD, including CHD, stroke, and peripheral vascular disease, as well as severe damage

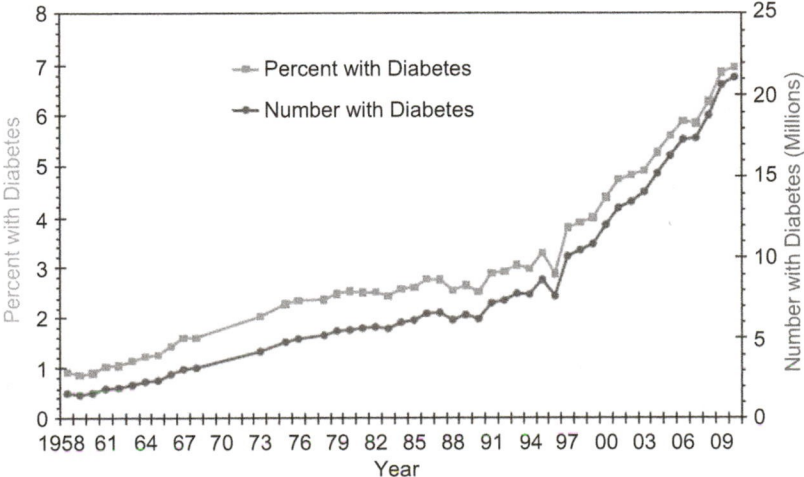

FIGURE 5.22   Number and percentage of US population with diagnosed diabetes mellitus, 1958–2010. *Source: Centers for Disease Control and Prevention. Division of Diabetes Translation. National Diabetes Surveillance System. Long-term trends in diagnosed diabetes. Atlanta, GA: CDC; October 2011. Available at: http://www.cdc.gov/diabetes/statistics/slides/long_term_trends.pdf [Accessed 22 October 2012].*

to nerves, kidneys, eyes, and other organs. There are two major types of diabetes: type 1 or insulin-dependent diabetes mellitus (IDDM), and type 2 or non-insulin-dependent diabetes mellitus (NIDDM). Type 2 diabetic patients represent approximately 95 percent of all diabetes patients. NIDDM was previously known as "adult-onset diabetes", and is closely related to diet and other lifestyle factors; while in the past, IDDM was regarded as "juvenile diabetes". However, this distinction has become somewhat blurred in recent years, as NIDDM is detected more and more often in children as well (in conjunction with the increasing prevalence of obesity in children) and as the mean age of NIDDM patients is declining steadily.

Diabetes prevalence has increased annually in the USA, rising to 11.3 percent by 2010. Over 1.9 million new cases of diabetes are diagnosed annually. About 215,000 people under the age of 20 are diabetic. Prevalence rates were approximately twice as high among African Americans as among whites in the USA. The number of diagnosed diabetics among the black population increased from some 0.9 million to 4.9 million in 2008. Among adults, the risk of diagnosis of diabetes is 77 percent higher among non-Hispanic blacks, while it is 66 percent higher among Hispanics compared to non-Hispanic white people. Among Asian Americans it is 66 percent higher. Native Americans also have high rates of diabetes and its complications. Among young people, the rate of new cases for type 1 diabetes among those over 10 years of age is 19.7 per 100,000, and for older youth 18.6 per 100,000. Furthermore, young people over 10 years of age have a rate of 8.5 per 100,000 of type 2 diabetes.

Diabetes can lead to CVD, blindness, kidney disease, and lower limb amputations. It is the leading cause of end-stage renal disease (ESRD or kidney failure) requiring dialysis or transplantation, which in the USA accounts for 44 percent of new cases of treated ESRD in 2007. In 2005, care for patients with kidney failure, mostly due to diabetes, cost the USA nearly US$32 billion. Visual impairment is approximately twice as common among diabetics compared to non-diabetics over the age of 50. Lower extremity peripheral arterial disease and neuropathy are chronic conditions mainly affecting the elderly as well as people with diabetes. They can result in ulcers, infections, gangrene, or amputation of the lower limbs. These disabling, crippling and painful conditions affect diabetics approximately twice as often as the elderly without diabetes.

The Canadian Community Health Survey 2002–2003 showed that the prevalence of self-reported hypertension, heart disease, and stroke among diabetics was four to six times higher than among similar age and gender groups without diabetes. The CDC estimates that the direct cost of medical care for diabetes may be in the order of US$50 billion per year. The addition of indirect costs may double that figure. Figure 5.23 shows diabetes prevalence by age group

in 1988–1994 and 2005–2008 (see companion web site at http://booksite.elsevier.com/9780124157668).

As measured by hemoglobin A_{1c} (HbA_{1c}, i.e., glucose attached to the hemoglobin of red blood cells) levels above 9 percent, poor diabetes control in the USA declined in all three age groups shown in Figure 5.24 between 1998–1994 and 2005–2008. For adults, the prevalence of physician-diagnosed diabetes increased from 8 percent in 1988–1994 to 11 percent in 2005–2008. This increase may be mainly attributed to increased physician and public consciousness of diabetes. The prevalence of undiagnosed diabetes is considered to be stable at 3 percent. Diabetes among US children and adolescents is reported with increasing frequency. Almost 200,000 people under 20 years of age had diabetes in 2007 (Health United States 2010).

The discovery of insulin and its use in 1921 by Frederick Banting and Charles Best in Toronto (Nobel Prize 1923 to Banting, Best, and McLeod) gave patients with type 1 diabetes the opportunity to live full lives provided their condition was carefully managed. The more recent developments of simpler monitoring, improved insulin preparations, and insulin pump devices have made management easier and more effective for people living with type 1 diabetes. Insulin preparations that may be inhaled are under clinical trial. Oral hypoglycemic medications provide effective help for type 2 diabetics in combination with weight reduction, physical fitness, and careful dietary management.

The WHO estimates the worldwide prevalence of diabetes to be 10 percent in adults; however, prevalence varies by income group and country, ranging between 8 and 11 percent. Estimates indicate that there are some 340 million people with diabetes globally, 90 percent with type 2 diabetes. Non-insulin-dependent and "adult-onset" type 2 diabetes is appearing at younger ages, including in children, and

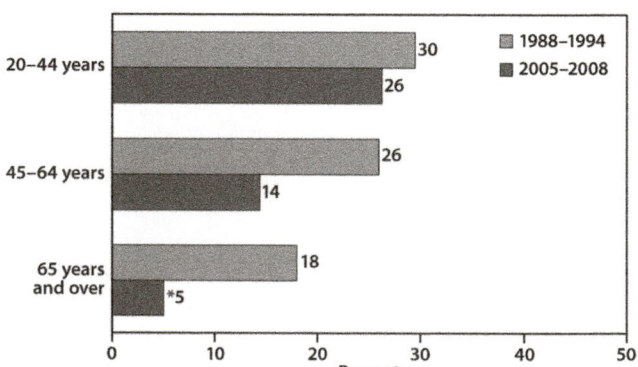

FIGURE 5.24   Diabetes control by age group, USA, 1988–2008. Note: Poor diabetes control = hemoglobin A1c > 9%. *Unreliable estimate (standard error 20–30%). *Source: Centers for Disease Control and Prevention. Health, United States, 2010: with special feature on death and dying. Hyattsville, MD: US DHHS, NCHS; February 2011. Available at: http://www.ncbi.nlm.nih.gov/books/NBK54380/#morbidity.s3 and http://www.ncbi.nlm.nih.gov/books/NBK54380 [Accessed 21 October 2012].*

also often requires insulin for adequate control. This phenomenon is due to the rapid increase in obesity as a major public health problem.

Globally, there were 3.4 million deaths in 2004 resulting from high blood sugar and its consequences. Diabetics are at very high risk for ischemic heart disease, stroke, kidney, eye, and peripheral vascular disease, as well as the direct effects of diabetes. Globally, the WHO reports that 50 percent of people with diabetes die of CVD (primarily heart disease and stroke). After 15 years of diabetes, approximately 2 percent of people become blind, and about 10 percent develop severe visual impairment. Diabetes often leads to kidney failure and 10–20 percent of people with diabetes die of kidney failure. Diabetic neuropathy affects up to 50 percent of people with diabetes and the overall risk of dying among people with diabetes is at least double that of their peers without diabetes (WHO, Diabetes fact sheet no. 312, August 2011).

Controlling high blood pressure in people with diabetes could reduce stroke death rates by 75–90 percent and CHD by 25–50 percent. Diabetic eye disease is the most common cause of blindness, with 25,000 new cases annually, 90 percent of which could be prevented by early diagnosis and treatment. Diabetic lower extremity disease causes more than 65,700 leg, foot, and toe amputations per year in the USA (in 2006); more than half of these could be prevented, with cost savings of US$600 million annually.

Costs for diabetes in the USA since 2002 have risen more than US$8 billion annually. In 2010, the Hastings Center reported that 10 percent of health care dollars in the USA are spent on overall direct costs related to diabetes, amounting to US$92 billion a year (1.5 times the amount spent on stroke or heart disease). The CDC predicts that spending on diabetes care will reach US$192 billion in 2020. Medicare reported spending US$1.4 billion for in-patient services (US$7383 per discharge).

Community-based intervention programs include improved prevention, education, and services designed to reduce complications of diabetes. Such programs should utilize epidemiological monitoring to identify high-prevalence populations. Using national mortality and hospitalization databases, it is possible to compare regional patterns of diabetes and its complications including CHD, cerebrovascular disease, eye disease, and reduced blood supply to lower limbs. Identifying population groups with a higher incidence/prevalence of these conditions permits the development of targeted community-based intervention programs to prevent obesity and diabetes through improved early diagnosis and management of diabetes and related conditions. Obesity among children and young adults is becoming a pandemic constituting a great threat to the health situation in the coming decades, in developed as well as developing countries.

Prevention of Diabetes and Its Complications

Prevention is feasible for diabetes at the health promotion as well as primary, secondary, and tertiary preventive levels:

- Health promotion involves education of the public to a high level of awareness of diabetes, its risk factors, and its complications, reduction in dietary intake of excess calories, fats, and sugars, with physical activity to promote fitness and weight appropriate for height (i.e., BMI <25). This includes national and local authorities' policies to regulate and work with food manufacturers and marketing to promote healthy food consumption and avoid factors contributing to unhealthy diets for schoolchildren and in work settings, promoting substitution of health food products in marketing practices.

- Primary prevention identifies obesity as a high-risk condition for preclinical diabetes, and promotes its prevention through proper nutritional practices and regular exercise to reduce the risk of type 2 diabetes.

- Secondary prevention to prevent complications of diabetes is aided by early case finding and management. Routine screening for diabetes will uncover early cases for whom treatment can reduce the severity of complications. Screening is recommended:
  - for people with a family history of diabetes
  - for patients with cardiovascular, renal, and eye diseases
  - during pregnancy
  - as follow-up for women with glucose intolerance in pregnancy, or a history of infants weighing over 4000 g
  - for obese people.
  - For people with abnormal blood glucose levels, follow-up management includes instruction and monitoring to prevent serious complications. Diet, exercise, regular urine and blood sugar testing, personal hygiene, and foot care should be followed closely in ambulatory and home care. In addition, medication, usually oral hypoglycemic agents, is sometimes needed to control blood sugar levels in type 2 diabetes. Insulin is always needed for control of type 1 diabetes. Social and medical support may be needed for those who are unable to care adequately for themselves, especially elderly, poor, and poorly educated people.

- Tertiary prevention aims to restore function and prevent further deterioration. For those with established peripheral vascular disease and potential gangrene, foot care can delay amputations; for amputees, rehabilitative care is essential to prevent total dysfunction. Close follow-up of diabetics by ophthalmologists promotes prevention of diabetic retinopathy by early treatment with photocoagulation (Box 5.12); regular professional foot care can help to prevent the onset of diabetic ulcers, which if not treated well

---

BOX 5.12 Cases of Diabetes Management in the Community

Case 1

A Los Angeles study of self-care behaviors of 200 adults with diabetes in two medically underserved communities, one urban and one rural, found that about 40 percent of the sample did not manage their illness. Problem areas were physical activity in the rural community and smoking in the urban community. The study focus was support needs of adults with diabetes in these communities by estimating their rates of various self-care behaviors, the amount of support provided by key sources, and the associations between support from these sources and adherence to recommended diabetes self-care behaviors. The conclusion was that there is a need to strengthen the social environmental resources in the community.

Case 2

In Winnipeg, Manitoba, Canada, a community diabetes prevention worker (CDPW) program of the Yellowquill Community College prepares students to provide health services to their home community in the First Nation reserves under the guidance of diabetes educators. This college is a First Nations postsecondary educational center founded by the Dakota Ojibway Tribal Council in 1984 to train adults for work in the health field and other sectors. The college operates with federal, provincial, tuition, and public support, and has produced over 950 graduates.

The CDPW helps those with diabetes and their families to manage the disease and promote awareness and prevention strategies among the community population. The CDPWs, with grade 12 education or equivalent and/or work experience in the health field, are selected from the communities to be served. The training program consists of 12 modules, each lasting one week, followed by a practicum.

The effectiveness of community-based interventions has become highly appreciated for addressing important lifestyle factors, such as dietary habits and physical activity, along with careful and continuing medical supervision. The ability of adults with diabetes to manage their illness properly and prevent complications is, in part, a function of support provided by the people and institutions surrounding them.

Sources: Centers for Disease Control and Prevention. Prevention research centers: building the public health research base with community partners at a glance, 2011 [updated 4 May 2011]. Atlanta, GA: CDC. Available at: http://www.cdc.gov/chronicdisease/resources/publications/aag/prc.htm [Accessed 22 October 2012].
Shaw BA, Gallant MP, Riley-Jacome M, Spokane LS. Assessing sources of support for diabetes self-care in urban and rural underserved communities. J Community Health 2006;31:393–412. Available at: http://www.ncbi.nlm.nih.gov/pubmed/17094647 [Accessed 24 October 2012].
Community Diabetes Prevention Worker Certificate (Modular). Manitoba: Yellowquill College. Available at: http://yellowquill.org/community-diabetes-prevention-worker-certificate [Accessed 24 October 2012].

The St. Vincent Declaration was developed at a WHO–sponsored conference in 1990, setting targets for reducing the complications of diabetes mellitus for the European Region of the WHO member countries. The European Region of the WHO has promoted development of national diabetes programs (NDPs) in countries from Western Europe to those in Central Europe and the Central Asian Republics and Azerbaijan. Each is created by broad-based working groups in each country with representatives of ministries of health, diabetes nursing, and patients' groups, as well as medical experts in the field. In 2009, about half of European Union (EU) member states lacked national diabetes plans, and in many that did have plans, implementation has been weak.

Similar programs in the USA, initially launched in 1997, are promoting the National Diabetes Education Program, sponsored by the CDC and National Institutes of Health (NIH), with participation of the American Diabetes Association and other NGOs. Such programs are intended to promote public and health care provider awareness of the seriousness of the problem and promote integrated approaches to care and improved access to diabetes care. Diabetes is a serious public health issue globally and increasing steadily. Its complications are serious, costly, and life shortening. The challenges include a wide array of public health measures combined with good medical care, active outreach programs to raise levels of understanding and compliance, and measures to improve the length and quality of life.

CANCER

Cancer is a leading cause of death worldwide, with about 30 percent of deaths attributable to behavioral and lifestyle risk factors, especially related to diet, low physical activity, and alcohol/tobacco consumption patterns. Medical costs for cancer care in the USA were nearly US$125 billion in 2010. National expenditures reflect the prevalence of disease, treatment patterns, and costs for different types of care.

Globally, cancer deaths include 4.2 million men and 3.3 million women (rounded), a total of 7.5 million (Box 5.13). For men the common cancers are lung, prostate, colorectal, stomach, and liver. For women the common cause of cancer deaths are breast, colorectal, cervix, lung, and stomach. About 30 percent of cancer deaths are due to the five leading behavioral and dietary risks: high BMI, low fruit and vegetable intake, lack of physical activity, tobacco use, and alcohol use. The main types of cancer death are: lung, 1.37 million; stomach, 736,000; liver, 695,000; colorectal, 608,000; breast, 458,000; and cervix, 275,000. Prostate cancer deaths are also of increasing concern, especially in the Americas and Europe.

Cancer is the second leading cause of death in the industrialized countries, representing about 13 percent of all deaths in 2008. Cancer is also increasingly recognized as

can and do lead to amputations; close monitoring for renal disease is equally important, and the monitoring of blood sugar and more importantly by HbA_{1c}, which is a measure of average blood sugar levels over a 3 month period.

BOX 5.13 WHO on How to Reduce the Global Burden of Cancer

Cancer is a leading cause of death worldwide, accounting for 7.6 million deaths (around 13 percent of all deaths) in 2008. This disease can be reduced and controlled by implementing evidence-based strategies for cancer prevention, early detection of cancer, and management of patients with cancer. The WHO estimates that more than 30 percent of cancer could be prevented by modifying or avoiding key risk factors with current preventions and treatment methods.

Prevention Strategies

1. Stop tobacco use and secondary exposure.
2. Control body weight from overweight or obese.
3. Diet – low fat, and high fruit, nuts, and vegetable intake.
4. Take regular frequent physical activity – moderate to active 150 minutes per week.
5. Moderate alcohol consumption.
6. Practice safe sex.
7. Avoid exposure to excessive air pollution.
8. Avoid indoor smoke and fumes from household use of solid fuels.
9. Immunization against human papillomavirus (HPV) and hepatitis B virus (HBV).
10. Treat infections of HIV, *Helicobacter pylori* (stomach), and schistosomiasis (bladder).
11. Control occupational and environmental carcinogenic hazards.
12. Moderate exposure to sunlight.

Early Detection

1. About one-third of the cancer burden can be decreased by early detection and treatment when the cancer is localized (before metastasis). This applies to skin, cervical, colorectal, breast, prostate, and lung cancer.
2. Education is imperative to help people recognize early signs of cancer and seek prompt medical attention for symptoms, which might include: lumps, sores, persistent indigestion, persistent cough, and bleeding.
3. Screening programs to identify early cancer or precancer before signs are recognizable include mammography for breast cancer, and cytology (Pap smear) for cervical cancer.

Treatment and Care

1. Treatment aims to cure, prolong life, and improve quality of life for patients. Some common cancer types (breast, cervix, and colorectal cancer) have high cure rates when detected early and treated according to best clinical practices.
2. Relief from pain and other problems can be achieved in over 90 percent of cancer patients through palliative care, which can also be provided by families in low-resource settings.

World Health Organization, European Region. Cancer: *prevention of cancer. Available at: http://www.euro.who.int/en/what-we-do/health-topics/noncommunicable-diseases/cancer/policy/prevention-of-cancer* [Accessed 6 August 2013].
Sources: Adapted from World Health Organization. Cancer. Fact sheet no. 297 [updated February 2012]. Geneva: WHO. Available at: http://www.who.int/mediacentre/factsheets/fs297/en/ [Accessed 22 October 2012].

a major factor in the epidemiology of developing countries, with nearly 70 percent of global cancer deaths occurring in low- and middle-income countries.

Cancer survival is improving and death rates are declining for total cancers and the four most common cancers (prostate, breast, lung, and colorectal) are declining in high-income countries owing to primary prevention, improved screening, and treatment. However, malignant melanoma of the skin, non-Hodgkin's lymphoma, childhood cancer, leukemia, and cancers of the kidney and renal pelvis, thyroid, pancreas, liver and intrahepatic bile duct, testis, and esophagus are all increasing. Liver cancer is associated with a world pandemic of hepatitis C, while hepatitis B levels are declining because of a highly successful global immunization program.

The WHO estimates that nine environmental and behavioral risks along with seven infectious causes are responsible for 45 percent of global cancer deaths. Known or probable causes of cancer include more than 150 chemical and biological agents, many of which are found in the work setting and are largely preventable. Worldwide, these work-related exposures cause an estimated 8 percent of lung cancer.

In the USA, the lifetime probability of developing cancer is one in three. From 1930 to 1990, there was an increase in age-adjusted cancer mortality rates, but the rate has been declining since then. In high-income countries breast cancer is the leading cause of death (one in every 10) among women in the 20–59-year age group. Moreover, global breast cancer mortality has remained relatively constant. In low- and middle-income countries cervical cancer mortality is about twice that in high-income countries (Figure 5.25).

Globally, cervical cancer, linked to genital HPV infection, is the second most common type of cancer among women, with almost 80 percent of the cases concentrated in low-income countries. Successful reduction of cervical cancer mortality has been achieved in many countries with the implementation of excellent preventive measures such as screening with the Pap smear and more recently the availability of the HPV vaccine. In Africa approximately 53,000 women die of cervical cancer each year. In Europe, Romania has mortality rates from cervical cancer six times higher than the rates of the Western European countries, and in countries of Eastern Europe the rates are three times higher than those in Western Europe.

Breast cancer mortality rates in the USA were steady from the 1950s through the 1980s and even rose in 1990. However, when comparing age-adjusted death rates from breast cancer between 1990 and 2008 (Figure 5.26), the reduction in deaths by 2008 was 32 percent. Although black women have lower rates of breast cancer incidence than whites or other US ethnic groups over the period 1999–2010, mortality rates for black women were some 1/3 higher than death rates for white woman and even more so for other ethnic groups in the US, suggesting later presentation and lack of preventive care for black women (CDC, 2013) (Figure 5.27).

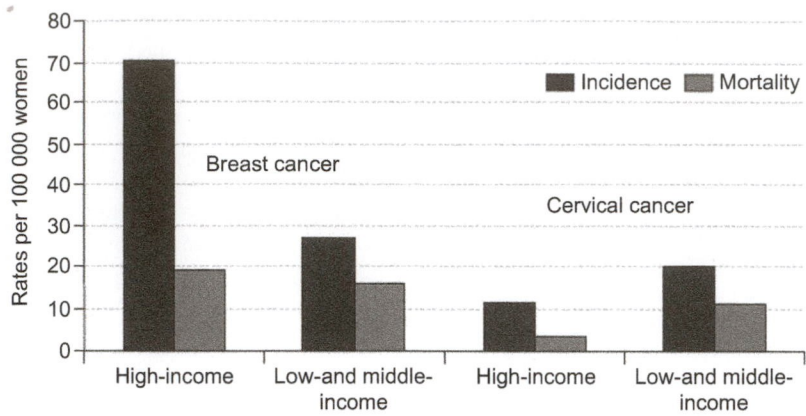

FIGURE 5.25 Age-standardized global breast and cervical cancer rates (per 100,000 women) by country income group, 2004. *Source: World Health Organization. Women and health, today's evidence tomorrow's agenda. Geneva: WHO; 2009. Available at: http://whqlibdoc.who.int/publications/2009/9789241563857_eng.pdf [Accessed 22 October 2012].*

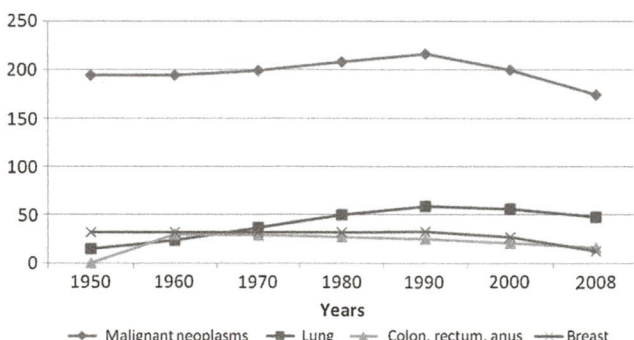

FIGURE 5.26 Death rates from malignant neoplasms for selected years, USA, 1950–2008. *Sources: Derived from: Centers for Disease Control and Prevention. Health United States. 2011, with special feature on socioeconomic status and health. Hyattsville, MD: US DHHS, NCHS; 2011. Available at: http://www.cdc.gov/nchs/hus.htm [Accessed 21 October 2012]. Trend table 32, pages 158–162. American Cancer Society. Global facts and figures 2007. Atlanta, GA: American Cancer Society; 2007. Available at: http://www.cancer.org/acs/groups/content/@nho/documents/document/globalfactsandfigures2007rev2p.pdf [Accessed 22 October 2012]. Centers for Disease Control and Prevention. NCHS Data. Hyattsville, MD: NCHS; 2010 July. Available at: http://www.cdc.gov/nchs/data/dvs/MORTFINAL2007_WorkTable210R.pdf [Accessed 29 September 2012].*

Cancer develops because of changes of genes responsible for cell growth and repair. These changes are the result of the interaction between genetic host factors and external agents, and accordingly, can be classified into the following categories: physical carcinogens such as ultraviolet (UV) and ionizing radiation; chemical carcinogens such as asbestos and tobacco smoke; biological carcinogens such as infections by viruses (hepatitis B virus and liver cancer, HPV and cervical cancer), bacteria (*H. pylori* and gastric cancer), and parasites (schistosomiasis and bladder cancer); and contamination of food by mycotoxins such as aflatoxins (products of *Aspergillus* fungi) causing liver cancer.

Cancer is responsible for much suffering and heavy economic demands on society in terms of health services, loss of work, and premature mortality. The potential for prevention

of cancer is so important that every public health program must include it as part of its duties. In many jurisdictions, laws require reporting of cancer to the public health department (cancer registry), which is vital for epidemiological monitoring of cancer and potential localization of environmental or occupational causes of specific cancers (Ginsberg, 1995; Mathers et al., 2002) or differences in case finding and management.

The WHO reports that *H. pylori* is the causative agent for 63 percent of stomach cancer deaths, 73 percent of liver cancer deaths are caused by viral hepatitis, and HPV causes 100 percent of cervical cancer deaths. These infections together with nine environmental and behavioral causes account for 45 percent of cancer deaths.

Many advances have been made in clinical management and in understanding the causes of some cancers. There is strong and increasing evidence of specific risk factors that are amenable to public health intervention. The most striking example is smoking, which is directly related to lung cancer, and is a factor in bladder and cervical cancers. Exposure to chemical carcinogens, such as asbestos, is of major public health interest, as is prevention of liver cancer by immunization against hepatitis B. Recent research indicates that dietary factors are important contributors to colorectal, breast, and possibly lung and cervical cancers as well. The findings linking *H. pylori* infection to peptic ulcers, gastritis, and stomach cancer have been instrumental in the declining incidence of these conditions internationally. The possibility of an effective vaccine for *H. pylori* would greatly enhance this benefit, joining hepatitis B and HPV as vaccines to prevent cancers.

In the USA, colorectal cancer death rates declined between 1998 and 2007 for both genders, from 25.6 to 20.0 per 100,000 for men and from 18.0 to 14.2 per 100,000 for women (annual decline of 2.8 percent and 2.7 percent, respectively) (Box 5.14). Breast cancer and cervical cancer death rates declined by 2.2 percent and 2.4 percent, respectively, during this period.

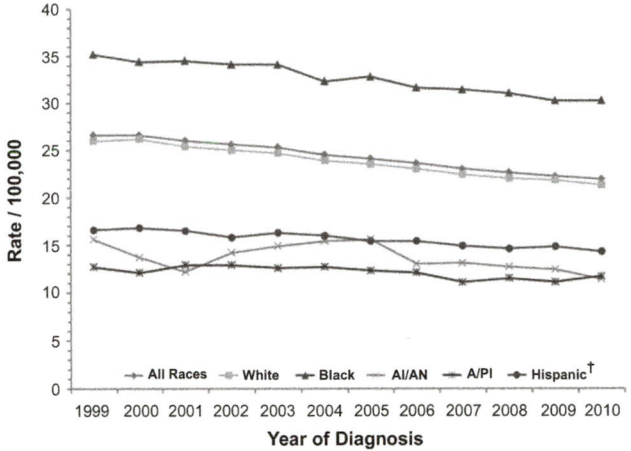

FIGURE 5.27   Female Breast Cancer Death Rates* by Race and Ethnicity, U.S., 1999–2010. Note: Mortality source: U.S. Mortality Files, National Center for Health Statistics, CDC. *Rates are per 100,000 and are age-adjusted to the 2000 U.S. standard population. Death rates cover 100% of the U.S. population. *Source: Centers for Diseases Control and Prevention. Breast Cancer Rates by Race and Ethnicity. Posted 12 August 2013 at: http://www.cdc.gov/cancer/breast/statistics/race.htm (accessed 11.2.2014).*

BOX 5.14 Colonoscopy for Prevention of Cancer of Colon

Colorectal cancer (CRC) is the second leading cause of cancer deaths in the USA, with over 53,000 deaths annually. Furthermore, in the USA, CRC is the leading cause of cancer deaths among non-smokers.

Screening for CRC is effective in reducing incidence and mortality by removal of premalignant polyps and through early detection and treatment. CRC screening prevalence has improved over the past decade. CRC incidence decreased by 3.4 percent per year, and the death rate decreased by 3.0 percent per year from 2003 to 2007 in the USA. These decreases in CRC incidence and mortality represent approximately 66,000 fewer new cases and 32,000 fewer deaths than expected from 2003 to 2007, compared with 2002. However, in 2006, records indicate that some 30 percent of eligible US residents had never been screened for CRC.

Most CRCs occur in people without a family history of CRC. However, as many as one in five people who develop CRC have other family members who have been affected by this disease. People with a history of CRC or adenomatous polyps in one or more first degree relatives are at increased risk. The risk is about doubled in those with one affected first degree relative. It is even higher if that relative was diagnosed with cancer when they were young, or if more than one first degree relative is affected.

The death rate from CRC has been declining in both men and women for more than 20 years. There are several reasons for this: polyps are being found by screening and, thus, removed before they develop into cancers; more CRCs are found earlier when the disease is easier to cure; and treatment for CRC has improved. As a result, there are now more than 1 million survivors of CRC in the USA.

It is generally recommended that *all average-risk adults should begin CRC screening at the age of 50 years*, with colonoscopy every 10 years. Individuals with a first degree relative diagnosed with colon cancer or adenomas when younger than 60 years, or with multiple first degree relatives diagnosed with colon cancer or adenomas, should undergo screening colonoscopy every 3–5 years initiated at an age 10 years younger than the youngest affected relative. Screening may be stopped after the age of 75.

Sources: *Centers for Disease Control and Prevention. Vital signs: colorectal cancer screening among adults aged 50–75 years – United States, 2008. MMWR Morb Mortal Wkly Rep 2010;59:1–5. Available at: http://www.cdc.gov/mmwr/preview/mmwrhtml/mm59e0706a1.htm [Accessed 22 October 2012].*
Centers for Disease Control and Prevention. National Program of Cancer Registries Cancer: United States cancer statistics: 1999–2008. Cancer incidence and mortality web-based report. Atlanta, GA: US DHHS. Available at: http://www.cdc.gov/uscs [Accessed 22 September 2012].
Winawer SJ, Zauber AG, Ho MN, O'Brien MJ, Gottlieb LS, Sternberg SS, et al. Prevention of colorectal cancer by colonoscopic polypectomy. The National Polyp Study Workgroup. N Engl J Med 1993;329:1977–81. Available at: http://www.ncbi.nlm.nih.gov/pubmed/8247072 [Accessed 22 October 2012].
US Preventive Services Task Force. Screening for colorectal cancer [updated March 2009]. Rockville, MD: US Preventive Services Task Force. Available at: http://www.uspreventiveservicestaskforce.org/uspstf/uspscolo.htm [Accessed 22 October 2012].
US Department of Health and Human Services. Healthy people 2020. Washington, DC: US DHHS; November 2010. Available at: http://www.healthypeople.gov/2020/TopicsObjectives2020/pdfs/HP2020_brochure_with_LHI_508.pdf [Accessed 22 October 2012].
American Cancer Society. Colorectal cancer [updated 1 October 2012]. Atlanta, GA: American Cancer Society. Available at: http://www.cancer.org/Cancer/ColonandRectumCancer/DetailedGuide/colorectal-cancer-key-statistics [Accessed 24 October 2012].
Clinical Guidelines: Screening for colorectal cancer: a guidance statement from the American College of Physicians. Ann Intern Med 2012;156:I-30. Available at: https://annals.org/article.aspx?articleID=1090694 [Accessed 22 October 2012].

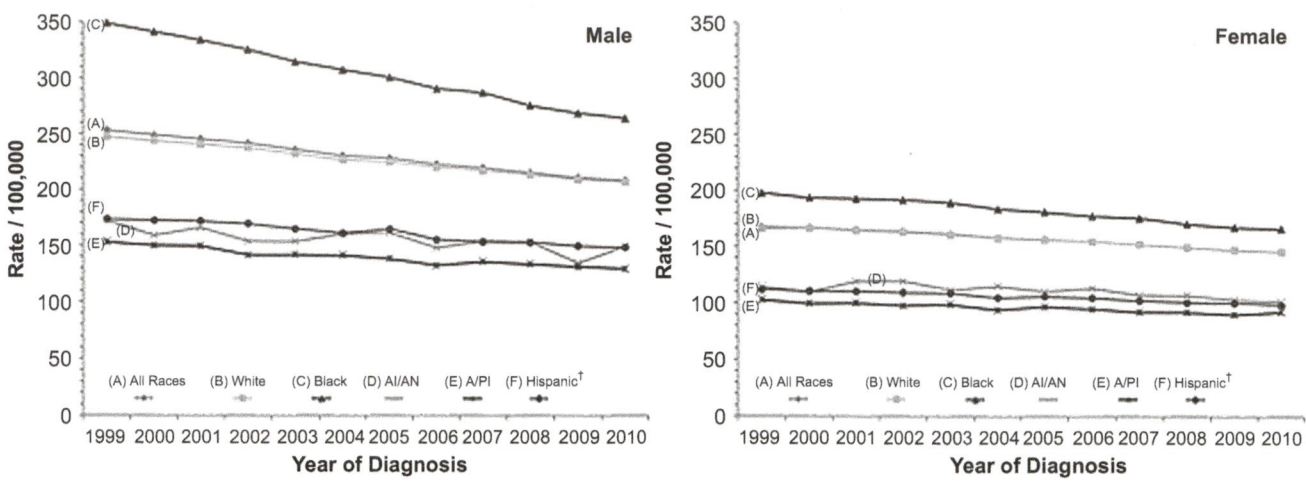

FIGURE 5.28   Age-adjusted cancer death rates by gender and race, USA, 1999–2010. *Source: Centers for Disease Control. Available at: http://www.cdc.gov/cancer/dcpc/data/race.htm (accessed 20 November 2013).*

Since 1975, total cancer mortality rates in the USA have declined for white and African American men and women (Figure 5.28). Cancer rates also vary widely among other ethnic groups, with Japanese, Filipino, Native American, and Mexican Americans having low age-adjusted rates. Japanese Americans have much higher cancer rates than Japanese from Japan, strongly suggesting that lifestyle differences, such as diet and smoking, influence or trigger events.

There are important social class differences in cancer mortality in the USA. The relative risk (RR) for African American men to die of lung cancer was 0.8 compared to whites in 1950, but by 1990 the RR had increased to 1.5. These differences, like those of higher rates of CHD mortality among African American males, were thought to be race related; however, the differences are now considered to be largely due to socioeconomic or social class differences, possibly related to different smoking, occupational, and dietary patterns as well as less access to medical care. This is borne out by studies showing differences in mortality rates in different social classes in the UK and the Scandinavian countries.

For males, total cancer age-adjusted mortality in the USA showed a slight decrease in rate between 1950 and 2008 (from 182 per 100,000 in 1950 to 153.2 in 2008). Figure 5.29 shows mortality rates for men by type of cancer from 1930 to 2008, with a steep rise in lung cancer death rates peaking in 1990–1991 and then declining by nearly one-third by 2008. Colorectal cancer remained stable but began to decline in the 1980s, by 35 percent up to 2008, and prostate cancer declined by nearly 50 percent from 1994 to 2008 (American Cancer Society, 2012).

For the USA, women's cancer mortality rates between 1930 and 2008 are shown in Figure 5.30. In 1987–1988, lung cancer surpassed breast cancer as the leading cause of cancer death among women, but remained steady between 1995 and 2008. Mortality rates from breast cancer declined by 43 percent from 1990–1991 to 2008, but it remains a greater cause of potential years of life lost, since it affects younger women than lung cancer. Female mortality from stomach cancer has been declining steadily since the 1930s and colorectal cancer mortality since the 1940s.

In 2011, CDC reported that lung cancer was the leading cause of death from cancer for both men (28 percent of all cancer deaths among men) and women (26 percent of all cancer deaths among women). The second leading cause for women was breast cancer (15 percent) and for men was prostate cancer (11 percent). Colon cancer was the third leading cause of death from cancer for both men (8 percent) and women (9 percent).

The single largest preventable cause of cancer in the world today is tobacco. It causes 71 percent of all lung cancer deaths, and about 80 percent of all cancer deaths in developing countries. It is currently responsible for the death of one in 10 adults worldwide (about 6 million deaths each year). Half of the people who smoke today (about 1 billion people) will eventually die of tobacco-related diseases. Thus, strong legislative and taxation policies and other efforts to reduce tobacco consumption and exposure are vital elements of public health.

Cancer incidence rates vary widely between countries and within countries. A comparison of total cancer mortality between some countries in the WHO European Region shows very marked differences, as seen in Figure 5.31, with declining rates since the mid-1970s in some countries (e.g., Finland) and since the late 1980s in others (e.g., the UK), and rising and then stabilizing rates in some (e.g., Poland). Even neighboring Scandinavian countries show quite marked differences in total, lung, and breast cancer standardized death rates.

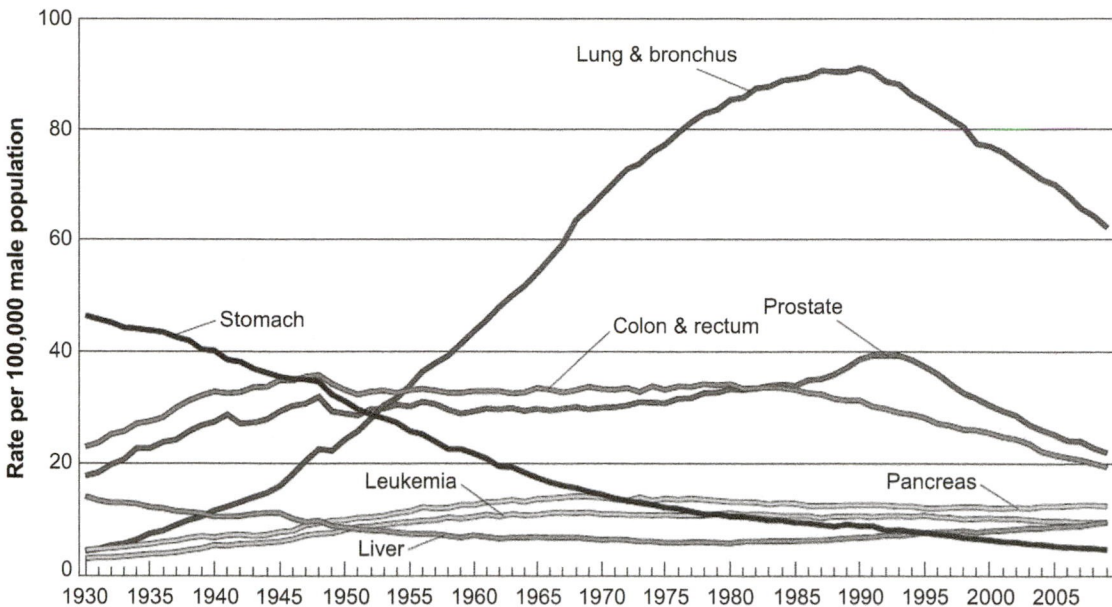

FIGURE 5.29   Age-adjusted cancer death rates, males by site, USA, 1930–2009. *Source: American Cancer Society. Cancer Facts & Figures 2013. Atlanta: American Cancer Society, Atlanta GA; 2013. Available at: http://www.cancer.org/acs/groups/content/@epidemiologysurveilance/documents/ document/acspc-036845.pdf (accessed 20 November 2013).*

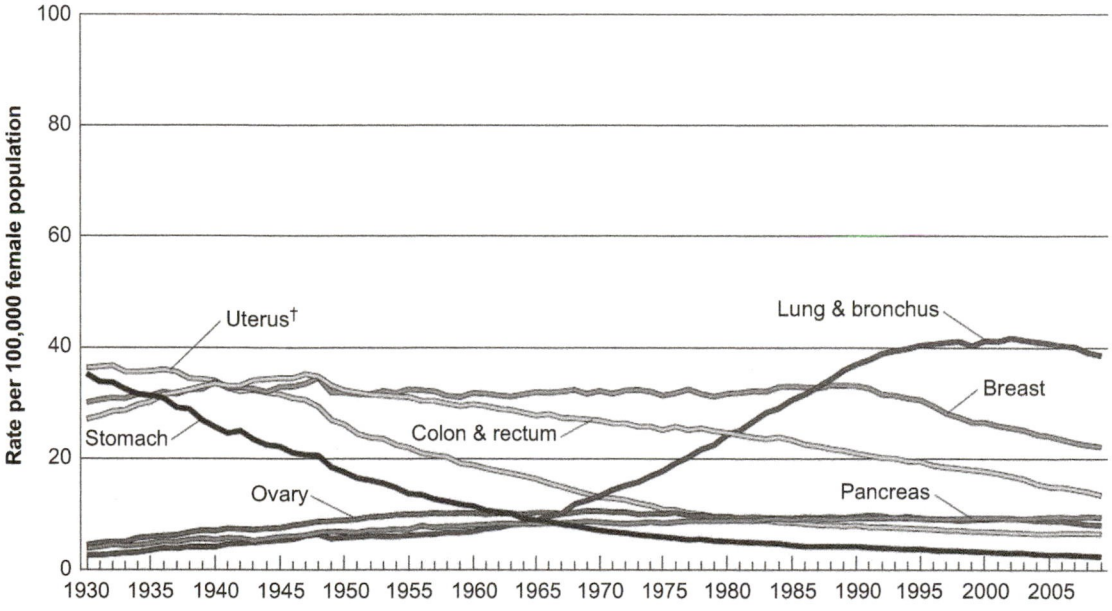

FIGURE 5.30   Age-adjusted cancer death rates, females by site, USA, 1930–2009. Note: †Uterus = cervix and corpus combined. *Source: American Cancer Society. Cancer Facts & Figures 2013. American Cancer Society. Atlanta, GA: Available at: http://www.cancer.org/acs/groups/content/@ epidemiologysurveilance/documents/document/acspc-036845.pdf. (accessed 20 November 2013).*

Differences in death rates from cancer of the trachea, bronchus, and lung are shown in Figure 5.32 for selected European countries. The explanations for such differences are not at all clear, but there may be differences both in risk factors (e.g., diet, environment, and smoking) and in treatment. Low rates in Israel and Spain have been attributed to high fruit and vegetable consumption but could also be due to sun exposure and vitamin D levels. Falling levels in Finland and the UK are likely to be related to decreased smoking. Such international comparisons can generate hypotheses for further investigation.

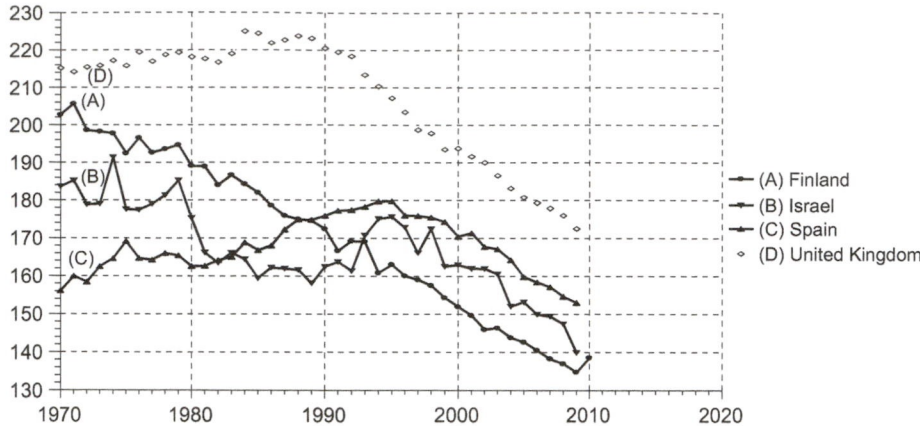

FIGURE 5.31 **Standardized mortality rates (per 100,000 population) from all malignant neoplasms, selected European countries, 1970–2010.** *Source: World Health Organization European Region. Health for All database (HFA-DB). Copenhagen: WHO Regional Office for Europe; January 2012. Available at: http://data.euro.who.int/hfadb/ [Accessed 22 October 2012].*

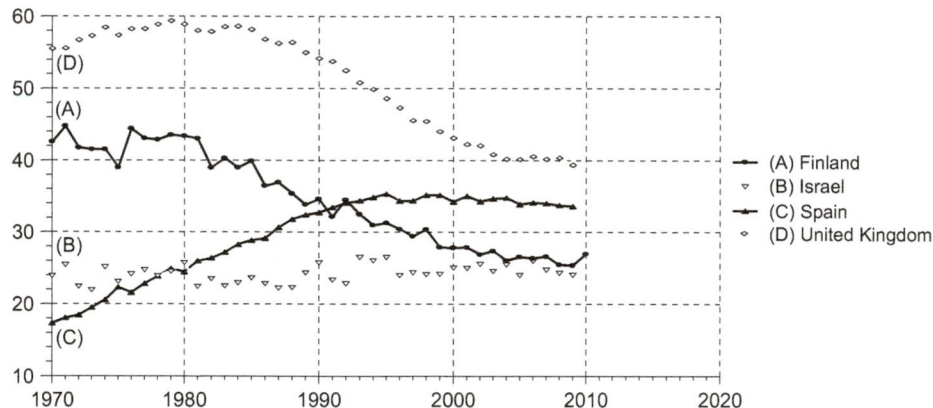

FIGURE 5.32 **Standardized mortality rates (per 100,000 population) from cancer of lung, trachea, and bronchus, selected European countries, 1970–2010.** *Source: European Health for All database (HFA-DB). Copenhagen: WHO Regional Office for Europe; January 2012. Available at: http:// data.euro.who.int/hfadb/ [Accessed 22 October 2012].*

Prevention of Cancer

Primary prevention of cancer requires a reduction of major risk factors: smoking (lung and bladder), fatty diet (breast and colorectal), carcinogenic chemical exposures (mesothelioma, lung, leukemia, lymphoma), exposure to infectious carcinogenic agents (hepatic), multiple sexual partners (cervical cancer), and excess sunlight exposure (melanoma). Lung and skin cancer, occupation-related cancers, and some gastrointestinal cancers are most amenable to primary prevention by reduced smoking and exposure to sunlight, asbestos, radon, and other carcinogens. Immunization against hepatitis B is protective against most liver cancers. Prevention also includes adequate levels of physical activity.

Secondary prevention in clinical services or in screening programs for high-risk groups focuses on early case finding. The medical care system has a major responsibility, emphasizing screening for breast, prostate, cervical, and colorectal cancer. Occupational, environmental, and social factors all play important roles in cancer. As a result, intersectoral preventive activities are needed for cancer prevention.

Cancer prevention involves reaching the total population to increase awareness of risk factors and promote change. Prevention is effective in reducing cancer risk for cancer of the cervix, liver, stomach, breast, and skin, but requires coverage, equity, quality control, and above all awareness among the public, policy makers, and providers of care. Health care programs and service providers have vital roles to educate the public, to set standards as to when and how to screen for disease, and to cross socioeconomic barriers in reaching patients.

More research will bring important breakthroughs in this field. Identification of genes that increase risk or cause cancer, infective agents associated with carcinogenesis, medications that stop or delay the progress of cancer, and other topics of research appear very promising. The evidence for a causal relationship between nutrition and cancer is increasingly convincing.

Key components of early detection efforts require attention by each individual and the health system. These include education to help people to recognize early signs of cancer and seek prompt medical attention for symptoms, screening programs to identify early cancer or precancer before signs are recognizable, and treatment aimed to cure, prolong life, and improve quality of life for patients (Table 5.11).

TABLE 5.11 Prevention of Cancer

Modality	Method
Education	Community and individual education – on cancer facts, prevention and control strategies, early signs of cancer, and seeking prompt medical attention
Self-care	Tobacco and smoking cessation Maintaining healthy body weight Regular physical exercise – 150 minutes per week – moderate to very active Diet – minimizing fat and red meat intake; increasing fruit, vegetable, nut, and grain consumption Protection from excess sun exposure Daily sun exposure of half an hour recommended Safe sex practices (e.g., reducing sexual risk behavior of multiple partners, promoting use of condoms) Moderating alcohol consumption Breast self-examination
Vaccination	Hepatitis B immunization and catch-up Vaccination against human papillomavirus (HPV)
Medical management/ screening, early detection	Cancer of cervix (Papanicolaou smears), colorectal cancer (fecal occult blood, rectal and colonoscopy) Breast cancer mammography (e.g., screening/ diagnostic)
Medical management/ treatment	Early detection and rapid treatment are important in cancer management Treatment for HIV and *Helicobacter pylori* (e.g., stomach – peptic ulcers, gastric cancer) and schistosomiasis (bladder cancer), STIs Palliative care provided by families in low-resource settings
Toxic/acrogenic exposures	Controlling occupational and environmental carcinogenic and toxic exposure; environmental and occupational health standards and enforcement Avoiding pesticides, herbicides, asbestos waste, and radon Controlling occupational and environmental carcinogenic hazards

Source: Adapted from World Health Organization. Cancer. Fact sheet no. 297 [updated February 2012]. Geneva: WHO. Available at: http://www. who.int/mediacentre/factsheets/fs297/en/ [Accessed 22 October 2012].

TRAUMA, VIOLENCE, AND INJURY

Injuries and violence are important and rising global problems. Globally, approximately 5.8 million lives are lost each year as a result of injuries, which represents 10 percent of total global deaths. This number is 32 percent more than the number of deaths caused by malaria, HIV/AIDS, and tuberculosis combined (approximately 4.2 million). Injury deaths are primarily from suicide and homicide (one-quarter of the total) and road traffic (also one-quarter of the total). Other causes include falls (8 percent), poisoning (6 percent), fire (6 percent), and war (3 percent). The major causes of death from injuries are expected to increase in rank when compared to other causes of mortality, so that by 2030 they are likely to emerge among the 20 leading causes of death globally. By this time, road accidents are predicted to become the fifth leading cause of death. Furthermore, experts predict that suicide and homicide will present as the 12th and 16th top causes of death. Many who survive have short- or long-term care needs, are at risk for other health consequences, and must adapt to changes in lifestyle (WHO, 2010). As illustrated in Figure 5.33, road accidents, homicide, and suicide combined account for 49 percent of the injury deaths that occurred in 2004.

Injury death rates place a heavy toll on society. Every year, globally 1.45 million people die from road crashes and almost 1 million people from suicide among the 5.8 million total injuries deaths. Twice as many men (traffic, suicide, and homicide) die from injuries and violence as compared to women (traffic, suicide, and fire-related deaths). Many survivors of trauma have lifelong physical and mental disabilities.

In the USA, injuries are the leading cause of death for people aged 1–44 years. In 2005, injuries cost an estimated

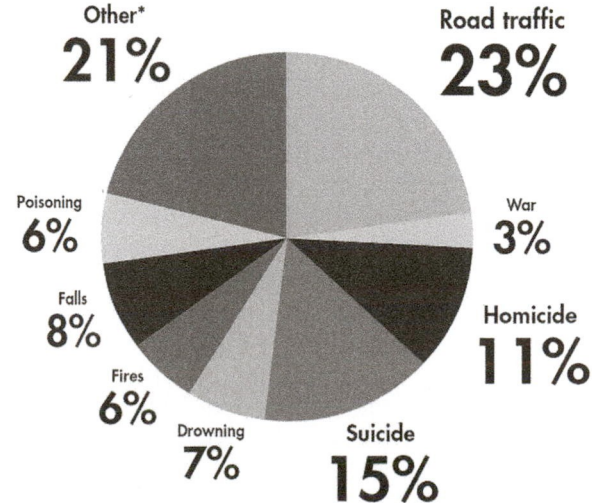

FIGURE 5.33   Global distribution of injury deaths by cause, 2004. *Source: World Health Organization. Injuries and violence – the facts. Geneva: WHO; 2010. Available at: http://whqlibdoc.who.int/publications/2010/9789241599375_eng.pdf [Accessed 22 October 2012].*

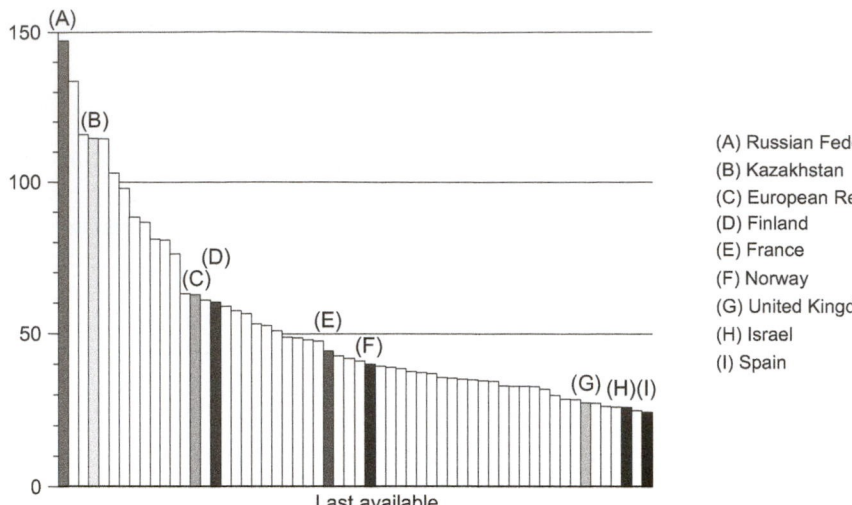

FIGURE 5.34 **Standardized mortality rates from injuries (external causes), European countries, 2008–2009.** *Source: World Health Organization, European Region. Health for All database. Copenhagen: WHO Regional Office for Europe; January 2013. Available at: http://data.euro.who.int/hfadb/ [Accessed 7 August 2013].*

US$406 billion for medical care costs, including deaths, hospitalizations, and emergency department visits and work loss. In 2007, 182,479 people died, 2.9 million were hospitalized, and 29.8 million were treated for non-fatal injuries in US hospital emergency departments and released (CDC, 2012). Injury is the leading cause of death of children in the USA, with over 9000 deaths in 2009, a decline of 30 percent over the previous decade. The causes include car crashes, suffocation, drowning, poisoning, fires, and falls.

In the European Region of the WHO, rates in the low- and middle-income countries are triple those of the higher income countries (Figure 5.34). Mortality from all trauma varies across the continent, with the highest rates in Russian Federation, closely followed by other former Soviet countries including the Central Asian Republics at five times higher than average EU rates. The leading cause of death of people aged 15–29 years in the European Region is road crashes, with the fourth and fifth causes due to homicide and suicide. Among the elderly, falls are the most common cause of deaths from injuries.

Trauma, or external injury, is a broad category that includes accidents, poisonings, suicide, homicide, and violence. In many countries, trauma is the leading cause of death because of its greater frequency among the young and the middle-aged. Motor vehicle injuries are among the five leading causes of death for teenage girls and women of reproductive age in all WHO regions with the exception of South-East Asia, where burns are the third leading cause of death. Trauma is a leading cause of YPLL in most developed countries and has become a major focus of intervention in modern public health program development. There are many possible interventions, requiring programmatic approaches as recommended in Healthy People 2020 targets. There is

high potential for beneficial effects of reduced injury and death, primarily among young people (Box 5.15).

Trauma morbidity can be reduced by public health measures including primary, secondary, and tertiary prevention. Primary prevention reduces risk factors that are associated with trauma and includes measures such as enforcement of laws against alcohol or cell phone use with driving, requiring motorcycle helmets, car seat belts, and speed limitation laws. Secondary prevention involves early and adequate medical care at the scene of an accident and rapid transportation with well-trained professionals to a hospital trauma center. Prevention of consequences of the trauma includes cardiopulmonary resuscitation, maintaining an airway, stopping bleeding, and treatment of shock at the accident site, which can greatly reduce case fatality rates. Tertiary prevention involves effective and early rehabilitation by which the degree of disability can be reduced and long-term management made more effective, as in cases of head injury.

An accident is a sudden, unintended event that may be associated with human injury. The term does not imply that the event is not potentially anticipated, since there may be neglect of fundamental safety and preventive procedures and therefore increased risk of the event occurring. If a driver is drunk and/or exceeding the speed limit, the crash and deaths constitute a criminal liability. If a plant operator allows or requires employees to work in dangerous conditions without adequate safety measures, then the event is not accidental but one that could have been anticipated and probably prevented, thus also constituting criminal liability.

Injuries and deaths inflicted intentionally include homicide, rape, assault, battery, child or vulnerable person abuse, and suicide. Public health is concerned with both intentional

BOX 5.15 Interventions to Prevent or Mitigate Motor Vehicle Injuries

Strategic program involving all levels of government, judiciary, ministries of transport, local authorities, and police:

1. Mandatory seat belt (front and rear seats) legislation and enforcement
2. Driver education and licensing control over risk groups, e.g., under age 25, previous offenders
3. Testing and enforcement of alcohol standards for drivers
4. Administrative suspension of driving licenses
5. Motorcycle and bicycle helmets mandatory and enforced
6. Enforcement of speed limits of 55 miles/hour (90 km/hour) on intercity roads (50 mph or 80 kph for trucks)
7. Marking and enforcement of safe driving near schools, hospitals, crosswalks (crossings), and construction crews
8. Enforcement of minimum age drinking laws
9. Mandatory child safety seats
10. Driver and passenger air bags
11. Mandatory vehicle safety inspection
12. Improved vehicle design and standards
13. Improved road design and standards
14. Education and public policy commitment
15. Graduated licensing for teenagers
16. Enhanced pedestrian safety, especially for elderly people
17. Public transport for disabled and mobility needs of elderly people
18. Develop public health surveillance systems of dangerous roads, emergency department visits, hospitalizations, disabilities, and deaths
19. Promote research on motor vehicle-related injury and death and contributory factors such as emergency care on site, transportation, and specialized trauma centers
20. Pedestrian crosswalks and traffic circles (roundabouts)
21. Speed cameras
22. Center lane barriers for highways
23. Seek and enforce severe fines and incarceration for repeat offenders
24. Improved public transport
25. Define, monitor, and police dangerous roads
26. Promote insurance rates by risk categories and individual experience.

Source: Adapted from Centers for Disease Control and Prevention. Achievements in public health, 1900–1999. Motor-vehicle safety: A 20th century public health achievement. MMWR Morb Mortal Wkly Rep 1999;48:369–74. Available at: http://www.cdc.gov/mmwr/preview/mmwrhtml/mm4818a1.htm [Accessed 22 October 2012].

public should be offered. Resuscitation training and emergency rescue measures are vital and emergency call systems should be well staffed and trained as well as being widely publicized.

Intervention to prevent violence is designed to break specific cycles (Table 5.12). Violence prevention involves increased awareness by teachers, police, social workers, health professionals, and the public at large in spotting potential and actual signs of violence, especially of abused children or women. Other forms of violence prevention include gun control, preventing weapons from entering schools, "hotlines" for victims to telephone and seek help, shelters for potential and actual victims, self-defense training, rapid response of police, enforcing drinking restrictions, and promotion of supervised teenage recreational activities. The patterns of mortality from various causes of trauma in the USA from 1950 to 2010 are seen in Figure 5.35.

Motor Vehicle Accidents

The most common cause of loss of life from trauma is motor vehicle accidents. Worldwide the leading cause of death among those 15–29 years of age is motor vehicle accidents, with the fourth and fifth leading causes ranked as homicide and suicide.

The CDC reports, "Motor-vehicle-related injuries kill more children and young adults than any other single cause in the USA and are the leading cause of death from unintentional injury for people age 5–34 years. Approximately 44,000 people in the USA die in motor vehicle crashes each year. Moreover, crash injuries result in approximately 500,000 hospitalizations and 4 million emergency department visits annually". In 2007 crashes were the third leading cause of potential years of life lost before 65 years of age. There have been large declines in pedestrian deaths (49 percent) and cyclist deaths (58 percent) even with the increased number of vehicle miles. Seat-belt usage increased to 84 percent in 2009; however, this important safety device is not used by millions of vehicle occupants, with the highest percent of non-use by teenagers.

In 2011, the United Nations launched a Decade of Action for Road Safety 2011–2020 to stimulate country and local political leadership in implementing evidence-based measures to reduce premature death and disability from road crashes. Dr Margaret Chan, head of the WHO, indicated that worldwide only 15 percent of countries have comprehensive legislation on the key risk factors.

Motor vehicle injuries are a problem not only in developed countries, but also in developing countries where road crash rates are extremely high. Alcohol use may be less restricted than in developed countries, road safety less advanced, cars poorly maintained, and drivers less experienced. Worldwide, the annual death toll on roads is over 1.2 million people with over 90 percent of deaths occurring in

and unintentional injury. Attempts to minimize their occurrence and effects require complex interactive programs that identify and target high-risk groups, such as for motor vehicle accidents and suicides, through preventive-oriented programs, public education, and rescue operations. Training of first responders including police, fire and rescue workers, ambulance personnel, schoolchildren, and the general

TABLE 5.12 Classification of Injuries and Primary Prevention Measures

Form of Injury	Regulatory Control Measures	Prevention/Education
Motor vehicle accidents	Mandatory seat belts, child safety seats, and speeding laws; police enforcement of speeding and drunk driving laws; lower permissible blood alcohol levels; raise age for driving permits; lower speed limits on interurban highways; cancellation of driving privileges in high-risk groups (e.g., repeat offenders); car safety examination, enforcement, air bags, structural standards; pedestrian-safe crosswalks, traffic circles (roundabouts); speed cameras; mandatory helmets for motorcyclists and bicyclists; regulation of insurance premiums related to risk factors	Driver education; alcohol and drug awareness campaigns; pedestrian safety education; police enforcement
Falls	Safety devices for children and elderly people; non-skid carpets, rails in homes of elderly, bathroom and stairs railings; hip protectors; monitor medication use	Education and awareness campaigns
Burns	Regulators on home heating systems, regulation for manufacturers of appliances; flame-resistant toys, children's clothing, bedding products; building code standards for electrical wiring, smoke detectors, doors opening outwards	Promote fire prevention awareness; occupational safety guidelines; education for care of thermal injuries
Poisoning	Manufacturers' labeling; childproof lids on medication and household chemicals	Education on dangers of household medications, chemicals, safe storage, labeling, and closure
Domestic violence	Mandatory reporting by medical and social workers of injuries to children and women that may originate in domestic violence and abuse; police and medical alertness; shelters for abused people; imprisonment and therapy for abusers	Education in reporting of domestic violence
Occupation-related injury	Enforcement of safety standards, employer criminal liability for unsafe conditions and injuries; monitoring small workshops and large industries, building sites, fisheries, lumber sites	Worker and employer education
Sports injury	Mandatory use of safety equipment, helmets for motorcycle, bicycle riders, in sports, professional and amateur; Consumer Product Safety Commission monitoring recreational sports/play equipment; licensing of coaches and sports facilities	Teaching safe sports practices in gymnastics and contact sports; good coaching and adjudication
Suicide	Raising awareness of the general population, especially health workers, teachers, social workers, risk groups, parents; reporting of high-risk persons to school and other authorities	Hotlines, counseling, group and individual therapy, group homes
Drowning	Water safety commission enforces guidelines for supervision of recreational swimming places, fencing of swimming pools; Lifeguards trained in rescue and CPR at public beaches and swimming pools	Swimming and boating safety awareness; swimming education in schools, summer camps; resuscitation training
Cardiac arrest	Defibrillator equipment and CPR training for personnel in sports facilities and events; police firemen, ambulance personnel, and the general public	Training of first responders and general public

Note: CPR = cardiopulmonary resuscitation.
Source: Adams PF, Lucas JW, Barnes PM. Summary health statistics for the US population: National Health Interview Survey, 2006. Atlanta, GA: NCHS; 2008. Available at: http://www.cdc.gov/nchs/data/series/sr_10/sr10_236.pdf [Accessed 22 October 2012]

low- and middle-income countries. These countries, with 48 percent of the registered vehicles, have double the fatality rates of high-income countries. Globally about 50 million people are injured annually.

Minimizing the effects of trauma from car crashes involves a complex of legislation, enforcement, education, and technical development initiatives. Passenger restraints and helmets for motorcycle and bicycle riders

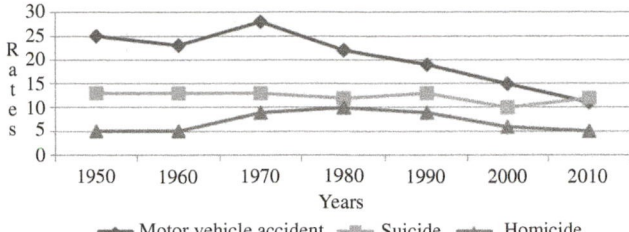

FIGURE 5.35 Age-adjusted mortality rates (per 100,000 population) by selected causes of trauma, total population, USA, 1950–2010. *Source: Kochanek KD, Xu J, Murphy SL, Minino A, Kung HC. National vital statistics reports. Deaths: Final data for 2009. Hyattsville, MD: NCHS; December 2011. Available at: http://www.cdc.gov/nchs/data/nvsr/nvsr60/nvsr60_03.pdf [Accessed 21 October 2012].*

have demonstrated great potential for reducing death and injury from transportation-related crashes. Other preventive approaches focus on reducing the number of unlicensed drivers and unlicensed vehicles on the roads, which serves as an ongoing challenge. Providers of insurance may cooperate with governments to devise measures for common initiatives.

Prevention of crashes caused by speeding, alcohol abuse, fatigue, or negligence involves dealing with the human factor, based on strict enforcement and education. Laws and enforcement should ensure that vehicles are in proper working order, especially lights and brakes. Governments must ensure well-maintained roads, adequate road lighting, and elimination of barriers to visibility. In the WHO European Region, mortality rates have fallen dramatically since 1970. Many factors contributed to this reduction, including seat-belt use, child safety measures, improved car design, improved policing and public attitudes leading to reduced drunk driving, and improved roads and lighting.

Driver testing, especially for people under the age of 25 and those over 75, should be strictly enforced. Statistically, women are safer drivers than men and should be encouraged to drive buses, trucks, and military transports. Restricting speed limits to less than 55 miles per hour (90 kilometers per hour; 50 mph or 80 kph for trucks) reduces accident fatality rates, as do non-rigid central and side road barriers.

Law enforcement is a major factor in reducing road crash deaths and injuries. Compulsory seat-belt laws are major contributors to reduced severity of injuries, and enforcement increases compliance. Enforcement of speed limits and the incarceration of drunk drivers have a strong deterrent effect and reduce death and severe injury rates. Strong police enforcement, heavy fines, criminal liability, and loss of driving permits for offenders should be implemented.

The physical relationship between speed and extent of injury is based on laws of mechanical energy or momentum. A person traveling at 55 miles per hour (90 kph) has four times the kinetic energy of a person traveling at 30 miles per hour (45 kph). Injury is more severe at higher speed. Permitting speed limits over 55 miles per hour (90 kph) or non-enforcement of the speed limits results in more serious injuries. This has been documented in studies of mortality patterns, which rose when speed limits were raised and, similarly, declined when speed limits were lowered. Low-resource countries in particular should reduce speed limits and invest in public transport and improved police enforcement, including prison for serial offenders, road and vehicle improvements, and driver education. The widespread use of speed cameras, with significant fines for speeding and using cell phones in cars, is also vital in controlling the tragic waste of life from motor vehicle crashes.

Emergency care has made important strides in the past several decades and has undoubtedly contributed to declining mortality rates from trauma. Central trauma units serving larger populations are more likely to have a broad multiservice potential and achieve better results from a larger volume of cases and experience. Organized ambulance services with well-trained and supervised paramedics taking patients to central trauma centers have been shown to be effective in reducing mortality and collateral damage rates. These programs are of vital importance in achieving goals for injury control and reducing deaths and serious complications of trauma. Improved care during the early minutes following violence, whether from terrorism or motor vehicle crashes, or at the earliest stages of AMI, can be improved by continuous training of first responders.

Domestic and Sexual Violence

Family or domestic violence includes spousal and child abuse, and sexual violence. Cases are more readily identified and brought to public attention than in previous generations, so that apparent increases may be due to better reporting. Some elements that determine these intentional injuries include socioeconomic status, alcohol use, and the family history of members who may have been abused themselves.

The National Intimate Partner and Sexual Violence Survey is a continuing telephone survey system in the USA conducted by the CDC. The 2010 survey findings reported that 1.3 million women were raped during the previous year, nearly one in five women (and one in 71 men) had experienced rape during their lifetime, and one in four women experience severe physical violence by an intimate partner. This survey found that over 80 percent of women who had experienced rape, stalking, or physical violence by an intimate partner reported short- or long-term emotional stress, including symptoms of post-traumatic stress disorder (PTSD). Men who experienced violent sexual assault also reported similar short- and long-term effects.

Prevention requires strong public concern, police and court intervention with enforced therapy, and/or imprisonment for repeat offenders. Measures to reduce violence against women are vital to achieving Millennium Development Goals relating to women's and children's health (MDGs 3, 4, 5, and 6) (see Chapters 2 and 16). Reports by girls and women of intimate partner violence indicate that prevalence varies widely between and within countries, with higher rates occurring in lower income countries. In some countries, "dating violence" or "date rape" is a frequent occurrence. Sexual, physical, and emotional maltreatment or abuse during childhood, by perpetrators including parents and caregivers, increases the likelihood of both sexual and partner abuse in later years. Population surveys in high-income countries show that many men report abuse, although the physical and emotional effects may be different from those in women. It is estimated that during childhood 20 percent of girls are sexually abused compared to 5–10 percent of boys. Child sexual abuse is a significant risk factor in depression, alcohol and drug use and dependence, panic disorder, PTSD, and attempted suicide. In the USA, rates of neglect, physical abuse, and sexual abuse of children declined by 10 percent, 55 percent, and 61 percent, respectively, between 1992 and 2009.

Suicide

Suicide is an important public health problem globally. Worldwide, suicide is a leading cause of death for women between the ages of 20 and 59 years. Internationally, there are wide variations in suicide rates, with the majority of cases occurring among adolescent men and elderly people. The WHO reports that by 2012 suicide had increased by 60 percent in the past 45 years, with a mortality rate of 16 per 100,000 worldwide. From 1981 to 2008, suicide rates in Canada among young adults (15–19 years) declined from 21.2 to 12.1 (42.9 percent) for men; in contrast, rates rose from 3.8 to 6.2 (63.2 percent) per 100,000 population for

women. Male suicide rates in most Central and South American countries are below 10 per 100,000 population. In the USA and Canada suicide rates are less than 20 per 100,000.

Suicide was ranked as the 11th leading cause of death in the USA in 2007. It increased slightly in 2010 to an age-adjusted death rate of 12.2, essentially no change from 2006. Among all age groups, age-adjusted mortality from suicide declined from 13.2 per 100,000 in 1950 to 10.8 in 2003 (Health United States, 2005). However, this masks an age/gender difference, with a major decline in all age groups over 45 years of age, and an increase in both the 15–24-year and 25–44 age groups.

For every 20 suicide attempts there is one successful suicide, with men more common in the latter group. It is estimated that 30 percent of suicides are the result of mental disorders, with the remainder due to decisions regarding life circumstances, low self-esteem, binge drinking, and situational depression. Suicide among young people can become a communicable condition by emulation among people who are alienated from home and society, and can appear attractive, as awareness of irreversibility is low.

In Europe, suicide rates vary widely, with the former Soviet countries the highest, at four times higher than those of the UK, Spain, Italy, and other low-rate countries; Scandinavian countries are in the middle range and Western Europe and the Mediterranean countries have the lowest suicide death rates. Overall suicide rates are lowest in the Mediterranean countries (Italy, Spain, Greece, and Israel) and above 40 per 100,000 in France, Switzerland, Sweden, Denmark, Finland, Austria, and Belgium. Suicide rates in Hungary are 40 per 100,000 population for men and 10.6 per 100,000 for women; among men in the former Soviet countries rates range between 28.1 and 61.3 per 100,000 population. Figure 5.36 shows declining rates of suicide for selected countries in the European Region from 1970 to 2009. Spain, Israel, and the UK have consistently low rates; Sweden has medium; and Russia and Hungary have high but declining suicide rates over this period.

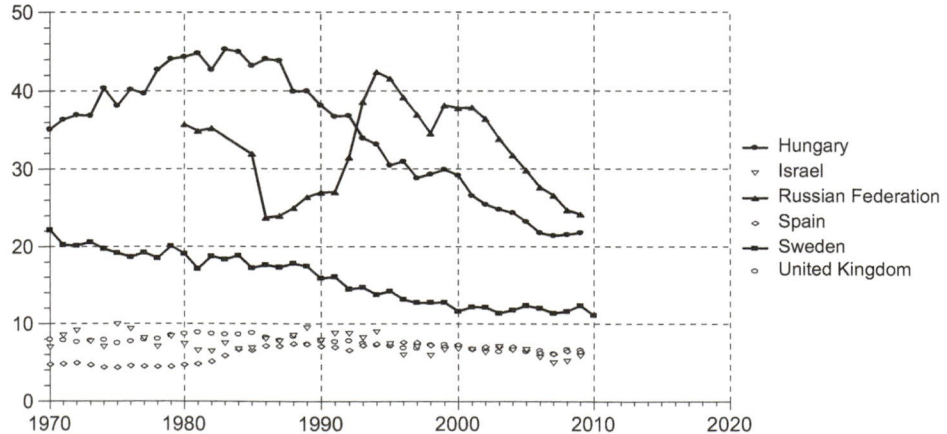

FIGURE 5.36 Rates of suicide (per 100,000 population) for selected countries in the European Region, 1970–2009. *Source: World Health Organization, European Region. Health for All database. Copenhagen: WHO Regional Office for Europe; January 2012. Available at: http://data.euro. who.int/hfadb/ [Accessed 22 October 2012].*

A recent review of suicide literature indicated that as many as 90 percent of suicide cases meet criteria for a psychiatric disorder, particularly major depression, substance use disorders, cluster B personality disorders, and schizophrenia. Transient factors that reflect an imminent risk of suicide crisis and therefore require immediate intervention include unbearable mental pain and related experiences of depression and hopelessness. Problems with help-seeking, social communication, and self-disclosure also pose a suicide risk, as do personality traits of aggression and impulsivity (Gvion and Apter, 2012).

An estimated 370,000 suicides (more than one-third) worldwide are caused by ingestion of pesticides, mostly in low- and middle-income countries. This places an enormous burden on health services in developing countries. In 2004, treatment for self-poisoning in Sri Lanka was an estimated US$1 million. In a Sri Lanka hospital during 1995–1996, 41 percent of intensive care beds were used by people poisoned with organophosphates. Sri Lanka has the most widely studied pesticide ban, which has been successful in reducing suicide rates (see Box 5.16 on companion web site at http://booksite.elsevier.com/9780124157668). Restricting access to the means of suicide is one approach needed to prevent unplanned impulsive suicide.

Studies of suicide patterns across the world showing unplanned, impulsive suicide patterns vary across countries, with poisoning by drugs in some countries, guns in others, and hanging or jumping from high-rise buildings in others. It is imperative that threats of suicide are taken seriously; health care providers, teachers, counselors, and religious leaders should be instructed in suicide prevention and how to assist people through periods of depression. Mental health and supportive counseling must be part of any health care system, as the suicidal individual requires immediate attention and care. Telephone counseling by volunteers on hotlines has proved to be valuable in suicide prevention.

Poisoning

Drug overdose death rates in the USA have more than tripled since 1990, to the highest level recorded. In 2008, more than 36,000 people died from drug overdoses, and most of these deaths were caused by prescription drugs. In 2008, there were 14,800 prescription painkiller deaths, amounting to more than cocaine and heroin deaths combined. Misuse and abuse of prescription painkillers was responsible for more than 475,000 emergency department visits in 2009, a number that nearly doubled in just 5 years. Moreover, opioid analgesics were involved in almost 40 percent of all poisoning deaths in 2006 (CDC).

In 2009, 41,592 poisoning deaths occurred in the USA, 76 percent of them unintentional (CDC, 2012). In 2010, unintentional poisoning generated 831,295 emergency room visits, with one-quarter of these being hospitalized. Poison Control Centers reported receiving calls about 2.4 million poisoning calls. Childhood poisonings from prescription or nonprescription drugs are common events. In 2004, an estimated 71,999 children were seen in hospital emergency departments for such events. In 2005, poisonings resulted in medical and productivity costs of US$33.4 billion in the USA.

Medication errors can result in serious risks of unintended drug interactions. Each year in the USA, adverse drug events result in over 700,000 visits to hospital emergency departments. Hospitals are liable for medical events resulting from drug errors or adverse events, mostly pertaining to litigation for damages. Hospitals and other health facilities are therefore acutely aware of their responsibility to prevent errors and this is incumbent on all health professionals, particularly prescribing doctors.

Poisoning in the work setting is also a major public health issue (see Chapter 9). Poisoning by pesticide was common in many Asian countries and in Latin America; poisoning by drugs was common in Nordic countries and the UK. Hanging was the preferred method of suicide in Eastern Europe, and firearm suicide in the USA. Suicide is a leading cause of death for women in the age group 20–59 years globally and the second leading cause of death in the low- and middle-income countries of the WHO Western Pacific Region.

Restriction of access to methods of suicide, such as firearms, is believed to reduce successful suicide, and restriction of publicity about suicides is thought to reduce suicide rates. More than restriction of access to weapons, however, is the importance of addressing the underlying issues triggering the desire or thoughts of suicide.

Pesticide storage, bans, and replacement by less toxic pesticides could prevent many of the currently estimated 370 thousand pesticide-related suicides each year. Rural communities in low- and middle-income countries have high rates of suicide related to pesticides. Controlling access to pesticides is not only critical in reducing self-directed violence, but also key to preventing unintentional poisoning and terrorism. There are international conventions towards restricting and managing highly toxic substances; however, many are still used widely. Chemical bans need to be considered alongside research and development of safer but effective pest control measures to reduce the availability of the highly toxic versions (WHO). Dangerous chemicals designed for chemical warfare are held in military storehouses in many countries that may become unstable and vulnerable to terrorist organizations or individuals and used for mass terrorism.

Among the majority of countries participating in a WHO survey, the most common form of deliberate self-harm among those 15–16 years of age is cutting. Self-cutting is three times more common in females than in males.

Homicide

Homicide has become one of the major causes of death in some countries, such as in Colombia. In young adult males between the ages of 15 and 24 years in the USA, homicide is the third leading cause of death. The epidemiological analysis of murders shows a relation to drug traffic, involving both rich drug competitors and street-level violence for control of street trafficking. Random violence among schoolchildren is a frequent event, as are drive-by or "road-rage" shootings, often resulting in child deaths. Murders associated with rival gangs and random violent crime with murder are now common in many former Soviet countries. Gang violence in US cities is matched by concern in rural areas, where murderous rampages by adolescents with access to and training in the use of deadly weapons occur with increasing frequency. Gun control legislation has made some minor gains, but weapons remain accessible to large sectors of the US population.

The WHO estimates that homicide and suicide, respectively, take the lives of 600,000 and 844,000 people annually worldwide, exceeding deaths from war and conflict (184,000). There are 875 million firearms in the world owned by private citizens and less than 10 percent of these firearms are registered. At least one-third of Americans are civilian owners of firearms. The proportion of homicides involving firearms ranges from 19 percent in Western and Central Europe to 77 percent in Central America. Countries in which firearms cause the vast majority of homicides (with a range of 70–90 percent) include Montenegro, Colombia, the USA, Brazil, and Yemen.

Firearms usage as a cause of death in male suicide ranges from 0.2 percent in Japan to 61 percent in the USA and 35.7 percent among women in Uruguay and the USA. Among European males aged 15–24 years, firearm use for suicide deaths ranges from 2.3 percent in England to 43.6 percent in Switzerland.

Health costs for hospital treatment of major abdominal firearms injuries are estimated at 4 percent of national health spending in South Africa. In England and Wales, the estimated annual societal cost of homicide is £1.5 million.

Evidence suggests that limiting access to guns, knives, and pesticides saves lives, prevents injuries, and saves costs to society, including psychological and medical, as well as loss of economic productivity in high-risk areas. Youth violence is so common in some societies that it affects efforts to achieve the MDGs.

Regulation of firearms can include bans, licensing schemes, minimum age for buyers, background checks, mandatory waiting time, and safe storage requirements. Such measures have been successfully applied in some countries such as Austria, Brazil, and some states in the USA. Introducing national legislation can be difficult politically but much can be done at local level. Rigorous

enforcement ("zero tolerance"), and improved security at youth attractions such as discos and concerts, and for state-owned firearms, are also helpful in the broad campaign against youth violence.

In the UK, knives and glass or bottles are the weapon in 10 percent of violence against adults. Sharp objects are commonly used in homicides in Malaysia, Scotland, Nigeria, and Australia. Control measures have focused on guns, but recent legislative measures have increased penalties for banned flick-knives, implemented a minimum age for the purchase of knives, and enacted "stop and search" initiatives; however, their impact is still unclear. Reducing the demand for guns and dangerous sharp objects requires measures for diverting vulnerable youth from gang membership; thus, governments need broad strategies to improve the socioeconomic factors underlying the violent atmosphere in areas defined by police and social agencies.

Sources of data for epidemiological analysis are available from national and local police, but hospital emergency department records may be more useful, as more than 50 percent of violent crime goes unreported to police. Murder rates have declined in the USA since 1990 (from 9.4 per 100,000 in 1990 to 5.3 in 2010). This decline is often attributed to improved economic conditions, low unemployment, stricter punishment laws, and successful community policing. The USA remains well above other industrialized countries in homicide rates, but well below rates in Russia.

Prevention of Violence

In many countries, violence is one of the leading causes of death, especially among teenagers and young adult males. Domestic violence leading to homicide is one of the most common causes. Prevention of violence and violence-related injuries is a major public health concern because of the large scale of loss of life and personal injury, as well as the long-term damage to society. Interventions involve the whole of society, not the health system alone. Nevertheless, public health has an advocacy role to play.

Violence inspired by religious, nationalistic, or other political motives is a fact of life in both developed and developing countries, sometimes occurring with shocking ferocity. These events, whether bombs in buses, subways, aircraft, or buildings, or "ethnic cleansing" warfare, cause enormous physical and psychological trauma that must concern health care providers and public health personnel. Reports of violence in the workplace are an increasing phenomenon.

Dramatic events involving school violence and mass homicide by teenagers occurred in the late 1990s in small communities all over the world, which are for the most part low-crime areas, such as in schools or university campuses, in the USA and also in Australia, Canada, and Scotland. A 2007 CDC report of suicide rates between 1990 and 2003 showed a decline of 28.5 percent in rates among the 10–24 age group, but an increase among boys and girls aged 15–19 between 2003 and 2004. In 2009, 650,843 teens and young adults received emergency medical care for non-fatal assault injuries.

CDC reports in a 2009 survey that 17.5 percent of US students have carried weapons to school and 7.7 percent of students reported being threatened with weapons during the past 12 months. Weapon use is often attributed to a high level of violence portrayed in the media, through films, television, or the Internet. The Internet has become a forum for politically extreme and neo-Nazi ideals, as well as an avenue for illegal weapons dealing. The consensus is that this gives troubled youth the opportunity to act out fantasies of murder and mayhem with the goal of satisfying a need for revenge, acting on an ideological hatred, or achieving notoriety. Within the context of high rates of risk-taking behavior in adolescent males, greater vigilance on the part of parents, health care providers, the educational system, and the community in seeking out warning signs, along with significant restrictions on gun sales, would help to rectify this serious problem.

A 2007 CDC review of school-based programs for the prevention of violent and aggressive behavior concludes that many different strategies documented are effective at all school levels. They have beneficial effects, beyond the benefits for reduced violent or aggressive behavior, including reduced truancy, and improvements in school achievement, "problem behavior", activity levels, attention problems, social skills, and internalizing problems (e.g., anxiety and depression).

There is an increasing prevalence of bullying, homophobic teasing, and sexual harassment among school students in the USA. Bullying others is reported by 12 percent of boys and girls, homophobic teasing directed at friends is reported by 26 percent of boys and 24 percent of girls, and 28–39 percent of both genders report making sexual comments or calling others names.

CHRONIC LUNG DISEASE

Chronic lung disease (CLD) is an important, diverse, and mostly preventable group of diseases which cause extensive morbidity and mortality. In 2007, CLD was the fourth leading cause of death in the USA. CLD can largely be prevented with good primary care and education for self-care.

When associated with acute respiratory infection, such as influenza, bronchitis, or pneumonia, CLD can result in lengthy hospitalization and premature death. Cough, shortness of breath, restricted exercise tolerance (e.g., climbing stairs), and difficulty sleeping are frequent symptoms, with impaired clearance of sputum and reduced lung capacity.

When assessing 2006–2009 COPD mortality for the age group 35–74 among 16 industrialized countries, the USA ranked highest for male mortality and second highest for female.

Chronic bronchitis and asthma can lead to emphysema, with fixed expansion and rigidity of lung tissue and reduced oxygen exchange capacity. Increasing shortness of breath, cough, loss of exercise ability, sleeplessness, repeated infections, hospitalizations, and death are all possible outcomes. The triad of emphysema, asthma, and bronchitis produces respiratory cripples who may function with limitations until a respiratory infection causes hospitalization and ultimately death.

Asthma

Asthma is an intermittent, reversible condition of airway obstruction in response to various stimuli, resulting in wheezing and shortness of breath due to variable airflow. Usually first appearing in children up to the age of 5 years, it affects an estimated 25.7 million people (2012 data) in the USA, including 7.1 million of those aged 0–17 years (28.9 percent). It is the most common chronic disease among children. During 2006–2008, asthma affected between 7.5 and 15 percent of people over the age of 18 years. Among women, the rates were 18.2 percent for Hispanics, 9.3 percent for white non-Hispanics, 19.1 percent for mixed race, and 9.5 percent for black non-Hispanics; while rates for men were 7, 5.9, 11.2, and 5.7 percent, respectively. Reportedly, only about one-third of children or adults were using long-term control medicine such as inhaled corticosteroids. The medical costs of asthma are estimated at US$50.1 billion.

Asthma prevalence increased in the USA from the 1980s to the 1990s, growing from 3.5 percent to 5.5 percent from 1980 to 1996; it continued to rise in 2010, with an annual increase from 2001 to 2009 of 1.2 percent. In 2009 asthma prevalence was reported to affect 8.2 percent of the US population (i.e., 24.6 million people). The mortality rate has remained stable at approximately 1.1 per 100,000 population since 1990. Asthma mortality from 2005 to 2007 averaged 1 per 100,000 population. Prevalence is highest among black and Puerto Ricans and those below the poverty line (11–15 percent). Comparisons of the burden of asthma in the USA are seen in Figure 5.37, showing ethnic, gender, and age differences for emergency room visits, hospitalizations, and mortality. In 2008, people with asthma missed 10.5 million school days and 14.2 million work days due to their asthma. In 2007, there were 1.75 million asthma-related emergency department visits and 456,000 asthma hospitalizations.

Although the specific etiology of asthma is unknown, it is associated with familial, infectious, allergic, environmental, and psychosocial factors. Risk factors include animal allergens (usually from pets), household dusts and mites, primary and secondary tobacco smoke, outdoor allergens, and pollutants. Air pollution may be a factor in

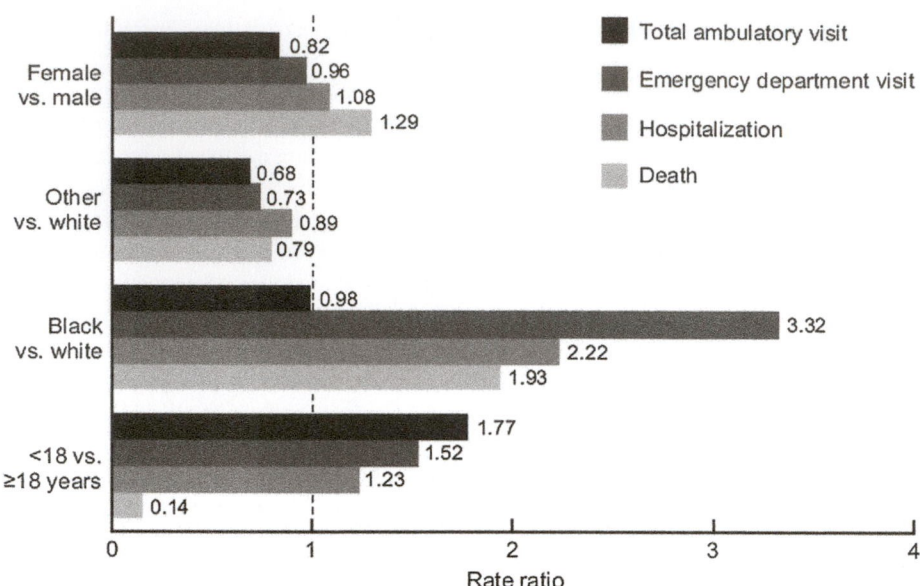

FIGURE 5.37   Relative burden of asthma health care use and mortality, adjusted for current prevalence by gender, race, and age group, USA, annual average 2005–2007. Note: Rate ratio of 1.0 (dashed line) = equal rates between the groups being compared. *Source: Akinbami L, Moorman J, Liu, X. Asthma prevalence, health care use, and mortality: United States, 2005–2009. National health statistics reports, no. 32. Hyattsville, MD: NCHS; 2011. Available at: http://www.cdc.gov/nchs/data/nhsr/nhsr032.pdf [Accessed 24 October 2012].*

the increase in asthma; 63 percent of cases live in areas where pollution exceeds recommended levels. Ozone pollution, the result of hydrocarbons and nitrogen oxide emissions from motor vehicles or other sources, mixed in the presence of sunlight, causes increased wheezing, coughing, and chest tightness, especially among susceptible children who play outdoors in polluted environments. Asthma prevalence is higher among those with family income below the poverty level, with 11.2 percent of those with incomes less than 100 percent of the poverty level during the period 2008–2010. The higher rates of asthma among African Americans may be related to lower socioeconomic conditions and residence in inner city communities. Conversely, keeping children indoors for safety reasons in poorly ventilated older homes may contribute to increasing asthma.

Identification and removal of antigens, the use of bronchodilators and corticosteroids, and treatment of infections as they occur are the major methods of management. Education plays a key role in asthma control for patients, their families, school personnel, and the general public. Mortality may be increased by medication overuse, substance abuse, and tobacco use. Health care providers need continuing education regarding this condition and its management, especially during pregnancy, and regarding the possibilities and problems associated with medication usage. CDC reports that despite the high and increasing prevalence of asthma, management strategies based on clinical guidelines for the treatment of asthma remained below the targets set by the Healthy People 2010 initiative.

Chronic Obstructive Pulmonary Disease

The term chronic obstructive pulmonary disease (COPD) represents advanced stages of chronic respiratory disease with airflow impairment due to chronic bronchitis affecting the smaller airways. This includes a variety of conditions resulting from damage to lung tissue, chronic narrowing of the respiratory tract, and obstruction of airflow. The term includes chronic bronchitis, emphysema, and other causes of COPD. Chronic bronchitis affects 20 percent of the adult male population of the USA. CLD increased from 100,000 deaths in 1994–1995 to 124,816 deaths in 2002, and more than 125,000 deaths in 2007. The WHO estimates that moderate to severe COPD affects more than 65 million worldwide, and that COPD mortality was more than 3 million or 5 percent of all deaths in 2005. Estimates also show that by 2030 COPD will be the third leading cause of death.

Tobacco use is certainly the greatest cause of COPD. Reduction of smoking is crucial to prevention, as is reduction of indoor pollution in the workplace and home, and outdoor pollution (see Chapter 9). Secondary and tertiary prevention should include annual influenza vaccination and pneumococcal vaccine usage. Careful monitoring of the patient at home for changing symptoms and early signs of infection would prevent long and costly hospitalization.

Tobacco use, which is increasing globally, is responsible for about 6 million deaths annually: of 5 million ex-users and 600,000 non-users. About 80 percent of the world's 1 billion users live in low- and middle-income countries. The WHO estimates that deaths could reach 8 million annually. Children of poor households working in tobacco farming may absorb nicotine from the wet leaves, leading to green tobacco disease. Tobacco use in the Russian Federation is among the highest in the world, with 43 million users and three times as many male smokers as females. The Health Ministry had no mandate for tobacco control until 2010. The WHO reports that health promotion educational measures such as warning labels, preferably graphic, on tobacco products are effective in targeting all age groups but have been found to be especially effective in reaching young people (Box 5.17).

Mortality rates from COPD have not declined markedly in the USA owing to aging of the population. Smoking reduction with increasingly aggressive legislation restricting smoking in public places including bars and restaurants, and wider use of influenza and pneumococcal pneumonia vaccines, are also major factors in improving the quality and duration of life of people with COPD.

Restrictive Lung Diseases

People with restrictive lung diseases have a reduced lung volume, either because of an alteration in lung parenchyma or because of a disease of the pleura, chest wall, or neuromuscular apparatus. Reduced total lung capacity, vital capacity, or resting lung volume result from diseases affecting the lung tissue, i.e., parenchymal lung disease, such as scarring of lung tissue, or to general or localized disease processes in the lung tissue, e.g., sarcoidosis, fibrosis, or other connective tissue diseases. Restrictive lung disorders are accompanied by reduced gas transfer, which may be marked clinically by respiratory distress on exercise.

Other causes may include drugs and other treatments such as cancer drugs and radiation. Inorganic dust exposure (e.g., silicosis, asbestosis, talc, pneumoconiosis, berylliosis, hard metal fibrosis, coal worker's pneumoconiosis) can also cause restrictive lung disease, as can exposure to organic dust (e.g., farmer's lung, bird fancier's lung, and mushroom worker's lung, with hypersensitivity pneumonitis).

Occupational Lung Diseases

Occupational lung diseases are related to particular occupational exposures in two main categories: diseases of lung tissue and diseases of the airway. Pulmonary fibrosis with restricted lung volume decreases lung diffusion capacity on pulmonary function testing, showing increased interstitial pulmonary markings on chest X-rays. Examples include silicosis and pneumoconiosis, with increased risk of tuberculosis, characterizing a major part of the burden of respiratory disease in the developing world (see Chapter 9).

Obstructive airways disease is also a common pattern of occupational lung disease, which may be reversible (occupational asthma) or become irreversible (chronic bronchitis with or without obstruction or emphysema or COPD). Obstructive airway diseases cause disturbed pulmonary function. The global burden of diseases related to occupational factors was estimated at 4–10 million cases per year, with approximately 3–9 million cases per year in developing countries. An estimated 12 percent of COPD deaths are from occupational exposure to airborne particulates. A further 29,000 deaths are caused by silicosis, asbestosis, and pneumoconiosis (WHO).

Occupational lung diseases are a group of conditions associated with workplace exposures to dusts and vapors, which act as irritants, carcinogens, or immunological agents. Microscopic airborne particles at work sites can cause lung cancer, COPD, silicosis, asbestosis, and pneumoconiosis (Table 5.13).

BOX 5.17 Progress in Banning Tobacco Advertising, Promotion, and Sponsorship

*"People have a fundamental right to information about the harms of tobacco;
countries have a legal obligation to provide it".*

- About half (3.8 billion) of the world population is covered by effective tobacco control.
- Nearly 2 billion people live in 23 countries that have at least one mass media campaign.
- Effective comprehensive tobacco programs combine education, communication, and training.
- Legislation that requires large geographic health warnings on every cigarette pack covers 1 billion people.
- Warning labels, at almost no government cost, are supported by 85–90 percent of the public, including smokers.
- Monitoring tobacco usage and tobacco control policy achievements are key to understanding and control.
- More than 739 million people are protected by comprehensive, national smoke-free laws, twice as many as in 2008.
- Nineteen countries with a population of 425 million (6 percent of the world population) are fully protected against tobacco industry marketing, with nearly all in low- or middle-income countries, mandating best practice health warning labels.
- A further 101 countries ban national television, radio, and print tobacco advertising,
- There are 74 countries (38 percent of all countries) with minimal or no restrictions on tobacco advertising.
- Low- and middle-income countries have been in the forefront of developing mass campaigns.
- Smokeless tobacco products are less likely to have warning labels or be covered in media campaigns.

Sources: *World Health Organization. WHO report on the global tobacco epidemic, 2011: warning about the dangers of tobacco. Geneva: WHO; 2011. Available at: http://whqlibdoc.who.int/publications/2011/9789240687813_eng.pdf [Accessed 22 October 2012]. Akinbami L, Moorman J, Liu, X. Health statistics reports: asthma prevalence, health care use, and mortality: United States, 2005–2009. Hyattsville, MD: NCHS; January 2011. Available at: http://www.cdc.gov/nchs/data/nhsr/nhsr032.pdf [Accessed 24 October 2012].*

TABLE 5.13 Occupational Disease of Lung

Disease	Disease Progression	Epidemiology
Coal worker's pneumoconiosis (CWP); also known as black lung disease, anthracosis	Prolonged exposure to coal dust for 10 years or more; diagnosis by chest X-ray or biopsy Asymptomatic accumulation of coal pigment without cellular reaction Inhaled coal dust becomes a problem when natural defense and processing of the dust becomes overwhelming and overreactive	Declined markedly in USA as a result of reduced mining workforce and improved regulation and legislation (Federal Coal Mine Health Safety Act). Similar to accumulation found in varying degrees among most urban dwellers and tobacco smokers
Silicosis	Due to chronic inhalation of crystalline silica; progresses over 20–40 years from cough and sputum production to crippling COPD due to massive fibrosis of lung tissue	Some 2000 cases reported in USA annually. People suffering from silicosis are at increased risk for tuberculosis and should undergo routine testing
Asbestosis	Caused by asbestos exposure; fibrous deterioration of lung tissue	High rates of lung cancer (especially among smokers); mesothelioma, a rapidly lethal form of cancer specific to occupational or community exposure to asbestos products
Byssinosis	Acute and chronic disease caused by exposure to cotton dust, flax, or hemp; results in shortness of breath, chest tightness, and chronic cough	Exposure over periods of 10 years or more causes reduced pulmonary function and COPD
Occupational asthma	Bronchial restriction following exposure to agents to which the individual has become sensitized. Asthma with airflow obstructive symptoms may be severe and chronic	Wide variety of occupations at risk; electronic workers, hairdressers; people exposed to dusts, chemical agents, and animal antigens

Note: COPD = chronic obstructive pulmonary disease.

Primary prevention by reducing exposure levels, and secondary prevention by close medical follow-up of exposed workers and ex-workers for many years after exposure, are integral parts of occupational health. The risk of such exposures causing serious disease is accentuated by cigarette smoking and environmental pollution.

In 2000, the WHO estimated that worldwide risk factors in the workplace were responsible for 13 percent of COPD, 11 percent of asthma, and 9 percent of lung cancer. Exposure to occupational airborne particulates caused an estimated 386,000 deaths globally including pneumoconiosis and almost 6.6 million DALYS. Occupational lung diseases are increasing in developing countries that lack worker safety guidelines and which attract mining and other industries from wealthy countries in search of mineral resources, lax regulation, potential corruption, and low-cost labor.

END-STAGE RENAL DISEASE

End stage renal disease (ESRD) is defined as reduced renal function (to less than 10 percent of normal capacity) requiring dialysis or kidney transplantation for survival. ESRD follows severe kidney damage from infection, glomerulonephritis, hypertension, drug reactions, or diabetes. Chronic kidney disease (CKD) is an important risk factor for CVD, requiring careful attention to the risk factors for CVD, and leads to ESRD. Treatment of CKD with medication and lifestyle changes as

well as dialysis and transplantation can alter the progression and outcomes of the disease. Nearly 15 percent of American adults – 20 years or older – (20 million people) have CKD, with variation of 8–14 percent by ethnicity (Figure 5.38).

About 37 people per 100,000 in 2009 in the USA have ESRD. Almost 100,000 are on chronic dialysis, with more than 20,000 having had a renal transplantation. Almost half of these are due to diabetes mellitus and most follow a long period of chronic renal failure. The US incidence of ESRD increased by 40 percent in a 10-year period, with 106,902 new cases and 85,790 deaths in 2005. The number of people undergoing treatment for ESRD increased from 26 to 35 per 100,000 population in 2004, and 110,000 patients started treatment in 2007.

Prevalence varies widely between different countries and among ethnic groups in the USA. High rates occur among some ethnic groups such as Mexican and Hispanic Americans, African Americans, Native Americans, Maoris, and Australian Aborigines, probably owing to high rates of diabetes and hypertension in these groups. A notable disparity in ESRD incidence is exemplified by data showing 277 per 1,000,000 new cases annually in whites compared to 976 per 1,000,000 annual new cases in African Americans (2009 data). The higher rates among African Americans compared to whites may be mostly attributed to socioeconomic status and associated risk factors, such as poor infant nutrition, chronic stress, high BMI, low levels of physical

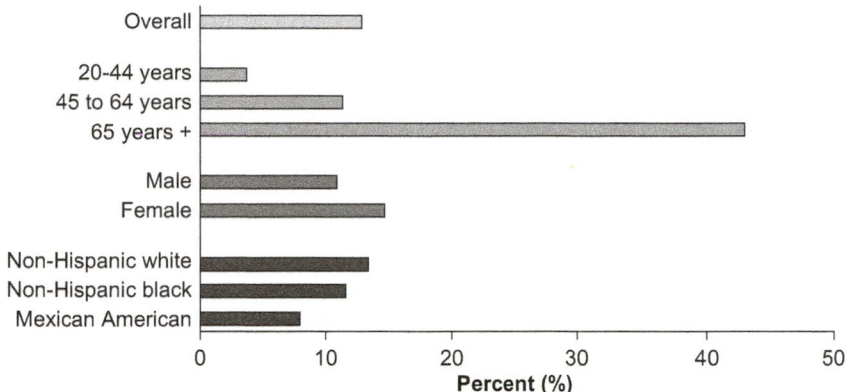

FIGURE 5.38 Chronic kidney disease among adults by age, gender, and race/ethnicity, USA, 2010. *Source: Centers for Disease Control and Prevention. National chronic kidney disease fact sheet [updated 2 April 2010]. Atlanta, GA: CDC. Available at: http://www.cdc.gov/diabetes/pubs/fact-sheets/kidney.htm [Accessed 22 October 2012].*

activity at work and recreationally, hypertension, lack of health insurance coverage, and lack of access to preventive health care.

Poor control of hypertension and diabetes are the main risk factors for ESRD. Sufficient control, as in diabetes, lowers the rate of ESRD; however, hypertension control is less rigorous in the USA. Consequently, ESRD resulting from hypertension is increasing (by 8 percent between 2000 and 2007), while diabetes-associated ESRD has risen by only 1 percent. Furthermore, ESRD due to glomerulonephritis fell by 21 percent in the same period. The USA estimates the cost of ESRD in 2009 at US$29 billion. Healthy People 2020 estimates that 25 per cent of the Medicare budget is used for treatment of ESRD and CKD (Healthy People 2020, Chronic Kidney Diseases, April 2013).

Prevention efforts to reduce the prevalence of ESRD should include the following:

- identification and effective treatment of streptococcal throat infections to prevent glomerulonephritis
- careful use of medications with potential for renal damage
- prompt treatment of urinary tract infections
- screening for the early detection of diabetes mellitus and hypertension
- proper ongoing monitoring and control of diabetes mellitus and hypertension.

(ESRD) is an important public health issue because it is partly preventable and is a large consumer of health care resources. The total cost of the US program for financing treatment of ESRD through Medicare was US$17.9 billion in 1999. It reached US$29.0 billion in 2009 and US$33.0 billion in 2010. Kidney transplantation is the most cost-effective method of treatment. It is approximately one-third of the cost of long-term hemodialysis; however, it is seriously restricted by a lack of donors. Continuous ambulatory peritoneal dialysis is a cost-effective option, especially where donors are limited. Usually done by the patient or the family at home, peritoneal dialysis allows the patient to engage in normal daily activities, and eliminates costly hospitalization and hemodialysis equipment needs. The proportion of dialysis patients waiting for a transplant grew from 1.4 percent in 1985 to 17.3 percent in 2009, while transplant rates declined from 8.5 to 4.1 per 100 dialysis patient years. The annual age-adjusted mortality rate per 1000 patient years for ESRD patients declined from 228.2 in 1985 to 146.4 in 2009. The overall mortality rate for dialysis patients declined from 245.5 to 180.2, adjusted between 1985 and 2009.

Diabetes mellitus is the leading cause of ESRD (i.e., kidney failure requiring dialysis or transplantation) in the USA. CDC reports that it accounted for 44 percent of new cases of treated ESRD in 2007. Between 1985 and 2002, the crude and the age-adjusted incidence of treatment for ESRD attributable to diabetes (ESRD-DM) per 1,000,000 population increased dramatically. Incidence increased from 35.8 to 149.5 per 1,000,000 population (a 409 percent increase). Similarly, the age-adjusted incidence increased from 40.9 to 155.0 per 1,000,000 population (a 374 percent increase) (US Renal Data System 2011). The rates rose to 160.3 in 2006, but in 2009 declined to 154.1 per million (age adjusted).

Ethical challenges pertaining to ESRD occur in both developed and developing countries. Some developing countries with specialty medical centers use donor kidneys purchased from poor people, providing kidneys for medical tourists. In developed countries, ESRD prevalence is increasing as the population ages, as this is an age-related condition. Hemodialysis is a life-saving procedure, and as costs of health care increase, important ethical issues arise surrounding its use. The use of hemodialysis is increasing, particularly in relation to the effective alternative of transplantation, which is limited mainly by short supplies of organ donors.

DISABLING CONDITIONS

As the population ages, many disabling conditions place a greater burden on the individual, on the health system, and on the need for adaptation of social and health policies to meet the needs of this population group. Figure 5.39 shows the growing burden of such conditions in the US population by age group.

The public health issues related to disabling conditions are of tertiary care, or preventing further deterioration and maximizing quality of life affecting all disabling conditions, involving families, caregivers, hospitals, nursing homes, social and employment services, disability benefits, as well as preventive care such as nutrition, immunizations, and mental health support. Prevention in multiple sclerosis (MS) focuses on good case management to reduce complications and promote rehabilitation. There are effective medications to reduce frequency and severity of relapses with increased length of remissions. Immunizations for influenza, pneumonia, and varicella are recommended.

The general issues relating to disabling conditions include:

- promotion of greater awareness and understanding of MS among the general public, employers, and health care professionals
- investing more in diagnostic tools and techniques
- investing in education and training of health professionals
- stimulating support for expansion of research into disabling conditions
- developing and strengthening initiatives and structures for health services offering treatment and rehabilitation equally available and accessible to all people with MS with a view to keeping them in employment
- investing in and supporting the development of the capacity of societies and patient groups to advocate and support public, private, and non-profit initiatives for public policy, service provision, and support.

All stakeholders in societal health need to invest in sustained efforts to improve the quality of life of people with disabling conditions, and to reduce the long-term financial impact on them, their families, and society as whole. This chapter focuses on physical disabilities, but mental health and related disabling conditions, discussed in Chapter 7, face the same challenges. The tight interaction of physical and mental health is very much present in these two sectors of health at both the individual and the population levels.

Arthritis and Musculoskeletal Disorders

Arthritis and musculoskeletal conditions are among the most common causes of physical disability, visits to doctors, and hospitalizations. Arthritis is the leading cause of disability in the USA, affecting as much as half of the population over the age of 65. It was associated with total direct and indirect costs of US$128 billion in the last Medical Expenditure Panel Survey (2003). Arthritis and other rheumatic conditions (bursitis, lupus, fibromyalgia) are among the most common chronic conditions, affecting 46.4 million people (21.6 percent of the adult population). Prevalence was higher among women (25.4 percent) than men (17.6 percent), and in older age groups (50 percent for age ≥65 years and 29.3 percent for 45–64-year-olds). By 2030, these conditions are expected to affect 60 million people. Prevalence is higher among obese and inactive people.

A higher proportion of black and Hispanic populations have work limitations resulting from arthritis, possibly reflecting variations in types of work as well as racial or ethnic differences and factors such as being overweight. Substantial increases in the cost of arthritis are expected in the coming years due to the aging of the population and the increasing use of expensive treatments and prescription drugs. In the USA, two-thirds of adults suffering from heart disease and arthritic conditions are physically inactive. There has been a steep increase in knee replacement rates among the older population in comparison to hip replacements (Figure 5.40).

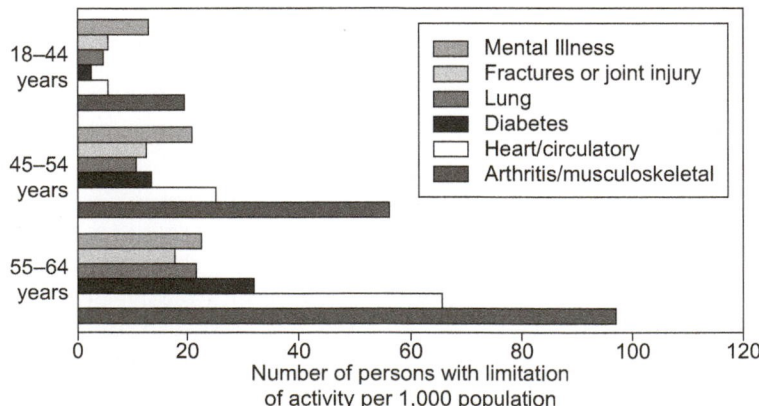

FIGURE 5.39 **Activity limitation among adults due to chronic conditions, USA, 2004–2005.** *Source: Centers for Disease Control and Prevention. Health, United States 2007, with chartbook on trends in the health of Americans. Hyattsville, MD: US DHHS, NCHS; 2007. Available at: www.cdc.gov/nchs/data/hus/hus07.pdf [Accessed 21 October 2012].*

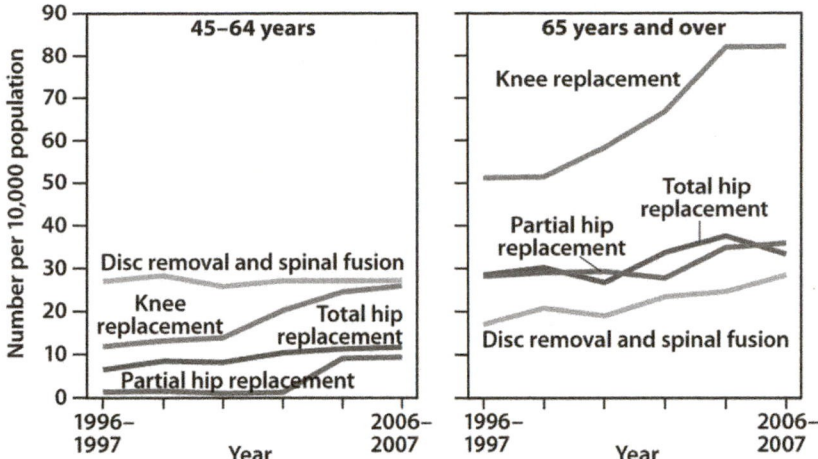

FIGURE 5.40   Back and joint procedures (per 100,000 population), USA, 1997–2007. *Source: Centers for Disease Control and Prevention. Health, United States, 2010: with special feature on death and dying. Hyattsville, MD: US DHHS, NCHS; February 2011. Available at: http://www.ncbi.nlm.nih. gov/books/NBK54380/#morbidity.s3 and http://www.ncbi.nlm.nih.gov/books/NBK54380 [Accessed 21 October 2012].*

Osteoporosis

Osteoporosis is a bone disorder resulting from a reduction of bone tissue density due to mineral and vitamin deficiency (calcium, vitamin D, and fluoride). The deficiency leads to weakening of the skeleton in older adults, thus causing fractures, most commonly of the spinal cord, hip, and wrists. It is a major cause of disability and death, as well as costly institutional care. It appears to be more common in northern countries and in cultural groups where women are completely covered for religious reasons and have darker skin coloring, reducing sun exposure and vitamin D production.

According to the US National Osteoporosis Foundation, osteoporosis is a major health threat for approximately 48 million Americans over 50 years of age and 10.7 million will have clinical osteoporosis by 2020. By 2030, the numbers will reach 11.9 million with clinical disease and 64.3 million with low bone mass. Osteoporosis results in more than 2 million fractures annually and is predicted to rise to more than 3 million by 2025, with an associated cost of US$25.3 billion in the USA (National Osteoporosis Foundation, April 2013). National surveys in the USA show adequate levels of folic acid, and vitamins A and D in most of the population, likely to be related to fortification of basic foods, but low vitamin D levels in black Americans, and low iodine levels as well (CFC, 2012).

Postmenopausal women are the primary group at risk, owing to low bone mass. The lifetime risk of a woman suffering an osteoporosis-related fracture is about 40 percent, and the chance of developing a hip fracture is equivalent to her combined risk of developing cancer of the breast, uterus, or ovary combined (US National Osteoporosis Foundation: http://www.nof.org/). The number of osteoporotic women is estimated to have been 3.8 million in 2005 and is projected to increase to 5.3 million by 2025.

Screening of those at risk by X-ray assessment of bone density is now being replaced by a relatively inexpensive, portable ultrasound instrument for use in ambulatory clinics. Postmenopausal women are at high risk for vitamin D deficiency and osteoporosis with factures of hip, spine, and other bones.

Age, gender, genetics, lifestyle (especially nutrition), and menopausal status are the major risk factors for osteoporosis. Other important factors include lack of adequate calcium and vitamin D intake, lack of appropriate exercise, smoking, and excessive alcohol intake. Fracture of the hip remains a serious threat to life despite improvements in surgical management. Osteoporosis management includes lifestyle issues as well as new advances in medical therapeutics. Osteoporosis is amenable to primary prevention and reducing risk factors. Primary prevention is directed towards adolescent, young adult, and perimenopausal women to ensure adequate physical activity, avoidance of smoking, adequate dietary intake of calcium and vitamin D, reduction of excess alcohol consumption, and prevention of falls. For postmenopausal women, prevention should also include home safety measures, bone density screening, and hormone replacement therapy to inhibit bone resorption.

In countries that do not mandate fortification of milk with vitamin D, as required for enriched milk in the USA, Canada, and other countries, vitamin D levels may be low in many parts of the population. Fortification is an essential public health measure, but not sufficient to prevent deficiency. This is important for children and, as a result, the American Academy of Pediatrics recommends supplements for children until they reach adolescence. This recommendation is associated with the shift in child play practices, from playing sports outdoors to using computers inside the home.

Vitamin D adequacy is increasingly linked to prevention of CVD and cancers such as breast and colorectal cancer. Fortification of a basic food with vitamin D, along with other fortificants, is a vital public health measure and should be augmented by vitamin D supplements for children, adolescents, adults, and elderly people, especially if confined to institutional care. Promotion of moderate sun exposure daily for 20–30 minutes should also be recommended.

Degenerative Osteoarthritis

Osteoarthritis is an important and growing public health problem. It is a degenerative disorder, and similar to many of the other conditions mentioned, it increases in prevalence as a population ages. Degenerative osteoarthritis is especially common in the knees in women and the hips in men. It is strongly correlated with both obesity and increasing age. CDC's analysis of NHIS data from 2007–2009 reported that 22.2 percent of adults (49.9 million) aged 18 years or older had self-reported doctor-diagnosed arthritis, and 9.4 percent (21.1 million or 42.4 percent of those with arthritis) had arthritis-attributable activity limitation (AAAL). This places a heavy burden on the economy owing to medical costs and reduced productivity (MMWR, 2010).

Reduction of obesity is the major preventive modality, requiring personal counseling and a climate of community attitudes that promotes healthy and sensible weight reduction. Treatment focuses on relief of symptoms and increasing mobility. Surgical replacement of the hip or knee followed by physiotherapy and other rehabilitation measures are well-established procedures to prevent deterioration and dependency and improve quality of life.

Rheumatoid Arthritis and Gout

Rheumatoid arthritis is an autoimmune disease causing chronic inflammation of joints with stiffness, pain, deformity, and limitations of ADLs, affecting as many as 1 percent of adults. It is two to three times more common in women than men, and can also occur in children. Rheumatoid arthritis is a generalized disease primarily affecting the joints but also other body systems, including the respiratory and gastrointestinal tracts. Furthermore, it is associated with excess mortality from infectious diseases and lymphomas. Supportive medical and other care should include immunization (e.g., influenza and pneumococcal pneumonia) as well as treatment of the primary condition, mainly with anti-inflammatory agents.

Gout is a metabolic disorder, causing deposition of uric acid crystals in and around joints, especially those in the foot. Gout is also associated with high lead exposure in certain occupational groups, such as painters, plumbers, and ship builders. National health survey data in the USA report a prevalence rate of 6 and 13 per 1000 for women and men,

respectively. In the USA, there were an estimated 2.6 million cases in 2008 and this number is estimated to rise to 3.6 million in 2025. Reduction of lead exposure and early diagnosis and management of gout improve outcome and quality of life. Follow-up care is essential in management.

Low Back and Neck Pain Syndromes

Low back and neck pain from muscle injury, and abnormalities of the vertebrae, discs, or joints, with or without compression of the lumbar nerves (sciatica) are both problematic and painful. It is estimated that between 12 percent and 15 percent of the US population will visit their physician with a complaint of back pain in one year, and this rate has seen a steady increase. In 2006, more than 44.4 million patients visited a physician with a complaint of back pain. The estimated number of sufferers from low back pain is projected to be 48.6 million in 2025.

In 2004, it was estimated that annual costs to the US economy of US$30.3 billion were directly related to spinal pain. Reports based on the US NHANES studies (2002 survey) indicate prevalence rates of back and/or neck pain among adults of 31 percent (low back pain 34 million, neck pain 9 million, both back and neck pain 19 million). Adults with low back and/or neck pain reported more comorbid conditions and more psychological distress (including mental illness), and engaged in more risky health behaviors than pain-free comparison adults.

Back and neck pain is especially common among industrial workers, equally for men and women and mainly in the 20–40-year age group. About 37 percent of back pain is due to occupational factors, a major cause of absence from work. Recovery is usual (90 percent), but chronic states often lead to surgical interventions which are expensive and often inappropriate. Prevention involves industrial engineering and education to reduce back strain, especially in heavy lifting work (see Chapter 9).

NEUROLOGICAL DISORDERS

Neurological disorders account for 1434.3 DALYs per 100,000 population or 6.3 percent of the global disease burden, ranging from 1150.1 to 1514.3 by World Bank income category. Among neurological disorders, cerebrovascular disease constituted nearly half of the total DALYs in 2005; it is projected to increase slightly by 2030. The percentage of total DALYs attributable to Alzheimer's and other dementias is estimated to increase from 0.75 percent to 1.2 percent by 2030. Higher rates occur among Caucasians in Europe and North America, with lower rates for Asians in China and Japan.

The WHO reports that neurological disorders contribute to 10.9 percent, 6.7 percent, 8.7 percent, and 4.5 percent of the global burden of disease in high, upper middle, lower middle, and low-income countries. In 2005, DALYs per 100,000 population for neurological disorders were the highest for lower and low-income countries (1514 and 1448, respectively).

Neurological disorders are an important burden on the affected individual and society in terms of disability, loss of productivity, premature mortality, and health costs. According to the National Institute of Neurological Disorders and Stroke, neurological disorders strike an estimated 50 million Americans each year, exacting an incalculable personal toll and an annual economic cost of hundreds of billions of dollars in medical expenses and lost productivity. This group of diseases includes stroke, epilepsy, multiple sclerosis (MS), Parkinson's disease, Alzheimer's disease, other neurodegenerative disorders, and autism.

Causes of neurological disorders include acute and chronic trauma to the brain and peripheral nerves, infection, and chemical poisoning. Projected changes in worldwide prevalence of neurological disorders are shown in Table 5.14. The burden of these disabling chronic conditions with increasing lifespan presents a multiple challenge to public health systems and costs. New innovations and scientific findings may present methods of reducing their severity and thus restoring better quality of life. The health system will need to adapt to these issues. As an example, nutrition-based disabilities may be reduced as food security improves and as fortification of basic foods becomes more widespread (see Chapter 8).

TABLE 5.14 Global Prevalence of Selected Neurological Disorders: WHO, per 1000 Population, Projected Rates for 2005, 2015, and 2030

	2005	2015	2030
Population (billions)	6.442	7.103	7.917
Disorder	Rate per 1000		
Epilepsy	6.2	6.3	6.4
Alzheimer's and other dementias	3.8	4.4	5.6
Parkinson's	0.8	0.8	0.9
Multiple sclerosis	0.4	0.4	0.4
Migraine	50.7	51.3	52.2
Cerebrovascular disease	9.6	9.5	9.7
Neurological infections	2.3	2.2	1.7
Nutritional and neuropathies	54.7	45.3	36.0
Neurological injuries	26.5	27.8	30.7

Source: World Health Organization. Neurological disorders: public health challenges. Geneva: WHO; 2006. Available at: http://www.who.int/mental_health/neurology/neurological_disorders_report_web.pdf [Accessed 24 October 2012].

Alzheimer's/dementia and cerebrovascular disease death rates are estimated to increase by 28.4 percent and 10.3 percent, respectively, by 2030. Stroke mortality is declining rapidly in the high-income countries, but remains high in former Soviet and in low-income countries.

Alzheimer's Disease

About 13 percent of people older than 65 years (and 43 percent of those over 85) are affected by dementia, with an estimated 5.4 million cases in the USA. The WHO estimates an annual incidence of 7.7 million new cases and 35–36 million people living with dementia. Alzheimer's disease is a brain disorder occurring later in life, possibly related to a genetic disorder. It is the leading cause (60–80 percent) of dementia among adults. It is estimated that a further 220–640,000 people under age 65 have Alzheimer's. It usually occurs after the age of 50 years, and more commonly in women than in men (1.6:1). There is no primary prevention as yet, but there is some evidence of benefit from hormone replacement and vitamin supplement therapy. Worldwide estimated costs of dementia reached US$604 billion in 2010. In high-income countries informal care accounts for 45 percent of the costs. Nearly 40 percent of those with dementia live in middle-income countries.

The Institute of Medicine reports that there are about 15 million caregivers providing care valued at US$202 billion in the USA, with family caregivers providing 80 percent of in-home care. Case management requires support for the family caregivers from community health resources. Other dementias, or organic brain syndromes, are due to cerebrovascular disease, Parkinson's disease, CJD, and AIDS. The WHO estimates that most of the 24 million people affected by dementia are elderly, and up to two-thirds of those with dementia live in low- and middle-income countries.

Parkinson's Disease

Parkinson's disease is a progressive neurological disorder that results from degeneration of neurons in a region of the brain that controls movement. It is common in men after the age of 60, more than in women, with a characteristic tremor, stiff walking gait, slowness of movement, and muscular rigidity. Age-adjusted rates are between 9.7 and 13.8 per 100,000, with higher rates for Caucasians in Europe and North America, lowest rates for black Africans, and intermediate rates for Asians in Japan and China. Incidence rates vary from 16 to 19 per 100,000.

In the USA, an estimated 500,000 people are believed to suffer from Parkinson's disease with about 50,000 new cases reported annually (NINDS). These figures are expected to increase as the average age of the population increases; the average age of onset is about 60. The disorder appears to be slightly more common in men than women. Genetic susceptibility is suspected. Treatment aimed at improving functional status is important along with support for caregivers, again to promote maximum independent function and avoid institutionalization. As the muscular rigidity affects chest muscles, respiratory infections are common, and therefore immunizations for influenza and pneumonia are valuable.

Multiple Sclerosis

Multiple sclerosis (MS) is a disorder of the myelin sheath of neurons, leading to impairment of vision, motor weakness, tremor, loss of coordination, dysfunction, and loss of sensation and bladder and bowel control. It most commonly occurs between the ages of 20 and 50, affecting more women than men (2:1 ratio) and is primarily found in areas distant from the equator, for example, Canada, northern Europe, and Australasia.

The WHO 2008 Atlas of MS reports that the global median estimated prevalence of MS is 30 per 100,000 population (with a range of 5–80). The highest rates were in Europe (80 per 100,000), followed by the Eastern Mediterranean (14.9), the Americas (8.3), the Western Pacific (5), South-East Asia (2.8), and Africa (0.3). The prevalence rates are highest in high-income countries. There are an estimated 1.1–2.5 million cases worldwide: the total estimated number of people reported with a diagnosis of MS is approximately 1.3 million. The cases are distributed globally as follows: 630,000 in Europe, 520,000 in the Americas, 66,000 in the Eastern Mediterranean, 56,000 in the Western Pacific, 31,500 in South-East Asia, and 11,000 in Africa.

There are several guidelines for diagnosis and staging of MS. The McDonald criteria are the diagnostic criteria typically used in most countries, followed by the Poser criteria.

Despite possible reporting biases, there are strong geographic patterns with frequency of MS variation, i.e., MS rates increasing with greater distance from the equator in both hemispheres. There are many theories as to the cause of the disease, with supportive evidence that it may be genetic, but other evidence points to the possibilities of an infectious agent origin and vitamin D deficiency due to poor exposure to UV sunlight. Medical care has improved with new medications, and it is hoped that continuing research into this disease will provide answers and effective prevention in the coming decades.

Epilepsy or Seizures

Epilepsy is characterized by uncontrollable convulsions starting abruptly, with or without warning symptoms, and with or without loss of consciousness. These are due to disturbances of cerebral function resulting from abnormal electrical activity in the brain. Isolated seizures can occur in anyone with brain hypoxia or hypoglycemia, and in children with fevers.

The WHO estimates that epilepsy affects one in every 130 people, or 50 million people worldwide, 90 percent of whom are in developing countries, with some 2 million new cases per year. The WHO estimates that the number of individuals living with active epilepsy (receiving or needing treatment) is 4–10 per 1000 population. Genetic factors, infections, and brain injuries are among the major causes of epilepsy. These include infection and toxicity in the prenatal period (e.g., maternal cocaine use), asphyxia and trauma during birth, postnatal infections causing febrile convulsions, infections of the central nervous system (e.g., meningitis and encephalitis), parasitic disease (e.g., malaria, schistosomiasis), and brain damage by alcohol, as well as trauma and toxic substances (e.g., lead, pesticides). Prevention of epilepsy is an important reason for good quality prenatal care, safe delivery, complete infant and childhood immunization, control of fever in children, control of infectious and parasitic diseases, reduction of brain injury (e.g., home and automobile accidents), and genetic counseling.

Brain and Spinal Cord Injury

About 1.7 million Americans sustain a traumatic brain injury (TBI) each year. TBIs are caused by a bump, blow, or jolt to the head or a penetrating head injury that disrupts the normal function of the brain. Of those injuries, about 75 percent are concussions or other forms of mild TBI; although termed mild, those injuries may exhibit long-term and even permanent effects.

TBI is associated with 52,000 deaths, 275,000 hospitalizations, and 1.3 million emergency room visits annually. There are an estimated 80,000 long-term or lifelong disability cases each year in the USA. The economic impact, including direct and indirect costs, is estimated to be over US$56 billion. Head injury causes prolonged hospitalization and often irreversible brain damage. The main causes in the USA are automobile injuries (46 percent), falls (22 percent), violence (16 percent), and sports or recreation injuries (12 percent).

If the injury is at the cervical spine level it causes quadriplegia and if lower in the spine, paraplegia. This condition is most common among young adults aged 15–35, often related to alcohol and motor vehicle crashes. There are about 200,000 cases living in the USA, with an estimate of 12,000–20,000 new cases per year. Each requires extensive hospital and rehabilitation services. Estimated lifetime cost per year is US$15,000, with lifetime costs of US$500,000 to more than US$3 million, depending on injury severity.

Brain and spinal cord injuries are prevented by safer motor vehicles with mandatory seat belts, air bags, helmets for bicycle (adult and child) riders, and full helmets for motorcycle drivers and riders. Work and home environments contain hazards that can result in serious falls and other injuries resulting in spinal cord damage. Helmets for football and hockey players, and other sports with frequent body contact, can reduce the incidence and seriousness of head trauma.

Safe transportation of the injured and strong management in emergency departments of hospitals are essential to reduce the extent of brain damage from the trauma. Advances in medical treatment have improved survival and recovery, but the damage to the patient's life can be severe and long-lasting. Rehabilitation plays a major role in returning the patient to an active lifestyle with promotion of employment, sports, and other normal life activities within the limits of the condition.

VISUAL DISORDERS

Blindness is defined as visual impairment sufficient to prevent an individual from performing work for which sight is essential. Vision loss is increasing owing to population growth and increased life expectancy. The WHO estimates global visual impairment at 285 million people and blindness at 39 million (Table 5.15). The vast majority (90 percent) of blind people live in developing countries where blindness prevalence rates commonly exceed 1–2 percent of the population.

Globally, the major causes of blindness are cataract (39 percent), uncorrected refractive errors (18 percent), glaucoma (10 percent), age-related macular degeneration (7 percent), corneal opacity (4 percent), diabetic retinopathy (4 percent), trachoma (3 percent), eye conditions in children (3 percent), and onchocerciasis (0.7 percent). Visual impairment has decreased over the past 20 years, largely as a result of the reduction in infectious eye diseases. Onchocerciasis (river blindness) is responsible for 1 million blind inhabitants in Africa and a smaller number in Latin America, but is responding to improved control measures, with important economic as well as health benefits in many countries in West Africa. Ghana (2010) and Morocco (2007) have reported elimination of trachoma.

TABLE 5.15 Blind and Visually Impaired People (Millions) by WHO Region

WHO Region	Impaired Vision	Blind
African	26.3	5.9
American	26.6	3.2
Eastern Mediterranean	23.5	4.9
European	28.2	2
South-East Asian	90.5	12
Western Pacific	90.2	10.6

Source: World Health Organization. Prevention of blindness and visual impairment. Geneva: WHO. Available at: http://www.who.int/blindness/table/en/index.html [Accessed 24 October 2012].

Cataracts are the principal cause of blindness in low- and middle-income countries. Each year 2.5 million older women go blind, with much of the disability burden avoidable if they have access to treatment such as surgery for the cataracts. In low-income countries trachoma is a significant and preventable cause of blindness affecting women more than men.

Vitamin A deficiency is suffered by 190 million preschool children globally, according to WHO estimates. It is a common cause of visual impairment in children under the age of 5, especially in developing countries, causing blindness in 250,000–500,000 children per year. Half of these children die within 12 months of becoming blind, and visual impairment occurs in millions more. International efforts are being made to prevent this by vitamin A supplements in conjunction with other public health interventions such as sick child visits, routine immunizations or special vaccination campaigns such as for poliomyelitis eradication or measles control. These supplements are estimated to have prevented 1.25 million child deaths and countless cases of blindness in 40 countries since 1998.

Untreated gonorrhea, syphilis, measles, cataracts, and glaucoma are also important causes of blindness in developing countries. Trachoma, which when untreated leads to marked conjunctivitis and eyelid deformities causing conjunctival abrasions, can lead to blindness. This disease is widespread in the Middle East, Africa, and some parts of Latin America. It is a disease of poverty, crowded living conditions, and lack of sanitation. A high percentage of cases of blindness are completely preventable by basic public health measures. Simple, inexpensive treatments are cost-effective and readily applied on a wide-scale basis where there are planned governmental or NGO programs.

The World Health Assembly of the WHO adopted a resolution endorsing the Action Plan for 2009–2013, focused on the prevention of avoidable blindness and visual impairment. Strategies were based on core elements of strengthening disease control, human workforce development, and infrastructure and technology.

The action plan called for increased political, financial, and technical commitment by member states, improved national and international cooperation, research, and monitoring of progress. Preventing avoidable visual impairment requires strengthening primary eye care services, developing skilled human resources, and technology and infrastructure development.

In the USA, blindness is defined as "the best-corrected visual acuity of 6/60 or worse (=20/200) in the better-seeing eye. Low vision is defined as the best-corrected visual acuity less than 6/12 (<20/40) in the better-seeing eye" (National Eye Institute, NIH). The prevalence of cataracts is 17.2 percent of the US population, with 42.8 percent among populations aged 70–79. Eye problems for people over 65 years of age include acute macular degeneration (AMD)

(47 percent), cataracts (20 percent), glaucoma (12 percent), diabetic retinopathy (2 percent), and others (19 percent). A study in England showed male rates increasing from 3 per 100,000 population in the 15–29 year age group to over 400 per 100,000 in the 75 and over group. For females, the rate increased from 2 to 475 per 100,000 in these same age groups. In 2003, 214,000 AMD cases of blindness were reported, and between 172,000 and 245,000 people have other forms of AMD. In England, 16,000 people annually become legally blind from AMD. The estimate is 394,000 women and 285,000 men with late AMD by 2020.

Prevention of blindness requires careful treatment of diabetes, screening of and treatment for glaucoma, cataract removal, and care of eyes using sunglasses in high sunlight areas. Public health measures include the use of safety glasses, shatterproof windows, seat belts and air bags in cars, prevention of infectious causes (STIs, measles, rubella), early treatment of eye diseases, and proper control of oxygen in incubators to prevent congenital retinal atrophy.

Blindness and visual impairment are among the 10 most common causes of disability in the USA and are associated with shorter life expectancy and lower quality of life. Prevention of blindness is a target for Healthy People 2010, as summarized in Box 5.18.

HEARING DISORDERS

Hearing loss is an important disabling condition. Those who are deaf without speech can learn to communicate by hand and finger signs or writing. Those with minimal hearing may learn to lip-read and to speak. Hard-of-hearing people have some useful hearing but require supplemental lip-reading. The psychological stress of deafness on the individual, family, and the community should be considered in developing prevention programs. Detection and correction of hearing loss can have a profound effect on an individual's well-being. Training manuals have been developed for health workers to aid in reducing the burden of deafness in developing countries.

The WHO estimates that worldwide in 2005, about 278 million people had moderate to profound hearing impairment, 80 percent them residing in low- and middle-income countries. Half of all cases of deafness and hearing impairment are avoidable through prevention, early diagnosis, and management. In developing countries, fewer than one out of 40 people who need a hearing aid actually have one. Despite representing one of the most common disabilities and a major factor in the global burden of disease, deafness and hearing impairment have been neglected in the public health field.

Screening infants for hearing loss is an accepted and integral part of well child care. They should be screened for hearing ability before the age of 3 months, with appropriate

---

BOX 5.18 CDC Strategic Approach to Assess, Evaluate, and Act in Prevention of Blindness

1. Define prevention of blindness as a national health target.
2. Set intervening subtargets such as regular eye care for all diabetics.
3. Mobilize key national partners, including NGOs, for collaboration with state and local health authorities.
4. Develop a national plan of action and evaluation.
5. Establish baseline and continuing monitoring of blindness.
6. Identify and alleviate health disparities in access to and use of needed services.
7. Focus interventions on high-risk populations, seeking untreated diabetics and hypertensives with follow-up referral and care.
8. Identify population-based health initiatives for education, screening, and care (e.g., infants and children).
9. Inform and educate vulnerable groups regarding self-care and medical care needed.
10. Collaborate with professional and community groups to promote common program.
11. Develop efforts to improve awareness and competency of providers and at-risk persons to provide and utilize needed care.
12. Apply public health research methods to follow up on prevalence and new innovations in prevention and care of visual disability.

Sources: *Centers for Disease Control and Prevention. Improving the nation's vision health: a coordinated public health approach. Atlanta, GA: CDC; 2006. Available at: http://www.cdc.gov/visionhealth/pdf/improving_nations_vision_health.pdf [Accessed 24 October 2012].*
Prevent Blindness America. A summary report of a five-year cooperative agreement with the vision health initiative of the Centers for Disease Control and Prevention (2003–2008). Chicago, IL: Prevent Blindness America; August 2009. Available at: http://www.cdc.gov/visionhealth/pdf/pba_complete.pdf [Accessed 24 October 2012].

---

follow-up management of treatment and education if necessary. Hearing impairment in children can delay the development of language and cognitive skills, thus hindering progress in school. In adults, hearing impairment often makes it difficult to obtain, perform, and sustain jobs. Hearing-impaired children and adults are often stigmatized and socially isolated.

In the USA, 10 million people have noise-induced hearing loss; another 20 million are exposed to hazardous noise levels in their place of employment. About 16 percent of adult-onset hearing loss worldwide is attributable to occupational noise exposure. A global data bank is under development by the WHO to track deafness and hearing impairment worldwide. Prevention programs can reduce the burden of this problem.

Hearing loss may be conductive or sensory due to a neurological defect. Conductive loss is due to obstruction in the ear canal or in the middle ear. These cases can be treated mechanically or surgically. Neurosensory hearing loss is caused by damage to specialized hearing cells in the inner ear due to aging, noise trauma, infection (e.g., measles or mumps), birth defects, metabolic disorders, autoimmune disorders, side-effects of medications, or unknown causes. Some causes of hearing loss can be prevented by adequate vaccination of children and limiting the use of medications that can cause hearing loss to situations in which there are no sufficient alternatives.

Noise control, especially in the workplace, is important in the prevention of hearing loss. Noise measurement in occupational settings includes decibel levels, frequency of sound waves (cycles per second or hertz), and loudness as perceived by the listener, and time or duration of exposure. Community noise levels of aircraft near airports, motor vehicle traffic, gardening equipment, and rock music are difficult to control, but should be considered in urban planning requirements. Preventive programs include modifying machinery, erecting sound barriers, and using protective ear devices. Public health programs should be implemented in schools with the use of mobile hearing units as well as education in methods of reducing ear damage from excess noise.

NON-COMMUNICABLE DISEASES AND THE NEW PUBLIC HEALTH

The burden of chronic conditions is an important factor in the health status of increasing aging populations, with many implications for the needs of health promotion, prevention, and health service systems. Health promotion, and primary, secondary, and tertiary levels of prevention make up a network of clinical, public health practices in a widely contextual New Public Health. They are an integral part of a framework in which social action provides health access and outreach to special needs groups for the individual benefit and well-being of society. The linkage between clinical care and public health is vital, one that requires strengthening of incentives and administrative changes, and one that will foster wide implementation of existing methods of care to sustain the chronically ill in their own homes and delay institutional care. Economics are an inevitable element of health policy, and the search for cost-effective ways to prevent disease and disability and to care for patients is central to progressive public health policy.

Chronic conditions often lead to acute crises resulting in hospitalization, or long-term dependency care or death. Improved preventive care can alleviate or defer crises exacerbating a long-standing disease process. The health care system has a responsibility to prevent these acute events to give those with chronic diseases good quality of life. Likewise, the health care system must avoid filling hospital intensive care units with patients whose health stability was upset by a medical or social crisis that was not addressed early enough. The onus for prevention and management falls on all components of a health system. Achieving control of the NCD pandemic requires coordinated and concerted efforts and sustained leadership, particularly by international agencies and organizations, national and subnational governments, and at the community level with medical practices and public health initiatives. The international level of activity, as exemplified in the WHO Framework Convention on Tobacco Control, is vital to reducing the terrible burden of NCDs now and in the coming decades for developing and middle-income countries.

The burden of chronic disease falls heavily on all, but especially on the poor in industrialized countries and in low-income countries. Obesity, diabetes, CVD, and cancer all have socioeconomic contributory causes related to poor nutrition and self-care. Even in countries with universal access systems, there are gradients in utilization, especially of preventive care. It may be more complex to prevent CVD, accidents, or diabetic complications than to treat their results, but the benefits to the individual and the community are far greater through prevention. The New Public Health includes health promotion and care of the ill in a context of limited resources and advancing medical technology, preserving individual dignity and rights, and ethical concerns.

INTEGRATED NON-COMMUNICABLE DISEASE PREVENTION AND CONTROL

Globally, the world's economies face a staggering risk of an estimated US$47 trillion in the next few decades. By 2015, the WHO projects economic losses from preventable NCDs to reach US$558 billion in China, US$303 billion in Russia, US$337 billion in India, US$49 billion in Brazil, and US$2.5 billion in Tanzania. Up to 80 percent of heart disease, stroke, and type 2 diabetes could be eliminated by reducing their common, underlying risk factors: smoking, obesity, lack of exercise, and lack of diabetes and hypertension control. A reduction of 36 million premature deaths (17 million in the under 70 age group) would reverse the tide and result in accumulated economic growth of US$26 billion in China, $15 billion in India, and US$20 billion in the Russian Federation.

The rise and spread of NCDs as the major health challenge of the twenty-first century was highlighted by the United Nations first global conference on healthy living and NCD control, held in Moscow in April 2011 (Box 5.19). Much of the excess mortality is caused by NCDs and injuries. The potential for global and national interventions that have shown some measure of success has emerged from the realization by government, public health, and people-based community social action that both the social components of causality in NCDs and the biological risk factors are major factors in the success of tobacco control programs. Poverty, unemployment, poor diet, smoking, and obesity are socially related and direct causative factors for NCDs.

BOX 5.19 Healthy Lifestyles and Non-Communicable Disease Control: Rationale for Action

1. NCDs, principally cardiovascular diseases, diabetes, cancers, and chronic respiratory diseases, are the leading causes of preventable morbidity and disability, and currently cause over 60 percent of global deaths, 80 percent of which occur in developing countries. By 2030, NCDs are estimated to contribute to 75 percent of global deaths.

2. Other NCDs such as mental disorders also significantly contribute to the global disease burden.

3. NCDs have substantial negative impacts on human development and may impede progress towards the Millennium Development Goals (MDGs).

4. NCDs impact significantly on all levels of health services, health care costs, and the health workforce, as well as national productivity in both emerging and established economies.

5. Worldwide, NCDs are important causes of premature death, striking hard among the most vulnerable and poorest populations. Globally, they impact on the lives of billions of people and can have devastating financial impacts that impoverish individuals and their families, especially in low- and middle-income countries.

6. NCDs can affect women and men differently, hence prevention and control of NCDs should take gender into account.

7. Many countries are now facing extraordinary challenges from the double burden of disease: communicable disease and non-communicable diseases. This requires adapting health systems and health policies, and a shift from disease-centered to people-centered approaches and population health measures. Vertical initiatives are insufficient to meet complex population needs. Integrated solutions that engage a range of disciplines and sectors are needed.

Strengthening health systems can improve capacity to respond to a range of diseases and conditions.

8. Evidence-based and cost-effective interventions exist to prevent and control NCDs at global, regional, national, and local levels. These interventions could have profound health, social, and economic benefits throughout the world.

9. Examples of cost-effective interventions to reduce the risk of NCDs, which are affordable in low-income countries and could prevent millions of premature deaths every year, include measures to control tobacco use, reduce salt intake, and reduce the harmful use of alcohol.

10. Particular attention should be paid to the promotion of healthy diets (low consumption of saturated fats, trans-fats, salt, and sugar, and high consumption of fruits and vegetables) and physical activity in all aspects of daily living.

11. Effective NCD prevention and control require leadership and concerted "whole of government" action at all levels (national, subnational, and local) and across a number of sectors such as health, education, energy, agriculture, sports, transport and urban planning, environment, labor, industry, industry and trade, finance and economic development.

12. Effective NCD prevention and control require active and informed participation and leadership by individuals, families and communities, civil society organizations, private sector where appropriate, employees, health care providers, and the international civil society.

Source: *Moscow Declaration Preamble. First Global Ministerial Conference on Healthy Living and Non-Communicable Disease Control, 28–29 April 2011, Moscow, Russian Federation. Geneva: WHO. Available at: http://www.un.org/en/ga/president/65/issues/moscow_declaration_en.pdf [Accessed 24 October 2011].*

A WHO 2010 survey of 185 countries showed that 48 percent of the countries had population-based mortality data for NCDs, but only 23 percent had population-based morbidity data as part of their national reporting systems. Risk factor surveillance data were reported by 59 percent of countries. Fewer than half of the countries (43 percent) had population-based cancer registries and even fewer had birth defect registries. NCD plans often are not integrated into the overall health planning for program implementation, priorities, policies, guidelines, and standards. Operational integrated policies and dedicated funding for NCD control ranged from 30 percent of lower-middle-income and high-income countries to 22 percent in low-income and 43 percent of upper- to middle-income countries (WHO, 2010).

The role of international actors in policy and knowledge transfer, which has been vital to the progress made to date in industrialized countries, urgently needs to be extended to those nations not yet able to limit the activities of the tobacco industry in spreading its wares. The WHO Framework Convention on Tobacco Control has done much to internationalize the struggle to control tobacco use and reduce the huge cost in shortened lives that tobacco causes (see Chapter 2). This requires national and local participation and is one of the great achievements of health promotion in the past several decades. A second great achievement of globalization of integrated NCD prevention and control is in the massive reduction in mortality from CVD. Here, the efforts and credits involve the medical community in its role of providing patient care with recommended lifestyle changes, and promoting the use of effective and inexpensive medications such as aspirin, statins, and antihypertensive for those at risk. Public health and health promotion factors play a huge role in changing public perceptions of the risk factors and importance of self-care.

A healthy public policy strategy with an emphasis on assessing new policies is key to reversing the current situation. This should include national programs focusing on salt reduction in consumable food products, fortification of basic foods such as iodization of salt and folic acid in flour through legislative measures, and policies and programs to reduce sedentary behaviors and increase physical activity.

The basis for integrated prevention and control of NCDs is to foster a comprehensive approach for preventable burden of disease. Globally, high blood pressure accounts for 13 percent of total deaths, followed by tobacco use (9 percent), high glucose levels (6 percent), physical inactivity (6 percent), and overweight/obesity (5 percent). Urbanization, global trade, population aging, and inadequate health and social systems are also influencing factors.

SUMMARY

Chronic conditions are major public health challenges in most industrialized countries, and are rapidly becoming problematic in developing countries as well. CVD and cancer are the major causes of death in most western countries, but the leading cause of years of potential life lost is trauma, as it predominantly affects young people. Increasing longevity, improved nutrition, social support, and medical care are creating an increasingly elderly population living longer and more healthily than previous generations. The public health challenge is to promote healthy middle-aged and elderly populations by reducing risk factors through health promotion and effective medical care.

NCDs are the most common causes of death and disability in the high-, medium-, and low-income countries. They are associated with many risk factors and contributing causes. Some are caused by specific infections. Most are related to poverty and lifestyle, and are responsive to medical and public health interventions through primary, secondary, and tertiary prevention.

The effective combination of medical and public health interventions has been demonstrated in powerful ways in the reduction of mortality from CVD, cancer, and other causes. The complex of applications available in the New Public Health makes the area of NCDs a prime candidate for concentration while attending to the other major public health issues simultaneously. Health promotion strategies have been shown to be extremely effective in reducing risk factors, just as biomedical interventions such as immunization and infection management have contributed to reductions in stomach and liver cancers, HIV prevalence, and CVD.

The key issues are national, social policy and health targets, creating a healthy living environment, access to quality foods, essential supplements, adequate physical activity, moderate alcohol use, and moderate sun exposure. However, prevention requires close coordination with development of adequate medical care for pregnancy, delivery, and follow-up care, screening for newborn diseases and growth patterns of children, screening for risk factors for NCDs, and proper management of long-term conditions such as hypertension, diabetes, and osteoporosis. Moreover, screening for cancers of the colon, cervix, and breast, attention to depression and other signs of mental distress, reaching out to groups with special risks for NCDs, economics of health care, and the obligation to reduce social inequalities are integral issues that must be addressed to improve the health of the public at large.

The New Public Health is a network of efforts combining biomedical and health promotion models needed for the future in meeting challenges to reduce morbidity and mortality, and to improve quality of life across the lifespan and around the globe. The New Public Health depends on working partnerships between public policy, health financing and organization, clinical standards, and organized public health programs to prevent chronic conditions, and to prevent or delay the onset of their complications. Cancer and trauma are also amenable to prevention. Dramatic reductions in mortality and morbidity from cerebrovascular and coronary heart diseases have been accomplished by this approach. Accordingly, the potential for prevention to increase the well-being of individuals affected by these conditions should be a central element of local, national, and global health policy.

NOTE

For a complete bibliography and guidance for student reviews and expected competencies please see companion web site at http://booksite.elsevier.com/9780124157668

BIBLIOGRAPHY

Recommended Reading

Bloom, B., Cohen, R.A., Freeman, G., December 2011. Summary health statistics for US children: National Health Interview Survey, 2010. US DHHS, NCHS, Hyattsville, MD. Available at: http://www.cdc.gov/nchs/data/series/sr_10/sr10_250.pdf (accessed 21.10.12).

Centers for Disease Control and Prevention, 1999. Achievements in public health, 1900–1999: Decline in deaths from heart disease and stroke – United States, 1900–1999. MMWR Morb. Mortal. Wkly. Rep. 48, 649–656.

Centers for Disease Control and Prevention. Chronic disease prevention and health promotion [updated 13 August 2012]. Atlanta, GA: CDC. Available at: http://www.cdc.gov/chronicdisease/overview/index.htm (accessed 21.10.12).

Centers for Disease Control and Prevention, 2011. Health United States, 2011, with special feature on socioeconomic status and health. US DHHS, NCHS, Hyattsville, MD. Available at: http://www.cdc.gov/nchs/hus.htm (accessed 21.10.12).

Centers for Disease Control and Prevention, National report on biochemical indicators of diet and nutrition in the US Population Second Nutrition Report 2012. Available at: http://www.cdc.gov/nutritionreport/ (accessed 14.8.13).

Centers for Disease Control and Prevention, July 2010. NCHS Data. NCHS, Hyattsville, MD. Available at: http://www.cdc.gov/nchs/data/dvs/MORTFINAL2007_WorkTable210R.pdf (accessed 29.9.12).

Centers for Disease Control and Prevention, 2011. Public health then and now: celebrating 50 years of MMWR at CDC. MMWR Morb. Mortal. Wkly. Rep. 60, 1–112. Available at: http://www.cdc.gov/mmwr/pdf/other/su6004.pdf (accessed 24.10.12).

Centers for Disease Control, 2013. CDC health disparities and inequalities report- United States. Morbid Mortal Wky Rep MMWR Supplement 3, 1–186.

Centers for Disease Control and Prevention, 2011. Ten great public health achievements – United States, 2001–2010. MMWR Morb. Mortal. Wkly. Rep. 60, 619–623. Available at: http://www.cdc.gov/mmwr/preview/mmwrhtml/mm6019a5.htm (accessed 8.8.13).

Centers for Disease Control and Prevention, 2010. The National Center for Health Statistics. US DHHS, NCHS, Hyattsville, MD. Available at: http://www.cdc.gov/nchs/data/about/nchs_50th_brochure.pdf (accessed 21.10.12).

Kochanek, K.D., Xu, J., Murphy, S.L., Minino, A., Kung, H.C., December 2011. National vital statistics reports. Deaths: final data for 2009. NCHS, Hyattsville, MD. Available at: http://www.cdc.gov/nchs/data/nvsr/nvsr60/nvsr60_03.pdf (accessed 21.10.12).

Moscow Declaration Preamble, April 2011. First Global Ministerial Conference on Healthy Living and Non-communicable Disease Control, 28–29. WHO, Moscow, Russian Federation. Geneva. Available at: http://www.un.org/en/ga/president/65/issues/moscow_declaration_en.pdf (accessed 24.10.11).

Murphy, B.S., Xu, J., Kochanek, K.D., January 2012. National vital statistics reports. Deaths: preliminary data for 2010. NCHS, Hyattsville, MD. Available at: http://www.cdc.gov/nchs/data/nvsr/nvsr60/nvsr60_04.pdf (accessed 21.10.12).

National Prevention Council, 2011. National prevention strategy, America's plan for better health and wellness. Office of the Surgeon General. US DHHS, Washington, DC. Available at: http://www.healthcare.gov/prevention/nphpphc/strategy/report.pdf (accessed 21.10.12).

US Department of Health and Human Services, November 2010. Healthy People 2020. US DHHS, Washington, DC. Available at: http://www.healthypeople.gov/2020/TopicsObjectives2020/pdfs/HP2020_brochure_with_LHI_508.pdf (accessed 22.10.12).

World Health Organization, June 2011. Action plan for implementation of the European strategy for the prevention and control of noncommunicable diseases 2012–2016. WHO Regional Office for Europe, Copenhagen. Available at: http://www.euro.who.int/__data/assets/pdf_file/0003/147729/wd12E_NCDs_111360_revision.pdf (accessed 26.10.12).

World Health Organization, Genes and human disease. Geneva: WHO. Available at: http://www.who.int/genomics/public/geneticdiseases/en/ (accessed 21.10.12).

World Health Organization, 2012. Global health observatory. Country statistics. WHO, Geneva. Available at: http://www.who.int/gho/countries/en/ (accessed 26.10.12).

World Health Organization, Human genetics areas of work genetics and common diseases. Geneva: WHO. Available at: http://www.who.int/genomics/about/commondiseases/en/index.html (accessed 21.10.12).

World Health Organization. The top 10 causes of death. Fact sheet no. 310 [updated June 2011]. Geneva: WHO. Available at: http://www.who.int/mediacentre/factsheets/fs310/en/index.html (accessed 21.10.12).

World Health Organization, 2013. World Health Statistics. . Available at: http://www.who.int/gho/publications/world_health_statistics/EN_WHS2013_Full.pdf. (accessed 08.08.13).

World Health Organization, January 2013. Regional Office for Europe. Health for All database (HFA-DB). WHO, Copenhagen. Available at: http://data.euro.who.int/hfadb/ (accessed 08.08.13).

Global Burden of Non-Communicable Diseases

Abegunde, D.O., Mathers, C.D., Adam, T., Ortegon, M., Strong, K., 2007. The burden and costs of chronic diseases in low-income and middle-income countries. Lancet 370, 1929–1938. Available at: http://www.ncbi.nlm.nih.gov/pubmed/18063029 (accessed 21.10.12).

Bodenheimer, T., Chen, E., Bennett, H.D., 2009. Confronting the growing burden of chronic disease: can the US health care workforce do the job? Health Aff. 28, 64–74. Available at: http://www.ncbi.nlm.nih.gov/pubmed/19124856 (accessed 22.10.12).

Bovet, P., Paccaud, F., 2012. Cardiovascular disease and the changing face of global public health: a focus on low and middle income countries. Public Health Rev. 33, 397–415. Available at: http://www.publichealthreviews.eu/upload/pdf_files/10/00_Bovet.pdf (accessed 21.10.12).

Institute of Medicine, 2010. Promoting cardiovascular health in the developing world: a critical challenge to achieve global health. National Academies Press, Washington, DC. Available at: http://www.ncbi.nlm.nih.gov/books/NBK45693/pdf/TOC.pdf (accessed 24.10.12).

Lopez, A.D., 2005. The evolution of the global burden of disease framework for disease, injury and risk factor quantification: developing the evidence base for national, regional and global public health action. Globalization and Health 1, 5. Available at: http://www.globalization-andhealth.com/content/1/1/5 (accessed 22.10.12).

Lopez, A.D., Mathers, C.D., Ezzati, M., Jamison, D.T., Murray, C.J., 2006. Global and regional burden of disease and risk factors, 2001: systematic analysis of population health data. Lancet 367, 1747–1757. Available at: http://www.ncbi.nlm.nih.gov/pubmed/16731270 (accessed 21.10.12).

Mathers, C.D., Loncar, D., 2006. Projections of global mortality and burden of disease from 2002 to 2030. PLoS Med. 3, e442. Available at: http://www.plosmedicine.org/article/citationList.action;jsessionid=DD3CB595FFE702CAD9B157C273CFC21D?articleURI=info%3Adoi%2F10.1371%2Fjournal.pmed.0030442 (accessed 21.10.12).

World Health Organization, 2012. Assessing national capacity for the prevention and control of noncommunicable diseases: report of the 2010 global survey. WHO, Geneva. Available at: http://www.who.int/cancer/publications/national_capacity_prevention_ncds.pdf (accessed 12.08.13).

World Health Organization, Change in rank order of disability adjusted life years (DALYs) for the 15 leading causes of death, worldwide, 1990–2020. Available at: http://www.wri.org/publication/content/8476 (accessed 21.10.12).

World Health Organization, 2005. Facing the facts #1: Chronic diseases and their common risk factors. WHO, Geneva. Available at: http://www.who.int/chp/chronic_disease_report/media/Factsheet1.pdf (accessed 21.10.12).

World Health Organization, 2009. Global health risks: mortality and burden of disease attributable to selected major risks. WHO, Geneva. Available at: http://www.who.int/healthinfo/global_burden_disease/GlobalHealthRisks_report_full.pdf (accessed 21.10.12).

World Health Organization, April 2011. Global status report on noncommunicable diseases 2010. WHO, Geneva. Available at: http://www.who.int/nmh/publications/ncd_report2010/en/ (accessed 21.10.12).

World Health Organization, 2011. Noncommunicable disease: country profiles 2011. WHO, Geneva. Available at: http://www.who.int/nmh/publications/ncd_profiles2011/en/ (accessed 21.10.12).

Social Determinants, Environment, and Economics of Non-Communicable Diseases

Asaria, P., Chisholm, D., Mathers, C., Ezzati, M., Beaglehole, R., 2007. Chronic disease prevention: health effects and financial costs of strategies to reduce salt intake and control tobacco use. Lancet 370, 2044–2053. Available at: http://www.ncbi.nlm.nih.gov/pubmed/18063027 (accessed 22.10.12).

Beaglehole, R., Ebrahim, S., Reddy, S., Voûte, J., Leeder, S., 2007. Chronic Disease Action Group. Prevention of chronic diseases: a call to action. Lancet 370, 2152–2157. Available at: http://www.ncbi.nlm.nih.gov/pubmed/18063026 (accessed 22.10.12).

Canadian Guide to Clinical Preventive Health Care [updated 16 August 2012]. Ottawa: Public Health Agency of Canada. Available at: http://www.phac-aspc.gc.ca/publicat/clinic-clinique/index-eng.php (accessed 22.10.12).

Centers for Disease Control and Prevention. Chronic diseases and health promotion [updated 13 August 2012]. Atlanta, GA: CDC. Available at: http://www.cdc.gov/chronicdisease/overview/index.htm (accessed 22.10.12).

Centers for Disease Control and Prevention. Health United States 2010: list of trend figures [updated 19 September 2011]. Atlanta, GA: CDC. Available at: http://www.cdc.gov/nchs/hus/contents2010.htm#chartbookfigures (accessed 22.10.12).

Cobiac, L.J., Magnus, A., Lim, S., Barendregt, J.J., Carter, R., Vos, T., 2012. Which interventions offer best value for money in primary prevention of cardiovascular disease? PLoS ONE 7, e41842. Available at: http://www.plosone.org/article/info%3Adoi%2F10.1371%2Fjournal.pone.0041842 (accessed 22.10.12).

Crosson, F., Madvig, P., 2004. Does population management of chronic disease lead to lower costs of care? Health Aff. 23, 76–78. Available at: http://www.ncbi.nlm.nih.gov/pubmed/15537587 (accessed 21.10.12).

Frank, J.W., Cohen, R., Yen, I., Balfour, J., Smith, M., 2003. Socioeconomic gradients in health status over 29 years of follow-up after mid-life: the Alameda county study. Soc. Sci. Med. 57, 2305–2323. Available at: http://www.ncbi.nlm.nih.gov/pubmed/14572839 (accessed 21.10.12).

Institute of Medicine, 2012. Living well with chronic illness: a call for public health action. National Academies Press, Washington, DC.

Lang, T., Lepage, B., Schieber, A.C., Lamy, S., Kelly-Irving, M., 2011. Social determinants of cardiovascular diseases. Public Health Rev. 33, 601–622. Available at: http://www.publichealthreviews.eu/upload/pdf_files/10/00_Lang.pdf (accessed 21.10.12).

Luepker, R.V., 2011. WHO MONICA project: what have we learned and where to go from here? Public Health Rev. 33, 373–396. Available at: http://www.publichealthreviews.eu/upload/pdf_files/10/00_Luepker.pdf (accessed 22.10.12).

Marmot, M., 2005. Social determinants of health inequalities. Lancet 365, 1099–1104. Available at: http://www.ncbi.nlm.nih.gov/pubmed?term=Marmot%20M.%20Social%20determinants%20of%20health%20inequalities.%20Lancet.%202005%3B%20365%3A1099-1104 (accessed 26.10.12).

Marmot, M.G., Bosma, H., Hemingway, H., Brunner, E., Stansfeld, S., 1997. Contribution of job control and other risk factors to social variations in coronary heart disease incidence. Lancet 350, 235–239. Available at: http://www.thelancet.com/journals/lancet/article/PIIS0140-6736(97)04244-X/abstract (accessed 22.10.12).

Tunstall-Pedoe, H., Connaghan, J., Woodward, M., Tolonen, H., Kuulasmaa, K.for the WHO MONICA project, 2006. Pattern of declining blood pressure across replicate population surveys of the WHO MONICA project, mid-1980s to mid-1990s, and the role of medication. BMJ 332, 629–635. Available at: http://www.bmj.com/content/332/7542/629 (accessed 22.10.12).

US Preventive Services Task Force, 2010. The guide to clinical preventive services – 2010–2011. Agency for Health Care Research and Quality, Rockville, MD. Available at: http://www.ncbi.nlm.nih.gov/books/NBK56707/ (accessed 22.10.12).

Risk Factors, Health Promotion, and Prevention of Non-Communicable Diseases

Bayer, R., Feldman, E., 2011. Tobacco control in industrialized nations: the limits of public health achievement. Public Health Rev. 33, 553–568. Available at: http://www.publichealthreviews.eu/upload/pdf_files/10/00_Bayer.pdf (accessed 22.10.12).

Bochud, M., Marques-Vidal, P., Burnier, M., Paccaud, F., 2012. Dietary salt intake and cardiovascular disease: summarizing the evidence. Public Health Rev. 33, 530–552. Available at: http://www.publichealthreviews.eu/upload/pdf_files/10/00_Bochud.pdf (accessed 22.10.12).

Brownstein, J.N., 2008. Addressing heart disease and stroke prevention through comprehensive population-level approaches. Prev. Chronic Dis. 5, 1–5. Available at: http://www.cdc.gov/pcd/issues/2008/apr/pdf/07_0251.pdf (accessed 22.10.12).

Centers for Disease Control and Prevention, 1999. Achievements in public health, 1900–1999. MMWR Morb. Mortal. Wkly. Rep. 48, 986–993. Available at: http://www.cdc.gov/mmwr/preview/mmwrhtml/mm4843a2.htm (accessed 22.10.12).

Centers for Disease Control and Prevention. Heart disease facts – America's heart disease burden [updated 16 October 2012]. Atlanta, GA: CDC. Available at: http://www.cdc.gov/HeartDisease/facts.htm (accessed 22.10.12).

Evans, A.S., 1976. Causation and disease: the Henle–Koch postulates revisited. Yale J. Biol. Med. 49, 175–195. Available at: http://www.ncbi.nlm.nih.gov/pmc/articles/PMC2595276/pdf/yjbm00143-0072.pdf (accessed 21.10.12).

Fuster, V., Kelly, B.B. (Eds.), 2010. Committee on Preventing the Global Epidemic of Cardiovascular Disease: meeting the challenges in developing countries; Institute of Medicine. Promoting cardiovascular health in the developing world: a critical challenge to achieve global health. National Academies Press, Washington, DC. Available at: http://www.nap.edu/catalog.php?record_id=12815. (accessed 22.10.12).

Hill, A.B., 1965. The environment and disease: association or causation? Proc. R. Soc. Med. 58, 295–300 PMC 1898525; PMID 14283879.

Institute of Medicine, 2012. Accelerating progress in obesity prevention: solving the weight of the Nation. National Academies Press, Washington, DC. Available at: http://www.nap.edu/catalog.php?record_id=13275 (accessed 22.10.12).

Institute of Medicine, 2001. Crossing the quality chasm: a new health system for the 21st century. National Academies Press, Washington, DC. Available at: http://www.nap.edu/openbook.php?record_id=10027&page=R1 (accessed 22.10.12).

Kromhout, D., 2012. Preface: a career in nutrition and cardiovascular disease: from research to results to public health policy. Public Health Rev. 33, 351–362. Available at: http://www.publichealthreviews.eu/upload/pdf_files/10/01_Kromhout.pdf (accessed 22.10.12).

Prospective Studies Collaboration, 2002. Age-specific relevance of usual blood pressure to vascular mortality: a meta-analysis of individual data for one million adults in 61 prospective studies. Lancet 360, 1903–1913. Available at: http://www.thelancet.com/journals/lancet/article/PIIS0140-6736(02)11911-8/fulltext (accessed 22.10.12).

Schober, S.E., Carroll, M.D., Lacher, D.A., Hirsch, R., 2007. High serum total cholesterol – an indicator for monitoring cholesterol lowering efforts; US adults, 2005–2006. NCHS, Hyattsville, MD. Available at: http://www.ncbi.nlm.nih.gov/pubmed/19389314. and http://www.cdc.gov/nchs/data/databriefs/db02.pdf (accessed 22.10.12).

World Health Organization, Joint effects of risk factors. Geneva: WHO. Available at: http://www.who.int/healthinfo/global_burden_disease/GlobalHealthRisks_report_part3.pdf (accessed 22.10.12).

World Health Organization, 2005. Preventing chronic diseases. a vital investment: WHO global report. WHO, Geneva. Available at: http://www.who.int/chp/chronic_disease_report/contents/en/index.html (accessed 22.10.12).

World Health Organization, 2009. Women and health, today's evidence tomorrow's agenda. WHO, Geneva. Available at: http://whqlibdoc.who.int/publications/2009/9789241563857_eng.pdf (accessed 22.10.12).

World Heart Federation, Cardiovascular disease risk factors. Geneva: World Heart Federation. Available at: http://www.world-heart federation.org/cardiovascular-health/cardiovascular-disease-risk-factors/stress/ (accessed 22.10.12).

Smoking

Centers for Disease Control and Prevention, 1999. Tobacco use – United States, 1900–1999. MMWR Morb. Mortal. Wkly. Rep. 48, 986–993. Available at: http://www.ncbi.nlm.nih.gov/pubmed/10577492 (accessed 22.10.12).

Centers for Disease Control and Prevention, 2010. Vital signs: current cigarette smoking among adults aged ≥ 18 years – United States, 2009. MMWR Morb. Mortal. Wkly. Rep. 59, 1135–1140. Available at: http://www.cdc.gov/mmwr/preview/mmwrhtml/mm5935a3.htm (accessed 22.10.12).

Doll, R., Peto, R., Boreham, J., Sutherland, I., 2004. Mortality in relation to smoking: 50 years' observations on male British doctors. BMJ 328, 1519–1523. Available at: http://www.bmj.com/content/328/7455/1519 (accessed 13.8.13).

Nikogosian, H.W.H.O., 2010. Framework convention on tobacco control: a key milestone. Bull. World Health Organ. 88 (2), 83.

Perry, C., Kelder, S.H., Murray, D.M., Klepp, D.I., 1992. Community-wide smoking prevention: long-term outcomes of the Minnesota Heart Health Program and the Class of 1989 Study. Am. J. Public Health 82, 1210–1216. Available at: http://www.ncbi.nlm.nih.gov/pmc/articles/PMC1694332/ (accessed 22.10.12).

Roemer, R., Taylor, A., Lariviere, J., 2005. Origins of the WHO Framework Convention on Tobacco Control. Am. J. Public Health 95, 936–938.

US Public Health Service, 1964. The Surgeon General's report on smoking and health 1964. US Department of Health, Education, and Welfare, Washington, DC.

World Health Organization, 2011. WHO report on the global tobacco epidemic, 2011: warning about the dangers of tobacco. WHO, Geneva. Available at: http://whqlibdoc.who.int/publications/2011/9789240687813_eng.pdf (accessed 22.10.12).

Diet, Physical Activity, and Obesity

Baker, J.L., Olsen-Lina, W., Sorensen, T.I.A., 2007. Childhood body-mass index and the risk of coronary heart disease in adulthood. N. Engl. J. Med. 357, 2329–2337. Available at: http://www.ncbi.nlm.nih.gov/pmc/articles/PMC3062903/ (accessed 22.10.12).

Bibbins-Domingo, K., Coxson, P., Pletcher, M., Lightwood, J., Goldman, L., 2007. Adolescent overweight and future adult coronary heart disease. N. Engl. J. Med. 357, 2371–2379. Available at: http://www.nejm.org/doi/full/10.1056/NEJMsa073166 (accessed 22.10.12).

Brownson, R.C., Ballew, P., Brown, K.L., Elliott, M.B., Haire-Joshu, D., Heath, G.W., et al., 2007. The effect of disseminating evidence-based interventions that promote physical activity to health departments. Am. J. Public Health 97, 1900–1907. Available at: http://www.ncbi.nlm.nih.gov/pubmed/17761575 (accessed 22.10.12).

Cecchini, M., Sassi, F., Lauer, J.A., Lee, Y.Y., Guajardo-Barron, V., Chisholm, D., 2010. Tackling of unhealthy diets, physical activity, and obesity: health effects and cost-effectiveness. Lancet 376, 1775–1784. Available at: http://www.thelancet.com/journals/lancet/article/PIIS0140-6736(10)61514-0/abstract (accessed 22.10.12).

Centers for Disease Control and Prevention, Second National Report on Biochemical Indicators of Diet and Nutrition in the US Population 2012. Available at: http://www.cdc.gov/nutritionreport/pdf/ExeSummary_Web_032612.pdf#zoom=100 (accessed 14.8.13).

Davey-Smith, G., Shipley, M.J., Batty, G., Morris, J.N., Marmot, M., 2000. Physical activity and cause-specific mortality in the Whitehall study. Public Health 114, 308–315. Available at: http://www.ncbi.nlm.nih.gov/pubmed/11035446 (accessed 22.10.12).

Dumanovsky, T., Huang, C.Y., Bassett, M.T., Silver, L.D., 2010. Consumer awareness of fast food calorie information in New York City after implementation of a menu labeling regulation. Am. J. Public Health 100, 2520–2525. Available at: http://www.ncbi.nlm.nih.gov/pubmed/20966367 (accessed 22.10.12).

Fox, M.K., Dodd, A.H., Wilson, A., Gleason, P.M., 2009. Association between school food environment and practices and body mass index of US public school children. J. Am. Diet. Assoc. 109, S108–S117. Available at: http://www.ncbi.nlm.nih.gov/pubmed/19166665 (accessed 22.10.12).

Glickman, D., Parker, L., Sim, L., Del Valle Cook, H., Miller, E., 2012. Accelerating progress in obesity prevention, solving the weight of the nation. National Academies Press, Washington, DC. Available at: http://books.nap.edu/openbook.php?record_id=13275 (accessed 22.10.12).

Guh, D.P., Zhang, W., Bansback, N., Amarsi, Z., Birmingham, C.L., Anis, A.H., 2009. The incidence of comorbidities related to obesity and overweight: a systematic review and meta-analysis. BMC Public Health 9, 88. Available at: http://www.biomedcentral.com/1471-2458/9/88 (accessed 22.10.12).

Institute of Medicine, 2012. Accelerating progress in obesity prevention: solving the weight of the nation. National Academies Press, Washington, DC. Available at: http://www.nap.edu/catalog.php?record_id=13275 (accessed 22.10.12).

Lovasi, G.S., Grady, S., Rundle, A., 2011. Steps forward: review and recommendations for research on walkability, physical activity and cardiovascular health. Public Health Rev. 33, 484–506. Available at: http://www.publichealthreviews.eu/upload/pdf_files/10/00_Lovasi.pdf (accessed 22.10.12).

National Osteoporosis Foundation, April 2013. NOF releases new data detailing the prevalence of osteoporosis. Available at: http://www.nof.org/news/1009 (accessed 13.8.13).

Ogden, C.L., Carroll, M.D., McDowell, M.A., Flegal, K.M., 2007. Obesity among adults in the United States – no change since 2003–2004. NCHS data brief no 1. NCHS, Hyattsville, MD. Available at: http://www.cdc.gov/nchs/data/databriefs/db01.pdf (accessed 22.10.12).

Park, S., Sappenfield, W.M., Huang, Y., Sherry, B., Bensyl, D.M., 2010. The impact of the availability of school vending machines on eating behavior during lunch: the Youth Physical Activity and Nutrition Survey. J. Am. Diet. Assoc. 110, 1532–1536. Available at: http://www.ncbi.nlm.nih.gov/pubmed/20869493 (accessed 22.10.12).

US Department of Health and Human Services, Health United States 2010. List of figures. Adult participation in physical exercise, by age group, United States, 1999–2009. Available at: http://www.cdc.gov/nchs/hus/contents2010.htm#chartbookfigures (accessed 22.9.12).

World Health Organization, 15 May 2012. Global database on body mass index: an interactive surveillance tool for monitoring nutrition transition. WHO, Geneva. Available at: http://apps.who.int/bmi/index.jsp (accessed 22.9.12).

World Health Organization, August 2011. New online nutrition initiative can help protect lives and health of millions of children. WHO, Geneva. Available at: http://www.who.int/mediacentre/news/releases/2011/nutrition_20110810/en/ (accessed 26.10.12).

Infectious Diseases and Non-Communicable Diseases

Centers for Disease Control and Prevention. *Helicobacter* and peptic ulcer diseases [updated 28 September 2006]. Atlanta, GA: CDC. Available at: http://www.cdc.gov/ulcer/history.htm (accessed 22.10.12).

De Martel, C., Ferlay, J., Franceschi, S., Vignat, J., Bray, F., Forman, D., et al., 2012. Global burden of cancers attributable to infections in 2008: a review and synthetic analysis. Lancet Oncol. 13, 607–615. Available at: http://www.thelancet.com/journals/lanonc/article/PIIS1470-2045(12)70137-7/fulltext (accessed 22.10.12).

Feinstein, L.B., Holman, R.C., Yorita Christensen, K.L., Steiner, C.A., Swerdlow, D.L., 2010. Trends in hospitalizations for peptic ulcer disease, United States, 1998–2005. Emerg. Infect. Dis. 16, 1410–1418. Available at: http://wwwnc.cdc.gov/eid/article/16/9/pdfs/09-1126.pdf (accessed 22.10.12).

National Cancer Institute. *Helicobacter pylori* and cancer [updated 16 November 2011]. Bethesda, MD: National Cancer Institute. Available at: http://www.cancer.gov/cancertopics/factsheet/Risk/h-pylori-cancer (accessed 22.10.12).

Watson, C., Alp, N.J., 2008. Role of *Chlamydia pneumoniae* in atherosclerosis. Clin. Sci. (Lond.) 114, 509–531. Available at: http://www.ncbi.nlm.nih.gov/pubmed/18336368 (accessed 22.10.12).

Williams, R., 2006. Global challenges in liver disease. Perspect. Clin. Hepatol. 44, 521–526. Available at: http://onlinelibrary.wiley.com/doi/10.1002/hep.21347/pdf (accessed 21.10.12).

Hypertension and Cardiovascular Diseases

Centers for Disease Control and Prevention. Heart disease facts: America's heart disease burden [updated 16 October 2012]. Atlanta, GA: CDC. Available at: http://www.cdc.gov/heartdisease/facts.htm (accessed 22.10.12).

Centers for Disease Control and Prevention, 2011. Prevalence of coronary heart disease – United States, 2006–2010. MMWR Morb. Mortal. Wkly. Rep. 60, 1377–1378. Available at: http://www.cdc.gov/mmwr/preview/mmwrhtml/mm6040a1.htm (accessed 22.10.12).

Centers for Disease Control and Prevention, 2012. Vital signs: awareness and treatment of uncontrolled hypertension among adults – United States, 2003–2010. MMWR Morb. Mortal. Wkly. Rep. 6, 703–709. Available at: http://www.cdc.gov/mmwr/preview/mmwrhtml/mm6135a3.htm?s_cid=mm6135a3_w (accessed 22.10.12).

Centers for Disease Control and Prevention, 2011. Vital signs: prevalence, treatment, and control of high levels of low-density lipoprotein cholesterol – United States, 1999–2002 and 2005–2008. MMWR Morb. Mortal. Wkly. Rep. 60, 109–114. Available at: http://www.cdc.gov/mmwr/preview/mmwrhtml/mm6004a5.htm?s_cid=mm6004a5_w (accessed 22.10.12).

Centers for Disease Control and Prevention, Division of Heart Disease and Stroke Prevention. Interactive atlas. Atlanta, GA: CDC. Available at: http://apps.nccd.cdc.gov/dhdspatlas/ (accessed 22.10.12).

Chiuve, S.E., Rexrode, K.M., Spiegelman, D., Logroscino, G., Manson, J.E., Rimm, E.B., 2008. Primary prevention of stroke by healthy lifestyle. Circulation 118, 947–954. Available at: http://circ.ahajournals.org/content/118/9/947.full (accessed 22.10.12).

Ford, E.S., Ajani, U.A., Croft, J.B., Critchley, J.A., Labarthe, D.R., Kottke, T.E., et al., 2007. Explaining the decrease in US deaths from coronary disease, 1980–2000. N. Engl. J. Med. 356, 2388–2398. Available at: http://www.nejm.org/doi/full/10.1056/NEJMsa053935 (accessed 22.10.12).

Framingham Heart Study, A timeline of milestones from the Framingham Heart Study. Framingham, MA: Framingham Heart Study. Available at: http://www.framingham.com/heart/timeline.htm (accessed 21.10.12).

Global Cardiovascular Infobase. Epidemiological profiles of cardiovascular and cerebrovascular diseases in the world [updated 3 March 2006]. Ottawa: Global Cardiovascular Infobase. Available at: http://www.cvdinfobase.ca/ (accessed 22.10.12).

Hall, M.J., Levant, S., DeFrances, C.J., May 2012. Hospitalization for strokes in US hospitals 1989–2009. NCHS data brief no. 95. CDC, Atlanta, GA. Available at: http://www.cdc.gov/nchs/data/databriefs/db95.pdf (accessed 22.10.12).

International Heart Health Society. The Victoria declaration on heart health. Ottawa: Health and Welfare, Canada; 1992. Available at: http://www.internationalhearthealth.org/Publications/victoria_eng_1992.pdf (accessed 22.10.12). Law M, Wald N, Morris J. Lowering blood pressure to prevent myocardial infarction and stroke: a new preventive strategy. Health Technol Assess 2003;7:1–94. Available at: http://www.ncbi.nlm.nih.gov/pubmed/14604498 (accessed 22.10.12).

Lenfant, C., Chobanian, A.V., Jones, D.W., Roccella, E.J., 2003. Seventh Report of the Joint National Committee on the Prevention, Detection, Evaluation, and Treatment of High Blood Pressure (JNC 7): resetting the hypertension sails. Circulation 107, 2993–2994. Available at: http://circ.ahajournals.org/content/107/24/2993.full.pdf (accessed 22.10.12).

Lim, S.S., Gaziano, T., Gakidou, E., Reddy, K.S., Farzadfar, F., Lozano, R., et al., 2007. Prevention of cardiovascular disease in high-risk individuals in low-income and middle-income countries: health effects and costs. Lancet 370, 205–206. Available at: http://www.thelancet.com/journals/lancet/article/PIIS0140-6736(07)61699-7/abstract (accessed 22.10.12).

Lloyd-Jones, D., Adams, J., Brown, T., Carnethon, M., Dai, S., De Simone, G., et al., 2010. Heart disease and stroke statistics 2010 update: a report from the American Heart Association. Circulation 121, e46–215. Available at: http://circ.ahajournals.org/content/121/7/e46.full.pdf (accessed 22.10.12).

Lloyd-Jones, D.M., Hong, Y., Labarthe, D., Mozaffarian, Appel LJ., Van Horn, L., et al., 2010. Defining and setting national goals for cardiovascular health promotion and disease reduction. The American Heart Association's strategic impact goal through 2020 and beyond. Circulation 121, 586–613. Available at: http://www.ncbi.nlm.nih.gov/pubmed/20089546 (accessed 22.10.12).

Mendis, S., Puska, P., Norrving, B. (Eds.), 2011. Global atlas on cardiovascular disease prevention and control. WHO, Geneva. Available at: http://whqlibdoc.who.int/publications/2011/9789241564373_eng.pdf. (accessed 22.10.12).

National Heart, Lung and Blood Institute, February 2012. Morbidity and mortality: 2012 chartbook on cardiovascular, lung and blood diseases. National Heart, Lung and Blood Institute, Bethesda, MD. Available at: http://www.nhlbi.nih.gov/resources/docs/2012_ChartBook_508.pdf (accessed 22.10.12).

Oppenheimer, G.M., Blackburn, H., Puska, P., 2011. From Framingham to North Karelia to US community-based prevention programs: negotiating research agenda for coronary heart disease in the second half of the 20th century. Public Health Rev. 33, 450–483. Available at: http://www.publichealthreviews.eu/upload/pdf_files/10/00_Oppenheiner.pdf (accessed 22.10.12).

Pająk, A., Kozela, M., 2012. Cardiovascular disease in Central and East Europe. Public Health Rev. 33, 416–435. Available at: http://www.publichealthreviews.eu/upload/pdf_files/10/00_Pajak.pdf (accessed 22.10.12).

Petrukhin, I.S., Lunina, E.Y., 2011. Cardiovascular disease risk factors and mortality in Russia: challenges and barriers. Public Health Rev. 33, 436–449. Available at: http://www.publichealthreviews.eu/upload/pdf_files/10/00_Petrukhin.pdf (accessed 22.10.12).

Roger, V., Go, A., Lloyd-Jones, D., Benjamin, E., Berry, J., Borden, W., et al., 2012. American Heart Association statistical update: heart disease and stroke statistics – 2012 update. Circulation 125, e2–220. Available at: http://circ.ahajournals.org/content/125/1/e2.full.pdf+html (accessed 22.10.12).

Schober, S.E., Carroll, M.D., Lacher, D.A., Hirsch, R., 2007. High serum total cholesterol – an indicator for monitoring cholesterol lowering efforts; US adults, 2005–2006. NCHS data brief. 1–8. Available at: http://www.ncbi.nlm.nih.gov/pubmed/19389314. (accessed 22.10.12).

Winkleby, M.A., Taylor, C.B., Jatulis, D., Fortmenn, S.P., 1996. The long-term effects of cardiovascular disease prevention: the Stanford Five-City Project. Am. J. Public Health 86, 1773–1779. Available at: http://www.ncbi.nlm.nih.gov/pubmed/9003136 (accessed 22.10.12).

World Health Organization, September 2011. Cardiovascular diseases (CVDs). Fact sheet no. 317 [updated. WHO, Geneva. Available at: http://www.who.int/mediacentre/factsheets/fs317/en/index.html (accessed 22.10.12).

World Heart Federation, Stress and cardiovascular disease. Geneva: World Heart Federation; Available at: http://www.world-heart-federation.org/cardiovascular-health/cardiovascular-disease-risk-factors/stress/ (accessed 22.10.12).

Young, F., Capewell, S., Ford, E.S., Critchley, J.A., 2010. Coronary mortality declines in the US between 1980 and 2000 quantifying the contribution of primary and secondary prevention. Am. J. Prev. Med. 39, 228–234. Available at: http://www.ncbi.nlm.nih.gov/pubmed/20709254 (accessed 22.10.12).

Metabolic Syndrome, Diabetes Mellitus, and End-Stage Renal Disease

American Diabetes Association, 26 January 2011. Diabetes statistics. Alexandria, VA: American Diabetes Association. . Available at: http://www.diabetes.org/diabetes-basics/diabetes-statistics/. (accessed 22.10.12).

Centers for Disease Control and Prevention, 2009. Estimated county-level prevalence of diabetes and obesity – United States, 2007. MMWR Morb. Mortal. Wkly. Rep. 58, 1259–1263. Available at: http://www.ncbi.nlm.nih.gov/pubmed/19940830 (accessed 22.10.12).

Centers for Disease Control and Prevention, 2010. Incidence of end-stage renal disease attributed to diabetes among persons with diagnosed diabetes – United States and Puerto Rico, 1997–2007. MMWR Morb. Mortal. Wkly. Rep. 59, 1361–1366. Available at: http://www.cdc.gov/mmwr/preview/mmwrhtml/mm5942a2.htm?s_cid=mm5942a2_w (accessed 22.10.12).

Centers for Disease Control and Prevention, 2005. Lower extremity disease among persons aged ≥40 years with and without diabetes – United States, 1999–2002. MMWR Morb. Mortal. Wkly. Rep. 54, 1158–1160. Available at: http://www.ncbi.nlm.nih.gov/pubmed/16292250 (accessed 22.10.12).

Centers for Disease Control and Prevention, Division of Diabetes Translation. Diabetes data and trends.. Atlanta, GA: CDC. Available at: http://www.cdc.gov/diabetes/statistics (accessed 22.10.12).

Centers for Disease Control and Prevention. National chronic kidney disease fact sheet [updated 2 April 2010]. Atlanta, GA: CDC. Available at: http://www.cdc.gov/diabetes/pubs/factsheets/kidney.htm (accessed 22.10.12).

Centers for Disease Control and Prevention, 2011. National diabetes fact sheet. National Center for Chronic Disease Prevention and Health Promotion, Atlanta, GA. Division of Diabetes Translation; 2011. Available at: http://www.cdc.gov/Diabetes/pubs/pdf/ndfs_2011.pdf (accessed 22.10.12).

International Diabetes Federation, 2011. Action plan 2011–2021. Global diabetes plan 2011. International Diabetes Federation, Brussels. Available at: http://www.idf.org/sites/default/files/Global_Diabetes_Plan_Final.pdf (accessed 22.10.12).

Signorello, L.B., Schlundt, D.G., Cohen, S.S., Steinwandel, M.D., Buchowski, M.S., McLaughlin, J.K., et al., 2007. Comparing diabetes prevalence between African Americans and whites of similar socio-economic status. Am. J. Public Health 97, 2260–2267. Available at: http://www.ncbi.nlm.nih.gov/pmc/articles/PMC2089102/ (accessed 22.10.12).

United States Renal Data System, 2012. USRDS 2012 annual data report: atlas of chronic kidney disease in the United States.. National Institutes of Health, National Institute of Diabetes and Digestive and Kidney Diseases, Bethesda, MD. Available at: http://www.usrds.org/2012/pdf/v1_00intro_12.pdf (accessed 22.10.12).

World Health Organization. Diabetes. Fact sheet no. 312 [updated September 2011]. Geneva: WHO. Available at: http://www.who.int/mediacentre/factsheets/fs312/en/index.html (accessed 22.10.12).

Cancer

American Cancer Society, 2012. Age adjusted cancer death rates, females by site, 1930–2008. American Cancer Society, Atlanta, GA. Available at: http://www.cancer.org/acs/groups/content/@epidemiologysurveilance/documents/document/acspc-032006.pdf (accessed 22.10.12).

American Cancer Society, 2011. Breast cancer facts and figures 2011–2012. American Cancer Society, Atlanta, GA. Available at: http://www.cancer.org/acs/groups/content/@epidemiologysurveilance/documents/document/acspc-030975.pdf (accessed 22.10.12).

American Cancer Society, 2012. Cancer facts and figures 2012. American Cancer Society, Atlanta, GA. Available at: http://www.cancer.org/acs/groups/content/@epidemiologysurveilance/documents/document/acspc-031941.pdf (accessed 22.10.12).

American Cancer Society. Colorectal cancer [updated 1 October 2012]. Atlanta, GA: American Cancer Society. Available at: http://www.cancer.org/Cancer/ColonandRectumCancer/DetailedGuide/colorectal-cancer-key-statistics. (accessed 22.10.12).

American Cancer Society, 2008. Global facts and figures, second ed. American Cancer Society, Atlanta, GA. Available at: http://www.cancer.org/acs/groups/content/@epidemiologysurveilance/documents/document/acspc-027766.pdf (accessed 22.10.12).

Centers for Disease Control and Prevention, 1993. Mortality trends for selected smoking related cancers and breast cancer – United States, 1950–1990. MMWR Morb. Mortal. Wkly. Rep. 42, 857–866. Available at: http://www.ncbi.nlm.nih.gov/pubmed/8232168 (accessed 22.10.12).

Centers for Disease Control and Prevention, National Program of Cancer Registries. Cancer: United States cancer statistics: 1999–2008. Cancer incidence and mortality web-based report. Atlanta, GA: US DHHS. Available at: http://www.cdc.gov/uscs (accessed 22.9.12).

Centers for Disease Control and Prevention, National Program of Cancer Registries (NPCR). United States Cancer Statistics (USCS). 1999–2009 cancer incidence and mortality data. Available at: http://apps.nccd.cdc.gov/uscs/ (accessed 14.8.13).

Centers for Disease Control and Prevention, 2010. Vital signs: colorectal cancer screening among adults aged 50–75 years – United States, 2008. MMWR Morb. Mortal. Wkly. Rep. 59, 1–5. Available at: http://www.cdc.gov/mmwr/preview/mmwrhtml/mm59e0706a1.htm (accessed 22.10.12).

Clinical Guidelines Committee of the American College of Physicians, 2012. Clinical guidelines: screening for colorectal cancer: a guidance statement from the American College of Physicians. Ann. Intern. Med. 156, I–30. Available at: https://annals.org/article.aspx?articleID=1090694 (accessed 22.10.12).

Ginsberg, G.M., Tulchinsky, T.H., 1992. Regional differences in cancer incidence and mortality in Israel: possible leads to occupational causes. Isr J. Med. Sci. 28, 534–543.

Mathers, C.D., Shibuya, K., Boschi-Pinto, C., Lopez, A., Murray, C.J.L., 2002. Global and regional estimates of cancer mortality and incidence by site: I. Application of regional cancer survival model to estimate cancer mortality distribution by site. BMC Cancer 2, 36. Available at: http://www.biomedcentral.com/content/pdf/1471-2407-2-36.pdf (accessed 14.8.13).

National Cancer Institute, Cancer trends progress report 2011–2012. Bethesda, MD: National Cancer Institute. Available at: http://progressreport.cancer.gov/ (accessed 22.10.12).

Qaseem A, Denberg TD, Hopkins RH, Humphrey LL, Levine J, Sweet DE, Shekelle P, for World Health Organization. Cancer. Fact sheet no. 297 [updated February 2012]. Geneva: WHO. Available at: http://www.who.int/mediacentre/factsheets/fs297/en/ (accessed 22.10.12).

Trauma, Violence, Injury, and Suicide

Ajdacic-Gross, Y., Weiss, M.G., Ring, M., Hepp, U., Bopp, M., Gutzwiller, F., Rössler, W., 2008. Methods of suicide: international suicide patterns derived from the WHO mortality database. Bull. World Health Organ. 86, 726–732. Available at: http://www.ncbi.nlm.nih.gov/pmc/articles/PMC2649482/ (accessed 22.10.12).

Centers for Disease Control and Prevention, 1999. Achievements in public health, 1900–1999. Motor-vehicle safety: a 20th century public health achievement. MMWR Morb. Mortal. Wkly. Rep. 48, 369–374. Available at: http://www.cdc.gov/mmwr/preview/mmwrhtml/mm4818a1.htm (accessed 22.10.12).

Centers for Disease Control and Prevention. Fast stats: accidents or unintentional injuries [updated 19 October 2012]. Atlanta, GA: CDC. Available at: http://www.cdc.gov/nchs/fastats/acc-inj.htm (accessed 22.10.12).

Centers for Disease Control and Prevention, 2007. Increases in age-group-specific injury mortality – United States, 1999–2004. MMWR Morb. Mortal. Wkly. Rep. 56, 1281–1284. Available at: http://www.cdc.gov/mmwr/preview/mmwrhtml/mm5649a1.htm (accessed 22.10.12).

Centers for Disease Control and Prevention. Injury center: connection between bullying and sexual violence perpetration [updated 11 January 2012]. Atlanta, GA: CDC. http://www.cdc.gov/ViolencePrevention/youthviolence/bullying_sv.html (accessed 22.10.12).

Centers for Disease Control and Prevention. Injury prevention and control. Injury: the leading cause of death among persons 1–44 [updated 9 June 2011]. Atlanta, GA: CDC. Available at: http://www.cdc.gov/injury/overview/leading_cod.html (accessed 22.10.12).

Centers for Disease Control and Prevention, National Intimate Partner and Sexual Violence Survey. Available at: http://www.cdc.gov/ViolencePrevention/pdf/NISVS_FactSheet-a.pdf (accessed 24.10.12).

Centers for Disease Control and Prevention. Poisoning in the United States: fact sheet [updated 29 June 2012]. Atlanta, GA: CDC. Available at: www.cdc.gov/homeandrecreationalsafety/poisoning/poisoning-factsheet.htm (accessed 22.10.12). Centers for Disease Control and Prevention, Task Force on Community Preventive Services. Motor-vehicle occupant injury: strategies for increasing use of child safety seats, increasing use of safety belts, and reducing alcohol-impaired driving: a report on recommendations of the Task Force on Community Preventive Services. MMWR Recomm Rep 2001;50:1–13. Available at: http://www.cdc.gov/mmwr/preview/mmwrhtml/rr5007a1.htm (accessed 22.10.12).

Fingerhut, L.A., February 2008. NCHS health e-stat: increases in poisoning and methadone-related deaths: United States, 1999–2005. NCHS, Atlanta, GA. Available at: http://www.cdc.gov/nchs/data/hestat/poisoning/poisoning.pdf (accessed 22.10.12).

Finkelhor, D., Jones, L., Shattuck, A., 2009. Updated trends in child maltreatment, 2009. Crimes Against Children Research Center, Durham, NH. Available at: http://unh.edu/ccrc/pdf/Updated_Trends_in_Child_Maltreatment_2009.pdf (accessed 22.10.12).

Garcia-Moreno, C., Watts, C., 2011. Violence against women: an urgent public health priority. Bull. World Health Organ. 89, 2. Available at: http://www.who.int/bulletin/volumes/89/1/10-085217.pdf (accessed 22.10.12).

Gvion, Y., Apter, A., 2012. Suicide and suicidal behavior. Public Health Rev. 34 (2). Available at: http://www.publichealthreviews.eu/upload/pdf_files/12/00_Gvion.pdf (accessed 22.10.12).

Kruse, M., Sørensen, J., Brønnum-Hansen, H., Helweg-Larsen, K., 2011. The health care costs of violence against women. J. Interpers Violence 26, 3494–3508. Available at: http://jiv.sagepub.com/content/early/2011/04/26/0886260511403754.abstract?rss=1 (accessed 22.10.12).

Stevens, J., Olson, S., 2000. Reducing falls and resulting hip fractures among older women. MMWR Recomm. Rep. 49, 1–12. Available at: http://www.cdc.gov/mmwr/preview/mmwrhtml/rr4902a2.htm (accessed 22.10.12).

Waller, P.F., 2002. Challenges in motor vehicle safety. Annu. Rev. Public Health 23, 93–113. Available at: http://www.annualreviews.org/doi/abs/10.1146/annurev.publhealth.23.100901.140522 (accessed 22.10.12).

World Health Organization, Global plan for the Decade of Action for Road Safety 2011–2020. Geneva: WHO. Available at: http://www.who.int/roadsafety/decade_of_action/plan/en/index.html (accessed 22.10.12).

World Health Organization, 2009. Global status report on road safety: time for action. Geneva: WHO, Department of Violence and Injury Prevention and Disability. . Available at: http://whqlibdoc.who.int/publications/2009/9789241563840_eng.pdf. (accessed 22.10.12).

World Health Organization, 2010. Injuries and violence – the facts. WHO, Geneva. Available at: http://whqlibdoc.who.int/publications/2010/9789241599375_eng.pdf (accessed 22.10.12).

World Health Organization, International Association for Suicide Prevention (IASP) – resources. World Health Organization. Geneva: WHO. Available at: http://www.iasp.info/resources/World_Health_Organization__WHO__/ (accessed 22.10.12).

World Health Organization. Mental health: suicide prevention (SUPRE) [updated 31 August 2012]. Geneva: WHO. Available at: http://www.who.int/mental_health/prevention/suicide/suicideprevent/en/ (accessed 22.10.12).

Chronic Lung Disease

Akinbami, L.J., Liu, X., 2011. Chronic obstructive pulmonary disease among adults aged 18 and over in the United States, 1998–2009. NCHS data brief. NCHS, Hyattsville, MD. Available at: http://www.cdc.gov/nchs/data/databriefs/db63.pdf (accessed 22.10.12).

Akinbami, L., Moorman, J., Liu, X., January 2011. Health statistics reports: asthma prevalence, health care use, and mortality: United States, 2005–2009. NCHS, Hyattsville, MD. Available at: http://www.cdc.gov/nchs/data/nhsr/nhsr032.pdf (accessed 24.10.12).

Bousquet, J., Khaltaev, N. (Eds.), 2007. Global surveillance, prevention and control of chronic respiratory diseases: a comprehensive approach. WHO, Geneva. Available at: http://whqlibdoc.who.int/publications/2007/9789241563468_eng.pdf. (accessed 24.10.12).

Centers for Disease Control and Prevention, 2011. Vital signs: asthma prevalence, disease characteristics, and self-management education – United States, 2001–2009. MMWR Morb. Mortal. Wkly. Rep. 60, 547–552. Available at: http://www.cdc.gov/mmwr/preview/mmwrhtml/mm6017a4.htm (accessed 24.10.12).

Centers for Disease Control and Prevention, March 20, 2013. High Blood Pressure. America's High Blood Pressure Burden. Available at http://www.cdc.gov/bloodpressure/facts.htm (accessed 21.11.13).

Disease Control Priorities Project, October 2008. Breathing easier: preventing chronic respiratory disease in adults. Disease Control Priorities Project, Washington, DC. Available at: http://www.dcp2.org/file/221/dcpp-respiratorydisease-web.pdf (accessed 24.10.12).

Halbert, R.J., 2006. Natoli JL, Gano A, Badamgarav E, Buist AS, Mannino DM. Global burden of COPD: systematic review and meta-analysis. Eur. Respir. J. 28, 523–532. Available at: http://erj.ersjournals.com/content/28/3/523.abstract (accessed 24.10.12).

Krishnamurthi, R.V., Feigin, V.L., Fourouzanfar, M.H., Mensah, G.A., Connor, M., Bennett, D.A., et al., 2013. Global and regional burden of first-ever ischaemic and haemorrhagic stroke during 1990—2010: findings from the Global Burden of Disease Study 2010. Lancet Global Health 1 (5), e259–e281. Available at http://www.thelancet.com/journals/langlo/article/PIIS2214-109X(13)70089-5/fulltext?elsca1=EMAIL-TLGHNov&elsca2=email&elsca3=2NPCT4F (accessed 21.11.13).

Mathers, C.D., Loncar, D., 2006. Projections of Global Mortality and Burden of Disease from 2002 to 2030. PLoS Med 3 (11), e442. http://dx.doi.org/10.1371/journal.pmed.0030442 (accessed 21.11.13).

Moorman, J.E., Zahran, H., Truman, B., Molla, M.T., 2011. Current asthma prevalence – United States, 2006–2008. MMWR Morb. Mortal. Wkly. Rep. 60, 84–86. Available at: http://www.cdc.gov/mmwr/preview/mmwrhtml/su6001a18.htm (accessed 24.10.12).

National Heart, Lung and Blood Institute. High Blood Pressure Education Program. Prevention, Detection, Evaluation, and Treatment of High Blood Pressure: The Seventh Report of the Joint National Committee on JNC 2003. Available at: http://www.nhlbi.nih.gov/guidelines/hypertension/express.pdf (accessed 20.11.2013).

Rafey, M.A., 2013. Cleveland Clinic Center for Continuing Education. Hypertension. Available at http://www.clevelandclinicmeded.com/medicalpubs/diseasemanagement/nephrology/arterial-hypertension/ (accessed 21.11.13).

World Health Organization, Chronic respiratory diseases: burden of COPD. Geneva: WHO. Available at: http://www.who.int/respiratory/copd/burden/en/index.html (accessed 24.10.12).

World Health Organization, 2011. Global status report on noncommunicable diseases 2010. Geneva: World Health Organization. Available at http://whqlibdoc.who.int/publications/2011/9789240686458_eng.pdf (accessed 21.11.13).

World Health Organization. Tobacco. Fact sheet no. 339 [updated May 2012]. Geneva: WHO. Available at: http://www.who.int/mediacentre/factsheets/fs339/en/index.html (accessed 24.10.12).

Disabling Conditions

Bailey, R.N., Indian, R.W., Zhang, X., Geiss, L.S., Duenas, M.R., Saaddine, J.B., 2006. Visual impairment and eye care among older adults – five states. MMWR Morb. Mortal. Wkly. Rep. 55, 1321–1325. Available at: http://www.cdc.gov/mmwr/preview/mmwrhtml/mm5549a1.htm (accessed 24.10.12).

Bolen, J., Murphy, L., Greenlund, K., Helmick, C.G., Hootman, J., Brady, T.J., et al., 2009. Arthritis as a potential barrier to physical activity among adults with heart disease – United States, 2005 and 2007. MMWR Morb. Mortal. Wkly. Rep. 58, 165–169. Available at: http://www.cdc.gov/mmwr/preview/mmwrhtml/mm5807a2.htm (accessed 24.10.12).

Centers for Disease Control and Prevention, 2011. Self-reported visual impairment among persons with diagnosed diabetes – United States, 1997–2010. MMWR Morb. Mortal. Wkly. Rep. 60, 1549–1553. Available at: http://www.cdc.gov/mmwr/preview/mmwrhtml/mm6045a2.htm (accessed 24.10.12).

Centers for Disease Control and Prevention. Spinal cord injury (SCI): fact sheet [updated 4 November 2010]. Atlanta, GA: CDC. Available at: http://www.cdc.gov/TraumaticBrainInjury/scifacts.html (accessed 24.10.12).

Dawodu ST. Traumatic brain injury (TBI) – definition, epidemiology, pathophysiology [updated 10 November 2011]. New York: Medscape. Available at: http://emedicine.medscape.com/article/326510-overview#aw2aab6b3 and http://emedicine.medscape.com/article/326510-overview# (accessed 24.10.12).

Gohdes, D.M., Balamurugan, A., Larsen, B.A., Maylahn, C., 2005. Age-related eye diseases: an emerging challenge for public health professionals. Prev. Chronic Dis. 2, A17. Available at: http://www.ncbi.nlm.nih.gov/pubmed/15963319 (accessed 24.10.12).

Halfon, N., Houtrow, A., Larso, K., Newachek, P.W., 2012. The changing landscape of disability in childhood. Children with Disabilities. Future Child. 22 (1). Available at: http://futureofchildren.org/futureofchildren/publications/journals/article/index.xml?journalid=77&articleid=559&sectionid=3866 (accessed 21.10.12).

National Osteoporosis Foundation, 2011. Why bone health is important. National Osteoporosis Foundation, Washington, DC. Available at: http://www.nof.org/node/150 (accessed 22.9.12).

Owen, C.G., Jarrar, Z., Wormald, R., Cook, D.G., Fletcher, A.E., Rudnicka, A.R., The estimated prevalence and incidence of late stage age related macular degeneration in the UK. Br J Ophthalmol 2012. Available at: http://bjo.bmj.com/content/early/2012/02/02/bjophthalmol-2011-301109.fullhttp://dx.doi.org/10.1136/bjophthalmol-2011-301109 (accessed 03.08.13).

Prevent Blindness America, August 2009. A summary report of a five-year cooperative agreement with the vision health initiative of the Centers for Disease Control and Prevention (2003–2008). Prevent Blindness America, Chicago, IL. Available at: http://www.cdc.gov/visionhealth/pdf/pba_complete.pdf (accessed 24.10.12).

World Health Organization. Deafness and hearing impairment. Fact sheet no. 300 [updated February 2012]. Geneva: WHO. Available at: http://www.who.int/mediacentre/factsheets/fs300/en/index.html (accessed 24.10.12).

World Health Organization. Epilepsy fact sheet [updated October 2012]. Geneva: WHO. Available at: http://www.who.int/mediacentre/factsheets/fs999/en/ (accessed 24.10.12).

World Health Organization, Micronutrient deficiencies – vitamin A deficiency. Geneva: WHO. Available at: http://www.who.int/nutrition/topics/vad/en/ (accessed 24.10.12).

World Health Organization, 2006. Neurological disorders: public health challenges. WHO, Geneva. Available at: http://www.who.int/mental_health/neurology/neurological_disorders_report_web.pdf (accessed 24.10.12).

World Health Organization, Prevention of blindness and visual impairment. Geneva: WHO. Available at: http://www.who.int/blindness/table/en/index.html (accessed 24.10.12).

World Health Organization, 2012. Alzheimer's Disease International. Dementia: a public health priority. WHO, Geneva. Available at: http://whqlibdoc.who.int/publications/2012/9789241564458_eng.pdf (accessed 24.10.12).

Chapter 6

Family Health

Learning Objectives

Upon completion of this chapter, the student should be able to:

1. Identify specific health issues of family members in their different stages of life;
2. Describe family health as part of the New Public Health;
3. Formulate public health strategies to address these issues.

INTRODUCTION

This chapter provides an introduction to health in the family context. The family structure provides an important foundation for the physical and emotional health of the individual and the community. Marital and family status and interactions among family members affect each person's health and the well-being of the community and nation. The family exists and functions within the context of cultural, economic, legal, and social patterns unique to each society, with important commonalities of the role of the family in health. However, each person is a unique individual who passes through life stages with changing health needs and support systems not only in the family, but also in peer groups and society more widely.

Family health issues relate to phases involving marital status, fertility and pregnancy, infancy, childhood, adolescence, adulthood, and old age, as well as the relationships among family members. Each phase has specific health risks in which prevention and other health services play an important role. This chapter addresses the health needs of family members at different stages of life. Poor pregnancy outcome affects the mother, child, family, and community. The family of the person who is chronically ill, injured, or killed at work or on the road suffers economically, emotionally, and socially. The public health and medical care systems must be sensitive to the special needs of the family by providing appropriate health promotion, disease prevention, medical care, and support programs for each member of the family and the family as a whole.

The New Public Health includes traditional elements of public health, such as maternal and child health, but within a larger context of total family needs. In developed and developing countries, single teenage pregnancies entail a multitude of family and societal problems. In societies where women's rights and education are restricted or repressed, the health of women and children suffers measurably. In many developing societies, the male and female adult in the family may be forced into dangerous or health-damaging work resulting in premature death. Unemployment, absence, or death of the father in a family setting has heavy consequences economically, emotionally, and functionally, often impoverishing the family as the male is in most societies is the sole or main income earner, in addition to providing a partner for the mother and a male role in child rearing. The role of public health is then to advocate societal change and provide direct responses to the health needs generated by societal patterns, whether this relates to high rates of morbidity for women and children due, for example, to lack of primary care services, or whether the problem is lack of health insurance coverage for children.

THE FAMILY UNIT

The family is the basic social support unit in virtually all human societies, providing the basis for childbearing and child rearing. It has important roles in the stability of basic physiological and psychological needs as well as the economic basis of the members of the unit. The family provides the key environment for the emotional needs, socialization, mutual help, and nurturing needed by adults as well as children.

The family is usually a group of two or more people related by marriage, common agreement, birth, or adoption who reside together in the same household, and may consist of one, two, or more generations. The family unit includes people living together, including single-parent families and same-gender couples, with children and parents and other relations. Families include sexual relations between adult parents, fertility, birth, and rearing of children through the many stages of development before they reach independent adulthood. It also includes caring for elderly parents and relatives, as well as maintaining close contact with adult siblings and children, themselves in the process of childbearing and child rearing.

The nuclear family typically includes a male and female couple related by marriage, or living together by common consent, with or without children from current and/or previous relationships. Increasingly, family units include single or divorced mothers, or fathers, living alone with children, or with a parent or another person related or unrelated by marriage. The definition is being widened to include couples of the same gender, with or without children from previous heterosexual relationships, from artificial insemination, or through adoption. The extended family is multigenerational and consists of the nuclear family and relatives of both parties, whether or not they are living in close geographic proximity. The extended family provides a broader basis of mutual support. It still exists in western societies, but frequently in an altered form.

Multigenerational families consisting of single, divorced, or widowed women or men with children are now common, in association with high divorce rates, single parenthood, and increasing longevity, especially for women. Same-gender households with children are becoming more common and more accepted, although they remain controversial. Single-parent families, most commonly headed by a single mother, are widespread in western societies owing to early age pregnancies or marriage breakdowns. The adolescent mother may have family support systems, but often is struggling alone to raise a child or children with or without social support. A 2006–2007 survey of Canadian new mothers showed that over 35 percent of lone mothers experienced abuse, three times the rate of all new mothers (Daoud, 2012). Table 6.1 shows the changes in US households. Following the 2008 recession, economic stress on the young adult population has increased, with high unemployment among the young especially, so that 14.2 percent of US adults in the 25–34 age group in 2011 are living in their parents' household, compared to less than 12 percent before the recession.

In developing countries, families often consist of large numbers of children born to poorly educated parents living in poverty. The father, or less commonly the mother, may be absent for long periods while working in a distant place. This set-up can create serious health hazards for all family members. In societies where death of adults occurs from civil wars, famine, accidents, or infectious diseases such as acquired immunodeficiency syndrome (AIDS), raising of children by single parents, neighbors, older siblings, or grandparents is common. Abandonment of children is also common in such situations and many children orphaned by human immunodeficiency virus (HIV/AIDS) are left to very limited public or charitable agencies for care in orphanages or to survive on the street, with potential for sexual exploitation and other violence.

Divorce and single parenthood are often associated with relative poverty, creating additional stresses in the functioning of the family unit. The age group under 18 years has the highest rates of poverty (Figure 6.1). The absence of one parent, usually the male, also strains the family. Abusive parenting, including psychological and sexual abuse and other violent behaviors, creates a burden on children in their most vulnerable years, with long-term psychological damage. Social and economic trends since the onset of recession in 2008 show important difficulties hitting young adults hard; more so for men than women because of long-term improved access of women to higher education in many professions, while a large sector of young men is not advancing for complex reasons that are not fully understood. There is an emerging trend of grandparents assuming parental responsibility for child rearing in the breakup of the nuclear family.

In the USA, there has been a decline in the percentage of children in two-parent family settings. In 1980, 86 percent of all family households in the white population were two-parent families, but this had fallen to 65.4 percent in 2010 (US Census Bureau, 2011). Among the African American population, 56 percent of all families were in two-parent family homes in 1980, declining to 28.5 percent in 2010, while female-headed black households increased from 30 percent 40 percent in the same period. For family groups with children under the age of 18 years, the percentage of one-parent family groups in black homes was 20.3 percent in 2010, compared to 24 percent in white homes and 31 percent in Hispanic homes. This is a continuing trend. A high percentage of African American homes and children are at high risk for poverty, insecurity, and inadequate health insurance. Health insurance coverage and access to services declined between 2003 and 2007 for Americans. This has resulted in lack of access to family doctors and regular care, poor utilization of immunization, and frequent use of hospital emergency departments.

In Canada, single-parent families make up 15.9 percent of all families, but all are included in universal health care run by the provinces with federal cost-sharing and guidelines. In the UK, 47 percent of all births occur to unmarried mothers, including cohabiting parents, with 2 million single-parent families, the highest proportion of children brought up in one-parent families of any major European country, and 48 percent of single mothers in the UK are unemployed. However, all are included in the National Health Service (NHS) and other social support systems so that society ensures that children have access to care regardless of their family circumstances.

Changes in family structure affect the health status of adults as well as children. Increases in longevity of women, more than men, together with high divorce rates, produce an excess of elderly women in relatively good health who live alone. The capacity of women to adapt to this condition is positively affected by the strong social support systems in the industrialized countries where universal health insurance, national pensions, pensions from prior work of spouses, and legal protection of economic rights in divorce have produced a relatively well-protected elderly population group in most industrialized countries.

TABLE 6.1 US Households (Millions) by Type and Percentage Change 2000 and 2010

Household Type	2000		2010		% Change 2000– 2010	
	Number	%	Number	%	Number	%
Total households	105.5	100.0	116.7	100.0	11.2	10.7
Family households	71.8	68.1	77.5	66.4	5.8	8.0
Husband–wife households	54.5	51.7	56.5	48.4	2.0	3.7
With own children	24.8	23.5	23.6	20.2	−1.2	−5.0
Without own children	29.7	28.1	32.9	28.2	3.2	11.0
Female householder, no spouse present	12.9	12.2	15.3	13.1	2.4	18.2
With own children	7.6	7.2	8.4	7.2	0.8	10.6
Without own children	5.3	5.1	6.9	5.9	1.5	29.0
Male householder, no spouse present	4.4	4.2	5.8	5.0	1.4	31.5
With own children	2.2	2.1	2.8	2.4	0.6	27.3
Without own children	2.2	2.1	3.0	2.6	0.8	35.6
Non-family households	33.7	31.9	39.2	33.6	5.5	16.3
Male householder	15.6	14.7	18.5	15.8	2.9	18.7
Living alone	11.8	11.2	13.9	11.9	2.1	18.1
Not living alone	3.8	3.6	4.6	3.9	0.8	20.5
Female householder	18.1	17.2	20.7	17.8	2.6	14.2
Living alone	15.5	14.6	17.3	14.8	1.8	12.0
Not living alone	2.7	2.5	3.4	2.9	0.7	27.3
Unmarried couple households[a]	5.5	5.2	7.7	6.6	2.3	41.4
Opposite-sex partners	4.9	4.6	6.8	5.9	2.0	40.2
Same-sex estimate[a]	0.6	0.6	0.9	0.8	0.3	51.8
Preferred estimates	0.4	0.3	0.6	0.6	0.3	80.4
Average household size	2.59	NA	2.58	NA	−0.01	NA
Average family size	3.14	NA	3.14	NA	0.00	NA

Note: Numbers rounded to one decimal place; NA=not applicable.

[a]Unmarried couple households can be family or non-family households depending on the relationship of others in the household to the householder. In this table, it is the sum of opposite-sex partners and same-sex partners from Summary File 1 counts. Summary File 1 counts in this table are consistent with Summary File 1 counts shown in American Fact Finder.

Source: US Census Bureau, Census 2000 Summary File 1 and 2010 Census Summary File 1. Available at: http://www.census.gov/prod/cen2010/briefs/c2010br-14.pdf (Accessed 20 March 2013).

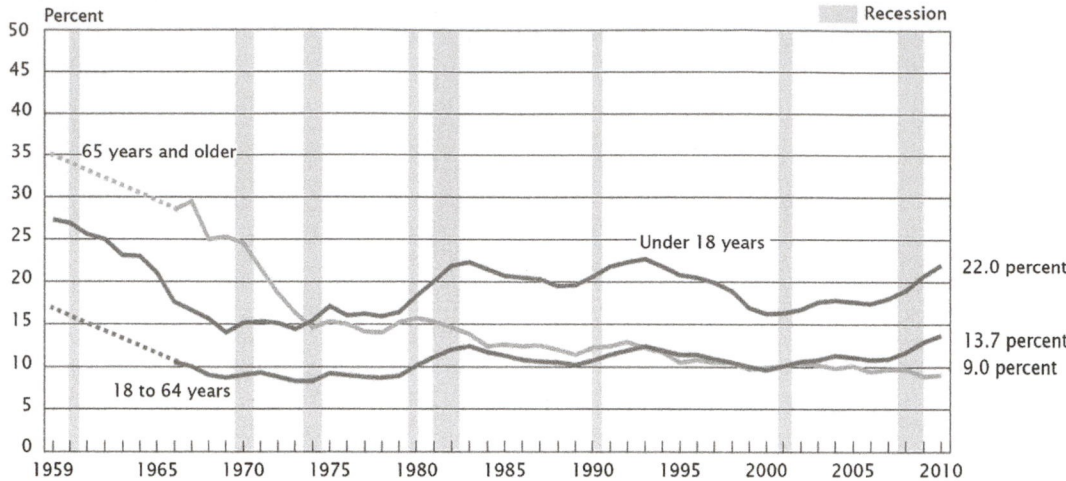

FIGURE 6.1   Poverty rates by age group, USA, 1959–2010. *Source: US Census Bureau. Income, poverty, and health insurance coverage in the United States 2010. Figure 5, p 17. Current population reports, September 2011. Available at: http://www.census.gov/prod/2011pubs/p60-239.pdf [Accessed 20 December 2012].*

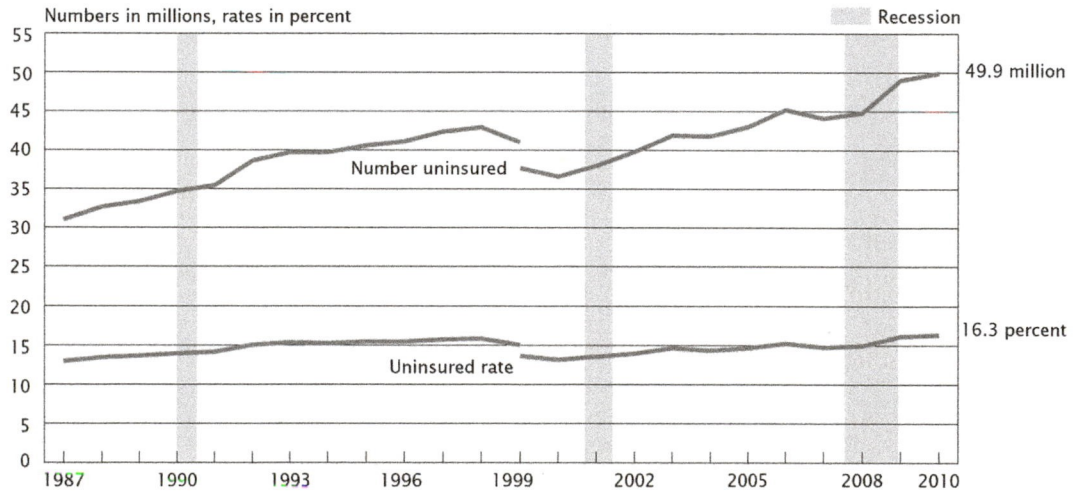

FIGURE 6.2   Uninsured population by age group, USA, 1987–2010. *Source: US Census Bureau. Income, poverty, and health insurance coverage in the United States: 2010. Figure 7, p 23. Current Population Reports 17.112. Available at: http://www.census.gov/prod/2011pubs/p60-239.pdf [Accessed 20 December 2012].*

In 2010, following the recession of 2008, the number of Americans without health insurance was nearly 50 million (Figure 6.2). Health insurance coverage for people born outside the USA and those living in poverty has increased in recent years. In 2010, about two and a half times as many of the foreign-born US population had no health coverage compared to the native-born population. Between 2000 and 2005 the number of children in poverty households without insurance declined from 32.5 percent to 30.5 in 2005, but rose to 34.8 percent in 2010, with 9 percent of all American children under the age of 6 years not covered by health insurance. For children aged under 18 years living in poverty, 15.8 percent were uninsured compared to 9.8 percent of all children. In 2010, slightly more than 16 percent of the American population had no health insurance. The Centers

for Disease Control and Prevention (CDC) reports that 18.2 percent of people under the age of 65 had no health insurance in 2011, and there was decreased employment-based health insurance (see Chapters 10 and 13).

Diseases associated with aging from middle age onwards include cardiovascular and other degenerative conditions, affecting men at an age nearly 10 years younger than women. Cancers, a leading cause of death worldwide, are the second leading cause of death in the USA. Epidemiological and demographic transitions affect the family in many ways. Care of the elderly has become an increasing factor in family life, not only in provision of care, but in inter-family relations, economic and other stresses for the family and society. Where health insurance fails to fully cover long-term care, serious financial burdens are placed on middle-aged children of elderly parents. It is not uncommon for middle-aged or elderly adults to care for very elderly parents. Where society does bear the cost, prevention and control of chronic disease are essential to avert breakdown of the health system.

Homelessness occurs in all societies and has impacts on the family unit, in particular women and children (Box 6.1). In the USA, African Americans comprised 12 percent of the population in 2008 but 39 percent of the sheltered homeless population. Public schools reported a 20 percent increase to nearly 1 million homeless students between the 2007–2008 and the 2008–2009 school years. Individual adults comprised 63 percent of the 983,835 people using shelters and transitional housing programs in 2009. The age group 50 and above represents 25 percent of homeless people, and 14 percent of individuals were in institutional settings the night before becoming homeless. The increase in homelessness is the result of a convergence of a number of key factors: loss of affordable housing and foreclosures; low wages and public assistance that have not kept pace with the cost of living; rising housing costs; job loss, underemployment and resulting debt; and the closing of state psychiatric institutions without the concomitant creation of community-based housing and mental health services.

PREVENTIVE HEALTH THROUGH LIFE STAGES

Monitoring and addressing health problems for a population are the key function of public health (health promotion, primary, secondary, and tertiary prevention). This applies to all stages of life. The health care system of health insurance and medical care services is an integral part of this process. Improving population health requires the identification and prioritization of effective and efficient preventive health services for the population, delivered by a coordinated team of service providers in the clinical and public health streams.

Health insurance systems have long learned from evidence-based policies that prevention is cost-effective in maintaining health and reducing costly burdens of disease. This requires the identification of priority preventive and screening services for an asymptomatic population as distinct from diagnostic and clinical services which focus on evaluation and diagnosis of a patient with specific health problems. Screening is a controversial area of policy. It is costly and in some cases the evidence is weak or negative for cost-effectiveness. Some screening measures are well accepted as "best practices", and the patient and the population deserve to have such screening. Levels of primary prevention and guidelines for prevention-oriented services are grouped under three levels based on efficacy and are important in ensuring a broad range of efficacious and efficient health services across the population for a cost-effective health system. The guidelines are evidence based

---

BOX 6.1 Homelessness

Homelessness has grown in North America since the onset of the recession in 2008. It affects families with children, veterans, mentally ill people, and others. Homelessness is associated with increased health problems, limited access to health care, higher rates of disease: chronic disease, poor mental health, substance abuse, and also violence.

In the USA, Medicaid covers 60 million low-income individuals and provides health insurance for 45 percent of those under the federal poverty levels, including one in three children, four in 10 births, and 8 million people with disabilities. The number of individuals and families affected by homelessness is increasing in both rural and urban areas, with a 9 percent increase between 2007 and 2008. As many as 3.5 million Americans are estimated to be homeless at sometime during the year and in 2009 an estimated 1.56 million spent at least one night in a shelter.

Women and families are the fastest growing segment of the homeless population; domestic and sexual violence is the leading cause, and the direct cause for up to 50 percent fleeing from domestic violence. Women make up 36 percent of those in shelters, and 81 percent of adults in family shelters; more than half the children are under 6 years of age.

Homeless women in the 18–44-year age group are more likely to die than women in the general population. Those in their mid-fifties are psychologically aged and affected by chronic illness comparable to women in their seventies in the general population. After adjusting for risk factors, homeless women are 2.9 times more likely to have a preterm delivery, 6.9 times more likely to have a low birth-weight infant ($<2000\,g$) and 3.3 times more likely to have a small for gestational age newborn.

The US government has plans to set ambitious but measurable goals: (1) end chronic homelessness in 5 years; (2) prevent and end homelessness among veterans in 5 years; (3) prevent and end homelessness for families, youth, and children within a decade; and (4) put the country on a path to ending all types of homelessness

In developing countries homelessness because of social determinants is an enormous societal problem; India has an estimated homeless population greater than 78 million.

Sources: Cheung AM, Hwang SW. Risk of death among homeless women: a cohort study and review of the literature. CMAJ 2004;170:1243–7. Available at: http://www.cmaj.ca/content/170/8/1243.full http://www.cmaj.ca/content/170/8/1243.full.pdf+html
Hwang SW, Wilkins R, Tjepkema M, O'Campo PJ, Dunn JR. Mortality among residents of shelters, rooming houses, and hotels in Canada: 11 year follow-up study. BMJ 2009;339:b4036. Available at: http://www.bmj.com/content/339/bmj.b4036 [Accessed 24 March 2013].
Centers for Disease Control and Prevention. National homeless persons' memorial day [updated 17 December 2012]. Available at: http://www.cdc.gov/Features/Homelessness/ [Accessed 19 May 2013].
National Healthcare for the Homeless Council, 2013. Available at: http://www.nhchc.org/ [Accessed 19 March 2013].

TABLE 6.2 US Preventive Screening Services Guidelines for Children, Adolescents and Adults

Level	Priority	Examples
Level I preventive screening services	Highest priority services for providers and care systems; these have the highest value and should be offered at every opportunity including outreach	Full immunization of infants, children, adolescents, and adults Maternal and child care, counseling, breastfeeding, folic acid, prenatal care, and nutritional and exercise counseling Newborn screening and child development care, safety, breast feeding, vitamin K injections, eye care, Apgar score, vitamin D and iron supplements Counseling adults re risk factors: obesity, smoking, inactivity, hypertension, cardiovascular diseases, diabetes, blood lipids, cancer, trauma, alcohol excess, sexual safety Screening for breast, cervical, skin, and colorectal cancer Body mass index, blood pressure, cholesterol, blood sugar Depression
Level II preventive screening services	Clinicians and care systems should assess and recommend to every patient whenever possible because of their demonstrated value	Motor vehicle safety, oral health Screening: hearing for newborns, children age 0–3 years, and adults; folic acid, obesity, hearing and vision impairment, depression, alcohol and tobacco use
Level III preventive screening services	Services recommended based on consideration of costs and health benefits. Evidence of effectiveness may be equivocal, incomplete or may result in harm	Screening: blood lead, hyperbilirubinemia, iron deficiency, cancer of skin, prostate, urinalysis, infectious disease, clinical breast examination, skin Counseling for domestic violence, child maltreatment, alcohol use
Level IV preventive screening services	Services that are not supported by evidence and should not be routinely recommended. Evidence of effectiveness may be lacking or clearly not effective with potential for harm without benefit for the individual	Screening: blood chemistry, ovarian cancer, hemoglobin, coronary heart disease, diabetes Child maltreatment screening Anemia screening at ages 5 and older Tuberculin screening (for average risk)

Sources: Agency for Health Care Research and Quality. Preventive services for children and adolescents. Available at: http://www.guideline.gov/content.aspx?id=38451.
Agency for Health Care Research and Quality. Preventive services for adults. Available at: http://www.guideline.gov/search/search.aspx?term=preventive+services+for+adults
Institute for Clinical Systems Improvement. Preventive services for adults, 2013. Available at: https://www.icsi.org/_asset/gtjr9h/PrevServAdults-Interactive0912.pdf
Institute for Clinical Systems Improvement. Preventive services for children and adolescents. Available at: https://www.icsi.org/_asset/x1mnv1/PrevServKids-Interactive0912.pdfs_2531.html
US Preventive Services Task Force. Recommendations for adults; Recommendations for children. Available at: http://www.uspreventiveservicestaskforce.org/recommendations.htm [All accessed 13 February 2013].
Wilkinson J, Bass C, Diem S, Gravley A, Harvey L, Hayes R, et al. Institute for Clinical Systems Improvement. Preventive services for adults [updated September 2012]. Available at: http://bit.ly.PrevServAdults0912. Available at: https://www.icsi.org/_asset/gtjr9h/PrevServAdults-Interactive0912.pdf [Accessed 15 February 2013].

and applicable to all providers in each health care setting across the health system. Table 6.2 categorizes the levels of priority which are based on reviews of available evidence. They are ranked by class of evidence (high, medium, and low quality), and by strength of recommendation (strong or weak) based on their clinical effectiveness and cost-effectiveness and the needs of the population. These are sometimes controversial and subject to change but represent the consensus of best practices.

THE FAMILY PHYSICIAN CANNOT DO ALL THIS ALONE

Clinical guidelines produced by many professional organizations are available online and are updated regularly by task forces, special study groups, and wide consultation. The CDC, American Academy of Pediatrics (AAP),

American Association of Family Physicians, and other professional organizations, after considerable preparation, adopted the World Health Organization (WHO) growth charts for child growth (up to age 2) based on an international standard (see Chapter 8). On 3 February 2013, the US Agency for Healthcare Research and Quality (AHRQ) listed nine new clinical guidelines from the USA (American Heart Association, Colorado Division of Worker's Compensation), Canada, the UK, and Spain. The Canadian Task Force on Preventive Health Care example updated a 2005 recommendation on screening adults for type 2 diabetes. The Catalan Agency for Health Information, Assessment Quality posted clinical guidelines on acute bronchiolitis. The British Committee for Standards in Haematology issued guidelines for screening and diagnosis of hemoglobinopathies (e.g., thalassemia and sickle-cell disease). The US National Collaborating Center for

Women's and Children's Health posted guidelines on antibiotics for prevention and treatment of early-onset neonatal infection. The fields of microbiology, nutrition, medicine, pediatrics, and family medicine, and the public health aspects of these fields continue to evolve as new evidence is provided by basic biological and genetic scientists, epidemiologists, social scientists, and many others.

The family physician is crucial in this process but cannot do it all alone. The health insurance or health service systems need to promote and even provide incentives and computer support to the primary care system to ensure that immunizations are completed, weight, height, body mass index (BMI), and blood pressure (BP) are measured, and periodic checkups for cervical, breast and colorectal cancer and many other preventive functions are implemented. The doctor's message regarding risks of smoking, alcohol excess, high blood pressure, and the need for screening for breast, cervical, and colorectal cancer has a powerful influence on patients and can be life saving. But this needs reinforcement from nurse practitioners, home visitors, community health workers (CHWs), health promotion, advocacy groups (e.g., American Cancer Society, March of Dimes, Mothers Against Drunk Driving), and culturally accepted norms. The latter can be influenced by professional organizations, government agencies, and the media, including the electronic world of communication by Facebook, Twitter, and many others. The US Preventive Services Task Force (USPSTF) issued guidelines for screening for excessive alcohol consumption using electronic screening and brief intervention (e-SBI), which has been shown to be effective. Telephone surveys are used by many research projects to monitor or evaluate community attitudes, knowledge, and practices relevant to the mission.

MATERNAL HEALTH

Women's health issues relate to their many roles: as family caregivers, individuals, workers, wives, grandmothers, mothers, and daughters. These roles demand lifelong responsibilities for knowledge, self-care, and family leadership in health-related issues, such as nutrition, hygiene, education, exercise, safety, fertility, child care, and care of the elderly. Changes in the social roles of women create extra demands and risks to health.

Fertility

Fertility is the natural potential and ability to conceive and have children through normal sexual activity. Infertility is defined as inability to conceive after a year of regular intercourse without the use of contraceptives. Childbirth and care of infants and children have been traditional concerns of public health because of the related vulnerability and historically large-scale loss of life. Improved nutrition and

living conditions as well as care during pregnancy and at delivery have combined to reduce maternal mortality globally and in all regions (see Table 6.5). Maternal and child survival and health are among the Millennium Development Goals (MDGs) selected by the United Nations (UN) in 2001 as global health targets for the year 2015. Child mortality reductions have been moving towards the targets, but maternal mortality is falling well behind.

The tragedies of high maternal mortality and morbidity are founded in poverty, poor nutrition, and unsanitary conditions, as well as lack of adequate professional health care. They are also the result of the poor general status of women, including low education and early age at fertility, previous number of births, the time or space between pregnancies, anemia and poor nutritional status, and lack of adequate facilities and standards of the health system. In developing countries, adolescence, low educational levels, and poor health prior to pregnancy are common. With little or no prenatal care and a lack of professional care during delivery, maternity is hazardous, especially for the rural population and the urban poor. These conditions are widespread in the developing world.

The maternal mortality rate (MMR) in sub-Saharan Africa fell from a 1990 rate of 850 deaths per 100,000 live births to a regional average of 500 by 2010 (41 per cent) (UNFPA, 2012). Equatorial Guinea reduced its MMR from 1200 to 240 per 100,000 (81 per cent) from 1990 to 2010, one of 10 countries worldwide that has achieved MDG5. Eritrea is also moving to achieve MDG5 with a reduction from 880 to 240 deaths per 100,000 live births in this period. However, MMR remains high in some countries; MMRs in 2010 in Chad were 1100, Somalia 1000, and Afghanistan 460 per 100,000 live births (reduced from 1300 in 1990) (WHO/UNICEF, 2012); in these countries the percentages of births with skilled birth attendance were 23, 34, and 23 percent, respectively (UNICEF, 2013).

India's MMR declined from 600 per 100,000 live births in 1990 to 200 in 2010 (−66 percent), with wide variation between the states and between urban and rural populations. Skilled birth attendance was reported as 53 percent in 2010. The Minister of Health and Family Welfare of India promised to addressed these issues in February 2013 (see Chapter 13): he promised new maternal and child health wings in district hospitals and medical colleges; comprehensive medical check-ups for children across the country; and to create networks of ambulances to cover remote and inaccessible areas. Providing free drugs for pregnant women and sick newborns, funds would be given for purchase of drugs for all types of illness for patients in government health facilities. The minister said that improving the health of mother and child was vital to increasing equity and reducing poverty. The poor performance of India in many of the basics of primary care despite the impressive economic growth of the country is a political failure which will hamper the development of the country in comparison to its competitors.

Female education, delayed marriage until adulthood, nutritional adequacy, and communicable disease control are all vital factors to ensuring that females reach pregnancy physically and intellectually prepared for childbirth and child rearing. Literacy for women contributes most to improved infant survival rates. The literate mother has a greater ability to address health issues, using written material instead of depending solely on community traditions, and is better able to cope with the complexities of a health care system. Greater education is also likely to lead to a better chance of employability and greater family income.

Fertility-related health issues include preparation for and timing of pregnancy, and professional care during and following pregnancy. Using health education and prevention services, public health provides the necessary resources through which these issues may be addressed. Cessation of smoking, alcohol, drug use, and risk-taking behavior at least for the duration of the pregnancy is an important part of protection of the fetus from harm.

Fertility patterns are affected by economic, social, cultural, religious, and other factors, including the technology available for birth control. With available and accessible safe and effective birth control measures and increasing education and job opportunities, fertility patterns have changed in the industrialized countries. Ethnic differences can be substantial within a country. For African American women, the crude birth rate fell from 22.4 per 1000 in 1990 to 16.6 in 2008, compared to 15.8 and 13.4 per 1000, respectively, for white women. The crude birth rate for Hispanic/Latino women declined from 26.7 to 22.2 during the same period. The USA has seen a decline in rates of teenage pregnancies to 67.8 per 1000 women aged 15–19 in 2008, a reduction of 42 percent from 116.9 per 1000 in 1990. But US rates are high compared to other industrialized countries, and disparities in rates among black and Hispanic American teens remain high compared to whites (CDC, Teen Pregnancy, 2012).

Fertility rates in many developed countries have declined considerably since the peak "baby boom" that occurred in the years following World War II. This has generated concern that the birth rate is lower than that needed to sustain the population even at current levels. Developing countries have the opposite problem, in that high birth rates with declining morbidity are placing a heavy economic, social, and medical burden on the community while the cultural and religious ethos produces little support for family planning.

Globally, despite declining birth rates, the world's population grew from 1 billion in 1804 to 2 billion in 1927, 3 billion in 1960, and to 7.1 billion in 2011. Family planning and spacing of pregnancy are vital issues in developing countries where the burden of frequent pregnancies contributes to high maternal and infant mortality rates. The traditional means of fertility limitation in these countries was through prolonged breastfeeding, which has declined with the promotion of commercial baby formulas. Improvements in socioeconomic conditions, with female education, infant and child survival, and the strengthening of family planning services, are likely to provide the needed impetus for increased contraceptive use and fertility decline in developing countries.

Public Health Challenges of Fertility

Provision of birth control along with prenatal, delivery, and postpregnancy care are among the central roles of any health care service. The high levels of maternal mortality of the past in the industrialized countries are still present in developing countries. Traditions of high fertility rates, unattended deliveries, unsafe practices of traditional birth attendants (TBAs), and female genital mutilation greatly contribute to the poor health status of women in developing countries.

Safe prenatal care and birth practices, along with better nutrition and general health status of women, reduced maternal morbidity and mortality dramatically in the twentieth century. But the health gap remains enormous between the industrialized and least developed countries. Fertility rates have been falling in most parts of the world as individual education levels improve. Economic incentives in rural areas for more children are replaced by incentives for fewer children, as urbanization fostered by a search for better standards of health and living conditions occurs. As the technology of birth control becomes widely available, better educated women have the power to control their own fertility. There is an encouraging pattern of fertility decline in all parts of the world. In sub-Saharan Africa total fertility rates above 6 until the 1980s declined to just over 4 in 2010 and are expected to continue declining in the coming decades. There is a global trend in that having reached their lowest fertility level with sub-replacement fertility, some countries including the USA and the UK are experiencing slight increases in birth rates. Fertility in sub-Saharan Africa and in the least developed countries remains high (Table 6.3), and is a major factor in poverty and high mortality rates, especially for children and mothers.

The UN Population Division estimates that fertility rates in most high- and medium-fertility countries will decline to a fertility level below 2.1 children per woman by 2020 but with little variation across world regions. Public health promotes spacing of pregnancy to improve health outcomes for both the mother and the newborn, with wider implications for family health. The least developed countries are expected to reduce total fertility rates to under 3 per woman by 2030 (Figure 6.3). Global population growth is estimated as follows by the UN (Department of Economic and Social Affairs, World Population Prospects Revision 2010):

- 3 billion – 20 October 1959
- 4 billion – 27 June 1974
- 5 billion – 21 January 1987
- 6 billion – 5 December 1998

- 7 billion – 31 October 2011
- 8 billion – 15 June 2025
- 9 billion – 18 February 2043
- 10 billion – 18 June 2083.

In many societies, infertility is a problem associated with significant personal distress and social stigma.

TABLE 6.3 Total Fertility Rates by World Demographic Regions,[a] Selected Years, 1980–2010

Region/Group	1980	1990	2000	2005	2010
Sub-Saharan Africa	6.7	6.3	5.6	5.4	4.9
India	4.9	3.7	3.1	2.9	2.6
China	2.8	2.2	1.8	1.7	1.6
South Asia	5.2	4.0	3.4	3.1	2.7
Latin America and Caribbean	4.2	3.2	2.6	2.5	2.2
Middle East and North Africa	5.9	5.1	3.7	3.1	2.8
CIS/CEE and Baltic States	2.2	2.3	1.6	1.7	1.8
Industrialized countries	1.9	1.7	1.6	1.6	1.7
Developing countries	3.9	3.1	3.0	2.8	2.6
Least developed countries	6.5	5.7	5.3	4.9	4.2
World	**NA**	**3.2**	**2.7**	**2.6**	**2.5**

Note: [a] Total fertility rate = average number of children who would be born per woman if she lived to the end of her childbearing years and bore children according to prevailing age-specific fertility rates (see Chapter 3 and Glossary).
CIS = Commonwealth of Independent States; CEE = Central and Eastern Europe.
Source: United Nations Children's Fund. The state of the world's children, 2012. Available at: http://www.unicef.org/sowc/files/SOWC_2012-Main_Report_EN_21Dec2011.pdf [Accessed 6 February 2013].

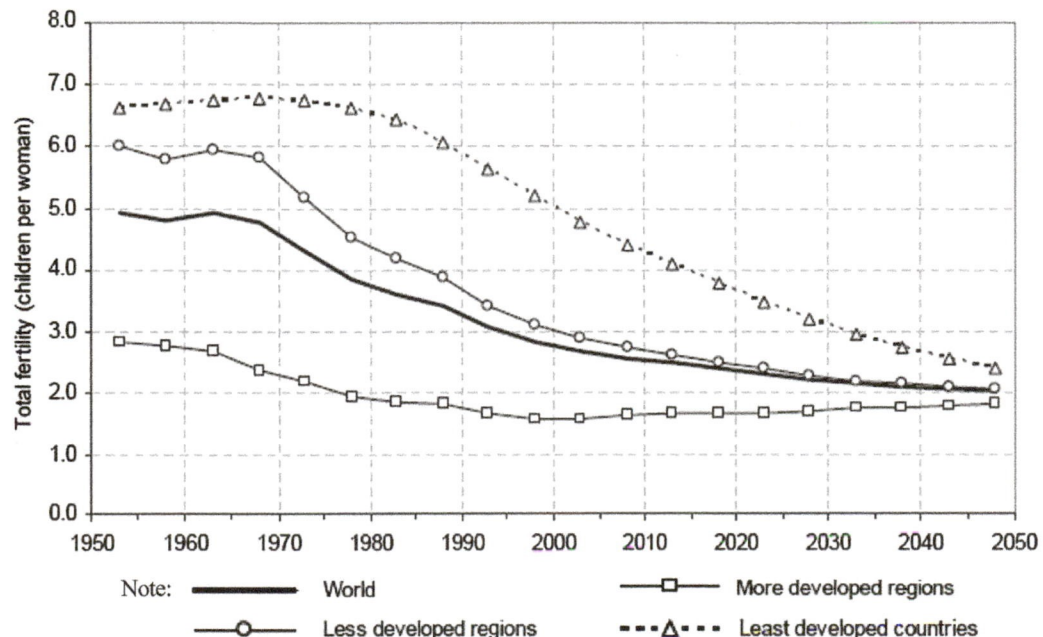

FIGURE 6.3   Total fertility trajectories for the world and major country development groups, 1950–2050 (medium variant). *Source: Population Division of the Department of Economic and Social Affairs of the United Nations Secretariat. World population prospects: the 2008 revision. New York: United Nations; 2009. Available at: http://www.un.org/esa/population/publications/wpp2008/wpp2008_highlights.pdf [Accessed 27 November 2012].*

TABLE 6.4 Percentage of Married Women Using Some Form of Contraception, World Regions, 1990–1997, 1997–2005, 2006–2010

Region	1990–1997	1997–2005	2006–2010
Sub-Saharan Africa	15	24	23
Middle East and North Africa	46	53	45
South Asia	38	46	51
Latin America and Caribbean	64	71	74
East Asia and Pacific	74	79	78

Source: United Nations Children's Fund. The state of the world's children, 1998, 2007, and 2012. Available at: http://www.unicef.org/sowc/ (Accessed 25.10.2013).

Sexually transmitted infections (STIs) are a major cause of infertility, so the prevention and treatment of STIs are important aspects of managing infertility. Treatment of infertility is associated with high cost, not only in expenditure, but also in emotional trauma. Modern services to treat infertility, including stimulation of ovulation, in vitro fertilization (IVF), and surrogate parenting, raise many ethical and financial issues. Despite these and other problems of multiple births and high rates of very low birth-weight infants with associated perinatal and developmental problems, infertility treatment is very much a part of modern health care systems.

Family Planning

Family planning enables a woman to determine the time, spacing, and frequency of pregnancy with a range of methods for preventing or expelling a conception while maintaining a normal sex life. Adoption is another aspect of family planning. The technology for birth spacing has been revolutionized, so that there are now safe and effective contraceptive methods widely available at reasonable cost in industrialized and developing countries (Table 6.4). Birth control is a public health issue and responsibility.

Between 2006 and 2010, reported contraceptive use ranged from 23 to 78 percent with a world average of 63 percent. Low levels of awareness or inaccessibility to birth control contribute to high birth rates and high maternal mortality. They also lead to the use of abortion for birth control, which may complicate subsequent pregnancies. High fertility rates or pregnancies late in life pose extra hazards to both the mother and the newborn, and place a burden on the family unit. In sub-Saharan Africa and the Caribbean, one in five and one in four women, respectively lack access to contraception.

The projected increased demand for contraceptive use will include an increased unmet need of 6.8 percent in developing countries by 2015. The unmet need, on average, is highest in East, Middle, and West Africa, with rates of 26.3, 26.1, and 25 percent, respectively. Little change

occurred between 1990 and 2010 in most WHO regions, particularly in sub-Saharan Africa, where estimated unmet need exceeded contraceptive use prevalence. In 2010, of the 42 countries with national estimates of unmet needs greater than 25 percent, 29 were in Africa, but there were also high rates (24.6 percent) in the Pacific island nations of Melanesia, Micronesia, and Polynesia. The absolute number of married women of reproductive age with a demand for contraception is projected to increase substantially by 2015, both worldwide in most developing countries. Increased investment is needed to overcome the stagnation in funding since the 1990s to improve reproductive health.

Modern birth control methods include pharmacological and chemical prevention of conception (hormone pills, hormonal implants or injections, spermicides) and physical methods [male and female condoms and intrauterine devices (IUDs)]. Traditional methods of prevention, such as breastfeeding, the rhythm method, and coitus interruptus, are less reliable than current methods, mainly the pill, tubal ligation, female condoms, and the IUD. Birth control measures for use by males include condoms or vasectomy. Condoms promote safer sexual contact, but both methods rely heavily on education and awareness levels.

Spacing of pregnancies by modern methods of birth control is a fundamental right of women. In many areas of the world, birth control remains a religious and political issue as well as a power struggle between men and women. The Roman Catholic Church is strongly opposed to "artificial" birth control, although it accepts the less reliable rhythm and withdrawal methods. Many political regimes oppose abortion for national demographic reasons. However, in many countries with a predominantly Catholic population, the birth rate has fallen and in some is below the population replacement level. Use of birth control has increased dramatically over recent decades in Latin America and East Asia.

Family planning has not been well understood or available in the former Soviet republics. They lack both a supply and widespread awareness of modern contraception, relying on frequent abortion instead. The political aspects of birth control were and remain highly controversial and their introduction was the result of technological advances, strong advocacy, and the will of women to control their own fertility practices (see Chapter 12).

While the absolute number of abortions has declined in Eastern Europe, abortion rates remain four times higher than Western European rates (43 and 12 per 100,000 live births, respectively), and a primary means of fertility control. Mortality due to abortion is highest in the Eastern Mediterranean Region, at a rate of 40 per 100,000 live births. In the Russian Federation, the 2006 abortion rate was almost 1.5 times the number of live births; however, abortion rates are falling in Commonwealth of Independent States (CIS) countries, including Russia.

The difference in abortion rates between Western and Eastern Europe partially reflects differences in grounds for legal abortion. Globally, nearly half of all abortions are unsafe and 98 percent of unsafe abortions occur in Africa; an estimated 21.6 million unsafe abortions took place in 2008 and caused 13 percent of all maternal deaths with a rate of 220 per 100,000 procedures, 350 times higher than the rate of 0.6 per 100,000 procedures for legal induced abortions in the USA.

The WHO reports that in 2008 unsafe abortions in the Eastern Mediterranean Region occurred at the rate of 24 per 1000 women (15–44 years); 20 per 1000 live births compared to 3 and 5, respectively, in the European Region. Abortion continues to be an important cause of preventable morbidity and mortality. The risk of death from unsafe abortion is 30 per 100,000 live births, but 80 per 100,000 in the least developed countries. In the sub-Saharan African Region the case death rate is 800 times higher, at 520 deaths per 100,000 procedures. Unsafe abortions increased from 2003 to 2008 by 2 million, almost entirely in developing regions. Among women of childbearing age in the developing world, an estimated 23 percent of maternal disability-adjusted life years (DALYs) lost was due to unsafe abortion, and worldwide a further 6 million DALYs are lost from postabortion morbidity and mortality. An estimated 3 million of the 8.5 million women in developing regions needing care after unsafe abortion do not receive it. Stability in abortion rates between 2003 and 2008 coincides with the plateau in UN funding of contraceptive uptake.

Despite the wide availability of contraception, teenage pregnancies account for a relatively large percentage of total births in the USA. This is a serious public health problem, being more prevalent in lower socioeconomic groups with problems of single mothers with welfare dependency and associated with poor health outcomes. Teenage unmarried pregnancies are a complex of social, educational, and labor force problems that cannot easily be addressed. They are a growing problem in other industrialized countries as well. In developing countries, teenage and even child marriage produces a wide range of maternal health problems of physical and emotionally immature mothers trapped in a life role with no chance of becoming educated or employed.

In the USA, progress is being made towards the health target for a reduction in teen pregnancies: the rate declined in the age group 15–19 from 61 per 1000 females in 1990 (a peak year) to 42 in 2006 and further declined to 34.4 in 2010. Sexual abstinence among females before 15 years of age and those aged 15–17 years has increased. African American and Hispanic youth comprise 35 percent of this age group but account for 57 percent of births in the age group. Both abstinence and use of condoms have increased among teen males. Recent emphasis on abstinence education programs focusing on prevention of a first pregnancy and the emerging availability of emergency contraception along with fear of HIV have contributed to this phenomenon, as indicated in achievements of the Midcourse Review of Healthy People 2010 and in the new Healthy People 2020.

Hormonal contraceptive implants and IUDs offer advantages over other contraceptive methods such as abstinence, withdrawal, condoms, injectable contraceptives, and combined oral contraceptives, all of which are dependent on user compliance. The first contraceptive implant system became available in 1983, after which two others were developed. Implants are approved and increasingly popular in more than 60 developed and developing countries and are currently being used by millions of women worldwide. They provide high contraceptive efficacy almost immediately after insertion, a low rate of complications and side-effects, and long-term effectiveness (for 3–5 years) after a single intervention. They allow rapid return of fertility after removal. Both insertion and removal take only a few minutes and prices are being lowered for use in developing countries. The WHO reviewed this among alternative methods and concluded that contraceptive implants are good candidates for inclusion in the choices offered by family planning programs, especially in underresourced settings, provided they are implanted by well-trained staff under sanitary conditions, with full explanation of their effects and the opportunity for removal if a return to fertility is desired (Bahamondes, 2008; Jacobstein and Stanley, 2013). The 2013 approval by the US Food and Drug Administration (FDA) of over-the-counter access for "Plan B", 24-hour postcoital contraception, will be important in reducing unwanted pregnancies, especially among teenagers.

Maternal Mortality and Morbidity

A maternal death as defined by the WHO is death of a woman while pregnant or within 42 days following termination of pregnancy from any cause related to or aggravated by the pregnancy or its management, i.e., early maternal mortality, but not from accidental or incidental causes. Late maternal mortality includes deaths from 42

TABLE 6.5 Maternal Mortality Rates, Selected Population Groups and Years, USA, 1970–2007

Group	1970	1980	1990	2000	2005	2007	Decline 1970–2007
White	14.4	6.7	5.1	6.2	9.1	7.7	46.5
Black	65.5	24.9	21.7	20.1	31.7	23.5	64.1
Total	21.5	9.4	7.6	8.2	12.4	10.2	52.6

Note: Age-adjusted rates per 100,000 live births. Data from 1999 according to ICD 10. Health target for 1990, not more than 5/100,000 overall and for all local and ethnic group rates. Black are African American.
Sources: Centers for Disease Control and Prevention. Health United States 2010. Table 36. Available at: http://www.cdc.gov/nchs/data/hus/hus10.pdf [Accessed 8 October 2012].
Centers for Disease Control and Prevention. Pregnancy related mortality in the United States. March 2012. Available at: http://www.cdc.gov/reproductivehealth/MaternalInfantHealth/Pregnancy-relatedMortality.htm [Accessed 3 February 2013].

days to 1 year after termination of the pregnancy. Maternal deaths are subdivided into direct causes, i.e., deaths resulting from obstetric complications of the pregnant state; and indirect causes, i.e., deaths resulting from preexisting disease or conditions not directly due to obstetric causes.

In past centuries, childbirth was highly life threatening. In 1664, a report of the Hôtel-Dieu Hospital in Paris indicated a maternal death rate of 33 percent among hospitalized parturient women. In England between 1660 and 1680, one woman in 44 was reported to have died in childbirth. Ignaz Semmelweiss in Vienna in the 1850s reported death rates of 2 percent in women during or following delivery by midwives as compared to rates of 10 percent in medically supervised wards. This was due to lack of precautions against transmission of puerperal fever through unhygienic practices of the medical staff. The work of Semmelweiss led to the control of streptococcal septicemia, a major cause of maternal death, by promoting hand washing by doctors. The work of Florence Nightingale and Joseph Lister in the Crimea and England made hygiene practices major issues in hospital and surgical facilities (see Chapter 1). Health care facility-associated infections remain an issue in preventing infections in maternal, neonatal, and many other components of health care facilities. Great stress is required for basic hygiene by health care providers, support staff and visitors, with careful and frequent hand washing in hospitals and in maternity–newborn care, along with other infection control measures (see Healthy People 2020 and Chapters 1 & 4).

In the USA, the MMR in 1900 was close to 850 per 100,000 live births, and in 1930 was 670 per 100,000, rates similar to those in very low-income countries today. Most maternal deaths were due to sepsis, hemorrhage, and eclampsia, associated with poverty and poor nutrition, poorly trained practitioners, excessive intervention, and home deliveries with low standards of hygiene. A 1933 White House Conference on Child Health Protection, Fetal, Newborn, and Maternal Mortality and

Morbidity identified the major problems and initiated hospital-based delivery standards and guidelines, defining physician qualifications for hospital delivery privileges, and increasing hospital deliveries during 1938–1948, from 55 percent to 90 percent. The transfer of deliveries to maternity wards of general hospitals contributed to the improved safety of maternity along with improved standards of hospitals across the country (see Chapter 10). The ratio declined substantially during the 1940s and 1950s, and continued to decline until 1982. The decline in maternal mortality was recognized by the CDC as "one of the great achievements of public health of the twentieth century" (CDC, 1999). MMR fell from 670 per 100,000 live births in 1930 to 21.5 in 1970, and 7.6 in 1990, rising to 8.2 in 2000, 12.4 in 2005, and declining to 10.2 in 2007 (Table 6.5).

During 1982–1996, the annual US MMR were between approximately 7 and 8 maternal deaths per 100,000 live births. Rates for black women generally fluctuated between 18 and 22 per 100,000 births and for white women between 5 and 6 per 100,000 live births. MMRs for African Americans remained more than three times higher than those of the white population. Access to and use of health care services for early diagnosis and effective treatment if complications develop, may be a factor. Maternal morbidity is also important in long term sequelae due to inadequate prepregnancy, antenatal, and delivery and post partum care. During the 2006–2008 period, the pregnancy-related mortality ratios were 11.3 deaths per 100,000 live births for white women and 34.8 deaths per 100,000 live births for black women. In 2010, the US rate overall was reported as 21 per 100,000, compared to 12 in Canada and 14 in the UK.

Figure 6.4 shows trends in MMR and infant mortality between 1960 and 2002 with a rise in MMR in 1995, continuing to 2007 (Table 6.5), and climbing to a rate of 21 per 100,000 in 2010 (APHA, 2011). The pregnancy-related mortality ratio was 15.2 deaths per 100,000 live births for the period 2006–2008.

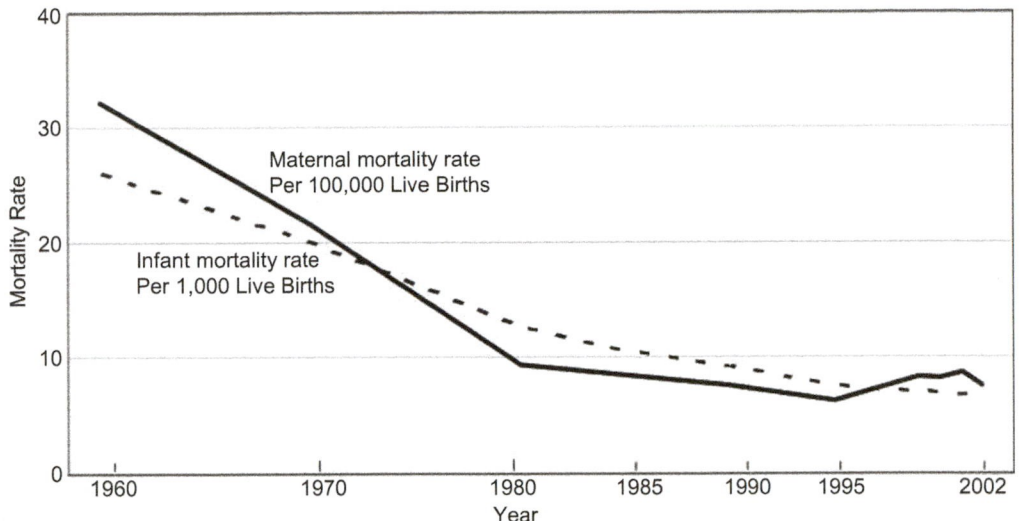

FIGURE 6.4   Maternal and infant mortality rates, USA, 1960–2002. *Source: Atrash HK, Johnson K, Adams M, Cordero JF, Howse J. Preconception care for improving perinatal outcomes: the time to act. Matern Child Health J 2006;10(Suppl 1):3–11. Published online 2006, June 14. http://dx.doi. org/10.1007/s10995-006-0100-4 PMCID: PMC1592246. Available at: http://www.ncbi.nlm.nih.gov/pmc/articles/PMC1592246/figure/Fig1/ [Accessed 6 February 2013].*

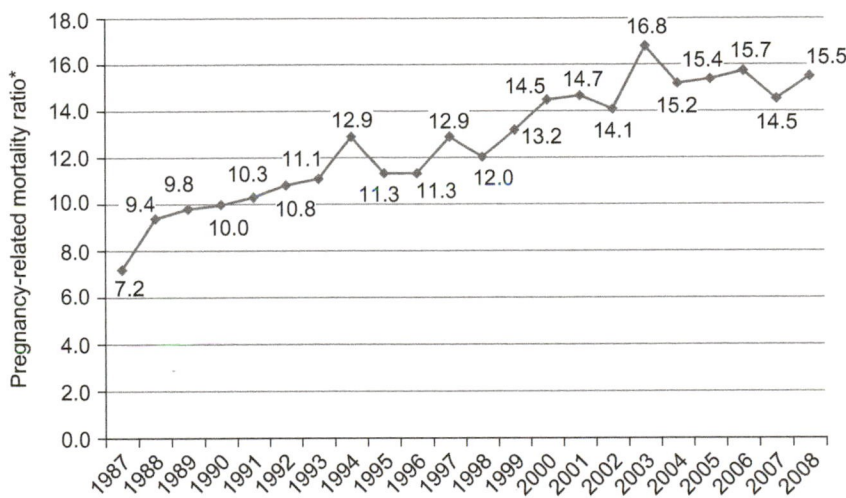

FIGURE 6.5   Trends in pregnancy-related mortality, USA, 1987–2008. Note: This is intended as a model for primary care early identification of high risk and prompt referral on basis of history and basic findings. Full prenatal care recommendations are available at American College of Obstetrics and Gynecology. *Number of pregnancy-related deaths per 1,000,000 live births per year. *Source: Centers for Disease Control and Prevention. Reproductive health: pregnancy mortality surveillance system [updated 7 March 2013]. Available at: http://www.cdc.gov/reproductivehealth/MaternalInfantHealth/ PMSS.html [Accessed 25 March 2013].*

CDC surveillance of maternal mortality includes deaths up to 1 year following delivery, as opposed to the 42 days after delivery used by the WHO, and includes medical conditions that are worsened by pregnancy. This may account for some of the excess US mortality compared to other countries. Nevertheless, more than half of all maternal deaths can be prevented through early diagnosis and appropriate medical care of pregnancy complications. Hemorrhage, pregnancy-induced hypertension, infection, and ectopic pregnancy continue to account for most (59 percent) maternal deaths (Figure 6.5). Pregnancy-related deaths in the USA in 2006–2008 were caused by:

- cardiovascular diseases – 14.6 percent
- cardiomyopathy – 12.4 percent
- non-cardiovascular diseases – 11.9 percent
- hemorrhage – 11.5 percent
- infection/sepsis – 11.1 percent
- hypertensive disorders of pregnancy – 10.5 percent
- thrombotic pulmonary embolism – 10.3 percent
- amniotic fluid embolism – 5.9 percent
- cerebrovascular accidents – 5.7 percent
- anesthesia complications – 0.6 percent
- unknown causes – 5.5 percent.

TABLE 6.6 Maternal Mortality Rates by World Health Organization Regions and Income Groups, 1990, 2000, and 2010

	1990	2000	2010	% Change 1990–2010
WHO region				
Africa	820	720	480	−41.5
Americas	100	80	63	−37.0
South-East Asia	590	370	200	−66.1
European	44	29	20	−54.5
Eastern Mediterranean	430	360	250	−41.9
Western Pacific	140	77	49	−65.0
Income group				
Low income	810	630	410	−49.4
Lower middle income	560	420	260	−53.6
Upper middle income	120	76	53	−55.8
High income	16	13	14	−12.5
Global	**400**	**320**	**210**	**−47.5**

Source: World Health Organization. World health statistics, 2012. Derived from Table 2. Cause specific mortality and morbidity, pages 80–82. Available at: http://www.who.int/healthinfo/EN_WHS2012_Part3.pdf [Accessed 21 March 2013].

The US Healthy People 2020 targets reduction of maternal deaths to 11.4 per 100,000 live births. The use of the International Classification of Diseases, 10th revision (ICD-10) and including maternal deaths up to a year following termination of pregnancy show an upward trend in MMR in the USA (CDC Reproductive Health, 2013).

During the period 1970–2007 in the USA, maternal mortality ratios declined by 45.6 percent for the white population and by 64.1 percent among blacks. However, the rate remains over three times higher among blacks than among white Americans. There remains a wide gap related to age and pregnancy preparation as well as adequacy of care, which continues to be a serious public health problem, especially for poor, often single, African American pregnant teenagers. The rise in maternal mortality may be due in part to lack of universal health insurance, increases in the uninsured population, and lower standards of health insurance benefits for many of the insured.

The MMR in the USA remains high compared to other leading industrialized countries. In 2010, the USA was placed 47th in the world for maternal mortality, near Iran and Hungary, 14 places lower than Canada, and 10 places below the UK. The US MMR is higher than almost all European countries, as well as several countries in Asia and the Middle East (Association of Reproductive Health Professionals, 2011). This dramatic decline in maternal mortality trends in most industrialized countries can be attributed to many factors, including better standards of living and nutrition, improved medical care, and advances in obstetric knowledge and training.

Internationally, an estimated 287,000 women died in 2010 from complications of pregnancy and childbirth, 99 percent of whom (284,000 or 200 per 100,000) were in developing countries. Sub-Saharan Africa (162,000: MMR 500) and Southern Asia (83,000: MMR 220) have the highest rates of maternal deaths, accounting for 85 percent of maternal deaths in 2010 (Table 6.6). Although the global MMR declined from 400 to 210 per 100,000 from 1990 to 2010, progress has been insufficient to meet the targeted reduction of 75 percent (the reach the target of a global MMR of 100) by 2015. The 2010 UN Global Strategy for Women's and Children's Health calls for renewed actions in all countries to achieve MDG5 and improve maternal health.

In 2001, the MDG5 target was for a 75 percent reduction in the MMR from 1990 to 2015 to a 2015 global MMR of 100. Maternal mortality levels are still far from the 2015 target and progress is mixed (see Table 6.6 and UN MDG Report 2012). Only 23 countries are on track to achieve the targeted 75 percent decrease in MMR by 2015 (Hogan et al., 2010). The WHO, United Nations Children's Fund (UNICEF), United Nations Population Fund (UNFPA), and the World Bank estimates (2012: Trends in Maternal Mortality 1990 to 2010) report that the number of women dying of pregnancy and childbirth-related complications has almost halved in 20 years. From 1990 to 2010, the annual number of maternal deaths dropped from more than 543,000 to 287,000, a decline of 47 percent. While substantial progress has been achieved in almost all regions, many countries, particularly in sub-Saharan Africa, will fail to reach the MDG target of reducing maternal death by 75 percent from 1990 to 2015. On an annual basis, sub-Saharan Africa

TABLE 6.7 Women and Child Health Indicators by Level of Country Development

	Unmet Need for Family Planning 2005–2010 (%)	Contraceptive Prevalence 2005–2010 (%)	Antenatal Visits 2006–2011, 1–4 visits (%)	Births by Skilled Personnel as % of Total Live Births 2005–2011 (%)	Under-5 Mortality Rates/1000 Live Births		
					1990	2000	2008
Region							
Africa	25	24	74–43	48	172	154	119
Americas	9	75	93–87	93	42	28	18
South-East Asia	13	58	76–52	59	111	80	57
European	–	71	NA	98	33	22	14
Eastern Mediter-ranean	20	42	72–43	59	100	80	68
Western Pacific	4	80	93–NA	91	48	33	19
Income group							
Low income	23	35	74–36	46	164	136	107
Lower middle income	14	52	76–53	58	112	89	69
Upper middle income	4	80	95–NA	96	49	33	19
High income	–	71	NA–96	99	12	8	6
Global	**11**	**63**	**81–55**	**69**	**88**	**73**	**57**

Note: Crude birth rate=number of live births per 1000 population; maternal mortality rate=maternal deaths per 100,000 live births. Rates for 2000 were adjusted by UNICEF.
Source: World Health Organization. World health statistics, 2012. Derived from Table 2. Cause specific mortality and morbidity, page 61. Available at: http://www.who.int/healthinfo/EN_WHS2012_Part3.pdf [Accessed 21 March 2013].

and Oceania regions have made the least progress. In 2010 the UN Secretary General launched the Global Strategy for Women's and Children's Health for renewed action towards improvement of women's and children's health to accelerate progress toward achieving MDG5.

The leading causes of maternal deaths globally are hemorrhage, followed by eclampsia, sepsis, and unsafe abortions. The World Bank and WHO/UNICEF estimate that the extension of prenatal, delivery, and postpartum care to 80 percent of the world's population would reduce by 40 percent the burden of disease associated with unsafe childbirth. They suggest that an appropriate intervention program should include information, education, communication, transportation, community-based obstetrics, and district hospital facilities. High MMRs are associated with high birth rates and delivery by untrained personnel (Table 6.7). Sustained and intensified effort is required for acceleration of progress. Sub-Saharan Africa and Southern Asia, with the highest maternal mortality, also have less than half of births attended by skilled personnel. The risk of a sub-Saharan African woman dying from complications of pregnancy and childbirth during her life is 1 in 31, compared to 1 in 4300 in the developed world.

More than 50 percent of all maternal deaths occurred in only six countries in 2008 (India, Nigeria, Pakistan, Afghanistan, Ethiopia, and the Democratic Republic of the Congo). Maternal mortality in India declined from 600 to 200 between 1990 and 2010. Nigeria's rate declined from 1100 in 1990 to 630 in 2010; Ethiopia's MMR declined from 950 to 350 per 100,000 live births. During 1990–2010, average rates of yearly decline in the MMR varied widely within the regions. In developing regions, Eastern Asia had the lowest MMR level, at 37 deaths per 100,000 live births, with 92 percent of the women seen at least once prenatally and 99 percent of births attended by skilled health personnel. Among countries variation ranged from a high average annual rate of decline of 8.8 percent (range 8.7–14.1) in the Maldives (from 830 per 100,000 in 1990 to 60 per 100,000 in 2010) to an average annual increase of 5.5 percent (from 450 to 570 per 100,000 in Zimbabwe).

Reaching MDG5 is feasible, with examples of countries that have achieved average annual declines of 5.5 percent including China, Egypt, Poland, Iran, Turkey, and Vietnam. The slow achievement in reducing maternal mortality is attributed by some to a lack of political commitment and related program planning. The MDG target may be extended and achievable globally, if not by 2015 then by 2020. But achieving this target in countries that have made "insufficient" or "no" progress requires more effort at the national level with country health plans, a comprehensive, integrated package of essential interventions, strengthening of health systems, coordinated research and innovation, and perhaps above all health workforce capacity building. International donors have focused on infectious disease control and especially HIV control, with much justification, but the MMR target requires development of human and capital infrastructure with massive training efforts (see Chapter 14) (WHO, 2012).

Maternal mortality reviews are a common method of investigating maternal deaths in order to ascertain preventable causes. Such reviews can help to identify factors leading to maternal deaths and provide important lessons for medical, obstetric, and community health staff. They are requirements for hospital accreditation in Canada and the USA (see Chapters 13 and 16).

The basic issue is that lowering maternal mortality requires preventing teenage pregnancies by education and access to birth control, universal access to reproductive health by 2015, cultural and political changes, with investment in training and infrastructure for safe delivery and postneonatal care. Without appropriate political commitment at the national and international levels to this as a priority issue the MDG cannot be achieved. Despite agreement on the MDGs, failure to progress in reducing maternal mortality is in part because other issues such as HIV and tuberculosis have received much more donor attention from the international community (see Chapter 16).

PREGNANCY CARE

The goals of prenatal, delivery, neonatal, and infancy care are to provide the mother and child with the optimal conditions and supervision to ensure the best possible outcome of the pregnancy, preserving the health and well-being of the mother and providing the newborn with the greatest chance of survival and optimal development. Since pregnancy is fraught with potential problems for the mother and the fetus, professional prenatal care and equally important self-care by the pregnant woman are necessary.

Prenatal care should be built around principles suggested by the American Congress (formerly College) of Obstetricians and Gynecologists (ACOG) and a Delphi Group Panel, US Department of Health and Human Services, in the late 1980s and periodically updated up to 2013, including:

- preparation for pregnancy
- early and continuous risk assessment
- health promotion
- medical and psychological intervention as needed.

Public health programs have the responsibility to assure prenatal care for the entire population either through direct provision of care or by obstetric services in managed care or private practice settings. Pregnancy is often a planned event, so that preparation for pregnancy is feasible. Pre-pregnancy and prenatal care should be early and complete for the health of the mother and the child. Preparation for pregnancy includes a general examination; inquiry as to possible allergies, infections, past obstetric history, and genetic problems; STI and HIV testing; nutrition status assessment and counseling; folic acid and iron supplements before and during pregnancy; smoking, alcohol, and drug use cessation; and mental health and social support systems with counseling as required. Other issues to explore are: undiagnosed, untreated, or poorly controlled medical conditions; immunization history, medication and radiation exposure in early pregnancy, nutritional issues, family history and genetic risk; tobacco, alcohol and substance use and other high-risk behaviors, occupational and environmental exposures; and social and mental health issues. Perinatal care standards recommended by ACOG are seen in Table 6.8.

The parents should be provided with user-friendly reading material on pregnancy and parenting. Pre-existing medical conditions and obstetric history are important to determine hypertension, diabetes, previous fetal or infant deaths, previous exposure to hepatitis C, and other medical or social risk factors. Good prenatal care presupposes a basic program of visits with extra care for those at special risk.

Prenatal care in the community includes early presentation, high-risk assessment and referral, and continuous care throughout the pregnancy. Early diagnosis of pregnancy is important in permitting the woman to attend prenatal care as early as possible, hopefully in the first trimester. Early presentation for prenatal care provides the opportunity to assess the health of the mother and to advise her on appropriate nutrition and self-care. In addition, it establishes a working relationship between the mother and the caregiver. Both the mother and the father should become involved in prenatal preparation, including prenatal classes and exercises. Early detection of potential complications offers better outcomes of care and genetic disorders can be diagnosed early. Great success has been achieved in the early twenty-first century with antiretroviral treatment for HIV-positive women, which protects against transfer of the virus perinatally to their newborns and in the first weeks of life, reducing the vertical transmission rate from 25 percent to 2 percent or less. Even instituting maternal prophylaxis during labor and delivery, or neonatal prophylaxis within 24–48 hours of delivery, or both, can substantially decrease rates of infection in infants.

ACOG also recommends pre-pregnancy consultation to discuss pregnancy and infant care issues, especially related to counseling on the hazards to the fetus of smoking, alcohol and other drugs, the need for folic acid and multivitamin

TABLE 6.8 American College of Obstetrics and Gynecology-Based Perinatal Care Guidelines Summary

Initial Prenatal Visit	First–Second Trimester	Third Trimester	Postpartum Visit
History and physical examination			
Medical history	Blood pressure	Blood pressure	Blood pressure
Obstetric/gynecological history	Weight	Weight	Weight
Family history	Fundal height	Fundal height and clinical evaluation of fetal weight	Breast, abdomen, and pelvic examinations
Demographic history	Fetal movement, heart rate, Doppler (begin 10–12 weeks)		Interim history
Psychosocial assessment	Interim history for problems/concerns	Interim history for problems/concerns/exercise and nutrition/preparation for delivery	Screen women with gestational diabetes mellitus for persistent diabetes 6–12 weeks postpartum
Physical examination	Fetal movement, contractions, vaginal bleeding, and leakage of fluid,	Fetal heart rate and movement, contractions, vaginal bleeding, and leakage of fluid	
Testing			
HCT/HGB	Urine screen for sugar and protein	Urine screen for sugar and protein	When indicated:
Pap test (if indicated)	Nuchal fold translucency test (first trimester)	GBS screen (35–37 weeks)	Pap smear
Urine culture/screen	When indicated		Other tests as indicated
Blood type and Rhesus screen	CVS (10–12 weeks)		
Antibody screen	Amniocentesis		
VDRL and RPR	Multiple marker serum/MSAFP (14–20 weeks)		
Rubella	Ultrasound for fetal anomaly (18–20 weeks)		
Chlamydia	HCT/HGB (24–28 weeks)		
HBsAg for hepatitis B carrier state	D(Rh) antibody screen (26–28 weeks)		
HIV education and screening (recommended with patient consent)	Diabetes screen (24–28 weeks)		
Optional labs			
Counseling			
Nutrition (iron and folic acid, or multivitamins)	Signs and symptoms of premature labor	Signs and symptoms of labor	Methods of birth control
Medication use	Medication use	Signs and symptoms of pregnancy-induced hypertension	Breastfeeding and mastitis
Physical and sexual activity	Exercise and nutrition	Exercise and nutrition	Postpartum depression
Avoidance of substance use – alcohol, drugs	Avoidance of alcohol and substance use	Avoidance/cessation of smoking	Restrictions and limitations
Avoidance/cessation of smoking	Avoidance/cessation of smoking	Infant care: breast or bottle feeding, infant car seat, circumcision	Exercise and nutrition
Expected prenatal care	Education courses available	Identifying a pediatrician	

Continued

TABLE 6.8 American College of Obstetrics and Gynecology-Based Perinatal Care Guidelines Summary—cont'd

Initial Prenatal Visit	First–Second Trimester	Third Trimester	Postpartum Visit
Safety belts and travel	Safety belts and travel	Post-term counseling	
Environmental hazards	Environmental hazards	Encourage the use of Adacel in addition to Rubella while in the hospital, postdelivery, to ensure immunizations up to date, include influenza and pneumococcal pneumonia	
Signs and symptoms requiring physician notification	Monitor fetal activity		
Domestic violence	Signs and symptoms requiring physician notification		
Influenza vaccine, if flu season			

Note: Visiting schedule for an uncomplicated pregnancy: every 4 weeks for week 4–28; every 2–3 weeks for week 39–35; every week for week 36 to delivery. Frequency of visits may vary with individual needs and risks. Postpartum visit: 4–6 weeks after delivery for uncomplicated delivery; visit advisable 7–14 days after delivery for complicated delivery or caesarean.
HCT = hematocrit; HGB = hemoglobin; Pap = Papanicolaou; VDRL = Venereal Disease Research Laboratory; RPR = rapid plasma reagin; HBsAg = surface antigen of hepatitis B virus; HIV = human immunodeficiency virus; CVS = chorionic villus sampling; MSAFP = maternal serum α-fetoprotein; GBS = group B *Streptococcus*.
Source: American College of Obstetricians and Gynecologists (ACOG). Guidelines for perinatal care, 6th edition (2007). Revised January 2013. Developed by American Academy of Pediatrics (AAP) and the American College of Obstetricians and Gynecologists (ACOG). Available at: http://www.anthem.com/provider/noapplication/f2/s2/t0/pw_ad094972.pdf?refer=ahpprovider&state=in (accessed 26.11.2013)

supplements as well as healthful diet, activity and rest levels. Prenatal care for a normal pregnancy should include a total of 13–15 visits to trained health providers, prenatal classes on physical care, and preparation for delivery, breastfeeding, and infant care. The mother's routine checkups should include gynecological, medical and family history, occupational and environmental exposures, tests for sexually transmitted diseases, prior pregnancies and infant health, nutritional status, weight gain, alcohol consumption, smoking, alcohol use, emotional well-being, partner or family support systems and potential complications; physical examination includes weight, height, blood pressure, fetal heart tones, fundal height, and fetal activity; and tests for anemia, blood type, Rhesus (Rh) factor, rubella, hepatitis B antibodies, syphilis, and HIV status; and ultrasound scans.

The frequency of visits and the content of normal prenatal care for healthy women vary widely among countries and are sometimes considered "excessive". However, good care should not be taken for granted. The achievement of low maternal mortality and morbidity along with low rates of risk for the newborn from low birth weight (LBW) and associated developmental problems should reinforce the importance of close supervision throughout a pregnancy. ACOG also recommends routine HIV screening for all pregnant patients unless they decline. Counseling is an important part of prenatal care, especially relating to lifestyle such as smoking alcohol and drug use, good nutrition with vitamin and mineral supplements, and preparation for motherhood with breastfeeding and infant care (ACOG

Clinical Practice Guidelines for Prenatal and Postpartum Care, 2010). The CDC recommends that unvaccinated pregnant women receive a dose of tetanus toxoid, reduced diphtheria toxoid, and acellular pertussis vaccine (Tdap) to protect the newborn until the baby is old enough to receive routine vaccination.

In some countries, such as Israel, prenatal care is separated from regular primary care and is offered in maternal and child health centers (MCHs) or women's clinics with nursing supervision, medical examinations, and free hospital delivery, mainly by midwives; public health nurses are the major providers of prenatal care. In France attendance is mandatory before 14 weeks and during the remainder of pregnancy to receive full maternity benefits including maternity and family allowances and postpregnancy leave from work. Special care services are provided to LBW or other risk factors for the mother and infant, including social and financial assistance if needed. In some countries, such as the Netherlands and the UK, family practitioners provide prenatal and obstetric services. In others, such as the USA and Canada, specialist obstetricians and general practitioners both provide prenatal and delivery services. Whether prenatal care is part of primary care services or a separate public health service, the goals must include universal coverage starting before or very early in pregnancy, risk assessment with referral and care, and ready access to specialized care. Health education provided in prenatal classes for the pregnant woman and her partner offers additional counseling and peer group support.

The pregnant woman should be provided with a complete medical record of her prenatal care during scheduled visits, so that when she arrives in the maternity unit, the provider of care will have adequate information. Therefore, even if she has not been seen previously by the new caregiver, the record grants a measure of continuity beneficial to both the woman, who knows that her record of previous care is available when needed, and the provider, who has the use of all her previous documentation. A public health system should ensure adequate records systems for maternity care to reduce unnecessary complications and mortality in the birth process.

In developing countries, improved access to prenatal care and increased use of midlevel health workers in "baby-friendly" birth centers, coupled with training and supervision of traditional birth attendants, spacing of pregnancies, risk assessment, and referral, are all needed to lower the present high rates of maternal and perinatal mortality. As discussed in the context of maternal mortality, progress in this regard has been extremely disappointing over the past several decades, even in countries with substantial economic growth and increased governmental revenues such as Nigeria and India.

High-Risk Pregnancy

Low-risk pregnancies are those in healthy women between the ages 18 and 34, who present at least once in the first trimester, who have had no more than three previous normal live births, no previous stillbirths or obstetric complications such as gestational diabetes or pre-eclampsia, and who have no history of drug or alcohol abuse, and no major medical conditions such as hypertension or kidney disease. Such women should be followed in a routine prenatal care program.

High-risk pregnancies (HRPs) are defined as those pregnancies with pre-existing or current conditions that put the mother, the fetus, and the newborn baby at higher than normal risk for complications during or after the pregnancy and birth. These include very young and older women, those with low levels of education and nutrition, as well as those with previous or current medical and obstetric complications and those in poverty or unstable living arrangements. HRPs should be identified as early as possible so that the patient can be given special care for her benefit and especially for the well-being of the fetus and newborn. Identification and management of high-risk factors initially and throughout pregnancy improve pregnancy outcomes for the mother and the newborn. Some predictors of HRP are maternal age (too young or too old), primiparas or grand multiparas, previous obstetric difficulties, other medical conditions (e.g., HIV, hypertension, heart disease, diabetes, kidney disease, or mental illness such as depression), malnourishment, poverty, women who attend STI clinics, and

use of cigarettes, alcohol, or other drugs. Risk factors may include social and economic factors such as adverse family circumstances, housing, financial status, and working conditions.

The medical and obstetric history provides evidence of previous risks such as frequent abortion, complications in pregnancy, or medical conditions that could affect the mother during the pregnancy or at the time of delivery. Pregnancy under the age of 16 or 17 or over the age of 35 should automatically define the pregnancy as being at higher than normal risk. Grand multiparity (i.e., more than five previous births) or a first pregnancy (primigravida) should also be considered as an extra risk for the mother, but more so for the newborn.

A scoring system provides a set of standards or guidelines for risk assessment to assist the primary care provider in early detection and referral of patients on the basis of a reasonably objective set of criteria for high-risk factors. Detailed guidelines are needed to implement this kind of standard for HRP and monitoring is of value to improved pregnancy care. The form and guidelines developed should take into account local risk factors, such as high consanguinity rates in some societies or chronic malnutrition in the population. Scoring systems produce a cumulative risk assessment by adding those factors that by themselves would not mean the pregnancy is high risk, but taken together indicate potential problems. The actual cut-off points for age, parity, and education levels can be adjusted to conditions in each country, but the principle of national standards is important. Fetal and neonatal monitoring is vital in developing countries, as neonatal mortality is a major part of total infant mortality. Low-cost, technologically adequate respirators and incubators are now being made and distributed, and training to support such efforts is crucial.

The HRP assessment and referral form outlined in Table 6.9, developed and used in a rural primary care setting, was adopted throughout a governmental health system in a developing area (West Bank and Gaza) for maternal and child health nurses and CHWs to promote early referral of HRPs (see Chapter 14). This record summarizes medical, obstetric, and prenatal care, especially parity, date of last menstrual period, risk factors, and prenatal care findings. In this case, A = low risk, B = medium risk, and C = high risk. Clear definitions and staff training and instruction are required. In some cases one C (e.g., hypertension) requires referral, or several Cs or Bs mean mandatory referral. The format and scoring system can vary, such as scoring each topic from 1 to 10 with systematic compilation of the score. Referral to HRP clinics needs not only a clear indication or reasons for referral, but also thorough assessment by specialists and feedback to the referring physician or other primary care provider. The HRP clinic should send the patient back to the referring center with a report of findings and a clear recommended plan of action. The

TABLE 6.9 High-Risk Pregnancy Assessment and Referral Form

I. Personal Data	II. Social/Personal	III. Obstetric History
Name	Age (<17, >35 years)	Gravida (1 or >5)
Identity Number	Education (<6 years)	Para (0 or >5)
	Marital status	Abortions (>2)
Date of Birth	Economic status	Miscarriages (1+)
Address	Consanguinity	Fetal deaths (1)
	Smoking	Bleeding in T3
City/Town/Village	Alcohol	Stillbirths (1)
	Drug use	Previous caesarean (1)
Clinic	Home conditions	Preterm deliveries
Date last menstrual period	Summary	Birth weights
Date of first visit	_____	Infant deaths (1)
	_____	Toxemia (1)
		Birth defects (1)
	_____	Summary
	A B C	
		_____
		A B C
IV. Medical and Family History	V. Present Pregnancy	VI. Summary of Risk Factors
Diabetes	Last menstrual period	I. Personal A B C
Hypertension	Number of present pregnancy	
Renal disease	Time of first visit T1 T2 T3	II. Social A B C
Heart disease	Weight before pregnancy <50 kg	
Chronic chest disease	Height <145 cm	III. Obstetric history A B C
Blood disorder	Blood pressure	
Endocrine disease	Bleeding	IV. Medical history A B C
Phlebitis	Pre-eclampsia	
STDs	Rh	V. Present pregnancy A B C
Other	Multiple pregnancy	
Summary __	Abnormal presentation	VI. Total score A B C
_____	Summary _____	
A B C	A B C	

Summary of Reasons for Referral: _____

_____

Date: _____ Signature: _____

Position: _____

Report of High-Risk Clinic: _____

_____

_____

Date: _____ Signature: _____

Position: _____

Note: This is intended as a model for primary care early identification of high risk and prompt referral on the basis of history and basic findings. Full prenatal care recommendations are available from the American College of Obstetrics and Gynecology.

HRP clinic may continue to follow the patient because of the risk factors, but the referring provider should have this information and assist in its implementation. A mother-carried record is desirable, since access to prenatal clinic records may be limited and may not include care given by other providers, such as emergency departments or private physicians.

A well-developed HRP assessment, referral and follow-up system contributes to improved outcomes for both mothers and newborns, preventing costly long-term consequences of maternal and infant morbidity and mortality. Its role in preventing complications during and following delivery is well justified on medical, public health, and economic grounds.

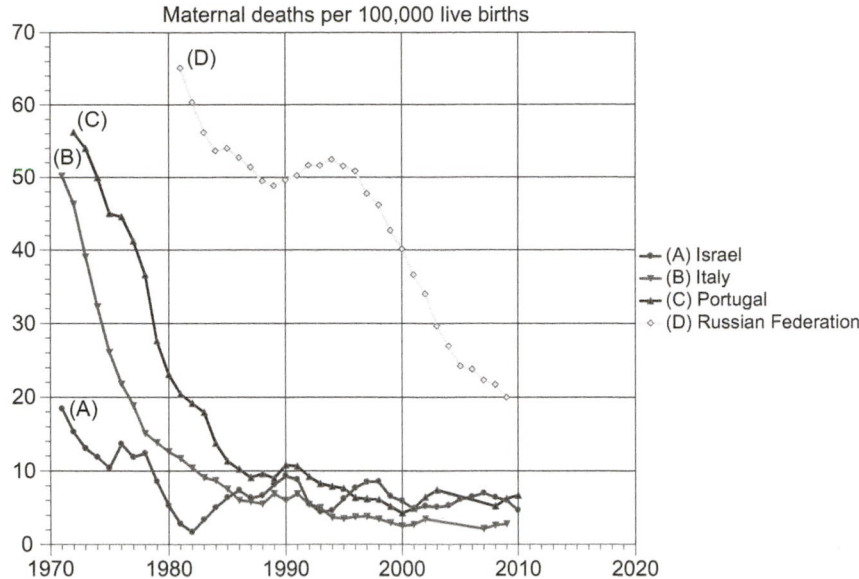

FIGURE 6.6   Maternal mortality (maternal deaths per 100,000 live births) in selected countries of the WHO European Region, 1970–2010. *Source: World Health Organization. European Region. Health for All database; July 2013. Available at: http://data.euro.who.int/hfadb/ [Accessed 22 November 2013].*

The issues may be different in developing and developed countries, but the principles are similar. In the USA, there are substantial population groups that cannot obtain prenatal care for financial or bureaucratic reasons, including those who may be at highest risk. A comparison of several European region countries (Figure 6.6) shows dramatic progress in Romania and considerable progress in Russia, with reduced maternal mortality from 74 (in 1990) to 16.9 per 1000 live births in 2010.

Maternal mortality does respond to improved access to medical care, as seen in its rapid decline in Italy and Portugal during the 1970s following introduction of their national health services. Countries with both universal access and well-developed obstetric care and risk assessment have low rates of maternal and infant mortality (e.g., Australia, Belgium, Denmark, Israel, and Norway). Despite great progress in reducing maternal mortality during the twentieth century (CDC, 1999), the USA has a comparatively poor showing of maternal mortality, being placed well behind almost all industrialized countries of Europe and Asia. This not only is due to a lack of universal health insurance, but also relates to many social and economic issues of a society that is prejudiced against mothers and children of minority groups and others living in poverty.

LABOR AND DELIVERY

Labor and delivery of a baby is a highly personal service but with important public health aspects. Assurance of safe delivery is important to both mother and newborn. Over the past century, the place of delivery has evolved from the private home to maternity or lying-in hospitals to maternity units in general hospitals, and to "mother and baby friendly" birth centers (in hospital settings). Delivery in maternity units in hospitals in developed countries has made an important contribution to reduced maternal and neonatal mortality and should be considered a primary goal in public health and primary care programs.

Hospital delivery is encouraged in many countries by giving maternity grants paid by national insurance or social security to the mother for covering the costs of delivery in a hospital. This was done previously in France and Israel, before the advent of national health insurance, to promote hospital delivery. France continues this tradition by providing grants to women who attend prenatal care early and frequently during pregnancy. In developing countries, the place of delivery needs to be addressed, making use of existing resources with a goal of reducing the current extremely high rates of maternal morbidity and mortality.

Dangerous emergencies can occur at any stage of labor and may necessitate caesarean section, blood transfusions, general anesthesia, fetal resuscitation, or treatment of respiratory distress syndrome. Delivery in a general hospital obstetric department is safer than at home or in a free-standing maternity home. Electronic fetal monitoring during labor does not necessarily reduce fetal intracranial damage, retardation, and cerebral palsy, compared to clinical fetal auscultation, but provides reassurance to the anxious mother and assists the harried staff of a busy obstetric department.

Intrapartum care should be made as homelike as possible and include the partner or other person desired by the pregnant woman. Care during labor and delivery requires information from the prenatal period, by transmission of a pregnancy care record, carried by the woman to the delivery room.

As noted above, some countries, including some with very low MMRs, use qualified midwives to perform routine deliveries while obstetricians are responsible for supervision and complex deliveries. Licensing of qualified midwives may be necessary in the USA to meet current obstetric needs, especially where there is a shortage of obstetricians. Qualification for midwives is usually an additional year of study for registered professional or academic level nurses (see Chapter 15). In developing countries with severe lack of physicians living in rural areas, training of community midwives is essential to improve maternal care and reduce maternal mortality.

Fetal monitoring during labor (intrapartum) by auscultation has been standard practice since the work of Evory Kennedy in Dublin in the 1830s. Since the late 1950s, electronic fetal monitoring has become widespread, offering reassurance to women and help to staff on a busy obstetric ward. Hospital maternity wards should be "mother, family, and baby friendly", promoting a homelike atmosphere with both midwives and medical support.

Attendance by trained anesthetists and pediatricians should be considered for routine cases and mandatory for High Risk Pregnancies (HRPs). The presence of a pediatrician or pediatric nurse practitioner at the time of birth gives greater assurance of appropriate early care for the newborn, which is especially important in cases of premature births, low birth weight babies (LBWs), respiratory distress, or other complications at birth. In countries with low overall infant mortality, early neonatal deaths are now a major element in that mortality. Further reduction in infant mortality rates requires greater attention to this risk period, when resuscitation and other immediate life-saving procedures may be needed.

In the USA in recent years, there has been a modest trend towards a return to home deliveries with highly trained midwives and guaranteed backup services. Home deliveries are declining in the UK with an increase in specialist obstetric departments in general hospitals. Home births are becoming popular among educated young mothers but should not be encouraged as an adequate alternative to delivery in the hospital for low-risk pregnancies, but only when carried out by well-trained professional midwives with the backup of emergency medical teams (and hospital) able within minutes to deal with the complications that can occur with any delivery. Home birth is a trend that could lead to increases in maternal and infant death as an unfortunate side-effect. Delivery in maternity homes or free-standing birth centers is an alternative to maternity units in general hospitals, but should be avoided in developed countries. The homes and centers lack full facilities in the event of complications unanticipated before the onset of labor. The place and methods of delivery may vary but should take into account not only the wishes of the pregnant woman but also the safety of the mother and the newborn. High-risk mothers, especially teenage primiparas, should always give birth in a skilled birth attended maternity center or preferably a hospital setting.

Safe Motherhood Initiatives

Safe motherhood is an initiative of the UN launched in 1987 with the goal of ensuring that women go through pregnancy and childbirth safely, and give birth to healthy children and reinforced by inclusion of reducing maternal mortality in the MDGs of 2000-2015. Yet, every year, hundreds of thousands of women die or suffer serious complications from pregnancy and childbirth. Safe motherhood begins before conception with proper nutrition and a healthy lifestyle. Planned pregnancy, appropriate prenatal care, prevention of complications when possible, and early and effective treatment of complications when they occur are all essential elements of maternal care. Labor at term with adequate care but without unnecessary interventions, followed by delivery of a healthy infant and provision of a healthy postpartum environment is vital for the physical and emotional needs of the woman, infant, and family. This is a process which requires the support of many diverse sectors of the nation, the family, educators, employers, health care providers, community groups, and others. Good maternity care is an investment in the physical, emotional, social, and economic well-being of women, their children, and their families and, by extension, the nation. Reducing inequalities in accessing professional care is key to attaining the goal for good quality and compassionate maternal health. The challenges are how to deliver services and scale up interventions, particularly to those who are vulnerable, hard to reach, marginalized, and excluded. Within the USA as well as in developing countries the challenges are reducing the inequalities between rich and poor in professional delivery care to achieve safe motherhood and reach the MDG for maternal health.

Surrogate motherhood is increasing as a method of bearing children on behalf of other parents, and can be a benefit or an exploitation of poverty. It should be carried out with great care ensuring the health of the mother and especially the well-being of the child. Surrogate mothering done internationally can become exploitive of poor women and a dangerous practice unless it includes adequate precautions of screening and counseling. The CDC reports that between 1998 and 2007 over 57,000 babies were conceived by in vitro fertilization (IVF) in the USA and many of these involved surrogate mothers.

Home births are still common in developing countries. They are unattended or attended by traditional birth attendants (TBAs). A home birth may be the only choice the mother has, in which case the public health authorities should train and supervise the TBAs. Special attention should be given to hygiene, prevention of tetanus, and identification and referral of high-risk cases. The WHO promotes guidelines for integrating TBAs into the health care system in developing countries. Setting standards and supervising TBAs are vital to reduce the large-scale loss of mothers and newborns in developing countries. High-risk cases, including primiparas and grand multiparas, women over the age of 35, multiple pregnancies, late presentation for care, and those with previous complicated deliveries, malnutrition, or chronic illness, as well as those with previous infant deaths, should be referred to a district hospital for delivery. Home deliveries have become fashionable among highly educated women in wealthy countries, albeit with well-qualified midwives, but this potentially dangerous trend seems to be modest and hopefully will remain so.

Public health authorities should ensure hospital delivery wherever possible, and make certain that hospitals are staffed with well-trained personnel. Equally important is assurance that early and consistent prenatal care is utilized by all pregnant women. District hospital facilities should be equipped for obstetric and newborn emergencies, including surgical, anesthesia, blood bank, laboratories, resuscitation staff and facilities for normal and caesarean deliveries. All care by community and district care centers and by TBAs should be documented in appropriate record systems developed by the public health authorities.

District and community-based obstetric service centers should be developed to provide prenatal risk assessment and care, working with TBAs in both a supportive and supervisory capacity. Community-based delivery facilities should be equipped and staffed to advise on breast feeding and good infant care practices, along with safe contraceptive use, provide counseling for pre-pregnancy and antenatal care, assess risk pregnancies and referrals, carry out normal deliveries and safe abortions, and handle emergencies such as pre-eclampsia, hemorrhage, and manual removal of placenta. TBAs should be trained to arrange evacuation of high-risk mothers and babies to district hospitals. Professional midwives or nurses should be appointed to supervise TBAs, including visits to villages and direct one-on-one training. TBAs performing deliveries on high-risk patients, except in dire emergencies or non-preventable situations, should be subject to a penalty enforced by the public health authority.

Supervision of TBAs involves a process of regulation, education, and incentives to meet higher standards of practice. The health authority is empowered to license and regulate all health providers and to discipline offenders who practice without a license or do not meet current standards of practice. TBAs can be motivated to participate in training programs with incentives including provision of essential equipment and supplies. TBAs should be required to keep a record of deliveries, with basic details and outcomes recorded in a state-issued log book. If TBAs are illiterate, they can be required to engage a literate person to keep such records, for regular inspection by the TBA supervisor. Training should include prenatal and delivery danger signs, hygiene in delivery, neonatal resuscitation and infancy care, and high-risk situations. TBAs can also be trained in regard to STIs, HIV, anemia, iodine deficiency, and obstetric or medical risks.

Care of the Newborn

The newborn is totally dependent on and vulnerable to many factors that may inhibit optimal growth and development. This dependency begins before conception and continues in utero during the prenatal period, the delivery, and up to the end of the first year of life. Care immediately following delivery should be given to provide warmth and suction to reduce the chance of aspiration and assure breathing. Many of the risk situations are preventable, and the drastic decline in infant mortality seen over the past century indicates this potential.

Key issues in immediate care of the newborn include:

- assuring respiration and suction if needed
- Apgar evaluation to assess functional status of the newborn (Table 6.10)
- eye care with silver nitrate or antibiotic ointment or drops to prevent gonococcus and chlamydia infection of the eyes, possibly acquired from the mother during the birth process
- cord care to ensure maximum sterility and safety
- vitamin K by injection to prevent hemorrhagic disease of the newborn

TABLE 6.10 Apgar Score

Sign/Score	0	1	2
Heart rate (beats/minute)	Absent	Slow (<100)	>100
Respiratory effort	Absent	Slow, irregular	Good, crying
Muscle tone/motion	Limp	Some flexion of extremities	Active motion
Reflex irritability	No response	Grimace	Cough or sneeze
Body color	Blue, pale	Body pink, extremities blue	Completely pink

Note: This is a widely used system of scoring an infant's physical condition 1 and 2 minutes after birth. The maximum score is 10; those with low scores require immediate attention. The test may be repeated at 5 or more minutes in order to monitor recovery, but the long-term significance is under debate.
Source: Committee on Fetus and Newborn. 2006. Policy statement: the Apgar score. American Academy of Pediatrics. Pediatrics 117:1444–7 [reconfirmed November 2012]. Available at: http://pediatrics.aappublications.org/content/123/5/1421.full [Accessed 3 February 2013].

- newborn blood drop screening to detect 29 diseases, especially phenylketonuria, congenital hypothyroidism, thalassemia, sickle-cell anemia, and many metabolic disorders
- hearing test
- hepatitis B vaccine to protect against maternal blood transmission
- complete check-up to record weight, length, and head circumference, and to detect other physical abnormalities, e.g., congenital heart disease, congenital dislocated hip, undescended testes, and other physical disorders.

Newborns weighing less than 2500 g (5 pounds 8 ounces) are considered low birth weight (LBW), with normal birth weight between 2500 and 4000 g (5 pounds 8 ounces to 8 pounds 13 ounces). In developed countries the LBW rate is usually below 8 percent and in developing countries often over 15 percent. Babies born weighing less than 1500 g (3 pounds 5 ounces) are considered very low birth weight (VLBW) and those under 1000 g (2 pounds 3 ounces) extremely low birth weight (ELBW), and have lower rates of survival and are often left with permanent disabling conditions even with excellent care. Survival of newborns falls gradually as birth weight declines. A major purpose of good prenatal care is to ensure continuance of the pregnancy beyond 36 weeks and as close to full term as possible in order to promote full maturity and development of the fetus in utero and ensure a greater chance of survival of the newborn. To reduce the factor of multiple births, the term low birth weight is sometimes used for single births only.

The newborn requires immediate assessment and immediate response to resuscitation needs or other necessary interventions. The Apgar score, developed by Virginia

Apgar in the early 1950s, provides an important standard method of evaluation of the newborn and its need for investigation and intensive care (Table 6.10). Scores of less than 6 at birth may indicate a need for resuscitation and specialized attention, and low scores after 2 minutes indicate an infant at risk. The Apgar score is a useful tool in assessing the newborn, but should not be considered an absolute marker for subsequent outcome and status of the infant.

CDC reports that uniform newborn screening for a panel of diseases has led to earlier life-saving treatment and intervention for over 3400 additional newborns each year with selected genetic and endocrine disorders. In 2003, most states in the USA were screening for only six of these disorders. By April 2011, all states reported screening for at least 26 disorders of an expanded and standardized uniform panel. Newborn screening for hearing loss increased from 46.5 percent in 1999 to 96.9 percent in 2008. Standardized terminology and reporting as well as national oversight of state screening programs has been recommended. Reduction in the number of newborns with neural tube defects (NTDs) has been seen as a result of increasing use of folic acid among women of fertile age. Mandatory folic acid fortification of enriched cereal grain products in the USA since 1998 contributed to a 36 percent reduction in NTDs from 1996 to 2006 and prevented an estimated 10,000 NTD-affected pregnancies from 2001 to 2010, with savings of US $4.7 billion in direct costs (CDC, 2011). The 10 leading causes of infant deaths in the USA are shown in Table 6.11. Birth defects and low birth weights are the leading causes and are particular areas where improvements can be made, as emphasized in professional organizations and the work of the March of Dimes. Folic acid taken before and

TABLE 6.11 Ten Leading Causes of Infant Deaths and Mortality Rates, United States, 2011

	Number	Rate
All causes	**23,907**	**604.7**
1 Congenital malformations, deformations and chromosomal abnormalities	4,984	126.1
2 Disorders related to short gestation and low birth weight, not elsewhere classified	4,116	104.1
3 Sudden infant death syndrome	1,711	43.3
4 Newborn affected by maternal complications of pregnancy	1,578	39.9
5 Accidents (unintentional injuries)	1,089	27.5
6 Newborn affected by complications of placenta, cord and membranes	992	25.1
7 Bacterial sepsis of newborn	526	13.3
8 Respiratory distress of newborn	514	13.0
9 Diseases of the circulatory system	496	12.5
10 Neonatal hemorrhage	444	11.2
All other causes	7,457	188.6

Note: Preliminary data. Rate are per 100,000 live births.
Source: Hoyert DL, Xu J. Deaths: Preliminary Data for 2011. National Vital Statistics Reports. Table 8. 2012;61 (6):1-52. Available at: http://www.cdc.gov/nchs/data/nvsr/nvsr61/nvsr61_06.pdf (accessed 26.11.2013).

during pregnancy and presence in all fortified flour will continue to help in reducing the burden of birth defects. The continued high rate of single motherhood and poverty in the USA is a barrier to improvement in this important health measure.

Low Birth Weight/Preterm Birth

Preterm births are births at less than 37 weeks; for comparison, the full term of pregnancy is 40 weeks. Preterm births are categorized as: extremely preterm, <28 weeks; very preterm, 28 to <32 weeks; and moderate/late preterm, 32 to <37 weeks; and number approximately 780,000, 1.6 million, and 12.6 million, respectively, worldwide per annum. About 80 percent occur between 32 and 37 weeks and survive with essential newborn intensive care. In 2010 preterm births made up 12 percent of births in the USA. In terms of birth weight, the normal is between 2500 and 4000 g. LBW (i.e., <2500 g), and especially VLBW (i.e., <1500 g), is a serious risk for the child's survival and future development. ELBW infants are those born weighing less than 1000 g and are at very high risk of survival and long-term developmental defects such as cerebral palsy, developmental delay and retardation, visual defects including blindness, hearing impairment, sudden infant death syndrome (SIDS), and chronic lung diseases.

UNICEF reports the percentage of births under 2500 g by country and region. The global average of LBW between 2006 and 2010 was 15 percent, with regional rates varying from 27 percent in South Asia, 13 percent in sub-Saharan Africa, and 11 percent in the Middle East and North Africa, to 7 percent in countries of Eastern Europe and the CIS; the US average for 2006–2010 was 8 percent. The US Institute of Medicine describes the risks of LBW and especially VLBW as follows:

"The spectrum of neurodevelopmental disabilities includes cerebral palsy, mental retardation, visual and hearing impairments, and more subtle disorders of central nervous system function. These dysfunctions include language disorders, learning disabilities, attention deficit-hyperactivity disorder, minor neuromotor dysfunction or developmental coordination disorders, behavioral problems, and social–emotional difficulties. Preterm infants are more likely to have lower intelligence and academic achievement scores, experience greater difficulties at school, and require significantly more educational assistance than children who were born at term. Preterm infants have an increased risk of rehospitalization during the first few years of life and increased use of outpatient care. Among the conditions leading to poorer health are reactive airway disease or asthma, recurrent infections, and poor growth. The smallest and most immature infants have the highest risk of health problems and neurodevelopmental disabilities. Limited evidence of the impact of prematurity on families suggests that caring for a child born preterm has negative and positive effects that change over time, that these effects extend to adolescence and are influenced by different environmental factors over time, and that many areas of family well-being are affected. The prevalence of neurodevelopmental disabilities and health impairments varies." (Behrman and Butler, IOM, 2012)

Risk factors for LBW include smoking, alcohol and drug use; maternal age (<15 and >35); low income and educational level; stress, domestic violence, or other abuse; unmarried status; previous preterm birth; and exposure to air pollution. Maternal health is a key factor, with regard to pre-existing medical problems (diabetes, renal disease, hypertension), maternal infections (rubella, cytomegalovirus), lack of weight gain, placental insufficiency, and others.

Of the 1.1 million deaths of preterm newborns, globally, 75 percent are preventable with feasible, cost-effective care, but without intensive care. More than 60 percent of preterm births occur in Africa and South Asia, but Brazil, India, Nigeria and the USA are among the 10 countries with the highest rates of preterm births. The US annual socioeconomic cost in 2005 was at least US $26.2 billion, including medical, educational, and lost productivity, with costs for inpatient and outpatient care 10 times greater for preterm than for full-term babies.

Preterm birth is a risk factor in as many as 50 percent of neonatal deaths; complications are responsible for 35 percent of the 3.1 million neonatal deaths worldwide. Widespread availability of prevention and care interventions focused on the multiple risk factors and with proven benefit such as kangaroo mother care would reduce the burden by saving nearly 1 million babies annually (Box 6.2).

Essential interventions for reproductive, maternal, newborn, and child health include a wide range of community and personal care by a well-structured and well-staffed health system at the primary care level, backed by hospital or medical supervised maternity care facilities. Education of women, promotion and access to birth control, spacing of pregnancies, and preparation for pregnancy, especially with iron and folic acid supplements and dietary adequacy, are fundamental. Full prenatal care and safe delivery, is required, preferably in hospitals but otherwise in maternity centers staffed and equipped for the needs of mothers and newborns. Training and supervision to ensure high risk identification, referral, and timely access to adequate facilities are all fundamental to saving lives of mothers and babies (Partnership for Maternal, Newborn & Child Health, 2011; March of Dimes, 2012).

Where home births take place, the potential for transport of the newborn with difficulties to a neonatal care center should be part of the preparation for LBW, respiratory distress, meconium inhalation, and other conditions for which the newborn requires highly trained medical and nursing personnel and equipment to improve survival and quality of outcomes. Early detection and intervention reduce unnecessary morbidity and mortality of the newborn.

In countries with high levels of neonatal mortality, such as those in the Central Asian Republics, there may be high levels of attendance for prenatal care, as was emphasized in the Soviet health system (see Chapter 13), but women's

BOX 6.2 Kangaroo Mother Care

Kangaroo mother care (KMC), developed by Rey and Martinez in Columbia, has been shown to be effective for thermal control (continuous skin-to-skin contact), promoting breastfeeding, and bonding in all preterm low birth-weight (LBW) stable newborns. It is a low-tech innovation that is equivalent to incubator care in safety and thermal protection. KMC has been successfully applied in many countries in Africa, Asia, and the Americas.

This method is effective in both low–medium- and high-income countries and is associated with a 51 percent reduction in neonatal mortality for stable babies born with weight under 2000 g if started in the first week as compared to incubator care. Systematic review and meta-analysis of several randomized controlled trials of KMC showed a decreased risk of mortality and severe infection sepsis. KMC should be continued at home but there are no studies on the home setting. KMC is an effective alternative care to incubators, especially in low-income countries, and is increasingly accepted in both high- and low-income countries.

A Cochrane review concluded that the evidence of the updated review supports the use of KMC in LBW infants especially in resource-limited settings. Further study is need on its effectiveness and safety for unstable LBW infants, long-term neurodevelopmental outcomes, and costs of care. KMC joins other low-tech solutions to health problems such as oral rehydration solution (ORS), impregnated bed nets to prevent malaria transmission, and simpler diagnostic tests for malaria, tuberculosis, and HIV (see Chapter 15).

Sources: USAID and MCHIP. Maternal and child integrated program (MCHIP): implementation guide, 2012. Available at: http://www.mchip.net/sites/default/files/MCHIP%20KMC%20Guide.pdf
Lawn JE, Mwansa-Kambafwile J, Horta BL, Barros FC, Cousens S. "Kangaroo mother care" to prevent neonatal deaths due to preterm birth complications. Int J Epidemiol 2010;39(Suppl 1):i144–54. Available at: http://ije.oxfordjournals.org/content/39/suppl_1/i144.full
Conde-Agudelo A, Belizán JM, Diaz-Rossello J. Kangaroo mother care to reduce morbidity and mortality in low birth weight infants. Cochrane Database Syst Rev 2011;(3):CD002771. Available at: http://www.ncbi.nlm.nih.gov/pubmed/21412879 and http://www.nichd.nih.gov/cochrane/conde-agudelo/conde-agudelo.htm
Howson CP, Kinney MV, Lawn JE, editors. Born too soon: the global action report on preterm birth. Geneva: WHO; 2012. Available at: http://www.who.int/pmnch/media/news/2012/preterm_birth_report/en/index1.html [All accessed 7 February 2013].

BOX 6.3 Vitamin K and Hemorrhagic Disease of Newborn or Vitamin K Deficiency Bleeding

Hemorrhagic disease of newborn, now called Vitamin K Deficiency Bleeding (VKDB), caused by vitamin K deficiency, and its prevention in newborns was the subject for which Henrick Dam (Denmark) and Edward Doisy (USA) were jointly awarded the Nobel Prize for Medicine in 1943. In 1961, the American Academy of Pediatrics (AAP) recommended vitamin K by imtramuscular injection routinely for all newborns. Routine use is widely practiced, but became mandatory in New York State after identification of cases of the disease that had not received vitamin K. It was then made mandatory as a public health measure adopted by most states in the USA. Vitamin K remains controversial in Europe and other parts of the world and is not included in perinatal management in WHO and UNICEF documents. The AAP recommendation was reaffirmed in 2003 to prevent early and late HDN [vitamin K deficiency bleeding (VKDB)] in newborns.

Early and late VKDB occurs in about 1 percent of births, with late HDN mainly in breastfed infants (rate 4.4–7.2 per 100,000 live births) due to immaturity of the liver and inefficient production of prothrombin due to lack of vitamin K production in the gut.

The problem of VKDB is neglected internationally due to misinterpretation of a now long-refuted study linking vitamin K and lymphoma in children. The problem of early and late VKDB is being reported in developing countries, such as Turkey and India, where this preventive measure is not practiced routinely, and where low birth weight and sole breastfeeding, important risk factors for late VKDB, are common. In 2013, 4 cases or late VKDB were reported in Tennessee due to refusal of injected vitamin K by mothers.

Sources: Presentation Speech, Nobel Prize in Physiology or Medicine, 1943. Available at: http://www.nobelprize.org/nobel_prizes/medicine/laureates/1943/press.html [Accessed 7 February 2013].
Committee on Fetus and Newborn. Policy statement. Controversies concerning vitamin K and the newborn. Pediatrics 2006;118:1266.
Tulchinsky TH, Patton, MM, Randolph L, Meyer MR, Linden JV. Mandating vitamin K prophylaxis for newborns in New York State. Am J Public Health 1993;83:1166–8.
Zipursky A. Prevention of vitamin K deficiency in newborns. Br J Haematol 1999;104:430–7. Available at: http://www.ncbi.nlm.nih.gov/pubmed/10086774 [Accessed 7 February 2013].

health may be compromised by poor nutrition, anemia, iodine deficiency, and delivery in maternity homes detached from hospitals. Neonatal deaths may be largely due to respiratory causes, with infection a likely cause, so that prophylactic antibiotics may be indicated. Vitamin K for newborns is also indicated on a routine basis.

All newborns should receive well-trained professional care including resuscitation if needed and routine eye care to prevent ophthalmic infection with gonococcus. Vitamin K should be given routinely either by injection to all newborns to prevent hemorrhagic disease of the newborn now called Vitamin K Deficiency Bleeding (VKDB). As many as 20 percent of newborns have low levels of prothrombin, a blood clotting factor, and are therefore subject to potentially

serious and even fatal bleeding, which is preventable by intramuscular vitamin K (Box 6.3). Vitamin K by injection is mandatory in all US states and other countries. In many countries it remains at the discretion of medical staff, often given only to LBW newborns. This is an issue often neglected in discussions of high neonatal mortality rates, especially in low-income countries where most deliveries take place at home especially in rural areas, but should be part of efforts to reduce neonatal mortality globally.

Screening for HIV infection if the mother's HIV status is unknown and for signs of drug or alcohol effects in newborns is important in high-risk settings. Screening for HIV status allows preventive treatment which has been successful in many developing countries in reducing maternal-fetal transmisson of the virus. It is up to modern public health agencies to establish the norms and training systems needed to foster high standards of care.

Complete examination by a pediatrician searches for birth defects and normal development and lays the basis for continued care in early infancy. The examination should include measurement of the length, head circumference, and birth weight, gestational assessment, inspection for jaundice, and assessment of the heart, vascular, and neurological systems, with results recorded. Screening for congenital disease or abnormalities such as congenital cardiac defects, Down syndrome, congenital dislocated hips (CDH), club feet, undescended testes, phenylketonuria (PKU), hypothyroidism, congenital hemolytic anemias (e.g., thalassemia and sickle-cell disease), galactosemia, and other inborn errors of metabolism should occur in this period with follow-up of any suspicious or positive findings. Early diagnosis of abnormal conditions allows the child to benefit from timely and suitable management, which can save the child and family from the burdens of serious handicapping conditions, painful and frequent hospitalizations, costly medical care, other complications, and early death.

The recommendations of the Genetics Committee of the AAP for newborn screening were updated in 2006 to take into account technological advances and greater knowledge of inherited diseases as well as ethical issues, and reaffirmed in 2011 (Box 6.4). The newborn screening system consists of newborn testing and timely follow-up of abnormal results, timely diagnostic testing and disease management. It is important to coordinate with the medical provider, to provide genetic counseling, and to conduct continuous evaluation and improvement of the newborn screening system. The March of Dimes reports that 90 percent of newborns in the USA are screened for inherited genetic disorders.

Establishing and sustaining breastfeeding is vital for mother and newborn and should be strongly encouraged by all health personnel as the sole source of feeding for the first 6 months, and continued with the addition of an increasing range of foods from 6 months onwards. The hospital setting, where almost all deliveries take place in the USA, should provide strong support to the mother to choose breastfeeding, which is important for bonding between mother and infant and for the optimal development of the infant. Staff support is essential to help the mother to choose breastfeeding and to initiate it well. It should continue for at least the first half year of life, with continued adequate supplementation by solid foods. Breastfeeding and follow-up nutrition are discussed in the chapters on nutrition (Chapter 8) and globalization of health (Chapter 16).

Postdelivery Care (Puerperium)

The puerperium is the 6–8-week period from the time of delivery of the placenta up to the resumption of normal ovulation. In the nursing mother, the puerperium may be prolonged because of the effects of lactation on hormonal balance. This is an important time for the new mother and

---

BOX 6.4 Recommended Standards for Routine Care of the Newborn: The American Academy of Pediatrics

- Resuscitation by trained personnel should be available at all high-risk pregnancy deliveries, transfer to neonatologist if necessary.
- Suction mouth and nose and establish airway.
- Clean, warm, and cover newborn, give to mother.
- Stimulate the newborn by rubbing back and soles of feet.
- Oxygen via Ambu bag with face mask, if needed; examine chest for adequate ventilation.
- Apgar score at 1 and 2 minutes following delivery.
- Apgar score at 5 minutes following delivery.
- If prolonged resuscitation is needed, further Apgar scores should be taken and recorded.
- Establish bonding and breastfeeding by giving newborn to mother as soon as possible.
- Vitamin K is routinely given to prevent Vitamin K deficiency bleeding.
- Eye care is given with antibiotic ointment to prevent chlamydia or gonococcal eye infection.
- Cord blood is taken for serology.
- Umbilical cord is cut and tied, applying antiseptic or antibiotic.
- Hepatitis B vaccine is given, the first of three doses.
- Newborn should be loosely wrapped, not swaddled.
- Newborn should preferably stay with mother ("rooming in").
- Father's visits and participation in care are to be encouraged.
- Breastfeeding is strongly encouraged.
- Complete examination by a pediatrician or primary care practitioner.
- Family and sibling visits are desirable.

Sources: *American Academy of Pediatrics/American College of Obstetrics and Gynecology. Guidelines for perinatal care. 7th ed. Washington, DC: APA/ACOG; 2012.*
Kaye CI. Introduction to the newborn screening factsheets. Technical report: newborn screening fact sheets. Pediatrics 2006;118(3):e034=963. 2006 policy. Available at: http://pediatrics.aappublications.org/content/118/3/1304.full [8 February 2013].
American Academy of Pediatrics. Hospital stay for healthy term newborns. Revised 2010. Pediatrics 2010;125:405–9. Available at: http://pediatrics.aappublications.org/content/125/2/405.full [Accessed 3 February 2013].

---

newborn to recover from the birth process and for their physiological and psychological adjustment. Adequate follow-up of the mother and infant soon after delivery helps to prevent complications and promotes optimal health for both.

Immediate postpartum concerns focus on assurance of complete expulsion of the placenta, contraction of the uterus, absence of bleeding from any site, or phlebitis. Basic examinations to protect the new mother's health include a breast examination, determination of the presence of any cervical or perineal tears or infections, and a hemoglobin test. Management of any residual problems should be part of the care process. The mother is provided with information on a variety of different topics including breastfeeding, nutrition, spacing for the next pregnancy, and birth control.

She is also given any needed support and counseling to address the potential for postpartum depression, which can become a debilitating disorder.

The mother's recovery from the pregnancy and birth process should be aimed at restoring her strength and normal family life. Postpartum depression is common and may need to be addressed by the care provider. Concern to lose weight rapidly may contradict the mother's nutritional demands, which are increased if she is breastfeeding. Iron supplements should continue in the puerperium and while breastfeeding to restore the iron lost to her during pregnancy, delivery, and breastfeeding. If necessary, Rh immune globulin to prevent future maternal–fetal blood immune reactions should be given by obstetricians. Also if necessary, rubella immunization should be given to the postpartum mother to prevent possible future congenital rubella syndrome (see Chapter 4).

Postnatal care should include a home visit by a public health nurse to assess and assist the adaptation of mother and infant and the home setting. A medical examination towards the end of the puerperium is an important element of pregnancy care because it is the first opportunity for complete assessment of the newborn after discharge from the hospital. The key concerns for the totally dependent newborn in this period are breastfeeding and bonding, hygiene and cleanliness, weight gain, vitamin supplements, beginning of immunization, and acceptance into the family.

First immunizations for hepatitis B, tuberculosis (with BCG, where used), and poliomyelitis begin as close to birth as possible and the routine immunization program begins in the first 6 weeks (see Chapter 4). For home deliveries, postnatal home visits should assess and weigh the newborn, administer vitamin K, initiate the vaccination program and routine screening tests, as well as counsel on breastfeeding and infant care, including immunization, additional feeding, safety and sleeping arrangements, and other aspects of care.

Counseling by the health care provider is very important to the new mother whether this is her first pregnancy or one of many. Family planning to promote spacing of pregnancies should be discussed as part of the health provider's attention and educational efforts. The need for support and advice is high in this period to reduce anxiety and to address many issues relating to maternal and infant health. Training of health providers working with delivery and newborn care should stress this role along with the professional skills of managing the delivery and immediate neonatal care.

GENETIC AND BIRTH DISORDERS

Worldwide, an estimated 7.9 million or 6 percent of all babies are born with serious genetic defects and millions of others are born with postconception defects. About 3.3 million children under the age of five die each year from birth defects. For those who survive, these disorders can cause lifelong mental, physical, auditory, or visual disability. Hundreds of thousands more are subject to conditions acquired during birth, such as HIV and hepatitis B and C, and tetanus and sepsis from unclean delivery procedures. They are also subject to the physiological condition of vitamin K deficiency, which can cause serious bleeding and death, and is also entirely preventable. They can acquire serious birth defects postconception, as the result of exposure to environmental factors, including alcohol, rubella, syphilis, and iodine deficiency (Table 6.12). The World Health Assembly

TABLE 6.12 Causes of Birth Defects and Conditions

Cause	Conditions
Genetic	Phenylketonuria, congenital hypothyroidism, hemoglobin disorders (thalassemia, sickle-cell disease), cystic fibrosis, Down syndrome, autistic disorder syndrome, fragile X syndrome, Tay–Sachs disease, other inborn errors of metabolism
Toxic	Thalidomide, fetal alcohol syndrome, drug abuse babies
Nutritional	Folic acid deficiency and neural tube defects, iodine, iron, zinc deficiency
Infection	Rubella syndrome, HIV, gonococci conjunctivitis congenital syphilis, cytomegalovirus, group B streptococcal infections
Physiological	Vitamin K deficiency and other hemorrhagic diseases, Rhesus hemolytic anemia
Multiple causation or unknown cause	Low birth weight, mental handicap and developmental disability (MRDD)
Trauma	Violence, accidents, burns

Source: March of Dimes Foundation. Global report on birth defects: the hidden toll of dying and disabled children, 2006. Available at: http://www.marchofdimes.com/downloads/Birth_Defects_Report-PF.pdf [Accessed 5 February 2013].

resolutions in 2010 called for global action for birth defect surveillance, prevention, and treatment.

The WHO estimates that there were some 270,000 neonatal deaths from birth defects worldwide (about 9 percent of all neonatal deaths) in 2010. Birth defects can result in lifelong disability with great impact on the child, the family, the health system, and the economy. The most common congenital anomalies are congenital heart defects and neural tube defects (NTDs). Congenital anomalies have genetic, infectious, nutritional, and probably environmental causes; many of these are preventable and also treatable. In 2010, over 100,000 infants were born with rubella syndrome, a condition that is entirely preventable by vaccination of the mother. The most common causes of birth defects globally in 2010 were:

- congenital heart defects – 1,040,835 births (10 per 1000 live births)
- neural tube defects (NTDs) – 323,904 births (1 per 1000 live births)
- hemoglobin disorders (thalassemia and sickle-cell disease) – 307,897 births
- Down syndrome (trisomy 21) – 217,293 births
- glucose-6-phosphate dehydrogenase (G6PD) deficiency – 177,032 births (March of Dimes, 2012)
- orofacial clefts – 1 per 700 births (WHO, 2011)
- hearing impairment – 2–3 per 1000 with permanent hearing loss in the USA with 95 percent neonate screening.

The World Health Assembly in 2010 adopted a resolution calling on all member states to promote primary prevention and health of children with birth defects, specifically by:

- developing and strengthening registration and surveillance systems
- developing expertise and capacity
- strengthening research and studies of etiology, diagnosis, and prevention
- promoting international cooperation.

The adoption of best practices in preventing toxic (e.g., exposure to alcohol, smoking, and drugs during pregnancy), infectious (e.g., maternal rubella and tetanus immunization) and nutritional causes of birth defects (e.g., folic acid fortification of flour) should be urgent on the global health agenda.

Diseases resulting from modifications in a single gene (monogenic) affect millions of people worldwide, with over 10,000 human diseases caused by a single error in a single gene in the human DNA. The disease depends on the functions performed by the modified gene. Monogenic diseases can be classified into three main categories: dominant, recessive, and X-linked.

All human beings have two sets or copies of each gene; one copy on each side of the chromosome pair. Dominant diseases are monogenic disorders that involve damage to only one gene copy. Recessive diseases are single-gene disorders due to damage in both copies of the gene. X-linked diseases are monogenic disorders that are linked to defective genes which can also be dominant or recessive on the X chromosome, which is the sex chromosome, i.e., more so in men, who carry only one copy of X chromosome (XY) while women carry two (XX). Monogenic diseases are responsible for a heavy loss of life. The global prevalence of all single-gene diseases at birth is approximately 10 per 1000. The major monogenic diseases include thalassaemia, sickle-cell anemia, hemophilia, cystic fibrosis (CF), Tay–Sachs disease, fragile X syndrome, and Huntington's disease (WHO, 2012). A Canadian estimate indicated that together, monogenic diseases may account for up to 40 percent of the work of hospital-based pediatric practice (Chesney, 2011).

In the USA, an estimated 120,000 infants are born with birth defects annually, with US$2.56 billion spent for hospital care in 2003, and 5800 such defects have a serious adverse effect on health, development, or functional ability. Birth defects are leading causes of pediatric hospitalization, medical expenditure, and infant mortality. Prevention and management of birth defects and genetic disorders are therefore of public health concern. In the USA, birth defects account for over 139,000 hospital stays during a single year, resulting in nearly US$2.6 billion in hospital costs alone, but also carry a financial and social cost to the family and community.

First trimester screening can detect immune status for rubella and hepatitis B, and Down syndrome using ultrasound, while second trimester ultrasound and blood tests for alpha-fetoprotein (AFP) from amniotic fluid can detect NTDs and other birth defects, but these are not considered essential for all pregnancies. Screening for maternal infections that can affect the newborn is also important for HIV, while hepatitis B and TORCH (toxoplasma, rubella, cytomegalovirus, herpes) are not considered necessary for all pregnancies, except for toxoplasmosis. Group B streptococcal (GBS) disease screening often focuses on mothers with defined risk factors (around 15 percent of all pregnancies) to identify candidates for intrapartum antibiotic prophylaxis (IAP) of GBS. A recent Israeli cost–utility analysis of implementing an alternative strategy to expand screening all pregnant women at 35–37 weeks' gestation based on vaginal culture for GBS found it to be cost-effective (Ginsberg et al., 2013). Hearing screening of newborns is now accepted as necessary for all.

Prevention of birth defects requires screening for genetic diseases, congenital diseases, and Rhesus incompatibility, and education and folic acid supplementation before and during pregnancy. The CDC estimates that the birth rates in 2005 for two of the most common NTDs, spina bifida and anencephaly, were 18 and 11 per 100,000 live births, respectively. In 2005, the American College of Medical

Genetics (ACMG) developed a list of 29 primary conditions with the aim of developing uniformity in screening (HSRA, 2011). Prevention also requires reduced exposure to teratogenic chemicals at home, at work, and in the community, and prevention of some infections in pregnancy (Box 6.5). Rhesus incompatibility prevention depends on screening of pregnant women and management where incompatibility between the maternal and fetal blood can develop. Screening for congenital diseases such as PKU and hypothyroidism must be followed up for confirmation of suspected cases and managed appropriately with suitable standards of care. Congenital hip dislocation, which may occur in as many as 50 per 1000 live births, and congenital cataracts should be sought on clinical examination of the newborn. Hearing testing of all newborns should take place within 3 months. Findings should be referred for follow-up care in pediatric specialized services.

The range of birth defects is wide and includes chromosomal and developmental abnormalities as well as metabolic, hematological, musculoskeletal, cardiovascular, and neurological disorders. Genetic disorders are transmitted by abnormal chromosomes from one or both parents. Dominant genes are those which cause the genetic disorder in a heterozygote; that is, a person who has one copy of the abnormal and one copy of the normal gene. If only one parent is affected, there is a 50 percent probability of offspring being affected. In recessive or homozygous conditions, both parents are carriers of the abnormal gene. Each child has a 25 percent chance of being affected, a 50 percent chance of being a carrier, and a 25 percent chance of being unaffected. If only one parent is a carrier of the homozygous gene, the disorder is not passed on although the child may also be a carrier. Screening in many states in the USA now includes 29 conditions and many more are technically feasible, but the benefits and effectiveness are not established for all of them.

Rhesus Hemolytic Disease of the Newborn

Rhesus hemolytic disease of the newborn is a serious condition of breakdown of red cells in the newborn due to an incompatibility between fetal and maternal blood. It occurs when the mother is Rh negative and the fetus Rh positive (inherited from the father). Fetal red blood cells can mingle with the maternal blood during delivery, causing the mother to produce antibodies to the fetal blood. This may affect subsequent pregnancies, causing hemolysis of fetal blood, producing anemia, jaundice, brain damage, or death. This disease was the cause of many stillbirths and neonatal deaths until the development of exchange transfusions in newborns reduced the death rate. In the 1970s, anti-D immunoglobulin was introduced. This is given to Rh-negative women following birth of an Rh-positive baby to prevent the mother from developing antibodies that would

BOX 6.5 Preventing Birth Defects

Community-wide Prevention
- Promote healthy pregnancy concepts – no teen pregnancy, no smoking, alcohol or avoidable drugs, and get prenatal care.
- Iodine fortification of salt to prevent iodine-deficiency brain damage.
- Folic acid fortification of flour to prevent neural tube defects.
- Folic acid supplements for all women of fertile age.
- Rubella immunization before pregnancy to prevent congenital rubella syndrome.
- Treatment of STIs to prevent congenital syphilis, gonococcal, CMV, and HIV infections.
- Hepatitis B immunization to prevent maternal–fetal transmission.
- Confirm full vaccination.

Prenatal Prevention
- Cessation of smoking, alcohol, and avoidable drug intake to prevent fetal damage.
- Pre-pregnancy and pregnancy nutrition counseling to promote optimal fetal development.
- Folic acid supplements, 400 µg per day, for at least 3 months before planned pregnancy and recommended for all women of fertile age.
- Genetic counseling for parents at risk (e.g., consanguineous marriages).
- Avoidance of unnecessary medication.
- Iron and multivitamin supplements (and iodine if salt is not fortified).
- Risk assessment and referral to a high-risk care program.
- Immunization with Tdap vaccine (adult tetanus toxoid, reduced diphtheria and acellular pertussis).
- Prenatal care protocol starting in first trimester.
- Fasting blood sugar to screen for gestational diabetes.
- Management of chronic diseases (e.g., diabetes, cardiac disease) before, during and after pregnancy.
- Prevention of exposure to teratogenic agents before and during pregnancy (e.g., anesthetic agents).
- Prevention of exposure to lead and mercury ingestion.
- Hemoglobin, Rhesus testing, Coombs' test, urinalysis.
- Adequate weight gain (11–14 kg).
- Sonography for suspected multiple gestation, structural, or placental abnormalities.
- Fetal monitoring, fetal size, position, and heart rate.
- Safe delivery with minimal sedation and early resuscitation to prevent brain damage at birth.

Note: STI = sexually transmitted infection; CMV = cytomegalovirus; HIV = human immunodeficiency virus.
Sources: *Centers for Disease Control and Prevention. Guidance for preventing birth defects. 24 February 2011. Available at: http://www.cdc.gov/ncbddd/birthdefects/prevention.html [Accessed 7 February 2013]. Agency for Healthcare Research and Quality. Guide to clinical preventive services, 2012. AHRQ publication no. 12-05154; October 2012. Rockville, MD: AHRQ. Available at: http://www.ahrq.gov/clinic/pocketgd.htm [Accessed 7 February 2013].*

TABLE 6.13 Estimated Proportion of Congenital Abnormality Deaths and Numbers of Neonatal Deaths Due to Neural Tube Defects (NTDs) by Region, 2005

Region	Neonatal Deaths Due to Congenital Abnormalities/Year	Neonatal Deaths Due to NTDs/year	Neonatal Deaths from Congenital Abnormalities Attributed to NTDs (%)
Asia	156,500	97,000	62
North Africa/Middle East	29,500	8,500	29
Sub-Saharan Africa	79,500	17,500	22
Other regions	41,000	8,000	20
World	306,500	131,000	43

Note: Provisional estimates by the Child Health Epidemiology Reference Group for 193 countries.
Sources: Blencowe H, Cousens S, Modell B, Lawn J. Folic acid to reduce neonatal mortality from neural tube disorders. Int J Epidemiol 2010;39(Suppl. 1):i110–21. Available at: http://www.ncbi.nlm.nih.gov/pmc/articles/PMC2845867/ [Accessed 24 March 2013].

affect the next pregnancy. Treatment with anti-D immunoglobulin should eliminate this disease, a major victory of preventive care.

Neural Tube Defects

Neural tube defects (NTDs) are defects of the neural tube, the precursor of the brain and spinal cord, which forms in the first 28 days of a pregnancy, before most women even know they are pregnant. This group is the second most common form of serious birth defects. NTDs include a range of abnormalities from anencephaly, a markedly defective development of the brain which usually results in death within a few hours of birth, to spina bifida, a defective closure of the vertebral column which is compatible with survival.

Each year, more than 300,000 infants worldwide are born with spina bifida and anencephaly. Approximately 4500 pregnancies every year in Europe result in a baby or fetus affected by an NTD, and in the USA 2500 live births are affected by NTDs each year. In China, 100,000 infants are born annually with NTDs. Evidence indicates that 50–70 percent of cases could be prevented by the appropriate consumption of folic acid before conception and during early pregnancy.

During the 1980s, Hungarian and British investigators confirmed that folic acid given before pregnancy greatly reduces the chance of this abnormality developing, by 50–70 percent. Pre-pregnancy care is not common, and compliance with recommended folic acid supplementation is under 30 percent, so the addition of folic acid to flour was adopted by the FDA in 1996 and elsewhere for primary prevention of this disorder.

The economic burden of NTDs falls on health care systems and on wider society, as well as on the family. Studies have helped to elucidate the cost-effectiveness of folic acid for the prevention of NTDs, including 12 studies from the USA, one each from Canada and Spain, and 10 economic evaluations on prevention with folic acid, four from the USA, two each from Chile and the Netherlands, and one each from Australia/New Zealand and South Africa. The benefits of preventing NTDs and spina bifida are very high compared to the cost of food fortification and folic acid supplements to women of fertile age (Yi et al., 2011).

Between 85 and 90 percent of those with spina bifida now survive to adulthood. NTDs vary in degree of severity but many require extensive surgical and medical care. In the USA, some 3000 pregnancies are affected by NTDs (spina bifida and anencephaly), and globally more than 300,000 such births occur annually. The lifetime cost of care for a child with spina bifida is estimated to be US$560,000 (2012). In the USA the CDC estimates the rate of NTDs as spina bifida 18 per 100,000 and anencephaly 11 per 100,000.

Use of folic acid daily among American women by age groups in childbearing years is highest (47 percent) among those age 25–34 years, compared to 30 percent of women age 18–24 years and 40 percent of women age 35–45 years. Because compliance with folic acid supplementation prior to pregnancy was under 50 percent of the population, fortification of flour (along with iron and vitamin B complex) was mandated in 1996 in the USA, Canada, Oman, and Chile, and then followed by over 50 more countries (Table 6.13). A 2008 study estimated that current folic acid fortification produces an annual savings of about US$300 million, or US$100 for each US$1 invested in fortification. Fortification also has resulted in substantial cost savings globally. Chile has demonstrated a saving of US$11 (in international dollars) for each US$1 invested in fortification.

Many countries do not have the surveillance systems to monitor the prevalence of NTDs and other birth defects. The incidence has declined as a result of screening in pregnancy and primary prevention through folic acid supplementation before pregnancy and in early pregnancy. Screening for NTDs became possible in the early 1970s with amniocentesis and testing of amniotic fluid, and later in blood tests for

BOX 6.6 Preventing Neural Tube Defects and Fortification of Flour: Global Experience

Neural tube defects (NTDs) such as spina bifida and anencephaly affect some 4000 births and pregnancies per year in the USA (1500 births with spina bifida and 1000 with anencephaly) and some 1500 NTD abortions are performed per year. In 2011, in the USA the affected pregnancies had declined to about 3000 with 840 deaths. NTDs are associated with 1.3 percent of all infant deaths, second only to cardiac conditions as a cause of neonatal deaths. NTD survivors suffer from spinal cord abnormalities with serious physical and mental disabilities. They may reach adulthood, but lifetime direct costs are estimated at US$560,000 (2003 dollars) per case of spina bifida and a total economic burden of nearly US$500 million.

Mandatory flour fortification began in Canada in 1998, along with the USA and several other countries. In Canada, spina bifida and NTDs were highest in Newfoundland with mean of rates between 1976 and 1997 of 3.40 per 1000 births. There was no significant change in the average rates between 1991–1993 and 1994–1997. The rates of NTDs fell by 78 percent after the implementation of folic acid fortification, from an average of 4.36 per 1000 births during 1991–1997 to 0.96 per 1000 births during 1998–2001 (relative risk 0.22). The average dietary intake of folic acid due to fortification increased in women aged 19–44 years and in seniors. There were significant increases in serum and red blood cell folate levels for women and seniors after mandatory fortification. Among seniors, there was no evidence of vitamin B_{12} deficiencies, and no evidence of masking hematological manifestations of vitamin B_{12} deficiency.

There is broad consensus that women of fertile age should take daily doses of 400–800 µg of folic acid to protect a possible pregnancy since about half of pregnancies are unplanned (ACOG, US Preventive Task Force). Over 70 countries in the Americas (USA, Canada, Chile, and others) and other parts of the world have adopted mandatory folic acid fortification of flour and cereal grain products as of 2012. Australia has added, and Nigeria is proceeding to add folic acid to mandatory fortification of flour. The WHO recommends flour fortification, but the European Union members have thus far failed to adopt flour fortification as an essential public health program, and no country in Western or Eastern Europe had implemented it by the end of 2012.

Sources: *Committee on Genetics, American Academy of Pediatrics. Folic acid for the prevention of neural tube defects. Pediatrics 1999;104:325–7. Available at: http://pediatrics.aappublications.org/content/104/2/325.full*
American College of Obstetrics and Gynecology. Periconceptional folic acid and food fortification in the prevention of neural tube defects. Scientific Advisory Committee, Opinion Paper 4. April 2003. Reaffirmed in 2008. Available at: http://www.guidelines.gov/content.aspx?id=3994
Wald NJ. Folic acid and the prevention of neural-tube defects. N Engl J Med 2004;350:101–3.
DeWals P, Tairou F, Van Allen MI, Soo-Hong U, Lowry RB, Sibbald B, et al. Reduction in neural-tube defects after folic acid fortification in Canada. N Engl J Med 2007;357:135–42.
Berry RJ, Bailey L, Mulinare J, Bower C. Folic Acid Working Group. Fortification of flour with folic acid. Food Nutr Bull 2010;31(Suppl. 1):S22–35. Available at: http://www.ncbi.nlm.nih.gov/pubmed/20629350
US Preventive Services Task Force. Folic acid for the prevention of infant neural tube defects: US Preventive Services Task Force Recommendation. Ann Intern Med 2009;150(9):1–50. Available at: http://annals.org/article.aspx?articleid=744495
US Preventive Services Task Force. Folic acid to prevent neural tube defects (May 2009) recommends between 400–800 micrograms/day. Available at: http://www.uspreventiveservicestaskforce.org/uspstf09/folicacid/folicacidrs.htm [All accessed 15 February 2013].

alpha-fetoprotein (AFP). Ultrasound can also detect NTDs. The CDC estimates that in 2005 the rates for two of the most common NTDs, spina bifida and anencephaly, were 17.96 and 11.11 per 100,000 live births, respectively.

Direct medical costs for patients with NTDs are significant, with the majority of costs being for inpatient care, for treatment at initial diagnosis in childhood, and for comorbidities in adult life. The lifetime indirect cost for patients with spina bifida, including caregiver time, is even greater owing to increased morbidity and premature mortality costs (Yi et al., 2011). An estimated 29 percent of neonatal deaths in low-income countries are related to NTDs, and folic acid fortification of basic foods could prevent 13 percent of neonatal deaths from congenital abnormalities (Blencowe et al., 2010).

In April 1998, the FDA and CDC recommended pre-pregnancy and pregnancy supplementation of folic acid, 400–800 µg/day, to augment the folic acid in fortified bread flour, which provides an intake of about 100–150 µg/day of folic acid, to ensure 100 percent recommended dietary allowance (RDA) intake (see Chapter 8) (Box 6.6). Mandatory fortification of flour with folic acid (along with iron and vitamin B complex in most cases) is practiced in over 50 countries and should now be considered "best practice" for prevention of NTDs and adopted globally to reduce the burden of this serious birth defect.

Following mandatory fortification of flour and cereal grain products not only with folic acid but also with vitamin B complex and iron, studies show a decline in NTDs, and the subject of increasing the level of fortification to reduce NTDs even further is actively being debated. CDC expects fortification of bread and supplements for women of fertile age to substantially reduce the annual incidence of NTDs in the USA. Since mandatory fortification, there have been reports of large reductions in the NTD incidence rate in Canadian provinces (Newfoundland, Nova Scotia) and the USA as well as Chile, and the subject is under active development in many other countries (e.g., the UK and Israel).

The WHO has issued a consensus statement endorsing fortification of flour with iron, folic acid, vitamin B_{12}, vitamin A, and zinc. CDC reports that flour fortification allows these vitamins and minerals to be included in bread, noodles, and other wheat or maize products, offering a cost-efficient strategy for addressing nutrient deficiencies.

Fortification is now mandatory practice in 57 countries and saves 22,000 fatal or disabling NTDs annually. Countries are reporting reductions of 30–70 percent in NTD new births (CDC, 2011).

Cerebral Palsy

Cerebral palsy is a group of neurological disorders occurring as a result of an injury to parts of the brain, or as a result of a problem with development in utero. Cerebral palsy causes motor disabilities, affecting a person's ability to move, and maintain balance and posture. Incidence rates are 1.4–4 per 1000 live births. It may be associated with mental retardation, seizure disorders, motor spasticity, or sensory problems. Cerebral palsy is related to LBW (<2500 g), and especially VLBW (<1500 g), as well as intracranial hemorrhage, Rhesus incompatibility, intrauterine and birth trauma, maternal exposure to heavy metals such as mercury, and other unidentified factors. About 20 percent of cerebral palsy cases are due to intrauterine fetal hypoxia.

In Europe and Australia, cerebral palsy rates have been reported of 5–8 per 1000 in babies of 32–36 weeks gestation and 35–80 per 1000 in those born between 28 and 32 weeks, compared to 1.1–1.7 per 1000 among those born at 37 or more weeks of gestation. In a study in Atlanta, Georgia, USA, the prevalence of cerebral palsy was reported as 6.2 per 1000 among LBW babies (1.5–2.5 kg), and nearly 60 per 1000 among VLBW babies (<1.5 kg), compared to 1.1 per 1000 among babies born over 2.5 kg.

Preventive measures include reducing LBW births by improving maternal nutrition, smoking and alcohol cessation throughout pregnancy, improving care prenatally and during labor and delivery, giving vitamin K at birth, and reducing infant trauma. The use of professionally trained midwives reduces the risk of cerebral palsy. Prevention is limited by a lack of identification of many causative factors. LBW increases the risk of neurological damage, including cerebral palsy, so prevention is focused on optimal maternal health and prenatal care.

Low Birth Weight and Developmental Disability

The WHO defines a birth weight of less than 2500 g as a public health concern because of its association with fetal and neonatal morbidity and mortality, impaired cognitive development, and association with chronic disease during the child's lifetime. LBW is a leading cause of neonatal deaths (i.e., from birth up to 28 days).

Preterm births in the USA are 12 percent of all births and the LBW rate in 2010 was just over 8 percent of total births, a slight decline from the peak in 2006. Rates vary by ethnicity, with rates of 13.5 percent for black non-Hispanics and 6.9 percent for Hispanics (US DHHS, 2011). This variation is largely related to social and economic factors and inadequate health care coverage, all of which constitute public health challenges. In the USA in 2005, the annual cost was more than US$26 billion, with medical care accounting for two-thirds of the costs; special education costs were US$1.1 billion. The prevalence of major disabilities includes cerebral palsy, mental retardation, hearing and vision, communication disabilities, behavioral and social challenges (IOM, 2007).

Intellectual disability, formerly called retardation, and more recently developmental disability (DQ) (Lichtenberger, 2005), is one of the most common birth and developmental disorders in industrialized countries and may be due to a wide variety of causes. These include Down syndrome, fetal alcohol syndrome (FAS), drug abuse, genetic conditions and infections (e.g., congenital cytomegalovirus), or defects that affect the brain (e.g., hydrocephalus or cortical atrophy), which occur before birth. LBW and asphyxia during delivery or soon after birth, and conditions occurring during childhood such as serious head injury, stroke, or infections such as meningitis are also causes of developmental disability.

Severe intellectual disability or intelligence quotient (DQ is now the preferred term) of less than 50 occurs in 3–4 per 1000 newborns. It may be related to genetic and prenatal factors and is often associated with cerebral palsy and seizures. More than one-third of cases are attributed to chromosomal abnormalities. Prenatal diagnosis, widespread use of amniocentesis in pregnant women over the age of 35, termination of affected pregnancies, but mostly improved birth weight have all contributed to reducing the incidence.

PKU, congenital rubella, congenital hypothyroidism, and maternal infections with toxoplasmosis and cytomegalovirus are all preventable or treatable (primary or secondary prevention) but untreated can all cause severe disability. Pregnancy complications such as toxemia, urinary tract infections, and anemia increase the risk but are treatable. Mild disability (IQ of 50–70) may relate to genetic, prenatal, perinatal, and postnatal factors, including LBW or asphyxia at birth. Prevention depends on healthy mothers with full participation in well-organized prenatal, perinatal, and postnatal care.

Down Syndrome

Down syndrome is a relatively common genetic disorder, with a combination of birth defects, including some degree of mild to moderate intellectual disability, characteristic facial features and frequently congenital heart disease and other health issues of varying severity. The baby has 47 instead of the normal 46 chromosomes, which causes the physical signs and additional problems of Down syndrome.

The WHO estimates a worldwide prevalence of 1 in 1000 births. Each year about 6000 babies, or 1 per 700 live births, are born with Down syndrome in the USA. The risk increases rapidly with increasing maternal age; for mothers over the age of 45 it occurs in 1 per 30 births; however, 80 percent of babies with Down syndrome are born to mothers younger than 35 years of age. Thus, although this condition occurs more commonly in older mothers, the majority of cases are in younger women. It is the most common cause of mental disability in industrialized countries. Prevalence at birth has been rising in the USA, from 9.5 per 10,000 live births in the period 1979–1983 to 11.8 in 1999–2003, with rates as much as five times higher in older mothers, and with lower rates among black Americans but higher among Hispanics (CDC, 2012).

In cases at risk, prenatal screening is done by amniocentesis, chorionic villus sampling, and fetal blood sampling, allowing for the parental choice of termination of pregnancy if the chromosomal abnormality is detected. Chorionic villus sampling can be performed in the first trimester, and amniocentesis in the second. Chorionic villus sampling has itself been reported to be associated with minor birth defects. Simpler and safer blood tests are soon expected to be available for screening. Screening for Down syndrome has not prevented an increase in cases and is associated with false positives, and therefore screening policies are currently under review.

Pregnant women over the age of 35 should be screened as early as possible in pregnancy. Biochemical markers will help to increase screening with less invasive procedures. In England and Wales, between 1989 and 2008, prenatal diagnoses of Down syndrome increased by 76 percent, from 1075 to 1843, but the incidence of live births with Down syndrome decreased by 1.5 percent, from 755 to 743, while there would have been a 48 percent increase in Down syndrome births if there had been no prenatal screening and subsequent terminations, with many parents choosing to delay starting a family.

Down syndrome symptoms can range from mild to severe; babies have varying degrees of intellectual disability, and often congenital cardiac defects and gastrointestinal obstruction. These people now survive well into their thirties, and 80 percent to their fifties, with many approaching their sixties and even seventies. The estimated lifetime costs of Down syndrome are US$1.8 billion in the USA. People with Down syndrome can lead happy and productive lives but age prematurely, and have an increased risk for Alzheimer's disease.

Emphasis has shifted in recent years from institutional care to care in the family and self-care in the community, along with education and training for Down syndrome children. With vocational and life skills training, many are able to live independently or in groups in the community. Those who are integrated into the community often live longer and have a better quality of life compared to those raised in institutions.

Cystic Fibrosis

Cystic fibrosis (CF) is a genetic disorder that affects mostly the lungs, pancreas, liver, and intestine. It is is the most common lethal genetic disease in the USA and the European Union (EU) white population, occurring in 1 per 2000 to 3500 for North Americans of Northern European descent, and the same rate in the EU. It is rarer among blacks, 1 per 17,000 and 1 per 11,500 for Hispanic newborns. The disease is a recessive genetic defect. Prenatal testing of couples of Caucasian background is helpful for early diagnosis. The gene is present in 5 percent of the white population and 2 percent of African Americans. The defect in homozygotes causes production of abnormally thick mucus in the lungs, intestines, and glands, causing chronic lung disease and malabsorption.

The clinical disturbance is chronic obstructive lung disease, repeated infections, and destruction of lung tissue. Early diagnosis is important to maintain aggressive preventive and care for infection and diet with vitamin supplements, especially vitamin D. CF leads to frequent hospitalization and death, even with the best of care. In 2005 in the USA, the predicted median age of survival rose from 20 years in 1970 to 32 years in 2000 and to 36.5 years in 2005. With careful management of lung, sinus, and other infections, nearly 40 percent of CF patients survive and many now live to their fifties or sixties. Prenatal screening can be done with chorionic villus sampling or amniocentesis.

Newborn screening is now mandatory in all states in the USA, with over 98 percent coverage of babies born in hospital. Screening is carried out in Canada and other industrialized countries because of demonstrated benefits of early treatment on growth, nutritional status, cognitive development, and reduced lung infections and hospitalization. Early diagnosis and support and education of parents can significantly improve the duration and quality of life of the affected person. Treatment is complex and costly, involving a multidisciplinary approach with a focus on lung and sputum management. Gene therapy techniques, new vaccines, and other therapies including lung transplantation are important areas of new development in case management for CF. Studies have shown that children with CF have low levels of vitamins K and D, which may adversely affect bone health. Since 1991, 2800 CF patients (150–200 annually) have received lung transplants in the USA.

Sickle-Cell Disease

Sickle-cell disease (SCD) is a common genetic disorder of the red cells in the blood. It is passed to a newborn when both parents are carriers of the affected gene, mainly among those whose ancestors are from sub-Saharan Africa, Spanish-speaking regions in the Western Hemisphere, Saudi Arabia, India, and Mediterranean countries. SCD is a major public health concern in the USA among African Americans. A registry surveillance system (RuSH) is coordinated by the National Institutes of Health (NIH) and several states. CDC estimates that SCD affects 90,000–100,000 Americans; 1 in 500 black African American infants, and 1 in 36,000 Hispanic American babies. Child mortality among children under the age of 4 years fell by 42 percent between 1999 and 2002, attributed to the introduction of vaccine for invasive pneumococcal disease (CDC Sickle Cell Disease 2011). In 2005, about 40 percent of SCD children were hospitalized at least once, with average medical care costs of US$11,702–14,772. The WHO estimates that there are about 300,000 SCD children born annually worldwide, and that about 25 percent of some regional populations are carriers.

SCD is an amino acid defect of red blood cells which affects 1 in every 500 African American newborns and is even more common in Africa, where it may have a protective effect against malarial parasites. The disease is caused by a recessive gene affecting hemoglobin structure, usually with a benign course in the carrier state, but with episodic pain due to small blood vessel closure in the homozygous state. Clinical symptoms appear in the second half of the first year of life. The patient develops moderately severe anemia, with an increased susceptibility to severe bacterial infections (meningitis, pneumonia, septicemia) and failure-to-thrive (growth retardation).

Screening of newborns mandated in all 50 US states and the District of Columbia identifies 50 carriers for every infant diagnosed with the disease. Identification of cases and carriers is important to ensure prompt care in crises, for preventive use of penicillin which reduces infection rates, and to ensure that they receive influenza and pneumococcal vaccines. Carriers should receive genetic counseling related to marriage and pregnancy. Healthy People 2020 objectives are to increase the use of prophylactic penicillin for children aged 4 months to 5 years and to reduce hospital admissions for preventable complications among children 9 years of age and younger.

Thalassemia

Beta thalassemia is a recessive genetic disorder of hemoglobin structure. Beta thalassemia minor is usually without clinical significance. Beta thalassemia major, the homozygous state when the gene is inherited from both parents, is characterized by hemolytic anemia (i.e., early breakdown of red blood cells). Beta thalassemia major, also known as Cooley's anemia or Mediterranean anemia, is widespread throughout the Middle East, Southern Europe, Iran, and across southern India and South-East Asia. Globally, nearly 75 percent of thalassemia cases and 72 percent of 22,522 deaths due to lack of transfusion occur in the Eastern Mediterranean and South-East Asia. It is ultimately fatal for those afflicted, but with current standards of treatment, including blood infusions and chelating agents (i.e., iron binding for

excretion) to reduce iron overload and hemochromatosis, patients survive into their thirties (Box 6.7).

The WHO reports that some 5 percent of the world population are carriers of gene traits for hemoglobinopathies, mostly SCD and thalassemia. Hemoglobin disorders are common in most countries with a high prevalence of malaria. Over 300,000 babies are born annually with hemoglobinopathies, about 275,000 of whom have SCD, mainly in Africa. Thalassemia major is responsible for 5500 deaths in the perinatal period, and a further 30,000 of the 56,000 with beta thalassemia major need regular blood transfusions and expensive chelating agent care to reduce iron overload from many transfusions and early breakdown of red blood cells. Hemoglobin disorders are widespread, originally localized in the Mediterranean areas, spreading to South-East Asia for thalassemia and sub-Saharan Africa for SCD. Because of migration, these diseases are now spread worldwide, with 10 percent of the population at risk in the USA.

Prevention of thalassemia is one of the success stories of genetic public health. Preventive approaches to this disease since the 1970s produced dramatic results in reducing the number and rate of new cases of beta thalassemia major in Sardinia, Cyprus, Greece, the UK, Canada, and other locations where this disease has been endemic among people of Mediterranean origin. Premarital screening, health education in schools, and access to prenatal screening are all part of a program to reduce new cases of this disease. The WHO stresses the public health importance of this disease and recommends adoption of demonstrably successful preventive approaches to member states that have this problem.

Screening for congenital anemias should take place at birth and at school age, because clinical cases may not appear until several years after birth. Gene carriers need to know at an age when they can understand the limitations this may place on them. In each pregnancy when both parents are carriers, chorionic villus sampling or amniocentesis should be carried out to determine if the fetus is affected. Abortion is currently recommended if the fetus is affected by thalassemia major, but bone marrow transplantation is showing promise in the treatment of new cases. The success of primary prevention in reducing new cases of thalassemia provides a model for application to disorders affecting other population groups, such as sickle-cell anemia.

Phenylketonuria

PKU is an inborn error of metabolism transmitted by a recessive gene. PKU involves a mutation in DNA which causes inadequate production of an enzyme needed to metabolize phenylalanine, an amino acid. In utero, the fetus is unaffected because the maternal metabolism handles the excess phenylalanine. However, if the newborn lacks the enzyme needed, phenylalanine accumulates in the blood, leading to brain and neurological disorders and severe retardation.

BOX 6.7 Eradication of Thalassemia Major in Cyprus and Sardinia

Between 4.5 and 7 percent of the world population are estimated to be carriers of hemoglobinopathy genes and an estimated 300,000–400,000 babies are born with major hemoglobinopathies. The problem is growing with improved treatment and better survival of cases. World Health Organization (WHO) approaches and recommendations for preventive measures have been established in different countries in all WHO regions and achieved success in reducing the burden of hemoglobinopathy, such as in the virtual eradication of beta-thalassemia major in formerly endemic areas in the Mediterranean region.

In Cyprus, 14 percent of the population were carriers of beta-thalassemia and nearly 1 percent of the Cyprus population (1 in 158 births) had the disease. Owing to an intervention program initiated in the 1970s, only rarely are new cases of beta-thalassemia major seen in infants born in Cyprus and other Mediterranean locations such as Sardinia, Sicily, and Greece. This was achieved by a long-term preventive program consisting of public education, screening for carriers, and genetic counseling. Marriage between carriers is reduced by a community education program, and, when marriage does occur, careful screening of all pregnancies and termination of affected pregnancies reduces the number of thalassemia major births. The success of the Cyprus approach provided a model of control of a genetic disorder via a combination of health education, screening, and community support. Other countries including Nigeria, Bahrain, Cuba, Brazil, Iran, Saudi Arabia, Tunisia, and Pakistan have similar models.

Education and planning for routine newborn screening to detect hematological diseases, inborn errors of metabolism, and other congenital conditions are part of modern care of newborns in developed countries. The benefits are in early case detection and treatment to prevent the damage done to brain (by phenylketonuria and congenital hypothyroidism), lungs (by cystic fibrosis), many inborn errors of metabolism, and other body systems including the hematological system, as in the case of thalassemia and sickle-cell anemia.

With rapidly increasing knowledge and science associated with the Human Genome Project, the potential in this field will grow. There are social and ethical issues associated with screening and ethnic focus of educational programs, but the success of thalassemia control is a great achievement of public health in the twentieth century and its application in the twenty-first century, and is being applied to other important genetic disorders such as Tay–Sachs and more recently cystic fibrosis. In the future it may be even more important in relation to genetic factors in cancer, cardiovascular disease, and other common conditions. This will raise many ethical and political issues that have to be addressed in the context of different settings, religious beliefs, and cultural norms.

Sources: *Angastiniotis M, Modell B, Engelzos P, Boulyjenkov V. Prevention and control of haemoglobinopathies. Bull World Health Organ 1995;73:375–86. Available at: http://www.ncbi.nlm.nih.gov/pmc/articles/PMC2486673/pdf/bullwho00407-0102.pdf [Accessed 27 June 2013].*
Ginsberg G, Tulchinsky T, Filon D, Godlfarb A, Abramov L, Racmilevitz EA. Cost–benefit analysis of a national thalssemiua prevention programme in Israel. J Med Screen 1996;5:120–6. Available at: http://www.ncbi.nim.gov/pubmed/9795870 [Accessed 27 June 2013].
Weatherall DJ, Clegg JB. Inherited haemoglobin disorders: an increasing global health problem. Bull World Health Organ 2001;79:704–12. Available at: http://www.ncbi.nlm.nih.gov/pmc/articles/PMC2566499/ [Accessed 27 June 2013].
World Health Organization. Genomics and world health. Geneva: WHO; 2002. Available at: http://whqlibdoc.who.int/hq/2002/a74580.pdf
Centers for Disease Control and Prevention. Public health genomics. Available at: http://www.cdc.gov/genomics/

The disease was first identified by Norwegian physician Ivar Følling in 1934. The famous PKU Guthrie test, developed in the USA by Robert Guthrie in 1961, permitted mass screening by a simple heel prick with blood on a filter paper sent to a specialized laboratory. If discovered early, PKU can be managed with a special low-phenylalanine diet with no significant retardation.

PKU is reported in 1 in 15,000 births in the USA. Higher rates are reported in Ireland (1 per 4500 births) and Turkey (1 per 2600 births), with low rates in Japan (1 per 120,000 births) and Finland (1 per 100,000 births). PKU has an incidence rate of 1 per 300,000 in African Americans and also a very low prevalence in Africa. In Europe, the prevalence is about 1 case per 10,000 live births. In Latin America, PKU varies from about 1 case per 25,000 to 1 per 50,000 births; prevalence is generally higher in southern regions.

Screening of all newborns for this condition is widespread in developed countries and reduces the number of cases requiring long-term institutional care. Where screening is not universal, pregnant women with previous PKU children should be tested during pregnancy and the newborn tested at birth. Women with PKU should maintain a special diet before and during pregnancy to help prevent mental retardation and birth defects. CDC considers PKU screening a major public health success, with an estimated 3000–4000 reproductive-aged women with PKU identified by newborn screening and placed on a diet that prevented severe mental retardation. People with PKU should stay on the special low-alanine diet throughout their life, but this severely restricted diet is often discontinued during adolescence. Babies born to women with PKU not following the diet are at high risk for mental retardation and birth defects, due to the mother's condition. These effects can be prevented if women follow the PKU diet before and during pregnancy (CDC, 2002, 2008).

Congenital Hypothyroidism

Congenital hypothyroidism is a relatively common congenital disorder in which the thyroid gland is deficient in development and function. It occurs in about 1 per 3500–4000 live births in North America, Europe, Japan, and Australia, with a range of 1–20 per 10,000 live births in different population groups, depending on genetic makeup and consanguinity rates. This genetic disorder affects females twice as often as males. It causes an inefficient development of the thyroid and may be confused with iodine deficiency disorders common in areas with deficient iodine in water and soil. Therapy with thyroid replacement in congenital hypothyroidism cases prevents the severe intellectual impairment that otherwise occurs. Screening within 48 hours of birth should uncover cases for long-term follow-up and management and is a must for neonatal care globally.

Congenital Syphilis

The adverse impact of syphilis on child health has been known for over 500 years, yet it was only in 2007 that the WHO targeted the eradication of congenital syphilis. Syphilis in pregnancy is more common than HIV in pregnancy. Untreated maternal HIV infection is transmitted to infants in about one-third of the cases. In contrast, untreated maternal syphilis nearly always results in adverse pregnancy outcome. There were 1.9 million pregnant women with syphilis in 2008 compared to 1.49 in 2010. Probably less than 10 percent of pregnant women are detected and treated. Most syphilis infections can be identified with laboratory-based and rapid point-of-care tests, effectively treated by a single dose of benzathene penicillin to prevent adverse pregnancy outcomes. The combined cost of testing and treatment is less than US$1 (2013). The available evidence is that elimination is achievable with community advocacy, political will, and private donor investment (Klausner, 2013).

Fetal Alcohol Syndrome

Alcohol use during pregnancy is a significant public health issue. Fetal alcohol syndrome (FAS) and fetal alcohol spectrum disorder (FASD) are caused by the toxic influence of alcohol on the fetus via the placenta. Alcohol consumption before and during pregnancy increases the risk of behavioral and cognitive disorders. Exposure of the fetus to maternal alcohol consumption, especially binge drinking (i.e., more than five drinks on one occasion), is a cause of fetal growth retardation and anomalies of the central nervous system. FAS is completely preventable, as it only occurs when the pregnant mother drinks alcohol.

The prevalence of FAS and FASD is estimated at 0.5–2 and 1.5–6 per 1000, respectively. Alcohol use and binge drinking among American women of childbearing age did not change markedly between 1991 and 2005 and the Healthy People 2010 targets were not met. Alcohol use levels prior to pregnancy indicate a risk of continued use during pregnancy, partly because many women do not realize they are pregnant in the early weeks of gestation. Alcohol use during pregnancy continues to be an important public health concern and should be addressed in health promotion and clinical care settings.

Tay–Sachs Disease

Tay–Sachs disease is an inborn error of metabolism associated with progressive mental deterioration and loss of vision by 4–8 months of age and death by 3–4 years. It affects 1 in 2500 births among Jews of Eastern European origin. The rate of carriers is 1 in 27 among American Jews of European descent as well as among French Canadians and Cajun Americans, and 1 in 50 among Irish Americans compared to 1 in 250 for Jews of North African or Middle Eastern origin (Sephardic Jews) and the general population.

Tay–Sachs disease can be prevented by screening these risk groups before marriage, at the time of marriage, or early in pregnancy. When both partners are carriers of the disorder, they should be informed of the disease and referred for genetic counseling and/or prenatal diagnosis. Prenatal screening by amniocentesis or chorionic villus sampling will confirm whether the fetus is affected. In such cases, termination of pregnancy is recommended. These screening programs have reduced the incidence of new cases in these risk groups. Implementation of such screening and prevention has virtually eliminated the disease among Ashkenazi Jews.

Glucose-6-Phosphate Dehydrogenase Deficiency

Glucose-6-phosphate dehydrogenase (G6PD) deficiency is a genetic disorder common in Mediterranean populations, including Sephardic Jews, Greeks, southern Italians, South-East Asians, and southern Chinese. The condition occurs in 400 million people worldwide; about 1 in 10 African American males are affected. It results in episodes of hemolytic anemia due to an infection, reactions to certain foods (e.g., fava beans), or reactions to oxidant drugs such as sulfonamides, antipyretics, and antimalarials. The degree of hemolysis varies with the agent and the degree of the enzyme deficiency. Identifying the condition helps the patient to avoid exposure to hemolytic inducing agents and start prompt treatment in crises.

Familial Mediterranean Fever

Familial Mediterranean fever is a recessive hereditary genetic condition when the child inherits the gene from both parents who carry the gene but not the illness. The condition is found in Arabic, Armenian, Turkish, and Sephardic Jewish populations. It results in periodic fevers and pains in the chest, abdomen, and joints. Control is by genetic counseling. The rate is 1 in 250 to 1 in 1000 people in these populations.

INFANT AND CHILD HEALTH

Public health has long played a major leadership role in improving the health of children by provision of care and regulation of conditions to prevent disease, provide early and adequate care of illness, and promote health. Pediatrics developed as a clinical specialty under the leadership of Abraham Jacobi, the acknowledged "founder of pediatrics". He opened the first pediatric clinic in New York City in 1860 based on German models, and was later appointed professor of infantile pathology and therapeutics at New York Medical College, coauthoring a major textbook on diseases of women and children. The first children's hospital in the USA was opened in 1865 in Philadelphia and continues to operate to this day.

The American Medical Association, recognizing that women and children had health needs apart from those of the general population, established a section to address those needs in 1879, leading to the founding of the American Pediatric Society in 1888. Pediatrics emerged as a separate specialty from general medicine with emphasis on the treatment of children's diseases and birth disorders, the prevention of infectious diseases, and infant nutrition. Well-child care was pioneered in the USA based on milk stations, adapted from those in France (*gouttes de lait*), in the mission houses providing home nursing to care to poor immigrant neighborhoods in New York City, leading to the development of public health nursing and the visiting nurse function.

The first textbook of pediatrics was written in 1869 by J. L. Smith, professor of children's diseases at Bellevue Hospital Medical College in New York. It was followed by Holt's *Diseases of Infancy and Childhood* in 1896. The US federal government established the Children's Bureau in 1913 to collect data on maternal and infant mortality. This later became the Bureau of Maternal and Child Health, which was subsequently empowered by federal legislation to provide grants to states for maternal and child health services. The AAP has come to the forefront of advocacy for improved child health standards, pioneering the development of clinical guidelines, and promoting professional standards in a wide range of child health topics, from breastfeeding to CF screening, in parallel to ACOG.

In the wider context of health internationally, maternal and child health are among the major priorities, with special focus on primary health care. Primary health care, promoted since the Alma-Ata Conference, has placed emphasis on infant immunization, diarrheal disease control, breastfeeding and nutrition practices, and the prevention of deficiency disorders. The WHO, UNICEF, and many non-governmental organizations (NGOs) have provided leadership in the development of primary health care in the developing world. The return to basics in health care has been of benefit to the industrialized world as well, as health care costs soared during the

1970s and 1980s and are increasing with advances in diagnostic technology. In the USA, public health also has to cope with the lack of health insurance for a substantial portion of the population, and even in the early twenty-first century, federal initiatives in health are focused on providing health benefits to about 16 percent of the population who lack health insurance, out of the total US child population of 75 million. In the USA in 2010, 9.8 percent of children under the age of 18 (7.3 million children) were living in poverty (reduced from 12 percent in 2009), and 20.8 percent of black children were without health insurance.

The global focus of the MDGs (discussed extensively in Chapter 16), launched by the UN in 2001 with targets set for 2015, includes MDG4: reduce by two-thirds, between 1990 and 2015, the under-five mortality rate, including objectives of reduced infant mortality and improved immunization coverage with measles-containing vaccines. The great dangers for maternal and infant mortality are poor preparation for pregnancy, unskilled deliveries, and inadequate nutrition of both mother and child. Community-based interventions with multivitamin supplements and low technology such as providing solar panels to provide lights for night deliveries would increase the effectiveness of services accessible to poor rural women (see Chapter 16).

Fetal and Infant Mortality

A live birth is defined by the WHO and the US National Center for Health Statistics as a completed expulsion or extraction from its mother of a product of conception, irrespective of the duration of the pregnancy, which after separation, breathes or shows any other evidence of life such as heartbeat, umbilical cord pulsation, or definite movement of voluntary muscles, whether or not the umbilical cord has been cut or the placenta is attached. Each product of such a birth is considered live born. The delivery can be described as spontaneous vaginal, forceps, vacuum extraction, caesarean section, or vaginal delivery after previous caesarean.

A fetal death or stillbirth is a death prior to the complete expulsion or extraction from its mother of the product of conception, irrespective of the duration of pregnancy. The death is indicated by the fact that after such separation, the fetus does not breathe or show any other evidence of life, such as beating of the heart, pulsation of the umbilical cord, or definite movement of voluntary muscles. For statistical purposes, tabulations for fetal deaths are for a stated or presumed gestation of 20 weeks or more and for 28 weeks or more gestation; the latter are known as late fetal deaths. Other indicators of infant mortality are seen in Box 6.8.

The infant mortality rate (IMR) is a generally accepted indicator of the health status of a population for internal regional and international comparisons, because it represents the cumulative effect of many socioeconomic, environmental, and health service factors. Most industrialized

BOX 6.8 Derivation of Rates in Infant Mortality

$$Total\ Stillbirth\ Rate = \frac{Number\ of\ stillbirths}{Total\ stillbirths + Live\ births\ (per\ annum)} \times 1000$$

$$Perinatal\ Mortality\ Ratio = \frac{Number\ of\ stillbirths + Number\ of\ deaths\ in\ first\ week\ of\ life}{Total\ live\ births + Stillbirths\ (per\ annum)} \times 1000$$

$$Early\ Neonatal\ Mortality\ Rate = \frac{Number\ of\ deaths\ in\ first\ week\ of\ life}{Total\ live\ births\ (per\ annum)} \times 1000$$

$$Late\ Neonatal\ Mortality\ Rate = \frac{Number\ of\ deaths\ between\ first\ 7\ and\ 28\ days}{Total\ live\ births\ (per\ annum)} \times 1000$$

$$Neonatal\ Mortality\ Rate = \frac{Number\ of\ deaths\ in\ first\ 28\ days\ of\ life}{Total\ live\ births\ (per\ annum)} \times 1000$$

$$Postneonatal\ Mortality\ Rate = \frac{Number\ of\ deaths\ between\ 28\ and\ 364\ days\ of\ life}{Total\ live\ births\ (per\ annum)} \times 1000$$

$$Infant\ Mortality\ Rate = \frac{Number\ of\ deaths\ from\ birth\ to\ 364\ days}{Total\ live\ births\ (per\ annum)} \times 1000$$

countries have IMRs under 5 per 1000 live births and many as low as 4 or even 2 per 1000. In these countries, most infant deaths are a result of congenital anomalies and perinatal conditions associated with prematurity and the neonatal period, usually occurring during the first week of life (early neonatal deaths).

In developing countries, most infant deaths are due to acute respiratory infections, diarrheal diseases, and prematurity (LBW), with measles and tetanus remaining large-scale causes of infant death (Box 6.9). Developing countries generally have IMRs over 30 and up to 90 per 1000, with high neonatal and postneonatal mortality. Neonatal mortality is largely due to low birth weight and poor standards of care at and following delivery. Postneonatal mortality is largely preventable by immunization, management of respiratory and diarrheal episodes, and breastfeeding, vitamin and mineral supplements with addition of solid foods after 6 months.

Between 1990 and 2010, IMRs declined by 23 percent in developing and 35 percent in least developed countries. Preventive health measures such as immunization, breastfeeding and nutritional supplementation, HIV treatment for infected pregnant women, and use of insecticide-treated bed nets, along with good management of acute respiratory infections and oral rehydration for diarrheal diseases, are highly effective in saving lives in the postneonatal period and up to age 5 years.

Neonatal mortality rates can be reduced by reducing the number of LBW births, providing good maternal nutrition and prenatal care, minimizing birth injury, and giving good care immediately following birth. Together, these constitute the child survival package of services that are one of

BOX 6.9 Leading Causes of Infant Mortality, USA and Developing Countries

US 2011 – Infant Mortality Rate Under 10 per 1000
1. Birth defects
2. Low birth weight – prematurity
3. Sudden infant death syndrome
4. Maternal complications
5. Accidents
6. Complications of placenta, cord, and membranes
7. Bacterial sepsis of newborns
8. Respiratory distress of newborn
9. Diseases of circulatory system
10. Neonatal hemorrhage

Developing Countries – Infant Mortality Rate Over 20 per 1000
1. Neonatal deaths due to low birth weight/prematurity, asphyxia, birth trauma: 40 percent of all infant deaths with 26–45 percent in the first 24 hours and 75 percent in the first week of life
2. Pneumonia – acute respiratory infections
3. Diarrheal diseases
4. Malaria
5. Measles
6. Newborn effects of maternal complications
7. Neonatal tetanus
8. Undernutrition: contributes to more than one-third of these deaths.

Sources: *National Vital Statistics Reports, Vol. 61, No. 6, October 10, 2012* http://www.cdc.gov/nchs/data/nvsr/nvsr61/nvsr61_06.pdf
World Health Organization. Newborns: reducing mortality. Geneva: WHO; May 2012. Available at: http://www.who.int/mediacentre/factsheets/fs333/en/index.html [Accessed 22 February 2013].

TABLE 6.14 Global Infant Mortality Rates by Region/Development Level, 1990–2010

Region/Development Level	1990	2010	% Change 1990–2010
Sub-Saharan Africa	105	76	−27.6
Middle East and North Africa	56	31	−44.6
South Asia	86	52	−39.5
East Asia and Pacific	41	19	−53.7
Latin America and Caribbean	43	18	−58.1
CEE/CIS	41	19	−53.7
Industrialized countries	9	5	−44.4
Developing countries	67	44	−34.3
Least developed countries	106	71	−33.0
World	61	40	−34.4

Note: Infant deaths from birth to one year per 1000 live births.
CIS = Commonwealth of Independent States; CEE = Central and Eastern Europe.
Source: United Nations Children's Fund. State of the world's children 2012. Available at: http://www.unicef.org/sowc2012/pdfs/SOWC%202012-Main%20 Report_EN_13Mar2012.pdf [Accessed 22 February 2013].

the main thrusts of public health. Undernutrition is a factor in 35 percent of child deaths; underweight is the leading risk factor in the burden of disease. The WHO estimates that 20 million children under five worldwide are severely malnourished, which leaves them more vulnerable to illness and early death. Poor maternal nutrition and failure to breastfeed in the early months of life create great danger from respiratory and diarrheal diseases.

LBW in industrialized countries ranges between 4 percent (Finland) and 8 percent (UK, USA, Japan). The use of fertility treatments and nutritional and lifestyle issues raise concerns because of the much higher rates of morbidity and mortality among LBW infants, especially in the neonatal period. LBW has become a major target program for the March of Dimes, an important NGO in the USA.

The MDGs initiated in 2001 and targeted to 2015 include reduction of child mortality by two-thirds by 2015. The WHO reports that the number of deaths of children below 5 years of age fell from 12 million in 1990 to 6.9 million in 2011. The percentage of births attended by skilled birth attendants increased globally, but in sub-Saharan Africa this is so for less than 50 percent of births. Progress has been made in developing infrastructure and in specific disease control measures to reduce infant mortality in low- and medium-income countries. Oral rehydration is life saving and can be distributed to mother and education provided by CHWs, who can also distribute and promote the use of insecticide-impregnated bed nets to prevent malaria, and rapid diagnostic tests for HIV and, hopefully in the near future, tuberculosis (see Chapter 4). Kangaroo carrying of newborns is also a life saver (see Box 6.2) and can be promoted by the same CHWs who can be trained to screen for HRPs and infants to refer for medical care.

Table 6.14 shows changes in infant mortality by WHO country group categories, with a worldwide decline in IMR from 61 per 1000 live births in 1990 to 40 per 1000 in 2010, a reduction of 34.4 percent. The most striking changes are seen in East Asia, Latin America and the Caribbean, and in Central and Eastern Europe and the Commonwealth of Independent States (CEE/CIS), with over 50 percent reduction in each case. Despite the impressive progress, the MDG for two-thirds reduction in child mortality is unlikely to be met by 2015 (see Chapter 16), but holds out prospects for dramatic changes in the coming decade.

The USA emphasizes targeted public health programming with attempts to provide care to people without access to prepaid medical care. Federal health officials set a goal to reduce the country's IMR to no more than 9.0 by 1990 and expanded access to prenatal and infant care in order to reach this goal, which was achieved. From 1983 to 2010, the IMR of the USA fell from 10.9 to 6.1 per 1000 live births, a decline of 44 percent.

Although mortality rates have declined for all ethnic groups, the African American IMRs remain more than double the rates for white infants, reflecting socioeconomic differences and less access to medical care for the African American population (Table 6.15). The trend among Hispanics is for marginally lower rates than white Americans.

INFANCY CARE AND FEEDING

The new infant is dependent on a healthy, caring mother with support of the family and health providers. Physical and emotional warmth, cleanliness, and bonding with the mother are important in care and feeding (see Chapter 8). Healthful feeding in infancy is a vital issue in infant care and for

TABLE 6.15 Infant Mortality Rates by Race and Percentage Change, USA, 1983–2010

Group	1983	1990	2000	2005	2006	2010	% Change 1983–2010
Neonatal white	6.1	4.8	3.8	3.8	3.7	3.5	−43
Neonatal black	12.5	11.6	9.4	8.9	8.7	7.5	−40
Post neonatal white	3.2	2.8	1.9	2.0	1.9	1.7	−47
Post neonatal black	6.7	6.4	4.7	4.3	4.2	4.1	−39
Total infant, white	9.3	7.3	5.7	5.7	5.6	5.2	−44
Total infant, black	19.2	16.9	13.5	13.3	12.9	11.6	−39
Total infant mortality	10.9	8.9	6.9	6.9	6.7	6.1	−44

Note: Data are rates per 1000 live births.
Sources: National Center for Health Statistics. Health, United States, 2010: with special feature on death and dying. Hyattsville, MD: NCHS; 2011. Available at: http://www.cdc.gov/nchs/data/hus/hus10.pdf; Chartbook: http://www.cdc.gov/nchs/data/hus/hus10.pdf#listfigures and National Vital Statistics Reports 2012;60(4):1–5. Available at: http://www.cdc.gov/nchs/data/nvsr/nvsr60/nvsr60_04.pdf

TABLE 6.16 Recommended Infant Feeding and Supplementation

					Age (Months)				
	Birth	2	4	6	8	10	12	16	
Breastfeeding	+	+	+	+	+	+	+	– optional	
Vitamin K injection	+								
Supplementary foods[a]	–	–	+	+ Add[a]	+	Eat at family table			
Vitamin A, D, or multivitamins	+	+	+	+	+	+	+	+	
Iron syrup	–	+ for LBW	+ For all	+	+	+	+	+	
Immunizations[b]	+	+	+	+	–	–	+	+	
Monitor development markers	+ Apgar, length, weight, head circumference	+ Physical measurements, responsiveness, movements, hearing	++	++	++	++	++	++	

Note: [a] Add gradually: cereals, vegetables, fruit, meat, eggs
LBW = low birth weight.
bSee childhood immunization table (Table 4.8) in Chapter 4.

primary care services, and is as important to infant survival and well-being as infectious disease control and immunization. Therefore, it has an important place in maternal education and in health provider orientation (Table 6.16).

Vitamin K should be given by injection within 6 hours of birth to all infants. Breastfeeding should be established early and encouraged as the sole feeding for 4–6 months, because the composition of breast milk is ideal nutritionally and for immunological protection of the infant. Breastfeeding should ideally continue to 1 year with complementary

feeding rich in iron and zinc beginning at 4–6 months. Vitamins A and D should be given daily after birth to prevent clinical and subclinical micronutrient deficiency conditions for both breastfed and formula-fed infants.

Supplementation of oral iron drops should be given from 4 months, and from 2 months for LBW babies, to support iron stores. Premature infants should receive both a multivitamin preparation and an oral iron supplement and this should continue into the second year. A completely mixed diet should be achieved in the second half of the first year of life. Fluoride supplements should be given from 6 months of age unless the community water supply is fluoridated (see Chapter 7).

Reports of clinical rickets and evidence of vitamin D deficiency in children and teenagers led the American Academy of Pediatrics (AAP) in 2003 to recommend routine vitamin D (400 IU) supplements for children up to the teenage years, in addition to the vitamin D fortification of milk. Immigrant children and women who are covered for religious reasons are particularly subject to vitamin D deficiency, but children spending more time on computers indoors and less time playing outside are also at risk.

Prevention of iron-deficiency anemia (IDA) is important because of its effects on the infant between 6 and 24 months of age, when rapid brain growth and psychomotor development are at their peak. Iron supplementation is given as iron-fortified formula, supplemental iron, and/or iron-enriched foods. If the child is not receiving iron-enriched formula, iron syrup, preferably with vitamin C, is given daily from 4 to 12 months at a level of 7–15 mg/day according to weight. Secondary prevention is by screening all infants at the age of 9–12 months to determine whether hemoglobin is normal or if further care or investigation is needed.

Following cessation of breastfeeding, infant formula enriched with iron should be given, preferably up to 1 year. Whole cow's milk should not be used until the infant is close to 1 year of age. Cereals fortified with iron and vitamins are also important in infant nutrition, along with fruit, vegetables, meat, and eggs for optimal growth and development and prevention of vitamin and mineral deficiencies. Later in the first year, the infant should be eating at the family table.

The physical and emotional love of the mother, father, and family provides the security and stimulus needed by the infant to develop physically, intellectually, and emotionally. Stimulus activities such as playing with and speaking to the baby help to promote the normal development of the infant.

Immunization is discussed in detail in Chapter 4, where the current US pediatric immunization schedule is presented. While there are many vaccination schedules and much controversy on this topic, coverage of the infant with vaccination on schedule is of paramount importance for a successful child health program, including both infancy and school-age follow-up.

ANTICIPATORY COUNSELING

An important element of the role of the pediatrician, family physician, or other health provider is counseling parents on what is to be expected at each stage of the child's development. This includes warnings of potential difficulties that may be experienced and how to cope with them. The AAP publishes detailed guideline questions for the practitioner addressing these issues from early infancy to the teenage years (see http://www.aap.org/ and http://pediatrics.aappublications.org/site/aappolicy/index.xhtml).

DOCUMENTATION, RECORDS, AND MONITORING

Standardized, user-friendly child health records for the primary care clinic help to promote national standards of child care practice, especially for preventive health care (Table 6.17). They help to guide busy practitioners to improve legal protection and professional group standards. The design of records systems and training of staff in their use should be given careful attention in any health care system, including private medical practice.

Well-constructed child health records set standards of care expected of the health provider. They furnish continuity of care when the contact with the health system may be episodic, or when the child is seen by different providers. They provide the caregiver with important time trends that may have clinical or epidemiological importance. Some child records are designed to continue through to adolescence, while others focus on the child up to school age, followed by a different record. Parent-carried child health records ("road to health cards") provide the parents with all data needed for care during childhood and should continue through to adolescence. They should be designed to include all basic information on one folded hard-paper record, preferably with a plasticized envelope or cover pages. Many designs are possible, but timelines and visually easy formats enhance ease of use by providers. Maintenance of a child health record by the parents is important for its successful use. A survey of US pediatricians indicated that most are generally satisfied with current ways of providing well-child care, but a majority think that a system that was less reliant on physicians, with other health workers providing counseling at other locations such as day care centers, would improve effective and efficient care (Coker et al., 2006).

Physical development in terms of weight and height for age and psychosocial developmental indicators should be carefully documented during infancy at the time of visits for immunization. In 2006, the CDC with the NIH and the AAP expert panel reviewed the scientific evidence and potential use of the new WHO growth charts in clinical settings in the USA. The WHO curves adopted for children aged less than 2 years were adjusted to include the 2.3rd and 97.7th percentiles as appropriate for use. Several factors are taken into consideration, including the recognition that breastfeeding is the recommended standard for infant feeding. The CDC growth charts based on the National Center for Health Statistics will continue to be used for the assessment of growth for children and adolescents aged 2–19 years.

The WHO growth standards/curves developed in 2006 for children under 5 years of age have been adopted by the UN and the International Pediatric Association standing

TABLE 6.17 Components of a Child Health (Preschool and School Entry) Health Record

I. Personal Data	II. Pregnancy/Birth	III. Risk Factors
Mother: name, age, family status, education, occupation	Gestation Place of birth	LBW Previous care
Father: name, age, family status, education, occupation	Date of birth Birth weight	Birth order Family status
Address, telephone no.	Delivery by	Social situation
Social security (ID) number	Condition at birth and at 1 and 5 minutes of age: Apgar	Home conditions
Home situation: parent(s), sibs		Family medical history
		Overall risk assessment

IV. Growth Patterns	V. Immunizations Completed	VI. Nutrition
WHO growth curves chart form	Dose, date, reactions	Age when added
Weight for age – WHO chart	Hepatitis B × 3	Breastfeeding: birth to 6 months plus
Height for age – WHO chart	DPT × 4	Formula
Weight for height – WHO chart	IPV × 3 (USA and industrialized countries)[a]	Vitamins A+D or multi vitamin
Head circumference – WHO chart	OPV × 4 (developing countries)[a] MMR × 2 Hib Varicella Influenza annual Pneumonia Rotavirus	Iron Supplements: solid foods, cereals, fruit, vegetables, eggs, meat: starting at 4–6 months

VII. Intercurrent Illnesses	VIII. Development	IX. Medical Examinations
Diarrheal: age, duration	Hearing	Birth
Respiratory: age, duration	Vision	6 weeks
Hospitalizations	Responsiveness	3 months
Surgery	Relates with family	6 months
Others	Grasp, smile, grab Roll over Sit, crawl Stand, walk Talk: words, sentences Other milestones	12 months 24 months 36 months 48 months Preschool School ages

X. Summary of Risk Factors	XI. Laboratory Results	XII. Continuing Notes
Social	PKU, hypothyroidism	Document, developments, routine examinations, illnesses, and care given
Family	Hemoglobinopathies: thalassemia, sickle cell	
Genetic	Hemoglobin	
Medical	Inborn errors of metabolism	
Other	Others: cystic fibrosis	

Note: [a] The USA and other developed countries have adopted an inactivated polio vaccine (IPV)-only policies. Developing countries use oral poliomyelitis vaccine (OPV) only. Some countries use combined OPV and IPV. See text for discussion.
DPT = diphtheria, pertussis, and tetanus; MMR = measles, mumps, and rubella; Hib = *Haemophilus influenzae* type b; PKU = phenylketonuria.

committees on nutrition and more than 90 countries. In 2010 the US CDC adopted a modified version of the new WHO growth charts for children up to 2 years of age. These curves have been developed and are available at http://www.cdc.gov/growthcharts. Training tools for clinicians are being developed and will also be available at this website.

The charts provide an optimal or gold standard for use internationally, as opposed to local standards which may represent less than optimal child health and development. Most importantly, they provide an external standard for comparison of change over time in the local setting.

Previous controversy on use of international or local standards (discussed in Chapter 8) has diminished, and there is wide agreement on the need for international growth standards for monitoring and early intervention where an infant or a group of infants shows signs of growth faltering or failure to thrive (FTT). Growth faltering occurs not only owing to food insufficiency, but also as a result of intercurrent illnesses, which themselves can be more severe when nutritional status is compromised, such as the relationship of high case fatality rates from measles and other infections when vitamin A deficiency is present. FTT is a medically urgent situation requiring full evaluation and intervention as rapidly and consistently as possible.

PRESCHOOL CHILDREN (AGE 1–5 YEARS)

The preschool child is undergoing rapid growth and development. The AAP recommends health assessments at 12, 15, and 18 months, and at 2, 3, 4, and 5 years (Steering Committee on Quality Improvement and Management and Committee on Practice and Ambulatory Medicine, 2008). The examination includes assessment of development status and physical examination as well as parental concerns and skills. Hearing should be checked along with signs of strabismus, and a general medical examination conducted. The examination includes behavioral assessment based on both interview and observation. The care provider should use the visits to assess parental concerns and convey information on the child's health needs. Abnormal findings on physical or developmental examination should lead to referral to a clinical or child development assessment service. Hemoglobin (or hematocrit) and urinalysis should be checked at least once from 18 months to 5 years. Millions of preschool children are in child day care, which provides an ideal setting for health promotion activities.

Childhood obesity is a global problem which requires attention in primary care. Child care settings should also be active in obesity prevention: regulations related to nutrition, physical activity, and media use for preschool-aged children in the USA should be incorporated in state licensing standards. CDC estimates that 25 percent of American children aged 2–5 years have a high BMI and nearly 11 percent are obese. The WHO estimates a doubling of the global pattern of obesity since 1980 and that over 40 million children under the age of 5 years were overweight in 2010.

Counseling on common child care problems (e.g., nighttime crying), safety issues, and stimulation of the child should be discussed with the parent at each of these visits. The dangers of accidents are serious at these ages. Prevention of accidental falls, poisoning, aspiration of food, electrocution, burns, and scalding requires diligence and safety planning in the home. Parents should be educated on how to explain safety to their children as well as how to childproof their homes (using childproof latches on cabinets and stairways, covering electrical outlets, safe storage of common household chemicals, and removal of poisonous houseplants). Signs of child abuse or neglect, especially injuries without adequate explanation, and reporting to police and welfare agencies is now required by law in many countries. Assessment and counseling regarding appetite problems, toilet training, separation anxiety, demanding or obstinate behavior, relationship with parents and siblings, nutrition, self-feeding, and psychomotor development are all part of well-child care.

The physical and emotional well-being of the child up to school age is a key determinant of his or her health and capacities afterwards. Completion of immunization, adequate nutrition with essential vitamin and mineral supplementation, and prompt and adequate treatment of intercurrent infections are essential. Warmth and responsiveness, including speaking and reading to the child to promote interest in books, pictures, and music, are part of needed stimulation and play. Even warm and caring parents need guidance, as well as advice and support from the health provider, to promote optimal development for the child.

Organizations at both community and national levels have a substantial role to play in the promotion of safety and health for toddlers. Mandatory safety practices for children can save many lives and reduce child morbidity. These include such factors as correct use of safe child car seats, childproof medication or poisonous material containers, lead abatement programs, regulation of child clothing standards to prohibit flammable materials, toy safety standards, supervision of nursery schools, and teacher qualifications.

Violence against infants and children is an increasingly recognized hazard at all ages of childhood and in all socioeconomic groups. Punishments of abusive parents or guardians, including imprisonment and mandatory treatment, are increasingly applied, but hidden physical, mental, or sexual abuse is far more common than previously recognized. The exposure of church and religion-related sexual abuse and its cover-up has caused widespread shock. Child abuse carries with it major risk factors for lifelong sequelae such as depression, anxiety disorders, smoking, alcohol and drug abuse, aggression and violence towards others, risky sexual behaviors, post-traumatic stress disorder, suicide, and elevated risk for chronic diseases. Preventive measures include public awareness through education and now mandatory reporting of signs of abuse by teachers, social workers, and doctors. Home, school, and playground safety involves a wide range of community activities ranging from mandatory use of helmets for bicycle riding and many sports, through proper use of car seats and seat belts for children in motor vehicles, to measures to reduce children's access to firearms.

In 2008, US state and local child protective services received 3.3 million reports of children being abused or neglected. They estimated that 772,000 (10.3 per 1000) of children were victims of maltreatment in 2008; in 2009 approximately 700,000 children were abused or neglected, with 1500 associated deaths. Approximately three-quarters of them had no history of prior victimization. In 2008, an estimated 1740 children aged 0–17 died from abuse and neglect (2.3 per 100,000 children); in 2009, child abuse deaths were estimated at 1500. The abuse rate is 16.6 per 1000 among black Americans. Family members are the principal perpetrators, and only 2 percent of perpetrators are strangers. The total cost of child maltreatment is US$124 billion annually. Young children who are abused are more likely to engage in health-risk behaviors such as smoking, heavy alcohol use, and substance abuse. Long-term abuse affects the nervous and immune systems, increasing vulnerability to chronic diseases of the heart, lung, and liver. It is a major risk factor for psychiatric disorders and suicide and long-term susceptibility to chronic diseases. The WHO estimates that the prevalence of forced intercourse and other types of sexual violence is higher in girls (14 percent – 150 million), but also widely prevalent in boys (7 percent – 73 million) among children and youths.

In most industrialized societies, a large proportion of women work outside the home, so that child care includes preschool centers in which health factors can be both positive and negative. The potential for spread of infectious diseases among child care center attendees is high, and outbreaks of foodborne disease can also occur. Children attending well-supervised preschool day care centers benefit from interacting with teachers and other children to develop social skills, and receiving nutritious meals and professional attention. Preschool care centers should provide education in hygiene, nutrition, and health services including updating of immunization and developmental assessment. Services such as these may help parents to assess and meet the emotional and health needs of their children.

The value of preschool education, especially for children from socioeconomically deprived situations, was demonstrated by follow-up studies of children who had been in the US Head Start program during the 1960s' War on Poverty program. Disadvantaged children participating in Head Start were found to perform significantly better in school and were less likely to have a criminal record or to be on social welfare than matched children not in the program.

This resulted in benefits from reduced costs for special education and public assistance.

Preschool examination and screening for developmental, cognitive, and behavioral problems can be helpful to parents and teaching personnel. A long list of screening tests is available and used widely in many US states, but they are controversial as to their validity and predictive value. They may be culture specific and not necessarily valid in all societies without developed management and support systems.

SCHOOL AND ADOLESCENT HEALTH

Youth risk behavior is a global major public health and societal problem. Globally, more than 2.6 million people aged 10–24 die annually mostly from preventable causes and 40 percent of all new HIV infections are in this group. Some 16 million girls aged 15–19 give birth annually. Estimates are that 430 people aged 10–24 die through violence and 700 from traffic injuries each day (WHO, 2011). Teen pregnancy and HIV are among the health issues addressed in the MDGs. In the USA, 72 percent of all deaths among people aged 10–24 years result from four causes: motor vehicle crashes, other unintentional injuries, homicide, and suicide.

The home environment is of major importance in the development of healthy behaviors in preschool children and young people, but teenagers gradually assume responsibility for their own behaviors. Many community settings, including school, should be utilized to foster health-promotion behaviors and incorporating screening, counseling, and confidentiality. School-based services are needed, with multidisciplinary teams providing a range of readily accessible services at no cost to the user. Teenagers account for 25 percent of all STI cases in the USA; 25 percent consume five or more alcoholic drinks at one time during a 30-day period, only 20 percent meet the recommended daily fruit and vegetable intake, and most do not meet the recommendations for physical activity.

Violence among and against school-aged children and adolescents is increasingly recognized as widely prevalent within schools and is becoming a much more important issue. Violence including bullying, pornography, and pedophilia sometimes takes place within church-related and family settings. Youth violence and risk behavior are major public health issues.

The hazards that schoolchildren face include many public health issues. A central goal of public health regarding this young population is life skill training to promote healthful practices and avoidance of risk behavior. Public health efforts to reduce the toll of teenage premature loss of life (Table 6.18) must be in concert with educational, social, and police authorities. Efforts to convey a message to these risk groups should involve popular figures of the sports and popular music fields recruited to stress a positive image of teen adjustment and transition to adulthood. Research is needed

TABLE 6.18 Total Mortality Rates for Children and Young Adults,[a] and Percentage Change, USA, 1980–2011

Age Group (years)	1980	1990	2000	2006	2011	% Change 1980–2011
1–4	64	47	32	28	27	−58
5–14	31	24	18	15	13	−58
15–24	115	99	82	82	68	−41

Note: [a] Rates per 100,000 population in each age group (rounded).
Source: Hoyert DL, Xu JQ. Deaths: preliminary data for 2011. National vital statistics reports 2012;61(6). Hyattsville, MD: NCHS; 2012. Available at: http://www.cdc.gov/nchs/data/nvsr/nvsr61/nvsr61_06.pdf [Accessed 26 June 2013].

BOX 6.10 Content of a School Health Program

- Parental and community participation in school programs
- Medical and nursing services at school
- First aid and emergency care
- Healthful and safe school environment
- Health consciousness and health-oriented attitude and curriculum in all age groups
- Risks of unhealthy diet, physical inactivity, smoking alcohol, driving, unsafe sex and drug use
- Early diagnosis of physical and psychological problems and follow-up
- Physical education/fitness
- Psychological services
- School age immunization at entry and in adolescence – assurance of completion
- Education for special or disabled children within regular school program
- Training and supervision of teaching staff in health content of teaching program
- Nutrition education and eating disorders prevention and case detection
- Healthful school lunch and breakfast programs
- Family life and sex education/parenthood/birth control
- STI and HIV education and prevention
- Prevention of bullying, violence, and sexual assault; and reporting
- Safety orientation – safety and prevention of fires; trauma as pedestrians; car, motorcycle, and bicycle operators

  or passengers; accidents at home; swimming and water safety
- Awareness of hazards of substance abuse, including smoking
- Personal hygiene and grooming
- Physical education and fitness habits
- Training in first aid/cardiopulmonary resuscitation
- Communicable disease control
- Mental and emotional health – suicide and violence prevention
- Understanding the health of a community
- Understanding health in the family – e.g., care of handicapped and elderly
- Understanding the environment and health
- Promotion of voluntarism in school health.

Note: STI = sexually transmitted infection; HIV = human immunodeficiency virus.

Sources: *Adapted from: American Academy of Pediatrics, Council on School Health. Available at: http://www.aap.org and http://www2.aap.org/sections/schoolhealth/ [Accessed 15 February 2013].*
World Health Organization. Child and adolescent health: school health services. Copenhagen: WHO European Region; posted 2013. Available at: http://www.euro.who.int/en/what-we-do/health-topics/Life-stages/child-and-adolescent-health/adolescent-health/school-health-services [Accessed 15 February 2013]. American School Health Association. What is school health? Posted 2013. Available at: http://www.ashaweb.org/i4a/pages/index.cfm?pageid=3278 Centers for Disease Control and Prevention. Adolescent and school health. Youth Risk Behavior Surveillance System (YRBSS) [updated 16 January 2013]. Available at: http://www.cdc.gov/HealthyYouth/yrbs/index.htm [Accessed 15 February 2013].

---

to develop educational approaches for high-risk teenagers and young adults. Prevention of bullying, violence, dangerous and risk behavior with alcohol, smoking, drugs, sexuality, and the societal determinants of these issues are cardinal public health and civil governmental responsibilities, not easily fulfilled, but not to be avoided.

A school health program should ensure a safe and healthful environment for the children, including sanitation, safety from violence, temperature control, and physical facilities for study and recreation. School-aged children need physical activity for their healthful development as an

integral part of a comprehensive school health education program. School meal services contribute to good child and adolescent nutrition, learning, and performance (Box 6.10).

Health monitoring includes ensuring full immunization at school entry and maintenance of its adequacy by appropriate booster doses at the elementary, secondary, and postsecondary stages. Monitoring of growth, vision, hearing, scoliosis, and skin testing for tuberculosis are frequent elements of school health programs.

The potential for learning about health in the school setting should be exploited by the public health system by

TABLE 6.19 Suicide, Homicide, and Motor Vehicle Accidents (MVAs), White and Black Males Aged 15–24, USA, 1980–2008

Group	1980	1990	2000	2007	2008	% Change 1980–2008
White males						
MVA	74	53	40	37	32	−57
Homicide	15	15	10	10	10	−33
Suicide	21	23	18	17	17	−19
Black males						
MVA	35	36	30	27	23	−34
Homicide	83	137	85	85	78	−6
Suicide	12	15	14	10	12	0

Note: Rates per 100,000 population. Homicide includes "legal intervention".
Source: Centers for Disease Control and Prevention. Health United States, 2011: With special features in socioeconomic status and health. Available at: http://www.cdc.gov/nchs/data/hus/hus11.pdf [Accessed 27 June 2013].

encouraging the educational authorities to include health as part of the regular curriculum and preparing the teachers on how and what to teach in this area. During the early school years, the areas of study can include safety (fire and traffic), oral health (fluoride, brushing, and flossing), healthy eating habits, hygiene, how the body functions, and healthy living (sleep and exercise). Topics for continuing education of teachers include assessment of child abuse, suicidal or violent tendencies, nutritional status, mood disorders, and substance abuse. Violence prevention has become an essential topic in educational systems at the primary as well as secondary levels.

Secondary school health programs include attention to personal behavior issues, including personal and family communication, sexual relationships, the right to say no (to peer pressure in situations ranging from alcohol and drug use to sexual activity), parental responsibilities, birth control, and STI and HIV prevention. Accident and violence prevention is also vital in this population, including safety in sports and work, use of vehicles, and overall safety as a life habit.

Preadolescent and adolescent children are subject to social pressures that promote bullying, early sexual activity, smoking, substance abuse, eating disorders, accidents, suicide, and violence. The school is an important part of their preteen and adolescent life and has the potential to serve a socially educational role for attitudes and practices that can be lifelong. An activist approach to school health promotion should be integral to the school curriculum. School health committees of students, teachers, and parents can serve to promote awareness and program development, with community participation.

Special needs for those with physical or mental disabilities should be identified and recognized as part of the regular duties of the school. Teaching staff should be trained and facilities modified to accommodate children with special problems and to be able to recognize physical and social ill-health, in a wide range of areas from nutrition to child abuse. The health of school-age children, adolescents, and young adults has improved in most countries as sanitation, vaccination, and nutrition programs have developed. Mortality rates overall have declined for all these age groups, but more so for preschoolers and primary school levels. Teenage and young adult health has also improved, but the problem of trauma in its various forms asserts itself and causes much loss of young life.

Adolescence is a difficult and dangerous period of transition from childhood to adulthood. The teenager feels insecure, yet demands independence and responsibility for his or her own well-being. The mix provides for a stormy personal and family transition, with important physical and mental health dangers. Violence, accidents, and suicide are major causes of death in this age group, especially for males. Sexual activity causes dangers of teenage pregnancy, hepatitis C, STIs, and HIV infection, especially when illicit drug use is involved. Developing individual responsibility in each of these categories is required of each individual, with the support of the family, the educational system, the public health system, and the community.

Table 6.19 shows a decline in mortality from motor vehicle crashes, homicide, and suicide of adolescents and young adults (ages 15–24) in the USA, particularly since 1990 for African American and white people in this age group. A sharp rise in mortality from homicide occurred in African American youths in the 1990s, which then declined dramatically up to 2008. While homicide and suicide decreased among white males, the biggest decline was in motor vehicle accidents. Rates of mortality indicate that the factors of motor vehicle accidents, homicide, and suicide still constitute major health problems associated with risk-taking behavior of teenage and young adult males (see Chapter 5).

Health targets for the USA updated for 2020 renewed the focus on reduction of societal determinants of health disparities. They include reducing suicide mortality to 10.2 per 100,000 and reducing suicide attempts among adolescents to 1.7 percent, primarily by health promotion and education, but also by community policing, and stricter weapons prevention programs in schools. Suicide is the fourth leading cause of death among American children aged 10–14 years. Suicide is the seventh leading cause of death of males in Canada, less than half that of motor vehicle deaths and nearly as high as deaths from diabetes. Male young adults are the highest risk group for suicide in North America. Worldwide, suicide is one of the top five leading causes of death among young people aged 15–34 years of both genders. In Canada in 2005, suicide was the second leading cause of death among individuals aged 15–34 years.

Youth risk behavior is an important aspect of monitoring the health status of a population, as reported by periodic survey data (CDC, 2010). The Midcourse Review of Healthy People 2010 and Healthy People 2020 provide an indication of the progress towards achieving the health targets. There are persistent variations between segments of the population. Of the Healthy People youth-related objectives for the first decade of the twenty-first century, only the objectives to reduce physical fighting and reduction of those riding with a person who had been drinking were achieved. Deaths from drug unintended overdosage have been increasing as a mortality risk in young adults.

The WHO estimates that almost one million people a year die from suicide; a global mortality rate of 16 per 100,000, or one death every 40 seconds. In terms of DALYs, i.e., the number of healthy years of life lost to an illness or event, the burden of suicide is about 20 million DALYs and is equal to the burden of all wars and homicides throughout the world.

In Europe, suicide rates have been falling since the 1980s (WHO report). Smoking, alcohol, and drug abuse are among the most common contributory factors for disease and early death in modern societies. These personal behaviors or "lifestyle factors" which cause personal ill-health and early death constitute an enormous burden on society. They usually begin in late childhood, preadolescence, or adolescence, and their health effects continue into adulthood (Figure 6.7).

Smoking

Smoking is a major contributor to cardiovascular diseases, cancer of the lung, and other diseases. The US Surgeon General's Report on Smoking and Health issued in 1964 highlighted the accumulating evidence of the central role of smoking in the causation of chronic diseases. Subsequent reports addressed the continuing consequences of smoking and the prevalence of smoking among different population groups. Nearly all first use of tobacco occurs before the age of 18, and most adolescent smokers are addicted to nicotine.

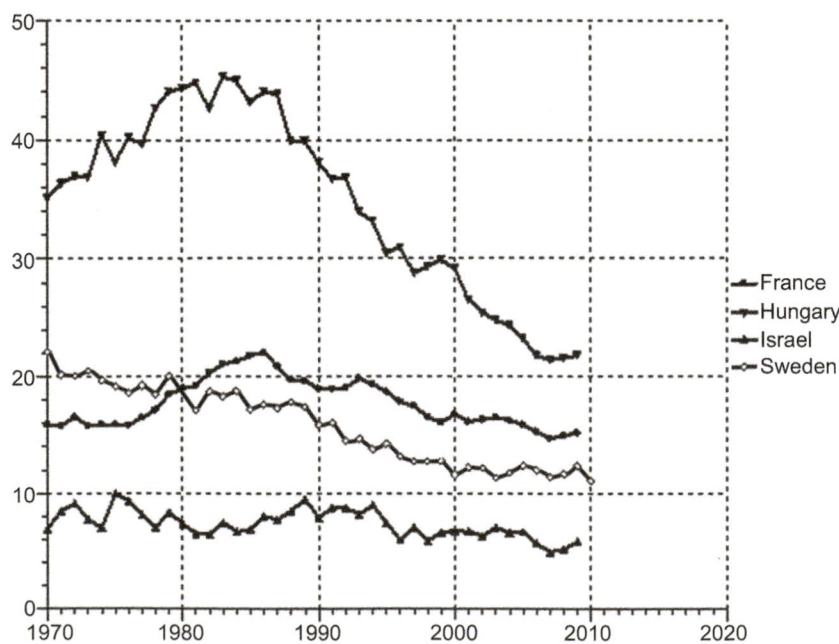

FIGURE 6.7   Suicide rates in selected European countries, 1970–2010. The graph shows standardized death rates for suicide and self-inflicted injury, all ages per 100,000. *Source: World Health Organization. European Region. Health for All database; July 2012. Available at: http://data.euro.who.int/hfadb/ [Accessed 22 February 2013].*

Students in Central and Eastern European countries reportedly have higher rates of smoking. Smoking on a daily basis starts at the age of 13 or younger. In 2009, 19.5 percent of students in grades 9–12 in the USA were regular cigarette smokers, 13 percent smoked at age 13, and 51 percent of smokers had tried to quit. There are identifiable psychosocial risk factors in smoking.

Community-wide projects show that public health efforts can successfully reduce adolescent use of tobacco. Many health-risk behaviors among US high-school students nationwide have decreased steadily from 1990 to 2009 (CDC, Youth Risk Behavior Surveillance System, 2010). However, many high-school students continue with risk behaviors, with similar rates across the country. Such behaviors include violence; tobacco, alcohol, and other drug use; risky sexual behaviors; unhealthy dietary behaviors; and low levels of physical activity. Some 74 percent of all deaths among youth and young adults result from motor vehicle accidents (30 percent), other unintentional injuries (16 percent), homicide (16 percent), and suicide (12 percent). Nearly 20 percent of students have been bullied on school property, according to a recent CDC survey. Other consequences include some 757,000 pregnancies among 15–19-year-olds, 9.1 million cases of STIs, and some 6610 cases of human HIV/AIDS occurring among people aged 15–24 years (CDC, Youth Risk Behavior Surveillance System, 2009).

Alcohol Abuse

Alcohol use has been part of human culture since early civilization. Its use in excess causes serious health problems to the individual and society. The effects of alcohol abuse can be acute or chronic, or relate to the dependence itself. Binge drinking (five or more drinks per occasion) is a threat to the drinker, their friends, and the community by associated violence, suicide, and motor vehicle and work accidents (Box 6.11). Mortality from all causes is elevated in alcoholics. In the USA, alcohol is the third leading lifestyle-related cause of death with 79,000 deaths yearly (4700 in underage youth) representing 2.3 million years of potential life lost (YPLL); 1.6 million hospitalizations and 4 million emergency room visits cost US$223.5 billion in 2006. In the USA, underage youth (12–20 years) drink 10 percent of all alcohol, with 190,000 alcohol-related emergency room visits in 2008.

Alcohol affects every organ in the body, and alcohol use is a compounding risk factor for a wide variety of diseases including hypertension, stroke, coronary heart disease, and cancers of the liver, esophagus, larynx, lung, stomach, large colon, and female breast. It increases reproductive disorders including amenorrhea, anovulation, early menopause, and poor outcome of pregnancy with LBW or FAS in newborns. The seventh and eighth grades are the peak time for starting to drink alcohol.

In many European countries binge drinking occurs in up to 60 percent of youth aged 15–16 years, about three times higher than in the USA. Intoxication rates are lower among American teens than those in most European countries. In 2009, 72.5 percent of high-school students reported drinking alcohol, 40.2 percent were current drinkers with 24.2 percent at a heavy level of consumption (five or more drinks in a row, once or more during the previous 2 weeks). In the USA, more than 12 percent of youth aged 18–20 have the highest dependence on alcohol. Between 1998 and 2004, increasing numbers of young people disapproved of substance use and abuse. Exposure of children and adolescents to alcohol advertising on US television increased by 71 percent between 2001 and 2009, more than advertising to adults. Among high-school seniors, 20.3 percent report never using alcohol. Alcohol use among pregnant teens aged 15–17 increased from 14.5 percent to 16.7 percent from 2002–2003 to 2008–2009 compared to a drop from 28.7 to 22.8 percent among a non-pregnant age-matched cohort.

Drug Abuse

Drug abuse, while not a new problem, has become extremely widespread in most industrialized countries. In the USA, drug overdose was responsible for 830,652 YPLL before age 65, comparable to the rate for car crashes. Adolescence is a period of experimentation, intense social pressure, and risk-taking behavior. With widespread experience and availability of illicit drugs and other intoxicants in the USA, drug-related deaths among young people aged 15–24 increased from 3.2 per 100,000 in 1999 to 8.2 per 100,000 in 2008 (Figure 6.8).

Cocaine-related emergency department episodes in the USA increased from 80,000 in 1990 to 142,000 in 1995, and to 422,696 in 2009, of which 5294 visits were for 11–17-year-olds and 53,015 for 18–24-year-olds (Substance Abuse and Mental Health Services Administration, Drug Abuse Warning Network, 2009). Adolescents use such drugs as glue (or other inhalants), marijuana, and "hard drugs" including heroin and crack cocaine. In 2009, 20.8 percent of US senior high-school students and 6.5 percent of eighth graders had used marijuana at least once in the previous month. Other drug use reported by high-school students in the past month in this survey included cocaine (2.8 percent) and inhalants (11.7 percent); 4.1 percent used methamphetamines. In the USA, rates of opioid pain reliever (OPR) overdose death, treatment admissions, and kilograms of OPR sold during 1999–2008 all increased substantially (MMWR, 2011).

In the USA in 2009, the use of prescription drugs without a prescription occurred among 20 percent of teens and continues to increase, partly because they believe these drugs to be safer than "street" drugs. In

BOX 6.11 Alcoholism: Global Health and Clinical Psychiatric Aspects

Harmful use of alcohol is a worldwide problem resulting in millions of deaths, including hundreds of thousands of young lives lost. Alcohol abuse is not only a causal factor in many diseases, but also a precursor to injury and violence with impact throughout a community or a country (WHO, 2011). The US Centers for Disease Control and Prevention reports 80,000 deaths attributable to excessive alcohol use each year in the USA. This makes excessive alcohol use the third leading lifestyle-related cause of death, with increased risk of health problems such as injuries, violence, liver diseases, and cancer (US CDC, 2012).

Cloninger and Svrakic (2009) describe two different types of alcoholism: one starting later in life, related to life events and easier to cure because of the major social competence of the affected person and better resilience factors. There is typically a less visible drinking pattern and more harm to the body than to the social environment (type I). A second form, starting early during adolescence, is often associated with personality disorders and abuse of other substances, so more difficult to cure (type II).

Both forms show previous emotional symptoms such as fear, anger, feelings of senselessness, shame or guilt, mainly due to complex interactions between life events and personal vulnerability. If the traumatic events are strong single major life events (type I addiction), such as the sudden death of a parent (e.g., by car accident, AIDS, or suicide), being a victim of or witness to acute sexual or other physical violence, the psychological reaction can be a protective emotional "freezing", guaranteeing nearly normal general functioning for years or decades.

Type I onset of an addiction later in life is often due to a second smaller life event (loss of a job, normal death of an elderly parent, separation from a partner, retirement). The "freezing" is lost and the alcohol abuse is a sort of failed attempt to control the painful emotions. Recent neuroscientific findings explain the post-traumatic overwhelming pain as if it is "imprisoned" in the right hemisphere of the brain. As the right hemisphere has nearly no linguistic capacity, non-verbal interventions such as art therapy, music therapy, bioenergetic approach, yoga, and eye movement desensitization and reprocessing (EMDR) are useful for the emotional reactivation of the original traumatic material for integration of this process to the left hemisphere of the brain with its linguistic competencies. Thus, a traumatized person can re-establish a new integrated reaction to the precursor traumatic event without being painfully overwhelmed by emotions.

The early onset of type II addiction has to be mainly interpreted as a failed general attempt of an emotional cure and control. This form is mostly caused by the social changes of recent decades: frequent divorces, translocations, mobility for work, loss of framing social or religious values, so that even smaller events become traumatic. This form is clinically more often characterized by binge drinking, violent episodes, and poor social integration of the affected person. In Italy, for example, social stability until recently was much better than in other Western countries, so this type of alcoholism is a relatively new phenomenon.

New neuroscientific findings explain better the catastrophic consequences of environmental changes on a neurohormonal and epigenetic level influencing emotional regulation, but they also explain why and how psychotherapy can work (Hinnenthal and Cibin, 2011).

Addiction therapy must include pharmacological support, motivational methods, self-help groups, relapse prevention, and other methods. Complete abstinence from alcohol and illegal substances is fundamental. The issues of alcohol abuse and frank alcoholism with their consequences on non-communicable disease and trauma are global in scale. Global, national, and local, as well as clinical policies are required to deal with this world pandemic.

Sources: *Hinnenthal IM, Health Economics Laboratory, Institute of Hygiene, Catholic University, Rome; University of Siena, Italy; Nante N, Department of Public Health, University of Siena, Italy; Ricciardi W, Department of Public Health, Catholic University, Rome; Cibin M, Addiction Department, Dolo, Venice, Italy; scientific board: Therapeutic community: Villa Soranzo, Venice. Personal communication.*

World Health Organization. Global status report on alcohol and health. Geneva: WHO; 2011. Available at: http://www.who.int/substance_abuse/publications/global_alcohol_report/msbgsruprofiles.pdf [Accessed 8 December 2012].

Centers for Disease Control and Prevention. Alcohol and public health. Available at: http://www.cdc.gov/alcohol/ [Accessed 8 December 2012].

Cloninger CR, Svrakic DM. Personality disorders. In: Sadock BJ, Ruiz P, editors. Comprehensive textbook of psychiatry. Philadelphia, PA: Lippincott, Williams and Wilkins; 2009.

Cibin M. Generalità sui trattamenti residenziali brevi. In: Hinnenthal IM, Cibin M, editors. Il trattamento residenziale breve delle dipendenze da alcol e cocaina: il Modello Soranzo. Turin: SEED; 2011.

2010, non-medical opiates were used by 4.8 percent of children under 12 years of age. As a baseline, 27.7 percent of high-school seniors reported never using alcoholic beverages and 53.3 percent of high-school seniors reported never using illicit drugs in 2009; the Healthy People 2020 report objective target set was for 10 percent improvement in these rates.

In Europe, marijuana is the most frequently used illicit drug, at 14 percent, but with a downward trend between 2003 and 2007, and usage has since stabilized; cocaine remains the most widely used stimulant. In the USA, the non-medical use of prescription drugs is epidemic; for example, one-fifth of seniors abuse oxycontin. In Europe, the trend in non-medical use of prescription drugs, highest among girls, is static, with the highest usage in Poland, France, and Lithuania (Trautmann et al., 2013). The percent of eighth graders who thought that it was a great risk to use marijuana declined from 79 percent to 49 percent between 1981 and 2010. The EU reports that new trends in drug use are occurring through expanded worldwide use of the Internet, which allows new patterns and trends in drug use to rapidly transcend geographic boundaries.

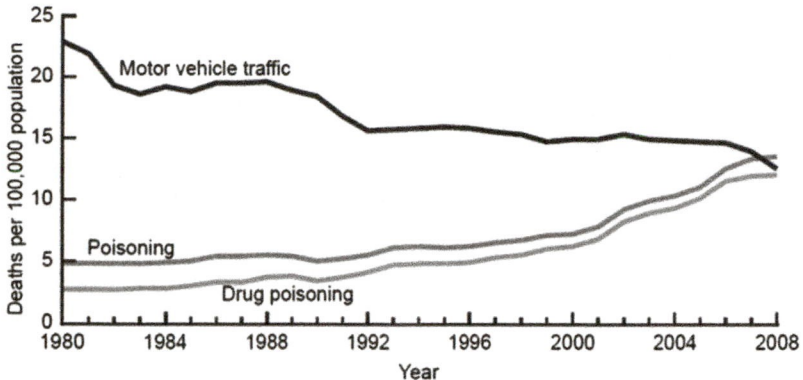

FIGURE 6.8   Poisoning and motor vehicle traffic death rates, USA, 1980–2008. *Source: Warner M, Chen LH, Makuc DM, Anderson RA, Minino AM. Drug poisonings deaths in the United States, 1980–2008. NCHS data brief no. 81. Hyattsville, MD: US DHHS, CDC, National Center for Health Statistics; 2011. Available at: http://www.cdc.gov/nchs/data/databriefs/db81.htm [Accessed 22 February 2013].*

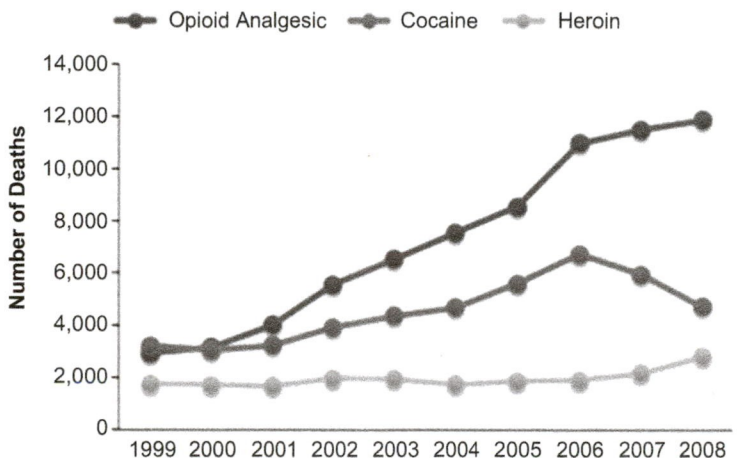

FIGURE 6.9   Deaths from unintentional overdose, by major types of drug, USA, 1999–2008. *Source: National Institute of Drug Abuse. Prescription drug abuse. Revised December 2011. Available at: http://www.drugabuse.gov/publications/topics-in-brief/prescription-drug-abuse [Accessed 23 February 2013].*

Prescription drug abuse has become a greater cause of death from overdosage than cocaine and heroin. Much of this is due to addictive opioid pain relievers (OPRs), tranquilizers, stimulants, and sedatives (Figure 6.9).

Drug addiction is a compulsion to use a substance and obtain it by any means with a need to increase the dosage to obtain the desired effect. There can be a physiological and/or psychological dependence on the effects of the substance. The addiction has a detrimental effect on the family and the community through increased, stress, violence, and crime. Drug use behavior may be classified as experimental, recreational, circumstantial, intensified, and compulsive. The experimental version is most common and motivated by curiosity and a desire to experience an altered mood state. Recreational use is voluntary but patterned in social settings where the user is not dependent on the drug. Circumstantial use may be related to certain situations, such as amphetamine use by students during examinations or by truck drivers on long drives. Intensified

drug use occurs daily, motivated by stress and/or the wish to maintain levels of performance, by the use of sedatives, barbiturates, or tranquilizers. The individual remains integrated within the social and economic context. Compulsive use is of high frequency, with a dominant psychological and/or physical dependence, by hard-core drug users whose lives are controlled by their dependence on and financing of the habit. This dependence leads to even more serious health problems, caused by exchange of sex for drugs, lack of condom use, sharing of needles, poor hygiene, and poor nutrition. Transmission of hepatitis, STIs, and HIV is common. Arrest, imprisonment, injury, and death by violence in drug-related crimes are also common.

Sexual Risk Behavior

Teenage sexual risks include unplanned pregnancy and parenthood, STIs, and inadequate preparation for adulthood. The educational role in prevention of these is a vital element of a school health program. National efforts are required to prevent exploitation of children for prostitution; to provide care for runaway, homeless, and street children; and to regulate standards of care for orphans and disabled children in institutions. These problems are particularly acute in South American and South-East Asian countries. The HIV/AIDS epidemic is taking an enormous toll among teenagers and young adult sex workers, truck drivers, military personnel, and increasingly in the general community. Almost half of the 19 million new cases of STIs each year are among young people aged 15–23 years, and 12.7 percent of students have been tested for HIV.

In the USA, 46 percent of high-school students report having had sexual intercourse at least once in their lives (5.9 percent before the age of 13), 13.8 percent having had four or more sexual partners, and 34.2 percent being currently sexually active (Youth Risk Behavior Survey, 2009). Among these, 61.1 percent used condoms and 19.8 percent used birth control pills. By 2009, more students were using condoms; the percentage of those engaged in sex, and sex with multiple partners had declined. Risks of STIs and HIV are commonly associated with high-risk sexual behavior and drug abuse. The Healthy People 2020 project called for efforts to reduce the rate of sexual intercourse among the 18–19 age group from the 2005 rate of 117.7 per 1000 to a rate of 105.9 by 2010 (Healthy People 2020). Although the reasons are not clear, there has been a decline in early-onset sexual activity among US teenagers and increased use of birth control, associated with a decline in teenage pregnancies (Martinez et al., 2011). Birth rates for women aged 15–17 years fell by 11% and for women aged 18–19 the decline was 7% from 2010 to 2011.

Dietary Risk Behavior

A 2010 US survey found that only 15 percent of students had eaten the recommended dietary allowance (RDA) of fruit and vegetables during the previous 24 hours. Nearly 30 percent consumed one or more servings of a sweetened variety of soda pop at least once a day. In addition, 28 percent thought of themselves as fat and 44 percent were currently trying to diet. Over 5 percent of girls and nearly 3 percent of boys vomited or used laxatives to lose weight. Dietary deficiency conditions, especially iron deficiency anemia (IDA), are common in adolescent girls owing to dieting behavior and extra need for iron to compensate for menstrual losses.

The Midcourse Review of Healthy People 2010 and Healthy People 2020 show that the increasing prevalence of overweight and obesity among children and adolescents aged 6–19 in the 1990s plateaued by 2007. In 2008, 19.6 percent of children aged 6–11 and 18 percent of young people aged 12–19 were obese (Health United States, 2010). In a 2009 survey, 12.0 percent of US teenagers were obese and 15.8 percent overweight. Nearly twice as many males (15.3

percent) as females (8.3 percent) were obese. The Healthy People targets are for obesity in less than 10 percent. During the National Health and Nutrition Examination Survey (NHANES) 1999–2008 periods, the age-adjusted proportion of adults aged 20 years and older at a healthy weight decreased from 35.5 percent to 31.7 percent, while the proportion of adults who were obese increased from 30.5 percent to 33.9 percent. Increased obesity is a worldwide phenomenon and is associated with an increasing prevalence of diabetes with occurrence at younger ages than in the past.

Addressing the increase in overweight and obesity requires both public health and individual approaches. Growing food abundance and sedentary lifestyles and working conditions contribute to this pandemic. The factors contributing to overweight and obesity are complex, including psychological, behavioral, cultural, and socioeconomic, as well as genetic, metabolic, and environmental issues. Slowing this trend requires attention by the health system in partnership with other sectors of society providing accessible and affordable healthy food choices, opportunities for regular physical activity, and a supportive environment to facilitate individual behavior change. For obese adults, even modest weight loss (e.g., 10 pounds or approximately 5 kg) has health benefits (CDC, Healthy People US Midterm Review, 2010).

In the USA, the prevalence of childhood obesity doubled from the 1960s to the 1980s and continued a rapid increase until the 1990s, when obesity prevalence among children and teens tripled from nearly 5 percent to approximately 15 percent. Obesity affects approximately 12.5 million children and 17 percent of teens (Figure 6.10) (MMWR, 2011). CDC reports that one-third of US children and adolescents were overweight or obese in 2008 and were likely to continue to become obese as adults, a trend opposite to the US Healthy People target for 2010 for 6–19-year-olds, which was a reduction of 5 percent in overweight and obesity. Efforts to reduce this include changes in school lunch menus, removing high-sugar drinks from school environments, and increased health promotion activities focusing on healthy diets and physical activity levels at school and at home.

Eating disorders are a common behavioral problem, especially among teenage girls influenced by the fashion-model body type. They enter a vicious cycle of anorexia nervosa or bulimia which can be fatal, or in less severe forms extremely traumatic to the teen and family. These conditions may affect as many as 5 percent of college-age females in the USA, and they are an occupationally associated condition among both male and female athletes and professional ballet dancers or models. Overeaters who take up obesity as a lifestyle may also suffer the melancholies of this vulnerable age group. Prevention includes education and early case finding with suitable counseling and group therapy.

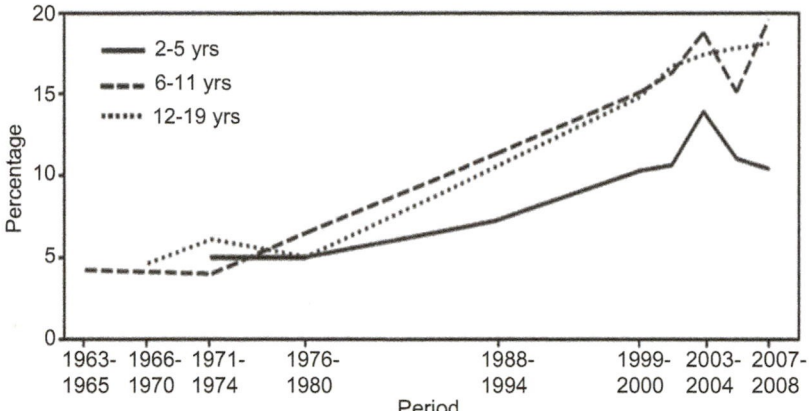

FIGURE 6.10   Prevalence of obesity among children and adolescents, by age group, USA, 1963–2008. *Source: Centers for Disease Control and Prevention. CDC grand rounds: childhood obesity in the United States. MMWR Morb Mortal Wkly Rep 2011;60:42–6. Available at: http://www.cdc.gov/mmwr/preview/mmwrhtml/mm6002a2.htm [Accessed 24 February 2013].*

Physical Activity

Among US adults in the age group 18–44 years, about 18 percent meet the guidelines for physical activity. Nearly 50 percent of US students meet the US guideline, and had engaged in strenuous physical activity on 5 or more days of the previous week, 52 percent were enrolled in physical education classes, and 58 percent were involved in team sports. As of 2012, less than half (48%) of all US adults meet the 2008 Physical Activity Guidelines. Men are slightly more active than women. Less than three in 10 high school students take at least 60 minutes of physical activity every day (CDC, 2012). Rates of inactivity are higher in southern states, where obesity and stroke mortality are high.

The prevalence of overweight and obesity among American children and adolescents is increasing, with 17 percent being obese by 2008, compared to the Healthy People target of 14.5 percent. Total fat consumption and sodium consumption remain high. Prevention-oriented policies and programs are the preferred means of intervention, addressing the social acceptability of physical activity and availability of healthy foods in school settings in place of junk foods and high sugar-content beverages.

Obesity has become a major issue in public health, which needs to encourage increased activity levels and reduced dietary intake of high-sugar drinks and fried foods. Institutional changes are required, such as banning of sugary drinks in dispensers in schools, and in some locations banning of large-size dispensers of popular sugar drinks.

Violence and Gang Behavior

Teenage violence in developing and industrial countries is a growing public health hazard. Teenage alienation, availability of drugs, weapons, and violent cultural icons provide breeding grounds for gangs and violence, or individual violent outbursts which can lead to random or organized shootings or terrorist outrages.

Youth violence refers to harmful behaviors that may start early and continue into young adulthood. These range in severity and include bullying, slapping, punching, sexual harassment, ethnic slurs, victimization, weapon use, rape, and incidents of mass murder in school or university settings. In 2009, 20 percent were bullied and 8 percent were threatened or injured with a weapon at school. Nearly 32 percent were involved in a physical fight; 11 percent were in a fight at school.

School authorities, students, and teachers are increasingly aware of these acts and the threat they pose. Preventive factors include security services, alert mental health services, and cooperation between students and school authorities. In some jurisdictions there is mandatory reporting of this type of maltreatment.

The impact of violence on US women's and men's health was reported in the CDC's National Intimate Partner and Sexual Violence Survey (NISVS), which included the following findings.

- Women are disproportionally affected by sexual violence, intimate partner violence, and stalking.
- 1.3 million women were raped during the year preceding the survey.
- Nearly one in five women have been raped in their lifetime; one in 71 men have been raped in their lifetime.
- One in six women have been stalked during their lifetime; one in 19 men have experienced stalking in their lifetime.
- One in four women have been the victim of severe physical violence by an intimate partner; one in seven men experienced severe physical violence by an intimate partner.
- Approximately 80 percent of female victims experienced their first rape before the age of 25 and almost half experienced the first rape before the age of 18 (30 percent between 11 and 17 years old and 12 percent at or before the age of 10).

- About 35 percent of women who were raped as minors were also raped as adults, compared to 14 percent of women without an early rape history.
- 28 percent of male victims of rape were first raped when they were 10 years old or younger.

Many other forms of violence constitute a public health hazard. These range from schoolyard or internet bullying and physical violence to mass murder by political adventurers or disturbed people bearing grudges. Severe politically motivated and genocidal threats and behaviors use rape and starvation as tools of ethnic cleansing, as well as actual mass slaughter of men, women, and children. Such acts are not spontaneous but always preceded by brutal verbal violence and threats. Incitement and delegitimization of others are tools used to incite and prepare for genocide, and serve as warning signs of intent and motivation to commit horrendous terrorist or genocidal acts.

ADULT HEALTH

The health of adults between 25 and 64 years of age is important to the well-being of both the family and society. The higher rate of death for males in this age group creates a social imbalance with a predominance of females in the elderly age group. Americans miss 2.5 million days of work each year due to chronic disease, resulting in lost productivity of more than US$1 trillion (US Surgeon General). Premature death or disability of males often results in loss of a major part of family income, as well as productivity in the society. Much of the excess loss of males can be prevented by currently available knowledge and techniques in clinical medicine and public health, as has already been seen in reduced age-specific mortality from heart disease and stroke. Adult males should be targeted by coordinated efforts of both preventive and curative services to reduce risk and promote health. In 2008, 18.1 percent of adults met guidelines for aerobic physical activity and muscle strengthening. Among adults aged 18 and over 55 percent had undergone a comprehensive eye examination in 2008. Visual impairment due to retinopathy occurs at a rate of 34.1 per 1000 people with diabetes. Adults experience noise-induced hearing loss in both ears at a rate of 121 per 1000 population (Healthy People 2020).

Risk factors that contribute to disease and mortality are both intrinsic and extrinsic. Intrinsic factors include age, gender, and genetics. Extrinsic factors include personal lifestyle and the environment. Improved health for the adult population requires a wide range of primary and secondary prevention programs focused on extrinsic factors that can reduce personal risks and rates of mortality from chronic diseases and trauma (see Chapter 5). Programs to reduce morbidity and mortality from chronic diseases include preventive activities including counseling by primary and other health professionals and public–private networks in clinics, health systems, according to age- and gender-recommended guidelines, and the following:

- counseling for smoking cessation, regular exercise, and dietary guidance
- a healthful diet (i.e., low in fat; moderate in carbohydrates; high in fruit, vegetables, and fiber)
- vitamin and mineral supplementation (e.g., calcium, iron, vitamins A, B, C, D and others)
- daily physical activity (i.e., 30 minutes of moderately intense physical exercise 5 days per week)
- road safety (e.g., seat belts, moderate speed enforcement, non-use of alcohol with driving)
- workplace safety (e.g., safety helmets and shoes, preventive work site inspections)
- reducing violence (e.g., gun control, adolescent recreation activity programs)
- medical examinations for selective screening (e.g., diabetes, glaucoma, mammograms, Pap smears)
- general screening for and management of hypertension
- general and high-risk group screening for and management of cancer (e.g., cervix, colorectum, and breast)
- management of chronic diseases such as diabetes mellitus and glaucoma
- screening and support services for mental health (e.g., depression, alcohol and drug use, social stress)
- good management and prevention of infectious diseases
- immunization with dT for all adults, influenza and pneumococcal vaccines for high-risk groups (i.e., people over 65 and those with chronic diseases)
- social and recreational activity and support groups
- employment, economic opportunity, and social equity
- fundoscopy to detect macular degeneration.

There are major gaps in evidence-based preventive care. For many decades, the debate on optimal methods for the prevention of cerebrovascular diseases has focused on low-fat diets, but recent research suggests that the "Mediterranean diet", with a high intake of olive oil, fish, vegetables, and fruit, is superior to the standard low-fat diet. The US Affordable Care Act includes provision of preventive care for many who were previously uninsured. The cost and effectiveness of screening is a continuing issue for epidemiologists and health economists and will continue to be so as new measures enter the armamentarium of public health and clinical practice. Screening for colorectal cancer has been shown to be cost effective, while prostate-specific antigen (PSA) screening for prostate cancer has not. Screening for lung cancer was until recently considered ineffective and costly, and thus a low priority, but recent evidence suggests that it may indeed be life saving and cost effective. Topics that deserve further research include screening for chronic kidney diseases and prostate cancer, and advice such as sun protection creams use in children. Screening for cervical cancer has achieved enormous success in reducing

BOX 6.12 Adult Health Screening Recommendations[a]

- *Counseling* – risk factors: smoking, obesity, inactivity, dietary habits, drug abuse, domestic violence/abuse, alcohol abuse, traffic safety
- *Evaluation* – obesity, hypertension, depression, hearing and visual impairment, iron, vitamin D, and folic acid deficiency, sexually transmitted diseases
- *Prevention* – immunization including childhood vaccines adult boosters (tetanus, diphtheria, pertussis: Td, TdTap), human papillomavirus, and pneumonia and influenza (annual), herpes zoster
- *Risk assessment*
- *Screening* – cancer of cervix, skin, colon, breast, and prostate
- *Initial and regular physical examinations and preventive medical visits* are recommended for all adults:
  - health behavior: smoking, exercises, diet, mood, birth control and sexual practice, employment life satisfaction
  - body mass index
  - dental: annual examination and dental hygiene care
  - vision: initial screening and regular glaucoma testing after age 64 or 40 for minorities
  - hypertension: initial and biannual blood pressure screening
  - diabetes: for all especially those with chronic disease, personal or family history of elevated blood glucose, overweight (body mass index >25) or other risk factors including cardiovascular diseases, and underserved population groups
  - cholesterol: initial screening and every 5 years after age 35 for men or 45 for women
  - immunization status: re influenza, pneumonia, dT, herpes zoster
  - osteoporosis: bone densitometry for women once after age 60
  - thyroid screening test
  - sexually transmitted infections: regularly for sexually active people
  - tuberculosis: purified protein derivative (PPD) skin test screening for people with risks, including chronic disease, residence, or employment in institutional settings and travel to areas with elevated prevalence
  - colorectal cancer: initial colonoscopy at age 50 and at 5-year intervals with yearly fecal occult blood test
  - pelvic and cervical cancer: examination and Pap test every 1–3 years for sexually active women[b]
  - breast cancer: regular breast self-examination and bi-yearly mammography starting at age 40
  - prostate cancer: digital rectal examination yearly beginning at age 50 or earlier with risk factors
  - skin cancer: annual.

Note: [a] People with chronic disease, personal or family history, and other risk factors should receive preventive services earlier and more often, based on medical recommendations.
[b] Immunization against common strains of human papillomavirus (HPV) is recommended for females ages 11–26, to greatly reduce future risk of cervical cancer.
Sources: *Agency for Healthcare Quality and Research. Guidelines: preventive services for adults. Available at: http://www.guideline.gov and http://www.guideline.gov/content.aspx?id=38450 [Accessed 14 March 2013]. Centers for Disease Control and Prevention. Vaccines and preventable diseases. Available at: http://www.cdc.gov/vaccines/vpd-vac/default.htm American Cancer Society. Guidelines for the early detection of cancer. Available at: http://www.cancer.org/Healthy/FindCancerEarly/CancerScreeningGuidelines/american-cancer-society-guidelines-for-the-early-detection-of-cancer Agency for Healthcare Quality and Research. Guide to preventive clinical services. Available at: http://www.ahrq.gov/professionals/clinicians-providers/guidelines-recommendations/guide/abstract.html Agency for Healthcare Quality and Research. Guide to clinical preventive services, 2012. AHRQ publication no. 12-051454; October 2011. Available at: http://www.ahrq.gov/professionals/clinicians-providers/guidelines-recommendations/guide/guide-clinical-preventive-services.pdf [Accessed 14 March 2013].*

cancer mortality, and the new human papillomavirus (HPV) vaccine will enhance control measures. Newer methods to identify the presence of HPV in the cervix are now in use and may make screening more effective.

Public health is a dynamic set of programs and activities to address population health protection and individual care with the resources made available by political decisions. It is vital to constantly review the evidence based on research and international best practices as they evolve to carry out these functions.

Detailed recommendations and clinical guidelines by disease classification are continuously updated and posted at http://www.guideline.gov/ and are referred to in various chapters of this book (Box 6.12). The health care system has the responsibility of promoting the primary prevention and screening programs needed to reduce the burden of disease as it affects the adult population.

WOMEN'S HEALTH

Women's health is a matrix of many factors relating to fertility, sexuality, and societal conditions that affect health. Traditionally, in the years between menarche and menopause, women's health needs have largely related to fertility. With higher levels of education and participation in the workforce and other societal changes, women's health increasingly consists of issues beyond those of fertility (Table 6.20).

Since the 1980s, women's health needs have received increased attention from the public health sector. Women are the largest consumers of health care, as well as the largest group of health providers. On average, women live longer than men; they visit physicians 25 percent more than men, and are hospitalized 15 percent more frequently. The three leading *causes of cancer* among women are

TABLE 6.20 Adult Women, Health Risk and Prevention by Age Group, 18–64 Years

Age (years)	Major Health Risk Factors	Screening/Counseling/Preventive Activities
18–34	Nutrition, iron-deficiency anemia, calcium, and micronutrient deficiencies	Nutrition counseling, supplements
	Eating disorders, body self-image	Education, family and social support
	Risk behavior: smoking, risky sex, drugs	Counseling
	Marriage and parenthood	Birth control education and access
	Pregnancy and child care	Safe sex education
	Single parenthood	Counseling, prenatal care and high-risk assessment
	Physical fitness	Safe delivery
	Self-reliance, self-determination	Nutrition, exercise, weight control
	Sexuality and birth control	Self-defense
	STIs and HIV	Smoking cessation
	Safety, unwanted sexual advances, assault, rape	Premarital counseling
	Safety from domestic abuse, physical and mental	Immunization: HPV, influenza, DTaP
	Physiological demands of menstrual cycle, PMS	Physical exercise
	Smoking, alcohol and drug use	Job training, academic work
	Poverty and economic status	Support groups; shelters for abused and raped women; police, social service, and teacher sensitization
35–44	Late pregnancy	Counseling and support groups
	Child-rearing problems	Physical exercise, weight control, fitness
	Marital stress, separation, divorce	Screening for obesity, hypertension, cholesterol
	Diet	Diet counseling, weight control
	Physical fitness	Preparation for menopause (counseling, hormonal therapy)
	Domestic abuse	Immunization (DT)
	Obesity	Screening for breast cancer, cervical cancer
	Cancers (breast, cervix, lung)	
45–64	Menopause	Periodic physical examinations
	Cancer (lung, breast, cervix)	Screening for diabetes, hypertension, bone density
	Diabetes, hypertension	Screening for breast, cervical, colonic cancer
	Physical fitness	Screening for thyroid and vitamin status
	Depression	Counseling for life crises
	Change in employment status	Immunization (DT), influenza, pneumococcal pneumonia
	Cardiovascular diseases	Physical activity
	Widowhood, divorce	Retirement planning

Note: STI=sexually transmitted infection; HIV=human immunodeficiency virus; PMS=premenstrual syndrome; HPV=human papillomavirus; DTaP=diphtheria, tetanus, acellular pertussis; DT=diphtheria and tetanus.
Sources: Adapted from US Preventive Services Task Force, 2012. Updates of US recommendations are posted at: http://www.ahrq.gov/clinic/uspstfix.htm and http://www.ahrq.gov/clinic/pocketgd.htm [Accessed 16 February 2013].
Breast cancer. Available at: http://www.canadiantaskforce.ca/recommendations/2011_01_eng.html
Guide to Clinical Preventive Services includes US Preventive Services Task Force (USPSTF) recommendations on screening, counseling, preventive medication topics, and clinical considerations for each topic. Available at: http://www.ahrq.gov/clinic/pocketgd.htm, http://www.uspreventiveservicestaskforce.org/and http://www.uspreventiveservicestaskforce.org/adultrec.htm

breast (123.1 per 100,000), lung (54.1 per 100,000), and colorectal cancer (37.1 per 100,000). However, the leading *causes of cancer death* are lung (38.6 per 100,000), breast (22.2 per 100,000), and colorectal (13.1 per 100,000). Although breast cancer is more than twice as common as lung cancer, lung cancer is a higher cause of death (CDC, 2013). Lung cancer has replaced breast cancer as the first cause of women's cancer mortality in many industrialized countries.

Despite declining mortality rates, cardiac disease in women is more common than previously thought. Indeed, stereotypes regarding women as having symptomatic coronary heart disease less frequently than men may lead to the underassessment of the severity of this problem.

The life cycle of women between menarche and menopause involves psychosocial and physiological functions of fertility. Menarche, premenstrual syndrome (PMS), and menopause are all associated with physical and psychological stress of clinical importance as well as public health significance.

When education and awareness of women's physiological needs are accepted by the family, co-workers, and society in general, there is a greater likelihood of a more appropriate response by women suffering from these effects. Other health issues of public health significance in women include IDA, eating disorders, and nutrition (see Chapter 8).

Women's health is related both to biological and social factors. Postmenopausal women are subject to diseases for which they had previously been protected by hormones, such as cardiovascular diseases and osteoporosis with its accompanying fractures. The higher incidence of mental illness among women, in particular anxiety, depression, and eating disorders, relates to self-image, economics, and society's varying pressures on women. Postpartum and menopausal depression, the pressures of childrearing and work, and long periods of loneliness as a result of family breakdowns and widowhood all contribute to an excess of mental disorders. As women have become a substantial part of the workforce, employment and career are major sources of income, self-esteem, and stress. Loss of employment can have negative effects on health for both financial and psychological reasons.

Important public health programs involving screening for specific female cancers such as breast, ovary, uterus, and cervix are leading to early diagnosis and treatment. Sexual abuse, violence, single parenting, and widowhood are major factors that determine the content of health care services needed for women in the community. Females have been underrepresented in chronic disease research, which has tended to center on male subjects. Generalizations regarding treatment or prevention of a disease may not always be appropriate when they are based on male subjects. This has been especially problematic in heart disease, which often presents without clear symptoms, and stroke-related issues, with hypertension being undermanaged.

Women's health issues relate to their role in society in terms of education, employment, and equal pay for work of equal value. Attitudes, customs, and laws and their enforcement regarding women's rights are extremely variable in different societies around the world, and the public health implications are widespread. In developed countries, the majority of people living in poverty are women and their children.

A WHO multicountry study found that 15–71 percent of women report intimate partner physical or sexual violence at some point in their lives. Violence can be mental and abusive in many forms. During 2013, acts of gang rape occurred in many locations, and the murder of polio workers also highlights the backward nature of this form of social control. Risk factors include poverty, low education levels, and exposure to violence during childhood. In upper income settings, violence is also present but the forms may be different: "date rape" and domestic murder and suicide are widely reported. So-called honor killings are widespread in regions such as the Middle East. Violence and mass rape are used as weapons of intimidation in civil conflicts and genocidal incitement, for example in Darfur and South Sudan (see Chapter 15).

The UN Universal Declaration of Human Rights and the Declaration on the Elimination of Violence against Women have recognized women's fundamental human right to live free from violence.

Addressing violence against women costs billions annually and is central to the achievement of the MDGs, primarily Goal 3 but also Goals 4, 5, and 6. Resource allocation is needed for all levels of prevention: primary, secondary, and tertiary. In the USA, intimate partner violence costs more than US$5.8 billion annually for direct health care, with a further US$1.8 billion in lost productivity. In the UK, a 2004 study estimated annual total costs for domestic violence at £23 billion.

In developing countries, women's health involves a wide range of problems including high rates of pregnancy and maternal mortality, STIs and AIDS, and lack of access to family planning. In some parts of Africa, up to one-third of pregnant women are HIV positive.

In 2010, the UN reported that 79 percent of women experience violence during their lifetime, with physical violence experienced by up to 59 percent of women depending on where they live; for example, Australia and Mozambique (48 percent), the Czech Republic (51 percent), and Zambia (59 percent). Among female murder victims, 40–70 percent in Australia, Canada, Israel, South Africa, and the USA are killed by their partners. Some 5000 women annually are victims of honor killings. Awareness is growing that violence occurs between women as well as being perpetrated by men. The WHO estimates that about one-quarter of the world's women are subject to violence and abuse in their own homes, with rates over 50 percent in Thailand, Papua New Guinea, and Korea, and as high as 80 percent in Pakistan and Chile.

The WHO estimates that between 100 and 140 million women in developing countries have undergone female genital mutilation (FGM) and some 3.3 million young girls are at risk of this procedure annually. FGM is a procedure that involves partial or total removal of the external female genitalia (clitoridectomy, excision of labia, narrowing of vaginal opening, or other injury to the female genital organs) for non-medical reasons. It is mostly carried out by physicians, nurses or midwives, but often by untrained village midwives. Even with hygienic practices complications are common. FGM can cause hemorrhage, shock, and infection in the short term. Long-term effects include pelvic inflammatory disease, infertility, psychological damage, and sexual dysfunction. It may also contribute to obstructed labor, a common contributor to maternal mortality.

FGM is a widespread practice in 28 African and Arab countries, approaching 100 percent of women aged 15–49 years in Guinea, Eritrea, and Egypt, and reaching rates of more than 80 percent of those over the age of 15 years in countries such as Sudan, Somalia, and Yemen, while lower rates (18–38 percent) are seen in sub-Saharan African countries such as Kenya and Tanzania. FGM is also practiced by immigrants to western countries.

This procedure has serious medical, obstetric, and psychological effects, as well as representing a societal contempt for women. It is internationally recognized as a violation of the human rights of girls and women, reflecting deep-rooted inequality between men and women, and a violation of the rights of children. Prevention of FGM will require basic changes in social attitudes towards women, their education, and their place in society.

Other outrageous violations include trafficking and commercial sex enslavement or keeping domestic servants in abusive conditions. Forms of serious abuse are present in the form of mass rape as a weapon of war or political weapon, and gang rape is tolerated in some societies. Honor killings of children or women for alleged forbidden liaisons are common in the Middle East and other parts of the world including western countries with immigrants from areas condoning these practices. The sale of preteen children into forced marriages, often with older men who are HIV positive, is another widespread form of abuse. The most common societal abuse of women is by denial of education and right of free movement. These issues are commonly decried at international political events and are present in the MDGs, but remain strongly entrenched in many societies and are especially linked with Taliban and Al Qaida-type religious–revolutionary extremism. Intimate partner violence and murder is common through all societies; on the other side of the coin, female abuse of older partners, both physical and verbal, is also widespread.

The Beijing Conference on Women in 1995 (reaffirmed in 2005) included health and reproduction among those issues of most concern to women. Despite major differences in points of view between conservative religious representatives and liberal delegates on such issues as birth control and sexuality, a consensus emerged that women have rights to control their own bodies in these crucial areas. Violence and lack of education and economic development were also highlighted and recognized as impinging on the health status of women in both developed and developing societies; with multiyear work plans focused on MDGs for 2010 and on violence and women for 2013.

Women's health is affected by socioeconomic status, education, information, societal equality, job opportunities, and women's input into health and social policies. Women are the primary caregivers in the family unit in most societies, but societal patterns may adversely affect a woman's health. Women's place in society affects many issues, ranging from fertility to public support for child care for working mothers. Some of these issues are politically contentious, with political, religious, and other social implications. Public health is closely involved in these issues, with responsibilities of advocacy, documentation, and innovative leadership (WHO, 2012).

MEN'S HEALTH

Health care for the adult male has not received as much attention as women's health issues. Women live an average of 4–5 years longer than men. Men are more vulnerable than women to non-communicable diseases, and have a higher risk for early disease of a wide variety of conditions and disabilities from cardiovascular disease, cancers of specific sites, violence and trauma, suicide, and occupational hazards (Table 6.21). Health systems often stress women's health issues more than men's. Men are less likely than women to visit their doctor, and more likely to be hospitalized for pneumonia and congestive heart failure and for complications of diabetes (AHRQ). Men are more likely to smoke and engage in risky sexual activity than women. While intimate partner violence is most common against women, there is an emerging awareness that men may also be victims of such violence.

In the USA, men under the age of 65 suffer 2.5 times more heart attacks than women. By age 65, one in three men suffers from high blood pressure, a primary risk for stroke and heart attacks, yet men are less likely than women to have their blood pressure checked. Each year, men in every age group make 150 million fewer trips to doctors than women. One in nine men will be diagnosed with prostate cancer, yet few will have a rectal examination and PSA blood test. Routine PSA screening is controversial and prostate cancer seems to have slow-growing or rapid-growing types. The latter need surgical care whereas the slow-growing type may have no effect on longevity. Men are at greater risk of stress-related illnesses than women, and 30 percent more likely to have a stroke. One out of three male strokes occurs before the age of 65. Each year, over 50,000 men die of emphysema, one of the most preventable diseases. It has been estimated that more than 3 million men are walking around with unrecognized early type 2 diabetes yet men are less likely than women to have their blood sugar checked.

Factors contributing to the lower life expectancy for males than females may include less attention to self-care and professional care, less social and psychological adaptability, more risk-taking behavior, and difficulty in the transition from middle age to elderly status. Emotional and psychosocial stresses on the middle-aged male (ages 45–64) may be very important in the morbidity and mortality patterns of a population. Hospitalizations for chronic illness are more likely among men than women; 24 percent more likely for congestive heart failure, 32 percent more often for lengthy hospitalization for diabetes, and more than twice as likely as women to have diabetes limb amputations due to insufficient attention to care measures. Pneumonia and hospitalizations occur more frequently among men owing to not being vaccinated.

Employment and career traditionally are central to the male's self-esteem. Feelings of competition with more aggressive younger colleagues, male and female, in a youth-oriented society may adversely affect the middle-aged male, as well as women. Personal failure in a job, or unemployment due to obsolescent industries, skills, or

TABLE 6.21 Adult Men, Health Risk Assessment and Prevention by Age Group, 18–64 Years

Age Group	Major Health Risk Factor Assessment	Screening/Counseling/Preventive Activities
18–34	Trauma, accidents, homicide, violence Suicide Alcohol, drug abuse Sexual problems, STIs Emotional problems Smoking Fatty diet – cholesterol Lack of exercise Marriage and parenthood	Counseling, community resources, law enforcement of drinking–driving, seat belt, motorcycle and bicycle helmet laws Mental health services Counseling Risky sexual practices education, HIV, HCV tests Peer group support Antismoking education; reminders of health risks Dietary counseling Exercise consciousness and regular activity, facilities Premarital screening/counseling for genetic disorders; preparation for marriage and fatherhood
35–44	Diet Hypertension Elevated cholesterol Lack of exercise	Screening for obesity, inactivity, BP, cholesterol Screening and continuing management Diet, exercise Exercise, diet, and weight control
45–64	Life crises Cardiovascular disease risk factors Hypertension, diabetes Colorectal cancer and prostate Diabetes and glaucoma Benign prostatic hypertrophy Infectious diseases Change in employment status Depression	Counseling for occupational, marital, family issues Diet, exercise, aspirin, statins Screening for hypertension, CVD, diabetes, glaucoma Screening for colorectal cancer from age 50, check prostate Periodic tests and eye examination Counseling Immunization: dT, influenza Counseling; referral, early retirement planning Counseling and treatment Physical activity

Note: STI = sexually transmitted infection; HIV = human immunodeficiency virus; HCV = hepatitis C virus; BP = blood pressure; CVD = cardiovascular disease; DT = diphtheria and tetanus.
Source: US Department of Health and Human Services. The guide to clinical preventive services 2012. Washington, DC: AHRQ; 2012. Available at: http://www.ahrq.gov/professionals/clinicians-providers/guidelines-ecommendations/guide/guide-clinical-preventive-services.pdf [Accessed 27 June 2013].

major economic change may result in stagnation at work, loss of self-esteem, and intrafamily strife that may lead to the breakdown of physical and mental health. Social and economic changes lead to loss of jobs and high rates of mortality among middle-aged men. Soaring rates of mortality from all types of chronic disease in Russia and other states of the former Soviet Union in the early 1990s reflect the social impact on physical health in a society in rapid transition. Further research in the psychological, sociological, and anthropological spheres is required.

Public health and clinical services should be aware of and ready to address self-image, social role, and occupational issues related to men's health. The health care provider should utilize health education and promotion within the community to reduce risk factors in the male population.

HEALTH OF OLDER ADULTS

Improved health of the population leads to increased longevity and an increase in the elderly population in good physical, mental, social, and financial condition in industrialized countries. With the changes in disease patterns in the population, more people are living and remaining relatively disease free for longer. Many life-threatening conditions that occurred during middle age are now postponed to much later in life. As a result, it is not uncommon for older adults to be well and free from major disease processes. Healthy aging is part of any public health program, with longer disease-free and functional years of disability-free life. At the same time, there is an inevitable burden of age-related conditions such as non-communicable diseases, and Alzheimer's and other dementias that impinge on the quality of life. These conditions are the subject of much effort in brain research to find causes and cures, and to promote a healthy lifestyle for elderly people.

Adults 65 years and older comprise the fastest growing group in the population of the industrialized countries. As the mortality rates in adult and middle-aged years have declined, more people survive into the over-65, the over-75, and the over-85-year-old groups, and even into the over-90 age group. These groups constitute 3.5–6 percent of the population in developing countries and up to 20 percent in developed countries. In the developed countries, this proportion is growing and will reach an average of 28 percent in European countries, with the group over 80 years of age being the most rapidly growing segment of the population.

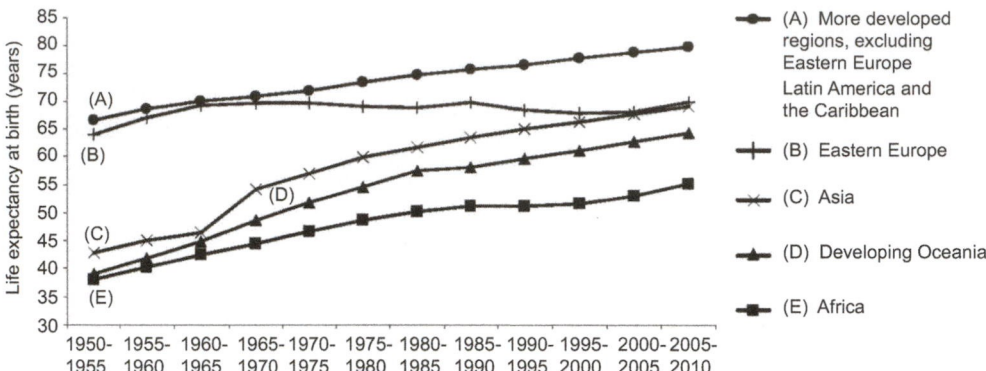

FIGURE 6.11   **Life expectancy at birth by selected global selected regions, 1950–1995 to 2005–2010.** World population prospects, 2010 revision. *Source: United Nations. Population facts no. 2012/2. Toward global equity in longevity. April 2012. Available at: http://www.un.org/esa/population/publications/popfacts/popfacts__2012-2.pdf [Accessed 27 June 2013].*

The UN uses the term "elderly" for the over-60 population, which in 2012 was 11 percent globally, 22 percent in the more developed countries, 9 percent in the less developed countries, and 5 percent in the least developed countries.

Less developed countries are experiencing rapid increases in the older age groups, often within a single generation, placing a strain on health and social services. In some countries negative population growth is occurring, such as in those with high rates of HIV/AIDS as in South Africa, which is expected to lose millions of its population over the next decade primarily due to HIV coupled with multidrug-resistant tuberculosis. Russia has lost millions of population owing to low birth rates and high death rates, which may be partly reversible in the coming decades. In most industrialized countries low death and birth rates combine to produce an aging population with a shrinking working age population. These trends have important impacts on health care as well as on economic and other aspects of society. Increasing longevity, even when associated with healthier aging, results in a growing elderly population which inevitable requires more health care, with its associated costs. These issues weigh on social security and health systems. Figure 6.11 shows the increase in life expectancy in different regions of the world.

Sub-Saharan Africa, despite the enormous burden of HIV/AIDS, other infectious disease, and rising mortality rates from non-communicable diseases, is forecast to have a rapid rise in the elderly population. In 2005, there were 34 million people aged 60 and over in sub-Saharan Africa, and this number is projected to double to over 67 million by 2030, a growth rate of older people more rapidly in sub-Saharan Africa than in the developed world. The proportion of the population over aged 60 years is projected to increase by annual rates of 2.5 percent rising to over 3.5 percent by 2050 in sub-Saharan Africa, while in developed countries the rate of increase will decline from just under 2.4 percent annually currently to under 0.5 percent in 2050. At that time about one in 30 people will reach the age 80-plus group

in the less developed regions and one in 100 in the least developed regions. Aging of the population has important implications for economics in the context of a shrinking workforce able to provide health care for the elderly, who are major consumers of health services (National Research Council, 2006).

Internationally, there are wide variations in life expectancy (Table 6.22) and in care programs for the elderly. The tradition of caring for frail elderly people in the family context is changing in industrialized societies where both men and women work, housing conditions may be crowded, and tolerance for older adults may be reduced. These factors, together with a growing elderly population, have increased the reliance on long-term care in institutional settings. Offsetting this trend in western countries has been the development of adequate pensions, social benefits, universal health care, home care, and a delay in the onset of debilitating effects of disease to a later time in life.

Biological aging is measured by various functional abilities and performance of the individual, not necessarily reflected by chronological age. Aging carries with it many social, occupational, psychological, and financial, as well as physical changes. These all directly affect health and the appropriate health and support needed by elderly people to sustain themselves. These factors can also interact with one another, causing or compounding health problems.

Physical and mental deterioration is associated with aging, frequently resulting in a heavy burden on the family unit and on social and health support systems. Seniors are at increased risk of abuse, including psychological and financial abuse by others including caregivers and family. Elder abuse by maltreatment at home is estimated at 4–6 percent in high-income countries and is likely to increase globally with the aging of the population, especially where resources are limited. Cooperation and leadership is needed between the education and social services and the health sector to respond to the challenge as the global population of elderly people increases to an estimated 1.2 billion in 2025 (WHO, 2011).

TABLE 6.22 Life Expectancy at Birth and at Age 65, by Gender, Selected Countries, 2009

	Male Life Expectancy (years)			Female Life Expectancy (years)	
Country	At Birth	At Age 65	Country	At Birth	At Age 65
Switzerland	79.9	19.0	Japan	86.4	24.0
Israel	79.7	18.9	Spain	84.9	22.4
Iceland	79.7	18.3	Switzerland	84.6	22.2
Japan	79.6	18.9	Italy	84.5	n/a
Sweden	79.4	18.0	France	84.4	n/a
Australia	79.3	18.7	Australia	83.9	21.8
Italy	79.1	18.2	Republic of Korea	83.8	21.5
New Zealand	78.8	18.6	Finland	83.5	21.5
Norway	78.7	18.0	Sweden	83.4	21.0
Spain	78.6	18.3	Luxembourg	83.3	21.4
Netherlands	78.5	17.4	Iceland	83.3	20.6
Canada	78.5	18.1	Austria	83.2	21.2
UK	78.3	18.1	Norway	83.2	21.1
Luxembourg	78.1	17.6	Canada	83.1	21.3
Germany	77.8	19.6	Belgium	82.8	21.1
Greece	77.8	18.1	Germany	82.8	20.8
France	77.7	18.2	New Zealand	82.7	21.1
Austria	77.6	17.7	Netherlands	82.7	20.8
Ireland	77.4	17.2	Greece	82.7	20.2
Belgium	77.3	17.5	Portugal	82.6	20.5
Denmark	76.9	16.8	UK	82.5	20.8
Republic of Korea	76.8	17.1	Ireland	82.5	20.6
Finland	76.6	19.3	Israel	82.4	21.2
Portugal	76.5	17.1	Slovenia	82.3	20.1
Slovenia	75.8	16.3	Denmark	81.1	19.5
USA	75.7	17.3	Chile	80.9	19.9
Chile	75.6	17.0	USA	80.9	20.3
Czech Republic	74.2	15.2	Czech Republic	80.5	18.8
Mexico	72.9	17.0	Estonia	80.1	18.3
Poland	71.5	14.7	Poland	80.0	19.1
Turkey	71.4	14.0	Slovak Republic	78.7	17.6
Slovak Republic	71.3	13.9	Hungary	77.9	17.6
Hungary	70.0	13.7	Mexico	77.6	18.3
Estonia	69.8	14.4	Turkey	76.1	15.9
Russian Federation	62.8	12.0	Russian Federation	74.7	16.5
South Africa	50.3	11.9	South Africa	63.1	15.8

Sources: Centers for Disease Control and Prevention. Health United States, 2011. Table 21. Available at: http://www.cdc.gov/nchs/data/hus/hus11.pdf [Accessed 27 June 2013].
The World Bank. Life expectancy at birth. 2011. Available at: http://data.worldbank.org/indicator/SP.DYN.LE00.MA.IN [Accessed 14 December 2012 and 27 June 2013].

Chronic diseases, such as diabetes, cardiovascular diseases, and cancer, are increasingly common with advancing age. Physical limitations can affect social interaction and mental status, with depression affecting physical abilities. Alzheimer's disease causes serious mental deterioration in people over the age of 50, with rates increasing with age. Parkinson's disease produces progressive physical muscular rigidity and limitations of movement (see Chapter 5).

Health Maintenance for Older Adults

Public health attempts to promote well-being by encouraging a healthful lifestyle and access to good health care. This involves nutrition, physical fitness, recreation, work or daily activities, a positive family life, social and religious participation, and an active sex life, all in keeping with the physical and emotional capability of the individual (Box 6.13). With rapid growth of the population of older adults, health targets for the USA and European countries have

addressed prevention and health maintenance for this group with specific targets as outlined in Box 6.14. Injuries from falls occur in about one-third of elderly adults, with fractures and head injuries occurring in about 20–30 percent of these situations. The annual death rate from falls is 45.3 per 100,000 people in the USA. Healthy People 2020 states that 46.3 percent of men age 65 or over and 47.9 percent of women meet the core set of clinical preventive services in 2008 and has targeted a 10 percent improvement as objectives to increased physical activity.

Good nutrition, vital for a healthy old age, can be impaired by financial, social, and psychological factors. In many industrialized countries, the elderly are protected by good pensions from their employment and social security, but there are also those with inadequate pensions or other

---

BOX 6.13 Health Challenges of the Elderly

- Social/occupational/economic status:
  - poverty
  - isolation
  - retirement
  - widowhood and bereavement
  - relocation
  - loss of friends and family
  - loss of financial security
  - loss of professional status and self-esteem
  - poor nutrition
  - physical inactivity.
- Physiological:
  - hormonal changes
  - onset of non-insulin-dependent diabetes
  - hypertension
  - thyroid dysfunction
  - osteoporosis
  - decreased absorption (e.g., vitamin B).
- Pathophysiological:
  - chronic disease of one or more organ systems
  - medical/surgical conditions, medication, or support services
  - disabilities limiting mobility, activities of daily living.
- Mental:
  - fear of death
  - loneliness and depression
  - loss of memory
  - senility and agitation
  - isolation from children, family, friends.

Sources: *Derived from National Institute of Aging. Improve health and quality of life of older people. Available at: http://www.nia.nih.gov/ [Accessed October 2007; confirmed 4 May 2008].*
National Institute on Aging. Healthy aging. Available at: http://nihseniorhealth.gov/category/healthyaging.html [Accessed 15 March 2013].

BOX 6.14 European and US Health Objectives for the Elderly

WHO Europe
- Ensure equity in health by reducing the gap in health status between countries and groups within countries.
- Add life to years by ensuring full development and use of physical and mental capacity to derive full benefit and to cope with life in a healthy way.
- Add health to years by reducing disease and disability.
- Add years to life by reducing premature deaths, thus increasing life expectancy for adults over age 65.

USA
- Vigorous exercise – 20 percent will engage in vigorous exercise three times a week for 30 minutes each occasion.
- Muscle tone and endurance – 50 percent will participate in physical activities that promote and develop muscle tone and endurance.
- Flexibility – 50 percent will regularly participate in physical activities that promote flexibility.
- All will have more knowledge, positive attitude, and greater practice of physical exercise on a regular basis.

Sources: *Agency for Healthcare Research and Quality. Guide to clinical preventive services, 2012: Recommendations of the US Preventive Services Task Force. Available at: http://www.ahrq.gov/professionals/ clinicians-providers/guidelines-recommendations/guide/guide-clinical-preventive-services.pdf and http http://www.ahrq.gov/professionals/ clinicians-providers/guidelines-recommendations/guide/index.html [Accessed 17 March 2013].*
Stahl T, Wismar M, Ollila E, Lahtinen E, Leppo K, editors. Health in all policies: prospects and proposals. Finland: Ministry of Social Affairs and Health; 2006. Available at: http://www.euro.who.int/__data/assets/pdf_ file/0003/109146/E89260.pdf [Accessed 17 March 2013].
Healthy People 2020. Older adults. Available at: http://www.healthy-people.gov/2020/topicsobjectives2020/pdfs/OlderAdults.pdf and http://www.healthypeople.gov/2020/topicsobjectives2020/overview. aspx?topicid=31 [Accessed 17 March 2013].
World Health Organization, Regional Office Europe. Health 2020: the European policy for health and well-being. Available at: http://www. euro.who.int/en/what-we-do/health-topics/health-policy/health-2020 [Accessed 17 March 2013].

financial support who live in poverty. Isolation, loneliness, passivity, and malnutrition become a lifestyle that produces illness. Many elderly people live on very limited budgets allowing little for food purchases, and many have poor dental health. These are common contributing factors to the "tea and toast" syndrome of semi-starvation that becomes a way of life for this population group. Even a well-off, physically capable elderly person, living alone, may lack the appetite and incentive to prepare a properly balanced diet. The situation is especially difficult when the person is restricted in activity and unable to move about. Vitamin deficiency conditions are common, especially associated with vitamins B and D, the latter mainly in winter months. Individual assessment by any caregiver should bear in mind the potential for low-level malnutrition among the elderly, especially those who are physically or mentally frail, or have severe dental problems. Community or voluntary programming to assist the elderly in nutrition and activity is a crucial contributor to maintaining the person in an independent life situation. The balanced diet for the elderly person is the same as for the younger adult, but must take into account the lower energy output in activity and less exposure to the sun, especially in winter.

Vitamin and mineral supplements are recommended for older adults. Fortification of basic foods is perhaps the most effective public health intervention for prevention of micronutrient deficiency in this and other vulnerable population groups. Food fortification not only makes essential nutrients readily available to the entire population but also supplies a significant portion of the RDAs for the elderly, including iron, iodine, and vitamins B, C, D, and E. Daily vitamin supplementation is increasingly common and is gaining support in geriatric and nutritional professional circles. In 1998, the US FDA recommended routine vitamin B complex supplements for the elderly. Annual immunization for influenza and periodic immunization against pneumococcal pneumonia (every 6 years) and diphtheria–tetanus (dT, every 10 years) should be standard care for the elderly. There is also a growing consensus that older adults should be taking routine aspirin (low dose) and possibly statins for cholesterol control.

Physical fitness is an important preventive measure in preparation for advanced age and to maintain health when reaching that state. Regular physical activity, individually or in groups, tailored to the individual's capacity, is helpful in promoting good appetite, sound sleep, and good physical appearance. It can help to prevent lethargy, apathy, and atrophy. Mutual support in group physical activity is part of regular socializing and mental well-being. Individuals, including disabled elderly people, can also carry out physical exercise seated or lying down.

Social and family networks are important to the elderly and contribute to their sense of acceptance and well-being. Familial and social relations have direct health benefits, including reduced hospitalization and long-term institutionalization. Mental health is strongly affected by the older person's self-perception of his or her role in the family and society. Recreation and social activity, including sexual relations, are part of the life of an older person. Recreational and social facilities designed to stimulate often socially isolated older people to participate in recreational and support activity are dependent on the development of a complex of adequate support systems such as rehabilitation, social security, transportation, and recreation facilities, all user friendly and readily accessible to the elderly person (Box 6.15).

Prevention services for the elderly begin well before the age of 65. A health maintenance approach to preserve well-being involves preparation for a healthy old age, and requires self-care and a preventive approach begun in earlier years. This involves early case finding of chronic disease and care of existing diseases to prevent debilitating complications. Continuous contact with and support of the elderly by health providers can prevent or alleviate complications from medication errors, misunderstanding of health needs, inadequate nutrition, and social isolation.

Communication is vital to the life and health of the older adult. The availability of an emergency communication system may be life saving, as well as providing the frail person with independence and confidence in activities, personal security, and familial and other social relations. A telephone can be used for personal contact and social support as well as for emergencies and contact with medical personnel, providing a sense of security as well as contact with family and others outside the home.

Transportation for seniors and disabled people enables them to have access to medical care, social activities, shopping, and other activities of daily living. Bus companies operated by municipalities often make special arrangements. They may provide special routes including home calls and lower costs of public transportation for older people. Ramps allow easy access for wheelchairs and walkers to public, residential, and commercial buildings and public transportation, making continued participation in community life possible for disabled and elderly people. Newer buses have ramps for easier access and are referred to as kneeler buses.

Relocation and transition are part of the life of the older person. Retirement from work brings with it the potential for rest and recreation or the possibility of isolation and depression. Training people for transition in life may be as important as their physical well-being. Death and bereavement are also part of this process of transition and require organized community, as well as family, support.

Organized community systems for assistance and interventions, based on community or social networks, require public health activities and support. Identification of needs, professional support for resource development, as well as

BOX 6.15 Community Health Needs of Seniors

Community programs for seniors should promote a wide range of knowledge and self-care and support services needed to prevent premature onset and progression of debilitation due to chronic disease and deterioration of general functioning of the older person, with the goal of helping them to function as independently as possible, consistent with safe and healthful conditions of a caring society.

- Preventive self-care: healthy nutrition, regular exercise, and exposure to the sun.
- Social contact: regular contact with family, friends, and social support systems (e.g., church, ethnic, recreation, and social clubs).
- Health education: to promote community awareness, and knowledge among the elderly.
- Medical care services: preventive, diagnostic, treatment, hospitalization, and rehabilitation care.
- Nutritional support programs: to provide assistance with counseling, home-delivered "meals-on-wheels", and group meals in senior citizen centers to promote socialization.
- Injury prevention programs: inspect homes and provide safety devices such as non-slip carpets, railings, shower aids, and bath mats.
- Medical devices loan service: a service to provide and maintain medical aid devices, such as wheelchairs and kitchen and bathroom assistance devices.
- Home care: organized nursing, physiotherapy, shopping, cleaning, and other services to assist disabled or frail elderly people to remain at home.
- Hospital care: accessible but kept as short as possible to avoid infections and other complications.

- Nursing homes: accredited facilities providing nursing and other care for elderly people not able to live independently and requiring daily nursing care.
- Supervised housing: group housing with supervisory nursing care, communal meals, and recreational activities.
- Recreation and occupational therapy: provided at community centers or in the home.
- Volunteer work programs: volunteer traffic control, public garden maintenance, mutual help organizations, and community health workers.
- Emergency call service: beeper service in case of emergency.
- Security and safety measures installations: bathtub grips, smoke detectors, safety locks, and window bars.
- Home help service: volunteer services to assist in maintaining independent homes by regular shopping, cleaning, and maintenance.
- Home health aide service: cleaning, bathing, light housework, shopping, laundry, meals-on-wheels.
- Mutual call service: telephone monitor services to maintain contact with seniors living alone.
- Security: against abuse and violence.
- Financial security: pensions and social support systems.

Sources: *World Health Organization. Community health needs assessment: an introductory guide for the family health nurse in Europe; Part 1: A pack for practitioners; Part 2: A pack for trainers.*
Healthypeople.gov. Older adults [updated 10 April 2013]. Available at: http://www.healthypeople.gov/2020/topicsobjectives2020/overview. aspx?topicid=31 [Accessed 27 June 2013].

direct provision of services are all part of public health. Public health plays an advocacy and promotion role, but many services will be provided by other agencies, such as comprehensive community-based home care programs or voluntary NGOs that provide support services for the elderly.

Finding methods of assisting seniors to remain functional and adjusted in their own homes and in the community is essential to preserve the capacity of the health system to meet the needs of this population group. Such measures can be simple helping services, such as shopping or house cleaning. They may be devices to help in activities of daily living such as adaptations to stoves, wall grips for bathtubs and toilets, or safety measures such as banisters, slip-free carpets, alarm systems, heating devices, safety locks, and police protection from vandals and burglars.

Finances in old age may be a serious burden for the younger generation. Children may need to provide financial assistance to the older members of their family who may have an inadequate pension or national social security system allocation. This can affect family relationships, living conditions, nutrition, medical care, social contact, and many other

aspects of life. National social security systems have been developed in most industrialized countries to provide income security for seniors, through contributions from wages during the working years. However, many social security systems are under pressure financially and politically. Increased lifespan and fewer births have created a situation in which the labor force may become smaller than the dependent members of society. Crises in social security systems may jeopardize the standards of living and security of the elderly in countries which today provide good income and other support systems for this population group.

Older adults in many countries comprise a significant proportion of the population and constitute a powerful political force. This has created a situation in which there is pressure on politicians to take into account the special needs of this population group. Seniors are a formidable group, strong in the political process, in professional organizations, and in municipal, provincial, and national political organizations. This "gray power" is a factor in Western European and North American politics, where the over-65 age group constitutes 13 percent or more of the total population, and an even larger percentage of the adult and politically active population.

The economic aspects of aging of the population are their pension support and increased health care needs. On the other hand, the older adults are consumers whose accumulated savings or social benefits employ many people as producers and care providers. They also constitute a social asset to a country, not only as part of extended family networks, but also in their potential for volunteer work in the community and the transmission of cultural heritage. Grandparents are important to the cultural and social well-being of their grandchildren and to their society.

Despite, or perhaps because of, their use of health resources, seniors are healthier than ever before, with many continuing in the labor or volunteer workforce. Furthermore, they are an important consumer group and tend to have the time, money, and health to play important economic and social roles. A healthy older population should be seen not as a burden but as a vibrant contributing part of the family and society generally. The New Public Health seeks to improve the health status of older adults and to ensure the provision of adequate support and health care services to assist them to function independently, for as long as possible, in their own homes. Society and the New Public Health should promote volunteer and self-help networking among seniors to provide support during critical times such as illness or injury, following hospitalization, or periods of emotional stress and depression.

The most rapid increase in older populations is occurring in the less developed world, with a projected increase of 140 percent between 2006 and 2030. The proportion of those living alone is rising in most countries, and as many as half of European women live alone. Living alone carries with it associated risks of social isolation, nutritional inadequacy, physical inactivity, and passivity in seeking helping services or compliance with needed medical regimens, and should thus be considered a risk factor for ill-health. Health care should address this with outreach contact by telephone and home visits to ensure that such conditions do not lead to preventable health consequences. A healthy lifestyle is a factor in the economic and social burden placed on society by Alzheimer's disease, which results in total costs of more than US$100 billion annually in the USA.

SUMMARY

Individuals live much of their lives in some form of family unit, but they also pass through stages of vulnerability as individuals and as population groups with common health problems. Traditionally, public health has paid great attention to some groups because of their particular vulnerability, as in maternal and child health. The benefits to society as a whole are great where such programs are well developed. Middle-aged men and women are important target groups for primary and secondary preventive programming

in preparation for old age. The elderly also need special attention in public health, because of increasing numbers, changes in society, and the need to find effective ways of promoting their healthy living and preventive care in place of costly services for unattended health problems. Preventive care for the elderly to sustain health can prevent unnecessary or premature dependency on high-cost medical or nursing care facilities.

Each age group, from the newborn to the elderly, has specific problems and concerns that need to be addressed by the health and social service systems. There are many medical, economic, and ethical issues involved in these aspects of public health. Public health needs to continuously monitor the health and social condition of the family as a key part of its overall individual care and population-oriented responsibility. Failure to do so leads to excess premature mortality and a costly burden on medical and hospital care to repair damage already done. The growth in the longevity of the population is a public health success story but much needs to be done to meet the new challenges, especially achieving the aims of the MDGs.

The New Public Health approaches the family unit both as a resource and as a target group needing preventive and curative services at different stages of life. Family members may be cared for by different service providers, but there is a functional and economic relationship among those services. Inadequate prenatal care will increase the chance of poor results in infant health that can have long-term irreversible effects on the potential of the child. Inadequate support for the family struggling to cope with a chronically ill child, spouse, or parent can result in unnecessary and damaging institutionalization. Lack of health promotion in nutrition, safety, and other community health issues may cause the premature death of a parent, most often the male, with serious economic and social as well as emotional consequences for remaining members of the family.

Children's chances for a healthy life depend on adults who often are unable to provide them with everything that they need for nutrition, health, and education. These children are likely to face handicaps in school, in health care, or in coping skills in adult life. Fewer children die as infants or toddlers where births are attended by skilled personnel, where children are fully immunized, where households have access to clean piped water and sanitation facilities, and where girls are better educated. Steps taken by governments make a difference when they invest in pregnancy care to make pregnancy safe for mother and newborn and when all children are immunized and have access to free and quality education. Pensions and minimum wages, set high enough so that working parents can support their family, enable children from the poorest families to go to school instead of work, regardless of ethnicity and social conditions. These factors set the health conditions that affect all the lives of each generation (Heymann and McNeill, 2013).

The holistic approach of the New Public Health when applied to family health also addresses social and economic issues that affect or prejudice family function. Unemployment, underemployment, and poverty promote family distress and crises with long-term consequences for all members. Planning and resource allocation need to address this complex matrix of health in the family and societal context.

This approach is based on unifying factors, such as integration of various service systems and new kinds of linkages between records or a new health provider, to assist the family to cope with normal family health events and the additional burdens of chronic disease. The family physician should ideally be supported in a team approach with a family nurse or health guide to help monitor and support families. Together, they can assist the family in coping with health problems of individuals within the family.

Health promotion and prevention services are all part of this complex. Their application varies with the societal organization to address health issues and the priorities placed on resource allocation and public policy. National health targets help to define specific program content needs. The MDGs similarly set targets for the global health community, but progress in many areas has been much less than hoped for, especially with no improvement in reducing horrendous maternal mortality rates. Progress is uneven and new challenges emerge, such as the growing prevalence of non-communicable diseases, obesity and diabetes, with potential to reverse progress made in public health in the twentieth and the early part of the twenty-first centuries. Important challenges in family health remain essential issues of the New Public Health.

NOTE

For a complete bibliography and guidance for student reviews and expected competencies please see companion web site at http://booksite.elsevier.com/9780124157668

BIBLIOGRAPHY

Aboderin, I., 2010. Understanding and advancing the health of older populations in sub-Saharan Africa: policy perspectives and evidence needs. Public Health Rev. 32, 357–376. Available at: www.publichealthreviews.eu (accessed 20.03.13).

Agency for Healthcare Research and Quality, Guide to clinical preventive services, 2012: Recommendations of the US Preventive Services Task Force. Available at: http://www.ahrq.gov/professionals/clinicians-providers/guidelines-recommendations/guide/guide-clinical-preventive-services.pdf. and http://www.ahrq.gov/professionals/clinicians-providers/guidelines-recommendations/guide/index.html (accessed 11.02.13).

Agency for Healthcare Research and Quality, 2011. Men: stay healthy at 50+. AHRQ publication no. 10(11)-IP005-A. Available at: http://www.ahrq.gov/ppip/men50.htm. (accessed 11.12.12).

Agency for Healthcare Research and Quality, 2010. Men: stay healthy at any age. AHRQ publication no. 10-IP004-A. Available at: http://www.ahrq.gov/ppip/healthymen.htm. (accessed 11.12.12).

Agency for Healthcare Research and Quality, National Guideline Clearinghouse. Available at: http://www.guidelines.gov/ (accessed 12.12.12).

Agency for Healthcare Research and Quality, September 2010. Women: stay healthy at any age. AHRQ publication no. 10-IP002-A. Available at: http://www.ahrq.gov/ppip/healthywom.htm. (accessed 11.02.13).

Alkema, L., Kantorova, K., Menozzi, C., Biddlecom, A., 2013. National, regional, and global rates and trends in contraceptive prevalence and unmet need for family planning between 1990 and 2015: a systematic and comprehensive analysis. Lancet, 12. March: doi: 10.1016/S0140-6736(12)62204-1. Available at: http://www.thelancet.com/journals/lancet/article/PIIS0140-6736(12)62204-1/fulltext (accessed 15.03.13).

American Academy of Pediatrics, 2003. Controversies concerning vitamin K and the newborn. Pediatrics 112, 191–192. Available at: http://pediatrics.aappublications.org/content/112/1/191.full (accessed 25.11.13).

American Academy of Pediatrics, 2008. Newborn screening authoring committee. Newborn screening expands: recommendations for pediatricians and medical homes – implications for the system. Pediatrics 121, 192–217. Available at: http://pediatrics.aappublications.org/content/121/1/192.full (accessed 23.02.13).

American Academy of Pediatrics, 1998. Committee on Environmental Health. Screening for blood lead levels. Pediatrics 101, 1072–1078.

American Academy of Pediatrics, 2006. Committee on Fetus and Newborn. The Apgar score. Pediatrics 117, 1444–1447. Reconfirmed November 2012. Available at: http://pediatrics.aappublications.org/content/123/5/1421.full (accessed 08.10.12).

American Academy of Pediatrics, 2013. Committee on Infectious Diseases. Recommended childhood and adolescent immunization schedule – United States, 2013. Pediatrics 131, 397–398. Available at: http://pediatrics.aappublications.org/content/131/2/397.full (accessed 25.11.13).

American Academy of Pediatrics, 1989. Committee on Nutrition. Iron fortified formulas. Pediatrics 84, 1114–1115. Available at: http://pediatrics.aappublications.org/content/84/6/1114 (accessed 25.11.13).

American College of Preventive Medicine, 1999. Public policy statement: folic acid fortification of grain products in the US to prevent neural tube defects. Am. J. Prev. Med. 16, 264–267. Available at: http://c.ymcdn.com/sites/www.acpm.org/resource/resmgr/policy-files/polstmt_folic.pdf (accessed 25.11.13).

American College/Congress of Obstetricians and Gynecologists, 2011. Prenatal and perinatal human immunodeficiency virus testing: expanded recommendations; reaffirmed. Available at: http://www.acog.org/Resources%20And%20Publications/Committee%20Opinions/Committee%20on%20Obstetric%20Practice/Prenatal%20and%20Perinatal%20Human%20Immunodeficiency%20Virus%20Testing%20-%20Expanded%20Recommendations.aspx. (accessed 26.06.13).

American College/Congress of Obstetricians and Gynecologists, 2009. The importance of preconception care in the continuum of women's health care: reaffirmed. Available at: http://www.acog.org/Resources_And_Publications/Committee_Opinions_List. (accessed 10.10.12).

American College/Congress of Obstetricians and Gynecologists, July 2011. Well woman and prenatal visits should include alcohol abuse screening. Available at: http://www.acog.org/About_ACOG/News_Room/News_Releases/2011/Well_Woman_and_Prenatal_Visits_Should_Include_Alcohol_Abuse_Screening. (accessed 10.10.12).

American Public Health Association, January 2011. Policy statement database. Reducing US maternal mortality as a human right. Database 11. Available at: http://www.apha.org/advocacy/policy/policysearch/default.htm?id=1430 (accessed 02.02.13).

American Public Health Association. Policy Statement Database 201114. 11.1.2011. Reducing maternal mortality as a human right. Available at: http://www.apha.org/advocacy/policy/policysearch/default.htm?id=1430 (accessed 25.11.13).

Anderson, R.J., March 2011. Dynamics of economic well-being: poverty 2004–2006. Current Population Reports P170-123. US Census Bureau, Washington, DC. Available at: http://www.census.gov/hhes/www/poverty/publications/dynamics04/P70-123.pdf (accessed 31.07.13).

Angastiniotis, M., Modell, B., Engelzos, P., Boulyjenkov, V., 1995. Prevention and control of haemoglobinopathies. Bull. World Health Organ. 73, 375–386. Available at: http://www.ncbi.nlm.nih.gov/pmc/articles/PMC2486673/pdf/bullwho00407-0102.pdf (accessed 31.07.13).

Antonucci, T.C., Wong, K.M., 2010. Public health and the aging family. Public Health Rev. 32, 512–531. Available at: www.publichealthreviews.eu (accessed 20.03.13).

Institute of Medicine (US) Committee on Understanding Premature Birth and Assuring Healthy Outcomes. Neurodevelopmental, health, and family outcomes for infants born preterm. Chapter 11. In: Behrman, R.E., Butler, A.S. (Eds.), Preterm birth, causes, consequences, and prevention. 2007. National Academies Press, Washington, DC, pp. 346–397. Available at: http://www.ncbi.nlm.nih.gov/books/NBK11356/#a20012272ddd00286. and book at http://www.nap.edu/openbook.php?record_id=11622&page=R1. (accessed 02.02.13).

Caravella, S., Clark, D., Dweck, H.S., 1987. Health codes for newborn care. Pediatrics 80, 1–5. Available at: http://pediatrics.aappublications.org/content/80/1.t (accessed 25.11.13).

Centers for Disease Control and Prevention, March 2012. About teen pregnancy: teen pregnancy in the United States. Available at: http://www.cdc.gov/teenpregnancy/aboutteenpreg.htm (accessed 06.10.12).

Centers for Disease Control and Prevention, 1999. Achievements in public health, 1900–1999: Healthier mothers and babies. MMWR. Morb. Mortal. Wkly. Rep. 48, 849–858. Available at: http://www.cdc.gov/mmwr/preview/mmwrhtml/mm4838a2.htm (accessed 25.11.12).

Centers for Disease Control and Prevention, Cancer prevention and control. Cancer among women [updated 2 January 2013]. Available at: http://www.cdc.gov/cancer/dcpc/data/women.htm (accessed 25.06.13).

Centers for Disease Control and Prevention, 2010. CDC grand rounds: additional opportunities to prevent neural tube defects with folic acid fortification. MMWR. Morb. Mortal. Wkly. Rep. 59, 980–984. Available at: http://www.cdc.gov/mmwr/preview/mmwrhtml/mm5931a2.htm (accessed 14.08.13).

Centers for Disease Control and Prevention, Chronic disease prevention and health promotion. Available at: http://www.cdc.gov/chronicdisease/index.htm (accessed 12.12.12).

Centers for Disease Control and Prevention, National Center on Birth Defects and Developmental Disabilities (NCBDDD). Available at: http://www.cdc.gov/ncbddd/ (accessed 12.12.12).

Centers for Disease Control and Prevention, National hospital discharge survey [updated 12 December 2012]. Available at: http://www.cdc.gov/nchs/nhds.htm (accessed 21.03.13).

Centers for Disease Control and Prevention, Pregnancy mortality surveillance system [updated 13 March 2013]. Available at: http://www.cdc.gov/reproductivehealth/MaternalInfantHealth/PMSS.html (accessed 24.03.13).

Centers for Disease Control and Prevention, 1999. Prevalence of selected maternal and infant characteristics. Pregnancy risk assessment monitoring system (PRAMS). MMWR. Surveill. Summ. 48 (SS-5), 1–43. Available at: http://www.cdc.gov/mmwr/PDF/ss/ss4805.pdf (accessed 31.07.13).

Centers for Disease Control and Prevention, Public Health Genomics. Available at: http://www.cdc.gov/genomics/ (accessed 31.07.13).

Centers for Disease Control and Prevention, 1998. Recommendations to prevent and control iron deficiency anemia in the United States. MMWR. Morb. Mortal. Wkly. Rep. 47 (RR-3), 1–30. Available at: http://www.cdc.gov/mmwr/preview/mmwrhtml/00051880.htm (accessed 25.11.13).

Centers for Disease Control and Prevention. Sickle Cell Disease, 2011. Available at: http://www.cdc.gov/ncbddd/sicklecell/data.html. (accessed 27.11.2013).

Centers for Disease Control and Prevention, 2012. Sickle cell disease (SCD): data and statistics. Available at: http://www.cdc.gov/ncbddd/sicklecell/index.html. (accessed 31.07.13).

Centers for Disease Control and Prevention, 2004. Spina bifida and anencephaly before and after folic acid mandate – United States, 1995–1996 and 1999–2000. MMWR. Morb. Mortal. Wkly. Rep. 53, 362–365. Available at: http://www.cdc.gov/mmwr/preview/mmwrhtml/mm5317a3.htm (accessed 25.11.13).

Centers for Disease Control and Prevention, 2011. Strategies to prevent obesity and other chronic diseases: the CDC guide to strategies to increase physical activity in the community. US Department of Health and Human Services, Atlanta, GA. Available at: http://www.cdc.gov/obesity (accessed 21.03.13).

Centers for Disease Control and Prevention, Teen pregnancy. Parent and guardian resources [updated 2 April 2013]. Available at: http://www.cdc.gov/TeenPregnancy/Parents.htm (accessed 25.06.13).

Centers for Disease Control and Prevention, 2008. Use of supplements containing folic acid among women of childbearing age – United States, 2007. MMWR. Morb. Mortal. Wkly. Rep. 57, 5–8. Available at: http://www.cdc.gov/mmwr/preview/mmwrhtml/mm5701a3.htm (accessed 11.12.12).

Centers for Disease Control and Prevention, Youth risk behavior surveillance system (YRBSS) [updated 27 February 2013]. Available at: http://www.cdc.gov/HealthyYouth/yrbs/index.htm (accessed 24.03.13).

Centers for Medicare and Medicaid Services (CMS). Available at: http://www.cms.gov/ (accessed 12.12.12).

Conde-Agudelo, A., Belizán, J.M., Diaz-Rossello, J., Kangaroo mother care to reduce morbidity and mortality in low birthweight infants. Cochrane Database Syst Rev 201116;(3):CD002771. Available at: http://www.ncbi.nlm.nih.gov/pubmed/21412879. and http://www.nichd.nih.gov/cochrane/conde-agudelo/conde-agudelo.htm (accessed 31.07.13).

Estimates of incidence of and mortality due to unsafe abortion, 2008 Unsafe abortion case–fatality (deaths per100 000 unsafe abortions) rounded Department of Reproductive Health and Research World Health Organization© World Health Organization 2012 http://apps.who.int/iris/bitstream/10665/75173/1/WHO_RHR_12.01_eng.pdf

Fried, L.P., Paccaud, F., 2010. Editorial: The public health needs for an ageing society. Public Health Rev. 32, 351–355. Available at: www.publichealthreviews.eu (accessed 20.03.13).

Gakidou, E., Cowling, K., Lozano, R., Murray, C.J.L., 2010. Increased educational attainment and its effect on child mortality in 175 countries between 1970 and 2009: a systematic analysis. Lancet 376 (9745), 959–974. Available at: http://www.ncbi.nlm.nih.gov/pubmed/20851260 (accessed 25.11.13).

Gartner, L.M., Greer, F.R., 2003. American Academy of Pediatrics. Prevention of rickets and vitamin D deficiency: new guidelines for vitamin D intake. Pediatrics 111, 908–911. Available at: http://www.ncbi.nlm.nih.gov/pubmed/12671133 (accessed 25.11.13).

Ginsberg, G., Tulchinsky, T.H., Filon, D., Goldfarb, A., Abramov, L., Rachmilevitz, E.A., 1998. Cost-benefit analysis of a national thalassemia prevention programme in Israel. J. Med. Screen 5, 120–126. Available at: http://msc.sagepub.com/content/5/3/120.full.pdf+html.

Healthy People 2020, Adolescent health. Available at: http://www.healthypeople.gov/2020/topicsobjectives2020/overview.aspx?topicid=2 (accessed 24.03.13).

Healthy People 2020, Health care associated infections [updated 6 December 2012]. Available at: http://www.healthypeople.gov/2020/topicsobjectives2020/overview.aspx?topicid=17 (accessed 22.03.13).

Healthy People 2020, Substance abuse. Available at: http://healthypeople.gov/2020/topicsobjectives2020/overview.aspx?topicid=40 (accessed 24.03.13).

Healthy People 2020, What's new for 2020 [updated 6 February 2013]. Available at: http://www.healthypeople.gov/2020/about/new2020.aspx (accessed 20.03.13).

Hilgartner, M.W., 1993. Vitamin K and the newborn [editorial]. N. Engl. J. Med. 329, 957–958. Available at: http://www.nejm.org/doi/full/10.1056/NEJM199309233291310 (accessed 25.11.13).

Hogan, M.C., Foreman, K.L., Naghavi, M., Ahn, S.Y., Wang, M., Makela, S.M., et al., 2010. Maternal mortality for 181 countries, 1980–2008: a systematic analysis of progress towards Millennium Development Goal 5. Lancet 375, 1609–1623. Available at: http://cdrwww.who.int/pmnch/topics/maternal/20100402_ihmearticle.pdf (accessed 24.03.13).

Howson, C.P., Controlling birth defects: reducing the hidden toll of dying and disabled children in low-income countries. Disease Control Priority Project. Available at: http://www.dcp2.org/file/230/dcpp-twpcongenitaldefects_web.pdf (accessed 31.07.13).

Institute of Medicine, 2007. Committee on Understanding Premature Birth and Assuring Healthy Outcomes. Preterm birth, causes, consequences, and prevention. National Academies Press, Washington, DC. Available at: http://www.nap.edu/openbook.php?record_id=11622 (accessed 31.07.13).

Kendig, J.W., 1992. Care of the normal newborn. Pediatr. Rev. 13, 262–268.

King, A.C., King, D.K., 2010. Physical activity for an aging population. Public Health Rev. 32, 401–426. Available at: www.publichealthreviews.eu (accessed 20.03.13).

Klausner, J.D., 2013. The sound of silence: missing the opportunity to save lives at birth. Bull. World Health Organ. 91. 158–A. Available at: http://www.ncbi.nlm.nih.gov/pmc/articles/PMC3590629/ (accessed 25.06.13).

Mladovsky, P., Allin, S., Masseria, C., Hernández-Quevedo, C., McDaid, D., Mossialos, E., 2009. Health in the European Union: trends and analysis. World Health Organization on behalf of the European Observatory on Health Systems and Policies. Available at: http://apps.who.int/bookorders/anglais/detart1.jsp?codlan=1&codcol=34&codcch=95&content=1 (accessed 21.03.13).

Modell, B., Darlison, D., 2008. Global epidemiology of haemoglobin disorders and derived service indicators. Bull. World Health Organ. 86 (6), 480–487. Available at: http://www.who.int/bulletin/volumes/86/6/06-036673/en/. and http://www.who.int/bulletin/volumes/86/6/06-036673-table-T1.html (accessed 27.06.13).

National Center for Health Statistics, 2011. Health, United States, 2010: with special feature on death and dying. CDC, Hyattsville, MD. Available at: http://www.cdc.gov/nchs/data/hus/hus10.pdf (accessed 24.03.13).

National Institute on Aging, Improve our ability to reduce health disparities and eliminate health inequities among older adults. Available at: http://www.nia.nih.gov/about/living-long-well-21st-century-strategic-directions-research-aging/research-goal-e-improve-our (accessed 31.07.13).

National Newborn Screening and Genetics Resource Center (NNSGRC), 2006. Newborn screening: toward a uniform screening panel and system: executive summary. Genetics Medicine. Available at: http://www.acmg.net/resources/policies/NBS/NBS_Exec_Sum.pdf. (accessed 12.12.12).

Nolte, E., McKee, M., 2008. Caring for people with chronic conditions: a health system perspective. European Observatory on Health Systems and Policies Series. McGraw Hill Open University Press, Maidenhead. Available at: http://www.euro.who.int/__data/assets/pdf_file/0006/96468/E91878.pdf (accessed 25.11.13).

Tulchinsky, T.H., Patton, M.M., Randolph, L.A., Meyer, M.R., Linden, J.V., 1993. Mandating vitamin K prophylaxis for newborns in New York State. Am. J. Public Health 83, 1166–1168. Available at: http://www.ncbi.nlm.nih.gov/pmc/articles/PMC1695173/ (accessed 25.11.13).

United Nations, 2010. Global strategy for women's and children's health. UN, New York. Available at: http://www.who.int/pmnch/topics/maternal/201009_globalstrategy_wch/en/index.html (accessed 08.10.12).

United Nations, 2012. The millennium development report. Available at: http://www.un.org/millenniumgoals/pdf/MDG%20Report%202012.pdf. (accessed 24.03.13).

United Nations, 2012. The millennium goals report 2012. UN, New York. Available at: http://www.un.org/millenniumgoals/goals/pdf (accessed 08.10.12).

United Nations, 2010. The world's women: Chapter 6. Violence against women. In: The world's women. Available at: http://unstats.un.org/unsd/demographic/products/Worldswomen/WW2010%20Report_by%20chapter(pdf)/Violence%20against%20women.pdf.

United Nations Children's Fund, Health: maternal and newborn health. Available at: http://www.unicef.org/health/index_maternalhealth.html (accessed 08.10.12).

United Nations Children's Fund, 2012. The state of the world's children. children in an urban world. Available at: http://www.unicef.org/media/files/SOWC_2012-Main_Report_EN_21Dec2011.pdf. (accessed 24.03.13).

United Nations Population Fund, May 2012. Sub-Saharan Africa's maternal death rate down 41 per cent 17. Available at: http://esaro.unfpa.org/public/public/cache/offonce/news/pid/10767 (accessed 15.08.13).

US Cancer Statistics Working Group, 2013. United States cancer statistics: 1999–2009 incidence and mortality web-based report. Department of Health and Human Services, Centers for Disease Control and Prevention, and National Cancer Institute, Atlanta, GA. Available at: http://www.cdc.gov/uscs (accessed 25.06.13).

US Department of Health and Human Services. Physical activity guidelines for Americans, 2008. Available at: http://www.health.gov/PAGuidelines/guidelines/default.aspx.

US Cancer Statistics Working Group, 2013. United States cancer statistics: 1999–2009 incidence and mortality web-based report. Department of Health and Human Services, Centers for Disease Control and Prevention, and National Cancer Institute, Atlanta, GA. Available at: http://www.cdc.gov/uscs (accessed 25.03.13).

US Census Bureau, 2009. America's families and living arrangements, 2009. Current population survey. Available at: http://www.census.gov/population/www/socdemo/hh-fam/cps2009.html. (accessed 12.12.12).

US Preventive Services Task Force, 2008. Screening for congenital hypothyroidism. Available at: http://www.ahrq.gov/clinic/uspstf08/conhypo/conhyprs.htm. (accessed 12.12.12).

US Preventive Services Task Force, 2008. Screening for phenylketonuria. Available at: http://www.ahrq.gov/clinic/uspstf08/pku/pkurs.htm. (accessed 12.12.12).

Warner, M., Chen, L.H., Makuc, D.M., Anderson, R.N., Miniño, A.M., December 2011. Drug poisoning deaths in the United States, 1980–2008. NCHS data brief no. 81. Available at: http://www.cdc.gov/nchs/data/databriefs/db81.pdf (accessed 24.03.13).

Weisberg, P., Scanlon, K.S., Li, R., Cogswell, M.E., 2004. Nutritional rickets among children in the United States: review of cases reported between 1986 and 2005. Am. J. Clin. Nutr. 80 (Suppl. 6) 1697–705S.

World Health Organization, 2009. 2008–2013 Action plan for the global strategy for the prevention and control of noncommunicable diseases. Available at: http://www.who.int/nmh/publications/9789241597418/en/index.html. (accessed 28.01.11).

World Health Organization, October 2012. Congenital anomalies. Fact sheet no. 370. Available at: http://www.who.int/mediacentre/factsheets/fs370/en/index.html. (accessed 15.03.13).

World Health Organization, 2002. Genomics and world health. WHO, Geneva. Available at: http://whqlibdoc.who.int/hq/2002/a74580.pdf (accessed 31.07.13).

World Health Organization, 2012. Guidelines on basic newborn resuscitation, 2012. WHO, Geneva. Available at: http://apps.who.int/iris/bitstream/10665/75157/1/9789241503693_eng.pdf (accessed 21.03.13).

World Health Organization, 2013. Standards for maternal and neonatal care. WHO, Geneva. Available at: http://www.who.int/reproductive-health/publications/maternal_perinatal_health/a91272/en/index.html (accessed 15.03.13).

World Health Organization. Trends in Maternal Mortality: 1990 to 2008. Estimates Developed by WHO, UNICEF, UNFPA and The World Bank. Geneva, Switzerland: World Health Organization, UNICEF, UNFPA, and The World Bank; 2010. Available at: http://whqlibdoc.who.int/publications/2010/9789241500265_eng.pdf. Available at: http://hqlibdoc.who.int/publications/2010/9789241500265_eng.pdf (accessed 25.11.13).

World Health Organization, 2011. Unsafe abortion: global and regional estimates of the incidence of unsafe abortion and associated mortality in 2008, sixth ed. WHO, Geneva. Available at: http://whqlibdoc.who.int/publications/2011/9789241501118_eng.pdf (accessed 08.02.13).

World Health Organization, 2009. Women and health: today's agenda, tomorrow's evidence. WHO, Geneva. Available at: http://whqlibdoc.who.int/publications/2009/9789241563857_eng.pdf (accessed 24.03.13).

World Health Organization, Women's health. Available at: http://www.who.int/topics/womens_health/en/ (accessed 12.12.12).

World Health Organization. World health statistics, 2012. Derived from Table 2, Cause specific mortality and morbidity. Available at: http://www.who.int/healthinfo/EN_WHS2012_Full.pdf. (accessed 08.10.12).

World Health Organization, Regional Office Europe, Health 2020: the European policy for health and well-being. Available at: http://www.euro.who.int/en/what-we-do/health-topics/health-policy/health-2020 (accessed 23.03.13).

Ylli, A., 2010. Health and social conditions of older people in Albania. Baseline data from a national survey. Public Health Rev. 32, 549–561. Available at: www.publichealthreviews.eu (accessed 20.03.13).

Zipursky, A., 1999. Prevention of vitamin K deficiency bleeding in newborns. Br. J. Haematol. 1064, 430–437.

Special Community Health Needs

INTRODUCTION

The New Public Health emphasizes the importance of seeing the individual with special needs, the special needs of groups, as well as the total population in the context of the community and of national health systems. No population is totally homogeneous in terms of health needs. Special community health needs may be defined according to the health risks, morbidity, mortality, economic and social burden related to specific social, professional, or economic conditions, and the presence of disabling chronic illnesses. Groups targeted for special community health needs may be defined according to particular population groups, conditions, and health risks. These may be related to characteristics of age, gender, and socioeconomic status, location of residence, occupation, ethnicity, religion, or disabling conditions. Examples include all children, as well as children with special needs, such as patients with human immunodeficiency virus/acquired immunodeficiency syndrome (HIV/AIDS), diabetes, or end-stage renal disease (ESRD). In the case of these groups, individual care may be supplemented by community, state and national efforts.

Other special needs include population-wide problems such as mental illness and oral health problems; exposure or potential exposure to terrorism, natural disasters, warfare, or genocide; and particular population subgroups such as the military, aborigines or native people, refugees, migrant workers, and prisoners. These issues may affect part or all of a population: minority groups interact with the general population, and are thus part of the total community. Their needs must be addressed with general and specific health promotion and community approaches based on planning and strategic targets. They require a wide array of programs in addition to medical care, such as education, outreach, screening, and risk reduction interventions. Research and evaluation require quantitative and qualitative research methods.

Special community health needs may change over time or be consistently present through lifespans or generations. Special communities include a range of vulnerable groups that require special attention in public health (Table 7.1). There are also special needs related to dental health and emergency health conditions which require organization of special services for the population as a whole. Traditionally, these special needs were dealt with through separate services which segregated them from the general health care system. Even if the initial reasons for separation are no longer relevant, tradition or vested interests of the systems sometimes continue this separation, but there is a growing trend to integrate such services within primary and secondary care services.

By the fact of birth in a special group or at any point in life, every person can become a member of a community with special health needs. The existing health system may be too slow to address emerging health needs. Many circumstances may prevent large sectors of a population from equitable access to services including economic, geographic, prejudice within the system, and informational barriers with lack of awareness. People may also have misinformation with fears of important preventive care, healthy life habits or even seeking medical advice when symptoms indicate it is needed. Special needs groups and services are essential to the evaluation of health service delivery for access, outreach, effectiveness, and efficiency for community leaders and health managers.

This chapter addresses these special health needs, affecting particular subgroups and the population as a whole. Some of the interventions needed to protect the health of particular groups are directed specifically towards the at-risk group. Other activities are directed at the general population because all are at risk, such as for the development of mental health problems. The New Public Health advocates attention to the needs of these often less privileged populations, and seeks to assure adequate attention to their health care through the framework and proactive role of the community.

The New Public Health. http://dx.doi.org/10.1016/B978-0-12-415766-8.00007-0

TABLE 7.1 Vulnerable Groups and Factors of Mental Dysfunction

Children	Genetic factors, birth injury, fetal alcohol syndrome, low birth weight, prenatal drug exposure Prenatal nutritional deficiency Nutritional deficiency: iodine, iron, vitamins A, D; vitamin K at birth Poverty and psychosocial deprivation Abuse, violence: actual or exposure to it in the home or environment Infectious (e.g., viral encephalitis, rubella) Toxic exposure (e.g., lead)
Adolescents	Sexual maturation and associated stress Family stress, abuse, violence Peer pressure, bullying School, career, occupational expectations Fear of failure Body image: fear of obesity, eating disorders Poverty, parental abuse, neglect Smoking, alcohol, drug exposure
Adults	Poverty Poor diet, lack of exercise, smoking Lack of preventive health care Women in relation to fertility, pregnancy Parenthood, especially single status Abuse, physical, sexual, and psychological violence against women Occupation-related stress Fear of aging, menopause Loss of reproductive function and virility Job and status loss, loss of self-esteem
Elderly	Loss of spouse, friends, home Poverty and isolation Retirement and loss of occupational status Deterioration of mental and physical powers in activities of daily life Memory loss Poor nutrition: "tea and toast syndrome" Abuse, violence Loss of independence Loneliness Fear of dying and long process of deterioration

MENTAL HEALTH

Mental health is defined by the World Health Organization (WHO) as a state of well-being in which every individual realizes his or her own potential, copes with the normal stresses of life, works productively and fruitfully, and is able to make a contribution to her or his community. Yet, there is no single consensus on the definition of mental illness or mental disorder, and the phrasing used depends on the social, cultural, economic, and legal contexts in different countries. Mental health issues typically have a variety of causes, and are determined by the interaction between each individual's socioeconomic, biological, and environmental conditions.

Some 450 million people suffer from mental disorders (WHO, 2010). The US National Institute of Mental Health

(NIMH) estimated that debilitating mental illness occurs in 13 million adults and the total mental health care expenditures were US$100 billion in 2008, an increase from US$52 billion in 2004. Mental disorders affect not only the patient but also the family and the society, and have associated comorbidity with other diseases, such as HIV, tuberculosis (TB), cardiovascular and other chronic diseases. Along with the growing recognition of the widespread nature of mental illness in all societies, there is progress in understanding biological, behavioral, neurological, and sociological factors in this group of conditions. As a result, new methods are being developed of preventing and managing mental and neurological illness with effective medications and other methods of intervention, so that this area of health has become an important element of the New Public Health.

The *Lancet* published a Special Series on Mental Health in September 2011, which noted that:

"Neuropsychiatric disorders comprise a substantial share of disease-related burden and disability – approaching 14 percent, with depression the leading global cause of disability – but receive a disproportionately low resource allocation – the average across countries is under 4 percent of the overall healthcare budgets. Resources for mental health care research are scarce and gaps exist. Alongside a shortfall in trained mental health professionals, these deficits are the backdrop to a disconcerting treatment gap for neuropsychiatric disorders in low-income countries, with over 75 percent of patients untreated." (Ravilia et al, Lancet 2011)

Short- or long-term mental and emotional problems may affect everyone to some degree during their life (Box 7.1). These include a wide spectrum of conditions: anxiety, depression, psychotic states, obsessive–compulsive disorders, eating disorders, drug abuse, delinquency, suicide and self-harm, alcoholism, and intrafamily physical and mental abuse. These conditions affect the physical and social well-being of the patient, the family, and the community. Services have often been based on waiting for the presentation with full clinical manifestation of psychotic disorders and dysfunction and acute suicide risk. In the past, long-term institutionalization was common, but emphasis has gradually evolved towards health promotion, early crisis intervention, and community-based approaches.

Psychiatric conditions have traditionally been classified as psychotic or neurotic and by mode of presentation. Psychoses are major mental illnesses, including schizophrenia, characterized by severe symptoms such as delusions and hallucinations. These are divided into organic and functional psychoses, the organic being caused by a demonstrable physical abnormality, while the functional has no physical disease demonstrated. Mood disorders include affective psychosis, severe or psychotic depression, and bipolar disorders, which can be psychotic in the manic phase.

DSM-5 is the fifth edition of the American Psychiatric Association's (APA) *Diagnostic and Statistical Manual of Mental Disorders*. In the USA the DSM serves as a

BOX 7.1 Global Prevalence of Mental Illness

- Hundreds of millions of people worldwide are affected by mental, behavioral, neurological, and substance use disorders.
- Mental, neurological, and behavioral disorders are common to all countries and cause immense suffering. People with these disorders are often subjected to social isolation, poor quality of life, and increased mortality. These disorders are the cause of staggering economic and social costs.
- Estimates of WHO, 2005:
  - 154 million people globally suffer from depression
  - 25 million people suffer from schizophrenia
  - 91 million people are affected by alcohol use disorders
  - 15 million are affected by drug use disorders
  - 50 million people suffer from epilepsy
  - 24 million suffer from Alzheimer's and other dementias
  - 170 million suffer from neurological sequelae of injury, migraine, cerebrovascular diseases, neuroinfections, or neurological sequelae of infections
  - 352 million suffer from nutritional disorders and neuropathies
  - 877,000 people die by suicide every year.
- One in four patients visiting a health service has at least one mental, neurological, or behavioral disorder but most of these disorders are neither diagnosed nor treated.

- Mental illnesses affect and are affected by chronic conditions such as cancer, heart and cardiovascular diseases, diabetes, and HIV/AIDS. Untreated, they bring about unhealthy behavior, non-compliance with prescribed medical regimens, diminished immune functioning, and poor prognosis.
- Cost-effective treatments exist for most disorders and, if correctly applied, could enable most of those affected to become functioning members of society.
- Barriers to effective treatment of mental illness include lack of recognition of the seriousness of mental illness and lack of understanding about the benefits of services. Policy makers, insurance companies, health and labor policies, and the public at large – all discriminate between physical and mental problems.
- Most middle- and low-income countries devote less than 1 percent of their health expenditure to mental health. Consequently, mental health policies, legislation, community care facilities, and treatments for people with mental illness are not given the priority they deserve.

Sources: *World Health Organization. Mental health. Available at: http://www.who.int/mental_health/management/en/ [Accessed 27 June 2013]. Kessler RC, Aguilar-Gaxiola S, Alonso J, Chatterji S, Lee S, Ormel J, et al. The global burden of mental disorders: an update from the WHO World Mental Health (WMH) surveys. Epidemiol Psichiatr Soc 2009;18:23–33. World Health Organization. Mental health atlas 2011. Geneva: WHO; 2011. Available at: http://whqlibdoc.who.int/publications/2011/9799241564359_eng.pdf [Accessed 28 June 2013].*

BOX 7.2 The New American Psychiatric Association (APA) Diagnostic and Statistical Manual of Mental Disorders (DSM-5), May 2013

The previous APA classification of mental disorders included the major categories of somatization. The current classification (DSM-IV) was developed in cooperation with the World Health Organization and is close to the ICD-10 classification, modified in 2000. The new DSM-5 was released in May 2013 and contains clarifications and definitions in keeping with changing experience and practice in mental health.

- *Intellectual disability (intellectual developmental disorder)* – the term mental retardation used in DSM-IV is replaced with intellectual disability, now in common use. Diagnostic criteria are based on both cognitive capacity (IQ) and adaptive function. Severity is determined by adaptive functioning rather than IQ score.
- *Communication disorders* – includes language disorder, speech sound disorder (a new name for phonological disorder), and childhood-onset fluency disorder (a new name for stuttering).
- *Autism spectrum disorder (ASD)* – ASD is a new term reflecting scientific consensus that four previously separate disorders are actually a single condition with different levels of symptom severity. ASD now includes the previous DSM-IV autistic disorder (autism), Asperger's disorder, childhood disintegrative disorder, and pervasive developmental disorder not otherwise specified.
- *Attention-deficit/hyperactivity disorder (ADHD)* – the diagnostic criteria for ADHD in DSM-5 are similar to those in DSM-IV based on the same symptoms, but divided into two symptom domains (inattention and hyperactivity/impulsivity).
- *Specific learning disorder* – combines the DSM-IV diagnoses of reading disorder, mathematics disorder, disorder of written expression, and learning disorders.

- *Motor disorders* – includes the DSM-5 neurodevelopmental disorders: developmental coordination disorder, stereotypic movement disorder, Tourette's disorder, persistent (chronic) motor or vocal tic disorder, and other specified and unspecified tic disorder.
- *Schizophrenia spectrum and other psychotic disorders* – some changes were made to DSM-IV definitions for schizoaffective disorder, delusional disorder, and catatonia.
- *Bipolar and related disorders* – diagnostic criteria were enhanced with several new categories and criteria added. Include major depressive disorders, bereavement-associated depression severity, and suicide risk.
- *Anxiety disorders* – agoraphobia, specific phobia, and social anxiety disorder, panic attack, specific phobias, social anxiety disorder, hoarding disorder, hair-pulling disorder.
- Substance/medication-induced obsessive–compulsive disorders.
- Other obsessive compulsive and related disorders.
- Dissociative disorders, somatic symptoms and related disorders, medically unexplained symptoms.
- Pica (i.e., persistent craving and compulsive eating of non-food substances) and ruminative (i.e., repeated regurgitation of binge eating food) disorders.
- *Avoidance/restrictive food intake disorders* – anorexia nervosa, bulimia nervosa, binge-eating disorder.
- *Disruptive, impulse control, and conduct disorders* – problems of emotional and behavioral self-control.
- *Opposition and defiant disorders* – conduct disorder, intermittent explosive disorder.

BOX 7.2 The New American Psychiatric Association (APA) Diagnostic and Statistical Manual of Mental Disorders (DSM-5), May 2013—cont'd

- *Neurological disorders* – delirium, neurocognitive disorders, dementias due to medical conditions, e.g., Alzheimer's disease.
- *Personality disorders* – long-term pattern of behaviors, emotions, thoughts interfering with ability to function in relationships, work and other settings.

- *Paraphilia and paraphilic disorders* – atypical sexual practices — with danger to others e.g., pedophilia

Sources: *American Psychiatric Association. DSM-5 implementation and support. Available at: http://www.dsm5.org/Pages/Default.aspx [Accessed 29 June 2013].*

American Psychiatric Association. Highlights of changes from DSM-IV-TR to DSM-5. Available at: http://www.psych.org/File Library/Practice/DSM/DSM-5/Changes-from-DSM-IV-TR--to-DSM-5.pdf [Accessed 27 June 2013].

TABLE 7.2 International Classification of Diseases, 10th Revision (ICD-10) Diagnostic Categories for Mental and Behavioral Disorders, F00–F99[a]

Diagnostic Category	ICD-10 Code	Clinical Features
Organic, symptomatic, mental disorders	F00–09	Mental disturbance due to brain damage, toxicity, or trauma (e.g., Alzheimer's, vascular dementia)
Mental and behavioral disorders due to psychoactive substance use	F10–19	Alcohol, opiates, sedatives, cannabis, cocaine, hallucinogens, volatile substances, multiple drug use, and other psychoactive substances
Schizophrenia, schizotypal, and delusional disorders	F20–29	Delusional psychotic disorders and schizoaffective disorders
Mood (affective) disorders	F30–39	Manic, bipolar, and depressive disorders
Neurotic, stress-related, and somatoform disorders	F40–49	Phobias, anxiety disorders, obsessive compulsive disorders, stress reactions, dissociation, somatoform, and other disorders
Behavioral syndromes associated with psychological disturbances and physical factors	F50–59	Eating, sleeping, sexual, behavioral, and other disorders
Disorders of adult personality and behaviors	F60–69	A variety of specific personality disorder syndromes, including habit and impulse disorders, gender disorders, sexual preference disorders (pedophilia, voyeurism, etc.)
Intellectual disability	F70–79	Mild, moderate, severe, and profound
Disorders of psychological development	F80–89	Speech and language, scholastic, motor, and development disorders (e.g., autism)
Behavioral and emotional disorders with onset usually occurring in childhood or adolescence	F90–98	Hyperkinetic conduct, emotional and social dysfunction disorders
Unspecified mental disorders	F99	

Note: [a] The 1994 American Psychiatric Association (APA) (DSM-IV) classification of mental disorders included the major categories alcohol or drug abuse or dependence, phobias, major depression, obsessive–compulsive disorders, antisocial personality, panic disorders, cognitive impairment, schizophrenia, mania, and somatization. DSM-IV was developed in cooperation with the WHO and is close to the ICD-10 classification, modified in 2000. The new DSM-5 was released in May 2013.
Sources: American Psychiatric Association. DSM-IV: diagnostic and statistical manual of mental disorders. 4th ed. APA in cooperation with WHO; 1994. Available at: http://www.google.co.il/books?hl=iw&lr=&id=w_HajjMnjxwC&oi=fnd&pg=PP1&dq=American+Psychiatric+Association.+1994.+DSM-IV:+Diagnostic+and+Statistical+Manual+of+Mental+Disorders,+Fourth+Edition&ots=i7ST5jcJ9H&sig=R7FvZnpy7ALHhIXg5bS2wbeKINY&redir_esc=y [Accessed 17 July 2013].
American Psychiatric Association. DSM-5 development implementation and support, 2012. Available at: http://www.dsm5.org/Pages/Default.aspx [Accessed 17 July 2013].

---

universal authority for the diagnosis of psychiatric disorders. DSM-5 was published on 18 May 2013 (Box 7.2), superseding DSM-IV-TR, published in 2000. Guidance in the International Classification of Diseases, 10th revision (ICD-10) (Table 7.2), developed in coordination with the WHO, will change with ICD-11, planned for release in 2015 (Box 7.3).

Neurotic conditions can vary in severity, but they generally reflect an exaggerated response, such as anxiety or obsessive thoughts, to normal life events. These are sometimes divided into anxiety neuroses, obsessive–compulsive neuroses, hysteria, and depression. Intellectual disability and personality disorders have traditionally been considered separately from mental illness because they begin

BOX 7.3 International Classification of Diseases, 11th Revision (ICD-11)

The International Classification of Diseases (ICD) is the world's standard tool to capture mortality and morbidity data. It organizes and codes health information that is used for statistics and epidemiology, health care management, allocation of resources, monitoring and evaluation, research, primary care, prevention and treatment. It helps to provide a picture of the general health situation of countries and populations.

The 11th version is now being developed through an innovative, collaborative process. For the first time, the World Health Organization is calling on experts and users to participate in the revision process through a web-based platform. The outcome will be a classification that is based on user input and needs. It is planned for release in 2015.

Source: *World Health Organization. International classification of diseases (ICD), 11th revision. Available at: http://www.who.int/classifications/icd/revision/icd11faq/en/ [Accessed 27 June 2013].*

in early life or adolescence, whereas mental illness has a recognized onset after a period of normal functioning in adult life. However, they are increasingly seen as part of a whole spectrum of community health problems and program needs. The WHO articulates a broad intersectoral approach to preventing and managing mental illness, as seen in Box 7.4.

Historical Changes in Methods of Treatment

Traditionally, mental illness has been stigmatized with superstition, brutal management, and treatment in isolation from the community. These methods of treating the mentally ill consisted of removal from the community into long-term institutionalization, using legal and physical restraints, and various severe forms of physical shock treatment. In the late eighteenth century, pioneering reforms by Vincenzo Chiarugi in Italy, William Tuke (and the Quakers) in England, and most influentially by Philippe Pinel in France, led to stopping the practice of chaining, starving, and beating patients in mental asylums. Psychiatric asylums, however, grew to be large isolated facilities for institutional care of the insane, usually under appalling conditions, and were the standard care of the mentally ill well into the twentieth century.

In the UK during the nineteenth century, local authorities were encouraged to build mental asylums and to supervise the notorious private asylums. In the Mental Treatment Act of 1930, community psychiatric clinics were established, providing some alternative to the grimness of mental hospitalization. In 1948, UK mental hospitals came under the National Health Service (NHS).

By the 1950s, the number of mental hospital beds in the USA and many other countries equaled the number of acute care beds. Hospitalization for a mental illness was long term (often lifelong), with repeated readmissions and custodial treatment. There was little optimism for improvement or release. Therapeutic measures relied heavily on custodial care with heavy sedation, insulin, and lobotomy as common forms of therapy, which are fortunately no longer practiced. Electroconvulsive therapy (ECT) is still in use. Reduction of tertiary syphilis and other organic causes of mental illness and the development of psychotropic drugs made major changes in custodial policies possible. A British classification in the 1970s divided hospital admissions into depressive illness, schizophrenia, personality disorder, neurosis, and mania.

Mental health services based mainly on custodial care were felt to be costly, inhumane and inefficient. The large custodial mental hospitals of the past consumed significant health resources. In many countries they were the mainstay of mental health services, with over 5 beds per 1000 population and large total expenditures. They had few cures and produced much long-term damage in the form of institutionalized patients unable to return to normal society.

The growing strength of advocates for mental health reform, called the mental hygiene movement, and passage of the National Mental Health Act in 1946 in the USA instigated new directions in mental health care. The UK Mental Health Act of 1959 encouraged a rapid reduction in mental hospital beds from 152,000 in 1952 to 98,000 in 1975, and to 59,000 in the early 1990s to 18,924 in 2012. During this period, there was a rapid increase in the number of psychiatric units in general hospitals for short-term admissions.

Since the 1960s, release from institutional care and the return of large numbers of patients to the community has necessitated supportive care in the form of effective medication, follow-up services in the community, and other forms of therapy, as well as back-up hospitalization for short- or even long-term care. Psychotropic drugs do not cure mental illness but can moderate or control symptoms, enabling the person to function despite mental problems and coping difficulty and to benefit from psychotherapy. The objective of management is relief of symptoms and restoration of maximum possible function, as in other chronic conditions such as diabetes.

Where community services are inadequate, this policy can contribute to increasing numbers of homeless mentally ill people who are unable to cope in modern society. Hospital admission rates for mental health in the USA during the period 1969–1993 declined in state and county mental hospitals but increased in short-term admissions to acute care general hospitals and private mental hospitals. The bed-to-population ratio in the USA has declined (Table 7.3), as it has in most industrialized countries. The concern that mentally ill people are contributing to the increase in the homeless population in the UK and the USA is leading to a reappraisal of mental health policies.

TABLE 7.3 Mental Health Beds for 24-Hour Hospital and Residential Treatment, by Type of Organization, Rates per 100,000 Civilian Population, USA, Selected Years 1990–2008

Type of Organization	1990	2000	2004	2008
All organizations, total beds	128.5	74.8	71.2	78.6
State and county mental hospitals	40.4	21.6	19.1	12.6
Private psychiatric hospitals	18.1	9.2	9.5	8.4
Non-federal general hospital psychiatric services	21.2	14.1	13.9	17.9
Department of Veterans' Affairs medical centers	9.8	3.1	–	3.9
Residential treatment centers for emotionally disturbed children	13.9	11.7	11.4	16.5
All other organizations	25.2	15.0	17.3	19.6

Source: Health United Sates 2011. With special features on socioeconomic status and health. Available at: http://www.cdc.gov/nchs/data/hus/hus11.pdf [Accessed 28 June 2013].

Mental health care reform initiated in 1999 in the UK increased funding for specialist mental health services with emphasis on supervised community treatment. Fewer hospitalizations were needed owing to the establishment of over 700 new mental health teams in the community providing early intervention or intensive support using modern drug treatments that had previously been rationed. National patient surveys show that 77 percent of community patients rate their care as good, very good, or excellent. The UK community mental health program has been regarded by the WHO as a successful new approach, and as one of the most progressive mental health services in Europe.

Mental Health Epidemiology

Mental disorders at the diagnosable level are common in the USA, affecting an estimated one in four adults, or 57.7 million people (2004). Nearly half of the population of the USA will develop a mental illness at some point throughout their lifetime. The main burden of illness is concentrated in about 6 percent of the population who suffer from a serious mental illness. Mental disorders are the leading cause of disability in the USA for people in the age group 15–44. Many suffer from more than one mental disorder, with nearly half (45 percent) of those with any mental disorder meeting criteria for two or more disorders. Comorbidity between mental and physical illness constitutes a major issue for health services and for human suffering.

In the mid-1950s, the Midtown Manhattan study defined the population as follows: well, 19 percent; mild symptoms, 36 percent; moderate to marked symptoms, 45 percent; and severe to incapacitated, 10 percent. The estimates of "any disorder" of 23.4 percent of lifetime mental health burden was similar to later studies in the US (Insel et al 2013). Studies carried out in other western countries show rates of clinical depression of between 4.5 percent and 7.2 percent in Finland and in cities such as Athens (Greece), Canberra

(Australia), and Camberwell (UK). In 1985, mental illness accounted for 29 percent of all hospital bed occupancy in the UK and 4 percent of hospital admissions. Studies in the UK indicate that between 25 and 30 percent of patients visiting general practitioners have important or exclusively psychiatric causes for their presenting condition, even if the symptoms are primarily somatic.

The National Comorbidity Survey (NCS, 2001-2002) studied lifetime and 12-month prevalence of psychiatric disorders (DSM-III) on a national probability sample of the adult population of the USA. Nearly 50 percent of respondents reported at least one lifetime mental disorder and at least 30 percent in the previous 12 months; however, this did not distinguish between "any disorder" (estimated at over 30 percent) and "serious disorder" (i.e., serious impairment, estimated at under 10 percent). The common disorders were major depressive episodes, alcohol dependence, and social or simple phobias. More than half of all people reporting lifetime disorders had three or more disorders, accounting for 14 percent of the total sample. Less than 40 percent of those with lifetime disorders were ever treated professionally. Women had higher rates of affective and anxiety disorders; men had higher rates of substance abuse and antisocial personality disorders.

The prevalence of psychiatric disorders was higher than expected, with a high proportion not receiving professional care. These findings suggest a need for widened outreach and integration of care for psychiatric needs within general primary care. This study is being followed up by the NIMH and linked to the global initiative on the epidemiology of mental disorders in 28 countries, coordinated through the WHO. This is a landmark study which will be important for policy development in mental health in the coming decade (Weinberg et al., 2013).

There were 136 million emergency department visits in the USA in 2009, of which 4.7 million had a primary psychiatric diagnosis (National Hospital Ambulatory Medical

Care Survey 2007, National Center for Health Statistics, CDC). Among all visits resulting in admission to hospital, 4 percent were admitted to a mental health or detoxification unit. In 2009, the number of US hospital discharges with psychoses listed as a first diagnosis was 1.6 million (National Hospital Discharge Survey, 2009). The average length of stay for mental disorders was 7.5 days.

The US Centers for Disease Control and Prevention (CDC, 2007) notes that depression takes an enormous toll on functional status, productivity, and quality of life, and is associated with elevated risk of heart disease and suicide. In addition, the cost of treatment for depression is increasing dramatically in the USA. The annual economic burden of depression alone (including direct care, mortality, and morbidity costs) totaled US$83.1 billion in 2000. In 2008, the cost of direct care for serious mental illness was US$100 billion and loss of income earnings was estimated at US$194 billion (National Institute of Mental Health, 2012). The increasing burden and cost of depression and its societal costs have stimulated numerous investigations into population-based strategies to prevent the occurrence of major depression. The goal of these interventions is to encourage more effective treatment to limit the course of depression while preventing recurrence. The estimated lifetime prevalence of anxiety was estimated to be 15 percent in the USA in 2011.

Survey methods used in prevalence studies are fraught with difficulties, such as recall and other sources of bias. There are also difficulties with response rates, diagnostic criteria, and representative sampling, as well as in survey instrument design. Mental health epidemiology draws on data from national health surveys and registries of ambulatory and hospital care. Studies require collaboration among social scientists, epidemiologists, biostatisticians, anthropologists, geneticists, and other disciplines in order to increase knowledge of factors contributing to mental illness and to promote the search for causes and treatments.

In a 2007 report of an updated comorbidity study, the prevalence of any personality disorder in the USA was reported at 9.1 percent of the population. Specific prevalence rates for borderline personality disorder and antisocial personality disorder were estimated at 1.4 percent and 0.6 percent, respectively. Thirty-nine percent of respondents with a personality disorder had received treatment for problems related to mental health or substance use at some time during the previous 12 months. On average, respondents made two visits seeking mental health treatment. Even though the majority of cases were seen by a psychiatrist or another mental health professional, respondents were more likely to receive treatment from general medical providers than mental health specialists. People with personality disorders are very likely to have co-occurring major mental disorders, including anxiety disorders [e.g., panic disorder, post-traumatic stress disorder (PTSD)], mood disorders

(e.g., depression, bipolar disorder), impulse control disorders (e.g., attention deficit hyperactivity disorder), and substance abuse or dependence. The association between personality disorders and major mental disorders affects functioning and help-seeking behaviors.

A World Mental Health Survey Initiative of the WHO coordinated the implementation and analysis of general population epidemiological surveys of mental, substance use, and behavioral disorders in countries in all WHO regions. This study fielded the same instrument assessing incidence, prevalence, and correlates in 30 or so countries, and validation studies were conducted on the instrument. The survey estimated that at least 27 percent of the adult population (aged 18–65) in European Union (EU) countries had experienced "any" or "serious" mental disorders in the past year. This figure represents approximately 83 million people affected and is likely to be an underestimation, since data were not collected for individuals over 65, who constitute a high-risk group.

Western European countries have seen radical changes in psychiatric care since the 1980s and 1990s, with a process of "deinstitutionalization" that combines discharge of patients previously kept in long-term institutions with their reintegration into society. Development of community-based treatment and support services to promote independent living, vital social and employment skills, and back-up support services have been part of this process. However, reforms in former Soviet countries are in the early stages of implementation, with lack of suitable resources for needed developments, and modern mental health services remain largely out of reach for most people in the near future (Petrea 2013).

The US NIMH (http://www.nimh.gov) estimated the lifetime prevalence of depression to be 16.5 percent among US adults in 2008. An estimated 2 percent of the US population suffered from serious if not severe depressive disorder, with a much higher lifetime prevalence. Women are 70 percent more likely than men to experience depression over their lifetimes. The average age of onset for depression is 32.0 years old. An estimated 51 percent of those suffering from depression are receiving treatment. This number is substantially lower in developing nations. The 12-month prevalence of schizophrenia among Americans is 1.1 percent of the US adult population. Bipolar (manic–depressive) disorders are estimated to affect 2.6 percent of the US adult population, with 83 percent of those cases being classified as severe.

Other conditions (phobias, PTSDs, and obsessive–compulsive disorders) together affect some 30 million, and a similar number is affected by Alzheimer's disease and other brain disorders. Eating disorders probably affect millions of teenagers, while substance abuse and associated comorbidity affect millions more. With a growing trend towards Accountable Care Organizations (ACOs) in the USA, and integration of mental health into general health systems under district

TABLE 7.4 Change in Rank Order of Depression Among the Leading Causes of Disability-Adjusted Life Years, World, 1990–2020

1990 Disease or Injury	2020 Disease or Injury
1. Lower respiratory infections	1. Ischemic heart disease
2. Diarrheal diseases	2. Unipolar major depression
3. Conditions arising in perinatal period	3. Road traffic accidents
4. Unipolar major depression	4. Cerebrovascular disease

health services in other countries, cost-effectiveness in care is increasingly a topic of concern for health economists and health systems managers (see Chapters 10 and 12).

The mission of the NIMH is to conduct research on the brain, behavior, and genetics, to develop new diagnostic and treatment methods, testing them in real-world settings. It focuses on the neurosciences, especially at the molecular level. Julius Axelrod, along with Sir Bernard Katz of University College London and Ulf von Euler of the Karolinska Institute, was awarded the Nobel Prize in Medicine in 1970 for work on neurotransmitters and drug therapies for treatment of affective, neurological, and cardiovascular disorders. Their research suggested that mental states were the result of complicated physiology and brain chemistry, rather than solely the result of psychological or environmental factors. This led to an era of pharmacological drugs designed to inhibit or stimulate neurotransmitters

in the nervous system. The presynaptic reuptake of dopamine is prevented by the use of drugs known as tricyclic antidepressants. Reserpine is an antischizophrenia drug that reduces the release of neurotransmitters. Even today, antidepressant drugs rely on discoveries made by Axelrod and others. The interaction of biological and developmental factors in mental illness is the subject of continuing study, with great importance for the development of new forms of therapy and diagnostic–prognostic instruments. The shift in emphasis from inpatient to ambulatory care has been controversial but the vast scope of mental ill-health requires strengthening of community care, including residential care for the severely disabled (Box 7.4).

The WHO regards mental disorders as one of the major global public health problems, with tens of millions of cases accounting for approximately 10 percent of years of healthy life lost, or disability-adjusted life years (DALYs). The projected change in the rank order of DALYs related to depression is shown in Table 7.4 below. The WHO reports that in 2002, 154 million people globally suffered from depression and 25 million people from schizophrenia; 91 million people were affected by alcohol use disorders, and 15 million by drug use disorders. WHO indicates that in 2005, 50 million people suffered from epilepsy and 24 million from Alzheimer's disease and other dementias. Millions of others suffer from neurological disorders, with 326 million people suffering from migraine, 61 million from cerebrovascular diseases, 18 million from neurological infections or their neurological sequelae, nutritional disorders and neuropathies (352 million), and neurological effects secondary to injuries (170 million). The cross-relationship between chronic physical illness and mental illness must also be considered important in the burden of disease and disability for both types of illness (Box 7.5).

Mental ill-health is as relevant in developing as in industrialized societies, as a result of increasing life expectancy, and because of complex interactions among the biological, psychological, and social factors (e.g., aging, poverty, war and trauma, human rights violations, limited education, gender discrimination, and malnutrition).

BOX 7.5 Chronic Physical Illness and Depression

The relationship between physical chronic illness (heart disease, stroke, and cancer) and depression is complex. People suffering from depression are at higher risk for developing chronic physical illness, or depression may result from or be worsened in people with chronic diseases. Furthermore, depression may be a risk factor for poor prognosis when diagnosed with a chronic condition.

The US Centers for Disease Control and Prevention states that mental illness such as depression is associated with increased occurrence of chronic diseases such as cardiovascular disease, diabetes, obesity, asthma, epilepsy, and cancer. In addition, mental illness is associated with lower use of medical care, reduced adherence to treatment therapies for chronic diseases, and higher risks of adverse health outcomes.

Many chronic illnesses are associated with mental illnesses, and treatment of mental illnesses associated with chronic diseases can reduce the effects of both conditions and result in improved outcomes. Studies of the effectiveness of traditional therapies such as antidepressant medications in populations with cancer, heart disease, and other chronic conditions are needed.

Researchers at Emory University in Atlanta, Georgia, USA, looked at a link between depression and heart disease (Vaccarino et al., 2001). The study found an increasing number of depressive symptoms to be a negative prognostic factor for patients with heart failure. The relationship between heart disease and depression may be especially relevant to women. Women are more susceptible than men to depression, and heart disease, often perceived as a predominantly male problem, is the leading killer of women, albeit at later ages. The relationship between depression and chronic disease may not be straightforward: a recent survey of 8000 patients in the UK found disability (i.e., difficulties in activities of daily living) to be associated with depression even after adjusting for physical ill-health.

Sources: *Silver MA. Depression and heart failure: an overview of what we know and don't know. Eur J Heart Fail 2010;12:389–96.*
Vaccarino V, Kasl SV, Abramson J, Krumholz HM. Depressive symptoms and risk of functional decline and death in patients with heart failure. J Am Coll Cardiol 2001;38:199–205.
World Health Organization. Women's health. Fact sheet no. 334; November 2009. Available at: http://www.who.int/mediacentre/factsheets/fs334/en/index.html [Accessed 2 July 2012].
Meltzer H, Bebbington P, Brugha T, McManu SS, Rai D, Dennis MS, Jenkins R. Physical ill health, disability, dependence and depression: results from the 2007 national survey of psychiatric morbidity among adults in England. Disabil Health J 2012;5:102–10.
Centers for Disease Control and Prevention. US Adult Mental Illness Surveillance Report [updated 7 September 2011]. Available at: http://www.cdc.gov/Features/MentalHealthSurveillance/ [Accessed 13 May 2012].
Beresford TP. Book forum: depression and cancer; depression and diabetes; and depression and heart disease. Am J Psychiatry 2012;169:102–3.

physical illness. The burden of mental health, as measured by DALYs and social and economic cost to families and society through illness, lost productivity, and personal financial outlays, is ranked among the highest of disease entities, although the mortality rates are lower. Therefore, it is important to assess population health via indicators that reflect well-being, and not only longevity and mortality. And there are some exceptions. People with schizophrenia have substantially elevated mortality, largely due to poorly managed chronic medical illness; the proximate cause of death is, for example, heart disease, but people with schizophrenia are both more likely to have heart disease and less likely to manage it well. In addition, death rates are elevated for bipolar disorder because of the high suicide risk. A hidden burden is the social stigma and loss of human rights so commonly associated with mental illness. Rejection by society, families, employers, and even caregivers compounds the isolation, humiliation, and pain suffered by the mentally ill. Loss of earning power and independence also play major roles in a cycle of dysfunction, depression, acting out, and the burden on caregivers and health systems. Health systems, especially at the primary care level, are vital elements in maintaining those with mental illness in functioning states in the community to prevent deterioration and descent to institutional care settings such as mental hospitals or prisons.

A Global Mental Health Resources and Services survey of 184 countries, conducted by the WHO in 2011, provides the most recent estimates on available resources for the treatment and prevention of neuropsychiatric disorders, covering 98 percent of the world's population. Resources are defined in terms of governance, financing, mental health care delivery, human resources, essential medicines, and information systems. The results indicate that 60 percent of countries have a dedicated mental health policy, 71 percent have a mental health plan, and 59 percent report having dedicated mental health legislation. Median mental health expenditures per capita are US$1.63, with large variation among countries of differing income levels, ranging from US$0.20 in low-income countries to US$44.84 in high-income countries. Globally, 67 percent of these financial resources are directed towards mental hospitals. High-income countries typically report more facilities, more human resources, and higher admission and utilization rates. Three-quarters of patients admitted to mental hospitals stay for less than 1 year. Globally, nurses represented the most prevalent professional group in the mental health sector. Mental health provider, advocacy and affected family associations are present in about two-thirds of the countries, more so in higher income countries. Results from the Mental Health Atlas 2011 reinforce the urgent need to scale up resources within countries in order to meet the high and growing burden of mental health needs (Morris et al., 2012). This study helped formulate WHO's Mental Health Action Plan 2013 - 2020. Resources for mental health services are shown in Box 7.6.

The challenge of mental illness is partly one of changing priorities in health policy and systems development to include this set of conditions and give them equal priority to

BOX 7.6 Resources for Mental Health: World Health Organization Mental Health Atlas 2011

The World Health Organization *Mental Health Atlas*, published in 2011, aimed to provide the latest estimates on resources available for treatment and prevent of neuropsychiatric disorders globally. The first version of the report was published in 2005.

Key Messages from the Executive Summary

- Resources to treat and prevent mental disorders remain insufficient.
- Almost half of the world's population lives in a country where, on average, there is one psychiatrist or less to treat 200,000 people.
- Globally, spending on mental health is less than US$2 per person, per year, and less than 25 cents in developing countries.
- Resources for mental health care are inequitably distributed globally.
- Only 36 percent of people living in low-income countries are covered by mental health legislation.
- The corresponding rate for high-income countries is 92 percent.
- Dedicated mental health legislation can legally reinforce goals and policies in line with international human rights and practice standards.
- Outpatient mental health facilities are 58 times more prevalent in high-income than in low-income countries.
- User/consumer organizations are present in 83 percent of high-income countries and 49 percent of low-income countries.
- Resources for mental health care are inefficiently utilized.
- Globally, 63 percent of psychiatric beds are located in mental hospitals, and 67 percent of global mental health spending is directed towards these institutions.
- Institutional care for mental disorders may be slowly decreasing worldwide.
- Although resources remain concentrated in mental hospitals, a modest decrease in mental hospital beds was found from 2005 to 2011 at the global level and in almost every income and regional group.

Source: *World Health Organization. Mental health atlas. Geneva: WHO; 2011. Available at: http://whqlibdoc.who.int/publications/2011/9799241564359_eng.pdf [Accessed 21 May 2012].*

Mental Disorder Syndromes

Mental disorders present with a wide range of symptoms including personality change, confused thinking, abnormal anxiety, fear or suspiciousness, withdrawal from social contact, suicidal thoughts or actions, sleeplessness, change in eating patterns, outbursts of anger and hostility, alcohol or drug abuse, or simply incapability of dealing with daily activities, such as school, job, or personal needs. Mental illness includes a heterogeneous group of disorders ranging from exaggerated responses to stressful events to altered

mental activity from specific neurological or genetic abnormalities (US Department of Health and Human Resources, 1999). Chronicity is a problem that has, in the past, required long-term hospitalization. However, with improved medications and community management, hospitalization has been reduced as the primary method of treatment.

Organic Mental Syndromes

Preventive care is vital to prevent prenatal damage to the fetus from exposure to infectious diseases (e.g., syphilis, rubella, toxoplasmosis), as well as reducing exposure to toxicity from alcohol, drugs, smoking, lead, and nutritional deficiencies (e.g., iodine, iron, vitamins B, D and folic acid). Medical and public health interventions targeting low birth weight are central to health-promoting pregnancy care and have major potential for the prevention of brain damage. Screening of newborns for phenylketonuria (PKU), congenital hypothyroidism, and many genetic defects of metabolism; infants for iron deficiency; and children for blood lead levels provides important public health measures for the prevention of organic brain damage (see Chapter 6).

Organic mental disorders in older adults can produce a range of symptoms, such as decline of memory, comprehension, learning capacity, language abilities, and judgment, including the ability to think and calculate, or severe dementia. Alzheimer's disease is the prominent condition in this category, but other causes of organic origin include traumatic and toxic brain damage, strokes (cerebrovascular accidents), Parkinson's disease, alcoholism, Creutzfeldt–Jakob disease, HIV, postencephalitic disorders, syphilis and other dementias, and mental disorders due to physical brain disease. The WHO estimates that 24.3 million people suffer from such conditions internationally, with 4.6 million new cases annually. Cognitive impairment in the elderly is less than 5 percent under the age of 75, but over 40 percent for those above age 80.

Prevention of brain injury, strokes, and encephalitis due to vaccine-preventable diseases is essential, as is management of alcoholism, and adequate nutrition reduces the prevalence of the dementias. Awareness and recognition of cognitive impairment in the elderly is an important function in primary care and specialized geriatric and psychiatric services. Because of increasing longevity, organic brain syndrome cases can place a great burden on families and the health care system. Support services for families caring for relatives affected by organic brain syndromes should be part of a comprehensive health program, and should include short-term respite care, home care, and long-term care services. Research into organic brain syndromes should be of high priority because of the long-term effects on an increasingly large group of the population, and the resultant effects on the individual and the family, as well as the costs of health services for this group.

Substance Abuse

Substance abuse (mental and behavioral disorders due to psychoactive drug use) is intoxication with a substance that causes physical or psychological harm, impaired judgment, or dysfunctional behavior, leading to disability and harming interpersonal relationships. Dependent syndromes are characterized by the presence of three of the following: a compulsion to use the substance, physiological withdrawal symptoms, tolerance of its effects, preoccupation with the substance, and persistence in using the substance despite negative effects. Substance abuse is associated with death from overdosage, crime to support the habit, sexually transmitted infections (STIs), AIDS and hepatitis transmission, imprisonment, social ostracism, and long-term brain damage and with long-term stress effects on the family.

Drug dependency syndromes are estimated by the WHO to affect at least 15.3 million people. Between 100,000 and 200,000 deaths occur from overdose annually. Inhalation of volatile solvents (i.e., sniffing of glue, paint thinners, gasoline, and aerosols) among preadolescents is widespread and causes death and serious brain damage. Cannabis use is extremely widespread. New chemical formulations ("designer drugs") are appearing, resulting in brain damage and deaths. Use of opiates, cocaine, and psychotropic drugs affects all levels of society in developed and developing countries. Their use is influenced by urbanization and other social stresses and promoted by powerful economic and political interests in the international drug trade. Attempts to eliminate or control drug traffic have many similarities to international efforts to ban the slave trade in the late eighteenth and early nineteenth centuries, in that with both instances some governments covertly foster the trade, while others attempt to stop it.

The WHO defines polydrug abuse as the concurrent or sequential abuse of more than one type of drug, with dependence on at least one. This type of abuse has been increasingly reported in emergency room admissions. A 2002 DHHS survey in the USA found that 56 percent of all admissions to publicly funded drug treatment facilities in Tennessee were for multiple substance abuse conditions: 76 percent abused alcohol, 55 percent marijuana, 48 percent cocaine, 27 percent opiates, and 26 percent abused other drugs (Kedia et al., 2007).

Prevention should target vulnerable groups, especially young people, street children, and female drug users. Methadone substitution is used widely and is succesful in stabilizing drug users; more recently suboxone (a combination of buprenorphine and naloxone) is being used to stabilize drug dependent persons, in place of methadone, but this is controversial. Needle exchange programs and condom distribution along with education have been successful in reducing the spread of HIV and hepatitis B and C among intravenous drug users, but are sometimes criticized for potentially encouraging drug use. Detoxification and long-term treatment and follow-up programs are costly and frustrating, but they are better than the alternatives of disease, crime, social breakdown, and imprisonment now common in this group.

Alcohol abuse affects 120 million people internationally. The WHO estimates that the harmful use of alcohol results in 2.5 million deaths each year (WHO, 2012). Worldwide, 320,000 young people between the ages of 15 and 29 die from alcohol-related causes, resulting in 9 percent of all deaths in that age group. Alcohol abuse, chronic alcoholism, diseases of cirrhosis, cancers of various sites, and social breakdown are widespread in many countries, with resulting morbidity and mortality from trauma, violence, and child and spouse abuse. Alcohol abuse during pregnancy is associated with stillbirth, prematurity, low birth weight, and Fetal Alcohol Syndrome Disorder (FASD), which is an increasing problem worldwide and is estimated to affect 1 percent of all births in the USA (CDC, 2012). Alcohol and drug abuse continues to be a major public health concern in the USA.

In the Department of Health and Human Services (DHHS) 2005 national survey of Americans aged 12 or older, 6.6 percent (16 million people) reported heavy drinking, 22.7 percent (55 million) reported binge drinking, and 8.1 percent (19.7 million) reported taking illicit drugs within the month prior to the survey.

Strategies for reducing alcohol abuse include raising the price and reducing the availability of alcohol, especially to adolescents, setting a minimum age for alcohol purchase, and legislation and enforcement to curb driving while under the influence of alcohol. These measures should be combined with restrictions on promotion, marketing, and advertisement of alcohol; public education and awareness programs; individual counseling; group therapy; and inpatient, outpatient, and rehabilitation programs.

Schizophrenia

Schizophrenia is a group of chronic conditions with episodes of psychotic illness and delusional hallucinatory thought or behavior disorders. It usually manifests itself around the age of 20, with distorted thinking, perception, and judgment. Symptoms may include excitement, withdrawal, or a catatonic state. Men tend to manifest symptoms earlier than women. A combination of factors can predict schizophrenia in up to 80 percent of young people who are at high risk of developing the illness. Hallucinations can be visual or auditory.

The disorder appears in episodes lasting for a few months with interval periods of normality, but it is a chronic illness, with periodic need for hospitalization. It occurs in both genders and all social classes, and affects about 1 percent of the adult population. Management with medication improves the outlook for many patients.

WHO estimates that there are 45 million schizophrenics in the world, including 33 million in developing countries. In the European Region mental ill-health accounts for some 20 percent of the burden of disease and mental health problems affect one in four people at some time in their life. Nine of the 10 countries with the highest rates of suicide in the world are in the European Region (WHO, 2007).

The number of schizophrenia discharges in US short-stay non-federal hospitals for those with a first listed diagnosis of schizophrenia increased from 262,000 in 1997 to 331,000 in 2004 but declined to 266,000 in 2010. The average length of stay declined slightly between 1997 (11.9 days) and 2004 (10.8 days), increasing to 19.6 days in 2010.

In the USA, the direct cost of treatment for schizophrenia is estimated to be close to 0.5 percent of the gross national product (GNP). While recognized as biological in origin, this condition is affected in course and outcome by social and cultural conditions. Acute care and long-term follow-up require well-integrated community and hospital services. Antipsychotic drugs such as chlorpromazine, introduced in the early 1950s, greatly reduce symptoms and enable patients to function in the community, especially with family support and suitable community-based services, with periodic short-term hospitalization if needed.

Mood Disorders

Elevated or depressed mood states affect some 340 million people around the world at any given time, ranking as the fourth leading cause of the total burden of disease in developing counties. In the USA, the yearly cost of depression is estimated at US$44 billion, equivalent to the cost of care of all cardiovascular diseases (CDC, 2007). WHO projects depression as second among three leading causes of burden of disease globally in the coming decades (Table 7.4).

Mood disorders range from manic to depressed conditions, often involving both mania and depression in sequence (manic–depressive or bipolar disorders). Depression is probably the most common affective disorder, affecting approximately 5 percent of the population at any one time.

Diagnosis of depression involves four or more of the following symptoms and signs: loss of interest or pleasure in normal activities, lack of emotional response, sleep disturbances (early waking, sleeplessness, or excessive sleepiness), depression that is worse in the morning, loss of appetite, loss of 5 percent of body weight in a month, loss of libido, and a psychomotor retardation (or agitation). Severity ranges from mild to severe.

Depressive episodes may be recurrent and are a major risk factor for suicide and social breakdown. Severe depression is common in the elderly. Seasonal affective disorder (SAD) is a syndrome related to darkness during winter months and is common in northern countries with long winter nights and an indoor lifestyle. It is associated with high rates of alcohol abuse, heart disease, and suicide.

A 2005–2008 survey found that over 1 in 20 Americans aged 12 and older had current depression, and 11 percent of Americans aged 12 and over used antidepressant medications. In 2010, the number of US hospital discharges with major depressive disorder listed as the first diagnosis was 395,000, and the average length of stay for major depressive disorder was 6.5 days (National Hospital Discharge Survey, 2010). The median age of onset for mood disorders is 30 years. Depressive disorders are often associated with anxiety disorders and substance abuse.

Economic distress, unemployment, discrimination, and practices that limit women's rights are all contributors to mood disorders. Women are more commonly affected by mood disorders, with depression especially common among married women with children, related to social isolation and social devaluation of the role of the housewife. Postpartum depression can be a precursor to chronic depressive illness with a need for support and referral to prevent the condition from becoming chronic.

Antidepressant drugs provide an important advance in the management of mood disorders, but should be accompanied by professional monitoring. In 1970, lithium was approved by the US Food and Drug Administration (FDA) for treatment of manic episodes, based on NIMH research, bringing major benefits to many people suffering from bipolar disorders. Lithium and many newer antidepressant treatments are available but require close monitoring and supportive care by health providers. They help to reduce the disabling quality of mood disorders, reducing the economic burden of these conditions on society as well as the personal suffering of patients and families.

Primary care providers need to work jointly with mental health specialists in the same setting, to ease referral and consultation with patients unlikely to present to a separate mental health setting. Awareness of mood disorders among primary caregivers is of paramount importance in addressing this problem with understanding, patience, supportive therapy, and referral if chronic or leading to social breakdown at work or at home. Psychotherapeutic skills are important for primary care, but require the support of specialized services. Community recognition is needed to address the widespread prevalence of domestic violence associated with mood disorders.

Neurotic (Anxiety and Dissociative) Disorders

This group includes a wide range of symptomatology and degrees of severity, including panic disorders, phobias, obsessive–compulsive disorders, anxiety conditions, and PTSDs. Specific phobias include fear of crowds, public places, traveling, social situations, objects, animals, and enclosed spaces. Panic conditions are discrete episodes of intense fear, starting abruptly with physical symptoms that are inconsistent with the perceived threat of a specific situation or trigger event. Obsessive–compulsive disorders are repetitive, unpleasant obsessions, causing distress or interfering with normal functioning. They include compulsive behavior such as frequent hand washing, hair brushing, cleaning, and counting. Body-based repetitive behaviors, such as compulsive nail picking, skin picking, or hair pulling, are disorders involving impulse control, which are often treating using cognitive behavioral therapy (CBT). Some obsessive–compulsive patients find relief with the new generation of medications. Stress reactions may be acute or occur long after the trigger event.

Post-Traumatic Stress Disorders

PTSDs are characterized by flashbacks and dreams of the event reawakening painful memories. These disorders usually occur within 6 months of the stressful event or time, and may include manifestations of depression or other affective or behavioral disturbances. PTSD, originally described as related to combat experience in the Vietnam and Iraq wars, is now also recognized as occurring in reaction to catastrophic events, such as violence, genocide, torture, disasters, and sexual abuse.

Public and professional awareness and sensitivity are part of health system response, in preparation for psychological support as part of disaster planning or in response to catastrophic events such as hurricanes, earthquakes, terrorist bombings (e.g., World Trade Center, New York City, 2001; Madrid, 2004), or mass murder by deranged individuals (e.g., Virginia Tech, 2007; or in Denmark, 2011, mass murder of teenagers at a summer camp). Veterans constitute a large population group at high risk for PTSD. About 19 percent of Vietnam veterans experienced PTSD at some point after the war and this continues with similar prevalence among veterans of the Iraq and Afghanistan wars. The disorder also frequently occurs after violent personal assaults such as rape, mugging, or domestic violence; terrorism; natural or human-caused disasters; accidents; and possibly social deprivation such as in early child care. PTSD can develop at any age, including childhood, with a median age of onset of 23 years. Approximately 7.7 million American adults aged 18 and older, or about 3.5 percent of people in this age group in a given year, have PTSD. Some estimates suggest that over 30 million US residents have struggled with PTSD.

Dissociative (Conversion) Disorders

Dissociative (conversion) disorders involve amnesia, stupor, or trance-like conditions, which have no physical cause but are related to a specific trigger event. Preoccupation with a variety of physical symptoms, not explainable by detectable physical disorders, and refusal to accept medical reassurance is called a *somatoform disorder*. Symptoms vary but usually involve gastrointestinal, cardiovascular, dermatological, genitourinary, or pain symptoms.

Behavioral Syndromes with Physiological Disturbance

Behavioral syndromes include eating disorders, anorexia nervosa (self-starvation) and bulimia nervosa (self-induced vomiting or purging), abuse of non-addictive substances (e.g., vitamins, antacids), sleep disorders, sleepwalking, and night terrors. Sexual dysfunction syndromes include loss of sexual desire or enjoyment, sexual aversion, failure of sexual response (male and female), orgasmic dysfunction, premature ejaculation, and painful intercourse.

Prevention requires public discussion, in the media, and awareness among family and caregivers concerning teenage psychological adjustment problems and life stress situations. The potential for conversion of normal anxieties into serious, even life-threatening disorders, such as bulimia and anorexia nervosa, is especially important in adolescent care. Social norms, such as the promotion of extremely thin models, who are themselves prone to eating disorders, encourage teenage emulation. Middle-aged men, who may be under stress from insecurity of employment or loss of status, may present with sleep or sexual dysfunction and are at risk for physical disease such as premature coronary heart disease events. Unemployment, loss of marriage partner, or financial distress may trigger excessive psychological and physical responses which may be life-threatening. Primary care services need to be oriented to detect and cope with potentially serious behavior syndrome situations in vulnerable groups and provide continuing support and referral services.

Personality Disorders

Personality disorders include deviations of perception and interpretation of people and events, self-images, and affect (i.e., mood or responsiveness). They are associated with difficulty in impulse control, gratification of needs, and manner of relating to people and situations. The symptoms are persistent and inflexible, and cause distress not explainable by other mental disorders. Personality disorders range from the paranoid, or excessively suspicious, to the schizoid (i.e., emotional detachment, flattened affect, unresponsiveness, solitary life with fantasy and introspection). Emotionally unstable personality disorder involves impulsive behavior with anger and violence and difficulty maintaining a course of action not offering immediate rewards.

Borderline personality disorder includes impulsive qualities with disturbed self-image, emotional crises, threats or acts of self-harm, and chronic feelings of emptiness. The health system has a role in recognition and support of this group of disorders as it does for any significant physical illness; the consequences of such conditions can be serious for the individual, the family, and society. Early recognition may help to initiate professional care and support to help affected individuals through critical life periods. Mutual and family support groups with professional assistance can be of value.

Disorders of Psychological Development

Psychological developmental disorders include some level of impairment of speech, language, sound categorization, visual perception, attention, and activity control. They include childhood autism; that is, abnormal social communication, attachments, and play appearing before the age of three, along with lack of spontaneity and social and emotional reciprocity, and failure to develop according to the mental age of peers. Early recognition and referral for specialist help require awareness and cooperation between education and health systems.

Autism spectrum disorders (ASDs) are a group of developmental conditions characterized by impaired social interaction and communication. Symptoms include restricted, repetitive and stereotyped patterns of behavior typically seen in the first few years of life and is life-long. ASD includes three subtypes of: autistic disorders, Asperger syndrome, and other unspecified developmental disorders. There is marked variability in the pattern and severity of symptoms. ASD incidence in the United States is estimated at about one percent of children, similar to estimates for other industrialized countries.

Behavioral and Emotional Disorders of Childhood and Adolescence

Hyperkinetic disorders include inattention, overactivity, and impulsivity. They include a variety of attention disorders such as attention deficit disorder (ADD) and attention deficit hyperactivity disorder (ADHD). Conduct disorders are characterized by aggressive behavior, repetitive behavior with temper tantrums, lying, stealing, use of dangerous weapons, and other unacceptable behavior.

Estimates suggest that 10–12 percent of children and adolescents suffer from mental disorders, including autism, hyperactivity, depression, developmental delay, behavior disorders, and emotional disturbances. A high percentage of children and adolescents who suffer from dysfunctional due to mental disorders do not receive appropriate therapy.

Intermittent explosive disorder (IED, listed in DSM-IV) is a highly prevalent and seriously impairing disorder beginning in late childhood and continuing to adolescence and later ages. Anger attacks in adolescents are common and often associated with heavy drinking of alcohol and outbursts of violence, which can be devastating for victim and the attacker as well, such as in weekend teen violence, road rage, and motor vehicle crashes.

Suicide

Suicidal behavior is a major public health issue. Suicide is the most serious outcome of mental disorders. In the USA suicide is the ninth leading cause of death among men, and the second leading cause of death among 15–24-year-olds. The total of completed suicides in 2011 was 38,285 or 12.3 suicides per 100,000 population (National Vital Statistics Reports, 2012).

According to the latest WHO Mortality Database, for the age group 15–19 suicide accounted for 9.1 percent of the 132,423 deaths reported from 90 of 130 member states. In these 90 countries, suicide was the fourth leading cause of death for males and the third for young females (see Chapter 6).

The global death toll from suicide – almost 1 million people per year – accounts for half of all violent death worldwide, and estimates suggest that self-inflicted fatalities could rise to 1.5 million by 2020. Suicide rates fluctuate and are closely associated with economic trends. Suicide rates in Western Europe (including Sweden) are low; those in eastern Europe were high in the 1980s and 1990s but falling rapidly since the mid 1990s (HFA Data Base, 2013). Rates are low in Latin America, Muslim countries, and some Asian nations, according to the Violence Prevention Alliance of WHO. More men than women complete suicide attempts, with the exception of rural China and some parts of India.

Suicide is a behavior that is probably the end result of the interaction of a complex involving biological, genetic, and environmental risk factors. It associated with severe mental, physical, financial, and emotional stress for family members, friends, society, and public resources, in terms of caring for people who have made suicide attempts.

An emotional crisis usually precedes suicide attempts and is often recognizable and treatable. Although most depressed people are not suicidal, most suicidal people are depressed. Serious depression can be manifested in obvious sadness, but often it is rather expressed as a loss of pleasure or withdrawal from usually enjoyed activities. One can help to prevent suicide through early recognition and treatment of depression and other psychiatric illnesses. Suicide is also a high risk among individuals with borderline personality disorder. Most people (90 percent) who die by suicide have a diagnosable and treatable psychiatric disorder at the time of their deaths, according to "psychological autopsy" studies with the survivors of suicide decedents, who are asked about the decedent's mental state in the period preceding suicide, although this could be subject to recall bias. Provision of easily accessible specialist and family support services is needed at crucial times in the history of a disorder or disease that can lead to emotional breakdown or suicide.

There are other high-risk groups for suicide, including older (white) men in the USA; and it varies across countries. In the USA, suicide attempts are 10–30 times more common than suicide deaths; and women are more likely to attempt, but less likely to die, than men. Some of this is due to differences in disease and suicidal intent, but some of it is due to differences in means, e.g., men in the USA are more likely than women to use guns.

School systems and health care providers must be especially alert to the potential for teenage suicide, while elderly care programs need to be alert to the potential for suicide among live alone and isolated elderly men. Suicidal behavior has been studied especially for trends, international variations, and underlying risk factors (e.g., aggression, impulsivity, suicidal intent), which remain unclear. Reducing the increasing trend of suicide rates among the most vulnerable populations will require further research and development of preventive initiatives. Primary care systems need to be sensitized to the dangers associated with depression and suicide risk within the mental health needs of a population they serve.

Learning Disabilities

Learning disabilities are defined by the US National Institute for Neurological Disorders and Stroke as disorders which affect the ability to understand or use spoken or written language, perform mathematical calculations, coordinate movements, or sustain direct attention. Such disabilities occur in very young children, but they may be recognized only when the child reaches school age. This requires special education with specially trained educators able to carry out diagnostic educational evaluation assessing the child's academic and intellectual potential and level of academic performance. Then, the basic approach is to teach learning skills by building on the child's abilities and strengths, while correcting and compensating for disabilities and weaknesses. Other professional services may be needed, including speech and language therapists, psychological therapy and medications to enhance attention and concentration, with wide degrees of severity from an isolated and mild learning difficulty to complex multiple problems.

The US DHHS reports mixed progress and regression on some health targets for Healthy People 2010, including expansion of primary care facilities providing mental health treatment, juvenile justice residential facilities that provide mental health screening, and state plans addressing mental health for elderly people. Three objectives moved away from their targets: suicide, adolescent suicide attempts, and homeless people with mental health problems who received services. Disordered eating behaviors among adolescents in grades 9–12 and state tracking of consumer satisfaction demonstrated no change towards or away from their targets.

Controversies in Mental Health Policies

During the 1960s, with the availability of improved medications for mental disorders, the population of mental hospitals and traditional psychiatric hospitals fell in the USA, Canada, and the UK. Canada reduced its psychiatric bed capacity by 32,000 beds while increasing general hospital psychiatric beds. The general hospital units tended to treat mild psychiatric conditions, such as mild depression, while resources were not put into needed community-based programs to serve the more severely disabled.

Inadequate follow-up in the community increases the risk of the psychiatric patient suffering exacerbation of symptoms, despondency, homelessness, alienation, imprisonment, and suicide. Community mental health programs should include case management, rehabilitation, housing programs, and other support services. Well-organized systems are required to avoid poor coordination, losing patients to follow-up, lack of accountability, ineffective case management, high readmission rates, and inadequate linkages between hospital and community care systems.

Reviews of mental health programs in the past several decades have raised a number of issues, including concern for the civil rights of the mentally ill, the destructive effects of long-term institutional care, the availability of knowledge from the neurosciences, new medications that enable treatment in the community, and the high cost of institutional care. The search for evidence-based medical practice, managed care systems, and capitation payment for mental health care are part of the continuing debate on mental health policy.

Mental health policy promoting reduced hospitalization and increased community care has come under much criticism. The major concern is that there has been a low level of investment in community-based services and residential help for the mentally ill, so they live in poverty and neglect, in nursing homes, in prisons, or on the streets. The "revolving door" or "dumping" of patients from one kind of institution to another or to hospital emergency departments is a dramatic social issue in every country. This phenomenon is a reminder of the need to provide sustainable support systems for the mentally ill in the community and not in institutions, but with institutional access and professional support when needed. It is not sufficient to seek transfer of funds from other health or social programs. This issue will continue with no prospect of an early solution except for a complex of well-coordinated service systems meeting the multiple needs of this major public health problem.

In 2012, the Strategic Mental Health Commission of Canada reported that some 30 percent of short- and long-term disability claims were related to mental health problems, with associated costs of US$6 billion in lost productivity. Many are in the criminal justice system. Follow-up by family physicians is much less than for patients with physical illness. Some studies indicate that 74 percent of homeless people have mental health problems and the percentage of people in federal prisons with mental health problems has increased by 60–70 percent since 1997.

Community-Oriented Mental Health

Advances in drug therapy, together with concern for the negative effects of institutionalization and the low cure rates of large psychiatric hospitals, prompted the development of community-oriented mental health (COMH) programs. During the 1960s, this model for delivery of community mental health services took the form of community mental health centers, but more recently there has been a trend towards increasing the involvement of primary care providers in mental health care.

Mental health services should be integral parts of other health and social services in the community. Patients may seek help at any one of the services, and appropriate care of the patient requires intervention at multiple service levels. This requires the development of intersectoral cooperation in health services and staff training in the elements of community mental health needs and services (Box 7.7).

COMH is a programmatic approach providing links between primary health care, hospitals, and long-term services, and includes social support, rehabilitation, and preventive services. These services need to be developed and function as a network in order to provide the individual and the community with the appropriate level of care needed at a particular point or particular stage of an illness.

The development of COMH may require different approaches in a metropolitan area as compared to a small city or rural area. Services should be tailored to the setting and be comprehensive in scope to help the patient in acute need as well as one in need of long-term support services, preferably within the community. Crises in the form of patient or family breakdown may require short-term hospitalization in a community hospital setting, as an integral part of the service.

Community mental health involves primary care providers willing and able to recognize and manage mental health problems. It requires back-up services of psychiatrists, psychologists, social workers, and perhaps paraprofessional community health workers trained to serve the mentally ill. This is particularly important in minority groups or others with special needs. Active educational and organizational efforts are an essential link between traditional mental health services and primary care.

Primary care physicians and other health personnel must be trained and oriented in the major mental dysfunctional conditions, such as mood and anxiety symptoms and disorders. Primary care providers and managed care programs need to assess the needs of people in vulnerable at-risk groups, such as menopausal women and people approaching retirement, as well as supply modalities of care that cannot be provided by mental health services alone. Traditional preventive and treatment services and back-up mental health services function in an interdependent manner. Building a community-oriented network approach to mental health may require finding new methods of collaboration, with multidisciplinary staffing either in the same setting or with different services in close coordination with one another.

BOX 7.7 Community Mental Health Policy

- Target at-risk populations, including a commitment to care for the significantly disabled in the community to the maximum extent possible.
- General hospital psychiatric units linked to community support, with greater diversification to include holding beds (e.g., 48-hour observation), day beds, crisis intervention, respite care services, and detoxification units.
- Psychiatric hospitals provide for major psychiatric and behavior disorders (e.g., dementia, brain damage, paranoid, and severely regressive schizophrenia); the supply of such beds is continuing to be reduced, but this requires well-funded and well-coordinated community-based services.
- Continuity of care is essential to ensure adequate maintenance of the chronically mentally ill with highly individualized care, needs assessment, planning, and monitoring.
- Attention to comorbidity (e.g., drug abuse with underlying psychopathology).
- Integration of mental health services with linkage to primary and other health care.
- Consumerism and advocacy groups help to define the needs of the mentally ill in peer and family support, self-help groups, income maintenance, retraining and job placement, adequate housing, and social support.
- Replace dependency with independence-promoting programs including group homes, training and employment.
- Psychiatric epidemiology provides greater evidence of prevalence of mental ill-health and can compare the costs and benefits of different treatment approaches.
- Community orientation includes outreach programs, supported housing, home services, patient and family support services, with attention to the cultural and organizational needs of minority and other special needs groups.
- Case management with multidisciplinary health care teams, with specialist and hospital back-up.
- Long-stay accommodation in group residences with therapeutic services in the community in home-like units, day care.

Source: *World Health Organization. The WHO mental health policy and service guidance package, 2013. Available at: http://www.who.int/mental_health/policy/essentialpackage1/en/ [Accessed 28 June 2013].*

Prevention and Health Promotion

Prevention requires working with at-risk groups and with groups with common problems where self-help or a support group may be the most effective form of therapy. This will encompass a wide range of resources that take into account the interaction of social, physical, and mental health issues.

Identification of people at risk for breakdown, such as suicide and violence, requires high levels of awareness by teachers, doctors, police, social workers, military personnel, employers, and the general public. Support groups take many forms, including Alcoholics Anonymous, Al-Anon for family members of an alcoholic, drug rehabilitation programs, overeater and binge eater groups, Schizophrenics Anonymous, National Association of the Mentally Ill (NAMI) for families of the mentally ill, and bereavement groups.

There are benefits to early detection and intervention. People with a first psychotic episode respond to dosages of antipsychotic medication that is one-quarter to one-half the strength of that needed by people with established schizophrenia (and antipsychotic medications can have nasty side-effects). Coordination is required between primary care systems and mental health services.

Primary prevention may be the most productive approach but is still not well defined. Ideally, programs will be designed to reduce the possibility that mental dysfunction will proceed to more severe forms in which serious functional breakdown occurs. Mental health promotion across the lifespan can mediate positive health outcomes for people in scarce-resource contexts. Given the potential to break the intergenerational cycle of poverty and mental ill-health and promote human and broader socioeconomic development, especially in resource-poor countries, mental health promotion can no longer be ignored in these contexts. This is a challenge that requires support and action across several sectors. The close connections between mental health and other aspects of health and productivity mean that promoting mental health is a necessity in both low- and high-income countries.

Social factors are vital in community mental health practices. Prevention in mental health involves many aspects of patient/client care on the primary, secondary, and tertiary levels. Health promotion in the mental health field involves encouraging public awareness of a healthy lifestyle, including activity, rest, recreation, and socializing, and the dangers of substance abuse. Specific programs for activity and socialization should be targeted at the young and elderly. Secondary prevention involves early detection and supportive treatment, and tertiary prevention requires the management of long-term mental illness with adequate community support systems.

Primary prevention includes prevention of nutritional deficiencies (i.e., iodine, iron, and vitamins), reduction of environmental contamination (e.g., lead), reducing social and educational deprivation (e.g., education and recreation for youth in high crime/drug abuse neighborhoods), promoting toddler and preschool enrichment programs, promoting family support systems (e.g., for single-parent families), providing vocational and employment assistance (e.g., for teenage mothers), providing social support for the aged and disabled (e.g., social security), targeting family and social violence (e.g., domestic, anti-female, and school violence),

and raising public and professional awareness and cooperation (e.g., education and health systems).

Secondary prevention requires crisis intervention by trained professionals or paraprofessionals at the primary care level, social relief and support for abuse victims (e.g., shelters); early diagnosis of mental health disorders including screening, early referral to adequate care by police, courts, schools, hospitals, armed forces, and employers, effective treatment and follow-up including outreach and use of paraprofessionals in COMH teams; and increased supply and accessibility of crisis intervention services at the community level, use of existing facilities such as hospitals for detoxification and crisis intervention, defining treatment goals and quality of care, and ensuring continuity of follow-up.

Tertiary prevention requires maintaining contact with clients to monitor medications, mood, function, social relations; providing support, referral, and assistance; assuring continuity in care and follow-up; prisons and the streets are not acceptable residences for those who should be housed in health care facilities. Community services should provide: facilitating client contact with social support networks; protecting/rehabilitating the patient/client, family, and community regarding the effects of the disorder (e.g., drug addiction); providing support groups and structured rehabilitation; funding community-based group transitional residences, halfway houses; promoting independence and self-support; and training and providing continuing education to caregivers, clients, families, and the community.

Mental health research is opening new approaches to management of clinical psychiatric conditions, along with a wider recognition in the community that mental illnesses are real and treatable conditions. Psychotherapeutic medications and psychotherapies are effective, helping people with mental illness by correcting abnormal brain function. More research is needed to further define the relationship between the brain and behavior. New knowledge also helps to establish the role of genetic and environmental factors in shaping brain and behavioral function. Great progress being made in the fields of neurosciences and genetics is helping biomedical scientists to study the normal and pathological function of the brain and devise new methods of treating mental disorders. This will bring new hope to millions of people and their families suffering the debilitating effects of mental illness.

Intellectual Disability

Intellectual disability is defined as an individual having impairment of general mental abilities that impact adaptive function in three or more areas e.g. coping abilities. It describes a range of disabilities, including:

- intellectual impairment: low levels of intelligence as measured by developmental tests [intelligence quotient (IQ)]

- learning disability or dysfunction: specific learning disorders, such as dyslexia, which are unrelated to intelligence, requiring specialist assessment and educational support
- mental disability: graded by social maladaptation, which relates to acceptance and accommodation of individuals with disabilities in society.

The term *mental disability* includes a wide range of conditions and their associated causes. Among them are organic neurological impairment, genetic endowment, and developmental or educational deprivation. Other factors that greatly influence the social adjustment of the mentally disabled include structure and orientation of services, professional attitudes and training, employment and training practices, social and cultural expectations, family and kinship structures, as well as legislation in relationship to health, welfare, education, and employment.

Many conditions resulting in mental disability are preventable. Prevention can be categorized into the following three areas.

- Prevention before conception includes good health habits for the future mother with nutrition, fitness, and other preparation. This includes taking folic acid tablets (along with fortification of flour), iron if anemic; screening and genetic counseling for genetic disorders as indicated by risk groups such as ethnicity, intrafamilial marriage, and previous birth defects, multiple spontaneous abortions or infant deaths; and smoking cessation and abstaining from alcohol and drug use. In addition, possible occupational, toxic, radiation or chemical or other environmental hazards should be taken into account. Immunization for rubella, and screening for HIV/TORCH (toxoplasma, rubella, cytomegalovirus, herpes), and STIs are part of preparation for pregnancy, as is spacing between pregnancies (see Chapter 6).
- Preventive processes during fetal life and birth include prevention targeted at the nutritional status of pregnant women, including folic acid and iron supplements, prevention of iron-deficiency and macrocytic anemia, and iodine deficiency before and during pregnancy; minimizing harm to the fetus by eliminating alcohol intake, drugs, and smoking during pregnancy; screening to detect fetal anomalies associated with intellectual disability; good prenatal and obstetric care to avoid fetal damage, anoxia, and birth trauma; good neonatal care to avoid anoxia and vitamin K deficiency-induced intracranial hemorrhages; and fetal monitoring in labor.
- Preventive processes after birth include vaccination against communicable diseases that may cause brain damage; screening for PKU, congenital hypothyroidism and other inborn errors of metabolism, and followed by case management; good infancy care and nutrition to avoid nutritional and emotional deprivation that can reduce psychomotor development; and strategies to reduce accidents and their impact.

BOX 7.8 Intellectual Disability: Decision-Making Rights

Epidemiological studies worldwide estimate the prevalence rate of intellectual disabilities (ID) to be 1–3 percent among the general population. Individuals with ID need specialized, integrated treatment within the health, education, and social welfare sectors with respect to care and decision making. According to Article 12 of the Convention on the Rights of Persons with Disabilities, all individuals should have the right to legal capacity. In order to exercise this basic human right of autonomy, these individuals must be allowed to make their own decisions and communicate these decisions to others. Services should support decision-making approaches to promote change in the way in which families, professionals, service providers, employers, and the general community perceive and act in reference to people with ID to respect these rights.

Sources: *United Nations General Assembly. Convention on the Rights of Persons with Disabilities A/RES/61/106: Resolution (24 January 2007). Available at: http://www.unhcr.org/refworld/docid/45f973632.html [Accessed 6 June 2012].*
Werner S. Individuals with intellectual disabilities: a review of the literature on decision-making since the Convention on the Rights of People with Disabilities (CRPD). Public Health Rev 2012;34. Available at: www.publichealthreviews.eu [Accessed 15 August 2012].

Better knowledge of nutrition and risk factors in pregnancy and technological developments are providing improvements in such areas as antenatal detection, treatment of fetal abnormalities, and genetic risk identification, so that intellectual disability from preventable causes can be reduced in frequency. Down syndrome can be reduced in terms of new cases by family planning and reduced late maternal age pregnancies and use of currently available screening methods of amniocentesis. People with Down syndrome can also achieve higher levels of learning and social adaptation than was thought possible even a decade ago.

The adaptation of mentally disabled people to living and functioning in the community is part of rehabilitation. Political, community, and parental acceptance and support are essential to establish and operate successful community group living and working arrangements for this population group. Public health can address issues relating to adaptation in non-institutional surroundings within the community. This requires advocacy and education among employers and the general public, and active promotion by the health community. The rights of people with intellectual disability to take decisions are shown in Box 7.8.

ORAL HEALTH

Oral health as defined by the WHO is "a state of being free from chronic mouth and facial pain, oral and throat cancer, oral sores, birth defects such as cleft lip and palate, periodontal (gum) disease, tooth decay and tooth loss, and other diseases and disorders that affect the oral cavity. Risk factors for oral diseases include unhealthy diet, tobacco use, harmful alcohol use, and poor oral hygiene" (WHO, 2008).

The term *stomatology* is used for dental health in some countries, derived from *stoma*, the Greek word for mouth. Oral health is an important element of general health status, affecting everyone. Poor dental health can cause pain and interfere with nutrition. Oral disease can progress to loss of teeth and costly treatment, yet much is preventable. In extreme cases, dental illness can result in osteomyelitis, brain abscesses, systemic infection, and death. Public health has a key role to play in oral health because it has important preventive methods available to it. These include health education for oral hygiene, fluoridation of community water supplies, and availability and assurance of quality standards in the dental and paradental professions.

Ancient societies faced dental problems and sought explanations and treatments based on practical experience, with explanations that "nematodes" grew in teeth. Egyptian mummies show evidence of dental treatment. Hippocrates and later Ibn Sinna (Avicenna) discussed hygiene for prevention of dental disease.

At the end of the seventeenth century dental care separated into its own field. Pierre Fauchard, a French surgeon (1678–1761), described 130 diseases of the mouth and teeth and is considered the father of modern dentistry. In 1736, artificial golden dental caps were first used. Silver amalgam was first used for filling cavities in 1819. Horace Wells introduced nitrous oxide for dental anesthesia in 1844, which, along with ether, was quickly adopted by the medical community for general surgery. Special cements for filling cavities were developed in 1855. The dental drill was invented in 1870. Four years later, fluoride was found to prevent tooth decay.

In 1942, the US Dental Health Officer reported on studies in 13 cities which determined a relationship between fluoride use and reduced caries, but with excessive levels of fluoride causing fluorosis. In 1946, a study was carried out simultaneously in Grand Rapids, Michigan, Kingston, New York, and Brantford, Ontario, whereby fluoride was administered through the community water systems. The results were compared to cities wherein fluoride was not dispensed. The study showed that fluoridated drinking water reduced caries by 48–78 percent. By 1999, 144 million Americans were using fluoridated community water, with fluoride levels adjusted to the optimal 0.7–1.2 parts per million. The CDC considers fluoridation to be one of the great public health achievements in the USA in the twentieth century.

In developing countries, as standards of living rise, increased use of sugars causes deterioration in children's dental health. Use of sugar in water or tea as a baby pacifier is common in some parts of the world and should be discouraged. Education of parents to prevent this should be part of health promotion.

The major issues in dental public health are dental caries, periodontal disease, malocclusion, and oral cancer. All of these contribute to loss of dentition with effects on general health status. The important interventions to reduce caries and periodontal disease are fluoridation of community water supplies, education in oral hygiene, reduction of sugar intake in children, regular dental care, and dental sealants. Feeding babies sugared water or tea and allowing babies to go to sleep with a bottle in the mouth can cause serious tooth decay. Self-care should include regular toothbrushing after meals, dental flossing, and preventive dental check-ups. These three aspects of proper dental health should be part of school and home health education.

The emerging allied dental profession of dental therapy shows promise to improve quality of care. Dental therapists receive specialized training in treating the teeth of children and adults, performing local anesthesia, restorations, cleaning, and radiography. Clinical outcomes worldwide indicate that dental therapists offer safe, effective dental care to children. In 2012, the W.K. Kellogg Foundation reviewed over 1100 reports regarding dental therapists and their work in various countries. According to the review, dental therapists can effectively expand access to dental care, especially for children, and the care they provide is technically competent, safe, and effective. The review indicated widespread public support for the role of dental therapists. In the USA, dental therapists practice in Alaska and Minnesota, but there is movement in other states to use these providers to expand access to needed dental care. Canada, New Zealand, Australia, and the Netherlands employ dental therapists in their oral health workforce. Adding dental therapists to the oral health care teams may be a new reality in dental health care. Numerous federal reports have recommended exploring using midlevel providers as a way to solve the current dental access crisis.

Fluoridation

Fluoridation of community water supplies reduces the incidence of caries and number of extractions in both children and adults by some 60 percent. Fluoride is naturally present in most water supplies. When adjusted to a level of 1 part per million, it prevents dental decay. Fluoridation is one of the most effective public health interventions available. It must be a controlled procedure because excessive fluoride in drinking water can cause fluorosis, which results in staining and fragility of the teeth. Good public health practice should therefore ensure adequate fluoride levels to reduce dental caries but avoid excessive levels of fluoridation.

Other methods of giving fluoride include tablets, rinses, and toothpaste. Salt and milk may also be enriched with fluoride. Fluoride tablets are a useful preventive adjunct to child health in areas without adequate fluoride in the water, but the most cost-effective method is by fluoridation of water supplies. The cost of fluoridation of water supplies depends on the size of the community served, but in the USA this varies from US$0.12–0.21 per person per year in communities of over 200,000 population to US$0.60–5.41 for smaller communities.

Fluoridation has been the subject of political and emotional controversy over many decades and continues to be so, with organized opposition in many countries. This has interfered with the institution of this preventive health measure in many locations, but over time fluoridation is being more widely implemented; it has been adopted by most of the major cities in the USA.

Reduction in the prevalence of dental caries has been progressing over the past several decades in countries where fluoridation has been adopted on a wide scale. Worldwide, approximately 370 million people consume drinking water with optimal levels of fluoride. Health education, fluoride tablets or rinses when fluoridation is not available, and access to dentists for assessment, cleaning, and treatment are also an essential part of public health. More recently, topical application of fluoride and the use of plastic dental sealants, developed in the 1970s, to reduce exposure of pits and fissures in teeth to caries-causing organisms has been shown to be highly effective.

The CDC has been the lead agency in promoting fluoridation since 1975. In 2011, CDC recommended reducing the recommended fluoride level from a range of 0.7–1.0 parts per million (ppm) to 0.7 ppm in order to reduce the possibility of excess fluoride from multiple sources, including natural levels in water and toothpaste, with consequent dental fluorosis or tooth staining. In 2012, the CDC announced that 73.0 percent of Americans on public water systems have fluoridated water (approximately 204 million people), compared to 67 percent in 2002. For US communities, every dollar spent on community water fluoridation results in a saving of US$38.00 in costs to repair a decayed tooth. Healthy People 2020 calls for an increase from the current 74 percent to 80 percent population coverage in the USA by 2020.

The WHO considers fluoridation of water supplies as the most effective public health measure for the prevention of dental decay. It requires a multidisciplinary team of engineers, chemists, physicians, nutritionists, and dentists. The efficiency of fluoridation is proven to be safe and cost-effective. Acceptability to the communities can be controversial, and antifluoridation as an organized movement has been very effective in slowing and even stopping governments and local authorities from employing this measure which is most beneficial to the poor part of a population. Even free dental care is not a substitute for continuous fluoride supply in community water systems. Nonetheless, free access to dental care is itself important to protect the dental and oral health of young and old people in the population.

The CDC considers community water fluoridation to be a safe and effective method of preventing tooth decay, and includes fluoridation among its selection of the 10 great achievements of public health of the twentieth century (CDC, 2007). Opposition to fluoridation remains very active and prevents or even reverses fluoridation of community water supplies, which causes much suffering to the poor who have less access to other methods of regular fluoride supplementation. Regrettably, some countries in Europe, for example the Netherlands, have banned fluoridation. In Israel, mandatory fluoridation was introduced a decade ago and supported by the Supreme Court when challenged, with great benefit shown in surveys of school children. However, a new minister of health appointed in 2013 immediately announced cancellation of mandatory fluoridation despite the strong opposition of her senior departmental staff and academics in the field.

Periodontal Disease

Disease of the oral tissues that support the teeth is described as periodontal disease, a bacterial infection that destroys the attachment fibers and supporting bone that hold the teeth in the mouth. Left untreated, these diseases can lead to tooth loss. The accompanying inflammation of the gums or gingival tissue extends to the periodontal ligament and causes the loss of the supporting bone. If control is not imminent, teeth become loose and must be extracted. The major factor in the development of periodontal disease is poor oral hygiene, especially plaque formation.

Research suggests that up to 30 percent of the population may be genetically susceptible to gum disease. Despite aggressive oral care habits, these people may be six times more likely to develop periodontal disease. Identifying these people with a genetic test before they even show signs of the disease and early intervention may help them to keep their teeth for a lifetime. Prevention of this disease should include such measures as tooth-brushing and flossing, along with use of antiplaque rinses. These measures improve oral hygiene and are protective against chronic infection. Chronic infections may have systemic effects, including possible association with cardiovascular diseases.

Oral health is a major issue among the elderly, who may have poor or no dentition, which adversely affects nutrition, as well as localized gum disease. Eating fruit, vegetables, and meat may be difficult. With aging of populations this will be an increasing health issue, especially where fluoridation has not been implemented.

Dental Care

In 2010, an estimated US$108 billion was spent on dental services in the USA. Expenditure for dental services increased from less than US$2 billion in 1960 to US$70.3 billion in 2002, a more than 35-fold increase. Each year, Americans make about 500 million visits to dentists. The US Centers for Medicare and Medicaid Services (formerly the Health Care Financing Administration) have projected the level of expenditure to 2013 to US$126.3 billion, an average annual increase of 5.5 percent since 2002. Expenditure per capita for dental services increased from US$10.86 in 1960 to US$244.20 in 2002, an average annual increase of 7.7 percent. However, the percentage share of dental care in total health expenditure declined from 7.1 percent in 1960 to 5.2 percent in 1980 and to 4.3 percent in 2004.

With the dramatic decline in dental caries in the child population of industrialized countries since the 1960s, there has been a tendency towards increased use of costly dental procedures. At the same time there has been a reassessment of the numbers of dentists required, whereas in the 1960s there was a continuous chorus of demand to increase the size and number of dental schools in western countries. Dental care is costly, and oversupply of personnel does not necessarily lower prices by competition.

The supply of dentists is affected by demand for services, and between 1985 and 1996 the number of dental schools in the USA declined from 60 to 53; the number of dental school enrollments of new students in the USA declined from 6132 in 1960 to 4612 in 2005, and remained static at 4688 in 2010.

Visits to dentists for preventive care, cleaning, or restorative work is limited by economic factors and personnel shortages in many countries. In the USA, rates of annual dental visits for people over age 25 increased from 54 percent in 1963 to 61 percent in 1993; in 2010, this figure was also 61 percent. Tooth decay affects more than one-quarter of US children aged 2–5 years and half of those aged 12–15 years. Advanced gum disease affects 4–12 percent of US adults. One-quarter of US adults aged 65 or older have lost all of their teeth (CDC, 2011).

Many countries with universal health plans do not include dental care as a covered benefit, largely because of its high cost. The UK NHS includes dental care, although most adults pay for dental check-ups and treatment, and conducts periodic surveys of dental health status by age groups and regions of the country, with steady improvement in indicators among children and young adults especially (NHS, 2011). Dental care is not required to be covered by the Canadian government national health insurance system.

Dental care programs in schools and through community organizations should include education, fluoride rinse programs where the water supply is inadequately fluoridated, and regular dental check-ups and care. The application of dental sealants to children's teeth after eruption of the first (age 6–8) and second molars (age 12–14) is a safe and highly effective means of preventing caries and should be used especially in areas without fluoridation.

In countries with dental health as a part of the national health program, such as the UK, giving priority to prevention by fluoridation of community water supplies and school fluoride rinse programs supplemented by dental sealants could reduce the need for restorative dentistry and allow dental human resources and costs to be reduced. Dental nurses providing treatment, oral hygiene, and education in school-based programs have been successfully used in New Zealand and parts of Canada, bringing dental care to children who would otherwise go unchecked and overcoming inaccessibility due to the costs of private dentistry.

Oral Cancer

In the USA, more than 7800 people, mostly elderly, die from oral and pharyngeal cancers each year and 36,500 new cases of oral cancer are diagnosed yearly (CDC, 2011). Control of oral cancer is primarily through early detection, as well as educating the public and training professional dentists and other health providers with regard to its epidemiology and presenting symptoms.

All suspicious lesions should be biopsied, especially in high-risk patients such as males above the age of 40 who smoke and/or drink heavily. Health promotion emphasizing the elimination of risk factors is most important. Risk factors include smoking (especially pipe and cigar smokers), heavy drinking of alcohol, use of chewing tobacco, and poor oral hygiene.

Human papillomavirus (HPV) causes oral cancers and is a risk for boys and for girls who may acquire the infection and cancer from oral sex with HPV-infected boys. Thus, the HPV vaccine is now recommended for boys as well as for girls.

PHYSICAL DISABILITY AND REHABILITATION

The United Nations (UN) estimates that there are 650 million people living with disabilities in the world, with 80 percent of them in developing countries. They constitute some 10 percent of any given country's population, the equivalent in size of the population 65 years of age and over in developed countries, but they receive much less attention in health service and social service systems. Despite some progress, only 45 countries have disability discrimination laws. The UN and WHO have undertaken promotion of political and professional awareness of the scope of the problem and unmet needs in the area, with a stress on prevention and rehabilitation.

A 2004 US survey found that only 35 percent of working-age people with disabilities are in fact working compared to 78 percent of those without disabilities. The numbers of adults and their percentages among the adult population in the USA in 2010 are shown in Box 7.9.

BOX 7.9 Disability and Functioning Among Non-Institutionalized Adults (18 Years and Older), USA, 2010

- Adults with hearing trouble: 37.1 million (16.2 percent of total)
- Adults with vision trouble: 21.5 million (9.4 percent)
- Adults unable (or able with great difficultly) to walk a quarter of a mile: 16.7 million (7.3 percent)
- Adults with any physical functioning difficulty: 35.8 million (15.6 percent)
- Adults with difficulty in at least one basic action or complex activity limitation: 73.7 million (32.8 percent).

Source: *US National Center for Health Statistics. Summary health statistics for US adults: National Health Interview Survey, 2010. Vital and health statistics. Series 10, No. 252. Available at: http://www.cdc.gov/nchs/data/series/sr_10/sr10_252.pdf [Accessed 28 June 2013].*

Many people with disabilities need long-term care with support services and devices. Such needs are only partially met in a hospital environment. Most care will be delivered by families within the community. Professional health care providers such as physiotherapists, occupational and speech therapists, and medical and nursing personnel should be part of the support services. The provision of high-quality rehabilitation services in a community should include the following:

- conducting a full assessment of people with disabilities and suitable support systems
- establishing a clear care plan
- providing measures and services to deliver the care plan.

Owing to the wide range of disabling conditions, this field is naturally extensive and complex. In most countries, services and legislation for individuals with disabilities have developed in a piecemeal fashion, often as the result of lobbying by social activists and reformers, or by organized groups of individuals living with disabilities.

The rights of individuals with disabilities were addressed internationally in a series of international declarations including the 1975 UN Declaration of the Rights of Disabled Persons, WHO's 1981 International Year of Disabled Persons, and the UN's Decade of Disabled Persons 1983–1992. Reviews of legislation and programs for individuals with disabilities in member countries of the WHO European Region set a benchmark in progress in the field. The UN Convention on the Rights of Persons with Disabilities declared its 2006 mission as follows: "to promote, protect and ensure the full and equal enjoyment of all human rights and fundamental freedoms by all persons with disabilities, and to promote respect for their inherent dignity". The Convention, in effect since 2008, had 153 signatories in 2012. This Convention is the first human rights treaty negotiated in the twenty-first century by the UN. This binding treaty addresses issues for disabled people that have been seriously neglected internationally and in most countries.

The Convention defines individuals with disabilities as:

"those who have long-term physical, mental, intellectual or sensory impairments which in interaction with various barriers may hinder their full and effective participation in society on an equal basis with others. The Guiding Principles of the Convention include promotion of measures to address needs of disabled people with:

- *respect for inherent dignity, individual autonomy including the freedom to make one's own choices, and independence of persons*
- *non-discrimination*
- *full and effective participation and inclusion in society*
- *respect for difference and acceptance of persons with disabilities as part of human diversity and humanity*
- *equality of opportunity*
- *accessibility*
- *equality between men and women*
- *respect for the evolving capacities of children with disabilities and respect for the right of children with disabilities to preserve their identities."*

(UN Convention on the Rights of Persons with Disabilities, 2008)

The Convention recognizes that prevention of disability is one of the most essential elements of a comprehensive approach, incorporating the classic public health definitions: primary prevention (i.e., stopping the disease or injury from occurring at all), secondary prevention (i.e., if the event occurs, stopping or reducing the severity of complications which lead to disability), and tertiary prevention (i.e., restoring the injured person to maximum feasible physical, psychological, and social functioning). Prevention of physical and mental disabilities from birth injury, prematurity, failure to thrive, fetal alcohol syndrome (FAS), and work, road, and other accidents is among the most important issues for public health to address.

Many countries have developed a comprehensive range of laws, activities, and programs for individuals with disabilities that include prevention of disability, reduction of the severity of complications, and physical, social, and employment rehabilitation, as well as support or compensation systems. In most countries there are difficulties coordinating policies, services, and support systems. The disabilities commonly dealt with in most countries include those due to medical, accidental, or occupational ill-health or injury: veterans and civilian war victims; those who are hard of hearing, blind, mentally disabled, or mentally ill, and others.

In 2011, the WHO jointly with the World Bank released a World Report on Disability. This report indicates that more than 1 billion people in the world today suffer from disability and this is growing as populations are aging and with increasing prevalence of chronic health conditions, such as diabetes, cardiovascular diseases, and mental disorders. People with disabilities have generally poorer health, lower

BOX 7.10 Principles of a National Public Health Program for Disabilities

- Review mainstream and disability-specific policies, systems, and services to identify gaps and barriers and to plan actions to overcome them.
- Develop a national disability strategy and action plan.
- Ensure political commitment of national, state, local governments.
- Encourage multisectoral involvement of relevant government agencies, NGOs, media, professional, judiciary, church, academic, private sector.
- Review and revise existing legislation and policies needed for legislative, regulatory, and moral persuasion, compliance and enforcement mechanisms.
- Comprehensiveness of approach.
- Prevention-oriented approach.
- Restoration-oriented approach for physical, psychological, and social function.
- Promote full participation and equality of disabled people in society.
- Promote acceptance of individuals with disabilities in society.
- Promote community versus institutional care.
- Reallocate resources and expenditures to community-based services.
- Local service responsibility.
- National coordination, support, and financing.
- Encourage participation of individuals with disabilities in decision making.

- Promote education, employment, housing, and support services.
- Define legal, financial, and social rights.
- Enable social benefits and compensation without litigation.
- Establish clear lines of responsibility and mechanisms for coordination, monitoring, and reporting across sectors.
- Regulate service provision by introducing service standards and by monitoring and enforcing compliance.
- Allocate adequate resources to existing publicly funded services and fund the implementation of the national disability strategy and plan of action.
- Adopt national accessibility standards and ensure compliance in new buildings, in transport, and in information and communication.
- Introduce measures to ensure that people with disabilities are protected from poverty and benefit adequately from mainstream poverty alleviation programmes.
- Include disability in national data collection systems and disability-disaggregated data wherever possible.
- Implement communication campaigns to increase public knowledge and understanding of disability.
- Establish channels for people with disabilities and third parties to lodge complaints on human rights issues and laws that are not implemented or enforced.

Source: *Adapted from World Health Organization. World report on disability. WHO and World Bank; 2011. Available at: http://whqlibdoc.who.int/publications/2011/9789240685215_eng.pdf [Accessed 28 June 2012].*

educational achievements, fewer economic opportunities, and higher rates of poverty than people without disabilities. There is a lack of services available to them and they face many obstacles in their everyday lives. The report provides the best available evidence about what works to overcome barriers to health care, rehabilitation, education, employment, and support services, and to create environments that enable people with disabilities to access and benefit from a normal lifestyle to their maximum possibilities, and to flourish. The report ends with a concrete set of recommended actions for governments and their partners (WHO Disability Report, 2011). Box 7.10 presents key issues in a national public health program for disabilities.

National programs in addition to medical, rehabilitative, and preventive health services include income maintenance and compensation, legal rights, normal and special education, vocational training, work in assisted or normal settings, housing and the physical environment, communication and transport, leisure and sport, and self-help groups.

Prevention of motor vehicle accidents, occupational hazards, and toxic exposure is an essential part of a national public health agenda. Every year, 255,000 people in the 25 EU member states die as a result of an accident or violence (2009 EU Report). Annually, more than 60 million

people receive medical treatment for an injury, of which an estimated 6.8 million are admitted to hospital. Injury presents the fourth major cause of death in Europe, after cardiovascular diseases, cancer, and respiratory diseases. In children and adolescents, it is the number one killer.

The UK stresses not only extensive social security legislation and support systems, but also the vital role of prevention. This ranges from adequate nutrition and vitamin/mineral enrichment of basic foods to smoking reduction, prevention of low birth weight and perinatal morbidity/mortality, infectious disease control, workers' health and safety, and accident prevention. Legislation and compensation for mental health and disability have also been given attention. The NHS has moved towards decentralized management through district health systems with national support. Organization of services for individuals with disabilities has evolved in this direction as well, reflecting a return to the principles of the local administration of the Elizabethan Poor Laws.

The Danish program for individuals with disabilities has evolved through social legislation over the past 100 years, especially the Public Assistance Act of 1933. A wide range of national legislation and programs has been implemented for individuals with disabilities. As in the UK, the Danish trend has also been to return responsibility for management of social and health programs from central government to the county and local governmental levels, with support and service back-up from national organizations and the national government.

The Finnish low birth weight rate of 4 percent, down from 5 percent over the past 10 years, is due in part to the 99 percent attendance rate at maternal and child centers and other social benefits. Finnish pregnant women average 12 prenatal and postnatal visits. A maternity benefit of US$125 is paid only to those who attend prenatal care before the end of the fourth month of pregnancy.

France has an impressively comprehensive maternal health program which also emphasizes prevention. As an example, there is much stress on prevention of low birth weight and associated morbidity by requiring early prenatal care for maternity benefits and 20 obligatory examinations for children from birth up to the age of 6 years, including nine during the first year of life. Screening for metabolic disorders at birth is now augmented by screening for thalassemia, hemophilia, and sickle-cell disease. Great importance is placed on integration of children with disabilities in regular schools and the social integration of disabled adults. Income support systems are tailored to promote these objectives. For example, in order to qualify for birth and child financial benefits, prenatal and infant visits to preventive care services are obligatory in France.

In many countries such as Russia, individuals with disabilities have traditionally been isolated from the mainstream of society. With over 13 million disabled people (2006), only 15 percent of disabled people in Russia are involved in any kind of productive work or business, compared to 35 percent in the USA and 80 percent in China. The responsibility for individuals with disabilities in Russia is shared jointly by the Ministry of Labor and Social Affairs and the Ministry of Health. The major issues in disability prevention relate to the high rates of binge drinking, violence, suicide, homicide, poisonings, and industrial and road accidents, with associated injury, disability, and mortality. FAS is a serious national problem due to widespread drinking among young women including during pregnancy.

Progress is being made in developing national programs in prevention, care, and services for individuals with disabilities. Sharing experiences between countries can help in defining how health and social services can be more effectively arranged both to prevent disability and to integrate the people with disabilities into society. Review of legislation and service organizations as well as the incorporation of new medical, therapeutic, and technological innovations are part of that process.

SPECIAL GROUP HEALTH NEEDS

In every society there are groups of people with special health needs. Indeed, a society is often judged on how it cares for its native peoples, its prisoners, its homeless, and others. They may be set aside from the mainstream of the society as a whole by historic, ethnic, legal, or economic circumstance. They may require special attention to deal with their health problems because they are more dependent, more vulnerable, or less able to access services based on traditional medical practice, or simply because their needs are greater or more specialized than those of the general population. The New Public Health has an advocacy and pioneering role here, as it does for special needs of the whole population.

Gay and Lesbian Health

Gay, lesbian, or bisexual orientation was in the past considered a mental disorder, but was removed from this category by the American Psychiatric Association in 1973, and is estimated to include some 10 percent of the population of the USA. The rights of gay men and women are still severely limited in some cultures and religions, but have gained growing acceptance in many secular and civil societies, although gains are made slowly, such as in the right to same-sex marriage, child adoption, parental rights, and spousal inheritance rights.

Accurate rates are difficult to determine: societal stigmatization of homosexuality may lead to underreporting, whereas non-random convenience studies (e.g., patients at STI clinics) may overestimate the true prevalence. Furthermore, sexual orientation may not always correlate with sexual behavior (e.g., 70 percent of gay men report having sex with married men and 45 percent of lesbians report having sex with men). Regardless of the exact figures, gay men, lesbians, and bisexuals represent a significant proportion of the population. In heterogeneous societies, such as the USA, the gay/lesbian population is part of the diversity of the population at large, in terms of ethnicity, religion, socioeconomic status, and geography. Thus, health care needs deriving from homosexual behavior must be addressed in an appropriate social context.

Specific health problems related to same-sex behavior include high incidence of STIs, gastrointestinal infections, and hepatocellular and anal cancer among gay men, related to oral–anal, oral–genital, and receptive sexual intercourse. Health problems of particular concern to lesbians include high incidence of breast, ovarian, and endometrial cancer related to low rates of parity, breastfeeding, and oral contraceptive use. Lesbians tend to visit gynecologists less frequently, and may therefore not receive important tests such as Papanicolaou (Pap) smears or mammograms. Anti-gay violence, discrimination, and stigmatization is another serious health and social issue for both gays and lesbians. Rates of smoking, alcoholism, substance abuse, depression, suicide, and cardiovascular disease are also higher in gay populations, all possibly related to the stress of living in a homophobic society.

The lack of preventive services and early treatment due to stigmatizing experiences with the health care establishment is a significant issue. Studies have shown that only 10–40 percent of primary care physicians routinely take a sexual history from a new adult patient and only 30 percent feel comfortable with gay patients. Over 60 percent of Gay and Lesbian Medical Association members surveyed felt that homosexuals who revealed their sexual orientation risked substandard care as a result. The overall outcome is a failure to screen, diagnose, and treat important medical problems of gay and lesbian patients.

The AIDS epidemic has brought sexual behavior issues to the forefront, especially for the homosexual population. However, with only an estimated 10 percent of US students receiving comprehensive sex education and fewer than 5 percent of primary care physicians routinely obtaining information regarding high-risk activities, gay and lesbian organizations have often had to fill the gap. Significant successes have been recorded (e.g., widespread adoption of condom use among gay men in the 1980s), but educational programs face enormous ongoing challenges.

One target group of particular concern is the gay and lesbian adolescent population. According to the American Academy of Pediatrics, these teens are severely hindered by societal stigmatization and prejudice, limited knowledge of human sexuality, a need for secrecy, a lack of opportunity for open socialization, and limited communication with healthy role models. As gay and lesbian adolescents are subject to overt rejection and harassment at the hands of family members, peers, school officials, and other community members, they may seek, but not find, understanding and acceptance. Such rejection may lead to isolation, runaway behavior, homelessness, domestic violence, depression, suicide, substance abuse, and professional failures. Heterosexual or homosexual promiscuity may occur, including involvement in prostitution (often by runaway youths) as a means to survive.

Thus, with the new generation as with the previous one, ostracism by society leads to those very behaviors that put both homosexuals and heterosexuals at increased risk for social isolation and the spread of diseases, including hepatitis B and AIDS. Breaking this vicious cycle will require a concentrated multidisciplinary effort, in which public health workers, primary care providers, and gay and lesbian organizations all have pivotal advocacy roles to play. Although there has been notable change in social attitudes to gay and lesbian people, which may reduce barriers to preventive and curative services, exposure to sexually transmitted diseases is still a significant health risk due to unsafe sexual lifestyles, particularly related to multiple partners and failure to use condoms and other contraceptives. Gay rights movements strongly advocate societal change such as same-sex marriages, including the right of same-sex couples to adopt children. Legal rights such as inheritance for marriage partners and parental benefits after adoption of children are also important issues. In some religious faiths and communities rejection of this movement is very strong; however, the past decade has seen enormous gains for this movement. In 2010, US President Barack Obama signed into law a bill that allows gay and lesbian Americans to serve openly in the military. In 2013, the US Supreme Court ruled that same-sex couples are entitled to federal benefits and declined a case from California against same-sex marriage law. This, in effect, legalizes same-sex marriage in states that allow it. Same-sex marriages are now allowed in many countries, including Argentina, Belgium, Brazil, Canada, Denmark, France, Iceland, the Netherlands, Norway, Portugal, Spain, South Africa, and Sweden.

Native Peoples' Health

The USA, Canada, Australia, and other countries provide health care for native (indigenous) populations as special categories under direct provision of care by federal government agencies. These populations are often segregated from the general population, with different schools, legal systems, and health care systems.

The Canadian Indian population in the sixteenth century was approximately 222,000 people, but disease and famine accompanying European immigration reduced this figure to 102,000 by 1867 (the year of Canada's confederation). By 1941, the number of aboriginal people was between 100,000 and 122,000, but by 1988 had reached 443,884. Population growth in aboriginal communities, with high birth and falling mortality rates, has resulted in an increase in the Indian population of Canada. According to the 2006 Canadian Census, there was a total of 1,172,785 aboriginal people in Canada, including First Nations people, Métis (part Indian), Inuit (Eskimo), and others, comprising 3.8 percent of the Canadian population (Statistics Canada, Canada at a Glance, 2012).

New wealth has benefited many reserves. They have shopping centers, gambling casinos, hotels, and other successful enterprises servicing nearby white communities. However, this has not reduced the flow to the cities, where native people are subject to severe stresses with alcohol and drug abuse, child neglect, family abuse and violence, and many acute and chronic diseases. The rate of heart disease remains 1.5 times higher among aboriginal peoples, type 2 diabetes is three to five times higher among first nations, and TB rates are eight to 10 times higher (Health Canada, 2008).

Health status indicators among Canada's native peoples show widespread alcoholism in the form of binge drinking among all age groups from children and adolescents upward. Glue sniffing, diabetes, hypertension, and hospitalization rates are about twice the Canadian averages. Over the past several decades, there has been a decline in mortality rates among Canadian Indian and Inuit people, but the gap with the overall population remains high.

A nutrition survey carried out in 1973 and subsequent follow-up studies showed marked levels of micronutrient deficiency conditions among Canadian Indians and Inuit, with inadequate intakes of vitamins A and D, iron, and calcium, and high levels of iron-deficiency anemia and rickets.

In 1990, the relative mortality rates compared to the total population varied from 5:1 in infancy to 1.1:1 for those over the age of 65. In 2000, the age-standardized mortality rate for the general population in Western Canada was 2.4 per 1000 population, less than half the rate for the Registered First Nations population (5.3 per 1000). In other words, the First Nations population experienced approximately three more deaths per 1000 population than it would have experienced if it had the same age-specific mortality rates as the general Canadian population. The major causes of death were, in order of decreasing frequency, injury and poisoning, diseases of the circulatory system, cancer, and diseases of the respiratory system. Suicide rates for those aged 15–24 were five to six times the general population rates for this age group.

Infant mortality in the Canadian aboriginal population in 1997 was twice the national average, and life expectancy at birth was 8 years less than the national level. However, there have been modest improvements: in 2001, life expectancy for the Registered First Nations population was lower than that of the general Canadian population by 6.6 years for males and 6.5 years for females. During the period 1980–2001, life expectancy for aboriginal peoples rose by approximately 10 years for males and 8 years for females. During this period, life expectancy gains in the general Canadian population were smaller. Chronic illness such as diabetes, cardiovascular disease, cancer, and ESRD are all far more common among native peoples than among the general population. The same pattern is seen for infectious diseases including TB, STIs, and hepatitis. Poverty, physical violence, and alcohol and drug abuse are widely prevalent.

The health of Indian and Inuit peoples comes under direct federal responsibility, but is also part of hospital and medical services operated by each province under Canada's universal health insurance program. Services not insured under provincial health plans are provided by the federal department, Health Canada. Recent trends to decentralize management of services to the tribal council have been matched by proposals for constitutional change making aboriginal communities recognized levels of government. The social problems of unemployment, poor education, alcohol abuse, family breakdown, violence, and welfare syndromes plague Canadian Indians who move to the cities, perhaps more so than those who remain on the reserves.

In the USA, wars against the Native Americans in the nineteenth century were followed by forcible concentration of these populations on poor land reservations and lethal administrative practices far harsher than the Canadian record. However, provision of care and prevention of social breakdown may have been even less effective in Canada than in the USA in the latter part of the twentieth century.

The history of US government provision of health care to Native Americans and Alaskan natives is equally troubling. From the early nineteenth century, care for Native Americans was provided by military doctors up to 1849, when it began being provided by the Bureau of Indian Affairs as part of the Department of the Interior, before being moved to the US Public Health Service in 1954. The Indian Health Service (IHS), formed as a national health service for Native Americans, is administered through regional offices, providing primary and specialty care with increasing emphasis on community participation and control.

The US population of Native Americans and Eskimos in 1996 totaled 2.3 million people and is expected to reach 4.3 million and just over 1 percent of the population by 2050 (US Census Bureau, 2008). The IHS provides care to approximately 60 percent of this population. Despite gains in health status, such as reduced infant and general mortality, native peoples remain in poorer health than the general US population. Poor nutrition, unsafe water supplies, isolation, poor transportation, inadequate waste disposal, and high rates of obesity, alcohol abuse, and violence shorten life expectancy. The IHS operates a network of 43 hospitals and over 110 health centers and health stations. Expenditures for Native American health care, however, are 60–65 percent of per capita national average expenditures.

Life expectancy for Native Americans improved from about 60 years in 1950 to 75.1 years in 1989–2011. Improved infant mortality (9.2 per 1000 live births in 2008) and a reduction in deaths due to TB, gastroenteritis, and acute respiratory disease have been accompanied by a reduction in alcohol- and violence-related mortality. Rates of diabetes (type 2) are high and increasing, compared to the non-native population. In 2008, the major causes of death in the Native American population were heart disease, malignant neoplasms, and unintentional injuries. Fetal Alcohol Syndrome (FAS) is a large problem for Native Americans, with an estimated incidence of 2.7 per 1000 live births, a higher rate than that for any other racial or ethnic group in the USA. Increasing tribal management of health services has been accompanied by increasing tribal activity to reduce alcohol abuse and violence.

The aboriginal populations, devastated initially by famine and acute infectious disease and currently by alcohol abuse, violence, and diabetes-related conditions, have a health status that remains well below the health of both the Canadian and American general populations. Federal administration and separateness from general health services have been subject to much criticism in both countries. Integration with other health services and decentralization of management with community involvement are current trends. Federal financing of this service has stabilized, while increasing local management leads to increased expenditures by this population group in the private medical market.

The prospects of the native populations of the USA, Canada, and elsewhere will depend on social and economic development, with a large measure of self-government. In 1999, the Canadian government established a self-governing Inuit (Eskimo) territory which has authority over health, social services, education, taxation and economic rights (e.g., minerals), and many other areas accorded to Canadian provinces. This will provide an important test for this concept.

Similar problems appear among aboriginal populations and minority groups in other countries, including Australia, New Zealand, and Peru and other South American countries. Aboriginal populations in Papua New Guinea, Taiwan, central Africa, Russia, and Australia suffer from high rates of cardiovascular disease, diabetes, ESRD, alcoholism, glue sniffing and drug abuse, rheumatic fever, infant and child mortality, and general social breakdown. The poor health of native peoples is a blight on the records of many countries, with no clearly defined health responsibility, authority, or programming available for this population. The issue of health of aboriginal peoples is a serious challenge to the New Public Health in finding ways to provide adequate preventive and curative care, and more importantly to reduce social gaps engendering apathy and social decay.

Prisoners' Health

Since the mid-1980s, there have been extraordinary changes in the condition of jails and the health of prison inmates in the USA. The increase in urban decay, illicit drug use, poverty and associated epidemics has had a great impact on incarcerated Americans. Prison medical services have been transformed into outposts faced with the challenge of meeting near-impossible demands.

Imprisonment grew in the USA from 1.6 million prisoners in 1997 to 2.3 million people held in federal or state prisons or in local jails in December 2006 (Bureau of Justice Statistics). This figure represents an increase of 2.9 percent from 2005, but that was less than the average annual growth of 3.4 percent since 1995. Incarceration rates rose from 411 prison inmates per 100,000 US residents at the end of 1995 to 501 at the end of 2006. There has been some improvement. In 2010, the incarceration rate decreased to 242 per 100,000 US residents, a percentage change of −2.5 percent since 2008. During 2010, the number of prisoners declined by 1.3 percent (91,700 offenders) to 7.1 million at the year end. The average annual percentage change from 2000 to 2009 was 1.9 percent, while that from 2009 to 2010 was −1.1 percent.

Incarceration rates show an increase in the number of women in state or federal prisons by 4.5 percent from 2005, reaching 112,498, while the number of men rose by 2.7 percent, totaling 1,458,363 in 2006. In 2010, black men were incarcerated at a rate of 3074 per 100,000 residents; Latinos were incarcerated at 1258 per 100,000, and white men were incarcerated at 459 per 100,000. Incarceration rates have fallen slightly in the USA for the first time since 1972. In 2000, 41.9 percent of US inmates were African American, but the percentage had declined to 37.8 percent in 2010. Most prisoners are young and poorly educated.

In 1970, concern for the poor state of prisoner health led the American Medical Association and the American Public Health Association to conduct reviews of prison health services, recommending guidelines which provided state prison services with standards to meet and resulting in improved services in many states. Prison health services took on a more professional role, with less reliance on part-time practitioners. Voluntary accreditation of services by the American Council on Health Services serves as a valuable external review procedure promoting transparency in the quality of care.

People in correctional institutions are at increased risk for TB because of the high prevalence of HIV and hepatitis B infections and latent TB, overcrowding, poor ventilation, and the frequent transfer of inmates within and between institutions. The recent emergence of multidrug-resistant TB as an important opportunistic infection of HIV-infected people underscores the need for improvement of infection control practices. The increase in TB in Russia since 1990 is in part related to the large-scale release of prisoners following perestroika in the 1980s. Specific control measures in correctional facilities should include the following:

- regular and systematic screening of inmates and staff for HIV and TB: those who test positive should be eligible for preventive therapy
- rapid identification, isolation, and treatment of suspected cases of TB
- directly observed therapy as well as rigorous follow-up and record-keeping to ensure treatment completion
- follow-up to ensure continuity of care both inside and outside the correctional facility.

Prisoners are also at increased risk for STIs. Both male and female prisoners may be subject to physical or sexual assault or intimidation by fellow prisoners and staff. Prison medical services should have screening and treatment capacity as well as powers to protect vulnerable prisoners, such as segregation of young prisoners from potentially violent or long-term inmates.

In the 1990s, US federal legislation mandating life sentences for repeat offenders ("three strikes" law) increased the number of elderly prisoners ineligible for parole. As larger numbers of inmates spend longer periods in prison, there is a need to regard prisoners' health as part of that of society as a whole.

TABLE 7.5 Comparative Prevalence of Psychiatric Disorders in Prisoners and in People Living in the Community

Disorder	Prevalence (%)	
	In Prisoners	In Community
Any psychiatric disorder	80	31
Psychosis	7	0.7
Affective disorder	23	9
Anxiety disorder	38	11
Substance abuse disorder	66	18
Personality disorder	43	9

Source: Butler T, Allnutt S, Cain D, Owens D, Muller C. Mental disorder in the New South Wales prisoner population. Aust N Z J Psychiatry 2005;39:407–13.

Short- and long-term health risks for prison staff require the attention of prison health services. Alcoholism, smoking, obesity, and job stress are linked to family breakdown and early mortality from cardiovascular disease, as well as other chronic conditions. Helping prison staff to deal with the stress and latent violence on the job should work to reduce personal risk and propensity to staff-initiated violence.

The need for mental health services in prison is great and growing. The deficiency of community mental health services has accelerated the movement of individuals from the street to the penitentiary. A study by the National Association for Mental Health found that 25 percent of prisoners admitted regularly experienced psychotic episodes, with another 14 percent manifesting some psychotic symptoms. Mental illness may be responsible for a large part of violent behavior and violent crime. There is a need to combine jail treatment facilities with rehabilitation programs. A study in New Zealand compared the prevalence of mental disorders among prisoners with people living in the community, as shown in Table 7.5.

Ethical issues in prisoner health include torture, rape, murder, starvation, unethical medical experimentation, and inadequate medical and psychiatric care. The Nuremberg and Helsinki codes of medical conduct in medical experimentation are internationally accepted standards. The Geneva Convention standards relating to war and military occupation, while often ignored, are also accepted and are still a basis for judgment. Abuse of prisoners in civil war situations, such as in Bosnia in the mid-1990s, reached genocidal levels with massacres, starvation, and rape applied widely. The health community in any country or conflict situation has a professional obligation to monitor the care of prisoners and play an advocacy role in times of peace, civil war, and genocide.

Migrant Population Health

In many countries, large numbers of people and families move with seasons for farm or other temporary work. They lack stable social environments and support systems and are often dependent on exploitative employers willing to provide only subsistence wages, and a possibly hostile surrounding community. This group includes migrant and seasonal farm workers, miners, foresters, and construction workers who often live away from their families for extensive periods. There are an estimated 1 billion migrants in the world today, of whom 214 million are international migrants and 740 million are internal migrants (WHO, 2012).

In the USA, this phenomenon is widespread. The Migrant Health Program of the US Department of Health and Human Services (DHHS) estimated the size of the population of immigrants (legal and illegal) living in the USA in 2005 to be 35.2 million. The majority are married with children, so the total size of the population including dependants is much larger. Most of these families are living below the poverty line. Many are illegal immigrants, subject to exploitation and abuse by employers or contractors. They are exposed to poor sanitation and housing conditions, pesticides, and infectious diseases. The Health Resources and Service Administration of the DHHS has established HRSA-funded health centers, which served 862,775 migrant or seasonal farm workers and their families in 2010.

Migratory workers in the USA are of mixed ethnic origin, including white Americans, Hispanics, Haitians, and Jamaicans. Some immigrant groups, such as Haitians, Mexicans, and Filipinos, have high rates of TB, and their work in agriculture may lead to disease transmission to other farm workers. Migrant farm workers are approximately six times more likely to develop TB than the general population of employed adults.

Migrant farm workers may bring their families with them to poor housing encampments operated by their employers with unsanitary conditions and limited access to food and other necessary supplies. Arrangements for health care, schooling, and recreation are likely to be inadequate. Migrant workers in urban settings are frequently single men, who may have left families at home and who migrate from declining rural regions in the same country or are brought into a foreign country for specific work in agriculture, construction, or mining. Women are often brought in for work as housekeepers, personal caregivers, or sex workers. Professional migrant workers in nursing medicine, computers, and other professions are also common in many parts of the world. People living in a foreign culture as contract workers are subject to abuse and exploitation as well as social isolation with associated mental stress, violence, and STIs.

Migrant workers are more likely to live in poverty, suffer unsanitary and crowded living conditions, and lack medical care. In addition, they generally lack the necessary documentation for establishment of legal residency status. Access to health insurance and regular medical care, especially for children and women, is often one of the most pressing issues, especially for illegal immigrants. Nutritional problems of iron deficiency, stunted growth, and micronutrient deficiencies are common. Dental disease is common in migrant families, and access to dental care is usually lacking. Injury and deaths from work accidents are high among farm workers. Agricultural workers have one of the highest mortality rates (37 per 100,000 agricultural workers) from work-related injuries of any occupational group (see Chapter 9).

The high prevalence of syphilis, HIV infection, and TB among migrant workers underscores the need for public health professionals who are trained to respond to health care needs within the migrant worker population. These studies led to development of cross-training for public health workers on STIs (including HIV infection), TB, and other communicable diseases among migrant farm workers.

Migration for work or to seek a better quality of life is an international phenomenon. In Europe, large numbers of "guest workers" provide unskilled labor in agriculture and the building trades, primarily to the wealthy economies which over several decades imported workers from poor countries in north and sub-Saharan Africa or Eastern Europe. This has created long-term social and political problems for workers and their families, who are often subject to ethnic resentment, violence, and culture shock. Public health agencies should supervise living conditions and arrange preventive and treatment services with screening for common diseases in this population group.

Migrants are a vulnerable group with many health needs, ranging from infectious diseases and chronic diseases to mental health problems. They may face discrimination, language difficulties, and cultural barriers, and work at menial jobs in agriculture, building trades, and caregiving. They leave their homes, moving from rural to urban centers in their own countries or to new countries to improve their economic conditions, sending money home to help poor families and hoping to stay or return with savings to re-establish themselves in a better condition. This process can be rewarding, tragic, or fall somewhere in between. Some conditions are akin to slavery, such as the trafficking of women for sex work. Exposure to disease and abuse is common.

Addressing the health needs of migrant populations is in the self-interest of host countries to prevent the spread of infectious diseases, such as HIV/AIDS and multidrug-resistant TB, usually acquired in the host country. Migrants are an important population group requiring special attention in the New Public Health.

Homeless Population Health

Homelessness due to poverty and lack of low-rental permanent housing has become a common problem in urban centers in industrialized countries and a long-standing issue in developing countries. It is increasingly capturing attention in the media because of the widening gap between the rich and the poor in society. The UN Commission for Human Rights estimates that there are approximately 100 million homeless people worldwide, with 1.6 billion inadequately housed (2005). An estimate suggests that as much as 7.4 percent of the US population has experienced homelessness at some time in their lives; however, this number is likely to be higher due to the effect of millions of home foreclosures in the late 2000s. Long-term street dwellers are at great risk for health problems.

Health care, especially preventive, is not a priority for homeless people. Instead, many of their resources, such as time and money, go towards securing a safe, warm, and dependable place to sleep, towards purchasing food, and often towards drugs and alcohol.

The global pandemic of homelessness is the result of the lack of affordable housing, unemployment, and the deterioration of social services and support. Homeless people do not choose to live on the streets; rather, they are often victims of circumstance. It is often argued that homelessness is the result of deinstitutionalization from mental institutions, and some estimates of the percentage of homeless individuals suffering from mental illness range from 25 to 50 percent. It is difficult to gauge whether homelessness causes mental illness or vice versa, but it is clear that the two combine to produce a vicious cycle in which individuals face social isolation, inadequate nutrition, and hygiene barriers to housing and employment, which together serve to perpetuate illness and insecurity. Released prisoners and the mentally ill are overrepresented in the homeless population, but this is also an issue among US veterans.

Surveys of health problems faced by the homeless show that rates of both chronic and acute diseases are extremely high. With the exception of stroke, cancer, and obesity, homeless people are far more likely to suffer from every category of chronic health complication. Because of the transient nature of the population, conditions that require vigilant care and attention, such as TB, are rampant. This, of course, has grave dangers for the public at large. Along with prisons, homeless shelters account for a large percentage of TB cases in the industrialized countries. Other health problems include death by freezing or violence, frostbite, leg ulcers, burns, respiratory infections, STIs including HIV, and trauma from muggings, beatings, and rapes. Homelessness precludes good nutrition, shelter, warmth, security, personal hygiene, and basic first aid. The homeless have poor access to care until medical conditions become dire.

As the members of this group are often uninsured, access to health care services is compromised by cost and distance. One of the most effective means of serving this population is through mobile health units. In Washington,

DC, among other US cities, "health vans", coupled with mobile "soup kitchens", operated by charitable organizations, provide nutritional and medical services to the homeless. If conditions are severe enough to warrant more substantial care, the individual is encouraged to enter a city or private homeless facility, or may be referred for medical care. These vans provide medical care, a hot shower, a place to stay the night, and hot meals. The role of non-governmental organizations (NGOs) and charitable organizations is important in helping the homeless to survive. A host of NGOs offers services to this population, but the process of obtaining these may be difficult to navigate, frustrating, or too complex. Even when it is feasible for the homeless to access these services, they are only temporary solutions to the millions of homeless people worldwide living in dangerous conditions.

Homelessness of female-headed families with young children may occur from family breakdown due to an abusive spouse, the mother's use of drugs and alcohol, pregnancy, migration, or the collapse of an economic base of the family with eviction from a stable home. Homeless runaway children and teenagers are vulnerable to serious social and health consequences, including violence, sexual abuse, substance abuse, suicide, and high-risk behavior, resulting in unwanted pregnancies, STIs, or HIV infection. In many developing countries, homelessness and poverty provide a source for the sale of babies and organs, and for child prostitution.

The UN has estimated the population of street children worldwide at 150 million, with the number rising daily. These young people are more appropriately known as community children, as they are the offspring of our communal world. Ranging in age from 3 to 18, about 40 percent of those are homeless – as a percentage of world population, this is unprecedented in the history of civilization. The other 60 percent work on the streets to support their families. They are unable to attend school and are considered to live in difficult and dangerous circumstances. Increasingly, these children are the defenseless victims of brutal violence, sexual exploitation, abject neglect, chemical addiction, and human rights violations.

Homelessness is a public health problem, mainly for the homeless themselves, but also due to the effects on the community around them. Solutions, like those for many other social problems, are complex and costly. The root issues lie at the crossroads of underlying societal issues such as poverty and inadequate supply of permanent housing. Reintegrating the homeless into mainstream society requires community-based treatment and social services, including client engagement, case management, housing options, long-term follow-up, and support. Vocational training and employment opportunities may be the key. Until such help is available, homelessness will continue to cause much suffering and subsequently strain the charitable

and public health systems aiming to provide care for this population.

Refugee Health

The UN definition, which the USA accepts, considers a refugee "any person who is outside any country of such person's nationality … and who is unwilling or unable to return … because of persecution or a well-founded fear of persecution on account of race, religion, nationality, membership in a particular social group, or political opinion". At the end of 2006, an estimated 35 million people had fled their countries as refugees and 32.9 million had been displaced internally. Of some 20 million cross-border refugees, about 14 million are in Africa, South-West Asia, or the Middle East. The mass exodus of people from their homelands can be the result of ethnic or religious persecution, political conflict, war, civil strife or political conflict, environmental degradation, or conflict over economic resources. Annually, millions of people abandon their homes, farms, regions, or countries in search of food and water, employment, or security.

The ability of a country to cope with a sudden mass refugee situation depends on the function of the state infrastructure and basic services, the security situation, and prior planning for disasters. Appropriate and timely response depends on local strengths in these areas, with international assistance as a back-up for needed supplies and services. Effective relief services must be primarily based on the establishment of security, shelter, clothing, sanitation, safe water and food supplies, family reintegration, personal preventive services, and treatment services. Education, family planning, and longer term health services are part of any relief program when the refugee situation continues.

The United Nations High Commissioner for Refugees (UNHCR) monitors and coordinates refugee assistance needs and relief services. The 1951 Geneva Refugee Convention established the UNHCR to protect the human rights of vulnerable people displaced by war and civil strife from being forced to return to a country where they may be persecuted. The UNHCR helps in civilian repatriation, integration in countries of asylum, or resettlement in third countries. Its world network provides shelter, food, water, and medical care in the immediate aftermath of refugee situations (UNHCR, 2008). At the end of 2006, UNHCR estimated the global refugee population at 9.9 million people.

The acute phase of a refugee situation, discussed later in the section on Health in Disasters, may become a long-term situation which imposes a different set of problems. The drama of the acute phase may become an international media event with many countries and NGOs offering assistance, relief supplies, and financial contributions. Longer term refugee situations require help which places more responsibility on the UNHCR and NGOs such as the

International Red Cross, Médicins sans Frontières (awarded the Nobel Peace Prize for 1999), Catholic Relief Services, and others. Long-term refugee centers should provide all the basic services of any population group, but there is the additional factor of the risk of collapse of fragile sanitation, nutrition, or health services following political change or further disaster. The Kosovo crisis of 1999 brought together international efforts of the UNHCR, the North Atlantic Treaty Organization (NATO), and many international NGOs, with bilateral and public assistance with finances and essential survival material. However, the immediate solution to the massive displacement was resolved through military and political action to force the home country to allow repatriation of the refugees, although long-term political solutions are hard to foresee.

The UN Environmental Program Report in 1985 defined environmental refugees as "people forced to leave their traditional habitat, temporarily or permanently, because of marked environmental disruption (natural and/or triggered by people) that jeopardized their existence and seriously affected the quality of their life". The Framework Convention on Climate Change (Bali, 2007) estimated the possibility for 50 million people being environmentally displaced by 2010, with a long-term projection of about 200–250 million people by 2050 (Biermann and Boas, 2007).

MILITARY MEDICINE

Armed forces are part of the national responsibility for defense and security. Preventive and medical care must be provided for the military and their families. Armed forces seek to recruit people in apparent good health, but conditions of service may cause disease both in peace and in wartime. Military medicine emphasizes prevention to maintain the health of personnel who will be placed in hazardous conditions that increase their risk for disease and injury.

Military medicine has played an important role in the development of surgery, related skills, and public health in general. Roman military success was aided by skill in preventive health measures through hygiene and camp discipline, no less than in care of wounds, basic military organization, and discipline. Armies and navies depend on medical and nutritional support to maintain the health of their forces and their ability to perform missions. The mandatory use of limes for British sailors following the epidemiological breakthrough of James Lind established the necessity of nutritional support and discipline to maintain the function and competence of the individual and the unit. The American army adopted mandatory vaccination soon after Jenner's method became known. Innovations in public health by military personnel during the nineteenth century were numerous, from Ronald Ross's work on the malaria parasite to the conquest of yellow fever by the US Army in Cuba in 1901.

TABLE 7.6 Number and Percent of Deaths from Disease and Battle Injury, US Army, 1860–2010

War	Disease Deaths (D)	Battle Injury Deaths (BI)	D/BI (%)
Civil War (North), 1861–1865	199,720	138,154	145
Spanish American War, 1898	1,939	369	525
Philippines, 1899–1902	4,356	1,061	410
World War I, 1917–1918	51,447	50,510	102
World War II, 1941–1945	15,779	234,874	7
Korean War, 1950–1953	509	27,704	2
Vietnam War, 1961–1975	1,433	30,900	5
Iraq War, 2003–2010	800	34,400	2.4

Sources: Adapted from Legters LJ, Llewellyn CH. Military medicine. In: Last JM, Wallace JB, editors. Public health and preventive medicine. London: Prentice Hall; 1992.
Goldberg M. Death and injury rates of US military personnel in Iraq. Mil Med 2010;175:220–6.

Death from disease outweighed deaths in battle in most armies until the early part of the twentieth century, as seen in data of the US Army from the Civil War to the Vietnam War (Table 7.6). Awareness of the predominance of disease in army casualties was a major contribution of Florence Nightingale from the Crimean War. Military medicine has contributed to the development of emergency medical care, bringing enormous benefit for the civilian sector, including innovations in medical adaptations from military technology, such as ultrasound.

Protection of troops from disease requires assurance of immunization to prevent diseases that could easily be transmitted in barracks or shipboard living conditions. This may include updating childhood vaccinations with boosters of diphtheria and tetanus and vaccinating against hepatitis B, influenza, polio, diphtheria, tetanus, measles, mumps, rubella, meningococcal meningitis, hepatitis A, anthrax, and others. Antimalarials and other preventive measures are used depending on location of service.

Nutrition protection from foodborne, waterborne, and vectorborne diseases; and prevention of training and motor vehicle accidents, suicides, exposure to toxic materials, contact with STIs, and HIV exposure are among the many issues of peacetime armed forces. Violence and brutality are frequent issues in the training period, as are suicides. Officer and non-commissioned officer vigilance and accountability must be emphasized, along with medical and psychological surveillance.

Prevention of war is the surest method of preventing related military and civilian deaths and injuries. In war, battle casualty prevention depends on armaments and skill in their application, discipline, and leadership. Medical support in the field, skilled evacuation and triage, and rapid transfer with life support systems to medical centers are crucial to keep fatality to injury ratios low. Using these methods may save many young lives. Organization of medical services from the medic on the battlefield to the evacuation post and the base hospital requires skilled management with assurance of adequate supplies of everything from water and food to diagnostic and treatment resources.

Casualty management to conserve fighting strength is based on prevention and collection, treatment, and evacuation of casualties in a manner that supports the morale of the troops and prevents complications or death following injury. Triage or sorting is based on providing the best care one can for the most patients under the circumstances. Triage categories include the following: urgent cases require airway clearance, chest tubes, hemorrhage control, and replacement therapy to prevent immediate death; immediate cases are life-threatening wounds temporarily stable but requiring surgery within a short time and a good chance of recovery; delayed cases are injuries that can be cared for successfully 8–16 hours after the injury; minimal, or superficial, wounds require minor surgery, fracture setting, or observation; and expectant cases are mortal wounds with little chance of survival.

Special US veteran hospitals and health services are operated by the Department of Veterans Affairs (VA). The VA deals with long-term disabilities from consequences of injuries and trauma, Agent Orange exposure during the Vietnam War, Gulf War syndrome, and PTSD: although each successive war brings renewed attention to this syndrome, it was not until the Vietnam War that PTSD was first identified and named. PTSD has gained increased recognition as a result of the Iraq and Afghanistan Wars, from which a high proportion of veterans have returned with serious mental health issues, including suicide. Mental health issues in combat service and in veterans has become a major issue in military medicine.

Conditions of potential atomic, biological, and chemical warfare require special preparation and, above all, prevention. New phases of historical and political developments bring with them the potential for such warfare, and preparation for its effects on the health of combat units and civilian populations are part of this special field of military medicine and, indirectly, public health. Military medical systems work with civilian public health to reduce the threats of bioterrorism by strengthening public health preparedness and biodefense research.

HEALTH IN DISASTERS

Natural and human-caused disasters are frequent occurrences (Table 7.7) and have important implications for public health. Natural disasters are naturally occurring extreme events that can cause excess morbidity and mortality among the population, and damage to the physical environment. They include earthquakes, floods, hurricanes, droughts, blizzards, and volcanic eruptions. The term disaster also includes a wide range of events including war, industrial accidents, terrorist incidents, and others. Disasters are often classified into "man-made" (human-caused), "natural", "sudden", or "slow onset". The distinctions are often blurred, since a natural disaster may be the result of inappropriate policy decisions such as a drought occurring in an area made vulnerable by political or policy actions. In semi-arid areas with a delicate food balance, chronic undernutrition, and epidemic infectious diseases such as measles and cholera, a moderate natural disaster may tip the balance and cause wide-scale suffering and frank malnutrition, which may continue over an extended period.

The accumulated experience of disaster relief management has been enhanced in the past several decades by improved technology in communications and transportability of air-mobile supplies of tents, blankets, food, chlorinators, generators, heavy equipment for lifting debris, and field hospitals, in addition to medical supplies such as sterilizing and trauma equipment, antibiotics, vaccines, and oral rehydration therapy.

Disasters require a highly professional and aggressive response based on intersectoral cooperation. This can be a matter of life and death for large groups of people. Disasters can occur in the midst of an urban metropolis or in a remote jungle. The details will differ, but meeting basic human needs is common to both settings. Limiting further death and injury requires attention to protection and security as well as the prioritization of safe water, food, and shelter. Organized prevention in the form of sanitation and management of the most common health problems such as diarrheal diseases, acute respiratory diseases, measles immunization, and other diseases associated with poor sanitation (e.g., hepatitis, typhoid, cholera, and gastroenteritis) is needed. Outbreaks of dysentery, malnutrition, and respiratory infection can kill large numbers of debilitated children and elderly people. PTSD can be dealt with as part of a community-oriented support program with involvement of the population affected as health aides.

Epidemiological monitoring of death, injury, and disease is usually difficult because of the chaotic events associated with disasters. Monitoring of the process and epidemiological patterns provides informs interventions which can be important in the planning and management of future crises. Monitoring hospital admissions data can indicate morbidity patterns for specific diseases such as typhoid fever and

TABLE 7.7 Selected Human-Caused and Natural Disasters, 1976–2011

Location	Type	Effects
Serveso, Italy, 1976	Chemical factory explosion	17,000 evacuated
Cambodia, 1979	Genocide, political	1–2 million deaths
Bhopal, India, 1984	Chemical leak	2000 deaths, 70,000 evacuated
Mexico, 1985	Earthquakes	10,000 deaths, 60,000 homeless
Colombia, 1985	Volcano	23,000 deaths, 200,000 homeless
Chernobyl, Ukraine, 1986	Nuclear reactor meltdown	30 deaths, 100,000 evacuated; long-term effects (cancer, birth defects) unknown
Bosnia, 1993–1995	Civil war, genocide	Tens of thousands of casualties, mass rape, genocide, breakdown of civilian services
Rwanda, 1994	Genocide, tribal	Up to 500,000 deaths
Kobe, Japan, 1995	Earthquake	6000 deaths
Caribbean, 1995	Hurricane Gordon	11,000 deaths in Haiti, Cuba, Jamaica, Dominican Republic
China, 1998	Flood	4150 deaths, 18.4 million displaced, 180 million affected
Nicaragua, 1998	Hurricane Mitch	10,000 deaths, 120,000 homeless
Kosovo, 1999	Ethnic cleansing, genocide	1 million people forcibly displaced, with mass murder, community destruction
Indian Ocean, 2005	Tsunami/tidal wave of 11 countries bordering Indian Ocean	>225,000 killed and missing
World Trade Center 9/11/2001, New York City, USA	Terrorist destruction	2,977 victims killed
New Orleans, USA, 2005	Hurricane Katrina	1836 deaths, most of New Orleans destroyed
Darfur, 2002–2008	Genocide	200,000–400,000 deaths, 2.5 million displaced
DR Congo, 2006–2008	Civil war	Hundreds killed monthly
Kenya, 2008	Political–ethnic conflict	350 deaths, 250,000 displaced
Brazil, 2008	Floods	49,506 people affected
Myanmar, 2008	Flood	>100,000 deaths, 1.6 million homeless
Haiti, 2010	Earthquake	46,190–84,961 deaths
Pakistan, 2010	Flood	17,000 deaths
Fukushima, Japan, 2011	Tsunami/earthquake/nuclear meltdown	15,859–18,880 deaths

Sources: Relief Web Open Forum. http://www.reliefweb.int/rw/dbc.nsf/doc100?Open Forum [Accessed 10 May 2008]; UN News Center. http://www.un.org/apps/news/ [Accessed 4 June 2012].

viral hepatitis. An example of such data indicating morbidity patterns is the 1980 earthquake in southern Italy, which caused water contamination. Cholera is also a serious danger in disaster situations, as occurred in the Rwandan refugee disaster. Haiti has suffered from recent earthquake disasters, particularly in 2010, and a massive cholera epidemic with over 216,000 people infected and nearly 6000 deaths. Haiti had been cholera free for 100 years even though cholera was reintroduced into South America in 1991 (see Chapter 1). It is possible that cholera was introduced after the Haiti earthquake by UN security forces from Bangladesh. Contamination of limited water supplies makes this one of the most urgent aid issues, with pumps and portable chlorinators

as basic equipment needs. Refugee camps, toilet, water and other facilities, especially food distribution facilities, need to be protected by armed guards.

Intervention by aid agencies should include the potential for vaccination against measles and poliomyelitis and sustaining regular diphtheria–tetanus–pertussis (DTP) vaccination, along with vitamin and iron supplements. Insecticide-impregnated bed nets should be used if the return to normal conditions is delayed over months. Typhoid and salmonellosis are combated by sanitary and case management approaches. The absolute necessity of potable water and oral rehydration therapy on a large scale was demonstrated in the Rwandan disasters of 1993–1994.

Over longer periods, nutritional monitoring of child weight and height for age may be done on sample populations to assess the effects and changing situation. Disaster relief should include the possibility for vitamin and iron supplements for children to offset the damage of food deprivation. Provision of food staples, shelter and water is a high priority, but the bulk shipment required may overwhelm even the ability of developed countries to cope with local disasters.

Secondary damage from unsanitary conditions can spread infectious disease. This situation may be aggravated by food deprivation and a lack of water and sanitary facilities. The large numbers of refugees from civil war and ethnic massacres in the 1994 Rwanda situation resulted in massive loss of life. International or local relief efforts may be overwhelmed and hampered by lack of coordination. Planning refugee camps in disaster situations must take into account natural drainage, water sources, access to roads, separation of sanitary facilities including latrines and garbage disposal, with safe areas for food and water distribution.

Better coordination of storm warnings could reduce deaths at sea. In addition, warnings to remain off the roads and curb the use of vehicles in dangerous areas may reduce road crash deaths. Emergency information to promote safe evacuation, safe havens, risk avoidance, and assurance of support services are part of public health action in emergency situations, in conjunction with other local, state, and federal agencies.

In late 1996, a massive refugee crisis emerged in Rwanda and Democratic Republic of Congo, with complex ethnic and political origins, which resulted in death on a massive scale due to genocidal military actions as well as dehydration, disease, and starvation. International intervention involving military field hospitals, aid, and health organizations was sporadic and insufficient. NGOs such as Médicins sans Frontières and UN aid organizations are often the first on the ground, but without strong international political and military backing, all efforts will be limited. Following the Indian Ocean tsunami of 2005, US armed forces provided immediate aid with troops, helicopters, and medicine. Bilateral governmental aid is important and often brings locally unavailable essential services and supplies. Medical interventions with field hospitals capture media attention, but more fundamental efforts to promote basic organization for shelter, safe water, and sanitation receive fewer resources and less support.

Disasters during 2005 caused 99,425 deaths globally, of which 84 percent were due to October's South Asian earthquake. Hurricane Katrina, which hit America's Gulf Coast in 2005, killed around 1300 people and was followed by Hurricane Stan, which killed over 1600 people in Guatemala. In 2005, natural disasters globally affected 161 million people and cost around US$160 billion. Between 1996 and 2005, disasters killed over 934,000 people, with 2.5 billion people affected around the globe. Earthquakes, floods, heat waves, and other natural and technological hazards cause thousands of deaths and hundreds of billions of dollars in economic losses each year.

In 2011, the US Department of Homeland Security declared 99 major disasters and 114 major fire assistances. In the same year, the Fukushima accident occurred in Japan following a 15 meter tsunami, resulting in a human death toll of over 20,000 dead or missing and enormous damage to coastal ports and towns, with over a million buildings destroyed or partly collapsed. It disabled the power supply, causing a disruption of the cooling system and meltdown of three nuclear plants at the Fukushima nuclear facility. In July 2012, a national inquiry into the episode ruled that this was a "man-made" disaster due to poor governmental regulation and standards in the nuclear industry.

In the WHO European Region between 1990 and 2006, 1483 disaster events were recorded, killing 98,119 people and affecting more than 42 million others, with an estimated economic loss of over US$168 billion (UN Human Habitat Global Report on Human Settlements, 2009). Increasing evidence indicates that climate change, environmental changes such as the depletion of stratospheric ozone, and increasing interconnectedness resulting from changes in trade, travel, and technology may threaten human health through complex and interdependent mechanisms. Health systems will need to include these evolving threats in comprehensive efforts to protect the public health.

Prevention of human injury and death in disasters must take into account the history of natural disasters in a given area. Earthquakes in California are a serious threat and have led to strict building codes that reduce the damage and secondary loss of life. Monitoring can give warning of potential disasters such as hurricanes, and the preparation of evacuation plans and facilities makes a major difference in disaster management. Zoning laws and building code enforcement can prevent many deaths when earthquakes strike in urban areas. Flood control measures in areas traditionally threatened are worthwhile investments to prevent the massive damage to property and homes. Storage of appropriate equipment and supplies and training of personnel for such emergencies is part of prudent public administration, and is important for public health as well.

Wars, genocide, and terrorism are part of past, recent, and current experience in many parts of the world. These too are public health disasters requiring interventions to provide security and basic needs of victims and refugees, and to limit the spread or continuation of human tragedies. Incitement to violence and terror are early warning signs of potential ethnic conflict or genocide and by themselves constitute a public health call to duty.

Documentation of disaster experiences is essential to improve efforts in future situations, and although lessons may be absorbed slowly, they are part of the development of public health. An action plan for addressing a disaster should include the following.

- Treat and evacuate the injured.
- Limit further death, injury, and disease.
- Assure safety/protection/security/public order to coordinate interventions of police, army, official health agencies, local and international NGOs.
- Provide shelter, safe water, food, and warmth.
- Provide sanitary facilities and prevent environmental hazards.
- Promote epidemic control/prevention and treatment of communicable diseases (e.g., diarrheal disease, acute respiratory infections, measles, hepatitis, malaria).
- Provide ongoing medical care for the injured and sick.
- Mobilize and coordinate all official and voluntary local, state, national, and international aid.
- Prevent malnutrition, provide micronutrient supplements.
- Provide maternal and child care for pregnancy, delivery, infancy, and childhood.
- Monitor disease and epidemiological surveys.
- Mobilize health aides among the affected population.
- Promote early restoration of normal functions (e.g., family, health care, work).
- Prevent PTSD.
- Assess, evaluate, monitor, and report the process and lessons learned.
- Survey, document, publish, and follow up.
- Educate (e.g., provide temporary preschool and school activities).
- Employ able-bodied people (e.g., promote participation in refugee care activities).
- Promote rapid return to homes and rehabilitation.
- Review experience and plan for potential future disasters.

Disaster planning is an essential component of public health agencies at the national, state, provincial, regional, and local health authority levels. Coordination with police, army, civil defense, fire service, local and state disaster planning, hospitals, and many other local agencies, as well as international relief agencies, is fundamental to coping with disasters. Preparation of protocols for operating procedures with drills may make a large difference in outcomes. Although disaster situations are unique, they have common features relevant to planning, adapted from documentation of trial, error, and experience.

SUMMARY

A society is often judged on how it treats its most vulnerable groups, including its minorities, its poor, its prisoners, and its refugees, as much as on how it cares for the main population groups. All such groups need special attention because they are people in need, but also because they can affect the health of others. Prime examples of this include the spread of TB to prison guards in the USA, and its spread in the general population by ex-prisoners in the former Soviet Union.

Public health agencies are key advocates and pioneers in implementing programs for such groups, to meet the special needs of the whole population or the needs of special groups in the population. The public health approach emphasizes defining groups at special risk prevention in all of its phases, and preparation for unforeseen and unexpected emergency health situations, while coordinating not only with health service systems but also with many other agencies in society.

Preparation for handling emergencies is an important challenge to all elements of a health system, from sanitation and pest control to tertiary care neurosurgery. In such a situation, there will be little question of the need for complete coordination to treat the injured and prevent further damage. This requires preparation as well as improvization at the time of the event. Mental and dental health are two areas where community care and prevention are established as effective measures but generally not well linked to mainstream organization and funding activities.

Mental health is increasingly recognized as a major burden on individuals, families, and the community as a whole. The interaction between mental illness and chronic disease has been described as both a contributing cause and a major effect. Planning for improved health conditions must take various factors into consideration and include the societal barriers that perpetuate inequalities, and the implications of the burden of cost and human suffering.

At-risk groups such as refugees, prisoners, minorities, migrants, or victims of natural and anthropogenic disasters provide living proof of John Donne's famous phrase, "No man is an island unto himself alone". Today's prisoner or refugee population can become tomorrow's free citizens infected with HIV or TB. Migrant workers may bring a disease from their home country or acquire it in the host nation and bring it back to the original place of residence. Disasters or terrorism can affect anyone. The role of health care providers and public health agencies is to prepare for and respond to such challenges.

Special community health needs are often stigmatized or neglected on social and political levels. Failure to address such needs can endanger the public health, and only an adequate response to a special community need can be a safeguard to population health. The New Public Health seeks to use all potential public and private activities, interventions, and resources of prevention and care effectively for what have previously been treated as marginal populations or issues.

NOTE

For a complete bibliography and guidance for student reviews and expected competencies please see companion web site at http://booksite.elsevier.com/9780124157668

BIBLIOGRAPHY

Mental Health

American Psychiatric Association, 1994. In: DSM-IV: diagnostic and statistical manual of mental disorders. fourth ed. APA, Washington, DC. [updated 2003] Available at: http://allpsych.com/disorders/dsm.html. (accessed 04.07.12).

American Psychiatric Association, DSM5: diagnostic and statistical manual of mental disorders. fiveth ed. Available at: http://www.psych.org/ (accessed 02.08.13).

Buka, S.L., 2008. Psychiatric epidemiology: reducing the global burden of mental illness. Am. J. Epidemiol. 168, 977–979. Available at: http://www.ncbi.nlm.nih.gov/pmc/articles/PMC2572566/ (accessed 08.08.12).

Butcher, J., Samarasekera, U., Wilkinson, E., Shetty, P., 2007. Special report. Lancet Series on Global Mental Health 370, 117–124. Available at: http://www.thelancet.com/journals/lancet/article/PIIS0140-6736(07)61069-1/fulltext (accessed 25.11.13).

Centers for Disease Control and Prevention, 1999. Achievements in public health, 1900–1999. Motor-vehicle safety: a 20th century public health achievement. MMWR 1999; 48(18): 369–374. Available at: http://www.cdc.gov/mmwr/preview/mmwrhtml/mm4818a1.htm (accessed 25.11.13).

Centers for Disease Control and Prevention, 2012. Alcohol use and binge drinking among women of childbearing age – United States, 2006–2010. MMWR. Morb. Mortal. Wkly. Rep. 61, 534–538. Available at: http://www.cdc.gov/mmwr/preview/mmwrhtml/mm6128a4.htm?s_cid=mm6128a4_w (accessed 25.11.13).

Centers for Disease Control and Prevention, Community mental health services. [updated 2012] Available at: http://www.thecommunityguide.org/mentalhealth/index.html (accessed 22.06.12).

Centers for Disease Control and Prevention, Mental illness surveillance among adults in the United States. [updated 2011] Available at: http://cdc.gov/mmwr/preview/mmwrhtml/su6003a1.htm (accessed 22.05.12).

Centers for Disease Control and Prevention, 2010. National Hospital Discharge Survey, 2010, first-listed diagnostic categories: United States. Available at: http://www.cdc.gov/nchs/data/nhds/2average/2010ave2_firstlist.pdf. (accessed 24.08.13).

Centers for Disease Control and Prevention. National Vital Statistics Reports. Deaths: preliminary data for 2011. Vol. 61, No. 6, 10 October 2012, Table 7. Available at: http://www.cdc.gov/nchs/data/nvsr/nvsr61/nvsr61_06.pdf (accessed 23.08.13).

Chilvers, R., Macdonald, G.M., Hayes, A.A., 2006. Supported housing for people with severe mental disorders. Cochrane. Database Syst. Rev. 18 (4), CD000453. Available at: http://www.update-software.com/pdf/CD000453.pdf. (accessed 25 November 2013).

Crosby, A.E., Ortega, L., Stevens, M.R., 2013. Suicides — United States, 2005–2009. MMWR. Morb. Mortal. Wkly. Rep. Suppl. 62 (3), 179–183. Available at: http://www.cdc.gov/mmwr/preview/mmwrhtml/su6203a31.htm?s_cid=su6203a31_e (accessed 22.11.13).

Diagnostic and Statistical Manual of Mental Disorders (DSM-5): Intellectual disability involves impairments of general mental abilities that impact adaptive functioning in three domains, or areas. These domains determine how well an individual copes with everyday tasks http://www.dsm5.org/Documents/Intellectual%20Disability%20Fact%20Sheet.pdf

Durkin, P.W., 1996. Beyond mortality – residential placement and quality of life among children with intellectual disability. Am. J. Public Health 86, 1359–1360.

Fleischman, A., Lurie, I., 2013. Cardiovascular mortality and related risk factors among persons with schizophrenia: a review of the published literature. Public Health Reviews, 34. epub ahead of print. Available at: http://www.publichealthreviews.eu/show/a/119 (accesed 22.11.13).

Guerino, P., Harrison, P.M., Sabol, W.J., 2011. Prisoners in 2010 (revised). Bureau of Justice Statistics, Washington, DC. Available at: http://www.bjs.gov/content/pub/pdf/p10.pdf (accessed 03.08.13).

Gvion, Y., Apter, A., 2012. Suicide and suicidal behavior. Public Health Rev. 34, 1–20. [Epub ahead of print]. Available at: www.publichealthreviews.eu (accessed 02.08.13).

Horton, R., 2007. Launching a new movement for mental health. Lancet 370, 806. Available at: http://www.thelancet.com/online/focus/mental_health/collection (accessed 14.06.12).

Insel TR, Fenton WS. It's not just about counting anymore. National Institute of Mental Health, 2013. Available at: http://www.nimh.nih.gov/about/director/publications/psychiatric-epidemiology.shtml and http://www.nimh.nih.gov/health/publications/the-numbers-count-mental-disorders-in-america/index.shtml#Intro (accessed 22 November 2013).

Institute of Medicine, 2002. Crossing the quality chasm: report for behavioral health. US Academies of Science, National Academies Press, Washington, DC. Available at: http://www.iom.edu/Global/News%20Announcements/Crossing-the-Quality-Chasm-The-IOM-Health-Care-Quality-Initiative.aspx (accessed 25.11.13).

International Helsinki Federation for Human Rights, 1997. Annual Report on United States. IHF 1999 Vienna. Available at: http://www.refworld.org/type,ANNUALREPORT,LVA,3ae6aa7214,0.html (accessed 25.11.13).

Kedia S, Sell MA, Relyea G. 2007. Mono- versus polydrug abuse patterns among publicly funded clients. Substance Abuse Treatment and Policy, 27 (33), Available at: http://www.substanceabusepolicy.com/content/2/1/33 (accessed 23.11.2013).

Kessler, R.C., Berglund, P., Chiu, W.T., Demler, O., Heeringa, S., Hiripi, E., et al., 2004. The US National Comorbidity Survey Replication (NCS-R): design and field procedures. Int. J. Methods Psychiatr. Res. 13, 69–92 (accessed 25.11.13).

Lancet, Series on Global Mental Health 2011 collection [updated 2012]. Available at: http://www.thelancet.com/online/focus/mental_health/ (accessed 08.05.12).

McLaughlin, K.A., Green Greif, J., Hwang, I., Sampson, N.A., Zaslavsky, A.M., Kessler, R., 2012. Intermittent explosive disorder in the National Comorbidity Survey Replication Adolescent Supplement. Arch. Gen. Psychiatry. 69, 1131–1139. Available at: http://archpsyc.jamanetwork.com/article.aspx?articleid=1206777 (accessed 03.07.12).

Mechanic, D., 2007. Mental health services then and now. Health Aff. 26, 1548–1550.

Menken, M., Munsat, T.L., Toole, J.F., 2000. The Global Burden of Disease Study Implications for Neurology. Neurology and Public Health. Arch Neurol 57, 419–420. Available at: http://wfneurology.org/pdfs/publications/menken.pdf (accessed 25.09.13).

Morris, J., Lora, A., McBain, R., Saxena, S., 2012. Global mental health resources and services: a WHO survey of 184 countries. Public Health Rev. 34 (2). [Epub ahead of print]. Available at: http://www.-publichealthreviews.eu/upload/pdf_files/12/00_Morris.pdf (accessed 27.06.13).

National Institutes of Health, National Institute of Mental Health. The numbers count – mental disorders in America [updated 2012]. Available at: http://www.nimh.gov/health/publications (accessed 08.05.12).

National Institute of Mental Health, Mental illness exacts heavy toll, beginning in youth. Available at: http://www.nimh.nih.gov/news/science-news/2005/mental-illness-exacts-heavy-toll-beginning-in-youth.shtml (accessed 17.07.13).

National Institute of Mental Health, 2008. National Strategic Plan. Available at: http://www.nimh.nih.gov/about/strategic-planning-reports/index.shtml. (accessed 24.08.13).

Nobelprize.org, 2013. Nobel lecture: Julius Axelrod: noradrenaline: fate and control of its biosynthesis. Nobel Media. Available at: http://www.nobelprize.org/nobel_prizes/medicine/laureates/1970/axelrod-lecture.html (accessed 17.07.13).

Petrea, I. Mental health in former Soviet countries: from past legacies to modern practices. Public Health Reviews. 2013;34: epub ahead of print. Available at: http://www.publichealthreviews.eu/show/a/121 (accessed 28.11.2013).

Raviola, G., Becker, A.E., Farmer, P., 2011. A global scope for global health - including mental health. Lancet (378), 1613–1614. Available at: http://www.thelancet.com/journals/lancet/article/PIIS0140-6736(11)60941-0/fulltext (accessed 23.11.13).

Rice, C.E., Rosanoff, M., Dawson, G., Durkin, M.S., Croen, L.A., Singer, A., Yeargin-Allsopp, M., 2012. Evaluating changes in the prevalence of the autism spectrum disorders (ASDs). Public Health Reviews. 34: epub ahead of print. Available at: http://www.publichealthreviews.eu/upload/pdf_files/12/00_Rice.pdf (accessed 24.11.13).

UK Department of Health, Mental Health Services [updated 2012]. Available at: http://www.dh.gov.uk/en/Publicationsandstatistics/Publications/PublicationsPolicyAndGuidance/DH_072730 (accessed 08.05.12).

UN General Assembly, Convention on the Rights of Persons with Disabilities A/RES/61/106: Resolution. Available at: http://www.unhcr.org/refworld/docid/45f973632.html (24 January 2007) (accessed 06.06.12).

United States Department of State, 1998. U.S. Department of State Country Report on Human Rights Practices 1997 - Latvia 30. January. Available at: http://www.refworld.org/docid/3ae6aa894.html (accessed 27.11.13).

Ustün, T.B., 1999. The global burden of mental disorders. Am. J. Public Health 89, 1315–1318. Available at: http://www.ncbi.nlm.nih.gov/pubmed/10474545 (accessed 08.08.12).

Werner, S., 2012. Individuals with intellectual disabilities: a review of the literature on decision-making since the Convention on the Rights of People with Disabilities (CRPD). Public Health Rev. 34. Available at: www.publichealthreviews.eu (accessed 15.08.12).

WHO, 2013. Mental Health Action PLan 2013-2020. WHO, Geneva. Available at: http://apps.who.int/iris/bitstream/10665/89966/1/9789241506021_eng.pdf (accessed 22.11.13).

World Health Organization, 2011. Mental health atlas 2011. WHO, Geneva. Available at: http://whqlibdoc.who.int/publications/2011/9799241564359_eng.pdf (accessed 08.08.12).

World Health Organization, Mental health program [updated 2012]. Available at: http://www.who.int/mental_health/en/ (accessed 22.06.12).

World Health Organization. Mental Health Action PLan 2013-2020. WHO, Geneva, 2013. http://apps.who.int/iris/bitstream/10665/89966/1/9789241506021_eng.pdf (accessed 22.11.2013).

Oral Health

American Dental Association, 2009–10 Survey of dental education. Available at: http://www.agd.org/files/webuser/website/membership/vol.%201_academic%20programs_enrollment_graduates.pdf (accessed 23.08.13).

British Fluoridation Society, November 2012. Fluoride Action Network. Countries that fluoridate their water. Available at: http://www.bfsweb.org/onemillion/onemillion2012.html. (accessed 23.08.13).

Centers for Disease Control and Prevention, 1999. Achievements in public health, 1900–1999: fluoridation of drinking water to prevent dental caries. MMWR. Morb. Mortal. Wkly. Rep. 48, 933–940. Available at: http://www.cdc.gov/mmwr/preview/mmwrhtml/mm4841a1.htm (accessed 25.11.13).

Centers for Disease Control and Prevention, Community water fluoridation [updated 25 July 2013]. Available at: http://www.cdc.gov/fluoridation/guidelines/index.htm (accessed 23.08.13).

Centers for Disease Control and Prevention, 2011. Oral health. [updated 2012]. Available at: http://www.cdc.gov/chronicdisease/resources/publications/AAG/doh.htm. (accessed 14.06.12).

Thornton-Evans, G., Eke, P., Wei, L., Palmer, A., Moeti, R., Hutchins, S., Borrell, L.N., 2013. Periodontitis Among Adults Aged ≥30 Years — United States, 2009–2010.

MMWR Morb, 2013. Mortal. Wkly Rep. Suppl. 62 (3), 129–135. Available at: http://www.cdc.gov/mmwr/preview/mmwrhtml/su6203a21.htm?s_cid=su6203a21_e (accessed 22.11.13).

World Health Organization, Oral health. [updated 2012]. Available at: http://www.who.int/topics/oral_health/en/ (accessed 14.06.12).

Disability and Rehabilitation

Drum, C.E., Krahn, G.L., Bersani, H. (Eds.), 2009. Disability and public health. American Public Health Association Press, Washington, DC.

United Nations, Convention on the Rights of Persons with Disabilities. Available at: http://www.un.org/disabilities/convention/conventionfull.shtml (accessed 12.06.12).

Gay and Lesbian Health

American Academy of Pediatrics, 1993. Homosexuality and adolescence. Pediatrics 92, 631–634. Available at: http://pediatrics.aappublications.org/content/92/4/631.abstract (accessed 25.11.13).

American Medical Association Council on Scientific Affairs, 1996. Health care needs of gay men and lesbians. JAMA. 275, 1354–1359. Available at: http://www.amsa.org/programs/barriers/jama96.pdf (accessed 25.11.13).

Mayer, K.H., Bradford, J.B., Makadon, H.J., Stall, R., Goldhammer, H., Landers, S., 2008. Sexual and gender minority health: what we know and what needs to be done. Am. J. Public Health 98, 989–995. Available at: http://www.ncbi.nlm.nih.gov/pmc/articles/PMC2377288/ (accessed 25.11.13).

Native Peoples' Health

Centers for Disease Control and Prevention, Native peoples' health [updated 2012]. Available at: http://www.cdc.gov/ and http://www.aaip.com/ (accessed 14.06.12).

Health Canada, A statistical profile on the health of First Nations in Canada: Vital Statistics for Atlantic and Western Canada; 2001/2002 [updated 2012]. Available at: http://www.hc-sc.gc.ca/fniah-spnia/pubs/aborig-autoch/stats-profil-atlant/index-eng.php (accessed 14.06.12).

US Department of Health and Human Services, Indian health service [updated 2012]. Available at: http://www.ihs.gov/ (accessed 14.06.12).

Prisoners' Health

Freudenberg, N., Daniels, J., Crum, M., Perkins, T., Richie, B.E., 2005. Coming home from jail: the social and health consequences of community reentry for women, male adolescents, and their families and communities. Am. J. Public Health 95, 1725–1736. Available at: http://www.ncbi.nlm.nih.gov/pubmed/16186451 (accessed 25.11.3).

Gatherer, A., Moller, L., Hayton, P., 2005. The World Health Organization European Health in Prisons project after 10 years: persistent barriers and achievements. Am. J. Public Health 95, 1696–1700. Available at: http://www.ncbi.nlm.nih.gov/pmc/articles/PMC1449422/ (accessed 25.11.13).

Migrant and Refugee Health

American Academy of Pediatrics, Migrant health. [updated 2012]. Available at: http://aappolicy.aappublications.org/ and http://www.bphc.hrsa.gov (accessed 14.06.12).

Centers for Disease Control and Prevention, Immigrant, refugee and migrant health. [updated 2012]. Available at: http://www.cdc.gov/ncidod/dq/refugee/ (accessed 14.06.12).

United Nations Refugee Agency. http://www.unhcr.org/cgi-bin/texis/vtx/home (accessed 14.06.12).

Military Medicine

Hoge, C.W., Castro, C.A., Messer, S.C., McGurk, D., Cotting, D.I., Koffman, R.L., 2004. Combat duty in Iraq and Afghanistan, mental health problems, and barriers to care. N. Engl. J. Med. 351, 13–22. Available at: http://www.ncbi.nlm.nih.gov/pubmed/15229303 (accessed 25.11.13).

Pols, H., Oak, S., 2007. War and military mental health: the US psychiatric response in the 20th century. Am. J. Public Health 97, 2132–2142. Available at: http://www.ncbi.nlm.nih.gov/pmc/articles/PMC2089086/ (accessed 25.11.13).

US Department of Veteran's Affairs. http://www.va.gov/ (accessed 14.06.12).

Disasters and Health

Blum, R., Stanton, G.H., Sagi, S., Richter, E.D., 2008. "Ethnic cleansing" bleaches the atrocities of genocide. Eur. J. Public Health 18, 204–209. Available at: https://www.ncbi.nlm.nih.gov/m/pubmed/17513346/?i=4&from=/21361895/related (accessed 25.11.13).

Chin, C.S., Sorenson, J., Harris, J.B., Robins, W.P., Richelle, C.C., Jean-Charles, R.R., et al., 2011. The origin of the Haitian cholera outbreak strain. N. Engl. J. Med. 364, 33–42. Available at: http://www.ph.ucla.edu/epi/snow/nejm364_33_42_2011.pdf (accessed 25.11.13).

Centers for Disease Control and Prevention, 2006. Assessment of health-related needs after Hurricanes Katrina and Rita – Orleans and Jefferson Parishes, New Orleans Area, Louisiana, October 17–22, 2005. MMWR. Morb. Mortal. Wkly. Rep. 55, 38–41. Available at: http://www.cdc.gov/mmwr/preview/mmwrhtml/mm5502a5.htm (accessed 25.11.13).

Fukushima Accident. http://www.world nuclear.org/info/fukushima_accident_inf129.html (accessed 04.07.12).

Leaning, J., 2004. Diagnosing genocide – the case of Darfur. N. Engl. J. Med. 351, 735–738. Available at: http://www.nejm.org/doi/full/10.1056/nejmp048206 (accessed 25.11.13).

Waring, S.C., Brown, B.J., 2005. The threat of communicable diseases following natural disasters: a public health response. Disaster Manag. Response 3 (2), 41–47. Available at: http://www.ncbi.nlm.nih.gov/pubmed/15829908 (accessed 25.11.13).

World Health Organization, Tsunami. [updated 2012]. Available at: www.searo.who.int/en/Section23/Section1108/Section1835.htm (accessed 14.06.12).

Chapter 8

Nutrition and Food Safety

Learning Objectives

Upon completion of this chapter, the student should be able to:

1. Describe the relationship of nutrition to disease and health;
2. Describe the relationship of nutrition to national and local socioeconomic development;
3. Describe monitoring and maintaining good nutrition for the individual and the community;
4. Define the role of nutrition policy in the New Public Health.

INTRODUCTION

Nutrition has a direct effect on growth, development, reproduction, and both physical and mental well-being. It is one of the most important factors for the health of an individual or a community and is, consequently, a fundamental issue in modern public health (Box 8.1). People require nutrients such as carbohydrates, fats, and protein to provide heat and energy, water, minerals, fiber, vitamins, proteins, and essential amino acids to regulate body processes and to build and renew body tissue (water, proteins, and mineral salts). The nutritional status of an individual and society is influenced by the supply, quality, distribution, access to, and cost of foods (Box 8.1). It is also affected by knowledge, attitudes, beliefs, and practices regarding essential and balanced nutrition. Food policy at the individual and governmental levels, eating habits, as well as economic and technical factors all contribute to the nutritional state of the public and the individual person.

Improved nutrition has made a major contribution to better health in recent ages. In the twentieth century, nutrition emerged as a basic and applied science. Knowledge of the elements of proper nutrition and its role in prevention of deficiency diseases or non-communicable diseases has played a vital part in the development of modern public health. And, despite rapid population growth, food production and average food consumption have improved steadily worldwide.

Nevertheless, malnutrition is widely prevalent throughout the world. Developed countries struggle with problems both of insufficiency (malnutrition or undernutrition) and of excess (overnutrition) so that communicable and non-communicable diseases are both rampant, along with diseases directly related to nutritional deficiencies. Rising standards of living in many developing countries have brought diseases of modern living to prominence while communicable diseases are increasingly brought under control. Subpopulations within rich and poor nations alike suffer from a broad spectrum of nutritional diseases.

Public health attempts to ensure that all groups in the population have adequate, but not excessive, intake of the basic food groups, essential vitamins and minerals for growth, health maintenance, and physical activity. It does this by recommending daily human needs for nutrition and energy, which vary according to age, gender, body size, level of activity, individual health status, and environmental conditions. It also requires monitoring of the nutritional status of the population and its subgroups.

DEVELOPMENT OF NUTRITION IN PUBLIC HEALTH

The steady improvement in life expectancy seen in Europe and North America in the seventeenth to nineteenth centuries probably had as much to do with improved nutrition as with improved sanitation. The pioneering epidemiological studies of James Lind in the mid-eighteenth century, the first recorded clinical epidemiological experiment, and of Joseph Goldberger in the early twentieth century, opened up the field of nutritional epidemiology. They each established proof of deficiency conditions that met the Koch–Henle criteria of causation in epidemiology.

Just as Snow's work on cholera preceded Koch's discovery of the *Vibrio cholerae* organism by 30 years, so Lind's work on scurvy preceded the isolation of ascorbic acid by more than a century. Antoine Lavoisier (1743–1794) in Paris developed basic concepts of metabolism, measuring oxygen consumption and carbon dioxide production, and is called the "father of the science of nutrition". Justus von Liebig (1803–1873) demonstrated that fat, protein,

The New Public Health. http://dx.doi.org/10.1016/B978-0-12-415766-8.00008-2

BOX 8.1 Food Security – Nutritional Adequacy

Food security is a complex, multifaceted, challenging and politically driven global and local issue. Food security was defined at the 1996 World Food Summit as: "when all people at all times have access to sufficient, safe, nutritious food to maintain a healthy and active life". Food security encompasses physical as well as economic access to food; it involves the concept of sustainable economics, trade, and environment development. Food security can be characterized as a concept built upon food availability, food access, and safe food use. Food availability is adequate amounts of food supply on a consistent basis. Food access refers to having resources to acquire nutritious food, while food use focuses on sufficient water and sanitation as well as one's knowledge and understanding of basic nutrition and care.

Hunger and starvation continue to ravage many parts of the world, leading to disease and death from malnutrition. Simultaneously, many populations are overnourished, leading to obesity, diabetes, cardiovascular diseases, and cancer, with premature disease and death. There are many debates associated with food security, the distribution of food, the impact of globalization, poverty, and malnutrition among rural populations in low-income countries, and disadvantaged populations in industrialized countries. International trade and agriculture agreements and policies impact on a country's food security and the nutritional status of the population.

In November 2009, world leaders and experts came together at the Food and Agriculture Organization World Summit on Food Security, where it was stated that over 1 billion people in the world suffer from hunger. The Summit stressed the importance of producing food in regions where the poor and hungry reside and stimulating agricultural investment in these areas. Experts at the summit discussed proactive strategies to combat climate problems affecting food security. Leaders focused on approaches to increasing agricultural investment, technology, and productivity in low-income countries.

Differences in food security exist not only between countries, cities, and neighborhoods but also between households. Problems associated with food insecurity cross borders, a global issue, but are also a challenge on a smaller scale such as within the local community and home environment. Global policies in human nutrition cross boundaries of agriculture,

health, and other agencies at national and international levels. The WHO and FAO joint policies adopted in 1992 will be under review in 2014 with a global perspective, but focusing mainly on nutrition challenges in developing countries; this review will address all forms of malnutrition, recognizing the nutrition transition and its consequences; and will seek to improve nutrition throughout the life cycle, focusing on the poorest and most vulnerable households, and on women, infants, and young children in deprived, vulnerable, and emergency contexts.

The USDA classifies food security into four ranges. High food security refers to households that have no problems obtaining sufficient food. This is followed by marginal food security and low food security. Very low food security describes households in which, at instances throughout the year, the eating patterns of at least one member were disrupted; as a result of economic difficulties or a lack of other resources, one's food intake was reduced. A household's food security level is determined based on a questionnaire.

Nutritional adequacy includes other factors, especially related to micronutrient deficiencies which can be present when the macronutrient supply, distribution, and intake are inadequate. The deficiencies are often called hidden hunger and affect many hundreds of millions of people, having deleterious effects on health, particularly in vulnerable groups such as pregnant women, infants and children, and elderly people. These are discussed extensively in this chapter because they are so important for public health and well-being in high-income countries as well as in middle- and low-income countries. Deficiencies are great challenges for public health nutrition because they are mostly unapparent clinically and require complex, but inexpensive interventions.

Sources: *World Health Organization. Trade, foreign policy, diplomacy and health, food security. Geneva: WHO. Available at: http://www.who.int/trade/glossary/story028/en/ [Accessed 2 April 2013].*
Food and Agriculture Organization of the United Nations. World summit in food security. Rome: FAO. Available at: http://www.fao.org/wsfs/en/ [Accessed 2 April 2013].
United States Department of Agriculture, Economic Research Service. Food security in the US [updated 4 September 2012]. Washington, DC: USDA. Available at: http://www.ers.usda.gov/topics/food-nutrition-assistance/food-security-in-the-us/measurement.aspx#measurement [Accessed 2 April 2013].

and carbohydrates are burned in the body and developed methods to analyze the composition of foods, body tissues, urine, and feces. In the mid-nineteenth century, scientists in France, Germany, and later in Britain and the USA made rapid advances in the chemistry of oxygen, carbon dioxide, calcium, iodine, and iron. In 1897, Christian Eijkman demonstrated the origin of beriberi by showing the disease in populations eating polished rice and its absence when rice was eaten with its husk (Nobel Prize, 1929).

In the early years of the twentieth century, animal studies showed that diets consisting of only pure protein, carbohydrates, and fats led to failure to thrive, sickness, and

death. These studies led Kasimir Funk to articulate the idea of "vital amines" missing in the diet, later termed *vitamins*, and the isolation of chemical substances with antineuritic and antiberiberi properties. The delineation of the vitamins necessary to prevent disease and promote health produced immeasurable improvement in human health where the knowledge was applied.

The development of applied nutritional public health interventions in the USA was led by the federal US Department of Agriculture (USDA) and Department of Education; the Department of Health and Human Services (DHHS) also contributed greatly to this evolution.

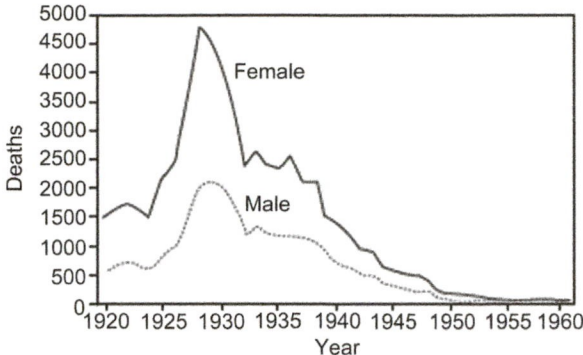

FIGURE 8.1   Reported pellagra deaths, by gender and year, USA, 1920–1960. *Source: Centers for Disease Control and Prevention. Achievements in Public Health, 1900–1999: safer and healthier foods. MMWR Morb Mortal Wkly Rep 1999;48:905–13. Available at: http://www.cdc.gov/mmwr/preview/mmwrhtml/mm4840a1.htm [Accessed 28 November 2013].*

A severe epidemic of pellagra occurred in the USA between 1907 and 1940. It was especially severe in the southern states among inmates of mental hospitals and prisons, poor sharecropping farmers, and millworkers, with an estimated 3 million cases and over 100,000 deaths. Pellagra is characterized by dermatitis, diarrhea, dementia, and death (the "4 Ds"). Pellagra is common where corn is the common base of a diet and may also develop as a consequence of gastrointestinal diseases or alcoholism. Joseph Goldberger of the US Public Health Service was sent to investigate the pellagra epidemic in 1914 (see Chapter 1). Pellagra was generally thought to be infectious in origin, but Goldberger concluded that the disease was linked to a deficiency in the diet and that dietary change eliminated the disease in institutional residents. Goldberger's findings aroused great opposition in the south. The specific deficiency was later shown to be of niacin (vitamin B_3), corrected and prevented by dietary changes. The epidemic curve of deaths, which were more common among women than men, rose rapidly during economic stress due to falling cotton prices and wages in the early 1920s, but declined rapidly as a result of thiamine fortification of flour and improved nutrition as well as the decline of the Great Depression (Figure 8.1).

The investigation of pellagra and subsequent dietary recommended food allowances, national food supplementation, and school lunch programs established nutrition as a central issue in public health in the USA. In many parts of the world, application of the knowledge of the fundamental importance of essential vitamins and trace minerals has still to be implemented as applied nutritional public health measures. This experience and evidence of poor nutrition among draftees to the US army both in World War I and in the lead-up to World War II led to the calling of White House Conferences on Nutrition, the establishment of recommended dietary allowances of nutrients in 1941, and the issuing of US War Order Number One mandating fortification of flour, salt with iodine, and milk with vitamin D to protect civilian health during times of austerity in the coming war (MMWR, 1999).

NUTRITION IN A GLOBAL CONTEXT

Worldwide, malnutrition affects one in three people. It affects all age groups, but is most devastating among children. It is associated with poverty, and inadequate access to health education, clean water, and good sanitation. Malnutrition contributes to half of all childhood deaths due to poor feeding practices, contaminated water, diarrheal and respiratory illnesses, as well as food deficiencies. Developing countries also suffer from a growing burden of chronic diseases associated with changing dietary patterns, including increased consumption of fats, particularly saturated fat, and refined carbohydrates (Box 8.2). These patterns are combined with an increasingly sedentary lifestyle (WHO and FAO, 2005).

The people most affected by poverty are women, children, and the elderly; poor nutrition is one of the key mediators of poverty and poor health. The effects of maternal malnutrition on newborns can be lifelong. Micronutrient deficiencies, despite their preventability by low-cost public health measures, are still widespread. Yet, while millions are starving or in a chronic state of undernutrition, there is enough food in the world to feed everyone. National strategies for combating the effects of poverty are needed based on good governance, suitable legislation, participation by all in society, and provision of effective basic services. Poverty, unequal distribution of income and food, ignorance, low education levels, large family size, acute and chronic illness, and government inaction are major contributors to world hunger (Table 8.1). The proportion of people living below the poverty line is as high as 80 percent in some countries (Chad and Haiti); and industrialized countries such as the USA, Germany, Belgium, and Chile have rates of 15 percent.

The cumulative effects of poverty, population, environmental degradation, overfarming of land, erosion of topsoil, harmful agricultural practices, and inadequate storage and transportation are vital in global health. Malnourished people, especially children, are susceptible to diseases including diarrhea and acute respiratory infections, many of which are vaccine preventable. Severe malnutrition resulting in wasting and stunting is most severe in the least developed countries (Table 8.2).

Micronutrient deficiencies are increasingly recognized as fundamental public health issues in low- and medium-income and also in high-income countries. Households in developing countries using iodized salt have increased markedly from 20 percent in the 1990s. Globally, for children under 5 years, nearly three-quarters of households have iodized salt, but high rates of use of non-iodized salt are seen in sub-Saharan Africa, the Middle East, North Africa, and South Asia (largely India). Nigeria has had excellent success in implementing and enforcing mandatory iodization of salt, as has Georgia since 2006. Even in Europe, control of iodine deficiency is highly variable.

BOX 8.2 Undernutrition, Malnutrition, and Overnutrition

Malnutrition, globally, causes about half of all deaths. China and Brazil have reduced malnutrition in children by more than half in less than 20 years. The world prevalence of hunger has increased since the late 2000s with the economic pressures facing all countries, including in high-income countries with high levels of unemployment and inadequate social safety nets.

Micronutrient malnutrition or "hidden hunger" affects around 2 billion people (over 30 percent of the world population), with serious public health consequences. Globally, the WHO estimates the prevalence of underweight in 18.3 percent (115 million) and stunting in 20.7 percent (186 million) of the population. Worldwide, 43 percent of children under age 5 years are chronically malnourished. About 80 percent of the developing world's children with stunting live in 24 countries, primarily in Asia and Africa. The combined effects of prolonged underinvestment in nutrition, economic downturns and events such as the increase in natural disasters have led to increased hunger and poverty in developing countries, with negative impacts on progress in meeting the Millennium Development Goals.

Undernutrition, micronutrient deficiencies, and overnutrition are a triple challenge for preventive health with impacts on the incidence and mortality of both communicable and non-communicable diseases. This triad is prevalent worldwide, related to poverty in low-, medium-, and high-income countries. In highly industrialized societies it usually affects particular segments of the population such as the poor, elderly, and unemployed, but may be found among those who self-starve, resulting in bulimia and anorexia.

The effects of undernutrition and malnutrition are staggering. About half of the 4–5 billion iron-deficient people are anemic. Those most at risk are pregnant women and children in developing countries. The WHO estimates that one-third of the 10 million deaths among children under the age of 5 years are a result of undernutrition. About one-third of these children in developing countries are stunted due to the effects of undernutrition and 148 million children are underweight.

At the same time, 43 million children under 5 years of age are overweight. More than 1.4 billion adults are overweight. Obesity has doubled since 1980, affecting some 500 million adults, and is increasingly prevalent in low- and middle-income countries, placing these people at risk for reduced quality of life and premature death from micronutrient deficiencies as well as comorbid conditions of communicable and non-communicable diseases.

Sources: Adapted from United Nations Children's Fund. Tracking progress on child and maternal nutrition. New York: UNICEF; 2009. Available at: http://www.unicef.org/publications/files/Tracking_Progress_on_Child_and_Maternal_Nutrition_EN_110309.pdf [Accessed 13 April 2013].
World Health Organization. Obesity and overweight. Fact sheet no. 311 [updated March 2013]. Available at: http://www.who.int/mediacentre/factsheets/fs311/en/ and http://www.who.int/nutrition/EB128_18_Backgroundpaper1_A_review_of_nutritionpolicies.pdf
Food and Agriculture Organization. Food insecurity table. 2009. Available at: ftp://ftp.fao.org/docrep/fao/012/i0876e/i0876e05.pdf
Central Intelligence Agency. World fact book; 2012. Available at: https://www.cia.gov/search?q=malnutrition&site=CIA&output=xml_no_dtd&client=CIA&myAction=%2Fsearch&proxystylesheet=CIA&submitMethod=get&x=0&y=0

NUTRITION AND INFECTION

The infectious disease–malnutrition cycle causes millions of deaths of children yearly from preventable causes. The control of measles and wide use of oral rehydration therapy alone would save hundreds of thousands of lives and improve the nutritional status of millions of children. Malnutrition and infection interact, each exacerbating the other. Infections such as measles may have a case fatality rate as high as 2 per 1000 in an industrialized country. Measles is one of the leading causes of death among young children, even though a safe and cost-effective vaccine is available. In 2011, there were 158,000 measles deaths globally – about 430 deaths every day or 18 deaths every hour. Up to 10 percent of measles cases result in deaths, usually in developing countries where there are widespread deficiencies of essential nutrient elements such as vitamin A. Vitamin A supplements have been shown to reduce the number of deaths from measles by 50 percent.

All children in developing countries diagnosed with measles should receive two doses of vitamin A supplements, given 24 hours apart. The case fatality rate for young children may be as much as 5–6 percent, with deaths clustered primarily in India and Africa, but 50 or more per 1000 in a developing country where there are widespread deficiencies of essential nutrient elements such as vitamin A. Conversely, even a relatively common infection can adversely affect a child's nutritional state, growth pattern, and resistance to further infection. The time required to compensate for nutrient losses from infection following recovery may be two to three times as long as the duration of the infection itself.

Normal growth in weight and height indicate that a child is more likely to resist infection or prolongation of infections that occur. Child health survival strategies rest on the twin pillars of infectious disease control and nutritional adequacy. Nutritional deficiency may also be a factor in reduced resistance to infection in elderly and immunocompromised people, as both groups may be socially and economically marginalized in industrialized societies.

FUNCTIONS OF FOOD

Consumed food substances provide varying levels of energy and essential requirements for growth and maintenance of body functions. Exercise and moderate eating habits maintain body weight and reduce the risk for chronic diseases

TABLE 8.1 Nutritional Conditions in Developing and Developed Countries

Issues	Factors in Developing Countries	Factors in Developed Countries
Poverty	Lack of resources or access to healthy foods; traditional crops and foods inadequate for healthy diet	General prosperity but poverty in significant minorities; supplementation programs; e.g., WIC, food stamps, school lunches offset deprivation
Educational deprivation	Lack of knowledge of good nutrition	High awareness of good nutrition, but obesity common, especially among the poor
Micronutrient deficiencies	Widespread deficiencies of iron, iodine, and vitamins A, B, C, and D; inadequate food processing technology	Food fortification with iodine, iron, and vitamins A, B, C, and D
Malnutrition–infection cycle	Poor sanitation; lack of control of vaccine-preventable diseases, parasitic, diarrheal and respiratory diseases; HIV, TB, and malaria all harm nutrition status	High levels of sanitation, vaccination coverage, and hygiene with benefit to nutritional state; high-risk subgroups: HIV, TB, drug abusers, the homeless
Food security	Harmful agriculture practices; overgrazing and overfarming; excessive rain or drought, soil erosion, small plots; inadequate and wasteful storage and distribution facilities; food costly relative to incomes	Science-based technological agriculture; high productivity, abundant produce, variety, food cheap relative to family income
Unhealthy diets and non-communicable diseases	Undernutrition with poverty, inadequate calories per person; middle-class overnutrition with high rates of CVD, diabetes	Overnutrition with excess animal fat intake; high rates of CVD; pockets of poverty and nutritional inadequacy; CVD mortality falling
Breastfeeding (capacity to care)	Common and prolonged but with poor supplementation practices	Increasing and well supplemented
Food quantity, quality, variety, and cost	Contamination, waste and destruction common; lack of supply and variety of vegetables, fruit, protein; costly, wasteful	Good supply and quality; good distribution, and marketing systems; labeling, regulation, and supervision; comparatively inexpensive
Monitoring	Monitoring supply, distribution, intake of food; growth, anemia, and intake studies needed	Nutrition surveys, including anthropometric, biochemical, and dietary intake studies
Need for national policies and objectives	Prevent deforestation and land loss; rural poverty, lack of education; lack of credit and agricultural support; poor and harmful farming practices; pesticides used dangerously	Government promotes agriculture, research, support systems, transport, and marketing; limit use of pesticides; industrialized farming

Note: WIC=Special Supplemental Nutrition Program for Women, Infants, and Children; HIV=human immunodeficiency virus; TB=tuberculosis; CVD=cardiovascular disease.
Source: Adapted from Committee on World Food Security. Social protection for food security: a report by the High Level Panel of Experts on Food Security and Nutrition of the Committee on World Food Security, Rome 2012. Available at: http://www.fao.org/fileadmin/user_upload/hlpe/hlpe_documents/HLPE_Reports/HLPE-Report-4-Social_protection_for_food_security-June_2012.pdf [Accessed 16 July 2013].

TABLE 8.2 Percentages of Children Under Age 5 by Type of Malnutrition, Vitamin A Supplement Coverage, and Households With Iodized Salt, Developing and Least Developed WHO Regions, 2006–2010

	Underweight	Wasting	Stunting	Vitamin A Supplement	Iodized Salt (households)
	2006–2010	2006–2010	2006–2010	2010	2006–2010
Africa	19	9	38	86	55
Sub-Saharan	20	9	39	86	53
Eastern and Southern	15	6	39	80	65
West and Central	23	11	40	90	–
Middle East and North Africa	11	9	28	–	48
Asia	27	13	34	56[a]	74
South Asia	42	19	47	50	55
East Asia and Pacific	10	6	19	84[a]	88
Latin America and Caribbean	4	2	15	–	89
Developing countries	18	10	29	66[a]	71
Least developed countries	25	10	41	88	61
World	16	10	27	86[a]	73

Note: [a]Excludes China.
Source: United Nations Children's Fund. The state of the world's children, 2012. Table 2 Nutrition. Available at: http://www.unicef.org/sowc2012/pdfs/SOWC-2012-TABLE-2-NUTRITION.pdf and http://www.unicef.org/sowc2012/pdfs/SOWC- [Accessed 16 July 2013].

associated with excess body fat, such as diabetes, hypertension, cardiovascular disease (CVD), and some cancers.

Nutrients have specific roles within the body, but their functions are interdependent. The diet of the individual determines nutrient availability. Consequently, it is important to identify the sources of proteins, carbohydrates, fats, vitamins, and minerals available in common foods in a target community. The six important nutrient groups are macronutrients (carbohydrates, fats, and proteins), micronutrients (trace minerals and vitamins), and water. The macronutrients provide energy, essential and other amino acids, and essential fatty acids. Micronutrients are required for utilization of that energy.

The body processes foods into simpler forms in order to absorb them by digestion through a continuous mechanical and chemical process in the digestive tract. Foods are first ground up by chewing, requiring good dentition. Mixing of the food with saliva and swallowing brings the partially digested food to the stomach and small intestine, where it is acted on by enzymes. These enzymes break the food down into smaller and smaller fragments which can then be absorbed through the walls of the small intestine to enter the bloodstream. Disease of the gastrointestinal tract can interfere with this process.

Pancreatic enzymes are released into the small intestine as proteases (which split proteins), amylases (which split polysaccharides), and lipases (which break down fats). The pancreas also produces insulin, vital for control of blood sugars. Carbohydrates are absorbed as sugar and stored to provide energy in the liver and muscles as glycogen, which is released into the bloodstream to sustain sugar levels. The liver also stores fat-soluble vitamins and manufactures enzymes, cholesterol, proteins, vitamin A, blood coagulation factors, and bile salts which are released into the intestine to help in absorption.

The human body is composed of approximately 62 percent water, 17 percent protein, 13 percent fat, 6 percent minerals, and 2 percent carbohydrates, by weight. Body composition can vary with stress and nutritional status. Body stores are used during periods of food deprivation. During starvation, depletion of carbohydrates is made up by synthesis from reserves of fat and protein. Depletion of up to 10 percent of total body water can occur without serious risk, but in small children the margin of safety is smaller.

HUMAN NUTRITIONAL REQUIREMENTS

Calculation of the appropriate amounts of nutrients for age and gender groups is a complex task. Energy from food is converted into mechanical work (up to 25 percent) and growth, dissipated as heat, and used in maintaining the basic functions and temperature of the body. The international unit of energy is a joule, but the more commonly used measure in nutrition is the kilocalorie, or calorie

(1 kcal = 1 cal = 4.1868 kilojoules, kJ). Technically, one calorie is the amount of energy required to raise the temperature of one gram of water by one degree centigrade. The calorie measure used commonly to discuss the energy content of food is actually a kilocalorie or 1000 real calories. This is the amount of energy required to raise the temperature of one kilogram of water (about 2.2 pounds) by one degree centigrade.

Need for energy intake varies with body size and is increased by activities of work and recreation. People with a sedentary lifestyle and average body size will need less food intake than those with moderate or high levels of activity or greater body size to maintain the status quo.

Both deficiency and excessive dietary intake of any nutrient can cause disease or death. The range of intake needed for optimal physiological function depends on age, body size, gender, and activity level, as well as on pregnancy, disease, or injury. The range of intake needed for optimal health emphasizes that there are subclinical phases of basic undernutrition and overnutrition. Table 8.3 lists the essential nutrients and their functions in the human body.

For nutrients and energy sources, there is a range of intake that confers optimal physiological function. Below this range, deficiencies can cause disease or death. Excessive intake, in some cases, also can lead to toxicity. The optimal range varies for each nutrient and is affected by many individual and environmental factors.

Carbohydrates

Carbohydrates are a major source of energy (4 kcal/g) used for metabolic processes and for producing cellular substances including enzymes and cell membranes. Carbohydrates are classified as monosaccharides or disaccharides (simple carbohydrates) or as polysaccharides (complex carbohydrates).

Monosaccharides are found as glucose and fructose in fruits and honey. They are simple sugars that can be absorbed in the gut without any digestive process. Disaccharides are made up of two monosaccharides and are commonly found in fruits and vegetables, including sugar beets and sugar cane, and as lactose in dairy products.

Polysaccharides are larger molecular structures of monosaccharides linked together. Disaccharides and polysaccharides must be broken down during digestion into monosaccharides before they can be absorbed. Excess glucose is stored as glycogen in the liver and muscles (not fat tissue). Simple monosaccharides are less healthy in that they promote rapid absorption, higher blood sugar, obesity, and diabetes. Diets rich in complex carbohydrates (polysaccharides) and fiber are associated with a reduced risk of cancer and CVD. Limiting the intake of simple carbohydrates is fundamental to a healthy diet.

TABLE 8.3 Essential Nutrients and Their Functions

Dietary Components	Types	Functions
Carbohydrates	Energy (4 kcal/g); sugar and starches	Provides efficient source of energy; water soluble, easily transported and available in tissue fluids; should comprise 40–85% of energy intake
Proteins	Energy (4 kcal/g); essential amino acids	Provides amino acids and building material for all body cells, especially muscle and bone; should comprise 10–15 percent of energy intake
Fats and oils	Energy (9 kcal/g); essential fatty acids (linoleic and linolenic acids)	Concentrated energy source; transports fat-soluble vitamins; enhances flavor; should comprise 25 percent of diet, mostly non-animal sources
Minerals	Arsenic, boron, calcium, chromium, copper, fluoride, iodine, iron, magnesium, manganese, molybdenum, nickel, phosphorus, selenium, silicon, vanadium, zinc	Essential for building healthy body tissue and fluids, blood, hormones, electrolyte balance; bones and teeth
Vitamins	Fat soluble (A, D, E, K); water soluble	Healthy body tissues of bone, muscle, blood, central nervous, and immunological systems
Fiber	Vegetable matter	Food bulk and prevention of cancer
Water		Fluid and body tissue balance; conveyer of food and water-soluble vitamins

Sources: Allen L, de Benoist B, Dary O, Hurrell R, editors. Guidelines on food fortification with micronutrients. Geneva: WHO; 2006. Available at: http://www.who.int/nutrition/publications/micronutrients/9241594012/en/ and http://whqlibdoc.who.int/publications/2006/9241594012_eng.pdf [Accessed 16 July 2013].
Institute of Medicine. Dietary reference intakes, 2005. Available at: http://www.nap.edu/openbook.php?isbn=0309085373/ [Accessed 29 April 2013].
National Research Council. Dietary reference intakes for energy, carbohydrate, fiber, fat, fatty acids, cholesterol, protein, and amino acids (macronutrients). Washington, DC: National Academies Press; 2005. Available at: http://www.nap.edu/catalog.php?record_id=10490 [Accessed 19 April 2013].

Proteins

Proteins are large molecules made up of chains of amino acids that are broken down by the digestive process into their component units (1 g protein yields 4 kcal). There are 20 common amino acids in biological materials required by the body. Humans lack the ability to synthesize nine of these amino acids, thus it is essential to obtain them by consuming protein from animal sources or combinations of foods, such as legumes and cereals. Young children and adolescents require protein for their growth spurts. Proteins function in the body as structural components of cells and tissues, enzymes which act as catalysts for chemical reactions, and hormones which act as chemical messengers. Lack of protein and calories in the diet is called protein–energy malnutrition (PEM).

Fats and Oils

Foods of animal and plant origin include a variety of substances known as fats and oils (lipids) that are soluble in organic solvents but not in water. Dietary fats are broken down in the gut for absorption into the body to provide energy and fatty acids needed for many physiological functions. They provide a concentrated form of energy (9 kcal/g as compared to 4 kcal/g for proteins and carbohydrates, and 7 kcal/g for alcohol). They also provide essential fatty acids

needed for production of hormones, cell membranes, and other substances. Linoleic and linolenic acids are essential fatty acids which cannot be synthesized in the body but can be retrieved from animal fats and plant sources such as walnuts and flaxseed oil.

Fats or lipids stored in body fat tissue insulate and protect vital organs, insulate the body against heat loss, and provide energy during periods of reduced consumption or greater body need of energy in periods of growth, illness, or injury. Fats provide for production of bile acids needed for absorption of fat-soluble vitamins (A, D, E, and K).

Dietary fats consist of mixtures of saturated, mono-unsaturated, and polyunsaturated fats (typically liquid at room temperature or when chilled), depending on their chemical structure. Most of the healthy unsaturated fats that should be eaten should come from polyunsaturated fats and monounsaturated fats. In general, nuts, seeds, vegetable oils – such as olive, canola, and safflower – and fish are excellent sources of unsaturated fats. Solid fats should be cut away from meats and those used in cooking should be replaced by liquid oils, such as olive and canola oils. All fats are high in calories, but replacing saturated fats with unsaturated fats is a step towards a healthier diet.

The degree of saturation of fatty acids is based on the number of double bonds in the side chains of molecules made up of carbon, hydrogen, and oxygen atoms. Fats from animal sources (meat, poultry, fish, and dairy products) are mostly saturated fats (i.e., contain no double bonds). Fats from plant sources such as sunflower, olive, or peanut oils are monounsaturated (i.e., contain side chains with one double bond) and are preferable to saturated or polyunsaturated fats. Whereas coconut and palm oils are high in saturated fats, fish are an excellent source of unsaturated fats, including omega-3 fatty acids, known to reduce harmful cholesterol levels.

Foods of animal origin contain high amounts of saturated fats and cholesterol. Cholesterol, which can also be synthesized in the body, is needed for the synthesis of sex hormones, vitamin D, and cell membranes. Excess dietary intake of saturated fats and cholesterol increases the risk of atherosclerosis, CVD, diabetes, and some forms of cancer.

Vitamins

Vitamins are organic compounds that are essential in small amounts for specific functions of the body for health, growth, reproduction, and resistance to infection (Table 8.4). They differ in physical and chemical properties and in biological functions. Vitamins function in highly specialized metabolic processes. They cannot be synthesized in sufficient quantity by the body alone, and must be obtained from the diet, or from fortified foods and from supplements. Many diets are deficient in these needed elements and along with trace minerals should be added to foods commonly eaten by most people, such as cooking oils, flour, salt, and sugar. Fortification of basic foods reaches almost all of the population and is justified in science and in practice over the past century. Supplements are essential for certain groups in the population who need more than that which can be made available by food fortification.

Vitamins are classified based on their solubility, either in fat or in water. Fat-soluble vitamins (A, D, E, and K) are found in high concentrations in the fatty portions of food. Excretion of excess intake of this type of vitamin is minimal. Vitamin C and those of the B-complex group are water soluble and should be supplied in adequate amounts in the daily diet, as they are easily excreted.

Storage of the water-soluble vitamins in the body is limited, so regular sources are even more essential than for the fat-soluble vitamins, which are stored in body fat and the liver. Deficiencies of either in the diet lead to depletion of body stores, followed by non-specific symptoms (fatigue, confusion, weakness, neuritis, and reduced resistance to infection) before classic deficiency can be recognized clinically. A deficiency condition of even one vitamin can jeopardize health.

Vitamins are present in natural foods, and an appropriate diet should supply most vitamin needs; however, since eating such a balanced diet is often problematic, food enrichment or supplementation is necessary, especially for vulnerable groups (e.g., children, elderly, adolescents, and institutionalized patients). Enrichment means replacing nutrients in foods to levels found in the natural product before processing. For example, enrichment of white flour should replace the 22 natural elements that are normally present in whole grains but are removed in processing, including B vitamins, vitamin D, calcium, and iron salts.

Minerals and Trace Elements

Minerals are distributed in a variety of foods, but are usually present in limited amounts. Diets must contain a sufficiency and variety of foods to meet daily requirements. Eighteen known minerals are required for body maintenance and regulatory functions. Of these, dietary reference intakes (DRIs) have been established for seven: calcium, iodine, iron, magnesium, phosphorus, selenium, and zinc. Other active minerals in the body include sodium, chloride, potassium, chromium, cobalt, copper, fluoride, manganese, molybdenum, sulfur, and vanadium. Sodium, potassium, chloride, and calcium are particularly crucial for electrolyte balance in the blood and body tissues. Essential trace minerals also include boron, silicon, nickel, and arsenic for optimal growth and membrane function. The body needs a small but continual intake of these elements for its structure and function. If metabolic needs are not met, a deficiency ensues. Deficiency disorders vary with the mineral element involved, the duration and extent of dietary intake deficiency, and depletion of body stores (Table 8.5).

GROWTH

Growth is not a steady progression, but a process during which nutrition requirements are determined by a genetic timetable, affected by nutritional intake and health status. Optimal growth occurs only if the organs and tissues receive the nutrients needed for synthesis of proteins and other molecules. Insufficient energy and protein (i.e., PEM) is common in developing countries or deprived populations and causes failure to thrive, stunting, and wasting. Deficiencies of essential minerals and vitamins also adversely affect growth and development. Iodine deficiency slows thyroid hormone production and causes adverse developmental effects. Lack of micronutrients such as vitamin A or D or minerals such as iron, iodine, calcium, and phosphorus adversely affects the growth and development of epithelial cells, bone, and red blood cells. Measurement of growth and development is one of the most important health status indicators of individual and population health.

MEASURING BODY MASS

The body mass index (BMI) is a standard method of measuring body size. It encompasses height and weight in one

TABLE 8.4 Essential Vitamins

Vitamins	Source and Activity in Body	Deficiency Condition
Water-soluble	Absorbed in intestines, excreted in urine, so very large amounts are required to produce overdosage. Body needs adequate daily intake or tissue depletion occurs within weeks or months; they are essential for enzymes to catalyze biochemical reactions in energy production, biosynthesis, and nervous system development and maintenance	
Vitamin B complex	Thiamine, riboflavin, niacin, pyridoxine, cobalamin, and folic acid; sources are whole-grain cereals, legumes, leafy vegetables, meat, and dairy foods	Loss of memory, mental confusion; occurs with chronic illness, alcoholism, dietary restriction; may lead to serious clinical conditions, such as amnesia, dementia, heart failure, neurological disorders, death
Vitamin B_1 (thiamine)	Part of enzyme systems for release of energy from carbohydrates	Beriberi, anorexia, emotional lability, depression, fatigue, constipation; cardiomyopathy, cardiac failure; polyneuritis; Wernicke's encephalopathy; Korsakoff's psychosis, amnesia, dementia, death
Vitamin B_2 (riboflavin)	Enzymes for metabolism of protein and carbohydrates	Dry skin and mucous membrane disorders, stomatitis, photophobia, blurred vision, polyneuritis
Vitamin B_3 (niacin, nicotinamide, nicotinic acid)	Maintains normal gastrointestinal and nervous system	Pellagra; gastrointestinal, skin, and neurological changes; depression, psychosis, neuropathy, dermatitis, diarrhea, dementia, death
Vitamin B_5 (pantothenic acid)	Metabolism of carbohydrates, proteins and fats	
Vitamin B_6 (pyridoxine)	Part of enzyme process in protein metabolism	Irritability, depression, muscle weakness, cardiomyopathy, liver damage; prevents neuropathy in isoniazid therapy for tuberculosis
Vitamin B_7 (biotin)	Skin condition	Acne
Vitamin B_9 (folic acid, folate)	Red blood cell formation; homocysteine metabolism	Megaloblastic anemia of pregnancy; neural tube defects (spina bifida and anencephaly); elevated homocysteine, possible link to coronary heart disease and mental deterioration with aging
Vitamin B_{12} (cobalamin)	Found only in foods of animal origin; essential for red blood cell formation	Macrocytic anemia; peripheral neuropathy, pernicious anemia, mental retardation
Vitamin C (ascorbic acid)	Source: fruits and vegetables; needed to form and maintain intercellular substances	Scurvy, poor bone and cartilage formation, anemia, stunting, infections, bleeding
Fat-soluble	Found mainly in fat component of food; absorbed, transported along with fat; requires bile and dietary fats; stored in body fats and takes longer to deplete than water-soluble vitamins	
Vitamin A (retinol)	Found in yellow vegetables; essential for epithelial cells of mucous membranes; regulation of vision in dim light	Night blindness, loss of color vision, dryness, corneal ulceration and scarring, blindness; poor bone and tooth formation; susceptibility to infection and poor survival rates from infectious diseases
Vitamin D (calciferol)	Produced in skin by exposure to sun; found in enriched foods (milk products); enhances calcium and phosphorus utilization, bone growth, thickness, and density	Rickets, stunting, soft bones, bowed legs, carious teeth, impaired development of bone length; osteomalacia, soft bones, fractures, muscle pain in adults; contributes to osteoporosis
Vitamin E (tocopherol)	Found in leafy green vegetables, legumes, nuts; protects fat from oxidation (antioxidant) and red cell breakdown	Low birth weight; hemolysis of the newborn; degenerative disorders
Vitamin K; phylloquinone (vitamin K_1), menanquinone (vitamin K_2)	Spinach, cabbage, cauliflower; formation of prothrombin	Hemorrhagic disease of newborn, prolonged clotting time

Note: Coenzymes including folate (vitamin B_{90}), pantothenate (vitamin B_5), cobalamin (vitamin B_{12}), biotin (vitamin B_8), and molybdenum cofactor (Moco) play essential roles in energy transfer and many vital metabolic processes.
Source: Adapted from Allen L, de Benoist B, Dary O, Hurrell R, editors. Guidelines on food fortification with micronutrients. Geneva: WHO; 2006. Available at: http://www.who.int/nutrition/publications/micronutrients/9241594012/en/ and http://whqlibdoc.who.int/publications/2006/9241594012_eng.pdf [Accessed 9 April 2013].

TABLE 8.5 Essential Minerals

Minerals	Activity in Body	Effect of Deficiency	Food Source
Calcium	Builds and maintains bone structure and teeth; muscle and cardiac function; blood coagulation; neuromuscular irritability	Poor bone and teeth formation; rickets in children, osteoporosis in elderly	Fortified milk, hard cheese, egg yolk, cabbage, clams, cauliflower, soybeans, spinach
Iron	Constituent of hemoglobin, muscle and bone; carries oxygen in red blood cells; between 4 and 5 billion people suffer from iron deficiency and an estimated 2 billion are anemic	Iron-deficiency anemia, fatigue, poor linear growth in infants, psychomotor deficiency affecting school and work performance. 50 percent of pregnant women and 40–50 percent of children under 5 in developing countries are iron deficient	Liver, red meat, turkey, legumes, egg yolk, peaches, apples, raisins, prunes, molasses
Iodine	Constituent of thyroxine, normal thyroid function needed for growth and mental development	Stunting, retardation, cretinism	Iodized salt, seafood; UNICEF reports households in the developing world with adequately iodized salt rose from <20 percent in 1990 to >70 percent today
Phosphorus	Builds and maintains bones, teeth, cells, body fluids	Poor teeth, stunting, rickets	Milk, cheese, egg yolk, meat, legumes, cereals, nuts, vegetables
Chloride	Needed to maintain acid–base and osmotic balance in body fluids	Lost in diarrhea, with dehydration; shock and death in children or elderly from diarrheal diseases	Milk, salt, fish, cheese
Copper	Needed for the central nervous system, hemoglobin formation	Anemia, liver function, metabolism of ascorbic acid	Seafood, nuts, whole-grain cereals, liver, meat, legumes, vegetables
Fluoride	Strengthens tooth enamel, bone formation	Dental caries and osteoporosis in the elderly	
Magnesium	Constituent of bones and teeth, enzymes; needed for cardiac and neurological functions	Cardiac arrhythmia, nervous irritability	Same as diet for calcium and phosphorus
Sodium	Intracellular and extracellular fluid balance, muscle and nerve irritability	Fluid loss, circulatory collapse	Table salt, meat, milk
Potassium	Electrolyte balance, cardiac arrhythmia	Fluid loss, circulatory collapse, and muscle irritability	Vegetables, cereals, fruits, bananas, melons
Selenium	Antioxidant	Changes in biochemical systems, cardiomyopathy, carcinogenesis, liver damage	Meat, seafood, cereals, grains

Source: National Research Council, Food Nutrition Board. Dietary reference intakes for energy, carbohydrate, fiber, fat, fatty acids, cholesterol, protein, and amino acids (macronutrients). Washington, DC: National Academies Press; 2005. Available at: http://www.nap.edu/openbook.php?record_id=10490&page=R2 [Accessed 27 November 2013] and http://www.nap.edu/openbook.php?isbn=0309085373 [Accessed 21 April 2013].

number, summarizing nutrition status. BMI is useful clinically for the individual patient and as a description of nutritional status of a community population based on survey data. It is calculated as follows:

$$BMI = \text{Body weight in kilograms}/(\text{Height in meters})^2$$

or

$$BMI = \text{Body weight in pounds} \left[\times 703 (\text{Height in inches})^2 \right]$$

Obesity is defined for males and females as a BMI above 30, overweight as a BMI between 25 and 30, and undernutrition as a BMI below 18.5. Obesity is reaching pandemic proportions globally, with increasing prevalence in developing and least developed countries. More than 1.4 billion adults and 40 million children are overweight, with an annual death toll of overweight of 2.8 million. The World Health Organization (WHO) reports that between 1980 and 2008 obesity doubled to 11

TABLE 8.6 World Health Organization Interpretation of Body Mass Index (BMI)

Classification	BMI: weight (kg)/ height (m²)
Underweight	<18.50
Severe thinness	<16.00
Moderate thinness	16.0–16.99
Mild thinness	17.0–18.49
Normal range	18.50–24.99
Overweight	>25.00
Preobese	25.0–29.99
Obese	>30.00
Obese class 1	30.00–34.99
Obese class II	35.00–39.99
Obese class III	≥40.00

Source: World Health Organization. Global database on body mass index. BMI classification [updated 5 January 2013]. Available at: http://apps.who.int/bmi/index.jsp?introPage=intro_3.html [Accessed 21 April 2013].

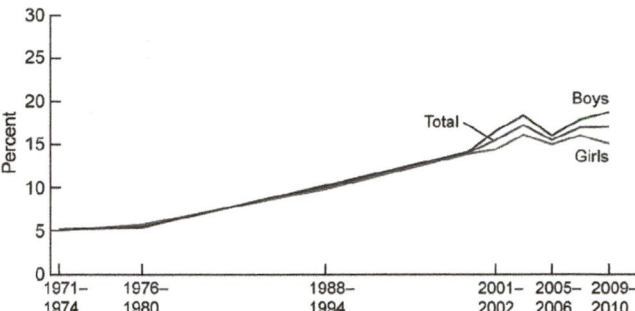

FIGURE 8.2   Trends in child and adolescent overweight, USA, 1971–2010. *Source: National Center for Health Statistics. Prevalence of overweight and obesity among children and adolescents, United States, 1971–2010. Available at: http://www.cdc.gov/nchs/data/hestat/obesity_child_09_10/obesity_child_09_10.pdf [Accessed 28 April 2013].*

percent of the population. The WHO Nutrition Landscape Information System (NLIS) provides country profiles on obesity.

Body weight is taken as usual weight and expressed as a percentage of desirable weight. The BMI is a convenient measure. Table 8.6 shows the WHO interpretation of the BMI ranges. These provide a useful categorization which may need to be augmented by clinical and other anthropometric measures and have become a guideline for proper nutrition for the health-conscious individual.

According to the WHO criteria, 31.2 percent of adult Americans are of normal body weight. More than two-thirds of US adults (68.5 percent), 63.9 percent of women and 73.3 percent of men, are overweight or obese (BMI ≥ 25) (National Institute of Diabetes and Digestive and Kidney Diseases, 2010). More than one-third of US adults aged 20 and older (35.7 percent, over 72 million people) and approximately 17 percent (or 12.5 million) of children and adolescents aged 2–19 years are obese (BMI ≥ 30). During 1980–2008, obesity rates doubled for adults and tripled for children.

The prevalence of extreme obesity in women (8 percent) is double that of men. Extreme obesity affects more than 1 in 10 blacks (13.1 percent), and about 1 in 20 whites (5.7 percent) and Hispanics (5 percent). The prevalence has steadily increased over the years among both genders, all ages, all racial and ethnic groups, and all educational levels.

From 1960 to 2010, the prevalence of overweight increased from 45 percent to 69 percent in US adults aged 20–74 [National Center for Health Statistics (NCHS), 2010]. The prevalence of obesity during the same period

more than doubled among adults aged 20–74, from 13 to 36 percent (extreme obesity increased from 0.9 to 6.6 percent), with most of this rise occurring since 1980. Figure 8.2 shows the increase in overweight and obesity rates among children in the USA from 1970 to 2010. Health costs are at least US$147 billion for adults and children (Flegal et al., 2012).

In 2010, children consumed fewer calories than in 2009. The distribution of protein, carbohydrate, and fat consumption varied by racial and ethnic group but within the recommended micronutrient ranges. However, on average, about 12 percent of caloric intake was from saturated fats. Obese preschoolers are more likely to have cardiovascular risk factors of high blood pressure, high cholesterol, and type 2 diabetes. Of the 27.5 million low-income preschool children 15 percent are obese and 2 percent have extreme obesity, with the highest rates among Native Americans and Alaskan natives (NCHS, Prevalence of overweight and obesity, 2012).

Many studies show measurable increases in mortality associated with obesity. Individuals who are obese have a 10–50 percent increased risk of death from all causes, compared with healthy weight individuals (BMI 18.5–24.9). Most of the increased risk is due to cardiovascular causes. In the USA, obesity is associated with over 162,000 excess deaths per year (2004) in the population relative to healthy weight individuals and may be as high as 300,000. The total economic burden of overweight and obesity in the USA and Canada due to excess medical costs, mortality, and disability is approximately US$300 billion per year (US Surgeon General, 2013).

RECOMMENDED DIETARY INTAKES

In 1941, the Committee of Food Nutrition of the National Research Council in the USA developed recommended daily allowances (RDAs) in response to a request from the

US Council of National Defense, which was concerned over possible food shortages and nutritional ill-effects on the health of the population during World War II.

The RDAs were developed by the Food and Nutrition Board of the Institute of Medicine of the US Academies of Sciences, established in 1940 to study adequacy and safety of food supply, and to recommend standards of adequate nutrition. The RDAs represent the levels of essential nutrients needed to adequately meet the nutritional needs of virtually all healthy people. They are used for planning for national emergencies and for the needs of institutionalized or socially deprived people. The RDAs are modified in light of expanding knowledge of nutrition.

These serve to remind us of what should be included in our regular diet, and assist in preventing excessive intake of essential nutritional elements that could be potentially harmful. The US National Academy of Sciences (NAS) revised the RDAs in 1989 and again in 1997.

From 1941 until 1989, the primary goal of RDAs was to prevent diseases caused by nutrient deficiencies. They were established and used to evaluate and plan menus that would meet the nutrient requirements of groups. Other applications of the RDAs included interpreting food consumption records of populations, establishing standards for food assistance programs, and establishing guidelines for nutrition labeling. Technically speaking, the RDAs were not intended to evaluate the diets of individuals, but they were often used in this way.

In the early 1990s, the Food and Nutrition Board of the NAS undertook the task of revising the RDAs, and a new set of nutrient reference values was born in 1997: the dietary reference intakes (DRIs). With the creation of DRIs, the NAS changed the way nutritionists and nutrition scientists evaluate the diets of healthy people. There are four types of DRI values: the estimated average requirement (EAR), the recommended dietary allowance (RDA), the adequate intake (AI), and the tolerable upper intake level (UL). (See Box 8.3 for definitions of these values.) The primary goal of having new dietary reference values was not only to prevent nutrient deficiencies but also to reduce the risk of non-communicable diseases such as osteoporosis, cancer, and CVD.

RDA values were reviewed and published as DRI values and released in stages. The first report, *Dietary Reference Intakes for Calcium, Phosphorus, Magnesium, Vitamin D, and Fluoride*, was published in 1997. Since then, additional reports have been released which address folate and other B vitamins, dietary antioxidants (vitamins C and E, selenium, and the carotenoids); micronutrients (vitamins A and K, and trace elements such as iron, iodine, copper, selenium, and zinc); macronutrients such as dietary fat and fatty acids, protein and amino acids, carbohydrates, sugars, and dietary fiber; as well as energy intake and expenditure.

The DRIs serve as nutritional guidelines and are useful in standardizing food manufacturing, package labeling, and education practices. They have become an essential part of both clinical nutrition and public health nutrition standards internationally. The DRIs are average figures and, although age and gender specific, they do not take into account periods of illness, injury, or physical stress, or the needs of the elderly population. There are, as in many other areas of public health, different points of view on the importance of DRIs. Intake of enriched foods along with additional vitamin or mineral supplements may have the potential for excessive intake if multiple sources of intake are used, as in the case of vitamin A. Recommendations for regular diet and for prevention of hypertension (see Chapter 5) are included in Tables 8.7 and 8.8.

The NAS sets DRIs, which are used by government, industry, researchers, and clinicians for individual and public health purposes. As an example, the NAS recently recommended folic acid and vitamin B supplements for routine use to prevent birth defects and micronutrient deficiency conditions among vulnerable groups in the US population.

DISORDERS OF UNDERNUTRITION

Undernutrition includes at least 25 different deficiency diseases resulting from lack of one of the essential nutrients, proteins, vitamins, fats, or minerals (Table 8.9). A person is more likely to suffer from multiple deficiencies than to be deficient in only one nutrient. Deficiency disorders may be subclinical, and characteristic clinical features may not be apparent.

The WHO and the United Nations Children's Fund (UNICEF) together estimate that severe acute malnutrition affects an estimated 20 million children under the age of 5 worldwide (WHO, 2007). Malnourished children can suffer developmental disability of varying degree, including reduced physical and/or mental ability, often associated with reduced strength (measured as hand-grip strength), impaired cognitive function, and reduced occupational activity.

Malnourished children are subject to illness and complications or death from diseases that would otherwise be less dramatic. Much of this is due to early cessation of breast-feeding, lack of adequate food supplies, ignorance, and failure of public health systems to fortify foods or provide supplements to vulnerable groups. Despite these integral factors, in many situations, the main underlying cause is political chaos or repression, with indifference to poverty and its consequences.

Failure of governments to sustain or promote conditions to ensure adequacy of food supply or distribution systems may cause widespread malnutrition or famine. Political and military conflict, often associated with drought and famine, may produce the conditions in which food production, storage, and marketing break down and prove inadequate for

BOX 8.3 Dietary Reference Intakes: Derivation and Uses

- *Dietary reference intake (DRI)* – qualitatively defines nutrient requirements as the lowest continuing intake level of a nutrient that will maintain a specific indicator of adequacy in an individual, as defined by a respected national governmental agency such as the FDA in the USA, UK, Canada, or other government.
- *Recommended dietary allowance (RDA)* – the average daily dietary intake level that is sufficient to meet the nutrient requirement of nearly all (97–98 percent) healthy individuals in a particular life stage and gender group.
- *Adequate intake (AI)* – a recommended intake value based on observed or experimentally determined approximations or estimates of nutrient intake by a group (or groups) of healthy people, that are assumed to be adequate; used when an RDA cannot be determined.
- *Tolerable upper intake level (UL)* – the highest level of daily nutrient intake that is likely to pose no risk of adverse health effects for almost all individuals in the general population. As intake increases above the UL, the potential risk of adverse effects increases.
- *Estimated average requirement (EAR)* – a daily nutrient intake value that is estimated to meet the requirement of half of the healthy individuals in a life stage and gender group; used to assess dietary adequacy and as the basis for the RDA.

Since their introduction in 1943, RDAs have been the accepted source in nutrient allowances for healthy people. DRIs have been developed for use across the USA and Canada to replace the former RDAs in the USA and the parallel Canadian RNIs to provide recommended nutrition policy for both countries. DRIs provide guidelines for nutrition policies, health practitioners, and educators, and represent a broad set of values to prevent poor nutrition (overnutrition and undernutrition), promote healthy lifestyles, and prevent chronic diseases including developmental handicaps in children.

In 1938, the Canadian Council on Nutrition developed recommended nutrient intakes (RNIs). In 1941, the US National Research Council issued the RDAs for Americans. Both systems, used for dietary assessment and planning diets for individuals and populations, have also served as a standard for nutrition labeling and food fortification.

In 1994, the Food and Nutrition Board of the US Institute of Medicine (National Academy of Sciences), supported by the governments of Canada and the USA, harmonized nutrition reference intakes for North America. In 2005, the Food and Nutrition Board and Health Canada developed DRIs as a practical reference guide for health practitioners and educators.

In the USA, the DRIs are used for food and nutrition policies and programs by the federal, state, and local governments, such as in the Special Supplemental Nutrition Program for Women, Infants, and Children (WIC) program of the Department of Agriculture and Healthy People 2020. They are also used by non-governmental organizations, private industry, and clinicians.

In 2011, the Institute of Medicine concluded that scientific evidence supports a key role of calcium and vitamin D in skeletal health, and determined intake requirements and DRIs for at least 97.5 percent of the population. For calcium the range is 700–1300 mg/day until at least 1 year of age; for vitamin D; DRIs of 600 IU/day for ages 1–70 years and 800 IU/day for age 71 years and older, are needed to maintain a serum 25-hydroxyvitamin D level of at least 20 ng/ml (50 nmol/liter). The American Academy of Pediatrics recommends vitamin D supplements for children from the first days after birth, especially for breastfed babies, continuing to the age of 18 years; this includes adding cod liver oil to the child's diet.

Sources: *Otten JJ, Hellwig JP, Meyers LD, editors. DRIs intakes: the essential guide to nutrient requirements. Washington, DC: National Academies Press; 2006. Available at: http://www.nap.edu/catalog.php?record_id=11537 [Accessed 10 April 2013].*
US Department of Agriculture. National Agricultural Library. Dietary reference intakes [updated 1 May 2013]. Available at: http://fnic.nal.usda.gov/dietary-guidance/dietary-reference-intakes [Accessed 3 May 2013].
Ross CA, Manson JE, Abrams SA, Aloia JF, Brannon PM, Clinton SK, et al. The 2011 report on dietary reference intakes for calcium and vitamin D from the Institute of Medicine: what clinicians need to know. J Clin Endocrinol Metab 2011;96:53–8. Available at: http://www.ncbi.nlm.nih.gov/pmc/articles/pmc3046611/ [Accessed 29 April 2013].
National Research Council. Dietary reference intakes for calcium and vitamin D. Washington, DC: National Academies Press; 2011. Available at: http://www.nap.edu/catalog.php?record_id=13050 [Accessed 2 May 2013].
American Academy of Pediatrics. Healthy living. Vitamin D on the double [posted 28 May 2013]. Available at: http://www.healthychildren.org/English/healthy-living/nutrition/pages/Vitamin-D-On-the-Double.aspx?nfstatus=401&nftoken=00000000-0000-0000-0000-000000000000&nfstatusdescription=ERROR%3a+No+local+token [Accessed 24 July 2013].

growing populations. Hunger is particularly common in populations living in sub-Saharan Africa, the Indian subcontinent, and South-East Asia, with chronic poverty of undernourished populations in all communities throughout the world.

Malnutrition is a pathological state caused by a relative or an absolute deficiency or excess of one or more essential nutrients, with the clinical results being detectable by physical examination or biochemical, anthropometric, or physiological tests. The types of malnutrition identified by Derrick and Patrice Jelliffe included underweight/starvation, overweight/obesity, specific deficiency, and imbalance. Fortifying staple foods that most of the population eats and in a production process that accommodates fortification is cost-effective; sugar fortification has been successful in Central American countries such as Guatemala and Honduras, and in many other countries. Food fortification, risk group supplementation, and agricultural and economic development are all part of the fight against malnutrition (WHO/World Food Programme/United Nations Committee on Nutrition/UNICEF, 2007).

TABLE 8.7 Recommended Dietary Reference Intakes by Age and Gender, USA, 2010

Gender	Age	Activity level[b,c,d]		
	(years)	Sedentary[b]	Moderately Active[c]	Active[d]
Child	2–3	1000–1200	1000–1400[e]	1000–1400[e]
Female[f]	4–8	1200–1400	1400–1600	1400–1800
	9–13	1400–1600	1600–2000	1800–2200
	14–18	1800	2000	2400
	19–30	1800–2000	2000–2200	2400
	31–50	1800	2000	2200
	51+	1600	1800	2000–2200
Male	4–8	1200–1400	1400–1600	1600–2000
	9–13	1600–2000	1800–2200	2000–2600
	14–18	2000–2400	2400–2800	2800–3200
	19–30	2400–2600	2600–2800	3000
	31–50	2200–2400	2400–2600	2800–3000
	51+	2000–2200	2200–2400	2400–2800

Note: [a]*These levels are based on estimated energy requirements (EER) from the Institute of Medicine Dietary Reference Intakes macronutrients report, 2002, calculated by gender, age, and activity level for reference-sized individuals. "Reference size", as determined by IOM, is based on median height and weight for ages up to 18 years and median height and weight for that height to give a BMI of 21.5 for adult females and 22.5 for adult males.*
[b]*Sedentary means a lifestyle that includes only the light physical activity associated with typical day-to-day life.*
[c]*Moderately active means a lifestyle that includes physical activity equivalent to walking about 1.5–3 miles per day at 3–4 miles per hour, in addition to the light physical activity associated with typical day-to-day life.*
[d]*Active means a lifestyle that includes physical activity equivalent to walking more than 3 miles per day at 3–4 miles per hour, in addition to the light physical activity associated with typical day-to-day life.*
[e]*The calorie ranges shown are to accommodate needs of different ages within the group. For children and adolescents, more calories are needed at older ages. For adults, fewer calories are needed at older ages.*
[f]*Estimates for females do not include women who are pregnant or breastfeeding.*
Source: US Department of Health and Human Services. Dietary guidelines for Americans; 2010. Available at: http://www.cnpp.usda.gov/Publications/DietaryGuidelines/2010/PolicyDoc/PolicyDoc.pdf [Accessed 21 April 2013].

Underweight: Protein–Energy Malnutrition

The state of being underweight, or protein energy malnutrition (PEM), is a nutritional deficiency resulting from either inadequate energy (caloric) or protein intake; it is manifested in either marasmus or kwashiorkor. The main characteristics are weight loss and wasting of body fat and muscle mass, low weight-for-height, and low height-for-age. Severe forms include a spectrum of failure to thrive, marasmus, or kwashiorkor in children and starvation in adults. Weight loss in adults may be the result of loss of appetite, fasting, anorexia nervosa, persistent vomiting, inability to swallow, incomplete absorption, and increased base metabolic rate such as in prolonged fever, hyperthyroidism, cancer, diabetes mellitus, or other medical conditions. Chronic underweight in developed countries can occur in high-risk population groups because of poverty, illness, or ignorance of appropriate diets.

In young children, PEM may be due to an infection, a lack of food, or both. In developing countries, undernutrition due to poverty or failure of the food supply is the most prevalent public health problem, particularly in infants and young children. This is especially risky because a malnourished child is more vulnerable to infection, with lowered resistance and reduced immunity. A child with an infection then becomes even more undernourished and may suffer long-term growth retardation as a result. In starvation there is a compensatory reduction in metabolic rate, slowed and weak pulse, lowered blood pressure, loss of body fat, muscle wasting, decreased muscle tone, loss of skin elasticity, mental dullness, and easy fatigue. The symptoms of specific deficiencies in vitamins and minerals are likely to be minimal. Recovery of weight loss after the start of feeding is slower in adults than in children. Starvation is more likely to affect children, women, and the elderly; infants and toddlers are especially vulnerable.

Failure to Thrive

Failure to thrive, or growth retardation, describes the failure of growth in keeping with age compared to a standard growth pattern. It is particularly common in inadequately fed babies in developing countries. Failure to recover growth following an illness such as diarrhea, acute respiratory infection, or measles is also common.

Marasmus

Marasmus is a severe failure-to-thrive condition due to marked deficiency of caloric and protein intake and general deprivation. It is characterized by wasting of body tissues, particularly muscles and subcutaneous fat, and is usually a result of severe restrictions in energy intake. It occurs in the 3–9-month age group, commonly due to early weaning and inadequate feeding with resultant starvation of the

TABLE 8.8 Comparison of Selected Nutrients in the Dietary Approaches to Stop Hypertension (DASH) Eating Plan, the USDA Food Guide, and Nutrient Intakes Recommended per Day by the Institute of Medicine (IOM)

Nutrient	DASH Eating Plan (2000 kcal)[a]	USDA Food Guide (2000 kcal)[b]	IOM Recommendations for Females 19–30[c]
Protein (g)	108	91	RDA: 46
Protein (% kcal)	21	18	AMDR: 10–35
Carbohydrate (g)	288	271	RDA: 130
Carbohydrate (% kcal)	57	55	AMDR: 45–65
Total fat (g)	48	65	–
Total fat (% kcal)	22	29	AMDR: 20–35
Saturated fat (g)	10	17	–
Saturated fat (% kcal)	5	7.8	ALAP
Monounsaturated fat (g)	21	24	–
Monounsaturated fat (% kcal)	10	11	–
Polyunsaturated fat (g)	12	20	–
Polyunsaturated fat (% kcal)	5.5	9.0	–
Linoleic acid (g)	11	18	AI: 12
α-Linolenic acid (g)	1	1.7	AI: 1.1
Cholesterol (mg)	136	230	ALAP
Total dietary fiber (g)	30	31	AI: 28[d]
Potassium (mg)	4706	4044	AI: 4700
Sodium (mg)	2329[e]	1779	AI: 1500, UL: <2300
Calcium (mg)	1619	1316	AI: 1000
Magnesium (mg)	500	380	RDA: 310
Copper (mg)	2	1.5	RDA: 0.9
Iron (mg)	21	18	RDA: 18
Phosphorus (mg)	2066	1740	RDA: 700
Zinc (mg)	14	14	RDA: 8
Thiamine (mg)	2.0	2.0	RDA: 1.1
Riboflavin (mg)	2.8	2.8	RDA: 1.1
Niacin equivalents (mg)	31	22	RDA: 14
Vitamin B_6 (mg)	3.4	2.4	RDA: 13
Vitamin B_{12} (µg)	7.1	8.3	RDA: 2.4
Vitamin C (mg)	181	155	RDA: 75
Vitamin E (AT) (mg)	16.5	9.5	RDA: 150
Vitamin A (mg) (RAE)	851	1052	RDA: 700

Note: [a]DASH nutrient values at the 2000 kcal level, based on a 1 week menu of the DASH Eating Plan (NIH publication no. 03-4082; www.cnhlbi.nih.gov).
[b]USDA nutrient values at the 2000 kcal level, based on population-weighted averages of typical food choices within each food group or subgroup.
[c]Recommended intakes for an adult female, 19–30 years of age.
[d]Amount listed is based on 14 g dietary fiber/1000 kcal.
[e]The DASH Eating Plan can also be followed at 1500 mg sodium per day.
AT = D-α-tocopherol; RAE = retinol activity equivalents; RDA = recommended dietary allowance; AMDR = acceptable macronutrient distribution range; ALAP = as low as possible while consuming a nutritionally adequate diet; AI = adequate intake; UL = upper limit.
Source: US Department of Health and Human Services Dietary Guidelines for Americans, 2010 Report. Healthy eating index [modified 22 March 2013]. Available at: http://www.cnpp.usda.gov/HealthyEatingIndex.htm and http://www.nutrition.gov/smart-nutrition-101/dietary-guidelines-americans [Accessed 28 April 2013].

TABLE 8.9 Terms and Syndromes of Undernutrition

Term/Syndrome	Condition
Protein–energy malnutrition (PEM) Kwashiorkor	Person does not get enough food, calories, or protein needed for normal growth and energy or normal human activities; low weight and height for age; fatigue, poor work or school capacity Calorie, vitamin, and protein deficiency commonly affecting children aged 2–4 when the next child is born and insufficient food, or triggered by an infection (e.g., measles); depigmented, flaky skin and hair; poor appetite; low serum albumin; edema or swelling of abdomen; enlarged liver in a child who is thin in trunk, arms, and legs
Marasmus	A wasting disease due to lack of energy and protein; usually occurs between 6 and 18 months with combination of maternal neglect and non-breastfeeding; occurs in urban slums; infant looks like a wizened old person, skin and bones, with recurrent infections; causes permanent impairment of brain development; needs warmth, attention, adequate nutrition, maternal education, and prevention of infectious diseases
Micronutrient deficiency conditions	Vitamin A deficiency – xerophthalmia, comorbidity with infectious diseases Vitamin B deficiency – pellagra, beriberi, polyneuropathy, dementia Folic acid deficiency – macrocytic anemia and congenital birth defects Vitamin C deficiency – scurvy Vitamin D deficiency – rickets, osteoporosis Vitamin K deficiency – hemorrhagic disease of newborn, bone disorders Iron-deficiency anemia (IDA) – common comorbid debilitating condition Iodine-deficiency disorders (IDDs) – cretinism, development disability, goiter
Low birth weight (LBW)	Birth weight of ≤ 2500 g; %LBW is a good indicator of nutrition status of a population
Low weight for age (wasting)	Indicator of current calorie and protein undernutrition (chronic or acute or both)
Low height for age (stunting)	Low height for age compared to reference population indicates chronic undernutrition at some time during the growth years, but may not be currently undernourished
Low weight for height	Thin person whose weight is low for his or her height, indicating current undernutrition
Infection–malnutrition cycle	Interaction of nutrition status, susceptibility to infection with effects of infectious disease on nutrition status

Source: World Health Organization and United Nations. WHO child growth standards and the identification of severe acute malnutrition in infants and children. A Joint Statement by the World Health Organization and the United Nations. Geneva: WHO; 2009. Available at: http://www.who.int/nutrition/publications/severemalnutrition/9789241598163_eng.pdf [Accessed 21 April 2013].

infant. The child appears wasted and irritable with depletion of subcutaneous fat and muscle tissue, which are burned up to maintain blood glucose.

Kwashiorkor

Kwashiorkor, a severe form of undernutrition, is usually the result of severe restrictions in protein intake and is characterized by edema (particularly ascites or abdominal swelling). It occurs in infants and children up to about the age of 6. This widespread protein deficiency syndrome occurs in young children who have been weaned, often after the birth of a new child, to a diet high in carbohydrates and low in protein. The condition is often aggravated by an infectious disease. Kwashiorkor is characterized by retarded growth and development, apathy, gastrointestinal irritability, depigmentation of hair, edema resulting in a swollen abdomen, fatty infiltration of the liver, and dry skin. Treatment consists of establishing adequate dietary intake and balance. Untreated, the mortality of this condition is high. Both types of

malnutrition can be present simultaneously (marasmic kwashiorkor) and the presence of edema can mask malnutrition.

Micronutrient Deficiencies

Malnutrition is often a synergistic factor underlying deaths in children in many developing countries but can also directly result in death. Deficiency conditions for one or more of the essential nutrients at a clinically detectable or subclinical level are crucial and undermanaged in public health. Worldwide about 1.62 billion (30 percent) of the population have anemia, with the lowest prevalence in men (12.7 percent). The prevalence of anemia is highest among the combined group of preschool children, pregnant and non-pregnant females (age 15–49 years) (Table 8.10). In developing countries, incidence varies according to urban/rural residence and social class, but among the majority of rural and urban poor they serve as major factors in excess morbidity and mortality. In developed countries, the predominant nutritional

TABLE 8.10 Prevalence of the Three Major Micronutrient Deficiencies by World Health Organization (WHO) Region[a]

WHO Region	Anemia[b]		Iodine Intake Deficiency[c]		Insufficient Vitamin A (Preschool Children)[d]	
	No. (millions)	Range (%)	No. (millions)	% Total	No. (millions)	% Total
Africa	170.6	47.5–67.8	321.1	40.0	56.4	44.4
Americas	66.0	17.8–29.3	125.7	13.7	8.7	15.6
South-East Asia	315.4	45.7–65.5	541.3	31.6	91.5	49.9
Europe	54.5	19.0–21.7	393.3	44.2	5.8	19.7
Eastern Mediterranean	47.7	32.4–46.7	199.2	37.4	13.2	20.4
Western Pacific	132	21.5–23.1	300.8	17.3	14.3	12.9
Total	786.2	30.2–47.4	1881.2	28.5	190	33.3

Note: [a]Based on the proportion of the population with hemoglobin concentrations below established cut-off levels by age group.
[b]Preschool, pregnant and non-pregnant women age 15–49 years.
[c]Based on the proportion of the population with urinary iodine < 100 μg/l.
[d]Based on the proportion of the population with clinical eye signs and/or serum retinol.
Sources: Allen L, de Benoist B, Dary O, Hurrell R. Guidelines on food fortification with micronutrients. Geneva: WHO; 2006.
de Benoist B, McLean E, Egl I, Cogswel M, editors. Worldwide prevalence of anaemia 1993–2005: WHO global database on anaemia. Geneva: WHO; 2008. Available at: http://whqlibdoc.who.int/publications/2008/9789241596657_eng.pdf [Accessed 13 April 2013].
Andersson M, Karumbunathan V, Zimmermann MB. Global iodine status in 2011 and trends over the past decade. J Nutr 2012;142:744–50. Available at: http://asn-cdn-remembers.s3.amazonaws.com/b837703fba319d397cf4169f3cdb6f05.pdf [Accessed 21 April 2013].
Andersson M, de Benoist B, Darnton-Hill I, Delange F. Iodine deficiency in Europe: a continuing public health problem. Geneva: WHO; 2007. Available at: http://whqlibdoc.who.int/publications/2007/9789241593960_eng.pdf [Accessed 21 April 2013].
World Health Organization. Global prevalence of vitamin A deficiency in populations at risk 1995–2005. WHO global database on vitamin A deficiency. Geneva: WHO; 2009. Available at: http://whqlibdoc.who.int/publications/2009/9789241598019_eng.pdf [Accessed 21 April 2013].
World Health Organization. Iron deficiency anaemia: assessment, prevention, and control. A guide for programme managers. WHO/NHD/01.3. Geneva: WHO; 2001. Available at: http://www.who.int/nutrition/publications/micronutrients/anaemia_iron_deficiency/WHO_NHD_01.3/en/ [Accessed 21 April 2013].

problem is obesity, but micronutrient deficiencies such as iron, iodine, zinc, selenium, and vitamins A and D deficiencies are common public health issues (Table 8.10).

Even in developed countries, vitamin and mineral deficiency can be widespread. In the early 1970s, the Canadian National Nutrition Survey showed vitamin and mineral deficiencies in different parts of the country, especially in native people, teenagers, women, and the elderly. Many developed countries have taken steps to reduce these problems through improved standards of living and the implementation of policies of vitamin and mineral enrichment of basic foods. The history of pellagra in the southern USA in the early part of the twentieth century (see Chapter 1) illustrates the serious damage that vitamin deficiencies can cause, and the remedy of vitamin B_1 fortification of flour demonstrates the effectiveness of public health interventions.

Vulnerable population groups include pregnant and lactating women, infants and toddlers, and the elderly, especially those in poverty groups. Other groups at risk include alcoholics, people with chronic or frequent infection such as people with acquired immunodeficiency syndrome (AIDS) and chronic diseases, and those who restrict themselves to certain foods, such as vegetarians. Vegetarianism is compatible with good health, provided it is practiced with adequate nutritional advice. However, when coupled with pregnancy and breastfeeding, it carries the risk of anemia and other deficiency conditions, especially for the breastfed infant. The largest vulnerable populations are the poor populations in sub-Saharan Africa, South-East Asia, and Latin America.

According to WHO Guidelines on Food Fortification (WHO: Allen et al., 2006), 0.8 million deaths (1.5 percent of the total) can be attributed to iron deficiency each year, and a similar number to vitamin A deficiency. The loss of healthy life, expressed as disability-adjusted life years (DALYs), for those with iron-deficiency anemia (IDA) results in 25 million DALYs lost (2.4 percent of the global total). For vitamin A deficiency, 18 million DALYs (1.8 percent of the global total), and for iodine deficiency, 2.5 million DALYs are lost (0.2 percent of the global total). These are important public health issues, particularly because they are preventable by currently accepted inexpensive interventions.

Vitamin A Deficiency

Vitamin A is essential for normal vision and ocular function, because of its primary role in forming visual pigment. Dietary sources of vitamin A include animal products, such as egg yolk, liver, dairy products, and breast milk. Plant sources include plants containing carotenoids, such as dark green leafy vegetables, yellow and reddish fruits, and red palm oil. Symptoms of vitamin A deficiency include growth retardation, alterations in the differentiation and morphology of epithelial and mesenchymal tissues, and impaired vision. The WHO estimates that 3 million children have some form of xerophthalmia (drying out of the eyes, leading to scarring and blindness) and, on the basis of blood levels, another 250 million preschool children are vitamin A deficient. It is likely that in vitamin A-deficient areas, a substantial proportion of pregnant women are vitamin A deficient; it is presumed that 250,000–500,000 vitamin A-deficient children become blind every year, with half of them dying within 12 months of losing their sight. The WHO and UNICEF estimate that vitamin A deficiency contributes to the following conditions and the estimated number of children affected:

- deficient intake (subclinical) – 562 million
- deficient intake (clinical susceptibility to infection) – 231 million
- night blindness – 13.5 million
- xerophthalmia – 3.1 million
- severe eye damage/blindness – 0.5 million.

Vitamin A deficiency decreases resistance to infections and increases the severity, complications, and risk of death from various diseases. It also results in night blindness and xerophthalmia. This nutritional deficiency is widespread among children in developing countries, and the problem is exacerbated by a tendency by some to withhold vegetables from children for cultural or other reasons. Vitamin A deficiency is associated with increased mortality from measles, as vitamin A has both protective and therapeutic roles in measles treatment. High doses of vitamin A should be given to susceptible populations and children during measles outbreaks.

Vitamin A deficiency is a public health problem in more than half of all countries, especially in Africa and South-East Asia, hitting young children and pregnant women in low-income countries the hardest. For children, lack of vitamin A causes severe visual impairment and blindness, and significantly increases the risk of severe illness, and even death, from common childhood infections such as diarrheal disease and measles. For pregnant women in high-risk areas, vitamin A deficiency occurs especially during the last trimester when demand by both the unborn child and the mother is highest. The mother's deficiency is demonstrated by the high prevalence of night blindness during this period. The impact of vitamin A deficiency on mother-to-child human immunodeficiency virus (HIV) transmission warrants further investigation.

Preventive measures for populations at risk for early or marginal vitamin A deficiency can markedly reduce the risk of mortality from intestinal and respiratory infections and measles. Methods for improving vitamin A status include periodic distribution of large-dose capsules appropriate for age, fortification of readily consumed dietary staples, and increased intake of vitamin A-rich foods. Vitamin A, along with vitamin D, was added to margarine, milk, and dairy products in the USA, Canada, and the UK during World War II. The practice is mandatory in Canada and nearly universal in the USA.

The basis for lifelong health begins in childhood. Vitamin A is a crucial component. Since breast milk is a natural source of vitamin A, promoting breastfeeding is the best way to protect babies from vitamin A deficiency. The WHO has a goal of worldwide elimination of vitamin A deficiency and its tragic consequences, including blindness, disease, and premature death. To successfully combat vitamin A deficiency, short-term interventions and proper infant feeding must be backed up by long-term sustainable solutions. The combination of breastfeeding and vitamin A supplementation, coupled with promotion of vitamin A-rich diets and food fortification, holds the promise of achieving this goal.

For vitamin A-deficient children, the periodic supply of high-dose vitamin A in swift, simple, low-cost, high-benefit interventions has also produced remarkable results, reducing mortality by 23 percent overall and by up to 50 percent among acute measles sufferers. Planting these "seeds" between 6 months and 6 years of age can reduce overall child mortality by 25 percent in areas with significant vitamin A deficiency. However, breastfeeding is time limited and the effect of vitamin A supplementation capsules lasts for only 4–6 months. Thus, rather than representing long-term solutions, they serve only as initial steps towards ensuring better overall nutrition. Food fortification takes over where supplementation leaves off. Fortification maintains vitamin A status, especially for high-risk groups and needy families.

Cultivating leafy vegetables in a home garden is a community-based phase necessary to achieve long-term results. For vulnerable rural families, for instance, in Africa and South-East Asia, growing fruits and vegetables in home gardens complements dietary diversification and fortification and contributes to better lifelong health.

In 1998 the WHO and its partners – UNICEF, the Canadian International Development Agency, the US Agency for International Development, and the Micronutrient Initiative – launched the Vitamin A Global Initiative. In addition, over the past few years, WHO, UNICEF, and others have provided support to countries in delivering vitamin A supplements. Linked to sick-child visits and national poliomyelitis immunization days, these supplements have averted an estimated 1.25 million deaths since 1998 in 40 countries.

Cumulative evidence since the mid-1980s has reinforced the importance of vitamin A in combating infections. Thus, food fortification is recommended especially in developing countries as well as populations subject to high infection rates, such as HIV-positive people. Routine vitamin A supplementation for children is now recommended by the WHO in conjunction with immunization programs (EPI-plus; see Chapter 4). Massive doses of vitamin A are used to treat children with measles, with great benefit in reducing mortality and complication rates.

Fortification of sugar with vitamin A has been implemented in a number of countries in South and Central America (e.g., Guatemala). The Philippines has required fortification of margarine with vitamin A since 1998. Indonesia is using a variety of techniques to increase supplementation and fortification to reduce deficiency conditions. Supplementation policies providing vitamin A supplements to children and postpartum mothers, carried out in 78 countries in 1996, are continuing and spreading with WHO and UNICEF promotion.

Although the major focus is on vitamin A *deficiency*, vitamin A toxicity may occur rarely when more than three times the RDA is consumed, usually from medication overdose, especially among pregnant women, alcoholics, or people with chronic liver conditions. This can result in

hyperkeratinosis (i.e., orange skin), nausea, vomiting, birth defects, and neurological, gastrointestinal, and dermatological symptoms.

Vitamin D Deficiency (Rickets and Osteomalacia)

Rickets from vitamin D deficiency was a common disorder in many parts of the developed world well into the twentieth century (Box 8.4). It constitutes one of the important diseases of infancy because of its serious complications, including disorders of long bone growth, bowing of legs, pelvic deformity and, in extreme forms in infants, tetany and convulsions. The vitamin D content of human milk is extremely low; the exclusively breastfed infant not adequately exposed to sunlight may develop clinical rickets. In adults, malabsorption or poor dietary intake of vitamin D can result in osteomalacia and osteoporosis with fragile bones and frequent fractures. Among the elderly, who

---

BOX 8.4 Rickets and Vitamin D Deficiency

Rickets is described in ancient medical writings from the first and second centuries CE. Soranus and Galen, Roman physicians of that era, described rickets as bone deformities in infants. In 1650, Francis Glisson, a Cambridge physician, published a treatise on rickets, describing its clinical features and suggesting treatments.

In 1870, as many as one-third of the poor children in cities such as London and Manchester suffered from obvious rickets. At the beginning of the twentieth century, rickets was rampant among infants living in industrialized, polluted cities of North America and Europe. In 1919, Edward Mellanby, an English physician, clearly established the role of diet in the cause of rickets via animal experiments. Elmer McCollum, an American nutritional biochemist, developed a method of biological analysis of nutritive value of foods and eventually discovered vitamin D. In 1919, investigators in Germany showed that exposure to sunlight cured rickets and that it acted by altering fats to produce vitamin D.

As late as 1921, McCollum claimed that probably half of the children in the USA had rickets. Rickets was, at that time, the most common nutritional disease of children, affecting approximately 75 percent of infants in New York City. Cod liver oil and irradiated foods were introduced to prevent rickets and widely used in Europe and North America.

Rickets remained widespread until fortification of milk was introduced in the 1940s in Britain and North America. Rickets prevalence in the UK, especially in the industrial cities in northern England and Scotland, declined dramatically. In Canada, fortification of milk with vitamin D was routine during the 1940s, but was stopped in the 1950s and 1960s. Significant deficiencies in certain population groups led to mandatory fortification in Canada starting from 1979, with subsequent disappearance of rickets.

Rickets in infants continues to be reported in the USA and other countries due to inadequate vitamin D intake and

decreased exposure to sunlight. In 2008, because of the reappearance of rickets and low levels of vitamin D, the American Academy of Pediatrics recommended that all infants, especially breastfed, have a minimum supplement of 400 IU of vitamin D per day from the first 2 months of life. They also recommend that 400 IU vitamin D per day be continued in childhood and adolescence because sunlight exposure is not easily determined for a given person. New vitamin D intake guidelines are based on US National Academy of Sciences recommendations.

Osteoporosis has its highest occurrence among postmenopausal women and results in the loss of bone mass, often leading to fractures, including those of the hip and spine. Hypovitaminosis D and related abnormalities in bone metabolism and strength are common in elderly people in Europe, but are reported in all elderly populations. In 1995, there were 382,000 hip fractures in 15 countries of the European Union with an estimated total care cost of about US$9 billion. In the UK, hip fractures were found to place the greatest demand on resources and have the greatest impact on elderly patients because of increased mortality, long-term disability, and loss of independence.

The problem of vitamin D deficiency should be addressed with a multiple strategy approach, including both supplementation (with calcium) targeted to higher risk groups and food fortification to reach the general population.

Sources: Centers for Disease Control and Prevention. Achievements in public health, 1900–1999: Safer and healthier foods. MMWR Morb Mortal Wkly Rep 1999;48:905–13.
American Academy of Pediatrics. Section on Breastfeeding and Committee on Nutrition. Prevention of rickets and vitamin D deficiency in infants, children, and adolescents. Pediatrics 2008;122:1142–52.
Holick MF. The vitamin D deficiency pandemic: a forgotten hormone important for health. Public Health Rev 2010;32:267–83. Available at: www.publichealthreviews.eu [Accessed 18 April 2013].

are closed in especially during winter months, and among homebound or institutionalized patients, vitamin D deficiency is common.

There are few natural sources of vitamin D, but deficiency is preventable by exposing the skin to the ultraviolet rays of the sun. Prevention became common through the practice of giving children cod liver oil, a successful antirachitic measure, until replaced by use of vitamin D supplements for infants. In colder climates or in foggy locations where exposure to the sun may be limited, rickets was common. However, seasonal deficiency of vitamin D can occur even in locations with ample sunlight. Cod liver oil was widely used for prevention of rickets in the early and middle decades of the twentieth century, and rickets prevalence declined, but remained a serious public health problem in poorer populations.

Rickets remained widespread until fortification of milk was introduced in the 1940s during World War II in Britain and North America. Rickets prevalence in the UK, especially in the industrial cities in northern England and Scotland, declined dramatically. The 1971 Canadian National Nutrition Survey found significant vitamin D deficiencies in certain age, gender, ethnic, and geographic population groups. Although antirachitic preventive measures (e.g., cod liver oil and adding vitamin D to milk) were routine during the 1940s, these practices waned in the 1950s and 1960s. An increase in hospitalizations for rickets in Montreal followed the abandonment of vitamin D milk enrichment.

Addition of vitamin D to milk, nearly universal in the USA, was made mandatory in Canada in 1979: rickets disappeared and no incidents of vitamin D toxicity were reported. Vitamin D toxicity may occur rarely owing to human error in formula preparation or from multiple sources of supplementation, with failure to thrive, nausea, vomiting, and weakness. Such events led to stopping wartime fortification of milk in Canada. Canada's mandatory milk and other food fortification policies of 1979 continue and were renewed in 2006.

Vitamin D deficiency is considered a worldwide pandemic occurring in countries as diverse as Great Britain, Austria, Germany, Finland, New Zealand, and India. Australia reports 30–50 percent of children and adults as being vitamin D deficient. The age groups included in studies of these countries include young adults and the elderly, with wide variation by country of residence. Even where vitamin D fortification of milk has been in place for many decades, as in the USA and Canada, there are still major deficiencies related to season, latitude of residence, age, gender, and social conditions (Holick, 2010). Even in a sunny climate, there are seasonal variations in vitamin D availability, and social customs of mothers keeping infants overly wrapped and out of the sun may cause rickets, especially in dark-skinned immigrants from southern climates whose religious customs are associated with complete covering of women

and infants. Studies of vitamin D levels among institutionalized elderly people found low levels, and among elderly people in the community vitamin D levels were reported to be low in the winter months. In climates with long winters and cloudy skies exposure to the sun's rays is limited and vitamin D levels are predictably low. Many European countries do not fortify their milk and people are thus at risk for this vitamin deficiency. Finland fortified its milk in 2005 and subsequent studies show increased vitamin D levels in young adults and others.

Vitamin D deficiency is linked to an elevated risk of falling, osteoporosis, fractures of the hip and vertebrae, cardiovascular disease, stroke, congestive heart failure, breast and colorectal cancer, type 1 diabetes, and multiple sclerosis. Vitamin D supplements have been shown to be effective in reducing the risk of these conditions. People living in northern climates with long winters and little sun exposure are at risk for vitamin D deficiency. Fortification of milk or other basic foods is protective against deficiency, but supplements are essential for specific groups such as all children (from age 0 to 18 years), women who work indoors and have little regular sun exposure, and middle-aged and elderly people, in particular. Multiple sclerosis is more common in upper and lower latitudes, suggesting vitamin D deficiency as a causative factor.

Prevention of vitamin D deficiency conditions should include routine supplements of vitamins A and D for infants (400 IU per day) and children up to adolescence (600 IU per day) (American Academy of Pediatrics, 2008; Institute of Medicine, 2011). Fortification of baby formulas and cereals is common. Standard textbooks of pediatrics and public health take the same position with respect to rickets (vitamin D deficiency): vitamin D deficiency is a problem in childhood and in the elderly that cannot be addressed by providing vitamins to high-risk groups, and therefore milk fortification is needed. Vitamin D fortification of milk products is one of the most important public health nutrition measures used. Studies in Finland before and after fortification of milk and milk products with vitamin D showed a high prevalence of wintertime deficiency in adolescents living above the 51.9° N latitude. In northern latitudes, the short length of daylight in winter increases vitamin D deficiency, while extended sun exposure during summer months reduces 1,25-dihydroxyvitamin D deficiency. Fortification of milk and margarine implemented in Finland has increased vitamin D intake in adolescents to close to the Nordic countries' reference dietary intake (Tylavsky, 2006).

There are increasing reports of vitamin D deficiency at the level of clinical rickets, and low vitamin D levels in older children. Risk factors include prolonged sole breastfeeding, and immigrant mothers with dark skin in northern climates, especially in winter. A further risk of vitamin D deficiency occurs among women completely covered for religious reasons, with no skin exposure, precluding them from sun exposure. Vitamin D deficiency conditions are being reported in the literature in countries, including the UK, which do not fortify milk or recommend routine vitamin D supplements.

Deficiency of vitamin D remains a major public health problem that requires attention in regard to both food fortification and supplementation for groups at risk of deficiency. These groups include infants, children, teenagers, adults, and especially the elderly. Failure to promote these public health measures widely has been an important lapse of the international health community. Table 8.11 indicates current recommended vitamin D intake for different age groups, based on minimal sunlight exposure and where the milk is nearly universally fortified with vitamin D. Concern about skin cancer has led to recommendations to avoid sun exposure, and lifestyle changes for people of all ages working or spending time indoors and with little spare time for even modest sun exposure lead to vitamin D deficiency as a widespread public health problem.

Vitamin C Deficiency

Scurvy is a deficiency disease due to a lack of vitamin C (ascorbic acid) in the diet. It was common among seamen (see Chapter 1) and others deprived of fresh fruits and vegetables. Vitamin C deficiency causes skin lesions, weakness, fatigue, weight loss, muscle pains, susceptibility to infection, hemorrhage, debility, and even death. It can occur at any age as the result of an inadequate diet. Infantile scurvy formerly appeared in bottle-fed infants, but with the advent of vitamin-fortified infant formulas this has become rare in industrialized countries. Infants should have a source of vitamin C from the first month of life, such as in orange juice.

Fortification of fruit juices and flavored drinks is a common practice to ensure adequate intake of vitamin C. As a water-soluble vitamin, there is no risk of excess intake from multiple sources in the course of a normal diet.

The use of vitamin C in prevention and treatment of the common cold and respiratory infections remains controversial, with ongoing research. For cold prevention, more than 30 clinical trials including over 10,000 participants have examined the effects of taking daily vitamin C (200mg or more). Overall, no significant reduction in the risk of developing colds has been observed. Other uses for vitamin C have been proposed, but few have conclusively shown evidence of benefit in scientific studies. In particular, research in asthma, cancer, and diabetes remains inconclusive, while no benefits have been found in the prevention of cataracts or heart disease.

Vitamin K Deficiency (Hemorrhagic Disease of the Newborn)

A deficiency of vitamin K occurs in the normal newborn and may cause impaired production of prothrombin, a key blood clotting factor. Lack of prothrombin shortly after birth may occur in up to 50 percent of newborns and can cause hemorrhagic disease of the newborn (HDN) or vitamin K deficiency bleeding disorder, so that the AAP recommends universal provision of vitamin K by injection (AAP, 2003) (see Chapter 6). Secondary HDN can occur weeks later. To prevent the deficiency, a single injection of vitamin K is given by injection within 6 hours of birth to all infants. Oral vitamin K is preferred in some countries, but this requires at least 3 separate administrations over the first 6 weeks of life and is not as effective as by injection (see Chapter 6). Vitamin K is added to fortified baby formulas and cereals. Administration of vitamin K to newborns is now instituted in most US hospitals and should be a priority for the prevention of neonatal mortality globally.

Vitamin B Deficiencies

Vitamin B₁ (Thiamine) Deficiency (Beriberi)

Thiamine deficiency causes derangement of carbohydrate metabolism. "Dry beriberi" results in neurological symptoms and death, whereas "wet beriberi" is characterized by cardiac symptoms and disorders, including heart failure and death. This disease was common among prisoners of war in

TABLE 8.11 Recommended Dietary Allowances (RDAs) for Vitamin D

Age	Male	Female	Pregnancy	Lactation
0–12 months[a]	400 IU (10 µg)	400 IU (10 µg)		
1–13 years	600 IU (15 µg)	600 IU (15 µg)		
14–18 years	600 IU (15 µg)	600 IU (15 µg)	600 IU (15 µg)	600 IU (15 µ)
19–50 years	600 IU (15 µg)	600 IU (15 µg)	600 IU (15 µg)	600 IU (15 µg)
51–70 years	600 IU (15 µg)	600 IU (15 µg)		
>70 years	800 IU (20 µg)	800 IU (20 µg)		

Note: [a]Adequate intake (AI).
Source: National Institute of Health Office of Dietary Supplements. Dietary supplement fact sheet: Vitamin D. Available at: http://ods.od.nih.gov/factsheets/VitaminD-HealthProfessional/ [Accessed 24 August 2013].

Japanese camps during World War II owing to diets consisting mainly of polished rice. Prevention requires a diet containing liver, glandular organs, yeast, wheat germ, whole wheat or cereals, unpolished rice, milk, legumes, soybeans, and peanuts.

Alcoholic dementia (Korsakoff's syndrome) and nutritional dementia (Wernicke's encephalopathy) may result from B complex deficiency caused by alcoholism and malnutrition, respectively, and constitute important public health problems. Wernicke's encephalopathy together with Korsakoff's syndrome is termed cerebral beriberi. These diseases are of particular concern for people who consume excess alcohol, i.e., more than 4 oz (113 g) per day. Habitual alcoholics may develop severe vitamin B deficiency leading to dementia and hemorrhagic stroke. Enrichment of bread, breakfast cereals, and other flour-based products is practiced in Canada and in many countries in Latin America on a mandatory basis, and in the USA, where it is mandatory when the term "enriched" is used.

Vitamin B$_3$ (Niacin) Deficiency (Pellagra)

Niacin (nicotinic acid) is essential for specific oxidation–reduction reactions in the body. A deficiency of niacin causes diarrhea, dermatitis, and dementia. Pellagra was established as a nutritional deficiency (and infectious) condition in investigation of the condition in orphanages and hospitals in the southern USA in 1917–1922 (see Chapter 1). Pellagra is prevented with adequate dietary intake of niacin or niacin substitutes contained in vitamin-enriched bread, liver, meat, fish, poultry, potatoes, green vegetables, peanuts, and cereals. Niacin also exerts beneficial effects by reducing blood lipids and raising high-density lipoprotein (HDL), slowing atherosclerosis and coronary artery lesions.

In 1998, the National Science Foundation issued a recommendation for multiple vitamin supplements for all adults.

Vitamin B$_9$ (Folate) Deficiency

Folate (folic acid) is required for normal blood formation and neurological health. Its deficiency is common among low socioeconomic groups, especially during pregnancy, infancy, and childhood (Box 8.5). Deficiency is common in alcoholics, owing to malnutrition or impaired absorption. Alcoholics with sufficient diets are less likely to develop folate deficiency than those with poor eating habits. Folate deficiency constitutes a major health risk for alcoholics to develop severe neurological damage in the spinal cord or in the optic or peripheral nerves.

Folic acid deficiency has been demonstrated to be a cause of neural tube defects (NTDs), an important congenital disorder (see Chapter 6). This condition ranges in severity from anencephaly to disabling defects of the spinal cord, and is largely preventable by prenatal supplements of folic

---

BOX 8.5 Folic Acid and Neural Tube Defects

In 1930, British physician Dr Lucy Willis successfully demonstrated in India that folate was a nutrient required to prevent macrocytic anemia of pregnancy; this was achieved using brewer's yeast and spinach leaves. The term folate was derived from its natural source in leafy green foliage. Folic acid was first synthesized in 1946. It is a water-soluble B vitamin, important in new cell formation via DNA synthesis. It is available in leafy green vegetables such as lettuce and spinach, as well as beans, peas, and liver. Folic acid levels in population studies in the USA showed widespread deficiency in the 1990s NHANES studies.

In the 1990s, international studies showed that folic acid supplements taken before pregnancy resulted in reduced rates of neural tube defects (NTDs) among newborns. When this was confirmed by the British Medical Council, recommendations for folic acid supplements among women of fertility age were adopted by many governmental and professional organizations. Recommendations for women of fertile age to take folic acid daily have been accepted practice by the Centers for Disease Control and Prevention, the American Congress of Obstetricians and Gynecologists, the American Academy of Pediatrics, and many others, since the 1990s. However, compliance with the use of routine folic acid supplements is variable and in many countries is still an unknown practice.

In 1996, the US Food and Drug Administration and the Canadian Federal Department of Health adopted mandatory fortification of flour, implemented in 1998, as did Chile. This has now become widespread, including over 50 countries. Some controversy along with apathy obstructs fortification with claims that potential harmful effects may occur by masking vitamin B$_{12}$ deficiency and neurological damage, so that vitamin B$_{12}$ is increasingly being added to fortified flour as well. Flour fortification has become widely adopted in other countries; however, no country in the European Union and only Kazakhstan in the European region (WHO) has yet adopted this measure.

Studies have shown reduced rates of NTDs in Canada and the USA. The benefit of folic acid fortification along with supplementation for high-risk groups (including women and infants) to achieve an intake of 400 µg per day has been demonstrated. Moreover, cumulative evidence is mounting regarding other important health benefits such as prevention of cardiovascular disease, dementia, and cancer.

This is a case in which evidence of benefit far outweighs theoretical possibilities of harm, and delayed implementation causes significant damage to newborns and others. The case for action is strong but reaction to implementation is also strong. The knowledge–action gap is still wide.

Sources: *National Institutes of Health. Dietary supplement fact sheet: Folate [updated 14 December 2012]. Office of Dietary Supplements, NIH Clinical Center, National Institutes of Health. http://ods.od.nih.gov/factsheets/folate.asp [Accessed 17 April 2013].*

De Wals P, Fassiatou T, Van Allen M, Uh SH, Lowry RB, Sibbald B, et al. Reduction in neural tube defects after folic acid fortification in Canada. N Engl J Med 2007;357:135–42.

acid or food fortification. Mandatory food fortification in the USA, Canada, and Chile in 1998 led to a decline in NTD incidence. In many countries, including in Europe, folic acid supplementation does not receive adequate attention. As a major method of preventing birth defects, all women of reproductive age should receive folic acid supplements. Low folate levels are found in 9.4 percent of black American women aged 15–49, twice the rate of other American women in the same age group. In addition, adequate intake is important for children, men, and women of all ages in maintaining proper homeostasis and health.

Folic acid supplementation may also be important in prevention of coronary heart disease by lowering homocysteine levels, but this requires further substantiation (see Chapter 5).

Vitamin B$_{12}$ Deficiency

Vitamin B$_{12}$ deficiency causes enlargement of red blood cells with poor hemoglobin content (macrocytic anemia). Vitamin B$_{12}$ deficiency may be due to malabsorption in the stomach (pernicious anemia) or from long-term deficiency of vitamin B$_{12}$ intake from vegetarian diets that exclude eggs and dairy products. Vitamin B$_{12}$ deficiency and folic acid deficiency are often comorbid, with serious deleterious effects for women of childbearing age, elderly people, and alcoholics. Pregnant women and lactating mothers, especially if vegetarian, need vitamin B$_{12}$ supplements. Vitamin B$_{12}$ deficiency can result in degeneration of the spinal cord, optic nerves, cerebral tissue, and peripheral nerves. Prevention is by promotion of healthy nutrition with foods rich in B-complex vitamins and supplementation for infants and for vegetarians in fortified cereals.

Iron-Deficiency Anemia

Iron deficiency is the most common and widespread nutritional disorder worldwide. It affects a large number of children and women in developing countries and is also highly prevalent in industrialized countries. The problem is staggering, with nearly 2 billion people (over 30 percent of the world's population) anemic, many through iron deficiency, and frequently exacerbated by infectious diseases. Poor and less educated people are most affected by iron deficiency and anemia. The reduced work capacity of affected people and populations has serious national economic consequences, especially in less developed countries. Improved population health with a priority on the reduction of micronutrient conditions of anemia and iron deficiency are at the core of the Millennium Development Goals (MDGs) of reducing poverty and hunger. Malaria, HIV/AIDS, tuberculosis, and worm infestations such tapeworm, hookworm, and schistosomiasis are also major factors in the high prevalence of anemia among some populations. Iron deficiency, affecting more people than any other condition, is a public health

challenge of epidemic proportions, with the heaviest overall toll in terms of ill-health, premature death, and lost earnings (WHO, 2013).

Iron plays a critical role in the transport of oxygen in human blood. Iron exists in two major forms in food. The first is heme iron, which is found only in animal sources. It is readily available, as absorption is not hindered by other constituents of the diet. The second type is inorganic iron, whose absorption is strongly influenced by factors present in foods ingested at the same time. The composition of a meal can affect the amount of iron absorbed. Consumption of foods containing heme iron will improve the absorption of non-heme iron. Vitamin C enhances the absorption of non-heme iron, but substances such as tannin from tea combine with the iron in the intestine to inhibit its absorption.

IDA as a public health problem has received considerable attention since the 1980s from international organizations such as UNICEF and the WHO. The issue is especially vital for developing countries, but even in developed countries iron enrichment of staple foods is an important public health measure. IDA particularly affects infants, pregnant women, adolescent and adult populations, as well as the elderly.

In developing countries, two-thirds of children and women of childbearing age are estimated to suffer from iron deficiency; one-third or more of them have the more severe form of the disorder, anemia. Symptoms include listlessness and fatigue. Low levels of iron are associated with often irreversible damage to brain development. Studies of the effects of iron deficiency and anemia have shown diminished psychomotor performance, suggestive of low levels of brain dysfunction or damage resulting from iron deficiency, even in the absence of apparent anemia. The intellectual development of children and the physical activity of both adults and children are impaired in those suffering from anemia. This may be complicated by lead toxicity, which is common in iron-deficient children, who are more susceptible to lead absorption. There is a direct correlation between low hemoglobin levels and the prevalence of diarrheal and respiratory diseases, which result from an impaired immune system. Lack of recovery in many children even after iron supplementation underscores the importance of preventing iron deficiency through dietary manipulations and health education.

The International Society for Prevention of Iron Deficiency Anemia promotes preventive approaches, such as iron fortification of basic foods (bread, sugar, salt) and routine supplementation for infants and pregnant women, to prevent what it defines as the most widespread nutritional deficiency in both developed and developing countries. The WHO reports that globally nearly 2 billion people are anemic and 3.6 billion are iron deficient. In many developing countries, IDA is prompted by worm infections, malaria, and other infectious diseases such as HIV and tuberculosis. The major health consequences include poor pregnancy outcome, impaired physical and cognitive development, increased risk of morbidity in children, and reduced work productivity in adults. Anemia is considered to contribute to 20 percent of all maternal deaths. Iron deficiency is the most common and widespread nutritional disorder in the world. As well as affecting a large number of children and women in developing countries, it is the only nutrient deficiency that is also significantly prevalent in industrialized countries (WHO, 2013).

Iodine-Deficiency Disorders

Iodine is an essential element in nutrition. Insufficient iodine in natural sources causes clinical or subclinical thyroid disorders. Deficient supply of iodine damages fetal development and produces fetal hypothyroidism. Low levels of circulating thyroid hormones, which can be measured in urinary iodine, cause varying degrees of brain damage in infants, including cretinism. A high prevalence of goiter, or enlargement of the thyroid gland, indicating suboptimal thyroid function, is present in large areas of the world where there are low levels of ground and surface water iodine.

In 2003, some 2 billion people suffered from iodine deficiency: 436 million in Europe, 624 million in South-East Asia, 365 million in the Western Pacific, 260 million in Africa, 229 million in the Eastern Mediterranean, and 75 million in the Americas (Andersson et al., 2007). In the USA, during World War I, high percentages of draft-eligible men were rejected because of goiter and these came primarily from mid-western prairie states. Drs David Marine and David Cowie, following a series of studies from 1910 onward, pioneered the idea of prevention of iodine deficiency by iodization of commercial table salt in 1924 (Morton's Iodized Salt) (see Chapter 1). By 1930, most of the salt consumed in the USA was iodized, and goiter had largely disappeared even in previously endemic areas.

Iodization of salt as a preventive measure has become standard public health practice in many countries since World War I. Iodization has been compulsory in Canada since 1979 and is widespread in Western Europe, although it is not universal and not always at effective levels.

In the 1980s, the WHO expressed growing concern over the widespread nature of iodine-deficiency disorders (IDDs) in large areas of the world, affecting an estimated 2.3 billion people, especially in China, the former Soviet Union, South-East Asia, and many developing countries. In 1986, the World Health Assembly called on all nations to introduce iodization of salt or other appropriate technology to reduce this silent pandemic.

Iodine deficiency is one of the main causes of impaired cognitive development in children. Australian scientist and public health advocate Basil Hetzel demonstrated that insufficient iodine during fetal development adversely affected brain development, prompting an international response to the issue of IDDs. In 1995, the WHO, in association with many international and donor organizations, called for universal salt iodization (USI). This program has been implemented gradually and incompletely, but has effectively reduced the prevalence of iodine deficiency conditions globally and can be seen as a substantial achievement of the global health community.

Iodine deficiency was described as follows in a 1993 editorial in the *New England Journal of Medicine*: "The most important effects of iodine deficiency are on the developing central nervous system, and they form a continuum from mild intellectual impairment to full blown cretinism." Prevention of iodine deficiency is best achieved through the iodization of salt on a national scale at a level of 1 part iodine to 10,000–20,000 parts salt.

UNICEF, the International Council for Control of Iodine Deficiency Disorders, the European Thyroid Association, Kiwanis International, some countries (e.g., Canada), and the World Bank have called for national and international action to control this widespread public health problem. The World Summit for Children called for USI, with a target of 95 percent iodization in each country by 1995. By 1994, 94 countries had national plans for iodizing salt, with 58 countries, including almost 60 percent of the world's children, on schedule.

In a 2004 report on iodine status, the WHO stated that the number of countries where iodine deficiency is a public health problem had halved over the previous decade. Therefore, the main strategy (USI) has been successful. The number of countries where iodine deficiency is a public health problem was reduced from 110 in 1993 to 54 in 2003. Of the 54 iodine-deficient countries, 40 are mildly iodine deficient and 14 moderately or even severely iodine deficient. Of the 126 countries for which data were available in 2003, iodine intake was adequate in 43. In 2013, WHO estimated that globally 66 percent of households have access to iodized salt. Progress is being made, but inertia and complacency remain barriers to achieving this goal.

China has achieved success in eliminating iodine deficiency. The percentage of Chinese households protected by iodized salt rose from 30 percent in 1995 to over 90 percent by 2000 and has been sustained at over 95 percent, but in remote areas of the country some 2 million children continue to be born each year unprotected from iodine deficiency. Nigeria has also achieved over 95 percent iodized salt consumption through enforcement of mandatory fortification of domestic and imported salt. Georgia has achieved success in mandatory iodization of salt since 2005, in a country traditionally endemic for IDD (Box 8.6).

Still, sustained efforts are required to strengthen salt iodization programs worldwide. Many countries, particularly in the Central Asian Republics and in South America, have difficulties with smuggling of non-iodized salt and poor compliance with national programs of iodization. Europe

BOX 8.6 Iodine Deficiency and Iodization of Salt in the Republic of Georgia, 2005

Traditionally, the Republic of Georgia has been a highly endemic country for iodine-deficiency disorders (IDDs), with the highland areas of the country most affected. Geographic and environmental factors such as low iodine content in the soil and water were the leading determinants of high IDD prevalence. Since the early 1990s, as the result of economic and political turmoil, as well as environmental changes (Chernobyl accident), the population has become especially vulnerable to iodine deficiency.

The situation became critical in 2000 when the Ministry of Health revealed alarming trends in high mountainous regions and remote areas of the country. In 2001–2002, as an emergency intervention, UNICEF supported the distribution of iodized oil capsules among children and pregnant/lactating women in the 35 worst affected districts of the country. In 1996, 64 percent of school children were found to be affected by iodine deficiency. In 1997, United Nations Development Programme indices reported a loss of IQ among the child population as a result of high IDD prevalence. In 1998, 80 percent of urinary iodine excretion (UIE) levels were low.

In recent years, the government of Georgia has initiated active steps to improve the nutritional status of the population. In 2005, the parliament adopted a law on "Prevention of iodine, other microelement and vitamin deficiencies" that banned the importation and sale of non-iodized salt and put in place mechanisms for food fortification policy in the country. Importation of iodized salt to Georgia increased 20-fold from 2017 tons in 1999 to more than 40,000 tons in 2005, meeting national requirements for edible iodized salt.

With heightened public awareness of IDD, consumption of iodized salt at household level increased eight-fold from 8 percent in 1999 to 67 percent in 2003, meeting the primary WHO criteria for IDD elimination. Goiter prevalence among children decreased from 54 percent in 1997 to 39 percent in 2003 and to 32 percent in 2005, while the total population iodine-deficiency indicator dropped from 58 percent in 1997 to 44 percent by 2003.

However, these positive developments may not have fully reversed the adverse impact on nutritional status of the difficult political situation and socioeconomic changes. According to the Georgia National Nutrition Survey (GNNS 2009), salt consumed by Georgian household members is currently well iodized; almost all specimens collected had added iodine. However, salt storage practices in many households expose salt to humid ambient air, which may result in dampening of the salt and seepage of iodine to the bottom of the container. Nonetheless, the results of iodine testing of household salt demonstrate show that, even if this occurs, it has little effect on the iodine content of household salt.

The GNNS 2009 recommends maintaining and enhancing current monitoring and evaluation practices of salt iodization following international recommendations. Salt iodization programs, even when functioning well as in Georgia, need frequent monitoring to ensure that proper fortification is performed, good coverage is maintained, and desired outcomes are continuously achieved.

Sources: *Lela Sturua, Non-Communicable Diseases Division, National Center for Disease Control and Public Health; Levan Baramidze, Public Health Department, Tbilisi State Medical University. Personal communication; April 2013.*
Children and women in Georgia: a situation analysis 2003. Available at: http://www.unicef.org/georgia/SITAN2003FinalFINALReport.pdf
Report of the 2009 Georgia National Nutrition Survey. Available at: http://www.unicef.org/georgia/GNNS2009_eng_with_cover_edit.pdf
Multiple Indicator Cluster Survey – Georgia; 2005. Available at: http://www.unicef.org/georgia/resources.html
United Nations Children's Fund. Sustained elimination of IDD through universal salt iodization in Georgia 2001–2005. Annual progress report. Tbilisi: Government of Georgia/UNICEF; 2004. Available at: http://www.ceecis.org/iodine/03_country/geo/03_08_geo.html [Accessed 30 May 2007].
Suchdev PS, Jashi M, Sekhniashvili Ze, Woodruff BA. Progress toward eliminating iodine deficiency in the Republic of Georgia. Int J Endocrinol Metab 2009;3:200–7.

still has many issues with iodization of salt and monitoring iodine deficiency (Zimmermann, 2011). A study carried out in northern France in 2003 reported 24 percent iodine deficiency in schoolchildren, which was mostly mild to moderate but still unsatisfactory. A study in the UK reported in 2013 that children of mothers with marginal and low levels of iodine had learning disabilities at age 8–9 years, showing the importance of iodine supplements even in a country considered "only mildly iodine deficient" (Bath et al, 2013).

Iodine deficiency continues to be a major public health problem in many parts of the world in the twenty-first century. The number of countries having iodine deficiency as a public health problem has halved over the past decade, yet 54 countries are still iodine deficient.

Osteoporosis

Osteoporosis is an important chronic condition, as discussed in Chapter 5, with serious health consequences, particularly among older women, in terms of fractures of the hip, spine, and forearm. Hip fracture mortality is between 12 and 20 percent, and many survivors are institutionalized as a result of complications. Osteoporosis is preventable to a large degree by adequate calcium and vitamin D supplementation and exposure to exercise and sunlight from the early years of life. Fluoridation of water may have beneficial effects as fluoride, in appropriate amounts, increases bone strength. Physical exercise, particularly weight bearing, helps to increase bone mass.

Exposure to sun produces vitamin D but this is countered by concerns that solar exposure increases skin cancer risk. Furthermore, use of sunscreens reduces ultraviolet light exposure of the skin, reducing vitamin D production. Sun exposure is also affected by seasonal variation and is dependent on geographic latitude, skin color, and total body coverage by clothing, particularly in dark-skinned people living in high-latitude countries, and in winter, when people tend to remain indoors and to cover up when they go outside.

Because of the changing demographics, migration from southern to northern countries, and increasing populations of elderly people, especially women, osteoporosis and its complications are important challenges for preventive, curative, and rehabilitative services as well as an issue for health promotion in terms of dietary and other self-care and preventive measures in individual medical practice as well as in public health. Food enrichment with calcium and vitamin D plays an important role in reducing the severity of this condition, which currently affects some 57 million Americans, 9 million with the disease and 48 million with low bone density and at risk for osteoporosis (National Osteoporosis Foundation, USA, 2013). The International Osteoporosis Foundation reports that, worldwide, more than 8 million fractures annually are caused by osteoporosis.

Eating Disorders

Eating disorders are an important health risk, primarily for teenage girls and young women, and an occupation-related health risk associated with sports, ballet, and modeling. These disorders can be communicated between people by group pressure of fashion, precedent, and close contact among teenage girls and young women. Fashions in a society have great influence on vulnerable adolescents who can enter a cycle of self-denial or purgation that may be extremely destructive and even fatal. In all eating disorders, a vital component of treatment and public health guidance should be behavioral and family therapy, as eating practices are highly influenced by family development and conceptions about image.

Eating disorders reportedly affect some 5 percent of US female college students. In the USA, among women the estimated lifetime prevalence of anorexia nervosa, bulimia nervosa, and binge-eating disorder is 0.9 percent, 1.5 percent, and 3.5 percent, respectively. The prevalence of eating disorders is lower in men than in women. Comorbid medical conditions are associated with anorexia and bulimia but binge eating is considered to be more linked to psychiatric disorders and severe obesity.

Anorexia Nervosa

Anorexia nervosa is a self-imposed severe dietary restriction that may be chronic or acute. This condition is most common among teenage girls and occupational groups including models and ballet dancers. Anorexia is often associated with laxative and diuretic abuse or excessive exercising. Specific features common to anorexia nervosa are abnormal sensitivity to being fat and fear of losing control over the amount of food eaten, severe restriction of food intake and refusal to accept food, marked weight loss, cessation of menstruation, damage to dentition and

chronic tooth pain. More serious consequences are liver and heart muscle damage, and not uncommonly death. The psychological features of rigidity, perfectionism, and fear of obesity preceding the condition are usually resistant to treatment. Hospitalization is commonly required with strict regimens of antidepressant and behavioral therapy. Anorexia nervosa often goes unrecognized and the risk that it may lead to death by self-starvation or suicide is not well known.

Bulimia and Binge-Eating Disorder

Bulimia nervosa is a condition most common among teenage girls who binge on food or who have regular episodes of overeating and feel a loss of control. The person then uses different methods, such as vomiting or abusing laxatives, to prevent weight gain. A binge eater has a compulsion to eat large quantities of food within a short period, usually two hours or less, with deliberate vomiting to expel the food to lose or maintain his or her weight. Many features are common to both bulimia and anorexia nervosa.

Binge-eating disorder is recurrent binge eating but without the characteristic compensatory behaviors of bulimia by vomiting or use of laxatives. This disorder is also more common in females than in males and is associated with severe obesity and/or a history of marked weight fluctuations. In the fifth edition of the *Diagnostic and Statistical Manual of Mental Disorders* (DSM-5), issued in May 2013, bulimia nervosa is designated as a feeding and eating disorder.

The physical effects of bulimia can be severe (such as esophageal damage from repeated vomiting) but usually less so than anorexia nervosa. The presence of depression may require antidepressant therapy supported by psychotherapy techniques or behavior modification therapy and hospitalization. Like anorexia, this disorder is most common in societies that promote images of beauty as thinness through advertising and related social pressures affecting psychologically vulnerable people with poor self-image.

DISEASES OF OVERNUTRITION

Although media attention tends to focus on the extreme forms of malnutrition, the more common nutritional health concerns in industrialized countries are those relating to specific deficiencies and dietary imbalances. In the USA, the *Surgeon General's Call to Action to Prevent and Decrease Overweight and Obesity, 2001* and *The Vision for a Healthy and Fit Nation, 2010* addressed the problems of nutrition in a developed country with an emphasis on a type of malnutrition which is often described as overnutrition. The reports also addressed the issues of developing a public health response with individual and community action to incorporate physical activity and healthy diet. Diseases of overnutrition, discussed in Chapter 5, are of increasing importance in developing countries as well, as the diseases of improving standards of living affect growing middle classes and as social patterns, food, and eating habits shift from traditional styles towards western diets including excess calories, especially from fatty foods and red meat.

Overweight/Obesity

Economic development leads to changes in a population's diet. Dietary habits identified with an affluent or western lifestyle are characterized by an excess of energy-dense foods rich in fat and simple sugars, with a relative deficiency of complex carbohydrates. Modest improvements in the economic state of a country are associated with an epidemiological shift, characterized by a rise in the incidence of chronic diseases of middle and later adult life. In developing countries, these typically coexist with the traditional and persistent problems associated with nutritional deficiencies. Obesity in the industrialized countries is often associated with poverty; its significance is increasingly recognized as a major public health issue not only for industrialized countries, but for developing countries as well.

Obesity is an excess of fat in the body caused when energy consumed is greater than the energy expended. Nutrient excess may occur in the short term or over a long period (chronic nutrient toxicity). Excessive weight for body size (BMI, see section entitled "Measuring Body Mass") is a public health problem because it is associated with premature death and is a risk factor for coronary heart disease, diabetes mellitus, hypertension, asthma, and gastrointestinal disorders. While there is evidence of a genetic predisposition for obesity, diet and other environmental factors, such as a sedentary lifestyle, play a major role in excess body fat. Increasing obesity at all ages is a widespread phenomenon in both industrialized and developing countries.

Restriction of caloric intake and increased exercise are strongly advised for individuals seeking to lose excess weight. Obese individuals tend to be inactive, and attempts to increase physical exercise have well-documented health benefits. Drugs and fad diets are treatment techniques that have the potential for damaging one's health, and any weight loss tends to be temporary. Surgical treatment may produce permanent weight loss but often with serious side-effects, such as chronic diarrhea. Nutritional counseling and assistance in the development of nutritious and low-fat diets will be more useful, especially when accompanied by exercise.

Obesity is extremely resistant to treatment, and thus primary prevention of obesity is a major public health target. Achieving this goal requires health education of mothers and very young children in proper nutritional practices. Eating and physical activity habits learned in childhood are very difficult, although not impossible, to change. Information about food choices should highlight the need to reduce fat and increase dietary fiber and complex carbohydrates. Food labeling can serve to inform the individual of the caloric and fat content of items intended for consumption. The individual person as well as governments, schools, parents, the media, and communities all have responsibilities in health promotion for the prevention of obesity.

Diabetes Mellitus

Diabetes mellitus, discussed in Chapter 5, is a chronic metabolic disorder found increasingly throughout the world associated with high caloric intake, increasing obesity, and sedentary lifestyle starting in early childhood. It develops in individuals who lack sufficient insulin production or whose insulin is impaired in function. As a result, they have reduced capacity to utilize glucose derived from carbohydrate foods or from body stores of glycogen. Type 1, also called insulin-dependent diabetes (IDDM) or juvenile diabetes, appears in childhood or adolescence, due to failure of insulin production. Type 1 diabetes is not diet generated; however, its management requires dietary controls as well as insulin.

Type 2, or mature-onset or non-insulin-dependent diabetes (NIDDM), typically develops in middle adulthood or old age, but is increasingly being seen in young people and even in children, so that the term type 2 diabetes is being used more often. It is mostly due to excess dietary caloric intake, with a high intake of saturated fats and low intake of dietary fiber. This results in decreased insulin sensitivity and abnormal glucose tolerance. Genetic predisposition and possibly fetal influences result in a high prevalence of type 2 diabetes in some ethnic groups such as Native Americans, Inuit, South Sea islanders, and Australian aborigines. However, prevalence is magnified by societal factors including poor diet, lack of physical exercise, and alcohol abuse. These interact with other risk factors, such as body weight, to instigate the onset of disease.

Type 1 and type 2 diabetes both require careful nutrition. Since the discovery of insulin by Frederick Banting and Charles Best in Toronto in 1921, type 1 diabetes has been managed by regulation of blood sugar with daily insulin injections, with monitoring of blood sugar, diet, and exercise. Type 2 diabetes is controllable by diet and exercise but may also require medication to lower the blood sugar. Approximately 80 percent of patients with type 2 diabetes are obese. Modern public health emphasizes weight reduction for obese diabetic patients to control their blood glucose and reduce their high risk of coronary heart disease and stroke. Nutrition plays a key role in the causation and management of type 2 diabetes.

Cardiovascular Diseases

As discussed in Chapter 5, a strong relationship exists among diet, lifestyle, and the risk of CVD, especially between saturated fat intake and incidence of this disease. Cholesterol levels can be reduced by lowering dietary intake of foods high in cholesterol and saturated fats, and by increasing foods containing fiber and monounsaturated fats, such as olive oil and avocado.

Mortality from coronary heart disease and strokes has declined dramatically in western countries, but strokes still occur in some 795,000 people (34 percent under age 65) in the USA each year, resulting in some 130,000 deaths. The American Heart Association estimates the direct and indirect costs for CVD in 2010 at US$445 billion (2008 dollars), rising by 61 percent to US$1094 billion by 2030. Estimates of the incidence of strokes based on the Framingham study underestimate the true incidence because the study primarily focuses on a white middle-class population; more representative population-based studies report higher rates because strokes are more common in African American and Hispanic populations.

Hypertension and diabetes are major risk factors for heart disease and stroke. Hypertension is associated with imbalanced nutrition, with excess fats and low intake of fruits and vegetables. Primary prevention is largely dietary, with weight loss, salt restriction, smoking cessation, and increased physical activity. Nutrition and diet play a major role in clinical care of the patient with CVD as well as in public health policy related to food, nutrition, and population awareness. Alcohol, especially red wine, in moderation, has a protective effect against coronary heart disease, but in excess contributes to increased rates. Clinical and health promotion intervention must also promote regular screening for elevated resting blood pressure, associated risk factors, and careful clinical management of hypertension to reduce the risk of CVD. Antihypertensive medications play an important role in preventing known complications of this condition.

Cancer

The relationship between specific nutrients and cancer is less well established than that between diet and CVD (see Chapter 5). The relationship between food consumption and cancer rates is supported by studies including the following: changing cancer rates in ethnic groups after migration with changes in dietary patterns, case–control studies of cancer patients and controls, prospective studies of populations with known dietary habits, and animal experimental data.

It is currently accepted that there is a relationship between diet and specific cancer sites. Cancers in which diet is implicated as an etiological factor include cancers of the oral cavity and pharynx, larynx, gastrointestinal tract, breast, liver, pancreas, lung, endometrium, cervix, and prostate. Dietary guidelines for prevention of cancer are shown in Box 8.7. The strongest dietary associations are high fat intake with cancers of the prostate and colon, high body weight with cancer of the endometrium, alcohol with cancer of the esophagus, and smoked, pickled, or salted foods with cancer of the stomach. Widespread consumption of betel nuts in India and other South Asian regions is a major risk factor for cancers of the alimentary tract. The protective effects of a diet high in fruit, whole grains, and vegetables on colon and other cancers are possibly due to vitamin A and C and their antioxidant effects.

Some epidemiologists attribute 30–40 percent of cancer in men and 60 percent in women as diet related. Since the 1960s, there has been a growing consensus that diet plays an important, although still undefined, role in carcinogenesis. The "Mediterranean diet", which is low in total and saturated fat, high in green and yellow vegetables and citrus fruits, and low in alcohol and salt-pickled, smoked, and salt-preserved foods, is consistent with a lower risk of many of the currently major cancers.

Public health nutrition includes a wide-ranging program of health education and health promotion programs as effective primary prevention. Governments, especially departments of agriculture and finance, should take steps to ensure fruit and vegetable supply at low cost to the consumer. Decisions affecting the supply and pricing of foods interact with the interests of farmers, producers, agribusiness, transport and storage, as well as marketers of food to the consumer and the food production industry. However, they exist within a social context in which public opinion and purchasing power affect the type of produce offered and available.

NUTRITION IN PREGNANCY AND LACTATION

Pregnant women and lactating women are, in effect, feeding two people. They must eat enough to meet their own requirements at a time of increased need, as well as provide for the growth of the fetus and infant. During pregnancy, a woman needs 300 kcal extra and an additional 20 g of protein per day. During breastfeeding, the mother needs an extra 500 kcal and 20 g of protein above her dietary needs based on height, weight, and activity levels (see Table 8.7). The needs for (many) vitamins and minerals are similarly higher during pregnancy and lactation.

Adequate nutrition and weight gain during pregnancy are important for the development of the fetus and for the woman's health. Weight gain of a mother is a good predictor of the birth weight of her infant. Since infant birth weight is a determinant of potential for survival and future development, it is important for the mother to achieve the recommended weight gain. Women who are of normal weight for their height or slightly overweight women typically have better pregnancy outcomes than those who are underweight.

Both caloric intake and nutritional quality need to be considered. A pregnant woman needs to increase her intake

BOX 8.7 Recommendations for Individual Diet and National Policies for Nutritional Security and Health

Recommended National Policies to Promote Nutritional Security

- National policies should promote legislation, regulation, and inspection to promote food security for preventing foodborne diseases.
- Food fortification is essential to prevent micronutrient deficiencies common in all countries; these include:
  - salt iodization with iodine, and monitoring of school-age children and pregnant women for iodine sufficiency
  - fortification of milk and milk substitutes, soft cheeses and yogurts with vitamin D
  - fortification of flour with iron, vitamin B complex including folic acid and vitamin B_{12}
  - fluoridation of community water supplies
  - fortification of breakfast cereals, pastas and other flour products, domestic and imported.
- Sentinel population groups such as pregnant women should be monitored for iron and vitamin D deficiency; infants for iron deficiency; and older groups for iron deficiency, excess lipids, thyroid deficiency, and vitamin D levels.
- Anthropometric monitoring should be undertaken of growth of children, and body mass index of children and adults.
- National health and nutrition surveys should be conducted.

Recommendations for Individual Nutritional Security

- Be as lean as possible without becoming underweight. Extra weight affects cancer risk. Excess fat around the waist is particularly harmful, releasing estrogen and other hormones into the bloodstream. This is strongly linked to colon cancer and probably to cancers of the pancreas and uterus, and breast cancer (postmenopausal).
- Be physically active for at least 30 minutes every day; moderate physical activity is beneficial at all ages.
- Avoid sugary drinks and energy-dense foods. Avoid fatty foods and baked goods with transfats.
- Eat more of a variety of vegetables, fruits, whole grains, nuts, olives, and legumes such as beans daily. Most meals should be based on plant foods. When preparing a meal, aim to fill at least two-thirds of the plate with vegetables, fruits, whole grains, and beans. A daily intake of vegetables and fruit is protective against a range of cancers, including mouth, pharynx, larynx, esophagus, stomach, lung, pancreas, and prostate.
- Limit consumption of red meats (such as beef, pork, and lamb) and avoid processed meats. Red meat contains substances linked to colon cancer. Frequent meals with fish and chicken are recommended.

- Drink water, tea, coffee, and milk; limit alcoholic drinks to two per day for men and one per day for women. Moderate alcohol intake is safe and heart healthy, but excess alcohol increases the risk of cancer of the mouth, pharynx, larynx, esophagus, breast, colorectum, and liver.
- Limit consumption of salty foods and foods processed with salt (sodium); salt and salt-preserved foods increase the risk of developing hypertension and stomach cancer.
- Vitamin and mineral supplements are essential for some groups and are recommended. The following are the most common situations when taking a supplement can be beneficial:
  - All women of childbearing age intending to conceive a child should take a folic acid supplement at least 3 months before conception, as well as during and following pregnancy, preferably with iron to reduce iron deficiency; in practice, women of fertile age should take folic acid daily as half of all pregnancies are unplanned.
  - Pregnant women and nursing mothers should also take a vitamin D supplement and possibly an iron supplement if their iron levels are low.
  - Children between 6 months and 5 years could benefit from taking drops containing vitamins A, C, and D; the American Academy of Pediatrics recommends vitamin D supplements up to the age of adolescence in view of reduced sun exposure lifestyles.
  - Middle-aged and older people should take a vitamin D supplement, especially people who rarely go outdoors, people who cover up all their skin when outdoors, and those who do not eat meat or oily fish.
  - Frail older people who have low calorie needs may benefit from a low-dose, balanced multivitamin supplement.
- New mothers should be strongly encouraged to breastfeed exclusively for up to 6 months and then add other liquids and foods, but not cow's milk until 1 year.
- After treatment, cancer patients should follow dietary recommendations for cancer prevention including daily vitamin D supplements.

Sources: *WCRF/AICR. Food, nutrition, physical activity, and the prevention of cancer: a global perspective. Available at: http://www.dietandcancerreport.org/expert_report/report_overview.php*
American Institute for Cancer Research. Recommendations for cancer prevention. Available at: http://preventcancer.aicr.org/site/PageServer?pagename=research_science_expert_report [Accessed 22 April 2013].

of folic acid, iron, and certain trace elements, through supplementation and food fortification. Other factors influencing weight gain include smoking, strenuous physical work, and chronic illness. The social pressure on women to be thin may make it difficult for some to allow themselves adequate weight gain. For further discussion on nutritional needs during pregnancy, see Chapter 6.

Mineral and vitamin supplementation is also important in pregnancy and lactation; iron, iodine, selenium, folic acid, and vitamins A, B, and C are especially important when these additional needs cannot be met from diet alone. As noted in Chapter 6, the need for folic acid supplements precedes pregnancy in order to prevent NTDs in the fetus, since the neural tube develops in the first weeks after conception, when a woman may not realize she is pregnant. As not all pregnancies are planned and prepregnancy preparation is not common, fortification of flour with folic acid has been made mandatory in Canada and the USA in order to prevent NTDs. Food enrichment is discussed later in this chapter.

Lactating women should continue iron, folate, and multiple vitamin supplementation. Return to pre-pregnancy weight is often a major preoccupation but should be secondary to meeting the calorie, micronutrient, and fluid needs of lactation and energy requirements of caring for a newborn.

PROMOTING HEALTHY DIETS AND LIFESTYLES

Nutrition plays a central role in health of an individual and a population, making this a major function of public health. Nutrition is also a very personal matter and requires understanding on the part of each individual as well as the society which deals with food and agricultural policies, costs and infrastructure, cultural standards, food enrichment, and many other aspects of nutrition. Education of the public in nutrition is a part of creating a social awareness of nutrition and its role in health (USDA).

Nationwide and community-based education to promote healthy eating patterns is an essential part of health promotion. Education for health can be promoted by departments or ministries of health, education, and agriculture, as well as by non-governmental organizations (NGOs). Availability, quality, variety, and cost of foods depend on national policy, economics, and personal preferences, knowledge, and community patterns. A nutrition program should provide consumers with information regarding food selection and should consider why people make their choices.

Healthy eating behavior is part of all stages of life. Beginning in kindergarten and grade school, children can be taught about the nutritional values of foods. School curricula and teacher training should provide sufficient background in nutrition to provide children with guidance in food selection and to promote desirable food habits. Child and adult education programs offer opportunities to emphasize the value of nutritious, balanced, and adequate diets for child and maternal health.

Dietary diversification programs need to take into account cultural beliefs about appropriate foods for different age groups, economic barriers to a healthy diet, and the availability of food types. The use of mass media to inform populations of food choices should not be overlooked.

Food support programs should be used to promote healthful nutrition (see Box 8.7). Community promotion of healthful nutrition can be implemented through schools and through health and social services, especially for vulnerable groups, including women at the age of menopause, single-parent families, and elderly, homeless, and immunocompromised people.

DIETARY GUIDELINES

Dietary guidelines provide a basis for individual and community education regarding healthful nutrition. The USDA's *My Food Plate* provides a personal eating guide with the foods and amounts suitable for each person. The food patterns are designed for the general public, ages 2 and over. They are not therapeutic diets for specific health conditions or for pregnancy or lactation. Those with a chronic health condition should consult with a health care provider to find a dietary plan appropriate for them. Special dietary requirements should be taken into account for adolescents, pregnant and lactating women, and the elderly. The recommended dietary guidelines are revised periodically. The Harvard adaptation, *Healthy Eating Plate*, is shown in Figure 8.3.

VITAMIN AND MINERAL ENRICHMENT OF BASIC FOODS

Health education regarding good nutrition practices alone is not sufficient to prevent significant manifestations (borderline or clinical) of vitamin and mineral deficiency states in vulnerable sections of the population, even in wealthy countries. It is a public health responsibility to ensure that all people receive an adequate basic vitamin/mineral daily intake even if their food budget or access to knowledge is limited. The best way to achieve this is through the appropriate vitamin and mineral enrichment of basic foods, such as bread, milk, and salt. Food enrichment to provide basic essential trace elements is standard practice widespread in North America and the UK. It must become a basic component of modern public health in nations where it has not yet been implemented.

The CDC includes improved nutrition as one of the great achievements of public health of the twentieth century, and "The discovery of essential nutrients and their roles in disease prevention has been instrumental in almost eliminating nutritional deficiency diseases such as goiter, rickets, and pellagra in the United States" (CDC, 1999). A policy of fortification of food began in the USA in 1924 with the fortification of salt with iodine. In the 1930s vitamin D was added to milk and in the 1930s there was voluntary fortification of flour with the B vitamins thiamine (B_1), riboflavin (B_2), and niacin (B_3), as well as iron. This became mandatory in 1943.

In the UK, since the 1940s there has been mandatory fortification of white flour with calcium, iron, and vitamins B_1 and B_3, and of margarines with vitamins A and D. These measures have helped to reduce the burden of many previously common deficiencies. Other commonly fortified foods in the UK include:

- breakfast cereals (B group vitamins and iron)
- soymilk (calcium, vitamin D, and some also have vitamins A and B_2 and iodine added)
- infant formula milks and many baby foods (a comprehensive spectrum of vitamins and minerals).

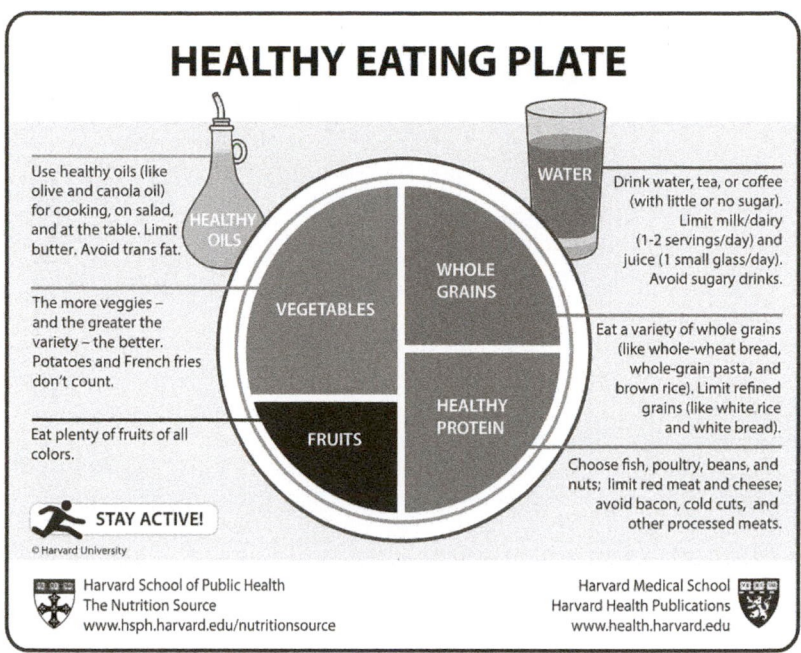

FIGURE 8.3   The Harvard Healthy Eating Plate. *Source: The Healthy Eating Plate: Copyright © 2011 Harvard University. For more information about The Healthy Eating Plate, please see The Nutrition Source, Department of Nutrition, Harvard School of Public Health. Available at: http://www. thenutritionsource.org and Harvard Health Publications, health.harvard.edu [Accessed 21 April 2013]; with permission.*

The UK Food Standards Agency recommends that a variety of foods will shortly be fortified with folic acid on a mandatory basis.

In Canada, as a result of a 1971 national nutrition survey, the federal government's Food and Drug Directorate passed regulations that made it illegal to sell milk products without vitamins A and D, to sell bread without iron or vitamin B complex, and to sell salt without iodine, each to specified levels. In the USA, flour as defined in FDA regulations for "enriched bread" is required to be fortified with iron, with vitamin B complex since 1941, and with folic acid since 1998. Folic acid enrichment is very widely practiced, and most breads are enriched in keeping with these regulations.

In the early 1990s, studies in the UK and elsewhere showed the protective effects of folic acid, if taken prior to and during early pregnancy (periconceptional), against NTDs, primarily anencephaly and spina bifida. This important finding raised new possibilities for prevention of birth defects by nutritional means. For planned pregnancies, this can be done through education and physician advice, and taking folic acid supplements before and during pregnancy. However, this is not practical for the majority of women as approximately half of all pregnancies are unplanned.

The alternative or supplementary approach is addition of folic acid to a common food such as bread and other flour- and wheat-based foods to reach the population at risk to prevent these fatal, severe, and difficult to manage birth defects (Figure 8.4). As a result, folic acid fortification of enriched flour is mandatory in the USA, Canada, all other countries in the Americas, and more recently in Australia. Over 50

countries have mandatory folic acid food fortification; however, most European countries do not.

Controversy in Food Enrichment

Food enrichment, a well-established practice in the USA, Canada, and Latin America, is increasingly being adopted in developing countries, but remains controversial in Europe. Opposition to food enrichment is largely based on interpretation of food fortification as an invasion of personal rights and a lack of professional support.

Food fortification is a cost-effective method of ensuring that nutrients reach large segments of the population without requiring changes in food consumption. Fortification has been used since the 1920s in industrialized countries to restore micronutrients lost in food processing, especially the B vitamins, and to add elements absent in the environment, such as iodine. Fortification, inexpensive and harmless to others, has played a key role in the eradication or control of diseases associated with deficiencies of these vitamins.

Is food enrichment an infringement of personal rights, or a manifestation of "paternalism versus liberalism"? It is the most liberal countries, including Canada and the USA, that have led in food enrichment for the common good, whereas the tradition-bound, more centralistic, less liberal countries have persisted in ignoring this issue. Chlorination and fluoridation of community drinking water are relevant precedents with positive experience in the western world. The philosophy of health promotion is based not only on prevention of disease, but also on ensuring an optimal

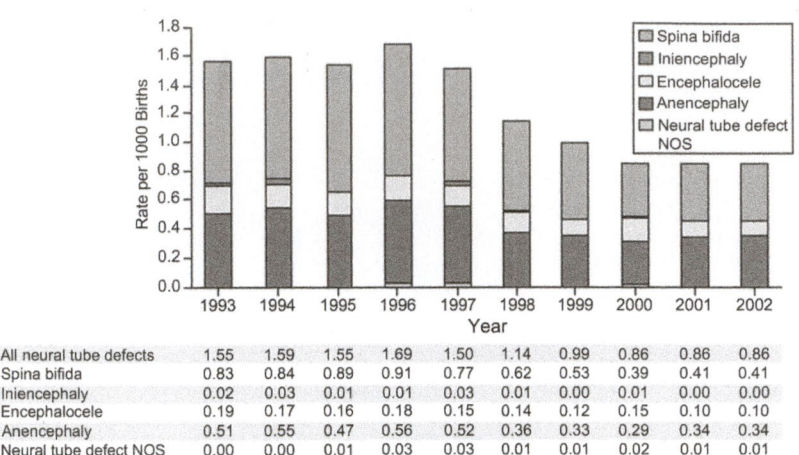

FIGURE 8.4   Decline in neural tube defects in Canada before and following fortification of flour with folic acid. *Source: De Wals P, Fassiatou-Tairou F, Van Allen MI, Uh SH, Lowry B, Sibbald B, et al. Reduction in neural-tube defects after folic acid fortification in Canada. N Engl J Med 2007;357:135–42; with permission.*

	1993	1994	1995	1996	1997	1998	1999	2000	2001	2002
All neural tube defects	1.55	1.59	1.55	1.69	1.50	1.14	0.99	0.86	0.86	0.86
Spina bifida	0.83	0.84	0.89	0.91	0.77	0.62	0.53	0.39	0.41	0.41
Iniencephaly	0.02	0.03	0.01	0.01	0.03	0.01	0.00	0.01	0.00	0.00
Encephalocele	0.19	0.17	0.16	0.18	0.15	0.14	0.12	0.15	0.10	0.10
Anencephaly	0.51	0.55	0.47	0.56	0.52	0.36	0.33	0.29	0.34	0.34
Neural tube defect NOS	0.00	0.00	0.01	0.03	0.03	0.01	0.01	0.02	0.01	0.01

health environment, particularly for the vulnerable groups in society (see Chapters 7 and 15).

Mandatory enrichment should be supported by mandatory labeling requirements, monitoring of levels of vitamin and mineral enrichment, and continuing health education on appropriate eating habits. National monitoring of the nutrition status of the population should be carried out in sentinel population groups through appropriate measurements (i.e., hematological, biochemical, and anthropometric).

Although malnutrition is widespread throughout the world, it is most common in the form of micronutrient deficiencies. These not only are health problems as defined by the WHO and UNICEF affecting well-being, but also affect the economic growth potential of a population. The World Bank has increasingly accepted the importance of public health nutrition as a basis for infrastructure and economic growth, with nutrition interventions being among the most cost-effective public health measures. These include nutrition education (including the promotion of breastfeeding), micronutrient fortification, micronutrient supplementation, food supplementation, food price subsidies, and control of parasitic diseases (World Development Report, 1993).

GENETICALLY MODIFIED FOODS

Genetically modified (GM) foods have become a major factor in food production in the USA and other countries, but are controversial (Box 8.8). They have important potential for improving production in water-scarce regions with less use of pesticides and potentially with enriched nutritional value in terms of quality of protein, vitamin, and mineral contents. A good example is the golden rice crop developed in Europe.

Herbicide-tolerant (HT) crops were developed to survive specific herbicides that previously would have destroyed the crop along with the targeted weeds, providing options for effective weed control in farming. USDA survey data show that HT soybeans increased from 17 percent of US soybean acreage in 1997 to 68 percent in 2001 and 93 percent in 2012. HT cotton expanded from about 10 percent of US acreage in 1997 to 56 percent in 2001 and 80 percent in 2012. Adoption of HT corn, slower in previous years, has accelerated, reaching 73 percent of US corn acreage in 2012.

Insect-resistant crops containing the gene from the soil bacterium *Bacillus thuringiensis* (Bt) have been available for corn and cotton since 1996. The bacteria produce a protein toxic to specific insects, protecting the plant over its entire life. Plantings of Bt corn grew from about 8 percent of US corn acreage in 1997 to 26 percent in 1999, then fell to 19 percent in 2000 and 2001, before climbing to 29 percent in 2003 and 67 percent in 2012, as shown in Figure 8.5. Commercial introduction in 2003–2004 of a new Bt corn variety that is resistant to the corn rootworm, a pest that may be more destructive to corn yield than the European corn borer, has promoted wider usage. Plantings of Bt cotton expanded from 15 percent of US cotton acreage in 1997 to 37 percent in 2001 and 77 percent in 2012.

FOOD AND NUTRITION POLICY

A national food policy aimed at preventing conditions related to nutritional deficiency or excess can only succeed with continuous intersectoral cooperation (Box 8.9). Regulation of food products by public health authorities involves national and provincial governments, as well as the local health authority. The food processing industry must be educated to include compliance and self-regulation as part of a corporate culture working within adequate government regulation and educated consumer demand, with the competition of the marketplace serving a positive role. Active

BOX 8.8 Genetically Modified Foods Progress and Controversy

The UN Food and Agriculture Organization (FAO) views genetically modified (GM) farming as "here to stay" despite controversy and opposition from many groups raising concerns of the unknown effects on humans, safety, the potential spread of GM seeds, and effects on traditional agriculture. FAO states: "Scientists in both public and private sectors clearly regard genetic modification as a major new set of tools, while industry sees GMOs [genetically modified organisms] as an opportunity for increased profits. Yet the public in many countries distrusts GMOs, often seeing them as part of globalization and privatization". A FAO consultation of experts convened at the FAO in January 2005 recommended that responsible deployment of GM crops needs to encompass a complete process, from development, through prerelease risk assessment, biosafety considerations, and postrelease monitoring.

There are common accusations of commercial interests promoting GM products with patents on the one hand, and protectionism of highly subsidized European agriculture on the other. Adoption of GM methods may, however, provide farmers in Africa and other developing areas with better crop yields, lower herbicide needs, and improved income, and could reduce malnutrition. The reduced dependency of GM crops on water during dry spells, alone, could make an enormous difference to public health in marginal farming areas in poor countries.

The scientific evidence regarding GMOs is still emerging, but so far there is no conclusive information on the definitive negative impacts of GMOs on health or the environment. Nevertheless, public perceptions about GMOs in food and agriculture are divided, with a tendency towards avoiding GM food and products in many developed and developing countries. The Cartagena Protocol on Biosafety came into force in 2003, and by October 2011 had been ratified by 161 countries. Labeling of GM foods is a current political issue in the USA.

Sources: *Food and Agriculture Organization of the United Nations. Agriculture and Consumer Protection Department; March 2005. Available at: http://www.fao.org/ag/magazine/0111sp.htm [Accessed 30 April 2013]. FAO Statistical Yearbook, World Food and Agriculture, 2012. Rome: FAO; 2012. Available at: http://www.fao.org/docrep/015/i2490e/i2490e00.htm and http://www.fao.org/docrep/015/i2490e/i2490e04d.pdf [Accessed 30 April 2013].*
US Department of Agriculture. Recent trends in GE adoption [posted 5 July 2012]. Available at: http://www.ers.usda.gov/data-products/adoption-of-genetically-engineered-crops-in-the-us/recent-trends-in-ge-adoption.aspx [Accessed 30 April 2013].

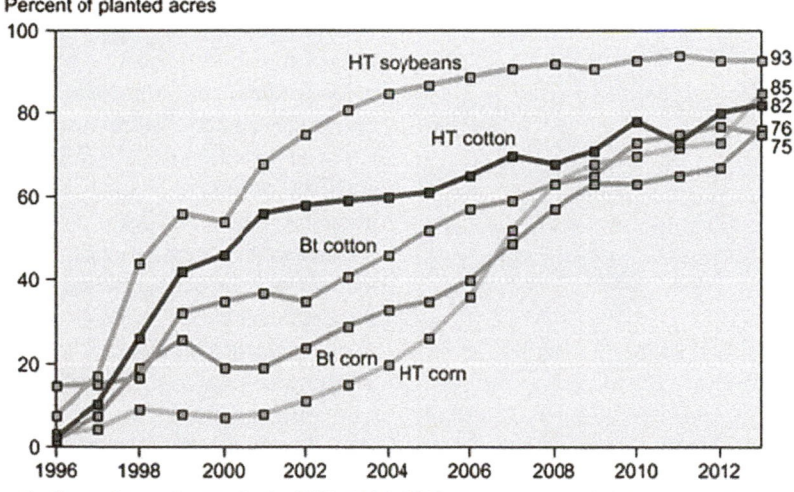

FIGURE 8.5   Growth of genetically engineered crops in the USA, 1996–2010. Note: HT=Herbicide-tolerant, Bt=*Bacillus thuringiensis. Source: US Department of Agriculture, Economic Research Service. Recent trends in GE adoption [posted 5 July 2012]. Available at: http://www.ers.usda.gov/data-products/adoption-of-genetically-engineered-crops-in-the-us/recent-trends-in-ge-adoption.aspx#.UpirQ8RDt8E [Accessed 29 November 2013].*

support is needed from government departments or ministries, food manufacturers and marketers, community leaders, health professionals, educators, women's groups, and the media in informing the public of the key issues.

Evolution of a Federal Role

The US federal government has been a force for improved nutrition for over a century. It has played a role in extension or outreach education programs as well as in food and agricultural research and development and in food supplementation programs. The federal role in nutrition research began in 1887 with the establishment of a nutrition laboratory which was the forerunner of the National Institutes of Health. The USDA instituted food monitoring programs in 1893 and extension services to promote nutrition education in 1914.

In 1917, national concern for nutrition increased as a result of research findings that large numbers of draftees were being rejected from military service because of poor nutritional status (e.g., goiter). The US Food Administration

BOX 8.9 Food and Nutrition Responsibilities of National and International Authorities

- *International agencies* – Establish and promote the science and practice of food production, processing, distribution, and marketing for adequacy of nutrition internationally. Agencies include the United Nations Food and Agriculture Organization (FAO), the WHO, UNICEF, United Nations Development Programme, the World Bank, and many bilateral governmental aid agencies such as USAID (USA), DFID (UK), CIDA (Canada), and many others (see Chapter 10).

- *National authority* – Establish national standards and regulatory requirements in such areas as food safety, labeling, fortification, and food content; responsible for regulating imported foods; establishing standards and enforcement of safety, packaging, and labeling; promotion of nutrition education and policies by government; coordination with the other departments of government (e.g., agriculture, trade, industry, and international agencies).

- *State/provincial authority* – Legislate ordinances which reinforce national standards; licensing of food establishments; supervisory and regulatory functions. Coordination with agricultural and industry departments of government; nutrition education; liaison with private food production, processing, and marketing industry.

- *Local authority* – Supervision of local food producers; inspection of meat and dairy products and certain processed foods; monitor school lunch programs; provision of nutrition services and education; periodic inspection of public eating places. Promotion of nutrition education, breastfeeding, and healthy weaning practices. Provision of well-child care including growth monitoring and nutrition education. Good nutrition in schools and control of unhealthy and fattening foods in schools and other child activity settings.

was established to supervise food supply during World War I. National initiatives in nutrition increased during the Great Depression (1929–1936) with the use of surplus agricultural commodities for food relief. This initiative later developed into routine school lunch programs on a national level continuing to the present time.

The US Food and Drug Administration (FDA) was established in 1927. During the 1930s, the USDA conducted the first nationwide food consumption surveys, and in 1939 a federal food stamp program was established. In the 1950s and 1960s, food stamp and lunch programs were expanded. In 1972, the USDA established the Special Supplemental Nutrition Program for Women, Infants, and Children (WIC), providing food supplements to needy pregnant women and children at risk. Programs to provide meals for the elderly were also developed and the school lunch program was expanded to include breakfasts in needy communities.

A 1940 report on malnutrition in the USA led to a national nutritional conference and to recommended dietary allowances

(RDAs). In 1941, the FDA issued enrichment standards for flour and bread with vitamin B complex and iron, and in 1942 addition of vitamins A and D to milk and margarine was mandated. Following World War II, in 1946 a national School Lunch Program was established. In 1965, as a part of the War on Poverty, the Head Start program was initiated. The Food Stamp Act was passed by Congress, followed by the Child Nutrition Act and School Breakfast Program in 1966. The National Health and Nutrition Examination Surveys, carried out periodically since 1971 (NHANES I, 1971–1974; NHANES II, 1976–1980; and NHANES III, 1988–1994) and annually since 1999, are vital to the documentation of the nutritional status of the US population (Table 8.12).

The NHANES provides ongoing studies that are important for US and global standards. The studies have a cross-sectional design and provide national estimates but no geographic or seasonal information, or data for specific population groups, or sources and uses of chemicals, and only limited data for children younger than 6 years of age. NHANES can provide ongoing data on exposure of the US population to environmental chemicals, such as lead. NHANES provides biomonitoring, clinical, and nutritional parameters that can be linked to evaluate health outcomes (Calafat, 2012).

The WIC supplementary food program for women, infants, and children and the elderly (Box 8.10) provides federal grants to states for supplemental foods, health care referrals, and nutrition education for low-income pregnant, breastfeeding, and non-breastfeeding postpartum women, and to infants and children up to the age of 5 who are found to be at nutritional risk. WIC distribution sites are also active in promoting immunizations and education regarding pregnancy and child nutrition.

Labeling of food products with nutritional content information and food claim statements are required by the FDA for domestic and imported foods to inform the consumer and to minimize legal action and delays (FDA 2013). Enrichment of infant cereals and formulas with vitamins B_1, B_2, B_6, and iron, with vitamin C in drinks and dessert powders, was required. In 1979, a Nutrition Policy Board and Surgeon General's publication *Healthy People* established nutritional objectives for the nation. In 1988, Dr Everett Koop, the Surgeon General of the USA, issued the *Surgeon General's Report on Nutrition and Health*, a landmark document on nutrition in national health policy. Federal regulations were issued in 1993 under the Nutrition Labeling and Education Act, and in 1994, regulations were issued for labeling of meat and poultry. This was followed in 1996 by federal requirements for addition of folic acid to enriched flour, and in 1998 federal agencies recommended vitamin B supplements for older people.

Nutrition Issues in Development Policies

Countries or regions that are poor and suffer from high rates of borderline or overt malnutrition must address the

TABLE 8.12 Examination and Laboratory Components of National Health and Nutrition Examination Surveys (NHANES), USA, 1999–2012

Examination Component	Sample Description	Laboratory Component	Sample Description
Arthritis body measures	20–69 years	Acrylamide	3 years and older
Audiometry	½ sample 20–69 years	Albumin	6 years and older
Balance	½ sample 20–69 years	Albumin (home urine collection)	
Bioelectrical impedance analysis	8–49 years	Apolipoprotein (B)	½ sample 12 years and older
Blood pressure	8 years and older	Arsenic	⅓ sample 6 years and older
Body measurements	Birth and older	Body alkaline phosphatase	8 years and older
Cognitive function	60 years and older	Bacterial vaginosis/trichomonas	Females 15–49 years
Cardiovascular (CV) fitness	12–49 years	Cadmium	1 year and older
Dermatology	20–59 years	Caffeine exposure	6 years and older
Dietary	Birth and older	CD4/CD8	18–49 years/HIV+ and controls
Dietary supplement	Birth and older	Celiac disease (TTG-EMA)	6 years and older
Dual-energy X-ray absorptiometry		Chemistry panel	12 years and older
Body composition	8 years and older	Chlamydia	14–39 years
Bone density: hip and spine	8 years and older	Cholesterol (total)	3 years and older
Food frequency	2 years and older	High-density lipoprotein (HDL)	3 years and older
Grip test	6 years and older	Low-density lipoprotein (LDL)	Subsample 3 years and older
Lower extremity disease		Complete blood count	1 year and older
Peripheral neuropathy	40 years and older	Colinine	3 years and older
Peripheral vascular disease	40 years and older	C-reactive protein (CRP)	3 years and older
Mental health	8–39 years	Creatine phosphokinase	12 years and older
Attention deficit hyperactivity disorder (ADHD) – youth	Parent of 8–15 years	Creatinine	6 years and older
Anxiety – youth	8–19 years	Creatinine (home urine collection)	6 years and older
Conduct disorders – youth	Parent of 8–15 years	Cryptosporidium	6–49 years
Depression – Youth	8–19 years	Cytomegalovirus	1–5 years
Eating disorders – youth	8–19 years	Dust allergens	1 year and older
Elimination disorders – youth	Parent of 8–11 years	Environmental phenols	⅓ sample 6 years and older
Panic disorder – youth	8–19 years	Erythrocyte protoporphyrin	1 year and older
Depression – adult	1/2 sample 20–39 years	Ferritin Fibrinogen Folate (RBC)	1 year and older 40 years and older 3 years and older

Note: TTG= tissue transglutaminase; EMA=endomysial antibodies; RBC=red blood cells.
Calafat AM. Infrastructures and procedures for a sustainable HBM applied in the US National Health and Nutrition Examination Survey (NHANES). Available at: http://www.eu-hbm.info/cophes/InfrastructuresandproceduresforasustainableHBMappliedintheU.S.NationalHealthandNutritionExaminationSurvey_NHANES.pdf [Accessed 21 April 2013].
Sources: Centers for Disease Control and Prevention. National Health and Nutrition Examination Survey, 1999–2012. Available at: http://www.cdc.gov/nchs/data/nhanes/survey_content_99_12.pdf [Accessed 21 April 2013].

nutritional needs of the population as an essential part of planning for development. Economic development and nutritional status are directly related, as a malnourished population is not productive and cannot learn well. Efficient planning and policy making require all governments to establish systematic assessment, intervention, and monitoring of the existing nutritional situation. This includes growth and body mass index (BMI) as well as biochemical

and hematological monitoring, such as with nutrition intake recall studies, urinary iodine levels, hematological, vitamin and trace element surveys, and others.

Piped or clean water and public sanitation and improved food safety help to reduce diarrheal disease which, especially in children, aggravates malnutrition (see Chapters 4 and 9). Microfinancing (small loans), cell phones, solar energy for electricity production, and technical support for

BOX 8.10 Special Supplemental Nutrition Program for Women, Infants, and Children (WIC)

WIC is a federal supplementary food program of the US Department of Agriculture for needy women, infants, and children, initiated in 1972. WIC provides federal grants to states for supplemental nutritious foods, and education for needy families. It is available in all 50 states, 34 Indian Tribal Organizations and American possessions, provided through 1836 local agencies and some 9000 clinic sites.

WIC also provides health care referrals, and nutrition education for low-income pregnant, breastfeeding, and non-breastfeeding postpartum women, and to infants and children up to age 5, based on income levels and state residency, and deemed to be at "nutritional risk" by a health professional, based on anemia, low weight for age, history of pregnancy complications and outcomes. Dietary risk means poor compliance with current dietary guidelines for Americans. WIC distribution sites are also active in promoting immunizations and education regarding pregnancy care and child nutrition.

Checks or vouchers are usually given to people to purchase food, but the program is gradually moving towards an electronic card system. In some states, food warehouses are provided and in some home deliveries as well.

Since 1992, WIC has supported a farmer's market nutrition support program in 46 states and US territories. WIC beneficiaries are enabled to purchase locally grown fresh fruit and vegetables and other produce at low prices.

The US FDA also supports school lunch programs serving free or subsidized nutritious meals meeting federal nutrition standards in over 100,000 public and non-profit private schools and residential institutions. The program provides cash subsidies and food donated by the Department of Agriculture.

In 2012 the program provided nutritional support for 8.9 million beneficiaries at a cost of US$6803 million, an increase from 4.5 million people and US$2122 million in 1990.

Sources: *US Department of Agriculture. Women, infants and children (WIC). Available at: http://www.fns.usda.gov/, http://wicworks.nal.usda.gov/ and http://www.fns.usda.gov/pd/wisummary.htm [All accessed 20 April 2013].*

The cost to the health system of sickness associated with overnutrition or undernutrition alone justifies expenditures on educational effort to reduce the burden of disease or disability associated with inappropriate nutritional status.

Role of the Private Sector and Non-Governmental Organizations

The private sector grows, processes, and markets food and thus plays a vital role in nutrition and food safety. This includes the farmer–producer, but equally the food processor and marketing industry. The private sector responds to the consumer but can also create demand for new products with a health component by labeling, advertising, and marketing practices.

Manufacturers of salt initiated iodization of salt in the USA, with widespread manufacture and marketing of iodized salt to prevent goiter in 1924 (Morton's Iodized Salt) (see Chapter 1). Baby food manufacturers enriched their products with vitamins and minerals without being required by law. Where governments are reluctant to mandate food enrichment, private industry may already have made the decision to do so. This can be a valuable marketing tool, and where government regulations stipulate permissible enrichment, private industry fills an important public health function by pursuing enrichment policies.

NGOs, especially women's organizations, have played important roles in nutrition. For example, in the early part of the twentieth century rural women's organizations in North America promoted better nutrition through education and consciousness raising. Lobbying of government and private industry for constructive nutrition policies is also important for promoting better nutrition for vulnerable groups. Concern for nutrition of the elderly is an area where NGO support for nutritional assistance programs can be very effective. Women's magazines have also played an important role in disseminating information on healthful diets and raising interest in the relationship between nutrition and health.

Role of Health Providers

Providers of care play a pivotal role in the education of patients surrounding appropriate nutrition during normal healthy periods and certainly during illness. The provider of health care has the trust of the patient and the professional knowledge to convey information to the patient in a time and manner that may be more effective than any other form of education. Such a patient may be someone at risk for development of ischemic heart disease, hypertension, and other diseases associated with nutritional factors, before and even more so after the disease process has become clinically apparent.

development of small agriculture are transforming village life in many deprived areas. This can help in food production; marketing businesses in rural and poor urban areas can improve the food supply and food quality.

Successful nutrition policy ultimately depends on consumer knowledge and access to quality foods at costs that are affordable. Fortification is an important supplement, but education of the producer, processor, marketer, and consumer is no less vital. Consideration must be given not only to the supply of food and nutrients but also to the short- and long-term effects of agricultural policies. Food security policies must aim to guarantee all families access to their minimum food requirements.

Malnutrition is associated with an increase in morbidity from infectious and non-communicable diseases. Conversely, disease, whether infectious or chronic, can cause a worsening of underlying malnutrition. Health providers need to be sensitive to these interactions and aware of their consequences. Prevention and management of infectious diseases can have an important effect on the nutritional status of children at risk for PEM, growth faltering, or failure to thrive. Identification of children at risk for malnutrition can result in additional attention and instruction of the mother regarding appropriate nutrition. Counseling and education for healthful nutrition according to age and needs comprise a key element of all patient care, as nutrition is a public health issue.

NUTRITION MONITORING AND EVALUATION

Ongoing monitoring of the nutritional status of representative samples of the target population is an essential part of evaluating the health of a population. Monitoring may be done through population samples or at selected sentinel centers. Nutrition monitoring systems, such as the US NHANES, enable governments to monitor trends of food intake of individuals sampled and trends in nutritional status and general health of the population. Study results can direct policy makers in dietary standards and recommendations for manufacturers, state and local governments, and the general public (Table 8.13).

Surveys of food purchases and consumption through direct observation, and interviews of family members can provide valuable nutrition monitoring data, as can interviews

with food marketing networks (Table 8.14). Food supply and consumption tables can be calculated from total national data to estimate average consumption. Anthropometric measurements of height and weight, skinfold, and BMI provide valuable data on nutrition status, as do hematological (iron, ferritin) and biochemical indicators (cholesterol, lipids, fasting blood sugar, vitamin levels). Large-scale surveys are costly, thus sentinel center studies in selected representative locations may be more practical and can provide time trends in successive cross-sectional studies.

Standard Reference Populations

Anthropometric measures were first developed in the nineteenth century by Richer using skinfold thickness as an index of fatness. During World War I, Matiega developed ways of measuring the composition of the human body to assess the physical efficiency of soldiers. Anthropometry includes composition of the body at the atomic, molecular, cellular, tissue, and whole body levels. Whole body studies involve measurements of body weight and stature, including skinfolds, circumferences, and bone measures. These studies have clinical as well as public health importance. The association between weight loss and disease outcome has become a useful predictor of prognosis in individual patient care.

Reference populations are used for comparison purposes and have been found to reflect the "optimal" growth of children who are not excessively ill and who receive what is currently thought to be "good nutrition". The data may be presented in chart form and as percentiles. Recorded weights or heights at specific ages for the individual or group of children are seen graphically at different ages.

The US National Center for Health Statistics (NCHS) growth pattern data, which were derived from a white middle-class population sample, have been increasingly used internationally as the standard reference population since 1977. The data were updated and reissued in 2000. They are based on revised curves and BMIs for children, including about half of breastfed children, with data from NHANES and new statistical analysis. The original NCHS standard was adopted and has been used by the WHO since the late 1970s as the international standard reference population. The WHO concluded that child growth potential is similar across national and ethnic lines if nutrition is appropriate and intercurrent illness is not excessive. The standards have been used internationally for several decades.

The NCHS/WHO reference standard was criticized, however, as the data used to construct the reference covering birth to 3 years of age came from a longitudinal study of children of European ancestry from a single community in the USA. Furthermore, measurements were taken every 3 months, which is inadequate to describe the rapid and changing rate of growth in early infancy. The statistical methods available at the time the NCHS/WHO growth curves were

---

TABLE 8.13 Standards and Samples for Monitoring the Nutrition Status of a Population

Standards/Samples	Factors
Reference standards	International standards preferable to local standards to compare and monitor changes over time with optimal "ideal growth" population (gold standard) as a standard measure for comparison
Representative samples	Epidemiologically representative sample of total population at risk
Sentinel populations	Monitor population in particular program such as food supplement program (e.g., WIC) or attendees of a mother and child preventive care program
Sentinel centers	Selected places of contact with population in which monitoring can be added to basic program such as maternal and child health centers, emergency departments, and doctors' offices

Note: WIC = Special Supplemental Nutrition Program for Women, Infants, and Children.

TABLE 8.14 Measures for Monitoring the Nutrition Status of a Population

Assessment	Indicators
Indirect measures	
Food supply and consumption	Total available food supply by categories of foods divided by the total population
Education levels and cultural practices	Literacy, years of schooling, and especially female education levels are directly and indirectly related to adequacy of infant and child care mediated through improved family incomes and competency in dealing with the challenges of child care; traditional dietary patterns and individual choices
Household income and food purchase surveys	Per capita GNP, average household incomes, and other indicators of purchasing power are derived from census data and special surveys, including purchasing practices
Food balance sheets	An indirect measure of nutrition state by aggregating national data on amounts of food products adding imports, subtracting exports; the balance is converted to calories, deriving per capita consumption, providing reasonable estimates of actual average intake
Infant mortality, low birth weight	Where these rates are high, it is assumed that malnutrition is widespread
Parasitosis	High rates of parasite infestation contribute to malnutrition
Direct measures	
Clinical assessment	Hospitalizations with primary or secondary diagnoses serving as index cases of extent of nutritional problems such as rickets and diabetes
Anthropometric measures and surveys	Weight and height for age and weight for height using WHO/NCHS standards now recommended for use as the international gold standard; skinfold thickness of upper arm is also widely used to indicate protein/calorie status of individual child or the population of children; body mass index is another widely used measure; WHO issued new child health growth curves in 2007
Biochemical measures and surveys	Examinations of body fluids including blood and urine are measures of nutrients or indicators of complex metabolic processes; anemia is measured by hemoglobin, hematocrit, serum iron, or ferritin; thyroid function tests and iodine levels in urine measure iodine status; vitamin C in cells and serum, serum levels of vitamins B and D also used; these are expensive tests not readily available for surveys; surrogate cheap field tests such as zinc protoporphyrin can serve as indicators of blood lead and/or anemia; routine fasting blood sugars in pregnant women can give an idea of prevalence of glucose intolerance
Dietary assessment	Dietary surveys in institutions or households are usually based on recall of intake over past 24 hours or 7 days, or on dietary records weighing food served and uneaten

Note: GNP = gross national product; WHO = World Health Organization; NCHS = National Center for Health Statistics.

constructed were too limited to correctly model the pattern and variability of growth. WHO experts agreed with the principle and the importance of international growth standards and undertook development of new growth charts consistent with "best" health practices that would show optimal child growth in all countries rather than in one country.

The WHO Multicentre Growth Reference Study (MGRS) produced the new international standards based on healthy children of different ethnic origins living under conditions likely to favor achievement of their full genetic growth potential. The mothers of the children selected for the construction of the standards engaged in fundamental health-promoting practices; namely, breastfeeding and not smoking. The new WHO standards were derived from an international study of 8440 children from a diverse set of countries, including Brazil, Ghana, India, Norway, Oman, and the USA, providing a sample with ethnic or genetic variability in addition to cultural variation in how children are nurtured, and thus strengthening universal applicability of the standards. The new WHO standards explicitly identify breastfeeding as the biological norm and the breastfed child as the normative model for growth and

development. They also include markers for six gross motor developmental milestones. WHO released the new growth charts in 2006, reviewed by a joint committee of the US Centers for Disease Control and Prevention (CDC) and the American Academy of Pediatrics (AAP), which recommended adoption of the WHO curves for children up to the age of 2 years. The WHO curves are available at http://www.who.int/childgrowth/standards/en/ and include the following indicators:

- length/height-for-age
- weight-for-age
- weight-for-length
- weight-for-height
- body mass index-for-age (BMI-for-age)
- head circumference-for-age
- arm circumference-for-age
- subscapular skinfold-for-age
- triceps skinfold-for-age
- motor development milestones
- weight velocity
- length velocity
- head circumference velocity.

The WHO curves have been endorsed by the European Childhood Obesity Group, the International Pediatric Association, the United Nations (UN) Standing Committee on Nutrition, and the International Union of Nutrition Sciences, and adopted for use in the UK, Canada, and other countries.

Combinations such as these indicators are used to calculate the fat and fat-free content of body mass, which is useful for clinical assessment and special studies of groups. Fat-free mass (FFM) indicates the muscle mass; this figure may be used in nutrition studies, for patients enrolled in long-term nutrition follow-up, and for teaching purposes. The CDC adopted the WHO growth curves for infants and children 0–2 years of age after expert committee deliberations. For children aged 2 years and older, CDC recommends the CDC growth curves because these charts can be used continuously up to 19 years of age.

Measuring Deviation from the Reference Population

Comparing the growth pattern of a study population to a standard provides a comparison similar to measuring blood cholesterol and comparing it to a normal range. A simple method of surveying a population for comparison with a standard, such as the NCHS/WHO reference population, is to record heights and weights of children presenting for immunization on a standard growth chart and compare the clustering of the observed measurements to the international standard. These records provide personnel in primary care with a method of monitoring the population they serve. The NCHS growth curves for boys aged 0–2 years for length-for-age and weight-for-age are shown in Figure 8.6, and BMI percentiles for girls aged 2–19 years in Figure 8.7 as examples. Weight-for-height and head circumference curves and other charts are available at the WHO website

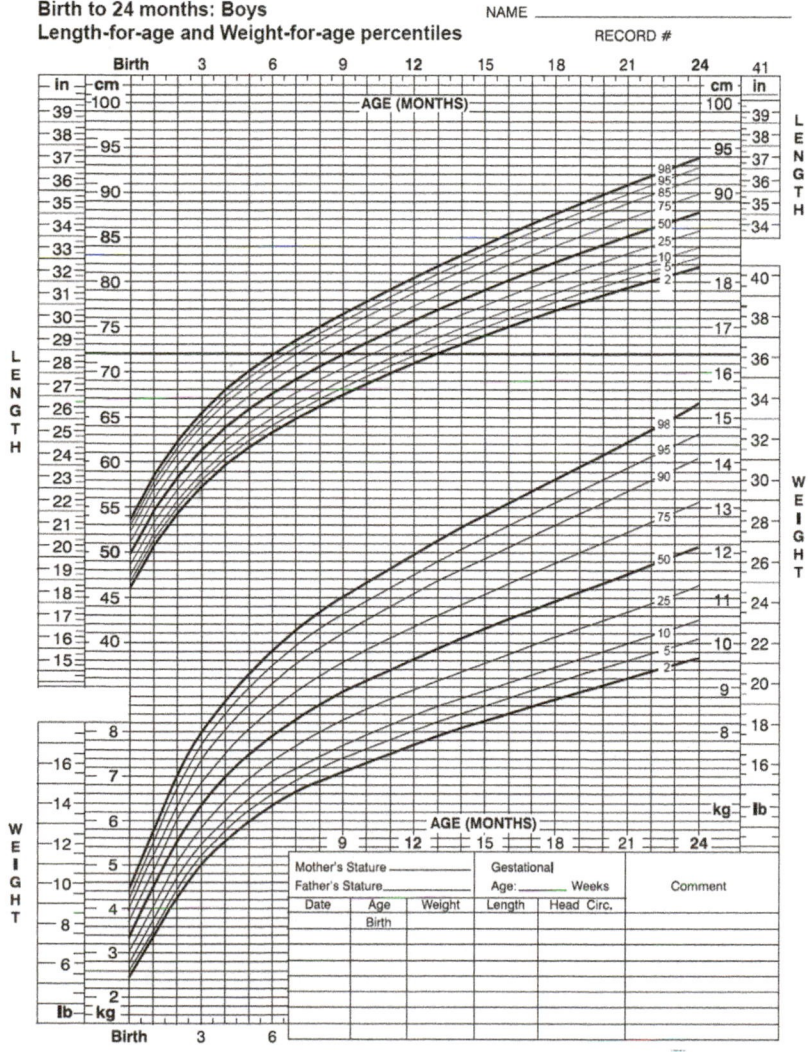

FIGURE 8.6   Standard growth curve (WHO), boys age 0–24 months, length-for-age and weight-for-age percentiles. *Source: Centers for Disease Control and Prevention. 1 November 2009. Available at: http://www.cdc.gov/growthcharts/data/who/grchrt_boys_24lw_9210.pdf [Accessed 19 April 2013].*

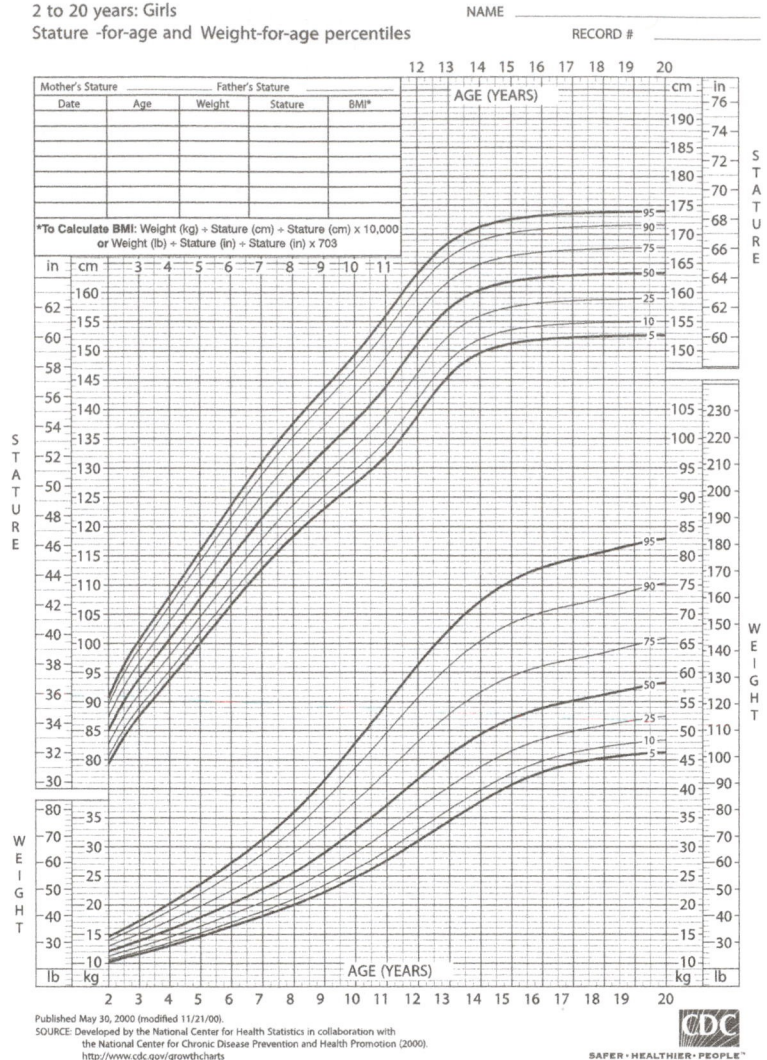

FIGURE 8.7 **Recommended growth curve, body mass index percentiles, girls age 2–19 years.** *Source: Centers for Disease Control and Prevention. Available at: http://www.cdc.gov/growthcharts/data/set1clinical/cj41c022.pdf [Accessed 29 November 2013].*

(http://www.who.int/childgrowth). The development of growth curve classification systems is shown in Box 8.11.

Z-scores are a widely used method of showing anthropometric data compared to a standard. A Z-score shows deviation from the mean in terms of standard deviations. Thus, the birth weight or growth pattern of children may be compared to a norm in standard deviations and not absolute numbers. Observations representing 68 percent of a normal distribution curve are included within one standard deviation, while 95 percent are included within 1.96 standard deviations (see Chapter 3). Z-scores simplify graphic summation of the data in anthropometric studies.

GLOBAL NUTRITION TARGETS FOR 2025

Following the MDGs of 2001–2015, the WHO has set new targets for maternal infant and young child nutrition for the year 2025. This move represents global consensus on

the centrality of nutrition in combating disease and death among women and children. Reaching these targets will involve complex issues of poverty reduction, education in general and in nutrition in particular, food fortification and micronutrient supplements for women, infants, and young children.

Box 8.12 shows post-MDG health targets recommended by WHO for 2025 to improve maternal, infant, and young child nutrition. These recognize the importance of nutrition in advancing child health and should help to direct international aid projects.

FOOD QUALITY AND SAFETY

Foodborne disease is a major public health issue in all societies. In developing countries, it is one of the major causes of morbidity and mortality, and even in the most advanced countries great care must be taken by governments, the

BOX 8.11 Child Growth Curves as the Standard to Monitor Child Growth Globally

The development of classification systems to monitor child growth was pioneered in 1956, by the Mexican physician Federico Gomez. He developed a system of simple anthropometric weight-for-age measurements to develop clinical profiles linking child malnutrition with risk of mortality.

By the 1970s, height was recognized as an important factor. The work of Waterlow resulted in using weight as the indicator for current nutrition status and height as the indicator for past nutrition status, thus distinguishing between acute and chronic malnutrition. The use of the two measures became the leading indicator for screening and identifying severely malnourished children at increased risk of death.

The Z-score system is accepted as the best statistical representation of population-based studies, with -2 and -3 standard deviation cut offs. For clinical care of the individual child, the percentile growth curves are used most widely.

By April 2011, 125 countries had adopted the WHO growth curves, but some 13 European and 7 African countries prefer to continue use of local growth standards (deOnis, 2012).

The WFLH (weight–height–length) indicator essential for assessing severe acute malnutrition, overweight, and obesity is used in many countries. The BMIFA (body mass index for age) indicator is used to monitor the child obesity epidemic (overweight and obesity) for the age group over age 2 years. The WHO growth curves provide a global gold standard measure for child health.

Sources: *de Onis M, Onyango A, Borghi E, Siyam A, Blössner M, Lutter C, WHO Multicentre Growth Reference Study Group. Worldwide implementation of the WHO child growth standards. Public Health Nutr 2012;15:1603–10. Available at: http://www.who.int/childgrowth/publications/global_implementation.pdf [Accessed 19 April 2013].*
de Onis M. Measuring nutritional status in relation to mortality. Bull World Health Organ 2000;78:1271–4. Available at: http://www.who.int/bulletin/archives/78(10)1271.pdf [Accessed 20 April 2013].
de Onis M, Onyango AW, Borghi E, Garza C, Yang H, WHO Multicentre Growth Reference Study Group. Comparison of the World Health Organization (WHO) child growth standards and the National Center for Health Statistics/WHO international growth reference: implications for child health programmes. J Nutr 2007;137:144–8. Available at: http://jn.nutrition.org/content/137/1/144.full.pdf [Accessed April 2013].

BOX 8.12 Post-Millennium Development Goal Global Targets 2025: To Improve Maternal, Infant, and Young Child Nutrition

- 40 percent reduction in the number of children aged <5 years who are stunted.
- 50 percent reduction of anemia in women of reproductive age.
- 30 percent reduction in low birth weight.
- No increase in childhood overweight.
- Increase the rate of exclusive breastfeeding in the first 6 months to at least 50 percent.
- Reduce and maintain childhood wasting to less than 5 percent.

Source: *World Health Organization. Nutrition. Global targets 2025: to improve maternal and infant and young child nutrition. Available at: http://www.who.int/nutrition/topics/nutrition_globaltargets2025/en/index.html [Accessed 28 April 2013].*

gastroenteric disease and contributes to the background level of diarrheal diseases not necessarily identified as outbreaks. In the USA, produce (often leafy green vegetables) is the source of nearly half of all foodborne illnesses. Annually in the USA, 48 million people suffer a foodborne illness, with 128,000 hospitalizations and 3000 deaths. Norovirus causes most foodborne illness and although usually a mild disease, is a leading cause of deaths. Leading causes of hospitalizations are non-typhoidal *Salmonella*, norovirus, *Campylobacter* and *Toxoplasma*. *Listeria* is a leading cause of death.

Food and water as sources of exposure to chemical and biological hazards place large health risks and economic burdens on individuals, communities, and countries. The WHO notes that new foods developed from biotechnology, along with animal husbandry, antibiotics and food additives, are factors in foodborne illness and food safety and security. The UN Food and Agriculture Organization (FAO) and WHO conduct a periodic review program for toxicity and residues of pesticides. Waterborne diseases and their prevention are discussed in Chapter 9.

Animals used for human food can be a source of transmission of communicable disease as well as chronic disease. Infectious diseases can spread when food is contaminated with pathogenic organisms during growth, transportation, preparation, storage, or handling before consumption. Food poisoning or diarrhea can result from *Salmonella* or *Campylobacter* on chicken meat or eggs. Animals raised for food can harbor diseases such as brucellosis that can be transmitted to humans through consumption of unpasteurized milk products from infected animals. People can become seriously affected by toxins in food from animal origin, including meat, milk, and seafood. Diseases such as brucellosis, anthrax, and Rift Valley fever can be spread from infected domestic animals by direct contact with people handling

food industry, and the consumer to prevent foodborne disease. Both government and the food industry have a vested interest in providing nutritious and safe food to the public at reasonable cost. Other concerns of private industry are attractiveness of food and its packaging, long shelf-life, low rates of loss of food or spoilage, contamination, and deterioration. Because the food producer and the processing/marketing industry wish to maximize their profit, self-policing is not sufficient, and a strong governmental regulatory agency is necessary to protect the public interest.

Fruits and vegetables contaminated with enteric pathogens are common causes of disease. Ingestion of contaminated products results in frequent outbreaks of

meat products. Table 8.15 lists various foodborne agents capable of affecting human health.

New agricultural products developed to be pest and disease resistant are reducing the use of pesticides and promoting agricultural productivity and economy. GM crops are now used in the production of many common foods, including most corn, soybeans, and canola for vegetable oil, as well as cotton. A GM tomato was first produced in 1994. In 2000, rice with enhanced nutritional value was developed using genetic modification methods. The papaya industry in Hawaii was saved from a deadly papaya-destroying virus by an engineered resistant papaya plant. GM foods are used in animal feeds and in vitamin enrichment of rice and potentially for cassava production. GM organisms are used in cheese production and GM animals such as goats for increased milk productive capacity have been approved by the FDA. Potential benefits include disease- and drought-resistant plants that require less water, fertilizer, and pesticides, increased productivity and supply of food with reduced cost and longer shelf-life, faster growing plants and animals, foods with medicinal properties that could include vaccines and essential vitamins (e.g., vitamin A) and minerals (e.g., iron), and more nutritious and tastier food (e.g., cassava plants in Africa). Controversy over GM

foods is especially strong in Europe and is a factor in global agriculture as it affects agriculture in sub-Saharan Africa, where farmers may fear exclusion from European markets for their products.

Labeling of GM foods is a controversy in the USA as the industry fears boycotts of GM foods due to public concerns over their safety, although scientific consensus on GM safety is widespread. Anti-GM groups are active in opposing expansion of GM crops and agricultural production methods on the grounds of suspicion of possible "contamination" of food with potentially dangerous consequences and owing to concerns that large manufacturers will take over agriculture. However, success in GM crops to increase production with less dependence on herbicides and water needs will be essential to meet growing demands for food from a growing global population.

Animal husbandry practices are part of public health concerns, along with food processing, transportation, and cooking. A large number of cases of dangerous *Escherichia coli* infections resulting from contaminated meat in the USA, Japan, and many other countries show the potential danger even in technologically advanced industrial countries. Use of animal materials rendered for animal feed is common practice instead of using more expensive

TABLE 8.15 Foodborne Disease Agents, Organisms, Chemicals, and Vectors

Agent	Organism/Chemical	Food Vector
Prions	Bovine spongiform encephalitis (BSE)	Food of animal origin for cattle transmitting Creutzfeldt–Jakob disease (CJD) to humans
Viruses	Poliomyelitis, hepatitis A and E	Vegetables, fruit irrigated with sewage; shellfish
Bacteria	*Campylobacter, Clostridia, E. coli* 0157, *Salmonella, Vibrio cholerae, Listeria*	Raw and processed foods contaminated with human, bird, or animal excreta
Molds	Mycotoxins, aflatoxins, ochratoxins, marine biotoxins and glycocides	Naturally occurring toxins; stored nuts and cereals
Protozoa	*Amoebae, Giardia, Cryptosporidia*	Vegetables, fruit, water, unpasteurized milk
Parasites	*Trichomonas*	Pigs
Helminths, parasites	*Ascaris, Fasciola, Taenia, Trichinella, Trichura, Opisthorchiasis*	Vegetables, undercooked meat, raw fish from areas with contaminated water; untreated sewage or irrigation water
Chemicals, agricultural, and veterinary	Pesticides, fungicides, fertilizers, hormones, antibiotics; food additives; persistent organic pollutants such as dioxins and PCBs	Contaminated food from abuse of agents; widespread antibiotic use may engender allergies and organism resistance
Heavy metals	Lead, mercury, cadmium, tin	Fish, seafood from water contaminated by industrial waste
Radionuclides	Radioactive materials	Foods contaminated from atmospheric fallout

Note: PCB = polychlorinated biphenyl.
Sources: Adapted from: Pesticide residues in food 2012. Joint FAO/WHO Meeting on Pesticide Residues. Geneva: Food and Agriculture Organization, WHO; 2013. Available at: http://www.fao.org/agriculture/crops/core-themes/theme/pests/jmpr/en/ [Accessed 29 November 2013].
World Health Organization. Health impact assessment: determinants of health – agricultural production issues and manufacturing [posted 2013]. Available at: http://www.who.int/hia/evidence/doh/en/index3.html [Accessed 24 April 2013].
World Health Organization. Chemical risks in food. Available at: http://www.who.int/foodsafety/chem/en/ [Accessed 24 April 2013].
Centers for Disease Control and Prevention. Estimates of foodborne illness in the United States. Atlanta, GA: CDC; 2011. Available at: http://www.cdc.gov/features/dsfoodborneestimates/ [Accessed 24 April 2013].
Centers for Disease Control and Prevention. Foodborne Diseases Active Surveillance Network (FoodNet). http://www.cdc.gov/foodnet/ [Accessed 24 April 2013].

alfalfa, which raises serious health risks because of the *Salmonella* and *Campylobacter* such materials contain. In August 1997, tainted beef found in products of one company resulted in tens of thousands of tons of hamburger (minced beef) meat being condemned in the USA. A ban was imposed on British beef by the European Union (EU) in the late 1990s as a result of the bovine spongiform encephalopathy (BSE or mad cow disease) linked to Creutzfeldt–Jakob disease cases in humans (see Chapters 4 and 5) due to animal feed using contaminated animal parts. In 2013, widespread concern in Europe focused on horse meat found in food products, raising concerns over the efficacy of food inspection practices in the EU. Modern methods of management of dairy cattle, which are raised in shelters instead of open fields with direct sunlight, probably contribute to low levels of vitamin D, even in whole milk. Recent experience of foodborne disease in the EU is shown in Box 8.13.

Hazards associated with foods require constant vigilance on the part of governments, the food industry, and consumers. Several agencies with overlapping authority have responsibility for food safety, creating challenges for a unified strategy for human health protection. Cooperation between government and food producers, processors, and manufacturers is one of mutual interest as well as a regulatory relationship. Private industry has a major role to play in improving food supply and nutritional quality for both commercial and legal liability reasons. The consumer is also responsible for safe food handling practices, in terms of hygiene, storage, and usage. Sources of health-related problems exist at the points of production, processing, transporting, storing, sale, preparation, and serving of food products. Outbreaks of foodborne diseases serve as a reminder of the need for careful inspection of fresh produce, meat and dairy products, and processed foods on both a national and a local basis.

The CDC classification of pathogens for foodborne illnesses includes three categories: viruses, bacteria, and parasites. Transmission of pathogens may be at the source, such as poultry contaminated with *Salmonella* during processing. Shellfish may be contaminated by sewage in the water. Ground crops may be contaminated with bacteria or spores. Food and water may cross-contaminate. Failures in cleaning, employee infection and poor hygiene (e.g., hand washing, utensil cleanliness), inadequate cooking and temperature control during storage or transportation, and methods of serving can all contribute to contamination.

Most foodborne diseases in the USA have been caused by pathogenic bacteria, but norovirus (Norwalk virus) of fecal origin, mostly due to contaminated seafood, unhygienic food handling origins, or person-to-person spread, has recently become the leading pathogenic organism. Hepatitis A has similar origins and outbreaks can occur with serious

---

BOX 8.13 Foodborne Disease Surveillance in the European Union

The European Food Safety Authority and the European Centre for Disease Prevention and Control have responsibility for surveillance of foodborne disease outbreaks. In 2011, 27 member states reported zoonoses and foodborne outbreaks.

Most of the 5648 reported foodborne outbreaks were caused by *Salmonella*, bacterial toxins, *Campylobacter*, and viruses, and the main food sources were eggs, mixed foods, and fish and fishery products.

- *Campylobacter* – the most common of the zoonoses with 220,209 confirmed cases.
- *Salmonella* – salmonellosis in 95,548 confirmed human cases; *Salmonella* is most often found in meat and poultry products but is declining, with progress in meeting the targets for reduction.
- *Listeria* – human listeriosis declined to 1476 cases, and was usually within the safety limit for ready-to-eat foods.
- *Escherichia coli* – 9485 confirmed human cases of verotoxigenic *E. coli* (VTEC), a major increase of 160 percent from 2010, occurring primarily in one country; VTEC was also found in food and animals.
- *Yersinia* – 7017 human yersiniosis cases; *Yersinia* enterocolitica laboratory confirmed in pig meat and pigs.
- *Mycobacterium bovis* – 132 human tuberculosis cases; bovine tuberculosis (in cattle) increased.
- *Brucella* – 330 cases of brucellosis in humans; prevalence of brucellosis decreased in cattle, sheep, and goats.
- *Echinococcus* – 781 human echinococcosis cases; number of alveolar and cystic echinococcosis cases varies.
- *Trichinella* – 268 human cases of trichinellosis.

Source: European Food Safety Authority (EFSA). Rise in human infections from Campylobacter and E. coli, whilst Salmonella cases continue to fall: EFSA and ECDC 2011 zoonoses report. Available at: http://www.efsa.europa.eu/en/press/news/130409.htm [Accessed 24 April 2013].

---

effects on adults, whereas young children usually have mild illness with this virus. The availability and gradual spread in use of hepatitis A vaccine will over time reduce this hazard, but its adoption worldwide has been slow.

Bacterial contamination is usually due to the toxins produced by ingested bacteria being spread by severe vomiting and diarrhea; for instance, *E. coli* 0157H7 can be deadly, especially in children. *Clostridium botulinum* can produce a fatal toxin which is readily destroyed by adequate heating of the food, so proper cooking and reheating are important preventive measures. *Staphylococcus aureus* is a common organism that produces a toxin which is heat stable but usually non-lethal, and best controlled by good hygiene and temperature control. Outbreaks of *S. aureus* usually begin with contamination of salads by meat, and in cream pastries. *Campylobacter* and *Salmonella* usually originate in contaminated poultry and unpasteurized milk.

Foodborne diseases may also be classified by the extent of the outbreak as:

- *small, local outbreak* – among a group of people who have had meals in common
- *regional or state-wide* – a contaminated common food source leading or illness in several areas but with the contaminated product sold through local food stores or from local suppliers
- *national* – at a national level with hundreds of people exposed to the contaminated product and many becoming sick
- *international* – due to cross-border transport of contaminated foods such as lettuce, tomatoes, salads, or meat products.

Contamination by pathogens occurs in many agricultural regions, causing incidents of foodborne disease ("food poisoning"). It is not uncommon in developing countries for food to be exposed to pathogens by the use of untreated or inadequately treated sewage water for irrigation, malfunctioning sewage systems, or deliberate use of "nightsoil" for fertilization. Such fruits and vegetables eaten raw can expose the consumer to a host of infectious diseases transmitted via the fecal–oral route. Destruction and contamination of food by ruminants, rodents, or other pests during growth, harvesting, storage, transportation, or processing of food can be major factors in the safety and adequacy of a food supply. Even in highly developed agriculture, contamination of food can occur at many stages in its processing. Imported and domestic food contamination caused 6647 reported outbreaks of foodborne diseases among 128,370 people in 1998–2002 in the USA.

In 2011, a major outbreak of enterohemorrhagic *Escherichia coli* (EHEC) occurred, with 39 deaths and over 3517 sick, 839 of them with life-threatening kidney disease. The origin was eventually traced to German sprouts grown on one farm but transported widely in Europe. In April 2013, an outbreak of enteric *Salmonella* serotype Saintpaul infection was linked to imported cucumbers from Culiacén, Mexico. CDC worked with local health officials to identify cases and the causative organism. The national subtyping network of public health and food regulatory agency laboratories coordinated by CDC provided DNA identification to verify cases of illness that may be part of this outbreak, with a total of 84 people infected, 28 percent hospitalized, but no reported deaths.

Another major concern in food safety is the hazard posed by ingestion of chemicals used in food production. Modern agriculture has come to rely on extensive use of chemical fertilizers and pesticides. These accumulate in the soil over time and form a residue on the produce grown. Pesticide residues on orchard and field crops and in processed foods include chlorinated hydrocarbons,

dieldrin, and dichlorodiphenyltrichloroethane (DDT). Pesticides can cause serious problems of toxic exposure to farm workers, contamination of rain runoff to surface waters or groundwater, and contamination of inadequately washed farm produce. Safety requires consumer self-protection by selection of foods and careful washing before use. Education in seeking effective alternatives and reducing the use of pesticides is important for primary prevention strategies (see Chapter 9).

Food additives are substances are used to enhance food taste, color, texture, nutritional value (vitamin and mineral enrichment), appearance, and resistance to deterioration. Some maintain product consistency or palatability and some retard spoilage. Other additives provide leavening and prevent acidity or alkalinity. They may be used at various stages of the production process, such as treatment, packaging, transportation, or storage of food. Food additives include simple additives such as salt, baking soda, vanilla, or yeast, as well as complex ones.

Many important natural additives are used in processed foods as an important method of ensuring nutritional adequacy for the population at large. Additives such as vitamins and minerals have been successfully used without safety concerns for the past century, and have had important public health benefits. Food fortification restores natural elements that may be removed from the food during processing; for example, the coating of wheat contains healthful vitamins but is removed during the preparation of white flour. In some cases the mineral/vitamin complex is added to provide basic levels in commonly used foods that would be difficult to acquire through a normal diet. This is true of iron, iodine, folic acid, and other B vitamins including B_{12}, especially for the poor whose diet is less than optimal from a health point of view. Adding vitamin B complex, folic acid, iron, and vitamin B_{12} to flour reduces substantially birth defects, anemia, and other deficiency conditions that especially affect fetuses in utero, infants, and pregnant women, but other people as well. Regulatory mechanisms are essential to ensure that the additives or micronutrients are within permitted ranges and as indicated on mandatory labeling requirements.

All additives are subject to FDA regulation, and some have been banned because of their harmful effects, including cyclamates (sweetener), cobalt salts (beer), tar derivative used for food coloring, some pesticides, fungicides, and herbicides, and polyvinyl chloride in plastic containers. Some additives are of concern owing to their potential toxic or carcinogenic effects, and more so when these additives are for cosmetic value only. In 1958, the Food Additives Amendment was added to the US Food, Drug, and Cosmetics Act. This amendment requires food manufacturers to comply with FDA rulings on food additives. In the same year, the Delaney Clause banned any additive that has been shown to be carcinogenic in any amount from use as a food additive. As a result, large-scale testing of additives was carried out. Additives already in use were categorized as "generally regarded as safe" (GRAS), but continued testing with improved methods found some to be potentially carcinogenic and they were removed from use by the FDA ruling.

Growth enhancers, including antibiotics and hormones to promote rapid and vigorous animal growth and milk production, can adversely affect milk and meat products. Prophylactic use of antibiotics may contribute to increasing the numbers of resistant organisms, which could potentially endanger public health.

Regulatory agencies in each country are responsible for ensuring the safety of food products. The *Codex Alimentarius* Commission (Codex) is the international food standards setting body recognized by national agencies. Codex also provides a reference point for international agreements such as the World Trade Agreements on Sanitary and Phytosanitary Measures and Technical Barriers to Trade as food standards in international trade, with the objectives of protecting the health of consumers and ensuring fair practices in the global food trade.

Methods of preserving foods, some as old as civilization itself, were developed to protect the quality of foods and ensure supply beyond the immediate period of harvesting of crops or foods of animal origin. They are meant to prevent microbiological and chemical deterioration, without harmful chemical, toxic, or carcinogenic effects. Some of the major categories of food preservation technology are shown in Table 8.16. Other methods are being investigated but show significant drawbacks and are not employed in the food industry.

TABLE 8.16 Major Food Processing and Preservation Technologies

Method	Description
Pasteurization	Heat to destroy microorganisms
Canning	Process and pack in airtight metal can
Drying	Radiant (solar), spray, air, freeze- or vacuum-drying/packing
Fermentation	Yogurt, cheese, soy sauce, wine and beer, vinegar, sauerkraut
Chemical treatment	Salting, pickling, sugaring, smoking, fumigating, or by food additives
Thermal treatment	Cooking (many forms), blanching, pasteurizing, aseptic filling with high-temperature short-time, retorting
Cold treatment	Refrigerating, freezing
Radiation	Microwave, ionizing radiation

While government and industry have major roles in ensuring safe food, the final responsibility rests with the consumer. Public education, knowledge, attitudes, and practices determine the safety of food served to the family. The consumer in a market economy can influence and even determine which goods to purchase, creating demands felt in the marketplace that guide producers and processors to change the content, packaging, labeling, advertising, and prices. Consumers can affect the market by refusing to purchase goods known to be problematic or lacking nutritional value, an effective part of quality control. Consumer advocacy can also play a role in food quality, as manufacturers and governments are sensitive to public expression of concern over food quality through the media. The Food and Drug Act, which set up regulatory mechanisms in the USA, was enacted in 1906 in response to newspaper articles and novels (such as *The Jungle*, by Upton Sinclair) exposing the quality of food in the USA. Consumer boycotts have also made producers very sensitive to public concerns.

Concern for the public health aspects of food have increased with serious incidents of food poisoning from new variants of *E. coli* and *Salmonella* resistant to available antibiotics. Large sectors of the population are vulnerable to infection, especially immunocompromised people such as HIV carriers and splenectomized patients. The US FDA reported that in 2010, total recalls totaled just over 34 million pounds (15.4 million kg), mostly of beef products. In 2011, over 36 million pounds (16.3 million kg) of fresh and frozen ground turkey *Salmonella*-contaminated products were recalled, out of a total of 39 million pounds of total food recalls. In 2012, the total recall was just under 3.5 million pounds (1.6 million kg).

Despite safe water supplies, good hygiene, and food technology, food poisoning has increased in many industrialized countries. Current estimates suggest that between 5 and 10 percent of the population of industrial countries are affected annually. CDC estimates that each year roughly one in six Americans (or 48 million people) get sick, 128,000 are hospitalized, and 3000 die of foodborne diseases (CDC, 2012).

Foodborne and waterborne diseases are perhaps the most important public health problems of developing countries, with 1.5 billion cases and 3 million deaths annually from diarrheal disease. Consumer resistance to use of irradiation of food remains high, despite its being widely regarded as probably the safest and most reliable method of protecting food supplies from contamination. This resistance may decline as more instances of food contamination are reported in the public media.

Irradiation to improve food safety, genetic engineering to produce food crops of higher quality, and freedom from the need for pesticides and chemical fertilizers are producing a new revolution in agriculture and food safety. They are also controversial and under critical review with regard to safeguards and labeling requirements to protect consumer freedom to choose. Nevertheless, these are important new technologies which will be vital to meeting the needs of developing countries, as well as those of the industrial world in promoting good nutrition and food safety.

NUTRITION AND THE NEW PUBLIC HEALTH

Population and individual health are deeply affected by nutrition status. The individual chronically ill patient may experience a serious decline in his or her condition as a result of inadequate daily nutrition from loss of appetite, funds, or energy to purchase and prepare adequate foods. The cost of hospitalization alone for such people may outweigh the support of pension and services that could assist that person in maintaining adequate nutrition and independent living.

Overt or subclinical deficiency conditions can cause widespread health damage if national health policies fail to take into account current knowledge and practices for their prevention. Failure to implement such policies engenders undernutrition, with important health consequences, such as anemia, pregnancy outcome including NTDs, osteoporosis from vitamin D deficiency, and serious developmental delay in children affected by iodine deficiency.

The success of many developed countries in controlling micronutrient deficiency conditions has been achieved partly by social and economic development and improved pediatric care, but also by enlightened food policy with food fortification. The adoption of population-oriented food policies is crucial for developing countries to cope with basic undernutrition for large vulnerable parts of the population.

The complexity of food and nutrition policy is demonstrated in Figure 8.8. The interaction of food supplies and prices with governmental policy and other related factors impinges on individual family food purchases and individual consumption, such as in promoting fruit and vegetable availability and consumption especially in low income urban neighborhoods. Education and culture play important roles, as do the economics of food and global factors, such as the use of grains for fuel production that leads to severe shortages of corn and grain for animal production and for even more basic foods.

Overnutrition and its associated conditions, such as obesity, diabetes, CVD, and cancer, are the cause of widespread pathology and high rates of use of costly health services, whereas education and other programs of intervention can have important preventive effects to reduce these effects. These conditions cannot be addressed solely as individual patient problems. They require a parallel population approach which involves a broad set of food and nutrition policies. The New Public Health takes a holistic approach to the prevention and management of disease, as well as the promotion of optimal health. Nutrition is central to that task, requiring well-planned interventions at the national, community, and individual levels.

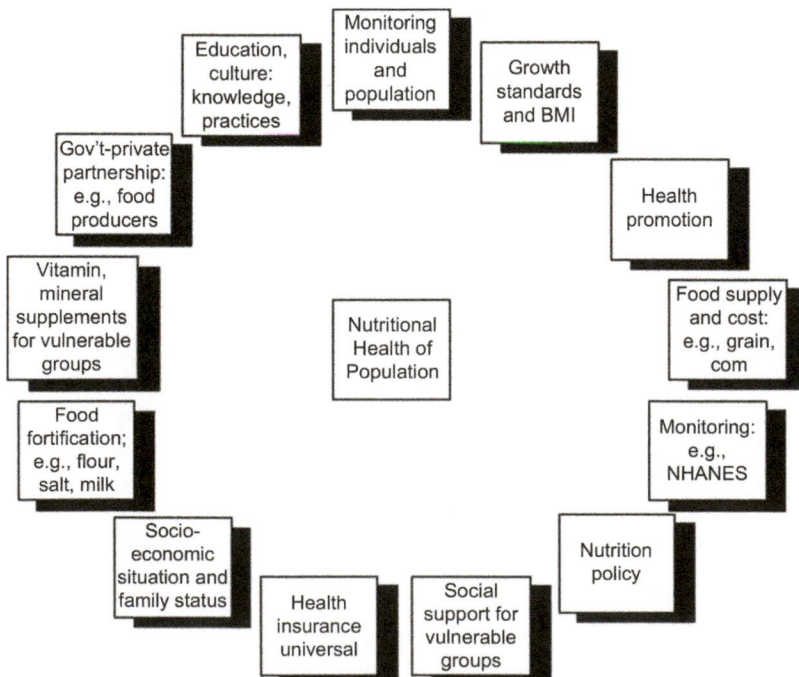

FIGURE 8.8 Factors in the nutritional health of a population. Note: BMI=body mass index; NHANES=National Health and Nutrition Examination Survey.

SUMMARY

While much of the glamour and emphasis in public health achievements over the past century have involved sanitation and control of communicable diseases, the contribution of improved nutrition and food safety to better health has been enormous. Mortality from all causes, including infectious diseases, began to decline in the seventeenth century, well before organized infectious disease control emerged, and that was, in part, a result of the agricultural revolution and improved food production. The advent of pasteurization, refrigeration, canning, and other food processes increased food safety and is a great achievement of public health, although foodborne disease remains a continuing challenge into the twenty-first century.

Nutrition has seen enormous advances over the past century, with improved standards of food and nutritional security a highlight of public health successes in many parts of the world. Yet, nutrition remains one of the major challenges of public health issues, in both developed and developing countries. In high- and middle-income countries the problems are largely of overnutrition or inappropriate balance and excessive caloric intake, yet deprivation among the poor remains a stain on societal responsibility and equity. In developing countries, mass nutritional deprivation is perhaps the most important public health problem, resulting in widescale deficiencies of calories, protein, and essential vitamins and minerals. Mid-level developing countries are experiencing "nutrition transition".

Cardiovascular mortality increased rapidly during the 1940s to the 1960s in the industrialized countries owing to excessive animal fat intake and other lifestyle issues such as smoking, as living standards increased for the major part of the population. This cause of mortality has since declined rapidly, with more awareness of healthy diet and reduced cigarette smoking as well as improved medical care (see Chapter 5). In developing countries, however, rapid lifestyle and nutritional change is affecting large sections of the population, while undernutrition remains widespread for the majority.

Since the 1960s, people in most industrialized countries have paid a great deal of attention to their personal nutrition, prompted by concern in the media and by the health professions. The dietary approach to preventive health care has contributed to the dramatic drop in death rates from coronary heart disease, stroke, and cancer of the stomach in most western countries. However problems of overnutrition and undernutrition have consequences for other non-communicable and infectious diseases. At the same time, illness, both acute and chronic, can have serious effects on nutrition status.

The success of folic acid fortification of flour in reducing NTDs has renewed interest in the issue of food fortification.

The WHO has come forward with important new initiatives in this field and in setting new growth standards for application internationally. The rapid growth of the interrelated problem of obesity and diabetes is one of the challenges to public health now and in the coming decades.

Monitoring nutrition status and developing national policies to assure adequate and high-quality food are major governmental functions. The economic, agricultural, and marketing systems all have roles to play in ensuring population nutritional health. Food fortification and promotion of healthful dietary habits are vital to prevent disorders due to deficient and excess nutrition. The supply and quality of food for growing populations with rising expectations will depend on new science and technology, such as genetic engineering and food irradiation, despite current controversy and manipulation of legitimate public concerns. The clinician works with the individual patient, while public health focuses on the individual and the community. This duality is the New Public Health in which nutrition plays a central role.

NOTE

For a complete bibliography and guidance for student reviews and expected competencies please see companion web site at http://booksite.elsevier.com/9780124157668

BIBLIOGRAPHY

Recommended Reading

Centers for Disease Control and Prevention, 1999. Achievements in public health, 1900–1999: safer and healthier foods. MMWR. Morb. Mortal. Wkly. Rep. 48, 905–913. Available at: http://www.cdc.gov/mmwr/preview/mmwrhtml/mm4840a1.htm (accessed 07.04.13).

Centers for Disease Control and Prevention, National Health and Nutrition Examination and Survey, 1999–2012. Available at: http://www.cdc.gov/nchs/data/nhanes/survey_content_99_12.pdf (accessed 21.04.13).

Dwyer, J., Piccian, M.F., Raiten, D.J., 2003. Members of the Steering Committee, National Health and Nutrition Examination Survey. Estimation of usual intakes – what we eat in America – NHANES. J. Nutr. 133. 609–23S. Available at: http://jn.nutrition.org/content/133/2/609S.full.pdf (accessed 09.04.13).

Flegal, K.M., Graubard, B.I., Williamson, D.F., Gail, M.H., 2005. Excess deaths associated with underweight, overweight, and obesity. JAMA 293, 1861–1867. Available at: http://jama.jamanetwork.com/article.aspx?articleid=200731 (accessed 10.04.13).

Food and Agriculture Organization, 2012. The state of food insecurity in the world 2012. FAO, Rome. Available at: http://www.fao.org/docrep/016/i3027e/i3027e00.htm (accessed 14.04.13).

Food and Agriculture Organization/World Health Organization, Second International Conference on Nutrition (ICN2), 2013. Rome: FAO. Available at: http://www.fao.org/food/nutritional-policies-strategies/icn2/en/ (accessed 07.04.13).

King, D.E., Mainous III, A.G., Geesey, M., 2007. Turning back the clock: adopting a healthy lifestyle in middle age. Am. J. Med. 120, 598–603. Available at: http://download.journals.elsevierhealth.com/pdfs/journals/0002-9343/PIIS0002934306011855.pdf (accessed 10.04.13).

Leathers, H.D., Foster, P., 2004. The world food problem: tackling the causes of under-nutrition in the third world, third ed. Lynne Rienner, Boulder, CO.

Mokdad, A.H., Ford, E.S., Bowman, B.A., Dietz, W.H., Vinicor, F., Bales, V.S., Marks, J.S., 2003. Prevalence of obesity, diabetes, and obesity-related health risk factors, 2001. JAMA 289, 76–79. Available at: http://www.ncbi.nlm.nih.gov/pubmed/12503980.

Nathoo, T., Holmes, C.P., Ostry, A., 2005. An analysis of the development of Canadian food fortification policies: the case of vitamin B. Health Promot. Int. 20, 375–382. Available at: http://www.ncbi.nlm.nih.gov/pubmed/15964882.

National Academy of Sciences, 1989. National Research Council. Recommended dietary allowances, tenth ed. National Academies Press, Washington, DC. Available at: http://www.nap.edu/catalog.php?record_id=1349 (accessed 27.11.13).

National Institutes of Health, 1998. Clinical guidelines on the identification, evaluation, and treatment of overweight and obesity in adults: the evidence report. National Heart, Lung, and Blood Institute, Bethesda, MD. Available at: www.nhlbi.nih.gov/guidelines/obesity/ob_gdlns.htm (accessed 16.07.13).

Ogden, C.L., Carroll, M.D., Curtin, L.R., McDowell, M.A., Tabak, C.J., Flegal, K.M., 2006. Prevalence of overweight and obesity in the United States, 1999–2004. JAMA 295, 1549–1555. Available at: http://www.ncbi.nlm.nih.gov/pubmed/16595758 (accessed 17.07.13).

Otten, J.J., Hellwig, J.P., Meyers, L.D. (Eds.), 2006. Dietary reference intakes: the essential guide to nutrient requirements. National Academies Press, Washington, DC. Available at: http://www.nap.edu/catalog.php?record_id=11537. (accessed 10.04.13).

Park, Y.K., Sempos, C.T., Barton, C.N., Vanderveen, J.E., Yetley, E.A., 2000. Effectiveness of food fortification in the United States: the case of pellagra. Am. J. Public Health 90, 727–738. Available at: http://ajph.aphapublications.org/doi/pdf/10.2105/AJPH.90.5.727 (accessed 16.07.13).

Roger, V.L., Go, A.S., Lloyd-Jones, D.M., Adams, R.J., Beny, J.D., Brown, T.M., et al., 2011. Heart disease and stroke statistics – 2011 update: a report from the American Heart Association Statistics Committee and Stroke Statistics Subcommittee. Circulation 123, e18–209. Available at: http://circ.ahajournals.org/content/123/4/e18.full.pdf (accessed 16.07.13).

Scrimshaw, N., 1995. The new paradigm of public health nutrition. Am. J. Public Health 85, 622–624. Available at: http://ajph.aphapublications.org/doi/pdf/10.2105/AJPH.85.5.622 (accessed 16.07.13).

Sidel, V., 1997. The public health impact of hunger. Am. J. Public Health 87, 1921–1922. Available at: http://ajph.aphapublications.org/doi/pdf/10.2105/AJPH.87.12.1921 (accessed 16.07.13).

United Nations Children's Fund, The state of the world's children, 2012. Table 2: Nutrition. Available at: http://www.unicef.org/sowc2012/pdfs/SOWC- (accessed 13.04.13).

Willett, W., 2012. Nutritional epidemiology, third ed. Oxford University Press, New York.

World Health Organization. Nutrition. Global targets, 2025: to improve maternal and infant and young child nutrition. Available at: http://www.who.int/nutrition/topics/nutrition_globaltargets2025/en/index.html (accessed 28.04.13).

World Health Organization, Vitamin and mineral nutrition information system (VMNIS). Available at: http://www.who.int/vmnis/en/index.html (accessed 28.04.13).

World Health Organization, 2012. World health statistics 2012, Part I: Health-related Millennium Development Goals. WHO, Geneva.

Available at: http://www.who.int/gho/publications/world_health_statistics/EN_WHS2012_Part1.pdf (accessed 04.04.13).

World Health Organization/United Nations, 2009. WHO child growth standards and the identification of severe acute malnutrition in infants and children. A joint statement by the World Health Organization and the United Nations. WHO, Geneva. Available at: http://www.who.int/nutrition/publications/severemalnutrition/9789241598163_eng.pdf (accessed 21.04.13).

Foodborne Illness, Nutrition, and Infectious Disease

Centers for Disease Control and Prevention. Feature: 2011 Estimates of foodborne illness in the United States [updated 15 April 2011]. Available at: http://www.cdc.gov/features/dsfoodborneestimates/ (accessed 24.04.13).

Centers for Disease Control and Prevention. Food safety and raw milk [updated March 2012]. Available at: http://www.cdc.gov/foodsafety/rawmilk/raw-milk-index.html (accessed 07.04.13).

Centers for Disease Control and Prevention, 2013. Incidence and trends of infection with pathogens transmitted commonly through food – Foodborne Diseases Active Surveillance Network, 10 US sites, 1996–2012. MMWR. Morb. Mortal. Wkly. Rep. 62, 283–287. Available at: http://www.cdc.gov/mmwr/preview/mmwrhtml/mm6215a2.htm?s_cid=mm6215a2_w (accessed 22.04.13).

Centers for Disease Control and Prevention, 2013. Surveillance for foodborne disease outbreaks – United States, 2009–2010. MMWR. Morb. Mortal. Wkly. Rep. 62, 41–47. Available at: http://www.cdc.gov/mmwr/preview/mmwrhtml/mm6203a1.htm?s_cid=mm6203a1_w (accessed 07.04.13).

Langer, A.J., Ayers, T., Grass, J., Lynch, M., Angulo, F.J., Mahon, B.E., 2012. Nonpasteurized dairy products, disease outbreaks, and state laws – United States, 1993–2006. Emerg. Infect. Dis. 18. Available at: http://wwwnc.cdc.gov/eid/article/18/3/11-1370_article.htm (accessed 07.04.13).

Painter, J.A., Hoekstra, R.M., Ayers, T., Tauxe, R.V., Braden, C.R., Angulo, F.J., et al., 2013. Attribution of foodborne illnesses, hospitalizations, and deaths to food commodities by using outbreak data, United States, 1998–2008. Emerg. Infect. Dis. 19, 407–415. Available at: http://wwwnc.cdc.gov/eid/article/19/3/pdfs/11-1866.pdf (accessed 04.04.13).

Monitoring Growth and Body Mass

Centers for Disease Control and Prevention, September 2010. WHO growth standards are recommended for use in the US for infants and children 0 to 2 years of age. Available at: http://www.cdc.gov/growthcharts/who_charts.htm. or http://www.cdc.gov/growthcharts/who_charts.htm. (accessed 21.04.13).

Grummer-Strawn, L.M., Reinold, C., Krebs, N., 2010. Centers for Disease Control and Prevention. Use of World Health Organization and CDC growth charts for children aged 0–59 months in the United States. MMWR. Recomm. Rep. 59 (RR9), 1–15. Available at: http://www.cdc.gov/mmwr/preview/mmwrhtml/rr5909a1.htm (accessed 04.04.13).

de Onis, M., Onyango, A., Borghi, E., Siyam, A., Blössner, M., Lutter, C., 2012. for the WHO Multicentre Growth Reference Study Group. Worldwide implementation of the WHO child growth standards. Public Health Nutr. 15, 1603–1610. Available at: http://www.who.int/childgrowth/publications/global_implementation.pdf (accessed 19.04.13).

Wang, Y., Moreno, L.A., Caballero, B., Cole, T.J., 2006. Limitations of the current World Health Organization growth references for children and adolescents. Food Nutr. Bull. 27, S175–S188. Available at: http://www.foodandnutritionbulletin.org/fnbhome.php (accessed 28.11.13).

World Health Organization, Child growth standards: the WHO child growth standards. Growth reference data for 5–19 years, 2007. Geneva: WHO. Available at: http://www.who.int/growthref/en (accessed 26.03.13).

Nutrient Requirements, Recommended Dietary Intakes, and Dietary Guidelines

American Academy of Pediatrics. Healthy living. Vitamin D on the double [posted 28 May 2013]. Available at: http://www.healthychildren.org/English/healthy-living/nutrition/Pages/Vitamin-D-On-the-Double.aspx (accessed 28.11.13).

Bath, S.C., Steer, C.D., Golding, J., Emmett, P., Rayman, M.P., 2013. Effect of inadequate iodine status in UK pregnant women on cognitive outcomes in their children: results from the Avon Longitudinal Study of Parents and Children (ALSPAC). Lancet 382 (9889), 331–337. http://www.thelancet.com/journals/lancet/article/PIIS0140-6736(13)60436-5/fulltext (accessed 29.11.13).

Doet, E.S., deWit, L.S., Dhonukshe-Rutten, R.A., Cavelaars, A.E., Raats, M.M., Timotijevic, L., et al., 2008. Current micronutrient recommendations in Europe: towards understanding their differences and similarities. Eur. J. Nutr. 47, 17–40. Available at: http://www.ncbi.nlm.nih.gov/pubmed/18427858 (accessed 26.03.13).

Ervin, R.B., Ogden, C.L., February 2013. Trends in intake of energy and macronutrients in children and adolescents from 1999–2000 through 2009–2010. NCHS data brief no. 113. Available at: http://www.cdc.gov/nchs/data/databriefs/db113.pdf (accessed 15.04.13).

Institute of Medicine, National Academy of Sciences. Food and Nutrition Board [updated 29 November 2011]. Available at: http://www.iom.edu/About-IOM/Leadership-Staff/Boards/Food-and-Nutrition-Board.aspx (accessed 02.05.13).

National Research Council, 2011. Dietary reference intakes for calcium and vitamin D. National Academies Press, Washington, DC. Available at: http://www.nap.edu/catalog.php?record_id=13050 (accessed 02.05.13).

National Research Council, 2005. Food Nutrition Board. Dietary reference intakes for energy, carbohydrate, fiber, fat, fatty acids, cholesterol, protein, and amino acids (macronutrients). National Academies Press, Washington, DC. Available at: http://www.nap.edu/openbook.php?isbn=0309085373 (accessed 09.04.13).

The Nutrition Source, Department of Nutrition, Harvard School of Public Health. Healthy eating. Available at: http://www.thenutritionsource.org and Harvard Health Publications, health.harvard.edu (accessed 24.04.13).

Ross, C.A., Manson, J.E., Abrams, S.A., Aloia, J.F., Brannon, P.M., Clinton, S.K., et al., 2011. The 2011 report on dietary reference intakes for calcium and vitamin D from the Institute of Medicine: what clinicians need to know. J. Clin. Endocrinol. Metab. 96, 53–58http://www.ncbi.nlm.nih.gov/pmc/articles/pmc3046611/ (accessed 29.04.13).

US Department of Agriculture. Dietary reference intakes [updated 27 March 2013]. Washington, DC: USDA. Available at: http://fnic.nal.usda.gov/nal_display/index.php?info_center=4&tax_level=2&tax_subject=256&level3_id=0&level4_id=0&level5_id=0&topic_id=1342&placement_default=0 (accessed 09.04.10).

US Department of Agriculture, 2010. US Department of Health and Human Services. Dietary guidelines for Americans, 2010, seventh ed. US Government Printing Office, Washington, DC. Available at: http://www.health.gov/dietaryguidelines/dga2010/DietaryGuidelines2010.pdf (accessed 07.04.13).

World Health Organization, Vitamin and mineral nutrition information systems (VMNIS). Geneva: WHO. Available at: http://www.who.int/vmnis/en/ (accessed 26.03.13).

World Health Organization/Food and Agriculture Organization, 2004. Vitamin and mineral requirements in human nutrition, second ed. WHO/FAO, Geneva. Available at: http://whqlibdoc.who.int/publications/2004/9241546123.pdf (accessed 26.03.13).

Micronutrient Deficiency Conditions

Abrams, S.A., 2011. Pediatrics dietary guidelines for calcium and vitamin D: a new era. Dietary guidelines for calcium and vitamin D: a new era. Pediatrics 127, 566. Available at: http://pediatrics.aappublications.org/content/127/3/566.full (accessed 24.08.13).

Allen, L., de Benoist, B., Dary, O., Hurrell, R., 2006. Guidelines on food fortification with micronutrients. WHO/FAO, Geneva. Available at: http://whqlibdoc.who.int/publications/2006/9241594012_eng.pdf (accessed 09.04.13).

Andersson, M., de Benoist, B., Darnton-Hill, I., Delange, F., 2007. Iodine deficiency in Europe: a continuing public health problem. WHO, Geneva. Available at: http://whqlibdoc.who.int/publications/2007/9789241593960_eng.pdf (accessed 07.04.13).

Andersson, M., Karumbunathan, V., Zimmermann, M.B., 2012. Global iodine status in 2011 and trends over the past decade. J. Nutr. 142, 744–750. Available at: http://jn.nutrition.org/content/early/2012/02/28/jn.111.149393.full.pdf (accessed 04.04.13).

Backstrand, J.R., 2002. The history and future of food fortification in the United States: a public health perspective. Nutr. Rev. 60, 15–26. Available at: http://onlinelibrary.wiley.com/doi/10.1301/002966402760240390/abstract (accessed 09.04.13).

Bath, S., Walter, A., Taylor, A., Rayman, M.P., 2008. Iodine status of UK women of childbearing age. J. Hum. Nutr. Diet. 21, 379–380. Available at: http://onlinelibrary.wiley.com/doi/10.1111/j.1365-277X.2008.00881_9.x/abstract (accessed 27.11.13).

Centers for Disease Control and Prevention, 1999. Achievements in Public Health, 1900–1999: Safer and healthier foods. MMWR. Morb. Mortal. Wkly. Rep. 48, 905–913. Available at: http://www.cdc.gov/mmwr/preview/mmwrhtml/mm4840a1.htm (accessed 23.08.13).

Centers for Disease Control and Prevention, 2002. Iron deficiency – United States, 1999–2000. MMWR. Morb. Mortal. Wkly. Rep. 51, 897–899. Available at: http://www.cdc.gov/mmwr/preview/mmwrhtml/mm5140a1.htm (accessed 09.04.13).

DeMaeyer, E.M., Dallman, P., Gurney, J.M., Hallberg, L., Sood, S.K., Srikantia, S.G., 1989. Preventing and controlling iron deficiency anemia through primary health care a guide for health administrators and program managers. WHO, Geneva. Available at: http://www.who.int/nutrition/publications/micronutrients/anaemia_iron_deficiency/9241542497.pdf (accessed 09.04.13).

Dunn, J.T., 1992. Iodine deficiency – the next target for elimination? [editorial]. N. Engl. J. Med. 326, 267–268. Available at: http://www.nejm.org/doi/pdf/10.1056/NEJM199201233260411 (accessed 09.04.13).

Hanley, D.A., Davison, K.S., 2005. Vitamin D insufficiency in North America. J. Nutr. 135, 332–337. Available at: http://jn.nutrition.org/content/135/2/332.long (accessed 10.04.13).

Harrison, G.G., 2010. Public health interventions to combat micronutrient deficiencies. Public Health Rev. 32, 256–266. Available at: http://www.publichealthreviews.eu/show/f/29 (accessed 04.04.13).

Hetzel, B.S., 2005. Towards the global elimination of brain damage due to iodine deficiency – the role of the International Council for Control of Iodine Deficiency Disorders. Int. J. Epidemiol 34, 762–764. Available at: http://ije.oxfordjournals.org/content/34/4/762.full.pdf (accessed 02.08.13).

Lazarus, J.H., Smyth, P.P., 2008. Iodine deficiency in the UK and Ireland. Lancet 372, 888. Available at: http://download.thelancet.com/pdfs/journals/lancet/PIIS0140673608613902.pdf (accessed 29.11.13).

National Institutes of Health, Office of Dietary Supplements. Dietary supplement fact sheet: Vitamin D. Available at: http://ods.od.nih.gov/factsheets/VitaminD-HealthProfessional/ (accessed 24.08.13).

Rayman, M.P., Sleeth, M., Walter, A., Taylor, A., 2008. Iodine deficiency in UK women of child-bearing age. Proc. Nutr. Soc. 67 E399. Available at: http://journals.cambridge.org/download.php?file=%2FPNS%2FPNS67_OCE8%2FS0029665108000736a.pdf&code=9dc39729be804045533e435a6523a1b7 (accessed 29.11.13).

Tulchinsky, T.H., 2010. Micronutrient deficiency conditions: global health issues. Public Health Rev. 32, 243–255. Available at: http://www.publichealthreviews.eu/show/f/28 (accessed 02.04.13).

World Health Organization, Micronutrient deficiencies: vitamin A deficiency. Geneva: WHO. Available at: http://www.who.int/nutrition/topics/vad/en/ (accessed 26.03.13).

World Health Organization, Nutrition: micronutrient deficiencies: iron deficiency anemia, 2013. Geneva: WHO. Available at: http://www.who.int/nutrition/topics/ida/en/index.html (accessed 26.03.13).

World Health Organization, 2000. Regional Office for Europe. Comparative analysis of progress on the elimination of iodine deficiency disorders. WHO, Copenhagen. Available at: http://www.euro.who.int/__data/assets/pdf_file/0009/119772/E68017.pdf (accessed 25.03.13).

Zimmermann, M.B., Andersson, M., 2011. Prevalence of iodine deficiency in Europe in 2010. Ann. Endocrinol. 72, 164–166. Available at: http://www.sciencedirect.com/science/article/pii/S0003426611000503 (accessed 04.04.13).

Vitamin D and Bone Health

Abrams, S.A., 2011. Dietary Guidelines for Calcium and Vitamin D: A New Era. Pediatrics 127, 566–568. Available at: http://pediatrics.aappublications.org/content/127/3/566.full.html (accessed 29.11.13).

Harvard School of Public Health, The Nutrition Source. Vitamin D and health. Available at: http://www.hsph.harvard.edu/nutritionsource/vitamin-d/ (accessed 16.07.13).

Hintzpeter, B., Mensink, G.B., Thierfelder, W., Müller, M.J., Scheidt-Nave, C., 2007. Vitamin D status and health correlates among German adults. Eur. J. Clin. Nutr. 62, 1079–1089. Available at: http://www.nature.com/ejcn/journal/v62/n9/abs/1602825a.html (accessed 25.03.13).

Holick, M., 2010. Vitamin D deficiency pandemic: a forgotten hormone important to health. Public Health Rev 32, 267–283. Available at: http://www.publichealthreviews.eu/upload/pdf_files/7/15_Vitamin_D.pdf (accessed 02.04.13).

Holick, M.F., 2004. Sunlight and vitamin D for bone health and prevention of autoimmune diseases, cancers, and cardiovascular disease. Am. J. Clin. Nutr. 80. 1678–88S. Available at: http://www.ncbi.nlm.nih.gov/pubmed/15585788 (accessed 25.03.13).

Rajakumar, K., 2003. Vitamin D, cod-liver oil, sunlight and rickets: a historical perspective. Pediatrics 112, e132–e135. Available at: http://www.ncbi.nlm.nih.gov/pubmed/12897318 (accessed 25.03.13).

Rovner, A.J., O'Brien, K.O., 2008. Hypovitaminosis D among healthy children in the United States: a review of the current evidence. Arch. Pediatr. Adolesc. Med. 162, 513–519. Available at: http://www.ncbi.nlm.nih.gov/pubmed/18524740 (accessed 03.04.13).

Sachan, A., Gupta, R., Das, V., Agarwal, A., Awasthi, P.K., Bhatia, V., 2005. High prevalence of vitamin D deficiency among pregnant women and their newborns in northern India. Am. J. Clin. Nutr. 81, 1060–1064.

Available at: http://www.ncbi.nlm.nih.gov/pubmed/15883429 (accessed 25.03.13).

Wagner, C.L., Greer, F.R., 2008. Prevention of rickets and vitamin D deficiency in infants, children, and adolescents. Pediatrics 122, 1142–1153. Available at: http://pediatrics.aappublications.org/content/122/5/1142.full.pdf (accessed 09.04.13).

Nutrition, Obesity, and Chronic Disease

American Heart Association, 2011. AHA Policy Statement: Forecasting the future of cardiovascular disease in the United States. Circulation 123, 933–944. Available at: http://circ.ahajournals.org/content/123/8/933.long (accessed 18.04.13).

American Institute for Cancer Research, Recommendations for cancer prevention. Available at: http://preventcancer.aicr.org/site/PageServer?pagename=research_science_expert_report (accessed 22.04.13).

Centers for Disease Control and Prevention, 2011. CDC grand rounds: childhood obesity in the United States. Morb. Mortal. Wkly. Rep. MMWR. 60, 42–46. Available at: http://www.cdc.gov/mmwr/preview/mmwrhtml/mm6002a2.htm (accessed 07.04.13).

Centers for Disease Control and Prevention, 2011. Obesity – halting the epidemic by making health easier, at a glance 2011. CDC, Atlanta, GA. Available at: http://www.cdc.gov/chronicdisease/resources/publications/aag/pdf/2011/Obesity_AAG_WEB_508.pdf (accessed 07.04.13).

Centers for Disease Control and Prevention, 2012. Second national report on biochemical indicators of diet and nutrition in the US Population 2012. National Center for Environmental Health, Atlanta, GA. Available at: http://www.cdc.gov/nutritionreport/pdf/Nutrition_Book_complete508_final.pdf#zoom=100 (accessed 07.04.13).

Centers for Disease Control and Prevention. Stroke facts: America's stroke burden [updated 2013]. Available at: http://www.cdc.gov/stroke/facts.htm (accessed 21.04.13).

Ervin, R.B., Ogden, C.L., February 2013. Trends in intake of energy and macronutrients in children and adolescents from 1999–2000 through 2009–2010. NCHS data brief no. 113. Available at: http://www.cdc.gov/nchs/data/databriefs/db113.pdf (accessed 21.04.13).

Estruch, R., Ros, E., Salas-Salvadó, J., Covas, M.I., Corella, D., Ardos, F., et al., 2013. Primary prevention of cardiovascular disease with a Mediterranean diet. N. Engl. J. Med. 368, 1279–1290. Available at: http://www.nejm.org/doi/full/10.1056/NEJMoa1200303 (accessed 04.04.13).

Flegal, K.M., Carroll, M.D., Kit, B.K., Ogden, C.L., 2012. Prevalence of obesity and trends in the distribution of body mass index among US adults, 1999–2010. JAMA 307, 491–497. Available at: http://www.ncbi.nlm.nih.gov/pubmed/22253363 (accessed 28.04.13).

Fryar, C.D., Carroll, M.D., Ogden, C.L., 2012. Prevalence of obesity among children and adolescents: United States, trends 1963–1965 through 2009–2010. CDC, Atlanta, GA. Available at: http://www.cdc.gov/nchs/data/hestat/obesity_child_09_10/obesity_child_09_10.pdf (accessed 07.04.13).

US Department of Health and Human Services. The Surgeon General's vision for a healthy and fit nation. Rockville, MD: US Department of Health and Human Services, Office of the Surgeon General, January 2010. Available at: http://www.ncbi.nlm.nih.gov/books/NBK44660/ (accessed 21.04.13).

US Surgeon General, Overweight and obesity: health consequences. Available at: http://www.surgeongeneral.gov/library/calls/obesity/fact_consequences.html (accessed 24.04.13).

Maternal/Child Nutrition and Survival

Bhutta, Z.A., Ahmed, T., Black, R.E., Cousens, S., Dewey, K., Giugliani, E., et al., 2008. Maternal and Child Undernutrition Study Group. What works? Interventions for maternal and child undernutrition and survival. Lancet 371, 417–440. Available at: http://www.thelancet.com/journals/lancet/article/PIIS0140-6736(07)61692-4/fulltext (accessed 04.04.13).

Black, R.E., Allen, L.H., Bhutta, Z.A., Caulfield, L.E., de Onis, M., Ezzati, M., et al., 2008. Maternal and Child Undernutrition Study Group. Maternal and child undernutrition: global and regional exposures and health consequences. Lancet 371, 243–260. Available at: http://www.thelancet.com/journals/lancet/article/PIIS0140-6736(07)61690-0/abstract (accessed 04.04.13).

Blencowe, H., Cousens, S., Modell, B., Lawn, J., 2010. Folic acid to reduce neonatal mortality from neural tube disorders. Int. J. Epidemiol. 39, i110–i121. Available at: http://ije.oxfordjournals.org/content/39/suppl_1/i110.full (accessed 04.04.13).

Getachew I. Community-based nutrition programme targets children at risk in Ethiopia [updated 25 August 2011]. New York: UNICEF. Available at: http://www.unicef.org/infobycountry/ethiopia_59657.html (accessed 04.04.13).

United Nations Children's Fund, 2009. Tracking progress on child and maternal nutrition. UNICEF, New York. Available at: http://www.unicef.org/publications/files/Tracking_Progress_on_Child_and_Maternal_Nutrition_EN_110309.pdf (accessed 13.04.13).

US Department of Agriculture, Women, infants and children (WIC). Available at: http://www.fns.usda.gov/. and http://wicworks.nal.usda.gov/ (accessed 20.04.13).

Food Fortification

Allen, L., de Benoist, B., Dary, O., Hurrell, R., 2006. Guidelines on food fortification with micronutrients. WHO, FAO, Geneva. Available at: http://www.who.int/nutrition/publications/guide_food_fortification_micronutrients.pdf (accessed 02.04.13).

Bentley, T.G.K., Willett, W.C., Wenstein, M.C., Kuntz, K.M., 2006. Population-level change in folate intake by age, gender, and race/ethnicity after folic acid fortification. Am. J. Public Health 96, 2040–2047. Available at: http://www.ncbi.nlm.nih.gov/pubmed/17018833 (accessed 27.03.13).

Flour Fortification Initiative, Nigeria adds folic acid to wheat fortification standard [July 2012]. Available at: http://www.ffinetwork.org/about/stay_informed/releases/NigeriaFolicAcid.html (accessed 04.04.13).

Canada, Health, Addition of vitamins and minerals to foods, 2005 Health Canada's proposed policy and implementation plans. Available at: http://www.hc-sc.gc.ca/fn-an/alt_formats/hpfb-dgpsa/pdf/nutrition/foritfication_final_doc-eng.pdf (accessed 09.04.13).

Oakley, G.P., Tulchinsky, T., 2010. Folic acid and vitamin B_{12} fortification of flour: a global basic food security requirement. Public Health Rev. 32, 284–295. Available at: http://www.publichealthreviews.eu/upload/pdf_files/7/16_Folic-acid.pdf (accessed 02.04.13).

Persad, V.L., Van den Hof, M.C., Dube, J.M., Zimmer, P., 2002. Incidence of open neural tube defects in Nova Scotia after folic acid fortification. CMAJ 167, 241–245. Available at: http://www.cmaj.ca/content/167/3/241.full.pdf (accessed 29.11.2013).

Piirainen, T., Laitinen, K., Isolauri, E., 2007. Impact of national fortification of fluid milks and margarines with vitamin D on dietary intake and serum 25-hydroxyvitamin D concentration in 4-year-old children. Eur. J. Clin. Nutr. 61, 123–128. Available at: http://www.ncbi.nlm.nih.gov/pubmed/16885927 (accessed 04.04.13).

World Health Organization, 2009. Recommendations on wheat and maize flour fortification meeting report: interim consensus statement. WHO, Geneva. Available at: http://www.who.int/nutrition/publications/micronutrients/wheat_maize_fort.pdf (accessed 04.04.13).

Food and Nutrition Policy, and Food Security

Brosco, J.P., 2012. Navigating the future through the past: the enduring historical legacy of federal children's health programs in the United States. Am. J. Public Health 102, 1848–1857. Available at: http://www.ncbi.nlm.nih.gov/pubmed/22897550 (accessed 04.04.13).

Food and Agriculture Organization of the United Nations, 2012. International Fund for Agricultural Development and World Food Programme. The state of food insecurity in the world 2012. Economic growth is necessary but not sufficient to accelerate reduction of hunger and malnutrition. FAO, Rome. Available at: http://www.fao.org/publications/sofi/en/ (accessed 07.04.13).

Food and Drug Administration, January 2013. A Food Labelling Guide: Guidance for Industry. Department of Health and Human Services. Available at: www.fda.gov/FoodLabelingGuide (accessed 29.11.2013).

Institute of Medicine, 2013. Challenges and opportunities for change in food marketing to children and youth: workshop summary. National Academies Press, Washington, DC. Available at: http://www.iom.edu/Reports/2013/Challenges-and-Opportunities-for-Change-in-Food-Marketing-to-Children-and-Youth.aspx (accessed 07.04.13).

Traoré, M., Thompson, B., Thoma, G., 2012. Sustainable nutrition security: restoring the bridge between agriculture and health. FAO, Rome. Available at: http://www.fao.org/fileadmin/user_upload/agn/pdf/ME785-E.pdf (accessed 07.04.13).

World Health Organization, Health impact assessment: determinants of health – agricultural production issues and manufacturing [posted 2013]. Available at: http://www.who.int/hia/evidence/doh/en/index3.html (accessed 24.04.13).

Nutrition Monitoring and Evaluation

Centers for Disease Control and Prevention. National Health and Nutrition Examination Survey [updated 22 March 2013]. Available at: http://www.cdc.gov/nchs/nhanes.htm (accessed 27.03.2013).

Lynch, M., Painter, J., Woodruff, R., Braden, C., 2006. Surveillance for foodborne-disease outbreaks, United States, 1998–2002. MMWR. Morb. Mortal. Wkly. Rep. 55, 1–34. Available at: http://www.cdc.gov/mmwr/preview/mmwrhtml/ss5510a1.htm (accessed 27.03.13).

Chapter 9

Environmental and Occupational Health

Learning Objectives

Upon completion of this chapter, the student should be able to:
1. Recognize and describe environmental health hazards;
2. Recognize and describe work-related disease and injury;
3. Formulate strategies for their prevention;
4. Recognize and describe the interaction between environmental and occupational exposure on health of the community.

ENVIRONMENTAL HEALTH

INTRODUCTION

Environmental challenges are receiving a growing level of public, private, and governmental attention. This varies from recognition of the vital importance of safe water, through waste disposal (nuclear, medical, industrial, household) and sewage disposal, to global climate change. The direct health effects of poor sanitation are horrendous in the cost to lives and health. Although progress is being made, high levels of danger to rural populations remain, especially in sub-Saharan Africa. In the industrialized countries, the sanitary revolution of the nineteenth century was a fundamental achievement of public health, resulting in a doubling of life expectancy.

The dramatic challenges of environmental degradation have come more into focus in the twenty-first century and the issues are by no means resolved. Public concern over environmental issues is higher than in previous decades. While the application of environmentally friendly policies is improving in many countries, it is still not an issue where reduced standards of living are an acceptable part of the solution, and technical solutions are slow in coming to substantive fruition. The international and national political levels address the environment in positive terms with many targets for action but progress remains slower and less comprehensive than the challenge requires. The United Nations Environment Programme (UNEP) defined themes and issues for the twenty-first century as shown in Box 9.1.

Of the eight Millennium Development Goals (MDGs) adopted by the United Nations (UN) in 2001 and accepted by virtually all member countries, one is: "to ensure environmental sustainability", with the following specific targets:

- *Target 9* – Integrate the principles of sustainable development into country policies and programs; reverse loss of environmental resources.
- *Target 10* – Reduce by half the proportion of people without sustainable access to safe drinking water.
- *Target 11* – Achieve significant improvement in lives of at least 100 million slum dwellers by 2020 (UNDP, 2008).

The MDGs call for international cooperation to prevent environmental degradation resulting in global warming. Progress in MDGs in terms of drinking water and sanitation is shown in Box 9.2.

A safe environment is fundamental to health; clean water is as important as shelter and food in a hierarchy of health and survival needs. Access to safe water has increased, but globally 19 percent of the burden of disease among children aged 0–1 years is from diarrheal disease largely due to contaminated water, while 10 percent is due to malaria and another 10 percent to malnutrition, intestinal infestation, and childhood disease clusters all related to poor environmental conditions.

Overshadowing other environmental issues are climate change and global warming as a result of both natural and human-caused phenomena. The result could be massive threats to public health through the spread of diseases related to climate, such as malaria and cholera with flooding and stagnant waters, desertification of highly vulnerable zones of the world, and disruption of safe drinking water and food supplies. Wide-scale natural disaster phenomena of rising sea levels with permanent flooding of coastal areas, hurricanes, and ecological changes of unpredictable severity are anticipated. A wide consensus of scientific opinion raises the level of concern over such disastrous effects such that governments and the public seem to be ready to act to reduce fossil fuel consumption and other root causes of greenhouse gases.

Safe water supplies and waste management are fundamental and still problematic aspects of public health and community hygiene. Incidences of contamination by biological, chemical, physical, or other disease-causing agents in the external environment and the workplace are major public health and political concerns of the twenty-first century. Since the 1960s, a high degree of consciousness has developed regarding these problems. Air, water, ground,

The New Public Health. http://dx.doi.org/10.1016/B978-0-12-415766-8.00009-4

BOX 9.1 The 21 Emerging Issues in Environmental Health

Cross-cutting Issues

1 Governance aligned with the challenges of global sustainability
2 Global environmental challenges and moving towards a green economy
3 Reconnecting science and policy
4 Catalyzing rapid and transformative changes in human behaviour towards the environment
5 Coping with creeping changes and imminent thresholds
6 Migration caused by new aspects of environmental change

Food, Biodiversity, and Land Issues

7 Ensuring food safety and food security for 9 billion people
8 Integrating biodiversity across the environmental and economic agendas
9 Boosting urban sustainability and resilience
10 New rush for land: responding to new national and international pressures

Freshwater and Marine Issues

11 Water–land interactions: shift in the management paradigm?
12 Degradation of inland waters in developing countries
13 Potential collapse of oceanic systems requires integrated ocean governance
14 Coastal ecosystems: addressing increasing pressures with adaptive governance

Climate Change Issues

15 Climate change mitigation and adaptation: managing the unintended consequences
16 Changing frequency of extreme events
17 Managing the impacts of glacier retreat

Energy, Technology, and Waste Issues

18 Accelerating implementation of environmentally friendly renewable energy systems
19 Minimizing risks of novel technologies and chemicals
20 Solving the impending scarcity of strategic minerals and avoiding electronic waste
21 Environmental consequences of decommissioning nuclear reactors

Source: United Nations Environment Programme. 21 Issues for the 21st century: results of the UNEP foresight process on emerging environmental issues. Nairobi: UNEP; 2012. Available at: http://www.unep.org/publications/ebooks/foresightreport/Portals/24175/pdfs/Foresight_Report-21_Issues_for_the_21st_Century.pdf [Accessed 15 August 2012].

BOX 9.2 Progress Towards the Millennium Development Goal on Drinking Water and Sanitation, 2012

Safe water and sanitation is vital to improving health and well-being. The 2001–2015 Millennium Development Goals (MDGs) set a target (7C) of reducing by half the proportion of people without sustainable access to safe drinking water and basic sanitation.

The MDG progress monitoring report by the WHO/UNICEF Joint Monitoring Programme for Water Supply and Sanitation 2012 states that the drinking water target for sustainable access to safe drinking water between 1990 and 2015 was met in 2010, 5 years ahead of schedule.

The 2012 report indicates that that more than 2 billion people have gained access to improved drinking water sources since 1990, but there are challenges of great disparities, with 780 million people remaining without access to improved drinking water sources and 2.5 billion lacking improved sanitation. Sub-Saharan Africa lags behind in this measure for rural dwellers and the urban poor in particular, with the burden of poor water supply mostly falling on girls and women.

Nevertheless, the global achievement indicates commitment of government leaders, public and private sector entities, communities and individuals. The international community (UN Rapporteur on Human Rights to Water and Sanitation) will maintain its surveillance on water and sanitation even following the 2015 MDG target date.

Source: World Health Organization/United Nations Children's Fund. Progress on drinking water and sanitation: 2012 update. Geneva: WHO/UNICEF; 2012. Available at: http://www.unicef.org/media/files/JMPreport2012.pdf [Accessed 16 August 2012].

Occupational health developed as a separate area of concern from environmental health, but in recent years there has been an increasing recognition of the interaction between workplace and community health hazards such as in asbestos and cancer. Occupational health is included as the second part of this chapter because of common advocacy, professionalism, technology, and regulatory approaches. The level of public response to environmental threats is illustrated by the groundswell of public opinion against environmental decay.

Issues have become more complex and go beyond the prevention of disease and traditional public health. While the resources needed to reduce the environmental neglect from inadequate sanitation and high levels of pollutants in the air, water, and soil are costly, the burden to society of environmental decay can be even greater in the long term.

Twentieth-century advocacy groups and reformers have made major contributions to public policy, which are akin to the achievements of their predecessor reformers of the eighteenth and nineteenth centuries in the areas of abolition of slavery, humane treatment of prisoners and the mentally ill, improvements in working conditions in factories and mines, and public health sanitary improvements (see Chapter 1).

and workplace pollution are issues of concern to the public, the business sector, the media, and governmental and non-governmental organizations, and are part of the general culture of our times. The growth of the concepts of right-to-know, consumerism, and advocacy in public health has led to greater sensitivity to these issues in many countries.

Globalization, industrialization, and fossil fuel dependency have become an accelerated threat to the global environment. Not only the scientific community, but also governments, the business community, and the general public, are increasingly accepting that human society must order its affairs so that its use of natural resources does not deplete or overwhelm the self-sustaining capacity or natural regenerative powers of the environment. Environmental health is a central issue in the New Public Health in that it is the root cause of much disease and death that is preventable and degrades the environment with irreversible loss to society.

GLOBAL ENVIRONMENTAL ISSUES

A World Summit on Sustainable Development called for world leaders: "aiming to achieve by 2020, that chemicals are used and produced in ways that lead to the minimization of significant adverse effects on human health and the environment". Specific recommendations for both technical and financial assistance will be needed for developing countries and economies in transition to build their capacity (Johannesburg, 2002).

The US Institute of Medicine in 2007 called for cooperative participation of industry in "green chemistry" and voluntary compliance both at home and internationally, eliminating double standards in industrialized and developing countries, and complying with a robust regulatory environment to achieve less industrial, air, and global environmental pollution (Harrison and Coussens, 2007).

The *World Health Report 2007* addressed the threats of increasing risk of disease epidemics, industrial accidents, natural disasters, and other health emergencies and their effects on global public health security. The International Health Regulations of 1995 (see Chapter 16) were an important asset to the process of international collaboration to identify risks and act to contain them. Pandemics of severe acute respiratory syndrome (SARS) and influenza H1N1 showed the dangers of spread of disease from animals and birds to humans and then their spread to distant parts of the world within hours. At the same time, natural disasters and human-caused catastrophes revealed both the potential and weaknesses of international cooperation to protect the public health.

International consensus on global warming has called for action to raise awareness, wide-scale preventive measures, and preparedness for the consequences of global climate change. Reviews by international agencies have confirmed the warnings and call for coordinated international and local action (Box 9.3). These warnings are associated with projections of attributable and avoidable burdens of disease associated with environmental degradation, global warming, and associated climate changes. The continuing struggle to reduce climate-damaging practices is meeting strong economic and political resistance, but progress is being made on

---

BOX 9.3 Global Environmental Challenges and Health Impact

- Global warming/climate instability
- Air pollution/carbon dioxide emissions with atmosphere/ozone depletion
- High population growth in low-income countries
- Inequalities between industrial and non-industrial countries
- Social, economic, and political inequalities nationally and internationally
- Deforestation/forest clearing and biodiversity loss
- Drought/flooding – lack of safe water/water shortage – management and technological advances
- Rural poverty, conservative agriculture and animal husbandry – food insecurity, passage of infectious diseases from animals to humans
- Food production, prices increase, poor food security – distribution and improved production methods
- Energy and resource depletion – search for alternative cost-effective energy sources
- Soil erosion/pesticide pollution/desertification – regional famine, migration, refugees
- Chemical/toxic wastes – long-term carcinogenic and teratogenic effects
- Wars and civil strife/terrorism/incitement and genocide/nuclear threats/armament costs
- Economic growth and economic decline
- Disease emergence and transference from local habitats to world exposure – HIV, SARS, West Nile fever, Lyme disease, Rift Valley fever, Lassa fever

Sources: *World Health Organization. The world health report 2007 – a safer future: global public health security in the 21st century. Geneva: WHO; 2007. Available at: http://www.who.int/whr/2007/en/ [Accessed 17 August 2012].*
Corvalin C, Hales S, McMichael A. Ecosystems and human wellbeing: health synthesis: a report of the millennium ecosystem assessment. Health impacts of ecosystem impairment due to environmental changes. Geneva: WHO; 2005. Available at: http://www.who.int/globalchange/ecosystems/ecosys.pdf [Accessed 15 August 2012].
United Nations Environment Programme. 21 Issues for the 21st century: results of the UNEP foresight process on emerging environmental issues. Nairobi: UNEP; 2012. Available at: http://www.unep.org/publications/ebooks/foresightreport/Portals/24175/pdfs/Foresight_Report21_Issues_for_the_21st_Century.pdf [Accessed 15 August 2012].

---

specific issues and technological advances such as in water and wastewater management, and searches for cost-effective energy sources. A vast complex of environmental issues having major health impacts is shown in Box 9.3.

The environment and human society interact and are mutually dependent. The ecological issues that face the world include those that can be addressed locally and nationally and others that require concerted international cooperation. Local action is part of global responsibility. Local issues require close cooperation among different agencies of government at all levels, with local authorities, supported at state and national levels. Non-governmental organizations (NGOs), the media, the private sector, and voluntary groups all have important roles in promotion of a healthful environment. Unrestrained population growth and rising standards of living in many developing countries with attendant demand for consumption standards of developed countries undermine local and international efforts to maintain a balance between nature and human society. At the same time, the industrialized countries are beginning efforts to reduce polluting standards, but the time available to prevent runaway global warming is very short.

Global society must face simultaneously those environmental, social, and health issues that relate to poverty and high population growth in the poorest countries. Urbanization, demographic, and epidemiological shifts are associated with growing populations with health needs related to long-term diseases and conditions. These can be aggravated by environmental pollution, which presents a severe challenge in rapidly developing societies with urban crowding, air and other pollution, and fossil fuel issues. Increasing consumption associated with increasing per capita incomes creates demands for increased food supply that will be difficult to meet. Agricultural reform with improved water management and smallholder methods including better transportation, marketing, and wider use of genetically modified foods will be essential to meet these demands. Such demands are occurring in an environment of decreasing water supply, uncertainty as to climate effects, and the continuing struggle to lift people from poverty and its adverse health effects. Some successes are impressive, with many mid-level income countries achieving better education and health, whereas in some newly wealthy countries the rapid increases in productivity and income coexist with large sectors of the population remaining in rural poverty and decay.

Among the long-range issues confronting many countries are water supplies and their quality, which are endangered by overuse and the pollution of groundwater sources. Air and soil pollution, deforestation, and desertification require local, national, and international multisectoral cooperative planning and intervention. Water alone will become a cause of technological change and possible conflict between countries. At present, nearly 700 million people suffer from water shortage, and recurrent droughts are expected to spread with global warming to affect global food supplies and prices, pushing more people into moderate and severe food insecurity.

Public consciousness regarding these issues has increased during the past several decades. Environmental concern has become an essential part of accepted public philosophy in many developed countries. Its place in developing countries is often of low priority, coming after the struggle to expand economically as well as the severe problems of population growth and basic services. Economic growth and health status are closely related to agriculture, food supplies, and distribution systems, as well as preservation of agricultural land and rational use of energy. As was the case in numerous countries during their industrial development and urbanization, many Eastern European countries prioritized industrialization over all other issues and subordinated environmental concerns, so that accumulated environmental degradation is part of the long-range burden of post-Soviet societies.

GEOGRAPHIC AND ENVIRONMENTAL EPIDEMIOLOGY

Geographic epidemiology is defined as the description of spatial patterns of disease incidence and mortality. It is part of descriptive epidemiology that generally describes the occurrence of disease according to demographic characteristics of the population at risk and in terms of place and time. Snow's description of cholera in London in 1854 and many other observational studies supported hypotheses that turned out in practice to be the case, even though the direct causal relationships were not demonstrable at the time.

Geographic epidemiology helps to generate hypotheses that can then be tested by rigorous methods. Environmental and occupational epidemiology applies a wide range of research methods to the study of disease in relation to environmental or work-related conditions. In practical everyday public health, the findings of a common point source of disease, injury, or death may lead directly to contaminated water, toxic exposure at a worksite, a risk condition, or polluted air of a city. While these may need case–control or other more formal studies for confirmation, the findings of known risk factors on routine surveillance should be sufficient to lead to adequate public health intervention by the appropriate regulatory authorities.

Epidemiological studies may describe in quantitative terms the relationship between the frequency of disease and the degree of exposure to a particular agent. Such studies are subject to errors in the measurement of exposure. Measurement of exposure by place of residence or work is only an approximation. Moreover, within the same community there will be wide variation in actual exposure levels to the toxic agent. The agent may affect different populations or subgroups differently. There may be genetic and social factors at play as well. In cases where there is a long time lapse between exposure and resultant disease, and many independent variables, it may be very difficult to attribute the disease to a specific exposure, as in the case of asbestos exposure and mesothelioma. The risk of asbestos exposure was compounded by cigarette smoking, affecting the workers directly exposed. But asbestos also affected the families of workers, as well as people in the community secondarily exposed to inhalation of the carcinogenic fibers, with residual asbestos material having been discarded or still present in roofing and many other applications. The cumulative evidence established the causal relationship and justified action to eliminate asbestos use in work settings and the wider environment.

TABLE 9.1 World Health Organization Environmental Burden of Disease: Health Targets and Issues

Target	Issues
Multisectoral policies to protect the environment	Coordination between agencies at international, national, regional, and local levels
Raising public awareness of global climate and the health effects of environmental health	Promotion of public/private consortia for environmentally friendly policies to reduce greenhouse effects, and promote alternative fuels, chemicals, construction, agriculture policies
Monitoring and control mechanisms for environmental hazards	Chemicals, ionizing radiation, noise, biological agents, consumer goods, risk assessment
Adequate supplies of safe drinking water	Quantity, quality of water; international, national programs, ground and surface water surveillance, quality control; water management standards
Protection against air pollution	Legislative administrative and technical measures to control indoor and outdoor pollution
Reduced risk of food contamination including harmful additives	Legislative administrative and technical measures to control food contamination and additives, and production, storage, transport, sale, and use
Eliminate risks of hazardous wastes	Effective legislative, administrative, and technical measures for surveillance and control of dumped wastes
Healthy and safe urban environment	Housing and urban planning standards, waste disposal, potable water supply, recreation, open spaces, traffic control, waste disposal, and sanitation
Protection against work-related risks	Protection against biological, chemical, physical hazards; worker education, industry self-monitoring and government regulation

Source: Campbell-Lendrum D, Woodruff R. Climate change: quantifying the health impact at national and local levels. Prüss-Üstün A, Corvalán C, editors. Geneva: WHO; 2007. Available at: http://whqlibdoc.who.int/publications/2007/9789241595674_eng.pdf [Accessed 15 August 2012].

ENVIRONMENTAL TARGETS

In 1962, publication of Rachel Carson's *Silent Spring* on the environmental effects of indiscriminate use of pesticides was a signal event galvanizing public opinion in the USA and elsewhere. The US Environmental Protection Agency (EPA) was established by President Nixon in 1970 in response to a growing public concern over environment issues. The EPA mandate was given broadly: "To declare a national policy which will encourage productive and enjoyable harmony between man and his environment", "To promote efforts which will prevent or eliminate damage to the environment and biosphere and stimulate the health and welfare of man", and "To enrich our understanding of the ecological systems and natural resources important to the Nation".

The World Health Organization (WHO) *Commission on Health and Environment Report* (1992) developed a consensus documentation of international environmental health issues. This commission, chaired by Simone Weil of the European Parliament, included many distinguished scientists, professional leaders, and international organizations. The report represented a strong international consensus on joint action to prevent and clean up environmental degradation that had occurred in Europe over several decades. The European Region of the WHO consensus statements on health targets that emerged represented a broad societal commitment to stop environmental degradation. These have been reissued in various forms and increasingly represent a wide commitment to action to slow global warming, as this has become the center-stage issue of environmental health in recent years.

The WHO (2006) defines the environment, in relation to health, as: "all the physical, chemical, and biological factors external to a person, and all the related behaviors"; "Environmental health consists of preventing or controlling disease, injury, and disability, related to the interactions between people and their environment" (Healthy People 2020; Last, 2001).

Public health has traditionally placed high priority on sanitation, housing, and urban planning in the battle to reduce the burden of infectious disease. The sanitary movement of the nineteenth century had an enormous impact on the control of communicable diseases. In the twenty-first century, environmental health issues are still an enormous challenge for public health and society in general (Table 9.1).

The Centers for Disease Control and Prevention (CDC) in the USA declared advances in public health organization, infectious disease control, and occupational health as being among the Ten Great Achievements of Public Health in the USA in the twentieth century (CDC 1999, 2011). The US Healthy People 2020 vision for environmental health in the USA includes six theme topics:

- outdoor air quality
- surface and ground water quality
- toxic substances and hazardous wastes
- homes and communities
- infrastructure and surveillance
- global environmental health.

Over the years, the EPA has addressed many specific issues setting national standards in the field. In 2012, the EPA focused on:

- air
- water
- green living
- health and safety
- land and cleanup
- pesticides, chemicals, toxics
- waste
- water issues.

In 2012, the EPA issued new regulations on fuel economy standards, pollution from power plants, and renewable fuels (EPA 2012), which provide hope for significant reductions in pollution in urban areas in the USA.

ENVIRONMENT AND INFECTIOUS DISEASES

Climate and environmental change can have important impacts on infectious disease. Diseases such as dengue, Lyme disease, West Nile fever, chikungunya, and Legionnaire's disease can move from their original sources to become endemic in new locations as environmental and vector conditions change.

The lessons learned in swamp and still-water drainage, disinfection of potable water supplies, and treatment of solid and liquid wastes are still not applied universally by either the less developed countries or middle-level developing and industrialized countries. The threat of local, national, and international disasters, including the re-emergence of cholera in South America and Russia and *Giardia* as a major outbreak in the USA, have returned the classic issues of water quality to center stage in modern public health. Plague broke out in India in 1994 and Rift Valley fever struck Egypt in 1977 and later affected various parts of the Middle East (Saudi Arabia, Yemen), where it is now likely to be endemic.

The resurgence of malaria and dengue fever in large parts of the world highlights the problems of vector control and damaging effects of environmental degradation on the burden of disease in modern public health. The USA experienced its biggest spike in human cases of West Nile virus since 2003, with 5,674 cases of the disease reported nationwide in 2012, including 286 deaths (up to 31 December 2012). Infections were reported in people, birds, or mosquitoes in 48 US states, 62 percent of these in California, Louisiana, Michigan, Mississippi, Oklahoma, South Dakota, and Texas. CDC (2013) stated that one-third of the total cases were in Texas.

Lyme disease (Box 9.4) exemplifies another vectorborne disease, representing an important and current public health problem affecting populations and substantially burdening health departments. In the 1990s, West Nile fever appeared

---

BOX 9.4 Lyme Disease

Lyme disease, which results from the bite of an infected blacklegged tick, is an example of a vectorborne disease caused by the bacterium *Borrelia burgdorferi*. Humans who acquire the disease are likely to experience a fever, headache, and a distinctive pink/red circular skin rash surrounding the area of the bite. The rash typically develops 3–30 days following the tick bite, and soon afterwards fatigue and other flu-like symptoms ensue. If untreated, symptoms worsen in severity and the infection may progress, affecting the joints, heart, and nervous system. This may include temporary paralysis of facial muscles.

The most common vectorborne disease in the USA, there are an estimated 20,000 new cases of Lyme disease documented annually. In 1991 Lyme disease grew to become nationally recognized in the USA. Since then, cases documented per year have more than doubled. Improved laboratory techniques may play a role in the increase in reported cases, but there appears to be a true rise in Lyme disease incidence. The UK Health Protection Agency (HPA) estimates that there are 2000–3000 cases of Lyme disease in England and Wales each year and, furthermore, the disease occurs widely in northern Europe and Asia.

In the USA, most cases are concentrated in north-eastern, mid-Atlantic, and north-central regions and mainly affect the age groups 5–14 and 45–54 years. The CDC established 10 reference states where the disease is endemic for surveillance purposes; of approximately 64,000 cases of Lyme disease between 2003 and 2005, 93 percent were from these 10 states. Between 2003 and 2005, the average yearly rate in these states was 29.2 cases per 100,000 population. The Healthy People 2010 objective was to lower the yearly incidence to 9.2 cases per 100,000 population, one-third of the current rate.

Prevention requires safety precautions: avoiding regions in which ticks are highly concentrated, applying insect repellents and carrying out body examinations and removal of ticks within the first 24 hours of attachment are strategies that aid in lowering the chances of bacterial transmission to the person. Landscaping maneuvers can substantially reduce the infestation of ticks in certain regions for people living in high-risk woodland areas. Antibiotics are used in treatment and various vaccines have been developed.

Sources: *Centers for Disease Control and Prevention. Lyme disease, United States 2003–2005. MMWR Morb Mortal Wkly Rep 2007;56:573–6. Available at: http://www.cdc.gov/mmwr/preview/mmwrhtml/mm5623a1. htm [Accessed 15 August 2012].*
National Health Service. Lyme disease [updated 15 August 2011]. UK: NHS. Available at: http://www.nhs.uk/Conditions/Lyme-disease/Pages/ Introduction.aspx [Accessed 15 August 2012].

---

for the first time in the north-east USA and it is now endemic in many parts of the USA. In 2003, SARS was spread from China to Toronto, causing the city to be placed under virtual quarantine. Chikungunya fever appeared in France and Italy in 2007, and in Hong Kong in 2008, brought by travelers from Asia. Spread by *Aedesalbopictus* and *Aedesaegypti* mosquitoes, which are common in many parts of Western and Central Europe, chikungunya is a viral disease characterized by fever, severe joint pain, nausea, muscle pain, and rash, often of sudden onset. Typically, the joint pain lasts for a few days or weeks, during which time the pain is severe and incapacitating. Most cases recover, but some individuals experience long-lasting joint pain or eye, neurological, or heart complications, which severely debilitate a patient. In regions where dengue is endemic, chikungunya is often misdiagnosed for dengue owing to their similar clinical manifestations. Infected female mosquitoes bites cause human-to-human virus transmission, with risk linked to the distance between mosquito breeding areas and human habitation. Preventive efforts focus on environmental measures to minimize breeding areas and control measures include the use of insecticides, repellents, protective clothing, and insecticide-impregnated mosquito bed nets.

The concept of environmental health has been widened in recent decades by the spectrum of global changes to the environment as a result of environmental pollution by humans and natural events such as volcanic eruptions. The greenhouse effect is the warming of the global environment through retention of solar heating of the Earth by increasing the greenhouse gases in the Earth's surrounding atmosphere. Disposal of toxic and radiological waste constitutes a very difficult public health challenge in many countries. Land degradation, loss of topsoil, deforestation, groundwater depletion, and acidification of water and soil are all challenges in environmental health in the twenty-first century. The effects of global environmental changes cannot be predicted with certainty, but there is scientific consensus on the serious and imminent dangers to the environment and human society which require both global and local preventive action and public health crisis response capacities.

Poverty, low levels of education, and rapid population growth in the poorest countries with limited food production potential stand in contrast to high levels of consumption and energy use and low rates of population growth in the industrialized countries. Many environmental issues involve more than one country, partly because of the transportation of waste products or hazardous materials from one country to another, by wind, water, or deliberately by people. Economic concerns include the destruction of fishing stocks, damage to forests, and more global concerns of ozone depletion, global warming, and ocean pollution. Intersectoral cooperation within a country, and international cooperation and regulation to reduce pollution of common waters in seas, lakes, and rivers shared by more than one country are part of a broad New Public Health agenda.

Global Climate Change

There is widespread consensus that the warming of the Earth is a result of human activities. It is an emerging risk factor for health, the spread of infectious diseases, and disruption of food and freshwater supplies. The effects of global warming are expected to include serious weather disruptions and changes in ecology that could threaten plant, human, and other animal life on Earth. The policy-making uses of estimates of health impacts include the identification of groups at risk for specific diseases and the use of scarce resources, which help to target measures needed for controlling the emissions of greenhouse gases.

The Human Development Report of 2007/2008 saw climate change as the defining human development challenge of the twenty-first century. The more recent Human Development Report in 2013 highlights the importance of sustainability and equity, two imperative elements strongly intertwined within the issue of climate change. Failure to respond to this challenge will stall and reverse international efforts to reduce poverty. The poorest countries and most vulnerable people will suffer the most damaging setbacks, but no country will be immune to the impact of global warming.

The 2013 Human Development Report indicates that it is impossible for the extraordinary advances made in human development to continue unless major steps are taken to diminish environmental risks and inequality. Increased exposure to droughts, floods, and storms is already destroying opportunity and reinforcing inequality. Overwhelming scientific evidence indicates that the world is moving towards the point of irreversible ecological catastrophe. Avoiding the impact of the most damaging climate changes requires global action in the decades ahead. The financial resources and technological capabilities exist, but implementation requires a sense of urgency, public interest, and political will to make deep cuts in greenhouse gas emissions. Achievement of MDG7 (to ensure environmental sustainability) rests very much on addressing the issues of global warming.

The WHO highlights major global environmental dangers that impact life globally. These effects include climate change, ozone depletion, reduction in biodiversity resulting in changes in ecosystems, land degradation, and pressure and strain on food-generating systems. Figure 9.1 provides examples of health effects emerging from major environmental and ecosystem changes. Severe droughts and prairie fires in Russia and Ukraine during 2011 and across North America in 2012 have had a serious impact on global production of corn and grain, almost immediate effects on food supply and prices, and an especially harsh impact on poor countries and poor people in rich countries.

An increasingly warmer climate warns us of the possibility of devastating consequences, including increased air pollution, more disease transmission via unsafe, unclean water and contaminated food, more natural disasters, and extreme weather conditions. Furthermore, agriculture and crops are likely to be affected. Of particular importance to

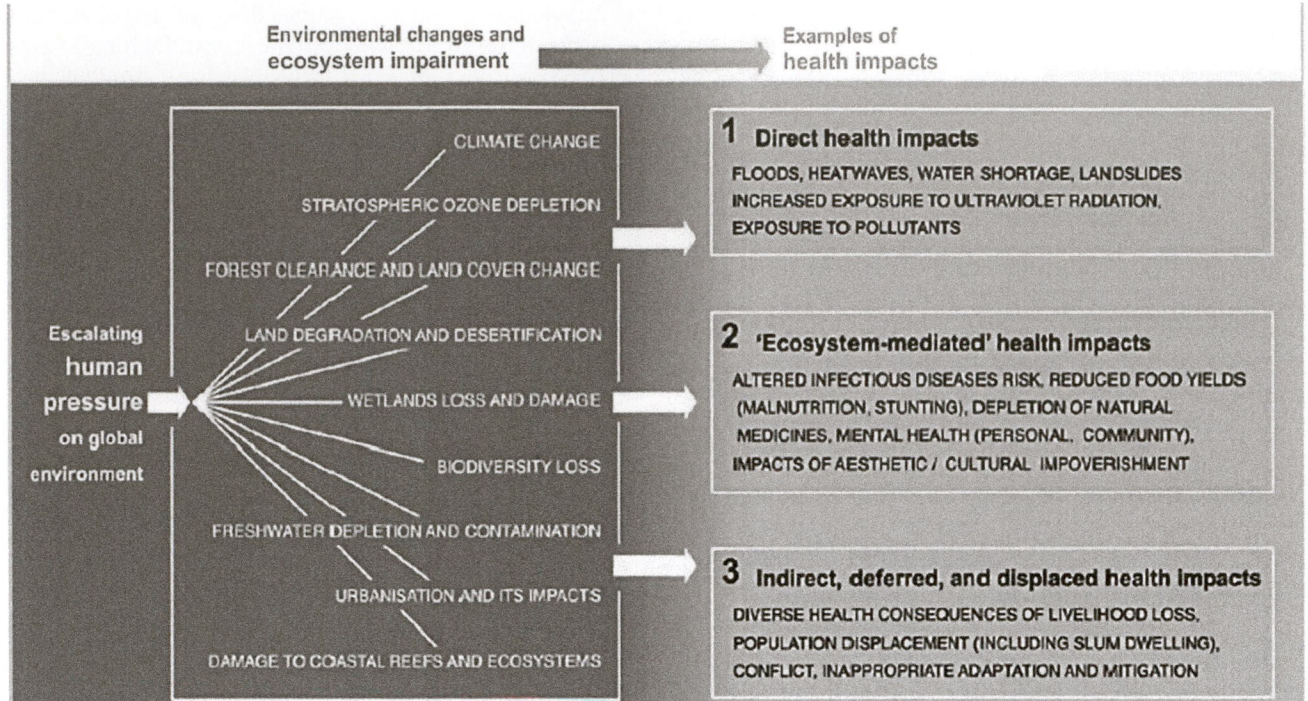

FIGURE 9.1   Environmental change, ecosystem effects, and human health. *Source: Corvalin C, Hales S, McMichael A. Ecosystems and human wellbeing: health synthesis: a report of the Millennium Ecosystem Assessment. Health impacts of ecosystem impairment due to environmental changes. Geneva: WHO; 2005. Available at: http://www.who.int/globalchange/ecosystems/ecosys.pdf [Accessed 15 August 2012].*

underdeveloped countries, climate change may bring about problems related to the containment and control of infectious diseases. Major causes of mortality, such as malaria, cholera, diarrheal diseases, dengue fever and other vector-borne diseases, are extremely sensitive to climate and influenced by temperature, humidity, and rainfall.

A special risk with regard to climate change is the effect on coastal populations worldwide of rising sea levels due to melting polar ice caps. The melting of taiga and tundra may release trapped methane gas in enormous qualities, further exacerbating global warming. Nearly 60 percent of the global population and 50 percent of the US population live in coastal counties.

Climate change is demonstrated through the elevation in global average air and ocean temperatures, the extensive melting of ice and glaciers, and a rise in sea levels. The World Meteorological Organization has annual updates on the status of climate change and greenhouse gas emissions (Box 9.5). There are documented changes in the frequency and degree of extreme weather conditions including hurricanes, heavy rainfall and flooding, forest fires, and heatwaves, all of which can create public health emergencies.

In situations of extreme weather, deaths resulting from cardiovascular or respiratory disease can be attributable to heatwaves. Heatwaves can cause high numbers of deaths among the elderly in particular, such as occurred in Chicago in 1995 with an estimated 700 deaths, and a 2003 heatwave in Europe which caused an estimated 50,000 deaths.

Floods, droughts, and unclean water increase the risk of disease, and when freshwater is unavailable, proper hygiene is seriously jeopardized. These devastating events cause serious damage to homes, communities, and much needed health facilities. These situations can be associated with cholera and other diarrheal diseases, such as occurred in Haiti in recent years (see Chapter 10).

Pre-existing medical conditions or malnutrition, a long-term contributor to high mortality in developing countries, may be worsened with widespread problems due to climate change affecting crop harvests. The WHO and the majority of scientists and environmental activists view climate change as a threat to the basic needs and primary determinants of health: air, water, food, shelter, and freedom from disease. The dramatic consequences of climate change underscore the need for international standards and cooperation to change the way in which we use energy and the way in which we live our daily lives.

Environmental Impact on Health Burden of Disease

Environmental hazards contribute to a wide range of diseases. The WHO reports that as much as 25 percent the burden of disease worldwide is from preventable environmental exposures, including more than 13 million deaths annually and nearly one-third of mortality and morbidity in low-income countries. Environmental factors are responsible for more

BOX 9.5 Climate Change

The World Meteorological Organization annual updates on the status of climate change and greenhouse gas emissions indicate that the decade 2001–2010 was the warmest on record since 1880, in terms of average global temperatures. These warmer temperatures surpassed the previous record decade, 1991–2000.

In 2008, the latest year for which data are available at the time of writing, global carbon dioxide emissions continued to rise, reaching 30.1 billion tonnes, an increase of 1.7 percent from the previous year. This change was smaller than in the period 2006–2007 (2.9 percent) owing to the economic crisis.

The December 2010 UN Climate Change Conference in Cancún, Mexico, was a step forward in international negotiations under the UN Framework Convention on Climate Change. A set of decisions known as the "Cancun Agreements" was adopted by the international community to address collectively and comprehensively the long-term challenges of climate change with a global drive of national actions to mitigate greenhouse gas emissions.

The Montreal Protocol for international reduction in ozone-depleting substances is a success story in protection of the ozone layer. At the end of 2009, the consumption of 98 percent of all ozone-depleting substances controlled under the listed in this Protocol had been phased out.

Biodiversity of the world's forests remains imperiled by the still high rate of global deforestation and forest degradation as well as a decline in primary forests. The problem remains, however, the rate of deforestation, and loss of forest from natural causes is slowing down.

Sources: *Climate change and human health: global environmental change [updated 2012]. Geneva: WHO. Available at: http://www.who.int/globalchange/environment/en/ [Accessed 15 August 2012].*
United Nations Development Programme. Human Development Report 2011, sustainability and equity: a better future for all. New York: UNDP; 2011. Available at: http://hdr.undp.org/en/reports/global/hdr2011/download/ [Accessed 11 August 2012].
World Health Organization. Protecting health from climate change: connecting science, policy and people. Geneva: WHO; 2009. Available at: http://whqlibdoc.who.int/publications/2009/9789241598880_eng.pdf [Accessed 2 August 2012].

BOX 9.6 Preventing Disease Through Healthy Environments

- Environmental hazards are responsible for about a quarter of the total burden of disease globally, and as much as 30 percent in regions such as sub-Saharan Africa. Worldwide, 13 million deaths could be prevented every year by making our environments healthier.
- In children under the age of five, one-third of all disease is caused by environmental factors such as unsafe water and air pollution.
- Every year, the lives of 4 million children under 5 years, mostly in developing countries, could be saved by preventing environmental risks such as unsafe water and polluted air.
- In developing countries, the main environmentally caused diseases are diarrheal disease, lower respiratory infections, unintentional injuries, and malaria.
- Better environmental management could prevent 40 percent of deaths from malaria, 41 percent of deaths from lower respiratory infections, and 94 percent of deaths from diarrheal disease – three of the world's biggest childhood killers.
- In the least developed countries, one-third of death and disease is a direct result of environmental causes.
- In developed countries, healthier environments could significantly reduce the incidence of cancers, cardiovascular diseases, asthma, lower respiratory infections, musculoskeletal diseases, road traffic injuries, poisonings, and drownings.
- Environmental factors influence 85 out of the 102 categories of diseases and injuries listed in the *World Health Report*.
- Much of this death, illness and disability could be prevented through well-targeted interventions such as promoting safe household water storage, better hygiene measures, and the use of cleaner and safer fuels.
- Other interventions that can make environments healthier include increasing the safety of buildings, promoting safe, careful use and management of toxic substances at home and in the workplace, and better water resource management.

Sources: *World Health Organization. Protecting health from climate change: connecting science, policy and people. Geneva: WHO; 2009. Available at: http://whqlibdoc.who.int/publications/2009/9789241598880_eng.pdf [Accessed 2 August 2012].*
World Health Organization. Ten facts on preventing disease through healthy environments. Geneva: WHO. Available at: http://www.who.int/features/factfiles/environmental_health/en/index.html [Accessed 11 August 2012].
Bridge JW, Oliver DM, Chadwick D, Charles H, Godfray J, Heathwaite AL, et al. Engaging with the water sector for public health benefits: waterborne pathogens and diseases in developed countries. Bull World Health Organ 2010;88:873–5. Available at: http://www.who.int/bulletin/volumes/88/11/09-072512.pdf [Accessed 18 August 2012].

than 33 percent of illness in children under 5 worldwide and as many as 4 million lives in this age group could be saved, mostly in developing countries, by preventive environmental measures. Safe household water storage and hygienic measures, cleaner and safer fuels, better built environment, less air pollution, better home and workplace management and use of toxic substances, and better water resource management would reduce diseases such as diarrhea, respiratory infections, malaria, dengue and West Nile Fever virus (Table 9.1). Knowledge regarding disinfection of drinking water supplies and treatment of solid and liquid wastes is not always applied, either in the less developed countries or in middle-level developing and industrialized countries (Box 9.6).

Environmental factors affect the developing countries most, as they suffer from poor water supplies, low levels of sanitation, low housing standards, poor education (especially of girls), and high rates of poverty. These topics are addressed in the MDGs. Most of the diseases with large numbers of deaths are amenable to change with available policies, technologies, and preventive and public health measures; these could result in 2.6 million fewer deaths annually from cardiovascular diseases, 1.7 million fewer deaths from diarrhea, 1.5 million fewer from respiratory diseases, 1.4 million fewer from cancers, and close to 1 million fewer from external injuries (motor vehicle accidents, poisonings, and others).

Sulfur and nitrogenous oxides from fossil fuel electric power plants can travel long distances after being released from tall chimneys. Pollutants falling as acid precipitation have led to the destruction of forests in countries of Central and Eastern Europe. Acid rain generated in one European country may fall in another, affecting bodies of water, animal life, and forests. Acid rain was reduced during the 1980s in North America by greater selectivity in fossil fuels, and the result is reduced damage to forests and water sources.

The release of various organic solvents, called chlorofluorocarbons (also known as freons or CFCs), used in cooling systems, refrigerators, and consumer aerosol products, causes damage to the Earth's ozone layer. This permits entry of ultraviolet (UV) light that was formerly excluded, into the Earth's atmosphere. UV light causes a rise in skin cancer and cataracts in humans. Substitution for freons is vital to reduce damage to the ozone layer, and can be achieved on an individual level by use of water-based paints and chemical products in daily life. The search for substitutes for refrigerants and toxic chemicals to replace those that damage the environment and exposed workers has, along with regulation, become the hallmark of environmental and occupational health.

Greenhouse gases are built up in the atmosphere by carbon dioxide emissions, largely due to increasing carbon dioxide and other gases produced from excessive and inefficient use of fossil fuels along with wide-scale destruction of forests, which are protective through natural conversion of carbon dioxide to water. These gases block infrared radiation from the Earth's surface, leading to trapping of heat. This effect resembles the use of glass or plastic covers to retain heat in a greenhouse. This global warming effect may have long-term serious consequences for the Earth's thermal balance. The effects on the polar ice caps can lead to global changes in the level of oceans. Reduction of the greenhouse effect requires international, national, and individual effort, and especially environmental consciousness and action by governments, the media, the scientific and business communities, as well as the general public.

Hazardous wastes are being exported from developed to developing countries. Box 9.7 discusses e-waste, a phenomenon in which unwanted, old electronic equipment ends up and accumulates in various African countries. This situation is potentially solvable by heightened national awareness and stronger international conventions, with publicity and fines imposed by international courts against offending firms or nations. In a global economy, all of these factors link up with effects on the physical environment as well as on working conditions and many social and political factors, such as the widening gap between rich and poor.

COMMUNITY WATER SUPPLIES

Freshwater is vital for all living organisms and is becoming an increasingly scarce resource. Waterborne diseases are among the major causes of death in developing countries, which often lack adequate supplies of water. In both developed and developing countries, pollution control, reuse of wastewater, and water planning are vital to the national economy and public health.

The International Decade for Drinking Water and Sanitation in the 1970s and early 1980s promoted national, binational, and international efforts to improve community water supplies, sanitation, drainage, education, and hygiene. Implementation of appropriate technology for maintaining water and sanitation infrastructure was emphasized. Safety of community drinking water, as defined by the WHO, requires a combination of standards and protection of raw water sources from contamination. Treatment of community water supplies requires sedimentation, coagulation, filtration, chlorination, and continuous monitoring. High standards of construction and maintenance of water distribution systems are needed, whether at the village well or in the municipal water supply system. Filtration removes solid and suspended particles, improving the quality of surface source water, and disinfection by chlorination effectively kills most microorganisms.

Covering and protection of reservoirs and canals is also beneficial in improving the security of water sources and in preventing contamination from natural sources, including birds, animals, and vegetation. Community water regulation and enforcement require both physical treatment and disinfection to protect the public against microbiological, chemical, and other health hazards. Agricultural runoff of pesticides and animal wastes are also important contaminants of water sources.

The Federal Water Pollution Control Act of 1948 was the first major US law to address water pollution. The Clean Water Act (CWA) of 1977 amended the 1972 Federal Water Pollution Control Act to address severe pollution of the Great Lakes and many of the major rivers of the USA. The CWA set new US national standards and regulatory mechanisms at federal, state, and local levels of government. It increased regulatory powers to "restore and maintain the chemical, physical, and biological integrity of the Nation's waters". It included provisions for cooperation with Canada in cleaning up the Great Lakes.

The CWA gave the EPA the authority to implement pollution control programs such as setting wastewater standards for industry and water quality standards for all contaminants in surface waters. It made it unlawful for to

BOX 9.7 E-Waste

Africa is beginning to benefit from electronic modernization, such as cellular phones where no telephone landlines exist, and further electronic gains as development advances. In another respect, however, it has fallen victim to the information and technology revolution. As a continent, it currently serves as a disposal ground for unwanted, discarded, and outdated electrical equipment, which typically possess toxic properties and release hazardous chemicals into the environment. This major global issue is referred to as e-waste, "a generic term encompassing various forms of electrical and electronic equipment that are old, end-of-life electronic appliances and have ceased to be of any value to their owner".

Examples of e-waste include televisions, cell phones, air conditioners, and refrigerators, which are transported from the developed world. The list of devices has greatly expanded through the years. The result is harmful pollutants into the atmosphere and underground water. E-waste disposal methods generally used (burning or landfill) contribute to the release of a mixture of toxic chemicals, harming the environment and humans, with leaching of poisons into groundwater aquifers following disposal.

Electronic devices contain a myriad of toxic compounds, and of particular concern is lead, characteristically found in high concentrations in many electronic products. Moreover, certain components such as polybutleneterephthalates (PBTs)

can accumulate to harmful amounts in living organisms. This may result even when limited quantities are dispersed, and consequently, PBTs place humans at risk for nerve damage, cancer, and reproductive complications.

Massive amounts of e-waste pose major problems in developing countries, which typically lack any form of infrastructure, defined protocols, or legislation to safely dispose of the e-waste and to cope with the overflow. Countries in West Africa, particularly Ghana and Nigeria, are especially burdened by electronic equipment waste. E-waste products are quite diverse and vary in complexity owing to the rapid evolution of new product design.

The lack of waste management protocols and enforcement in low-income countries allows this practice to continue. In contrast, regulations surrounding the disposal of unwanted devices are well established and growing in developed countries. The US Environmental Protection Agency informs consumers on ways to recycle or donate used electric devices, and many states have created specific laws pertaining to the disposal and recycling of electronic equipment.

Sources: *Orisakwe OE, Frazzoli C. Electronic revolution and electronic wasteland: the West/waste Africa experience. J Nat Envi Sci 2010;1:43–7. Available at: http://www.asciencejournal.net/asj/index.php/NES/article/viewFile/27/ORISAKWE [Accessed 17 August 2012].*
Environmental Protection Agency. E-cycling [updated 24 July 2012]. Washington, DC: EPA. Available at: http://www.epa.gov/epawaste/conserve/materials/ecycling/index.htm [Accessed 15 August 2012].

discharge any pollutant from a point source into navigable waters, unless a permit was obtained under its provisions. It provided grants for construction of sewage treatment plants. It also recognized the need for planning to address problems of non-point source, i.e., generalized pollution (EPA, Laws and Regulations: History of the Clean Water Act; 2013).

CWA regulations permitted effective action against industrial and other polluters and for controls to be established where multiple municipalities were involved in a river or regional water system. This has led to steady improvement in water quality of lakes, rivers, and groundwater sources throughout the country. However, a 2013 EPA report of a survey in 2008–2009 showed that: "21 percent of the nation's river and stream length is in good biological condition, 23 percent is in fair condition, and 55 percent is in poor condition according to commonly used measurements" (EPA, 2013). A Clean Water Restoration Act proposed in the US Congress in 2007 is intended to clarify federal jurisdiction and standards of water supervision. However, it is controversial because of alleged federal infringement of state responsibilities (Box 9.8).

Concern about the potential carcinogenic effects of trihalomethanes may cause withdrawal of mandatory chlorination of surface waters. The absence of adequate disinfection with chlorine increases the risk of serious waterborne disease outbreaks such as the wave of cholera epidemics in South America during the 1990s. New standards may

require time to be implemented because of prevailing conservative professional and public attitudes and the cost of treatment plants. In Israel, for example, opinion gradually shifted towards a mandatory chlorination policy. This was due to a number of factors: increased public and news media awareness of drinking water quality, a greater recognition at the leadership level of the Ministry of Health of the importance of preventive and environmental factors in enteric disease, an increasing presence of younger, better trained sanitary engineers willing to challenge previously accepted dogmas, and persuasive documentation of the impact of contaminated community water supplies on the infectious disease burden of the country. Principles of water quality regulation are shown in Box 9.9.

Waterborne Diseases

Despite the long-standing success in reducing mortality and morbidity in the industrialized countries by emphasis on safe water supplies and careful monitoring, waterborne disease remains a serious challenge to public health in the twenty-first century. Waterborne diseases are among the most common causes of death in developing countries and remain an important public health issue even in high-income countries. They may be so common as to escape detection in point outbreak form. This seems to be the case in many countries, where hepatitis (especially hepatitis A

BOX 9.8 Clean Water Restoration Act

The Clean Water Act in 1972 made great progress in reducing pollution of waters and establishing standards for water quality in the USA and was a successful environmental law, but has fallen behind in recent years. The Act prohibits designated point source discharges, an oil spill prevention program, and the impaired waters cleanup program. The federal Environmental Protection Agency and Army Corps of Engineers regulations implement the law to protect all of America's waters, including tributaries, wetlands, and intrastate waters with linkages to interstate commerce. These rules were upheld by the vast majority of courts. In 1993, a large epidemic of waterborne cryptosporidium, a parasitic disease spread by contaminated water, occurred in Milwaukee with 400,000 cases and more than 100 deaths, which led to national concern regarding water safety in the USA.

The EPA regulates cryptosporidium in drinking water, and the regulatory process was strengthened by the Safe Drinking Water Act Amendments of 1996 providing funds to upgrade water treatment plants and requiring public drinking water suppliers to inform customers about chemicals and microbes in the water. The Great Lakes, the largest surface freshwater system in the world, provide drinking water to over 30 million people, some 10 percent of the US population and 30 percent of Canada's (EPA 2012). The International Joint Commission advises the national governments on issues regarding joint stewardship of the Lakes. The Great Lakes Cleanup of 1997 agreement between the US Environmental Protection Agency (EPA) and the Government of Canada agreed to a joint project to remove toxic substances from the Great Lakes by 2006.

In 2001, the Supreme Court ruled that non-navigable, intrastate waters are not protected by the Clean Water Act, weakening the regulation of national water systems. The Clean Water Restoration Act of 2009, introduced by US Senator Russell Feingold, D-WI, would accomplish these important goals and has been endorsed by Clean Water Action. An EPA Hudson River Cleanup project in 2002 moved to clean up a 40 mile (64 km) stretch of the Hudson River of polychlorinated biphenyl (PCB) contamination, removing approximately 2.65 million cubic yards (3.5 million cubic meters) of contaminated sediment. The Clean Water Restoration Act of 2009 and 2010 clarifies the jurisdiction of the federal government over US waters.

In 2012, the 1624 miles (2614 km) of California coastline, from Mexico to Oregon, and surrounding major islands were deemed a no-discharge zone by the EPA as a ban was placed on large cruise ships and other large ocean-going ships from discharging any sewage into California's marine waters. Over 20 million gallons (75 million liters) of vessel sewage is kept off the state's beaches, a boon to the state's economy, helping to protect marine species, fisheries, residents, and tourists.

Sources: *Clean Water Action. Overview: Clean Water Restoration Act of 2009. Available at: http://www.cleanwateraction.org/mediakit/overview-clean-water-restoration-act-2009 and http://www.govtrack.us/congress/bills/111/s787/text [Accessed 27 June 2013].*

US Environmental Protection Agency. Water: Clean Water Act 40th anniversary. Protecting and restoring our nation's waters. Available at: http://water.epa.gov/action/cleanwater40/cwa101.cfm [Accessed 27 June 2013].

US Environmental Protection Agency. Great Lakes. Basic Information, geography & hydrology. Available at: http://www.epa.gov/greatlakes/basicinfo.html [Accessed 16 July 2013].

US Congress. S. 787 (111th): Clean Water Restoration Act. Text as of 10 December 2010. Available at: http://www.govtrack.us/congress/bills/111/s787/text [Accessed 28 June 2013].

US Environmental Protection Agency. Water is worth it. Available at: http://water.epa.gov/action/cleanwater40/ [Accessed 27 June 2013].

US Environmental Protection Agency. Laws and regulations. Available at: http://www2.epa.gov/laws-regulations [Accessed 27 June 2013].

and E) is endemic and where the incidence of gastroenteritis from *Shigella*, *Escherichia coli*, and rotavirus remains high. In industrialized countries, waterborne disease outbreaks have become uncommon events because of high levels of water management (Box 9.10).

Water contamination and enteric disease can also occur from organisms for which routine testing is not currently practiced. For example, testing for rotaviruses (which cause enteric disease) and organisms such as *Campylobacter* and *Giardia* is not done routinely; however, water is tested for these if there is a suspicion of contamination. Widespread infection of up to half of the world's population via contaminated water may have caused asymptomatic or symptomatic infection with *Helicobacter pylori*, the major cause of chronic peptic ulcer disease and gastric cancer (see Chapter 5). Safe water requires physical treatment as well as disinfection of all community water sources to ensure its safety during distribution to homes for private use. Effective water management can reduce the burden of gastroenteric disease even in a relatively developed country (Box 9.11). The WHO water guidelines, which also incorporate chemical and radiation hazards, are periodically updated.

Israel, developing rapidly in the 1960s, built a national water carrier to distribute unfiltered, chlorinated water to communities and for agriculture. Local groundwater was not necessarily chlorinated, and sewage system development was inadequate. In the 1970s and 1980s, Israel experienced large numbers of waterborne disease outbreaks. A 1985 outbreak resulted from the contamination of groundwater sources by a sewage pipe which accidentally broke during roadwork, resulting in 9000 cases of shigellosis, 49 cases of typhoid fever, and one death. The introduction of mandatory chlorination in 1988 in Israel resulted in a substantial improvement in the quality of community water supplies and greatly reduced the incidence of waterborne disease outbreaks and the total burden of diarrheal disease (Tulchinsky et al., 2001).

An Environmental Protection Ministry is responsible for ensuring that all industries comply with the standards.

BOX 9.9 Principles of Water Quality Regulations

- There is an explicit link between drinking water quality regulations and protection of public health.
- Regulations are designed to ensure safe drinking water from source to consumer, using multiple barriers.
- Regulations are based on good practices proven to be appropriate and effective over time.
- A variety of tools are in place to build and ensure compliance with regulations, including education and training programs, incentives to encourage good practices and penalties, if enforcement is required.
- Regulations are appropriate and realistic within national, subnational and local contexts, including specific provisions or approaches for certain contexts or types of supplies, such as small community water supplies.
- Stakeholder roles and responsibilities, including how they should work together, are clearly defined.
- "What, when and how" information is shared between stakeholders – including consumers – and required action is clearly defined for normal operations and in response to incidents or emergencies.
- Regulations are adaptable to reflect changes in contexts, understanding and technological innovation and are periodically reviewed and updated.
- Regulations are supported by appropriate policies and programs.
- The aim of drinking water quality regulations should be to ensure that the consumer has access to sustainable, sufficient, and safe drinking water.
- Enabling legislation should provide broad powers and scope to related regulations and include public health protection objectives, such as the prevention of waterborne disease and the provision of an adequate supply of drinking water.

Sources: *World Health Organization. A conceptual framework for implementing the guidelines. In: Guidelines for drinking-water quality. 4th ed. Geneva: WHO; 2011. Available at: http://www.who.int/water_sanitation_health/publications/2011/9789241548151_ch02.pdf [Accessed 15 August 2012].*
Bridge JW, Oliver DM, Chadwick D, Charles H, Godfray J, Heathwaite AL, et al. Engaging with the water sector for public health benefits: waterborne pathogens and diseases in developed countries. Bull World Health Organ 2010;88:873–5. Available at: http://www.who.int/bulletin/volumes/88/11/09-072512.pdf [Accessed 18 August 2012].

BOX 9.10 Control of Waterborne Disease as a Crucial Element in Reducing Morbidity and Mortality from Infectious Diseases, Twentieth Century, USA

"The 19th century shift in population from country to city that accompanied industrialization and immigration led to overcrowding in poor housing served by inadequate or nonexistent public water supplies and waste-disposal systems. These conditions resulted in repeated outbreaks of cholera, dysentery, TB, typhoid fever, influenza, yellow fever, and malaria.

By 1900, however, the incidence of many of these diseases had begun to decline because of public health improvements, implementation of which continued into the twentieth century. Local, state, and federal efforts to improve sanitation and hygiene reinforced the concept of collective 'public health' action (e.g., to prevent infection by providing clean drinking water).

By 1900, 40 of the 45 states had established health departments. The first county health departments were established in 1908. From the 1930s through the 1950s, state and local health departments made substantial progress in disease prevention activities, including sewage disposal, water treatment, food safety, organized solid waste disposal, and public education about hygienic practices (e.g., food handling and hand washing).

Chlorination and other treatments of drinking water began in the early 1900s and became widespread public health practices, further decreasing the incidence of waterborne diseases." (CDC, 1999)

Sources: *Centers for Disease Control and Prevention. Achievements in public health, 1900–1999: control of infectious diseases. MMWR Morb Mortal Wkly Rep 1999;48:621–9. Available at: http://www.cdc.gov/mmwr/preview/mmwrhtml/mm4829a1.htm [Accessed 16 August 2012]. Centers for Disease Control and Prevention. Achievements in public health, 1900–1999: changes in the public health system. MMWR Morb Mortal Wkly Rep 1999;48:1141–7. Available at: http://www.cdc.gov/mmwr/preview/mmwrhtml/mm4850a1.htm [Accessed 27 June 2013].*

seawater, and building reservoirs to store rainwater. These measures have improved the water situation in the country, and thus a large industry of water systems technology has developed for international needs.

Waterborne Disease Surveillance and Prevalence, USA

During the nineteenth and early twentieth centuries, cholera and typhoid were major causes of waterborne disease outbreaks. Data collection on waterborne disease outbreaks in the USA dates back to 1920. Decades later, in 1971, the CDC, EPA, and Council of State and Territorial Epidemiologists collaborated to establish the Waterborne Disease and Outbreak Surveillance System (WBDOSS). This system allows experts to identify the prevalence and sources of outbreaks from water exposures which, in turn, helps to determine the epidemiology and the etiology of waterborne diseases. In many cases, the investigations reveal outbreaks resulting from pathogens or engineering problems within water systems. WBDOSS-documented outbreaks include those resulting from drinking water, recreational water, and

Recycling of water for agricultural use has become a widespread practice, although recycled waste is not be used for household purposes. The shortage of water supplies has led to large-scale desalination along with improved water-pipe maintenance and conservation methods including reservoir construction.

There has been a marked reduction in the overall burden of enteric diseases in Israel, including hepatitis A; however, foodborne salmonellosis continues to be a public health problem. Initiatives to deal with overall water shortages include reclaiming sewage water for irrigation, drip irrigation, reducing water demand by one-third, desalination of

---

BOX 9.11 International, National, State, and Local Water Management Standards

- *International* – The United Nations and World Health Organization promoted the International Decade for Drinking Water and Sanitation and promulgated clear standards of water quality for community water supplies (1958, 1963, 1971, 1984, and 1997).
- *National, state, and local authorities* – policy commitment, funding, and professional departments for supervision of community water systems.
- *Municipal water systems* – water management and testing vary according to the quality of the source water and methods of treatment, including:
  - high standards of acceptability of source surface water
  - physical treatment: coagulation and filtration
  - disinfection by chlorination: routine and mandatory
  - maintaining and monitoring of residual chlorine
  - construction and maintenance of water storage and distribution systems
  - monitoring of enteric disease
  - investigation of suspected waterborne disease outbreaks
  - continuous monitoring by bacteriological and chemical testing
  - assurance of safe distance between sewage and water pipes
  - integrity of water distribution systems against inflow.
- *Village wells*
  - protection of wells from human and animal wastes
  - regular or periodic chlorination
  - supervision by trained and supervised village health workers.
- *Sanitary education* – at all levels of society including governments, non-governmental organizations, intersectoral cooperation, public, medical and other professional communities, and schools.

Source: *World Health Organization. Guidelines for drinking-water quality. 4th ed. Geneva: WHO; 2011. Available at: http://whqlibdoc.who.int/publications/2011/9789241548151_eng.pdf [Accessed 15 August 2012].*

---

other forms of water exposure. Samples from the patient and samples of water are examined and recorded, as are environmental components and water disinfection procedures. Surveillance is passive and reporting voluntary, so the figures obtained are not the complete incidence of waterborne outbreaks, as reporting depends on public awareness and resources accessible to local health departments.

Pathogens that have yet to be identified may play a crucial role in the future, owing to the changing characteristics of waterborne pathogens. Waterborne pathogens that require more of scientists' efforts include those that may be resistant to modern water treatment procedures. Recently, treatment protocols have been enhanced and upgraded in an effort to kill *Giardia* and *Cryptosporidium*, since outbreaks of these organisms in the USA have raised concerns,

as they are not efficiently eliminated by standard water treatment and are not routinely tested for in regular water-sampling monitoring. In addition, these waterborne organisms, among others, constitute a special risk for immunocompromised people, including cancer patients treated with chemotherapy, human immunodeficiency virus (HIV)-positive people, and patients on immunosuppressants following organ transplantation.

During 1991 to 2002 there were 2007 waterborne disease outbreaks reported in the USA, with 433,947 cases of illness reported. Problems in the distribution system were the most commonly identified deficiencies under the jurisdiction of a water utility, underscoring the importance of preventing contamination after water treatment. Most notably, in 1993, the largest reported waterborne disease outbreak in US history occurred in Milwaukee, Wisconsin, resulting in approximately 403,000 ill people, with 4400 requiring hospitalization. The organism responsible for the thousands of illnesses was *Cryptosporidium parvum*, a protozoan parasite that causes gastrointestinal problems. Transmission occurs by ingesting oocytes that have been eliminated from the body via animal or human feces. This has sparked increased focus on water quality assurance and epidemiological follow-up of waterborne disease.

Cryptosporidium can be transmitted from person to person, from animal to person, and by ingesting contaminated food or water. It can also be transmitted in swimming pools, and accordingly, this parasite is reportedly present in 65–87 percent of surface water samples tested in the USA. The degree of illness from this outbreak ranged from mild to severe among residents of Milwaukee, and attack rates were as high as 50 percent in some parts of the city. Most affected individuals experienced diarrhea, dehydration, and fever. Some cases resulted in deaths, which generally occurred among the immunocompromised or the elderly. The underlying cause of this massive Milwaukee outbreak was inefficiency of the water filtration process, causing contamination of a public water source. This resulted in the insufficient elimination of *Cryptosporidium* oocytes in one of the two major municipal treatment plants.

In 2011 the CDC's *Morbidity and Mortality Weekly Report* (MMWR) highlighted waterborne disease outbreaks in the USA, reporting 36 outbreaks documented during 2007–2008 with over 4000 cases of illness and three deaths. The four major illness recorded included outbreaks of acute gastrointestinal illness, acute respiratory illness, hepatitis and skin irritation related to a chemical exposure. Almost 60 percent of the 36 outbreaks were bacterial in nature, with a significant portion attributed to *Legionella*. The remaining etiologies include viruses, parasites, and chemicals. The underlying reasons for water contamination include untreated groundwater, treatment deficiencies, and substandard distribution systems. This report indicates the need for improved strategies to target, remove, and control

(a)

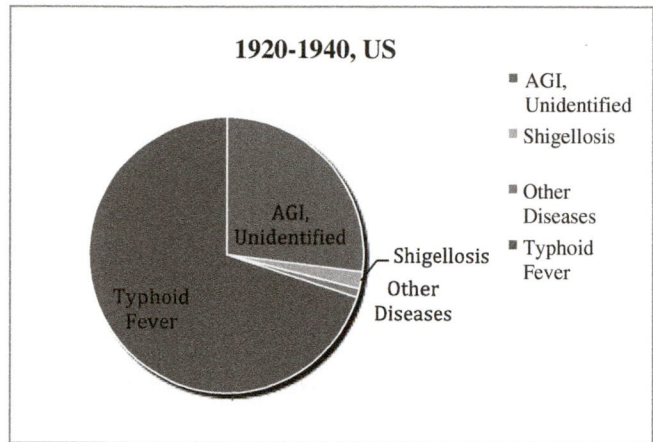

(b)

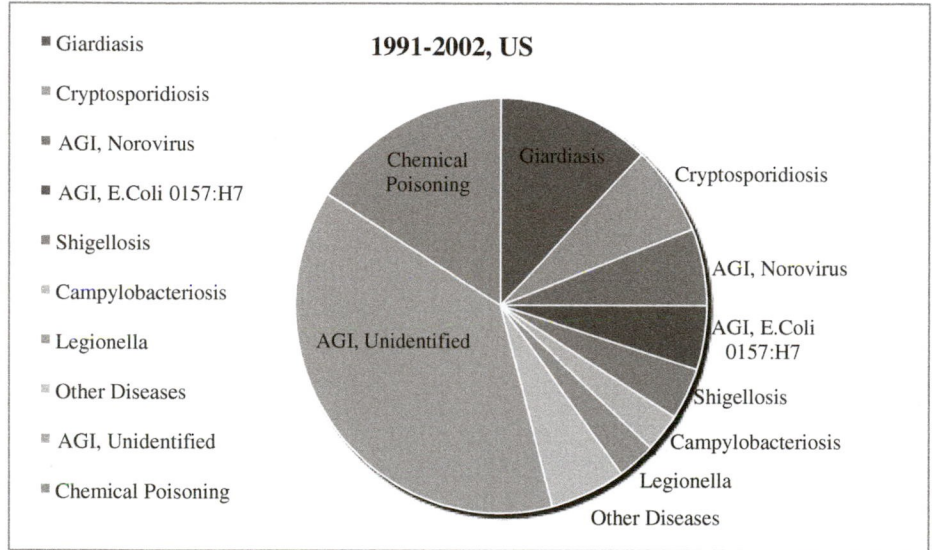

FIGURE 9.2   Etiology of waterborne diseases, USA, 1920–1940, 1991–2002. Note: AGI = Acute gastrointestinal illness. *Sources: Adapted from Craun MF, Craun GF, Calderon RL, Beach MJ. Waterborne outbreaks reported in the United States. J Water Health 2006;2:19–30. Available at: http://courses. washington.edu/h2owaste/group1.pdf [Accessed 16 August 2012].*

Lang IA, Galloway TS, Scarlett A, Henley WE, Depledge M, Wallace RB, et al. Association of urinary bisphenol A concentration with medical disorders and laboratory abnormalities in adults. JAMA 2008;300:1303–10. Available at: http://jama.jamanetwork.com/article.aspx?articleid=182571 [Accessed 16 August 2012].

Melzer D, Rice NE, Lewis C, Henley WE, Galloway TS. Association of urinary bisphenol A concentration with heart disease evidence from NHANES 2003/06. PLoS ONE 2010;5:e867. Available at: http://www.plosone.org/article/info%3Adoi%2F10.1371%2Fjournal.pone.0008673 [Accessed 16 August 2012].

Centers for Disease Control and Prevention. Surveillance for waterborne disease outbreaks associated with drinking water – United States, 2007–2008. MMWR Surveill Summ 2011;60(ss12):38–68. Available at: http://www.cdc.gov/mmwr/preview/mmwrhtml/ss6012a4.htm [Accessed 19 August 2012].

Legionella, which is the most common documented etiology among outbreaks associated with drinking water in the USA.

The graphs in Figure 9.2 illustrate an interesting comparison and historical perspective, showing the major etiologies of US waterborne outbreaks from 1920–1940 and from 1991–2002. Looking at the first graph, it is evident that the recorded waterborne diseases each fall into one of four categories. When examining the graph representing

more the recent decades, we see the various diseases falling into several more classifications. Thus, the number of pathogens that have been identified as causative agents in waterborne disease outbreaks has expanded.

Early detection by laboratory diagnosis requires preparation of laboratories for identification of these organisms. Regular testing of the community water supply at its origin and within the supply system is essential to monitor water safety. The presence of coliform bacteria indicates fecal contamination and potential hazards, warning sanitation officials that other more dangerous organisms, such as dysentery bacilli or enteric viruses such as hepatitis, may be present. Testing for *Cryptosporidium*, *Giardia*, and viruses is difficult, costly, and insensitive; therefore, routine testing is not done. Chlorination and filtration may not be sufficient to prevent waterborne disease transmission of these organisms. This is a problem for sanitary control, thus new methods of testing and disinfection of water supplies must be devised. At present, filtration and chlorination remain the basic methods of ensuring safe community water supplies, supplemented by boiling of suspect water during outbreaks of disease.

Standard water treatment processes (Figure 9.3) remove solid and suspended material, bacteria, and odors from water and have been outstandingly successful in reducing waterborne disease. New concerns over chemical contamination of community water supplies have become prominent in recent decades. Heavily polluted waters have been linked to neurological damage and cancers of the bladder, intestinal tract, liver, and kidney.

The US Safe Drinking Water Act of 1974, as amended in 1996, establishes criteria for monitoring of public water systems for microbiological, chemical, and other contaminants (Box 9.9). The act defines maximum contaminant levels (MCLs) for specified chemical pollutants. The EPA sets MCLs for pollutants, out of hundreds of organic, inorganic, biological, and radiological contaminants detected in water supplies around the country. This area of public health concern still requires much epidemiological and sanitary engineering research.

Right-to-know laws, a critical investigative press, and an environmentally conscious public are fundamental to prevent serious ecological degradation. Environmental activism has made important contributions to public health, but such activism can be a two-edged sword. One example is the excessive zeal focused on the environmental impact of chlorination and its byproducts, in particular, trihalomethanes. This class of disinfectant byproducts is produced when natural organic and inorganic substances present in the water are combined with the disinfectants chorine and chloramine. Total trihalomethanes include chloroform, bromoform, bromodichloromethane, and dibromochloromethane. The current MCL, which includes the combined concentrations of each of these four trihalomethanes, is 0.080 mg/l (EPA, 2012), although this level has changed throughout recent years. Similar to the case for other disinfectants, the MCL is set as a yearly average. People who drink water in which total trihalomethane levels exceed the

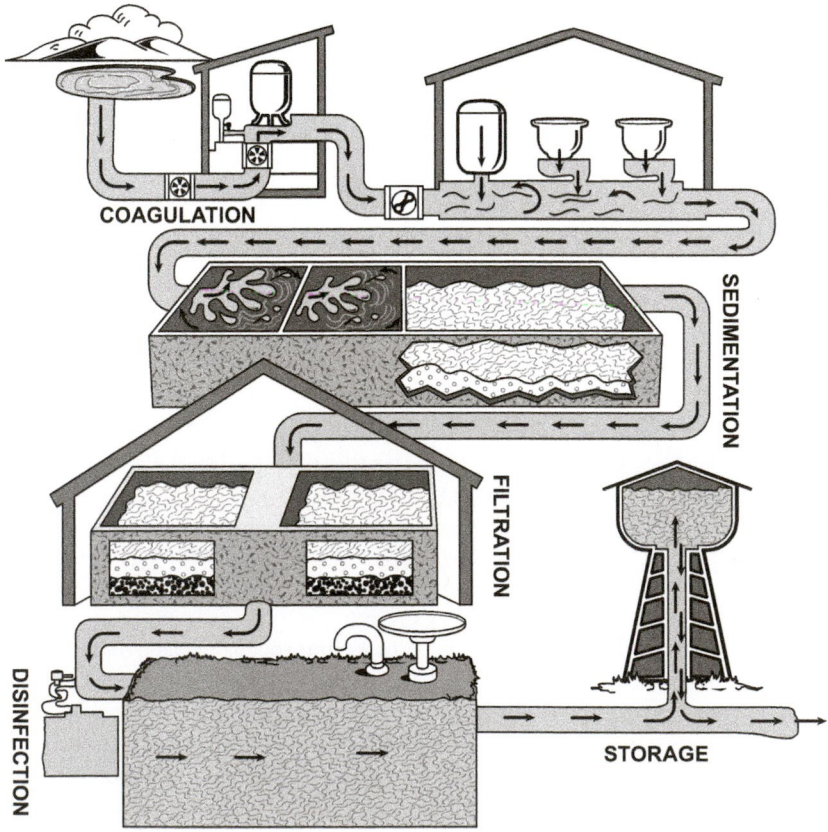

FIGURE 9.3 Process of community water treatment. *Source: Centers for Disease Control and Prevention. Healthy water: community water treatment figure, courtesy of EPA [updated 3 April 2009]. Atlanta, GA: CDC. Available at: http://www.cdc.gov/healthywater/drinking/public/water_treatment_fig.html [Accessed 16 August 2012].*

BOX 9.12 Water Contaminants Under US Environmental Protection Agency Regulation

- *Microorganisms* – turbidity, total coliforms, viruses, *Giardia lamblia*, *Cryptosporidium*, *Legionella*
- *Disinfectants* – chloramines and chlorine (as Cl_2), chlorine dioxide (as ClO_2)
- *Disinfection by-products* – bromate, chlorite, haloacetic acids, total trihalomethanes
- *Organic chemicals* – acrylamide, atrazine, tetrachloroethylene, carbon tetrachloride, vinyl chloride, benzenes, lindane, endrine, carbon tetrachloride, carbofuran, chlorobenzene, dichloromethane, dichloropropane, dinoseb, dioxin, diquat, endrin, epichlorohydrin, ethylbenzene, ethylenedibromide, glyphosate, heptachlor and epoxide, hexachlorobenzene, methoxychlor, oxamyl, polychlorinated biphenyls (PCBs), pentachlorophenol, simazine, styrene, terachloroethylene, toluene, toxaphene, trichloroethylene, vinyl chloride, xylene
- *Inorganic chemicals* – antimony, arsenic, asbestos, barium, beryllium, cadmium, chromium, copper, cyanide, fluoride (>4 mg/l), lead, mercury, nitrates, nitrites, selenium, thallium
- *Radionuclides* – alpha particles, beta particles and photon emitters, uranium, radium-226 and radium-228 (combined)

Note: Maximum contamination levels (MCLs) are set by the EPA for the listed contaminants.
Source: *Environmental Protection Agency: Water: drinking water contaminants [updated 5 June 2012]. Washington, DC: EPA. Available at: www.epa.gov/safewater/contaminants/index.html#micro [Accessed 16 August 2012].*

prevent greater health damage than benefit. Box 9.12 shows major waterborne disease contaminants that are defined by EPA regulations.

WATER AS A GLOBAL RESOURCE

Created in 2003, UN-Water is the United Nations interagency collaboration organization for all issues pertaining to freshwater. Its purpose is to assess and document the conditions and utilization of freshwater resources on a global scale. UN-Water's important functions include evaluating a geographic region's past and determining whether and how much progress has been made. Notable advances have been achieved through WASH (water, sanitation, and hygiene) services.

UNICEF reports that MDG target 7c, of reducing by half the proportion of people without sustainable access to safe drinking water and basic sanitation, was reached by 2010. This means that the percentage of individuals who have no access to improved drinking water sources was reduced by more than half since 1990. The figure dropped from 24 percent to 11 percent, as more than 2 billion individuals obtained access to improved water.

This is a significant achievement, but the benefits of clean water are not available equally or uniformly. Only minimal advancement has been made in expanding access to drinking water among those living in poverty in sub-Saharan Africa and in access to sanitation facilities among the poor in South Asia, especially in rural regions. The primary challenges impeding continued progress in making safe drinking water available to everyone are issues of disparities and insufficient human and financial resources.

In March 2011, the United Nations Development Programme (UNDP) stated that, "the right to water emphasizes the importance of water-related development for marginalized and vulnerable groups, who are commonly socially excluded" (UNDP, 2011). In a 2012 WHO document discussing environmental health inequalities in Europe, rural populations in particular are identified as a vulnerable group. In examining the prevalence of inadequate water supply among several European countries, data demonstrate that rural populations are significantly more affected by this than urban populations. Those residing in rural areas are especially exposed to sources of inadequate drinking water, and this puts them at increased risk for waterborne diseases. Underlying reasons for this disparity include distance to water sources, minimal access to a water distribution network, and cost, as the price associated with water services is substantially higher in rural regions than in urban areas. The availability of safe, potable water for vulnerable populations would lead to marked improvements in the health status of many, as the risk of diarrheal diseases would drop and better hygiene would be maintained. Moreover, access to improved water may result in less of a need for both storing water in the home and transporting water. This is an

established MCL over a period of several years can develop liver, kidney, or central nervous system damage and may be at increased risk for cancer.

Opposition to disinfection by use of chlorination led to the spread of cholera in South America during the 1990s (see Chapter 4). The offset of benefits against risks has resulted in current professional consensus that this is not a justification to cease chlorination. Rather, it provides additional justification for physical treatment of raw water before chlorination to reduce the nitrogenous material content and thereby reduce the combination with chlorine which produces trihalomethanes, improving water potability and clarity. Residual chlorine within the water distribution system is a protector for cross-contamination from sewage sources where maintenance of pipes and valves may be inadequate in aging distribution systems.

Developmental programs including local and large-scale dam projects can have negative health effects by providing a hospitable environment for vectors for diseases such as malaria, schistosomiasis, and onchocerciasis, resulting in the resurgence of diseases once controlled. Planning of development projects must take into account the potential ecological effects and the needed control measures to

BOX 9.13 Population Growth, Food Security, Water Scarcity, Science and Technology

The global population of some 7 billion people will increase by another 2 billion by 2050. The highest rates of population growth are occurring in water-scarce sub-Saharan Africa. Water security is essential for human life, socioeconomic development, and healthy ecosystems. Urban development requires increased utilization of groundwater and surface water sources for domestic, agricultural, and industrial needs. Population and economic growth requires more food production, which demands large amounts of water (e.g., 1 kg (2.2 lbs) of wheat consumes 1500 liters (US 396.3 gal) of water).

Water use has grown at more than twice the rate of population increase in the past century. Although there is enough freshwater on the planet for 6 billion people, it is distributed unevenly and much is wasted, polluted, and unsustainably managed. Thus, pressure on water resources intensifies, with conflicts among users and damage to the environment. For example, Ethiopia is building a dam on the upper Nile, which Egypt views as a threat to its water supply.

Water scarcity affects every continent. Around 1.2 billion people, or almost one-fifth of the world's population, live in areas of scarcity, with another 500 million people close to this situation. A further 1.6 billion people (one-quarter of the world's population) face water shortage due to lack of infrastructure to harvest, transport, and utilize water from surface and aquifer sources. Water scarcity limits agricultural production, leads to population relocation from regions impacted by droughts, and adversely affects food security in developing countries.

Recent evidence indicates that the notoriously dry continent of Africa has vast reservoirs of deep aquifers with 100 times the amount of surface water. But the challenges of secure recovery systems require good stable governance, massive investment, and improved distribution and utilization systems. Developed countries such as the USA also face serious national and regional water shortages.

Important innovations developed to meet the growing challenge of water scarcity include:

- safe water recovery from wells and surface water with protection from contamination
- safe water processing – filtration, chlorination, and new innovations in microfilters
- improved distribution systems
- monitoring and replacement of leaking distribution systems that waste large amounts of treated water
- improved irrigation techniques – drip irrigation, sprinklers, microsprinklers
- new plant varieties that can survive drought conditions
- home and industrial wastewater recovery and treatment – now over 75 percent utilized in Israel
- storage of runoff waters
- rainwater collection and storage
- desalination – now provides half of total water usage in some countries
- filtration and treatment technologies – self-cleaning filtration systems
- decontamination – from oil and chemical contamination
- community education to reduce overusage of scarce water – reduced garden irrigation, toilet flushing using "gray water" (e.g., from showers, kitchen sink).

Safe water in adequate quantities for a growing world population with potential for serious climate change will make water conservation and innovative management a crucial public issue in the coming decades.

Note: Maximum contamination levels (MCLs) are set by the EPA for the listed contaminants.
Sources: *Environmental Protection Agency. Water: drinking water contaminants [updated 5 June 2012]. Washington, DC: EPA. Available at: www.epa.gov/safewater/contaminants/index.html#micro [Accessed 16 August 2012].]
United Nations Department of Economic and Social Affairs. International Decade for Action: Water for life, 2005–2015. New York: UNDESA. Available at: http://www.un.org/waterforlifedecade/scarcity.shtml [Accessed 16 August 2012].
UN-Water. Coping with water scarcity, a strategic issue and priority for system-wide action. New York: United Nations; 2006. Available at: ftp://ftp.fao.org/agl/aglw/docs/waterscarcity.pdf [Accessed 18 August 2012].
Organisation for Economic Co-operation and Development. Biodiversity, water and natural resource management: Global Forum on Environment: Making Water Reform Happen. OECD Conference Centre, Paris 25–26 October 2011. Available at: http://www.oecd.org/env/resources/globalforumonenvironmentmakingwaterreformhappen.htm [Accessed 1 December 2013].
World Health Organization/United Nations Children's Fund. Progress on drinking water and sanitation: 2012 update. Geneva: WHO/UNICEF; 2012. Available at: http://www.unicef.org/media/files/JMPreport2012.pdf [Accessed 16 August 2012].
Mekorot: Israel National Water Company. Wastewater treatment and reclamation. Tel Aviv: Mekorot. Available at: http://www.mekorot.co.il/Eng/Activities/Pages/WastewaterTreatmentandReclamation.aspx [Accessed 18 August 2012].*

important and favorable outcome, as typically, it is these actions that play a role in water contamination.

The MDG to improve access to clean drinking water has made good progress since 2001. Globally, population coverage with safe water increased from 77 percent in 1990 to 87 percent in 2008. An estimated 1.1 billion people in urban areas and 723 million people in rural areas gained access to improved drinking water sources over the period 1990–2008. However, the most vulnerable populations still lack such access (UN, 2013; Millennium Development Goals Report 2013; UNICEF, 2012). Consequently, water scarcity is identified as a major global problem (Box 9.13).

SEWAGE COLLECTION AND TREATMENT

Sewage collection and treatment, along with filtration and disinfection of drinking water, have made enormous contributions to improved public health, perhaps even more than the use of modern medicines and vaccines. Collection of sewage prevents surface environmental contamination as well as seepage into groundwater and contamination of local water sources. Sewage contains bacteria, viruses, protozoa, and other pathogens that can cause serious disease; treatment entails killing the pathogenic organisms present in the sewage. The purpose of sewage treatment is to improve the

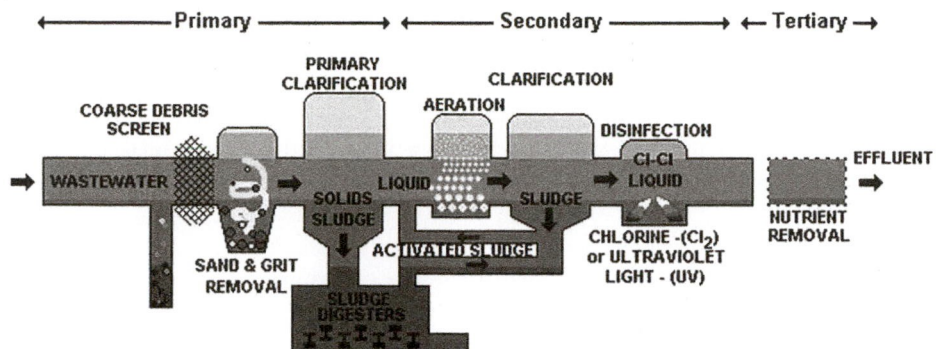

FIGURE 9.4   Wastewater treatment process. *Source: Water Treatment Process. Waste water treatment process. Available at: http://watertreatment-process.net/waste-water-treatment-process/waste-water-treatment-process/ [Accessed 16 August 2012].*

quality of wastewater to a level where it can be discharged into a waterway or prepared for reuse for agriculture without damaging the aquatic environment or causing human health problems in the form of waterborne disease.

Figure 9.4 illustrates each level involved in the wastewater treatment process. Primary treatment of community wastewater begins with the removal of solids from the wastewater, through several mechanical processes of screening and sedimentation. The wastewater is passed through screens to remove large solid objects and then through grinders to further break up the solid wastes. The wastewater then flows at reduced velocity through a grit chamber where sand, gravel, and other inorganic materials settle out. Air is injected into the tank to remove trapped gases and to maintain an aerobic environment. The wastewater then flows into secondary settling tanks where further sedimentation of solid particles takes place. Primary treatment removes just over half of the suspended material and particles in preparation for secondary treatment.

Secondary treatment of wastewater is based on biological treatment assisted by mechanical methods, accelerating the natural decomposition of organic wastes. Aerobic microorganisms are used in the presence of an abundant oxygen supply to decompose the organic material into carbon dioxide, water, and minerals. The wastewater is sprayed over trickling filters or beds of crushed stone covered with a slime containing various types of microbes. These microbes absorb the organic material and act to break it down into its various components. The sewage is then processed by the activated sludge method, carried out by introducing bacteria-containing sludge into a tank of wastewater along with compressed air. The waste is then agitated and mixed for 4–10 hours. The microbes are adsorbed to suspended particles and oxidize the organic material. After this process, the sludge, consisting of masses of bacteria, settles out into the tank. The sludge is then removed and recycled into the next tank of wastewater.

Following primary and secondary treatment, the suspended material and the biochemical oxygen demand (BOD) are reduced by approximately 90 percent. This process depends on temperature, which affects the metabolic rate and activity of the organisms needed to break down the suspended organic material. Secondary treatment is most effective in removing protozoa, worms, and bacteria, but less effective against viruses, heavy metals, and other chemicals. Since 1988, all sewage plants in the USA have been required by federal regulations to provide at least secondary treatment.

Tertiary treatment is required if the wastewater is to be recycled for the purposes of agricultural irrigation, recreation, or community use. Tertiary treatment includes a combination of physical, chemical, and biological processes to reduce the particles and BOD to less than 1 percent of those of the original wastewater. The process includes chemical coagulation, filtration, sedimentation, activated carbon adsorption, oxygenation ponds and aerated lagoons, osmosis, ion exchange, foam separation, and land application. All of these processes remove different pollutants present in the wastewater, especially tiny particles of suspended organic matter. They also remove synthetic chemicals, ammonia, nitrates, phosphates, and dissolved organic materials. Recycled wastewater is an important source of water in a world running short of water. Desalination is also becoming an attractive option as the costs of treatment are reduced and become competitive with other forms of water management. Another important potential in new technology is to use evaporated water in the air as a source of household water supply.

Disposal of the sludge remaining after sewage treatment by incineration or ocean dumping is environmentally problematic. Use of the sludge for compost in agriculture or gardening is increasing, but contamination may enter the food chain and create another hazard. Sludge disposal should be carefully regulated.

Reuse of treated sewage water has become important in national water management systems to address widespread shortages of high-quality water supplies from surface or groundwater sources.

Disinfection is the final stage, accomplished by introducing chlorine into the water so that there is a residual level of chlorine to protect the water from contamination in the water storage and distribution system. In many countries or regions, lack of sufficient local water supplies for community agriculture and industrial uses necessitates recycling of wastewater as part of the process of water conservation. Supplementation of water sources by desalination and recycling will be increasingly important as population growth, increasing standards of living, and pressures of agricultural and industrial contamination of water sources increase. New technology in membrane filtration offers hope to improve the efficiency and economics of this sector of the ecological sciences. The post MDG 2015 period will require renewed effort to improve sanitation with infrastructure development with new methods of water and sewage management to reach the poor people so far bypassed.

SOLID WASTE

The disposal of solid waste has been a challenge from prehistoric to modern times which will only increase in the future. With the growth of cities, the disposal of refuse took on a more significant health importance. In biblical times, Jerusalem burned its garbage in a valley outside the city walls (the valley of *Gehennam*, a term later adopted for "Hell"). The Greek city-states had ordinances against dumping refuse in or near cities, providing waste disposal sites for this purpose. In medieval European cities, garbage as well as human and animal wastes were discarded into the streets and areas surrounding the home. In the thirteenth century, Parisians were forbidden to throw waste on the streets and had to dump it outside the walls of the city. In 1388, the English parliament prohibited waste disposal in public waterways. During the industrial revolution, medieval cities evolved into working-class slums. Crowding, poor housing, and poor sanitation forced municipal governments to organize measures to reduce the nuisance and health hazards of solid waste.

Waste management continues to be a problem as greater amounts are generated by the affluent lifestyles of the population of industrialized countries. In developing countries, where rural to urban population shift is under way on a massive scale, rapid population growth, crowding, and slums increase the burden of solid waste disposal. Since the 1980s, return, recycling, and reuse of waste products have entered the popular culture in many countries, and these practices are beginning to have an impact on reducing landfill needs. Recycling of paper, plastic, glass bottles, and metals contributes to reduced solid waste for disposal and has become an economically attractive activity. Biogas methods are improving so that animal wastes can be used for production

of methane gas to be used for energy produced for home or general use.

Waste management is becoming more controversial. There is confusion about the differences between issues such as hospital waste, industrial waste, and toxic household waste, and a lack of trust by the community in governments and scientific communities. Hence, there is great need for continuing community education and communication by government agencies and community leaders.

In the USA, 95 percent of solid wastes comes from agriculture, mining, and industry. The remainder is from household waste, which generates 150–180 million tons (approximately 140–160t) of solid waste annually. This is the equivalent of 4 pounds (1.8kg) of refuse per person per day. Municipal waste collection and disposal are serious problems involving high costs and a serious public health burden if not done well.

Waste management involves a variety of techniques, including reusing and recycling, composting, incineration, and land refill. Each has its advantages and disadvantages. These techniques are part of the engineering of community infrastructure. Seawater dumping is still practiced in some countries, but increasing global concern about the effects of such practices on the ecology of the lakes and oceans makes this solution unacceptable. Landfill is the most prevalent method of solid waste disposal. It involves spreading garbage in layers 8–10 feet (2–3 m) deep and covering them with a thin layer of soil. This method is adequate if well planned and supervised and has the benefit that methane gas produced by anaerobic decomposition can be recovered for use. The problems of seepage of toxic materials and potentially explosive gas accumulation require careful assessment of landfill sites and limit the potential of landfills to serve as a sustainable, feasible option. Limited possibilities for suitable landfill locations in large urban concentrations make this method of disposal a serious urban planning problem. Sanitary landfill is expensive because of the cost of collection and transportation, the land value, and the human resources required. Landfill under sanitary conditions requires compaction of waste and covering by well-spread yet compacted earth far from ground and surface water. The site must be fenced to prevent scavenging by people, animals, and off-hours dumpers. The landfill should be located away from residential areas, be well maintained and tidy, and have well-paved and well-drained access roads. It should be seeded, in completed areas, with grass and trees to control erosion, and must be maintained by well-trained sanitarians.

Composting or conversion of waste products into topsoil can be applied at the household and municipal levels. Byproducts of wood and food processing can be composted and used to reduce soil pollution from petroleum-based products. This process involves separation of non-biodegradable from biodegradable materials and their treatment to break

TABLE 9.2 Hazardous Waste Disposal Sites: Summary of Epidemiological Evidence

Health Effect	Level of Evidence	Health Effect	Level of Evidence
Early fetal deaths (spontaneous abortion)	Inadequate	Orofacial birth defects	Inadequate
Late fetal deaths (stillbirths)	Inadequate	Musculoskeletal birth defects	Inadequate
Intrauterine growth retardation	Inadequate	Genitourinary birth defects	Limited
Small for gestational age	Inadequate	Gastrointestinal birth defects	Inadequate
Birth weight adjusted for gestation length	Inadequate	Chromosomal abnormalities (structural)	Inadequate
Term birth weight	Inadequate	All childhood cancers	Inadequate
Low birth weight (not adjusted for gestation length)	Limited	Leukemia	Inadequate
Preterm birth, gestation length	Inadequate	Lymphoma	Inadequate
Total birth defects	Limited	Reproductive system development	Inadequate
Central nervous system birth defects	Limited	Thyroid function	Inadequate
Cardiovascular birth defects	Limited	Kidney function	Inadequate

Sources: World Health Organization. Population health and waste management: scientific data and policy options. Report of a WHO Workshop, 2007, March 29–30, Rome, Italy. Copenhagen: WHO, European Region; 2007. Available at: http://www.euro.who.int/__data/assets/pdf_file/0012/91101/E91021.pdf [Accessed 17 August 2012].
Agency for Toxic Substances and Disease Registry [updated 14 August 2012]. Atlanta, GA: ATSDR. Available at: http://www.atsdr.cdc.gov/ [Accessed 15 August 2012].
Environmental Protection Agency. Household hazardous wastes [updated 24 July 2012]. Washington, DC: EPA. Available at: http://www.epa.gov/osw/conserve/materials/hhw.htm [Accessed 15 August 2012].

down organic waste. Decomposition at high temperatures (140°F, i.e., 60°C) kills flies, weed seeds, and potentially pathogenic organisms. In closed systems with forced draft aeration, this process can be accomplished in a few days; however, with passive methods it takes many months. After further treatment of "curing" and screening or grinding, an excellent soil conditioner can be produced that can be used to enhance agricultural or horticultural work such as in nurseries, public gardens, and parks. Incineration is attracting wide interest, but its use is limited by high capital cost and the possible release of potentially toxic materials such as dioxin and heavy metals into the atmosphere.

Meticulous maintenance is needed to thoroughly mix the materials for clean burning at high temperatures. In addition, there is the residual problem of disposal of the ash, which is toxic. Waste-to-energy incineration reduces the volume of waste products by 80–90 percent and produces energy that can generate electricity and replace fossil fuels. In Japan and Western Europe, 30–40 percent of solid waste is incinerated in waste-to-energy plants.

Using garbage as feed for pigs is no longer acceptable because of the problem of meat contamination with trichinosis (pork tapeworm). However, the practice is returning on an experimental basis with scraps ground and steamed before being used as animal feed. Control of the use of animal parts for animal feed is now being re-evaluated and more intensively regulated following the bovine spongiform encephalopathy (BSE) experience in the UK and Europe in the 1990s (see Chapter 4).

Recycling and waste reduction are methods gaining wide support. Reducing the use of disposables (e.g., packaging materials, disposable diapers) requires an ecologically conscious public, and municipal, non-governmental, or volunteer collection systems. Scrap metal, paper, glass, and plastic recycling can be commercially successful. Industry and commercial enterprises can be convinced to reduce the use of bulky packaging materials and to adopt "ecologically friendly" practices. Plastics and rubber tires are also recyclable in economically valuable ways. Ecological consciousness is fundamental to the success of such practices.

The potential harmful effects of hazardous waste sites are shown in Table 9.2. The location and management of landfills requires professional management with a fair and transparent process and involvement of the community, with a focus on replacing poor-quality landfill practices. Epidemiological surveillance programs should only be undertaken after a feasibility analysis and with suitable protocols. The chemical exposure pathways and the effects on at-risk segments of the population should be considered. The adverse effects on health from factors such as noise and odor, negative impacts on property values, and the views of the community should all be considered. A 2007 WHO review of the epidemiological evidence of hazardous waste effects on health showed such evidence to be weak or limited (Table 9.2).

Production of both steel and aluminum from virgin ore is very polluting and energy intensive. Therefore, recycling of iron, steel, and aluminum in the USA makes up a substantial part of total new production of these metals.

Use of recycled iron and steel reduces air pollution by 86 percent, water pollution by 76 percent, and solid wastes by 105 percent, compared to production from new ores. Similar benefits accrue from recycling of aluminum scrap. In North America, steel is the most recycled material. During the first 6 months of 2013, recycling of steel was 37.6 million tons compared to recycling of paper, 25.5 million tons; aluminum, 2.3 million tons; glass, 1.6 million tons; and plastics, 1.2 million tons. Community waste collection and recycling has become widely practiced in many countries. In the USA, the rate of recycling municipal waste products has doubled to over 32 percent of total waste, saving some 64 million tons of household waste from landfill or incinerator disposal. In 2011, recycling and composting of trash was about 250 million tons, equivalent to a rate of 34.7 percent of total waste in the USA (EPA 2011).

TOXINS

A toxin is a substance in the environment with the potential for causing human disease or injury. Toxicology is the study of such substances and their effects on humans. All chemicals are toxic under some conditions, depending on the dose, concentration, and threshold or sensitivity of a given species for that substance. The range of chemical toxins and methods of classifying them are shown in Box 9.14.

The factors that affect the toxicity of an agent, in addition to the extent and duration of exposure, include host factors (e.g., age, gender, fitness level, previous exposure), environmental factors (e.g., temperature, air flow), and the

---

BOX 9.14 Hazardous Substances: Chemical Classifications According to their Structure, Properties, or Use

Chemical classes are groupings of those related by similar features: by their structure (e.g., hydrocarbons), uses (e.g., pesticides), physical properties [e.g., volatile organic compounds (VOCs)], radiological properties (e.g., radioactive materials), or other factors, as used by the US Agency for Toxic Substances and Disease Registry (ATSDR) to address hazardous substances.

Chemical Classification

- Benzidines/aromatic amines
- Dioxins, furans, polychlorinated biphenyls (PCBs) (contain phenyl rings of carbon atoms)
- Hydrocarbons (contain hydrogen and carbon atoms)
- Inorganic substances
- Metals/elements (the simplest forms of matter)
- Nitrosamines/ethers/alcohols
- Organophosphates and carbamates
- Pesticides (chemicals used for killing pests, such as rodents, insects, or plants)
- Phenols/phenoxy acids
- Phthalates
- Radionuclides (radioactive materials)
- Volatile organic compounds
- Warfare and terrorism agents (used in acts of war or terror)

Most Viewed Toxic Substances

- Aluminum
- Ammonia
- Arsenic
- Asbestos
- Benzene
- Cadmium
- Chromium
- DDT, DDE, and DDD
- Formaldehyde
- Lead
- Mercury
- PCBs
- Polycyclic aromatic hydrocarbons (PAHs)
- Toluene
- Trichloroethylene (TCE)

Source: *Agency for Toxic Substances and Disease Registry [updated 14 August 2012]. Atlanta, GA: ATSDR. Available at: http://www.atsdr.cdc. gov/ [Accessed 15 August 2012].*

BOX 9.15 Basic Concepts of Toxicology

- *Bioavailability* – the ability of a substance that enters the body to be liberated from its environmental matrix (water, tissue, soil) and to enter the circulation of the host.
- *Dose–response relationship* – the relationship between the quantity of a toxicant received by the host and the probability of an effective concentration at the vulnerable site.
- *Intermediary metabolism* – the metabolic changes that a chemical undergoes once it reaches the cells of the body, usually in the liver. The substance may be detoxified to benign compounds, or may be converted to biologically harmful metabolites. The toxic substance acts on a cellular or subcellular level to disrupt the living organism. Some toxic agents are metabolic poisons; others act on cell membranes, interfere with chemical reactions, or bind to nucleic acids.
- *Susceptibility* – the ability of a living thing to be harmed by an agent, which may be influenced by age, gender, genetic disposition, nutrition, prior exposure, immune state or general health, stress, location at work, airflow, temperature, and humidity.
- *Threshold* – the lowest dose of a chemical that has a detectable effect.
- *Toxic effect* – damage to an organism as measured in terms of loss, reduction, or change of function, clinical symptoms, or signs. Effects may be adverse in one person and not in others.

nature of the toxic agent (e.g., physical and chemical properties) (Box 9.15). Toxicology is an important part of environmental and occupational health; further reference will require a specialized text and appropriate Internet websites (see Bibliography).

Toxic Effects on Fertility

Toxins can adversely affect fertility, pregnancy, and early or later child development. Reproductive potential can be adversely affected by reduced male reproductivity, such as by exposure to the pesticide dibromochloropropane (DBCP). Other chemicals have been implicated in increased abortion rates among exposed pregnant women; for example, birth defects or teratogenesis occurred with exposure to thalidomide. Other chemicals relate to low birth weight and toxicity in newborns. Exposures to chemicals such as lead produce brain damage in children.

Teratogens are substances that cause birth defects, diseases, or abnormalities in the embryo or fetus either by disturbing maternal homeostasis or by acting directly on the fetus. Birth defects historically were attributed to retribution for sin, witchcraft, or moral or physical defects in the mother. Scientific knowledge of genetic disorders has grown since the 1940s, and many agents have been shown to cause birth defects. Such agents act on fetal development and not on genetic DNA, so that a threshold effect is assumed; that is, the effect occurs only if the causative exposure is above a certain threshold. Some currently known teratogenic agents and their effects are shown in Table 9.3.

TABLE 9.3 Teratogens and Their Effects on the Fetus and Newborn

Teratogen	Effects on Fetus and Newborn
Maternal Infections	
Rubella	Congenital rubella syndrome, deafness, cataracts, heart defects
Syphilis, herpes simplex	Mental retardation, microcephaly
Cytomegalovirus	Infected kidney, liver, lungs
Toxoplasmosis	Central nervous system lesions
HIV	HIV neonatal transmission
Others – varicella, mumps, parvovirus	Nerve deafness
Nutritional deficiency	
Protein deficiency	Abortion, prematurity, low birth weight
Folic acid deficiency	Anencephaly, spina bifida
Ionizing radiation	
X-rays or nuclear radiation or fallout	Central nervous system disorders, microcephaly, mental retardation
Drugs	
Alcohol	Mental retardation, microcephaly, facial defects
Cocaine	Prematurity, retardation, addiction
Thalidomide	Phocomelia (i.e., small deformed limbs)
Dilantin, valproic acid	Heart malformations, cleft palate, retardation, microcephaly
DES (diethylstilbestrol)	Vaginal cancer in girls, genital deformities in boys
Anesthesia	Miscarriages, structural deformities
Barbiturates	Heart defects, microcephaly, retardation
Chemicals and heavy metals	
Methyl mercury, lead, cadmium	Miscarriages, mental retardation, neurological disorders
Dioxin	Physical deformities, miscarriage
Cigarette smoke – direct and "secondhand smoke"	Miscarriage, prematurity, low birth weight

Sources: Nadakavukaren A. Our global environment: a health perspective. 7th ed. Prospect Heights, IL: Waveland Press; 2011.
Chung W. Notes: teratogens and their effects. New York: Columbia University. Available at: http://www.columbia.edu/itc/hs/medical/humandev/2004/Chpt23-Teratogens.pdf [Accessed 18 August 2012].
University of New South Wales. Abnormal development teratogens [updated 5 November 2011]. Sydney: UNSW, Embryology. Available at: http://php.med.unsw.edu.au/embryology/index.php?title=Abnormal_Development_-_Teratogens [Accessed 18 August 2012].

Toxic Effects of Lead in the Environment

In the USA in the 1920s, the use of tetraethyl lead in fuel was promoted to improve automobile performance. This led to a long struggle between public health regulatory agencies and the automobile industry. Industry won, and leaded gasoline was used well into the 1960s and is still available in many parts of the world. Alice Hamilton investigated the widespread use of lead in industry during the 1920s and successfully lobbied for legislative changes to increase surveillance and improve safety by reduced exposure (Box 9.16). Community exposure to lead was identified as a public health problem in the 1960s when trace quantities were found in food, beverages, soil, and air. The main sources of community exposure were from leaded fuels for cars and lead-based paints manufactured from the 1920s to the 1960s.

Children are especially vulnerable to these environmental contaminants. Clinical effects appeared particularly in children and at lower blood concentration levels than previously thought to be significant. "Acceptable" levels were lowered and lead abatement programs introduced. These programs were especially needed in urban slum areas, where children were exposed to lead-based paints in older homes and heavy urban traffic. Consequently, many children were found to have high blood lead levels (BLLs), placing them at risk of brain damage.

Since 1991, the recommended standard of lead exposure necessitating follow-up action was $10\,\mu g/dl$ in blood samples. Between 1991 and 1994 in the USA, 4.4 percent of children 1–5 years of age had elevated BLLs ($>10\,\mu g/dl$). In 1992, the American Academy of Pediatrics adopted a lower BLL as a danger sign of lead toxicity sufficient to cause brain damage in children. Current professional opinion is that there is no safe level of blood lead and that levels under $5\,\mu g/dl$ are also harmful to the brains of young children (Box 9.17). In 2012, the CDC Advisory Committee on Childhood Lead Poisoning recommended adoption of a BLL of $5\,\mu g/dl$ as indicating an exposure risk. The new standard is considered to be a risk for impairment of cognitive ability and risks for cardiovascular, immunological, and endocrine disorders.

There may be no BLL which is harm free. Current recommendations include routine testing of infants and young

---

BOX 9.16 Alice Hamilton and Tetraethyl Lead

Alice Hamilton, a pioneering researcher and public health advocate in the 1910s and 1920s, demonstrated workplace hazards and toxic substances such as white phosphorus used in match production, lead additives to gasoline, and radium in watch dials. Tetraethyl lead (TEL) was produced and promoted by DuPont, despite being identified as hazardous.

Despite strenuous opposition from Hamilton and others, the use of TEL use expanded, and with it her research on behalf of state and federal government commissions. Environmental lead toxicity increased until the 1970s, when further research revealed the extent of the problem and its public health effects, especially on children. Hamilton's work set standards for toxicology research in occupational and environmental health that led to the regulatory successes of the 1970s in the USA.

Sources: Rossner D, Markowitz GA. "Gift of God"? The public health controversy over leaded gasoline in the 1920s. Am J Public Health 1985;75:344–52. Available at: http://www.ncbi.nlm.nih.gov/pmc/articles/PMC1646253/ [Accessed 17 August 2012].
Centers for Disease Control and Prevention. Achievements in public health, 1900–1999: Improvements in workplace safety – United States, 1900–1999. MMWR Morb Mortal Wkly Rep 1999;48:461–9. Available at: http://www.cdc.gov/mmwr/preview/mmwrhtml/mm4822a1.htm [Accessed 17 August 2012].

BOX 9.17 Lead Abatement in the USA, 1977–2009

Reduction of elevated blood lead levels (BLLs $>10\,\mu g/dl$) was one of the targets of Healthy People 2010. Studies in the USA based on the National Health and Nutrition Examination Survey (NHANES II) showed elevated BLLs in 88.2 percent of children aged 1–5 years in 1976–1980 (13.5 million children). The rate of elevated BLL fell to 8.6 percent in 1988–1991 (1.7 million children), 4.4 percent in 1991–1994 (850,000 children), and 2.2 percent in 1999–2000 (434,000 children). The lower level accepted as a marker by the Centers for Disease Control and Prevention (CDC) was reduced to $<5\,\mu g/l$ in May 2012.

The reduction in prevalence of elevated BLL was due to a number of factors, including:

- reduction in use of lead in gasoline since 1976
- reduced use of food and soft drink cans containing lead solder
- reduced use of lead-based house paint
- national standards for lead exposure in industry
- ban on use of lead soldering on household plumbing
- screening of children as part of routine child care and intervention where elevated BLLs are found
- lead abatement by county health departments through removal of lead-based paint in older housing
- increased provider and parental awareness of lead-induced permanent brain damage hazard
- strong positions of the CDC, American Academy of Pediatrics, state and county health departments, and child advocacy organizations
- increased public awareness
- regulation and elimination of lead paint on child toys
- reduction of house lead paint exposure by removal, painting over, and aluminum covers. Lead prevention has high priority in US worker and child preventive care but is not emphasized in Europe or international programs.

Sources: Centers for Disease Control and Prevention. Lead [updated 10 August 2012]. Atlanta, GA: CDC. Available at: http://www.cdc.gov/nceh/lead/ [Accessed 17 August 2012].
Agency for Toxic Substances and Disease Registry [updated 14 August 2012]. Atlanta, GA: ATSDR. Available at: http://www.atsdr.cdc.gov/ [Accessed 15 August 2012].

children as well as exposed workers, along with environmental measures to reduce emission levels and industrial or home use of lead or lead-containing products. This topic has received a lot of attention in the USA and Canada, but less in other countries. CDC lead control programs have targets of eliminating elevated BLLs in the USA. This program focuses on assisting states and municipalities in lead poisoning prevention programs. The WHO recommends preventive measures including:

- environmental standards that remove lead from gasoline, paint, and plumbing
- replacement of lead pipes; where they cannot be removed, cold water should be flushed through in the morning before drinking

- enforcement of occupational health standards
- surveillance of potentially exposed population groups, especially vulnerable ones (young children, pregnant women, workers)
- water treatment
- removing lead solder from food cans
- use of lead-free paint in homes
- screening of children for BLLs over acceptable limits and referral for medical care as necessary.

The health-based guideline for lead in drinking water is 0.1 mg/l (WHO, 1993). If high levels are detected in a supply, alternative supplies or bottled water may be necessary to protect young children.

Scientists continue to discover new chemicals with potential for harmful effects on human health. Many hazardous substances are found in common everyday items, as exemplified by bisphenol A and its possible association with heart disease (Box 9.18).

BOX 9.18 Bisphenol A and Heart Disease

Increasing evidence suggests that bisphenol A (BPA) may play a role in the development of heart disease and diabetes. Most commonly recognized as a component in food and beverage packaging such as plastic, BPA is problematic as it is classified as an endocrine-disrupting chemical. Results from the 2003–2004 and 2005–2006 NHANES, or the National Health and Nutrition Examination Survey conducted in the USA, indicate that elevated urinary BPA levels are correlated with heart disease.

In addition to food packaging, further sources of BPA include drinking water, dental sealants, and household dust. Polycarbonate, commonly used to strengthen plastics and reusable bottles, also contains BPA. When heat is applied to the plastic (such as containers or water bottles), the BPA typically is released. A UK researcher on this topic estimates that globally, 5 billion people are ingesting BPA.

A causal association exposure and health outcome of BPA and coronary artery disease cannot be established based on NHANES, which are cross-sectional in nature. A case–control study in the UK investigated baseline urinary BPA levels of participants diagnosed with cardiovascular disease, and compared them with the BPA levels of participants who did not have the outcome (i.e., did not develop heart disease). This study indicated an 11 percent increase in the likelihood of acquiring heart disease with each standard-deviation elevation in levels of urinary BPA.

The direct risk of BPA to humans is still unclear and controversial. However, increasing evidence suggests that environmental BPA exposure may be harmful to behavioral and other effects in children (Rochester 2013).

Sources: *Lang IA, Galloway TS, Scarlett A, Henley WE, Depledge M, Wallace RB, Melzer D. Association of urinary bisphenol A concentration with medical disorders and laboratory abnormalities in adults. JAMA 2008;300:1303–10. Available at: http://jama.jamanetwork.com/article.aspx?articleid=182571 [Accessed 17 August 2012].*
Melzer D, Rice NE, Lewis C, Henley WE, Galloway TS. Association of urinary bisphenol A concentration with heart disease evidence from NHANES 2003/06. PLoS ONE 2010;5:e867. Available at: http://www.plosone.org/article/info%3Adoi%2F10.1371%2Fjournal.pone.0008673 [Accessed 16 August 2012].

AGRICULTURAL AND ENVIRONMENTAL HAZARDS

Pesticide and herbicide use to increase agricultural production is a worldwide phenomenon. Resistance to widely used chemicals has developed, and consequently, there is a continuing search for new chemicals. Excess use affects the ecosystem by the buildup of pesticides in the food chain and in groundwater, and the long-term effects may be serious.

Short-term exposure to agricultural chemicals may result in acute poisoning, especially in developing countries, where it is estimated to affect some 3 million people with 220,000 deaths annually. Suspected concentration of pesticides in breast fat tissues may be linked with excess breast cancer risk. Pesticide use in North America and the former Soviet countries is high but has declined since the 1980s, while it is increasing in Western Europe. Widespread use of pesticides in developing countries is often poorly supervised, and pesticide poisoning episodes are common.

The use of herbicides and pesticides within the recommended limits of the *Codex Alimentarius* (joint foods standards manual of the Food and Agriculture Organization, or FAO, and the WHO) and methods recommended by the International Code of Conduct on the Distribution and Use of Pesticides are considered safe. Current recommended practice is to reduce the amounts of pesticide and herbicide use, accompanied by care and safe use and storage practices to reduce the chance of acute poisonings. Alternative agricultural methods, using few or no chemicals, are the subject of wide research and experimentation.

AIR POLLUTION

The External Environment

Air pollution is contamination of the air by smoke, solid material, or chemicals that cause health and ecological damage to the community and the environment. It includes the oxides of sulfur and nitrogen spread locally and over long distances, domestically and internationally. The effects are increasingly important as the demand for and use of fossil fuels have grown for the internal combustion engine, heating, and power generation. Coal fuel used in homes created the terrible air pollution of nineteenth- and early twentieth-century London. In his classic novel *The Jungle*, Upton Sinclair described air pollution in Chicago in the early part of the twentieth century (Box 9.19). Such levels of pollution have subsided since the 1950s with the reduction of brown soft coal use in individual homes in most industrialized countries, but remain common in many mid-level developing countries. Use of coal-fueled energy plants in Central and Eastern Europe has created a gray zone of air pollution carried over long distances, destroying forests, creating serious damage to the human environment, and presenting health hazards to large population groups. Similarly, extensive damage has occurred in Canadian forests from acid rain originating in the USA.

Large modern fossil fuel plants built near population centers may use high chimneys to disperse the effluent. This reduces exposure of the adjacent population but contributes to long-distance pollutant effects, carrying sulfur and nitrogen oxides to forests and bodies of water, and creating sulfuric and nitric acid or acid rain. Acid precipitation affects rivers, streams, and lakes, many already burdened with sewage effluent and pesticide runoff, damaging the ecosystem and animal and plant life. The effects on human health are not easily measurable in a directly attributable way, but environmental damage affects the quality of life. International transmission of environmental damage is seen in nearly a quarter of Europe's forests from acid rain originating in Eastern European countries with poor emission control standards.

The range of damage measured by the percentage of dead and dying trees varies from over 24 percent in Central and Western European countries (Denmark, Norway, the Netherlands, and Germany) to over 50 percent in some Eastern European countries (the Czech Republic and Poland). Environmental studies, primarily conducted in the early and mid-1970s, demonstrated that air pollutants are capable of traveling several thousands of kilometers. Thus, the harmful pollutants may accumulate and cause damage at a distance far from the original source of air pollution. International collaboration was imperative in developing solutions for large-scale environmental issues.

In the 1979 Convention on Long-Range Transboundary Pollution, European states agreed to reduce emissions that could cross international boundaries by 30 percent by 1993. This convention strengthened the development of international environmental law with a foundation to control and minimize the risk to human health as a result of transboundary air pollution. This represents the first legal tool concerning air pollution on an extensive regional level. Further, it provides a mechanism to limit and gradually reduce and prevent local and long-range transboundary air pollution. Cooperation includes developing policies and strategies to combat the discharge of air pollutants through exchanges of information, consultation, research, and monitoring. Studies are underway on the applicability of these standards to countries of Eastern Europe and Central Asia.

Air pollutants can enter the food chain by contaminating fish, fowl, and livestock. Changes in the acidity of water can create further harmful effects by corrosion of water pipes, affecting the lead, mercury, aluminum, cadmium, or copper content of drinking water. Acidified metals may cause chronic conditions such as chronic obstructive pulmonary disease (COPD) and asthma as well as specific chemical toxicity. Toxicity levels are difficult to measure epidemiologically; regulation of source emissions is set as a proxy measure for preventable exposure to unhealthy contaminants.

Particulate matter in air pollution has both physical and chemical effects on the nasopharynx and respiratory tract. Excess cancer of the respiratory tract and COPD can be demonstrated in exposed populations. A variety of syndromes is associated with specific respiratory irritants, such as coal dust (miner's lung) and cotton dust (byssinosis), among exposed occupational groups. A study of regional cancer rates in Israel in the 1980s showed an excess of cancers of the nasopharynx and respiratory tract in people living in an area exposed to high levels of silicate materials in emissions from a local cement plant. Geographic cancer epidemiology in the UK shows higher levels of many diseases in terms of standardized mortality rates (SMRs) in urban or other polluted areas in Britain, correlating excess morbidity with excess air pollution (see Chapter 3).

The London "killer fog" incident in 1952, implicated in up to 4000 deaths, raised international concern over the

---

BOX 9.19 Upton Sinclair – *The Jungle*

"A full hour before the party reached the city they had begun to note the perplexing change in the atmosphere. It grew darker all the time, and upon the earth the grass seemed to grow less green. Every minute as the train sped on, the colors of things became dingier; the fields were grown parched and yellow, the landscape hideous and bare. And along with the thickening smoke, they began to notice another circumstance, a strange, pungent odor …. It was now no longer something far off and faint, that you caught in whiffs; you could literally taste it as well as smell it."

Source: *Sinclair U. The jungle. New York: Airmont; 1905.*

deadly effects of critical levels of pollution as well as the long-term effects. In Britain, this led to controls on the use of soft coals for home fires and a gradual reduction in the Victorian levels of smog that had fouled ambient air quality in British industrial and commercial centers. A similar inversion in 1948 in Donora, Pennsylvania, affected over 40 percent of the population of 14,000, with 20 deaths. A smog crisis in New York City in 1966 occurred a month before the third National Conference on Air Pollution, followed by a series of smog crises in California.

Localized air pollution is largely generated by automobiles and general industry. Pollution of urban areas with lead, sulfur dioxide (SO_2), and nitric oxide (NO) has been reduced where catalytic converters and unleaded gasoline are compulsory. However, the beneficial effect is reduced simply by the increase in the number of automobiles, as shown in the southern California experience. During the 1960s and 1970s, there was a growing sense of crisis in environmental pollution in the USA. Until the 1970s, solid and liquid industrial wastes were dumped or discharged indiscriminately, with volatile chemicals contaminating water sources and the air. Pollution of lakes and rivers and poor air quality in the cities led to a series of federal legislative acts, including the Motor Vehicle Air Control Act of 1967, the Air Quality Act of 1967, the more effective Clean Air Act of 1970, the Clean Water Act of 1977, the Safe Drinking Water Act of 1974 (amended 1996), and the Water Quality Act of 1987, as well as the establishment of the EPA in 1970.

Traffic congestion in modern cities exposes car occupants and pedestrians to exhaust fumes containing particulate matter and air pollutants. Pollutants may act to compound the ill-effects of other risk factors such as smoking. Los Angeles is subject to heavy pollution and temperature inversions, producing harsh conditions for those prone to chronic bronchitis, asthma, and COPD. A study in Los Angeles showed that an increase of 10 parts per million of carbon monoxide (CO) levels in the air was associated with a 37 percent increase in hospital admissions. Action by state and local authorities to enact more stringent car and industrial emission standards, and replace high emission trucks and buses, has brought Los Angeles' air pollution levels down and continue to work to further the provisions of the US federal Clean Air Act. The problems of air pollution are widespread in urban areas of mid-level and low-income level countries; increasingly crowded and automobile-oriented cities such as Mexico City, Beijing, and Mumbai have poor standards of pollution control and serious air pollution levels.

Where the numbers of cars and trucks increase, but measures to control air quality standards are not implemented, pollution can have alarming effects on adult and child health. In children, asthma exacerbation and high BLLs can have serious detrimental effects on health. Among adults with a predisposition to cardiovascular disease and respiratory tract damage, chemical and particulate pollutants can cause increased mortality. Carbon monoxide blocks the uptake of oxygen by red blood cells and can reduce the oxygen carrying capacity of blood. In vulnerable groups, such as children, the elderly, pregnant women, and the immunosuppressed, this can have serious deleterious effects on psychomotor function. Polycyclic hydrocarbons released from car emissions and other sources are carcinogens. Nitrogen oxides (NO_x) affect the terminal respiratory tract alveoli, increasing susceptibility to lower respiratory tract infection in children. Ozone (O_3) and secondary pollutants affect UV light absorption, increasing skin cancer incidence. Ozone can travel hundreds of kilometers, causing clinical effects close to and well away from the site of the traffic. Carbon dioxide affects global warming with potentially important effects on world climate and water supplies. The health and environmental effects of air pollutants pose severe challenges to the global community. The Organisation for Economic Co-operation and Development (OECD, 2012) reports that by 2050, air pollution will become the leading environmental cause of early death worldwide.

Emission control through regulation and new technology should be seen in the context of overall transportation policy. Policy in transportation has long-term effects in determining degrees of air pollution, land use, and trauma from motor vehicle crashes. A full accounting of the costs of morbidity and mortality associated with air pollution and traffic accidents should be included in cost-effectiveness studies of rail versus road transport, especially in crowded urban communities and in countries with limited land space.

The US Federal Clean Air Act of 1970 established air quality standards for major pollutants such as NO_x, CO, SO_2, O_3, asbestos, dioxin, and other toxic air contaminants. Improving enforcement, especially of automobile emissions, has led to improved air quality in many parts of the country. Though federally legislated, implementation is at the state level. Standards are set for ambient air quality, automobile emissions, and emission by stationary facilities, such as power plants and factories. Such standards are also being implemented in many other countries.

The Clean Air Act Amendments of 1990 listed 189 hazardous air pollutants (HAPs) for which Congress mandated the EPA to issue standards. These include asbestos, dioxin, diesel, and many other potentially toxic agents, including latex, which has been identified as a factor in causing asthma. The US Food and Drug Administration (FDA) is continuing to develop standards for other HAPs. The Clean Air Act as amended provides for state agencies to regulate local air districts.

The California Air Quality Management Board regulates regional air quality management boards (e.g., southern California) that carry out a certification process of local industry. This board has powers to sanction changes in industrial practices in any given industry by attributing its component of ambient air pollution, with the potential for closing down an offending industry. As a result, California has been able to reduce air pollution dramatically since the mid-1980s, with only one major smog alert occurring in 1997 compared to 66 in 1987 in Los Angeles. The California Air Resources Board estimates that fine particle pollution causes 9000 excess premature deaths annually through its effects on pre-existing ischemic heart and chronic respiratory disease, and indicates the benefits of meeting national ambient air quality standards. Air pollutants are linked with excess hospitalizations of children for asthma-related conditions.

Beginning in the late 1990s, technological innovations have become standard in some new automobiles, thus further reducing emissions and increasing the mileage per gallon of gasoline. Other innovations incorporate hydrogen fuel cells and hybrid and electric vehicles that will release nearly zero pollutants at the point of use, and may contribute to a reduction in cardiovascular and repiratory disease deaths due to fine particle exposure.

The goal of the US Clean Air Act is to reduce the proportion of people exposed to air that does not meet the standards. Ozone affects 43 percent, particulate matter 12 percent, carbon monoxide 20 percent, nitrogen dioxide and sulfur dioxide 2 percent, and lead less than 1 percent of the population. It is estimated that improvements in healthy air have prevented 160,000 premature deaths. The estimated benefits of reduced air pollution are as many as 230,000 fewer deaths, 200,000 fewer heart attacks, and 2.4 million fewer asthma attacks in the next decade (Welker-Hood et al., 2011). Collaboration among multiple countries and responses to air pollution on an international level are exemplified in the Montreal Protocol (Box 9.20) and the United Nations Framework Convention on Climate Change (UNFCCC) (Box 9.21).

Diesel air pollutants became the subject of scrutiny by the state Air Resources Board, which carried out a meta-analysis and defined diesel pollution as a health hazard. This decision requires use of best available control technology (BACT) to reduce emissions from the defined "acceptable risk" of 10 excess cases of cancer per million population. In the case of diesel emissions the excess rate was determined to be 100 times in excess of the acceptable rate. Industry opponents raise the specter of the tremendous economic effects of such decisions, but the BACT approach minimizes this potential harm to the economy. At the same time, the process of identifying the issue spurs industry to seek out technological solutions that are compatible with greater efficiency in the long run. Box 9.22 discusses the development of US regulations designed to reduce fine particle emission air pollution.

BOX 9.20 International Treaties to Protect the Ozone Layer in the Atmosphere

The high-altitude or stratospheric ozone layer of the air acts as a shield in the atmosphere that protects life on Earth by blocking the sun's harmful ultraviolet (UV) radiation, which affects humans and ecosystems. During the 1980s, scientists observed that the stratospheric ozone layer was getting thinner over Antarctica (the "ozone hole"). A series of conventions and amendments produced a high degree of international acceptance and implementation of measures to reduce the use of chemical agents that adversely affected the ozone layer.

The Vienna Convention on the Protection of the Ozone Layer in 1985 was the precursor to the Montreal Protocol. The Montreal Protocol on Substances that Deplete the Ozone Layer was adopted in 1987. This treaty is the basis on which Title VI of the US Clean Air Act was established. The Montreal Protocol was updated by a series of conferences: London Amendment (1990), Copenhagen Amendment (1992), Montreal Amendment (1997), Beijing Amendment (1999), and Vienna (2000).

All UN recognized nations have ratified the treaty and continue to phase out the production of chemicals that deplete the ozone layer while searching for ozone-friendly alternatives. A broad coalition developed and implemented effective approaches to ensure stratospheric ozone layer protection. By agreements of an Executive Committee, the key agencies involved were the United Nations Environment Programme (UNEP), United Nations Development Programme (UNDP), United Nations Industrial Development Organization (UNIDO), and the World Bank.

Specific chemicals declared in the process include chlorofluorocarbons (CFCs), hydrochlorofluorocarbons (HCFCs), and hydrofluorocarbons (HFCs), but not all are banned or being reduced. Atmospheric concentrations of many of the important CFCs and related chlorinated hydrocarbons have either leveled off or decreased. Banning many chemicals and finding better alternatives and technologies has contributed to the process. The US Environmental Protection Agency has approved over 300 safe alternative chemicals to those which contribute to ozone depletion. The ozone layer has not grown thinner since 1998, but continued enforcement of the phasing out of ozone-depleting substances is vital to return the ozone layer to normal levels over the coming decades.

The Montreal Protocol process led to a broad coalition of governments, scientists, and others who work to develop effective approaches to protect human health and the global environment. Its adoption and implementation provide an approach to global cooperation in order to achieve global environmental protection. However, implementation by member nations is slow and variable.

Sources: *Sarma K, Bankobeza G. The Montreal Protocol on substances that deplete the ozone layer. Nairobi: United Nations Environment Programme; 2000. Available at: http://ozone.unep.org/pdfs/Montreal-Protocol2000.pdf [Accessed 17 August 2012].*

BOX 9.21 The Kyoto Accord and UNFCCC: The International Response to Climate Change

The United Nations Framework Convention on Climate Change (UNFCCC) was created in 1992, when countries collaborated in an effort to join an international treaty to define actions needed to control increases in average global temperature and the resulting climate change.

By 1995, it became apparent to many country leaders that emission reduction provisions in the Convention were insufficient. Negotiations to strengthen the global response to climate change led to the adoption of the Kyoto Protocol.

The Kyoto Protocol legally binds developed countries to emission reduction targets. The Protocol's first commitment period started in 2008 and ran until 2012. In Durban, government Parties to the Kyoto Protocol made a decision to implement a second commitment period, from 2013 onwards, extending the first commitment by either 5 or 8 years.

There are currently 195 parties participating in the Convention. The UNFCCC supports all bodies involved in the international climate change negotiations, particularly the Conference of the Parties (COP) and each of its supporting institutions.

A complex global environmental issue, climate change carries with it consequences relating to all spheres of existence on this planet. It impacts and influences many underpinning continuous global challenges such as poverty, economic development, population growth, sustainable development, and resource management. Therefore, solutions to help mitigate these large-scale challenges require assistance from all disciplines and fields of research, and above all political commitment.

The central issue in the response to climate change lies in the need to reduce greenhouse gas emissions. In 2010, multiple governments reached an agreement that emissions must be reduced in order for global temperature increases to be limited to below 2°C.

Rio+20, the United Nations Conference on Sustainable Development held in June 2012, planned to address climate change, oceanography, fisheries, resource management (water, land), food security, and further issues of this nature. However, the conference representatives came to an agreement to switch the key issue to economic development. Thus, a major emphasis was put on poverty eradication, an underlying and complex issue involving a myriad of global environmental challenges.

Sources: *Adapted from United Nations Meetings. Bonn: UN Framework Convention on Climate Change. Posted 2012. Available at: http://unfccc. int/2860.php [Accessed 15 August 2012].*
United Nations. The future we want. Rio+20 United Nations Conference on Sustainable Development; 2012. Available at: http://www.un.org/en/ sustainablefuture/ and http://www.uncsd2012.org/futurewewant.html [Accessed 11 August 2012].

Methyl Tertiary Butyl Ether

Methyl tertiary butyl ether (MTBE) is a synthesis of methanol and isobutylene, developed as an additive to gasoline to improve octane performance. It was widely adopted in the USA, especially in California, in place of ethanol used in other states. Ethanol is a farm product that was encouraged by the US Department of Agriculture as a new economic opportunity for farmers and a relatively non-toxic agent environmentally. Some gasoline producers opted to use MTBE instead, as it is produced and promoted by the chemical industry.

MTBE is a volatile, ether-based chemical agent that when present in drinking water imparts a bad taste. MTBE was adopted widely without adequate testing for potential toxic effects and has come under scrutiny. Evidence of carcinogenesis in rats raised concerns that MTBE may have the same effect in humans, especially drivers, gasoline station attendants, and refinery workers exposed to high levels. The effectiveness of MTBE in promoting clean burning of gasoline and reducing exhaust pollutants has also been questioned. MTBE has been found in 3.4 percent of water districts in California. Some 50 percent of drinking water wells in Santa Monica, California, were closed owing to MTBE contamination in 1995.

The MTBE case is an example of EPA-sanctioned use of a harmful chemical substance widely used in industrial settings as a gasoline additive, and therefore present in automobile emissions. It replaced a more environmentally safe substance from the farming industry even before environmental concerns brought the issue under public scrutiny. Since 1996, the American Public Health Association (APHA) has called on the FDA to ban MTBE as a hazardous chemical, to place restrictions on the use of gas-powered boats on lakes and rivers, to return to ethanol-based fuel additives, and to inspect underground storage facilities for gasoline to reduce leakage and contamination of groundwater. In 2000, the EPA included MTBE under the Toxic Substances Control Act and called for efforts to reduce MTBE contamination of ground and surface water supplies (EPA, 2007).

INDOOR POLLUTION

Effects of Indoor Air Pollution

Contaminants within private dwellings may be a greater health hazard than external pollution. Housing is understood as a crucial factor influencing population health, as people generally spend a large portion of time inside their dwelling. According to studies, people spend about two-thirds of their time inside their house; however, this figure differs by populations and subgroups. Women, as well as vulnerable groups such as the elderly, children, and those who are ill, are particularly affected by indoor air pollution. Others, who are unemployed or rarely participate in external affairs, are also more affected. According to WHO data, use of household solid fuel contributes to indoor pollution, causing serious illness resulting in the premature death of almost

BOX 9.22　Reduction in Fine Particle Emissions: Air Pollution Standards in the USA

Since the Clean Air Act was enacted more than in 1970, pollution in the USA has been reduced while the population and economy have grown. The health benefits far exceed the costs of reducing pollution. Less pollution lower the risks of premature death and other serious health effects. Environmental damage from air pollution is reduced. New cars, trucks and non-road engines use state-of-the-art emission control technologies. New plants and factories install modern pollution control technology. Power plants have cut emissions that cause acid rain and harm public health. Interstate air pollution has been reduced. Mobile and industrial pollution sources release much less toxic pollution to the air than in 1990. Actions to protect the ozone layer are saving millions of people from skin cancers and cataracts. National parks are clearer due to reductions in pollution-caused haze. EPA has taken initial steps to limit emissions that cause climate change and ocean acidification. The Act prompted deployment of clean technologies, with impetus for technology innovations that reduce emissions and control costs (EPA, 2012, 2013).

These regulations require states to significantly improve air quality by reducing power plant emissions that are partially responsible for ozone and/or fine particle pollution in other states. CSAPR identifies 28 states that are required to reduce annual SO_2 emissions, annual nitrogen oxide (NO_x) emissions and/or ozone-season NO_x emissions. Together, these substantial reductions play a meaningful role in attaining the 1997 ozone and fine particle and 2006 fine particle NAAQS, thus replacing EPA's 2005 Clean Air Interstate Rule (CAIR). The new regulations will improve air quality throughout the eastern half of the USA, helping states to achieve national clean air standards.

Similar to all health regulations, the purpose of CSAPR and its implementation is to improve a population's health outcomes. Expected health benefits directly resulting from CSAPR include reduced premature mortality by approximately 13,000–34,000 cases. Moreover, experts predict that the following number of cases will be averted due to CSAPR: 15,000

non-fatal heart attacks, 19,000 emergency room visits, 420,000 cases of upper and lower respiratory illnesses, and 400,000 cases of asthma attacks. The benefits include avoiding loss of 1.8 million days of work or school. This will result in achieving hundreds of billions of dollars in public health benefits, and will yield US\$120–280 billion in annual health and environmental benefits in 2014. Scientists predict that the new air pollution ruling will improve visibility in national and state parks, as well as tighten protection for sensitive ecosystems, including Adirondack lakes and Appalachian streams, coastal waters and estuaries, and forests.

The US\$800 million annual projected costs of this rule in 2014, as well as the approximately US\$1.6 billion per year in capital investments already advancing due to CAIR, are making progress to improve air quality for over 240 million Americans. Moreover, the emission reductions anticipated to result from EPA's newly finalized Mercury and Air Toxics Standards (MATS) are not included in the estimated emission reductions from the CSPAR; once those standards are implemented, SO_2 emissions from the power sector are likely to be reduced even further. More recently, since April 2012, the EPA has issued new standards under the Clean Air Act. These regulations pertain to hydraulic fracturing oil and gas systems, thus affecting the oil and gas industry. Anticipated outcomes of the standards include not only a reduction in groundwater pollution, but also a reduction in hazardous methane, benzene, and other volatile organic compound emissions by 95 percent.

Sources: *Environmental Protection Agency. Air pollution and the Clean Air Act. Last updated 16 August 2013. Available at: http://www.epa.gov/air/caa/ [Accessed 22.12.2013].*
Environmental Protection Agency. Progress Cleaning the Air and Improving People's Health http://www.epa.gov/air/caa/progress.html [Accessed 22.12.2013].
Environmental Protection Agency. Oil and natural gas air pollution standards [updated 18 April 2012]. Washington, DC: EPA. Available at: http://www.epa.gov/airquality/oilandgas/index.html [Accessed 17 August 2012].

2 billion people. Likewise, COPD, resulting from exposure to indoor air pollution, is responsible for killing over 1 million people annually.

Increased insulation, window layers, sealed doors, and smoking all contribute to increased concentrations of indoor pollutants, including benzene, formaldehyde, carbon monoxide, and radon gas, as well as bacteria, fungi, and viruses. Smoking is a widespread habit, and passive smoking, or inhalation of smoke generated by other people, is a long-term health hazard.

Wood and its waste products, vegetable matter, and animal dung are sometimes referred to as *bamboo fuels*. These are less efficient than fossil fuels in terms of heat produced per unit mass. Approximately half of the world's population depends on such fuels for their daily needs. These fuels are used extensively in rural areas of developing countries because they are cheap and widely available, but they require much time to gather. Moreover, they lead

to deforestation with other damage to the environment. Primitive stoves are often used, creating fire hazards and high levels of continuous daily indoor pollution due to poor ventilation. The dangers associated with use of bamboo fuels include fires, smoke inhalation, and chronic indoor pollution. These fuels release many chemical compounds including suspended particulate matter, carbon monoxide, nitrogen and sulfur oxides, aldehydes, hydrocarbons, benzene, phenols, and complex hydrocarbons.

Women in India show high rates of right heart failure (cor pulmonale) from the fumes released from cooking stoves. Technological development of more efficient wood stoves would reduce the problem; however, other forms of energy are more efficient and less damaging to health in the home and to the environment. Currently, approximately 3 billion people are still using solid fuels in open fires or leaky stoves to cook or warm their houses. These methods generate pollutants and soot pieces that can enter deep into a person's lungs. Serious health consequences of these outdated methods and the impure air they produce include pneumonia, COPD, lung cancer, and lung and airway inflammation. Further research suggests that indoor air pollution may be associated with low birth weight, the development of tuberculosis, and ischemic heart disease. As one would expect, indoor air pollution primarily affects poor populations and those living in developing countries. The following statement demonstrates the severity of the problem: in the homes of some families living in developing countries, women who cook for three hours per day can be exposed to equivalent levels of benzo(a)pyrene as one would obtain from smoking two packs of cigarettes daily.

The UN Foundation, in collaboration with the WHO and other agencies, has established the Global Alliance for Clean Cookstoves. The Alliance is working to encourage better quality biomass cookstoves to significantly diminish indoor air pollution. To advance in this program, WHO established a household energy database, used to evaluate the world's progress in the conversion to cleaner fuels and safer stoves. This database will allow for monitoring and evaluation of disease burden and important health outcomes. Progress made in this area of health will influence the success of multiple MDGs. It will play a role in reducing child mortality (MDG4), and improving maternal health (MDG5) and gender equality (MDG3); furthermore, improved sources of household energy will influence environmental sustainability (MDG7).

Research conducted by the WHO on housing conditions in Europe validates the notion that insufficient housing standards are correlated with increased risk of respiratory diseases, such as asthma, lung infections, and allergies. These health outcomes are likely to be due to dampness within a house, which offers a welcoming environment for agents that may trigger respiratory diseases, such as roaches, mites, viruses, and molds. Indoor pollution from materials used in construction is a serious health problem. Asbestos in the home may contribute to mesothelioma and lung cancer. Lead paint in the home increases the hazard of lead toxicity among young children, which is associated with brain damage. Unsafely packaged household chemical solvents and mold in the home contribute to poisonings as well as asthma morbidity and mortality. Overcrowding and insufficient hygiene are further hazards within the home. Studies demonstrate an association between low indoor temperature and suboptimal health status, specifically in relation to cardiovascular disease. Past interventions focused on modifying housing through thermal improvement have proven to substantially improve mental health. Furthermore, the temperatures in households of low-income populations tend to be more extreme and unbearable owing to inadequate insulation or no air conditioning. Consequently, health professionals have recognized a link between substandard living conditions (specifically defined by thermal inefficiency) and elevated winter or summer mortality rates.

Indoor air pollution is a very serious, widespread public health problem, affecting some populations more than others. Globally, it is the most impoverished people, those susceptible to countless other health issues who suffer the most. Interventions, policy changes, and global support should be a priority in establishing and achieving the goal of reducing exposure to indoor air pollution, while finding alternative, cleaner, safer, methods to meet a population's cultural needs and energy requirements.

Radon Gas

Radon is a very heavy gas that produces harmful alpha particles as a byproduct. Radon originates in the natural radioactive decay of uranium from soil and rocks such as granite, shale, and phosphate, and is present as a gas in ground crevices, dissolved water, or dispersed open air. It seeps into homes via basement cracks and into well water and point sources. Radon was first detected in homes in the USA in 1984 near Philadelphia. Early investigations showed in-home radiation exposure as high as the equivalent of 455,000 chest X-rays. Further investigation revealed that sections of eastern Pennsylvania, New Jersey, and New York lie over uranium-rich geological formations that result in high levels of radon contamination.

The US EPA in 1988 advised that all homes be checked for radon levels. Inexpensive home radon detectors are available that meet EPA standards. In 1988, the EPA estimated that radon contributes to between 7000 and 30,000 cases of lung cancer per year, or up to 10 percent of all lung cancer deaths in the USA. In 2009, the WHO estimated that 5–15 percent of all lung cancers are caused by radon, which is the primary cause of lung cancer among non-smokers. Radon-induced cancers are caused by low and moderate levels of radon as in the home environment. Hundreds of thousands of US citizens receive as much radiation as did people living near the Chernobyl plant at the time of the nuclear accident in 1986. Cigarette smoking has a synergistic effect, enhancing the radon-related risk of lung cancer by a factor of 10. Radon reduction can be carried out in high-risk homes by carefully planned sealing of identified sources, ventilation, and fans for high radon basements.

Outdoor–Indoor Pollutants

Carbon monoxide, nitrogen oxides, chemicals, and particulate matter are common outdoor pollutants that can accumulate in homes with kerosene and wood stoves, attached automobile garages, or cigarette use. Passive smoking can expose the non-smoker to benzene and other carcinogens. Formaldehyde is produced from insulation material, plywood, and floor coverings, especially in mobile homes. Chemical fumes from household products, such as disinfectants, solvents, hair sprays, furniture polish, and dry cleaning solvent, also pollute the home atmosphere and can potentially cause childhood poisonings.

Carbon monoxide poisoning from home heaters where there is inadequate ventilation causes 100 deaths per year in the UK. Some deaths from carbon monoxide poisoning may be attributed to heart disease and can only be diagnosed affirmatively by measurement of carbon monoxide in the air or blood carboxyhemoglobin levels.

Biological Pollutants

Bacteria and fungal spores can enter a building and infect its inhabitants, usually through the air conditioning or ventilation system, as is the case with Legionnaires' disease (see Chapter 4). Occupants of a building may suffer from allergies due to fungal spores, mites, animal dander, and feces of roaches or mites. These allergies are more likely to occur in buildings using humidifiers or vaporizers with stagnant water, which favor bacterial and fungal growth. Sick building syndrome is discussed in Box 9.23.

Built Environment and Health

A Healthy City, as defined in the Zagreb Declaration (WHO, 2009), is a city for all its citizens: inclusive, supportive, sensitive and responsive to their diverse needs and expectations. It provides conditions and opportunities that encourage, enable and support healthy lifestyles for people of all social groups and ages. It offers a physical and built environment that encourages, enables and supports health, recreation and well-being, safety, social interaction, accessibility and mobility, and a sense of pride and cultural identity, and is responsive to the needs of all its citizens.

BOX 9.23 Sick Building Syndrome

Recurrent respiratory infections, wheezing, fatigue, dizziness, headache, eye and nose irritation: when these symptoms affect people working inside a specific building, the term used to describe this condition is *sick building syndrome*.

Although studied for decades, it is still a poorly understood condition, as scientists have been unable to pinpoint one single cause. The consensus is that it is a combination of various risk factors that lead to people experiencing symptoms of sick building syndrome. Potential factors include poor ventilation, low humidity, dramatic changes in temperature throughout the day, airborne pollutants (dust, fungal spores), and chemical pollutants, such as cleaning materials. Poor levels of cleanliness in the work environment may play a role in employees developing symptoms of sick building syndrome. Poor ventilation systems may fail to provide adequate fresh air relief from microbiological pollution, formaldehyde in furniture, ozone emissions from photocopying machines, and cigarette smoke.

The common symptoms listed above typically improve or disappear after an affected person has left the building, that is, once the person is no longer surrounded by the exposure. While sick building syndrome is an increasing occupational problem, there are preventive and control measures that can and should be implemented. Some of the many strategies include elimination or adaptation of the pollutant sources, maintaining adequate waterproofing, removing water-stained ceiling and flooring, and keeping products such as paints, solvents, and adhesives tightly enclosed and in highly ventilated areas. Measuring air distribution and ventilation levels is also critical. Building codes should specify minimum levels of outside air admission; acceptable levels of oxygen, carbon monoxide, and carbon dioxide; odor dilution; and adequacy of ventilation equipment. Further crucial factors include education and communication to advance the process of air quality programs, as well as prohibition of smoking in the workplace.

Although the condition is not well understood and an affected individual can experience a wide variety of symptoms, a set of elements is associated with a higher prevalence of sick building syndrome. Host factors include being female and working in a job that is considered lower in building hierarchy (a more menial job), which increases risk of being affected by sick building syndrome. Exposure to paper dust, office dust, cigarette smoke, and increased use of computers are common factors. The likelihood of developing this syndrome increases with high indoor temperature ($>23°C$ in air-conditioned buildings), low fresh air ventilation, poor individual control of lighting and temperature, excessive use of air conditioning, water damage, inadequate building service maintenance, and insufficient levels of cleanliness.

Depending on the severity of a worker's symptoms, employees suffering from sick building syndrome are likely to have economic consequences. Maintaining a healthy, safe, comfortable environment comes at a cost; however, so does low worker productivity and satisfaction. A study conducted in Switzerland showed that, compared with workers in air-conditioned offices, those who work in naturally ventilated buildings typically take less sickness absence. Moreover, a study carried out in the Netherlands examined the number of days of missed work due to sickness and whether a worker was given the opportunity to control his or her own office environment. The researchers found that among office workers who were given the freedom to control their own environment, 34 percent less sick days were taken.

It is crucial that employers respond efficiently and effectively to their workers' complaints, as each individual deserves to work in a safe, healthy, clean environment. A substantial portion of many people's lives is spent in an office building; therefore, an individual's level of health is very much influenced by his or her working conditions.

Sources: National Health Service. Sick building syndrome [updated 29 October 2010]. UK: NHS. Available at: http://www.nhs.uk/conditions/Sick-building-syndrome/Pages/Introduction.aspx [Accessed 17 August 2012].
Joshi SM. The sick building syndrome. Indian J Occup Environ Med 2008;12:61–4. Available at: http://www.ijoem.com/article.asp?issn=0019-5278;year=2008;volume=12;issue=2;spage=61;epage=64;aulast=Joshi [Accessed 17 August 2012].

The public health aspects of the built environment include the physical parts of places where people live and work. These include homes, buildings, streets, open spaces, and infrastructure to provide recreation, commerce, facilities for physical activities, such as jogging, bicycle paths, workout equipment to promote physical activity and leisure, and access to well-stocked grocery and fruit and vegetable supplies. These amenities are meant to promote physical activity, reduce sedentary habits, and promote healthful eating and recreation. Roads, industry, and commerce zoning regulation are part of Healthy Cities, and public health can be improved by far-sighted local authority planning and implementation. Alleviation of industrial and air pollution from power plants and roads is vital to improving habitation in urban settings. In developing countries, the environment of urban slums, without adequate safe water supply or sewage and garbage disposal, promotes illness and inequality that healthful urban environments can do much to alleviate.

In recent years this healthy urban environment approach has expanded to include open space and rooftop gardening for urban dwellers to grow fresh vegetables, and for recreational and economic benefit. This is now common practice in cities such as New York City, London, and Hong Kong. It may be especially important for the urban poor with limited access to stores selling fresh fruit and vegetables, making healthy eating difficult; and private gardening can make a huge difference to a struggling family.

HAZARDOUS OR TOXIC WASTES

Toxic materials used in industrial processes can cause ill-effects in workers exposed to the material at the site of production and in storage, transport, and use of the materials. They can also cause harmful effects to people living near the material, as well to the environment. Case studies of serious environmental pollutants and their effects on health demonstrate the problems involved.

Hazardous wastes are defined as any discarded material that may pose a substantial threat to human health or the environment when improperly handled. They include toxic wastes such as arsenic, heavy metals, and pesticides which can cause acute or long-term health problems. Ignitable wastes include organic solvents, oils, plasticizers, paint waste, and corrosive wastes (with a pH of <2 or >12.5) which can eat away metal containers or living tissue. Reactive wastes include obsolete munitions and acids that react with water or air to produce explosions or toxic fumes. Radioactive and infectious wastes from hospitals are also hazardous to public health. Hospital wastes took on new importance with the dangers of transmission of hepatitis B, HIV, and drug-resistant microorganisms in contaminated materials. The problem caught worldwide attention in the late 1980s when waste material from hospitals washed on

to beaches in the USA. Box 9.24 presents information on hospital and health care facility waste.

Prevention and waste site remedies have gained wide attention by industry as well as federal, state, and local government. Media and public concern was generated by episodes such as the Love Canal in the late 1970s, which served to mobilize public awareness of environmental health in the USA. In the 1890s, Mr. William T. Love built a canal bypassing Niagara Falls with the intent of building an industrial city using inexpensive hydroelectric power. The project failed and the canal was abandoned and the land sold at public auction. In 1942, the Hooker Chemical Company (subsidiary of the Occidental Petroleum Co.) received permission to use the canal to dump chemicals from its several plants in the area. Up to 1953, when it was covered by landfill, 21,000 tons of chemical wastes (acids, alkalis, solvents, chlorinated hydrocarbons, etc.) were disposed of at the site. Despite warnings, the land was sold and over 1000 homes, apartments, and schools were constructed along the covered canal.

Beginning in the 1950s, local residents complained of foul odors and chemicals oozing from the covered canal. In 1978, pressure from local congressmen and news media prompted investigation by the EPA and the New York State Department of Health. Over 200 different chemicals were identified, including dioxin and 12 known or suspected carcinogens, mutagens, and teratogens. The New York State Commissioner of Health proclaimed an imminent health peril and called for evacuation of pregnant women and children under the age of 2. Over 1000 families were evacuated and 300 homes demolished at public expense. Work at the site to contain the chemicals and prevent seepage and groundwater contamination cost over US$180 million. An initial investment of US$2 million by the Hooker Chemical Company at the time of the disposal could have prevented the damage to health and associated costs.

Epidemiological studies of the exposed residents showed that they experienced statistically significant elevated rates of miscarriage, birth defects, and chromosomal abnormalities, but the studies' methods and conclusions remain controversial. This episode focused national concern on the approximately 16,000 hazardous waste sites throughout the USA. In 1980, Congress established a superfund program, funded by federal tax on the chemical and petroleum industries, to locate, investigate, and clean up the worst sites in the country.

Minimata Disease

Minimata disease is a chronic neurological disorder caused by methyl mercury, a heavy metal with many industrial uses. The disease was first reported near Minimata Bay in Japan in 1968, when mercury oxide was being discharged from a chemical plant into the waters

BOX 9.24 Hospital and Health Care Waste

Medical facilities and health providers use countless tools, drugs, vaccines, dressings, needles, and syringes in patient care and performing procedures, generating massive amounts of waste. Waste and its by-products can be classified into eight different categories: infectious waste, pathological waste, sharps, chemicals, pharmaceuticals, genotoxic waste, radioactive waste, and heavy metals. The World Health Organization (WHO) estimates that 80 percent of the total waste produced from health care functions is general waste, similar to domestic waste, while 20 percent is classified as hazardous material, which may be infectious or toxic, or possess radioactive properties.

Infectious waste includes material contaminated with blood, cultures of infectious agents, and contaminated products such as bandages and swabs. Pathological waste consists of identifiable body parts, while sharps are syringes, needles, and disposable tools, such as scalpels and blades. Common chemical forms of waste include mercury and various disinfectants, and pharmaceutical waste consists of expired or contaminated drugs, as well as vaccines and sera. Certain forms of waste can be especially dangerous and require very careful handling, in particular, genotoxic waste, as well as mutagenic, teratogenic, and carcinogenic materials, such as cytotoxic drugs used for cancer treatment, and radioactive diagnostic material.

High-income countries produce on average 0.5 kg of hazardous waste per hospital bed each day, while low-income countries typically produce 0.2 kg. In low-income countries it is often the case that no distinction is made between hazardous and non-hazardous waste generated from health care; thus, the true amount of hazardous waste is likely to be significantly greater than the estimated figure. Globally, approximately 16,000 million injections are delivered annually. Not all syringes are disposed of appropriately, thus the risk of infection and possibility of reuse emerges. The WHO in 2000 estimated that 21 million cases of hepatitis B virus, 2 million cases of hepatitis C, and 260,000 HIV infections resulted worldwide from injections using contaminated syringes. Reuse of disposable syringes is a dangerous and common practice, especially in some African, Asian, and Central and Eastern European countries.

In the USA, most infectious medical waste (over 90 percent) is disposed of through incineration. In 1997, the Environmental Protection Agency (EPA) announced regulations to control levels of emissions released from medical waste incinerators. Although incineration is the primary form of disposing of medical and health care waste, when incinerators are unavailable the waste can be taken to solid waste landfill depots, where safe packaging and proper precautions are taken to avoid contact with the public. Other alternatives to incineration include thermal treatment, steam sterilization, and chemical–mechanical systems.

Substandard management of medical waste can result in health workers, waste handlers, and patients being exposed to the risks of infection, toxic materials, and injuries. Further hazards include environmental damage, which affects the entire community. As WHO has declared, it is imperative that products of health care waste, including unused medications, are segregated immediately after use, adequately treated, and cautiously disposed of. Home health care waste should also be attended with safety in mind for caregivers, patients, families, neighbors, and the community.

Sources: Environmental Protection Agency. Medical waste frequent questions [updated 24 July 2012]. Washington, DC: EPA. Available at: http://www.epa.gov/osw/nonhaz/industrial/medical/mwfaqs.htm [Accessed 17 August 2012].
Environmental Protection Authority. Guidelines: medical waste – storage, transport and disposal. South Australia: Environmental Protection Authority; 2003. Available at: http://www.epa.sa.gov.au/xstd_files/Waste/Guideline/guide_medical.pdf [Accessed 15 August 2012].
World Health Organization. Waste from health care activities [updated November 2011]. Geneva: WHO. Available at: http://www.who.int/mediacentre/factsheets/fs253/en/index.html [Accessed 17 August 2012].
World Health Organization. Medical waste [updated 2012]. Geneva: WHO. Available at: http://www.who.int/topics/medical_waste/en/ [Accessed 17 August 2012].

---

of the bay. It was converted to an organic form, methyl mercury, by organisms in the mud and slime of the bay floor. Mercury poisoning of fish is a recurrent phenomenon where industrial wastes discharged into rivers, lakes, and the sea enter the food chain, and humans are affected through fish consumption. As of March 2001, 2265 victims had been officially recognized (1784 of whom had died) and over 10,000 had received financial compensation. Minimata disease is one of four major pollution diseases of Japan caused by environmental pollution due to improper handling of industrial wastes by Japanese corporations. Compensation, cleanup, and damages cost hundreds of millions of dollars. This episode also served to mobilize international public opinion to the dangers of toxic waste disposal as a health hazard. In 1999, people living in remote areas of Brazil were found to have methyl mercury poisoning, probably from fish contaminated by methyl mercury used to purify gold.

Toxic Waste Management

Pollution prevention in the workplace has become part of management processes as industry responds to increasing federal and state regulation and as the public demand for greater corporate responsibility leads to increasing punitive litigation. In 1986, the Federal Office of Technology Assessment published a comprehensive work on the topic entitled *Serious Reduction of Hazardous Waste*. A 1992 publication by the OECD called on workers to play a greater role in pollution prevention. The chemical industry responded with the idea of total quality environmental management (TQEM), adopting pollution prevention as integral to industrial management. Companies such as 3M, Monsanto, and Rhone-Poulenc, and industrial associations (the Chemical Manufacturers' Association) undertook environmental prevention policies. Community activism helped industry to respond positively and openly to environmental hazards in their communities. The search for safe alternatives to toxic chemical waste management can be costly, but inevitably saves a company large expenditures in fines, litigation, and damage to corporate image. The issue is now very broadly shared among government, private industry, workers, and community, involving planners, scientists, engineers, regulators, residents, as well as environmental organizations and consumers.

The EPA is currently promoting waste minimization of persistent, bioaccumulative, and toxic (PBT) chemicals from industrial sources. This includes source reduction by development of new materials for packaging and biodegradable products, recovery of waste electronics, and many other social and public activities for environmental improvement, particularly recycling aimed at reducing hazardous waste products (EPA, 2008). The OECD indicates that the quantity of municipal waste generated in the OECD area exceeded 658 million tonnes in 2010, or 540 kg per inhabitant, a reduction from 2008 despite population growth, possibly related to improvements in waste management and recycling in member countries and to the economic downturn.

In Côte d'Ivoire in August 2006, over 500 t of chemical waste was unloaded from a cargo ship and illegally dumped by trucks at several sites. One month later nearly 85,000 consultations were recorded at various medical facilities, with 69 hospital admissions and eight deaths. A mixture of sodium hydroxide, phenols, mercaptanes, hydrogen sulfide, hydrocarbons, and other chemicals used to clean oil transporters' tanks was found in the waste. Serious consequences ensued, as people inundated medical facilities, and many were concerned about damage to the food supply chain and pollution of rivers and lakes. Concern grew since the ship had sailed from northern Europe with calls at a number of other ports, thus making the tracing of sources very difficult.

RADIATION

Radiation occurs when electromagnetic energy travels in waves or subatomic particles through space with a spectrum of varying degrees of energy. The two primary classifications are non-ionizing and ionizing radiation. Non-ionizing radiation indicates radiation in which there is sufficient amount of energy to allow atoms in a molecule to vibrate, but there is not enough to remove electrons. The properties of non-ionizing radiation are used for everyday applications, such as microwaves, radio broadcasting, and infrared heat lamps. Ionizing radiation represents higher

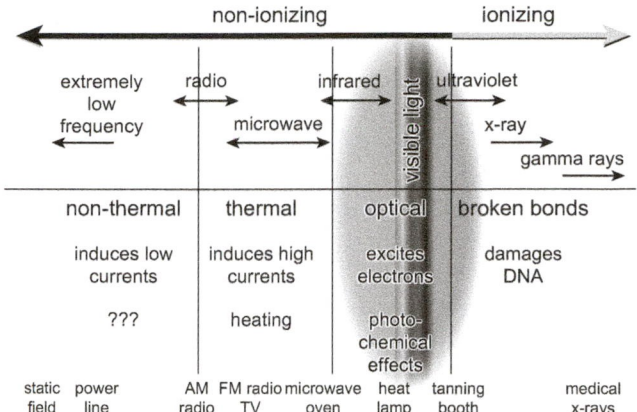

FIGURE 9.5 Spectrum of radiation levels. *Source: Environmental Protection Agency. Radiation: non-ionizing and ionizing [updated 7 August 2012]. Washington, DC: EPA. Available at: http://www.epa.gov/rpdweb00/understand/index.html#nonionizing [Accessed 17 August 2012].*

levels of energy. Consequently, a charged part of the molecule has the ability to break free from the atom, thus creating particles with a charge, or ions. Ionizing radiation includes particulate radiation of alpha and beta particles, as well as electromagnetic X-rays and gamma rays. Alpha particles are easily stopped by a thin sheet of paper, while beta and gamma radiation can penetrate barriers both inside and outside the body. Ionizing radiation can dislodge atoms or parts of atoms and destroy chemical bonds. This can adversely affect living organisms, especially vulnerable fetal cells, resulting in mutations or carcinogenesis.

Figure 9.5 illustrates radiation energy levels on a spectrum. The degree of energy increases from left to right, non-ionizing radiation to ionizing, as frequency also increases.

Ionizing Radiation

The US National Cancer Institute defines ionizing radiation as follows: "Ionizing radiation is a type of radiation made or given off by x-ray procedures, radioactive substances, rays that enter the Earth's atmosphere from outer space, and other sources. At high doses, ionizing radiation increases chemical activity inside cells and can lead to health risks, including cancer."

Ionizing radiation is characterized by having enough energy to be able pull away tightly attached electrons from their atoms. It includes high-energy electromagnetic radiation, such as X-rays and gamma rays, which are of shorter wavelength and higher energy than UV or visible radiation. It also includes high-energy particles such as electrons, neutrons, protons, and alpha particles. Excessive exposure to these forms of radiation has early and late effects depending on dose and the tissue exposed. Early effects of exposure to high doses of radiation may be fatal owing to acute damage to the gastrointestinal, erythropoietic (blood-forming), and central nervous systems. Late effects include malignant disease such as leukemia and birth defects.

Background radiation is composed of both natural and artificial sources of radiation. Approximately 81 percent comes from nature, while the remaining 19 percent can be attributed to non-natural sources. The principal sources of radiation exposure for the general public can be broken down as follows: radon (55 percent), external (15 percent), internal (11 percent), medical (15 percent), consumer (3 percent), and other (1 percent). Exposure to ionizing radiation from artificial sources is largely derived from medical procedures. Moreover, natural background radiation is composed of cosmic radiation, terrestrial radiation, and internal radiation. Cosmic radiation is attributed to the sun and stars, which transmit a continuous flow of radiation to Earth. The terrestrial radiation that affects humans comes from the Earth itself, as radioactive substances are naturally present in soil, rock, air, and water. Lastly, humans are exposed to internal radiation, which originates inside the body and is present from birth; people are, in fact, a source of exposure to those around them.

The primary forms of ionizing radiation consist of alpha particles, beta particles, gamma rays, and X-rays. Fairly heavy and high energy, alpha particles have a positive charge. The process of alpha emission can be hazardous to health. This is characterized by a change occurring in the nucleus and a particle being released, resulting in a decay product. The majority of alpha emission takes place naturally in the environment. When soil, minerals, or rock formations are disrupted (such as during mineral extraction), this creates a risk of environmental and human exposure. For instance, high levels of uranium and radium are found in uranium mining wastes. When they rise to the surface, they may contaminate surface water or become airborne. The severity of health effects is highly dependent on the route of exposure. Alpha particles do not have the energy to penetrate the outer skin layer, so exposure external to the body is significantly less harmful than internal exposure. However, if alpha emitters are inhaled or swallowed, or enter the bloodstream, this is a cause for concern. The radiation affects living tissue and may increase the risk of developing cancer, more specifically lung cancer if alpha emitters have been inhaled.

Beta particles have a negative charge, and while their emission can result in serious health problems, their properties are used for various important medical uses. They are important in diagnosis, imaging, and treatment, such as treatment of thyroid disorders and drug metabolism research. Both human-made and naturally occurring beta emitters exist, with some present in our bodies. Health may be compromised when energetic particles are released. Sources of beta particle exposure vary, and may result from a nuclear reactor accident or as a consequence of a

patient taking radioactive iodine. The patient will release beta particles and therefore must adhere to a rigid protocol to prevent exposure to family members. Similar to alpha particles, the route of exposure is important in determining the severity of the health outcome. Inhalation and ingestion are very serious, as particles in contact with living tissue can harm the molecules and disturb cell function. Beta particles are significantly smaller than alpha particles. As a result, they can travel deeper into tissues, causing more widespread cellular damage. In addition, if released from a strong enough source, when beta particles come into contact with skin, they can cause irritation, reddening, or burning. Although radiation from beta particles can result in both acute and chronic health problems, chronic effects are significantly more common. The major health outcome from this radiation is cancer, which typically develops from a relatively low degree of exposure over a long period. It follows a dose–response relationship, in which a higher dose is associated with a higher risk of cancer. While some beta emitters disperse throughout the body, others concentrate in particular organs.

The third class of ionizing radiation is gamma rays. This form of radiation is extremely high energy. Gamma photons are characterized as having no mass and no charge, and being able to travel at the speed of light. They have the power to penetrate many types of objects and materials, including human tissue. As a result, dense, heavy materials are often used as shield to retard or halt gamma photons. Gamma rays and X-rays present similar health risks, their difference lying in the portion of the atom from which they originate. Radionuclides that emit gamma photons are the most commonly used sources of radiation. Some of their uses include treatment of cancer, pasteurization of foods, sterilization of medical tools, and measuring soil density in construction areas. Radioactive isotopes are used in diagnostic examinations, for instance, to obtain images of bone, liver, or brain. For the general population, the majority of gamma exposure comes from natural radionuclides, typically present in soil and water. Further sources include meats and foods that contain high levels of potassium, such as bananas. The majority of gamma and X-ray exposure is external, and the danger lies in the rays' ability to travel far through air and deep into human tissue, putting all organs as risk. For this reason, when X-rays are needed for dental or medical purposes, they are typically carried out in a controlled, well-planned environment. Although less common, radionuclides can release gamma rays that may be ingested with water or food, or inhaled, resulting in internal risks. Determined by the source, the radionuclide can follow different pathways, as it is either absorbed in body tissue or excreted through urine or feces.

Short exposures at high dosage are far more serious than long-term, low-dose exposure. Radiation sickness in people exposed to radiation from the atomic bomb explosions at Hiroshima and Nagasaki and nuclear accidents ranged in severity, with a variety of short- and long-term responses. The long-term responses have been less severe than originally feared. Ionizing radiation of humans can act as a mutagen, a carcinogen, and a teratogen. It can cause cataracts, impaired fertility, premature aging, and skin damage. Radiation-induced cancer can occur as little as 2–5 years after exposure, or following a latency period of up to 25 years after exposure. Greater risk occurs for those exposed in utero. X-ray-induced disease from excess exposure, faulty equipment, or human error is a hazard of medical care. There is perhaps no safe exposure to ionizing radiation beyond atmospheric background, and any extra exposure should be limited, with prudent exposure to X-rays and limited exposure to atomic radiation from domestic or military uses.

Non-Ionizing Radiation

There are two types of non-ionizing radiation: optic and some electromagnetic fields. Optic radiation includes ultraviolet and infrared. Electromagnetic fields, such as those induced by microwave or radio frequencies, are described in terms of wavelengths or frequency. The harmful effects of non-ionizing radiation are of three main types: photochemical (sunburn or snow blindness), thermal, and electrical.

The health effects of UV radiation include increasing incidence of squamous and basal cell carcinoma and melanoma of the skin, a highly malignant cancer. This kind of radiation is associated with excess exposure to the sun, which in addition to these skin cancers, causes skin and eye burns, cataracts, reduced immunity, and damage to blood vessels. Infrared radiation exposure over long periods is associated with increased risk of cataracts, impaired fertility, and tissue damage. UV radiation is generally classified as non-ionizing radiation; however, in Figure 9.5, it appears to be on the cusp of non-ionizing and ionizing radiation. While most UV radiation is non-ionizing, radiation of higher frequency and higher energy is more powerful, and thus can be considered ionizing.

Long-term exposure to cellular phone use, high-voltage power lines, and radio and radar transmitters is suspected to be associated with increased risk of cancer, but this has not yet been proven. Microwave exposures at high levels can damage vulnerable tissues, but the level of dangerous exposure has not yet been conclusively determined. Lasers are pulsed electromagnetic waves used increasingly in medicine and industry. Excessive use of magnetic resonance imaging (MRI), computed tomography (CT) scans, airport total body scanners and tasers may also contribute to excess radiation exposure, as may the use of cell phones by children as well as adults. Safer technology with reduced radiation should be a high priority as these useful instruments are increasingly used globally.

Low-dose irradiation is used in the production, processing, and handling of foods to prevent food hazards, and is widely supported by professional organizations. It provides an important adjunct to sanitation and good manufacturing practices to reduce morbidity and mortality associated with foodborne diseases, even in industrialized countries. More than 40 years of research and use in the USA and many other countries have demonstrated the effectiveness and safety of low-dose irradiation. This is rapidly becoming an essential part of public health protection from foodborne disease in the USA and internationally, although public acceptance is still problematic.

When discussing different degrees of radiation energy, the issue of both lasers and tasers arises. They are both increasingly used in various fields for a myriad of purposes. Lasers (an acronym for light amplification by stimulated emission of radiation) are used in many electrical devices, and play an important role in law enforcement. They are also used for military, medical, and surgical purposes, skin treatments, hair removal, and in research. Lasers not intended for medical use (and misused medical lasers) can cause irreparable retinal damage and severe burns. The potential for damage to the eye is the principal issue behind laser application, standards, safety, and control measures. The severity of eye damage is determined by both the wavelength and the part of the eye exposed to the laser. If the laser burn affects peripheral vision, this will have minimal or no effect on vision; however, if the fovea is exposed to a laser beam, reading vision may be destroyed.

Tasers are characterized as weapons that function by use of electric currents, which stun and cause temporary debilitation. Those exposed to tasers lose control of their muscles, as the electricity causes involuntary muscle contractions. As these weapons incapacitate normal muscle function, they are widely used by police forces on potentially dangerous individuals to make it easier to arrest or restrain them. According to the US Department of Justice, the use of tasers by law enforcement officials was adopted in an effort to provide stronger control over problematic or aggressive suspects, while resulting in less severe injuries. Despite this, tasers have become very controversial, as there have been cases in which the use of these instruments has resulted in head injuries, broken bones, and even deaths. Because of their potential for misuse and danger, legal restrictions and caution in use are imperative.

ENVIRONMENTAL IMPACT

The US National Environmental Policy Act (NEPA), passed in 1970, made protection and restoration of the environment matters of national policy. NEPA required all federal agencies to take environmental considerations into account in decision-making processes and program implementation. *Environmental impact statements* are required for major construction and public works programs, delineating positive impact, possible adverse effects, alternatives, and any irreversible effects. This legislation resulted in changes in many national projects and promoted a governmental regulatory approach to supervision, control, and prevention of pollution with materials and processes that could harm human health and the environment.

Emergency Events Involving Hazardous Substances

Since World War II, there has been a rapid increase in the number of chemicals developed and used worldwide. More than 60,000 chemicals are available, with some 600 new substances produced every year, an unknown number of which are hazardous. The health effects resulting from the release of a hazardous substance are often unknown. A hazardous substance release is defined as the uncontrolled or illegal release or threatened release of chemicals or their hazardous byproducts.

Reportable events are defined as those events in which the substances need to be removed or cleaned up. Plant management is liable for damages due to negligence in both civil and criminal law. Where community exposure occurs from negligence, accident, or natural disaster, a public health emergency response is required, based on prior preparation.

The environment is a factor in more than 80 percent of diseases, 23 percent of deaths (premature mortality), 24 percent of the global burden of disease, and more than a third of the burden of disease in children (WHO).

Environmental contaminants such as chemicals and industrial products are responsible for causing approximately 10 percent of all birth defects. There are over 4 million chemicals present in the home and work environments. Some 114 million Americans live in areas where concentrations of air pollution are over the standards, and 41 million live within a 4-mile (6.4 km) radius of the 1270 most hazardous sites. In the USA annually, more than 2 billion pounds (1 billion kg) of toxic pollutants are released into the air, and a similar amount into surface water, on to land, or underground. There are at least 10,000 accidental and illegal releases of hazardous chemicals each year, totaling 4.1 billion pounds (1.9 billion kg) of toxic chemicals. Between 2005 and 2013, there were 113,307 hazardous chemical events, with 319 fatalities and a further 11,728 people injured as a result of these incidents (ATSDR, 2013). Volatile organic substances, other inorganic substances, mixtures of more than one chemical, and acids account for more than 51 percent of the hazardous substances released.

European Union (EU) legislation requires member countries to identify high-risk industrial sites, to prevent major accidents, and to limit effects on the population and the environment. The goal is a high level of protection for the population across countries. Even in industrialized countries, monitoring for heavy metals is problematic. The US Agency for Toxic Substances and Disease Registry (ATSDR) was established by Congress to monitor the effects on public health of hazardous substances in the environment. This includes "public health assessments of waste sites, health consultations concerning specific hazardous substances, health surveillance and registries, response to emergency releases of hazardous substances, applied research in support of public health assessments, information development and dissemination, and education and training concerning hazardous substances" (ATSDR, 2013).

Biomonitoring is a public health a standard for assessing human exposure to toxic substances for responding to serious environmental issues. The US national biomonitoring program is based on National Health and Nutrition Examination Surveys (NHANES) of blood, urine, breast milk, and saliva samples to determine population prevalence of toxic chemicals above a known toxicity level (see Chapter 8). The US goal is to increase the number of states and territories that monitor for diseases from heavy metal environmental hazards. Since the publication of the Fourth Report, 2009, tables for 117 chemicals have been updated and 34 chemicals have been added, making a total of 151 chemicals in 2013.

All US states monitor for lead poisoning, while 20 monitor for pesticides, 14 for mercury, 10 for arsenic, 10 for cadmium poisoning, and 35 for birth defects. The goal is to monitor exposure to pesticides in humans by measuring urine concentrations of metabolites. Linked health effect, exposure, and hazard data for environmental public health surveillance were used by 15 states in 2004 (CDC, National Biomonitoring Program, 2013).

Widespread programs for lead exposure reduction and restrictions on lead use have reduced lead exposure in the environment. As a result, lead poisonings have decreased and become less severe, but still occur. CDC estimates that some 500,000 US children aged 1–5 years have blood lead levels (BLLs) greater than 5 micrograms of lead per deciliter of blood (μg/dl), the level at which CDC recommends public health interventions (CDC, Child Lead Poisoning, 2013).

Internationally, a number of major disasters involving radiation has occurred in recent decades (Box 9.25). In Seveso, Italy, in 1976, an explosion in a chemical factory resulted in 17,000 people being evacuated and many terminations of pregnancy among exposed women. An elevated risk of hemopoietic cancers and some elevation of breast cancer risk was found after 20-year follow-up of the exposed population, although no consistent pattern with time since the accident was evident.

In 1984, a sudden release of highly toxic methyl isocyanide from a chemical plant in Bhopal, India, caused thousands of deaths, and blinding and permanent injury of several thousands more, requiring evacuation of an estimated 300,000 people living in adjacent neighborhoods. While

BOX 9.25 Nuclear Accidents: Three Mile Island (1979), Chernobyl (1986), and Fukushima (2011)

Three Mile Island (1979)

In 1979, the nuclear plant at Three Mile Island in Pennsylvania suffered a near disaster that devastated the plant but did not release nuclear material. It led to a review of safety procedures and heightened public concern as to the overall safety of nuclear energy facilities. There was no loss of life or radiation release, but the incident led to major changes in emergency response planning, operator training, radiation protection, and other safety measures of nuclear power plant operations. It also caused the US Nuclear Regulatory Commission to tighten and heighten its regulatory oversight, resulting in changes in the nuclear power industry and its regulation, with much improved safety.

Chernobyl, Ukraine (1986)

In 1986, a nuclear energy plant located at Chernobyl in the Ukraine (then in the USSR) suffered a meltdown that breached the integrity of the containment vessel, resulting in a massive explosion of the reactor. Design problems and a series of staff errors led to loss of control of the reactor with power levels soaring to 120 times the normal, rupturing the fuel rods, and vaporizing the cooling system. A steam explosion then blasted open the 100 ton concrete slab covering the reactor, starting uncontrollable fires. Despite valiant attempts by emergency personnel and staff, the fires could not be controlled immediately. Air-dropping of sand, lead, clay, and limestone controlled the fire, but the heat of the reactor and radiation could not be reduced for many days. Immediate deaths numbered 33 individuals, mostly among the firefighters, with 237 suffering acute radiation poisoning. Around 135,000 people were evacuated from a 19 square mile ($49\,km^2$) area.

The nuclear fallout material carried in a 600 m plume, including iodine-131, cesium-137, and xenon isotopes, spread across much of Europe. Fallout reached some 20 countries, and an international public health threat of major proportions occurred. Ten years after the incident, there was a highly significant increase in thyroid cancer cases in children in the three affected countries: Ukraine, Belarus, and Russia. The long-term effects in terms of increased cancer and birth defects are hard to assess, but current estimates are of 500 (1–2 percent) additional cancer cases among 100,000 people exposed to 10–20 rads. The actual increase in incidence of thyroid cancer, other cancers, and birth defects and the general impact on health will only be determined by careful epidemiological follow-up of the exposed population over many years. The economic impact of the disaster is estimated at over US$19 billion, and replacement of the plant reaches a similar sum. Close to the tenth anniversary of the Chernobyl disaster, a second nuclear leak nearly occurred due to human error. The Ukrainian government reopened the second reactor in 1999. International assistance in technical and financial aspects of nuclear energy in the Ukraine is in process, but a large area around the plant remains uninhabitable.

Fukushima, Japan (2011)

On 11 March 2011, the Great East Japan Earthquake, with a magnitude of 9.0 on the Richter scale, struck Japan's eastern coast, initiating a string of deadly events. Soon afterwards, the earthquake caused a major, 15 m tsunami tidal wave, which swept the coast of Japan and caused hurricane-scale damage, with many aftershocks. The loss of life was measured in the thousands, with extensive damage to homes, buildings, transportation, and many other community resources. By February 2012, there were 15,534 confirmed deaths and 7092 missing people among 12 prefectures, including Fukushima Prefecture.

The force of the tsunami destroyed the power supply and cooling capabilities of three Fukushima Daiichi reactors, and a nuclear emergency became apparent. The main priority in alleviating the disaster was preventing the reactors from overheating. A second crucial objective was the prevention of radiation leakage, specifically in contaminated water released from the three reactors.

The Fukushima nuclear accident did not cause any deaths or cases of radiation illness; however, this favorable outcome was only achieved through the rapid evacuation of over 400,000 people from within a specified evacuation zone. Despite no fatalities from the nuclear accident, a more tragic outcome shadowed both the earthquake and the tsunami. Owing to the amount of radioactive material released into the air, sea, and land, the Fukushima nuclear accident, caused by the earthquake and tsunami, is considered to be one of the most devastating nuclear accidents globally. Declared a public health emergency requiring worldwide support, the disaster was given the highest rating, level 7, on the International Nuclear Event Scale.

One month after the earthquake, leakage of radioactive material from the power plant remained substantial, and the reactors' inadequate cooling capabilities still posed a serious threat. Cooling operations such as water injection and high-level water spraying were carried out by fire trucks and helicopters. Policies were established, issuing levels of radiation exposure in Japan's ports and in various areas. An area occupying a 20 km radius surrounding the Fukushima Daiichi power plant was pronounced a "No-Entry Zone", and the area is left uninhabited.

Sources: *World Nuclear Association. Chernobyl accident 1986 [updated April 2012]. London: WNA. Available at: http://www.world-nuclear.org/info/chernobyl/inf07.html [Accessed 15 August 2012].*
Washington State Department of Health. Background radiation: natural vs. man-made. Olympia, WA: Washington State Department of Health, Division of Environmental Health Office of Radiation Protection; 2002. Available at: http://www.doh.wa.gov/Portals/1/Documents/Pubs/320-063_bkvsman_fs.pdf [Accessed 15 August 2012].
World Nuclear Association. Fukushima accident 2011 [updated 3 August 2012]. London: WNA. Available at: http://www.world-nuclear.org/info/fukushima_accident_inf129.html [Accessed 17 August 2012].
World Health Organization, Western Pacific Region. The Great East Japan Earthquake. Geneva: WHO; 2011. Available at: http://www.wpro.who.int/publications/9789290615682/en/index.html [Accessed 17 August 2012].

BOX 9.26 The Bhopal Gas Disaster, 1984

In Bhopal, India, a city of some 900,000 people, an accidental release of methyl isocyanate (MIC) gas from the Union Carbide pesticide plant occurred on the night of 3 December 1984. The gas spread rapidly across the sleeping city. Some 3800 people died within minutes and many others were left with lifelong illnesses.

Failure of inadequate safety systems allowed an estimated 27 tons of MIC gas to escape. A warning siren was turned on, but quickly turned off again so as to not cause panic; however, choking and coughing panic led to chaos and deaths by trampling.

Estimates of the death toll vary; at least 3700 people died from immediate exposure to the gas, with other estimates of up to 8000. In the two decades following the disaster, an estimated 20,000 additional people died from effects of the gas. Estimates of long-term illness vary, with some 120,000 people living with the effects from the gas, including blindness, extreme shortness of breath, cancers, birth deformities, and early onset of menopause. Chemicals from the pesticide plant and from the leak have infiltrated the water system and the soil near the old factory and thus continue to cause poisoning in the people who live near it.

Three days after the disaster, the chairman of Union Carbide, Warren Anderson, was arrested. Released on bail, he fled the country. He is still wanted in India for culpable homicide, but is apparently living in the USA.

Extradition procedures have been obstructed by political issues. Union Carbide claims that the incident was due to sabotage, but acknowledged responsibility and paid a modest compensation sum of US$470 million to the families of those who died, although refusing to pay for the long-term effects. The case is still before the courts in India. Regulation of chemical industries remains weak. The Bhopal gas tragedy is remembered as "one of the ugliest industrial disasters that have ever taken place" (Broughton, 2005).

Environmental cleanup has been hampered by a lack of governmental regulatory capacity. In June 2010, seven former employees of the Union Carbide subsidiary, all Indian nationals and many in their seventies, were convicted of causing death by negligence and each sentenced to 2 years' imprisonment and fined. All were released on bail shortly after the verdict. Damages for personal injury, medical monitoring, and injunctive relief in the form of cleanup of the drinking water supplies for residential areas near the Bhopal plant are being sought in an appeal before the federal district court in New York. The legal and political implications of this disaster are still ongoing.

Sources: *Sharma DC. Bhopal's health disaster continues to unfold. Lancet 2002;360:859.*
Broughton E. The Bhopal disaster and its aftermath: a review. Environ Health 2005;4:6. Available at: http://www.ncbi.nlm.nih.gov/pmc/articles/PMC1142333/ [Accessed 16 July 2013].
Maps of India. Bhopal gas tragedy [updated 23 July 2012]. Available at: http://www.mapsofindia.com/bhopal/gas-tragedy.html [Accessed 16 August 2012].
About.com. Twentieth century history. 1984 – Huge poison gas leak in Bhopal, India. Available at: http://history1900s.about.com/od/1980s/qt/bhopal.htm [Accessed 20 August 2012].

the transfer of hazardous occupations and industries to less developed areas is a growing issue, the tragedy at Bhopal led to greater recognition by policy makers and the public that toxic accidents can potentially happen at any time and in any place, not just in developed countries (Box 9.26).

Nuclear and chemical disasters have become a major element in disaster planning for corporate, investor, and occupational and environmental health agencies, as well as for communities adjacent to chemical production, storage, or transportation. Emergency responses to chemical, radiation, or biological catastrophes involve specialized expertise, based on common principles of prevention, monitoring, and crisis management. These include prior emergency planning, speed, coordination of civil and military resources, skilled professional teams providing information to the public, logistic, medical, and laboratory support, on-site case management and evacuation, investigation of causes, and continuous teamwork among all involved agencies.

Human-Caused Disasters

The *Exxon Valdez*, a large oil tanker that ran aground in Alaska in 1989, spilling large amounts of crude oil in Prince William Sound, served as a precedent case in acknowledging that personal and fiscal responsibility for cleanup and other costs to reduce the environmental damage lies with the company that owns the ship. Cleanup efforts required enormous amounts of money, and the spill became a *cause célèbre* for the environmental movement. The incident highlighted the importance of monitoring seagoing chemical and fuel vessels. New cleanup techniques have been researched and applied in recent years. In response, the Coalition for Environmentally Responsible Economics (CERES) of investment fund advisors and social advocates established the "CERES principles" demanding environmental monitoring of corporations regarding energy use, public disclosure, damage compensation, sustainable use of natural resources, and environmental representatives on boards and in management of corporations. The EPA monitored air, water, sediment, and wastewater generated from the 2010 BP oil spill in the Gulf of Mexico. The response to this disaster was coordinated with four states and the oil company and the cleanup is still being monitored.

War, Terrorism, and Genocide

The man-made disaster of war has used chemical, biological, and nuclear methods of destruction as well as traditional methods of warfare including economic blockade. War's offspring, terrorism, has used chemical armamentaria and may use biological or even nuclear destruction sooner or later. When disasters occur, lessons can be learned to improve services for future disasters, be they natural or human-made (see Chapter 7).

Although the Hague Convention of 1899 specifically opposed the use of gas in warfare, poison gas has been used as a weapon against both frontline troops and civilian populations since World War I. This practice continues to this day. Gas warfare was banned by international treaty in 1997 under the Chemical Weapons Convention, which has so far been signed by 188 states. The Organization for the Prohibition of Chemical Weapons (OPCW), located in The Hague, oversees the implementation of the guidelines.

In World War I, gas warfare was used by German forces and in retaliation by the Allied armies, primarily with chlorine, mustard gas, and diphosgene. Other gases were also used, including hydrogen cyanide and cyanogen chloride. Gas is estimated to have killed between 300,000 and 900,000 soldiers, mainly in the Russian army, during the war and affected the health of 1 million others. Nerve gases (e.g., sarin, tabun, soman) were developed by the Germans up to World War II. Poison gas was used with deadly efficiency by the Nazis in the Holocaust, but not in warfare because of fear of retaliation.

In the Vietnam War, the US armed forces used napalm widely and Agent Orange as a mass defoliant, with long-term effects on exposed military personnel and Vietnamese civilians, causing large-scale loss of life and birth defects. Egypt used poison gas in its war in Yemen in the 1960s, and in the 1980s Iraq targeted Kurdish villages, killing thousands of civilians.

During the Gulf War of 1991, the potential use of poison gas in long-range rockets on civilian population targets was narrowly averted. Several years later, thousands of US service personnel reported a variety of neurological symptoms and general fatigue. By 1996, these cases were acknowledged by the Department of Defense as possible long-term sequelae of accidental exposure by troops to toxic agents following destruction of Iraqi chemical weapons, or due to antidotes taken for potential gas warfare exposure (soman). In 1995, a chemical attack with a very dangerous chemical warfare agent (sarin) was carried out by an extremist cult in Japan on subway passengers in Tokyo, resulting in 12 deaths and 3000 injuries, and hundreds of hospitalizations.

In 2013, in a bitter civil war in Syria, a neurotoxin (probably sarin) was launched by rockets on to a suburb of Damascus held by rebel forces, killing hundreds of men, women, and children. This gas warfare, suspected to have been carried out by the Syrian regime against civilians of their own country, has provoked international outrage. The issue of chemical, biological, and atomic weapons will reverberate in the Middle East for years to come and may provide precedents for the use of nuclear and biological weapons directed at civilian populations.

Terrorist bombing incidents occurred in many parts of the world during the 1990s. The Lockerbie airplane bombing incident of a Pan Am flight to New York City killed 270 people in 1988. The 1995 bomb detonated by domestic terrorists in a federal building in Oklahoma City in the USA killed over 160 people. Terrorist bombings of a US military housing complex in Saudi Arabia in 1996, US embassies in Africa in 1998, Moscow apartment buildings in 1999, and Israeli public bus lines and restaurants by suicide bombers killed some 1,000 persons during the second Intifada (2000-2005), and numerous highly lethal suicide bombings in Iraq, Yemen and over 120,000 killed in a civil war in Syria from 2010 to 2013 as well as many terrorist incidents in other parts of the world, caused large numbers of deaths and injuries. Each incident resulted in national concern over the threat of terrorist action causing mass casualties. Destruction of pipelines and oil fields caused extensive environmental damage in the aftermath of the Gulf War in 2001. The infamous September 11, 2001 terrorist attacks in New York City and Washington, DC killed thousands of people and caused many more casualties. They created a new world struggle against terrorism and many major man-made catastrophes. This terrorist event resulted in an unspeakable amount of irreparable damage. Refer to Chapter 10 for a more detailed account of this day and the ensuing health outcomes of this as well as other major disasters.

Unmarked landmines cause huge loss of life and limbs, often among farmers and children. Millions of landmines are present in many areas of conflict, and cleanup is dangerous and costly. Between 2003 and 2005, there were an estimated 7000 landmine deaths and casualties a year worldwide. Most were concentrated in Iraq, Afghanistan, Cambodia, and Colombia. Landmines limit land and water use and have serious economic consequences for farmers. An international movement to ban the use of landmines gained international prominence with support from Princess Diana and by the awarding of the 1997 Nobel Peace Prize to Jody Williams, founder of this movement. Prevention is achieved by raising awareness and political action to prevent landmine use and support efforts for landmine clearance.

The potential for intentional, negligent, or accidental disasters, whether caused by humans or natural, is a real and present danger requiring health officials to coordinate with civil defense and military authorities to prepare disaster plans for such events. Planning can greatly reduce the number and severity of casualties of toxic chemical disasters.

Incitement to organized mass murder of ethnic or political groups has continued into the twenty-first century and constitutes a grave danger to public health as well as to peace. The Syrian crisis of 2011–2013 is a good example of massive use of heavily armed forces against civil uprisings with religious and ethnic overtones and the potential for use of poison gas by the government against the rebel population.

Preventing and Managing Environmental Emergencies

Public health has an important role to play in prevention, management, and mitigation of the effects of human-caused and natural disasters. The US Congress passed the Emergency Planning and Community-Right-to-Know

Act (EPCRA) of 1986 following the 1984 Bhopal disaster (Box 9.26). This legislation established state and local agencies for managing chemical emergencies. It requires facilities that handle hazardous chemicals to make information available to the public and preparedness for possible chemical accidents. This involves a holistic approach integrating technology, procedures, and management practices. The first responsibility lies with management, which must have a high level of awareness and commitment to accident prevention and safe practices. The range of industries at risk is very broad in modern societies; it includes local dry cleaners and furniture manufacturers as well as the chemical industry. The right to know extends from governments, professional societies, trade associations, labor unions, the research community, the news media, and environmentalists as well as the general public. The right to know has become the need to know. Box 9.27 discusses the International Health Regulations and the importance of strong national frameworks capable of managing public health events and situations in which global health is threatened.

Environmental emergencies occur from release of chemicals or radiation into the air. Inhalation and fallout effects downwind of the site depend on weather conditions and dispersal of the smoke plume. Clinical management of exposed civilians and emergency personnel is an activity of health management that involves organizing triage and transportation services at the site of the disaster. The decision to evacuate civilians is often made with limited information but must take into account the potential for exposure during evacuation, weighed against the protective effect of sealing homes and staying indoors (Box 9.28).

The approach to management of environmental health problems requires a continuum of interrelated activities and phases ranging from prevention, through preparedness, detection, and response, to recovery. It includes measures to rebuild infrastructure and also lives and livelihoods affected by the emergency. Prevention is ultimately the most cost-effective and cost-beneficial means of dealing with potential environmental health problems. The principles of disaster and environmental preparedness include:

- planning
- coordination with sister agencies
- preparation
- research
- adaptation with science and technology
- training
- monitoring
- supply
- detection
- prevention
- event
- response
- revision based on lessons learned.

BOX 9.27 Progress of International Health Regulations National Core Capacities

Adopted at the 58th World Health Assembly in 2005, the International Health Regulations (IHR) are an important tool in managing the growing globalized nature of disease transmission and health challenges. Implemented in most countries by June 2007, the IHR are a public health global framework, for consistency in national and international public health programs.

The revised 2005 IHR, a legally binding agreement, includes chemical threats to public health, requiring countries to strengthen capacities on the effects of chemical events on human health. In 2012, a significant 5-year milestone was reached in global efforts to promote national capacities to investigate, evaluate, and take action towards public health events. This requires strategies to create, strengthen, and sustain routine and emergency public health capabilities at identifiable points of entry.

Chemical safety, including prevention and preparedness globally, is promoted through the Strategic Approach for International Chemicals Management (SAICM). The comprehensive approach and global action plan sets out the scope, principles, objectives, financial aspects, implementation, and review arrangements.

The World Health Organization *Manual for Public Health Management of Chemical Incidents* provides a comprehensive overview of the public health management of chemical incidents and emergencies, and outlines the steps to support implementation, with an emphasis on prevention. These include national legislation, policy, and financing; coordination and national focal point (NFP) communications; surveillance; response; preparedness; risk communication; and human resources and laboratory capacity.

Public health has an essential role to play in preventing and minimizing adverse effects of chemical on humans and the environment. With the complexity of environmental challenges, fragmentation of roles and unclear responsibilities are common among the many functional centers, not only across but also within sectors.

Sources: *Hardiman MC. World Health Organization perspective on implementation of International Health Regulations. Emerg Infect Dis 2012;18(7). Available at: http://wwwnc.cdc.gov/eid/article/18/7/12-0395_article.htm [Accessed 29 June 2012].*
World Health Organization. Inter-Organization Programme for the Sound Management of Chemicals (IOMC). Available at: http://www.who.int/iomc/en/index.html [Accessed 6 July 2013].
World Health Organization. WHO manual for the public health management of chemical incidents. Geneva: WHO; 2009. Available at: http://whqlibdoc.who.int/publications/2009/9789241598149_eng.pdf [Accessed 6 July 2013].

The appropriate team to handle such a situation involves public health, occupational health, and epidemiology investigators as well as police, fire services, civil defense, armed forces, chemical warfare units, and psychological staff. While it is crucial that experts from a wide range of disciplines are involved in rescue and relief efforts, it must be recognized how to apply the most effective model of management. The

BOX 9.28 Emergency Procedures for Hazardous Substances, Chemical, or Radiation Disasters

- Plan and prepare local public health, hospital, and other first responders for possible chemical or radiation disasters or attacks.
- Establish early warning reporting and communication procedures.
- Contain and reduce spread of the toxin.
- Inform the community to remain inside homes or other buildings.
- Notify municipal, state, and federal emergency organizations.
- Minimize exposure by ensuring the potentially exposed population remains inside, with the affected areas sealed off and quarantined or by limited evacuation.
- Identify, decontaminate, and triage exposed persons.
- Measure exposure and reaction.
- Determine causative agents and antidotes.
- Initiate antichemical procedures for exposed persons including removal of clothing, showers, and antidote.
- Coordinate on-site triage and evacuation for medical care.
- Ensure medical or hospital care for exposed persons.
- Provide accurate information to the public.
- Promote health and supportive care at evacuation sites.
- Investigate – professional and criminal.
- Compensate the injured or displaced.
- Pursue civil and criminal charges against negligent management persons and corporations.
- Provide documentation and recommendations from lessons learned.
- Review procedures and revise disaster plan operation.
- Promote public and professional discussion.

Sources: *Khan AS, Levitt AM, Sage MJ. Biological and chemical terrorism: strategic plan for preparedness and response: recommendations of the CDC Strategic Planning Workgroup. MMWR Recomm Rep 2000;49:1–14. Available at: http://www.cdc.gov/mmwr/preview/mmwrhtml/rr4904a1. htm [Accessed 17 August 2012].
Environmental Protection Agency. Emergency management [updated 3 August 2012]. Washington, DC: EPA. Available at: http://www.epa.gov/emergencies/index.htm [Accessed 15 August 2012].*

most desirable outcomes (lowest possible rates of deaths, injuries and least amount of damage) will result from use of a chain-of-command system. Past public health emergencies, such as Hurricane Katrina and the BP oil spill off the Gulf of Mexico, illustrate system weaknesses and inadequate management of personnel and responsibilities. Many experts agree that effective management will only occur once society chooses to adopt a military approach of command, with a single chain of authority and accountability.

Moreover, relief efforts do not end once the surroundings of a public health disaster have been cleaned up; post-disaster recovery planning is part of the planning process. (Disaster planning is discussed in Chapter 7.) Long-term effects include post-traumatic stress disorder (PTSD), which can result in serious psychological dysfunction in affected individuals. PTSD can be alleviated by early psychological support for victims of mass disasters at the site and at evacuation or follow-up centers, and should be part of emergency care planning.

Rapid risk assessment involves weighing the hazard, exposure potential, dose–response, and both short- and long-term risks. Command centers and designated leaders are needed to maintain control of the multitude of needs for information, coordination between agencies, and the distribution of resources to areas of greatest need. Long-term epidemiological assessment may be necessary for legal and compensation purposes, as well as for training and preparation for future events.

Advocacy is a key public health function, and environmental and safety issues are areas where advocacy can bring important public benefit. Leadership in defining public health problems and in defining necessary action to reduce risk factors, or short- or long-term ill-effects, requires skill in interpretation of epidemiological events and studies, providing perspective for policy makers addressing those issues.

ENVIRONMENTAL HEALTH ORGANIZATION

The World Health Organization Commission on Health and Environment Report (1992) developed a consensus documentation of international environmental health issues. This commission, chaired by Simone Weil of the European Parliament, included many distinguished scientists, professional leaders, and international organizations. The report represented a strong international consensus on joint action to prevent and clean up environmental degradation that had occurred in Europe over several decades.

National organization for environmental health can take various forms. In the past, it was common for ministries of health to have environmental health departments, but in recent years this has increasingly moved to ministries of the environment.

Since 9/11, there has been increased governmental and public concern regarding possible emergencies along with environmental decay and incidents of food or waterborne disease. The possibilities of natural and human-made disasters in the environment call for renewed efforts to prepare emergency plans, and conduct suitable training and exercises with public health, hospital, and primary care centers, as well as ambulance, fire, police, and military services.

Terrorist incidents may involve microbiological (bacterial, viral) pathogens, nerve gas (e.g., sarin), and lethal plant toxins (e.g., ricin), as well as explosive or firearm attacks. Preparing "homemade" agents or explosive devices can be technically simple. Weaponization of biological agents for localized or mass dissemination is feasible and such agents have been used. Many potential biological agents prepared in military or secret warfare laboratories could possibly reach terrorist groups. Such agents may be highly contagious, causing public panic and high mortality rates. Public health and other local first responders including police, fire services, emergency evacuation ambulance, medical and hospital providers, volunteers and social services may all be required in rapid mobilization. Large-scale attacks with chemical or microbiological agents require rapid procurement and distribution of large quantities of drugs and vaccines, which must be available quickly.

Natural and other environmental disasters require at least as much mobilization as local terrorist acts, and may cause damage on a massive scale due to droughts, floods, hurricane, tsunamis, oil spills, and forest fires, all of which happen on a frequent basis in many especially vulnerable parts of the world. The challenges include massive evacuation, damage control, security, provision of food, water and shelter, prevention of epidemic disease, reconstruction, and rehabilitation of refugees (see Chapter 10 and http://www. epa.gov/emergencies/index.htm).

Because of concern over environmental decay and fragmentation of government regulation efforts, the EPA was established in the USA in 1970 as the head federal agency reporting to the president to coordinate the administration of a wide range of environmental health problems. The EPA sets standards and regulations for a variety of legislation pertaining to the environment, such as air and water pollution, solid and hazardous waste management, noise, public water supplies, pesticides, and radiation. Despite the growth of the EPA and its control of a superfund to reduce toxic and other waste sites, interagency coordination is complex. In the US federal government, a wide variety of agencies located in different government departments has responsibilities related to the environment (Box 9.29). The substantial environmental progress made in the USA in the past 25 years is outlined in Table 9.4.

BOX 9.29 US Federal Government Agencies with Environmental Responsibilities

- Environmental Protection Agency (Independent)
- Council on Environmental Quality (Executive Office)
- Nuclear Regulatory Commission (Independent)
- Office of Environmental Safety and Health (Department of Energy)
- Office of Environmental Management (Department of Energy)
- Office of Environmental Policy and Assistance (Department of Energy)
- Office of Surface Mining Reclamation and Enforcement (Department of the Interior)
- Bureau of Land Management (Department of the Interior)
- Center for Environmental Health, CDC (Department of Health and Human Services)
- National Institute of Occupational Safety and Health (NIOSH) (CDC)
- Consumer Product Safety Commission (Independent)
- Public Health Service (Department of Health and Human Services)
- Centers for Disease Control and Prevention (Department of Health and Human Services)
- Food and Drug Administration (Department of Health and Human Services)
- Agency for Toxic Substances and Disease Registry (Department of Health and Human Services)
- Occupational Safety and Health Administration (OSHA) (Department of Labor)
- Mine Safety and Health (Department of Mines)
- Fish and Wildlife Service (Department of the Interior)
- Soil Conservation Service (Department of Agriculture)
- Department of Homeland Security

Note: *The department under which each agency falls is listed in parentheses.*

TABLE 9.4 Environmental Milestones in the USA, 1970–2006

1970	President Richard Nixon creates EPA to protect the environment and public health. Congress amends the Clean Air Act to set national air quality, auto emission, and antipollution standards
1971	Congress restricts use of lead-based paint in residences and on cribs and toys
1972	EPA bans DDT, a cancer-causing pesticide, and requires extensive review of all pesticides USA and Canada agree to clean up the Great Lakes, which contain 95 percent of America's freshwater Congress passes the Clean Water Act, limiting raw sewage and other pollutants flowing into rivers, lakes, and streams Only 36 percent of the nation's assessed stream miles are safe for fishing and swimming; in 2006 about 60 percent are safe for such uses
1973	EPA begins phasing out leaded gasoline; OPEC oil embargo triggers energy crisis, stimulating conservation and research on alternative energy sources. EPA issues its first permit limiting a factory's polluted discharges into waterways. Endangered Species Preservation Act passed
1975	Congress establishes fuel economy standards and sets tailpipe emission standards for cars, resulting in the introduction of catalytic converters
1976	Congress passes the Resource Conservation and Recovery Act, regulating hazardous waste from its production to its disposal President Gerald Ford signs the Toxic Substances Control Act to reduce environmental and human health risks; EPA begins phase-out of cancer-causing PCB production and use
1977	Clean Air Act Amendments to strengthen air quality standards and protect human health

TABLE 9.4 Environmental Milestones in the USA, 1970–2006—cont'd

1978	Residents discover Love Canal, New York, is contaminated by buried leaking chemical containers Federal government bans CFCs as propellants in aerosol cans; CFCs destroy the ozone layer which protects the Earth from harmful ultraviolet radiation
1979	EPA demonstrates scrubber technology for removing air pollution from coal-fired power plants. This technology is widely adopted in the 1980s Three Mile Island nuclear power plant accident near Harrisburg, Pennsylvania, increases awareness and discussion about nuclear power safety. EPA and other agencies monitor radioactive fallout
1980	Congress creates a superfund to clean up hazardous waste sites. Polluters are made responsible for cleaning up the most hazardous sites. Agency for Toxic Substances and Disease Registry created
1981	National Research Council report finds acid rain intensifying in the north-eastern USA and Canada
1982	Nuclear Waste Repository act for safe disposal of nuclear waste Dioxin contamination forces the government to purchase homes in Times Beach, Missouri; federal government and responsible polluters share the cleanup costs A PCB landfill protest in North Carolina begins the environmental justice movement
1983	Cleanup actions begin to rid the Chesapeake Bay of pollution stemming from sewage treatment plants, urban runoff, and farm waste EPA encourages homeowners to test for radon gas, which causes lung cancer; more than 18 million homes tested for radon. Approximately 575 lives are saved annually due to radon mitigation and radon-resistant new construction
1984	Bhopal disaster in India stirs public opinion on chemical industrial hazards globally
1985	Scientists report that a giant hole in the Earth's ozone layer opens each spring over Antarctica
1986	Congress declares the public has a right to know when toxic chemicals are released into air, land, and water. Superfund Amendment Act promotes hazardous site cleanup
1987	USA signs the Montreal Protocol, pledging to phase out production of CFCs Medical and other waste washes up on shores; beaches closed in New York and New Jersey
1988	Congress bans ocean dumping of sewage sludge and industrial waste
1989	Exxon Valdez spills 11 million gallons of crude oil in Alaska's Prince William Sound
1990	Clean Air Act Amendments require states to show progress in improving air quality EPA Toxic Release Inventory of pollutants released from specific facilities in their communities Number of chemicals listed in EPA's Toxic Release Inventory nearly doubled, from 328 in 1990 to 644 in 1999 Pollution Prevention Act signed, emphasizing importance of preventing, not just correcting, environmental damage National Environmental Education Act signed, for educating the public to ensure scientifically sound, balanced, and responsible decisions about the environment
1991	Federal agencies begin using recycled content products. EPA launches voluntary industry partnership programs for energy-efficient lighting and for reducing toxic chemical emissions
1992	EPA launches the Energy Star® Program to help consumers identify energy-efficient products
1993	EPA reports that secondhand smoke contaminates indoor air, with serious health risks to non-smokers *Cryptosporidium* outbreak in drinking water in Milwaukee, Wisconsin, sickens 400,000 people and kills more than 100 Federal government uses its US$200 billion annual purchasing power to buy recycled and environmentally preferable products
1994	EPA Brownfields Program to clean up abandoned, contaminated sites to return them to productive community use EPA issues new standards for chemical plants to reduce toxic air pollution by more than half a million tons each year, equivalent to removing 38 million vehicles annually
1995	EPA launches an incentive-based acid rain program to reduce SO_2 emissions EPA requires municipal incinerators to reduce toxic emissions by 90 percent from 1990 levels
1996	Public drinking water suppliers required to inform customers about chemicals and microbes in their water; funding made available to upgrade water treatment plants Vast majority of American households now have safe drinking water EPA requires that home buyers and renters be informed about lead-based paint hazards Food Quality Protection Act signed to tighten standards for pesticides used to grow food, with special protections to ensure that foods are safe for children to eat
1997	Executive Order issued to protect children from environmental health risks, including childhood asthma and lead poisoning EPA issues tough new air quality standards for smog and soot, an action that would improve air quality for 125 million Americans Chemical Weapons Convention banning chemical warfare, now signed by 188 nations
1998	Clean Water Action Plan announced to continue making America's waterways safe for fishing and swimming

Continued

TABLE 9.4 Environmental Milestones in the USA, 1970–2006—cont'd

1999	New emissions standards for cars, sport utility vehicles, minivans and trucks, requiring them to be 77–95 percent cleaner in the future EPA announces new requirements to improve air quality in national parks and wilderness areas
2000	EPA establishes regulations requiring more than 90% cleaner heavy-duty highway diesel engines and fuel
2002	Small Business Liability Relief and Brownfields Revitalization Act signed to reclaim and restore thousands of abandoned properties
2003	Healthy Forests Restoration Act signed to prevent forest fires and preserve nation's forests Over 4000 school buses to be retrofitted through the Clean School Bus USA program, removing 200,000 pounds of particulate matter from the air over next decade Clear Skies legislation and alternative regulations proposed to create a cap and trade system to reduce SO_2 emissions by 90 percent and NO_x emissions by 65 percent below current levels
2004	New, more protective, 8-hour ozone and fine particulate standards go into effect across the country. Clean Air Rules of 2004 are proposed that will make people healthier EPA requires cleaner fuels and engines for off-road diesel machinery such as farm or construction equipment
2005	EPA issues the Clean Air Interstate Rule and the Clean Air Mercury Rule
2006	Program to raise awareness about the importance of water efficiency and water-efficient products, and provide good consumer information. Clean Water Restoration Act of 2007 to clarify federal jurisdiction in surface waters in process in Congress
2010	California Desert Protection Act for conservation areas
2012	EPA issues new emission standards for industrial sites

Note: EPA=Environmental Protection Agency; DDT=dichlorodiphenyltrichloroethane; OPEC=Organization of the Petroleum Exporting Countries; PCB=polychlorinated biphenyl; CFC=chlorofluorocarbon; SO_2=sulfur dioxide; NO_x=nitrogen oxides.
Sources: Environmental Protection Agency. Earth day and EPA history [updated 12 July 2011]. Washington, DC: EPA. Available at: http://www.epa.gov/earthday/history.htm [Accessed 15 August 2012].
Environmental Protection Agency. Water sense [updated 20 July 2012]. Washington, DC: EPA. Available at: http://www.epa.gov/watersense/ [Accessed 16 August 2012].

OCCUPATIONAL HEALTH

INTRODUCTION

One of the main functions of the WHO was mandated in Article 2 of its 1946 Constitution: to promote improvement of working conditions and other aspects of environmental hygiene. This mandate led to international recognition that occupational health is closely linked to public health and health systems development. All determinants of workers' health, including risks for disease and injury in the occupational environment, social and individual factors, and access to health services, need to be addressed.

The CDC considers the improvement in workers' health and safety as one of the 10 great achievements of public health in the USA in the twentieth century. The US National Safety Council reports from 1933 to 1997 indicate that deaths from unintentional work-related injuries declined by 90 percent, from 37 per 100,000 workers to 4 per 100,000, a "reduction of the number of deaths from 14,500 to 5,100; during this same period, the workforce more than tripled, from 39 million to approximately 130 million" (CDC, 1999).

Occupational health is the promotion and maintenance of the highest levels of physical, mental, and social well-being of workers in all occupations by preventing departures from health, controlling risks, and adapting of work to people and people to their jobs (International Labour Organization and WHO, 1950). Diseases related to occupations, always an essential part of public health, increasingly relate to environmental health, but to other fields as well. The worker is also a member of a family and a breadwinner, so the health of the worker is related to family health. The worker is concerned not only with what happens at the place of employment but also with hazardous agents that he or she might accidentally bring home. The retired or laid-off worker is worried about well-pensioned and honorable retirement. Occupational health in this wider context has an important place in the New Public Health.

DEVELOPMENT OF OCCUPATIONAL HEALTH

Occupational health is one of the oldest sectors of public health, dating back to Roman times. Documentation of occupational diseases began in 1700 by Bernardino Ramazzini (1633–1714) (see Chapter 1). Historic examples of work-related health hazards and diseases include scurvy among sailors, cancer of the scrotum specific to chimney sweeps in nineteenth-century England, black lung in coal miners, mercury poisoning in hat makers, byssinosis in cotton mill workers, and mesothelioma in asbestos workers. The list is long and extends to musculoskeletal injuries and hepatitis B in hospital workers, spinal disorders in typists, and medial neuritis (carpal tunnel syndrome) in computer users. Interventions vary widely, from the banning of asbestos use to modifying the office work environment through better chairs, exercise breaks, and ergonomic training of workers.

During the early part of the nineteenth century, the harsh working conditions of children, women, and other workers led to parliamentary action to regulate mines and factories, improving conditions generally. The first factory inspectors in the UK were appointed in 1833 to administer the provisions of the Factory and Workshops Acts. In 1898, Thomas Legge became the first medical doctor appointed to the post of Chief Factory Inspector in the UK. He articulated the basic public health approach to workers' health and established the principle that management is responsible for the health of the employees. These issues are termed Legge's axioms and are still relevant to the field of occupational health today (Table 9.5).

Government responsibility for setting standards, monitoring, intervening, and regulating compensation grew slowly over the past century. Case reports, epidemiological studies, and advocacy regarding the effects of lead, asbestos, vinyl chloride, silica, and dust fibers led to steps to reduce the hazards to workers and provided the professional support for legislative initiatives. International standards developed by the League of Nations, the International Labour Organization, and other international organizations promoted development of this field.

THE HEALTH OF WORKERS

Workers are subject to normal health threats for the adult population, but there are specific threats to health associated with the work situation. Workers have lower death rates than the general population because they are demographically different from the general population and even epidemiologically different from a population matched for age and gender. This is due to the fact that there is a process of selection of workers that excludes the severely ill and disabled from employment. The selection process continues with attrition of unhealthy people from the workplace. This is termed the *healthy worker effect* and is a factor that needs to be considered in occupational health studies and practice. Death rates or other population-based norms from the general population may be inappropriate for comparison if this effect is not taken into account. Case matching or control studies may be needed to accommodate this phenomenon. Other population groups such as immigrants or refugees go through similar selection, where only the healthy may be included or survive.

THE BURDEN OF OCCUPATIONAL MORBIDITY AND MORTALITY

In the USA, the workforce is made up of 154 million people, with 66.9 percent between the ages of 25 and 54. The age group 25–54 years is projected to decline to 63.7 percent of the total population in 2020. Workforce participation declined during the 2007–2009 recession by 2.4 percent and is projected to decline further by 2.2 percent by 2020.

Premature disease, injury, and death related to occupational exposures are a major burden on the economy and the health system. The number of fatal and non-fatal injuries in the USA in 2007 was estimated to be more than 5600 and almost 8,559,000, respectively, at a cost of US$6 billion and US$186 billion. The number of fatal and non-fatal illnesses was estimated at more than 53,000 and nearly 427,000, respectively, with cost estimates of US$46 billion and US$12 billion. National costs of occupational injury and illness among civilians in the USA for 2007 were estimated at US$67 billion, and indirect costs were almost US$183 billion, equivalent to the costs of cancer (Leigh, 2011).

The largest numbers of deaths occur in the following industries: construction, transportation/communications/public utilities, and manufacturing (14.0 percent). The decrease in occupation-related deaths from 1980 to 2005 is related to the cumulative effect of increased awareness and regulation of worksite dangers and toxins, as well as new technology and mechanization, changes in the economy, and workforce distributions (CDC MMWR, 1999).

In 2001 in the USA, there were 5.7 non-fatal injuries with lost workdays per 100 employees in the private sector, a reduction of 34 percent from 1992. There was a trend towards substantial improvements in the more dangerous occupations such as agriculture, fishing and forestry, mining, construction, and manufacturing during the 1990s and early twenty-first century, as seen in Table 9.6.

The 10 most frequent work-related diseases and injuries in the USA are:

- lung disease
- musculoskeletal injuries
- cancers
- severe trauma
- cardiovascular disorders
- disorders or reproduction
- neurotoxic disorders
- noise-related hearing loss
- dermatological conditions
- psychological strain and boredom.

TABLE 9.5 Thomas Legge's Axioms on Workers' Health and Modern Equivalents

Legge's Axiom	Modern Version
Unless and until the employer has done everything – everything means a good deal – the workman can do next to nothing to protect himself.	Don't blame the victim; the health of workers is the responsibility of management.
If you can bring an influence to bear, external to the worker, that is one over which he can exercise no control, you will be successful; if you cannot or do not, you will never be wholly successful.	Structural change is best.
Practically all industrial lead poisoning is due to inhalation of dust and fumes.	If you stop the exposure, you stop the poisoning.
All workmen should be told something of the danger of the material with which they come into contact and not be left to find out for themselves – sometimes at the cost of their lives.	Workers have the right to their health at their place of employment.

Sources: Harrington JM. 1998 and beyond – Legge's legacy to modern occupational health. Ann Occup Hyg 1998;43:1–6. Available at: http://www.sciencedirect.com/science/article/pii/S0003487898000751 [Accessed 17 August 2012].
Waldron T. Thomas Morison Legge (1863–1932): the first medical factory inspector. J Med Biogr 2004;12:202–9. Available at: http://www.ncbi.nlm.nih.gov/pubmed/15486616 [Accessed 18 August 2012].

TABLE 9.6 Occupational Injury Death Rates (per 100,000 Workers) by Industry, USA, Selected Years 1985–2010

Industry	1985	1990	1993	2003	2005	2010	% Change 1985–2010
Total civilian workforce	5.8	4.6	4.2	4.0	4.0	3.5	−39.6
Mining	30.0	30.0	25.4	26.9	25.6	19.8	−34.0
Agriculture, fishing, forestry	23.7	18.0	18.5	31.9	32.5	26.8	+11.6
Construction	16.6	14.0	11.8	11.7	11.1	9.5	−42.7
Transportation, communication, public utilities	15.7	10.4	10.1	17.8	17.7	17.1	+8.9
Public administration	6.4	3.8	4.2	2.7	2.4	2.2	−65.6
Manufacturing	4.0	4.0	3.6	2.5	2.4	2.2	−45.0
Wholesale trade	2.8	3.6	3.6	4.2	4.6	4.8	+74.1
Retail trade	2.7	2.8	2.9	2.1	2.4	2.2	−18.5
Services	1.8	1.5	1.4	NA	NA	NA	–

Sources: National Center for Health Statistics. Health, United States, 2007, with chartbook on trends in the health of Americans. Hyattsville, MD: US Department of Health and Human Services; 2007. Available at: http://www.cdc.gov/nchs/data/hus/hus07.pdf [Accessed 17 August 2012].
National Center for Health Statistics. Health, United States, 1998, with socioeconomic status and health chartbook. Hyattsville, MD: US Department of Health and Human Services; 1998. Available at: http://www.cdc.gov/nchs/data/hus/hus98.pdf [Accessed 17 August 2012].
Bureau of Labor Statistics. National census of fatal occupational injuries in 2010 (preliminary report). Washington, DC: US Department of Labor; 2011. Available at: http://www.bls.gov/news.release/cfoi.nr0.htm [Accessed 17 August 2012].

A preliminary total of 4547 fatal work injuries was recorded in the USA in 2010, about the same as the final count of 4551 fatal work injuries in 2009. The rate of fatal work injury for US workers in 2010 was 3.5 per 100,000 full-time equivalent (FTE) workers (Table 9.6), as was the final rate for 2009 (US Bureau of Labor Statistics, National Census of Fatal Occupational Injuries, 2010).

Occupational Health Priorities in the USA

Priorities for research in occupational health are focusing on work-related anxiety and neurotic disorders, hearing loss, musculoskeletal disorders, injuries (fatal and non-fatal), poisoning, respiratory conditions, and skin disorders.

These are the most common occupational health issues and the costliest to the economy (CDC, Worker Health Chartbook, 2004). Injury surveillance in the USA is maintained by the CDC's National Institute of Occupational Safety and Health.

INTERNATIONAL ISSUES IN OCCUPATIONAL HEALTH

The WHO defines occupational health as including: "all aspects of health and safety in the workplace", and occupational health "has a strong focus on primary prevention of hazards. The health of the workers has several determinants, including risk factors at the workplace leading to cancers,

BOX 9.30 Occupational Health Issues in the Global Economy

- Technology transfer from industrial to developing countries or areas within a country
- Child labor, sexual exploitation and slave-like conditions in low income countries
- Pesticide overuse, toxicity, and food contamination
- Ecological damage from toxic waste spills and waste disposal
- Toxic waste transfer from industrial to developing countries
- High-technology industrial toxic wastes
- Nuclear energy, accidents, and wastes
- Technological and professional common interest between occupational and environmental health
- Poor safety and control standards in former Soviet and developing countries
- Poor wages, psychological stress, boredom, and shift work
- Management negligence and lack of accountability for workplace safety
- Governmental negligence and corruption in developing regulatory role
- Dangerously inadequate health and safety measures in developing countries, e.g., Bangladesh clothing workers disasters 2013
- Widening income gap between upper and lower income groups

accidents, musculoskeletal diseases, respiratory diseases, hearing loss, circulatory diseases, stress related disorders and communicable diseases and others. Employment and working conditions in the formal or informal economy embrace other important determinants, including, working hours, salary, workplace policies concerning maternity leave, health promotion and protection provisions" (WHO, 2012).

WHO is implementing a Global Plan of Action on Workers' Health 2008–2017, endorsed by the World Health Assembly in 2007, with the following objectives:

- devising and implementing policy instruments on workers' health
- protecting and promoting health at the workplace
- improving the performance of and access to occupational health services
- providing and communicating evidence for action and practice
- incorporating workers' health into other policies.

Occupational health has become an international issue as the global economy enables the transfer of manufacturing from one country to another with great speed and ease (Box 9.30). This is often motivated by lower wages, and also by lower occupational and environmental regulatory controls and less stringent or non-existent legal protection against toxic exposures and child labor in developing

countries. Transfer of occupational hazards from industrialized to non-industrialized countries has become an issue in international cooperation and trade agreements. Developed countries have stricter environmental regulations and worker organization than developing countries that are anxious for job-producing industry at any price.

In 2000, occupational risk factors were responsible worldwide for 37 percent of back pain, 16 percent of hearing loss, 13 percent of chronic obstructive lung disease (COPD), 11 percent of asthma, 8 percent of injuries, 9 percent of lung cancer, and 2 percent of leukemia. These risks at work caused 850,000 deaths worldwide and resulted in the loss of about 24 million years of healthy life. Needle-stick injuries accounted for about 40 percent of hepatitis B and hepatitis C infections and 4.4 percent of HIV infections in health care workers. Exposure to occupational hazards accounts for a significant proportion of the global burden of disease and injury, which could be substantially reduced through application of proven risk prevention strategies.

NATIONAL AND MANAGEMENT RESPONSIBILITIES

In the USA, workers' health benefits cost more than the steel to make a car. As a result, there is a growing interest on the part of both management and workers in promoting workers' health through improved nutritional monitoring of canteens and cafeterias, antismoking activities, and physical fitness programs. The management interest in a healthier workforce to contain rising health care costs is part of the modern corporate culture. The primary responsibility, however, legally and morally, lies with management, in addition to protecting the worker by monitoring risks, providing a safe environment, and providing care at the time of injury (Box 9.31).

More than 23 million Americans have asthma and a further 13.6 million have COPD. The workplace environment often results in exposure to many chronic respiratory hazards among workers. Examples of respiratory hazards identified in the past 10 years include butter-flavouring chemicals; minifibers (flock) of nylon, rayon, and polypropylene; and dust from the World Trade Center. Analysis of workplace asthma in the USA showed that among adults an annual average of 1.4 million cases of workplace-related asthma could be prevented (Box 9.32).

Occupational injuries and illnesses are social as well as engineering and management concerns. Compensation, litigation, class-action suits, and union action are all associated with increasing awareness of toxic and trauma effects on workers, and court decisions regarding management liability. The field is made more complex because some occupational illness may occur long after the exposure: silicosis, asbestos-related mesothelioma, and asbestosis may develop after a long latency of up to 20–30 years following exposure.

BOX 9.31 Principal Tasks of Occupational Health

- *Anticipation* – dealing with potential disease and injury to include preparation for prevention as facilities are planned or renovated.
- *Surveillance and monitoring* – assuring timely and accurate identification, reporting, and recording of occupational disease and injury; medical surveillance: passive or active and industrial hygiene and safety.
- *Right to know* – for workers, health professionals, community at large.
- *Epidemiological analysis* – analyzing collected data; linking exposure to outcome data helps to locate trends, clusters, associations, and causes of disease and injury for more in-depth investigation and prevention.
- *Exposure reduction* – minimizing toxic exposure, to prevent approaching or exceeding established limits.
- *Substitution* – substituting less toxic substances.
- *Awareness* – promoting awareness at government, management, community, worker, and consumer levels.
- *Government regulation* – on-site supervision by regulatory agencies; publication of standards of exposure and "good practices".
- *Compensation* – compensating for illness and loss of life related to work accidents, toxicity, and stress.
- *Management–worker cooperation* – recognizing that worker participation in health and safety is of mutual benefit.

Source: *Weeks JL, Levy BS, Wagner GR, editors. Preventing occupational disease and injury. Washington, DC: American Public Health Association; 2005.*

BOX 9.32 Work-Related Asthma, USA (38 States and District of Columbia), 2006–2009

The CDC reports that work-related asthma (WRA) includes exacerbation of pre-existing asthma concurrent asthma worsened by factors related to the workplace environment. Occupational asthma is new-onset asthma attributed to the workplace environment. WRA is a preventable occupational lung disease associated with serious adverse health and socioeconomic outcomes.

Among workers with similar occupational exposures, WRA diagnosis offers unique opportunities for prevention. The American Thoracic Society estimated that for 15 percent of US adults with asthma the condition is attributable to occupational factors.

To estimate current asthma prevalence and the proportion of asthma that is WRA, CDC analyzed data from the 2006–2009 Behavioral Risk Factor Surveillance System (BRFSS) from 38 states and the District of Columbia (DC).

Among ever-employed adults with current asthma, the overall proportion of WRA was 9.0 percent, ranging from 4.8 to 14.1 percent in different states. Proportions of WRA were highest among people aged 45–64 years (12.7 percent), blacks (12.5 percent), and people of other races (11.8 percent).

Source: *Centers for Disease Control and Prevention. Work-related asthma – 38 states and District of Columbia, 2006–2009. MMWR Morb Mortal Wkly Rep 2012;61:375–8. Available at: http://www.cdc.gov/mmwr/preview/mmwrhtml/mm6120a4.htm [Accessed 17 August 2012].*

Follow-up of exposed workers may be difficult, and issues such as compensation may also be complicated. Occupational health involves a governmental regulatory function and legislated responsibility to protect workers from toxic or physical risks at the worksite and those exposed to the workers and the products found in the environment (e.g., asbestos fibers on workers' clothing). In 2012, preliminary findings were reported of excess cancer rates among people working with asphalt roofing, a reminder of Percivall Pott's findings of cancer of the scrotum among young chimney sweeps in the late eighteenth century.

Standards and Monitoring

Monitoring of occupational health involves a set of activities designed to increase the safety and protection of the workers. It involves a number of parallel services to promote the health of the individual worker and the safety of the work environment, and should be coordinated in an overall strategy.

In the USA prior to 1970, prevention of occupational injuries, death, and disease was governed by state and local government or market forces. Federal initiatives to raise standards of occupational health and safety were mandated in the Occupational Safety and Health Act of 1970, which established two government agencies to implement the Act, the Occupational Safety and Health Administration (OSHA) and the CDC's National Institute of Occupational Safety and Health (NIOSH). OSHA is responsible for promulgation and enforcement activities, within the US Department of Labor. OSHA sets standards based on consensus derived from professional organizations in consultation with labor, industry, and health authorities, meant to promote safety and reduce risk for employees and set performance standards for employers. NIOSH was established to conduct research related to the objects of the act for occupational disease, particularly those derived from exposure to toxic physical and chemical agents.

The Act provides an environment for regulation and study of occupational health issues including public petitions, court decisions, and new research findings used to formulate priorities for standards development. Monitoring is done by a combination of federal, state, and local health authorities with participation of professional and industrial organizations. The legal responsibility for worker safety and health is placed with the employer (Table 9.7), but worker awareness and participation in safety programs are vital to a successful approach.

TABLE 9.7 Management and Governmental Responsibilities in Workers' Health

Management Responsibility	Governmental Responsibility
Substitute less dangerous materials	Legislation – substitute, ban, define legal responsibility (civil and criminal compensation)
Enclose/separate dangerous materials	Regulation to set and enforce standards for toxic emissions and controls
Process exhaust fumes	Litigation – civil suits vs. compensation
General ventilation	Test environment and workers with notification of test results
Good housekeeping	Designation of hazardous materials, labeling, disposal and reporting requirements
Monitor health of workers	Monitor health of workers
Personal protection	Educate managers and workers
Investigate potential unhealthy events	Research – scientific and operational
GMP (good manufacturing practices)	Regulate compensation for income loss and health damage
Payment for workers' compensation protection	Compensation for injury or disease

OCCUPATIONAL HEALTH TARGETS

The US Surgeon General's report *Healthy People 2020* formulated a number of targets for occupational health and safety issues (Table 9.8). These are national targets that are also being adopted by state departments of health and have organizational as well as legal implications. The Midcourse Review (2010) identifies the progress made since establishment of the goals and targets. The review found difficulties in the systematic evaluation of actions and in establishing that some medical conditions are related to the job. The review also found a lack of awareness of prevention methods and risks for some conditions that are otherwise preventable. There is renewed effort in activities such as research, surveillance, new preventive measures, and information dissemination and training.

OCCUPATIONAL INJURY

The majority of employed healthy individuals who are able to contribute to society on a daily or weekly basis spend a large portion of their lives at work. Spending extensive amounts of time in the workplace allows for the possibility of occupational injury; however, risks and type of injury are, as one would expect, influenced by work category or the task being carried out. New and interesting technological tools are being developed in an effort to prevent or limit an employee's physical distress (Box 9.33). Thus, the same tasks can be achieved with significant less risk of injury, which in turn reduces workers' compensation claims as well as lost workdays. In the USA, major progress has been made in workplace safety. While this is a great achievement, the importance of this public health measure crosses all borders, and there are advances to be made in occupational health globally. Progress in reduced mortality from occupational injuries is shown in Figure

9.6. The introduction of federal workplace safety initiatives was associated with a reduction of the deaths in miners; this is one of the greatest public health achievements of the twentieth century (Figure 9.7).

TOXICITY AT THE WORKPLACE AND IN THE ENVIRONMENT

Toxic substances are widely used in industry, not only in manufacturing but also in services such as laboratories, and they constitute a major concern in both occupational and environmental health. Extensive information on toxic substances is published by the WHO and the CDC.

Much of the concern in occupational health has been on detection, prevention, and reduction of exposure to toxic materials in the workplace, but more recently concern has increased with regard to contamination of the surrounding environment. The scientific knowledge of toxins used in occupational settings and their sources, uses, effects, actions, and target organs is extensive.

Factors that affect the toxicity of an agent include the extent and duration of exposure, and host factors such as age, gender, fitness, previous exposure, and compounding risk factors such as smoking and nutritional status. Environmental factors include temperature and air flow, as well as the physical and chemical properties of the toxic agent. The following examples of toxic substances and the history of measures to control them illustrate the complexity of this problem.

Lead

Lead is a mineral with thousands of applications because of its plasticity and softness. Lead poisoning has been a worker hazard since ancient times. Lead enters the body

TABLE 9.8 US Health Targets in Occupational Safety and Health for 2020

Subject	Previous (Baseline)	Target
OSH-1: Reduce death from work-related injuries	4 work-related injury deaths/100,000 workers (2007)	3.6/100,000
OSH-2: Reduce work-related injuries (non-fatal), resulting in medical treatment, lost time from work, or restricted work activity	4.2 injuries/100 full-time workers (2008)	3.8/100
OSH-3: Reduce the rate of injury and illness cases involving days away from work due to overexertion or repetitive motion	29.6 injury and illness cases/10,000 workers	26.64/10,000
OSH-4: Reduce pneumoconiosis deaths	2430 pneumoconiosis deaths (2005)	2187 deaths
OSH-5: Reduce deaths from work-related homicides	628 work-related homicides (2007)	565 deaths
OSH-6: Reduce work-related assaults	8.4 assaults/10,000 workers (2007)	7.6/10,000
OSH-7: Reduce the proportion of people who have elevated blood lead concentrations from work exposures	22.5/100,000 employed adults have elevated blood lead concentrations	20.2/100,000
OSH-8: Reduce occupational skin diseases or disorders	4.4 occupational skin diseases/100,000 workers (2008)	4/10,000
OSH-9 (Developmental): Increase the proportion of employees who have access to workplace programs that prevent or reduce employee stress		
OSH-10: Reduce new cases of work-related, noise-induced hearing loss	2.2 new cases of work-related hearing loss/10,000 workers (2008)	2/10,000

Note: OSH= occupational safety and health.
Source: Healthy People: Occupational health and safety [updated 26 July 2012]. Washington, DC: US Department of Health and Human Services. Available at: http://www.healthypeople.gov/2020/topicsobjectives2020/objectiveslist.aspx?topicId=30 [Accessed 18 August 2012].

through inhalation and ingestion, affecting the gastrointestinal, nervous, hematological, and circulatory systems. It is associated with intestinal colic, encephalopathy, delirium, and even coma in its acute forms. Chronic forms of plumbism or lead poisoning cause mental dullness, headache, memory loss, neurological defects (wrist drop), anemia, and a blue line on the gums.

Lead toxicity has been a traditional health problem of glaziers and potters because of lead use in the manufacturing process. Wines or rum produced and stored in lead containers or in pewter (lead–tin alloy) utensils were known to be associated with the "dry gripes" in the seventeenth and eighteenth centuries. The Devonshire colic, described in 1776 by George Baker, was widespread for more than 100 years in parts of England where cider was made and stored in lead containers.

Lead toxicity and excess exposure in the workplace remain problems in the USA. Lead-induced hypertension, neuropathy, carcinogenesis, reproductive damage for men, and abortion for women are the major toxic effects. In 1995, CDC's NIOSH Adult Blood Lead Epidemiology and Surveillance Program, which monitors elevated blood levels among adults, reported a continuing hazard of work-related exposures as an occupational hazard in the USA. Studies of lead exposure in industrial settings in the USA have shown widespread exposure above permissible exposure limits, in the traditional high-exposure industries such as primary and secondary lead smelting, battery and pigment manufacturers, brass/bronze foundries, and 47 other industries.

Workers with the highest exposure jobs throughout industry were painters.

Occupational exposure continues to be an important source of lead toxicity. OSHA standards promulgated in 1978 came at a time when lead prices dropped, reducing the number of producers and the degree of compliance overall.

Concern over lead toxicity evolved from strictly occupation-related to environmental toxicity in which both the exposed worker and the general population are adversely affected by this widely used metal. In 1997, the CDC adopted a BLL standard of less than 10 μg/dl, a level at which a negative effect on cognitive development is recorded. The US NHANES found that between 1976 and 1980, and 1980 and 1991, geometric mean BLLs of people aged 1–74 in the USA declined from 12.8 μg/dl to 2.9 μg/dl, and even further in 1991–1994 to 2.3 μg/dl. The geometric mean BLL for the US population aged above 1 year decreased by 30 percent from 1991–1994 to 1999–2002; the prevalence of elevated BLLs decreased by 68 percent overall and by 64 percent for children aged 1–5 years, largely due to very active regulatory and inspection services of low-income housing and lead paint removal efforts in housing rehabilitation (CDC, 2005).

Despite major improvements (see Box 9.17), some 1.7 million children aged 1–5 in the USA still have BLLs above 10 μg/dl. Further progress in BLLs will require reduction in lead hazards in housing and reduced contact with lead-contaminated dust, house paint lead, and worksite exposure. Work-related and environmental lead exposures continue to

The Centers for Disease Control and Prevention designated workplace safety as one of the ten great achievements of public health in the USA in the twentieth century. In the early years of that century, workers in the USA faced high health and safety risks at work. Large decreases in work-related deaths and injuries resulted from the combined efforts of government regulators, unions, employers, scientists, individual workers, and others. Despite reductions in injury, disease, and death from these improved work conditions, "much work remains, with the goal for all workers being a productive and safe working life and a retirement free from long-term consequences of occupational disease and injury", especially in mining and forestry (CDC, 1999).

Since 2000, progress has continued to improve working conditions, to reduce the risk of workplace-associated injuries. Patient lifting has been a substantial cause of lower back injuries among the 1.8 million US health workers in nursing care and residential facilities. Patient-handling programs introduced mechanical patient-lifting equipment, reducing by 66 percent rates of workers' compensation injury claims and lost workdays. Investment in lifting equipment can be recovered in less than 3 years. Following widespread dissemination and adoption of these practices by the nursing home industry, US Bureau of Labor Statistics data showed a 35 percent decline in lower back injuries in residential and nursing care employees between 2003 and 2009.

The annual cost of farm-associated injuries among young people has been estimated at US$1 billion annually. A comprehensive childhood agricultural injury prevention initiative was established to address this problem. Among its interventions was the development by the National Children's Center for Rural Agricultural Health and Safety of guidelines for parents to match chores with their child's development and physical capabilities. Follow-up data have demonstrated a 56 percent decline in youth farm injury rates from 1998 to 2009 (National Institute for Occupational Safety and Health, unpublished data, 2011).

In the mid-1990s, crab fishing in the Bering Sea was associated with a rate of 770 deaths per 100,000 full-time fishermen, mostly due to overloading of vessels which overturned because of their heavy loads. In 1999, the US Coast Guard implemented Dockside Stability and Safety Checks to correct stability hazards. Since then, one vessel has been lost and the fatality rate among crab fishermen has declined to 260 deaths per 100,000 full-time fishers (see Plimsoll line, Chapter 1).

Sources: *Centers for Disease Control and Prevention. Achievements in public health, 1900–1999: Improvements in workplace safety – United States, 1900–1999. MMWR Morb Mortal Wkly Rep 1999;48:461–9. Available at: http://www.cdc.gov/mmwr/preview/mmwrhtml/mm4822a1.htm [Accessed 17 August 2012].*
Centers for Disease Control and Prevention. Ten great public health achievements – United States, 2001–2010. MMWR Morb Mortal Wkly Rep 2011;60:619–23. Available at: http://www.cdc.gov/mmwr/preview/mmwrhtml/mm6019a5.htm [Accessed 17 August 2012].

be public health problems in the USA, requiring continued diligence on the part of pediatricians and internists as well as occupational and public health workers.

The Upper Silesia region of Poland, with a population of 4 million, is the site of many non-ferrous metal plants, especially using lead and zinc. In the Katowice district there are four such plants, two of which are more than a century old and have a high output of atmospheric lead, and two built in the 1960s with inadequate pollution control equipment. Although emissions of lead and cadmium from one plant reportedly fell during the late 1980s, high levels of blood lead and cadmium are found in children, and soil contamination is extensive, including high levels of contamination of vegetables. This problem is widespread in Eastern Europe.

Reduction in lead exposures has been achieved in the USA by a combination of legislation and professional and social pressures, resulting in the adoption of lead-free gasoline, removal of lead from paints, and its substitution in many industrial practices. Awareness and active lobbying by public health-minded groups have had a beneficial effect in reducing lead toxicity in the community and workplace. The CDC in May 2012 announced a change in the "level of concern" from $10\,\mu g/dl$ to $5\,\mu g/dl$. The APHA welcomed this change, which will increase the number of children in the USA needing follow-up from 100,000 to 450,000. However, the 2011 budget for the US lead screening program was reduced from US$29 million to US$2 million: the APHA is calling for restitution and even expansion of the budget. The APHA is promoting a wide-ranging program of further abatement of lead paint hazards including litigation against manufacturers, lead-safe workplaces, home renovations, and community-based prevention and health education programs (APHA, 2012).

Asbestos

Asbestos is a commercial name for six different fibrous materials that have many features making them attractive for many diverse uses in commercial and military products for fireproofing, automotive brakes, textiles, cement products, and wallboard material. The fibers fragment easily and remain in the air, where they can be inhaled readily by workers or others exposed to the asbestos-containing material, during its use or disposal.

Asbestos-related disease is an occupational and public health problem that grew from the rapid increase in the use of asbestos during World War II. It left a legacy of death and disease that only became apparent many years later. Fibrotic lung disease resulting from asbestos exposure was called *asbestosis* by W. E. Cooke in 1927. A subsequent British government investigation of the subject reported to Parliament that inhalation of asbestos dust over a period of years results in the development of a serious type of fibrosis of the lung, and recommended dust suppression measures. This report was followed by many

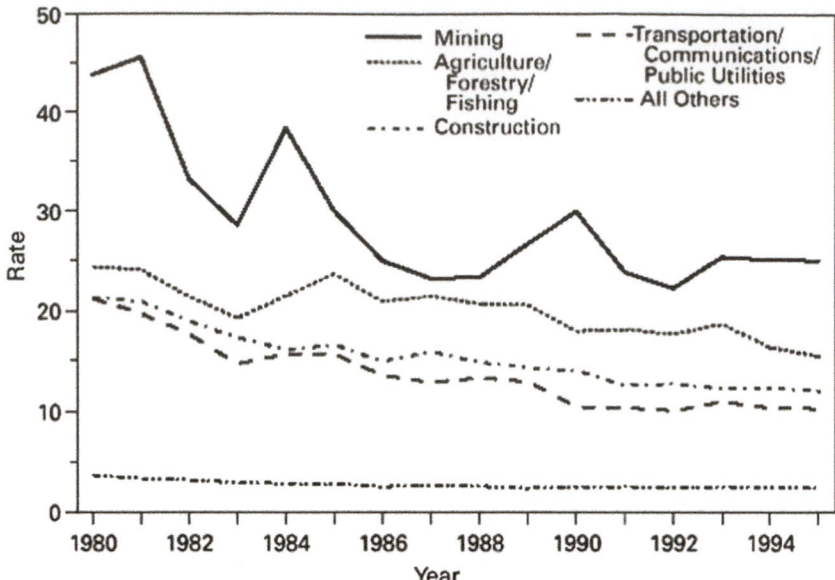

FIGURE 9.6 Occupational injury death rates per 100,000 workers, by industry, USA, 1980–1995. Note: "All Others" includes public adminis-tration, manufacturing, wholesale trade, retail trade, services, and finance/insurance/real estate. *Source: Centers for Disease Control and Prevention. Achievements in public health, 1900–1999: Improvements in workplace safety – United States, 1900–1999. MMWR Morb Mortal Wkly Rep 1999;48:461–9. Available at: http://www.cdc.gov/mmwr/preview/mmwrhtml/mm4822a1.htm [Accessed 17 August 2012].*

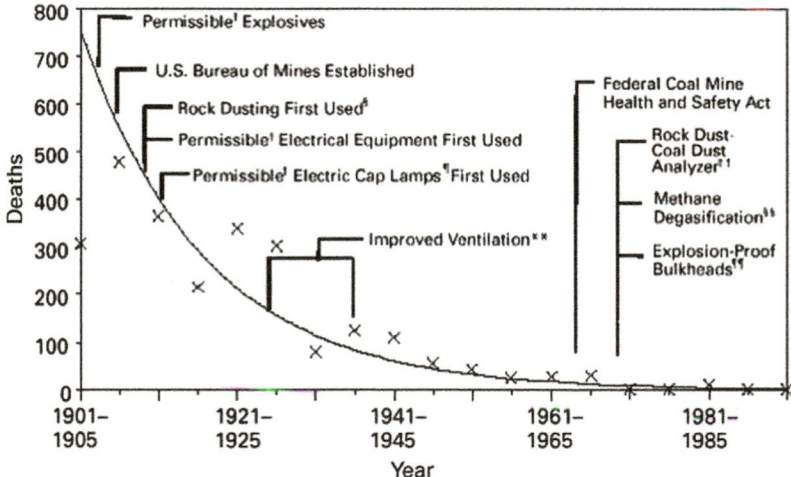

FIGURE 9.7 Public health achievements: reduction in deaths of miners with safer equipment and improved ventilation 1901–1995. Note: Each X represents the 5-year average of the number of deaths resulting from explosions; the line is a smoothed regression line through the 5-year averages. †Explosives and equipment that can be used in an explosive methane-rich environment without causing a methane explosion. §The process of applying a layer of rock dust over the coal dust, which creates an inert mixture and inhibits a coal dust explosion. ¶Lamps worn on miners' caps. ** Ventilation improvements, including the use of reversible fans, reduce the concentration of methane and remove the explosive gas from the mine. ††A handheld monitor that provides instantaneous readings of the rock-to-coal dust mixture to ensure that it is inert. §§Techniques to remove methane from the coal bed before mining the coal. ¶¶Explosion-proof walls used to seal abandoned (mined-out) areas to protect workers in active parts of the mine. *Source: Centers for Disease Control and Prevention. Achievements in public health, 1900–1999: Improvements in workplace safety – United States, 1900–1999. MMWR Morb Mortal Wkly Rep 1999;48:461–9. Available at: http://www.cdc.gov/mmwr/preview/mmwrhtml/mm4822a1.htm [Accessed 27 June 2013].*

case reports and the wide recognition of the health hazards associated with asbestos exposure. During World War II, the US Navy issued minimum requirements for safety in shipyards contracting for naval work, involving some 1 million workers.

The first reports of an association between asbestos and lung cancer began to appear in the 1930s. Studies by Irving Selikoff in 1965 in New York reported high rates of lung cancer

in several large population groups of ex-shipyard workers. Selikoff and colleagues also showed a synergistic relationship between asbestos exposure and cigarette smoking (Table 9.9); namely, a greater risk of lung cancer with heavier smok-ing and a reduction in risk following cessation of smoking. Similar dose-related findings for lung cancer among smoking ex-asbestos workers were reported for the period 1985-90 in Stockholm Sweden (Gustavvson et al., 2002).

TABLE 9.9 Lung Cancer Death Rates (Age Standardized) for Workers Exposed to Asbestos Dust and Cigarette Smoking with Controls

Group	Asbestos Exposure	Cigarette Smoking	Death Rate	Mortality Difference	Mortality Ratio
Control	No	No	11.3	0	1.0
Asbestos workers	Yes	No	58.4	+47	5.2
Control	No	Yes	122.6	+111	10.9
Asbestos workers	Yes	Yes	601.6	+590	53.2

Note: Rates per 100,000 person-years, age standardized, on 12,051 asbestos-exposed workers followed prospectively between 1967 and 1976. Controls included 73,763 similar men in a prospective American Cancer Society study for the same decade. The number of lung cancer deaths is based on death certificate information.
Sources: Hammond EC, Selikoff IJ, Seidman H. Asbestos exposure, cigarette smoking, and death rates. Ann N Y Acad Sci 1979;330:473–90. Available at: http://www.ncbi.nlm.nih.gov/pubmed/294198 [Accessed 17 August 2012].
Frank AL. Public health significance of smoking–asbestos interactions. Ann N Y Acad Sci 1979;330:791–794. Available at: http://www.ncbi.nlm.nih.gov/pubmed/294223 [Accessed 18 August 2012].
Selikoff IJ. Asbestos-associated diseases. In: Last JM, editor. Maxcy-Rosenau: Public health and preventive medicine. 12th ed. Norwalk, CT: Appleton-Century-Crofts; 1986.
Case BW. Asbestos, smoking, and lung cancer: interaction and attribution. Occup Environ Med 2006;63(8):507-508. doi: HYPERLINK "http://dx.doi.org/10.1136%2Foem.2006.027631" \t "pmc_ext" 10.1136/oem.2006.027631PMCID: PMC2078131
Gustavvson P, Nyberg F, Pershagen P, Patrik-Schele P, Jakobson R, Plato N. Low-Dose Exposure to Asbestos and Lung Cancer: Dose-Response Relations and Interaction with Smoking in a Population-based Case-Referent Study in Stockholm, Sweden. Am J Epidemiol 2002;155:1016–22. Available at: http://aje.oxfordjournals.org/content/155/11/1016.full.pdf+html (accessed 22.12.2013)

The US Toxic Substances Control Act of 1976 placed the responsibility for harmful chemicals, including asbestos, on those who would profit from their sale. The long time-lag between the first reports of asbestos-related disease followed by definitive studies and implementation of control measures raised questions as to the way in which occupational health functions. As a result of these studies and the regulatory responses by federal legislators, there was a fourfold reduction in asbestos use in the USA from 1972 to 1982. In 1986, the US Asbestos Hazard Emergency Response Act reinforced federal regulation of asbestos use.

Asbestos exposure is accepted as the cause of mesothelioma, a highly malignant cancer of the chest or abdominal lining. The latency period may be 20–30 years or more, and the risk of this disease as well as lung cancer is related to the extent of exposure and to cigarette smoking. The exposure may occur in asbestos-cement production, shipyard workers, garage workers exposed to brake linings, plumbers, and construction workers using asbestos-based products.

During the 1980s, concern was expressed that asbestos was being exported to developing countries lacking the regulatory mechanisms of the developed world. In the 1990s, there was still some concern that asbestos products manufactured in developing countries were being imported to developed countries. In 1999, the EU effectively banned the use of asbestos products. In March 2012, CDC's NIOSH stated: "Although the use of asbestos and asbestos products has dramatically decreased in recent years, they are still found in many residential and commercial settings and continue to pose a health risk to workers and others".

Silica

Silica is a non-metallic element occurring in rocks, clay, and sand. It is widely used in the manufacture of glass and clay products. It is inert but becomes biologically active when inhaled as dust. It causes silicosis, a progressive disabling pulmonary fibrosis.

Silicosis is one of the oldest known occupational diseases, affecting miners in particular. It was described in ancient Greece and Rome as the "fatal dust". Silica occurs in minerals and rocks throughout the world either as free silica or combined in quartz, flint, or sandstone. Mining, tunneling, stone cutting, quarrying, iron and steel works, sandblasting, brick making, polishing of stone, glass, and metals, and many other industries expose workers to inhalation of silica dust (Box 9.34).

Silicosis is a condition of massive fibrosis of the lungs resulting from prolonged inhalation of silica dust. It is classified as a pneumoconiosis, a general inflammatory fibrotic lung condition caused by inhalation of dust particles. This condition can progress through mild symptoms to shortness of breath, with radiological evidence of pulmonary consolidation and concomitant tuberculosis. Silica is also associated with lung cancer, tuberculosis, and bronchial airway diseases. It is also associated with development of autoimmune disorders, chronic renal disease, and other adverse health effects.

Studies of hard coal miners in the nineteenth century documented the effects of silicosis. By 1918, English workers could receive disability compensation for silicosis and tuberculosis. In the 1920s and 1930s in the USA, studies showed silicosis in cement production workers, anthracite miners, tunnel workers, lead–zinc miners, and other hard rock miners. In the mid-1930s, an estimated 700 US workers died as a result of construction of the Hawk's Nest Tunnel in Gauley Bridge, Fayette County, Virginia, leading to compensation laws covering workers with silicosis. At present, there is still controversy over legally enforceable standards, and the problem remains difficult to prevent.

Cotton Dust (Byssinosis)

Cotton dust has been a common cause of COPD among long-term workers in textile industries, widespread in the USA until the 1960s. OSHA promulgated new standards in 1978 based on assessment of the potential of improved ventilation and filtration, and improved machinery use. The industry at that time was in the process of replacing old equipment with modern and more automated machines, which gave improved production speed, more effective use of floor space, reduced labor input, and a higher quality product, along with lower dust levels. The technical and economic feasibility of the higher standard was correct, and compliance by industry exceeded early expectations at about one-third of anticipated costs.

Vinyl Chloride

Vinyl chloride is a colorless, flammable gas with a faintly sweet odor. It is an important component of the chemical industry because of its flame-retardant properties, low cost, and many end-product uses. It is also a carcinogen causing liver, brain, and lung cancer, as well as spontaneous abor-tion. Vinyl chlorides are dangerous primarily when inhaled or ingested. Vinyl chloride usage increased from the 1930s and more dramatically after the end of World War II until the 1970s. In the 1960s, polyvinyl chloride (PVC) was shown to be associated with Raynaud's phenomenon and later with malignancies, including hemangiosarcoma of the liver.

The carcinogenicity of PVC was established as a result of the review of all evidence in 1974 by the US Office of Tech-nology Assessment and OSHA. Scientists concluded that there was no safe level of exposure to vinyl chlorides. OSHA adopted 1 part per million as the maximum possible dose. While the risk assessment issues are still controversial, reduc-tion of exposure to workplace carcinogens such as vinyl chlo-ride is the accepted standard of modern occupational health.

Despite the industry's vigorous opposition to this reduc-tion in permissible emission level, full compliance was achieved within 18 months by improving ventilation, reduc-ing leaks, modifying reactor designs and chemical path-ways, and using greater automation of the process. Even more effective was a major improvement in the production of PVC using less vinyl chloride. The costs to industry of reducing exposure levels were less than 25 percent of expected costs because of unanticipated innovations in the production process.

Agent Orange

The herbicide Agent Orange, which was used widely by US armed forces in the 1960s and early 1970s during the Vietnam War to defoliate large areas of the country, contains dioxin and is carcinogenic. High levels of dioxin have been found in the breast milk, adipose tissue, and blood of the Vietnamese population. Even though sampling has not been systematic, studies carried out between 1984 and 1992 found high levels of dioxin-like contaminants (2,3,4,8-tetrachloro-dibenzo-*p*-dioxin, TCDD) in blood samples of the Vietnam-ese population exposed to Agent Orange during the war.

Studies of effects among US veterans of the Vietnam War have not produced convincing evidence of long-term effects. Additional studies will be needed to verify effects such as increased cases of cancer or birth defects. However, court and compensation decisions have been made in favor of veterans exposed to Agent Orange despite inconclusive epidemiological evidence of its ill-effects on health.

WORKPLACE VIOLENCE

Violence is endemic in many societies and affects many organizations and institutions. Violence has become a lead-ing cause of fatal injuries in the workplace. Violence in the health setting has an extensive history, with the first docu-mented case in 1849 when a patient fatally assaulted a psy-chiatrist in a mental health care facility. Since then there have been many other studies reporting assaults, hostage taking, rapes, robbery, and other violent acts in the health care and community settings. During the 1990s, homicide became the leading occupational cause of death for females and the second leading cause, after motor vehicle accidents, for men in the USA.

Small businesses have special risk factors. Murder of small business employees in the USA and other countries has become a major occupational problem. Some work-places such as convenience stores with few employees are at particular risk. Between 1992 and 2010 there were 13,827 workplace homicides in the USA. More than half of these violent deaths occurred in three workplace classifications, with 28 percent in sales and related occupations. Protective measures are being introduced and the OSHA provides small businesses with free safety and health guidelines and consul-tations on worker protection. In 2010 these covered 30,000 small businesses with 1.5 million workers (OSHA, 2013).

Shocking incidents of violence and homicide have occurred in which bombs and handguns were used in assassinations of health workers in clinics carrying out abortions, while health workers have also been assaulted and murdered in hospitals and other settings. The US target of work-related homicides for the year 2010 was 0.4 per 100,000 workers over 16 years of age.

Homicide at work has only recently been addressed as an occupational hazard, and research in this area is in its infancy. No universal standards exist to protect workers from work-related violence, and no policy has been created to protect workers. Preventing violence in the workplace is essential and must be addressed at the national level. The California Occupational Safety and Health Authority promulgated guidelines with an emphasis on preventing violence before it occurs, by developing an effective policy to ensure workplace safety. Management and workers' organizations as well as the health system share responsibility. Prevention of drug, alcohol, and sexual abuse or exploitation at work is vital to eliminate workplace violence.

OCCUPATIONAL HEALTH IN CLINICAL PRACTICE

The clinical physician should be aware of the patient's occupation and previous work history. The inclusion of questions related to workplace factors of current or past employment (Table 9.10) may be crucial in the investigation of patients; without this information it may be impossible to determine the cause of the disease. The health care provider should be aware of industries in the community and their potential hazards. The clinician is particularly important because he or she may be the first to see index cases of toxicity. The clinician should ask simple questions, such as the following: What is your job or hobby? What do you do at work? Are you exposed to any chemicals at work or at home? Are there others at work with similar exposure and similar symptoms? How long have you been exposed to these chemicals? Clinical suspicion is the key to finding a potential toxic cause to a set of symptoms, and may uncover a wider public health problem.

INSPECTING THE PLACE OF WORK

The public health authority responsible for health at the place of work may be under the authority of a Ministry of Labor or under a public health authority. Site inspection provides a guide to management and workers for safety and health issues. Non-compliance with federal, state, or local standards should lead to regulatory action to correct deficiencies and should include, if necessary, punitive damages to management. Examination of the worksite involves on-site observations as listed in Table 9.10. The inspection should be documented and made available to management, workers, and follow-up inspections.

RISK ASSESSMENT

Identifying and quantifying occupational and environmental risks may be difficult, but clinical or public health observations, supplemented by epidemiological analysis, can identify toxic or carcinogenic factors that can be reduced or eliminated by public health intervention. High levels of awareness by clinicians of potential health effects from environmental or occupational exposures can help in the identification of index cases just as in infectious disease, leading to an investigation and removal of the cause. Similarly, epidemiological small area analysis can identify populations at high risk for cancers or other toxic effects, providing localization for further investigation.

The establishment of dose–response relationships requires well-conducted observational studies. Some studies may be so insensitive as to dismiss risks that are at low levels of statistical significance, but still represent preventable risks that can be sufficient to warrant compensation. This was the case for veterans in the USA who were exposed to Agent Orange in Vietnam in the 1960s and those suffering from effects attributed to toxic exposures in the Gulf War in 1991 and Iraq War (Second Gulf War) of 2003–2008.

Regulatory and compensation decisions must often be made in the face of inconclusive or contradictory evidence from epidemiological studies. In the 1960s, the FDA used the Delaney Clause applied to food additives or coloring in which any degree of ill-effect noted in animal studies was enough to disqualify a drug from acceptability, but this has not become an accepted legal standard. The topic remains one of controversy and contradiction, with cases providing precedents that affect future court and regulatory decisions. The contribution of epidemiology to resolving such issues also remains controversial.

PREVENTING DISASTERS IN THE WORKPLACE

A disaster in a workplace can affect the workers and the surrounding community. The major responsibility for prevention lies with management, but the worker and society also have roles in the process. Prevention involves education of workers and management, and constant vigilance. Government has the overall responsibility to legislate and enforce standards, safe conditions of work, and control of toxic materials, and to ensure fair compensation for injury or disease. The simple qualitative observations listed in Table 9.11 can provide a useful picture of the disaster management capacity of a worksite. These observations can be made by management, health professionals, and workers' representatives to monitor and promote improved worker health and safety.

The principle of "good worksite practice" is parallel to good manufacturing practices required by food and drug authorities. It is based on the concept that current standards of acceptable safety involve standards of facilities, staffing,

TABLE 9.10 Factors for Walk-Through Inspection of Worksites

Marker	Observations of Conditions, Safety Arrangements, and Effects on Workers
Sensory effects	Eye irritation, poor lighting, noise levels, metallic taste in air, visible fumes, exhaust, temperature (heat/cold)
Safety devices	Use of hard helmets, welding masks, safety shoes and clothing, ear protectors, eye and face protectors, first aid facilities, respirators, monitoring procedures
Storage	Hazardous chemical substances closets; unlabeled bottles, containers
Toilets	Cleanliness, fixtures, soap, toilet paper, waste disposal bins
Worker hygiene	Changing place, showers, lockers, clothing change
Eating place	Separate tables, cleanliness, wash-up facilities
Workers' ages	Children, teenagers, elderly, pregnancy
Workers' complaints	Headache, fatigue, dizziness, nausea, breathlessness, skin problems
Worker morale	Reflected in turnover and absenteeism
Worksite layout	Safety in movement of supplies, products, ventilation
Medical service	On-site staff, first aid, evacuation procedures
Emergency procedures	Spills, contamination, terrorist attack, communications, reporting, evacuation, staff training
Hazard control	Labeling, process recording, worker records, periodic screening
Cleanliness	Removal of waste products, oil or chemicals on floors, machines, tables
Vents, fans	Exhaust of fumes, odors, dust
Worker–management cooperation	Mechanisms for worker and management to consult and share responsibility to reduce hazards and improve performance

TABLE 9.11 Markers and Indicators of Disaster Management Capability in an Industrial Setting

Marker	Indicator
Engage key stakeholders	Ongoing consultative mechanism to develop and implement plan
Administrative	Occupational health disaster plan; access to first aid; frequent disaster drills; close supervision of subcontractors
Investigation	Thorough investigation of complaints, leaks, and spills
Monitoring workers	Monitor worker injuries, illnesses, toxic levels; use of safety measures
Technological	Fail-safe monitoring devices; real-time monitoring; minimal on-site storage; automatic alarm/shut-down devices; local incineration/neutralization
Transportation	Vehicle and container standards; driver training, fatigue, alcohol and drug abuse, traffic offenses
Information/feedback	Workers' information; right-to-know of workers and community; community disaster plan

Source: Koh D, Aw T. Surveillance in occupational health. Occup Environ Med 2003;60:705–10. Available at: http://www.ncbi.nlm.nih.gov/pmc/articles/PMC1740637/ [Accessed 17 August 2012].

and operational criteria. The healthful and safe worksite should be maintained and accredited on that basis.

OCCUPATION AND THE NEW PUBLIC HEALTH

Social class, often defined by occupation and education, is a key determinant of health status. A population of unskilled workers has much higher rates of coronary heart disease, strokes, and cancer, and their children have much higher rates of mortality and morbidity than higher skilled workers or business and professional people. The evidence points to a feeling of having less control over one's own life as a major consideration. The worker who has little say in determining his or her own activities may be subject to higher stress at work, such as on the production line, or in job security, advancement, and wages. Loss of work is a key factor in increasing the vulnerability of men in particular to a variety of life-threatening conditions, including suicide, alcoholism, violence, and cardiovascular disease. The phenomenon of downsizing, or reducing the workforce, affects production workers disproportionately, but also reaches middle- and upper-management levels, so that the danger of losing a position at an age when finding new employment is unlikely may become a real health hazard. Awareness of and responsiveness to a variety of risks associated with employment and occupation are part of health responsibility. Prevention may predominate in some situations, screening for case finding in others, and clinical management in yet others.

SUMMARY

Environmental health and occupational health are interrelated topics that are becoming increasingly prominent in the New Public Health, along with concern for global ecology. The 2013 Bangladesh incident of over 370 clothing workers burned to death in grossly unsafe and locked premises highlight the dangers of globalized industry and poor occupational health amidst poverty and poor safety measures. Safe water and waste disposal, so fundamental to public health and so successful over the past century in reducing morbidity and mortality, remain serious problems in low-income countries. Their inclusion in the MDGs, challenges that are accepted by almost all countries in the world, attests to this recognition, and progress is being made.

The problems of the environment have become more complex in recent decades as scientific consensus on global ecological concerns has emerged. These include global warming, hazards associated with nuclear accidents on the scale of Chernobyl and Fukushima, and frequent chemical disasters. Water safety is compounded by water shortages and threats of desertification in large parts of the world where droughts are common. Concern for the environment and the worker often clashes with a desire for economic growth, especially in poorer countries as they try to cope with rapidly increasing populations and increasing expectations of a better life.

Air pollution, climate change, deforestation, desertification, water shortages, and population growth, especially in low-income countries, can lead to environmental and human tragedies on a vast scale. In higher income countries, environmentally friendly lifestyles, urban planning, land and water pollution cleanup, and oil spills are continuing issues at the forefront of public health in collaboration with environmental and occupational health protection agencies.

Important progress has been made in the management of water, waste products, toxic wastes, and air quality standards, especially since the 1970s. Workers' health and safety have improved dramatically over the past century in the industrialized countries. However, some of these gains are at the cost of moving hazardous materials and working conditions to newly industrializing or developing countries in the global economy, along with widespread outsourcing of low-skilled work in dangerous conditions to countries rife with corruption.

Even a vigilant health sector is, by itself, incapable of dealing with the problems of the environment and of occupational health. It requires many levels and agencies of government as well as the support of public opinion. The role of the public health community is to act in the professional and advocacy roles with intersectoral cooperation to address these complex and vital issues. Epidemiology provides tools to measure mortality, morbidity, and physiological changes that may occur as a result of environmental damage, but these may not be sufficiently rapid or sensitive. Both epidemiology and testing technology are improving steadily, providing hope for standards that one would expect to contribute to a cleaner, safer, and more aesthetically pleasing environment. Technical advances in water and waste management are becoming available and can play an important role in global health progress.

The environment affects everyone, but the poor more so, for many reasons. Work, or lack of satisfactory work, occupies a large portion of a person's time and energy. The workplace is also the location of many activities of daily life, including diet and physical activity. Policy makers, employers, and workers all need to take this into account in developing worksite conditions, management, access to health services, life habits, nutrition, and planned activities, along with safety and risk reduction for the benefit of both employer and worker, to protect the health of workers. This is increasingly important as more people are employed in knowledge-based industries, and with the aging of society, fewer workers are available to perform unskilled jobs. For these reasons, health targets cut across all aspects of society, including environment and workplace health.

The New Public Health includes long-standing public health issues of the environment and occupational health, but widens the field to include clinical services, the community, and the individual. Everyone needs to be involved in healthy public policy, in case finding, and in documenting the results of workplace and environmental risks. For a society there are choices to be made in creating a less toxic and hazardous environment, including private versus public transportation, jobs in industries with toxic emissions, and producing energy from fossil fuels or from nuclear sources.

Substitutes for toxic materials and an increased level of social consciousness are needed to reduce the gross pollution that was the price of industrialization over the twentieth century. Equally challenging is the need to prepare and deal with natural and human-caused disasters that may involve conventional explosives or biological, chemical, and even nuclear methods of destruction.

Achievements made with the MDGs to reduce poverty and ensure environmental sustainability will need to be carried beyond the current target of 2015. Avoiding the impact of the most damaging changes to the climate requires global action in the decades ahead. The financial resources and technological capabilities exist but implementation requires a sense of urgency, public interest, and political will to make deep cuts in greenhouse gas emissions. How global society addresses the issues of global warming will be crucial to our future as unrestrained pollution and anthropogenic destruction will be too great a burden to bear. Investment in a healthy environment and work settings is a health, economic, and quality-of-life issue for each community, each country, and the entire planet.

NOTE

For a complete bibliography and guidance for student reviews and expected competencies please see companion web site at http://booksite.elsevier.com/9780124157668

BIBLIOGRAPHY

Environmental Health

Agency for Toxic Substances and Disease Registry, 2010. National Toxic Substance Incidents Program (NTSIP) annual report. Available at: http://www.atsdr.cdc.gov/ntsip/docs/ATSDR_Annual%20Report_031413_FINAL.pdf. (accessed 01.07.13).

Bolte, G., Braubach, M., Chaudhuri, N., Deguen, S., Fairburn, J., Fast, I., et al., 2012. Environmental health inequalities in Europe. WHO, Copenhagen. Available at: http://www.euro.who.int/__data/assets/pdf_file/0010/157969/e96194.pdf (accessed 16.08.13).

Bruce, N., Perez-Padilla, R., Albalak, R., 2000. Indoor air pollution in developing countries: a major environment and public health challenge. Bull. World Health Organ. 1078–1092. Available at: http://www.who.int/bulletin/archives/78(9)1078.pdf (accessed 17.08.12).

Brunkard, K., Ailes, E., Roberts, V., Hill, V., Hilborn, E., Craun, G., et al., 2011. Surveillance for waterborne disease outbreaks associated with drinking water – United States 2007–2008. MMWR. Surveill. Summ. 60. 38–8. Available at: http://www.cdc.gov/mmwr/preview/mmwrhtml/ss6012a4.htm (accessed 16.08.12).

Burge, P.S., 2004. Sick building syndrome. Occup. Environ. Med. 61, 185–190. Available at: http://oem.highwire.org/content/61/2/185.full (accessed 17.08.12).

California Air Resources Board, 2010. California Environmental Protection Agency. Estimate of premature deaths associated with fine particle pollution (PM2.5) in California using a US Environmental Protection Agency methodology. 31 August. Available at: http://www.arb.ca.gov/research/health/pm-mort/pm-report_2010.pdf. (accessed 24.08.13).

Centers for Disease Control and Prevention, 1999. Achievements in public health, 1900–1999: control of infectious diseases. MMWR. Morb. Mortal. Wkly. Rep. 48, 621–629. Available at: http://www.cdc.gov/mmwr/preview/mmwrhtml/mm4829a1.htm (accessed 18.08.12).

Centers for Disease Control and Prevention, 1999. Achievements in public health, 1900–1999: fluoridation of drinking water to prevent dental caries. MMWR. Morb. Mortal. Wkly. Rep. 48, 933–940. Available at: http://www.cdc.gov/mmwr/preview/mmwrhtml/mm4841a1.htm (accessed 18.08.12).

Centers for Disease Control and Prevention, 1999. Achievements in public health, 1900–1999: motor-vehicle safety: a 20th century public health achievement. MMWR. Morb. Mortal. Wkly. Rep. 48, 369–374. Available at: http://www.cdc.gov/mmwr/preview/mmwrhtml/mm4818a1.htm (accessed 18.08.12).

Centers for Disease Control and Prevention, 2000. Biological and chemical terrorism: strategic plan for preparedness and response. MMWR. Recomm. Rep. 49, 1–14. Available at: http://www.cdc.gov/mmwr/preview/mmwrhtml/rr4904a1.htm (accessed 15.08.12).

Centers for Disease Control and Prevention, 2012. Childhood lead poisoning. Available at: http://ephtracking.cdc.gov/showChildhoodLeadPoisoning.action. (accessed 16.07.13).

Centers for Disease Control and Prevention, March 2013. National report on human exposure to environmental chemicals. DHHS, Atlanta, GA. Available at: http://www.cdc.gov/exposurereport/index.html (accessed 16.07.13).

Centers for Disease Control and Prevention, 2006. Rapid community needs assessment after Hurricane Katrina – Hancock County, Mississippi, September 14–15, 2005. MMWR. Morb. Mortal. Wkly. Rep. 55, 234–236. Available at: http://www.cdc.gov/mmwr/preview/mmwrhtml/mm5509a3.htm (accessed 15.08.12).

Centers for Disease Control and Prevention, 2011. Ten great public health achievements – United States, 2001–2010. MMWR. Morb. Mortal. Wkly. Rep. 60, 619–623. Available at: http://www.cdc.gov/mmwr/preview/mmwrhtml/mm6019a5.htm (accessed 02.08.13).

Centers for Disease Control and Prevention, 1995. Update: *Vibriocholerae* O1 – western hemisphere, 1991–1994, and *V. cholerae* O139 – Asia, 1994. MMWR. Morb. Mortal. Wkly. Rep. 44, 215–219. Available at: http://www.cdc.gov/mmwr/preview/mmwrhtml/00036609.htm (accessed 18.08.12).

Craun, M.F., Craun, G.F., Calderon, R.L., Beach, M.J., 2006. Waterborne outbreaks reported in the United States. J. Water. Health. 2, 19–30. Available at: http://courses.washington.edu/h2owaste/group1.pdf (accessed 16.08.12).

Dalbokova, D., Krzyzanowski, M., Lloyd, S. (Eds.), 2007. Children's health and the environment in Europe: a baseline assessment. WHO, European Region, Copenhagen. Available at: http://www.euro.who.int/__data/assets/pdf_file/0009/96750/E90767.pdf. (accessed 18.08.12).

Edling, C., 1985. Radon daughter exposure and lung cancer. Br. J. Ind. Med. 42, 721–722. Available at: http://www.ncbi.nlm.nih.gov/pmc/articles/PMC1007566/ (accessed 17.08.12).

Environmental Protection Agency, E-cycling [updated 24 July 2012]. EPA, Washington, DC. Available at: http://www.epa.gov/epawaste/conserve/materials/ecycling/index.htm (accessed 15.08.12).

Environmental Protection Agency, Earth day and EPA history [updated 12 July 2011]. EPA, Washington, DC. Available at: http://www.epa.gov/earthday/history.htm (accessed 15.08.12).

Environmental Protection Agency, Emergency management [updated 3 August 2012]. EPA, Washington, DC. Available at: http://www.epa.gov/emergencies/index.htm (accessed 15.08.12).

Environmental Protection Agency, Household hazardous wastes [updated 24 July 2012]. EPA, Washington, DC. Available at: http://www.epa.gov/osw/conserve/materials/hhw.htm (accessed 15.08.12).

Environmental Protection Agency, Laws and regulations: a history of the Clean Water Act [updated 17 April 2013]. Available at: http://www2.epa.gov/laws-regulations/history-clean-water-act (accessed 25.08.13).

Environmental Protection Agency, Medical waste: frequent questions [updated 24 July 2012]. EPA, Washington, DC. Available at: http://www.epa.gov/osw/nonhaz/industrial/medical/mwfaqs.htm (accessed 17.08.12).

Environmental Protection Agency, Municipal solid waste generation, recycling, and disposal in the United States: facts and figures for 2011. Available at: http://www.epa.gov/epawaste/nonhaz/municipal/pubs/MSWcharacterization_508_053113_fs.pdf (accessed 28.06.13).

Environmental Protection Agency, February 2013. National rivers and streams assessment 2008–2009: a collaborative survey. EPA, Washington, DC. Available at: http://water.epa.gov/type/rsl/monitoring/riverssurvey/upload/NRSA0809_Report_Final_508Compliant_130228.pdf (accessed 26.08.13).

Environmental Protection Agency, Radiation: non-ionizing and ionizing [updated 7 August 2012]. EPA, Washington, DC. Available at: http://www.epa.gov/rpdweb00/understand/index.html#nonionizing (accessed 17.08.12).

Environmental Protection Agency, Water: drinking water contaminants [updated 5 June 2012]. EPA, Washington, DC. Available at: www.epa.gov/safewater/contaminants/index.html#micro (accessed 16.08.12).

Environmental Protection Agency, Water sense [updated 20 July 2012]. EPA, Washington, DC. Available at: http://www.epa.gov/watersense/ (accessed 16.08.12).

Europa, Major accidents involving dangerous substances. Available at: http://europa.eu/legislation_summaries/environment/civil_protection/l21215_en.htm (accessed 01.07.13).

Europa, Summaries of EU legislation. Chemical products. Available at: http://europa.eu/legislation_summaries/internal_market/single_market_for_goods/chemical_products/index_en.htm (accessed 25.07.13).

Gofin, R., 2005. Preparedness and response to terrorism: a framework for public health action. Eur. J. Public Health 15, 100–104. Available at: http://eurpub.oxfordjournals.org/content/15/1/100.full (accessed 18.08.12).

Grandjean, P., Landrigan, P.J., 2006. Developmental neurotoxicity of industrial chemicals. Lancet 368, 2167–2178. Available at: http://www.ncbi.nlm.nih.gov/pubmed/17174709 (accessed 18.08.12).

Greenberg, M.R., 2007. Contemporary environmental and occupational health issues: more breadth and depth. Am. J. Public Health 97, 395–397. Available at: http://www.ncbi.nlm.nih.gov/pubmed/17267707 (accessed 18.08.12).

Greenberg, M.R., 2009. Water, conflict and hope. Am. J. Public Health 99, 1928–1930. Available at: http://www.ncbi.nlm.nih.gov/pmc/articles/PMC2759781/ (accessed 16.08.12).

Hardiman, M.C., 2012. World Health Organization perspective on implementation of international health regulations. Emerg. Infect. Dis. 18, 7. Available at: http://wwwnc.cdc.gov/eid/article/18/7/12-0395_article.htm (accessed 17.08.12).

Hernberg, S., 2000. Lead poisoning in a historical perspective. Am. J. Ind. Med. 38, 244–254. Available at: http://www.rachel.org/files/document/Lead_Poisoning_in_Historical_Perspective.pdf (accessed 17.08.12).

Khan, A.S., Levitt, A.M., Sage, M.J., 2000. Biological and chemical terrorism: strategic plan for preparedness and response: recommendations of the CDC Strategic Planning Workgroup. MMWR. Recomm. Rep. 49, 1–26. Available at: http://www.cdc.gov/mmwr/preview/mmwrhtml/rr4904a1.htm (accessed 18.08.12).

Kleinman, M.T., Sioutas, C., Froines, J.R., Fanning, E., Hamade, A., Mendez, L., et al., 2007. Inhalation of concentrated ambient particulate matter near a heavily trafficked road stimulates antigen-induced airway responses in mice. Inhal. Toxicol. 19, 117–126. Available at: http://informahealthcare.com/doi/abs/10.1080/08958370701495345 (accessed 18.08.12).

Kouznetsova, M., Huang, X., Ma, J., Lessner, L., Carpenter, D.O., 2006. Increased rate of hospitalization for diabetes and residential proximity of hazardous waste sites. Environ. Health Perspect. 115, 75–79. Available at: http://ehp03.niehs.nih.gov/article/fetchArticle.action?articleURI=info%3Adoi%2F10.1289%2Fehp.9223 (accessed 18.08.12).

Last, J.M., 2001. A dictionary of epidemiology, fourth ed. Oxford University Press/International Epidemiological Association, New York.

MacKenzie, W.R., Hoxie, N.J., Proctor, M.E., Gradus, M.S., Blair, K.A., Peterson, D.E., et al., 1994. A massive outbreak in Milwaukee of Cryptosporidium infection transmitted through the public water supply. N. Engl. J. Med. 331, 161–167. Available at: http://www.nejm.org/doi/full/10.1056/NEJM199407213310304 (accessed 18.08.12).

Maisonet, M., Correa, A., Misra, D., Jaakkola, J.J., 2004. A review of the literature on the effects of ambient air pollution on fetal growth. Environ. Res. 95, 106–115. Available at: http://www.ncbi.nlm.nih.gov/pubmed/15068936 (accessed 17.08.12).

Mekorot, Israel National Water Company. Wastewater treatment and reclamation. Tel Aviv: Mekorot. Available at: http://www.mekorot.co.il/Eng/Activities/Pages/WastewaterTreatmentandReclamation.aspx (accessed 18.08.12).

Morris Jr., J.G., 2011. Cholera – modern pandemic disease of ancient lineage. Emerg. Infect. Dis. 17, 2099–2104. Available at: http://www.ncbi.nlm.nih.gov/pmc/articles/PMC3310593/ (accessed 16.08.12).

National Health Service, Lyme disease [updated 15 August 2011]. NHS, UK. Available at: http://www.nhs.uk/Conditions/Lyme-disease/Pages/Introduction.aspx (accessed 15.08.12).

National Health Service, Sick building syndrome [updated 29 October 2010]. NHS, UK. Available at: http://www.nhs.uk/conditions/Sick-building-syndrome/Pages/Introduction.aspx (accessed 17.08.12).

Organisation for Economic Co-operation and Development, 2003. OECD guiding principles for chemical accident prevention, preparedness and response. OECD, Paris. Available at: http://www.oecd.org/chemicalsafety/riskmanagementofinstallationsandchemicals/2789820.pdf (accessed 17.08.12).

Organisation for Economic Co-operation and Development, 2012. The OECD environmental outlook to 2050. OECD, Paris. Available at: http://www.oecd.org/env/indicators-modelling-outlooks/oecdenvironmentaloutlookto2050theconsequencesofinaction-keyfactsandfigures.htm (accessed 25.08.13).

Organisation for Economic Co-operation and Development, The water challenge: OECD's response. OECD, Paris. Available at: http://www.oecd.org/document/47/0, 3746, en_2649_37465_36146415_1_1_1_3 7465,00.html (accessed 18.08.12).

Patz, J., Campbell-Lendrum, D., Gibbs, H., Woodruff, R., 2008. Health impact assessment of global climate change: expanding on comparative risk assessment approaches for policy making. Annu. Rev. Public. Health. 29, 27–39. Available at: http://www.ncbi.nlm.nih.gov/pubmed/18173382 (accessed 15.08.12).

Pesatori, A.C., Consonni, D., Rubagotti, M., Grillo, P., Bertazzi, P.A., 2009. Cancer incidence in the population exposed to dioxin after the "Seveso accident": twenty years of follow-up. Environ. Health 8, 39. Available at: http://www.ehjournal.net/content/8/1/39 (accessed 15.12.13).

Resnik, D.B., Wing, S., 2007. Lessons learned from the children's environmental exposure research study. Am. J. Public Health 97, 414–418. Available at: http://www.ncbi.nlm.nih.gov/pmc/articles/PMC1805023/ (accessed 18.08.12).

Rochester, J.R., 2013 Dec. Bisphenol A and human health: A review of the literature. Reprod Toxicol. 42, 132–55. doi: 10.1016/j.reprotox.2013.08.008. Epub 2013 Aug 30. Available at: http://www.ncbi.nlm.nih.gov/pubmed/23994667.

To, T., Shen, S., Atenafu, E.G., Guan, J., McLimont, S., Stocks, B., Licskai, C., 2013. The air quality health index and asthma morbidity: a population-based study. Environ. Health. Perspect. 121(1), 46–52. Available at: http://ehp.niehs.nih.gov/wp-content/uploads/121/1/ehp.1104816.pdf (accessed 15.12.13).

Tulchinsky, T.H., Burla, E., Brown, A., Goldberger, S., Clayman, M., Sadik, C., 2000. Safety of community drinking-water and outbreaks of waterborne enteric disease: Israel, 1976–97. Bull. World. Health. Organ. 78, 1466–1473. Available at: http://www.ncbi.nlm.nih.gov/pmc/articles/PMC2560668/ (accessed 16.08.12).

United Nation. We can end poverty 2013 Fact Sheet. Available at: http://www.un.org/millenniumgoals/environ.shtml (accessed 22.12.2013).

United Nations, Background on the UNFCCC: the international response to climate change. Bonn: United Nations Framework Convention on Climate Change. Available at: http://unfccc.int/essential_background/items/6031.php (accessed 15.08.12).

United Nations, 2012. The future we want. Rio+20 United Nations Conference on Sustainable Development. Available at: http://www.un.org/en/sustainablefuture/. and http://www.uncsd2012.org/futurewewant.html. (accessed 11.08.12).

United Nations Children's Fund, Progress on drinking water and sanitation. 2012 update. Available at: http://www.unicef.org/media/files/JMPReport2012.pdf. (accessed 25.08.13).

United Nations Meetings, 2012. Bonn: United Nations Framework Convention on Climate Change. Available at: http://unfccc.int/2860.php. (accessed 15.08.12).

United Nations Department of Economic and Social Affairs, International decade for action: water for life, 2005–2015. UNDESA, New York. Available at: http://www.un.org/waterforlifedecade/scarcity.shtml (accessed 16.08.12).

United Nations Development Programme, 2011. Human Development Report 2011, sustainability and equity: a better future for all. UNDP, New York. Available at: http://hdr.undp.org/en/reports/global/hdr2011/download/ (accessed 11.08.12).

United Nations Environment Programme, 2012. 21 Issues for the 21st century: results of the UNEP foresight process on emerging environmental issues. UNEP, Nairobi. Available at: http://www.unep.org/publications/ebooks/foresightreport/Portals/24175/pdfs/Foresight_Report21_Issues_for_the_21st_Century.pdf (accessed 15.08.12).

United Nations Development Report 2013. The Rise of the South: Human Progress in a Diverse World. Available at: http://hdr.undp.org/sites/default/files/reports/14/hdr2013_en_complete.pdf (accessed 22.12.2013).

US Nuclear Regulatory Commission, Backgrounder on Chernobyl nuclear power plant accident [updated 18 April 2012]. NRC, Washington, DC. Available at: http://www.nrc.gov/reading-rm/doc-collections/fact-sheets/chernobyl-bg.html (accessed 17.08.12).

Valent, F., Little, D., Bertollini, R., Nemer, L., Barbone, F., Tamburlini, G., 2004. Burden of disease attributable to selected environmental factors and injury among children and adolescents in Europe. Lancet 363, 2032–2039. Available at: http://www.ncbi.nlm.nih.gov/pubmed/15207953 (accessed 11.08.12).

Vandentorren, S., Bretin, P., Zeghnoun, A., Mandereau-BrunoL, L., Croisier, A., Cochet, C., et al. 2006. August 2003 heat wave in France: risk factors for death of elderly people living at home. Eur. J. Public. Health. 16, 583–591. Available at: http://eurpub.oxfordjournals.org/content/16/6/583.full.pdf (accessed 15.12.13).

World Health Organization, Climate change and human health: global environmental change [updated 2012]. WHO, Geneva. Available at: http://www.who.int/globalchange/environment/en/ (accessed 15.08.12).

World Health Organization, 2006. Preventing disease through healthy environments. WHO, Geneva. Available at: http://www.who.int/quantifying_ehimpacts/publications/preventingdisease/en/index.html (accessed 26.08.13).

World Health Organization, Radon and cancer. Fact sheet no. 291 [updated September 2009]. Available at: http://www.who.int/mediacentre/factsheets/fs291/en/index.html (accessed 16.08.12).

World Health Organization, Ten facts on preventing disease through healthy environments. WHO, Geneva. Available at: http://www.who.int/features/factfiles/environmental_health/en/index.html (accessed 11.08.12).

World Health Organization, Waste from health-care activities [updated November 2011]. WHO, Geneva. Available at: http://www.who.int/mediacentre/factsheets/fs253/en/index.html (accessed 17.08.12).

World Health Organization, Water sanitation health, water-related diseases: malnutrition [updated 2012]. WHO, Geneva. Available at: http://www.who.int/water_sanitation_health/diseases/malnutrition/en/ (accessed 16.08.12).

World Health Organization, 2011. Western Pacific Region. The Great East Japan Earthquake. WHO, Geneva. Available at: http://www.wpro.who.int/publications/9789290615682/en/index.html (accessed 17.08.12).

World Health Organization/United Nations Children's Fund, 2012. Progress on drinking water and sanitation: 2012 update. Geneva, WHO/UNICEF. Available at: http://www.unicef.org/media/files/JMPReport2012.pdf (accessed 16.08.12).

World Nuclear Association, Chernobyl accident 1986 [updated April 2012]. WNA, London. Available at: http://www.world-nuclear.org/info/chernobyl/inf07.html (accessed 15.08.12).

World Nuclear Association, Fukushima accident 2011 [updated 3 August 2012]. WNA, London. Available at: http://www.world-nuclear.org/info/fukushima_accident_inf129.html (accessed 17.08.12).

Occupational Health

Centers for Disease Control and Prevention, 1999. Achievements in public health, 1900–1999: improvements in workplace safety – United States, 1900–1999. MMWR. Morb. Mortal. Wkly. Rep. 48, 461–469. Available at: http://www.cdc.gov/mmwr/preview/mmwrhtml/mm4822a1.htm (accessed 27.06.13).

Centers for Disease Control and Prevention, 2006. Advanced pneumoconiosis among working underground coal miners – Eastern Kentucky and Southwestern Virginia, 2006. MMWR. Morb. Mortal. Wkly. Rep. 56, 652–655. Available at: http://www.cdc.gov/mmwr/preview/mmwrhtml/mm5626a2.htm (accessed 17.08.12).

Centers for Disease Control and Prevention, Asbestos [updated 16 March 2012]. CDC, Atlanta, GA. Available at: http://www.cdc.gov/niosh/topics/asbestos/ (accessed 17.08.12).

Centers for Disease Control and Prevention, 2007. Fatal occupational injuries and illnesses – United States, 2005. MMWR. Morb. Mortal. Wkly. Rep. 56, 297–301. Available at: http://www.cdc.gov/mmwr/preview/mmwrhtml/mm5613a1.htm (accessed 17.08.12).

Centers for Disease Control and Prevention, 2007. Nonfatal occupational injuries and illnesses – United States, 2004. MMWR. Morb. Mortal. Wkly. Rep. 56, 393–397. Available at: http://www.cdc.gov/mmwr/preview/mmwrhtml/mm5616a3.htm (accessed 17.08.12).

Centers for Disease Control and Prevention, April 2013. Occupational violence. CDC, Atlanta, GA. Available at: http://www.cdc.gov/niosh/topics/violence/ (accessed 14.06.13).

Centers for Disease Control and Prevention, Workplace safety and health [updated 30 November 2011]. CDC, Atlanta, GA. Available at: http://www.cdc.gov/workplace/ (accessed 17.08.12).

Concha-Barrientos, M., Nelson, D.I., Fingerhut, M., Driscoll, T., Leigh, J., 2005. The global burden due to occupational injury. Am. J. Ind. Med. 48, 470–481. Available at: http://www.ncbi.nlm.nih.gov/pubmed/16299709 (accessed 18.08.12).

Environmental Protection Agency, The birth of EPA [updated 5 July 2012]. Washington, DC, US, EPA. Available at: http://www.epa.gov/aboutepa/history/topics/epa/15c.html (accessed 17.08.12).

Gochfeld, M., 2005. Chronologic history of occupational medicine. J. Occup. Environ. Med. 47(2), 96–114. Available at: http://journals.lww.com/joem/Abstract/2005/02000/Chronologic_History_of_Occupational_Medicine.2.aspx (accessed 15.12.13).

Gochfeld, M., 2005. Occupational medicine practice in the United States since the Industrial Revolution. J. Occup. Environ. Med. 47 (2), 115–131.

Healthy People 2020, Occupational safety and health. Available at: http://www.healthypeople.gov/2020/topicsobjectives2020/overview.aspx?topicid=30 (accessed 17.07.13).

Herbert, R., Landrigan, P, 2000. Work-related death: a continuing epidemic. Am. J. Public Health 90, 541–545. Available at: http://www.ncbi.nlm.nih.gov/pmc/articles/PMC1446189/pdf/10754967.pdf (accessed 22.12.13).

Lahiri, S., Levenstein, C., Nelson, D.I., Rosenberg, B.J., 2005. The cost effectiveness of occupational health interventions: prevention of silicosis. Am. J. Ind. Med. 48, 503–514. Available at: http://www.ncbi.nlm.nih.gov/pubmed/16299711 (accessed 18.08.12).

Marsh, S.M., Menéndez, C., Baron SL Steege, A.L., Myers, J.R., November 22, 2013. Fatal Work-Related Injuries — United States, 2005–2009. CDC Health Disparities and Inequalities Report — United States, 2013. Supplement Vol. 62 (No. 3). Available at: http://www.cdc.gov/mmwr/pdf/other/su6203.pdf (accessed 22.12.03).

Pesatori, A.C., Consonni, D., Rubagotti, M., Grillo, P., Bertazzi, P.A., 2009. Cancer incidence in the population exposed to dioxin after the "Seveso accident": twenty years of follow-up. Environ Health 8, 39. Available at: http://www.ehjournal.net/content/8/1/39 (accessed 15.12.13).

Rosenstock, L., Cullen, M.R., Fingerhut, M., February 2005. Advancing worker health and safety in the developing world. J. Occup. Environ. Med. 47 (2), 132–136.

Toossi, M., 2012. Labor force projections to 2020: a more slowly growing workforce. Mon. Labor. Rev. 43–64. Available at: http://www.bls.gov/opub/mlr/2012/01/art3full.pdf (accessed 25.08.13).

Tulchinsky, T.H., Ginsberg, G.M., Shihab, S., Goldberg, E., Laster, R., 1992. Mesothelioma mortality among former asbestos-cement workers in Israel, 1953–90. Isr. J. Med. Sci. 28, 543–547. Available at: http://www.ncbi.nlm.nih.gov/pubmed/1428808 (accessed 17.08.12).

Health, U.K., Executive, Safety, Mesothelioma mortality in Great Britain 1968–2010. Available at: http://www.hse.gov.uk/statistics/causdis/mesothelioma/mesothelioma.pdf (accessed 10.07.13).

US Department of Health and Human Services, Healthy People: occupational health and safety [updated 26 July 2012]. US DHHS, Washington, DC. Available at: http://www.healthypeople.gov/2020/topicsobjectives2020/objectiveslist.aspx?topicid=30 (accessed 18.08.12).

World Health Organization, Occupational health. Available at: http://www.who.int/topics/occupational_health/en/ (accessed 18.08.12).

Organization of Public Health Systems

INTRODUCTION

Formal structures to ensure public health evolved over the centuries as local authorities addressed fundamental societal needs for sanitation, safe water and food safety business licensing and other issues. These structures developed in response to the challenges of industrialization and urbanization along with growing scientific and applied methodologies for disease prevention and health promotion. Non-governmental charitable, religious, and advocacy organizations pioneered many services that were part of addressing the broad spectrum of public health needs. With the widening range of public responsibilities, state and national governments took on increasing roles of leadership. These included financial support and professional development of public health and, in parallel, medical care systems to meet the growing public expectation for good health. These challenges remain important for current and future needs of both individual and population health. In the USA the high and rising cost of health care, and lack of universal insurance coverage are continuing political and public health issues, while many other industrialized countries have better health outcomes such as longer life expectancy (see Chapters 11 and 13).

This chapter examines the organization of public health and health care delivery services, illustrating how separate systems of service coexist and interact. Each system evolved in its own organizational and financing format, yet they come together, as medical care and prevention become more mutually interdependent. Traditional public health systems must increasingly develop intersectoral cooperation with other components of the health care industry, as well as with government and related fields, such as agriculture, business, and social welfare, education, police, and community organizations.

Governments have legislative, regulatory, and taxation powers set out in constitution and law for common action for the public good, including powers to promote health and to restrict individual actions that may jeopardize the health of others. City-states in ancient Greece provided sanitation for the entire community and medical care for the poor. The Elizabethan Poor Laws in Britain in the early seventeenth century established the responsibility of the local authority for health and welfare. Subsequent developments brought local, state, and national government into sanitation, disease control, and other aspects of public health and health planning. Later this extended to assuring provision of comprehensive health care on a social-equity basis for all or to meet the specific needs of vulnerable groups within a society.

Societies have learned to prevent disease by social action and have learned that individual health depends on such action. Governments are involved in that process, whether the governmental structure is based on democratic and free market principles, or is centrally managed with a command economy. Society has accepted some limitations on individual rights for the public good. These limit the individual from attacking and harming another person, or damaging goods, whether private or public. A person is restricted from throwing garbage in the street, and industry is prohibited from polluting the environment or endangering its workers.

Public health policy, legislation, and action involve common measures to protect the individual and the community. Such measures may take the form of mandatory reporting of an infectious disease, chlorinating and fluoridating community water systems, sanitary waste disposal, regulating food and drug industries, requiring children to be immunized before entry to school, or fining or imprisoning industry managers whose negligence causes death and injury, or whose industry pollutes the environment.

Achievement of public health goals requires organization (Box 10.1). Public health organization requires a formal structure for a defined population in which finance, management, scope, and content are defined in law and regulations. It includes services contributing to people's health as well as health care to be delivered in many settings, such as homes, communities, educational institutions, workplaces, hospitals, and clinics. Public health also addresses

The New Public Health. http://dx.doi.org/10.1016/B978-0-12-415766-8.00010-0

> ### BOX 10.1 What Is a Public Health System?
>
> *"A network of public, private, and voluntary entities that contribute to the health and well-being of a community."*
>
> **Source:** *World Health Organization. World health report 2004. Geneva: WHO. Available at: http://www.who.int/whr/2004/en/*

the policy, legislative, and regulatory functions of societal health, including the physical and psychosocial environment. A health system is organized at various levels, starting at the most peripheral, the community or primary level. It includes district, regional, state, and national levels as well as international aspects. International and national strategies for health and national health systems should be seen as investments that produce health gain rather than merely management of existing medical care institutions and services.

Function and structure are interdependent. Structure should evolve from the desired function; that is, to achieve national goals and objectives for health. This aim is fulfilled through legislative, regulatory, financing, and service functions, which provide the underpinnings to meet health needs in any country. Some countries provide universal health care through a governmental system. Others legislate financing of health care, while another approach focuses on financing for certain population subgroups, such as the elderly and the poor, placing greater emphasis on provision of facilities and research in health care.

This chapter describes public health organization primarily using examples from the USA, including federal, state, and local public health authorities. In contrast to most industrialized countries, the USA lacks universal health care. As a result, health care is provided through a mix of independent, private, and public agencies. While this is sometimes described as a "non-system", it is in fact a complex network of interactive services. Yet, it lacks universality, leaving many individuals without access to even basic private health care. As a result, public health organizations in the USA play a very important role in providing essential services for people or needs not otherwise met. Yet there are socioeconomic, ethnic, and regional variations and inequalities in insurance coverage and resource allocation, leaving substandard access and outcomes for many in the US health system. Public health has played a leadership role particularly in advocacy, development, and achievement in promoting health, partly to compensate for this fragmentation of health care in the USA. In many ways medium and low income countries are similar in lacking universal care, so that the institutions of US public health provide examples of infrastructure development needed to meet deficiencies in any health system.

GOVERNMENT AND HEALTH OF THE NATION

Public health involves a wide variety of issues that should be directly under governmental responsibility as they require legislation, enforcement, and taxing powers. These include environment, nutrition, food and drug control, sanitation, immunization, traffic laws, firearms control, and health education. Many of these functions are promoted by non-governmental organizations (NGOs), with delegated governmental regulatory powers.

Financing and allocation of public funds for health care provide important means of influencing health activities, which may mean direction of public funds to support research, teaching facilities, and provision of services. National governments may directly provide services, but increasingly this is being decentralized to lower levels of government (regional, district, municipal) or to non-governmental health care providers. Academic, professional, and public advocacy organizations play important roles in the New Public Health, such as in personnel training, education, research, and professional standards setting. These functions can be diffused to a variety of professional, consumer, and academic institutions, enabling governments to act through direct regulatory functions. Governments can also act indirectly, setting standards and norms through financial and other incentives or sanctions, and involving an organized system of accountability, accreditation, licensing activities, and quality guidelines.

Federal and Unitary States

Public health requires a basis in law, public administration, and financing. The constitution, law and form of government may differ from country to country, some being federal, others unitary.

In a federal system, three levels of government – federal, state, and local – have separate but overlapping responsibilities for public health. Federal states have constitutions conceived and written in a historical period when state rights were emphasized and health care was perceived mainly as a private activity between patient and doctor. Consequently, primary responsibility for health rested with a combination of the state, provincial, regional, or local levels of government. However, because of greater resources at the national level, federal government roles have increased in the health field over the years. National governments have a responsibility to ensure equity of social policy. A growing federal or national role has been a historical process common to many countries. At a minimum, the federal level is responsible for national health policy, planning, and setting national health targets. The USA, Canada, Russia, Argentina, and Nigeria are examples of countries with federal forms of government.

A unitary state is a form of government that has a central national level and local governments, but no intermediary legislating level. The UK despite having four constituent units of England, Scotland, Northern Ireland and Wales with semi independant health services is constitutionally a unitary state. Countries with governments based on the French Napoleonic Code, including most Spanish-speaking countries, are examples. In these countries, the central government has great responsibility for health, but here,

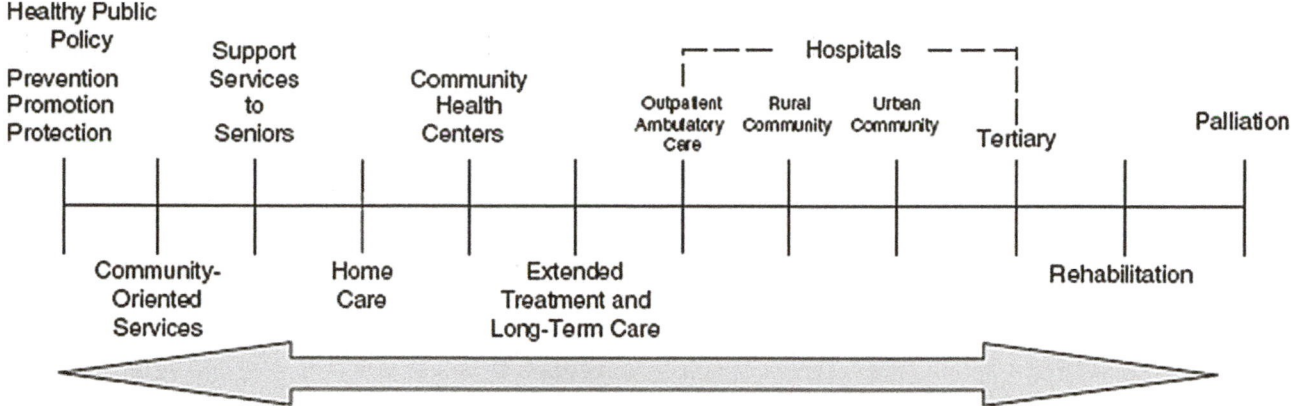

FIGURE 10.1   The continuum of health services.

too, local government is still a major factor in sanitation and local public health. The powers of regional and local authorities are derived from the national structure. Public health grew initially at the local level with regulations for sanitation, business premises and product licensing, food safety, and the like. In the UK, the national government promoted local public health organization, later organizing personal health services programs for the entire population in the centrally controlled National Health Service (NHS) (UK National Audit Office 2012).

Diffusion of authority is common to all health systems to differing degrees, mainly based on historical precedents. In recent years, national health authorities have largely been responsible for overall policy, law, financing, standards, monitoring, research, and assurance of services to meet national health goals. Management of services is generally decentralized, with responsibility at the state, regional, and local health authority or institutional level. Diffusion or sharing of responsibility from each level of authority is common in current planning to cope with the wide range of activities and interests that make up the health sector of a society. Non-governmental agencies (NGOs) often precede governmental authority in the field, and their presence and participation make up important elements of the health complex, whether as providers of services, as advocates, or as fundraisers for programs that a government cannot manage to include in its "basket of services".

Local authorities often delegate administration of services to independent institutions or other public or private agencies. Diffusion of responsibilities occurs to different degrees in administration of services, in education, in training, as well as registry of health professionals including the related professional and accreditation organizations. Diffusion also occurs in research, in intersectoral cooperation between governmental agencies, along with NGOs, or advocacy groups, and in academic as well as research facilities. Legislation may initiate and direct changes in health programs using regulatory and financing measures, but implementation also requires a broad spectrum of participation of individuals and organizations of consumers, providers, and other health interest groups. Health is not an isolated service, but a reflection of the social values and standards and economic development of a society, with a large degree of interdependence and interaction between health agencies and other governmental and non-governmental elements of that society.

Checks and Balances in Health Authority

The balance between government intervention and private organization, between regulation and self-governance, is not easy to define or to achieve in health. Historically, elements of health care developed at different times and with different degrees of political, economic, and public support. The accumulated experience of modern public health indicates that all elements of health need to be considered as part of a spectrum of services (Figure 10.1). Weakness in one area threatens the well-being of the totality. Poor levels of nutrition and sanitation breed disease, for which treatment is more expensive and less effective than prevention. At the same time, low medical care standards due to inadequate training, motivation, resources, and supervision can lead to low standards of health among large segments of the population.

Public health services have developed separately from curative services by providing care for special needs populations such as maternal and child care, primarily for the poor. However, there is growing recognition that health promotion, health protection, and preventive care are interwoven and at least partially integrated with curative service systems. Where health care is provided by private services, a public–private mix is essential in delivery of specific preventive services, such as screening and immunization. This leaves out part of the population, so that health promotion and outreach services are also required. When such services are provided by private medical care services, there will always be a need for special provision to the uninsured, those lacking financial and physical access, and those lacking information or awareness of the need for such services.

Organization for public health services, whether integrated into a total care system or separate from curative service systems, requires a combination of centralized and decentralized responsibilities. The overriding national responsibility requires political leaders to set policy goals and standards, including measures to promote regional and social equity in health. Decentralization in public health allows local authorities to take direct operational responsibility, with resources and accountability, in some cases such as in Scandinavian countries with direct management of health services, and most commonly in sanitation, business licensing, and disaster planning and management.

Diffusion of responsibility means that many agencies operate at different levels of a nation state, with some overlapping functions and some gaps. Each level has its own sphere of responsibility, working in cooperation and under regulation set by the higher level of government, but linking together to form a working whole, with checks, balances, and cooperation among them. Even in highly centralized organizations of health services, cooperation with other governmental agencies such as social welfare, education, environmental, and other agencies is essential to modern public health.

A centralized health organization that controls policy, administration, financing, services, personnel training, research, and regulation may lack checks and balances needed to prevent authoritarian control. Formerly highly centralized health systems are now seeking decentralization as a means of infusing additional funding, local identification, pride, privacy, and quality in their health systems. They are combining this with universal access and regional, ethnic, and social equity. Comprehensiveness and cost constraint are the challenges of organization of public health systems.

A federal structure of government divides health responsibilities with the senior level of government as the overall policy level with financing and regulatory roles. The state or provincial level is responsible for public health and, in many cases, health insurance systems serving its population, while the local government is responsible for public health at the community level. National and state or provincial funding, regulations, support services, and policy direction guidelines and accountability provide backing to promote community health interests.

The New Public Health is a population health model which seeks a balance and cooperation between government-operated health services and the diffused network of private, often competing, organizations, working together to use resources effectively to achieve common health targets that meet the needs of the individual and the population as a whole.

Government and the Individual

Conflicting ideas as to the overall role that government should play affect public health in many ways. In 1869, John Stuart Mill, the founder of modern economics, wrote in the introduction to *On Liberty*, "The only purpose for which power can rightfully be exercised over any member of a civilized society, against his will, is to prevent harm to others. His own good, either physical or moral, is not a sufficient warrant". This philosophy has been adapted to a recognized and essential role for government in public health, as in education and other essential services.

The institutions of basic sanitation and community hygiene have had to contend with such individualistic ideas. The issue of governmental interference in "private matters", such as in health, is not new and is actively debated in industrialized western societies, in the post-Soviet countries, and in developing nations alike. *Laissez-faire* economists promote the idea of minimal governmental involvement in all economic affairs including social services such as health.

During the nineteenth and increasingly in the twentieth centuries, it became apparent and imperative for protection and promotion of health that the state intervene to set and enforce public health measures in all societies. At the other extreme, disillusionment occurred when governments assumed total responsibility for health and total central management of health services. Most countries have their own balance between the two extremes. Paradoxically, the most decentralized and privatized of all national health systems, that of the USA, has been proactive in, and has emphasized development of, national and professional standards, monitoring, setting national targets and regulation in health, and is in the process of profound change from individual care towards managed care systems.

FUNCTIONS OF PUBLIC HEALTH

The American Public Health Association (APHA), founded in 1872, periodically issues policy statements on the mission and essential services of public health organizations. These guidelines help government to provide or assure provision of services through other agencies. The 1994 APHA statements of the overall vision and the mission of public health in the USA were endorsed by the Association of State and Territorial Health Officials, the National Association of County and City Health Officials, the Institute of Medicine, the Association of Schools of Public Health, the US Public Health Service, and others. Periodic review and revision, with consensus among the many professional organizations concerned with public health, help to maintain relevance for local and central public health organizations. The mission and essential services of public health in the USA as published by the Centers for Disease Control and Prevention (CDC) are shown in Box 10.2.

For many of the responsibilities legislated for public health agencies at the national, state, provincial, or local health authority levels, a combination of methods and approaches is needed. Regulatory functions are those based on the legal authority of a public health agency to set and enforce standards. Setting health targets, policies and financing, and national or state standards is important

BOX 10.2 Mission and Essential Services of Public Health, USA

Public Health Responsibilities or Mission

- Prevent epidemics and spread of disease.
- Protect against environmental hazards.
- Prevent injuries.
- Promote and encourage healthy behaviors.
- Respond to disasters and assist communities in recovery.
- Assure quality and accessibility of health services.

Essential Public Health Services

- Monitor health status to identify and solve community health problems.
- Diagnose and investigate health problems and health hazards in the community.
- Inform, educate, and empower people about health issues.
- Mobilize community partnerships and action to solve health problems.
- Develop policies and plans that support individual and community health efforts.
- Enforce laws and regulations that protect health and ensure safety.
- Link people to needed personal health services and assure provision of health care when otherwise unavailable.
- Assure an expert public health workforce.
- Evaluate effectiveness, accessibility, and quality of health services.
- Research for new insights and innovative solutions to health problems.

Source: *Centers for Disease Control and Prevention. National public health performance standards program. 10 essential public health services [updated December 2010]. Available at: http://www.cdc.gov/nphpsp/essentialservices.html [Accessed 26 October 2012].*

BOX 10.3 Examples of Regulated Aspects of Public Health in the USA

- *Regulation and processing data* from birth and death certificates and other data sources from local, state, and national authorities – National Center for Health Statistics (NCHS)
- *Business premises and product licensing approval* – local health authorities
- *Building code compliance* – local health authorities under state and federal codes
- *Sanitation and environmental health* – municipal, state, and national agencies, such as the Environmental Protection Agency (EPA)
- *Regulation of health professionals* – state boards
- *Licensing and certification of health facilities* – local, state, and federal authorities
- *Communicable disease control* – local, state, and federal authorities with the Centers for Disease Control and Prevention (CDC)
- *Food safety* – local, state, and federal standards and inspections by the Food and Drug Administration (FDA)
- *Pharmaceutical standards* – safety, efficacy, labeling, and manufacturing standards by the FDA
- *Occupational health and safety* – local, state, and federal standards and inspections within the Occupational Safety and Health Administration (OSHA), for standards, regulation, and enforcement, and the National Institute for Occupational Safety and Health (NIOSH), for research.

in promoting new program initiatives. Health promotion includes not only direct and formal teaching, but also promotion of awareness of public health problems to the general public, health care providers, and other agencies. Services may be provided directly or may be funded and supervised by the public health agency. Direct service is the provision of services to the public, especially useful in areas where universal coverage is essential (e.g., immunizations), or for high-risk groups not able to access other services (e.g., prenatal care for the poor).

Intersectoral cooperation is the coordination with other agencies of government, NGOs, or service providers to work towards common objectives that will improve public health. This is an area where public health advocacy is important in that the public health authority tries to engage other agencies, as in the development of water and sewage systems or the policing of highways, to reduce road accident deaths and related morbidity. NGOs, voluntary organizations, and advocacy groups have in the past and will in the future play a vital role in developing health programs.

Regulatory Functions of Public Health Agencies

Regulatory function in public health is based on a legal mandate to protect the public from health hazards and to assure certain standards for provision of care. Whatever degree of decentralization occurs, there are key central standards in public health that must be maintained at the federal level in essential areas such as nutrition, sanitation, food and drug control, and others over which the individual citizen or health provider has no direct control. The regulatory function covers a wide range of public health activities (Box 10.3).

Investigative Functions of Public Health

Public health reporting in addition to vital statistics includes communicable diseases (see Chapter 4). The purpose of such reporting is to monitor health and to investigate unusual events such as infectious disease outbreaks, which may be due to many sources, including food or water contamination, hospital-acquired infections, sexually transmitted infections (STIs), and tuberculosis (TB). Investigation of disease in a population includes non-communicable diseases, cancers, injuries, birth defects, suicides, and suspicious deaths. Cardiovascular diseases and cancers as the leading causes of death are discussed in Chapter 5, along with many other health risk factors.

Many examples of such reporting and investigation have led to the identification of new diseases, including human immunodeficiency virus/acquired immunodeficiency syndrome (HIV/AIDS) in the 1980s, Lyme disease, Legionnaires' disease, West Nile fever, and the spread of dengue and chikungunya to new parts of the world. There has been a recurrence of measles, pertussis, and other vaccine-preventable disease thought to have been brought under control; these diseases and others are identified by reporting to local health departments and supported by national and international epidemiological reporting and investigation. Smallpox eradication was achieved by a combination of mass immunization and, in later stages, outbreak identification and rapid vaccination of local communities to stop the spread of the disease.

The Program for Monitoring Emerging Diseases (ProMED) presents an Internet site reporting daily on global disease outbreak monitoring. ProMED obtains information from local reports including newspaper stories, and is supported by professional investigations that provide fresh data on emerging and re-emerging diseases. The US CDC publishes *Morbidity and Mortality Weekly Report* and the European Centre for Disease Prevention and Control publishes *Eurosurveillance*. These are highly professional, regular investigative reporting systems, supported by *Emerging Infectious Diseases* and other journals to bring rapid sharing of information of epidemiological importance to public health systems. (See Chapter 4 for references. URLs are available in support material at the *New Public Health* website.)

In 2012, a diffuse outbreak of fungal meningitis/encephalitis occurred in the USA from the use of contaminated injectable medications used for pain control. Initial cases were diagnosed by an astute physician in Tennessee and traced to a pharmaceutical company in Massachusetts, with cases then appearing widely across the country. Investigation showed contamination in the production site; the offending material was withdrawn from pharmacies and the company's operation was closed (Box 10.4).

In 2013, the emergence of a new Middle East respiratory syndrome corona virus (MERS-CoV) in Saudi Arabia and other Middle Eastern countries is a new episode in disease identification. This emerging disease has the potential to become a widespread epidemic with cases among visitors to the area, although control measures are limited to patient identification and isolation techniques.

Prevention of Injuries

Prevention of road crash injuries is a major public health challenge which requires networking with other governmental agencies, policy makers, and public opinion. A public policy of allowing high speed limits or increasing the speed limit accounted for an estimated 12,545 deaths in the USA over a 10-year period of follow-up. The US Department of Transportation estimated in 2002 that the comprehensive cost of each fatality was US$977,000 and the cost for each critically injured person was US$1.1 million, so that the 10-year cumulative cost for fatalities alone of repealing the 55 mph speed limit was approximately US$12 billion. The department reports that in 2010 there were 3092 deaths and 416,000 injuries in distraction-affected motor vehicle accidents. Distracted driving includes the use of cell phones or performance of other tasks while driving (see Chapter 5). Local and state legislation regarding the enforcement of drink–driving laws, mandatory use of seat belts, safe car seats for children, helmets for motorcyclists and bicycle riders, education, and enforcement against use of cell phones while driving are all necessary to reduce the toll of death and disability from road crashes (see Chapter 5).

Repeal of the National Maximum Speed Law and its aftermath show that policy decisions that appear harmless can have long-term repercussions. Reduced speed limits lower crash rates, case fatality, and injury severity, thus saving lives as well as reducing fuel consumption, emissions, and air pollutants; save valuable years of productivity; and reduce the societal cost of motor vehicle crashes. Coupled with mandatory seat belts and child safety seats, air bags, road and car safety measures, lowering legal speed limits on rural and urban highways, improved enforcement and use of speed cameras could reduce traveling speeds and fatalities immediately (Friedman et al., 2009).

Methods of Providing or Assuring Services: Direct or Indirect?

Whether a governmental agency provides or assures the provision of services varies from country to country. Canada's health insurance program is operated by the provinces, with federal cost sharing. In Scandinavian countries, the counties, which have many of the characteristics of provinces, operate most local health services. In centrally managed economies, such as former Soviet countries, health services have been operated with a high degree of central control. The international movement towards decentralization of management of services is under critical review, and a mix of centrally managed and decentralized services is likely to be the trend in coming decades.

Only government can perform many public health functions because certain services require legislative, taxing, and regulatory powers, or because they are directed at the total population. Central coordination is required for key public health functions such as epidemiology and disease control, monitoring population health, nutrition, sanitation, and food and drug control.

BOX 10.4 US Iatrogenic Fungal Meningitis Outbreak 2012: Lessons Learned from the National Distribution with State-Based Regulation

Detection and control of infectious disease outbreaks require a concerted effort by frontline and specialty clinicians, local and regional and/or state public health professionals, and national level scientists and regulators. The 2012 outbreak of fungal meningitis in the USA due to contamination of preservative-free methyl prednisolone acetate illustrates several of the strengths and weaknesses of this approach.

Meningitis is an infection of the central nervous system, specifically of the fluid and materials that surround the brain and spinal cord. This system is normally sterile, but injections of medications near or into the spinal column can cause contamination. Agents that can cause meningitis are typically viruses or bacteria, but meningitis due to fungi can occur.

In early September 2012, patients began to present to emergency departments and other sites with symptoms consistent with meningitis but without common causal agents. An infectious disease expert at a Tennessee academic medical center identified one of the initial patients and, given the unusual etiological agent, reported the case via email to the State Health Department of Tennessee (TDOH). There, the Director of the Healthcare Associated Infections, among others, began to look for other patients. As noted by the *New York Times* report on the investigation, while physicians and other clinicians on the frontlines are most likely to detect initial cases, "only health departments and other governmental agencies have the ability and authority to track down additional cases to document disease outbreaks and warn those at risk. It is work that private groups seldom can do".

Within 48 hours of the first report, the TDOH notified the US federal Centers for Disease Control and Prevention (CDC); the key point was the unusual fungal agent involved. After consultation with the CDC, the TDOH inspected the facility where the index case received treatment, and identified potential causes, such as local contamination due to environmental contamination, mishandling of equipment, or contamination at the compounding source. Similar to case identification and tracking those potentially exposed, inspection of health care facilities requires authority usually reserved only for state and national regulatory agencies.

While narrowing the possible causes, the TDOH reached out directly to the out-of-state compounding source, as well as the state health department in that state (Massachusetts, MDOH).

The company voluntarily recalled potentially tainted lots of the medications. The US agency that oversees most drug manufacture in the USA, the Food and Drug Administration (FDA), was also informed. Here, the lines of authority and responsibility overlap, and the outbreak reveals potential weaknesses in state and federal regulation. Compounding companies are not routinely regulated by the FDA, as such businesses are technically pharmacies. In the USA, pharmacies are routinely overseen by professional boards, often including volunteer professionals.

Clinicians in many other states began to report fungal meningitis cases to their state health departments, and the CDC coordinated formal national surveillance. Finally, within 8 days of the initial email report to the TDOH, the MDOH conducted a detailed inspection of the production facility involved. The company voluntarily recalled all of its products and shut down. A month later, a preliminary report by MDOH noted contamination of floors, floors mats, and a leaking boiler, all near sterile mixing areas. An FDA report found that many of the drug vials contained foreign matter and that the "clean" compounding rooms had either mold or bacterial overgrowth, or both.

By 15 November 2012, over 400 cases of fungal infection due to the contaminated drug had been found, with 32 deaths, and over 14,000 potential exposed patients. In reaction to the situation, hearings at the national level were scheduled to discuss the need for regulatory changes. The director of the Massachusetts Board of Pharmacy has been fired, while new cases are still occurring, and recurrence in some of the initial cases is being documented.

Sources: DiFerdinando GD. Personal communication; November 2012.
Perfect JR. Iatrogenic fungal meningitis: tragedy repeated. Ann Intern Med 2012;157:825–6. Available at: https://annals.org/article.aspx?articleid=1384984 [Accessed 10 November 2012].
Lyon JL, Gireesh ED, Trivedi JB, Bell R, Cettomai D, Smith BR, et al. Fatal Exserohilum meningitis and central nervous system vasculitis after cervical epidural methylprednisolone injection. Ann Intern Med 2012;157:835–6. Available at: https://annals.org/article.aspx?articleid=1384432 [Accessed 10 November 2012].
Altman LK. Chasing clues to detect outbreak. N Y Times 2012; 5 November. Available at: http://www.nytimes.com/2012/11/06/health/doctors-chased-clues-to-identify-meningitis-outbreak.html?pagewanted=all [Accessed 10 November 2012].
Outterson K. Regulating compounding pharmacies after NECC. N Engl J Med 2012;367:1969–72. Available at: http://www.nejm.org/doi/full/10.1056/NEJMp1212667?query=featured_meningitis#t=references [Accessed 11 November 2012].

In keeping with specific health targets formulated by national or international public or professional bodies, local, state, or national health authorities directly provide certain basic public health services, such as those of specialized laboratories. In the USA, public health agencies provide services not otherwise available to high-risk or otherwise underserved population groups. Many of these developed under special funding by higher levels of government to promote specific programs such as immunization, lead abatement, prenatal care, and HIV testing. They

are generally services that are often not adequately covered by health insurance systems or by private practitioners and health care systems.

Immunization may be provided as a governmental service, which is the case in Israel, or by private or managed care providers, but the state retains overall responsibility for policy and implementation of an adequate immunization schedule and level, as in the UK and the USA. Even in countries with well-developed primary care systems, there may be a need for additional special services, such as screening for cancer of the cervix, hypertension, or congenital disease. Health education, a function of all levels of government and non-government health services, involves those activities centered on raising consciousness and knowledge in the health professions, the public, or vulnerable target groups, cutting across virtually all public health activities.

Collaboration may take place with parallel departments of government including education, social welfare, agriculture, urban planning, and voluntary agency groups. Healthy Cities can be an important vehicle for promoting public health-related interests when civic authorities place health on the agenda for urban development in particular. With regard to departments of education, issues of the school health curriculum for education program content, quality of nutrition, and obesity reduction programs are of vital importance.

Financial incentives in the form of grants or other categorical funding may be directed to programs to promote specific public health services, research, or education. Financial incentives are used widely in seeking solutions to particular problems, such as incentive payments to physicians for achieving performance indicators or national health targets such as full immunization, or Papanicolaou (Pap) smears and mammography for target population groups in the UK. National goals may be set in a consultative process, taking into account their importance to the health of the nation. They must also address economic and human resource capacity to organize and deliver relevant programs to meet goals stated with the potential impact evaluated. Incentive or categorical funding is often a useful method to introduce a new set of activities, to strengthen a weak area of public health, or to promote a shift in emphasis in the health system.

NON-GOVERNMENTAL ROLES IN HEALTH

Both the government and the private sector, including not-for-profit and for-profit service systems, have vital roles to play in public health and health care. The private sector includes service providers; professional organizations; universities; and consumer, volunteer, and advocacy groups. Because of the private sector's contribution to service delivery, professional standards, and education of health personnel, it can make a major contribution to any health system.

NGOs may be able to innovate through voluntary action and programming to meet areas of need with which formal health systems may have difficulty. In the USA, the March of Dimes (Box 10.5) is an outstanding example of a volunteer organization and its contribution in the development of the Salk polio vaccine in the 1940s, subsequently in the care of people affected by poliomyelitis and, currently, in the prevention of birth defects. There are many organizations raising funds for promotion of research and services for specific health concerns, ranging from diabetes to multiple sclerosis.

BOX 10.5 The March of Dimes

Founded in 1938 to address the issue of poliomyelitis by President Franklin Delano Roosevelt, a 1921 victim of polio himself, the March of Dimes (MOD) played a major role in providing care for polio-stricken children and the search for a vaccine to prevent the disease. Thousands of volunteers helped to raise funds and to organize widescale clinical trials of the breakthrough vaccine developed by Jonas Salk in 1955. Following the eradication of polio in the USA, the March of Dimes shifted its focus to major health problems of children: birth defects, low birth weight, infant mortality, and lack of prenatal care.

The MOD's 2005 Global Report on Birth Defects states: "Every year an estimated 8 million children – 6 percent of total births worldwide – are born with a serious birth defect of genetic or partially genetic origin. Additionally, hundreds of thousands more are born with serious birth defects of postconception origin due to maternal exposure to environmental agents. At least 3.3 million children less than 5 years of age die annually because of serious birth defects and the majority of those who survive may be mentally and physically disabled for life".

The organization promotes and funds activities to reduce birth defects and infant mortality by measures to prevent low birth weight (to 5 percent or less), and to increase the number of women receiving prenatal care in the first trimester (to 90 percent). It funds work to promote genetic research including gene therapy, testing, counseling, and gene mapping. MOD promotes work on the Human Genome Project with genes related to immune disorders, mental retardation, leukemia, improved blood tests for newborn screening, and improved perinatal care for cerebral palsy and respiratory distress of the newborn. MOD works actively to promote use of folic acid among women in the age of fertility to reduce risks of neural tube defects, and supports comprehensive newborn screening for all babies, for at least 29 conditions for which there are good screening capacity and management of affected children.

Source: *March of Dimes. A history of the March of Dimes. August 2010. Available at: http://www.marchofdimes.com/mission/history_indepth. html [Accessed 26 October 2012].*

Voluntary organizations can often initiate services that the public sector cannot. Examples are numerous, but the following may suffice. In Jerusalem, a father and son established a voluntary organization in memory of the wife and mother (Yad Sarah) in 1976 to provide a wide range of free, loaned medical devices and services, from wheelchairs, through home meals, to day care centers and emergency call systems. The mission of Yad Sarah is to help the elderly and handicapped to function in their own homes. Subsequently, branches were established in 70 cities all over Israel. Other organizations established similar projects in over 25 cities of the former Soviet Union, and plans are in progress for a similar organization in New York City.

In international efforts to reduce the burden of diseases in low-income countries, bilateral governmental aid, such as the work of the US Agency for International Development (USAID), is important, but international agency and donor aid is equally or even more important. The idea of public–private partnership has achieved much in the global arena, with agencies such as the World Health Organization (WHO), United Nations Children's Fund (UNICEF), Joint United Nations Programme on HIV/AIDS (UNAIDS) and many others, along with private foundations such as the Bill and Melinda Gates Foundation. These are discussed in Chapter 16 on global health.

DISASTERS AND PUBLIC HEALTH PREPAREDNESS

After September 11, 2001, preparedness for terrorism became a high priority for federal, state, and local governments (Box 10.6). With federal funding and other support, communities have strengthened their ability to respond to public health emergencies. Collaborative relationships developed for bioterrorism preparedness have proven useful in addressing other threats, such as health impacts of natural disasters and infectious disease outbreaks. The primary role in disaster response is increasingly recognized as a local responsibility. Funding constraints, inadequate surge capacity, public health workforce shortages, competing priorities, and jurisdictional issues all continue to hamper adequate preparation and response, as witnessed by the aftermath of Hurricane Katrina in New Orleans in 2005. The US federal and many state governments have responded with an investment of some US$5 billion since 2001 to upgrade the public health system's ability to prevent and respond to large-scale public health emergencies, whether caused by terrorism or by natural agents.

As most natural disasters affect many communities and require major resource support, state and federal agencies are necessarily involved. Non-governmental and bilateral aid support is vitally important but basically subsidiary to the governmental agency responsibility and coordination. There has been some criticism of governmental agencies placing too much responsibility on NGOs, such as the American Red Cross following both Hurricane Katrina in 2005 and Storm Sandy in 2012, in terms of location of support supplies and speed of response, but governmental agencies also came under criticism for their poor preparedness and slow speed of effective response.

In 2012, the WHO and World Meteorological Organization published an *Atlas of Health and Climate*, which provides an excellent review of the effects of climate on infections (malaria, diarrhea, meningitis, and dengue fever), emergencies (floods and cyclones, drought, and airborne dispersion of hazardous materials), and emerging

---

BOX 10.6 Planning Assumptions for Emergency Mass Critical Care

- Mass casualties from bioterrorist attacks or accidental, chemical, or biological releases may occur without warning and could result in hundreds, thousands, or more critically ill victims.

- National, state, and local health authorities should prepare, direct, and coordinate activities in planning and managing such critical situations as illness due to pandemic, natural disaster, and other human-caused or natural disaster situations, utilizing all public and private resources for such events.

- Prehospital care by first responders trained in first care measures of triage, and in chemical contamination, is a vital part of public health systems. They should include well-trained personnel with standard protocols for bleeding, blast injury, or compromised airway care with oxygen and intubation. Ambulances or other transportation to well-organized emergency departments in hospitals are also crucial to life saving in disaster situations.

- Mass illness (or injury) from a pandemic may produce large numbers of critically ill patients requiring acute respiratory care.

- Mass critical illness will place great stress on local community hospitals, which will have a key role in decreasing morbidity and mortality rates after a bioterrorist attack or pandemic disaster situation.

- Surge capacity pre-event planning is required for mass critical care with new approaches to triage and care, fluid infusion, and rapid transport to the nearest hospital.

- Any hospital will have limited ability to divert or transfer patients to other hospitals in such an event.

- Currently deployable medical and epidemiological teams of the US federal government will have a limited potential for increasing a hospital's immediate ability to provide critical care to large number of victims of a bioterrorist attack.

- Hospitals will need to depend on non-federal sources or reserves of medications and equipment necessary to provide critical care to the seriously ill for the first 48 hours following discovery of the bioterrorist attack, or during a pandemic.

Sources: Khan AS, Levitt AM, Sage MJ. Biological and chemical terrorism: strategic plan for preparedness and response. Recommendations of the CDC Strategic Planning Workgroup. MMWR Morb Mortal Wkly Rep 2000;49(RR-04):1–14.
Rubinson L, Nuzzo JB, Talmor DS, O'Toole T, Kramer BR, Ingelsby TV. Augmentation of hospital critical care capacity after bioterrorist attacks or epidemics: recommendations of the working group on emergency mass critical care. Crit Care Med 2005;33:E1–13.
Centers for Disease Control and Prevention. Emergency preparedness and response. Mass casualty information for emergency medical services (EMS) providers. Available at: http://www.bt.cdc.gov/masscasualties/ems.asp [Accessed 2 November 2012].
National Center for Injury Prevention and Control. Updated: In a moment's notice: surge capacity for terrorist bombings. Atlanta, GA: Centers for Disease Control and Prevention; 2010. Available at: http://www.bt.cdc.gov/masscasualties/pdf/cdc_surge-508.pdf [Accessed 2 November 2012].

environmental challenges (heat stress, ultraviolet radiation, pollen, and air pollution).

The public health system will continue to face demands for emergency preparedness and health protection in the face of natural disasters and terrorism. The challenges are to use focused, risk-based resource allocation, regional planning, technological upgrades, workforce restructuring, and improved monitoring.

Disaster preparedness requires activities and readiness at all levels of government, and by first responders (police, firefighting, and ambulance services) as well as by health care institutions. Activities include preparation of essential supplies, organizational guidelines, staff training and orientation, as well as adequate funding to meet these needs. Since disasters with mass casualties may appear in many forms, the response teams need flexibility and capacity for improvization. Coordination between different levels of government can be difficult, with lines of command and lateral communication unclear and potentially disastrous. Preparation for treatment of mass casualties of bioterrorism requires similar resources to a situation of pandemic and mass illness due to a new variant of severe acute respiratory syndrome (SARS) or avian influenza (Box 10.7).

The US Federal Emergency Measures Agency (FEMA) was established by President Jimmy Carter in 1978. It acts on the request of a state governor, who declares a state of emergency and requests federal assistance. FEMA provides experts in specialized fields of disaster management, and funds for reconstruction, emergency relief, and support services. FEMA has assisted state and local authorities in many instances of hurricanes, floods and other disasters, including the Love Canal toxic chemical waste site in New York State and the Three Mile Island nuclear near-meltdown threat in the late 1970s.

FEMA was attached to the new US Department of Homeland Security created in 2002, but suffered from reduced budgetary and restricted definition of functions, so that when Hurricane Katrina struck in Louisiana and the Gulf states, with devastating effects on New Orleans, the municipal, state, and federal responses were seriously lacking. Because of the bitter legacy of Hurricane Katrina, FEMA was strengthened in its terms of reference and budgetary support.

In late October 2012, Superstorm Sandy, and its associated snowstorms, reached a wide sector of the US eastern seaboard, with the overwhelming power of a hurricane, high waves from the sea sweeping inland, widespread flooding, and destruction of everything in its path. It led to flooding of major parts of New York City and New Jersey, millions of people being affected by fires and power and transportation outages, and some 100 deaths, mainly from falling trees and drowning. The responses of city, state, and federal authorities were impressive in their initial disaster management and provision of public information. FEMA played a

vital supportive role, and continues to provide support in the reconstruction phase. But the first responders were local city employees of the police, fire, and ambulance services, who helped to coordinate health services, and evacuated patients from facilities threatened with flooding and fires, and the loss of electricity, food supplies, and other essentials. The public health impact is likely to be immense, from floodwaters as well as the potential for carbon monoxide poisoning from misuse of generators. Relief efforts by local, state, and national authorities to alleviate the immediate impact on millions of people will be followed by reconstruction that may take years and cost an estimated US$50 billion. Agencies involved in relief include the American Red Cross (http://www.redcross.org). The federal government has a number of useful websites containing valuable information, including:

- US Government – www.ready.gov
- US DHHS – www.phe.gov/emergency
- CDC – http://www.cdc.gov/Features/AfterAFlood/index.html
- EPA – www.epa.gov/hurricanes
- FEMA – www.fema.gov/response-recovery

MEDICAL PRACTICE AND PUBLIC HEALTH

Public health and clinical services are interactive and mutually supportive. Both have important roles to play in individual and population health. Ready access to high-quality health care services is a basic right and a requirement of good public health. This calls for high-quality organization and the availability of professionals to provide both clinical and preventive care. The phenomenon of private payment to physicians working in public sector health systems is widespread, as is that of physicians in public service who practice privately after official working hours. Under-the-table payments are common in many countries and difficult to stop, but regulated private services in public or voluntary hospitals can be regulated allowing onsite private services with a portion of the funds remaining with the hospital.

In Canada, the Supreme Court of Quebec ruled in 2005 that delays in the health system for medically justified procedures were in contravention of the Quebec Charter of Human Rights. This caused national controversy over the integrity of Canadian provincial health plans, supported on one side by the public and all political parties, and on the other side by medical associations and opponents of public medical care systems.

In the UK, private practice by specialists employed by hospitals is permitted and encouraged, allowing faster access to hospital care for private patients. This situation is often seen as a built-in injustice in the NHS. In Israeli teaching hospitals, a private medical service is organized using

BOX 10.7 Lessons from Recent Disasters and Threatened Pandemics

The twenty-first century began with the 9/11 massive terrorist attack on New York City's World Trade Center in Manhattan using hijacked civilian aircraft, causing over 2500 deaths and many injuries. This event stirred worldwide repercussions and was followed by deadly terrorist strikes in Madrid, London, Bali, Mumbai, and many other parts of the world. These attacks caused national and international reactions including calls for disaster preparedness with stress on local capacity for response to human-caused and natural disasters, with emphasis on basic "first responder" service capacity.

During 2003, a threatened pandemic of severe acute respiratory syndrome (SARS) started in China and, in a short time was transmitted via an infected person to Toronto, Canada. The Canadian provincial and municipal authorities were taken by surprise and lacked adequate federal mechanisms for addressing the problem. Provincial and municipal authorities managed the epidemic by hospitalization and isolation of all suspected cases with quarantining of hospitals involved. As a result of review of this experience, Canadian governmental authorities developed new federal institutions, in part modeled on the US Centers for Disease Control and Prevention, establishing a federal Public Health Agency whose director was also a deputy minister in the federal Department of Health, with direct authority to increase the federal presence in epidemic control.

In 2004–2005, three huge natural disasters occurred in different parts of the world, showing the crucial importance of disaster preparedness and response organization, preparation, and intergovernmental coordination. The tsunami in Thailand and surrounding regions, Hurricane Katrina in Louisiana and especially New Orleans, and the earthquake in northern Pakistan showed the crucial need for coordination and speed as well as preparation for natural disasters by all levels of government working with voluntary organizations for rescue and relocation needs.

In 2006, the H5N1 influenza virus, also called "avian flu", threatened to become a new world pandemic of a scope similar to the influenza pandemic of 1917–1918. National and world public health organizations mobilized under the leadership of the WHO, implementing monitoring and control measures. These largely rest on identification of cases among wild and domestic birds, and the rapid identification, isolation, and treatment of human cases. Culling of domestic agricultural birds took place to restrict transmission of the H5N1 virus, which could produce a human pandemic of epic proportions if transmitted from birds to humans and then by human-to-human transmission.

In May 2008, a cyclone disaster in Burma (Myanmar) killed many tens of thousands of people, and left some 1.5 million homeless, destitute, and vulnerable to secondary disasters from new floods, exposure, famine, and infectious diseases. The response from the military government has been alleged as criminally negligent, preventing foreign aid reaching the people in need. China was struck by a massive earthquake and series of aftershocks which killed an estimated more than 100,000 people and devastated many cities, towns, and villages. The governmental response was immediate and effective, accepting limited foreign assistance, which was unable to cope with the calamity, but limited the secondary effects of famine and infectious diseases.

Hurricane Sandy in 2012 resulted in one of the largest disaster areas affecting the Caribbean and six states in the USA, including New York City. The damage in the USA included over 100 deaths, and an estimated US$50 billion of damage to property and public facilities. With power outages, the unsafe use of home generators and indoor use of charcoal grills resulted in fatal carbon monoxide poisonings.

In August 2013, in a civil war in Syria with over 100,000 deaths and millions of refugees, a large-scale use of a neurotoxic chemical weapon (probably sarin) caused many hundreds of deaths and casualties. This caused international outrage and possible military response by the USA, the UK, and France. The intervention is legally based on the precedent of NATO's Kosovo intervention in the 1990s to prevent continued genocide and the Hague Convention on the use of chemical weapons in warfare (see Chapter 9).

These experiences and threatened pandemics have brought public health organizations and key public health functions into the spotlight of national thinking in many countries, after many years of financial cutbacks and administrative neglect or outsourcing to private providers. This public awareness may be fleeting, and should be used to help strengthen public health infrastructure capacity and workforce development.

Sources: *Centers for Disease Control and Prevention. Public health emergency response guide for state, local, and tribal public health directors. Available at: http://www.bt.cdc.gov/planning/responseguide.asp [Accessed 8 November 2012].*
World Health Organization. Myanmar disaster. Available at: http://www.searo.who.int/LinkFiles/Myanmar-Cyclone_sitrep_170508.pdf [Accessed 8 November 2012].
US Department of Health and Human Services. Public health emergency preparedness and recovery. Hurricane Sandy and response 2012. Available at: http://www.phe.gov/emergency/events/Pages/sandy-midatlantic-2012.aspx [Accessed 8 November 2012].
Centers for Disease Control and Prevention. Emergency preparedness and response. Hurricanes. Hurricane Sandy. Available at: http://emergency.cdc.gov/disasters/hurricanes/index.asp [Accessed 8 November 2012].

senior physicians on the hospital premises, with a percentage of the generated funds going to the hospital.

Fee-for-service payment practice of medicine is still common in the USA and Canada, even though each of these countries has different methods of financing services. Canada's national health insurance program is based on private fee-for-service practice of medicine. Fee schedules are negotiated between each province and their respective medical associations. Federal legislation bans extra billing by physicians, which could threaten equity of access for all population groups, as part of federal criteria for the support of provincial health plans.

The USA has a mixed situation of private health coverage, mainly through employer-subsidized insurance, Medicare for those over age 65, and Medicaid for the poor and people with disabilities. This combined system has proven inadequate on a societal level; some 48.6 million people (or 15.7 percent of the US population in 2012, increasing to 16.3 percent in July 2013) lack health insurance and another 15 million have poor levels of coverage, with further difficulties for those who change jobs and lose their health insurance coverage. In 2010 nearly 26 percent of people in the USA had at least one month without health insurance coverage. Growth of managed care plans is occurring as private medical practice is declining in the USA. Operated as for-profit or as not-for-profit programs, managed care plans provide lower cost and more comprehensive coverage than traditional insurance plans (US Census Bureau).

Medical care outside hospitals was reviewed in eight countries where health care financing is based predominantly on social health insurance and in others funded through taxation (Ettelt et al., 2006). This and another study pointed out wide variation in patterns of organization, use of computerized medical records, insurance restrictions, quality incentives, and other factors (Schoen et al., 2009). Common issues that are emerging are the increasing burden of chronic conditions, the tendency to move services out of hospitals, the use of information technology, and group practices with ancillary health workers. Reforms in various countries encouraging multispecialist and general practitioner networks with integration into single centers providing medical service are becoming an increasing trend. In the USA, accountable care organizations (ACOs) are linking primary care with hospitals for comprehensive care and this will be fostered by elements of the Patient Protection and Affordable Care Act (PPACA, "Obamacare") being introduced in 2014.

Health care is being reformed in many countries. Such reform requires incentives to promote ambulatory and community outreach services, through incentives and integration of hospital and long-term care. Managed care is important in the USA, and the model is relevant in other countries because of the link with reducing unnecessary use of hospital and unreferred specialist services, placing emphasis on primary care and preventive care (see Chapters 11–13).

INCENTIVES AND REGULATION

Incentives and disincentives are important tools in health policy and management. Governments are responsible for assuring adequate supplies and quality of health facilities and personnel to meet the needs of the population. They also are responsible for assuring that financing of the system is adequate and efficient. These responsibilities include the use of public authority to ensure a balanced and high-quality system of care equitably available to people of all regions and social classes. Whether services are owned and administered by government, non-profit agencies, or private

auspices, the public authority is responsible and accountable for ensuring that the health needs of the population are met.

The appropriate balance among different elements of health systems serving the same regional or district population is an important public health planning issue. Health facilities such as hospitals and long-term and community care facilities are licensed and regulated by the appropriate public health authority. This regulatory power is necessary, but not sufficient without financing arrangements to combine incentives and disincentives (Box 10.8).

The ratios of hospital beds and medical personnel per thousand population are crucial determinants of health economics, so that national and state health authorities must use their regulatory powers to contain supply and distribution. Excess labor supply of medical specialists is a problem in many mid-level developing countries, such as in Latin America. Regulatory or financial powers, as well as financial controls, can be used to reduce the oversupply of specialists and to redirect doctors to underserved areas of a country and primary care.

A federal government authority can act to promote health programs by setting financial incentives and disincentives. The categorical grant approach provides funds for a specific purpose or cost sharing for a program that meets defined guidelines. Canadian health insurance is based on provincial plans meeting federal guidelines to qualify for a share of the costs. The Canadian national health insurance system is based on provincial plans with federal cost

---

BOX 10.8 Regulation and Incentives: Carrots and Sticks

"Carrot and stick" is a phrase used to refer to the act of simultaneously rewarding "good" behavior while punishing "bad" behavior. An older interpretation is the use of a carrot dangling on a stick in front of an uncooperative mule, so that the encouragement is constant, but the satisfaction is permanently elusive.

The combination provides financial mechanisms, and limiting the supply of, for example, hospital beds by regulation or financial incentives is meant to encourage health facilities to develop, in keeping with national, state, or local needs. In developed countries, this may mean closure of excess hospital beds and reallocation of resources to community-based health services, as in the UK, many European countries, Canada, the USA, and others. In Russia and many former Soviet countries, the incentives and requirements produced a heavily hospital-oriented health system with lower priorities to community-based services.

Pay for performance (P4P) is being adopted in other countries. The US Patient Protection and Affordable Care Act of 2010 (Obamacare) includes incentives to institutions to improve quality of care and rural care, and incentives to provide free preventive care for breast and cervical cancer screening and other preventive care services.

sharing and conditions. The first public health insurance plan was enacted in 1947 by the province of Saskatchewan, and led to passage in 1957 of the federal Hospital Insurance and Diagnostics Services Act, which ensured universal coverage for in-hospital services in provinces that met federal criteria. By 1961, all of Canada's 10 provinces had signed on. In 1962, the government of Saskatchewan passed an act requiring doctors to collect fees solely through the government-run plan. Thus, the Canadian system is based on provincial responsibility and administration, but with federal cost-sharing incentives that helped to induce the provinces to participate. Federal conditions for funding in health include universal coverage, comprehensiveness, portability, and public administration as criteria for the provincial plans. When the federal government moved from a fixed percentage of expenditures to block grants, it lost some control over detailed management of provincial plans, but it retains a strong voice in requirements for equity, portability (i.e., transferability of insured benefits from one province to another), public administration, and prohibiting extra billing by providers for insured services. As federal shared cost program funding declined as a share of total provincial health costs, the provinces were under pressure to reform, mainly by reducing the hospital bed supply and promoting community-wide health service organization. The federal parliament unanimously passed the Canadian Medical Care Act of 1966, giving the national framework a stronger legislative base, setting standards for provincial plans, disallowing extra payment for medical services, and ensuring a standard across the country of which Canadians are very proud.

In the USA, national health insurance was included in the proposed social security legislation during the Roosevelt administration but excluded from the Social Security Act of 1935 because of severe opposition to the major elements of the act by the medical association and the insurance industry. During World War II, the Emergency Maternity and Infant Care Program (EMIC) was established by the federal government to help state governments to provide wives and infants of lower grades of servicemen with generous obstetric and pediatric care. Thus, to meet the needs of military families, the government became involved in health care. This was the first national health services program for a significant sector of the US population.

Following the end of World War II, in 1946, the proposed Wagner–Murray–Dingell Bill for national health insurance failed to reach the floor of Congress, dying in committee, under severe pressure from the American Medical Association and the health insurance industry. A portion of that proposal emerged, however, as the Hill–Burton Act (HBA) to provide federal assistance to local agencies to build or upgrade hospitals. The Hill–Burton model is a relevant approach to problem solving in a federal state using a categorical grant mechanism to promote what is

seen as a health priority. Such an approach may be useful to strengthen a weak health program such as immunization and maternity care in a developing country. It may be used to change the balance in supply of services and resources. A system of incentives or cost-sharing arrangements can provide capital funding; for example, to reduce total bed capacity and to promote integration of maternity, mental health, geriatric, and TB facilities into general hospitals. A "downsize and upgrade" conditional grant would provide for renovation and transition to an approved program of facilities to modernize hospital services. The federal grant system, pioneered by the HBA (Box 10.9), would encourage the local authority to apply for and match part of the funding, and meet federal criteria and guidelines for this process.

The HBA is relevant today as a model for top–down health services development based on transfer of federal funds to promote state and local health services development, and may be applied to many targeted needs such as in financing community-based networks of primary and secondary care services. In some ways it is a component of the 2010 Obamacare plan now being introduced in the USA to extend insurance coverage and to control the costs of public and private insurance by ACOs. These are basically networks of service systems with financial as well as administrative linkages (see Chapters 11 and 13).

The 1965 Medicare and Medicaid titles under the Social Security Act enacted Medicare, which provides health insurance for people over 65 years and those with major disabilities. Medicaid, also established under the Social Security Act, provides a system of federal assistance to state health insurance for the poor (see Chapter 13). Subsequent attempts to introduce various forms of national health insurance failed in Congress, with some exceptions, until the Obama administration passed the PPACA in 2010, extending health insurance coverage to millions of Americans, with many cost savings and incentives to improved preventive care coverage in the US population.

In countries where health systems were highly centralized, such as the UK NHS and in former Soviet health systems, decentralization and diffusion of power were promoted by financing mechanisms. These are discussed in Chapters 13 and 15.

Unregulated chronic care facilities operated by private interests resulted in proliferation of poor-quality facilities and sometimes extremely low levels of care in many communities in the USA. Public health authorities were powerless to interfere except in cases of gross neglect or poor sanitary facilities. The introduction of Medicare for the elderly and Medicaid for the poor provided federal and state agencies with the power to set minimum standards for care facilities, by requiring all facilities serving Medicare patients to be accredited by a non-governmental agency accepted by the federal health authorities. This

BOX 10.9 The Hill–Burton Act

The Hill–Burton Act (HBA), adopted by the US Congress in 1946 as the Hospital Survey and Construction Act, provided a federal–state–local partnership that channeled large federal grants to assist the development of hospitals and standards for construction (Hospital Survey and Construction Act, 1946, Title VI of the Public Health Service Act). This affected 4000 communities in 6800 projects to modernize hospitals suffering from a lack of investment from the depression and World War II period. Initially it covered hospitals, but later was expanded to extended care, rehabilitation facilities, and public health centers. In 1975, this was further expanded to grants, loan guarantees, and interest subsidies for health facilities. Facilities assisted under Title XVI were required to provide uncompensated services in perpetuity. The HBA gave hospitals, nursing homes, and other health facilities grants and loans for construction and modernization. The HBA required facilities which benefited with federal grants to provide a "reasonable volume of services to persons unable to pay and to make their services available to all persons residing in the facility's area". Although the program stopped providing funds in 1997, approximately 170 US health care facilities still have to provide free or reduced-cost care.

This Act of Congress brought national standards and financing to local hospitals. The program helped to raise standards of medical care throughout the USA in the 1950s and 1960s. It led to an increase in numbers of hospitals in underserved areas and the renovation of obsolete facilities. It promoted

desegregation in the southern USA and provided a mechanism for treatment of the uninsured in the nation's hospitals.

The program also succeeded in limiting the buildup of an excess of hospital beds, setting standards at 4–4.5 acute care hospital beds per 1000 population (more for rural areas), without an increase in the total supply of beds. While it favored middle-class communities because it required local financial contributions, it also channeled federal monies to poor communities, thus raising standards of hospitals and equity in access to quality care. In setting upper limits on hospital beds, it limited hospital expansion and contributed to a continuing process of improvement of diagnostic and patient care shortening hospital stays. Limiting hospital bed supply over time influenced medical ideology and helped to promote community-based health services.

The program had a number of basic failings, including the promotion of the hospital as the main center of health care, leaving community care out of the main flow of added funds. It led to an increase in the proportion of health expenditures going to hospital care. Expenditures for hospital care in the USA as a percentage of total health expenses increased from 34.5 percent in 1960 to a high of 41.5 percent in 1980, but declined to 35.4 percent in 1995. In the 1980s, the HBA was expanded to promote clinic and primary care facilities.

Source: *Department of Health and Human Services. Human Services and Resources Administration (HSRA). Hill–Burton free and reduced-cost health care. Available at: http://www.hrsa.gov/gethealthcare/affordable/hillburton/ [Accessed 1 November 2012].*

has become a standard requirement throughout the USA. The Canadian provincial health insurance plans also apply economic sanctions on unaccredited hospitals or other inpatient facilities.

Another measure to increase regulation of health care facilities was the requirement for any hospital proposing expansion or renovation to seek state approval through a Certificate of Need (CON). The CON, as used in the USA under state health legislation, makes approval by the state contingent on demonstrating need and sources of funding which comply with state regulations. This measure can be linked with incentive grants but can also be used as a simple regulatory mechanism. The CON approach by state departments of health was only partially successful in limiting unbridled ambitious expansion of hospital facilities. In the 1980s and especially the 1990s, competition and changes in payment systems have resulted in hospital closures and downsizing in the USA.

Promotion of Research and Teaching

Research and education are the basis for future developments in health care. They foster new health scientific developments in health, such as diagnostic devices, vaccines, and medications. The Human Genome Project has already generated

new diagnostic and treatment for genetic and chronic diseases. Research contributes to the development of medical schools, but also safeguards, guarantees, and increases their quality, raising standards of care. Research in public health depends on the basic and clinical sciences, but equally on epidemiology and documented experience of field programs.

In the USA, the National Institutes of Health (NIH), starting with the National Cancer Institute in the 1930s, have done much to encourage high-quality medical education and research. The NIH granting system has been a major factor in promoting standards of medical education by financing research and teaching faculties in medical schools throughout the USA. NIH funding has played a major role in moving the USA to the forefront of the biomedical sciences since World War II. There are currently 27 separate National Institutes of Health including centers and divisions (Box 10.10).

A combination of professional competition, the free publication and exchange of research studies, views in peer-reviewed journals, and professional meetings in government agencies promotes scientific and applied progress in the medical sciences. Clinical guidelines and recommended practices contribute to quality of care. The private sector manufacture of drugs and medical devices contributes to

---

BOX 10.10 US National Institutes of Health, Centers and Divisions, and Internet Addresses, 2012

- Institutes home page – http://www.nih.gov/
- National Cancer Institute (NCI) – http://www.cancer.gov/
- National Eye Institute (NEI) – http://www.nei.nih.gov/
- National Heart, Lung, and Blood Institute (NHLBI) – http://www.nhlbi.nih.gov/
- National Human Genome Research Institute (NHGRI) – http://www.genome.gov/A/
- National Institute on Ageing (NIA) – http://www.nia.nih.gov/
- National Institute of Alcohol Abuse and Alcoholism (NIAA) – http://www.niaaa.nih.gov/
- National Institute of Allergy and Infectious Diseases (NIAID) – http://www.niaid.nih.gov/Pages/default.aspx
- National Institute of Arthritis and Musculoskeletal and Skin Diseases (NIAMS) – http://www.niams.nih.gov/
- National Institute of Biomedical Imaging and Bioengineering– http://www.nibib.nih.gov/
- National Institute of Child and Human Development (NICHD) – http://www.nichd.nih.gov/
- National Institute of Deafness and Other Communication Disorders (NIDCD) – http://www.nidcd.nih.gov/Pages/default.aspx
- National Institute of Dental and Craniofacial Research (NIDCR) – http://www.nidcr.nih.gov/
- National Institute of Diabetes and Digestive and Kidney Diseases (NDDK) – http://www2.niddk.nih.gov/
- National Institute of Drug Abuse (NIDA) – http://www.drugabuse.gov/
- National Institute of Environmental Health Sciences (NIEHS) – http://www.niehs.nih.gov/
- National Institute of General Medical Sciences (NIGMS) – http://www.nigms.nih.gov/
- National Institute of Mental Health (NIMH) – http://www.nimh.nih.gov/index.shtml
- National Institute of Neurological Disorders and Stroke (NINDS) – http://www.ninds.nih.gov/
- National Institute of Nursing Research (NINR) – http://www.ninr.nih.gov/
- National Institutes of Health Clinical Center (NIHCC) – http://clinicalcenter.nih.gov/
- Center for Information Technology (CIT) – http://www.cit.nih.gov/
- National Library of Medicine (NLM) and MEDLARS – http://www.nlm.nih.gov/
- National Institute on Minority Health and Health Disparities (NIHHD) – http://www.nimhd.nih.gov/
- National Center for Research Resources (NCRR) – http://www.nih.gov/about/almanac/organization/NCRR.htm
- National Center for Complementary and Alternative Medicine (NCCAM) – http://nccam.nih.gov/
- John Fogarty International Center (FIC) – http://www.fic.nih.gov/Pages/Default.aspx
- Center for Scientific Review (CSR) – http://public.csr.nih.gov/Pages/default.aspx

MEDLARS = Medical Literature Analysis and Retrieval System.
Source: *National Institutes of Health, Bethesda, MD. Sites confirmed: 8 November 2012.*

---

the continued development of medical and public health sciences. National centers of excellence in public health in other countries include the Pasteur Institute in France and Cambridge Laboratories in the UK. They receive national funding and have a critical mass of high-quality researchers. Federal funding of medical teaching centers supports development and maintenance of academic standards for undergraduate medical education.

Federal or external granting mechanisms can be used to promote schools of public health and health administration that are needed to prepare the next generation of health leaders, academics, and researchers. Research may be initiated in response to requests for proposals by scientists in university or research institutes, or in the governmental or private sector. A competitive peer-reviewed grant system can be useful to upgrade medical education and university academic standards by promoting research and graduate education, as developed by the US NIH since 1946.

Accreditation and Quality Regulation

Public health authorities have sufficient powers to regulate health facilities. However, in practice, accreditation based on professional guidelines and systems outside the governmental structure (see Chapter 15) plays an important role in quality of health care provider organizations, as an important adjunct to the official regulatory approach of health departments.

The Joint Commission on Hospital Accreditation (JCHA) started in the USA in 1913, and included Canada from 1951 until 1959, when the latter established its own accreditation system. The JCHA was established by a consortium of the American College of Surgeons, the American Hospital Association, and other voluntary professional bodies. It carries out voluntary peer review of hospitals throughout the USA. The commission established minimum standards in 1918, and has gone on to develop extensive guidelines based on physical, organizational, and professional criteria, to protect the safety and rights of the patient, standards of care, and efficient organization of services. Accreditation involves a process of external review of the facilities, organization, staffing, and related functions including staff qualifications, continuing education, medical records, and quality assurance (see Chapter 15).

The JCHA review was initially conducted on the basis of a voluntary request by the institution, but accreditation has become virtually mandatory for the economic survival of a hospital in the USA and Canada. Since 1965, Medicare and Medicaid accept accreditation as compliance with federal standards for the purpose of payment, and refuse to pay for services in an unaccredited hospital. The renamed Joint Commission for Accreditation of Healthcare Organizations (JCAHO) has gone on to develop standards for accreditation of facilities for the mentally retarded (1969), psychiatric facilities (1970), long-term care facilities (1971), ambulatory facilities (1975), hospices (1983), managed care programs (1989), and home care and ambulatory care (1990). There is a growing emphasis on action plans for quality improvement for rural hospitals, health care networks, laboratories, and public health programs. The JCAHO has become active in promoting accreditation organizations in other countries such as the UK and Australia.

The New York State Department of Health has its own mandatory regulatory system for hospitals and long-term care facilities. Regulation of hospitals and other health care institutions or programs including public health organizations is essential to the maintenance of quality standards and prevention of professional and human rights abuses. Accreditation by non-governmental agencies such as the Joint Commission may be accepted in lieu of state inspection. The New York State Department of Health has a collaborative agreement with the Joint Commission. In that agreement, the Department will waive a routine onsite survey of a facility if that facility requests accreditation by the Joint Commission. Israel, during the 1990s, established a national system of inspection of private long-term care facilities, which has improved standards of facilities and care. While opponents may see this as excessive state interference, in principle accreditation is for the protection of patients' rights in public service facilities, even under private auspices. Resultant improvements in quality of care measures have justified prudent regulation and oversight of health care facilities. These models could be useful for raising standards in other health care systems.

NATIONAL GOVERNMENT PUBLIC HEALTH SERVICES

National governments can use their financial power to promote programs directly to the state, provincial, or local governmental level or indirectly through non-governmental agencies. The latter include universities, voluntary teaching hospitals, and private NGOs. Direct or indirect funding may be used to diffuse and promote national standards, such as in medical education and research. Both federal and unitary governments often try to ensure regional equity of services by the use of cost sharing or grants that favor poorer regions of the country. National governmental health agencies are responsible for external relations, including those with international bodies such as the United Nations, the WHO, the Food and Agriculture Organization, and the International Labour Organization (see Chapter 16), as well

> **BOX 10.11 Key Functions of a Federal or National Ministry or Department of Health**
>
> - National health planning
> - National health financing
> - National health insurance
> - Assurance of regional equity
> - Defining goals, objectives, and targets
> - Setting standards and quality of care
> - Promotion of research in quantity and quality
> - Operating or delegating professional standards/licensing
> - Environmental protection
> - Food and drug standards, licensing
> - Epidemiology of acute and chronic disease
> - Health status monitoring
> - Medical/pharmaceutical industrial development
> - Health promotion
> - Nutrition and food policy
> - National reference laboratories
> - Social assistance
> - Social security
> - Identification of reportable diseases
> - Immigration health requirements

as with parallel ministries of health in other countries, and other national agencies in the same country.

Before and after World War II, most western industrialized countries developed some form of national health program. In North America, health care was provided through private insurance, largely union-negotiated, employment-based health plans. Attempts by US President Harry Truman to bring in a national health insurance plan in 1946 were unsuccessful. As a result, federal support for health was channeled into many categorical programs by funding state and county public health services and research and teaching facilities, and the CDC and the NIH were established. This promoted high levels of competitive, peer-reviewed programs throughout the country, but failed to ensure universal access to health care (see Chapter 13).

In all forms of government, the national responsibility for health has led to specialized public health services as well as supervisory and regulatory functions (Box 10.11). These include provision of vital support services, such as public health reference laboratories, epidemiology and communicable disease control activities (e.g., national epidemiological publications, airport, and port surveillance), national health statistics, approval and supervision of drugs and biologicals, research and teaching facilities, and cooperation among federal, state, and local authorities. Standards bureaux and agencies, such as the Food and Drug Administration (FDA) and Environmental Protection Agency (EPA) in the USA and the National Institute for Health and Care Excellence (NICE) in the UK (see Chapter 13), while created by governments, need to have a high degree of semi-autonomy to provide regulations, enforcement, guidelines, monitoring, and/or supervision of health care at the lower levels of government and in the non-governmental and private sectors.

The federal government entered the public health arena in areas where only a national jurisdiction could function. The Marine Hospital Service was established in 1798 to provide care for US and foreign seamen, becoming the United States Public Health Service (USPHS) in 1889. Under the organizational structure of the US Department of Health and Human Services (DHHS), the USPHS provides direct care in many areas of US society, including Native American reservations, areas of physician shortage, the US Coast Guard, and penal institutions. The federal Food and Drug Act of 1906, which has been updated frequently, protects the consumer from adulterated foods and ineffective or dangerous medicines. The Social Security Act has provided pensions for elderly and disabled people since 1935. In 1965, the Social Security Act was extended to include Medicare as a federal program providing health insurance for the elderly. In the same year, Medicaid was also established, providing health care for the poor, set up as a cost-sharing program with state and local authorities. The history of development of public health in the USA reflects advancing scientific knowledge, societal demands for better health, and the evolution of interactive organization at federal, state, and local levels. In some respects, public health in the USA has provided professional leadership in the field internationally; in other respects, the USA has lagged behind other industrialized countries.

The Department of Health, Education, and Welfare was established in 1953 under a cabinet-level officer of the executive branch of the Eisenhower administration. This brought together a variety of federal agencies and programs, and subsequent reorganization led to the emergence of the DHHS. The present organizational structure of the DHHS is presented in Figure 10.2. The federal role in direct regulation and funding of projects deemed to be in the national interest helps to promote state and local health authority response to public health problems. The categorical grant system has been instrumental in advancing specific areas of activity, such as maternal and child health, which remain a major activity of both state and local public health departments. The initiatives of the Health Care Financing Administration (HCFA) in promoting changes in methods of paying for hospital care through diagnosis-related groups (DRGs; discussed in Chapters 12 and 13) helped to reduce hospital lengths of stay, days of care, and the hospital bed to population ratio.

The Surgeon General of the Public Health Service is also the Assistant Secretary for Health and provides important professional leadership to the public health movement in the USA. Dr C. Everett Koop, an outstanding surgeon general, who served from 1982 to 1989 during the Reagan administration, exemplified this kind of leadership role. As a pediatric cardiac surgeon, he was initially poorly accepted by the public health community as an "outsider", but came to be a highly respected leader and advocate for public health, responsible for many accomplishments, most notably increased awareness of the deadly effects of tobacco use and for HIV/AIDS research and treatment funding.

The CDC plays a continuous role in dispersing epidemiological data and evaluation throughout the country and the world (see Chapter 4). The training program of Epidemic Intelligence Service (EIS) officers for federal, state, and local health departments continues to provide high-quality medical epidemiologists capable of developing leadership in this field.

Other agencies of the federal government control health-related programs, including the Departments of Agriculture, Defense, the Environment, the Interior, Labor, and Transportation. The Department of Agriculture operates a National School Lunch program and a food stamp program to supplement food purchasing power for the working poor. The Department of Labor operates the Occupational Safety and Health Administration. The EPA is an independent federal agency responsible for air and water quality, pollution control, pesticide regulation, solid waste control, radiation and toxic substance hazard control, and noise abatement.

STATE GOVERNMENT PUBLIC HEALTH SERVICES

State or provincial governments have leading roles in health in most federal countries, as constitutions written in the eighteenth or nineteenth century. These left health to state or provincial responsibility for ensuring adequacy in organization, setting standards and targets, assisting financially, and providing professional and technical support services to local health departments. State functions, such as financing and in some cases direct services and monitoring health status, are listed in Box 10.12. In Canada, the provinces are responsible for universal health insurance programs within federal standards and financial support. In the USA, states are responsible together with local welfare authorities for operating Medicaid programs within the federal funding and guidelines, but the access and support levels vary widely by state.

State or provincial departments of health are complex organizations with many responsibilities for financing, regulating, inspecting, and assuring health-related issues. In the USA, responsibilities include administration of health insurance for the poor under Medicaid; in Canada, the provinces administer universal health insurance plans. States may initiate programs that are shared with local health authorities and with federal cost-sharing, or respond to federal initiatives and seek funding for a wide variety of programs through federal requests-for-proposals for maternal and child health or other categorical grants.

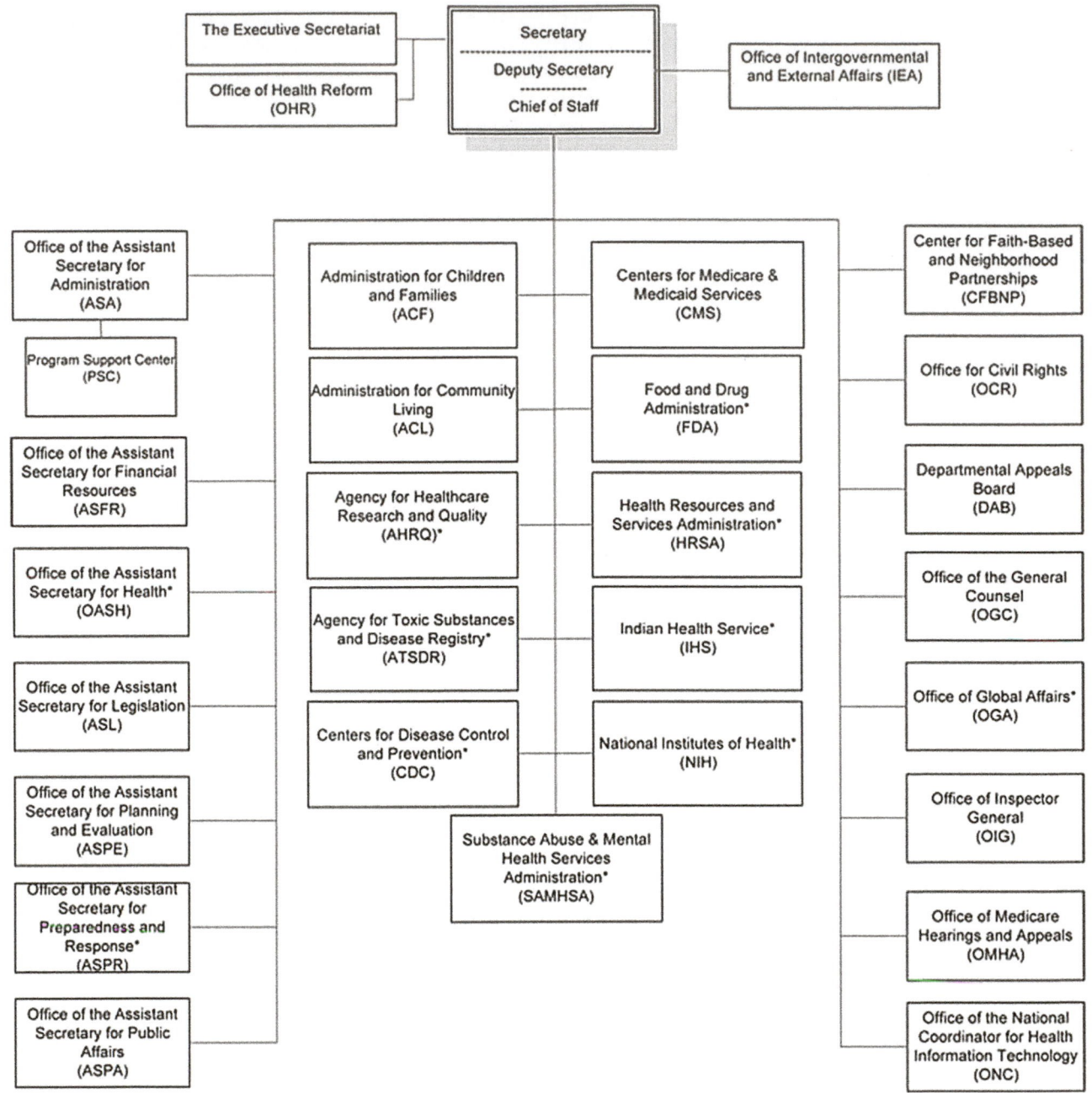

FIGURE 10.2   US Department of Health and Human Services. Note: The Assistant Secretary for Health is also the Surgeon General of the United States. *Designates a component of the US Public Health Service. *Source: US Department of Health and Human Services. Available at: http://www.hhs. gov/about/orgchart.html [Accessed 8 November 2012].*

The New York State Department of Health (DOH) has a strong tradition of regulation in chronic care facilities, laboratories, and hospitals, and in environmental health, including arrangements with the JCAHO. The various regulatory functions of the department make it a powerful determinant of the operation of health care in the state. Among its functions are granting certificates of need, regulation of reimbursement methods for hospital care (see Chapter 13), establishing health standards and surveillance systems, rural health systems, and many other activities. This state DOH is active in screening programs for congenital and infectious diseases of the newborn, laboratory certification, and quality assurance. An AIDS Institute is responsible for prevention, screening, and AIDS care programs. The Center for Community Health operates a wide range of public health programs, from epidemiological surveillance of infectious diseases, to prenatal and newborn care among the underserved, to community health worker programs, to nutrition monitoring and many other intervention programs focused on high-risk groups or topics. Environmental epidemiology

BOX 10.12 Functions of a State/Provincial Ministry or Health Department

- Coordinate with other government departments: governmental planning and priorities; education, social welfare, labor, agriculture, mental health, and financing of universities.
- Establish standards; finance, develop, advise, and supervise local health departments.
- Legislate and regulate health-related matters: preparation, assistance, and enforcement.
- Plan and set health priorities and targets.
- Provide epidemiological and laboratory services to local health departments and conduct biological surveys.
- Maintain and publish vital statistics, epidemiology, and health information systems.
- Develop standards and monitor quantity, quality, and distribution of diagnostic and treatment services.
- Ensure occupational health supervision.
- Ensure environmental health monitoring and supervision.
- License and discipline health professionals and health care institutions.
- May provide occupational and personal health services to state employees.
- Coordinate with related state services: social services, mental retardation, drug and rehabilitation, and prison services.
- Ensure mental health services are part of mainstream health.
- Coordinate with national and other state/provincial health authorities.
- Monitor health status indicators of state/province and local authorities.
- Provide health education.
- Promote quality of care in long-term care and hospitals, and in primary care.
- Ensure communicable and infectious disease control.
- Prepare and train for natural and human-made disasters as well as health emergencies, including potential mass epidemics and bioterrorism.
- Legislate for and promote positive health behaviors, such as smoking restriction and environments in schools, workplaces, and public spaces.

Source: *Turnock BJ, Atchison C. Governmental public health in the United States: the implications of federalism. Health Aff 2002;21(6):68–78.*

and monitoring are also strong in the state, which experienced the Love Canal incident (see Chapter 9). Figure 10.3 shows the 1996 configuration of the New York State DOH. This arrangement is not necessarily typical but does show the wide range of activities, including state, federal, and local initiatives.

In New York State, selected public health functions are the responsibility of other government departments or agencies. Table 10.1 displays the range of public health responsibilities in other agencies. The New York State

"prevention agenda" is an important and valuable initiative, similar to Healthy People objectives for the nation. As part of this agenda in New York State, local health departments work with community partners, hospitals in particular, in a collaborative effort to promote community health.

The New York State Department of Health is unique in that it is a cosponsor with the State University of New York (SUNY) of a School of Public Health at Albany, which involves departmental personnel as faculty and students in internships in branches of the DOH. While not necessarily representative of other states, this health department represents the broad scope of public health at the state level of government (Table 10.1).

LOCAL HEALTH AUTHORITIES

Historically, the local health authority (LHA) was responsible for sanitation and the provision of direct care to the poor and high-risk population groups. Boards of Health were established in Philadelphia in 1794 and in New York City in 1796 for these purposes.

The city or county local public health department is the official public health agency closest to the population served. The LHA provides a range of direct supervisory sanitation functions to ensure compliance with local, state, and federal sanitary codes. The local public health department may also provide direct services, usually personal preventive services, such as those for uninsured pregnant women, funded by the local government authority or by higher levels of government. In the USA, the local public health department is the agency attempting to ensure services to people inadequately served by voluntary or federal and state insurance plans. Programs may be funded by cost sharing or may be based on categorical or block grants from state or federal governments.

Even though there has been massive growth in the involvement of higher levels of government in public health, the LHA remains the major force for public health at the community level (Box 10.13). In the USA and Canada, the LHA is organized in the form of city or county/municipal health departments. In Quebec, the community level of government operates Local Community Service Centers (CLSCs). In Scandinavian countries, the county is the key operating level for public health as well as hospital and medical services. Current reforms in the UK are moving in this direction as well (see Chapter 13).

In new health initiatives, such as Health for All, district health systems, and Healthy Cities, the LHA is involved in a wider set of programs for the health of its population. In recognition of the objectives of these programs, formerly highly centralized systems, such as those of the UK, the Scandinavian countries, developing nations, and republics of the former Soviet Union, are being decentralized to

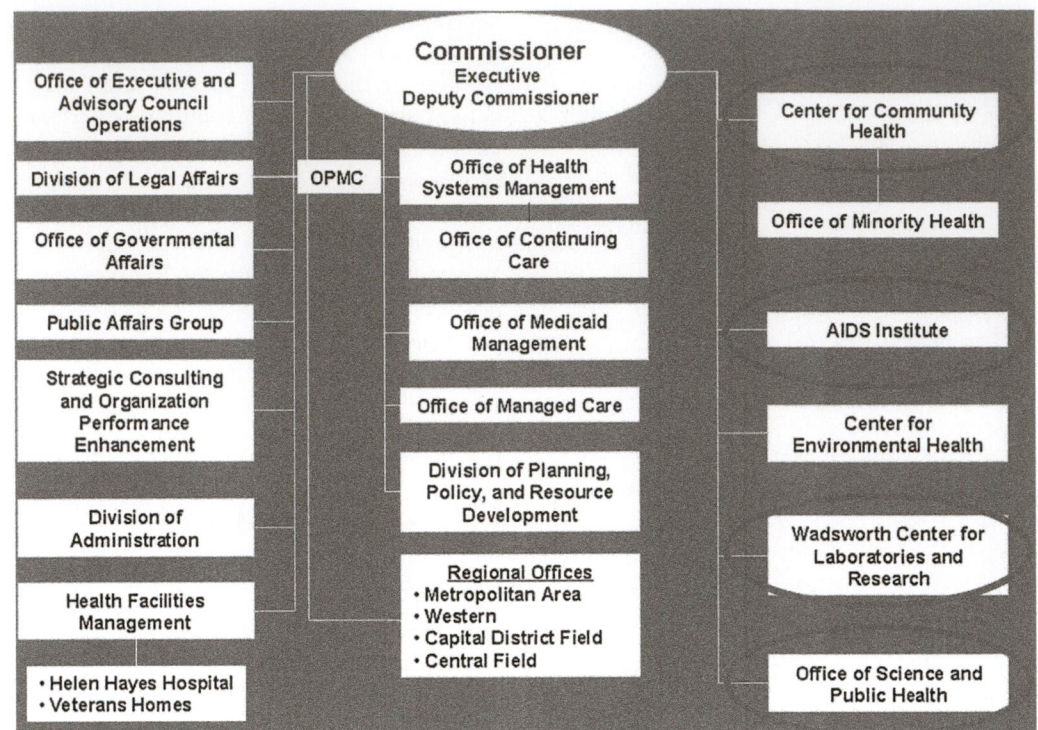

FIGURE 10.3   New York State Department of Health: Organization Chart. *Source: New York State Department of Health. Available at: http://www. cdc.gov/nchhstp/programintegration/attachments/G-PCSItheNewYorkExperience/G-PCSItheNewYorkExperience_03.pdf [Accessed 2 November 2012].*

TABLE 10.1 Agencies with Public Health Responsibilities (New York State)

Agency	Responsibilities
Department of Education	School sanitation, health education, licensure of physicians and other health professionals
Department of Labor	Health and safety of workers, in-plant pollution and radiation control
Department of Environmental Conservation	Control of pesticides, rabies control, air pollution, sewage and solid waste control
Department of Social Services	Medicaid (program for the poor)
State University of New York	School of Public Health, student health services
Department of Mental Hygiene	Mental institutions and community services
Narcotics Addiction Control Commission	Treatment facilities, research, education
Department of Agriculture	Licensure of meat dealers and slaughterhouses, inspection of restaurants, school-meal regulation of food additives
Department of Corrections	Operation of prison hospitals and clinics, tuberculosis case finding
Department of Motor Vehicles	Highway safety promotion

Source: New York State. Prevention agenda toward the healthiest state. Available at: http://www.health.ny.gov/prevention/prevention_agenda/ [Accessed 15 November 2012].

BOX 10.13 Health Responsibilities of a Local (Health) Authority (LHA)

- Registration and vital statistics
- Epidemiology of infectious diseases
- Maintaining documentation and reports as required by the government, e.g., fiscal records, reportable diseases, inspection and laboratory reports
- Health education and health promotion
- Environmental protection and sanitation
- Control of communicable diseases, sexually transmitted infections, human immunodeficiency virus, tuberculosis
- Preventive prenatal, infant, and toddler care
- Coordination and cooperation with Departments of Education, Social Welfare, Agriculture, Environmental Protection, Urban Planning, and others
- Allocation of resources
- Planning and management of services
- Licensing and supervision of health facilities
- Hospitals and home care
- Care of disabled
- Rehabilitation and long-term care
- Coordination of health services
- Intersectoral cooperation
- Mental health
- Emergency and disaster preparedness
- Social assistance
- Nutrition, including licensing of food establishments
- Community participation advocacy

LHAs, with varying degrees of central funding, planning, and direction.

In 1940, the APHA adopted a recommended standard of six basic responsibilities of the LHA, known in the public health community as the Haven Emerson Six:

- vital statistics
- communicable disease control: childhood diseases, TB, STIs, and tropical diseases
- environmental sanitation: water, food processing and marketing, sewage, garbage, sanitary condition of places of business, public eating places, and workplaces
- laboratory services
- maternal, child, and school health
- health education.

In 1950, the APHA adopted an expanded list of program of responsibilities for the LHA, which included the above plus the following:

- non-communicable and chronic disease control
- housing and urban planning
- accident prevention
- coordination with other agencies
- surveillance of total health status; births, deaths, chronic disease, morbidity data, surveys, reporting of morbidity, and evaluation of community needs
- education of the public and professional community regarding health status and needs
- supervisory and regulatory activities including health services providers
- personal health services: direct provision and supportive services, varying from comprehensive service programs to services for those in need
- planning of health facilities, urban planning and renewal
- special diagnostic services, including STIs, TB, cancer, child development, and dental care.

Cooperation between the different levels of government is vital to define and achieve national health objectives. Each level of government has a unique role to play. There is growing emphasis on responsibilities for emergency and disaster preparedness. Decentralized administration of public health without national financing and policies will not achieve the full potential of public health and will produce inequities between different regions of a country. National governments are responsible for setting policies, priorities, and goals with definable health targets. State and provincial governments are direct providers and supervisors of public health standards, while local authorities are those directly responsible for sanitation, local planning, and direct services to reduce public health risks. As an example, the programs of the Albany, New York Department of Health are summarized in Box 10.14, and an organizational chart of departmental activities in 2009 and 2010 is provided in Figure 10.4.

ACCREDITATION OF PUBLIC HEALTH DEPARTMENTS

Accreditation of public health departments has been promoted on a national level in the USA in recent years by the American Public Health (APHA) and other national associations of public health professionals. A Public Health Accreditation Board was established and has published guidelines and standards for conducting accreditation. The objective is to raise standards and assist health departments to achieve excellence in performance.

Standards and measures were developed based on many years of state-based public health accreditation programs, a National Public Health Performance Standards Program, and operational definition of a local health department. The standards and measures, developed by a working group comprising public health professionals, experts, and researchers, can be used to advance public health practice, strengthen the role of public health, and demonstrate accountability, and apply to all health departments and all forms of governance. As of 27 August 2013, 126 local health departments, 18 state health departments, and one tribal health department have successfully undergone accreditation. They include the award of accreditation to five departments in August 2013 (Central Michigan, Chicago, El Paso County, Kansas City, and Tulsa Oklahoma) serving communities ranging from 45,000 to millions in Chicago city (Public Health Accreditation Board, 2013).

This is seen as a method of improving quality and performance standards in local, state, and other departments of public health. It is a trend which has gained momentum in the USA and will become a standard in other countries as well. Standards for public health services are a component of Accreditation Canada developed in response to the need for public accountability and the organizational changes in health care delivery. The comprehensive program addresses the five core functions of public health service systems: health surveillance, health assessment, health promotion, health protection, and disease and injury prevention (Accreditation Canada, Public Health Services, 2013).

MONITORING HEALTH STATUS

As discussed in Chapter 3, public health depends on information and evidence, just as an army depends on intelligence in order to modify approaches in accordance with changing circumstances and need. Collection, collation, and analysis of this information are vital for informed health policy, and the information must be available to all concerned with health for analysis and policy debate. All levels of government are engaged in health status monitoring, with the geographic information system (GIS), a multisource database related to health indicators for the

BOX 10.14 The Albany, New York Department of Health

Programs of the Albany County Department of Health include:

- Public health emergency preparedness (PHEP).
- Preparing for a widespread natural disease outbreak since SARS, and the potential threat of avian influenza in 2011 and the global H1N1 influenza pandemic.
- New and revised programs:
  - Kids: Growing Healthy, Growing Strong!
  - Lyme disease monitoring
  - sanitation: individual sewage disposal systems, individual water supply
  - mobile home parks
  - nuisance and housing complaints
  - pesticide notification: enforcing law requiring notification of commercial and residential lawn pesticide use
  - investigation and control of outbreaks of communicable diseases
  - public water supply
  - realty subdivision
  - swimming pools and beaches
  - toxic exposures, indoor air, and chemicals
  - animal rabies/bites
  - children's camps
  - food service and vendors
  - hotels and motels
  - schools and day care centers
  - investigation and information for the Clean Indoor Air Act (smoking law)
  - implementation and enforcement of the Adolescent Tobacco Use Prevention Act
  - supervision of tattoo and piercing sites
  - West Nile virus surveillance and emergencies

- community health worker program: providing in-home health education and assisting families in getting basic needs for healthy living (medical care, food, clothing, and shelter); preventive care and dental treatments for children up to age 18.
- Anonymous and confidential HIV counseling and testing; informational sessions and programs targeted at high-risk populations and the general public.
- Home care – registered nurses, social workers, and public health nurses design a patient specific plan of care, coordinate needed health and support services, and provide ongoing follow-up and treatment under the orders of the patient's physician.
- Influenza vaccination – everyone 6 months and older should have a flu vaccination each year.
- Testing for lead poisoning for uninsured children aged 6 months to 5 years; nursing visits to assist with education and treatment; home inspections to find and correct lead problems.
- New York State Smokers' Quitline – a free and confidential telephone-based counseling service that provides effective stop-smoking services.
- Identifying hepatitis B-positive mothers; ensuring hepatitis B vaccine series for infants.
- Investigations of potential contacts of humans with rabid animals.
- Residential public health programs (water, sewage, pesticides).
- Free confidential STI diagnosis and treatment to all age groups; medical care for active and inactive TB, skin testing, and medications.

Source: *Crucetti J. Personal communication; October 2012.*

population of a geographic region, helping to identify localized or national problems for intervention.

The responsibility for gathering vital statistics lies largely at the local government level, as does the reporting of infectious diseases and other events. Initial collation of the data occurs at this level, and information is then sent to state health authorities and subsequently to the national level. The gathering of information is a strongly developed tradition in the industrialized countries, and the USA has in many ways done this effectively. In the USA, the CDC serves as a national leadership and reference center, not only for infectious diseases, but also for chronic diseases such as cardiovascular disease, nutrition, diabetes, perinatal epidemiology, and many other conditions.

Health statistics provide the ongoing data needed for monitoring the health status of populations. They provide routine diagnostic and population-based monitoring data that supply valuable epidemiological information on congenital conditions, STIs, TB, and HIV infection. Centers of excellence of all kinds, funded or administered directly by federal or state government or by the NIH mechanism, provide tertiary level medical care and conduct biomedical and epidemiological research, making important contributions to the information pool needed to promote quality analysis and health care.

The national health authority is responsible for the central collation and analysis of health information on the epidemiology of infectious and chronic diseases, vital statistics, utilization of services, and monitoring of national and regional variations in health. This information is only of value if gathered, processed, and published so that it is readily available to health administrators, planners, epidemiologists, care providers, and the public. Census data provide the population denominators for calculation of rates of death and disease incidence or prevalence.

Inexpensive technology of personal computers with modems, as well as telephones and facsimiles, enables local public health agencies to receive real-time information through Internet connections for continuous health profiles of their communities. Sources of data include the following:

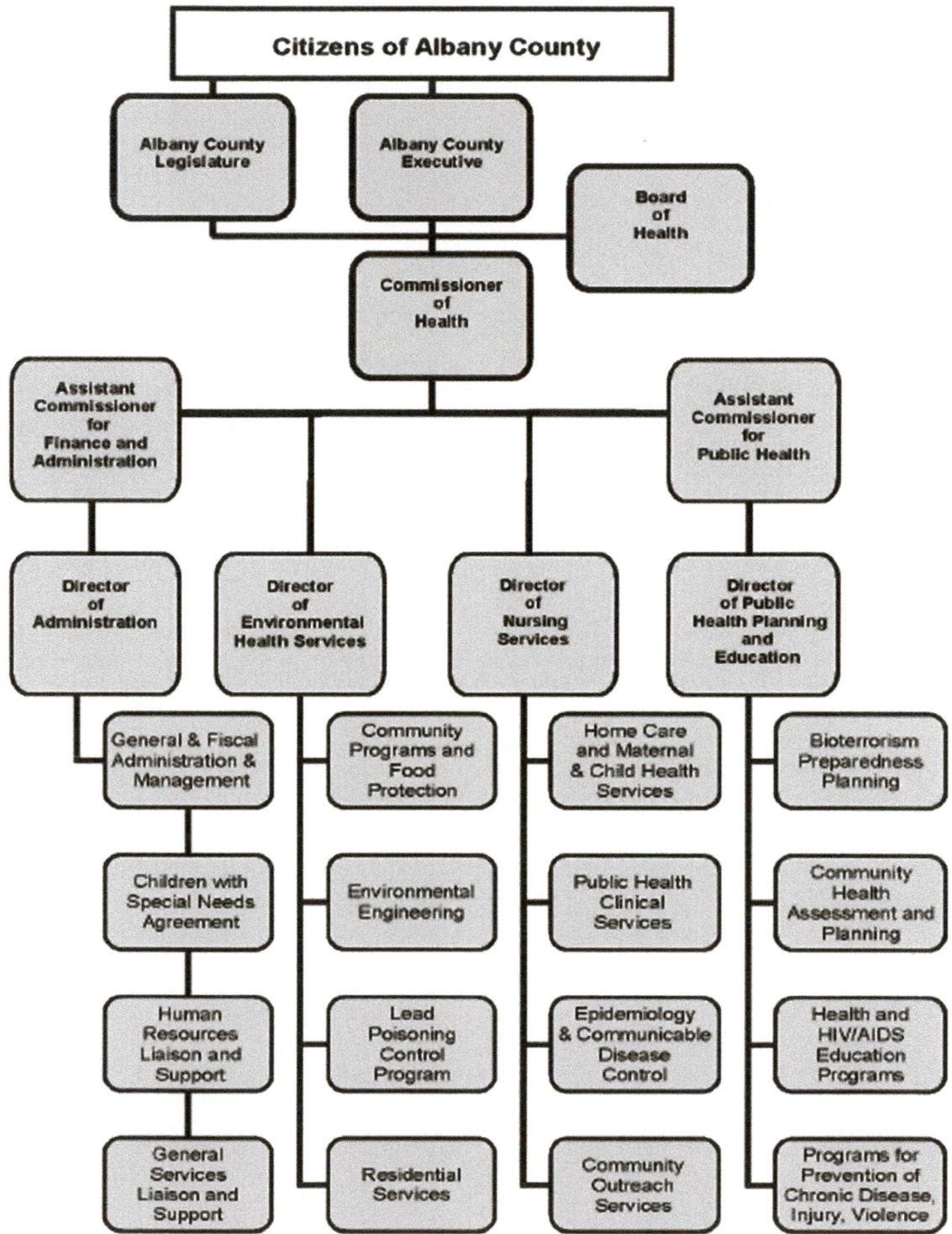

FIGURE 10.4 Albany County Department of Health Annual Report 2009-2010. Programs and Services Report. *Courtesy James B Crucetti, Commissioner, personal communication 7.11.2012. Department website: http://www.albanycounty.com/Government/Departments/DepartmentofHealth. aspx [Accessed 28 December 2013].*

- vital statistics and national centers for health statistics
- epidemiological reports of infectious and reportable diseases, including STIs
- state, national, and international reporting centers for disease control
- census data
- special disease registries (e.g., cancer)
- hospital and health and residential facilities discharge information systems
- public health laboratories

- poison control centers
- central medical libraries with Medline
- registries of medical, nursing, and dental professionals.

Geographic epidemiology has been important in the history of public health. Fragmentation of information systems has delayed the application of modern information technology to multiphasic evaluation and the integration of data from multiple sources. Pooled information can be used to identify a basic framework of standards and policies for both public and private sector participation. A common framework of policies and standards would be strengthened by information sharing among regional and other health networks including academic institutions, service sectors, and community organizations, as well as the media. Hospital discharge information systems and disease registries provide monitoring of sentinel events which can have important public health implications (see Chapter 3).

An outstanding example of such an information system is the Health for All database, provided and updated by the WHO European Region. This database is also available for use within countries to show interprovincial or intracounty variations in health status indicators (available at http://www.who.dk/hfadb). It is uniquely user friendly and can be adapted to other regions, countries, and states, such as the USA, Canada, the UK, and others wishing to understand the regional variations in health status of their populations as a public benefit. It is also especially useful for teaching purposes, as well as for policy and research background material, and should be included in all public health teaching programs.

Increased financial and human resources in local health departments in the USA show a relationship with lowered rates of infectious diseases (AIDS, hepatitis A and B, and TB) and higher rates of reduction for cardiovascular mortality than in comparable local health departments which had reduced or static resources between 1997 and 2005 (Erwin et al., 2011).

As an example of targeted public health issues, bicycle traffic deaths in the USA declined from 830 in 1995 to 677 in 2011 or 2 percent of total traffic fatalities while bicycle injuries declined from 61,000 to 38,000 in the same time period. A large proportion of deaths and injuries from bicycling can be prevented with helmets, but helmet use among cyclists in the USA remains low, particularly among adults. A legal requirement of use of bicycle helmets may be seen as the "nanny state" interfering with personal liberty, but the injuries cost the community many preventable deaths and cost the health system large amounts of money for hospitalization. In 2009, there were 418,700 emergency department visits and 27,900 inpatient community hospital stays for injuries related to bicycle accidents as well as loss of life. State regulation and local enforcement are part of legitimate public health activity (Stranges et al., 2012).

NATIONAL HEALTH TARGETS

The US Public Health Service has set national health targets since 1979. These are increasingly accepted at all levels of the national public health complex. Targets highlight areas of concern that require effort by all levels of government and the health care system. They also serve an educational role for health providers and the community.

Some of the progress made in reducing morbidity and mortality from epidemiologically important diseases is the result of that wider awareness and a growing concept of "self-care". Healthy People 2010 is a set of health objectives for the USA. It is important as a guideline for states, communities, professional organizations, and others to help them to develop programs to improve health. This initiative began in 1979 with the Surgeon General's Report, Healthy People, and Healthy People 2000: National Health Promotion and Disease Prevention Objectives. These were developed through a broad consultation process, incorporating available scientific knowledge, and are monitored by measurable indicators over time. The publication *Health, United States* provides annual updating of a wide range of health statistics.

Healthy People 2010 Midcourse Review, issued by the National Center for Health Statistics and the CDC, showed progress being made towards over 450 separate objectives in 28 focus areas designed to prevent disease and injury and to promote health in the USA. Of the 281 objectives with tracking data, some 10 percent of the goals have been met and progress has been made in another 49 percent. Midcourse reviews showed that progress in their implementation was not uniform: for 20 percent of targets there were regressions; for 20 percent mixed results or no change. The Leading Health Indicators are composed of 26 indicators organized under 12 topics. The Healthy People 2020 Leading Health Indicators are shown in Table 10.2. Each of these 26 indicators listed under the 12 topics is being tracked, measured, and reported on regularly throughout the decade.

Another approach to national health promotion developing in Europe relates to decision making in public health (Box 10.15). The European Union (EU) lacks many of the institutions available to a federal state such as the USA. It is attempting to find ways to compensate, such as by establishing the European Centre for Disease Prevention and Control (ECDC) to promote pan-European cooperation in communicable and control of other diseases with guidelines and common policies of health promotion. This effort is in its early stages, but has been advanced by concern over the threats of pandemics such as SARS and avian influenza.

The concepts of prevention and health promotion are integral to setting and attaining health targets. The methods of public health are increasingly moving towards wider responsibilities in terms of health monitoring and

TABLE 10.2 Healthy People 2020: Leading Health Indicator Topics

Topic	Indicators
Access to health services	People with medical insurance People with a usual primary care provider
Clinical preventive services	Adults who receive a colorectal cancer screening based on the most recent guidelines Adults with hypertension whose blood pressure is under control Adult diabetic population with an A1c value >9% Children aged 19–35 months who receive the recommended doses of DTaP, polio, MMR, Hib, hepatitis B, varicella, and PCV vaccines
Environmental quality	Air quality index >100 Children aged 3–11 years exposed to secondhand smoke
Injury and violence	Fatal injuries Homicides
Maternal, infant, and child health	Infant deaths Preterm births
Mental health	Suicides Adolescents who experience major depressive episodes
Nutrition, physical activity, and obesity	Adults who meet current federal physical activity guidelines for aerobic physical activity and muscle-strengthening activity Adults who are obese Children and adolescents who are considered obese Total vegetable intake for people aged ≥2 years
Oral health	People aged ≥2 years who used the oral health care system in the past 12 months
Reproductive and sexual health	Sexually active females aged 15–44 years who received reproductive health services in the past 12 months People living with HIV who know their serostatus
Social determinants	Students who graduate with a regular diploma 4 years after starting ninth grade
Substance abuse	Adolescents using alcohol or any illicit drugs during the past 30 days Adults engaging in binge drinking during the past 30 days
Tobacco	Adults who are current cigarette smokers Adolescents who smoked cigarettes in the past 30 days

Note: A1c=glycosylated hemoglobin; DTaP=diphtheria, tetanus, acellular pertussis; MMR=measles, mumps, and rubella; Hib=*Haemophilus influenzae* type b; PCV=pneumococcal conjugate vaccine.
Source: US Department of Health and Human Services. 2020 LHI topics [updated 12 March 2012]. Available at: http://www.healthypeople.gov/2020/LHI/2020indicators.aspx [Accessed 29 October 2012].

organization to reach the stated goals and objectives. The New Public Health provides a conceptual basis for this process.

UNIVERSAL HEALTH COVERAGE AND THE NEW PUBLIC HEALTH

Because the USA lacks a universal coverage national health insurance program, it is commonly cited in the literature that the USA has a "non-system". This is misleading: the USA has a very complex and unfinished health system, with a major deficiency in lack of universal access health insurance. Yet, the USA is a world leader in public health, not only in the development of new vaccines, but in implementation of important advances in prevention and health promotion, such as fluoridation of community water supplies. The USA has the costliest health system, with total expenditures reaching nearly 18 percent of gross domestic product (GDP) in 2011, but it lags behind many other countries in important indicators of health status (see Chapter 13). Still, the USA has other indirect public health programs that support poverty groups, including a universal school lunch program and the Special Supplemental Nutrition Program for Women, Infants, and Children (WIC), which provides food supplementation for pregnant women and toddlers in need. Furthermore, the US health system is a complex interactive set of organizations, subject to system changes, that has pioneered many innovations in health sciences, health care administration, and public health.

Publicly administered universal access elements exist, even if they are underfunded. The middle class is protected by employment-based health insurance, the elderly by Medicare, and the poor by the federal–state–locally administered Medicaid program. The failure to adopt national

BOX 10.15 Effective Decision Making for Public Health Policy

Public health does not take place in a vacuum. It requires a societal commitment that places health in a high social priority for funds and public policy. Allin and colleagues examined public health policy in eight countries (Denmark, Finland, France, Germany, the Netherlands, Sweden, Australia, and Canada). The authors discussed the following key issues for strong public health policy:

- political commitment and support at all governmental levels (national, state, local)
- intersectoral cooperation between government agencies and with non-governmental organizations
- preparation of the population (e.g., societal acceptance of smoking restriction legislation)
- health law developed and codified with appropriate enforcement capacity
- promotion of individual and population behavior changes consistent with "healthy lifestyle" and supportive socioeconomic context such as in alleviating poverty and inequities in health
- adequate infrastructure and resources for organized public health structures at all levels of government with sufficient, well-trained personnel and programs
- independence from political control so that the voice of public health can operate to identify and meet challenges in population health and not be submerged under a clinically oriented health system
- organization, funding, and support for research to provide the skills and material to evaluate health of the population and identify new risk factors and associations
- health policies that are realistic and targeted to measurable goals with identification of priorities and feasible programs to meet these objectives
- development of training and research environments and capacities consistent with the standards and culture of public health at the highest international standards.

Source: Modified from Allin S, Mossalio, McKee M, Holland W. Making decisions in public health: a review of eight countries. Copenhagen: World Health Organization on behalf of the European Observatory; 2004. Available at: http://www.euro.who.int/document/E84884.pdf [Accessed 8 November 2012].

health insurance providing equitable access to health care continues to be a major obstacle to improving health of the vulnerable poor and marginalized sectors of society. Public health services at all levels of government spend much energy and resources trying to cover deficiencies resulting from inequities in access to services.

Managed care plans in which financial incentives are in play to promote ambulatory and preventive care and decreasing use of hospital care increasingly cover the US population. Collaboration between organized public health and medicine, long-standing antagonists in the USA, took a new direction in the mid-1990s with development of a "new paradigm" of cooperation. The American Medical Association and the APHA agreed to work together to promote networking in the form of collaborative local programs to resolve unmet health needs of the community. This mutual awareness represents recognition of the importance of both clinical medicine and public health. Intersectoral dialogue helps to identify the potential for cooperation in the context of the dramatic changes taking place in the USA in health care organization. Health insurance coverage of people aged 18–44 years and 45–64 years in the USA (Figure 10.5) shows a decline in private coverage in both age groups, with an increase in uninsured and Medicaid insured people.

In other countries such as the UK and the Scandinavian countries, organization of health services moved to district health systems in which public health is a full partner with clinical services, and where prevention is integral to the economics and function of a population-based program. The managed care evolution in the USA since the 1990s may well promote a new level of cooperation between clinical medicine and public health. Integration of services financed by Medicare and Medicaid, with federal waivers of eligibility conditions for age and poverty, may allow a new approach based on residence in areas of need. Expanding Medicaid will occur largely through enabling enrollment into managed care programs of large numbers of eligible people who are not currently enrolled.

Downsizing the hospital sector, constraining health costs, increasing enrollment in managed care, focusing on health targets, and increasing coverage through managed care will constitute a national health program evolving towards some form of the New Public Health. The USA has been very innovative in financing systems to promote efficiency in use of services, and other countries have begun to apply those lessons in their national health insurance plans. The USA will benefit from examining the reforms going on in many countries, including Canada and European countries, as their health systems also evolve. The US public health community, including the schools of public health, has capacity and experience with professional leadership and advocacy, and it can make a great contribution towards adaptation of the New Public Health.

In 2010, expenditures for governmental public health services in the USA were 3.2 percent of total health expenditures, an increase from 2.8 percent in 1990. Personal health care accounted for 84.3 percent of expenditures, including 31.4 percent for hospital care and 26.6 percent for professional services including dental care (Health United States 2012). Thus, most expenditure by state health departments was for personal care services, mostly for people ineligible for health insurance or with benefits excluding preventive care. This represents the predominant priority for hospital and ambulatory care services based on insured or personal outlay for services. While much of ambulatory care involves preventive services, the relatively low expenditure

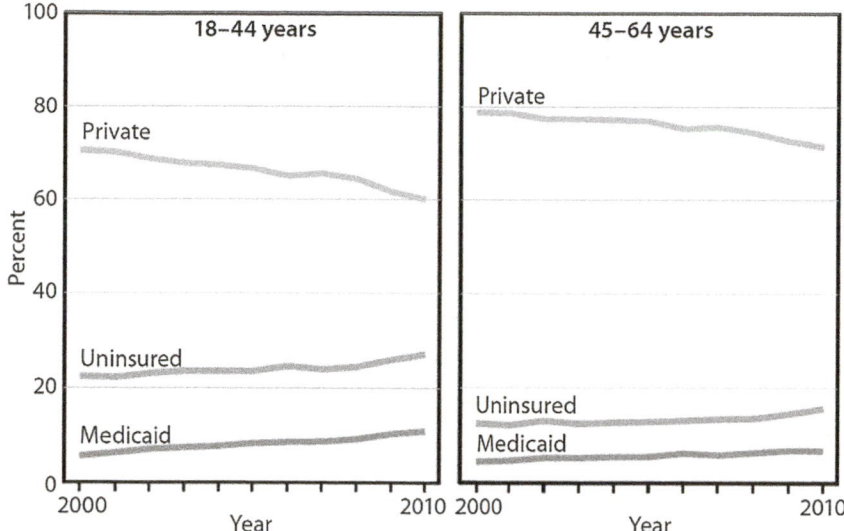

FIGURE 10.5   Health insurance coverage among young and older adults, USA, 2000–2010. *Source: National Center for Health Statistics. Health, USA, 2011: with special feature on socioeconomic status and health. Hyattsville, MD: NCHS; 2012. Available at: http://www.cdc.gov/nchs/hus/contents2011.htm#fig15 [Accessed 8 November 2012].*

for community-oriented public health activities reflects traditional values and underevaluation of the potential impact of community-oriented approaches to health promotion. The health reforms going on in most countries, especially those in transition from the Soviet system, require a shift of priorities for expenditure from a hospital orientation to a community orientation. This is a difficult process with many political implications, especially loss of jobs in many communities.

HOSPITALS IN THE NEW PUBLIC HEALTH

Hospitals evolved under municipal, religious, voluntary, governmental, university, private, or other sponsorship. Hospitals have traditionally been separate administrative units from other health services, although often with a strong connection to medical and paramedical training programs. The organizational structure is often based on the history of the organization, and may need adaptation to address the facility's mission, resources, and role as part of a larger community health system.

The hospital is an important element of the New Public Health. Inpatient health care facilities characterized as hospitals include many different types of facilities with important roles in a health system. They include general or specialized hospitals, rehabilitation centers, nursing homes, mental and other special hospitals. Each has a defined role, administrative structure, funding sources, operating and capital budgets, and modus operandi as a unique service-providing organization. They are mutually dependent even if entirely independent administratively and financially. This is a key issue in cost control in public health insurance or service systems.

Hospitals are often the largest employers in a community. They employ some three-quarters of all health personnel and, depending on the country and its traditions and reform processes, between 38 percent and 75 percent of total health expenditure. The magnitude of the hospital sector and the key role it plays in the health service system make it vital to rationalize its services, preventing duplication, bed surpluses, overemphasis on specialized services versus primary care, and depersonalization of patients and workers. Hospital spending in the US between 2003 and 2006 grew by an average of 7.4 percent, slowing to 5.5 percent annual growth between 2007 and 2010, reaching US$814.0 billion in 2010.

The modern hospital is the most costly and visible element of a health system to the public; it employs the most personnel and it provides care for the seriously ill. Hospital management is therefore an important factor in managing the total health system. While health care is an organizational system, the component facilities such as hospitals are also living organizational entities that require structure, management, and planning.

The supply and utilization of beds in community general hospitals in the USA have declined over the past three decades from 4.4 acute care beds per 1000 population in 1980 to 2.6 beds per 1000 in 2009 (OECD, 2012). Occupancy rates of community hospitals declined from 75 percent in 1960 to 65 percent in 2009. A trend of reducing hospital bed supply has also occurred in most industrialized countries, and more recently in some of the former Soviet countries, although with rates still well above those in Western Europe (see Chapters 11 and 13).

Under managed care systems, the hospital will try to satisfy two parties: the patient and the managed care system, with its economic constraints. These two parties may have different objectives and methods of assessment of the functioning of the institution and the community it serves. The insured patient, in his or her role as a hospital patient and with the option to change health plans, will be able to exert some influence on the quality of care he or she receives. Similarly, the managed care system can judge the quality of care rendered by a hospital and express dissatisfaction by choosing an alternative provider.

The mission of a hospital is to provide high-quality care and service to the patient within the limits of current standards of knowledge and resources. In addition, there are many other objectives of the hospital as an organization, including professional and economic survival as an institution, teaching functions, research, and publication. The hospital makes an important contribution to the community, providing employment, financial stability and solvency, prestige, education, research, and a system of access to health care.

To meet these diverse goals and objectives, hospitals have become complex organizations with an extensive division of labor (see Chapter 12). The organization involves many different professional areas, as well as "hotel services and facilities" such as the provision of food, laundry, housekeeping, supplies, and financial and personnel administrative functions. As a large organization of great complexity, a hospital must have a formal, quasi-bureaucratic structure with clear lines of authority and responsibility. However, the modern hospital cannot function under a traditionally authoritarian, paternalistic pattern of administration. Coordination of the many complex skills brought together in a hospital requires lateral coordination between departments and staff at all levels or the machine simply will not function. As a result, the hospital is highly dependent on the motivation and integrity of its staff, and their ability to network with others in different departments or professional levels freely and without excessive bureaucratic constraints.

Nevertheless, basic teamwork, acceptance of authority, professional standards and clinical guidelines, and quality assurance on a continuous basis to maintain standards of care are still essential to hospital function and predictability of performance. A great demand on hospitals is efficiency, so that waste, duplication of service, poor maintenance and function of facilities and equipment, corruption, negligence, or theft cannot be tolerated by the organization. The modern hospital has formal bureaucratic lines of authority, and hundreds or perhaps thousands of examples of informal networks and sometimes formal organizations to carry out the daily work of patient care, while meeting the other needs of the hospital and good standards of care with efficiency in use of resources. There are many checks and balances in the structure with multiple lines of authority and responsibility, and sometimes even tension between administrative and professional elements.

Hospital Classification

Hospitals are institutions whose primary function is to provide diagnostic and therapeutic medical, nursing, and other professional services for patients in need of care for medical conditions. Hospitals have at least six beds, an organized staff of physicians, and continuing nursing services under the direction of registered nurses. The WHO considers an establishment a hospital if it is staffed continuously by at least one physician, can offer inpatient accommodation, and can provide active medical and nursing care.

Any hospital bed that is set up and staffed for care of inpatients is counted as a bed in a facility. A bed census is usually taken at the end of a reporting period. The WHO defines a hospital bed as one that is regularly maintained and staffed for the accommodation and full-time care of inpatients and situated in a part of the hospital that provides continuous medical care. A bed is measured functionally by the number and quality of staff and support services that provide diagnostic and treatment care for the patient in that bed.

Hospitals include those operated on a not-for-profit and those on a for-profit basis. Most are operated as not-for-profit facilities as public services provided by government, municipalities, religious organizations, or voluntary organizations. In the UK, hospitals formerly operated by the NHS have been transformed into public trusts to operate as not-for-profit public facilities. In the Scandinavian countries, hospitals and other local health services are operated by the county health department. Private, for-profit hospitals, though increasing, are still a minority of general hospitals but include a large proportion of chronic care facilities.

In the USA, Canada, and Israel, long-term care for the elderly and infirm is largely provided by private for-profit facilities. In these countries, private facilities arose because of inadequate public resources for direct provision of services. As payment systems evolved, private operators were encouraged to enter the field. Government supervision and regulation have diminished the abuses and exploitation that occurred in the 1960s, but the standards of care can be compromised by the profit motive. There are, however, good examples of large-scale operations of long-term care facilities run by private organizations that are efficient and provide good standards of care. As illustrated in Box 10.16, hospitals are also defined by the types of services provided, the population served, and average length of stay.

Supply of Hospital Beds

The supply of hospital beds is measured in terms of hospital beds per 1000 population, a ratio which varies widely between and within countries. Historically, hospital development was initiated by church or religious groups, municipalities or voluntary charitable societies, or by local, state,

BOX 10.16 Types of Hospital

- *Short-stay hospitals* are those in which more than half of the patients are admitted to units in the facility with an average length of stay of fewer than 30 days. These include teaching, general, community, and district hospitals providing a broad range of services, as well as specialized hospitals that focus on special categories of patients by age, gender, or medical condition.

- *Long-stay hospitals* are those in which more than half of the patients are admitted to units in the facility with an average length of stay of more than 30 days. These may include special hospitals and may be jointly managed with short-stay hospitals.

- *Nursing homes* are establishments with three or more beds that provide nursing or personal care to the aged, infirm, or chronically ill. They employ one or more registered or practical nurses and provide nursing care to at least half of the residents.

- *Skilled nursing homes* provide more intensive nursing care, as defined by nursing care hours per patient day.

- *Hostels* are residential facilities attached to a medical center for overnight stay of patients undergoing outpatient investigation or care.

- *Hospices* are facilities related to a medical center especially organized to provide a humane, personalized, and family-oriented setting for care of dying patients.

- *Non-profit hospitals* are operated by a government, voluntary, religious, university, or other organization whose objectives do not include financial profit.

- *Proprietary hospitals and nursing homes* are operated for profit by individuals, partnerships, or corporations.

- *General hospitals* provide diagnoses and treatment for patients with a variety of medical conditions or for more than one category of medical discipline (e.g., general medicine, specialized medicine, general surgery, specialized surgery, and obstetrics). This excludes hospitals which provide a more limited range of care.

- *Community hospitals* serve a town or city and are usually short-stay (fewer than 30 days average length of stay) general hospitals.

- *District hospitals* are general hospitals that serve a population of a defined geographic district and have, as a minimum, four basic services: general medicine, surgery, obstetrics and gynecology, and pediatrics.

- *Teaching hospitals* are those operated by or affiliated with a medical faculty in a university or institute.

- *Special hospitals* are single-category inpatient care facilities such as a children's, maternity, psychiatric, tuberculosis, chronic disease, geriatric, rehabilitation, or alcohol and drug treatment center which provide a particular type of service to the majority of their patients.

- *Tertiary care hospitals* are referral and teaching hospitals; a secondary level hospital is a community or district hospital providing a wide range of services; and a primary level hospital is a limited service community hospital in a rural area.

Source: *American Hospital Association, 2006; Health, United States.* Available at: *http://www.aha.org/about/index.shtml* [Accessed 15 November 2012].

---

or national governments without national planning criteria. In all health systems, regardless of administration and financing methods, the supply of hospital beds and their utilization are fundamental to health economics and planning.

The hospital bed is often a political issue. In some countries, the hospital has been traditionally regarded as a center of refuge from the harsh conditions of life, climate, and social conditions. This is especially the case in rural areas with lesser access to health care. Pressures for more beds may come from physicians or from the public. Political figures tend to favor more hospitals because they provide jobs in a community, signify access to medical care, and create a public sense of well-being. The addition or closing of hospital beds is one of the difficult and controversial issues in health planning and health politics. However, if politicians are responsible for paying the hospital operational costs, they must take into account that operational costs will equal capital costs in about 2 years. It is also difficult to close redundant or uneconomic hospital beds, because this means a loss of jobs in the community unless combined with transfer of personnel to other services, a painful procedure itself.

The hospital bed is a functional economic unit with accompanying staff and fixed costs, so it has important economic implications for the health system. The cost per bed is measured by the total expenditure of the hospital divided by the number of beds. Building and operating costs, on average, are such that the cost of construction of a hospital unit is usually equal to the cost of operating the bed over 2–3 years. The decision to build a bed obliges the health system to indefinitely fixed costs even if that bed is unused as a result of regulation or reduced utilization from professional or economic incentives. Hospital planning is no longer left to the initiative of the facility itself, even in the most competitive, market economy-oriented health system.

The tendency to build excess hospital beds and the resultant costs of maintaining them were common to both developed and developing countries in the 1950–1980s. Excess supply is associated with high utilization rates and long lengths of stay. Most non-emergent diseases may be better treated on an outpatient basis, as hospital-associated infections and disease, such as deep vein thrombosis (DVT), increase length of stay, morbidity, and mortality, and raise health costs dramatically. Where there is no incentive for the hospital or physician to increase efficiency, patients tend to linger in the hospital. This situation results in higher overall costs of health care and is associated with medical mishaps, including falls in the hospital, errors in care, drug errors, anesthetic mishaps, and nosocomial infections. Excess bed capacity can be managed in a number of ways. Essentially, it requires conversion of bed stock and staff to other purposes or closure of obsolete facilities.

Especially since the 1980s, many countries have been reducing excess hospital bed utilization by shortening the length of stay, increasing the efficiency in diagnostic procedures, decreasing unwarranted surgical procedures, and adopting less traumatic procedures (e.g., breast-conserving surgery for breast cancer, and endoscopic surgery). Ambulatory services replace inpatient care for many types of surgery, including most eye, ear, nose, and throat surgery, and for medical care in oncology, hematology, mental health, and many internal medical problems (see Chapter 11).

Alternatives to hospital care, such as organized home care, assist in earlier discharge of patients from acute care hospitals by providing services to the patient at home, such as nursing, physiotherapy, intravenous care, changing dressings, or removal of stitches following surgery. Rehabilitation facilities provide appropriate low-cost alternatives to lengthy recovery periods after surgery such as hip or knee replacements. Long-term care facilities provide services for geriatric patients requiring extensive nursing care. These patients may not benefit from lengthy stays in acute care hospitals, and need access to alternatives to hospital care. Closure or reduction of beds is important to assure that savings in one area of service are transferred to a common financing system to provide funding for those alternative services. Investment may be required in these extended community services before savings are realized from reduced hospital utilization. While hospitals are vital for acute care in life-threatening disease, preventive capacity is optimized by decentralizing and taking medical care to the community. Hospital size, number, and beds must

be balanced using an economic- and public health-focused approach.

The capitation system of payment provides incentives for district health or managed care systems to limit admissions and lengths of stay. Sweden succeeded in reducing the percentage of gross national product (GNP) spent on health care during the 1980s by reducing hospital bed supplies, while maintaining the improvement of health status indicators. Managed care systems and diagnostic related groups (DRGs) have the same effect in the USA. District health system capitation is leading to reduced hospital bed supplies in the UK. This is a complex and controversial issue, but managing the numbers of hospital beds is essential especially in view of aging populations with chronic diseases, and the highly intensive and expensive kinds of care needed by many patients (see Chapter 11). Figure 10.6 shows general hospitalization rates (adjusted by population age) from 1998 to 2010. The decline in utilization is part of a long-term trend to reduce hospitalizations and length of stay with improved diagnostic and treatment methods and a stronger emphasis on ambulatory and primary care.

The Changing Role of the Hospital

Hospitals are technologically oriented and costly to operate. Under the influence of rising costs, incentives for alternative forms of care have led to the development of home care, ambulatory services, and linkages with long-term care. Forces acting on the hospital as an organization and economic unit place the hospital in a context where community-based care is an essential alternative that requires organizational and funding linkage to promote integration.

As a key element of any health system, the hospital will undergo changes as technology and health management sciences advance. Managing health systems with fewer hospital days requires reorganization within the hospital to provide the support services for ambulatory, diagnostic, and treatment services, as well as home care. The

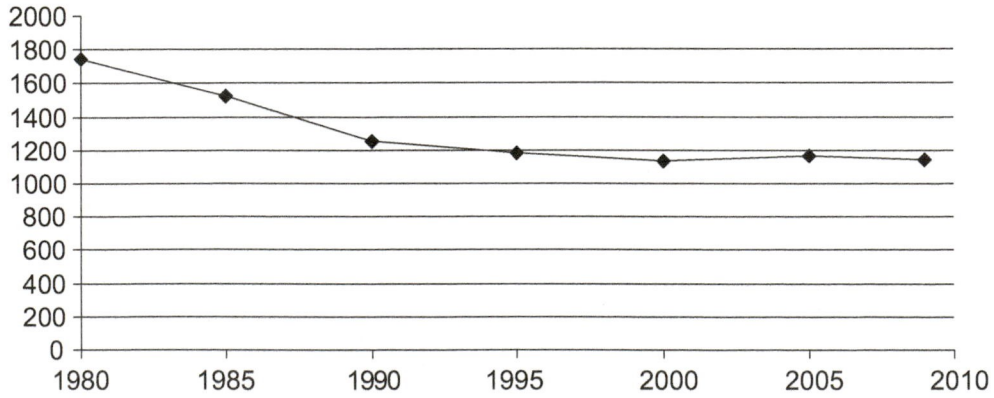

FIGURE 10.6 Total age-adjusted discharges from general hospitals per 10,000 population, USA, 1980–2009. *Source: Centers for Disease Control and Prevention. Health United States, 2011. Table 123. Available at: http://www.cdc.gov/nchs/data/hus/hus11.pdf [Accessed 9 November 2012].*

interaction between the hospital- and community-based services requires changes in the management culture and community-oriented approaches. Involvement of all staff in ensuring the quality of the service has become part of this management (Box 10.17).

In countries that operate hospitals as part of the Ministry of Health or National Health Service, there is a growing tendency to transfer hospital ownership and operation to not-for-profit agencies, or trusts as free-standing economic

BOX 10.17 Hospital Mergers in Los Angeles County: The University of California, Los Angeles (UCLA), Health Sciences Center and Community Outreach

Los Angeles is a large, multi-ethnic, and rapidly growing metropolitan city of over 9 million people in southern California. The hospital bed-to-population ratio was 3.5 per 1000 population during the 1980s and 1990s. Payment by diagnosis-related group in the 1980s, and growing membership in managed care, led to reduced hospital bed occupancy, with 45 percent of beds occupied in 1996. In 1998, the vast majority of insured Angelinos belonged to managed care programs. As a result, many for-profit hospitals are being sold to for-profit hospital chains, or are under threat of closure, some being converted to long-term or ambulatory care facilities.

As an example, the UCLA network includes the Santa Monica Hospital, a 337-bed acute care facility serving the health care needs of Los Angeles and Santa Monica since 1926. The UCLA network includes community clinics (Brentwood, Malibu, Santa Monica, Westwood, and others). This provides a wide population for the tertiary care center in competition with other tertiary care centers in Los Angeles. The UCLA Health Sciences Center is a teaching hospital owned and operated by the university.

In the mid-1990s this center developed contracts to provide hospital care to many managed care programs. In order to broaden its community service base, the center purchased several community hospitals and established affiliation agreements with medical group practices in adjacent areas of the city. This enabled the center to ensure its catchment population in a highly competitive market. The emphasis is increasingly on developing contractual arrangements with primary care medical services. The Health Sciences Center is replacing the hospital owing to damage in the 1994 earthquake and will do so with a substantially lower number of beds. This is the survival strategy adopted to ensure its continuing role as a major teaching and community service hospital in the changing medical market in the twenty-first century.

The UCLA Medical Center is also linked to many educational facilities including the Faculties of Medicine and Nursing, and the UCLA Fielding School of Public Health.

Source: UCLA Health. Ronald Reagan UCLA Medical Center. Available at: http://www.uclahealth.org/homepage_med.cfm [Accessed 2 November 2012].

units, or integrated within service programs of district health authorities. Competition for patients and payment for services such as by a DRG system will increase competition and the need for excellence in hospital care and its management for the financial survival of the facility. There is a trend in the UK, Israel, and many countries in transition from the Soviet and postcolonial health systems towards less centralized management and greater competition in health care. The trend to include hospitals in district health authorities, as in the Nordic countries, as part of geographic managed care programs is another important policy direction of health reform. Some Nordic countries, however, are reversing this trend and re-establishing centralized management of district hospitals.

In the USA, hospital networks are developing in the for-profit and not-for-profit sectors with integration of management and other cost savings in scale of purchasing and operation. Integration of health services can be "lateral", integrating related services and the medical providers of these services, or "vertical", integrating different types of services and different levels of health prevention, such as acute with long-term care, and community care services (Figures 10.7 and 10.8). Contracts with managed care

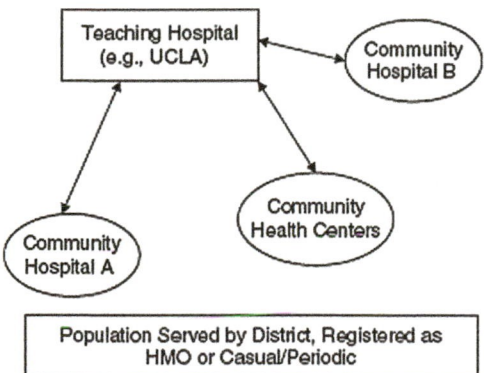

FIGURE 10.7   Integration of health services, University of California, Los Angeles (UCLA) Medical Center. Note: HMO=health maintenance organization.

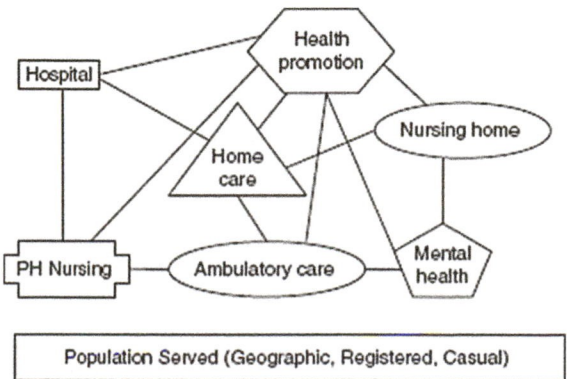

FIGURE 10.8   Vertical integration of health services. Note: PH=public health.

organizations for hospital care have replaced the previous system under which the insured patients' hospitalization depended on whether the attending doctor had privileges or worked on staff. The for-profit hospital corporations, along with similar managed care organizations, have brought health care to the stock market with profits larger than many other sectors of the private economy. Hospital mergers may then be seen in the context of any other business merger or corporate development.

For example, the UCLA health system includes the Ronald Reagan UCLA Medical Center; UCLA Medical Center, Santa Monica; Resnick Neuropsychiatric Hospital at UCLA; Mattel Children's Hospital UCLA; and the UCLA Medical Group, with its wide-reaching system of primary-care and specialty-care offices throughout the region. The links with community and rehabilitation hospitals and group medical practices provide a strong referral system and access to a top medical center for primary care physicians and their patients.

State governments have the responsibility and authority to assure standards of health for the population. Licensing of health facilities is a traditional method used to ensure public safety and prevent harmful practices in patient care facilities. State licensing is the basis for regulation of quantity as well as quality and the content of the service, and is essential for controlling health care expenditures (see Chapter 11).

Governments have a number of methods to regulate hospitals. One method is through control of the funding mechanism; this allows room for negotiation and influence on standards and level of satisfaction with care. The second is regulation of the number of hospital beds as the licensing and standards authority. The third is control of capital expenditures. A fourth method is to link payment for insured patients to accreditation of the hospital. The level of government responsibility for regulation varies from country to country, usually depending on the constitutional division of responsibility between the different levels of government and the size of the country. In general, the state and local authorities have the greatest influence because of their proximity. Where government agencies operate hospitals directly, there is a conflict of interest in the form of self-regulation.

The combination of governmental roles of financing, operating, and regulating hospitals in a highly centralized health system may appear to have some advantages, but separation of these conflicting functions is important in promoting a high-quality service. The separation of financing and regulation from operation of services is a widening trend in national health systems.

Governmental regulation may be augmented by the use of non-governmental accreditation systems, making them virtually mandatory by conditioning payment on accreditation. Accreditation agencies' standards are accepted by government as a requirement for hospitals and long-term care facilities in the USA. This use of an NGO inspection system as a proxy for governmental standards frees the government from the need to establish large-scale regulatory and inspection systems. National accreditation by Ministries of Health is standard, but external transnational accreditation offers a wider and perhaps more objective system using international standards of organization, facilities, management, quality, and ethical standards. Accreditation systems outside routine governmental licensing have been developed in many countries including Canada, Australia, the UK, Malaysia, Taiwan, and Norway, while most countries have governmental or semi-autonomous accreditation processes for hospitals and other health care programs.

INNOVATIONS IN HEALTH CARE DELIVERY

Health care provided by physicians has traditionally been on a fee-for-service basis in the USA. Since the 1930s demonstration programs called prepaid group practice developed the idea of a group of physicians contracting to provide care for construction sites in remote communities, such as the Hoover Dam or for mining communities, to registered clients, including workers and families lacking access to other arrangements for medical care.

Prepaid group practice came to prominence in the USA during World War II to provide care for war industry workers and families. The Kaiser Permanente system grew to cover millions of people in many states and other similar programs developed with doctors having incentives to promote preventive care and reduce hospitalization and unnecessary interventions. This model later developed into health maintenance organizations (HMOs) and more recently into accountable care organizations (ACOs), which are becoming increasingly common methods of organization of health care for Americans and will be fostered by Obamacare in the coming years (see Chapter 13).

The link between medical care and public health has been a distant goal for those who see a need to link prevention and curative services, including health promotion and long-term support systems for patients with chronic illnesses and problems of aging.

The development of HMOs in the USA since the 1990s has been accompanied by a decrease in acceptability of private for-profit programs and a sense of substandard services. Nevertheless, the principles of organized group practice with the emphasis on preventive care came to be recognized as vital to controlling costs and reducing inequities in care. The introduction of the Patient Protection and Affordable Care Act (PPACA) will promote new approaches to medical care with group practice, social and preventive support systems, and the ideas of community-oriented primary care. Innovations under development include the patient-centered medical home (PCMH), ACO, and population health management system (PHMS). They include new payment arrangements that reward health outcomes achieved rather than payment of a fee for each service rendered. Evidence on the performance of such innovations will be needed to promote their wider adoption (Shortell et al., 2010).

THE UNINSURED AS A PUBLIC HEALTH CHALLENGE

While most industrialized countries have some form of national health insurance or national health service, low- and medium-income countries usually have very mixed systems which do not have such guaranteed access to health care for the majority of people. At the same time, public health systems in those countries are weak, with shortages of trained personnel and organized infrastructure. People without insurance or entitlements in a national health service lack access to regular medical care, including preventive services that are taken for granted in the industrialized countries. These countries also have low levels of national expenditure for health from all sources, generally under 5 percent of GDP. As a result, maternal and child health care are weak, with high maternal, neonatal, and postneonatal mortality, and high child death rates, often from diseases that could be prevented or treated inexpensively. The Millennium Development Goals (MDGs), discussed in several chapters, are only partially being reached, although significant progress has been made in many countries. The burden of non-communicable disease is also high and access to medical care is crucial for the management of hypertension, early cancer discovery and treatment, and management of malaria, tropical diseases, and TB. HIV and hepatitis C are at pandemic levels and these too require access to care which, if available, mainly comes from foreign donor sources.

The WHO has been calling for progress in health systems development in medium- and low-income countries as essential to achieving health goals. Western experience with national health insurance is, however, not necessarily appropriate as it tends to favor the middle and wealthy classes as opposed to the urban and rural poor majority, so the infrastructure development of public health services may be a more suitable approach. Private insurance is developing for the urban middle class in employment settings such as the civil service, commercial and industrial enterprises, and the military. There is a wide gap between currently available medical and public health technology and its implementation. More investment in health is needed to bridge that gap. In 2004, the WHO concluded that "much more investment is needed for a new, innovative approach to research on health systems; health research must be managed more effectively if it is to strengthen health systems and build public confidence in science; [and] stronger emphasis should be placed on translating knowledge into action to improve public health by bridging the gap between what is known and what

is actually being done". National health insurance or service systems need to be developed that reach the rural and urban poor, who are most at risk for high morbidity and mortality from preventable diseases.

In the USA, national health insurance has been only slowly and partially achieved for the elderly and the poor, but the country may take a large step forward as a result of the PPACA of 2010, known widely as "Obamacare". Historically, the USA has been pioneering in many scientific, medical, and public health achievements. The US health system functions adequately, albeit with major handicaps of the uninsured and underinsured. County and municipal health departments are well developed and focus a great deal of their activities and attention on this population, who are largely poor and in need of health care. The coverage of the elderly and the very poor under Medicare and Medicaid has given a base of protection to these groups, but the near-poor and the near-elderly are still highly vulnerable, especially when job layoffs are a major part of the economic condition. While this problem is becoming more acute with a growth in the number of uninsured following the failure to enact national health insurance in 1994, there are increasing federal and state initiatives to widen coverage for Medicaid and especially to cover children who are uninsured. The USA, despite still being the only industrialized country lacking universal health insurance, has established and led in the development of public health programs that have had positive health effects, such as expanding the content of routine immunization of children and adults, school lunch programs, a wide range of categorical health programs to promote prenatal care, lead screening and exposure reduction, mammography, Pap smears, and other preventive services.

The delay in establishing a universal health insurance program remains a continuing burden on the full realization of America's national health potential, for its individual citizens and for the nation as a whole. To improve health in the USA in the twenty-first century, the political echelons at federal and state levels will need to find a suitable formula for the implementation of universal health coverage. The USA will adopt national health insurance, or alternatively state programs to mandate health insurance coverage for all in stages. Implementation of the 2010 act is proceeding as a result of its being declared constitutional by the US Supreme Court and the re-election of President Barack Obama in 2012. Public health professionals have to engage the public, the business community, and public policy makers to promote this process towards achievement of individual and community health as well as a healthy workforce.

In contrast to HMOs, which were largely led by for-profit insurance companies, ACOs are led by medical provider groups such as hospitals (e.g., Beth Israel-Deaconess Medical Center in Boston), clinics, physicians, and other health care providers. ACOs may also integrate with health departments, social security departments, safety net clinics, and home care services. The various providers within an ACO need to work with one another to provide coordinated care to the beneficiary population, to adjust financial incentives, and to lower overall health care costs. Primary targets for enrollment are Medicare beneficiaries, but may include private insurance or employer-purchased insurance. Payers may play several roles in helping ACOs to achieve higher quality care and lower expenditures. Payers may collaborate with one another to align incentives for ACOs and create financial incentives for providers to improve the quality of health care. The Obama health insurance plan is complex; it establishes federal support for state health insurance initiatives to expand Medicaid coverage and a federal–state program of mandatory private or public health insurance.

US Patient Protection and Affordable Care Act (Obamacare)

The US Patient Protection and Affordable Care Act (PPACA), commonly referred to as "Obamacare", was passed by the US Congress and signed into law on March 2010. The PPACA was challenged as to its constitutionality, but in 2012 the US Supreme Court ruled in favor of most clauses of the act. It is the most fundamental reform in US health care since the 1965 introduction of Medicare and Medicaid under the Social Security Act. Obamacare is aimed primarily at decreasing the number of uninsured Americans, recently approximately 16 percent of the population, but rising as chronic unemployment has increased. Features of the plan are shown in Box 10.18. The plan also focuses on reducing the overall costs of health care. The PPACA is highly controversial in the USA, and its application will depend on political events in the coming years.

In terms of preventive care, the Act has expanded access to vaccination for influenza, diabetes screenings, and mammograms. The list of preventive services covered contains over 100 services for adults and children, including a range of preventive services for adults such as screening for breast, cervical, and colorectal cancer. The care provided includes screening for chronic and infectious diseases, including mental health conditions such as depression. Counseling is

---

BOX 10.18 Features of Obamacare: Patient Protection and Affordable Care Act 2010

- Most Americans will be required to have health insurance coverage in 2014.
- Those unable to obtain affordable health coverage through employers will be able to purchase insurance through a Health Insurance Exchange with premiums and cost sharing to those who cannot afford the insurance on their own.
- This will extend coverage to some 32 million Americans previously lacking health insurance under a public insurance plan (Health Care for America Plan).
- Regulations will prevent private health insurance plans from denying coverage for pre-existing conditions or charging higher premiums based on health or gender; this mandates insurance companies to cover all applicants and offer the same rates regardless of pre-existing conditions or gender and bans punitive exclusions of insurance benefits with denial of coverage by private insurance plans.
- Reduces costs of premiums to millions of families and small businesses.
- Caps out-of-pocket expenses under private insurance and eliminates co-payments, co-insurance, and deductibles for benefits defined as part of an "essential benefits package" of preventive care, and protects patients' rights.
- Encourages small businesses to provide health insurance to employees by mandates, subsidies, and tax credits to promote coverage.
- Introduces reforms aimed at improving health care outcomes and streamlining the delivery of health care by bundling payments to organized health networks (Accountable Care Organizations) as opposed to fee-for-service.
- Reduces fraud and abuses in private insurance plans.
- Requires insurance to promote free preventive care.

- Encourages young adult coverage on parents' insurance up to age 26, extending coverage to half a million young people.
- Expands Medicaid with child health benefits expansion (CHIP) and simplifies enrollment.
- Encourages state initiatives to improve care for Medicaid and Medicare beneficiaries.
- Reduces Medicare spending.
- Waives co-payments for preventive measures for seniors and many services for women.
- Provides incentives to institutions to improve quality of care and rural care.
- Promotes prevention and wellness programs; provides nutritional information to reduce costs to patients for preventive care measures to keep people well and reduce costs.
- Promotes women's preventive health measures, including mammography.
- Regulates and provides incentives to improve quality in nursing homes.
- Provides incentives to states to improve legal tort reforms, protect patients' safety, and improve liability insurance law.
- Promotes cutting-edge medical and health-related research.
- Promotes development of community health centers.
- Provides scholarships for young people training in health professions.

Sources: *US Department of Health and Human Services. Washington, DC. Key features of the law: preventative care [updated 11 October 2012]. Available at: http://www.healthcare.gov/law/features/rights/preventive-care/index.html [Accessed 25 October 2012].*
Obamacare. The Patient Protection and Affordable Care Act summary. Available at: http://obamacarefacts.com/affordable-care-act-facts.php [Accessed 28 August 2013].

available in various areas: breastfeeding counseling for new mothers, therapy to treat alcohol misuse, STI prevention, and dietary counseling for those at risk for chronic disease. The Act provides coverage for certain vaccinations and all approved contraceptive methods.

Many services for children and newborns are covered, including fluoride supplements for children without fluoride-fortified water sources, screening for autism, and behavioral assessments for children of all ages. In addition, a large component of the plan is a US$15 billion "prevention and public health fund", which invests in proven prevention and public health programs including smoking cessation and antiobesity programs. Included in these programs are "well-woman visits", which focus solely on preventive care for women, free of charge, including human papillomavirus testing, screening for gestational diabetes, and free breastfeeding equipment rental.

The PPACA expands access to private insurance plans, by offering a Pre-Existing Condition Insurance Plan (PCIP) to individuals who have been uninsured owing to pre-existing conditions or other factors. States have the option of running this program. By 2014, all discrimination against pre-existing conditions will be prohibited. The idea of lifetime limits on coverage was problematic for young children incurring high costs early in life. Under the new law, insurance companies are prohibited from imposing lifetime dollar limits on benefits such as hospital stays. Young Americans have benefited from staying on their parents' plan until they turn 26 years old. In addition, the law gives small businesses tax credits to provide insurance benefits for their employees.

The plan makes prescription drugs more affordable for eligible seniors by sending rebates to those who fell into the "doughnut hole", or those seniors who had to pay expensive premiums for prescription drugs because they had reached a limit. This system was implemented in 2010, when each eligible senior received a one-time, tax-free US$250 rebate check.

The PPACA also contains a program called New Exchanges, which will be fully implemented in 2014. This US$5 billion program is intended to provide financial assistance to employment-based plans for the provision of health insurance coverage to people who retire between the ages of 55 and 65 (Box 10.18).

SUMMARY

Public health is organized at local, state, and national levels to define and work towards a healthy population with achievement of health targets. A balanced health care system requires resources to be rationally allocated to the different preventive, curative, or environmental elements of health. Resources must be directed to all vulnerable groups in the population, recognizing that some groups have greater needs than others. At the same time, issues that affect everyone, such as nutrition, sanitation, housing, and socioeconomic conditions, affect the poor and the elderly disproportionately. Sound public policy must also take into account the need to ensure adequate quality of care by health care providers and institutions, through developing and regulating standards, licensing procedures, and quality assurance mechanisms.

Impressive progress has been made in public health in the USA over more than two centuries since the federal government established the US Marine Hospital Service in 1792, and public health systems have continued to evolve in the twentieth and twenty-first centuries. Despite lacking a national health system, the USA has been a leader in formulating administrative mechanisms to improve the efficiency of health care.

The evolution of health care in the USA since the 1990s towards managed care is causing a large-scale reorganization of hospitals with both vertical and lateral integration; that is, the formation of networks of hospitals and linkage of hospitals with primary care and other care facilities and programs. Adjustment to meet the health care organization environment of the twenty-first century requires further changes for hospitals, including downsizing, development of ambulatory and home care services, and linkages with primary care services to ensure a catchment population. The competitive factors in which primary care providers and the community have a role in determining a hospital's utilization, occupancy, and ultimately its survival will help to build a more community-oriented health system.

In the USA, public health has been separated from and is poorly funded compared to medical services. The advent of managed care for a large portion of the population creates a professional and economic challenge for both sides. Organized public health in the USA needs to seek a closer liaison with managed care to promote a more comprehensive New Public Health approach. Managed care organizations need to develop health promotion and at the same time ensure the interests of the patient to successfully promote their long-term economic interests, and vice versa. If public health remains outside the issues of organization and financing of personal care services, the isolation of public health in the USA will deepen.

The New Public Health is a comprehensive approach to health care, stressing the interdependence of medical and hospital services with prevention and health promotion. Clinical medicine, management of health services, and community health approaches are interactive in many forms, in the USA and elsewhere. In northern Europe and the UK, district health systems incorporating public health are responsible for and are budgeted on a per capita basis to ensure community health and the availability of all levels of personal care services to the catchment population. In the USA, the lack of universal health access and central payment systems for all has, paradoxically, promoted development of managed care systems linking all levels of health care. However, public health remains detached from this process, being organized and financed separately.

Health impact assessment applied systematically at the community level is important for determining priorities and use of evidence-based public health. Health impact assessment is an approach to assessing both the health burden from conditions in sectors other than health and the potential of health improvements by modifying those conditions. It combines procedures, methods, and tools by which a policy, program, or project may be judged as to its potential effects on the health of a population and the distribution of those effects within the population.

Systematic review is a formal process that identifies all of the relevant scientific studies on a topic, assesses their quality, individually and collectively, and sums up their results. Systematic reviews make it easier for practitioners and policy makers to understand all of the relevant information that is available, how it was collected and assembled, and how the conclusions and recommendations relate to the information that was reviewed. This range of techniques and tools can serve to ensure that an intervention or policy will be appropriate and feasible in particular settings.

Intersectoral collaboration is vital between health and with other sectors of government such as agriculture, education, economic policy, transportation, and housing, as well as with non-governmental sectors including industry, community, advocacy and donor groups, and the media. Working with political leadership is just as crucial. Ongoing and long-term support for public health is overshadowed by clinical medicine, not only in funding but also in public perception. However, the medical community is increasingly aware of the vital importance of prevention and organized public health activity. Understanding the bond between curative medicine and public health is the foundation for addressing non-communicable diseases and conditions as well as communicable diseases.

The Obamacare plan introduced in the USA in 2012 creates a new dynamic towards national health insurance by covering many uninsured Americans, promoting preventive care, and regulating private insurance to remove many exclusions, co-payments, and caps on coverage. This plan is controversial and full implementation will depend on political evolution in the coming years, but is a great step forward towards universal coverage in the USA.

The development of universal coverage in low-income countries is still a major challenge, and great care must be taken to protect the rights of rural and poor people in the population from plans that would mainly benefit the middle and upper classes. At the same time, the development of public health infrastructure and training of large cadres of public health workers at the bachelor's and community health worker levels should be of the highest priority for both national governments and international agencies.

While public health and health protection of the population is largely a governmental function, it is vitally linked to many sectors of society to be effective. The New Public Health approach seeks to link those activities of local, state, and national government with public awareness and health systems organization, all vital to protect and promote the health of a population, including the provision of personal care in hospitals, community, and long-term care settings. The New Public Health approach also seeks to link those activities of local, state, and national governments with non-governmental agencies and sectors that are related to achieving such goals.

NOTE

For a complete bibliography and guidance for student reviews and expected competencies please see companion web site at http://booksite.elsevier.com/9780124157668

BIBLIOGRAPHY

Agencies and Organizations

Agency for Health Care Policy Research. http://www.ahcpr.gov/ (accessed 11.11.12).

American Hospital Association. http://www.aha.org/about/index.shtml (accessed 11.11.12).

American Public Health Association. http://www.apha.org/ (accessed 11.11.12).

Bureau of Primary Care (DHHS). http://www.bphc.hrsa.gov/ (accessed 11.11.12).

Centers for Disease Control and Prevention (DHHS). http://www.cdc.gov/ (accessed 11.11.12).

Centers for Disease Control and Prevention, Executive summary. Available at: http://www.healthypeople.gov/2010/data/midcourse/html/execsummary/introduction.htm (accessed 11.11.12).

Centers for Disease Control and Prevention, Healthy People 2010. Available at: http://www.healthypeople.gov/2010/ (accessed 11.11.12).

Centers for Disease Control and Prevention, Healthy People 2010, Midcourse review. Available at: http://www.healthypeople.gov/2010/data/midcourse/default.htm (accessed 11.11.12).

Health Care Financing Administration (DHHS). http://www.thebody.com/content/art13091.html (accessed 11.11.12).

Health Care Indicators, United States. http://www.cms.gov/ or http://www.cms.gov/site-search/search-results.html?q=health%20care%20indicators (accessed 11.11.12).

Hill–Burton Obligated Facilities, Health Resources and Services Administration. http://www.hrsa.gov/gethealthcare/affordable/hillburton/facilities.html (accessed 11.11.12).

Indian Health Services (DHHS). http://www.ihs.gov/ (accessed 11.11.12).

Institute of Medicine, National Academy of Science. http://www.iom.edu/ (accessed 11.11.12).

US Department of Health and Human Services (DHHS). http://www.hhs.gov/ (accessed 11.11.12).

Recommended Reading

Accreditation Canada, Public Health Services. Available at: http://www.accreditation.ca/accreditation-programs/qmentum/standards/public-health-services/ (accessed 28.08.13).

Albany County Department of Health, Programs and services. Available at: http://www.albanycounty.com/departments/health/programs_services.asp?id=235 (accessed 02.11.12).

Allin, S., Mossalios, E., McKee, M., Holland, W., 2004. Making decisions in public health: a review of eight countries. European Observatory on Health Systems and Policies. WHO, Copenhagen. Available at: http://www.euro.who.int/__data/assets/pdf_file/0007/98413/E84884. pdf (accessed 11.07.13).

Baker, S.L., Beitsch, L., Landrum, L.B., Head, R., 2007. The role of performance management and quality improvement in a national voluntary public health accreditation system. J Public Health Manag. Pract. 13, 427–429. Available at: http://journals.lww.com/jphmp/Fulltext/2007/07000/The_Role_of_Performance_Management_and_Quality.18.aspx (accessed 20.12.13).

Baum, F.E., Begin, M., Houweling, T.A.J., Taylor, S., 2009. Changes not for the fainthearted: reorienting health care systems toward equity through action on the social determinants of health. Am. J. Public Health 99, 1967–1974. Available at: http://ajph.aphapublications.org/doi/abs/10.2105/AJPH.2008.154856 (accessed 20.12.13).

Benjamin, G.C., 2006. Putting the public in public health: new approaches. Health Affairs 25, 1040–1043. Available at:http://content.healthaffairs.org/content/25/4/1040.full (accessed 20.12.13).

Beth Israel-Deaconess Physician Groups at BIDMC. Available at: http://www.bidmc.org/AboutBIDMC/AffiliatesandPartnerships/PhysicianGroupsat-BIDMC.aspx (accessed 15.11.12).

Brown, L.D., 2010. The political face of public health. Public Health Rev. 32, 155–173. Available at: http://www.publichealthreviews.eu/show/f/34 (accessed 20.12.13).

Centers for Disease Control and Prevention, 1999. Achievements in public health, 1900–1999: changes in the public health system. MMWR. Morb. Mortal. Wkly. Rep. 48, 1141–1147. Available at: http://www.cdc.gov/mmwr/preview/mmwrhtml/mm4850a1.htm (accessed 20.12.13).

Centers for Disease Control and Prevention, 2012. CDC's vision for public health surveillance n the 21st century. MMWR. Morb. Mortal. Wkly. Rep. Suppl. 61, 1–39. Available at: http://www.cdc.gov/mmwr/pdf/other/su6103.pdf (accessed 20.12.13).

Centers for Disease Control and Prevention, Emergency preparedness and response. Hurricanes. Hurricane Sandy. Available at: http://emergency.cdc.gov/disasters/hurricanes/index.asp (accessed 08.11.12).

Centers for Disease Control and Prevention, Emergency preparedness and response. Mass casualty information for Emergency Medical Services (EMS) providers. Available at: http://www.bt.cdc.gov/masscasualties/ems.asp (accessed 02.11.12).

Centers for Disease Control and Prevention, 1997. Estimated expenditures for essential public health services – selected states, fiscal year 1995. MMWR. Morb. Mortal. Wkly. Rep. 46, 150–152. Available at: hhttp://www.cdc.gov/mmwr/preview/mmwrhtml/00046330.htm (accessed 20.12.13).

Centers for Disease Control and Prevention, Health United States 2012, Table 113. Available at: http://www.cdc.gov/nchs/data/hus/hus12.pdf#113 (accessed 11.07.13).

Centers for Disease Control and Prevention, Healthy People 2020. Available at: http://www.cdc.gov/nchs/healthy_people/hp2020.htm (accessed 11.11.12).

Centers for Disease Control and Prevention, 1996. Historical perspectives of CDC. MMWR. Morb. Mortal. Wkly. Rep. 45, 526–530. Available at: http://www.cdc.gov/mmwr/preview/mmwrhtml/00042732.htm (accessed 20.12.13).

Centers for Disease Control and Prevention, National public health performance standards program. 10 essential public health services [updated December 2010]. Available at: http://www.cdc.gov/nphpsp/essential Services.html (accessed 26.11.12).

Centers for Disease Control and Prevention, 2001. Recommendations and Reports. Updated guidelines for evaluating public health surveillance systems. Recommendations from the Guidelines Working Group. MMWR. Morb. Mortal. Wkly. Rep. 50 (RR-13), 1–35. Available at: http://www.cdc.gov/mmwr/PDF/rr/rr5013.pdf (accessed 20.12.13).

Centers for Disease Control and Prevention, 1999. Ten great public health achievements – United States, 1900–1999. MMWR. Morb. Mortal. Wkly. Rep. 48, 241–243. Available at: http://www.cdc.gov/mmwr/preview/mmwrhtml/00056803.htm (accessed 20.12.13).

Centers for Medicare and Medicaid Services, National health expenditures 2010 highlights. Available at: http://www.cms.gov/Research-Statistics-Data-and-Systems/Statistics-Trends-and-Reports/NationalHealthExpendData/downloads/highlights.pdf (accessed 10.11.12).

Centers for Medicare and Medicaid Services, 2012. Department of Health and Human Services. Medicaid program; eligibility changes under the Affordable Care Act of 2010. Final rule, Interim final rule. Fed. Regist. 77, 17144–17217. Available at: http://www.gpo.gov/fdsys/pkg/FR-2012-03-23/pdf/2012-6560.pdf (accessed 20 December 2013).

Commission on Social Determinants of Health, 2008. Closing the gap in a generation: health equity through action on the social determinants of health. WHO, Geneva. Available at: http://www.who.int/ social-determinants/thecommission/finalreport/en/index.html (accessed 07.04.10).

DeNavas-Walt, C., Proctor, B.D., Smith, J.C., 2012. Income, poverty, and health insurance coverage in the United States: 2011. US Census Bureau. Current population reports, P60–243. US Government Printing Office, Washington, DC. Available at: http://www.census.gov/prod/2012pubs/p60-243.pdf (accessed 08.11.12).

Erwin, P.C., Greene, S.B., Mays, G.P., Ricketts, T.C., Davis, M.V., 2011. The association of changes in local health department resources with changes in state-level health outcomes. Am. J. Public Health 101, 609–615. http://www.ncbi.nlm.nih.gov/pmc/articles/PMC3052341/ (accessed 20.12.13).

Ettelt, S., Nolte, E., Mays, N., Thomson, S., McKee, M., 2006. International Healthcare Comparisons Network. Policy brief: Health care outside hospital: accessing generalist and specialist care in eight countries. European Observatory on Health Systems and Policies. WHO European Region, Copenhagen. Available at: http://www.euro.who.int/__data/assets/pdf_file/0009/108963/E892592.pdf (accessed 11.07.13).

Fielding, J.E., Briss, P.A., 2006. Promoting evidence-based public health policy: can we have better evidence and more action? Health Aff. 25, 969–978. Available at: http://health-equity.pitt.edu/518/1/BETTER_EVIDENCE_MORE_ACTION.pdf (accessed 20.12.13).

Fielding, J.E., Teutsch, S., Breslow, L., 2010. A framework for public health in the United States. Public Health Rev. 32, 174–189. Available at: www.public healthreviews.eu (accessed 02.11.12).

Frederic, E., Shaw, F.E., Kohl, K.S., Lee, L.M., 2011. Public health then and now: celebrating 50 years of MMWR at CDC. MMWR. Morb. Mortal. Wkly. Rep. Suppl. 60, 1–129. Available at: http://www.cdc.gov/mmwr/pdf/other/su6004.pdf (accessed 20.12.13).

Gebbie, K.M., Turnock, B.J., 2006. The public health workforce, 2006: new challenges. Health Affairs 25, 923–933. Available at: http://faculty.unlv.edu/ccochran/HCA452_652/Articles/Public%20Health%20Workforce.pdf (accessed 20.12.13).

Gee, R.E., 2012. Preventative services for women under the Affordable Care Act. Obstet Gynecol. 120, 12–14.

Green, L.W., 2001. From research to "best practices" in other settings and populations. Am. J. Health Behav. 25, 165–178. Available at: http://png.publisher.ingentaconnect.com/content/png/ajhb/2001/00000025/00000003/art00002 (accessed 20.12.13).

Greenough, P.G., Kirsch, T.D., 2005. Hurricane Katrina. Public health response – assessing needs. N. Engl. J. Med. 353, 1544–1546. Available at: http://www.nejm.org/doi/full/10.1056/NEJMp058238 (accessed 20.12.13).

Halpin, H.A., Morales, M.M., Martin-Moreno, J.M., 2010. Chronic disease prevention and the new public health. Public Health Rev. 32, 120–154. Available at: http://www.publichealthreviews.eu/show/f/24 (accessed 20.12.13).

Henry, J., Kaiser Family Foundation. Health care costs: a primer – key information on health care costs and their impact. May 2012. Available at: http://www.kff.org/insurance/upload/7670-03.pdf (accessed 08.11.12).

Institute of Medicine, 2011. For the public's health: revitalizing law and policy to meet new challenges. National Academies Press, Washington, DC. Available at: http://www.iom.edu/Reports/2011/For-the-Publics-Health-Revitalizing-Law-and-Policy-to-Meet-New-Challenges.aspx (accessed 20.12.13).

Institute of Medicine, 2003. Who will keep the public healthy? Educating public health professionals for the 21st century. National Academies Press, Washington, DC. Available at: http://www.nap.edu/openbook.php?isbn=030908542X (accessed 20.12.13).

Institute of Medicine, 1988. The future of public health. National Academies Press, Washington, DC. Available at: http://iom.edu/Reports/1988/The-Future-of-Public-Health.aspx (accessed 20.12.13).

Institute of Medicine, 2002. The future of the public's health in the 21st century. National Academies Press, Washington, DC. Available at: http://www.iom.edu/Reports/2002/The-Future-of-the-Publics-Health-in-the-21st-Century.aspx (accessed 20.12.13).

Joint Commission, About the Joint Commission. Available at: http://www.jointcommission.org/about_us/about_the_joint_commission_main.aspx (accessed 08.11.11).

Joint Commission, Improving America's hospitals – the Joint Commission's Annual Report on Quality and Safety 2012. State recognition details: New York. Available at: http://www.jointcommission.org/state_recognition/state_recognition_details.aspx?ps=100&s=NY (accessed 08.11.12).

Lurie, N., Wasserman, J., Nelson, C.D., 2006. Public health preparedness: evolution or revolution? Health Aff. 25, 935–945. Available at: http://content.healthaffairs.org/content/25/4/935.long (accessed 20.12.13).

Macinko, J., Silver, D., 2012. Improving state health policy assessment: an agenda for measurement and analysis. Am. J. Public Health 102, 1697–1714. Available at: http://www.ncbi.nlm.nih.gov/pubmed/22813417 (accessed 20.12.13).

National Audit Office. Healthcare across the UK: A comparison of the NHS in England, Scotland, Wales and Northern Ireland. Report by the comptroller and auditor general, HC 192session 2012-13, 29 June 2012. Available at: http://www.nao.org.uk/wp-content/uploads/2012/06/1213192es.pdf (accessed 27.12.13).

National Center for Health Statistics, 2012. Health, United States, 2011: with special feature on socioeconomic status and health. NCHS, Hyattsville, MD. Available at: http://www.cdc.gov/nchs/hus/contents2011.htm#fig15 (accessed 08.11.12).

National Center for Injury Prevention and Control, 2010. Updated: In a moment's notice: surge capacity for terrorist bombings. CDC, Atlanta, GA. Available at: http://www.bt.cdc.gov/masscasualties/pdf/cdc_surge-508.pdf (accessed 02.11.12).

New York State Department of Health. Available at: http://www.cdc.gov/nchhstp/programintegration/attachments/G-PCSItheNewYorkExperience/G-PCSItheNewYorkExperience_03.pdf (accessed 02.11.12).

Nicklin, W., March 2013. The value and impact of health care accreditation: a literature review: driving quality health services. Accreditation Canada. Available at: http://www.accreditation.ca/uploadedFiles/Value%20of%20Accreditation_EN.pdf (accessed 28.08.13).

Obamacare, Affordable Care Act Summary. Available at: http://obamacarefacts.com/affordablecareact-summary.php (accessed 27.10.12).

Oberle, M.W., Baker, E.L., Magenheim, M.J., 1994. Healthy people 2000 and community health planning. Annu. Rev. Public Health 15, 259–275. Available at: http://nciph.sph.unc.edu/cvs/baker/pdfs/084.pdf (accessed 20.12.13).

Organisation for Economic Cooperation and Development, June 2012. Health data 2012 OECD. Available at: http://www.oecd.org/health/healthdata. (accessed 12.11.12).

Public Health Administration Board, Accreditation overview. Available at: http://www.phaboard.org/ (accessed 28.08.13).

Rubinson, L., Nuzzo, J.B., Talmor, D.S., O'Toole, T., Kramer, B.R., Ingelsby, T.V., 2005. Augmentation of hospital critical care capacity after bioterrorist attacks or epidemics: recommendations of the working group on emergency mass critical care. Crit. Care Med. 33(10):2393–2403. Available at: http://www.upmchealthsecurity.org/website/resources/publications/2005/2005-10-15-hospcritcareafter-bioterror.html (accessed 20.12.13).

Salinsky, E., 2010. Governmental public health: an overview of state and local public health agencies. Background paper no. 77. National Health Policy Forum. George Washington University. Available at: http://www.nhpf.org/library/background-papers/BP77_GovPublicHealth_08-18-2010.pdf (accessed 20.12.13).

Saltman, R., Bankkauskaite, V., Vrangbaek, K. (Eds.), 2007. Decentralization in health care. European Observatory on Health Systems and Policies Series. Open University Press, Buckingham. Available at: http://www.euro.who.int/en/about-us/partners/observatory/studies/decentralization-in-health-care.-strategies-and-outcomes (accessed 20.12.13).

Schoen, C., Osborn, R., Doty, M.M., Squires, D., Peugh, J., Applebaum, S., 2009. A survey of primary care physicians in eleven countries, 2009: perspectives on care, costs, and experiences. Health Aff. 28, w1171–w1183. Available at: http://www.commonwealthfund.org/Publications/In-the-Literature/2009/Nov/A-Survey-of-Primary-Care-Physicians.aspx (accessed 20.12.13).

Shortell, S.M., Gillies, R., Wu, F., 2010. United States innovations in healthcare delivery. Public Health Rev. 32, 190–212. Available at: http://www.publichealthreviews.eu/show/f/26 (accessed 20.12.13).

Sommers, B.D., Kronick, R., 2012. The Affordable Care Act and insurance coverage for young adults. JAMA 307, 913–914. Available at: http://jama.jamanetwork.com/data/Journals/JAMA/22492/jlt0307_913_914.pdf (accessed 20.12.13).

Stuckler, D., Basu, S., McKee, M., 2010. Public health in Europe: power, politics, and where next? Public Health Rev. 32, 213–242. Available at: http://www.publichealthreviews.eu/show/f/27 (accessed 20.12.13).

Task Force on Community Preventive Services, 2005. The guide to community preventive services. Oxford University Press, New York. Available at: http://www.thecommunityguide.org/library/book/FrontMatter.pdf (accessed 20.12.13).

Thacker, S.B., Simon, P.A., Fielding, J.E., 2006. Public health and business: a partnership that makes cents. Health Aff. 25, 1029–1039. Available at: http://content.healthaffairs.org/content/25/4/1029.full (accessed 20.12.13).

Tulchinsky, T.H., 2010. It is not just the Broad Street pump. J. Public Health 32, 134–135. Available at: http://intl-jpubhealth.oxfordjournals.org/content/32/1/134.full (accessed 20.12.13).

Tulchinsky, T.H., Varavikova, E.A., 2010. What is the "New Public Health"? Public Health Rev. 32, 25–53. Available at: http://www.publichealthreviews.eu/show/f/23 (accessed 20.12.13).

Turnock, B.J., 2004. Public health: what it is and how it works, third ed. Jones and Bartlett, Boston, MA. Available at: http://books.google.ca/books?id=tLk0K127ZQgC&printsec=frontcover#v=onepage&q&f=false (accessed 20.12.13).

UCLA Health System, Ronald Reagan UCLA Medical Center. Available at: http://www.uclahealth.org/homepage_med.cfm (accessed 02.11.12).

US Census Bureau. Health insurance highlights 2011. Available at: http://www.census.gov/hhes/www/hlthins/data/incpovhlth/2011/highlights.html (accessed 08.11.12).

US Department of Health and Human Services, 1992. Healthy People 2000: national health promotion and disease prevention objectives. DHHS publication PHS 91–50212. US DHHS, Washington, DC. Available at: http://www.cdc.gov/nchs/healthy_people/hp2000.htm (accessed 11.11.12).

US Department of Health and Human Services, Hill–Burton facilities compliance and recovery. Available at: http://www.hrsa.gov/gethealthcare/affordable/hillburton/compliance.html (accessed 11.11.12).

US Department of Health and Human Services, Human Services and Resources Administration (HSRA). Hill–Burton free and reduced-cost health care. Available at: http://www.hrsa.gov/gethealthcare/affordable/hillburton/index.html (accessed 11.11.12).

US Department of Health and Human Services, Key features of the law: Patients' Bill of Rights. Available at: http://www.ama-assn.org/ama/pub/category/1736.html (accessed 02.11.12).

US Department of Health and Human Services, Key features of the law: preventative care. Available at: http://www.healthcare.gov/law/features/rights/preventive-care/index.html (accessed 02.11.12).

World Health Organization, May 2008. Integrated health services: what and why? Technical brief no. 1. WHO, Geneva. Available at: http://www.who.int/healthsystems/technical_brief_final.pdf (accessed 08.11.12).

World Health Organization, 2008. World Health Report 2008: primary care now more than ever. WHO, Geneva. Available at: http://www.who.int/whr/2008/en/ (accessed 20.12.13).

World Health Organization, 2010. World Health Report 2010: health systems financing, the path to universal coverage. WHO, Geneva. Available at: http://www.who.int/whr/2010/en/index.html (accessed 12.11.12).

Chapter 11

Measuring Costs: The Economics of Health

Learning Objectives

Upon completion of this chapter, the student should be able to:

1. Explain the relationship between economic issues and priority selection in health;
2. Apply economic analysis to alternative methods of providing health services;
3. Explain a comprehensive approach to the evaluation of health status of a population;
4. Apply economic and evaluative approaches to the New Public Health.

INTRODUCTION

Management of health care requires an understanding of the use of resources, priorities, and trade-offs in health. Monetary resources for health are limited in all countries, and difficult choices have to be made in their allocation. Health economics is an important element of health policy, at both the strategic (macroeconomics) and tactical levels (microeconomics). Macroeconomics is defined as the aggregate of economic activity in health and deals with overall financing and allocation of health resources. Microeconomics, in contrast, is the theory of how individual organizations (e.g., suppliers) and consumers behave, and focuses on comparing alternative approaches to dealing with specific health issues.

More recent economic theory and applications apply operational research methods and theory of complex systems when there are many decision makers with limited choices and external factors such as regulation of supply and demand. Game theory has come to be recognized as an approach that is relevant to economic analysis of many issues in health care such as allocation of staff, election of hospitals for medical residencies, and allocation of limited kidneys for transplantation. The Nobel Prize in Economics in 2012 was awarded to Lloyd Shapley and Alvin Roth for cooperative game theory to study and compare different matching methods and their practical application in health and other fields.

All professional health care providers and planners need a working knowledge of the fundamentals of health economics and how regulation and economic incentives and disincentives affect the supply, demand, and ultimately the cost of health services. This knowledge helps one to understand and appreciate how health care, while beneficial in terms of reduced morbidity and mortality, also has a cost in terms of resources used, and how health can be improved while facing constraints of limited resources.

Health economic analysis provides a set of tools for management and decision making in the selection of priorities. It can add a measurable empirical element to policy formation as a necessary, though not sufficient, instrument for health policy decisions. Sometimes, there is a conflict between health economics and professional, ethical, and moral issues in solving the everyday problems of preventive and curative services.

With high and rising costs, health systems are increasingly coming under public scrutiny and economic analysis. The total expenditures and how they are allocated are central issues in all health systems. The achievement of a balance between these issues is part of everyday health management, and therefore of the New Public Health.

ECONOMIC ISSUES OF HEALTH SYSTEMS

Health care expenditures vary widely among different countries, ranging from under 4 percent to nearly 18 percent of gross domestic product (GDP), and are rising faster than general inflation in the industrialized and many mid-level developing and transitional countries. In many developing nations total health expenditures are far below levels needed to produce a sustainable infrastructure and meet the ongoing crises of human immunodeficiency virus/acquired immunodeficiency syndrome (HIV/AIDS), malaria, tuberculosis (TB), maternal and child health, as well as rising mortality from cardiovascular diseases and diabetes.

In the USA, total expenditures for health care rose from 7.2 percent of GDP in 1965 to 17.9 percent of GDP in 2010, while the total GDP declined in 2008–09 and increased only modestly in 2010 (Table 11.1). Public expenditures for health increased to over 50 percent of total expenditures for the first time in 2010. Despite this high rate of expenditure on health, and the strength of US leadership in medical sciences and research, a substantial portion of the population lacks any or adequate health insurance and the USA is well behind many other less wealthy countries in terms of key health indicators (see Chapters 10 and 13).

The New Public Health. http://dx.doi.org/10.1016/B978-0-12-415766-8.00011-2

TABLE 11.1 National Health Expenditures and Average Change, USA, Selected Years, Annual Percent, 1960–2010

	1960	1970	1980	1990	2000	2005	2008	2009	2010
Amount									
GDP (billion US$)	526	1,039	2,790	5,803	9,952	12,623	14,292	13,939	14,527
NHE per capita (US$)	148	356	1,110	2,854	4,878	6,868	7,911	8,149	8,402
Percent Distribution									
NHE as % of GDP	5.2	7.2	9.2	12.5	13.8	16.1	16.8	17.9	17.9
Average % increase in NHE from previous year shown	NA	10.6	13.1	11.0	6.6	8.1	5.8	3.8	3.9
Private expenditures (as % NHE)	75.3	62.4	58.1	59.8	55.9	54.6	54.0	52.3	46.9
Public expenditures (as % NHE)	24.7	37.6	41.9	40.2	44.1	45.4	46.0	47.7	53.1

Note: NHE=national health expenditures; GDP=gross domestic product; NA=not available.
Sources: National Center for Health Statistics. Health, United States, 2011. Table 125. Available at: http://www.cdc.gov/nchs/data/hus/2011/125.pdf [Accessed 6 November 2012]
World Bank. World Bank data. Available at: http://data.worldbank.org/indicator/SH.XPD.PUBL/countries [Accessed 6 November 2012].

Comparisons of health expenditure and health indicators between the USA and other high-income countries have been much discussed in the literature, with the USA ranking tenth among Organisation for Economic Co-operation and Development (OECD) countries in health status measures while being by far the largest spender on health (Figure 11.1). Life expectancy (LE) in the USA in 1960 was 1.5 years *above* the OECD average, but in 2010 the US LE of 78.7 years was more than 1 year *below* the OECD average of 79.8 years. In Japan, Switzerland, Italy, and Spain, all of which spend less than 11.4 percent of GDP on health, life expectancies exceeded 82 years, while the USA spends 17.9 percent (2010) (OECD Health Data, 2012).

The Commonwealth Fund reported in 2007 that health spending in the USA has continued to climb, but the rate of increase has slowed, and remained at a nearly constant percentage of GDP per capita from 2003 to 2013. This report called for steps to reduce the rate of cost increase to the level of growth of the economy, through: "increasing transparency and public reporting of cost and quality information, rewarding quality and efficiency, and expanding the use of information technology and systems of health information exchange".

National health expenditures remained stable at 17.9% of GDP in 2009-2013, but are expected to rise to just over 18% annually from 2014-2018 (Centers for Medicare & Medicaid Services, Office of the Actuary, 2013). The rate of increase in annual spending on health in the USA has declined since 2007 to rates of 3.9 percent annually from 2009 to 2011 with 4% predicted for 2013, partly related to the economic recession, but expected to increase to 6 percent in the following years (Cuckler GA et al, Health Affairs 2013). Medicare Part D drug coverage in 2006 produced a shift in spending but had little net effect on aggregate

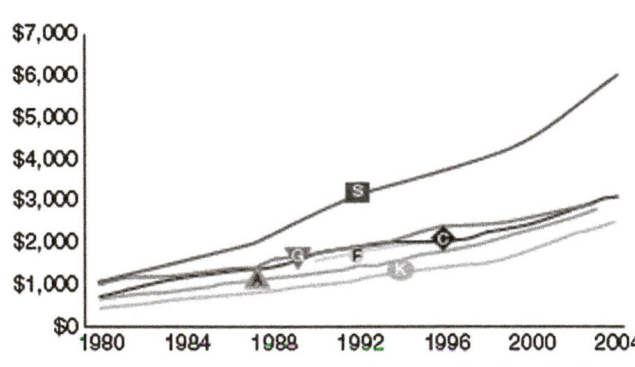

Average spending on health per capita ($US PPP)

FIGURE 11.1   Trends in international comparisons in spending per capita on health, 1980–2004. *Source: Davis K, Schoen C, Stremikis K. Mirror, Mirror on the wall: how the performance of the US health care system compares internationally, 2010 Update. Commonwealth Fund; June 2010. Available at: www.commonwealthfund.org [Accessed 31 October 2012].*

spending growth. Health spending is now more than four times US spending on defense, and is expected to grow faster than the economy. Projections indicate that health spending will reach 20 percent of GDP by 2015. At the same time, the severe problem of nearly 47 million uninsured Americans remains a political and social dilemma for the USA.

Investing in Health

Basic economic problems in health affect different countries through underinvestment, overinvestment, and misallocation of health resources. In 1987, the United Nations (UN) published the *Report of the World Commission on Environment and Development: Our Common Future*, which addressed health within the broader issues of agricultural, economic, and environmental issues of sustainable development. In 1993, the now classic World Bank's *World Development Report: Investing in Health* directed itself to economic aspects of health development seeking identification of priority needs, mainly, but not exclusively, in developing countries (Box 11.1). This report justified economic investment in health care focusing on less costly and more effective interventions. This approach has influenced international donors and the World Bank as a lending agency. The World Bank has been widely criticized both externally and internally for health project lending policies that have been oriented towards medical services without adequate emphasis on population health issues, promoting privatization in health care and inadequate external evaluation of projects. In 2006, the World Bank Group, a consortium supported by the World Health Organization (WHO), the Gates Foundation, and other donor agencies, published its second edition of *Disease Control Priorities in Developing Countries*. This publication reviewed progress made and guidelines for addressing current public health issues in developing countries, again with an economic justification that a healthier population is essential for economic growth.

The UN-approved Millennium Development Goals (MDGs) for 2015 and their ongoing review provide a globally accepted framework linking poverty reduction and health indicator progress, such as reducing child and maternal mortality. Progress made and the complexity of achieving these goals are discussed elsewhere, with little chance of achieving all the targets by 2015. Nevertheless, they provide a universally accepted set of defined goals and specific targets to monitor progress. Global indicators and follow-up global health goals are currently under discussion for post-2015 follow-up based on progress, failures, and new issues that need global health approaches, especially in relation to non-communicable conditions (see Chapters 2, 5, and 16).

The Commission on Macroeconomics and Health (CMH) launched by WHO Director-General Dr Gro Harlem Brundtland in 2000, authored by Jeffrey Sachs, then at Harvard University, analyzed the impact of health on development. The commission recommended specific interventions for health sector investments that could have a positive impact on economic growth and equity in developing countries through poverty reduction and economic development.

The UN MDGs, along with other international initiatives and many national programs, strongly promote the concept that health is a prerequisite for development and economic growth. They call for international and national policy responses and financing mechanisms to address global problems such as poverty, AIDS, malaria, TB, environmental sanitation, and the threat of avian influenza, as well as chronic diseases such as diabetes and obesity.

Professional and institutional capacity for health economics analysis is important to examine health intervention alternatives and priorities. There is a need for accessible data systems and people trained and authorized to analyze such data for planning purposes, ranging from mortality and morbidity to hospital utilization, special disease registries, and more (see Chapter 3). The importance of health as an international issue also requires the attention of non-health sector bodies, including addressing global warming, water shortages, and agricultural and trade policies in the interests of the world as a whole.

The wide acceptance of the vital role of health in economic development is based on the concept that a healthy population is not only a well-meaning social goal based on human rights but, like an educated population, good health is essential to the development of a strong economy. Health is not only a large budgetary cost to society but health status improvement is an added value. Healthier populations are better workers and contributors to economic growth. Healthier children learn better in schools and, thus, also have prospects for contributing to the economic development of their country.

These reports address investment in health as follows:

- Good health is a crucial part of well-being.
- Spending on health can be justified on purely economic grounds.

BOX 11.1 Essential Cost-Effective Health Services for Developing Countries: World Bank

Public Health
- Immunizations – EPI-plus=DPT, polio, measles, hepatitis B, yellow fever
- Nutritional supplements for women and children e.g. iodine, iron, vitamin A and D, zinc, folic acid
- School-based health services
- Information for family planning and nutrition
- Programs to reduce tobacco and alcohol consumption
- AIDS prevention

Clinical Services
- Pregnancy-related care
- Family planning
- Tuberculosis control
- STI control
- Management of diseases of children – ARI, diarrheal diseases, measles, malaria, malnutrition

---

Note: *EPI=Expanded Programme on Immunization (see Chapter 4); DPT=diphtheria–pertussis–tetanus; ARI=acute respiratory infection; AIDS=acquired immunodeficiency syndrome; STI=sexually transmitted infection (see Chapter 4).*
Sources: World Bank. World development report, 1993: investing in health: world development indicators. New York: Oxford University Press; 1993. p 117.
Disease Control Priority Project, World Bank Group. Available at: http://www.dcp2.org/pubs/DCP [Accessed 6 November 2012].

- Improved health contributes to economic growth by:
  - reducing production loss by worker illness
  - permitting use of natural resources that have been inaccessible because of disease
  - increasing the enrollment of children in school and improving their ability to learn
  - freeing resources for alternative uses that would otherwise have to be spent on treating illness.
- Sound policies in financing and resource allocation are essential to achieve good health.

The World Bank report calls on governments to increase spending on health and to foster an atmosphere in which families and communities do the same. It calls for competition and diversity in health care, based on an economic rationale, with health development based on "a basket of essential public health and clinical services" (Box 11.1).

This report is a landmark document in international health, as important as the Declaration of Alma-Ata, Health for All 2000, and the more recent MDGs (see Chapter 2). These reports continue to press the policy that investment in health is an efficient contributor for economic development, and thereby places health among the priorities for national and international financial investment. Health is essential for national productivity and economic growth, and allocation of limited resources to health should promote quality and efficiency in care using fewer relatively unproductive services, such as excessive hospital services, placing greater emphasis on quality, health promotion, and primary care.

In 2007, the prestigious journal *Science* ran a lead article describing the technology–implementation gap, in which "many evidence-based innovations fail to produce results when transferred to communities in the global south, largely because their implementation is untested, unsuitable, or incomplete. For example, rigorous studies have shown that appropriate use of insecticide-treated bed nets can prevent malaria, yet on average fewer than 10 percent of children in 28 sub-Saharan African countries regularly sleep with this protection" (Madon et al., 2007). This article and many others call for implementation sciences to address this "know–do gap" with promotion of research to identify barriers to the adoption of currently available technology. The United Nations Children's Fund (UNICEF) reports that insecticide-treated bed nets are "critical to eliminating deaths from malaria". The use of insecticide-treated bed nets has increased from less than 5 percent in 2000 to more than 30 percent by 2010, yet "malaria still kills 660,000 people every year, most of them African children" (UNICEF, 2013).

The use of antiretroviral therapy (ART) to stop the epidemic of HIV in low- and medium-income countries has progressed to 50–60 percent, with sub-Saharan Africa reaching 56 percent, while countries of Eastern Europe and Central Asia have reached only 23 percent. ART has been shown to save countless lives with affordable quality drugs now available to those countries, but using public health strategies. According to estimates by WHO and UNAIDS, 35.3 million people were living with HIV at the end of 2012. In the same year, some 2.3 million people became newly infected, and 1.6 million died of AIDS-related causes (WHO 2013). There has been remarkable progress in the global fight against HIV/AIDS, but the struggle is far from over. The highest prevalence regions still suffer from weak basic public health infrastructure with very high rates of mortality from preventable maternal and child deaths, TB, and malaria, as well as HIV/AIDS.

BASIC CONCEPTS IN HEALTH ECONOMICS

All societies have limited resources and must, according to politically determined priorities, provide funds for health care in competition with funds for education, defense, agriculture, and others. The availability of limited funds requires making choices. These choices reflect the overall political commitment to health and should, as far as possible, be based on an objective assessment of the costs and benefits of the available options.

The components of economic evaluation in health care are seen in Figure 11.2. Expenditure of resources (in terms of financial and human resources), both direct and indirect, is targeted to a health program. This is expected to produce health and economic benefits, which can also be both direct and indirect. Health benefits may be expressed in terms of a direct reduction in morbidity and mortality, or as improved productivity and quality of life.

Measurement of both input and output is an essential part of health management. Health inputs include resources such as expenditures on buildings, hospital or nursing home beds, equipment, personnel, home care, ambulatory care, and preventive programs. Other elements of health costs, not directly related to the provision of health services but resulting from it, include patients' travel time, loss of work time for patients and for caregivers, loss of full functioning

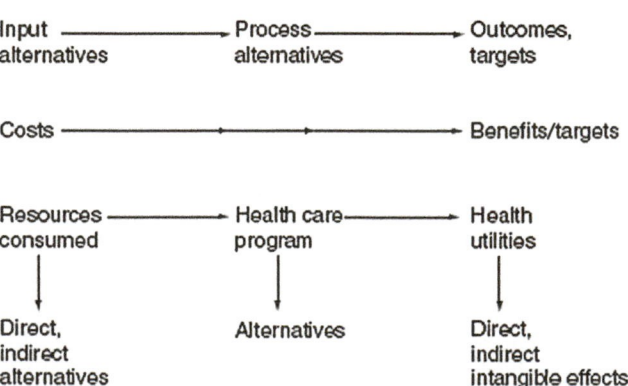

FIGURE 11.2 Economic models of health evaluation: resources–programs–benefits. *Source: Commonwealth Fund. www.commonwealth-fund.org/ [Accessed 22 May 2008].*

years of life, and loss of quality of life. The "input–output" theory of health economics may sound simplistic, but it provides a useful marker when examining the benefits and costs of a specific health intervention. Alternatives can be examined and their cost-effectiveness analyzed, in order for decision makers to select the most suitable ones.

SUPPLY, NEED, DEMAND, AND UTILIZATION OF HEALTH SERVICES

Supply and demand are fundamental concepts of economics, particularly of a market economy. *Demand* refers to the quantity of a product or service wanted by buyers, and this relates to the price of the service and the availability of supply. *Supply* is the quantity of the service or product available. Demand is affected also by price, which if too high will reduce demand for that specific product. In a traditional market situation, promotion, supply, price, and demand are interactive.

In health care, price and demand are not the same as products on a free market. Supply may be limited by government regulation and price is offset by third party payment through insurance mechanisms. In most industrialized countries, these are established by government and cover the entire population. In the USA, health insurance coverage is through a combination of governmental and private insurance arrangements, with a substantial population having inadequate or no such coverage. Therefore, the supply of services and the method of payment are important economic factors in demand and utilization of health care.

Need and demand for medical services are not necessarily the same. Need for medical care exists when an individual has a disorder or risk of such, with symptoms, illness, or disability, for which he or she believes there to be an effective, acceptable, and beneficial treatment or cure. Need also refers to preventive care which may not be a pressing issue for the individual (e.g., immunization, smoking reduction). Demand for medical care exists when the individual considers that he or she has a need and is willing to spend resources including money, time, and energy, and to incur loss of work, traveling costs and time, and inconvenience to receive care. Utilization of services occurs when the individual acts on this demand or need and receives health services.

Normative Needs

Normative needs are those services determined by experts to be essential for a specific need or condition for a particular population group. These services include many standard guidelines for both preventive and clinical health care, such as prenatal care, immunization, child care for infants and toddlers, management of diabetes and hypertension, and screening for breast and prostate cancer. "Evidence-based public health" consists of summation of the published literature and reports from countries with successful application of public health "best practices". There are very often legitimate differences of opinion about public health issues, based on alternative interpretations of information, incomplete evidence, or lack of access to international sources and literature. Despite great successes in public health there is often public skepticism and resistance to basics such as immunization, hospital deliveries, drinking pasteurized milk, where negligent behavior often leads to important health crises. Public opinion and disinformation are important challenges to public health and health promotion with many anti-public health opinions voiced on easily accessed internet non-professional lobby group sources.

As scientific knowledge advances, new information is not absorbed into decision-making processes as rapidly as needed, especially in developing and transitional countries. Professional value judgments may be traditions or biases in medical opinion, and not adequately responsive to advances in clinical, technological, and epidemiological evidence, and evidence from best practices in leading countries. Normative needs, such as cancer screening by Papanicolaou (Pap) smears and mammograms, should be under continuous review by professional panels, with representatives from epidemiological, clinical, and public health services, as well as managers and consumers of health care. These reviewers must consider the available literature and experience with such programs worldwide. Disciplines such as health economics, sociology, health education, and urban planning add to the understanding of factors contributing to a disease, its presence and effects on a population, and how to address it. A current problem requiring much multidisciplinary consideration is the obesity epidemic, with its related problems of diabetes and its complications. Each professional field can contribute to interpretation and decision making about standards for addressing such problems in the health system.

The individual characteristics of people seeking care, including factors such as age and gender, help to determine the type and amount of health services needed. For example, a woman of 40 may not need a mammography as frequently as a woman over 50 years of age. An infant may need to be seen for preventive care assessment more often than a 3-year-old. A male aged 45 needs his blood pressure checked more often than a 25-year-old, and a teenager needs more attention paid to prevention of risk-taking behavior than a 35-year-old.

Clinical guidelines are in common use in many countries, providing norms for care based on consensus by professional groups or health insurance providers, and have been adopted in countries such as the USA, the UK, and Israel. Norms are also used for payment purposes, but this method, as used in the Soviet Semashko health system, has proved to be rigid and difficult to alter when health conditions change. A norm system for hospital beds was a major barrier to adaptation to changing scientific information and population needs, as well as the expectations of the clients, beneficiaries, or patients. Where economic incentives (and disincentives) can be used to promote selected health priorities and resources are limited, as occurs in all countries, then choices must be made. This often comes into conflict with public expectations in health care systems.

Felt Need

Felt need is the subjective view of the patient or the community, which may or may not be based on actual physiological needs. Although subjective, felt need is a prerequisite to whether a person actually undertakes to seek care. There is growing recognition of the importance of sharing health information with the population to increase the possibility that rational choices will be made (the health-belief model). Greater public knowledge is vital to the acceptance of preventive programs such as immunization and compliance with treatment regimens for chronic diseases. Felt needs also affect health planning.

A community or donor may, for example, feel that a community needs a new hospital, whereas the same resources may be better spent on developing primary care or health educational services that have a greater impact on the health of the population. Even in an authoritarian society, public opinion may direct decision makers to make irrational choices, such as placing an excessive portion of health expenditure on high levels of hospital bed supply, rather than adequately addressing non-communicable diseases (NCDs) or nutrition issues.

Expressed Need

Expressed need is a felt need that is acted on, by visiting a clinic or general practitioner, for example. Felt needs may not be acted on because economic, geographic, social, or psychological barriers may inhibit a person from seeking or receiving care. Accessibility may be limited because the individual cannot afford to pay the fee. A service may be free, but not readily accessible owing to such obstacles as distance, language, religious or cultural barriers, difficulty in arranging an appointment, or a long waiting period. As a result, the person seeking care may not be able to receive it and may delay interfacing with the health system until a more urgent, and often more costly, problem arises. Distance, time, and cost of travel, inconvenience, and loss of wages may affect the seeking of services, more so for preventive care than urgent surgical conditions, for example, even if the service is free of charge. Elderly people may sometimes avoid turning their felt needs into actions as they may not feel comfortable with the fact that they are ill, or may not wish to become a burden. Altering the supply, location, type of service, and its availability can change these factors, thus improving equity of access.

Unexpressed need may be the result of lack of knowledge, awareness, or access, or of taboos from religious, cultural, or even political factors which, for example, may prevent a woman from using birth control even when further pregnancies may jeopardize her life. Lack of knowledge may also interfere with appropriate use of available clinical or preventive health care. Therefore, outreach services may be necessary to access such people as migrant laborers, immigrants, refugees, intravenous drug users, commercial sex workers, and other groups whose social circumstances place them at risk of disease, but whose access to appropriate preventive and curative services may be very limited.

Comparative Need

Comparative need is a term that relates the needs of similar population groups, as in two adjacent regions with the same mix of age, gender, ethnicity, and socioeconomic status. One region may have a certain service, such as fluoridation of the community water supply, while the comparison community does not. The population of the second community is objectively in need of that service according to the best current professional and scientific evidence. There are no definable absolutes in the extent of demand for health care, but there are accepted basic standards that are part of world standards at a particular point in time for health promotion, prevention, or health care. These standards are derived from trial and error as much as from science and must be continuously re-examined in the light of new information, as well as the measurable benefits and costs derived from them.

Demand

Demand is based on individual and community expectations (Figure 11.3). Economists consider this to be a part of the economic demand theory of *laissez-faire*, in which the individual is seen as the best judge of his or her need. The individual may feel that he needs a service, but expert opinion may say that this is not a reasonable demand. A patient may ask a physician for an antibiotic to treat a viral infection, which would not help and may even cause harm. A community hospital may wish to increase its bed supply or purchase a highly technical piece of medical equipment because of consumer demand. A patient may feel that denial of a referral by a doctor in a managed care plan is infringement of his rights, but there may be a legitimate and ethical reason for the refusal. Doctors may wish to have the prestige and convenience of certain equipment locally, but economic and planning assessments may say that this is not justified on economic or medical grounds. Such analyses, however, are not immune to change, as the costs of a procedure or technology and clinical experience may change.

Cost is a major factor in adding new vaccines or drugs to those services covered by a governmental or other insurance system. Wide-scale adoption of hepatitis B vaccine was greatly influenced by the fall in price of the vaccine, originally costing US$100 per immunization and currently

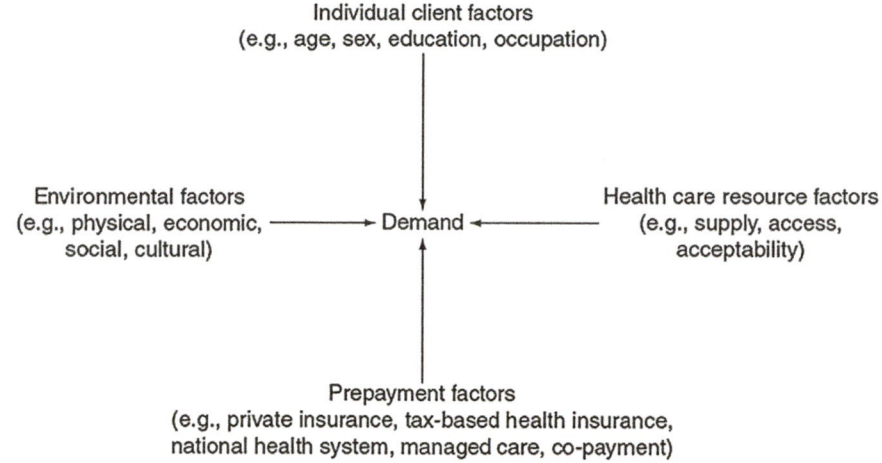

FIGURE 11.3   **Factors in demand for health services.**

less than US$5 per dose. What was once difficult to include in budgets for free immunization has now become highly cost-effective even in poor countries. The human papillomavirus (HPV) vaccine is relatively new, and costs about US$130 per dose or US$390 for a full series (CDC, 2012). Most US health insurance plans cover the cost of vaccines, but if it does not cover vaccines, or for those without insurance, the Vaccines for Children (VFC) program may help. HPV should be included in the routine basket of services in all countries, but the cost is still prohibitive for many; however, it may be reduced in developing countries if purchased by international agencies. ART medications were not accessible in developing countries until prices were lowered, but international aid programs have made HIV control a major priority and encouraged their use to reduce the mortality and spread of the disease; the use ART of to prevent HIV transmission from mother to child is especially encouraging. Priorities are therefore very much affected by prices and such considerations are unavoidably part of health care planning.

Where government or private insurance pays for services, the costs of a service are totally or largely borne by the insuring agency. The patient or consumer may be more willing to obtain the services in this case, but the insurer must determine the relative value of covering the service. In the case of vaccinations, the insurer may be regulated and required to provide this service. The insurer may initiate new preventive services because they are cost-effective and save on hospitalizations or other costly care.

Supply

Demand may also be induced by the supply or provision of care. Making available more hospital beds may increase their use beyond justifiable need, or it may lead to an expectation or demand by patients or their families for an unnecessarily long stay in the hospital. Providing some services at no cost to patients may induce people to use those services

more than they really objectively need to for health reasons according to current best standards. An inappropriate or excessively frequent use of a service may be promoted, and used by the upper middle class, while there may be a lack or scarcity of other important services for the poor owing to selection of priorities and inequitable allocation of resources. Sometimes the interest of the health care providers is such that they may act to promote the use of services because payments are received for each service rendered (fee-for-service). This occurs, for example, in situations where a greater supply of surgeons results in unnecessary cases and rates of surgery being performed.

Since the 1970s there has been a general movement in the industrialized countries to reduce hospital bed supply and utilization. Length of stay in hospitals has been reduced as supply has gone down, but the process was encouraged by many factors. Improved treatment methods such as laparoscopic surgery, outpatient surgical care for previously inpatient procedures, improved early management of acute coronary syndromes, and many others have reduced the need for lengthy or any hospitalization for many common conditions. In some cases, diseases that once resulted in hospitalization and surgical treatment, such as chronic peptic ulcer diseases, have been virtually eliminated by advances in science and clinical practices.

Grossman's Demand Model

A frequently used economic demand model is that described by Michael Grossman in 1972. This method looks at health within the framework of a production function; that is, health status (output) is a result of health care activities (input) by the environment, the individual, and the health services system. Individual demand for health care is affected by many factors, such as socioeconomic, educational, and cultural barriers or incentives to health care, as well as age, gender, and health status.

In this model, everyone inherits a stock of health when they are born. Health depreciates over time, however, and investment is required to sustain it. As people age, there is an increase in the rates of illness and death and in utilization of health services. The rate at which a person's stock of health depreciates over time is represented by a health depreciation–time curve. The stock of health can be sustained by investment to maintain health, such as investment in health-promoting activities (e.g., recreation and fitness facilities) and health services.

Change in health is thus a function not only of medical care received, but also of exercise, good housing, nutrition, smoking restraint, and societal factors that are difficult to quantify. Over their life cycle, people will attempt to offset to an increasing extent the rate of depreciation in their health by increasing their expenditures or use of medical care services. The production function depends on environmental and behavioral variables, such as education, that alter the efficiency of the production process. Personal choices affect health, depending on the way the individual allocates resources to its production; for example, the amount of leisure time dedicated to jogging or the decisions whether to eat fatty foods or to smoke cigarettes.

Health is also an investment good. Being unhealthy brings discomfort, a reduced sense of well-being, and a measurable loss of income from reduced work hours or performance. Health as a consumption good means that a health-related activity will improve the quality and enjoyment of life, prevent discomfort or illness, or improve appearance, as in cosmetic surgery. Older people use more hospital and ambulatory medical care and other services than younger people as their health declines, and they are subject to more disease. There are also factors within the health services system or in an insurance system related to the way individuals act. Personal lifestyle may influence the process of services delivery and provision, including access to services and quality of care. For example, an insurance plan may refuse to accept an enrollee on grounds of age, personal habits (e.g., smoking), or pre-existing medical conditions, or may not cover preventive services.

In Grossman's model of health demand, rising income may adversely affect health because of an increase in unhealthy or risk-taking behavior. Together with the rise in per capita income, excessive consumption of fatty foods, smoking, alcohol abuse, and motor vehicle accidents also increased, and led to rising death rates from cardiovascular diseases and trauma in the industrial countries in the 1940s and 1950s. This also occurred in the 1980s and 1990s in the growing middle class of developing countries. Very poor populations, whose basic health problems are inadequate food intake and shelter, as well as infectious diseases, benefit from rising incomes. In developing countries, as family incomes increase as a result of improved agricultural practices, for example, this allows for more expenditure on food

and better nutrition, with positive effects on health status, especially of children, but also of adults. However, rising incomes in poor populations may also increase obesity, smoking, and poor health habits, leading to a rise in NCDs, including cardiovascular diseases and cancer, and injuries.

A consumption good is defined by economists as use of resources in a manner that most benefits the individual and society, based on the best currently available evidence. Free market economists, and institutions such as the World Bank, once considered health care as ineffective consumption, and recommended investment of resources in economically "productive" development such as industry. At the other end of the political spectrum, Marxist economics also regarded spending on health as a non-productive consumption burden, in comparison to investments in heavy industry or infrastructure. In contrast, social democratic governments, such as the British Labour Party and those in the Nordic countries, have long held the view that investment to improve health is justified on both social and economic grounds, with health, like education, being a social right and foundation of civil society.

The concept of health as a human right and societal responsibility is in keeping with the idea that it is a social and economic investment. This approach gained international acceptance with the Health for All program for fundamental social justice and human rights, as expressed in the Alma-Ata Declaration of 1978. The World Bank's 1993 report *Investing in Health* and the WHO's *Macroeconomics and Health* in 2004 take a utilitarian view that spending money on health and education is a sound investment for economic growth, and not a drain on the economy. Despite the view that investment in health is a benefit for economic health, budgetary battles are still fought over the idea that health expenditure is a burden on society, especially when much of it is dedicated to illness and the terminal phases of life.

COMPETITION IN HEALTH CARE

Health services were historically established as private, charitable, religious, or governmental public services. In recent years, even in countries in which government finances the services, a market-oriented approach has evolved, giving the consumer the choice to use competing health services (Box 11.2).

Reforms in the UK's National Health Service (NHS) in the 1990s included the primary care provider who held the funds for the people who selected him or her for their primary care. Fundholding by general practitioners was stopped and transferred to primary care trusts (PCTs). These are established in all parts of England and receive budgets directly from the Department of Health, and since 2002, PCTs have taken control of provision of local health care, under the supervision of standards and performance monitoring of regional health authorities.

BOX 11.2 Trends in Health Reforms

- Universal health coverage
- Health protection – safe food, water, environment, occupational health
- Food security and quality with fortification to prevent micronutrient deficiencies, birth defects, poor child health, and deficiencies of iodine, folic acid, iron, vitamin D and vitamins B
- Promote healthy environment and lifestyle, e.g., tackle air pollution, smoking, high fat content of diet, lack of physical exercise
- Downsize the hospital sector
- Develop alternatives to hospitalization
- Home care to assist patients for early return home
- Ambulatory medical and surgical care for mental health, cancer, and many chronic conditions
- Palliative care
- Infection and error control to reduce long stays
- Close supervision of chronically ill patients with congestive heart failure, diabetes, and chronic obstructive pulmonary disease
- Preventive care, e.g. colorectal, cervix, and breast cancer screening; pneumonia and influenza vaccination for all
- Develop primary health care including outreach services
- Linkage between insurance and service to promote efficiency and preventive orientation of services
- Linkage between services – vertical and lateral integration
- Define basket of services, i.e., benefits
- Generic drugs to reduce costs and maintain safety
- Clinical guidelines
- Technology assessment
- Cost-effectiveness analysis
- Computerization and linkage of medical records

BOX 11.3 Obligations of Health Service Provider Systems

- *Availability* – provides continuous service coverage, 24 hours every day and on holidays, e.g., urgent care centers in the community
- *Accessibility* – the client can readily reach the service within reasonable traveling and waiting time
- *Accountability* – the system and provider explain programs, decisions, and actions to the client
- *Affordability* – services are provided at reasonable cost to consumer and insurer
- *Acceptability* – user friendly in style and manner of staff to the patient and family
- *Accredited* – undergoes external evaluation and implements recommendations of accrediting agencies
- *Equality* – provides fair and equivalent access to services needed regardless of age, gender, ethnic origin, religions, social, or political identification, place of residence, prior medical condition, or ability to pay
- *Efficiency* – produces effective results with a minimum of waste, expense, or unnecessary effort in use of human, financial, and other resources
- *Consumer's rights* – ensures the patient and consumer are informed of their rights and alternatives of care, before medical or administrative decisions are made
- *Financial soundness* – is able to meet financial obligations
- *Goals and objectives* – defined, written, reviewed, and used as a basis for planning and monitoring
- *Innovativeness* – open to new methods in clinical and preventive approaches to health
- *Quality promotion* – sets high standards of facilities and services in accordance with current professional criteria and standards of leading providers
- *Community acceptability* – meets community expectation with participation in promoting health standards
- *Comprehensive* – includes community, home care, hospital, long term care
- *Health Promotion* – includes healthy lifestyle, health protection legislation

---

In the USA, managed care health plans receive insurance payments per capita and select hospitals, physicians, or other services on a competitive basis. In some other countries (e.g., Israel, Colombia, Philippines, Jordan), clients select health plans or sick funds that are then responsible for all care and costs, seeking competitive benefits from providers such as hospitals or medical practitioners.

Health care organizations, whether hospitals, nursing homes or primary care services, must provide quality service to the community in order to meet their mission and to survive as institutions (Box 11.3). The patient or consumer is, in principle, at the center of the process. However, market analysis regards health care program costs, contents, level of satisfaction of individuals and the community, as well as health care providers' (individual or institutional) satisfaction and performance indicators. Since 2000, many countries have tried to introduce free market ideas of competition into their health systems to move away from state-run health systems. The introduction of free market mechanisms in a national health program compromises universal entitlement to health care, creating different levels of services for those who can afford to pay and those who cannot. A two-tier system introduces greater degrees of inequity between the urban middle class and the urban and rural poor.

To fulfill the potential of the US health system, it is important to continue to seek savings and better value for the high investment in health. More efficient and effective health promotion, health care, and insurance systems with universality and equity are required as major goals. Strategies with the potential to achieve savings, slow growth in spending, and improve health system performance include:

- increased public information and greater competition between health care organizations
- state-mandated universal health coverage, access, affordability, and equity

- reduction of insurance administrative overheads
- financial incentives to promote efficient and effective care
- expanded efforts in health promotion and training in public health
- patient- and risk group-centered primary care
- health information technology
- categorical grants to improve equity in health.

However, health systems are very costly and choices need to be made in services provided or available. Regulation of the supply and utilization of costly services is an inevitable result. This has led to the development of new models of health care organization. Managed care systems depend on reducing unnecessary hospital care and increasing the use of ambulatory and home care services. This has become a part of the entire system of health care in the USA because of reduced hospital bed supply and payment by diagnosis-related groups (DRGs). Services previously provided on an inpatient basis are being provided equally or even more successfully on an outpatient basis. The manager or chief executive officer of a health facility needs to guide its transition from a passive receiver of the sick to an institution with a strong ability to meet health needs in the community, while at the same time controlling costs to remain competitive and financially viable. The US federal Patient Protection and Affordable Care Act (PPACA or "Obamacare") includes many features that make health insurance available to millions of previously uninsured Americans, and introduces incentives for preventive care and new models of accountable care systems that promote quality and cost-saving care approaches.

ELASTICITIES OF DEMAND

Classically, *elasticity of demand* means that demand is related to the price of a product. Demand may increase more rapidly than the fall in price (elastic), or may not change even if the price falls (inelastic), or may fall in direct relation to price change (unitary). But demand in health care is influenced by many other factors determined by the consumer, the provider, the supply, the location of services, and the knowledge and motivation of the consumer. Cost to the consumer is a factor in choosing to purchase goods or seek services. If the price goes up, then demand will decline, and vice versa. In other words, demand is not an absolute but can be affected by supply, price, and type of payment required for the service.

Classical market mechanisms fail in health care because there are so many intervening factors, including the high degree of information asymmetry between provider and the consumer, the role of third party payers such as national or private health insurance, and regulatory roles of government. Universal access through national systems of coverage is also not a guarantee of equity, as access is often unequal and because the consumer may not be sufficiently informed of the potential benefit of care or prevention before making

his or her choice. Cost containment can be addressed using many "market" factors in health care, such as the supply of facilities that affect demand; reducing costs of a specific vaccine, drug, or surgical procedure; and method of payment for doctors and hospital services. All of these affect the economics of health. However, where consumers have no choice of service provider, there is a risk of exploitation by providers and consumers alike.

In classic market economic theory, the individual is seen as the best judge of his or her own needs and decides what to buy (consumer sovereignty). It is assumed that consumers purchase services or a health plan based on factors such as cost and quality, as one would do when purchasing a refrigerator. Individual decisions are made on the basis of personal perception, information, priorities, and resources. Proponents of the market approach in health care suggest that it provides the consumer with more control and choice, and indirectly raises the competition for consumer demand and the quality of services while lowering costs.

Opponents of this view claim that this market approach fails in medical care because the issue is more complex than purchaser and provider. The market mechanism in health care is as much determined by supply, access, and method of payment as by consumer choice. Supply of services creates demand, as does prepayment through insurance or a national health service. Consumers rely on their doctors to advise them; this is known as "the agency relationship". Physicians make or recommend decisions for their patients on the basis of both patient needs and the supply of services. Payment for services by a third party, such as an insurance plan or the government, where the consumer may provide little or no direct payment, can lead to the provision of unneeded services, particularly if the doctor has incentives of the fee-for-service payment system.

Someone may want to purchase additional or different services according to one's wishes for expected benefits (marginal utility), such as reduced waiting time for an elective surgical procedure. The marginal cost is the cost of an extra unit of the commodity used. In classic economic theory, the consumer decides to purchase a service when the marginal utility is equal to or greater than the marginal cost; that is, the added benefit is worth the additional cost. This free market approach in health care may result in underprovision of vital preventive health services, especially to the population in greatest need. If medical care services operated in a free market, then consumers would not necessarily take into account benefits to people other than themselves (externalities). For example, if vaccines are available purely on a free market basis, many people would not purchase the vaccinations, owing to a lack of resources or lack of awareness of the importance of the preventive measure. This would increase the risk to the population at large by reducing herd immunity. Externalities should be taken into account in public policy decisions, valuing the benefits to society as a whole.

Where there are many providers, they will compete with each other in principle by offering services at lower prices in order to attract clients. However, this rarely occurs in medical services where fees or salaries are set by collective bargaining. In terms of hospital and insurance services, the existence of monopolies (i.e., only one provider) or oligopolies (i.e., too few providers) prevents price competition and often results in collusion to fix prices. A monopsony is a situation when there is only one purchaser, so that the provider may be subjected to pressures to lower prices or meet additional demands of the buyer, such as a managed care organization (MCO) purchasing hospital services.

In a highly privatized system of health care, such as that in the USA up to 2008, demand for care is rationed through fees, co-payment by the consumer, or limitations set by the indemnity insurance plan, and lack of insurance benefits for many. Indemnity insurance plans require co-payments and deductibles, so that the insured person has to take out additional insurance to cover expenditures or pay part of the charges. Medicare beneficiaries in the USA have two alternatives: to pay part of their health care expenditures and have a greater variety of service providers to choose from, or to join a managed care option which covers their service costs, but with a restricted list of physicians. Therefore, the beneficiary must choose between extra payments and a limited choice of service providers.

In a public system of health care, demand is rationed through limitations on supply of services by a requirement that the patient wishing to see a specialist be referred by a general practitioner who plays the role of a "gatekeeper", so that the consumer's choice is limited to the options decided by the general practitioner. This arrangement has been adopted by many health systems, including managed care plans in the USA. This is a controversial limitation on the consumer, who may wish to consult other specialists, but it may be essential to avoid frivolous "shopping" for care that drives up the cost of health care.

The market approach is based on choices being made by consumers on the basis of anticipated benefits in terms of improved health or health care, or reassurance. The employer must offer the employee several options for health insurance. In a fee-for-service indemnity plan, allowing the consumer greater choice of physicians, the employee pays additional monthly premiums, as compared to a managed care option. This carries with it a measure of inequality because of differences within the population, both in health needs and in the ability to purchase or utilize needed services.

Insurance for health care means sharing the risk and paying from your place of employment to protect yourself against the costs of illness which are unpredictable. The insuring body, whether public or private, can predict needs based on population data. When governments provide an insurance system, they seek to share the risks among the total population, and not only the highest risk groups such as the elderly and the poor; this is done in the USA, where the government has taken responsibility for these groups, leaving the general public, who are generally a healthier age group and lower risk population group, to arrange insurance privately or through collective bargaining at their workplace. Where the government insures the highest risk groups it is a form of adverse selection of the highest risk population groups. Yet other risk groups are left without coverage or basic information on health insurance. Sharing risk is the basis of insurance, but those left outside insurance protection or with inadequate coverage are at risk for major economic loss due to illness. Those with insurance may also seek extra coverage or co-insurance to protect themselves against the costs of uninsured health needs, such as drugs or dental care.

Often those with the greatest need are those least able to access the desired or required services. People who have no source of income cannot be consumers or make decisions to purchase health services on the free market. Rather, they are dependent on free charitable services. This problem may be addressed by a number of economic alternatives, such as providing low-income individuals with free health insurance under Medicaid or with vouchers to purchase services.

Decisions or actions of patients are influenced not only by cost and access, but also by knowledge and attitudes towards care. A person unaware of modern birth control, or living in a society that discourages or limits its use on religious or political grounds, is not able to make an informed decision about its use. Market mechanisms in health work in different ways in different health systems. Even where care is a free service and it is seen as a right for everyone, there are financial and human resource limitations in the supply of services. Market mechanisms have a major role in the reform of many health systems by giving consumers the choice of provider, even if this means a choice of health care plan such as with managed care systems. Financial incentives are used to promote quality care, such as immunization or preventive procedures such as Pap smears, or disincentives to reduce unnecessary or wasteful services, as in the limitation of hospital bed supply.

MEASURING COSTS

Costs in health can be analyzed in various ways. Direct costs of services are those paid by the patient or by the insurer or sick fund on behalf of the patient, including costs of the hospital or other provider. Indirect costs of an illness to the patient, his or her family, and society include loss of income due to time off work or lowered productivity as expressed in work or school absence, and poor quality of product or learning capacity.

Opportunity cost refers to the resources used that could have been applied to other uses. Hospital land and building costs, for example, could be allocated for other purposes, such as primary health care facilities or facilities outside the health sector such as after-school programs for children. Increasing the proportion of gross national product (GNP) spent on health care may limit society's ability to spend money on education and other important social programs.

Social costs include indirect expenditures for health and illness, such as the total value of lost production and costs of social support for a person whose health and work capacity have been impaired by illness. Private costs include out-of-pocket expenditures that an individual makes to purchase health care plus related expenses such as payments for health insurance, loss of wages, purchase of pharmaceuticals, and co-payments for health services.

ECONOMIC MEASURES OF HEALTH STATUS

Economic analysis assesses not only input, as in costs and resources, but also output, as in extension of years of life and reduction of disability, morbidity, and mortality. Greater functional levels that improve the quality and quantity of life are output measures of health care, and should be part of an economic evaluation of the use of national or personal resources. Disability-adjusted life years (DALYs) and quality-adjusted life years (QALYs) are measures of the total burden of disease as a guide to population health status (of death and disability).

DALYs are calculated as the present value in years of disability-free life that might be lost as a result of premature death and disability occurring due to a disease in a particular year. QALYs measure life expectancy adjusted by changes in quality of life, measured by assessing two or more aspects of health, such as pain, disability, mood, or capacity to perform self-care or socially useful activities such as paid employment or housework. DALYs and QALYs are constructed using expert evaluation to estimate the degree of impairment (normal, impaired, or incapacitated) from specific diseases. These include impairments such as loss of ability to communicate; sleep disturbance; pain; depression; and sexual, eating, and mobility dysfunctions.

The value of the health status of an individual can be expressed in numerical values for comparisons. The values of the total scores are then added together and the overall score is calculated out of a maximum value for comparison. This allows a measure of health status and may be used for the purpose of comparing the effectiveness of alternative interventions. Such assessments are subjective, depending on the perception of the assessor, and interobserver variability may be high. While such measures do not include all factors that may contribute to development of a disease, they contribute towards using the economic impact of disease as part of health planning by pooling mortality and economic indicators.

DALYs and QALYs are useful economic measures to provide a common basis for comparison of different cases, settings, and time changes by using mortality, disability, and quality of life as measures. They are used as proxy health status indicators to analyze different approaches to health policy, to justify specific interventions, and to determine priorities. The World Bank, WHO, and other organizations are examining alternative indicators to link health and its underlying determinants of the total burden of disease and disability, and to refine the process of establishing priorities for research and decision making for interventions.

Gains in life expectancy from preventive or curative medical interventions may be measured from published data sources. The gain in life expectancy calculated for a patient who survived cardiac arrest by placement of an implantable pacemaker is calculated as 36–46 months, or who had bone marrow transplantation for recurrence of non-Hodgkin's lymphoma is calculated at 72 months. For preventive measures, the average gains for the total population appear smaller. Cervical cancer screening, for example, increases the life expectancy of all women by 3 months, but for a woman whose cancer of the cervix was detected early, the gain is an average of 25 years. This methodology relies on published studies, and can be of great value in the comparison and analysis of alternative strategies and health care priorities.

COST–BENEFIT ANALYSIS

Cost–benefit analysis (CBA) compares the expense of a specific program to its expected monetary yield or savings. Costs include direct expenditures as well as the indirect costs of loss of productivity and loss of contribution to society. Direct benefits include prevention of premature loss of life through reductions in morbidity and mortality and the associated savings in medical care costs, such as hospitalization, doctors' services and drugs used, and the attachment of economic value to this. Indirect benefits include savings to the patient's family in terms of expenditures to visit the patient (transportation costs) or time away from work to look after a sick child or relative. Other indirect benefits also accrue to society, in terms of savings in reduction of lost work time by the patient or his or her family during an illness.

The assessment of costs and benefits involves three stages: enumeration, measurement, and explicit valuation. Assessing a particular treatment, or enumeration, requires measurement of change in health status, the cost of use of resources, and the patient's productive output. Economic appraisal depends on determination of the many factors needed in managing a public health problem and its expected outcomes. Explicit valuation, or estimation of the cost of a variable, is based on determination of the economic value of these factors. Many factors need to be taken into account and simplified; consequently, the estimations of the costs are approximations rather than exact figures.

A CBA study of a phenylketonuria (a congenital metabolic disorder) program in the USA showed the total cost, including screening 660,000 newborns, confirmation tests, the special diet for those affected, and administration of the program, to be US$1.39 million. The benefits gained were US$1.26 million for medical and other services, and US$1.05 million for prevented loss of productivity, making a total of US$2.31 million. The benefit/cost ratio was $2.31/1.39 = 1.66$. Therefore, for each dollar invested, the gain to society was US$1.66. According to CBA studies on the benefit-to-cost ratios of vaccines, the addition of a second dose of measles vaccine was found to have a higher benefit-to-cost ratio (CBA = 4.5/1) in both developed and developing countries compared to one dose of hepatitis B or *Haemophilus influenzae* type b (Hib) vaccine on an immunization schedule. Hepatitis B vaccinations were found to have high benefit-to-cost ratios, even in countries with intermediate levels of endemicity (CBA = 4.5/1). For Hib vaccine, the social benefits were found to exceed the costs to society; these include indirect social benefits such as reduced need for special education for brain-damaged children. The benefit-to-cost ratio, if viewed solely from the point of view of the health sector, was lower. Such a result might prevent the health system from adopting a beneficial program if the benefit-to-cost study is too narrowly applied.

The decision to adopt a specific program may include a CBA but is often made on other grounds, including public and professional opinion as well as political factors. A CBA can give a prioritized ranking to alternative interventions, and thereby help in the decision-making process. Ranking according to the relative costs and benefits can help a health ministry to choose among putting resources into a high-technology hospital, home care, expansion of an immunization program, or investment in primary care services.

Both CBA and cost-effectiveness analysis (CEA) include initial as well as ongoing costs, but they must take into account that the future value of money will be less than the present value, referred to as *discounting*. The costs as well as benefits to be derived from the project must be calculated as they accrue, so that a portion of the effect is observed the next year, and a portion the following year. The cumulative discounted value is called the net present value (NPV). Cost–utility analysis (CUA) (Box 11.4) is commonly used. CUA, CEA, and CBA are used for different purposes. CUA has been criticized by some economists because this method is based on the incremental cost calculation, and it does not take into account the overall health system and population. It is, however, widely used under the pressing and frequent need for CEA of health projects. Each health project could be evaluated on the basis of its incremental cost per extra QALY

provided to patients, as in the case of new drugs for the treatment of advanced cancers such as of the breast or colon.

In recent years, the gold standard for assessing the feasibility and the ranking of potential interventions has been the cost per QALY. The costs and QALYs lost under a null hypothesis scenario, when no intervention is provided, are used as a baseline. Then, a calculation is made as to the costs and QALY losses associated with the current scenario of provision (e.g., Pap smears once every 5 years). This enables the average cost-effectiveness ratio in terms of costs per QALY to be calculated (from the differences in costs and QALYs between the current and null scenarios). Finally, a calculation is made as to the costs and QALY losses associated with the various proposed scenarios (e.g., Pap smears annually and HPV vaccination). Thus, the incremental cost-effectiveness ratio can be calculated in terms of costs per QALY comparing the costs and QALYs between the proposed and the current scenarios.

COST-EFFECTIVENESS ANALYSIS

CEA in health care is the net gain in health or in reducing the burden of disease from a specific intervention in relation to its cost. It is used to determine the least expensive way of achieving the goal, by comparing alternative methods of intervention to make a choice. The most cost-effective method is the one that achieves the same goal using the fewest resources. A low cost per DALY gained indicates a high degree of cost-effectiveness, and therefore an intervention that should be of high priority, given limited resources.

Alternative methods of treatment may also be compared, such as use of medication compared to surgery, day surgery compared to inpatient care, or treatment in the community compared to hospital inpatient care. Expanded programs of immunization, maternity care are the most cost-effective services for developing countries where vaccine-preventable diseases and early childhood deaths are the major causes of loss of DALYs for the population (as seen in Table 11.2 at the companion web site: http://booksite.elsevier.com/9780124157668). Other cost-effective programs include preventing iodine and vitamin A deficiency and treating intestinal worms, even though these are relatively minor causes of lost DALYs. While these calculations are from an earlier period the issues identified remain valid. The number of deaths of children up to the age of 5 has been reduced from the 1990s to 2010, but at current rates of decline will not meet the MDGs by 2015. The main interventions are vaccination (through the Expanded Programme on Immunization, EPI), nutritional support to prevent malnutrition and micronutrient deficiencies (iron, iodine, folic acid), family planning, malaria control (vector reduction, case management, and bed nets), management of serious infections in children with antibiotics, rotavirus, pneumonia, and Hib vaccines, promoting breastfeeding, and vitamin/mineral supplements.

BOX 11.4 Why Do We Need Cost-Effectiveness and Cost–Utility Analysis?

In the field of public health, there are numerous potential interventions, including immunizations against infectious diseases (e.g., measles, diphtheria, HPV infections), health promotion programs for reducing risk factors (e.g., smoking, overweight, sedentary life styles), screening programs for early diagnosis of diseases (e.g., thalassemia, colorectal and cervical cancer), and various treatment options (e.g., surgical or pharmaceutical, hospital or home care).

All societies (whether capitalist, socialist, communist, or mixed economies) have limited resources that they can allocate to any sector, including health. So decisions have to be made about which of the many potential public health programs should be adopted (or not) and which existing programs should be discontinued.

Often such decisions are made through struggles between political groups, pressure groups, and interest groups (professional, public, private, and industrial). A more rational, scientific, objective, and evidence-based approach to prioritizing programs was made towards the end of the twentieth century with the use of cost–benefit analysis (CBA), where expected monetary benefits and monetary costs of interventions were compared. However, CBA did not directly measure (in non-monetary terms) the day-to-day health and functional benefits (e.g., ability to walk, lack of pain) that individuals could receive from a successful intervention.

Cost-effectiveness analysis (CEA) searches for the cheapest way of achieving a given goal. CEA compares the costs of different interventions with their outcomes, often measured in terms of cost per case prevented, cost per averted death, or cost per life year saved. Since the late twentieth century, the major tool used for prioritization has been a form of CEA where the outcomes are measured in terms of quality-adjusted life years (QALYs), known as cost–utility analysis (CUA).

CUA is a method of analysis in which the outcome of a program or an intervention is measured by outcomes such as the cost per QALY. CUA integrates economic information and epidemiological information as to the proposed intervention. A QALY is the decrease in morbidity and mortality as measured by quality-adjusted life years gained as a result of the project. Sometimes outcomes are measured in terms of costs per disability-adjusted life year (DALY).

A CUA calculates the cost per QALY (or DALY) as defined by the following formula:

$$Cost\ per\ QALY = \frac{Cost\ of\ intervention - Savings\ from\ intervention}{QALY\ added\ from\ decreased\ mortality\ and\ morbidity}$$

Both basic and advanced explanations on how to carry out CUA calculations are available. The World Health Organization has applied CUA using standardized methodologies to prioritize different interventions in various regions of the world in order of their cost per QALY.

Sources: *Courtesy: Ginsberg GM. Personal communication; November 2012.*
Ginsberg GM. Assessing evidence – the strengths and weakness of various ways of assessing outcomes, cost-effectiveness analysis. In: Aceljas C, editor. Assessing evidence to improve population health and well-being. Exeter: Learning Matters; 2011. p 91–111.
Tan-Torres Edejer T, Baltussen R, Adam T, Hutubessy R, Acharya A, Evans DB, Murray CJ, editors. Making choices in health. WHO guide to cost-effectiveness analysis. Geneva: WHO; 2003.
Chisholm D, Baltussen R, Evans DB, Ginsberg G, Lauer JA, Lim S, et al. What are the priorities for prevention and control of non-communicable diseases and injuries in sub-Saharan Africa and South East Asia? BMJ 2012;344:e586.

The Disease Control Priorities Project (2006) examined 25 priority health conditions in low- and middle-income countries, assessing their public health significance and the cost-effectiveness of various clinical and public health interventions. Such analyses help a basket of essential services to be constructed on the basis of comparative cost-effectiveness. Highly cost-effective interventions include vitamin A supplementation, measles control, and directly observed chemotherapy for TB. A high-cost but highly effective intervention is chemotherapy for leukemia in children under the age of 15. This intervention is justified as benefits are high. The same chemotherapy in a 75-year-old would present a low DALY value. CEA studies examine issues such as day surgery versus inpatient surgery, operations versus medications (e.g., for peptic ulcers and coronary heart disease), public versus individual dental prevention (e.g., fluoride versus dental hygienist care), and community versus institutional care.

The 2010 child mortality estimates published in 2012 show that infectious diseases still account for nearly two-thirds of deaths. Neonatal deaths account for 40 percent of all deaths under the age of 5. The study report's global trend analysis reveals a marked decline in the annual number of child deaths between 2000 and 2010. The rate of decline is far from fast enough to reach the UN MDG of reducing under-five mortality by two-thirds between 1990 and 2015 (Liu et al., 2012). Addressing causes of death is not sufficient for a program approach, which relates to underlying causes or predisposing conditions, such as:

- education of girls and preventing early-age maternity
- promoting birth control and spacing between pregnancies
- augmenting maternal malnutrition
- breastfeeding and vitamin support
- vaccinations
- malaria control by vector control, bed nets, and case management
- safe delivery
- maternal tetanus immunization
- newborn vitamin K
- education of parents in child care.

A comparison of life years gained for patients with end-stage renal disease in the USA found that renal transplantation was less expensive (US$3600 per year of life gained) than home dialysis (US$4200 per life year gained) and hospital dialysis (US$116,000 per life year gained). Moreover, transplantation provides a higher quality of life. This study was perhaps the first example of cost–utility analysis, where life years gained was weighted according to quality of life. This can be expressed as the cost-effectiveness per QALY. Policy decisions based on such findings are constrained by difficulty in obtaining sufficient donor kidneys, and a lack of personnel and facilities suitable to carry out transplantation effectively.

Surgical removal of the gallbladder and even cancers of the colon are now performed with endoscopy instead of the traditional abdominal procedures. Endoscopy is less traumatic and leads to faster recovery, so the patient is discharged from the hospital on the next day and returns to work within a day or two, whereas the patient who has undergone the conventional abdominal procedure requires a much longer hospital stay and recuperation at home before returning to work. The newer procedure is easier on the patient and safer. CBA must take into account not only the medical and hospital costs, but also the social costs of lost work time for the patient and caregivers. Computed tomography (CT), magnetic resonance imaging (MRI), and positron emission tomography (PET) scanning, once considered costly and for special use, have become a valuable part of the investigation of many conditions and are used frequently, CT in particular, in place of the costly, dangerous, and less effective procedures previously used to investigate many conditions, such as head and neck injuries.

Care of an infirm elderly person, up to a certain level of disability, in a private home with outside help, including meal preparation and delivery, nursing, physiotherapy, and social worker visits, is less costly than care of the same patient in an institution. Home care promotes earlier hospital discharge and recuperation at home. These assessments must take into account social costs and transfer of costs of services provided in an institution, such as food, laundry, heating, and electricity, to the patient's family. Home settings promote improved recovery, avoidance of hospital infections, and a general feeling of well-being of the patient. For people with severe illness or many disabilities, requiring a higher degree of nursing and/or medical care, institutions are more cost-effective than home care. Respite care sometimes provides support for a family in caring for a patient with multiple disabilities, delaying more costly institutional care.

Studies comparing the treatment of psychiatric patients in a large mental hospital, in the psychiatric ward of a general hospital, and in a day treatment center show the day treatment center care to be the least costly, but some measure of the severity of illness and need of care must be added to this assessment. Planning mental health services and facilities with reduced hospitalization requires adequate resources for mental health care in the community in order to prevent patients with chronic mental health problems from becoming part of the homeless population, as has happened in many large cities.

> **BOX 11.5 Management for Cost-Effectiveness**
> - Promotion of healthy lifestyle and healthy cities
> - Prevention of micronutrient deficiencies by fortification of basic foods (salt, flour, milk, soya)
> - Cost containment and moderation (of increase) in health expenditures
> - Priorities and resources shift towards health promotion and disease prevention
> - Cost-effective health initiatives
> - Decentralized management
> - National policy, monitoring, and standards
> - Information systems/monitoring
> - District health profiles with comparisons of key markers or tracer conditions indicating quality of care
> - Increase in primary care resources, funding, and linkage with hospitals as accountable care organizations
> - Outreach services in follow-up of non-attenders with diabetes, hypertension, HIV, and other dangerous conditions
> - Increase home care
> - Increase in long-term care facilities linkage with accountable care organizations
> - Increase non-admission surgery and long-term care
> - Health information systems
> - Managed care or accountable care organizations
> - Diagnosis-related groups for payment of hospital care

Sometimes the least costly method is the least effective. For example, a study showed that prevention of pregnancy by the withdrawal method is least costly but is far less effective than use of the birth control pill. Abortion as a method of birth control may be less costly than use of the pill, but, in addition to the ethical issues, it produces complications and contributes to excess morbidity and mortality in subsequent pregnancies, for both the mother and the newborn.

CEA takes into account both the cost and effectiveness of interventions, as a measure of value for cost, but does not answer the question of whether or when the intervention should be carried out. Management of health systems is part of the New Public Health. It reflects the priorities and capacity development of a broad spectrum from health promotion through prevention to institutional care. Box 11.5 summarizes management issues related to operational aspects of cost containment, and substitution of high and inappropriate costs by lower cost services that address actual patient and community needs.

Projects whose cost per QALY exceeds three times the GNP per capita are deemed not to be cost-effective. Those whose cost is between one and three times the GNP per capita are cost-effective, and those whose cost is less than the GNP per capita are very cost-effective. Projects whose net cost is negative, as a result of savings in treatment costs exceeding the project costs, are deemed to be cost saving.

The choice of a cost-effectiveness threshold is a value judgment that depends on several factors:

- who the decision maker is and what the purpose is of the project
- how a decision maker values health outcomes and money, how he or she is willing to substitute one for the other, and what his or her attitude is about risk
- the resources available.

On a societal basis, the use of ART for HIV infection illustrates vividly the dependence of the cost-effectiveness threshold on resources: although ART may be considered cost-effective in the USA, a cost-effectiveness threshold of US$50,000 per life year gained is completely implausible in the developing world, where per capita health spending may be less than US$10 annually. Resources and the cost-effectiveness threshold tend to rise and fall together, all other factors being equal.

Economic analysis must be part of deliberations on how to develop health systems to ensure good value for money but equal access and financial help for those in need, especially poor and vulnerable people (Chisholm et al., 2012). While economic analysis is important, it cannot be the sole determinant of health policy. It provides a tool for policy makers to help in efficient allocation of resources, but other factors must be taken into account, such as poverty reduction and epidemiological factors.

BASIC ASSESSMENT SCHEME FOR INTERVENTION COSTS AND CONSEQUENCES

Assessments of effectiveness and costs of an intervention have become a basic part of policy making in health. An approach called the Basic Assessment Scheme for Intervention Costs and Consequences (BASICC) has been widely promoted by the US Centers for Disease Control and Prevention (CDC). BASICC is a complex approach that looks at the efficacy of the intervention and the cost, including direct outlays, productivity costs (e.g., loss of time for work or recreation), and intangible costs (e.g., pain and suffering).

Costs include fixed costs, or those that do not vary according to the quantity of the service provided but also include a portion of rent, utilities, and equipment allocated to the program. The average costs are the total cost of a program divided by the total units of output produced. Variable costs are those that vary according to the level of service provided, such as the number of nursing visits required for a home care patient. Marginal costs are those additional costs to basic program costs, such as expansion of staff or facilities to accommodate extra activities.

BASICC focuses on intervention costs and direct cost savings in terms of medical care. Net costs can be summarized as the cost of the intervention and its side-effects for n people minus the direct costs of the expected number of cases averted for the same n people, calculated as follows:

$$\text{Net cost} = (\text{Cost of program} + \text{Cost of side-effects}) \\ - \text{Cost of averse health outcomes adverted}$$

The steps of BASICC include the following:

1. Describe the program, its objectives, target population, effectiveness of intervention, external constraints, resources required, management of the program, implementation strategy, and scientific evidence of effectiveness.
2. Define the burden of the disease, its incidence, and prevalence without the program.
3. Define outcomes anticipated in terms of improved quality of life, reduced incidence or severity of the disease, and prevention of premature death.
4. Measure efficacy of the intervention, taking into account that interventions are rarely 100 percent successful in practice because of compliance and effectiveness of the intervention.
5. Measure intervention costs per unit.
6. Measure direct medical costs of outcome averted by the intervention.
7. Assess resources required for the intervention, which include fixed, variable, and total, as well as unit costs.

THE VALUE OF HUMAN LIFE

The main anticipated benefit of a health intervention is the saving of human life. Placing an economic value on life is useful in calculating the benefits of specific interventions or perhaps for compensation to the family of a person who loses his or her life as a result of, for example, negligence by a doctor or plant manager.

The value of a human life in economic terms was first calculated by William Petty in 1699 while developing his idea of political arithmetic. In 1876, William Farr used life tables to calculate economic equivalents. More recently, economists have introduced other methods of quantifying the value of human life, such as calculating the value of human capital, willingness to pay for services, years of life saved, and more precise evaluations involving QALYs and DALYs saved.

Ethical and political conflicts surround the issue of calculating the economic value of human life. A materialistic approach would evaluate human life based solely on the value of contributions that the individual might make to society. A humanistic approach would place virtually unlimited value on a human life according to the ethical principle that saving one life is as important as saving all human beings (sanctity of human life; see Chapter 1).

By placing infinite value on human life, society provides doctors with precious resources to save one human, without considering that this may be at the expense of other lives. For example, the cost of a heart transplant that may add quality and years to one person's life may be equivalent to the cost of a program on prevention of heart disease that might save many more lives. International agencies spend hundreds of millions of dollars to eradicate poliomyelitis, a much feared, crippling, but usually non-lethal disease, while measles, often thought to be a common benign disease, kills over hundreds of thousands of children annually. The valuing of human life is not meant to fuel ethical argument, but rather to provide a measurement tool for the planning of priorities and litigation needs of health planners.

In health economics some arbitrary measures are used to demonstrate alternative ways of using limited resources. The implicit social value (ISV) of life (Box 11.6) rates a program by the lives it saves, and assumes that, in a democratic society, all lives have the same intrinsic value. Governments have made some decisions showing inconsistency in the appraisement of ISV. A UK government decision not to introduce childproof drug containers implied an estimated value of a life saved at less than US$5000, while the same government decided to change a building code that implied a valuation of US$50 million per life saved. Estimated costs per person year of life saved may vary among specific public health interventions: annual mammography for women aged 40–49 is estimated to cost US$62,000 per life year saved, compared to US$2700 for a program of mammography every 3 years for women aged 60–65. A smoking cessation advice program for men aged 50–54 yields a cost-effectiveness of US$990 for a year of life saved (Brownson et al., 1998). The question of routine mammograms in reducing cancer mortality has been questioned and the issue is again being reviewed, but until better screening methods are developed it continues to be recommended by leading professional organizations. Smoking cessation has a demonstrated strongly positive benefit-to-cost ratio. Economic analyses such as the ISV are part of decision making in health systems management.

Early economists valued life in terms of loss of net output to society, or the future loss of earnings minus the future loss in consumption resulting from the premature death of an individual. This human capital method is still widely used because of the simplicity of its calculations. However, it does not take into account the grief of the family. It places a negative value on the life of a pensioner who is no longer a worker and "producer" in society, and gives no value to work done in the household, such as cooking, home maintenance, and rearing children. Nor does it give value to the intangible social and psychological benefits of the multi-generational family for all its members.

Another approach to valuation of life is based on court awards for compensation. It is a highly subjective method, often based on the court's interpretation and judgment of degree of contributory negligence, such as whether the injured person in a car accident was wearing a seat belt at the time, or may be based on assumed earning power of the individual in the years of work lost.

A major method is the willingness-to-pay approach, where valuations of life are based on what individuals are willing to pay for reductions in their probability of dying. For example, how much would people pay for new car tires, or how much extra would they pay in order to travel on an airline with a better safety record? How much would a patient be ready to pay above his or her insurance coverage to have a world-famous surgeon operate on him or her as opposed to accepting the surgeon available within the health service? Such measurement is difficult to perform and is often based on asking questions about hypothetical situations. Answers are also influenced by the income level of the respondent, their attitude towards risk, and the probability of death.

The issue is not only theoretical. If the cost to prevent HIV transmission to newborns is US$3000, and the number of cases of HIV-positive pregnant women who may transmit the virus in a developing country is such that a very large part of the national budget for health may go to this purpose, while there are insufficient funds for basic immunization, then choices need to be made, and they may be painful

---

BOX 11.6 Implicit Social Value of Life

The implicit social value (ISV) of life is summarized in the following equation:

$$ISV = \frac{Sum\ of\ costs - Sum\ of\ benefits}{Sum\ of\ life\ years\ saved}$$

The World Health Organization (WHO) has a rule of thumb for the ISV of a human life: three times the average per-person income per quality-adjusted life year (QALY) gained is a cost-effective intervention. In the USA, per-person income is about US$40,000, so an intervention that costs less than US$120,000 per QALY would be considered cost-effective according to the WHO rule. A Harvard economist has suggested US$100,000 as a reasonable value.

If a doctor prescribes a beta-blocker for a high-risk patient after a heart attack, it costs about US$5000 to buy that person one QALY. If a doctor gives a patient with HIV combination antiretroviral therapy, it costs US$20,000 to buy one QALY. Dialysis for end-stage kidney failure costs US$50,000–60,000 per QALY, which is still good value in the USA (Weinstein, 2010).

Sources: Salonen T, Reina T, Oksa H, Rissanen P, Pasternack A. Alternative strategies to evaluate the cost-effectiveness of peritoneal dialysis and hemodialysis. Int Urol Nephrol 2007;39:289–98.
Harvard Public Health. Can cost effective = better health care? Interview with Harvey Weinstein, Henry J. Kaiser Professor of Health Policy and Management. Harvard School of Public Health; Winter 2010. Available at: http://www.hsph.harvard.edu/news/magazine/winter-2010/winter-10assessment.html [Accessed 28 November 2012].

ones. All societies must make choices in priorities and in allocation of resources. Choosing to build large superhighways and neglecting public mass transit is a decision which assumes certain social values, but will cost lives and health because of downstream effects such as increased pollution, motor vehicle injuries, and deaths.

HEALTH FINANCING: THE MACROECONOMIC LEVEL

Financing health care has evolved from personal payment at the time of service delivery to financing through health insurance (prepayment) by the employer and employee at the workplace. This has progressed in most industrialized countries towards governmental financing through social security or general taxation, supplemented by private and non-governmental organizations (NGOs) (Table 11.3), and personal out-of-pocket expenditures. Ultimately, every country faces the need for governmental funding of health care either for the total population or at least for vulnerable groups such as the elderly and the poor, as in the USA, where governmental funding comes to nearly 50 percent of total health expenditures. Government funding is necessary also for services that insurance plans avoid or are inefficient in reaching, including as community-oriented services and groups at special risk, such as infants and women (see Chapter 13).

Health financing involves not only methods of raising money for health care, but also allocation of those funds. National health expenditures are derived from government and non-government sources and are used to finance a wide array of programs and services. There is competition for funds in any system, and the way in which money is allocated affects not only the way the services are provided but also setting of priorities, as indicated in the "laws" of health economics in Box 11.7.

The economic consequences of decisions made in resource allocation are major determinants of health care economics. Each country has to cope with similar issues in reforms to adjust for changing health needs and the economic results of previous decisions (see Chapter 13). A comparison of total national health expenditures is seen in Table 11.4. The USA has consistently been the highest spender on health care, but succeeded in reducing the rate of cost increase in the 1990s. Canada also experienced high rates of cost increase in health during the 1970s and 1980s, but managed to reduce the rate of increase and moved from

TABLE 11.3 Sources of Financing Health Services

Public	Private	International Aid
Federal, state and local government general revenues; mainly from taxes: income, excise, resources, business, inheritance, value added, capital gains, property, special taxes	Private health insurance	United Nations affiliates
Social security payroll tax	Personal expenditures	Foundations
Compulsory health insurance	Private donations, bequests	Religious organizations
Lotteries	Private foundations	Other non-governmental organizations
Dedicated taxes: cigarettes, alcohol, gambling	Voluntary community service	World Bank
	User fees	Government bilateral aid

BOX 11.7 "Laws" of Health Resource Allocation

- *Sutton's law* – Willy Sutton was a bank robber and when asked by a reporter why he robbed banks, he replied: "Well, that's where the money is". This expression is used to indicate that health services emphasize those aspects which are better financed. If more funds are available for treatment services, and preventive care is relatively underfunded, then treatment will have greater emphasis than prevention.
- *Capone's law* – Al Capone, a well-known gangster, planning the division of Chicago among his colleagues, said: "You take the north side and I'll take the south side", i.e., let's divide things up according to our mutual interest. This expression in the health context is taken to mean that planning may reflect interests of providers, as opposed to that of the general public. An alternative use of the concept is that macroeconomics planning may serve a general interest at the expense of the individual patient.
- *Roemer's law* – "Hospital beds, once built and insured, will be filled". The supply of hospital beds is a key determinant of utilization, especially where the public has health insurance benefits covering hospitalization. This "law" was modified by the experience of changing payment systems with incentives to reduce utilization. Following the introduction of the diagnosis-related group (DRG) method of payment in the USA in the 1980s, there was a reduction in hospital bed supply and occupancy. Incentives to control both hospital bed supply and utilization are crucial elements of health planning in most industrialized countries.
- *Bunker's law* – "More surgeons; more surgery". A greater supply of surgeons generates more surgery. This has also been modified as managed care and gatekeeper functions limit referrals and self-referral to specialists, and as professional organizations and governments limit training positions and licensing for such specialists.

the second leading country in per capita health expenditures to fourth place after the USA, Germany, and France.

Health care expenditure involves money spent from all sources for the entire health sector, regardless of who operates or provides the services. The methods of financing health care include tax supported, social security supported, employer–employee financed, charitable organizations, or consumer payment at the time of service. The total of expenditures for health care and how those funds are spent are the most fundamental issues in health economics and planning. Allocation of resources requires a skillful planning process to balance spending on different subsectors of the system and to ensure equity between regions and various socioeconomic groups in society.

What is the "right" amount of health care financing? This is a political decision which reflects the social and economic value placed on health by a nation. These attitudes affect such issues as how well medical and other health care staff are paid in comparison to other professions, and the supply of physical and human resources for health care in a given society. Virtually all developed countries have recognized the importance of national health and the role of financing systems to make health care universally available. Some basic principles and recommendations for successful health care financing policies are outlined in Box 11.8.

These rules are not absolute and solutions vary from country to country, as discussed in Chapter 13, and they change as experience, technology, science, and resources change. But it is important to stress that the system of financing greatly affects the services provided. The UK continues to operate its NHS at a relatively low percentage of GDP, as does Japan (Table 11.4). There are large differences in levels of expenditure on health between countries. In the established market economies, on average 9.3 percent of GDP goes to health, while the former socialist economies expend 3.6 percent, and developing countries generally under 4.5 percent.

Per capita health expenditures also vary widely. The total per capita expenditure on health, whether as a percentage of GDP or as dollars per capita, does not reflect the efficiency with which the resources are used. Many countries not only have low overall levels of health expenditure but also allocate those meager resources inefficiently.

Regardless of how efficiently money is allocated, countries spending less than 4 percent of GNP on health will have poorly developed health care. Those spending between 4 and 5 percent of GNP may try to have universal coverage, but often achieve this through low staff salaries, inadequate equipment, and spreading limited resources too thinly. Problems are

BOX 11.8 Health Care Financing

- Incentives for improving health performance measures, including:
  - preventive health measures
  - health promotion
- Universal coverage through social security or tax-based system
- Financing within national means for social benefits
- Adequate overall financing (>6 percent GNP)
- Shift from supply-side planning to costs per capita
- Performance or output measures
- Categorical grants to promote national objectives and specific health target programs
- Increase financing at national, state, and local government levels (7–10 percent GNP)
- Health insurance as a supplement
- Define "basket of services" and consumer rights
- Reduce acute care beds to <3 per 1000 population
- District health authorities with capitation funding
- Disincentives for excess hospitalization, surgery
- Incentives for integration of services

Note: *GNP=gross national product.*

TABLE 11.4 Total Health Expenditures as a Percentage of Gross Domestic Product and Per Capita Health Expenditures in Dollars, by Selected Countries: Selected Years, 1970–2011

Country	1970	1980	1990	2000	2005	2006	2007	2008	2009	2010	2011
USA	7.1	9.0	12.4	13.7	15.7	15.8	16.0	16.4	17.4	18	18
France	5.4	7.0	8.4	10.1	11.1	11.0	11.0	11.1	11.8	12	12
Germany	6.0	8.4	8.3	10.3	10.7	10.6	10.5	10.7	11.6	12	11
Canada	6.9	7.0	8.9	8.8	9.8	10.0	10.0	10.3	11.4	11	11
Sweden	6.8	8.9	8.2	8.2	9.1	8.9	8.9	9.2	10.0	10	9
Denmark	–	8.9	8.3	8.7	9.8	9.9	10.0	10.3	11.5	11	11
UK	4.5	5.6	5.9	7.0	8.2	8.5	8.4	8.8	9.8	10	9
Japan	4.5	6.4	5.9	7.7	8.2	8.2	8.2	8.5	–	9	9

Note: Data for 2010 and 2011 are rounded as from WHO Global Health Expenditure Data. NHA Indicators. Available at: http://apps.who.int/nha/database/DataExplorer.aspx?ws=1&d=1 (accessed 2.1.2014)
Source: National Center for Health Statistics. Health, United States, 2011: with special feature on socioeconomic status and health. Hyattsville, MD: NCHS; 2012. Available at: http://www.cdc.gov/nchs/hus/contents2011.htm#124 [Accessed 7 November 2012].

accentuated when a disproportionately large hospital system and excessive supply of physicians create a siphoning effect on health care spending, or when resources are concentrated in cities although most of the population is rural.

Developed countries that spend between 8 and 18 percent of GNP on health care have made a value judgment. They have placed health care among the vital priorities in their societies. In those countries with high health care expenditures, such as the USA, physicians' incomes are very high, even when compared with other highly paid professionals. Where financing is centralized in a single paying agency, administrative costs are less than in countries with multiple funding sources. Canada's provincial health insurance plans operate with administrative overheads of less than 5 percent, compared to some 30 percent in US private health insurance.

The WHO issued a Global Strategy for Health Development, which stressed the importance of efficiency in use of resources as a vital element of health development. The WHO recommends preferential allocation to primary and intermediate care services, especially for currently underserved rural populations. In most countries, reallocation of resources is necessary to strengthen primary care and to adopt new technology and health programs, shown to be cost-effective in terms of costs as well as anticipated benefits.

Where there are multiple sources of health financing, it is more difficult to develop effective national planning. Regulation and supplemental funding by government are needed and required to prevent inequity between socioeconomic groups and between urban and rural populations. When multiple agencies are involved in health insurance or direct government granting systems for specific services, there are gaps (inadequate coverage or access) in services, usually for politically, geographically, and socially disadvantaged sectors of the population, who may have the greatest needs. Under such circumstances, public health services very often become oriented to provision of basic services for people excluded from health benefits because of lack of health insurance. This places a great financial burden on public health services, which are generally underfunded in comparison to clinical services. Such countries often bring in national health insurance for the disadvantaged groups (e.g., the elderly and the poor). These insurance plans may pay less well than private insurance for the middle class and organized workers. This situation applies to the USA and to many mid-level developing countries (see Chapter 13). The USA addresses this issue by promotion of national (and state) health targets, guidelines, accreditation systems, and strong professional organizations and medical centers with high levels of research capacity, but still lacks universal coverage for health care. However, the USA provides some US$100 billion in "uncompensated care" for the poor and large families, and these costs are built into the budgets. Hence, this care is subsidized by the insured person.

Where financing of health care is centralized, a potential exists for rational allocation of resources. But this depends on adequacy of total financing and rational allocation policies to promote equitable access to services and a balance between one service sector and another. Allocation of monies within the total health expenditures means selection from many alternatives. Misallocation of resources between sectors within the health sector can lead to a wasteful and even counterproductive health system, such as excessive funding of tertiary care while primary care is lacking (Box 11.9).

Where funds are allocated to regional or local health authorities, the potential for shifting resources to meet local needs should be greater. But this may be limited by lack of data or lack of analysis on a local or district basis to highlight priority areas of need. Where there is a highly decentralized management system, some centralized functions are essential to promote national health needs and equity between regions of the country. These functions include setting policy and standards, monitoring health status indicators, and determining health targets with funding to promote national priorities.

BOX 11.9 Major Categories of Health Expenditures

- *Institutional care* – teaching hospitals, general hospitals, mental and other special hospitals, long-term nursing care, residential care, hospices
- *Pharmaceuticals and vaccines*
- *Ambulatory care* – primary care, family practice, pediatric, prenatal, and medical specialist; medical, diagnostic, and treatment; ambulatory and day hospital clinics; surgical, medical, geriatric, dialysis, mental, oncological, drug and alcohol treatment
- *Home care*
- *Elderly support activity/service centers*
- *Categorical programs* – immunization, maternal and child health, family planning, mental health, STIs, HIV, tuberculosis, screening for birth defects, cancer, diabetes, hypertension
- *Dental health*
- *Community health activities* – healthy communities, health promotion in the community for risk groups; smoking restriction, promotion of physical fitness and healthy diet; environmental and occupational health; nutrition and food safety, safe water supplies, special groups
- *Research*
- *Professional education and training*

Note: *STI = sexually transmitted infection; HIV = human immunodeficiency virus.*
Sources: Testimony of A. Bruce Steinwald before the Subcommittee on Health, Committee on Energy and Commerce, House of Representatives; 2006.
US Government Accountability Office. Medicare physician payments: trends in service utilization, spending, and fees prompt consideration of alternative payment approaches. Statement of Director, Health Care. Available at: http://www.gao.gov/new.items/d061008t.pdf [Accessed 7 November 2012].

The range of services or programs requiring funding for a population group is indicated in Box 11.9. Medicaid includes those enrolled in the State Children's Health Insurance Program (SCHIP). Medicaid and SCHIP covered 37.5 million low-income individuals (12.9 percent of the US population), primarily children, pregnant women, elderly, and disabled people. Medicare, which insures the elderly, the disabled, and patients with end-stage renal disease. Both Medicare and Medicaid are parts of the Social Security Act (SSA) Amendments of 1965, Medicare (SSA Title XVIII) is under federal administration, while Medicaid (SSA Title XIX) is shared among federal, state, and local administration. Figure 11.4 shows the breakdown of health insurance coverage in the USA. In 2010, Medicare covered 47.5 million Americans; in 2012, Medicare covered 16 percent of the US population. In 2011 more than 48.6 million Americans lack health insurance, and another 16 million have low insurance coverage.

The US Institute of Medicine reports that lack of adequate insurance is the cause of an estimated 20,000 excess deaths per year in the USA (IOM, 2008). The Commonwealth Fund reports that the USA is the highest spender on health care but ranks below many other nations in terms of health indicators (Commonwealth Fund, 2012; OECD, 2012, 2013).

COSTS OF ILLNESS

Direct expenditures for health care in the USA by type of illness are measured in periodic National Medical Expenditure Surveys of the civilian, non-institutionalized population, covering thousands of people and homes for self-reported expenditures. The largest items of health care expenditure are for cardiovascular disease, followed by injury, then neoplasms.

Following especially high rates of increase in health expenditures during the 1980s in the USA, measures were taken to restrain growth in health care costs, leading to a slower rate of increase. In part this was due to growth of managed care and incentives for lower hospital utilization, and shifts in payment schedules for hospital and ambulatory

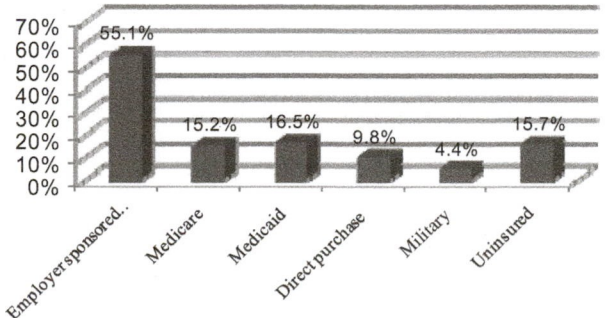

FIGURE 11.4   Sources of insurance coverage in the USA, 2011. *Source: ASPE Issue Brief. Overview of the uninsured in the United States: a summary of the 2012 current population survey report. Available at: http://aspe.hhs.gov/health/reports/2012/UninsuredInTheUS/ib.pdf [Accessed 7 November 2012].*

care. In the USA, out-of-pocket expenditures as a percentage of total health expenditures declined from 55 percent in 1960 to 15 percent in 2005, while private health insurance increased from 21 percent in 1960 to 36 percent in 2005; government expenditures for health grew from 24 percent in 1960 to 45 percent (34 percent federal and 11 percent state and local) in 2005 (Health, United States, 2007, Table 125).

From 1990 to 2004, acute care hospital beds in the USA declined from 3.7 to 2.7 per 1000 population, while occupancy rates were stable at 67 percent. The rate of growth of health expenditures slowed from annual increases of 9.1 percent in 2002 to 6.6 percent in 2005; however, health expenditures are expected to reach 20 percent of GDP by 2015. The economic effect of adoption of any possible universal health plan in the coming decade will be difficult to forecast, but it will not be entirely additional to current expenditures, as much episodic health care is provided by hospitals to the uninsured.

Costs and Variations in Medical Practice

Increasing costs of health care, waste, variations, and fraud in medical practice inevitably come under scrutiny whether prepayment is in the private or in the public sector. Variations due to the different needs of various population groups may be justified. However, if an epidemiological analysis reveals no apparent reasons for the variations, then they become administrative problems that require other approaches. Comparing the quantity and quality of services between population groups is part of epidemiological and administrative health practice. This approach, when supported by review of relevant current literature on methods of treatment, provides a basis for what is termed *evidence-based medical practice*.

Analysis of medical practice by examination of medical and hospitalization data may show quite startling differences between different cities, regions, and countries. What has come to be called "small area analysis" looks at patterns of practice and tries to determine what may be the cause of such differences. For example, no evidence exists of benefit from higher rates of some types of surgery, such as hysterectomy, cholecystectomy, and tonsillectomy. Furthermore, there is a cost attached to a surgical procedure that includes a certain mortality rate from anesthetic mishaps and other iatrogenic complications; that is, caused by medical care itself. For example, cholecystectomy rates in the early 1990s in Canada were 600 per 100,000 population, 370 in the USA, and 122 in the UK. These studies concluded that an excess supply of surgeons and the fee-for-service type of payment may lead to an excess of unnecessary and potentially harmful surgical procedures (Bunker's law). The cost implications for a health care system are high and can be calculated. Since that time, reduced hospital bed capacity, and increased oversight within hospitals and externally by insurers have resulted in reductions in these procedures, which are no longer regarded as routine operations.

In the USA, health maintenance organizations (HMOs) have shown the capacity to provide comprehensive care over long periods to large population groups with relatively low hospital utilization. HMOs and for-profit managed care coverage increased dramatically in the 1990s. For the period between 1993 and 1994, hospital admissions, average length of hospital stay, and days of care for HMO members (non-Medicare) as well as for Medicare beneficiaries, were well below comparable rates for fee-for-service-based insurance plans.

Technological innovations using simpler, less costly, less invasive, and less risky procedures have led to important changes in health care standards. Continuous evaluation of criteria for "good practice" leads to change based on new knowledge, experience, consensus or leading opinion, and meta-analysis, and is essential to quality promotion in health care (see Chapter 15).

Cost Containment

High public and professional expectations from health care can, along with increasing demands of an aging population, costly medical technology, and oversupply of highly technological medical services, lead to a rapid rise in health care costs and a fiscal crisis in many countries. Cost containment became important as the costs in all health systems increased at rates well above economic growth during the 1970s and 1980s. Governments everywhere sought ways to restrain cost increases. Cost-effectiveness and cost–benefit analyses have become a part of the planning and management review of ongoing or new interventions in health for both operational and capital expenditures as critical tools of health service planning for rational decision making to restrain health cost increases. Because hospitals are the major consumers of health care expenditures (between 40 and 60 percent in different countries), the emphasis on cost containment has been placed on reducing hospital utilization and developing alternative services or programs of ambulatory and community care.

Cost containment and high-quality health care can coexist. Indeed, cost-containment measures (Table 11.5) are associated with greater precision in care and more appropriate use of resources than previous patterns of care. Some measures relate to substitution of lower cost care for more costly services, such as home care for acute hospital care. Others relate to changes in professional services; for example, outpatient surgery in place of inpatient care, and shorter hospital length of stay following myocardial infarction.

Countries with public funding of health care systems are especially concerned with establishing cost containment to reduce the rate of increase in health costs. In Canada, governments have shifted their concern from assurance of access to care to cost containment.

TABLE 11.5 Example Health Service Programs Promoting Cost Containment

Program	Mode of Operation
Home care	Reduces length of stay following medical or surgical hospital treatment; reduces the incidence of nosocomial (hospital-acquired) infections; helps elderly or chronically ill to remain at home rather than enter a long-term care facility
Long-term care facilities	For people unable to be cared for in the family setting, reduce length of hospital stay
Regulatory limitations	Limit supply of beds; limit medical services
Ambulatory or day care surgical, medical, mental	Reduces stay in hospital, with fewer secondary infections and iatrogenic complications
Prevention	Primary, secondary, and tertiary prevention reduces hospitalization for vaccine-preventable disease, cardiovascular disease, diabetes, and their complications
Environmental health	Chlorination of community water supplies prevents diarrheal diseases and hospitalizations; fluoridation reduces dental disease
Road safety measures	Improved emergency care and transportation; interventions to reduce trauma from motor vehicle accidents; restriction of smoking leads to less lung cancer and coronary heart disease
Diagnosis-related groups	Payment promotes reduced length of hospital stay
Health maintenance organizations and managed care organizations	Promote alternatives to hospitalization and long stays, lowers hospital utilization; incentives for employers, employees, and governments to enroll beneficiaries in less costly managed care organizations; capitation provides incentive to prevent illness and institutional care, strengthen ambulatory and preventive care
District health system or regionalization	Promotes rationalization of services, elimination of excess facilities and duplication; promotes greater community orientation in service complex

Health expenditure grew by 8-10 percent annually in the 1980s in Canada, well above the growth of the economy. Canada's cost-containment approaches include controls on fees, regionalization to reduce duplication, excess hospital supply and utilization, and increased oversight in utilization of medical care. During the 1990s, these measures succeeded in slowing the rate of cost increase (see Chapter 13).

MEDICAL AND HOSPITAL CARE: MICROECONOMICS

Resource allocation policy, made at the national, regional, health insurance, or sick fund level, must address many specific factors affecting the way services are provided and paid for. Incentives and disincentives for efficient care include how doctors and hospitals are paid, and how services are organized. Payment for doctor's services includes fee-for-service, case payment, capitation, salary, or a combination of these methods. Each has its historical roots, its advantages and disadvantages, and its proponents and opponents.

Payment for Doctors' Services

Fee-for-service is payment for each unit of service, such as a visit or surgical procedure. Payment for a complete service covering the whole period of an illness or another type of care such as obstetric care including prenatal care and delivery, or other, is called case payment. Fee-for-service is historically the common method of paying for doctors' services and is still the norm in Canada, Germany, and other countries. In some places, payment may be according to a fixed-fee schedule negotiated between the insurance mechanisms, whether public or private, and the doctors' representatives. Fee schedules are often weighted towards medical specialists, who have greater prestige than primary care physicians.

Fee-for-service tends to promote an overabundance of the more expensive kinds of care, including surgery, often without real need. This is especially so when the patient is fully covered by health insurance and is therefore better able to pay for the service than the person without insurance. Some insurance systems require participation of the user in the co-payment or user fees or charges. This is often promoted by the idea that it restrains the consumer from seeking unnecessary care, as well as helping to cover costs, while opponents justly reply that user fees affect the poorer sector of any population disproportionately and discourage preventive care.

Capitation is payment of doctors by a fixed sum of money for the individual registered for care for a specified period. This can apply to a comprehensive health service, as in MCOs, as well as to general practitioner services, as

in the UK. Compared with salaried service, this method allows a greater degree of personal identification of the patient with the doctor. It has been in use in the UK since the introduction of national health insurance in 1911. The recent introduction of incentive fees for full immunization or screening programs has improved performance in these areas.

The UK's budget holder system initiated in the late 1980s, which pays a group of general practitioners for their registered patients, has been replaced by per capita funding of primary care trusts. Hospitals are being increasingly paid on a DRG basis.

Salary payments for doctors and other health workers are common in hospitals even where fee-for-service or capitation is the prominent method of payment. This system has advantages for the physician in predictability of income, with less incentive to promote unnecessary services. Salary payment may be combined with incentive payments for additional services.

Pay-for-performance is a relatively new approach, which has been implemented in the UK NHS since 2004. It is meant to provide incentives for doctors to focus more on preventive care by measuring 146 quality indicators resulting in increases in existing income according to performance with respect to covering clinical care for 10 chronic diseases, organization of care, and patient experience. This approach is also being promoted in the USA in HMOs and accountable care organizations (ACOs) coming into being under the PPACA (Obamacare) plan, being implemented among other objectives to promote preventive care practices.

The method of payment for doctors has an important impact on the way in which medical services are used. Empirical evidence indicates that fee-for-service promotes excessive use of the system, including unnecessary surgical procedures, while salaried services are often criticized for diminished identification with patients and, perhaps, underservicing. Mixed systems of payment are increasingly emerging, with capitation as a predominant method.

Payment for Comprehensive Care

Per capita budgeting is a system of payment based on a defined population registered for care with a specific health service system providing a comprehensive range of services, such as a district health system or a managed care organization (MCO). Capitation payment covers responsibility for total care, so that economies in hospital care can be applied to cost-effective alternatives such as strong ambulatory care, home care, and long-term care. The population may be enrolled either on a voluntary basis, as in HMOs, prepaid group practice systems and managed care systems, or on a geographic basis, as in regional or district health systems.

In some financing systems, the per capita payment takes into account the age and gender distribution of the region, locality, or registered population. It applies national hospital utilization rates for different categories by age and gender. The capitation method provides an incentive against unnecessary admissions and decreases length of hospital stay, but it is not in a hospital's best interest to discharge a patient prematurely because of the potential for litigation and because the patient may later return in need of more care, adversely affecting hospital costs.

Capitation values may be adjusted by applying regional standard mortality rates (SMRs), as in the UK, to account for age, gender, and morbidity differences. The UK NHS is paying many of its general practitioners by a combination of capitation and DRG systems, as discussed later in this chapter.

HEALTH MAINTENANCE AND MANAGED CARE ORGANIZATIONS

HMOs are integrated health insurance and provider systems, responsible for hospital, ambulatory, and preventive care for an enrolled population. The HMO is a system of prepaid health care in which the insured person joins or becomes an enrolled member of a health plan that has received a fixed per capita payment from the insurer to provide comprehensive health care for a defined period. This approach, which was developed in the USA, creates non-profit organizations sponsored by industry, unions, and cooperative groups. Formerly called prepaid group practice, these plans were developed by Kaiser Permanente in California during World War II and later in many other parts of the country.

Since the 1973 HMO Act, the HMO has become part of the accepted mainstream of health care in the USA. Some large HMOs operate their own hospitals, utilizing 1.5 beds per 1000 population, well below US averages, even when adjusting for age and selection factors. They operate with 1.2 doctors per 1000 enrollees, compared to 4.5 per 1000 for fee-for-service health care systems. Doctors working in HMOs may be paid by salary or capitation in a staff and group HMO or on a fee-for-service basis in an independent practice association (IPA) or a preferred provider organization (PPO) (Box 11.10).

Health care in the USA has been influenced by the HMO experience and that of other health insurers using HMO-like cost-control measures which limit unrestricted fee-for-service practice. The HMO or managed care approach to health care organization is less costly, largely because of better management of patients in the community and lower hospital utilization patterns.

The major increase in enrollment in managed care took place in the 1990s, much of it in for-profit managed care. Managed care was successful in taking a large part of the market share of health insurance because of its advantages of lesser cost and more comprehensive coverage than traditional fee-for-service health insurance.

BOX 11.10 Managed Care Organization Models

Managed care plans are health insurance plans that contract with health care providers and medical facilities to provide care at reduced costs. They provide a network of services and are responsible for the quality of care and comprehensiveness of services according to the contract with the insured people.

- *Health maintenance organization (HMO)* – a health system providing insurance and service to enrolled members. The traditional group model HMO is based on the prepaid group practice in which the HMO employs or contracts with physician groups to provide comprehensive care. Payment is on a capitation payment basis to enrolled members, usually in health centers operated by the HMO and in hospitals owned or contracted with the HMO. Group HMOs may be partnerships that share in the incentive payments. Staff model HMOs are plans which employ physicians and other providers in HMO-owned facilities. Network model HMOs contract with multiple physician groups including single or multi-specialty medical groups.
- *Preferred provider organization (PPO)* – a formally organized entity, usually of physicians, hospitals, pharmacies, laboratories, or other providers, which contracts to provide care to HMO members on an agreed (discounted) fee schedule or capitation basis. Each provider works independently but agrees to contracted conditions, including utilization review. The beneficiary has a choice of providers within the panel.
- *Individual practice association (IPA)* – providers may include individual practice physicians who contract to provide services to HMOs, and may also provide services to members of other health insurance plans.
- *Point of service (POS) plan* – this type of plan allows a choice of HMO or PPO services at any time.
- *Accountable care organization (ACO)* – a group of health care providers organized to give coordinated care and chronic disease management, and thereby improve the quality of care that patients receive. The organization's payment is tied to achieving health care quality goals and outcomes that result in cost savings.

Sources: *US National Library of Medicine and National Institutes of Health. Managed care. Available at: http://www.nlm.nih.gov/medlineplus/managedcare.html [Accessed 12 July 2013]*
National Center for Health Statistics. Health, United States, 2007. Available at: http://www.cdc.gov/nchs/data/hus/hus07.pdf [Accessed 12 July 2013].
Healthcare.gov. Accountable care organizations. Posted 30 November 2012. Available at: http://www.healthcare.gov/glossary/a/accountable.html [Accessed 12 July 2013].

Managed care health plans undertake responsibility for the comprehensive care of enrolled members. Managed care systems are being promoted by private employers, by insurance companies, by states for Medicaid beneficiaries, and by the federal government PPOs.

Since the 1970s, managed care has become the predominant form of health care in most parts of the USA.

More than 70 million Americans are enrolled in HMOs and almost 90 million are part of PPOs. Enrollment in HMOs peaked in 2001 and has declined substantially since, but managed care remains a dominant type of health care and coverage. Medicaid managed care grew rapidly in the 1990s. In 1991, 2.7 million beneficiaries were enrolled in some form of managed care, and by 2004, that number had grown to 27 million. Of the total Medicaid enrollment in the USA in 2005, some 63 percent receive Medicaid benefits through managed care. All states (except for Alaska, New Hampshire, and Wyoming) have all, or a portion of their Medicaid population enrolled in a MCO. States can make managed care enrollment voluntary, or require certain populations to enroll in a MCO. For 2006, the breakdown of enrollment by plan type was as follows: 20 percent HMO, 60 percent PPO, 13 percent point of service (POS) providers, 4 percent high-deductible health plan (HDHP), and 3 percent conventional indemnity plans. The US Health Care Financing Administration (HCFA) regulates HMOs and has instituted guidelines for reporting and quality assessment in an accreditation approach to quality assurance (see Chapter 15). There has been some backlash against managed care, with negative publicity regarding restrictions in referrals and other client concerns.

Managed care, especially in the for-profit sector, is under criticism in medical and public health organizations and journal editorials, as well as in the media and state and federal legislatures. It is alleged that the system promotes denial of access to specialists and other needed care because of the economic incentives built into the capitation system, especially when administered by for-profit companies. The economic benefits are generally accepted. The controversy focuses on the incentives to underservice and on loss of choice by the consumer in for-profit managed care systems. The quality and ethical issues of managed care are discussed further in Chapter 15. Legislative efforts at state and federal levels to define patients' rights, grievance procedures, and minimum baskets of service have been under way in Congress, with a narrow (50–47) defeat in late 1998. President Bill Clinton then actively promoted a Patient's Bill of Rights which was strongly opposed by the health insurance lobby, but contributed to the concepts of Obamacare introduced in 2010.

Opponents of the managed care approach argue that lower HMO hospital utilization may in part be attributable to lower costs due to a bias by selection of healthy members, and that HMOs may underservice patients in order to reduce costs, or increase physician incomes or profits.

Available evidence supports HMO experience as providing high-quality medical care at lower cost than competing open-ended, fee-for-service insurance systems. The leveling off of expenditure for health in the USA during the 1990s is largely attributable to the move from fee-for-service care plans to managed care of a large percentage of the population. Managed care is also emerging in other countries, in the Sick Funds in Israel and in some European countries, in Latin America (Argentina, Brazil, Mexico, Chile, Peru, and others), as well as in the Philippines, all seeking to restrain cost increases while extending health care to a greater part of their populations.

Under the PPPACA (Obamacare), ACOs are to be established within the Medicare program encouraging primary care physicians to join together with other providers, such as hospitals, taking responsibility for the full continuum of their primary care patients' care. They must commit to reporting comprehensive measures of the quality and outcomes of care. The ACOs receive additional payments for demonstrably improved quality of care and reduced overall costs, as a share of the savings achieved. They are expected to provide a "medical home" for registered patients, strengthening primary care and improving care coordination (*Dartmouth Atlas*, 2012). An example is the Beth Israel–Deaconess Hospital initiative sponsored by the Center for Medicare and Medicaid Services (CMS) Innovation Center, which provides Medicare beneficiaries with higher quality care, while reducing growth in Medicare expenditures through enhanced care coordination. The Harvard University-affiliated Beth Israel–Deaconess Physician Organization in Boston has more than 1700 providers including over 1300 specialists and 400 primary care physicians. This organization provides integrated primary care as well as secondary and highly specialized services to its enrolled population. It is defined as an ACO with links to community hospitals in the Boston area. It provides services to managed care with incentives for efficient and high-quality services to private insured populations as well as assisting participating medical providers with administrative help.

DISTRICT HEALTH SYSTEMS

In the UK and the Scandinavian countries, a comprehensive service model has existed in the form of district health systems for many years. The residents of a district have their health benefits provided by or contracted out by the district. In principle, the geographic unit of service allows for efficiency in transfer of resources and patients from one service to another, based on need, and not on the financial interests of the insurance system or the provider.

The Scandinavian countries have a long tradition of management of health facilities at the county level with budgets derived from a combination of local taxation and national grants. In reforms since the 1980s, integration of various services into district health systems with reduced hospital bed supplies has resulted in a leveling off of cost increases for health.

In the UK, methods of budget allocation to health regions were the subject of a long and detailed study by the Regional Allocation Working Party (RAWP) in the 1980s. The decision was made that the optimal method of allocating funds to district health authorities would be by per capita grants with adjustment by the SMRs of the district. This adjustment takes into account age and morbidity differences between different districts and promotes an equitable approach to resource allocation. Reforms since the 1990s have focused on building workforce and physical capacity, and issues such as waiting times, which have fallen, clinical outcomes for cancer and heart disease, which have improved, and NHS facilities, which have been modernized. Since 2003, a new stage of reform has promised more choice for patients: more freedom to innovate and improve services; competition on quality with financial incentives to improve care and promote sound financial management; and national standards and regulation to guarantee quality, safety, and equity, along with improved information management and technology to support reforms and deliver better, safer care.

Health reform in some provinces in Canada includes reducing the per capita hospital bed supply along with regionalization and integration of services in regional or district health systems. Many provinces have adopted district health boards, which amalgamated hospital, nursing home, and public health boards. Per capita funding allows transfer of funds from hospital care to other sector services such as home and community care. The provinces have managed to level off health expenditure increases to rates less than the growth of GDP.

Regionalization of hospital and other services is another approach to rationalization of health care and cost control. In communities with excessive hospital beds and competing services, regionalization provides a method of rationalization, with voluntary or mandatory elimination of wasteful, competing departments or investigative units such as in vitro fertilization, cardiac surgery units, advanced imaging devices (e.g., MRI), or excess bed capacity. In the USA, efforts were made to regionalize certain services, such as with the introduction of perinatal care systems in the 1980s, but did not lead to wider application of this approach. In the 1990s, hospital networks in both the for-profit and not-for-profit sectors have expanded aggressively to increase market share and vertical integration for service and management cost-efficiency as part of the managed care dominance of the US health insurance market.

The UK health reforms initiated in 2012 (Health and Social Care Act 2012) place more responsibility for public health with the local authorities, which will employ a director of public health, and will be supported by a block budget. The act requires directors of public health to publish annual reports that can chart local progress. The idea is that many of the wider determinants of health (e.g., housing, economic development, transport) are more easily impacted by local authorities, with responsibility for improving the local area for their populations. This reform is discussed in Chapter 13.

PAYING FOR HOSPITAL CARE

Hospitals are the most costly component of a health service. Traditionally, hospitals were paid on a per diem or flat rate per patient-day. The per diem may be determined by using actual costs or by national, state, or regional averages. The daily operating costs are divided by the number of beds, with possible adjustment for teaching or research functions. The per diem based on actual costs per patient in specific units in a hospital, such as intensive care, may be higher or lower than the budget provides for that specific service.

The per diem method of payment encourages long lengths of stay, rewards hospitals with low technology, and if based on national or regional averages may penalize hospitals with high levels of staffing and technology, such as teaching hospitals. When the service is insured, there is no financial incentive for shortening the patient's hospital stay. The per diem method is associated with inefficient use of facilities, such as admission to the hospital for diagnostic tests or prolonging a stay for additional testing or care that could be provided in alternative and less costly ways. The provider has an incentive to hospitalize and provide prolonged care to a relatively healthy patient, while the sicker patient is a financial liability, as are teaching and research functions, unless funded separately. This system lacks incentives to improve efficiency by developing alternative ambulatory or day care services, and it punishes more efficient hospitals which reduce length of stay or occupancy rates.

Fee-for-service payment for each service supplied in a hospital favors unnecessary marginal care, long lengths of stay, high admission rates, and the provision of duplicative or unnecessary services. This method was common in the USA with its multiple insurance systems but is increasingly being replaced by DRG payment (discussed later in this section). Fee-for-service payment provides incentives to overservice with no incentive to reduce costs or admissions of length of stay.

Historical budgeting is remuneration based on the previous year's budget, adjusted for inflation and the cost of new services. The budget may be reviewed line by line by the paying authority or be on a global or block budget basis, which frees the hospital to make internal reallocations within the overall allotment, but sometimes the funder may use a combination with some programs on a line-by-line system, especially in the case of new programs such as bone marrow transplants. Payment can include a capital fund for renovation. This method is often used when a hospital is directly operated by the Ministry of Health. As distinct from the per diem payment system, this method should theoretically provide some incentive to reduce length of stay and to search for efficiency in the use of hospital resources.

Payment by norms means financing according to nationally fixed standards of numbers of beds, staffing, and other measures. This method, as practiced in the Soviet health system, provided national incentives to maintain high hospital bed to population ratios, long length of hospital stay, little investment in improving ambulatory care, low salaries, and generally low quality of care. Reform in post-Soviet countries required cancellation of these historic norms, reducing excess hospital bed capacity, and adoption of incentives for efficiency in health care (see Chapter 13).

As a result of concern over high costs and utilization rates, alternative methods of payment have developed in the USA since the 1960s. The DRG system was adopted in 1983 by the US HCFA as the basis for payment for hospitalization of Medicare patients. The DRG system has been the basis for paying for hospital care in the USA since 1999, and it is increasingly being used in other industrialized countries, such as the UK and Israel, and some developing countries, such as the Philippines.

The DRG system is a prospective payment system for hospital care reimbursement, and the system pays the hospital according to 495 treatment classifications of diagnoses or procedures, each with a fixed hospital payment rate. This system provides an incentive to reduce length of stay, make more efficient use of diagnostic and treatment services, and reduce overall bed capacity. As a result, hospital outpatient services increased rapidly in the USA while bed occupancy rates and the ratio of hospital beds to population declined steadily over the 1990s. The DRG system does not lead to fewer admissions and may encourage falsification of diagnostic criteria or increasing the diagnostic severity of case definition to increase revenues ("DRG creep").

Different hospital budgeting methods have advantages and disadvantages. Payment by DRGs is most likely to promote rational use of hospital care. Regional budgets allocated on a per capita basis with hospital payment by DRGs may be the most effective way of achieving a balance between ambulatory and hospital care, combining regional equity and incentives for efficient use of diagnostic and treatment services. Prospective payment systems must be associated with quality assurance mechanisms, a vital issue in health management (see Chapter 15).

CAPITAL COSTS

The capital cost to build or renovate a health facility is based on long-term considerations but has important effects on current operating costs. The cost of operating a new health care facility may equal the capital cost in 2–3 years. Capital costs may be financed by public or private donations, risk-capital investment, or government-guaranteed loans. Government regulatory agencies may approve a capital project for construction or equipment of a hospital under a certificate of need (CON) procedure and then agree to a grant mechanism to provide funds to match local contributions or to budget or adjust rates to include repayment of long-term

loans for capital costs. This occurred both in the USA under the Hill–Burton Act (see Chapter 10) and in Canada under the National Health Insurance System. Where hospitals are operated independently of government, they may borrow or raise money privately through long-term bonds or low-interest loans. Repayment can be built into the operating costs and amortization of the loan over many years.

When government finances capital costs, it has greater control over the direction, distribution, and supply of hospital facilities. Government norms may encourage an increased bed supply by encouraging hospital construction, or maintenance of high numbers of beds that may not be used or may be of poor quality. Norms may also be used to set upper limits or provide incentives to reduce bed supply. One of the common elements of cost-containment strategies in many industrialized countries is a reduction in hospital bed supply, which is occurring without apparent harm to the quality of care. Hospital bed reduction is partly offset by transfer of long-stay patients to home care programs or to nursing homes with a transfer of capital and operating costs. Overall, maintaining quality of care is not compatible with supporting a large and costly number of beds to population ratio (the converse of the aforementioned Roemer's law, Box 11.7) because of the excessive resources required to maintain these beds at the expense of other needed services in the community.

Health expenditures by category of service in the USA are shown over the period 1960–2010 in Table 11.6. In 2005, hospital care accounted for about 30 percent of total expenditures, a reduction from nearly 40 percent in 1980. Physician and other clinical services were 21 percent of total expenditures, a small decline from 1990, prescription drugs 13 percent, dental care 4.4 percent, nursing homes just over 6 percent, governmental and private health insurance administration 7.2 percent, governmental public health activities 2.8 percent, research 2.0 percent, and structures and equipment 4.4 percent.

The USA spent far more per capita on health care in 2008 than any other developed country – US$7538 per person, compared to US$3129 in the UK, US$4079 in Canada, and US$5003 in Norway (Kaiser Family Foundation, 2008). This is in part due to high administrative costs associated with multiple insurance plans with different policies, networks, and reimbursement rates, and exclusions.

HOSPITAL SUPPLY, UTILIZATION, AND COSTS

Acute care hospital bed to population ratios in the USA increased from the 1940s to the 1980s and declined thereafter. The supply and utilization of hospital beds are changing as economic incentives increase pressure to find less costly forms of care, and as ambulatory and community-oriented care is perceived to be more effective in many instances. In

TABLE 11.6 National Health Expenditures, Percentage Distribution, USA, 1960–2010

	1960	1970	1980	1990	2000	2005	2008	2009	2010
NHE (billions US$)	28	75	255	724	1,377	2,029	2,403	2,495	2,593
Percentage Distribution									
Total NHE	100.0	100.0	100.0	100.0	100.0	100.0	100.0	100.0	100.0
Investment (structure and capital equipment)	6.8	7.8	5.7	5.0	4.5	4.2	6.4	5.9	5.7
Research	2.5	2.6	2.1	1.8	1.8	2.0	1.8	1.8	1.9
Personal Health Care									
Hospital care	32.8	36.3	39.3	34.6	30.2	30.0	30.3	31.1	31.4
Physician and clinical services	20.6	19.1	18.7	21.9	21.1	20.5	20.2	20.1	19.9
Dental care	7.3	6.3	5.2	4.4	4.6	4.3	4.3	4.1	4.0
Other personal care	3.6	2.7	2.7	3.8	5.6	5.8	4.7	4.9	5.0
Nursing homes	2.9	5.4	7.5	7.3	7.0	6.1	5.5	5.6	5.5
Home health care	0.2	0.3	0.9	1.8	2.3	2.4	2.6	2.6	2.7
Prescription drugs and medical products	18.0	14.0	10.1	10.4	12.6	13.0	13.4	13.4	13.2
Government administration and net cost of private health insurance	4.4	3.7	4.8	5.5	6.0	7.2	6.9	6.6	6.8
Public health (governmental)	1.5	1.9	2.5	2.8	3.2	2.8	3.0	3.1	3.2

Note: NHE = national health expenditures.
Source: National Center for Health Statistics. Health, United States, 2011. Tables 124 and 125. Available at: http://www.cdc.gov/nchs/data/hus/2011/125.pdf and http://www.cdc.gov/nchs/hus/contents2011.htm#124 [Accessed 7 November 2012].

the USA, hospital utilization, average length of stay, and percentage occupancy showed a decline mainly during the period 1980–2004. Hospital staff per 1000 patient days increased from 226 in 1960 to 583 in 1991, reflecting increased support and technical services and greater severity of illness of those hospitalized. Increased staffing, technological innovations, and expensive medications increased the cost of patient care in hospitals. Table 11.7 shows the trend in hospital bed supply and percentage occupancy in acute care non-federal general hospitals in the USA from 1970 to 2009.

There has been a trend towards decreasing the hospital bed supply and utilization in the USA during the 1980s, 1990s, and 2000s. Despite aging of the population, the trend towards lower overall hospital utilization has resulted from the following: changing morbidity patterns, ambulatory services in place of inpatient care, adoption of the DRG system of payment reducing length of stay, greater stress on health economics and cost containment in medical considerations, more efficient methods of care, greater health consciousness in the general population, and improved self-care and prevention.

TABLE 11.7 Acute Care Hospital Bed Supply and Utilization, USA, 1970–2009

Facilities	1970	1980	1990	2000	2005	2009
Beds per 1000 population	4.3	4.5	3.7	2.9	2.8	2.6
Discharges per 1000 population	NA	175	125	113	116	115[a]
Average length of stay	NA	7.5	6.5	4.9	4.8	4.8[a]
Total days of care per 1000 population	NA	1303	819	558	554	556
Percent occupancy	77	75	67	64	67	66

Note: Includes community hospitals; does not include federal hospitals.
[a]2008–09; figures rounded; Tables 103 and 104. NA = data not available.
Source: National Center for Health Statistics. Health, United States, 2011. Available at: http://www.cdc.gov/nchs/data/hus/hus11.pdf [Accessed 29 November 2012].

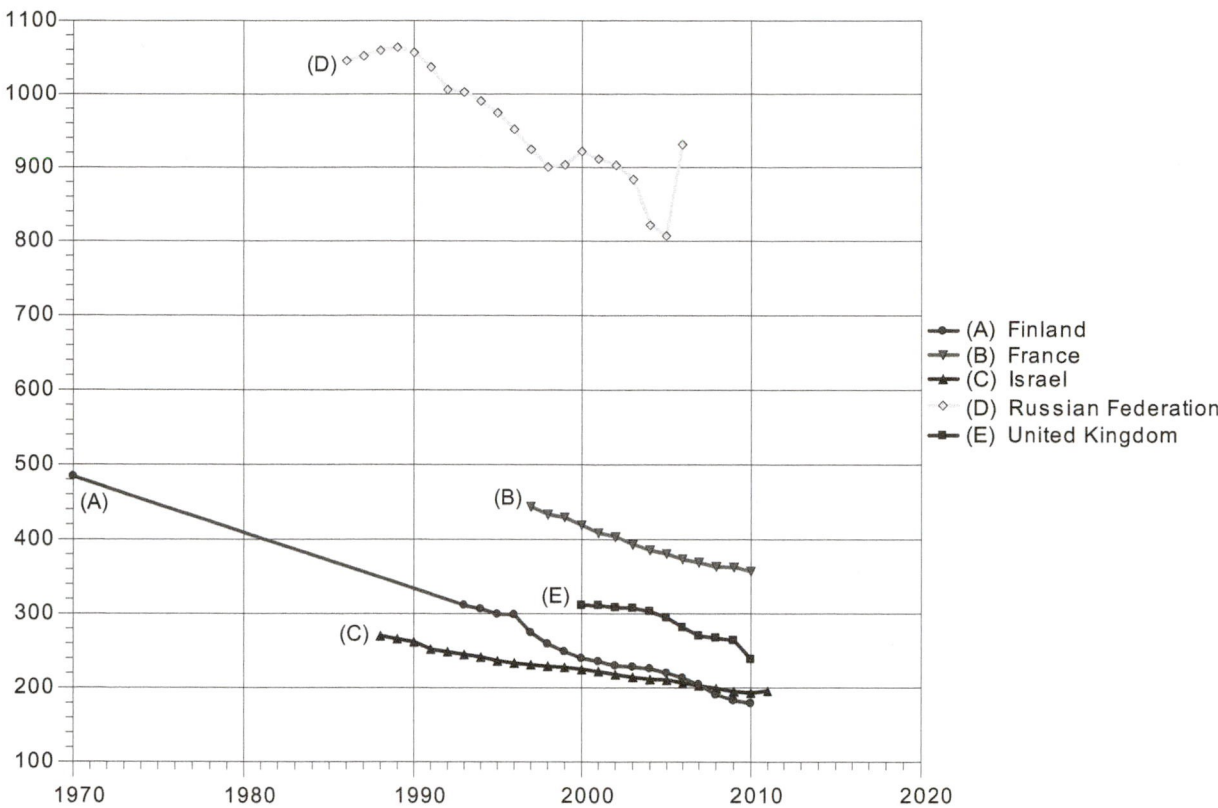

FIGURE 11.5   Acute care hospital beds per 100,000, selected countries, WHO European Region, 1970–2010. *Source: World Health Organization, Regional Office for Europe. Health for All database; Copenhagen: WHO; August 2012.*

Mortality from coronary heart disease has decreased markedly during this period; however, admission rates for heart disease overall have not declined, while total days of care fell by 38 percent. This is in part due to changing patterns of care, with shorter length of stay and a more aggressive rehabilitation approach to myocardial infarction, and emphasis on ambulatory care. Medical treatment during the acute myocardial infarction stage is more effective than previous treatments, with technology such as streptokinase, angioplasty, stents, and other interventions. All of this has been accompanied by a steady fall in mortality rates (see Chapter 5).

Many Western European countries have been reducing their hospital bed ratios (i.e., beds per 1000 population) since the 1980s, as seen in Figure 11.5, by over 50 percent in some cases. Countries of Eastern Europe and the former Soviet Union have high but declining hospital bed to population ratios and hospital inventories. Some have increased health spending, but some (e.g., Russia) still have relatively low overall expenditures on health care per capita and have spent much of their new funds for health on hospitals in recent years.

Reducing hospital utilization and bed supply creates a problem and an opportunity for staff and resource reallocation. Hospital facilities can sometimes be converted to other purposes, as outlined in Box 11.11. Often, the most constructive use of obsolete hospital facilities is to transfer them out of the health sector, since the land may be of greater value than its continuing use for health care purposes.

MODIFIED MARKET FORCES

Classically, market forces are seen as a means of empowering the purchaser to seek the least expensive and/or the best goods or services from competing providers. The classical market forces in health systems are presented in Box 11.12.

In health care, there are modifying factors that affect market forces. Understanding the modifiers of market forces summarized in Table 11.8 is part of the preparation of a manager, provider, and policy planner for a strategic role in health systems. Some of these modifiers are governmental regulatory factors, such as supply of hospital beds. Others are related to access to services and the amount and method of payment and other factors that affect needs for care, as well as the quality and efficiency of a service.

Market mechanisms are modified by regulations, incentives, and other factors used to promote a balance of preventive, curative, and rehabilitative services, including health promotion to improve health and help the individual to seek and find the most appropriate care at any point in time. These factors are summarized in Box 11.13.

BOX 11.11 Managing Excess Hospital Bed Capacity

- *Convert to long-term care (LTC) facility* – extended care, rehabilitation, chronic care, or elderly people's housing.
- *Close maternity homes* – replace with maternity units in district general hospitals.
- *Develop acute psychiatric units in general hospitals* – conversion of excess acute care beds, closure of long-term psychiatric beds, development of community services and group residential facilities.
- *Develop acute geriatric units in general hospitals* – short-term care, with home and LTC facilities.
- *Develop tuberculosis (TB) units in general hospitals* – short-term investigation and therapy, with closure of long-term beds in TB hospitals and strengthening community care systems.
- *Develop detoxification units for alcohol and drug abuse* – general hospitals with community facilities support.
- *Convert to special needs shelters* – homeless people, abuse or rape crisis shelters.
- *Convert to ambulatory care facilities* – use inpatient facilities and staff for outpatient, day hospital services.
- *Develop hospice care* – for terminally ill patients.
- *Convert to other socially useful functions* – community centers, schools, or vocational training.
- *Demolish and dispose of obsolete facilities* – land value may pay for part of new health programs.

BOX 11.12 Classical Market Factors in Health

- Insurance
- Supply and its regulation
- Demand
- Competition in cost, quality
- System macroefficiency
- Vertical integration (see Chapter 10)
- Lateral integration (see Chapter 10)
- System microefficiency
- Incentives
- Disincentives
- Reputation
- Accessibility, proximity to population
- Primary care
- Home care
- Taxation on alcohol and cigarettes
- Health promotion to reduce risk factors, promote health

BOX 11.13 Regulatory Factors in Economics of Health

- Regulate supply
- Regulate demand – gatekeeper, user fees
- Regulate price
- Regulate benefits
- Regulate method of payment
- Health promotion issues
- Accreditation of providers and facilities

ECONOMICS OF PREVENTION

The market approach to health economics is commonly based on the assumption that individuals are in the best position to judge their own welfare and choose to make use of health services to improve their well-being, especially if insured for those services. However, the health care market is not efficient in changing lifestyle choices. Use of services and consumption behaviors may not be the best or only way to improve individual and societal welfare. OECD studies indicate that government intervention may be needed, such as in mandatory fortification of flour with folic acid, iron and vitamins B, and banning use of transfats in baking and food production or reducing salt content of foods, where "market forces" fail to meet public health needs. Consumers may lack information to make informed choices on the consequences of consumption decisions, or individual behavior may affect the welfare of others, or consumers are unable to affect the environment in which they live or work, or to make sufficiently rational and informed choices (as is generally the case among children, for example). There may also be confusing information available to those who seek guidance from anonymous Internet or other media sources which have dissonant views on health topics. Prevention policies provide opportunities for increasing social well-being, and can be used to reduce health disparities by reaching out to population groups at special risk to themselves and others.

The OECD reports that the obesity epidemic has slowed down in several countries since 2009, with school and institutional intervention programs, improved menus, and reduced access to junk foods in schools (see Chapter 6). Rates grew less than previously projected or did not grow at all in 10 OECD countries. Child obesity rates stabilized in England, France, South Korea, and the USA. However, rates remain high and social disparities in obesity are crucial in the epidemiology of NCDs globally.

Governments are increasing their efforts to tackle the causes of obesity, embracing increasingly comprehensive strategies and involving communities and key stakeholders such as food producers and marketers, including restaurants. Strategies include new labeling of calorie and fat content of foods in worldwide chain restaurants such as McDonalds, the banning of transfats in baking in New York City, followed by many locations in the USA and other countries, and the banning of sugary drinks in schools and other public places. Governments are also interested in differential taxes on foods rich in fat and sugar, and in 2011 new laws to this effect were passed by several governments, including Denmark, Finland, France, and Hungary (OECD, 2012).

OECD studies show that preventive interventions can improve health at a lower cost than many treatments offered today by health systems. Studies in Canada show that

TABLE 11.8 Modifying Factors on Market Forces in the Economics of Health

Determinants in Health Demand	Modifiers of Demand	Examples
Classical market factors	Supply	Hospital beds per 1000; providers combine to restrict supply
	Demand	Prepayment increases effective demand
	Competition in cost, quality	Managed care versus fee-for-service plans
	System macroefficiency	District health systems and HMOs
	Vertical integration	Multiservice systems for defined populations
	Lateral integration	Multihospital networks for efficiency
	System microefficiency	Quality improvement, computerization, staff attitudes, scheduling, and service hours
	Incentives: disincentives	Budget, method of payment: block, per diem, DRG
	Reputation	Consumer, community, provider satisfaction
Regulatory factors	Regulate supply	Reduce hospital beds per 1000 and personnel
	Regulate demand	Gatekeeper functions
	Regulate price	User fees
	Regulate method of payment	Fee control, income capping, salary or capitation payment for doctors; DRGs, block budgets for hospitals
	Health promotion issues	Food enrichment, safety factors, seat belts
Health and societal factors	Differing population needs	Demographic and epidemiological transitions
	Social inequities	Reduce social gaps, universal access, risk groups
	Improve infrastructure to reduce needs	Sewage, water treatment, road safety
	Socioeconomic improvements	National and family incomes
	Public social policies	Social security, pensions, compensation
	Health as a national and local priority	Expenditures and resource allocation; health system reform
	Health promotion	Involve community and providers in prevention; outreach to risk groups
	Improve KABP (knowledge, attitudes, beliefs, and practices)	Providers, beneficiaries, consumers' rights, needs, responsibilities
System determinants	Shift in resource allocation	Balance of institutional and community care; infection and error control
	Technological innovations	New vaccines, drugs, diagnostic equipment, ORS, community health workers
	Regulation	Home care, generic drugs, nurse practitioners
	Accreditation	External accreditation, internal review systems, patient choice, continuous quality improvement
	Substitution	
	Total quality management	

Note: HMO = health maintenance organization; DRG = diagnosis related group; ORS = oral rehydration salts.

prevention programs, such as hypertension, cervical cancer, mammography, and colorectal cancer screening, could avoid up to 25,000 deaths (i.e., *avoidable mortality*) from NCDs every year. The number of deaths avoided could increase to 40,000 if different interventions were combined in a comprehensive prevention strategy. For example, an organized program of counseling of obese people by their family doctors would also lead to an annual gain of 40,000 years of life in good health (Figure 11.6) (OECD, 2012). Such programs require policy and resource commitment to health promotion, and organized efforts to orient and support such interventions, including training of family doctors, public health and nurse practitioners, dieticians, dental hygienists, pharmacists and, in low-income countries, community health workers (see Chapter 14) and other health providers in dietary counseling as a priority in their practices.

GAME THEORY, ECONOMICS, AND HEALTH POLICY

Game theory has become part of health economics and management, in part as a replacement of market approaches to health care, which are not helpful in systems of universal health care and regulation and limitation of supply of services affecting the "free market". According to Aumann, game theory was first applied in Talmudic examination of practical problems such as bankruptcy. Operations research was initiated in Britain in the nineteenth century to resolve problems of sorting and delivering the "penny post" of the then new public mail system.

Operations research became an important aspect of the British strategic and tactical warfare planning efforts during World War II. The Battle of the Atlantic involved complex planning based on calculations of interactive thinking in

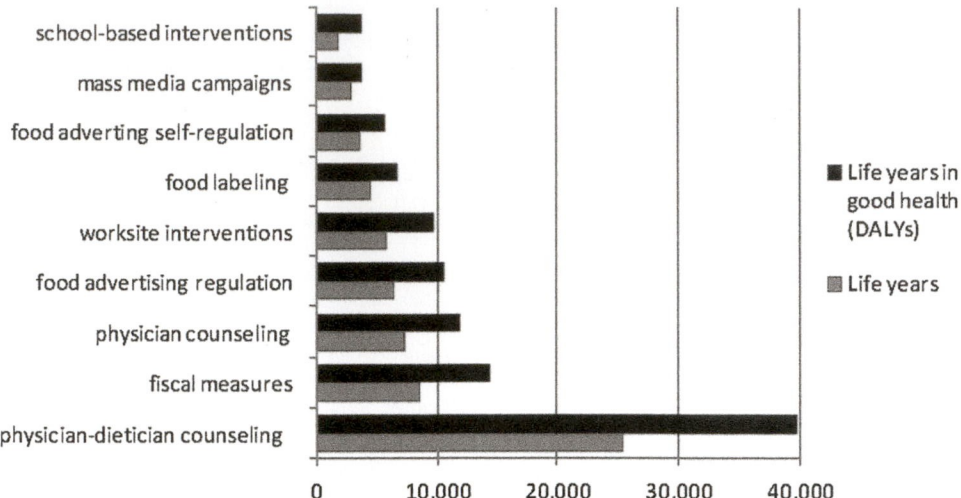

FIGURE 11.6   Deaths potentially avoided by selected prevention programs, Canada, 2012. Note: DALYs=disability-adjusted life years. *Source: Organisation for Economic Cooperation and Development. Obesity and the economics of prevention: fit for fat: key facts – Canada, update 2012. Available at: http://www.oecd.org/els/healthpoliciesanddata/49712071.pdf [Accessed 1 December 2012].*

war games in the struggle between the highly skilled Nazi submarine arm and the relatively new and untrained Allied antisubmarine warfare system, to move convoys from North America to Britain safely. Operations research contributed to important technical developments and tactical issues at sea or in the air, as applied to planning the bombing campaign of Germany during the same war, and in many cases made the difference between success and failure (Table 11.9).

Thomas Schelling and Robert Aumann were awarded the Nobel Prize in Economic Sciences in 2005; and in 2012, Alvin Roth and Lloyd Shapley were also awarded Nobel Prizes in Economics for basic theory and applications of game theory to real-life situations where market mechanisms alone would not explain decision-making interactions. Schelling served in the Economic Cooperation Administration in Europe during the period of the Marshall Plan. He published work based on game theory models on military strategy and arms control, energy and environmental policy, climate change, nuclear proliferation, terrorism, organized crime, foreign aid and international trade, conflict and bargaining theory, racial segregation and integration, the military draft, health policy, tobacco and drugs policy, and ethical issues in public policy and in business. Schelling's co-laureate, Aumann, was the first to analyze "infinitely repeated games", which helped to explain why some people or communities cooperate better than others over time.

Tarrant et al. (2004) addressed game theory and its potential for value in increasing the understanding of doctor–patient relationships. Game theory is based on assumptions of rational choice and interactive decision making. It may provide "a new conceptual and theoretical basis for future empirical work on the interaction between doctors and their patients", in which the patient, or a mother facing immunization of her child, may make decisions in predictable ways, and thus may be useful in helping to determine policies to maximize

positive responses to, for example, the continuation of blood pressure management for life or the acceptance of vaccines.

Roth studied practical issues such as student allocation in the New York City school system, and resident physician selection for hospital residencies. Both the individual and the system have roles in the choices, and the complexities of each system were worked out based on game theory.

Game theory is a field of analysis with great potential in the health field. It was applied by Anderson and May in predicting epidemic patterns of infectious diseases and will be used in non-communicable and movement issues in health care in years to come.

ECONOMICS AND THE NEW PUBLIC HEALTH

The New Public Health has a vital interest in methods of financing of health services, the economics of health care, and allocation of resources, whether in terms of money, human resources, or fixed assets. No other approach has taken the broad view of health promotion, preventive, curative, and long-term care services. The balance and interdependence among these elements of the health system are influenced by financing systems and choices among alternative ways of expending resources.

Economic analysis, like epidemiological assessment, is a vital tool in health planning and management, particularly in evaluating allocation of resources within a health system. Failure to carry out such assessments results in inefficiency in health planning. Methods of economic analysis, described briefly in this chapter, are part of the armamentarium of the New Public Health. Both the new and "veteran" health professionals need a basic understanding of the economic issues involved in setting priorities; service organization; utilization of services; and the complex of related ethical, political, social, and management issues.

TABLE 11.9 Operations Research and Game Theory

Operations research before and in 19th century	Talmudic solutions to bankruptcy issues and other practical problems "Penny post" management in Britain in mid-19th century: Operations research developed mathematical modeling of transport and sorting of Royal Mail
Operations research during World War II	World War II, Battle of Atlantic: convoy system and escorts, convoy speed, placement and attack tactics of escorts, depth charge settings and kill ratios; tactical changes, aircraft search patterns, camouflage and submarine sinking ratios RAF bombing of Nazi Germany: allocation, spacing, routing, escorts, rescue, evaluation of effects
Post-World War II: applications in business, education, health	Critical path analysis for project planning: identifying components in a complex project which affect the overall duration, cost, and effectiveness of the project Facility planning: location, size, design, materials, lighting and heating costs, layout of equipment in a factory or components on a computer chip to reduce manufacturing, shipping time, costs, and marketing Network optimization: setup of telecommunications networks to maintain quality of service, supplies, storage, ordering and sales; collaboration with other agencies, companies Budget and staff: allocation, training, advancement Evaluation: costs and benefits, utility
Game theory and economics	Game theory: study of strategic decision making, through mathematical models of conflict, or diseases and rational decision making via interactive decision theory Game-theorists: eight Nobel Prizes in Economics to 2011 Nobel Prize Economics 2005: Robert Aumann (Hebrew University, Jerusalem) and Thomas Schelling (University of Maryland), "for having enhanced our understanding of conflict and cooperation through game-theory analysis" Nobel Prize Economics 2012: Lloyd Shapely (UCLA), Alvin Roth (Harvard/Berkeley), economics of market and non-market mechanisms Strategic allocation algorithms for practical problems, e.g., strategic activities, arms control negotiation policies, cooperation and conflict resolution, bargaining and limited war, strategy of conflict Algorithms for practical issues: dating and marriage, college selection, NYC high-school student allocation to city schools, medical doctors to hospitals for residency training, kidney transplant exchange systems Applications in biology: infectious disease modeling (Anderson and May); and in molecular/genetic biology

Note: RAF = Royal Air Force; UCLA = University of California, Los Angeles; NYC = New York City.
Sources: Aumann RJ, Maschler M. Game theoretic analysis of a bankruptcy problem from the Talmud. J Econ Theory 1985;36:195–213. Available at: http://www.sciencedirect.com/science/article/pii/0022053185901024 [Accessed 20 October 2012].
Gale D. Book review of: Games and decisions: introduction and critical survey. By Luce RD, Raiffa H. John Wiley and Sons; 1957. Available at: http://projecteuclid.org/DPubS/Repository/1.0/Disseminate?view=body&id=pdf_1&handle=euclid.bams/1183522326 [Accessed 20 October 2012].
Nobel Prize in Economic Sciences. Robert J. Aumann and Thomas C. Schelling, 2005. Available at: http://www.nobelprize.org/nobel_prizes/economics/laureates/2005/aumann.html# and http://www.nobelprize.org/nobel_prizes/economics/laureates/2005/schelling.html [Accessed 20 October 2012].

SUMMARY

The basic view of Health for All promoted by the WHO at the Alma-Ata Conference in 1978 was adopted anew in the *Macroeconomic Health Report* of 2001, and more recently on the thirtieth anniversary of Alma-Ata. It was also adopted in the UN MDGs of 2001. The incorporation of an economic rationale for investment in health, expounded in the World Bank Human Development Report of 1993 and later updated in the Disease Control Priorities Project of 2006, has strengthened the concept that investment in health is a part of and necessary for economic development. This is the principle that health is a human right for everyone, and that management of resources is crucial to achieve improved health. This combination of overall public health commitment with an economic justification and management is the essence of the New Public Health.

Advancement in its application requires political commitment, and adequate funding for public health programs including its oversight and policy formulation for a balanced program including health promotion, primary health care, hospital care, and long-term care, all essential elements of an effective health system. Health promotion and primary care are the most cost-effective interventions in improving the health status of the population. Where there has been an excessive emphasis on institutional care, there is real potential for transfer of resources and emphasis within the health system as part of the process of raising primary care and health promotion standards. This is the essential direction of health reform in many countries which had overemphasized hospitalization for health care.

Innovations in health care financing and administration, such as Accountable Care Organizations, managed care, capitation payment, fundholding general practitioners, district health systems, and DRGs, are all part of the search for more efficient ways of using resources and limiting cost increases. Innovations in achieving better individual and population health include a wide array of technologies, organizational, and other improvements in medical care and prevention care, health promotion including legislation and regulatory public health functions, smoking reduction and greater awareness of healthy lifestyle, and outreach and home care services for people at high risk. Many new technologies and innovations are making an impact on health systems costs, supply, training and research, and, most importantly, improving successful prevention and medical care. Examples include endoscopic, robotic and outpatient surgery, early and more effective care and rehabilitation for cardiovascular disease, simple and inexpensive diagnostics/screening/laboratory tests, cure for chronic peptic ulcers, new vaccines and screening for prevention of cancer, electronic medical records and many more. Basic science leads to new methods of prevention and care that are best applied in universal access systems with special care for risk groups and those with special needs.

The effectiveness of new health innovations and the economics of health in society are critically important elements of the New Public Health. As populations age and as technology advances, costs will inevitably increase, but this can be moderated by resource allocation to prevent or delay the onset of complications among chronically ill people in the community. This includes health promoting activities and services for the population to avoid or deter the emergence of acute illness or complications of existing conditions such as diabetes, hypertension, chronic lung and cardiovascular diseases, and risk factors such as smoking, obesity, lack of exercise and unhealthy dietary habits.

Health must compete with other government programs for resource allocation. In the New Public Health, health care is an investment in human capital for the development of a country, as well as an ethical obligation of a society to its individual members. Classical public health protection, preventive care, health promotion, health and social policy, and systems management together make up a New Public Health that is involved in management of health in all its aspects. An understanding of basic issues in health economics is as vital to public health practice as is an understanding of communicable disease or any other element of the broad panorama of health.

NOTE

For a complete bibliography and guidance for student reviews and expected competencies please see companion web site at http://booksite.elsevier.com/9780124157668

BIBLIOGRAPHY

Agency for Healthcare Research and Quality, Managed care. Available at: http://www.ahrq.gov/research/managix.htm (accessed 01.11.12).

American Public Health Association, Policy Statement 9615(PP). Supporting national standards of accountability for access and quality in managed health care, advocacy/policy statements. Available at: http://www.apha.org/ (accessed 30.10.12).

American Public Health Association, Policy Statement 9716(PP). The issue of profit in health care, advocacy/policy statements. Available at: http://www.apha.org/ (accessed 30.10.12).

Aumann, R.J., Maschler, M., 1985. Game theoretic analysis of a bankruptcy problem from the Talmud. J. Econ. Theory 36, 195–213. Available at: http://www.sciencedirect.com/science/article/pii/0022053185901024 (accessed 20.10.12).

Beth Israel Deaconess Care Organization, http://www.bidpo.org/aboutus/ (accessd 12.07.13).

Beth Israel Deaconess Medical Center, Beth Israel Deaconess physician organization named a Medicare pioneer accountable care organization. Available at: http://www.bidmc.org/News/AroundBIDMC/2011/December/BIDPOACO.aspx (accessed 26.11.12).

Borger, C., Smith, S., Truffer, C., Keehan, S., Sisko, A., Poisal, J., et al., 2006. Health spending projections through 2015: changes on the horizon. Health Aff. 25, 61–73. Available at: http://content.healthaffairs.org/content/25/2/w61.full (accessed 21.12.13).

Casalino, L., Nicholson, S., Gans, D., Hammons, T., Morra, D., Karrison, T., et al., 2009. What does it cost physician practices to interact with health insurance plans? Health Aff. 28 (4), w533–w543. Available at: http://content.healthaffairs.org/content/28/4/w533.long (accessed 21.12.13).

Center for Medicare & Medicaid Services, http://www.cms.gov/ (accessed 01.11.12).

Center for Medicare & Medicaid Innovation, 24 May 2012. Selected participants in the pioneer ACO model. . Available at: http://innovations.cms.gov/Files/x/Pioneer-ACO-Model-Selectee-Descriptions-document.pdf.

Centers for Disease Control and Prevention, Advisory Committee on Immunization Practices: Guidance for health economics studies: guidance for health economics studies presented to the ACIP. Available at: http://www.cdc.gov/vaccines/acip/committee/guidance/economic-studies.html (accessed 24.11.12).

Centers for Disease Control and Prevention, 1995. Assessing the effectiveness of disease and injury prevention programs: costs and consequences. MMWR Recomm. Rep. 44 (RR10). Available at: http://www.cdc.gov/mmwr/preview/mmwrhtml/00038592.htm (accessed 29.11.12).

Centers for Disease Control and Prevention, 2012. CDC Global Health Strategy 2012–2015. CDC, Atlanta, GA. Available at: http://www.cdc.gov/globalhealth/strategy/pdf/CDC-GlobalHealthStrategy.pdf (accessed 29.11.12).

Centers for Disease Control and Prevention, 1995. Economic costs of birth defects and cerebral palsy – United States, 1992. MMWR Morb. Mortal Wkly. Rep. 44, 694–699. http://www.cdc.gov/mmwr/preview/mmwrhtml/00038946.htm (accessed 21.12.13).

Centers for Disease Control and Prevention, Vaccine costs. Available at: http://www.cdc.gov/vaccines/ programs/vfc/cdc-vac-price-list.htm (accessed 31.10.12).

Centers for Medicare and Medicaid Services, Office of the Actuary. National Health Expenditure Projections 2011-2021 Available at: http://www.cms.gov/Research-Statistics-Data-and-Systems/Statistics-Trends-and-Reports/NationalHealthExpendData/Downloads/Proj2011PDF.pdf (accessed 2.1.2013).

Chernew, M.E., Goldman, D.P., Pan, F., Shang, B., 2005. Disability and health care spending among Medicare beneficiaries. Health Aff. 24 (Suppl. 2) W5R42–52. Available at: http://content.healthaffairs.org/content/early/2005/09/26/hlthaff.w5.r42.long (accessed 21.12.13).

Chisholm, D., Baltussen, R., Evans, D.B., Ginsberg, G., Lauer, J.A., Lim, S., et al., 2012. What are the priorities for prevention and control of non-communicable diseases and injuries in sub-Saharan Africa and South East Asia? BMJ 344, e586. Available at: http://www.bmj.com/content/344/bmj.e586 (accessed 21.12.13).

Centers for Medicare and Medicaid Services. Office of the Actuary. National Health Expenditure Projections, 2011-2021. Available at: http://www.cms.gov/Research-Statistics-Data-and-Systems/Statistics-Trends-and-Reports/NationalHealthExpendData/Downloads/Proj2011PDF.pdf (accessed 2.1.2013).

Collins, S.R., Robertson, R., Garber, T., Doty, M.M., 26 April 2013. Insuring the future: current trends in health coverage and the effects of implementing the Affordable Care Act. Commonwealth Fund, New York. Available at: http://www.commonwealthfund.org/Publications/Fund-Reports/2013/Apr/Insuring-the-Future.aspx (accessed 12.07.13).

Commonwealth Fund Commission, 10 January 2013. Confronting costs: stabilizing US health spending while moving toward a high performance health care system. Commonwealth Fund, New York. Available at: http://www.commonwealthfund.org/Publications/Fund-Reports/2013/Jan/Confronting-Costs.aspx (accessed 12.07.13).

Cuckler GA, Sisko AM, Keehan SP, Smith SD, Madison AJ, Poisals JA, et al. National health expenditure projections, 2012–22: slow growth until coverage expands and economy improves. Health Aff September 2013 10.1377/hlthaff.2013.0721 Published online before print September 2013, doi: 10.1377/hlthaff.2013.0721

Dartmouth Institute of Health Care and Practice, 2012. Dartmouth atlas of health care, 2012. Dartmouth College, NH. Available at: http://www.dartmouthatlas.org/ (accessed 25.10.12).

Dartmouth Institute of Health Care and Practice, Dartmouth atlas of health care, 2012. Accountable care. Available at: http://www.dartmouthatlas.org/keyissues/issue.aspx?con=2943 (accessed 26.10.12).

Davis, K., Schoen, C., Stremikis, K., June 2010. Mirror, mirror on the wall: how the performance of the US health care system compares internationally, 2010 update. Commonwealth Fund.. Available at: www.commonwealthfund.org (accessed 31.10.12).

Desvarieux, M., Landman, R., Liautaud, B., Girard, P.M., 2005. The INTREPIDE initiative in global health. Antiretroviral therapy in resource-poor countries: illusions and realities. Am. J. Public Health 95, 1117–1122. Available at: http://www.ncbi.nlm.nih.gov/pmc/articles/PMC1449328/ (accessed 21.12.12).

Detsky, A.S., Laupacis, A., 2007. Relevance of cost-effectiveness analysis to clinicians and policy makers. JAMA 298, 221–224. Available at: http://ehealthecon.hsinetwork.com/Detsky_JAMA_2007_221.pdf (accessed 21.12.13).

Doran, T., Fullwood, C., Gravelle, H., Reeves, D., Kontopantelis, E., Urara, E., et al., 2006. Pay-for-performance programs in family practices in the United Kingdom. N. Engl. J. Med. 355, 375–384. Available at: http://content.nejm.org/cgi/content/full/355/4/375 (accessed 01.12.12).

Drummond, M.F., O'Brien, B., Stoddart, G.L., Torrance, G.W., 2005. 3rd edition. Methods for the economic evaluation of health care programmes. Oxford University Press, New York. Available at: http://www.amazon.com/Methods-Economic-Evaluation-Health-Programmes/dp/0198529457 (accessed 21.12.13).

Drummond, M.F., Stoddart, G., Labelle, R., Cushman, R., 1987. Health economics: an introduction for clinicians. Ann. Intern. Med. 107, 88–92. Available at: http://annals.org/article.aspx?articleid=701989 (accessed 21.12.13).

Evans, R.G., Lomas, J., Barer, M.L., Labelle, R.J., Fooks, C., Stoddart, G.L., et al., 1989. Controlling health expenditures: the Canadian reality. N. Engl. J. Med. 320, 571–577. Available at: http://www.nejm.org/doi/full/10.1056/NEJM198903023200906 (accessed 21.12.13).

Foote, S.B., Halaas, G.W., 2006. Defining a future for fee-for-service Medicare. Health Aff. 25, 864–868. Available at: http://content.healthaffairs.org/content/25/3/864.full (accessed 21.12.13).

Friede, A., Taylor, W.R., Nadelman, L., 1993. On-line access to a cost–benefit/cost–effectiveness analysis bibliography via CDC WONDER. Med. Care 31 (Suppl.) JS12–7. Available at: http://www.ncbi.nlm.nih.gov/pubmed/8392124 (accessed 21.12.13).

Ginsberg, G.M., 2011. Assessing evidence – the strengths and weakness of various ways of assessing outcomes, cost-effectiveness analysis. In: Aceljas, C. (Ed.), Assessing evidence to improve population health and well-being. Learning Matters, Exeter, pp. 91–111.

Ginsberg, G.M., Lauer, J.A., Zelle, S., Baeten, S., Baltussen, R., 2012. Cost effectiveness of strategies to combat breast, cervical, and colorectal cancer in sub-Saharan Africa and South East Asia: mathematical modeling study. BMJ 344, e614. Available at: http://www.ncbi.nlm.nih.gov/pmc/articles/PMC3292522/pdf/bmj.e614.pdf (accessed 21.12.13).

Ginsberg, G.M., Tulchinsky, T.H., 1990. Costs and benefits of a second measles inoculation of children in Israel, the West Bank, and Gaza. J. Epidemiol. Commun. Health 44, 274–280. Available at: http://jech.bmj.com/content/44/4/274.full.pdf+html (accessed 21.12.13).

Ginsberg, G., Tulchinsky, T., Filon, D., Goldfarb, A., Abramov, L., Rachmilevitz, E.A., 1998. Measuring costs: the economics of health benefit analysis of a national thalassemia programme in Israel. J. Med. Screen 5, 120–126. Available at: http://www.ncbi.nlm.nih.gov/pubmed/9795870 (accessed 21.12.13).

Goetghebeur, M.M., Wagner, M., Khoury, H., Levitt, R.J., Erickson, L.J., Rindress, D., 2008. Evidence and value: impact on decision making – the EVIDEM framework and potential applications. BMC Health Serv. Res. 8, 270. Available at: http://www.biomedcentral.com/1472-6963/8/270 (accessed 21.12.13).

Goldie, S.J., Gaffikin, L., Goldhaber-Fiebert, J.D., Gordillo-Tobar, A., Levin, C., Mahe, C., et al., 2005. Alliance for cervical cancer prevention cost working group. Cost-effectiveness of cervical-cancer screening in five developing countries. N. Engl. J. Med. 353, 2158–2168. Available at: http://www.nejm.org/doi/full/10.1056/NEJMsa044278 (accessed 21.12.13).

Goldman, L., 2005. Cost-effectiveness in a flat world – can ICDs help the United States get rhythm? N. Engl. J. Med. 353, 1513–1515. Available at: http://www.nejm.org/doi/full/10.1056/NEJMe058214 (accessed 21.12.13).

Gonzalez-Perez, J.G., Vale, L., Stearns, S.C., Wordsworth, S., 2005. Hemodialysis for end-stage renal disease: a cost-effectiveness analysis of treatment-options. Int J. Technol. Assess. Health Care 21, 32–39. Available at: http://journals.cambridge.org/action/displayAbstract?fromPage=online&aid=285789&fulltextType=RA&fileId=S026646230505004X (accessed 21.12.13).

Jacobs, P., Ohinmaa, A., 2010. A comparison of the use of economics in vaccine expert reviews. Vaccine 28, 2841–2845. Available at: http://www.ncbi.nlm.nih.gov/pubmed/20153796 (accessed 21.12.13).

Jamison, D.T., 2002. Cost-effectiveness analysis: concepts and applications. In: Detels, R.G., McEwen, J., Beaglehole, R., Tanaka, H. (Eds.), Oxford textbook of public health. fourth ed. Oxford University Press, Oxford, pp. 903–919. Available at: http://medtextfree.wordpress.com/2011/11/06/7-6-cost-effectiveness-analysis-concepts-and-applications/ (accessed 21.12.13).

Jamison, D.T., Breman, J.G., Measham, A.R., Alleyne, G., Claeson, M., Evans, D.B. (Eds.), 2006. Priorities in health. World Bank, Washington, DC. Available at: http://www.ncbi.nlm.nih.gov/books/NBK10257/. (accessed 03.11.12).

Kahn, J.G., Kronick, R., Kreger, M., Gans, D.N., 2005. The cost of health insurance administration in California: estimates for insurers, physicians, and hospitals. Health Aff. 24, 1629–1639. Available at: http://content.healthaffairs.org/content/24/6/1629.long (accessed 21.12.13).

Kaiser Family Foundation, Health care spending in the United States and OECD countries. Available at: http://www.kff.org/insurance/snapshot/chcm050206oth2.cfmhttp://www.kff.org/insurance/snapshot/oecd042111.cfm. and http://www.kff.org/insurance/snapshot/chcm010307oth.cfm (accessed 01.11.12).

Kaiser Family Foundation, Employer health benefits: 2013 Summary of findings. Available at: http://kaiserfamilyfoundation.files.wordpress.com/2013/08/8466-employer-health-benefits-2013_summary-of-findings1.pdf (accessed 28.08.13).

Kelley, J.E., Burrus, R.G., Burns, R.P., Graham, L.D., Chandler, K.E., 1993. Safety, efficacy, cost and morbidity of laparoscopic versus open cholecystectomy; a prospective analysis of 228 consecutive patients. Am. Surg. 59, 23–27. Available at: http://www.ncbi.nlm.nih.gov/pubmed/8480927 (accessed 21.12.13).

Kuttner, R., 2008. US Market based failure – a second opinion on US health costs. N. Engl. J. Med. 358, 549–551. Available at: http://www.nejm.org/doi/full/10.1056/NEJMp0800265 (accessed 21.12.13).

Leroy, J.L., Habicht, J.P., Pelto, G., Bertozzi, S.M., 2007. Current priorities in health research funding lack an impact on the number of child deaths per year. Am. J. Public Health 9, 219–223. Available at: http://www.ncbi.nlm.nih.gov/pmc/articles/PMC1781402/ (accessed 21.12.13).

Lieu, T.A., Cochi, S.L., Black, S.B., Halloran, E., Shinefield, H.R., Holmes, S.J., et al., 1994. Cost-effectiveness of a routine varicella vaccination program for US children. JAMA 271, 375–381. Available at: http://jama.jamanetwork.com/article.aspx?articleid=363837 (accessed 21.12.13).

Lord, J., Thomason, M.J., Littlejohns, P., Chalmers, R.A., Bain, M.D., Addison, G.M., et al., 1999. Secondary analysis of economic data: a review of cost–benefit studies of neonatal screening for phenylketonuria. J. Epidemiol. Commun Health 53, 179–186. Available at: http://jech.bmj.com/content/53/3/179.abstract (accessed 21.12.13).

Madon, T., Hoffman, K.J., Kupfa, L., Glass, R.I., 2007. Implementation science. Science 318, 1728–1729. Available at: http://www.sciencemag.org/content/318/5857/1728.short (accessed 21.12.13). http://www.sciencemag.org/content/318/5857/1728.summary.

Managed Care Digest, http://www.managedcaredigest.com (accessed 01.11.12).

McKee, M., Edwards, N., Atun, R., 2006. Public–private partnerships for hospitals. Bull. World Health Organ 84, 890–896. Available at: http://www.who.int/bulletin/volumes/84/11/06-030015.pdf (accessed 21.12.13).

Meltzer, M.I., 2000. Economic consequences of infectious diseases. In: Lederberg, J. (Ed.), Encyclopedia of microbiology. second ed. Academic Press, San Diego, CA.

Meltzer, M.I., 2001. Introduction to health economics for physicians. Lancet 358, 993–998. Available at: http://www.thelancet.com/journals/lancet/article/PIIS0140-6736(01)06107-4/abstract (accessed 21.12.13).

Neumann, P.J., Rosen, A.B., Weinstein, M.C., 2005. Medicare and cost-effectiveness analysis. N. Engl. J. Med. 353, 1516–1522. Available at: http://www.nejm.org/doi/full/10.1056/NEJMsb050564 (accessed 21.12.13).

Neumann, P.J., Stone, P.W., Chapman, R.H., Sandberg, E.A., Bell, C.M., 2000. The quality of reporting in published cost–utility analyses, 1976–1997. Ann. Intern. Med. 132, 964–972. Available at: http://annals.org/article.aspx?articleid=713548 (accessed 21.12.13).

Nobelprize.org, Nobel Prize in Economic Sciences. Robert J. Aumann and Thomas C. Schelling, 2005. Available at: http://www.nobelprize.org/nobel_prizes/economics/laureates/2005/aumann.html# (accessed 01.11.12).

Nohynek, H., 2008. The Finnish decision-making process to recommend a new vaccine. J. Public Health 16, 275–280. Available at: http://link.springer.com/article/10.1007%2Fs10389-008-0204-y#page-1 (accessed 21.12.13).

Nord, E., Daniels, N., Kamlet, M., 2009. QALYs: some challenges. Value Health 12 (1), S10–S15. Available at: http://onlinelibrary.wiley.com/doi/10.1111/j.1524-4733.2009.00516.x/abstract (accessed 01.11.12).

O'Donnell, O., van Doorslaer, E., Wagstaff, A., Lindelow, M., 2007. Analyzing health equity using household survey data. World Bank, Washington, DC. Available at: https://openknowledge.worldbank.org/bitstream/handle/10986/6896/424800ISBN978011OFFICIAL0USE0ONLY10.pdf?sequence=1 (accessed 21.12.13).

Ortegón, M., Lim, S., Chisholm, D., Mendis, S., 2012. Cost effectiveness of strategies to combat cardiovascular disease, diabetes, and tobacco use in sub-Saharan Africa and South East Asia: mathematical modeling study. BMJ 344, e607. Available at: http://www.who.int/choice/publications/tobacco.pdf (accessible 21.12.13).

Sachs, J.D., Macroeconomics and health: investing in health for economic development; 2001. Available at: http://www.cid.harvard.edu/archive/cmh/cmhreport.pdf (accessed 01.11.12).

Sackett, D.L., Rosenberg, W.M., Gray, J.A., Haynes, R.B., Richardson, W.S., 1996. Evidence based medicine: what it is and what it isn't. BMJ 312, 71–72. Available at: http://www.bmj.com/content/312/7023/71 (accessed 21.12.13).

Scanlon, W.J., 2006. The future of Medicare hospital payment. Health Aff. 25, 70–80. Available at: http://content.healthaffairs.org/content/25/1/70.long (accessed 21.12.13).

Schoenbaum, S.C., Davis, K., Holmgren, A.L., 2007. Health care spending: an encouraging sign? Commonwealth Fund, New York. Available at: http://www.commonwealthfund.org/~/media/Files/Publications/Data%20Brief/2007/Jan/Health%20Care%20Spending%20%20An%20Encouraging%20Sign/Schoenbaum_hltcarespendingencouragingsign_databrief_993%20pdf.pdf (accessed 21.12.13).

Shillcutt, S.D., Walker, D.G., Goodman, C.A., Mills, A.J., 2009. Cost-effectiveness in low- and middle-income countries: a review of the debates surrounding decision rules. Pharmacoeconomics 27, 903–917. Available at: http://www.ncbi.nlm.nih.gov/pmc/articles/PMC2810517/ (accessed 21.12.13).

Smith, S., Freeland, M., Heffler, S., McKusick, D., 1998. Health Expenditures Projection Team. The next ten years of health spending: what does the future hold? , Health Aff., 128–140. Available at: http://content.healthaffairs.org/content/17/5/128.long (accessed 21.12.13).

Tan-Torres, T., Baltussen, R., Adam, T., Hutubessy, R., Acharya, A. (Eds.), 2003. Making choices in health. WHO guide to cost-effectiveness analysis. WHO, Geneva. Available at: http://www.who.int/choice/publications/p_2003_generalised_cea.pdf (accessed 21.12.13).

Tarrant, C., Stokes, T., Colman, A.M., 2004. Models of the medical consultation: opportunities and limitations of a game theory perspective. Qual. Saf. Health Care 13, 461–466. Available at: http://www.ncbi.nlm.nih.gov/pmc/articles/PMC1743922/ (accessed 21.12.13).

Thorpe, K.E., 2005. The rise in health care spending and what to do about it. Health Aff. 24, 1436–1445. Available at: http://content.healthaffairs.org/content/24/6/1436.abstract (accessed 21.12.13).

United Nations Children's Fund, August 2013. Despite progress, 1,500 African children die daily from malaria. UNICEF 28. Available at: http://www.unicefusa.org/news/releases/world-malaria-day.html (accesssed 30.08.13).

Wagner, E.H., Sandhu, N., Newton, K.M., McCulloch, D.K., Ramsey, S.D., Grothaus, L.C., 2001. Effect of improved glycemic control on health care costs and utilization. JAMA 285, 182–189. Available at: http://jama.jamanetwork.com/article.aspx?articleid=193448 (accessed 21.12.13).

Weinstein, M.C., Stason, W.B., 1977. Foundations of cost-effectiveness analysis for health and medical practices. N. Engl. J. Med. 296, 716–721. Available at: http://www.nejm.org/doi/full/10.1056/NEJM197703312961304 (accessed 21.12.13).

World Bank, 1993. World Development Report, 1993: Investing in health. World development indicators. Oxford University Press, New York. Available at: http://econ.worldbank.org/external/default/main?pagePK=64165259&theSitePK=469372&piPK=64165421&menuPK=64166093&entityID=000425962_20130227111342 (accessed 21.12.13).

World Bank Group, 2006. Disease control priorities in developing countries, second ed. World Bank, Washington, DC. Available at: http://www.ncbi.nlm.nih.gov/books/NBK11728/ (accessed 21.12.13).

World Bank Group, Global burden of disease and risk factors 2006. Available at: http://www.dcp2.org/pubs/GBD (accessed 03.11.12).

World Health Organization, April 2004. Macroeconomics and health: an update. WHO, Geneva. Available at: http://www.who.int/macrohealth/action/mh_and_country_update.pdf (accessed 12.07.13).

World Health Organization, 2010. WHO guide for standardization of economic evaluations of immunization programmes. Vaccine 11, 2356–2359 Available at: Science Direct.

World Health Organization, 2010. World health report: Health systems financing: the path to universal coverage. WHO, Geneva. Available at: http://www.who.int/whr/2010/en/index.html (accessed 24.11.12).

World Health Organization, 1996. Regional Office for Europe. European health care reforms: analysis of current strategies, summary. WHO, Copenhagen. Available at: http://www.euro.who.int/__data/assets/pdf_file/0005/111011/sumhecareform.pdf (accessed 21.12.13).

Chapter 12

Planning and Managing Health Systems

Learning Objectives

Upon completion of this chapter, the student should be able to:

1. Define the role of management at all levels of health service and public health organization;
2. Apply management theory to health planning and the New Public Health;
3. Continue preparation for leadership roles in health service organizations.

INTRODUCTION

Health systems are complex organizations and their management is an important concept in the New Public Health. Health is a major sector of any economy and often employs more people in the industrialized countries than any other industry. Health has complex networks of services and provider agencies, including funding through public or private insurance or through national health service systems. Whether insurance is provided by the state or through private and public sources combined, skilled management is required at the macro- or national and the micro- or local level, including the many institutions that make up the system. Management training of public health professionals and clinical services personnel is a requisite and not a luxury.

Planning and management are changing in the era of the New Public Health with advances in prevention and treatment of disease, population health needs, innovative technologies such as genetic engineering, new immunizations that prevent cancers and infectious diseases, prevention of non-communicable diseases, environmental and nutritional health, and health promotion to reduce risk factors and improve healthful living for the individual and the community. Modern and successful public health also must address social, economic, and community determinants of health and the promotion of public policies and individual behaviors for health and well-being. The social capital and norms that promote cooperation among people are the basis of a "civil society" (i.e., the totality of voluntary, civic, and social organizations and institutions of a functioning society alongside the structures of governmental and commercial institutions). Health systems are ideally knowledge- and evidence-based in using technologies available in medicine and the environment to promote the health and well-being of a population, including security against the effects of threatened terrorism, growing social isolation, and inequities in health. Management in health can learn much from concepts of business management that have evolved to address the economic and human resource aspects of a health system at the macrolevel or an individual unit of service at the microlevel.

The New Public Health is not contained within one organization, but rather reflects the collective efforts of national, state, regional, and local governments, many organizations in the public and non-governmental sectors, and finally efforts of individual or group advocates and providers and the public itself. The political level is crucial for adequate funding, legislation, and promotion of health-oriented policy positions and in public health management. The responsibility for health management is shared across all parts of society, including individuals, communities, business, and all levels of government.

The New Public Health identifies and addresses community health risks and needs. Planning is critical to the process of keeping a health system sustainable and adaptable and in creating adequate responses to new health threats. Monitoring, measurement, and documentation of health needs are vital to design and adapt an effective program and to measure impact. Data on the targeted issues must be accessible while protecting individual privacy.

Health is a hugely expensive and expansive complex of services, facilities, and programs provided by a wide range of professional and support service personnel making up one of the largest employers of any sector in a developed country. Services are increasingly delivered by organized groups of providers. But all health systems operate in an environment of economic constraints, imposing a need to seek efficiency in the use of resources. How organizations function is of great importance not only for their economic survival, but also, and equally important, for the well-being of the clients and providers of care.

An organization is two or more people working together to achieve a common goal. Management is the process of defining the goals and making effective use of an organization to attain those goals. Even very small units of a human organization require management. Management of human resources is vital to the success of an organization, whether in a production or service industry. Health systems may

The New Public Health. http://dx.doi.org/10.1016/B978-0-12-415766-8.00012-4

vary from a single structure to a network of many organizations. No matter how organizations are financed or operated, they require management.

Management in health care has much to learn from approaches to management in other industries. Elements of theories and practices of profit-oriented sector management can be applied to health services even if they are operated as non-profit enterprises. Physicians, nurses, and other health professionals will very likely be involved in the management of some part of the health care system, whether a hospital department, a managed care system, a clinic, or even a small health care team. At every level, management always means working with people, using resources, providing services, and working towards common objectives.

Health providers require preparation in the theory and practice of management. A management orientation can help providers to understand the wider implications of clinical decisions and their role in helping the health care system to achieve goals and targets. Students and practitioners of public health need preparation in order to recognize that a health care system is more complex than the direct provision of individual services. Similarly, policy and management personnel need to be familiar with both individual and population health needs and related care issues.

HEALTH POLICY AND PLANNING AS CONTEXT

Health has evolved from an individual one-on-one service to complex systems organized within financing arrangements, mostly under government auspices. As a governmental priority, health may be influenced by political ideology, sometimes reflecting societal attitudes of the party in power and sometimes apparently at odds with its general social policy. Following Bismarck's introduction in Germany in 1881 of national health insurance for workers and their families, funded by both workers and their employers, most countries in the industrialized world introduced variants of this national health plan. Usually, this has been at the initiative of socialist or liberal political leadership, but conservative political parties have preserved national health programs once implemented. Despite the new conservatism since the 1990s with its pre-eminent ideology of market forces, the growing roles for national, state, and local authorities in health have led to a predominantly government role in financing and overall responsibility for health care, even where there is no universal national health system, as in the USA. The UK's National Health Service (NHS), initiated by a Labour government in 1948, has survived through many changes of government, including the conservative Margaret Thatcher period in which many national industries and services were privatized.

Health policy is a function of national (government) responsibility overall for health, but implementation is formulated and met at state, local, or institutional levels. The division of responsibilities is not always clear cut but needs to be addressed and revised both professionally and politically within constitutional, legal, and financial constraints. Selection of the direction to be taken in organizing health services is usually based on a mix of factors, including the political view of the government, public opinion, and rational assessment of needs as indicated through epidemiological data, cost–benefit analysis, the experience of "good public health practice" from leading countries, and recommendations by expert groups. Lobbying on the part of professional or lay groups for particular interests they wish to promote is part of the process of policy formulation and has an important role in the planning and management of health care systems. There are always competing interests for limited resources of funding, by personnel within the health field itself and in competition with other demands outside the health sector.

The political level is vitally involved in health management in establishing and maintaining national health systems, and in determining the place of health care as a percentage of total governmental budgetary expenditures, in allocating funds among the competing priorities. These competing priorities for government expenditures include defense, roads, education, and many others, as well as those within the health sector itself. Traditionally, there are competing priorities between the hospital and medical sector and the public health and community programs sector. A political commitment to health must be accompanied by allocation of resources adequate to the scope of the task. Thus, health policy is largely determined by societal priorities and is not a prerogative of government, health care providers, or any institution alone.

As a result of long struggles by trade unions, advocacy groups, and political action, well-developed market economies have come to accept health as a national obligation and essential to an economically successful and well-ordered society. This realization has led to the implementation of universal access systems in most of the industrialized countries. Once initiated, national health systems require high levels of resources, because the health system is labor intensive with relatively high salaries for health care professionals. In these countries, health expenditures consume between 7 and nearly 18 percent of gross domestic product (GDP). Some industrialized countries, notably those in the former Soviet bloc, lacking mechanisms for advocacy, including consumer and professional opinion, tended to view health with a political objective of social benefits, and also as a "non-productive" consumer of resources rather than a producer of new wealth. As a result, budget allocations and total expenditures for health as a percentage of GDP were well below those of other industrialized countries (Figure 12.1). Salaries for health personnel in the Semashko system were low compared to industrial workers

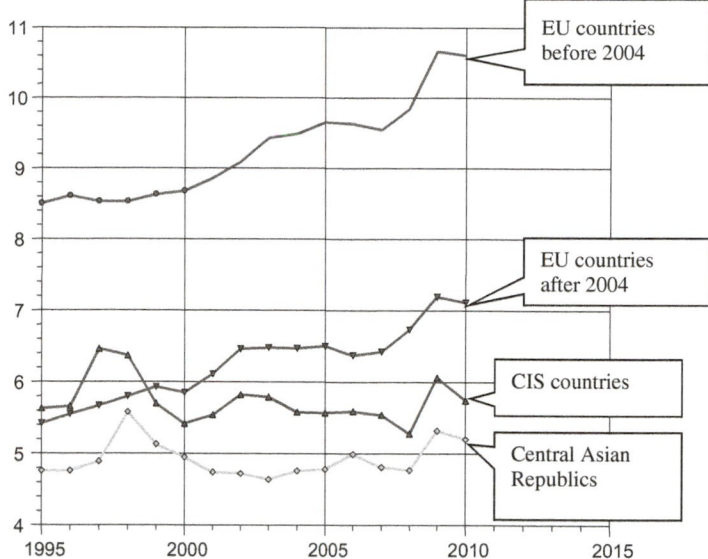

FIGURE 12.1   Total health expenditure as percentage of gross domestic product (GDP), World Health Organization estimates.

in the "productive" sectors. Furthermore, industrial policy did not promote modern health-related industries, compared to the military or heavy industrial sectors.

The former socialist countries of Eastern Europe which have joined the European Union (EU) have gradually increased allocation to health from 5.44 percent of GDP in 1995 to 7.1 percent in 2010, while the pre-2004 members of the EU increased their expenditures from 8.5 percent of GDP to 10.6 percent. The average spend in the Commonwealth of Independent States (Russia, Ukraine, and others) increased from 5.6 percent in 1995 to 5.74 percent in 2010, and in the Central Asian Republics (Kazakhstan, Uzbekistan, and others) from 4.8 percent in 1995 to 5.2 percent in 2010 (WHO Health for All database, January 2013). However, Russian health expenditure in 2011 was still only 6.2 percent of GDP and there is a lingering idea of health being a non-productive investment. The developing countries generally spend under 4 percent of GNP on health, because health is addressed as a relatively low political priority, and they depend very much on international donors for even the most basic of public health programs such as immunization.

Financing of health care and resource allocation requires a balance among primary, secondary, and tertiary care. Economic assessment, monitoring, and evaluation are part of determining the health needs of the population. Regulatory agencies are responsible for defining goals, priorities, and objectives for resulting services. Targets and methods of achieving them provide the basis for implementation and evaluation strategies. Planning requires written plans that include a statement of vision, mission objectives, target strategies, methods, and coordination during the implementation. Designation and evaluation of responsibilities, resources to be committed, and

participants and partners in the procedure are part of the continuous process of management.

The dangers of taking a "wrong" direction may be severe, not only in terms of financial costs, but also in terms of high levels of preventable morbidity and mortality. Health policy is often as imprecise a science as medicine itself. The difference is that inappropriate policy can affect the lives and well-being of very large numbers of people, as opposed to an individual being harmed by the mistake of one doctor. There may be no "correct" answer, and there are numerous controversies along the path. Health policy remains more an "art" than the more quantitative and seemingly precise field of health economics. Societal, economic, and cultural factors as well as personal habits have long been accepted as having an important impact on vulnerability to coronary heart disease. But other factors such as the degree of control over one's life, as suggested in studies of British civil servants, religiosity, and the effects of migration on families left behind are part of the social gradients and inequalities seen in many disease entities, with consequent excess morbidity and mortality in some contexts, such as in Russia and Ukraine.

Health policy, planning, and management are interrelated and interdependent. Any set goal should be accompanied by planning how to attain it. A policy should state the values on which it is based, as well as specify sources of funding, planning, and management arrangements for its implementation. Examination of the costs and benefits of alternative forms of health care helps in making decisions as to the structure and the content of health care services, both internal structures (within one organization) and external linkages (intersectoral cooperation with other organizations). The methods chosen to attain the goals become the applied health policy.

The World Health Organization's (WHO's) 1977 Health for All strategy was directed at the political level and intended to increase governmental awareness of health as a key component of overall development. To some degree it succeeded despite its expansive aspirations, and even after nearly 40 years, its objectives remain worthwhile even in well-developed health systems. Within health, primary care was stressed as the most effective investment to improve the health status of the population. In 1993, the World Bank's *World Development Report* adopted the Health for All strategy and promoted the view that health is an important investment sector for general economic and social development. However, economic policies promoting privatization and deregulation in the health sector threaten to undermine this larger goal in countries with national health systems.

In the USA, major steps are being taken to increase coverage of health insurance for all as the number of uninsured Americans declined from 50 million people uninsured in 2010 to 48.6 million in 2011, edging down from 16.3 to 15.7 percent of the total population. Further decline in the uninsured population is expected as the Patient Protection and Affordable Care Act (PPACA, or "Obamacare") comes into effect in the coming years, bringing many millions of Americans into health insurance and meeting federal standards of fair practices such as eliminating exclusion for pre-existing conditions by private insurers. The PPACA comes into effect on 1 January 2014 and will guarantee coverage for pre-existing conditions, and ensure that premiums cannot vary based on gender or medical history. It will subsidize the cost of coverage, and new state-based health insurance exchanges will help consumers to find suitable policies. It will introduce many preventive care measures into public and private insurance plans, and will promote efficiencies in the health systems including reduction in fraudulent claims and wasteful funding systems. All of this will require skilled management in the components of the health system (see Chapters 10 and 13).

In the New Public Health, health promotion, preventive care, and clinical care are all part of public health because the well-being of the individual and the community requires a coordinated effort from all elements of the health spectrum. Establishing and achieving national health goals require planning, management, and coordination at all levels. The achievement of health advances depends on organizations and structured efforts to reach health goals such as those defined above, and more recently by the United Nations (UN) in the Millennium Development Goals (MDGs) (see Chapter 2), and requires some understanding of organizations and how they work.

THE ELEMENTS OF ORGANIZATIONS

The study of organizations developed within sociology, but has gradually become a multidisciplinary activity involving many other professional fields, such as economics, anthropology, individual and group psychology, political science, human resources management, and engineering. Organizations, whether in the public or private sector, exist within an external environment, and utilize their own structure, participants, and technology to achieve goals. For an organization to survive and thrive, it must adapt to the physical, social, cultural, and economic environment.

Organizations participating in health care establish the connection between service providers and consumers, with the goal of better health for the individual and the community. The factors for this include legislation, regulation, professionalism, instrumentation, medications, vaccines, education, and other modalities of intervention for prevention and treatment. The social structure of an organization may be formal (structured stability), natural (groupings reflecting common interests), or open (loosely coupled, interacting, and self-adjusting systems to achieve goals).

Formal systems are deliberately structured for the purposes of the organization. Natural systems are less formal structures where participants work together collaboratively to achieve common goals defined by the organization. Open systems relate elements of the organization to coalitions of partners in the external environment to achieve mutually desirable goals. In the health system, structures should focus on prevention and treatment of disease and improvement in health and well-being of society. The social structure of an organization includes values, norms, and roles governing the behavior of its participants.

Government, business, or service organizations, including health systems, require organizational structures, with a defined mission and set of values, in order to function. An organizational structure needs to be tailored to the size and complexity of the entity and the goals it wishes to achieve. The structure of an organization is the way in which it divides its labor into distinct tasks and coordinates them. The major organizational models, which are not mutually exclusive and may indeed be complementary, are the pyramidal (bureaucratic) and network structures. The bureaucratic model is based on a hierarchical chain of command with clearly defined roles. In contrast, the matrix or network organization brings together professional or technical people to work on specific programs, projects, or tasks. Both are vital to most organizations to meet ongoing responsibilities and to address special challenges.

SCIENTIFIC MANAGEMENT

Some classic organization theory concepts help to set the base for modern management ideas as applied to the health sector. *Scientific management* was pioneered by Frederick Winslow Taylor (1856–1915). His work was pragmatic and based on empirical engineering, developed in observational studies carried out for the purpose of increasing worker, and therefore system, efficiency. Taylor's industrial engineering studies of scientific management were based on the concept that the best way to improve worker productivity was by designing improved techniques or methods used by workers. This theory viewed workers as instruments to be manipulated by management, and assumed that efficient, rationally planned methods would produce better industrial results and industrial peace as the tasks of managers and workers would be better defined.

Time and motion studies analyzed work tasks to seek more efficient methods of work in factories. Motivation of workers was seen to be related to payment by piecework and economic self-interest to maximize productivity. Taylor sought to improve the productivity of each worker and to make management more efficient in order to increase earnings of employers and workers. He found that the worker was more efficient and productive if the worker was goal oriented rather than task oriented. This approach dominated organization theory during the early decades of the twentieth century.

Resistance to Taylor's ideas came from both management and labor; the former because it seemed to interfere with managerial prerogatives and the latter because it expected the worker to function at top efficiency at all times. However, Taylor's work had a lasting influence on the theory of work and organizations.

BUREAUCRATIC PYRAMIDAL ORGANIZATIONS

The traditional pyramidal bureaucratic organization is classically seen in the military and civil services, but also in large-scale industry, where discipline, obedience, and loyalty to the organization are demanded, and individuality is minimized. This form of organization was analyzed by sociologist Max Weber between 1904 and 1924. Leadership is assigned by higher authority, and is presumed to have greater knowledge than members lower down in the organization. This form of organization is effective when the external and internal environments, the technology, and functions are relatively well defined, routine, and stable.

The pyramidal system (Figure 12.2) has an apex of policy and executive functions, a middle level of management personnel and support staff, and a base of the people who produce the output of the organization. The flow of information is generally one way, from the bottom to the top level, where decisions are made for the detailed performance of duties at all levels. Lateralizing the information systems so that essential data can be shared to help staff at the middle and field or factory-floor levels of management is generally discouraged because this may promote decentralized rather than centralized management. Even these types of organization have increasingly come to emphasize small-group loyalty, leadership initiative, and self-reliance.

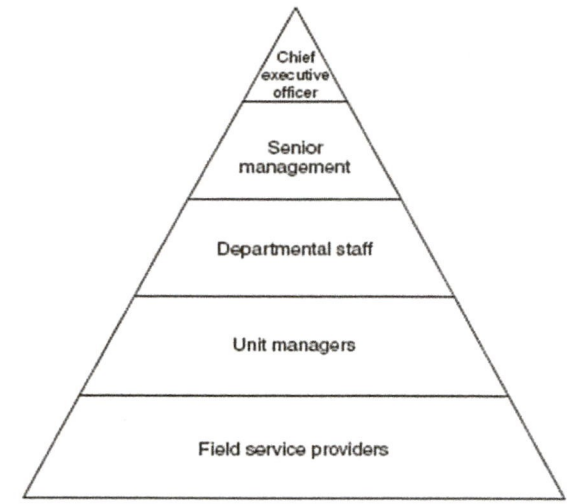

FIGURE 12.2   Pyramid structure of organizations.

The bureaucratic organization has the following characteristics:

- There is a fixed division of labor with a clear jurisdiction and based on assignments, which are subject to change by the leader.
- There is a hierarchy of offices, with each lower functionary controlled and supervised by a higher one.
- A documented, stable set of rules governs decisions and actions.
- Property and rights belong to the office, not the person in the office.
- Officials are selected on the basis of qualifications; salaries and benefits are based on technical competencies.
- Employment is viewed as a tenured career for officials, after an initial trial period.

The bureaucratic system, based on formal rationality, structure, and discipline, is widely used in production, service, and governmental agencies, including military and civilian departments and agencies.

ORGANIZATIONS AS ENERGY SYSTEMS

Health systems, like other organizations, are dynamic and require continuous management, adjustment, and systems control. Continuous monitoring and feedback, evaluation, and revision help to meet individual and community needs. The input–process–output model (Figure 12.3) depends on feedback systems to make the administrative or educational changes needed to keep moving towards the selected objectives and targets.

Organizations use resources or inputs that are processed to achieve desired results or outputs. The resource inputs are money, personnel, information, and supplies. Process is the accumulation of all activities taken to achieve the results intended. Output, or outcome, is the product, its marketing,

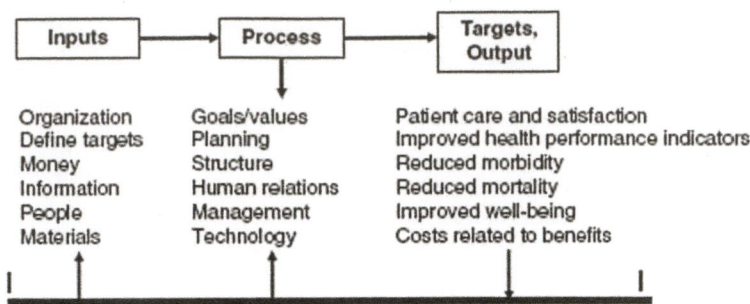

FIGURE 12.3   Organizations as energy systems.

its reputation and quality, and profit. In a service sector such as health, output or impact can be measured in terms of reduced morbidity and/or mortality, improved health, or number of successfully treated and satisfied patients at affordable costs. The management system provides the resources and organizes the process by which it hopes to achieve the established goals.

Program implementation requires systematic feedback for the process to work effectively. When targets are set and strategy is defined, resources, whether new or existing, are placed at the service of the new program. Management is then responsible for using the resources to achieve the intended targets. The results are the outcome or output measures, which are evaluated and fed back to the input and process levels.

Health systems consist of many subsystems, each with an organization, leaders, goals, targets, and internal information systems. Subsystems need to communicate within themselves, with peer organizations, and with the macro (health) system. Leadership style is central to this process. The surgeon as the leader of the team in the operating room depends on the support and judgment of other crucial people on the team, such as anesthesiologists, operating room nurses, pathologists, radiologists, and laboratory services, all of whom lead their own teams. Hospital and public health directors cannot function without a high degree of decentralized responsibility and a creative team approach to quality development of the facility.

Health systems management includes analysis of service policy, budget, decision-making in policy, as well as operation, regulation, supervision, provision, maintenance, ethical standards, and legislation. Policy formulation involves a set of decisions made in pursuit of a course of action for achieving selected health targets, such as those in the MDGs or continuing to update *Healthy People 2020* health targets in the USA (see Chapter 2).

Cybernetics and Management

Cybernetics, a term coined by Norbert Wiener, refers to systems or organizations which are dependent on each other to function, and whose interdependence requires flexibility of

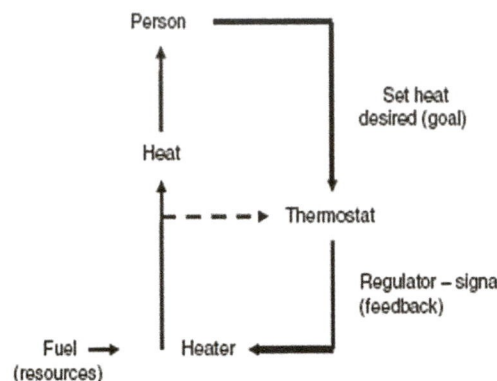

FIGURE 12.4   Cybernetic feedback control organization.

response. Cybernetics gained wide credence in engineering in the early 1950s, and feedback systems became part of standard practice of all modern management systems. Its later transformations appeared in operating service systems, as information for management. Application of this concept is entering the health sector. Rapid advances in computer technology, by which personal computers have access to Internet systems and large amounts of data, have already enhanced this process. In mechanistic systems, the behavior of each unit or part is constrained and limited; in organic systems, there is more interaction between parts of the system. The example used in Figure 12.4 is the use of a thermostat to control the temperature and function of a heater according to conditions in the room. This is also described as a feedback system.

Cybernetics opens up new vistas on the use of health information for managing the operation of health systems. A database for each health district would allow assessment of current epidemiological patterns, with appropriate comparisons to neighboring districts or regional, state, and national patterns.

Data would need to be processed at state or national levels in comparable forms for a broad range of health status indicators. Furthermore, the data should be prepared for online availability to local districts in the form of current health profiles. Thus, data can be aggregated and disaggregated to meet the management needs of the service, and may be used to generate real targets and measure progress towards meeting them. A geographic information system may demonstrate high rates of a disease in a region due to local population risk factors, and thus become the basis for an intervention program.

In the health field, the development of reporting systems based on specific diseases or categories has been handicapped by a lack of integrative systems and a geographic reporting approach. The technology of computers and the Internet should be used to process data systems in real time and in a more user-friendly manner. This would enable local health authorities and providers to respond to actual health problems of the communities.

Health is a knowledge-based service industry, so that knowledge management and information technology are extremely important parts of the New Public Health, not only in patient care systems in hospitals, but also in public health delivery systems in the community, school, place of work, and home. Mobilization of evidence and experience of best practices for policies and management decision-making is a fundamental responsibility of health leaders. The gap between information and action is wide and presents an ethical as well as a political challenge. Regions with the most severe health problems lack trained personnel in assessment and exploitation of current state-of-the-art practices and technology in many practical public health fields, including immunization policy and in management of risk factors for stroke.

Knowledge and evidence are continuously evolving, but the capacity to access and interpret information is commonly poorly implemented in many countries so that very large numbers of people die of preventable diseases even when there are, overall, sufficient resources to address the challenges. International guidelines are vital to help countries to adopt current standards and make use of the available knowledge for public policy. Political support and openness to international norms are crucial to this process of technology diffusion and building the physical and human resource infrastructure needed to achieve better population health with current best practices. Development of health standards in low-income countries is progressing but is seriously handicapped by low levels of funding, lack of emphasis on training sufficient and appropriate human resource personnel and administrative support to promote measures which can save millions of lives. In high-income countries, the slow adoption of best international health standards can have harsh effects on population health, such as in the long delay in adopting national health insurance in the USA. In the European context, the EU has failed to adopt a harmonized recommended immunization program, which is badly needed for the new and potential members, as well as the older member countries. In countries of the former socialist bloc, mortality rates from stroke and coronary heart disease are slowly declining but remain two to four times higher than in countries of Western Europe (see Chapter 5). Systems management requires access to and the use of knowledge to bridge these gaps.

Adoption and adaptation of knowledge to address local problems are essential in a globalized world, if only to prevent the international spread of threatened pandemics or adoption of unhealthy lifestyles (diet, smoking, and lack of exercise) to middle-income countries, which are developing a growing middle class alongside massive poverty. The application of knowledge and experience that has been successful in leading countries can foster innovation and create experience that may generate a local renewal process. Management is crucial to address the complex "strategy areas for improving performance of health organizations: standards and guidelines, organizational design, education and training, improved process, technology and tool development, incentives, organizational culture, and leadership and management" (Bradley et al., 2012). Managing a knowledge-based service industry or facility relies on leadership, collaboration to realize the potential of technology, professional skills, and social capital to the address the health problems faced by all countries.

TARGET-ORIENTED MANAGEMENT

The management of resources to achieve productivity and measurable success has been characterized and accompanied by the development of systems of organizing people to create solutions to problems or to innovate towards defined objectives.

Operations Research

Operations research is a concept developed by British scientists and military personnel in search of solutions for specific problems of warfare during World Wars I and II. The approach was based on the development of multidisciplinary teams of scientists and personnel. The development of the Anti-Submarine Detection Investigation Committee for underwater detection of submarines during World War I characterized and pioneered this form of research. The famous Bletchley Park Enigma code-breaking success in Britain and the Manhattan Project, in which the USA assembled a powerful research and development team which produced the atomic bomb, are prime World War II examples.

Team- and goal-oriented work was very effective in problem solving under the enormous pressure of wartime needs. It also influenced postwar approaches to developmental needs in terms of applied science in such areas as the aerospace and computer industries. The computer hardware and software industries are characterized by innovation conceived and developed through informal working groups with a high level of individual competence, peer group dynamism, and commitment to problem solving. Thus, the "nerds" of Macintosh and Microsoft beat the "suits" of IBM in innovation and introduction of the personal computer. Similar startup groups, such as Google and Facebook, successfully took the Internet to startling new levels of global applications, showing the capacity of innovation from California's Silicon Valley and its counterparts in other places in the USA and worldwide.

In the health field, innovation in organization developed prepaid group practice which became the health maintenance organization (HMO), and later the managed care organization (MCO), now a major, if controversial, factor in health care provision in the USA. Other examples may be found in multidisciplinary research teams working on vaccines or pharmaceutical research, and in the increasingly multidisciplinary function of hospital departments and especially highly interdependent intensive care or home care teams.

Management by Objectives

The business concept of management by objectives (MBO), pioneered in the 1960s, has become a common theme in health management. MBO is a process whereby managers of an enterprise jointly identify its goals, define each individual's areas of responsibility in terms of the results expected, and use these measures as guides for operating the unit and assessing the contributions of its members.

The common goals and then the individual unit goals must be established, as well as the organizational structure developed to help achieve these goals. The goals may be established in terms of outcome variables, such as defined targets for reduction of infant or maternal mortality rates. Goals may also be set in terms of intervening or process variables, such as achieving 95 percent immunization coverage, prenatal care attendance, or screening for breast cancer and mammography. Achievements are measured in terms of relevancy, efficiency, impact, and effectiveness.

The MBO approach has been subject to criticism in the field of business management because of its stress on mechanical application of quantitative outcome measures and because it ignores the issue of quality. This approach had great influence on the adoption of the objective of "Health for All" by the WHO, and on the US Department of Health and Human Services' 1979 health targets for the year 2000, later as Healthy People 2010, and now, based on these experiences and new evidence, renewed as Healthy People 2020. Targeting diseases for eradication may contribute to institution building by developing experience and technical competence to broaden the organizational capacity.

However, categorical programs or target-oriented programs can detract from the development of more comprehensive systems approaches. Addressing the MDGs of reducing child and maternal mortality is at odds to some extent with targeting poliomyelitis for eradication and reliance on national immunization days, which distract planning and resource allocation for the buildup of the essential public health infrastructure for the basic immunization system so fundamental to child health. Immunization and human immunodeficiency virus (HIV) control draw the major part of donor resources in developing countries, while education for strengthening human resources and infrastructure draw less donor attention. A balance between comprehensive and categorical approaches requires very skilled management. The MDGs agreed to by the UN in 2001 as targets for the year 2015 provide a set of measurable objectives and a formula for international aid and for national development planning to help the poorest nations, with the wealthy nations providing aid, education, debt relief, and economic development through fairer trade practices. They are now being reviewed for extension to 2020 based on experience to date, with successes and failures, and recognizing the vital importance of non-communicable diseases as central to the health burden of low- and middle-income countries.

HUMAN RELATIONS MANAGEMENT

Management is the activity of coordinating and integrating organizational resources, including people, money, materials, time, and space. The purpose is to achieve defined/stated objectives as effectively and efficiently as possible. Whether in terms of producing goods and profits or in delivering services effectively, management deals with human motivation and behavior because workers are the key to achieving goals. Knowledge and motivation of the individual client and the community are also essential for achieving good health. Thus, management must take into account the knowledge, attitudes, beliefs, and practices of the consumer as much as or more than those of the people working within the system, as well as the general cultural and knowledge level in the society, as reflected in the media, political opinions, and organizations addressing the issues.

Management, like medicine, is both a science and an art. The application of scientific knowledge and technology in medicine involves both theory and practice. Similarly, management practice involves elements of organizational theory, which, in turn, draws on the behavioral and social sciences and quantitative methodologies. Sociology, psychology, anthropology, political science, history, and ethics contribute to the understanding of psychosocial systems, motivation, status, group dynamics, influence, power, authority, and leadership. Quantitative methods including statistics, epidemiology, survey methods, and economic theory are also basic to development of systems concepts. Comparative institutional analysis helps principles of organization and management to develop, while philosophy, ethics, and law are part of understanding individual and group value systems.

Organizational theory, a relatively new discipline in health, as an academic study of organizations, addresses health-related issues using the methods of economics, sociology, political science, anthropology, and psychology. The application of organizational theory in health care has evolved and become an integral part of training for, and the practice of, health administration. Related practical disciplines include human resources, and industrial and organizational psychology. Translation of organizational theory into management practice requires knowledge, planning, organization, mobilization of professional and other staff support for evidence-based best practices, assembly of resources, motivation, monitoring and control. Health organizations have become more complex and costly over time, especially in their mix of specializations in science, technology, and professional services.

Organization and management are particularly crucial for successful application of the principles of the New Public Health, as it involves integration of traditionally separate health services. Delegation of responsibilities in health systems, such as in intensive care units, is fundamental to success in patient care, with nurses taking increasing responsibility for the management of the severely ill patient suffering from multiple system failure. Delegation or devolution of health care responsibilities to non-medical practitioners has been an ongoing development affecting nurse practitioners, physician assistants, paramedics, community health workers and others, as discussed in Chapter 14. It is a vital process to provide needs not met by physicians because of shortages and inappropriate location or specialty preferences that leave primary care or other medical specialties unable to meet community and patient needs.

The Hawthorne Effect

Elton Mayo of the Harvard School of Business carried out a series of observational studies at the Hawthorne, Illinois, plant of the Western Electric Company between 1927 and 1932. Mayo and his industrial engineer, along with psychologist colleagues, made a major contribution to the development of management theory. Mayo began with industrial engineering studies of the effect of increased lighting on production at an assembly line. This was followed by other improvements in working conditions, including reduced length of the working day, longer rest periods, better illumination, color schemes, background music, and other factors in the physical environment. These studies showed that production increased with each of these changes and improvements. However, the researchers discovered, to their surprise, that production continued to increase when the improvements were withdrawn. Furthermore, in a control group where conditions remained the same, productivity also grew during the study period. These results led Mayo to conclude that the performance of workers improved because of a sense that management was interested in them, and that worker participation contributes to improved production.

Traditionally, industrial management viewed employees as mechanistic components of a production system. Previous theory was that productivity was a function of working conditions and monetary incentives. What came to be known as the Hawthorne effect showed the importance of social and psychological factors on productivity. Formal and informal social organizations among management and employees were recognized as key elements in productivity, now called *industrial humanism*. Research methods adapted from the behavioral sciences contributed to scientific studies in industrial management. Traditional theories of the bureaucratic model of organization and management were modified by the behavioral sciences. This led to the emergence of the systems approach, or scientific analysis to analyze complex structures or organizations, taking into account the mutually interdependent elements of activities, interactions, and interpersonal relationships between management and workers.

Some revisits to the Hawthorne studies suggest that the data do not support the conclusions, and offer a different interpretation. One is that informal groups such as workers on a production line themselves set standards for work which assert an informal social control outside the authority system of the organization. The informal cohesive group can thus control the norms of the amount of work acceptable to the group, i.e., not "too much" and not "too little". Others point out that the effects were temporary and that there were extraneous factors, but the added value of the Hawthorne effect remains part of the history of and had a culture-changing effect on management theory.

The Hawthorne effect in management is in some ways comparable to the placebo effect in clinical research and health care practice. It is also applied to clinical practice, whereby medical care provided by doctors is measured for specific "tracer conditions" to assess completeness of care according to current clinical guidelines. Review of clinical records has been shown to be a factor in improving performance by doctors in practice, such as in treatment of acute myocardial infarction, management of hypertension, or completeness of carrying out preventive procedures such as screening for cancer of the cervix, breast, or colon (see Chapters 3 and 15). Awareness of being studied is a factor in improved performance or response to an intervention. Studies of clinical practice-based research or public health interventions need to consider whether different types of studies and outcomes are more or less susceptible to the Hawthorne effect (Fernald et al., 2012).

Maslow's Hierarchy of Needs

Abraham Maslow's hierarchy of human needs made an important contribution to management theory. Maslow

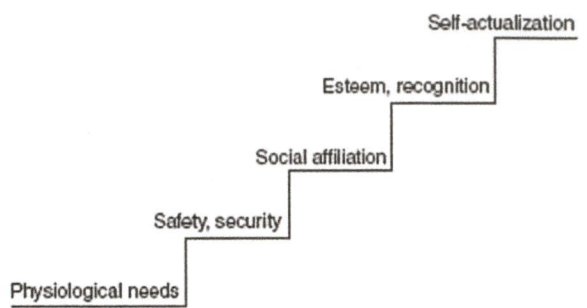

FIGURE 12.5   Maslow's hierarchy of needs.

(1908–1970) was an American psychologist, considered "the father of humanism" in psychology. Maslow defined a prioritization of human needs (Figure 12.5), starting with those of basic physical survival; at higher levels, human needs include social affiliation, self-esteem, and self-fulfillment. Others in the hierarchy include socialization and self-realization; later revisions include cognitive needs.

The survival needs of an employee include a base salary and benefits, including health insurance and pension; the safety and security needs include protection from injury, toxic exposure or excess stress; social needs at work include an identity, pride, friendships, union solidarity, company social activities and benefits; esteem and recognition include job titles, awards, and financial rewards for achievement by individuals, groups, or all employees; and self-actualization includes promotion to more challenging jobs with benefits, both financial and in terms of recognition.

This concept is important in terms of management because it identifies human needs beyond those of physical and economic well-being. It relates them to the social context of the work environment with needs of recognition, satisfaction, self-esteem, and self-fulfillment. Maslow's conclusions opened many positive areas of management research, not only in the motivation of workers in production and service industries, but also in the motivation of consumers.

Maslow's hierarchy of human needs contributed to the idea that workers' sense of well-being is important to management. His theories played an important role in application of sociological theory to client behavior, just as the topic of personal lifestyle in health became a central part of public health and clinical management of many conditions, such as in risk factor reduction for cardiovascular diseases. This concept fits well with the epidemiological studies referred to in the Introduction, such as those showing strong relationships with sociopolitical factors as well as socioeconomic conditions.

Theory X–Theory Y

Theory X–Theory Y (Table 12.1), developed by clinical psychologist and professor of management Douglas McGregor

in the 1960s, examined two extremes in management assumptions about human nature that ultimately affect the operations of organizations. Organizations with centralized decision-making, a hierarchical pyramid, and external control are based on certain concepts of human nature and motivation. McGregor's theory, drawing on Maslow's hierarchy of needs, describes an alternative set of assumptions that credit most people with the capacity for self-direction.

Traditional approaches to organization and management stress direction and external control. Theory X assumes that workers are lazy, unambitious, uncreative, and motivated only by basic physiological needs or fear. Theory Y places stress on integration and self-control. This model provides a more optimistic leadership model, emphasizing management development programs and promoting human potential, assuming that, if properly motivated, people can be self-directed and creative at work, and that the role of management is to unleash this potential in workers with performance appraisal. Many other theories of motivation and management have been developed to explain human behavior and how to utilize inherent skills to produce a more creative work environment, reduce resistance to change, reduce unnecessary disputes, and ultimately create a more effective organization.

Variants of the human motivation approach in management carried the concept further by examining industrial organization to determine the effects of management practices on individual behavior and personal growth within the work environment. They describe two contrasting models of workforce motivation. Theory X assumes that management produces immature responses on the part of the worker: passivity, dependence, erratically shallow interests, short-term perspective, subordination, and lack of self-awareness. In contrast, at the other end of the immaturity–maturity spectrum was the mature worker, with an active approach, an independent mind capable of a broad range of responses, deeper and stronger interests, a long-term perspective, and a high level of awareness and self-control. This model has been tested in a variety of industrial settings, showing that giving workers the opportunity to grow and mature on the job helps them to satisfy more than basic survival needs and allows them to use more of their potential in accomplishing organizational goals. This model became widely influential in human resource management theory of organizational behavior, organizational communication, and organizational development, and in the practical management of business and service enterprises.

In *The Motivation to Work* (1959), US clinical psychologist Frederick Herzberg wrote of his motivation–hygiene theory. He developed this theory after extensive studies of engineers and accountants, examining what he called *hygiene factors* (i.e., administrative, supervisory, monetary, security, and status issues in work settings). His motivating factors included achievement, recognition of

TABLE 12.1 Theory *X*–Theory *Y*

Theory *X*	Theory *Y*
Work is inherently distasteful to people	Work is as natural as play in favorable conditions
Most people are not ambitious, have little desire for responsibility, and prefer to be directed	Self-control is indispensable in achieving goals
Most people have little capacity for creatively solving organizational problems	The capacity for creativity in solving organizational problems is widely distributed in the population
Motivation occurs only at physiological and safety levels	Motivation exists at the social, esteem, and self-actualization levels, as well as the physiological and security levels
Most people must be closely controlled and often coerced to achieve organizational objectives	People can be self-directed and creative at work if properly encouraged

Source: Adapted from McGregor D. The human side of enterprise. New York: McGraw-Hill; 1960.

accomplishment, challenging work, and increased responsibility with personal and collective growth and development. He proved that the motivating factors had a substantial positive effect on job satisfaction.

These human resource theories of management helped to change industrial approaches to motivation from "job enrichment" to a more fundamental and deliberate upgrading of responsibility, scope, and challenge of work, by letting workers develop their own ways of achieving objectives. Even when the theories were applied to apparently unskilled workers, such as plant janitors, the workers changed from an apathetic, poorly performing group into a cohesive, productive team, taking pride in their work and appearance. This approach gave members of the team the opportunity to meet their human self-actualization needs by taking greater responsibility for problem solving, and it resulted in less absenteeism, higher morale, and greater productivity with improved quality.

Rensis Likert, with McDougal and Herzberg, helped to pioneer the "Human Relations School" in the 1960s, applying human resource theory to management systems and styles. Likert classified his theory into four different systems, as follows.

- *System 1* – Management has no confidence or trust in subordinates, and avoids involving them in decisions and goal setting, which are made from the top down. Management is task oriented, highly structured, and authoritarian. Fear, punishment, threats, and occasional rewards are the principal methods of motivation. Worker–management interaction is based on fear and mistrust. Informal organizations within the system often develop that lead to passive resistance of management and are destructive to the goals of the formal organization.
- *System 2* – Management has a condescending relationship with subordinates, with some degree of trust and confidence. Most decisions are centralized, but some decentralization is permitted. Rewards and punishments

are used for motivation. Informal organizations become more important in the overall structure.

- *System 3* – Management places a greater degree of trust and confidence in subordinates, who are given a greater degree of decision-making powers. Broad policy remains a centralized function.
- *System 4* – Management is seen as having complete confidence in subordinates. Decision-making is dispersed, and communication flows upward, downward, and laterally. Economic rewards are associated with achieving goals and improving methods. Relationships between management and subordinates are frequent and friendly, with a sense of teamwork and a high degree of mutual respect.

Case studies showed that a shift in management from Likert system 1 towards system 4 radically changed the performance of production, cut manufacturing costs, reduced staff turnover, and increased staff morale. Furthermore, workers and managers both shared a concern for the quality of the product or service and the competitiveness and success of their business. The health industry includes highly trained professionals and paraprofessional workers who function as a team with a high degree of cohesion, mutual dependence, and autonomy, such as a surgical or an emergency department team.

NETWORK ORGANIZATION

The network, or task-oriented working group, is basically a more democratic and participatory form of organization meant to elicit free interchange of concerns and ideas. This is a more organic form of organization, best suited to be effective for adaptation when the environment is complex and dynamic, when the workforce is largely professional, and when the technology and system functions change rapidly. Complexities and technological change require information, expertise, flexibility, and innovation, strengths best promoted in free exchange of ideas in a mutually stimulating environment.

In a network organization, leadership may be formal or informal, assigned to a particular function, which may be temporary, medium term, or permanent, to achieve a single defined task or develop an intersectoral program. The task force is usually for a short-term specific assignment; a working group, often for a medium-term project, such as integrating services of a region; and a committee for permanent tasks such as monitoring an immunization program.

Significant advantages of this form of organization are the challenge and the sharing of information and responsibility, which give professionals responsibility and job satisfaction by providing the opportunity to demonstrate their creativity. Members of the task force may each report within their own pyramidal structure, but as a group they work to achieve the assigned objective. They may also be interdisciplinary or interagency working groups to review the state of the art in this particular issue as documented in reports and professional literature, and to coordinate activities, review previous work, or plan common future activities.

An ongoing network organization may be a government cabinet committee to coordinate government policy and the work of various government departments, or a joint chiefs of staff to coordinate the various armed services. This approach is commonly used for task groups wherein interdisciplinary teams of professionals meet to coordinate functions of a department in a hospital, or where a multidisciplinary group of experts is established with the specified task of a technical nature.

Network organizational activity is part of the regular functions of a health professional. Informal networking is a day-to-day activity of a physician in consultations with colleagues and also a part of more formalized network groups. The hospital department must, to a large extent, function as a network organization with different professionals working as a team more effectively than would be possible in a strictly authoritarian pyramidal model. A ministry of health may need to develop a joint working group with the ministry of transport, the police, and those responsible for standards of motor vehicles to seek ways to reduce road accident deaths and injuries. If a measles eradication project is envisioned, a multidisciplinary and multiorganizational team, or a network, should be established to plan and carry out the complex of tasks needed to achieve the target (Figure 12.6).

In a public health context, a task group to determine how to reduce obesity rates in school-aged children, or to eradicate measles locally, might be chaired by the deputy chief medical officer or senior health promotion person; if the project is reduction of obesity among school children, the lead agency may be the department of education, perhaps jointly with the local department of health; if reduction in road traffic deaths is the topic, the lead may be the police department with participation of emergency transportation and hospital emergency room lead personnel. Members

Chairperson/Facilitator/Coordinator

A	B	C	D
E	F	G	H
I	J	K	L

FIGURE 12.6   Network organization structure. Note: The letters represent the participation by individuals or organizations as a task group to achieve a defined goal, as set out in terms of reference and a time-frame.

may include the chief district nurse, an administrative and budget officer, a pharmacist, the chief of the pediatric department of the district hospital, a primary school administrator, a health educator, a medical association representative, the director of laboratories, the director of the supply department, a representative of the department of education, representatives of voluntary organizations interested in the topic, and others as appropriate.

Most organizational structures are mixed, combining elements of both the formal pyramidal and the less structured network structure with a task-oriented mandate. It is often difficult for a rigid pyramidal structure to deal with parallel bodies in a structured way, so the network approach is necessary to establish working relations with outside bodies to achieve common goals. A network is a democratic functional grouping of those professionals and organizations needed to achieve a defined target, sometimes involving people from many different organizations. The terms of reference of the working group are crucial to its function as well as its composition, time-frame, and access to relevant information. The application of this concept is increasingly central in health care organization as multilevel health systems evolve in the form of managed care or district health systems. These are vertically integrated management systems involving highly professional teams and units whose interdependence for patient care and financial responsibility are central elements of the New Public Health.

TOTAL QUALITY MANAGEMENT

In the USA during World War II, W. Edwards Deming, a physicist and statistician, developed a system of economic and statistical methods of quality control in production industries. Following the war, Deming was invited to teach in Japan and moved from the university to the level of industrial management. Japanese industrialists adopted his principles of management and introduced quality management into all industries, with astonishingly successful results within a decade. The concept, later called total quality management (TQM), has since been adopted widely in production and service industries.

BOX 12.1 Traditional Management Theory

- Quality is expensive.
- Inspection is the key to quality, and control experts and inspectors can assure this.
- Systems are designed by outside experts – no input is needed from workers.
- Work standards, quotas, and targets can help productivity.
- People may be hired when needed and laid off when not needed.
- Rewards and punishments will lead to greater productivity and creativity.
- Buy at the lowest cost.
- Change suppliers frequently, based on price alone.
- Profits are based on keeping costs down and revenue high.
- Profit is the most important indicator of a company.

BOX 12.2 Total Quality Management

- Quality leads to lower costs and less waste.
- Inspection is after the damage is done; worker involvement in quality services eliminates defects.
- Quality is determined by management.
- Most defects are caused by the system, not the worker.
- Eliminate work standards and quotas in industry as sole criteria of performance.
- Fear leads to disaster.
- Make workers feel secure in their jobs.
- Judgment, punishment, and reward for above- or below-average performance destroy teamwork essential for quality production.
- Work with suppliers to improve quality and costs.
- Profits are generated by loyal customers – running a company for profit alone is like driving a car by looking in the rearview mirror.

In the Deming approach to company management, quality is the top priority and is the key responsibility of management, not of the workers. If management sets the tone and involves the workers, quality goes up, costs come down, and both customer satisfaction and loyalty increase. Having their ideas listened to, and avoiding a punitive inspection approach, enhances the pride of the workers. It is the responsibility of leadership to remove fear and build mutual participation and common interest. Training is one of the most important investments of the organization. The differences between traditional management and the TQM approach are shown in Boxes 12.1 and 12.2. In societies with growing economies, the role of an educated workforce becomes greater as information technology and services, such as health, become larger parts of the economy and require professionalism and self-motivating workers.

The TQM approach integrates the scientific management and human relations approaches by giving workers credit for intellectual capacity and expects them to use it to analyze and improve the tasks they perform. Even more, this approach expects workers at all levels to contribute to better quality in the process of design, manufacture, and even marketing of the product or the service.

The TQM ideas were revolutionary and successful when applied in business management in production industries. The TQM concept is much in discussion in the service industries. The WHO has adapted TQM to a model called continuous quality improvement (CQI), with the stress on mutual responsibilities throughout a health system for quality of care. The application of TQM and CQI approaches is discussed in Chapter 15, including the external regulatory and self-development TQM approaches.

In the health sector, issues such as prevention of health facility-acquired infections require staff dedicated to promoting a culture of cleaning, frequent and thorough hand washing, sterilization, isolation techniques, intravenous and intratracheal catheter and tube care technique, and immunization of hospital personnel. These and many other cross-disciplinary measures promote patient safety and prevent the costly and frequently deadly effects of serious respiratory or urinary tract injection acquired in hospitals or other health care facilities.

CHANGING HUMAN BEHAVIOR

Human behavior is individual but takes place in a social context. Changes to individual behavior are needed to reduce risk factors for many diseases. Change can be threatening; it requires alteration, substitution, transformation, or modification of purposes, procedures, methods, or style. The implementation of plans usually requires some change, which often meets resistance. The resistance to change may be professional, technical, psychological, political, emotional, or a mix of all of these. The manager of a health facility or service has to cope with change and gather the support of those involved to participate in creating or implementing the change effectively.

The behavior of the worker in a production or service industry is vital to the success of the organization. Equally important is the behavior of the purchaser or consumer of the product or service. Diagnosing organizational problems is an important skill to bring to leadership in health systems. Even more important is the ability to identify and alter the variables that require change and adaptation to improve the performance of the organization. High expectations are essential to produce high performance and improved standards of service or productivity. Conversely, low expectations not only lead to low performance, but produce a downward spiraling effect. This applies not only within the organization, but to the individuals and community served, whether in terms of purchase of goods produced or in terms of health-related behavior.

People often resist change because of fear of the unknown. Participation in the process of defining problems, formulating objectives, and identifying alternatives is needed to bring about changes. Change in organizational performance is complex, and this is the test of leadership. Similarly, change at the individual level is essential to achieve the goals of the group, whether this is in terms of the functioning of a health care service unit, such as a hospital, or whether it is an individual's decision to change from smoking to non-smoking status. The health of both an individual and a population depends on the individual health team member's motivation and experience.

The behavior of the individual is important to his or her personal and community health. Even small steps in the direction of a desirable change in behavior should be rewarded as soon as possible (i.e., reinforcing positive performance in increments). Behavior modification is based on the concept that change of behavior starts with the feelings and attitudes within the individual, but can be influenced by knowledge, peer pressure, media coverage, and legislative standards. Change involves a number of elements to define a current or previous starting point:

- *Knowledge* – What is the level of adequate health information?
- *Attitudes* – What is the person's perception of that information?
- *Behavior of the individual* – What does the individual actually do?
- *Behavior of the group* – What are the social norms and acts?
- *Behavior of the organization* – What does the health system do to change these factors?
- *Behavior of society* – What do legislation, regulations, and enforcement say about harmful acts endangering individuals and the public?
- *Preparation for emergencies* – What organizations are in place and organized to meet local or national emergencies, and how are public perception and participation affected by messages from authorities, such as in evacuation of hurricane or tsunami danger zones?
- *Behavior of the media* – How do the media convey public health messages and warnings, and how does this affect behavior or responses?

Change in behavior is vital in the health field: in the organization, in the community, in individual behavior, and in societal regulation and norms. The health belief model (Chapter 2) is widely influential in psychology and health promotion. The belief intervention approach involves programs meant to reduce risk factors for a public health problem. It may require change in the law and in organizational behavior, with involvement and feedback to the people who determine policy, those who manage services, and the community being served.

Obesity in school-aged children is being fought by many measures including healthier menus and banning the sale of high sugar drinks on school property. High cholesterol is being fought on many fronts including dietary change and banning the use of transfats in food processing. Deaths from bulimia are not uncommon and may stem from teenage identification of beauty with ultrathin body image. Banning television and modeling agencies from using models with a very low body mass index is an intervention in advertising which encourages harmful practices that are a danger to health and life. Banning cigarette advertising and smoking in public places promotes behavioral change, as does raising the taxes on cigarettes. Gun control laws are meant to prevent disturbed individuals or political fanatics having easy access to firearms to commit mass murder. Strict enforcement of drinking and driving laws can prevent drunk driving and reduce road traffic deaths (see Chapter 15).

EMPOWERMENT

In the 1980s, major industries in the USA were unable to compete successfully with the Japanese in the consumer electronics and automobile industries. Management theory began to place greater emphasis on empowerment as a management tool. The TQM approach stresses teamwork and involvement of the worker in order to achieve better quality of production. Comparatively, empowerment went further to involve the worker in operation, quality assessment, and even planning of the design and production process. Results in production industries were remarkable, with increased efficiency, less absenteeism, and greater searching for ideas to improve quality and quantity of production, with the worker as a participant in the management and production process.

The concept of empowerment entered the service industries with the same rationale. The rationale is that improvements in quality and effectiveness of service require the active physical and emotional participation of the worker. Participation in decision-making is the key to empowerment. This requires management to adopt new methods that allow the worker, whether professional or manual, to be an active participant. Successful application of the empowerment principles in health care extends to the patient, the family, and the community, emphasizing patients' rights to informed participation in decisions affecting their medical care, and the protection of privacy and dignity.

Diffusion of powers occurs when management of services is decentralized. Delegation of powers to professional groups, non-governmental organizations (NGOs), and advocacy organizations is part of empowerment in health care organizations. Governmental powers to govern or promote areas such as licensure, accreditation, training, research, and service can be devolved to local authorities or NGOs by delegation of authority or transfer of funds. Organizational change may involve decentralization. Institutional changes such as amalgamation of hospitals, long-term care facilities, home care programs, day surgery, ambulatory care, and public health services are needed to produce a more effective use of resources. Integration of services under community leadership and management should encourage transfer of funds within a district health network from institutional care to community-based care. Such changes are a test of leadership skills to achieve cultural change within an organization, which requires behavioral change and involvement of health workers in policy and management of the change process.

STRATEGIC MANAGEMENT OF HEALTH SYSTEMS

Strategic management emphasizes the importance of positioning the organization in its environment in relation to its mission, resources, consumers, and competitors. It requires development of a plan of action or implementation of a strategy to achieve the mission or goal of the organization within acceptable ethical and legal guidelines. Articulation of these is a key role of the management level of an organization. Defining the mission and goals of the organization must take into account the external and internal environment, resources, and operational needs to implement and evaluate the adequacy of the outcomes. The strategy of the organization matches its internal approach with external factors, such as consumer attitudes and competing organizations. Strategy is a set of methods and skills of the health care manager to attain the objectives of a health organization, including:

- providing high-quality care at current professional standards
- innovating to avoid obsolescence
- developing good internal and external professional relationships
- utilizing human resources effectively
- ensuring accountability and accreditation within the local and national environment
- promoting the service to improve market share
- managing financial, human, and other resources efficiently
- promoting the public and professional reputation of the institution.

Policy is the formulation of objectives and priorities. *Strategy* refers to long-range plans to achieve stated objectives, indicating the problems to be expected and how to deal with them. Strategy does not identify all actions to be taken, but it includes evaluation of progress made towards a stated goal. While the term has traditionally been used

---

BOX 12.3 The Strategic Management Process

1. Policy and planning:
   (a) Define mission, goals, and objectives.
   (b) Undertake surveillance.
   (c) Analyze external environment.
   (d) Analyze internal environment.
   (e) Assess capabilities.
   (f) Evaluate strategic choices, short-range.
   (g) Develop strategic planning, long-range.
   (h) Guide the implementation process.
   (i) Communicate policy direction.
2. Implementation:
   (a) Motivate: clearly communicate the goals and plans of the organization.
   (b) Differentiate between short- and long-term goals.
   (c) Ensure that staff understand their responsibilities.
   (d) Ensure provision of adequate resources.
   (e) Promote sense of staff involvement.
   (f) Modify structure to meet needs.
   (g) Delegate authority, assign responsibility.
   (h) Promote interdepartmental coordination and interpersonal relations.
   (i) Promote capacity to deal with change.
   (j) Review policies in keeping with progress towards goals.
   (k) Promote understanding of change and resistance to change.
3. Monitoring
   (a) Evaluate effectiveness.
   (b) Evaluate outcome, lessons learned.
   (c) Revise strategic plan.
   (d) Redeploy resources in keeping with lessons learned.

---

in a military context, it has become an essential concept in management, whether of industry, business, or health care. Tactics are the methods used to fulfill the strategy. Thus, strategic MBO is applicable to the health system, incorporating definitions of goals and targets, and the methods to achieve them (Box 12.3).

Change in health organizations may involve a substantial alteration in the size or relationships between existing, well-established facilities and programs (Table 12.2). A strategic plan for health reform in response to the need for cost containment, redefined health targets, or dissatisfaction with the status quo requires a model or a vision for the future and a well-managed program. Opposition to change may occur for psychological, social, and economic reasons, or because of fear of loss of jobs or changes in assignments, salary, authority, benefits, or status. Downsizing in the hospital sector, with buildup of community health services, is one of the major issues in health reforms in many countries. It can be accomplished over time by naturally occurring vacancies or attrition due to retirement, or by retraining and reassignment, all of which require skilled leadership.

TABLE 12.2 Transformation of Health Care Paradigms

Old Paradigm	New Paradigm
Emphasis on inpatient care	Emphasis on continuum of care
Emphasis on treating illness	Emphasis on maintaining and promoting wellness
Responsibility for the individual patient	Accountable for defined population
Specialists rewarded more than generalists	Greater economic parity between providers
Surgery rewarded more than medical services	Prevention rewarded versus surgery
Goal to fill beds	Provision of care at appropriate level of care
Separate organization, funding of hospital and other services	Integrate health delivery system
Managers run an organization or a department	Managers promote market share
Managers coordinate services	Managers promote intersectoral cooperation

Source: Adapted from Shortell SM, Kaluzny AD. Health care management, organization, design and behavior. 5th ed. Albany, NY: Delmar; 2005.

The introduction of new categories of health workers in hospitals such as phlebotomists, hospitalist doctors, and technicians of all kinds has improved hospital efficiency and safety. Community health has benefited from home care and in many situations community health workers to assist and supervise patient care in remote rural villages and in urban centers, even in high-income countries, with health guides trained to help people to function with chronic illnesses and dementias (see Chapter 14).

HEALTH SYSTEM ORGANIZATION MODELS

The New Public Health is an integration or coordination of many participating health care facilities and health-promoting programs. It is evolving in various forms in different places as networks with administrative and financial interaction between participating elements. Each organization provides its own specific services or groups of services. How they function internally and how they interact functionally and financially are important aspects of the management and outcomes of health systems. The health system functions as a network with formal and informal relationships; it may be very broad and loosely connected as in a highly decentralized system, with many lines of communication, payment, regulation, standards setting, and levels of authority.

The relationship and interchange between different health care providers have functional and economic elements. As an example, an educated adult woman is more likely than an uneducated woman to prepare herself for the requirements of pregnancy by smoking and alcohol or drug cessation, folic acid intake, healthful diet, and attending professional antenatal care. A pregnant woman who is healthy and prepared for pregnancy physically and emotionally, and who receives comprehensive prenatal care, is less likely than a woman whose health is neglected to develop complications and require prolonged hospital care

as a result of childbirth. The cost of good prenatal care is a fraction of the economic cost of treating the potential complications and damage to her health or that of the newborn. A health system is responsible for ensuring that a woman of reproductive age takes folic acid tablets orally before becoming pregnant, has had access to family planning services so that the pregnancy is a desired one, ensures that the space between pregnancies is adequate for her health and that of her baby, and receives adequate prenatal care. An obstetrics department should be involved in assuring or providing the prenatal care, especially for high-risk cases, and delivery should be in hygienic and professionally supervised settings.

Similarly, for children and elderly people, there is a wide range of public health and personal care services that make up an adequate and cost-effective set of services and programs. The economic burden of caring for the sick child falls on the hospital. When there is a per capita grant to a district, the hospital and the primary care service have a mutual interest in reducing morbidity and hence mortality. This is the principle of the HMOs and district health systems discussed elsewhere. It is also a fundamental principle of the New Public Health.

Health care organizations differ according to size, complexity, ownership, affiliations, types of services, and location. Traditionally, a health care organization provides a single type of service, such as an acute care hospital providing episodic inpatient care, or a home health care agency. In present-day health reforms, health care organizations, such as an HMO or a district health system, provide a population-based, comprehensive service program. Each organization must have or develop a structure suited to meet its goals, in both the internal and external environments. The common elements that each organization must deal with include governance of policy, production or service, maintenance, financing, relating to the external environment, and adapting to changing conditions.

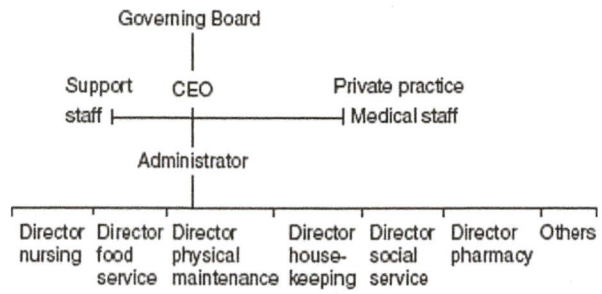

FIGURE 12.7   Functional model of organization.

Functional Model

A functional model of an organization perhaps best suited to the smaller hospital is the division of labor into specific functional departments; for example, medical, nursing, administration, pharmacy, maintenance, and dietary, each reporting through a single chain of command to the chief executive officer (CEO) (Figure 12.7). The governing agency, which may be a local non-profit board or a national health system, has overall legal responsibility for the operation and financial status of the hospital, as well as raising capital for improvements.

The medical staff may be in private practice and work in the hospital with their own patients by application for this right as "attending physician", according to their professional qualifications, or the medical staff may be employed by the hospital in a similar way to the rest of the staff. Salaried medical staff may include physicians in administration, pathology, anesthesia, and radiology, so that even in a private practice market system many medical staff members are hospital employees. Increasingly, hospitals are employing "hospitalists", who are full- or part-time physicians whose work is in the health facility, to provide continuity of inpatient and emergency department services, augmenting the services of senior or attending staff or private practice physicians. This shift is in part related to the increasing numbers of female physicians who run their homes and families as well as practice medicine and who find this mode of work more attractive than full-time private practice.

This model is the common arrangement in North American hospitals. The governing board of a "voluntary", nongovernmental, not-for-profit organization with municipal and community representatives may be appointed by a sponsoring religious, municipal, or fraternal organization.

Corporate Model

The corporate model in health care organization (Figure 12.8) is often used in larger hospitals or where mergers with other hospitals or health facilities are taking place. The CEO delegates responsibility to other members of the senior management team who have operational responsibility for major sectors of the hospital's functioning.

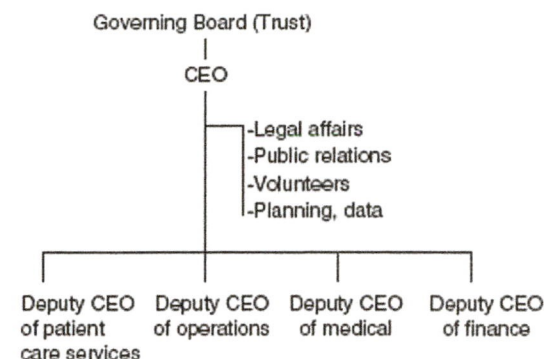

FIGURE 12.8   Corporate model of health care organization.

A variation of the corporate model is the divisional model of a health care organization based on the individual service divisions allowing middle management a high degree of autonomy (Figure 12.9). There is often departmental budgeting for each service, which operates as an economic unit; that is, balancing income and expenditures. Each division is responsible for its own performance, with powers of strategic and operational decision-making authority. This model is used widely in private corporations, and in many hospitals in the USA. With increasing complexity of services, it is also employed in corporate health systems in the USA, with regional divisions.

Matrix Model

The matrix model of a health care organization is based on a combination of pyramidal and network organization. This model is suited to a public health department in a state, county, or city. Individual staff people report in the pyramidal chain of command, but also function in multidisciplinary teams to work on specific programs or projects. A nutritionist in the geriatric department is responsible to the chief of nutrition services but is functionally a member of the team on the geriatric unit. In a laterally integrated health maintenance organization or district health system, specialized staff may serve in both institutional (i.e., hospital) and community health roles (Figure 12.10).

The organizational structure appropriate to one set of circumstances may not be suitable for all. Whether the payment system is by norm (i.e., by predetermined numbers of staff, their salaries, and fixed costs for all services), per diem (i.e., payment of a daily rate times the number of days of stay), historical budget, or per capita in a regional or district health system structure (see Chapters 10 and 11), the internal operation of a hospital will require a model of organization appropriate to it. Hospitals need to modify their organizational structure as they evolve, and as the economics of health care change.

SKILLS FOR MANAGEMENT

Leadership in an organization requires the ability to define the goals or mission of the organization and to develop a

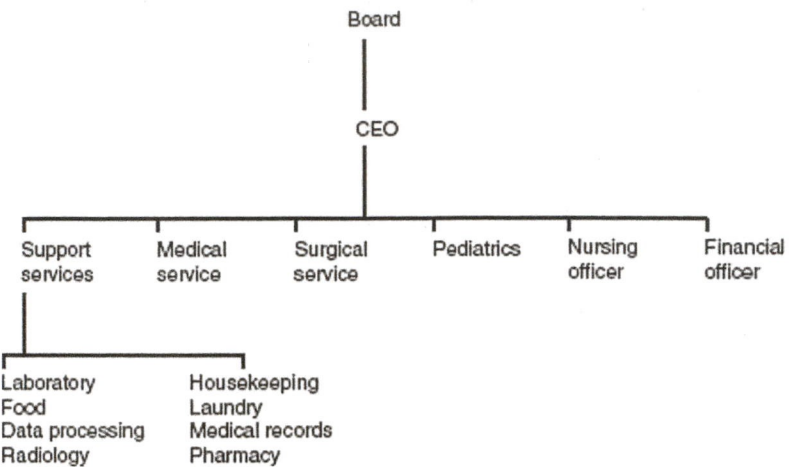

FIGURE 12.9   Divisional model of health care organization.

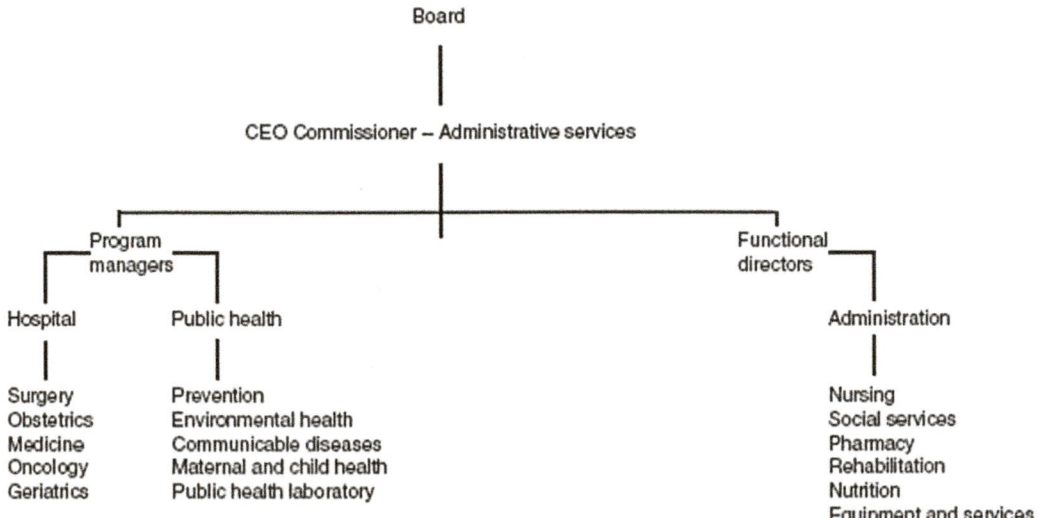

FIGURE 12.10   Matrix model of health care organization.

strategy and define steps needed to achieve these goals. It requires an ability to motivate and engender enthusiasm for this vision by working with others to gain their ideas, their support, and their participation in the effort. In health care as in other organizations, it is easier to formulate plans than to implement them. Change requires the ability not only to formulate the concept of change, but also to modify the organizational structure, the budgeted resources, the operational policies and, perhaps most importantly, the corporate culture of the organization.

Management involves skills that are not automatically part of a health professional's training. Skilled clinicians often move into positions requiring management skills in order to build and develop the health care infrastructure. In some countries, hospital managers must be physicians, often senior surgeons. Clinical capability does not transfer automatically into management skills to deal with personnel, budgets, and resources. Therefore, training in management is vital for the health professional.

The manager needs training for investigations and fact-finding as well as the ability to evaluate personnel, programs, and issues, and set priorities for dealing with the short- and long-term issues. Negotiating with staff and outside agencies is a constant activity of the manager, ranging from the trivial to major decisions with wide implications. Perhaps the most crucial skill of the manager is communication: the ability to convey verbal, written, or unwritten messages that are received and understood and to assess the responses as an equal part of the exchange.

Interpersonal skills are a part of management practice. The capable manager can relate to personnel at all levels in an open and equal manner. This skill is essential to help foster a sense of pride and involvement of all personnel in working towards the same goals and objectives, and to show that each member of the team is important to meeting the objectives of the organization. At the same time, the manager needs to communicate information, especially as to how the organization is doing in achieving its objectives.

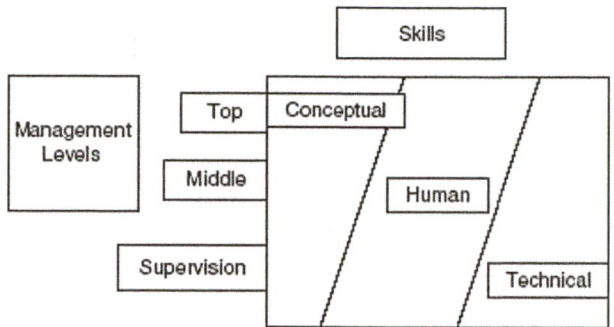

FIGURE 12.11   Distribution of skills within an organization.

The manager is responsible for organizing, planning, controlling, directing, and motivating. Managers assume multiple roles. A role is an organized set of behaviors. Henry Mintzberg described the roles needed by all managers: informational, interpersonal, and decisional roles. Robert Katz (1974) identified three managerial skills that are essential to successful management: technical, human, and conceptual: "Technical skill involves process or technique knowledge and proficiency. Managers use the processes, techniques and tools of a specific area. Human skill involves the ability to interact effectively with people. Managers interact and cooperate with employees. Conceptual skill involves the formulation of ideas. Managers understand abstract relationships, develop ideas, and solve problems creatively". Technical skill deals with things, human skill concerns people, and conceptual skill has to do with ideas. The distribution of these skills between the levels of management is shown in Figure 12.11.

THE CHIEF EXECUTIVE OFFICER OF HEALTH ORGANIZATIONS

Hospital directors in the past were often senior physicians, often called *superintendents*, without training in health management. The business manager CEO has become common in hospital management in the USA. During the 1950s, the CEO was called an *administrator*, and worked under the direction of a board of trustees who raised funds, set policies, and were often involved in internal administration.

Where the CEO was a non-physician, the usual case in North American hospitals, a conflict often existed with the clinical staff of the hospital. In some settings, this led to appointment of a parallel structure with a full-time chief of medical staff with a focus on clinical and qualitative matters. In European hospitals, the CEO is usually a physician, often by law, and the integration of the management function with the role of clinical chief is the prevalent model.

Over time, as the cost and complexity of the health system have increased, the CEO role has changed to one of a "coordinator". The CEO is now more involved in external

relations and less in the day-to-day operation of the facility. The CEO is a leader/partner but *primus inter pares*, or first among equals, in a management team that shares information and works to define objectives and solve problems. This de-emphasizes the authoritarian role and stresses the integrative function.

The CEO is responsible for the financial management of operational and capital budgets of the facility, which is integral to the planning and future development of the facility. Budgets include four main factors: income, fixed or regular overhead, variable or unpredictable overhead, and capital or development costs, all essential to the survival and development of the organization.

The key role of top management is to develop a vision, goals, and targets for the institution, to maintain an atmosphere and systems to promote the quality of care, financial solidity, and to represent the institution to the public. The overall responsibility for the function and well-being of the program is with the CEO and the governing board of directors.

COMMUNITY PARTICIPATION

Community participation in management of health facilities has a long-standing and constructive tradition. The traditional hospital board has served as a mechanism for community participation and leadership in promoting health facility development and management at the community level. The role of hospital boards evolved from primarily a philanthropic and fund-raising one to a greater overall responsibility for policy and planning function working closely with management and senior professional staff. This change occurred as operational costs increased rapidly, as government insurance schemes were implemented, and as court decisions defined the liability of hospitals and reinforced the broadened role of governing boards in malpractice cases and quality assurance. Centrally developed health systems such as the UK's NHS have promoted district and county health systems with high degrees of community participation and management, both at the district level and for services or facilities.

The role of local authorities, as well as state and national governments, is crucial to the functioning of public health in its traditional issues such as safe water supply, sanitation, business licensing, social welfare, and many others, as discussed in Chapter 10. These functions have not diminished with the greater roles of state and federal or national governments in health. In healthful living environments the local authority functions are of continuing and indeed expanding importance, as in urban planning and transportation, promoting easy access to commercial facilities for shopping and healthy food sources for poorer sections as well as those available to prosperous members of the community.

BOX 12.4 Changing the Law Banning Birth Control in Massachusetts: The Role of Advocacy

In 1942 and 1948, referenda were held on a Massachusetts law which banned dissemination of birth control devices and information; both were defeated. Massachusetts and Connecticut alone of all the states continued to ban birth control. But with the advent of the birth control pill in the 1960s, the issue was reopened. Richard Cardinal Cushing, head of the Archdiocese of Boston, no longer opposed a change in the state law, although still opposed the practice of birth control.

Some Catholic doctors, including Dr John Rock, the gynecologist who conducted the key clinical studies of the birth control pill, favored a change in the law. An article published by a young Catholic doctor, later specializing in public health, in the prestigious New England Journal of Medicine in 1964, called for changing the Massachusetts birth control law. This drew the ire of some of the hierarchy of the Church.

However, the article served to stimulate the Legislature to revisit the law, leading to its repeal in 1966, thus allowing use of all methods of birth control. The controversy subsided and women were free to control their own fertility as a result of this advocacy.

Sources: *Dorsey JL, Emeritus Clinical Professor of Medicine, Harvard Medical School, Boston, Massachusetts. Personal communication; December 2012.*
Dorsey JL. Changing attitudes toward the Massachusetts birth control law. N Engl J Med 1964;271:823–7.
Meehan S. From patriotism to pluralism: how Catholics initiated the repeal of birth control restrictions in Massachusetts. Catholic University of America Press. Catholic Hist Rev 2010;96(3):470–98. http://dx.doi.org/10.1353/cat.0.0864.

Advocacy has always been an important part of public health. An illustration of this is seen in Box 12.4 in changing the law banning birth control in Massachusetts in the 1960s. The issue of birth control still casts a heavy burden on women globally owing to religious objections, so this example from the 1960s is still relevant as a political issue both in the USA and in many other countries.

Community participation can be crucial to the success of an intervention to promote community health. Sensitivity to local, religious, or ethnic concerns is part of planning any study or intervention in public health. This does not mean that the national, state, and local health authorities must continuously canvass public opinion, but there is advantage in holding referenda on some issues compared to governmental fiat. The USA has higher rates of fluoridation than most countries, and this is implemented after referenda in each municipality (see Chapter 7). In Portland, Oregon, the City Council profluoridation vote in 2012 (*New York Times*, 12 September 2012) was later rejected in the public referendum. Portland is the only major American city without fluoridation (*Portland Tribune*, 21 May 2013).

INTEGRATION: LATERAL AND VERTICAL

Rationalization of health facilities increasingly means organizational linkages between previously independent facilities. Mergers of health facilities are common events in many health systems. In the USA, there are frequent mergers between hospitals, or between facilities linked to HMOs or managed care systems. Health reform in many countries is based on similar linkages. Governmental approval and alteration to financing systems are needed to promote linkages between services to achieve greater efficiency and improve patient care (see Chapters 10 and 11).

Lateral integration is the term used for amalgamation among similar facilities. Like a chain of hotels, in health care this involves two or more hospitals, usually meant to achieve cost savings, improve financing and efficiency, and reduce duplication of services. Urban hospitals, both not-for-profit as well as for-profit, often respond to competition by purchasing or amalgamating with other hospitals to increase market share in competitive environments. This is often easier for hospital-oriented CEOs and staff to comprehend and manage, but it avoids the issues of downsizing and integration with community-based services.

Vertical integration describes organizational linkages between different kinds of health care facilities to form integrated, comprehensive health service networks. This permits a shift of emphasis and resources from inpatient care to long-term, home, and ambulatory care, and is known as the managed care or district health system model. Community interest is a factor in promoting change to integrate services, which can be a major change for the management culture, especially of the hospital.

The survival of a health care facility may depend on integration with appropriate changes in concepts of management. In the 1990s, a large majority of California residents moved to managed care programs because of the high cost of fee-for-service indemnity health insurance and because of federal waivers to promote managed care for Medicare and Medicaid beneficiaries. Independent community hospitals without a strong connection to managed care organizations (MCOs) were in danger of losing their financial base.

Hospital bed supplies were reduced in the USA from 4.5 beds per 1000 population in 1980 to 2.9 in 2000 and 2.6 in 2009. Occupancy rates also fell, from 75 percent in 1980 to 64 percent in 2000 and rose slightly to 66 percent in 2009. Hospital discharges also fell during these years, from 173 per 1000 population in 1980 to 113 in 2005 and 112 in 2007, while days of care fell from 1297 to 558 and 540, respectively (Health United States, 2011). These data are monitored by the National Hospital Discharge Survey and the Centers for Disease Control and Prevention's (CDC's) National Center for Health Statistics. The lower hospital bed supply and utilization since the 1980s and 1990s reflect the adoption of insurance system payments by diagnosis-related group (DRG), rather than on a per diem basis. Similar trends are seen in European countries, although in the Commonwealth of Independent States the number of hospital beds declined between 1990 and 2005-2011 but stabilized at high and inefficient levels (8 beds per 1000 population) compared to the number in Western Europe, which fell from 5 beds per 1000 in 1990 to 3.4 in 2011, and in some countries to 2 per 1000 population despite increased longevity and aging of the population.

There was a shift to stronger ambulatory care, as occurred throughout the industrialized countries despite an aging of the population. These trends were largely due to greater emphasis on ambulatory surgery and other care, and major medical centers responded with strategic plans to purchase community hospitals and develop affiliated medical groups and contract relationships with managed care organizations to strengthen their "market share" service population base for the future. The new payment environment and managed care also promoted hospital mergers (lateral integration) and linkages between different levels of service, such as teaching hospitals with community hospitals and primary community care services (vertical integration).

Vertical integration not only is important in urban areas, but can serve as a basis for developing rural health care in both developed and developing countries. The district hospital and primary care center operating as an integrated program can provide a high-quality program. Hospital-centered health care, common in industrialized countries, has traditionally channeled a high percentage of total health expenditures into hospital services. Over recent years, there has been a reduction in hospital bed supply in most industrialized countries, with shorter length of stay, more emphasis on ambulatory care, improved diagnostic facilities, and improved outcomes of care (see Chapter 3).

Expenditures on the hospital component of care have come down to between 40 and 45 percent of total health expenditures in many countries, with a growing proportion going to ambulatory and primary care, and increased percentages to public health. This shift in priorities has been an evolutionary process that will continue, but requires skilled management leadership, grounded in health systems management training and epidemiological knowledge, and skilled negotiating skills to foster primary care and health promotion approaches both within the organization and in relation to outside services, especially preventive services. This shift in policy direction will be fostered in implementation of the PPACA (Obamacare), discussed in Chapters 10 and 13. Managed care systems or accountable care organizations (ACOs) will integrate hospital and community care and try to limit hospital care by strengthening ambulatory and primary care, and especially preventive care. This will have both economic and epidemiological benefits, but will depend on skilled management to understand and lead in their implementation.

Much of the rationale for these changes is discussed in the literature and summarized in a 2012 report from the US Institute of Medicine, entitled "Best care at lower cost". This report calls for overhauling the health system in a continuous evolution based on evidence and lessons learned from decades of innovative care systems and research into their workings. The health system needs to relate to other community services with a shared population orientation (Institute of Medicine, 2012).

NORMS AND PERFORMANCE INDICATORS

Norms are useful to promote efficient use of resources and promote high standards of care, if based on empirical standards proved by experience, trial and error, and scientific observation. Norms may be needed even without adequate evidence, but should be tested in the reality of observation, experience, and experiment. This process requires data for selected health indicators and trained observers free to examine, report, and publish their findings for open discussion among colleagues and peers in proceedings open to the media and the general public.

Normative standards of planning are the determination of a number per unit of population that is deemed to be suitable for population needs; for example, the number of beds or doctors per 1000 population or length of stay in hospital. Many organizations based on the bureaucratic model used norms as the basis for planning and allocation of resources including funding (see Chapter 11). This led to payment systems which encouraged greater use of that resource. If a factory is paid by the number of workers and not the number and quality of the cars produced, then management will have no incentive to introduce efficiency or quality improvement measures. If a district or a hospital is paid by the number of beds, or by days of care in the hospital, there is no incentive to introduce alternative services such as same-day or outpatient surgery and home care.

Performance indicators are measures of completion of specific functions of preventive care such as immunization, mammography, Pap smears, and diabetes and hypertension screening. They are indirect measures of economy, efficiency, and effectiveness of a service and are being adopted as better methods of monitoring and paying for a service, such as by paying a premium. General practitioners in the UK receive additional payments for full immunization coverage of the children registered in their practices. A block grant or per capita sum may be tied to indicators that reflect good standards of care or prevention, such as low infant, child, and maternal mortality. Incentive payments to hospitals can promote ambulatory services as alternatives to admissions and reduce lengths of stay. Limitations of financial resources in the industrialized countries and even more so in the developing countries make the use of appropriate performance indicators of great importance in the management of resources.

Pay-for-performance is a system of paying for health services developed in the UK for paying general practitioners, with apparently satisfactory results. It is now widely used in the USA. It is defined as "a strategy to improve health care delivery that relies on the use of market or purchaser power. Agency for Healthcare Research and Quality (AHRQ) Resources on Pay for Performance (P4P), depending on the context, refers to financial incentives that reward providers for the achievement of a range of payer objectives, including delivery efficiencies, submission of data and measures to payer, and improved quality and patient safety" (Agency for Healthcare Research and Quality, 2012). More than half of commercial HMOs are using pay-for-performance. Recent legislation requires the Medicare and Medicaid programs to adopt this approach for beneficiaries and providers. As commercial programs have evolved during the past 5 years, the categories of providers (clinicians, hospitals, and other health care facilities), number of measures, and dollar amounts at risk have increased. This method of payment is likely to be promoted in the Affordable Care Act implementation to improve quality and control cost increases in US health care (see Chapters 10, 11, and 13). Pay-for-performance has also been adopted in other countries trying to improve quality of care, such as Macedonia (Lazarevik and Kasapinov, 2012).

HEALTH PROMOTION AND ADVOCACY

Social marketing is the systematic application of marketing alongside other concepts and techniques to achieve specific behavioral goals for a social good. Initially focused on commercial goals in the 1970s, the concept became part of health promotion activities to address health issues where there was no current biomedical approach, such as in smoking reduction and in safe sex practices to prevent the spread of HIV.

Social marketing was based initially on commercial marketing techniques but now integrates a full range of social sciences and social policy approaches using the strong customer understanding and insight approach to inform and guide effective policy and strategy development. It has become part of public health practice and policy setting to achieve both strategic and operational targets. A classic example of the success is seen with tobacco reduction strategies in many countries using education, taxation, and legislative restrictions. Other challenges in this field include risk behavior such as alcohol abuse through binge drinking, unsafe sex practices, and dietary practices harmful to health.

PHILANTHROPY AND VOLUNTEERISM

Philanthropy and volunteerism have long been important elements of health systems through building hospitals, mission houses, and food provision, and other prototype initiatives on a demonstration basis. This approach has been instrumental in such areas as improved care and prevention of HIV, immunization in underdeveloped countries, global health strategies, and maternal and child health services.

During the late twentieth and early twenty-first centuries, a new "social entrepreneurship" was initiated and developed by prominent reform-minded former US President Bill Clinton, Microsoft's Bill Gates, and the Open Society Institute of George Soros. The Rotary Club International has been a major factor in funding and promoting the global campaign to eradicate poliomyelitis. This has promoted integration and consortia for the promotion of acquired immunodeficiency syndrome (AIDS) prevention and malaria control in many developing countries. The Global Alliance for Vaccine and Immunization (GAVI) is a US-based organization which links international public and private organizations and resources to extend access to immunization globally. It includes the United Nations Children's Fund (UNICEF), WHO, bilateral donor countries, the vaccine industry, the Gates Foundation, and other major donors. GAVI has made an important contribution to advancing vaccine coverage and adding important new vaccines in many developing countries and regions. These organizations focus funds and activities on promoting improved care and prevention of HIV, tuberculosis, and malaria, along with improved vaccination for children, reproductive health, global health strategies, technologies, and advocacy. These programs generate publicity and raise consciousness at political levels where resource allocations are made. A central feature of these programs is the promotion of "civil society" as active partners in a globalized world of free trade, democracy, and peace.

Specific initiatives included promoting improved large-scale marketing of antiretroviral drugs for the treatment of HIV infection, including price reduction so that developing countries can offer antiretroviral treatment, especially to reduce mother-to-infant transmission. Programs have branched out into the distribution of malaria-preventing bed nets, provision of low-cost pharmaceuticals, marketing drugs for the poor, desalination plants, solar roof units, low-cost small loans, and cell phones, mainly in Africa.

Another form of social entrepreneurship that has gained support in the private sector is proactiveness in environmental consciousness to address issues raised by the environmental movement, and public interest for environmental accountability. The automobile industry is facing both public concern and federal legal mandates for improved gas mileage as opposed to public demand for larger cars. Hybrid cars using less fuel have been successfully introduced into the market for low-emission, fuel-efficient cars, and electric cars are gradually entering the field. Public opinion is showing signs of moving towards promoting environmentally friendly design, marketing, and purchasing practices in energy consumption, conservation practices, and public

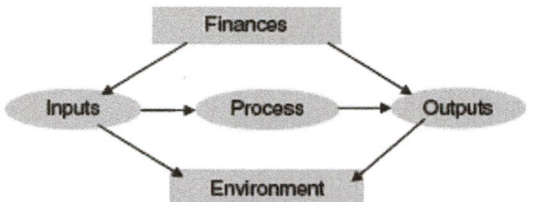

FIGURE 12.12 Basic elements of an organizational system. *Source: Modified from Gibson JL, Ivancevich J, Donnelly JH, Konopaske R. Organizations: behavior, structure, process. New York: McGraw-Hill/Irwin; 2003.*

policy. Public opinion and the price of fuel will play a major part in driving governments to legislate energy and conservation policies to address global warming and damage to the environment, with their many negative health consequences. However, such changes must work with public opinion because of the sensitivity of consumers to the price of fuel. In addition, when food crops, such as corn, are used to produce ethanol for energy to replace oil, then food prices rise and consumers suffer and respond vigorously.

Corporations adopt policies of environmental responsibility in part because of public relations and partly because of potential liability claims. Much of the planning and financial costs of offshore petroleum and gas drilling is spent on safety measures to protect the environment. The explosion in 2010 at a British Petroleum site in the Gulf of Mexico, off the coast of Texas and Louisiana, caused massive pollution and environmental damage, and resulted in the US government being awarded US$4.5 billion against BP for cleanup and damages. The reputation of the corporation suffered and some executive officers lost their positions. Thus, corporate social responsibility can be seen as self-interest.

NEW ORGANIZATIONAL MODELS

New models of health care organization are emerging and developing rapidly in many countries. This is partly a result of a search for more economical methods of delivering health care and partly the result of the target-oriented approach to health planning that seeks the best way to define and achieve health objectives. The developed countries seek ways to restrain cost increases, and the developing countries seek effective ways to quickly and inexpensively raise health standards for their populations. New organizational models that try to meet these objectives include district health systems, managed care organizations (MCOs) and accountable care organizations (ACOs), described in greater detail in Chapter 11. Critical and basic elements of a health system organization are shown in Figure 12.12.

NEW PROJECTS AND THEIR EVALUATION

New initiatives are part of the growth and development of any organization or health service system, as needs,

technologies, resources, and public demand change. Identification of issues and decisions to launch new endeavors or projects to advance the state of the art, to address unmet needs, or to meet competition are part of organizational responsibility, in the public sector to meet needs, and in the private sector to remain competitive.

In developing and developed countries, many NGOs provide funding from abroad for essential services that a government may be unable to provide. Such projects focus on issues directed from the head offices in the USA or Europe of the funding source or management offices for specific vertical programs which are often not fully integrated with national priorities and programs. However, these need coordination and approval by the local national government agency responsible for that sector of public service. New projects run by NGOs may run in parallel to each other, or to state health services as uncoordinated activities. Governmental public health agencies have responsibility for oversight of health systems and can play a leadership and regulatory role in coordinating activities and directing new programs to areas of greatest national need.

The public health agency may also seek funding to launch new pilot or specific needs programs. The agency may introduce a new vaccine into a routine immunization program in phases, pending government approval and funding to incorporate it as a routine immunization program based on evaluation of the initial phase. An example is the introduction of *Haemophilus influenzae* type b vaccine in Albania in 2006, which was funded by GAVI for 5 years based on a study and proposal including a cost-effectiveness study (Bino S, Ginsberg G, personal communication, 2007).

Proposals for health projects by NGOs or private agencies need to be prepared in keeping with the vision, mission, and objectives of the responsible governmental agency, with ethics review and community participation. A project proposal should include why the project is important, its specific goals and objectives, available or new resources, and the time-frame required to achieve success (Box 12.5). It should describe the means proposed to accomplish the goals, and how the proposed program will impact the community, providing recommendations for follow-up and/or further action.

The introduction of the project proposal outlines the current state of the problem and the case for action. It should describe existing programs which address that issue, with proposed collaboration, and expansion or improvement of programs, but avoiding duplication of services. Background information needs to relate the project to the priorities of the prospective funding organization. The objectives should follow the acronym "SMART": specific, measurable, achievable, relevant, and time-based. This term, originally used for computer disc self-management, has been adapted as a current form of MBO from the 1950s and 1960s.

The project objectives should be feasible and the expected results of the project should be based on the stated objectives.

BOX 12.5 Program Evaluation Information Needs

The following utility standards ensure that an evaluation will serve the information needs of intended users:

- Identify and engage stakeholders, including relevant government agencies, people or communities involved in or affected by the evaluation, so that their needs and concerns can be addressed.
- Develop and describe the program.
- Focus the evaluation design with ethical standards and review requirements respected.
- Gather credible evidence – The people conducting the evaluation should be trustworthy and competent in performing the evaluation for findings to achieve maximum credibility and acceptance. Information collected should address pertinent questions regarding the program and be responsive to the needs and interests of clients and other specified stakeholders.
- Justify the conclusions – The perspectives, procedures, and rationale used to interpret the findings should be carefully described so that the bases for value judgments are clear.
- Ensure sharing and use of information and lessons learned – Evaluation reports should clearly describe the program being evaluated, including its context and the purposes, procedures, and findings of the evaluation so that essential information is provided and easily understood. Substantial interim findings and evaluation reports should be disseminated to intended users so that they can be used in a timely fashion to encourage follow-through by stakeholders, to increase the likelihood of the evaluation being used.
- Standards of a project should focus on scientific justification, utility, feasibility, propriety, and accuracy.
- A program in this context includes:
  - direct service interventions
  - community mobilization efforts
  - research initiatives
  - surveillance systems
  - policy development activities
  - outbreak investigations
  - laboratory diagnostics
  - communication campaigns
  - infrastructure building projects
  - training and education services
  - administrative systems and others.

Sources: Centers for Disease Control and Prevention. Framework for program evaluation in public health. MMWR Morb Mortal Wkly Rep 1999;48(RR-11):1–40.
Centers for Disease Control and Prevention. A framework for program evaluation. Office of the Associate Director for Program – Program Evaluation; 12 September 2012. Available at: http://www.cdc.gov/eval/framework/index.htm [Accessed 30 December 2012].

BOX 12.6 Project Proposal Summary

- *Title page* – Name of project; principal people and implementing organizations; contact person(s); timeframe; country (state, region); target group of project; estimated project cost; date of submission.
- *Introduction* – Provides project background including the health issue(s) to be addressed, a situational analysis of the health problem, the at-risk and target populations, and existing programs in the community; includes an international and national literature review of the topic with references.
- *Aim of the project* – Intended accomplishment of the project; what will be evaluated (i.e., what is "the program" and in what context does it exist?); what aspects of the program and evidence will be used to indicate how the program has performed?
- *Objectives* – Specific, measurable, achievable, relevant, and time-based (SMART).
- *Expected results* – Based on the objectives: what will be produced and delivered.
- *Activities* – Actions and time-frame in keeping with the objectives and expected results.
- *Work plan* – Timeline of all activities, including preparation, training, pilot, and implementation stages.
- *Budget* – Estimated cost of expenditures, including human resources, activities, running costs, and overheads for project and evaluation.
- *Monitoring and evaluation* – What evidence will be used to indicate how the program has performed? What plan is recommended for periodic follow-up of project activities (including timeline and measures) to implement lessons learned from positive or negative outcomes, and use of resources? How efficient and effective is the project?
- *Conclusions* – What conclusions regarding program performance may be drawn? What conclusions regarding program performance are justified by comparing the available evidence to the selected standards?
- *Reporting* – Report the project to the key stakeholders and public bodies; publication in peer-reviewed journal if possible.
- *Justification* – Why is this project important and timely, and how will implementation benefit health of the community?

Sources: Adapted from Centers for Disease Control and Prevention. Framework for program evaluation in public health. MMWR Morb Mortal Wkly Rep 1999;48(RR-11):1–40.
Centers for Disease Control and Prevention. A framework for program evaluation. Office of the Associate Director for Program – Program Evaluation; 12 September 2012. Available at: http://www.cdc.gov/eval/framework/index.htm [Accessed 30 December 2012].

The funding organization will want to know what will be the expected product of the program in measurable process (e.g., immunization coverage) or outcome indicators (e.g., reduced child mortality). Projections will be based on the intended activities and known outcomes of other past programs with similar goals in the same or other countries (environmental scan), and should be supported by a review of local and international literature on the topic. The activities section of a proposal should include a timeline of the intended actions and a description of activities based on best practices. The expected outcomes, monitoring and evaluation, and justification are all part of the presentation (Box 12.6).

The proposed funding agency expects convincing evidence of how this program will be effective, efficient, practical, and realistic. This information is presented in the activities section, which also needs to address the resources that will be needed to implement the program such as the budget for staff, supervison, training, management, materials (vaccines, syringes, equipment, ongoing supplies and others), transportation, and costs of premises. After completing the activities section, a realistic and achievable work plan and time-frame are required.

Well-planned projects have monitoring and evaluation criteria. Monitoring follows the performance of the program, documenting successes, failures, and lessons learned, as well as expenditures. Evaluation guidelines of the program define the methods used to assess the impact of the project and whether the project was carried out in an effective and efficient manner, and may be required periodically throughout the life of the project.

The most difficult issue is sustainability. A project funded by an NGO is usually time limited to 3–4 years and the survival of the program usually depends on its acceptability and the capacity of government to continue it. Thus, evaluation becomes even more crucial for the follow-up of even successful short-term projects. Harm reduction programs include tackling HIV in drug users, reducing maternal–child HIV transmission, tobacco control programs, and reducing levels of obesity in schoolchildren. Sustainability and diffusion of positive findings to wider application are important challenges, especially to global health. Even in high-income countries, diffusion of best practices is often slow and fraught with controversy and inertia. Examples of this slow or non-diffusion of evidence-based public health include the failure of most European countries to harmonize salt fortification with iodine or total indifference to flour fortification with folic acid to prevent neural tube defects (see Chapters 6 and 8).

Public health work within departments or ministries of health or local health authorities operates at a disadvantage in comparison with other health activities, especially hospitals, pharmaceuticals, diagnostics, and medical care. The competition for resources in a centrally funded system is intense, and the political and bureaucratic battles for funds may pit new immunization agents or health promotion programs against new cancer treatment drugs or scanners, and this is very often a difficult struggle. The presentation of program proposals for new public health interventions requires skill, professionalism, good timing, and the help of informed public and professional opinion. Allocation of resources is decided at the political level in a tax-based universal system, while even in a social security (Bismarckian) system where funding is through an employee–employer payroll deduction, additional funding from government is essential to keep up with the continuing flow of new modalities of treatment or prevention.

Public health is handicapped in portraying the costs and benefits of important interventions, leaving new programs with insufficient resources, including the staffing and administrative costs (e.g., office space, phone service, transportation costs), which are essential parts of any public health program. Portraying the cost of the new proposed program should be based on the total population served, not just the specific target population for a new program; that is, it should be represented as a per capita cost. Similarly, projected benefits should extrapolate the results from other areas, such as pandemic or avian flu or severe acute respiratory syndrome (SARS), and the likely impact on the target geographic area and its population.

COMPETENCIES IN HEALTH POLICY AND MANAGEMENT

Public health has prime responsibility for monitoring the health status of the population as well as in preventing infectious and non-communicable diseases and injuries, preparing for disasters, and many other functions. This role requires an adequate multidisciplinary workforce with high levels of competencies. This topic is discussed extensively in Chapter 14.

Canada's experience with the SARS epidemic in 2003 led to a reappraisal of public health preparedness and standards. This, in turn, led to the establishment of the national Public Health Agency of Canada, which is mandated to develop standards and practices to raise the quality of public health in the country and especially to prepare for possible pandemics. The Agency issued standards of competency for public health personnel and fostered the development of regional laboratories, and schools of public health were developed across Canada. Core competencies for program planning implementation and evaluation are seen in Box 12.7.

Health care systems throughout the world are being scrutinized because of their growing costs in relation to national wealth. At the same time, techniques for evaluating health care with respect to appropriateness, quality, and resource allocation are being developed. These techniques are multifactorial since they must relate to all aspects of health care, including the characteristics of the population being served; available health care resources; measures of the process and utilization of care; measures of health care outcomes; peer review, including quality assessment of health care providers; consumer attitudes, knowledge, and compliance; care provided for "tracer" or sample conditions; and economic cost–benefit studies.

Evaluation in health care assumes that a health care system and the providers of health care within that system are responsible and accountable for the health status of the population. It must, however, recognize that health services are not the sole determinants of health status; social, economic, and cultural factors also play key roles. A comprehensive approach to evaluation in health care

BOX 12.7 Core Competencies for Program Planning, Implementation and Evaluation

Core competencies are essential knowledge, skills, and attitudes necessary for the practice of public health. They transcend the boundaries of specific disciplines and are independent of program and topic. They are the building blocks for effective public health practice, and the use of an overall public health approach.

Generic core competencies provide a baseline for what is required to fulfill public health system core functions. These include population health assessment, surveillance, disease and injury prevention, health promotion, and health protection.

The core competencies are needed to effectively choose options, and to plan, implement, and evaluate policies and/or programs in public health, including the management of incidents such as outbreaks and emergencies.

A public health practitioner is able to:

● describe selected policy and program options to address a specific public health issue
● describe the implications of each option, especially as they apply to the determinants of health and recommend or decide on a course of action
● develop a plan to implement a course of action taking into account relevant evidence, legislation, emergency planning procedures, regulations, and policies
● implement a policy or program and/or take appropriate action to address a specific public health issue
● demonstrate the ability to implement effective practice guidelines
● evaluate an action, a policy, or a program
● demonstrate an ability to set and follow priorities, to maximize outcomes based on available resources
● demonstrate the ability to fulfill functional roles in response to a public health emergency.

Source: Public Health Agency of Canada. Core competencies for public health in Canada. Available at: http://www.phac-aspc.gc.ca/php-psp/ccph-cesp/pdfs/cc-manual-eng090407.pdf [Accessed 28 December 2012].

is described in Chapter 3. Many of the components that are available in health care systems exist, while others that remain to be developed are discussed. Evaluation is an integral part of a comprehensive health care system, in that the components of evaluation must be built into any national system. As long as rationality is expected of health care, evaluation is an essential element of the overall system (Tulchinsky, 1982) (see Chapter 3).

SYSTEMS APPROACH AND NATIONAL PLANNING

The purpose of management in health is the improvement of health, and not merely the maintenance of an institution. Separate management of a variety of health facilities serving a community has derived from different historical development and funding systems. In competition for

public attention and political support, public health suffers in comparison to hospitals, new technology and drugs, and other competitors for limited resources. The experience of successes in reducing mortality from both non-infectious and infectious conditions comes largely from public health interventions. Medical care is also an essential part of public health, so that management and resource allocation within the total health sector are interactive and mutually dependent. The New Public Health looks at all services as part of a network of interdependent services, each contributing to health needs, whether in hospital care or in enforcing public health law regarding; for example, motor vehicle safety and smoking restriction in public places.

Separate management and budgeting of a complex of services results in disproportionate funds, staff, and attention being directed towards high-cost services such as hospitals, and fails to redirect resources to more cost-effective and patient-sensitive kinds of services, such as home and preventive care. However, reducing the supply of hospital beds and implementing payment systems with resources for early diagnosis and incentives for short stays have changed this situation quite dramatically in recent decades. The effects of incentives and disincentives built into funding systems are central issues in determining how management approaches problem solving and program planning, and are therefore important considerations in promoting health.

The management approach to resolving this dilemma is professional vision and leadership to promote the broader New Public Health. Thus, managers of hospitals and other health facilities need broad-based training in a New Public Health in order to understand the interrelationships of services, funding, and population health. Managers who continue to work with an obsolescent paradigm with the traditional emphasis, regardless of the larger picture, may find the hospital non-competitive in a new climate where economic incentives promote downsizing institutions and upgrading health promotion. Defensive, internalized management will become obsolete, while forward-looking management will be the pioneers of the New Public Health. This may be seen as a systems approach to improve population and individual health, based on strategic planning for immediate needs and adaptation of health systems in the longer term issues in health.

Examples of national planning that cut across health and social services include national insurance policies and the provision of new services to meet rising needs, as shown for Alzheimer's disease, in France since 2001 (Box 12.8) and in the USA since 2011 (Box 12.9).

SUMMARY

Health care is one of the largest and most important industries in any country, consuming anywhere from 3 to nearly 18 percent of GNP, and still growing. It is a service, not a production industry, and is vital to the health and well-being

---

BOX 12.8 France – National Dementia Plans 2001–2012

First National Dementia Plan 2001–2005

An estimated 600,000 French people lived with dementia; half were diagnosed and one-third were receiving treatment; 75 percent of people with Alzheimer's disease were living at home; 50 percent of all nursing home residents lived with some form of dementia; a day's care cost €60 while full-time residency in a nursing home ranged between €3000 and €4600.

- Identify the early symptoms of dementia and refer people to specialists.
- Create a network of "memory centers" to enable earlier diagnosis.
- Produce ethical guidelines for families and care homes.
- Provide financial support for people with dementia.
- Establish day care centers and create local dementia information centers.
- Build new residential care homes and improve existing homes.
- Provide support for research and clinical studies.

Second Alzheimer Plan 2005–2007

By 2004 nearly 800,000 French people lived with dementia; a growing proportion of women and 18 percent of all people over 75; over 165,000 new cases of dementia diagnosed annually with an associated life expectancy of 8 years; Alzheimer's disease now recognized as a chronic disease by the French social security system, with the need for continuing support.

- Eligibility of dementia for 100 percent insurance coverage.
- Identify and support the needs of younger people with dementia.
- Provide training and support to professional and volunteer workers.
- Develop emergency housing resources.

Third Alzheimer Plan 2008–2012

- The growing need is clear.
- Improve diagnosis.
- Strengthen coordination between providers.
- Provide better treatment and support for caregivers.
- Provide supportive home help more effectively.
- Speed up research.
- Provide public information.

Fourth Alzheimer Plan

- Pan European.
- In preparation.

Source: *Alzheimer Europe. France – national plans for Alzheimer and related diseases. Available at: http://www.alzheimer-europe.org/Policy-in-Practice2/National-Dementia-Plans/France#fragment-1 [Accessed 24 December 2012].*

BOX 12.9 US National Alzheimer's Disease Project Act

Vision Statement

"For millions of Americans, the heartbreak of watching a loved one struggle with Alzheimer's disease is a pain they know all too well. Alzheimer's disease burdens an increasing number of our Nation's elders and their families, and it is essential that we confront the challenge it poses to our public health."

US President Barack Obama

Action Plan

On 4 January 2011, President Barack Obama signed into law the National Alzheimer's Project Act (NAPA), requiring the Secretary of the US Department of Health and Human Services (HHS) to establish the National Alzheimer's Project to:

- Create and maintain an integrated national plan to overcome Alzheimer's disease (AD).
- Coordinate Alzheimer's disease research and services across all federal agencies.
- Accelerate the development of treatments to prevent, halt, or reverse the course of AD.
- Improve early diagnosis and coordination of care and treatment of AD.
- Improve outcomes for ethnic and racial minority populations that are at higher risk for AD.
- Coordinate with international bodies to fight AD globally.

The law also establishes the Advisory Council on Alzheimer's Research, Care, and Services and requires the Secretary of HHS, in collaboration with the Advisory Council, to create and maintain a national plan to overcome AD.

Goals and Strategies for 2025

Research funds are being allocated towards that end. Education for health providers, strengthening of the workforce, for direct care and for public health guidelines for management of AD, education and support for caring families, addressing special housing needs for AD patients and many other initiatives are proposed in this comprehensive approach to a growing public health problem. Enhancing public awareness is crucial to achieve the goals set out in this plan.

Source: *Department of Health and Human Services. National plan to address Alzheimer's disease. Available at: http://aspe.hhs.gov/daltcp/napa/NatlPlan.pdf [Accessed 29 December 2012].*

---

of the individual, the population, and the economy. Because health care employs large numbers of skilled professionals and many unskilled people, it is often vital to the economic survival of small communities, as well as for a sense of community well-being.

Management includes planning, leading, controlling, organizing, motivating, and decision-making. It is the application of resources and personnel towards achieving targets. Therefore, it involves the study of the use of resources, and the motivation and function of the people involved, including the producer or provider of service, and the customer, client, or patient. This cannot take place in a vacuum, but is based on the continuous monitoring of information and its communication to all parties involved. These functions are applicable at all levels of management, from policy to operational management of a production or a service system. Creative management of health systems is vital to the functioning of the system at the macrolevel, as well as in the individual department or service. This implies effective use of resources to achieve objectives, and community, provider, and consumer satisfaction. These are formidable challenges, not only when money is available in abundance, but even more so when resources are limited and difficult choices need to be made.

Modern management includes knowledge and skills in identifying and measuring community health needs and health risks. Critical needs are addressed in strategic planning with measurable impacts and targets. Public health managers should have skills gained in marketing, networking, data management, managing human resources and finance, engaging community partners, and communicating public health messages.

Many of the methods of management and organization theory developed as part of the business world have become part of public health. These include defining the mission, values and objectives of the organization, strategic planning and management, MBO, human resource management (recognizing individual and professional values), incentives–disincentives, regulation, education, and economic resources. The ultimate mission of public health is the saving of human life and improving its quality, and achieving this efficiently with high standards of professionalism and community involvement.

The scope of the New Public Health is broad. It includes the traditional public health programs, but equally must concern itself with managing and planning comprehensive service systems and measuring their function. The selection of targets and priorities is often determined by the feasible rather than the ideal. The health manager, either at the macrolevel of health or managing a local clinic, needs to be able to conceptualize the possibilities of improving the health of individuals and the population in his or her service responsibility with current and appropriate methods. Good management means designing objectives based on a balance between the feasible and the desirable. Public health has benefited greatly from its work with the social sciences and assistance from management and systems sciences to adapt and absorb the new challenges and technologies in applied public health. The New Public Health is not only a concept; it is a management approach to improve the health of individuals and the population.

NOTE

For a complete bibliography and guidance for student reviews and expected competencies please see companion web site at http://booksite.elsevier.com/9780124157668

BIBLIOGRAPHY

Electronic Resources

Agency for Healthcare Research and Quality. http://www.ahrq.gov/ (accessed 28.12.12).

American College of Healthcare Executives. http://www.ache.org/ (accessed 28.12.12).

American College of Medical Quality. http://www.acmq.org/ (accessed 28.12.12).

American Hospital Association. http://www.aha.org/ (accessed 28.12.12).

Centers for Disease Control and Prevention, Management and Analysis Services Office. http://www.cdc.gov/maso/ (accessed 28.12.12).

Glossary of Managed Care Terms. http://www.pohly.com/terms.html and http://www.thci.org/other_resources/glossary.htm (accessed 28.12.12).

Joint Commission. Updated 27 December 2012 at: http://www.jointcommission.org/ (accessed 28.12.12).

National Association of Public Hospitals and Health Systems. http://www.naph.org/(accessed 28.12.12).

National Center for Health Statistics, Centers for Disease Control and Prevention. http://www.cdc.gov/nchs/ (accessed 28.12.12).

World Health Organization, The health manager's website. http://www.who.int/management/en/ (accessed 28.12.12).

Recommended Reading

Agency for Healthcare Research and Quality, 2012. Pay for performance (P4P): AHRQ resources. AHRQ, Rockville, MD. Available at: http://www.ahrq.gov/professionals/quality-patient-safety/quality-resources/tools/pay4per.html (accessed 02.01.14).

Aguayo, R., 1990. Dr. Deming: the American who taught the Japanese about quality. Simon & Schuster, New York. Available at: http://www.amazon.com/Dr-Deming-American-Japanese-Quality/dp/0671746219 (accessed 02.01.14).

American College of Physicians, 2008. Health and Public Policy Committee. Achieving a high performance health care system with universal access: what the United States can learn from other countries. Ann. Intern. Med. 148, 55–75. Available at: http://annals.org/article.aspx?articleid=738556 (accessed 02.01.14).

Bar-Yam, Y., 2006. Improving the effectiveness of health care and public health: a multi-scale complex systems analysis. Am. J. Public Health 96, 459–466. Available at: http://www.ncbi.nlm.nih.gov/pmc/articles/PMC1470498/ (accessed 02.01.14).

Centers for Disease Control and Prevention, 1999. Framework for program evaluation in public health. MMWR Recomm. Rep. 48 (RR-11), 1–40. Available at: http://www.cdc.gov/mmwr/pdf/rr/rr4811.pdf (accessed 02.01.14).

Centers for Disease Control and Prevention, 2000. Biological and chemical terrorism: strategic plan for preparedness and response. Recommendations of the CDC strategic planning workgroup. MMWR Recomm. Rep. 49 (RR-04), 1–14. Available at: http://www.cdc.gov/mmwr/preview/mmwrhtml/rr4904a1.htm (accessed 02.01.14).

Centers for Disease Control and Prevention, 12 September 2012. A framework for program evaluation. Office of the Associate Director for Program – Program Evaluation. CDC, Atlanta, GA. Available at: http://www.cdc.gov/eval/framework/index.htm (accessed 30.12.12).

CEO online, Developing leadership skills. Available at: http://www.ceoonline.com/expert_talk/leadership_management/leadership_skill.aspx (accessed 24.12.12).

Dutton, D.B., 1979. Patterns of ambulatory health care in five different delivery systems. Med. Care 17, 221–243. Available at: http://journals.lww.com/lww-medicalcare/toc/1979/03000 (accessed 02.01.14).

Ettelt, S., Nolte, E., Thomson, S., Mays, N., 2008. Capacity planning in health care: a review of the international experience. WHO, on behalf of the European Observatory on Health Systems and Policies, Copenhagen. Available at: http://www.euro.who.int/__data/assets/pdf_file/0003/108966/E91193.pdf (accessed 02.01.14).

Ginter, P.M., Duncan, W.J.P., Swayne, L.E., 2013. Strategic management of health care organizations. 7th edition, Jossey-Bass, Available at: http://ca.wiley.com/WileyCDA/WileyTitle/productCd-1118466462.html (accessed 02.01.14).

Heldman, A.B., Schindelar, J., Weaver III, J.B., 2013. Social media engagement and public health communication: implications for public health organizations being truly "social". Public Health Reviews, 35: epub ahead of print. http://www.publichealthreviews.eu/show/a/129 (accessed 04.01.13).

Institute of Medicine, 2001. Crossing the quality chasm: a new health system for the twenty-first century. National Academies Press, Washington, DC. Available at: http://www.nap.edu/catalog.php?record_id=10027 (accessed 02.01.14).

Institute of Medicine, 2003. Future of the public's health in the 21st century. National Academies Press, Washington, DC. Available at: http://www.nap.edu/catalog.php?record_id=10548 (accessed 02.01.14).

Institute of Medicine, 2012. Best care at lower cost: the path to continuously learning health care in America. National Academies Press, Washington, DC. Available at: http://www.nap.edu/download.php?record_id=13444 (accessed 13.07.13).

Johnson, W.D., Diaz, R.M., Flanders, W.D., Goodman, M., Hill, A.N., Holtgrave, D., et al., 2008. Behavioral interventions to reduce risk for sexual transmission of HIV among men who have sex with men. Cochrane Database Syst. Rev. (3), CD001230. Available at: http://onlinelibrary.wiley.com/doi/10.1002/14651858.CD001230.pub2/abstract;jsessionid=2CEB240AD4112A2A5696505CD03A7CDD.f04t04 (accessed 02.01.14).

Katzenbach, J.R., Smith, D.K., 1993. The wisdom of teams: creating the high-performance organization. Harvard Business School Press, Cambridge, MA. Available at: http://www.amazon.com/The-Wisdom-Teams-High-Performance-Organization/dp/0060522003 (accessed 02.01.14).

Lee, N.R., Kotler, P.A., 2011. Social marketing: Influencing behaviors for good [fourth edition]. Sage, Thousand Oaks, CA. Available at: http://www.amazon.com/Social-Marketing-Influencing-Behaviors-Good/dp/1412981492 (accessed 01.02.14).

Levesque, J.-F., Breton, M., Senn, N., Levesque, P., Bergeron, P., Roy, D.A., MD, 2013. The interaction of public health and primary care: functional roles and organizational models that bridge individual and population perspectives. Public Health Reviews, 35: epub ahead of print. http://www.publichealthreviews.eu/show/a/130 (accessed 04.01.13).

Maynard, A., 2012. The powers and pitfalls of the payment for performance. Health Econ. 21, 3–12. Available at: http://avym.com/wp-content/uploads/2013/07/Powers-and-Pitfalls-of-Fee-For-Service-payment-models.pdf (accessed 02.01.14).

Molinsky, A.L., Davenport, T.H., Iyer, B., Davidson, C., 2012 (January–February). Three skills every 21st-century manager needs. Harv. Bus. Rev. Available at: http://hbr.org/2012/01/three-skills-every-21st-century-manager-needs (accessed 23.12.12).

Powel, T.C., 1995. Total quality management as competitive advantage: a review and empirical study. Strateg. Manage. J. 16, 15–37. Available at: http://www.thomaspowell.co.uk/article_pdfs/TQM_as_CA.pdf (accessed 02.01.14).

Ryan, A., Blustein, J., 2012. Making the best of hospital pay for performance. N. Engl. J. Med. 366, 1557–1559. Available at: http://www.nejm.org/doi/pdf/10.1056/NEJMp1202563 (accessed 02.01.14).

Satcher, D., Higginbotham, E.J., 2008. The public health approach to eliminating disparities in health. Am. J. Public Health 98, 400–403. Available at: http://www.ncbi.nlm.nih.gov/pmc/articles/PMC2253560/ (accessed 02.01.14).

Scutchfield, F.D., Ingram, R.C., 2013. Public health systems and services research: building the evidence base to improve public health practice. Public Health Reviews, 35. epub ahead of print http://www.publichealthreviews.eu/show/a/126 (accessed 04.01.13).

Shortell, S.M., Gillies, R., Wu, F., 2011. United States innovations in healthcare delivery. Public Health Rev. 32, 190–212. Available at: http://www.publichealthreviews.eu/show/f/26 (accessed 02.01.14).

Smith, J.M., Topol, E., 2013. A call to action: lowering the cost of health care. Am. J. Prev. Med. 44 (1S1), S54–S57. Available at: http://www.ajpmonline.org/article/S0749-3797(12)00637-X/fulltext (accessed 02.01.14).

Sutton, M., Nikolova, S., Boaden, R., Lester, H., McDonald, R., Roland, M., 2012. Reduced mortality with hospital pay for performance in England. N. Engl. J. Med. 367, 1821–1828. Available at: http://www.nejm.org/doi/full/10.1056/NEJMsa1114951 (accessed 02.01.14).

Trochim, W.M., Cabrera, D.A., Milstein, B., Gallagher, R.S., Leischow, S.L., 2006. Practical challenges of systems thinking and modeling in public health. Am. J. Public Health 96, 538–546. Available at: http://www.ncbi.nlm.nih.gov/pmc/articles/PMC1470516/ (accessed 02.01.14).

Turnock, B.J., 2012. Public health: essentials of public health, second ed. Jones and Bartlett, Sudbury, MA. Available at: http://www.amazon.com/Essentials-Of-Public-Health-Essential/dp/1449600220 (accessed 02.01.14).

US Department of Health and Human Services, 2011. Health United States, 2011, with chartbook on trends in the health of Americans. DHHS, Washington, DC. Available at: http://www.cdc.gov/nchs/data/hus/hus11.pdf (accessed 02.01.14).

Chapter 13

National Health Systems

Learning Objectives

Upon completion of this chapter, the student should be able to:

1. Describe major types of national health insurance and health services systems;
2. Assess factors in health reform policies in various countries, including developing countries and the former Soviet countries;
3. Apply the experience of different countries to current health reform in the USA;
4. Formulate public health reforms in the context of the New Public Health.

INTRODUCTION

Assuring access to quality health care for all is a basic principle of the New Public Health. There are many personal or community risk factors which affect health status, and medical care is a vital aspect of the broad spectrum of health needs. Despite its value, medical care by itself is not sufficient to produce a high standard of population health. In order to promote optimal health, effective population-level prevention methods as described in previous chapters, availability of and access to care must be seen in the context of the individual and of societal conditions that increase the risk of disease, and application of appropriate measures to reduce those risks to prevent disease and promote health. Some of those interventions are provided by medical care and its preventive aspects. Other key aspects include social, sanitary, environmental, legal, economic, and educational factors. This interrelates with human resources for health (Chapter 14), financing and economics (Chapter 11), organization (Chapter 10), technology, law, and ethics (Chapter 15), and global health (Chapter 16).

The World Health Organization (WHO) defines a health system as: "The people, institutions and resources, arranged together in accordance with established policies, to improve the health of the population they serve, while responding to people's legitimate expectations and protecting them against the cost of ill-health through a variety of activities whose primary intent is to improve health. It is a set of elements and their relationship in a complex whole, designed to serve the health needs of the population. Health systems fulfill three main functions: health care delivery, fair treatment to all, and meeting health expectations of the population" (WHO, 2000). The WHO also addresses six basic building blocks for health systems: service delivery; health workforce; information; medical products, vaccines, and technologies; leadership and governance; and financing and a growing emphasis on universal access and reducing inequalities in health (WHO, 2013).

Most industrialized countries have implemented national health programs such as health insurance systems or national health services. Each system developed in the political, social, and historical context of the country and continues to evolve. Developing countries are also struggling to achieve universal access to care and health for all by expanding primary health care and social security plans which provide benefits to workers and for certain vulnerable populations, primarily mothers and children. As they move up the scale of economic development, developing countries also address the problem of how to decrease morbidity and mortality, achieve equity in access to health care, and expand the funding basis for health care through national health insurance. Some countries are experiencing rapid economic development but lag behind in directing increased national wealth towards improving health status. This is often due to a lack of focused political commitment, trained policy analysts, and trained public health professionals (see Chapters 14 and 16).

Each national health system has its own characteristics and challenges. System management requires continuous evaluation based on well-developed information systems, trained health management personnel, and societal involvement through professional organizations and advocacy groups. There is no defined "gold standard" plan for providing universal access to health care that is suitable for all countries. Each country develops and modifies a program of national health appropriate to its own cultural needs and available resources. However, there are evolving sets of patterns in health care, so that countries can and do learn from one another (Box 13.1).

Barriers to care can be geographic, cultural, social, and psychological as well as financial. Removing financial barriers to care is necessary but not sufficient for optimal health and to address the health problems of an individual and of a society. Equity in financial access with universal coverage is vital to population and individual health since

BOX 13.1 Key Elements of National Health Systems

- A tradition of government and non-governmental initiatives to improve the health of the population
- Health targets
- Demographic, epidemiological, economic monitoring
- Public health programs including health promotion
- Universal access by public insurance or service system
- Access to a broad range of health services
- Strategic planning for health and social policies
- Monitoring of health status indicators
- Outreach to special needs of high-risk groups and related issues
- Portability and accessibility of benefits when changing employer or residence
- Efforts to reduce inequality in regional and sociodemographic accessibility and quality of care
- Adequacy of financing
- Cost containment
- Efficient use of resources for a well-balanced health system
- Consumer satisfaction and choice of primary care provider
- Provider satisfaction and choice of referral services
- Public administration and regulation
- Promotion of high-quality service
- Promote patient and staff safety
- Comprehensive primary, secondary, and tertiary levels of care
- Well-developed information and monitoring systems
- Continual policy and management review
- Promote standards and accreditation of services, professional education, training, and research
- Governmental and private provision of services
- Decentralized management and community participation
- Assurance of ethical standards of care for all
- Conduct health systems research
- Preparation for mass casualties from disasters, terrorism and genocide

anyone can have serious illness at any time, and long-term preventive care is essential to good public health standards and quality health care. Inequalities exist in all societies but many have successfully reduced them by poverty reduction, job creation, education, and many other systems that reduce interregional, socioeconomic, and demographic differences in health. Special attention to high-risk groups in the population is essential. Groups may be based on age, gender, occupation, risky lifestyle, location of residence, ethnicity, religion, sexual orientation, economic status, or other factors that increase susceptibility to disease, premature death, or disability. Services should be based on need and not only demand, which can escalate costs by overservicing or, in effect, selective servicing of those with insured access and the knowledge, time, and capacity to make use of a health

system. Health systems planning needs to promote access on patient assessment, but also those services that reach the entire population, especially people at high risk who are often least able to seek appropriate care.

A program that provides equal access for all may not achieve the objective of better health for its population unless it is accompanied by other important governmental activities. These include enactment and enforcement of environmental and occupational health laws; food safety, nutrition, and water standards; improved rural care; higher educational levels; and provision of health information to the public. Additional national programs are needed to promote health generally and to reduce specific risk factors for morbidity and mortality. Responsibility for health lies not only with medical and other health professionals, but also with governmental and voluntary organizations, the family, the individual, and the community.

Individual access to an essential basket of services as a prepaid insured benefit is integral to a successful national health program. Each country addresses this issue according to its means and traditions, but the most cost-effective method of meeting the country's epidemiological and demographic needs should be employed. Payments for heart transplantation may be beyond the means of a health system, but early and aggressive management of acute myocardial infarction is an effective method of saving lives at modest cost and containing the need for more intrusive personal health interventions. Improved diets, smoking reduction, and physical fitness are even more effective and less costly. Prevention is cost-effective and should be integral to the development of service priorities within the basket of services.

Globalization affects health systems around the world not only in the ease of spread of infectious diseases, but in increased access to modern preventive, diagnostic, and treatment modalities. Access to antiretroviral therapy (ART) is changing the face of human immunodeficiency virus/acquired immunodeficiency syndrome (HIV/AIDS) in many developing countries with support of international and bilateral donors. Adoption of vaccines, such as *Haemophilus influenzae* type b (Hib), rotavirus, and pneumococcal pneumonia vaccines, will save the lives of many hundreds of thousands of children, especially in the low- and middle-income countries. Information technology, migration of medical professionals, and internalization of educational standards are all global health issues affecting national health systems (see Chapters 14–16). Health systems are facing similar problems in population health, with rising population age, obesity and diabetes prevalence, and health care costs. Health systems research capacity is important in each country as it attempts to cope with rapid changes in population health and individual health needs with limited resources.

In this chapter, selected national health systems are presented representing major models of organization and different parts of the world. These organizational models influence health care system formulation in both developing and developed countries, as well as in countries restructuring their health services. Health care systems and financing are under pressure everywhere, not only to assure access to health for all citizens, but also to keep up with advancing medical technology, and contain the cost increase to sustainable levels. Because a health system is judged by more than its costs and measures of medical services, this chapter includes indicators of the health status of the population, including morbidity and mortality. This topic has developed a complex terminology of its own. Some of the key words are defined in this and other chapters in this text.

Finally, health systems are meant to improve health and quality of life, as measured by quantitative and qualitative methods (see Chapter 3). Since 2000, the Human Development Index (HDI) has provided a standard method of comparison which combines many health indices, including life expectancy at birth, gross domestic product (GDP) per capita, and child mortality, into a summary figure. Table 13.1 provides some of the key indicators discussed in this chapter for some industrialized as well as mid-level and other developing countries.

Selected leading causes of mortality among adults are shown in Table 13.2 for selected countries discussed in this chapter. Mortality rates vary not only within income level but also between income levels. Cancer mortality rates are similar among countries, but mortality rates for cardiovascular and diabetes-related diseases on average are more than three times higher in low-income than in high-income countries. Countries with different health systems have different outcome indicators (see Chapter 3).

In this grouping, Australia and Japan have an outstanding record with low mortality rates in all classes of cases of deaths, with Israel and Sweden coming close, followed by Norway, Canada, and the Netherlands. Nigeria has the highest rates in this group, followed by Russia, suffering from extremely high mortality from non-communicable diseases (NCDs), principally cardiovascular diseases (CVDs). India suffers from high riates of both non-communicable and communicable diseases. The poor performance of the USA in standardized adult mortality rates is a subject for continuing debate politically and in professional discussions, but largely rests on a lack of universal health insurance and poor dietary patterns, especially in some parts of the country. Canada does much better than the USA and somewhat better than the UK, although the UK is doing well in rates of

TABLE 13.1 Human Development Index (HDI) Ranking, Selected Countries, 2000–2012

	HDI Rank 2012	LE at Birth 2011	GDP Spent on Health 2011	Under-5 Mortality Rate per 100 live births		Maternal Mortality Ratio per 1,000 live births	
		(years)	(%)	2000	2011	2000	2010
Canada	11	82.0	11.2	6	6	70	12
China	101	76.0	5.2	35	15	61	37
Colombia	91	78.0	6.1	25	18	130	92
Denmark	15	79.0	11.2	6	4	8	12
Finland	21	81.0	8.9	4	3	5	5
Germany	5	81.0	11.1	5	4	7	7
Israel	16	82.0	7.7	7	4	9	7
Japan	10	83.0	9.3	5	3	10	5
Netherlands	4	81.0	12.0	6	4	13	6
Nigeria	153	53.0	5.3	188	124	970	630
Norway	1	81.0	9.1	5	3	8	7
Russian Federation	55	69.0	6.2	21	12	57	34
Sweden	7	82.0	9.4	4	3	5	4
UK	26	80.0	9.3	7	5	12	12
USA	3	79.0	17.9	9	8	14	21

Note: LE = life expectancy; GDP = gross domestic product.
Sources: Human Development Report 2013. Available at: http://hdrstats.undp.org/indicators/ and http://hdrstats.undp.org/en/indicators/53906.html [Accessed 14 April 2013].
World Health Organization. World Health Statistics 2013. Available at: http://apps.who.int/iris/bitstream/10665/81965/1/9789241564588_eng.pdf [Accessed 14 April 2013].
World Bank. Health expenditure total (% of GDP). Available at: http://data.worldbank.org/indicator/SH.XPD.TOTL.ZS/countries [Accessed 14 April 2013].

TABLE 13.2 Cause-Specific Age-Standardized Mortality Rates, per 100,000 Population Aged 30–70 Years, Selected Countries, 2008

Countries	All Causes	Cancer	Cardiovascular Diseases and Diabetes	Non-Communicable Respiratory Diseases
Australia	278	125	65	11
Canada	320	138	82	11
China	568	179	199	49
Colombia	493	112	152	21
Denmark	411	170	92	18
Finland	395	113	112	8
France	360	169	65	8
Germany	362	150	102	11
India	1002	108	328	139
Israel	289	125	72	12
Japan	281	119	68	6
Netherlands	323	165	77	12
Nigeria	1632	148	377	90
Norway	315	138	74	15
Russian Federation	1172	180	517	21
Sweden	293	121	71	9
UK	359	144	91	20
USA	460	143	137	24
Low-income countries	1354	154	375	77
Mid-level countries	808	150	273	73
High-income countries	375	141	104	14
World	764	150	245	52

Source: World Health Organization. World Health Statistics 2013. Table 2. Available at: http://www.who.int/gho/publications/world_health_statistics/2013/en/ [Accessed 14 June 2013].

cardiovascular mortality. The differences between all countries grouped by development level are very clear in this comparison; notably, mortality rates from CVDs are highest in the poorest countries, while respiratory causes are high in both the poorest and the mid-level countries.

HEALTH SYSTEMS IN THE INDUSTRIALIZED COUNTRIES

EVOLUTION OF HEALTH SYSTEMS

The tradition of prepayment of health care goes back to ancient times, when municipal doctors were employed by local authorities to provide care for the poor and slaves. In the Middle Ages, the Church provided charitable care for the poor. In the medieval and Renaissance periods, guilds provided prepaid health care to members and their families. These later evolved into the "friendly societies", as mutual benefit programs that provided for burials, pensions, and payment for health services for members and their survivors (see Chapter 1).

In the twentieth century, these programs developed through collective bargaining into health insurance plans with private or professionally sponsored insurers, and labor union-sponsored health plans. Governmental responsibility for health systems evolved in public health and health protection systems in the nineteenth and twentieth centuries, and continues to evolve to face new challenges and preventive and treatment capacities.

Social Insurance

Otto von Bismarck, Chancellor of Germany, introduced the first national health insurance plan for workers. It followed previous legislation in Germany establishing workmen's compensation on railroads (1838) and compulsory miners' benevolent societies (1854). Workmen's compensation and other benefits were extended in 1871 to many workers in other industries, such as those in domestic service, workers in mines, factories, and quarries, and seamen. Bismarck's 1883 compulsory health insurance legislation was intended to improve the health of workers and their families, and especially of potential army recruits, as well as to stave off the political advancement of the social democratic parties. The program was based on the principle of social insurance, involving payroll deductions at the workplace with contributions from the employer and employee, to cover medical care, unemployment benefits, and pensions for workers.

The Bismarckian model established state social insurance with prepayment by workers and their employers. It utilized Sick Funds (Krankenkassen) as insurers to provide payment to the physician, hospital, or other provider. In the years before World War I, many countries in Central and Eastern Europe implemented similar health plans. In the period between the world wars, national health insurance programs were developed in many countries in the industrialized world. In Europe, most countries developed models based on the Bismarckian approach, with compulsory contributions by workers and their employers to a national social security system, which then finances approved Sick Funds that pay for services usually paid through private medical practice with fee-for-service payment. The Bismarckian model is a successful model used widely in Europe and Israel. This model has also influenced post-Soviet health reforms and countries of Eastern Europe.

In 1911, the Liberal government of Great Britain, initiated by Chancellor of the Exchequer David Lloyd George and influenced by the German compulsory health insurance scheme, introduced the National Health Insurance Act. It was compulsory for all wage earners between the ages of 16 and 70. These workers made payments along with their employers and a state contribution. This two-part plan provided a contributory system for unemployment insurance and for medical care against illness for workers and their families. General practitioners (GPs) were paid on a capitation basis rather than a salary, preserving their status as self-employed professionals. Initially this plan covered one-third of the population, but coverage increased to one-half by 1940. Administration was through approved mutual benefit societies (friendly societies), some based on insurance companies and others founded by professional associations and trade unions. European countries and Japan gradually developed compulsory health insurance following World War I, and completed universal coverage following World War II.

The social security model of health insurance for urban workers also became prominent in many countries in Latin America. Social security plans are financed by mandatory contributions of workers and employers, and administered by the state. The Social Security Act of 1935 in the USA was instituted to alleviate the social distress of the Great Depression. This "New Deal" social experiment of President Franklin Roosevelt provided cash benefits for widows, orphans, and disabled people, as well as pensions for the elderly, and provided a base for future reform including health insurance. Since 1965, this legislation has provided the basis for US medical and hospital coverage of the elderly under Medicare and the poor under Medicaid. Later proposals for national health insurance in the USA have also largely been based on the federal social security funding system.

National Health Service

In some countries, the state directly assumed responsibility for both social security and health care. The welfare state took on measures such as unemployment and disability insurance, and special disability benefits for the blind, widows, orphans, and the elderly through pensions. Several states also instituted child benefits to raise levels of child care and nutrition through general governmental revenues from taxation and other sources.

In 1918, following the Russian Revolution, the new Soviet Union (USSR) introduced its national health plan for universal coverage within a state-run system of health protection. The Soviet model, designed and implemented by Nikolai Semashko, provided free health care for all as a government-financed and -organized service. It developed health services across the vast underdeveloped regions of the USSR with free health services to the population, with a system of primary and secondary care based on the principles of universal and equitable access to care through district organization of services. It achieved control of epidemic and endemic infectious diseases and expanded services into the most remote areas of the country.

In the early days of World War II, the British government established a national Emergency Medical Service to operate hospitals in preparation for the large-scale civilian casualties expected. The plan established national health planning and rescued many hospitals from near bankruptcy resulting from the effects of the Great Depression in the UK. During World War II, a postwar social reconstruction program was developed by William Beveridge, at the behest of the wartime prime minister Winston Churchill. The Beveridge Report of 1942, *Social Insurance and Allied Services*, outlined the nature of the future welfare state including a national health service, placing medical care in the context of general social policy for the total population.

The wartime coalition government approved the principle of a national health service, which had wide public support, despite opposition from the medical profession. In 1948, the Labour government of Prime Minister Clement Attlee under the leadership of the Minister of Health, Aneurin Bevan, implemented the National Health Service (NHS), a nationally financed, universal coverage system providing free care by GPs, specialists, hospitals, and public health services. The NHS is one of the major successful international models for national health systems and continues to this day, albeit with many challenges and periodic reforms.

National Health Insurance

The Canadian system of tax-based national health insurance is based on provincial health plans meeting federal government requirements for cost-sharing. The program evolved from provincial initiatives led by Tommy Douglas, Premier of the Province of Saskatchewan. Initiated in 1946, provincial plans provided universal insured hospital services under provincial public administration, later followed by medical and other services.

Developed over the period 1946–1971, the plans were promoted by federal governmental cost-sharing, political support, and national standards. The plans were initially financed by taxation and premiums, but later by general tax revenues alone in most provinces; Alberta, British Columbia, and Ontario also have premiums. The Canadian "Medicare" plans are publicly administered by the provinces with federal standards, cost-sharing, and comprehensive coverage. Care is provided by private medical practitioners on a fee-for-service basis under negotiated medical fee schedules. Hospitals may be operated by nonprofit voluntary, regional health, or municipal authorities, with payment by block budgets. This Medicare-type plan was later adopted in a number of other countries including Australia. Figure 13.1 indicates the distribution of governmental and private funding of health expenditures by the member countries of the Organisation for Economic Co-operation and Development (OECD).

THE UNITED STATES

The US population in 2012 was 313 million, with a per capita GDP of US$48,387 (in 2011), an increase from US$43,800 in 2006. Health expenditures rose from 15.8 to 17.4 percent of GDP between 1990 and 2010 (Table 13.3). In 2011, health expenditures reached 17.9 percent of GDP, US$8233 per capita, the highest among the OECD countries. The US child mortality rate of 8 per 1000 live births in 2010 was 41st in world ranking. The US life expectancy (total) at birth was 79 years in 2009 (51st in world ranking). In 2011, the infant mortality ranked 48th with a rate of 6 per 1000 live births.

TABLE 13.3 Health Expenditures, USA, 1990–2010

Health Expenditures	1990	2000	2005	2010
Total expenditure per capita, US$ PPP	2851	4791	6728	8233
Annual growth rate of total expenditure on health per capita, in real terms from previous or to next year	NA	6.0	3.3	2.7
% of GDP spent on health	15.8	16.0	16.4	17.4
Health expenditures (% distribution)	100	100	100	100
Private (%)	60.6	57.0	55.8	51.8
Public (%)	39.4	43.0	44.2	48.2

Note: PPP = purchasing power parity; GDP = gross domestic product; NA = not available.
Source: Organisation for Economic Co-operation and Development. OECD database, 2012. Available at: http://www.oecd.org/els/health-systems/oecdhealthdata2012-frequentlyrequesteddata.htm [Accessed 21 April 2013].

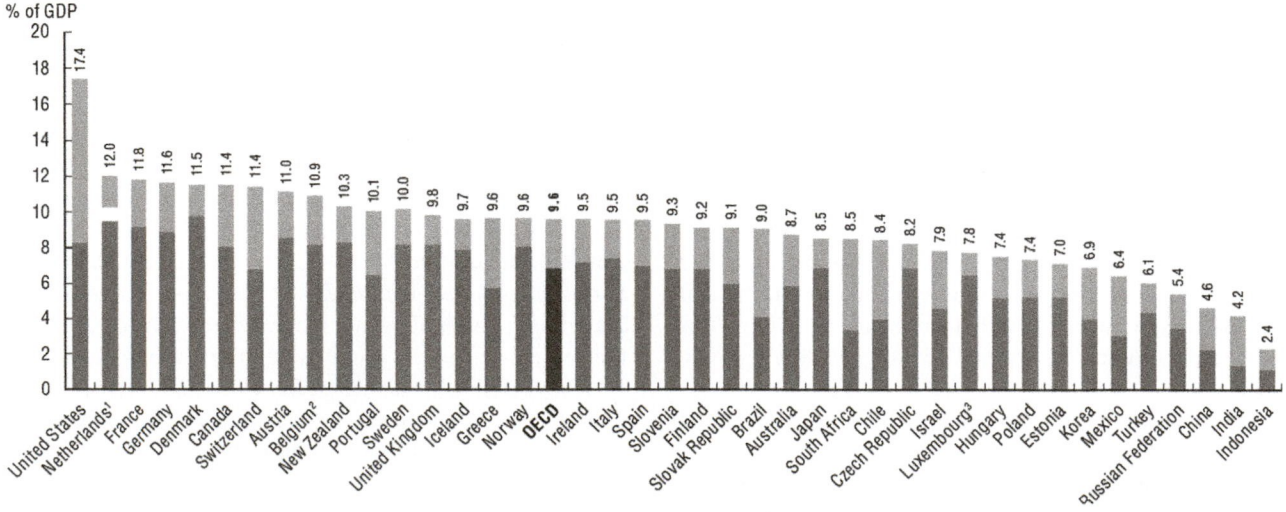

FIGURE 13.1   Total health expenditures as a share of gross domestic product (GDP), Organisation for Economic Co-operation and Development countries, 2009. Note: Some 2010 and 2011 data are available but incomplete. *Source: OECD Report Health at a Glance 2011. Available at: http://www. oecd.org/els/health-systems/49105858.pdf [Accessed 2 January 2014].*

In 2011, the USA stood third among the leading nations in the HDI. It has steadily improved in HDI since 1975 and ranks above the average for OECD countries. In health indicators, however, the USA is near the middle of the 32 OECD nations in death rates from all causes, mainly heart disease, cancer, and stroke. With declining birth rates and increasing longevity, the population is aging, with more than 12 percent aged 65 and over. Nearly 40 percent of the population is now in non-working dependent age categories (over 65 or under 15).

The USA has a federal system of government, with each of the 50 states having its own elected government with legislative, judicial, enforcement, and taxing powers. The US Constitution gives primary responsibility for health and welfare to the states, but direct federal services are provided to the armed forces, veterans, and Native Americans. However, the federal government has established a major leadership role in health by developing national standards, regulatory powers, and information systems. It also serves as a major agency for financing research, health services, and training programs.

Federal Health Initiatives

In 1798, the federal government under President John Adams established the US Marine Hospital Service to provide hospitals in the major port cities to provide prepaid care for sick and disabled merchant seamen. This later became the Marine Hospital Service and then the US Public Health Service Commissioned Corps as a uniformed service headed by the US Surgeon General (1873). It then became the location for the United States Public Health Service (USPHS) services for Native Americans, military personnel and their families, the National Institutes of Health (NIH), the Centers for Disease Control and Prevention (CDC). and other federal programs of research, service, and teaching.

In the late nineteenth and early twentieth centuries the US Federal Department of Agriculture Extension Service promoted nutrition and hygiene education throughout the rural areas of the country. Later legislation provided federal grants to establish state, municipal, and county health departments. Health hazards caused by poor food and drug standards, lack of care for the elderly and the poor, dangerous automobiles, environmental pollution, and health service deficiencies led to government intervention to protect the public interest. The Food and Drug Control Act of 1906 was promulgated to regulate and control commerce. In 1921, the Sheppard–Towner Act established the federal Children's Bureau that administered grants to assist states to operate maternal and child health programs, which were later incorporated into the Social Security Act.

In 1927, the Committee on the Costs of Medical Care, a commission funded by several private foundations, recommended that the USA implement a universal national health program based on medical group practices with voluntary prepayment. From the 1920s, labor unions won health insurance benefits through collective bargaining, which became the main basis for prepayment for health care in the USA until today. These initiatives were slowed due to the Great Depression from 1929 to 1939, but resumed during and after World War II. The Social Security Act of 1935 increased social support for millions of individuals living with disabilities or occupational injuries, as well other vulnerable groups such as widows, orphans, and elderly people. This act successfully alleviated some of the Depression's most devastating effects.

With the coming of war and implementation of compulsory registration for conscription, significant percentages of eligible males were considered unfit for the draft because of physical unfitness and reasons such as the lack of six adjacent teeth due to poor dental health, just as a high percentage of draftees had been rejected in World War I owing to goiter from iodine deficiency. In 1941, before the USA had actually joined World War II, President Roosevelt initiated mandatory fortification of "enriched" salt with iodine, flour with iron and vitamin B complex, and milk with vitamin D which became nearly universal national standards.

During World War II (1941–1945) millions of Americans in the armed forces and their dependants, previously with limited access to prepaid health care, were enrolled in a national plan for free health care (Emergency Maternity and Infant Care for the Wives and Children of Servicemen or EMIC). At the same time, health benefits through voluntary insurance for workers were vastly expanded in place of wage increases, which were forbidden by federal wartime regulation. At the end of the war, millions of veterans were eligible for health care through the Veterans Health Administration, which established a national network of federal hospitals and primary care services for this sole purpose.

In 1946, President Truman attempted to bring in national health insurance but the legislation (the Wagner–Murray–Dingell Bill) failed in the US Congress. One section of the bill was approved, enabling the federal government to initiate a program of categorical grants to upgrade countrywide hospital facilities under the Hill–Burton Act. Another section provided massive federal funding for health to strengthen the NIH, established after World War II. The NIH promotes research and strengthens public and private medical schools, teaching hospitals, and research facilities. In the 1950s, the federal government also established the Centers for Disease Control (CDC) and increased public health grants providing assistance for state and local public health activities.

From the 1940s through the 1960s, voluntary health insurance became the major method of prepayment for health care needs, mostly through employment contracts. The private insurance industry developed rapidly, with minimal governmental regulation to ensure fair pricing and payment. During the 1970s and 1980s, employers grew concerned about health insurance costs for their workers and pressed the government to restrain health care costs. Federal initiatives included public insurance for the elderly and the poor, promoting efficiency in payment for hospital care. Later on, the promotion of health maintenance organizations (HMOs) and managed care was also emphasized.

Medicare and Medicaid

In the mid-1960s, despite the growth of voluntary and employment-based health insurance, a large percentage of elderly and poor Americans lacked health insurance. In 1965, President Lyndon Johnson introduced Medicare for the aged (over age 65), disabled people, and people on renal dialysis as Title XVII of the 1935 Social Security Act. This brought some 10 percent of the population under a limited form of national health insurance.

Medicaid, Title XIX of the Social Security Act, also enacted in 1965, provided federal cost-sharing for acceptable state health plans for the poor, with local authority participation. These two plans brought some 25 percent of Americans into public systems of health insurance. Limitations included variable definitions of poverty in each state, and co-payments for Medicare beneficiaries. Medicare covers hospitalization, skilled nursing home care, medical appliances, and other benefits with co-payments. In 2006, a drug benefit program was added.

In 1997, Title XXI of the Social Security Act the State Children's Health Insurance Program (SCHIP) was initiated to provide federal funds to assist approved state plans to extend health insurance for children. This program provides health coverage for families that are ineligible for Medicaid owing to their income status but cannot afford to purchase independent insurance. While funding for SCHIP is provided by both federal and state governments, each state runs its own SCHIP program under the broad guidelines of the federal government and the specific guidelines created by each state. Congress initially authorized SCHIP for 10 years, from 1998 to 2007. It was vetoed by President George W. Bush in October 2007. A congressional effort to override the veto failed by 13 votes (273 to 156, with two-thirds approval required) 15 days later. In 2009, President Barack Obama signed the Children's Health Insurance Program Reauthorization Act of 2009 (CHIPRA) expanding the health care program to an additional 4 million children and pregnant women.

In 2006, 67.9 percent of the US population was covered under private health insurance, mostly employment based, 13.6 percent under Medicare, and 12.9 percent under Medicaid, while 15.8 percent were uninsured. In 2010, 55.3 percent of the US population was covered by employer-sponsored insurance, 14.5 percent under Medicare, 15.9 percent under Medicaid, 9.8 percent under other private coverage, and 4.2 percent under military plans, while 16.3 percent were uninsured. Medicare and Medicaid brought many previously uninsured people under health insurance

coverage. Public funding for health care in the USA includes Medicare, Medicaid, and SCHIP, research and medical education, and promotion of community health centers and services in impoverished or underserved areas (see Table 13.3). The percentage of public funding in the USA rose from under 25 percent of total health expenditures in 1960 to approximately 45 percent of total health expenditures in the years 1995–2004. This figure was 47.7 percent in 2009.

The population enrolled in Medicare increased from 19 million in 1966 to 49.4 million or 16 percent of the US population in 2013, including over 9 million disabled people under the age of 65. The Medicaid-enrolled population increased from 28.2 million in 1991 to 72.6 million in 2012 or some 23 percent of the US population. This contributed to growth of health expenditures in the public sector, a concern for both critics and supporters of public health care programs. Medicaid beneficiaries must be US citizens or legal permanent residents, and may include low-income adults, their children, and people with certain disabilities. Poverty alone does not necessarily qualify someone for Medicaid. Enrollment is projected to reach 78.0 million in 2019.

A federal–state Children's Health Insurance Program (CHIP) provides health coverage to nearly 8 million children in families with incomes too high to qualify for Medicaid. Each state receives an annual allotment of federal funds, available as a federal match based on the state's expenditures. In general, states have 3 years to use each fiscal year's allotment, after which unspent federal funds may be redistributed.

In 1997, Congress created the Medicare+Choice plan (now known as Medicare Advantage), which gives Medicare enrollees the choice of various health plans. It was also created in the hope of controlling Medicare costs. The Medicare Prescription Drug Improvement and Modernization Act of 2003, signed into law by President George W. Bush, included prescription drugs for Medicare enrollees. The Patient Protection and Affordable Care Act (PPACA, widely known as the Affordable Care Act or "Obamacare") will expand both eligibility for and federal funding of Medicaid beginning on 1 January 2014. Despite a seriously flawed start-up, this plan will allow all US citizens and legal residents with income up to 133 percent of the poverty line, including adults without dependent children, to qualify for coverage.

Medicare costs are increasing at a faster rate than the economy, especially since the start of the recession which began in 2008. Medicare spending grew 6.2 percent to US$554.3 billion in 2011, or 21 percent of total national health expenditures. Financing of Medicare comes from two trust funds: the Hospital Insurance and the Supplementary Medical Insurance (SMI). Taxes paid by employees and employers support the Hospital Insurance trust fund, which finances inpatient care. This trust fund is expected to be depleted by the year 2019. The SMI is supported by general income tax revenues and enrollee premiums, and covers physician services, outpatient and hospital services, and prescription drugs. The federal government faces the challenge of making appropriate reforms in Medicare in order to avoid consuming more federal revenues and taking from other federal programs, especially as the postwar "baby boom" generation becomes eligible for Medicare benefits.

The Changing Health Care Environment

From the 1960s through the 1990s, rapid cost increases were attributed to many factors, including an increasing elderly population, high levels of morbidity in the poor population, the spread of AIDS, rapid innovation and costly medical technology, specialization, high laboratory costs, and large-scale public investment in medical education and research and health facility construction. Other equally important factors were high levels of preventable hospitalizations, the institutional orientation of the health system, high administrative costs due to multiple private billing agencies in the private insurance industry, high incomes for physicians, especially for specialists, and high medical malpractice insurance costs. The pressure for cost constraint came from government, industry, and the private insurance industry.

Most hospitals are owned and operated by non-profit agencies, including federal, state, and local governments, voluntary organizations, and religious organizations. Privately owned hospitals operating for profit increased from 7.8 percent of community, short-term hospital beds in 1975 to 12.7 percent in 1996 and to 20.6 percent in 2013. Private medical practice, with payment by fee-for-service, was the major form of medical care until the 1990s. HMOs and other forms of managed care have grown rapidly to become the predominant method of organizing health care in the USA.

Prepaid group practice (PGP) originated from company-provided contract medical care, especially in remote mining camps. The Community Hospital of Elk City, Oklahoma, established in 1929, is considered the first real medical cooperative or prepaid group practice. Later, many rural cooperatives were formed to provide prepaid medical care. Union-sponsored health services were developed to provide medical care in poor mining areas in the Appalachian Mountains, as well as in an urban cooperative in Washington, DC in 1937. In the 1940s, New York City sponsored the Health Insurance Plan of Greater New York to provide prepaid medical care for residents of urban renewal and low-income housing areas. This was later supported by organized union groups such as municipal employees and garment industry workers.

PGP became best known in the Kaiser Permanente network developed for workers of Henry J. Kaiser Industries, at the Boulder Dam and Grand Coulee Dam construction sites in the 1930s. This experience was applied in Kaiser's rapidly growing industries in the San Francisco Bay area during World War II when salaries were frozen but health benefits were expanding. Kaiser Permanente health plans expanded rapidly in many other states and now provide care for millions of Americans. Initially opposed by the organized medical profession and the private insurance industry, PGP gained acceptance by providing high-quality, less costly health care. It became attractive to employers and unions alike, and later to governments seeking ways to constrain increases in health costs.

Since the 1970s, the generic term health maintenance organization (HMO) has been used, especially by the federal government seeking to promote this concept. The HMO model links health insurance and medical care in the same organization, and the concept was promoted through the HMO Act by President Richard Nixon in 1973. The HMO including both HMOs and other forms of prepaid insurance plans, later called "managed care", has become an accepted, if often criticized, part of medical care in the USA and an important alternative to fee-for-service, private practice medicine (Figure 13.2). In 2011, 70.2 million Americans, or 22.5 percent of the total US population, were registered in HMO plans.

In recent years, the terms accountable care organization (ACO), patient-centered medical home (PCMH), and population health management system (PHMS) have come into wide use to denote organizations that take responsibility for comprehensive care for enrolled patients, with payment based on a form of capitation rather than fee-for-service. The ACO comes in different models, but many include a hospital base and may be linked to independent practice associations (IPAs), which may include specialty groups, or hospital medical staff organizations, or a network of hospitals linked with other providers as organized delivery systems.

This approach to health reform in the USA is based on evidence of cost-effective care with emphasis on prevention and reduced hospitalization as given to millions of Americans by well-established care systems such as Kaiser Permanente and the Cleveland Clinic. These are not for profit, based on group practice, led by doctors who are salaried rather than fee-for-service, and subject to rigorous annual professional review. It may provide a set of models adaptable on a wider scale to improve quality and cost-effective care to improve the health of Americans (Devers and Berenson, 2009; Shortell, 2010).

In order to encourage more efficient use of hospital care, the method of payment was changed during the 1980s. In 1983, a prospective payment system, called diagnosis-related groups (DRGs), was adopted for Medicare, with payment by categories of diagnosis (HCFA, 1998). This replaced the previous system of paying by the number of hospital days, or per diem. DRGs encourage hospitals to diagnose and treat patients effectively and expeditiously and to discharge them as quickly as their condition allows.

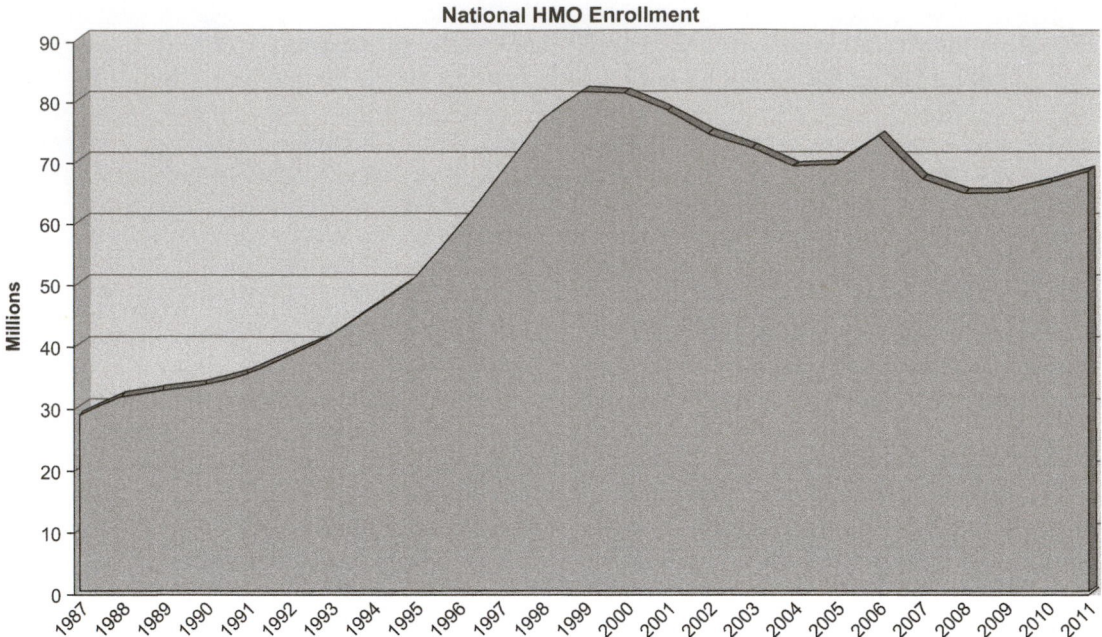

FIGURE 13.2 National health maintenance organization (HMO) enrollment, 1987–2011. *Source: Managed Care Fact Sheets. National HMO enrollment graph 1987–2011. Available at: http://www.mcol.com/factsheet_hmo_enrollment_graph [Accessed 31 March 2013].*

Payment for Medicare and Medicaid patients was shifted to this method. In many states this has also become standard for patients with private health insurance. Between 1980 and 1990, because of the DRG payment system and HMOs or managed care systems, which promote alternatives such as home and ambulatory care, hospital utilization was reduced in the USA. While total costs of health care increased during this period, without the reduction in hospital utilization the increase would have been considerably higher.

During the late 1980s, managed care expanded from non-profit HMOs of the Kaiser Permanente type to include both non-profit and for-profit systems operated by the insurance industry. Managed care plans of the HMO type operate their own clinics and staff (i.e., the staff model). Other managed plans operate on a not-for-profit or a for-profit basis. These are IPAs, which operate with physicians in private practice, or preferred provider organizations (PPOs), which cover care with doctors and other providers associated with the plan providing services to the enrolled members or beneficiaries at negotiated prices (see Chapter 12).

Following the failure of the Clinton national health insurance proposal in 1993, managed care experienced tremendous growth as employers sought to provide their employees with comprehensive coverage at reasonable costs. Managed care systems have been able to cut costs in health care in ways that governments could not. In 1996, 74 percent of insured American workers were enrolled in managed care plans, compared to 55 percent in 1992. In California, with a long tradition of HMOs such as Kaiser Permanente, enrollment at the end of 2006 was 65 percent of the total state population. In the USA as a whole, in addition to the nearly 58 million people enrolled in HMOs, another 91 million people are enrolled in PPOs, with 25 percent of Medicaid and 10 percent of Medicare beneficiaries in managed care plans.

The search for cost containment led to the development of a series of important innovations in health care delivery, payment, and information systems. HMOs have demonstrated that good care provision can be operated efficiently with lower hospital admission rates than care provided on a fee-for-service basis. The managed care systems brought about profound changes in health care organization in the USA. The number of plans declined from 572 in 1990 to 412 in 2004 as a result of mergers. A total of 149 million people (Table 13.4) or 51 percent of the insured population and 49 percent of the total US population are enrolled in managed care. More than 70 million Americans have been enrolled in HMOs and almost 90 million have been part of PPOs (National Conference of State Legislatures, 2013).

Managed care coverage peaked in 2001 and has subsided slightly since owing to negative publicity of the private for-profit insurance operators apparently making tight restrictions on access to care to reduce costs; thus, the capitation payment method is criticized by supporters of open-ended fee-for-service. Proponents of managed care point to high-quality programs such as the Harvard Pilgrim Managed Care plan and the ACOs, which have been very successful in growing and sustaining high-quality care within reasonable cost parameters. Such programs have pioneered

TABLE 13.4 Enrollees in Managed Care Coverage by Type of Health Insurance Plan, USA, 2011

Numbers of Enrollees by Insured System (millions)	Total US (millions)	US (%)	Managed Care (millions)	Managed Care (%)
Medicare	48.0	15.4	12.2	25.5
Medicaid	45.8	14.7	32.6	71.2
Military	4.0	1.3	4.0	100.0
Commercial	162.8	52.3	161.1	99.06
Uninsured	50.7	16.3	0.0	0.0
Total	311.3	100.0	210.0	67.5

Source: MCOL. Managed care fact sheet. Managed care penetration 2011; September 2011. Available at: http://www.mcol.com/managed_care_penetration [Accessed 31 March 2013].

computerization of medical records, utilization review, preventive practices as part of regular medical care, and quality promotion. The topic of managed care remains a central issue for the federal government in the search for universal coverage health insurance at affordable costs.

Hospitals and other specialty services are competing for contracts with managed care organizations and establishing community service systems of their own in order to compete for "market share" of insured clients. In many locales, excess hospital beds have become an economic burden, forcing many hospitals to downsize or become part of larger hospital chains or a local multihospital of vertically integrated health networks (see Chapter 10). Hospitals have responded by establishing contracts with managed care organizations and by reducing bed capacity; others have closed as they were unable to compete for sufficient patient flow.

Federal and state legislative initiatives are attempting to define patients' rights in managed care because of public complaints with limitations of managed care. In response to widespread criticism regarding managed care restricting access to specialty services and shortened hospital stays, in 1998 the US Congress passed a bipartisan-sponsored law that requires minimum 48-hour maternity stays. Many other pieces of legislation to protect consumers' rights and choice of doctor have been proposed in Congress and in state legislatures.

Health Information

The USA has developed extensive information systems of domestic and international importance. The CDC publishes the *Morbidity and Mortality Weekly Report (MMWR)*, which sets high standards in disease reporting and policy analysis. The US National Center for Health Statistics (NCHS), Health Care Financing Administration (HCFA), USPHS, Food and Drug Administration (FDA), NIH, and many non-governmental organizations (NGOs) carry out

data collection, publication, and health services research activities important for health status monitoring. National nutrition surveillance, via the National Health and Nutrition Examination Survey (NHANES; see Chapter 8), and other systems of health status monitoring are reported in the professional literature and in publications of the CDC. National monitoring of hospital discharge information facilitates the understanding of patterns of utilization and morbidity. These information systems are vital for epidemiological surveillance and managing the health care system.

The Surgeon General in the USA has a high rank akin to the military services, but reports to the Department of Health and Human Services (DHHS) and is head of the USPHS. Periodic Reports of the Surgeon General have an important influence on health systems not only in the USA but also internationally. The 1965 US Surgeon General's *Report on Smoking*, linking smoking and lung cancer, had a major impact on public knowledge and behavior. This classic report made the issue of smoking a major public health challenge due to the very strong evidence of links with lung cancer, cardiovascular and other diseases. Other reports that have made important contributions to the evolution of public health include:

- 1988 – Nutrition and Health
- 1996 – Physical Activity and Health
- 1998 – Tobacco Use Among US Racial/Ethnic Minority Groups
- 2001 – Mental Health: Culture, Race, and Ethnicity
- 2001 – Women and Smoking
- 2001 – Youth Violence
- 2004 – Bone Health and Osteoporosis
- 2004 – The Health Consequences of Smoking
- 2006 – The Health Consequences of Involuntary Exposure to Tobacco Smoke
- 2007 – Children and Secondhand Smoke Exposure
- 2010 – How Tobacco Smoke Causes Disease: The Biology and Behavioral Basis for Smoking-Attributable Disease

- 2011 – Call to Action to Support Breastfeeding
- 2012 – National Strategy for Suicide Prevention: Goals and Objectives for Action.
- 2012 – Preventing Tobacco Use Among Youth and Young Adults.

The Surgeon General continues to promote awareness of knowledge on important public health issues including physical activity and health, mental health, oral health, youth violence, bone health and osteoporosis, underage drinking, and sexual health. Reductions in smoking and obesity continue to be priority issues. Media coverage of health-related topics is extensive, and is important to promote health consciousness in the public. The sheer volume of information may make it difficult to discern which information is most relevant, and much misinformation appearing on Internet sites can also create trends counter to public health such as refusals to vaccinate children. Public levels of health knowledge are growing steadily but vary widely with social class and educational levels.

The CDC created the National Center for Public Health Informatics (NCPHI) in 2005 to provide leadership and coordination of shared systems and services, to build and support a national network of integrated, standards-based, and interoperable public health information systems. This is meant to strengthen capabilities to monitor, detect, register, confirm, report, and analyze, as well as to provide feedback and alerts on important health events. This will enable partners to communicate evidence that supports decisions that impact health. Electronic medical and personal health records are now widely used. They both protect patient privacy and confidentiality and serve legitimate clinical and public health needs. US health costs are rising, increasing from 17.4 percent of GDP in 2010 to 17.9 percent of GDP in 2011. At the same time, acute and chronic health threats challenge the US and global capacity to address them with efficient and effective disease prevention.

Health Targets

Despite rapid increases in health care expenditures during the 1970s and 1980s, improved health promotion activities, and rapidly developing medical technology, the health status of the US population has improved less rapidly than that in other western countries. Infant mortality in the USA remains high in comparison to other OECD countries and ranked 34th among all countries in 2012 (estimated). Even the rate of infant mortality of the white population of the USA was higher than that of 16 countries that spent much less per person and a lesser percentage of gross national product (GNP) per capita on health care.

The 1979, the US Surgeon General's Report *Healthy People* set forth a series of national health targets for a wide variety of public health issues. The program defined 226 objectives in 15 program areas within the three categories of prevention, protection, and promotion. These goals and objectives were formulated based on research and consultation by 167 experts in different fields who participated in a conference by the USPHS. Consensus was based on position papers, studies, and conferences involving the national governmental health authority, the National Academy of Sciences' Institute of Medicine (IOM), and professional organizations, such as the American Academy of Pediatrics (AAP) and the American Congress of Obstetricians and Gynecologists (ACOG). Many private individuals and organizations contributed to this effort, including state and local health agencies, representatives of consumer and provider groups, academic centers, and voluntary health associations.

These targets (Table 13.5) are periodically assessed as performance indicators of the US health system and then updated. Progress made during the 1980s included major reductions in death rates for three of the leading causes of death: heart disease, stroke, and unintentional injuries. Infant mortality decreased, as did the incidence of vaccine-preventable infectious diseases.

TABLE 13.5 Healthy People 2020

Framework	Specific groups or activities with measurable targets
Vision	A society in which people lead healthy, long lives
Mission	Identify nationwide health improvement priorities Increase public health and awareness of the determinants of health, disease, and disability and the opportunities for progress Provide measurable objectives and goals that are applicable at the national, state, and local levels Engage multiple sectors to take action to strengthen policies and improve practices that are driven by the best available evidence and knowledge Identify critical research, evaluation, and data collection methods
Overarching goals	Attain high-quality, longer lives free of preventable disease, disability, injury, and premature death Achieve health equity, eliminate disparities, and improve the health of all groups Create social and physical environments that create good health for all Promote quality of life, healthy development, and healthy behaviors across all life stages
Progress indicators	General health status Healthy related quality of life and well-being Determinants of health Disparities

Source: United States Department of Health and Human Services. Healthy People 2020 Framework. Available at: http://www.healthypeople.gov/2020/Consortium/HP2020Framework.pdf [Accessed 19 May 2013].

Healthy People 2000, published in 1992 by the Surgeon General, detailed 332 specific health targets, in six groups, for the year 2000, in the areas of health promotion, health protection, preventive services, surveillance and data systems, and age-related and special population groups (see Chapter 11). The final reviews of Healthy People 2000 showed significant decreases in mortality from coronary heart disease (CHD) and cancer. *Healthy People 2020* is renewing this effort to establish national targets which are adopted by state-level governments and strongly influence policy in health insurance systems (Table 13.5). The 2010 "Obamacare" program will include over 30 million previously uninsured Americans in health insurance within better regulated private insurance or in state-run Medicaid plans (see Chapter 10).

In 2000, the DHHS released *Healthy People 2010*, with two main goals: to "increase the quality and years of healthy life" and to "eliminate health disparities". These goals focus on 28 specific areas developed by over 350 national membership organizations and 250 state health, mental health, substance abuse, and environmental agencies. A midcourse review of Healthy People 2010 shows that 60 percent of the objectives are either being met or moving forward. The USA is moving towards the goal to "increase the quality and years of healthy life", although there are still clear gender, race, and ethnic discrepancies. Reducing health disparities continues to be a challenge in the USA.

Many states have adopted these targets as their own measures of health status and performance. Annual publications by the USPHS, in cooperation with the NCHS, make available a wide set of data for updating health status and process measures relating to these national health goals. The value of working towards health targets is widely accepted. Healthy People 2020 has defined similar overarching goals (see Table 13.5).

Health promotion has received wide public, governmental, and professional support in the USA over the past decades. In part, this reflects a long tradition of education on health matters in the rural agricultural sector and school health education. Nutrition and antismoking consciousness has grown in part because of wide media attention to many important epidemiological studies.

Consumer advocacy has been a potent factor for change in the USA in the twenty-first century, and especially since the 1960s. It has contributed to strengthened governmental regulation in a wide area of public health-related fields (see Chapter 2). These include automobile safety features and emission control, environmental standards, Mothers Against Drunk Driving, nutritional labeling, vitamin and mineral fortification of basic foods, and legal action against cigarette manufacturers. Food fortification, pioneered in the USA, is not mandatory as in Canada, but is nevertheless nearly universal, and mandatory for those foods labeled "enriched" (see Chapter 8). This is accepted in the general population based on advocacy, informed public opinion, and an innovative, highly competitive food industry. Despite much public controversy, fluoridation of community water supplies covers 67 percent of the population, a higher coverage than in most industrialized countries.

Advocacy groups can also promote regression in public health measures, as with groups currently fighting against immunization on the grounds of disinformation and opposition to vital vaccination programs. Some opposed to abortion have greatly affected public policy and promote sometimes violent activities against proponents and providers of abortions. Groups opposed to hospital births have sparked a widespread home birthing movement, which may lead to dangerous complications. Research and wide media coverage of health issues encourage a high level of individual and community consciousness of health-related issues and a climate receptive to health promotion.

Social Inequalities

Lack of universal access and the lack of empowerment it brings encourages an alienation or non-engagement with early health care. This promotes inappropriate reliance on emergency department care and hospitalization in response to undertreated health needs. With large numbers of uninsured people and many lacking adequate health insurance, access to and utilization of preventive care are below the levels needed to achieve social equity in health. This is especially true for maternal and child health and for NCDs such as diabetes, cancer, and heart disease.

Infant mortality rates in the USA vary greatly by race and ethnicity. As measured by the infant mortality rate in 2007, the rate among non-Hispanic black mothers was 2.4 times as high as the rate for white non-Hispanic mothers. A significantly higher rate of infant mortality exists among Puerto Rican and American Indian populations compared to the national average. This is primarily due to higher levels of preterm births and preterm-related causes of death. Efforts to improve immunization coverage of US infants to meet national health targets have been partially successful with efforts directed towards poor population groups. Vaccination against measles was 90 percent in the USA in 2009. The rate has been steadily decreasing since 2004, when it was 93 percent. Vaccination against diphtheria, tetanus, and pertussis (DTP) has also decreased, from 85.5 percent in 2004 to 83.3 percent in 2009. Vaccination against influenza has increased, from 58.7 percent in 2005 to 67.7 percent in 2009. In 2002, a program called the Racial and Ethnic Adult Disparities in Immunization Initiative was introduced to tackle the low levels of influenza and pneumococcal vaccinations among minorities aged 65 and over. In 2009, President Obama allocated US$2.3 billion in Recovery Act funds to improve preventive health care for children and vulnerable groups. Of that amount, US$300 million was directed towards vaccination efforts.

The US Department of Agriculture's Special Supplemental Nutrition Program for Women, Infants, and Children (WIC) enables millions of poor Americans to have good nutritional security (see Chapter 8). The WIC program covers pregnant women (through pregnancy and up to 6 weeks after birth or after pregnancy ends), breastfeeding women (up to the infant's first birthday), non-breastfeeding postpartum women (up to 6 months after the birth of an infant or after pregnancy ends), infants (up to their first birthday), and children (up to their fifth birthday). WIC serves 53 percent of all infants born in the USA. The benefits include supplemental nutritious foods; nutrition education and counseling at WIC clinics; and screening and referrals to other health, welfare, and social services such as completion of immunization and special needs counseling.

School lunch programs and nutrition support for pregnant women and children in need have reduced some of the ill-effects of poverty in the USA, but a lack of health insurance affects these groups severely. NCD and trauma are also diseases of poverty, with higher rates of morbidity and mortality in virtually all categories compared to higher income groups.

Health disparities are a complex problem that goes beyond the issue of uninsured Americans. Low-income and illegal immigrants face challenges accessing medical insurance. New immigrants to the USA who obtained citizenship after August 1996 must wait 5 years before they are eligible for Medicaid. The structure of the medical system plays an important role in an individual's ability to obtain medical care. This includes convenience of appointment making and office hours, waiting times, and transportation. A lack of health literacy also plays a role in an individual's ability to seek medical attention. Individuals not fluent in English experience communication gaps. In 2003, it was estimated that an excess of US$58 billion a year is spent on health care in the USA as a result of low health literacy. In certain areas of the country, medical facilities are scarce. Minorities are underrepresented in medical professions. Black, Latino, and Native American populations make up approximately 6 percent of the physician workforce, although these populations represent over 26 percent of the population in the USA.

Health disparities remain an important social and political issue in the USA. The Office of Minority Health of the DHHS was established in 1986 to address issues of health disparities among racial and ethnic minorities. One of the main goals of Healthy People 2020 is to eliminate health disparities.

Health care reform was a contentious issue in the debates surrounding the 2012 presidential election, with proposals for the introduction of the PPACA (Obamacare). The US Supreme Court declared the legislation constitutional in 2012. Republican legislators, the majority in the House, continue efforts to repeal the legislation, which will come into effect in 2014 with the addition of millions of Americans to health insurance coverage and much improved protection for those insured under private insurance. From 1 January 2014, insurers will no longer be permitted to deny coverage for pre-existing conditions, and all Americans will be required to have health insurance under the PPACA. Insurance rates began to fall in 2013, and state insurance regulators indicate rates for 2014 over 50 percent lower on average than those currently available. With competitive pricing and federal subsidies health insurance costs should be even lower. When the health insurance marketplaces were established (in October 2013), consumers are able to shop among alternative insurance plans that meet federal standards, either through state-established insurance bureaux or through the federal alternatives registry to seek their best advantage coverage. Consumer education campaigns are being conducted to educate the public.

Important health disparities exist in America in relation to region of residence, with the southern states having high rates of obesity, stroke, and CHD mortality, which are thought to be due to customary diets rich in fat and salty foods. State health departments will need to address these issues in order to reduce gaps in life expectancy due to lifestyle factors which are grounded in tradition and poverty as well as a lack of health insurance.

The Dilemma of the Uninsured

Universal access is widely accepted as essential to reduce the social inequalities in health even when income gaps are high. Conversely, increasing family disposable income for the poor is an effective way of reducing health inequalities. The two are complementary and equally important in social policy in the USA.

High percentages of the population are without any, or have inadequate, health insurance. Loss of health coverage with change of place of employment and the rapidly increasing cost of private health insurance generated widespread pressure for a national health program. The business community, too, had lost confidence in voluntary health insurance as costs of health insurance mounted rapidly as a cost of employment in an increasingly competitive international business climate.

The Clinton health care plan (1994) was based on federally administered compulsory universal health insurance through the place of employment, with alternative plans available to choose from at different costs. A state could opt to form its own health insurance program and even designate its own department of health to fulfill this function. Physicians could contract with health insurance plans to provide care on a fixed-fee schedule, or in HMOs, whether based on group or individual practice.

The Clinton health plan failed in Congress. Apathy or frank opposition was widespread among the majority of the population who already had good insurance benefits under their employment-based health insurance plans or Medicare. Their interest was in the status quo, and the insurance industry and organized medical community used this to defeat the bill. Federal legislation protecting workers' health rights under collective bargaining prevented states from mandating health insurance benefits. Federal assistance and waivers for state health insurance allow states to opt for managed care for Medicaid beneficiaries. Medicare and Medicaid waivers also allow states to include these beneficiaries in state health plans, but universal access to care would require enabling legislation in Congress. At the same time, conservative attacks on public programs such as Medicare keep the issue of national health insurance on the public agenda.

Many employers have switched to promoting managed care coverage, while offering indemnity plans as options to the employees but with additional premiums. The movement to managed care became an avalanche in the 1990s, with a high percentage of the population insured at their workplace becoming members of HMOs or other forms of managed care. The swing to managed care produced major effects in the health care system, not only for doctors increasingly pressed to join HMOs or PPOs, but also for hospitals and for the consumer who had to adjust to the rules of managed care. Restrictions on access to specialists and new procedures generated public and political criticisms leading to a decline in enrollment from 1999 to 2005, with an increase in 2006, but this economically driven changeover has had profound effects on the US health system.

In 2008, President Clinton's wife Hillary Rodham Clinton promoted a health care plan in her bid for the Democratic presidential nomination. When Obama took office, Democratic-controlled Congress spent a year crafting legislation to require most companies to cover their workers; mandate that everyone have coverage or pay a fine; and require insurance companies to accept all comers, regardless of any pre-existing conditions, and assist people unable to afford insurance. Congress passed the measure in 2010.

In 1996, many states introduced legislation to regulate HMOs, of which 56 laws were enacted in 35 states. Criticisms of for-profit HMOs are appearing frequently in the popular media, and there is a growing backlash of opinion against imposed limitations on specialist referrals, emergency department visits, hospitalization, and some therapeutic interventions (e.g., bone marrow transplants for terminal cancer cases). Some of these have also generated legal suits for malpractice, with large settlements. A 1998 Commission on Health Quality appointed by President Clinton produced a bill of rights for patients that called for additional information on health plans and for the right of appeal to an independent panel on health plan decisions regarding denials of coverage for emergency care or access to specialists.

The non-profit PGP type of HMO uses over 90 percent of premiums for patient care, whereas the for-profit plans spend higher proportions of premiums for administration, including very high salaries for executive staff. The growth trend of managed care will certainly continue, but perhaps with greater regulation of for-profit HMOs to ensure access to services based on medical criteria in the patient's interest and quality assurance.

In 2010, US health care spending increased to nearly US$2.6 trillion, or US$8402 per person, which was 17.9 percent of GDP. Prescription drug spending growth increased to 10.0 percent of total expenditures, in part due to Medicare Part D covering prescription drugs for older adults. Most other major health care services experienced slower growth since 2008 than in previous years.

The 2010 national health expenditures of US$2.6 trillion or for hospital care were 31.4 percent; for nursing home care, 5.5 percent; and for physician and clinical services 19.9 percent of total expenditures (Health United States, 2012 Table 113). This represents a long term shift in distribution of expenditures from acute hospital care toward ambulatory and long term care services and a slowing of the rate of growth in total expenditure especially since 2008 and the great recession.

Summary

The USA has managed to achieve many of the targets set by the 1979 Surgeon General's *Healthy People* report. At the same time, average annual increases in health care expenditures in the USA slowed markedly from the 1986–1990 period, which had average annual increases of 11 percent, falling to under 8.1 percent annually in 1995 and to 3.9 percent in 2010. This is due partly to lower general inflation rates (below 3 percent), but also to cost-containment measures being adopted by government insurance (Medicare and Medicaid), the health insurance industry, the growth of managed care, and rationalizing the hospital sector by downsizing and promoting lower cost alternative forms of care.

National health insurance was delayed by congressional rejection of the Clinton health plan, but President Obama's struggles to pass a health care bill resulted in the Affordable Care Act (PPACA). Several possibilities exist to extend health insurance coverage, including state health insurance initiatives with federal waivers and cost-sharing, or a federal plan based on Medicare or the Veterans Health Administration program. In the mid- to late 1990s, employers promoted managed care options for their workers, so that managed care grew rapidly through market mechanisms. State governments are acting to regulate this by legislation, such as requiring minimum hospital stays for obstetrics cases, limiting managed care programs from certain kinds of contracts for services, and establishing appeals mechanisms for managed care members. Increased access to Medicaid may be fostered by states raising the income levels defining poverty to increase health insurance coverage under Obamacare.

The term *non-system* is often applied to health care in the USA. There are many stakeholders and providers, high costs, and poorer results than health systems in other industrialized countries. Much of this implied criticism is justified. The US health system is a diffuse and incomplete system with good to outstanding quality of care for the majority with insurance but very inadequate care for the over 30 percent with none or poor levels of health insurance. Social and regional inequalities in health status are still present, but not necessarily greater than in some countries with universal access to health care. Furthermore, there are many parallel programs in the USA that have important positive public health content, such as universal school lunch programs; nutrition support for poor women, infants, and children (the WIC program); food stamps for the working poor; fortification of basic foods; free care in emergency departments, urgent hospital care for the poor, Medicare for the elderly, and Medicaid for the poor.

Nevertheless, equitable universal access is lacking, and the system is the costliest in the world. In 2012, costs were an estimated US$2.7 trillion but without the best levels of health as measured by process indicators, such as immunization and prenatal care coverage, nor the best in outcome measures such as infant and other mortality rates. Life expectancy at birth in the USA increased by 7.9 years between 1960 and 2004, substantially less than the increase of over 14 years in Japan or 8.9 years in Canada. In 2009, the life expectancy increased to 78.2 years, well below the 2009 OECD average of 81.6 years (OECD, 2012).

Social inequalities in these health status indicators are further evidence of failures of the US health system to reach its full potential, despite its being the most expensive system in the world and its high quality for those with access (Davis, 2008). The advent of Obamacare in 2014 will, over some years, bring affordable health insurance to millions of Americans and is expected to level off the increase in health expenditures. It is not the universal health plan of Canadian or European tradition, but it is a huge step forward in the USA, where the working poor are in large measure excluded from health protection. However, the struggle for universal coverage and cost containment remain formidable challenges in the USA.

CANADA

Canada is a federal country with 10 provinces and three northern territories, a population of 34.5 million, and a gross national income (GNI) per capita of US$45,550 in 2011. Life expectancy in 2011 was 82.0 years. The International Human Development Index (HDI) rating for Canada was 11th in 2009, a drop from its position as 4th in 2005–2006, after having increased steadily since 1975, but it remains well above the OECD average. Canada's total health expenditure as a percentage of GDP is almost two percentage point higher than the average of 9.5 percent in OECD countries. Total health expenditures as a percentage of GDP have been steadily increasing since 2005, reaching 11.2 percent in 2011. The percentage of total expenditures from the public sector has remained relatively constant, at 70 percent, from 2005 to 2013.

The Canadian constitution sets responsibility for health at the provincial level of government, except for the aboriginal Indian and Inuit populations, armed forces, prisoners, Royal Canadian Mounted Police, and veterans. Despite many geographic, historic, cultural, and political similarities to the neighboring USA, Canada developed its own unique national health insurance program.

Starting in the 1930s, federal grants-in-aid were given to the provinces for categorical health programs, such as cancer and public health programs. Based on this precedent, Canada's national health program is a system of provincial health insurance with federal government financial support and standards. It developed in stages between 1946 and 1971, first with hospital and diagnostic services and subsequently with medical care insurance, now collectively known as Medicare. It brought all Canadians into a system of publicly financed health care, while retaining the private practice model of medical care. Hospital care is provided mostly through non-profit, non-governmental hospitals.

The Canadian health program differs markedly from those of the UK and the USA. Each national health system is an important part of the political and cultural traditions of the country. Each within its own tradition is attempting to constrain the rate of cost increases and preserve, or develop universal coverage. Comparisons are attempted using various health indicators and can be controversial but the Canadian universal health service or insurance coverage seems to have improved the health status of the population more rapidly than similar indicators for the total US population, but not necessarily for the insured US population. After decades of emphasis on developing national health insurance, Canada became a leading innovator in health promotion since the 1970s.

Initiatives for national health insurance in Canada go back to the 1920s, but definitive action occurred only after World War II. The development of national health insurance was partly the result of the experience of the Great Depression of the 1930s, a strong agrarian cooperative movement, and the collective wish for a better society following the war. In 1946, the recently elected social democratic government of Saskatchewan, a large wheat-growing province of 1 million people on the western prairies, under the leadership of Tommy Douglas, the founder of Canada's Medicare program, established a hospital insurance plan. This plan provided free hospital care for all residents of the province on a prepaid basis under public administration. Within several years, other provinces developed similar plans, and in 1956, the federal government passed legislation (the Hospital Insurance and Diagnostic Services Act) to provide a cost-sharing plan for provinces, adopting universal, publicly administered hospital insurance plans. By 1961, all 10 provinces and the (then) two territories had implemented hospital insurance plans meeting federal criteria in a two-tiered national health insurance plan; that is, universal provincial health plans with federal standards and cost-sharing.

In 1961–1962, Douglas and the province of Saskatchewan again led the way by implementing a universal plan for medical services (Medicare). This was opposed by a bitter, 23-day doctors' strike which resulted in some compromises, but the universal plan came into effect, paying doctors' bills on a fee-for-service basis. Again, this was based on the principles of universal coverage, comprehensive benefits, and public administration.

Following the controversies over this plan, a federal Royal Commission on Health Services (the Hall Commission) recommended the establishment of similar plans with federal cost-sharing. In 1966, the federal government introduced its Medicare Act, providing federal cost-sharing of approved provincial plans. Federal reimbursement to the provinces included 25 percent of national average medical care expenditures per capita and 25 percent of the actual expenditures by each individual province. This provided higher than national average rates of support to poorer provinces as well as portability between provinces. By 1971, all provinces had implemented such plans.

Reform Pressures and Initiatives

The Canadian health program established universal coverage for a comprehensive set of health benefits without changing the basic practice of medicine from individual medical practice on a fee-for-service basis. Poorer provinces were able to use the cost-sharing mechanism to raise standards of health services, and a high degree of health services equity was achieved across the country.

Rapid increases in health care costs led to a review of health policies in 1969 (the Federal–Provincial Committee on the Costs of Health Services). The resulting report stressed the need to reduce hospital bed to population rations and develop lower cost alternatives to hospital care, such as home care and long-term care. Federally-led initiatives during this period extended coverage to include home care and long-term nursing home care, while restricting federal participation in cost-sharing to the rate of increases in GNP. Since then, many provincial and federal reports have examined the issues in health care and recommended changes in financing, cost-sharing, hospital services, and development of primary care and other community services.

In 1974, a new approach to health was outlined by the Federal Minister of Health, Marc Lalonde, in a landmark public policy document, *New Perspectives on the Health of Canadians*. This report described the health field theory, in which health was seen as a result of genetic, lifestyle, and environmental issues, as well as medical care itself (see Chapter 2). As a result, health promotion became a feature of Canadian public policy, with the objective of changing personal lifestyle habits to decrease cross-cutting risky behaviors such as smoking, obesity, and physical inactivity. The pioneering work in nutrition from the National Nutrition Survey published in 1971 led to the 1979 adoption of federal mandatory enrichment regulations for basic foods with essential vitamins and minerals (see Chapter 8). This and other initiatives in the 1980s led to the Ottawa Charter on Health Promotion (see Chapter 2).

In the mid-1980s, physicians' organizations pressed for the right to bill patients above the rates paid for by Medicare, but this was forbidden by national legislation (the Medical Care Act), passed unanimously by the Federal Parliament. This act penalizes provincial governments which allow extra billing by physicians by withholding federal funding. The Canada Health Act was passed by the Federal Parliament in 1984. This act outlines specific principles and requirements for all Canadian provinces and territories on health care, in order to qualify for federal public funding. Annual reports are published outlining the status of health care for provinces and territories. Those who do not adhere to the requirements are subject to withholding of transfers or penalties.

Canadian health expenditures showed high rates of cost increase, approximately 4 percent annually, during the 1980s (Figure 13.3). GDP grew at 3 percent per year but

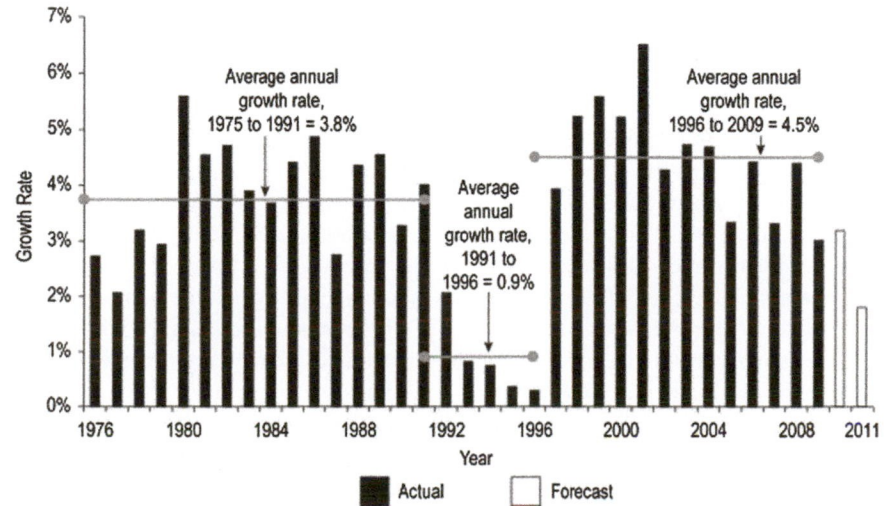

FIGURE 13.3   Health expenditures, Canada, 1976–2011. *Source: CIHI Canadian Institute Health Information. National Health Expenditures 1975– 2011. Available at: https://secure.cihi.ca/free_products/nhex_trends_report_2011_en.pdf [Accessed 30 May 2013].*

declined in 2011 to 2.0 percent. National expenditures on health rose from 5.4 percent of GDP in 1960 to 7.0 percent in 1980, 8.9 percent in 1990, stabilizing at 8.8 percent in 2000, but again rising to 9.8 percent in 2005, and reaching 11.3 percent in 2011 (Figure 13.3).

During the late 1990s, the rate of increase in health care costs was reduced by politically painful measures of retrenchment, especially in hospitals. In the period 1975–1991, when the rate of growth of health expenditures was averaging 4 percent (7.4 percent 1998–2008), Canada was second only to the USA in percentage of GNP expended on health. In 1998 it was the fourth highest in the world (9.2 percent of GNP in 1996), after the USA, Germany, and France. In 2005, Canada's percentage of GDP spent on health was above the OECD average, but well below the USA (16 percent) and seven other OECD countries. In 2009 Canada's health care spending, at 11.4 percent of GDP, was below that of five other OECD countries but remained above the OECD average of 9.4 percent. However, the current figure, 11.3 percent, is still well below that of the USA (17.9 percent).

In 2006, Stephen Harper, leader of the Conservative Party, was elected Prime Minister of Canada. Many criticize his leadership for failing to address the lack of family physicians and long waiting times. Canada spent $191.6 billion on health care in 2010, up from an estimated $182.1 billion in 2009 and $171.8 billion in 2008, growing by an estimated $9.5 billion or 5.2 percent since 2009, according to the Canadian Institute for Health Information (CIHI).

Hospitals, drugs, and physician services, in that order, continue to account for the largest share of health dollars. In 2010, spending on hospitals was expected to reach $55.3 billion, spending on drugs $31.1 billion, and spending on physicians $26.3 billion. For the past two decades, there has been an increase in the share of spending on drugs and a decrease in the share of spending on hospitals. However, more recent trends show that spending patterns may be shifting.

In 2010, the total spending on health care in Canada represented an increase of $216 per Canadian, bringing total health expenditure per capita to an estimated $5614. After removing the effects of inflation, Canada's health care spending per person increased by 1.4 percent in 2010, the lowest annual growth rate seen in 13 years.

Financing of total health expenditures in 2005 was 70 percent from public sector sources, including federal, provincial, and municipal governments and workers' compensation, and has held at a stable rate since 1996. Federal government cost-sharing in health expenditures has gradually declined since the 1970s, so that provincial governments are facing difficulty with continued financing at current levels, and are under pressure to control rates of increase. This has led many provinces to reduce the hospital bed supply, from 6.9 beds per 1000 in 1979 to 4.7 in 1995 and to 3.5 in 2003. This figure steadily declined until it dropped to 3.3

beds per 1000 population in 2008 and to 1.8 acute beds per 1000 population in 2011.

The Canada Health and Social Transfers (CHST), established in 1997, is the mechanism for federal transfer of money to provinces and territories through cash contributions and tax transfers. As long as the provinces and territories adhered to the ideology of the Canada Health Act, money could be allocated to various social programs. In 2004, the CHST was split into the Canada Health Transfer (CHT) and the Canada Social Transfer (CST). By creating the CHT and the CST, the federal government can allocate cash contributions and tax transfers in a way that is accountable and transferable to provinces and territories in order to maintain the goals and obligations set forth by the Canada Health Act.

The Public Health Agency of Canada and the position of Chief Medical Officer were established in 2004, following severe criticism of public health organization in Canada over the confused management of the 2003 outbreak of severe acute respiratory syndrome (SARS). From 2006 onward as part of the review of public health deficiencies, fifteen MPH programs opened universities across Canada (Massé 2012, see Chapter 14).

Provincial Health Reforms

In the 1970s, a growing emphasis on health promotion and development of alternatives to acute hospital care led the province of Manitoba to institute reforms in the delivery of services. It established district health systems in rural areas in order to strengthen services and attract health personnel and their families to remain in these areas rendering more comprehensive and integrated services. This model integrated the organization of hospitals, nursing homes, home care, preventive services, and medical practice, reaching many of the rural areas of the province over the next decades. This evolved into the regional health systems model. Following the 1970s' Castonguay–Nepveau Commission Report, Quebec implemented Community Health Centers (CLSCs) throughout the province. In the 1990s, Saskatchewan began development of similar integrated district health service and regionalized hospital systems. Other provinces have since followed suit, each using a unique formula to regionalize and consolidate services, and increase the provision of services in the community setting. Ontario, the economically largest province, but the last to regionalize, recently introduced 14 local health integrated networks to manage services to its 12 million inhabitants.

In most provinces, regional health authorities (RHAs) are autonomous health care organizations responsible for health administration within a defined geographic region of a province or territory. The regions have appointed or elected boards of governance responsible for funding and delivering community and institutional health services within the regions. RHAs fund and provide core services including public health, home care and community-based services, mental health services and long-term care institutions, alcohol and substance abuse programs, and hospitals within provincial standards of the Ministry (Department) of Health. Provincial and territorial ministries collect taxes to finance the health care system and develop regional funding envelopes; regional health boards allocate funds to service organizations based on their own needs assessments and policy priorities.

Continuing care, including home and nursing homes (long-term care), and prescription drugs are integral components of provincially insured health systems. Services provided are based on residence in the province and are not transferrable between provinces. Manitoba Continuing Care, implemented in 1974, is an integrated community-based public sector funded program with a single entry point for home care and nursing home care. Assessment for admission is based on assessed health need; there are no income-based or ability-to-pay restrictions. Home care has developed as an appropriate and cost-effective alternative to

both hospital and long-term care (Box 13.2). Services above the assessed health need are the responsibility of the client and not funded through the program. The program serves as an important adjunct to earlier hospital discharge. The range of services includes short-term (medical and postsurgical), long-term, and palliative care.

Health Status

Criticism of the Canadian health system focuses on long waiting times (in comparison to the USA) for diagnostic and surgical procedures, lesser access to high-tech equipment and procedures, and reduction in hospital staff positions. Such comparisons, however, are not substantiated by objective analyses or in measurable health indicators. Waiting times have reportedly been reduced in recent years and the supply of high-tech equipment such as computed tomography (CT) and magnetic resonance imaging (MRI) scanners has increased in comparison to other OECD countries.

BOX 13.2 Manitoba Home Care Program

The Manitoba Home Care Program was established in 1974, as an integral component of the publicly funded health system enabling people to remain at home as an appropriate community-based alternative to care in hospital or nursing homes. Home Care is provided free of charge to permanent Manitoba residents who are Canadian citizens or landed immigrants. Services are based on assessed need, taking into account other resources available to the individual including families, and other community resources/programs.

Regional health authorities are responsible for program management and deliver the services including assessment for and coordination of nursing home admissions within the policy framework and standards of the Department of Health. Accreditation Canada is the established body for standards and the accrediting process for home care programs of the regional health authorities.

In 2011/2012, of the $4.7 billion ($3800 per capita) provincial health expenditures for insured health services, 6.5 percent ($300 million) or $244 per capita was for home care. In comparison, long-term care (nursing home) and acute care hospital costs were $482 and $1754 per capita, respectively. The cost of insured health services in 2011–2012 was $4000 per capita. The percentage distribution of costs for acute care, long-term care, and mental health was 46 percent, 13 percent, and 5 percent, respectively. However, the federal Canada Health Act for hospital and medical services is not applicable to extended health care or home care programs; therefore, there is no portability of services between provinces.

Eligibility and admission to the program are based on a functional needs assessment by the assigned health professional (case manager) with reassessments at predetermined intervals. The program requires the designation of a relative or close friend as the primary contact person as needed from

time to time. In some situations a family member living in another province is the contact person. Services may include nursing, personal care (bathing, dressing, feeding), physiotherapy, occupational therapy, and homemaking and respite care. Other services such as speech therapy, social work, and dietitian services may be accessible. Supportive services may include day care in a funded health facility; other community services include meals on wheels and friendly visitor services.

In 2011, of the 17,202 assessments for eligibility, 15,481 were admitted and 14,710 were discharged. There is a monthly average of nearly 24,000 individuals receiving services, with the majority of services provided by a home care attendant, and about one-third of the clients receive registered nurse services.

The Manitoba Health Care Appeal Board was established to ensure that residents of Manitoba have access to an independent arms-length appeal process for publicly funded insured health services including home care. If a person is not satisfied with certain decisions regarding a financial or service delivery matter, an appeal may be made to the Board.

Since 1974 results of periodic extensive external reviews by internationally recognized management consultants have consistently shown the program to be a cost-effective appropriate alternative to facility care. Shortly after its implementation, the program was recognized as an important model for health systems by the World Health Organization.

Sources: *Joan Bickford MSN, Former Chief Public Health Nurse, Manitoba Health, Province of Manitoba, Canada. Personal communication; 2013. Government of Manitoba. Health expenditures. Available at: http://www.gov.mb.ca/finance/budget12/papers/r_and_e.pd Manitoba Health. Annual statistics 2010–2011. Available at: http://www.gov.mb.ca/health/annstats/as1011.pdf Accreditation Canada. http://www.accreditation.ca/about-us/ Health Canada. Home and community care. Available at: http://www.hc-sc.gc.ca/hcs-sss/home-domicile/commun/index-eng.php*

Since the implementation of Medicare, Canada's position in major health status indicators has improved in comparison to other countries. Infant mortality rates were higher than those in the USA until the 1960s (28 versus 22 per 1000), but lower in the 1990s (6 versus 7 per 1000 in 1997). Canada's infant mortality rate was 5.3 deaths per 1000 live births in 2004, but increased to 6.0 deaths per 1000 live births in 2009 and declined to 5 per 1000 live births in 2011, lower than in the USA (6.5) and the OECD average (5.4). Canada ranked 42nd and the USA 50th in international ranking of infant mortality rates in 2013 (estimated) for 224 countries.

In 2005, Canada's maternal mortality rate (MMR) was 7 per 100,000 live births, compared to 7 for the UK and 11 for the USA. In 2012, the Canadian MMR increased to 12 per 100, 000 live births. In 2007, Canada's life expectancy at birth was 80.7 (81 years in 2011) compared to 78 years in the USA; the Canada–USA gap decreased from 2.7 years in 1993 for men to 1.8 years in 2007, and the difference among women stayed at 2.9 years. Canada ranks among the 15 OECD countries with the lowest total mortality rate.

In Canada, public health is generally identified with the following discrete functions: population health assessment, health promotion, disease and injury control and prevention, health protection, surveillance, and emergency preparedness and epidemic response. Immunization coverage for infants was reported in 2011 as more than 95 percent. Canada has made major progress in reducing tobacco consumption. The rate of daily smokers among adults fell by half, from 34.4 percent in 1980 to 17.3 percent in 2011, with the second lowest percent of the population consuming tobacco (14.3 percent) among the OECD countries. Smoking prevalence in 2011 among those aged 15–19 years was 12 percent compared to 21 percent for the age group 20–24 years; in 2000 the prevalence had been 25 percent and 32 percent, respectively.

Canada is fifth lowest among the OECD countries in rates of hospitalization for circulatory system disease, as cerebrovascular disease is lower in Canada than in many OECD countries (Table 13.6), reflecting both prevention programs and outpatient and home management of stroke patients. The main causes of death in Canada are cancer, and circulatory, respiratory, digestive, and infectious diseases. Obesity is a growing health concern, with approximately 24.2 percent of the population obese in 2008. The aboriginal Canadian populations suffer higher rates of poor health status from immunization-preventable diseases and alcohol- and tobacco-related illness (see Chapter 7).

Summary

Canada's health system successfully established universal tax-supported national health insurance in North America. Prior to the advent of universal coverage, Canada was on a similar track to the USA but the country has since bypassed

TABLE 13.6 Hospital Discharge Rates, per 1000 Population, Life Expectancy at Age 65, Selected Countries, 2009[a]

	Circulatory Diseases	Cancers	Total	Life Expectancy at 65	
				Females	Males
Canada	10.9	6.1	84	21.3	18.1
Denmark	19.9	13.9	170	19.1	16.3
France	22.2	19.9	263	20.8	17.4
Germany	36.0	24.5	237	20.8	17.6
Israel	13.3	6.1	146	21.2	17.7
Netherlands	16.7	10.7	117	20.8	17.4
Sweden	25.4	14.1	166	20.8	17.1
UK	13.2	9.4	138	20.8	18.1
USA	19.5	6.7	131	20.0	17.3

Note: [a]Or nearest year.
Source: Organisation for Economic Co-operation and Development. Health at a glance 2011: OECD indicators. Available at: http://www.oecd-ilibrary.org/social-issues-migration-health/health-at-a-glance-2011_health_glance-2011-en [Accessed 13 June 2013].

the USA in increased longevity and lower cause-specific mortality rates for CVD, cancer, and stroke, as well as child mortality, all at considerably lower per capita expenditures.

The Canadian health program has important lessons for health care reform internationally. Universal health insurance was implemented without changing the basic mix of services or the way they are funded, but with inadequate attention to the public health portion of the system until the SARS episode of 2003, when serious deficiencies in public health organization and training were revealed. These are all issues that need to be considered in the Canadian experience.

Canada pioneered the idea of health promotion from the 1970s in such areas as Healthy Cities, fitness, and food enrichment. The primary functions that are the focus of the Canadian public health system include population health assessment and surveillance, health promotion, prevention-oriented services for disease control, health protection, and emergency readiness. These have helped to achieve positive results in reduced smoking, falling rates of cardiovascular morbidity and mortality, food fortification, and increased health consciousness of the political leadership and the general population over a number of decades. Pioneering reform in integrated regional health management, provincial insured home care and nursing home programs is now widespread across the country, based on the Manitoba model developed in the 1970s, which has received international recognition.

The Canadian health insurance model can be regarded as a success, although with some drawbacks. It has tended to freeze the medical private practice model, and was slow in implementing reform measures. Despite transfer of tax points from federal to provincial governments, the provinces have difficulty coping with health costs, which are the largest item of provincial budgets. The principle of universal health insurance delivered by provincial plans under federal regulation and cost-sharing has been preserved. The quality of care is high, and Medicare is one of the most popular public institutions in Canada, of which most Canadians, including physicians, are proud. Reforms carried out during recent decades appear to be succeeding in controlling the rate of increase in costs.

The health status of Canadians is rated among the highest in the world. Despite the financial burden and the need for economic analysis with priority selection, the Medicare program remains highly popular with the Canadian public, the federal and most of the provincial political leadership remains committed to universal, publicly administered health care, and this is likely to continue. The Canadian model is a success from many points of view and is of importance as a working model for reform for other countries, particularly the neighboring USA.

THE UNITED KINGDOM

The UK's population in 2011 was 62.6 million and the GDP per capita was US$35,441 [purchasing power parity (PPP)], ranking it economically 16th, well below Scandinavia, North America, and many European countries. Health expenditures as a percentage of GDP (9.3 percent, 2009) parallel the OECD trend for selected countries (Figure 13.4). In 2010–2011, the UK ranked 28th among OECD countries in the HDI, a large decrease from 16th position in 2005–2006. In 2011, the total life expectancy was 80.2 years and the child mortality rate was 6 per 1000 live births.

The UK is a parliamentary democracy with a royal head of state, House of Commons and House of Lords, which unites England, Wales, Scotland and Northern Ireland. The UK is a national entity with a central unitary state government but with decentralized authorities for Scotland, Wales and Northern Ireland, each of which has a parliament or house of assembly with limited autonomy and governing powers.

National health initiatives have evolved slowly since the mid-nineteenth century. The NHS has developed and maintained high professional and technical standards, despite modest levels of funding of the service. Immunization coverage in 2008–2009 was 93 percent for DTP, 93 percent for poliomyelitis (polio), and 85 percent for measles [through the measles–mumps–rubella (MMR) vaccine], with a second dose of measles vaccine at school entry.

The UK developed a unique and important model of health care as a tax-financed public service that is widely influential in other national health systems. The NHS

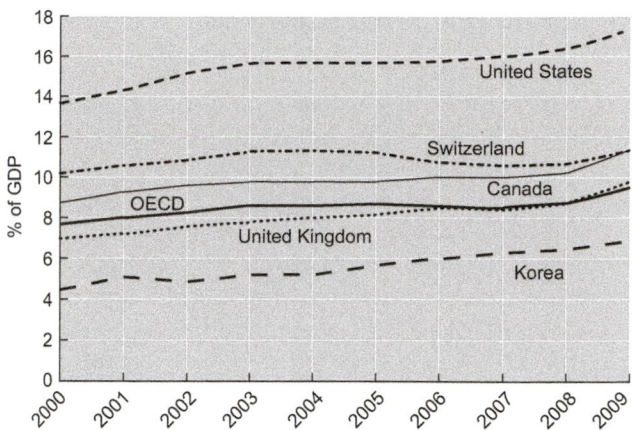

FIGURE 13.4 Health expenditures, percentage of gross domestic product (GDP), selected Organisation for Economic Co-operation and Development countries, 2000–2009 *Source: OECD Indicators. Health at a glance, 2011. Available at: http://www.oecd.org/health/health-systems/49105858.pdf [Accessed 14 June 2013].*

is popular with the British public and was successful in achieving its initial goals. It has undergone periodic reforms since its inception in 1948, surviving many changes of government and political philosophy and continues with a current reform process.

The National Health Service

As described earlier, the UK developed its present NHS over many decades. This program evolved from previous milestones including reform of the Poor Laws of the eighteenth to nineteenth centuries, the friendly societies, the National Health Insurance Act of 1911 for workers and their families, the Emergency Medical Service of World War II, and the (William) Beveridge Report of 1942. In 1946, under the Labour government of Clement Attlee, Parliament approved the National Health Service Act, with implementation in 1948, under the leadership of Aneurin Bevan.

THE NHS operates through four systems which are funded from allocations by the British Parliament to the NHS England which gives block grants to the independently administered individual systems: National Health Service (England); Health and Social Care in Northern Ireland (HSCNI); NHS Scotland; and NHS Wales. These separate organizations operate free health services with many common features.

The NHS is financed through general tax revenues to provide a comprehensive service to the entire population. The NHS was originally organized as three parallel services: the hospital service with salaried doctors, the GP (and dental) services provided by independent practitioners with capitation payment, and the public health service with salaried staff. The hospital and general practice services were operated by separate public boards or councils; the public health service was administered by the local authorities.

Reforms in the 1970s and 1980s

During several stages of reform in the 1970s and 1980s, the NHS was reorganized, reducing the number of administrative levels in an attempt to achieve integration and coordination between highly specialized and fragmented services. The 1974 reform established regional health authorities, and integrated area health authorities (AHAs) beneath them to replace the previous multiplicity of hospital management committees, boards of governors, and local health authority committees. The AHAs were non-elected lay bodies that absorbed public health and hospital management functions, consolidating many previously overlapping jurisdictions. Multidisciplinary management teams were introduced at the AHA and district levels, with decision-making by consensus, stressing professional managerial competency.

A further reorganization in 1982 abolished the AHAs, placing the managerial responsibility at the district health authority (DHA) level, with a stress on further decentralization of management authority to hospital and community service structures. Reviews of the NHS were conducted during the Conservative government led by Margaret Thatcher, focusing on managerial efficiency, government and business viewpoints, the growth of the private sector, consumer group advocacy issues, and protection of consumer rights.

Despite the aging of the population, the number of acute care hospital beds in the UK has been steadily decreasing from 3.0 per 1000 population in 2004, to 2.7 in 2007, and 2.4 in 2010. The average length of stay in acute care hospitals fell from 8.5 days in 1980 to 5.0 days in 1996. However, this figure grew to 8.0 in 2000, gradually decreasing to 7.0 in 2007, and reaching 6.6 in 2010. While the supply of physicians is below that of other European OECD countries, the supply of nurses is above OECD levels. Numbers of psychiatric beds were also reduced in these years. Overall geriatric and nursing care beds decreased slightly between 2004 and 2008. There has been an increase in medical practitioners (from 2.3 per 1000 population in 2004 to 2.49 in 2007 and 2.79 in 2011) and nurses (from 9.2 per 1000 population in 2004 to 9.6 in 2010) in the UK.

Access to advanced technology such as CT scanners (8.9 per million in 2011 versus 12.5 in France and 40.9 in the USA) and MRI scanners (5.9 versus 7.5 in France and 34.5 in the USA in 2012) has increased in the past decade, but remains well below OECD averages. Community care services of all kinds increased during this period. Health expenditures increased from 3.9 percent GDP in 1960 to 5.6 percent in 1980, 6.0 percent in 1990, 7.3 percent in 2000, 8.3 percent in 2005, and 9.4 percent in 2011.

Reforms Since 1990

In 1990, the National Health Service and Community Care Act attempted to further rationalize management of the NHS. Three types of statutory health authority were redefined: regional health authorities (RHAs), district health authorities (DHAs), and family health service authorities (FHSAs). The RHAs and DHAs became the primary administrative levels, while the FHSAs manage contracts with GPs. The NHS operations in England, Wales, Scotland, and Northern Ireland operate under similar arrangements.

The 14 RHAs assess health needs, set strategic directions for service development, monitor quality of management and care, and allocate resources to promote cost-effective services. They also promote medical audits and specific program development (e.g., transplantation services), and provide assistance to health providers such as hospitals with management problems. The RHAs do not provide services. The DHAs operate under the authority of boards similar to those of the RHAs, and are the major purchasers of services from hospitals and other providers. They contract with hospitals for services based on assessed need and on satisfaction with hospital performance. They may also operate NHS hospitals or other services, such as ambulances.

Reforms since 1990 include the introduction of competition between providers, the development of community health services, and further reduction of the supply of hospital beds. These were intended to introduce greater choice for the patient and the primary care provider (the GP), with incentives for efficiency and quality of care.

FHSAs under the 1990 Act are governed by boards similar to the RHAs and DHAs. The FHSA is responsible for contracting with GPs, general dental practitioners, optometrists, and community pharmacists. The role of FHSAs expanded to include formulation of policies, supervision of facilities and services, and remuneration of contracting providers. Patients register with GPs and are referred to hospitals and specialists in accordance with medical needs. GPs have traditionally been paid on a capitation basis for the patients registered with them, and the patient has the right to change GP. Capitation is the allocation of funds per person registered as a service beneficiary for a specified period to cover care for a range of services. Weighted capitation is allocation per person, with adjustments made for factors such as age, gender, and regional standardized mortality rates (SMRs), which reflect both need and demand for health services. SMRs are used as a proxy for morbidity in capitation allocation (see Chapter 3). GPs are paid extra premiums for performance indicators; for example, specific preventive services such as immunization, Papanicolaou (Pap) smears, and mammogram screening. In 2008, a pay-per-performance program called Advancing Quality was introduced in 24 NHS hospitals in the Northwest of England (population 6.8 million), and a clinically significant reduction in hospital mortality was observed compared to other hospitalized populations in other regions; the largest reduction was observed for heart diseases and especially for pneumonia (1.9 percentage points) (Sutton et al., 2013).

A major innovation was been allowing the FHSAs to administer budgets for fundholder GPs, in the form of per capita payments including both GP and hospital services. GPs are increasingly working in health centers, along with district public health nurses. By 1995, about one-third of GPs worked as fundholders with per capita payment by the NHS for ambulatory and hospital care. This empowered GPs to negotiate with the hospitals, reduce waiting times, and improve other health care conditions for their patients, placing the hospital in the position of having to compete for the referral work of the GP. Experiments with financing of hospital care through the GPs were designed to raise the quality of care and promote cost containment. The GP fundholder movement seems to be a successful program, although it has not been well evaluated.

Hospitals are encouraged to become NHS trusts, which are non-profit public corporations governed by boards of trustees and appointed by the national government, usually representing the local authorities. Hospital trusts must demonstrate management capacity and viability to operate as economic units. They must compete for referrals, striving for patient and GP satisfaction. Hospitals are no longer funded directly by the NHS, but derive their income from providing services to the health authorities, fundholding GPs, private insurance, and self-paying patients, paid for services by a DRG system. This permits them to operate as independent economic units, enabling them to charge for services, determine staff conditions, raise capital by borrowing money and, within limits, buy or sell land or facilities.

Financing of the NHS continues to be through governmental allocations from general tax revenues. Some revenues come from other sources, including user fees, such as for prescription drugs or dental services. Operating budgets are allocated to RHAs to cover costs of hospital, community health, and primary care services. The allocation is determined on the basis of population size and adjusted by SMRs, with some local weighting factors based on service utilization. The DHAs are, in turn, funded by the RHAs, using similar criteria. RHAs administer GP fundholding units, whereas FHSAs are funded to pay for contracting primary care services. Capital allocations to replace and modernize facilities and equipment are based on long-term planning at the RHA level.

Market reforms in the UK are still developing. Despite being the subject of continuous critical scrutiny in the press and at political levels, the NHS continues to have support of the general public and all political parties and has provided universal access and maintained high quality at reasonable costs. Health expenditures increased under Tony Blair's government. Widespread criticisms, focused on the idea that the NHS was being operated at lower levels of expenditures than in most industrialized countries, and of underfunding of important areas such as hospital bed supply, led the Blair government to increase funding modestly and increase the bed supply.

In 2005, the cabinet officer responsible for the NHS put forward a series of proposed changes, including greater funding for disadvantaged areas of the country, changes in hospital financing methods from a block budget to an incentive-based budget, greater flexibility in funding for primary care trusts (PCTs) and GPs to innovate in developing programs, and greater choice of hospitals by patients.

Public support for the NHS is an important element in its durability. The values of the NHS were described in a 2007 review by the Nuffield Trust as including:

- universalism – compulsory coverage
- equity – social justice, fairness
- democracy – accountability, answerability
- choice – autonomy, freedom
- respect for human dignity – honesty, consideration, fair dealing
- public service – public service ethos, altruism, non-commercial motives
- efficiency – cost-effectiveness, waste avoidance
- promotion of desirable outcomes and processes
- accountability.

Social and Regional Inequalities

Social and Regional inequalities in the health status of the population of Britain, which were part of the justification for the establishment of the NHS in 1946, have persisted. The Black Report (by Douglas Black in 1980) documented this problem, and subsequent reports indicate the persistence and even worsening of social inequalities into the 2010s. There are sharp differences in mortality rates from preventable diseases including cancers, stroke, coronary heart disease, lung and liver diseases, with local authorities such as Manchester, Liverpool, and Blackpool having high rates, while local authorities in southern England have very much lower low rates after age adjustment. Many of these differences originate in poverty-related reasons, such as poor diet, obesity, alcohol consumption, and smoking. The NHS has not been able to reduce the inequalities, in part because of slow adoption of population-based strategies of intervention (NHS, 2013).

Changes in definitions and distribution of the population in the different social classes may explain some of the differences; however, there is a widening of the gap between the social classes, with continuous increase in the SMR of class V and continuous decline in SMRs for classes I and II (see Chapter 4). Many studies show higher mortality from all causes by social class, and health profiles for every local authority and region across England are now published by the Department of Health and public observatories. Higher cause-specific mortality in lower socioeconomic classes is seen especially for CVD, trauma, and cancer.

This social gap is not easily explained on the grounds of the classic health risk factors alone. The health gap correlating to economic disparities may be due to poor diet, high rates of smoking, lower rates of physical activity, and social and working conditions offering less reward, personal satisfaction, and control of life events than for the higher social classes. There are also regional differences in SMRs in the UK; the reasons for these are not always well understood but may relate to a variety of social, economic, lifestyle, and environmental risk factors.

In 1998, Donald Acheson, a senior professor of public health in the UK, reported on the Blair government-initiated inquiry into social disparities in health in Britain. His report confirmed the findings of the Black Report and evaluated findings of the many studies of social gradients in health status since that report was issued. The Acheson Report has been a factor in government policies in tax and welfare reform, preschool child care programs, and tobacco legislation, as well as in some aspects of NHS reform.

Health Promotion

During the 1950s and 1960s, mortality from CVD increased in the UK, as in most industrialized countries. These rates began to decline in the 1970s in the USA, Canada, and other European countries, but remained high in the UK for another decade, with CHD mortality declining substantially only since 1985. This delay in the reduction of CVD mortality may be explained by then prevailing conservative attitudes towards treatment of acute myocardial infarction in the UK, such as aggressive, intervening methods of treatment and intensive care units. The NHS was also slow in responding to changing health promotion and risk factor reduction approaches. The UK continued to have much higher mortality rates from CVD as well as lung and cervical cancer than Western European countries. These and other public health issues, including relatively low immunization coverage levels, led to the formulation of health promotion strategies in the Department of Health.

In the late 1980s and early 1990s, a number of major initiatives sought to improve prevention and health promotion activities in the UK, including greater public awareness of healthy nutrition and smoking risks. Mortality rates from stroke declined from 1970 to 2010 by 72 percent, from 151 to 42 per 100,000 population. Ischemic heart disease fell from a peak in 1972 of 277 per 100,000 to 77 per 100,000 in 2010 (a decline of 72 percent). Cancer of cervix and lung mortality rates, although still higher than western European rates, have been declining precipitously since the mid 1980s.

Regional variation remains high, with rates in England and Wales for men and women under age 75 ranging from 7 to 25 per 100,000 population by county or borough in London. Higher mortality rates are seen in Scotland, Northern Ireland, and the Midlands than in southern England.

CVD standardized mortality rates have continued to decline across the UK, but in 2010 at 164 per 100,000 population remained above those of France (119 per 100,000 in 2009) and Israel (119 per 100,000 in 2010) although below that of Sweden (164 per 100,000) (WHO European Region, Health For All database, January 2013). Much of this decrease is attributed to declining tobacco consumption and improvements in medical care. Incentive payments to GPs resulted in a sharp increase in preventive care practices such as blood pressure and cholesterol control and immunization rates. Local authorities are required to have specialized staff to promote motor vehicle safety, which contributed to a reduction in road crash mortality from over 24 per 100,000 population in 1970 to just over 3 per 100,000 in 2010. A new Water Act and Environmental Protection Act of 1990 increased the supervisory and regulatory role of the national government in these areas of public health.

The Health of the Nation report (Secretary of State for Health, 1991) placed health promotion and national health targets as a major focus of a national health program. Declining mortality from the major causes of death (CVD, cancer, and trauma) may reflect an increasing effectiveness of health promotion activities in the UK. Rates of stroke mortality have decreased steadily since 1980, and the number of deaths from diseases of the cardiovascular system has also decreased steadily from 2000 to 2010, but rates of overweight and obesity continue to increase.

Health Reforms

Between 1991 and 1997, the Conservative government introduced the option of holding budgets for general practices for prescribing and elective secondary care. The Labour government of Tony Blair, elected in 1997, undertook further reform in the NHS, especially in methods of financing primary care and the market forces of GP fundholding. The government increased NHS funding by 4 percent above inflation over the period 1999–2003 to strengthen clinical service sectors, which had suffered from excessive cutbacks in the previous decade, mainly going towards improvement in salaries and medical equipment.

In 1999, the Blair government initiated a new reform of the NHS, establishing primary care groups (PCGs) throughout the country with GP groups serving population groups of between 30,000 and 250,000 people. The PCGs replaced the purchasing of services previously performed by the fundholding GPs and the health authorities.

NHS policy in England is directed from the center by the Department of Health. The Department of Health has provided new tools to improve monitoring with the Commission for Health Improvement and the National Institute for Clinical Excellence (NICE) established in 1999, now called the National Institute for Health and Care Excellence. GPs gained online appointment systems and patients have access to a free 24-hour telephone nurse consulting service, improving access and patient contact.

NHS funding was increased in 2000; this mainly went towards increasing staff salaries and capitation payments. Family doctors have benefited most, with more than doubling of their incomes. In 2004, an additional payment was added to the basic capitation payment to family practitioners based on their performance on 146 quality indicators relating to clinical care for 10 NCDs, organization of care, and patient experience. Fundholding was reintroduced in 2005.

Hospital reform became the prominent issue in public assessment of the NHS, particularly in reducing waiting times. This was to be achieved through a combination of targets and, since 2003, financial incentives to promote an incentives-based system. Payment by results started in 2006 and provides for competition for hospital trusts, increasing local control and the range of non-hospital treatment options.

The NHS had been relatively underfunded but increased funding during the Blair government brought overall health expenditures of the NHS to above the European Union (EU) average from 2002. The growth in expenditures between 1997 and 2009 declined from 9.9 percent to 9.4 percent of GDP in 2011. Private health care expenditures decreased by 5.7 percent between 2008 and 2010 but in 2011 increased to slightly above the usual annual growth of 6 percent.

Primary Care Trusts

Primary care trusts (PCTs) are local health organizations charged with providing and commissioning services to a geographic population. They are supervised by 10 regional Strategic health authorities (SHAs). The number of PCTs was reduced to 152 in 2006 to match the geographic division of other social services bodies. They are based on the registered populations of enrolled GPs in the geographic area and are responsible for primary care, hospitalizations, community health promotion, dental care, and health promotion.

The NHS health has used trusts to manage NHS hospitals as well as for community care and mental health services. A new system is in process of development, and all NHS trusts are expected to become foundation trusts by 2014 under an NHS trust development authority.

PCTs are budgeted through a capitation formula with many factors taken into consideration, with adjustments for age, mortality, in- and out-migration between geographic areas, ethnic mix, prison populations, army personnel and their dependants, and other factors. Incentive payments for immunization and preventive procedures such as Pap smears and mammography have become part of the capitation base budget. Only HIV/AIDS is identified separately.

In 2006, the Department of Health published its Seventh White Paper, promising a "fundamental shift" towards integrated services provided in local communities. It was intended to improve access and local coordination among services, provide cost-effectiveness in reducing hospitalization, improve quality of care, and save money in the long term through an on emphasis of prevention to avoid costly illnesses. It abolished the original fundholding scheme (where GPs held budgets and bought services on behalf of their patients) and set up PCTs with the aim of improving the quality of local services. The PCTs set up GP contracts with new quality incentives built in, and from 2000 promoted a policy very similar to GP fundholding. Further health reforms occurred in 2012 with new legislation under the Health and Social Care Act.

PCTs pay for hospital care from the allocations for geographic areas with resident populations and for community hospital and office visits, and so on. DRGs are used but are called case-mix groups. Community health services are part of the Hospital and Community Health Service (HCHS). Chiropody, family planning, and screening are part of this component. Community health services include district nurses, community psychiatric nurses, health promotion programs, community dental health, and health visitors (i.e., public health nurses).

Electronic medical records are now used by almost all GPs in the UK. This has contributed to implementation of the performance measurement system to evaluate practice on a national level system, called The Quality and Outcomes Framework (QOF). This system is used both to calculate payments and as a public source of quality of care information, providing a base for comparison against individual GP previous performance and comparison to other practices locally and nationally, with data accessible at http://qof.hscic.gov.uk/index.asp. These also affect GP payments for the performance of specified services; 83 percent of incentive payments claimed in the first years of the program.

The indicators, particularly those in the clinical areas, represent a mixture of process measures and intermediate outcome measures. In general, intermediate outcome indicators are more difficult to achieve and so represent a greater workload. Most clinical measures are process in nature (registers, improving systems), but many include intermediate measures such as lowering blood pressure, lipid and glucose levels in heart disease, stroke, hypertension, diabetes, and kidney disease patients.

Successive governments of the different political parties have supported the NHS, and despite criticism, it remains a popular institution with the British public, surviving many changes in political leadership over the past 65 years. Reform has enabled the NHS to evolve with experience and to meet the changing economic and health needs of the country.

Changing epidemiological patterns have also led the UK Department of Health and the NHS to develop health promotion strategies. These have helped to reduce high rates of mortality from CVD and trauma, and may help to reduce the social and regional inequalities in health still present after over half a century of universal access. NHS reforms in the early 1990s and in the 2000s have promoted local community participation and clients' rights with the PCTs, and GP satisfaction with much higher incomes and control (as "gatekeepers") over use of secondary and tertiary care.

Numerous innovations in organization, incentive funding, information technology, and quality promotion with clinical guidelines appear to have had beneficial effects on access to care, shortening waiting times for primary and specialty care, and probably improving the quality of care as well. Health promotion activities such as smoking regulation, physical activity, and dietary change to combat obesity have been active elements of the Department of Health and NHS for over a decade. Modest progress has been made in smoking cessation. The percentage of the population aged 15 and over who are daily smokers has been decreasing steadily since 2000, from 27 percent in 2000 to 21.5 percent in 2009. However, obesity has not been sufficiently reduced. In terms of body weight, the proportion of the population that is overweight steadily increased from 29 percent in 1980 to 40 percent in 2001, and then gradually decreased to 36.7 percent in 2010. The percentage of the population that is obese was 7 percent in 1980, followed by a steady increase to 24.5 percent in 2008, dropping to 23 percent in 2009, and increasing to a record high of 26 percent in 2010.

In March 2012, the Health and Social Care Act was passed, an Act of Parliament in the UK. This act was one of the most extensive reforms of the NHS. It is controversial and has been criticized for being too costly. The mandate is to give local authorities a stronger role in shaping services, and widen the focus on education, research, and training. Previously, all NHS planning and delivery was done by the Department of Health, strategic health authorities (SHAs), and PCTs. Under this act, NHS providers are no longer performance managed by the SHAs. In 2013 this plan includes the abolition of PCTs and SHAs, and the introduction of clinical commissioning groups (CCGs). These CCGs are to replace the PCTs and SHAs, which will become foundation trusts by 2014. The NHS planning and delivery functions of the Department of Health, the PCTs, and the SHAs will be merged into the authority of a commissioning board called NHS England. This reform is still in process and its outcome and effects will be much studied in the coming decade. This reform is meant to introduce cross-cutting themes (Box 13.3). Ministers in the Department of Health are still accountable to the NHS.

The UK's constitutional arrangements allow a powerful executive to implement wide-ranging reforms at great speed, often with limited consultation. After seven or more major reorganizations since the NHS was founded in 1948 and a widespread feeling of "reform fatigue", the Conservative Party pledged, prior to the 2010 general election, "no more top–down reorganizations of the NHS". However, this pledge did not survive the election. The 2012 reform has created an extremely complex system with many unclear lines of accountability. The legislation is extremely permissive and the eventual result is uncertain. At some stage it

> **BOX 13.3 Cross-Cutting Themes of the Health and Social Care Act 2012**
> - Improving quality of care
> - Tackling inequalities in healthcare
> - Promoting better integration of health and care services
> - Choice and competition
> - Role of the Secretary of State
> - Reconfiguration of services
> - Establishing new national bodies
> - Embedding research as a core function of public service
> - Education and training
>
> **Source:** *UK Department of Health. Health and Social Bill explained. 17 February 2012. Available at: http://webarchive.nationalarchives.gov. uk/20130805112926/ http://healthandcare.dh.gov.uk/factsheets/ [Accessed 3 January 2014].*

will have to be revised significantly to create a workable system (see Box 13.4). An animated description of the NHS in England, setting out the extent of the confusion, can be viewed at the King's Fund website.

Studies of deaths from potentially avoidable causes include CVDs in men and lung cancer in women. They accounted for approximately 24 percent of all deaths registered in England and Wales in 2011, but the rates fell by 28 percent (from 243.2 to 175.8 per 100,000 population) between 2001 and 2011. Avoidable mortality rates were significantly higher in Wales than in England throughout this period and rates varied across the regions of England, with higher rates in the North of England and lowest rates in the South and East of England (Office for National Statistics, 2013).

Summary

The NHS has succeeded in its mission of providing universal access in a tax-financed and relatively economical service. It has guaranteed access to health care for all, but has failed to alleviate social class inequalities in health status. This has fostered new efforts and resource allocation to needier geographic areas and to health promotion efforts as in other industrialized countries to reduce the burden of CVD, cancer, and other diseases that disproportionately affect the poor. Health promotion activities have been successful in reducing tobacco use, but work remains to be done in combating overweight and obesity.

The Beveridge model NHS has been influential in the Nordic countries since the 1950s and in countries of southern Europe (Greece, Italy, Portugal, Spain, and Turkey) in their various reform programs since the 1970s. The NHS continues to evolve, and is an important and successful international model of health care systems, and one of the most cost-effective. It has adopted many measures of health promotion and increased service efficiency. Despite many criticisms, the NHS remains one of the most important and respected social

---

BOX 13.4 National Health Service Reform of 2013: Clarity Needed

The proposed new NHS reform promised to hand power to general practitioners (GPs) who, it was claimed, knew best what their patients needed as the previous primary care trusts purchased services on behalf of a defined population but were criticized as bureaucratic and remote. Draft legislation expanded this simple idea into an incomprehensible 300 page document. But barely concealed within it was a major drive to open delivery of care to the full force of the market, implementing the plans of Conservative politicians of two decades previously.

This was presented as enabling small non-governmental organizations (NGOs) to provide services such as mental health and palliative care. Ministerial reassurances attracted support from some general practitioners and heads of NGOs, although these were contradicted by the draft legislation. Given the confusion, the government instituted a "listening pause", and then reintroduced the legislation with a few cosmetic changes. Regulations published a few days before the legislation came into force in April 2013 required that almost all health care delivery be subject to competitive tendering, but lacked clarity as to what this meant in practice.

The resulting system is one of remarkable complexity with confusion about responsibility. Clinical commissioning groups (CCGs), now with only token involvement of GPs, are responsible for purchasing some aspects of care. However, they are subject to detailed oversight by a new body, NHS England, which is responsible for purchasing general practice and more complex services. CCGs obtain technical support from soon to be privatized clinical support units, seen by some as the basis for future competing insurers. They are also advised by

clinical senates, designed to provide advice on complex clinical issues. This system is overseen by two regulatory bodies, the Care Quality Commission, responsible for quality of care, and Monitor, responsible for competition (overlapping with the Office of Fair Trading, which covers all sectors of the economy).

Delivery of care is becoming a very mixed economy. As predicted, most contracts (mainly community services so far) have gone not to niche NGOs but to large multinational corporations. NHS hospitals are increasingly becoming independent foundation trusts, responsible for their own budgets. Since 2010, cuts in NHS funding have left many with severe financial problems, with debts from earlier private finance initiatives for capital development. Many have responded by cutting staffing, with resulting problems of quality and accountability.

The demise of primary care trusts means that the public health function no longer has a home, and is fragmented among a new body, Public Health England, and departments of local government. Some public health staff involved in purchasing care have moved to CCGs.

Sources: *Martin McKee, Professor of European Public Health, London School of Hygiene and Tropical Medicine. Personal communication; 19 July 2013.*
McKee M. Does anyone understand the government's plan for the NHS? BMJ 2012;344:e399.
Pollock A, Price D, Roderick P, Treuherz T, McCoy D, McKee M, Reynolds L. How the Health and Social Care Bill 2011 would end entitlement to comprehensive health care in England. Lancet 2012;379:387–9.
Reynolds L, McKee M. Opening the oyster: the 2010-2011 NHS reforms in England. Clin Med 2012;12:128–32.
Reynolds L, McKee M. "Any qualified provider" in the NHS reforms: but who will qualify? Lancet 2012;379:1083–4.

---

institutions of the UK. Recent reforms such as the Health and Social Care Act of 2012 remain to be evaluated and the impact on the NHS hospital system is yet to be determined.

THE NORDIC COUNTRIES

The Nordic countries share common principles of a "Nordic welfare model" with features of universality (right to social protection), a strong public sector, and tax funding based on legislative rights of citizens, equal treatment, and high social benefits. Church-based philanthropy and charity have not played much of a role in welfare provision. The roots of the municipal welfare model roots go back to the early eighteenth century, long before the emergence of organized philanthropy and charity.

The Nordic countries have working committees to focus on joint cooperative projects in the health sector and many institutions. The work concerns the common interests between these countries and related matters with the EU.

Each of the health care systems in the Nordic countries has its own characteristics, and reforms are in progress in each country. Denmark, Finland, Norway, and Sweden, with social democratic governments both before and after World

War II, in many ways pioneered the welfare state. They were later influenced by the UK's NHS, but with strong regional or local governmental organization and taxation have had more emphasis on a decentralized program of health services. Their achievements in social welfare and health care over many decades have been widely acclaimed successful models for social protection in prosperous industrial economies. In 2010, the HDI ranks for the Nordic countries – Denmark 15, Finland 21, Norway 1, and Sweden 8 – were among the highest in the world, with a steady increase in life expectancy (Figure 13.5). Total health expenditures per capita have increased moderately on an annual basis. Expenditures on health have remained less than 11 percent, and are similar among the Nordic countries (Table 13.7).

Most (74–85 percent) of the health system revenues are from public sources (Table 13.8). Commonly, between 50 and 70 percent of health system revenues are generated from personal income taxes levied at the regional (Sweden, Norway, Denmark) or municipal (Finland) levels of government. Most of the remainder comes from general revenues raised by the national government through value-added or excise taxes and personal or corporate income taxes. The national funds are distributed as block grants to minimize

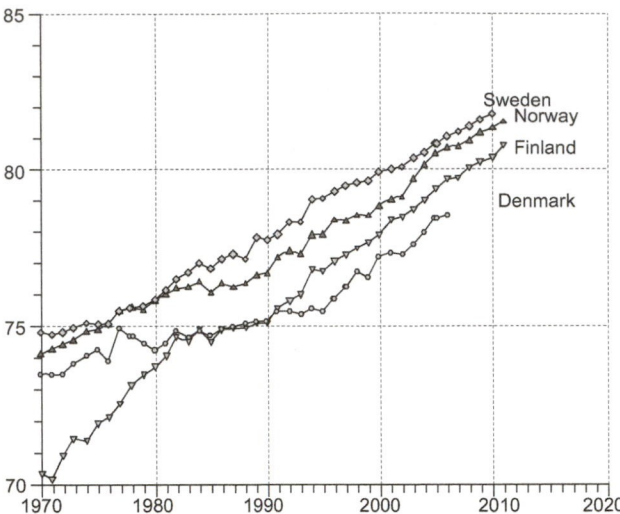

FIGURE 13.5 Life expectancy at birth in years, Nordic countries, 1970–2011. *Source: World Health Organization, European Region. Health for All database; January 2013. Available at: http://data.euro.who. int/hfadb/*

TABLE 13.7 Total Expenditures on Health as Percentage of Gross Domestic Product, Nordic Countries, Canada, USA, and UK, Selected Years, 1985–2010

	1985	1990	1995	2000	2005	2010
Denmark	8.5	8.3	8.1	8.3	9.1	11.1
Finland	7.1	7.7	7.5	6.6	7.5	8.9
Norway	6.6	7.6	7.9	8.4	9.1	9.4
Sweden	8.6	8.3	8.1	8.4	9.1	9.8
Canada	8.1	8.9	9.4	8.8	9.8	11.4
UK	5.8	5.9	6.8	7.0	8.2	9.6
USA	10.4	12.4	13.7	13.7	15.8	17.6

Source: Organisation for Economic Co-operation and Development. OECD health data 2012. Available at: http://www.oecd.org/health/healthdata

TABLE 13.8 Public Expenditure on Health as Percentage of Health Expenditures, Nordic Countries, Canada, USA, and UK, 1985–2010

	1985	1990	1995	2000	2005	2010
Denmark	85.6	82.7	82.5	82.4	84.1	85.1
Finland	78.6	80.9	71.7	71.3	75.4	74.5
Norway	85.8	82.8	84.2	82.5	83.6	85.5
Sweden	90.4	89.9	86.6	84.9	81.2	81.0
Canada	75.5	74.5	71.2	70.4	70.2	71.1
UK	85.8	83.6	83.9	78.8	81.7	83.2
USA	39.6	39.4	45.1	43.0	44.2	48.2

Source: Organisation for Economic Co-operation and Development. OECD health data 2013. Available at: http://www.oecd.org/health/healthdata

TABLE 13.9 Acute Care Beds per 1000 Population, Nordic Countries, Canada, USA, and UK, Selected Years, 1985–2010

	1985	1990	1995	2000	2005	2010
Denmark	4.7[a]	4.1[a]	3.9[a]	3.5	3.1	2.9
Finland	4.8	4.3	3.0	2.4	2.2	1.8
Norway	4.7	3.8	3.3	3.1	2.9	2.4
Sweden	4.6	4.1	3.0	2.5	2.2	2.0
Canada	4.5	4.1	3.9	3.2	2.8	1.7[b]
UK	NA	NA	NA	3.1	2.9	2.4
USA	4.2	3.7	3.5	3.0	2.7	2.6[b]

Note: [a]Reported in previous OECD reports; [b]2009 data.
NA = not available.
Source: Organisation for Economic Co-operation and Development. OECD health data 2012. Available at: http://www.oecd.org/health/healthdata

interregional inequalities, with additional grants for medical education. National sickness funds pay for ambulatory visits. Municipal governments pay for long-term care for the elderly. Patient co-payments provide 2–3 percent of Sweden's county health expenditures and were introduced in Finland in 1993. The user fees are not a significant hardship because of the widespread prosperity and well-established social security systems of the Scandinavian countries.

The Nordic countries have traditionally emphasized maternal and child health and have achieved very low rates of infant mortality. Immunization coverage in 2011 for Hib vaccine ranged from 91 to 98 percent. They have higher rates of mortality from CVD than do countries in southern Europe. This is thought to be related to traditional dietary patterns with high-fat diets, along with smoking and heavy alcohol usage. These risk factors have been the subject of much successful effort at health promotion and are slowly declining. All Nordic countries have greatly reduced the supply of acute care hospital beds, as shown in Table 13.9.

Sweden

In 2011, Sweden's population was 9.5 million, and the 2011 gross national income (GNI) per capita was US$36,143 purchasing power parity (PPP). The 2011 infant mortality rate was 2 per 1000 live births, and life expectancy at birth was 82.0 years. In 2005, crude birth rate was 11 per 1000 population, with 100 percent of births taking place in medical facilities, and the maternal mortality rate averaged 4 per 100,000 between 2000 and 2010. Immunization coverage in infancy in 2011 was over 95 percent.

Sweden's health insurance system evolved over many decades, and with the Health Care Act became compulsory and universal in 1955, covering compensation for medical clinics, hospital services, and private ambulatory care. Swedish health care is tax financed, with funding mainly from employers and government, but patients are charged a co-payment for services. Health expenditures as a percentage of GDP ranged between 8.1 and 8.6 percent from 1985 to 2000, increasing to 9.6 percent in 2011. The publicly financed health system covers public health and preventive services, inpatient and outpatient hospital care, primary health care, inpatient and outpatient prescription drugs, mental health care, dental care for children and young people, rehabilitation services, disability support services, patient transport support services, home care, and long-term and nursing home care (Commonwealth Fund, 2011).

The county or municipality is the principal level of government responsible for management of health care. There are 20 county councils and three large municipalities with populations ranging from 60,000 to 1.5 million. The counties or municipalities, which have an income tax base of financing, provide over 70 percent of funding for health care, with 11 percent coming from the national government, 5 percent from national insurance, 2–3 percent from patient fees, and the remainder from miscellaneous sources. Current reforms include improved primary care coupled with a reduction in the hospital bed supply. Primary care is provided in health centers staffed by salaried GPs, nurses, and other staff serving about 15,000 clients, but about 12 percent is provided by private physicians. Sweden has a system of economic equalization to compensate for uncontrollable factors such as age differences and rate differences for certain costly disease conditions.

Sweden has traditionally had a very high ratio of hospital beds to population. In 1985, this included 4.6 acute care beds, 6.2 long-term care beds, and 2.4 mental hospital beds per 1000 population. The acute care beds were reduced to 2.0 per 1000 by 2010. Hospitalization was a common form of care, especially in areas with a sparse population and long distances to hospitals and doctors. Reduction in hospital bed supplies has been a long-term strategy in Sweden since the 1940s, and emphasized since the 1960s, with a steady reduction in medical, surgical, and community care beds, as well as psychiatric beds. Long-term social care for the elderly has been transferred to social service agencies. This was accomplished while maintaining high-quality service and improving national health indicators, such as infant mortality rates and maternal mortality rates, which are among the lowest in the world.

Recent reforms in Sweden allowed contracting out for public sector services. This strengthened the role of primary care providers, who are now able to select more efficient and user-friendly services. Hospitals operate as economic units, balancing revenues and expenditures, and must compete for patients in the new public market for health care. Public institutions must also compete with the private sector and, in some instances, purchase services from private providers. This has helped to reduce waiting times for operations and led to bankruptcy of inefficient or unacceptable hospitals.

Sweden, like other Nordic countries, has refocused health planning on the principle that all people should have equal access to the same conditions for good health, with a renewed emphasis on vulnerable groups such as immigrants and single parents and their children. It includes a focus on avoidable hospital days for non-communicable long-term conditions (e.g., asthma, diabetes, heart failure, and hypertension) and acute conditions (bleeding ulcers, diarrhea, and inflammatory conditions).

Denmark

In 2010, Denmark's population was 5.5 million; in 2011 the GNI per capita was US$33,518 (PPP). The 2011 infant mortality rate was 3 per 1000 live births, life expectancy at birth was 79.0 years, and the HDI was 0.01, ranking 15th highest of 177 countries. The crude birth rate was 11 in 2005, with 100 percent of births occurring in medical facilities. The maternal mortality rate was 12 per 100,000 (2010); in 2000 it was 8.0. Immunization coverage in infancy in 2011 was 91 percent for polio and DTP, and declined from 99 percent in 2000 to 87 percent in 2011 for measles.

In 1803, the predecessor of the National Board of Health was established; from 1858 local boards of health began to be set up. There is a long history of decentralized health services, which have been the responsibility of local towns and municipalities since the early years of the eighteenth century.

Reforms focus on ensuring continuity of care across administrative sectors, with easy access to unified prevention, primary care, and rehabilitation services. The focus is on improved service for multiproblem situations, the disadvantaged, the chronically ill, and at-risk children. Denmark has not built any institutional accommodation since 1987 but has developed subsidized housing and extensive home care services for older people. The percentage of GDP spent on health increased from 8.1 percent in 1980 to 11.1 percent in 2010. Between 1990 and 2010, acute care beds declined from 4.1 to 2.9 per 1000 population.

Norway

In 2012, Norway's population was 4.96 million; the GNI per capita was US$48,688 (PPP 2005). The 2011 under-five mortality rate was 3 per 1000 live births, and life expectancy at birth was 81.3 years. The HDI for Norway of 0.955 in 2012 made it the number one country in the world. The crude birth rate was 12 in 2011. In 2011, 99 percent of all births were in medical facilities. The maternal mortality rate was 7 per 100,000 in 2010 and immunization coverage in infancy in 2011 was over 93 percent.

Norway, with a GDP 43 percent above the average in the EU, is one of the richest countries in the world. Health expenditures in 2011 were US$8967 (PPP) per capita. The percentage of GDP spent on health rose from 6.6 percent of GDP in 1985 to 9.4 percent in 2010. The proportion of government expenditures spent on health stayed stable at just over 86 percent and is similar to that of the other Nordic countries. Norway is the only Nordic country where central government is directly involved in the decision-making process for tertiary care services. Sweden, Denmark, and Finland delegate this to regional authorities or municipalities.

The trends in health reform over the past few decades may be summarized as:

- 1970s – reducing inequalities and building up health services
- 1980s – cost containment and decentralization
- 1990s – efficiency and leadership
- 2000s – structural changes in delivery and organization with a focus on reducing inequalities.

Primary care is the responsibility of the local municipalities; five regional health authorities are responsible for specialist care; with ownership of hospitals transferred to central government. Hospital services are organized as enterprises, with day-to-day operations run by a general manager and an executive board. National reforms have focused on responsibility for providing service, priorities, patient rights, and cost containment.

Finland

Finland is a republic with a population of 5.4 million people in 2011, a GNI per capita of US$32,510 (PPP), and an HDI of 0.892, ranking 21st out of 177 countries. Finland has achieved one of the lowest infant mortality rates in the world, declining from 22 per 1000 live births in 1960 to 2 per 1000 in 2011. Maternal mortality averaged 5 per 100,000 live births (2000–2012). Child care is provided free by the municipalities. Immunization coverage rates include 97 percent for polio, 99 percent for DTP, and 97 percent for MMR for 1-year-old children (2011). Despite high immunization coverage, Finland experienced an outbreak of polio due to use of an inadequately immunizing inactivated polio vaccine (IPV) in the 1980s. Longevity increased by 5.5 years for men and 5.1 years for women from 1971 to 1991, with a life expectancy of 81.0 years overall in 2011, an increase from 68 years in 1960.

Finland has three tiers of government. Strong municipal governments provide primary, secondary, and tertiary care services, as well as public health, education, and other social services. The states subsidize municipalities to provide these services, with management by locally elected officials. Taxes on income are shared between the municipal

and national governments. Universal access to care is guaranteed.

Health policy is determined at the level of the national government, which regulates capital investment in health facilities and subsidizes municipalities, which are responsible for providing health and social services. State and municipal governments together collect approximately half of total taxation, which is high compared to other countries, reaching 46 percent of GDP. The economy was in recession during the early 1990s with a decline in GDP; as a result, the percentage of GDP spent on health care rose sharply, from 6.3 percent in 1980 to 7.7 percent in 1990, falling to 6.6 percent in 2000 and rising to 8.9 percent in 2010.

The constitution provides social protection for the people made up of preventive social and health policy, social welfare and health services, and sickness, unemployment, old age, and other benefits.

Public health care services consist of primary care provided by municipal health centers and specialized hospital care. A health center can be run by more than one municipality on a cooperative basis. Primary care includes well child care, school health care, medical rehabilitation, and dental care. Services may be purchased from private providers. Finland has 20 districts which provide specialized hospital care and includes a central and a regional hospital. There are five university hospitals. There is a fee paid at the time of visit to the health center for municipal services but with a cap on the annual amount the person is charged, while fees for long-term care are based on the person's income.

High rates of mortality from CVD, injury, and suicide affect middle-aged men disproportionately. The widely

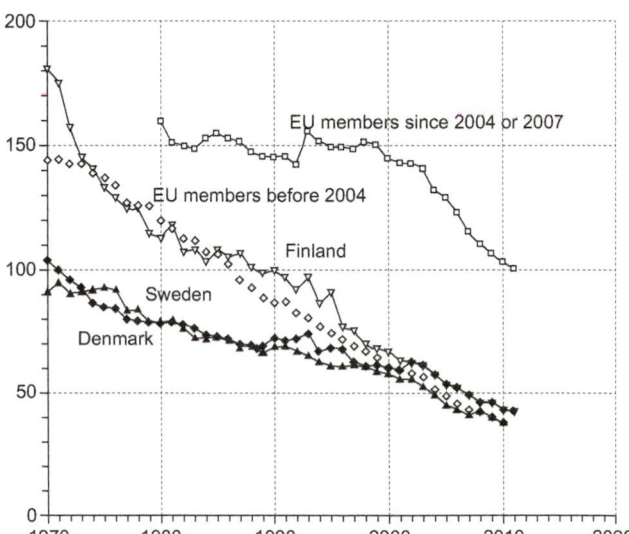

FIGURE 13.6 Nordic countries, standardized death rate, cerebrovascular diseases, all ages per 100,000 population, 1970–2011. Note: EU=European Union. *Source: World Health Organization, European Region. Health for All database; June 2013. Available at: http://data.euro. who.int/hfadb/*

known North Karelia project (see Chapter 5) to promote reduction in risk factors for heart disease stimulated national efforts and contributed to substantial reductions in mortality rates from these diseases (Figure 13.6). Cardiovascular mortality rates declined 52 percent from 1970 to 1996, in part because of changes in diet with less meat and greater vegetable consumption. Hospital discharge rates for CVDs declined significantly (Table 13.10). Smoking rates for men in the early 1970s reached 50 percent but declined to 22 percent in 2009, with 16 percent of women smoking (Table 13.11). Overall alcohol consumption is low, but binge drinking is common and relates to the high suicide and trauma rates.

Finland had high hospital bed ratios up until the 1980s when it changed health policy, recognizing the limitations of hospital care and placing greater emphasis on primary care, preventive and social services, and health promotion. Hospital bed supplies are still being reduced, with shorter lengths of stay and increasing ambulatory care and outpatient care. Mental hospital beds were decreased by 50 percent during the 1980s. The total hospital bed to population ratio declined from 15.6 in 1980 to 9.3 per 1000 population

in 1995 and to 6.2 in 2011. Acute care hospital beds per 1000 decreased from 4.9 to 4.0 from 1980 to 1995 and 2.9 in 2005 (OECD data report, 2007).

Reform in primary care services during the 1980s reduced inefficiency, bureaucracy, and waiting times, and raised consumer satisfaction. A combination of capitation and fee-for-service payment is used. During 1993, reforms in health care financing converted national support for municipal health services to block grants based on capitation formulae to the municipalities, which now fund both the hospitals and primary care services. This allows the municipalities greater freedom in seeking a new balance of services and redirecting resources from the hospital to the primary care sectors. Local health centers provide most medical and health-related services, including rehabilitation and addiction services. Recent health reform activities have emphasized guaranteed access to care within maximum time-frames with uniform criteria for non-emergency care. Oral health care is supported by public funding that covers the total population.

Hospital-based physicians are permitted to practice privately. Over 90 percent of GPs work in publicly operated health facilities, but nearly one-third also conduct private practices in their off-duty time. GP satisfaction with the changes in the health system, with the combination of capitation and fee-for-service, is reportedly high.

The search for greater efficiency now includes a mix of planned and market economies in health. The strong tradition of publicly operated health services will continue despite introduction of market elements, but regional inequalities may be an undesired result. Health reform in Finland continues with decentralized service management and central planning and financial support. Finland emphasizes *Health in All* policies, whereby health and social issues are included in all local and national planning (Ministry of Social Affairs and Health, Finland and European Observatory on Health Systems and Policies, 2006).

TABLE 13.10 Hospital Discharge Rates, Cerebrovascular Diseases, Nordic Countries and European Union (EU), Selected Years, 1990–2005

	1990	1995	2000	2005	2010
Denmark	430	394	452	384	362
Finland	681	820	658	561	NA
Norway	Na	382	319	342	306
Sweden	613	617	506	451	NA
EU before 3004	Na	339	348	351	NA

NA = not available.
Source: World Health Organization, European Region. Health for All database; January 2013. Available at: http://data.euro.who.int/hfadb/

TABLE 13.11 Tobacco Consumption, Daily Smokers 15 Years and Older, Percentage of Population, Selected Years, Nordic Countries, 1980–2009

	1980	1990	1995	2000	2004–2005	2009
Denmark	50.5	44.5	35.5	30.5	26.0	19.0
Finland	26.1	25.9	24.0	23.4	21.8	18.6
Norway	36.0	35.0	33.0	32.0	25.0	21.0
Sweden	32.4	25.8	22.8	18.9	15.9	14.9

Source: Organisation for Economic Co-operation and Development. OECD health data 2011 – Version: October 2007, data available from 1980–2009. Available at: http://www.oecd.org/els/health-systems/oecdhealthdata2012-frequentlyrequesteddata.htm [Accessed 14 June 2013].

WESTERN EUROPE

The countries of continental Western and Central Europe pioneered national health insurance through place of employment, with the national government regulating conditions of insurance, establishing fee schedules, and setting national health policies. The generic type is termed the Bismarckian national health insurance program, and is characteristic of Germany, France, the Netherlands, Belgium, Luxembourg, Austria, and Switzerland, each having distinct characteristics and mixed features of social insurance with national service elements. These have been termed "sickness insurance", based on the solidarity principle of workers' benefits, including old-age pensions, disability benefits, and compensation for loss of working capacity. The funds have maintained a treatment-oriented approach, and only under exceptional circumstances have they undertaken disease prevention, much less health promotion.

Germany

Germany is a federal state with a century-old tradition of social protection legislation. Most aspects of management are delegated to self-governing insurers and associations of providers. The population of Germany in 2010 was 81.8 million, and life expectancy at birth in 2011 was 81.0 years; infant mortality (2011) was 3.0 per 1000 live births, and maternal mortality was 7.0 per 100,000. In 2011, the GNI per capita was US$35,431 (PPP) and Germany ranked 22nd of 177 countries on the HDI (0.920), just above the OECD average. Immunization coverage for infants was over 90 percent in 2011.

Bismarckian Health Insurance

Germany's system of national health insurance is based on Chancellor Otto von Bismarck's plan, which introduced care for low-income workers financed through a social security system by employer and employee contributions. The Sickness Insurance Act of 1883 provided that all workers earning below a designated level be insured by a Sick Fund, with employer–employee contributions. This is also known as statutory health insurance (SHI) or as the Bismarckian system, based on making health insurance mandatory for certain employees.

The Sick Funds (*Krankenkassen*) might be owned by unions or employer associations, which can operate their own health services to provide comprehensive medical and hospital services for enrolled members and their families. The Sick Funds or mutual benefit societies may also provide cash benefits for accidental injuries, burial benefits, and widows' pensions. This plan was later extended to cover virtually the entire population and remains the foundation of Germany's health and social insurance up to the present time.

In 1911, a framework for social insurance was introduced with adoption of the Imperial Insurance Regulation. In 1923 the Imperial Committee of Physicians and Sickness Funds (later known as the Federal Joint Committee) was created as the authority responsible for decisions regarding benefits and the delivery of outpatient care. Later, the Sick Funds became obliged by law to provide hospital care not only to their members but also to family dependants, and coverage extended to include health care benefits for pensioners. Health care benefits were gradually extended further and in 2004 the unemployed, students, disabled, and recipients of social welfare were incorporated into the statutory health plan.

The statutory health insurance system is characterized by three main principles: solidarity – the willingness of the healthy people to pay for the sick and availability of a universal and comprehensive benefit package; decentralization and organization of the health care system from the bottom up; and the principle of corporatist organization, namely representation of employees and employers on the management boards of Sick Funds.

In Germany, health care is governed at the national level by the Federal Assembly, the Federal Council, and the Federal Ministry for Health and Social Security as the key authorities liable for passing health reforms concerning statutory health insurance. The federal government is responsible for setting the health policy for delivery of medical services. The corporatist level consists of 292 non-profit, quasi-public Sick Funds and associations of SHI-contracted physicians and dentists on the provider side. The 16 *Länder* are accountable for planning and management of the hospital sector, policy development, and implementation for social and nursing care services, including prevention and monitoring of transmissible diseases, pharmaceuticals and drugs, and environmental hygiene.

The entire German population is entitled to health care services; in 2003, 88 percent were covered by SHI, 10 percent by private health insurance companies, and the remaining 2 percent by specific governmental schemes (military, police, social welfare, and assistance for immigrants seeking asylum). Thirty-seven percent of SHI insured were members of general regional funds (AOK), 33 percent were insured by substitute funds, 21 percent were members of company-based sickness funds (BKK), and 6 percent were covered by guild funds (IKK).

Statutory health insurance is the core of the German health care system. Outpatient care is provided by private for-profit care providers characterized by a monopoly and no gatekeeping functions. Physicians and other health professionals working in hospitals or institutions for nursing care or rehabilitation are paid salaries. Private physicians and dentists are paid on a fee-for-service basis with the fee schedule determined by the Federal Ministry of Health and Social Security. Inpatient care is delivered by a mixture of public and private providers. The Sick Funds represent the collectors, purchasers, and payers of SHI and long-term care insurance. Sick Funds are self-governed and based on mandatory membership.

As a result of amalgamations, the number of Sick Funds decreased from 1200 in 1993 to 292 in 2004. By law, they have the right to raise contributions, and to negotiate prices and quality assurance with providers of care with whom they contract. Sick Fund membership is mandatory for employees whose gross income does not exceed a specified upper level of the gross salary per month (in 2005) in order to prevent high-earning voluntary members from leaving the SHI. Contributions for SHI are dependent on income, and not on risk. From 1949 to 2004, contributions were shared equally between employees and their employers. In

2005, the contribution rate for employees was increased to 54 percent, with employers obliged to pay the remaining 46 percent. For people earning below a threshold minimum salary, the employers pay a standard rate of 11 percent contributions for all Sick Funds. Since 2004, pensioners have had to pay the full contribution rate. In 1995, mandatory insurance for long-term care was introduced. The long-term care insurance scheme is run by the Sick Funds and private health insurers. There is a uniform co-payment for outpatient services and products and co-payment of ₡10 per inpatient day for a maximum of 28 days.

Since reunification of East and West Germany in 1990, several health care reforms have been launched with the main focus on expenditure control and improving technical efficiency by enhancing managed competition and taking measures to avoid adverse effects on equity and quality. In 2004, the total government expenditure as a percentage of GDP was 47 percent, whereas in 2010 the total health spending as a percentage of GDP was 11.6 percent, placing Germany among the OECD countries with highest expenditure on health. Only the USA (15.3 percent), Switzerland (11.6 percent), and France (11.1 percent) allocated more of their GDP to health than Germany in 2005 (OECD).

Health Insurance Reform

Since 2009, universal health insurance (SHI) has been mandatory for all citizens and permanent residents. It covers preventive services, inpatient and outpatient hospital care, physician services, mental health care, dental care, optometry, prescription drugs, medical aids, rehabilitation, hospice and palliative care, and sick leave compensation. Preventive services include regular dental checkups, well-child checkups, basic immunizations, checkups for NCDs, and cancer screening at certain ages. A separate mandatory insurance scheme, LTCI, covers long-term care in the whole population.

This is a public–private health care system. In 2008, public expenditures covered 76.8 percent of total health expenditures, private spending 13 percent, and out-of-pocket expenditures 10.2 percent. Although the insurance system pays for these services and prescription drugs, user fees are charged for both medical visits and prescription drugs.

Hospitals are paid on a per diem basis, including salaried physician services. Hospital bed supply and discharge rates are high (227 per 1000 population compared to the OECD average of 163 per 1000), with low hospital occupancy rates.

Traditionally, the German citizen had no right to choose the Sick Fund and was assigned to the appropriate fund based on geographic and/or job characteristics. However, since 2002, every SHI member has a choice of Sick Fund membership at any time of the year, but a minimum membership period of 18 months is required before being able to switch to another Sick Fund. The company-based funds (BKK) and the guild funds (IKK) have the right to remain closed, but if they decide to open, they are obliged to contract with all applicants. Only the farmers', sailors', and miners' funds remained closed with assigned membership.

In order to assure competition and to balance income-level differences in contribution rates among the funds, a risk structure compensation scheme was launched in two stages during 1994–1995. In 2001, disease management programs were introduced as a new instrument to avoid "cream-skimming" among the Sick Funds, as well as providing incentives for care of the insured chronically ill. Since 2004, the Sick Funds have been obliged to receive a fixed amount from the federal budget for several benefits relevant to family policies, such as maternity benefits, sick pay for parents caring for sick children, and in vitro fertilization, and in 2007 the scheme became "morbidity oriented".

State governments have the authority to plan hospitals. By 1985 legislation, hospital capital costs were funded by state and local governments through a certificate of need. In 2002, 54 percent of hospitals were public; most were operated by municipalities, with 38 percent by non-profit NGOs, and 8 percent by for-profit corporations. Until 2003, hospitals, except for university hospitals, traditionally provided inpatient care only. Since then, hospitals have been able to treat patients with diseases requiring highly specialized treatment on an outpatient basis. In 2005, Germany's acute care hospital bed supply had declined from 8.4 per 1000 in 1991 to 6.3, compared to 3.7 in France and 3.9 among the original members of the EU in 2004. The average length of stay for acute care hospitals in 2004 was 8.6 days and bed occupancy rate was 75.6 percent, in comparison with most other EU countries which had an average of 6.7 days and 75.5 percent bed occupancy. Acute care bed supply declined to 5.7 and average length of stay declined to 7.5 days in 2009.

One of the major reforms in the German health care system concerns the hospital payment system. Operating costs were paid on a per diem basis by the Sick Funds at standard rates for all patients but differing among hospitals. There were no incentives for hospitals to reduce the costs of utilization. In 1986, global budgeting was introduced for hospitals, intended to promote cost-effective services, outpatient treatment, and hospital financing for greater ambulatory care and coordination of medical care. Germany has a high hospital bed supply and low occupancy rates. In 1988, and again in 1993, health reform laws were passed trying to restrain health cost increases. These included a law limiting fee increases, the supply of physicians, and use of expensive technologies in ambulatory care. Since 2004, hospitals have been reimbursed on the basis of DRGs; in 2005, the acute hospital cases were classified in 878 DRGs. Mandatory quality assurance carried out by external authorities was initiated in 2004 to provide transparency and improve quality of care.

The professional associations and hospitals have had a strong role in determining the costs of health care by negotiating high salary levels and promoting an emphasis on high technology, high levels of surgery, and overlapping services. Patients have a choice of physician but may be obliged to join one of the 294 Sick Funds according to the choice of their employer or their professional grouping.

Germany pioneered social security-based health insurance. Its health system coped well with the challenge of integrating the former East German health system and population. Germany's health care standards are among the highest in the world; life expectancy rates are improving steadily but continue to be below those of France and the EU, although well above neighboring countries of Eastern Europe (Figure 13.7). Mortality rates from cerebrovascular disease and heart disease are well above those of France and the original EU members, while cancer mortality is slightly lower. Health promotion approaches are not part of Sick Fund responsibilities, but are being developed in recent health reforms.

The Netherlands

In 2011, the Netherlands had a population of 16.7 million, with a GNI per capita of US$37,282 (2011) and life expectancy of 81.0 years (2011), among the world's highest. Health expenditures in 2011 were 9.2 percent of GDP, just above the OECD average of 9.0 percent. The crude birth rate was 11.5 live births per 1000 population in 2011. The infant mortality rate declined from 18 in 1980 to 3 per 1000 live births in 2011, compared to US rates of 26 and 6 per 1000, respectively. Maternal mortality was 6 per 100,000 live births in 2000–2010. Immunization coverage in infancy in 2011 was 96–97 percent for DTP, polio, and measles. The Netherlands experienced two outbreaks of polio among non-immunized religious groups from imported polio virus in 1987 and 1992 and a large mumps outbreak in 2008.

The health care system of the Netherlands is a combination of public and private financing, with private delivery of care. The system evolved from medieval guilds and mutual benefit associations to health insurance through employer–employee payments to non-profit Sick Funds or private insurance plans. By 1933, health insurance offered by such groups covered 41 percent of the population. National health insurance was introduced in 1941 (by Germany). Sick Funds were established on a geographic basis covering a majority of the population. Physicians are paid on a fee-for-service basis for insurance patients and by capitation for Sick Fund patients.

A new health insurance system was created in 2006, replacing the former fragmented insurance system, and includes occupational disease and workplace injuries. It is a private insurance system with statutory safeguards covering the total population, covering long-term nursing care, acute care, and supplementary insurance. It is described as a hybrid model between public and private insurance. The medical insurers (30 companies) are required to accept all applicants and offer the same insurance coverage under the same terms and conditions. The insured person pays a nominal premium and an income-related contribution. At the year's end, those who made little use of the system receive a rebate of part of the premium. The tax system levies the income-related contribution through the employers.

Municipal health services are responsible for public health services on behalf of local and regional authorities (governments). Lifestyle factors are seen as important aspects of public health policy on smoking, alcohol abuse, physical activity, nutrition, diabetes, and mental depression.

Preventive and health promotion targets for improving health include smoking, problem drinking, overweight, diabetes, and depression as key areas of reducing health inequalities. More than 70 percent of care expenditures are for treatment of those with NCDs. The private insurance system for personal health services limits opportunities

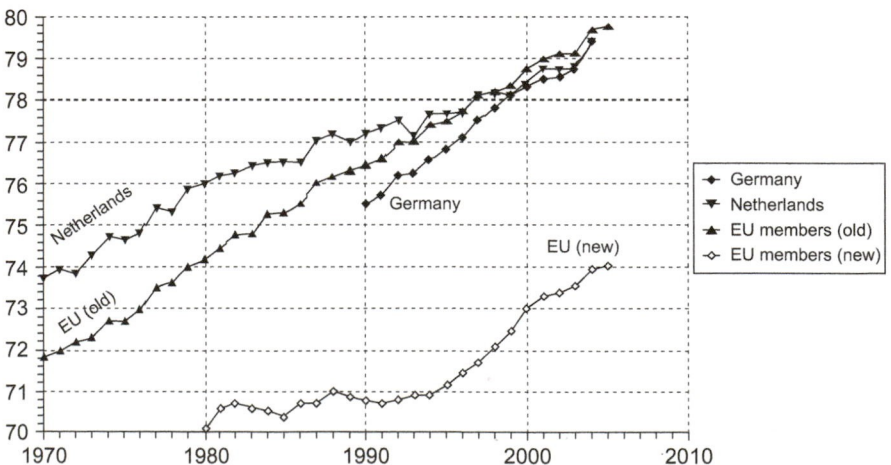

FIGURE 13.7 Life expectancy at birth in years, Germany, the Netherlands, and European Union (EU), 1970–2005. *Source: World Health Organization, European Region. Health for All database; November 2007. Available at: http://data.euro.who.int/hfadb/*

for prevention-oriented activities for lifestyle-related conditions. In the absence of objectives and targets, providers and insurers determine the types and levels of preventive health services.

Patients must have a referral from their GP before seeing a specialist (i.e., the GP as gatekeeper). This helps to prevent unnecessary referrals, strengthening the role of the GP and helping to control health care costs. Most specialists are hospital based and are paid on a fee-for-service basis. Most hospitals are not-for-profit and paid on a block budget negotiated with the private insurers. The supply of hospital beds is closely regulated by the government, as is technology investment, restraining cost increases for the hospital sector.

Reform of the health system in the Netherlands emphasizes competition and market-based approaches to private insurance. Health expenditures as a percentage of GDP increased from 6.7 percent in 1972 to 8.4 percent in 1982, remained relatively stable until 1990 at 8.0 percent, and subsequently increased to 12.0 percent in 2011. The acute care hospital bed-to-population ratio was reduced from 5.5 in 1970 to 3.1 beds per 1000 population in 2011.

Mortality patterns show the Netherlands population to be at relatively high risk for cancer, but at lower risk than most northern European countries for CVD. The Dutch health system has been successful in restraining cost increases compared with the USA, while providing universal coverage, preserving primary care medical services, and achieving health status measures among the best in the world.

RUSSIA

The Russian Federation is the largest country in the world, stretching from Europe to the Pacific Ocean. Russia has a highly urbanized (74 percent) and educated (99.6 percent), multiethnic population of 143.0 million people (2012) and abundant natural resources. Following the collapse of the former Soviet Union, the Russian Federation went through tumultuous times but then developed an economic growth pattern based mainly on oil and other resources. The Russian Federation GDP per capita grew at an average 6.6 percent between 2001 and 2008, with inflation and widespread poverty, especially in rural areas. After the economic crisis in 2008 there was a drop in GDP but during 2010–2012 GDP stabilized at an average growth rate of 4.2 percent. GNI is estimated at US$14,461 (PPP) for 2012. Owing to the recession the unemployment rate rose to 8.4 percent but decreased to 6 percent in 2012 (equal to the 2007 level). Immigration from neighboring countries such as the Central Asian Republics helps to moderate the depopulation trend to some extent. In 2012 the HDI of 0.788 placed Russia in 55th position, with a life expectancy at birth of 69.8 years (females 75.6 and males 64.0 years).

Russia's population decline since the beginning of 1990s was largely due to low birth rates and premature deaths from stroke, CHD, violence, traffic accidents, and alcoholism. Life expectancy is slowly improving with the improving social conditions. The government took measures to influence birth rates in the country through financial support for parents after they give birth to a second child (this Maternal Capital Program has been implemented since 2007). CVDs are the most frequent cause of death in Russia. SMRs are high compared to Western European countries, which have been experiencing declining mortality rates especially since the 1960s.

CVDs cause 57 percent of all deaths in Russia, of which 49.3 percent are from CHD and 35.4 percent from cerebrovascular disease; most of these deaths occur among people of working age (Petrukhin and Lunina, 2012). The high incidence of CHD among men reflects the gap between the life expectancy of men and women. Stroke mortality in Russia declined by one-third from 2003 to 2010 (from 317 to 215 per 100,000), while CHD mortality fell by only 16 percent (from 415 to 349 per 100,000).

According to the National Statistic Committee of Russia, the total number of deaths due to CVD in 2011 was 1.1 million, whereas the total number of births was 1.7 million. CVD mortality among Russian men remains five times higher than rates in Western Europe and more than double the rates in Central European countries (Health for All database, January 2013). Up to 70 percent of men and 30 percent of women smoke in Russia. Mortality rates from CVD vary in different regions of the country: rates are higher in the north-west regions of Russia (over 1200 deaths per 100,000 in 2009), whereas in southern regions the rates are significantly lower (309 per 100,000 in Chechnya and 167 per 100,000 in Ingushetia in 2009). Figure 13.8 compares CVD SMRs in the Russian Federation with the EU and countries of Central Asia.

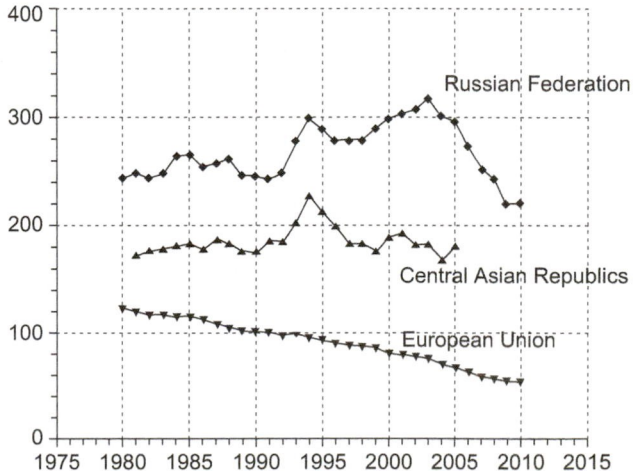

FIGURE 13.8   Standardized death rate, cerebrovascular diseases, all ages, per 100,000 population, Russia, European Union members, and Central Asian Republics (CARK), 1980–2010. Note: Three-year moving averages. *Source: World Health Organization, European Region. Health for All database; 2013. Available at: http://data.euro.who.int/hfadb/*

Excessive alcohol consumption and binge drinking also result in high motor vehicle death rates and other trauma, and homicide and suicide rates which are among the highest in the world. The situation with HIV infection is not improving. The number of officially registered HIV-positive cases (in 2011) was 695,484, for an HIV prevalence rate of 393.9 cases per 100,000 population. The rates doubled in 5 years. The most affected age group is between 18 and 24 years old. The HIV epidemic in the Russian Federation is concentrated in high-risk population groups with the principal driving forces being injection drug use and unsafe sex. But there has been a four-fold increase (from 10 to 41 percent) in cases of heterosexual HIV transmission since 2001, whereas the number of HIV transmissions via unsafe drug use has been decreasing (from 95 to 56 percent). Tuberculosis (TB) case notification rates more than doubled from 34.2 to 90.4 per 100,000 population between 1990 and 2000, but decreased by 8.5 percent from 2001 to 2003, and have been decreasing since 2008 (to 77.4 per 100,000 in 2010). Multidrug-resistant strains of TB are present in as much as 20 percent of cases in some regions of Russia.

This crisis in health is not only related to the period of economic transition in the 1990s, but goes deep into the former Soviet health system. The "old" state-operated service provided free universal health care with ample, indeed excessive, resources in medical personnel, hospital beds, polyclinics, and other services, but with quantity compromising quality since the epidemiological transition to a predominance of NCDs and changes in the health profile of the population. The system operated as a state monopoly, with the central government controlling budgets, setting mandatory norms, and totally controlling personnel training and research. The system lacked mechanisms for epidemiological or economic analysis and accountability to the public. Medical standards, research, and education were very isolated from the outside world with poor access to literature and professional contacts. The epidemiological transition from predominance of infectious to non-infectious diseases was addressed by further increases in the quantity of services. Policy and funding favored hospitals over ambulatory care and individual routine checkups over community-oriented preventive approaches.

Health expenditures have increased from 2.5 percent of GDP in 1992 to 5.1 percent of GDP in 2010, while other industrialized European countries expend an average of 9.9 percent of GDP on health. The Russian per capita GNI declined from US$3220 in 1991 to US$2410 in 1996, but subsequently increased to US$14,461 (PPP) in 2011. Russia has traditionally maintained a very high hospital bed-to-population ratio, which has been declining since 1990.

After the 1991 breakup of the Soviet Union, the Russian Federation entered a period of political, economic, and social reform with important effects on the national health system and health of the population. In 1993, a compulsory (mandatory) national health insurance (MHI) plan was adopted to augment funding and promote decentralized management of health care and movement towards a market economy in health. The health issues are, however, complex, and changing methods of financing medical care services alone may worsen the health situation by reducing access to care.

The Soviet Model

Before the 1917 revolution, Russia was a largely rural country with higher mortality rates than European countries. Public medical care and other social services for the rural poor majority were established in Czarist Russia in 1864 under the local district assemblies (*Zemstvos*) providing tax-financed services for medical and hospital care. Health insurance was established in 1912 based on the Bismarckian social security model, covering about 20 percent of industrial workers.

Following World War I, the 1917 October Revolution, and the Civil War, Russia was racked by mass epidemics and starvation. In 1918, reconstruction planning included the Soviet concept of health care formulated by Nikolai Semashko, based on the principles of government responsibility for health; universal access to free services; a preventive approach to the "social diseases"; quality professional care; a close relation between science and medical practice; continuity of care between health promotion, treatment of the sick, and rehabilitation; and community participation.

The state undertook to provide free medical services for all, through a governmental unified health system. The "social diseases" referred to all diseases related to the poor living and working conditions of the workers, mainly infectious and occupational diseases as well as maternal and child health problems, and were the focus of special attention and measures of prevention and control. Epidemic control was successfully implemented on an urgent basis, especially for TB, typhoid fever, typhus, malaria, and cholera. Community prevention approaches were enforced, often with use of punishment measures. Prophylactic measures such as quarantine were implemented, urban sanitation and hygiene improved, and malarial swamps drained in the huge territories of the USSR, resulting in the elimination of malaria by 1960.

Medical prevention of social diseases focused on routine checkups for the working population. From the 1920s, emphasis was placed on prevention and control of infectious diseases. In order to meet the needs of the system of providing health care throughout the country, increases in the supply of hospitals, polyclinics, doctors, and nurses were a national priority. In 1937, all insurance and hospital-based Sick Funds were closed, and hospitals and other health facilities nationalized and organized under district health management. Virtually all health personnel became public employees. Parallel services were provided within industries and for special categories, especially party leadership, some ministries, defense and security personnel, miners, workers in heavy industries, and transport workers.

General government revenues provided financing of health services as part of national plans for social and economic development. The central administration directly employed staff, paid salaries, and provided supplies for all health care facilities and research and training institutes. Directors of health facilities therefore administered their allotted resources, supplies, and human resources with no opportunity for program management or internal accounting of service costs. The health system was developed, financed, and managed under strong central government control, with payments based on norms such as for hospital beds and staffing. Mandatory norms for facilities and personnel were enacted by the Commissariat (later Ministry) of Health, under strict regulation of the central authorities of the Communist Party, and later by the Ministry of Finance. These norms were revised periodically at Party Congresses, with expansion of services being the major policy orientation. The policy of continuing to increase the supply of hospital beds and medical personnel was reiterated in the mid-1980s and continued into the 1990s, but has been reduced since 2000.

During World War II, the Soviet health system was mobilized for the war effort, effectively providing care for huge numbers of military and civilian casualties. Some 20 million Soviet military personnel and civilians were killed in World War II. Despite the harsh conditions for both military and civilian populations, no mass epidemics occurred. External observers including Garrison in the 1920s, Sigerist in the 1940s, Field in the 1960s, and more recently Roemer (1991, 1993), as well as Russian medical historians Yeravinski, Smirnov, and others, noted the remarkable achievements of reducing epidemic diseases, meeting wartime demands, and bringing health care to the whole country. Postwar stabilization allowed health services to be restored and trained personnel lost in the conflict to be replaced.

In order to assure equal access, each province (*Oblast*) operates a complete health system including medical institutes for training and for research, laboratories, and specialty services. Each district (*Rayon*) also has a health system with sanitary epidemiological stations (*Sanepid*), hospitals, polyclinics, and specialized treatment facilities according to national norms based on population size. The *Sanepid* supervises water, sewage, air, and ground quality; conducts epidemiological investigations of infectious disease outbreaks; and monitors child health and nutrition status. *Medsanchast* clinics located in industrial plants provide on-site medical and occupational health services, and prophylactic health centers provide a variety of medical rehabilitation services, sanatoria, and vacation benefits. Originally, polyclinics in each district were linked as outreach facilities

to the district hospital with staff rotation between them to promote continuity of care and improve professional education. However, this became impractical because of rapid expansion of the number of polyclinics. Prevention of disease continued to be based on routine screening checkups for workers and other specified groups.

With universal access of the population to preventive and curative care, control of infectious disease was achieved and the health status of the population dramatically improved. A strong system of epidemiological surveillance and control evolved and successfully defended the huge population through the challenges of Russian history of the twentieth century with the social disruption, starvation, migration, mass imprisonment, and executions in the gulags in the 1920s and 1930s. The enormous losses of soldiers and civilians during World War II (13.7 percent of the total population) were followed by the dramatic return to society of millions of prisoners from the *Gulag* prison system after Stalin's death in 1953. During the 1950s, the Soviet model of a state-operated health system was widely promoted and emulated in countries of Eastern Europe, Central Asia, newly independent countries in Africa, Asia, and the Middle East, and in Latin America. It also influenced the development of the Alma-Ata approach of Health for All based on universal access to primary health care.

As a response to the increasing prevalence of NCDs in the mid-1960s, the Communist Party Plenum in 1983 decided to implement annual *Dispanserizatzia* or checkups as a uniform program for the general population, provided in polyclinics, hospitals, and specialized clinics. The checkups and treatment involved clinical care, follow-up ambulatory or hospital care, sanatoria, and a change of work if necessary. The screening program increased demands for hospitalization because of limited ambulatory diagnostic resources, placing the major focus of care on hospitalization and institutional care. In the mid-1980s, the Ministry of Health enunciated the continued direction of health policy as concentrating on "development of preventive medicine and improvement of health care facilities through a program for building general and specialized hospital establishments". With central control of financing, the state set mandatory norms for personnel and hospital beds, and controlled medical education to produce the human resources to operate the system. The state monopoly on health, however, led to stagnation with a bias of the system towards hospital care, without financial or epidemiological accountability for efficiency and effectiveness. The focus on hospitalization and institutional care has begun to change and the per capita acute hospital bed supply has declined since the mid-1980s to under 8 per 1000 by 2005 (Figure 13.9).

In 2005, President Vladimir Putin established priority projects in education, health, housing, and agriculture. Priority health projects were intended to improve the health status of the population, increase accessibility, and improve the

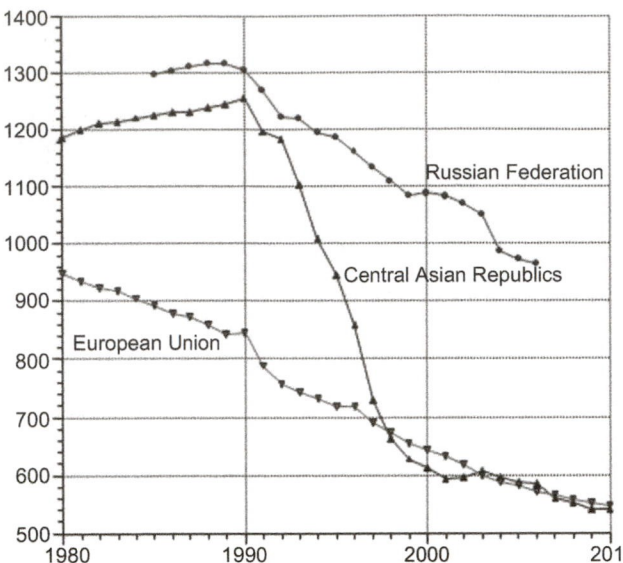

FIGURE 13.9 Acute care hospital beds per 1000 population, Russia, European Union, and Central Asian Republics (CARK), 1980–2010. *Source: World Health Organization, European Region. Health for All database; March 2013. Available at: http://data.euro.who.int/hfadb/*

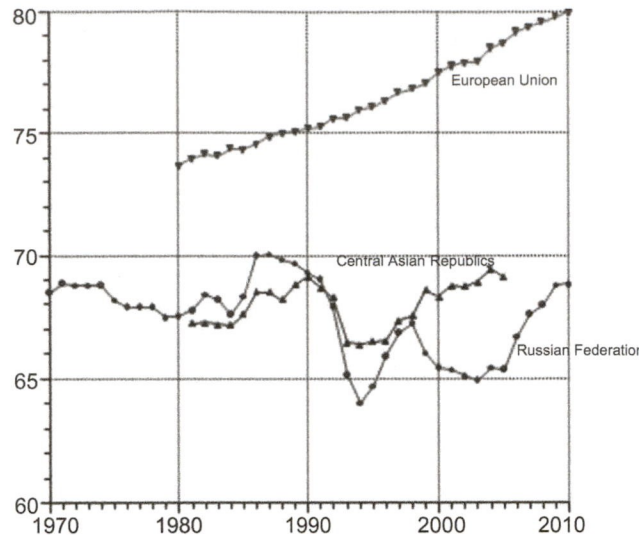

FIGURE 13.10 Life expectancy at birth in years, Russia, European Union, and Central Asian Republics (CARK), 1970–2010. *Source: World Health Organization, European Region. Health for All database; March 2013. Available at: http://data.euro.who.int/hfadb/*

quality of medical care. There was an emphasis on strengthening primary care as well as health promotion and disease prevention activities, and projects to improve accessibility to tertiary care. These included upgrading ambulatory care, additional immunization programs, new check-up programs for infants and pregnant women, and AIDS prevention and treatment. Primary care centers are being re-equipped with cardiograph and ultrasound equipment. Salaries for GPs and nurses have been improved to attract young staff.

Epidemiological Transition

Despite major improvements during the Soviet period (1917–1991), mainly due to control of infectious diseases, the health status of the population dramatically deteriorated in the last quarter of the twentieth century.

Life expectancy improved up to the 1960s, but has since lagged well behind other countries (Figure 13.10). Very high mortality rates from CVDs and trauma are primarily responsible for low and declining life expectancy. CVD mortality is twice as high as in OECD countries, and mortality rates from transport accidents in the Russian Federation were twice as high as in countries of the European Region in 2010. By 2000, life expectancy at birth for males had fallen to less than 55 years, almost 14 years fewer than in 1990. However, there has been a slight increase in life expectancy and decline in mortality in Russia since 2005 (Table 13.12).

Even before the impact of the collapse of the Soviet system was felt in 1991, mortality rates in Russia were much higher than those in other industrialized countries. SMRs in Russia were 1.5 times higher for total mortality, and Table 13.13 shows even higher rates in categories such as

TABLE 13.12 Life Expectancy at Birth, by Gender, Russian Federation, Selected Years, 1970–2011

Year	Males	Females	Total	European Region Total
1970	63.1	73.6	68.8	
1980	61.5	73.1	67.6	
1990	63.8	74.3	69.2	72.0
2000	50.0	72.5	65.3	
2011	63.0	75.0	69.0	76.0

Sources: World Health Organization. World health statistics, 2012 and 2013. Part 3. Statistical indicators. Available at: http://www.who.int/gho/publications/world_health_statistics/EN_WHS2013_Full.pdf [Accessed 18 March 2013].
World Health Organization, European Region. Health for All database; January 2013. Available at: http://data.euro.who.int/hfadb/

cerebrovascular disease, trauma, and infectious diseases, with alcohol binge drinking and violence as major factors.

The crude birth rate declined from 17.2 per 1000 in 1987 to 8.4 per 1000 in 1999, rising to 12.4 in 2010. The total fertility rate declined from 2.0 in 1989 to 1.1 in the period 2000–2005 and then increased to 1.6 in 2010. Infant mortality rates fell from 22 per 1000 in 1980 to 16 in 1998, and 10.0 in 2011, still twice the rates in Western European countries. Abortion is the main method of birth control, and modern methods are not widely available or trusted. Maternal mortality in Russia declined from 68 per 100,000 live births in 1980 to 44 in 1998, and to 34.0 per 100,000 live

TABLE 13.13 Death by Cause, Rates per 100,000 Population, Russian Federation, Selected Years, 1992–2011

	1992	1995	2000	2005	2007	2008	2009	2010	2011
Deaths – all causes	1217	1498	1529	1610	1464	1462	1417	1419	1348
Circulatory system	647	791	846	908	834	836	801	806	749
Neoplasm	202	203	205	201	203	204	207	205	203
Accidents, poisonings and injuries:	173	237	219	221	183	172	158	152	132
Transport injuries	30	26	27	28	28	25	21	20	20
Alcohol poisonings	18	30	26	29	18	17	15	13	8
Drowning	9	14	11	10	9	8	7	8	6
Suicides	31	41	39	32	29	27	27	23	21
Homicides	23	31	28	25	18	17	15	13	12
Respiratory system	58	74	70	66	55	56	56	52	51
Digestive system	33	46	44	66	62	64	63	64	61
Infectious/parasitic diseases	13	21	25	27	24	24	24	24	23

Source: State Committee of the Russian Federation, Statistics (GOSKOMSTAT). Available at: http://www.gks.ru/bgd/regl/b12_12/IssWWW.exe/stg/d01/05-08.htm

TABLE 13.14 Age-Standardized Mortality Rate per 100,000 Population for Selected Causes of Death, Russia and Other Countries, 2008

Country	Non-Communicable Diseases (Total)	Cardiovascular Diseases and Diabetes (Total)	Cancer (Total)	Trauma (Total)
Russia	797	517	180	159
Poland	546	219	219	54
Germany	394	102	150	25
Denmark	440	92	170	53
USA	418	137	143	53
UK	401	91	144	25
Israel	337	72	125	24
Sweden	358	79	121	32
Canada	346	82	138	32
France	336	65	169	38
Japan	273	68	119	36

Note: Data are standardized to the world population.
Source: World Health Organization. World health statistics 2013. Part 3. Statistical indicators. Available at: http://www.who.int/gho/publications/world_health_statistics/EN_WHS2013_Full.pdf [Accessed 18 March 2013].

births in 2010, compared with rates under 15 in the industrialized countries (WHO, 2013).

The decline in health status since 1990 cannot be blamed solely on the current economic crisis, or entirely on the health care system. The worsening mortality pattern is due to a combination of factors: stress, alcohol abuse, smoking, violence, lack of a balanced diet, lack of modern health care technology, environmental pollution, and a general mood of anxiety and depression related to the dramatic decline in economic and political stability since 1990 (Tables 13.13 and 13.14). On average, males are surviving only slightly beyond pension age at 60 years.

A combination of factors encouraged a medical bias towards care of individual patients and failure to apply the successful experience of the 1930s to the control of epidemics of NCDs. The concept of prevention took on a primarily medical orientation, stressing routine checkups. Health policy continued to promote increased supplies of doctors, polyclinics, and an emphasis on hospital beds. The number of medical graduates per 100,000 population was declining

TABLE 13.15 Human Resources for Health (Selected) per 10,000 Population, Selected Countries and Years, 2005–2010

	Physicians	Midwives and Nurses	Dentists	Pharmacists
Canada	19.8	104.3	12.6	9.2
China	14.2	13.8	0.4	2.5
Israel	35.6	51.8	8.8	6.7
Japan	21.4	41.4	7.4	13.6
Russian Federation	43.1	85.2	3.2	80.5
UK	27.4	101.3	5.3	6.6
USA	24.2	98.2	–	–

Source: World Health Organization. World health statistics 2013. Part 3. Statistical indicators. Available at: http://www.who.int/gho/publications/world_health_statistics/EN_WHS2013_Full.pdf [Accessed 18 March 2013].

but has been stable since 2000. Table 13.15 shows a comparison of human resources in Russia to other countries.

A hospital sector with a passive strategy of treatment and long hospital stays was unable to keep up with technological advances and consumed a large share of the very limited amount of funds allocated to health care.

Post-Soviet Reform

The Russian Federation continues to provide basic social security and health care for all citizens. Until 1993, when compulsory health insurance was established, all social benefits were funded from the general budget of the government. The health insurance scheme was based on mandatory payment by employers to regional health insurance funds.

Since the early 1990s, the World Bank has played a significant role in influencing policy makers in Russia promoting health insurance to partially replace state funding of health services, as well as decentralization and privatization of health and social services. Decentralization of management of most state health services and financing of health care increased regional and local health autonomy. The sudden and nearly complete decentralization of authority and funding hampered central management by the Ministry of Health in its capacity to develop new policies for public health issues. Central management of the sanitary epidemiological service was hampered by limited funding to expand the immunization program or to promote nutrition and other health promotion initiatives.

Decentralization of compulsory health insurance through regional systems allowed local change in health management issues and shifting from obsolescent inflated national norms for hospital beds and personnel. Epidemiological, economic, and cost-effectiveness analysis is vital to reform in health care, especially in harsh economic conditions. Calculation of the cost of a service is fundamental. Regional and municipal authorities now have more financial responsibility and power to reallocate funds and shift priorities from institutional treatment to prevention and ambulatory care, but lack trained health management personnel to challenge old assumptions, such as the norms for hospital beds and human resources, still used as guidelines.

The deteriorating health situation is a part of a health care system in decline. The structure continued to focus on inpatient services, with less attention to ambulatory and primary health care, disease prevention, and health promotion. The per capita hospital bed supply and average length of stay are much greater than in the EU. In addition, public funding of health care has declined considerably, and the collection of informal user fees by public health providers has reduced the access of the poor to health care. Public sector funding for health care is through the federal, regional, and municipal levels and the 3.1 percent payroll tax. International reports, including by those of the WHO, OECD, and United Nations (UN), show that the federal level allocates 3.2 percent of expenditures to health.

Development of information systems, training of leadership personnel in modern management theory and practice, and reduction of the hospital sector with transfer of resources to primary care are needed to improve health care efficiency and quality. Access to international literature and the Internet will help to improve the quality of continuing education in the health sector. Reforms based, in part, on reallocation of existing resources, will require additional funding to meet the cost of the transition and raise the quality of care.

Health reforms are essential to preserve universal access and to raise population health status, medical care, and public health to international standards. Changes in financing of health care, adoption of international health targets, and changes in workforce development programs are needed. But these depend on a new set of priorities and new standards at the national, oblast (province/state in Russia), and local health authority levels of government. Decentralization and diffusion of the overly centralized system require epidemiological information and dialogue on health issues to raise health awareness and management practices to meet the health needs of the people.

The sanitary epidemiological stations are a force with the potential to expand their traditional roles to lead health promotion activities at the community and district levels, raising public awareness and knowledge of health issues. This will mean a change in attitude from being defenders of the old system to responders to community needs. That means redefining objectives, instituting training programs for new personnel, modernizing technology for laboratories, and environmental quality and enforcement issues.

Training for modern behavioral epidemiological data collection, analysis, and distribution is essential to promote knowledge of risk factors and their control. Policy, provider, and community levels need to define cost-effective programs to meet local conditions. Wide distribution of relevant data to government and the general public is needed to help change knowledge, attitudes, beliefs, and practices related to risk factors.

Future Prospects

The Russian health system has important assets with potential for change. Health expenditures per capita, traditionally low in the USSR, but despite a modest increase in 2009-2010 to just over 6 percent, remain well below western country standards, and declined in 2011. Allocation of resources will need to represent a reordering of priorities and allocating funds to promote primary care, upgrading salaries and equipment, and revision of the role of the existing polyclinic system. A system of grants to local authorities specifically directed towards strengthening primary care and health promotion would help to downsize and upgrade the hospital sector in terms of equipment and upgrade primary care, with the government of Russia developing new approaches towards reallocating rising expenditures and more effective management of financing the health system. There is a multichanneled system of financing in place, based on the state budgets, the compulsory insurance system, household expenditures, services by other governmental and industrial complexes, and voluntary insurance.

Bureaucratic segmentation of services presents a formidable barrier to reform. Privatization is not a solution and will not help to reduce the current burden of excess mortality in the Russian Federation. Instead, a reform process should build on the main existing structure of sanitary epidemiological stations and polyclinics but with major revision of content and quality standards.

OECD recommendations (2012) for reform in the Russian health system include the following: "The first objective concerns ensuring that patients can access the care that they need under the Government Guarantee Package on a timely basis. The second concerns the quality of care and whether it is adapted to patient needs. The third key goal concerns the resources allocated to the public health care system and whether this is sustainable over the longer haul. The final key issue concerns the scope for easing any overall resource constraints on the public health care system through improved efficiency of health care provision."

Necessary health reforms include major refocusing on a number of key issues:

- preserving universal access to health care for the population
- control of privatization in health care

- health promotion regarding smoking reduction, alcohol abuse, nutrition, physical activity, trauma prevention, and NCDs
- sustained increased level of funding for health
- pooling of regional health budgets and health insurance
- national standards and guidelines for a regional "basket of services" for all
- replacement of obsolescent financing norms of the Ministry of Finance
- reduced hospital bed supply with upgrading of hospital quality and greater emphasis on ambulatory and home care
- development of polyclinics with preventive and curative services for defined populations, with capitation funding and incentives for improved quality and efficiency of services
- financing patient care on a capitation basis
- control of corruption and under-the-table payments
- raising standards of care and quality of services
- increased contact with the international community through health literature and professional meetings.

Summary

The Soviet health system brought health care to a vast, underdeveloped rural country. This system provided universal access, within a totally state-operated system of service. It was a source of pride to the Soviet state, and was recognized internationally as an important model because of its successes from the 1930s to the 1960s. During the 1950s, the Soviet model, a state-operated health system, was widely promoted and emulated in Eastern Europe and in newly independent countries in Africa, Asia, and the Middle East, as well as in Latin America. This model also influenced development of the Alma-Ata approach of Health for All based on universal access to primary health care.

From the 1960s to the 1990s, an epidemiological transition occurred in varying degrees in the different republics and ethnic populations of the Soviet Union. This transition was characterized by sharply declining mortality from infectious diseases and rising death rates from non-infectious diseases. Life expectancy remained static during the 1970s and 1980s. In the 1990s, life expectancy declined dramatically, especially for men, during the economic and social crisis following the breakup of the Soviet Union.

This crisis in health not only related to the period of economic transition in the 1990s, but went deep into the former Soviet health system with quantity compromising quality since the epidemiological transition and changes in health profile of the population. The system operated as a state monopoly, with the central government in control. It lacked mechanisms for epidemiological or economic analysis and accountability to the public. The epidemiological transition from predominance of infectious to non-infectious diseases was addressed by further increases in the quantity of services. Policy and funding favored hospitals. Individual health was deeply affected by stress associated with great uncertainty, economic collapse, and the breakdown of social safety nets. Levels of alcohol consumption, homicide, and suicide in Russia are among the highest in the world. The challenges of health promotion and adequate prevention and treatment are not met by the existing health system. A decrease in alcohol consumption in Russia is critical for the long-term improvement in the demographic crisis.

Reform since 1991 has centered on compulsory national health insurance and decentralized management of services. In order to free resources to address health needs more effectively, reforms aimed at rationalizing the health care delivery system are needed. However, the reform movement was lacking a broad national health strategy to address the fundamental public health problems and especially the present enormous excess of preventable mortality. A new national health project reported 138,000 fewer deaths in 2006 compared to 2005 and this decline is attributed to new national initiatives in social and health policy in the Russian Federation. Rising national income and standards of living in recent years will foster this improvement, but structural and content reform of Russia's health system is important to reduce the continuing dreadful toll of preventable deaths in the country.

A World Bank and WHO joint report of 2007 pointed out the urgent need for increased funding, and for fundamental reforms of the Russian health system. The system is still primarily based on the Semashko model with the addition of national compulsory health insurance. The reforms suggested are wide in scope, from primary care to hospital reformation. The wide gaps between Russia and other former socialist countries, and especially countries of Western Europe, are primarily in the sector of NCDs and injuries which share common risk factors, underlying social and cultural determinants and opportunities for intervention. These are, in particular: high blood pressure, tobacco consumption, alcohol binge drinking, obesity, and especially low fruit and vegetable consumption and physical inactivity. These factors account for most of the disease burden, with highest rates among the poor, rural, and vulnerable. Improving health for the people of the CIS countries requires an emphasis on the detection and management of hypertension and risk factor reduction. Increased funding in health care should be directed to primary care and the hospital sector should be downsized and upgraded to adapt to the underlying determinants and risk factors of NCDs and injuries. Since 2006 there have been trends towards improvement but dedicated reform is needed to close the gap with other countries in Europe.

ISRAEL

Israel had a population of 7.8 million in 2011 (Israel Central Bureau of Statistics). The GDP per capita rose from US$23,340 in 2005 to US$28,809 in 2011, as compared to US$30,272 and US$35,831 for the countries of Western Europe. Health indicators for Israel show an advanced state of health but with important ethnic, regional, and gender inequalities. Life expectancy at birth was 82.0 years in 2011 (fifth in the OECD). Israel's ranking in the 2012 HDI of 0.900 was 16th. Infant mortality in 2011 was 4.0 per 1000 live births (18th out of 34 OECD countries), and maternal mortality was 4–7 per 100,000 (2010), similar to the median in the OECD.

In terms of NCD mortality among the OECD countries, Israel recorded the seventh lowest rate of ischemic heart disease, the lowest stroke mortality rate, and the second lowest male cancer mortality rate, but ranked only 18th out of 34 with respect to female cancer mortality. On the other hand, it was only 28th out of 40 in cancer incidence. Israel ranked in the upper third of countries with the lowest transport mortality rate (2009 or nearest year data). Total health expenditures as a percentage of GDP were 7.7 percent in 2011, with 61.5 percent coming from the earmarked health tax and other public sources in 2011 (Table 13.16).

Immunization rates are high, with 2011 rates of 98 percent for DTP, 95 percent for polio, and 99 percent for MMR. The content of the publicly funded program has been gradually expanded to include hepatitis A and B, Hib, pneumococcal pneumonia, varicella, rotavirus, and influenza vaccines for children. The human papillomavirus (HPV) vaccine for 12-year-old girls is the latest vaccine to be funded. Hib and hepatitis B (three doses) immunization rates were 93 and 99 percent, respectively, in 2011.

Despite high immunization rates, an epidemic of polio with 15 cases occurred in 1988, and measles epidemics in 1991, 1994, and 2007–2008, leading to the adoption of improved immunization policies. In 2013, wild poliovirus was identified in sewage in several parts of southern Israel. The immunization rate is over 90 percent with inactivated polio vaccine (IPV), but the previous system of combined oral poliomyelitis vaccine (OPV) and IPV has been reintroduced (Israel Ministry of Health, 2013).

Origins of the Israeli Health System

Israel's health system evolved gradually over the past century. Palestine under the Ottoman Turkish Empire was a poor, disease-ridden, remote province rife with malaria, dysentery, and other infectious diseases. Immigration of Jews from Eastern Europe and Arabs from surrounding countries since the 1880s led to the initiation of charitable hospitals to provide care for the urban poor.

Jewish immigrants from Eastern Europe formed labor brigades and mutual aid associations. Sick Funds were initiated in 1912 based on mutual benefit principles derived from European models, associated with the union movement, and later with other political organizations. The Sick Funds grew to provide medical care insurance and services to over

TABLE 13.16 Health Expenditures, Hospital Resources and Utilization, Israel, 1970–2011

Resources/Utilization	1970	1980	1990	2000	2005	2011
Acute care beds/1000 population	3.2	3.0	2.6	2.3	2.1	2.0
Hospitalization (acute) days/1000 population/year	1148	997	834	785	761	NA
Discharges (acute)/1000 population/year	129	139	156	155	155	NA
Average length of stay (acute care) (days)	8.9	6.8	5.3	4.3	4.2	4.0[a]
Mental health beds/1000 population	2.4	2.2	1.5	0.9	0.8	0.5
Mental diseases days/1000 population	631	721	496	379	220	NA
Nursing and elderly beds/1000 population	NA	1.4	2.0	2.9	3.1	2.4

Note: [a]2009 data.
NA = not available.
Sources: Rosen B, Samual H, Merkur S. Israel: health system review. Health Syst Transit 2009;11(2):1–226. European Observatory on Health Systems and Policies. Available at: http://www.euro.who.int/__data/assets/pdf_file/0007/85435/E92608.pdf [Accessed 21 July 2013].
World Health Organization, European Region. Health for All database; July and August 2012. Available at: http://www.euro.who.int/en/what-we-do/data-and-evidence/databases/european-health-for-all-database-hfa-db2 [Accessed 20 July 2013].

95 percent of the population. They provide services through neighborhood and specialized clinics, or affiliated doctors in their own clinics, purchasing hospital care from government or NGO-operated hospitals in areas where they lack their own.

Preventive care originated in 1911 by nurses from the USA sponsored by Hadassah, an international women's organization. Following the conquest of the area by British forces from the Turks in 1917, Hadassah sent the American Zionist Medical Unit from the USA to help establish a network of health facilities in Palestine. This consisted of 44 doctors, nurses, dentists, and other personnel with equipment and financial support from Hadassah and the Joint Distribution Committee. The unit opened hospitals in many urban centers, and established nursing training and preventive care programs for immigrants and schoolchildren, as well as mother and child health stations (*Tipot Halav* or "drop of milk" stations). These were gradually located in towns, villages, and neighborhoods throughout the country, providing prenatal care and child care for infants and toddlers. They provided immunization, child development monitoring, and nutrition counseling to almost all the infants in the country, and prenatal care for most women in the country, with others going to private doctors.

The British Mandate from 1917 to 1948 brought successful colonial administrative experience and development of basic public health law and systems, licensing of medical professions, sanitation, food and drug laws, as well as public health laboratories, malaria control, and many other features of public health standards of the time.

From 1912 to 1948, the health system grew based on primary health care through the *Tipot Halav* and the labor movement's Sick Fund clinics in towns and villages throughout the country, providing ready access to primary care treatment and referral services.

Following the establishment of the State of Israel in 1948, massive immigration from post-Holocaust Europe and the Middle East brought an enormous burden of health problems to the country. The new Ministry of Health established regional hospitals throughout the country in abandoned British army camps, providing acute care, rehabilitative, mental health, and long-term care services. Other hospitals are owned by the major Sick Funds and by NGOs. Reliance on ambulatory and primary care with regional medical and hospital centers is the basis of the Israeli health system.

Health Resources and Expenditures

Israel spent US$2185 per capita on health in 2009, compared to the OECD average of US$3233. The rate of growth in health expenditures was one of the lowest in the OECD countries, at 1.5 percent annual average over the period 2000–2009, compared to the OECD annual average of 4.0 percent. Expenditures on hospital care increased from 34 to 41 percent of total health expenditures from 1975 to 1990, dropping to 35 percent in 2000 and to 34 percent in 2008 (Table 13.17).

Health care expenditures as a percentage of GDP increased from 7.9 percent in 1990 to 9.3 percent in 2002,

TABLE 13.17 Health Expenditures (Percent) by Type of Service, Israel, 1985–2009

Category/Year	1985	1990	1995	2000	2005	2009/10
Population (millions)	4.23	4.66	5.55	6.29	6.93	7.63
Life expectancy at birth (years)	75.4	76.8	77.5	79.0	80.2	82.1
Total expenditure on health per capita (US$ PPP)	784	1028	1435	1765	1829	2081
Total health expenditures (% of GDP) (WHO estimate)	NA	NA	7.4	7.4	7.7	7.6
Hospital costs (% of total health expenditures)	42.8	39.9	39.4	35.4	34.0	33.1
Public clinics/prevention	32.5	32.6	34.5	38.5	41.2	NA
Other	24.7	26.7	24.9	23.6	20.4	NA
Total health expenditures	100	100	100	100	100	100
Public sector health expenditures (% of total health expenditures)	NA	NA	69.2	64.0	60.5	60.3
Social security (% of total government expenditure)	NA	NA	47.1	48.5	50.0	NA

Note: PPP=purchasing power per capita; GDP=gross domestic product; NA=not available.
Sources: World Health Organization, European Region. Health for All database; August 2012. Available at: http://www.euro.who.int/en/what-we-do/data-and-evidence/databases/european-health-for-all-database-hfa-db2 [Accessed 20 July 2013].
Organisation for Economic Co-operation and Development. OECD health data 2012. How does Israel compare? Available at: http://www.oecd.org/els/healthpoliciesanddata/BriefingNoteISRAEL2012.pdf [Accessed 1 February 2013].
Organisation for Economic Co-operation and Development. OECD health data 2013. Frequently requested data. Available at: http://www.oecd.org/els/health-systems/oecdhealthdata2013-frequentlyrequesteddata.htm [Accessed 31 July 2013].

and declined to 7.7 percent in 2011. Acute care hospital beds per 1000 population were reduced in Israel from 3.0 in 1980 to 2.6 in 1990, 2.3 in 2000, and 2.0 in 2011. During this period, psychiatric beds were reduced from 2.2 per 1000 in 1980 to 1.5 in 1990, 0.9 in 2000, and 0.5 in 2011. Nursing and elderly care beds increased from 1.4 per 1000 in 1980 to 2.0 in 1990, 2.9 in 2000, and 2.4 in 2011 (Tables 13.16 and 13.17).

Along with the decline in acute care beds, average length of hospital stay (acute care) fell from 6.8 days in 1980 to 5.8 days in 2009, and bed occupancy increased from 90 percent to 96 percent, which was the highest among the OECD-25 countries. Ambulatory care and community health consume about 40 percent of total expenditures, an increase since 1975. Salaries in the health sector have been low compared to other sectors in the society, as has been capital investment. Physicians and nurses have successfully negotiated for salary increases during the 2011–2012 period. The physician–patient ratio has declined over recent years. Various initiatives have been developed to increase this ratio. These include the opening of a fifth medical school, developing a post-baccalaureate 4-year medical school program to complement the existing postsecondary education 6-year programs, and working to ease the certification of foreign-trained Israeli physicians. Cost restraint, improving the physical infrastructure, and keeping up with technological advances in medicine are major challenges for the future.

Health Reforms

After many years of debate, several national commissions on health, and gradual reform of health services, Israel's national health insurance (NHI) plan was implemented on 1 January 1995. It covers the total population through the universal National Insurance social security system. The individual pays for this through a 3 percent deduction from his or her salary along with an equivalent employer's contribution to a mandatory NHI program, which also covers old age and disability pensions, workers' compensation, and other social benefits. Each family must select membership in a Sick Fund which functions as a health maintenance organization. Each individual is entitled to change Sick Funds semi-annually. The National Insurance Institute transfers funds to the Sick Fund and HMOs according to a per capita formula, with a larger per capita payment for the elderly and for populations in the periphery of the country.

The Ministry of Health supervises the Sick Funds, which are required by the 1995 NHI law to provide a basic basket of services that is very comprehensive. This basket is updated on an annual basis by a multispecialty committee through a comprehensive prioritization method. The Sick Funds are obliged to provide all specified services or to arrange for those services that they cannot provide. They provide comprehensive care, either through their own neighborhood clinics or through affiliated private physicians who are paid on a capitation basis. Additions to the obligatory basket of services, such as new medications and diagnostic tests, are made by a multiprofessional team coordinated by the Ministry of Health on an annual basis. Co-payments for specialists, diagnostic testing, and pharmaceuticals are a source of cost-sharing. People receiving welfare do not pay co-payments and the elderly population pays 50 percent of the quarterly ceilings. Attendance at well-baby clinics is free. However, there are no subsidies to low-income groups, resulting in inequalities in access to health care.

The Sick Funds are accountable for the services rendered. Most hospital beds are operated directly by the Ministry of Health, although several large hospitals are run by the largest Sick Fund, Clalit Health Services. The government wishes to transfer government hospitals to independent trusts, to operate as economic units able to allocate funds internally and compete for clients, with payment on a DRG basis, but various economic and other considerations have delayed this transition. Regionalization of services will be difficult to achieve in the present configuration of the NHI law because each Sick Fund has its own regional organization.

Health promotion is gaining strength in Israel, and health awareness has generally increased (Table 13.18). The compulsory seat belt law has met with compliance by a large majority of car drivers, and similar legislation requiring use of helmets for motorcycle drivers is also generally implemented. Similar requirements were passed in 2007 for bicycle users. Increases in permitted speed limits on major highways have been followed by a rise in motor vehicle deaths and case fatality rates. Studies by the Israeli Road Safety Authority show widespread non-adherence of drivers with speed limits. On the positive side, the creation of speed-reducing roundabouts/traffic circles and increased police enforcement have led to traffic calming and increased safety in cities such that, on the whole, Israel has reduced fatal motor vehicle accidents by nearly 50 percent over the past two decades, from 525 in 1990 to 287 in 2012.

Non-smoking legislation banning smoking from public buildings and workplaces has helped to reduce smoking in the adult population, especially among males. As of 2011, 20.6 percent of adults smoked, which is a substantial reduction from the 34 percent rate in 1991. In 2007, smoking was banned by law in bars, restaurants, and cafés, with the owners held responsible and fined for violations. This was expanded in 2012 to include several other locations, such as bus and train stops/platforms, swimming pools, cultural/entertainment performances and others, and compliance is reportedly good. In 2012 legislation was passed to increase cigarette and nargilah (water-pipe) tobacco prices, and proposals were introduced in the Knesset to limit tobacco product advertising and to increase the graphic content of anticigarette information on cigarette containers.

Low-fat foods are now commonly available in supermarkets. Private food manufacturers fortify baby formula and cereals with vitamins and minerals. Breakfast cereals are also enriched, but food fortification of bread, milk, and salt with essential minerals and vitamins is not practiced. The Ministry of Health, in collaboration with food manufacturers, is working to reduce the salt content of processed foods, as well as to reduce the price of wholegrain bread. Another initiative aims at eliminating advertising of unhealthful foods ("junk food") on television during children's peak viewing hours. Nearly 35 percent of the population is overweight and roughly 15 percent obese. This is especially so in Arab women. While consciousness of the importance of physical fitness is increasing, it is still not at an acceptable level. Only 8 percent of adults meet the recommendations for weekly physical activity of at least 150 minutes per week of moderate physical activity or 75 minutes of intensive physical activity. In 2011, the Ministries of Health and of Finance authorized an incentive program to encourage improved assessment and counseling of the overweight and obese in the Sick Funds. Legislation to increase the physical activity infrastructure and ease requirements for medical certification for all, prior to exercising in health clubs has been proposed.

Mortality from stroke and CHD has declined dramatically over the past three decades, largely as a result of improved treatment of hypertension and myocardial events, but also because of a decline in smoking and greater interest in self-help to maintain health. CVDs have now fallen below cancer as leading causes of mortality, although prevalence rates remain high. Obesity and diabetes are growing as national health problems.

In terms of mortality amenable to prevention compared to rates in 20 European countries, Israel ranked eighth lowest for males and twelfth lowest for females in 2008. However, Israel ranked higher in the decrease in amenable mortality rates between 2001 and 2007 for females than males in a 19-country comparison (Goldberger and Haklai, 2012).

Regional, social, and ethnic disparities are still important in Israel's health status; the Arab population has higher rates of infant mortality than the Jewish population, 6.9 versus 2.7 per 1000 in 2010. Large-scale immigration of Russians and Ethiopians since the 1990s brings together people with different risk factors. This trend has continued, albeit on a more modest scale, with nearly 17,000 immigrants coming to Israel in 2011. The traditional distribution of health resources favors the more concentrated population centers, while the more rural areas receive fewer resources per capita.

The health agenda has paid increasing attention to health promotion as well as to the structure of health services in health reform. The Ministry of Health has concentrated in the past on reform in health services and NHI, but has expanded the breadth of its vision with development of the multiyear "Pillar of Fire" strategy, whose goals include the

TABLE 13.18 Health Promotion Initiatives in Israel

Topic	Action	Effects
Smoking	Non-smoking legislation; restricted advertising; non-smoking promotion by NGOs	Increased awareness of health effects of smoking; ban on smoking in public places, extended in 2007 to include bars and restaurants with responsibility of owners, and in 2012 to include certain open areas such as bus and train platforms
Cancer prevention	Promoting mammography and colon cancer screening, reduced sun exposure; restricting smoking, encouraging physical activity	Improving public awareness and adherence with screening tests; overall cancer incidence stable in Jewish population but increasing in Arab population, partly due to improved availability and awareness of screening programs, higher than OECD average, relatively high incidence of breast cancer
Nutritional health education	Mediterranean diet; high fruit and vegetable intake; low consumption of animal fats; incentivization of Sick Fund clinical teams to screen for and counsel obesity, especially in pediatric population	Increasing public awareness of healthy nutrition contributes to rapidly declining cardiovascular mortality and low rates of cancer; further efforts needed to slow and reverse increasing obesity rates
Motor vehicle accidents	Mandatory seat belt use implemented; highway speed limits raised; adequate highway patrols; urban roundabouts, speed bumps, improved highway infrastructure, increased interurban speed cameras	Improved emergency care and transport; police activity in driver licensing, road monitoring; electronic monitoring improved; overall fatality rates have declined substantially
Water quality	Mandatory chlorination since 1988; filtration plan operational for main surface water source from 2006	Less waterborne diarrheal disease
Sewage	Increasing treatment	Reuse of wastewater increasing; now >75 percent recycled for agriculture
Enrichment of basic foods	Law permits but does not require private manufacturers' initiatives	Breakfast cereals enriched; infant formulae and cereals enriched; basic foods (bread, salt, milk) not enriched
Food quality	Food supervision strengthened; standards at international levels; clearer and more comprehensive labeling to be implemented	Public awareness increased; low-fat foods now widely available; salt reduction campaign underway
AIDS/STI prevention	School health education, free confidential testing, AIDS hotline, campaigns for the general public and gay populations, availability of rapid combination antibody–antigen tests	Widespread information on use of condoms and avoidance of transmission in drug use, stable annual incidence – lower than most Western European and North American countries
Preventive health care	Universal health insurance; Sick Funds become health maintenance organizations; health professional schools increase preventive curricula	Prevention increasing in primary care for all ages; national quality indicators in community health care stress disease prevention, computerized medical records, and regularly updated clinical guidelines
Healthy Cities	Active association and networking of healthy cities, incorporation in National Active and Healthy Lifestyle Program	31 cities and regional councils have Healthy Cities programs; health profiles and sustainable strategy
Health promotion	MOH and Sick Funds act to raise professional and public consciousness of health and lifestyle, MOH incentivizes Sick Funds to increase the number of health promotion personnel	Growing public consciousness of diet, fitness, and smoking as health factors
Healthy aging	Community centers for elderly, quality markers and updated guidelines for lay and clinical prevention	Municipal and NGO sponsorship of activity centers and programs
Healthy Israel 2020	Define health targets for 2020 with measurable indicators and recommend evidence-based interventions	Increased awareness by health professionals, stimuli for translational research, development of multiorganizational implementation efforts

Note: AIDS = acquired immunodeficiency syndrome; STI = sexually transmitted infection; NGO = non-governmental organization; MOH = Ministry of Health; OECD = Organisation for Economic Co-operation and Development.
Source: Adapted from Israel Center for Disease Control. Health status in Israel, 1999; and Donchin M, Shemesh AA, Horowitz P, Daoud N. Implementation of the Healthy Cities' principles and strategies: an evaluation of the Israel Healthy Cities network. Health Promot Int 2006;21:266–73.
Updated from relevant Ministry of Health, Healthy Cities, Road Safety Authority, and OECD websites: http://www.who.int/gho/publications/world_health_statistics/EN_WHS2013_Full.pdf and Rosenberg E, Lev B, Bin-Nun G, McKee M, Rosen L. Healthy Israel 2020: a visionary national health targeting initiative. Public Health 2008;122:1217–25.

strengthening of public health topics such as health promotion and disease prevention. Reducing regional disparities in health resources and health status is another important goal of the Ministry of Health and is being addressed proactively by the Sick Funds.

Mental Health

Mental health reform has been a controversial subject since the establishment of NHI. In 2006, the Ministry of Finance and the Ministry of Health agreed to transfer mental health care to the Sick Funds. The Sick Funds will be responsible for including mental health care in the basic basket of services. A debate between various stakeholders delayed the process, but in 2012 the economic cabinet of the Knesset voted to implement the transfer and it will be completed by 2015.

Healthy Israel 2020

The *Healthy Israel 2020* initiative was created by the Ministry of Health to define Israeli policy in the areas of disease prevention and health promotion (Box 13.5 and Table 13.18). It has established and prioritized objectives, quantitative targets, and evidence-based interventional strategies necessary to improve health and reduce health inequalities. The initiative is similar Healthy People 2020 of the US DHHS and has served as one of the templates for the WHO's *Health 2020*.

The initiative involves collaboration between a broad spectrum of individuals and organizations, including representatives of government ministries, health care organizations (the Sick Funds or "Kupot Holim"), academic researchers, local government, NGOs, and the Knesset. An international panel of experts provided consultation to each committee. Twenty focus areas were established,

---

BOX 13.5 Healthy Israel 2020 Initiative

Healthy Israel 2020 (HI2020) is a national health targeting initiative coordinated by the Israeli Ministry of Health. It was conceived in 2005 along the lines of other international efforts such as Healthy People 2010 of the US Department of Health and Human Services. HI2020 is meant to provide a preventive health blueprint for the country to improve life expectancy and quality of life, and at the same time to reduce health inequalities.

Twenty broad domains were chosen, which were further subdivided into specific topics, bringing the total number to 30. These included health determinants such as lifestyle behaviors, nutrition, injury and violence prevention, and enhancement of occupational and environmental health; health states such as oral health, mental health, non-communicable and infectious disease prevention; and age-related topics such as maternal and child health and geriatric health. Infrastructure topics such as education and training of the preventive workforce, as well as data development, and utilitarian/implementation topics such as health communications and marketing were also addressed. Committee members were selected from relevant government ministries, the academic community, the four health maintenance organizations or "Sick Funds", and non-governmental organizations. Committees were asked to generate reports containing an epidemiological overview describing the health and economic burden of their respective topics, craft health objectives and target values to reach by the year 2020, and to prioritize evidence-based interventions to achieve them.

The first reports, published in 2011, dealt with three main lifestyle health behaviors: physical activity enhancement, obesity control, and healthful nutrition. These served as the basis for a broad implementation program entitled the National Healthy Lifestyle Program (NHLP) led by the Ministry of Health in partnership with the Ministry of Education and the Ministry of Culture and Sports. The NHLP focuses on legislative initiatives, implementation efforts in a variety of healthy cities, and

incentive packages for health care organizations to enhance preventive screening and counseling by clinic personnel countrywide. It was adopted by the Israeli government in late 2011.

Expert workshops are another means of implementing committee recommendations. To date, workshops have been held on topics such as curbing excessive alcohol use, geriatric health, and alertness enhancement, to develop specific, real-world recommendations. The introduction of Internet-based, interactive information is planned in 2013 for both health professionals and the lay public.

The HI2020 initiative has received international recognition as one of the templates for the Health 2020 of the European office of the World Health Organization. That said, various challenges loom on the horizon: Can implementation efforts already underway stay true to the guiding recommendations? How should resource appropriation be optimized among the large number of current and future recommendations? Will governmental and municipal funding for the interventions be sustained? How will new interventional, health economic, and health services research findings be integrated to upgrade existing programs? Proven Ministry commitment to evidence-based public health to date augurs well for the continued integration of this approach in developing government health promotion policies.

Sources: *Elliot Rosenberg MD, MPH, National Coordinator, Healthy Israel 2020 and Director, Department of Occupational Health, Israel Ministry of Health. Personal communication; 2013.*
Rosenberg E, Lev B, Bin-Nun G, McKee M, Rosen L. Healthy Israel 2020: a visionary national health targeting initiative. Public Health 2008;122:1217–25.
Rosen L, Rosenberg E, McKee M, Gan-Noy S, Levin D, Mayshar E, et al. Healthy Israel 2020 Tobacco Control Subcommittee. A framework for developing an evidence-based, comprehensive tobacco control program. Health Res Policy Syst 2010;8:17.
Ginsberg GM, Rosenberg E. Economic effects of interventions to reduce obesity in Israel. Isr J Health Policy Res 2012;1:17.

on topics ranging from health behaviors such as obesity control and healthful nutrition to injury prevention and preventive health education. Interventions from the Healthy Israel 2020 report on tobacco control helped to frame new national legislation spearheaded by the Ministry of Health. Another large-scale implementation program is the National Program for an Active and Healthy Lifestyle, which is a tri-ministerial effort (together with the Ministries of Education and Culture and Sports), focusing on legislative, clinical, and community-oriented interventions. Over the coming years, it is expected that many such initiatives will be created using the scientific framework developed by Healthy Israel 2020 and should be expected to expand their reach via facilitation by increasingly accessible and sophisticated social media.

Quality Assurance

In 1995, Israel implemented its NHI law providing a universal coverage standardized basket of medical services for all residents of the country through four Sick Fund health plans (Box 13.6). The law specifies that health care should reflect "justice, equality, and mutual assistance", with medical services provided on a timely basis at reasonable quality as close as possible to the insured person's home. The Ministry of Health supervises implementation of the law and external organizations for the purposes of evaluating the effect of the law on health services' quality, efficiency, and expenditure.

Summary

Israel has achieved high standards of health care and health status indicators. The Israeli health system has been a quasi-national health service for many decades, with over 95 percent coverage through Sick Funds. The NHI in 1995 brought universal coverage to all residents of the country. The health system has helped the Israeli population to achieve low rates of mortality from infectious and non-infectious diseases and life expectancies among the highest in the world.

Medical and paramedical professional education, research, and the medical and drug industries have reached high levels of excellence. The 1995 implementation of NHI through Social Insurance provides greater equity in financing and reduces political manipulation of the health system. Primary care services still separate community-based preventive and treatment facilities, but the Sick Funds have become increasingly prevention oriented with improved standards of primary care. New Ministry of Health initiatives, including Healthy Israel 2020, and the Pillar of Fire strategy, as well as other national programs such as the national quality marker in community care project, promise to reduce further the disease and mortality burden and to reduce regional

BOX 13.6 Quality Indicators for Community Health in Israel

In 2004, the Ministry established the National Program for Quality Indicators in Community Healthcare (QICH), begun as a joint research project with Ben-Gurion University and Israel's four health plans; since 2010, the project has been under the direction of the Braun School of Public Health (Hebrew University, Jerusalem). The program provides annual reports of a national set of quality indicators for community health care (available at: http://healthindicators.ekmd.huji.ac.il).

The purpose is to evaluate community-based medical care in Israel, and the variations in quality of care between subgroups in the population. The indicators collected are population measures to enable evaluation of the development of quality medical care, identification of areas that require intervention, improved data collection, effectiveness of care, and comparison of Israel's indicators to those of other countries. Indicators are based on a consensus of Israel's four health plans and national and international guidelines.

The community health care indicators in the QICH report are based on the computer databases of each of the health plans without personal identifiers; missing data are a small percentage (0.6 percent) of the population. QICH indicator data are harmonized to produce national rates which undergo a data audit by each health plan, the program management team, and a certified external auditor.

The QICH 2008–2010 report comprised 35 indicators of community health care covering six health topics: asthma, cancer screening, immunizations for elderly, children's health, cardiovascular health, and diabetes care. Indicator domains comprise primary prevention, disease management, and effectiveness of care. Rates are available for year, gender, and a proxy for socioeconomic status based on exemption from national health insurance.

The quality indicators show continuing improvements in health promotion in the general population and disease control, increasing quality of care over time for some measures, and maintenance of the existing high levels of quality for others. Annual reports with longitudinal assessment of quality measures for community health, along with data on financial performance and patient satisfaction, provide policy makers with data for making informed decisions and health policy.

Sources: *Dena Jaffe PhD, Israel national Program for Quality Indicators in Community Healthcare. Personal communication; 2013.*
Jaffe DH, Shmueli A, Ben-Yehuda A, Paltiel O, Calderon R, Cohen AD, et al. Community healthcare in Israel: quality indicators 2007–2009. Isr J Health Policy Res 2012;1:3.
Manor O, Shmueli A, Ben-Yehuda A, Paltiel O, Calderon R, Jaffe D. National Program for Quality Indicators in Community Healthcare in Israel Report, 2008–2010. Jerusalem: Hebrew University, Hadassah, Israel Ministry of Health and Israel National Institute for Health Policy and Health Services Research; 2012.

and ethnic inequalities to promote care for the whole population.

HEALTH SYSTEMS IN DEVELOPING COUNTRIES

In most developing countries, health services were inherited from colonial regimes and subsequently influenced by the Soviet model of health care in the 1950s and 1960s. The development of primary health care was neglected and underfunded, with excessive allocation of resources to teaching hospitals in the main population centers, leaving little for the rural majority. As a result, most developing countries are facing the need to reform their health systems.

During the 1980s, emphasis slowly moved towards primary care under the influence of the *Health for All* initiatives sponsored by the WHO. Achievements during the 1980s and 1990s included greatly improved immunization coverage, widescale use of oral rehydration therapy (ORT), and better sanitation. There has been a decline in birth rates in most regions of the world, including sub-Saharan Africa, which had until the 1990s seemed totally resistant to birth control. National health programs emphasize primary care with immunization, oral rehydration therapy (ORT), promotion of breastfeeding, supplemental feeding for infants, and birth spacing. Progress has been made in working towards the Millennium Development Goals (MDGs), although the targets for 2015 remain out of reach.

Productivity and per capita GDP are rising in many countries in the developing world so that a combination of universal primary education and improved economic status is furthering the potential for continuing improvement in health standards. In sub-Saharan Africa, low and declining levels of economic activity reduce the likelihood of increasing funds for health care. AIDS, malaria, TB, measles, other infectious diseases, poverty, malnutrition, and high birth rates with high child mortality aggravate a poverty–population–environment cycle, which impairs national growth potential. Improving the living conditions would mean that one-quarter of the people in sub-Saharan Africa would not be undernourished, and one-third of African children would not have stunted growth. World Bank data show that the poverty rate (people living on less than US$1.25 per day) has fallen from 58 percent in 1999 to 47 percent in 2008. Two-thirds of Africans are estimated to be deficient in vitamin A or iodine and half of the children are deficient in more than one micronutrient. As developing countries absorb western diets, lifestyles, and technology, they face a dramatic increase in NCDs such as hypertension, diabetes, stroke, CHD, and motor vehicle accidents. This, along with increasingly costly technology, places new burdens on health services.

Most developing countries spend less than 4 percent of their low national incomes on health and much of that on costly hospitals in the capital cities. Lack of adequate government budgetary funding raises political interest in national health insurance, especially in the mid-level developing countries. The purpose is to bring more of the population into the health care system and raise additional funds for health care beyond the little that is provided through government allocations.

This section, following brief regional overviews, gives examples of developing countries actively working to reform their health care systems. Health insurance is needed to increase funding for health care and provide for the growing urban employed and middle-class health needs, but at the same time, ministries of health must provide direct services to the rural poor majorities. As in developed countries, there will be no uniform approach, but sharing of lessons learned will be helpful.

Sub-Saharan Africa includes 40 countries with a total population of 874.8 million people in 2012 (expected to reach 1284 million by 2030) and a GNI per capita of US$2010. Annual births of 32.1 million in 2010 represent a decline in total fertility rates from 6.6 in 1960 to 5.9 in 1997, to 5.4 in 2005, and 4.8 in 2012. Life expectancy increased from 44 years in 1970 to 50 years in 1990, declining to 46 in 2005, but rising to 54.6 by 2011. The age group 15–49 years has an HIV prevalence rate slightly below 5 percent. Mortality of children under the age of 5 declined from 244 per 1000 in 1970 to 188 per 1000 in 1990, then to 169 per 1000 in 2005, and 121 per 1000 in 2010, and the infant mortality rate was 107 per 1000 live births in 2011.

In 2010, the WHO estimated that 287,000 women died of maternal causes worldwide; of these deaths, 85 percent occurred in sub-Saharan Africa and south Asia and less than 1 percent in more developed countries. Maternal mortality remains very high, with a rate of 475 per 100,000 in 2011, and with only 49 percent of births attended by a trained attendant (2005–2012; MDG5).

Immunization rates for the major childhood diseases have improved markedly in recent years and range between 61 and 75 percent, with tetanus immunization of pregnant women at 39 percent. Malnutrition (acute and chronic) is a crucial factor in child health status, with a high prevalence of stunting (estimated 41 percent) and wasting (UNICEF/WHO), contributing to high mortality from otherwise transient diseases with low mortality rates. TB, malaria, AIDS, measles, and other infectious diseases are major contributors to high rates of morbidity and mortality. Africa, with 10 percent of the global population, had 60 percent of the people living with HIV and nearly 30 million people with HIV/AIDS in 2005. Some 57 percent of those with advanced HIV receive ART, while the treatment success rate for TB increased from 71 percent to 82 percent between 2000 and 2010 (MDG6). Only 32 percent of young children sleep under insecticide-treated nets. In addition, NCDs and trauma are increasingly important contributors to the total burden of disease.

In the face of economic decline, political chaos in many countries, and the aforementioned health problems, resource allocations for health budgets have been jeopardized in many countries. Some countries in the region devote less than US$2 per capita to health budgets. Despite these challenges, progress has been made in efforts to improve sanitation and expand primary care services to the underserved rural areas and urban slums, giving hope for effective public health in the twenty-first century.

The dramatic effects of HIV/AIDS, with the accompanying epidemic of TB and multidrug-resistant forms, have created a public health crisis of great severity in sub-Saharan Africa, largely overwhelming the nascent health infrastructure. Yet progress is being made in immunization coverage for children and successful external assistance programs to widen the impact of the Expanded Programme on Immunization (EPI), directly observed treatment, short-course (DOTS), ORT, ART, and other modalities of public health hold out hope for a better future. The funding in low-income countries has, however, relied excessively on aid from international organizations and not placed health as a high priority in their budget allocations. Furthermore, the loss of skilled health personnel to wealthy countries in Europe and North America is a serious deficit, although signs of hope in political, economic, and social development are being seen.

Dramatic progress has been made in the eradication of polio, dracunculiasis, and onchocerciasis. Less progress is seen in TB, schistosomiasis, and malaria control. The WHO recommends wide-ranging new efforts to prevent cancer by immunization against hepatitis B, screening transfusion blood for hepatitis C, schistosomiasis control, smoking and alcohol control, reducing risk factors for CVD and diabetes, mental health, and oral health. The effects of civil war, the collapse of governments in some areas, and the refugee situation have had dreadful effects on public health. Since the mid-1990s, signs of stabilization in the governments of some countries of sub-Saharan Africa (such as Nigeria, discussed below) and significant economic progress offer new hope for the future of this potentially wealthy continent.

FEDERAL REPUBLIC OF NIGERIA

The Federal Republic of Nigeria, located in sub-Saharan Africa, is the 14th largest country in Africa and the eighth most populous country in the world, with a population of 166.6 million in 2012, with 49.7 percent living in more than 90,000 rural villages. Nigeria has more than 250 ethnic groups, with varying languages and customs. Religious groups are Muslim 50 percent, Christian 40 percent, and indigenous beliefs 10 percent. The number of children born in one year is around 6.4 million. Nigeria has 55 percent of its population living in poverty, and about 63 percent of primary-age children attend school.

In 2005, 4.4 percent of the population was infected with HIV, rising to 4.6 in 2008 and then declining to 4.1 percent in 2010, based on serosurveys. There are almost 3.5 million people now living with HIV and an estimated 1.5 million require ART. In 2011, about 390,000 new infections occurred and there were over 200,000 AIDS-related deaths. Between 2008 and 2011 those requiring ART nearly doubled, increasing from 0.86 million to 1.5 million. There are 17.5 million vulnerable children, some 7.3 million have lost one or both parents, 2.23 million are orphans because of AIDS, and an estimated 260,000 children have HIV/AIDS. About 20.3 percent of these millions of children do not attend school regularly and 18 percent are victims of sexual abuse.

The 2012 GNI per capita was US$2102 (PPP). The HDI ranked Nigeria as 153 (as shown in Table 13.19) out of 177 countries in 2012. Life expectancy at birth was 51.3 years, with a difference between males and females of 6.5 years. Nigeria's child mortality rate fell from 230 in 1990 to 143 in 2012, and infant mortality was reduced from 120 to 88 in the same period (UNICEF, 2012). Table 13.19 provides some vital statistics indicators for Nigeria.

In 1960, Nigeria gained independence from the UK, becoming a republic in 1963. It has a federal system of government with 36 states and the federal capital territory and 774 local government areas. It has had a turbulent political history since then, full of violence and instability, resulting in a slow rate of development. Despite the vast oil wealth discovered during the 1970s, Nigeria [a member of the Organization of the Petroleum Exporting Countries (OPEC) and the sixth largest producer of oil in the world], long ruled by repressive military regimes, saw corruption erode all levels of government functioning. The military in Nigeria has played a major role in the country's history since independence.

Africa's attainment of the MDGs depends on Nigeria's success, as one in every five Africans is a Nigerian. There is

TABLE 13.19 Vital Statistics, Selected Indicators, Nigeria, 2010

Life expectancy at birth (years), males	52
Life expectancy at birth (years), females	54
Children under-5 mortality rate/1000 live births	124
Infant mortality rate/1000 live births	78
Neonatal mortality rate/1000 live births	40
Maternal mortality ratio/100,000 live births	630

Sources: United Nations Children's Fund. Statistical report 2012. Available at: http://www.unicef.org/sowc2012/pdfs/SOWC-2012-TABLE-1-BASIC-INDICATORS.pdfWorld Health Organization. World health statistics 2013. Part III. Global health indicators. Available at: http://www.who.int/gho/publications/world_health_statistics/EN_WHS2013_Part3.pdf [Accessed 25 May 2013].

a big discrepancy between stated health strategy, developed with modern knowledge and good understanding of New Public Health, and existing progress with the health status of the population. Table 13.20 provides some HDI comparisons with other countries in sub-Saharan Africa.

In Nigeria, adult literacy rates for men increased from 47 percent in 1960 to 67 percent in 1995, and 76 percent in 2005, declining to 72 percent in 2010; respective rates for women rose from 23 to 47 percent and 61 percent, declining to 50 percent. According to the UN, Nigeria has experienced very rapid population growth; it has one of the highest fertility rates in the world with annual growth rates of 2.5 percent. By the UN projections, Nigeria will be one of the countries in the world that will account for most of the world's total population increase by 2050.

The birth rate has declined among the educated urban population, but remains high in the Muslim northern half of the country and the southern primarily Christian rural areas, with an overall total fertility rate of 6.0 children per woman in 1997. In 2010, estimated fertility rates were 5.5 children born per woman and 40.2 births per 1000 population. Only 39 percent of infants are delivered by trained health personnel; 57 percent are born at home, with considerable variation by region. Infant mortality declined from 122 per 1000 in 1960 to 112 in 1997, and to 88 (male 78; female 68) in 2010 per 1000 live births. Child mortality in Nigeria in 2010 was the 12th highest in the world at 143 per 1000 live births, down from 207 in 1960.

The maternal mortality ratios were reported to be between 550 and 840 in the period 2006–2010 (630 in 2010). Maternal mortality rates showed geographic disparity from 166 in the south-west to 1549 in the north-west per 100,000 live births (Nigerian Demographic and Health Survey, 2003). In 2010, the WHO estimated that 287,000 women died of maternal causes worldwide; of these deaths, 85 percent occurred in sub-Saharan Africa and south Asia and less than 1 percent in more developed countries. Some 14 percent (40,000 women, estimated in 2010) of the worldwide losses by maternal mortality occurred in Nigeria, despite the fact that the country contains only 2.5 percent of the world's population. Large regional differences in maternal deaths demonstrate that most of these deaths are preventable. The WHO estimates that for each woman who dies from childbirth in Nigeria, another 30 suffer long-term damage to urogenital organs, often with vesicovaginal fistula (with continuous leakage of urine through the vagina), tubal damage (resulting in infertility and ectopic pregnancy), and chronic pelvic pain. The United Nations Population Fund (UNFPA) estimates that some 2 million women are affected by fistulae in the developing world, of whom 800,000 (40 percent) are in Nigeria, especially in the northern part of the country where early marriage and frequent pregnancies are promoted.

The most common causes of maternal mortality and morbidity in Nigeria are bleeding immediately after delivery (postpartum hemorrhage, 23 percent), prolonged obstructed labor, eclampsia (hypertensive disease of pregnancy, 11 percent), postpartum infection (17 percent), and unsafe abortion (11 percent), along with anemia (11 percent), malaria (11 percent), and other causes (5 percent). As the result of the restrictive abortion law in the country, women often use dangerous methods to produce abortion, with high rates of complications, often resulting in death; every day, 160 pregnant Nigerian women die from the complications of pregnancy. It has been estimated that nearly 610,000 women resort to induced abortion each year, and of

TABLE 13.20 Human Development Index (HDI), Selected African Countries, 2013

Country	HDI Rank 2012	Life Expectancy at Birth (years) 2012	Under-5 Mortality Rate/1000 Live Births 2010	Maternal Mortality Ratio/100,000 Births 2010
Ghana	135	64.6	74	350
Cameroon	150	52.1	136	690
Togo	159	57.5	103	300
Kenya	145	57.7	85	360
Nigeria	153	52.3	143	630
Benin	166	56.5	123	350
Ivory Coast	168	56.0	123	400
Chad	184	49.9	173	1100
Sierra Leone	176	40.6	283	890
Niger	177	44.3	259	590

Source: United Nations Development Programme. Human Development Report 2013. Available at: http://hdr.undp.org/en/reports/global/hdr2013/download/

this number 10,000 die. Low levels of use of contraception in the 15–25-year age group result in 60 percent of pregnancies being unwanted, with 80 percent of women with such pregnancies resorting to unsafe and illegal abortion.

About one-fifth of children who are born in Nigeria die before reaching 5 years, twice as high as in Ghana. Life expectancy at birth increased from 43 years in 1970 to 52 in 1997, dropped to 43.3 in 2005 and rose to 52.3 in 2010. The quality of health and vital statistics is low. Of the 39 percent of births with a skilled birth attendant there is a wide variation between urban (63 percent) and rural (28 percent).

Female Genital Mutilation

Female genital mutilation (female circumcision) is among traditional practices that are deeply entrenched in Nigeria. This practice has received global attention and condemnation over the years because of its many serious physical, mental, social, economic, and political implications. The Nigerian government observes the International Day for Zero Tolerance to Female Genital Mutilation (FGM). The fight against this harmful practice is to be marked on 6 February each year. Nigeria is one of the 28 countries in Africa where FGM is still practiced. In Nigeria among women and girls under 15, there are slightly more than 10 million (11 percent) of the 91.5 million African women undergoing FGM (WHO, 2011). It is also estimated that some 25 percent of the 140 million women living with FGM are in Nigeria, and that 40–60 percent of Nigeria's women are victims to the practice of FGM, with the level being over 90 percent in some regions. This practice has no health benefits and harms girls and women in many ways. It involves removing and damaging healthy and normal female genital tissue, and hence interferes with the natural function of girls' and women's bodies. It poses a great burden on the women of the country and on the health system, including its adverse effects in transmission of HIV and sexually transmitted infections.

There is no federal law prohibiting the practice of FGM in Nigeria. Although the Nigerian federal government has publicly condemned FGM as a harmful practice, it has not taken any legal action against it. FGM is done by largely untrained women with crude implements, with no anesthesia or antibiotics; there is usually bleeding which sometimes leads to death or anemia. Besides the direct consequences of bleeding, there is the ever-present risk of infection, especially tetanus or HIV/AIDS. Many NGOs have been established by Nigerian women to fight this issue.

A WHO multicountry study in which more than 28,000 women participated confirmed that women who had undergone genital mutilation had significantly increased risks for adverse events during childbirth, with high rates of caesarean section and postpartum hemorrhage. Genital mutilation of mothers has negative effects on their newborn babies. The consequences of genital mutilation are even more severe for the majority of Nigerian women who deliver outside a hospital setting.

Communicable Diseases

The World Bank Country Status Report on Nigeria (2005) notes that communicable diseases, often in association with malnutrition, are the major causes of mortality among children, predominantly malaria, measles, meningitis, pneumonia, yellow fever, dysentery, TB, and AIDS. The United Nations Children's Fund (UNICEF, 2013) notes that Nigeria ranks as the second largest contributor to the under-five mortality rate in the world. Immunization coverage in Nigeria is among the lowest in Africa. Malaria causes the largest number of child deaths in Nigeria, estimated at 172,000 (2010). Other major causes of childhood deaths are foodborne and waterborne diseases, bacterial and protozoal diarrhea, hepatitis A, typhoid fever, respiratory disease, meningococcal meningitis, and aerosolized dust or soil contact disease. Some parts of Nigeria are highly endemic for Lassa fever.

Highly pathogenic H5N1 avian influenza was identified among birds in this country, or the surrounding region, in January 2006. The potential devastation from emergence of a pandemic strain in Africa has led to a sudden shift in disease control to a public health focus with international aid funding available for pandemic preparedness, but this has led to concern over the possible distortion of priorities and damage to critical basic public health programs.

Immunization rates increased during the early 1980s but declined in the latter part of the decade. The level of immunization of pregnant women against tetanus was 23 percent in 1995–1997; infant immunization with bacille Calmette–Guérin (BCG) was 29 percent, 21 percent for DTP, 25 percent for polio, and 38 percent for measles. Coverage with DTP (three doses) was 47 percent, measles 71 percent, hepatitis B 50 percent (WHO 2013), and polio 79 percent (2011, WHO Regional Office, Africa). Measles accounts for 12 percent of child deaths and AIDS is a major public health issue in Nigeria, as in other sub-Saharan countries.

Since 2007, Nigeria has been the only polio-endemic country in the African region, and one of only three in the world. There was an 80 percent reduction in wild poliovirus in 2007; however, a total of 62 wild poliovirus cases was detected in 2011, an increase from 21 cases in 2010. There is also a significant antivaccination movement with the murder by Islamic extremists of some vaccination service workers in 2013. Polio control and eradication measures are ongoing with advocacy and support of community and religious leaders. This support, especially in the vaccine-averse north, along with efforts to control measles and other childhood killer diseases, creates awareness of acute flaccid paralysis and disease surveillance, and intersectoral cooperation of governmental, private sector, and community financial and logistic support for immunization activities. Immunization-plus days have been helpful but have led to deterioration in the routine immunization program in Nigeria.

TB rates are declining in Nigeria, as in many countries, but it remains among the high-prevalence TB countries, with more than 280,000 people infected. The WHO estimates that there were over 84,000 new cases of TB in 2011, an incidence of 181 per 100,000 total population, and a prevalence of 171 per 100,000 population compared to 282 per 100,000 in 1990 (WHO, 2011). Multidrug-resistant TB accounts for 3.1 percent of new TB cases.

Non-Communicable Diseases

Although communicable diseases are major causes of mortality and morbidity in the country, NCDs represent a fairly large share of the burden of disease among Nigerians, representing 47 percent of total mortality in 2011. Half of the deaths are due to CVDs, a quarter due to cancers, and about a tenth due to respiratory diseases. Sickle-cell anemia is the most common genetic disorder affecting Nigerians. Hypertension affects an estimated 11.2 percent (4.3 million) of Nigerians over 15 years of age. In 2011, about 4.14 million Nigerians over the age of 15 years were smokers. Diabetes prevalence is also high with over 1 million Nigerians estimated to be suffering from this disease and its complications.

Nigeria's Health System

The federal, state, and local governments support works in a three-tier system of health care. The essential features of the system are its comprehensive nature, multisectoral inputs, community involvement, and collaboration with non-governmental providers of health care. The system is based on the 1979 constitution of the country, which put health care on the concurrent legislative list of responsibilities of all three levels of government. International health quarantine and control of drugs and poisons are exclusively the responsibility of the federal government. A national health policy based on the philosophy of social justice and equity was developed in 1984 and adopted in October 1988. The policy was revised in 2004 as the "Nigerian Health System on Primary Health Care".

The health system inherited from the British colonial period included limited hospital care in the urban centers, and some medical training facilities. Following independence in 1960, the state-operated health system began to develop a widened network of primary care services, in parallel with state primary education. Health care expenditures in 1992 were US$1.50 per capita, and health constituted 5 percent of the national budget.

The present health system is seriously underfunded and covers less than two-thirds of the population, with large parts of the rural population outside the system. UNICEF estimates access to health services at 85 percent for the urban population, 62 percent for the rural population, and 66 percent for the total population. Curative services in hospitals and primary care clinics receive the major share of the fiscal resources. Proposed changes in allocation will divide health resources in Nigeria as follows: 15 percent to federal government-operated specialty hospitals; 25 percent to state government-operated district hospitals; and 60 percent to local government-operated primary health care clinics, including maternal and child health, school health, and other aspects of primary health care.

Spending on health is low, with total expenditures reported as 5.3 percent of GDP including 1.9 percent from government sources (World Bank, UNHDR, 2011). According to the Central Bank of Nigeria, federal government health spending in 2011 decreased by 14.6 percent compared to 2010. Most of the federal health spending goes to teaching and specialized hospitals and federal medical centers. Tertiary health care institutions receive more than two-thirds of the total budget allocated to health, of which about two-thirds is spent on personnel and administrative overheads. Out-of-pocket expenditure accounts for 70 percent of Nigeria's total health expenditure and represents more than 9 percent of total household expenditure.

Health care insurance is through social security and state national assistance, together with special group coverage for members of the armed forces and organized urban groups such as those working in the transport sector. The public health services suffer from low salaries, lack of supplies, and inefficient administration. Private practice is common in the urban centers, serving mainly the middle class. Drugs are expensive and imported in an unrestricted fashion. The National Health Care Fund receives funds from federal government general revenues, rural cooperative health insurance premiums, and employed people's health insurance.

Despite a high standard of medical training, the overall quality of care and efficiency in health management are low. The federal government is undertaking initiatives to broaden health insurance in order to raise revenues for health care and to increase equity of access to services. Although the data are incomplete, available information from the Federal Ministry of Health Record for 2005 reported the following federal government-operated hospitals: 19 teaching and specialist hospitals, eight psychiatric hospitals, three orthopedic hospitals, and 24 federal medical centers. In addition, there are 59 tertiary health facilities operated by the states. The number of tertiary and specialized hospitals suggests that there is a relatively good average availability of high-level services. In 2000, there were 3275 secondary care-level facilities in the public sector, a population-to-facility ratio of around 135,000 people per facility supplemented by

TABLE 13.21 Health Professionals, Nigeria, 2003 and 2008

| | 2003 | | | 2008 | |
Personnel	No.	Rate/10,000 Population	Personnel	No.	Rate/10,000 Population
Physicians	34,923	2.8	Physicians	55,376	4.0
Nurses	127,580	1.03	Nurses and midwives	224,943	16.1
Midwives	82,726	0.67			
Community health workers	115,761	9.0	Community health workers	19,268	1.4

Note: Data estimated as of 2003 and 2008. Nurses and midwives are reported as one category in 2008.
Source: World Health Organization. World health statistics 2008, 2012. Available at: http://www.who.int/gho/publications/world_health_statistics/EN_WHS08_Part2.pdf, http://www.who.int/whosis/data-base/core/core_select_process.cfm, and http://www.who.int/gho/publications/world_health_statistics/EN_WHS2012_Full.pdf [Accessed 25 May 2013].

3000 facilities in the private sector. Primary care was based in over 21,585 public sector and almost 7000 private primary health care facilities in 2003. As of 2008, responsibility for tertiary care services is with the federal government, secondary health care services are the responsibility of the state governments, and local governments are responsible for primary care services. This is expected to expand coverage for basic health care to a large part of the rural and urban poor population.

In 2003 human resources included 115,761 community health workers. There has been a steady increase in the numbers of health professionals trained in Nigeria to meet the health care needs. There were 34,923 physicians in 2003, 127,580 registered nurses, and 82,726 registered midwives. In 2011 human resources included 55,376 physicians or 4.0 per 10,000 population, but most physicians are located in the urban areas. There were 224,943 registered nurses and midwives as of 2008 (Table 13.21).

Medical education has been given high priority; there are 18 fully and five partially accredited medical schools in the country and many more awaiting accreditation. They graduated about 2000 doctors, 5000 nurses, and 800 pharmacists in 2002/2003. Some of the universities/teaching hospitals have fully developed departments of public health where public health physicians are trained in conjunction with the National Postgraduate Medical College of Nigeria and West African College of Physicians.

A master's degree in public health programs is also offered, but there is no school of public health. The priority given to curative services largely fails to address the basic health problems of the country, which require the application of well-known and cost-effective public health programs. Increasing death rates from no-infectious diseases and trauma require attention in planning preventive and curative services for the future.

Nigeria is one of several major health staff-exporting countries in Africa, with nurses and physicians emigrating, both legally and illegally, mainly to Britain, which is a threat to sustainable health care delivery in Africa's most populous country. About 20,000 health professionals are estimated to emigrate from Africa annually. Data on Nigerian doctors legally migrating overseas are scarce and unreliable, but estimates are that hundreds of Nigerian-trained doctors continue to migrate annually. Internal migration from state and rural posts is a major threat to the achievement of the MDGs. Doctors are attracted to university teaching hospitals rather than employment by states because the salaries are far higher in the federal establishments than in state employment. Almost half of Nigerian doctors choose to work in Lagos in federal health institutions. A unified salary scale for doctors in both state and federal government employment should be implemented to motivate doctors to stay in the state of origin or local governments to render services, because that is where they are most needed. The Nigeria Medical Association has for many years been advocating for a unified salary scale, the Medical Salary Scale, to counter the maldistribution of medical doctors in the country.

Millennium Development Goals

Nigeria's MDG achievements in the past few years include the extension of primary health care services to over 20 million people, provision of safe water to over 8 million people, a six-fold increase in the distribution of insecticide-treated nets to protect the under-fives from malaria, and a 98 percent reduction in the incidence of polio, albeit with a resurgence of cases which remains a challenge in pockets in the north of the country.

The under-five mortality has fallen by over 20 percent in 5 years, from 201 deaths per 1000 live births in 2003, to 157 deaths per 1000 in 2000 and 143 per 1000 in 2010. In the same period, the infant mortality rate fell from 100 to 88 deaths per 1000 live births in 2010. Recent interventions including Integrated Management of Childhood Illnesses that reflect the underlying causes of child deaths have contributed to these successes. However, these need to be rapidly expanded and accelerated if Nigeria is to achieve MDG4. Nigeria has had striking success in almost eradicating polio, reducing the number of cases by 98 percent between 2009 and 2010. However, a climate of insecurity and violence in parts of Nigeria threatens the solid programmatic advances in polio eradication in Nigeria made in recent years.

Maternal mortality fell by 32 percent, from 800 deaths per 100,000 live births in 2003 (then one of the highest maternal mortality rates in the world) to 545 deaths per 100,000 live births in 2008, but rose to 630 in 2010. However, the proportion of births attended by a skilled health worker has remained low and threatens to hold back further progress. An innovative Midwives Service Scheme is expected to contribute substantially to ongoing shortfalls but its impact has yet to be reflected in the data. If the scheme is expanded in proportion to the national gap in the number of midwives, this will further accelerate progress. In addition, more mothers will be covered by antenatal care as access to quality primary health care improves and incentives attract health workers to rural areas (Nigeria MDG Report, 2010).

While the level of violence against Nigerian women in the home remains poorly mapped, pilot studies conclude that it is "shockingly high". Up to two-thirds of women in certain communities in Nigeria's Lagos State are believed to have experienced physical, sexual, or psychological violence in the family; in other areas, around 50 percent of women say that they are victims of domestic violence.

In the absence of official studies, research into the prevalence of violence in the family has been conducted by individuals and organizations. In a recent small-scale study of gender inequality in Lagos and Oyo states, 40 percent of the women interviewed said that they had been victims of violence in the family, in some cases for several years. The widespread practice of FGM may be a further indicator of the level of violence against women and children. According to the UN Committee on the Rights of the Child, acceptance of domestic violence is high even among law enforcement officers and court personnel.

As a means of promoting gender equality, the Strategic Implementation Framework and Plan sets out the objectives, targets, and monitoring framework needed to work towards eliminating gender discrimination and improving the participation of women in national life.

Cancer

The National System of Cancer Registries was established in 2009 in collaboration with the US University of Maryland, various Nigerian health institutions, and the Nigerian Ministry of Health. It works cooperatively with other international cancer agencies and the CDC in Atlanta, USA, to develop a strong population-based cancer data system for the most common cancers as a basis for health policy and research.

Data from the National Cancer Registry show that 100,000 new cases of cancer are currently diagnosed each year in Nigeria. The most common cancers are of the cervix, liver, breast, and lymph glands.

HIV/AIDS

Nigeria has the third largest number of people infected with HIV/AIDS in the world. Around 26,000 children have HIV/AIDS. The need for ART increased from 0.85 million to 1.5 million people between 2008 and 2011. As a result of the HIV/AIDS epidemic, 7.3 million children have lost one or both parents, with 2.3 million orphans. Violations of women's rights escalate the rate of HIV infections throughout Africa. Sexual oppression combined with a high biological receptiveness of viral transmission due to FGM puts women at risk. As a consequence, the violence against women threatens to destroy whole communities.

Success in reducing the prevalence of HIV among pregnant young women aged 15–24 led to a decline from a prevalence of 5.8 percent in 2001 to 4.1 percent in 2010. Nationally, Nigeria has already achieved this MDG target, although some states still have high prevalence rates. Success depends on better awareness and use of contraceptives. There has been a sharp decrease in malaria prevalence rates. Nationwide distribution of 72 million long-lasting insecticide-treated bed nets, although only in its initial stages, protected twice as many children (10.9 percent) in 2009 as in 2008 (5.5 percent). Similar progress has been made with TB: with sustained attention, TB is expected to be a limited public health burden by 2015.

Food Fortification/Malnutrition

The initiative to control and reduce micronutrient deficiency disorders in Nigeria goes back to 1990. Iodination of salt, begun in 1993, reduced the prevalence of goiter to 11 percent at the pilot sites and household consumption of iodized salt increased to 98 percent. Nigeria was the first African country to receive a certificate of achievement.

In 2002, the government adopted a new strategy: the fortification of staple foods with vitamin A, with published mandatory standards, in flour, sugar, and vegetable oil. By 2004, 70 percent of the sugar, 100 percent of wheat flour, and 55 percent of vegetable oil were fortified with vitamins. Wheat flour is also fortified with iron. Food fortification with folic acid has been extended to common staple foods such as margarine, pasta, popular drinks, and some brands of powdered milk. However, cassava, the most commonly consumed food in Nigeria, is not fortified.

The National Policy on Food and Nutrition launched in 2002 set specific targets for 30 percent reduction in malnutrition (acute and chronic) among under-fives by 2010, and a 50 percent reduction in micronutrient deficiencies (vitamin A, iodine, and iron) by 2010. The strategy for reducing malnutrition includes both the agricultural and non-agricultural sectors.

Health Reform

The Nigerian Colonial Development Plan in the 1940s had a limited framework for a unitary health service. In the 1950s, regional governments ran independent and sometimes parallel health systems to the federal government; in the 1960s, the Second National Development Plan in the postindependence era did not articulate a system with clear responsibilities for each level of government.

Between 1986 and 1992, progress was made in the development of primary health care, focusing on local government areas. This was supervised by the National Primary Health Care Development Agency (NPHCDA), established in 1992 as a body reporting to Nigeria's Federal Ministry of Health. In 2007 this agency was merged with the national immunization program with the mandate to improve access to care and control preventable diseases.

Since 2005, ongoing health service reforms have continued in areas of food fortification under the national response to malnutrition, repositioning of the NPHCDA, establishment of the National Health Insurance Scheme (NHIS), as well as the proposed National Health Bill. The NHIS, launched in 2005, provides services to enrollees through 5949 health care provider plans, 24 bank plans, five insurance companies, and three insurance brokers.

In May 2011, the new National Health Bill sought to establish stable funding for health and basic services for certain vulnerable groups including young children, pregnant women, the elderly and those with disabilities, as well as those living in hard-to-reach rural areas. However, the accompanying measure, the Primary Healthcare Development Fund, has yet to be established.

The first major development policy framework introduced by the federal government after the Millennium Declaration was the National Economic Empowerment and Development Strategy (NEEDS) in 2004. The State Economic Empowerment and Development Strategy (SEEDS) was the corresponding strategy at state level. NEEDS was a medium-term development strategy implemented between 2004 and 2007. It laid down the overall framework and strategic direction for the sector policies that followed. NEEDS and SEEDS formed the basis for policy coordination in programs and projects between the federal and state governments.

The NEEDS is based on three pillars:

- empowering people and improving social service delivery
- growing the private sector and focusing on non-oil growth
- changing the way government works and improving governance.

Nigeria Vision 20:2020 was developed as a longer term growth and development framework for the country. It foresees Nigeria being among the 20 largest economies by the year 2020. The growth prospects assumed by the Vision, quite apart from the policy interventions for the MDGs, are expected to make a substantial contribution to poverty reduction.

Summary

The evolution of health care in Nigeria from a very limited colonial health service to a centrally managed service with serious underfunding, and then to a more universal system, reflects postindependence trends in many countries. Ethnic violence over the oil-producing Niger Delta region, interreligious relations, corruption, and inadequate infrastructure are basic issues in the country. Facing a population explosion and contracting economies, African countries went through a very difficult transition in the 1980s and again in the first decade of the twenty-first century. In natural resource-rich countries health expenditures per capita declined. Health information systems are inconsistent with limited reliability for policy and decision making within the country and regionally. Recent activities and reports of international organizations (UN, WHO, UNICEF) have highlighted these challenges.

The primary care system needs strengthening to meet the challenge of preventable diseases, which have been exacerbated by a decline in immunization coverage in recent years. Decentralization of organization to increase the role of the state and local government authorities may improve community participation and efficiency of services. It may also increase revenues by providing a mechanism for local financial input.

There has been a lack of prioritization of maternal and child health in terms of resource allocation and systematic programming. The low rate of political attention given to maternal and child health in the country in part reflects the continuing adverse affects of some harmful traditional, religious, and cultural practices. The Minister of Health, Professor Adenike Grange, in November 2007 reported to the National Health Council on a seven-point agenda that placed a high premium on the development of human capital, and recognized that health and education are the twin engines that drive national development by developing human capital.

Since the 1990s, the Nigerian health sector has proposed reform targeted towards improving health service delivery and quality of care, but "these programmes have fallen short of making a significant impact towards improving health service delivery, due to a relatively poor emphasis on implementation, monitoring and evaluation". Professor Grange promised that the administration would move health sector reform forward, building on the policies and frameworks that have been developed, focusing on implementation, integration, monitoring, and evaluation. Current and new legislative initiatives and service program require adequate funding levels to ensure availability and accessibility of services across populations, especially the most vulnerable, in both rural and urban areas.

LATIN AMERICA AND THE CARIBBEAN

The Latin American region includes 29 countries, with a population of more than 589 million. It experienced rapid economic growth until the world financial crisis of 2008, but is burdened with widespread poverty and inequality in incomes, health, and well-being. GDP per capita ranged from a low of US$700 in Haiti to US$12,280 in Chile in 2011. GNI per capita in 2006 averaged US$8571, compared to US$32,217 for the industrialized countries.

Life expectancy at birth increased from 60 years in 1970 to 68 in 1990, to 73 years in 2006 and 74 years in 2011. There are over 11 million births annually in the region (2006), but the crude birth rate fell from 37 to 27 to 20 to 18.2 per 1000 population in 1970, 1990, 2006, and 2011, respectively. Crude mortality declined in the same years from 10 to 7 and to 6 per 1000 population. From 1970 to 1990 and 2006, the child (under 5 years) mortality rate fell from 123 to 55 to 27 per 1000 live births, with average annual reductions of 4 percent and 4.4 percent in the periods 1970–1990 and 1990–2006. Infant mortality fell from 106 to 43 to 22 per 1000 live births from 1970 to 1990 and 2006, respectively. Maternal mortality in the period 2000–2006 was still high, 130 per 100,000 live births, with 86 percent of deliveries taking place in a health facility (UNICEF, 2008).

These figures indicate impressive economic and health care progress for the region. Continent-wide eradication of wild poliovirus and control of measles and other vaccine-preventable diseases have been achieved. However, violence and trauma, CVDs, TB, malaria, dengue fever, Chagas' disease, and cholera are still major public health problems.

Despite the impressive but uneven progress in health, inequalities of income and health status between and within countries are also dramatic, with widespread poverty in rural and urban slums. Sustained economic growth and higher quality work will be needed to achieve the MDGs of reducing poverty and hunger. The health sector can help very much, however, by furthering the successes to date in control of infectious diseases, paying attention to malaria, TB, and other endemic diseases, improving sanitation, extending immunization, and improving maternal and child health care. Colombia is presented as an example of the progress and challenges facing the health sector in Latin America.

Colombia

Colombia is located in the north-western region of South America and has a land area of 1.141 million square kilometers divided into 32 departments (states), and further subdivided into 1076 municipalities. It is a mid-level developing nation with 47.6 million inhabitants, with a per capita GDP of US$8861 (PPP) in 2012. The population is 76 percent urban (2012), with 93.4 percent literate and 52.7 percent living below the national poverty line (8.2 percent living on less than US$1.25 per day) (World Bank data, 2011).

Life expectancy at birth increased from 57 years in 1960 to 71 in 1997, 72.3 in 2005 and 73.9 years in 2012. From 1970 to 1990 and 2000, the crude mortality rate fell from 9 to 6 to 5 per 1000 and remained at 5 between 2005 and 2012. The total fertility rate in 2000–2005 was 2.5 per woman and declined to 2.1 in 2011. The infant mortality rate decreased from 82 per 1000 live births in 1960 to 25 in 1997, 17 in 2005 and 15 in 2011. Child mortality rates declined from 130 per 1000 live births in 1960 to 30 in 1997, 21 in 2005 and 19 in 2010. Maternal mortality remains high at 92 per 100,000 in 2010. Primary school enrollment is universal, and adult literacy rates are high (93.4 percent in 2010).

Total expenditures on health were 6.1 percent of GDP in 2011. Colombia's HDI in 2012 was 0.719, placing the country in 91st position out of 187 countries surveyed. Furthermore, Colombia has a Gini index (a measure of inequality of wealth distribution) of 65.9 (on a scale from 0 to 100, 0 being total equality and 100 total inequality).

The internal civil conflict has caused the displacement of nearly 4 million Colombians since 1985, with a devastating impact on the health profile of the Colombian population.

In Colombia in 2002, the leading causes of death included CVDs and diabetes (27 percent), violence and trauma (19 percent), and chronic and lower respiratory infections (8 percent). CVDs are increasing as the associated risk factors of smoking, fatty diet, inactivity, hypertension, and diabetes are more prevalent than in the past throughout the area.

The main causes of mortality in 2010 were cancer, external causes, CVDs, and communicable disease. Smoke-free legislation applies to all public places with national laws and fines levied against both violators for smoking and the establishments. Homicides among males are an important cause of mortality, with a rate as high as 109.2 per 100,000 men. Road traffic deaths among men declined from 35.9 to 29.9 per 100,000 population between 2000 and 2005. There is wide variability in reported cases of malaria but these declined from 120,096 in 2006 to 64,309 in 2011.

Communicable and infectious diseases are highly prevalent in Columbia. In 2010, TB incidence was 24.7 per 100,000 population, with 15.2 laboratory-confirmed positive sputum smears. In contrast, AIDS incidence was 3.1 per 100,000. It is estimated that 8 million Colombians live in high-risk areas for Chagas disease, with 1.2 million cases. There was a large outbreak of dengue in 2010, with 157,152 cases and a case fatality rate of 2.3 percent. Ten million people are at risk of leishmaniasis, mainly in rural areas, with 14,000 cases reported on average between 2000 and 2010. Annually, there are 140,000 malaria cases.

Cancer and CVD are the leading causes of NCDs, with mortality rates of 120.7 and 101.7 per 100,000, respectively, in 2010. Cerebrovascular diseases account for 51.2 deaths per 100,000 population and diabetes mellitus mortality was 24.2 per 100,000 population in 2010.

External causes are an important group mainly related to homicide (including illegal and legal interventions, and war operations) with 44.6 deaths per 100,000, police reported intentional homicides rates of 66.7, 42.1 and 34 per 100,000 for 2000, 2005 and 2010 respectively, while road traffic accident deaths are 12 per 100,000 (2010). Violence is the leading cause of death in the 15–45 year age group for men and women. Deaths from violence are the second leading cause of male deaths, with a rate of at 3.1 per 100,000, twice the rate for women of 1.5 per 100,000; violence is the fourth leading cause of death among Colombian women.

During the 1990s, Colombia's health system experienced a major reform that replaced the previous national health system and the Bismarckian social security system with a new social security system that covers standards governing the general system of pensions, professional risks, and complementary social services. The reform of the 1970s' National Health System attempted to respond to the global initiative promoted by the World Bank in 1987 that aimed to consolidate health systems in different nations. As a result, Colombia implemented Law 100 in 1990, by which territorial entities became financially and administratively autonomous to operate the public hospitals circumscribed to their area and to execute free public health activities within the frame of their local plans.

This process of decentralization was favored by the new Colombian constitution of 1991, which conferred more power to the territorial entities and defined social security as a mandatory public service that should be coordinated and controlled by the state. This mandate was enacted in 1993 under Law 60, which governs matters relating to the authority and resources of the various territorial entities (today Law 715/2001), and Law 100, which created a new scheme for the General Social Security System for Health. Based on the concept of universal access through a demand-oriented model, the reform seeks to implement equity of access, free choice of HMOs (Entidades Promotoras de Salud), institutional autonomy, decentralized administration, and national regulatory mechanisms by assigning each person a per capita unit adjusted by risk.

To assure universality and financial solidarity, the reform intends to cover all individuals under both contributory and subsidized systems based on a partnership scheme of income redistribution. The law stipulates that employed people contribute 12 percent of their salary (two-thirds of which is paid by their employer), while the self-employed pay 12 percent of their declared income. However, the subsidized system is financed with the resources of the municipalities, one-twelfth of the resources collected through the compulsory system, fiscal allocations to the departments, national income assigned to the departments, resources from gambling taxes, voluntary contributions from the municipalities and departments, royalties from new oil wells, contributions from the compensation funds, value-added tax destined for social programs, tax on firearms and ammunition, and co-payments and prorated fees from members and their families.

Funding resources are collected in the National Solidarity and Guaranty Fund. The system is directed, standardized, regulated, and controlled by the National Council of Social Security for Health, a body of the Ministry of Social Protection (previously called the Ministry of Health), composed of a professional group of the main participants in the system and the sectional health services in each state. The legal framework is supervised and evaluated by Committee VII of the Senate and the House of Representatives. The HMOs administer the provision of the services and the health provider institutions (*Instituciones Prestadoras de Salud*) provide the services. The Superintendancy of Health controls and monitors the system.

The contributory system offers a comprehensive compulsory health plan (*Plan Obligatorio de Salud*, POS) under the social security system that includes initiatives to benefit the individual, the family, and the community in general. The compulsory health plan of the subsidized program (*Plan Obligatorio de Salud del Regimen Subsidiado*, POS-S) is territorially based, composed mainly of actions in the area of health promotion and disease prevention, and provides only 70 percent of the services offered through the contributory system. To select the subsidized population, the municipal authorities apply annually a survey that combines criteria of the Poverty Line Index and the Index of Unsatisfied Basic Needs. The HMOs mobilize financial resources, organize health promotion activities, arrange complementary health plans, provide the POS and the POS-S for affiliated individuals, and provide other medical services for people with disabilities or those who have an occupational disease or a work-related accident. These medical services are provided either by the HMOs through their own health provider institutions or through other health provider institutions (public hospitals, independent health service centers, individual or groups of health professionals) that are contracted by the HMOs.

During 2005–2012, Colombia's supply of doctors was 14.7 per 10,000 and 6.2 nurses per 10,000, compared to US levels of 24.2 doctors and 98.2 nurses per 10,000 population (WHO, 2013). In order to respond to the human resources needed by the health sector, the government implemented Law 30 and Law 115 of 1994, which authorized educational institutions to create new programs. Consequently, Colombia is experiencing an uncontrolled and hazardous growth of study programs and private vocational schools at the technical and auxiliary levels. The National Council on Human Resources Development regulates the basic formation of the health technicians, such as health promoters (promatores de salud), family, community health workers, and nursing assistants. Training of health technicians in rural areas recruited from the population served constitutes a great asset, because it guarantees intensive outreach and culturally sensitive health educational functions.

Use and quality control of pharmaceutical products have been supervised since 1995 by the National Institute for the Surveillance of Drugs and Food (INVIMA), which follows the good manufacturing practices (GMP) guidelines of the WHO. At the same time, the Bureau of Pharmaceutical and Laboratory Services of the Ministry of Social Protection develops strategies to promote the development of services for pharmaceutical care and the rational use of drugs, and also designs policies related to this area.

After more than a decade of implementation and the recognition by the WHO as one of the most responsive models in Latin America, diverse evaluations have criticized the performance of the Colombian health system. As with other countries, and especially since the world economic crisis of 2008, the Colombian health system has been under stress, with consequences of access for the unemployed and the poorest segment of the population. Some 4.3 percent of Colombians are not covered by the General Social Security System for Health. Private expenditures by families have been impacted without an increase in coverage rates or improvements in the quantity and quality of services. There is evidence of a deterioration in public health services due to the lack of commitment of the HMOs to fulfill their obligations in regard to public health. In addition, local governments and local health authorities have been unable to ensure adequate levels of public health services. Immunization coverage for DTP and measles decreased to 85 percent and 88 percent, respectively, in 2011, while morbidity and mortality from malaria, TB, and other communicable diseases have increased.

Expenditure for health care increased from 7 percent of GDP in 1990 to 10.5 percent in 1999, then fell to 7.8 percent in 2004 and to 6.1 percent in 2011. Between 2008 and 2011, public expenditures increased from 68.1 to 74.8 percent of total expenditures. Private expenditures have increased, but with no change in coverage rates or the quantity and quality of services, indicating that resources are being diverted from social objectives by the HMOs. In the market-driven system, reform promoted privatization and minimal state involvement in care delivery. Owing to the imbalanced competition between private and public providers and the enormous debt that the government has with the public hospitals, five of the largest national public hospitals have closed and 10 more are in the process of liquidation. In practice, the reform has promoted privatization and minimal state involvement in care delivery.

In an attempt to overcome most of these problems, in 2006 the government enacted Law 52, which constitutes the first reform of Law 100. This law is intended to increase the level of coverage from the current 47 percent to 85 percent and to equalize the mandatory health plan for both contributory and subsidized systems. It is also meant to diminish access barriers such as co-payments and prorated fees for the subsidized system, and the waiting period required to treat chronic conditions for the contributory system. The government also pays the debts of the subsidized system and fortifies the provision of public health services through the implementation of the National Plan of Public Health. This law also proposes the creation of the Health Regulatory Commission and the Colombian Territorial Fund, to define new regulations within the system, control the use of resources, and monitor the quality of the services provided by the HMOs. The Colombian health system is in a process of continuing change meant to improve quality of health care, universal coverage, and equity.

A new health reform is being debated in 2013, motivated by the barriers to access, failures in health promotion and disease prevention, and failures of the HMOs.

Health reform is intended to introduce universal coverage and consumer choice of HMO-like organizations based on the concept of universal access to a market-oriented set of service alternatives. They will provide care paid on a capitation basis and be subject to accreditation and quality control with GPs as gatekeepers. Direct service development of primary health care continues as a responsibility of the Ministry of Health with some assistance by NGOs. *Promatoras* (community health workers) are an important part of that strategy. These reforms will provide important experience in health care reorganization in a mid-level developing country, as well as a major step forward for Colombia's social security, but will not resolve the problem of providing care for the underserved rural population.

Summary

Colombia faces continuing struggles such as armed conflict with rebel groups, drug trafficking, poverty, unemployment, and poor sanitation and nutrition in many sectors of the country. Rural populations are at serious social and health disadvantage, and the inaccurate health statistics with poor public health surveillance systems in these underserved areas, due to the lack of resources, training, lack of awareness about the policies, do not reveal the real magnitude or characteristics of the problems, or the impact of the health programs and policies. Promoting health to achieve the MDGs for the country will be a serious challenge in the coming years. The health system is an important factor in this process but is concentrated in the cities and requires a strengthening of health promotion activities with prioritization of improved sanitation, maternal and child health, and communicable disease control, as well as facing the growing burden of NCDs.

ASIA

UNICEF divides Asia into two groups: (1) South Asia, and (2) East Asia and the Pacific (Table 13.22). The former includes India and has a total population of 1.5 billion people, while the latter includes China and has 2.0 billion people. Japan is excluded, being linked to the industrialized countries.

TABLE 13.22 Countries of South and East Asia, Demographic and Health Indicators

Indicator	South Asia	East Asia and Pacific
Population 2006	1.54 billion	1.97 billion
Annual births 2006	37.9 million	29.7 million
Total fertility rate		
1970	6.8	5.8
1990	5	4.3
2005	3.1	3.1
Under-5 mortality rate		
1970	206	122
1990	128	58
2005	84	33
Maternal mortality rate		
2005	500	150
Life expectancy at birth		
1970	48	58
1997	61	68
2005	64	91

Source: United Nations Children's Fund. State of the world's children, 1999 and 2007. Available at: http://www.unicef.org/sowc/

The countries of South Asia have progressed less rapidly than those of East Asia in terms of economic, demographic, and health status indicators. Each is a diverse group of nations, but many have common problems, including infectious diseases (e.g., AIDS, TB, and malaria), poor nutrition for the majority, problems related to rapid urbanization, and the growing problem of non-infectious diseases.

India

Located in the South Asia region, the Republic of India is one of the oldest civilizations in the world. India is a federal constitutional republic under a parliamentary system of government. This system is subdivided into 28 states and seven union territories administering 629 districts in their respective areas.

The seventh largest country by area, India's population increased from 700 million to 1.26 billion from 1980 to 2012. India has 17 percent of the world's population, making it the second most populous country in the world.

The HDI in 2012 ranks India 136th among the nations. Life expectancy at birth has reached 65.8 years. The 2011 Census showed that 68.8 percent of the people live in rural areas and 31.2 percent in urban areas. The overall literacy rate in the country is 74.0 percent (rural 68.9 percent, urban 84.9 percent) with a mean of 4.4 years of schooling for adults. There is a huge difference in female literacy rate between urban (79.9 percent) and rural (58.7 percent) areas. The WHO ranked India's health system 112th in the world in 2000, with an HDI rank of 134 in 2011. Over the past few decades, India has emerged as one of the fastest growing economies in the world, transforming the country from a traditionally agrarian to an increasingly industrialized economy.

India has one of the most ancient and richest civilizations, known as the Indus Valley, dating back to 3000 BCE. The existence of basic infrastructure for drainage and bathing highlights the hygienic and environmental sanitation practices during that period. The *Ayurveda* (or science of life) and *Siddha* system of medicine with broad concepts of health came into existence in 1400 BCE. Medical education was initiated in the ancient universities of Nalanda and Taxila during the post-Vedic period (600 BCE to 600 CE). During the period of Muslim rule (650–1850 CE), the Arabic system of medicine, *Unani*, was widely adopted in India.

In the mid-eighteenth century, the British established their rule in India, which lasted until 1947. The British mandate in India brought some successful initiatives in the development of public health laws and systems. The significant events of public health history included several acts passed or promulgated under British rule. In 1896, India faced a severe epidemic of plague which led to urgent action to improve public health.

The first major achievement for state health administration came in 1919, when states attained autonomy from the central government under the Montague–Chelmsford constitutional reforms. This change led to the decentralization of health administration and the creation of basic public health organizations in all states by 1921–1922. The Government of India Act 1935 provided further independence. All health activities were grouped in three categories: federal, concurrent, and provincial.

The Bhore Committee Report of 1946, based on a survey of health conditions and organizations, became the foundation for most of the planning and measures taken after India gained independence from Britain in 1947. The Committee recommendations included short- as well as long-term plans to improve the health services in the country.

With the Constitutional Amendment Acts (1992), the local bodies were assigned development activities, which have direct and indirect impacts on health. These include health and sanitation, family welfare, drinking water, women and children's development, the public distribution system, and poverty alleviation programs.

About two-thirds (65 percent) of the total population are aged 15–64 years. The World bank reports India's expenditure on health to be US$44 per capita in 2009 and US$59 in 2011. Out-of-pocket private expenditure accounts for nearly 86 percent of health expenditures.

Health Status Indicators

Life expectancy at birth increased from 42.2 years for males and 43.9 years for females in 1961 to 63.9 years for males and 67 years for females in 2011. Only 34 percent of the population have access to improved sanitation facilities. The crude death rate is 8 per 1000 population. The total fertility rate declined from 3.8 in 1990 to 2.6 in 2009, with marked differences between rural and urban areas (Figure 13.11). The child mortality rate (up to age 5 years) declined by 45.2 percent from 115 per 1000 live births in 1990 to 63 in 2010 (UNICEF, 2012); this decline is less than needed to achieve the targets set by the MDGs of a 75 percent reduction by 2015. Infant mortality declined from 80 per 1000 live births in 1990 to 50 in 2009, a reduction of 37.5 percent (Figure 13.12). The maternal mortality ratio fell by 35.2 percent, from 327 per 100,000 in 1999 and 2001 to 212 in 2007–2009, also falling short of the MDG targets (Figure 13.13).

Historically, the gender ratio in India has not been favorable to females, and there has been a steady fall in the ratio of females to males since the pre-independence period. In 1901, 972 females were recorded per 1000 males, the highest in the past century; the lowest ratio of females to males was recorded at 927 per 1000 males in 1991. The current ratio is 940 females per 1000 males, the highest since 1971

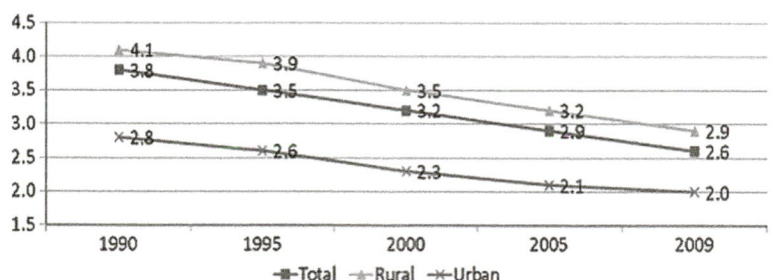

FIGURE 13.11   Fertility rate, India, 1990–2009. *Source: Office of Registrar General, India. Maternal and child mortality and fertility rates; 7 July 2011. Available at: http://censusindia.gov.in/vital_statistics/SRS_Bulletins/MMR_release_070711.pdf [Accessed 16 April 2013].*

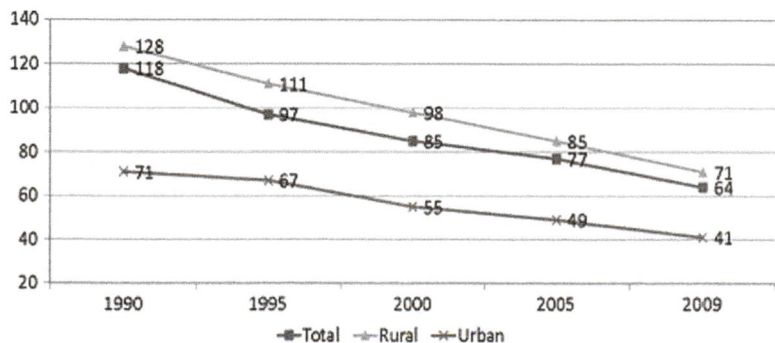

FIGURE 13.12   Under-five child mortality, rate per 1000 live births, India, 1990–2009. *Source: Office of Registrar General, India. Maternal and child mortality and fertility rates; 7 July 2011. Available at: http://censusindia.gov.in/vital_statistics/SRS_Bulletins/MMR_release_070711.pdf [Accessed 16 April 2013].*

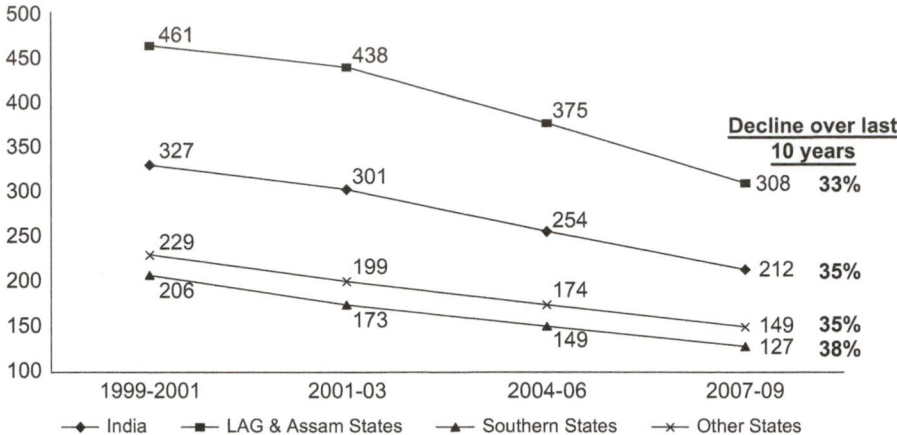

FIGURE 13.13 Maternal mortality rates, India, trend 1999–2009. Note: EAG=Empowered Action Group states. *Source: Office of Registrar General, India. Maternal and child mortality and fertility rates; 7 July 2011. Available at: http://censusindia.gov.in/vital_statistics/SRS_Bulletins/MMR_release_070711.pdf [Accessed 16 April 2013].*

but far below the overall the global picture of 984 females per 1000 males. The improvement in the gender ratio has largely taken place in urban areas. The population of children aged 0–6 years is 158.8 million, with a gender ratio of 914 girls per 1000 boys. The child population of India declined by 5 million between 2001 and 2011 because of declining birth and still high child mortality rates.

Health System Organization

The public health system in India uses both modern (allopathic) medicine and traditional Indian systems of medicine woven together to attempt to provide the envisioned goal of universal health care. Modern medicine makes up the majority of the health system; however, recent debates on strengthening the forms of Indian medicine have helped their integration at various levels within the system.

Major weaknesses of Indian health are the lack of universal access, poor levels of immunization coverage, poor maternal and child care, lack of access to prenatal and delivery care, and weak newborn care. In 2008, 53 percent of births were attended by skilled health workers, and the prevalence of contraceptive use among women aged 15–49 years was 54 percent. In 2010, the infant mortality rate was 48 per 1000 live births. Infant mortality rates vary widely within the country, from Madhya Pradesh with 62, Uttar Pradesh 61, and Odisha 61 per 1000 live births; to Kerala with 13, Goa 10, and Manipur 14 per 1000 live births. India is one of 10 countries worldwide with the highest rate of preterm births, with 3.5 million annually, accounting for 60 percent of preterm births. The maternal mortality ratio is 200 per 100,000 live births.

India is in the midst of an epidemiological and demographic transition, with declining mortality and fertility rates, an increasing burden of NCDs, and an increasing elderly population. The major health problems are communicable diseases including TB, HIV/AIDS, and diarrheal diseases, road accidents, vectorborne diseases, and NCDs.

With 2.4 million people living with HIV/AIDS, India accounted for more than 60 percent of Asia's estimated HIV infections. The prevalence rate of 0.34 percent in 2007 declined to 0.22 percent in 2010. The incidence of TB is 181 per 100,000 population. The burden of NCDs is rising and estimated to account for 53 percent of all deaths. In 2007, approximately 140,000 people in India lost their lives in road accidents.

The health system in India has three tiers: central, state, and local. India's constitution places the responsibility for the delivery of health care largely on state governments. Each state, therefore, has developed its own system of health care delivery, independent of central government. The central government plays a guiding, supporting, and coordinating role to strengthen the efforts of the state governments, and to ensure coverage of every area of the country for coordination of health activities and programs.

At the national level, there are three main organizations: the Union Ministry of Health and Family Welfare, headed by a cabinet minister with functions set out in the constitution; the Directorate General of Health Services, which provides technical advice to the union government on medical and public health issues; and the Central Council of Health, which provides continuous guidance, mutual understanding, and cooperation regarding a large number of health matters between the center and the states.

At the state level, the management comprises two organizations: the State Ministry of Health, headed by a minister at state level; and the State Health Directorate, which performs the role of technical advisor to the ministry for medicine and public health issues. Each state is responsible for all health services for the people in its jurisdiction.

The district level is further subdivided into six main types of administrative areas (subdivisions, talukas,

TABLE 13.23 Supply of Doctors and Nurses per Hospital Bed and per 1000 Population, India and Selected Countries, 2009

Indicator	India	USA	UK	Brazil	China
Average number of doctors per bed	0.6	0.81	0.53	0.69	0.46
Average number of nurses per bed	1.27	3	0.16	1.18	3.02
Number of doctors/1000 population	0.6	2.7	2.1	1.7	1.4
Number of nurses/1000 population	1.3	9.8	0.6	2.9	1

Sources: Organisation for Economic Co-operation and Development. http://www.oecd.org/World Health Organization. Available at: http://www.whoindia.org

community development blocks, municipalities and corporations, villages, and panchayats). The district is headed by a collector, who is responsible for various administrative offices including health. Primary care is provided by teams of health workers, trained volunteers and dais; secondary care in district hospital or community health centers; and tertiary care by regional or central level institutions.

A large percentage of the population of India lives in rural areas but health and care facilities are concentrated in urban areas. Moreover, 74 percent of physicians are located in urban areas serving approximately one-quarter of the population. Most of the people living in rural areas rely on local or traditional–cultural remedies.

Human resources and infrastructure capacities are limited compared to some developing countries and globally. India has a ratio of six physicians, 13 nurses, and nine hospital beds per 10,000 people (Table 13.23). In addition to allopathic care, various alternative and traditional systems of medicine are practiced. Annually, on average 26,499 allopathic doctors, 9865 Ayurvedic graduates, 1525 Unani graduates, 320 Siddha graduates, and 12,785 Homeopathic graduates are produced in the country.

The health care system in India is a mixed system in which the government provides health care at the primary, secondary, and tertiary levels. There is also a strong private sector infrastructure. The health insurance schemes are generally basic and inaccessible, with only 11 percent of the population having any form of health insurance coverage. Since 2005, the National Rural Health Mission has attempted to encourage state governments to join a central sponsored scheme that seeks to quickly increase the delivery of good-quality health care, especially to poor people living in rural areas. In 2009, a national health insurance (Rashtriya Swasthya Bima Yojana) for people living below the poverty line was initiated, with joint federal (75 percent) and state (25 percent) financing. In general, most of the southern states are better organized for immunization and other primary care services than the northern states,

demonstrating the high variability among them in health financing, outputs, and outcomes.

The National Urban Health Mission was launched in 2013 to improve access to health care services in urban parts of India, to connect with the National Rural Health Mission program, and to provide better health services with a focus on the health needs of the massive population of urban poor.

Medical tourism is on the rise in India owing to the low cost and high-quality health care facilities offered by the private health sector compared to the high costs in the western world. The participation of private sector in health care has risen significantly in the early twenty-first century. In the absence of resources and development in the governmental sector, the private sector seems to offer some hope for improving access to and quality of health care in India. India is one of the major suppliers of several bulk drugs produced at lower cost than in other countries, and many drug companies source their products from Indian manufacturers.

The first National Family Health Survey was carried out in 1992–1993, with subsequent surveys in 1998–1999 and 2005–2006 providing detailed information on health indicators. The plan is to repeat the survey every 5 years.

Various initiatives in India have included a goiter control program established in 1962, followed by a trachoma program in 1963. Government legislation includes: the Water Act (1974) for prevention and control of pollution, the Cigarettes Regulation (Of Production, Supply and Distribution) Act (1975), the Prevention of Food Adulteration (Amendment) Act (1976), and the Air (Prevention and Control of Pollution) Act (1981). Policy initiatives include the National Health Policy (1983), National Nutritional Policy (1993), National Population Policy (2000), National Health Policy (2002), National AIDS Policy (2002), and National Urban Sanitation Policy (2008). Program initiatives and events include: the Kartar Singh Committee recommending multipurpose community health workers in 1973, India becoming smallpox free in 1975, Integrated Child Development Services (1975), the Bhopal gas tragedy (1984), universal salt iodization (USI) efforts launched in 1992; Revised National Tuberculosis program with DOTS (1993), National Vector Borne Diseases Control Programme (2003), Integrated Diseases Surveillance Project (2004), a safe motherhood scheme (2005), National Rural Health Mission (2005), National Family Health Survey-3 (2006), NCD programme (2007), National Health Insurance for the poor below the poverty line (2009), National Urban Health Mission (2012), many more programs, and many institutions being built. The sustainability of these programs is critical to development of India at a time when its economic growth is high and a large middle class is emerging. As of 2012, only 61 percent of Indians are using iodized salt and iodine deficiency is still widespread.

Summary

A country as huge in size and population as India with such a wide range of socioeconomic settings requires health programs designed with enough elasticity to meet differing population needs. India's economy has grown enormously in the past decade, but health sector development has not kept pace with the country gaining middle-income status and a large part of the population, both rural and urban, continues to live in dire poverty with poor or no sanitation and little access to health care.

There is a need to reduce out-of-pocket expenditures by encouraging and providing nationalized or social or private insurance to minimize the financial burden on people. With low per capita health expenditures, even compared with other developing countries, India must focus more on strengthening basic public health needs. Similarly, more allocation of resources is needed in building the primary health care facilities in rural areas. A culture of professionalism needs

to be established in urban as well as rural areas by ensuring the quality of care through robust quality assessment and evaluation systems.

Many diseases in India are from preventable causes and sincere efforts are required to improve hygiene, water and sanitation facilities, nutrition, and education. Instead of merely focusing on treating diseases, trained public health professionals need to prevent disease by prioritizing health promotion.

One of the major constraints in achieving universal access to health services is limited or non-availability of skills and trained human resources. By international standards, India represents an unfortunate scenario and needs strong remedial actions. There is a gigantic need to develop a skilled workforce at each level in the area of modern health care as well as in traditional medicine systems, and to strengthen the role of civil society and other community-based organizations in developing community health programs. Population health should be a top priority by focusing more on the health sector to build and maintain the basic health infrastructure even at the village level.

There are opportunities to create new models by focusing on the strengths of diverse sectors. India has a good opportunity to tackle its health and care challenges by learning from, and avoiding, the expensive errors of industrialized economies. The improvement in strategies and policies will not only affect those who live in the country but also help other countries in the region that are struggling with their health care needs. As India is now included in the BRIC group of countries (Brazil, Russia, India, and China), representing mid-level and rapidly developing countries, its health and social systems need a major overhaul to keep pace with the rising population and international expectations. Table 13.24 shows life expectancy for BRIC countries in comparison to other selected countries.

TABLE 13.24 Life Expectancy at Birth, 1980–2012, BRIC Countries and Selected Countries

Country	1980	1990	2000	2005	2012	HDI Rank 2012
Brazil	62.5	66.3	70.1	71.6	73.5	85
Russian Federation	67.5	68.0	65.0	66.1	68.1	55
India	55.3	58.3	61.6	63.3	65.8	136
China	67.0	69.4	71.2	72.1	73.7	101
South Africa	56.9	61.5	54.0	51.1	53.4	121
Nigeria	45.5	45.6	46.3	49.0	51.9	153
Israel	74.1	76.5	79.0	80.1	81.6	16
Egypt	56.2	62.0	69.1	71.6	73.2	112

Note: BRIC = Brazil, Russia, India, and China; HDI = Human Development Index.

China

The People's Republic of China, with a population of more than 1.34 billion people (2010), is 47 percent urbanized and in the process of very rapid change and economic growth. China's GDP has grown at the extraordinary annual rate of 8 percent during the past 25 years, and its economy is now among the world's largest and most rapidly expanding. The GDP per capita increased from US$300 in 1988 to US$6757 in 2005 (PPP). WHO reports China's per capita expenditure on health in 2011 was US$432 and 5.2 percent of GDP, an increase from 4.7 percent of GDP in 2004. The HDI for China rose by 2.0 percent annually, from 0.407 in 1980 to 0.699 in 2012, when China ranked 101 out of 187 countries (HDI Index China, 2013), with health being the strongest positive indicator, as compared to education and income. In 2011, China's total public expenditures on health were 5.2 percent of GDP.

Life expectancy at birth increased from 63 in 1970 to 69 in 1990 and 73.5 in 2012. Infant mortality fell from 140 per 1000 live births in 1960 to 85 in 1970, 38 in 1990 and 16 in 2010. The under-five child mortality rate fell from 185 per 1000 live births in 1970 to 48 in 1990 and 18 in 2010, ranking 108th among all countries.

China has traditionally placed a strong social value on health and education, with major achievements in the development of a health care infrastructure during the twentieth century. Primary school education is nearly universal (96 percent for boys and 95 percent for girls). Youth literacy was 99 percent for both boys and girls in 2005–2010, with total adult literacy at 94 percent. As a result of falling birth rates and mortality patterns, the population pyramid is becoming similar to that of developed countries, with a rapidly aging population. The demographic transition is contributing to China's health challenges, with increasing longevity and declining mortality rates.

Health in Pre-Revolutionary China

Ancient China had a rich tradition of medical care and vital statistics. The Confucian and Taoist streams of Chinese culture supported a "high-order" medical system, emphasizing both preventive and curative services. Classical medical texts documented an empirical base of pharmacopoeias and therapeutic traditions. The yin–yang principle of resonant harmonies between alternative structures was in contrast to the single causation emphasis of western culture. Ancient Chinese medicine was based on treatment with herbal medicines, and at the same time included a holistic, psychosomatic perspective. Preventive medicine included attention to diet, rudimentary sanitation, personal hygiene, destruction of rabid animals, inoculation against smallpox, and an orientation towards the well-being of the individual as essential to health. However, this high-order medicine was available only to the elite of a rigid feudal–bureaucratic society. The vast bulk of the rural population relied on folk medicine based on herbal and other traditional practices.

Western medicine was introduced to China with the advent of missionary activities in the nineteenth century. It was accepted as another eclectic element of medicine, and medical schools were opened in the early twentieth century to train medical personnel in western medicine. In the period 1911–1949, medicine and public health advanced with the establishment of the national Ministry of Public Health (1927), 30 medical colleges, municipal public health departments, rural district hospitals, military medical services, a factory inspection service, and an array of public health professional departments including maternal and child health, and a large number of provincial medical centers. This brought vaccination, ophthalmic and other forms of surgery, western hospitals, clinics, and medical schools to the provinces and rural areas. The Japanese invasion and civil war that ravaged China from 1936 to 1948 halted this progress.

The Maoist Period

With the establishment of the People's Republic of China under Mao Tse Tung in 1949, the improvement of living and health conditions among the rural population became a high national priority. Between 1949 and 1965, China's national government, acting with advisors from the Soviet Union, emphasized the rapid expansion of training of mid-level health personnel – nurses, midwives, dispensers, and feldshers (see Chapter 14) – as well as doctors, whose numbers increased from 13,000 in 1945 to 150,000 in 1966. Hospital bed supply was also expanded rapidly so that by 1965 every county had at least one modern hospital.

The People's Republic of China established a centrally directed health system with strong emphasis on primary care and combating infectious diseases. Life expectancy rose as infant mortality was reduced and infectious diseases came under control. Under the slogan "away with all pests", vector control and sanitation works helped to eradicate previously endemic parasitic and other infectious diseases. The success of these programs contributed a solid base for later rapid development of the country, but the urban–rural gap is wide and a large sector of the Chinese population remains in poorer health than the rapidly increasing middle class (Hillier and Shen, 1996).

Rural Health Care

In 1966, as part of the Cultural Revolution, a new policy placed emphasis on developing rural health care by combining traditional medicine and self-sufficiency in health care at the community level. Western medical training was reduced in scope and duration. Auxiliary or "barefoot doctors" were trained briefly in a mixture of western and traditional Chinese medicine. The barefoot doctors brought health care to the rural population living in 27,000 communes, as well as to urban neighborhoods, focusing on sanitation, family planning information, immunization, and treatment of common illnesses.

The rural population of China then constituted some 80 percent of the total population. Rural health care was based on cooperative medical services (CMS) funded by the rural communes using barefoot doctor and referral services. Some medical staff were transferred to rural regions. The quality of care was questionable, but the program brought expanded access to the health care system to the rural population as part of the socialist program, and provided effective preventive and curative services to the vast bulk of the rural population of China during the 1960s and 1970s.

Market Reforms

Economic reforms were launched from 1978, including in agriculture, as part of the transition to a market economy. These reforms were meant to end the endemic problem of low productivity and management chaos. The communal agricultural system was replaced by individual farm units which were contracted with state agencies and sold excess products on private markets. This in effect, abolished the rural communes, virtually dismantling their apparently successful health care and public health system overnight, and putting nothing in its place. As a result, the CMS system, with no organizational or financial basis, was replaced with fee-for-service practice by the former barefoot doctors, who became private medical practitioners during the 1980s (Blumenthal and Hsaio, 2006).

From 1979, with the adoption of market-oriented reforms and new economic policies, a new focus on modernization replaced the ideological zeal and violence of the Cultural Revolution, and has since been associated with a period of rapid economic growth. Barefoot doctors were retrained and examined for licensing as village doctors. China's earlier high health standards have played a key role in the country's economic success, but the benefits of economic growth have not been shared equally, with a wide gap in socioeconomic indicators between different regions and communities, between urban and rural, and migrant and resident communities within cities. Surveys show that 30–50 percent of poor people in China indicate that health is the single biggest factor in their poverty, with reduced earning capacity and unaffordable medical care costs. As the Chinese economy boomed, largely by emulating western economic methods, its health care system nearly collapsed as a result of radical health care privatization.

By 1986, only 9.5 percent of the rural population was still covered by the CMS system, in comparison to 90 percent in 1978. This has resulted in greater use of emergency services and hospitalization, with less diligence in performance of preventive health services. In some areas, the CMS model is being restored as cooperative measures under local initiatives. The national Ministry of Health and provincial/regional or municipal departments of public health are responsible for health services in their jurisdictions, with a high degree of local autonomy.

Health Achievements

In the 1990s, a national campaign to eradicate poliomyelitis was conducted, showing good results with a reduction of cases from 5065 in 1990 to 1191 in 1992, through supplemental OPV national immunization days for children up to the age of 4 years. China has since joined the polio-free nations of the world. Crude mortality rates fell from 25 per 1000 in 1949 to 7.6 in 1970 and 6.8 per 1000 in 2006. Maternal mortality fell from 1500 per 100,000 births in 1949 to 95 per 100,000 in 1990 and 37 in 2011. Life expectancy increased from 44 years in 1960 to 63 years in 1970 and 76 in 2011. These health indicators are shown in Tables 13.25 and 13.26.

Health System Development

The WHO reports total health expenditures in China in 2005 at 4.7 percent of GDP; national or provincial governments covered 39 percent of total health expenditures, with private expenditures covering 61 percent (WHO, 2007). A national health survey in 2003 indicated a decline in access to health care, especially in rural areas, with a falling level of coverage in private or public health insurance systems.

TABLE 13.25 Health Indicators, China, 1970–2011

Indicator	1970	1990	2011	% Change 1990–2011
Infant mortality rate/1000 live births	140	39	13	−66.7
Child mortality rate/1000 live births	209	49	15	−69.4
Maternal mortality ratio/100,000 live births	NA	120	37	−69.2
Life expectancy at birth	63	69	76	+10.1

Note: NA = not available.
Sources: United Nations Children's Fund. State of the world's children 1999 and 2012. Available at: http://www.unicef.org/sowc2012/ [Accessed 21 July 2013].
World Health Organization. World Health Statistics 2013. Available at: http://www.who.int/gho/publications/world_health_statistics/2013/en/ [Accessed 14 June 2013].

TABLE 13.26 Vital Statistics, People's Republic of China, 1970–2010

Indicator	1970	1980	1990	2000	2010
Crude birth rate/1000	36	18	21	14	12
Crude mortality rate/1000	9	6.3	6.7	6.5	7.1
Natural annual increase (%)	1.7	1.7	0.8	0.8	0.8

Source: China's Statistical Yearbook 2012. Table 3-2. Available at: http://www.stats.gov.cn/tjsj/ndsj/2012/indexeh.htm [Accessed 15 May 2013].

Hospital bed ratios in China increased from 4.6 beds per 1000 population in 1985 to 6.1 in 1989 in urban areas, and went from 1.5 to 1.4 beds per 1000 in rural areas during the same period. Similarly, in 1989, the number of health professionals increased to 12.6 per 1000 urban residents compared to 2.3 per 1000 rural residents. In 2006, polyclinics and sanitary epidemiological stations were established throughout the country; patients are charged fees for services to support the health system. In the period 2006–2013, the health workforce included 14.6 doctors and 16.1 nurses per 10,000 population, hospital beds were 3.9 per 1000 population and psychiatric hospital beds were 0.14 per 1000 (WHO, World Health Statistics 2013).

Government continued to exert tight controls over the amount that publicly owned hospitals and clinics could charge for routine visits and services such as surgeries, standard diagnostic tests, and routine pharmaceuticals. However, it permitted facilities to earn profits from new drugs, new tests, and technology, with profit margins of 15 percent or more. The government modified its salary-based system of compensating hospital physicians with bonuses determined according to the revenue the physicians generate for their hospitals. Between 1990 and 2002, while total national spending on health care of all types (including public health) rose from 3.0 percent to nearly 5.5 percent of the GDP, public funding as a proportion of local public health revenues fell from nearly 60 percent to 42 percent.

During recent decades, one of the country's priorities was achieved with respect to human resources for health, namely an increase in the quantity of health personnel with 2–6 years of professional training. Consequently, the availability of health services has expanded rapidly, particularly in cities and better-off rural areas. Privatization of health services has, however, created a difficult situation in that half the population is unable to afford health services; only 25 percent of the urban and 10 percent of the rural population have any form of health insurance.

Health workers are not evenly distributed, and the poor rural areas suffer from shortages. There are also concerns about the quality of public health professional and clinical standards in education, training, and practice. The rising costs of health services create a paradox of increased numbers of health personnel and decreasing use of health services.

Health facilities for profit increased during 2000–2003, despite a decline in the number of patients; it is estimated that only 25 percent of the urban population and 10 percent of the rural population use any form of health care. Preventive and health promotion services are more cost-effectively delivered by nurses and other health disciplines. A national strategy for human resources planning will be needed to redefine the roles of health care practitioners, and to meet the needs of rural areas.

Emerging Infectious Diseases

At the end of 2005, there were 650,000 HIV-infected people reported, including 75,000 clinical cases, with 25,000 previous deaths from AIDS. China had an estimated 120 million people infected with hepatitis B in 2004. The public health challenges of SARS in 2003 provided a shock to governmental health authorities, revealing the weaknesses in national and provincial epidemiological and laboratory systems of health monitoring. Monitoring is especially important for frequently avian and domestic animal-borne infectious diseases which can cause major epidemics with international importance. Both HIV and syphilis have increased dramatically since 2000 and are predicted to become major epidemics fueled by millions of migrant workers with poor levels of sex education working in China's booming megacities. Many workers, far from their restrictive rural home environments, tend to access commercial sex workers, in part because of a shortage of young women in the population.

China's achievements in the control of vaccine-preventable and other infectious diseases have been matched by success in birth control and in arranging access to medical care for a population of over 1.3 billion people. Sixty-nine percent of the urban population and 28 percent of the rural population live with good sanitary conditions; however, some 100 million of the urban and most of the rural populations did not have access to safe water in 2008 (Carlton et al., 2012).

Emerging infectious diseases, such as SARS and avian influenza, are increasingly important because of their potential to become epidemics and pandemics. In addition to illness and death, they can cause social instability. China is a source of dangerous emerging viral diseases because of its enormous population living close to animal populations, intensive animal farming practices, enormous global trade, and poor infrastructure of veterinary and human health services. The SARS epidemic in 2003 affected 5327 people in mainland China, with 348 deaths, and spread to other countries via air transport. This is a precursor of more pandemics developing in the Chinese epicenter of newly emerging infectious disease.

In 1996, the H5N1 virus was identified in Guangdong province in China and later became an emerging global threat (WHO, H5N1 avian inflenza timeline, 2012). Since 2003, there have been 25 reported human cases of H5N1 in China, with 16 deaths. In 2013, a new threat in the form of H7N9 serotype avian influenza virus appeared in China, causing severe respiratory symptoms and a high death rate. It is primarily transferred from chickens to humans, but human-to-human infection may soon occur, with the threat of H7N9 avian influenza spreading locally and globally into a new pandemic.

Maternal and Child Health

China has achieved better outcomes in terms of infant and child mortality and life expectancy with lower health expenditures (3.5–4.7 percent of GDP) than many other developing countries. The transition to a market economy left many, especially in rural areas, with no medical care. About 90 percent of children in the rural areas have serious health problems and low vaccination rates. The collapse of state medicine led to a decline in the health of children in the rural areas of China at the beginning of the twenty-first century.

Since the 1960s, an emphasis on family planning resulted in a slowing of population growth. The "one child" policy adopted since the 1960s is enforced with many sanctions. This policy has led to widespread use of illegal ultrasound prenatal testing, promoted the abortion of female fetuses and female infanticide, and resulted in a high male to female population ratio in young age groups. There is now a large-scale deficit of marriage-age women, and important societal problems, particularly in the rural population.

Fertility declined in China with the crude birth rate declining from 36 per 1000 population in 1970 to 21 in 1990 and to 12 into 2010. The total fertility rate decreased from 2.4 births per woman in 1980 to 1.6 in 2010. Contraceptive prevalence reached 85 percent and institutional delivery rates reached 96 percent in 2006–2010. Maternal mortality declined from 60 per 100,000 in 1997 to 38 per 100,000 in 2008 (UNICEF, 2012). Immunization coverage in 2010 was reported as 99 percent for BCG, DTP, measles, polio, and hepatitis B (three doses), but no Hib vaccination was reported.

Non-Communicable Diseases

Serious health problems in China include high rates of CVDs, and lung cancer in polluted industrial cities, with very high rates of smoking. The leading causes of death are similar to those in developed countries, but regional disparities are apparent, with rural populations having higher death rates

in all categories. Urban health care has always been at an advantage in China. The epidemiological transition brought NCDs to the fore, with an increase during the period 1973–2009 from 53 percent to 85 percent of deaths due to NCDs and injuries. The Third National Death Survey reported the four leading causes of death as cerebrovascular disease, cancer, respiratory system diseases, and heart disease, and the total mortality rate for NCDs has reached 503 per 100,000. Cerebrovascular diseases, malignant neoplasms, and heart disease account for more than 50 percent of all deaths.

Aging of the population and the one-child-per-family policy create a situation where the tradition of family care of the elderly will be by a couple who will have sole responsibility for four parents. This will be compounded by the rapid movement of young people to the cities for economic opportunity, so that care of the elderly will be a major problem in the coming decades. Demographic projections suggest that there will be close to 350 million people older than 65 years (24.5 percent) in China in 2050. With economic growth, and dietary and lifestyle changes, vascular-related diseases are increasing rapidly. NCDs cause about 80 percent of deaths and are projected to result in US$550 billion of lost productivity between 2005 and 2015 due to associated deaths and disabilities.

As the country rapidly expands its economic potential, national health insurance is in an advanced stage of preparation. The Chinese experience in health status improvement for its huge population during a chaotic period is an enormous achievement considering the economic level of development in China. The country has successfully reduced fertility rates in an attempt to limit population growth and reduce infant, child, and general mortality rates, but it faces challenges not only in transforming the health system to a market economy but also from the effects of the profound demographic shift.

Obesity and smoking are major health problems in China. Currently, 23 percent of the population is overweight and 150 million people are suffering from hypertension. Diabetes prevalence is projected to double by 2030 to more than 42 million cases. In 2010, there were an estimated 301 million current smokers in China, 53 percent of men and just over 2 percent of women, increasing their risk of developing related NCDs. The WHO Framework Convention on Tobacco Control (FCTC) came into force in 2006 in China. However, the tobacco industry continues to grow with consent of the Chinese government. Cigarette production grew by 25 percent since the FCTC came into effect. Comparison of mortality rates for China, Japan, India, and several western countries shows that China's total mortality rate was just over half that of India and twice that of Japan. The cardiovascular mortality rate for China was three times that of Japan. China's cancer mortality rate was about 70 percent higher than that of India (2008 data from WHO World Health Statistics 2013).

Millennium Development Goals

China has made good progress towards achieving the MDGs since 2000, particularly in reducing childhood maternal and childhood mortality, stunting, and malaria. Progress is, however, markedly varied among the provinces, with the rural population and poorer provinces at a significant disadvantage. Progress in control of TB has been successful where DOTS was implemented, but China still lags behind in this MDG. This is in part due to fee-for-service payments required by the current health system.

- MDG1. *Eradicate extreme poverty and hunger* – extreme poverty and hunger reduced, but rural poverty still a problem; some regions lag well behind urban and industrialized parts of the country.
- MDG2. *Achieve universal primary education* – has been achieved.
- MDG3. *Promote gender equality and empower women* – target to reduce gender inequality in education achieved at primary school level and improved at higher levels; target likely to be achieved.
- MDG4. *Reduce child mortality* – target to reduce child mortality by two-thirds has been met; from 1990 to 2011 child mortality rate reduced by 69.4 percent (from 49 to 15 per 1000 live births).
- MDG5. *Improve maternal health* – targets to reduce maternal mortality by three-quarters and achieve universal access to reproductive health almost met; from 1990 to 2011 maternal mortality reduced by 69.2 percent (from 120 to 37 per 100,000).
- MDG6. *Combat HIV/AIDS, malaria and other diseases* – HIV spread halted; access to treatment for all in need achieved; malaria and other major infectious diseases being reduced.
- MDG7. *Ensure environmental sustainability* – some improvement, but massive air pollution problem in rapidly growing urban areas, and poor sanitation in rural areas; reforestation progressing; safe water supplies to rural areas increased coverage by an additional 220 million people.
- MDG8. *Develop a global partnership for development* – rated by UNDP as "ongoing" with increase in projects in southern countries; Internet use increased from 2.1 million people in 1998 to 420 million in 2010 (UNDP, MDG Report 2010).

Health Reforms (2006–2015)

The Chinese government announced a new wide-ranging health initiative for the period 2006–2010 and 2011–2015 of increasing government investment in health, improving the public health and clinical service delivery system, and establishing a medical safety net for the poor. Measures taken included improving capacity in disease prevention and control, including improved control of HIV/AIDS, schistosomiasis, and hepatitis B. The Health Ministry has undertaken action to prevent occupational and endemic diseases, strengthen maternal and child health care, and promote development of community health services. Other measures include deepening health system reform and allocating health resources rationally, better regulating pharmaceutical production/products and the market, and fostering a modern traditional Chinese medicine industry (WHO, 2005).

In 2001 insurance coverage was very low, with some 60 percent of total health expenditure being out of pocket. Health care is mostly on a fee-for-service basis, but fees and salaries are set artificially low so that drug sales and tests provide alternative income for facilities and providers. Hospitals are managed as profit-making enterprises, rather than for the public good. Public health and information systems are particularly weak, as seen during the SARS crisis (World Bank, 2011). For the tens of millions in the countryside, health provision is patchy, with poor access and rampant corruption causing social discontent. China's Health Ministry has announced a plan to reform the health system and provide a national service for all citizens, including the rural population. The *Healthy China 2020* program would provide a universal national health service and promote equal access to public services, with some comparisons being made with the NHS in Britain.

Between 2003 and 2011, national insurance coverage reportedly increased from 23 percent to 90 percent, accompanied by increased service utilization, particularly in rural areas: the participation rate in the New Rural Cooperative Medical Scheme had reached 96 percent. Employee health insurance, medical insurance for urban residents, and rural cooperative medical and hospitalization cost insurance have increased their reimbursement levels. However, benefits are not portable across regions, which is a concern for migrant populations and migrant workers (WHO, 2011).

China's recent 11th and 12th Five-Year Plans (2006–2010 and 2010–2015) respectively stress rebalancing the economy from export at all costs to promotion of domestic consumption and promoting quality of life and reduced inequalities. There is also stress on protecting the environment. The 12th plan envisions a GDP growth rate target of 7 percent, promoting domestic consumption over investments and exports, closing the income gap through minimum wage improvements and strengthened safety nets. Three sectors designated to receive a major boost are health care, energy and technology (*China's 12th Five Year Plan, 2010*).

In 2009, China launched a health care reform plan that included expanding access to basic medical coverage for citizens, modernizing the country's health care infrastructure and improving grassroots health care delivery which will continue in the 2010-2016 national goal to improve living standards for the rural population in particular.

The New Rural Cooperative Medical Scheme, launched in 2003, increased rural population health insurance to move away from a 25-year-old system in which out-of-pocket payments dominated health spending, and from the previous multiple insurance agencies towards unified payer systems. Public spending on health is increasing, along with an increased role of government in the direction of health services, a growing emphasis on NCDs and their prevention, and rising standards of training and performance in health facilities (Wagstaff et al., World Bank, 2009).

Summary

China, the country with the largest population on Earth, has experienced very rapid and sustained growth over the past three decades. Very great progress has been achieved in health and education indicators. The national health system developed in the Maoist period provided a base for health care, but has undergone massive changes from a governmental health plan to a largely privatized one.

National health insurance coverage is reported to have reached 90 percent of the population by 2011. New health reforms are in process, bringing critical changes to the current health system building on ongoing health care organization and financing, with the objective of reducing NCDs in the coming decades by 50 percent. Service delivery reforms will place emphasis on primary care and raising standards of training.

The rise of NCDs as major health issues presents an enormous challenge to the Chinese health system as the population ages, and as the increasing middle class in the rapidly growing urban population adopts unhealthy diet and lifestyle patterns; for instance, China continues to have a heavily smoking population.

Progress in achieving the MDGs has been impressive, especially in those related to health and universal primary education. However, China remains well behind in the HDI at 101st place and is ranked as a medium human development country. The social and health gap between urban and rural populations remains very high.

Despite over three decades of very high rates of industrialization and economic growth and a large and rapidly growing wealthy urban population, China continues to have a large poor rural population and a severe urban/rural divide in health, social and economic indicators. The severe earthquake of 2008, with its effects of tens of thousands of deaths and millions displaced from shattered homes, towns, and villages, revealed the weak infrastructure of the country. China is on the road to becoming an economic and political superpower. A great effort is required to ensure that the health system can meet this challenge.

Japan

Japan is a centralized industrialized democratic country with a 2010 population of 126.5 million and a GNI per capita of US$42,150. Longevity is among the world's highest, with a combined male and female life expectancy increase from 72 years in 1970 to 80 in 1997 and 83 years in 2010, and infant mortality of 2.3 in 2010. Japan ranks high (eighth) in the HDI, well above its 17th GDP ranking in 2005.

The OECD reports that total health spending increased from 4.8 percent of GDP in Japan in 1972 to 6.8 percent in 1982, 8 percent in 2005, and 9.6 percent in 2010, similar to the OECD average of 9.5 percent. Japan's public percentage of public expenditures as a percentage of total health expenditures increased from 77.6 percent in 1990 to 82.1 percent in 2010. Expenditures per capita in Japan were US$3120.4 in 2010, compared to US$8232.9 for the USA, US$5257.4 for Canada, and US$3434.3 for the UK (PPP).

In terms of health resources, Japan has fewer physicians per capita than most other OECD countries. In 2010, it had 2.2 practicing physicians per 1000 population, well below the OECD average of 3.1. Government policies limit the number of new entrants to medical schools. In the same year, Japan had 10.1 nurses per 1000 population, above the OECD average of 8.7. Japan had the highest number of acute care hospital beds of all OECD countries, with 8.1 beds per 1000 population in 2010, more than twice the OECD average of 3.4. Japan had by far the highest number of MRI units, with 46.9 per million population (2011) compared to 12.5 per million in OECD countries, and 101.3 CT scanners per million population, which is four times the OECD average of 22.6 (2010) (see Chapter 15).

Following World War II, the Japanese placed emphasis on maternal and child health, providing free maternal and child care services. Pregnant women receive maternity bonuses to encourage early prenatal care; child care services include an extensive immunization program, screening for diseases of the newborn, developmental testing, and special care for low birth weight or disabled newborns.

In 2011, Japan had the highest life expectancy among OECD countries at 83.0 years. Improved longevity has been largely due to declining death rates from heart diseases (the lowest of all OECD countries for both males and females). The birth rate and infant mortality rates in Japan have both fallen dramatically in recent decades. In 2010, the birth rate was 1.4 births per woman, and the 2011 infant mortality rate was one of the lowest in the world at 2.3 per 1000 live births, about half of the OECD average of 5.4 per 1000. Maternal mortality (adjusted) was 6 per 100,000 live births in 2008. Immunization coverage in 2006 was 97–99 percent for DTP, polio, and measles vaccines (OECD, 2012).

Japan has very low rates of heart disease, diabetes, and malignant disease mortality, but relatively high rates of stroke and trauma (motor vehicle accidents and suicides). CHD death rates in Japan are low, 25 per 100,000 for men as compared to 118–164 in Canada, the USA, Sweden, and the UK. However, stroke death rates are higher than in these and other countries. OECD reports that Japan has one of the lowest case fatality rates for stroke, with less than 2 percent of patients dying within 30 days after ischemic stroke. However, 10 percent of Japanese patients die within 30 days of having a heart attack, compared to 8 percent in Singapore, 6 percent in the Republic of Korea, and 3 percent in New Zealand.

Stomach cancer rates are higher, but lung and breast cancer mortality are lower than OECD averages. The Japanese diet is low in animal fat and cholesterol, which may relate to the low CVD mortality rates, but high in smoked and salty foods, perhaps explaining the higher cerebrovascular disease and stomach cancer mortality rates.

Japanese policy makers stress the importance of prevention and wellness to control health care costs. Current priorities include reducing smoking and improving blood pressure management. Since 2008, annual checkups have been obligatory for those between the ages of 40 and 74.

National Health Insurance

The basic health insurance program was enacted in Japan in 1922 as an extension of the employment-related social insurance law of 1874. In 1935, health insurance was extended to all manual workers, and further expanded in 1938 to self-employed people. By the mid-1960s, virtually the entire population was covered by a health insurance plan, through employers, local government, or trade associations. Government-managed health insurance covers employees of small businesses of fewer than 300 employees, which include some 29 percent of the population. Large companies, or groups of companies, with more than 700 employees, as an alternative to the government health insurance plan, can set up independent insurance plans for their employees. These currently cover some 25 percent of the population. Mutual aid associations provide coverage for civil servants, educators, and others (approximately 10 percent of the population).

Two laws promulgated in 1972 and in 1992 provide coverage for the elderly and low-income earners (32 percent of the population). Insurance for these groups is administered by local authorities or trade associations. There are also many health laws governing a wide range of issues including nutrition, TB prevention, communicable disease control, mental health, environmental sanitation, and health planning.

Financing and Services

In Japan, 81 percent of total health expenditures are from the public sector. Japan's health service is financed by a payroll tax with rates fixed by law at 3.6–4.5 percent for employees and 4.1–4.7 percent for employers. Government subsidies for health insurance cover 65 percent of health costs, with control of costs by national obligatory fee schedules for a basket of covered services. Co-payments by patients include 10 percent for employees and 30 percent for their dependants for hospital care and outpatient care.

Health plan benefits include medications, long-term care, dental care, and some preventive services, as well as medical and hospital services. Preventive care is provided free of charge through a nationwide network of health centers, with costs shared by the central and local governments.

Japan has a very high hospital bed-to-population ratio, with 8.1 acute care hospital beds per 1000 population, more than double the OECD average of 3.9, with few beds designated for long-term or nursing care. Hospital utilization rates are therefore high, with average lengths of stay much longer than in western countries. Over 55 percent of hospital beds are in private non-profit hospitals. Hospitals, generally small with an average size of 166 beds, include both acute and chronic nursing care patients. Since 2000, all patients aged 65 and older and some disabled between 40 and 64 have been covered under the national long-term insurance program, administered by the municipalities, financed half by taxation and half through premiums.

Patients have a free choice of doctors, two-thirds of whom work as private practitioners in both public and private hospitals. About one-third of physicians are solo GPs, paid on a fee-for-service basis, which favors primary care. National fee schedules promote primary care by financial incentives. Physicians also dispense medicines in their private clinics, so that the Japanese consume more medications than most industrialized populations. Physician contact rates are at least double those in western countries, at 12.9 contacts per capita per year, compared to 2.8 in Sweden and 5–7 in Canada, the USA, and the UK.

Japan has had very low birth and fertility rates since the 1950s. This fact, coupled with low mortality rates and increasing longevity, contributes to an aging of the population, posing problems for the health services in the years ahead. These include a need for geriatric facilities, nursing homes, home care, and support services for family care of the elderly. Proliferation of medical technology is a problem in the health system, and cost containment is now a major issue, with government regulation in health care likely to increase. Obesity rates have increased over recent years, but at 3 percent are well below rates in the USA (32 percent).

Smoking prevalence in Japan is one of the highest in the OECD countries (especially among males, at 46 percent), with 30 percent of all adults reporting smoking, compared to the OECD average of 24 percent. Japanese life expectancy for both men and women is among the world's highest (OECD, 2012; HDI, 2012). Japan has one of the largest proportions of elderly people in the world, with 22.7 percent of the population in 2009 over the age of 65 (compared to the USA's 13.0 percent and the OECD median of 15.5 percent). Yet per capita expenditure on health care of US$2878 was well below the USA (US$7960) and the OECD median (US$3128). This is due largely to strict regulation of the prices paid for all health care services included in the national benefit package. All insurers adhere to a national fee schedule, which is revised every 2 years, and providers are banned from charging above that fee.

In 2010, the government announced a new health information initiative, including patient electronic medical records accessible to all providers; telehealth to link patients with doctors and nurses in underserved areas; monitoring pharmaceutical prescriptions and adverse events to improve patient safety and monitoring; and a claims database of all conditions and interventions. This initiative is hindered by a lack of unique identifiers, and information exchange between providers and linking various databases come with privacy and data security issues.

Summary

The Japanese health system is highly decentralized, but regulated by the national authorities. It has achieved success in lowering mortality rates for most ages and conditions to among the lowest in the world, while restraining health care expenditures. Incentives for primary care seem to have been successful, despite the promotion of excess use of medication. Japan has a high total hospital bed ratio, in part because it has a high percentage of elderly people in its population and lacks alternative facilities for long-term care. The problem of caring for the elderly will be a challenge in the years ahead. The massive earthquake, tsunami, and nuclear plant disaster of 2011 damaged a large proportion of the health system facilities and placed great stress on the health system. Massive investment is required to restore the infrastructure, which is difficult during a time of economic recession.

COMPARING NATIONAL HEALTH SYSTEMS

The major participants in national health insurance networks include governments, employers, insurers, consumers, providers, and the public. Governments have increasingly come to recognize the economic and social value of improving the health of the population (Box 13.7). They carry this out through public health measures to ensure the basic health of the nation, as well as through legislation regarding the nature of health insurance, whether it is provided through private or public insurance mechanisms. In both the original UK Beveridge and the Soviet Semashko models, the government directly finances and provides health care. Services in the UK are provided by independent contractors, GPs, and hospitals operated by free-standing hospital boards (now trusts). The Semashko model was a totally state-financed and -operated service, with national norms and decentralized management. It brought health care to the far reaches of the Soviet Union, but failed to adjust to changing epidemiological and technological standards and thus the population health fell far behind that of the advanced countries.

In the Bismarckian model, health insurance is financed through social insurance, paid at the place of employment,

> **BOX 13.7 Stakeholders in National Health Systems**
> - The public, society, community, the nation, the regional and global community
> - Individual members of society
> - Government – national, state, region, and local authorities (town, county, city)
> - Employers – through negotiated health benefits for employees
> - Insurers – public, not-for-profit, and private for-profit
> - Patients, clients, or consumers – as individuals or groups
> - Risk groups – people with special risk factors for disease (e.g., age, poverty, occupational, or social groups)
> - Providers – hospitals, managed care plans, medical, dental, nursing, laboratories, others
> - Not-for-profit provider institutions
> - For-profit institutions, individual providers, and groups
> - Teaching and research institutions – universities, hospitals, institutes
> - Professional associations, societies, academies, colleges
> - Social security systems – with employer and employee contributions
> - The public, the community, public opinion
> - Political parties, philosophies, and social agendas
> - Advocacy groups – age, disease, poverty, or public interest groups
> - The media – advocacy and watchdog roles
> - Economies – national, regional, and local
> - International health organizations and movements
> - Pharmaceutical and medical technology industries.

with Sick Funds paying for services of private medical practice and non-government hospitals. The Canadian plan finances health services by provincial governments funded by general tax revenues with federal government financial support, but care is provided by private practitioners and not-for-profit community-based hospitals. In all variations of health insurance systems, the place of the government as provider and insurer is important to the care received by the consumer and the general state of public health.

There are many variations in methods of assuring national access to health care. Different approaches taken in the development and current structure of health systems in the USA, Canada, the UK, European and Nordic countries, Japan, Russia, Israel, and the developing countries are given as examples in this chapter. Improved health, as measured by outcome indicators such as increased longevity and reduced morbidity, mortality, or social and physiological dysfunction, is the major underlying objective of a national health system. This is sometimes forgotten in debates that may reflect interests of groups such as insurers, providers, institutions, governments, professional groups, or even political philosophies.

A typology of national health systems based on methods of financing and administration of health services provides a framework for their classification and for

TABLE 13.27 Typology of Financing and Administration of National Health Systems

Type	Financing Source	Administration
Bismarckian health insurance through social security, e.g., Germany, Japan, France, Austria, Belgium, Switzerland, Israel	Compulsory employer–employee tax payment to Sick Funds or through social security	Germany – governments regulate Sick Funds which pay private services; strong Sick Fund and doctors' syndicates; Israel's Sick Funds compete as HMOs with per capita payments for a mandatory "basket of services"
Beveridge National Health Service, e.g., UK, Norway, Sweden, Denmark, Italy, Spain, Portugal, Greece	Government – taxes and revenues; UK national financing; Nordic countries combine national, regional, and local taxation	Central planning, decentralized management of hospitals, GP service, and public health; integrated district health systems with capitation financing in UK
Semashko national health systems, e.g., former USSR	Government – taxes and revenues; post-Soviet national health insurance	Strong central government planning and control; financing by fixed norms per population; allocation of facilities and human resources promote increase in hospital beds and medical staff; post-1990 reforms emphasize decentralization with capitation and compulsory health insurance (i.e., payroll taxation)
Douglas national health insurance through government, e.g., Canada, Australia	Taxation – cost-sharing between provincial and federal governments	Provincial government administration; federal government regulation; medical services paid by fee-for-service; hospitals on block budgets; reforms to regionalize and integrate services
Mixed private/public system, e.g., USA, Latin America (e.g., Colombia), Asia (e.g., Philippines), and African countries (e.g., Nigeria)	Private insurance through employment and public insurance through social security for specific population groups	Strong government regulation (USA); mixed private medical services, public and private hospitals, state/county preventive services; DRG payment to hospitals, rapid increase in managed care; extension of Medicaid coverage

Note: HMO = health maintenance organization; GP = general practitioner; DRG = diagnosis-related group.

comparisons (Table 13.27). Mixed models have also developed as the dynamics of health system reform evolves in many countries.

Economic Issues in National Health Systems

As discussed in Chapter 11 and earlier in this chapter, health expenditures and costs of health care are major issues in national health systems. This is in part due to the rising costs of technology in medicine and the increasing age of the population with the associated increasing importance of NCDs, but it is also due to the traditional emphasis on institutional care. Health spending per capita since 2000 has increased more than twice as fast as economic growth on average across OECD countries (4.0 percent versus 1.6 percent), resulting in an increasing share of the economy being devoted to health in most countries, but there is some slowing in the rate of increase. This is particularly harsh in countries most affected by the economic recession, such as Greece and Ireland. In contrast, many developing countries are experiencing good levels of economic growth with a rising middle class, wider poverty reduction, and health benefits.

Health expenditures for preventive care, health promotion, and environmental health are generally not well financed or analyzed in routine economic data reporting. This makes economic analysis and comparison of interventions difficult, thereby handicapping the search for cost-effective interventions such as smoking reduction, hypertension management, obesity reduction measures, and promotion of physical exercise. These require greater emphasis in health systems development to reach out to populations at greatest risk, including the poor and disadvantaged ethnic groups, and education to promote greater public support of population health issues such as immunization, food fortification, fluoridation, and wider issues of environment and climate change.

National expenditures on health care are usually expressed in terms of US dollars as a percentage of GNP or GDP. The two economic figures are expressions of the total goods and services in a country, but GDP excludes international transfer of funds. Health care costs are also expressed directly as expenditures per capita (per person, per year), and indirectly as resources such as the number of hospital beds or medical personnel per 1000 (or 10,000) population (Table 13.28). The percentage of GNP spent on health care often is not necessarily directly related to health indicators, such as infant mortality or longevity, as funds may be allocated to or spent on less effective and more costly care. This said, countries with low GNP per capita that spend less than

TABLE 13.28 Population, Gross Domestic Product (GDP) per Capita, Health Facilities and Health Indicators, Selected Countries and Years, 2010–2011

Country	Population (millions) 2011	GDP per capita (US$) 2011	% GDP for Health 2010	Acute Care Beds/1000 2010	Average Length of Acute Care Hospital Stay (days) 2010	Discharges/ 1000 2010	Infant Mortality/ 1000 2010	Life Expectancy 2011
USA	313.1	48,043	17.6	2.6	5.4	126	6.2	78.7
Canada	34.4	40,470	11.4	1.7	7.7	83	4.9	80.0
Sweden	9.4	56,927	9.6	2.0	6.0	162	3.8	81.9
Germany	81.8	39,852	11.6	5.7	9.5	240	3.4	80.8
Finland	5.4	49,391	9.0	3.8	11.6	181	2.3	80.6
Denmark	5.6	59,683	11.2	2.9	4.6	181	3.4	79.9
Israel	7.8	31,282	7.6	1.9	5.8	198	3.7	81.8
UK	62.4	38,818	9.6	2.4	7.4	138	4.2	81.1
Russian Federation	142.8	13,089	5.0	9.3	11.3	216	9.8	69.0

Sources: World Health Organization, European Region. Health for All database; January 2013. Available at: http://data.euro.who.int/hfadb/Organisation for Economic Co-operation and Development. OECD health data 2013. Available at: http://stats.oecd.org/Index.aspx?DataSetCode=HEALTH_REAC (Accessed 5.1.2014).

4 percent on health have poorer health indicators because there are insufficient resources to provide a basic health level for all. Underfinancing and inappropriate allocation of funds have been severe problems in most post-Soviet health systems and even more so in most developing countries.

The supply of health care services remains one of the difficult and controversial topics in health planning. Economic analysis usually focuses on methods of financing in health care, and on methods of reimbursement or payment for services, placing less emphasis on the supply and quality of services. The World Bank's 1993 *World Development Report*, discussed in previous chapters, places major emphasis on the economic benefits of prevention and cost-effective measures to reduce the burden of disease. Excessive hospital utilization is not cost-effective.

Roemer's law (see Chapter 11) states that hospital utilization under insurance varies directly with bed supplies. Despite its essential validity, subsequent evidence shows that payment systems for hospital care can be modified so that there are incentives to prevent unnecessary admissions and to shorten hospital stays. As health costs increased rapidly, the concept of providing health care with fewer hospitalizations and more emphasis on ambulatory service has become one of the essentials of health policy in many countries since the 1970s.

Health resources indicators are quite variable among the developed market economy countries. Acute care bed ratios represent the number of general short-term beds per 1000 population. A hospital bed is not only a piece of furniture; it represents a service unit with staffing, services, maintenance, food, laundry, and other services. It is therefore an economic unit with fixed and variable costs when in use or even empty. Total hospital beds per 1000 population includes all institutional beds utilized for inpatient medical care, but not geriatric custodial care. Acute care hospital beds per 1000 is a more precise and comparable indicator (see Table 13.28). Many countries have reduced or are actively reducing hospital bed supplies (UK, the Nordic countries, most Western European countries, the USA, and Israel), developing alternatives to hospital care, using incentive payments to promote ambulatory or day-hospital treatments.

The hospital bed supply (i.e., the acute care bed-to-population ratio) of a country reflects historical patterns, medical practice traditions, concepts, medical technology, and the ability of an organization to adjust to changing circumstances and needs. It is also a function of financial incentives or disincentives. Reduced hospital bed supply has become part of standard health reforms in industrialized countries as more efficient care is achieved through better diagnostic facilities, ambulatory care, and other community-based services and facilities, including not-for-admission outpatient surgery, home care, and day care. There is also a wide recognition that hospitals are vital for short-term acute care, but themselves are health risks from infections and there are incidents that relate to errors, infections, disorientation of patients, and the discomfort of being away from the family environment. Because the elderly are greater consumers of health services than the young, another major factor that influences this ratio is the age distribution of the population. Investment in alternatives to hospital care and health promotion to reduce morbidity is essential to help control the rate of increase in costs of health care. This requires investment in education, legal action, screening, nutrition education, group counseling, selective home support services, and many other elements of the broad concept of health promotion.

Important factors in determining costs of national health systems include the salary or income of providers, levels of technology in the service, health planning criteria (norms), and hospital bed supply and utilization. Other factors are availability of home care and comprehensive community care services, use of integrated or regionalized models of health care delivery, methods of paying for hospital services, use of incentive payment systems to promote more efficient use of resources, and emphasis on prevention and health promotion. All of these are issues in the reform of national health systems. Table 13.28 shows a comparison of expenditure, resource utilization, and outcome indicators for selected industrialized countries. Globalization of economics and weakening of public services with trends towards privatization in health care are accompanied by technological advances, aging, and migration, all creating new challenges for a New Public Health.

No analysis of a health system can be complete without addressing the importance of poverty as a major contributing factor to morbidity and mortality. Poverty is associated with high rates of mortality from stroke, CHD, trauma, asthma, and cancer. Poverty is also related to many specific risk factors for illness, including low educational levels, poor housing conditions, poor nutrition, psychological depression, cigarette smoking, alcohol and drug abuse, teenage pregnancies, single parenthood, early bereavement or abandonment, lack of prenatal care, low birth weights, and family and neighborhood violence. Universal access to traditional medical care may alleviate some of these effects, but it fails to address the core issues. Social policy and health programs are interdependent, each contributing to improving the quality and length of life. Health planning, including economic indicators, must take this factor into account.

REFORMING NATIONAL HEALTH SYSTEMS

Health care systems are developed in the historical and political context of each country and continue to evolve slowly to meet the challenges of demographic, economic, and epidemiological change, public awareness and expectation, and changing technology in health. Impetus for reform of a health system may derive from a need for cost restraint, universal coverage, or efficiency in use of resources, or an effort to improve the satisfaction of consumers or providers (Table 13.29). The aim of improving the health of the population is the overall objective, but this is often expressed as "process indicators" such as improved access, equity, efficiency, and quality of care, as well as outcome measures of reduced morbidity, mortality, or loss of function.

Political and philosophical considerations for health reform often stress issues such as universal access, social solidarity, and equity in resource distribution, human resources, and hospital beds, but it is equally important to focus on targets for improving the health of the general population and special groups at risk. Philosophical and historical issues and arguments for national health insurance have included the need for social protection as a matter of national honor, but a system that fails to improve national health in terms of international outcome indicators does not meet this objective.

Debates and reforms in organizing health systems continue and are increasing in intensity as the political objectives of *Health for All* or *Health in All* policies meet the reality of rising costs, aging populations, and new health challenges such as unanticipated epidemics or technological breakthroughs. Efficiency in use of resources and satisfaction of the public and providers are major issues in all health systems. There is no single best means, despite claims by proponents of state-operated systems and equally ideological claims by market-force proponents. Direct importation of a total health system model is not feasible, because there are many factors contributing to the development of a health system relating to the political, social, and professional cultures of each country.

The assumption that market forces produce a better quality of health care is commonly expressed. This point of view has merit if taken in the sense that personal management of finances and choices in health care empower the individual to choose. This may be an advantage for a better educated urban population living near specialized services unavailable to others. Choice for consumers and freedom of choice (autonomy) for providers are different aspects of the market-force issue. Taken together, they provide a measure of protection of the rights of the consumer and provider to choose health systems. However, they diminish the responsibility and ability of the system to reach out and provide care and preventive services, or to manage resources effectively, so that important programs, particularly in health promotion and public health (such as care for high-risk groups, immunization, prenatal care, and care of the elderly), may suffer as a result. This set of rights is also sometimes in conflict with the imperative of cost control and the rapid increase in availability of new innovations in diagnosis and management whose benefits may be limited and costly, preventing other proven measures from being implemented. They may also have the undesired effect of promoting excess services such as unnecessary surgery, which has costly and potentially harmful consequences. Market mechanisms that

TABLE 13.29 Goals, Issues, Strategies, and Tactics for National Health Policies

Goal	Issues	Strategies/Tactics
National political commitment to improved health for all	Health as a government responsibility Universal access Adopt international standards Regional and social equity in access Rights to choose within health system Healthy lifestyle as national policy	Health promotion as policy Law/regulations Regulate consumers' rights in health Public information on health Advocacy groups – public, professional
Financing within national means for social benefits	Adequate overall financing (>6 percent GNP) Shift from supply planning to cost per capita per output Categorical grants to promote national objectives	Increase financing at national, state, and local government levels Health insurance as supplement Define "basket of services" and consumer rights Reduce acute care beds to <3/1000 District health authorities with capitation funding
Management for cost-effectiveness	Cost containment Cost-effective health initiatives Decentralized management National policy, monitoring, and standards Information systems/monitoring District health profiles	Incentives for primary care and outreach services Incentives for home care, long-term care facilities Increase home care, non-admission surgery, and long-term care facilities Health information systems Managed care and DRGs
Defining national health targets	Define leading causes of morbidity, mortality, and YPLL, hospitalization Regional, socioeconomic, ethnic analysis Health promotion vs treatment philosophy Prioritization for use of available resources Use relevant international standards	Social factor analysis in health Improve health KABP Community attitudes to health promotion Promote public health, nutrition, environment, immunization policies
Monitoring health status	Reporting, data systems, information technology	Computerization of medical records, IT and public access to population-based statistics

Note: GNP=gross national product; YPLL=years of potential life lost; DRG=diagnosis-related group; KABP=knowledge, attitudes, beliefs, and practices.

promote individual as well as health system responsibility can make important contributions in health.

Government responsibility to implement health promotion initiatives may limit individual rights. These include adding chlorine and fluoride to community water supplies; iodine to salt; and vitamin B, iron, and folic acid to flour. This is part of the substance of public health and requires people who may not directly benefit to accept this social solidarity in the interests of the need of others in the community and the community at large. A local, state, or national health authority may close a business that is hazardous to health, such as an unhygienic restaurant or a manufacturer of lead-contaminated toys. Management of health care systems must address macroeconomic and microeconomic issues for efficiency. Communities and regions will often address health planning in terms of its impact on business, jobs, and prestige in the community, as opposed to national or regional plans and priorities.

Since the 1970s, there has been a growing stress on health promotion as a way of reducing the burden of NCDs and the cost of health care for these diseases. This was stimulated and promoted by the Health Field Concept (Marc Lalonde, 1974), the Alma-Ata Conference on primary health care (1978 and 2000), and the WHO's Health for All concept (1978). Specific health targets in the USA (Healthy People 2010 and Healthy People 2020) and in the European Region of the WHO (1985 and 2005) place emphasis on measurable objectives as the basis for health planning, affecting the planning process (see Chapter 2). Even the most developed countries have substantial population groups living in poverty, with poor health conditions.

The 1990s was a decade of major reforms in national health systems. Industrialized countries attempted to restrain cost increases while retaining universal access. Sweden has brought down its health expenditures by reducing hospital bed supplies. The USA, building on its social security-based health insurance plan for the elderly and the poor since 1965, has brought health care to a large sector of the elderly and poor population. Many attempts have failed to bring in universal coverage as national health insurance, but the country is undergoing dramatic changes in the managed care revolution, with health insurance coverage by public and private insurance systems working under regulations to protect individual rights, and propelled by the need to control the rate of cost increases. In the USA about three-quarters of working-age adults on low incomes, an estimated 40 million people, lack any or adequate health insurance (Commonwealth Fund, 2012). The Obamacare plan, implemented over the years 2013–2015, is expected to lower private insurance costs and improve conditions for many who were excluded from medical insurance coverage by prior medical conditions. This is a major new health reform. Understanding the international experience of health care systems is essential for policy development to promote international standards and criteria for health systems development.

The Canadian Medicare program is about 70 percent funded from the public sector through general taxation of the federal and provincial governments. As the federal government has withdrawn from its earlier levels of participation, the provinces struggle to support the comprehensive range of services and increasing costs associated with an aging population, increasing professional fees and costly health technology. The federal Medical Care Act is limited to sharing provincial costs for physician and hospital services, leaving many essential service programs to provincial funding alone.

Israel has moved from voluntary Sick Funds to national health insurance, with the Sick Funds as managed care systems. The Eastern European countries are in a state of transition away from the pre-1990 Soviet model, adopting national health insurance and decentralized administration of services. In developing countries there is concern that directly financing services through the government will hinder the development of health services, so that there is a tendency to look towards national health insurance as a way to improve funding of services and bring more people into care. China has moved towards fee-for-service in its rural health care for some 70 percent of its population. All countries are struggling to develop adequate prevention models to reduce the burden of disease that can bankrupt a national health system.

Universal access to health care does not necessarily address social inequalities in health. Removal of financial barriers by itself does not guarantee good health. Many social, cultural, and environmental health risk factors are not correctable or preventable by medical or hospital care. They may be of greater importance than the medical care provided (see Chapter 3). The models presented may serve as examples for other countries, and will continue to do so. It is therefore useful to understand how they evolved, their successes and failures, and how they are continuing to develop.

There are two basic directions for reform, which are sometimes in conflict. One is the primary health care approach, which is based on tackling the basic health problems of developing countries by promoting primary health care as a public service through decentralized delivery and administration. The alternative approach, based on the market economy theory, is to promote access to health care

by national health insurance, funded through employer–employee contributions or through general taxation.

The fundamental differences in these two approaches present a dilemma for the developing countries and in many ways for the developed countries as well, as they struggle to control health care costs. A health insurance approach may increase funds available for health care, but it invites increases in expenditures for care, inequalities in access to care, and an emphasis on curative as opposed to preventive service. This is decidedly a medical approach, promoting hospital and physician services, with public health inadequately addressed and left to the care of private medical practitioners.

The market approach assumes that promoting competition will increase the quality of care and attention to consumer needs, but it is often associated with overutilization of costly services and drives health costs to very high levels. It is a luxury available only to the very wealthiest countries and still not providing all citizens with equal access to services. Developing countries may not have adequate funds to provide health care for all. At the same time, developed economies may not be able to fund health services on demand at levels that consumers and providers might consider ideal. This has led many countries to restrict access to specialist services and place other limitations on services, and is the basis for the managed care approach in the USA.

The public service model often leaves a national program underfunded, leading to problems of quality and morale for the provider as well as the consumer. However, a national health policy is still essential for vulnerable population groups or areas, whether in a developing or developed country. Even countries with universal national health insurance or service systems have population groups living in poverty, with poor health conditions. All countries have difficulties with health care in rural areas ill-served by collapsed rural health services.

The health sector is under great pressure to constrain costs. Employer–employee contribution systems are implementing changes to control costs because health costs are partly responsible for making their industry non-competitive in the global market. At the same time, there are inflationary pressures of the aging of the population, medical technological innovation, and high professional and public expectations. Health system reform includes downsizing the hospital sector and building up community health care.

SUMMARY

National health systems throughout the world are in a process of change, seeking restraint in increasing costs, universal coverage, equity in access and quality, as well as efficiency and effectiveness in use of resources to achieve health targets. Many countries are looking for ways to provide universal and equitable care, while controlling costs and improving efficiency. There is no single answer to the search for a health system that works.

Social security and social welfare systems took up the task of assuring access to health services during the twentieth century. National health systems evolved to provide access to medical, hospital, preventive, and community health services. Financing of services through general taxation based on progressive income tax, resource taxes, and excise taxes may be the most equitable way of raising funds. Many countries use social security systems based on employer–employee contributions to pay for health services. Universal access is a means of assuring that the economic barrier is removed for the total population and may lead to increased access to medical and hospital services for those previously excluded. It does not, in itself, guarantee achievement of important health targets. Allocation of resources is an even more fundamental problem.

NGOs serve many purposes such as testing out new ways of doing things before government makes a commitment, identifying gaps, and raising consciousness of important issues, such as advocacy groups and fund raising for cancer, and specialized conditions that require innovation and non-governmental support. However, governments often leave important issues to NGOs and do not absorb them into the total health system programming.

Beyond financing and resource allocation, there are many "non-tariff" barriers to health. Even in highly developed national health systems, such as that of the UK, social class, place of residence, education level, and ethnicity play important roles in morbidity and mortality rates. Factors other than medical or hospital care are vital, as classic risk factors for disease, such as diet, smoking, and physical fitness. Partly, however, social class differences in morbidity and mortality are the result of less well-defined aspects of poverty, such as depression, fear, insecurity, and lack of control over one's life. These are issues that are important to the achievement of national health goals and equity.

Health systems must be continuously evaluated. Traditional outcome indicators, such as infant, child and maternal morbidity and mortality, and disease-specific mortality rates, are important but not sufficient. Information on the incidence of vaccine-preventable diseases, immunization rates for infants, anemia rates in infancy and pregnancy, and disabling conditions is also necessary. Newer measures such as DALYs and QALYs (see Chapters 3 and 11) may help to change the emphasis from mortality to quality of life measures as part of the evaluation. National health systems require data systems that generate information needed for this continuous process of monitoring. High-quality academic centers for epidemiological, sociological, and economic analysis are needed to train health leaders and managers and to carry out the studies and research vital for health progress.

Despite the structural diversity and underlying philosophical differences in national health systems, there are important common elements. They are large employers and among the biggest economic sectors in their respective countries. All face problems of financing, cost constraint, overcoming structural inefficiencies, and, at the same time, funding incentives for high quality and efficiency. Funding in health care still predominantly goes to biomedical aspects including research, so that community-oriented health promotion aspects are less well supported despite social inequities being so widespread, in even the high-income countries. The long-standing struggle between the germ theory and the miasma theory orientation is still present, and a new balance needs to be found to deal with the issues of aging, mental health, and health promotion in all its aspects, as well as providing medical and hospital care.

A national health system is a complex with many parts that includes but goes well beyond medical care. The quality of the health protection community infrastructure (sewage, water, roads, and communication), the quantity and quality of food, levels of education, and professional organization are all parts of this continuum. Narrow planning for health systems ignores this message at the risk of missing its targets of improved health indicators, such as those adopted by the UN as the MDGs, and control of the burden of NCDs and injuries. National health systems are not only a matter of adequacy and methods of financing and assuring access to services; they address health promotion, national health targets, and adaptation to changing needs of the population, the environment, and a broad intersectoral approach to health of the population and the individual. The structure, content, and quality of a health system play a vital role in the social and economic development of a society and its quality of life.

Since the end of the Cold War in 1991, a new movement of globalization with economic and political dimensions has taken place with greater stress on human rights with direct application to health. The former socialist countries have gone through painful periods of transition. Many countries have developed free-market systems with dynamic growth in national economies. Health systems have struggled to adapt but great gains in longevity and reduced mortality from preventable diseases have been made in many countries. Public and private donor partnerships have emerged to help the poorest countries to cope with overwhelming health problems of HIV, TB, malaria, diarrheal and respiratory diseases, and the vaccine-preventable diseases.

The MDGs represent an international consensus on reducing poverty and preventable mortality, especially of women and children. The potential for achieving these goals depends on developing infrastructures of health systems which provide access for all and distribution to meet geographic and social inequalities in health. Each country needs to develop its own system, but can learn from the experience of others. The purpose of this chapter is to highlight the unique and common features of national health systems.

Universal access is a means of ensuring that the economic barrier is removed for the total population and may lead to increased access to medical and hospital services for those previously excluded. It does not, in itself, guarantee achievement of important health targets. A system of national health must be able to allocate resources to meet the needs of those with the highest risk of early disability or death, and not simply be a payment system for doctors and hospitals. The issue of changing demographics and epidemiological challenges must also be addressed. For global health, universal access to health care despite all its difficulties is a basic goal that must be achieved.

NOTE

For a complete bibliography and guidance for student reviews and expected competencies please see companion web site at http://booksite.elsevier.com/9780124157668

BIBLIOGRAPHY

International

American College of Physicians, 2008. Achieving a high-performance health care system with universal access: what the United States can learn from other countries. Ann. Intern. Med. 148, 1–21. Available at: http://annals.org/article.aspx?articleid=738556 (accessed 05.01.14).

Anderson, G.F., Markovich, P., 2011. Multinational comparisons of health systems data. Commonwealth Fund, New York. Available at: http://www.commonwealthfund.org/Publications/Chartbooks/2011/Jul/Multinational-Comparisons-of-Health-Systems-Data-2010.aspx (accessed 05.01.14).

Anderson, G.F., Hussey, P.S., Frogner, B.K., Waters, H.R., 2005. Health spending in the United States and the rest of the industrialized world. Health Aff. 24, 903–914. Available at: http://content.healthaffairs.org/content/24/4/903.long (accessed 05.01.14).

Banks, J., Marmot, M., Oldfield, Z., Smith, J.P., 2006. Disease and disadvantage in the United States and in England. JAMA 295, 2037–2045. Available at: http://jama.jamanetwork.com/article.aspx?articleid=202788 (accessed 05.01.14).

Brown, L.D., 2003. Comparing health systems in four countries: lessons for the United States. Am. J. Public Health 93, 52–56. Available at: http://www.ncbi.nlm.nih.gov/pmc/articles/PMC1447691/ (accessed 05.01.14).

Cylus, J., Anderson, G.F., Multinational comparisons of health systems data, 2006. Commonwealth Fund; May 2007. Available at: http://www.commonwealthfund.org/Publications/Chartbooks/2007/May/Multinational-Comparisons-of-Health-Systems-Data--2006.aspx (accessed 06.01.14).

DeWitt, L., Historical background and development of social security. Social Security Administration, Historian's Office; 2003. Available at: http://www.ssa.gov (accessed 24.05.08).

Figueras, J., McKee, M. (Eds.), 2012. Health systems, health, wealth and societal well-being: assessing the case for investing in health systems. Open University Press. WHO. on behalf of the European Observatory on Health Systems and Policies. Available at: http://www.euro.who.int/__data/assets/pdf_file/0007/164383/e96159.pdf (accessed 05.01.14).

Hussey, P.S., Anderson, G.F., Osborn, R., Feek, C., McLaughlin, V., Millar, J., et al., 2004. How does the quality of care compare in five countries? Health Aff. 23, 89–99. Open University Press. WHO. on behalf of the European Observatory on Health Systems and Policies. Available at: http://content.healthaffairs.org/content/23/3/89.full.pdf (accessed 05.01.14).

International Monetary Fund, 2011. World Economic Outlook. IMF, Washington, DC 2011. Available at: http://www.imf.org/external/pubs/ft/weo/2011/02/pdf/text.pdf (accessed 05.01.14).

Lasser, K.E., Himmelstein, D.U., Woolhandler, S., 2006. Access to care, health status, and health disparities in the United States and Canada: results of a cross-national population-based survey. Am. J. Public Health 96, 1300–1307. Available at: http://www.ncbi.nlm.nih.gov/pmc/articles/PMC1483879/ (accessed 05.01.14).

Nolte, E., McKee, C.M., 2008. Measuring the health of nations: updating an earlier analysis. Health Aff. 27, 58–71. Available at: http://content.healthaffairs.org/content/27/1/58.full.pdf (accessed 05.01.14).

Or, Z., 2000. Determinants of health outcomes in industrialised countries: a pooled, cross-country, time-series analysis. OECD Economic Studies No. 30, 2000/I. OECD, Paris. Available at: http://www.oecd.org/eco/growth/2732311.pdf (accessed 05.01.14).

Sarti, C., Rastenyte, D., Cepaitis, Z., Tuomilehto, J., 2000. International trends in mortality from stroke, 1968 to 1994. Stroke 31, 1588–1601. Available at: http://stroke.ahajournals.org/content/31/7/1588.long (accessed 05.01.14).

Schoen, C., Osborn, R., Doty, M.M., Squires, D., Peugh, J., Applebaum, S., 2009. A survey of primary care physicians in 11 countries, 2009: perspectives on care, costs, and experiences. Health Aff. 28, w1171–w1183. Available at: http://content.healthaffairs.org/content/28/6/w1171.full.pdf (accessed 05.01.14).

Squires, D., November 2012. Multinational comparisons of health systems data, 2012. Commonwealth Fund. Available at: http://www.commonwealthfund.org/~/media/Files/Publications/In%20the%20Literature/2012/Nov/PDF_2012_OECD_chartpack.pdf (accessed 05.01.14).

United Nations Children's Fund, 2013. The state of the world's children 2013. Children with disabilities. UNICEF, New Yorkhttp://www.unicef.org/sowc2013/ (accessed 04.08.13).

Walshe, K., McKee, M., McCarthy, M., Groenewegen, P., Hansen, J., Figueras, J., Ricciardi, W., 2013. Health systems and policy research in Europe: Horizon 2020. Lancet, 382. Available at: http://dx.doi.org/10.1016/S0140-6736(12)62195-3 (accessed 24.08.13).

World Bank, Health expenditure total (% of GDP). Available at: http://data.worldbank.org/indicator/SH.XPD.TOTL.ZS/countries (accessed 14.04.13).

World Bank, 2011. World development indicators 2011. Washington, DC: World Bank. Available at: http://data.worldbank.org. (accessed 15.05.11).

World Health Organization, 2007. Everybody's business: strengthening health systems to improve health outcomes: WHO's framework for action. WHO, Geneva. Available at: http://www.who.int/healthsystems/strategy/everybodys_business.pdf (accessed 16.08.13).

World Health Organization, 2012. Global TB report. Available at: http://www.who.int/tb/publications/global_report/en/ (accessed 12.06.13).

World Health Organization, 2000. The world health report 2000 – Health systems: improving performance. WHO, Geneva. Available at: http://www.who.int/whr/2000/en/whr00_ch1_en.pdf (accessed 15.08.13).

World Health Organization, 2013. World health statistics 2013. WHO, Geneva. Available at: http://www.who.int/gho/publications/world_health_statistics/EN_WHS2013_Full.pdf (accessed 14.04.13).

United States

America's Health Insurance Plans, http://www.ahip.org/ (accessed 24.03.13).

Centers for Disease Control and Prevention, Health United States 2012. Health expenditures Table 111. Available at: http://www.cdc.gov/nchs/fastats/hexpense.htm (accessed 06.01.14).

Center for Health Statistics. http://www.cdc.gov/nchs/ (accessed 24.03.13).

Fielding, J.E., Teutsch, S., Breslow, L., 2010. A framework for public health in the United States. Public Health Rev. 32, 174–189. Available at: http://www.publichealthreviews.eu/show/f/25 (accessed 05.01.14).

Fuchs, V.R., 2013. How and why US health care differs from that in other OECD countries. Viewpoint January 2, 2013. JAMA 309, 33–34. http://dx.doi.org/10.1001/jama.2012.125458. Available at: http://users.phhp.ufl.edu/jharman/healthecon/Fuchs%20JAMA%202013.pdf (accessed 05.01.14).

Health Care Financing Administration, http://www.os.dhhs.gov/about/opdivs/hcfa.html (accessed 24.03.13).

Ibrahim, S.A., 2007. The Veterans Health Administration: a domestic model for a national health care. Am. J. Public Health 97, 2124–2126. Available at: http://www.ncbi.nlm.nih.gov/pmc/articles/PMC2089116/ (accessed 05.01.14).

Iglehart, J.K., 2007. The battle over SCHIP. N. Engl. J. Med. 357, 957–960. Available at: http://www.nejm.org/doi/pdf/10.1056/NEJMp078156 (accessed 05.01.14).

Kaiser Family Foundation, http://www.kff.org/ (accessed 24.03.13).

Kerr, E., Fleming, B., 2007. Making performance indicators work: experience of the US Veteran's Health Administration. BMJ 335, 971–973. Available at: http://www.ncbi.nlm.nih.gov/pmc/articles/PMC2072029/ (accessed 05.01.14).

Kuttner, R., 2008. Market-based failure – a second opinion on US health care costs. N. Engl. J. Med. 358, 549–551. Available at: http://www.nejm.org/doi/pdf/10.1056/NEJMp0800265 (accessed 05.01.14).

Laugesen, M.J., Glied, S.A., 2011. Higher fees paid to US physicians drive higher spending for physician services compared to other countries. Health Aff. 30, 1647–1656. Available at: http://www.ncbi.nlm.nih.gov/pubmed/21900654 (accessed 05.01.14).

Lindenauer, P.K., Rothberg, M.B., Pekow, P.S., Kenwood, C., Benjamin, E.M., Auerbach, A.D., 2007. Outcomes of care by hospitalists, general internists, and family physicians. N. Engl. J. Med. 357, 2589–2600. Available at: http://www.nejm.org/doi/pdf/10.1056/NEJMsa067735 (accessed 05.01.14).

McGlynn, E., Asch, M., Adams, A., Keesey, J., Hicks, J., DeCristofaro, A., Kerr, E.A., 2007. The quality of health care delivered to adults in the United States. N. Engl. J. Med. 348, 2635–2645. Available at: http://www.nejm.org/doi/pdf/10.1056/NEJMsa022615 (accessed 05.01.14).

Medicare. http://www.medicare.gov/ (accessed 24.03.13).

National Center for Health Statistics, 2011. Health, United States, 2010: With special feature on death and dying. NCHS, Hyattsville, MD. Available at: http://www.cdc.gov/nchs/data/hus/hus10.pdf (accessed 05.01.14).

National Center for Health Statistics, 2013. Health, United States, 2012: With special feature on emergency care. Hyattsville, MD. Available at: http://www.cdc.gov/nchs/data/hus/hus12.pdf. (accessed 15.06.13).

National Conference of State Legislatures, May 2013. Managed Care, Market Reports and the States. Available at: http://www.ncsl.org/issues-research/health/managed-care-and-the-states.aspx (accessed 17.08.13).

Shaffer, E.R., 2013. The Affordable Care Act: the value of systemic disruption. Am. J. Public Health 103, 969–972.

Shortell, S.M., Gillies, R., Wu, F., 2010. United States innovations in healthcare delivery. Public Health Rev. 32, 190–212. Available at: www.publichealthreviews.eu (accessed 15.07.13).

United Nations Development Programme, Human Development Index – United States. Available at: http://hdrstats.undp.org/countries/country_fact_sheets/cty_fs_USA.html (accessed 24.03.13).

US Census Bureau, http://www.census.gov/# (accessed 21.07.13).

US Department of Health and Human Services, 2011. CDC health disparities and inequalities – United States. MMWR. Morb. Mortal. Wkly. Rep., 60.

US Department of Health and Human Services, Healthy People. Available at: http://www.healthypeople.gov/ (accessed 24.03.13).

US Department of Health and Human Services, 2006. Healthy People 2010. Midcourse review. Executive summary. DHHS, Washington, DC. Available at: http://www.cdc.gov/mmwr/pdf/other/su6001.pdf (accessed 05.01.14).

US Department of Health and Human Services, Healthy People 2020 framework: the vision, mission, and goals of Healthy People 2020. Available at: http://www.healthypeople.gov/2020/consortium/HP2020Framework.pdf (accessed 24.03.13).

US Department of Health and Human Services, Surgeon General. Available at: http://www.surgeongeneral.gov/ (accessed 06.01.14).

Welch, W.P., Miller, M.E., Welch, H.G., Fisher, E.S., Wennberg, J.E., 1993. Geographic variation in expenditures for physicians' services in the United States. N. Engl. J. Med. 328, 621–627. Available at: http://www.nejm.org/doi/full/10.1056/NEJM199303043280906. (accessed 05.01.14).

Canada

Bensimon, C.M., Smith, M.J., Pisartchik, D., Sahni, S., Upshur, R.E., 2012. The duty to care in an influenza pandemic: a qualitative study of Canadian public perspectives. Soc. Sci. Med. 75, 2425–2430. Available at: http://www.ncbi.nlm.nih.gov/pubmed/23089615 (accessed 05.01.14).

Buhr, K., 2013. Access to medical care: how do women in Canada and the United States compare? Prev. Med. 56, 345–347. Available at: http://www.ncbi.nlm.nih.gov/pubmed/23462478 (accessed 05.01.14).

Butler-Jones, D., 2007. The health of the public is the foundation of prosperity: the work of the Public Health Agency of Canada at home and around the world. CMAJ 177, 1063–1064. Available at: http://www.cmaj.ca/content/177/9/1063.full.pdf (accessed 05.01.14).

Canadian Institute for Health Information. http://www.cihi.ca/; (accessed 24.03.13).

Deber, R.B., 2003. Health care reform: lessons from Canada. Am. J. Public Health 93, 20–24. Available at: http://www.ncbi.nlm.nih.gov/pmc/articles/PMC1447685/ (accessed 05.01.14).

Evans, R.G., 1989. Controlling health expenditures – the Canadian reality. N. Engl. J. Med. 320, 571–577. Available at: http://www.nejm.org/doi/full/10.1056/NEJM198903023200906 (accessed 05.01.14).

Guyatt, G.H., Devereaux, P.J., Lexchin, J., Stone, S.B., Yalnizyan, A., Himmelstein, D., et al., 2007. A systematic review of studies comparing health outcomes in Canada and the United States. Open Med. 1, e27–e36. Available at: http://www.openmedicine.ca/article/viewFile/8/15 (accessed 05.01.14).

Hermus, G., Stonebridge, C., Thériault, L., Bounajm, F., Home and community care in Canada: an economic footprint. Available at: http://www.conferenceboard.ca/temp/42707adf-5444-4dbd-8cc6-0d4bfdcd84ca/12-306_homeandcommunitycare_prt.pdf.

LaLonde, M., 1974. New perspectives on the health of Canadians. Department of National Health and Welfare, Ottawa. Available at: http://www.phac-aspc.gc.ca/ph-sp/pdf/perspect-eng.pdf (accessed 05.01.14).

Marchildon, G.P., 2013. Canada: Health system review. Health Systems in Transition. European Observatory on Health Systems and Policies., Copenhagen 15(1), 1–179. Available at: http://www.euro.who.int/__data/assets/pdf_file/0011/181955/e96759.pdf (accessed 05.01.14).

Mental Health Commission of Canada, February 2013. Why investing in mental health will contribute to Canada's economic prosperity and to the sustainability of our health care system: the key facts. Available at: http://www.mentalhealthcommission.ca/node/742. (accessed 24.08.13).

Organisation for Economic Co-operation and Development, OECD health data 2007. How does Canada compare? Available at: http://www.oecd.org/health/health-systems/36956887.pdf; http://www.oecd.org/els/health-systems/oecdhealthataglance2009keyfindingsforcanada.htm (accessed 24.03.13).

Roos, N.P., Brownell, M., Shapiro, E., Roos, L.R., 1998. Good news about difficult decisions: the Canadian approach to hospital cost control. Health Aff. 17, 239–246. Available at: http://content.healthaffairs.org/content/17/5/239.full.pdf. (accessed 05.01.14).

United Nations Development Programme. http://hdrstats.undp.org/countries/country_fact_sheets/cty_fs_CAN.html (accessed 24.03.13).

United Kingdom

Acheson, D., 2000. Health inequalities: impact assessment. Bull. World Health Organ. 78, 75–77. Available at: http://www.scielosp.org/pdf/bwho/v78n1/v78n1a08.pdf. (accessed 05.01.14).

Boyle, S., 2011. United Kingdom (England): health system review. Health Syst. Transit. 13, 1–486. Available at: http://www.euro.who.int/__data/assets/pdf_file/0004/135148/e94836.pdf (accessed 05.01.14).

Campbell, S., Reeves, D., Kontopantelis, E., Middleton, E., Sibbald, B., Roland, M., 2007. Quality of primary care in England with the introduction of pay for performance. N. Engl. J. Med. 357, 181–190. Available at: http://www.qualitymed.org.il/vault/files/p4p%20in%20uk%20nejm.pdf (accessed 05.01.14).

Coulter, A., 1995. Evaluating general practice fundholding in the United Kingdom. Eur. J. Epidemiol. 5, 233–239. Available at: http://eurpub.oxfordjournals.org/content/5/4/local/back-matter.pdf (accessed 05.01.14).

Dixon, T., Shaw, M.E., Dieppe, P.A., 2006. Analysis of regional variation in hip and knee joint replacement rates in England using Hospital Episodes Statistics. Public Health 120, 83–90. http://www.ncbi.nlm.nih.gov/pubmed/16198381 (accessed 05.01.14).

Doran, T., Fullwood, C., Gravelle, H., Reeves, D., Kontopantelis, E., Hiroeh, U., Roland, M., 2006. Pay-for-performance programs in family practices in the United Kingdom. N. Engl. J. Med. 355, 375–384. Available at: http://www.nejm.org/doi/full/10.1056/NEJMsa055505 (accessed 05.01.14).

Dusheiko, M., Gravelle, H., Yu, N., Campbell, S., 2007. The impact of budgets for gatekeeping physicians on patient satisfaction: evidence from fundholding. J. Health Econ. 26, 742–762. Available at: http://www.ncbi.nlm.nih.gov/pubmed/17276530 (accessed 05.01.14).

HM Government, 2010. Health Lives, Healthy People. Our strategy for public health in England. Available at: https://www.gov.uk/government/publications/healthy-lives-healthy-people-our-strategy-for-public-health-in-england. (accessed 24.03.13).

Klein, R., 2006. The troubled transformation of Britain's National Health Service. N. Engl. J. Med. 355, 409–415. Available at:. http://www.nejm.org/doi/pdf/10.1056/NEJMhpr062747 (accessed 05.01.14).

Marmot, M., 2007. Commission on Social Determinants of Health. Achieving health equity: from root causes to fair outcomes. Lancet 370, 1153–1163. Available at: http://www.thelancet.com/journals/lancet/article/PIIS0140-6736(07)61385-3/abstract (accessed 05.01.14).

McKee, M., 2012. Does anyone understand the government's plan for the NHS? BMJ 344 e399 Available at: http://www.bmj.com/content/344/bmj.e399 (accessed 05.01.14).

Pocock, S.J., Shaper, A.G., Cook, D.G., Phillips, A.N., Walker, M., 1987. Social class differences in ischemic heart disease in British men. Lancet 330(885): 197–201. Available at: http://www.thelancet.com/journals/lancet/article/PIIS0140-6736(87)90774-4/abstract (accessed 05.01.14).

Pollock, A., Price, D., Roderick, P., Treuherz, T., McCoy, D., McKee, M., Reynolds, L., 2012. How the Health and Social Care Bill 2011 would end entitlement to comprehensive health care in England. Lancet 379, 387–389. Available at: http://www.thelancet.com/journals/lancet/article/PIIS0140-6736(12)60119-6/fulltext (accessed 05.01.14).

Townsend, N., Wickramasinghe, K., Bhatnagar, P., Smolina, K., Nichols, M., Leal, J., Luengo-Fernandez, R., Rayner, M., 2012. Coronary heart disease statistics 2012 edition. British Heart Foundation, London (accessed 05.01.14.). Available at: http://www.bhf.org.uk/plugins/PublicationsSearchResults/DownloadFile.aspx?docid=508b8b91-1301-4ad7-bc7e-7f413877548b&version=-1&title=Coronary+Heart+Disease+Statistics+2012+&resource=G608 (accessed 08.01.14).

UK National Health Service, September 2012. Acute hospital summary and NHS beds information. Quarter ending. Available at: http://data.gov.uk/dataset/annual_acute_hospital_activity_and_nhs_beds_information. (accessed 24.03.13).

UK National Health Service, Quality and Outcomes Framework. at: http://www.qof.hscic.gov.uk/index.asp (accessed 06.01.14).

UK Office for National Statistics, Statistical bulletin. Avoidable mortality in England and Wales. Available at: http://www.ons.gov.uk/ons/dcp171778_311826.pdf (accessed 17.08.13).

Europe

Greer, S.L., Hervey, T.K., Mackenbach, J.P., McKee, M., 2013. Health law and policy in the European Union. Lancet 381, 1135–1144. Available at: http://www.thelancet.com/journals/lancet/article/PIIS0140-6736(12)62083-2/fulltext (accessed 30.03.13).

Joossens, L., Raw, M., 2011. The tobacco control scale 2010 in Europe. Association of the European Cancer Leagues, Brussels. Available at: http://media.hotnews.ro/media_server1/document-2011-03-23-8428154-0-raportul-tobacco-control-scale-2010-europa.pdf (accessed 05.01.14).

Judge, K., Platt, S., Costongs, C., Jurczak, K., 2005. Health inequalities: a challenge for Europe. A background paper on health inequalities commissioned by the UK Presidency to coincide with Presidency Conference on Inequalities in Health. Available http://ec.europa.eu/health/ph_determinants/socio_economics/documents/ev_060302_rd05_en.pdf. (accessed 06.01.14).

Kringos, D.S., Boerma, W., van der Zee, J., Groenewegen, P., 2013. Europe's strong primary care systems are linked to better population health but also to higher health spending. Health Aff. 32, 686–694. Available at: http://content.healthaffairs.org/content/32/4/686.abstract (accessed 05.01.14).

Leppo, K., Ollila, E., Pena, S., Wismar, M., Cook, S. (Eds.), 2013. Health in All policies: seizing opportunities, implementing policies. Ministry for Social Affairs and Health. Available at: http://www.euro.who.int/__data/assets/pdf_file/0007/188809/Health-in-All-Policies-final.pdf (accessed 05.01.14).

LSE Health, 2008. Health system snapshots: perspectives from six countries. Eurohealth 14 (1). Available at: http://www.lse.ac.uk/collections/LSEhealth/pdf/eurohealth/vol14no1.pdf (accessed 21.08.13).

Mackenbach, J.P., Karanikolos, M., McKee, M., 2013. The unequal health of Europeans: successes and failures of policies. Lancet 381, 1125–1134. Available at: http://www.thelancet.com/journals/lancet/article/PIIS0140-6736(12)62082-0/fulltext (accessed 30.03.13).

Mackenbach, J.P., McKee, M., 2013. A comparative analysis of health policy performance in 43 European countries. Eur. J. Public Health 23, 195–344. Available at: http://eurpub.oxfordjournals.org/content/23/2/195.full.pdf (accessed 05.01.14).

Nolte, E., McKee, C.M., 2008. Measuring the health of nations: updating an earlier analysis. Health Aff. (Millwood) 27, 58–71 Available at: http://content.healthaffairs.org/content/27/1/58.full.pdf (accessed 05.01.14).

Organisation for Economic Co-operation and Development, 2012. Health at a glance: Europe. Available at: http://www.oecd-ilibrary.org/social-issues-migration-health/health-at-a-glance-europe-2012_9789264183896-en. (accessed 21.07.13).

Wismar, M., McQueen, D., Lin, V., Jones, C.M., Davies, M., 2013. Rethinking the politics and implementation of health in all policies. Isr. J. Health Policy Res. 2, 17. Available at: http://www.ijhpr.org/content/2/1/17/abstract (accessed 17.08.13).

World Health Organization, 2012. Health 2020: a European policy framework supporting action across government and society for health and well-being. WHO, Geneva. Available at: http://www.euro.who.int/__data/assets/pdf_file/0009/169803/RC62wd09-Eng.pdf (accessed 17.08.13).

World Health Organization, European Region, The European health report 2012: charting the way to well-being. Available at: http://www.euro.who.int/en/what-we-do/data-and-evidence/european-health-report-2012/the-european-health-report-2012-charting-the-way-to-well-being (accessed 14.06.13).

Nordic Countries, Germany, and the Netherlands

Allin, S., Mossialos, E., McKee, M., Holland, W., 2004. Making decisions on public health: a review of eight countries. European Observatory of Health Systems, Brussels. Available at: http://www.euro.who.int/__data/assets/pdf_file/0007/98413/E84884.pdf (accessed 05.01.14).

Anell, A., 2011. Choice and privatisation in Swedish primary care. Health Econ. Policy Law 6, 549–569. Available at: http://www.ncbi.nlm.nih.gov/pubmed/20701829 (accessed 05.01.14).

Bennema-Broos, M., Groenewegen, P.P., Westert, G.P., 2001. Social democratic government and spatial distribution of health care facilities. The case of hospital beds in Germany. Eur. J. Public Health 11, 160. Available at: http://eurpub.oxfordjournals.org/content/11/2/160.full.pdf (accessed 05.01.14).

Brown, L.D., Amelung, V.E., 1999. "Manacled competition": market reforms in German health care. Health Aff. 18, 76–91. Available at: http://content.healthaffairs.org/content/18/3/76.full.pdf (accessed 05.01.14).

Busse, R., 2001. Risk adjustment compensation in Germany's Statutory Health Insurance. Eur. J. Public Health 11, 174–177. Available at: http://eurpub.oxfordjournals.org/content/11/2/174.full.pdf (accessed 05.01.14).

Busse, R., Riesberg, A., 2004. Health care systems in transition: Germany. WHO Regional Office for Europe on behalf of the European Observatory on Health Systems and Policies, Copenhagen. Available at: http://www.euro.who.int/Document/E85472.pdf (accessed 24.05.08).

European Observatory on Health Systems and Policies – Sweden, 2005. Copenhagen. WHO, European Region. Available at: http://www.euro.who.int/Document/E88669sum.pdf. (accessed 24.05.08).

Finland, Ministry of Social Affairs and Health. Available at: http://www.stm.fi/en/frontpage (accessed 04.08.13).

Garcy, A.M., Vagero, D., 2013. Unemployment and suicide during and after a deep recession: a longitudinal study of 3.4 million Swedish men and women. Am. J. Public Health 103, 1031–1038. Available at: http://ajph.aphapublications.org/doi/pdf/10.2105/AJPH.2013.301210 (accessed 05.01.14).

Germany, Federal Ministry of Health. https://www.bundesgesundheitsministerium.de/ministerium/english-version.html (accessed 04.08.13).

Glenngård, A.H., Halte, F., Svensson, M., Anell, A., Bankauskaite, V., 2005. Health systems in transition: Sweden. WHO Regional Office for Europe on behalf of the European Observatory on Health Systems and Policies, Copenhagen. Available at: http://www.euro.who.int/__data/assets/pdf_file/0010/96409/E88669.pdf (accessed 05.01.14).

Harrison, M., Calltorp, J., 2001. The reorientation of market-oriented reforms in Swedish health care. Health Policy 50, 219–240. Available at: http://www.ncbi.nlm.nih.gov/pubmed/10827309 (accessed 05.01.14).

Johnsen, J.R., 2006. Health Systems in Transition: Norway. WHO Regional Office for Europe on behalf of the European Observatory on Health Systems and Policies, Copenhagen. Available at: http://www.euro.who.int/__data/assets/pdf_file/0005/95144/E88821.pdf (accessed 05.01.14).

Russia

Andreev, E., Nolte, E., Shkolnikov, V., Varavikova, E., McKee, M., 2003. The evolving pattern of avoidable mortality in Russia. Int. J. Epidemiol. 32, 437–446. Available at: http://ije.oxfordjournals.org/content/32/3/437.full (accessed 05.01.14).

Balabanova, D., Falkingham, J., McKee, M., 2003. Winners and losers: the expansion of insurance coverage in Russia in the 1990s. Am. J. Public Health 93, 2124–2130. Available at: http://www.ncbi.nlm.nih.gov/pmc/articles/PMC1448163/ (accessed 05.01.14).

Balabanova, D., Roberts, B., Richardson, E., Haerpfer, C., McKee, M., 2012. Health care reform in the former Soviet Union: beyond the transition. Health Serv. Res. 47, 840–864. Available at: http://onlinelibrary.wiley.com/doi/10.1111/j.1475-6773.2011.01323.x/pdf (accessed 05.01.14).

Bobak, M., Murphy, M., Rose, R., Marmot, M., 2007. Societal characteristics and health in the former communist countries of Central and Eastern Europe and the former Soviet Union: a multilevel analysis. J. Epidemiol. Community Health 61, 990–996. Available at: http://www.ncbi.nlm.nih.gov/pmc/articles/PMC2465607/ (accessed 05.01.14).

Centers for Disease Control and Prevention, 1992. Public health assessment – Russian Federation. MMWR. Morb. Mortal. Wkly. Rep. 41, 89–91. Available at: http://www.cdc.gov/mmwr/preview/mmwrhtml/00016056.htm (accessed 05.01.14).

Danichevski, K., McKee, M., Balabanova, D., 2008. Prescribing in maternity care in Russia: the legacy of Soviet medicine. Health Policy 85, 242–251. Available at: http://www.ncbi.nlm.nih.gov/pubmed/17854946 (accessed 05.01.14).

Leon, D.A., Saburova, L., Tomkins, S., Andreev, E., Kiryanov, N., McKee, M., Shkolnikov, V.M., 2007. Hazardous alcohol drinking and premature mortality in Russia: a population based case–control study. Lancet 369, 2001–2009. Available at: http://www.thelancet.com/journals/lancet/article/PIIS0140-6736(07)60941-6/abstract (accessed 05.01.14).

Organisation for Economic Co-operation and Development. 2012. Available at: http://www.oecd-ilibrary.org/social-issues-migration-health/oecd-reviews-of-health-systems-russian-federation-2012_9789264168091-en;jsessionid=1ijla4yyng5uc.x-oecd-live-02 (accessed 05.01.14)

Popovich, L., Potapchik, E., Shishkin, S., Richardson, E., Vacroux, A., Mathivet, B., 2011. Russian Federation: health system review. Health Syst. Transit. 13, 1–190. Available at: http://www.euro.who.int/__data/assets/pdf_file/0006/157092/HiT-Russia_EN_web-with-links.pdf (accessed 05.01.14).

Rechel, B., Kennedy, C., McKee, M., Rechel, B., 2011. The Soviet legacy in diagnosis and treatment: implications for population health. J. Public Health Policy 32, 293–304. Available at: http://www.ncbi.nlm.nih.gov/pubmed/21808248 (accessed 05.01.14).

Rechel, B., Roberts, B., Richardson, E., Shishkin, S., Shkolnikov, V.M., Leon, D.A., et al., 2013. Health and health systems in the Commonwealth of Independent States. Lancet 381, 1145–1155. Available at: http://www.thelancet.com/journals/lancet/article/PIIS0140-6736(12)62084-4/fulltext (accessed 30.03.13).

Tulchinsky, T.H., Varavikova, E.A., 1996. Addressing the epidemiologic transition in the former Soviet Union: strategies for health system and public health reform in Russia. Am. J. Public Health 86, 313–320. Available at: http://www.ncbi.nlm.nih.gov/pmc/articles/PMC1380508/ (accessed 05.01.14).

United Nations Development Programme. Russian Federation: National Human Development Report. Available at: http://www.undp.ru/index.php?lid=1&cmd=publications1&id=48 (accessed 8.1.14).

World Health Organization, Health profile: Russian Federation. Available at: http://www.who.int/gho/countries/rus.pdf (accessed 15.04.13).

Israel

Busch, S.H., Hodgkin, D., 2013. Declines in psychiatric care in inpatient settings in Israel mirror global trend. Isr. J. Health Policy Res. 2, 30. Available at: http://www.ijhpr.org/content/2/1/33/abstract (accessed 15.08.13).

Chernichovsky, D., Navon, G., August 2012. Private expenditure for medical services, distribution of incomes and poverty in Israel. Bank of Israel Discussion Paper Series. Bank of Israel, Jerusalem. Available at: http://www.boi.org.il/en/Research/DiscussionPapers1/dp1213e.pdf (accessed 24.08.13).

Chernichovsky, D., Regev, E., 2012. Israel's healthcare system. In: Ben-David, D. (Ed.), State of the Nation Report 2011–2012. Society, economy and policy in Israel: Taub Center for Social Policy Studies. pp. 507–558. Available at: http://taubcenter.org.il/tauborgilwp/wp-content/uploads/State-of-the-Nation-Report-2011-2012-21.pdf. (accessed 17.08.13).

Ginsberg, G., Tulchinsky, T., Filon, D., Goldfarb, A., Abramov, L., Rachmilevitz, E.A., 1998. Cost–benefit analysis of a national thalassaemia prevention programme in Israel. J. Med. Screen 5, 120–126. Available at: http://msc.sagepub.com/content/5/3/120.full.pd (accessed 05.01.14).

Ginsberg, G.M., Rosenberg, E., 2012. Economic effects of interventions to reduce obesity in Israel. Isr. J. Health Policy Res. 1, 17. Available at: http://www.ijhpr.org/content/1/1/17 (accessed 17.08.13).

Goldberger, N., Haklai, Z., 2012. Mortality rates in Israel from causes amenable to health care, regional and international comparison. Isr. J. Health Policy Res. 1, 41. Available at: http://www.ncbi.nlm.nih.gov/pmc/articles/PMC3502080/ (accessed 05.1.14).

Gross, R., 2003. Implementing health care reform in Israel: organizational response to perceived incentives. J. Health Pol. Policy Law 28, 659–692. Available at: http://jhppl.dukejournals.org/content/28/4/659.short (accessed 05.01.14).

Israel Center for Disease Control, Jerusalem: Ministry of Health. Available at: http://www.health.gov.il/English/MinistryUnits/HealthDivision/Icdc/Pages/default.aspx (accessed 14.06.13).

Israel Central Bureau of Statistics, 2011. Survey on Research and Development in Hospitals (2009) and Physicians in Hospitals Survey. Available at: http://www1.cbs.gov.il/www/hodaot2012n/12_12_308e.pdf. (accessed 15.06.13).

Kahan, N.R., Waitman, D.A., Blackman, S., Chinitz, D.P., 2006. Suboptimal pneumococcal pneumonia vaccination rates among patients at risk in a managed care organization in Israel. J. Manag. Care Pharm. 12, 152–157. Available at: http://www.ncbi.nlm.nih.gov/pubmed/16515373 (accessed 05.01.14).

Kranzler, Y., Davidovich, N., Fleischman, Y., Grotto, I., Moran, D.S., Weinstein, R., 2013. A health in all policies approach to promote active, healthy lifestyle in Israel. Isr. J. Health Policy Res. 2, 16. Available at: http://link.springer.com/article/10.1186/2045-4015-2-16#page-2 (accessed 17.08.13).

Levav, I., Levinson, D., Radomislensky, I., Shemesh, A.A., Kohn, R., 2007. Psychopathology and other health dimensions among the offspring of Holocaust survivors: results from the Israel National Health Survey. Isr. J. Psychiatry Relat. Sci. 44, 144–151. Available at: http://doctorsonly.co.il/wp-content/uploads/2011/12/2007_2_9.pdf (accessed 05.01.14).

Manor, O., Shmueli, A., Ben-Yehuda, A., Paltiel, O., Calderon, R., Jaffe, D.H., 2011. National program for quality indicators in community healthcare in Israel report, 2007–2009. School of Public Health and Community Medicine, Hebrew University-Hadassah, Jerusalem, Israel. Available at: http://healthindicators.ekmd.huji.ac.il/reports/QICH%20Report%202007-2009%20English.pdf (accessed 08.01.14).

Rabinovich, M., Wood, F., Shemer, J., 2007. Impact of new medical technologies on health expenditures in Israel 2000–2007. Int. J. Technol. Assess Health Care 23, 443–448. Available at: http://www.ncbi.nlm.nih.gov/pubmed/17937832 (accessed 05.01.14).

Rosen, B., Goldwag, R., Thomson, S., Mossialos, E., 2003. Health care systems in transitions – HiT summary Israel. European Observatory on Health Systems and Policies. Available at: http://www.euro.who.int/document/E81826.pdf. (accessed 24.05.08).

Rosenberg, E., Lev, B., Bin-Nun, G., McKee, M., Rosen, L., 2008. Healthy Israel 2020: a visionary national health targeting initiative. Public Health 122, 1217–1225. Available at: http://www.ncbi.nlm.nih.gov/pubmed/18672257 (accessed 05.01.14).

Sharon, A., Levav, I., Brodsky, J., Shemesh, A.A., Kohn, R., 2009. Psychiatric disorders and other health dimensions among Holocaust survivors 6 decades later. Br J Psych 195, 331–335. Available at: http://bjp.rcpsych.org/content/195/4/331.full.pdf (accessed 08.01.14).

United Nations Children's Fund, Israel at a glance. Available at: http://www.unicef.org/infobycountry/israel_statistics.html (accessed 15.06.13).

United Nations Development Programme, Human Development Index – Israel. Available at: http://hdrstats.undp.org/en/countries/profiles/ISR.html (accessed 24.05.13).

Weil, L.G., Bin Nun, G., McKee, M., 2013. The recent physician strike in Israel: a health system under stress? Isr. J. Health Policy Res. 2, 33. Available at: http://www.ijhpr.org/content/pdf/2045-4015-2-33.pdf.

World Health Organization, Health policy monitor. Available at: http://www.hpm.org/en/Country_Facts/Country_Selection/Middle_East/Israel.html; jsessionid=DBA0686151FD7A62200533BE7AB47E59 (accessed 24.05.08).

World Health Organization, Israel statistics. Available at: http://www.who.int/countries/isr/en/ (accessed 15.06.13).

World Health Organization, 2009. European Observatory. Health systems in transition. Israel vol. II. Available at: http://www.euro.who.int/__data/assets/pdf_file/0007/85435/E92608.pdf (accessed 24.05.13).

Yosefy, C., Dicker, D., Viskoper, J.R., Tulchinsky, T.H., Ginsberg, G.M., Leibovitz, E., Gavish, D., 2003. The Ashkelon Hypertension Detection and Control Program (AHDC Program): a community approach to reducing cardiovascular mortality. Prevent. Med. 37, 571–576. Available at: http://www.ncbi.nlm.nih.gov/pubmed/14636790 (accessed 05.01.14).

Developing Countries

Anyangwe, S.C., Mtonga, C., 2007. Inequities in the global health workforce: the greatest impediment to health in sub-Saharan Africa. Int. J. Environ. Res. Public Health 4, 93–100. Available at: http://www.ncbi.nlm.nih.gov/pmc/articles/PMC3728573/ http://www.ncbi.nlm.nih.gov/pmc/articles/PMC3728573/pdf/ijerph-04-00093.pdf (accessed 05.01.14).

Beaglehole, R., Sanders, D., Dal Poz, M., 2003. The public health workforce in sub-Saharan Africa: challenges and opportunities. Ethn. Dis. 13, S24–S30. Available at: http://www.ishib.org/ED/journal/ethn-13-02s-24.pdf (accessed 06.01.14).

Johansen, R.E.B., Diop, N.J., Glenn Laverack, G., Leye, E., 2013. What works and what does not: a discussion of popular approaches for the abandonment of female genital mutilation. Obstet Gynecol. Int. Available at: http://www.hindawi.com/journals/ogi/2013/348248/ (accessed 06.02.24).

Pauly, M.V., Zweifel, P., Schleffer, R.M., Preker, A.S., Basset, M., 2006. Private health insurance in developing countries: voluntary private insurance could fill in the gaps that limited public resources cannot cover. Health Aff. 25, 369–379. Available at: http://content.healthaffairs.org/content/25/2/369.full.pdf (accessed 06.01.14).

Ron, A., Abel-Smith, B., Tamburri, G., 1990. Health insurance in developing countries: the social security approach. International Labor Office, Geneva. Available at: http://www.amazon.com/ (accessed 06.01.14).

Schieber, G., Maeda, A., 1993. Health care financing and delivery in developing countries. Health Aff. 18, 135–143. Available at: http://content.healthaffairs.org/content/18/3/193.full.pdf (accessed 06.01.14).

Stilwell, B., Diallo, K., Zurn, P., Vujicic, M., Adams, O., Dal Poz, M., 2004. Migration of health-care workers from developing countries: strategic approaches to its management. Bull. World Health Organ. 82, 595–600. Available at: http://www.who.int/bulletin/volumes/82/8/595.pdf (accessed 06.01.14).

World Health Organization, 2011. An update on WHO's work on female genital mutilation (FGM). Progress report. WHO. Available at: http://whqlibdoc.who.int/hq/2011/WHO_RHR_11.18_eng.pdf. (accessed 04.08.13).

World Health Organization, 2013. The African Regional Health Report: The Health of the People. Bull. World Health Org., 91. Available at: http://www.who.int/bulletin/africanhealth/en/. and http://whqlibdoc.who.int/afro/2006/9290231033_rev_eng.pdf (accessed 31.07.13).

Nigeria

Central Bank of Nigeria, Annual Report: 2011. Section 5. Available at: http://www.cenbank.org/Out/2012/publications/reports/rsd/arp-2011/Chapter%205%20-%20Fiscal%20Policy%20and%20Government%20Finance.pdf (accessed 02.08.13).

Editorial, 2008. Nigeria still searching for right formula. Bull. World Health Organ. 86, 663–665. Available at: http://www.who.int/bulletin/volumes/86/9/08-020908.pdf (accessed 06.01.14).

Obalum, D.C., Fiberesima, F., 2012. Nigerian National Health Insurance Scheme (NHIS): an overview. Niger. Postgrad. Med. J. 19, 167–174. Available at: http://www.npmj.edu.ng/sample-link/74-nigerian-national-health-insurance-scheme-nhis-an-overview (accessed 06.01.14).

Okeke, T.C., Anyaehie, U.S.B., Ezenyeaku, C.C.K., 2012. An overview of female genital mutilation in Nigeria. Ann. Med. Health Sci. Res. 2, 70–73. Available at: http://www.ncbi.nlm.nih.gov/pmc/articles/PMC3507121/ (accessed 03.08.13).

United Nations Children's Fund, June 2006. Nigeria includes folic acid in fortification standard. Food fortification initiative. FFI Newslett. Available at: http://www.unicef.org/wcaro/WCARO_Nigeria_Factsheets_Nutrition.pdf (accessed 06.01.14).

United Nations Children's Fund, Universal salt iodization in Nigeria: process, successes and lessons. Available at: http://www.unicef.org/nigeria/ng_publications_USI_in_Nigeria_Report.pdf. (accessed 14.04.13).

United Nations Development Programme, 2013. Human Development Report. Nigeria. Available at: http://www.ng.undp.org/content/nigeria/en/home.html. (accessed 06.01.14).

World Health Organization, 2012. Regional Office for Africa. Eradicating polio in the African Region: 2011 Annual Report. WHO, Brazzaville. Available at: http://www.afro.who.int/en/downloads/doc_download/7539-eradicating-polio-in-the-african-region-2011-annual-report.html (accessed 06.01.14).

Latin America and the Caribbean

Pan American Health Organization. Available at: http://www.paho.org/ (accessed 06.01.14).

United Nations Development Programme, 2010. Achieving the Millennium Development Goals with equality in Latin America and the Caribbean: progress and challenges. Available at: http://www.eclac.cl/mdg/default.asp?idioma=IN. (accessed 06.01.14).

United Nations Development Programme, 2013. Available at: http://www.undp.org/content/undp/en/home/librarypage/hdr/human-development-report-2013/ (accessed 06.01.14).

Colombia

De Vos, P., De Ceukelaire, W., Van der Stuyft, P., 2006. Colombia and Cuba, contrasting models in Latin America's health sector reform. Trop. Med. Int. Health 11, 1604–1612.

Hernandez, M., 2002. Health reform, equity and the right to health in Colombia. Cad. Saude. Publica. 18, 991–1001.

Ministry of Health and Social Protection, 2011. Colombia's basic health indicators. Available at: http://www.minsalud.gov.co/salud/Paginas/INDICADORESBASICOSSP.aspx.

Ministry of Health and Social Protection, February 2013. Health system reform. (Spanish). Available at: http://www.minsalud.gov.co/Documents/General/Reforma-salud-febrero-2013.pdf. (accessed 04.08.13).

National Department of Statistics, March 2013. Press communication. Released 13. (Spanish). Available at: http://www.dane.gov.co/files/investigaciones/condiciones_vida/calidad_vida/cp_ECV_2012.pdf. (accessed 04.08.13).

United Nations Development Programme, 2013. Human Development Report. Available at: http://hdr.undp.org/en/content/human-development-report-2013. (accessed 06.01.14).

India

All India Institute of Hygiene and Public Health. Available at: http://www.aiihph.gov.in/html/about_us.htm (accessed 18.11.10).

Bangdiwala, S.I., Tucker, J.D., Zodpey, S.M., Griffiths, S., Li, L.-M., Reddy, K.S., et al., 2011. Public health education in India and China: history, opportunities, and challenges. Public Health Rev. 33, 204–224. Available at: http://www.publichealthreviews.eu/upload/pdf_files/9/Bangdiwala.pdf (accessed 06.01.14).

Garg, B.S., 2004. Accreditation of public health courses in India – the challenge ahead. Indian J. Commun. Med. 29, 52. Available at: http://www.indmedica.com/journals.php?journalid=7&issueid=56&articleid=682&action=article (accessed 06.01.14).

Kumar, A.K.S., Chen, L.C., Choudhury, M., Ganju, S., Mahajan, V., Sinha, A., Sen, A., 2011. Financing health care for all: challenges and opportunities. Lancet 377, 668–679. Available at: http://www.thelancet.com/journals/lancet/article/PIIS0140-6736(10)61884-3/abstract (accessed 06.01.14).

Negandhi, H., Sharma, K., Zodpey, S.P., 2012. History and evolution of public health education in India. Indian J. Public Health 56, 12–16. Available at: http://www.ijph.in/text.asp?2012/56/1/12/96950 (accessed 21.07.13).

Park, K., 2011. Park's textbook of preventive and social medicine, twenty-first ed. Banarsidas Bhanot, Jabalpur. Available at: (ebook) http://www.e-bookspdf.org/download/preventive-and-social-medicine-k-park-edition.html (accessed 21.12.13).

Rao, M., Rao, K.D., Shiva Kumar, A.K., Chatterjee, M., Sundararaman, T., 2011. Human resources for health in India. Lancet 377, 587–598. Available at: http://www.thelancet.com/journals/lancet/article/PIIS0140-6736(10)61888-0/abstract (accessed 06.02.14).

Sharma, K., Zodpey, S., 2010. Need and opportunities for health management education in India. Indian J. Public Health 54, 84–91. Available at: http://www.ijph.in/article.asp?issn=0019-557X;year=2010;volume=54;issue=2;spage=84;epage=91;aulast=Sharma (accessed 06.01.14).

World Health Organization, India. Available at: http://www.who.int/countries/ind/en/ (accessed 06.01.14).

China

Akin, J.S., Dow, W.H., Lance, P.M., Loh, C.P., 2005. Changes in access to health care in China, 1989–1997. Health Policy Plann. 20, 80–89. Available at: http://heapol.oxfordjournals.org/content/20/2/80.full.pdf (accessed 06.01.14).

Bloom, G., Xingyuan, G., 1997. Health sector reform lessons from China. Soc. Sci. Med. 45, 351–360. Available at: http://www.ncbi.nlm.nih.gov/pubmed/9232730 (accessed 06.01.14).

Blumenthal, D., Hsiao, W., 2005. Privatization and its discontents – the evolving Chinese health care system. N. Engl. J. Med. 353, 1165–1170. Available at: http://journal.9med.net/qikan/article.php?id=216829 (accessed 15.05.13).

Carlton, E.J., Liang, S., McDowell, J.Z., Li, H., Luo, W., Remais, J.V., 2012. Regional disparities in the burden of disease attributable to unsafe water and poor sanitation in China. Bull. World Health Organ. 90, 578–587. Available at: http://www.who.int/bulletin/volumes/90/8/11-098343/en/ (accessed 24.08.13).

Chinas 12th Five Year Plan, October 2010. How it actually works and what's in store for the next five years. Available at: http://apcoworldwide.com/content/PDFs/Chinas_12th_Five-Year_Plan.pdf. (accessed 06.01.14).

Griffiths, S.M., 2010. Leading a healthy lifestyle: the challenges for China. Asia Pac. J. Public Health 22, 110S. Available at: http://www.ncbi.nlm.nih.gov/pubmed/20566542 (accessed 06.01.14).

Grogan, C.M., 1995. Urban economic reform and access to health care coverage in the People's Republic of China. Soc. Sci. Med. 41, 1073–1084. Available at: http://www.ncbi.nlm.nih.gov/pubmed/8578330 (accessed 06.01.14).

Hesketh, T., Zhu, W.X., 1997. Health care in China: maternal and child health in China. BMJ 314, 1898–1899. Available at: http://www.ncbi.nlm.nih.gov/pmc/articles/PMC2126969/ (accessed 06.01.14).

Hillier, S., Shen, J., 1996. Health care systems in transition: People's Republic of China: an overview of China's health care system. J. Public Health Med. 18, 258–265. Available at: http://jpubhealth.oxfordjournals.org/content/18/3/258.full.pdf (accessed 15.05.13).

Hipgrave, D., Guo, S., Mu, Y., Guo, Y., Yan, F., Scherpbier, R.W., et al., 2012. Chinese-style decentralization and health system reform. PLoS Med. 9 (11) http://dx.doi.org/10.1371/journal.pmed.1001337 (accessed 06.01.14).

Hsiao, W.C.L., 1995. The Chinese health care system: lessons for other nations. Soc. Sci. Med. 41, 1047–1055. Available at: http://www.ncbi.nlm.nih.gov/pubmed/8578327 (accessed 06.01.14).

Li, L.M., 2004. The current state of public health in China. Annu. Rev. Public Health 25, 327–339. Available at: http://www.annualreviews.org/doi/abs/10.1146/annurev.publhealth.25.101802.123116 (accessed 06.01.14).

Li, L.M., Tang, J.L., Lu, L., Jiang, Y., Griffiths, S.M., 2011. The need for integration in health sciences sets the future direction for public health education. Public Health 125, 20–24. Available at: http://www.sciencedirect.com/science/journal/00333506/125/1 (accessed 06.01.14).

Li, Q., Hsia, J., Yang, G., 2011. Letter to the Editor. Prevalence of smoking in China in 2010. N. Engl. J. Med. 364, 2469–2470. Available at: http://www.nejm.org/doi/full/10.1056/NEJMc1102459 (accessed 23.07.13).

Lim, M.K., Yang, H., Zhang, T., Feng, W., Zhou, Z., 2004. Public perceptions of private health care in socialist China. Health Aff. (Millwood) 23, 222–234. Available at: http://content.healthaffairs.org/content/23/6/222.full.pd (accessed 06.01.14).

Liu, Y., 2004. China's public health-care system: facing the challenges. Bull. World Health Organ. 82, 532–538. Available at: http://www.ncbi.nlm.nih.gov/pmc/articles/PMC2622899/f (accessed 21.12.13).

Liu, Y., Rao, K., Fei, J., 1998. Economic transition and health transition: comparing China and Russia. Health Policy 44, 103–122. Available at: http://www.ncbi.nlm.nih.gov/pubmed/10180676 (accessed 06.01.14).

National Bureau of Statistics of China, 2012. China statistical yearbook. China Statistics Press, Beijing. Available at: http://www.stats.gov.cn/tjsj/ndsj/2012/indexeh.htm (accessed 15.05.13).

United Nations Development Programme, 2010. MDGs in China. Available at: http://www.undp.org.cn/modules.php?file=article&catid=32&sid=6. (accessed 16.05.13).

Wagstaff, A., Lindelow, M., Wang, S., Zhang, S., 2009. Reforming China's rural health system. Directions in human development. World Bank. Available at: http://elibrary.worldbank.org/doi/abs/10.1596/978-0-8213-7982-0. (accessed 06.01.14).

World Bank, 2005. Rural health in China: China's health sector – why reform is needed. World Bank. Available at: http://siteresources.worldbank.org/INTEAPREGTOPHEANUT/Resources/502734-1129734318233/BN3whyreformfinal.pdf. (accessed 06.01.14).

World Health Organization, January 2012. H5N1 avian influenza: timeline of major events, 25. Available at: http://www.who.int/influenza/human_animal_interface/H5N1_avian_influenza_update.pdf. (accessed 23.07.13).

Japan

Hashimoto, H., Ikegami, N., Shibuya, K., Izumida, N., Noguchi, H., Yasunaga, H., et al., 2011. Cost containment and quality of care in Japan: is there a trade-off? Lancet 378, 1174–1182. Available at: http://www.thelancet.com/journals/lancet/article/PIIS0140-6736(11)60987-2/abstract (accessed 06.01.14).

Ikeda, N., Saito, E., Kondo, N., Inoue, M., Ikeda, S., Satoh, T., et al., 2011. What has made the population of Japan healthy? Lancet 378, 1094–1105. Available at: http://www.thelancet.com/journals/lancet/article/PIIS0140-6736(11)61055-6/abstract (accessed 06.01.14).

Ikegami, N., Campbell, J.C., 1999. Health care reform in Japan: the virtues of muddling through. Health Aff. 18, 56–75. Available at: http://content.healthaffairs.org/content/18/3/56.full.pdf (accessed 06.01.14).

Ingelhart, J.K., 1988. Health policy report: Japan's medical care system, parts 1 and 2. N. Engl. J. Med. 319, 807–812 1166–72. Available at: http://www.nejm.org/doi/pdf/10.1056/NEJM198810273191731 (accessed 06.01.14).

Nishimura, S., 2007. Promoting health during the American occupation of Japan: the public health sections, Kyoto Military Government Team, 1945–1949. Am. J. Public Health 98, 424–434. Available at: http://www.ncbi.nlm.nih.gov/pmc/articles/PMC2253585/ (accessed 06.01.14).

Organisation for Economic Co-operation and Development, Health policies and data. Asia/Pacific Region. Improving quality of its healthcare. Available at: http://www.oecd.org/els/health-systems/asiapacificregionimprovingthequalityofitshealthcare.htm (accessed 27.05.13).

Organization of Economic Cooperation and Development. Country statistical profile: Japan, 15 Novemeber 2013. Available at: http://www.oecd-ilibrary.org/economics/country-statistical-profile-japan_20752288-table-jpn (accessed 8.1.14).

Tatara, K., Okamoto, E., 2009. Japan health system review. Health Syst. Transit. 11 (5), 1–164. Available at: http://www.euro.who.int/__data/assets/pdf_file/0011/85466/E92927.pdf (accessed 06.01.14).

United Nations Development Programme, Human Development Index – Japan. Available at: http://hdrstats.undp.org/countries/country_fact_sheets/cty_fs_JPN.html (accessed 24.05.08).

Human Resources for Health

Upon completion of this chapter, the student should be able to:

1. Define criteria for determining the supply of personnel in the health sector;
2. Define criteria for evaluation of health personnel training programs;
3. Define criteria for development of new training facilities and new health workers;
4. Define licensing, regulatory, and accreditation functions in the New Public Health.

INTRODUCTION

The health workforce can be defined as "all people engaged in actions whose primary intent is to enhance health" (Dal Poz et al., 2009). These people include clinical care providers, nurses, doctors, pharmacist, and many others, as well as public health, management, and support staff. All are vital to the work of maintaining and improving health. There are no absolute standards of need and there is wide variation in the health workforce, even in the industrialized countries. But there are clearly needs for supplies of health personnel to meet the needs of a population and methods to train them, employ them, and retain them. Those countries with exceedingly low levels of numbers of health workers will have difficulty achieving the Millennium Development Goals (MDGs) by the year 2015 and beyond. Developing adequacy in the quality and quality of the health workforce is one of the essential aspects of public health.

Development and sustainability of the New Public Health and its ability to respond to old and emerging threats depend on the quantity and quality of human resources of the total and especially the public health workforce. The great achievements of public health of the twentieth century led to doubling of life expectancy in the industrialized countries and the emergence of important health advances even in the least developed countries. New challenges of globalization of diseases such as human immunodeficiency virus/acquired immunodeficiency syndrome (HIV/AIDS), severe acute respiratory syndrome (SARS), persisting tropical diseases and other emerging communicable diseases, and disasters, terrorism, genocide and pandemics of avian influenza present great challenges in the twenty-first century. But the

main causes of death are the non-communicable diseases (NCDs), where prevention has shown truly amazing positive results in cardiovascular diseases, cancer, and injuries. However, much remains to be done in the high-income as well as in middle- and low-income countries for primary, secondary, and tertiary prevention to reduce morbidity and avoidable deaths, i.e., deaths that may be prevented by risk reduction or preventive measures, such as lung cancer from cigarette smoking and exposure. The key to many of the future achievements in reducing the burden of disease lies in health promotion along with advances in the biomedical side of prevention with, for example, new vaccines to prevent or modify infectious diseases and NCDs, especially cancers. Despite improved prevention, NCDs, including mental and oral disease, are still a major public health problem in high-income countries and growing rapidly in low- and middle income countries (see Chapters 6 and 7).

It is vital to address the issues of human resources for health in all countries. While there has been much progress in this field, the challenges ahead are daunting. There is an increasing flow of trained health professionals from poor to rich countries, from rural to urban locations, from public to private sector services, and from poor countries to wealthy ones providing attractive financial and professional rewards. In low-income countries, development of infrastructures capable of providing community health services in poor urban and rural areas is a challenge requiring new approaches such as training and deploying well-supervised community health workers (CHWs) and university- or college-level trained public health managers and health promoters. Similarly, health promoters with various training backgrounds will be required in high-income countries to meet the growing demand from aging populations in need of extensive support systems in their homes and in the community.

The New Public Health is concerned with the total health system and related issues. It requires an understanding of issues related to the training, supply, distribution, and management of many kinds of human resources, including the balance between personnel working in institutions and in the community. It was for these reasons that the World Health Organization (WHO) World Health Report of 2006 noted "an estimated shortage of almost 4.3 million doctors, midwives, nurses, dental and optometric and other health professions and support workers worldwide" and

The New Public Health. http://dx.doi.org/10.1016/B978-0-12-415766-8.00014-8

recommended a 10-year program to address this fundamental issue, particularly for the developing world. The developed nations are facing many shortages in critical areas such as nursing personnel, but also in other skills needed to care for an increasingly elderly population, as well as rising tides of diabetes and obesity, and their long-term sequelae.

Health systems require adequate numbers of well-trained, well-remunerated, and up-to-date providers working with adequate facilities and support systems. Health professions are made up of many disciplines working in a complex network of facilities and programs. Their range of activities includes provision of patient care in the community and in hospital or other institutional settings. They are also needed to promote health, prevent disease, treat illness, and rehabilitate in a compassionate, ethical, professional, and cost-effective manner. Health care providers at all levels must not only be educated for competence and humaneness in clinical functions, but also be continuous learners, knowledgeable in the medical and social sciences of health including the economic aspects of health care. They must be aware of and able to synthesize knowledge from related fields such as epidemiology, economics, and management, as well as the social and behavioral sciences. The quality of the practitioner depends on the recruitment of socially motivated and talented people, on education, training, and professionalization as providers, as well as on the structure, content, and quality orientation of the health system in which they work.

Determining policy, need, and allocation of human resources is an important health planning issue. A relative oversupply or undersupply of one or more health professions creates a bias or imbalance in the health system and its economics. Mid-level practitioners and CHWs are being recognized as essential to ensure access to appropriate levels of service and to provide for unmet service needs in both developed and developing countries. The ongoing pandemics of AIDS, tuberculosis (TB), malaria, and comorbidities of infectious diseases with micronutrient deficiencies stretch beyond the limited capacities of existing resources. At the same time, public and political attention is unable to focus on health needs sufficiently to provide for developing and sustaining the human resources necessary to meet such challenges as the MDGs. Many countries will be unable to meet such goals, particularly in relation to child and maternal mortality, with the financial and human resources available, even with assistance from donor countries and organizations.

OVERVIEW OF HUMAN RESOURCES

Globalization has increased our awareness and understanding of the importance of building the essential health system governance, infrastructure, organization, funding, universal access with associated workforce needs. Threats and experience have shown the dangers of high levels of premature mortality from communicable and non-communicable diseases, environmental and natural disasters, violence and terrorism, ongoing pandemics, with movements of diseases from one habitat crossing oceans via rapid transportation from one end of the Earth to the other. Unprecedented migration of the skilled health workforce from developing countries to developed countries is happening owing to perceptions of more attractive incomes, professional settings, and way of life. At the same time, with aging populations and new technologies revolutionizing medicine, new generations of health professionals and health workers are needed to meet rising demands and expectations.

In seeking efficient and effective ways of improving health, health systems have opened many new professional roles in new organizational frameworks. As definitions of health service were widened to include health maintenance, new health professions were added to the total health service spectrum. Continuing education is vital to maintain and upgrade quality in a health care system. Registration systems and databases are important to provide basic information on all relevant aspects of health personnel.

The World Health Report 2006 states the issue as follows:

"The world community has sufficient financial resources and technologies to tackle most of these health challenges; yet today many national health systems are weak, unresponsive, inequitable – even unsafe. What is needed now is political will to implement national plans, together with international cooperation to align resources, harness knowledge and build robust health systems for treating and preventing disease and promoting population health. Developing capable, motivated and supported health workers is essential for overcoming bottlenecks to achieve national and global health goals." (WHO, 2006)

This report indicates that the world's poorest countries, mostly in sub-Saharan Africa, have deficits of 4.3 million health workers, especially doctors, nurses, and midwives. Scaling up training programs as realistically suggested over a 10–20-year period is costly and is associated with major loss of graduates to migration to high-income countries, which actively recruit doctors and nurses from poor countries. Additional costs of training and employing the professional workers needed for low- and medium-income countries will require large-scale investment by the developing countries themselves and international donor aid. However, many policy makers, donors and health service agencies do not place infrastructure development on their agendas, so that the estimate by the WHO at hundreds of millions of dollars per country-year, and even more in incremental annual salary costs, is well beyond current levels of national and donor expenditures in developing countries.

In 2010, the WHO sharpened the focus on building sustainable health systems in low- and medium-income countries by calling for more intensive national and donor emphasis on a combination of issues: service delivery, health workforce, health information systems, access to essential medicines, financing, and leadership-governance in a context of universal access to basic primary care and public health services to achieve the MDGs and their follow-up after 2015 (WHO: Monitoring the Building Blocks of Health Systems, 2010).

This chapter examines the importance of human resources for the New Public Health and the elements essential for training in relation to the quantity, quality, and changing interaction among the health professions. The global problem of human resources for health is not only a severe problem for developing countries but an ongoing issue in the industrialized countries as well. The numbers, types, and distribution of personnel supply are major determinants of access, availability, appropriateness, and costs of health care. The training, quality, and performance of health personnel and the technology they use are all important health planning issues. Every health professional needs knowledge of the principles and current standards of public health in order to perform his or her functions, as all of health care now routinely involves prevention, teamwork, management, quality assurance, cost containment, and related ethical issues.

In many countries, the major focus of education of health personnel has been to prepare clinicians, without equal emphasis on preparation of public health policy analysts, health managers, and public health professionals. Yet the latter are especially important when health reforms are under way and when health promotion and prevention are needed to cope with changes in the health needs of a society.

The principal problems in human resources development are not the same for high-income countries as for low-income countries, which also have low rates of expenditures on health (<5 percent of gross domestic product). While there is variation between countries, the following are common to the human resources issues generally:

- inadequate funding, training positions, salaries, incentives, safety, and support systems for health workers
- imbalance in training of health professionals; severe shortages of nurses and other health professionals, compared to the physician workforce
- insufficient training for medical and nursing personnel in developing countries; possibly excess capacity for medical training in post-Soviet countries
- excess of medical subspecialists, and insufficient incentives and training of primary care physicians, inflating health costs and compromising access to care
- geographic maldistribution of vital professional categories with concentration in urban centers and poor supply in rural areas
- underfinancing for public systems of health care in comparison to private services, fostering poor work conditions, low remuneration, and indifferent career opportunities, with low staff morale, performance, and patient satisfaction or compliance with needed care

- insufficient standards and length of training of specialist physicians to produce well-qualified professional leaders
- lack of public health orientation of policy makers and health providers with overmedicalization in the health field, and excessive influence of the pharmaceutical industry on medical education, practice, and health priorities
- lack of postgraduate accredited academic centers for research and training of public health specialists in epidemiology, health-related social sciences, health system policy analysis, or health system management, compromising the ability of a health system to monitor its outcomes and resource allocation, or to evaluate program effectiveness
- licensing of health providers by the government, which may allow for compromises in quality to ensure adequate numbers of graduates; conversely, delegation of licensing to professional syndicates may result in a protectionist approach, placing the interests of the profession above those of the public
- compromising the quality of human resources by inadequate recruitment and educational standards, inadequate continuing examination and recertification
- conflicts of academic, professional, government, or insurer interests with public and individual patient interests in training policies
- poor coordination and communication between government and managerial sectors involved in health policy
- developing countries, where they are most needed
- inadequate development of community health workforces as front-line access and outreach personnel in rural and underserved urban areas as integral participants in the health system, including for underserved high-need populations in developed countries
- resistance to addition of new health workers and transfer of tasks and responsibilities
- in underserviced communities or health guides for high risk populations such as diabetics to supplement limited medical and nursing services and health needs to community-based workerstrained and supervised to provide for unmet needs
- inter-professional rivalries over distribution of tasks and skills for future professional workforce training needs
- addition of new health professional roles to improve efficiency in organized health care systems, such as accountable care organizations.

From the 1950s to the 1970s, few new medical schools were opened in the USA, while existing schools expanded to meet problems of access to care and the perceived shortage of doctors. However, none opened during the 1980s and 1990s. The number of medical schools in 1980–81 was 125 and 14 osteopathy schools, and in 2008–09, these increased to 131 and 26, respectively. Since 2007, more than a dozen schools have started working with the Liaison Committee on Medical Education accreditation process; 10 more are under discussion, and five osteopathic medical colleges have opened. These will not meet the need for 125,000 physicians by 2025, an estimate that is debatable; however, a shortage does exist with an aging society and more Americans gaining health care entitlement coverage under the Affordable Care Act of 2010. Reliance on immigration of doctors from poor countries is problematic in other ways.

It was thought that increased numbers of doctors would increase competition and lower doctors' incomes; however, medical incomes continued to rise and problems of access to care were unresolved. With growing emphasis on health economics and health promotion and disease prevention, there was a realization that excess medical personnel would not contribute to the national health, and that in some countries the excess of medical personnel had become a liability. In both fee-for-service medicine and salaried health service, increases in physician supply generate increases in health expenditures. Supply and demand market forces do not adapt well to health care, because the consumer demand is to a large extent generated by provider decisions (e.g., for return visits, investigation, or hospitalization). Fees may be fixed arbitrarily or by negotiation with a public insurance mechanism; the service is paid by a third party, and the consumer is less knowledgeable about the medical condition and needs than the provider.

Each country addresses the issue of how many and what kinds of human resources to train for its own needs, related to the design and operation of its health system. During the 1970s, the province of Alberta, Canada, had relatively stable expenditures for medical services and physician-to-population ratios. During the 1980s, the province experienced a marked economic downturn and zero population growth, but the supply of physicians and services per physician increased by some 20 percent. The reduced numbers of clients per physician led to an increase both in fees and in volume of services per capita so that physician incomes were sustained. As a result, total and per capita expenditures for health care increased sharply. In many countries during the 1980s, policies were reformulated to reduce the size of medical school training entry classes.

Oversupply of medical specialists can also be a serious problem for a health system, promoting a bias towards a specialized medical orientation in health care at the expense of other more basic needs of public health, primary care, and fundamental support systems for vulnerable groups in society. An excessively specialized medical orientation fosters misallocation of limited resources by creation of tertiary care and high physician density in central cities, leaving rural and primary care underdeveloped. This situation is widely prevalent in developing countries such as India, Mexico, Colombia, and other Latin American countries. In some countries the problem is often compounded by an inability of the health budget to employ needed numbers of physicians. Unemployment among young physicians is a substantial problem in many countries.

In developing countries, health workforce shortages are already at crisis levels. As both the populations and workforces of industrialized nations increase in age, these societies also face an increasing demand for health workers across the professional spectrum that outstrips supply. In 1996, the Association of American Medical Colleges (AAMC) in cooperation with the Council on Graduate Medical Education issued a call for a 30 percent increase in medical school capacity over the following decade. As the current workforce nears retirement age it becomes clear that even this may be insufficient to maintain basic health care services. The shortages encompass nearly every field, but are most pronounced for rural areas, primary care, and public health; fewer than 32 percent of physicians in the USA practice as generalists. The American Association of Family Physicians predicts a need for an increase of at least 40 percent of physicians in primary care alone by 2020. Increasing the percentage of female medical graduates has only partially met increased workforce needs as many work part time during parts of their careers because of family obligations, while there is a higher tendency for male graduates to leave the medical profession for other fields, and specialized medical practice makes a more attractive career than primary care.

The ratio of physicians in active medical practice per 10,000 population in the USA increased from 13.5 in 1975 to 18.0 in 1985, 21.3 in 1995, and 23.8 in 2005 (Health, United States, 2011), and is expected to rise to 29.2 by the year 2020. Concerns of shortages of both primary care and specialist doctors are based on current geographic and specialty training distribution, but also the changing demography of the aging population. These concerns are, however, based on current medical practice organization and should take into account more efficient methods of practice, such as prepaid group practice, full-time salaried physician staffing in hospitals, and increased potential for preventive care by nurse practitioners (NPs) and CHWs.

The debate on increasing graduate training positions by removing the current ceiling on their funding by Medicare is in full swing. Many argue that there has been a steady increase in medical personnel, but the geographic distribution favors the major population centers over rural and underserved remote areas. Furthermore, some medical specialties, such as anesthesia and primary care, have severe shortages. They further argue that increasing the output of medical schools increases current social disparities and only major reforms including national health insurance would reduce such inequities.

In the USA, mid-level health care providers fill more than half the supply of primary care clinicians. Compounded by a generalized nursing workforce shortage, NP graduation rates are decreasing at a rate of 4.5 percent each year. Policy studies in family medicine and primary care in the USA are needed to help determine future professional workforce training needs, and the addition of new health professions to more effectively organized health care systems, such as accountable care organizations (ACOs) linking hospitals, primary care, and outreach services (see Chapters 10 and 12). There is a predicted similar decline in physician assistant (PA) graduations, reaching a 25 percent loss by 2020. Mid-level practitioners have grown in influence and have proven to be outstanding public health and primary care professionals, bringing needed health care to rural and other marginalized populations. Enhancement and support of the NP and PA professions are essential for successful efforts to meet health care workforce demands. As graduation rates for mid-level practitioners decrease, a crisis of unmet need arises. Current educational programs must be expanded, new programs developed, and incentives created to attract workers to public health and primary care.

While nearly all health professions face projected shortages, the situation for nurses is most severe. Throughout the world, the nursing workforce is far from sufficient to meet the public's needs. In sub-Saharan Africa, simply to accomplish immediate health intervention goals, an additional 600,000 nurses would be required. In the USA, the number of registered nurses (RNs) increased by 8 percent between 2000 and 2004 to a new high of 2.9 million, but this is increasingly a group approaching retirement age. The picture varies from country to country; however, every region of the world faces dramatic nursing shortages. A 2006 report by the International Council of Nurses presents policy guidelines for recruiting new workers, curtailing migration trends which have stripped impoverished regions of nurses, and improving working conditions and labor strength, a priority in the retention of the current workforce.

Table 14.1 highlights the imbalance between nursing and medical professions for several developed nations. While total physician supply varies greatly, there is a global pattern of severely unequal distribution of human resources, with greater than half of providers practicing in subspecialties and metropolitan areas. In terms of public health need, the shortage of both doctors and nurses is at crisis levels for rural and marginalized populations.

The achievement of the goal of Health for All through primary health care requires the effective and coordinated services of many types of health personnel within a national health system designed to reach this goal. Government policy is crucial for the preparation, composition, and work patterns of the health workforce. National expenditures on health are dependent on the political priority given to health compared to other issues that may be equally or more pressing to the governing power. A strong national health policy can nevertheless be constructed, even in a poor country, by well-defined health programs. Community and rural health policy in China during the 1950s was based on a number of elements: development of a 3-year family doctor training program for rural service, upgrading training of

TABLE 14.1 Physician and Nurse Density per 1000 Population for Selected Countries, 2010

Country	Physicians	Nurses
UK	2.7	9.6
Sweden	3.8	11.0
Germany	3.7	11.3
USA	2.4	11.0
Russian Federation[a]	4.3	8.1
Israel	3.5	4.8
Greece	6.1	3.3
Mexico	2.0	2.5
France	3.3	8.5
Canada	2.4	9.3

Note: Rounded to one decimal place.
[a]*From the WHO Health for All database.*
Sources: Organisation for Economic Co-operation and Development. Health data, 2012. Health policies and data: frequently requested data. Available at: http://www.oecd.org/els/health-systems/oecdhealthdata2013-frequentlyrequesteddata.htm [Accessed 15 July 2013].
World Health Organization, European Region. Health for All database; January 2013. Available at: http://data.euro.who.int/hfadb/ [Accessed 3 July 2013].

village doctors to physician assistant level, and incentives to encourage work in the countryside and at a grassroots level. This program was successful in raising health standards in China beyond those that might have been expected from its economic level. With recent economic reforms, this system is going through profound changes (see Chapter 13).

HUMAN RESOURCES PLANNING

The health infrastructure of a country includes the resources available and their organization. Human resources are essential to any health system. The supply of personnel and facilities, economic support of the system, management and policy, methods of payment of providers, and organization of the services are therefore vital in health planning (Box 14.1).

Resources available and needed for health systems include facilities, personnel, and financial resources for health care. The organizational and financial structure of a health system determines how these resources are allocated or expended, in the public as well as the private health care sectors. Both structure and methods of payment affect how services are provided. Health systems require economic support sufficient for basic and continuing education of high-quality human resources, as well as managing their appropriate and optimum use.

Regulation of health personnel includes licensure and discipline and is an important governmental function. Measures to control or limit the supply of medical practitioners, along with incentives to promote more efficient health care, are important issues in rationalizing health care systems.

BOX 14.1 The Problem Statement of the Independent Commission: Education of Health Professionals for the Twenty-First Century

The "Frenck report", published in *The Lancet* in 2010 as the report of an independent commission on education of health professionals for the twenty-first century, reviewed the history of medical education since the Flexner report of 1910 (see Box 14.3). The Flexner report initiated a rapid change-over from the poor-quality private medical schools common in the USA at that time to science- and university-based faculties modeled after successful medical education in then leading German schools. The new model medical faculties stressed the education of health professionals based on the integration of modern science into the curricula at university-based schools.

Following World War II, these reforms were reinforced by massive US federal funding of the National Institutes of Health, which promoted research and raised standards of medical education and research to leading levels globally. This helped to create the basis for medical, nutrition, and many other research fields, which took giant steps forward and contributed to public health successes that helped to double the lifespan during the twentieth century. Although raising general levels of health, these changes did not eliminate the gaps and inequities in health within the USA and between countries.

Equity in health is still far from having been achieved, and new health challenges are still unmet and unexpected. New infectious diseases, and the recurrence of those thought to have been controlled, such as measles, and demographic and epidemiological transitions, with potentially overwhelming climate and environmental risks, threaten health security of everyone. Behavioral risk factors and loss of support for public health initiatives create heavy burdens for health promotion and classical public health.

The commission noted that medical professional education is slow to change, with inadequate attention being paid to patient and population needs, and poor teamwork. There is gender stratification of professional status and a narrow technical focus lacking contextual understanding, with episodic rather than continuous care. Training is primarily in hospitals with little in primary care, and there is a lack of attention to improvement in health-system performance.

Efforts to change have been limited in their effect. Educational reforms include the introduction of science-based curricula and problem-based instructional innovations. A new reform movement should now address core professional competencies to specific contexts, based on evidence from global knowledge.

Source: Frenk J, Chen L, Bhutta ZGA, Cohen J, Crisp N, Evans T, et al. Health professionals for a new century: transforming education to strengthen health systems in an interdependent world. a global independent commission: education of health professionals for the 21st century. The Lancet Commissions. Lancet 2010;376:1924–58. Available at: http://www.thelancet.com/journals/lancet/article/PIIS0140-6736(10)61854-5/fulltext?_eventId=login [Accessed 14 July 2013].

BOX 14.2 Issues in Health Personnel Planning

- Current and projected demographic changes, i.e., population growth and aging of the population.
- Current and projected supply of practitioners and their geographic distribution by specialty.
- Technological advances requiring new professions.
- Immigration and emigration, effects on personnel supply.
- Costs/benefits of increasing professional-to-population ratios versus prevention, health promotion measures.
- Changing epidemiological patterns, e.g., reduction of dental service needs by fluoridation of community water supplies, aging of population with increasing prevalence of chronic disease.
- Health system shift from institutional to ambulatory and preventive care.
- Shift of tasks from higher level to other personnel specifically prepared for needed health services, increasing range of health personnel, e.g., optometrists, psychologists, social workers, midwives, dental nurses, nurse practitioners, and community health workers.
- Migration of doctors and nurses from low income to high income countries and from rural to urban areas.

Continuing education is a vital part of health personnel planning. The rapid and continuous development of medical sciences (e.g., genetics and nanotechnology) and technology (e.g., new vaccines and health promotion techniques) requires health workers to have access to continuing education to keep up with new developments. The methods of doing this should include short courses, longer formal training periods, such as the Master of Public Health (MPH) degree for health managers, and development of distance learning with wide access to Internet resources. Increased access to well-developed consensus guidelines for both clinical care and public health policy is vital for human resource planning.

Fundamental to the process of determining labor needs is knowledge of the current personnel situation (Box 14.2). Essential for this are data systems based on periodic registration or census-taking of people practicing a health profession. Practitioners may retire, die, migrate, or leave the profession, and should be deleted from registries of those actively practicing. An accurate, up-to-date picture of actual human resources provides information on specialty, geographic distribution, age, gender, and current work activities. International comparisons of professional personnel help to place a national pattern in the context of other countries with similar socioeconomic and health standards. Human resources supply should be matched to the targets and resources of a country. Alternative approaches may be needed if the supply of workers is insufficient or inappropriate to meet health needs and targets.

Assessing current personnel supply and determining future needs are specific tasks of a government agency concerned with comprehensive national socioeconomic planning. They may be assigned to a planning agency, board, commission, or committee empowered by authorities working with education systems, consistent with general health planning. Academic training centers play an important role not only in training but also in implementation of national human resource policies, so they are integral to determining policy.

Supply and Demand

A common form of quantitative human resources planning, or non-planning, is a market-oriented approach, based on the needs of the training institution and demands of trainees. The demand for training as physicians may be high, and the medical schools have an interest in training more students for financial or prestige reasons. The creation of new private medical schools, if unregulated, will be based on a profit motive and not take into account the needs or capacity of the country to absorb new graduates. This leads to the creation of excessive training capacity and a surplus of poorly trained doctors with little prospect of professional employment, as is happening in some mid-level developing countries.

During the 1950s and 1960s, planners in many countries thought that universal access to medical care would solve most health problems and more doctors would be needed to fulfill that dream. However, increasing the supply of medical graduates is costly to society and results in oversupply, especially in major urban centers, with increased utilization and subspecialization of medical care.

Increasing the supply of physicians was expected to increase access to health care and to increase the numbers of doctors entering less popular fields of practice, such as primary care, and moving to underserved geographic areas. This approach has been less accepted since the 1980s; even in free market societies, it inflates the costs of health care and fails to meet needs in underserved populations or specialties. Immigration and emigration of medical personnel, or departures from active practice, are also factors in the supply and distribution of health personnel. There were also concerns during the 1980s of a possible oversupply of doctors, and medical school enrollments were limited as a policy. This approach was promoted as part of the concern for rapid increases in health care costs and partially as a result of concerns in the medical profession of excess supply bringing more competition and possibly lowering the incomes of physicians. In the 1980s, health care costs rose rapidly, associated with increasing specialization and a search for new organizational patterns of health care such as the health maintenance organizations (HMOs). There was also a growing realization that population and personal health needs depend more on prevention and health promotion than on increased supply of physicians. This led to a trend to reduce numbers of new students entering medical schools, a decrease in subspecialty training positions, and

a greater reliance on immigration of doctors to the USA, Canada, and the UK. This created a growing migration of medical and nursing personnel and reliance of developed countries on importation of doctors and nurses from poorer countries, often with serious and destructive effects on poorer countries in Europe and especially in severely work-force-deficient countries in sub-Saharan Africa.

Comparing Canada to the USA and other countries in the Organisation for Economic Co-operation and Development (OECD) shows that in 2010 Canada and the USA each had a ratio of 2.4 practicing physicians per 1000 population, compared to the OECD average of 3.1 per 1000. In 2009 the USA had 10.8 nurses per 1000 population, compared to 9.4 for Canada and an OECD average of 8.4. As seen in Chapter 13, Canadians have better life expectancy than their US counterparts (80.8 versus 78.7 years), lower mortality rates, and better access to physicians. The difference seems to be in universal coverage still lacking in the USA, despite higher health expenditures there.

Medical and health profession schools are costly to establish and operate; they can generate high costs to a health system if they produce excess graduates. Founding new schools and maintaining existing schools at present levels of enrollment require careful consideration of the effects of the numbers of medical graduates on the health system. In either a regulated environment or a free market situation, the supply of human resources can be powerful in driving up health care costs. A period of restraint in health expenditures calls into question the wisdom of continuous increases in personnel and unlimited service as direct public service insured benefits. Even in free market settings such as the USA, government funding and regulatory powers are used to reduce the number of training positions in the specialties in favor of increased incentives and openings in primary care.

Trends in medical education and physician supply in the USA between 1970 and 2009 are shown in Table 14.2. The number of medical graduates increased by 84 percent between 1970 and 1980 but has grown at a slower rate since 1990. Medical personnel per population increased by over 26 percent from 1970 to 1980, and by over 36 percent from 1980 to 2000, with only a 2.2 percent increase from 2000 to 2009. Some 25 percent of practicing physicians in the USA are graduates of international medical schools. The number of US medical graduates per 100,000 population during the 1980s was between 7.0 and 7.5, but declined to an average of 6.3 per 100,000 population in the 2000–2010 period. Countries such as Sweden and Switzerland average between 8 and 10 per 100,000 in recent years (OECD, 2012). Owing to population expansion and aging, the American Medical Association (AMA) and American Association of Medical Colleges (AAMC) now predict a growing shortage of physicians in the USA, particularly in primary care and underserved areas. This shortage is partially being addressed by the expansion of osteopathic schools and immigration

TABLE 14.2 Schools, Graduates, and Physician Supply, USA, 1970–2009

	1970	1980	1990	2000	2005	2008–09
Medical and osteopathy schools	110	140	142	144	145	157
Graduates (thousands)[a]	8.8	16.2	16.9	18.0	18.5	20.1
Doctors/10,000 population[b]	15.5	19.6	23.2	26.8	26.9	27.4

Note:
[a]Includes all graduating allopathic and osteopathic physicians.
[b]Includes all practicing and non-practicing physicians.
Source: National Center for Health Statistics. Health, United States, 2011: With special feature on socioeconomic status and health. Tables 109 and 114. Hyattsville, MD: NCHS; 2012. Available at: http://www.cdc.gov/nchs/data/hus/hus11.pdf [Accessed 15 July 2013].

of foreign graduates; however, the AMA and AAMC have called for a rapid increase in US medical school admissions, by 30 percent. Immigration of physicians from developing countries is providing an important source of medical personnel, but is contributing to the deficiency of physicians in the source countries.

A normative approach uses standards (or norms) derived in some systematic, arbitrary way. The standards may be based on empirical criteria of the number of physicians, nurses, or other health personnel required. This approach may be excessively rigid and unresponsive to changes in disease prevalence and technological changes in health service needs. Standards may also be adopted from ratios found in other countries or in other successful or "gold standard" areas of the same country.

Human resources planning may set certain goals intended to produce personnel in numbers maintaining or increasing the current supply-to-population ratio by a selected percentage, for example, by 5 or 10 percent within a chosen period. A country wishing to increase this ratio will need to take into account new training needs and loss due to emigration. This approach is less likely to lead to an oversupply but may maintain an arbitrarily high level of human resources despite changes in epidemiological patterns or increased efficiency of the services. For example, as TB declined, fewer TB specialists were needed, but as the disease recurs, there is a demand to improve the training and numbers of specialists in the field. Hospitals are becoming less the center of health care, and reduction in hospital beds has become part of restructuring of services. This should lead to a shift of personnel from institutional to community-based services, with provision for retraining and skilled system management.

Many countries require medical graduates to serve one or several years in rural or underserved areas. Young graduates are thus exposed to the realities of primary care as part of their professional development and, it is hoped, infused with concern for the harsh realities of the living conditions of rural poverty. However, this system places inexperienced young professionals in isolated locations without adequate collegial support or supervision, where they are unlikely to remain beyond a compulsory period of service. Efforts to require young graduates to work in rural areas are temporary solutions, generally frustrated by the desire of doctors and nurses to live in urban areas and practice in clinical subspecialties.

Partly in search of methods to constrain cost increases and partly to find ways to improve access to care for high-risk groups, mid-level health provider training is increasingly accepted in human resources planning. Human resources planning should take into account the many different disciplines needed for both clinical care and public health, taking into account changing patterns of need, technology, and spread of health care responsibility among many professions.

Organization of care affects the numbers and types of different health workers required. Independent private practice and free choice of physician or specialist promote higher utilization patterns and create waiting lists, rapid cost increases, and an apparent shortage of personnel. Centrally controlled health systems such as the Soviet health system created inflated norms of staff-to-population ratios and low-efficiency health services (see Chapter 13).

Qualitative methods are as important as quantitative planning. Quality of training programs at the undergraduate and graduate levels, accreditation, licensure procedures, and ongoing quality assurance measures are important elements in the quality of national health systems.

The supply of medical doctors varies widely in European countries (Table 14.3) and has little relationship to health indicators, for example in comparing Austria and Belgium, or Greece and Israel. Greece has a very high doctor to population ratio (as well as acute care hospital bed to population ratio) compared to all other European countries and the supply is increasing even while the country is in a severe economic crisis. Greek life expectancy was above most European countries until the 1970s but its rate of increase has slowed to levels below those countries since 1990.

MEDICAL EDUCATION

The education of medical doctors and the training of specialists are, in principle, national commitments. Governments

TABLE 14.3 Physicians per 100,000 Population, Selected European Countries, 1980–2010

Country	1980	1990	2000	2010
Austria	222	299	381	478
Belgium	231	327	283	297
Czech Republic	226	271	337	358
Germany			326	373
Greece	243	338	433	610
Israel			339	350
Norway	197		340	407
Portugal	189	274	310	383
UK	132	162	196	273
European Region	247	291	301	329
EU	198	259	283	334
EU members before May 2004	193	260	290	356
EU members since 2004 or 2007	212	254	263	270
CIS	355	387	378	381

Note: Numbers rounded.
CIS = Commonwealth of Independent States (Russian Federation, Ukraine and others).
Source: World Health Organization. Health for All database; July 2012. Available at: http://data.euro.who.int/hfadb/ [Accessed 3 December 2012].

BOX 14.3 The Flexner Report, 1910

"For twenty-five years there has been an enormous overproduction of uneducated and ill-trained medical practitioners in absolute disregard and without serious thought to the interests of the public.

Taking the USA, physicians are four or five times as numerous in proportion as in older countries like Germany. Over-production is due to the very large numbers of commercial schools.

Colleges and universities have failed to appreciate the great advance in medical education and the increased cost of teaching it along modern lines.

A hospital under complete educational control is as necessary to a medical school as is a laboratory of chemistry or pathology.

Trustees of hospitals, public and private, should, therefore, go to the limit of their authority in opening hospital wards to teaching. Progress for the future would seem to require a very much smaller number of medical schools, better equipped and better conducted and the needs of the public would equally require fewer physicians graduated each year better educated and better trained."

Sources: *Pritchett HS. Introduction. In Flexner A. Medical education in the United States and Canada: a report to the Carnegie Foundation for the Advancement of Teaching; 1910. Reprinted New York: Arno Press and New York Times; 1972.*
Frenk J, Chen L, Bhutta ZA, Cohen J, Crisp N, Evans T, et al. Health professionals for a new century: transforming education to strengthen health systems in an interdependent world. Education for health professionals for the 21st century. The Lancet Commissions. Lancet 2010;376:1924–58.

have a responsibility to ensure an adequate number of well-trained health professionals to provide services. This is a combined function of health and education authorities, carried out by providing financial support and standards for the universities or medical training institutes where education occurs. Funding support and accreditation of educational institutions provides mechanisms for applying national or state policy for both quantity and quality of educational programs. National or provincial departments of education set guidelines and standards for funding through a university grants mechanism or commission, often based on enrollment. Standards may be set for curriculum, faculty, basic sciences, and clinical training, as part of approval for funding or through non-governmental accreditation structures organized by the medical schools themselves (see Chapter 15).

The long tradition of multifaculty, university-based medical education is widespread in the industrialized countries and their former colonies, now independent states. Medical training gains from an environment that promotes research and service in an academic atmosphere with its associated standards. This tradition of linking research with education and service is important in promoting quality education. Having a research climate of peer-reviewed work raises the aspirations of the institution and its faculty and sets

a standard for students for their life's work. A university degree confers prestige to a profession, encourages the pursuit of peer recognition of excellence, and academic criteria for student selection, curriculum, and faculty standards. This is widely the case for medical schools, and increasingly for schools of nursing and other health professions. However, a university degree is not required for all health professions. Community colleges may more appropriately provide a multifaculty educational environment and a broad education base for some health care jobs.

In the nineteenth century, medical training in the USA was primarily carried out by private, commercial schools of medicine with poor facilities, staffing, and standards. The Carnegie Foundation sponsored a study of medical education in the USA and Canada, carried out by Abraham Flexner, a non-physician educator, who reported in 1910 on the poor quality of these commercial schools (Box 14.3). This report promoted university-based medical schools modeled on the Johns Hopkins University, which itself was based on successful, scientifically oriented German medical schools, combined with the strong clinical orientation of British teaching hospital medical schools. Most of the 450 commercial schools in the USA closed soon after this report and were replaced by the present 126 university-based medical schools with high standards of medical

BOX 14.4 Syllabus for Medical Education in Epidemiology, Clinical Epidemiology and Biostatistics

- Lectures
- Measures of disease frequency, morbidity, and mortality
- Rates and standardization
- Morbidity and mortality in Israel
- Research design I – cohort studies
- Measures of association
- Statistical inference
- Research design II – case–control studies
- Sample size
- Occupational epidemiology
- Clinical trials
- Analysis of clinical trials (multivariate models)
- Survival analysis
- Diagnostic tests
- Screening
- Meta-analysis
- Evidence-based medicine
- Class interactive exercise on diagnosis
- Causal and non-causal associations
- Summary and questions
- Small group sessions: prognosis; therapy; prevention
- Final written examination: short answers, multiple choice based on lectures and journal articles

Source: Paltiel O, Brezis M, Lahad A. Principles for planning the teaching of evidence-based medicine/clinical epidemiology for MPH and medical students. Public Health Rev 2002;30:261–70.

education and academic research. Since the 1950s, US medical schools have been stimulated by large amounts of federal funds channeled into research and training through the National Institutes of Health (NIH), as well as from non-governmental sources, including private and foundation donations.

Medical schools are resources for service to the community as well as being centers of training and academic research excellence. Their goal should be to provide a balanced education in an academic environment where teaching, research, and service interact to produce medical graduates competent and oriented to meet the needs of the population (Box 14.4). This requires a balance among the biomedical, psychological, population-based, and sociological perspectives on health care. Teaching methods should be designed to promote the objectives of the program. Many medical schools teach primarily by lecture to very large classes, with limited supervised clinical experience, which reduces the chance for the student to develop patient-oriented and problem-solving skills. It also promotes a didactic approach to medicine, and minimizes the opportunity for the student to work with multidisciplinary teams, or to see medical care as part of a diverse team. Working with students of other sciences and professions in a collegial fashion helps the medical student to understand the team role of workers in the health care system.

The purpose of training medical practitioners is to have skilled professionals providing patient care and the professional leadership needed to develop and maintain high-quality health care and public health systems. In order to meet these goals, high standards are required in selection of candidates. Medical schools in the USA are graduate schools requiring a prior university degree for candidates. In other countries such as Canada and the UK, medical education includes 2 years of premedical studies followed by 4 years of medical school. Quality medical education requires continuous curriculum development and review, as well as highly qualified teachers, library access, clinical training, and examination during and at completion of training. The nature of undergraduate training will be a key factor in determining the lifelong practice habits of the providers, but equally important are the specialization period and ongoing education throughout their professional lives.

In most industrialized countries, enrollment of women and minority groups has increased dramatically in recent decades as part of social policy. There are social and political reasons to promote access to professional schools for all segments of a population, but this should be without compromising academic standards, and should not adversely affect the quality of services provided to the patient or the population as a whole. Private medical schools are a highly lucrative business in some developing countries, which, if unregulated, may contribute to overproduction of inadequately trained doctors, compromising national efforts to promote quality of training.

Where the language of instruction is not one used internationally for scientific literature, the local medical community may be limited in access to current textbooks and peer-reviewed professional literature, domestic and especially international journals, as well as Internet access. The language of instruction in most schools of medicine is the national language, but English is increasingly required as a second language in many European schools. Lack of English-language training prevents or hinders access to the world literature and participation in international exchanges and effectively holds back scientific progress in many countries.

During the transition period of the post-Soviet era, medical schools sought enrollment of foreign students to increase revenues, reducing the numbers of local students. However, they still produce graduates at levels well above the capacity of the system to absorb them if salaries of physicians are to rise above tradesperson levels. In some developing countries, private medical schools have sprung up with inadequate facilities and faculty producing large numbers of poorly trained doctors with little chance of employment in the profession in their own countries.

Curriculum reform, as in the days of Flexner's recommendations, must be an ongoing process to meet the health needs of the population, in keeping with current international standards. Adequate attention must be paid to basic medical sciences, clinical experience and patient care, hospital and community-based training, and research. Access to libraries with an adequate supply of current international literature, textbooks, and computers with Internet services is essential to maintain acceptable standards.

Reform in medical education is focusing on producing practitioners to meet the needs of both primary care and specialized medical services including orientation to public health. In recent years, there has been growing concern that there has been an overemphasis on science and specialization in undergraduate training of US physicians, to the detriment of primary care. All medical students should be exposed to patient contact earlier in their training than in the past, in different health care settings, including teaching hospitals, outpatient clinics, and community clinics, as well as public health programs. They should also be familiar with community-based resources for the infirm, disabled, and poor. Training should include multidisciplinary components so the student is familiar with the professional elements of other disciplines including those in public health, health-related economics, and social sciences.

International conferences on medical education sponsored by the World Federation for Medical Education (WFME) in 1988, 1993 and in 2003 attempted to define a new direction for education of physicians to promote their role in promotion of health as well as treatment and prevention of illness (Box 14.5). This included a global perspective on damage done by migration of doctors and nurses from low income to high income countries. Quality and uniformity in standards for basic medical education were stressed as priorities at the global level. Sponsored by the WHO, United Nations Children's Fund (UNICEF), United Nations Educational, Scientific and Cultural Organization (UNESCO), United Nations Development Programme (UNDP), and the World Bank, these conferences established an international forum for re-evaluation of medical education in the twenty-first century in the context of changes in medical and public health technology, organization of health care, and needs of the population. Change in medical education is often difficult because of competing concepts of what medical students should know and a lack of focus on how practicing physicians, especially for primary care, should be trained and encouraged to remain in this vital social role.

The costs of medical education are high and require public subsidies. University grant commissions are semi-autonomous bodies with financial grants from governmental education departments. Thus, both financial and regulatory powers are used to set criteria for standards and accreditation of faculties of medicine. This represents an important diffusion of power and responsibility from direct control

---

BOX 14.5 Medical Education Issues: The Edinburgh Declaration (1988) and the World Summit on Medical Education, 1993 (Reaffirmed in 2005, Standards Revised 2012)

- Conducted in relevant educational settings – hospital, community, workplace, homes
- Curriculum based on national health needs
- Emphasis on disease prevention and health promotion
- Lifelong active learning
- Competency-based learning
- Teachers trained as educators
- Integration of science with clinical practice
- Selection of entrants for social commitment, intellectual attributes
- Coordination of medical education with health care services
- Balanced production of categories of doctors
- Multiprofessional training
- Continuing medical education requirements
- Students involved in planning and evaluation of medical education
- A multi science-based medical graduate
- Ethical and moral basis of medical practice
- Curriculum options for dealing with information overload
- Postgraduate education in relation to community needs
- Health teams and multidisciplinary education
- Community participation in medical education
- Population-based education – care for individual patients in context of needs for a defined population

Sources: Adapted from World Federation for Medical Education. World Summit on Medical Education: The changing medical profession; Edinburgh, August 1993, reaffirmed at Copenhagen in 2005. Promotion of accreditation of basic medical education: a programme within the framework of the WHO/WFME strategic partnership to improve medical education. Copenhagen: WFME; revised 2012. The Panum Institute, Faculty of Health Sciences, University of Copenhagen; November 2005.
World Federation for Medical Education. Basic medical education. WFME global standards for quality improvement. Copenhagen: WFME; 2003. Available at: http://www.wfme.org and http://www.wfme.org/news/general-news/263-standards-for-basic-medical-education-the-2012-revision [Accessed 19 November 2012].

---

by government. Regulation by accreditation of schools is also strengthened by national organizations which promote national standards of medical education.

The Medical Council of Canada (MCC) grants a qualification in medicine known as the Licentiate of the Medical Council of Canada (LMCC) to graduate physicians of accredited Canadian and US medical schools who have satisfied the eligibility requirements and passed the MCC Qualifying Examination Parts I and II. The MCC registers candidates who have been granted the LMCC in the Canadian Medical Register. Graduates of medical schools outside Canada and the USA are required to first pass the MCC Evaluating Examination. Physician licensing to practice in a province is delegated to the provincial medical associations or a provincial college of physicians upon successful completion of national examinations.

In 2011, Canada had 2.4, the USA 2.5 and the UK 2.8 physicians per 1,000 population (OECD 2013). These countries augment local production of doctors by promoting immigration of foreign graduates, mostly from low- and medium-income countries, often through postgraduate education followed by retention processes fostered by teaching institutions. This policy causes grave concerns among developing nations and international organizations, which are calling for ethical practices to stop this promotion of a notorious "brain drain" from countries that are severely deficient in medical and other health professionals, pushed by poor conditions at home and pulled by wonderful conditions in the wealthy countries: the UK, Canada, the USA, Australia, and others. These issues are discussed extensively in the World Health Report 2006, and a Commission under the auspices of the Lancet (see Box 14.1) (Frenk et al., 2010).

Medical education needs should include exposure to the principles of evidence-based medicine and methodological training in epidemiology, biostatistics, economics, and research methods in order to cope with the explosion of medical information, and to appraise, interpret, and perform clinical research (see Chapters 3 and 15). There is growing interest in MD/MPH programs in the USA and Israel. There has also been an increase in clinical specialists taking MPH and other special programs, for example at the Harvard School of Public Health, to gain knowledge in these areas needed to advance clinical research capacity, in scientific writing, and providing potential leadership roles in the health system.

POSTGRADUATE MEDICAL TRAINING

Undergraduate medical training provides an educational base, but is not adequate preparation for a medical practitioner. Postgraduate training of high quality and adequate duration is essential to assuring quality in health care services (Box 14.6). Specialty training requirements should be regulated by a national or state authority, or a professional body (college or association) delegated the legal right to license practitioners. This includes designation of facilities accredited for training, academic, and research areas within the curriculum, clinical experience, duration of training, and requirements for examination at several stages during the training period.

National standards are needed to ensure equivalent quality and permit freedom of movement for professionals. However, this may put some areas at a disadvantage by promoting a brain drain, or loss of professionals, usually from rural to urban areas or from poor countries to wealthy ones. The rights of an individual practitioner to select place and type of practice are limited by open positions in training centers or in practice settings.

BOX 14.6 Standards for Postgraduate Medical Training

- Regulated by national board with professional, governmental, and public representation for quality, admissions/enrollment, and clinical and community-based training
- Duration of training of 4–6 years, depending on specialty
- Supervised independent clinical experience
- Accreditation of training centers based on academic and service criteria of licensing body
- Supervised research period in basic science laboratory or epidemiological study
- Required familiarity with relevant international literature
- Rotation with part of training in a different medical center
- Demonstrated high levels of clinical ability, responsibility, knowledge, and ethical standards
- Examinations in mid-training with written examinations based on international standards
- Examinations at end of training; clinical and written examinations based on international standards
- State board or professional college setting examinations and certification
- Recertification requirements

Source: *World Federation for Medical Education. Post graduate medical education. WFME global standards for quality improvement. Copenhagen: WFME; 2003. Available at: http://www.wfme.org/standards [Accessed 29 November 2012].*

Licensing of medical specialists is a state responsibility, but in some countries this is delegated to a professional association. In the USA, postgraduate training is under the control of state and national boards, made up of state-appointed officials and public and professional representatives. In Canada, postgraduate examinations and certification are under the authority of a professional body, the Royal College of Physicians and Surgeons of Canada. In the UK, licensing of physicians is under a state-appointed body, the General Medical Council (GMC), while specialty recognition is by a series of Royal Colleges, including a Faculty of Public Health.

Standards for specialty training must reflect the views of the specialty practitioners as well as the public interest. The public interest is best protected by a combination of state and professional supervisory systems with the force of law, including the regulatory and disciplinary measures needed to maintain professional and ethical standards demanded by the public interest. The specialist trainee requires supervised time and experience to mature as a professional. Supervised clinical experience, research, publication in peer-reviewed journals, and continuing peer review are all essential in the training process to produce motivated specialists who are capable of keeping up with the rapidly evolving standards of modern medicine. Clinical specialization time requirements vary widely from country to country. Eligibility for specialty boards in the

USA is generally 3–4 years of recognized training after graduation, with examination by member boards of the American Board of Medical Specialties.

SPECIALIZATION AND FAMILY PRACTICE

Good medical care depends on access to primary care and appropriate referral for specialty care. Most systems utilize the primary care physician as a gatekeeper for referral to specialty care. Insured service systems allowing non-referred access to specialty care face the difficulty of maintaining primary care medicine and continuing pressures on physicians to select specialty training as their career choice. Still too few graduating physicians choose a career in family practice, but specialists in internal medicine, pediatrics, and other fields are essentially providing primary care. The trend seen in Table 14.4 shows a relatively stable pattern of doctor to population ratio and nearly half of the medical workforce practicing primary care.

Uneven distribution of medical practitioners is widespread, with rural and urban poverty areas often suffering from a lack of access to primary care. In the USA, regional variation in practicing physicians in 2009 ranged from high ratios in Massachusetts (34.4 per 10,000 population) to a low rate of 18.6 in Nevada. Specialist physicians are less likely than generalists to live in rural areas. Distribution of physicians by specialty is another problem in medical resource planning. Regulations to redistribute the number of training positions are now operational in the USA and common in many countries. Medical teaching centers in the USA now come under regulations which require them to include primary care in their postgraduate training programs.

National health systems address these problems with regulations to mandate and financial incentives to attract physicians to underserved areas and understaffed specialties. In the UK, as in many other European countries, the National Health Service (NHS) uses the general practitioner (GP) as the key family practitioner (FP), primary care provider for all beneficiaries, with specialty access through the GP. Managed care programs in the USA also stress and require patients to see primary care physicians. The changing economic environment of health care will be associated with changes in medical specialization more easily than the urban–rural inequities. These issues are leading to greater role delegation to nursing and new kinds of health workers.

TRAINING IN PREVENTIVE MEDICINE

Preventive medicine is defined by the American Board of Preventive Medicine (ABPM) as: "the specialty of medical practice that focuses on the health of individuals, communities, and defined populations. Its goal is to protect, promote, and maintain health and well-being and to prevent disease, disability, and death. Preventive medicine specialists have core competencies in biostatistics, epidemiology, environmental and occupational medicine, planning and evaluation of health services, management of health care organizations, research into causes of disease and injury in population groups, and the practice of prevention in clinical medicine".

Preventive medicine is recognized as a clinical specialty in the USA. Promoted since the 1970s, this specialty attempts to bring public health and clinical medicine closer together. Preventive medicine training is one of 24 accredited clinical specialties in the USA, with physicians becoming board certified in one or more subspecialties: general preventive medicine and public health, occupational medicine, and aerospace medicine. These programs are part of the postgraduate training program system of the AMA, in conjunction with the ABPM. Master's or doctoral degrees are earned in graduate programs situated in departments of community or preventive medicine within a medical faculty.

Preventive medicine is a specialized field of medical practice composed of distinct disciplines that utilize skills focusing on the health of defined populations to maintain and promote health and well-being and prevent disease, disability, and premature death. The ABPM requires trainees to have core competencies in biostatistics, epidemiology, administration, planning, organization, management, financing, and evaluation of health programs. Training also includes environmental and occupational health, and social and behavioral factors in health and disease and the practice of prevention in clinical medicine. It applies primary,

TABLE 14.4 Physician Workforce and Percentage of Primary Care Doctors, USA, Selected Years, 1950–2009

	1985	1990	1995	2000	2005	2009
Active physicians/10,000 population	20.7	NA	24.2	25.8	23.8	27.4
% in primary care	NA	44.6	45.6	47.3	46.2	48.1

Note: Primary care includes general primary care specialists and primary care subspecialists.
NA=not available.
Source: National Center for Health Statistics. Health, United States, 2011: With special feature on socioeconomic status and health. Tables 109 and 111. Hyattsville, MD: NCHS; 2012. Available at: http://www.cdc.gov/nchs/data/hus/hus11.pdf [Accessed 15 July 2013].

secondary, and tertiary prevention measures within clinical medicine.

Graduates in this field provide a supply of potential health planners, administrators, and teachers of preventive medicine, researchers, and clinicians applying preventive medicine in health care settings of practice. They may also serve in governmental (local, state, national, and international) public health departments, educational institutions, organized medical care groups, in industry, other employment settings, and the community, voluntary health agencies, and professional and related health organizations. Requirements include a graduate year of training and experience in a clinical area of medicine; a year of academic training in a fundamental aspect of preventive medicine; and a practicum or year of supervised practical experience (e.g., occupational health). Training of clinicians in health services research and clinical epidemiology also provides a potential career path for physicians entering one of the many fields of public health.

Preventive medicine specialty is not common in countries outside the USA as a combined clinical and public health specialty. UK and Canadian specialization in public health is also largely divorced from clinical practice. Social medicine is widely used as a specialty in Eastern Europe but is primarily non-clinical in orientation and function.

NURSING EDUCATION

Nursing is the backbone profession in hospital and community health care. The place of nursing in a health system reflects the cultural values of the society and has an important effect on the health system. Whereas medicine is generally a high-prestige profession, in many countries nursing is of low social status, with strong cultural biases against women entering nursing. Germany and Canada have more than four nurses per physician, while developing countries such as Pakistan, Nepal, and India have between 0.7 and 1.5 nurses per physician (2009 figures). This disparity reflects a widespread overemphasis on medical training and an underemphasis on training of nurses in developing countries.

The health system thus suffers from a lack of personnel to develop and operate primary care services, with biases towards high-cost secondary and tertiary care services. Furthermore, the lack of high-level professional nursing personnel prevents full development of quality secondary and tertiary care services. Lack of nursing at the professional level may be one of the biggest factors in retarding the development of health services in many countries. The OECD data in Table 14.5 show the wide range of nurse to population ratios in selected European countries, with the highest ratios in Belgium, Germany, Norway, and Ireland, and much lower ratios in Israel, Italy, and the Czech Republic. There is wide variation in the nursing workforce between registered nurses (RNs) and practical nurses. More than 75 percent of

TABLE 14.5 Nurse Density in Countries of the Organisation for Economic Co-operation and Development, 2000

	Practicing Nurses per 1000 Population	% Registered Nurses
Ireland	14.0	96
Australia	11.7	NA
Switzerland	10.7	84
Canada	9.9	76
Denmark	9.5	15
USA	9.1	NA
Denmark	9.5	15
Sweden	8.8	42
UK	8.8	81
New Zealand	9.6	88
Portugal	3.7	NA
Korea	3.0	47
Mexico	1.9	NA

NA=not available.
Source: Derived from Simoens S, Villeneuve M, Hurst J. Tackling nurse shortages in OECD countries. OECD health working papers no. 19. Paris: OECD; 2005. Available at: http://www.oecd.org/health/healthpoliciesanddata/34571365.pdf [Accessed 2 December 2012].

the workforce in Ireland, Belgium, New Zealand, Switzerland, the UK, and Canada are registered nurses. Practical nurses are the predominate group in Sweden, Norway, Netherlands, Denmark, and Korea with the proportion ranging from 53 to 85 percent of the nursing workforce.

In the USA, the number of nursing graduates increased by nearly 85 percent from 1970 to 2005 (Table 14.6). From 1996 to 2000, there was a decline in nursing graduates by 31 percent overall when nursing education moved from diploma to baccalaureate training as the basic qualification. But the demand is growing: the American Association of Colleges of Nursing (AACN) reports that nearly 300,000 jobs were added to the health care sector in 2011, with the largest demand for RNs. With more than 3 million members, the nursing profession is the largest segment of the US health care workforce. Adequate numbers and quality of nurses will be vital in implementing the 2010 Patient Protection and Affordable Care Act ("Obamacare"), the most important change in US health care in the USA since the creation of the Medicare and Medicaid programs in 1965.

The promotion of the academic aspects of nursing is seen in the growth of baccalaureate nursing education from 13 percent of all nursing graduates in 1960 to 26.5 percent in 2000. The decline in the number of nursing schools in the 1950s was due to closure or consolidation of individual hospital schools of nursing. North American schools for

TABLE 14.6 Nursing Schools and Graduates in the USA, 1970–2010

	1970	1980	1990	1996	2000–01	2003	2005	2010
Nursing schools	1340	1385	1470	1508	NA	1370	1446	1691
Total nursing graduates (thousands)	43.1	75.5	66.1	94.8	79.7	76.6	84.9	NA
Graduates with BA/BSc (thousands)	9.1	25.0	18.6	32.4	26.5	31.4	28.0	NA
Registered nurses/10,000 population	35.6	56.0	69.0	79.8[a,b]	102	101	104	110

[a]Includes bachelor of sciences degrees.
[b]Data for 1995.
Note: Some schools have more than one program so the number of programs is larger than the number of schools. In 2011, there were very few (63) diploma programs, 1060 associate degree (AD) programs, and 677 baccalaureate programs for a total of 1800 programs in 1712 schools.
NA = not available.
Sources: US Department of Health and Human Services. Health, United States, 1998, 2006.
National League for Nursing. 2012. Nursing education statistics: annual survey of schools of nursing, academic year 2010–2011. Available at: http://www.nln.org/researchgrants/slides/index.htm [Accessed 8 December 2012].
Organisation for Economic Co-operation and Development. Health policies and data: OECD health data 2012 – frequently requested data. Available at: http://www.oecd.org/health/healthpoliciesanddata/oecdhealthdata2012-frequentlyrequesteddata.htm [Accessed 8 December 2012].
American Nurses Association. Nursing fact sheet. Registered nurses in the US 2011. [Accessed 8 December 2012].

nursing education are now largely associated with university or associate degree programs in community colleges. University-based schools in the USA provide academic degree programs at the bachelor, master, and doctorate levels. Nursing education at the master and doctorate levels provides the teaching, research, and management cadres needed for a progressive health care system. In 2008, 50.0 percent of the RN workforce in the USA held a baccalaureate or graduate degree while 36.1 percent earned an associate degree and 13.9 percent a diploma in nursing.

The first baccalaureate program in nursing was established at the University of Minnesota in 1909. By 1980 there were 377 and in 1995 521 bachelor's programs for RNs. Upgrading educational standards for existing professions, such as nursing or midwifery, involves consideration of the costs and effects on personnel supply as well as the desirability of raising professional standards. The advent of degree programs in nursing raised the level of prestige, leadership, research, teaching, and service of the profession. The transition from hospital apprenticeship training to university-based education (i.e., "academization") was opposed by traditional interests such as hospital management and the medical profession, but this resistance subsided with the demonstration of greater capacity in the nursing profession to take responsibility and incorporate rapid scientific and technological advances. Table 14.7 shows a decrease in diploma-level training and increases in associate and bachelors (or higher) in the USA. The trend towards increasing nursing graduates at the baccalaureate level continues to the present time, with most other RNs being trained in 2-year community college programs. In recent years there has been an increasing flow of people entering the profession in their late twenties and early thirties, as well as in their early twenties; nursing is attracting interest from different age groups and the entry classes are limited by capacity and not by shortage of qualified applicants. Expansion of the number of nursing educational programs and their capacity will require investment in facilities and faculty training. It is estimated that the shortages of nurses in the USA in 2020 will range from 240,000 to 600,000, but probably closer to the smaller number.

The scope of activities that professional nurses are authorized to carry out by law and custom has gradually broadened over the past several decades to include procedures previously performed only by physicians in the USA. This change is partly associated with increasing academization of the nursing profession and with the emphasis on bachelor's and master's degrees for nurses and PhDs for nursing teachers.

NPs are trained to diagnose and treat illness, usually under authorization from a supervising physician. In some developing countries, especially in rural areas, auxiliary nurses, as well as professional nurses, are expected to diagnose and treat common ailments, in addition to conducting health education and primary and secondary prevention. This role is vital, especially in areas without medical practitioners, and should be under the supervision and guidelines of the ministry of health or other public health agency.

Nursing specialization may be at a certificate or master's level. Certificate courses are in fields where the nursing role involves highly skilled practice crucial to patient outcomes such as in intensive care or emergency department nursing. Master's programs in areas such as pediatrics, geriatrics, or adult health produce a more broadly based and independent practitioner, researcher, or educator. Nursing specialties include ambulatory care, cardiac, critical care, education,

TABLE 14.7 Estimated Population of Registered Nurses by Graduation Cohort, USA

Initial Education	Graduated Before 2001 (%)	Graduated 2001–2004 (%)	Graduated 2005–2008 (%)
Diploma	24.6	3.5	3.1
Associate	42.7	56.4	56.7
Bachelor's and higher	32.7	40.1	40.3

Source: US Department of Health and Human Services, Health Resources and Services Administration. The registered nurse population: findings from the 2008 national sample survey of registered nurses, 2010. Available at: http://bhpr.hrsa.gov/healthworkforce/rnsurveys/rnsurveyfinal.pdf [Accessed 6 December 2012].

emergency, flight, forensic, geriatrics, holistic, home health, hyperbaric, management, maternal–child, medical–surgical, midwifery, military, neonatal, obstetrics, occupational health, oncology, orthopedics, pediatrics, perianesthesia, perioperative, psychiatric and mental health, private duty, and public health (AACN, 2012).

The AACN cites the US Bureau of Labor Statistics' Employment Projections 2010–2020 indicating that the RN workforce is the top occupation in terms of job growth to 2020. The number of employed nurses is expected to grow from 2.74 million in 2010 to 3.45 million in 2020, an increase of 712,000 or 26 percent. There is an additional need for 495,500 replacements in the nursing workforce, bringing the total number of job openings for nurses due to growth and replacements to 1.2 million by 2020 (AACN, 2012). A 2011 AACN survey found that total enrollment in all nursing programs leading to the baccalaureate degree was 259,100, an increase from 238,799 in 2010. Within this population, 169,125 students are enrolled in entry-level baccalaureate nursing programs. In graduate programs, 94,480 students are enrolled in master's programs, 4907 in research-focused doctoral programs, and 9094 in practice-focused doctoral programs in nursing (AACN, 2012). Figure 14.1 shows the growth of the US nursing workforce and the distribution between registered and employed from 1980 to 2008. The nurse-to-population ratio increased from 56 per 10,000 in 1980 to 69 in 1990 and 78.5 in 1994. The number of employed RNs increased during the period 1999–2005 by an average of 1.2 percent per annum, from 2.2 million to 2.4 million (Health United States, 2007).

In 2009, the USA had 10.8 nurses per 1000 population compared to the OECD average of 8.4 per 1000. Nursing colleges and universities in the USA are expanding enrollment levels to meet the current nursing shortage and the rising demand for nursing personnel. The number of qualified applicants not accepted in nursing baccalaureate programs in the USA reached over 75,000 in 2011, raising the issue of providing more training capacity and qualified faculty to meet the growing need for nurses to meet anticipated shortages and expanded roles in nursing practice. The shortage of schools, qualified faculty personnel, and general funds makes it difficult to meet the challenge of nursing shortage

through private–public initiatives with governmental support. In the USA, 833 nursing schools offered baccalaureate and graduate programs, with a large increase in the number of available seats in entry-level baccalaureate programs, from 20,000 in 2000 to nearly 65,000 in 2010.

The hospital bed supply in 2010 in the USA was 3.1 beds per 1000 (a decline from 6.0 beds in 1980), compared to the 2010 OECD average of 4.9 (OECD Health Data, 2012). Acute care hospital bed supplies have been reduced and long-term beds for nursing care have increased (see Chapter 11), so the need for nurses does not decrease. Auxiliary personnel staff, including ward clerks, nursing aides, and personal care workers in nursing homes, are important in the team in institutional care, and should be strengthened in community health care as well.

In-hospital patient care focuses on more severely ill patients and intensive care, requiring increased nursing staff ratios. In 2010, the number of nurses employed in the USA was 2.74 million and this figure is expected to grow to 3.45 million in 2020 (AACN, 2012). In addition, there is an increasing range of specialties in nursing, with more career options than in the past. Demand for RNs is expected to grow by 2–3 percent each year. Some estimates indicate a need for 30,000 additional nurses to graduate annually in order to meet the nation's health care needs and that more than one million new and replacement nurses will be needed in the USA by 2016.

As health care copes with both an aging and a healthier total population, people with chronic diseases require care at the primary level, with increased roles (and needed retraining) for physicians and for nurses in home care and other outreach programs of care for people at high risk (e.g., secondary prevention for hypertension or diabetes). Expanding roles of nurses in the changing health care environment will be vital to meet the challenges. There is a need to address retention of nurses and prevention of burnout, as well as to produce greater training capacity to absorb qualified applicants. Nursing has become an attractive profession because of the expansion of its roles and because of academization with greater recognition for the profession, yet the profession remains underpaid in comparison to the crucial role of nurses in health systems and to other professions with similar

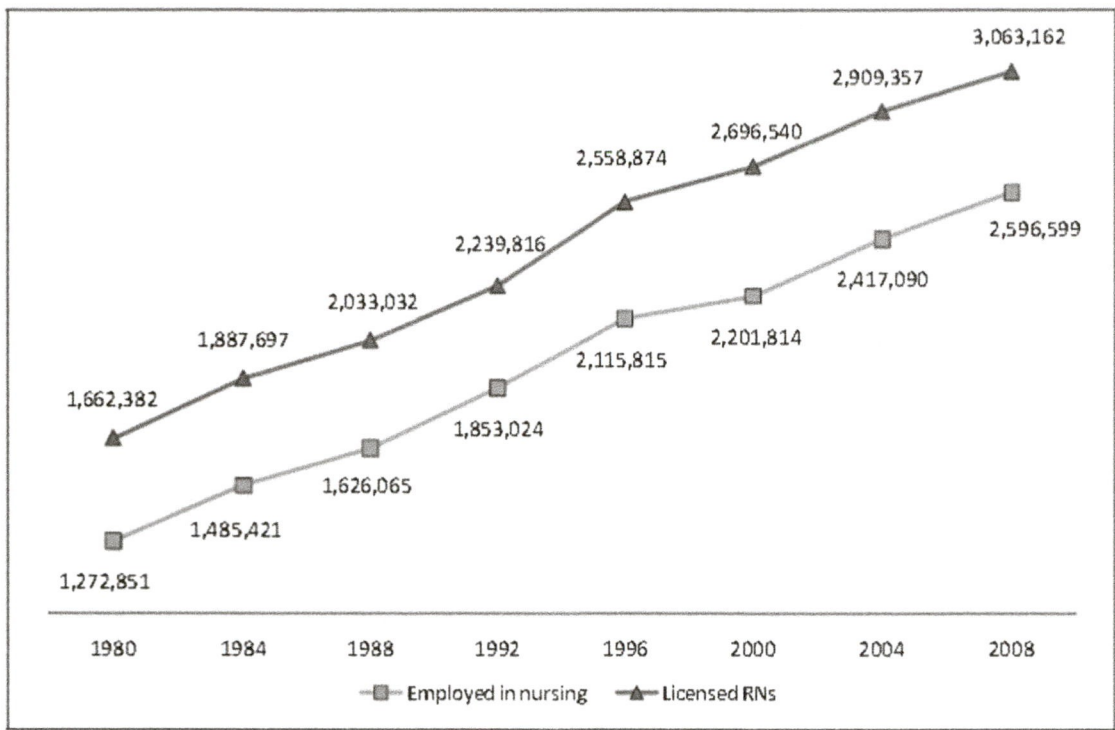

FIGURE 14.1   Nursing workforce, USA, 1980–2008. *Source: US Department of Health and Human Services, Health Resources and Services Administration. The registered nurse population: findings from the 2008 national sample survey of registered nurses, 2010. Available at: http://bhpr.hrsa. gov/healthworkforce/rnsurveys/rnsurveyfinal.pdf [Accessed 6 December 2012].*

qualifications. However, as economies shift from primary production to service-based industries, this may change the relative economic remuneration for this professional group.

National health authorities need to take professional views into account, but balance the vested interests of each profession with other factors such as the hospital bed supply and utilization, alternative forms of care (which can be labor intensive), aging of the population, changing disease patterns, and technological changes in prevention and health care. The development of training capacity for doctors, nurses, and all other health professions requires long-term planning and investment not only in building institutions for education, but also in developing faculty, research capacity, and scholarships for students along with modest tuition fees. Government has the key responsibility in this but private organizations and corporate, organizational, and individual donors can play crucial roles as well.

The medical and nursing professions have been in conflict over the numbers and roles of nurses. The nursing profession has struggled to establish greater autonomy and academic quality while the medical association has fought to maintain large numbers of nurses and a subordinate role for nursing. The conflict between need and demand is a matter of definition, viewpoint, and priority. Reorientation of the health system with greater emphasis on community care will provide more employment opportunities for nurses with expanded professional responsibilities.

IN-SERVICE AND CONTINUING EDUCATION

Rapid changes in all fields of medical science and practice make in-service and continuing education a necessity of any health program to maintain professional standards. In-service education increases the sense of self-esteem of workers and motivates staff to improve their performance. It serves to reinforce knowledge and introduce new information, and is essential to facilitate change in an institution. It also provides opportunities for the supervisory staff to reinforce and raise quality standards. The introduction of new programs and technologies should be accompanied by staff orientation as part of an ongoing in-service education program.

Continuing education refers to ongoing professional education in the form of courses, conferences, workshops, and literature. Medical graduates who complete requirements for specialization must continue to upgrade their training with periodic courses in specialty areas, where rapid advances are continuous. In public health, staff may take summer courses in epidemiology at schools of public health or departments of clinical medicine. Many medical, nursing, and other professional organizations require proof of continuing education for continued licensure and for professional advancement.

Recent advances in genetics, biotechnologies, patient safety, and many other important findings and discoveries require continuing education for all health workers. Science is changing our health paradigms and professional education needs to provide continuous updating in new knowledge and professional skills.

Governments, educational authorities, professional associations, provider organizations, non-governmental health agencies, and the general public all have strong interests in continuing education for the health professions. In-service and continuing education should be part of the working schedule of a health institution and included in budgetary planning for all levels of health personnel, from laundry room staff to hospital managers and from CHWs to medical officers of health.

ACCREDITATION OF HEALTH PROFESSIONAL EDUCATIONAL OR TRAINING FACILITIES

All facilities training health professionals should be accredited to do so by the national or provincial authority or by an agency recognized by them for this purpose. In Canada, accreditation is carried out by the Medical Council of Canada (MCC), which is also the examining body for graduates of all medical schools. Provincial licensing bodies accept the Licentiate of the MCC (LMCC) as the basic requirement for licensure. In the USA, the AAMC provides guidelines and accreditation of existing schools and reviews applications for new schools wishing to be recognized. Medical schools are subject to the state educational boards governing higher education facilities. State boards are responsible for examination of graduates and their licensure. Similar accreditation agencies exist for nursing, dentistry, pharmacy, and other health professions, in addition to governmental licensing bodies.

Universities or colleges establishing schools for other health disciplines are subject to the requirements of the authorities governing postsecondary education. In the USA, the Accreditation Commission for Education in Nursing (ACEN) is the recognized body for accreditation of all nursing education programs. A university wishing to establish a medical, dental, nursing, pharmacy, or other professional school would need prior approval showing the need for the facility, financial resources, and a complete proposal including curriculum, staffing, facilities, organizational affiliations, and objectives. Recruitment standards and policies, clinical affiliations, quality of library and basic sciences facilities, and budget would be scrutinized. Staff qualifications, tenure procedures and requirements, publications and research, access to international professional literature, availability of textbooks, and students' ability to read them (e.g., in a foreign language) should be part of the accreditation process.

Many new medical schools have followed patterns set at schools such as McMaster University in Hamilton, Ontario, Canada, and Ben Gurion University in Beersheva, Israel,

with a focus on preparing primary care physicians, but it is not clear to what extent they have succeeded in this objective. Curriculum review has become widespread in schools of medicine, with a concern that there may be an excessive emphasis on basic sciences and specialty clinical services so that the graduate has little orientation towards family and community practice or public health.

THE RANGE OF HEALTH DISCIPLINES

New professions such as the NP are developed from graduates of degree programs and require a master's level of training in an accredited program. Establishing or recognizing new health professional roles, such as NPs, optometrists, or CHWs, is dependent on and related to the needs of the health system. Development of curricula, criteria for enrollment, and site of the training program should be governed by the objectives of the program, but also should ensure wide acceptance of the new profession and potential for career advancement. Acceptance by the community is important, especially in programs intended to improve services for people in high-risk groups with the health and social services systems. Traditional birth attendants and CHWs are categories of personnel providing health care where cultural adaptation is especially important.

Clinical medicine has evolved from primarily a medical and nursing service to involve a highly complex team of professionals. Similarly, in public health the range of professions is broad. Interdisciplinary training is important for adequate functioning of a department or service increasingly dependent on teamwork.

The complexity of modern public health and clinical services is shown in the number of different professions listed in Table 14.8. This broad range of professions in public health requires graduate studies with an interdisciplinary approach to the preparation of leaders, teachers, and researchers for the field. Public health professionals work in a variety of settings. They need a wide base of training to understand the broad professional aspects of public health that relate to complex and rapidly changing professions and practices.

LICENSURE AND SUPERVISION

All countries have legal or regulatory systems by which newly trained health personnel are permitted to practice their profession. Requirements differ from country to country and for various types of personnel within a country. In some countries, health personnel must pass licensing examinations in addition to completing the prescribed training. In others, registration by the government is more or less automatic after the prescribed training, including the examinations, has been successfully completed. For some disciplines, such as medicine,

TABLE 14.8 Health Workforce

Category	Type
Physicians	Generalist and all specialist medical practitioners, including epidemiologists, laboratory, imaging, pathology, genetics, health systems analysis, policy and emergency care and systems specialists, researchers and academics, global health specialists
Nursing and midwifery personnel	Professional nursing, midwifery professionals, nursing associate and midwifery associate professionals
Dentistry personnel	Generalist and specialist dentists, dental hygienists, assistants, technicians and related occupations
Pharmaceutical personnel	Pharmacists, pharmaceutical assistants, pharmaceutical technicians, industrial pharmacists, quality assurance technicians and related occupations
Laboratory health professions and technologists	Laboratory scientists, biochemists, microbiologists, food scientists, toxicologists, geneticists, laboratory assistants, laboratory technicians, radiographers and related occupations
Public health professions and technologists	Occupational/industrial, environmental and public health officers, sanitary engineers, epidemiologists, public health nurses, geriatric, school, military, prison and other institutional health staff, health promotion specialists, public health nutritionists, environmental and public health technicians, sanitarians, hygienists, district health officers, public health inspectors, food sanitation and safety inspectors and related occupations, statisticians, medical anthropologists and sociologists
Community and traditional health workers	Community health workers, education and nutrition workers, family health workers, traditional and complementary medicine practitioners, traditional birth attendants and related occupations
Allied health professions and technicians	Medical assistants, dieticians, nutritionists, occupational therapists, medical imaging and therapeutic equipment technicians, optometrists, ophthalmic opticians, physiotherapists, personal care workers, speech pathologists and medical trainees, radiology, audiology and imaging technicians, occupational health therapists
Health management and support	Managers of health and personal-care services, health economists, health statisticians, demographers, lawyers, public relations, vital and medical records technicians, health information technologists, ambulance drivers, building maintenance staff, and other general management and support staff (kitchen, laundry, technical, janitorial and cleaning staff, many others)

Source: Modified from Technical Notes – Global Health Workforce Statistics database. Available at: http://www.who.int/hrh/statistics/TechnicalNotes.pdf [Accessed 2 February 2013].

dentistry, nursing, or pharmacy, the legal requirements for the license may be delegated to professional colleges or to state or national boards. Certification and relicensing of medical and other health care practitioners have become standard practice in the USA and some other jurisdictions to ensure that the health care provider meets the accepted professional standards of the day and public expectations.

Examination of undergraduate students is generally by the teaching institution itself, but examination at completion of training for licensing to practice medicine should be by external examination, preferably at a national or even an international level. National examinations are formulated and supervised by professional and governmental authorities, which establish and maintain the standards of medical graduates. In the USA, state boards govern medical licensure and specialty certification.

Licensing of health professions in some countries such as Canada allows the health professions self-government to set standards and govern the discipline within the profession as a form of peer review. National examinations and limitations of foreign graduates are spelled out in regulation or by decisions of the governing body of the profession. Foreign schools may be accepted for equivalent status or examination requirements may be established. As many as 30 percent of doctors working in the UK NHS obtained their primary qualifications from a country outside the European Union (EU).

In Canada, the MCC, a consortium of provincial professional bodies, establishes and supervises medical graduation examinations, while licensure for medical practice is by a provincially authorized medical body. Other countries regulate medical licensure directly but delegate specialty training supervision to professional organizations. Many countries have developed national examinations for medical, dental, nursing, pharmacy, and other professional licensure to promote high-level requirements and avoid the conflict of interests of a school examining its own graduates.

In the USA, medical graduates of accredited medical schools are licensed by state boards. Graduates of US medical schools in one state are accepted in other states for postgraduate training but not necessarily for medical practice, although some states have agreements of reciprocity.

Canadian provinces used to accept graduates of British medical schools, but this was restricted in the 1970s to reduce the flow of immigrant doctors. In the UK, the GMC is the legislated body empowered to license local graduates and immigrant physicians.

Licensing of physicians, nurses, midwives, psychologists, optometrists, NPs, or other professionals must be based on legislation or regulation under public statutes to designate the scope of permissible functions in each profession, licensing and examination procedures, as well as a code of ethics. Diffusion of power in governance of medical practice has contributed towards setting high standards of practice.

Control of education and licensing by the same authority that operates the national service may compromise standards. The development of multiple systems of accountability in a previously totally state-controlled system, as in Russia, will require many changes in existing practices of medical education, examinations, licensure, specialty training, and examination and discipline, as well as the development of independent professional organizations, accreditation bodies and standards of care (see Chapter 16).

It is now widespread practice for high-income countries to be very open to accepting foreign medical graduates for training positions and then facilitating their remaining in the host country with family visas and highly attractive work or research positions. This results in the loss of a skilled workforce from needy countries and represents a very exploitive aspect of globalization.

CONSTRAINTS ON THE HEALTH CARE PROVIDER

Maintaining standards requires organized supervision of performance by public bodies, in written guidelines, based on accepted current standards of care. This is often based on a consensus of professional views and practices, as well as recommended guidelines of professional bodies. Care should be taken to avoid penalizing legitimate innovations or differences of professional opinion, such as whether simple lumpectomy is sufficient care for cancer of the breast as opposed to radical mastectomy. This is part of quality assurance, discussed in Chapter 15.

Ideally, the constraints that impinge on the health care providers are the sum of training, licensing, practice, collegial relationships, and self-governance of ethical, humanitarian, and professional standards. These constraints are under scrutiny and potential disciplinary procedure from a variety of sources, including legal responsibility and standards of care expected by the employer or institution in which the provider functions.

The provider is also under scrutiny in the eyes of the public or consumers where consumer choice is part of the

system. The total effect of peer review on a continuous basis, encouraging good standards of practice in the community, helps to assure basic standards, but at the same time may promote conformity, limiting medical innovation, especially in the areas of organization of health care. Recognition of new professions can lead to conflicts of interest with self-governing professions, as has happened in the case of optometrists and NPs. Similarly, professional groups may oppose changes in health care financing and organization, both for the public good and sometimes for professional self-interests.

Application of standard medical curricula, national standard examinations, national licensing and disciplinary boards, and standards of medical practice is essential for maintaining and improving health care. External peer review is consistent with the current emphasis on total quality management or continuous quality improvement. When entering medical practice, doctors seek access to hospitals where they apply for "hospital privileges" and are assessed by professional peers for experience and qualifications. Individuals are expected to be members of professional organizations in their specialty and participate in departmental staff meetings, quality improvement committees, infection control, error and unusual event investigations, and programs of continuing education organized by professional associations, hospitals, and medical schools.

Health insurance plans monitor the billing or practice patterns of physicians and investigate aberrant practice or potential fraud. Investigations may be followed by administrative action against the offending physician or, rarely, by criminal procedures for fraud. Monitoring of surgical procedures may point out poor practice that may lead to disciplinary procedures, while criminal conviction means a suspension of one's license to practice. Malpractice insurance is vital to protect any physician or other health care provider against litigation and may be very costly depending on the specialty. The National Vaccine Injury Compensation Program passed by the US federal government in 1986 is designed to compensate individuals quickly, easily, and generously so as to prevent litigation, which damaged the uptake of immunization in the USA. The no-fault insurance for injury protects the provider and the injured party and may be a model for a more rational system than court litigation against the provider and the manufacturer.

Professional accountability is specific to each country. In the UK, the GMC is empowered by the state to issue medical licenses and discipline practitioners. A Patient's Charter sets out the rights, entitlements, and standards of service that the citizen may expect in health care. This, coupled with the right to change GP, empowers the patient to seek redress of grievances. Complaints regarding hospital care are investigated and can be pursued through stages of investigation. Consumer satisfaction is a factor in the recent innovation of GP fundholding in the UK (see Chapter 13), where the patient may, with the GP, select among hospitals or other support services. Health care is complex and requires a skilled and integrated team functioning with mutual trust, based on a common set of professional and ethical goals and standards. This is clear in the hospital dramas seen in popular television programs, but applies equally in the larger, real-world scale of health system organization and interaction among institutions, insurers, and public health networks. In addition to oversight by financial authorities and accreditation bodies, the scrutiny of the media, the political sector, the consumer, and the public at large is important. In short, the health provider and the health system are, and should be, under scrutiny, internally and externally.

OTHER HEALTH PROFESSIONS

There are other health professionals providing patient care that can be part of health promotion and health education. Pharmacists, for example (Box 14.7), can help to educate patients with long-term care needs, such as those with hypertension, diabetes, and obesity management, about risk reduction, complications, and compliance with medical management. Dental hygienists can help to reinforce health promotion messages such as smoking cessation, mouth care, healthful nutrition, and weight control, during routine dental hygiene care. The same can be said for physiotherapists, hearing technicians, and many other specialized ancillary care providers. Public health orientation should be included in the basic training of all health professionals and technicians in messages of how a healthful lifestyle is important for the prevention of many chronic conditions and their complications, for reinforcement of basic primary care messages for self-care.

Because health self-awareness is crucial to the successful prevention of many common diseases, and the promotion of healthy lifestyles is basic to self-care, it is more important than ever that all health professionals be participants in passing the key messages of health to their clients or patients on all possible occasions. Nurses, nutritionists, social workers, and many other caregivers should be enabled and encouraged to communicate dietary messages during care or routine follow-up. This is especially important for elderly patients, where nutrition and compliance with medication may be compromised by isolation, clouded memory, and confusion. It will require the introduction of public health topics or courses in the basic education of a broad range of health and social care giving professions in the New Public Health concept to recognize the cross-relationship of many social and health problems with health-promoting messages.

Cross-disciplinary work to promote health can be seen in many examples. One is the Special Supplemental Nutrition Program for Women, Infants, and Children (WIC), a federal program of the US Department of Agriculture to provide states with support for nutritional support for eligible pregnant, postpartum, and breastfeeding women, infants, and children up to 5 years of age. Although primarily meant for nutrition support, since 2000 state WIC programs have also checked immunization status and promoted the completion of immunization, and played a key role in raising the immunization levels of low-income children in the USA (see Chapter 8).

Health promotion is a strong force in public health and its range is wide and broadening. It is part of the roles of all health professions and technical support services, but has a cross-disciplinary orientation to create the societal and legal basis of modern public health with understanding and capability to address individual and community concerns, knowledge, attitudes, and practice that harm health and hinder efforts to remediate the health risks of different communities and risk groups (Carter et al., 2012).

EXPANDING ROLES IN PRIMARY CARE

New professional roles have emerged and continue to do so as the health needs of the community evolve. Public health and health management professionals as well as health care providers are all essential, tailored to the community they serve. The health system needs to assure that traditional health workers are available in numbers sufficient to provide for community and national care needs. In addition, there has been a growing realization of the need for new levels of health workers, which will affect the planning for medical personnel. Perceived shortages of doctors may be addressed by training more health providers, including NPs and professional midwives, and expanding the roles of public health and health promotion personnel. Improved computer and telecommunication usage may also help to relieve shortages of highly trained health professionals, and extend the outreach capacity to underserved areas or populations.

Mid-level health worker experience in Russia with the feldsher (see below) was important in the provision of primary care in rural, underserved areas. In many developing settings, experience with CHWs has been growing, and this has also been applied in some developed countries. In the USA, NPs and PAs have emerged as new health professional roles to augment the medical workforce, provide health care in underserved areas, and in some cases provide health care for targeted underserved population groups such as the elderly and diabetics.

Nurse Practitioners

In the 1960s health care providers and planners in the USA became aware of the growth of specialization in medicine and the decline in the availability of general practitioners. The nursing profession promoted expansion of nursing roles to fill this gap, including examinations and initiation of treatment and follow-up without direct supervision of physicians.

BOX 14.7 Pharmacists as Health Promoters in Management of Cardiovascular and Other Non-Communicable Diseases

Compliance and safety of medical management is a major issue, especially in relation to health care for non-communicable diseases such as hypertension and cardiovascular diseases (CVDs), the leading cause of death in adults worldwide, or in the management of diabetes, a major and growing public health problem. Pharmacist skills are complementary to those of primary care physicians and nurses, and can contribute to health promotion and improved safety of care of chronic patients. On the basis of their knowledge of medications, communication skills, and high accessibility to the public, pharmacists are valuable resources, for example in the management of CVDs.

Pharmacists are well positioned to provide health promotion counseling, such as in smoking cessation, safe medication management and instructions to patients, and formulating and implementing medication plans with computerized records. Pharmacists can help to analyze reasons for non-adherence, and provide electronic reminders to support essential medication adherence, screening, and monitoring services, as well as vaccination or home care services.

Pharmacists can also serve the patient by conducting medication reviews and can act in liaison with the community care team (primary care physician, nurse) as well as hospital care. Pharmacists can therefore play an increased role in health care of chronic patients by assisting physicians and other health care professionals in patient care and by participating in prevention and management programs.

A recently published systematic review and meta-analysis showed substantial benefit of pharmacist interventions in the management of major CVD risk factors among outpatients in North America, Asia, Australia, and Europe. Patients allocated to pharmacist interventions achieved greater reductions in systolic and diastolic blood pressure, total cholesterol, and low-density lipoprotein cholesterol, and in the risk of smoking compared to the usual group.

The types of intervention that can be conducted by pharmacists, alone or in collaboration with physicians or nurses, include:

- patient education and counseling on medications, lifestyle or medication adherence
- recommendations to physicians regarding medication changes or problems of medication adherence
- medication management – reviewing patient medications directly from patient interviews or from medical records, assessing medication adherence or adjustment of medications
- scrutinizing and improving the quality of medication prescription among elderly people in the primary care setting to prevent common inappropriate medication prescriptions and multiple specialist consultations with risks for adverse events
- measurement of biomarkers and risk factors such as blood pressure and total blood cholesterol for screening and monitoring
- health care professionals' education.

Pharmacists can play a valuable role in the long-term care of patients with other chronic diseases, e.g., diabetes, heart failure, asthma, chronic obstructive lung disease, or cancer. Greater integration of pharmacists as part of a health care team is a potentially valuable resource for improving management of CVDs and other chronic diseases. The role of pharmacists in the community is slowly changing and, with appropriate training, pharmacists can become more effective and relevant health promoters.

Sources: Courtesy of: Valérie Santschi, Institute of Social and Preventive Medicine (IUMSP), Lausanne University Hospital, Lausanne, Switzerland. Personal communication.

Santschi V, Chiolero A, Burnand B, Colosimo AL, Paradis G. Impact of pharmacist care in the management of cardiovascular disease risk factors: a systematic review and meta-analysis of randomized trials. Arch Intern Med 2011;171:1441–53.

Santschi V, Chiolero A, Paradis G, Colosimo AL, Burnand B. Pharmacist interventions to improve cardiovascular disease risk factors in diabetes: a systematic review and meta-analysis of randomized controlled trials. Diabetes Care 2012;35:2706–17.

Collins C, Limone BL, Scholle JM, Coleman CI. Effect of pharmacist intervention on glycemic control in diabetes. Diabetes Res Clin Pract 2011;92:145–52.

Carter BL, Ardery G, Dawson JD, James PA, Bergus GR, Doucette WR, et al. Physician and pharmacist collaboration to improve blood pressure control. Arch Intern Med 2009;169:1996–2002.

Opondo D, Eslami S, Visscher S, de Rooij SE, Verheij R, Korevaar JC, et al. Inappropriateness of medication prescriptions to elderly patients in the primary care setting: a systematic review. PLoS ONE 2012;(8):e43617. doi: 10.1371/journal.pone.0043617.

These factors and increasing costs of medical care fostered the wider role of nurses to provide medical services.

In 1965, the first program to train pediatric NPs was developed at the University of Colorado. Nurses were taught specific medical functions, not as a doctor, but as a logical extension to the nurse's traditional function of assisting patients to regain health and independence. In 1971, the federal government adopted the Nurse Training Act, which defined the role of NPs.

A nurse practitioner is a registered nurse who has advanced training and education in the medical field. NPs perform detailed physical examinations, order laboratory tests for diagnostic purposes, assess the results of such tests, write prescriptions for medications, undertake investigations, make referrals, and perform other delegated medical functions within authorized limitations and medical supervision. An NP can provide care in hospitals or the community as part of a health team. The 2008 survey of practicing nurses in the USA reports that there were 158,348 NPs in practice in the USA in that year, of whom 88 percent held master's degrees.

According to studies commissioned by the US Agency for Healthcare Research and Quality (AHRQ) in 2010, the estimated number of NPs in the USA totaled 106,073 or 3.8 percent of the estimated 3 million RNs currently licensed, with 52 percent practicing in primary care and the others in subspecialty care. The NPs were 44 percent of all advanced practice nurses (APNs), who represent some 6.3 percent of the total RN population. Applicants and enrollees more than doubled in the USA between 1993 and 1995, and the number of graduates nearly doubled, increasing by a factor of 1.8. More recent surveys show that this expansion has slowed, however, and NP graduation rates are currently in decline. The number of NPs may reach 10 percent of all RNs in the next decade as primary care roles for NPs increase. NPs serve in family practice; women, adult, and school health; pediatrics; nurse midwives, and gerontology. Specialty care tracks include neonatal and acute care, occupational therapy, psychiatric care, and others in institutional care settings.

In each state, the practice of nursing is established and regulated by nurse practice acts and common law. These acts establish educational and examination requirements. The acts also provide for licensing or regulation of individuals who have met these requirements, and define the functions of the professional nurse in general and specific terms. The criteria establish parameters within which an NP may practice. The Commission on Collegiate Nursing Education, an accreditation organization, has made great strides in standardizing NP training programs to ensure the quality of graduates.

Nurses constitute the largest single group of professionals among health personnel. The expansion of nursing roles is essential to improved quality of health care, especially in medically underserved areas. Equally important, NPs in underserved professional areas such as geriatrics and primary care have innovated programs and provided a research and theoretical basis for new directions in health care. With the growing use of clinical guidelines, the roles of the NP may be expected to increase in the twenty-first century. To meet this goal, NP programs must work to expand and attract nurses to the profession. The 2008 US National Sample Survey of Registered Nurses reported that there were 59,242 practicing specialist nurse midwives.

"Nurse navigators" are RNs, usually with a bachelor's or master's degree, who provide liaison and coordination among patients, particularly cancer patients, and their many doctors and other medical and related professionals, home care and other services. Patient safety nurses are increasingly appointed to develop safety measures, monitoring, and education for all health care staff, with duties expanded beyond in-service training, including implementation of safety, prevention of drug errors, and infection control. With increasing educational levels of bachelor's, master's and PhD levels in nursing, the expansion of roles will continue to be a positive force for meeting population and patient health needs as medicine and public health continue to evolve. Nurses specializing in public health usually take master's degrees in this field.

Physician Assistants

In 1923, 89 percent of US physicians were general practitioners. By the mid-1960s, the figure had declined to about 25 percent. Concern over the shortage of primary care physicians in the country led Eugene Stead, at Duke University in Durham, North Carolina, to develop the first physician assistant (PA) 2-year training program in 1965 to provide care in rural areas without on-site general practicing physicians. PA training programs were developed for several reasons: to help alleviate a perceived shortage of primary care physicians, to compensate for the geographic and specialty maldistribution of physicians, to help control escalating health care costs, and to accommodate medics from the military returning to civilian life and seeking a continued role in health care.

PA programs are not part of nursing, and this has engendered conflict with NP programs and nursing authorities. There are currently 52 accredited PA training programs in the USA. The majority of PAs in the USA work outside public health and primary care; however, the profession is very adaptable. In addition to medical science, PAs are taught preventive health care, patient education, utilization of community health and social service agencies, and health maintenance. The main focus in basic PA training is patient care in the primary care practice setting. The first 6–12 months of training is devoted to preclinical studies and clinical laboratory procedures, followed by 9–15 months of clinical training. The curricula are reviewed regularly and modifications made in keeping with changes in the health care setting.

On completion of a PA training program, an entry-level competence examination is given by the National Commission of Certification of Physicians' Assistants. When passed, it allows PAs to append the title PA-C (physician assistant – certified) to their names. PAs must register every 2 years, documenting 100 hours of approved continuing medical education. To ensure clinical competency, PAs must take a recertifying examination every 6 years.

PAs perform tasks such as history taking, physical examination, simple diagnostic procedures, data gathering, synthesis of data for a physician, formulation of diagnoses, initiation of basic treatment, management of common acute and emergent conditions, management of stable chronic conditions, patient and family counseling, supportive functions, and prescribing privileges throughout the country. Task delegation of PAs is determined by the State Board of Medical Examiners within each state. The PA is not a substitute for the physician or for an independent provider like an NP. The PA is not licensed for independent practice, and the physician must assume all responsibilities and bear all the professional and legal consequences of the PA's actions.

Of the estimated 70,383 PAs practicing in the USA in 2010, 43.4 percent were working in primary care (AHRQ, 2011). The 2008 physician assistant census reported that some two-thirds had taken various baccalaureate degrees prior to PA training. Some one-third of respondents were employed in hospitals, another one-third in solo or group practice offices, and 9 percent in community health centers.

PAs work in over 60 specialty fields, with over 50 percent in general/family practice. Employment opportunities are increasing for PAs in various specialties and practice settings. A PA may be able to handle one-half to three-quarters of the clinical services provided by the supervising physician, indicating that they are productive and cost-effective in their employment settings. PA salaries are one-quarter to one-third those of physicians, so the costs and benefits of PAs are an attractive option for organized health systems, but this discipline should not be seen as an independently practicing profession. The literature on this topic reports that the quality of care offered by PAs is comparable to that given by a physician in a physician-supervised practice. The Joint Commission on Healthcare Accreditation (JCHA) allows for licensed advance practice nurses and physician assistants to perform medical history and examinations on admissions within the limitations of state laws.

The PA, although initially controversial, is now a widely accepted role. The nursing profession has regarded this as a method of increasing physician incomes, an infringement of nursing roles, and a reaction on the part of physicians to the growing professionalization of nursing. The medical profession defends the PA as a way of extending the possibility of medical care to larger population groups and improving the economics of medical practice. These factors will become increasingly important as the Affordable Care Act comes into full effect with promotion of prevention of NCDs and new methods of organizing and paying for health care in the USA (see Chapters 10, 11, and 13), with potential employment of NPs on a wider scale. The use of medics and emergency medical technicians with college-level training as first responders is becoming more widespread in the USA and other countries. Many begin with military training and continue in civilian life in emergency care services. This is generally seen as improving survival possibility in serious trauma and medical conditions requiring rapid intervention during transportation to hospital under the guidance of emergency room physicians or other specialist physicians.

The Israeli Ministry of Health in 2013 announced that it would accept and credential PAs and NPs licensed in the USA.

Feldshers

The feldsher is a unique Russian mid-level health worker. The role originated as military company-level surgeons introduced by Peter the Great in the seventeenth century. Retired army feldshers returned to rural areas not served by physicians, becoming the sole providers of rural medical care. The feldsher was adapted to the Soviet health system to provide care in rural areas still underserved by doctors, despite the increase in medical personnel.

The feldsher is trained in a course lasting for 2–3 years following intermediate school graduation. Feldshers complement physicians in urban and rural practice, especially in small medical posts, in mobile emergency medical services, and in industrial health stations. Since the health reforms of the 1990s, the feldsher has become a declining profession, which may result in serious difficulties in maintaining rural health care.

Community Health Workers

The concept of the CHW is not new but has found new expression in health programs in many parts of the world as part of the primary health care initiatives springing from Alma-Ata. It is an adaptation of traditional village practice of midwives and healers to modern, organized public health services. CHWs were first developed to provide care in rural areas in developing countries without access to health care. More recently, there has been an interest in the CHW model for urban community health needs where access to health is limited for geographic or socioeconomic reasons. Another category of services providers comprises home care workers for patients not requiring inpatient services of hospital or nursing home, but needing assistance in the community and home setting. Paramedics and other emergency care technicians are also categories of health workers needed in a comprehensive care system. Training programs for such health care workers need to be supervised with state standards to assure capacity to provide quality of care.

CHWs include categorical or targeted health workers, and preventive CHWs. The exact functions of these personnel, the duration of training, and the framework within which they work have varied much more than their titles. The generalist village health worker programs in some areas lack close supervision, and may seek fee-for-service practice. CHWs, advisors, or *promatores* are commonly used in Latin American countries, often as volunteers. CHWs are recommended for rural locations in developing countries without supervisory or organized contact with professional health services, providing a wide range of diagnostic and treatment services (Box 14.8).

CHWs may provide services for categorical target diseases. These services include malaria and TB control, directly observed therapy, short-course (DOTS), providing medication under supervision to ensure compliance, support services, and counseling for families with multiple problems in inner-city poverty areas, follow-up of sexually transmitted infections (STIs), and promotion of immunization. Prototypes of the task-oriented CHW include malaria control CHWs in Colombia, and TB DOTS and AIDS case workers, TB case workers, and public health nurse assistants in New York State. In Africa, CHWs are crucial to programs for the eradication of guinea worm disease and river blindness.

A CHW program with a focus on preventive services was developed in 1985 in Hebron, in the West Bank, under Israeli jurisdiction, and later continued and expanded the

BOX 14.8 Community Health Worker (CHW) Program Models

Independent CHWs

- Feldsher in Russia for rural health care.
- Barefoot doctors in China for rural heath care.
- "Where There Is No Doctor" CHWs provide all health care in remote villages in Latin America.

Categorical CHWs

- Program-specific CHWs, e.g., providing malaria control in Colombia, Guinea worm disease and onchocerciasis in Africa; injectable contraceptives in Ethiopia, Kenya and Uganda; AIDS or tuberculosis care in New York City; immunization and discovery of unmanaged diabetes and hypertension in low-income housing in Los Angeles.
- Public health nurse extender CHWs, e.g., Albany County Health Department, New York State.

Preventive-Oriented CHWs

- Preventive village health workers provide on-site services as part of public health system, visiting medical-nursing services with continuing education and close supervision, e.g., Hebron, West Bank, Palestinian Authority.
- "Urban villages", i.e., urban poverty area CHWs in USA, with CHWs as part of county health department or community-based organization services such as Housing Authority of City of Los Angeles.

The American Public Health Association defines a CHW as: "a frontline public health worker who is a trusted member of and/or has an unusually close understanding of the community served. This trusting relationship enables the CHW to serve as a liaison/link/intermediary between health/social services and the community to facilitate access to services and improve the quality and cultural competence of service delivery. A CHW also builds individual and community capacity by increasing health knowledge and self-sufficiency through a range of activities such as outreach, community education, informal counseling, social support and advocacy."

An extensive experience is building up in the use of CHWs in many settings with specific tasks and training, or as primary care coordinators in locations or conditions with poor access or follow-up in meeting community health needs. There is wide application of the CHW model in Ethiopia. A CHW program for cardiovascular health implemented in Colorado in the USA showed statistically significant improvements in diet, weight, blood pressure, lipids, and risk scores, with the greatest effects among those with uncontrolled risk factors and successful telephone interaction by the CHW in lowering risk scores on retests.

Sources: *American Public Health Association. Community health workers. Available at: http://www.apha.org/membergroups/sections/aphasections/chw/ [Accessed 20 September 2012].*
Wittmer A, Seifer SD, Finocchio L, Leslie J, O'Neill EH. Community health workers: integral members of the health care work force. Am J Public Health 1995;85:1055–8.
Tulchinsky TH, Al Zir AM, Abu Munshar J, Subeih T, Schoenbaum M, Roth M, et al. A successful, preventive-oriented village health worker program in Hebron, the West Bank, 1985–1996. J Public Health Manage Pract 1997;3:57–67.
United Nations Children's Fund. The state of the world's children. 2009. Maternal and newborn health. Available at: http://www.unicef.org/sowc09/report/report.php [Accessed 22 December 2012].
Kong S, Brown M. USAID Knowledge Services Center. Community health workers: Ethiopia, 2008. Available at: http://pdf.usaid.gov/pdf_docs/PNADM019.pdf [Accessed 29 November 2012].
Krantz MJ, Coronel SM, Whitley EM, Dale R, Yost J, Estacio RO. Effectiveness of a community health worker cardiovascular risk reduction program in public health and health care settings. Am J Public Health 2013;103:e19–27.

under Palestinian Authority. CHWs act as preventive care workers as people of contact representing the government health service, providing on-site prenatal and child care, immunization, nutritional counseling, pregnancy care with a medical nurse support team, first aid, and emergency or non-urgent referrals to the district hospital or to nearby medical clinics. The CHW is trained and supervised to provide primary care and outreach services in a community as part of an organized health system. Visits by supervisory professional staff are vital to the function of this system of care. The emphasis may be on preventive and community services in small villages without on-site services, or as part of an outreach and support service in a large urban setting.

CHWs should be recruited from the community to be served and trained in settings with both didactic and field experience. Training of village CHWs may be very different from schools for training conventional health personnel. Training usually takes place in rural health centers or hospitals, to which classrooms and student living quarters have been added. The training is nearly always sponsored by the ministry of health, sometimes in collaboration with ministries of education, academic colleges, bilateral aid agencies, or international organizations. Candidates for such training are typically young people from rural families, who have been selected by their communities. During training, the student should be salaried, and on completion of training return to work in the community of origin.

In the USA, public health nurses have been traditional providers of home visiting programs or outreach programs to provide health care and education for families in need. The emergence of community-oriented primary health care broadens this strategy, selecting and training community residents as CHWs. The CHW concept was used to carry out many Great Society programs in the USA in the 1960s and 1970s. Volunteer and paid workers worked as lay home visitors or health guides in programs in selected areas or target populations, such as pregnant women or parenting families. Other CHW programs in the USA included Navajo communities, urban health centers, rural Texas, and Alaska.

CHWs can be trained to provide outreach and case management in the complex environment of New York State, going into the community to assist high-risk families in underserved areas of large cities and in AIDS, STI, and TB patient care, especially with DOTS for TB. CHW experiences working with the homeless and mental health patients have shown positive results. CHW projects have been developed in many urban settings such as in low-income housing unite or "poverty urban villages" in Los Angeles, with backup services of the County Health department, and clinics or tertiary care centers with follow-up management for health issues found to be untreated, such as diabetes and hypertension.

Evaluation and cost–benefit justification of CHW programs are difficult in terms of establishing population denominators and control groups, and in determining changes in outcome measures in mortality, morbidity, and physiological indicators, such as growth patterns. This limitation is shared with many health programs, not only in primary care but perhaps even more so in highly technological medicine. The village health worker concept has received criticism as well as advocacy in recent years. The CHW as a health promoter and provider of both preventive and treatment services may be impossible to sustain and is undesirable conceptually. But, training to tasks and programs in a selective approach may be more feasible and manageable, with the CHW showing the promising potential to provide a new parameter in health care. CHWs resident in the villages they serve can provide many services to the rural population with little access to medical services (Tulchinsky et al., 1997; Haines et al., 2007). Training to tasks and continuing supervision with ongoing support and training are essential to promote standards and quality of CHW village programs.

In Ethiopia, Uganda, and other countries, the CHW is enabled to provide injectable birth control medications. The Ethiopian Health Extension Program, a community-based nutrition program, has trained some 34,000 health extension workers providing a package of health nutrition and sanitation services to rural populations in village health posts supported by volunteer CHWs. This program, which began in 2004, trained people in assessing, analyzing, and acting to teach and supervise volunteers CHWs to work in the village to promote breastfeeding and other good nutrition practices, vaccinations, vitamin supplements, deworming, and other preventive health practices such as reducing the practice of female genital mutilation and exclusion of AIDS patients from the community. This program has reached over a million village children, helping to improve mothers' knowledge of good health and nutrition practices, and referring severely malnourished children to treatment centers to reduce child stunting and failure to thrive (UNICEF, 2010). In Rwanda, CHWs are active in promoting birth control and improved maternity care, and improving health insurance coverage in rural areas.

US experience with CHWs is also extensive. A study conducted in a statewide chronic disease prevention program in 34 rural counties in Colorado reported on successful CHW intervention to reduce cardiovascular risk factors and diabetes management (Krantz et al., 2013). With a growing elderly population along with rising levels of diabetes and obesity, chronic care will require a greater emphasis on task sharing. Medical and nurse practitioner providers will be a part of health teams with many different skills and levels of training whose common purpose will be to help patients to cope with their conditions and retain maximum independent living.

ALTERNATIVE MEDICINE

The use of alternative medical treatment has become a widespread phenomenon in industrial countries and remains a mainstay of health care in preindustrial societies. Alternative medical care is based on the belief that illness includes physical, mental, social, and spiritual factors. Alternative medicine views health as a positive state, rather than as the absence of disease, and believes in the natural healing capacity of the human body. Its interventions are generally non-invasive and less technological than conventional medical care.

The growth in popularity of alternative medicine in part relates to a widespread disillusionment with the depersonalization and technological orientation of medical care. Other factors include medicine's failure to treat the patient as a whole person, bias towards single causes, and "magic bullet" treatments for disease. Disappointment with medical outcomes is common, with failure of cures or complications of treatment itself as iatrogenic disease. In the USA, the number of visits to providers of unconventional therapies is greater than the number of visits to primary care doctors, with high levels of out-of-pocket expenditures. Users of alternative medicine are largely in the 25–49-year age group and are among better educated people in upper income categories. Alternative medicine categories include acupuncture, chiropractic, massage, commercial weight-loss programs, lifestyle diets, herbal medicine, megavitamin therapy, self-help groups, energy healing, biofeedback, hypnosis, homeopathy, and folk remedies.

Changing attitudes within conventional medical care are seen in a growing tendency to accept some previously excluded professions, including optometry, chiropractic, and acupuncture, in insured benefits or even multidisciplinary health care systems as complementary or supplementary to conventional medical care. Conventional medicine is itself in a process of change with the addition of many paramedical professions and awareness of the limitations of the biomedical model as the sole basis for health care.

Consumer demand plays an important role in this change, and so does willingness of medical practitioners to refer chronic health problems such as back and neck pain, stress and related problems, phobias and addictions, allergy and skin disorders, and hormonal and menstrual disorders. Conversely, alternative practitioners seem to be increasingly aware of limitations in treatment of conditions such as cancer, hypertension, and chronic and hereditary disorders, and the need for referral for treatment by conventional medical methods. The nursing profession in the USA has promoted the use of touch therapy as an independent healing modality over the past several decades.

Absorption of "holistic methods" by medical practitioners has become widespread. Many major medical centers now include departments of alternative or holistic medicine, with acupuncture, hypnosis, and other non-traditional approaches in the roster of services. This increases recognition, legitimacy, status, and incomes of practitioners, but is only achieved over long periods of conflict and opposition by orthodox medical practice. In 1992, the US Congress mandated establishment of an office of alternative medicine within the NIH, which in 1998 was expanded to become the National Center for Complementary and Alternative Medicine. This move indicates the growing acceptance of alternative medicine as a significant element of community health, with recognition by insuring agencies. Issues of cost containment and competition with traditional medicine are thus raised, but gradual acceptance is happening. European medicine and health agencies, with long traditions of healing baths and rest cures, have been more open to complementary and alternative medicine than their North American counterparts.

CHANGING THE BALANCE

Health systems are under pressure to change for a number of reasons; one is cost, and other major factors are changing demography and morbidity patterns. Populations are aging so that a growing sector of the population is reaching ages where chronic diseases are more prevalent, but healthier aging is also more common. The health needs for care of the elderly, however, are growing and this will require a changing mix of services with greater emphasis on outreach and supportive care by paramedical health workers with physicians in the clinical role of guidance. Furthermore, population health needs are heavily influenced by social and economic factors which require political and societal attention to alleviate poverty, unemployment, and deterioration of urban and rural settings, to reduce the inequities found in most countries, even those with well-organized health care systems.

Health systems will need to adapt to these forces by making changes to the organization and financing of health care and placing much greater stress on health promotion and preventive care. Home care and patient advisory services will help many patients with long-term health problems to be sustained in their own homes with medical support systems, such as "patient homes" for comprehensive oriented

care. Advances in health technology will bring growing capacity for prevention, early diagnosis, and management of serious infectious diseases, cancer, and cardiovascular diseases. Trauma, violence, and mental illness will also be affected by societal factors, and addressed by improved diagnostic and support systems.

As cost containment becomes more important and many countries cut back on health expenditures in order to control rates of increase, the stress of readjustment falls on health workers. If such measures are implemented on an emergency basis rather than over time as part of transition and realignment, the process will generate hostility, defensiveness, and political opposition. If, however, long-term planning takes into account the issue of human resources in a changing balance of services, then the burden of the individual or group of health workers who suffer from the downsizing can be minimized.

During the 1960s, Canada was embarking on its national health insurance program, and leading thinkers of the day called for rapid expansion of medical and nursing schools to meet future needs of the population. They assumed that national health insurance would bring a significant portion of the population who lacked access to care to the health services, overloading the medical and hospital services. Removing financial barriers does not, however, remove differences in use of medical services and incorporation of changes in lifestyle in daily life to reduce health risk factors. Education and outreach need to be incorporated in health plans to reduce social inequities in health status that are prevalent in all countries, including those with well-established health insurance or service systems.

In the mid-1990s, Canadian provinces cut back on the hospital bed supply, creating unemployment for nurses and maintenance staff. Provinces are attempting to restrict the numbers of practicing doctors and to modify the payment systems away from fee-for-service. Inflation in health care utilization and costs has resulted in a crisis management approach, rather than structural reform to promote a more integrated population-oriented health care organization. Such an approach can lead to a great deal of public and professional dissatisfaction within a health care system. Public education is vital to raise consciousness and participation rates in preventive care such as influenza vaccination, screening for colon and cervical cancer, and many other aspects of health promotion and disease prevention.

EDUCATION FOR PUBLIC HEALTH AND HEALTH MANAGEMENT

In 1915, the topic of public health education was addressed in the Welch–Rose Report, sponsored by the Rockefeller Foundation. It set up different models: one based on national schools of public health in the governmental sector, and another promoting schools of public health within universities.

Both models exist to this day. The Welch university-based model was promoted and financed by the Rockefeller Foundation to create the Johns Hopkins School of Public Health and Hygiene in 1916, and the Harvard School of Public Health in 1922. Most schools of public health in the USA followed the Welch model as independent faculties in universities.

Successful implementation of the New Public Health requires that many health disciplines work together. The training milieu should have a capacity for interdisciplinary training in a comprehensive program, including fundamental and applied research, as well as a relationship with service programs and community health assessments. To facilitate service standards and teamwork, all public health practitioners need a background in the medical sciences, epidemiology, economics, social sciences, environmental and occupational health, health systems analysis, and management theory. It is important that they be familiar with the terms and concepts of fields other than their own specialized discipline. This arrangement is more likely to be found or created in a university atmosphere and is difficult to foster in separate, categorical institutes.

University resources are essential for a school of public health to provide teachers and courses from other faculties such as schools of business administration and the social, physical, and biological sciences. Schools that are unaffiliated with a degree-granting university lack the broad academic atmosphere and requirements, as well as the connection with parent disciplines, such as economics, sociology, microbiology, and business management faculties, and public recognition as a quality reputable and accredited school.

Preparation of personnel for public health and health management should be at the graduate school level followed by continuing education. The US Institute of Medicine's 1988, 2002, and 2003 reports on public health defined the need for schools of public health to teach not only professional and technical skills, but also an understanding of how a particular discipline relates to public health as a whole, and the value system that is part of public health's coherence.

Training of public health physicians began in the UK in 1871 in Dublin's Trinity College, which granted the Diploma in Public Health (DPH). The program was designed to provide for the training of medical officers of health to lead the work of the boards of health, which were established under public health legislative statute and were required to have qualified medical practitioners as medical officers of health. Other universities later offered DPH or equivalent training, supervised by the College of Physicians and Surgeons, as public health became a recognized medical specialty. In 1924, the London School of Hygiene and Tropical Medicine brought together several institutes and produced a major center of training and research in public health. Since the 1991 Acheson Report on schools of public

health in the UK, there has been a rapid growth in schools of public health under various names in a number of universities. This growth has coincided with the increased interest by the NHS in working towards health targets as opposed to simply managing health services.

The tradition of schools of public health is especially strong in the USA, where the Johns Hopkins and Harvard Schools of Public Health were founded in the early twentieth century. In 1915, the Rockefeller Foundation sponsored a national program to promote public health education at the University of Michigan, Yale University, and the University of Pennsylvania, and thus established graduate training for public health professionals to meet the needs of the crowded urban industrial cities of the USA. These schools saw their mission as the training of, primarily, public health practitioners, and secondly academics, educators, and researchers. They attempted to develop and assimilate new knowledge into public health practice. The development of public health as a multidisciplinary field made it vital to be independent from but affiliated with a medical faculty. Schools of public health in the USA produced many generations of well-trained epidemiologists, social scientists, health educators, practitioners, and leaders who were crucial for development of the field.

In the 1960s, schools of public health were criticized because of their separateness from and lack of influence on clinical medical training. In some jurisdictions, this led to closure and replacement by departments of community medicine within faculties of medicine. This occurred in the UK and in British Commonwealth countries. Canada's two schools of public health were closed, replaced by departments of social and preventive medicine or community health, developed within medical schools. Departments of community health within a medical faculty serve as only one department among many clinical or basic science departments. Such departments may lack prestige in the hierarchy of medical schools, in an environment promoting a narrow, medically oriented approach to public health, with an insufficiently multidisciplinary program and faculty. This model provides training at the undergraduate, MPH, and doctoral levels. The full academic potential of a graduate school of public health is most suited to be in an independent, multidisciplinary, university-based academic center for public health research and training.

Periodically, threats of absorption of schools of public health into other sectors of the university arise. In 1994, a plan to close the University of California, Los Angeles (UCLA), School of Public Health by transfer to the School of Public Policy was halted by university and nationwide protests. Health administration is sometimes considered as better integrated within schools of business, as happened at the University of Minnesota School of Public Health.

In Canada, two long-standing traditional schools of hygiene were closed during the 1960s on the idea that integration into medical faculties would provide greater acceptance of public health in the medical community. Epidemiology and health management were taught in different departments in medical faculties. This had the effect of limiting the growth of the public health workforce and research capacity, despite the development of departments of epidemiology and health management in medical faculties. In part as a result of serious deficiencies in the management of the SARS epidemic in 2003 and subsequent reviews of education in public health, since 2008 seven schools of public health and fifteen MPH programs have been established across Canada (Massé and Moloughney, 2011).

Since the 1980s, there has been a renaissance of schools of public health in the USA with an expanding market for graduates. The Association of Schools of Public Health (ASPH) represents the 46 accredited schools in the USA and many with associate member programs or schools not yet accredited through a formal review process of the Council on Education for Public Health (CEPH). Many other programs also provide postgraduate education in public health. The core curriculum themes include:

- *biostatistics* – collection, storage, retrieval, analysis and interpretation of health data; design and analysis of health-related surveys and experiments; and concepts and practice of statistical data analysis
- *epidemiology* – distributions and determinants of disease, disabilities, and death in human populations; the characteristics and dynamics of human populations; and the natural history of disease and the biological basis of health
- *environmental health sciences* – environmental factors including biological, physical, and chemical factors that affect the health of a community
- *health services administration* – planning, organization, administration, management, evaluation, and policy analysis of health and public health programs
- *social and behavioral sciences* – concepts and methods of social and behavioral sciences relevant to the identification and solution of public health problems (CEPH, 2005).

Policy analysis, advocacy, and health promotion should be woven through the studies in all fields, along with law, ethics, economics, and other issues in public health policy. They should provide the new-entry public health worker with competencies in basic tools of social analysis, as outlined in Roemer's classic paper and Gebbie's renewal of this approach, as seen in Box 14.9. These include a broad education in the social sciences, history of public health, the statistical and epidemiological methods, as well as qualitative research, infectious and chronic diseases, nutrition, environment, risk groups, global ecology of disease, promotion of health and prevention of disease, accreditation and quality promotion in health care, information systems, monitoring, and research methods for management, and global health.

A combination of an MPH with medical doctor training has become widespread among schools of public health, as has public health training for clinicians in many specialties including family practice. MPH training gives medical clinicians a good background in epidemiology, biostatistics, and economics, which are important subjects for authors of clinical studies intended for publication. Many schools offer multiple graduated degree programs in health administration, epidemiology, and other fields in the broad area of public health.

The success model for schools of public health since the 1920s provides postgraduate training in a multifaculty academic setting; a multidisciplinary approach; teaches problem-oriented skills training to identify targets and problem-solving management approaches; and links education, research, and service in public health. The organization and stakeholders for a teaching program at graduate levels are illustrated in Figure 14.2, indicating the complexity of internal and external organizational relationships to achieve the support and environment to train and conduct important and responsible research, and to influence social and health policy. The complex of stakeholders and connections of a school of public health is part of the multiple roles of teaching, research, advocacy, and service.

Teaching programs include undergraduate (bachelor's), master's, and doctoral levels, with the MPH and PhD levels involved in the faculty research programs. School faculty members are frequently leaders in advocacy or consultants for topics in public health where their research and expertise are most relevant. They are able to speak publicly when public health officials of governmental authorities and nongovernmental organizations cannot. This is a powerful tool in public dialogue on frequently controversial topics, but often comes to the aid of public servants in special meetings of parliamentary committees and other policy-making forums.

Students in schools of public health come from many different backgrounds, including medicine, dentistry, nursing, engineering, economics, social sciences, statistics, mental health, and veterinary sciences. In 2008, ASPH member schools of public health in the USA, Puerto Rico, and Mexico consisted of a combined 4000 faculty, with 19,000 students, and 7000 graduates per year. The most popular specialties were international health and epidemiology. The career outlook for graduates of schools of public health now includes other traditional public sector positions, but also higher paying positions in the private sector, including consulting firms and managed care organizations. In the 1996–1997 academic year, 14,007 students were enrolled in schools of public health in the USA.

Core Elements for Master of Public Health Programs

The Council of Education in Public Health (CEPH) in the USA sets standards and accredits graduate schools of public health, which provide approximately 85 percent of the public health graduates in the USA.

The Association of Schools of Public Health (ASPH) in the USA identified core competencies for students, upon graduation, in Bachelor and Master of Public Health programs in response to the challenges of twenty-first century public health practice, with widening use of competency-based training in the field of public health, emphasis on accountability in higher education, growing incorporation of competencies into accreditation criteria, and voluntary credentialing examinations for public health graduates. The planning of curricula content tries to address the needs of the public health profession and discipline, which focuses on the role of the population and society in monitoring and achieving good health and quality of life.

The competencies are intended to serve as a *resource* and *guide* for improving the quality and accountability of public health education and training. They are not meant to prescribe the methods or processes for achievement, recognizing that implementation of the competencies may vary with the mission and goals of each school.

Five core discipline areas (biostatistics, environmental health sciences, epidemiology, health policy management, and social and behavioral sciences) are considered the foundation of public health education and competencies for graduates (Box 14.10). In addition, interdisciplinary, cross-cutting competency domains addressed in assessing educational content and competencies in public health education include communication and informatics, diversity and culture, leadership, professionalism, program planning, public health biology, and systems thinking. These seven areas, however, have become increasingly important to effective public health practice and, thus, are included along with the five discipline-specific competency domains in the ASPH model. Other disciplines that contribute to public health include ethics, economics, law, education, engineering, political science, psychology, business, and public administration.

In the USA, graduate schools of public health accredited by CEPH have grown from 28 in 1999 to 40 in 2008, with an additional 70 recognized MPH programs. They are all based on a multidisciplinary approach to training for public health, including technical and administrative leadership in epidemiology of communicable and non-infectious diseases, biostatistics, management of personal health services, environmental health, maternal and child health, health economics, health education, and other related fields. The interdisciplinary aspect of public health is emphasized by the wide range of backgrounds and experiences of the students and their many areas of specialization. Analysis

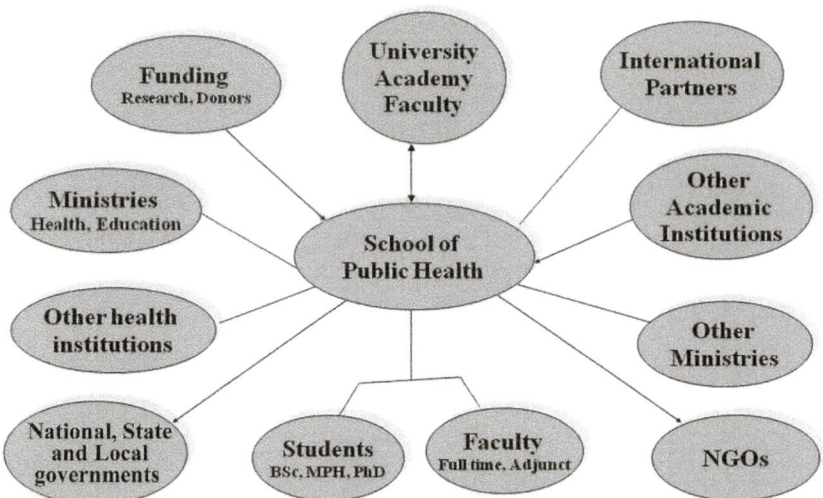

FIGURE 14.2   Structure and stakeholders in schools of public health. Note: NGO=non-governmental organization.

BOX 14.10 Council on Education for Public Health: Core Disciplines and Cross-Disciplines in Public Health Education

Core Disciplines

- Biostatistics
- Environmental health sciences
- Epidemiology
- Health policy and service management
- Behavioral science and health education

Cross-Disciplines

- Communication and informatics
- Diversity and culture
- Professionalism
- Program planning
- Systems thinking
- Public health biology
- Ethics (not yet included in APHA list)
- Law (not yet included in APHA list)

Note: APHA=Association of Schools of Public Health.
Source: *Association of Schools of Public Health. MPH core competency model [updated 9 September 2010]. Available at: http://www.asph.org/document.cfm?page=851 [Accessed 2 December 2012].*

---

of problems with the skills of epidemiology, sociology, and other related disciplines permits the graduate to enter practice with a problem-solving approach. Another 26 university graduate programs in community health education, community health, and preventive medicine are accredited to extend public health training to most parts of the country.

The importance of newer training models for the public health workforce is one of the most urgent keys to successful, well-functioning, and well-managed health systems. This is reinforced and evidenced by the events and outcomes of newly developing schools of public health in countries of Eastern Europe and Central Asia, as well as the new schools now in development in India and South-East Asia.

Accreditation involves external review of facilities, faculty, curriculum, student selection criteria, internships or field experience, and academic standards. An accredited school is better able to generate research and scholarship funds and is more attractive to students for future career advancement in an expanding job market. These schools, plus many other non-accredited schools or university departments, provide operating public health and health care agencies with well-trained personnel with a wide range of undergraduate and professional experience. This enriches the field not only with practitioners, but also with researchers, administrators, and policy analysts of high quality.

Schools of public health should have close working relations with state and local health agencies, recruiting part-time faculty from service agencies and conducting research in real public health problems that confront health agencies. Schools of public health can provide important services to departments of health in research, consultation, and assessment on public health issues.

The WHO addressed the "global crisis in health systems and health workforce needs" by choosing Human Resources for Health as the theme of the World Health Report 2006. The major stress was placed on the shortage of medical and nursing and other health providers, and the issues of training, conditions of work, and migration. There is also recognition that preparing a competent public health workforce is a key element for effective and sustainable health systems. There is no clear agreement on what the public health workforce is. Definitions vary for identifying the specific members of the public health workforce, and their corresponding roles within the health systems, as classifications and roles differ across countries.

The World Health Report 2006 states that the education of the health workforce requires attention to curricular content, pedagogical learning methods, training of teaching staff, research and service, and moreover, that "more schools of public health are needed".

Public health workforce development is a crucial element in increasing the capacity of national health systems, allowing them to address present and future population health challenges. The development of advanced-level programs of postdiploma public health education in the Central and Eastern European region is an important innovation to help countries cope with public health crises of low performance of their health systems.

Ranking Universities

University and college education is a large, highly competitive industry. Students and their families are faced with difficult choices and admission requirements as well as the high costs of higher education. At the same time, education is a national resource vital to sustaining competitiveness in a global economy increasingly based on highly educated human resources, highly technical engineering, and sociological and biomedical research. This applies to public health no less than to other health and technical professions. The idea of ranking of educational institutions is also applied to colleges and even to schools.

Graduate schools of public health are ranked annually by surveys of deans, top administrators, and senior faculty conducted as part of continuous surveys of graduate schools in the USA which were initiated in 1983 by the weekly news magazine *US News and World Report*. There are many ranking systems for higher education, with *The Times* [of London] *Good University Guide* and the Shanghai Jiao Tong's *Academic Ranking of World Universities* being the best known.

These surveys are based on annual assessments, including reputation among deans and senior faculty (40 percent), research activity (30 percent), student selectivity (20 percent), and faculty resources (10 percent). The rankings of US schools of public health are shown in Table 14.9. In 2011, the top ranked schools of public health included Johns Hopkins, the University of North Carolina (Chapel Hill), and Harvard.

While these surveys and rankings have no official status, they can be very important for the recognition of achievements and excellence, and are widely used as guides for student selection of graduate schools, possibly affecting research grants, fundraising, and faculty recruitment. The CEPH provides a recognized accreditation system in the USA with requirements for standardized public health curricula and core content, and the process of accreditation ensures quality graduate education across the country, and some abroad (e.g., Mexico, and Alberta, Canada).

TABLE 14.9 Ranking of US Schools of Public Health, 2011

Rank	School Name	Score
1	Johns Hopkins University, Baltimore, MD	4.8
2	University of North Carolina–Chapel Hill, Chapel Hill, NC	4.6
3	Harvard University, Boston, MA	4.5
4	University of Michigan–Ann Arbor, Ann Arbor, MI	4.3
5	Columbia University, New York, NY	4.2
6	Emory University, Atlanta, GA	4.1
6	University of Washington, Seattle, WA	4.1
8	University of California–Berkeley, Berkeley, CA	3.8
8	University of Minnesota–Twin Cities, Minneapolis, MN	3.8
10	University of California–Los Angeles, Los Angeles, CA	3.7
11	Boston University, Boston, MA	3.4
11	University of Pittsburgh, Pittsburgh, PA	3.4
13	Tulane University, New Orleans, LA	3.3
13	Yale University, New Haven, CT	3.3
15	University of Texas–Houston Health Sciences Center, Houston, TX	3.2
16	George Washington University, Washington, DC	3.0
16	University of Alabama–Birmingham, Birmingham, AL	3.0
16	University of Illinois–Chicago, Chicago, IL	3.0
16	University of Iowa, Iowa City, IA	3.0
20	Ohio State University, Columbus, OH	2.8
21	Drexel University, Philadelphia, PA	2.7
21	University of South Florida, Tampa, FL	2.7
23	University of Arizona (Zuckerman), Tucson, AZ	2.5
23	University of South Carolina, Columbia, SC	2.5
25	Texas A&M Health Science Center, College Station, TX	2.4

Source: US News and World Report. Education, Grad Schools, Public Health. Available at: http://grad-schools.usnews.rankingsandreviews.com/best-graduate-schools/top-health-schools/public-health-rankings [Accessed 4 December 2012].

Field Epidemiology

Field epidemiology is an essential component of effective public health practice, and developing such capacity is a critical step in a country's efforts to improve the health of its citizens. The US Centers for Disease Control and Prevention (CDC) was founded in 1946 and Alexander Langmuir established its Epidemic Intelligence Service (EIS) program in 1951. It has made important contributions to training in and field investigation of many public health issues. EIS trainees and graduates have played important leadership roles in many domestic issues such as AIDS, Lyme disease, legionnaires' disease, foodborne and waterborne diseases, occupational health, NCDs, and trauma. Many trainees have become state epidemiologists and international trainees have become leaders in their own countries in the continuing struggles against infectious diseases (CDC Timeline, August 2013).

The EIS has played important leadership roles in international infectious disease control including smallpox eradication, measles control, and cholera epidemics. Since 1980, the CDC has established a total of 37 field epidemiology training programs (FETPs) worldwide, including 2 years' full-time classroom and field supervised training. FETPs are based on partnerships among CDC, host country health agencies, the WHO, the US Agency for International Development (USAID), and others. Modeled on CDC's EIS, FETPs follow the EIS approach of combining service with training. FETPs also participate in training programs in the epidemiology and public health interventions network, which provides a venue for information sharing, program development, and quality improvement. FETPs have graduated approximately 2100 field epidemiologists, most staying in public service and many reaching leadership positions in their ministries of health (CDC, 2012).

Public Health Education in Europe

In Europe, there is a growing trend towards public health education at the master's, PhD, and increasingly the bachelor's levels. The Association of Schools of Public Health in the European Region (ASPHER) represents this movement and is promoting peer review, mission and values statement, competency standards, and an accreditation program. ASPHER now includes over 80 member schools or programs offering MPH programs. These are mostly housed in faculties of medicine as departments of social medicine, occupational health and others, with a health economics and planning orientation. ASPHER offers a constructive role in promoting core curricula development, competencies, and PhD program development. It also promotes studies in public health ethics and is currently working to develop bachelor's degree training in public health, as is already common in such countries as Germany, Sweden, and Albania.

In 2011, ASPHER initiated the establishment of an independent accreditation body, the Agency for Accreditation of Public Health Education in Europe, which has established an organization, criteria, and procedures for accreditation of MPH programs, initially with the London School of Hygiene and Tropical Medicine and the Kazakhstan School of Public Health. This new movement is beginning to work in alliance with the European Public Health Association (EUPHA) to promote public health policies, research, and educational standards. Similar organizations are working in other regions such as in South-East Asia, where new schools of public health are developing, for example in sub-Saharan Africa and India.

In former Soviet countries, public health training is a stream within medical training institutes or academies at the undergraduate level. The Sanepid doctors provide communicable disease control, and food and environmental hygiene, and are the main cadre of public health practitioners in the Semashko system (see Chapter 13). Postgraduate training is provided as a medical specialty through *Ordinatura*, *Aspirantura*, and *Candidat* levels in the medical academies. Research institutes in various fields of public health provide graduate-level training up to the doctoral (PhD and Doctor of Science) levels.

Most former Soviet countries provided specialization during basic medical training (i.e., internal medicine, pediatrics, or sanitary epidemiology). Specialized graduate training takes place in various departments of medical academies, with criteria including length of training and examinations established by the Ministry of Health. Newly established graduate schools or programs in public health are providing training at the MPH level in many countries of Eastern Europe, the former Soviet Union, and Central Asia, including Hungary, Poland, Romania, Macedonia, Bulgaria, Moldova, Albania, the Czech Republic, Ukraine, and in the Central Asian Republics of Kazakhstan, Uzbekistan, and Tajikistan, with several in the Russian Federation. developing in Russia.

The MPH degree is not yet recognized in all these countries, but is recognized in Western and Central Europe, in keeping with the Bologna Agreement on postgraduate education as agreed to by most of these countries. In many countries in transition from socialist systems, the model of public health continues to be one largely focused on infectious disease and is hospital oriented, although there is advent of compulsory health insurance and elements of privatization of services in some countries. Non-infectious disease issues are seen as clinical problems and left to medical practitioners to resolve; health promotion remains a vague concept. As a result, the populations of countries in transition suffer the tragic consequences of high rates of preventable morbidity and mortality from chronic diseases.

In continental Europe, training in public health was largely traditionally through a job-oriented, vocational approach carried out in government or independent institutes as courses for medical officers of health or hospital managers, except for several outstanding examples such the London School of Hygiene and Tropical Medicine, and counterparts in the Netherlands, Scandinavia, and others, as well as many medical school or university-based research departments.

In Europe there has been a burgeoning of schools of public health since the 1990s and increasingly this is also occurring in countries of Eastern Europe and Central Asia. Associations of schools of public health have promoted these new schools with help from the Open Society Institute (Soros Foundation) of New York. The economic and political consolidation of Europe has highlighted the need for uniformity, standardization, and reciprocity in the EU's systems of higher education.

Europe recognized a need to modernize its university education standards. This came at a time when the health workforce was becoming an increasingly urgent issue recognized by the WHO and many individual countries as crucial to health policy. Global threats to population health, such as HIV, SARS, threatened bioterrorism, and the avian flu pandemic, have increased public and political recognition of the vital importance of a trained cadre of public health experts.

Following a series of meetings initiated by the EU, in June 1999 the European ministers of education issued what has come to be known as the Bologna Declaration. By this agreement, signatory states agreed in principle to work towards a European Higher Education Area (EHEA), to improve the quality of higher (i.e., post-secondary) education, and to provide for enhanced movement of students and academics within the EU. The motivation was to harmonize and raise European standards of postsecondary education to competitive levels in an education-based world competitive economy. The three overall objectives of the Bologna Process have been from the start: introduction of the three cycle system (bachelor/master/doctorate), quality assurance, and recognition of qualifications and periods of study. The Bologna Declaration initiated a "series of reforms needed to make European Higher Education more compatible and comparable, more competitive and more attractive for Europeans and for students and scholars from other continents. Reform was needed then and reform is still needed today if Europe is to match the performance of the best performing systems in the world" (European Commission, 2012) (Box 14.11).

The implications for public health education are great. The creation of new schools of public health has been initiated in several countries, particularly in Eastern and South-Eastern Europe, where the need for trained health workers is perhaps greatest. The Bologna Declaration allows for widespread recognition of the MPH degree and clarification of the public health worker's role as a professional. There are similar consequences for nursing, medicine, and allied health professions, which will move towards common nomenclature, educational standards, and academic exchange.

Since the 1980s and early 1990s, European schools of public health of the broader model have been established in Germany, the Netherlands, France, Spain, Poland, and Romania, and in countries of the former Yugoslavia and South-Eastern European region (Box 14.12). ASPHER has developed and promotes the idea of standardization and reciprocity for MPH degrees in the EU.

The Bologna Agrement of 1999 was intended to promote equality, not only in the context of health standards, but also in educational standards, in recognition of qualifications and accreditation systems, and in improving the health of the peoples of Europe. Many European countries are still in transition or development with currently poor standards of public health. Training in professional aspects of public health is vital to the educational goals of participating members, but also to the public health movement as a whole, within basic principles of ethics and values for the modern public health movement,

---

BOX 14.11 Bologna Declaration: Unified Standards of Postgraduate Education in the European Union

The Bologna Declaration initiated reforms in the study structure of the European higher education area, for both undergraduate and graduate university education.

Accreditation is a key instrument to support the processes of changes in Europe's higher education systems. The Bucharest Communiqué builds on the previous Leuven Communiqué of 2009 to establish the following priorities for 2010–2020:

- Ensuring a quality higher education system in the European Union.
- Adopting a two- or three-cycle system of study (BA, MA, PhD).
- Promoting the mobility of students and academic and administrative staff.
- Introducing a credit system (ECTS) for the assessment of study performance.
- Recognition of levels: adopting a system of easily identifiable and comparable levels.
- Active involvement of higher education institutions, teachers and students in the Bologna Process.
- Student participation in the management of higher education.
- Promoting a European dimension in higher education.
- Promoting the attractiveness of the European higher education area.
- Lifelong learning.
- A European higher education area and a European research area – two pillars of a society based on knowledge.
- Accreditation or certification of a degree program requires a review of the standards for content and vocational relevance of the degree to be awarded for a limited period within the frame of a transparent, formal, and external peer review.

Sources: Declaration 19 June 1999 by European Ministers of Education in Bologna, Italy. Available at: http://ec.europa.eu/education/policies/educ/bologna/bologna.pdf [Accessed 4 December 2012].
European Commission. Education and training: the Bologna Process – towards the European higher education area [updated 29 May 2012]. Available at: http://ec.europa.eu/education/higher-education/bologna_en.htm [Accessed 29 November 2012] and http://www.bologna-berlin2003.de/en/glossary/index.htm [Accessed 3 December 2012].

BOX 14.12 Principles of Public Health Education Standardization for Europe

- Public health as an organized system is vital for societies for long-term improvement in health of the population and to meet emergency situations of pandemics, disasters, and bioterrorism.
- Public health depends on laws, organizational structure, resources, planning, and training.
- Many countries in Europe lack clear differentiation of powers between different levels of government.
- Trained workforce is crucial to an effective public health network.
- Training, licensure, and accreditation are all needed in Europe in the western and eastern countries, especially in their transition stages.
- Development of schools of public health as multidisciplinary settings with autonomy as separate faculties of semi-autonomous status within medical faculties is vital for them to succeed in their mission of training, research, and service.
- The WHO must take a leadership role in promoting public health reform and education, networking with organizations already active in this field such as ASPHER, EUPHA, and others.
- Ministries of Health should be proactive in developing robust structures and training programs for public health in conjunction with sister ministries and academic organizations.
- Europe-wide standards and health targets will help individual countries to cope with these challenges.
- Europe-wide funding to promote new organization and training capacities in member countries is essential to enable this process to happen and to implement the intent of the Bologna Declaration in higher education.

Note: WHO=World Health Organization; ASPHER=Association of Schools of Public Health in the European Region; EUPHA=European Public Health Association.
Sources: Dubois CA, McKee M, Nolte E. Resources for health in Europe. Buckingham: Open University Press; 2005.
Paccaud F, Weihofen A, Nocera S. Public health education in Europe: old and new challenges. Public Health Rev 2011;33:66–86.

TABLE 14.10 Core Subject Domains to be Included in Master of Public Health (MPH) Curricula as Required by Agency Public Health Europe Accreditation

Core Subject Areas	Curriculum Content	ECTS Credit Ranges
Introduction	Introduction to public health	2
Methods in public health	Epidemiological methods, biostatistical methods, qualitative research methods, survey methods	18–20
Population health and its determinants	Environmental sciences (including physical, chemical and biological factors), communicable and non-communicable disease, occupational health, social and behavioral sciences, health risk assessment, health inequalities along social gradient	18–20
Health policy, economics, and management	Economics, health care systems planning, organization and management, health policy, financing health services, health program evaluation, health targets	16–18
Health education and promotion	Health promotion, health education, health protection and regulation, disease prevention	16–18
Cross-disciplinary themes (mandatory and/or elective courses)	Biology for public health, law, ethics, aging, nutrition, maternal and child health, mental health, demography, information technology use, health informatics, leadership and decision-making, social psychology, global public health, marketing, communication and advocacy, health anthropology, human rights, program planning and development, public health genomics, technology assessment	21–23
Internship/final project resulting in thesis/dissertation/memoire	Supervised by faculty (full time and/or adjunct)	24–26

Note: ECTS is the European Credit Transfer System, allowing transfer of students between universities and some degree of standardization of education at bachelor's level in keeping with the Bologna Agreement.
Source: Otok R, Levin I, Sitko S, Flahault A. European accreditation of public health education. Public Health Rev 2011;33:30–8.

and of the highest standards of public health practice. In 2011, ASPHER established the Agency for Public Health Education Accreditation (APHEA) in association with other prominent public health organizations to accredit MPH programs in Europe. The core subject domain criteria for accreditation are shown in Table 14.10 and the first two European MPH programs underwent accreditation in 2012. At the same time, strong interest is evolving in defining competencies of graduates of MPH programs (Box 14.13).

Schools of public health are of special importance for developing countries because of the prime importance of public health approaches in meeting their health needs. Nigeria is a very populous country where the main health

issues are those in the public health sector, yet there are over 20 medical schools and no school of public health. Departments of social or community medicine, primarily teaching medical students, are common in most medical schools, but this fails to provide students at the graduate-school level with an academic environment and the multidisciplinary training and specialization that the field requires to provide the professional leadership needed to meet the health challenges of their societies.

Leadership positions in health systems at the local, state, and federal levels are now held by doctors trained in clinical medicine, often with added training in public health. Few, however, have training in management and many received training focused on subspecialties of public health. In contrast, schools of public health at the doctoral level focus on the preparation of scholars in research and teaching rather than health leadership and management. Some seek preparation in Master's of Business Administration (MBA) programs. Preparation at the PhD level of doctors in public health requires post-baccalaureate training with broad areas of knowledge: tools of social analysis, health and disease in populations, promotion of health and prevention of disease, and health care systems and their management (see Box 14.10). Some schools of public health are moving in this direction by providing special part-time programs for working health executives. These will be especially important in the preparation of leadership capable of coping with the complexities of managed care or district health programs which are developing in many countries facing the organizational, economic, and ethical aspects of individual and population health.

The impact of public health and preventive medicine on national health has gained prominence since the 1980s, with the idea that medical need can be reduced by decreasing the burden of illness. This led to an increase in demand for preventive medicine and public health training, as medical care costs were thought by economists to be a function of need and demand. Health education programs designed to reduce health risks and reduce costs were shown to have documented effectiveness, with reductions in claims of 20 percent in some health insurance systems in the USA. Specific program features including chronic disease self-management, risk reduction, and increased self-efficacy appear important. This concept was further supported by the review by the CDC of public health achievements in the twentieth century, in which reductions of the burden of disease were directly related in large measure to public health programs. In the late 20th and first decade of the 21st century, new challenges with important achievements were seen with HIV, SARS, bioterrorism, and natural pandemics heightened the sense of concern, need, and urgency to strengthen public health training in the USA and internationally.

Countries of Eastern Europe and the Commonwealth of Independent States are facing a combination of high rates of mortality from preventable diseases and pressures for reform of the health care systems. The development of schools of public health as independent schools within single- or multifaculty universities should be an important priority for international aid and for national authorities. The challenge lies in integrating experience from many countries in the industrialized world with local academic centers in the field of public health. Traditional departments of social hygiene within medical academies need to evolve to educate new generations of doctors and other health professionals to cope with challenges facing the health systems in these countries. Postgraduate centers of training are essential to provide the leadership and professional staffing to address the New Public Health.

For developing countries, such as India and Nigeria, with vast populations and poor health, the need for schools of public health is even more crucial for creating the infrastructure to meet the health needs. Achieving the MDGs in many countries will not be possible without developing and sustaining a strong workforce of well-trained public health analysts, leaders, and field workers. This will require academic centers capable of training, research, and service to prepare such public health workers and to advocate policies and priorities to achieve these targets.

The essential competencies for public health are those outlined by the American Public Health Association (APHA) and discussed in Chapter 10. These are skills acquired through training and experience and not part of the skills of physicians per se. They include what the APHA calls the essential public health services:

- Monitor health status to identify community health problems.
- Diagnose and investigate health problems and health hazards in the community.
- Inform, educate, and empower people about health issues.
- Mobilize community partnerships to identify and solve health problems.
- Develop policies and plans that support individual and community health efforts.
- Enforce laws and regulations that protect health and ensure safety.
- Link people to needed personal health services and assure the provision of health care when otherwise needed.
- Assure a competent public health and personal health care workforce.
- Evaluate effectiveness, accessibility, and quality of personal and population-based health services.
- Research for new insights and innovative solutions to health problems.

These are the elements essential for public health practice and a training curriculum to address those needs. A professional, qualified, and multidisciplinary workforce, in sufficient numbers, is vital to the organization and management of effective public health systems in Europe and around the world. Such a workforce is essential to evaluate and respond to growing threats to population health, to address health inequalities between and within countries, and to develop and implement scientifically based interventions in a timely and appropriate manner within the limits of available resources.

BOX 14.13 Competencies in Public Health Education and Practice

- Competencies denote the potential for specific, defined performance standards of knowledge and skills based on public health theory and science.
- Competencies must be observable and transparent through appropriate performance patterns required for their achievement, related to appropriate methods to meet existent public health challenges.
- In public health, competencies are needed throughout complex interactive systems for achievement of goals in the mission of public health.
- Patterns of competencies can be assigned to individuals, professional groups, and public health organizations – thus both to individual public health professionals and to accreditation of public health institutions, systems, and other organizations, e.g., those serving specific geographic areas or populations.
- European lists of public health competencies cover the education levels defined by the Bologna Process, i.e., bachelor's, master's, PhD and continuing professional development (CPD).
- Documentation of satisfactory demonstration of competencies can be used as effective indicators of progress through education and training programs, leading to certification of fitness to practice in either service or academic public health careers.
- Certification or licensing of public health professionals, especially in countries with established coherent public health systems and well-defined career structures, can be based upon effective completion of a systematic education and training program.
- Lists of public health competencies have been developed by public health organizations in the USA, Canada, Australia, and Europe, as well as by groups and associations concerned with public health in general or subspecialties of public health such as health promotion.
- Such lists have many practical applications, e.g., in planning, delivery, and evaluation of education, training, and service programmes, in demonstration of completion of training, or to support employment in construction of job descriptions and in annual reviews of performance.
- Even with similarities, there are variations in the structure of lists of public health competencies, e.g., between two UK lists (one designed for training, the other for employment), and the lists produced by the US Association of Schools of Public Health (ASPH).
- European lists produced by the Association of Schools of Public Health in the European Region (ASPHER) cover

population health as well as health systems and public health operations; the latter classifies and subdivides public health competencies into:
  - methods in public health
  - population health and its social and economic determinants
  - population health and its material (physical, radiological, chemical, and biological) determinants
  - health policy; economics; organizational theory and management
  - health promotion: health education, health protection and disease prevention
  - ethics.
- ASPHER's European Core Competencies Programme has, since its initiation in 2006, involved about 100 researchers/teachers at member schools, European Ministries of Health, civil servants, and public health practitioners. International conferences were devoted to discussion of competencies, workshops at European and global public health conferences, as well as practitioner-school workshops in a series of European countries, with support by the EU in collaboration with public health partner organizations.
- As of September 2012, ASPHER's European lists of expanded core competencies (population health, health systems, and public health operations) have been endorsed by member states of the World Health Organization's European Region.
- Development, implementation, and use of lists of competencies and their interaction with public health practice and research is a continuing process rooted in well-established organizational structures.

Sources: *Anders Foldspang, School of Public Health, Aarhus University, Denmark; Christopher Birt, Department of Public Health, University of Liverpool, UK. Personal communication; December 2012.*
Birt C, Foldspang A. The developing role of systems of competences in public health education and practice. Public Health Rev 2011;33: 134–47. Available at: www.publichealthreviews.eu [Accessed 12 December].
Faculty of Public Health. Learning outcomes for public health. Available at: http://outcomes.fph-groups.org.uk/learning_outcomes/ [Accessed 12 December 2012].
Calhoun G, Wrobel CA, Finnegan JR. Current state in US public health competency-based graduate education. Public Health Rev 2011;33:148–67. Available at: www.publichealthreviews.eu [Accessed 12 December].
Birt C, Foldspang A. European Core Competences for MPH Education (ECCMPHE). ASPHER publication no. 6. Brussels: ASPHER; 2011.
World Health Organization, European Region. European action plan for strengthening public health capacities and services. Copenhagen: WHO Europe; 2012. p 18.

HEALTH POLICY AND MANAGEMENT OF HUMAN RESOURCES

Preparation for policy and management roles in public health and health systems has become widespread in schools of public health and in business schools in the USA. This trend will undoubtedly increase as managed care increases, and as intersectoral mergers or other

functional arrangements become more common. Departments of management and policy or health services in schools of public health have the mission to study and seek methods of improving efficiency and effectiveness of personal and population-based health organizations. As academic fields, they share a population perspective that includes interdisciplinary faculty from economics, law, management, medicine, history, sociology, and policy analysis. They focus on societal, population, economic, and organizational perspectives.

Personnel and managerial costs are the largest single component of total health expenditures, so that management and utilization of this resource is of prime importance to a health system. Health personnel should be recruited, trained, and utilized in a manner appropriate to meet the health needs of the population. This means employing their skills under conditions that promote effective work. Human resource management includes determining which category of worker can best provide specific services, how many are required, and which organizational frameworks are required to provide needed care most effectively. This requires not only delegation of responsibility, but also the resources with accountability to carry out the tasks.

Human resources management includes determination of numbers and types of health personnel needed for the health system now and in the future. Responsibility for planning, management, financing constraints, licensing, discipline procedures, quality control measures, and so on, can be delegated from professional levels to appropriately qualified paraprofessionals. Accountability for performance is essential for any system dedicated to provision of quality care and to meeting its goals of improved health outcomes.

Restricting numbers of professionals is sometimes done in the self-interest of a professional group to restrain competition. Oversupply can be costly and destructive to the public interest by misdirecting health resources in non-productive or even harmful ways. An excess supply of surgeons generates higher rates of elective surgical procedures than necessary or safe, while shortages of primary care physicians prevent adequacy in basic health services. Poor supply or quality of nursing personnel compromises the quality of hospital care and primary care in the community. Long-term retraining and redeployment policies are required, which are developed and implemented over time, rather than spasmodic mass layoffs of nurses and other hospital workers.

Managed care in the USA and similar comprehensive service programs in other countries provide the opportunity to seek a new balance of services and introduce new roles in health care. Physicians, NPs, CHWs, and many other kinds of health care professionals and technicians will be part of the complex of health care provision when the economics of care necessitate cost-effectiveness, where prevention and treatment are part of the same complex, and where a health promotion approach is fundamental to the objectives of the organization. In turn, public health workers should take active roles informing and participating in health systems management.

The evolution of public health is discussed in Chapter 2 and its organization in Chapter 10. The preparation of public health leaders and professional staff is well described by the ASPH. The ASPH represents 46 graduate schools of public health, which provide approximately 84 percent of the public health graduates in the USA. ASPH identified core competencies for MPH students upon graduation. These focus on the role of the public health graduate, as a professional in a discipline that addresses population health and society's role in monitoring and achieving good health and quality of life. Public health professionals work in many settings, as defined by the ASPH, to achieve:

- optimal human growth, development, and dignity across the lifespan
- respect for community participation and preferences in health
- air, food, and water safety
- workplace, school, and recreation safety
- timely detection of disease outbreaks and public health threats
- science-based responses to public health problems
- health care access, efficiency, and effectiveness
- encouragement of healthy choices that prolong a high quality of life
- design and maintenance of policies and services to meet community and individual needs for physical and mental health.

Public health professionals also recognize the contributions of other disciplines, including but not limited to the health professions. Public health education should be included in all undergraduate education programs as part of preparation of students for fields as diverse as business, economics, education, engineering, law, political science, psychology, sociology, anthropology, urban planning, and public administration.

The CDC established a Prevention Research Centers Program, offering training courses for public health practitioners working in the field. These include training program offerings: evidence-based public health, physical activity and public health, and social marketing. These courses illustrate the commitment of the Prevention Research Centers Program to helping in creation of a better trained public health workforce. The Moldova Ministry of Health, working with the WHO local office, developed a program for public health workers in control of NCDs; and the Moldova School of Public Health in Chisinau provides distance learning technology to support the spread of this knowledge to all parts of this small and very poor country in Eastern Europe. The Braun School of Public Health in Jerusalem provided support materials for use in a "training of trainers" program, a distance learning initiative implemented by the local WHO office in Moldova.

The demographic and epidemiological transition from acute diseases to chronic health problems is now moving to what Breslow called "the third era of health" (see Chapter 2), which is basically a health promotion model. This will necessitate a shift in the health workforce education policies. The traditional emphasis on diagnosis and treatment will need to shift to producing educational and supportive health workers at different levels of training to help the elderly and people with chronic health problems such as cancer, diabetes, asthma, and congestive heart failure, not in place of medical providers, but augmenting what is now being called a "patients' home" (Shortell et al., 2010) to "a new set of core competencies (knowledge, skills, abilities, personal qualities, experience, or other characteristics", with a mix of professional and paraprofessionals providing comprehensive care over long periods of duration of the most prevalent health problems (Pruitt and Epping-Jordan, 2005).

New information technology and medical devices will change the face of health care as patient monitoring and educational support will become key to the long-term well-being of people even with chronic medical conditions, helping to improve life quality and preventing unnecessary hospitalization and disabling health status. The educational approaches will need to shift towards the creation of more mid-level and CHW capacity to be part of the teams that will enable a more comprehensive approach to health and patient care with efficiency and support.

The rapid decline in mortality from coronary heart disease, stroke, cancer, injury, and other leading causes of death has prolonged and improved quality of life over the past half century. The health workforce will need to change to cope with these and other new challenges of disease, trauma, mental health, and newly evolving technologies to improve disease control and healthy quality of life. New technologies such as robotics will change surgical practice, as genetics will change medical practice, but health promotion and disease prevention will still be the keys to a healthier and longer life. The health workforce needs to be prepared to deal with these challenges.

SUMMARY

Education and training of medical and allied health personnel are important issues in health care systems development, and include issues of both quantity and quality. Regular reassessment is needed lest the numbers of practitioners produced become larger or fewer than the needs of the services, and as a result health care standards may decline or the system may become excessively costly while needed health promotion is inadequate. Preparation of managers and planners skilled in data and program analysis and leadership is as important as training health care providers.

Training of health professionals should be accompanied by orientation to the broad sweep of the New Public Health, including its management and evaluation skills. New health professional roles will evolve based on individual patient and community health needs. This is crucial in low- and medium-income countries struggling with both communicable and non-communicable diseases with insufficient health workforce and educational capacity while international donor aid focuses mainly on specific diseases rather than on building capacity and political support.

The build up of an adequate supply, quality and distribution of health workers is a matter of life and death in developing countries which lose many of their trained personnel through migration to high income countries. The receiving countries absorb many such emigre health personnel to meet their own needs without compensation to the countries losing their badly needed health personnel trained at great cost. The need for training of health personnel at levels appropriate to their country needs such as community health workers and bachelor degree public health personnel. This is a vital issue to continue and expand progress toward the MDGs and their followup health goals.

Management in the New Public Health is confronted with many difficult challenges in human resource policy. These include not only the quantity and quality of training but also flexibility in utilization, including redeployment of personnel from institutional care settings to community health and health promotion activities. Personnel and management expenses constitute some 75 percent of costs of patient care; any program for reallocation of resources towards community and preventive care must necessarily involve health workers, not only as an economic issue but as a qualitative one. These professional and personnel issues must be treated with care and sensitivity.

A health care system depends on the quality, ethics, pride, and professional skills of its team members. The training and retraining of such personnel are therefore fundamental considerations of the New Public Health. The training and work situations of the people who make up the complex of the New Public Health and provide its monitoring and services are crucial to the workings of the system and in networking with the multiple partners in health for an individual and for a community. New human resource issues will need to be addressed in facing the challenges of changing demographics and epidemiological patterns and to overcome still enormous inherent regional and social inequalities in health globally.

NOTE

For a complete bibliography and guidance for student reviews and expected competencies please see companion web site at http://booksite.elsevier.com/9780124157668

BIBLIOGRAPHY

Adany, R., Villerusa, A., Bislimovska, J., Kulzhanov, M., 2011. Public health education in Central and Eastern Europe, and Central Asia. Public Health Rev. 33, 105–133. Available at: www.publichealthreviews.eu (accessed 02.02.13).

Agency for Healthcare Research and Quality, October 2011. Primary care workforce facts and stats no. 2: The number of nurse practitioners and physician assistants practicing primary care in the United States. AHRQ publication no. 12-P001-3-EF. AHRQ, Rockville, MD. Available at: http://www.ahrq.gov/research/pcwork2.htm (accessed 04.12.12).

Aiken, L.H., Buchan, J., Sochalski, J., Nichols, B., Powell, M., 2004. Trends in international nurse migration. Health Aff. 23, 69–77. Available at: http://content.healthaffairs.org/content/23/3/69.full.pdf (accessed 07.01.14).

Aiken, L.H., Cheung, R.B., Olds, D.M., 2009. Education policy initiatives to address the nurse shortage in the United States. Health Aff. 28, 646–656. Available at: http://content.healthaffairs.org/content/28/4/w646.full.pdf (accessed 07.01.14).

Allin, S., Mossalios, E., McKee, M., Holland, W., 2003. Making decisions on public health: a review of eight countries. WHO Observatory, Brussels. Available at: http://www.euro.who.int/__data/assets/pdf_file/0007/98413/E84884.pdf (accessed 07.01.14).

American Academy of Physician Assistants, 2008 AAPA physician assistant census report. Available at: http://www.aapa.org/uploadedFiles/content/Common/Files/ob_gyn08c.pdf (accessed 06.01.14).

American Association of Colleges of Nursing, 2012. Employment of new nurse graduates and employer preferences for baccalaureate-prepared nurses. AACN, Washington, DC. Available at: http://www.aacn.nche.edu/leading_initiatives_news/news/2011/employment11 (accessed 04.12.12).

American Association of Colleges of Nursing, 6 August 2012. Nursing shortage. Available at: http://www.aacn.nche.edu/media-relations/fact-sheets/nursing-shortage (accessed 03.12.12).

American Board of Preventive Medicine. What is preventive medicine? Available at: http://www.abprevmed.org/aboutus.cfm (accessed 04.12.12).

American Public Health Association, 2012. Community health workers. Available at: http://www.apha.org/membergroups/sections/aphasections/chw/ (accessed 20.09.12).

Anand, S., Barnighausen, T., 2004. Human resources and health outcomes: cross-country econometric study. Lancet 364, 1603–1609. Available at: http://www.thelancet.com/journals/lancet/article/PIIS0140-6736(04)17313-3/abstract (accessed 06.01.14).

Auerbach, D.I., Buerhaus, P.I., Staiger, D.O., 2011. Registered nurse supply grows faster than projected amid surge in new entrants ages 23–26. Health Aff. 30, 2286–2292. Available at: http://content.healthaffairs.org/content/30/12/2286.abstract (accessed 07.01.14).

Bangdiwala, S.I., Fonn, S., Okoye, O., Tollman, S., 2010. Workforce resources for health in developing countries. Public Health Rev. 32, 296–318. Available at: www.publichealthreviews.eu (accessed 12.12.12).

Beaglehole, R., Dal Poz, M., 2003. Public health workforce: challenges and policy issues. Hum. Resour. Health 1, 4. Available at: http://www.ncbi.nlm.nih.gov/pmc/articles/PMC179882/ (accessed 07.01.14).

Bender, D.E., Pitkin, K., 1987. Bridging the gap: the village health worker as the cornerstone of the primary health care model. Soc. Sci. Med. 24, 515–528. Available at: http://www.ncbi.nlm.nih.gov/pubmed/3589747 (accessed 07.01.14).

Benner, P., Sutphen, M., Leonard, V., Day, L., 2009. Educating nurses: a call for radical transformation: highlights. Jossey-Bass. Available at: http://www.carnegiefoundation.org/elibrary/educating-nurses-highlights. (accessed 06.01.14).

Bologna Agreement on the European Space for Higher Education: an Explanation. Available at: http://ec.europa.eu/education/policies/educ/bologna/bologna.pdf (accessed 8.1.14).

Bhutta, Z.A., Ali, S., Cousens, S., Ali, T.M., Haider, B.A., Rizvi, A., et al., 2008. Alma-Ata: rebirth and revision 6 interventions to address maternal, newborn, and child survival: what difference can integrated primary health care strategies make? Lancet 372, 972–989. Available at: http://www.thelancet.com/journals/lancet/article/PIIS0140-6736(08)61407-5/abstract (accessed 06.01.14).

Birt, C., Foldspang, A., 2011. The developing role of systems of competences in public health education and practice. Public Health Rev. 33, 134–147. Available at: www.publichealthreviews.eu (accessed 07.01.14).

Calhoun, G., Wrobel, C.A., Finnegan, J.R., 2011. Current state in US public health competency-based graduate education. Public Health Rev. 33, 148–167. Available at: www.publichealthreviews.eu (accessed 07.01.14).

Calhoun, J.G., Ramiah, K., Weist, E.M., Shortell, S.M., 2008. Development of a core competency model for the Master of Public Health degree. Am. J. Public. Health 98, 1598–1607. Available at: http://www.ncbi.nlm.nih.gov/pmc/articles/PMC2509588/ (accessed 07.01.14).

Cavallo, F., Rimpela, A., Normand, C., Bury, J., 2001. Public health training in Europe: development of European master's degrees in public health. Eur. J. Public. Health 11, 171–173. Available at: http://eurpub.oxfordjournals.org/content/11/2/171.full.pdf (accessed 07.01.14).

Centers for Disease Control and Prevention, 2005. Assessment of Epidemiologic Capacity in State and Territorial Health Departments – United States, 2004 May 13, 2005. MMWR. Morb. Mortal. Wkly. Rep. 54(18); 457–459. Available at: http://www.cdc.gov/mmwr/preview/mmwrhtml/mm5418a2.htm (accessed 07.01.14).

Centers for Disease Control and Prevention, CDC timeline. Available at: http://www.cdc.gov/about/history/timeline.htm (accessed 02.09.13).

Centers for Disease Control and Prevention, Field epidemiologic training programs (FETP). Available at: http://www.cdc.gov/globalhealth/fetp/ (accessed 04.12.12).

Centers for Disease Control and Prevention, 2006. Field epidemiology training program standard core curriculum. Available at: http://www.cdc.gov/globalhealth/FETP/pdf/FETP_standard_core_curriculum_508.pdf (accessed 19.02.13).

Centers for Disease Control and Prevention, 1999. Ten great public health achievements – United States, 1900–1999. MMWR. Morb. Mortal. Wkly. Rep. 48(50), 1141–1147. Available at: http://www.cdc.gov/mmwr/preview/mmwrhtml/mm4850a1.htm (accessed 07.01.14).

Centers for Disease Control and Prevention, 2011. Ten great public health achievements – worldwide, 2001–2010. MMWR Morb. Mortal. Wkly. Rep. 60, 814–818. Available at: http://www.cdc.gov/mmwr/preview/mmwrhtml/mm6024a4.htm (accessed 10.1.14).

Centers for Disease Control and Prevention, 2011. Ten great public health achievements – United States, 2001–2010. MMWR Morb. Mortal. Wkly. Rep. 60, 619–623. Available at: http://www.cdc.gov/mmwr/preview/mmwrhtml/mm6019a5.htm (accessed 10.1.14).

Chen, L.C., 2010. Striking the right balance: health workforce retention in remote and rural areas. Bull. World Health Organ. 88, 323A. Available at: http://www.who.int/bulletin/volumes/88/5/10-078477.pdf (accessed 07.01.14).

Chen, L.C., Boufford, J.I., 2005. Fatal flows – doctors on the move. N. Engl. J. Med. 353, 1850–1852. Available at: http://www.nejm.org/doi/pdf/10.1056/NEJMe058188 (accessed 07.01.14).

Chinn, P.L., 2006. The global nursing shortages and healthcare. ANS. Adv. Nurs. Sci. 29, 1. Available at: http://www.ncbi.nlm.nih.gov/pubmed/16495683?report=abstract (accessed 07.01.14).

Cole, K., Sim, F., Hogan, H., 2011. The evolution of public health education and training in the United Kingdom. Public Health Rev. 33, 87–104. Available at: www.publichealthreviews.eu (accessed 02.03.13).

Collins, C., Limone, B.L., Scholle, J.M., Coleman, C.I., 2011. Effect of pharmacist intervention on glycemic control in diabetes. Diabetes Res. Clin. Pract. 92, 145–152. Available at: http://www.ncbi.nlm.nih.gov/pubmed/20961643 (accessed 07.01.14).

Commission on Social Determinants of Health, 2008. Final report: Closing the gap in a generation: health equity through action on the social inequalities. WHO, Geneva. Available at: http://whqlibdoc.who.int/publications/2008/9789241563703_eng.pdf (accessed 02.03.13).

Dal Poz, M.R., Gupta, N., Quain, E., Soucat, A.L.B. (Eds.), 2009. Handbook on monitoring and evaluation of human resources for health with special applications for low- and middle-income countries. WHO, Geneva. Available at: http://www.who.int/hrh/resources/handbook/en/ (accessed 07.01.14).

Dubois, C.A., McKee, M., Nolte, E. (Eds.), 2006. Human resources for health in Europe. European Observatory on Health Systems and Policies. Open University Press, New York. Available at: http://www.euro.who.int/__data/assets/pdf_file/0006/98403/E87923.pdf (accessed 07.01,14).

Flexner, A., 1910. Medical education in the United States and Canada: a report to the Carnegie Foundation for the Advancement of Teaching. The Carnegie Foundation for the Advancement of Teaching, New York. Available at: http://www.carnegiefoundation.org/sites/default/files/elibrary/Carnegie_Flexner_Report.pdf (accessed 07.01.14).

Foldspang, A., 2008. Public health education in Europe and the Nordic countries: status and perspectives. Scand. J. Public Health 36, 113–116. Available at: http://www.ncbi.nlm.nih.gov/pubmed/18519274 (accessed 07.01.14).

Franks, A.L., Brownson, R.C., Bryant, C., Brown, K.M., Hooker, S.P., Pluto, D.M., et al., 2005. Prevention research centers: contributions to updating the public health workforce through training. Prev. Chronic. Dis. 2 (2). Available at: http://www.cdc.gov/pcd/issues/2005/apr/04_0139.htm (accessed 26.12.12).

Frenk, J., Chen, L., Bhutta, Z.A., Cohen, J., Crisp, N., Evans, T., et al., 2010. Health professionals for a new century: transforming education to strengthen health systems in an interdependent world. Education for health professionals for the 21st century. The Lancet Commissions. Lancet 376, 1924–1958. Available in expanded form at http://healthprofessionals21.org/images/healthprofnewcentreport.pdf (accessed 03.03.13).

Fried, L.P., Bentley, M.E., Buekens, P., Burke, M.E., Frenk, J.J., Klag, M.J., et al., 2010. Global health is public health. Lancet 375, 535–537. Available at: http://www.thelancet.com/journals/lancet/article/PIIS0140-6736(10)60203-6/fulltext (accessed 06.01.14).

Gebbie, K., Rosenstock, L., Hernandez, L.M. (Eds.), 2003. Committee on educating public health professionals for the 21st century: who will keep the public healthy? Educating public health professionals for the 21st century. National Academies Press, Washington, DC. Available at: http://www.nap.edu/catalog.php?record_id=10542 (accessed 07.01.14).

Gebbie, K.M., Potter, M.A., Quill, B., Tilson, H., 2008. Education for the public health profession: a new look at the Roemer proposal. Public Health Rep. 123 (Suppl. 2), 18–26. Available at: http://www.ncbi.nlm.nih.gov/pmc/articles/PMC2431092/ (accessed 07.01.14).

Goodman, J., Overall, J., Tulchinsky, T., April 2008. Public health workforce capacity building: lessons learned from quality development of public health teaching programmes in Central and Eastern Europe. ASPHER publication no. 3. ASPHER, Brussels (accessed 07.01.14).

Hayes, E., 2007. Nurse practitioners and managed care: patient satisfaction and intention to adhere to nurse practitioner plan of care. J. Am. Acad. Nurse. Pract. 19, 418–426. Available at: http://www.ncbi.nlm.nih.gov/pubmed/17655571 (07.01.14).

Institute of Medicine, 2010. The future of nursing: leading change, advancing health. National Academies Press, Washington, DC. Available at: http://www.iom.edu/Reports/2010/The-Future-of-Nursing-Leading-Change-Advancing-Health.aspx (accessed 02.12.12).

Institute of Medicine, 1988. The future of public health. National Academies Press, Washington, DC. Available at: http://iom.edu/Reports/1988/The-Future-of-Public-Health.aspx (accessed 08.03.13).

Institute of Medicine, 2002. The future of the public's health in the 21st century. National Academies Press, Washington, DC. Available at: http://www.iom.edu/Reports/2002/The-Future-of-the-Publics-Health-in-the-21st-Century.aspx (accessed 08.03.13).

Institute of Medicine, 2003. Who will keep the public healthy? Educating public health professionals for the 21st century. National Academies Press, Washington, DC. Available at: http://www.nap.edu/openbook.php?isbn=030908542X (accessed 08.03.13).

Jaskiewicz, W., Tulenko, K., 2012. Increasing community health worker productivity and effectiveness: a review of the influence of the work environment. Human Resour. Health 10, 38. Available at: http://www.human-resources-health.com/content/10/1/38 (accessed 09.03.13).

Kimberly, J.R., 2011. Preparing leaders in public health for success in a flatter, more distributed and collaborative world. Public Health Rev. 33, 289–299. Available at: www.publichealthreviews.eu (accessed 12.12 .13).

Krantz, M.J., Coronel, S.M., Whitley, E.M., Dale, R., Yost, J., Estacio, R.O., 2013. Effectiveness of a community health worker cardiovascular risk reduction program in public health and health care settings. Am. J. Public Health 103, e19–e27. Available at: http://www.ncbi.nlm.nih.gov/pubmed/23153152 (accessed 07.01.14).

Lee, L.M., Wright, B., Semaan, S., 2012. Expected ethical competencies of public health professionals and graduate curricula in accredited schools of public health in North America. Am. J. Public Health 103, 938–942. Available at: http://www.ncbi.nlm.nih.gov/pubmed/23153152 (accessed 07.01.14).

Lehmann, U., Sanders, D., 2007. Community health workers: what do we know about them? The state of the evidence on programmes, activities, costs and impact on health outcomes of using community health workers. WHO, Geneva. Available at: http://www.who.int/hrh/documents/community_health_workers.pdf (accessed 09.03.13).

Lehmann, U., Van Damme, W., Barten, F., Sanders, D., 2009. Task shifting: the answer to the human resources crisis in Africa? Hum. Resour. Health 7, 49 doi: 10.1186/1478-4491-7-49 PMID: 19545398. Available at: http://www.ncbi.nlm.nih.gov/pmc/articles/PMC2705665/ (accessed 07.01.14).

Massé, R., Moloughney, B., 2011. New era for schools of public health in Canada. Public Health Rev. 33, 277–288. Available at: www.publichealthreviews.eu (accessed 12.12.12).

Noack, H., 2011. Governance and capacity building in German and Austrian public health since the 1950s. Public Health Rev. 33, 264–276. Available at: http://www.publichealthreviews.eu/upload/pdf_files/9/Noack.pdf (accessed 12.12.12).

Oliver, R., Sanz, M., 2007. The Bologna Process and health science education: times are changing. Med. Educ. 41, 309–317. Available at: http://www.ncbi.nlm.nih.gov/pubmed/17316217 (accessed 07.01.14).

Organization of Economic Cooperation and Development. Health Policies and Data. OECD Health Data 2013 - Frequently Requested Data. Available at: http://www.oecd.org/health/health-systems/Table-of-Content-Metadata-OECD-Health-Data-2013.pdf (accessed 8.1.14).

Otok, R., Levin, I., Sitko, S., Flahault, A., 2011. European accreditation of public health education. Public Health Rev. 33, 30–38. Available at: http://www.publichealthreviews.eu/upload/pdf_files/9/03_Otok.pdf (accessed 12.12.12).

Paccaud, F., Weihofen, A., Nocera, S., 2011. Public health education in Europe: old and new challenges. Public Health Rev. 33, 66–86. Available at: www.publichealthreviews.eu (accessed 12.12.12).

Pendergrast, M., 2011. Inside the outbreaks: the elite detectives of the epidemic intelligence services. Houghton Mifflin Harcourt, Boston, MA. Available at: http://www.amazon.com/ (accessed 07.01.14).

Popova, S., Georgieva, L., Koleva, Y., 2011. Development of public health education in Bulgaria. Public Health Rev. 33, 323–330. Available at: http://www.publichealthreviews.eu/upload/pdf_files/9/Georgieva.pdf (accessed 12.12.12).

Pruitt, S.D., Epping-Jordan, J.E., 2005. Preparing the 21st century global healthcare workforce. BMJ 330, 637–639. Available at: http://www.bmj.com/content/330/7492/637?view=long&pmid=15774994 (accessed 07.01.14).

Roemer, M., 1999. Genuine professional doctor of public health the world needs. J. Nurs. Scholarsh. 31, 43–44. Available at: http://onlinelibrary.wiley.com/doi/10.1111/j.1547-5069.1999.tb00419.x/abstract (accessed 07.01.14).

Rosenstock, L., Helsing, K., Rimer, B.K., 2011. Public health education in the United States: then and now. Public Health Rev. 33, 39–65. Available at: http://www.publichealthreviews.eu/upload/pdf_files/9/Rosenstock.pdf (accessed 08.03.13).

Santschi, V., Chiolero, A., Paradis, G., Colosimo, A.L., Burnand, B., 2012. Pharmacist interventions to improve cardiovascular disease risk factors in diabetes: a systematic review and meta-analysis of randomized controlled trials. Diabetes Care 35, 2706–2717. Available at: http://care.diabetesjournals.org/content/35/12/2706.long (accessed 07.01.14).

Scheffler, R.M., Liu, J.X., Kinfu, Y., Dal Poz, M.R., 2008. Forecasting the global shortage of physicians: an economic- and needs-based approach. Bull. World Health Organ. 86, 516–523. Available at: http://www.ncbi.nlm.nih.gov/pmc/articles/PMC2647492/ (accessed 07.01.14).

Schneider, D., Evering-Watley, M., Walke, H., Bloland, P.B., 2011. Training the global public health workforce through applied epidemiology training programs: CDC's experience, 1951–2011. Public Health Rev. 33, 190–203. Available at: http://www.publichealthreviews.eu/upload/pdf_files/9/Schneider.pdf (accessed 12.12.12).

Southall, D., Cham, M., Sey, O., 2010. Health workers lost to international bodies in poor countries. Lancet 376, 498–499. Available at: http://www.thelancet.com/journals/lancet/article/PIIS0140-6736(10)61157-9/fulltext (accessed 07.01.14).

Tilson, H., Gebbie, K.M., 2004. The public health workforce. Annu. Rev. Public Health 25, 341–356. Available at: http://www.ncbi.nlm.nih.gov/pubmed/15015924 (accessed 07.01.13).

Tulchinsky, T.H., 2002. Developing schools of public health in countries of Eastern Europe and the Commonwealth of Independent States. Public Health Rev. 30, 179–200. Available at: http://www.ncbi.nlm.nih.gov/pubmed/12613705 (accessed 07.01.14).

Tulchinsky TH, Epstein L, Normand C. (Eds), 2002. Proceedings of the Jerusalem Conference on Developing New Schools of Public Health. Public Health Rev 30 (1–4), 1–392. To be archived at: www.publichealthreviews.eu

Tulchinsky, T.H., Goodman, J., 2012. The role of schools of public health in capacity building. J. Public Health (Oxf.) 34, 462–464. doi: 10.1093/pubmed/fds045. Available at: http://jpubhealth.oxfordjournals.org/content/34/3/462.full.pdf (.07.01.14).

Tulchinsky TH, McKee M. (Eds), 2011. Education for a public health workforce in Europe and globally. Public. Health. Rev. 33, 7–15. Available at: http://www.publichealthreviews.eu/upload/pdf_files/9/01_Tulchinsky_Editorial.pdf (accessed 12.12.12).

UK's Faculty of Public Health, Learning outcomes for public health. Available at: http://outcomes.fph-groups.org.uk/learning_outcomes/ (accessed 12.12.12).

World Health Organization, Technical Notes – Global health workforce statistics database. Available at: http://www.who.int/hrh/statistics/TechnicalNotes.pdf (accessed 02.03.13).

World Health Organization, 1995. The community health worker: working guide: guidelines for training: guidelines for adaptation. WHO, Geneva. Available at: http://apps.who.int/medicinedocs/en/d/Jh2940e/1.html (accessed 08.03.13).

World Health Organization, 2012. WHO country assessment tool on the uses and sources for human resources for health (HRH) data. WHO, Geneva. Available at: http://www.who.int/hrh/resources/HRH_data-online_version_survey_use_sources.pdf (accessed 04.12.12).

Xu, D., Sun, B., Wan, X., Ke, Y., 2010. Reformation of medical education in China. Lancet 375, 1502–1504. Available at: http://www.thelancet.com/journals/lancet/article/PIIS0140-6736(10)60241-3/fulltext (accessed 06.01.14).

Zhang, D., Unschuld, P.U., 2008. China's barefoot doctor: past, present, and future. Lancet 372, 1865–1867. Available at: http://www.thelancet.com/journals/lancet/article/PIIS0140-6736(08)61355-0/fulltext (06.01.14).

Chapter 15

Health Technology, Quality, Law, and Ethics

Learning Objectives

Upon completion of this chapter, the student should be able to:

1. Describe responsibility for and methods of assessing and regulating technological developments in health care;
2. Describe methods of health facility accreditation and peer review;
3. Describe the concept of total quality management;
4. Identify and discuss ethical and legal issues in national health systems;
5. Apply ethical considerations to health issues in his or her home setting.

INTRODUCTION

Management of a production or a service system requires attention to the quality of personnel as much as to the system in which they work. Their motivation and sense of participation, the scientific and technological level of the program, and the legal and ethical standards of individual providers and of the system as a whole, are all important to the quality of care provided and equity of health status achieved.

Quality is the result of input and process, and is measured by outcome or performance indicators as well as perception of the service by the patients, the staff, and the community as a whole. Input refers to the institutional and financial resources for education, human resources, supplies, medications, vaccines, diagnostic capacity, and services available. Process refers to the use of those resources, including peer group expectations of professionalism. Outcomes generally include measures of morbidity, mortality, and functional status of the patient and the population. Defining and measuring achievements of national health objectives and targets, the methods of financing services, and the efficiency of organization help to determine quality. Training, supply, and distribution of health personnel are all determinants of access to and quality of care. Continuous and adequate availability of essential preventive, diagnostic, and treatment services, as well as accountability and internal methods of promoting standards, are all elements of the quality of a health service for the individual, the population as a whole, and groups within the population with special needs.

The content and standards of service are assessed through organized review by professional peers within an institution, and from outside. Peer review within an institution and external evaluation by accreditation or governmental inspection, based on cumulative evidence and the recognized current "state of the art", contribute to accountability and improved quality of care. Continuous quality improvement (CQI) among health care teams and organizations includes regular practice assessments, evidence gathering, remediation, and re-evaluation, which will be discussed later in this chapter. The perception of the services by the community, along with the knowledge, attitudes, beliefs, and practices of health, are all vital to improvement of health status.

Health-related technology is also in a continuing state of change. Systematic review and absorption of new scientific knowledge, technology, and innovations are essential to promote and renew health care methods. Public health serves in a regulatory role to assure high-quality care to the individual and the community. New technology, whether in the form of diagnostic procedures, new drugs, devices, or vaccines, or new types of health personnel, requires evaluation for effectiveness and appropriateness to the system.

Technology assessment also involves epidemiological and economic aspects of effectiveness. Failure to continuously monitor developments and to assimilate those that are demonstrably successful is an ethical and management failure which tragically costs many millions of lives from preventable diseases yearly, such as in delayed adoption of well-proven vaccines or tobacco restriction legislation. This is due to political failure even more than professional weakness, and constitutes one of the saddest ethical dilemmas of public health: failure to convince policy makers of the prime importance of health promotion and disease prevention in the health sector.

Ethics and law in public health reflect the values of a society. They inevitably evolve as they face dramatic social, economic, demographic, and political changes; new health challenges; and new technological and scientific possibilities for improving health. Ethics are the foundation of the value systems of a society and thus of its health concepts.

The New Public Health. http://dx.doi.org/10.1016/B978-0-12-415766-8.00015-X

Biblical sources articulated values of the Ten Commandments, Sanctity of Human Life, Improve the World, along with the Hippocratic Oath of physicians to "do good and do no harm". Modern definitions of public health and bioethics emerged from lessons learned from the horrors of eugenics and genocide in the twentieth century with humanistic precepts of "Universal Human Rights" and "Health for All" in the recent era (see Chapters 1 and 2).

The law is both permissive and restrictive. It sets the basic responsibilities, powers, and limitations of public health practice, with legislation and court decisions. Innovations in the technology of medical care and public health are powerful forces contributing to increased longevity, quality of life, and economic growth, but they also bring challenges to implementation impeded by additional costs of the health system and slow adaptation in countries with the greatest need. These are challenges to national and international political,

organizational, and economic systems to address health with the full potential for saving lives. Determining standards of "good practice" is a continuing process with the rapid development of new knowledge, technology, and experience.

The law is a dynamic process involving old and new legislation, court decisions, and new issues not previously faced, often following rather than anticipating public health issues. Public health has had both positive and negative ethical experiences and continues to face new issues with changing population needs, technology, science, and economics.

INNOVATION, REGULATION, AND QUALITY CONTROL

Health care technology has advanced with an increasing stream of innovation since the seventeenth-century epidemiological discoveries of Lind on scurvy (1747) and smallpox

TABLE 15.1 Health Care Innovations from the Seventeenth to the Twenty-First Centuries

Period	Selected Highlights of Scientific, Technological, and Organizational Innovations in Health
17th century	Biological basis of disease (Descartes), circulation of blood (Harvey), microscope (Leeuwenhoek)
18th century	Thermometer, lime juice supplements (Lind, 1756), vaccination (Jenner, 1796), surgical anatomy (Hunter), clinical sciences (Sydenham)
19th century	Miasma theory vs germ theory; inventions of stethoscope (1816), blood transfusion (1818), anesthesia (1842), hypodermic syringe (1852), ophthalmoscope (1851), laryngoscope (1855), pasteurization of wine, beer, milk (1860s), cholera vaccine (1879), X-ray (1895), blood pressure cuff (1896); sanitation, municipal health departments, chlorination and filtration of community water supplies, antisepsis, Braille printing, hygiene in obstetrics, nursing, microscopic pathology, pathological chemistry, microbiology, vaccines, X-ray, national health insurance, syringes, well-child care, aspirin (1899), Bismarkian social insurance (1881)
1900–1930	Electrocardiogram (1901), Flexner report on medical education, salvarsan, insulin (1922), blood groups, vitamins, conquest of yellow fever, vitamin B, vaccine for diphtheria (1923), tetanus vaccine (1924), electroencephalogram (1924), iron lung respirator (1927), Social Security Act (1935), cost–benefit analysis, food fortification (iodized salt, flour with vitamin B complex), improved work safety
1931–1945	Mandatory fortification of milk, salt, and flour in USA (1941), Pap test (1942), penicillin (1928), streptomycin, randomized clinical trials, antimalarial drugs, vector controls, dialysis machine (1945)
1946–1960	Contact lens (1948), DNA double helix (1953), heart–lung bypass machine (1953), ultrasound (1955), cardiac pacemaker (1958), Salk polio vaccine (1955), kidney transplant (1959), advances in vaccines, antihypertensives, psychotropic drugs, cancer chemotherapy, prepaid group practice, UK National Health Service (1948), Medicare in Canada (1946–1971)
1961–1980	Oral polio vaccine (Sabin), hip replacement (1962), oral rehydration therapy, measles vaccine (1964), coronary bypass (1964), Medicare, Medicaid (1965), mammography (1965), portable defibrillator (1965), measles–mumps–rubella vaccine, cost-effectiveness analysis, open heart surgery, pacemakers, organ transplantation, computed tomography (CT), eradication of smallpox (1972), health maintenance organizations (HMOs), diagnosis-related groups (DRGs), district health systems
1981–2000	Health promotion (1987), magnetic resonance imaging (MRI), positron emission tomography (PET), endoscopic surgery, *Helicobacter pylori* and chronic peptic ulcer disease (1982), managed care, *Haemophilus influenzae* b (Hib) vaccine, statins (1987), poliomyelitis eradication campaign (1982), local eradication of beta-thalassemia, pandemic of HIV (1981 onward), AZT antiretroviral approved (1987), robotic surgery (2000)
2001–2013	Millennium Development Goals (MDGS 2000) with substantial progress achieved, managing emergencies of mass terrorism and natural disasters, new vaccines (HPV), managing epidemics of measles and influenza, new diagnostic technologies, flour fortification to prevent birth defects, HIV still deadly but effective treatment and control measures, new treatments for hepatitis C, robotic surgery, nanotechnology, scientific advances with great potential benefit, Affordable Care Act (2010), Accountable Care Organizations

Source: Adapted from Health United States 2009. Special Feature: Medical technology. Introduction and timeline. 2009. Available at: http://www.ncbi.nlm.nih.gov/books/NBK44737/#specialfeature.sec1 [Accessed 15 December 2012]. See Historical Markers in Chapter 1.

vaccination by Jenner (1796), to the dramatic innovations of the end of the twentieth century (Table 15.1). The pace of innovation is rapid, creating the need for regulation, quality control, and technology assessment.

National governments are responsible for assuring that pharmaceuticals, biological products, food, and the environment are regulated to protect the public. In some countries, these responsibilities are divided among ministries of trade, industry, commerce, health, and environment. In a federal system of government, there may be a division of responsibility among federal, state, and local government, but with the national government often providing national standards and leadership in this area.

Government regulation and control are meant to protect the public health. The US Food and Drug Administration (FDA) is responsible for enforcing the Food, Drug and Cosmetic Act, the Fair Packaging and Labeling Act, sections of the Public Health Services Act relating to biological products for control of communicable diseases, and the Radiation Control for Health and Safety Act. The FDA is a Division of the Department of Health and Human Services (DHHS). State governments have the authority to supervise pharmacies and their products, which may be marketed across different states. All national governments have departments responsible for conducting supervision of food, drugs, and medical devices, often relying on international standards.

Drugs and devices include all drugs, diagnostic products, blood and its derivatives, biologicals, veterinary medicines, and medicated premixed animal products. All manufacturers and distributors are required by law to register these products with the national authority in order to be allowed to market or import them. All countries need to govern the food, drugs, vaccines, and cosmetics regulated for production, importation, marketing, and use within their jurisdiction. Organizations within each government must be responsible for assuring the consumer that foods are pure (unadulterated) and wholesome, safe to eat, and produced under sanitary conditions; that drugs and medical devices are safe and effective for their intended uses; that cosmetics are safe and made from appropriate ingredients; and that labeling is truthful, informative, and not deceptive.

National authorities such as the FDA, under legislation and regulations, govern both domestic and imported products. They establish and enforce standards, or adopt external agency standards as a "gold standard", meaning that products meet high standards of safety and efficacy. The FDA also monitors and inspects contents manufacturing standards under good manufacturing practices (GMPs), which includes regular accreditation of a manufacturer's facilities, staffing, planning, and monitoring capacity. Testing of products is carried out to assess safety, potency, and toxicity using accepted reference laboratory procedures as published in the compendium *Official Methods of Analysis of the Association of Official Analytical Chemists.*

When federal, state, or local investigators, sometimes known as *consumer safety officers*, detect through laboratory monitoring or observe conditions that may result in a public health hazard, and violation of food and drug laws and regulations, they issue a written report to the manufacturers with recommendations for correcting the conditions. In more blatant cases, the authorities may issue urgent recall or seizure orders for products in violation of standards constituting a danger to public health, such as contaminated products, lead-painted children's toys, or contaminated foods causing foodborne disease outbreaks, which occur not infrequently in imported and domestically produced foods in the USA. The Los Angeles County Department of Health inspects restaurants regularly and places a prominent placard in the window giving a grade A, B, or C to the restaurant for sanitation and safety. Those given D ratings may be closed until specified faults are eliminated, or a restaurant may be closed permanently. State governments require restaurants to list calorie and salt content of foods on their menus as part of the public health efforts to reduce obesity.

Supervision of food standards may also fail, as occurred in Israel in 2004 when total absence of vitamin B_1 in a soy-based baby formula imported from Germany resulted in three deaths and permanent brain damage to other infants due to severe beriberi. This episode led to criminal charges in 2008 of negligence resulting in death against the owners of the company that imported or produced the foods and staff members of the Ministry of Health. Animal foods in 2007 and infant milk products imported from China in 2008 were found to be contaminated with melamine, which was meant to mimic protein content but was toxic in combination with other chemicals used. The infant formula caused serious illness in some 300,000 Chinese babies and six deaths.

The FDA and its counterparts in each country are responsible for regulation of:

- *food* – foodborne illness, nutritional content, labeling, dietary supplements
- *drugs* – prescription drugs and generics, over-the-counter products
- *medical devices* – pacemakers, stents, contact lenses, hearing aids
- *biologics* – vaccines, blood products
- *animal feed and drugs* – for livestock, pets
- *cosmetics* – safety, labeling
- *radiation-emitting products* – cell phones, lasers, microwaves
- *combination products.*

New drugs and biological products for human use are required to pass rigorous review before approval for marketing is granted. Applications are submitted by the manufacturer or sponsor with acceptable scientific data including test results to evaluate the safety and effectiveness of the product for the conditions under which it is being offered. All manufacturers of drugs are required to be registered with the FDA and to meet its requirements for each drug produced and marketed, including the reporting of adverse reactions and labeling criteria. Manufacturers are required to operate in conformity with current GMPs, which include stringent control over manufacturing processes, personnel training, computerized operations, and testing of finished products. The FDA publishes guidelines to help manufacturers to familiarize themselves with current standards. The *United States Pharmacopoeia*, *National Formulary*, and *WHO Model Formulary 2008* are the official listings of approved products.

Medical devices are also regulated by the FDA. Thousands of products for health care purposes require premarket approval, ranging from basic articles such as thermometers, tongue depressors, and intrauterine devices (IUDs), to more complex devices such as cardiac monitors, pacemakers, breast implants, and kidney dialysis machines. These products are subject to controls of GMPs, labeling, registration of the manufacturer, and performance standards.

Monitoring for efficacy and potential hazards has been strengthened since the 1970s as a result of findings of long-term carcinogenic and mutagenic effects of estrogens, and toxic effects of chloramphenicol on bone marrow. The drug thalidomide, widely used as an antinauseant and sleeping pill for pregnant women in Europe, Canada, and Australia in the 1960s, was not approved by the US FDA. This drug was found to cause large numbers of serious birth deformities leading to its being banned in most countries. Controls of blood and blood products have been strengthened since the transmission of human immunodeficiency virus (HIV), hepatitis B, and hepatitis C by contaminated blood products in the 1980s. The responsibility of this regulatory function is well illustrated by the 1995 criminal conviction of several senior health officials in France for failing to stop the use of blood products contaminated with HIV in the mid-1980s. Concern regarding possible carcinogenic effects of silicone breast implants led to legal action and greater controls of all implantable products. A balance between safety and well-regulated approval of new products requires a highly professional and motivated regulatory agency, well-developed procedures, and well-trained staff.

The concepts of standardization of GMPs for pharmaceutical products and written protocols for good medical practice or good public health practice are accepted norms based on best available evidence of current scientific knowledge and experience. Recommended immunization schedules, water quality, ambient air standards, food fortification, and screening programs for early stages of diabetes are examples of accepted practice that have become recommended standards of public health practice, paralleling qualitative measures developed in clinical care.

APPROPRIATE HEALTH TECHNOLOGY

The concept of intermediate technology pioneered by Dr Ernst Schumacher in the 1960s proposed the development of simple and inexpensive technology for developing countries such as India to promote local economic development. Environmentally sustainable development and sources of energy, energy conservation, and reductions in toxic and harmful emissions are encouraged. In recent years ideas have included small loan systems for rural entrepreneurs in developing countries, and the use of simple cell phones for communication, farm produce marketing, cash transfers in remote areas without banking services, and many others. Now called appropriate technology, this topic has gained adherence in the health field in the search for low-cost and simple techniques for preventing and managing common illnesses.

Appropriate technology is defined by the World Health Organization (WHO) as the level of medical technology needed to improve health conditions in keeping with the epidemiological, demographic, and financial situation of each country. All countries have limited resources and so must select strategies of health care and appropriate technology to use those resources effectively to achieve health benefits. Improved water pumps, solar energy, rainwater collection and water reservoirs, sanitary latrines, fly traps, insecticide-impregnated bed nets, biogas from animal waste, improved home cooking stoves, and many other simple devices can make enormous differences in local sustainable agriculture, economic growth, and living conditions. Cell phones are now used to monitor health conditions such as hypertension, diabetes control, weight and body mass index, and other non-communicable conditions, and to transmit imaging from remote areas to specialists in medical centers who can provide test readings online. Simple, affordable, portable information technology can effectively support public health programs, even in resource-poor environments.

The topics discussed in the growing literature and meetings of the International Society of Technology Assessment in Health Care represent the dynamic field of technology assessment. The issues range from economic evaluation of pharmaceuticals to modeling approaches, measures of quality of life, technology dissemination and impact, and outcomes measurement. The range of issues also includes finance and health insurance, health care in developing countries, informatics, telemedicine, technologies for the disabled, screening, and cost-effectiveness. Evaluations in the scrutiny of both high- and low-technology services based on a combination of clinical, epidemiological, and economic factors are necessary. As health costs rise, disabling conditions increase and populations age, medical innovation proceeds at a rapid rate, and both client and community expectations in health care continually rise.

In developing countries, the training and supervision of traditional birth attendants (TBAs) for prenatal preparation and normal deliveries are important ways to reduce maternal mortality in rural areas, as discussed elsewhere, and an important Millennium Development Goal (MDG) which will not be met by 2015. Community health workers (CHWs) in well supervised and supported programs are essential to provide preventive care to underserved rural poor populations with a defined package of services that can be tailored to meet specific local needs, such as immunization, child growth monitoring, nutrition counseling, and malaria and TB control.

A major example of appropriate technology has been the WHO initiatives to promote national drug formularies (NDFs) as a consensus list of essential drugs that are sufficient for the major health needs of a country, eliminating unnecessary duplication and combined products on the commercial market. The WHO calls on all member states to ensure the availability and rational use of drugs and vaccines, and supports states wishing to select an essential list of drugs for economic procurement. Assistance with drug regulatory agencies, legislation, quality control, information, supply, and training is offered to help the member countries. Standard reference laboratories, the *International Pharmacopoeia*, and the *WHO Drug Bulletin* promote international standards and provide guidance to member states. The *WHO Model List of Essential Drugs* is a valuable tool to improve quality and cost management in national health systems.

Cochlear implants are now routinely used for children with congenital or other loss of hearing, as well as in elderly people. In August 2013, a new cell phone application was announced which photographs the eye and can be used to diagnose cataracts, macular degeneration (AMD), and other eye pathology, for interpretation by experts far away and to enable arrangements to be made for appropriate intervention to prevent blindness, which is common in developing countries. Other applications allow for monitoring of blood sugar of diabetics, hypertension, exercise, dietary management, and other aspects of health. In the same month, a camera, computer, and auditory device allowing blind people to "see and read" was demonstrated. The costs of such devices are initially high but will fall with advances in computing and other technical developments.

In both developing and industrialized countries major causes of death include cardiovascular diseases (coronary heart disease and stroke), along with respiratory diseases, cancer and injuries, all amenable to preventive and curative medical care. The key preventive measures for these are: healthful diet, reduced obesity, smoking cessation, exercise and physical fitness, hypertension management, aspirin, immunizations and other low-cost and highly effective medications such as statins. These are all low-cost self-care measures that can be promoted by local, state, and national governments, private advocacy organizations, and individuals in their families and communities. The principles of low technology, cost-effectiveness, and sound health policy converge in addressing these fundamental issues.

Priority Interventions in Low- and Medium-Income Countries

Disease control priorities for low- and medium-income countries are an important challenge for public health. Selection has often been based on individual initiatives due to strong advocacy in international organizations by donor countries, organizations, or individuals. In 1993, two landmark documents attempted to apply a logical system to such considerations: one was the World Bank's now classic *World Development Report: Investing in Health* and the other was *Disease Control Priorities in Developing Countries*. The *World Development Report* defined cost-effective clinical and public health cluster programs essential to improving health outcomes for low- and middle-income developing countries. The programs focus on those diseases that contribute heavily to the burden of disease and are amenable to relatively inexpensive interventions. The report defined interventions most able to reduce the burden of disease in low- and middle-income countries using clinical and public health interventions, as summarized in Table 15.2.

The 1993 *World Development Report* provided policy makers and public health practitioners with a concept and tools for assessing cost-effectiveness of available interventions for the major health problems in the developing world. It also provided useful measuring tools in the form of disability-adjusted life years (DALYs) to calculate the burden of disease and the cost-effectiveness of interventions to address them. This World Bank report addressed clinical interventions that would reduce DALYs lost by 24 percent in low-income countries and 8 percent in middle-income countries, including treatment of TB, with directly observed therapy, short course (DOTS); integrated management of the sick child; prenatal and delivery care; family planning; treatment of STIs; and limited care for pain, infections, and trauma as resources permit. It also addressed public health interventions, which would reduce DALYs lost by 8.2 percent in low-income countries and 4 percent in middle-income countries, expanded immunization with vitamin A supplements; tobacco and alcohol control; AIDS prevention; and school health including deworming. Together, the total reductions would be 32 percent for low-income and 12 percent for middle-income countries (Table 15.2).

These estimates have been refined by numerous studies conducted over the subsequent two decades. The second edition of the *Disease Control Priorities in Developing Countries* (2006) incorporates important changes in the technologies available. The concept of viewing priorities with an economic epidemiology model is still applicable, and has increased in importance (Box 15.1).

TABLE 15.2 World Bank Model for Priority Cost-Effective Health Interventions in Low- and Middle-Income Developing Countries

Service Type	Burden of Disease Averted (%)	
	Low-Income Countries	Middle-Income Countries
Public health interventions		
EPI-plus immunization (DPT, polio, measles, BCG, hepatitis B, yellow fever, vitamin A)	6.0	1.0
Other public health programs (family planning, health, and nutrition education)	NA	NA
Tobacco and alcohol control programs	0.1	0.3
AIDS prevention program	2.0	2.3
School health program (including deworming)	0.1	0.4
Subtotal (public health)	8.2	4.0
Clinical interventions		
Treatment of tuberculosis (short course)	1.0	1.0
Integrated management of the sick child	14.0	4.0
Prenatal and delivery care	4.0	–
Family planning	3.0	1.0
Treatment of STIs	1.0	1.0
Limited care: pain, trauma, infection plus as resources permit	1.0	1.0
Subtotal (clinical care)	24.0	8.0
Total	32.2	12.0

Note: Low-income=<US$350 gross national product (GNP) per capita; middle income=>US$2500 GNP per capita. Cost per immunized child=US$14.60 (US$0.50 per capita) and US$27.20 (US$0.80 per capita) in low- and middle-income countries, respectively.
DPT=diphtheria–pertussis–tetanus; BCG=bacille Calmette–Guérin; AIDS=acquired immunodeficiency syndrome; STI=sexually transmitted infection.
Note: The World Development Report was an innovative basis for follow-up work, as reported in Jamison DT, Breman JG, Measham AR, Alleyne G, Claeson M, Evans DB, et al., editors. Disease control priorities in developing countries. 2nd ed. Disease Control Priorities Project. Washington, DC: World Bank; 2006.
Source: Adapted from World Bank. World development report. Investing in health. New York: Oxford University Press; 1993.

BOX 15.1 Disease Control Priorities in Developing Countries

- Average life expectancy in low- and middle-income countries increased dramatically since the 1960s, while cross-country health inequalities decreased.
- Improved health has contributed significantly to economic welfare since the 1960s.
- Five critical challenges face developing countries (and the world) at the beginning of the twenty-first century:
  - Rapid demographic growth
  - HIV pandemic improved but still rampant
  - Persistent malaria, TB, diarrhea, pneumonia
  - Micronutrient malnutrition for mothers and infants
  - NCDs
  - Possible pandemics.
- Cost-effective interventions include:
  - Interventions to reduce neonatal mortality (50 percent of total child deaths)
  - Treatment of HIV-positive mothers, treatment of sexually transmitted infections
  - Controlling tobacco use, particularly through taxation
  - Lifelong medical management of risk factors in individuals at high risk for heart attacks or strokes, using aspirin and other drugs, would benefit tens of millions of individuals.
- Reform of health services and systems is needed, including:
  - Provider incentives
  - Provider focus on selected intervention to gain experience
  - Strengthening surgical capacity at district hospitals
  - Targeting limited resources to diseases affecting the poor, e.g., TB in low-income countries
  - In middle-income countries, public finance (or publicly mandated finance) of a substantial package of clinical care for all.
- Generation and diffusion of new knowledge and products underpinned the enormous improvements in health in the twentieth century and need to be applied for the control of NCDs, HIV, TB, and neglected populations.

Note: HIV=human immunodeficiency virus; TB=tuberculosis; NCD=non-communicable disease.
Source: Adapted from Jamison DT, Breman JG, Measham AR, Alleyne G, Claeson M, Evans DB, et al., editors. Disease control priorities in developing countries. Chapter 1, Investing in health, Table 1.1. 2nd ed. Washington, DC: World Bank; 2006.

As the MDGs are reaching their endpoint in 2015, follow-up global health targets will need to recognize the vital importance of non-communicable diseases (NCDs) in developing countries. The global consensus on MDGs, set out by the United Nations (UN) in 2001, indicates progress in the epidemiological understanding of realities in low-income countries and the need for consensus over common targets. Since then, attention has been directed towards the epidemiological shift to NCDs, which are the most common causes of death in low- and medium-income countries. Thus there is a double burden of infectious, nutrition, maternal, and child priorities, alongside the NCDs. The increasing adoption of vaccines such as *Haemophilus influenzae* type b (Hib) and rotavirus alongside the standard diphtheria–pertussis–tetanus (DPT), poliomyelitis (polio), and measles–mumps–rubella (MMR) vaccines provides new possibilities to control the major infectious disease killers of children. New technologies such as the advent of antiretroviral treatment for HIV have led to startlingly successful improvements in the quality of life and longevity of HIV/AIDS patients, and the prevention of onward transmission of HIV from mothers to babies and sexual partners. The WHO and many other global health stakeholders continue this work and produce analyses to contribute to policy making based on economic epidemiological evidence. This work affects policy, slowly but importantly.

In 2003, the Bellagio Study Group on Child Survival estimated that the lives of 6 million children could be saved each year if 23 proven interventions were universally available in the 42 countries in which 90 percent of child deaths occurred in 2000. The MDGs set out in 2001 provided targets for economic, educational, and environmental improvements, with three specifically focused on health: reducing child mortality; reducing maternal mortality; and control of HIV, TB, and other diseases. While important progress is being made, some of these targets will not be achieved by 2015. The global public health infrastructure will need to be expanded in content and strengthened in order to implement lessons learned in childhood routine immunization, safe maternity care, and nutritional security (see Chapter 16).

In medium- and low-income countries the difficulties are much more severe because of limited resources for health and the weak infrastructure of facilities and human resources in many countries. The key issues relate to NCDs, as in developed countries, so the interventions most needed address cardiovascular diseases, cancer, and injury, as well as diarrheal diseases, malnutrition, vaccine-preventable diseases, HIV, TB, malaria, and neglected tropical diseases. Efforts should be focused on low-cost interventions such as smoking reduction, vitamin and mineral fortification of foods, HIV, TB, and malaria control, along with maternal and child health protection.

Priority Selection in High-Income Countries

As discussed in Chapters 5 and 13, high- and middle-income countries also face complex health challenges, including aging populations, health costs, rapid development of new drugs and technologies, high rates of NCDs, and the rising prevalence of obesity and diabetes. Selection of priorities for health care expenditure from public and private sources has become a major focus of managing health systems.

In the industrialized countries, technological advances in the medical and public health fields have been major contributors to increasing longevity but also rising health costs. This situation has led to pressures for greater selectivity in adopting costly innovations without adequate assessment of benefits and costs. Many countries have adopted more cautious policies with regard to financing high levels of expansion of new

technology in the field of medical equipment, clinical procedures, or medications. Organized assessment of technology is now an essential feature of health management at the international, national, and local levels of service delivery. The major responsibility for technology assessment is at the national level, even with decentralization of service management.

With available resources being limited, health systems must choose interventions to be selected and how health systems are to be organized for efficiency and effectiveness while meeting public expectations. The US Patient Protection and Affordable Care Act (PPACA, more generally called ACA or "Obamacare") is undertaking reform measures to promote efficiency and prevention to reduce per capita health costs and to include more people in prepaid health care (see Chapters 10 and 13). These include preventive measures as recommended by Healthy People 2020 and implementation committees for selection of cost-effective measures to reduce morbidity and mortality to reduce health costs. The range of services to be promoted includes smoking cessation, increased physical activity, weight loss, healthy dietary practices, cancer screening, and many others that have not been previously accessible to those living in poverty and with no or limited health insurance. There is an emphasis on vaccination for children and adults.

The WHO promotes the widespread use of basic radiological units (BRUs) to increase access to low-cost, effective, diagnostic X-rays, especially in rural areas in developing countries. BRUs are hardy, relatively inexpensive pieces of radiological examination equipment that can be used in harsh field conditions for simple diagnosis of fractures and respiratory infections. The WHO estimates that 80 percent of all diagnostic radiology can be performed adequately using simple, safe, and low-cost equipment, supported by training of local people to operate and maintain the equipment. This is a consensus view of leading radiologists and clinicians helping the WHO to develop model equipment and training material.

The WHO *World Health Report* of 2009 focused on health technology assessment, stating:

"Technology continues to transform the medical care system and to improve length and quality of life – but at substantial cost. It is almost inconceivable to think about providing health care in today's world without medical devices, machinery, tests, computers, prosthetics, or drugs. Medical technology can be defined as the application of science to develop solutions to health problems or issues such as the prevention or delay of onset of diseases or the promotion and monitoring of good health."

Appropriate technology in the health field is becoming increasingly complex, laden with economic, legal, and ethical issues. Professional and public opinion demands make this a highly sensitive area of health policy, but responsible management of resources requires decision making that includes consideration of the effectiveness, costs, and alternatives of any new technology (Box 15.2). Failure to adopt

BOX 15.2 Health Technology Assessment

Questions that form the basis of technology assessment for a medical innovation include the following:

- Is it safe and cost-effective for the stated purpose?
- Is it a new service, or does it replace a less efficient intervention which can be phased out of service?
- What is the need it addresses?
- Where is it in the order of priorities of development of the facility?
- Does it duplicate a service already available in the community?
- Does it make medical sense (i.e., does it help in diagnosis and treatment for the patient's benefit)?
- What are the alternatives?
- What are the resources needed in terms of supplies, staffing, and upkeep?
- Can the facility afford it?
- What could otherwise be done with the resources it requires?

Sources: *Adapted from Kass N. Public health ethics: from foundations and frameworks to justice and global public health. J Law Med Ethics 2004;32:232–42.*
Sullivan SD, Watkins J, Sweet B, Ramsey SD. Health technology assessment in health-care decisions in the United States. Value Health 2009;12:S39–44. Available at: http://www.ispor.org/htaspecialissue/Sullivan.pdf [Accessed 14 December 2012].
Velasco-Garrido M, Busse R. Policy brief: Health technology assessment: an introduction to objectives, role of evidence, and structure in Europe. Geneva: WHO on behalf of the European Observatory on Health Systems and Policies; 2005. Available at: http://www.euro.who.int/__data/assets/pdf_file/0018/90432/E87866.pdf [Accessed 22 October 2012].

new innovations can result in obsolescence, while excessive expenditures for hospitals and medical technology prevent a health system from developing more cost-effective preventive approaches, such as improved ambulatory care, or supportive care for the chronically ill.

HEALTH TECHNOLOGY ASSESSMENT

Technology adoption can be a highly emotional and controversial issue, in advocacy of new cancer treatments or in criticism of managed care or national regulatory agencies, but spending limited national resources on some devices or medications of unproven value or inappropriately long hospital stays denies resources needed for other aspects of health care. A society must be able and willing to pay for medical innovation or improving quality of life by medical and public health interventions. Underfunding of a health system can deny these benefits just as misallocation of resources does, and this is a political issue even more than a professional one.

Medical and health technology assessment is the process of determining the contribution of any form of care to the health of the individual and community. It is a systematic analysis of the anticipated impact of a particular technology in regard to its safety and efficacy as well as its social, political, economic, legal, and ethical consequences. The technology may be a machine, a vaccine, an operation, or a form of organization and management of services. Analysis should include cost–benefit and cost-effectiveness studies (see Chapter 11) as well as clinical outcomes and other performance indicators.

Pressures from medical professionals, manufacturers of new medical equipment, and the public for adoption of new methods can be intense and continuous. Care must be taken that the specialists involved in committees for assessment are not those who may directly or indirectly benefit from the exploitation of technology, and who therefore may have conflicts of interest. Assessment must be multidisciplinary, involving policy analysts, physicians, public health specialists, economists, epidemiologists, sociologists, lawyers, and ethicists. The available information needs to include evidence from clinical trials, critical analysis of the literature, and the economic effect of adopting the technology on allocation of resources.

Medical technology varies in complexity and cost, not only to produce but in its utilization. Medical technology that is inexpensive to supply and administer is known as low technology or *low-tech*, while high technology or *high-tech* refers to costly and complex diagnostic and treatment devices or procedures.

At the low-tech end of the technology scale, oral rehydration therapy (ORT) was developed in the 1960s for oral replacement of fluids and electrolytes lost in diarrheal disease, particularly in children. It has been described as one of the greatest medical breakthroughs of the twentieth century. The introduction and wide-scale use of ORT for prevention of dehydration from diarrheal diseases throughout the world has saved hundreds of thousands of lives. Use of insecticide impregnated bed nets and reintroduction of DDT in household spraying along with vector control and improving diagnostic tools are low-tech but effective and key tools in malaria control.

Advances in endoscopic surgical techniques since the 1990s, and in robotic surgery since 2000, have greatly improved patient care by reducing trauma, discomfort, and length of hospital stay and endoscopy has become the surgical approach of choice for many procedures. Since reports of the first 100 operations performed in France in 1990, endoscopy has spread rapidly to all parts of the world. It is now recognized by surgeons worldwide as a safer, less traumatic and more effective alternative to traditional invasive surgery. Although the operating time is longer, patients are discharged from the hospital within several days and return to work shortly thereafter, compared to the long hospital stays after more invasive surgical procedures in the past. Following traditional abdominal surgery, a patient may acquire infections and require intensive care initially and a recovery period of many weeks.

Endoscopic surgery for cholecystectomy and esophageal, colorectal, hernia repair, renal, orthopedic, and other forms of surgery which previously were carried out with the patient remaining in hospital for many days are now done on a not-for-admission basis. Not-for-admission surgery has become standard practice in hospitals, extending the range of outpatient surgery and the comfort of patients who can return to their own homes to recuperate and return to regular activities much sooner. Fewer complications arise and patient comfort and economic implications are important. As a result, fewer hospital beds are needed for postoperative care than previously thought necessary, while surgical and ambulatory care facilities may need expansion to accommodate the growing elderly populations needing surgical interventions but requiring shorter recovery. This innovation is now accepted as the standard of much of modern surgical care and shows that simple organizational changes can save money and improve patient safety and comfort.

The bacterium *Helicobacter pylori* was first identified as the cause of peptic ulcers of the stomach and duodenum in 1982 (Robin Warren and Barry Marshall, Nobel Prize 2005). This discovery led to effective diagnosis and rapid, inexpensive treatment of chronic peptic ulcer disease. This has resulted in elimination of a major component of surgical procedures for chronic peptic ulcer diseases as well as a reduction in gastric cancer (see Chapter 4). Surgery for gastrectomies, vagotomies, and other outdated forms of treatment are now virtually gone, contributing to a decreased need for hospital beds even for an aging population. This and many other innovations in medical care have led to a growth in the use of ambulatory care for many forms of surgical, medical, and mental health care, along with much shorter length of hospital stay than in previous times. All of these factors have led to greater emphasis on ambulatory, outpatient, and home care services.

The dissonance between high-tech and low-tech procedures may lead to serious consequences in any health system. Choices require well-informed analysis of benefits, costs, alternatives, ethical considerations, and political consequences before limited health care resources are allocated between hospital-based high-tech medicine and low-tech primary care.

High-tech procedures are usually applied in hospital settings in the context of other highly specialized care for seriously ill, often terminal, patients. Computed tomography (CT), invented in the 1960s, quickly proved to be an extremely valuable diagnostic tool. Advances in CT, magnetic resonance imaging (MRI), and subsequent imaging techniques have proven to be cost-effective and lifesaving, replacing less efficient and more dangerous invasive procedures. The CT and MRI scans allow the clinician to reach a rapid diagnosis of many lesions before they can be detected by other invasive and dangerous diagnostic techniques, at stages where the lesions are subject to earlier and more effective interventions. Imaging technology is advancing rapidly and promising inexpensive new systems for long-distance transmission of imaging to medical centers may provide enormous benefits to people living in rural or developing countries. Recent advances in low-intensity CT screening of long-term heavy smokers for lung cancer have recently been added to recommended and potentially effective and cost-saving practice and may change the outlook for this disease in the coming decade (US Preventive Services Task Force, 2012 Flahault and Martin Moreno, 2013).

Technology assessment also examines methods of preventing and managing medical conditions. Treatment protocols or clinical guidelines are based on decision analysis of accumulated weight of evidence. Published clinical studies are assessed in meta-analyses, using statistical methods to combine the results of independent studies, where the studies selected meet predetermined criteria of quality. This provides an overview from pooling of data, but also implies an evaluation of the studies and data used. Clinical guidelines are part of raising standards of care, but also contribute to cost containment. Many countries form professional study groups to carry out meta-analyses on important health policy issues and new technologies.

Technology Assessment in Hospitals

There is considerable variance among countries, hospitals consume between 40 and 70 percent of total national health expenditures, with pressures for increased staffing and novel medical technology being a continual inflationary factor. Industrialized countries have all reduced their acute care hospital bed supplies and length of stay so that their expenditures for hospital care have fallen to between 30-40 percent of total health expenditures. Shorter stays and older patients have resulted in a drift towards intensive care, especially for internal medicine patients. Medical innovation is a continuing process with new diagnostic and treatment modalities reaching the market.

Hospitals no longer live in splendid isolation in the medical economy. A national or state government needs regulatory procedures to rationalize distribution of medical technology. The "certificate of need" is a form of technology assessment that has been used in the USA since the 1960s to assess and regulate the development of hospital services to prevent oversupply and costly duplication of services. It attempts to establish and implement the use of rational criteria for diffusion of expensive new technology. Whether this has had a lasting impact on restraining the excesses of high-tech medicine is arguable. This regulatory approach was limited to the hospital setting and failed to stop the development of high-tech medical services such as ambulatory for-profit CT, imaging, and in vitro fertilization centers.

Many countries have adopted national technology assessment systems to review topics as far-ranging as guidelines for acute cardiac interventions; liver, heart, and lung transplantation; minimal access surgery; and beam and isotope radiotherapy. Other technology assessment guidelines include diagnostic ultrasound, sleep apnea, molecular biology, prostate cancer, MRI, and new medications for inclusion in a national health system's approved basket of services.

Despite the limitations of this approach, where governments do not directly operate health care services, governmental regulation is necessary to prevent inequities in services by excessive development in some geographic areas at the expense of others, or by overexpansion of the institutional sector of health care at the expense of primary care. Regulatory mechanisms are essential in health care planning to restrain excessive and inappropriate use of high-tech services, but need augmentation by fiscal incentives to promote other essential services.

Hospitals everywhere face serious problems of hospital-acquired infections, which occur in about 5 percent of all hospitalizations. Healthcare-associated infections (HAIs), including multidrug-resistant bacterial infections, cause long lengths of stay, high costs, and most importantly, unexpected deaths and serious disabilities. Prevention of hospital-acquired infection requires ongoing training, staffing, and organization. The Centers for Disease Control and Prevention (CDC) defines HAIs as "infections caused by a wide variety of common and unusual bacteria, fungi, and viruses during the course of receiving medical care". Some of the preventive measures are simply promoting frequent hand washing by caregivers and visitors, and immunization of staff members against influenza and pneumonia, which can be problematic if there is staff resistance to influenza vaccination.

Training and routine supervision of cleaning staff are also vital, as are strict infection control measures for isolation rooms, strict protocols for catheter care, surgical suite sterility, surgical site infections, central line associated bloodstream infections, ventilator-associated pneumonias, catheter-associated urinary tract infections, and *Clostridium difficile*-associated disease. Guidelines for their control in surgical dialysis, pediatric, outpatient, and other vulnerable departments are available from CDC. The benefits of preventive procedures for this problem include cost estimates ranging from US$5.7–6.8 billion (20 percent of infections preventable) to US$25.0–31.5 billion, yet 70 percent of HAIs are preventable by well-known methods such as frequent and careful hand washing by medical and nursing staff, catheter and infusion care, and other similar measures (CDC, 2012).

Technology Assessment in Prevention and Health Promotion

Technology assessment of preventive care programs includes evaluation of the methodology itself, along with the costs and measurable benefits, as in reduced burden of disease. DOTS is the standard management of sputum-positive and sputum-negative TB, at low cost for DALYs saved. The coexistence of HIV and other complications has created multidrug-resistant tuberculosis (MDR-TB), which is difficult and costly to treat and cases constitute a source of continuing spread of the disease. A 2012 meta-analysis of cost-effectiveness of MDR-TB treatment in Estonia, Peru, the Philippines, and Russia shows it to be cost-effective and best carried out on an ambulatory basis (Fitzpatrick and Floyd, 2012).

Wide use of available and effective vaccines such as Hemophilus influenza b (Hib), pneumococcal pneumonia, influenza and rotavirus reduce hospitalizations and mortality from respiratory and diarrhoeal diseases among children, the elderly and other age groups. Vaccine prices generally fall after their initial period of use as manufacturing costs are lessened by improved methods or by bulk purchase contracts, as occurs in the public sector. For example, in 2012 MMR vaccine cost US$19.33 per dose if purchased through the CDC, but US$52.73 per dose if purchased in the private sector in 10 packs of single-dose units of the vaccine. A combined diphtheria, tetanus, acellular pertussis (DTaP) vaccine cost US$15.00 when purchased through CDC, while the same vaccine purchased with hepatitis B and inactivated polio vaccine (IPV) cost US$52.10 per dose. But the combination saves repeated visits and loss of compliance for that reason. The new human papillomavirus (HPV) cervical cancer vaccine cost US$130.27 per dose for the series of three doses per person, while the vaccine against diarrhea-causing rotavirus, approved in 2006, cost US$106.57 per dose for the recommended three doses (CDC, 2012).

The WHO recommends the inclusion of rotavirus vaccination in a country's immunization program, but the costs of the current generations of rotavirus vaccines are high in comparison to the budgets for vaccines for prevention of childhood illnesses in many developing countries. Many cost-effectiveness studies have shown this vaccine to be highly beneficial and it could help to reduce the very high global burden of disease of over 500,000 child deaths and 2 million hospitalizations occurring annually (Tu et al., 2011).

Vaccine programs must take into account transportation and administrative costs and expenses of ordering, storing, inventory control, cold chain, insurance, wastage, and spoilage. Multiple vaccines in one dose are less costly and less inconvenient for all. Examples include DTaP plus polio and Hib, or MMR (see Chapters 4 and 6). There is a need for implementation of legal protection of manufacturers from excessive litigation judgments while protecting the interests of the public and individuals who may have reactions to vaccines.

In 2012, the reappearance of pertussis and diphtheria raised concerns about immunization coverage and efficacy. Public opinions on vaccination may not be as supportive as in previous years. Mothers who oppose pertussis immunization for their children, such as occurred in the UK during the 1980's, leave their children vulnerable to a serious and often deadly disease, which has recurred since 2010.

The WHO estimates the cost of all immunization activities in all 117 low- and middle-income countries for the period 2006–2015 to be US$75 billion, while low-income countries would need US$35 billion. The rate of adoption of currently available and new vaccines will be determined by governmental decisions in each country, although external aid – such as that of the Global Alliance for Vaccines and Immunization (GAVI), an international public–private consortium to promote vaccination – is a valuable resource. The United Nations Children's Fund (UNICEF) is concerned about supply problems as well as costs, but the key issue relates to political decisions, funding, and capacities of national immunization systems.

Despite an excellent vaccine having been available since the 1960s, measles epidemics continue to occur in the industrialized countries. In the 1900s global deaths from measles were in the order of 1 million people per annum. Two major epidemics of measles occurred in Canada in the early 1990s, despite high rates of immunization coverage. Following this, a 1993 Delphi conference of experts from 31 countries reached a consensus recommending a two-dose measles immunization policy. Measles eradication has been set as a goal by the WHO and 90 percent reduction in cases and fatalities has been achieved since the 1990s. However, measles elimination requires coverage of 95 percent of children and two doses of a measles-containing vaccine (preferably MMR).

Measles reappeared as a widespread disease in Europe in 2010–2013 with tens of thousands of cases, many hospitalizations, and some deaths. It spread to the Americas, brought by travelers, and resulted in modest sized outbreaks, including the UK in 2012–2013. Eradicating measles by 2020 is projected to cost an additional discounted US$7.8 billion and avert a discounted 346 million DALYs between 2010 and 2050. As new vaccines enter the field, it is important to evaluate their effectiveness, costs, and the benefits to be derived.

The cost of the hepatitis B vaccine initially was over US$100 for an immunization schedule of three doses but has come down dramatically to less than US$1 per dose in developing countries for bulk purchases. However, in the USA, the price of vaccination per dose is estimated at US$41 if given by a general practitioner, US$15 if administered through an existing childhood immunization program, and US$17 if given through the school medical system. This is a standard vaccine covered by public and private health insurance systems. The vaccine is a cost-effective method to prevent liver cancer and the long-term effects of chronic hepatitis.

Screening and education for thalassemia in high-prevalence areas have nearly eradicated the clinical disease but not its carrier status in Cyprus, southern Greece, and other countries. Newborn screening and case management for phenylketonuria, congenital hypothyroidism, Tay–Sachs disease, and many other genetic diseases have been shown to be far less expensive than post-facto treatment of severely developmentally delayed and dependent children born with these diseases (see Chapter 6).

The success of Papanicolaou (Pap) smear screening in reducing cancer of the cervix mortality since the 1960s has been dramatic. The discovery of causation of cancer of the cervix by HPV strains led to development of an effective vaccine, which has been in use since 2006. Recent evidence shows that male circumcision can reduce transmission of HPV as well as HIV and other sexually transmitted diseases, and it is being adopted as an effective intervention in countries with high rates of both HIV and cancer of the cervix, such as in sub-Saharan Africa.

The drastic reduction in cancer of the cervix provides a powerful demonstration of the effectiveness of public health screening and other measures to control this major malignant cause of death in women. Screening for cervical cancer by Pap smears is recommended annually for high-risk groups, and every 2 or 3 years for other adult women (Box 15.3). Screening will remain vital for many years to come as the HPV vaccine comes into general use, and as its cost is reduced, but its protective effect for individual and herd immunity will not replace the need for ongoing screening for this very common cancer. HPV vaccine is also being recommended for all boys to prevent oral and anogenital cancers and HPV transmission to girls.

Routine mammography screening for breast cancer every 1–2 years is recommended by the US National Cancer Institute for women over the age of 40 and for younger women with high-risk factors (e.g., previous cancer, family history, genetic markers). Cost-effectiveness analysis is now an essential part of decision making in health policy and priorities. While there is controversy over the frequency of routine testing, mammography remains a mainstay in women's health and contributes to early case finding and falling mortality rates from breast cancer. Figure 15.1 demonstrates differences in utilization of mammography among US women in the age group 50–64 years within the previous 2 years, by insurance status. US women with private insurance (mostly through place of employment) had over 70 percent compliance, those with public insurance (primarily Medicaid) averaged about 60 percent compliance, while those with health insurance had average compliance rates of about 45 percent during the period 1993–2010. The UK National Health Service (NHS) invites women between the ages of 50 and 70 for screening every 3 years;

BOX 15.3 Technology for Prevention of Cervical, Colorectal, Liver, Stomach and Lung Cancers

Cancer of the cervix is the second most common cancer among women worldwide, with about 500,000 new cases and 250,000 deaths worldwide annually. Approximately 80 percent of cases occur in low-income countries, where cervical cancer is the second commonest cancer in women (WHO, 2012).

In the USA, and other industrialized countries, the incidence and mortality of cancer of the cervix have been going down steadily since the introduction of Papanicolaou (Pap) smear testing. Cervical cancer incidence declined during the period 1999–2008 by 2.3 percent per year and mortality declined by 1.9 percent per year an estimated 12,170 cases of invasive cervical cancer diagnosed in the USA with 4220 deaths in 2012.

Prevention of cancer of the cervix has until recently mainly focused on Pap smears to detect the disease while still in a precancerous (cancer in situ) phase, and this procedure reduced rates dramatically over the latter part of the twentieth century. The newly developed and highly effective vaccines against key strains of human papillomavirus (HPV) is now being used in routine immunization of young girls and more recently boys as well. The high cost of the vaccine precludes its rapid diffusion to most parts of the world but its use is spreading and being included in immunization programs funded by donor agencies in sub Saharan Africa. The vaccine should, in principle, also be used by adult women, in addition to continuation of routine Pap smear testing.

In the past decade, evidence of HPV as the cause of cancer of the cervix and the presence of HPV in uncircumcised men has brought circumcision back to professional and public debate. Reports from Africa of reduced risk of acquiring HIV among circumcised men have brought new attention to adult male circumcision, which is now actively promoted many sub-Saharan African countries.

The technological breakthroughs of the Pap smear in the 1950s, HPV testing in the 1990s, and the HPV vaccine in the 2000s should also include prevention by male circumcision. Visual inspection of the cervix and cryotherapy can treat precancerous cervical lesions in areas of developing countries as part of community health worker programs.

Colorectal cancer, the 7th leading cause of death in high income countries, is amenable to prevention by early screening using colonoscopy and fecal occult blood (FOB) testing. Screening is recommended for all persons over age 50 at 5 year intervals along with annual FOB testing. Where there is a family history of colorectal cancer or polyps, routine screening should begin earlier. Increasing use of screening and improved medical care are resulting in improving survival and declining mortality rates.

Stomach cancer is 10th leading cause of death in upper middle income countries. Prevention relies on early treatment of chronic peptic ulcer disease caused by Helicobacter pylori infection. This is readily diagnosed by a simple breath test and completely cured by low cost antibiotics. Increased awareness and access to these services would enhance long term trends of reducing mortality from stomach cancer.

Liver cancer is 8th leading cause of death in upper middle income countries due to the global prevalence of hepatitis B and helaptitis C. Hepatitis B is now falling due to widespread vaccination in childhood. Hepatitis C is now the major cause of liver cancer affecting hundreds of millions of persons worldwide. There is still no vaccine currently available, but screening and treatment is now used in the industrialized countries and will become more widely used as simpler, less costly treatments with less side effects are becoming available.

Early detection of lung cancer with spiral low dose tomodensitometry for smokers is recently being recommended by many professional bodies.

More basic cancer preventive measures such as smoking cessation, healthy diets, regular exercise, and moderate alcohol use are discussed in chapter 5.

Sources: *World Health Organization. Sexual and reproductive health. Cancer of cervix. Available at: http://www.who.int/reproductivehealth/topics/cancers/en/ World Health Organization. The top 10 leading causes of death (2011). Available at: http://who.int/mediacentre/factsheets/fs310/en/index1.html*
Centers for Disease Control and Prevention. Cervical cancer trends 2012. Available at: http://www.cdc.gov/cancer/cervical/statistics/trends.htm [Accessed (13.12.2012)].

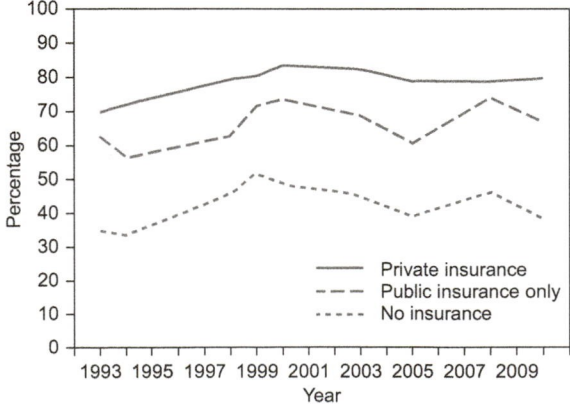

FIGURE 15.1 **Self-reported percentage of women aged 50–64 years receiving a mammogram in the past 2 years, by health insurance status, USA, 1993–2010.** *Source: Centers for Disease Control and Prevention. QuickStats from the National Center for Health Statistics: Percentage of women aged 50–64 years who reported receiving a mammogram in the past 2 Years, by health insurance status — National Health Interview Survey, United States, 1993–2010. MMWR Morbid. Mortal. Wkly. Rep. 2013;62:651. Available at: http://www.cdc.gov/mmwr/pdf/wk/mm6232.pdf (accessed 12.1.14).*

with an average of 75 percent respond to the invitation, of whom 4 percent are referred for further testing (UK Cancer Research, 2012) and as in most industrailized countries breast cancer mortality rates are falling impressively (see Chapter 5).

Health promotion in reducing exposure to HIV and cigarette smoking has been shown to be very cost-effective despite its low-tech or non-technological methodology, involving primarily group or mass education. Hypertension screening and case management is low-tech but highly effective in preventing strokes and blindness.

Low-tech innovations have had an important impact in reducing death and injury. These include mandatory use of car seat belts (introduced since the 1970s and 1980s in many countries), children's car seats, air bags, and bicycle and motorcycle helmets. Iodization of salt, vitamin A supplementation, and food fortification prevent large numbers of clinical cases of severe retardation, death, and blindness at low cost per child protected.

TABLE 15.3 Examples of High-Tech and Low-Tech Health Problem Solving

Problem	High Tech	Low Tech
Birth defects	Surgical repairs, rehabilitation	Folic acid fortification of flour, vitamin supplementation before and during pregnancy
Infectious diseases	Treatment – antibiotics	Vaccination, sanitation, handwashing, infection control in hospitals, health facilities, and nursing homes
Breast cancer	Screening – mammography	Nutrition, self-examination, routine medical examination
Colon cancer screening	Colonoscopy	Nutrition, vitamin D supplements; fecal occult blood testing
Acute myocardial infarction, primary, secondary prevention	Coronary angioplasty, stent, bypass surgery, heat transplantation	Antiplatelet thrombosis treatment (e.g., aspirin, intravenous streptokinase, beta-blocker); rehabilitation; diet, exercise, smoking cessation
Gallstones	Lithotripter, abdominal cholecystectomy	Endoscopic surgical removal
Head injuries	Intensive care	Helmets for bicycle riders and motorcyclists, seat belts in front and rear of motor vehicles
Thalassemia	Transfusions, chelating agents; prenatal diagnosis, amniocentesis, chorionic villus biopsy	Screening, education, counseling
Dehydration	Infusions	Oral rehydration
Neural tube defects	Surgery, pregnancy termination	Folic acid fortification of flour and grain products, supplements for women of fertile age
Liver cirrhosis, liver failure, cancer	Liver transplant	Hepatitis B vaccine, risk reduction activities among intravenous drug users, screening blood donors
Cancer of stomach	Surgery, chemotherapy	Dietary change, cure of *Helicobacter pylori*-generated gastric ulcers
Cancer of cervix		Pap smear screening; visual inspection and cryotherapy; human papillomavirus vaccine

Education for reducing risk factors for the cardiovascular disorders is far less costly than the premature deaths and high medical costs of patients suffering stroke and congestive heart failure. Health education, condom and needle supply, and screening of blood donations are the most important effective community health measures against the spread of HIV. Table 15.3 shows a comparison of high-tech and low-tech approaches, which often complement each other, to selected health problems.

Technology assessments represent the current consensus derived from reviews of published studies and exchange of views of highly qualified clinicians, epidemiologists, and economists within a context of technology assessment. They may change over time as new data or innovations are reported, and this possibility should be kept in mind in such discussions. Technology assessment mobilizes information and critically analyzes many aspects of medical technology to build a wide community consensus to influence policy decisions. Public opinion, political leadership, and administrative practice, as well as the scientific merit of a case are all factors in developing a consensus.

Technology Assessment in National Health Systems

Technology assessment requires an organization within the framework of national regulatory agencies. The FDA serves this purpose as a statutory body within the US Public Health Service. Sweden, Canada, Australia, the UK, the Netherlands, Spain, and other countries also have technology assessment advisory or regulatory agencies established by national governments to monitor and examine new technologies as they appear. Sweden has a widely representative national Swedish Council for Technology Assessment in Health Care which has an advisory role to the national health authorities.

The processes used in traditional systems to regulate food and drugs for efficacy, safety, and cost are more recently being applied to new medical devices and procedures. The unrestricted proliferation of new procedures presents serious dilemmas for national agencies concerned with financing health care and controlling cost increases. Non-governmental health insurance shares this concern, as does industry, which bears much of the cost of health insurance through negotiated, collective bargaining, "voluntary" health insurance in the USA. Most industrialized countries have national health services or national health insurance and are thus vitally interested in health costs and technology assessment. Many industrialized countries maintain technology assessment and cost-control activities. In the USA, the Agency for Healthcare Research and Quality (AHRQ) maintains oversight and studies related to clinical information, including evidence-based practice, outcomes and effectiveness, comparative clinical effectiveness, risks and benefits, and preventive services.

In Canada, the Health Protection Branch of the Federal Department of Health reviews medical devices and drugs and, with consent of the provincial governments, approves new medical procedures. Concern by governments over the cost implications of new procedures led to this practice. Since 1988, a network of government and professional bodies has formed a non-profit agency for technology assessment (Canadian Agency for Drugs and Technologies in Health, CADTH). This supports the provincial administration of health insurance in resisting professional, commercial, or political pressures to add untested technology or procedures to the health system as covered benefits. A comparison of rates of procedure performance between provinces shows very high discrepancies, as high as two-fold, in procedures such as coronary artery bypass graft or prostatectomy. Control of acquisition of high-tech equipment by national or state authorities is essential to prevent expenditures on high-cost equipment without adequate assessment.

DISSEMINATION OF TECHNOLOGY

The rapid spread of high-tech medical equipment has played a substantial role in escalating health costs. A comparison of the number of MRI scanners per million population in member countries of the Organisation for Economic Co-operation and Development (OECD) (Table 15.4) showed Japan and the USA with the highest number at 43.1 in 2010 (23.2 in 2000–01) and 31.6 (15.4 in 2000) per million, respectively, while the median was 10.5. Canada ranked fourteenth among the 20 OECD countries with 8.2 MRI scanners per million. Comparing CT scanners showed that Japan had the highest number at 97.3 while the median was 15.6. Canada was in eighteenth place among the 28 OECD countries reporting in 2010 with 14.2 CTs per million population. However, it should be noted that countries with the most machines have lower productivity per machine. The USA has far fewer examinations per machine than other countries (OECD, 2012).

The use of endoscopic surgery has spread worldwide since the 1990s. Health professionals become almost instantly aware of new developments from the news media as well as professional diffusion of information at conferences, in exchange visits, in published articles, and most dramatically via the Internet. National policy to foster the introduction of appropriate new technology requires a careful program of regulatory and financial incentives and disincentives to encourage or discourage diffusion of new methods of prevention as well as of treatment and community health care. Kidney transplantation has become a cost-effective and patient-friendly alternative to long-term dialysis both in hospital and at home. The key limitation is

TABLE 15.4 High-Tech Medical Equipment Units, Selected Organisation for Economic Co-operation and Development Countries and Years, 1986–2010 (Rate per Million Population)

Country	CTs				MRIs			
	1986	2000	2005	2010	1986	2000	2005	2010
Japan	27.5	84.4[a]	92.6	97.3[c]	0.1	23.2[a]	40.1	43.1[c]
USA	12.8	25.1[a]	32.3	40.7[d]	0.5	15.4[a]	26.7	31.6
Israel	–	5.7	6.6	9.2	–	1.4	1.7	2.0
Germany	6.9	12.2	15.4	17.7	0.7	4.9	7.1	10.3
UK	2.7	5.4	7.5	8.2	0.3	5.6	5.4	5.9
Canada	–	9.8[b]	11.6	14.2	–	2.5	5.7	8.2
France	4.7	7.0	10.0	11.8	0.5	1.7	4.8	7.0

CTs = computed tomography scanners per million population; MRIs = magnetic resonance imaging units per million population.
Notes:
[a]1999
[b]2001
[c]2008
[d]2011.
Source: Organisation for Economic Co-operation and Development. Health policies and data: OECD health data – 2012 data; frequently requested data. Available at: http://www.oecd.org/health/healthpoliciesanddata/oecdhealthdata2012-frequentlyrequesteddata.htm [Accessed 13 October 2012].

the shortage of donors. The same can be said for liver transplantation, which has also been shown to be cost-effective in terms of DALYs saved from chronic liver diseases such as cirrhosis and hepatitis B and C. Heart transplantation has spread among major centers in western countries.

The black market in organs for transplantation has become an international scandal of exploitation of poor people in low-income countries, and is the subject of police investigation in many countries, but it is difficult to control.

Bone marrow transplantation is now used widely and is effective in saving the lives of many people with malignant and non-malignant hematological disorders. Stem cell therapy, by introducing new adult and embryonic stem cells into damaged tissue to treat disease or injury, is becoming feasible for a wide variety of conditions including cancer, type 1 diabetes mellitus, Parkinson's disease, Huntington's disease, celiac disease, cardiac failure, muscle damage, and neurological disorders.

Limitation of new techniques or procedures to selected medical centers allows the passage of time to fully assess the merits and deficiencies of new technology before general diffusion into the health care system. Such limitation, however, is fraught with the danger of depriving the population of benefits of new medical technology, and the possibility of restraint of trade to the economic advantage of selected providers. Current advances in robotic-assisted surgery will expand during the coming decade for brain, bone, joint, prostate, and other surgery, and need to be assessed with regard to patient care benefits, costs, and the economics of capitalization of such innovations. Stem cell therapy is already widespread for some malignant conditions, and with advances in genomics and molecular biological technology it is very likely to become a major therapeutic intervention for many more conditions in the coming years. The effects of new technology on insurance and managed care systems are necessarily involved in decision making as to inclusion of new procedures in their service plans.

Publication in the professional literature is an accepted method of establishing the scientific merit of a treatment or an intervention. Too rapid diffusion of a medical practice can lead to disillusionment and confusion as to the merits of a particular medical procedure, as happened during the 1960s and 1970s with anticoagulant therapy for acute myocardial infarction and gastric freezing for peptic ulcers. Reviews of the literature should be critical and should assess the scientific merits of published data, as well as the sources of funding. Well-controlled large-scale clinical trials are vital to establish the relative values of alternative therapeutic approaches, as are meta-analyses of multiple studies.

Dissemination of information about new medical innovations in the popular media is almost immediate. Many major newspapers and television networks have well-informed medical reporters and commentators who have access to electronic medical journals as quickly as do medical specialists in each field. News magazines may carry special articles on new innovations, creating instant demand for them as benefits in a health program. This ready access to information has both benefits and dangers.

In the USA, health insurers have led the way in developing technology assessment and information synthesis, and in evaluating the costs and benefits of new procedures. The process is affected by public opinion, as well as by court decisions. A landmark decision against a health maintenance organization (HMO) in 1993 awarded US$29 million in damages to the family of a terminal breast cancer patient who died following refusal of the HMO to authorize a bone marrow transplant, which was at the time an experimental procedure. Denial of new technology may lead to increases in malpractice suits. In countries with limited financial resources, selection of technological innovations in health care that can benefit patient care or the public health requires a careful balance in order to use limited resources well, and to gain from the application of appropriate new health care technology.

Payment systems by national or private insurance systems are crucial to introduce and control diffusion of technology. Block budgets for hospitals have been more effective in Canada than in the USA in restraining the proliferation of high-tech equipment. This has led to criticism of the limited access of Canadians to medical technology, such as CT, MRI, and advanced cancer therapies. In the USA, universal application of the diagnosis-related group (DRG) payment system for Medicare, Medicaid, and most private insurance had the effect of increasing ambulatory surgery very dramatically, from 16 percent of all surgery in 1980 to 80 percent in 2008 of all surgical procedures in community hospitals (i.e., non-federal short-stay hospitals or 85 percent of all hospitals in the country). Inpatient surgical procedure rates declined from 85 percent in 1980 to 35 percent in 2008. Although the rate of visits to hospital-based surgery centers remained largely unchanged in the USA from 1996 to 2006, the rate of visits to private ambulatory surgery centers increased by about 300 percent.

HMOs and managed care organizations are paid on a per capita basis and have a strong incentive for cost containment. They have developed procedures and medical guidelines for investigation and intervention that seek to reduce unnecessary procedures. At the same time, HMOs are very active in promoting preventive care and non-hospital care insofar as this is compatible with good patient care.

Coronary bypass procedures decreased in frequency in the USA between 2001–02 and 2007–08. In the USA, such procedures are less frequently carried out in women and African Americans, because of lesser access to health insurance for African Americans and possibly because of biases in terms of case assessment criteria in women. Cardiac invasive procedures increased dramatically since the 1980s in most industrialized countries, but with wide variation in their use. The benefits of aggressive invasive management of cardiovascular diseases remain controversial, but many such procedures have proven beneficial in reducing mortality rates and improving quality of life.

Critical analysis of the need for surgery has resulted in lower tonsillectomy and radical mastectomy rates along with the increased use of outpatient procedures. Tonsillectomy, a routine procedure until the 1960s, is now performed infrequently since it was found to be of little medical value. Cataract surgery is now largely done on an ambulatory basis. The technology of home care has come to play an important role in early discharge of patients from the hospital, as has the wide use of cancer chemotherapy and radiation therapy on an outpatient basis.

DIFFUSION OF TECHNOLOGY

Innovations in health care through scientific and technological advances are continuing, with exciting breakthroughs being made in effective new treatments and public health interventions, and this requires health authorities, practitioners, and the public to maintain constant awareness of the current state of the art. Diffusion of new technology or adaptations from basic science advances may begin slowly, and then reach a "tipping point", at which time a dramatic change of trend occurs and it becomes the new standard or fashion.

Those with economic interests in the product try to advertise and promote sales, while practitioners are ready to try new methods to help their patients, but those who must pay for services may ask for evidence of effectiveness, safety, added value over present and known methods, and benefit to the length or quality of life of the individual. This can become a highly charged debate when those responsible for adopting new measures in national health plans must weigh one proposed addition against another, each with its ardent professional, community, or business promoters. The new HPV vaccine approved by the FDA in 2006 for prevention of cancer of the cervix is an example.

The HPV vaccine is recommended for preteen girls at the age of 11–12 years and also for females aged 13–26 to offset future sexual exposure to HPV-infected males and since 2012 recommended routinely for teenage boys as well. The two competing main manufacturers of HPV vaccine are naturally interested in increasing their market and market share, and willing to reduce prices. The cost has been lowered substantially for use in developing countries if purchased in bulk, but costs are still prohibitive unless funded by international donors. Competing low-cost manufacture in India has encouraged the two main manufacturers to lower prices to seek broader markets. In 2011, one manufacturer lowered its price dramatically to US$5 per dose, a 67 percent reduction in the current lowest public price. This has allowed GAVI to adopt an HPV strategy for developing countries, where 88

percent of cervical cancer deaths occur, with 275,000 deaths of relatively young women each year (GAVI, 2011, 2012).

Pioneering projects promoting visual examination of the cervix and local cauterization of abrasions by trained nurses and community health workers are meant to increase access to care in traditional villages remote from medical centers. The duration of immunity and whether booster doses will be required are still not known. Policy makers need to consider whether the same money would have greater benefit if used to provide pneumococcal pneumonia and rotavirus vaccine for children in developing countries, which would quickly save hundreds of thousands of lives. It is likely that the wonderful new public health technology that is the HPV vaccine will be absorbed quickly into public health practice at least in the industrialized countries, and is now being introduced by international donor agencies in sub-Saharan Africa.

QUALITY ASSURANCE

Quality assurance is an integral part of public health function and involves ensuring the quality of both health practitioners and facilities. It is an approach that measures and evaluates the proficiency or quality of services rendered. Hospital accreditation is a long-standing method of quality assurance, providing many generations of health providers in North America with first hand experience of quality assurance in community hospitals and long-term care facilities, as well as ambulatory and mental health services. Hospital accreditation has contributed to improvement in standards of facilities and patient care throughout Canada and the USA and has provided a working model for replication or adaptation internationally.

Adverse Events and Negligence

Iatrogenic diseases are adverse events that occur as a result of medical management and result in measurable disability. Negligent adverse events are those events caused by a failure to meet standards of care reasonably expected of the average physician or other provider of care. Hospital-acquired infections, anesthesia mishaps, falls, and drug errors are the most common iatrogenic events.

Iatrogenic disease is a major cause of morbidity, prolongation of hospitalization, and even death. Hospital-acquired (nosocomial) infections are estimated to occur in 7–10 percent of hospital cases in Britain and the USA. Primarily these are caused by urinary, respiratory tract, and wound infections. It is becoming more common that infections involving organisms previously responsive to antibiotics are now resistant to many antibiotics and difficult to treat. Infection control in hospitals is therefore an essential part of hospital organization. Because hospitals are increasingly being paid by DRGs, any secondary event prolonging hospital stays may have adverse financial effects on the hospital. In the USA, recent decreases in Medicare reimbursements for nosocomial infections reflect this trend to provide financial incentives to improve hospital infection control. There is, therefore, a strong financial as well as professional interest in reducing hospital-acquired infections.

A classic study of 32,000 hospitalizations in New York State carried out by a Harvard University team showed that 3.7 percent of hospitalized patients suffered adverse events or injuries caused by medical mismanagement which resulted in measurable disability. Of these, 28 percent were due to negligence, so that 1.03 percent of all hospitalizations involved medical negligence leading to measurable injury. Of the total of some 100,000 adverse events in the study group, 57 percent recovered within a month and 7 percent had severe injury. Some 14 percent or 14,000 people with adverse events died as a result; 51 percent of these deaths were due to negligence. A 1999 report of the US National Institute of Medicine estimated that between 44,000 and 98,000 people die annually in the USA from medical errors occurring in hospitals, but these data are considered to be overestimated in some studies. Adverse drug events (ADEs) result in 700,000 emergency department visits and 120,000 hospitalizations annually, with US$3.5 billion spent on extra medical costs. CDC estimates that at least 40 percent of the costs of ambulatory (non-hospital setting) ADEs are preventable.

A 2008 report by the Office of Inspector General of the US DHHS reported that 13.5 percent of Medicare beneficiaries experienced adverse events and that for 1.5 percent of beneficiaries, these adverse events contributed to their deaths. An additional 13.5 percent of beneficiaries in the sample experienced temporary harm as a result of their medical care, bringing the total percentage of beneficiaries experiencing instances of care-related harm to 27 percent. Nearly half (44 percent) of these adverse or temporary harm events were preventable.

Hospital-acquired infections cause 99,000 patients deaths in the USA every year (AHRQ, 2009). Higher rates are seen among the elderly and the poor. Rates are lower in teaching hospitals than in community hospitals. About 20 percent of the events were related to drug reactions or dosage errors. Less than 3 percent of those injured brought civil litigation for the negligence. The search for "bad apples" – that is, unethical, criminal, or incompetent health providers – is necessary, but not sufficient to stem the problems created by the health system itself. Prevention requires organized activity. Investigation of adverse events helps to identify methods of prevention and to protect the patient's rights. A program of measures to reduce hospital infection must be based on epidemiological analysis of recorded events in the search for common causes and preventable factors.

Organized surveillance and control requires a ratio of one infection control practitioner per 250 acute care beds, a trained hospital epidemiologist, and routine reporting of wound infections to practicing surgeons (CDC, Hospital Infection Program). Computer-aided medication dispensing, as well as automated and other safety systems are critical elements in minimizing morbidity and mortality resulting from preventable human errors. In response to the high frequency and cost of medical litigation, many states in the USA have enacted legislation to restrict court awards for medical negligence. Proposals for alternatives to the tort system of medical malpractice compensation include arbitration and mediation, an administrative system similar to that used for workers' compensation, and a no-fault system of compensation, such as exists in New Zealand, Sweden, and Finland. In a no-fault system the complainant need not prove negligence on the part of the provider, but only that he or she suffered an adverse event which is compensable at standard rates depending on the degree of disability. In the USA, federal legislation provides compensation for vaccine injuries, and three states have enacted restricted no-fault systems for birth-related neurological injuries.

In addition, there is greater emphasis on the adoption of failsafe mechanisms, such as introducing warning systems in anesthesia machines to alert the anesthetist if oxygen flow in the patient's tubing falls below a safe point. This system was tested in Boston hospitals and found to reduce adverse anesthetic events to zero cases over a 3-year period. Vitamin K injection was made mandatory for all newborns in New York State, as was already the case in some other states, when a study showed deaths from hemorrhagic disease of the newborn in cases where vitamin K was not administered.

Inappropriate medical practice patterns are an equal, or even larger problem for health systems. Comparisons of surgical rates within the USA for coronary bypass procedures, hysterectomies, and caesarean sections show wide variation between different areas of the country. The costs of excess surgery not only are economically wasteful but also involve risks for the patient from the surgery itself or anesthesia mishaps, infection, pain, and discomfort, with legal and ethical questions of unwarranted interventions not for the benefit of the patient. Health systems are increasingly required to evaluate and control excess surgical, investigative, or other medical procedures, not only for financial reasons but also for protection against litigation and infringement of patients' rights.

Licensure and Certification

The requirements that society establishes for allowing an individual to practice medicine, and any health profession, are vital to maintaining and improving the quality of care (see Chapter 14). These standards require defining the training and experience needed by the individual, examination procedures, and recognition for continued education and maintenance of competence. This requires a statutory base and national bodies operating under a national authority, separate from the agency operating the health system services. Separation of licensing from operation of the health service is essential in maintaining high professional standards.

The licensing authority is accountable to the state and the public. In some cases, this function is delegated to self-regulating professional bodies. In Canada, the licensing of the medical profession and specialty recognition are carried out by the medical profession with self-regulation. In the UK, medical licensing is by a state-appointed board and in the USA by state boards.

Medical schools, postgraduate training programs, and fellowships are all subject to periodic comprehensive assessments. Institutions that fail to meet the standard may have funding or licensure suspended until they have performed adequate remediation.

Health Facility Accreditation

Hospital accreditation in North America is by a voluntary grouping of professional associations, including the Canadian and American Colleges of Physicians and Surgeons, the hospital associations, and the Colleges of Nurses. The Joint Commission, originally operating in both Canada and the USA, carries out regular inspections of hospitals. In Canada, other organizations including the federal Department of Health, provincial ministries of health, the Canadian Diabetes Association, the Public Health Association, and the Standards Council of Canada participate in the Joint Commission as observers. Initially focusing on acute care hospitals, accreditation has been gradually extended to cover special hospitals, long-term facilities, home care programs, public health departments, and ambulatory care services.

Health facility accreditation is a systematic, multidisciplinary inspection of the physical and organizational structure of the facility or program and the functioning of its component parts. Factors measured include staff qualifications, facilities, organization, record keeping, and continuing education of staff.

The process of accreditation requires a request for accreditation from the board of governors of the hospital or health facility, implying acceptance of the standards of the commission. The accreditation process includes a self-assessment, an on-site survey, and follow-up action for correction of deficits and improvements. The commission is invited to conduct a survey, and resurvey as it sees fit. The hospital pays a fee and commits itself to provide all data requested and to cooperate with the site visit. The commission issues a confidential report, giving the accreditation rating and interim statement of deficiencies, and requests progress reports in correcting deficiencies. It is also empowered to carry out follow-up inspections and

resurveys. Box 15.4 lists the areas of a large community or teaching hospital, regional health authorities, hospitals, and community-based programs and services, from both private and public sectors, not only in Canada but around the world.

The assessment survey examines the goals and objectives of the organization and its administration, the direction and staffing of the facility, policies, and procedures. Review includes medical staff organization, credentials and review procedures, clinical privileges, selection of department chairpersons and their responsibilities, standing committees, schedule of meetings, bylaws, and the role of the governing board of the hospital. The presence and nature of quality assurance organization, records review procedures, and continuing educations are assessed. The quality of clinical records is assessed by examination of charts for the completeness of histories and documentation of the course of the hospital stay including laboratory reports.

Each section of the program being accredited is assessed in the following categories:

- statement of purposes, goals, and objectives
- organization and administration
- human and physical resources
- orientation, staff development, and continuing education
- patient care
- quality assurance.

These categories are also used in the programs covered by the contracts between Accreditation Canada, formerly the Canadian Council on Health Services Accreditation (CCHSA), and other health and social service agencies.

Hospital accreditation was established in the UK and Australia in the 1980s and is attracting interest in other countries seeking ways to maintain and promote standards. The procedure for accreditation of hospitals is still voluntary in Canada, but in effect has become universal for hospitals of medium and large size (over 75 beds) and common for smaller hospitals. It is seen as advantageous for the governing board and the community and also for the medical staff in terms of medicolegal protection. In the USA, hospital accreditation has become virtually universal since payment for federally funded health insurance (Medicare and Medicaid) beneficiaries is not allowed for non-accredited hospitals, and many private insurers make this requirement as well. In some states, accreditation is mandatory for all hospitals.

Since the 1990s, CCHSA's accreditation program has expanded to cover a diversity of health care and service areas, through contract arrangements with independent non-hospital facilities such as highly specialized programs as well as community health and social service organizations. In 2006, CCHSA introduced standards for child welfare, hospice, palliative and end-of-life care facilities, prison facilities, biomedical laboratories, and supplementary criteria for telehealth. In

BOX 15.4 Accreditation Canada Standards

- System wide:
  - Governance
  - Infection prevention and control
  - Leadership
  - Leadership for aboriginal health services
  - Leadership for assisted reproductive technology
  - Leadership for primary care
  - Managing medications
- Population based:
  - Cancer populations
  - Child and youth populations
  - Maternal/child populations
  - Mental health populations
  - Populations with chronic conditions
  - Public health services
- Service excellence:
  - Acquired brain injury services
  - Ambulatory care services
  - Ambulatory systemic cancer therapy services
  - Assisted reproductive technology
  - Case management services
  - Child welfare services
  - Community health services
  - Community-based mental health services and support standards
  - Critical care services
  - Developmental disabilities services
  - Diagnostic imaging services
  - Emergency department services
  - Health care staffing services
  - Home care and support services
  - Hospice palliative and end-of-life services

- Independent medical/surgical facilities
- Laboratory and blood services
- Long-term care services
- Medical imaging centers
- Medicine services
- Mental health services
- Obstetrics services
- Operating rooms
- Organ and tissue donation standards for deceased donors
- Organ and tissue transplant
- Organ donation standards for living donors
- Point-of-care testing
- Primary care services
- Rehabilitation services
- Reprocessing and sterilization of reusable medical devices
- Spinal cord injury acute services
- Spinal cord injury rehabilitation services
- Substance abuse and problem gambling services
- Surgical care services
- Telehealth services
- Service distinction:
  - Acute stroke services
  - Audit tool for reprocessing and sterilization of reusable medical devices
  - Inpatient stroke rehabilitation services
  - Providing an integrated system of services to people with stroke

Source: *Accreditation Canada. Available at: http://www.accreditation.ca/ en/content.aspx?pageid=54 [Accessed 14 December 2012].*

2008 CCHSA officially became Accreditation Canada, providing services to other countries. The ever-changing health and social environment now accommodates specialized needs in a diversity of service areas as an adjunct to the hospital accreditation process. Examples are shown in Box 15.5.

Licensing and regulation of health facilities are a government responsibility, but an independent accreditation authority has advantages. The national authority may fail to monitor its own facilities with the diligence or objectivity needed, and there may be a conflict of interest. Where there is a national system of organization, distinct departmentalization of the operating and certification functions may provide a greater measure of objectivity. Assistance from countries experienced in voluntary accreditation can help to establish accreditation mechanisms and provide technical and professional support to countries wishing to establish such programs.

In the current period of transition from central to decentralized management of health services in many countries, health facilities are being transferred from government operation to independent operation as not-for-profit or even for-profit facilities. Present methods of regulation by national or state levels of government will require review as decentralization and privatization take place. Regulation by governmental authorities and non-governmental professional bodies is mutually complementary in promoting accountability, standards, and quality of services.

Peer Review

A large part of the work of clinical and departmental managers in hospitals or other care settings relates to quality assurance. A major method of improving quality in a health program is through peer review by which the staff organizes systematic review of cases and records, using statistics on performance indicators. In hospitals, this includes review of deaths, maternal mortality and infant mortality cases, surgical rates, complications following surgery, and infection rates. Medical records and computer information systems permit users to review records

BOX 15.5 Accreditation Canada International Accreditation Program

The Canadian health services accreditation program began in 1917 in conjunction with the American College of Surgeons (ACS) with a hospital standardization program. The first Minimum Standard for Hospitals developed requirements of just one page. In 1918, on-site inspections of hospitals began, with 89 of 692 hospitals surveyed meeting the requirements of the Minimum Standard. In 1926, the first Standards Manual was issued.

In 1951, the American College of Physicians, the American Hospital Association, the American Medical Association, and the Canadian Medical Association joined with the ACS to create the Joint Commission on Accreditation of Hospitals (JCAH). It is an independent, not-for-profit organization whose purpose is to provide voluntary accreditation. In 1953, the Canadian Hospital Association (now the Canadian Healthcare Association), the Canadian Medical Association, the Royal College of Physicians and Surgeons, and l'Association des Médecins de Langue Française du Canada established the Canadian Commission on Hospital Accreditation. The Commission's purpose was to create a Canadian program for hospital accreditation, and in 1958 the Canadian Council on Hospital Accreditation (CCHSA) was incorporated.

In 2008 CCHSA became Accreditation Canada International. The accreditation program is used by all types of health facilities, from large and complex hospitals, to health systems, community health organizations, and residences providing long-term care. Its scope includes a wide range of programs, including standards on child welfare, hospice palliative and end-of-life care, biomedical laboratory services, blood banks, and supplementary criteria for Telehealth. The accreditation program covers a diversity of health care and service areas, service programs for brain injury, ambulatory care, assisted reproductive technology – clinical and laboratory services, Canadian Forces health services, cancer agencies, child welfare organizations, First Nations and Inuit addictions and community health services, the Federal Department of Veterans' Affairs, substance abuse and problem gambling treatment services.

The accreditation service is on a contract basis with specialized health programs, other federal government departments, for-profit health facilities, and community organizations across the provinces.

Accreditation Canada International works with other countries to develop national accreditation programs for their countries, and launched its first international program for acute care, primary care, ambulatory care, and clinical laboratories in 2010.

Source: Accreditation Canada International. Available at: http://www. internationalaccreditation.ca/Accreditation/AccreditationProgram.aspx [Accessed 12 September 2012].

by diagnosis. These records can be utilized to assess other events in hospitals, such as time from admission to surgery, lengths of stay by diagnosis, response to abnormal laboratory findings, and many other indicators of the process of care. Obstetric departments can review the frequency of and criteria for caesarean section deliveries. Surgical departments review their appendectomy rates to separate pathological findings from normal appendices. Organized peer review has also been called *medical audit* and essentially describes methods of self-policing and education to learn from mistakes and experience and to improve the quality of care.

In 1972, an amendment to the US Social Security Act required hospitals and long-term care facilities to monitor the quality of care given to Medicare and Medicaid patients through professional standards review organizations (PSROs). These were medical audit committees with specified tasks to conduct utilization review, medical care evaluation, and profile analysis of physician or institutional performance compared to accepted standards of the medical community. In 1982, peer review organizations (PROs) were created by federal statutes to replace PSROs. The PROs are non-profit corporations, staffed by physicians and nurses, to review medical necessity, quality, and appropriate level of care under the Medicare and Medicaid programs. The Centers for Medicaid and Medicare Services have an Office of Clinical Standards to conduct surveys, provide certification, and develop best practices guidelines, in a health care quality improvement program (HCQIP).

Hospitals have departmental clinical meetings, adverse incident or outcome committees, mortality rounds, and clinical pathology conferences to help staff to evaluate and learn from difficult cases. The presence of functioning peer review mechanisms indicates that quality is of concern to the professional and administrative network, raising the consumer's confidence in the system.

Maternal mortality committees have been widely used to assess preventable factors in deaths related to maternity and to point out areas of needed improvement in services. Identification of high-risk pregnancies emerged from this process and has become an important part of prenatal care. Infant mortality reviews by professional groups can similarly demonstrate areas of needed improvement in services. Death rounds are held to review cases of death following surgery or soon after admission, or "incidents", such as inappropriate medication given in error.

The successive waves of peer review initiatives in the USA represent attempts by the federal government to establish mandatory quality of care review by professional peers for facilities providing care to Medicare and Medicaid patients. The concept of requiring standards of care review has probably contributed to a greater awareness of the accountability of hospital-based practice. Frequent litigation may have contributed more to the sense that the physician is accountable for services and outcomes of care. PROs are a form of quality regulation that represent a commitment by funding agencies to accountability in care systems and to identification of organizational and administrative weaknesses in health care generally and not only in hospitals. The generation of US physicians and health systems managers trained since the 1970s accepts peer review as an integral part of health services. Other countries use this kind of mechanism to maintain and promote quality of care.

Tracer Conditions

Tracer conditions are common medical conditions (or procedures) for which diagnostic criteria are well established and clear, there are effective preventions or treatments, and a lack of treatment can cause significant harm to the patient. Examples of tracer conditions include otitis media, appendectomy, caesarean section, and hysterectomy. These conditions, if evaluated in terms of incidence and actual chart review, can provide useful insights into departmental medical standards. Incident reports by nursing staff and nosocomial infections are examples of the functioning of the tracer condition concept.

Incident reports in hospitals are designed to determine the causes of errors, so that remedial action can be taken and similar events prevented. Tracer condition studies have become such an accepted part of modern health management that the absence of an organized review system could be considered a serious structural flaw in a health service, requiring remedial action.

Setting Standards

Standards recommended by independent professional organizations or by advisory committees appointed by ministries of health can play important roles in defining standards of care for specified conditions. In addition, organized professional bodies can issue practice guidelines or help governments or health care agencies to develop standards or algorithms for management of specific topics and conditions.

Specifying standards for preventive care, such as for infants and adults, assists local health authorities in planning and evaluating services. The American Academy of Pediatrics (AAP) has an extensive professional committee structure that publishes periodic guidelines for pediatricians on a wide variety of infant and child topics including nutrition, immunization, prevention of anemia and lead toxicity, child safety, and school health. Mandatory preventive care for newborns includes eye care and vitamin K injection in the USA (see Chapter 6). Mandatory immunization requirements for school entry and for health care personnel are discussed in Chapter 4.

The American Public Health Association (APHA) publishes the *Control of Communicable Diseases Manual*, now in its nineteenth edition (2008). It is the authoritative US manual on this topic. The AAP's Red Book on infectious diseases is used across North America by pediatricians in clinical practice. These organizations and their counterparts in obstetrics and many other clinical fields directly relevant to public health continually update practitioners and policy personnel in the "state of the art" or "gold standard", discussed previously. This constitutes a professional self-guidance system in standards. Managed care and other health provider systems also issue guidelines for member practitioners that serve to maintain standards of service.

The wide use of treatment protocols and scoring systems in hospital medicine helps to define standards of care in a measurable way. The Apgar score for rating newborn status has been a standard in hospitals worldwide for decades, helping to standardize infant assessment and care. The APACHE (Acute Physiology And Chronic Health Evaluation) scoring system is used widely to assess the chances of survival of patients admitted to intensive care units and to compare outcomes, for example, between teaching hospitals and community hospitals. It is also used in assessing patient outcomes with different modes of treatment. Scoring systems are also used in community health care, as in risk scoring for pregnancy care (see Chapter 6).

Algorithms and Clinical Guidelines

Algorithms are decision trees or a systematic series of decisions based on the outcomes of previous decisions, tests, or findings. Derived from operations research, this approach applied to medicine identifies all available choices (e.g., exposed versus non-exposed) and follow-up decisions based on findings from each previous option substantiated by observation. It is often presented graphically like the branches of a tree, showing the alternatives and subsequent decisions to be made.

A clinical algorithm is a systematic process defining a sequence of alternative, logical steps depending on outcomes of previous ones, incorporating clinical, laboratory, and epidemiological information, applied to maximize benefits and minimize risks for the patient. It gives the provider a review of the relevant literature and recommended standards of practice on a particular topic for preventive care or case management. These guidelines are usually arrived at by consensus of multidisciplinary working groups taking into account published studies on the topic. The guidelines may suggest that some procedures should not be carried out routinely.

Clinical guidelines are meant to establish accepted standards of care and may have important economic implications. *Medical Letter*, published by the Consumers' Union, is a long-standing and useful publication that reviews

TABLE 15.5 Adult Health Maintenance Checklist by Age Group

Procedure	Age (years)		
	20–39	40–64	65+
Checkup visit	Every 3 years	Every 2 years	Annually
Cholesterol	With checkups	With checkups	With checkups
Fecal occult blood	Age 40–49 if high risk	Annually	Annually
Clinical breast examination	Every 1–3 years	Annually[a]	Annually[a]
Mammography	Baseline age 35	Age 40–49, every 1–2 years	Over 70, every 2 years
Pelvic examination	Every 1–3 years	Every 1–3 years	Every 1–3 years
Pap smear	From age 21-29 every 3 years; from 30-65 every 5 years with HPV DNA test	From age 30–65 every 5 years with HPV DNA test	If previously negative, may stop 3 years
Colonoscopy	No	From age 40 for those with family history of colon cancer or polyps. After age 50, every 3–5 years	After age 50, every 3–5 years
Prostate and PSA Immunizations	No	Annually[a]	Annually[a]
Tetanus–diphtheria	Every 10 years	Every 10 years	Every 10 years
Pneumococcal pneumonia	For high risk	For high risk	Every 6 years
Influenza	For high risk	For high risk	Annually
Skin cancer	Annually[a]	Annually[a]	Annually[a]
Bladder cancer	Annual routine urinalysis	Annual routine urinalysis	Annual routine urinalysis
Lung cancer	Routine examination[b]	Routine examination[b]	Routine examination[b]
Testicular cancer	Routine examination[b]	Routine examination[b]	Routine examination[b]
Oral cancer	Routine examination[b]	Routine examination[b]	Routine examination[b]
Ovarian cancer	Routine examination[b]	Routine examination[b]	Routine examination[b]
Pancreatic cancer	Routine examination[b]	Routine examination[b]	Routine examination[b]
Routine vitamin supplements	Routine[b]	Routine[b]	Routine[b]

Note:
PSA = prostate-specific antigen.
Agency for Healthcare Research and Quality. Rockville, MD: AHRQ. http://www.ahrq.gov [Accessed 13 September 2012].
[a]*Inconclusive*
[b]*negative recommendation. The topics are under continuing review, and recommendations are in some cases left to the opinion of the provider as the current cumulative evidence is not affirmative, e.g., clinical breast examination annually or breast self-examination.*
Sources: US Preventive Services Task Force Ratings: Strength of recommendations and quality of evidence. guide to clinical preventive services. 3rd ed. Periodic updates, 2000–2003. Available at: http://www.uspreventiveservicestaskforce.org/3rduspstf/ratings.htm [Accessed 13 September 2012].

therapeutic issues of everyday medical practice and the relevant studies. It represents a balanced, updated view of medical practice and summaries of current literature, reviewed by respected, experienced, and competent medical authorities. Clinical practice guidelines are produced by hundreds of professional, medical, and governmental agencies in order to standardize and improve medical care.

Clinical and preventive care guidelines are helpful in clinical practice and in preventive medicine. They are increasingly used in managed care environments to assure standards, quality of care, and cost-effectiveness as well as legal protection. Guidelines for preventive medicine and public health practice are also part of the process of promoting the quality of individual and community health, as discussed in Chapter 11. Annual revision of the infant

immunization program, discussed in Chapter 4, is a prime example, as is the set of guidelines for preventive care for adult health maintenance in Table 15.5.

The issue of application of current scientific knowledge for population health is a continuing struggle for recognition of the prime importance of health promotion and preventive care for health of a population. The selection of priorities in use of resources is vital especially in the many developing countries that are in various stages of economic development, or which have abundant income from natural resources such as oil and minerals. Implementation of programs designed to achieve the MDGs can help to serve this purpose.

Public health standards and clinical practice guidelines are an increasing part of quality improvement. It is important, however, that they are developed as best practices and influenced as little as possible by commercial interests of drug or vaccine manufacturers. The proliferation of such guidelines by health authorities or professional associations of the USA, the UK, Canada, Australia, and other countries indicates a wide consensus on the importance of such written standards, guidelines, or "best practice" statements. The recommended childhood immunization program put forward annually by the CDC in conjunction with the AAP and other professional organizations is an example of such best practices and is accepted by health insurers and providers as the gold standard in this field. The concept of promotion of quality in health care and the adoption of current scientific standards are global issues and an integral part of the New Public Health (Box 15.6).

The Canadian Province of Saskatchewan Health Services Utilization and Research Commission publishes periodic reports presenting consensus positions of panels of medical faculty, clinical specialists in pathology and physical medicine, and public health specialists in nutrition, community health, and epidemiology. Its reports are circulated widely and serve to update medical practitioners, reduce unnecessary testing, promote appropriate use of laboratory and other diagnostic procedures, and provide standards of care for individual patients and community services, such as long-term care facilities and home health agencies.

The Canadian Medical Association issued its *Handbook on Clinical Practice Guidelines* in 2007, based on a systematic review of the literature, interviews of key professionals, consensus conferences, and continuing evaluation of both process and content of such guidelines. The Guideline International Network (GIN) Fourth International Conference, held in Toronto in 2007, involved experts in national and international practice guidelines from 31 countries to share experience and concepts in this ongoing field. The GIN library contains more than 6600 (by October 2012) guidelines, evidence reports and related documents, developed or endorsed by GIN member organizations (GIN, 2012).

An Institute for Clinical Evaluation (ICES) organization at the University of Toronto, established in 1992 with core funding provided by Ontario's Ministry of Health and Long Term Care, is mandated to conduct research that contributes to the effectiveness, quality, equity, and efficiency of health care and health services in Ontario. ICES uses an interdisciplinary research approach to health care, health services, and health policy.

The American College of Cardiology (ACC) provides a framework of evidence-based clinical statements and guidelines developed by leaders in the field of cardiovascular medicine with continuing adoption of new scientific information and experience in many aspects of this field (ACC, 2012). Many professional organizations such as the AAP, American Congress of Obstetricians and Gynecologists (ACOG), UK Faculty of Public Health, and European Society of Cardiology produce clinical guidelines which are updated regularly to provide physicians and health systems managers with current consensus on state-of-the-art standards, such as the European Society of Hypertension Guidelines released in 2013 (i.e., less than 140 mm. systolic for all).

The US Health Care Financing Administration (HCFA), Center for Medicare & Medicaid Services (CMS), and National Institutes of Health (NIH) have consensus programs to develop guidelines that are widely disseminated and set standards of practice. In 1977, the NIH issued its first consensus paper on breast screening for cancer, and this has been followed by many other topics each year since. The AHRQ also produces research related to efficacy of current and new practices and training material to promote their diffusion across the US health system. Cochrane reviews and the Cochrane Library provide high standards of literature reviews and meta-analysis on many topics which serve to guide practitioners and policy makers in current standards. The US Healthy People 2020 project provides gold standards for preventive care which serve clinicians, public health practitioners, and health planner standards for their work. Evidence-based consensus guidelines were issued on the following topics: breast cancer screening for women aged 40–49, interventions to prevent HIV risk behavior, management of hepatitis C, genetic testing for cystic fibrosis, acupuncture, and effective medical treatment for heroin addiction.

Clinical guidelines are increasingly being promoted by professional, governmental, and managed care organizations with the purpose of promoting rational use of health care resources and at the same time promoting standards of care to incorporate good standards of clinical practice. Clinical practice guidelines are now common in the practice of primary care, mental health, and clinical specialties. The University of Southern California's list of clinical guidelines website (http://medicine.ucsf.edu/) provides access to hundreds of websites for such practice guidelines.

Clinical guidelines provide practicing doctors, peer review committees, health care managers, managed care companies, governmental bodies, and professional organizations with channels to set standards of practice and expectations of care standards. Legal aspects of health care also increasingly recognize the importance of clinical guidelines where committees of appropriate medical professionals convene and set out average or minimum standards of care for defined clinical entities. Thus, peer-reviewed guidelines set an appropriate standard (a silver if not a gold standard) for judging malpractice or adequate practice. Clinical guidelines should be under periodic review and subject to critical discussion and updating using the Cochrane review methods of literature review and analysis. Promotion by advocacy or special interest groups can be constructive, but the influence of drug companies can be insidious and reduce the professional objectivity of such reviews and their recommendations, a concern that must be carefully monitored and continuously kept in mind as a potential compromising bias.

The AAP produces policy statements, practice parameters, and model bills which have a wide distribution and influence; they are published in the academy's journal, *Pediatrics*. The AAP clinical practice guidelines issued include diagnosis and treatment of urinary tract infection in febrile infants and young children, long-term treatment of the child with simple febrile seizures, management of acute gastroenteritis in young children, management of otitis media with effusion in young children, and others. The policy statements of the AAP cover a wide range of topics including use of bicycle helmets, 55 mile per hour maximum speed

limits, folic acid for the prevention of neural tube defects, and ethics in the care of critically ill infants and children. AAP guidelines are valid for 5 years only and are reissued or reconfirmed in order to keep up to date and to incorporate new or revised knowledge into practice standards.

Empirically derived, peer-reviewed, regularly updated guidelines have become an appropriate standard for practice and for judging malpractice, as well as balancing quality and cost-effectiveness. Clinical guidelines may become restrictive, but they help to reduce practice by whim and unsubstantiated belief to improve the quality of care overall. In large health care organizations they provide a basis for continuing education for staff and advancement of standards of the organization.

The *Community Guide* produced by the CDC provides an excellent source of evidence-based advice for community programs. It serves the needs of public health professionals, health care providers, legislators and policy makers, researchers, community-based organizations, employer–employee groups, and other purchasers of health services. The guide covers a wide range of health issues including alcohol, cancer, diabetes, mental health, motor vehicle safety, nutrition and obesity, oral health, physical activity, pregnancy, sexual behavior, social environment, substance abuse, tobacco, vaccines, violence, and workplace health issues.

In 1999, the UK National Health Service (NHS) established the National Institute for Clinical Excellence (NICE) as an independent organization to provide guidelines for public health, health technologies, and clinical practice guidelines for specific conditions. The Health Development Agency of the NHS was included in the NICE organization in 2005. Now called the National Institute for Health and Care Excellence, NICE publishes guidelines that provide a helpful basis for clinical practice and public health as well as other areas in the NHS to update the services provided. Topics for public health include smoking and tobacco control, diet and obesity, exercise and physical activity, sexual and mental health, and alcohol.

ORGANIZATION OF CARE

Administrative and financing systems are essential elements of quality assurance. They can be designed to promote standards of care and to reduce fiscal incentives that foster excess supply and overservicing. The organization of financing health care has important implications for quality, technology, and ethical issues in the New Public Health.

Diagnosis-Related Groups

DRGs, discussed extensively in Chapter 11, were developed in the 1960s as an alternative way of paying for hospital care in order to encourage shortened lengths of stay. Experience with payment by days of care (per diem) showed that it promoted unnecessary, lengthy, and potentially dangerous use of hospital care, an important factor in the rapid escalation of costs in the health system. DRGs were adopted for payment for Medicare beneficiaries in the USA in 1983 and later became the standard method of payment for all insurance systems.

In the DRG system the insurer pays the provider hospital for a procedure or diagnosis rather than the number of days of stay in hospital. This has led to a large reduction in hospital days of care and a remarkable growth in the number of surgical procedures done on an outpatient basis. Since the introduction of DRGs, outpatient surgical procedures have grown from less than one-fifth to more than half of inpatient surgical cases. Outpatient surgery is safer for the patient and less costly to the insurer. DRGs have gradually been adopted as a case payment system for reimbursing hospitals in most developed countries.

The DRG system is widely considered to promote quality of care as an active process focusing on quickly addressing the diagnosis and management of the patient with rapid mobilization of treatment and return home. Critics of this system allege that DRGs encourage inappropriate early discharge of patients before optimal patient education and follow-up care have been provided, but long length of hospital stay has not been shown to improve patient outcomes. Critics also suggest that this may promote altering diagnoses to higher cost units of service. Others think that DRGs, by reducing length of stay, have turned hospitals into intensive care units with ultra-sick patients. Despite these issues, the trend towards short hospital stays and newer approaches to active treatment seems to be compatible with better care and improved outcomes, according to some measures. The rapid decline in mortality rates from coronary heart disease is thought to be due in large part to the activist treatment approach, with lengths of stay of 1 week or less for acute myocardial infarction compared to 6 weeks on average up to the 1970s.

Managed Care

Managed care systems developed in the USA in response to rapid cost escalation for health care and the successful experience of HMOs. Managed care is based on the concepts of resource management, and quality assurance with rationalized use of technology. The system developed over time with checks and balances to provide comprehensive care at lower cost than traditional fee-for-service systems by discouraging excessive utilization without compromising quality of service. Managed care systems include traditional HMOs and various other organizations which employ physicians or are made up of independent physicians working together who own or contract for hospital services (see Chapter 10).

HMOs, both for-profit and not-for-profit, and managed care itself, have been widely criticized as excessively limiting patient access to appropriate care in the interest of cost containment. The 2010 PPACA (Obamacare) is promoting development of newer innovations including patient-centered medical homes (PCMH), accountable care organizations (ACOs), and population health management systems (PHMSs), and early evidence shows that these models are quality management approaches for integrated primary and hospital patient care (see Chapter 11). Obamacare is a highly politicized and much debated topic in the USA; it seems likely to make a very big difference in coverage and fair practices of insurance with lower costs of private insurance.

District health systems in the UK, the Scandinavian countries, and the post-Soviet model of health care incorporate organizational and financial linkage between care systems and funding from tax sources. HMOs, sick benefit funds, and district health systems provide both prepayment and health services. Even in traditional private health insurance systems, the insurer is increasingly taking on the role of regulating reimbursement for medical services in order to contain costs and curb abuses by providers. In this context, emphasis is placed on maintaining health, preventive care, and financial incentives to efficiency in overall care. Clinical indications, utilization review, and organizational and professional standards are now becoming accepted parts of the health insurance milieu.

The competition between hospitals for referrals from managed care plans in the USA has created a market situation in which a high proportion of hospital beds are empty, and in which mergers or closures of hospitals are common. Closures or reductions in hospital bed supply are also occurring in the UK and in most industrialized countries of Europe.

PERFORMANCE INDICATORS

Performance indicators are measures such as morbidity, mortality, functional status, or immunization rates in a community, used to monitor the functioning of a health service. Routinely collected statistics are analyzed to compare performance against objectives, help monitor efficiency and effectiveness, point out problem areas within the service, and plan new health programs. This method is based on the use of the concept of management-by-objectives in health administration to promote achievement of national health targets.

The UK has a strong tradition of mapping diseases as a basis of epidemiological analysis and has applied this strategy to mapping of performance indicators to assess health care performance. The UK financing system is based on capitation adjusted by standardized mortality rates on the premise that mortality rates standardized and compared to the national average serve as indicators of need. In this way, the approach helps to promote equitable funding among wealthy and poorer regions of the country, and thereby improve services in areas of greater need.

Performance indicators were introduced into the NHS during reforms of the late 1980s, providing a series of outcome or performance measures that are used to adjust payments allocated on a per capita basis to district health authorities. These authorities can be penalized for low rates of immunization, whereas general practitioners receive incentive payments for full immunization coverage. The result was a rapid improvement in immunization coverage of infants and children compared to rates in the previous decade. Incentive payments in many countries encourage women to go to hospitals for delivery or to attend prenatal care by making social maternity grants conditional on seeking care.

Use of performance indicators requires the development of health information systems with district health profiles to provide ongoing monitoring of health indicators in a district, compared to regional and national rates and targets. Health profiles help to establish and monitor the prevalence of chronic disease and measure the impact of health services. This enables the study of the performance of preventive and curative services, such as managing hypertension to reduce the incidence of strokes and related conditions. There are criticisms of performance indicators alleging a potential for manipulation and abuse of health intervention measures when the financial incentives are used for a specific activity. However, financial incentives are part of the DRG system and have been successfully used in the UK to improve vaccination coverage and implementation of other preventive health practices by family physicians. In Israel in 2007, payments to hospitals provided a bonus for surgical interventions for hip fracture within 48 hours of the event, resulting in a marked rise in early intervention and a reduction in mortality from hip fractures.

CONSUMERISM AND QUALITY

With decentralization and the growth of managed care, health systems must increase their attention to the attitudes of the consumer. Quality is, in part, how the client perceives the system, and how the system meets client needs in an acceptable manner, where privacy, dignity, the right to know, and the right to a defined set of services are protected. However, the rights of the client are not unlimited. A public or private health plan has the duty to manage the basket of services responsibly, which includes limitations such as in access to specialist services.

Patients' rights and consumer protection in health care often (but not always) include the right to select and change a health care provider, as well as the right to receive high-quality care for a designated range of services. The UK NHS

> **BOX 15.7 Patients' Rights, European Union, 2009**
>
> A review of patients' rights in countries of the European Union in 2009 focused on the following:
> - Right to informed consent based on access to information for care or participation in research
> - Right to information concerning own health, diagnosis
> - Right to medical records
> - Right to confidentiality of personal and health information and physical privacy during care
> - Right to complain and compensation
> - Right of free choice of provider and of treatment
> - Respect of patient's time
> - Right to observance of quality standards access to high-quality health services
> - Right to safety and freedom from harm caused by the poor functioning of health services, medical malpractice and errors, and the right of access to health services and treatments that meet high safety standards
> - Right of access to innovative procedures, including diagnostic procedures, according to international standards and independently of economic or financial considerations.
>
> **Source:** European Patients' Forum. Patients' rights in the European Union. Available at: http://www.eu-patient.eu/Documents/Projects/Valueplus/Patients_Rights.pdf [Accessed 25 October 2012].

issued a patient's Charter of Rights during the 1990s, which is perhaps idealistic and may not be actualized in practice, but still outlines an ideal of value both for practical application and for legal rights. The consumer's formal protection includes the right to complain and to seek redress of grievance and compensation for injury suffered from neglect or incompetent care (Box 15.7). In North America and Europe, there are at least four models of defining the rights of patients: the paternalistic model, the informative model, the interpretive model, and the deliberative model (WHO, 2012). Many new charters have been established such as data protection, end-of-life care, mental health, access to health services, quality of care and care giving environment, nationally approved treatments, drugs and programs, respect, consent and confidentiality specific to the UK and in Canada, New Zealand and other countries. The new US federal Affordable Care Act of 2010 (PPACA) includes a large element of patient's rights protection, as discussed in Chapter 10.

The patient or consumer of health care needs to be informed and conscious of health care costs if efforts to restrain cost increases are to be effective. Public attitudes are vital in terms of self-care, demands on the health service, and limitations to the potential of health care and resources for health care. The media and consumer organizations can play important roles in advocacy for health, in raising public consciousness of self-care, and as watchdogs on abuses.

Consumer acceptance is manifested through choice of health plan and practitioner, or by seeking alternative care privately when service is unacceptable because of quality or style. Erosion of confidence in a public system of care can lead to a two-tier system with the public system serving the poor and a private parallel system serving the middle and wealthy classes. Such a division can seriously undermine a public system unless it is addressed by improving the quality and manner of the service and by establishing supervision and limitations on public and private practice.

The growing inequality caused by the rise of private practice outside a national health care system is a chronic problem in the UK's NHS, in Israel's health system, and in many countries developing their health systems through parallel public and private care. The issue is also surfacing in the USA in the transition to managed care with its inherent limitations of choice for people insured through their place of work or covered under the Medicare and Medicaid programs.

The PPACA requires insurance companies to accept anyone requesting cover without restrictions due to prior conditions or high expenses for serious conditions, and without other forms of discrimination common in the past. It also includes provisions for coverage of preventive care services and incentives for quality improvement. Extra billing, banned in Canada's national health insurance plan, is a recurring issue with the medical profession in some provinces.

Consumer knowledge, attitudes, beliefs, and practices are part of the health system, from health promotion to tertiary care. Informed and health-conscious consumers are stronger partners in the health system in achieving improved health than an ill-informed and apathetic public, so that health education and health promotion are fundamental to modern public health. The role of the consumer in health care is unique in that there is a significant information asymmetry between the consumer and provider. Health education programs and wide use of the Internet increase access to health and medical information, but this gap can never completely be eliminated. Patients may use their power as consumers to demand inappropriate care, such as unnecessary surgery or antibiotics when clearly not indicated, because of their preference for intervention and action over watchful waiting. However, there is an equal or perhaps greater danger of provider-induced demand for repeated and possibly unnecessary interventions that may be related to methods of paying the doctor or the hospital. The traditional doctor–patient relationship is still an important factor for the interests of patients and their health. A still effective method of having an individual quit smoking is a brief but stern lecture by the family physician.

THE PUBLIC INTEREST

Population-based interventions are often more effective and less costly ways to reduce morbidity and mortality than individual prevention or treatment services. A population-based preventive program may require behavior change by the individual, such as in mandatory seat belt and motorcycle helmet enforcement or banning smoking in public places. Fortification of flour, milk, and salt with essential micronutrients is a well-established public health measure. There is an element of compulsion in this, with the social gain usually considered to be sufficiently important to outweigh individual rights. Immunization is for the protection of individuals but also for the population, so that refusals to immunize children and adults can cause injury to others. Herd immunity is protective of people who are at high risk. Mandatory immunization for school entry in the USA has been effective in increasing coverage to levels akin to the most advanced health systems, over 95 percent coverage. Refusals and failure to harmonize immunization policies in Europe have resulted in mass epidemics of measles, rubella, and mumps in recent years.

There is often a delicate balance between community rights and individual rights which can lie at the heart of many controversies in modern public health and health care, ranging from chlorination or fluoridation of community water supplies to managed care systems for health services. Women's rights, gay rights and abortion are highly controversial and politicized in the USA, and in many other countries. The differences can become extreme and the source of international strife, such as in the movement to promote fundamentalist Sharia law in many countries that are severely restrictive of women's and minority rights.

In public health, issues should be examined on their merits, especially in terms of what is accepted as good public health practice, based on evidence from clinical trials, documented experience, and best practices in other countries. The evidence of successful public health measures in improving individual and collective health status is powerful, yet must always be balanced within the context of individual rights and the public interest. The ethical issues of individual and community rights of public health are discussed later in this chapter.

TOTAL QUALITY MANAGEMENT

Total quality management (TQM), as discussed in Chapter 12, was adapted from business management theory and practice to health care in the 1990s and provides a basis for promoting continuous improvement in health care systems. TQM involves everyone in the system, from all levels of management to production or service personnel and support staff, and thus helps to raise staff morale because of the shared involvement. Health is provided through multidisciplinary groups which need to approach problems with open and shared scientific inquiry and hypothesis formation, testing, and revision to find operational solutions to problems.

Electronic health records and information technology provide many new opportunities to improve patient care and data systems for monitoring the health status of population groups for process and outcome measures, or health targets, such as immunization coverage, or screening compliance for colon, cervical, or breast cancer, as measures of performance in primary care. Information technology adds a great deal of capacity for quality monitoring and improvement measures.

TQM incorporates statistical methods, comparing variations in patterns of service or use of resources. It employs epidemiological methods to draw conclusions for policy needs. It looks for continuous improvement, encouraging cooperation, and motivation to achieve common goals of service and client satisfaction. Psychological theory helps to foster higher levels of motivation, with early identification and resolution of conflict. Leadership is shared, and there is a basic need for cooperation. Cost and quality are interrelated, as poor quality leads to waste, inefficiency, and dissatisfaction of both clients and staff. High-quality, humane, and effective services are especially important in a competitive environment where clients have the right to choose and where costs and efficiency are factors in the well-being and indeed the survival of institutions.

Medical care is increasingly practiced in larger health care organizations. To provide technically competent medicine is not by itself sufficient. The patient's rights and sense of personal worth are also of great importance. Financial incentives can be effective in redirecting health care priorities, such as in reducing hospital length of stay and admissions, but may result in the patient or the family feeling that they are not receiving the best care. DRGs, HMOs, and other organizational and funding systems meant to increase efficiency of care may have the effect of alienating patients from a health care system. Staff attitudes towards patients are important for client satisfaction. The service must include ready access to a continuum of supportive services, such as home care and counseling, so that the patient and family do not feel abandoned by the system.

A byproduct of TQM is continuous quality improvement (CQI), by which institutions wishing to improve quality train and empower the staff to work in teams to assess their own performance and seek solutions to problems in their operational unit. People of different ranks and professions work in a network organization as well as in a traditional hierarchical organization in which rank and seniority provide authority. This community of practice is important for staff morale and a shared sense of responsibility for the patient and the institution.

CQI involves multidisciplinary approaches, not only to review problems but also to seek better ways of functioning and improving consumer satisfaction. The process includes all those involved in providing care, support services, and administration of a department, hospital, clinic, or community health program. This is not only professional self-policing but a method to find better ways of meeting needs and using resources. The involvement of all providers improves motivation and promotes a sense of common purpose in the organization.

Applying these principles in a health care setting can take many forms. Selection of topics by TQM/CQI committees in a hospital or another health facility may be based on surveys or interviews with staff, patients, or management. Satisfaction surveys among women following delivery in an obstetrics unit could point out remediable problems. An obstetrics department may be faced with issues related to high or low volume of deliveries, staff training, equipment and supplies, communication among staff, and among staff and patients and their families, cleanliness, sterile technique, staff satisfaction, client satisfaction, and many others. The team looking at such a problem should be multidisciplinary, and the emphasis should be on client attitudes and satisfaction.

Examination of the function of an emergency department in a hospital would similarly look at many functional and attitudinal aspects of the service including staff attitudes, training needs, waiting times, consultation services, and others. Addressing waiting times, for example, can lead to ways to reduce these substantially, improving both client satisfaction and the efficient management of the emergency department. Any service is there to serve patients and the community. A service is not primarily for the benefit of the staff, but staff satisfaction and morale are essential for successful service to clientele. CQI can also be applied to assessing and improving compliance with clinical guidelines or evidence. An example is assessing the proportion of diabetics whose hemoglobin A_{1c} (Hb_{A1c}) is measured at least twice annually, who have eye and feet examinations regularly, or whose blood pressure is managed with an angiotensin-converting enzyme (ACE) inhibitor.

The European Region of the WHO and the national medical associations in Europe agreed in 1995 that medical associations should take leading roles in programs of CQI to achieve better outcomes of health care in terms of functional ability, patient well-being, consumer satisfaction, and cost-effectiveness. This is in keeping with the European Region's *Health for All* targets: there should be structures and processes in all member states to ensure continuous improvement in the quality of care and appropriate development and use of health technologies.

The introduction in the 1990s of general practitioner fundholding for hospital care for patients on the general practitioners' roster in the UK encouraged the hospital to maximize patient satisfaction with the care system. This promotes application of CQI to improving the quality and acceptability of care. Similarly, performance indicators provide regional and district health authorities in the UK with tools for CQI approaches. The UK NHS established

BOX 15.8 The UK's National Institute for Health and Care Excellence (NICE)

The National Institute for Clinical Excellence (NICE), established in 1999, has a mandate to review health service treatments and effective therapies that should be commissioned and made available within the National Health Service (NHS) throughout England and Wales. The mission statement for NICE is that it "contributes to better health around the world through the more effective and equitable use of resources". In 2005 NICE was revised to include reviews of public health interventions, and its mandate was expanded to include quality standards for the English social care sector (English Health and Social Care Act of 2011). Now renamed the National Institute for Health and Care Excellence, NICE operates as a statutory independent special health authority in England and Wales. Commissioning bodies of the NHS are required to observe its recommendations. Guidance can be used by the NHS, local authorities, employers, voluntary groups, and anyone else involved in delivering care or promoting well-being.

NICE recommendations are respected elsewhere in the UK, but are not mandatory; in Scotland NICE recommendations are published after further review by NHS Quality Improvement Scotland (for health services issues) and by NHS Health Scotland (for public health recommendations). NICE recommendations are respected worldwide, including by the European Commission and by national governments; NICE International is a section of NICE established to meet non-UK needs (e.g., evaluating rural health programs in China).

An independent committee including lay representation advises on priorities for NICE consideration but final decisions on topics referred to NICE are made by the Department of Health. When making recommendations to the NHS on which services (e.g., treatments) should be provided routinely, it calculates the cost-effectiveness of treatment for each quality-adjusted life year (QALY) of health gain purchased.

NICE publications include guidance on 374 interventional procedures, 270 technology appraisals, 162 clinical guidelines, and 43 public health topics. From this latter group, some examples include:

- *Prevention of cardiovascular diseases (2010)* – provides evidence of effectiveness of population-based prevention programs as more effective than programs aimed at high-risk groups.
- *Alcohol dependence and harmful alcohol use (2011)* – summarizes all NICE guidance; designed to inform members of the public as well as health professionals.

- *Preventing type 2 diabetes through population and community interventions (2011)* – provides guidance to government departments, the commercial sector, health service organizations, and non-governmental organizations on integration of public policy to prevent obesity, and reduce diabetes prevalence and complications.
- *Preventing uptake of smoking by children and young people (2008)* – document to advise local health service commissioners; identifies target populations, reviews campaign messages, and provides recommendations for the mass media and retailers.
- *Promoting mental well-being at work (2009)* – guidance aimed at employers; reviews evidence in the field and recommends strategic approaches by firms, opportunities to promote well-being and assess risk, and systems of flexible working.
- *Preventing unintentional injuries among under-15s in the home (2010)* – reviews evidence and makes recommendations to local authorities and related agencies on training an appropriate workforce, advises government to fund curricula development, and indicates to the NHS appropriate surveillance and treatment services.

NICE is often criticized (especially by the pharmaceutical industry) for the time taken to carry out investigations of new treatments. It is also criticized by relatives of patients with "glamorous" conditions (e.g., cancer) for not approving drugs that might extend life by only 4–6 weeks, and perhaps approving instead new psychiatric therapies. The current government has recently sought to overrule some of these NICE recommendations in England. However, NICE methods and recommendations are held in high repute, within the UK and beyond.

Sources: *Christopher Birt FRCP FFPH, University of Liverpool, UK. Personal communication.*
National Institute for Health and Care Excellence. 2012. Available at: http://guidance.nice.org.uk [Accessed 24 December 2012].
O'Flaherty M, Flores-Mateo G, Nnoaham K, Lloyd-Williams F, Rayner M, Capewell S. Estimating potential cardiovascular mortality reductions with different food policy options in the UK. Bull World Health Organ 2012;90: 522–31.
National Institute for Health and Clinical Excellence. Promoting mental wellbeing through productive and healthy working conditions: guidance for employers. NICE; 2009. Available at: http://www.nice.org.uk or http://www.apho.org.uk/resource/item.aspx?RID=83868 [Accessed 18 August 2013].
Campbell B. Regulation and safe adoption of new medical devices and procedures. Br Med Bull 2013;1–14 [Epub ahead of print]. http://dx.doi.org/10.1093/bmb/ldt022.

NICE as an independent body to promote "national guidance on promoting good health and preventing and treating ill-health". NICE produces guidance in three areas:

- *public health* – guidance for those working in the NHS, local authorities and the wider public and voluntary sector on promotion of good health and the prevention of disease
- *health technologies* – guidance on use of new and existing medicines, treatments and procedures within the NHS

- *clinical practice* – guidance on appropriate treatment and care within the NHS of people with specific diseases and conditions.

NICE guidelines are recommended practices with the objective of reducing ineffective practices. During 2007, guidelines were issued on topics including asthma, dermatitis, caesarean section, chronic obstructive lung disease, depression (in children and adults), eating disorders, fertility, contraception, multiple sclerosis, post-traumatic stress disorder, and diabetic foot care (Box 15.8).

BOX 15.9 Organizations to Promote Quality in Health, USA

- *National Committee for Quality Assurance (NCQA)* – This non-profit organization, founded in 1979 by the managed care industry, conducts surveys among managed care plans to evaluate clinical standards, members' rights, and health service performance. It accredits over 550 managed care plans in the USA, and in 2007 published rankings of the "best" health plans. Website: http://www.ncqa.org/

- *Agency for Healthcare Research and Quality (AHRQ)* – This is part of the US Public Health Service. Founded in 1995, it was mandated to develop an evidence-based practice program in 12 centers in the USA. It conducts systematic reviews of the literature and publishes analyses and findings of these reviews. Website: http://www.ahrq.gov/

- *Centers for Medicare & Medicaid Services (CMS) 2001* – The CMS, previously the Health Care Financing Administration (HCFA, 1977), is the federal agency of the Department of Health and Human Services, responsible for administering the Medicare and Medicaid and the State Children's Health Insurance Program (CHIP) health plans. Its roles include quality assurance, the requirements for managed care organizations, and quality improvement. Website: http://www.cms.gov/

- *Institute for Healthcare Improvement (IHI)* – Non profit organization founded in 1991 as a global resource for health care improvement knowledge to improve health care by fostering collaboration among health care organizations. IHI examines office practices of physicians, educational reform, and promotes interdisciplinary team work in quality improvement. Website: http://www.ihi.org/ihi/

- *National Patient Safety Foundation (NPSF)* – Sponsored by the American Medical Association as a response to findings of high rates of injury and death from iatrogenic disease in the USA, the NPSF promotes research into human error among health care providers, seeking ways to reduce the frequency and effects of medical error, such as misdiagnosis, medication errors, and mistakes during procedures. Website: http://www.npsf.org/au/

- *Joint Commission on Accreditation of Healthcare Organizations (JCAHO)* – Originating in 1917 by the American College of Surgeons, it began accrediting hospitals in 1918. It developed in 1953 as the JCAHO, becoming a national voluntary accreditation organization focusing mainly on hospitals. Its mandate was broadened in 1987 and, as of 2007, had accredited more than 15,000 health care organizations. Accreditation is mandatory for Medicare and Medicaid payment. The JCAHO is changing its approach from standards-based assessment every 3 years to one of reviewing performance data quarterly as a continuous surveillance activity for risk reduction. Website: http://www.jointcommission.org/

Source: *Websites accessed 12 September 2012.*

The USA has a number of government and independent organizations dedicated to improving quality in health care systems. The CDC and the Institute of Medicine of the US National Academies of Science play active roles in promoting research quality and methods of CQI in the US health care system. Canada is also very active in this regard, having national and provincial institutes for the evaluation of clinical effectiveness and clinical guidelines, and so too are European countries (Box 15.9).

PUBLIC HEALTH LAW

Public health workers need knowledge of government structure and public health legislation as basic to their professional work to understand their responsibilities, powers and liabilities. Law consists of a system of rules, regulations, and orders that govern the behavior of individuals and of society. Law represents the consensus of a society, as enacted by an elected legislature, put into effect by the executive branch of government, and interpreted by the courts as need be from time to time. The legislative and executive branches are separate under the US Constitution, but the two are united in the parliamentary system (Box 15.10). The authority, responsibility, and power to provide for and protect the public health are basic functions of a sovereign government, which may be delegated to another level of government (higher or lower) or even a non-governmental agency. The constitution of a sovereign government states explicitly or implicitly that responsibility, but accepted practice and court decisions (i.e., the common law) define the powers of the national, state, or local government to monitor and protect the health of its citizens.

In the USA, national legislation is enacted under the powers of the federal government, namely to regulate interstate commerce and the power to tax and spend for the general welfare. State legislation is enacted under the basic power of the state to protect the health, welfare, and safety of its citizens. Under these federal and state powers, a wide range of health legislation and regulations is enacted affecting public health, labor, and occupational health and safety, environmental controls, public welfare, and the financing of health services, agriculture, food, drugs, cosmetics, and medical devices. Public health law relies on a wide range of constitutional, statutory, administrative, and judicial decisions in both civil and criminal actions. Appropriation of funds is a legal act of legislative bodies to achieve objectives directly or indirectly by financial incentives.

Categorical programs may be directed to specific issues such as combating TB and promoting immunization or for work to combat NCDs such as diabetes, or in improving standards of facilities, and in providing health care services. The regulatory, enforcement, policing, and punitive functions of public health laws have evolved over many decades and in many countries lack clear definition. In the USA,

BOX 15.10 Legal Structure of Federal and Unitary Countries

In federal nations, political authority is divided between two autonomous sets of governments, one national and the other subnational. Both operate directly with the people in their jurisdiction based on a constitutional division of power between the national government, which exercises authority over the whole national territory, and state or provincial governments with independent authority within their own territories. The constitution is the supreme law of a country. It sets out the divisions of governmental powers including statutory authority, administrative, natural resources, and taxation between federal and state levels of government.

A federal legislature or congress makes the law of the land, but is subject to rulings of a Supreme Court as are state and local governments. State or provincial governments in a federal system have functions set out in the Constitution. They also have elected legislatures, and executive branches with taxing, regulatory, and punitive powers. Local governments for county, municipal, or city governments also have delegated taxing and regulatory powers including those of public health.

Canada, the USA, Brazil, Australia, India, and Argentina are organized on a federal basis. Federal countries also include Austria, Germany, Malaysia, Mexico, Nigeria, Switzerland, and Venezuela. Russia is called a federation. Usually there is some overlapping or shared powers between national and state constitutions, legislatures, and court systems, and public agencies, taxing powers and regulatory functions, such as in interstate commerce and emergency response to natural or other disasters.

In a unitary government system, most or all of the governing power resides in a centralized government. This contrasts with a federal system. In unitary systems the central government commonly delegates authority to subnational units and channels policy decisions down to them for implementation. A majority of nation-states are unitary systems. They vary greatly. The UK includes England, Scotland, Wales, and Northern Ireland, each with legislatures, but the Westminster Parliament in London maintains national powers. In health, each of the four member entities of the UK conducts a National Health Service with autonomy but common features. The national government may delegate certain powers to self-governing regions/local authorities, and there is a growing tendency to devolve various governmental functions such as health to regional authorities. More than 150 countries are unitary states, including France, Italy, Spain, China, and Japan.

In both forms of government, local authorities are established under state law with governance by councils elected by the people, with taxing and regulatory powers within the state or provincial laws, with a high degree of autonomy but within state regulation, standards, and financial support. Local authorities have major responsibilities in public health such as in sanitation, licensing, and regulation of businesses and zoning, as well as many other areas, including social welfare.

Note: See also Chapter 10.
Source: *Differences between federal and unitary forms of government. Available at: http://www.preservearticles.com/201107139054/difference-between-unitary-and-federal-forms-of-government.html* [Accessed 15 December 2012].
Encyclopedia Britannica. Unitary government. Available at: http://www.britannica.com/EBchecked/topic/615371/unitary-system [Accessed 15 December 2012].
Encyclopedia Britannica. Political systems. Available at: http://www.britannica.com/EBchecked/topic/467746/political-system/36704/Federal-systems [Accessed 15 December 2102].

efforts are being made to update and reform laws in the public health sector. In 1988, the Institute of Medicine (IOM) in the USA (the Future of Public Health) called for codification of public health law as essential for the public good, while questioning the soundness of certain US public health laws. More recently, the Model State Emergency Health Powers Act in the USA, the Quarantine Act in Canada, and the revised International Health Regulations (2007) have sought to update century-old legislation. The revised international regulations provide for a global approach to control the spread of epidemics and public health emergencies while minimizing disruption to international activities such as travel, trade, and economics.

A combination of the regulatory, persuasive, and funding approaches is widely used in public health in control of communicable and non-communicable diseases, in improving standards of facilities, and in providing health services. The regulatory, enforcement, policing, and punitive functions of public health are important in health promotion and assurance of health care. The taxing power of government is essential for public health to ensure that adequate facilities and access to care are available to all members of the community, especially those in financial need and thus at greater risk for disease.

Medical officers of health and their staff have legal authority to issue formal orders for health protection of the public. Situations which require court proceedings are referred to the justice system. Situations that may require enforcement by court proceedings are referred to the justice system. Laws may be enacted to fund public health activities, whether provided by public health authorities or by acting through official or non-official agencies or providers. Public health authorities, namely medical officers of health, have the legislative power to issue orders to individuals or businesses where there is a threat to the health of the public such as food establishments. Administrative resources are needed to enforce laws, such as through the FDA and the Environmental Protection Agency, which come under the aegis of the Department of Health and Human Services. Other departments such as Agriculture, Education, or interdepartmental agencies (e.g., Homeland Security), also are key to public health activities, such as in disaster situations.

Other intergovernmental activities may require special legislation to empower, finance, and promote their cooperation, such as in the case of establishing an authority to manage long-term efforts to clean up a contaminated river or basin, which involves the cooperation and coordination of many local authorities.

Health protection of individuals and communities may require legal action to detain a person in order to prevent the spread of a reportable communicable disease, to protect a mentally ill patient, or to restrain a violent person. Such powers should be used as a last resort if voluntary compliance and education fail, and where the danger to the community or the individual is sufficient to convince a court of the public need to override the personal liberty of an individual. An example is a 2007 case of a person with MDR-TB who was taken into custody on arrival for compulsory treatment after traveling across the Atlantic Ocean on a commercial airline, against the specific instructions of his physician, thus endangering fellow passengers. Outbreaks of measles in the UK (2006–2007) and in Israel via imported cases among ultraorthodox Jews or conservative protestant groups in the Netherlands, with transmission among religious people who tend not to immunize their children, led to pressure by health authorities to immunize those placed at risk by such contacts at weddings or other large public events.

However, these measures are currently used less than voluntary isolation or quarantine and placarding homes for reportable infectious diseases such as measles. Powers are essential in extreme cases where refusal to comply with public health measures endangers others. Such powers should have been used more vigorously in the early years of the AIDS epidemic at a time when individual rights took precedence over protection of the population, including vulnerable high-risk groups. The severe acute respiratory syndrome (SARS) epidemic of 2003 led to sequestering hospital staff in Toronto, Canada, for lengthy periods to prevent spread of the disease, and subsequent influenza pandemic threats have raised questions as to whether hospital personnel should be required to be immunized to protect patients and their families from onward transmission of dangerous infections.

Recent cases in the USA, the UK, and Norway demonstrate the responsibility of governments to protect the public from incidents of violence by dangerous, mentally disturbed individuals who carry out mass killings. In Norway, 69 people, mostly teenagers, were killed by a radical ideologue while many others sustained serious injuries; and in the USA, Islamic terrorists at the Boston Marathon killed three and seriously injured more than 200 others; a 20-year old fatally shot his mother then killed 20 children and six adult staff members at Sandy Hook elementary school in Newtown, Connecticut, before killing himself; and an army psychiatrist who had become an increasingly devout and radicalized Muslim psychiatrist shot and killed 13 people and injured more than 30 others in a Texan army base. Background checks and other restrictions on gun sales are an important public health and political issue, especially in the USA. The wide availability of guns, including military-style assault weapons, presents a serious danger for impulsive or planned mass killings.

Public health has generally evolved with greater reliance on health promotion through voluntary cooperation of a patient or community than on compulsion. Enabling legislation may permit a local authority to fluoridate its water supply, but the enactment of local legislation and funding to implement it may also require a public referendum. In some states in the USA and in Israel, fluoridation of community water supplies is mandatory, which is also part of the health promotion approach to public health.

Appropriation of public funds to promote public health is through approval by the legislature for a specified program. Provision of public funds may take the form of categorical grants for specified services, such as immunization, prenatal care, school health, or specific disease management such as TB control, cancer control, or AIDS education. Programs may be designed to promote certain types and quality of services, such as the Hill–Burton Act, which provided federal grants for hospital construction in the 1950s to 1970s, conditioning these grants on certain requirements concerning hospital licensure and hospital planning. Such legislation has a "carrot and stick" effect of attracting lower levels of government to seek such funding but also requiring them to accept the conditions and regulations that accompany the grants. The Canadian federal government's cost sharing of provincial health (hospital and medical) insurance programs is based on federal criteria requiring public administration, portability between provinces, accessibility without payment, comprehensiveness, and banning extra billing by physicians (see Chapter 13).

Public funds are also appropriated in the context of legislated programs in which people are entitled to the services defined in the appropriation legislation, such as in the amendments to the Social Security Act providing Medicare and Medicaid programs, or national health insurance legislation in many countries. These and their regulations spell out categories and specified entitlement benefits.

Legislation and court decisions to protect the rights of the individual are part of public health. Public health law is meant to protect individuals and communities from potential abuse, of both individual and community human rights, as in the US Bill of Rights. Enforcement of public health law may infringe on individual rights by enforcing sanitation, food and drug safety, and supervision of restaurants and catering firms. Laws may allow restriction of civil rights, such as rarely used mandatory treatment of a person with a dangerous contagious disease or mental illness. Freedom of religion may come into conflict with other laws in public health where restrictive practices may deny the use of publicly supported health facilities, as when a religiously affiliated hospital may refuse an abortion procedure in a case of rape. Religious practices or other personal beliefs may endanger others in the community, such as in the refusal to immunize children so that an imported infectious disease may spread among non-immunized people and even affect those who are immunized, as occurs with imported measles cases even when domestic transmission of the disease has previously been eradicated. General legislative provisions applied to public health forbid misleading or unethical advertising. Legislative provisions may also ban advertising for products, such as tobacco, which are legal but may be harmful to health. These laws affect public health but are provisions in other statutes such as the regulation of business enterprises. Legislation may also make smoking in public places illegal, with fines for offenders and operators of places such as public bars.

Since the 1973 US Supreme Court decision of *Roe v. Wade*, the law has allowed women to seek safe and legal abortion. This remains a highly controversial political issue in the USA and several other countries. The potential conflict between community and individual interests and rights is part of the dynamics of public health law and public health practice. The issues involved are complex and highly politicized, and often involve ethical distinctions where "the greatest good for the greatest number" may limit the legitimate rights of individuals and vice versa. The PPACA in 2010 is a fundamental legislative initiative, upheld by the US Supreme Court to become the law of the land. It will bring millions of Americans into regulated health insurance with many protective elements to prevent abuse by private insurance company through arbitrary exclusions or limitations.

The legal aspects of public health are vital to its operation and are increasingly complicated by ethical issues, and by public and political debate. Health protective legislation and regulation for sanitation of food, water, and air are fundamental to public health, as is the control of drugs, cosmetics, vaccines, and biologicals, the manufacture of devices, and the licensing of health personnel and facilities. Limitations of legal suits (torts) against manufacturers of vaccines proved to be a successful measure in the USA with the introduction of the National Vaccine Injury Compensation Program (NVICP) in 1988. This is funded by a modest surcharge tax collected from vaccine manufacturers. It protects both public and private interests while providing a fair compensation system to ensure patients' rights but without jeopardizing immunization to prevent widespread disease, and also protects manufacturers from litigation with high legal costs and excessive compensation awards by the jury system. Promoting healthy behavior through the prudent use of the legal system of regulation and taxation is increasingly utilized to protect the health of the population. This is widely applied in promoting road safety, in tobacco control measures regarding banning of advertising, high taxes on alcohol, and banning smoking in public places including restaurants and bars.

Environmental Health

There is growing concern by the public and by governments over climate change, global warming, air and water pollution, and other noxious and harmful industrial and commercial processes. Environmental laws affecting the public health include legislation on clean air, clean water, toxic substances, solid waste control, and other noxious substances. Non-compliance with the legislative provisions can result in prosecution in the civil or criminal courts or both.

Infringement of public health laws and regulations may lead to criminal action as an increasingly common method of sanction. While such violations may not be seen as "truly" criminal and may be treated in the courts as misdemeanors, they can lead to fines or even jail. Such cases are increasingly being addressed seriously in the judicial system.

The CDC, in 1999, defined 10 great achievements of public health of the twentieth century. These achievements are identified as control of infectious disease, motor vehicle safety, fluoridation of drinking water, recognition of tobacco use as a health hazard, immunization, decline in deaths from coronary heart disease and stroke, safer and healthier foods, healthier mothers and babies, family planning, and safer workplaces (Goodman et al., 2006). Of the 10 great achievements in the twenty-first century (2001–2010) identified by CDC, seven of the 15 leading causes of death (largely NCDs) resulted in a decline in the age-adjusted death rate in the USA from 881.9 per 100,000 population in 1999 to 741.0 in 2009. This decline was a result of a combination of supportive laws and legal tools at the local, state, and federal levels. In other industrialized countries similar legislation has led to equal or greater achievements in public health over the past century.

Public Health Law Reform

Public health law is scattered through many legislative statutes and administrative documents which developed historically. Efforts to codify public health law may contribute to greater understanding and enforceability of the many separate pieces of legislation (Box 15.10). Such reform will enhance understanding in the legislative, judicial, and administrative branches of government as well as in business, non-governmental organizations, and the community. Box 15.11 suggests topics for model public health consolidation or compendia for states. The principles of this formulation may also apply to other countries at the national and state or provincial levels.

ETHICAL ISSUES IN PUBLIC HEALTH

The field of public health includes a wide range of activities and professional disciplines, ranging from health promotion to disease protection, epidemiology to environmental health, and financing to supervision or provision of clinical care. Each of these disciplines works within systems that face ethical dilemmas, and public health workers' understanding and motivation within the ethical guidelines of their professions and roles are important in their training and practice conduct. Ethical frameworks have evolved in part as the result of bitter experience with ethical failures which were later recognized and affect public health standards of practice for future generations (Box 15.12).

Ethics in health are based on the fundamental religious and humanistic values and concepts of a society. If the principle of saving a life is valued above all other considerations (i.e., Sanctity of Life or *Pikuah Nefesh*) (see Chapter 1), then all measures available are to be used, irrespective of the condition of the patient or the cost. If sickness and death are seen as acts of God, possibly as punishment for sin, then prevention and treatment may be considered to be interfering with the divine will, and the ethical obligation may be limited to relief of suffering. Humanism balances these two ethical imperatives: saving of life and relief of suffering. Materialistic political philosophies may view health care as primarily a function to preserve health for economic prosperity and social well-being. Secular humanism adopted many of the religious precepts of the worth and rights of the individual and these have become part of the standards of law and ethics in modern secular societies.

The role of society in protecting the health of the population grew during the nineteenth century with the sanitation movement, while medical care became an effective part of public health during the twentieth century. The astonishing successes of public health during the past century increased life expectancy in the high-income countries by some 30 years, mostly through improved living conditions and health protection, as well as societal and medical advances to make care available to all. In the 1970s the Lalonde concept that individual behavior was one of the key determinants of health (see Chapter 2) placed much of the onus of illness and its prevention on the individual, but fostered health promotion as an essential component of public health theory and practice. All these points of view are involved in the ethical issues of the New Public Health (Box 15.13).

Resources for health care are limited even in industrialized countries, so that priority setting and judicious allocation of scarce resources are always issues. Money spent on new technology with only marginal medical advantages is often at the expense of well-tried and proven lower cost techniques to prevent or treat disease. The potential benefits gained by the patient from more and more interventions are sometimes very limited in terms of length or quality of life. These are difficult issues when the physician's commitment to do all to preserve the life of the patient conflicts with the patient's concept of quality of life and his or her right to decline or terminate heroic measures of intervention. Many health systems use clinical guidelines that are mandatory for a health facility or a doctor in the clinic. Preparation for surgery requires a signature from the patient to consent to the procedure being carried out, careful preoperative procedures to ensure that the correct organ is addressed, antiseptic preparation of the site, and checking that all instruments are accounted

---

BOX 15.13 Study and Practice of Public Health Ethics

Ethics is a branch of philosophy that deals with distinctions between right and wrong, with the moral consequences of human actions. The ethical principles that arise in epidemiological practice and research include:

- informed consent
- confidentiality
- respect for human rights
- scientific integrity.

"As a field of study, public health ethics seeks to understand and clarify principles and values which guide public health actions. Principles and values provide a framework for decision making and a means of justifying decisions. Because public health actions are often undertaken by governments and are directed at the population level, the principles and values which guide public health can differ from those which guide actions in biology and clinical medicine (bioethics and medical ethics) which are more patient or individual-centered.

As a field of practice, public health ethics is the application of relevant principles and values to public health decision making. Public health ethics inquiry carries out three core functions:

(1) *identifying and clarifying the ethical dilemma posed,*

(2) *analyzing it in terms of alternative courses of action and their consequences, and*

(3) *resolving the dilemma by deciding which course of action best incorporates and balances the guiding principles and values." (CDC, 2001)*

Sources: *Last JM, editor. A dictionary of epidemiology. 4th ed. New York: Oxford University Press; 2001.*
Centers for Disease Control and Prevention. Science coordination and innovation. Public health ethics; 2001. Available at: http://www.cdc.gov/od/science/phec/ [Accessed 23 September 2012].

---

for. The checklist approach is well established for care in many settings and protects the patient from neglect or faulty follow-up, such as in the management of hypertension and diabetes.

The suffering that a terminally ill patient may endure during radical treatment, which may prolong life by only hours or days, clashes with the physician's ethical obligation to do no harm to the patient. The ethical value of sustaining the life of a terminally ill patient suffering extensively is an increasing medical dilemma. The issue is even more complex when economic values are included in the equation. There are potential conflicts among the economic issues, the role of the physician in preserving life, the physician's obligation to do no harm, the felt needs of the patient and his or her family, and the needs of the community as a whole. The complex issues involved in the "right to die" and end-of-life care raise many ethical and legal questions for the patient, the family, society, and caregivers.

The state represents organized society and has, among its responsibilities, a duty to promote healthful conditions and to provide access to health care and public health services. The conflict between individual rights and community needs is a continuous issue in public health. Application of accepted public health measures for the benefit of some people in society may require applying an intervention to everyone in a community or a nation. The majority thus are subject to a public health activity to protect a minority, without designating which individual's life may be saved. Furthermore, a society may in special cases need to restrict individual liberties to achieve the goal of reducing disease or injury in the population. Raising taxes on alcohol and tobacco products, mandatory speed limits, driving regulations, and seat belt usage laws are examples of public health interventions that interfere with individual liberty but protect individuals, and thereby the community at large, from potential harm.

Many public health measures originally criticized as interventions in private rights are generally accepted as essential for health protection and promotion to reduce the risk of disease in the population. Chlorination of community water supplies is a well-established, effective, and safe intervention to protect the public health. Fluoridation of drinking water to prevent tooth decay in children means that other people are also drinking the same fluoridated water, which is of less direct benefit to them. Fortification of foods with vitamins and minerals is also a cost-effective community health measure with advocates and opponents. The addition of folic acid to food as the most effective way to prevent neural tube defects in newborns is an intervention mandated by the US FDA since 1998.

Confidentiality to assure the right of the individual to privacy involves ethical issues in the use of health information systems. Birth, death, reportable conditions (not all reportable diseases are infectious), and hospitalization data are basic tools of epidemiology and health management. The use of detailed individual data is needed for case-finding and follow-up activities which are vital to good epidemiological management of diseases, including STIs. However, caution is needed in data use to avoid individual identification that could be used punitively, for example, in denial of access to health insurance for smokers, alcoholics, or AIDS patients because health damage may be attributable to a self-inflicted risk factor. Increasingly, however, reporting is also mandatory for physical or sexual abuse and criminally linked injuries as essential for the protection of individuals at risk or the general public from serious harm.

Individual and Community Rights

The protection of the individual's rights to privacy, and freedom from arbitrary and harmful medical treatments, procedures, or experiments, may come up against the rights of the community to protect itself against harmful health issues. This conflict comes into much of what is done in public health practice, which has both an enforcement basis in law and practice and a humanitarian and protective aspect based on education, persuasion, and incentives. Society permits its governments to act for the common good, but sets limits that are protected by the courts and administrative appeal mechanisms.

Society has the right to legislate the side of the road on which one is permitted to drive, the speed permitted, the wearing of seat belts, and the non-use of alcohol or drugs before driving or cell phones while driving. Offenders may be punished by significant fines or jail and are subject to strong educational efforts to persuade them to comply. Similarly, the community must ensure sanitary conditions and prevent hazards or nuisances from bothering neighbors or the public. Society must act to protect the environment against unlawful contamination or poisoning of food, drugs, the atmosphere, the water supply, or the ground.

Enforcement is thus a legitimate and necessary activity of the public health network to protect the community from harm and danger to health. Table 15.6 shows topics where individual rights and responsibilities predominate, and a second set of rights that are the prerogative of the community to protect its citizens against public health hazards. Sometimes the issues overlap and sometimes come to political, advocacy, or legal action, so that court decisions are needed to adjudicate precedents for the future.

The AIDS epidemic in the 1980s and 1990s raised a host of public health, ethical, and issues. Management of the AIDS epidemic is in some respects in conflict with the long-established role of society in contacting and quarantining people suffering from transmissible diseases. It is not acceptable or feasible in modern society to isolate HIV carriers. But failure or delay of public health authorities even in the late 1980s to close public bathhouses in New York and other cities in the USA, where exposure to multiple same-sex partners promoted transmission of the infection, could be interpreted as negligence. During the 1980s, the gay community in the USA centered its concern that HIV testing would be used in a discriminatory manner. AIDS was initially addressed as a civil liberties issue and not as a public health problem. Screening, reporting, and case contact follow-up were seen as an invasion of privacy and proved counterproductive by increasing resistance to and avoidance of testing. Protection of privacy and an educational approach were adopted as most feasible and acceptable. International opinion and national court decisions have emphasized the right to privacy with decriminalization of non disclosure of HIV status to sex partners (UNAIDS 2013).

The AIDS epidemic and public anxiety about contracting AIDS through casual contact reinforced the need for public education on safe sex. This has been raised as an ethical issue because such education may be construed as condoning teenage and extramarital relations. The issue of

HIV screening of pregnant women in general or in high-risk groups took on a new significance with the findings that treatment of the pregnant woman reduces the risk of HIV infection of the newborn, and that breastfeeding may be contraindicated. This issue is arising anew in the context of using the HPV vaccine for preteen girls to prevent the sexually transmitted infection, which is also controversial, and in the USA this vaccination will be mandatory for school entry.

A pre-eminent ethical issue in public health is that of assuring universal access to services, and/or the provision of services according to need. An important ethical, political, and social issue in the USA in the twenty-first century is how to achieve universal access to health care. The solidarity principle of socially shared responsibility for funding universal access to health care is based on equitable prepayment for health care for all by nationally regulated mechanisms through place of work or general revenues of government. A society may see universal access to health care as a positive value, and at the same time utilize incentives to promote the use of services of benefit to the individual, such as hospital care, immunization, and screening programs. Some services may be arbitrarily excluded from health insurance, such as dental care, although this is to the detriment of children and a financial hardship for many. Strategies for program inclusion are often based on historical precedent rather than cost-effectiveness or evidence. While efforts are being made to include more children in the program, the Medicaid system in the USA defines eligibility at income levels of 185 percent of the poverty line, thus excluding a high percentage of the working poor. Health is also a political issue in countries with universal health systems where funding may be inadequate or patient dissatisfaction common.

Choices in health policy are often between one "good" and another. Limitations in resources may make this issue even more difficult in the future, with aging populations, increasing population prevalence of physical disabilities, and rapid increases in technology and its associated costs. For example, the UK's NHS at one point refused to provide dialysis to people over the age of 65. When computed tomography was first introduced, Medicare in the USA refused to insure this service as an untested medical technique. Owing to a lack of facility resources such as incubators and poor prospects for the survivors, the Soviet health system considered newborns as living only if they weighed over 1000 g and survived for more than 7 days. Those under 1000 g, who would be considered living by other international definitions, would be placed in a freezer to die. At the opposite extreme, many western medical centers use extreme and costly measures to prolong life in terminally ill patients, preserving life temporarily but often with much suffering for the person and at great expense to the public system of financing health care.

In many countries, such as those in the former Soviet system of health care, spending for hospital services, in

TABLE 15.6 Individual and Community Rights and Responsibility in Health: Ethical/Legal Issues

Ethical/Legal Issues	Individual Rights and Responsibilities	Community Rights and Responsibilities
Sanctity of human life	Right to health care; responsibility for self-care and risk reduction	Responsible for providing feasible basket of services, equitable access for all
Individual vs community rights	Immunization for individual protection	Immunization for herd immunity and community protection; education; community may mandate immunization
Right to health care	All are entitled to needed emergency, preventive, and curative care	Community right to care regardless of location, age, gender, ethnicity, medical condition, and economic status
Personal responsibility	Individual responsible for health behavior, diet, exercise, and non-smoking	Community education to health-promoting lifestyles; avoid "blame the victim"
Corporate responsibility	Management accountability to criminal and civil action	Producer, purveyor of health hazard accountable for individual and community damage
Provider responsibility	Professional, ethical care and communication with patient	Access to well-organized health care, accredited to accepted standards
Personal safety	Protection from individual, family, and community violence	Public safety, law enforcement, protection of women, children, and elderly; safety from terrorism
Freedom of choice	Choice of health provider; limitations of gatekeeper functions; control costs while function; right to second opinion; right of appeal	Confidentiality; informed consent; birth control ensuring individual rights; limitations of self-referrals to specialist
Euthanasia	Individual's right to die; limitations by societal, ethical, and legal standards	Assure individual and community interests; prevention of abuse by family or others with conflict of interests
Confidentiality	Individual's right to privacy, limitation of information	Mandatory reporting of specified diseases; data for epidemiological analysis
Informed consent	Right to know, risks vs benefits; agree or disagree to treatment or participation i n experiment	Helsinki Committee approval of research; regulate fair practice in right to know; Patient's Bill of Rights
Birth control	Right to information and access to birth control and fertility treatment; woman's rights over her body	Political, religious promotion of fertility; alternatives to abortion; protection of women's rights to choose
Access to health care	Universal access, prepayment; individual contribution through workplace or taxes	Solidarity principle and adequate funding; right to cost containment, limitations on service benefits
Regulation and incentives to promote preventive care	Social security for hospital delivery, attendance for prenatal care; primary care, ambulatory care; home care	Incentive grants to assist communities for programs of national interest; limit institutional facilities
Global health	Human rights and aspirations; economic development, health, education, and jobs	Transfer of health risks; occupational hazards and environmental damage
Rights of minorities	Equality in universal access	Special support for high-needs groups
Prisoners' health	Human rights	Security and human rights; reduce inequalities in sentencing convicts, harsh dangerous conditions in prisons; prohibition of torture and execution
Allocation of resources	Lobbying, advocacy for equity and innovation	Equitable distribution of resources; targeting high-risk groups; cost containment

some cases grossly in excess of need, is accompanied by a lack of adequate funds for primary care or adding new vaccines to the immunization program for children. The majority of Americans have health insurance which increasingly includes preventive care services, but a substantial percentage lack such coverage which limits their access to routine preventive care. The Affordable Care Act brings an improvement in coverage and inclusion of preventive care with incentives (see Chapters 10 and 13). In many countries, including in Europe, delay in updating immunization programs may be due to a lack of funding or to delays in professional or governmental acceptance of "new" vaccines.

The closure or amalgamation of hospitals involves difficult decisions and is a source of friction between central health authorities, the medical professions, and local communities. Health reforms in many industrialized countries, such as reducing hospital bed supplies and managed care systems promoting cost containment and reallocation of resources, raise ethical and political issues often based on vested interests such as private insurance systems, hospitals, and private medical practitioners.

Where there is a high level of cumulative evidence from the professional literature and from public health practice in "leading countries" with a strong scientific base and case for action on a public health issue, when does it become bad practice or even unethical public health practice to ignore and fail to implement such an intervention? Such ethical failures occur frequently and widely. For example, is it "unethical" not to fortify grain products with folic acid, and salt with iodine? Should there be a recommended European immunization program; should milk be fortified with vitamin D; should vitamin and mineral supplements be given to women and children; should all newborns be given intramuscular vitamin K routinely? Other examples include the issues of fluoridation of water supplies and opposition to genetically modified crops or generic drugs in African countries. These issues are continuously debated and the responsibility of the trained public health professional is to review the international literature on a topic and formulate a position based on the cumulative weight of evidence. It is not possible to wait for indisputable evidence because in epidemiology and public health this rarely occurs. This is another reason for guidelines established by respected agencies and professional bodies, which are free from financial obligations to vested interest groups, being essential for review of the evidence which continues to accumulate on many issues thought to have been resolved or which reappear repeatedly despite strong evidence of effectiveness and public health benefit.

Tragic Deviations in Public Health Ethics

In the nineteenth century the germ and miasma theories both produced enormous gains in public health. The biomedical paradigm addressed alleviation of disease risk or manifest disease; the health paradigm addressed the improvement of social and environmental conditions for reducing disease. During the early part of the twentieth century, a segment of the social hygiene movement promoted ideas of Social Darwinism or racial improvement by sterilization of mentally ill, retarded, and other "undesirable" people.

The dominant biomedical model of public health and medical professionals adopted policies of eugenics in Sweden, the USA, and Canada, leading to policies and programs to force the sterilization of mentally handicapped or mentally ill patients. This distorted a socially oriented concept of public health. This euthenasia policy was adapted to a racially oriented policy with horrendous policies of mass murder in the name of racial purity as a public health policy in Nazi Germany with the near-total support and participation of a highly Nazified medical profession, and used in murder, by gassing or planned starvation, of half a million "undesirables" under the eugenics "T-4" program administered from Hitler's headquarters. Although this program was stopped after parental and Church protests in Germany, the methods used were adopted in newly occupied countries and for concentration camps organized for the mass extermination of Jews, Gypsies, and others in the Holocaust.

The eminent historian Sir Richard Evans (Regius Professor of History at Cambridge University), in his classic *The Third Reich at War*, wrote:

"At the heart of German history in the war years lies the mass murder of millions of Jews in what the Nazis called 'the final solution to the Jewish question in Europe'. This book provides a full narrative of the development and implementation of this policy of genocide, while also setting it in the broader context of Nazi racial policies toward the Slavs, and toward Gypsies, homosexuals, petty criminals and 'asocials'. … For many years, and not merely since 1933, the medical profession, particularly in the field of psychiatry, had been convinced that it was legitimate to identify a minority of handicapped as 'a life unworthy of life', and that it was necessary to remove them from the chain of heredity if all the many measures to improve the German race under the Third Reich were not to be frustrated. Virtually the entire medical profession has been actively involved in the sterilization programme, and from here it was but a short step in the minds of man to involuntary euthanasia."

The twentieth century was replete with mass murders, executions, and genocide, with nationalistic, ideological, and racist motives perpetrated by fascist, Stalinist, and radical xenophobic political or religious movements when gaining governmental power by election or by revolution, in some cases applying common public health terminology and concepts to uses of genocide and ethnic cleansing (Box 15.14).

An outline of genocides of the past 100 years is seen in Box 15.15. These include the Turkish genocide of the Armenians in 1917 followed by horrific genocides in which many millions of people were killed, carried out under the communist regime of the Soviet USSR in the 1920s and subsequently, in the People's Republic of China under Chairman Mao in the 1950s, and by the Khmer Rouge in Cambodia in the 1980s, and in the wars resulting from the breakup of the Yugoslav Republic in the 1990s.

The human and national cost of genocide lasts for generations. The hatred and fear may wane but the trauma goes deep. It lasts with the victims and their descendants, but

BOX 15.14 Values and Ethical Principles of Public Health

- *Sanctity of human life.*
- *Individual human rights* – liberty, privacy, protection from harm.
- *Solidarity* – sharing the burden of promoting and maintaining health.
- *Beneficence* – reduce harm and burdens of disease and suffering.
- *Non-malfeasance* – do no harm.
- *Proportionality* – restriction on civil liberties must be legal, legitimate, necessary, and use the least restrictive means available.
- *Reciprocity principle* – public responsibility to those who face disproportionate health and social burden.
- *Transparency principle* – honest and truthfulness in the manner and context in which decisions are made must be clear and accountable.
- *Precautionary principle* – decision makers have a general duty to take preventive action to avoid harm even before scientific certainty has been established.
- *Failure to act* – public health officials and policy makers have a duty to act and implement preventive health measures demonstrated to be effective, safe, and beneficial to population health. Failure to enforce public health regulations with resulting disease or deaths may constitute negligence on the part of responsible officials with civil or criminal penalties.
- *Equity* – reduce inequities.
- *Cost and benefits*
- *Stewardship* – responsibility of governance in a trustworthy and ethical manner.
- *Trust* between the many stakeholders in health.
- *Reasonableness* – decisions should be evidence based and revised based on new evidence.
- *Responsive to needs* and challenges as they may be anticipated and appear with close monitoring of health status.

Source: *Modified from Lee LM. Guest editorial: Public health ethics theory: review and path to convergence. Public Health Rev 2012;34(1). Available at www.publichealthreviews.eu [Accessed 17 December 2012].*

German medical community, one even being elected to head the World Medical Association, then discussing the Helsinki Declaration of Ethics in Biomedical Research, before being forced to resign. The Nuremberg Trials and the subsequent Helsinki Declaration laid the fundamentals of biomedical ethics for the following generations, regulated by requirements of ethical procedures and institutional research board approvals for funding, conducting, and publishing research involving human subjects (Table 15.7).

The United Nations Convention on Prevention and Punishment of the Crime of Genocide (UNGC) of 1948 defines acts committed with intent to destroy, in whole or in part, members of a national, ethnical, racial, or religious group as crimes against humanity. This convention specifies that incitement to genocide is itself a crime against humanity. Legal action should focus on state-sanctioned incitement as a recognized early warning sign. The UNGC defines genocidal acts to include the following as punishable under international law:

- genocide
- conspiracy to commit genocide
- direct and public incitement to commit genocide
- attempt to commit genocide
- complicity in genocide.

The reappearance of genocide in the late twentieth century in the Balkans and Rwanda, and in the twenty-first century by Sudanese in Darfur, highlights genocide as a public health concern and its prevention as a public health and international political responsibility. Incitement to genocide is a crime against humanity and was the basis for the trials and convictions of leaders of the Rwandan Tutsi tribe, as well as inciters to ethnic violence and the political leaders and perpetrators of mass murders in the former Yugoslav Republic. The threat and practice of genocide are still present, whether in the murderous raids of Sudanese Janjaweed militias in Darfur and South Sudan, the threats of genocide by Iran and associated terrorist organizations against Israel and Jews in general, or the killing of Christians in northern Nigeria and Egypt, of Muslims in Burma/Myanmar, and others. Incitement to genocide is now common as part of international discourse.

Genocide represents the most extreme assault on the right to life and respect for life. In the twentieth century, an estimated 200 million people perished through genocide. Totalitarian dictatorships, past wars, and ideologies of exclusiveness, ethnic purity, and religious fundamentalism increase the risks for genocide. Perpetrators use dehumanizing, demonizing, and delegitimizing hate language to desensitize or intimidate bystanders and to mobilize, order, and instruct followers.

Genocide prevention requires international surveillance networks for monitoring and reporting incitement and hate language in the media, textbooks, places of worship, and the Internet, which should monitor and identify their sources

also with the perpetrating country and its culture. The Nazi Holocaust has had downstream effects in public health in the German-speaking countries which last to the present time, seven decades since the events took place. The long-term damage done to public health in Germany and Austria is described in Box 15.16.

The Nuremberg Doctors' Trial in 1946–47 convicted many leading Nazi physicians of crimes against humanity and resulted in severe punishments including hanging or long prison terms. This trial was a seminal event in establishing the ethical standards required for medical research and human rights. However, many in the medical profession aligned with these horrors remained leading figures in the

BOX 15.15 Eugenics and Genocide: The Slippery Slope

Eugenics was a movement within the "social hygiene" concept of the early part of the twentieth century. It was widely promoted to reduce births among mentally ill and handicapped people in some states in the USA and was upheld in decisions of the Supreme Court. It was also practiced in Canada and Sweden. This idea was promoted by Hitler in *Mein Kampf* and adopted by the Nazi Party, which was legally elected to office in 1933 and began to implement it. Organized massacres of mentally ill and handicapped children and adults led to practices of organizing various modes of killing, including gas chambers, which were applied in concentration camps and in the Holocaust murder of 6 million Jews and millions of others.

Genocide represents the most extreme assault right to life and respect for life. In the twentieth century, an estimated 200 million have perished from genocide. Totalitarian dictatorships, past war and defeat, ideologies of exclusiveness, ethnic purity, and religious fundamentalism increase risks for genocide. Perpetrators use dehumanizing, demonizing, and delegitimizing hate language to desensitize or intimidate bystanders and to mobilize, order, and instruct followers.

- 1915–1917 Armenian genocide by Ottoman Turkish Empire – 1.2 million killed
- 1920s–1940s Eugenics movement in USA and Sweden
- 1920s Mass executions, deportations, and starvation as policy in Soviet Union Stalinist regimes
- 1930s–1940s Mass sterilization of "defectives" in the USA and Sweden
- 1930–1940s Mass murder of "defectives" in Nazi Germany – 750,000 killed
- 1940s Quarantining as pretext for ghettos by Nazis
- 1940s Concentration camps, human experimentation
- 1940s Holocaust of 6 million Jews and genocide in Nazi occupation of Poland and in Soviet Union
- 1947 Nuremberg Trials – convictions and capital punishment for war crimes and genocide by Nazi leaders and doctors
- 1950s Mass starvation in Maoist China – estimated deaths of 21 million people
- 1948 Convention on the Prevention and Punishment of the Crime of Genocide
- 1975–1979 Cambodian genocide – 1.7 million killed
- 1988 Iraqi genocide of Kurds in town of Halabja by poison gas
- 1988 Brazil genocide conviction of Tikuna people
- 1995 Serbian massacres in Srebrenica in Bosnia and Herzegovina
- 2004 Rwandan genocides
- 2003–2012 Sudanese genocide in Darfur – over 400,000 killed
- 2011 Sudanese genocide of Nuba people
- 2012 Iran incitement to genocide of Israel
- 2012 Syria: civil war and genocide
- 2012 Democratic Republic of Congo massacres of Kivu reported

Sources: *Richter ED, Genocide Prevention Center, Braun School Public Health, Hebrew University –Hadassah, Jerusalem, Israel. Personal communication.*
United Nations. Convention on the prevention and punishment of the crime of genocide. Available at: http://www.hrweb.org/legal/genocide.html and www.un.org/millennium/law/iv-1.htm [Accessed 16 December 2012].
Richter ED. Commentary. Genocide: can we predict, prevent, and protect? J Public Health Policy 2008;29:265–74.
Stanton G. The eight stages of genocide; 1998. Available at: www.genocide-watch.org/aboutgenocide/8stagesofgenocide.html
Genocide Watch. http://www.genocidewatch.org/ [Accessed 25 December 2012].

and map their distribution and spread. Dehumanization, demonization, delegitimization, disinformation, and denial are the danger signs of potential genocidal actions. Genocide results from human choice and bystander indifference. One lesson of the Holocaust is that silence in response to incitement to genocide makes one a complicit bystander. Public health professionals and institutions have a responsibility to speak out publicly on such dangerous early warning signs (Richter E, personal communication, 2012).

Human Experimentation

Human experimentation has been a subject of great concern since the Nazi and Imperial Japanese armed forces' experiments on prisoners and concentration camp victims during World War II. The Nuremberg Trials set forth standards of professional responsibility to comply with internationally accepted medical behavior (Table 15.7).

The Helsinki Declaration was first adopted by the World Medical Assembly in 1964, and amended in 1975, 1983, 1989, and 1996. It delineates standards of medical experimentation and requires informed consent from subjects of medical research. These standards have become an international norm for experiments, with national, state, and hospital Helsinki committees regulating research proposals within their jurisdiction. Funding agencies require standard approval by the appropriate Helsinki committee before considering any proposal, with informed consent on any research project.

The Tuskegee experiment (Box 15.17) was a grave and tragic violation of medical ethics, but in the context of the 1930s was consistent with widespread and institutionalized racism. It provides an important case study which has repercussions until the present time in suspicion of public health endeavors, particularly among the African American community in the USA.

BOX 15.16 The Rise, Fall, and Slow Recovery of German Public Health

In the German context the social–ecological health paradigm can be traced back to the late eighteenth and early nineteenth centuries when the country was a loose alliance of kingdoms or princedoms lagging behind the economic, cultural, and political developments in England and France. Ensuring population health was seen as the obligation of the state, while the family was responsible for caring for the health and well-being of its members. Organized health care and health maintenance was seen in the framework of *Medizinische Polizey*, as a model of the health systems. Leading scholars in law and medicine shared a normative perspective of promoting a healthy lifestyle (known as *dietetics*), and provision of shelter, food, and spiritual aid in asylums for the sick and disabled, or in private homes the for wealthy.

In the second half of the nineteenth century evidence from medical statistics and overwhelming practical experience indicated that widespread poverty was the critical factor explaining high rates of typhus or cholera epidemics in lower social classes among children and industrial workers. A social health movement fought for healthier living and working conditions, education, and democracy. The movement's prominent leaders were Salomon Neumann, a physician pioneer in medical statistics, and Rudolf Virchow, the renowned pathologist and outspoken political activist.

Between 1890 and 1930 the conceptual framework of public health was defined as "social hygiene" or "health science", an interdisciplinary field to conduct scientific research, academic teaching, and community-based activities aiming at the promotion of individual and collective health and the prevention of disease. In the 1920s the field was highly developed and pioneering the modern academic public health.

Social hygiene was a general framework open to different definitions. A group of academic teachers and publishing scientists sharing the social–ecological paradigm, among them a high proportion of German Jews, wanted to continue the social reform strategy and to strengthen local communities to take an active role in the formulation and implementation of health policies.

Public health activists sharing the biotechnological disease paradigm favored a more focused approach aiming at the control of disease through medical care. Although there was no supportive evidence, in the late nineteenth century a racial eugenic movement emerged widely in Europe and the USA. A conceptual model derived from the disease paradigm postulated racial factors to explain disease. A healthy population was assumed to be "free" of "racially contaminated" individuals and inferior groups. Health-related public policy was supposed to eliminate racially "unclean" members, e.g., by forced sterilization or murder. This was a central theme in Hitler's *Mein Kampf* and was enacted as basic policy by the Nazi Party in Germany as a fundamental ideological basis of racial theory and public health.

When the Nazis were legally elected in Germany in 1933, and later seized power in Austria, this policy provided fertile ground to open the door to euthanasia, leading to mass murder. This was implemented in the well-organized, medically directed execution of mentally and physically handicapped Germans and others in psychiatric facilities. This provided a working model for the industrialized murder of 6 million Jews in the Holocaust and millions of gypsies, homosexuals, communists, and others.

It took only 10 years to eradicate a 200-year tradition of German socially oriented public health grounded largely in the political philosophy of human rights and social justice. Most of those advocates were exiled or murdered. Many of the academic medical leaders after World War II remained in key positions in the German public sector for decades.

In contrast to many other countries, the two wealthy German-speaking countries, with over 90 million people, have few academic public health resources. In there is only one German School of Public Health, and a small number of institutes, far fewer in Austria than in Germany. More than half a century has passed since the Nazi period and the populations of these two countries are slow to build a new socially oriented public health system.

Sources: *Horst Noack MD, PhD, Professor Emeritus, Medical University of Graz, Austria. Personal communication; 24 December 2012.*
Flügel A. Public Health und Geschichte. Weinheim: Beltz Juventa; 2012.
Heinzelmann W. Sozialhygiene als Gesundheitswissenschaft. Bielefeld: Transcript Verlag; 2009.
Noack H. Governance and capacity building in German and Austrian public health since the 1950s. Public Health Rev 2011;33:264–76.

Ethics in Public Health Research

The border between practice and research is not always easy to define in public health, which has as one of its major tasks the surveillance of population health. This surveillance is mostly anonymous but relies on individually identifiable data needed for reportable and infectious disease control as well as for causes of death, birth defects, mass screening programs, and other special disease registries. It may also be necessary to monitor the effects of chronic disease, for example, to ascertain repeat hospitalizations of patients with congestive heart failure to assess the long-term effects of treatment, and the effects of strengthening ambulatory and outreach services to sustain chronic patients at a safe and functional level in their own homes.

Hospitalizations, immunizations, and preventive care practices (e.g., Pap smears, mammography, and colonoscopies) are all part of the New Public Health. Impact assessment of preventive programs may require special surveys and are important to assess smoking and nutritional status and other measures of health status and risk factors. Every effort must be made to preserve the anonymity and privacy of the individual but in some cases, where the disease is contagious, case contact is crucial. This can entail identifying people who attended an

TABLE 15.7 Ethical Issues of Medical Research Derived from the Nuremberg Trials, the Universal Declaration of Human Rights, and the Declaration of Helsinki

Nuremberg Doctors Trial, 1946–47	The voluntary consent of a human subject is absolutely essential, with the exercise of free power of choice without force, fraud, deceit, duress, or coercion
	Experiments should be such as to bear fruitful results, based on prior experimentation and the natural history of the problem under study. They should avoid unnecessary physical and mental suffering
	The degree of risk should not exceed the humanitarian importance of the experiment
	Persons conducting experiments are responsible for adequate preparations and resources for even the remote possibility of death or injury resulting from the experiment
	The human subject should be able to end his participation at any time
	The scientist in charge is responsible to terminate the experiment if continuation is likely to result in injury, disability, or death
Universal Declaration of Human Rights, 1948	Everyone has the right to a standard of living adequate for the health and well-being of himself and of his family, including food, clothing, housing, and medical care and necessary social services
United Nations covenants for protection of human rights	Covenant on Civil and Political Rights
	Optional Protocol to the Covenant on Civil and Political Rights
	Covenant on Economic, Social, and Cultural Rights
	Convention Against Torture
	Convention Against Genocide
	The Geneva Conventions
	Convention on the Rights of the Child
	Convention on Elimination of Discrimination Against Women
	Charter of the United Nations
Declaration of Helsinki, 1964	Research must be in keeping with accepted scientific principles, and should be approved by specially appointed independent committees
	Biomedical research should be carried out by scientifically qualified persons, only on topics where potential benefits outweigh the risks, with careful assessment of risks, where the privacy and integrity of the individual is protected, and where the hazards are predictable. Publication must preserve the accuracy of research findings
	Each human subject in an experiment should be adequately informed of the aims, methods, anticipated benefits, and hazards of the study. Informed consent should be obtained, and a statement of compliance with this code
	Clinical research should allow the doctor to use new diagnostic or therapeutic measures if they offer benefit as compared to current methods
	In any study, the patient and the control group should be assured of the best available methods. Refusal to participate should never interfere with the doctor–patient relationship. The well-being of the subject takes precedence over the interests of science or society

Source: Summarized from the Nuremberg Trials (1948) and World Medical Association, Declaration of Helsinki.
Website sources include: World Medical Association. Available at: http://www.wma.net/
Australian Government Department of Health and Ageing. Available at: http://www.nhmrc.gov.au/health-ethics/human-research-ethics-committees-hrecs/human-research-ethics-committees-hrecs/national
United Nations. A Summary of United Nations Agreements on Human Rights. Available at: http://www.hrweb.org/legal/undocs.html (accessed 10.1.14).
United Nations. Available at: http://www.un.org/en/events/humanrightsday/2007/hrphotos/declaration%20_eng.pdf (accessed 10.1.14).
US Food and Drug Administration. World Medical Association Declaration of Helsinki. Available at: http://www.fda.gov/ohrms/dockets/dockets/06d0331/06D-0331-EC20-Attach-1.pdf (accessed 10.1.14).

event or traveled on an airplane where an infected person may have been, so as to take appropriate preventive measures.

The general distinction between research and practice has to do with the intent of the activity. Clinical research uses experimental methods to establish the efficacy and safety of new interventions or unproved interventions; many drugs and procedures in common use have never been subjected to randomized controlled trials. In practice, many methods are devised that are held to be effective and safe by expert opinion and documented as such. Researchers comparing HIV or hepatitis B transmission rates among intravenous drug users not using needle-exchange programs would be conducting unethical research, according to accepted current standards, by giving needles to the experimental group

BOX 15.17 The Tuskegee Experiment

The Tuskegee experiment was carried out by the US Public Health Service between 1932 and 1972. It was meant to follow the natural course of syphilis in 399 already infected African American men in Alabama and 201 uninfected men. The men were not told that they were being used as research subjects. The experiment had been intended to show the need for additional services for those infected with syphilis. However, when penicillin became available, the researchers did not inform or offer the men treatment, even those who were eligible when drafted into the army in 1942. The experiment was stopped in 1972 as "ethically unjustified" when the media exposed it to public scrutiny.

The case is considered unethical research practice because, even at the time it was conducted, it did not provide the patients with available care and their well-being was put aside in the interest of the descriptive study. A similar experiment was conducted by the US Public Health Service in cooperation with the Guatamala Ministry of Health during the 1960s, in which syphilis was actually given to soldiers, prisoners, and others by sexual contact with prostitutes known to have the disease, but the study was terminated when it was discovered by a public health historian and reached public attention in the USA.

In 1997, President Bill Clinton apologized to the survivors and families of the men involved in the experiment on behalf of the US government. The Tuskegee experiment is the source of lingering widespread suspicion in the African American community to the present time.

Sources: Lombardo PA, Dorr GM. Eugenics, medical education and public health: another perspective on the Tuskegee syphilis experiment. Bull Hist Med 2006;80:291–316.
Centers for Disease Control and Prevention. US Public Health Service Syphilis Study at Tuskegee. Available at: http://www.cdc.gov/tuskegee/timeline.htm [Accessed 13 December 2012].

and withholding them from the control group. The scientific justification of an experiment must be made explicit and justifiable. Clinical equivalence is a necessary condition of all clinical and public health research and provision of standard of care treatment to control groups is a minimal requirement for most research ethics boards. Determination of the standard, and whether it should be place, time, and community specific, is an area of ongoing controversy.

In 1996 a US Public Health Service study, supported by the NIH and WHO, compared a short course of zidovudine (AZT) to a placebo given late in pregnancy to HIV-positive women in Thailand, measuring the rate of HIV infection among the newborns. The experiment was terminated when a protest editorial appeared in a prominent medical journal. This study confirmed previous findings that AZT given during late pregnancy and labor reduced maternal–fetal HIV transmission by half. When a study shows clearly positive results, it should be discontinued and reported so that the findings can be applied generally. The findings indicated that AZT should be used in developing countries, and the

manufacturers agreed to make it available at reduced costs. The result has been a major success in helping with more recent medications to reduce maternal–fetal transmission in many places in Africa with help from GAVI, and a slowing of the spread of HIV/AIDS-related deaths.

Public health may face the challenge of pandemic influenza, such as avian flu, with decisions regarding the allocation of vaccines, treatment of massive numbers of patients arriving at hospitals in acute respiratory distress with very limited resources available, coping with sick or absent staff, and many other issues requiring not only individual life and death situations, but mortality en masse. The ethical questions will be replaced by struggles to cope with such situations. Preparation for such potential catastrophic events will be a challenge to public health organizations and the health system in general.

An outstanding case of a breach of ethics in public health research occurred with the "Wakefield effect", as described in Chapter 4 and Box 15.18.

Ethics in Patient Care

Ethical issues between the individual patient and health care provider are important in the New Public Health. A doctor is expected to use diligence, care, knowledge, skill, discretion, and caution in keeping with practice standards accepted at the time by responsible medical opinion and to maintain the basic medical imperative to do no harm to the patient. Patients have the right to know their condition, available alternatives for treatment, and the risks and benefits involved. They also have a right to seek alternative medical opinions, but this right is not unlimited, as any insurance plan or health service may place restrictions on payment for further opinions and consultation without the agreement of a primary care provider.

Health care has a responsibility beyond that of the payment of health service bills and individual care by a physician, in institutions, or through services in the community or the home. The contract for service is becoming less between an individual physician and his or her patient, and more among a health system, its staff, and the client. This places a new onus on the physician to ensure that patients receive the care they require. Conversely, the US provider often faces the dilemma of knowing that a patient may not access needed services because of a lack of adequate health insurance.

Sanctity of Life Versus Euthanasia

The imperative to save a life is an important ethical and practical issue in health care. Advocates of physician-assisted suicide (euthanasia) argue for the right of the patient to die with dignity when the illness is terminal and the individual is suffering excessively. This is not a medical decision alone, and is an agonizing issue for society to address. The Nazi euthanasia program and its human experiments provided

BOX 15.18 The Wakefield Effect

In 1998, *The Lancet*, published an article by a number of well-known researchers headed by Dr Andrew Wakefield. The article reported on 12 cases of autistic children and alleged to show a connection to immunization with the MMR (measles–mumps–rubella) vaccine.

The immediate effect of this "revelation" was widespread alarm over the MMR vaccine and a fall off in immunization coverage by measles-containing vaccines in the UK and elsewhere with many mothers refusing to have their child vaccinated due to a "risk of autism". As a result, measles epidemics occurred in the UK and in many other countries, with measles again becoming endemic in many parts of Europe, especially England and France.

After a long series of investigative journalism in the British press, the article came under scientific scrutiny and withdrawal of many of the coauthors but a consistent insistence by the lead author of its authenticity.

Investigation by British medical authorities later found Dr Wakefield guilty of medical negligence and the UK General Medical Council withdrew his license to practice medicine. The coauthors were found to have been credulous and insufficiently vigilant in agreeing to coauthorship of the paper. In 2000, 12 years after the original publication, *The Lancet* formally withdrew the article.

The effect of this fraudulent scientific publication was a serious loss of credibility of immunization in general and especially regarding the MMR vaccine, one of the greatest life savers in public health technology.

The return of measles in Europe to large scale epidemics with frequent international transmission furthered the loss of confidence of mothers in immunizations and public health. Measles-containing vaccines were particularly strongly affected owing to the publicity given to the Wakefield case. The journal editors could be seen as irresponsible for failing to ensure the scientific integrity of lead authors and coauthors, and the journal for failing to retract a fraudulent article sooner than 12 years after the first publication.

In other public health issues, single publications of findings of small sample and poorly assessed studies published in haste without adequate inquisitive review occur with great frequency. The electronic media often include unscientific opinion blogs which appear larger than life which provoke great anxiety over accepted and successful public health interventions such as fluoridation or folic acid fortification of flour, with unsubstantiated claims that they cause cancer, asthma, and other ill-effects.

The interface between ethics, law, and science in public health requires continuous sensitivity to the downstream effects of "shouting fire in the theater".

Sources: Wakefield AJ, Murch SH, Anthony A, Linnell, Casson DM, Malik M, et al. Ileal lymphoid nodular hyperplasia, non-specific colitis, and pervasive developmental disorder in children [retracted]. Lancet 1998;351:637–41.

Office of Research Integrity. Definition of research misconduct. Available at: http://ori.hhs.gov/misconduct/definition_misconduct.shtml

General Medical Council. Andrew Wakefield: determination of serious professional misconduct 24 May 2010. Available at: www.gmc-uk.org/Wakefield_SPM_and_SANCTION.pdf_32595267.pdf

Murch SH, Anthony A, Casson DH, Malik M, Berelowitz M, Dhillon AP, et al. Retraction of an interpretation. Lancet 2004;363:750.

Godlee F, Jane Smith J, Harvey Marcovitch H. Editorial. Wakefield's article linking MMR vaccine and autism was fraudulent. BMJ 2011;342:c7452.

---

the direst of warnings to societies of what may follow when the principle of the sanctity of the individual human life is breached. The issue, however, returned to the public agenda in the 1980s and 1990s as advances in medical science have allowed the prolongation of human life beyond all hope of recovery. Legislation in the Netherlands, the USA ("assisted suicide" in the states of Washington, Oregon, and Montana), and northern Australia has legally sanctioned euthanasia with various safeguards in a variety of circumstances, such as long-term comas or terminal illnesses.

Doctors, patients, relatives, and health care organizations need clear guidelines, orientation, procedures, legal protection, and limitations where failure to take utmost steps to "save" the patient by intubation, resuscitation, or transplantation may cause legal jeopardy. Even though a distinction can be drawn theoretically between permitting and facilitating death, in practice, doctors in intensive care units face such decisions regularly where the line is often blurred. Hospital doctors routinely go to extreme measures to prolong the life of hopeless cases. Such decisions should not be considered for economic reasons alone, but in practice the costs of care of the terminally ill will be a driving force in debate of the issue. Living wills allow a patient to refuse heroic measures such as resuscitation, with "do not resuscitate" standing orders and assignment of power of attorney to family members to make such decisions. Family attitudes are important, but the social issue of redefining the right of a patient to opt for legal termination of life by medical means will be an increasingly important issue in the twenty-first century.

The Imperative to Act or Not Act in Public Health

As in other spheres of medicine and health, in public health the decision whether to intervene on an issue is based on identification and interpretation of the problem, the potential of the intervention to improve the situation, to do no harm, and to convince the public and political levels of the need for such intervention along with the resources to carry it out. This process requires patience and a longer timeframe than many other fields in health.

Some interpretations of ethics in health consider that the only purpose for which power can be rightfully exercised over any member of a democratic community, against his will, is to prevent harm to others. But this is not a dictum that is applied to public health, which is obliged to act to protect the public health in so many spheres such as food and drug safety and environmental health, on a spectrum that extends to banning smoking in public places, mandating food fortification, and many other areas of civil society.

Failure to act is an action, and when there is convincing evidence of a problem that can be alleviated or prevented entirely by an accepted and demonstrably successful intervention, then the onus is on the public health worker to advocate such action and to implement it as best as possible under the existing conditions. Failure to do so is a breach of "good standards of practice" and could be unethical. Inertia of the public health system in the face of evidence of a demonstrably effective modality such as adoption of state-of-the-art vaccines or fortification of flour with folic acid to prevent birth defects would come under this categorization and may even constitute neglect and unethical practice. This is not an easy categorization, because there is often disagreement and even opposition to public health interventions, as was the case with opposition to vaccination long after Jenner's crucial discovery of this procedure in the late eighteenth century. It is also true today with opposition to many proven measures such as fluoridation or fortification of basic foods. Box 15.19 shows the ethical standards of the APHA in 2006.

The use of ethical and high standards of practice in public health (Box 15.20) requires an ideological commitment to the advancement of health standards and use of best practices of international standards to the maximum extent possible under the local conditions in which the professional is working. This is not an easy commitment as there is often dispute and outright hostility to public health activities, in part because of ethical distortions of great magnitude in the past. But this is an optimistic field of activity because of the great achievements it has brought to humankind. Preparation for disasters and unanticipated health emergencies in addition to addressing current issues is a vital part of the New Public Health and our ethical and professional commitments.

BOX 15.19 Principles of Ethical Public Health Practice: American Public Health Association, 2006

- Public health should address principally the fundamental causes of disease and requirements for health, aiming to prevent adverse health outcomes.
- Public health should achieve community health in a way that respects the rights of individuals in the community.
- Public health policies, programs, and priorities should be developed and evaluated through processes that ensure an opportunity for input from community members.
- Public health should advocate and work for the empowerment of disenfranchised community members, aiming to ensure that the basic resources and conditions necessary for health are accessible to all.
- Public health should seek the information needed to implement effective policies and programs that protect and promote health.
- Public health institutions should provide communities with the information they have that is needed for decisions on policies or programs and should obtain the community's consent for their implementation.
- Public health institutions should act in a timely manner on the information they have within the resources and the mandate given to them by the public.
- Public health programs and policies should incorporate a variety of approaches that anticipate and respect diverse values, beliefs, and cultures in the community.
- Public health programs and policies should be implemented in a manner that most enhances the physical and social environment.
- Public health institutions should protect the confidentiality of information that can bring harm to an individual or community if made public. Exceptions must be justified on the basis of the likelihood of significant harm to the individual or others.
- Public health institutions should ensure the professional competence of their employees.
- Public health institutions and their employees should engage in collaborations and affiliations in ways that build the public's trust and the institution's effectiveness.

Source: *American Public Health Association. Public Health Leadership Society. Principles of the ethical practice of public health. APHA; 2002. Available at: http://www.apha.org/NR/rdonlyres/1CED3CEA-287E-4185-9CBD-BD405FC60856/0/ethicsbrochure.pdf [Accessed 13 December 2012].*

SUMMARY

In order to maintain and improve standards of care, health systems need quality assurance and technological assessment as part of their ongoing operation. Poor-quality care is costly in terms of iatrogenic diseases and prolonged or repeated hospitalization. If innovations such as endoscopic surgery are not introduced, then longer hospital stays are needed for the same operation, wasting the patient's time and productivity, while utilizing expensive health care resources, and incurring the risks associated with more invasive surgery.

Health care is provided by people, as well as by institutions with a range of devices and equipment. The people providing care, more than the technological facilities, set the quality of care. Nevertheless, progress on the technological side of medical care is vital to the continuing development of the field. Modern medications, monitoring equipment, laboratory services, and imaging devices have made enormous contributions to advances in medical care. Appropriate

BOX 15.20 The Ethics of Publication in Public Health

Publication in peer-reviewed journals is a key part of the advancement in science and a vital part of the development of the scientific basis for public health practice. The process of publication should promote rigorous standards of high quality ethical research and the wide dissemination of their findings. Codes of practice for editors and publishers of peer-reviewed journals have been developed by both the Committee on Publication Ethics (COPE) (Rees, 2011) and the World Association of World Editors (WAME).

Editors are subject to competitive pressures, and the over-arching metric of success is seen to be the impact factor, a measure of the frequency with which the "average article" in a journal has been cited in a particular year or period. Relevant, rigorous research of better quality will tend to be cited more frequently, and thus editorial strategies that look for quality and relevance in the given field will increase the impact factor. However, there can also be potential distorting factors. Publishing a highly controversial paper can result in high citation levels. Publishing studies which demonstrate negative findings may be less likely to attract large numbers of citations.

Key issues relate to conflicts of interest, and the potential for advertising and sponsorship to distort editorial decision making (Gray, 2012). A particular concern has been the pernicious influence of the tobacco industry in sponsoring, frequently covertly, research which has aimed to confuse or obfuscate key findings linking second hand exposure to tobacco to adverse impacts on health. Similar tactics are used in other areas where health and commercial interests collide. Clear statements of potential conflicts of interest are essential. Journal owners must not interfere in the evaluation, selection, or editing of individual articles, either directly or by creating an environment in which editorial decisions are strongly influenced.

Other challenging areas are plagiarism and research misconduct. The latter is extremely difficult both to detect and to deal with, and requires close working between institutions and editors who may suspect professional misconduct. In cases of fraud, the publishing journal should withdraw the article in a timely fashion (see Box 15.18: The Wakefield Effect).

There has been a rapid rise in open access publishing, in part underpinned by an ethical belief that research is a public good, and an increasing number of influential research funders now require that there should be unrestricted access to the published output of research. In addition, several publishers make their journals free to those in selected low-income countries, promoting dissemination to those who might not otherwise afford them.

In summary, publication in peer-reviewed journals remains a key method for establishing and progressing the evidence base for public health practice. The consequences of poor or frankly fraudulent science can have a substantial adverse impact both on health and on the use of resources. Editors must adhere to high ethical and professional standards and remain vigilant to avoid allowing external drivers to distort their decision-making processes. They must strive to maintain integrity and high scientific standards to advance the field of public health practice (Smith, 2007).

Sources: *Selena Gray, BSc, MBCHB, MD, FFPH, FRCP, Professor, University of West of England, Bristol, and Deputy Postgraduate Dean, Severn Deanery, Bristol, UK. Personal communication.*
Rees M. Code of conduct and best practice guidelines for journal editors. Committee on Publication Ethics; 2011. Available at: http://publicationethics.org/ [Accessed 21 August 2012].
Gray S. The ethics of publication in public health. Public Health Rev 2012;34. Epub ahead of print. Available at: www.publichealthreviews.eu [Accessed 20 December 2012].
Smith R. The trouble with medical journals. London: Royal Society of Medicine Press; 2007.

technology is a critical issue for international health, since the most advanced technology may be completely inappropriate in a setting that cannot afford to maintain it or lacks the trained personnel to operate it, or where it comes in place of more vital basic primary care services. Technology assessment needs to be seen in the context of the country and its resources for health care.

Ethical issues in public health are no less demanding than those related to **individual** clinical care. The rights of the individual and those of the community are sometimes in conflict. Technology, quality, the law, and ethics are closely interrelated in public health. Well-informed and sensitive analysis of all aspects of their development is a part of the New Public Health. The balance between individual and community rights is very sensitive and must be kept under continuous surveillance.

The New Public Health is replete with technological and ethical questions, especially in a time of cost restraint, increasing technological potential, the public expectation

of universal access to health care, and the assumption that everyone will live a healthy and long life. Health status has always been linked with socioeconomic status and, despite enormous gains, this remains true even in the most egalitarian countries. Expansion of market mechanisms, such as controlling the supply of hospital beds, doctors, and access to referrals, competition and incentives/disincentives in payment systems for hospital and managed care systems, contribute to a need for dynamic health policy management capacity. The New Public Health assumes a social responsibility for health for all, using community and personal care modalities as effectively as possible to achieve that overall goal.

NOTE

For a complete bibliography and guidance for student reviews and expected competencies please see companion web site at http://booksite.elsevier.com/9780124157668

BIBLIOGRAPHY

Health Technology

Agosti J.M., Goldie S.J. 2007. Introducing HPV vaccine in developing countries–key challenges and issues. N. Engl. J. Med. 356:1908–1910. Available at: http://www.nejm.org/doi/full/10.1056/NEJMp078053 (accessed 10.01.14).

Alliance, G.A.V.I., 2011. GAVI alliance secures lower price for rotavirus vaccine. Available at: http://www.gavialliance.org/library/news/press-releases/2012/gavi-secures-lower-price-rotavirus-vaccine/ (accessed 15.12.12).

American Cancer Society, 30 July 2013. US task force makes recommendations for lung cancer screening. Available at: http://www.cancer.org/cancer/news/us-task-force-makes-recommendations-for-lung-cancer-screening (accessed 12.08.13).

American Congress of Obstetricians and Gynecologists, 2012. ACOG practice bulletin no. 131: screening for cervical cancer. Obstet. Gynecol. 120, 1222–1238. http://www.acog.org/About%20ACOG/Announcements/New%20Cervical%20Cancer%20Screening%20Recommendations.aspx (accessed 22.5.13).

Brenzel L., Wolfson L.J., Fox-Rushby J., Miller M., Halsey N.A. 2006. Vaccine-preventable diseases. In: Jamison D.T., Breman J.G., Measham A.R., Alleyne G., Claeson M., Evans DB, et al., (Ed.) Disease control priorities in developing countries, second ed. World Bank Washington, DC. Available at: http://www.ncbi.nlm.nih.gov/pubmed/21250343 (accessed 7.10.2012).

Campbell, B., 2013. Regulation and safe adoption of new medical devices and procedures. Br. Med. Bull. 1–14. Available at: http://www.ncbi.nlm.nih.gov/pubmed/23896485 (accessed 10.01.14).

Canadian Council of Health Services Accreditation. http://www.accreditation.ca/ (accessed 10.01.14).

Centers for Disease Control and Prevention. Healthcare associated infections (HAI) [updated 23 September 2013]. Available at: http://www.cdc.gov/hai/ (accessed 10.01.14).

Centers for Disease Control and Prevention, 2009. The direct medical costs of healthcare-associated infections in US hospitals and the benefits of prevention. CDC, Atlanta, GA. Available at: http://www.cdc.gov/HAI/pdfs/hai/Scott_CostPaper.pdf (accessed 14.12.12).

Centers for Disease Control and Prevention. Vaccine price list. Available at: http://www.cdc.gov/vaccines/programs/vfc/awardees/vaccine-management/price-list/index.html (accessed 19.09.12).

Centers for Disease Control and Prevention, 2013. Percentage of women Aged 50–64 years who reported receiving a mammogram in the past 2 years, by health insurance status – National Health Interview Survey, United States, 1993–2010. MMWR. Morb. Mortal. Wkly. Rep. 62, 651. Available at: http://www.cdc.gov/mmwr/preview/mmwrhtml/mm6232a5.htm?s_cid=mm6232a5_w (accessed 10.01.14).

Danish Centre for Health Technology Assessment, 2008. Health technology assessment handbook, second ed. National Board of Health. Available at: http://sundhedsstyrelsen.dk/~/media/C0ED080616D7410E8B-6020B903AD0339.ashx (accessed 10.01.14).

Deyo, R.A., 2002. Cascade effects of medical technology. Annu. Rev. Public Health 23, 23–44. Available at: http://www.annualreviews.org/doi/abs/10.1146/annurev.publhealth.23.092101.134534 (accessed 10.01.14).

Fitzpatrick, C., Floyd, K., 2012. A systematic review of the cost and cost effectiveness of treatment for multidrug-resistant tuberculosis. Pharmacoeconomics 30, 63–80. Available at: http://www.ncbi.nlm.nih.gov/pubmed/22070215 (accessed 13.12.12).

Flahault, A., Martin-Moreno, JM., 2013. Why do we choose to address health 2020? Public Health Reviews. 2013;35: epub ahead of print. Available at: http://www.publichealthreviews.eu/upload/pdf_files/13/00_Editorial.pdf (accessed 04.01.14).

Greenberg, M.R., 2006. The diffusion of public health innovations. Am. J. Public Health 96, 209–210. Available at: http://www.ncbi.nlm.nih.gov/pmc/articles/PMC1470490/ (accessed 10.11.14).

Harris, R.P., Helfand, M., Woolf, S.H., Lohr, K.N., Mulrow, C.D., Teutsch, S.M., et al., 2001. Current methods of the US Preventive Services Task Force. A review of the process. Am. J. Prev. Med. 20, 21–35. Available at: http://www.ncbi.nlm.nih.gov/pubmed/11306229 (accessed 10.01.14).

Health Services/Technology Assessment (HSTAT). Available at: http://www.ncbi.nlm.nih.gov/books/NBK16710/ (accessed 17.08.12.).

Institute of Medicine, 1985. Assessing medical technologies. National Academies Press, Washington, DC. Available at: http://www.nap.edu/openbook.php?record_id=607 (accessed 10.01.14).

International Network of Agencies for Health Technology Assessment (INAHTA). http://www.inahta.org/ (accessed 17.08.12).

Ikeda, N., Sapienza, D., Guerrero, R., Aekplakorn, W.,et al. 2914, Control of hypertension with medication: a comparative analysis of national surveys in 20 countries. Bull. World Health Organ.; 92(1): 10–19. Available at: http://www.ncbi.nlm.nih.gov/pmc/articles/PMC3865548/ (accessed 10.01.14).

Lancet, 2011. Financing HPV vaccination in developing countries. Lancet 377, 1544. Available at: http://www.thelancet.com/journals/lancet/article/PIIS0140-6736(11)60622-3/fulltext (accessed 10.01.14).

Larson, E.L., Quiros, D., Lin, S.X., 2007. Dissemination of the CDC's hand hygiene guideline and impact on infection rates. Am. J. Infect. Control. 35, 666–675. Available at: http://www.ncbi.nlm.nih.gov/pubmed/18063132 (accessed 10.01.14).

Lehoux, P., Tailliez, S., Denis, J.L., Hivon, M., 2004. Redefining health technology assessment in Canada: diversification of products and contextualization of findings. Int. J. Technol. Assess 20, 325–336. Available at: http://www.ncbi.nlm.nih.gov/pubmed/15446762 (accessed 10.01.14).

Martelli, F., La Torre, G., Di Ghionno, E., Neroni, M., Cicchetti, A., Von Bremen, K., et al., 2007. Health technology assessment agencies: an international overview of organizational aspects. Int. J. Technol. Assess 23, 414–424. Available at: http://www.ncbi.nlm.nih.gov/pubmed/17937828 (accessed 10.01.14).

Murray, A., Lourenco, T., de Verteuil, R., Hernandez, R., Fraser, C., McKinley, A., et al., 2006. Clinical effectiveness and cost-effectiveness of laparoscopic surgery for colo-rectal cancer: systematic reviews and economic evaluation. Health Technol. Assess 10, 1–141. Available at: http://www.ncbi.nlm.nih.gov/books/NBK62293/ (accessed 10.01.14).

National Library of Medicine, 2007. National information center on health services research & health care technology. Etext on health technology assessment (HTA) information resources. Available at: http://www.nlm.nih.gov/archive//20060905/nichsr/ehta/ehta.ht (accessed 10.01.14).

Nixon, J., Stoykova, B., Glanville, J., Christie, J., Drummond, M., Kleijnen, J., 2000. The UK NHS economic evaluation database. Economic issues in evaluations of health technology. Int. J. Technol. Assess 16, 731–742. Available at: http://www.ncbi.nlm.nih.gov/pubmed/11028129 (accessed 10.01.14).

Noorani, H.Z., Husereau, D.R., Boudreau, R., Skidmore, B., 2007. Priority setting for health technology assessments: a systematic review of current practical approaches. Int. J. Technol. Assess 23, 310–315. Available at: http://www.ncbi.nlm.nih.gov/pubmed/17579932 (accessed 10.01.14).

Sassi, F., 2003. Setting priorities for the evaluation of health interventions: when theory does not meet practice. Health Policy 63, 141–154. Available at: http://www.ncbi.nlm.nih.gov/pubmed/12543527 (accessed 10.01.14).

Soto, J., 2002. Health economic evaluations using decision analytic modeling. Principles and practices – utilization of a checklist to their development and appraisal. Int. J. Technol. Assess 18, 94–111. Available at: http://www.ncbi.nlm.nih.gov/pubmed/11987445 (accessed 10.01.14).

Upshur, R.E.G., 2002. Principles for the justification of public health interventions. Can. J. Public Health 93, 101–103. Available at: http://journal.cpha.ca/index.php/cjph/article/download/217/217 (accessed 10.01.14).

Williams, I., Bryan, S., McIver, S., 2007. How should cost-effectiveness analysis be used in health technology coverage decisions? Evidence from the National Institute for Health and Clinical Excellence approach. J. Health Serv. Res. Policy 12, 73–79. Available at: http://www.ncbi.nlm.nih.gov/pubmed/17407655 (accessed 10.01.14).

Woods, K., 2002. Health technology assessment for the NHS in England and Wales. Int. J. Technol. Assess 18, 161–165. Available at: http://www.ncbi.nlm.nih.gov/pubmed/12053415 (accessed 10.01.14).

Quality

Agency for Healthcare Research and Quality. http://www.ahrq.gov/ (accessed 14.12.12).

AHRQ. U.S. Preventive Services Task Force USPSTF), an introduction. Available at: http://www.ahrq.gov/professionals/clinicians-providers/guidelines-recommendations/uspstf/index.html.

American College of Cardiology (ACC), http://www.cardiosource.org/science-and-quality.aspx (accessed 20.08.12).

Cancer Research UK, October 2012. Who is screened for breast cancer? Available at: http://www.cancerresearchuk.org/cancer-help/type/breast-cancer/about/screening/who-is-screened-for-breast-cancer (accessed 10.01.14).

Centers for Disease Control and Prevention, 2013. Quick stats. MMWR. Morb. Mortal. Wkly. Rep. 62, 651. Available at: http://www.cdc.gov/mmwr/pdf/wk/mm6232.pdf (accessed 17.08.13).

Reviews, Cochrane, http://www.cochrane.org/cochrane-reviews (accessed 14.12.12).

Dannenberg, A.L., Bhatia, R., Cole, B.L., Dora, C., Fielding, J.E., Kraft, K., et al., 2006. Growing the field of health impact assessment: an agenda for research and practice. Am. J. Public Health 96, 262–270. Available at: http://www.ncbi.nlm.nih.gov/pmc/articles/PMC1470491/ (accessed 10.01.14).

Deeks, J.J., 2001. Systematic reviews in health care: systematic reviews of evaluations of diagnostic and screening tests. BMJ 323, 157–162. Available at: http://www.ncbi.nlm.nih.gov/pmc/articles/PMC1120791/ (accessed 10.01.14).

European Society of Cardiology, Guidelines. Available at: http://www.escardio.org/GUIDELINES-SURVEYS/ESC-GUIDELINES/Pages/GuidelinesList.aspx (accessed 17.08.12).

Epstein, A.J., Polsky, D., Yang, F., Yang, L., Groeneveld, P.W., 2011. Coronary revascularization trends in the United States, 2001–2008. JAMA 305, 1769–1776. Available at: http://www.ncbi.nlm.nih.gov/pubmed/21540420 (accessed 13.08.13).

Fielding, J.E., Briss, P.A., 2006. Promoting evidence-based public health policy: can we have better evidence and more action? Health Aff. 25, 969–978. Available at: http://content.healthaffairs.org/content/25/4/969.full.pdf (accessed 10.01.14).

Institute of Medicine, 2007. State of quality improvement and implementation research: expert views. Workshop summary. Institute of Medicine of the National Academies of Science. National Academies Press, Washington, DC. Available at: http://www.nap.edu/catalog.php?record_id=11986 (accessed 10.01.14).

International Guideline Network Library. http://www.g-i-n.net/library/international-guidelines-library/ (accessed 14.12.12).

Kohn, L.T., Corrigan, J.M., Donaldson, M. (Eds.), 1999. Institute of Medicine. Committee on quality of health care in America. To err is human: building a safer health system. National Academies Press, Washington, DC. Available at: http://www.nap.edu/openbook.php?isbn=0309068371 (accessed 10.01.14).

Krech, L.A., El-Hadri, L., Evans, L., Fouche, T., et al., 2014. The medicines quality database: a free public resource. Available at: http://www.who.int/bulletin/volumes/92/1/13-130526.pdf (accessed 10.01.14).

National Council for Quality Assurance. http://www.ncqa.org/ (accessed 10.01.14).

Pilkington, H., Blondel, B., Drewniak, N., Jennifer Zeitlin, 2014. Where does distance matter? Distance to the closest maternity unit and risk of foetal and neonatal mortality in France. Eur. J. Public Health, 1–6. Available at: http://eurpub.oxfordjournals.org/content/early/2014/01/01/eurpub.ckt207.full (accessed 10.01.14).

Shortell, S.M., Gillies, R., Wu, F., 2010. United States innovations in health care delivery. Public Health Rev. 32, 190–212. Available at: http://www.publichealthreviews.eu/show/f/26 (accessed 26.10.12).

The National Audit Office, 2012. Healthcare across the UK: A comparison of the NHS in England. Wales and Northern Ireland, Scotland. Available at: http://www.officialdocuments.gov.uk/document/hc1213/hc01/0192/0192.pdf (accessed 10.01.14).

UK National Institute for Health and Care Excellence (NICE). Available at: http://www.nice.org.uk/ (accessed 17.08.12).

US Department of Health and Human Services, August 2008. Office of Inspector General. Adverse events in hospitals: overview of key issues. Available at: http://oig.hhs.gov/oei/reports/oei-06-07-00470.pdf (accessed 17.08.13).

US Food and Drug Administration, 2007. Department of Health and Human Services. Amendment to the current good manufacturing practice regulations for finished pharmaceuticals. Direct final rule. Federal Register 72, 68064–68070. Available at: http://www.ihi.org/Pages/default.aspx (accessed 17.08.12).

US National Library of Medicine. Available at: http://www.nlm.nih.gov/ (accessed 13.02.14).

US National Information Center on Health Services Research. https://www.nlm.nih.gov/nichsr/ (accessed 13.02.14).

US Preventive Services Task Force. Guide to preventive services. Report of the US Preventive Services Task Force (USPSTF). http://www.ahrq.gov/professionals/clinicians-providers/guidelines-recommendations/uspstf/index.htm. Available at: http://www.ahrq.gov/clinic/USpstfix.htm (accessed 13.02.14).

Law

American Society of Law, Medicine and Ethics. The public's health and the law in the 21st century: Second annual partnership conference on public health law. J. Law. Med. Ethics. Special Suppl. 31 (4). Available at: http://www.stacks.cdc.gov/view/cdc/6751/cdc_6751_DS1.pdf (accessed 10.1.14).

Annas, G., 1998. Human rights and health – the universal declaration of human rights at 50. N. Engl. J. Med. 339, 1778–1781. Available at: http://www.nejm.org/doi/full/10.1056/NEJM199812103392411 (accessed 10.01.14).

Burris, S., Wagenaar, A.C., Mello, M.M., 2010. Making the case for laws that improve health: A framework for public health law research. Milbank. Q. 88 (2), 169–210. Available at; http://www.ncbi.nlm.nih.gov/pmc/articles/PMC2980343/ (accessed 10.01.14).

Centers for Disease Control and Prevention, Collaborating Center for the Law and the Public's Health. Available at: http://www.publichealthlaw.net/ (accessed 17.08.12).

Fox, D., Kramer, M., Standish, M., 2003. From public health to population health: how can law redefine the playing field? J. Law Med. Ethics. 31, 21–29. Available at: http://www.stacks.cdc.gov/view/cdc/6751/cdc_6751_DS1.pdf (accessed 10.01.14).

Goodman, R.A., Moulton, A., Matthews, G., Shaw, F., Kocher, P., Mensah, G., et al., 2006. Law and public health at CDC. MMWR. Morb. Mortal. Wkly. Rep. 55, 29–33. Available at: http://www.cdc.gov/mmwr/preview/mmwrhtml/su5502a11.htm (accessed 10.01.14).

Gostin, L.O., 2000. Public health law in a new century. Part II: Public health powers and limits. JAMA 283, 2979–2984. Available at: http://www.ncbi.nlm.nih.gov/pubmed/10865277 (accessed 10.01.14).

Gostin, L.O., 2000. Public health law in a new century: Part III: Public health regulation: a systematic evaluation. JAMA 283, 3118–3122. Available at: http://www.ncbi.nlm.nih.gov/pubmed/10865307 (accessed 10.01.13).

Ruhl, S., Stephens, M., Locke, P., 2005. The role of NGOs in public health law. J. Law Med. Ethics. 31, 76–77. Available at: http://www.stacks.cdc.gov/view/cdc/6751/cdc_6751_DS1.pdf (page 76). (accessed 10.01.14).

Thiel, K.S., 2003. New developments in public health case law. J. Law Med. Ethics. 31, 86–87.

Zasa, S., Clymer, J., Upmeyer, L., Thacker, S., 2003. Using science-based guidelines to shape public health law. J. Law Med. Ethics. 31, 65–67.

Ethics

Aceijas, C., Brall, C., Schröder-Bäck, P., Otok, R., Maeckelberghe, E., Stjernberg, L., et al., 2012. Teaching ethics in schools of public health in the European Region: findings from a screening survey. Public Health Rev. 34. Epub ahead of print. Available at: www.publichealthreviews.eu (accessed 05.12.12).

American College of Epidemiology, 2000. Ethics guidelines. Ann. Epidemiol. 10, 487–497. Available at: http://acepidemiology.org/content/ethics.

American Public Health Association, Principles of the ethical practice of public health. Available at: http://www.apha.org/NR/rdonlyres/1CED3CEA-287E-4185-9CBD-BD405FC60856/0/ethicsbrochure.pdf (accessed 17.08.12).

Bachrach, S., 2004. In the name of public health – Nazi racial hygiene. N. Engl. J. Med. 351, 417–420. Available at: http://content.nejm.org/cgi/content/full/351/5/417?ijkey=8bee5b41cf11cadc3a826ad16fc2deef59aa8f32&keytype2=tf_ipsecsha (accessed 05.12.12).

Bayer, R., Fairchild, A., 2004. The genesis of public health ethics. Bioethics 18, 473–492. Available at: http://www.ncbi.nlm.nih.gov/pubmed/15580720 (accessed 10.01.14).

Birn, A.E., Molina, N., 2005. In the name of public health. Am. J. Public Health 95, 1095–1097. Available at: http://www.ncbi.nlm.nih.gov/pmc/articles/PMC1449322/ (accessed 10.01.14).

Carter, S.M., Cribb, A., Allegrante, J.P., 2012. How to think about health promotion ethics. Public Health Rev. 34. Epub ahead of print. Available at: www.publichealthreviews.eu (accessed 05.12.12).

Centers for Disease Control and Prevention, US Public Health Service syphilis experiment at Tuskegee. Available at: http://www.cdc.gov/tuskegee/timeline.htm (accessed 17.08.12).

Centers for Disease Control, Advancing excellence and integrity of CDC science. Available at: http://www.cdc.gov/od/science/integrity/phethics/ (accessed 10.01.14).

Coleman, C.H., Bouësseau, M.-C., Reis, A., 2008. The contribution of ethics to public health. Bull. World Health Org. 86, 578–579. Available at: http://www.ncbi.nlm.nih.gov/pmc/articles/PMC2649472/ (accessed 10.01.14).

Coughlin, S.S., Barker, A., Dawson, A., 2012. Ethics and scientific integrity in public health, epidemiological and clinical research. Public Health Rev. 34 (1). Available at: www.publichealthreviews.eu (accessed 05.12.12).

Coughlin, S.S., 2006. Ethical issues in epidemiologic research and public health practice. Emerg Themes Epidemiol 3, 3–16. Available at: http://www.ncbi.nlm.nih.gov/pmc/articles/PMC1594564/ (accessed 10.01.14).

Dawson, A., Paul, Y., 2006. Mass public health programs and the obligations of sponsoring and participating organizations. J. Med. Ethics 32, 580–583. Available at: http://www.ncbi.nlm.nih.gov/pmc/articles/PMC2563318/ (accessed 10.01.14).

Dawson, A., 2004. Vaccination and the prevention problem. Bioethics 18, 515–530. Available at: http://onlinelibrary.wiley.com/doi/10.1111/j.1467-8519.2004.00414.x/abstract?deniedAccessCustomisedMessage=&userIsAuthenticated=false (accessed 10.01.14).

El Amin, A.N., Parra, M.T., Kim-Farley, R., Fielding, J.E., 2012. Ethical issues concerning vaccination requirements. Public Health Rev. 34 (1). Available at: www.publichealthreviews.eu (accessed 05.12.12).

Kass, N., 2004. Public health ethics: from foundations and frameworks to justice and global public health. J. Law Med. Ethics. 32, 232–242. Available at: http://www.ncbi.nlm.nih.gov/pubmed/15301188 (accessed 10.01.14).

Lee, L.M., 2012. Guest editorial: Public health ethics theory: review and path to convergence. Public Health Rev. 34 (1). Available at: www.publichealthreviews.eu (accessed 05.12.12).

Lynch, H.F., 2012. Ethical evasion or happenstance and hubris? The US Public Health Service STD Inoculation Study. Hastings Cent. Rep. 42, 30–38. Available at: http://onlinelibrary.wiley.com/doi/10.1002/hast.17/abstract (accessed 10.01.14).

Nuremberg Code, 1949. Trials of war criminals before the Nuremberg military tribunals under control council law no. 10, Vol. 2. US Government Printing Office, Washington, DC. pp. 181–182. Organisation for Economic Co-operation and Development, 2007. Global Science Forum. Best practices for ensuring scientific integrity and preventing misconduct. OECD, Available at: http://www.oecd.org/dataoecd/37/17/40188303.pdf (accessed 19.5.2012).

Reverby, S.M., 2012. Ethical failures and history lessons: the US Public Health Service research studies in Tuskegee and Guatemala. Public Health Rev. 34. Epub ahead of print. Available at: http://www.publichealthreviews.eu/upload/pdf_files/11/00_Reverby.pdf (accessed 10.01.14).

Thieren, M., Mauron, A., 2007. Nuremberg Code turns 60. B World Health Organ 85, 573. Available at: http://www.publichealthreviews.eu/upload/pdf_files/11/00_Reverby.pdf (accessed 10.01.14).

Thomas, J.C., Sage, M., Dillenberg, J., Guillory, V.J., 2002. A code of ethics for public health. Am. J. Public Health 92, 1057–1060. Available at: http://www.ncbi.nlm.nih.gov/pmc/articles/PMC1447186/ (accessed 10.01.14).

US Presidential Commission for the Study of Bioethical Issues, 2011. Creation of the commissions. Available at: http://bioethics.gov/cms/creation-of-the-commission (accessed 05.12.12).

United Nations Programme on HIV/AIDS (UNAIDS), 2013. Ending overly broad criminalisation, non-disclosure, exposure and transmission: Critical scientific, medical and legal considerations. Available at: http://www.unaids.org/en/media/unaids/contentassets/documents/document/2013/05/20130530_Guidance_Ending_Criminalisation.pdf (accessed 20.1.14).

Zusman, S.P., 2012. Water fluoridation in Israel, 1968–2012. Public Health Rev. 34. Epub ahead of print. Available at: http://www.publichealthreviews.eu/upload/pdf_files/11/00_Zusman.pdf (accessed 10.01.14).

Global Health

INTRODUCTION

Globalization of health is the growth of international transfer of diseases and cooperation in combating societal conditions that create diseases and their effects on nations and the global community. The concern over communicable diseases goes back throughout history to epidemics and pandemics that cost countless lives. Global health includes cancers and other non-communicable diseases (NCDs) and conditions fostered by lifestyle, poor nutrition, and harmful personal habits, as well as adverse social conditions such as poverty and insecurity. International cooperation is essential to address disease threats and health conditions common to many countries in the world. Globalization of health includes the rapid movement of large numbers of people, foods, drugs, vaccines, medical education, and technology from place to place. It recognizes that health and economic development are interlinked, and that social equity in health is essential to achieve the newly reiterated goals of Health for All. In addition, many new factors including rapid mass travel, global communication, and entertainment promote broadening of the effect of communicable diseases and common risk factors leading to epidemics of chronic diseases. Transportation, colonization, and commerce have been responsible for the dissemination of disease throughout history. With rapid movement of large numbers of people by sailing ships, steamships, rail, and later air, the possibility of disease transmission by travelers has increasingly become a global public health problem.

More generally, globalization is a process of international trade liberalization, privatization, and domestic deregulation, by which the peoples of the world are becoming part of a single society. The globalization process has intensified since the 1990s with the removal of barriers to international trade and foreign direct investments. At the same time, a combination of economic, technological, environmental, social, cultural, and political forces came into play which increased the commonality of global health concerns and promoted shared attempts to address global warming, poverty, ill-health, and inequities. Globalization has a complex influence on health, both negative and positive. The Millennium Development Goals (MDGs) constitute an attempt by the United Nations (UN) to address global inequities of poverty, maternal and child health, communicable and other diseases, environmental degradation, and other markers of an increasingly interactive world (Box 16.1).

Events in any part of the world can affect the status of health of people in other parts of the globe. There is a dynamic interaction and interdependence so that a global approach is essential to achieve health targets, even locally. Without restating arguments elaborated throughout the text, current and future generations of public health professionals must be well aware of what is occurring outside their communities and countries. This means not only learning from the news media of outbreaks of exotic diseases, but also recognizing that the political, social, and economic upheavals that define everyday existence, across the street or halfway around the world, affect us all. Even the most remote communities in the world are not immune to the global impacts of distant military coups, civil wars, natural disasters, economic crises, or epidemics. In the spring of 2013, a new avian influenza virus, H7N9, appeared in China, with 131 cases and 36 deaths to the end of May 2013. Over 75 percent of the victims probably contracted the virus from poultry or in markets selling live chickens. Some caught the virus from family members, although no one seems to have been infected by breathing the same air as an infected person (Zhu et al., 2013). A new coronavirus, Middle East respiratory syndrome coronavirus (MERS-CoV), appeared in 2013 in Saudi Arabia and other countries of the Middle East, as well as among travelers from other countries including France and the UK. MERS-CoV causes severe respiratory illness with high mortality rates and may be transmitted from person to person, with 90 cases with 50 percent case fatality reported as of June 2013. Each of these two new threats raises the possibility of a wide epidemic or pandemic. Either

The New Public Health. http://dx.doi.org/10.1016/B978-0-12-415766-8.00016-1

BOX 16.1 Health in the Post-2015 Development Agenda: Need for a Social Determinants of Health Approach

The United Nations (UN) initiated the Millennium Development Goals approach, supported by almost all member states, which has had positive effects in mobilizing national and donor support for life-saving projects and has made real progress in achieving the set goals. Preparation for the post-2015 period will require a new commitment to reducing social inequalities in health.

The UN stresses social determinants of health based on the Constitution of the World Health Organization, which states that "health is a state of complete physical, mental and social well-being and not merely the absence of disease or infirmity" and that "the enjoyment of the highest attainable standard of health is one of the fundamental rights of every human being without distinction of race, religion, political belief, economic or social condition".

The UN also emphasizes that "the health of all peoples is fundamental to the attainment of peace and security and is dependent upon the fullest co-operation of individuals and states".

The responsibility for health lies with national governments, which requires the "provision of adequate health and social measures".

The Universal Declaration of Human Rights states that "everyone has the right to a standard of living adequate for the health and well-being of himself and of his family, including food, clothing, housing and medical care and necessary social services, and the right to security in the event of unemployment, sickness, disability, widowhood, old age or other lack of livelihood in circumstances beyond his control".

Health and social security are thus recognized as human rights. These include special care and social protection for children, compulsory elementary education for children, and care and protection of the rights of mentally ill and handicapped people.

Source: *United Nations. Health in the post-2015 development agenda: need for social determinants of health approach Joint statement of the UN Platform on Social Determinants of Health. Available at: http://www.who.int/social_determinants/advocacy/UN_Platform_FINAL.pdf [Accessed 29 May 2013].*

help as well as self-help. This inequality is not only between countries, but within even developed and much more so in the rapidly developing countries, with middle and upper classes living with luxuries and a comfortable lifestyle. Globalization has a complex influence on health. Some components of globalization such as trade liberalization and technology transfer could increase efficiency, welfare, and health. However, under the existing barriers to the international markets, the weaker countries need help to address the stagnation preventing them from improving their market position, thereby increasing wealth and health of the population. Financial aid alone can have destructive effects by creation of dependency and misuse of resources through corruption and promotion of military primacy use of funds, while fundamental poverty and health issues are not a high priority on national agendas.

The tools to prevent and control disease are available and widely successful, yet not well applied in many countries. In the health sector, there are inappropriate balances of health resource allocation so that important and effective preventive measures are underfunded and often left to international donors to provide. The common interest and social solidarity represented by international health efforts have achieved much but international public health efforts are necessary to combat disease and create a healthy world. Such achievements are shown in global initiatives to eradicate smallpox and great progress in reducing poliomyelitis (polio) from 125 countries to only three endemic countries in 2013, although spread to other countries is still being seen. Global polio surveillance data from 28 May 2013 report 41 polio cases from the three remaining endemic countries: Afghanistan, Nigeria, and Pakistan. In 2012, a total of 223 polio cases was reported from five countries: Afghanistan, Chad, Niger, Nigeria, and Pakistan. Of the 2012 polio cases, 97 percent (217 of the 223 cases) were reported from the three remaining endemic countries (CDC, 2013).

Previous generations of public health advocates have made tremendous contributions to understanding disease, how it spreads, and how it affects all forms of life. But mistakes have been made; human immunodeficiency virus (HIV) infection was not detected early enough, or its impact realized, until it had already reached pandemic proportions. In the twenty-first century, public health is globalized as the health of all humans is globally linked. Yesterday's local SARS, Ebola, measles, or cholera outbreak 10,000 miles away may manifest itself today in the arrival hall of your local airport. With this very real possibility in mind, the future of public health will require advocacy of international economic, political, and social justice policies, with the necessity for organized common efforts to improve health around the world.

NCDs have also been transferred to new populations with economic development and gross inequality by the adoption of lifestyle risk factors, such as smoking, automobiles, and western diets heavy in fats, bringing rising

or both of these could become a new pandemic on the scale of the severe acute respiratory syndrome (SARS) in 2003, or may fade from concern. The lessons learned from SARS may have made global public health more alert and prepared to address new pandemic threats, but huge risks still prevail despite its legacy (Braden et al., 2013). Only by careful professional monitoring and international cooperation will we know of or be able to face the challenge if one or both of H7N9 or MERS-CoV becomes a pandemic.

The world is, and has long been, global in political, economic, and health terms. Yet now there are methods of identifying and controlling potential and actual threats, even though inequities in health are still blatant and call for mutual

waves of mortality from these causes to areas with previously low rates. The impacts of economic, demographic, and epidemiological changes are not uniform, either within or between countries. Domestically and internationally, poor health and poverty affect the stability of us all. A popular late twentieth-century slogan, "think globally, act locally", expresses the interdependent realities of public health.

Another aspect of globalization of health is the increased mobility of health professionals, health consumers, and international organizations working or seeking help across boundaries. The resulting massive flow of trained health workers from developing to developed nations makes building the human resource infrastructure a serious problem. This aspect of globalization of professional training and mobility threatens poor countries with the loss of skilled professionals they so badly need to build their health infrastructure capacity. The development of bachelor's level training for public health managers and of training different levels of mid-level providers and community health workers (CHWs) becomes a necessary and increasingly implemented approach.

Previous chapters addressed demographic and epidemiological issues with examples from different countries, as well as regional and global trends. In this chapter, major trends and contemporary patterns of health and disease in the world are presented, along with policy issues for improving those patterns. Global trends can be analyzed by grouping countries by geographic region, by levels of economic development, and by political, cultural, or ethnic characteristics. Global health requires both official and unofficial international health organizations to stimulate and facilitate joint efforts to achieve common goals, such as preventing the transmission of communicable disease or, more generally, promoting *Health for All*.

THE GLOBAL HEALTH SITUATION

Global health status involves a wide diversity of social and economic standards, disease, disability, and mortality throughout the world. Environmental and socioeconomic factors and health interventions all play a role in health status. Differences between and among developed and developing countries in these factors are great, yet there are common concerns and shared interests in health development. Studies of countries classified by geographic region, such as the World Health Organization (WHO) regions, or by economic status, such as the Organisation for Economic Co-operation and Development (OECD) countries, European Union (EU), Central and Eastern Europe (CEE), and the former Soviet Union (Commonwealth of Independent States or CIS), help to provide an overall picture of demographic transitions and epidemiological shifts. Economic groupings of countries are usually measured by gross domestic product (GDP) per capita, a measure of national productivity, which in industrialized countries is more than 20 times greater than that of the developing countries.

Globally, average life expectancy at birth has increased by about 24 years, from 46.5 years in 1950–1955 to 64 years in 1990 and to 70 years in 2011 (WHO, 2013). The gain in African life expectancy from 1990 to 2011 is estimated to be 6 years (from 50 to 56), some 14 years lower than the global rate and 20 years below life expectancy in the Americas and European Region. For the Americas, the gain from 1990 to 2011 was 5 years (from 71 to 76); for the European Region 4 years (from 72 to 76); South-East Asia 8 years (from 59 to 67); and the Western Pacific 6 years (from 70 to 76) (Table 16.1). Life expectancy at age 60 shows much smaller differences between the regions and slower rates of improvement.

Table 16.2 shows health status indicators by levels of development of countries. From 1990 to 2009, life

TABLE 16.1 Life Expectancy at Birth and 60 Years by World Health Organization Region, 1990, 2000, and 2011

Region	Life expectancy at age 0 (years)			Life expectancy at age 60 (years)		
	1990	2000	2011	1990	2000	2011
Africa	50	50	56	15	15	16
Americas	71	74	76	20	21	22
South-East Asia	59	63	67	15	17	17
Europe	72	73	70	19	20	21
Eastern Mediterranean	61	65	68	17	18	18
Western Pacific	70	72	76	18	19	21
Global	64	66	70	18	19	20

Source: World Health Organization. Life expectancy by WHO region, 1990, 2000, 2011. Available at: http://apps.who.int/gho/data/view.main.690?lang=en [Accessed 29 May 2013].

TABLE 16.2 Health Status Indicators by Country Income Group, Selected Years, 1990–2009

	Low Income	Lower Middle Income	Upper Middle Income	High Income
GNI per capita 2009 (PPP, US$)	$1,457	4,696	12,447	36,708
Life expectancy at birth (years)				
1990	52	63	68	76
2009	57	68	71	80
Adolescent fertility rate (per 1000 girls aged 15–19 years), 2000–2008	115	37	53	21
Total fertility rate (per 1000 adult women)				
1990	5.4	3.4	2.8	1.8
2009	3.9	2.5	2.0	1.7
Adult literacy rate (%)				
1990–1999	40	63	87	NA
2000–2008	63	80	92	NA
Births attended by trained attendant 2000–2010 (%)	40	64	96	100
Maternal mortality ratio (confidence interval)				
1990	850 (590–1300)	400 (290–590)	120 (95–160)	15 (14–19)
2008	580 (420–840)	230 (170–330)	82 (66–110)	15 (14–18)
Infant mortality rate (per 1000 live births)				
1990	108	64	40	10
2009	75	42	19	6
Under-5 mortality rate (per 1000 live births)				
1990	170	90	50	12
2009	117	57	22	7
Low birth weight, 2000–2009 (%)	15	17	8	7
Immunization (%)				
Measles				
1990	57	73	78	84
2009	78	79	93	93
DPT3				
1990	58	78	74	88
2009	80	79	93	95

Note: GNI=gross national income; PPP=purchasing power parity; DPT3=diphtheria, pertussis, tetanus vaccine, three doses.
Source: World Health Organization. World health statistics. 2011. Geneva: WHO. Available at: http://www.who.int/whosis/whostat/EN_WHS2011_Full.pdf [Accessed 12 June 2013].

expectancy at birth for boys and girls gained 5 years in both the low-income developing and middle-income countries gained 5 years' life expectancy (from 52 to 57 and from 63 to 68, respectively), upper middle-income countries gained 3 years (from 68 to 71), and high-income countries gained 4 years (from 76 to 80). Between socioeconomic regions, life expectancy varies more widely. In 2011, life expectancy at birth ranged from 55 years in sub-Saharan Africa to over 80 years in industrialized countries. In the European Region,

average life expectancy for men varies widely from west to east; for men in the Commonwealth of Independent States (CIS) it is 65 years compared with 79 years in Western Europe. Most of the excess mortality in the CIS was due to cardiovascular disease (CVD) related to high rates of coronary heart disease, strokes, binge drinking, diet, and smoking, and a further 20 percent due to trauma (see Chapter 13).

The differences are due to many factors, including quality of water supply and sanitation, housing, education, nutrition, lifestyle, family planning measures, and effective public health measures such as immunizations against infectious diseases and access to health services. Life expectancy has increased in the developed world but also in the low-mortality developing countries. Countries such as China and India have shown great improvements in under-fives' mortality in the past 50 years, in contrast to most African countries. The under-five child mortality rate in India declined from 118 in 1990 to 66 per 1000 population in 2009; in China during the same period the mortality rate for this group declined from 46 to 19 (UNICEF, 2010).

Countries now considered developed had in the past disease patterns similar to developing countries today. It should be noted that infant and maternal mortality rates in many developing countries today are similar to those in the USA in the 1920s (Table 16.2). Furthermore, within industrialized countries, there are social, ethnic, or immigrant groups whose current health status is characteristic of developing countries. In many developing countries, the rising middle-class populations show epidemiological patterns similar to those in developed countries, such as rising rates of heart disease.

The enormous differences in GDP per capita and in fertility rates are reflected by differences in almost all health status indicators. Population growth in the developing countries, due to high fertility rates and declining child mortality, is a key factor in poverty and poor health status, and thus is a major health problem.

Trends in demographic and health indicators for countries classified as high income, upper middle income, and lower middle income all show positive changes in health status: fertility rates have declined in developing countries, but also more recently in the low-income countries. Adult literacy rates are increasing in the low-income and lower-middle-income countries. Immunization coverage has improved globally, as have infant, child, and crude mortality rates, so that life expectancy is generally rising. However, there remains a large discrepancy between rich and poor in health status indicators; in the low-income countries, reported maternal mortality in 2008 was more than 38 times higher, and infant mortality in 2009 was more than 12 times that of the high-income countries.

Despite the fact that health status indicators have improved for the least developed countries, on the basis of current trends, the UN Population Division estimates that the world infant mortality rate was 53.9 in 2000–2005 and projects that it will decline to 45.1 by 2015 and to 23.4 per 1000 population by 2030. Globally, child mortality rates have declined by almost half between 1990 and 2011. In 1990, some 12 million children under the age of 5 died, while in 2010, there were 7.6 million deaths in under-fives (Figure 16.1). In 2011, the number of these deaths declined to 6.9 million, still mostly in developing countries and largely from preventable or easily treatable diseases. Recent data show that some progress has been made in reducing maternal,

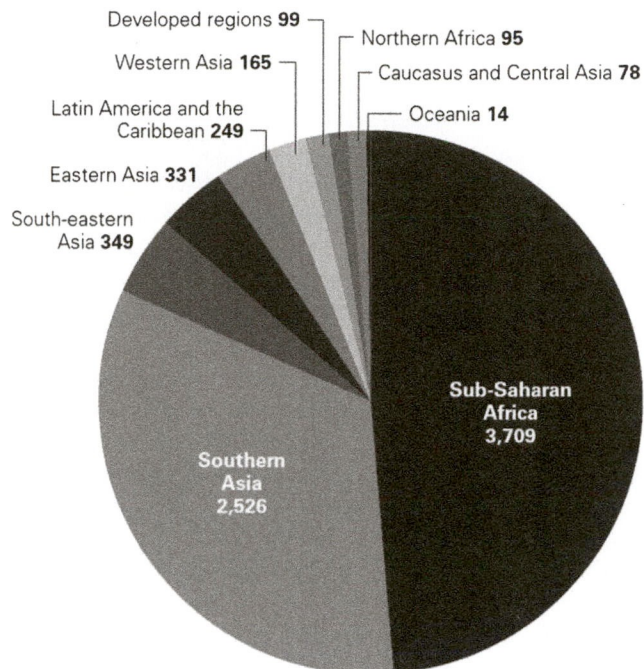

FIGURE 16.1   Distribution of global mortality (thousands) for children, 2010. *Source: United Nations Children's Fund. Levels and trends in child mortality report 2012. New York: UNICEF; 2012. Available at: http://apromiserenewed.org/files/UNICEF_2012_child_mortality_for_web_0904.pdf [Accessed 30 May 2013].*

newborn, and under-five mortality, but not swiftly enough to reach the targets by 2015. Of the eight goals, MDG4 (child mortality) and MDG5 (maternal mortality) are the farthest from being achieved by 2015. Failure to halve child mortality by 2015 would mean that "5 million would die, still largely from preventable diseases" (UNICEF, 2012).

While the gap between developed and developing countries in health remains very wide, the adaptation and dissemination of health technology are having profound effects in raising standards in the latter. Each can learn, for good or ill, from the other. A developing country may spend most of its health resources in central teaching hospitals, while primary care is neglected. Adoption of appropriate priorities, including adoption of available vaccines and other health technologies, can bring dramatic improvements in health in developing countries. Conversely, innovations in the health field from developing countries can also be applied in developed countries. For example, oral rehydration therapy (ORT) and CHWs, providing care in developing country conditions, can be applied to unmet needs in industrialized countries.

PRIORITIES IN GLOBAL HEALTH

Poverty and disease are interactive. Sick people have reduced or no capacity to perform well economically. Health is not only a development goal, but also a means to promote development. The eight MDGs agreed upon by all the countries of the world target poverty all over the world with a community-based bottom–up approach as well as national targeted programs which must be integrated at the community level (Table 16.3). Three of the eight MDGs directly address health, and all the others are intricately linked to health. Many other global issues in health include climate change, bioterrorism, global epidemiological surveillance, drug-resistant organisms for tuberculosis (TB), malaria and other infectious diseases, aging, obesity and diabetes, alcohol and drug abuse, violence, prostitution, and human trafficking. Achievements were unequally distributed across and within regions and countries. Moreover, progress has slowed for some MDGs after the multiple crises of 2008–2009. Inadequate levels of health spending by the world's poorest countries are the major factor in continuance of the vicious cycle of poverty and illness.

The MDGs represent an effort by the international community at the highest political level to address the various pressing issues in the world as a global village in which problems of one country cannot be seen in isolation. With 2015 approaching, an MDG monitoring system has been established to help track the progress of individual countries, to learn about new challenges, and to support organizations worldwide working on the MDGs. The first MDG has two targets: to reduce by half, between 1990 and 2015, the proportion of people whose income is less than US$1 a day; and to halve, between 1990 and 2015, the proportion of people who suffer from hunger (Table 16.3).

The WHO's World Health Reports of 2006 and 2007 addressed human resources for health and global health security as their main topics. The deficiency in the trained health workforce in developing counties and the movement of health workers from poor to richer nations are matters of global concern. The revised International Health Regulations (IHR) present a new code for public health based on long traditions derived from the plague and quarantine, cholera and sanitation, and smallpox and immunization. They include issues related to preparedness for chemical and bioterrorist emergencies, and attempts to provide a basis for localization

TABLE 16.3 Millennium Development Goals from 1990 to 2015

1. Eradicate extreme poverty and hunger	Reduce by half the proportion of people with income less than one dollar a day Achieve full and productive employment, decent work for all; including women and young people Halve the proportion of people who suffer from hunger
2. Achieve universal primary education	Ensure by 2015 that children everywhere, boys and girls alike, be able to complete primary schooling
3. Promote gender equality and empower women	Eliminate gender disparity in primary and secondary education, preferably by 2005, and in all levels of education no later than 2015
4. Reduce child mortality	Reduce by two-thirds, between 1990 and 2015, the under-five child mortality rate
5. Improve maternal health	Reduce by three-quarters, between 1990 and 2015, the maternal mortality ratio Achieve, by 2015, universal access to reproductive health
6. Combat HIV/AIDS, malaria, and other diseases	Have halted by 2015 and begun to reverse the spread of HIV/AIDS Achieve, by 2010, universal access to treatment for HIV/AIDS for all in need Have halted by 2015 and begun to reverse the incidence of malaria and other major diseases
7. Ensure environmental sustainability	Integrate principles of sustainable development into country policies and programs and reverse the loss of environmental resources Reduce biodiversity loss, by 2010, a significant reduction in the rate of loss of land area covered by forest; CO_2 emissions, total, per capita; consumption of ozone-depleting substances; proportion of fish stocks within safe biological limits; proportion of total water resources used; proportion of terrestrial and marine areas protected; proportion of species threatened with extinction Halve, by 2015, the proportion of people without sustainable access to safe drinking water and basic sanitation Proportion of population using an improved drinking water source Proportion of population using an improved sanitation facility By 2020, to achieve significant improvement in the lives of at least 100 million slum dwellers
8. Develop a global partnership for development	In cooperation with pharmaceutical companies, provide access to affordable essential drugs in developing countries In cooperation with the private sector, make available the benefits of new technologies, especially information and communications

Note: HIV/AIDS = human immunodeficiency virus; AIDS = acquired immunodeficiency virus; CO_2 = carbon dioxide.
Sources: United Nations, 2005. Available at: http://www.un.org/millenniumgoals/ and http://mdgs.un.org/unsd/mdg/Resources/Attach/Indicators/OfficialList2008.pdf [Accessed 30 April 2013].
United Nations. MDG indicators: all indicators should be disaggregated by sex and urban/rural as far as possible. Effective 15 January 2008. Available at: http://mdgs.un.org/unsd/mdg/host.aspx?Content=indicators/officiallist.htm [Accessed 11 June 2013].

and control of potentially highly dangerous infectious diseases, to prevent them spreading uncontrollably.

Poverty–Illness–Population–Environment

The process and outcome evaluation method is being used to track the work done in this regard by the World Bank, the United Nations Children's Fund (UNICEF), the WHO, and the UN Food and Agriculture Organization (FAO), using indicators such as underweight children under the age of 5 years and the proportion of the population with an income below US$1 per day. Progress is variable across regions and within countries. The global economic crisis since 2008 has been a factor in slower progress for some targets.

The decade 2000–2010 has seen a substantial decline in poverty all over the world, largely because of the economic growth in China, India, Indonesia, Brazil, Russia, and other mid-level income countries, leading to a spurt in the GDP of the countries in Asia, especially in the Indian subcontinent and South-East Asia, and in sub-Saharan Africa. The situation in Brazil, Russia, India, and China (BRIC countries),

South Africa and sub-Saharan Africa has improved, with many countries experiencing high rates of economic growth.

Globally, the number of people living in conditions of extreme poverty, defined as living on less than US$1.25 per person per day, has fallen by 700 million from 1990 to 2010. The first MDG target, of reducing extreme poverty by half (MDG1), has been met well before 2015, but 1.2 billion are still living in extreme poverty. Globally, 384 million workers lived below the $1.25 a day poverty line in 2011, a reduction of 294 million since 2001. Some 870 million people are estimated to be undernourished, including more than 100 million children under the age of 5.

Major economic growth patterns in China and India in recent decades, and more recently in Africa, have created strong middle classes, but rural and urban poverty remains high (UN Goal 1, Eradicate extreme poverty and hunger, 2013). The third target (MDG3), devoted to equality, has been achieved. For MDG2, increased enrollment of boys and girls in primary school has reached 90 percent in developing countries, so this target is also being reached. Box 16.2 indicates

---

BOX 16.2 The World Health Organization (WHO) Agenda, 2013

1. *Promoting development* – During the past decade, health has been recognized as a key driver of socioeconomic progress, and more resources than ever are being invested in health. Yet poverty continues to contribute to poor health, and poor health keeps large populations in poverty. Health development is directed by the ethical principle of equity: access to life-saving or health-promoting interventions should not be denied for economic or social reasons. Activities are aimed at health development with priority to health outcomes in poor and disadvantaged countries or vulnerable groups. The Millennium Development Goals addressing poverty reduction and other programs preventing and treating chronic diseases, and the neglected tropical diseases are cornerstones of the health and development agenda.

2. *Fostering health security* – Great threats to international health security arise from outbreaks of emerging and epidemic-prone diseases. These are occurring in increasing numbers, fueled by rapid transportation, urbanization, environmental mismanagement, the way food is produced and traded, and the way antibiotics are used and misused. Shared vulnerability to health security threats demands joint action to collectively defend against outbreaks. Collective security is crucial to identify and control potential pandemics. This was strengthened by newly revised International Health Regulations (2007), and absorption of lessons learned from the HIV, SARS, and subsequent pandemics.

3. *Strengthening health systems* – For health improvement to help reduce poverty, health services must reach poor and underserved populations, especially those in neglected rural and urban slums. Health systems in many parts of the world are unable to do this, so strengthening of health

systems is a high priority. This includes addressing the provision of adequate numbers of appropriately trained staff, sufficient financing, suitable systems for collecting vital statistics, with access to appropriate technology and essential drugs, with priority on local primary care.

4. *Harnessing research, information, and evidence* – Evidence and awareness of "best practices" provides the foundation for setting priorities, policies, defining strategies, and measuring results. WHO generates authoritative health information, in consultation with leading experts, to set norms and standards, articulate evidence-based policy options, and monitor the evolving global health situation.

5. *Enhancing partnerships* – WHO works with the support and collaboration of many partners, including UN agencies and other international organizations, donors, civil society and the private sectors. The role of donors in advancing health in developing countries must encourage national governments to increase their allocation and commitment to health as a priority. WHO uses the strategic power of evidence to encourage partners implementing programs within countries to align their activities with best technical guidelines and practices, as well as with the priorities established by member countries.

6. *Improving performance* – WHO participates in reforms to improve its efficiency and effectiveness, both at the international level and within countries. WHO aims to enhance its work in a motivating and rewarding environment at country, regional and global levels.

Note: HIV=human immunodeficiency virus; SARS=severe acute respiratory syndrome; UN=United Nations.
Source: *Adapted from World Health Organization. The WHO agenda. Geneva: WHO; 2013. Available at: http://www.who.int/about/agenda/en/index.html [Accessed 30 April 2013].*

current WHO thinking on the agenda for the post-MDG period, i.e., after 2015.

The seventh MDG has three targets which deal with sustainable environment, sanitation, and housing (for at least 100 million slum dwellers). This is a crucial goal as the great economic growth of many countries will be halted because of environmental degradation. Table 16.4 shows important indicators from developing countries on their track to achieve the MDGs.

The interactions among poverty, population growth, and environmental degradation combine to adversely affect many developing countries and hundreds of millions of people globally. In many developing countries, economic stagnation and political instability compound these issues, causing an inability to address basic human needs and condemning more generations to ill-health and early death. Although the effects of low income, lack of basic sanitation, and crowding in rural poverty or the slums of megacities cannot be overcome by health measures alone,

the potential for raising the quality of life and survival rates by public health measures is very great. The leading causes of disease and mortality among adults are shown in Tables 16.5–16.7. These tables indicate the leading causes of death globally are ischemic heart disease, stroke and respiratory diseases. There is a commonality between low and medium income countries with upper income countries so that cost-effective preventive measures are needed for NCDs for countries at all economic levels. Medium and even low income countries need to priorize prevention for NCDs as much as for HIV, vaccine preventable diseases and maternal-child health and nutrition issues.

Industrialized countries have made great progress in reducing mortality from infectious diseases as well as NCDs, but there is still a long way to go in coping with very substantial pockets of poverty, homelessness, violence, preventable disease, environmental degradation, and limited access to health care. The southern hemisphere is largely made up of developing countries with massive economic

TABLE 16.4 Global Progress in Reducing Child Mortality Rate: Average Annual Rate of Reduction in Under-Five Mortality Rate (Deaths per 1000 Live Births) for 1990–2011, and Percentage Decline During 1990–2011

Region	Under-5 Mortality Rate		MDG Target	Decline, 1990–2011 (%)	Average annual rate of reduction, 1990–2011 (%)	Progress Towards MDG Target 2011
	1990	2011				
Developed regions	15	7	5	55	3.8	On track
Developing regions	97	57	32	41	2.5	Insufficient progress
North Africa	77	25	26	68	5.5	On track
Sub-Saharan Africa	178	109	59	39	2.3	Insufficient progress
Latin America and Caribbean	53	19	18	64	4.8	On track
Caucasus and Central Asia	76	42	25	44	2.8	Insufficient progress
East Asia	48	15	16	70	5.7	On track
Excluding China	28	17	9	38	2.3	On track
South Asia	116	61	39	47	3.1	Insufficient progress
Excluding India	119	60	40	50	3.3	Insufficient progress
South-East Asia	69	29	23	58	4.1	On track
West Asia	63	30	21	52	3.5	On track
Oceania	74	50	25	33	1.9	Insufficient progress
World	87	51	29	41	2.5	Insufficient progress

Note: The table shows progress towards Millennium Development Goal 4 (MDG4), with countries classified according to the following thresholds. On track = under-five mortality was < 40 deaths per 1000 live births in 2011 or that the annual rate of reduction was at ≥ 4% over 1990–2011; insufficient progress = under-five mortality was ≥ 40 deaths per 1000 live births in 2011 and that the annual rate of reduction was ≥ 1% but < 4% over 1990–2011.
All calculations are based on unrounded numbers. These standards may differ from those in other publications by Inter-agency Group for Child Mortality Estimation members.
Source: United Nations Children's Fund. Levels and trends in child mortality report 2012. Estimates developed by the UN Inter-agency Group for Child Mortality Estimation. New York: UNICEF; 2012. Available at: http://apromiserenewed.org/files/UNICEF_2012_child_mortality_for_web_0904.pdf [Accessed 30 May 2013].

and social needs. The north–south socioeconomic divide is one that will shape global health and politics in the twenty-first century. Table 16.6 shows the prevalence of risk factors for causes of death in countries grouped by income levels. Table 16.7 indicates the disability-adjusted life years (DALYs) lost by disease risk factors which shows the effects on younger ages in the population (see Chapters 3 and 11).

During the 1990s, a number of developing countries entered a phase of rapid economic and industrial growth, combining the advantages of educated low-wage workforces with market economies. Some Asian countries moved ahead rapidly with economic development, creating strong rates of growth. The breakdown of traditional social patterns, traditional rural ways of life, and intergenerational family structure to attain better education, upward mobility, and small family units in South-East Asian economies will compound these problems.

The recognition that poverty and ill-health are interactive led the industrialized nations (G7 plus Russia) to take an important step in mid-1999 to relieve debt and provide aid-related loans to very poor countries by some US$118 billion. This helped many countries in sub-Saharan Africa by reducing their debt repayment by one-third to one-half. Despite this important step, most poor countries pay more on servicing debts than they do on health and education for their people.

Child Health

Almost 19 percent of total deaths in the world (10.5 million) are of children under 5 years of age and almost 98 percent are in the developing world. The figures seem to have improved in many countries since 1970 when the figure was 17 million, but in 14 African countries the present levels of under-five mortality are higher than they were in 1990. About 35 percent of Africa's children are at a higher risk of death than they were 10 years ago. Nineteen African countries are in the list of the top 20 countries in the world with the highest rates of under-five mortality. A baby born in an African country, for example, in Sierra Leone, is three and a half times more likely to die before its fifth birthday than a child born in India, and more than 100 times more likely to die than a child from a developed European country or even Singapore (WHO, 2003). The leading causes of under-five mortality in the poor countries are perinatal conditions, lower respiratory tract infections, diarrheal diseases, and malaria. Sub-Saharan Africa has undergone a severe onslaught by the HIV epidemic, wiping out an estimated 332,000 children in 2002 (WHO, 2003). There is a growing health inequality in children's health all over the world, with a higher rate of death if they are poor and undernourished. Table 16.8 shows progress made in MDG indicator 4 (child mortality) for 1990 and 2010, along with infant mortality by UNICEF regions.

Most deaths in children under the age of 5 are preventable, as can be seen from the list given in Table 16.9. The 10 leading causes listed in the table constitute 86 percent of all causes of under-five mortality. Five diseases – pneumonia, diarrhea, malaria, measles, and acquired immunodeficiency syndrome (AIDS) – account

---

TABLE 16.5 Ten Leading Causes of Mortality Worldwide, 2008

Cause	Deaths (thousands)	% of Deaths
Ischemic heart disease	7,250	12.8
Stroke and other cerebrovascular disease	6,150	10.8
Lower respiratory infection	3,460	6.1
Chronic obstructive pulmonary disease	3,280	5.8
Diarrheal disease	2,460	4.3
HIV/AIDS	1,780	3.1
Trachea, bronchus, and lung cancers	1,390	2.4
Tuberculosis	1,340	2.4
Diabetes mellitus	1,260	2.2
Road traffic accidents	1,210	2.1
Subtotal	29,580	52.0
Total estimated global deaths	57,000	100.0

Note: Total deaths globally estimated as 57 million, 2008.
HIV/AIDS = human immunodeficiency virus; AIDS = acquired immunodeficiency virus.
Source: World Health Organization. Top ten causes of death. Fact sheet no. 310 [updated June 2011]. Available at: http://www.who.int/mediacentre/factsheets/fs310/en/ [Accessed 30 May 2013].

TABLE 16.6 The 10 Leading Causes of Death by Income Group, 2011

Country Income Groupings	Leading Causes of Death	Deaths per 1,000 Population	Amenable to Preventive Care
High income countries	Ischemic heart disease	119	+++
	Stroke	69	++++
	Lung, trachea, bronchus cancer	51	++++
	Alzheimers disease, other dementias	48	+
	COPD	32	++
	Lower respiratory infections	32	+++
	Colorectal cancers	27	+++
	Diabetes mellitus	21	++
	Hypertensive heart disease	20	+++
	Breast cancer	18	+++
Upper middle income countries	Stroke	126	++++
	Ischemic heart disease	120	+++
	COPD	45	++
	Lung trachea, bronchus cancer	28	++++
	Lower respiratory infections	22	+++
	Road injury	21	+++
	Diabetes mellitus	20	++
	Liver cancer	19	++
	Hypertensive heart disease	18	+++
	Stomach cancer	18	+++
Lower middle income countries	Ischemic heart disease	93	+++
	Stroke	75	++++
	Lower respiratory infections	60	+++
	COPD	51	+++
	Diarrheal diseases	47	+++
	Prematurity	27	+++
	HIV/AIDS	24	++
	Tuberculosis	22	++
	Diabetes mellitus	20	++
	Road injury	19	+++
Low income countries	Lower respiratory infections	98	+++
	HIV/AIDS	70	++
	Diarrheal diseases	69	+++
	Stroke	56	+++
	Ischemic heart disease	47	+++
	Prematurity	43	++
	Malaria	38	++
	Tuberculosis	32	++
	Protein energy malnutrition	32	++++
	Birth asphyxia and birth trauma	30	++

Note: WHO Member States are classified according to the World Bank income categories for the year 2011 (World Bank list of economies, July 2012). Estimate of "amenable to preventive care" by text authors.
Source: Adapted from World Health Organization. The top 10 causes of death. Available at: http://www.who.int/mediacentre/factsheets/fs310/en/ (accessed 14.1.14.)

for about half of under-five deaths. Most of these lives can be saved by expanding coverage of existing interventions, especially among poor families using a bottom–up approach. Many vaccine-preventable diseases are listed in the top causes, and it has been seen, as with polio, that a concerted effort by health authorities in promoting immunization can dramatically reduce child mortality. With simple methods like ORT, mortality related to diarrhea can be reduced within a short time. This calls

for a range of health interventions, including greater emphasis on:

- contraceptive care – acceptability, availability, costs and use
- antenatal care – accessibility and quality
- skilled attendants at delivery – accessibility, hygiene and quality; vitamin K injection and antibiotics for newborns

TABLE 16.7 Ranking of Selected Risk Factors: 10 Leading Risk Factor Causes of Disability-Adjusted Life Years (DALYs) by Income Group, 2004

Rank	Risk Factor	DALYs (millions)	% of Total
(a) World			
1	Childhood underweight	91	5.9
2	Unsafe sex	70	4.6
3	Alcohol use	69	4.5
4	Unsafe water, sanitation, hygiene	64	4.2
5	High blood pressure	57	3.7
6	Tobacco use	57	3.7
7	Suboptimal breastfeeding	44	2.9
8	High blood glucose	41	2.7
9	Indoor smoke from solid fuels	41	2.7
10	Overweight and obesity	36	2.3
(b) Low-income countries			
1	Childhood underweight	82	9.9
2	Unsafe water, sanitation, hygiene	53	6.3
3	Unsafe sex	52	6.2
4	Suboptimal breastfeeding	34	4.1
5	Indoor smoke from solid fuels	33	4.0
6	Vitamin A deficiency	20	2.4
7	High blood pressure	18	2.2
8	Alcohol use	18	2.1
9	High blood glucose	16	1.9
10	Zinc deficiency	14	1.7
(c) Middle-income countries			
1	Alcohol use	44	7.6
2	High blood pressure	31	5.4
3	Tobacco use	31	5.4
4	Overweight and obesity	21	3.6
5	High blood glucose	20	3.4
6	Unsafe sex	17	3.0
7	Physical inactivity	16	2.7
8	High cholesterol	14	2.5
9	Occupational risks	14	2.3
10	Unsafe water, sanitation, hygiene	11	2.0
(d) High-income countries			
1	Tobacco use	13	10.7
2	Alcohol use	8	6.7
3	Overweight and obesity	8	6.5
4	High blood pressure	7	6.1
5	High blood glucose	6	4.9
6	Physical inactivity	5	4.1
7	High cholesterol	4	3.4
8	Illicit drugs	3	2.1
9	Occupational risks	2	1.5
10	Low fruit and vegetable intake	2	1.3

Note: [a]Countries grouped by gross national income per capita: low income=US$825 or less; high income=US$10,066 or more.
Source: Adapted from World Health Organization. 2009. Global health risks: mortality and burden of disease attributable to selected major risks. Geneva: WHO; 2009. Available at: http://www.who.int/healthinfo/global_burden_disease/GlobalHealthRisks_report_full.pdf [Accessed 28 May 2013].

TABLE 16.8 Infant and Child Mortality Rate, 1990–2010

	Infant Mortality		Child Mortality	
	1990	2010	1990	2010
World	61	40	88	57
Least developed countries	106	71	170	110
Developing countries	67	44	97	63
Industrialized countries	9	5	10	6
Sub-Saharan Africa	105	76	174	121
Latin America and the Caribbean	43	18	54	23
East Asia and Pacific	41	19	55	24
South Asia	86	52	120	67
CEE/CIS	41	19	50	23

Note: Deaths of children before reaching the age of 1 year, per 1000 live births.
CEE = Countries of Eastern Europe; CIS = Commonwealth of Independent States (Russia, Ukraine, Bylorussia, etc.); NA = not available.
Sources: United Nations Children's Fund. State of the world's children 2012. Table 1: Basic indicators. Available at: http://www.unicef.org/sowc/files/SOWC_2012-Main_Report_EN_21Dec2011.pdf [Accessed 6 June 2013].

TABLE 16.9 Global Causes of Death Among Children Under 5 Years, 2011

Cause of Death	% of Under-5 Deaths
Pneumonia	18
Preterm birth complications	14
Diarrhea	11
Birth asphyxia	9
Malaria	7
Other causes	41
Under-5 deaths, total causes	6.9 million

Note: More than 40% of child deaths occur during the neonatal period. Other causes include congenital anomalies, pertussis, tetanus, HIV/AIDS, and malnutrition.
Source: World Health Organization. Children: reducing mortality. Fact sheet no. 178. September 2012. Available at: http://www.who.int/mediacentre/factsheets/fs178/en/ [Accessed 1 June 2013].

- postnatal care for infant and mother – promoting breast-feeding, hygiene, immunization, bed-net use, and follow-up
- child health – nutrition, vitamin and mineral supplements, immunization, growth and development monitoring
- financial investments in maternal, newborn and child health with priority in planning
- equity of access, health systems, and policy in universal health care development (WHO, 2010).

Between 2000 and 2010, the global burden of deaths in children younger than 5 years decreased by 2 million, of which pneumonia, measles, and diarrhea contributed the most to the overall reduction (0.451 million, 0.363 million, and 0.359 million, respectively). However, only tetanus, measles, AIDS, and malaria (in Africa) decreased at an annual rate sufficient to attain MDG4 (Liu et al., 2012). Table 16.9 shows the leading causes of death of infants and children up to the age of 5 globally, with 40 percent being in the first month of life (neonatal). But progress is being made with deaths due to vaccine-preventable diseases (diphtheria, measles, neonatal tetanus, pertussis and poliomyelitis) estimated to have fallen from 0.9 million in 2000 to 0.4 million in 2010 (GAVI 2012).

MDG4 is to reduce the under-five mortality rate by two-thirds between 1990 and 2015. Although child mortality has fallen by more than one-third progress is still too slow to reach the target (Figure 16.2). The most recent 2012 Millennium Goals Report reveals rural–urban inequities, with children in rural areas at a greater disadvantage for higher mortality.

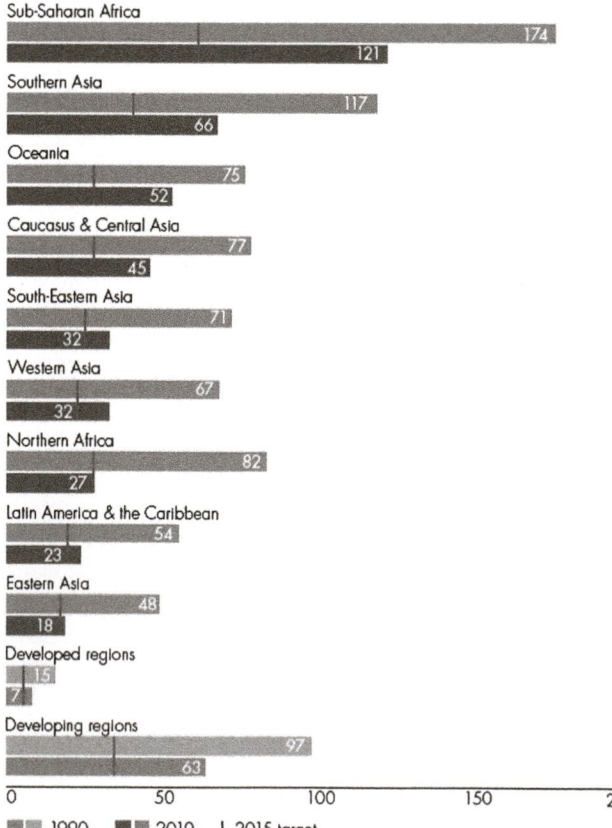

Under-five mortality rate, 1990 and 2010 (Deaths per 1,000 live births)

■ 1990   ■ 2010 | 2015 target

FIGURE 16.2 Progress in achieving reduction in child mortality in World Health Organization regions, 1990–2011. *Source: United Nations. Millennium Development Goals Report 2012. Goal 4: Reduce child mortality. Available at: http://www.un.org/millenniumgoals/pdf/MDG%20Report%202012.pdf [Accessed 2 June 2013].*

Maternal Health

The fifth target (MDG5), as seen in the previous section, is to reduce the maternal mortality ratio by three-quarters between 1990 and 2015. In developing countries, complications in the antepartum, intrapartum, and postpartum periods are leading causes of death among women. Every year, approximately 9 million women suffer some form of injury from pregnancy or childbirth which has lasting effects on their health. Africa and Asia have the majority of pregnancy-related deaths, accounting for 95 percent of the total maternal deaths in the world.

Globally, an estimated 287,000 women died during pregnancy and childbirth in 2010, a decline of 47 percent from levels in 1990. The main cause is poor access to skilled routine and emergency care. However, there is progress as some countries are on track to meet the MDG5 target of reducing maternal mortality by three-quarters by 2015. The global maternal mortality ratio in 2008 was 260 per 100,000 live births. There was wide variation among WHO regions and by level of GDP per capita country grouping. Maternal mortality ratios for the regions were: Africa 620, South-East Asia 240, Eastern Mediterranean 320, Western Pacific 51, and the Americas 66 per 100,000 live births. Rates among the countries by level of income were: low income 580 per 100,000, lower mid-level income 230, upper mid-level income 82, and high level of income 15 per 100,000 (WHO Statistical Information System, 2008). In 2005, 358,000 women

died of maternal causes, compared to 546,000 in 1990. Ninety-nine percent of these deaths occurred in developing countries. Maternal mortality was estimated as 287,000 maternal deaths in 2010 worldwide, a decline of 47 percent from 1990, but the levels in many countries were still far above the 2015 target. In the European Region progress has been seen in all sectors: EU members before 2004, new EU members, CIS countries, and Central Asian Republics (Figure 16.3).

Institutionalized delivery or the presence of a skilled attendant decreases the risk of maternal deaths to a great extent. The number of deliveries with a skilled attendant increased significantly between 1990 and 2003, from 41 to 57 percent, and increased to 66 percent between 2007 and 2012 (UN DESA, 2005; UNICEF, 2013).

The health of women in relation to fertility is fundamental to national health standards. Education and improved nutrition for girls and women, better access to modern birth control, spacing of pregnancies, and adequate care in all stages of pregnancy are the vital means to achieve improvement in women's reproductive health. Traditional birth attendants (TBAs) provide care during most deliveries in developing countries. There are no adequate substitutes for good prenatal medical care, but the work of TBAs can be improved by a strict program of licensing, training, and supervision (see Chapter 14).

Simple, inexpensive measures can improve outcomes: education about the right to care for safe pregnancy; routine iron and folic acid during pregnancy; prenatal care

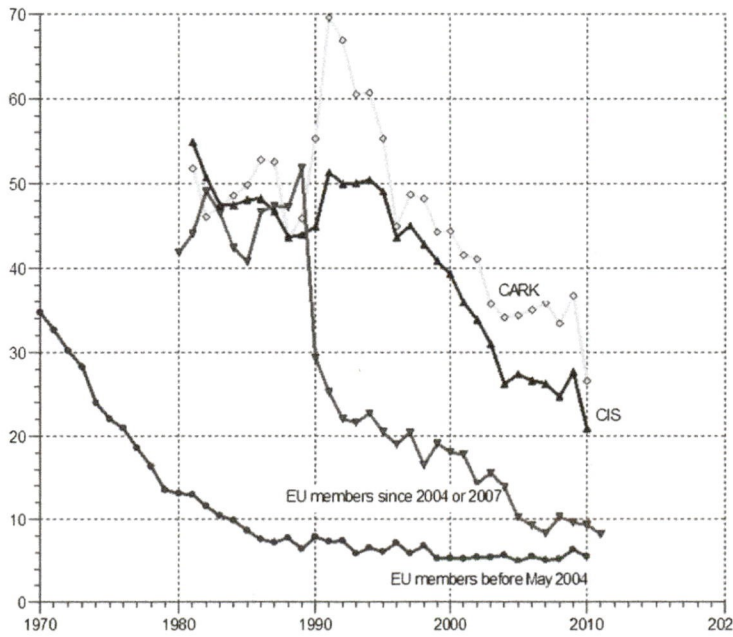

FIGURE 16.3   Maternal mortality ratios in sectors of the World Health Organization European Region, 1970–2011. Note: EU = European Union; CARK = Central Asian Republics; CIS = Commonwealth of Independent States. *Source: World Health Organization, European Region. Health for All database; January 2013. Available at: http://data.euro.who.int/hfadb/ [Accessed 30 May 2013].*

stations (maternal and child health); HIV and sexually transmitted infection (STI) screening and care; professionally supervised birth centers (in hospitals if possible); high-risk identification and referral systems (see Chapter 5); training, licensing, and supervision of traditional birth attendants; and deployment of well-trained CHWs for preventive health care (see Chapter 15) can all make a difference.

The failure to advance sufficiently in lowering maternal mortality in many countries is largely a failure of political will and initiative. The means (knowledge) to prevent most of these deaths is available, but applying the national and international will and investment of necessary resources to this task has been wanting. Safe motherhood has taken a back seat to HIV and child health in priorities in many countries with a high maternal mortality rate, and stagnation of effort and results has followed. A combination of international and national organization, improved databases, and program entrepreneurship is needed to reduce the staggering toll of maternal mortality, even in countries such as Nigeria with oil wealth, and India with rapid economic development.

Population Growth

Despite the fact that population growth is a religious and political controversy in many societies, falling birth rates are now seen in most parts of the world. Developing countries are increasingly recognizing that high fertility rates hinder economic development, perpetuating poverty, a fundamental cause of ill-health. The politics of population have traditionally rested on the assumption that population increase is essential for economic growth and national power. At the microlevel in traditional farming societies the assumption is that more children provide greater security for the family. In recent decades the expansion of family planning technology has been accompanied by a gradual shift to the view that unrestrained population growth is a barrier to economic development.

In many poor countries, high rates of population growth perpetuate poverty and ill-health for mothers and children. Improved child survival and reduced economic imperatives for more children to work farms or to contribute to family incomes have led most countries to lower overall birth rates. Higher education levels for women have increased knowledge and use of birth control. Religious injunctions against birth control no longer have the power to prevent use, so that birth rates have fallen worldwide, and in many countries to below replacement levels (i.e., negative population growth).

Despite the decline in fertility rates in most regions of the world, the world population has passed the 7 billion mark, is growing at an average annual rate of 1.73 percent, and will reach a predicted level of more than 9 billion in 2050. Asia

accounts for almost 60 percent of the world's population, and this will decline to 57 percent in 2050, while Europe is about 12 percent, but declining to an estimated 7.2 percent in 2050. Africa accounts for just over 13 percent of the world population but will reach a forecast 21.7 percent in 2050. Fertility rates are crucial for economic growth, as has been demonstrated in countries around the world. In many Asian countries, the birth rate has declined precipitously to rates close to those in developed countries. Over the period 1970–2011, crude fertility rates in East Asia and the Pacific fell from 36 per 1000 population, to 27 per 1000 population in 1990, and to 14 per 1000 population in 2011. In Latin America and the Caribbean crude birth rates declined from 36 to 18 per 1000 population from 1970 to 2011, and in the Middle East and North African Region from 44 per 1000 to 24 per 1000 in the same period. By 2011, the crude fertility rate declined from 47 to 37 per 1000 population. In CCE and CIS crude birth rates declined from 20 to 14 per 1000 population (UNICEF, 2013). Table 16.10 shows population trends from 1960 to 1910 and projections to 2020 and 2030.

Many countries in sub-Saharan Africa will see their population double within 20 years, although even here there has been what appears to be the beginning of a decline in total fertility rates. Population growth is below replacement level in most industrialized countries and falling in most developing countries as well, but the fertility gap is still high, with growth rates of 1.5 percent currently and projected to decline to 0.5 percent by 2030. In 2011 the world fertility rate was 2.5, and 4.2 for developing countries (UNICEF, 2013).

Governments have a crucial role in family planning. Distribution of information and promotion of family planning as a national policy and priority must be part of a new emphasis on primary health care. In China in the 1950s, Chairman Mao Tse Tung called birth control a new form of

TABLE 16.10 World Population in Billions, by Development Level, Medium Variant, 1965–2030

Level of Development	1960	1980	2000	2010	2020	2030
Least developed regions	0.2	0.4	0.7	0.8	1.0	1.3
Less developed regions	1.8	3.0	4.3	4.8	5.3	5.8
More developed regions	0.9	1.1	1.2	1.2	1.3	1.3
Total world population	3.0	4.5	6.1	6.9	7.7	8.3

Source: United Nations, Population Division of the Department of Economic and Social Affairs of the United Nations Secretariat. World population prospects: the 2010 revision. Available at: http://esa.un.org/unpd/wpp/unpp/panel_population.htm [Accessed 29 May 2013].

genocide of the developed countries against the developing countries. The legacy of this tragic pronouncement was no less destructive to public health than the pronouncements of religious bodies that still equate birth control with mortal sin. Both had the effect of promoting high fertility rates in those populations that can least afford the health and economic burden of raising large numbers of children. For the past four decades, birth control has been promoted in India and China, the two countries with the world's largest populations, but the momentum of population growth continues and is unlikely to level off in the next 20 years. In addition, China's one-child-per-family policy has reportedly led to female infanticide, forced abortions, and sterilization in a primarily rural society valuing male children.

A demographic transition occurs when the age makeup of the population shifts. As countries move from developing to a developed or industrialized status, population age patterns change. With greater life expectancy and declining birth rates, the ages of the population shift towards older age groups. Developed countries are experiencing a rapid growth of the "old-old" and "oldest-old" (i.e., over 75 and over 85 years old, respectively), more dependent population. These trends are of vital importance to the future of individual countries as they try to sustain or improve economic and social conditions. All countries need a working-age population sufficient to sustain the elderly and the young dependent groups.

High birth rates in developing countries still restrict the potential for economic growth, and the care and nurturing of children. Food supplies have been expanded by improved agriculture, but this may not be able to sustain high rates of population growth. In addition, rising standards of living and aspirations place further demands on natural resources and the environment, with great stress on the Earth's fragile ecology.

Malnutrition

Food production has increased in most parts of the world, but has steadily declined per capita in sub-Saharan Africa, although gross national product (GNP) per capita increased from US$751 in 2004 to US$1447 in 2011. Increased production in other parts of the developing world during the 1960s and 1970s slowed during the 1980s. In developing countries, the capacity to produce food faster than population growth is limited. The developed countries, with one-quarter of the world's population, produce over half of the world's food supply. They dominate food production but have low rates of population growth. Developing countries may purchase this surplus of food, but many lack the hard currency to do so. GNP alone cannot measure wealth; it must also be weighed in terms of the capacity to produce food, its affordability and ease of access, awareness of nutritional preferences for good health and food quality and

safety, and some basic foods fortified with essential minerals and vitamins.

Hunger, adaptation, and starvation are difficult to measure. Hunger is a subjective phenomenon; adaptation occurs when people adjust to lower energy intake; and when energy output exceeds intake, starvation occurs. Starvation may be acute or chronic. Hunger and famine are associated with natural disasters and war, but they also occur chronically in settings where food production cannot keep up with population growth. Although hunger and famine affect all ages and genders, the most vulnerable groups in the population are infants and children, pregnant women, women as a whole, and the elderly. Men are affected in terms of reduced capacity to work. The Chinese famine of 1959–1961, one of the most tragic disasters of the twentieth century, killed up to 36 million people.

Estimating the number of people lacking food is difficult because of limited data. Few countries maintain monitoring systems of national nutrition status because of a lack of financial and human resources. The nutritional status of the population, more specifically of children, is usually measured by birth weight, weight-for-age, and height-for-age. Low height for a given age, or stunting, is the most prevalent symptom of protein–energy malnutrition (PEM). Approximately 40 percent of all 2-year-olds in developing countries are stunted. The prevalence of stunting may be as high as 65 percent in India, about 40 percent in China and sub-Saharan Africa, and more than 50 percent in the rest of Asia. According to WHO standards, some 780 million people worldwide are estimated to be energy deficient or in a state of PEM. This is not always manifested by hunger, but rather represents long-term inadequate food intake, especially protein and calories for energy needs. Malnutrition may be so widespread among children that parents and health providers assume the children's lethargy and stunting to be normal.

Micronutrient deficiency conditions affect some 2 billion people worldwide, with serious sequelae including premature death, poor health, blindness, growth stunting, mental retardation, learning disabilities, and low work capacity. Iodine, iron, and vitamins A, B, C, and D are commonly deficient in diets in developing countries, adversely affecting the health of the whole population but especially vulnerable subgroups. Iron deficiency is the most common of these, affecting mainly women and children but also men and the elderly. In developing countries, children and women are especially vulnerable because of frequent childbirth and poor diets. Poor nutrition is the underlying contributor to excess deaths of children and women in relation to pregnancy and early life of the premature newborn, and it is underaddressed in global program implementation.

Iron deficiency and iron-deficiency anemia are, as discussed in Chapter 8, the commonest nutritional deficiencies globally. They are primarily associated with low dietary intake of iron rich foods, female gender and menstrual functions, single or multiple pregnancies and other deficiencies such as vitamin C, as well as concomitant infections or parasitic diseases or other chronic conditions such as renal disease. In industrialized countries, anemia of pregnancy affects 18 percent of pregnant women, but 40 percent are affected in China and Latin America and 88 percent in India. Iron deficiency in Russian women exceeds 50 percent, and iron supplementation is not routinely practiced; there is a lack of hemoglobin testing. Iron deficiency in infancy causes reduced growth (in height) and potential learning capacity in school. In adults, it reduces work potential. Distribution of inexpensive iron (ferrous sulfate) to pregnant and lactating women could largely prevent this onerous health burden, which is estimated by the WHO to affect 1.8 billion people.

Iodine-deficiency disorders (IDDs) have been identified as one of four key global risk factors for impaired child development. Iodine deficiency affects some 1.88 billion people: almost one-third of the global population, including 241 million schoolchildren, still lack sufficient dietary iodine intakes. Annually, 38 million or nearly 30 percent of the world's newborns born every year are unprotected from brain damage due to IDD. The severity of iodine deficiency varies from subclinical deficiency to cretinism and severe retardation. Even subclinical brain damage of the newborn handicaps the affected child for life. Iodine is deficient in soil and water in many parts of the world, and deficiency conditions at subclinical and clinical levels are widespread. Routine iodization of salt has been recommended by the WHO, UNICEF, and many other organizations for decades. It is widely used to prevent IDDs and has been adopted as a major objective of the 1990 World Summit of Children, along with elimination of vitamin A deficiency, but implementation remains problematic even in Western Europe, and more so in low- and medium-income countries. Global iodine fortification has markedly improved over the past decade and the number of iodine-deficient countries decreased from 54 in 2003 to 32 in 2011.

Iodine deficiency remains a major global health problem, even in countries with iodization, but is poorly monitored in Europe, the Eastern Mediterranean, and sub-Saharan Africa. Regions with the greatest proportions of children with inadequate iodine intake are reported to be in the European (52.4 percent) and Eastern Mediterranean (48.8 percent) Regions. The International Council for the Control of Iodine Deficiency Disorders (ICCIDD) works with a network for the Sustained Elimination of Iodine Deficiency, including UNICEF, WHO, Kiwanis International, and the US Centers for Disease Control and Prevention (CDC), and the global salt industry. The WHO estimates the cost of eradication of iodine deficiency by iodination of salt at US $0.05 per person per year (Salt Institute, 2011; Andersson, 2012). Progress is being made, with some outstanding successes in salt fortification in countries such as China, Nigeria, and Georgia in the past decade, but much more remains to be done.

Vitamin A supplements reduce mortality from measles and prevent xerophthalmia and blindness in children. This knowledge has created a major change in public health nutrition needs in developing countries by demonstrating nutritional comorbidity and the vital importance of nutritional interventions to prevent morbidity and mortality in vulnerable population groups. The extent of iodine and vitamin A deficiency conditions is enormous and entirely within the scope of current technology to prevent at low cost. Elimination of vitamin A deficiency can be achieved by giving vitamin A capsules to children over 6 months of age three times per year at a cost of US$0.10 per capsule, dietary modification to promote vitamin A-rich foods, and/or fortification of basic foods (oil, margarine, milk, or sugar).

Food fortification is now recognized as a major need and cost-effective intervention necessary in all countries, particularly those in the middle and lower levels of development. Since the 1990s, fortification of flour with folic acid has been implemented in a number of countries to prevent birth defects (neural tube defects). This has provided a new impetus to promote food fortification, and new deficiency conditions of public health importance are reported for vitamin D, vitamin K, vitamin B complex (including B_2), selenium, zinc, and others (see Chapter 8). The WHO issued new guidelines for food fortification in 2006, which have great importance in international aid and development policies for the second decade of the twenty-first century.

In many countries in the African, South-East Asian, and Eastern Mediterranean Regions, infectious and parasitic diseases occur in association with malnutrition and continue to be major public health problems. The infection–malnutrition comorbidity causes much of the mortality among infants and children and shortens life expectancy (see Chapter 6).

Nutritional security, including food adequacy and prevention of micronutrient deficiencies, along with the prevention of obesity and diabetes, are central issues in improving health and reducing mortality globally, and should be given a central place in post-MDG and global funding priorities.

The Fight Against HIV/AIDS and Other Communicable Diseases

Globalization of the spread of disease is as old as the migration of humans, animals, or disease vectors. The emergence of the AIDS pandemic has affected all countries of the world, regardless of their level of development. In 2008 the WHO lowered its estimates of HIV-infected people globally to between 30 and 37 million: it had previously been estimated that around 40 million people in the world are infected with HIV, of which Africa alone has 25 million. More than 10 million children in sub-Saharan Africa have been orphaned by AIDS. The prevalence rate of HIV infection among adults in sub-Saharan Africa was estimated to be more than 7 percent in 2004. Although the prevalence rate of HIV in countries in Asia, especially in India and China, seems to be low, there is a definite potential danger of epidemics if not controlled soon. The youngest population of the world resides in China and India and worldwide about one-third of those currently living with HIV/AIDS are between 15 and 24 years of age. This figure is rising and with the rising level of substance abuse there is an urgent need for concerted international action.

As discussed in Chapter 4, the emergence of "new" infectious diseases and the reemergence of well-known but still uncontrolled diseases pose great challenges for public health and clinical care. The problems of these diseases are compounded by the rise of resistant microbial strains. The basic priorities in control of infectious diseases remain the need for universal coverage with childhood immunization; high standards of food and water safety and sanitation; education to reduce the spread of HIV and STIs; improved primary care for prevention, diagnosis, and management of TB and malaria; and provision of antimicrobial therapy. Education and behavior are still crucial, while new interventions such as circumcision and condom use reduce transmission. It is to be hoped that an effective, safe, and inexpensive vaccine against HIV will soon be developed.

The HIV epidemic has engulfed the economies of many nations and has been the cause of the rising spread of poverty, reversal of human development, worsening health inequalities, and crippling government machineries in various parts of the world, thus reducing the provision of essential services. The very obvious health inequalities observed in the world led the WHO to declare a global health emergency to combat HIV/AIDS when it was found that only 5 percent of those in the developing world who require antiretroviral therapy (ART) receive it. It will be a pity if one part of the world is oblivious to the health situation in the other, not only for moral reasons but also for practical ones; as the world is a global village, ill-health in one part of the world will definitely affect the other parts.

Malaria remains endemic in many poor countries, especially in the tropical and subtropical regions of Africa, the Americas, and Asia. Although successful treatment and prevention methods for malaria have been available for a long time, there are still an estimated 154–289 million clinical malaria cases, and about 600,000 (uncertainty range of 490,000–836,000) deaths from malaria every year. Approximately 3.3 billion people, almost half the world's population, live in malaria-endemic areas, making them vulnerable to the disease. Increased prevention and control measures have led to a reduction in malaria mortality rates by more than 25 percent globally since 2000 and by 33 percent in the WHO African Region. Factors include the increasing availability of bed nets impregnated with insecticide, indoor spraying with insecticide, greater awareness of the importance of vector control, and improving diagnostic procedures. However, resistance to currently available drugs is a worrying factor in malariology and the long hoped-for vaccine still remains an unfulfilled dream. Genetic modification of mosquitoes to prevent their transmission of the parasite, and biological methods of vector control by larvae-eating bacteria are promising but still not game changers in malaria control.

Tuberculosis is a disease which experts thought could be eradicated in the 1970s, but it reemerged as one of the major killer diseases in the world, partly because of comorbidity with HIV. If TB is detected early and fully treated, people with the disease quickly become non-infectious and are eventually cured. Most are treated with directly observed therapy, short course (DOTS), which has been highly effective in completion of treatment that is needed even after symptoms recede. In 2011, an estimated 5.8 million new cases of TB occurred; there were 1.4 million deaths, including half a million HIV-associated TB deaths. TB is a global phenomenon but is mainly restricted to poor countries where it has been increasing, particularly in sub-Saharan Africa and South Asia, and is compounded by the additional problem of multidrug-resistant organisms. Multidrug-resistant tuberculosis (MDR-TB) and extremely drug-resistant tuberculosis (XDR-TB), HIV-associated TB, and weak health systems are major challenges. TB is high on the agenda of the MDGs, with the goal to reduce the burden of TB, and the target to reduce by half TB deaths and prevalence by 2015 has been achieved with the help of international donor agencies and national governments. New cases of TB have been falling since 2006 and fell at a rate of 2.2 percent in 2011.

Important policy and technological advances are needed before TB, HIV, and malaria can be controlled globally, but much can be accomplished with existing technology. Providing ART and risk-reduction measures at primary care levels, along with antimalarial activities, is essential to control current epidemics. New diagnostic methods and vaccines research may provide more effective preventive measures, but improved use of current methods can markedly reduce the toll of these diseases.

Preventable deaths in developing countries can be avoided by many measures of proven effectiveness at relatively low cost. Currently available vaccines for children can eliminate deaths from measles, and markedly reduce deaths from diarrheal and respiratory diseases, which are major causes of death and disability. Relatively new but well-established vaccines are being gradually incorporated in internationally supported vaccination programs in developing countries, including *Haemophilus influenzae* b (Hib), pneumococcal pneumonia, and rotavirus, but influenza, hepatitis A, and varicella vaccines are not yet widely used in developing countries. WHO with UNICEF, GAVI, national governments and many other partners have declared "the Decade of Vaccines (2011–2020) is of a world in which all individuals and communities enjoy lives free from vaccine-preventable diseases". This means not only increasing routine immunization to all children especially. But equally important is expanding the kinds of vaccines used to include those developed in recent decades and reducing the time lag from their proven success until their adoption globally (GAVI/WHO/UNICEF 2012).

Along with immunizations, nutrition counseling and monitoring of child development are vital, along with supplementation providing vitamin A, iron, and iodine, and deworming treatments. These give children greater resistance to infectious disease and encourage healthy development. Environmental control measures such as the widening use of insecticide-coated bed nets, DDT spraying of homes, and actions to reduce environmental conditions ideal for mosquito breeding are helping to reduce child mortality from malaria.

Currently available public health measures are saving millions of lives and could save many more if applied in organized and sustained programs. It is hoped that breakthroughs in the development of new antimicrobial therapies and vaccines for HIV, TB, dengue, and malaria to address the multidrug-resistant organisms will be achieved in the second decade of the twenty-first century and help to accelerate the revolution in global health that is now underway.

Improved food technology will be needed to prevent *Salmonella* and *Escherichia coli* from infecting food sources. Medical care will need to improve its methods of control of infectious diseases to avoid the emergence of resistant organisms through more restrained use of antibiotics. The achievements towards eradication of important infectious diseases such as smallpox, polio, measles, guinea worm, leprosy, and onchocerciasis provide a basis for cautious optimism even though tempered by realistic appraisal of unsolved and new challenges of infectious diseases.

Non-Communicable Diseases

Tobacco is one of the largest causes of preventable deaths globally, killing between one-third and half of those using it from ischemic heart diseases, stroke, and chronic lung disease, and accounting for between 5 and 8 million deaths per year (WHO, 2008). Progress in the industrialized countries in reducing all smoking, however, masks a large increase in teenage smoking. The tobacco interests promote smoking in developing and transition countries, which are less able to carry out the legal and other issues associated with prevention of smoking. Antismoking legislation has advanced in North America and in the EU, but smoking remains one of the great challenges of public health.

The WHO, in its World Health Report 2003, used the term *neglected global epidemics* to emphasize three important and growing threats to the world: CVDs, tobacco, and motor vehicle accidents (Figure 16.4). Developing countries suffer from the dual burden of communicable diseases and increasing rates of NCDs and injury. These conditions are now greater causes of morbidity and mortality than the communicable diseases affecting the poorest countries around the world and the poor in the developed countries. Risk factors for CVDs are indicative of future health status, and five of the top 10 risks worldwide are specific to NCDs, namely, elevated blood pressure, tobacco use,

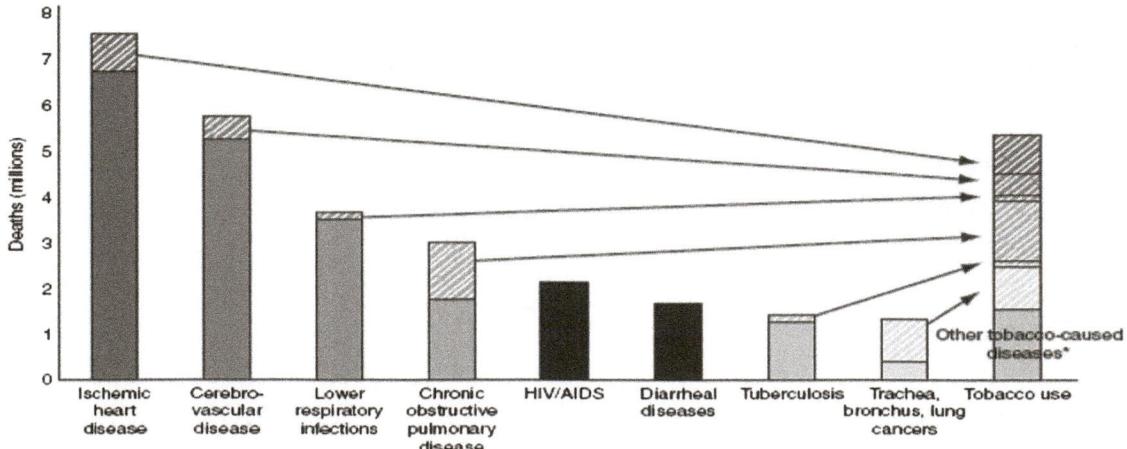

FIGURE 16.4   The eight leading causes of death worldwide and deaths attributable to tobacco use, 2005. Note: Hatched areas indicate proportions of deaths that are related to tobacco use and are marked according to the column of the respective cause of death. *Includes mouth and oropharyngeal cancer, stomach cancer, liver cancer, other cancers, cardiovascular diseases other than ischemic heart disease and cerebrovascular disease, diabetes mellitus, and digestive diseases. *Source: World Health Statistics 2008, Part 1, Ten highlights in health statistics. Available at: http://www.who.int/whosis/whostat/EN_WHS08_Part1.pdf [Accessed 16 January 2014].*

alcohol abuse, increasing fat consumption with choles-terol, and obesity (World Health Report 2000). An esti-mated 32 million deaths are attributable to NCDs and around 16.7 million are because of CVD (WHO, 2003 and 2013) (Figure 16.4).

In developing countries such as India, CVDs have become the leading cause of death, responsible for one-third of all deaths. Developing countries have double the number of deaths due to CVD in comparison to developed countries. Overall, in developing countries, CVD ranks third in disease burden (after injuries and neuropsychiatric disorders). Even in high-mortality developing countries, CVD morbidity is ranked very high. The major risk factors for CVDs have been identified, even while mortality rates are declining (see Chapter 6), and their effects through globalization can be seen all over the world with diets rich in saturated fats, sugar, and salt in vogue everywhere, along with increasing sedentary life habits. With the lack of fruits and vegetables in the diet, increase in tobacco use and lack of physical activity, the "global" diet leads to an increase in cases of CVD worldwide.

The epidemiological transition from predominance of infectious diseases to the chronic conditions that occurred in the industrialized countries by the mid-twentieth century is also occurring in the middle- and even low-income developing nations. CVDs, cancer, other degen-erative conditions and mental disorders, and trauma are already the major causes of death in many developing countries. Trauma also constitutes a vast public health issue, with serious individual, social, and economic con-sequences. The WHO reports that in 2008, 7.3 million deaths occurred from ischemic heart disease and another 6.2 million from strokes (WHO, CVD, 2013). More than 5 million deaths from injury and poisoning are reported

yearly, accounting for 9 percent of global deaths, with 90 percent occurring in low- and middle-income coun-tries, and they result in considerable loss of potentially productive years of life. Motor vehicle accidents rank first in causality, followed by domestic accidents, includ-ing falls, burns, poisonings, and drowning, all of which are particularly prevalent among young and elderly people.

The prevalence of obesity worldwide, highest in the Americas, more than doubled between 1980 and 2008. Obesity is a major and rapidly increasing phenomenon in high-, medium-, and low-income countries, due to dietary practices and sedentary lifestyles. Two billion people suffer from one or more micronutrient deficiencies, while 1.4 bil-lion are overweight, of whom 500 million are obese (FAO, June 2013). Twenty-six percent of all children under five are stunted and 31 percent suffer from vitamin A deficiency. In sub-Saharan Africa underweight DALYs declined from 694 in 1990 to 278 per 1000 population in 2010 (Table 16.11). Even in Africa, DALYs related to overweight increased.

Obesity has become a world pandemic. A survey in Eng-land predicted that more than 12 million adults and 1 million children would be obese by 2010 if no action was taken. In the USA the high and rising prevalence of overweight and obesity makes this perhaps the leading public health prob-lem. From 2000 to 2011, the prevalence of obesity increased from 27.5 to 35.5 percent in adults, and overweight in chil-dren and adolescents tripled. Currently, some 78 million, or more than one-third of US adults are obese, with almost half of black adults in that category. Obesity affects 12.5 million children and adolescents. The prevalence of obesity and overweight in the USA varies among the states, with a greater prevalence among the southern states. The US rate of overweight and obese categories combined continues to

TABLE 16.11 Disability-Adjusted Life Years (DALYs) by Malnutrition-Related Risk Factor, Population Group and Region, 1990 and 2010

	Child and Maternal Malnutrition		Underweight				Overweight and Obesity			
	Total DALYs (thousands)		Total DALYs (thousands)		DALYs per 1000 population (no.)		Total DALYs (thousands)		DALYs per 1000 population (no.)	
Region	1990	2010	1990	2010	1990	2010	1990	2010	1990	2010
World	339,951	166,147	197,774	77,346	313	121	51,613	93,840	20	25
Developed regions	2,243	1,731	160	51	2	1	29,956	37,959	41	44
Developing regions	337,708	164,416	197,614	77,294	356	135	21,657	55,882	12	19
Africa	121,492	78,017	76,983	43,990	694	278	3,571	9,605	15	24

Source: Food and Agriculture Organization. The state of food and agriculture, 2013. Available at: http://www.fao.org/docrep/018/i3300e/i3300e.pdf [Accessed 3 June 2013].

rise, with a prevalence of 69 percent among adults. Obesity contributes to risk factors for heart disease, hypertension, diabetes, stroke and cancer, which have substantial DALYs and impact on the economy.

Diseases related to smoking, overeating, and unhealthy diet are increasing in developing countries among the middle-class and working-class populations. Rising death rates from coronary heart disease and strokes in the former Soviet countries constitute an enormous contributor to premature death and a burden on underfinanced health systems. As infectious diseases are better controlled and as eating patterns shift in the urban middle and working classes to high meat and fat intake, patterns of CVD seen in the industrialized countries are occurring in developing nations. Public health practitioners need to prepare for this epidemiological transition with interventions such as antismoking campaigns, nutrition education, and other health promotion programs. Similarly, the mental, dental, and other health needs of societies are part of global health planning for developed countries in transition.

Mental health is gaining increasing recognition as a health issue of global proportions affecting hundreds of millions with moderate to severe disability, not only in the industrialized countries, but also in developing countries. As measures of the burden of disease include morbidity as well as mortality (such as in DALYs), major depression (unipolar), alcohol dependence, bipolar disorders, and especially schizophrenia come high on the list of causes of disability, particularly in young adults aged 15–24. These require attention in the health system and especially in primary care. Drug abuse and comorbidity with personality disorders and other mental illnesses are global problems, associated with related issues such as HIV, STIs, TB, hepatitis B and C, violence, crime, and other destructive behavior with great cost to society.

National and international bodies responsible for population health are aware of the magnitude of NCDs and the related risk factors. This constitutes a core issue for public health and social policy for the coming decade at least. Paradoxically, mortality from CVDs and cancer is falling in the high-income countries, but increasing in the middle- and low-income countries. The lessons learned in NCD control need to be central to bilateral and international aid programs along with the traditional emphasis on HIV, vaccine-preventable disease, and maternal and child health. Proposals for the post-MDG period are being widely discussed and will become part of policy for global health and for funding agencies. A modified outline is suggested in Table 16.12 (see Chapters 5 and 6).

Disaster Management

Tragic events leading to large-scale loss of property and life created by nature and by humans require organized international response to limit the damage, to reduce suffering, and to restore normality. These situations may be natural disasters such as hurricanes, floods, droughts, earthquakes, or volcanic eruptions, with severe consequences for public health. They may be larger scale events created by human initiatives, such as bi-national and civil wars, genocide, and civil strife or repression. Such events can take on enormous proportions as displacement, murder, and other forms of violence disrupt human norms and civil society.

The public health aspects of such events lie within the context of restoration of safety, provision of safe water, shelter, food, and sustenance, and efforts to restore civil life. Such events are now brought to the immediate attention of the world's community in television coverage. International action is forthcoming, but often inadequately coordinated as local civil and security authorities are overwhelmed. Preparation and organization for such disasters are important elements of global health. The 2004 tsunami in South-East Asia, with massive loss of life and property, repeated floods, and other disasters, resulted in large-scale displacement of people and public health challenges of the most basic kind. Provision of safety, shelter, safe water and food, disposal of the dead, and many other burdens can overwhelm local resources and require national and international assistance to reduce the scope of the tragedy and to help restore normal life (see Chapter 9).

Investment in infrastructure to ensure safety and prevent collapse of buildings during earthquakes, floods, tsunamis, civil strife, or war can save thousands of lives. Incitement of religious and political strife can lead to large-scale death and injury, so that prevention of incitement should be a political priority. Investment in preparation and training of first responders, such as police, firefighters, emergency medical teams, food and shelter supplies, and many other methods of reducing the scale of damage and injury, is a vital function of civic authorities, public health, emergency transportation, and hospital facilities. The twenty-first century will continue to experience natural and human-caused disasters with huge loss of life, breakdown in civil society, and loss of personal security from shelter, food, water, warmth, and family contact.

Environment

The environment is a global health concern, not only because it affects every country but also because its maintenance requires joint action. Air pollution caused by industry, power plants, and domestic use of coal is common in urban areas worldwide. The quality of air in the industrialized countries improved over the 1980s and 1990s, but in many developing countries and the former

TABLE 16.12 World Health Organization Proposals for Global Control of Non-Communicable Diseases (NCDs)

Target	Measures
Reduce premature mortality from NCDs by 25%	Education, regulation, taxation
Reduce harmful use of alcohol by 10%	Education, taxation, legal enforcement
Reduce physical inactivity by 10%	Education, school programs, public facilities in parks
Reduce sodium intake by 30%	Mandatory regulations for processed foods
Reduce tobacco use by 30%	Ban advertising and smoking in public places; increase taxation; implement international tobacco agreements
Reduce elevated blood pressure by 25%	Promote education of physicians, patients and their families, and the general public on dangers and control of elevated blood pressure
Diabetes/obesity: 0% increase	Promote school and public education; school restrictions on sweet beverages and dispensers; healthful diet in institutions
Mental health, drug therapy, and counseling: 50% coverage	Train GPs and primary care workers in mental health issues; increase access to drug treatment centers, and family doctor counseling management of methadone
Essential NCD medications and technologies: 80% coverage	Promote preventive medications such as aspirin, statins, and antihypertensive medications for prevention of development of ischemic heart disease with treatment of medicines and counseling (including glycemia control) to prevent heart attacks and strokes; promote use of generics for aspirin, statins, antihypertensive medications; increase capacity for PTCA and stent insertion for acute coronary events
Training, organization, and reporting systems for primary care	Strengthen reporting and data systems; develop medical and specialty training; train and deploy community health workers to actuate education at worksites, schools, hospitals, and clinics
Health protection measures	Promote sanitation, antimosquito vector control, and bed net use. Implement full immunization of children (DPT, polio, MMR, rotavirus, Hib, pneumococcal pneumonia, influenza) and adults (hepatitis B, HPV, tetanus, diphtheria)
Healthful diets; promote nutrition security, reduce malnutrition and micronutrient deficiencies	Ban transfats; reduce animal fats; increase vegetable, olives, nuts, and whole grain; promote fruit and vegetable production and availability at low cost; food fortification with folic acid and vitamin B_{12} in flour, vitamin D in milk, and iodine in salt, with monitoring; school meals; multivitamin supplements to women before and during pregnancy; cereal consumption
Promote screening for cancer	Pap smear, colonoscopy, breast examination
Healthful fertility, delivery and newborn care	Promote contraception, spacing of pregnancies, folic acid and iron for women in age of fertility, antenatal care and safe delivery; vitamin K and immunizations for newborns, antibiotics for newborns where neonatal mortality is high
Healthful aging	Community social centers for the elderly; home care and hospice care centers
Mental health	Educate primary care providers in diagnosis and management of mental health problems; develop community-based mental health capacity of trained personnel and accessible facilities

Note: GP = general practitioner; PTCA = percutaneous transluminal coronary angioplasty; DPT = diphtheria–pertussis–tetanus ; MMR = measles–mumps–rubella; Hib = *Haemophilus influenzae* type b; HPV = human papillomavirus.
Source: Adapted from World Health Organization. NCD global monitoring framework, 2013. Available at: http://www.who.int/nmh/global_monitoring_framework/en/ [Accessed 29 May 2013].

socialist countries, air quality has deteriorated because of poor quality in power generation and urban congestion. Excessive production of carbon dioxide by the use of fossil fuels is contributing to a global warming effect, and chemicals used in industrialized societies cause ecological damage, with potentially serious global consequences. Climate change is an overarching issue affecting health, economics, political developments, and human society generally (see Chapter 9).

Global Partnership for Development

The eighth goal in the MDGs calls for a partnership between the developed and the developing world, and official development assistance (ODA). The ODA includes measures to ensure debt sustainability on a long-term, rule-based basis, along with predictable and non-discriminatory multilateral trading, and financial systems to address the special needs of the least developed countries. The developed nations undertook to share responsibility for ensuring the global partnership. The UN General Assembly proposed an ODA target of 0.7 percent of the donors' national income but the OECD countries have contributed to just around half of the promised amount for many years, which fell to about one-third in the 1990s. World leaders met in 2002 at the International Conference on Financing for Development in Monterrey, Mexico, and established a new framework for a global development partnership (MDG8) (Table 16.3). Since then, the signatories have started to deliver on commitments made during the conference and aid has reached a record high of US$79 billion, with countries such as Denmark, the Netherlands, Norway, and Sweden honoring the initial commitment in the Declaration by the UN General Assembly.

It is vital for developing countries to increase their participation in the global economy, and this depends on reduction in trade barriers imposed by developed countries on imports from developing countries and tariffs on goods that are strategically important to developing economies, such as textiles, clothing, and farm products. Steps need to be taken to write off the debts on very poor countries, especially the economically stagnant African economies.

Economic growth in Africa since 2000 has been strong and grew by 4.3 percent in 2010 and 5.3 percent in 2011 based on high prices and growing investment in commodity production. However, this has not been translated into increased employment, industrial development, or investment in infrastructure, social, and health programs. The contrast with East Asian growth is startling: in East Asia, GDP grew from US$305 per capita in 1970 to US$8483 in 2010 (an average annual growth of 60.6 percent), while Africa's growth was from US$246 per capita to US$1701 in the same period, a respectable average annual increase of 14.8 percent. The total debt burden on African economies was estimated to be US$206 billion at the end of 2000, reaching US$300 billion in 2009 (UN Economic Commission for Africa, 2013).

DEVELOPMENT AND HEALTH

Environmental conditions are profoundly related to health. Some 1.3 billion people in developing countries lack access to clean water; nearly 2 billion lack adequate sanitation. Improved access to sanitation facilities globally within the MDG7 objective of environmental sustainability has resulted in more than 2 billion people gaining access to improved drinking water, increasing the global coverage from 76 percent in 1990 to 89 percent in 2010. Important progress has been made in urban slums. Despite this progress, 2.5 billion people in developing countries still lack access to improved sanitation facilities (UN, MDG7, 2013).

Poverty, low educational and job skills, poor nutrition, an unsanitary environment, and poor housing conditions all contribute to the enormous burden of disease in developing countries. Indoor pollution from the use of cooking fuels with inadequate ventilation in developing countries contributes to high rates of acute respiratory disease and deaths in children, as well as to chronic lung disease in the elderly.

Health status and economic development are interdependent, and the prevailing social and political philosophies have a vital impact on health, not only in terms of the amount of the funds allocated to health, but also in the form of the health care delivery system adopted. Rapid economic development also has its price. Environmental pollution and increases in occupational health hazards occur when new technology is transferred to developing countries. Further degradation also occurs with the tendency of the rural poor to move to cities, where basic sanitation and other infrastructures are often lacking.

Measurement of economic development by GNP alone is misleading. The distribution of wealth in a country is an important variable, along with other measures such as school enrollment. The Human Development Index (HDI) includes life expectancy, educational attainment, and measures of income (giving lower weight to income above the poverty level, since extra income for upper income groups is less important to survival). The HDI, along with DALYs and quality-adjusted life years (QALYs) (see Glossary and Chapters 3 and 11), adds an element of the quality of life to the usual economic indices.

Equally important to the amount of money spent on health is how the money is used. Some countries have succeeded in achieving marked improvements in health while remaining poor as measured by GNP per capita. Some countries have higher ratings in terms of HDI than their ranking by GNP. China, with a GNP per capita of US$410 in 1993 rising to US$4940 in 2011, has succeeded in attaining the infant mortality and life expectancy rates of mid-level developing countries by bringing primary care to the vast rural population. Sri Lanka, with a per capita GNP of US$498 in 1993 rising to US$2580 in 2011, has an infant mortality rate of 15 per 1000, comparable to well-advanced developing countries. The Indian state of Kerala is well above Indian national standards in HDI, even though economically poorer than the national average. On the other hand, some countries with high per capita GNP have lower HDIs; for example, Kuwait and Saudi Arabia have large GNP per capita but fewer public health achievements than much poorer countries such as Cuba, Costa Rica, and Jamaica. In some countries, this may be due to the large economic gap between the small, very wealthy ruling class and the large, poor population.

ORGANIZATION FOR INTERNATIONAL HEALTH

As seen in Chapter 1, from the decline of the Roman Empire in the fifth century CE, Europe passed into a millennium of scientific repression. Knowledge, including medical knowledge, passed into the hands of the Church, and the Greek and Roman writings that were preserved in the west survived in isolated monasteries of Ireland and Europe, and in Arab civilization, where during the next few hundred years Arabian, Byzantine, and Jewish scholars translated and preserved ancient medical knowledge in Europe. In the ninth century, a medical school was founded in Salerno near Naples, and medical schools spread to cities throughout Europe and the Arab world.

European colonial expansionism, beginning in 1415 with the Portuguese attack on Muslim settlements in nearby North Africa, had extremely important effects on international health. European ships brought smallpox and measles to the natives of the South Pacific and the Americas, decimating their populations. Syphilis, possibly originating from yams, may have been introduced into Europe by sailors returning from the Americas. European adventurers and settlers often suffered severely from the many endemic diseases to which they had scant resistance. In addition, the slave trade brought communicable diseases from Africa to favorable habitats in the Americas.

Colonialism led to near-eradication of many of the world's native peoples and changed the character of many populations, most dramatically in North America, Australia, New Zealand, South Africa, and Latin America. Colonial governments introduced western medical organization and practice, including public health and professional education systems, and influenced health with respect to concepts of causality and the treatment of diseases. Widespread education and medical training were important legacies in many developing countries that gained their independence in the mid-twentieth century.

The development of sanitation and later microbiology depended on the scientific and technological underpinning provided by the industrial revolution. In the latter half of the nineteenth century, repeated epidemics of cholera in Europe and continuing havoc from other communicable diseases were intense stimuli for researchers to identify the causal agent and means of transmission of almost every major bacterial and parasitic disease. Asiatic cholera arrived in Europe in 1832 and spread throughout the continent in repeated epidemics during the nineteenth century. This led to a convening of the International Sanitation Conference in Paris in 1851, with follow-up meetings held in 1874, 1881, and 1885. These conferences were held more frequently between 1892 and 1903 regarding maritime quarantine and control of international transmission of cholera, yellow fever, and typhus. In the early 1880s, a pioneering step in international public health occurred when, at the request of the International Cholera Commission, Robert Koch led a team to investigate cholera epidemics in Egypt. This resulted in identification of the *Vibrio cholerae* organism and recommendation of preventive procedures.

The Health Organization of the League of Nations (1921–1946), established in Geneva, was an attempt to develop the idea of international collective security for health. As part of its function, the Health Office provided an Epidemic Intelligence Service. The Health Office organized many expert committees on infectious diseases and other public health problems, including the establishment of standards for biologicals, maternal and child health, nutrition, health insurance, and medical education. Malaria, leprosy, and rabies control activities were promoted, as were the establishment of cancer registries and preparation for an international classification of disease; pharmacopoeias were coordinated and standards for housing and nutrition developed. The scope of organized international work was broadened from prevention of international transmission of disease to disease control and improved health conditions for vulnerable groups in the population. The collapse of peace in the late 1930s led to the League of Nations being disbanded.

During World War II, the United Nations Relief and Rehabilitation Agency was established by the allied powers to assist in the resettlement of millions of displaced people. This became part of the initiative to establish a new international health organization as part of an international consensus to build a better world after the war, in the context of a stronger, more coordinated United Nations. In the postwar period, international organization for health was seen to be a crucial need for world peace and progress.

WORLD HEALTH ORGANIZATION

The World Health Organization (WHO) was founded in 1948 as a UN agency in the spirit of cooperation and idealism following World War II. The WHO charter states that one of the fundamental rights of every human being is "the highest attainable standard of health", and the UN's Universal Declaration of Human Rights in 1948 stated, "Everyone has the right to a standard of living adequate for the health and well-being of himself and his family".

The WHO has made an enormous contribution to global health. It serves as the central, unified, intergovernmental organization representing all countries and covering all fields of health. The WHO consists of 193 member states working together and with other organizations towards the achievement of the highest possible level of health. It replaced previous organizations, especially the Health Organization of the League of Nations and the Pan American

BOX 16.3 Successful Areas of International Health Leadership

Successes and important initiatives of the international health movement led by the WHO and UNICEF include the following:

- Eradication of smallpox
- Massive increase in immunization coverage (EPI)
- Control and closing in on eradication of poliomyelitis
- Reduced measles incidence (decline in deaths from 535,300 in 2000 to 139,300 deaths in 2010)
- Massive reduction in incidence of tetanus, diphtheria, pertussis
- Tuberculosis control using DOTS and DOTS-plus for multidrug-resistant cases
- Improving control of diarrheal disease (CDD) and reduced death rates
- Improving control of acute respiratory illness (ARI)
- Improved control in neglected tropical diseases: onchocerciasis, leprosy, yaws; potential eradication of dracunculiasis
- Leadership in principles of primary health care: influence on national health programs in developing countries
- Raising public and political consciousness of health issues
- Health for All initiatives
- Health targets initiatives
- The Healthy Cities movement
- Health promotion, raising health in national priorities
- Increasing awareness of health information needs
- Intersectoral cooperation in vaccines and immunization purchase, distribution and delivery
- GAVI, Gates Foundation, and public–private sector cooperation in global health
- Health systems reforms
- Health human resources issues
- Global tobacco control
- Millennium Development Goals promotion and achievements
- International Health Regulations
- Post-2015 health targets.

Note: WHO=World Health Organization; UNICEF=United Nations Children's Fund; EPI=Expanded Programme on Immunization; DOTS=directly observed treatment, short-course; GAVI=Global Alliance for Vaccines and Immunization.

control. The 30th anniversary of the Alma-Ata Declaration renewed the call for primary care development. The WHO has also been effective in its technical services, epidemiology functions, statistics, standardized nomenclatures for disease and drugs, and publications. It has established good working relations with major donors and others in GAVI and in parallel efforts in a wide variety of fields (Box 16.3). The lack of progress in important issues such as women's health and maternal mortality in many large-population countries is primarily due to insufficient national political commitment.

The organizational structure of the WHO includes headquarters in Geneva and seven regional offices, including Europe (EURO-Copenhagen), the Eastern Mediterranean (EMRO-Alexandria), Africa (AFRO-Brazzaville), South-East Asia (SEARO-Delhi), western Pacific (WPRO-Manila), and the western hemisphere [the Pan American Health Organization (PAHO) in Washington, DC]. The central headquarters in Geneva has many offices dealing with a diversity of topics (Table 16.13).

The WHO has led in the formulation of a worldwide consensus on a new direction in health policy. It formulated a strategy that incorporated the principles of governments' responsibility for the health of their peoples, the right of people to take part in developing and controlling their health care, and equality in health. It helped to formulate and promote the concept that cooperative activity among different parts of the public and private sectors (intersectoral cooperation) is necessary to advance health causes. The concept of appropriate technology is also a WHO initiative (see Chapter 15).

The problems and limitations of the WHO are important to assess. The organization is part of the UN system and is governed by its membership countries. It cannot avoid being influenced by political conflicts such as during the Cold War period, regional wars such as in the Middle East, and genocides such as in Darfur. This politicization can be a detriment to its leadership and moral authority. It can also limit contacts of the WHO with the highest quality of professional leadership, impairing its ability to relate to the forefront of medical science, epidemiology, and public health practice. It has also led to inadequate leadership in areas where the response of the WHO to important issues may offend national pride.

The global struggle against tobacco as a major contributor to the spread of chronic diseases is being led by the WHO and regional and national bodies as well as non-governmental organizations (NGOs). Progress in many fields of health, as outlined in Box 16.3, has been a major contribution by the WHO and shows its important role in global health and the world community. Although hampered by its political nature, it exists as an international body in health representing all countries and dealing with health in a broad definition. The WHO's leadership in the Declaration of

Sanitary Bureau. A Technical Preparatory Commission developed the organization and, in the new optimism of the time, undertook the enormous task of dealing with global health problems. Its direction and coordinating functions are the primary assets of the organization, especially in the definition of health goals and in initiating international cooperation to achieve them.

The WHO carried this optimistic approach further in the Alma-Ata Conference and Declaration of 1978, in the successful pursuit of smallpox eradication and great success in nearing polio eradication, in major reduction of measles morbidity and mortality, and in many other fields of disease

TABLE 16.13 National and International Strategies for Coping with Emerging Infectious Diseases

Goal/Topic	Activities	Examples
Surveillance	Detect, promptly investigate, and monitor emerging pathogens, the diseases they cause, and the factors influencing their emergence	Monitoring in sentinel surveillance networks (e.g., blood banks, emergency departments, laboratories, sentinel settings) Population-based surveillance Increase field investigation of outbreaks Dissemination of epidemiological data locally and internationally using electronic media; Internet, websites, ProMED, MMWR, Eurosurveillance, and others Rapid laboratory diagnosis; Monitor vectorborne diseases
Applied research	Integrate laboratory science and epidemiology to optimize public health practices	Promote reporting by sentinel laboratories and clinical settings Improve laboratory diagnostic techniques, genotyping, subtyping, and mapping "fingerprinting" (e.g., *Escherichia coli*, cholera, poliomyelitis, measles, meningitis)
Prevention and control	Safe animal husbandry; vector control; safe water and sanitation; food control; immunization, rapid diagnosis, directly managed treatment (DOTS); preventive treatment for tropical diseases; promote cooperation between public health, clinical services, veterinary services and IT monitoring of infectious diseases and preventive activities, e.g., immunization	Enhance communication of public health information about emerging diseases and ensure rapid implementation of preventable strategies. Wide and immediate dissemination of health information on infectious diseases to health professionals, general public, groups at special risk Promote health education on prevention of spread of communicable diseases
Infrastructure	Strengthen local, state, and federal public health infrastructures to support surveillance and implement prevention and control programs	Improve laboratories, reporting, and training
International cooperation	Strengthen international effort and funding to promote immunization, nutrition interventions, maternal and child health	WHO, UNICEF, UNDP, WB, FAO, GAVI, Médecins sans Frontières, bilateral aid programs Rotary International, Gates Foundation, and many others working side by side and in growing cooperation to help national governments achieve MDGs
International Health Regulations (IHR 2005)	Approved by WHO and in effect in 2007, ratified by most countries	Legal obligations to report infectious or other public health emergencies (chemical, radiation) of international public health significance. All countries agreed to increase surveillance capacity for emergencies such as SARS or human influenza

Note: DOTS=directly observed treatment, short-course; IT=information technology; ProMED=Program for Monitoring Emerging Diseases; WHO=World Health Organization; UNICEF=United Nations Children's Fund; UNDP=United Nations Development Programme; WB=World Bank; FAO=Food and Agriculture Organization; GAVI=Global Alliance for Vaccines and Immunization; SARS=severe acute respiratory syndrome.
Sources: Modified from Centers for Disease Control and Prevention. Addressing emerging infectious disease threats. Atlanta, GA: US Public Health Service; 1994. National Institutes of Health. Understanding emerging and re-emerging infectious diseases, 2007. Available at: http://www.ncbi.nlm.nih.gov/books/NBK20370/ [Accessed 2 June 2013].
World Health Organization. International health regulations. 2005. Available at: http://www.who.int/features/qa/39/en/index.html [Accessed 17 January 2008].
National Institute of Allergy and Infectious Diseases. Emerging and re-emerging infectious diseases. Available at: http://www.niaid.nih.gov/topics/emerging/Pages/Default.aspx [Accessed 2 June 2013].

Alma-Ata and Health for All 2000 represented an important step forward in international health with its major commitment to primary health care (see Chapter 2).

Tropical disease work on malaria, bilharzia, filariasis, TB, onchocerciasis, leishmaniasis, schistosomiasis, helminthic diseases, and diarrheal disease control is of particular importance to the developing countries. The WHO's leadership in the eradication of smallpox and virtual eradication of guinea worm disease, onchocerciasis, and polio has made outstanding contributions to improved global health. Its

initiatives in reducing nutritional deficiency conditions, in chronic disease control, in defining health personnel needs, and in health services financing have also been important for both the developing and developed countries.

The WHO develops programs of work that guide its activities and its regional offices as well as member states. The WHO has defined 15 objectives and a number of targets for each objective. These involve a global strategy for health, including promotion of food production and distribution, social progress in literacy, poverty reduction, and economic growth. Also included are the following: intersectoral cooperation; development of health care systems with a stress on primary care and improved management skills and efficiency; community involvement; improved levels of health resources, including financial support by governments and universities involved in training health personnel; research, technology, and cooperation between countries; and environmental sanitation. All were included as areas for action within this program. WHO's policy framework stresses work with member governments, international organizations, banks, NGOs, and other organizations related to health, economic, and social development.

Many other UN agencies and other organizations play important roles in international health. These include UNICEF, the United Nations High Commissioner for Refugees (UNHCR), the United Nations Development Programme (UNDP), the International Labour Organization (ILO), the FAO, and the International Atomic Energy Commission (IAEC).

UNITED NATIONS CHILDREN'S FUND

Following World War II, the new UN General Assembly created the United Nations International Children's Emergency Fund (UNICEF, now the United Nations Children's Fund), principally to assist the children of war-torn Europe. The program gradually expanded to include other activities and other areas, particularly in developing countries.

This agency has spent large sums of money, especially on food and supplies, for the promotion of child and maternal health and welfare activities throughout the world. Beyond this, usually through partnership with the WHO, UNICEF has been carrying out large and significant programs of bacille Calmette–Guérin (BCG) vaccination and yaws and malaria control. The promotion of family planning in developing countries is one of its major activities. UNICEF plays an important leadership role in fostering primary care and community preventive approaches worldwide.

UNICEF's annual State of the World's Children reports provide thorough reviews of essential topics and valuable data presentations of key health indicators for all countries. UNICEF has been crucial in developing and promoting the UN Convention on the Rights of Persons with Disabilities (CRPD) and the Convention on the Rights of the Child (CRC). As of 2013, 127 countries and the EU have ratified the CRPD with commitments to promote full equality and participation of people with disabilities in society, and 76 countries have signed the CRC.

These conventions set new standards for integrating children with handicaps into general society and call for an end to separating children with disabilities from their families. Children have the right to be cared for by their parents unless this is deemed by a competent authority to be incompatible with the individual child's best interests. Making "public services, schools and health systems accessible and responsive to the needs of children with disabilities and their families" will reduce the pressure to send children to institutional care at all (UNICEF, State of the World's Children, 2013).

NON-GOVERNMENTAL ORGANIZATIONS

Numerous NGOs carry out specialized activities worldwide. They vary widely in content, funding, ideology, and modus operandi. Many provide important support for developing countries, often succeeding where international agencies have failed, precisely because they work outside the national political framework. This is particularly true in the case of emergencies and areas of conflict.

The earliest NGOs were those of the various church missions and sectarian organizations. Among the many that might be mentioned are the Unitarian Services Committee, the American Friends Services Committee, Catholic Relief Services, the American Jewish Joint Distribution Committee, the International Rotary Club, and the American Bureau for Medical Aid to China. The International Committee of the Red Cross (ICRC), Médecins sans Frontières (MSF), Terres des Hommes, and other European-based NGOs provide direct assistance in developing countries during crises. In 1999, MSF was awarded the Nobel Peace Prize in recognition of its worldwide health achievements.

Philanthropic foundations have made and continue to make major contributions to international health. Private foundations such as the Ford, Soros, and Rockefeller Foundations carry out important international health work within their own exclusive structures. They are important sources of grants to promote pilot programs and research in health care systems. In addition, they contribute extra governmental funding that can stimulate the development of innovative programs later affecting general health services. From its inception in 1999 to late 2007, the Gates Foundation donated some US$8.5 billion to the international child vaccination program, aid to small farmers, and other health and education programs, mainly in Africa.

Among the foundations, the Rockefeller Foundation is the best known in the field of international assistance in health. Since its inception in 1913, it has operated in almost every country worldwide. Its many significant contributions include support of control programs for malaria and yellow fever, the development of recognized centers of learning in medicine and public health, postgraduate fellowships, and the demonstration of sound methods of organization and operation of health programs.

Despite the many positive aspects of NGOs, they can be a source of distortion in health care services in both developed and developing countries. They tend to focus on one kind of service, are very proud of their independence from government, and can create pressure for services that will place a burden on the system of financing or provision of health care. NGOs or bilateral aid can promote hospital development in places where there is already an oversupply and limited primary care. They can provide a primary service but be unwilling to coordinate with the basic governmental program in immunization, so that no one agency is fully responsible.

Coordination of NGO services into a comprehensive population-based service program may be compromised by political and international sensibilities, which can create chaos in an emergency situation. The balance of services for a population requires inclusion of governmental, NGOs, and private services as a coordinated if not integrated whole. This may be impossible with highly independent NGOs, but the state public health authorities are responsible for overseeing the functions of NGOs, no matter how well meaning or charitable the cause.

THE WORLD BANK

The International Bank for Reconstruction and Development (IBRD), also known as the World Bank, is based in Washington, DC. It was established by the industrialized countries following the Bretton Woods Conference towards the end of World War II. It was an important financial institution to facilitate the reconstruction of postwar Europe. It has since become a major source of financing for development projects throughout the world. Traditionally, it focused on large-scale infrastructure, industrial, and farming development projects. Its policies in health development focused on promoting market mechanisms and privatization of health care in countries lacking infrastructure. This strategy fostered an inappropriate stress on medical and hospital care when a community health orientation was needed.

The World Bank examined the health sector and its importance for economic development and produced a new assessment of good health as an economic value contributing to economic growth. This assessment led to a growing emphasis on health in global and national economic planning. The 1993 World Development Report: Investing in Health was a landmark document in the development of global health. It examined the interaction of health status, health policy, and economic development, and stated that, contrary to the views held by many traditional economists, health is essential for economic growth, and not a burden on the economy (Box 16.4). The report advocated a four-pronged approach by governments to improve health in developing and former Soviet countries:

- Foster an economic environment that will enable households to improve their own health by promoting income gains for the poor and expanding social investment in raising standards of education, especially for girls.
- Redirect government spending away from specialized care towards low-cost and highly effective activities such as immunization, programs to combat micronutrient deficiencies, and the control and treatment of infectious diseases.
- Encourage greater diversity and competition in the provision of health services by decentralizing government services and promoting competitive procurement practices.
- Foster greater involvement of NGOs and private organizations, and regulate insurance markets.

World Bank assistance to health-related projects is growing steadily. The World Bank and the WHO have worked together on projects such as a special Program for Research and Training in Tropical Diseases and the Onchocerciasis Control Project in West Africa. Long-standing World Bank pro-privatization policies and practices, most notably structural adjustment programs, have led to reduced social welfare infrastructure of developing countries in areas such as housing, education, health services, subsidies, and family transfers. While the World Bank recognizes the importance of health to development, its advocacy of privatization of health services has exacerbated poor health outcomes by reducing access to health services for those unable to pay for care.

BOX 16.4 World Development Report: Investing in Health

The World Bank published the World Development Report in 1993. It was ground breaking in stating that good health is essential for human well-being and for national economies. It provided evidence that good health is an economic asset to a nation and that investment in health is a sound economic decision for governments. Health improvement contributes to economic growth by reducing worker absence due to illness; it allows economic development in areas previously too unhealthy for productive work. Good health helps children to attend and learn at school, and good health reduces the costs of health care for treating illness.

This document had a major influence on economists who previously had seen investment in health as a net economic drain on society. This report indicated the opposite. Investment in health contributes to economic growth and societal well-being. The report is widely quoted in current literature on economics of health and is a vital issue in discussions with ministries of finance, which are often staffed by market economics-oriented staff who fail to appreciate the positive contribution that health investment makes to economic growth.

Source: World Bank. World Development Report, 1993. Available at: http://wdronline.worldbank.org/worldbank/a/c.html/world_development_report_1993/abstract/WB.0-1952-0890-0.abstract1 [Accessed 25 July 2013].

The promotion of privatization mirrored the global trend towards more market-oriented economic policies. The World Bank promoted privatization and health insurance to replace direct provision of services by the state. This created problems in many developing countries, where most citizens earn less than US$2 a day, and private sector services limited access to care of acceptable quality at affordable prices. World Bank policies in health have also obligated many countries to large repayment loans for programs of questionable value while basic services such as adding more successful vaccines to immunization programs are delayed, so that many countries lag behind in the adoption of internationally proven vaccines.

TRENDS IN GLOBAL HEALTH

It is now widely understood that the socioeconomic environment is a basic determinant of the state of health of an individual or a population, even though the precise nature of intervening variables may not be sufficiently elucidated. The southern hemisphere has witnessed, along with its demographic explosion, the persistence of chronic problems plaguing the education, food, and housing sectors. In addition, more acute situations have emerged over the past few decades in relation to conflict, employment, migration, trade, and degradation of the physical environment. The northern hemisphere has enjoyed a rising level of affluence, with negative aspects that have made a sizable impact on public health: overeating, overdrinking, smoking, pollution, illicit drugs, and motor vehicle accidents.

Global changes in the twenty-first century hold the promise of improvements in diagnostic and therapeutic measures, such as targeting drug therapy with nanotechnology, improved diagnostic methods including long-distance imaging methods, and less invasive diagnostic measures. Climate change and the potential for spread of vectorborne diseases, and food supplies and the issues related to genetically modified foods are all issues of huge importance for public health in the coming decades.

Progress in control of vaccine-preventable diseases has been one of the most important public health achievements with eradication of smallpox, closing in on polio eradication, and increasingly effective control of measles since adoption of the two-dose policy. The development and increasingly wide use of vaccines with combinations such as measles–mumps–rubella (MMR), hepatitis B, Hib, rotavirus, varicella, pneumococcal pneumonia, and influenza vaccine is one of the vital issues for achieving the MDG of reducing child mortality. Coverage has improved globally, but expanding the content of child and adult immunization has been painfully slow, despite aid from GAVI and other international donors.

The WHO reports that the maternal mortality ratio in 2010 in developing regions, with 240 maternal deaths per 100,000 live births, is in stark contrast to the average rate of 16 in developed regions and 32 in the countries of the CIS. The drop in the global maternal mortality ratio reflects mainly the reduction that occurred including in countries with high levels of maternal mortality. Countries with the highest initial levels of mortality have made virtually no progress over the past 20 years. While gains are being made in developing and middle-income countries, the annual decline between 1990 and 2010 in both the sub-Saharan Africa and Latin America and Caribbean Regions was only 2.6 percent, and in Eastern Asia excluding China the annual decline was 0.8 percent. The North Africa Region achieved an annual 5.7 percent decline, exceeding the targeted 5.5 percent annual decline. During the same period, annual percentage declines for Southern Asia, South-Eastern and Western Asia were 4.4, 4.9, and 4.2 percent, respectively (Table 16.14). Overall, 51 countries are "making progress", while 14 countries have made "insufficient progress", and 11 are characterized as having made "no progress" and are likely to miss the MDG target. This MDG will remain an important health target in the post-2015 period.

The concept of primary health care as the basis of health system development has been almost universally accepted, yet evidence of public commitment to its implementation is still lacking. Problems reported include poor distribution of resources and inadequate orientation of health workers to primary health care, with continuing emphasis primarily on curative services. The community is often insufficiently

TABLE 16.14 Maternal Mortality Ratio, World Millennium Development Goal Regions, 1990–2010

Region	1990	2010	% Change
World	400	210	−47
Developed regions	26	16	−39
Developing regions	440	240	−46
North Africa	230	78	−66
Sub-Saharan Africa	850	500	−41
East Asia	120	37	−69
South Asia	590	220	−63
South-East Asia	410	150	−63
West Asia	170	71	−58
Caucasus and Central Asia	71	46	−35
Latin America and Caribbean	140	80	−43
Oceania	320	200	−38

Source: World Health Organization. Trends in maternal mortality: 1990 to 2010. Table 3. WHO, UNICEF, UNFPA and The World Bank estimates. Comparison of 1990 and 2010 maternal mortality ratio (MMR, maternal deaths per 100 000 live births) and number of maternal deaths, by United Nations Millennium Development Goal region. Available at: http://whqlibdoc.who.int/publications/2012/9789241503631_eng.pdf

aware of the role it should play and is frequently willing to accept competing demands for expensive secondary and tertiary care. A lack of resources to develop preventive services and health promotion is likely to erode the confidence and commitment of health workers and the community to primary health care.

The formulation and analysis of health personnel policy have emerged as growing concerns in the world (see Chapter 14). There is a consensus regarding the urgent need to ensure the relevance and quality of human resources to the requirements of the health system, and to avoid imbalances in the production of health professionals, especially with regard to physicians, nurses, and dentists. In most developing countries, health personnel development plans either do not exist or are in the process of being developed.

International organizations, such as the WHO, UNICEF, UNDP, and OECD, use their wealth of information on a broad range of topics to help governments to foster prosperity and fight poverty through economic growth and financial stability. The OECD helps to reinforce the work of the health-oriented organizations to promote the priority of health as an essential investment for economic development.

While there are rising expectations for better health for all, global changes also constitute challenges to continued progress in health. Population growth, aging of the population, increasing incidence of chronic diseases, high expectations of the public for health care, increasing costs and medical technology, economic recession, and limited resources for health have all contributed pressures for health system reforms to maintain universal coverage. During the latter part of the twentieth century, many industrialized countries developed health reforms that included reduction in hospital bed supply, financial incentives to promote development of community-based services, and a combination of decentralized management and integration of services in those countries with national health services (e.g., the UK). Control of oversupply and excess utilization of hospital beds is also a feature of reforms for cost containment. In the USA, rapidly rising costs led to the expansion of managed care systems seeking cost-effective health care combined with health promotion to reduce disease prevalence and dependency on treatment services (see Chapters 13 and 14).

The relationship between disease and society is such that many of the factors needed to reduce preventable diseases largely lie outside the biomedical framework of genetics, medical care, public health, and health promotion, but are determined by social preconditions that are in the realm of human rights. This, however, does not absolve governments or the health community from the imperative of applying known measures of prevention and curative services for all as a basic human right.

EMERGING INFECTIOUS DISEASE THREATS

International health began as an activity to prevent the spread of epidemics and communicable diseases. This involved the collection and dissemination of information in a timely fashion, preventive measures such as appropriate immunization campaigns to control the spread of a disease, and subsequent follow-up. Success in the eradication of smallpox and increasing control of the vaccine-preventable diseases led to enthusiastic assessments that epidemic diseases were under control. This optimism has been tempered by setbacks in malaria and TB control, along with a host of other emerging and re-emerging disease issues. The experience of SARS and the threat of a potential pandemic of avian influenza indicate a new scale of public health threat and the need for global preparation (see Chapter 4).

The globalization of food and medical products marketing has become an enormous worldwide phenomenon. This has been accompanied by the emergence and dissemination of new infections in human populations, with geographic spread of disease such as HIV, hepatitis C via blood products, and variant Creutzfeldt–Jakob disease (vCJD). Marketing and growth of global demand for beef and animal feed and for human anticoagulant factors for medical treatment contributed to the transmission of HIV, hepatitis C, and vCJD during the decade prior to the detection of these diseases in humans. The incubation period of bovine spongiform encephalopathy (BSE or mad cow disease) is in the order of years. Consumption of beef was declining in the 1980s in keeping with changing lifestyles. Changes in processing may have permitted prions to cross the species barrier from ruminant to human. Slaughterhouse practices in the UK in the 1990s were ineffective in deactivating prions, so that vCJD appeared and mad cow disease reappears periodically, requiring large-scale culling of cattle.

Commercial production of blood factors developed in the USA grew from approximately US$50 million in 1975 to US$325 million in 1988. During this period, it was estimated that half of the hemophiliacs in the USA were infected with HIV through this route, and an untold number of hemophiliacs were infected worldwide. In some countries, such as Japan, this was probably the primary route of transmission into the population. By the end of the 1990s, there were 400 commercial centers for plasmapheresis operating in the USA. These centers, which employ paid donors, provided 60 percent of the worldwide requirement for plasma.

By mid-1982 the possible link between AIDS and the blood supply was reported and became widely known and accepted even though AIDS was seen as primarily a sexually transmitted disease of gay men. In the following year the occurrence of cases in hemophiliacs living in geographically dispersed areas led to epidemiological investigations which identified contaminated blood as the source of infection. The blood factor industry used pooled serum largely from paid donors who were often from groups with high exposure to HIV, such as homosexual men. Self-exclusion was relied on for screening because of fear of potential lawsuits for discrimination by excluding high-risk donors, and the voluntary sector failed to stop blood collection in high-risk areas.

Viral inactivation methods had been under development since the early 1970s to reduce hepatitis transmission in blood. However, the industry leaders considered such steps proprietary information, so the work towards successful strategies was not shared across the corporate competitors. In 1984 the major producers had all been licensed to distribute heat-treated products to reduce the threat of hepatitis and HIV infection, but recall orders were not issued by the US Food and Drug Administration as soon as the risk of HIV transmission in blood factors became known in March 1983, a failure later criticized by the US Institute of Medicine.

By 1985, enzyme-linked immunosorbent assay (ELISA) screening tests for HIV and hepatitis C had come into use. This, coupled with heat treatment of the factor product, markedly reduced the risk of HIV transmission in blood. However, global spread of the virus, in part facilitated by the global trade in factor VIII, had already occurred in the 1980s, affecting the majority of Japanese hemophiliacs and AIDS-affected people during 1983–1985 by non-heat-treated factor concentrates imported from the USA, according to the World Federation of Hemophilia. Worldwide sales of manufactured blood factor products contaminated with HIV and HCV resulted in mass infections and deaths of thousands of hemophiliacs worldwide.

Continued sale of contaminated blood products to hemophiliacs in the mid-1980s led to legal action against French officials of the national blood bank and the ministry of health. Their trial, initially for "poisoning", resulted in convictions and jail terms for three including a former minister for their role in HIV transmission to some 4000 hemophiliacs in France. In Japan in 2000, three former drug company executives were convicted of selling blood products tainted with HIV and given prison terms.

The spectrum of infectious disease in a community evolves rapidly with changing conditions of the environment and society. Population growth, crowding in urban slums, homeless populations, massive migration, and travel contribute to the transmission of once-localized diseases internationally. Resistance to available antimicrobial medications is creating a new dilemma for modern medicine and public health. The synergism of infectious diseases such as AIDS with TB or *Cryptosporidium* causes deterioration in the patient and spread by secondary infection to other people. The "post-antibiotic era" is widely discussed as a serious threat to modern public health. New research and strategies are required to prevent the loss of some of the

important gains of the twentieth century in communicable disease control. Among the lessons learned from AIDS are that improvements in early warning systems and attention to new threats are vital tasks of public health.

In the USA, several new or resurgent infectious diseases are of increasing public health concern. These include HIV/AIDS, *E. coli* O157:H7 disease, cryptosporidiosis, coccidioidomycosis, multidrug-resistant pneumococcal disease, MDR-TB, vancomycin-resistant enterococci, influenza A Beijing/32/92, hantavirus infections, leishmaniasis in veterans of the 1991 Gulf War, legionnaires' disease, and Lyme disease.

Newly emerging and re-emerging diseases of concern internationally include HIV/AIDS, multidrug-resistant malaria, TB, cholera, *Shigella dysenteriae*, diphtheria, and *E. coli* O157:H7. Tropical diseases, such as yellow fever and dengue, are reappearing in Asia and Latin America, and Rift Valley fever in Egypt, Saudi Arabia, and Yemen; Lassa fever in West Africa; Ebola virus in the Democratic Republic of the Congo (formerly Zaire); Marburg virus via imported monkeys; Oropouche arbovirus and Sabia in Brazil; Junin virus in Argentina; and Machupo virus in Bolivia all are health concerns as they may spread from their natural habitat to other countries before the carrier shows symptoms.

West Nile fever spread from the Middle East to New York City and then to other parts of the USA in 1999, with 62 severe cases and seven deaths, but spread farther in 2000 and 2001, followed by a dramatic increase in 2002 with over 4000 human cases in North America and Europe. The disease has become endemic in mosquito populations and the cause of hundreds of deaths annually in the USA alone. The spread of new diseases was dramatically demonstrated during the SARS episode of 2003 and with avian influenza since 2005 (see Chapter 4). In 2007, chikungunya fever spread from South-East Asia to Italy and southern France, where it has become endemic.

It can no longer be assumed that such diseases will remain in their natural habitats; they can be transmitted all over the world via human or animal carriers and, in appropriate conditions, become serious local or even general public health concerns. The 1995 outbreak of Ebola virus in the former Zaire raised international concern of the very real possibility of widespread transmission of this deadly disease (Box 16.5). The public has been made aware of this kind of situation in graphic detail in the news media, novels, and movies. Again, the threat of avian influenza remains a looming public health disaster of global proportions.

In 1996, a large-scale epidemic of food poisoning in Japan involving *E. coli* O157:H7 (first described in 1982) spread through contaminated school lunches and caused over 3000 illnesses, hundreds of hospitalizations for bloody diarrhea, many with severe hemolytic uremic syndrome, and seven deaths. The identification of the source proved to

BOX 16.5 Ebola Virus

Ebola virus is named after a river in the Democratic of the Congo (DRC, formerly Zaire), where it was first recognized in 1976. It is a member of the family of ribonucleic acid (RNA) viruses called the Filoviridae with identified subtypes including Ebola-Zaire, Ebola-Sudan, and Ebola-Ivory Coast. It is the cause of a deadly hemorrhagic fever, with high mortality. An outbreak of Ebola virus in Zaire in 1995 was of major international concern, because of its 77 percent mortality rate and a fear that it could spread rapidly. This outbreak in Kikwit was limited to a total of 315 cases with 250 deaths (81 percent) were identified, many of whom were hospital workers. The organism has been isolated in specific species of monkeys and can be transmitted to humans via blood and secretions.

An international team mobilized by the World Health Organization (WHO) and Centers for Disease Control and Prevention (CDC) went to the site to assist the DRC public health staff. Stress on case detection, laboratory confirmation, isolation, and staff protection helped to limit the disease spread. The WHO Collaborating Center on Arboviruses and the Viral Hemorrhagic Fevers reference laboratory at the CDC in Atlanta, USA, played an important part in management of this epidemic. Rapid international response and heightened surveillance are part of the global concern for newly emerging infectious diseases, partly resulting from the lessons learned from the slow response to the AIDS epidemic. Outbreaks occurred in South Sudan in 2004, and in DRC in 2007.

More recent outbreaks occurred in DRC in 2008, with 32 cases and 15 deaths (47 percent), and in Uganda in 2011 with a single case who died. In 2013, evidence of Ebola virus in fruit bats in Bangladesh suggested a possible reservoir and source of infection of humans.

Sources: World Health Organization. World Health Report, 1996. Geneva: WHO.
Centers for Disease Control and Prevention. Questions and answers about Ebola hemorrhagic fever. CDC; 2005. Available at: http://www.cdc.gov/ncidod/dvrd/Spb/mnpages/dispages/ebola/qa.htm
World Health Organization. Ebola hemorrhagic fever. WHO; 2007. Available at: http://www.who.int/csr/disease/ebola/en/
World Health Organization. Ebola haemorrhagic fever. WHO; 2012. http://www.who.int/mediacentre/factsheets/fs103/en/index.html
Centers for Disease Control and Prevention, Special Pathogens Branch. Known cases and outbreaks of Ebola hemorrhagic fever in chronological order. Available at: http://www.cdc.gov/ncidod/dvrd/spb/mnpages/dispages/ebola/ebolatable.htm [Accessed 2 June 2013].
Olival KJ, Islam A, Yu M, Anthony SJ, Epstein JH, Khan SA, et al. Ebola virus antibodies in fruit bats, Bangladesh. Emerg Infect Dis 2013;19(2). Available at: http://dx.doi.org/10.3201/eid1902.120524 [Accessed 30 May 2013].

be very difficult. Milder epidemics have occurred in many other countries, including Australia, Canada, the USA, and various European countries, so that there is continued concern for recurrence of this potentially severe form of food poisoning. The ProMED website monitors infectious diseases globally, with email reports on virtually a daily basis. Other web-based infectious disease monitoring and information-sharing systems are invaluable public health

teaching and service programs that are available via the CDC website.

Food safety monitoring is vital in the control of food-borne disease outbreaks in both domestic and international trade. The WHO established the Global Public Health Intelligence Network (GPHIN), a web-based system that monitors news reports of infectious disease outbreaks around the world; Salm-Surv, a global network linking laboratories tracking the incidence of *Salmonella* and other food-borne diseases; the Global Outbreak Alert and Response Network (GOARN), which provides technical assistance within 24 hours to governments facing potential epidemics; and the International Food Safety Authorities Network (INFOSAN), which enables transborder collaboration and assistance among food safety officials. These systems supplement individual national surveillance and diagnostic functions.

In 2005, the WHO adopted new International Health Regulations (IHR), which have been adopted by most countries and came into effect in June 2007. The IHR address all diseases and health events that may constitute a public health emergency of international concern, to contain the threat of international spread of diseases such as SARS, or a new human influenza virus. The IHR includes threats of public health emergencies that may spread across borders, such as chemical spills, leaks, and dumping, or nuclear meltdowns. They replace the previous IHR (1969) which addressed only four diseases: cholera, plague, yellow fever, and smallpox, since then eradicated. The repealed IHR focused on the control at borders and relatively passive notification and control measures. The new IHR provides a legal basis for global disease surveillance, alert, and response. It defines the rights, obligations, and procedures in ensuring international health security without unnecessary interference in international traffic and trade. It requires all member states to strengthen their existing capacity for disease surveillance and response.

International standards of food, plant, and animal safety are addressed in the *Codex Alimentarius* ("Food Law") Commission (WHO and FAO), the International Plant Protection Commission (IPPC), and the Organization for Animal Health (Office International des Épizooties, or OIE). The World Trade Organization (WTO) works to break down – not erect – trade barriers; world exports of agricultural products increased by 21 percent in 2011 to US$1659 billion.

While the health and safety measures provided by the WHO and WTO are considerable, they are limited by the respective organizations' resources and priorities. Global trade in agricultural products in 2002 was US$583 billion, rising to US$1660 billion in 2011. Food scientists work in both industry and academic centers to discover new products, ensure quality assurance of specific products, and provide standards for the production, processing, marketing, and distribution of foods, with improved nutritional value and safety.

International cooperation to identify new infectious or other health threats to prevent global epidemics is an urgent priority for both international agencies and national health systems throughout the world. Control of communicable diseases requires medical, laboratory, and epidemiological intelligence services of a high order, with rapid means of communication, publication, and coordination, backed up by skilled professional services. Examples of international activity in the control of infectious disease are numerous (Box 16.6). The crowning achievement in this field was the eradication of smallpox. This great feat may soon be matched by the international eradication of polio. Such achievements can only be made with major efforts in international commitment and cooperation, working with international and national agencies, donor organizations, and ultimately with the people themselves.

Governmental agricultural policies have important impacts on health. Therefore, food safety monitoring remains an ongoing problem for the international agricultural and public health authorities, especially as market-driven agricultural forces become an increasingly important part of world trade, with growing populations and increasing economic power in rapidly developing economies. Traditionally, governmental subsidies on the production of meat, dairy products, and sugar have led to their increased consumption. At the same time, such subsidies penalize fruit and vegetable production and consumption to the health detriment of populations owing to their cost or limited supply. This adversely affects vulnerable groups in the population both within countries and globally. High intakes of fat and sugar with low consumption of fruits and vegetables are associated with obesity, hypertension, stroke, and early coronary heart disease. Domestic and international policies to reduce beef and other saturated fat consumption while increasing fruit and vegetable consumption are vital to a

---

BOX 16.6 The Global Alliance for Vaccines and Immunization (GAVI): An Innovative Partnership

The Global Alliance for Vaccines and Immunization (GAVI) was established in 2000 at the World Economic Forum at Davos as an alliance of different stakeholders from the private and public sectors, with the mission of saving children's lives through the worldwide expansion of mass vaccination programs. GAVI's partners include 16 donor countries, United Nations agencies and institutions (UNICEF, WHO, the World Bank), civil society organizations (International Pediatric Association), public health institutes (Johns Hopkins Bloomberg School of Public Health), donor and implementing country governments, the Bill & Melinda Gates Foundation, other private philanthropists, vaccine industry representatives, the financial community, and others whose collective efforts and expertise are enabling great progress to be made in this field.

The GAVI Alliance is a unique, multidimensional partnership of public and private sector resources with a shared focus: to improve child health in the poorest countries by extending the reach and quality of immunization coverage within strengthened health services.

These efforts are directed through the financing mechanisms of the GAVI Fund and the work of the Geneva-based GAVI Secretariat, which channels funding, optimizes product availability and market pricing, and coordinates the field support necessary to plan and implement programs in the world's poorest countries.

Funding is time limited. The intent is to enable countries to develop sustainable programs and progress towards integrating them into national health budgets. The sustainability of these gains depends on national governments placing financial and political priority on these programs and adopting them into their regular budget or financing systems. More than 40 countries now have multiyear immunization plans in keeping with GAVI objectives.

GAVI's efforts are vital to achieving the Millennium Development Goal on child health, which calls for reducing childhood mortality by two-thirds by 2015. By the end of 2010, GAVI had supported the full immunisation of 296 million additional children in 77 countries. Under-five mortality rate in GAVI-supported countries declined from 78 in 2010 to 73 per 1,000 live births in 2012, helping to achievement of the MDG target of 68 in 2015.

Introducing pentavalent vaccine (DPT, Hep B and Hib), rotavirus, pneumococcal pneumonia, and human papillomavirus vaccines to low-income countries was a distant dream a decade ago, but the combination of donors has made near-eradication of poliomyelitis and introduction of new life-saving vaccines possible. GAVI brings modern vaccines to low-income countries, quickening the pace of their introduction and reaching a growing percentage of vulnerable children. The problem of sustainability in times of economic distress is a real one, but economic growth in Africa and Asia has made it possible for national governments to provide increased resources to health and continue to save the lives of children and mothers.

Source: Global Alliance for Vaccines and Immunization (GAVI): an innovative partnership. Available at: http://www.gavialliance.org/about/in_partnership/index.php [Accessed 28 May 2013].
GAVI. Global vaccine action plan 2011-2020 as adopted by the World Health Assembly in 2012. Available at: http://www.who.int/immunization/global_vaccine_action_plan/GVAP_doc_2011_2020/en/index.html?utm_source=Self-Subscribe&utm_campaign=9a43426e42-WIW_Annual_Report_launch_letter4_22_2013&utm_medium=email (accessed 12.1.14.).
GAVI. Goal-level indicators updated October 2013. Available at: http://www.gavialliance.org/results/goal-level-indicators/ (accessed 12.01.14.).
GAVI. Global vaccine action plan 2011-2020 as adopted by the World Health Assembly in 2012. Available at: http://www.unicef.org/videoaudio/PDFs/GVAP_single_pages_PRINT.pdf (accessed 12.1.14.).

healthy nutrition program. This needs to be coupled with the fortification of commonly used foods to address micronutrient deficiencies common in developing and developed countries and as part of larger programs to prevent the rapid spread of CVD, obesity, diabetes, and other chronic conditions.

In 2006–2007, high prices for oil led to a demand for alternative energy sources for motor vehicles with increased use of grains and corn for ethanol fuel production, causing sharp rises in price of wheat, corn, and flour-based food products. Use of foods for energy is seen as inefficient and uneconomic, but the search for non-polluting energy sources will generate major economic growth in new sectors of technology and industry in the decades ahead. These agroeconomic and effective energy policies, along with poverty reduction, educational opportunities, and global and equitable economic growth, are issues of great public health importance internationally.

Despite progress and optimism, the tragic global toll of death and mental or physical disability continues, with preventable infectious or vitamin deficiency conditions numbering in the hundreds of millions. Other issues remain to be dealt with in global health, especially regarding women's health, including family planning, reduction of maternal mortality and morbidity, reduction of violence and abuse against women, and improved education and job opportunities. Child abuse, child labor, female-child murder, and sexual exploitation remain large-scale global health problems. Increasingly, non-infectious conditions affecting young adult and middle-aged males are becoming issues of global health; for example, CVD, diabetes mellitus, trauma, and cancer. Care of special needs groups in the population, such as the mentally ill, the disabled, and the elderly (see Chapters 6 and 7), are global health problems that require attention in each country and locality.

International, individual government and community action is vital to deal with these issues, not only in the poorest countries, but also in developed countries. Known public health measures applied effectively can reduce the burden of these conditions within a very few years. This is a challenge of historical importance and necessity.

EXPANDING NATIONAL HEALTH CAPACITY

The idea of a *cordon sanitaire* to protect a nation's health from invading epidemics is a form of passive defense that has not been effective in major epidemics in the latter part of the twentieth century. A forward defense is now part of the New Public Health. Countries need to reach out to other countries to improve international public health capacity as their own first line of defense. The tragedy of the late discovery of AIDS and inadequate early response was matched by the equally poor handling of the first phase of the 1991–1996 cholera epidemic in South America. The 1991 epidemic cost Peru some US$770 million in lost food exports and reduced tourism. Cholera in Haiti, possibly introduced by international UN troops following the earthquake of 2010, infected many thousands of people in the wreckage of sanitation and housing in the aftermath of the earthquake. In 2012 there were 112,076 cases of cholera in Haiti, with 894 deaths (WHO: Cholera 2012).

Building the first line of defense means strengthening the capacity of individual countries to detect, report, and request help in controlling potentially serious disease outbreaks. Help is available from the WHO, the CDC in Atlanta, USA, and newly strengthened counterparts in France and the UK, as well as international organizations such as the International Red Cross, MSF, and many others. Training in basic epidemiology, sterile techniques, and laboratory services can mean the difference between local containment and widespread infection of hemorrhagic fever viruses, with person-to-person transmission amplified in a hospital setting.

Even the industrialized countries are in need of strengthening of epidemiological capacity. Few have adequate information systems to collect hospitalization data that can provide vital measures of morbidity and the economics of health services. Few have the training capacity for public health epidemiologists, economists, sociologists, sexologists, psychologists, or anthropologists, let alone entomologists, geneticists, and the many other professionals making up the New Public Health team.

Many industrialized countries, satisfied with universal access to doctors and hospitals and the feeling that infectious diseases were going away under the power of sanitation, vaccination, and antibiotics, allowed their public health infrastructure to decline with poor pay, reward, recognition, and motivation, and lack of training capacity. The 1990s brought a different reality of emerging and re-emerging infectious diseases and other plagues such as violence and trauma, drugs, heart disease, cancer, and stroke. Failure to prepare public health professionals and support systems is an invitation to disaster, both epidemiologically and economically. No country can afford such laxity. Training of public health professionals requires graduate schools of public health, which are more essential than the excess of medical schools that already exist in most countries.

The 2006 World Health Report (*Working Together for Health*) focused on worldwide shortages of health personnel, especially in the countries with the most severe health problems. The supply of health workers ranges from 2.3 per 1000 population in Africa and 4.3 in South-East Asia to 18.9 and 24.8 in Europe and the Americas, respectively.

BOX 16.7 Achievements of the Millennium Development Goals 2001–2012

Working together, governments, the United Nations family, the private sector, and civil society have succeeded in saving many lives and improving conditions for many more. The world has met some important targets – ahead of the deadline.

- Extreme poverty is falling in every region. For the first time since poverty trends began to be monitored, the number of people living in extreme poverty and poverty rates fell in every developing region – including in sub-Saharan Africa, where rates are highest. The proportion of people living on less than US$1.25 a day fell from 47 percent in 1990 to 24 percent in 2008 – a reduction from over 2 billion to less than 1.4 billion people.

- The poverty reduction target was met. Preliminary estimates indicate that the global poverty rate at US$1.25 a day fell in 2010 to less than half the 1990 rate. If these results are confirmed, the first target of the Millennium Development Goals – cutting the extreme poverty rate to half its 1990 level – will have been achieved at the global level well ahead of 2015.

- The target of halving the proportion of people without access to improved sources of water has been met. The target of halving the proportion of people without sustainable access to safe drinking water was also met by 2010, with the proportion of people using an improved water source rising from 76 percent in 1990 to 89 percent in 2010. Between 1990 and 2010, over 2 billion people gained access to improved drinking water sources, such as piped supplies and protected wells.

- Improvements in the lives of 200 million slum dwellers exceeded the target. The share of urban residents in the developing world living in slums declined from 39 percent in 2000 to 33 percent in 2012, improving the lives of at least 100 million slum dwellers, well ahead of the 2020 deadline.

- Parity in primary education between girls and boys has been achieved and many more of the world's children are enrolled in primary school since 2000. Girls have benefited the most. The ratio between the enrolment rate of girls and that of boys grew from 91 in 1999 to 97 in 2010 for all developing regions. The gender parity index value of 97 falls within the plus-or-minus 3-point margin of 100 percent, the accepted measure for parity.

- Enrolment rates of children of primary school age increased markedly in sub-Saharan Africa, from 58 to 76 percent between 1999 and 2010. Many countries in that region succeeded in reducing their relatively high out-of-school rates even as their primary school age populations were growing.

- Child survival is gaining momentum. Despite population growth, the number of under-five deaths worldwide fell from more than 12.0 million in 1990 to 7.6 million in 2010. Progress in the developing world as a whole has accelerated. Sub-Saharan Africa, the region with the highest level of under-five mortality, doubled its average rate of reduction, from 1.2 percent a year over 1990–2000 to 2.4 percent during 2000–2010.

- Access to treatment for people living with HIV increased in all regions. At the end of 2010, 6.5 million people were receiving antiretroviral therapy for HIV or AIDS in developing regions. This total constitutes an increase of over 1.4 million people from 2009, and the largest one-year increase ever. The 2010 target of universal access, however, was not reached.

- Globally, tuberculosis incidence rates have been falling since 2002, and current projections suggest that the 1990 death rate from the disease will be halved by 2015. The world is on track to achieve the target of halting and beginning to reverse the spread of tuberculosis.

- Global malaria deaths have declined. The estimated incidence of malaria has decreased globally, by 17 percent since 2000. Over the same period, malaria-specific mortality rates have decreased by 25 percent. Reported malaria cases fell by more than 50 percent between 2000 and 2010 in 43 of the 99 countries with ongoing malaria transmission.

Note: *HIV=human immunodeficiency virus; AIDS=acquired immunodeficiency syndrome.*
Source: *The Millennium Development Goals Report 2012. Available at: http://www.un.org/en/development/desa/publications/mdg-report-2012.html [Accessed 18 February 2013].*

Issues relate not only to quantity but also to access, quality, and support systems for health workers. Migration of educated people in the population tends to drain doctors and nurses from poor countries to wealthy countries, exacerbating shortages and the problems of developing and sustaining standards of care (see Chapter 14).

The UN set out eight key targets for improved health education and social development, the MDGs, in 2001. The targets agreed to by 191 nations stimulated international efforts towards the achievement of improved population health, especially for the world's most disadvantaged people. The 2012 review of progress with important results is summarized in Box 16.7.

TOP–DOWN AND BOTTOM–UP DEVELOPMENT

Improved health requires commitment and leadership and funding from the top of government pyramid, while at the same time enabling local communities to act to improve education, sanitation, communication, and health. Exciting developments are taking place on a large scale in countries such as Ethiopia, Rwanda, and Swaziland.

Ethiopia is struggling to improve child health and nutrition. Stunting was reduced from 58 percent in 2000 to 44 percent in 2011, underweight rates decreased from 41 percent in 2000 to 29 percent in 2011, with improved treatment

TABLE 16.15 Ethiopian Health Extension Worker Program Priorities

Hygiene and Environmental Sanitation	Disease Prevention and Control	Family Health Services	Health Education and Communication
Proper and safe excreta disposal system	Prevention and control of HIV/AIDS	Maternal and child health	Health education and communication
Proper and safe solid and liquid waste management	Prevention and control of tuberculosis	Family planning	
Water supply safety measures	Prevention and control of malaria	Immunization	
Food hygiene and safety measures	First aid	Adolescent reproductive health	
Healthy home environment		Nutrition	
Arthropod and rodent control			
Personal hygiene			

Sources: Shimelis. Personal communication; May 2013.
McCord GC, Lio A, Singh P. Deployment of community health workers across rural sub-Saharan Africa: WHO financial considerations and operational assumptions. Bull World Health Organ 2012;91:244–53B.
Ethiopian Ministry of Finance and Economic Development and UNDP. Ethiopian MDG report: trends and prospective for meeting MDG by 2015. Addis Ababa, Ethiopia; 2010.

of around 300,000 severely malnourished children treated in eight drought-affected regions in 2011 (84 percent cure rate and 0.6 percent death rate). Although child and maternal mortality are significantly decreasing in Ethiopia, the rates remain the highest among the developing world. The infant mortality rate decreased from 97 per 1000 in 2000, to 77 per 1000 in 2005, and to 59 per 1000 in 2011. The under-five mortality rate was reduced from 166 to 123 then 82 per 1000 live births in these same years.

In 2007, Ethiopia expanded a health extension program (HEP), tested in 2004–2005, with training and deployment of 2800 female health extension workers (HEWs), later increased to 33,819 HEWs (a worker to population ratio of two per 5000 people), reaching 89 percent of health posts in rural subdistricts, and working with 2566 HEP supervisors deployed by the end of 2009. This preventive-oriented CHW model provides universal access to primary care with a community-based health promotion and disease prevention strategy. Some 1 million children and 700,000 pregnant or lactating women receive vitamin A supplementation and deworming tablets every 6 months. Iron–folate supplementation is targeted to reach 80 percent of pregnant women. The nutrition program is being revised to focus on children under 2 years, pregnant and lactating women, and adolescent females as a strategy to break the intergenerational cycle of malnutrition, and incorporates other government agencies and the private sector (Ferew and Matji, 2013). The HEWs are highly accepted by the community and work with volunteer CHWs in the villages to achieve specified goals with modest cost. They carry out a baseline survey of the village using a standardized tool, mapping households and the population by age category. They also prioritize health problems of the village, and set targets with respect to the four major areas and 17 service packages of services as seen in Table 16.15.

BOX 16.8 Lessons from Rwanda: Strategies for Strengthening Comprehensive Health Systems

- *National leadership* – high-level political commitment to equity and to service delivery as well as a clear plan for action.
- *Health systems approach* – harnessing funding for disease-specific or other "vertical" programs to build and strengthen platforms for integrated service delivery.
- *Country ownership* – health system spending managed by or in partnership with national and local government.
- *Community-based care* – for example, using community health workers to increase the effectiveness and efficiency of care delivery, especially for chronic diseases.
- *Evidence-based policy making* – a critical "feedback loop" linking research to service and training to promote accountability and improve the quality of care.
- *Cross-sector collaboration* – strengthening health systems with partnerships between the public and private sectors and also across sectors and ministries.

Source: Farmer PE, Nutt CT, Wagner CM, Sekabaraga C, Nuthulaganti T, Weigel JL, et al. Reduced premature mortality in Rwanda: lessons from success. BMJ 2013;346:f65. http://dx.doi.org/10.1136/bmj.f65.

The deployment of CHWs has helped to produce sustainable and low-cost interventions for common health problems in the developing countries. Progress in child health in Rwanda has been achieved by a combination of government and donor organization development of community-based preventive and treatment services with improved coverage of a widened range of vaccines, including rotavirus and HPV (Box 16.8). There have been increases in the number of married women using contraception, deliveries in medically supervised facilities, the use of bed nets for malaria control, and HIV treatment.

The results have been seen in major reductions in child and maternal mortality and HIV deaths (Farmer et al., 2013). Similarly, good progress has been made in Ethiopia in implementing CHW systems on a large scale, which have succeeded in improving nutrition and immunization, and reducing female genital mutilation (Ferew and Matji, 2013).

Famer and colleagues suggest applying systems-level analysis to the complex processes and interventions that must occur, across a health care system and over time, to deliver high-value care for patients with HIV/AIDS and co-occurring conditions, from TB to malnutrition. To deliver value, vertical or stand-alone projects must be integrated into a shared delivery infrastructure so that personnel and facilities are used wisely and economies of scale reaped. Two other integrative processes are necessary for delivering and assessing value in global health: one is the alignment of delivery with local context by incorporating knowledge of both barriers to good outcomes (from poor nutrition to a lack of water and sanitation) and broader social and economic determinants of health and well-being (jobs, housing, and physical infrastructure). The second is the use of effective investments in care delivery to promote equitable economic development, especially for those struggling against poverty and high burdens of disease (Kim et al., 2013).

Many CHW programs have demonstrated the effectiveness of this approach in many countries such as Ethiopia, and in a 2003 study in Colorado, USA, in which CHWs' efforts reduced cardiovascular risk factor prevalence in trial communities compared to control groups. These principles apply in Los Angeles, in New Delhi, and most certainly in rural India and sub-Saharan Africa (see Chapter 14). The Human Development Report entitled *The Rise of the South* (UNDP, 2013) indicates rapid economic and social development, with large population groups moving up the "ladder of development" as seen in Box 16.9. Health gains are major factors and are integral to economic development.

BOX 16.9 The Rise of the South

Countries in the southern hemisphere are showing remarkably rapid and widespread progress. This is bringing broad human development and "dramatic expansion of individual capabilities and sustained human development progress" in countries with the vast majority of the world's people. Dozens of countries and billions of people are moving up the development ladder, with direct impact on wealth creation and broader human progress in all countries and regions of the world. "This provides new opportunities for catch-up in less developed countries with creative policy initiatives that could benefit the most advanced economies as well."

Source: *United Nations Development Programme. Human development report 2013. The rise of the south: human progress in a diverse world. New York: UNDP; 2013. Available at: http://hdr.undp.org/en/media/ HDR2013_EN_Summary.pdf [Accessed 26 May 2013].*

GLOBAL HEALTH AND THE NEW PUBLIC HEALTH

The New Public Health is concerned with globalization of health in several senses. First, it includes all health activities in any one country, and second, what happens in the rest of the world, including the effects of globalization, is of direct interest to each country, no matter how wealthy, industrialized, or isolated. The lessons of the bubonic plague may seem to be remote history to the generation raised on concepts of the success of public health in the control of communicable diseases, but the lessons of HIV should surely be learned. John Donne's famous idea that "no man is an island unto himself" expresses the issue clearly. Global health means identifying and addressing the acute infectious and chronic diseases as early as possible before they spread or amplify by common risk factors.

In the twenty-first century, many developing countries are reaching an epidemiological transition that took place in the industrialized world in the mid-twentieth century. The resurgence of long-known diseases and the emergence of new and sinister infectious disease threats are occurring worldwide. The industrialized countries are again facing serious infectious disease challenges, including those imported from developing countries.

In the 1950s and 1960s, control of infectious diseases looked extremely promising. Vaccines and antibiotics seemed to provide the answer to age-old infectious diseases. But in the 1970s and 1980s new infectious organisms appeared, along with a frightening increase in resistance of microorganisms to therapeutic agents. Diseases spread from country to country, as did HIV in the 1980s, and cholera in Peru and diphtheria in Russia in the 1990s, and the plague outbreak in India in 1994.

The Global Burden of Diseases, Injuries, and Risk Factors Study (GBD 2010), headed by Professor Chris Murray of the Harvard School of Public Health, was a collaboration of 488 scientists from 303 institutions in 50 countries. The study documented the state of health indicators around the world using uniform methods. The burden of each disease, injury, or risk factor was calculated in terms of deaths, years of life lost due to premature mortality (YLLs), years lived with disability (YLDs), and disability-adjusted life years (DALYs). Age-specific mortality was analyzed for each of 187 countries for the years from 1970 to 2010 (Murray et al., 2012). This study provides important data for economic analysis studies such as those carried out by the World Economic Forum and the Harvard School of Public Health (2011) to define the economic burden associated with NCDs. This joint report focuses on low- and middle-income countries, which account for 84 percent of the world's population and 83 percent of the NCD burden. A 2011 WHO report (Scaling up action against non-communicable diseases: How much will it cost?), quantifying the expected costs of addressing projected national NCD mortality rates against the current and future economic output of a country, shows great economic gains with appropriate interventions (Table 16.16).

The chronic diseases associated with the risk factors of overnutrition and smoking are increasing in low- and medium-income countries just as the public health field is gaining momentum in controlling infectious and childhood diseases. In addition, all countries are facing the strains of health expenditures and the painful process of health reform. The legal, ethical, and technological challenges are increasingly important in managing health care systems (Box 16.10).

All health systems are obliged to face these challenges through the sharing of information and improved monitoring of the use of resources, as well as seeking effective ways of preventing diseases and managing them to promote early and complete return to function. All industrialized countries are facing serious problems financing health care in its traditional form, and reform is taking place amid aging of the population, increasing technology, and high expectations of health care. Reforms shifting emphasis and resources from hospital to ambulatory and primary care show a strong return to the idea of health promotion by regulation and education.

Some answers to unconquered infectious diseases have come from simple technology, such as the use of ORT to reduce morbidity and mortality from diarrheal diseases. The resurgence of TB and multidrug-resistant organisms has been successfully handled by another simple innovation of directly observed therapy by CHWs to ensure compliance and completion of treatment, especially in high-risk groups.

Malaria control, using specially trained CHWs, is another application of inexpensive, simple, appropriate technology.

Simpler technologies are also having a major impact on the chronic diseases. Cardiovascular mortality rates are falling in most industrialized countries as a result of healthier lifestyles and improved treatments such as antihypertensive treatments, low-fat diets, statins, aspirin, and physical activity. New screening techniques are being developed continuously. Lung cancer screening using low-dose helical computed tomography as compared to chest radiography among heavy-smoking older adults shows reduced mortality from lung cancer. New diagnostic tests for cancer will include refined robotic smell techniques. New urine tests will help in early detection of bladder cancer and help to continue the reduction in cancer mortality being seen in

---

TABLE 16.16 Economic Burden of Selected Non-Communicable Diseases: Economic Lost Output 2011–2025, (trillions of US Dollars in 2008)

Country Income Group	Diabetes	CVDs	Respiratory Diseases	Cancer	Total
Upper middle	0.30	2.52	1.09	1.20	5.12
Lower middle	0.91	1.07	0.44	0.26	1.85
Low income	0.02	0.17	0.06	0.06	0.31
Total of low and middle	0.42	3.76	1.59	1.51	7.28

Note: CVD = cardiovascular disease.
Sources: World Health Organization. Burden to "best buys": reducing the economic impact of non-communicable diseases in low- and middle-income countries. Available at: http://www.who.int/nmh/publications/best_buys_summary.pdf [Accessed 3 June 2013].
World Economic Forum and Harvard School of Public Health. The global economic burden of non-communicable diseases; 2011. Available at: http://www3.weforum.org/docs/WEF_Harvard_HE_GlobalEconomicBurdenNonCommunicableDiseases_2011.pdf [Accessed 29 July 2013].
World Health Organization. Scaling up action against non-communicable diseases: how much will it cost? WHO; 2011. Available at: http://whqlibdoc.who.int/publications/2011/9789241502313_eng.pdf [Accessed 29 July 2013].

BOX 16.10 Costs of Scaling Up a Core Intervention Package for Non-Communicable Diseases in Low- and Middle-Income Countries

The health and economic consequences of non-communicable diseases (NCDs) are staggering. NCDs are the leading causes of death globally and 80 percent occur in low- and middle-income countries. Half of these deaths occur in the productive years of life, affecting economic activity of the countries. In older people NCDs cause disability and health care needs that are costly to economies. In a "business as usual" scenario where intervention efforts remain static and rates of NCDs continue to increase, yearly loss to the economy is equivalent to approximately 4 percent of these countries' current annual output. On a per-person basis, the annual losses amount to an average of US$25 in low-income countries, US$50 in lower middle-income countries and US$139 in upper middle-income countries.

A World Health Organization study indicates that the price tag for scaled-up implementation of a core set of NCD "best buy" intervention strategies is comparatively low. Population-based measures for reducing tobacco and harmful alcohol use, as well as unhealthy diet and physical inactivity, are estimated to cost US$2 billion per year for all low- and middle-income countries – less than US$0.40 per person. Individual-based NCD "best buy" interventions, ranging from counseling and drug therapy for cardiovascular disease to measures to prevent cervical cancer, bring the total annual cost to US$11.4 billion. On a per-person basis, the annual investment ranges from under US$1 in low-income countries to US$3 in upper middle-income countries. Reducing mortality rates for ischemic heart disease and stroke by 10 percent would reduce economic losses in low- and middle-income countries by an estimated US$25 billion per year, which is three times greater than the investment needed for the measures to achieve these benefits.

Source: World Economic Forum and Harvard School of Public Health. The global economic burden of non-communicable diseases; 2011. Available at: http://www.weforum.org/EconomicsOfNCD [Accessed 11 June 2013].

TABLE 16.17 Basic Risks and Continuum of Care for Prevention of Child Morbidity and Mortality

Stage	Risks	Prevention/Treatment
Pre-pregnancy	Education, marriage before age 19; poor nutrition, female genital mutilation, HIV, STIs	CHW assessment, counseling, iron and folic acid supplements, teach and monitor hygiene, promote breastfeeding, screen and treat hypertension, renal disease, HIV, STIs
Pregnancy and lactation	Lack of prenatal care, abuse, smoking, alcohol, drugs, lack of adequate diet; anemia; low birth weight, neural tube defect, fetal alcohol syndrome	CHW care; good nutrition, multi-micronutrient supplements including vitamin A, iron, and folic acid; risk assessment, referral if high risk, antenatal care from earliest stage, HIV treatment, counseling
Delivery	Preterm birth, lack of trained midwifery care, poor hygiene, anemia, hemorrhage, eclampsia, infection, hypertension, prolonged and obstructed labor; maternal death; vesicovaginal fistula	Trained attendants, hygienic, safe delivery, referral to medical center/hospital for high-risk patients; telemedicine (cell phone or landline); transportation; family support
Neonatal	Half of neonatal deaths occur in first 24 hours and three-quarters in first week of life; non-breastfeeding, lack of trained care providers, asphyxia, respiratory infection, diarrheal diseases, tetanus, HDN	Appropriate care by trained attendant; Apgar score; initiate and sustain exclusive breastfeeding; HIV care; vitamin K and hepatitis B injections and eye care with antibiotic after birth; oxygen and antibiotic if respiratory distress; ORS for diarrhea; register births and birth weight and complications
Infancy	Inadequate milk formula and feedings, contaminated water and food; worms; malnutrition; lack of stimulus of developmental tasks; infectious diseases – diarrhea, respiratory, malaria; non-breastfeeding with risk of childhood obesity, type 1 and type 2 diabetes, sudden infant death syndrome	Exclusive breastfeeding for minimum 6 months; add complementary feeding gradually; for adequate nutrition use multi-micronutrient powders for home fortification; immunization with DPT, polio, MMR, Hib, pneumococcal pneumonia, rotavirus; height and weight monitoring and recording on WHO growth charts; ORS and respiratory care as needed; vitamin supplementation; insecticide in home and impregnated bed nets, vector control
1–5 years	Malnutrition – stunting, inadequate feeding, lack of iron and vitamin A supplements, and infectious diseases – pneumonia, diarrhea, malaria; childhood obesity	Monitoring and recording on growth chart; counseling, developmental assessment and support; ensure complete immunization; impregnated bed nets, multivitamins for children at risk; for adequate nutrition use multi-micronutrient powders for home fortification including iron supplements; add zinc with ORT for diarrhea; refer children with malnutrition, failure to thrive

Note: HIV = human immunodeficiency virus; STI = sexually transmitted infection; HDN = hemorrhagic disease of the newborn; CHW = community health worker; ORT = oral rehydration therapy; DPT = diphtheria–pertussis–tetanus; MMR = measles–mumps–rubella; Hib = *Haemophilus influenzae* type b; WHO = World Health Organization.
Sources: See Chapters 4 and 6.
World Health Organization. The integrated global action plan for the prevention and control of pneumonia and diarrhea. WHO/UNICEF; 2013. Available at: http://apps.who.int/iris/bitstream/10665/79200/1/9789241505239_eng.pdf [Accessed 11 June 2013].
World Health Organization. Essential nutrition actions: improving maternal, newborn, infant and young child health and nutrition. Geneva: WHO; 2013. Available at: http://apps.who.int/iris/bitstream/10665/84409/1/9789241505550_eng.pdf [Accessed 12 June 2013].
World Health Organization. Essential interventions, comorbidities and guidelines for reproductive, maternal, newborn and child health: a global review of the key interventions related to reproductive, maternal, newborn and child health. Available at: http://www.who.int/pmnch/topics/part_publications/201112_essential_interventions/en/index1.html [Accessed 12 June 2013].

high-income countries. Trauma death rates are falling as a result of mandatory improvements in car and road safety, as well as stringent policing to prevent alcohol and drug use among drivers. Occupational safety standards are continuing the decline in occupational mortality. Poisoning deaths from drug abuse are increasing and require improved treatment of substance abuse. The simpler technology of home care allows chronically ill patients to return to their homes with less lengthy, high-cost, and risky hospitalizations. Ambulatory care can be provided safely for many conditions previously requiring hospitalization, which incurred greater cost and danger from hospital-acquired infections.

New technology will emerge, such as synthetic vaccines, which as safer and cheaper than organically grown vaccines; vaccines genetically engineered in basic foods; affordable genetic sequencing, with genetically customized cancer treatments; remote patient monitoring, using Internet and wireless technology to enable patient monitoring and data sharing between health care systems; synthetic blood vessels for replacing damaged arteries; and laboratory-grown organs, to replace damaged organs such as livers. A biotechnological revolution is in the making which will provide a wider array of public health and medical interventions to prevent and cure diseases with safer and less expensive technology.

Preventing maternal and child deaths remains a huge challenge globally. Development of a community-based infrastructure is the fundamental challenge to reach the rural and urban poor where the risks are greatest. CHWs, trained, supervised, supported and preferably salaried, can make and are making huge differences in connecting these high-risk populations to basic health care. The support systems needed are evolving in many countries and should be fostered by the donor agencies, but mainly by the national governments to meet their obligations under the global MDG program and its subsequent iterations after 2015. The basic risks, their prevention and management are indicated in Table 16.17.

While care of the individual mother and child is crucial there are equally important initiatives by national governments such as in mandatory fortification of salt with iodine, fortification of flour with iron, vitamin B complex, including folic acid and vitamin B_{12} including those used in food manufacturing. Vitamin D deficiency should also be addressed even in sunny climates where dark skin tone and religious customs of total coverage of the body may reduce vitamin D production by sun exposure. In northern climates vitamin D fortification of milk and supplements are essential because of long winters and cloudy weather even in summer (see Chapter 8). Global nutrition programs have been characterized as disorganized and uncoordinated.

The future of public health and health care will see tremendous change and adoption of new modalities of preventing and managing disease: recombinant vaccines will reduce costs and introduce new vaccines, bringing more infectious diseases under control, including viral hepatitis and respiratory and diarrheal diseases. Vaccine technology for cancer and genetic disorders is evolving. Congenital disorders will be controlled by education, screening, and appropriate interventions. Dietary change will help in the control of cancer, as will screening and reduced exposure to carcinogens. It is now established that infectious agents can cause chronic disease, such as *Helicobacter pylori* and peptic ulcer, and cancers such as those of the stomach, liver, and cervix. There are synergies between micronutrient deficiency conditions with infectious and chronic disease, such as folic acid deficiency and birth defects, and these associations open many new vistas for research, preventive breakthroughs, and applied public health.

Health for All means access to care for everyone. This requires sound management of finances and other resources to provide the needed services efficiently and by reducing the waste and extravagances of unnecessary servicing. It also requires a social and physical environment that enables people to experience healthful, satisfying, and productive lives. To attain these lofty goals, broad partnerships or coalitions of health services and providers working with communities and an increasingly knowledgeable and participating general public must be achieved. This is especially important for compliance with immunization, healthful infant and child nutrition and care, self-care in pregnancy, and healthful adult nutrition. Paternalistic, traditional services of doctors dominating both the health systems and patients are not able to raise the level of patient and community participation needed.

The goal of better health requires a sharing of tasks and resources between the clinical and community levels, and between countries. Assisting countries in developing the staff and infrastructure of epidemiology in infectious and chronic disease is an investment in the frontline of public health protection and self-defense. This is the substance of work by international organizations and bilateral aid. In international partnerships in Europe, the industrialized countries help each other, and this model needs to be applied to promote public health infrastructure in developing countries as well.

SUMMARY

Health for All sounded like a hopeless, idealistic dream when first promulgated by the WHO in 1977. Yet the progress made since then in lowering mortality and birth rates, raising longevity, and improving quality of life has been dramatic. *Globalization of health* means that what happens anywhere is the concern of everyone everywhere, as the world learned with plague in the fourteenth century and AIDS in the late twentieth century. At the same time, globalization means all aspects of health for a population, because of the interaction of health care, economics, and the political priority given to health. Global partnership efforts to contain the HIV/AIDS pandemic advanced greatly with support and adoption of generic antiretrovirals manufactured at very low prices to treat 10 million people with HIV/AIDS in developing countries. Similar efforts with newly successful short course antiviral treatments could provide answers to the HCV pandemic for low and middle income countries. "Large-scale manufacture of treatment to cure Hepatitis C is feasible, with target prices of US$100-250 per 12 week treatment course" (Hill et al, 2014).

In the globalized world of the twenty-first century, public health of one country cannot be considered in isolation. Globalization has bridged countries together, intensified human interactions, and made international boundaries increasingly irrelevant in the control of disease. *Global health* is a very complex term which is influenced by actions or circumstances in countries other than the one affected directly. Today, the determinants of global health include poverty, environmental degradation, climate change, violence, terrorism, illegal drug trafficking, and international or bilateral trade laws. With all its drawbacks, globalization also has its benefits in the transfer of education, science, and technology, helping to provide the benefits of development from developed to developing countries. Many countries are emerging from economic stagnation with rapid development of industry and trade based on domestic and global markets. The middle- and low-income countries are experiencing rapid growth in their middle-class populations and trends in disease prevalence of heart disease, stroke, overnutrition, obesity, and diabetes, and growing gaps between rich and poor. The economic burden of these diseases in developing countries makes public health programs for intervention essential, to prevent them from undermining economic and social development.

Global action means that countries must be committed to health at all levels, including state and local governments as well as voluntary, educational, and many other elements of a society. The potential gain is enormous, and this requires systematic organization and information, with well-defined targets, strategies, and tactics. The WHO Framework Convention on Tobacco Control is a key global public health treaty and is meant to help developing countries to address the tobacco epidemic promoted by the tobacco industry.

Measuring disease (infectious and chronic), family health, special groups in the population, nutrition, environmental and occupational health, organization of public health, management of health systems, comparing with other national health systems, human resources, technology assessment, quality assurance, law, and ethics, all chapter topics in this book, are the substance of the New Public Health. Altogether, they are the subjects of day-to-day life in health systems.

The great achievement of smallpox eradication has been followed by other global disease eradication efforts, and great health and economic benefits have already been achieved. The eradication of poliomyelitis is progressing with coordinated global and country-level activities. In 2012, a total of 223 polio cases was reported from five countries: Afghanistan, Chad, Niger, Nigeria, and Pakistan. India's last case was reported in 2011, but continued control efforts are underway to ensure its polio-free status. As of 16 July 2013, 132 polio cases have been reported from the three remaining endemic countries in which wild poliovirus is still circulating: Afghanistan, Nigeria, and Pakistan. Continuous monitoring and special immunization efforts in still endemic areas are using type 1 oral poliomyelitis vaccines to reduce this most virulent strain, while also addressing type 2 and type 3 areas. In both Pakistan and Afghanistan, polio field workers have been murdered by Islamic terrorists who oppose immunization, which has made continued immunization and eradication efforts in those endemic countries problematic. However, polio eradication is within sight with the sustained efforts of the donors and national governments, as well as international organizations such as the WHO, UNICEF, UNDP, GAVI, Rotary International, and the Gates Foundation.

New emerging diseases continue to threaten to spread out from localized cases. West Nile fever has become endemic in North America and Europe. Dengue, Chagas' disease, Rift Valley fever, and chikungunya have spread to many countries far from their original habitat. In 2013, MERS-CoV emerged in Saudi Arabia with 90 cases and 45 fatalities (case fatality rate of 50 percent) in what seems to be animal-to-human transmission and subsequent close-contact human-to-human transmission, with the reservoir likely to be the horseshoe bat (CDC, 25 July 2013). This virus has similarities to the coronavirus which caused the SARS pandemic of 2003, from which the global health community learned many lessons and has adopted measures to take those lessons into account. MERS-CoV could become a problem associated with the Hajj pilgrimage to Mecca in 2013.

Lessons learned from the new and re-emerging infectious disease patterns help to strengthen the capacity to meet future challenges such as pandemic avian influenzas. The SARS epidemic helped Canada, for example, to develop a strong federal investigative and laboratory capacity to deal with national and global health threats, and China also improved its surveillance capacity.

The experience, skills, and infrastructure of infectious disease control will also bring changes in chronic disease control. New acute and chronic disease challenges will emerge; preparation will increase the chances of coping with them before they reach epidemic proportions. Addressing the global rising tide of CVD and cancer mortality, especially in the low-income countries and the former Soviet Union and many developing countries, is the central challenge of public health for the coming decades. Interventions to reduce poverty and poor nutrition, along with development of health systems that address social inequalities and human behavior, are the major challenges facing countries at all levels of development. These are no less complex and no less important than recognizing the resurgence of infectious diseases, as well as multidrug-resistant and non-communicable diseases, all of which are serious public health problems. The shared predominance of leading causes of death from cardiovascular diseases in high medium and low income countries should reinforce global approaches to the preventive measures that have achieved much success in the high income countries of smoking reduction, hypertension and lipid control, healthy diet and exercise with their greater application in the medium and low income countries.

The conceptual basis of the New Public Health provides an idealized yet practicable model for all countries, including developing countries. This concept has since grown with many influences, including health promotion, health targets, and factors outside the direct domain of public health organizations and health insurance. The important influence of poverty on health can be addressed by poverty alleviation, job creation, urban planning, and education, as well as the foundations of hygiene and environmental health. The New Public Health involves the actual management of health systems and integration of secondary and tertiary care services of hospitals, and the whole range of programs or services that relate to improving the health of the individual and the society.

Health inequalities within nations and around the globe present an important challenge to public health. Huge excesses in rates of mortality from preventable diseases are still prevalent in many countries, such as those in Eastern Europe, along with slow rates of adoption of current best practices in health protection and health care. Failure or delay in adoption of common policies on fundamentals such as immunization or essential vitamins and minerals in nutritional security are issues that public health and political leaders need to address. New disciplines such as health promotion helped to control HIV before there was a medical treatment. New innovations in public health offer the promise of control of chronic conditions that plagued earlier generations, such as chronic peptic ulcer disease, cancer of the cervix and colorectal cancer. The future will offer more breakthroughs but their absorption into common policies and practices will, as in the past, often be unconscionably slow (Tulchinsky and Varavikova, 2010).

The international health community, through the MDGs, has succeeded, in part, in changing the health agenda of many countries towards prevention, primary care, and health promotion. The development of goals and targets with international sanction helps each country to resist pressure to place most of its health care resources into curative and tertiary services. An international commitment to Health for All has taken on an important meaning in member countries. It has helped national and regional health leadership to tackle the difficult task of changing priorities to an emphasis on primary health care and modern public health.

Coalitions of forces are needed to take up the challenges that the health community cannot do alone. The isolation of health from other sectors, or of parts of the health spectrum from each other, lowers the capacity of all to reach common goals. Networks of international agencies, including the WHO, UNICEF, World Bank, FAO, UNDP, private donor organizations, the private sector, and many others, are needed to face the health challenges and tasks. Similarly, at the national, state, and local levels, globalized approaches and networks of organizations can help to define targets and mobilize the resources needed to achieve them. International partnerships such as WHO and GAVI have declared "the Decade of Vaccines (2011–2020) so that all can live free from vaccine-preventable diseases, both by increasing routine immunization but equally important reducing the time for incorporating the more recently developed and proven vaccines.

It is appropriate to end this book with a dedication pledge taken by graduates of a prominent school of public health in the USA, as a personal commitment for public health professionals graduating from their training programs (Box 16.11). This statement of personal mission and

---

BOX 16.11 A Public Health Graduate's Pledge to Public Health

- We as health professionals do hereby commit ourselves to advocacy and action to promote the health rights of all human beings.
- The enjoyment of the highest attainable standard of health is one of the fundamental rights of every human being. It is not a privilege reserved for those with power, money, or social standing.
- Health is more than the absence of disease, but includes prevention of illness, development of individual potential, and a positive sense of physical, mental, and social well-being.
- Health care should be based on dialogue and collaboration among citizens, professionals, communities, and policy makers.
- Health services should emphasize equity, accessibility, community, participation, prevention, and sustainability.
- Health begins with healthy development of the child and a positive family environment. Health must be sustained by the active role of men and women in health and development. The role of women and their rights must be recognized, respected, and promoted.
- Health care for the elderly should preserve dignity, respect, and concern for quality of life, and not merely extend life.
- Health requires a sustainable environment with balanced human population growth and preservation of cultural diversity.
- Health depends on more than access to health care. It depends on healthy living conditions and the availability to all people of basic essentials: food, safe water, housing, education, productive employment, protection from pollution, and prevention of social alienation.
- Health depends on protection from exploitation and discrimination on account of race, religion, political belief, ethnic group, national origin, gender, sexual preference, or economic or social status.
- Health requires peaceful and equitable development and collaboration of people.

Source: University of California Los Angeles (UCLA) School of Public Health. Graduation ceremony oath 2013. Confirmed by Dean Jody Heymann; July 2013.

---

values represents the best ideals of public health and is relevant to the many training programs in public health not only in the USA, but also in Europe and other parts of the world.

Healthy people are more productive than ill people, so investments in health and nutrition, along with education and sanitation, contribute to economic growth. Science, technology, and successful public health practice have shown great achievements in reducing and eliminating many previously devastating diseases while promoting longevity and healthy aging. This is seen in the high-income countries and increasingly in the middle- and low-income countries as well. Political priorities in resource allocation are crucial to support for health protection, nutritional security, universal health coverage, and societal commitment to reducing economic gaps in terms of relative and absolute poverty. Global organizations for health and the private sector donors are vital to achieving the current MDGs (by 2015). The global efforts to achieve the MDGs, even if only partly successful, show that political action can achieve major results; although not uniformly and satisfactorily in all countries, many have done well. Their follow-up targets will surely add a major focus on the prevention of NCDs, trauma, and mental health, which affect all nations.

Reduction in smoking alone will reduce lung and other cancers. Improved screening and management of hypertension will continue and accelerate the decline in mortality from CVDs. Increasing vaccination coverage and expanding the vaccines to include those against pneumonia and diarrheal diseases will reduce child mortality and save millions of lives. Failure to implement the reduction of lead; to fortify food to eliminate micronutrient deficiencies (the "silent hunger") of iodine, iron, and vitamin D, and other deficiencies; and to fortify flour with folic acid to prevent birth defects, has meant that the health needs of mothers and children have not been met. Furthermore, non use of alcohol during pregnancy will prevent fetal alcohol syndrome, while malnutrition and overnutrition are factors in much of the global morbidity that produces the most common causes of death and disability.

Public health's record of achievements should lend optimism to the serious challenges of continuing this progress and facing the issues of an aging population in times of economic slumps and with newly emerging diseases. The achievements of science and technology need support and implementation. New breakthroughs in genetics, nanotechnology, immunology, pharmacology, vaccinology, nanotechnology, and robotics will produce great advances in diagnostics, simpler screening methods, and improved treatments for chronic debilitating brain and neurological disorders. Societal efforts to reduce poverty may be the most important contributor to improvements in health.

The New Public Health is a conceptual framework and methodology for implementation of these lofty, but achievable goals. It addresses policy and management of health systems as well as health promotion and disease prevention so that changes in priorities can be implemented by appropriate shifts in resources to meet the health needs of individuals, vulnerable groups, and national and international communities. Globalization of health is more than ever vital to human well being requiring international consensus and coordination for reducing pollution and its fearsome effects of climate change. Similarly in health infrastructure and human resources global cooperation is crucial to achieve desired outcomes in reducing inequalities both within and between countries. This means reduced inequities, maximum use of available technologies and statistical, epidemiological, social, and basic sciences of public health with a renewed global commitment to the global ideal of Health for All.

NOTE

For a complete bibliography and guidance for student reviews and expected competencies please see companion web site at http://booksite.elsevier.com/9780124157668

BIBLIOGRAPHY

Global Health

Abegunde, D.O., Mathers, C.D., Adam, T., Ortegon, M., Strong, K., 2007. The burden and costs of chronic diseases in low-income and middle-income countries. Lancet 370, 1929–1938. Available at: http://www.thelancet.com/journals/lancet/article/PIIS0140-6736(07)61815-7/abstract (accessed 12.01.14).

Black, R.E., Allen, L.H., Bhutta, Z.A., Caulfield, L.E., de Onis, M., Ezzati, M., et al., 2008. for the Maternal and Child Undernutrition Study Group). Maternal and child undernutrition: global and regional exposures and health consequences. Lancet 371, 243–260. Available at: http://www.thelancet.com/journals/lancet/article/PIIS0140-6736(07)61690-0/abstract (accessed 12.01.14).

Bouwman, H., van den Berg, H., Kylin, H., 2011. DDT and malaria prevention: addressing the paradox. Environ. Health Perspect 119, 744–747. Available at: http://www.ncbi.nlm.nih.gov/pmc/articles/PMC3114806/pdf/ehp-119-744.pdf (accessed 12.01.14).

Brandt, A., 2013. How AIDS invented global health. N. Engl. J. Med. 368, 2149–2152. Available at: http://www.nejm.org/doi/full/10.1056/NEJMp1305297 (accessed 26.07.13).

Bryce, J., Coitinho, D., Darnton-Hill, I., Pelletier, D., Pinstrup-Andersen, P., 2008. for the Maternal and Child Undernutrition Study Group. Maternal and child undernutrition: effective action at national level. Lancet 371, 510–526. Available at: http://www.thelancet.com/journals/lancet/article/PIIS0140-6736(07)61694-8/abstract 12.01.14.

Centers for Disease Control and Prevention, 2013. Antismoking messages and intention to quit – 17 countries, 2008–2011. MMWR. Morb. Mortal. Wkly. Rep. 62, 417–422. Available at: http://www.cdc.gov/mmwr/preview/mmwrhtml/mm6221a2.htm 13.01.14.

Centers for Disease Control and Prevention, 2011. Chronic disease prevention and health promotion. Targeting the nation's leading killer at a glance. CDC, Atlanta, GA. Available at: http://www.cdc.gov/chronicdisease/resources/publications/aag/osh.htm (accessed 12.06.13).

Centers for Disease Control and Prevention, 31 May 2013. Global health: polio. Updates on CDC's polio eradication efforts. Available at: http://www.cdc.gov/polio/updates/ (accessed 01.06.13).

Centers for Disease Control and Prevention, Smoking and tobacco use: data and statistics [posted 5 June 2013]. Available at: http://www.cdc.gov/tobacco/data_statistics/fact_sheets/fast_facts/ (accessed 26.07.13).

Centers for Disease Control and Prevention, Special Pathogens Branch. Known cases and outbreaks of Ebola hemorrhagic fever in chronological order. Available at: http://www.cdc.gov/ncidod/dvrd/spb/mnpages/dispages/ebola/ebolatable.htm (accessed 02.06.13).

Chan, M., Lake, A., 2013. Integrated action for the prevention and control of pneumonia and diarrhea. Lancet 381, 1436–1437. Available at: http://www.thelancet.com/journals/lancet/article/PIIS0140-6736(13)60692-3/fulltext (accessed 12.01.14).

Denny, J., Boelaert, F., Borck, B., Heuer, O.E., Ammon, A., Makela, P., 2007. Zoonotic infections in Europe: trends and figures – a summary of the EFSA–ECDC annual report. Eurosurveill. Wkly. Rep. 12, 12. Available at: http://www.eurosurveillance.org/ViewArticle.aspx?ArticleId=3336 (accessed 12.01.14).

Eisenberg, J.N.S., Scott, J.C., Porco, T., 2007. Integrating disease control strategies: balancing water sanitation and hygiene interventions to reduce diarrheal disease burden. Am. J. Public Health 97, 846–852. Available at: http://www.ncbi.nlm.nih.gov/pmc/articles/PMC1854876/pdf/0970846.pdf (accessed 12.01.14).

El-Harrak, M., Martín-Folgar, R., Llorente, F., Fernández-Pacheco, P., Brun, A., Figuerola, J., et al., 2011. Rift Valley and West Nile virus antibodies in camels, North Africa [letter]. Emerg. Infect. Dis. 17 (12). Available at: http://dx.doi.org/10.3201/eid1712.110587 (accessed 02.06.13).

Ferew, L., Matji, J., 2013;6 June. Delivery platforms for sustained nutrition in Ethiopia. Lancet. Available at: http://press.thelancet.com/nutrition-ethiopia.pdf (accessed 07.06.13).

Food and Agriculture Organization, 2013. FAO urges end of malnutrition as priority. Available at: http://www.fao.org/news/story/en/item/176888/icode/. (accessed 12.01.14).

GAVI. Global vaccine action plan 2011-2020 as adopted by the World Health Assembly in 2012. Available at: http://www.unicef.org/video-audio/PDFs/GVAP_single_pages_PRINT.pdf (accessed 12.1.14).

Global Burden of Disease 2010 country results, 2013. a global public good. Lancet 381, 965–970. Available at: http://www.thelancet.com/journals/lancet/article/PIIS0140-6736(13)60355-4/abstract (accessed 12.01.14).

Gostin, L.O., 2007. The "tobacco wars" – global litigation strategies. JAMA 298, 2537–2539. Available at: http://scholarship.law.georgetown.edu/cgi/viewcontent.cgi?article=1479&context=facpub (accessed 12.01.14).

Hogan, M.C., Foreman, K.J., Naghave, M., Ahn, S.Y., Makela, S.M., Wang, S.M., et al., 2010. Maternal mortality for 181 countries, 1980–2008: a systematic analysis of progress towards Millennium Development Goal 5. Lancet 375, 1609–1623. Available at: http://www.thelancet.com/journals/lancet/article/PIIS0140-6736(10)60518-1/abstract (accessed 12.01.14).

Health in the Post-2015 Agenda. Report of the Global Thematic Consultation on Health, April 2013. Available at: http://www.worldwewant2015.org/file/337378/download/366802 (accessed 12.01.14).

Hill A, Khoo S, Fortunak J, Simmons B, Ford N. Minimum costs for producing Hepatitis C direct acting antivirals, for use in large-scale treatment access programs in developing countries. Clinical Infectious Diseases Advance Access published January 6, 2014. Available at: http://cid.oxfordjournals.org/content/early/2014/01/06/cid.ciu012.full.pdf+html (accessed 14.1.14).

Hutton, G., Bartram, J., 2008. Global costs of attaining the Millennium Development Goal for water supply and sanitation. Bull. World Health Organ. 86, 2–3. Available at: http://www.ncbi.nlm.nih.gov/pmc/articles/PMC2647341/ (accessed 12.01.14).

Jamison, D.T., 2006. Investing in health. In: Jamison, D.T. (Ed.), Disease control priorities in developing countries. second ed. Oxford University Press, New York, pp. 3–36. Available at: http://www.ncbi.nlm.nih.gov/books/NBK11754/ (accessed 12.01.14). Chapter 1..

Kikwete, J., Jenkins, K., Whitbread, J., 2013;12 April. Playing our part to save children's lives. Lancet. Available at: http://dx.doi.org/10.1016/S0140-6736(13)60719-9 (accessed 12.01.14).

Liu, L., Johnson, H.L., Cousens, S., Perin, J., Scott, S., Lawn, J.E., et al., 2012. Global, regional, and national causes of child mortality: an updated systematic analysis for 2010 with time trends since 2000. Lancet 379, 2151–2161. Available at: http://www.thelancet.com/journals/lancet/article/PIIS0140-6736(12)60560-1/abstract (accessed 12.01.14).

Lopez, A.D., Mathers, C.D., Ezzati, M., Jamison, D.T., Murray, C.J.L. (Eds.), 2006. Global burden of disease and risk factors. Oxford University Press/World Bank, Washington, DC. Available at: http://www.ncbi.nlm.nih.gov/books/NBK11820/ (accessed 12.01.14).

McCord, G.C., Liu, A., Singh, P., 2012. Deployment of community health workers across rural sub-Saharan Africa: WHO financial considerations and operational assumptions. Bull. World Health Organ. 91. 244–53B. Available at: http://www.who.int/bulletin/volumes/91/4/12-109660/en/ (accessed 09.06.13).

Morris, S.S., Cogill, B., Uauy, R., 2008. for the Maternal and Child Undernutrition Study Group. Effective international action against undernutrition: why has it proven so difficult and what can be done to accelerate progress? Lancet 371, 608–621. Available at: http://www.thelancet.com/journals/lancet/article/PIIS0140-6736(07)61695-X/fulltext (accessed 12.01.14).

Munos, M., Fischer Walker, C.L., Black, R.E., 2010. The effect of rotavirus vaccine on diarrhoea mortality. Int. J. Epidemiol. 39, i56–i62. Available at: http://ije.oxfordjournals.org/content/39/suppl_1/i56.long (accessed 12.01.14).

Murray, C.J.L., Richards, M.A., Newton, J.M., Fenton, K.A., Andersin, H.R., Atkinson, C., et al., 2013. UK health performance: findings of the Global Burden of Disease Study 2010. Lancet 381, 997–1020. Available at: http://www.thelancet.com/journals/lancet/article/PIIS0140-6736(13)60355-4/abstract (accessed 12.01.14).

Nolte, E., McKee, M., 2011. Variations in amenable mortality – trends in 16 high-income nations. Health Policy 103, 47–52. Available at: http://www.ncbi.nlm.nih.gov/pubmed/21917350 (accessed 12.01.14).

Olival, K.J., Islam, A., Yu, M., Anthony, S.J., Epstein, J.H., Khan, S.A., et al., 2013. Ebola virus antibodies in fruit bats. Bangladesh. Emerg. Infect. Dis. 19 (2). Available at: http://wwwnc.cdc.gov/eid/article/19/2/pdfs/12-0524.pdf (accessed 12.01.14).

Piot, P., Quinn, T.C., 2013. Response to the AIDS pandemic – a global health model. N. Engl. J. Med. 368, 2210–2218. Available at: http://www.nejm.org/doi/full/10.1056/NEJMra1201533 (accessed 12.01.14).

Rai, R.K., Tulchinsky, T.H., 2012;2 February. Addressing the sluggish progress in reducing maternal mortality in India. Asia Pac. J. Public Health http://dx.doi.org/10.1177/1010539512436883 [Epub ahead of print].

Salomon, J.A., Wang, H., Freeman, M.K., et al., 2012. Healthy life expectancy for 187 countries, 1990–2010: a systematic analysis for the Global Burden of Disease Study 2010. Lancet 380, 2144–2162. Available at: http://www.thelancet.com/journals/lancet/article/PIIS0140-6736(12)61690-0/abstract (accessed 12.01.14).

Shiffman, J., 2007. Generating political priority for maternal mortality reduction in 5 developing countries. Am. J. Public Health 97, 796–803. Available at: http://www.ncbi.nlm.nih.gov/pmc/articles/PMC1854881/ (accessed 12.01.14).

Stephenson, J., 2006. HIV and circumcision. JAMA 296, 759. Available at: http://jama.jamanetwork.com/article.aspx?articleid=203183 (accessed 12.01.14).

United Nations, 2009. World mortality report, 2009. UN, New York. Available at: http://www.un.org/en/development/desa/population/publications/pdf/mortality/worldMortalityReport2009.pdf (accessed 28.05.13).

United Nations, 2011. World population prospects. Social affairs, economics. The 2010 revision highlights and advance tables. UN, New York. Available at: http://esa.un.org/wpp/Documentation/pdf/WPP2012_HIGHLIGHTS.pdf (accessed 12.01.14).

United Nations Children's Fund, 2012. Levels and trends in child mortality report 2012. Estimates developed by the UN Inter-agency Group for Child Mortality Estimation. UNICEF, New York. Available at: http://apromiserenewed.org/files/UNICEF_2012_child_mortality_for_web_0904.pdf (accessed 30.05.13).

United Nations Children's Fund, 2013. The state of the world's children, 2013. Children with disabilities. Oxford University Press, New York. Available at: http://www.unicef.org/sowc2013/files/SWCR2013_ENG_Lo_res_24_Apr_2013.pdf (accessed 30.05.13).

United Nations Children's Fund/World Health Organization, 2012. Progress on drinking water and sanitation: 2012 update. UNICEF, New York. Available at: http://www.unicef.org/media/files/JMPreport2012.pdf (accessed 12.01.14).

United Nations Development Programme, 2011. Human Development Report 2011: Sustainability and equity: a better future for all. UNDP, New York. Available at: http://hdr.undp.org/en/media/HDR_2011_EN_Contents.pdf (accessed 30.05.13).

United Nations Economic Commission for Africa, Report on Africa, 2013. Making the most of Africa's commodities: industrializing for growth, jobs and economic transformation. Available at: http://www.uneca.org/sites/default/files/publications/unera_report_eng_final_web.pdf (accessed 02.09.13).

Victora, C.G., Adair, L., Fall, C., Hallal, P.C., Martorell, R., Richter, L., Sachdev, H.S., 2008. for the Maternal and Child Undernutrition Study Group. Maternal and child undernutrition: consequences for adult health and human capital. Lancet 371, 340–357. Available at: http://www.thelancet.com/journals/lancet/article/PIIS0140-6736(07)61692-4/abstract (accessed 12.01.14).

Wang, H., Dwyer-Lindgren, L., Lofgren, K.T., Rajaratnam, J.K., Marcus, J.R., Levin-Rector, A., et al., 2012. Age-specific and sex-specific mortality in 187 countries, 1970–2010: a systematic analysis for the Global Burden of Disease Study 2010. Lancet 380, 2071–2094. Available at: http://www.thelancet.com/journals/lancet/article/PIIS0140-6736(12)61719-X/abstract (accessed 12.01.14).

Wibulpolprasert, S., Tangcharoensathien, V., Kanchanachitra, C., 2008. Three decades of primary health care: reviewing the past and defining the future. Bull. World Health Organ. 86, 1–80. Available at: http://www.ncbi.nlm.nih.gov/pmc/articles/PMC2647359/ (accessed 12.01.14).

World Bank, World Databank. Millennium Development Goal, 2013. Available at: http://databank.worldbank.org/Data/Views/VariableSelection/SelectVariables.aspx?source=Millennium%20Development%20Goals (accessed 28.05.13).

World Economic Forum and Harvard School of Public Health, 2011. The global economic burden of non-communicable diseases. Available at: http://www3.weforum.org/docs/WEF_Harvard_HE_GlobalEconomicBurdenNonCommunicableDiseases_2011.pdf. (accessed 29.07.13).

World Health Organization, Cardiovascular diseases (CVDs). Fact sheet no. 317 [updated March 2013]. http://www.who.int/mediacentre/factsheets/fs317/en/ (accessed 02.09.13).

World Health Organization, September 2012. Children: reducing mortality. Fact sheet no. 178. WHO, Geneva. Available at: http://www.who.int/mediacentre/factsheets/fs178/en/ (accessed 01.06.13).

World Health Organization. Cholera, 2013. 2012 Wkly. Epidemiol. Rec. 88, 321–336. Available at: http://www.who.int/wer/2013/wer8831.pdf (accessed 02.09.13).

World Health Organization, 2 August 2013. Global epidemics and impact of cholera. WHO. Available at: http://www.who.int/topics/cholera/impact/en/. (accessed 02.09.13).

World Health Organization, Health statistics and information systems. Global burden of disease. Available at: http://www.who.int/healthinfo/global_burden_disease/en/ (accessed 02.06.13).

World Health Organization, 2005. International health regulations. Available at: http://www.who.int/ihr/en/ (accessed 11.06.13).

World Health Organization, Levels and trends in child mortality, report 2012. Estimates of under-five, infant and neonatal mortality by Millennium Development Goal region; by UNICEF region. Available at: http://apromiserenewed.org/files/UNICEF_2012_child_mortality_for_web_0904.pdf (accessed 30.05.13).

World Health Organization, November 2012. Millennium Development Goals (MDGs). Fact sheet no. 290. WHO, Geneva. Available at: http://www.who.int/mediacentre/factsheets/fs290/en/ (accessed 29.05.13).

World Health Organization, Trends in maternal mortality: 1990 to 2010. Available at: http://whqlibdoc.who.int/publications/2012/9789241503631_eng.pdf (accessed 11.06.13).

World Health Organization. World Health Report, 2007. A safer future: global public health security in the 21st century. WHO, Geneva. Available at: http://www.who.int/whr/2007/whr07_en.pdf (accessed 12.01.14).

World Health Organization. World Health Report, 2010. Health systems financing: the path to universal coverage. WHO, Geneva. Available at: http://www.who.int/whr/2010/en/ (accessed 12.01.14).

World Health Organization, 2013. World Health Statistics 2013. WHO, Geneva. Available at: http://www.who.int/gho/publications/world_health_statistics/EN_WHS2013_Full.pdf (accessed 12.01.14).

World Health Organization, 2012. World malaria report. Available at: http://www.who.int/malaria/publications/world_malaria_report_2012/en/. (accessed 10.06.13).

World Health Organization, 2009. Department of Violence and Injury Prevention and Disability. Global status report on road safety: time for action. WHO, Geneva. Available at: http://whqlibdoc.who.int/publications/2009/9789241563840_eng.pdf (accessed 22.10.12).

World Health Organization, Global Health Observatory. Life expectancy. Available at: http://www.who.int/gho/mortality_burden_disease/life_tables/situation_trends/en/ (accessed 02.06.13).

World Health Organization/United Nation Children's Fund, 2013. Progress on sanitation and drinking water, 2013 update. WHO/UNICEF, Geneva. Available at: http://apps.who.int/iris/bitstream/10665/81245/1/9789241505390_eng.pdf (accessed 12.01.14).

International Organizations

Institute for Health Metrics and Evaluation, 2013. The global burden of disease: generating evidence, guiding policy. IHME, Seattle, WA. Available at: www.healthmetricsandevaluation.org (accessed 25.05.13).

National Institute of Allergy and Infectious Diseases, Emerging and re-emerging infectious diseases. Available at: http://www.niaid.nih.gov/topics/emerging/Pages/Default.aspx (accessed 02.06.13).

National Institutes of Health, 2007. Understanding emerging and re-emerging infectious diseases. Available at: http://www.ncbi.nlm.nih.gov/books/NBK20370/. (accessed 02.06.13).

Pan American Health Organization, June 2013. Health in All Policies in the Americas. The 8th Global Conference on Health Promotion. Finland, Helsinki. 10–14. Available at: http://www.paho.org/hiap/index.php?option=com_docman&task=doc_view&gid=396&Itemid= (accessed 11.06.13).

United Nations, Health in the post-2015 development agenda: need for a social determinants of health approach. Joint statement of the UN Platform on Social Determinants of Health. Available at: http://www.worldwewant2015.org/node/300184 (accessed 08.06.13).

United Nations, Millennium Development Goal 7: Ensure environmental sustainability. Available at: http://www.un.org/millenniumgoals/environ.shtml (accessed 25.07.13).

United Nations, 2013. Millennium Development Goals Momentum. Available at: http://www.un.org/millenniumgoals/mdgmomentum.shtml. (accessed 25.07.13).

United Nations, 1948. Universal Declaration of Human Rights. UN, Geneva. Available at: http://www.un.org/en/documents/udhr/ (accessed 12.01.14).

World Bank, 1993. World Development Report 1993: Investing in health. Oxford University Press, New York. Available at: http://wdronline.worldbank.org/worldbank/a/c.html/world_development_report_1993/abstract/WB.0-1952-0890-0.abstract1 (accessed 12.01.14).

World Bank, 2006. World Development Report 2006: Equity and development. World Bank, Washington, DC. Available at: http://wwwwds.worldbank.org/external/default/WDSContentServer/IW3P/IB/2005/09/20/000112742_20050920110826/Rendered/PDF/322040World0Development0Report02006.pdf (accessed 12.01.14).

World Health Organization, 2009. Achieving Millennium Development Goal 5: target 5A and 5B on reducing maternal mortality and achieving universal access to reproductive health. WHO, Geneva. Available at: http://whqlibdoc.who.int/hq/2009/WHO_RHR_09.06_eng.pdf (accessed 11.06.13).

World Health Organization, 2011. Closing the gap: policy into practice on social determinants of health. Discussion paper of the World Conference on Social Determinants of Health, 19–21 October 2011. WHO, Geneva. Available at: http://www.who.int/sdhconference/Discussion-Paper-EN.pdf (accessed 12.01.14).

World Health Organization, 1946. Constitution of the World Health Organization. WHO, New York. Available at: http://whqlibdoc.who.int/hist/official_records/constitution.pdf (accessed 12.01.14).

World Health Organization, Health statistics and health information. WHO mortality database. Available at: http://www.who.int/healthinfo/mortality_data/en/index.html (accessed 02.06.13).

World Health Organization, WHO Framework Convention on Tobacco Control. Available at: http://www.who.int/fctc/text_download/en/ (accessed 28.05.13).

World Health Organization, WHO mortality database. Available at: http://apps.who.int/healthinfo/statistics/mortality/whodpms/ (accessed 02.06.13).

World Health Organization, 2011. WHO report on the global tobacco epidemic: warning about the dangers of tobacco. WHO, Geneva. Available at: http://whqlibdoc.who.int/publications/2011/9789240687813_eng.pdf (accessed 12.06.13).

Global Health in the Future

Health in the Post-2015 Agenda, April 2013. Report of the Global Thematic Consultation on Health. Available at: http://www.worldwewant2015.org/file/337378/download/366802 (accessed 12.01.14).

Hutton, G., Bartram, J., 2008. Global costs of attaining the Millennium Development Goal for water supply and sanitation. Bull. World Health Organ. 86, 2–3. Available at: http://www.ncbi.nlm.nih.gov/pmc/articles/PMC2647341/ (accessed 12.01.14).

Kim, J.Y., Farmer, P., Porter, M.E., 20 May 2013. Redefining global health-care delivery. Lancet, Early Online Publication. Lancet. 382: 1060–1069. (Epub 2013 May 20) Available at: http://press.thelancet.com/healthcaredelivery.pdf (accessed 12.01.14).

Kolan, J.P., Butler-Jones, D., Tsang, T., Yu, W., 2013; June. Public health lessons from severe acute respiratory syndrome – ten years later. Emerg Infect Dis. Available at: http://wwwnc.cdc.gov/eid/article/19/6/12-1426_article.htm (accessed 12.01.14).

Lagomarsino, G., Garabrant, A., Adyas, A., Muga, R., Otoo, N., 2012. Moving towards universal health coverage: health insurance reforms in nine developing countries in Africa and Asia. Lancet 380, 933–943. Available at: http://www.thelancet.com/journals/lancet/article/PIIS0140-6736(12)61147-7/abstract (accessed 12.01.14).

Levine, O.S., O'Brien, K.L., Knoll, M., Adegbola, R.A., Black, S., Cherian, T., et al., 2006. Pneumococcal vaccination in developing countries. Lancet 367, 1880–1882. Available at: http://www.thelancet.com/journals/lancet/article/PIIS0140-6736(06)68703-5/fulltext (accessed 12.01.14).

Mathers, C.D., Loncar, D., 2006. Projections of global mortality and burden of disease from 2002 to 2030. PLoS Med. Available at: http://www.plosmedicine.org/article/info:doi/10.1371/journal.pmed.0030442 (accessed 12.01.14).

McCord, G.C., Liu, A., Singh, P., 2012. Deployment of community health workers across rural sub-Saharan Africa: WHO financial considerations and operational assumptions. Bull. World Health Organ. 91. 244–53B. Available at: http://www.who.int/bulletin/volumes/91/4/12-109660/en/ (accessed 09.06.13).

Morris, S.S., Cogill, B., Uauy, R., 2008. for the Maternal and Child Undernutrition Study Group. Effective international action against undernutrition: why has it proven so difficult and what can be done to accelerate progress? Lancet 371, 608–621. Available at: http://www.thelancet.com/journals/lancet/article/PIIS0140-6736(07)61695-X/abstract (accessed 12 .01.14).

Munos, M., Fischer Walker, C.L., Black, R.E., 2010. The effect of rotavirus vaccine on diarrhoea mortality. Int. J. Epidemiol. 39, i56–i62.
. Available at: http://ije.oxfordjournals.org/content/39/suppl_1/i56.full (accessed 12.01.14).

Suba, E.J., Murphy, S.K., Donnelly, A.D., Furia, L.M., Huynh, M.L., Raab, S.S., 2006. Systems analysis of real-world obstacles to successful cervical cancer prevention in developing countries. Am. J. Public Health 96, 480–487. Available at: http://www.ncbi.nlm.nih.gov/pubmed/16449592 (accessed 12.01.14).

Tulchinsky, T.H., Varavikova, E.A., 2010. What is the new public health? Public Health Rev. 32, 25–53. Available at: http://www.publichealthreviews.eu/show/f/23 (accessed 12 .01.14).

UNICEF. State of the world's children, UNICEF, New York. Available at: http://www.unicef.org/sowc2013/files/SWCR2013_ENG_Lo_res_24_Apr_2013.pdf (accessed 14.1.14).

Wolfson, L.J., Gasse, F., Lee-Martin, S.P., Lydon, P., Magan, A., Tibouti, A., et al., 2008. Estimating the costs of achieving the WHO–UNICEF global immunization vision and strategy, 2006–2015. Bull. World Health Organ. 86, 27–39.

World Economic Forum, 2011. From burden to "best buys": reducing the economic impact of non-communicable diseases in low- and middle-income countries. WHO, Geneva. Available at: http://www.who.int/nmh/publications/best_buys_summary/en/ (accessed 12.01.14).

World Health Organization, 2013. Essential nutrition actions: improving maternal, newborn, infant and young child health and nutrition. WHO, Geneva. Available at: http://apps.who.int/iris/bitstream/10665/84409/1/9789241505550_eng.pdf (accessed 12.06.13).

World Health Organization, 2013. Global alert and response. Ebola hemorrhagic fever. WHO, Geneva. Available at: http://www.who.int/csr/disease/ebola/en/ (accessed 11.06.13).

World Health Organization, 2012. Global costs and benefits of drinking-water supply and sanitation interventions to reach the MDG target and universal coverage. WHO, Geneva. Available at: http://www.who.int/water_sanitation_health/publications/2012/global_costs/en/index.html (accessed 11.06.13).

World Health Organization, Global nutrition policy review: what does it take to scale up nutrition action? Available at: http://apps.who.int/iris/bitstream/10665/84408/1/9789241505529_eng.pdf (accessed 07.06.13).

World Health Organization, Global plan for the Decade of Action for Road Safety 2011–2020. Geneva: WHO. Available at: http://www.who.int/roadsafety/decade_of_action/plan/en/index.html (accessed 22.10.12).

World Health Organization, 2012. Global progress report on implementation of the WHO Framework Convention on Tobacco Control. WHO, Geneva. Available at: http://www.who.int/fctc/reporting/2012_global_progress_report_en.pdf (accessed 29.05.13).

World Health Organization, 2013. Monitoring the building blocks of health systems: a handbook of indicators and their measurement strategies. WHO, Geneva. Available at: http://www.who.int/healthinfo/systems/monitoring/en/index.html (accessed 12.01.14).

World Health Organization, 2013. Nutrition. Global targets 2025: To improve maternal, infant and young child nutrition. WHO, Geneva. Available at: http://apps.who.int/iris/bitstream/10665/84409/1/9789241505550_eng.pdf (accessed 12.01.14).

World Health Organization, 2011. Scaling up action against non-communicable diseases: how much will it cost? WHO, Geneva. Available at: http://whqlibdoc.who.int/publications/2011/9789241502313_eng.pdf (accessed 29.07.13).

World Health Organization, 2013. End preventable deaths: Global Action Plan for Prevention and Control of Pneumonia and Diarrhoea. WHO/UNICEF. Available at: http://apps.who.int/iris/bitstream/10665/79200/1/9789241505239_eng.pdf. (accessed 11.06.13).

World Health Organization, Millennium Development Goals (MDGs). Available at: http://www.who.int/topics/millennium_development_goals/en/ (accessed 12.01.14).

World Health Organization. World Health Report, 2006. Working together for health. WHO, Geneva. Available at: http://www.who.int/whr/2006/en/ (accessed 12.01.14).

World Health Organization, 2008. Commission on Social Determinants of Health. Closing the gap in a generation: health equity through action on the social determinants of health. Final report of the Commission on Social Determinants of Health. WHO, Geneva. Available at: http://whqlibdoc.who.int/hq/2008/WHO_IER_CSDH_08.1_eng.pdf (accessed 12.01.14).

World Health Organization, 2012. Regional Office for Europe. European Health Report. Charting the way to well-being. Available at: http://www.euro.who.int/__data/assets/pdf_file/0004/185332/The-European-Health-Report-2012,-Executive-summary-w-cover.pdf (accessed 12.01.14).

World Health Organization, 11 January 2013. Secretariat. Draft action plan for the prevention and control of non-communicable diseases 2013–2020. Report by the Secretariat. EB132/7, 132nd session. Available at: http://apps.who.int/gb/ebwha/pdf_files/WHA66/A66_9-en.pdf. (accessed 11.06.13).

World Health Organization/United Nation Children's Fund, 2010. Countdown to 2015: Decade report (2000–2010): taking stock of maternal, newborn and child survival. WHO, Geneva. Available at: http://www.childinfo.org/files/CountdownReport_2000-2010.pdf (accessed 12.06.13).

Note: Page Numbers followed by "f" indicate figures; t, tables; b, boxes.